Lecture Notes in Computer Science 10032

Commenced Publication in 1973
Founding and Former Series Editors:
Gerhard Goos, Juris Hartmanis, and Jan van Leeuwen

Jung Hee Cheon · Tsuyoshi Takagi (Eds.)

Advances in Cryptology – ASIACRYPT 2016

22nd International Conference on the Theory and Application of Cryptology and Information Security
Hanoi, Vietnam, December 4–8, 2016
Proceedings, Part II

Editors
Jung Hee Cheon
Seoul National University
Seoul
Korea (Republic of)

Tsuyoshi Takagi
Kyushu University
Fukuoka
Japan

ISSN 0302-9743 ISSN 1611-3349 (electronic)
Lecture Notes in Computer Science
ISBN 978-3-662-53889-0 ISBN 978-3-662-53890-6 (eBook)
DOI 10.1007/978-3-662-53890-6

Library of Congress Control Number: 2016956613

LNCS Sublibrary: SL4 – Security and Cryptology

Printed on acid-free paper

This Springer imprint is published by Springer Nature
The registered company is Springer-Verlag GmbH Germany
The registered company address is: Heidelberger Platz 3, 14197 Berlin, Germany

Preface

ASIACRYPT 2016, the 22nd Annual International Conference on Theory and Application of Cryptology and Information Security, was held at InterContinental Hanoi Westlake Hotel in Hanoi, Vietnam, during December 4–8, 2016. The conference focused on all technical aspects of cryptology, and was sponsored by the International Association for Cryptologic Research (IACR).

Asiacrypt 2016 received a total of 240 submissions from all over the world. The Program Committee selected 67 papers from these submissions for publication in the proceedings of this conference. The review process was made via the usual double-blind pier review by the Program Committee comprising 43 leading experts in the field. Each submission was reviewed by at least three reviewers and five reviewers were assigned to submissions co-authored by Program Committee members. This year, the conference operated a two-round review system with a rebuttal phase. In the first-round review the Program Committee selected the 140 submissions that were considered of value for proceeding to the second round. In the second-round review the Program Committee further reviewed the submissions by taking into account their rebuttal letter from the authors. The selection process was assisted by a total of 309 external reviewers. These two-volume proceedings contain the revised versions of the papers that were selected. The revised versions were not reviewed again and the authors are responsible for their contents.

The program of Asiacrypt 2016 featured three excellent invited talks. Nadia Heninger gave a talk on "The Reality of Cryptographic Deployments on the Internet," Hoeteck Wee spoke on "Advances in Functional Encryption," and Neal Koblitz gave a non-technical lecture on "Cryptography in Vietnam in the French and American Wars." The conference also featured a traditional rump session that contained short presentations on the latest research results of the field. The Program Committee selected the work "Faster Fully Homomorphic Encryption: Bootstrapping in Less Than 0.1 Seconds" by Ilaria Chillotti, Nicolas Gama, Mariya Georgieva, and Malika Izabachène for the Best Paper Award of Asiacrypt 2016. Two more papers, "Nonlinear Invariant Attack—Practical Attack on Full SCREAM, iSCREAM, and Midori64" by Yosuke Todo, Gregor Leander, Yu Sasaki and "Cliptography: Clipping the Power of Kleptographic Attacks" by Alexander Russell, Qiang Tang, Moti Yung, Hong-Sheng Zhou were solicited to submit full versions to the *Journal of Cryptology*.

Many people contributed to the success of Asiacrypt 2016. We would like to thank the authors for submitting their research results to the conference. We are very grateful to all of the Program Committee members as well as the external reviewers for their fruitful comments and discussions on their areas of expertise. We are greatly indebted to Ngo Bao Chau and Phan Duong Hieu, the general co-chairs for their efforts and overall organization. We would also like to thank Nguyen Huu Du, Nguyen Quoc Khanh, Nguyen Duy Lan, Duong Ngoc Thai, Nguyen Ta Toan Khoa, Nguyen Ngoc Tuan,

Le Thi Lan Anh, and the local Organizing Committee for their continuous supports. We thank Steven Galbraith for expertly organizing and chairing the rump session.

Finally we thank Shai Halevi for letting us use his nice software for supporting the paper submission and review process. We also thank Alfred Hofmann, Anna Kramer, and their colleagues at Springer for handling the editorial process of the proceedings. We would like to express our gratitude to our partners and sponsors: XLIM, Microsoft Research, CISCO, Intel, Google.

December 2016

Jung Hee Cheon
Tsuyoshi Takagi

ASIACRYPT 2016

The 22nd Annual International Conference on Theory and Application of Cryptology and Information Security

Sponsored by the International Association for Cryptologic Research (IACR)

December 4–8, 2016, Hanoi, Vietnam

General Co-chairs

Ngo Bao Chau	VIASM, Vietnam and University of Chicago, USA
Phan Duong Hieu	XLIM, University of Limoges, France

Program Co-chairs

Jung Hee Cheon	Seoul National University, Korea
Tsuyoshi Takagi	Kyushu University, Japan

Program Committee

Elena Andreeva	KU Leuven, Belgium
Xavier Boyen	Queensland University of Technology, Australia
Anne Canteaut	Inria, France
Chen-Mou Cheng	National Taiwan University, Taiwan
Sherman S.M. Chow	Chinese University of Hong Kong, Hong Kong, SAR China
Nico Döttling	University of California, Berkeley, USA
Thomas Eisenbarth	Worcester Polytechnic Institute, USA
Georg Fuchsbauer	École Normale Supérieure, France
Steven Galbraith	Auckland University, New Zealand
Sanjam Garg	University of California, Berkeley, USA
Vipul Goyal	Microsoft Research, India
Jens Groth	University College London, UK
Sylvain Guilley	Secure-IC S.A.S., France
Alejandro Hevia	Universidad de Chile, Chile
Antoine Joux	Foundation UPMC and LIP6, France
Xuejia Lai	Shanghai Jiaotong University, China
Hyung Tae Lee	Nanyang Technological University, Singapore
Kwangsu Lee	Sejong University, Korea
Dongdai Lin	Chinese Academy of Sciences, China
Feng-Hao Liu	Florida Atlantic University, USA
Takahiro Matsuda	AIST, Japan
Alexander May	Ruhr University Bochum, Germany

Florian Mendel	Graz University of Technology, Austria
Amir Moradi	Ruhr University Bochum, Germany
Svetla Nikova	KU Leuven, Belgium
Tatsuaki Okamoto	NTT, Japan
Elisabeth Oswald	University of Bristol, UK
Thomas Peyrin	Nanyang Technological University, Singapore
Rei Safavi-Naini	University of Calgary, Canada
Peter Schwabe	Radboud University, The Netherlands
Jae Hong Seo	Myongji University, Korea
Damien Stehlé	ENS de Lyon, France
Ron Steinfeld	Monash University, Australia
Rainer Steinwandt	Florida Atlantic University, USA
Daisuke Suzuki	Mitsubishi Electric, Japan
Mehdi Tibouchi	NTT, Japan
Yosuke Todo	NTT, Japan
Hoang Viet Tung	University of California Santa Barbara, USA
Dominique Unruh	University of Tartu, Estonia
Ivan Visconti	University of Salerno, Italy
Huaxiong Wang	Nanyang Technological University, Singapore
Meiqin Wang	Shandong University, China
Aaram Yun	UNIST, Korea

External Reviewers

Michel Abdalla
Aysajan Abidin
Shashank Agrawal
Shweta Agrawal
Ahmad Ahmadi
Mamun Akand
Saed Alsayigh
Joël Alwen
Abdelrahaman Aly
Daniel Apon
Muhammad Rizwan Asghar
Tomer Ashur
Nuttapong Attrapadung
Benedikt Auerbach
Saikrishna Badrinarayanan
Shi Bai
Razvan Barbulescu
Lejla Batina
Georg T. Becker
Christof Beierle
Fabrice Benhamouda
Begül Bilgin
Céline Blondeau
Tobias Boelter
Carl Bootland
Jonathan Bootle
Yuri Borissov
Christina Boura
Colin Boyd
Wouter Castryck
Dario Catalano
Andrea Cerulli
Gizem Cetin
Pyrros Chaidos
Nishanth Chandran
Yu-Chen Chang
Lin Changlu
Binyi Chen
Cong Chen
Jie Chen
Ming-Shing Chen
Yu Chen
Céline Chevalier
Chongwon Cho
Kyu Young Choi
HeeWon Chung
Kai-Min Chung
Eloi de Chérisey
Michele Ciampi
Craig Costello
Joan Daemen
Ricardo Dahab
Wei Dai
Bernardo David
Thomas de Cnudde
David Derler
Apoorvaa Deshpande
Christoph Dobraunig
Yarkin Doroz
Ming Duan
Léo Ducas

Dung Hoang Duong
Maria Eichlseder
Martianus Frederic Ezerman
Xiong Fan
Pooya Farshim
Serge Fehr
Max Fillinger
Dario Fiore
Victor Fischer
Marc Fischlin
Thomas Fuhr
Jake Longo Galea
David Galindo
Peter Gazi
Essam Ghadafi
Mohona Ghosh
Zheng Gong
Rishab Goyal
Hannes Gross
Vincent Grosso
Berk Gulmezoglu
Chun Guo
Jian Guo
Qian Guo
Divya Gupta
Iftach Haitner
Dong-Guk Han
Kyoohyung Han
Shuai Han
Goichiro Hanaoka
Christian Hanser
Mitsuhiro Hattori
Gottfried Herold
Felix Heuer
Takato Hirano
Shoichi Hirose
Wei-Chih Hong
Yuan-Che Hsu
Geshi Huang
Guifang Huang
Jialin Huang
Xinyi Huang
Pavel Hubacek
Ilia Iliashenko
Mehmet Sinan Inci
Vincenzo Iovino
Gorka Irazoqui
Ai Ishida
Takanori Isobe
Tetsu Iwata
Aayush Jain
Sune Jakobsen
Yin Jia
Shaoquan Jiang
Chethan Kamath
Sabyasachi Karati
Sayasachi Karati
Yutaka Kawai
Carmen Kempka
HeeSeok Kim
Hyoseung Kim
Jinsu Kim
Myungsun Kim
Taechan Kim
Paul Kirchner
Elena Kirshanova
Fuyuki Kitagawa
Susumu Kiyoshima
Jessica Koch
Markulf Kohlweiss
Vladimir Kolesnikov
Thomas Korak
Yoshihiro Koseki
Ashutosh Kumar
Ranjit Kumaresan
Po-Chun Kuo
Robert Kübler
Thijs Laarhoven
Ching-Yi Lai
Russell W.F. Lai
Virginie Lallemand
Adeline Langlois
Sebastian Lauer
Su Le
Gregor Leander
Kwangsu Lee
Gaëtan Leurent
Anthony Leverrier
Jingwei Li
Ming Li
Wen-Ding Li
Benoit Libert
Fuchun Lin
Tingting Lin
Meicheng Liu
Yunwen Liu
Zhen Liu
Zidong Lu
Yiyuan Luo
Atul Luykx
Vadim Lyubashevsky
Bernardo Magri
Mary Maller
Alex Malozemoff
Antonio Marcedone
Benjamin Martin
Daniel Martin
Marco Martinoli
Daniel Masny
Maike Massierer
Mitsuru Matsui
Willi Meier
Bart Mennink
Peihan Miao
Kazuhiko Minematsu
Nicky Mouha
Pratyay Mukherjee
Sean Murphy
Jörn Müller-Quade
Valérie Nachef
Michael Naehrig
Matthias Nagel
Yusuke Naito
Mridul Nandi
María Naya-Plasencia
Kartik Nayak
Khoa Nguyen
Ivica Nikolic
Ventzislav Nikov
Ryo Nishimaki
Anca Nitulescu
Koji Nuida
Maciej Obremski
Toshihiro Ohigashi
Miyako Ohkubo
Sumit Kumar Pandey
Jong Hwan Park

Seunghwan Park
Alain Passelègue
Christopher Patton
Bo-Yuan Peng
Rachel Player
Antigoni Polychroniadou
Bertram Pöttering
Sebastian Ramacher
Vanishree Rao
Shuqin Ren
Reza Reyhanitabar
Bastian Richter
Thomas Ristenpart
Mike Rosulek
Hansol Ryu
Akshayaram Srinivasan
Yusuke Sakai
Kochi Sakumoto
Amin Sakzad
Simona Samardjiska
Yu Sasaki
Pascal Sasdrich
Falk Schellenberg
Benedikt Schmidt
Tobias Schneider
Jacob Schuldt
Okan Seker
Nicolas Sendrier
Jae Hong Seo
Minhye Seo
Yannick Seurin
Masoumeh Shafienejad
Barak Shani
Danilo Sijacic
Alice Silverberg
Siang Meng Sim
Dave Singelee
Luisa Siniscalchi
Daniel Slamanig
Nigel Smart
Raphael Spreitzer
Douglas Stebila
Christoph Striecks
Takeshi Sugawara
Yao Sun
Berk Sunar
Koutarou Suzuki
Alan Szepieniec
Mostafa Taha
Somayeh Taheri
Junko Takahashi
Katsuyuki Takashima
Benjamin Tan
Jean-Pierre Tillich
Junichi Tomida
Yiannis Tselekounis
Himanshu Tyagi
Thomas Unterluggauer
Damien Vergnaud
Gilles Villard
Vanessa Vitse
Damian Vizar
Michael Walter
Han Wang
Hao Wang
Qiungju Wang
Wei Wang
Yuyu Wang
Yohei Watanabe
Hoeteck Wee
Wei Wei
Mor Weiss
Mario Werner
Bas Westerbaan
Carolyn Whitnall
Alexander Wild
Baofeng Wu
Keita Xagawa
Zejun Xiang
Hong Xu
Weijia Xue
Shota Yamada
Takashi Yamakawa
Hailun Yan
Jun Yan
Bo-Yin Yang
Bohan Yang
Guomin Yang
Mohan Yang
Shang-Yi Yang
Kan Yasuda
Xin Ye
Wentan Yi
Scott Yilek
Kazuki Yoneyama
Rina Zeitoun
Fan Zhang
Guoyan Zhang
Liang Feng Zhang
Liangfeng Zhang
Tao Zhang
Wentao Zhang
Yusi Zhang
Zongyang Zhang
Jingyuan Zhao
Yongjun Zhao
Yixin Zhong
Hong-Sheng Zhou
Xiao Zhou
Jincheng Zhuang

Local Organizing Committee

Co-chairs

Ngo Bao Chau — VIASM, Vietnam and University of Chicago, USA
Phan Duong Hieu — XLIM, University of Limoges, France

Members

Nguyen Huu Du	VIASM, Vietnam
Nguyen Quoc Khanh	Vietcombank, Vietnam
Nguyen Duy Lan	Microsoft Research, USA
Duong Ngoc Thai	Google, USA
Nguyen Ta Toan Khoa	NTU, Singapore
Nguyen Ngoc Tuan	VIASM, Vietnam
Le Thi Lan Anh	VIASM, Vietnam

Sponsors

XLIM
Microsoft Research
CISCO
Intel
Google

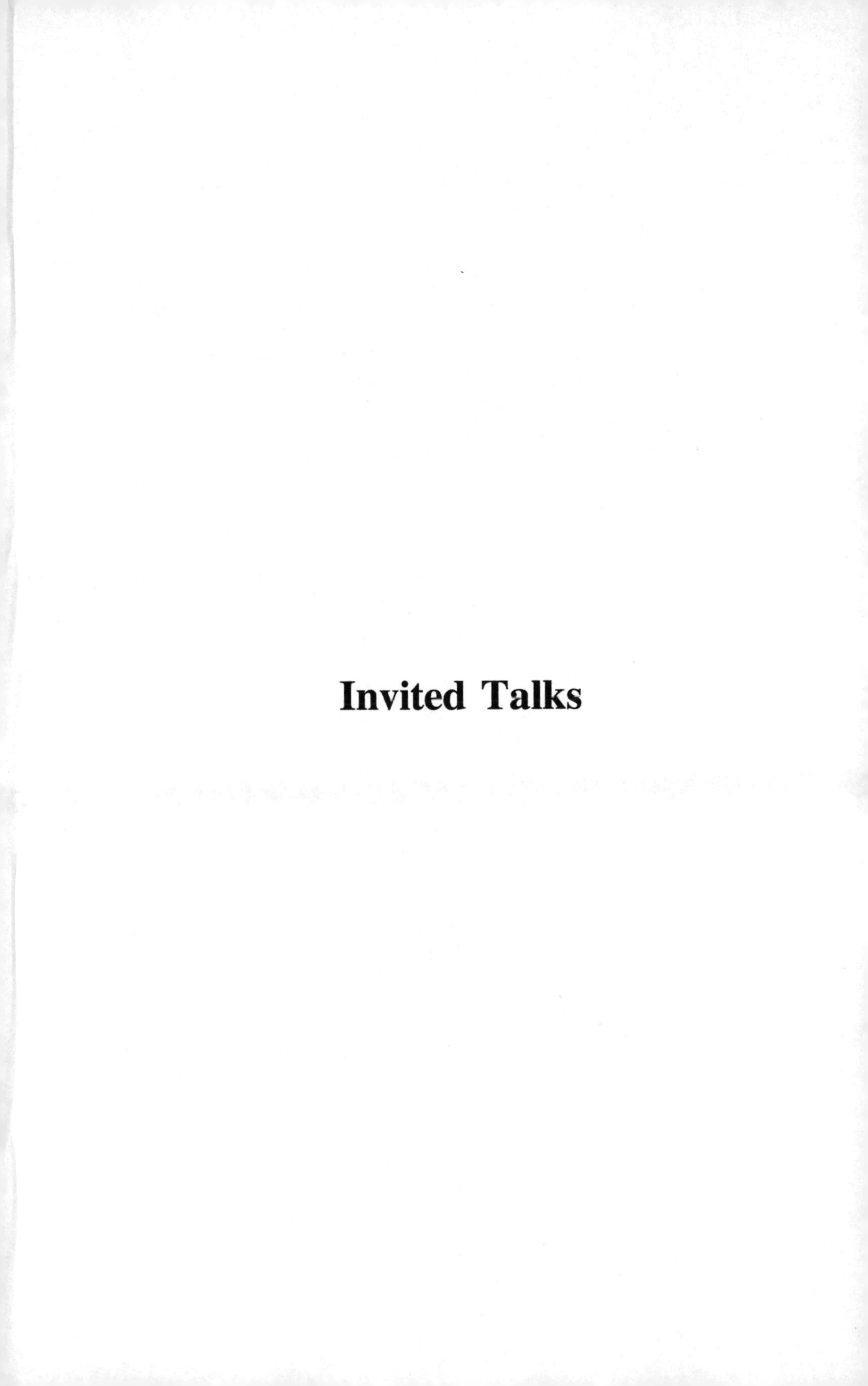

Invited Talks

Advances in Functional Encryption

Hoeteck Wee

ENS, Paris, France
wee@di.ens.fr

Abstract. Functional encryption is a novel paradigm for public-key encryption that enables both fine-grained access control and selective computation on encrypted data, as is necessary to protect big, complex data in the cloud. In this talk, I will provide a brief introduction to functional encryption and an overview of the state of the art, with a focus on constructions based on lattices.

CNRS, INRIA and Columbia University. Supported in part by ERC Project aSCEND (H2020 639554) and NSF Award CNS-1445424.

The Reality of Cryptographic Deployments on the Internet

Nadia Heninger

University of Pennsylvania, Philadelphia, USA

Abstract. Security proofs for cryptographic primitives and protocols rely on a number of (often implicit) assumptions about the world in which these components live. They assume that implementations are correct, that specifications are followed, that systems make sensible choices about error conditions, and that reliable sources of random numbers are present. However, a number of real world studies examining cryptographic deployments have shown that these assumptions are often not true on a large scale, with catastrophic effects for security. In addition to simple programming errors, many real-world cryptographic vulnerabilities can be traced back to more complex underlying causes, such as backwards compatibility, legacy protocols and software, hard-coded resource limits, and political interference in design choices.

Many of these issues appear on the surface to be at an entirely different level of abstraction from the cryptographic primitives used in their construction. However, by taking advantage of the structure of many cryptographic primitives when used at Internet scale, it is possible to uncover fundamental vulnerabilities in implementations. I will discuss the interplay between mathematical cryptanalysis techniques and the thorny implementation issues that lead to vulnerable cryptographic deployments in the real world.

Contents – Part II

Asiacrypt 2016 Award Papers

Zero Knowledge

Post Quantum Cryptography

Provable Security

Foundation

Cryptographic Protocol

Multi-party Computation

Contents – Part I

Asiacrypt 2016 Best Paper

Mathematical Analysis I

AES and White-Box

Hash Function

Randomness

Authenticated Encryption

Block Cipher I

SCA and Leakage Resilience I

Asiacrypt 2016 Award Papers

Nonlinear Invariant Attack

Practical Attack on Full SCREAM, iSCREAM, and Midori64

Yosuke Todo[1,3(✉)], Gregor Leander[2], and Yu Sasaki[1]

[1] NTT Secure Platform Laboratories, Tokyo, Japan
{todo.yosuke,sasaki.yu}@lab.ntt.co.jp
[2] Horst Görtz Institute for IT Security,
Ruhr-Universität Bochum, Bochum, Germany
gregor.leander@rub.de
[3] Kobe University, Hyogo, Japan

Abstract. In this paper we introduce a new type of attack, called *nonlinear invariant attack*. As application examples, we present new attacks that are able to distinguish the full versions of the (tweakable) block ciphers Scream, iScream and Midori64 in a weak-key setting. Those attacks require only a handful of plaintext-ciphertext pairs and have minimal computational costs. Moreover, the nonlinear invariant attack on the underlying (tweakable) block cipher can be extended to a ciphertext-only attack in well-known modes of operation such as CBC or CTR. The plaintext of the authenticated encryption schemes SCREAM and iSCREAM can be practically recovered only from the ciphertexts in the nonce-respecting setting. This is the first result breaking a security claim of SCREAM. Moreover, the plaintext in Midori64 with well-known modes of operation can practically be recovered. All of our attacks are experimentally verified.

Keywords: Nonlinear invariant attack · Boolean function · Ciphertext-only message-recovery attack · SCREAM · iSCREAM · Midori64 · CAESAR competition

1 Introduction

Block ciphers are certainly among the most important cryptographic primitives. Since the invention of the DES [1] in the mid 70's and even more with the design of the AES [2], a huge amount of research has been done on various aspects of block cipher design and block cipher analysis. In the last decade, many new block ciphers have been proposed that aim at highly resource constrained devices. Driven by new potential applications like the internet of things, we have witnessed not only many new designs, but also several new cryptanalytic results. Today, we have at hand a well established set of cryptanalytic tools that, when are carefully applied, allow to gain significant confidence in the security of a block cipher design. The most prominent tools here are certainly differential [5] and linear [21] attacks and their numerous variations [4,7,14,15].

J.H. Cheon and T. Takagi (Eds.): ASIACRYPT 2016, Part II, LNCS 10032, pp. 3–33, 2016.
DOI: 10.1007/978-3-662-53890-6_1

Despite this fact, quite some of the recently proposed lightweight block ciphers got broken rather quickly. One of the reasons for those attacks, on what is supposed to be a well-understood field of cryptographic designs, is that the new lightweight block ciphers are designed more aggressive than e.g. most of the AES candidates. Especially when it comes to the design of the key schedule, many new proposals keep the design very simple, often using identical round keys. While there is no general defect with such a key schedule, structural attacks become much more of an issue compared to a cipher that deploys a more complicated key schedule. In this paper we introduce a new structural attack, named *nonlinear invariant attack*. At first glance, it might seem quite unlikely that such an attack could ever be successfully applied. However, we give several examples of ciphers that are highly vulnerable to this attack.

1.1 Our Contribution

Given a block cipher $E_k : \mathbb{F}_2^n \to \mathbb{F}_2^n$, the general principle of the nonlinear invariant attack is to find an efficiently computable nonlinear Boolean function $g : \mathbb{F}_2^n \to \mathbb{F}_2$ such that

$$g(x) \oplus g(E_k(x))$$

is constant for any x and for many possible keys k. Keys such that this term is constant are called weak keys. The function g itself is called *nonlinear invariant* for E_k. Clearly, when the block cipher E_k has a (non-trivial) nonlinear invariant function g, $g(p) \oplus g(E_k(p))$ is constant for any plaintext p and any weak key k. On the other hand, the probability that random permutations have this property is about 2^{-N+1} when g is balanced. Therefore, attackers can immediately execute a distinguishing attack. Moreover, if the constant depends on the secret key, an attacker can recover one bit of information about the secret key by using one known plaintext-ciphertext pair.

For round-based block ciphers, our attack builds the nonlinear invariants from the nonlinear invariants of the single round functions. In order to extend the nonlinear invariant for a single round to the whole cipher, all round-keys must be weak keys. It may be infeasible to find such weak-key classes for block ciphers with a non-trivial key schedule. However, as mentioned above, many recent block ciphers are designed for lightweight applications, and they adopt more aggressive designs to achieve high performance even in highly constrained environments. Several lightweight ciphers do not deploy any key schedule at all, but rather use the master key directly as the identical round key for all rounds. In such a situation, the weak-key class of round keys is trivially converted into the weak-key class of the secret key. In particular, when all round keys are weak, this property is iterative over an arbitrary number of rounds.

(Ciphertext-Only) Message-Recovery Attacks. The most surprising application of the nonlinear invariant attack is an extension to ciphertext-only message-recovery attacks. Clearly, we cannot execute any ciphertext-only attack without some information on the plaintexts. Therefore, our attack is ciphertext-only attack under the following environments. Suppose that block ciphers which are vulnerable against the nonlinear invariant attack are used in well-known modes of operation, e.g., CBC, CFB, OFB, and CTR. Then, if the same unknown plaintext is encrypted by the same weak key and different initialization vectors, attackers can practically recover a part of the plaintext from the ciphertexts only.

Applications. We demonstrate that our new attack practically breaks the full authenticated encryption schemes SCREAM[1] [11] and iSCREAM [10] and the low-energy block cipher Midori64 [3] in the weak-key setting.

Table 1. Summary of the nonlinear invariant attack

	# of weak keys	Max. # of recovered bits	Data complexity	Time complexity
SCREAM	2^{96}	32 bits	33 ciphertexts	32^3
iSCREAM	2^{96}	32 bits	33 ciphertexts	32^3
Midori64	2^{64}	$32h$ bits	$33h$ ciphertexts	$32^3 \times h$

h is the number of blocks in the mode of operation.

We show that the tweakable block ciphers Scream and iScream have a nonlinear invariant function, and the number of weak keys is 2^{96}. Midori64 also has a nonlinear invariant function, and the number of weak keys is 2^{64}. Table 1 summarizes the result of the nonlinear invariant attack against SCREAM, iSCREAM, and Midori64. The use of the tweakable block cipher Scream is defined by the authenticated encryption SCREAM, and the final block is encrypted like CTR when the byte length of a plaintext is not multiple of 16. We exploit this procedure and recover 32 bits of the final block of the plaintext if the final block length ranges from 12 bytes to 15 bytes. We can also execute a similar attack against iSCREAM. Note that our attack breaks SCREAM and iSCREAM in the nonce-respecting model. Midori64 is a low-energy block cipher, and we consider the case that Midori64 is used by well-known modes of operation. As a result, we can recover 32 bits in every 64-bit block of the plaintext if Midori64 is used in CBC, CFB, OFB, and CTR.

Comparison with Previous Attacks. Leander et al. proposed invariant subspace attack on iSCREAM [19], which is a weak-key attack working for 2^{96} weak keys. The attack can be a distinguishing attack and key recovery attack in the chosen-message and chosen-tweak model. Guo et al. presented a weak-key attack on full Midori64 [12], which works for 2^{32} weak keys, distinguishes the cipher with 1 chosen-plaintext query, and recovers the key with 2^{16} computations.

[1] Note that throughout the paper SCREAM always refer to the latest version as SCREAM, i.e. SCREAM (v3).

Compared to [19], our attack has the same weak key size and we distinguish the cipher in the known-message and chosen-tweak model. Compared to [12], our weak-key class is much larger and the cipher is distinguished with 2 known-plaintext queries. In both applications, the key space can be reduce by 1 bit, besides a part of message/plaintext can be recovered from the ciphertext.

1.2 Related Work

The nonlinear invariant attack can be regarded as an extension of linear cryptanalysis [21]. While linear cryptanalysis uses a linear function to approximate the cipher, the nonlinear invariant attack uses a nonlinear function and the probability of the nonlinear approximation is one. When g is linear, ciphers that are resistant against the linear cryptanalysis never have a linear approximation with probabilistically one.

The use of the nonlinear approximation has previously been studied. This extension was first discussed by Harpes et al. [13], and Knudsen and Robshaw later investigated the effectiveness deeply [16]. However, they showed that there are insurmountable problems in the general use of nonlinear approximations. For instance, one cannot join nonlinear approximations for more than one round of a block cipher because the actual approximations depend on the specific value of the state and key. Knudsen and Robshaw demonstrated that nonlinear approximations can replace linear approximations in the first and last rounds only [16]. Unfortunately, nonlinear cryptanalysis has not been successful because of this limited application. Our attack can be seen as the first application of the nonlinear cryptanalysis against real ciphers in the past two decades.

Other related attacks are the invariant subspace attack [18,19] and symmetric structures [8,17,23]. Similar to the nonlinear invariant attack, those attacks exploit a cryptanalytic property which continues over an arbitrary number of rounds in the weak-key setting. While those attacks have to choose plaintexts, i.e. are chosen plaintext attacks, the nonlinear invariant attack does not need to choose plaintexts in general. This in particular allows us to extend the nonlinear invariant attack from a pure distinguishing attack to the (ciphertext-only) message-recovery attack.

1.3 Paper Organization

We explain the general ideas and principles of the new attack in Sect. 2. Section 3 explains how, in many cases, the attack can be constructed in an almost automatic way using an algorithmic approach that is for most ciphers practical. Moreover, we give a structural reason why some ciphers, more precisely some linear layers, are inherently weak against our attack and why our attacks are possible against those ciphers. In Sect. 4 we explain in detail our attacks on SCREAM and iSCREAM. Moreover, Sect. 5 details our nonlinear invariant attack on Midori64. Finally, in Sect. 6, we give some additional insights into the general structure of nonlinear invariant functions and outline some future work.

2 Nonlinear Invariant Attack

In this section, we describe the basic principle of the attack and its extension to (ciphertext-only) message-recovery attacks when used in common modes of operations. While our attack can be applied to any cipher structure in principle, we focus on the case of key-alternating ciphers and later on substitution permutation networks (SPN) ciphers to simplify the description. We start by explaining the basic idea and later how, surprisingly, the attack can be extended to a (ciphertext-only) message-recovery attack in many scenarios.

2.1 Core Idea

Let $F : \mathbb{F}_2^n \to \mathbb{F}_2^n$ be the round function of a key-alternating cipher and $F_k(x) = F(x \oplus k)$ be the round function including the key XOR. Thus, for an r-round cipher, the ciphertext C is computed from a plaintext P using round keys k_i as

$$\begin{aligned} x_0 &= P \\ x_{i+1} &= F_{k_i}(x_i) = F(x_i \oplus k_i) \quad 0 \leq i \leq r-1 \\ C &= x_r \end{aligned}$$

where we ignore post-whitening key for simplicity.

The core idea of the nonlinear invariant attack is to detect a nonlinear Boolean function g such that

$$g(F(x \oplus k)) = g(x \oplus k) \oplus c = g(x) \oplus g(k) \oplus c \quad \forall x$$

for many keys k, where c is a constant in $\mathbb{F}_2$. Keys for which this equality holds will be called *weak keys*. The function g itself is called *nonlinear invariant* in this paper.

The important remark is that, if all round-keys k_i are weak then

$$\begin{aligned} g(C) &= g(F(x_{r-1} \oplus k_{r-1})) \\ &= g(x_{r-1}) \oplus g(k_{r-1}) \oplus c \\ &= g(F(x_{r-2} \oplus k_{r-2})) \oplus g(k_{r-1}) \oplus c \\ &= g(x_{r-2}) \oplus g(k_{r-2}) \oplus g(k_{r-1}) \\ &\vdots \\ &= g(P) \oplus \bigoplus_{i=0}^{r-1} g(k_i) \oplus \bigoplus_{i=0}^{r-1} c. \end{aligned}$$

Thus, the invariant is iterative over an arbitrary number of rounds and immediately leads to a distinguishing attack.

Distinguishing Attack. Assume that we found a Boolean function g that is nonlinear invariant for the round function F_k of a block cipher. Then, if all round keys are weak, this function g is also nonlinear invariant over an arbitrary number of rounds.

Let (P_i, C_i) $1 \leq i \leq N$ be N pairs of plaintexts and corresponding ciphertexts. Then, $g(P_i) \oplus g(C_i)$ is constant for all pairs. If g is balanced, the probability that random permutations have this property is about 2^{-N+1}. Note that the case that g is unbalanced can be handled as well, but is not the main focus of our paper. Therefore, we can practically distinguish the block cipher from random permutations under a *known-plaintext attack*.

Suitable Nonlinear Invariants. We next discuss a particular choice of a nonlinear invariant g for which it is directly clear that weak keys exist. Imagine we were able to identify a nonlinear invariant g for F, i.e. a function such that

$$g(F(x)) \oplus g(x)$$

is constant, such that g is actually linear (or constant) in some of the inputs. In this case, all round keys that are zero in the nonlinear components of g, are weak.

More precisely, without loss of generality, assume that the nonlinear invariant g is linear in the last t bits of input (implying that g is nonlinear in the first s bits of input where $s = n - t$). Namely, we can view g as

$$g : (\mathbb{F}_2^s \times \mathbb{F}_2^t) \to \mathbb{F}_2$$

such that

$$g(x, y) = g(x, 0) \oplus g(0, y) = f(x) \oplus \ell(y)$$

where f is the nonlinear part of g, and ℓ is the linear part of g. As g is a nonlinear invariant for F, it holds that

$$g(x, y) \oplus g(F(x, y)) = c,$$

where c is a constant in $\mathbb{F}_2$. Now consider a round key $k \in \mathbb{F}_2^s \times \mathbb{F}_2^t$ of the form $(0, k')$. That is, we consider a round key such that its first s bits are zero. Then it holds that

$$\begin{aligned} g(F_k(x, y)) &= g(F(x, y \oplus k')) \\ &= g(x, y \oplus k') \oplus c \\ &= f(x) \oplus \ell(y \oplus k') \oplus c \\ &= f(x) \oplus \ell(y) \oplus \ell(k') \oplus c \\ &= g(x, y) \oplus g(0, k') \oplus c. \end{aligned}$$

In other words, all those round-keys that are zero in the first s bits are weak. Phrased differently, the density of weak keys is 2^{-s}.

Example 1. Let $g : \mathbb{F}_2^4 \to \mathbb{F}_2$ be a nonlinear invariant as

$$g(x_4, x_3, x_2, x_1) = x_4x_3 \oplus x_3 \oplus x_2 \oplus x_1.$$

Then, the function g can be viewed as

$$g(x_4, x_3, x_2, x_1) = f(x_4, x_3) \oplus \ell(x_2, x_1).$$

Now consider a round key $k \in \mathbb{F}_2^2 \times \mathbb{F}_2^2$ of the form $(0, k')$. Then, the function g is a nonlinear invariant for the key XOR because

$$g(x) \oplus g(x \oplus k) = g(x) \oplus g(x) \oplus g(0, k') = g(0, k').$$

On Key Schedule and Round Constants. Many block ciphers generate round keys from the master key by a key schedule. For a proper key schedule, it is very unlikely that all round keys are weak in the above sense. However, many recent lightweight block ciphers do not have a well-diffused key schedule, but rather use (parts of) the master key directly as the round keys. From a performance point of view, this approach is certainly preferable.

However, the direct XORing with the secret key often causes vulnerabilities like the slide attack [6] or the invariant subspace attack [18]. To avoid those attacks, round constants are additionally XORed in such lightweight ciphers. While dense and random-looking round constant would be a conservative choice, many such ciphers adopt sparse round constants because they are advantageous in limited memory requirements.

Focusing on the case of identical round keys, assume that there is a Boolean function g which is nonlinear invariant for the round function F. Now if all used round constants c_i are such that c_i is only involved in the linear terms of g, the function g is nonlinear invariant for this constant addition. This follows by the same arguments for the weak keys above. We call such constants, in line with the notation of weak keys from above, *weak constant*.

To conclude, given a key-alternating cipher with identical round-keys and weak round-constants, any master-key that is weak, is immediately weak for an arbitrary number of rounds. In this scenario, the number of weak keys is 2^t, or equivalently, the density of weak keys is 2^{-s}.

2.2 Message Recovery Attack

As described so far, the nonlinear invariant attack leaks at most one bit of the secret key. However, if a block cipher that is vulnerable to the nonlinear invariant attack is used in well-known modes of operation, e.g., CBC, CFB, OFB, and CTR, surprisingly, the attack can be turned into a *ciphertext-only message recovery attack*.

Clearly, we cannot execute any ciphertext-only attack without some information on the plaintexts. When block ciphers are used under well-known modes of operation, the plaintext itself is not the input of block ciphers and the input is rather initialization vectors. Here we assume that an attacker can collect several ciphertexts where the same plaintext is encrypted by the same (weak) key and different initialization vectors. We like to highlight that this assumption is more practical not only compared to the chosen-ciphertext attack but also to the known-plaintext attack. In practice, for instance, assuming an application sends secret password several times, we can recover the password practically. While the feasibility depends on the behavior of the application, our attack is highly practical in this case.

Attack Against CBC Mode. Figure 1 shows the CBC mode, where h message blocks are encrypted. Let P_j be the jth plaintext block, and C_j^i denotes the jth ciphertext block by using the initialization vector IV^i. The attacker aims at recovering the plaintext $(P_1, P_2, \ldots, P_h)$ by observing the ciphertext $(IV^i, C_1^i, C_2^i, \ldots, C_h^i)$. Moreover, we assume that the block cipher E_k is vulnerable against the nonlinear invariant attack, i.e., there is a function g such that $g(x) \oplus g(y)$ is constant, where x and y denote the input and output of the block cipher.

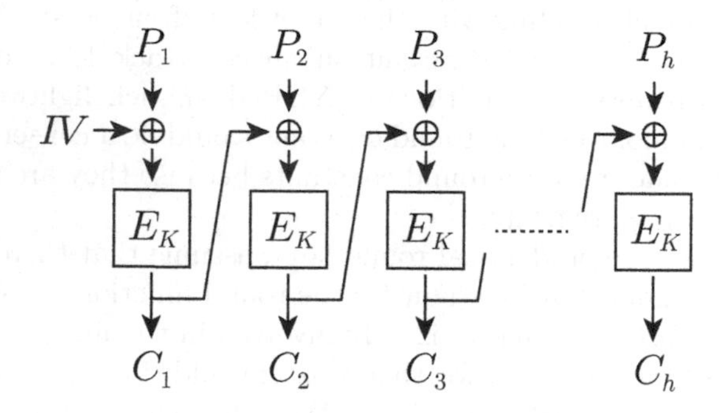

Fig. 1. CBC mode

First, we explain how to recover the plaintext P_1 by focusing on the first block. Since E_k is vulnerable against the nonlinear invariant attack, there is a function g such that $g(P_1 \oplus IV_1^i) \oplus g(C_1^i)$ is constant for any $i \in \{1, 2, \ldots, N\}$. If g would be a linear function,

$$g(P_1 \oplus IV_1^j) \oplus g(C_1^j) = g(P_1) \oplus g(IV_1^j) \oplus g(C_1^j)$$

is constant, and the attacker could only recover at most one bit of secret information. However, g is nonlinear in our attack. Therefore, we can guess and determine the part of P_1 that is involved in the nonlinear term of g. More precisely, assume as above – without loss of generality – that g is nonlinear in the first s inputs and linear in the last t inputs, i.e.

$$g : \mathbb{F}_2^s \times \mathbb{F}_2^t$$

such that

$$g(x,y) = f(x) \oplus \ell(y)$$

where f is any Boolean function, and ℓ is a linear Boolean function. Consider again a plaintext $P_1 = (x,y)$ with $x \in \mathbb{F}_2^s$ and $y \in \mathbb{F}_2^t$. The corresponding ciphertext C_1^i is split as $C_1^i = (c_i, d_i)$ and the IVs as $IV^i = (a_i, b_i)$. With this notation, we can rewrite the following

$$\begin{aligned} g(P_1 \oplus IV^i) &= (f(x \oplus a_i) \oplus \ell(y \oplus b_i)), \\ g(P_1 \oplus IV^j) &= (f(x \oplus a_j) \oplus \ell(y \oplus b_j)), \\ g(C_1^i) &= (f(c_i) \oplus \ell(d_i)), \\ g(C_1^j) &= (f(c_j) \oplus \ell(d_j)). \end{aligned}$$

Now, by using two distinct initialization vectors IV^i and IV^j

$$0 = g(P_1 \oplus IV^i) \oplus g(C_1^i) \oplus g(P_1 \oplus IV^j) \oplus g(C_1^j)$$

implies

$$f(x \oplus a_i) \oplus f(x \oplus a_j) = \ell(b_i \oplus b_j) \oplus g(C_1^i) \oplus g(C_1^j). \tag{1}$$

Assuming that the left side of Eq. (1) randomly changes depending on x, that is the left part of P_1, we can recover one bit of information on P_1 by using two initialization vectors. Similarly, we can recover $N-1$ bits of P_1 by using N initialization vectors. Note that we can usually efficiently recover these bits by solving linear systems if the algebraic degree of f is small [22]. We show the specific procedure for SCREAM and Midori64 in Sects. 4 and 5, respectively. The relationship among (P_1, IV, C_1) is equivalent to that among (P_i, C_{i-1}, C_i). Therefore, we can similarly guess and determine the part of P_i from C_{i-1} and C_i for any of the plaintext blocks. One interesting remark is that as long as we start to recover the message from the second block, the attack can be executed even without the knowledge of the IV.

Attacks Against Other Modes. We can execute similar attack against the CFB, OFB, and CTR modes.

In the CFB mode, the hth ciphertext block C_h is encrypted as

$$C_h = E_k(C_{h-1}) \oplus P_h,$$

where the initialization vector IV is used as the input of the first block. For simplicity, let C_0 be IV. Then, we can recover the part of P_h from two ciphertext blocks C_{h-1} and C_h.

In the OFB mode, the hth ciphertext block C_h is encrypted as

$$C_h = (E_k)^h(IV) \oplus P_h,$$

where $(E_k)^h(IV)$ is h times multiple encryption. Since the nonlinear invariant property is iterative over an arbitrary number of rounds, the multiple encryption is also vulnerable against the nonlinear invariant attack. Therefore, we can recover the part of P_h from IV and C_h.

In the CTR mode, the hth ciphertext block C_h is encrypted as

$$C_h = E_k(IV + h) \oplus P_h.$$

Therefore, we can recover the part of P_h from $IV + h$ and C_h.

3 Finding Nonlinear Invariants for SP-ciphers

We start by considering the very general problem of finding nonlinear invariants. Namely, given any function

$$F : \mathbb{F}_2^m \to \mathbb{F}_2^m,$$

our goal is to find a Boolean function

$$g : \mathbb{F}_2^m \to \mathbb{F}_2$$

such that

$$g(x) = g(F(x)) \oplus c \tag{2}$$

where c is a constant in $\mathbb{F}_2$.

The description so far is generic in the sense that it applies to basically any block cipher. For now, and actually for the remainder of the paper, we focus on key-alternating ciphers with a round function using a layer of S-boxes and a linear layer, so called substitution-permutation networks (SPN).

3.1 SPN Ciphers

In the following, we consider the un-keyed round function only. That is to say that we ignore the key schedule and also any round constants.

For simplicity, we focus on the case of identical S-boxes, but the more general case can be handled in a very similar manner. We denote by t the number of S-boxes and by n the size of one S-box. Thus, the block size processed is $n \cdot t$ bits. With this notation, we consider one round R of an SPN

$$R : (\mathbb{F}_2^n)^t \to (\mathbb{F}_2^n)^t$$

as consisting of a layer of S-boxes $\mathcal{S}$ with

$$\mathcal{S}(x_1, \ldots, x_t) = (S(x_1), \ldots, S(x_t))$$

where S is an n-bit S-box and a linear layer

$$L : (\mathbb{F}_2^n)^t \to (\mathbb{F}_2^n)^t$$

which can also be seen as

$$L : \mathbb{F}_2^{nt} \rightarrow \mathbb{F}_2^{nt}.$$

The round function R is given as the composition of the S-box layer and the linear layer, i.e.

$$R(x) = L \circ \mathcal{S}(x).$$

We would like to find nonlinear invariant g for R. However, computing this directly is difficult as soon as the block size is reasonable large. For any function F, let us denote by

$$U(F) := \{g : \mathbb{F}_2^m \rightarrow \mathbb{F}_2 \mid g(x) = g(F(x)) \oplus c\}$$

the set of all nonlinear invariants for F, and it holds that

$$g \in (U(\mathcal{S}) \cap U(L)) \subset U(R).$$

In other words, functions that are invariant under both $\mathcal{S}$ and L are clearly invariants for their composition R.

As we will explain next, computing parts of $U(\mathcal{S}) \cap U(L)$ is feasible, and sufficient to automatically detect the weaknesses described later in the paper.

The S-box Layer. We start by investigating the S-box-layer. Given the S-box as a function

$$S : \mathbb{F}_2^n \rightarrow \mathbb{F}_2^n$$

computing $U(S)$ is feasible as long as n is only moderate in size.

Note that, for any function F, $U(F)$ is actually a subspace of Boolean functions. To see this, note that given two Boolean functions $f, g \in U(F)$, it holds

$$\begin{aligned}(f \oplus g)(x) &= f(x) \oplus g(x) \\ &= (f(F(x)) \oplus c) \oplus (g(F(x)) \oplus c') \\ &= (f \oplus g)(F(x)) \oplus (c \oplus c')\end{aligned}$$

for any x. Thus the sum, $f \oplus g$, is in $U(F)$ as well. Moreover, the all-zero function is in $U(F)$ for any F. Therefore, any nonlinear invariant $g_S \in U(S)$ can actually be described by a linear combination of basis elements of $U(S)$. More precisely, let $b_1, \ldots, b_d : \mathbb{F}_2^n \rightarrow \mathbb{F}_2$ be a basis of $U(S)$, then any $g_S \in U(S)$ can be written s

$$g_S(x) = \bigoplus_{i=1}^{d} \gamma_i b_i(x)$$

for suitable coefficients γ_i in $\mathbb{F}_2$.

To identify a nonlinear invariant $g_S \in U(S)$, the idea is to consider the algebraic normal form (ANF) of g_S, that is to express g_S as

$$g_S(x) = \bigoplus_{u \in \mathbb{F}_2^n} \lambda_u x^u,$$

where $\lambda_u \in \mathbb{F}_2$ are the coefficients to be determined and x^u denotes $\prod x_i^{u_i}$. The key observation is that Eq. (2), for any fixed $x \in \mathbb{F}_2^n$, translates into one linear (or affine) equation for the coefficients λ_u, namely

$$\bigoplus_{u \in \mathbb{F}_2^n} \lambda_u (x^u \oplus S(x)^u) = c.$$

The ANF of $(x^u \oplus S(x)^u)$ is computed for all $u \in \mathbb{F}_2^n$, and we can easily solve the basis $b_1, \dots, b_d \in U(S)$ for n not too big. Appendix A shows the algorithm in detail. In particular, for commonly used S-box sizes of up to 8 bits, the space $U(S)$ can be computed in less than a second on a standard PC.

So far, we have considered only a single S-box, and it still needs to be discussed how those results can be translated into the knowledge of invariants for the parallel execution of S-boxes, i.e. for $\mathcal{S}$. Again, for a layer of S-boxes $\mathcal{S}$ computing $U(\mathcal{S})$ directly using its ANF is (in general) too expensive. However, we can easily construct many elements in $U(\mathcal{S})$ from elements in $U(S)$ as summarized in the following proposition.

Proposition 1. *Let $g_i \in U(S)$, for $i \in \{1, \dots, t\}$ be any set of invariants for the S-box S. Then, any function of the form*

$$g_{\mathcal{S}}(x_1, \dots, x_t) = \bigoplus_{i=1}^{t} \alpha_i g_i(x_i)$$

with $\alpha_i \in \mathbb{F}_2$ is in $U(\mathcal{S})$, that is an invariant for the entire S-box layer. The set of function form a subspace of $U(\mathcal{S})$ of dimension $d * t$ where d is the dimension of $U(S)$, and t is the number of parallel S-boxes.

We denote this subspace of invariants for $\mathcal{S}$ by $U_\ell(\mathcal{S})$, and $U_\ell(\mathcal{S}) \subset U(\mathcal{S})$.

It turns out that, in general, many more elements are contained in $U(\mathcal{S})$ than those covered by the construction above. We decided to shift those details, which are not directly necessary for the understanding of the attacks presented in Sects. 4 and 5 to the end of the paper, in Sect. 6.

The Linear Layer. For the linear layer computing $U(L)$ using its ANF seems again difficult. But, as stated above, we focus on

$$g \in (U(L) \cap U_\ell(\mathcal{S})) \subset (U(L) \cap U(\mathcal{S})) \subset U(R),$$

and computing $U(L) \cap U_\ell(\mathcal{S})$ is feasible in all practical cases.

Recall that any nonlinear invariant $g \in U(S)$ can actually be described by a linear combination of basis of $U(S)$ as

$$g_S(x) = \bigoplus_{i=1}^{d} \gamma_i b_i(x)$$

for suitable coefficients γ_i in $\mathbb{F}_2$.

As any f in $U_\ell(S)$ is itself a direct sum of elements in $U(S)$, it can be written as

$$f(x_1, \dots, x_t) = \bigoplus_{i=1}^{t} \bigoplus_{j=1}^{d} \beta_{i,j} b_j(x_i)$$

with $\beta_{i,j} \in \mathbb{F}_2$. Computing those coefficients $\beta_{i,j}$ can again be done by solving linear system, as any fixed $x \in (\mathbb{F}_2^n)^t$ results in a linear equation for the coefficients by using

$$f(x) = f(L(x)).$$

As long as the dimension of $U_\ell(S)$, i.e. the number of unknowns, is not too large, this again can be computed within seconds on a standard PC.

Experimental Results. When the procedure explained above was applied to the ciphers SCREAM and Midori, it instantaneously detected possible attacks. Actually, as we will explain next, there is a common structural reason why non linear invariant attacks are possible on those ciphers.

3.2 Structural Weakness with Respect to Nonlinear Invariant

Let us consider linear layers which are actually used in the LS-designs [9] (cf. Sect. 4) and also in any AES-like cipher that uses a binary diffusion matrix as a replacement for the usual MixColumns operation. Then, we consider a linear layer that can be decomposed into the parallel application of n identical $t \times t$ binary matrices M. The input for the first $t \times t$ matrix is composed of all the first output bits of the t S-boxes, the input for the second matrix is composed of all the second output bits of the S-boxes, etc.

Here, when M is an orthogonal matrix, that is if

$$\langle x, y \rangle = \langle xM, yM \rangle \quad \forall\, x, y,$$

any quadratic nonlinear invariant for the S-box can be extended to a nonlinear invariant of the whole round function as described in Theorem 1.

Note that from a design point of view, taking M as an orthogonal matrix seems actually beneficial. Thanks to the orthogonality of M, bounds on the number of active S-boxes for differential cryptanalysis directly imply the same bounds on the number of active S-boxes for linear cryptanalysis.

Theorem 1. *For the SPN ciphers whose round function follows the construction used in LS-designs, let $M \in \mathbb{F}_2^{t \times t}$ be the binary representation of the linear layer and M is orthogonal. Assume there is a nonlinear invariant g_S for the S-box. If g_S is quadratic, then the function*

$$g(x_1, \dots, x_t) := \bigoplus_{i=1}^{t} g_S(x_i)$$

is a nonlinear invariant for the round function $L \circ S$.

Proof. First, due to Proposition 1, it is immediately clear that g is a nonlinear invariant for the S-box layer $\mathcal{S}$.

Next, let us consider the linear layer L. Let $x \in (\mathbb{F}_2^n)^t$ and $y \in (\mathbb{F}_2^n)^t$ be the input and output of L, respectively. Moreover, $x_i[j]$ and $y_i[j]$ denotes the jth bit of x_i and y_i, respectively. For simplicity, let $x^T \in (\mathbb{F}_2^t)^n$ and $y^T \in (\mathbb{F}_2^t)^n$ be the transposed input and output, respectively, where $x_j^T \in \mathbb{F}_2^t$ denotes $(x_1[j], x_2[j], \ldots, x_t[j])$. Then, it holds $y_i^T = x_i^T \times M$ for all $i \in \{1, 2, \ldots, n\}$. Since the Boolean function g_S is quadratic, the function is represented as

$$g_S(x_i) = \bigoplus_{i_1=1}^{n} \bigoplus_{i_2=1}^{n} \gamma_{i_1,i_2} (x_i[i_1] \wedge x_i[i_2]),$$

where γ_{i_1,i_2} are coefficients depending on the function g. From the inner product $\langle x_{i_1}^T, x_{i_2}^T \rangle = \bigoplus_{i=1}^{t} x_i[i_1] \wedge x_i[i_2]$,

$$g(x) = \bigoplus_{i=1}^{t} g_S(x_i) = \bigoplus_{i_1=1}^{n} \bigoplus_{i_2=1}^{n} \gamma_{i_1,i_2} \langle x_{i_1}^T, x_{i_2}^T \rangle.$$

Then,

$$g(y) = \bigoplus_{i_1=1}^{n} \bigoplus_{i_2=1}^{n} \gamma_{i_1,i_2} \langle x_{i_1}^T M, x_{i_2}^T M \rangle$$

From the orthogonality of M,

$$\begin{aligned} g(y) &= \bigoplus_{i_1=1}^{n} \bigoplus_{i_2=1}^{n} \gamma_{i_1,i_2} \langle x_{i_1}^T, x_{i_2}^T \rangle \\ &= \bigoplus_{i=1}^{t} g_S(x_i) = g(x) \end{aligned}$$

Therefore, the function $g(x) = \bigoplus_{i=1}^{t} g_S(x_i)$ is a nonlinear invariant for L. □

Assuming that the matrix M is orthogonal, Theorem 1 shows that there is a nonlinear invariant for the round function $L \circ S$ if there is a quadratic function which is nonlinear invariant for the S-box.

4 Practical Attack on SCREAM

The most interesting application of the nonlinear invariant attack is a practical attack against the authenticated encryption SCREAM and iSCREAM in the nonce-respecting model. Both authenticated encryptions have 2^{96} weak keys, and we then practically distinguish their ciphers from a random permutation. Moreover, we can extend this attack to a ciphertext-only attack.

4.1 Specification of SCREAM

SCREAM is an authenticated encryption and a candidate of the CAESAR competition [11]. It uses the tweakable block cipher Scream, which is based on the tweakable variant of LS-designs [9].

LS-Designs. LS-designs were introduced by Grosso et al. in [9], and it is used to design block ciphers. We do not refer to the design rational in this paper, and we only show the brief structure to understand this paper. The state of LS-designs is represented as an $s \times \ell$ matrix, where every element of the matrix is only one bit, i.e., the block length is $n = s \times \ell$. The ith round function proceeds as follows:

1. The s-bit S-box S is applied to ℓ columns in parallel.
2. The ℓ-bit L-box L is applied to s rows in parallel.
3. The round constant $C(i)$ is XORed with the state.
4. The secret key K is XORed with the state.

Figure 2 shows the components of a LS-design. Let SB and LB be the S-box layer and L-box layer, respectively. Then, we call the composite function $(LB \circ SB)$ a LS-function. Let $x \in \mathbb{F}_2^{s \times \ell}$ be the state of LS-designs. Then $x[i, \star] \in \mathbb{F}_2^{\ell}$ denotes the row of index i of x, and $x[\star, j] \in \mathbb{F}_2^{s}$ denotes the column of index j of x. Moreover, let $x[i, j]$ be the bit in the $(i+1)$th row and $(j+1)$th column. The S-box S is applied to $x[\star, j]$ for all $j \in [0, \ell)$, and the L-box L is applied to $x[i, \star]$ for all $i \in [0, s)$.

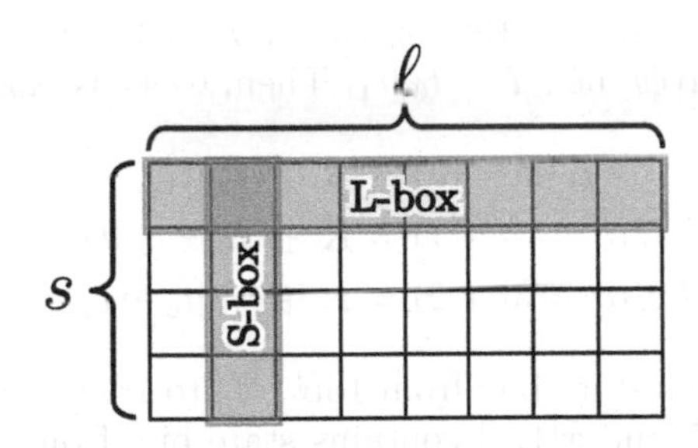

Fig. 2. The components of a LS-design

Tweakable Block Cipher Scream. Scream is based on a tweakable LS-design with an 8×16 matrix, i.e., the block length is $8 \times 16 = 128$ bits. Let $x \in \mathbb{F}_2^{8 \times 16}$ be the state of Scream, then the entire algorithm is defined as Algorithm 1. Here S and L denote the 8-bit S-box and 16-bit L-box, respectively. The round constant $C(r)$ is defined as

$$C(r) = 2199 \cdot r \bmod 2^{16}.$$

Algorithm 1. Specification of Scream

1: $x \leftarrow P \oplus TK(0)$
2: **for** $0 < \sigma \leq N_s$ **do**
3: **for** $0 < \rho \leq 2$ **do**
4: $r = 2(\sigma - 1) + \rho$
5: **for** $0 \leq j < 16$ **do**
6: $x_j^T = S[x[\star, j]]$
7: **end for**
8: $x \leftarrow x \oplus C(r)$
9: **for** $0 \leq i < 8$ **do**
10: $x_i = L[x[i, \star]]$
11: **end for**
12: **end for**
13: $x \leftarrow x \oplus TK(\sigma)$
14: **end for**
15: **return** x

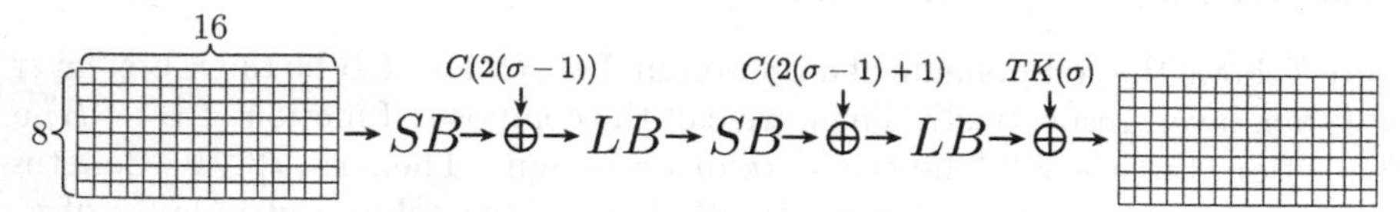

Fig. 3. The σth step function of Scream

The binary representation of $C(r)$ is XORed with the first row $x[0, \star]$. Scream uses an 128-bit key K and an 128-bit tweak T as follows. First, the tweak is divided into 64-bit halves, i.e., $T = t_0 \| t_1$. Then, every tweakey is defined as

$$\begin{aligned} TK(\sigma = 3i) &= K \oplus (t_0 \| t_1), \\ TK(\sigma = 3i + 1) &= K \oplus (t_0 \oplus t_1 \| t_1), \\ TK(\sigma = 3i + 2) &= K \oplus (t_1 \| t_0 \oplus t_1). \end{aligned}$$

Here, the $x[i, \star]$ contains state bits from $16(i-1)$ to $16i - 1$, e.g., $x[0, \star]$ contains state bits from 0 to 15 and $x[1, \star]$ contains state bits from 16 to 31. Moreover, Fig. 3 shows the step function, where SB and LB are the S-box layer and L-box layer, respectively.

Authenticated Encryption SCREAM. The authenticated encryption SCREAM uses the tweakable block cipher Scream in the TAE mode [20]. SCREAM consists of three steps: associated data processing, encryption of the plaintext block, and tag generation. Since our attack exploits encryption of the plaintext block, we explain the specification (see Fig. 4). Plaintext values are encrypted by using Scream in order to produce the ciphertext values, and all blocks use $T_c = (N \| c \| 00000000)$. If the last block is a partial block, its bitlength is encrypted to generate a mask, which is then truncated to the partial block size

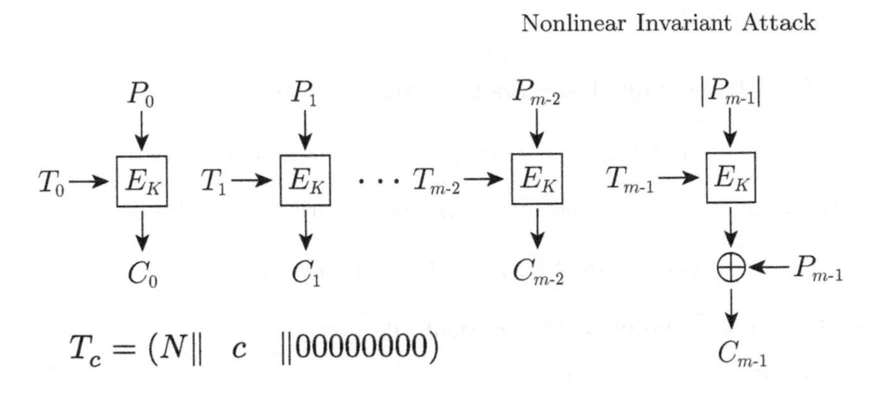

Fig. 4. Encryption of plaintext blocks

and XORed with the partial plaintext block. Therefore, the ciphertext length is the same as the plaintext length.

Security Parameter. Finally, we like to recall the security parameters of SCREAM, as described by the designers. Let n_b be the nonce bytesize, and it can be chosen by the user between 1 and 15 bytes. However, the designers recommend that $n_b = 11$, and we also use the recommended parameter in this paper.

SCREAM has three security parameters, i.e., lightweight security, single-key security, and related-key security. They are summarized as follows.

Lightweight security: 80-bit security, with a protocol avoiding related keys. Tight parameters: 6 steps, Safe parameters: 8 steps.

Single-key security: 128-bit security, with a protocol avoiding related keys. Tight parameters: 8 steps, Safe parameters: 10 steps.

Related-key security: 128-bit security, with possible related keys. Tight parameters: 10 steps, Safe parameters: 12 steps.
More precisely, designers order their recommended sets of parameters as follows:

- First set of recommendations: SCREAM with 10 steps, single-key security.
- Second set of recommendations: SCREAM with 12 steps, related-key security.

4.2 Nonlinear Invariant for Scream

The L-box of Scream is chosen as an orthogonal matrix. Therefore, there is a nonlinear invariant for the LS function from Theorem 1 if we can find quadratic Boolean function $g : \mathbb{F}_2^8 \to \mathbb{F}_2$ which is a nonlinear invariant for the S-box S.

Let $x \in \mathbb{F}_2^8$ and $y \in \mathbb{F}_2^8$ be the input and output of the S-box S, respectively. Moreover, $x[j] \in \mathbb{F}_2$ and $y[j] \in \mathbb{F}_2$ denote the jth bits of x and y, respectively. Then, the Scream S-box has the following property

$$(x[1] \wedge x[2]) \oplus x[0] \oplus x[2] \oplus x[5] = (y[1] \wedge y[2]) \oplus y[0] \oplus y[2] \oplus y[5] \oplus 1.$$

Let $g_S : \mathbb{F}_2^8 \to \mathbb{F}_2$ be a quadratic Boolean function, where

$$g_S(x) = (x[1] \wedge x[2]) \oplus x[0] \oplus x[2] \oplus x[5].$$

Then, the function g_S is a *quadratic* nonlinear invariant for S because

$$g_S(x) \oplus g_S(S(x)) = g_S(x) \oplus g_S(x) \oplus 1 = 1.$$

Therefore, due to Theorem 1, the Boolean function

$$g(x) = \bigoplus_{j=0}^{15} g_S(x[\star, j]) = \bigoplus_{j=0}^{15} \left(x[1,j] \wedge x[2,j] \oplus x[0,j] \oplus x[2,j] \oplus x[5,j] \right)$$

is a nonlinear invariant for the LS function. Note that this nonlinear invariant g is clearly balanced, as it is linear (and not constant) in parts of its input.

Next, we show that this Boolean function is also a nonlinear invariant for the constant addition and tweakey addition. The round constant $C(r)$ is XORed with only $x[0, \star]$. Since $C(r)$ linearly affects the output of the function g,

$$g(x \oplus C(r)) = g(x) \oplus g(C(r))$$

for any x. The tweakey $TK(\sigma)$ is defined as

$$\begin{aligned} TK(\sigma = 3i) &= K \oplus (t_0 \| t_1), \\ TK(\sigma = 3i+1) &= K \oplus (t_0 \oplus t_1 \| t_1), \\ TK(\sigma = 3i+2) &= K \oplus (t_1 \| t_0 \oplus t_1), \end{aligned}$$

where $T = t_0 \| t_1$. Therefore, if we restrict the key and tweak by fixing

$$\begin{aligned} &K[1, \star] = K[2, \star] = 0, \\ &T[1, \star] = T[2, \star] = T[5, \star] = T[6, \star] = 0, \end{aligned}$$

$TK(\sigma)[1, \star]$ and $TK(\sigma)[2, \star]$ are always zero vectors. Then, since the tweakey linearly affects the output of the function g,

$$g(x \oplus TK(\sigma)) = g(y) \oplus g(TK(\sigma)),$$

and all those keys are weak. Therefore, the density of weak keys is 2^{-32}, i.e., there are 2^{96} weak keys.

Let P and C be the plaintext and ciphertext of Scream, respectively. In N_s-step Scream, the relationship between p and c is represented as

$$\begin{aligned} g(P) &= g(C) \oplus \bigoplus_{r=1}^{2N_s} g(C(r)) \bigoplus_{\sigma=0}^{N_s} g(TK(\sigma)) \\ &= g(C) \oplus c \oplus g_T(N_s, T) \oplus g_K(N_s, K), \end{aligned}$$

where $c = \bigoplus_{r=1}^{2N_s} g(C(r))$, and $g_T(N_s, T)$ and $g_K(N_s, K)$ are defined as

$$g_T(N_s,T) = \begin{cases} g(t_0\|t_1) & N_s = 0 \bmod 6, \\ g(t_1\|0) & N_s = 1 \bmod 6, \\ g(0\|t_0 \oplus t_1) & N_s = 2 \bmod 6, \\ g(t_0\|t_0) & N_s = 3 \bmod 6, \\ g(t_1\|t_0 \oplus t_1) & N_s = 4 \bmod 6, \\ 0 & N_s = 5 \bmod 6, \end{cases}$$

and

$$g_K(N_s,K) = \begin{cases} g(K) & N_s = 0 \bmod 2, \\ 0 & N_s = 1 \bmod 2, \end{cases}$$

respectively. When the master key belongs to the class of weak-keys, $g(p) \oplus g(c) \oplus g_T(N_s, T)$ is constant for all plaintexts and a given key. When the key does not belong to the weak-key class, the probability that the output is constant is about 2^{-n+1} given n known plaintexts. Therefore, we can easily distinguish whether or not the using key belongs to the weak-key class. Note that all recommended numbers of rounds are even number. Therefore, from

$$g(K) = g(p) \oplus g(c) \oplus c \oplus g_T(N_s, T),$$

we can recover one bit of information about the secret key K.

4.3 Practical Attack on SCREAM

Known-Plaintext Attack. We exploit the encryption step of SCREAM (see Fig. 4). The nonlinear invariant attack is a chosen-tweak attack under the weak-key setting. First, let us consider the class of weak tweaks. In the encryption step, the tweak $T_c = (N\|c\|00000000)$ is used, where we assume that $n_b = 11$. Figure 5 shows the structure of T_c. From the condition of the nonlinear invariant attack, the following T_c

$$T_c[1,\star] = T_c[2,\star] = T_c[5,\star] = T_c[6,\star] = 0$$

are weak tweaks. Namely, we choose N whose 3rd, 4th, 5th, 6th, and 11th bytes are zero. Then, if the counter c is less than 256, i.e. from T_0 to T_{255}, the tweak fulfils the condition. Moreover, the actual nonce fulfils the needs of the tweak if the nonce is implemented as a counter increment, which seems to occur in practice. If the master key belongs to the weak-key class, we can recover one bit of information about the secret key by using only one known plaintext. Moreover, by using n known plaintexts, the probability that the output is constant is about 2^{-n+1} when the key does not belong to weak-key class. Therefore, an attacker can distinguish whether or not the used key belongs to the weak-key class.

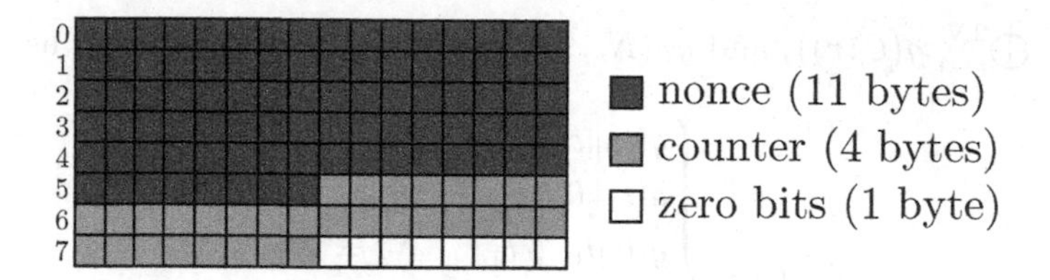

Fig. 5. Tweak mapping

Ciphertext-Only Message Recovery Attack. The interesting application of the nonlinear invariant attack is a ciphertext-only attack. This setting is more practical than the known-plaintext attack.

We focus on the procedure of the final block. The input of Scream is the bitlength of P_{m-1}, and the bitlength is encrypted to generate a mask. Then the mask is truncated to the partial block size and XORed with P_{m-1}. Therefore, the ciphertext length is the same as the plaintext length. In the ciphertext-only attack, we cannot know P_{m-1}. On the other hand, we know ciphertext C_{m-1} and the bitlength $|P_{m-1}|$ can be obtained from $|C_{m-1}|$. Therefore, we guess P_{m-1} and evaluate

$$g(|P_{m-1}|) \oplus g(P_{m-1} \oplus C_{m-1}) \oplus g_T(N_s, T),$$

and the above value is always constant for any weak tweaks T. Therefore, from two ciphertexts corresponding to the same final plaintext block encrypted by distinct tweaks, we create a linear equation as

$$g(P_{m-1} \oplus C_{m-1}) \oplus g(P_{m-1} \oplus C'_{m-1}) = g_T(N_s, T) \oplus g_T(N_s, T'). \quad (3)$$

We can compute the right side of Eq. (3). Moreover, we regard the function g as

$$g(X) = f(X) \oplus \ell(X),$$

where

$$f(X) = \bigoplus_{j=0}^{15} \Big(X[1,j] \wedge X[2,j] \Big),$$

$$\ell(X) = \bigoplus_{j=0}^{15} X[0,j] \oplus X[2,j] \oplus X[5,j].$$

Then,

$$g(P_{m-1} \oplus C_{m-1}) \oplus g(P_{m-1} \oplus C'_{m-1})$$
$$= f(P_{m-1} \oplus C_{m-1}) \oplus f(P_{m-1} \oplus C'_{m-1}) \oplus \ell(C_{m-1}) \oplus \ell(C'_{m-1})$$

$$= \bigoplus_{j=0}^{15} \left(C_{m-1}[1,j] P_{m-1}[2,j] \oplus P_{m-1}[1,j] C_{m-1}[2,j] \oplus C_{m-1}[1,j] C_{m-1}[2,j] \right)$$
$$\bigoplus_{j=0}^{15} \left(C'_{m-1}[1,j] P_{m-1}[2,j] \oplus P_{m-1}[1,j] C'_{m-1}[2,j] \oplus C'_{m-1}[1,j] C'_{m-1}[2,j] \right)$$
$$\oplus \ell(C_{m-1}) \oplus \ell(C'_{m-1}).$$

The equation above is actual a linear equation in 32 unknown bits, $P_{m-1}[1,j]$ and $P_{m-1}[2,j]$, as all other terms are known. Therefore, we can create t linear equations by collecting $t+1$ ciphertexts encrypted by distinct tweaks. We can recover 32 bits, $P_{m-1}[1,j]$ and $P_{m-1}[2,j]$ by solving this system as soon as the corresponding system has full rank. Assuming the system behaves like a randomly generated system of linear equations, we can expect that the system has full rank already when taking slightly more than 33 equations. The time complexity for solving this system is negligible.

Note that the system involves four 16-bit words, $C_{m-1}[0,j]$, $C_{m-1}[1,j]$, $C_{m-1}[2,j]$, and $C_{m-1}[5,j]$. Since the bitlength of C_{m-1} is equal to that of P_{m-1}, we cannot solve this system if $|P_{m-1}| < 96$. Therefore, the necessary condition of this attack is $96 \leq |P_{m-1}| < 128$.

Experimental Results. In order to verify our findings and in particular to verify that the system indeed behaves like a random system of linear equations, we implemented our ciphertext-only message recovery attack for SCREAM. In our experiment, the key is randomly chosen from the weak-key class. Moreover, we use N distinct nonces such that the corresponding tweak is weak, and collect N corresponding ciphertexts. We repeated our attack 1000 times. Table 2 summarizes the success probability of recovering the correct 32 bits. Moreover, in the table we compare the experimental success probability to the theoretically expected probability in the case of a randomly generated system of linear equations. As can be seen, the deviation of the experimental results to the theoretically expected results is very small.

Table 2. The success probability of recovering the correct 32 plaintext bits on SCREAM.

# nonces	33	34	35	36	37	38	39	40	41	42	43
Experimental	0.289	0.571	0.762	0.885	0.942	0.976	0.991	0.995	0.998	0.999	1
Theoretical	0.289	0.578	0.770	0.880	0.939	0.969	0.984	0.992	0.996	0.998	0.999

4.4 Application to iSCREAM

The authenticated encryption iSCREAM also has the similar structure of SCREAM. We search for the nonlinear invariant for the underlying tweakable block cipher iScream. As a result, the following quadratic Boolean function

$$g_S(x) = (x[4] \wedge x[5]) \oplus x[0] \oplus x[6].$$

is nonlinear invariant for the S-box[2], and it holds

$$g_S(x) \oplus g_S(S(x)) = g_S(x) \oplus g_S(x) = 0.$$

Therefore, from Theorem 1, the following Boolean function

$$g(x) = \bigoplus_{j=0}^{15} g_S(x[\star, j]) = \bigoplus_{j=0}^{15} \left(x[4,j] \wedge x[5,j] \oplus x[0,j] \oplus x[6,j] \right).$$

is nonlinear invariant for the LS function.

5 Practical Attack on Midori64

5.1 Specification of Midori64

Midori is a light-weight block cipher recently proposed by Banik et al. [3], which is particularly optimized for low-energy consumption. There are two versions depending on the block size; Midori64 for 64-bit block and Midori128 for 128-bit block. Both use 128-bit key. The nonlinear invariant attack can be applied to Midori64, thus we only explain the specification of Midori64 briefly.

Midori64 adopts an SPN structure with a non-MDS matrix and a very light key schedule. The state is represented by a 4×4-nibble array. At first the plaintext is loaded to the state, then the key whitening is performed. The state is updated with a round function 16 times, and a final key whitening is performed. The resulting state is the ciphertext. The overall structure is illustrated in Fig. 6. More details on each operation will be given in the following paragraphs.

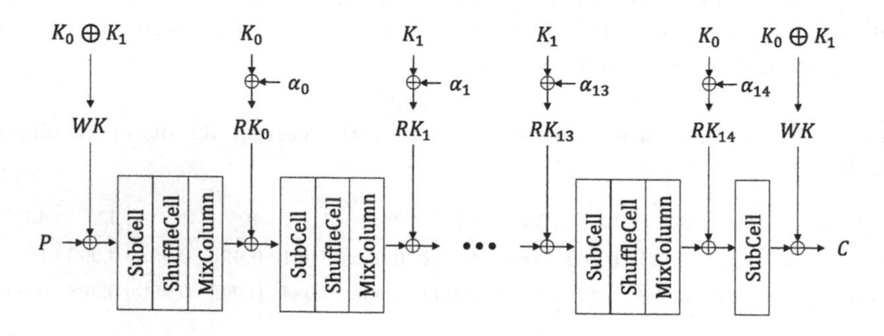

Fig. 6. Computation structure of Midori64

[2] In the round function of iScream with the constant addition, the equation, $g_S(x) = (x[5] \wedge x[6]) \oplus x[2] \oplus x[5] \oplus x[6] \oplus x[7]$, is another nonlinear invariant.

Key Schedule Function. A user-provided 128-bit key is divided into two 64-bit key states K_0 and K_1. Then, a whitening key WK and 15 round keys $RK_i, i = 0.1, \ldots, 14$ are generated as follows.

$$WK \leftarrow K_0 \oplus K_1, \qquad RK_i \leftarrow K_{i \bmod 2} \oplus \alpha_i,$$

where the α_i are fixed 64-bit constants. The round constant α_i are binary for each nibble, i.e. any nibble in α_i is either 0000 or 0001. Using such constants is beneficial to keep the energy consumption low. The exact values of the α_i are given in Table 3 for the first 6 rounds. We refer to [3] for the complete specification.

Table 3. Examples of round constant α_i

α_0	α_1	α_2	α_3	α_4	α_5
0010	0110	1000	0000	0001	1000
0100	1010	0101	1000	0011	1010
0011	1000	1010	1101	0001	0010
1111	1000	0011	0011	1001	1110

Round Function. The round function consists of four operations: SubCell, ShuffleCell, MixColumn, and KeyAdd. Each operation is explained in the following.

SubCell: The 4-bit S-box S defined below is applied to each nibble in the state.

x	0	1	2	3	4	5	6	7	8	9	A	B	C	D	E	F
$S(x)$	C	A	D	3	E	B	F	7	8	9	1	5	0	2	4	6

ShuffleCell: Each cell of the state is permuted as ShiftRows in AES. Let s_0, s_1, s_2, s_3 be four nibbles in the first row. Let $s_4, \ldots, s_{15}$ be the other 12 nibbles similarly defined. Then, the cell permutation is specified as follows.

$$(s_0, s_1, \ldots, s_{15}) \leftarrow (s_0, s_{10}, s_5, s_{15}, s_{14}, s_4, s_{11}, s_1, s_9, s_3, s_{12}, s_6, s_7, s_{13}, s_2, s_8)$$

Note that our nonlinear invariant attack would actually work in exactly the same way for any other cell permutation as well.

MixColumn: The following 4×4 *orthogonal* binary matrix M is applied to every column of the state.

$$M = \begin{pmatrix} 0\,1\,1\,1 \\ 1\,0\,1\,1 \\ 1\,1\,0\,1 \\ 1\,1\,1\,0 \end{pmatrix}$$

KeyAdd: The round key RK_i is xored to the state in round i.

In the last round, only SubCell (followed by the post-whitening) is performed.

5.2 Nonlinear Invariant for Midori64

The matrix used in MixColumn is a binary and orthogonal matrix. Thus, Theorem 1 implies that any quadratic Boolean function $g : \mathbb{F}_2^4 \to \mathbb{F}_2$, which is a nonlinear invariant for the S-box S, allows us to find nonlinear invariant for the entire round function. Similarly to the previous section, we use the notation $x[j] \in \mathbb{F}_2$ and $y[j] \in \mathbb{F}_2$ to denote the jth bits of 4-bit S-box input x and 4-bit S-box output y, respectively.

We search for g such that $g(x) = g(S(x))$. Different from Scream, the S-box of Midori64 is small, and many of such g usually exist. Actually, we found 15 choices of such g.

We then pick up ones that are also nonlinear invariant for the key addition RK_i, which is computed by $RK_i \leftarrow K_{i \bmod 2} \oplus \alpha_i$. Here, α_i takes 0 or 1 in each nibble, i.e. the 2nd, 3rd, and 4th bits are always 0. Thus we need to avoid g in which the first bit is included in the nonlinear component, i.e. g cannot involve $x[0]$ and $y[0]$ in their nonlinear component.

Among 15 choices, only one can satisfy this condition. The picked S-box property of Midori64 is as follows.

$$(x[3] \wedge x[2]) \oplus x[2] \oplus x[1] \oplus x[0] = (y[3] \wedge y[2]) \oplus y[2] \oplus y[1] \oplus y[0].$$

Then, the following $g_S : \mathbb{F}_2^4 \to \mathbb{F}$ is nonlinear invariant for S;

$$g_S(x) = (x[3] \wedge x[2]) \oplus x[2] \oplus x[1] \oplus x[0].$$

Here, ShuffleCell does not affect the nonlinear invariant. Therefore, from Theorem 1, the following Boolean function

$$g(x) = \bigoplus_{j=0}^{15} g_S(s_i)$$

is a nonlinear invariant for the round function of Midori64. Note, as for SCREAM the Boolean function g is actually balanced.

5.3 Distinguishing Attack

As mentioned in Sect. 2, the simple distinguishing attack can be mounted against a weak key. Let $\ell : \mathbb{F}_2^4 \to \mathbb{F}$ be a linear part of g, namely $\ell(x) = x[2] \oplus x[1] \oplus x[0]$. We have $g(p) \oplus g(c) = const$ and $const$ is a linear part of the values injected to round function during the encryption process;

$$\begin{aligned} const &= \ell(WK) \oplus \ell(RK_0) \oplus \ell(RK_1) \oplus \cdots \oplus \ell(RK_{14}) \oplus \ell(WK), \\ &= \ell(RK_0) \oplus \ell(RK_1) \oplus \cdots \oplus \ell(RK_{14}). \end{aligned}$$

Given $RK_i = K_{i \bmod 2} \oplus \alpha_i$, the above equation is further converted as

$$const = \ell(K_1) \oplus \ell(\alpha_0) \oplus \ell(\alpha_1) \oplus \cdots \oplus \ell(\alpha_{14}).$$

As $\alpha_i[2] = \alpha_i[1] = 0$ for any i, it can be simply written as

$$const = \ell(K_1) \oplus \bigoplus_{i=0}^{14} \bigoplus_{j=0}^{15} \alpha_{i,j},$$

where $\alpha_{i,j}$ is the jth nibble of α_i. We confirmed that the total number of 1 in all α_i is even, thus $\bigoplus_{i=0}^{14} \bigoplus_{j=0}^{15} \alpha_{i,j} = 0$. In the end, $g(p) \oplus g(c) = \ell(K_1)$ always holds for Midori64, while this holds with probability 1/2 for a random permutation.

5.4 Experimental Results

As mentioned in Sect. 2, the above property can reveal 32 bits (the two most significant bits from each nibble) of an unknown plaintext block in the weak-key setting when Midori64 is used in well-known block cipher modes.

We implemented our ciphertext-only message recovery attack for Midori64 in the CBC mode. In our experiment, the key and IV are chosen uniformly at random from the weak-key space and the entire IV space. We also choose a 64-bit plaintext block p, uniformly at random, and assume that p is iterated over b blocks, where $33 \leq b \leq 43$. We executed our attack of repeating 1000 times, and Table 4 summarizes the success probability of recovering the correct 32 bits.

Table 4. The success probability of recovering the correct 32 bits on Midori64-CBC.

# blocks	33	34	35	36	37	38	39	40	41	42	43
Experimental	0.279	0.574	0.753	0.883	0.931	0.968	0.988	0.991	0.999	0.997	1
Theoretical	0.289	0.578	0.770	0.880	0.939	0.969	0.984	0.992	0.996	0.998	0.999

Similarly to the case of SCREAM the system of equations behaves very much like a random system of equation in the sense that the probability that it has full rank is very close to the corresponding probability for a random system with the same dimensions.

6 Extensions and Future Work

In this section we outline some extensions to the previously described attacks. Furthermore, we give some additional insights in the structure of nonlinear invariants in general. Finally, we explain how invariant subspace attacks can be seen as a special, *chosen plaintext*, variant of nonlinear invariant attacks. It is important to point out that none of the observations in this section lead to any attacks. But we feel that those explanations provide good starting points for future investigations.

More General Nonlinear Invariant. We continue to use the notation that we fixed in Sect. 3. First recall Proposition 1, that allowed to construct nonlinear invariants for the whole S-box layer by linearly combining nonlinear invariants for each single S-box. This proposition can actually be easily extended. Instead of only linearly combining the nonlinear invariants for each S-box, any combination by using an arbitrary Boolean function results in an invariant for the whole S-box layer as well. The following proposition summarizes this observation.

Proposition 2. *Given any Boolean function $f : \mathbb{F}_2^t \to \mathbb{F}_2$ and t elements*

$$g_1, \ldots, g_t : \mathbb{F}_2^n \to \mathbb{F}_2$$

from $U(S)$ it holds that

$$\begin{aligned} g_{\mathcal{S}} &: (\mathbb{F}_2^n)^t \to \mathbb{F}_2 \\ g_{\mathcal{S}}(x_1, \ldots, x_n) &= f(g_1(x_1), \ldots, g_t(x_t)) \end{aligned}$$

is an element of $U(\mathcal{S})$

Note that the special case of f being linear actually corresponds to the choice made in Proposition 1.

While this generalization potentially allows a much larger variety of invariants, and therefore potential attacks, we like to mention that the restriction made in Proposition 1 has two crucial advantages. First, the choice is small enough, so that it can be handled exhaustively and second, the invariants generated by Proposition 1 are usually balanced, while this is not necessarily the case for the generalization.

At first sight, one might be tempted to assume that the above construction actually covers all invariants for the S-box layer. However, in general, this is not the case.

One counter-example, that is a nonlinear invariant not covered by this construction, can easily be identified as follows: For simplicity, consider an S-box layer consisting of two identical n bit S-boxes only. If the two inputs to those two S-boxes are equal, so are the outputs. Thus, the function

$$\begin{aligned} g &: \mathbb{F}_2^n \times \mathbb{F}_2^n \to \mathbb{F}_2 \\ g(x, y) &= \begin{cases} 1 & \text{if} \quad x = y \\ 0 & \text{else} \end{cases} \end{aligned}$$

is an nonlinear invariant of the S-box layer as

$$g(x,y) = 1 \Leftrightarrow x = y \Leftrightarrow S(x) = S(y) \Leftrightarrow g(S(x), S(y)) = 1.$$

Moreover, this nonlinear invariant can certainly not be generated by Proposition 2.

Cycle Structure. Actually, there is a nice, and potentially applicable way, of describing all nonlinear invariants for a given permutation F by considering its cycles. Recall that a cycle of F being a set

$$C_x := \{F^i(x) \mid i \in \mathbb{N}\}$$

for a value $x \in \mathbb{F}_2^n$. Actually, one can show that a mapping g is contained in $U(F)$ if and only if g is either constant on all cycles of F or alternating along the cycles of F. The later case corresponds to nonlinear invariants such that

$$g(x) + g(F(x)) = 1.$$

This is because $g(x) = g(F(x))$ implies

$$g(x) = g(F(x)) = g(F(F(x))) = \cdots = g(F^i(x)).$$

Thus, looking at the cycle structure of F, we can assign to each cycle one value the function g should evaluate to on this cycle. That view point also shows that the number of invariant functions g is equal to

$$|U(F)| = 2^{(\#\text{ cycles of } F)},$$

in the case where there exist at least one cycle of odd length or

$$|U(F)| = 2^{(\#\text{ cycles of } F)+1},$$

in the case where all cycles of F have even length. This perspective allows to actually compute a basis of $U(F)$ very efficiently. Consider, for simplicity, the case were not all cycles are of even length. Then, a basis of $U(F)$ clearly consists of the set of all indicator functions of C_x, i.e.

$$U(F) = \text{span}\{\delta_{C_a} \mid a \in \mathbb{F}_2^n\}.$$

Here, for a subset $A \subseteq \mathbb{F}_2^n$, the function δ_A denotes the indicator function of the set A, i.e.

$$\delta_A(x) = \begin{cases} 1 & \text{if} \quad x \in A \\ 0 & \text{else} \end{cases}$$

Example 2. Consider the function $F : \mathbb{F}_2^2 \to \mathbb{F}_2^2$ with

x	0	1	2	3
$F(x)$	1	2	0	3

The cycle composition of F is

$$(0,1,2)(3).$$

Thus we have two cycles of odd length. Following the above, any nonlinear invariant of F is constant on those cycles. In this case we have the following invariants

$$g_1(x) = \delta_{\{0,1,2\}}(x)$$
$$g_1(x) = \delta_{\{3\}}(x)$$

or, more explicitly

x	0	1	2	3
$g_1(x)$	1	1	1	0

and

x	0	1	2	3
$g_2(x)$	0	0	0	1

together with the trivial invariants, that is the identical zero or identical one functions. So in total F has 4 invariants. □

Relation to Invariant Subspace Attack. Along the same lines, one can also see the invariant subspace attack as a special case of a nonlinear invariant. Recall that a subspace $V \subseteq \mathbb{F}_2^n$ is called invariant under (a block cipher) F if

$$F(V) = V.$$

That is, the set V is mapped to itself by the function F. Note that the complement $\bar{V}$ is also mapped to itself because the function F is permutation. This means nothing else than that the nonlinear Boolean function $\delta_V(x)$ is a nonlinear invariant for F as

$$\delta_V(x) = 1 \Leftrightarrow x \in V \Leftrightarrow F(x) \in V \Leftrightarrow \delta_V(F(x)) = 1,$$
$$\delta_V(x) = 0 \Leftrightarrow x \in \bar{V} \Leftrightarrow F(x) \in \bar{V} \Leftrightarrow \delta_V(F(x)) = 0.$$

In other words, invariant subspace attacks are nonlinear invariant attacks where the support of the nonlinear invariant is a subspace of $\mathbb{F}_2^n$. And as such, nonlinear invariant attacks could be called *invariant set attacks*, as the function g splits in the inputs into two sets, the support of g and its complement, that are invariant under F.

Further Research. Other interesting directions for further research include the generalization of the nonlinear invariant to the case where one does not consider the same function g in every round, but rather a sequence of functions that can be chained together. In fact, we also found quadratic Boolean function $g' : \mathbb{F}_2^4 \rightarrow \mathbb{F}_2$ such that $g(x) = g'(S(x))$ for Midori64. Owing to the involution property of the S-box, $g(x) = g'(S(x))$ always implies $g'(x) = g(S(x))$. Combining with the alternative use of K_0 and K_1 in the key schedule, such g, g' may be exploited in the attack. Unfortunately, since such Boolean functions are not nonlinear invariant for the constant addition in Midori64, we cannot exploit them in real cryptanalysis. However, it is clearly worth discussing this extension. And last but not least, even so it seems notoriously difficult, it would be nice to be able to use a statistical variant of the attack described here, i.e. consider nonlinear functions such that $g(F(x)) = g(x)$ for many – but not necessarily for all – inputs x.

A Algorithm to Solve Basis of $U(S)$

Let $g_S \in U(S)$, and the algebraic normal form (ANF) is expressed as

$$g_S(x) = \bigoplus_{u \in \mathbb{F}_2^n} \lambda_u x^u,$$

where $\lambda_u \in \mathbb{F}_2$ are the coefficients to be determined and x^u denotes $\prod x_i^{u_i}$. From the definition of the nonlinear invariant, for any $x \in \mathbb{F}_2^n$, the following equation

$$\bigoplus_{u \in F_2^n} g_{S,u}(x) = \bigoplus_{u \in F_2^n} \lambda_u (x^u \bigoplus S(x)^u)$$

is constant. The ANF of $g_{S,u}$ is computed for all $u \in \mathbb{F}_2^n$, and the ANF is expressed as

$$g_{S,u}(x) = \bigoplus_{v \in \mathbb{F}_2^n} \lambda_{u,v} x^v.$$

Then, we prepare a matrix $[I \| M]$, where I is a $(2^n \times 2^n)$ identical matrix and coefficients of M is computed as

$$M[u, v] = \lambda_{u,v}$$

Then, by Gaussian elimination like computation, we compute matrix $M' = [M_1' \| M_2']$. If rows of M_2' are $[0, 0, \ldots, 0]$ or $[1, 0, 0, \ldots, 0]$, the corresponding row of M_1 is the basis of $U(S)$. In particular, for commonly used Sbox sizes of up to 8 bits, the space $U(S)$ can be computed in less than a second on a standard PC.

From our experiments, 4-bit S-boxes usually have quadratic nonlinear invariant. On the other hand, it is generally rare that 8-bit S-boxes have quadratic nonlinear invariant. However, as described in this paper, it is not always rare if low-degree S-boxes are applied like Scream or iScream for the efficiency.

References

1. DATA ENCRYPTION STANDARD (DES). National Bureau of Standards, federal Information Processing Standards Publication 46(1977)
2. Specification for the ADVANCED ENCRYPTION STANDARD (AES). U.S. DEPARTMENT OF COMMERCE/National Institute of Standards and Technology, federal Information Processing Standards Publication 197(2001)
3. Banik, S., Bogdanov, A., Isobe, T., Shibutani, K., Hiwatari, H., Akishita, T., Regazzoni, F.: Midori: a block cipher for low energy. In: Iwata, T., Cheon, J.H. (eds.) ASIACRYPT 2015. LNCS, vol. 9453, pp. 411–436. Springer, Heidelberg (2015). doi:10.1007/978-3-662-48800-3_17
4. Biham, E., Biryukov, A., Shamir, A.: Cryptanalysis of skipjack reduced to 31 rounds using impossible differentials. In: Stern, J. (ed.) EUROCRYPT 1999. LNCS, vol. 1592, pp. 12–23. Springer, Heidelberg (1999). doi:10.1007/3-540-48910-X_2
5. Biham, E., Shamir, A.: Differential cryptanalysis of DES-like cryptosystems. In: Menezes, A.J., Vanstone, S.A. (eds.) CRYPTO 1990. LNCS, vol. 537, pp. 2–21. Springer, Heidelberg (1991). doi:10.1007/3-540-38424-3_1
6. Biryukov, A., Wagner, D.: Slide attacks. In: Knudsen, L. (ed.) FSE 1999. LNCS, vol. 1636, pp. 245–259. Springer, Heidelberg (1999). doi:10.1007/3-540-48519-8_18
7. Bogdanov, A., Rijmen, V.: Linear hulls with correlation zero and linear cryptanalysis of block ciphers. Des. Codes Crypt. **70**(3), 369–383 (2014)
8. Bouillaguet, C., Dunkelman, O., Leurent, G., Fouque, P.-A.: Another look at complementation properties. In: Hong, S., Iwata, T. (eds.) FSE 2010. LNCS, vol. 6147, pp. 347–364. Springer, Heidelberg (2010). doi:10.1007/978-3-642-13858-4_20
9. Grosso, V., Leurent, G., Standaert, F.-X., Varıcı, K.: LS-Designs: bitslice encryption for efficient masked software implementations. In: Cid, C., Rechberger, C. (eds.) FSE 2014. LNCS, vol. 8540, pp. 18–37. Springer, Heidelberg (2015). doi:10.1007/978-3-662-46706-0_2
10. Grosso, V., Leurent, G., Standaert, F., Varici, K., Journault, A., Durvaux, F., Gaspar, L., Kerckhof, S.: SCREAM v1 (2014 b). submission to CAESAR competition
11. Grosso, V., Leurent, G., Standaert, F., Varici, K., Journault, A., Durvaux, F., Gaspar, L., Kerckhof, S.: SCREAM v3. submission to CAESAR competition (2015)
12. Guo, J., Jean, J., Nikolić, I., Qiao, K., Sasaki, Y., Sim, S.M.: Invariant subspace attack against full midori64. Cryptology ePrint Archive, Report 2015/1189 (2015)
13. Harpes, C., Kramer, G.G., Massey, J.L.: A generalization of linear cryptanalysis and the applicability of matsui's piling-up lemma. In: Guillou, L.C., Quisquater, J.-J. (eds.) EUROCRYPT 1995. LNCS, vol. 921, pp. 24–38. Springer, Heidelberg (1995). doi:10.1007/3-540-49264-X_3
14. Hermelin, M., Cho, J.Y., Nyberg, K.: Multidimensional linear cryptanalysis of reduced round serpent. In: Mu, Y., Susilo, W., Seberry, J. (eds.) ACISP 2008. LNCS, vol. 5107, pp. 203–215. Springer, Heidelberg (2008). doi:10.1007/978-3-540-70500-0_15
15. Knudsen, L.R.: Truncated and higher order differentials. In: Preneel, B. (ed.) FSE 1994. LNCS, vol. 1008, pp. 196–211. Springer, Heidelberg (1995). doi:10.1007/3-540-60590-8_16
16. Knudsen, L.R., Robshaw, M.J.B.: Non-linear approximations in linear cryptanalysis. In: Maurer, U. (ed.) EUROCRYPT 1996. LNCS, vol. 1070, pp. 224–236. Springer, Heidelberg (1996). doi:10.1007/3-540-68339-9_20

17. Le, T., Sparr, R., Wernsdorf, R., Desmedt, Y.: Complementation-like and cyclic properties of AES round functions. In: Dobbertin, H., Rijmen, V., Sowa, A. (eds.) AES 2004. LNCS, vol. 3373, pp. 128–141. Springer, Heidelberg (2005). doi:10.1007/11506447_11
18. Leander, G., Abdelraheem, M.A., AlKhzaimi, H., Zenner, E.: A cryptanalysis of PRINTcipher: the invariant subspace attack. In: Rogaway, P. (ed.) CRYPTO 2011. LNCS, vol. 6841, pp. 206–221. Springer, Heidelberg (2011). doi:10.1007/978-3-642-22792-9_12
19. Leander, G., Minaud, B., Rønjom, S.: A generic approach to invariant subspace attacks: cryptanalysis of robin, iSCREAM and zorro. In: Oswald, E., Fischlin, M. (eds.) EUROCRYPT 2015. LNCS, vol. 9056, pp. 254–283. Springer, Heidelberg (2015). doi:10.1007/978-3-662-46800-5_11
20. Liskov, M., Rivest, R.L., Wagner, D.: Tweakable block ciphers. J. Cryptology **24**(3), 588–613 (2011)
21. Matsui, M.: Linear cryptanalysis method for DES cipher. In: Helleseth, T. (ed.) EUROCRYPT 1993. LNCS, vol. 765, pp. 386–397. Springer, Heidelberg (1994). doi:10.1007/3-540-48285-7_33
22. Moriai, S., Shimoyama, T., Kaneko, T.: Higher order differential attack of a CAST cipher. In: Vaudenay, S. (ed.) FSE 1998. LNCS, vol. 1372, pp. 17–31. Springer, Heidelberg (1998). doi:10.1007/3-540-69710-1_2
23. Özen, M., Çoban, M., Karakoç, F.: A guess-and-determine attack on reduced-round Khudra and weak keys of full cipher. IACR Cryptology ePrint Archive 2015, 1163 (2015). http://eprint.iacr.org/2015/1163

Cliptography: Clipping the Power of Kleptographic Attacks

Alexander Russell[1], Qiang Tang[2(✉)], Moti Yung[3], and Hong-Sheng Zhou[4]

[1] University of Connecticut, Storrs, USA
acr@cse.uconn.edu
[2] New Jersey Institute of Technology, Newark, USA
qiang@njit.edu
[3] Snapchat Inc., Columbia University, New York City, USA
moti@cs.columbia.edu
[4] Virginia Commonwealth University, Richmond, USA
hszhou@vcu.edu

Abstract. Kleptography, introduced 20 years ago by Young and Yung [Crypto '96], considers the (in)security of malicious implementations (or instantiations) of standard cryptographic primitives that may embed a "backdoor" into the system. Remarkably, crippling subliminal attacks are possible even if the subverted cryptosystem produces output indistinguishable from a truly secure "reference implementation." Bellare, Paterson, and Rogaway [Crypto '14] recently initiated a formal study of such attacks on symmetric key encryption algorithms, demonstrating that kleptographic attacks can be mounted in broad generality against randomized components of cryptographic systems.

We enlarge the scope of current work on the problem by permitting adversarial subversion of (randomized) key generation; in particular, we initiate the study of cryptography in the *complete subversion model*, where *all* relevant cryptographic primitives are subject to kleptographic attacks. We construct secure one-way permutations and trapdoor one-way permutations in this "complete subversion" model, describing a general, rigorous immunization strategy to clip the power of kleptographic subversions. Our strategy can be viewed as a formal treatment of the folklore "nothing up my sleeve" wisdom in cryptographic practice. We also describe a related "split program" model that can directly inform practical deployment. We additionally apply our general immunization strategy to directly yield a backdoor-free PRG. This notably amplifies previous results of Dodis, Ganesh, Golovnev, Juels, and Ristenpart [Eurocrypt '15], which require an honestly generated random key.

We then examine two standard applications of (trapdoor) one-way permutations in this complete subversion model and construct "higher level" primitives via black-box reductions. We showcase a digital signature scheme that preserves existential unforgeability when *all* algorithms (including key generation, which was not considered to be under attack before) are subject to kleptographic attacks. Additionally, we demonstrate that the classic Blum–Micali pseudorandom generator (PRG), using an "immunized" one-way permutation, yields a backdoor-free PRG.

J.H. Cheon and T. Takagi (Eds.): ASIACRYPT 2016, Part II, LNCS 10032, pp. 34–64, 2016.
DOI: 10.1007/978-3-662-53890-6_2

Alongside development of these secure primitives, we set down a hierarchy of kleptographic attack models which we use to organize past results and our new contributions; this taxonomy may be valuable for future work.

1 Introduction

Consider conventional use of a cryptographic primitive in practice, such as an encryption scheme: To encrypt a desired plaintext, one simply runs an implementation (or an instantiation with particular parameters) of the encryption algorithm obtained from a hardware or software provider. Although the underlying algorithms may be well-studied and proven secure, malicious implementations or instantiations may cleverly "backdoor" the system or directly embed sensitive information—such as the secret key—into the ciphertext in a fashion that permits recovery by the provider/manufacturer but is undetectable to other parties. Notably, *such leakage is possible even if the implementation produces "functionally and statistically clean" output that is indistinguishable from that of a faithful implementation.* While the underlying concept of kleptography was proposed by Young and Yung two decades ago [27,28], striking recent examples—including those of the Snowden revelations [20]—have reawakened the security community to the seriousness of these issues [21]. As a result, the topic has received renewed attention; see, e.g., [1,2,4,10,18]. In particular, Bellare, Paterson, and Rogaway [4] studied algorithm substitution attacks—with a focus on symmetric key encryption—and demonstrated a devastating framework for such attacks that apply in broad generality to randomized algorithms. These results were later amplified [3] to show that such attacks can be carried out even if the adversarial implementation is *stateless.* Soon after, Dodis, Ganesh, Golovnev, Juels, and Ristenpart [10] formalized the subversion of Dual_EC pseudorandom generators (PRG) and studied backdoored PRGs in generality; they additionally studied methods for "immunizing" PRG in such hostile settings.

Our contributions. We continue this line of inquiry. Specifically, we are motivated to develop cryptographic schemes in a *complete subversion model,* in which *all* algorithms of a scheme are potentially subverted by the adversary. This model provides a conceptually simple abstraction of the adversary's power, and significantly amplifies previously studied settings, which rely on trusted key generation or clean randomness that is assumed private from the adversary.

In particular, motivated by the question of defending against the kleptographic attacks on **key generation** as demonstrated in the original paper of [27,28], we study two fundamental cryptographic primitives in the complete subversion model—one-way permutations (OWP) and trapdoor one-way permutations (TOWP)—and apply these primitives to construct other cryptographic tools such as digital signatures and PRGs. Along the way, we identify novel generic defending strategies and a hierarchy of attack models. We hope to stimulate a systematic study of "**cliptography**," the challenge of developing a broad class of familiar cryptograhpic tools that remain secure in such kleptographic settings. As mentioned above, prior to our work kleptographic attacks on various primitives have been addressed in weaker models; see the discussion of related work in Sect. 1. In detail, we show the following:

- We set down a hierarchy of security models that capture practical kleptographic settings. The models are characterized by three parties: an adversary, who may provide potentially subverted implementations of *all* cryptographic algorithms; a "watchdog," who either certifies or rejects the implementations by subjecting them to (black-box) interrogation;[1] and a challenger, who plays a conventional security game (but now using the potentially subverted algorithms) with the adversary. Armed with the "specification" of the cryptographic algorithms and oracle access to the implementations provided by the adversary, the watchdog attempts to detect any subversion in the implementations. Various models arise by adjusting the supervisory power of the watchdog; see Sect. 2.
- We study (trapdoor) one-way permutations in the presence of kleptographic attacks, introducing notions of subversion-resistance that can survive various natural kleptographic attacks. We first give a simple example of a OWP that can be proven secure in the conventional sense, but can be completely broken under the kleptograhic attack. This demonstrates the need for judicious design of cryptographic primitives to defend against kleptographic attacks.
  We then construct subversion-resistant (trapdoor) one way permutations via a general transformation that "sanitizes" arbitrary OWPs by *randomizing the function index*. This transformation clips potential correlation between the function and the possible backdoor that the adversary may possess.
  Additionally, we introduce a *split-program* model to make the general method above applicable using standard hash functions (see Sect. 3.3).
- In Sect. 4, we observe that subversion-resistant trapdoor OWPs give us a way to construct key generation algorithms (for digital signature schemes) against kleptographic attacks. We then showcase a concrete example of a digital signature scheme in the complete subversion model. More concretely, we achieve this result by (1) using the subversion-resistant trapdoor OWP directly as a key generation algorithm, and then (2) instantiating the unique signature generation mechanism via full domain hash (FDH). We stress that the reduction of the standard FDH signature scheme does not go through in the kleptographic setting. To resolve this issue, we slightly modify the FDH approach by hashing the message *together with the public key.* We remark that the original kleptographic attacks [27,28] were indeed applied to the key generation algorithm, while recent efforts [1,4] shift focus to other algorithmic aspects of encryption or digital signature schemes, assuming that key generation is honest. Our result is the first digital signature scheme allowing the adversary to sabotage all algorithms, including key generation.

[1] Without the watchdog, it is elusive to achieve interesting cryptographic functionalities in those stringent settings.

– We then turn our attention to PRGs. Previous work of Dodis et al. [10] investigated a notion of "backdoored PRG" in which the adversary sets up a PRG instance (i.e., the public parameter), and is able to distinguish the output from uniform with a backdoor. They then proposed powerful immunizing strategies which apply a keyed hash function to the output—*assuming the key is unknown to the adversary*—in the public parameter generation phase. Motivated by their success, we focus on constructing backdoor-free PRGs in the complete subversion model (where such clean randomness is not permitted). Our first construction is based on the classic Blum-Micali construction, using our subversion-resistant OWP and the Goldreich-Levin hardcore predicate. Dodis et al. [10] additionally show that it is impossible to achieve a public immunizing strategy for all PRGs by applying a public function to the PRG output. We sidestep this impossibility result via an alternative public immunizing strategy: Rather than randomizing the output of the PRG, we randomize the public parameter of PRG, which yields a general construction for PRG in the complete subversion model. See Sect. 5.

Finally, we remark that black-box constructions and reductions do not, in general, survive in the kleptographic model. However, two of the results above—the Blum-Micali construction and the signature scheme—give explicit examples of reductions that can be salvaged.

Remarks: Our techniques and the "nothing up my sleeve" principle; single use of randomized algorithms and subliminal channels. We remark that our general defending technique significantly differs from known methods: We use a—potentially subverted—hash function to "randomize" the index and public parameter of a (perhaps randomized) algorithm so that any correlation with some potential backdoor can be eliminated. This can be seen as an instance of the folklore wisdom of a "nothing up my sleeve number" [26] which has been widely used in practical cryptographic designs. The basic principle calls for constants appearing in the development of cryptographic algorithms to be drawn from a "rigid" source, like the digits of π; the idea is that this prevents them from possessing hidden properties that might give advantage to an attacker (or the designer). In our setting, the fact that a given value v is *supplied along with a preimage* x *so that* $h(x) = v$ (for a hash function h) is a evidence that v has "nothing up its sleeve." In fact, the situation is complicated: While this does effectively mean that v is generated by selecting x and computing $h(x)$ and, thus, severely restricts the possibility for tampering with v, it does not eliminate subliminal channels introduced by rejection sampling or entirely "clean" v. In particular, detailed analysis is still required to control the behavior of v.

Previous results either use a trusted random source to re-randomize the output of a randomized algorithm, or consider only deterministic algorithms. Permitting randomized algorithms in a kleptographic framework immediately invites the (devastating) general "steganochannel" attack of Bellare et al. [3,4]. Apparently, the prospect of full "immunization" for general randomized algorithms (in particular, generic destruction of a steganochannel) is a presumably challenging direction of future work. We note that our primitives here do permit randomized algorithms, although the security games we analyze invoke them only once (to, e.g., derive a key). Very interestingly, recent subsequent work of Russell et al. [23] addresses this major problem for a class of randomized algorithms; as a consequence, they can achieve the first IND-CPA secure public key encryption in the kleptographic setting.

For simplicity, we focus on (potentially subverted) algorithms that do not maintain "internal state" between invocations. We remark that typical steganographic attacks can indeed be in carried out in a stateless model [3]. Moreover, this restriction can be lifted for the constructions in the paper; see Remark 1.

Related work. The concept of *kleptography*—subverting cryptographic algorithms by modifying their implementations to leak secrets covertly—was proposed by Young and Yung [27,28] in 1996. They gave concrete examples showing that backdoors can be embedded into the public keys of commonly used cryptographic schemes; while the resulting public keys appear normal to every user, the adversary is nevertheless capable of learning the secret keys. Young and Yung have shown kleptographic backdoors in digital signature algorithms, key exchanges, SSL, symmetric crypto (e.g., block ciphers), composite key generation (e.g., RSA), and public key cryptosystems [27–32]. It may not be surprising that defending against such deliberate attacks is challenging and only limited feasibility results exist. We next briefly describe these existing results.

In [16], Juels and Guajardo suggested the following idea: the user and a trusted certificate authority (CA) jointly generate the public key; as a part of this process, the user proves to the CA that the public key is generated honestly. This contrasts markedly with our setting, where the the user does not have any secret, and every component is provided by the adversary.

Bellare et al. considered a powerful family of kleptographic attacks that they call *algorithm substitution attacks*, and explore these in both symmetric-key [3,4] and public-key [2] settings. They first proposed a generic attack, highlighting the relevance of steganographic techniques in this framework: specifically, a sabotaged randomized algorithm can leak a secret bit-by-bit by invoking steganographic rejection-sampling; then an adversary possessing the backdoor can identify the leaked bits from the biased output, which appears unmolested to other observers. The attack and analysis relies on the effectiveness of subliminal channels [15,24,25]. They then introduced a framework for defending against such attacks by focusing on algorithms that having a unique output for each input: relevant examples of such algorithms include unique ciphertext encryption algorithms. These results were later refined by [9]. Their defending mechanism does not, however, address the (necessarily randomized) process of key generation—it implicitly assumes key generation to be honest. This state of affairs is the direct motivation of the current article: we adopt a significantly amplified *complete subversion model* where *all* cryptographic algorithms—including key generation—are subject to kleptographic (i.e., substitution) attacks. The details of the model, with associated commentary about its relevance to practice, appear below.

Dodis et al. [10] pioneered the rigorous study of pseudorandom generators in such settings, developing an alternative family of kleptographic attacks on pseudorandom generators in order to formalize the notorious Dual_EC PRG subversion [7,19]. In their model, the adversary subverts the security of the PRG by opportunistically setting the public parameter while privately keeping some backdoor information (instead of providing an implementation). They then demonstrate an impossibility result: backdoored PRGs cannot be immunized by applying a public function—even a trusted random oracle—to the output. They also proposed and analyzed immunizing strategies obtained by applying a keyed hash function to the output (of the PRG). Note that the (hash) key plays a special role in their model: it is selected uniformly and is unknown to the adversary during the public parameter generation phase. These results likewise inspire our adoption of the amplified *complete subversion model*, which excludes such reliance on public randomness beyond the reach of the adversary. In particular, our general immunizing strategy (randomizing the public parameter of a backdoored PRG instead of randomizing the PRG output) permits us to bypass the impossibility result. Additionally, our results on subversion-resistant OWFs can be applied to construct a specific "backdoor-free" PRG following the classic Blum-Micali framework.

Other works suggest different angles of defense against mass surveillance. For example, in [11,18] the authors proposed a general framework of safeguarding protocols by randomizing the incoming/outgoing messages via a trusted (reverse) firewall. Their results demonstrate that with a trusted random source, many tasks become achievable. As they rely on a "subversion-free" firewall, these results require a more generous setting than provided by our *complete subversion model*.

Ateniese et al. [1] continued the study of algorithm substitution attacks on signatures and propose two defending mechanisms, one utilizes a unique signature scheme assuming the key generation and verify algorithms to be honest; the other adopts the reverse firewall model that assumes trusted randomness. We construct a signature scheme that can be proven secure in the complete subversion model which does not make assumptions on honesty or require trusted randomness. We remark that the strength of the "watchdog" that is required for the signature scheme is, however, stronger than that required for the other primitives; it must be permitted a transcript of the security game. See Sect. 4.

2 A Definitional Framework for Cliptography

2.1 From Cryptography to Cliptography

In this section, we lay down a definitional framework for cliptography. The adversary in this new setting is "*proud-but-malicious*": the adversary wishes to supply a subverted implementation in order to break security while keeping the subversion "under the radar" of any detection. Thus the basic framework should reflect the ability of the adversary to provide (potentially subverted) implementations of the cryptographic algorithms of interest, the ability of an efficient "watchdog" to interrogate such implementations in order to check their veracity, and a classical "challenger-adversary" security game. Specifically, the model considers an adversary that commences activities by supplying a (potentially subverted) implementation of the cryptographic primitive; one then considers two parallel procedures: a classical challenger-adversary security game in which the challenger must use only (oracle access to) the adversary's implementations, and a process in which the "watchdog" compares—also via oracle access—the adversary's implementations against a specification of the primitives. (For entertainment, we occasionally refer to the adversary as "big brother.")

Cryptographic games. We express the security of (standard) cryptographic schemes via *cryptographic games* between a challenger $\mathcal{C}$ and an adversary $\mathcal{A}$.

Definition 1. (Cryptographic Game [14]). *A cryptographic game* $\mathsf{G} = (\mathcal{C}, \delta)$ *is defined by a random system* $\mathcal{C}$*, called the challenger, and a constant* $\delta \in [0, 1)$*. On security parameter* λ*, the challenger* $\mathcal{C}(1^\lambda)$ *interacts with some adversary* $\mathcal{A}(1^\lambda)$ *and outputs a bit* b*. We denote this interaction by* $b = (\mathcal{A}(1^\lambda) \Leftrightarrow \mathcal{C}(1^\lambda))$*. The advantage of an attacker* $\mathcal{A}$ *in the game* G *is defined as*

$$\mathbf{Adv}_{\mathcal{A},\mathsf{G}}(1^\lambda) = \Pr\left[(\mathcal{A}(1^\lambda) \Leftrightarrow \mathcal{C}(1^\lambda)) = 1\right] - \delta\,.$$

We say a cryptographic game G *is* secure *if for all* PPT *attackers* $\mathcal{A}$*, the advantage* $\mathbf{Adv}_{\mathcal{A},\mathsf{G}}(1^\lambda)$ *is negligible in* λ*.*

The above conventional security notions are formulated under the assumption that the relevant algorithms of the cryptographic scheme are faithfully implemented and, moreover, that participants of the task have access to truly private randomness (thus have, e.g., truly random keys). In the kleptographic setting, these assumptions are relaxed.

The complete subversion model. A basic question that must be addressed by a kleptographic model concerns the selection of algorithms the adversary is permitted to subvert. We work exclusively in a setting where the adversary is permitted to provide implementations of *all* the relevant cryptographic elements of a scheme, a setting we refer as the *complete subversion* model. Thus, all guarantees about the quality of the algorithms are delivered by the watchdog's testing activities. This contrasts with all previous work, which explicitly protected some of the algorithms from subversion, or assumed clean randomness. Such a setting we refer to as *partial subversion* model.

Choosing the right watchdog. By varying the information provided to the watchdog, one obtains different models that reflect various settings of practical interest. The weakest (and perhaps most attractive) model is the *offline* watchdog, which simply interrogates the supplied implementations, comparing them with the specification of the primitives, and declares them to be "fit" or "unfit." Of course, we must insist that such watchdogs find the actual specification "fit": formally, the definition is formulated in terms of distinguishing an adversarial implementation from the specification.

One can strengthen the watchdog by permitting it access to the full transcript of the challenger-adversary security game, resulting in the *online* watchdog. Finally, we describe an even more powerful *omniscient* watchdog, which is even privy to private state of the challenger. (While we do not use such a powerful watchdog in our results, it is convenient for discussing previous work.)

We remark these various watchdogs reflect various levels of "checking" that a society might entertain for cryptographic algorithms (and conversely, various levels of tolerance that an adversary may have to exposure): the offline watchdog reflects a "one-time" laboratory that attempts to check the implementations; an online watchdog actually crawls public transcripts of cryptographic protocols to detect errors; the omniscient watchdog requires even more, involving (at least) individuals effectively checking their results against the specification.

2.2 A Formal Definition

Having specified the power of the big brother (the adversary) and that of the watchdog, we are ready to introduce *cliptographic games* to formulate security. To simplify the presentation, we here initially consider *complete* subversion with an *offline* watchdog. In the next section, we will discuss the other variants.

A cryptographic scheme Π consists of a set of (possibly randomized) algorithms $(F^1, \ldots, F^k)$. (In general, deterministic algorithms determine functions $F^i : (\lambda, x) \mapsto y$, whereas randomized algorithms determine distributions $F^i(\lambda, x)$ over an output set Y_λ.) For example, a digital signature scheme consists of three algorithms, a (randomized) key generation algorithm, a signing algorithm, and deterministic verification algorithm. The definition of Π results in a *specification* of the associated algorithms; for concreteness, we label these as $\Pi_{\text{SPEC}} = (F^1_{\text{SPEC}}, \ldots, F^k_{\text{SPEC}})$, when a scheme is (perhaps adversarially) implemented, we denote the implementation as $\Pi_{\text{IMPL}} = (F^1_{\text{IMPL}}, \ldots, F^k_{\text{IMPL}})$. If the implementation honestly follows the specification of the scheme, we overload the notation and represent them interchangeably with the specification as Π_{SPEC}.

In our definition, the adversary $\mathcal{A}$ will interact with both the challenger $\mathcal{C}$ and the watchdog $\mathcal{W}$. (In the offline case, these interactions are independent; in the online case, $\mathcal{W}$ is provided a transcript of the interaction with $\mathcal{C}$.) Following the definition of cryptographic game, we use $b_\mathcal{C} = (\mathcal{A}(1^\lambda) \Leftrightarrow \mathcal{C}^{F^1_{\text{IMPL}}, \ldots, F^k_{\text{IMPL}}}(1^\lambda))$ to denote the interaction between $\mathcal{A}$ and $\mathcal{C}$; $b_\mathcal{C}$ denotes the bit returned by the challenger $\mathcal{C}$. (Note that the challenger must use the implementation of Π provided by the adversary, while the interaction between $\mathcal{A}, \mathcal{C}$ is the same as in the classical cryptographic game.)

As for the watchdog $\mathcal{W}$, the adversary provides $\mathcal{W}$ his potentially subverted implementations Π_{IMPL} of the primitive (as oracles); $\mathcal{W}$ may then interrogate them in an attempt to detect divergence from the specification, which he possesses. On the basis of these tests, the watchdog produces a bit (Intuitively, the bit indicates whether the implementations passed whatever tests the watchdog carried out to detect inconsistencies with the specification.)

Definition 2. (Cliptographic Game). *A cliptographic game* $\widehat{\mathsf{G}} = (\mathcal{C}, \Pi_{\text{SPEC}}, \delta)$ *is defined by a challenger* $\mathcal{C}$*, a specification* Π_{SPEC}*, and a constant* $\delta \in [0, 1)$*. Given an adversary* $\mathcal{A}$*, a watchdog* $\mathcal{W}$*, and a security parameter* λ*, we define the* detection probability *of the watchdog* $\mathcal{W}$ *with respect to* $\mathcal{A}$ *to be*

$$\mathbf{Det}_{\mathcal{W},\mathcal{A}}(1^\lambda) = \left|\Pr[\mathcal{W}^{F^1_{\text{IMPL}},\ldots,F^k_{\text{IMPL}}}(1^\lambda) = 1] - \Pr[\mathcal{W}^{F^1_{\text{SPEC}},\ldots,F^k_{\text{SPEC}}}(1^\lambda) = 1]\right| ,$$

where $\Pi_{\text{IMPL}} = (F^1_{\text{IMPL}}, \ldots, F^k_{\text{IMPL}})$ *denotes the implementation produced by* $\mathcal{A}$ *. The* advantage *of the adversary is defined to be*

$$\mathbf{Adv}_{\mathcal{A}}(1^\lambda) = \left|\Pr\left[(\mathcal{A}(1^\lambda) \Leftrightarrow \mathcal{C}^{F^1_{\text{IMPL}},\ldots,F^k_{\text{IMPL}}}(1^\lambda)) = 1\right] - \delta\right| .$$

We say that a game is ***subversion-resistant*** *if for any polynomial* $q(\cdot)$*, there exists a* PPT *watchdog* $\mathcal{W}$ *such that for all* PPT *adversaries* $\mathcal{A}$*, either* $\mathbf{Det}_{\mathcal{W},\mathcal{A}}(1^\lambda)$ *is non-negligible, or* $\mathbf{Adv}_{\mathcal{A}}(1^\lambda)$ *is negligible, in the security parameter* λ*.*

Other watchdog variants. In the above definition, we chose the strongest model: the watchdog is *universal* and *offline.* In particular, primitives secure in this model are secure in any of the other models considered. The detection algorithm of the watchdog must be designed for a *given specification*, regardless of how the adversary subverts the implementation; furthermore, it may only carry out a one-time check on the implementation (and may not supervise the security game). To permit a broader class of feasibility results, it is possible to extend the basic model in both directions.

Swapping the quantifiers. It is also reasonable to consider a watchdog that may be tailored to the adversary, i.e., the quantifiers are changed to be $\forall \mathcal{A}, \exists \mathcal{W}$. Indeed, such quantification (or even more generous settings, see below) was considered implicitly in previous works, e.g., [3,4,10]. We remark that such a model is still highly non-trivial in that the adversary can be randomized by, e.g., selecting a random backdoor. (Thus knowing the code of the adversary does not necessarily help the watchdog to identify a faulty implementation which might based on a random backdoor that is only known to $\mathcal{A}$.) Note that such a model is particularly interesting for evaluating *attacks*, where one would like to guarantee that the attack is undetectable even by a watchdog privy to the details of the adversary: specifically, when establishing security, weak watchdogs are preferable; when establishing the value of an attack, strong watchdogs are preferable.

We develop one-way permutations and pseudorandom generators in the offline model. However, it appears that richer primitives may require qualitatively stronger watchdogs. Considering that an offline watchdog cannot ensure *exact* equality for deterministic algorithms, we remark that a clever adversary may be able to launch attacks by altering a deterministic function at a single location. Imagine a security game where the adversary supplies a string m to which the challenger is expected to apply one of the subverted algorithms; this takes place, e.g., in the typical signature security game. The adversary may now select a random string w and implement the deterministic algorithm in such a way that it diverges from the specification at (only) this preselected point. While such inconsistencies are (essentially) undetectable by an offline watchdog, the adversary can ensure that the subverted algorithm is indeed queried at w during the security game. Such "input-triggering attacks" in [1,4,9] motivated them to consider extra "decryptability condition" and "verifiability condition" assumptions.

An *online watchdog* can guard against this possibility; he is permitted to monitor the public interactions between users. More precisely, the online watchdog is permitted to certify both the implementations *and* the transcript between the challenger and adversary. The security game is then altered by considering $\mathcal{W}^{\Pi_{\text{IMPL}}}(1^\lambda, \tau)$, identical to the offline case except that the watchdog is provided the transcript τ of the security game $(\mathcal{C} \Leftrightarrow \mathcal{A})$.[2] (We use the shorthand notation Π_{IMPL} here to denote the collection of oracles $F^1_{\text{IMPL}}, \ldots, F^k_{\text{IMPL}}$.) The detection game must then be adjusted, guaranteeing that the transcripts produced when the challenger uses Π_{IMPL} are indistinguishable from those produced when the challenger uses Π_{SPEC}. Our results on digital signature schemes will require such a watchdog. We remark that previous work on subversion-resistant digital signatures [1] assumes a verifiability condition: every message-signature pair produced by the subverted sign algorithm (at least the responses to the signing queries) can pass the verification of the specification of the verify algorithm. This extra assumption can be guaranteed by an online watchdog (and, indeed, it demands either an absolute universal guarantee or an on-line guarantee for those pairs that appear in the security game).

An *omniscient watchdog* is even stronger. In addition to access to the transcript, the omniscient watchdog is aware of the entire internal state of the challenger (and can monitor the interactions between users and the subverted implementations). Similarly, by replacing $\mathcal{W}$ in Definition 2 above with an omniscient watchdog, we obtain cliptographic games with omniscient watchdog. As mentioned, omniscient watchdog has been considered in literature [4,9]. In those works, they assume the extra decryptability condition such that ciphertext generated by the subverted encryption algorithm decrypts correctly with the honest decryption algorithm. Again, without allowing the watchdog to input the whole transcript and the decryption key, this assumption cannot supported.

Discussion: The guarantees provided by an offline watchdog. We make some general observations about the guarantees that an offline watchdog provides.

Consider a *deterministic* algorithm implemented by the adversary; an offline watchdog cannot ensure that such an algorithm is perfectly implemented. However, it can ensure that the implementation agrees with the specification with high probability over a particular (sampleable) distribution of choice (by simply drawing from the distribution and checking equality). This frequently arises in our setting, where we are led to study the behavior of a deterministic algorithm on a particular "public input distribution."

[2] We remark that the transcript τ includes the final output bit of the challenger.

Lemma 1. *Consider an adversarial implementation* $\Pi_{\text{IMPL}} := (F^1_{\text{IMPL}}, \ldots, F^k_{\text{IMPL}})$ *of a specification* $\Pi_{\text{SPEC}} = (F^1_{\text{SPEC}}, \ldots, F^k_{\text{SPEC}})$ *in a cliptographic game, where* $F^1, \ldots, F^k$ *are deterministic algorithms. Additionally, for each security parameter* λ, *(sampleable) public input distributions* $X^1_\lambda, \ldots, X^k_\lambda$ *are defined respectively. If* $\exists j \in [k]$, $\Pr[F^j_{\text{IMPL}}(x) \neq F^j_{\text{SPEC}}(x) : x \leftarrow X^j_\lambda]$ *is non-negligible, then there is a* PPT *offline watchdog that can detect with a non-negligible probability.*

The above includes the cases that the deterministic algorithm is with a known input distribution (e.g., uniform distribution), or with an input distribution that is generated by other (adversarial) implementations. Jumping ahead, the evaluation function of a one way permutation takes a uniform input distribution; and a pseudorandom generator stretch function takes $\mathcal{K} \times \mathcal{U}$ as (public) input distribution, where $\mathcal{K}$ is the output distribution of a parameter generation algorithm implemented by the adversary and $\mathcal{U}$ is the uniform seed distribution.

In our analysis, we will use this simple observation extensively. In particular, when a hash specification is modeled as a random oracle we can check that the hash function has been faithfully implemented by the adversary (with high probability) for any particular sampleable distribution of choice; in many cases, these will be distributions generated by other adversarial implemented algorithms.

Next, consider a *randomized* algorithm (with fixed inputs) that is supposed to output a high-entropy distribution. The offline watchdog can provide a weak guarantee of min-entropy by simply running the algorithm twice to see whether there is collision. While this does not guarantee large entropy, it can guarantee a critical feature: the result is unpredictable to the adversary.

Lemma 2. *Consider an adversary* $\mathcal{A}$ *which prepares the implementation* F_{IMPL} *of a specification* F_{SPEC}, *where* F_{SPEC} *is a randomized algorithm that produces an output distribution with* $\omega(\log \lambda)$ *min-entropy. If* $\Pr[x = x' : x \leftarrow \mathcal{A}(\lambda), x' \leftarrow F_{\text{IMPL}}] \leq \mathsf{negl}(\lambda)$ *does not hold, then there is a* PPT *offline watchdog that can detect this with a non-negligible probability.*

Discussion: random oracles. In many settings, we establish results in the conventional *random oracle* model which requires some special treatment in the model above. In general, we consider a random oracle to be an (extremely powerful) heuristic substitute for a deterministic function with strong cryptographic properties. In a kleptographic setting with complete subversion, we must explicitly permit the adversary to tamper with the "implementation" of the random oracle supplied to the challenger. In such settings, we provide the watchdog—as usual—oracle access to both the "specification" of the random oracle (simply a random function) and the adversary's "implementation" of the random oracle, which may deviate from the random oracle itself. For concreteness, we permit the adversary to "tamper" with a random oracle h by providing an efficient algorithm $T^h(x)$ (with oracle access to the random oracle h) which computes the "implementation" $\widetilde{h}$—thus the implementation $\widetilde{h}(x)$ is given by $T^h(x)$ for all x. Likewise, during the security game, the challenger is provided oracle access only to the potentially subverted implementation $\widetilde{h}$ of the random oracle. As usual, the probabilities defining the security (and detection) games are taken over the choice of the random oracle. In this sense, the random oracle assumption used in our complete subversion model is weaker than the classical one, since we can allow even "imperfect" random oracles. Fortunately, when the random oracle is applied to a known input distribution, an offline watchdog can ensure that the implementation is almost consistent with its specification (see Lemma 1).

Remark 1. **Stateless/stateful implementations.** In principle, algorithms in the specification of a cryptographic scheme or implementations provided by an adversary could be stateful; for simplicity, we focus on stateless implementations in the above lemmas. However, to jump ahead a bit, those results still hold (with simple modifications) in natural stateful settings. To see this, (1) consider a randomized algorithm specified to produce a high-entropy output distribution: in the case of a stateful implementation (maintaining a local state), the unpredictability requirement can still be ensured by an offline watchdog who can *rewind* the implementation. The watchdog simply samples (rewinds to the same state and then samples) from the randomized algorithm to see whether there is a collision. (2) For deterministic algorithms with a state, as an example, we consider a stateful PRG, where the seed is updated in each iteration. In this case, the public input distribution is evolving during the iterations. Observe that the offline watchdog can indeed ensure the consistency of the implementation of the PRG when the input is chosen from a uniform distribution. This means the "bad" input set (on which the implementation deviates from its specification) could be at most negligibly small (in the uniform distribution). Note that starting from a uniform seed, any polynomially number of PRG iterations will yield poly-many pseudorandom bits. Thus the probability for any of them falls into the "bad" input set would still be negligible.

Schemes with augmented system parameter. Often, deployment of a cryptographic scheme may involve a *system parameter generation* algorithm $pp \leftarrow \mathsf{Gen}(1^\lambda)$. When we consider such an augmented scheme $\Pi = (\mathsf{Gen}, F^1, F^2, F^3)$ in our setting, we can treat the system parameter pp in two natural ways: (1) as in Definition 2, the adversary simply provides the implementation $\mathsf{Gen}_{\text{IMPL}}$ to $\mathcal{W}$ (and $\mathcal{C}$) as usual and the challenger computes pp by running $\mathsf{Gen}_{\text{IMPL}}$ during the security game; (2) the *adversary provides* pp directly to the watchdog $\mathcal{W}$ (and $\mathcal{C}$); we write $\mathcal{W}^{\Pi_{\text{IMPL}}}(1^\lambda, pp)$ to reflect this. By replacing $\mathcal{W}^{\Pi_{\text{IMPL}}}(1^\lambda)$ in Definition 2 with $\mathcal{W}^{\Pi_{\text{IMPL}}}(1^\lambda, pp)$, and suitably changing the security game so that the challenger does not generate pp, we can obtain the *adversarially chosen parameter model.* This model was used for studying pseudorandom generator under subversion in [10], we choose to present it as a general model that would be interesting to consider for any cryptographic primitive.

It is clear that if a primitive is secure in the adversarially chosen parameter model, then it is secure according to Definition 2. (Observe that the adversary is always free to generate pp according to the algorithm provided to the challenger.) We record this below.

Lemma 3. *If Π is secure in the adversarially chosen parameter model, then Π is secure according to Definition 2.*

Schemes with split-program. Randomized algorithms play a distinguished role in our kleptographic setting. One technique we propose for immunization may also rely on the decomposition of a randomized generation algorithm $y \leftarrow \mathsf{Gen}(1^\lambda)$ into two algorithms: a random string generation algorithm RG responsible for producing a uniform poly(λ)-bit random string r, and a deterministic output generation algorithm dKG that transforms the randomness r into an output y. Note that dKG is deterministic and is always applied to a public input distribution. In light of Lemma 1, we may assume that the maliciously implemented $\mathsf{dKG}_{\text{IMPL}}$ is consistent with the honest implementation $\mathsf{dKG}_{\text{SPEC}}$ with overwhelming probability. See results in this model in Sect. 3.3, and definition in the full version [22].

We remark that this perspective only requires a change in the specification of Π_{SPEC}. When we apply Definition 2 with a specification that has been altered this way, we say that a primitive is proven secure in the *split-program model.*

The split-program model is quite general and can be applied to most practical algorithms. To see this, the user is provided the source code of the implementation which makes calls to some API for generating randomness (e.g., `rand()`) whenever necessary. The user can hook up the calls to the randomness API with the separate program RG provided by the adversary. (In fact, full source code is not strictly necessary in this setting; object code that adopts a particular fixed convention to gather randomness would also suffice.)

3 Subversion-Resistant One-Way Permutations

In this section, we study one-way permutations (OWP) in our cliptographic framework. As mentioned before, this is motivated by the problem of defending against subverted key generation. In particular, we propose general constructions for subversion-resistant OWPs that require only the weakest (offline) watchdog with adversarially chosen parameters. Our "immunizing strategy" consists of coupling the function generation algorithm with a hash function that is applied to the function index—intuitively, this makes it challenging for an adversary to meaningfully embed a backdoor in the permutation or its index.[3] We prove that if the specification of the hash function is modeled as a random oracle, then randomizing the permutation index using the (adversarially implemented) hash function destroys any potential backdoor structure. We emphasize that the permutation evaluation algorithm, the name generation algorithm, and the hash function may all be subverted by the adversary.

The cliptographic model introduces a number of new perspectives on the (basic) notion of security for one-way permutations. We actually consider three different notions below, each of which corresponds to distinct practical settings: the first corresponds to the classical notion, where the challenger chooses the index of the function (using subverted code provided by the adversary)—we call this OWP^{C}; the second corresponds to a setting where the adversary may

[3] In concrete constructions, the hash function becomes a component of, e.g., the evaluation function, so that the syntax of the primitive is still the same.

choose the index—we call this $\mathrm{OWP^A}$; the last corresponds to our "split program model," discussed above—we call this $\mathrm{OWP^{SP}}$.

In many cases of practical interest, however, the permutation index may have special algebraic structure, e.g., RSA or DLP. In such cases, it would appear that the public hash function would require some further "structure preserving" property (so that it carries the space of indices to the space of indices). Alternatively, one can assume that the space of indices can be "uniformized," that is, placed in one-to-one correspondence with strings of a particular length. In order to apply our approach to broader practical settings, we apply the "split-program" model discussed above. This effectively "uniformizes" index space by insisting that the function generation algorithm is necessarily composed of two parts: a random string generation algorithm RG that outputs random bits r, and a deterministic function index generation algorithm dKG which uses r to generate the index. Hashing is then carried out on r; see Sect. 3.3 for details.

3.1 Defining Subversion Resistant OWP/TDOWP

In this subsection, following our general definitional framework, we define the security of one-way permutations and trapdoor one-way permutations. We first recall the conventional definitions.

One-way permutation (OWP). A family of permutations $\mathcal{F} = \{f_i : X_i \to X_i\}_{i \in I}$ is *one-way* if there are PPT algorithms $(\mathsf{KG}, \mathsf{Eval})$ so that (i) KG, given a security parameter λ, outputs a function index i from $I_\lambda = I \cap \{0,1\}^\lambda$; (ii) for $x \in X_i$, $\mathsf{Eval}(i, x) = f_i(x)$; (iii) $\mathcal{F}$ is one-way; that is, for any PPT algorithm $\mathcal{A}$, it holds that $\Pr[\mathcal{A}(i, y) \in f_i^{-1}(y) \mid i \leftarrow \mathsf{KG}(\lambda); x \leftarrow X_i; y := f_i(x)] \leq \mathsf{negl}(\lambda)$.

Trapdoor one way permutation (TDOWP). A family of permutations $\mathcal{F} = \{f_i : X_i \to X_i\}_{i \in I}$ is *trapdoor one-way* if there are PPT algorithms $(\mathsf{KG}, \mathsf{Eval}, \mathsf{Inv})$ such that (i) KG, given a security parameter λ, outputs a function index and the corresponding trapdoor pair (i, t_i) from $I_\lambda \times T$, where $I_\lambda = I \cap \{0,1\}^\lambda$, and T is the space of trapdoors; (ii) $\mathsf{Eval}(i, x) = f_i(x)$ for $x \in X_i$; (iii) $\mathcal{F}$ is one-way; and (iv) it holds that $\Pr[\mathsf{Inv}(t_i, i, y) = x \mid i \leftarrow \mathsf{KG}(\lambda); x \leftarrow X_i; y := f_i(x)] \geq 1 - \mathsf{negl}(\lambda)$.

Sometimes, we simply write $f_i(x)$ rather than $\mathsf{Eval}(i, x)$.

$\mathbf{Subversion-resistant^C\ one-way\ permutations: OWP^C}$. As described in Sect. 2, we assume a "laboratory specification" of the OWP, $(\mathsf{KG}_{\text{SPEC}}, \mathsf{Eval}_{\text{SPEC}})$, which has been rigorously analyzed and certified (e.g., by the experts in the cryptography community). The adversary provides an alternate (perhaps subverted) implementation $(\mathsf{KG}_{\text{IMPL}}, \mathsf{Eval}_{\text{IMPL}})$. We study OWP/TDOWP in the offline watchdog model; while the implementations may contain arbitrary backdoors or other malicious features, they can not maintain any state.

Intuitively, the goal of the adversary is to privately maintain some "backdoor information" z so that the subverted implementation $\mathsf{KG}_{\text{IMPL}}$ will output functions that can be inverted using z. In addition, to avoid detection by the watchdog, the adversary must ensure that implementations $(\mathsf{KG}_{\text{IMPL}}(z), \mathsf{Eval}_{\text{IMPL}}(z))$ are computationally indistinguishable from the specification ($\mathsf{KG}_{\text{SPEC}}, \mathsf{Eval}_{\text{SPEC}}$) given only oracle access. Formally,

Definition 3. *A one-way permutation family $\mathcal{F} = \{f_i : X_i \to X_i\}_{i \in I}$ with the specification $\mathcal{F}_{\text{SPEC}} = (\mathsf{KG}_{\text{SPEC}}, \mathsf{Eval}_{\text{SPEC}})$, is* ***subversion-resistant***^{C} *in the offline watchdog model if there exists a* PPT *watchdog $\mathcal{W}$, s.t.: for any* PPT *adversary $\mathcal{A}$ playing with the challenger $\mathcal{C}$ in the following game, (Fig. 1), either the detection probability* $\mathbf{Det}_{\mathcal{W},\mathcal{A}}$ *is non-negligible, or the advantage* $\mathbf{Adv}_{\mathcal{A}}$ *is negligible.*

Here the detection probability *of the watchdog $\mathcal{W}$ with respect to $\mathcal{A}$ is defined as*

$$\mathbf{Det}_{\mathcal{W},\mathcal{A}}(1^\lambda) = \left|\Pr[\mathcal{W}^{\mathsf{KG}_{\text{IMPL}},\mathsf{Eval}_{\text{IMPL}}}(1^\lambda) = 1] - \Pr[\mathcal{W}^{\mathsf{KG}_{\text{SPEC}},\mathsf{Eval}_{\text{SPEC}}}(1^\lambda) = 1]\right| ,$$

and the advantage *of the adversary $\mathcal{A}$ is defined as*

$$\mathbf{Adv}_{\mathcal{A}}(1^\lambda) = \Pr\left[(\mathcal{A}(1^\lambda) \Leftrightarrow \mathcal{C}^{\mathsf{KG}_{\text{IMPL}},\mathsf{Eval}_{\text{IMPL}}}(1^\lambda)) = 1\right] .$$

For convenience, we also say that such $\mathcal{F}_{\text{SPEC}}$ is a OWP^C in the offline watchdog model. On the other hand, we say that an OWP is subvertible *if:*
$\mathbf{Det}_{\mathcal{W},\mathcal{A}}(1^\lambda)$ *is negligible for all* PPT *$\mathcal{W}$, and* $\mathbf{Adv}_{\mathcal{A}}(1^\lambda)$ *is non-negligible.*[4]

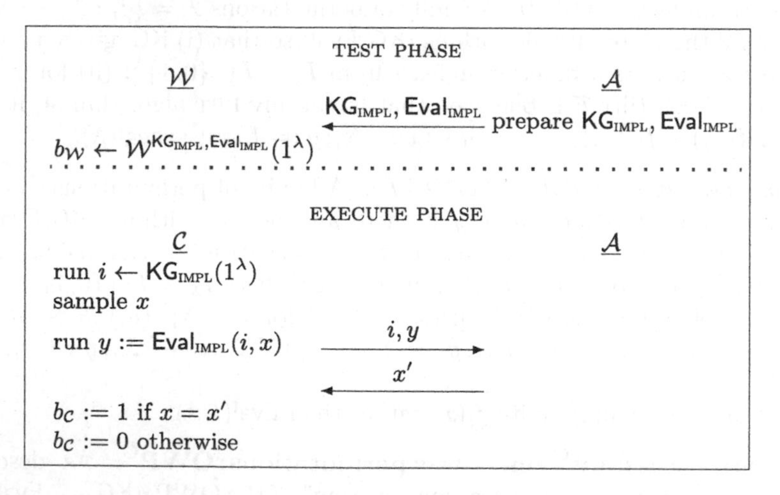

Fig. 1. Subversion-resistant^{C} security game: OWP^{C}.

Subvertible OWPs. Next we observe that it is easy for an adversary to break the security of a conventional OWP in the kleptographic setting. In particular, the following lemma shows that one can construct a *subvertible* OWP (so the

[4] We choose a stronger definition for subvertibility (swap the quantifiers of $\mathcal{A}, \mathcal{W}$) to describe attacks instead of directly negating the definition of OWP^{C}.

subverted implementation can evade detection by all watchdogs and the adversary can invert) using a conventional trapdoor OWP. In particular, if we wish to use public-key cryptography in a kleptographic setting, nontrivial effort is required to maintain the security of even the most fundamental cryptographic primitives.

Our construction of a subvertible OWP substantiates the folklore knowledge that sufficient random padding can render cryptosystems vulnerable to backdoor attacks, e.g., [27,28]. Specifically, the random padding in the malicious implementation can be generated so that it encrypts the corresponding trapdoor using the backdoor as a key. For detailed proofs, we defer to the full version [22].

Lemma 4. *One can construct a subvertible OWP from any TDOWP. In particular, given a TDOWP, we can construct a OWP that is not a* OWP^C.

We defer the question of the existence of a OWP^{C} to the next section, where we will construct permutations that satisfy a stronger property.

Subversion-resistant OWPs with adversarially chosen indices: $\mathbf{OWP^A}$. The notion of OWP^{C} formulated above defends against kleptographic attacks when the adversary provides a subverted implementation of the defining algorithms. In many cases, however, it is interesting to consider a more challenging setting where the adversary may directly provide the public parameters, including the function index. Indeed, this is the case in many real-world deployment settings, where a "trusted" agency sets up (or recommends) the public parameters. One notorious example (for a different primitive) is the Dual_EC PRG [7]. Note that, in general, this notion is not very suitable for asymmetric key primitives, e.g. TDOWP, since allowing the adversary to set up the public key gives him the chance to generate the trapdoor. We will focus on OWP^{A}.

Definition 4. *A one-way permutation family* $\mathcal{F} = \{f_i : X_i \to X_i\}_{i \in I}$ *with the specification* $\mathcal{F}_{\text{SPEC}} = (\mathsf{KG}_{\text{SPEC}}, \mathsf{Eval}_{\text{SPEC}})$, *is* ***subversion-resistant***$^{\mathbf{A}}$ *in the offline watchdog model, if there is a* PPT *watchdog* $\mathcal{W}$, *such that: for any* PPT *adversary* $\mathcal{A}$ *playing the following game (Fig. 2) with the challenger* $\mathcal{C}$, *either the detection probability* $\mathbf{Det}_{\mathcal{W},\mathcal{A}}$ *is non-negligible, or the advantage* $\mathbf{Adv}_{\mathcal{A}}$ *is negligible.*

Here the detection probability *of the watchdog* $\mathcal{W}$ *with respect to* $\mathcal{A}$ *is defined as:*

$$\mathbf{Det}_{\mathcal{W},\mathcal{A}}(1^\lambda) = \left|\Pr[\mathcal{W}^{\mathsf{Eval}_{\text{IMPL}}}(1^\lambda, i_\bullet) = 1] - \Pr[\mathcal{W}^{\mathsf{Eval}_{\text{SPEC}}}(1^\lambda, i) = 1]\right| ,$$

and the advantage *of the adversary* $\mathcal{A}$ *is defined as*

$$\mathbf{Adv}_{\mathcal{A}}(1^\lambda) = \Pr\left[(\mathcal{A}(1^\lambda) \Leftrightarrow \mathcal{C}^{\mathsf{Eval}_{\text{IMPL}}}(1^\lambda, i_\bullet)) = 1\right] ,$$

where $i \leftarrow \mathsf{KG}_{\text{SPEC}}(1^\lambda)$, *and* $i_\bullet$ *is chosen by the adversary.*

We also say that such $\mathcal{F}_{\text{SPEC}}$ *is a* OWP^A *in the offline watchdog model.*

Relating $\mathbf{OWP^C}$ and $\mathbf{OWP^A}$. Following Lemma 3, an adversary that successfully breaks the OWP^{C} game can be easily transformed into an adversary that breaks the OWP^{A} game; thus *any* OWP^A *is also a* OWP^C. As far as existence is concerned, then, it suffices to construct a OWP^{A}.

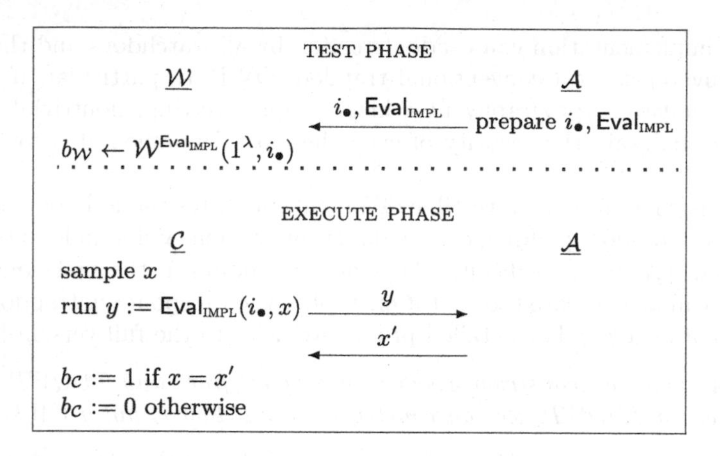

Fig. 2. Subversion-resistant$^\text{A}$ security game: OWP$^\text{A}$.

3.2 Constructing Subversion-Resistant$^\text{A}$ OWP

In this section, we discuss methods for safeguarding OWP against kleptographic attacks. We first present a general approach that transforms any OWP to a OWP$^\text{A}$ under the assumption that a suitable hash function can be defined on the index space. Specifically, we prove that randomizing the function index (via hashing, say) is sufficient to eliminate potential backdoor information. These results assume only the weakest (offline) watchdog. More importantly, we permit the hash function—like the other relevant cryptographic elements—to be implemented and potentially subverted by the adversary.

Note that we treat only the specification of the hash function in the random oracle model, assuming that the adversary may arbitrary subvert the (randomly specified) hash function; thus the watchdog is provided both the adversary's arbitrarily subverted "implementation" and the correct (random) hash function "specification."[5] Despite the adversary's control over the OWP and the hash function (which is partially constrained by the watchdog), it is difficult for the adversary to arrange a backdoor that works for a large enough target subset of function indices that these can be reliably "hit" by the hash function.

One remaining difficulty is to keep the "syntax intact," that is, to avoid changing the structure of the specification. For this purpose, we propose to treat the hash function only as a component of (jumping ahead) the evaluation algorithm (see Fig. 3). The adversary only implements the evaluation algorithm as a whole with the hash function built in (as the specification demands). In this case, the hash implementation (and specification) are not explicitly given to the watchdog anymore. However, we still manage to show the security by exploring the fact that

[5] Note that we place no a priori constraints on the subverted hash function provided by the adversary. The watchdog, of course, can ensure that the subverted function and the specification (which is just a random function, in this case) agree with high probability on slices of the space, or possess other common statistical properties.

both hash and the evaluation algorithm are deterministic algorithms with public input distribution, so that the offline watchdog can force the implementation of $\mathsf{Eval}_{\mathrm{IMPL}}$ to agree with the specification $\mathsf{Eval}_{\mathrm{SPEC}}$ with overwhelming probability when inputs are sampled according to the input generation distribution.

General feasibility results for OWP$^{\mathbf{A}}$. Let $\mathcal{F}$ be any OWP family with specification $\mathcal{F}_{\mathrm{SPEC}} := (\mathsf{KG}^{\mathcal{F}}_{\mathrm{SPEC}}, \mathsf{Eval}^{\mathcal{F}}_{\mathrm{SPEC}})$; while we assume, of course, that it is one-way secure (in the classical sense), it may be subvertible. We also assume that $\mathsf{KG}^{\mathcal{F}}_{\mathrm{SPEC}}(\lambda)$ outputs uniform i from the index set I_λ and that we have a public hash function with specification $h_{\mathrm{SPEC}} : I_\lambda \rightarrow I_\lambda$, acting on this set. Then we construct a subversion-resistant^{A} OWP family $\mathcal{G}$ with specification $\mathcal{G}_{\mathrm{SPEC}} := (\mathsf{KG}^{\mathcal{G}}_{\mathrm{SPEC}}, \mathsf{Eval}^{\mathcal{G}}_{\mathrm{SPEC}})$ defined as follows:

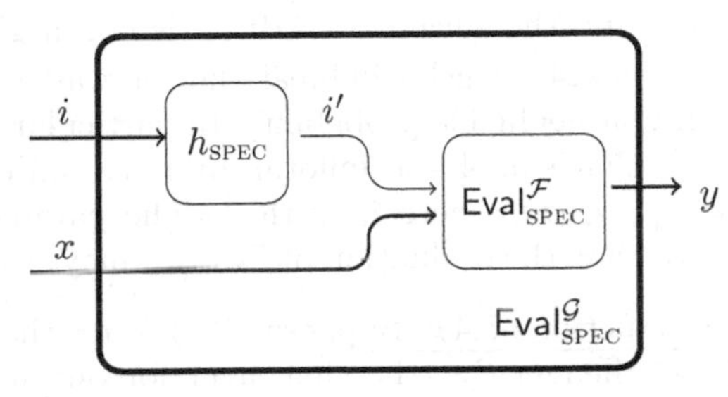

Fig. 3. New specification $\mathsf{Eval}^{\mathcal{G}}_{\mathrm{SPEC}}$.

- Function index generation $i \leftarrow \mathsf{KG}^{\mathcal{G}}_{\mathrm{SPEC}}$, where $\mathsf{KG}^{\mathcal{G}}_{\mathrm{SPEC}}$ is given by: Sample $i \leftarrow \mathsf{KG}^{\mathcal{F}}_{\mathrm{SPEC}}(\lambda)$; output i.
- Function evaluation $y \leftarrow \mathsf{Eval}^{\mathcal{G}}_{\mathrm{SPEC}}(i, x)$, where $\mathsf{Eval}^{\mathcal{G}}_{\mathrm{SPEC}}$ is given by: Upon receiving inputs (i, x), compute $i' = h_{\mathrm{SPEC}}(i)$ and compute $y = \mathsf{Eval}^{\mathcal{F}}_{\mathrm{SPEC}}(i', x)$; output y. See also the pictorial illustration for $\mathsf{Eval}^{\mathcal{G}}_{\mathrm{SPEC}}$ in Fig. 3.

Remark 2. Note that the specification of the hash function is "part of" of the specification of the evaluation function. In fact, an interesting property of the construction above is that it is secure even if the (subverted) hash function is not separately provided to watchdog.[6]

Security analysis. Roughly, the proof relies on the following two arguments: (1) any particular adversary can only invert a sparse subset of the one-way permutations; otherwise, such an adversary could successfully attack the (classical) security of the specification OWP. Thus, randomizing the function index will map it to a "safe" index, destroying the possible correlation with any particular backdoor. (2) The $\mathsf{Eval}_{\mathrm{IMPL}}$ (having the hash function h_{IMPL} built in) is a *deterministic* function that is only called on fixed public input distributions ($\mathcal{I} \times \mathcal{U}$, where $\mathcal{I}$ is the output distribution of $\mathsf{KG}_{\mathrm{IMPL}}$ and $\mathcal{U}$ is the uniform distribution over the input space, and both of them are known to the watchdog). Following Lemma 1 of Sect. 2, the watchdog can ensure that $\mathsf{Eval}^{\mathcal{G}}_{\mathrm{IMPL}}$ is consistent with its specification an overwhelming probability when inputs are generated according to $\mathcal{I} \times \mathcal{U}$. We remark that on all inputs for which the hash implementation (running inside $\mathsf{Eval}^{\mathcal{G}}_{\mathrm{IMPL}}$) is consistent with h_{SPEC}, random oracle queries have to be made.

[6] In general, development of secure primitives in the complete subversion model would presumably be easier if the watchdog can separately "check" the implementation of h even though we do not need this for the above construction.

Theorem 1. *Assume h_{SPEC} is random oracle, and $\mathcal{F}$ with specification $\mathcal{F}_{\text{SPEC}}$ is a OWP. Then $\mathcal{G}$ with specification $\mathcal{G}_{\text{SPEC}}$ defined above is a OWP^{A} in the offline watchdog model.*

Proof. Suppose that $\mathcal{G}$ is not subversion-resistant^{A}. Then for any watchdog $\mathcal{W}$, there is a PPT adversary $\mathcal{A}_{\mathcal{G}}$ so that the detection probability $\mathbf{Det}_{\mathcal{W},\mathcal{A}_{\mathcal{G}}}$ is negligible and the advantage $\mathbf{Adv}_{\mathcal{A}_{\mathcal{G}}}$ is non-negligible, say δ. We will construct an adversary $\mathcal{A}_{\mathcal{F}}$ which will break the one-way security of $\mathcal{F}_{\text{SPEC}} := (\mathsf{KG}^{\mathcal{F}}_{\text{SPEC}}, \mathsf{Eval}^{\mathcal{F}}_{\text{SPEC}})$ with non-negligible probability. In particular, we define a simple watchdog algorithm that samples a uniform input x, and compares whether $\mathsf{Eval}^{\mathcal{G}}_{\text{SPEC}}(i_{\bullet}, x) = \mathsf{Eval}^{\mathcal{G}}_{\text{IMPL}}(i_{\bullet}, x)$, where $i_{\bullet}$ is the public parameter chosen by the adversary $\mathcal{A}_{\mathcal{G}}$. (Note that the evaluation of $\mathsf{Eval}_{\text{SPEC}}$ may involve querying random oracle.)

Construction of $\mathcal{A}_{\mathcal{F}}$. Suppose (i^*, y^*) are the challenges that $\mathcal{A}_{\mathcal{F}}$ receives from the challenger $\mathcal{C}_{\mathcal{F}}$ (the challenger for one way security of $\mathcal{F}_{\text{SPEC}}$), where $y^* = \mathsf{Eval}^{\mathcal{F}}_{\text{SPEC}}(i^*, x^*)$ for a randomly selected x^*. $\mathcal{A}_{\mathcal{F}}$ simulates a copy of $\mathcal{A}_{\mathcal{G}}$. In addition $\mathcal{A}_{\mathcal{F}}$ simulates the subversion-resistant^{A} OWP game with $\mathcal{A}_{\mathcal{G}}$.

Before receiving the function index $i_{\bullet}$ and the implementation $\mathsf{Eval}^{\mathcal{G}}_{\text{IMPL}}$ from $\mathcal{A}_{\mathcal{G}}$, the adversary $\mathcal{A}_{\mathcal{F}}$ (also acting as the challenger in the OWP^{A} game playing with $\mathcal{A}_{\mathcal{G}}$) operates as follows: First, note that h_{SPEC} is random oracle; whenever $\mathcal{A}_{\mathcal{G}}$ wants to evaluate h_{SPEC} on some points (or implementing the component for $\mathsf{Eval}^{\mathcal{G}}_{\text{IMPL}}$ that is consistent with h_{SPEC} for those points), $\mathcal{A}_{\mathcal{G}}$ has to make random oracle queries. Without loss of generality, assume $\mathcal{A}_{\mathcal{G}}$ asks q random oracle queries $i_1, \ldots, i_q$ where $q = \text{poly}(\lambda)$. Here $\mathcal{A}_{\mathcal{F}}$ randomly chooses a bit b to decide whether to embed i^* in the answers of random oracle queries at this stage. If $b = 0$, $\mathcal{A}_{\mathcal{F}}$ randomly selects an index $t \in \{1, \ldots, q\}$, and sets i^* to the answer for $h_{\text{SPEC}}(i_t)$; for all others $j \in \{1, \ldots, q\} \setminus \{t\}$, $\mathcal{A}_{\mathcal{F}}$ uniformly samples i'_j from the index set I and sets $h_{\text{SPEC}}(i_j) = i'_j$. If $b = 1$, for all $j \in \{1, \ldots, q\}$, the adversary $\mathcal{A}_{\mathcal{F}}$ uniformly samples i'_j from the index set I and sets $h_{\text{SPEC}}(i_j) = i'_j$.

After receiving $i_{\bullet}, \mathsf{Eval}^{\mathcal{G}}_{\text{IMPL}}$ from $\mathcal{A}_{\mathcal{G}}$, if $b = 1$ the adversary $\mathcal{A}_{\mathcal{F}}$ sets i^* to $h_{\text{SPEC}}(i_{\bullet})$; otherwise, it chooses a random value and sets that to $h_{\text{SPEC}}(i_{\bullet})$. Next, $\mathcal{A}_{\mathcal{F}}$ gives y^* to $\mathcal{A}_{\mathcal{G}}$ as the challenge and receives an answer x' from $\mathcal{A}_{\mathcal{G}}$. Note that in this phase, whenever $\mathcal{A}_{\mathcal{G}}$ makes random oracle queries on i, if $i \in \{i_1 \ldots, i_q\} \cup \{i_{\bullet}\}$, it is provided the previous response as answer; otherwise, i' is randomly chosen in the index set I and is returned as the answer.

Last, $\mathcal{A}_{\mathcal{F}}$ checks whether $b = 0 \wedge i_{\bullet} \neq i_t$, or $b = 1 \wedge i_{\bullet} \in \{i_1, \ldots, i_q\}$ (in those cases, $\mathcal{A}_{\mathcal{F}}$ fails to embed i^* into the right value); if yes, $\mathcal{A}_{\mathcal{F}}$ aborts; otherwise, $\mathcal{A}_{\mathcal{F}}$ submits x' to challenger $\mathcal{C}_{\mathcal{F}}$ as his answer.

Probabilistic analysis. Now we bound the success probability of $\mathcal{A}_{\mathcal{F}}$. Suppose x^* is the random input chosen by $\mathcal{C}_{\mathcal{F}}$; let W denote the event that $\mathcal{A}_{\mathcal{F}}$ aborts, W_1 the event that $b = 0 \wedge i \neq i_t$, and W_2 the event that $b = 1 \wedge i \in \{i_1, \ldots, i_q\}$. Recall that $\Pr[x' = x^*] = \Pr[x' = x^* \mid \overline{W}] \Pr[\overline{W}]$. Let $Q = \{i_1 \ldots, i_q\}$.

We first bound $\Pr[\overline{W}]$. Note that $\Pr[\overline{W}] = 1 - \Pr[W]$, and $\Pr[W] = \Pr[W_1 \vee W_2] \leq \Pr[W_1] + \Pr[W_2]$. Assuming $\Pr[i_\bullet \in Q] = \eta$, it follows that:

$$\begin{aligned}\Pr[W_1] &= \Pr[b = 0 \wedge i_\bullet \neq i_t] = \Pr[b = 0] \cdot \Pr[i_\bullet \neq i_t] \\ &= (1/2)(\Pr[i_\bullet \neq i_t \mid i_\bullet \in Q]\Pr[i_\bullet \in Q] + \Pr[i_\bullet \neq i_t \mid i_\bullet \notin Q]\Pr[i_\bullet \notin Q]) \\ &= (1/2)\left[(1 - (1/q)) \cdot \eta + (1 - \eta)\right] = (1/2)(1 - \eta/q).\end{aligned}$$

While $\Pr[W_2] = \Pr[b = 1] \cdot \Pr[i \in Q] = \eta/2$, we have: $\Pr[W] \leq (1/2)(1 - (\eta/q) + \eta) = (1/2)(1 + \eta(1 - 1/q)) \leq (1/2)(1 + 1 \cdot (1 - 1/q)) = 1 - 1/(2q)$. Thus we can derive that $\Pr[\overline{W}] \geq 1/(2q)$.

Next, we bound $\Pr[x' = x^* | \overline{W}]$. From the assumption that $\mathcal{A}_\mathcal{G}$ breaks the security of $\mathcal{G}$, we have the following two conditions: (1) the detection probability $\mathbf{Det}_{\mathcal{W},\mathcal{A}_\mathcal{G}}$ is negligible; (2) the advantage $\mathbf{Adv}_{\mathcal{A}_\mathcal{G}}$ is non-negligible δ.

From condition (1), we claim: $\Pr[\mathsf{Eval}^{\mathcal{G}}_{\text{IMPL}}(i_\bullet, x) = \mathsf{Eval}^{\mathcal{G}}_{\text{SPEC}}(i_\bullet, x)] \geq 1 - \mathsf{negl}(\lambda)$. The probability is over the choices of x from uniform distribution over the input space. If not, the portion of inputs that $\mathsf{Eval}^{\mathcal{G}}_{\text{IMPL}}$ deviates from its specification is non-negligible (say, δ_1) in the whole domain. The watchdog $\mathcal{W}$ we defined (that simply samples an x and tests if the values $\mathsf{Eval}^{\mathcal{G}}_{\text{IMPL}}(i_\bullet, x)$ and $\mathsf{Eval}^{\mathcal{G}}_{\text{SPEC}}(i_\bullet, x)$ are equal) satisfies that $\Pr[\mathcal{W}^{\mathsf{Eval}_{\text{IMPL}}}(1^\lambda, i_\bullet) = 1] = 1 - \delta_1$. On the other hand, $\Pr[\mathcal{W}^{\mathsf{Eval}_{\text{SPEC}}}(1^\lambda, i) = 1] = 1$. This implies that $\mathbf{Det}_{\mathcal{W},\mathcal{A}_\mathcal{G}}$ is δ_1, which contradicts condition (1). Conditioned on $\overline{W}$, the equalities:

$$y^* = \mathsf{Eval}^{\mathcal{F}}_{\text{SPEC}}(i^*, x^*) = \mathsf{Eval}^{\mathcal{F}}_{\text{SPEC}}(h_{\text{SPEC}}(i_\bullet), x^*) = \mathsf{Eval}^{\mathcal{G}}_{\text{SPEC}}(i_\bullet, x^*) = \mathsf{Eval}^{\mathcal{G}}_{\text{IMPL}}(i_\bullet, x^*)$$

hold with an overwhelming probability. That said, conditioned on $\overline{W}$, from $\mathcal{A}_\mathcal{G}$'s view, the distribution of y^* is identical to what she expects as a challenge in the subversion-resistant^{A} game.

Recall now from condition (2) the advantage $\mathbf{Adv}_{\mathcal{A}_\mathcal{G}}$ is non-negligible δ; this means $\mathcal{A}_\mathcal{G}$ inverts challenge $y^* = \mathsf{Eval}^{\mathcal{G}}_{\text{IMPL}}(i_\bullet, x^*)$ and returns a correct $x' = x^*$ with probability δ. Combining the above, we can conclude that:

$$\Pr[x' = x^*] \geq \delta(1 - \mathsf{negl}(\lambda))\frac{1}{2q} = \frac{\delta}{2q} - \mathsf{negl}(\lambda)$$

which is non-negligible; note that $q = \text{poly}(\lambda)$. Thus $\mathcal{A}_\mathcal{F}$ breaks the security of $\mathcal{F}_{\text{SPEC}}$, which leads to a contradiction. □

3.3 Constructing Subversion-Resistant^{SP} OWP/TDOWP

We can define the notion of subversion-resistant^{C} TDOWP similar as OWP^{C}. (Note that a *subvertible* TDOWP means that the adversary can invert the TDOWP using a backdoor which may have no relation to the regular trapdoor.) We defer the formal definition to the full version [22].

Indices (names) of a OWP family may have structure. For example, for OWPs based on discrete logarithm, $f_{g,p}(x) = g^x \bmod p$, the function index consists of an algebraically meaningful pair (p, g), where p is a prime and g a random generator. Applying the immunization strategy above would then require a hash function that respects this algebraic structure, mapping meaningful pairs (g, p) to meaningful pairs (g', p'). Furthermore, for a TDOWP, we must assume there is a public algorithm that can map between (public key, trapdoor) pairs.

To address these difficulties, we propose a practical *split-program* model in which every function generation algorithm (and, in general, any randomized algorithm) is composed of two components: a "random string generation algorithm" RG that outputs a uniform ℓ-bit string r, and a deterministic function index generation algorithm dKG that transforms the randomness r into a function index i. In this model, dKG is deterministic and is coupled with a known public input distribution (the output distribution of RG). Following Lemma 1 and the elaboration in Sect. 3.1, a watchdog can ensure that the implementation $\mathsf{dKG}_{\text{IMPL}}$ is "almost consistent" with $\mathsf{dKG}_{\text{SPEC}}$ (the specification) over this input distribution, i.e., $\Pr[\mathsf{dKG}_{\text{IMPL}}(r) = \mathsf{dKG}_{\text{SPEC}}(r) : r \leftarrow \mathsf{RG}_{\text{IMPL}}] \approx 1$. Morally, this forces the adversary to concentrate his malicious efforts on subverting RG.

Since we already demonstrated how to analyze the immunizing strategy for OWP, in this section we present results for TDOWP^{SP}. It is straightforward to adapt the construction and analysis to OWP^{SP}. The standard TDOWP definitions can be easily adapted in the split-program model, where the challenge index is generated by running $\mathsf{dKG}_{\text{SPEC}}$ on a uniform string r generated by $\mathsf{RG}_{\text{SPEC}}$. It is easy to see that a standard TDOWP is also a TDOWP in the split program model. For detailed definition, we refer to the full version [22].

Next we define the notion of a $\text{subversion-resistant}^{\text{SP}}$ TDOWP in the split-program model by simply augmenting Definition 3. It is easy to see the same method applies to OWP^{SP} as well. For detailed discussions of OWP^{SP}, we defer to the full version.

Generic construction of $\textbf{TDOWP}^{\textbf{SP}}$. Consider a TDOWP family $\mathcal{F}$ with specification $\mathcal{F}_{\text{SPEC}} := (\mathsf{RG}^{\mathcal{F}}_{\text{SPEC}}, \mathsf{dKG}^{\mathcal{F}}_{\text{SPEC}}, \mathsf{Eval}^{\mathcal{F}}_{\text{SPEC}}, \mathsf{Inv}^{\mathcal{F}}_{\text{SPEC}})$, where $\mathsf{RG}^{\mathcal{F}}_{\text{SPEC}}$ outputs uniform bits. Assuming a public hash function with specification $h_{\text{SPEC}} : \{0,1\}^* \to \{0,1\}^*$, we construct a TDOWP^{SP} family $\mathcal{G}$ with specification $\mathcal{G}_{\text{SPEC}} = (\mathsf{RG}^{\mathcal{G}}_{\text{SPEC}}, \mathsf{dKG}^{\mathcal{G}}_{\text{SPEC}}, \mathsf{Eval}^{\mathcal{G}}_{\text{SPEC}}, \mathsf{Inv}^{\mathcal{G}}_{\text{SPEC}})$, defined below:

- Randomness generation $r \leftarrow \mathsf{RG}^{\mathcal{G}}_{\text{SPEC}}$: $\mathsf{RG}^{\mathcal{G}}_{\text{SPEC}}$ is the same as $\mathsf{RG}^{\mathcal{F}}_{\text{SPEC}}$. That is, $\mathsf{RG}^{\mathcal{G}}_{\text{SPEC}}$ runs $\mathsf{RG}^{\mathcal{F}}_{\text{SPEC}}$ to get r and outputs r.
- Index/trapdoor generation algorithm $(i, t_i) \leftarrow \mathsf{dKG}^{\mathcal{G}}_{\text{SPEC}}(r)$: Upon receiving inputs r, it computes $\tilde{r} \leftarrow h_{\text{SPEC}}(r)$, and outputs $(i, t_i) \leftarrow \mathsf{dKG}^{\mathcal{F}}_{\text{SPEC}}(\tilde{r})$.

– $\mathsf{Eval}^{\mathcal{G}}_{\text{SPEC}}, \mathsf{Inv}^{\mathcal{G}}_{\text{SPEC}}$ are the same as $\mathsf{Eval}^{\mathcal{F}}_{\text{SPEC}}, \mathsf{Inv}^{\mathcal{F}}_{\text{SPEC}}$.[7]

See also the pictorial description for $\mathsf{dKG}^{\mathcal{G}}_{\text{SPEC}}$ in Fig. 4:

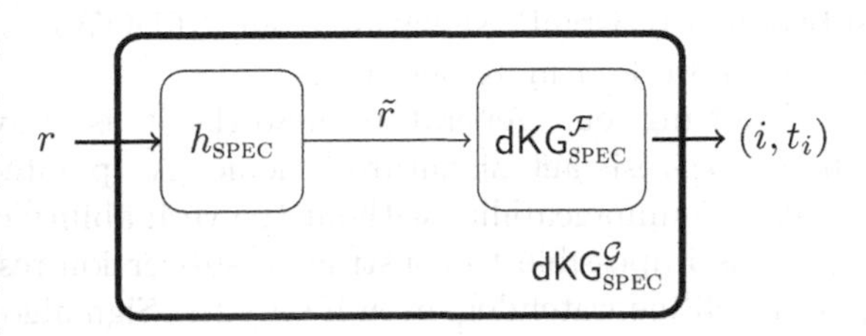

Fig. 4. New specification $\mathsf{dKG}^{\mathcal{G}}_{\text{SPEC}}$.

Security analysis. The security of $\text{OWP}^{\text{SP}}/\text{TDOWP}^{\text{SP}}$ is more subtle than it looks. Randomizing the function index directly indeed destroys any backdoor structure; however, simply randomizing the random coins for generating the function index might lead the adversary to another index/backdoor pair. It will be critical in the split-program model that, with an offline watchdog, the output of $\mathsf{RG}_{\text{IMPL}}$ is unpredictable even to the adversary who implements it.

A few words about the security proof: Recall that in the OWP^{A} proof, the reduction tries to "program the random oracle" so that the challenge of the specification can be embedded into the challenge to the adversary. In the split-program model, however, the reduction can directly embed the challenge if outputs of RG are unpredictable to the adversary; in this case, from the view of the adversary, any random index as challenge is possible to appear in the TDOWP^{SP} game. Therefore, we here do not need to program the random oracle. We defer the full proof to the full version [22].

Theorem 2. *Assume h_{SPEC} is random oracle, and $\mathcal{F}$ with specification $\mathcal{F}_{\text{SPEC}}$ is a TDOWP. Then $\mathcal{G}$ with specification $\mathcal{G}_{\text{SPEC}}$ defined above is a TDOWP^{SP} in the offline watchdog model.*

4 Subversion-Resistant Signatures

In this section, we consider the challenge of designing digital signature schemes secure against kleptographic attacks. Previously results [1,4] suggest that a unique signature scheme [13,17] is secure against subversion of the signing algorithm assuming it satisfies the *verifiability condition*: every message signed by the sabotaged $\mathsf{Sign}_{\text{IMPL}}$ should be verified via $\mathsf{Verify}_{\text{SPEC}}$. As mentioned in the introduction, in these constructions *the key generation and verification algorithms are assumed to be faithfully implemented* while, in practice, all implementations normally come together. Thus, our goal in this section is to construct a signature scheme secure in the *complete subversion* model.

[7] We remark that in the split-program model, the hash function applies to the random bits, and the hash function is implemented by the adversary inside $\mathsf{Eval}^{\mathcal{G}}_{\text{IMPL}}$. The specification of the hash function can be modeled as a random oracle so that replacing the random oracle with an explicit function like SHA256 may be heuristically justified.

We emphasize that, in general, bringing a reduction between two primitives in the classical cryptographic world into the kleptographic world turns out to be highly non-trivial. We will see that the well-known reduction for full domain hash does not go through in the kliptographic setting when we try to build a subversion-resistant^{C} signature from a TDOWP^{C}. (See Remark 3 and the proof of Theorem 3 for more details).

Following our general framework, it is easy to derive a definition for subversion-resistant signature scheme. As pointed out in [1], it is impossible to achieve unforgeability without the verifiability condition. Using our terminology, it is impossible to construct a subversion-resistant signature scheme with just an offline watchdog, even if only the Sign algorithm is subverted. So we will work in the online watchdog model where the watchdog can check the transcripts generated during the game between $\mathcal{C}$ and $\mathcal{A}$.[8] Next we define the security for digital signature schemes in the complete subversion model where all algorithms are implemented by the adversary, including the key generation algorithm.

Definition 5. *The specification* $\Pi_{\text{SPEC}} = (\mathsf{KG}_{\text{SPEC}}, \mathsf{Sign}_{\text{SPEC}}, \mathsf{Verify}_{\text{SPEC}})$ *of a signature scheme is* ***subversion-resistant***^{C} *in the online watchdog model if there exists a* PPT *watchdog* $\mathcal{W}$ *such that: for any* PPT *adversary* $\mathcal{A}$ *playing the following game (Fig. 5) with the challenger* $\mathcal{C}$*, either the detection probability* $\mathbf{Det}_{\mathcal{W},\mathcal{A}}$ *is non-negligible, or the advantage* $\mathbf{Adv}_{\mathcal{A}}$ *is negligible.*

Here the detection probability *of the watchdog* $\mathcal{W}$ *with respect to* $\mathcal{A}$ *is defined as*

$$\mathbf{Det}_{\mathcal{W},\mathcal{A}}(1^\lambda) = \left|\Pr[\mathcal{W}^{\Pi_{\text{IMPL}}}(1^\lambda, \tau) = 1] - \Pr[\mathcal{W}^{\Pi_{\text{SPEC}}}(1^\lambda, \widehat{\tau}) = 1]\right| ,$$

and the advantage *of the adversary* $\mathcal{A}$ *is defined as*

$$\mathbf{Adv}_{\mathcal{A}}(1^\lambda) = \Pr\left[(\mathcal{A}(1^\lambda) \Leftrightarrow \mathcal{C}^{\Pi_{\text{IMPL}}}(1^\lambda)) = 1\right] ,$$

where τ *is the transcript that generated when the challenger uses* Π_{IMPL} *and* $\widehat{\tau}$ *is the transcript generated when the challenger uses* Π_{SPEC}*.*

Discussion. To extend previous results to the complete subversion model, the main challenge is to protect the (randomized) key generation algorithm against subversion attacks. While the subliminal channel attacks of Bellare et al. [4] apply to arbitrary sabotaged randomized algorithms, we observe that the key generation algorithm will be run only once (as in the security definition) which provides some hope that the subliminal channel can be controlled.

Next, we will prove that a variant of the widely deployed full domain hash scheme [5,8] can achieve the security in the complete subversion model. More concretely, in this variant, the signing algorithm needs to hash m *together with* pk; we remark that this modification is critical for the security reduction (see Remark 3).

[8] Note that, for digital signature schemes, it seems far preferable to adopt an online watchdog rather than an omniscient watchdog as in [4,9]. Due to the nature of signature schemes, transcripts consist of message-signature pairs which could arguably be *publicly* verified, and an online watchdog is sufficient.

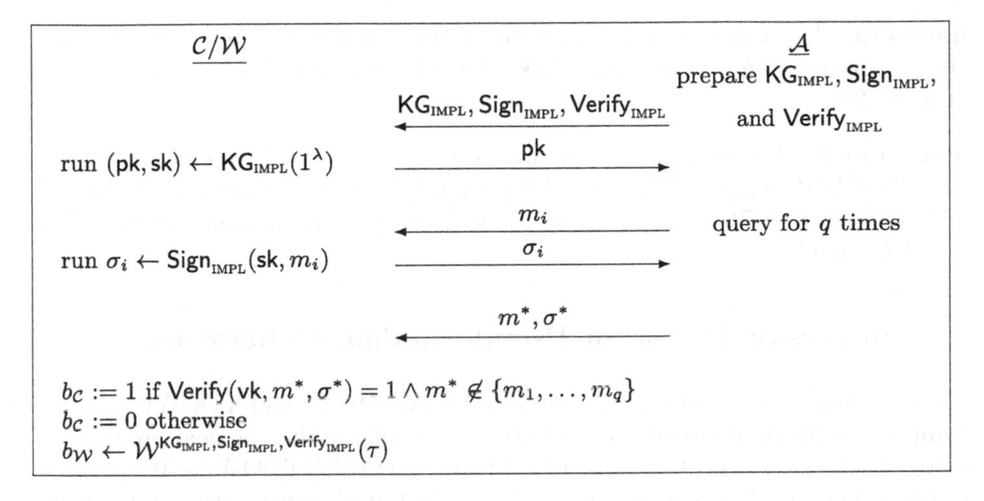

Fig. 5. Subversion-resistant$^\text{C}$ Signature Game, where $\tau := (\mathsf{pk}, \{m_i, \sigma_i\}_{i \in [q]}, m^*, \sigma^*)$

When instantiating its key generation with our subversion-resistant TDOWP$^\text{C}$, this variant gives a subversion-resistant signature scheme.

Constructing signature schemes with an online watchdog. Given a TDOWP$^\text{C}$, with specification $\mathcal{F}_{\text{SPEC}} := (\mathsf{KG}^{\mathcal{F}}_{\text{SPEC}}, \mathsf{Eval}^{\mathcal{F}}_{\text{SPEC}}, \mathsf{Inv}^{\mathcal{F}}_{\text{SPEC}})$, and a public hash function with specification $h_{\text{SPEC}} : \mathcal{PK} \times \mathcal{M} \to \mathcal{M}$, where $\mathcal{PK}$ is the public key space and $\mathcal{M}$ is the message space, we construct a subversion-resistant$^\text{C}$ signature scheme $\mathcal{SS}$ with specification $\mathcal{SS}_{\text{SPEC}} := (\mathsf{KG}^{\mathcal{SS}}_{\text{SPEC}}, \mathsf{Sign}^{\mathcal{SS}}_{\text{SPEC}}, \mathsf{Verify}^{\mathcal{SS}}_{\text{SPEC}})$ as follows:

- Key generation $(\mathsf{pk}, \mathsf{sk}) \leftarrow \mathsf{KG}^{\mathcal{SS}}_{\text{SPEC}}(\lambda)$, where $\mathsf{KG}^{\mathcal{SS}}_{\text{SPEC}}$ is given by: Compute $(i, t_i) \leftarrow \mathsf{KG}^{\mathcal{F}}_{\text{SPEC}}(\lambda)$, and set $\mathsf{pk} := i$ and $\mathsf{sk} := t_i$;
- Signature generation $\sigma \leftarrow \mathsf{Sign}^{\mathcal{SS}}_{\text{SPEC}}(\mathsf{pk}, \mathsf{sk}, m)$, where $\mathsf{Sign}^{\mathcal{SS}}_{\text{SPEC}} = (h_{\text{SPEC}}, \mathsf{Inv}^{\mathcal{F}}_{\text{SPEC}})$ is given by: Upon receiving message m, compute $\tilde{m} = h_{\text{SPEC}}(\mathsf{pk}, m)$, and $\sigma = \mathsf{Inv}^{\mathcal{F}}_{\text{SPEC}}(\mathsf{pk}, \mathsf{sk}, \tilde{m})$, where $\mathsf{pk} = i, \mathsf{sk} = t_i$.
- Verification algorithm $b \leftarrow \mathsf{Verify}^{\mathcal{SS}}_{\text{SPEC}}(\mathsf{pk}, m, \sigma)$, where $\mathsf{Verify}^{\mathcal{SS}}_{\text{SPEC}} = (h_{\text{SPEC}}, \mathsf{Eval}^{\mathcal{F}}_{\text{SPEC}})$ is given by: Upon receiving message-signature pair (m, σ), if $\mathsf{Eval}^{\mathcal{F}}_{\text{SPEC}}(\mathsf{pk}, \sigma) = h_{\text{SPEC}}(\mathsf{pk}, m)$ then set $b := 1$, otherwise set $b := 0$; here $\mathsf{pk} = i$.

Remark 3. We emphasize here that the specification of the Sign algorithm defines the hash and the inversion function separately, thus the adversary has to provide the implementation for each of them to the watchdog. Verify is treated similarly.

We also stress that using the full domain hash directly (without adding pk into the hash) results in the possibility that the random oracle query for m^* is asked before the implementations are prepared. In this case, the simulator has not yet received y^* from the TDOWP challenger, and the simulator has no way to embed y^* into the target. Including the pk in the hash essentially renders any random oracle queries made before the implementations are provided essentially useless (as they will be unrelated to any of the signatures), since the adversary cannot predict the actual value of pk. We defer the detailed proof to the full version [22].

Theorem 3. *Assume h_{SPEC} is random oracle, and $\mathcal{F}$ with specification $\mathcal{F}_{\text{SPEC}}$ is a TDOWP^{C} in the offline watchdog model. Then the signature scheme $\mathcal{SS}$ with specification $\mathcal{SS}_{\text{SPEC}}$ constructed above is subversion-resistant^{C} in the online watchdog model.*

5 Subversion-Resistant Pseudorandom Generators

Having studied the fundamental building blocks (OWPs and TDOWPs) in the complete subversion model, we now attempt to mimic the classical program of constructing richer cryptographic primitives from OWP/TDOWPs. We proceed in two different ways. The first is to carry "black-box" construction, the second is to generalize the immunizing strategy to broader settings. We remark that typical "black-box" constructions and reductions may not survive in the cliptographic model (indeed, even such basic features as the presence of multiple calls to a randomized algorithm can significantly affect security [4].) We begin by focusing on pseudorandom generators (PRG).

We first review the basic notions of PRG under subversion and then provide a specific construction that mimics the classical Blum-Micali PRG construction in this cliptographic context. We then examine how to extend the applicability of our general sanitizing strategy for OWP/TDOWPs to more general settings, demonstrating a strategy of public immunization for PRGs. Note that an impossibility result was shown in [10] that no public immunizing strategy is possible if it is applied to the output of the backdoored PRG, so a solution involves some trusted randomness is proposed. We also remark that all algorithms in our backdoor-free PRG construction—including the sanitizing function (which can be part of the KG algorithm in the specification)—can be subverted. Thus we provide the first PRG constructions secure in the complete subversion model.

We remark that since we follow the formalization of [10], the stretching algorithm is deterministic and stateful. In this case, the input distribution is evolving and not fixed, a universal watchdog cannot exhaust all those distributions. Fortunately, in the case of PRG stretching algorithm, we can still establish such a result, see security analysis of Theorem 4.

5.1 Preliminaries: The Definition of a Subversion-Resistant^{A} PRG

We adopt the definition from [10]: a pseudorandom generator consists of a pair of algorithms $(\mathsf{KG}, \mathsf{PRG})$, where KG outputs a public parameter pk and $\mathsf{PRG} : \{0,1\}^* \times \{0,1\}^\ell \to \{0,1\}^\ell \times \{0,1\}^{\ell'}$ takes the public parameter pk and an ℓ-bit random seed s as input; it returns a state $s_1 \in \{0,1\}^\ell$ and an output string $r_1 \in \{0,1\}^{\ell'}$. PRG may be iteratively executed; in the i-th iteration, it takes the state from the previous iteration s_{i-1} as the seed and generates the current state

s_i and output r_i. We use $q - \mathsf{PRG}$ to denote the result of q iterations of PRG with outputs $r_1, \ldots, r_q$ (each $r_i \in \{0,1\}^{\ell'}$).

They then define the notion of a *backdoored PRG* [10]: the adversary sets up a public parameter pk and may keep the corresponding backdoor sk. The output distribution $\mathsf{PRG}(pk, \mathcal{U})$ must still look pseudorandom to all algorithms that do not hold the backdoor sk (e.g., it fools the watchdog), where $\mathcal{U}$ is the uniform distribution; however, with sk, the adversary is able to distinguish the output from a uniform string, breaking the PRG.

The definition of a backdoored-PRG [10] is closely related to the subversion-resistant^A definition in our definitional framework, as the adversary is empowered to choose the "index" pk. Although there are several variants that all appear meaningful and interesting for PRG in the cliptographic settings, we will initially focus on the subversion-resistant^{A} PRG as the striking real world example of Dual_EC subversion is indeed in this model. Additionally, from Lemma 3, we remark that any PRG^{A} is a PRG^{C}.

We first reformulate the definition of [10] in the subversion-resistant^{A} cliptographic framework: There exist "specification" versions of the algorithms and an offline watchdog. The parameter generation algorithm $\mathsf{KG}_{\mathrm{SPEC}}$ has the requirement that the distribution of the adversarially generated public parameter must be indistinguishable from the output distribution of $\mathsf{KG}_{\mathrm{SPEC}}$. Additionally, as the PRG algorithm is deterministic, and its input distribution is public, an offline watchdog can ensure that it is consistent with its specification $\mathsf{PRG}_{\mathrm{SPEC}}$ on an overwhelming fraction of the inputs. The formal definitions are as follows:

Definition 6. *We say that a PRG (with the specification* $(\mathsf{KG}_{\mathrm{SPEC}}, \mathsf{PRG}_{\mathrm{SPEC}})$*) is* q**-subversion-resistant$^{\mathbf{A}}$** *in the offline watchdog model if, there exists a* PPT *watchdog* $\mathcal{W}$ *such that: for any* PPT *adversary* $\mathcal{A}$ *playing the following game (Fig. 6) with the challenger* $\mathcal{C}$*, either the detection probability* $\mathbf{Det}_{\mathcal{W},\mathcal{A}}$ *is non-negligible, or the advantage* $\mathbf{Adv}_{\mathcal{A}}$ *is negligible.*

Here the detection probability *of the watchdog* $\mathcal{W}$ *with respect to* $\mathcal{A}$ *is defined as*

$$\mathbf{Det}_{\mathcal{W},\mathcal{A}}(1^\lambda) = \left|\Pr[\mathcal{W}^{\mathsf{PRG}_{\mathrm{IMPL}}}(1^\lambda, pk_\bullet) = 1] - \Pr[\mathcal{W}^{\mathsf{PRG}_{\mathrm{SPEC}}}(1^\lambda, pk) = 1]\right| ,$$

and the advantage *of the adversary* $\mathcal{A}$ *is defined as*

$$\mathbf{Adv}_{\mathcal{A}}(1^\lambda) = \left|\Pr\left[(\mathcal{A}(1^\lambda) \Leftrightarrow \mathcal{C}^{\mathsf{PRG}_{\mathrm{IMPL}}}(1^\lambda, pk_\bullet)) = 1\right] - \frac{1}{2}\right| .$$

where $pk \leftarrow \mathsf{KG}_{\mathrm{SPEC}}(1^\lambda)$*, and* $\mathsf{PRG}_{\mathrm{IMPL}}, pk_\bullet$ *are chosen by the adversary.*

We say that such PRG is a PRG^{A} to stress that the public parameters are generated by the adversary.

5.2 Constructing q-PRG^A from a OWP^{A}

In this section, we provide constructions of a PRG^{A} based on a OWP^{A}. We start with a construction based on a (simplified) Blum-Micali PRG, and then extend

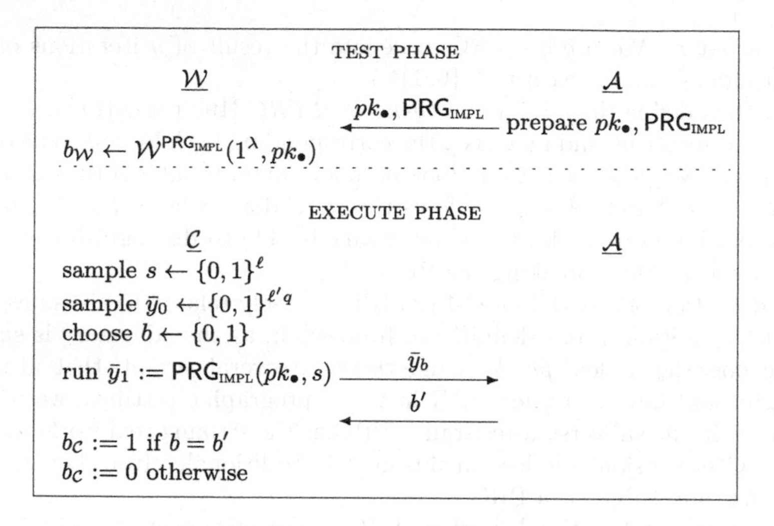

Fig. 6. Subversion-resistant[A] PRG Game

it to a full-fledged solution. We remark that a similar reduction can be used to construct a subversion-resistant[C] PRG from a subversion-resistant[C] OWP (where the challenger queries $\mathsf{KG}_{\text{IMPL}}$ to choose a public parameter).

Before describing the details of our construction, we recall the classic generic construction of Goldreich-Levin (GL), yielding a hardcore predicate [12] for any OWF f. We suppose the input x of f is divided into two halves $x = (x_1, x_2)$ and define the bit $B(x) = \langle x_1, x_2 \rangle$; $B(x)$ is hard to predict given $x_1, f(x_2)$, assuming that f is one-way. Moreover, if there is a PPT algorithm that predicts $B(x)$ with significant advantage δ given $x_1, f(x_2)$, then there is a PPT algorithm I that inverts f with probability poly(δ).

Basic construction. We will show that given a subversion-resistant[A] one-way permutation (OWP) family $\mathcal{F}$ with specifications and implementations $\mathcal{F}_{\text{SPEC}} := (\mathsf{KG}^{\mathcal{F}}_{\text{SPEC}}, \mathsf{Eval}^{\mathcal{F}}_{\text{SPEC}})$ and $(\mathsf{KG}_{\mathcal{F}}, \mathsf{Eval}_{\mathcal{F}})$ respectively, the classic Blum-Micali PRG [6] (using the GL hardcore predicate) is 1-subversion-resistant[A]. Our basic construction $\mathcal{G}$ with the specification $\mathcal{G}_{\text{SPEC}} := (\mathsf{KG}^{\mathcal{G}}_{\text{SPEC}}, \mathsf{PRG}^{\mathcal{G}}_{\text{SPEC}})$ is as follows:

- Parameter generation algorithm $pk \leftarrow \mathsf{KG}^{\mathcal{G}}_{\text{SPEC}}(\lambda)$: compute $i \leftarrow \mathsf{KG}^{\mathcal{F}}_{\text{SPEC}}(\lambda)$ and set $pk := i$;
- Bit string generation algorithm $(s', b) \leftarrow \mathsf{PRG}^{\mathcal{G}}_{\text{SPEC}}(pk, s)$: upon receiving s and pk, where $pk = i$, $s = s_1 || s_2$ and $|s_1| = |s_2| = \ell$, compute the following: $s'_1 := s_1$, $s'_2 := \mathsf{Eval}^{\mathcal{F}}_{\text{SPEC}}(i, s_2)$, and $s' = s'_1 || s'_2$, $b := \langle s_1, s_2 \rangle$.

Security analysis. We can show in the lemma below that, with a specification designed as above, the basic construction above is a 1-subversion-resistant PRG. The intuition is that in the (simplified) Blum-Micali PRG, a distinguisher can be transformed into an OWP inverter (following the GL proof); thus an adversary who can build a backdoor for this PRG violates the subversion-resistance of $\mathcal{F}$.

We present the lemma for its security, while due to lack of space, we refer the detailed proof to the full version [22].

Lemma 5. *If $\mathcal{F}$ with specification $\mathcal{F}_{\text{SPEC}}$ is a OWP^{A} in the offline watchdog model, then $\mathcal{G}$ with specification $\mathcal{G}_{\text{SPEC}}$ constructed above is a 1-subversion-resistant^{A} PRG in the offline watchdog model.*

Full – fledged PRG$^{\mathbf{A}}$. We can easily adapt our basic construction to the full-fledged PRG^{A} construction via the iteration as the BM-PRG and argue the security following the classic hybrid lemma. We refer the details of the construction and analysis to the full version [22].

5.3 A General Public Immunization Strategy for PRG$^{\mathbf{A}}$

An impossibility result concerning public immunization of a PRG (to yield a PRG^{A}) was presented in [10]. However, we observe that this impossibility result only applies to an immunization procedure that operates on the *output* of the PRG^{A}. The general construction of OWP^{A} shown above inspires us to consider an alternate general immunizing strategy for (potentially subvertible) PRGs. We establish that—similar to the procedure above for eliminating backdoors in OWPs—one can randomize the public parameter to sanitize a PRG.[9]

The intuition for this strategy to be effective in the setting of PRG is similar: considering a specification $\mathsf{KG}_{\text{SPEC}}$ that outputs a uniform pk from its domain, no single backdoor can be used to break the security for a large fraction of public parameter space; otherwise, one can use this trapdoor to break the PRG security of the specification. As above, while the adversary can subvert the hash function, an offline watchdog can ensure the hash function is faithful enough to render it difficult for the adversary arrange for the result of the hashed parameter to be amenable to any particular backdoor.

Consider a (potentially subvertible) PRG with specification $\mathcal{F}_{\text{SPEC}} = (\mathsf{KG}^{\mathcal{F}}_{\text{SPEC}}, \mathsf{PRG}^{\mathcal{F}}_{\text{SPEC}})$; we assume that $\mathsf{KG}^{\mathcal{F}}_{\text{SPEC}}$ outputs a uniform element of its range PP. Consider hash function with specification $h_{\text{SPEC}} : PP \to PP$. Then we construct a PRG^{A} $\mathcal{G}$ with its specification $\mathcal{G}_{\text{SPEC}} := (\mathsf{KG}^{\mathcal{G}}_{\text{SPEC}}, \mathsf{PRG}^{\mathcal{G}}_{\text{SPEC}})$:

- Parameter generation algorithm $pk \leftarrow \mathsf{KG}^{\mathcal{G}}_{\text{SPEC}}$: Compute $\mathsf{KG}^{\mathcal{F}}_{\text{SPEC}}$, resulting in the output pk;
- Bit string stretch algorithm $(s', r) \leftarrow \mathsf{PRG}^{\mathcal{G}}_{\text{SPEC}}(pk, s)$ which is given by: Upon receiving a random seed s and public keys pk as inputs, it computes $\widetilde{pk} = h_{\text{SPEC}}(pk)$ and it computes $\mathsf{PRG}^{\mathcal{F}}_{\text{SPEC}}(\widetilde{pk}, s)$ and obtains s', r as outputs, where r would be the actual output, while s' would be used as the seed for next iteration. See also the pictorial illustration for $\mathsf{PRG}^{\mathcal{G}}_{\text{SPEC}}$ in Fig. 7.

[9] To interpret this results, the solution of [10] is in a semi-private model which requires a trusted seed/key generation, thus part of the PRG algorithms can not be subverted. It follows that the construction of PRG in the complete subversion model was still open until our solution. In contrast, our sanitizing strategy does not require any secret, and even the deterministic hash function can be implemented by the adversary as part of the KG algorithm.

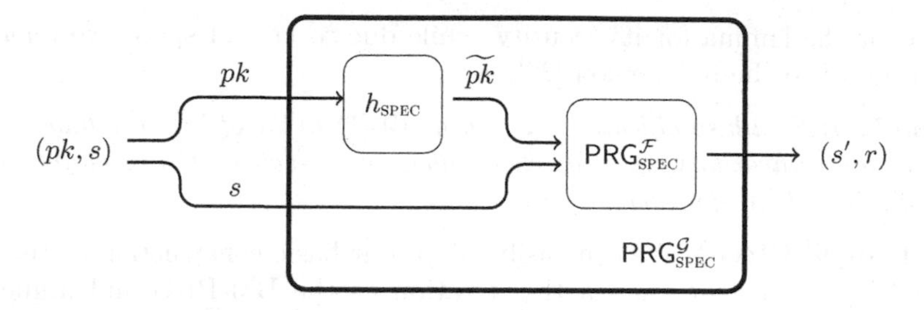

Fig. 7. Public immunization strategy for PRG.

Security analysis. If the above PRG only iterates once, the security analysis would be very similar to that of Theorem 1; since any potential backdoor embedded in the public parameter is now destroyed, and the stretch algorithm is a deterministic algorithm with a public input distribution, thus an offline watchdog can already ensure it to be (essentially) consistent with its specification.

Things become trickier when the PRG may be iterated with arbitrary number of times. For example, suppose the watchdog checks only for t iterations, $\mathsf{PRG}_{\text{IMPL}}$ might deviate from the $t+1$-th iteration. This might be indeed problematic for general deterministic algorithms. Fortunately, for this particular example of PRG, the watchdog simply checks for one uniform input and compares the output with that generated by the specification is enough to ensure almost-everywhere consistency. To see this, the adversary can create a subset of inputs $S = \{s\}$, such that: $\mathsf{PRG}_{\text{IMPL}}(pk, s) \neq \mathsf{PRG}_{\text{SPEC}}(pk, s)$, where pk is the adversarially generated public parameter. Observe that the probability that a randomly chosen input s falls in S would be negligible. Otherwise the watchdog can detect with a non-negligible probability. While the difference with a stateful stretching algorithm is that it offers the adversary more chances to hit the bad set S because of the iterations. Note that when $\mathsf{PRG}_{\text{IMPL}}(pk, s) = \mathsf{PRG}_{\text{SPEC}}(pk, s)$ for some randomly chosen s, then the output s' would also be pseudorandom; iterating on this input, the stretching algorithm yields a polynomially many pseudorandom strings, thus the probability of any of those hit the bad set S would be still negligible. With this observation, we can still claim that with an overwhelming probability, $\mathsf{PRG}_{\text{IMPL}}$ will be consistent with $\mathsf{PRG}_{\text{SPEC}}$ even after arbitrary number of iterations (polynomially bounded). We defer the proof to the full version.

Theorem 4. *Assume h_{SPEC} is random oracle, and $\mathcal{F}$ with specification $\mathcal{F}_{\text{SPEC}} = (\mathsf{KG}^{\mathcal{F}}_{\text{SPEC}}, \mathsf{PRG}^{\mathcal{F}}_{\text{SPEC}})$ is a pseudorandom generator, where $\mathsf{KG}^{\mathcal{F}}_{\text{SPEC}}$ outputs pk randomly from its range. Then $\mathcal{G}$ with specification $\mathcal{G}_{\text{SPEC}}$ in the above construction yields a q-subversion-resistant PRG^{A} for any polynomially large q.*

Remark 4. If the public parameter contains only random group elements, e.g., the Dual_EC PRG, we may simply encode them into bits and use a regular hash function like SHA-256, and convert the resulting bits back to a group element;

References

1. Ateniese, G., Magri, B., Venturi, D.: Subversion-resilient signature schemes. In: Ray, I., Li, N., Kruegel, C. (eds.), ACM CCS 15, pp. 364–375. ACM Press, October 2015
2. Bellare, M., Hoang, V.T.: Resisting randomness subversion: fast deterministic and hedged public-key encryption in the standard model. In: Oswald, E., Fischlin, M. (eds.) EUROCRYPT 2015. LNCS, vol. 9057, pp. 627–656. Springer, Heidelberg (2015). doi:10.1007/978-3-662-46803-6_21
3. Bellare, M., Jaeger, J., Kane, D.: Mass-surveillance without the state: Strongly undetectable algorithm-substitution attacks. In: Ray, I., Li, N., Kruegel, C. (eds.), ACM CCS 15, pp. 1431–1440. ACM Press, October 2015
4. Bellare, M., Paterson, K.G., Rogaway, P.: Security of symmetric encryption against mass surveillance. In: Garay, J.A., Gennaro, R. (eds.) CRYPTO 2014. LNCS, vol. 8616, pp. 1–19. Springer, Heidelberg (2014). doi:10.1007/978-3-662-44371-2_1
5. Bellare, M., Rogaway, P.: The exact security of digital signatures-how to sign with RSA and rabin. In: Maurer, U. (ed.) EUROCRYPT 1996. LNCS, vol. 1070, pp. 399–416. Springer, Heidelberg (1996). doi:10.1007/3-540-68339-9_34
6. Blum, M., Micali, S.: How to generate cryptographically strong sequences of pseudo random bits. In: 23rd FOCS, pp. 112–117. IEEE Computer Society Press, November 1982
7. Checkoway, S., Niederhagen, R., Everspaugh, A., Green, M., Lange, T., Ristenpart, T., Bernstein, D.J., Maskiewicz, J., Shacham, H., Fredrikson, M.: On the practical exploitability of dual EC in TLS implementations. In: Proceedings of the 23rd USENIX Security Symposium, San Diego, CA, USA, August 20–22, 2014, pp. 319–335 (2014)
8. Coron, J.-S.: On the exact security of full domain hash. In: Bellare, M. (ed.) CRYPTO 2000. LNCS, vol. 1880, pp. 229–235. Springer, Heidelberg (2000). doi:10.1007/3-540-44598-6_14
9. Degabriele, J.P., Farshim, P., Poettering, B.: A more cautious approach to security against mass surveillance. In: Leander, G. (ed.) FSE 2015. LNCS, vol. 9054, pp. 579–598. Springer, Heidelberg (2015). doi:10.1007/978-3-662-48116-5_28
10. Dodis, Y., Ganesh, C., Golovnev, A., Juels, A., Ristenpart, T.: A formal treatment of backdoored pseudorandom generators. In: Oswald, E., Fischlin, M. (eds.) EUROCRYPT 2015. LNCS, vol. 9056, pp. 101–126. Springer, Heidelberg (2015). doi:10.1007/978-3-662-46800-5_5
11. Dodis, Y., Mironov, I., Stephens-Davidowitz, N.: Message transmission with reverse firewalls–secure communication on corrupted machines. Cryptology ePrint Archive, Report 2015/548 (2015). http://eprint.iacr.org/2015/548
12. Goldreich, O., Levin, L.A.: A hard-core predicate for all one-way functions. In: 21st ACM STOC, pp. 25–32. ACM Press, May 1989
13. Goldwasser, S., Ostrovsky, R.: *Invariant* signatures and non-interactive zero-knowledge proofs are equivalent. In: Brickell, E.F. (ed.) CRYPTO 1992. LNCS, vol. 740, pp. 228–245. Springer, Heidelberg (1993). doi:10.1007/3-540-48071-4_16
14. Haitner, I., Holenstein, T.: On the (im)possibility of key dependent encryption. In: Reingold, O. (ed.) TCC 2009. LNCS, vol. 5444, pp. 202–219. Springer, Heidelberg (2009). doi:10.1007/978-3-642-00457-5_13
15. Hopper, N.J., Langford, J., Ahn, L.: Provably secure steganography. In: Yung, M. (ed.) CRYPTO 2002. LNCS, vol. 2442, pp. 77–92. Springer, Heidelberg (2002). doi:10.1007/3-540-45708-9_6

16. Juels, A., Guajardo, J.: RSA key generation with verifiable randomness. In: Naccache, D., Paillier, P. (eds.) PKC 2002. LNCS, vol. 2274, pp. 357–374. Springer, Heidelberg (2002). doi:10.1007/3-540-45664-3_26
17. Lysyanskaya, A.: Unique signatures and verifiable random functions from the DH-DDH separation. In: Yung, M. (ed.) CRYPTO 2002. LNCS, vol. 2442, pp. 597–612. Springer, Heidelberg (2002). doi:10.1007/3-540-45708-9_38
18. Mironov, I., Stephens-Davidowitz, N.: Cryptographic reverse firewalls. In: Oswald, E., Fischlin, M. (eds.) EUROCRYPT 2015. LNCS, vol. 9057, pp. 657–686. Springer, Heidelberg (2015). doi:10.1007/978-3-662-46803-6_22
19. NIST. Special publication 800-90: Recommendation for random number generation using deterministic random bit generators. National Institute of Standards and Technology (2012). http://csrc.nist.gov/publications/PubsSPs.html
20. Perlroth, N., Larson, J., Shane, S.: N.S.A. able to foil basic safeguards of privacy on web. The New York Times (2013). http://www.nytimes.com/2013/09/06/us/nsa-foils-much-internet-encryption.html
21. Rogaway, P.: The moral character of cryptographic work. Cryptology ePrint Archive, Report 2015/1162 (2015). http://eprint.iacr.org/2015/1162
22. Russell, A., Tang, Q., Yung, M., Zhou, H.-S.: Cliptography: Clipping the power of kleptographic attacks. Cryptology ePrint Archive, Report 2015/695 (2015). http://eprint.iacr.org/2015/695
23. Russell, A., Tang, Q., Yung, M., Zhou, H.-S.: Destroying steganography via amalgamation: Kleptographically cpa secure public key encryption. In Cryptology ePrint Archive, Report 2016/530 (2016). http://eprint.iacr.org/2016/530
24. Simmons, G.J.: The prisoners' problem and the subliminal channel. In: Chaum, D. (ed.) Advances in Cryptology, pp. 51–67. Springer, Heidelberg (1983)
25. Simmons, G.J.: A secure subliminal channel (?). In: Williams, H.C. (ed.) CRYPTO 1985. LNCS, vol. 218, pp. 33–41. Springer, Heidelberg (1986). doi:10.1007/3-540-39799-X_5
26. Wikipedia. Nothing up my sleeve. https://en.wikipedia.org/wiki/Nothing_up_my_sleeve_number
27. Young, A., Yung, M.: The dark side of "black-box" cryptography or: should we trust capstone? In: Koblitz, N. (ed.) CRYPTO 1996. LNCS, vol. 1109, pp. 89–103. Springer, Heidelberg (1996). doi:10.1007/3-540-68697-5_8
28. Young, A., Yung, M.: Kleptography: using cryptography against cryptography. In: Fumy, W. (ed.) EUROCRYPT 1997. LNCS, vol. 1233, pp. 62–74. Springer, Heidelberg (1997). doi:10.1007/3-540-69053-0_6
29. Young, A., Yung, M.: Monkey: black-box symmetric ciphers designed for MON*opolizing* KEY*s*. In: Vaudenay, S. (ed.) FSE 1998. LNCS, vol. 1372, pp. 122–133. Springer, Heidelberg (1998). doi:10.1007/3-540-69710-1_9
30. Young, A., Yung, M.: An elliptic curve backdoor algorithm for RSASSA. In: Camenisch, J.L., Collberg, C.S., Johnson, N.F., Sallee, P. (eds.) IH 2006. LNCS, vol. 4437, pp. 355–374. Springer, Heidelberg (2007). doi:10.1007/978-3-540-74124-4_24
31. Young, A.L., Yung, M.M.: Space-efficient kleptography without random oracles. In: Furon, T., Cayre, F., Doërr, G., Bas, P. (eds.) IH 2007. LNCS, vol. 4567, pp. 112–129. Springer, Heidelberg (2007). doi:10.1007/978-3-540-77370-2_8
32. Young, A., Yung, M.: Kleptography from standard assumptions and applications. In: Garay, J.A., Prisco, R. (eds.) SCN 2010. LNCS, vol. 6280, pp. 271–290. Springer, Heidelberg (2010). doi:10.1007/978-3-642-15317-4_18

Zero Knowledge

Zero-Knowledge Accumulators and Set Algebra

Esha Ghosh[1(✉)], Olga Ohrimenko[2], Dimitrios Papadopoulos[3], Roberto Tamassia[1], and Nikos Triandopoulos[4]

[1] Department of Computer Science, Brown University, Providence, USA
esha_ghosh@brown.edu, rt@cs.brown.edu
[2] Microsoft Research, Cambridge, UK
oohrim@microsoft.com
[3] University of Maryland, College Park, USA
dipapado@umd.edu
[4] Stevens Institute of Technology, Hoboken, USA
ntriando@stevens.edu

Abstract. Cryptographic accumulators allow to succinctly represent a set by an accumulation value with respect to which short (non-)membership proofs about the set can be efficiently constructed and verified. Traditionally, their security captures soundness but offers no privacy: Convincing proofs reliably encode set membership, but they may well leak information about the accumulated set.

In this paper we put forward a strong privacy-preserving enhancement by introducing and devising *zero-knowledge accumulators* that additionally provide hiding guarantees: Accumulation values and proofs leak nothing about a dynamic set that evolves via element insertions/deletions. We formalize the new property using the standard real-ideal paradigm, namely demanding that an adaptive adversary with access to query/update oracles, cannot tell whether he interacts with honest protocol executions or a simulator fully ignorant of the set (even of the type of updates on it). We rigorously compare the new primitive to existing ones for privacy-preserving verification of set membership (or other relations) and derive interesting implications among related security definitions, showing that zero-knowledge accumulators offer stronger privacy than recent related works by Naor et al. [TCC 2015] and Derler et al. [CT-RSA 2015]. We construct the first dynamic universal zero-knowledge accumulator that we show to be perfect zero-knowledge and secure under the q-Strong Bilinear Diffie-Hellman assumption.

Finally, we extend our new privacy notion and our new construction to provide privacy-preserving proofs also for an authenticated dynamic set collection—a primitive for efficiently verifying more elaborate set operations, beyond set-membership. We introduce a primitive that supports a *zero-knowledge verifiable set algebra*: Succinct proofs for union, intersection and set difference queries over a dynamically evolving collection of sets can be efficiently constructed and optimally verified, while—for the first time—they leak nothing about the collection beyond the query result.

J.H. Cheon and T. Takagi (Eds.): ASIACRYPT 2016, Part II, LNCS 10032, pp. 67–100, 2016.
DOI: 10.1007/978-3-662-53890-6_3

Keywords: Zero-knowledge dynamic and universal accumulators · Zero-knowledge updates · Zero-knowledge set algebra · Outsourced computation · Integrity · Privacy · Bilinear accumulators · Cloud privacy

1 Introduction

A cryptographic accumulator is a primitive that offers a way to succinctly represent a set of elements $\mathcal{X}$ by a single value acc referred to as the *accumulation value*. Moreover, it provides a method to efficiently and succinctly prove (to a party that only holds acc) that an element x belongs to $\mathcal{X}$, by computing a constant-size proof w, referred to as *witness*. The interaction is in a three-party model, where the trusted *owner* of the set runs the initial key generation and setup process to publish the accumulation value. Later an untrusted *server* handles queries on the set issued by *clients*, providing membership answers with publicly verifiable witnesses.

Accumulators were originally introduced by Benaloh and del Mare in [4]. Numerous constructions have been proposed since, operating in various models [2,3,7,9–12,19,23,44,54][1]. Most notably, the primitive was extended to support non-membership witnesses [2,19,41], and efficient updates [2,11], introducing the notion of *universal* and *dynamic accumulators*, respectively. At the same time, accumulators found numerous other applications in the context of public-key infrastructure, certificate management and revocation, time-stamping, authenticated dictionaries, set operations, authenticated database queries, anonymous credentials, and more.

Traditionally in the literature, the security property associated with accumulators is *soundness* (or *collision-freeness*), expressed as the inability to forge a witness for an element, i.e., if $x \notin \mathcal{X}$, it should be hard to prove $x \in \mathcal{X}$ (and vice-versa for universal accumulators). No notion of privacy was considered until recently [20,23], e.g., "does the accumulation reveal anything about the elements of $\mathcal{X}$" or "what can an adversarial client, that asks queries and is presented with the accumulation and witnesses, learn about the set $\mathcal{X}$". It is clear that such a property would be attractive, if not—depending on the application—crucial. For example, in the context of securing the Domain Name System (DNS) protocol by accumulating the set of records in a zone, it is crucial to leak no information about values in the accumulated set while responding to queries.[2] As additional examples, Miers et al. [49] developed a privacy enhancement for Bitcoin, that utilizes the accumulator from [11], while Hanser and Slamanig [35] used accumulators to build randomizable polynomial commitments for anonymous credentials. In such a context, it is very important to minimize what is leaked by accumulation values and witnesses in order to achieve anonymity (for individuals and transactions).

In this work, we propose the notion of *zero-knowledge* for cryptographic accumulators. We define this property via an extensive real/ideal game, similar to

[1] We refer interested readers to [23] for a comprehensive review of existing schemes.
[2] See for example, https://tools.ietf.org/html/rfc5155.

that of standard zero-knowledge [34]. In the real setting, an adversary is allowed to choose his challenge set and to receive the corresponding accumulation. He is then given oracle access to the querying algorithm as well as an update algorithm that allows him to request updates in the set (receiving the updated accumulation value every time). In the ideal setting, the adversary interacts with a simulator that does not know anything about the set or the nature of the updates, other than the fact that an update occurred. Zero-knowledge is then defined as the inability of the adversary to distinguish between the two settings.

We provide the first zero-knowledge accumulator construction and prove its security. Our construction builds upon the *bilinear accumulator* of Nguyen [53] and achieves *perfect* zero-knowledge. Our scheme falls within the category of *dynamic universal* cryptographic accumulators: It allows to prove both membership and non-membership statements (i.e., one can compute a witness for $x \notin \mathcal{X}$), and supports efficient updates in the accumulation value due to insertions and deletions in the set. It satisfies soundness under the q-Strong Bilinear Diffie-Hellman assumption (q-SBDH), introduced in [6]. In order to provide non-membership witness computation in zero-knowledge, we had to deviate from existing non-membership proof techniques for the bilinear accumulator [2,19]. We instead use the set-disjointness technique of [56], appropriately enhanced for privacy. From an efficiency perspective, we show that introducing zero-knowledge to the bilinear accumulator comes at an insignificant cost: Asymptotically all query overheads are either the same or within a poly-logarithmic factor of the construction in [53] that offers no privacy.

Zero-knowledge vs. indistinguishability. Recently, de Meer et al. [20] and Derler et al. [23] introduced an *indistinguishability* property for cryptographic accumulators. Unfortunately, the definition of the former was inherently flawed, as noted in [23].[3] The accumulator definition in [23], while meant to support changes in the accumulated set (i.e., element insertion or deletion), did not protect the privacy of these changes. In particular, any adversary suspecting a particular modification in the set could easily check the correctness of his guess. Our notion of zero-knowledge differs from the privacy notion of [23], by protecting not only the originally accumulated set but also all subsequent updates. In fact, we formally prove that, for cryptographic accumulators, zero-knowledge is a strictly stronger property than indistinguishability.

Relation to zero-knowledge sets. Our privacy notion is reminiscent of that of zero-knowledge sets [14,15,43,48,58] where set membership and non-membership queries can be answered without revealing anything else about the set. Zero-knowledge accumulators can be seen as a relaxation of zero-knowledge sets in an "honest-committer" setting. In Sect. 3.2 we discuss this relation in more detail, also looking into the dynamic setting, comparing with existing work on updatable zero-knowledge sets [45].

[3] Subsequently, the definition was strengthened in [61], but it is still subsumed by that of [23].

Relation to zero-knowledge authenticated data structures. A cryptographic accumulator can be viewed as a special case of an *authenticated data structure* (ADS) [51,63], where the supported data type is a set. Likewise, the zero-knowledge accumulator we introduce here, falls within the framework of *zero-knowledge authenticated data structures* (ZKADS) introduced recently in [29]. We discuss the relation in detail in Sect. 3.2.

Beyond set-membership. One question that arises naturally is how to build a "set-friendly" ZKADS with a supported functionality beyond set-membership. In particular, given multiple sets, we are interested in accommodating more elaborate set-operations: set union, intersection and difference.[4] We introduce the primitive of *zero-knowledge authenticated dynamic set collection* for the following setting. A party that owns a database of sets outsources it to an untrusted server that is subsequently charged with handling queries, expressed as set operations among the database sets, issued by multiple clients; at any point, the owner can make updates to the outsourced sets. We present the first scheme that provides not only integrity of computations but also privacy for the queried set (i.e., the provided proofs leak nothing beyond the answer). The basic building block is our zero-knowledge accumulator, together with a carefully deployed *accumulation tree* [57]. We note that if we restrict the security properties only to soundness—as is the case in the traditional literature of ADS—there are existing schemes (specifically for set-operations) by Papamanthou et al. [56] for the single-operation case, and by Canetti et al. [12] and Kosba et al. [40] for the case of multiple (nested) operations. However, none of these constructions offers privacy, thus our scheme is a natural strengthening of their security guarantees, while maintaining the same efficient performance. Preserving efficiency while maintaining integrity and zero-knowledge privacy turned out to be quite challenging. In particular, answering union and set difference queries for set collections required new techniques to be developed. At a high level, the efficiency of the proof techniques in [56] strongly relies on revealing much of the non-queried information and hence could not be extended to support privacy-preserving queries.

Contributions. Our contributions can be summarized as follows:

- We introduce the property of zero-knowledge for cryptographic accumulators and show that it is strictly stronger than existing privacy notions for accumulators.
- We give an overview of the connection between zero-knowledge accumulators and related cryptographic primitives in the area (e.g., we show that zero-knowledge accumulators imply primary-secondary-resolver systems proposed in [52]).
- We provide the first construction of a zero-knowledge dynamic universal accumulator. Our scheme is perfect zero-knowledge and secure under the q-SBDH assumption; it achieves these security properties with only a small

[4] We stress that, in the computational setting, these operations form a complete set algebra.

(or no) overhead. We compare efficiency with the accumulator of [53] in Fig. 3 in terms of number of cryptographic operations performed.
- Using our zero-knowledge accumulator as a building block, we construct the first protocol for zero-knowledge outsourced set operations. Our scheme is non-interactive and offers secure and efficient subset, intersection and union operations under the q-SBDH assumption. For set-difference queries, our construction is secure under q-SBDH assumption as well, but proof construction entails a Sigma protocol thus requiring interaction. This secure set-difference protocol can also be made non-interactive, albeit in the random oracle model, (in which case the construction is in the Common Reference String model). Our construction (except for the update cost) is asymptotically as efficient as the previous state-of-the-art construction from [56], that offered no privacy guarantees.

1.1 Other Related Work

Existing works (e.g., [2, 11, 41, 53]) equip some accumulators with zero-knowledge proof-of-knowledge protocols, such that a party that knows that value x is (or is not) in $\mathcal{X}$ can efficiently prove it to a third-party arbitrator, without revealing the value. While hiding x, all existing constructions trivially expose the accumulation value as part of the proven statement. This may itself reveal information about set $\mathcal{X}$. Our privacy goals are therefore different, yet the techniques are compatible. Developing zero-knowledge proof-of-knowledge protocols for membership and non-membership, that can work with a zero-knowledge accumulator will yield a strong tool that leaks nothing about either the set or the particular element in the proof.

Most widely-used accumulator constructions—including ours—are in the trusted-setup model, i.e., the party that generates the scheme parameters originally, holds some trapdoor information that is not revealed to the adversary. E.g., for the RSA-based constructions, any adversary that knows the factorization of the modulo can trivially cheat. An alternative body of work aims to build *trapdoorless* accumulators (also referred to as *strong* accumulators) [7, 9, 44, 54, 55, 62], where the owner is entirely untrusted (effectively the owner and the server are the same entity). Unfortunately, the earlier of these works are quite inefficient for all practical purposes, while the more recent ones either yield witnesses that grow logarithmically with the size of $\mathcal{X}$ or rely on algebraic groups that are not yet well-studied in cryptography. A straight-forward way to construct a strong accumulator is via a black-box reduction from zero-knowledge sets (with corresponding efficiency caveats). While a scheme without the need for a trusted setup is clearly more attractive in terms of security, it is safe to say that we do not yet have a practical scheme with constant-size proofs, based on standard security assumptions.

Recently, Naor et al. [52] introduced primary-secondary-resolver membership proof systems, a primitive that is also a relaxation of zero-knowledge sets in the three-party model, and showed applications in network protocols [33]. While our definitions have similarities, in Sect. 3.2 we show that zero-knowledge accumulators are a stronger primitive than primary-secondary-resolver systems.

Regarding related work for set operations, the focus in the cryptographic literature has been on the privacy aspect with a very long line of works (see for example, [5,27,36,37,39]), some of which focus specifically on set-intersection (e.g., [17,18,24,38]). The above works fit in the secure two-party computation model and most are secure (or can be made with some loss in efficiency) also against malicious adversaries, thus guaranteeing the authenticity of the result. However, this approach typically requires multi-round interaction or larger communication cost than our construction. On the other hand, our two security properties are "one-sided": Only the server may cheat with respect to soundness and only the client with respect to privacy; in this setting we achieve non-interactive solutions with optimal proof-size. There also exist works that deal exclusively with the integrity of set operations, such as [50] that achieves linear verification and proof cost, and [64] that only focuses on set-intersection but can be combined with an encryption scheme to achieve privacy versus the server.

Another work that is related to ours is that of Fauzi et al. [25] that presents an efficient non-interactive zero-knowledge argument for proving relations between committed sets. Conceptually this work is close to zero-knowledge sets, allowing also for more general set operation queries. From a security viewpoint, it is in the stronger two-party model and, from a functionality viewpoint, it works for (more general) multi-set operations. However, its security relies on non-falsifiable knowledge assumptions, and the construction trivially leaks an upper-bound on the committed sets. Moreover, it cannot be efficiently generalized for operations on more than two sets at a time and it does not explicitly support efficient modifications in the sets.

We also note that recently other instantiations of zero-knowledge authenticated data structures have been proposed, including lists, trees and partially-ordered sets of bounded dimension [29,32].

2 Preliminaries

We denote with λ the security parameter and with $\nu(\lambda)$ a negligible function. A function $f(\lambda)$ is negligible if for each polynomial function $poly(\lambda)$ and all large enough values of λ, $f(\lambda) < 1/(poly(\lambda))$. We say that an event can occur with negligible probability if its occurrence probability can be upper bounded by a negligible function. Respectively, an event takes place with overwhelming probability if its complement takes place with negligible probability. The symbol $\xleftarrow{\$} \mathbb{X}$ denotes uniform sampling from domain $\mathbb{X}$. We denote the fact that a party Adv (instantiated as Turing machine) is probabilistic and runs in polynomial-time by writing PPT Adv.

Bilinear pairings. Let $\mathbb{G}$ be a cyclic multiplicative group of prime order p, generated by g. Let also $\mathbb{G}_T$ be a cyclic multiplicative group with the same order p and $e : \mathbb{G} \times \mathbb{G} \rightarrow \mathbb{G}_T$ be a bilinear pairing with the following properties:

(1) Bilinearity: $e(P^a, Q^b) = e(P, Q)^{ab}$ for all $P, Q \in \mathbb{G}$ and $a, b \in \mathbb{Z}_p$; (2) Non-degeneracy: $e(g, g) \neq 1_{\mathbb{G}_T}$; (3) Computability: There is an efficient algorithm to compute $e(P, Q)$ for all $P, Q \in \mathbb{G}$. We denote with $pub := (p, \mathbb{G}, \mathbb{G}_T, e, g)$ the bilinear pairings parameters, output by a randomized polynomial-time algorithm $\mathsf{GenParams}$ on input 1^λ. For clarity of presentation, we assume for the rest of the paper a symmetric (Type 1) pairing e. We note though that both our constructions can be securely implemented in the (more efficient) asymmetric pairing case, with straight-forward modifications (see [16] for a general discussion on pairings). Our security proofs make use of the q-Strong Bilinear Diffie-Hellman (q-SBDH) assumption over groups with bilinear pairings introduced in [6].

Assumption 1 (q-Strong Bilinear Diffie-Hellman). *For any PPT adversary* Adv *and for q being a parameter of size polynomial in λ, there exists negligible function $\nu(\lambda)$ such that the following holds:*

$$\Pr\left[\begin{array}{c} pub \leftarrow \mathsf{GenParams}(1^\lambda); s \leftarrow_R \mathbb{Z}_p^*; \\ (z, \gamma) \in \mathbb{Z}_p^* \times \mathbb{G}_T \leftarrow \mathsf{Adv}(pub, (g^s, ..., g^{s^q})) : \gamma = e(g, g)^{1/(z+s)} \end{array}\right] \leq \nu(\lambda) .$$

Complexity model. For ease of notation, we measure the asymptotic performance of our schemes by counting numbers of operations and group elements, ignoring a, poly-logarithmic in λ, factor (e.g., an operation in $\mathbb{G}$ takes one unit time).

Characteristic polynomial. A set $\mathcal{X} = \{x_1, \dots, x_n\}$ with elements $x_i \in \mathbb{Z}_p$ can be represented by a polynomial following an idea introduced in [27]. The polynomial $\mathsf{Ch}_{\mathcal{X}}(z) = \prod_{i=1}^n (x_i + z)$ from $\mathbb{Z}_p[z]$, where z is a formal variable, is called the *characteristic polynomial* of $\mathcal{X}$. In what follows, we will denote this polynomial simply by $\mathsf{Ch}_{\mathcal{X}}$ and its evaluation at a point y as $\mathsf{Ch}_{\mathcal{X}}(y)$. Characteristic polynomials enjoy a number of homomorphic properties w.r.t. set operations. We use the following characterization of set intersection of the sets: Given a collection of sets $\mathcal{X}_{i_1}, \dots \mathcal{X}_{i_k}$ and their characteristic polynomial representation, we summarize a characterization of the intersection of the sets in the following lemma.

Lemma 1 ([56]). *A set* answer *that is a common subset of sets $\mathcal{X}_{i_1}, \dots \mathcal{X}_{i_k}$, is their intersection if and only if there exist polynomials $q_1[z], \dots q_k[z]$ such that $\sum_{j \in [i_1, i_k]} q_j[z] P_j[z] = 1$ where $P_j[z] = \mathsf{Ch}_{\mathcal{X}_j \setminus \mathsf{answer}}[z]$. Computing polynomials $q_j[z]$ where $j \in [i_1, i_k]$ has $O(N \log^2 N \log\log N)$ complexity where $N = \sum_{j \in [i_1, i_k]} n_j$ and $n_j = |\mathcal{X}_j|$.*

The following two lemmas characterize the efficiency of computing the characteristic polynomial of a set –note that there is no requirement for the existence of an n-th root of unity in $\mathbb{Z}_p$ for such an algorithm to exist– and the probability that two polynomials are equivalent at a randomly chosen point.

Lemma 2 ([59]). *Given a set $\mathcal{X} = x_1, ..., x_n \in \mathbb{Z}_p^n$, its characteristic polynomial $\mathsf{Ch}_{\mathcal{X}} := \sum_{i=0}^n c_i z^i \in \mathbb{Z}_p[z]$ can be computed with $O(n \log n)$ operations by FFT interpolation.*

Lemma 3 (Schwartz–Zippel–DeMillo–Lipton). *Let $p[z], q[z]$ be two d-degree polynomials from $\mathbb{Z}_p[z]$ with $p[z] \neq q[z]$, Then for $w \xleftarrow{\$} \mathbb{Z}_p$, the probability that $p(w) = q(w)$ is at most d/p, and the equality can be tested in time $O(d)$.*

If $p \in O(2^\lambda)$, it follows that the above probability is negligible, if d is $poly(\lambda)$.

Accumulation tree. Given a collection of sets $\mathbb{S} = \{\mathcal{X}_1, \mathcal{X}_2, \ldots, \mathcal{X}_m\}$, let $\mathsf{acc}(\mathcal{X}_i)$ be a succinct representation of $\mathcal{X}_i$ using its characteristic polynomial. We describe an authentication mechanism that does the following. A trusted party computes m hash values $h_i := h(\mathsf{acc}(\mathcal{X}_i))$ (using a collision resistant cryptographic hash function) of the m sets of $\mathbb{S}$. Then given a short public digest information of the current set collection $\mathbb{S}$, the authentication mechanism provides publicly verifiable proofs of the form "h_i is the hash of the ith set of the current set collection $\mathbb{S}$". A popular authentication mechanism for such proofs are Merkle hash trees [47] based on a single value digest can provide logarithmic size proofs and support updates. An alternative mechanism to Merkle trees, (specifically in the bilinear group setting) are *accumulation trees* [57]. Intuitively, an accumulation tree can be seen as a "flat" version of Merkle trees. In this work, we use our extension (for batch updates) of the accumulation tree in [56]. The detailed construction can be found in the full version.

3 Zero-Knowledge Universal Accumulators (ZKUA)

A *dynamic universal accumulator* (DUA) consists of five probabilistic polynomial time algorithms (GenKey, Setup, Witness, Verify, Update). It represents a set $\mathcal{X}$, with elements from domain $\mathbb{X}$, by an accumulation value $\mathsf{acc} \in \mathbb{A}$. It supports queries of the form "is $x \in \mathcal{X}$?" for $x \in \mathbb{X}$ and updates to the current set (e.g., using "insert x" or "remove x" operations). The algorithms of DUA, as described below, are run between three parties: the owner, the server and the client. We follow the definitional style of [23,26] where the accumulator is described as a tuple of algorithms. In the full version we provide a discussion regarding our chosen definitional style.

Definition 1 ***(Dynamic Universal Accumulator).*** *A dynamic universal accumulator is a tuple of five PPT algorithms,* DUA = (GenKey, Setup, Witness, Verify, Update) *defined as follows:*

$(sk, vk) \leftarrow \mathsf{GenKey}(1^\lambda)$
This probabilistic algorithm takes as input the security parameter and outputs a (public) verification key vk that will be used by the client to verify query responses and a secret key sk that is kept by the owner.

$(\mathsf{acc}, ek, \mathsf{aux}) \leftarrow \mathsf{Setup}(sk, \mathcal{X})$
This probabilistic algorithm is run by the owner. It takes as input the source set $\mathcal{X}$ and produces the accumulation value acc *that will be published to both server and client, and an evaluation key ek as well as auxiliary information* aux *that will be sent only to the server in order to facilitate proof construction.*

$(b, \mathsf{w}) \leftarrow \mathsf{Witness}(\mathsf{acc}, \mathcal{X}, x, ek, \mathsf{aux})$

This algorithm is run by the server. It takes as input the evaluation key and the accumulation value ek, acc *generated by the owner, the source set* $\mathcal{X}$*, a queried element* x*, as input. It outputs a boolean value* b *indicating whether the element is in the set and a witness* w *for the answer.*

$(\mathsf{accept}/\mathsf{reject}) \leftarrow \mathsf{Verify}(\mathsf{acc}, x, b, \mathsf{w}, vk)$

This algorithm is run by the client. It takes as input the accumulation value acc *and the public key* vk *computed by the owner, a queried element* x*, a bit* b*, the witness* w *and it outputs* $\mathsf{accept}/\mathsf{reject}$.

$(\mathsf{acc}', ek', \mathsf{aux}') \leftarrow \mathsf{Update}(\mathsf{acc}, \mathcal{X}, x, sk, \mathsf{aux}, \mathsf{upd})$

This algorithm takes as input the current set with its accumulation value and auxiliary information, as well as an element x *to be inserted to* $\mathcal{X}$ *if* $\mathsf{upd} = 1$ *or removed from* $\mathcal{X}$ *if* $\mathsf{upd} = 0$*. If* $\mathsf{upd} = 1$ *and* $x \in \mathcal{X}$*, (likewise if* $\mathsf{upd} = 0$ *and* $x \notin \mathcal{X}$*) the algorithm outputs* $\perp$ *and halts, indicating an invalid update. Otherwise, it outputs* $(\mathsf{acc}', ek', aux')$ *where* acc' *is the new accumulation value corresponding to set* $\mathcal{X} \cup \{x\}$ *or* $\mathcal{X} \setminus \{x\}$ *(to be published),* ek' *is the (possibly) modified evaluation key, and* aux' *is respective auxiliary information (both to be sent only to the server).*

To update existing witnesses efficiently (i.e., not recomputing them from scratch) after a change of the accumulation value, we define the WitUpdate functionality.

$(\mathsf{upd}, \mathsf{w}') \leftarrow \mathsf{WitUpdate}(\mathsf{acc}, \mathsf{acc}', x, \mathsf{w}, y, ek', \mathsf{aux}, \mathsf{aux}', \mathsf{upd})$

This algorithm is to be run after an invocation of Update*. It takes as input the old and the new accumulation values and auxiliary informations, the evaluation key* ek' *output by* Update *, as well as the element* x *that was inserted or removed from the set, according to the binary value* upd *(the same as in the execution of* Update *). It also takes a different element* y *and its existing witness* w *(that may be a membership or non-membership witness). It outputs a new witness* w' *for* y*, with respect to the new set* $\mathcal{X}'$ *. The output must be the same as the one computable by running* $\mathsf{Witness}(\mathsf{acc}', \mathcal{X}', y, ek', \mathsf{aux}')$.

We point out that the ability to update membership witnesses is inherently more important than that of non-membership witnesses. The former corresponds to the (polynomially many) values in the set whereas the latter will be exponentially many (or infinite). A server that wants to cache witness values and update them efficiently can thus benefit more from storing pre-computed positive witnesses than negative ones (that are less likely to be used again).

Untrusted vs. trusted setup. The way we formulated our definition, Setup and Update require knowledge of sk, Witness requires ek and Verify takes only vk. From a practical point of view, the owner is the party that is responsible for maintaining the accumulation value at all times (e.g., signing it and posting it to a public log); all changes in $\mathcal{X}$ should, in a sense, be validated by him first. On the other hand, in most popular schemes (e.g., the RSA construction of [11] and the bilinear accumulator of [53]), setup and update can be run by the server (without trapdoor sk) and the only distinction is that the owner can achieve this much faster. The same holds for our construction, but in the security definitions we adopt the more general framework where the adversary is given oracle access to these algorithms. It should be noted that for our construction, all security properties hold even if sk is empty –only the complexity analysis changes.

3.1 Zero-Knowledge Accumulators: Security Properties

The first property we require from a cryptographic accumulator is completeness, i.e., a witness output by any sequence of invocations of the scheme algorithms, for a valid statement (corresponding to the state of the set at the time of the witness generation) is verified correctly with all but negligible probability.

Definition 2 (Completeness). *Let $\mathcal{X}_i$ denote the set with elements from $\mathbb{X}$, constructed after i invocations of the* Update *algorithm (starting from a set $\mathcal{X}_0$) and likewise for ek_i, aux_i. A dynamic universal accumulator is complete if, for all sets $\mathcal{X}_0$ where $|\mathcal{X}_0|$ and $l \geq 0$ are polynomial in λ and for all $x_i \in \mathbb{X}$, for $0 = 1, \ldots, l$, there exists a negligible function $\nu(\lambda)$ such that:*

$$\Pr\left[\begin{array}{c}(sk, vk) \leftarrow \mathsf{GenKey}(1^\lambda); (ek_0, \mathsf{acc}_0, \mathsf{aux}_0) \leftarrow \mathsf{Setup}(sk, \mathcal{X}_0);\\ \{(\mathsf{acc}_{i+1}, ek_{i+1}, \mathsf{aux}_{i+1}) \leftarrow \mathsf{Update}(\mathsf{acc}_i, \mathcal{X}_i, x_i, sk, \mathsf{aux}_i, \mathsf{upd}_i)\}_{0 \leq i \leq l}\\ (b, \mathsf{w}) \leftarrow \mathsf{Witness}(\mathsf{acc}_l, \mathcal{X}_l, x, ek_l, \mathsf{aux}_l) : \mathsf{Verify}(\mathsf{acc}_l, x, b, \mathsf{w}, vk) = \mathsf{accept}\end{array}\right] \geq 1 - \nu(\lambda)$$

where the probability is taken over the randomness of the algorithms.

In the above we purposely omitted the WitUpdate algorithm that was introduced purely for efficiency gains at the server. In fact, recall that we restricted it to return the exact same output as Update (run for the corresponding set and element) hence the value w in the above definition might as well have been computed during an earlier update and subsequently updated by (one or more) calls of WitUpdate.

The second property is soundness which captures that fact that adversarial servers cannot provide accepting witnesses for incorrect statements. It is formulated as the inability of Adv to win a game during which he is given oracle access to all the algorithms of the scheme (except for those he can run on his own using ek, aux –see discussion on private versus public setup and updates above) and is required to output such a statement and a corresponding witness.

Definition 3 (Soundness). *For all PPT adversaries* Adv *running on input 1^λ and all l polynomial in λ, the probability of winning the following game, taken over the randomness of the algorithms and the coins of* Adv *is negligible:*

Setup. *The challenger runs $(sk, vk) \leftarrow \mathsf{GenKey}(1^\lambda)$ and forwards vk to* Adv. *The latter responds with a set $\mathcal{X}_0$. The challenger runs $(ek_0, \mathsf{acc}_0, \mathsf{aux}_0) \leftarrow \mathsf{Setup}(sk, \mathcal{X}_0)$ and sends the output to the adversary.*

Updates. *The challenger initiates a list* $\mathcal{L}$ *and inserts the tuple* $(acc_0, \mathcal{X}_0)$*. Following this, the adversary issues update* x_i *and receives the output of* $\mathsf{Update}(\mathsf{acc}_i, \mathcal{X}_i, x_i, sk, \mathsf{aux}_i, \mathsf{upd}_i)$ *from the challenger, for* $i = 0, \ldots, l$*. After each invocation of* Update*, if the output is not* $\perp$*, the challenger appends the returned* $(\mathsf{acc}_{i+1}, \mathcal{X}_{i+1})$ *to* $\mathcal{L}$*. Otherwise, he appends* $(\mathsf{acc}_i, \mathcal{X}_i)$*.*

Challenge. *The adversary outputs an index* j*, and a triplet* (x^*, b^*, w^*)*. Let* $\mathcal{L}[j]$ *be* $(\mathsf{acc}_j, \mathcal{X}_j)$*. The adversary wins the game if:*

$$\mathsf{Verify}(\mathsf{acc}_j, x^*, b^*, \mathsf{w}^*, vk) = \mathsf{accept} \wedge ((x^* \in \mathcal{X}_j \wedge b^* = 0) \vee (x^* \notin \mathcal{X}_j \wedge b^* = 1))$$

A discussion on the winning conditions of the game is due at this point. This property (also referred to as collision-freeness) was introduced in this format in [41] and was more recently adapted in [23] with slight modifications. In particular, Adv outputs set $\mathcal{X}^*$ and accumulation value acc^* as well as the randomness used (possibly) to compute the latter (to cater for randomized accumulators). It is trivial to show that the two versions of the property are equivalent.

An alternative, more demanding, way to formulate the game is to require that the adversary wins if he outputs two accepting witnesses for the same element and with respect to the same accumulation value (without revealing the pre-image set): a membership and a non-membership one. This property, introduced in the context of accumulators in [7], is known as *undeniability* and is the same as the privacy property of zero-knowledge sets. Recently, Derler et al. [23] showed that undeniability is a stronger property than soundness. However, existing constructions for undeniable accumulators are in the trapdoor-less setting (with the limitations discussed in Sect. 1.1); since our construction is in the three-party setting, we restrict our attention to soundness. This should come as no surprise, as undeniability allows an adversary to provide a candidate accumulation value, without explicitly giving a corresponding set. In a trusted-setup setting, the accumulation value is always maintained by the trusted owner; there is no need to question whether it was honestly computed (e.g., whether he knows a set pre-image or even whether there exists one) hence undeniability in this model is an "overkill" in terms of security (see also the related discussion in Sect. 3.2).

The novel property we introduce here is zero-knowledge. Informally, this property ensures that an adversarial party (i.e., the client) that sees the accumulation value as well as all membership and non-membership witnesses exchanged during the protocol execution, and has the ability to issue arbitrary queries, learns nothing about the set, *not even its size.* Zero-knowledge guarantees that nothing can be learned from the protocol except for the answer to a query itself. In other words, explicitly querying for an element is the only way to learn whether it appears in the set or not. We formalize this in a way that is very similar to zero-knowledge sets (e.g., see the definition of [15]) appropriately extended to handle not only queries but also updates issued by the adversary. In particular, we want the proofs to be ephemeral, i.e., proofs generated before an update should be invalidated after an update. We require that there exists a simulator such that no adversarial client can distinguish whether he is interacting with the algorithms of the scheme or with the simulator that has no knowledge of the set or the element updates that occur, other than whether a queried element is in the set and whether requested updates are valid. This information is given to the simulator as the output of a function D that checks the validity of a requested operation[5].

Definition 4 (Zero-Knowledge). *Let D be a binary function defined as follows. For queries, $D(\mathsf{query}, x, \mathcal{X})) = 1$ iff $x \in \mathcal{X}$. For updates $D(\mathsf{update}, x, c, \mathcal{X})) = 1$ iff $(c = 1 \wedge x \notin \mathcal{X})$ or $(c = 0 \wedge x \in \mathcal{X})$. Let $\mathsf{Real}_{\mathsf{Adv}}(1^\lambda), \mathsf{Ideal}_{\mathsf{Adv},\mathsf{Sim}}(1^\lambda)$ be games between a challenger, an adversary Adv and a simulator $\mathsf{Sim} = (\mathsf{Sim}_1, \mathsf{Sim}_2)$, defined as follows:*

$\mathsf{Real}_{\mathsf{Adv}}(1^\lambda)$:

Setup. *The challenger runs $(sk, vk) \leftarrow \mathsf{GenKey}(1^\lambda)$ and forwards vk to Adv. The latter chooses a set $\mathcal{X}_0$ with $|\mathcal{X}_0| \in poly(\lambda)$ and sends it to the challenger who in turn runs $\mathsf{Setup}(sk, \mathcal{X}_0)$ to get $(\mathsf{acc}_0, ek_0, \mathsf{aux}_0)$. He then sends acc_0 to Adv and sets $(\mathcal{X}, \mathsf{acc}, ek, \mathsf{aux}) \leftarrow (\mathcal{X}_0, \mathsf{acc}_0, ek_0, \mathsf{aux}_0)$.*

Query. *For $i = 1, \ldots, l$, where $l \in poly(\lambda)$, Adv outputs (op, x_i, c_i) where $\mathsf{op} \in \{\mathsf{query}, \mathsf{update}\}$ and $c_i \in \{0, 1\}$:*

If *$\mathsf{op} = \mathsf{query}$: The challenger runs $(b, \mathsf{w}_i) \leftarrow \mathsf{Witness}(\mathsf{acc}, \mathcal{X}, x_i, ek, \mathsf{aux})$ and returns the output to Adv.*

If *$\mathsf{op} = \mathsf{update}$: The challenger runs $\mathsf{Update}(\mathsf{acc}, \mathcal{X}, x_i, sk, \mathsf{aux}, c_i)$. If the output is not $\perp$ he updates the set accordingly to get $\mathcal{X}_i$, sets $(\mathcal{X}, \mathsf{acc}, ek, \mathsf{aux}) \leftarrow (\mathcal{X}_i, \mathsf{acc}_i, ek_i, \mathsf{aux}_i)$ and forwards acc to Adv. Else, he responds with $\perp$.*

Response. *The adversary outputs a bit d.*

$\mathsf{Ideal}_{\mathsf{Adv}}(1^\lambda)$:

Setup. *The simulator Sim_1, on input 1^λ, outputs a vk which he forwards to Adv. The adversary chooses a set $\mathcal{X}_0$ with $|\mathcal{X}_0| \in poly(\lambda)$. Sim_1 (without seeing $\mathcal{X}_0$) responds with acc_0 and maintains state state_S. Finally, let $(\mathcal{X}, \mathsf{acc}) \leftarrow (\mathcal{X}_0, \mathsf{acc}_0)$.*

Query. *For $i = 1, \ldots, l$ Adv outputs (op, x_i, c_i) where $\mathsf{op} \in \{\mathsf{query}, \mathsf{update}\}$ and $c_i \in \{0, 1\}$:*

If *$\mathsf{op} = \mathsf{query}$: The simulator runs $(b, \mathsf{w}_i) \leftarrow \mathsf{Sim}_2(\mathsf{acc}, x_i, \mathsf{state}_S, D(\mathsf{query}, x_i, \mathcal{X}))$ and returns the output to Adv.*

If *$\mathsf{op} = \mathsf{update}$: The simulator runs $\mathsf{Sim}_2(\mathsf{acc}, \mathsf{state}_S, D(\mathsf{update}, x_i, c_i, \mathcal{X}))$. If the output of $D(\mathsf{update}, x_i, c_i, \mathcal{X})$ is 1, let $\mathcal{X} \leftarrow \mathcal{X}_i \cup x_i$ in the case $c_1 = 1$ and $\mathcal{X} \leftarrow \mathcal{X}_i \setminus x_i$ in the case $c_1 = 0$ —i.e., $\mathcal{X}$ is a placeholder variable for the latest set version at all times according to valid updates, that is however never observed by the simulator. The simulator responds to Adv with acc'. If the response acc' is not $\perp$ then $\mathsf{acc} \leftarrow \mathsf{acc}'$.*

---

[5] Instead of using D with different arguments for checking the validity of query and update, we could make D work only for queries, i.e., $D(\mathsf{query}, \ldots)$, and express the validity of a requested update as $D(\mathsf{query}, \ldots) \oplus c_i$. We chose to use the former notation because we feel it is cleaner.

Response. *The adversary outputs a bit d.*

A dynamic universal accumulator is zero-knowledge if there exists a PPT simulator $\mathsf{Sim} = (\mathsf{Sim}_1, \mathsf{Sim}_2)$ *such that for all adversaries* Adv *there exists negligible function* ν *such that:*

$$|\Pr[\mathsf{Real}_{\mathsf{Adv}}(1^\lambda) = 1] - \Pr[\mathsf{Ideal}_{\mathsf{Adv}}(1^\lambda) = 1]| \leq \nu(\lambda).$$

If Adv *is PPT, then this defines computational zero-knowledge; perfect and statistical zero-knowledge can be defined similarly.*

Observe that, even though Adv may be unbounded (in the case of statistical or perfect zero-knowledge) the size of the set is always polynomial in the security parameter as in [15]; in fact it is upper bounded by $|\mathcal{X}_0| + l$. This ensures that we can have polynomial-time simulation, matching the real-world execution where all parties run in polynomial-time. Having computationally unbounded adversaries is still meaningful; such a party may, after having requested polynomially many updates, spend unlimited computational resources trying to distinguish the two settings.

As already observed in [20,21,23], when formulating a notion of privacy for cryptographic accumulators the fact that the accumulation value computation must be randomized becomes evident. If Setup (and similarly, Update) is a deterministic algorithm, then each set has a uniquely defined accumulation value (subject to particular sk) that can be reproduced by any adversary with oracle access to the algorithm.

In our definition, the server holds the evaluation key ek that is used to produce witnesses, and that is not available to the client. This is not a restriction of the model, but should rather be seen as a generalization, in order to capture zero-knowledge in all settings; if ek does not leak any information about the set, it may be included in the public vk. Specifically for our construction from Sect. 4, if we choose to make ek public, then what is leaked is an upper-bound on the set size, formally captured by the notion of *functional zero-knowledge* [52].

3.2 Relation to Other Primitives

There exist various cryptographic primitives that address the problem of secure set (non-)membership, in the same or related models, and it is imperative to compare the primitive of zero-knowledge accumulators with these.

We present a mapping of the research literature for the construction of cryptographic proofs for set-membership and non-membership, which has attracted significant attention lately; proofs can be found in the full version. This is far from a complete presentation of results in the area; we focus on the relation between those primitives that are most closely related to the problem, avoiding general approaches (e.g., general-purpose zero-knowledge protocols) or related models that address similar problems (such as group signatures, e.g., [1]). The overall picture for the static case (i.e., without assuming changes in the set) can be seen in Fig. 1. Arrows denote implication; an arrow from A to B translates to "B can be built in black-box manner from A". Double-sided arrows denote equivalence of definitions, i.e., both can be constructed in a black-box manner from each other.

The most prominent such primitive is zero-knowledge sets [14,15,43,48]. There, queries can be answered without revealing anything about the set, albeit at a stronger setting where the server and the owner are the same (untrusted) entity. In the same setting, we also discussed trapdoorless (or strong) accumulators (see Sect. 1.1). Zero-knowledge sets are a stronger primitive than accumulators; they satisfy the same soundness property with trapdoorless accumulators but they additionally offer privacy. Hence all other primitives in our mapping can be built from them. Additionally, if a scheme is a trapdoorless accumulator it is secure with an untrusted setup execution, therefore (and quite trivially) it is also secure with a trusted setup, hence it is a also an accumulator.

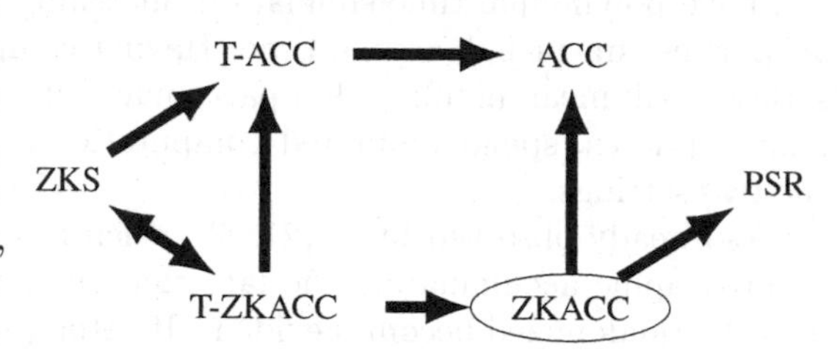

Fig. 1. Relations among cryptographic primitives for proof of membership and non-membership (static case). ZKS: zero-knowledge sets, T-ACC: trapdoorless accumulators, ACC: accumulators, T-ZKACC: trapdoorless zero-knowledge accumulators, ZKACC: our zero-knowledge accumulators (circled), PSR: primary-secondary-resolver membership proof systems.

As a mental exercise, let us now try to define the privacy-preserving counterparts of trapdoorless accumulators, i.e., *trapdoorless zero-knowledge accumulators*[6]. Quite informally, the completeness and zero-knowledge definitions remain the same but the soundness property is replaced by the, strictly stronger, property of undeniability (see, e.g., [44] for a concrete definition), which is the same as the soundness property of zero-knowledge sets: By "merging" the existing soundness guarantee of trapdoorless accumulators with our zero-knowledge property (which, for the static case, is identical to that of zero-knowledge sets) we –quite unsurprisingly– ended up with zero-knowledge sets. We stress that the latter exist in the common reference string model (or the trusted parameters model) hence this must also be true for trapdoorless zero-knowledge accumulators (e.g., a trusted authority runs the key-generation algorithm and publishes the result as a common reference string). On the contrary, this is not necessary for trapdoorless accumulators (without privacy) since the security game there is one-sided; the client can perform key-generation himself. As a final note, we point out, that zero knowledge (trapdoorless) accumulators

[6] It should be noted that, in the accumulators literature, the trapdoor refers to a secret value possibly used for efficiency purposes when computing accumulation values and witnesses by the trusted owner. This should not be confused with the trapdoor typically used in zero-knowledge protocols for simulation purposes.

imply (trapdoorless) accumulators since the former satisfy a strict superset of the security properties of the latter.

This equivalence of zero-knowledge sets and trapdoorless zero-knowledge accumulators can be useful in two ways: (i) more efficient (e.g., with smaller proof sizes) zero-knowledge sets may be achievable with techniques borrowed from the accumulators literature, and (ii) an impossibility result in one of the two models is translatable to the other. This holds, for example, in the case of the batch-update impossibility for accumulators of [8]. We want to stress that our construction in Sect. 4 is not trapdoorless; to the best of our knowledge, the best known way to construct trapdoorless zero-knowledge accumulators is via a black-box reduction from zero-knowledge sets.

Another related primitive is primary-secondary-resolver proof systems (PSR), introduced by Naor et al. [52]. Their privacy notion is a relaxation of zero-knowledge defined as *functional zero-knowledge*, i.e., the simulator may be allowed to learn some function of the set (typically its size). Also, the games in the PSR definition are non-adaptive in the following sense: Adv needs to declare its cheating set before he even receives the corresponding keys (ek, vk for soundness and only vk for zero-knowledge –using our terminology)[7]. For the above reasons, while it is trivial that zero-knowledge accumulators imply PSR (where the leaked function is void), the other direction is generally not true. We stress that the above distinction between adaptive and selective security does not hold in the dynamic setting. There an adversary may declare a cheating set originally, receive the keys, and then modify his choice via a series of update calls (see, however, our discussion for this setting in the next paragraph).

Our results here are complementary to the relations proven in [52]. There, the authors prove that PSR systems exist, if and only if, one-way functions exist, which in turn implies that zero-knowledge sets cannot be built in a black-box manner from PSR.

Dynamic setting. Once we move to the dynamic setting, where there exist efficient algorithms for modifications in the set, the relations are largely the same as in Fig. 1, but some clarifications are in order. The first work addressing updatable zero-knowledge sets was by Liskov [45], where two notions of privacy were introduced: opacity and transparency. Constructions of the latter form were presented in [13,45]. The above relations between definitions hold with respect to opacity. A construction for efficiently updatable opaque zero-knowledge sets (from standard assumptions) remains an open problem. However, when restricted to the three-party model (i.e., with trusted setup), it can be shown that our construction from Sect. 4 (with minor modifications) satisfies the opacity property. On the other hand, transparency is a weaker form of privacy, as it trivially leaks

[7] One could possibly modify the PSR model –and the security games– significantly to make them adaptive, by separating the key generation and setup algorithms. Indeed, to the best of our knowledge, the PSR construction of [33] would probably satisfy such a modified definition, assuming it was instantiated with an adaptively-secure signature scheme and an adaptively-secure verifiable random function.

whether a particular element, that has been previously queried, was affected by an update (but it otherwise allows parties to maintain cached witnesses).

Regarding the relation between zero-knowledge accumulators and PSR, matters are also straight-forward as the latter are explicitly defined only for the static case. In [52], the authors recommend the usage of techniques from certificate-revocation lists [51], as an additional external mechanism to accommodate updates. Contrary to this, our definitional approach is to make update-handling mechanisms explicitly part of the scheme. In this sense, zero-knowledge accumulators are a natural definitional extension of PSR in the dynamic setting. That said, we explicitly require that clients can at all times access the latest accumulation value, which would not be the case following the revocation scheme approach. We stress however that this does not necessitate active authenticated channels between owner and clients; in practice it is achievable with a "timestamp-sign-and-publish" from the owner.

We note that recently [42] introduced the general notion of functional commitments (which can capture accumulators as a special case). However, their construction handles only subset queries and it does not support updates on the committed set. On the other hand, [60] introduced the notion of asynchronous accumulators in a distributed setting and does not consider privacy.

Relation to zero-knowledge authenticated data structures. Another important observation is the relation of zero-knowledge accumulators with the framework of zero-knowledge authenticated data structures (ZKADS), recently introduced in [29].[8] ZKADS extend the well-known primitive of authenticated data structures (ADS) adding an additional zero-knowledge property. The setting is the standard three-party model but now the supported type may be any kind of data structure. The choice of data structure defines the kind of data stored and the type of supported queries. In [29,31,32], the authors provided constructions for various types of data structures, in particular for a zero-knowledge authenticated list (i.e., a data structure that supports "insert-after", "delete" operations, as well as "order" queries), a tree, and a partially-ordered set (poset) of bounded dimension and range queries. Consequently, a zero-knowledge accumulator is a type of ZKADS where the data structure is a set of elements supporting –unordered– insertions and deletions, and (non-)membership queries.

The above constructions are the only ZKADS instantiations in the literature so far. One natural way to extend zero-knowledge authenticated sets to accommodate more elaborate query types is by allowing for set-operations beyond (non-)membership. In particular, consider a data structure, called set collection, that consists of a collection of sets and accommodates operations among (an arbitrary selection among) them. We stress that a construction that accommodates set unions, intersection and differences, allows for a complete set algebra.[9] In the

[8] Though [29] uses the term Privacy-Preserving Authenticated Data Structures, we use ZKADS to fit our notation.

[9] In the computationally-bounded setting, a negation operation is infeasible unless the element domain is of polynomial size in the security parameter. In that case, a negation can be instantiated as a set difference from the set that contains the entire domain.

full version we provide a definition of *zero-knowledge authenticated set collection*, in the style of [29], and the corresponding construction (which naturally uses our zero-knowledge authenticated set construction from Sect. 4 as a building block).

3.3 Zero-Knowledge Implies Indistinguishability (for Accumulators)

The notion of zero-knowledge defined here is a strengthening of the indistinguishability property introduced in [23]. There the authors introduce a notion similar to ours that also requires the accumulation value produced by Setup to be randomized. If we restrict our attention to static accumulators, the effect of both notions is the same, i.e., the clients see a randomized accumulation value and corresponding "blinded" witnesses.

However, while the indistiguishability game entails updates, it inherently does not offer any privacy for the elements inserted to or removed from the set, as the Update algorithm is deterministic. At a high level, that notion only protects the original accumulated set and not subsequent updates. We believe this is an important omission for a meaningful privacy definition for accumulators, as highlighted by the following example. Consider, a third-party adversary that observes the protocol's execution before and after an insertion (or deletion) update. If the adversary has reasons to suspect that the inserted (or deleted) value may be y, he can always test that. A very realistic example of this behavior is a setting where the accumulator is used to implement a revocation list. In that case an adversary may want to know if his fake certificate (value y in the above case) has been "caught" yet. We provide the following result[10].

Theorem 1. *Every zero-knowledge dynamic universal accumulator is also indistinguishable under the definition of [23], while the opposite is not always true.*

Proof. We first show that every scheme that is zero-knowledge is also indistinguishable. Then we show that the construction of [23] is not zero-knowledge.

ZK ⇒ IND: We prove this direction by contradiction. Assume there exists an accumulator that is zero-knowledge but not indistinguishable. Then, there exists a PPT adversary Adv that wins the indistinguishability game. Adv gives two sets $\mathcal{X}_0, \mathcal{X}_1$ to a challenger who flips a coin b and provides oracle access to Adv for the algorithms with respect to $\mathcal{X}_b$. By assumption, Adv can

[10] In [23] the indistinguishability definition assumes that the adversary is also given access to the Setup algorithm arbitrarily many times. This makes sense in their model, since they explicitly require that Setup is randomized whereas Update is deterministic. Here this requirement is redundant since both processes may be randomized; any setup response can be emulated by a series of update calls that shape the required set. To simplify the process, we assume that the indistinguishability adversary only makes Update and Witness calls. We stress that this is not a limitation of the reduction. We could alternatively have chosen to define our zero-knowledge game giving the adversary access to Setup and the result would still hold.

output a bit b' correctly guessing b with non-negligible advantage ϵ over 1/2. The (natural) constraint is that Adv cannot issue a query (or update request) that is trivially revealing the chosen set (e.g., if $x \in \mathcal{X}_0$ and $x \notin \mathcal{X}_1$, Adv is not allowed to query for x). We defer interested readers to [23] for a formal definition of the indistinguishability game.

We will now construct a PPT adversary Adv′ that breaks the zero-knowledge property of the scheme as follows. Adv′ on input $1^\lambda, vk$ runs Adv with the same input and receives sets $\mathcal{X}_0, \mathcal{X}_1$. He then forwards $\mathcal{X}_1$ as the challenge for the zero-knowledge game and receives accumulation value acc_0, which he forwards to Adv. Consequently, he responds to all messages of Adv (queries and updates) with calls to the zero-knowledge game interface and forwards all responses back to Adv. Finally, he outputs the output bit b' of Adv.

First, observe that Adv′ is clearly PPT, since Adv is PPT. Now let us argue about his success probability in distinguishing between real and ideal interaction. Observe that, if Adv′ is interacting with the algorithms of the scheme (i.e., is playing the real game), the interface he is providing to Adv is a perfect simulation of the indistinguishability game for $b = 1$. On the other hand, if he is interacting with Sim, the view of the latter during this interaction is exactly the same independently of whether the set chosen by Adv′ is $\mathcal{X}_0$ or $\mathcal{X}_1$. Hence, the view offered to Adv is the same in both cases, and therefore $\Pr[b' = 1] = \Pr[b' = 0] = 1/2$. Let E be the event that the Adv′ is playing the real game (and likewise for the complement E^c). From the above analysis (recall that Adv′ outputs the bit b' returned by Adv), it holds that $\Pr[b' = 1|E] > 1/2 + \epsilon$ and $\Pr[b' = 1|E^c] = 1/2$. This implies that Adv′ can distinguish between the two executions with non-negligible probability, breaking the zero-knowledge property of the scheme. The claim follows by contradiction.

IND $\not\Rightarrow$ ZK: The main observation for this part of the proof is that in the construction of [23], given the accumulation acc of set $\mathcal{X}$, the new accumulation value after inserting or deleting an element is computed via a deterministic algorithm. Assume now an adversary Adv that operates as follows when playing the zero-knowledge game against the scheme of [23]. He initially plays the setup phase of Definition 4 choosing a set $\mathcal{X}_0$ and receiving acc_0 from the challenger. Then he chooses $e \xleftarrow{\$} \{0,1\}$. If $e = 0$ then Adv chooses x uniformly from $\mathbb{Z}_p \setminus \{\mathcal{X}_0\}$ and sends to the challenger first $(\mathsf{update}, x, 1)$, receiving acc_1, and then $(\mathsf{update}, x, 0)$, receiving acc_2. Else, if $e = 1$ he chooses x, y uniformly from $\mathbb{Z}_p \setminus \{\mathcal{X}_0\}$ with $y \neq x$, and sends to the challenger first $(\mathsf{update}, x, 1)$, receiving acc_1, and then $(\mathsf{update}, y, 1)$, receiving acc_2. Finally, if $(\mathsf{acc}_2 = \mathsf{acc}_0 \wedge e = 0)$ or $(\mathsf{acc}_2 \neq \mathsf{acc}_0 \wedge e = 1)$, he outputs $d = 1$. In all other cases he outputs $d = 0$.

Observe first that Adv is clearly PPT as all algorithms of the scheme are run in polynomial time. Regarding his success probability, we argue as follows. If Adv is playing the real game versus the challenger running the algorithms of [23], then we identify the following two probabilities $\Pr[\mathsf{acc}_2 = \mathsf{acc}_0|e = 0]$ and $\Pr[\mathsf{acc}_2 = \mathsf{acc}_0|e = 1]$. The first probability is equal to 1 whereas the second one is negligibly small; as explained above, the updates of the scheme are deterministic therefore adding and removing the same element will result in the same accumulation value, whereas adding two elements will always result in a different accumulation value, unless the latter happens to be the multiplicative inverse of the former. On the other hand, if Adv is playing the ideal game against the simulator, the latter is only given access to the information that two updates occurred (not even the nature of the update operations). Therefore, the simulator's view is the same, independently of the value of e, and $\Pr[\mathsf{acc}_2 = \mathsf{acc}_0 | e = 0] = \Pr[\mathsf{acc}_2 = \mathsf{acc}_0 | e = 1] = 1/2$. Let E be the event that the Adv′ is playing the real game (and likewise for the complement E^c). From the above analysis it follows that $Pr[d = 1|E] = 1 - \nu(\lambda)$, whereas $Pr[d = 1|E^c] = 1/2$ therefore Adv distinguishes the two games with non-negligible probability and the accumulator of [23] is not zero-knowledge. ∎

Other privacy notions. The indistinguishability property of [23] is a strengthening of a that of [20]. The latter was the first work to formally define a privacy property for cryptographic accumulators, however their definition had inherent problems, e.g., it was easy to prove that deterministic accumulators –that clearly were not private– satisfied it. Another technique for providing privacy to cryptographic accumulators was proposed earlier in [41], without a formalization. The idea is to simply produce a randomized accumulation value for a set $\mathcal{X}$ by choosing at random an element x from the elements universe during Setup and outputting the accumulation of set $\mathcal{X} \cup \{x\}$. This generic mechanism will work for any static accumulator, but will also not protect updates. Moreover it weakens soundness as an adversary could potentially produce a membership witness for the element $x \notin \mathcal{X}$. Out approach does not suffer from this as there is no additional element accumulated and the randomness r used to blind the accumulation value during Setup is explicitly given to the server without compromising soundness. Finally, Theorem 1 implies that our construction from Sect. 4, is also the only known algebraic construction of a universal indistinguishable accumulator. The two schemes of [23] are a black-box reduction from the stronger primitive of zero-knowledge sets, and a construction similar to ours that only offers membership witnesses.

4 A Zero-Knowledge Universal Accumulator Construction

In this section we present our construction for a zero-knowledge dynamic universal accumulator. It builds upon the bilinear accumulator of Nguyen [53], adopting some of the techniques of [23] that we further expand to achieve zero-knowledge. It supports sets with elements from $\mathbb{Z}_p \setminus \{s\}$ where p is prime and $p \in O(2^\lambda)$ and s is the scheme trapdoor. Note that, the fact that the elements must be of $\log p$ bits each, is not a strong limitation of the scheme; one can always apply a collision-resistant hash function that maps arbitrarily long strings to $\mathbb{Z}_p$. We now make several observations about our ZKUA construction in Fig. 2.

Notation: The notation $q[z]$ denotes polynomial q over undefined variable z and $q(s)$ is the evaluation of the polynomial at point s. All arithmetic operations are performed $\bmod p$. N is a variable maintained by the owner.

Key Generation $(sk, vk) \leftarrow \mathsf{GenKey}(1^\lambda)$

Run $\mathsf{GenParams}(1^k)$ to receive bilinear parameters $pub = (p, \mathbb{G}, \mathbb{G}_T, e, g)$. Choose $s \xleftarrow{\$} \mathbb{Z}_p^*$. Return $sk = s$ and $vk = (g^s, pub)$.

Setup $(\mathsf{acc}, ek, \mathsf{aux}) \leftarrow \mathsf{Setup}(sk, \mathcal{X})$

Choose $r \xleftarrow{\$} \mathbb{Z}_p^*$. Set value $N = |\mathcal{X}|$. Return $\mathsf{acc} = g^{r \cdot \mathsf{Ch}_{\mathcal{X}}(s)}$, $ek = (g, g^s, g^{s^2}, \ldots, g^{s^N})$ and $\mathsf{aux} = (r, N)$.

Witness Generation $(b, \mathsf{w}) \leftarrow \mathsf{Witness}(\mathsf{acc}, \mathcal{X}, x, ek, \mathsf{aux})$

If $x \in \mathcal{X}$ compute $\mathsf{w} = (\mathsf{acc})^{\frac{1}{s+x}} = g^{r \cdot \mathsf{Ch}_{\mathcal{X} \setminus \{x\}}(s)}$ and return $(1, \mathsf{w})$.

Else, proceed as follows:

- Using the Extended Euclidean algorithm, compute polynomials $q_1[z], q_2[z]$ such that $q_1[z]\mathsf{Ch}_{\mathcal{X}}[z] + q_2[z]\mathsf{Ch}_{\{x\}}[z] = 1$.
- Pick a random $\gamma \xleftarrow{\$} \mathbb{Z}_p^*$ and set $q_1'[z] = q_1[z] + \gamma \cdot \mathsf{Ch}_{\{x\}}[z]$ and $q_2'[z] = q_2[z] - \gamma \cdot \mathsf{Ch}_{\mathcal{X}}[z]$.
- Set $W_1 := g^{q_1'(s) r^{-1}}, W_2 = g^{q_2'(s)}$ and $\mathsf{w} := (W_1, W_2)$. Return $(0, \mathsf{w})$.

Verification $(\mathsf{accept}/\mathsf{reject}) \leftarrow \mathsf{Verify}(\mathsf{acc}, x, b, \mathsf{w}, vk)$

If $b = 1$ return accept if $e(\mathsf{acc}, g) = e(\mathsf{w}, g^x \cdot g^s)$, reject otherwise. If $b = 0$ do the following:

- Parse w as (W_1, W_2).
- Return accept if $e(W_1, \mathsf{acc})e(W_2, g^x \cdot g^s) = e(g, g)$, reject otherwise.

Update $(\mathsf{acc}', ek', \mathsf{aux}') \leftarrow \mathsf{Update}(\mathsf{acc}, \mathcal{X}, x, sk, \mathsf{aux}, \mathsf{upd})$

Parse aux as (r, N). If $(\mathsf{upd} = 1 \wedge x \in \mathcal{X})$ or $(\mathsf{upd} = 0 \wedge x \notin \mathcal{X})$ output $\perp$ and halt.

Choose $r' \xleftarrow{\$} \mathbb{Z}_p^*$. If $\mathsf{upd} = 1$:

- Compute $\mathsf{acc}' = \mathsf{acc}^{(s+x)r'}$.
- If $|\mathcal{X}| + 1 > N$, set $N = |\mathcal{X}| + 1$ and compute $\mathsf{ek}' = g^{s^N}$.

Else, compute $\mathsf{acc}' = \mathsf{acc}^{\frac{r'}{s+x}}$ and $ek' = \mathbf{0}$. In both cases, set $\mathsf{aux}' := (r \cdot r', N)$ and return $(\mathsf{acc}', ek', \mathsf{aux}')$.

Witness Update $(\mathsf{upd}, \mathsf{w}') \leftarrow \mathsf{WitUpdate}(\mathsf{acc}, \mathsf{acc}', x, \mathsf{w}, y, ek', \mathsf{aux}, \mathsf{aux}', \mathsf{upd})$

Parse $\mathsf{aux}, \mathsf{aux}'$ to get r, r'.

- If w is a membership witness:

  If $\mathsf{upd} = 1$ output $(1, \mathsf{w}' = (\mathsf{acc} \cdot \mathsf{w}^{x-y})^{r'})$. Else, output $(0, \mathsf{w}' = \mathsf{acc}'^{\frac{1}{(y-x)}} \cdot \mathsf{w}^{\frac{r'}{(x-y)}})$.
- If w is a non-membership witness:

  Let $\mathcal{X}'$ be the set produced after the execution of Update for element x (i.e., the current set). Run $\mathsf{Witness}(\mathsf{acc}', \mathcal{X}', y, ek', \mathsf{aux}')$ and return its output.

Fig. 2. Zero-knowledge dynamic universal accumulator construction.

The main property of our construction is that the algorithms do not reveal anything about the set in the units sent to the client. The key vk published from the key-generating algorithm reveals nothing about the set. The accumulation value produced by Setup is the standard bilinear accumulation value of [53] which is now blinded by a random value r, also revealed to the server. Witness generation also utilizes this randomness r.

For membership queries, the process is the same as in [19,53] with one additional exponentiation with r for privacy purposes. This technique proves that an element $x \in \mathcal{X}$ iff the degree-one polynomial $x + z$ divides $\mathsf{Ch}_{\mathcal{X}}[z]$. The major deviation occurs in the non-membership case. As previously discussed, there are existing works [2,19] that enhance the bilinear accumulator to provide non-membership witnesses. Their technique is a complement of the one used for the membership case. At a high level, it entails proving that the degree-one polynomial $x + z$ does not divide $\mathsf{Ch}_{\mathcal{X}}[z]$, by revealing the scalar (i.e., zero-degree polynomial) remainder of their long division. Unfortunately, using this approach here entirely breaks the zero-knowledge property: It reveals r (multiplied by an easily computable query-specific value) to any client. Instead, we adopt an entirely different approach. Our scheme uses the set-disjointness test, first proposed in [56]. In order to prove that $x \notin \mathcal{X}$, the server proves the statement $\mathcal{X} \cap \{x\} = \emptyset$. The different nature of the proved statement allows us to use fresh query-specific randomness γ together with r to prove non-membership in zero-knowledge. As a consequence, the verification for membership and non-membership is also different, but always efficient.

Finally, the way updates are handled is especially important as it is another strong point of divergence from previous schemes that seek to provide privacy. After each update, a fresh randomness r' is used to blind the new accumulation value. This re-randomization technique perfectly hides the nature of the change in $\mathcal{X}$ and lets us achieve zero-knowledge. Observe that, at all times, the owner maintains a variable N which is the maximum set-cardinality observed up to that point (through the original setup and subsequent insertions). If an insertion increases N (at most by one), the owner provides the server with an additional ek component that can be used by the server for subsequent witness generation. This is a slight deviation from our notation in Sect. 3 where the new key produced from Update replaces the previous ek. Instead the new evaluation key must be set to $ek \cup ek'$. This has no meaningful impact to the security of our scheme; we could always have Update output the entire old key together with the additional element. From an efficiency perspective though, that overly naive approach would require Update to run in time linear to N –the same holds for $\mathsf{WitUpdate}$. Regarding witness updates, for the (more meaningful, as discussed in Sect. 3) case of membership witnesses there indeed exists a fast method. On the other hand, for non-membership witness updates, our scheme resorts to re-computation from scratch.

We can now present our main result. We give the proof of security below and defer the asymptotic analysis to the full version [30].

Theorem 2. *The algorithms* $\{\mathsf{KeyGen}, \mathsf{Setup}, \mathsf{Witness}, \mathsf{Verify}, \mathsf{Update}, \mathsf{WitUpdate}\}$ *constitute a zero-knowledge dynamic universal accumulator with perfect completeness, soundness under the q-SBDH assumption (with* $q = N$ *set to the maximum set size observed during the soundness game) and perfect zero-knowledge. Let* N *be the cardinality of the set. Then, the runtime* *of* GenKey *is* $O(poly(\lambda))$ *where* λ *is the security parameter, the complexity of* Setup *is* $O(N)$*, that of* Witness *is* $O(N \log N)$ *for membership witnesses and* $O(N \log^2 N \log\log N)$ *for non-membership witnesses, that of* Verify *is* $O(1)$*, that of* Update *is* $O(1)$*, and that of* WitUpdate *is* $O(1)$ *for membership witnesses and* $O(N \log^2 N \log\log N)$ *for non-membership witnesses. Finally, witnesses consist of* $O(1)$ *bilinear group elements.*

Proof. Completeness follows by close inspection of the algorithms' execution. We proceed to prove soundness and zero-knowledge.

Proof of Soundness. Assume there exists PPT adversary Adv that on input 1^λ breaks the soundness of our scheme with non-negligible probability. We will construct a PPT adversary Adv′ that breaks the q-SBDH assumption for $q = N$, running as follows:

1. On input $(pub, (g^s, \ldots, g^{s^N}))$, run Adv on input $(g^s, pub, 1^\lambda)$.
2. Upon receiving set $\mathcal{X}_0$, choose $r_0 \xleftarrow{\$} \mathbb{Z}_p^*$. Use r_0 and $(g^s, \ldots, g^{s^N})$ to compute $\mathsf{acc}_0 = g^{r_0 \cdot \mathsf{Ch}_{\mathcal{X}_0}(s)} = g^{(\mathsf{Ch}_{\mathcal{X}_0}(s))^{r_0}}$ and respond with $(ek_0 = (g, g^s, \ldots, g^{s^{|\mathcal{X}_0|}}), \mathsf{acc}_0, r_0)$. Initiate list $\mathcal{L}$ and insert triplet $(\mathsf{acc}_0, \mathcal{X}_0, r_0)$ as $\mathcal{L}[0]$ (i.e., the first element of the list). The notation $\mathcal{L}[i]_j$ denotes the first part of the i-th element of the list (e.g., $\mathcal{L}[0]_0 = \mathsf{acc}_0$). Also set $n = |\mathcal{X}_0|$.
3. Initiate update counter $i = 0$. While $i \leq l$ proceed as follows. Upon receiving update upd_i, x_i, check whether this is a valid update for $\mathcal{X}_i = \mathcal{L}[i]_1$. If it is not, respond with $\perp$ and re-append $acc_i = \mathcal{L}[i]_0, \mathcal{X}_i, r_i$ to $\mathcal{L}$. Otherwise, pick $r' \xleftarrow{\$} \mathbb{Z}_p^*$ and set $r_{i+1} = r_i \cdot r'$. Update $\mathcal{X}_i$ according to upd_i, x_i to get $\mathcal{X}_{i+1}$. If $|\mathcal{X}_{i+1}| > n$, set $n = |\mathcal{X}_{i+1}|$ and $ek_{i+1} = g^{s^n}$. Else, $ek_{i+1} = \emptyset$. Use r_{i+1} and $(g^s, \ldots, g^{s^N})$ to compute $\mathsf{acc}_{i+1} = g^{r_{i+1} \cdot \mathsf{Ch}_{\mathcal{X}_{i+1}}(s)} = g^{(\mathsf{Ch}_{\mathcal{X}_{i+1}}(s))^{r_{i+1}}}$ and respond with $(ek_{i+1}, \mathsf{acc}_{i+1}, r_{i+1})$. Append triplet $(\mathsf{acc}_{i+1}, \mathcal{X}_{i+1}, r_{i+1})$ to $\mathcal{L}$. In both cases, increase i by 1.
4. Upon receiving the j-th challenge with triplet (x^*, b^*, w^*) proceed as follows:
   - If $b^* = 1$, then $x^* \notin \mathcal{X}_j$ yet $\mathsf{Verify}(\mathsf{acc}_j, x^*, 1, vk)$ accepts. Compute polynomial $q[z]$ and scalar c such that $\mathsf{Ch}_{\mathcal{X}_j}[z] = (x^* + z)q[z] + c$. Output $[x^*, (e(\mathsf{w}^*, g)^{r_j^{-1}} e(g, g^{-q(s)}))^{c^{-1}}]$.
   - If $b^* = 0$, then $x^* \in \mathcal{X}_j$ yet $\mathsf{Verify}(\mathsf{acc}_j, x^*, 0, vk)$ accepts. Parse w^* as (W_1^*, W_2^*). Compute polynomial $q[z]$ such that $\mathsf{Ch}_{\mathcal{X}_j}[z] = (x^* + z)q[z]$. Output $[x^*, (e(W_1^*, g^{r_j \cdot q(s)})e(W_2^*, g))]$.

First of all observe that Adv′ perfectly emulates the challenger for the DUA security game to Adv. This holds since all accumulation values and witness are computable without access to trapdoor sk in polynomial time. All the necessary polynomial arithmetic can be also run efficiently hence Adv′ is PPT. Regarding its success probability, we argue for the two cases separately as follows:

$\mathbf{b^* = 1}$ Since $x^* \notin \mathcal{X}_j$, it follows that $(x^* + z) \not| \mathsf{Ch}_{\mathcal{X}_j}[z]$ which guarantees the existence of $q[z], c$. Also observe that c is a scalar (zero-degree polynomial) since it is the remainder of the polynomial division and it must have degree less than that of $(x^* + z)$. Since verify accepts we can write

$$e(\mathsf{w}^*, g^{x^*} \cdot g^s) = e(\mathsf{w}^*, g)^{(x^*+s)} = e(\mathsf{acc}_j, g) = e(g^{r_j \cdot \mathsf{Ch}_{\mathcal{X}_j}(s)}, g) = e(g,g)^{r_j((x^*+s)q(s)+c)}$$

from which it follows that:

$$\begin{aligned} e(\mathsf{w}^*, g)^{r_j^{-1}(x^*+s)} &= e(g,g)^{(x^*+s)q(s)+c} \\ e(\mathsf{w}^*, g)^{r_j^{-1}} &= e(g,g)^{q(s)+c/(x^*+s)} \\ e(\mathsf{w}^*, g)^{r_j^{-1}} e(g,g)^{-q(s)} &= e(g,g)^{c/(x^*+s)} \\ [e(\mathsf{w}^*, g)^{r_j^{-1}} e(g,g)^{-q(s)}]^{c^{-1}} &= e(g,g)^{1/(x^*+s)}. \end{aligned}$$

$\mathbf{b}^* = \mathbf{0}$ Since $x^* \in \mathcal{X}_j$, it follows that $(x^* + z)|\mathsf{Ch}_{\mathcal{X}_j}[z]$ which guarantees the existence of $q[z]$. Since verify accepts we can write:

$$\begin{aligned} e(W_1^*, \mathsf{acc}_j)e(W_2^*, g^{x^*} \cdot g^s) &= e(g,g) \\ e(W_1^*, g^{r_j \cdot \mathsf{Ch}_{\mathcal{X}_j}(s)})e(W_2^*, g^{(x^*+s)}) &= e(g,g) \\ e(W_1^*, g^{r_j(x^*+s)q(s)})e(W_2^*, g^{(x^*+s)}) &= e(g,g) \\ [e(W_1^*, g^{r_j \cdot q(s)})e(W_2^*, g)]^{(x^*+s)} &= e(g,g) \\ [e(W_1^*, g^{r_j \cdot q(s)})e(W_2^*, g)] &= e(g,g)^{1/(x^*+s)} \end{aligned}$$

Observe that in both cases the left hand of the above equations is efficiently computable with access to *pub*, $(g^s, \ldots, g^{s^N}), r_j, \mathcal{X}_j, x^*, \mathsf{w}^*$. Hence, whenever Adv' succeeds in breaking the soundness of our scheme, Adv' outputs a pair breaking the q-SBDH assumption for $q = N$. By assumption the latter can happen only with negligible probability, and our claim that our scheme has soundness follows by contradiction. ∎

Proof of Zero-Knowledge. We define simulator $\mathsf{Sim} = (\mathsf{Sim}_1, \mathsf{Sim}_2)$ as follows. At all times, we assume state_S contains all variables seen by the simulator this far.

- Sim_1 runs $\mathsf{GenParams}$ to receive *pub*. He then picks $s \xleftarrow{\$} \mathbb{Z}_p^*$ and sends g, g^s, pub to Adv. After Adv has output his set choice $\mathcal{X}$, Sim_1 picks $r \xleftarrow{\$} \mathbb{Z}_p^*$ and responds with $\mathsf{acc} = g^r$. Finally, he stores r and initiates empty list $\mathcal{C}$.
- For $i = 1, \ldots, l$ upon input (op, x_i, c_i):
  - If $\mathsf{op} = \mathsf{query}$, the simulator checks if $x_i \in \mathcal{C}$.
    * If $x_i \notin \mathcal{C}$, then if $D(\mathsf{query}, x_i, \mathcal{X}) = 1$, he computes $\kappa = r \cdot (x_i + s)^{-1}$ and responds with $(b = 1, \mathsf{w} = g^\kappa)$. Else, if $D(\mathsf{query}, x_i, \mathcal{X}) = 0$ he computes q_1, q_2 such that $q_1 \cdot r + q_2 \cdot (x_i + s) = 1$, picks $\gamma \xleftarrow{\$} \mathbb{Z}_p^*$ and responds with $(b = 0, \mathsf{w} = (W_1 = g^{q_1 + \gamma(x_i+s)}, W_2 = g^{q_2 - \gamma r}))$. In both cases, the simulator appends (x_i, b, w) to $\mathcal{C}$.
    * If $x_i \in \mathcal{C}$ he responds with the corresponding entries b, w from $\mathcal{C}$.

- If op = update then the simulator proceeds as follows. If $D(\mathsf{update}, x_i, c_i, \mathcal{X}) = 0$ then he responds with $\bot$. Else, he picks $r' \xleftarrow{\$} \mathbb{Z}_p^*$ and responds with $\mathsf{acc} = g^{r'}$. Finally he sets $r \leftarrow r'$ and $\mathcal{C} \leftarrow \emptyset$.

The simulator $\mathsf{Sim} = (\mathsf{Sim}_1, \mathsf{Sim}_2)$ produces a view that is identically distributed to that produced by the challenger during $\mathsf{Real}_{\mathsf{Adv}}$. Observe that random values r are chosen independently after each update (and initial setup) in both cases. Once s, r are fixed then for any possible choice of $\mathcal{X}$ there exists unique $r^* \in \mathbb{Z}_p^*$ such that $g^r = g^{r^* \cdot \mathsf{Ch}_{\mathcal{X}}(s)}$. It follows that the accumulation values in $\mathsf{Real}_{\mathsf{Adv}}$ are indistinguishable from the (truly random) ones produced by Sim. For fixed s, r, given a set-element combination $(\mathcal{X}, x_i)$ with $x_i \in \mathcal{X}$, in each game there exists a unique membership witness w that satisfies verification. For negative witness $\mathsf{w} = (W_1, W_2)$, given a set-element combination $(\mathcal{X}, x_i)$ with $x_i \notin \mathcal{X}$, for each possible independently chosen value of γ, in both games there exists only one distinct corresponding pair W_1, W_2 that satisfies the verifying equation. It follows that the distributions in Definition 4 are identical and our scheme is perfect zero-knowledge. ∎

Efficiency comparison with the bilinear accumulator of [53]**.** Here we compare the efficiency of our accumulator with the bilinear accumulator of [53] –as extended in [2]– which is secure under the same assumption, but does not offer privacy. In Fig. 3, we show the number of necessary cryptographic operations for the constructions. We denote by ADD, MUL point addition and scalar multiplication in the elliptic curve group $\mathbb{G}$, by ADD_T point addition in $\mathbb{G}_T$ and by PAIR a bilinear pairing computation. We stress that we do not measure the number of "non-cryptographic" operations, i.e., additions and multiplications modulo p.

As can be seen, our construction requires the same number of cryptographic operations for setup and membership witness construction and verification.

	[53]	**This paper**
Setup	nMUL	nMUL
Update	1MUL	2MUL
Witness (Member)	nMUL + $(n-1)$ADD	nMUL + $(n-1)$ADD
Witness (Non-Member)	nMUL + $(n-1)$ADD	$(n+1)$MUL + $(n-1)$ADD
Verify (Member)	1(MUL + ADD + PAIR)	1(MUL + ADD + PAIR)
Verify (Non-Member)	2(MUL + ADD + PAIR)	1(MUL + ADD + ADD_T) + 2PAIR
Witness Update (Member)	1(MUL + ADD)	2MUL + 1ADD
Witness Update (Non-Member)	2MUL + 1ADD	$(n+1)$MUL + $(n-1)$ADD

Fig. 3. This table compares the number of cryptographic operations involved in each operation between our construction and that of [53] as extended in [2]. ADD, MUL denote point addition and scalar multiplication in the elliptic curve group $\mathbb{G}$, ADD_T point addition in $\mathbb{G}_T$ and PAIR a pairing computation, whereas n is the size of the set.

For all other algorithms, the additional number of operations is only a constant (at most one) highlighting that zero-knowledge is achieved in practice with only a very small overhead[11]. The only notable exception is the update of non-membership witnesses in which case our construction resorts to re-computation from scratch.

Proving (non-)membership in batch. Another important property of our construction is that it allows the server to efficiently prove statements in batch. Consider the case when a client wants to issue a query on every element of a set $(y_1, \ldots, y_m)$. One way to achieve this would be to provide a separate membership/non-membership witness. This approach would yield a proof that consists of $O(m)$ group elements. Instead, with our construction the server can produce a single membership witness for all $y_i \in \mathcal{X}$ and a single non-membership witness for those $\notin \mathcal{X}$. We will use this technique for our construction in Sect. 5.

5 Zero-Knowledge Authenticated Set Collection (ZKASC)

Zero-knowledge accumulators, as presented so far, can be viewed as zero-knowledge authenticated sets where authenticated zero-knowledge membership/non-membership queries are supported on an outsourced set. In this section, we generalize the problem of zero-knowledge authentication from a set to a collection of sets to support outsourced set algebra operations: *is-subset, intersection, union and set difference.* We refer to this primitive as *zero-knowledge authenticated set collection* (ZKASC) since it falls in the general model of zero-knowledge authenticated data structures [29].

We consider a *dynamic collection* $\mathbb{S}$ of m sets $\mathcal{X}_1, \ldots, \mathcal{X}_m$, with elements from $\mathbb{X}$, that is remotely stored with an untrusted server. $\mathbb{S}$ has two types of operations defined on it: immutable operations $Q()$ and mutable operations $U()$. $Q(\mathbb{S}, q)$ takes a set algebra query q (wrt the indices of $\mathbb{S}$) as input and returns an answer and a proof and it does not alter $\mathbb{S}$. $U(\mathbb{S}, u)$ takes as input an update request and changes $\mathbb{S}$ accordingly. An update $u = (x, \mathsf{upd}, i)$ is either an insertion (if $\mathsf{upd} = 1$) of an element x into a set $\mathcal{X}_i$ or a deletion (if $\mathsf{upd} = 0$) of x from $\mathcal{X}_i$.

ZKASC is a tuple of six probabilistic polynomial time algorithms ZKASC = (KeyGen, Setup, Update, UpdateServer, Query, Verify). Informally, ZKASC lets the owner outsource $\mathbb{S}$ and some auxiliary information and an evaluation key ek to the server (using KeyGen, Setup) and publish a verification key vk and public digest for $\mathbb{S}$. Then, the client can query $\mathbb{S}$ by sending queries to the server. For each query, the server generates answer and prepares its proof (using Query). The owner can also update his set collection and make corresponding changes

[11] Note however, that computing the coefficients of the polynomials that will be encoded in the exponents of the witnesses requires different types of polynomial arithmetic operations. In our construction the server runs an Extended Euclidean algorithm on input two polynomials of degree n and 1 respectively whereas in [2] he runs a polynomial division on the same inputs.

to digest (using Update) and the changes are propagated by the server to his copy of $\mathbb{S}$ and auxiliary information and ek (using UpdateServer). The client verifies the query answer against proof and the digest corresponding to the latest update using vk (in Verify). The security properties of ZKASC are: completeness, soundness and zero-knowledge. They are similar to those of ZKUA as described in Sect. 3, since both follow definition of ZKADS [29].

In the rest of the section we informally introduce our efficient construction of ZKASC, present the main theorem and compare the asymptotic complexity of the algorithms of our ZKASC scheme with that of [56] in Fig. 4. Our construction makes use of *zero-knowledge dynamic universal accumulator* introduced in Sect. 3 and *accumulation tree* described in Sect. 2. For the detailed algorithms and their security analysis we refer the reader to the full version.

5.1 Setup and Update Algorithms

The construction uses $pub = (p, \mathbb{G}, \mathbb{G}_T, e, g)$ as in Sect. 4. The owner runs Setup algorithm with the secret key s, the verification key (g^s, pub) (after generating them using KeyGen) and the set collection $\mathbb{S}$ as input and generates a short public digest for the client, the evaluation key ek and some authentication information of $\mathbb{S}$ for the server. The algorithm computes $\mathsf{acc}(\mathcal{X}_i)$ (zero-knowledge accumulation using Setup algorithm of Sect. 4) for each set $\mathcal{X}_i \in \mathbb{S}$. It then builds an accumulation tree on $\mathsf{acc}(\mathcal{X}_1) \ldots \mathsf{acc}(\mathcal{X}_m)$ and publishes the root of this tree as the public digest of $\mathbb{S}$. It sets the evaluation key to $(g, g^s, \ldots, g^{s^N})$ where $N = \sum_{i \in [1,m]} |\mathcal{X}_i|$. The auxiliary information for the server contains the randomness used for computing each $\mathsf{acc}(\mathcal{X}_i)$.

Update algorithm takes as input an update string u and updates the corresponding set in the set collection (using the Update algorithm in Sect. 4) and accordingly updates the authentication path in the accumulation tree and the auxiliary information, (possibly) the evaluation key and the public digest. As described so far, the update does not guarantee zero-knowledge. If a client queries wrt some set $j \neq i$ before and after u was performed, and sees that $\mathsf{acc}(\mathcal{X}_j)$ has not changed, then he learns that $\mathcal{X}_j$ is not affected by the update. This will also imply that the proofs that the client holds wrt $\mathcal{X}_j$ between updates are still valid. To achieve zero-knowledge, we require Update to re-randomize all the accumulation values that the client has seen (due to queries) since the last update. The update involves changes to authentication information stored with the server. To this end, the server runs UpdateServer algorithm to propagate owner's update on the set collection and authentication information. This algorithm updates the relevant set and updates all the authentication paths in the accumulation tree corresponding to the sets whose accumulation value has been changed or refreshed by the owner.

5.2 Set Algebra Query and Verify Algorithms

Query and Verify algorithms let the server construct a proof of a response to a set operation query and the client verify it, respectively. Since ZKASC supports several set operations, we describe each algorithm in terms of modular subroutines.

Is-subset query: A subset query $q = (\Delta, i)$ is parametrized by a set of elements Δ and an index i of a set collection. Given q, the subset query returns $\mathsf{answer} = 1$ if $\Delta \subseteq \mathcal{X}_i$ and $\mathsf{answer} = 0$ if $\Delta \nsubseteq \mathcal{X}_i$. This query is an efficient generalization of Witness (Sect. 4) where membership/non-membership query is supported for a batch of elements instead of a single element. The proof technique is similar to the membership and non-membership proof generation for a single element using Witness algorithm.

Set intersection query: Set intersection query $q = (i_1, \ldots, i_k)$ is parameterized by a set of indices of the set collection. The answer to an intersection query is a set of elements which we denote as answer and a simulatable proof of the correctness of the answer. If the intersection is computed correctly then $\mathsf{answer} = \mathcal{X}_{i_1} \cap \mathcal{X}_{i_2} \cap \ldots \cap \mathcal{X}_{i_k}$. We express the correctness of intersection with the two following conditions as in [56]:

Subset condition: $\mathsf{answer} \subseteq \mathcal{X}_{i_1} \wedge \ldots \wedge \mathsf{answer} \subseteq \mathcal{X}_{i_k}$. This condition ensures that the returned answer is a subset of all the queried set indices, i.e., every element of answer belongs to each set in the query.

Completeness condition: $(\mathcal{X}_{i_1} - \mathsf{answer}) \cap \ldots \cap (\mathcal{X}_{i_k} - \mathsf{answer}) = \emptyset$. This ensures that answer indeed contains *all* the common elements of $\mathcal{X}_{i_1}, \ldots, \mathcal{X}_{i_k}$, i.e., none of the elements have been omitted from answer.

To prove the first condition, we will use subset query as a subroutine. Proving the second condition is more tricky; it relies on the fact that the characteristic polynomials for the sets $\mathcal{X}_j - \mathsf{answer}$, for all $j \in [i_1, i_k]$, do not have common factors. In other words, these polynomials should be co-prime and their GCD should be 1 (Lemma 1). Since the proof units should be simulatable, we cannot directly use the technique as in [56]. To this end, we randomize the proof units by generalizing the randomization technique in Sect. 4 used to prove non-membership in a single set. The technique essentially adds noise in the exponent for each unit of the intersection proof such that they cancel out when used by the client in the bilinear map equality check.

Set union query: Set union query $q = (i_1, \ldots i_k)$ is parameterized by a set of indices of the set collection. The answer to a union query contains a set of elements, denoted as $\mathsf{answer} = \mathcal{X}_{i_1} \cup \mathcal{X}_{i_2} \cup \ldots \cup \mathcal{X}_{i_k}$, and a simulatable proof of the correctness of the answer. We *introduce* a technique for checking correctness of union operation based on the following conditions:

Superset condition: $\mathcal{X}_{i_1} \subseteq \mathsf{answer} \wedge \mathcal{X}_{i_2} \subseteq \mathsf{answer} \wedge \ldots \wedge \mathcal{X}_{i_k} \subseteq \mathsf{answer}$. This condition ensures that no element has been excluded from the returned answer.

Membership condition: $\mathsf{answer} \subseteq \tilde{U}$ where $\tilde{U} = \mathcal{X}_{i_1} \uplus \mathcal{X}_{i_2} \uplus \ldots \uplus \mathcal{X}_{i_k}$. $\uplus$ denotes multiset union of the sets, i.e., $\uplus$ preserves the multiplicity of every element in the union. This condition ensures that every element of answer belongs to *at least* one of the sets $\mathcal{X}_{i_1}, \ldots, \mathcal{X}_{i_k}$. We note that the trivial way (used in [56]) of proving this condition is to prove that each element of answer is a member of $\mathcal{X}_j$ for $j \in [i_1, i_k]$. This technique obviously does not support zero knowledge as it reveals which set the element comes from.

The first condition can be checked by using the subset proof as a subroutine.[12] The second condition should be proved carefully and not reveal (1) whether an element belongs to more than one of the sets in the query, and (2) which set an element in the union comes from. For example, returning $\tilde{U}$ in the clear trivially reveals the multiplicity of every element in answer. Instead, we request the server to return $\mathsf{acc}(\tilde{U})$ which equals $g^{\mathsf{Ch}_{\tilde{U}}(s)}$ blinded with randomness in the exponent. In order to prove that the server computed $\mathsf{acc}(\tilde{U})$ correctly, we introduce a *union tree*.

A union tree (UT) is a binary tree computed as follows. Corresponding to the k queried indices, $\mathsf{acc}(\mathcal{X}_{i_1}), \ldots, \mathsf{acc}(\mathcal{X}_{i_k})$ are the leaves of UT. The leaves are computed bottom up. Every internal node v is computed as follows. Let v_1 and v_2 be its two children. The (multi)set associated with v is the multiset $M = M_1 \uplus M_2$ where M_1 and M_2 are (multi)sets for v_1 and v_2 respectively. Let r_1 and r_2 be the blinding factors used in computing the accumulation values of v_1 and v_2, respectively. Then the node v stores value $a(v) = g^{r_1 r_2 \mathsf{Ch}_M(s)}$. Finally, the server constructs a proof of subset for answer in $\tilde{U}$.

The client can verify the correctness of each node of UT bottom up using a bilinear map as follows: $e(a(v), g) \stackrel{?}{=} e(a(v_1), a(v_2))$, where g is a part of the verification key. The membership proof verification of $\mathcal{X}_j \subseteq \mathsf{answer}, \forall j \in [i_1, i_k]$, and $\mathsf{answer} \subseteq \tilde{U}$ is done using subset verification subroutine.

Set difference query: Set difference query q is parameterized by two set indices of the set collection $q = (i_1, i_2)$. The answer to a set difference query is $\mathsf{answer} = \mathcal{X}_{i_1} - \mathcal{X}_{i_2}$ and a proof of correctness of the answer. We express the correctness of the answer using the following statement: $(\mathsf{answer} = \mathcal{X}_{i_1} - \mathcal{X}_{i_2}) \iff \mathcal{X}_{i_1} \setminus answer = \mathcal{X}_{i_1} \cap \mathcal{X}_{i_2}$. It ensures two conditions: (1) all the elements of answer indeed belong to $\mathcal{X}_{i_1}$ and (2) *all* the elements of $\mathcal{X}_{i_1}$ that are not in $\mathcal{X}_{i_2}$ are contained in answer. In other words, the union of answer and the intersection $I = \mathcal{X}_{i_1} \cap \mathcal{X}_{i_2}$ equals $\mathcal{X}_{i_1}$.

The second condition is tricky to prove for the following reasons. The server can reveal neither $\mathcal{X}_{i_1} - \mathsf{answer}$ nor $\mathcal{X}_{i_1} \cap \mathcal{X}_{i_2}$ to the client, since this reveals more than the set difference answer the client requested for (hence, breaking our zero-knowledge property).[13] Hence, we are required to provide blinded accumulators corresponding to these sets. Unfortunately, the blinded version of $\mathcal{X}_{i_1} \setminus answer =$

[12] We note that even the security proof does not assume the security proof for subset in a blackbox fashion since here it is the superset rather than the subset that is the known answer.

[13] We note that the sets are revealed to the client in [56] where privacy is not a concern.

$\mathcal{X}_{i_1} \cap \mathcal{X}_{i_2}$, even if the server computed them correctly, would be different. This is caused by different blinding factors used for these accumulators, even though the exponent that corresponds to the elements of the sets is the same. We use the latter fact and require the server to prove that the non-blinded exponents are the same. For this we use standard Schnorr proofs that can be made NIZKPoK in the Common Reference String model using standard techniques [22,28,46]. We describe the properties of a particular NIZKPoK protocol for discrete log in the full version [30]. We can now state the following result. The security proof and an efficiency analysis can be found in [30].

Theorem 3. *The scheme* ZKASC = (KeyGen, Setup, Update, UpdateServer, Query, Verify) *has perfect completeness, soundness under the q-SBDH assumption (with q set to the sum of maximum set sizes produced during the soundness game) and perfect zero-knowledge. Let* $\mathbb{S} = \{\mathcal{X}_1, \ldots, \mathcal{X}_m\}$ *be the original set collection. Define* $M = \sum_{i \in m} |\mathcal{X}_i|$, $n_j = |\mathcal{X}_j|$, *and* $N = \sum_{j \in [i_1, i_k]} n_j$. *Let k be the number of group elements in the query input (for the subset query, it is the cardinality of a queried subset, and for the rest of the queries it is the number of set indices). Let ρ be the size of a query answer, L be the number of sets touched by the queries between updates* u_{t-1} *and* u_t, *and* $0 < \epsilon < 1$ *be a constant chosen at the time of setup. We have:*

- KeyGen *has complexity* $O(1)$*;*
- Setup *has complexity* $O(M + m)$*;*
- Update *and* UpdateServer *have complexity* $O(L)$*;*
- Query *and* Verify *have the following complexity:*
  - *For is-subset, the complexity is* $O(N \log^2 N \log\log N + m^\epsilon \log m)$. *The proof size is* $O(k)$ *and the verification has complexity* $O(k)$.[14]
  - *For set intersection, the complexity is* $O(N \log^2 N \log\log N + km^\epsilon \log m)$. *The proof size is* $O(\rho + k)$ *and the verification has complexity* $O(\rho + k)$.
  - *For set union, the complexity is* $O(k\rho \log \rho + N \log N \log k + km^\epsilon \log m)$. *The proof size is* $O(\rho + k)$ *and the verification has complexity* $O(\rho + k)$.
  - *For set difference, the complexity is* $O(N \log^2 N \log\log N + m^\epsilon \log m)$. *The proof size is* $O(\rho)$ *and the verification has complexity* $O(\rho)$.

5.3 Efficiency Comparison with the Scheme of [56]

We compare the asymptotic complexity of the algorithms of our ZKASC scheme with that of [56] in Fig. 4, which provides *only authenticity* and *trivially reveals information about the set collection.* We show that *only* update algorithms are more expensive compared to that of [56]. The extra cost is to achieve zero-knowledge, which requires all the proofs be ephemeral, i.e., proofs should not hold good between updates. We defer a more detailed comparison to the full version.

---

[14] Note that if the subset query is of the form: is set at index i a subset of the set at index j, then the proof complexity can be made constant.

		[56] \ This paper
Setup		$M+m$
Update	Owner	$1 \backslash L$
	Server	$1 \backslash L$
Subset	Query	$N\log^2 N\log\log N + m^{\varepsilon}\log m$
	Verify/Proof size	k
Instersection	Query	$N\log^2 N\log\log N + km^{\varepsilon}\log m$
	Verify/Proof size	$\rho + k$
Union	Query	$kN\log N + km^{\varepsilon}\log m$
	Verify/Proof size	$\rho + k$
Difference	Query	$N\log^2 N\log\log N + m^{\varepsilon}\log m$
	Verify/Proof size	ρ

Fig. 4. This table compares the asymptotic complexity of each operation with that of [56]. When only one value appears in the last column, it applies to both constructions. We note that the complexity of Union Query was originally mistakenly reported as $O(N \log N)$ in [56]. Notation: $m = |\mathbb{S}|$, $M = \sum_{i \in m} |\mathcal{X}_i|$, $n_j = |\mathcal{X}_j|$, $N = \sum_{j \in [i_1, i_k]} n_j$, k is the number of group elements in the query input (for the subset query it is the size of a queried subset Δ and for the rest of the queries it is the number of set indices), ρ is the size of the answer, L is the number of sets touched by queries between updates u_{t-1} and u_t, and $0 < \epsilon < 1$ is a constant chosen during setup.

6 Conclusion

In this work, we introduced the property of zero-knowledge for cryptographic accumulators. This is a strong privacy property, requiring that witnesses and accumulation values leak nothing about the accumulated set at any given point in the protocol execution, even after insertions and deletions. We showed that zero-knowledge accumulators are located between zero-knowledge sets and the recently introduced notion of primary-secondary-resolver membership proof systems, as the they can be constructed (in a black-box manner) from the former and they can be used to construct (in a black-box manner) the latter. We also presented a construction of an accumulator that achieves computational soundness and perfect zero-knowledge. Using this construction as a building block, we have designed a zero-knowledge authenticated set collection scheme that handles set-related queries that go beyond set (non-)membership. In particular, our scheme supports set unions, intersections, and differences, thus offering a complete set algebra. Future directions in the area include developing constructions that support efficient witness update, constructions based on constant-size assumptions (such as RSA) and constructing an efficient non-interactive set-difference protocol that does not rely on NIZKPoK's. Another interesting future direction is to equip zero-knowledge accumulators with zero-knowledge proofs of knowledge for membership/non-membership.

Acknowledgments. We thank the reviewers for their insightful comments and suggestions. We also thank Markulf Kohlweiss, Leonid Reyzin and Asaf Ziv for helpful discussions. This research was supported in part by the U.S. National Science Foundation under CNS grants 1012798, 1012910, 1228485 and 1645661.

References

1. Ateniese, G., Camenisch, J., Joye, M., Tsudik, G.: A practical and provably secure coalition-resistant group signature scheme. In: Bellare, M. (ed.) CRYPTO 2000. LNCS, vol. 1880, pp. 255–270. Springer, Heidelberg (2000). doi:10.1007/3-540-44598-6_16
2. Au, M.H., Tsang, P.P., Susilo, W., Mu, Y.: Dynamic universal accumulators for DDH groups and their application to attribute-based anonymous credential systems. In: Fischlin, M. (ed.) CT-RSA 2009. LNCS, vol. 5473, pp. 295–308. Springer, Heidelberg (2009). doi:10.1007/978-3-642-00862-7_20
3. Barić, N., Pfitzmann, B.: Collision-free accumulators and fail-stop signature schemes without trees. In: Fumy, W. (ed.) EUROCRYPT 1997. LNCS, vol. 1233, pp. 480–494. Springer, Heidelberg (1997). doi:10.1007/3-540-69053-0_33
4. Benaloh, J., de Mare, M.: One-way accumulators: a decentralized alternative to digital signatures. In: Helleseth, T. (ed.) EUROCRYPT 1993. LNCS, vol. 765, pp. 274–285. Springer, Heidelberg (1994)
5. Blanton, M., Aguiar, E.: Private and oblivious set and multiset operations. In: ASIACCS (2012)
6. Boneh, D., Boyen, X.: Short signatures without random oracles. In: Cachin, C., Camenisch, J.L. (eds.) EUROCRYPT 2004. LNCS, vol. 3027, pp. 56–73. Springer, Heidelberg (2004). doi:10.1007/978-3-540-24676-3_4
7. Buldas, A., Laud, P., Lipmaa, H.: Accountable certificate management using undeniable attestations. In: CCS (2000)
8. Camacho, P., Hevia, A.: On the impossibility of batch update for cryptographic accumulators. In: Abdalla, M., Barreto, P.S.L.M. (eds.) LATINCRYPT 2010. LNCS, vol. 6212, pp. 178–188. Springer, Heidelberg (2010). doi:10.1007/978-3-642-14712-8_11
9. Camacho, P., Hevia, A., Kiwi, M., Opazo, R.: Strong accumulators from collision-resistant hashing. In: Information Security (2008)
10. Camenisch, J., Kohlweiss, M., Soriente, C.: An accumulator based on bilinear maps and efficient revocation for anonymous credentials. In: Jarecki, S., Tsudik, G. (eds.) PKC 2009. LNCS, vol. 5443, pp. 481–500. Springer, Heidelberg (2009). doi:10.1007/978-3-642-00468-1_27
11. Camenisch, J., Lysyanskaya, A.: Dynamic accumulators and application to efficient revocation of anonymous credentials. In: Yung, M. (ed.) CRYPTO 2002. LNCS, vol. 2442, pp. 61–76. Springer, Heidelberg (2002). doi:10.1007/3-540-45708-9_5
12. Canetti, R., Paneth, O., Papadopoulos, D., Triandopoulos, N.: Verifiable set operations over outsourced databases. In: Krawczyk, H. (ed.) PKC 2014. LNCS, vol. 8383, pp. 113–130. Springer, Heidelberg (2014). doi:10.1007/978-3-642-54631-0_7
13. Catalano, D., Fiore, D.: Vector commitments and their applications. In: Kurosawa, K., Hanaoka, G. (eds.) PKC 2013. LNCS, vol. 7778, pp. 55–72. Springer, Heidelberg (2013). doi:10.1007/978-3-642-36362-7_5
14. Catalano, D., Fiore, D., Messina, M.: Zero-knowledge sets with short proofs. In: Smart, N. (ed.) EUROCRYPT 2008. LNCS, vol. 4965, pp. 433–450. Springer, Heidelberg (2008). doi:10.1007/978-3-540-78967-3_25

15. Chase, M., Healy, A., Lysyanskaya, A., Malkin, T., Reyzin, L.: Mercurial commitments with applications to zero-knowledge sets. In: Cramer, R. (ed.) EUROCRYPT 2005. LNCS, vol. 3494, pp. 422–439. Springer, Heidelberg (2005). doi:10.1007/11426639_25
16. Chatterjee, S., Menezes, A.: On cryptographic protocols employing asymmetric pairings - the role of ψ revisited. Discrete Appl. Math. **159**(13), 1311–1322 (2011)
17. Cristofaro, E., Tsudik, G.: Practical private set intersection protocols with linear complexity. In: Sion, R. (ed.) FC 2010. LNCS, vol. 6052, pp. 143–159. Springer, Heidelberg (2010). doi:10.1007/978-3-642-14577-3_13
18. Dachman-Soled, D., Malkin, T., Raykova, M., Yung, M.: Efficient robust private set intersection. In: Abdalla, M., Pointcheval, D., Fouque, P.-A., Vergnaud, D. (eds.) ACNS 2009. LNCS, vol. 5536, pp. 125–142. Springer, Heidelberg (2009). doi:10.1007/978-3-642-01957-9_8
19. Damgård, I., Triandopoulos, N.: Supporting non-membership proofs with bilinear-map accumulators. Cryptology ePrint Archive, Report 2008/538 (2008)
20. de Meer, H., Liedel, M., Pöhls, H.C., Posegga, J.: Indistinguishability of one-way accumulators. In Technical Report MIP-1210, Faculty of Computer Science and Mathematics (FIM), University of Passau (2012)
21. de Meer, H., Pöhls, H.C., Posegga, J., Samelin, K.: Redactable signature schemes for trees with signer-controlled non-leaf-redactions. In: E-Business and Telecommunications (2014)
22. Santis, A., Crescenzo, G., Ostrovsky, R., Persiano, G., Sahai, A.: Robust non-interactive zero knowledge. In: Kilian, J. (ed.) CRYPTO 2001. LNCS, vol. 2139, pp. 566–598. Springer, Heidelberg (2001). doi:10.1007/3-540-44647-8_33
23. Derler, D., Hanser, C., Slamanig, D.: Revisiting cryptographic accumulators, additional properties and relations to other primitives. In: Nyberg, K. (ed.) CT-RSA 2015. LNCS, vol. 9048, pp. 127–144. Springer, Heidelberg (2015). doi:10.1007/978-3-319-16715-2_7
24. Dong, C., Chen, L., Wen, Z.: When private set intersection meets big data: an efficient and scalable protocol. In: ACM CCS (2013)
25. Fauzi, P., Lipmaa, H., Zhang, B.: Efficient non-interactive zero knowledge arguments for set operations. In: Christin, N., Safavi-Naini, R. (eds.) FC 2014. LNCS, vol. 8437, pp. 216–233. Springer, Heidelberg (2014). doi:10.1007/978-3-662-45472-5_14
26. Fazio, N., Nicolosi, A.: Cryptographic accumulators: Definitions, constructions and applications. In Technical report. Courant Institute of Mathematical Sciences, New York University (2002)
27. Freedman, M.J., Nissim, K., Pinkas, B.: Efficient private matching and set intersection. In: Cachin, C., Camenisch, J.L. (eds.) EUROCRYPT 2004. LNCS, vol. 3027, pp. 1–19. Springer, Heidelberg (2004). doi:10.1007/978-3-540-24676-3_1
28. Garay, J.A., MacKenzie, P., Yang, K.: Strengthening zero-knowledge protocols using signatures. In: Biham, E. (ed.) EUROCRYPT 2003. LNCS, vol. 2656, pp. 177–194. Springer, Heidelberg (2003). doi:10.1007/3-540-39200-9_11
29. Ghosh, E., Goodrich, M.T., Ohrimenko, O., Tamassia, R.: Verifiable zero-knowledge order queries and updates for fully dynamic lists and trees. In: Zikas, V., Prisco, R. (eds.) SCN 2016. LNCS, vol. 9841, pp. 216–236. Springer, Heidelberg (2016). doi:10.1007/978-3-319-44618-9_12
30. Ghosh, E., Ohrimenko, O., Papadopoulos, D., Tamassia, R., Triandopoulos, N.: Zero-knowledge accumulators and set operations. ePrint, 2015/404 (2015)

31. Ghosh, E., Ohrimenko, O., Tamassia, R.: Efficient verifiable range and closest point queries in zero-knowledge. In: Privacy Enhancing Technologies Symposium (PETs) (2016)
32. Ghosh, E., Ohrimenko, O., Tamassia, R.: Zero-knowledge authenticated order queries and order statistics on a list. In: Malkin, T., Kolesnikov, V., Lewko, A.B., Polychronakis, M. (eds.) ACNS 2015. LNCS, vol. 9092, pp. 149–171. Springer, Heidelberg (2015). doi:10.1007/978-3-319-28166-7_8
33. Goldberg, S., Naor, M., Papadopoulos, D., Reyzin, L., Vasant, S., Ziv, A.: NSEC5: Provably preventing DNSSEC zone enumeration. Cryptology ePrint Archive, Report 2014/582 (2014)
34. Goldwasser, S., Micali, S., Rackoff, C.: The knowledge complexity of interactive proof-systems (extended abstract). In: STOC (1985)
35. Hanser, C., Slamanig, D.: Structure-preserving signatures on equivalence classes and their application to anonymous credentials. In: Sarkar, P., Iwata, T. (eds.) ASIACRYPT 2014. LNCS, vol. 8873, pp. 491–511. Springer, Heidelberg (2014). doi:10.1007/978-3-662-45611-8_26
36. Hazay, C., Nissim, K.: Efficient set operations in the presence of malicious adversaries. J. Cryptology **25**(3), 383–433 (2012)
37. Huang, Y., Evans, D., Katz, J.: Private set intersection: are garbled circuits better than custom protocols? In: NDSS (2012)
38. Jarecki, S., Liu, X.: Efficient oblivious pseudorandom function with applications to adaptive OT and secure computation of set intersection. In: Reingold, O. (ed.) TCC 2009. LNCS, vol. 5444, pp. 577–594. Springer, Heidelberg (2009). doi:10.1007/978-3-642-00457-5_34
39. Kissner, L., Song, D.: Privacy-preserving set operations. In: Shoup, V. (ed.) CRYPTO 2005. LNCS, vol. 3621, pp. 241–257. Springer, Heidelberg (2005). doi:10.1007/11535218_15
40. Kosba, A.E., Papadopoulos, D., Papamanthou, C., Sayed, M.F., Shi, E., Triandopoulos, N.: TRUESET: faster verifiable set computations. In: USENIX (2014)
41. Li, J., Li, N., Xue, R.: Universal accumulators with efficient nonmembership proofs. In: Katz, J., Yung, M. (eds.) ACNS 2007. LNCS, vol. 4521, pp. 253–269. Springer, Heidelberg (2007). doi:10.1007/978-3-540-72738-5_17
42. Libert, B., Ramanna, S.C., Yung, M.: Functional commitment schemes: from polynomial commitments to pairing-based accumulators from simple assumptions. In: ICALP (2016)
43. Libert, B., Yung, M.: Concise mercurial vector commitments and independent zero-knowledge sets with short proofs. In: Micciancio, D. (ed.) TCC 2010. LNCS, vol. 5978, pp. 499–517. Springer, Heidelberg (2010). doi:10.1007/978-3-642-11799-2_30
44. Lipmaa, H.: Secure accumulators from euclidean rings without trusted setup. In: Bao, F., Samarati, P., Zhou, J. (eds.) ACNS 2012. LNCS, vol. 7341, pp. 224–240. Springer, Heidelberg (2012). doi:10.1007/978-3-642-31284-7_14
45. Liskov, M.: Updatable zero-knowledge databases. In: Roy, B. (ed.) ASIACRYPT 2005. LNCS, vol. 3788, pp. 174–198. Springer, Heidelberg (2005). doi:10.1007/11593447_10
46. MacKenzie, P., Yang, K.: On simulation-sound trapdoor commitments. In: Cachin, C., Camenisch, J.L. (eds.) EUROCRYPT 2004. LNCS, vol. 3027, pp. 382–400. Springer, Heidelberg (2004). doi:10.1007/978-3-540-24676-3_23
47. Merkle, R.C.: Protocols for public key cryptosystems. In: IEEE Symposium on Security and Privacy (1980)
48. Micali, S., Rabin, M.O., Kilian, J.: Zero-knowledge sets. In: FOCS (2003)

49. Miers, I., Garman, C., Green, M., Rubin, A.D.: Zerocoin: anonymous distributed e-cash from bitcoin. In: IEEE Symposium on Security and Privacy (2013)
50. Morselli, R., Bhattacharjee, S., Katz, J., Keleher, P.J.: Trust-preserving set operations. In: IEEE INFOCOM (2004)
51. Naor, M., Nissim, K.: Certificate revocation and certificate update. IEEE J. Sel. Areas Commun. **18**(4), 561–570 (2000)
52. Naor, M., Ziv, A.: Primary-secondary-resolver membership proof systems. In: Dodis, Y., Nielsen, J.B. (eds.) TCC 2015. LNCS, vol. 9015, pp. 199–228. Springer, Heidelberg (2015). doi:10.1007/978-3-662-46497-7_8
53. Nguyen, L.: Accumulators from bilinear pairings and applications. In: Menezes, A. (ed.) CT-RSA 2005. LNCS, vol. 3376, pp. 275–292. Springer, Heidelberg (2005). doi:10.1007/978-3-540-30574-3_19
54. Nyberg, K.: Commutativity in cryptography. In: 1st International Trier Conference in Functional Analysis (1996)
55. Nyberg, K.: Fast accumulated hashing. In: Gollmann, D. (ed.) FSE 1996. LNCS, vol. 1039, pp. 83–87. Springer, Heidelberg (1996). doi:10.1007/3-540-60865-6_45
56. Papamanthou, C., Tamassia, R., Triandopoulos, N.: Optimal verification of operations on dynamic sets. In: Rogaway, P. (ed.) CRYPTO 2011. LNCS, vol. 6841, pp. 91–110. Springer, Heidelberg (2011). doi:10.1007/978-3-642-22792-9_6
57. Papamanthou, C., Tamassia, R., Triandopoulos, N.: Authenticated hash tables based on cryptographic accumulators. Algorithmica (2015)
58. Prabhakaran, M., Xue, R.: Statistically hiding sets. In: Fischlin, M. (ed.) CT-RSA 2009. LNCS, vol. 5473, pp. 100–116. Springer, Heidelberg (2009). doi:10.1007/978-3-642-00862-7_7
59. Preparata, F., Sarwate, D., I. U. A. U.-C. C. S. LAB: Computational Complexity of Fourier Transforms Over Finite Fields. DTIC, 1976
60. Reyzin, L., Yakoubov, S.: Efficient asynchronous accumulators for distributed PKI. In: Zikas, V., Prisco, R. (eds.) SCN 2016. LNCS, vol. 9841, pp. 292–309. Springer, Heidelberg (2016). doi:10.1007/978-3-319-44618-9_16
61. Samelin, K., Pöhls, H.C., Bilzhause, A., Posegga, J., Meer, H.: Redactable signatures for independent removal of structure and content. In: Ryan, M.D., Smyth, B., Wang, G. (eds.) ISPEC 2012. LNCS, vol. 7232, pp. 17–33. Springer, Heidelberg (2012). doi:10.1007/978-3-642-29101-2_2
62. Sander, T.: Efficient accumulators without trapdoor. In: ICICS (1999)
63. Tamassia, R.: Authenticated data structures. In: Battista, G., Zwick, U. (eds.) ESA 2003. LNCS, vol. 2832, pp. 2–5. Springer, Heidelberg (2003). doi:10.1007/978-3-540-39658-1_2
64. Zheng, Q., Xu, S.: Verifiable delegated set intersection operations on outsourced encrypted data. IACR Cryptology ePrint Archive (2014)

Zero-Knowledge Arguments for Matrix-Vector Relations and Lattice-Based Group Encryption

Benoît Libert[1(✉)], San Ling[2], Fabrice Mouhartem[1], Khoa Nguyen[2], and Huaxiong Wang[2]

[1] École Normale Supérieure de Lyon, Laboratoire LIP, Lyon, France
benoit.libert@ens-lyon.fr

[2] School of Physical and Mathematical Sciences, Nanyang Technological University, Singapore, Singapore

Abstract. Group encryption (GE) is the natural encryption analogue of group signatures in that it allows verifiably encrypting messages for some anonymous member of a group while providing evidence that the receiver is a properly certified group member. Should the need arise, an opening authority is capable of identifying the receiver of any ciphertext. As introduced by Kiayias, Tsiounis and Yung (Asiacrypt'07), GE is motivated by applications in the context of oblivious retriever storage systems, anonymous third parties and hierarchical group signatures. This paper provides the first realization of group encryption under lattice assumptions. Our construction is proved secure in the standard model (assuming interaction in the proving phase) under the Learning-With-Errors (LWE) and Short-Integer-Solution (SIS) assumptions. As a crucial component of our system, we describe a new zero-knowledge argument system allowing to demonstrate that a given ciphertext is a valid encryption under some hidden but certified public key, which incurs to prove quadratic statements about LWE relations. Specifically, our protocol allows arguing knowledge of witnesses consisting of $\mathbf{X} \in \mathbb{Z}_q^{m \times n}$, $\mathbf{s} \in \mathbb{Z}_q^n$ and a small-norm $\mathbf{e} \in \mathbb{Z}^m$ which underlie a public vector $\mathbf{b} = \mathbf{X} \cdot \mathbf{s} + \mathbf{e} \in \mathbb{Z}_q^m$ while simultaneously proving that the matrix $\mathbf{X} \in \mathbb{Z}_q^{m \times n}$ has been correctly certified. We believe our proof system to be useful in other applications involving zero-knowledge proofs in the lattice setting.

Keywords: Lattices · Zero-knowledge proofs · Group encryption · Anonymity

1 Introduction

Since the pioneering work of Regev [49] and Gentry, Peikert and Vaikuntanathan (GPV) [23], lattice-based cryptography has been an extremely active research area. Not only do lattices enable powerful functionalities (e.g., [22,26]) that have no viable realizations under discrete-logarithm or factoring-related assumptions, they also offer a number of advantages over conventional number-theoretic techniques, like simpler arithmetic operations, their conjectured resistance to quantum attacks or a better asymptotic efficiency.

J.H. Cheon and T. Takagi (Eds.): ASIACRYPT 2016, Part II, LNCS 10032, pp. 101–131, 2016.
DOI: 10.1007/978-3-662-53890-6_4

The design of numerous cryptographic protocols crucially relies on zero-knowledge proofs [25] to prove properties about encrypted or committed values so as to enforce honest behavior on behalf of participants or protect the privacy of users. In the lattice settings, efficient zero-knowledge proofs are non-trivial to construct due to the limited amount of algebraic structure. While natural methods of proving knowledge of secret keys [31,40,42,44] are available, they are only known to work for specific languages. When it comes to proving circuit satisfiability, the best known methods are designed for the LPN setting [30] or take advantage of the extra structure available in the ring LWE setting [10,54]. Hence, these methods are not known to readily carry over to standard (i.e., non-ideal) lattices. In the standard model, the problem is even trickier as we do not have a lattice-based counterpart of Groth-Sahai proofs [28] and efficient non-interactive proof systems are only available for specific problems [48].

The difficulty of designing efficient zero-knowledge proofs for lattice-related languages makes it highly non-trivial to adapt privacy-preserving cryptographic primitives in the lattice setting. In spite of these technical hurdles, a recent body of work successfully designed anonymity-enabling mechanisms like ring signatures [2,31], blind signatures [50], group signatures [9,27,35,36,38,41,45] or, more recently, signature schemes with companion zero-knowledge protocols [37]. A common feature of all these works is that the zero-knowledge layer of the proposed protocols only deals with linear equations, where witnesses are only multiplied by public values.

In this paper, motivated by the design of advanced privacy-preserving protocols in the lattice setting, we construct zero-knowledge arguments for non-linear statements among witnesses consisting of vectors and matrices. For suitable parameters $q, n, m \in \mathbb{Z}$, we consider zero-knowledge argument systems whereby a prover can demonstrate knowledge of secret matrices $\mathbf{X} \in \mathbb{Z}_q^{m \times n}$ and vectors $\mathbf{s} \in \mathbb{Z}_q^n$, $\mathbf{e} \in \mathbb{Z}^m$ such that: (i) $\mathbf{e} \in \mathbb{Z}^m$ has small norm; (ii) A public vector $\mathbf{b} \in \mathbb{Z}_q^n$ equals $\mathbf{b} = \mathbf{X} \cdot \mathbf{s} + \mathbf{e} \bmod q$; (iii) The underlying pair $(\mathbf{X}, \mathbf{s})$ satisfies additional algebraic relations: for instance, it should be possible to prove possession of a signature on some representation of the matrix $\mathbf{X}$. In particular, our zero-knowledge argument makes it possible to prove that a given ciphertext is a well-formed LWE-based encryption with respect to some hidden, but certified public key. This protocol comes in handy in the design of *group encryption* schemes [33], where such languages naturally arise. In this paper, we thus construct the first construction of group encryption under lattice assumptions.

Group Encryption. As suggested by Kiayias, Tsiounis and Yung [33], group encryption (GE) is the encryption analogue of group signatures [19], which allow users to anonymously sign messages on behalf of an entire group they belong to. While group signatures aim at hiding the source of some message within a crowd administered by some group manager, group encryption rather seeks to hide its destination within a group of legitimate receivers. In both cases, a verifier should be convinced that the anonymous signer/receiver indeed belongs to a purported population. In order to keep users accountable for their actions, an opening authority (OA) is further empowered with some information allowing it to un-anonymize signatures/ciphertexts.

Kiayias, Tsiounis and Yung [33] formalized GE schemes as a primitive allowing the sender to generate publicly verifiable guarantees that: (1) The ciphertext is well-formed and intended for some registered group member who will be able to decrypt; (2) the opening authority will be able identify the receiver if necessary; (3) The plaintext satisfies certain properties such as being a witness for some public relation or the private key that underlies a given public key. In the model of Kiayias *et al.* [33], the message secrecy and anonymity properties are required to withstand active adversaries, which are granted access to decryption oracles in all security experiments.

As a natural application, group encryption allows a firewall to filter all incoming encrypted emails except those intended for some certified organization member and the content of which is additionally guaranteed to satisfy certain requirements, like the absence of malware.

GE schemes are also motivated by natural privacy applications such as anonymous trusted third parties, key recovery mechanisms or oblivious retriever storage systems. In optimistic protocols, GE allows verifiably encrypting messages to *anonymous* trusted third parties which mostly remain off-line and only come into play to sort out conflicts. In order to protect privacy-sensitive information such as users' citizenship, group encryption makes it possible to hide the identity of users' preferred trusted third parties within a set of properly certified trustees.

In cloud storage services, GE enables privacy-preserving asynchronous transfers of encrypted datasets. Namely, it allows users to archive encrypted datasets on remote servers while convincing those servers that the data is indeed intended for some anonymous certified client who paid a subscription to the storage provider. Moreover, a judge should be able to identify the archive's recipient in case a misbehaving server is found guilty of hosting suspicious transaction records or any other illegal content.

As pointed out by Kiayias *et al.* [33], group encryption also implies a form of hierarchical group signatures [53], where signatures can only be opened by a set of eligible trustees operating in a very specific manner determiner by the signer.

Related work. Kiayias, Tsiounis and Yung (KTY) [33] formalized the notion of group encryption and provided a modular design using zero-knowledge proofs, digital signatures, anonymous CCA-secure public-key encryption and commitment schemes. They also gave an efficient instantiation using Paillier's cryptosystem [46] and Camenisch-Lysyanskaya signatures [15].

Cathalo, Libert and Yung [18] designed a non-interactive system in the standard model under non-interactive pairing-related assumptions. El Aimani and Joye [3] suggested various efficiency improvements with both interactive and non-interactive proofs.

Libert *et al.* [39] empowered the GE primitive with a refined traceability mechanism akin to that of traceable signatures [32]. Namely, by releasing a user-specific trapdoor, the opening authority can allow anyone to publicly trace ciphertexts encrypted for this specific group member without affecting the privacy of other users. Back in 2010, Izabachène, Pointcheval and Vergnaud [29] considered the problem of eliminating subliminal channels in a different form of traceable group encryption.

As a matter of fact, all existing realizations of group encryption or similar primitives rely on traditional number theoretic assumptions like the hardness of factoring or computing discrete logarithms. In particular, all of them are vulnerable to quantum attacks. For the sake of not putting all one's eggs in the same basket, it is highly desirable to have instantiations based on alternative, quantum-resistant foundations.

OUR RESULTS AND TECHNIQUES. We put forth the first lattice-based realization of the group encryption primitive and prove its security under the Learning-With-Errors (LWE) [49] and Short-Integer-Solution (SIS) [4] assumptions. As in the original design of Kiayias, Tsiounis and Yung [33], the security analysis of our scheme stands in the standard model if we avail ourselves of interaction between the prover and the verifier. In the random oracle model [8], the Fiat-Shamir paradigm [21] readily provides a non-interactive solution based on the same hardness assumptions.

As a core ingredient of our GE scheme, we develop a new technique allowing to prove that a given ciphertext is a valid LWE-based encryption under some hidden but certified public key. Via a novel extension of Stern-like zero-knowledge arguments [31,52] in the lattice setting, we provide a method of proving quadratic relations between a secret certified matrix and a secret vector occurring in LWE-related languages. We believe our zero-knowledge arguments to be of independent interest as they find applications in other protocols involving zero-knowledge proofs in lattice-based cryptography.

It was shown by Kiayias *et al.* [33] that, in order to design a GE scheme, three ingredients are necessary: we need digital signatures, anonymous (i.e., key-private [7]) public-key encryption and zero-knowledge proofs. While the first two ingredients are available in lattice-based cryptography, suitable zero-knowledge proof systems are currently lacking. The underlying proof system should allow the sender to prove that the ciphertext is well-formed and is decryptable by some certified group member without betraying the latter's identity. Such statements typically involve equations of the form $\mathbf{b} = \mathbf{X} \cdot \mathbf{s} + \mathbf{e} \bmod q$, for which given integers n, m, q and vector $\mathbf{b} \in \mathbb{Z}_q^m$, the prover has to demonstrate possession of a certified matrix $\mathbf{X} \in \mathbb{Z}_q^{m \times n}$, vector $\mathbf{s} \in \mathbb{Z}_q^n$ and small-norm error vector $\mathbf{e} \in \mathbb{Z}^m$ satisfying the equation. Existing mechanisms of proving relations appearing in lattice-based cryptosystems belong to two main classes. The first one, which uses "rejection sampling" techniques for Schnorr-like protocols [51], was introduced by Lyubashevsky [42]. The second class, which was initiated by Ling *et al.* [40], appeals to "decomposition-extension-permutation" techniques in lattice-based extensions [31] of Stern's protocol [52]. These techniques mainly deal with *linear equations*, where each term is a product of a public matrix with a secret vector, which possibly satisfies some additional constraints (e.g., smallness) to be proven. Here, we are presented with *quadratic equations* where some terms $\mathbf{X} \cdot \mathbf{s}$ are products of two secret witnesses $\mathbf{X} \in \mathbb{Z}_q^{m \times n}$ and $\mathbf{s} \in \mathbb{Z}_q^n$ which are involved in other equations. Proving such quadratic equations thus requires new ideas.

To overcome the above hurdle, we employ a divide-and-conquer strategy. First, we consider the binary representations of $\mathbf{X}$ and $\mathbf{s}$, and view the product $\mathbf{X} \cdot \mathbf{s}$ as a bunch of bit-wise products $\{x_i \cdot s_j\}_{i,j}$. Now, although these bit-wise products still admit a quadratic nature, but to prove that each of them is well-formed, it suffices to demonstrate in zero-knowledge that $x_i \cdot s_j$ belongs to the set $B = \{0 \cdot 0, 0 \cdot 1, 1 \cdot 0, 1 \cdot 1\}$ of cardinality 4. This can be done with a Stern-like sub-protocol, using the following extending-then-permuting technique. We first extend $x_i \cdot s_j$ to vector $\mathsf{ext}(x_i, s_j) \stackrel{\mathsf{def}}{=} (\overline{x}_i \cdot \overline{s}_j, \overline{x}_i \cdot s_j, x_i \cdot \overline{s}_j, x_i \cdot s_j)^\top \in \{0,1\}^4$ whose entries are elements of B (here, $\bar{c}$ denotes the bit $1 - c$). We then employ a special permutation, determined by two random bits b_x and b_s, to the entries of $\mathsf{ext}(x_i, s_j)$, such that the permuted vector is exactly the correct extension $\mathsf{ext}(x_i \oplus b_x, s_j \oplus b_s)$, where $\oplus$ denotes the addition modulo 2. Seeing that a permutation of $\mathsf{ext}(x_i, s_j)$ has entries in the set B, the verifier should be convinced that $x_i \cdot s_j \in B$. Meanwhile, the bits b_x and b_s act as one-time pads that perfectly hide x_i and s_j. Furthermore, to prove that the same bits x_i and s_j are involved in other equations, we establish similar extending-then-permuting mechanisms for their other appearances, and use the same one-time pads b_x and b_s, respectively, as those places.

Having settled the problem of proving quadratic relations, we are able to realize the desired zero-knowledge layer by combining our proof system with the techniques of [37,41]. These help us demonstrate possession of a signature on the user's public key while proving that this key is encrypted under the OA's public key. Since users' public keys consist of a matrix $\mathbf{B}_\mathsf{U} \in \mathbb{Z}_q^{n \times m}$, we actually encrypt a hash value of this matrix under the OA's public key while the sender proves knowledge of a signature on the binary decomposition of $\mathbf{B}_\mathsf{U}$. By using a suitable lattice-based hash function [24], the Stern-like protocols of [37,41] make it possible to prove that the hashed matrix encrypted under the OA's public key coincides with the one for which the sender knows a certificate and which served as a public key to encrypt the actual plaintext.

The last issue to sort out is to determine the appropriate encryption schemes to work with in the two public-key encryption components. The CCA2-secure cryptosystem implied by the Agrawal-Boneh-Boyen (ABB) identity-based encryption (IBE) scheme [1] via the CHK transformation [16] is a natural choice as it is one of the most efficient LWE-based candidates in the standard model. For technical reasons, we chose to use a variant of the ABB cryptosystem based on the trapdoor mechanism of Micciancio and Peikert [43] because it allows dispensing with zero-knowledge proofs of public key validity. Indeed, the Kiayias-Tsiounis-Yung model [33] mandates that certified public keys be valid public keys (for which an underlying private key exists). This requirement is easier to handle using Micciancio-Peikert trapdoors [43] since, unlike GPV trapdoors [23], they are guaranteed to exist for any public matrix.

2 Background and Definitions

2.1 Lattices

In our notations, all vectors are denoted in bold lower-case letters while bold upper-case letters will be used for matrices. If $\mathbf{b} \in \mathbb{R}^n$, its Euclidean norm and infinity norm will be denoted by $\|\mathbf{b}\|$ and $\|\mathbf{b}\|_\infty$ respectively. The Euclidean norm of matrix $\mathbf{B} \in \mathbb{R}^{m \times n}$ with columns $(\mathbf{b}_i)_{i \le n}$ is denoted by $\|\mathbf{B}\| = \max_{i \le n} \|\mathbf{b}_i\|$. If $\mathbf{B}$ is full column-rank, we let $\widetilde{\mathbf{B}}$ denote its Gram-Schmidt orthogonalization.

When S is a finite set, we denote by $U(S)$ the uniform distribution over S and by $x \hookleftarrow D$ the action of sampling x according to the distribution D.

A (full-rank) lattice L is the set of all integer linear combinations of some linearly independent basis vectors $(\mathbf{b}_i)_{i \le n}$ belonging to some $\mathbb{R}^n$. We work with q-ary lattices, for some prime q.

Definition 1. *Let $m \ge n \ge 1$, a prime $q \ge 2$ and $\mathbf{A} \in \mathbb{Z}_q^{n \times m}$ and $\mathbf{u} \in \mathbb{Z}_q^n$, define $\Lambda_q(\mathbf{A}) := \{\mathbf{e} \in \mathbb{Z}^m \mid \exists \mathbf{s} \in \mathbb{Z}_q^n \text{ s.t. } \mathbf{A}^T \cdot \mathbf{s} = \mathbf{e} \bmod q\}$ as well as*

$$\Lambda_q^{\perp}(\mathbf{A}) := \{\mathbf{e} \in \mathbb{Z}^m \mid \mathbf{A} \cdot \mathbf{e} = \mathbf{0}^n \bmod q\},\ \Lambda_q^{\mathbf{u}}(\mathbf{A}) := \{\mathbf{e} \in \mathbb{Z}^m \mid \mathbf{A} \cdot \mathbf{e} = \mathbf{u} \bmod q\}$$

For any $\mathbf{t} \in \Lambda_q^{\mathbf{u}}(\mathbf{A})$, $\Lambda_q^{\mathbf{u}}(\mathbf{A}) = \Lambda_q^{\perp}(\mathbf{A}) + \mathbf{t}$ so that $\Lambda_q^{\mathbf{u}}(\mathbf{A})$ is a shift of $\Lambda_q^{\perp}(\mathbf{A})$.

For a lattice L, a vector $\mathbf{c} \in \mathbb{R}^n$ and a real $\sigma > 0$, define $\rho_{\sigma,\mathbf{c}}(\mathbf{x}) = \exp(-\pi\|\mathbf{x} - \mathbf{c}\|^2/\sigma^2)$. The discrete Gaussian distribution of support L, parameter σ and center $\mathbf{c}$ is defined as $D_{L,\sigma,\mathbf{c}}(\mathbf{y}) = \rho_{\sigma,\mathbf{c}}(\mathbf{y})/\rho_{\sigma,\mathbf{c}}(L)$ for any $\mathbf{y} \in L$. We denote by $D_{L,\sigma}(\mathbf{y})$ the distribution centered in $\mathbf{c} = \mathbf{0}$. We will extensively use the fact that samples from $D_{L,\sigma}$ are short with overwhelming probability.

Lemma 1 ([6, Lemma 1.5]). *For any lattice $L \subseteq \mathbb{R}^n$ and positive real number $\sigma > 0$, we have $\Pr_{\mathbf{b} \hookleftarrow D_{L,\sigma}}[\|\mathbf{b}\| \le \sqrt{n}\sigma] \ge 1 - 2^{-\Omega(n)}$.*

As shown in [23], Gaussian distributions with lattice support can be sampled from efficiently, given a sufficiently short basis of the lattice.

Lemma 2 ([14, Lemma 2.3]). *There exists a* PPT *(probabilistic polynomial-time) algorithm* GPVSample *that takes as inputs a basis $\mathbf{B}$ of a lattice $L \subseteq \mathbb{Z}^n$ and a rational $\sigma \ge \|\widetilde{\mathbf{B}}\| \cdot \Omega(\sqrt{\log n})$, and outputs vectors $\mathbf{b} \in L$ with distribution $D_{L,\sigma}$.*

Lemma 3 ([5, Theorem 3.2]). *There exists a* PPT *algorithm* TrapGen *that takes as inputs 1^n, 1^m and an integer $q \ge 2$ with $m \ge \Omega(n \log q)$, and outputs a matrix $\mathbf{A} \in \mathbb{Z}_q^{n \times m}$ and a basis $\mathbf{T_A}$ of $\Lambda_q^{\perp}(\mathbf{A})$ such that $\mathbf{A}$ is within statistical distance $2^{-\Omega(n)}$ to $U(\mathbb{Z}_q^{n \times m})$, and $\|\widetilde{\mathbf{T_A}}\| \le \mathcal{O}(\sqrt{n \log q})$.*

Lemma 3 is often combined with the sampler from Lemma 2. Micciancio and Peikert [43] recently proposed a more efficient approach for this combined task, which should be preferred in practice but, for the sake of simplicity, we present our schemes using TrapGen.

We rely on a basis delegation algorithm [17] which extends a trapdoor for $\mathbf{A} \in \mathbb{Z}_q^{n \times m}$ into a trapdoor of any $\mathbf{B} \in \mathbb{Z}_q^{n \times m'}$ whose left $n \times m$ submatrix is $\mathbf{A}$.

Lemma 4 ([17, Lemma 3.2]). *There exists a* PPT *algorithm* ExtBasis *that takes as inputs a matrix* $\mathbf{B} \in \mathbb{Z}_q^{n \times m'}$ *whose first* m *columns span* $\mathbb{Z}_q^n$*, and a basis* $\mathbf{T_A}$ *of* $\Lambda_q^{\perp}(\mathbf{A})$ *where* $\mathbf{A}$ *is the left* $n \times m$ *submatrix of* $\mathbf{B}$*, and outputs a basis* $\mathbf{T_B}$ *of* $\Lambda_q^{\perp}(\mathbf{B})$ *with* $\|\widetilde{\mathbf{T_B}}\| \leq \|\widetilde{\mathbf{T_A}}\|$.

Like [11,13], we use a technique due to Agrawal, Boneh and Boyen [1] that realizes a punctured trapdoor mechanism [12]. Analogously to [43], we will use such a mechanism in the real scheme and not only in the proof.

Lemma 5 ([1, Theorem 19]). *There exists a* PPT *algorithm* SampleRight *that takes as inputs matrices* $\mathbf{A} \in \mathbb{Z}_q^{n \times m}, \mathbf{C} \in \mathbb{Z}_q^{n \times \bar{m}}$*, a low-norm matrix* $\mathbf{R} \in \mathbb{Z}^{m \times \bar{m}}$*, a short basis* $\mathbf{T_C} \in \mathbb{Z}^{\bar{m} \times \bar{m}}$ *of* $\Lambda_q^{\perp}(\mathbf{C})$*, a vector* $\mathbf{u} \in \mathbb{Z}_q^n$ *and a rational* σ *such that* $\sigma \geq \|\widetilde{\mathbf{T_C}}\| \cdot \Omega(\sqrt{\log n})$*, and outputs a short vector* $\mathbf{b} \in \mathbb{Z}^{m+\bar{m}}$ *such that* $\left[\mathbf{A} \mid \mathbf{A} \cdot \mathbf{R} + \mathbf{C}\right] \cdot \mathbf{b} = \mathbf{u} \bmod q$ *and with distribution statistically close to* $D_{L,\sigma}$ *where* L *denotes the shifted lattice* $\Lambda_q^{\mathbf{u}}\big([\mathbf{A} \mid \mathbf{A} \cdot \mathbf{R} + \mathbf{C}]\big)$.

2.2 Computational Problems

The security of our schemes provably relies on the assumption that both algorithmic problems below are hard, i.e., cannot be solved in polynomial time with non-negligible probability and non-negligible advantage, respectively.

Definition 2. *Let* m, q, β *be functions of a parameter* n*. The Short Integer Solution problem* $\mathsf{SIS}_{n,m,q,\beta}$ *is as follows: Given* $\mathbf{A} \hookleftarrow U(\mathbb{Z}_q^{n \times m})$*, find* $\mathbf{x} \in \Lambda_q^{\perp}(\mathbf{A})$ *with* $0 < \|\mathbf{x}\| \leq \beta$.

If $q \geq \sqrt{n}\beta$ and $m, \beta \leq \mathsf{poly}(n)$, then $\mathsf{SIS}_{n,m,q,\beta}$ is at least as hard as standard worst-case lattice problem SIVP_γ with $\gamma = \widetilde{\mathcal{O}}(\beta\sqrt{n})$ (see, e.g., [23, Sect. 9]).

Definition 3. *Let* $n, m \geq 1$*,* $q \geq 2$*, and let* χ *be a probability distribution on* $\mathbb{Z}$*. For* $\mathbf{s} \in \mathbb{Z}_q^n$*, let* $A_{\mathbf{s},\chi}$ *be the distribution obtained by sampling* $\mathbf{a} \hookleftarrow U(\mathbb{Z}_q^n)$ *and* $e \hookleftarrow \chi$*, and outputting* $(\mathbf{a}, \mathbf{a}^T \cdot \mathbf{s} + e) \in \mathbb{Z}_q^n \times \mathbb{Z}_q$*. The Learning With Errors problem* $\mathsf{LWE}_{n,q,\chi}$ *asks to distinguish* m *samples chosen according to* $\mathcal{A}_{\mathbf{s},\chi}$ *(for* $\mathbf{s} \hookleftarrow U(\mathbb{Z}_q^n)$*) and* m *samples chosen according to* $U(\mathbb{Z}_q^n \times \mathbb{Z}_q)$.

If q is a prime power, $B \geq \sqrt{n}\omega(\log n)$, $\gamma = \widetilde{\mathcal{O}}(nq/B)$, then there exists an efficient sampleable B-bounded distribution χ (i.e., χ outputs samples with norm at most B with overwhelming probability) such that $\mathsf{LWE}_{n,q,\chi}$ is as least as hard as SIVP_γ (see, e.g., [14,47,49]).

2.3 Syntax and Definitions of Group Encryption

We use the syntax and the security model of Kiayias, Tsiounis and Yung [33]. The group encryption (GE) primitive involves a sender, a verifier, a group manager (GM) that manages the group of receivers and an opening authority (OA) which is capable of identitying ciphertexts' recipients. In the syntax of [33], a GE scheme is specified by the description of a relation $\mathcal{R}$ as well as a tuple $\mathsf{GE} = (\mathsf{SETUP}, \mathsf{JOIN}, \langle \mathcal{G}_r, \mathcal{R}, \mathsf{sample}_{\mathcal{R}} \rangle, \mathsf{ENC}, \mathsf{DEC}, \mathsf{OPEN}, \langle \mathcal{P}, \mathcal{V} \rangle)$ of algorithms or protocols. In details, SETUP is a set of initialization procedures that all take (implicitly or explicitly) a security parameter 1^λ as input. We call them $\mathsf{SETUP}_{\mathsf{init}}(1^\lambda)$, $\mathsf{SETUP}_{\mathsf{GM}}(\mathsf{param})$ and $\mathsf{SETUP}_{\mathsf{OA}}(\mathsf{param})$. The first one of these procedures generates a set of public parameters param (like the KTY construction [33], we rely on a common reference string even when using interaction between provers and verifiers). The latter two procedures are used to produce key pairs $(\mathsf{pk}_{\mathsf{GM}}, \mathsf{sk}_{\mathsf{GM}})$, $(\mathsf{pk}_{\mathsf{OA}}, \mathsf{sk}_{\mathsf{OA}})$ for the GM and the OA. In the following, param is incorporated in the inputs of all algorithms although we sometimes omit to explicitly write it.

$\mathsf{JOIN} = (\mathsf{J}_{\mathsf{user}}, \mathsf{J}_{\mathsf{GM}})$ is an interactive protocol between the GM and the prospective user. After the execution of JOIN, the GM stores the public key pk and its certificate $\mathsf{cert}_{\mathsf{pk}}$ in a public directory database. As in [34], we will restrict this protocol to have minimal interaction and consist of only two messages: the first one is the user's public key pk sent by $\mathsf{J}_{\mathsf{user}}$ to J_{GM} and the latter's response is a certificate $\mathsf{cert}_{\mathsf{pk}}$ for pk that makes the user's group membership effective. We do not require the user to prove knowledge of his private key sk or anything else about it. In our construction, valid keys will be publicly recognizable and users will not have to prove their validity. By avoiding proofs of knowledge of private keys, the security proof never has to rewind the adversary to extract those private keys, which allows supporting concurrent joins as advocated by Kiayias and Yung [34]. If applications demand it, it is possible to add proofs of knowledge of private keys in a modular way but our security proofs do not require rewinding the adversary in executions of JOIN.

Algorithm $\mathsf{sample}_{\mathcal{R}}$ allows sampling pairs $(x, w) \in \mathcal{R}$ (made of a public value x and a witness w) using keys $(\mathsf{pk}_{\mathcal{R}}, \mathsf{sk}_{\mathcal{R}})$ produced by $\mathcal{G}_r(1^\lambda)$ which samples public/secret parameters for the relation $\mathcal{R}$. Depending on the relation, $\mathsf{sk}_{\mathcal{R}}$ may be the empty string (as in the scheme [33] and ours which both involve publicly samplable relations). The testing procedure $\mathcal{R}(x, w)$ uses $\mathsf{pk}_{\mathcal{R}}$ to return 1 whenever $(x, w) \in \mathcal{R}$. To encrypt a witness w such that $(x, w) \in \mathcal{R}$ for some public x, the sender fetches the pair $(\mathsf{pk}, \mathsf{cert}_{\mathsf{pk}})$ from database and runs the randomized encryption algorithm. The latter takes as input w, a label L, the receiver's pair $(\mathsf{pk}, \mathsf{cert}_{\mathsf{pk}})$ as well as public keys $\mathsf{pk}_{\mathsf{GM}}$ and $\mathsf{pk}_{\mathsf{OA}}$. Its output is a ciphertext $\Psi \leftarrow \mathsf{ENC}(\mathsf{pk}_{\mathsf{GM}}, \mathsf{pk}_{\mathsf{OA}}, \mathsf{pk}, \mathsf{cert}_{\mathsf{pk}}, w, L)$. On input of the same elements, the certificate $\mathsf{cert}_{\mathsf{pk}}$, the ciphertext Ψ and the random coins $coins_\Psi$ that were used to produce Ψ, the non-interactive algorithm $\mathcal{P}$ generates a proof π_Ψ that there exists a certified receiver whose public key was registered in database and who is able to decrypt Ψ and obtain a witness w such that $(x, w) \in \mathcal{R}$. The verification algorithm $\mathcal{V}$ takes as input Ψ, $\mathsf{pk}_{\mathsf{GM}}$, $\mathsf{pk}_{\mathsf{OA}}$, π_Ψ and the description of $\mathcal{R}$ and outputs 0 or 1. Given Ψ, L and the receiver's private key sk, the output of DEC is either a witness w such that $(x, w) \in \mathcal{R}$ or a rejection symbol $\perp$. Finally, OPEN takes as input a ciphertext/label pair (Ψ, L) and the OA's secret key $\mathsf{sk}_{\mathsf{OA}}$ and returns a receiver's public key pk.

The model of [33] considers four properties termed correctness, message security, anonymity and soundness.

3 Warm-Up: Decompositions, Extensions, Permutations

This section introduces the notations and techniques that will be used throughout the paper. Part of the covered material appeared (in slightly different forms) in recent works [20,37,38,40,41] on Stern-like protocols [52]. The techniques that will be employed for handling quadratic relations (double-bit extension $\mathsf{ext}(\cdot,\cdot)$, expansion $\mathsf{expand}^{\otimes}(\cdot,\cdot)$ of matrix-vector product and the associated permuting mechanisms) are novel contributions of this paper.

3.1 Decompositions

For any $B \in \mathbb{Z}_+$, define the number $\delta_B := \lfloor \log_2 B \rfloor + 1 = \lceil \log_2(B+1) \rceil$ and the sequence $B_1, \ldots, B_{\delta_B}$, where $B_j = \lfloor \frac{B+2^{j-1}}{2^j} \rfloor$, $\forall j \in [1, \delta_B]$. As observed in [40], the sequence satisfies $\sum_{j=1}^{\delta_B} B_j = B$ and any integer $v \in [0, B]$ can be decomposed to $\mathsf{idec}_B(v) = (v^{(1)}, \ldots, v^{(\delta_B)})^\top \in \{0,1\}^{\delta_B}$ such that $\sum_{j=1}^{\delta_B} B_j \cdot v_j = v$. We describe this decomposition procedure in a deterministic manner:

1. $v' := v$
2. For $j = 1$ to δ_B do:
   (i) If $v' \geq B_j$ then $v^{(j)} := 1$, else $v^{(j)} := 0$;
   (ii) $v' := v' - B_j \cdot v^{(j)}$.
3. Output $\mathsf{idec}_B(v) = (v^{(1)}, \ldots, v^{(\delta_B)})^\top$.

Next, for any positive integers $\mathfrak{m}, B$, we define the decomposition matrix:

$$\mathbf{H}_{\mathfrak{m},B} := \begin{bmatrix} B_1 \ldots B_{\delta_B} & & & \\ & B_1 \ldots B_{\delta_B} & & \\ & & \ddots & \\ & & & B_1 \ldots B_{\delta_B} \end{bmatrix} \in \mathbb{Z}^{\mathfrak{m} \times \mathfrak{m}\delta_B}, \qquad (1)$$

and the following injective functions:

(i) $\mathsf{vdec}_{\mathfrak{m},B} : [0, B]^{\mathfrak{m}} \to \{0,1\}^{\mathfrak{m}\delta_B}$ that maps vector $\mathbf{v} = (v_1, \ldots, v_{\mathfrak{m}})^\top$ to vector $\big(\mathsf{idec}_B(v_1)^\top \| \ldots \| \mathsf{idec}_B(v_{\mathfrak{m}})^\top\big)^\top$. Note that $\mathbf{H}_{\mathfrak{m},B} \cdot \mathsf{vdec}_{\mathfrak{m},B}(\mathbf{v}) = \mathbf{v}$.

(ii) $\mathsf{vdec}'_{\mathfrak{m},B} : [-B, B]^{\mathfrak{m}} \to \{-1,0,1\}^{\mathfrak{m}\delta_B}$ that maps vector $\mathbf{w} = (w_1, \ldots, w_{\mathfrak{m}})^\top$ to vector $\big(\sigma(w_1) \cdot \mathsf{idec}_B(w_1)^\top \| \ldots \| \sigma(w_{\mathfrak{m}}) \cdot \mathsf{idec}_B(w_{\mathfrak{m}})^\top\big)^\top$, where for each $i = 1, \ldots, \mathfrak{m}$: $\sigma(w_i) = 0$ if $w_i = 0$; $\sigma(w_i) = -1$ if $w_i < 0$; $\sigma(w_i) = 1$ if $w_i > 0$. Note that $\mathbf{H}_{\mathfrak{m},B} \cdot \mathsf{vdec}'_{\mathfrak{m},B}(\mathbf{w}) = \mathbf{w}$.

We also define the following matrix decomposition procedure. For positive integers n, m, q, define the injective function $\mathsf{mdec}_{n,m,q} : \mathbb{Z}_q^{m \times n} \to \{0,1\}^{mn\delta_{q-1}}$ that maps matrix $\mathbf{X} = [\mathbf{x}_1 | \ldots | \mathbf{x}_n] \in \mathbb{Z}_q^{m \times n}$, where $\mathbf{x}_1, \ldots, \mathbf{x}_n \in \mathbb{Z}_q^m$, to vector

$$\begin{aligned} \mathsf{mdec}_{n,m,q}(\mathbf{X}) &= \big(\mathsf{vdec}_{m,q-1}(\mathbf{x}_1)^\top \| \ldots \mathsf{vdec}_{m,q-1}(\mathbf{x}_n)^\top\big)^\top \\ &= (x_{1,1}, \ldots, x_{1,mk}, x_{2,1}, \ldots, x_{2,mk}, \ldots, x_{n,1}, x_{n,mk})^\top \in \{0,1\}^{nmk}, \end{aligned}$$

where, for each $(i,j) \in [n] \times [mk]$, $x_{i,j} \in \{0,1\}$ denotes the j-th bit of the decomposition of the i-th column of $\mathbf{X}$.

Looking ahead, when proving knowledge of witnesses $(\mathbf{X}, \mathbf{s}) \in \mathbb{Z}_q^{m \times n} \times \mathbb{Z}_q^n$ satisfying $\mathbf{b} = \mathbf{X} \cdot \mathbf{s} + \mathbf{e} \bmod q$, we will have to consider terms of the form $x_{i,j} \cdot s_{i,t}$, where $\mathbf{s} = (s_1, \ldots, s_n)^\top \in \mathbb{Z}_q^n$ and $(s_{i,1}, \ldots, s_{i,k})^\top = \mathsf{idec}_{q-1}(s_i)$ for each $i \in [n]$.

3.2 Extensions and Permutations

We now introduce the extensions and permutations which will be essential for proving quadratic relations.

- For each $c \in \{0,1\}$, denote by $\bar{c}$ the bit $1 - c \in \{0,1\}$.
- For $c_1, c_2 \in \{0,1\}$, define the vector

$$\mathsf{ext}(c_1, c_2) = (\bar{c}_1 \cdot \bar{c}_2, \bar{c}_1 \cdot c_2, c_1 \cdot \bar{c}_2, c_1 \cdot c_2)^\top \in \{0,1\}^4.$$

- For $b_1, b_2 \in \{0,1\}$, define the permutation T_{b_1,b_2} that transforms vector $\mathbf{v} = (v_{0,0}, v_{0,1}, v_{1,0}, v_{1,1})^\top \in \mathbb{Z}_q^4$ to vector $(v_{b_1,b_2}, v_{b_1,\bar{b}_2}, v_{\bar{b}_1,b_2}, v_{\bar{b}_1,\bar{b}_2})^\top$.
  Note that, for all $c_1, c_2, b_1, b_2 \in \{0,1\}$, we have the following:

$$\mathbf{z} = \mathsf{ext}(c_1, c_2) \iff T_{b_1,b_2}(\mathbf{z}) = \mathsf{ext}(c_1 \oplus b_1, c_2 \oplus b_2), \tag{2}$$

where $\oplus$ denotes the bit-wise addition modulo 2.

Now, for positive integers n, m, k, and for vectors

$$\mathbf{x} = (x_{1,1}, \ldots, x_{1,mk}, x_{2,1}, \ldots, x_{2,mk}, \ldots, x_{n,1}, x_{n,mk})^\top \in \{0,1\}^{nmk}$$

and $\mathbf{s}_0 = (s_{1,1}, \ldots, s_{1,k}, s_{2,1}, \ldots, s_{2,k}, \ldots, s_{n,1}, \ldots, s_{n,k})^\top \in \{0,1\}^{nk}$, we define the vector $\mathsf{expand}^{\otimes}(\mathbf{x}, \mathbf{s}_0) \in \{0,1\}^{4nmk^2}$ as

$$\begin{aligned}
\mathsf{expand}^{\otimes}(\mathbf{x}, \mathbf{s}_0) = \big(&\mathsf{ext}^\top(x_{1,1}, s_{1,1}) \| \mathsf{ext}^\top(x_{1,1}, s_{1,2}) \| \ldots \| \mathsf{ext}^\top(x_{1,1}, s_{1,k}) \| \\
&\| \mathsf{ext}^\top(x_{1,2}, s_{1,1}) \| \mathsf{ext}^\top(x_{1,2}, s_{1,2}) \| \ldots \| \mathsf{ext}^\top(x_{1,2}, s_{1,k}) \| \ldots \\
&\| \mathsf{ext}^\top(x_{1,mk}, s_{1,1}) \| \mathsf{ext}^\top(x_{1,mk}, s_{1,2}) \| \ldots \| \mathsf{ext}^\top(x_{1,mk}, s_{1,k}) \| \\
&\| \mathsf{ext}^\top(x_{2,1}, s_{2,1}) \| \mathsf{ext}^\top(x_{2,1}, s_{2,2}) \| \ldots \| \mathsf{ext}^\top(x_{2,1}, s_{2,k}) \| \ldots \\
&\| \mathsf{ext}^\top(x_{2,mk}, s_{2,1}) \| \mathsf{ext}^\top(x_{2,mk}, s_{2,2}) \| \ldots \| \mathsf{ext}^\top(x_{2,mk}, s_{2,k}) \| \ldots \\
&\| \mathsf{ext}^\top(x_{n,1}, s_{n,1}) \| \mathsf{ext}^\top(x_{n,1}, s_{n,2}) \| \ldots \| \mathsf{ext}^\top(x_{n,1}, s_{n,k}) \| \ldots \\
&\| \mathsf{ext}^\top(x_{n,mk}, s_{n,1}) \| \mathsf{ext}^\top(x_{n,mk}, s_{n,2}) \| \ldots \| \mathsf{ext}^\top(x_{n,mk}, s_{n,k}) \big)^\top.
\end{aligned}$$

That is, $\mathsf{expand}^{\otimes}(\mathbf{x}, \mathbf{s}_0)$ is obtained by applying ext to all pairs of the form $(x_{i,j}, s_{i,t})$ for $(i,j,t) \in [n] \times [mk] \times [k]$.

Now, for $\mathbf{b} = (b_{1,1}, \ldots, b_{1,mk}, b_{2,1}, \ldots, b_{2,mk}, \ldots, b_{n,1}, b_{n,mk})^\top \in \{0,1\}^{nmk}$ and $\mathbf{d} = (d_{1,1}, \ldots, d_{1,k}, d_{2,1}, \ldots, d_{2,k}, \ldots, d_{n,1}, \ldots, d_{n,k})^\top \in \{0,1\}^{nk}$ we define the permutation $P_{\mathbf{b},\mathbf{d}}$ that transforms vector

$$\begin{aligned}
\mathbf{v} = \big(&(\mathbf{v}_{1,1,1}^\top \| \cdots \| \mathbf{v}_{1,1,k}^\top) \| (\mathbf{v}_{1,2,1}^\top \| \cdots \| \mathbf{v}_{1,2,k}^\top) \| \cdots \| (\mathbf{v}_{1,mk,1}^\top \| \cdots \| \mathbf{v}_{1,mk,k}^\top) \| \\
&(\mathbf{v}_{2,1,1}^\top \| \cdots \| \mathbf{v}_{2,1,k}^\top) \| (\mathbf{v}_{2,2,1}^\top \| \cdots \| \mathbf{v}_{2,2,k}^\top) \| \cdots \| (\mathbf{v}_{2,mk,1}^\top \| \cdots \| \mathbf{v}_{2,mk,k}^\top) \| \\
&(\mathbf{v}_{n,1,1}^\top \| \cdots \| \mathbf{v}_{n,1,k}^\top) \| (\mathbf{v}_{n,2,1}^\top \| \cdots \| \mathbf{v}_{n,2,k}^\top) \| \cdots \| (\mathbf{v}_{n,mk,1}^\top \| \cdots \| \mathbf{v}_{n,mk,k}^\top) \big)^\top \in \mathbb{Z}^{4nmk^2},
\end{aligned}$$

consisting of nmk^2 blocks of length 4, to vector $P_{\mathbf{b},\mathbf{d}}(\mathbf{v})$ of the form

$$\begin{aligned}
&((\mathbf{w}_{1,1,1}^\top\|\cdots\|\mathbf{w}_{1,1,k}^\top)\|(\mathbf{w}_{1,2,1}^\top\|\cdots\|\mathbf{w}_{1,2,k}^\top)\|\cdots\|(\mathbf{w}_{1,mk,1}^\top\|\cdots\|\mathbf{w}_{1,mk,k}^\top)\|\\
&(\mathbf{w}_{2,1,1}^\top\|\cdots\|\mathbf{w}_{2,1,k}^\top)\|(\mathbf{w}_{2,2,1}^\top\|\cdots\|\mathbf{w}_{2,2,k}^\top)\|\cdots\|(\mathbf{w}_{2,mk,1}^\top\|\cdots\|\mathbf{w}_{2,mk,k}^\top)\|\\
&(\mathbf{w}_{n,1,1}^\top\|\cdots\|\mathbf{w}_{n,1,k}^\top)\|(\mathbf{w}_{n,2,1}^\top\|\cdots\|\mathbf{w}_{n,2,k}^\top)\|\cdots\|(\mathbf{w}_{n,mk,1}^\top\|\cdots\|\mathbf{w}_{n,mk,k}^\top))^\top,
\end{aligned}$$

where for each $(i,j,t) \in [n] \times [mk] \times [k]$: $\mathbf{w}_{i,j,t} = T_{b_{i,j},d_{i,t}}(\mathbf{v}_{i,j,t})$.
Observe that, for all $\mathbf{b} \in \{0,1\}^{nmk}$, $\mathbf{d} \in \{0,1\}^{nk}$, we have:

$$\mathbf{z} = \mathsf{expand}^{\otimes}(\mathbf{x}, \mathbf{s}_0) \iff P_{\mathbf{b},\mathbf{d}}(\mathbf{z}) = \mathsf{expand}^{\otimes}(\mathbf{x} \oplus \mathbf{b}, \mathbf{s}_0 \oplus \mathbf{d}). \tag{3}$$

Next, we recall the notations, extensions and permutations used in previous Stern-like protocols [20,37,40,41] for proving linear relations.

For any positive integer t, denote by $\mathcal{S}_t$ the symmetric group of all permutations of t elements, by B_{2t} the set of all vectors in $\{0,1\}^{2t}$ having Hamming weight t, and by B_{3t} the set of all vectors in $\{-1,0,1\}^{3t}$ having exactly t coordinates equal to j, for each $j \in \{-1,0,1\}$. Note that for any $\phi \in \mathcal{S}_{2t}$ and $\psi \in \mathcal{S}_{3t}$, we have the following equivalences:

$$\mathbf{x} \in \mathsf{B}_{2t} \iff \phi(\mathbf{x}) \in \mathsf{B}_{2t} \quad \text{and} \quad \mathbf{y} \in \mathsf{B}_{3t} \iff \psi(\mathbf{y}) \in \mathsf{B}_{3t}. \tag{4}$$

The following extending procedures are defined for any positive integers t.

- $\mathsf{ExtendTwo}_t : \{0,1\}^l \to \mathsf{B}_{2t}$. On input vector $\mathbf{x}$ with Hamming weight w, it outputs $\mathbf{x}' = (\mathbf{x}^\top\|\mathbf{1}^{t-w}\|\mathbf{0}^w)^\top$.
- $\mathsf{ExtendThree}_t : \{-1,0,1\}^t \to \mathsf{B}_{3t}$. On input vector $\mathbf{y}$ containing n_j coordinates equal to j for $j \in \{-1,0,1\}$, output $\mathbf{y}' = (\mathbf{y}^\top\|\mathbf{1}^{t-n_1}\|\mathbf{0}^{t-n_0}\|(-\mathbf{1})^{t-n_{-1}})$.

We also use the following encodings and permutations to achieve fine-grained control over coordinates of binary witness-vectors.

- For any positive integer t, define the function encode_t that encodes vector $\mathbf{x} = (x_1,\ldots,x_t)^\top \in \{0,1\}^t$ to vector $\mathsf{encode}_t(\mathbf{x}) = (\bar{x}_1, x_1, \ldots, \bar{x}_t, x_t)^\top \in \{0,1\}^{2t}$.
- For any positive integer t and any vector $\mathbf{c} = (c_1,\ldots,c_t)^\top \in \{0,1\}^t$, define the permutation $F_{\mathbf{c}}^{(t)}$ that transforms vector $\mathbf{v} = (v_1^{(0)}, v_1^{(1)}, \ldots, v_t^{(0)}, v_t^{(1)})^\top \in \mathbb{Z}^{2t}$ into vector $F_{\mathbf{c}}^{(t)}(\mathbf{v}) = (v_1^{(c_1)}, v_1^{(\bar{c}_1)}, \ldots, v_t^{(c_t)}, v_t^{(\bar{c}_t)})^\top$.

Note that the following equivalence holds for all $t, \mathbf{c}$:

$$\mathbf{y} = \mathsf{encode}_t(\mathbf{x}) \iff F_{\mathbf{c}}^{(t)}(\mathbf{y}) = \mathsf{encode}_t(\mathbf{x} \oplus \mathbf{c}). \tag{5}$$

To close this warm-up section, we remark that the equivalences observed in (3), (4) and (5) will play crucial roles in our zero-knowledge layer.

4 The Supporting Zero-Knowledge Layer

In this section, we first demonstrate how to prove in zero-knowledge that a given vector $\mathbf{b}$ is a correct LWE evaluation, i.e., $\mathbf{b} = \mathbf{X} \cdot \mathbf{s} + \mathbf{e} \bmod q$, where the hidden matrix $\mathbf{X}$ and vector $\mathbf{s}$ may satisfy additional conditions. This sub-protocol, which we believe will have other applications, is one of the major challenges in our road towards the design of lattice-based group encryption. We then plug this building block into the big picture, and construct the supporting zero-knowledge argument of knowledge (ZKAoK) for our group encryption scheme (Sect. 5).

4.1 Proving the LWE Relation with Hidden Matrices

Let n, m, q, β be positive integers where $\beta \ll q$, and let $k = \lceil \log_2 q \rceil$. We identify $\mathbb{Z}_q$ as the set $\{0, 1, \ldots, q-1\}$. We consider a zero-knowledge argument system allowing prover $\mathcal{P}$ to convince verifier $\mathcal{V}$ on input $\mathbf{b} \in \mathbb{Z}_q^m$ that $\mathcal{P}$ knows secret matrix $\mathbf{X} \in \mathbb{Z}_q^{m \times n}$, and vectors $\mathbf{s} \in \mathbb{Z}_q^n$, $\mathbf{e} \in [-\beta, \beta]^m$ such that:

$$\mathbf{b} = \mathbf{X} \cdot \mathbf{s} + \mathbf{e} \bmod q. \tag{6}$$

Moreover, the argument system should be readily extended to proving that $\mathbf{X}$ and $\mathbf{s}$ satisfy additional conditions, such as:

- The bits representing $\mathbf{X}$ are certified by an authority, and the prover also knows that secret signature-certificate.
- The (secret) hash of $\mathbf{X}$ is correctly encrypted to a given ciphertext.
- The LWE secret $\mathbf{s}$ is involved in other linear equations.

Let $q_1, \ldots, q_k \in \mathbb{Z}_q$ be the sequence of integers obtained by decomposing $q - 1$ using the technique recalled in Sect. 3.1, and define the row vector $\mathbf{g} = (q_1, \ldots, q_k)$. Let $\mathbf{X} = [\mathbf{x}_1 | \ldots | \mathbf{x}_n] \in \mathbb{Z}_q^{m \times n}$ and $\mathbf{s} = (s_1, \ldots, s_n)^\top$. For each index $i \in [n]$, let us consider $\mathsf{vdec}_{m,q-1}(\mathbf{x}_i) = (x_{i,1}, \ldots, x_{i,mk})^\top \in \{0,1\}^{mk}$. Let $\mathsf{vdec}_{n,q-1}(\mathbf{s}) = (s_{1,1}, \ldots, s_{1,k}, s_{2,1}, \ldots, s_{2,k}, \ldots, s_{n,1}, \ldots s_{n,k})^\top \in \{0,1\}^{nk}$ and observe that $s_i = \mathbf{g} \cdot \mathsf{idec}_{q-1}(s_i) = \mathbf{g} \cdot (s_{i,1}, \ldots, s_{i,k})^\top$ for each $i \in [n]$. We have:

$$\begin{aligned} \mathbf{X} \cdot \mathbf{s} = \sum_{i=1}^{n} \mathbf{x}_i \cdot s_i &= \sum_{i=1}^{n} \mathbf{H}_{m,q-1} \cdot \mathsf{vdec}_{m,q-1}(\mathbf{x}_i) \cdot s_i \\ &= \mathbf{H}_{m,q-1} \cdot \Big(\sum_{i=1}^{n} (x_{i,1} \cdot s_i, \ldots, x_{i,mk} \cdot s_i)^\top \Big) \bmod q. \end{aligned}$$

Observe that, for each $i \in [n]$ and each $j \in [mk]$, we have

$$x_{i,j} \cdot s_i = x_{i,j} \cdot \mathbf{g} \cdot (s_{i,1}, \ldots, s_{i,k})^\top = (q_1, \ldots, q_k) \cdot (x_{i,j} \cdot s_{i,1}, \ldots, x_{i,j} \cdot s_{i,k})^\top.$$

We now extend vector $(q_1, q_2, \ldots, q_k)$ to $\mathbf{g}' = (0, 0, 0, q_1, 0, 0, 0, q_2, \ldots, 0, 0, 0, q_k) \in \mathbb{Z}_q^{4k}$. For all $(i, j) \in [n] \times [mk]$, we have:

$$x_{i,j} \cdot s_i = \mathbf{g}' \cdot (\mathsf{ext}^\top(x_{i,j}, s_{i,1}) \| \ldots \| \mathsf{ext}^\top(x_{i,j}, s_{i,k}))^\top.$$

Let us define the matrices

$$\mathbf{Q}_0 := \mathbf{I}_{mk} \otimes \mathbf{g}' = \begin{bmatrix} \mathbf{g}' & & & \\ & \mathbf{g}' & & \\ & & \ddots & \\ & & & \mathbf{g}' \end{bmatrix} \in \mathbb{Z}_q^{mk \times 4mk^2}, \quad (7)$$

and $\widehat{\mathbf{Q}} = [\overbrace{\mathbf{Q}_0 | \dots | \mathbf{Q}_0}^{n \text{ times}}] \in \mathbb{Z}_q^{mk \times 4nmk^2}$. For each $i \in [n]$, define

$$\begin{aligned} \mathbf{y}_i = \ & (\mathsf{ext}^\top(x_{i,1}, s_{i,1}) \| \dots \| \mathsf{ext}^\top(x_{i,1}, s_{i,k}))^\top \| \mathsf{ext}^\top(x_{i,2}, s_{i,1} \| \dots \| \mathsf{ext}^\top(x_{i,2}, s_{i,k}) \\ & \| \dots \| \mathsf{ext}^\top(x_{i,mk}, s_{i,1} \| \dots \| \mathsf{ext}^\top(x_{i,mk}, s_{i,k}))^\top \in \{0,1\}^{4mk^2}. \end{aligned}$$

Then, for all $i \in [n]$, we have: $(x_{i,1} \cdot s_i, \dots, x_{i,mk} \cdot s_i)^\top = \mathbf{Q}_0 \cdot \mathbf{y}_i$. Now, we note that

$$(\mathbf{y}_1^\top \| \dots \| \mathbf{y}_n^\top)^\top = \mathsf{expand}^\otimes(\mathsf{mdec}_{n,m,q}(\mathbf{X}), \mathsf{vdec}_{n,q-1}(\mathbf{s})),$$

and

$$\begin{aligned} & \sum_{i=1}^{n} (x_{i,1} \cdot s_i, \dots, x_{i,mk} \cdot s_i)^\top \\ & \qquad = \sum_{i=1}^{n} \mathbf{Q}_0 \cdot \mathbf{y}_i = \widehat{\mathbf{Q}} \cdot \mathsf{expand}^\otimes(\mathsf{mdec}_{n,m,q}(\mathbf{X}), \mathsf{vdec}_{n,q-1}(\mathbf{s})). \end{aligned} \quad (8)$$

Letting $\mathbf{Q} = \mathbf{H}_{m,q-1} \cdot \widehat{\mathbf{Q}} \in \mathbb{Z}_q^{m \times 4nmk^2}$ and left-multiplying (8) by $\mathbf{H}_{m,q-1}$, we obtain the equation:

$$\mathbf{X} \cdot \mathbf{s} = \mathbf{Q} \cdot \mathsf{expand}^\otimes\big(\mathsf{mdec}_{n,m,q}(\mathbf{X}), \mathsf{vdec}_{n,q-1}(\mathbf{s})\big) \bmod q.$$

This means that the task of proving knowledge of $(\mathbf{X}, \mathbf{s}, \mathbf{e}) \in \mathbb{Z}_q^{m \times n} \times \mathbb{Z}_q^n \times [-\beta, \beta]^m$ such that $\mathbf{b} = \mathbf{X} \cdot \mathbf{s} + \mathbf{e} \bmod q$ boils down to proving knowledge of $\mathbf{z} \in \{0,1\}^{4nmk^2}$, $\mathbf{x} \in \{0,1\}^{nmk}$, $\mathbf{s}_0 \in \{0,1\}^{nk}$ and a short $\mathbf{e} \in \mathbb{Z}^m$ such that

$$\mathbf{b} = \mathbf{Q} \cdot \mathbf{z} + \mathbf{I}_m \cdot \mathbf{e} \bmod q \qquad \text{and} \qquad \mathbf{z} = \mathsf{expand}^\otimes(\mathbf{x}, \mathbf{s}_0).$$

As the knowledge of small-norm $\mathbf{e}$ can easily be proved with Stern-like protocol (e.g., [40]), the challenging part is to prove in ZK the constraint of $\mathbf{z} = \mathsf{expand}^\otimes(\mathbf{x}, \mathbf{s}_0)$. To this end, we will use the following permuting technique inspired by the equivalence of Eq. (3). We sample uniformly random $\mathbf{d}_x \in \{0,1\}^{nmk}$ and $\mathbf{d}_s \in \{0,1\}^{nk}$, send $\mathbf{x}' = \mathbf{x} \oplus \mathbf{d}_x$ and $\mathbf{s}' = \mathbf{s}_0 \oplus \mathbf{d}_s$ to the verifier, and let the latter check that $P_{\mathbf{d}_x, \mathbf{d}_s}(\mathbf{z}) = \mathsf{expand}^\otimes(\mathbf{x}', \mathbf{s}')$. This will be sufficient to convince the verifier that the original vector $\mathbf{z}$ satisfies the required constraint. The crucial point is that no additional information about $\mathbf{x}$ and $\mathbf{s}_0$ is leaked, since these binary vectors are perfectly hidden under the "one-time pad" $\mathbf{d}_x$ and $\mathbf{d}_s$, respectively.

In the framework of Stern's protocol, the idea of using "one-time-pad" permutations further allows us to prove that $\mathbf{x}$ and $\mathbf{s}_0$ satisfy additional conditions, i.e., they appear in other equations. This is done by first setting up an equivalence similar to (3) in the places where these objects appear, and then, using the same "one-time pad" for each of them in all appearances. We will explain in detail how this technique can be realized in the next subsection.

4.2 The Main Zero-Knowledge Argument System

The zero-knowledge argument of knowledge used in our group encryption scheme (Sect. 5) will involve a system of 10 modular equations:

$$\begin{cases} \mathbf{v}_1 = \mathbf{M}_{1,1} \cdot \mathbf{w}_1 + \mathbf{M}_{1,2} \cdot \mathbf{w}_2 + \ldots + \mathbf{M}_{1,15} \cdot \mathbf{w}_{15} \bmod q, \\ \mathbf{v}_2 = \mathbf{M}_{2,1} \cdot \mathbf{w}_1 + \mathbf{M}_{2,2} \cdot \mathbf{w}_2 + \ldots + \mathbf{M}_{2,15} \cdot \mathbf{w}_{15} \bmod q, \\ \cdots\cdots\cdots\cdots\cdots\cdots\cdots\cdots\cdots\cdots\cdots\cdots \\ \mathbf{v}_{10} = \mathbf{M}_{10,1} \cdot \mathbf{w}_1 + \mathbf{M}_{10,2} \cdot \mathbf{w}_2 + \ldots + \mathbf{M}_{10,15} \cdot \mathbf{w}_{15} \bmod q, \end{cases} \tag{9}$$

where $\{\mathbf{M}_{i,j}\}_{(i,j)\in[10]\times[15]}$, $\{\mathbf{v}_i\}_{i\in[10]}$ are public matrices and vectors (which are possibly zero). Our goal is to prove knowledge of vectors $\mathbf{w}_1, \ldots, \mathbf{w}_{15}$, such that (9) holds, and that these vectors have the following constraints.

1. $\mathbf{w}_1 \in \{0,1\}^{n\bar{m}k}$, $\mathbf{w}_2 \in \{0,1\}^{nk}$ and $\mathbf{w}_3 = \mathsf{expand}^{\otimes}(\mathbf{w}_1, \mathbf{w}_2) \in \{0,1\}^{4n\bar{m}k^2}$. (Note that these vectors are obtained via the techniques of Sect. 4.1.)
2. $\mathbf{w}_4, \mathbf{w}_5, \mathbf{w}_6, \mathbf{w}_7$ are $\{0,1\}$ vectors.
3. Vectors $\mathbf{w}_8, \ldots, \mathbf{w}_{14}$ have bounded infinity norms.
4. Vector $\mathbf{w}_{15}$ has the form $\left(\mathbf{d}_1^\top \,\|\, \mathbf{d}_2^\top \,\|\, \tau[1]\cdot \mathbf{d}_2^\top \,\|\, \ldots \,\|\, \tau[\ell]\cdot \mathbf{d}_2^\top\right)^\top$, for some vectors $\mathbf{d}_1, \mathbf{d}_2 \in [-\beta, \beta]^m$ and $\tau = (\tau[1], \ldots, \tau[\ell])^\top \in \{0,1\}^\ell$.

Towards achieving the goal, we employ a 4-step strategy.

1. The first step transforms all the secret vectors with infinity norm larger than 1 into vectors with infinity norm 1. This is done with the decomposition technique of Sect. 3.1.
2. The norm-1 vectors is then encoded or extended into vectors whose constraints are invariant under random permutations. This is done with the techniques described at the end of Sect. 3.2. The public matrices $\{\mathbf{M}_{i,j}\}_{i,j}$ are transformed accordingly to preserve the equations.
3. The third step unifies all the equations into one of the form $\mathbf{M} \cdot \mathbf{x} = \mathbf{v} \bmod q$, where $\mathbf{x}$ is a concatenation of the newly obtained witness-vectors.
4. In the final step, we run a Stern-like protocol to prove the unified equation $\mathbf{M} \cdot \mathbf{x} = \mathbf{v} \bmod q$, where a composed permutation is employed to prove the constraints of vector $\mathbf{x}$.

Our strategy subsumes the central ideas underlying recent works on Stern-like protocols [37,40,41] for lattice-based relations: preprocessing secret witness-vectors to make them provable-in-zero-knowledge with random permutations, unifying them into just one vector for the sake of convenience, and then running Stern's protocol in a classical manner.

The first step is applicable to vectors $\mathbf{w}_8, \ldots, \mathbf{w}_{14}$ and $\mathbf{w}_{15}$. Suppose that $\mathbf{w}_i$ has dimension m_i and infinity norm bound β_i, for $i \in [8, 14]$. Then we compute vector $\mathbf{w}'_i = \mathsf{vdec}_{m_i,\beta_i}(\mathbf{w}_i) \in \{-1, 0, 1\}^{m_i \delta_{\beta_i}}$. Note that $\mathbf{H}_{m_i,\beta_i} \cdot \mathbf{w}'_i = \mathbf{w}_i$. To decompose $\mathbf{w}_{15}$, we compute $\mathbf{d}'_j = \mathsf{vdec}_{m,\beta}(\mathbf{d}_j) \in \{-1, 0, 1\}^{m\delta_\beta}$, for $j = 1, 2$.

The second step performs the following encodings and extensions.

- Encode $\mathbf{w}_1$ and $\mathbf{w}_2$: Let $\mathbf{w}''_1 = \mathsf{encode}_{n\bar{m}k}(\mathbf{w}_1)$ and $\mathbf{w}''_2 = \mathsf{encode}_{nk}(\mathbf{w}_2)$. Note that to prove knowledge of $\mathbf{w}''_1$ and $\mathbf{w}''_2$, we will employ the "one-time pad" permuting technique implied by (5). The same one-time pads are used to prove that $\mathbf{w}_3 = \mathsf{expand}^{\otimes}(\mathbf{w}_1, \mathbf{w}_2)$, as discussed in Sect. 4.1.
- Extend vectors $\mathbf{w}_4, \ldots, \mathbf{w}_7, \mathbf{w}'_8, \ldots, \mathbf{w}'_{14}$ and $\mathbf{d}'_1, \mathbf{d}'_2, \tau$.
  For $i \in [4, 7]$, suppose that the binary vector $\mathbf{w}_i$ has dimension m_i. Then we extend it to $\mathbf{w}''_i = \mathsf{ExtendTwo}_{m_i}(\mathbf{w}_i) \in \mathsf{B}_{2m_i}$. For $i \in [8, 14]$, we extend $\mathbf{w}'_i$ to $\mathbf{w}''_i = \mathsf{ExtendThree}_{m_i\delta_{\beta_i}}(\mathbf{w}'_i) \in \mathsf{B}_{3m_i\delta_{\beta_i}}$. It follows from (4) that, the knowledge of vectors $\{\mathbf{w}''_i\}_{i=4}^{14}$ can be proved in zero-knowledge using random permutations.
  Meanwhile, we need a more sophisticated treatment for the components of vector $\mathbf{w}_{15}$. For $j = 1, 2$, we let $\mathbf{d}''_j = \mathsf{ExtendThree}_{m\delta_\beta}(\mathbf{d}'_j) \in \mathsf{B}_{3m\delta_\beta}$. We also extend τ to $\tau'' = \mathsf{ExtendTwo}_\ell(\tau) = (\tau[1], \ldots, \tau[\ell], \tau[\ell+1], \ldots, \tau[2\ell])^\top \in \mathsf{B}_{2\ell}$. Then we form the vector:

$$\mathbf{w}''_{15} = \big((\mathbf{d}''_1)^\top \,\|\, (\mathbf{d}''_2)^\top \,\|\, \tau[1](\mathbf{d}''_2)^\top \,\|\, \ldots \,\|\, \tau[\ell](\mathbf{d}''_2)^\top \,\|\, \ldots \,\|\, \tau[2\ell](\mathbf{d}''_2)^\top \big)^\top.$$

  Next, we define CorMix as the set of all vectors in $\{-1, 0, 1\}^{(2\ell+2)3m\delta_\beta}$, that have the form $\big(\mathbf{z}_1^\top \,\|\, \mathbf{z}_2^\top \,\|\, \rho[1]\mathbf{z}_2^\top \,\|\, \ldots \,\|\, \rho[2\ell]\mathbf{z}_2^\top\big)^\top$ for some $\mathbf{z}_1, \mathbf{z}_2 \in \mathsf{B}_{3m\delta_\beta}$ and $\rho \in \mathsf{B}_{2\ell}$. Clearly, $\mathbf{w}''_{15} \in \mathsf{CorMix}$. Furthermore, as shown in [37,41], this set is closed under a special composition of 3 permutations $\phi_1 \in \mathcal{S}_{3m\delta_\beta}, \phi_2 \in \mathcal{S}_{3m\delta_\beta}, \phi_3 \in \mathcal{S}_{2\ell}$, which we denote by T_{ϕ_1,ϕ_2,ϕ_3}. That is, we have the equivalence:

$$\mathbf{w}''_{15} \in \mathsf{CorMix} \Longleftrightarrow T_{\phi_1,\phi_2,\phi_3}(\mathbf{w}''_{15}) \in \mathsf{CorMix}. \tag{10}$$

- As we have changed the dimensions of the witness-vectors, we also have to transform the public matrices $\{\mathbf{M}_{i,j}\}_{i,j}$ accordingly to preserve the equations. In short, this can be done through right-multiplying by the decomposition matrices (if needed), and then inserting zero-columns at suitable positions. We denote the transformed public matrices by $\{\mathbf{M}''_{i,j}\}_{i,j}$.

At the end of the second step, we are presented with the following system of equations, which is equivalent to (9).

$$\begin{cases} \mathbf{v}_1 = \mathbf{M}''_{1,1} \cdot \mathbf{w}''_1 + \mathbf{M}''_{1,2} \cdot \mathbf{w}''_2 + \ldots + \mathbf{M}''_{1,15} \cdot \mathbf{w}''_{15} \bmod q, \\ \mathbf{v}_2 = \mathbf{M}''_{2,1} \cdot \mathbf{w}''_1 + \mathbf{M}''_{2,2} \cdot \mathbf{w}''_2 + \ldots + \mathbf{M}''_{2,15} \cdot \mathbf{w}''_{15} \bmod q, \\ \ldots\ldots\ldots\ldots\ldots\ldots\ldots\ldots\ldots\ldots\ldots\ldots\ldots\ldots\ldots\ldots \\ \mathbf{v}_{10} = \mathbf{M}''_{10,1} \cdot \mathbf{w}''_1 + \mathbf{M}''_{10,2} \cdot \mathbf{w}''_2 + \ldots + \mathbf{M}''_{10,15} \cdot \mathbf{w}''_{15} \bmod q. \end{cases} \tag{11}$$

The third step involves only basic linear algebra. Let

$$\mathbf{M} = \left(\begin{array}{c|c|c|c} \mathbf{M}''_{1,1} & \mathbf{M}''_{1,2} & \cdots & \mathbf{M}''_{1,15} \\ \hline \mathbf{M}''_{2,1} & \mathbf{M}''_{2,2} & \cdots & \mathbf{M}''_{2,15} \\ \hline \cdots & \cdots & \cdots & \cdots \\ \hline \mathbf{M}''_{10,1} & \mathbf{M}''_{10,2} & \cdots & \mathbf{M}''_{10,15} \end{array}\right); \quad \mathbf{x} = \begin{pmatrix} \mathbf{w}''_1 \\ \mathbf{w}''_2 \\ \vdots \\ \mathbf{w}''_{15} \end{pmatrix}; \quad \mathbf{v} = \begin{pmatrix} \mathbf{v}_1 \\ \mathbf{v}_2 \\ \vdots \\ \mathbf{v}_{10} \end{pmatrix},$$

then we obtain the unified equation $\mathbf{M} \cdot \mathbf{x} = \mathbf{v} \bmod q$.

Given the above preparations, we now comes to **the final step** where we formally present our protocol. Let D be the dimension of vector $\mathbf{x}$. Denote by VALID the set of all vectors in $\{-1, 0, 1\}^D$, that have the form $\mathbf{z} = (\mathbf{z}_1^\top \| \ldots \| \mathbf{z}_{15}^\top)^\top$, where:

1. $\mathbf{z}_1 = \mathsf{encode}_{n\bar{m}k}(\mathbf{y}_1)$, $\mathbf{z}_2 = \mathsf{encode}_{nk}(\mathbf{y}_2)$ and $\mathbf{z}_3 = \mathsf{expand}^{\otimes}(\mathbf{y}_1, \mathbf{y}_2)$, for some $\mathbf{y}_1 \in \{0,1\}^{n\bar{m}k}$ and $\mathbf{y}_2 \in \{0,1\}^{nk}$.
2. For $i \in [4, 7]$, vector $\mathbf{z}_i \in \mathsf{B}_{2m_i}$. For $i \in [8, 14]$, vector $\mathbf{z}_i \in \mathsf{B}_{3m_i\delta_{\beta_i}}$.
3. Vector $\mathbf{z}_{15} \in \mathsf{CorMix}$.

It can be seen that our vector $\mathbf{x}$ is an element of this tailored set VALID. By construction, the task of proving knowledge of vectors $\mathbf{w}_1, \ldots, \mathbf{w}_{15}$ that have the required constraints, and that satisfy system (9) has boiled down to proving the possession of vector $\mathbf{x} \in \mathsf{VALID}$ such that $\mathbf{M} \cdot \mathbf{x} = \mathbf{v} \bmod q$. We will fulfill this task with a Stern-like zero-knowledge protocol, in which we hide $\mathbf{x}$ from the verifier's view by a random permutation and a random masking vector.

Let us determine the type of permutations to be applied for $\mathbf{x}$. Let

$$\mathcal{S} = \{0,1\}^{n\bar{m}k} \times \{0,1\}^{nk} \times \mathcal{S}_{2m_4} \times \ldots \times \mathcal{S}_{2m_7} \times \mathcal{S}_{3m_8\delta_{\beta_8}} \times \ldots$$
$$\ldots \times \mathcal{S}_{3m_{14}\delta_{\beta_{14}}} \times (\mathcal{S}_{3m\delta_\beta})^2 \times \mathcal{S}_{2\ell}.$$

We associate each element $\pi = (\mathbf{b}_1, \mathbf{b}_2, \phi_4, \ldots, \phi_{14}, \phi_{15}^1, \phi_{15}^2, \phi_{15}^3) \in \mathcal{S}$ with the permutation Γ_π that transforms vector $\mathbf{z} = (\mathbf{z}_1^\top \| \ldots \| \mathbf{z}_{15}^\top)^\top \in \mathbb{Z}^D$, where the length of block $\mathbf{z}_i$ equals to that of $\mathbf{w}''_i$ for all $i \in [15]$, into vector

$$\Gamma_\pi(\mathbf{z}) = \big(F_{\mathbf{b}_1}^{(n\bar{m}k)}(\mathbf{z}_1) \| F_{\mathbf{b}_2}^{(nk)}(\mathbf{z}_2) \| P_{\mathbf{b}_1,\mathbf{b}_2}(\mathbf{z}_3) \| \phi_4(\mathbf{z}_4) \| \ldots$$
$$\ldots \| \phi_{14}(\mathbf{z}_{14}) \| T_{\phi_{15}^1, \phi_{15}^2, \phi_{15}^3}(\mathbf{z}_{15})\big).$$

It is implied by the equivalences given in (3), (4), (5) and (10) that the following holds for all $\pi \in \mathcal{S}$:

$$\mathbf{x} \in \mathsf{VALID} \iff \Gamma_\pi(\mathbf{x}) \in \mathsf{VALID}.$$

Additionally, if $\mathbf{x} \in \mathsf{VALID}$ and π is uniformly random in $\mathcal{S}$, then $\Gamma_\pi(\mathbf{x})$ is uniformly random in VALID. In the framework of Stern's protocol, these facts allow us to prove in zero-knowledge the knowledge of $\mathbf{x} \in \mathsf{VALID}$.

Furthermore, proving that equation $\mathbf{M} \cdot \mathbf{x} = \mathbf{v} \bmod q$ holds can be done by sampling a uniformly random masking vector $\mathbf{r}_x \in \mathbb{Z}_q^D$, and demonstrating to the verifier that $\mathbf{M} \cdot (\mathbf{x} + \mathbf{r}_x) - \mathbf{v} = \mathbf{M} \cdot \mathbf{r}_x \bmod q$.

The interaction between prover $\mathcal{P}$ and verifier $\mathcal{V}$ is described in Fig. 1. Prior to the interaction, both parties obtain matrix $\mathbf{M}$ and vector $\mathbf{v}$ from the public input, while $\mathcal{P}$ construct witness-vector $\mathbf{x}$ from his secret input, as described above. The protocol employs the statistically hiding and computationally binding string commitment scheme COM from [31].

The properties of the given protocol are summarized in Theorem 1. The proof of the theorem employs standard simulation and extraction techniques for Stern-like protocols [31,40,41], and is detailed in the full version of the paper.

Theorem 1. *The protocol in Fig. 1 is a statistical* ZKAoK *with perfect completeness, soundness error 2/3, and communication cost* $\widetilde{\mathcal{O}}(D \log q)$. *Namely:*

- *There exists a polynomial-time simulator that, on input* $(\mathbf{M}, \mathbf{v})$, *outputs an accepted transcript which is statistically close to that produced by the real prover.*
- *There exists a polynomial-time knowledge extractor that, on input a commitment* CMT *and* 3 *valid responses* $(\mathrm{RSP}_1, \mathrm{RSP}_2, \mathrm{RSP}_3)$ *to all* 3 *possible values of the challenge* Ch, *outputs* $\mathbf{x}' \in \mathsf{VALID}$ *such that* $\mathbf{M} \cdot \mathbf{x}' = \mathbf{v} \bmod q$.

Note that, given vector $\mathbf{x}'$ outputted by the extractor, one can efficiently compute 15 vectors satisfying the conditions described at the beginning of this subsection, simply by "backtracking" the transformations conducted by our first 3 steps. In the group encryption scheme presented next, the constructed ZKAoK will be invoked by algorithm $\langle \mathcal{P}, \mathcal{V} \rangle$, while its simulator and extractor will come in handy in the proofs of security theorems, that are defined in the full version of the paper.

1. **Commitment:** Prover samples $\mathbf{r}_x \leftarrow U(\mathbb{Z}_q^D)$, $\pi \leftarrow U(\mathcal{S})$ and randomness ρ_1, ρ_2, ρ_3 for COM. Then he sends $\mathrm{CMT} = \big(C_1, C_2, C_3\big)$ to the verifier, where

$$C_1 = \mathsf{COM}(\pi, \mathbf{M} \cdot \mathbf{r}_x; \rho_1), \quad C_2 = \mathsf{COM}(\Gamma_\pi(\mathbf{r}_x); \rho_2), \quad C_3 = \mathsf{COM}(\Gamma_\pi(\mathbf{x} + \mathbf{r}_x); \rho_3).$$

2. **Challenge:** The verifier sends a challenge $Ch \leftarrow U(\{1,2,3\})$ to the prover.
3. **Response:** Depending on Ch, the prover sends RSP computed as follows:
   - $Ch = 1$: Let $\mathbf{t}_x = \Gamma_\pi(\mathbf{x})$, $\mathbf{t}_r = \Gamma_\pi(\mathbf{r}_x)$, and $\mathrm{RSP} = (\mathbf{t}_x, \mathbf{t}_r, \rho_2, \rho_3)$.
   - $Ch = 2$: Let $\pi_2 = \pi$, $\mathbf{y}_2 = \mathbf{x} + \mathbf{r}_x$, and $\mathrm{RSP} = (\pi_2, \mathbf{y}_2, \rho_1, \rho_3)$.
   - $Ch = 3$: Let $\pi_3 = \pi$, $\mathbf{y}_3 = \mathbf{r}$, and $\mathrm{RSP} = (\pi_3, \mathbf{y}_3, \rho_1, \rho_2)$.

Verification: Receiving RSP, the verifier proceeds as follows:

- $Ch = 1$: Check that $\mathbf{t}_x \in \mathsf{VALID}$ and $C_2 = \mathsf{COM}(\mathbf{t}_r; \rho_2)$, $C_3 = \mathsf{COM}(\mathbf{t}_x + \mathbf{t}_r; \rho_3)$.
- $Ch = 2$: Check that $C_1 = \mathsf{COM}(\pi_2, \mathbf{M} \cdot \mathbf{y}_2 - \mathbf{v}; \rho_1)$, $C_3 = \mathsf{COM}(\Gamma_{\pi_2}(\mathbf{y}_2); \rho_3)$.
- $Ch = 3$: Check that $C_1 = \mathsf{COM}(\pi_3, \mathbf{M} \cdot \mathbf{y}_3; \rho_1)$, $C_2 = \mathsf{COM}(\Gamma_{\pi_3}(\mathbf{y}_3); \rho_2)$.

In each case, the verifier outputs 1 if and only if all the conditions hold.

Fig. 1. Our zero-knowledge argument of knowledge.

5 Our Lattice-Based Group Encryption Scheme

To build a GE scheme using our zero-knowledge argument system, we need to choose a specific key-private CCA2-secure encryption scheme. The first idea is to use the CCA2-secure public-key cryptosystem which is implied by the Agrawal-Boneh-Boyen identity-based encryption (IBE) scheme [1] (which is recalled in Appendix A.2) via the Canetti-Halevi-Katz (CHK) transformation [16]. The ABB scheme is a natural choice since it has pseudo-random ciphertexts (which implies the key-privacy [7] when the CHK paradigm is applied) and provides one of the most efficient CCA2 cryptosystem based on the hardness of LWE in the standard model. One difficulty is that the Kiayias-Tsiounis-Yung model [33] requires that certified public keys be valid public keys (i.e., which have a matching secret key). When new group members join the system and request a certificate for their public key $\mathbf{B}_{\mathsf{U}} \in \mathbb{Z}_q^{n \times \bar{m}}$, a direct use of the ABB/CHK technique would incur of proof of existence of a GPV trapdoor [23] corresponding to $\mathbf{B}_{\mathsf{U}}$ (i.e., a small-norm matrix $\mathbf{T}_{\mathbf{B}_{\mathsf{U}}} \in \mathbb{Z}^{\bar{m} \times \bar{m}}$ s.t. $\mathbf{B} \cdot \mathbf{T}_{\mathbf{B}_{\mathsf{U}}} = \mathbf{0}^n \bmod q$). While the techniques of Peikert and Vaikuntanathan [48] would provide a solution to this problem (as they allow proving that $\mathbf{T}_{\mathbf{B}_{\mathsf{U}}} \in \mathbb{Z}^{\bar{m} \times \bar{m}}$ has full-rank), we found it simpler to rely on the trapdoor mechanism of Micciancio and Peikert [43].

If we assume public parameters containing a random matrix $\bar{\mathbf{A}} \in \mathbb{Z}_q^{n \times m}$, each user's public key can consist of a matrix $\mathbf{B}_{\mathsf{U}} = \bar{\mathbf{A}} \cdot \mathbf{T}_{\mathsf{U}} \in \mathbb{Z}_q^{n \times \bar{m}}$, where $\mathbf{T}_{\mathsf{U}} \in \mathbb{Z}^{m \times \bar{m}}$ is a small-norm matrix whose calms are sampled from a discrete Gaussian distribution. Note that, if $\bar{\mathbf{A}} \in \mathbb{Z}_q^{n \times m}$ is uniformly distributed, then [23, Lemma 5.1] ensures that, with overwhelming probability, any matrix $\mathbf{B}_{\mathsf{U}} \in \mathbb{Z}_q^{n \times \bar{m}}$ has an underlying small-norm matrix satisfying $\mathbf{B}_{\mathsf{U}} = \bar{\mathbf{A}} \cdot \mathbf{T}_{\mathsf{U}} \bmod q$. This simplifies the joining procedure by eliminating the need for proofs of public key validity.

In the encryption algorithm, the sender computes a dual Regev encryption [23] of the witness $\mathbf{w} \in \{0,1\}^m$ using a matrix $[\bar{\mathbf{A}} \mid \mathbf{B}_{\mathsf{U}} + \mathsf{FRD}(\mathsf{VK}) \cdot \mathbf{G}] \in \mathbb{Z}_q^{n \times (m+\bar{m})}$ such that: (i) $\mathsf{VK} \in \mathbb{Z}_q^n$ is the verification key of a one-time signature; (ii) $\mathsf{FRD} : \mathbb{Z}_q^n \to \mathbb{Z}_q^{n \times n}$ is the full-rank difference[1] function of [1]; (iii) $\mathbf{G} = \mathbf{I}_n \otimes [1|2|\ldots|2^{k-1}] \in \mathbb{Z}_q^{n \times \bar{m}}$ is the gadget matrix of [43]. Given that $\mathbf{G}$ has a publicly known trapdoor allowing to sample short vectors in $\Lambda_q^{\perp}(\mathbf{G})$, the user can use his private key $\mathbf{T}_{\mathsf{U}} \in \mathbb{Z}^{m \times \bar{m}}$ to decrypt by running the SampleRight algorithm of Lemma 5.

Having encrypted the witness $\mathbf{w} \in \{0,1\}^m$ by running the ABB encryption algorithm, the sender proceeds by encrypting a hash value of $\mathbf{B}_{\mathsf{U}} \in \mathbb{Z}_q^{n \times \bar{m}}$ under the public key $\mathbf{B}_{\mathsf{OA}} = \bar{\mathbf{A}} \cdot \mathbf{T}_{\mathsf{OA}} \in \mathbb{Z}_q^{n \times \bar{m}}$ of the opening authority. The latter hash value is obtained as a bit-wise decomposition of $\mathbf{F} \cdot \mathsf{mdec}_{n,m,q}(\mathbf{B}_{\mathsf{U}}^{\top}) \in \mathbb{Z}_q^{2n}$, where $\mathbf{F} \in \mathbb{Z}_q^{2n \times n\bar{m}\lceil \log q \rceil}$ is a random public matrix and $\mathsf{mdec}_{n,m,q}(\mathbf{B}_{\mathsf{U}}^{\top}) \in \{0,1\}^{n\bar{m}\lceil \log q \rceil}$ denotes an entry-wise binary decomposition of the matrix $\mathbf{B}_{\mathsf{U}} \in \mathbb{Z}_q^{n \times \bar{m}}$.

[1] This means that, for any two distinct one-time verification keys $\mathsf{VK}, \mathsf{VK}' \in \mathbb{Z}_q^n$, the difference $\mathsf{FRD}(\mathsf{VK}) - \mathsf{FRD}(\mathsf{VK}') \in \mathbb{Z}_q^{n \times n}$ is invertible over $\mathbb{Z}_q$.

By combining our new argument for quadratic relations and the extensions of Stern's protocol suggested in [37,41], we are able to prove that some component of the ciphertext is of the form $\mathbf{c} = \mathbf{B}_{\mathsf{U}}^{\top} \cdot \mathbf{s} + \mathbf{e} \in \mathbb{Z}_q^{\bar{m}}$, for some $\mathbf{s} \in \mathbb{Z}_q^n$ and a small-norm $\mathbf{e} \in \mathbb{Z}^{\bar{m}}$ while also arguing possession of a signature on the binary decomposition $\mathsf{mdec}_{n,m,q}(\mathbf{B}_{\mathsf{U}}^{\top}) \in \{0,1\}^{n\bar{m}\lceil \log q \rceil}$ of $\mathbf{B}_{\mathsf{U}}^{\top}$. For this purpose, we use a variant of a signature scheme due to Böhl *et al.*'s signature [11] which was recently proposed by Libert, Ling, Mouhartem, Nguyen and Wang [37] (and of which a description is given in Appendix A.1). At the same time, the prover $\mathcal{P}$ can also argue that a hash value of $\mathsf{mdec}_{n,m,q}(\mathbf{B}_{\mathsf{U}}^{\top})$ is properly encrypted under the OA's public key using the ABB encryption scheme.

5.1 Description of the Scheme

Our GE scheme allows encrypting witnesses for the Inhomogeneous SIS relation $\mathrm{R}_{\mathsf{ISIS}}(n, m, q, 1)$, which consists of pairs $((\mathbf{A}_R, \mathbf{u}_R), \mathbf{w}) \in (\mathbb{Z}_q^{n \times m} \times \mathbb{Z}_q^n) \times \{0,1\}^m$ satisfying $\mathbf{u}_R = \mathbf{A}_R \cdot \mathbf{w} \bmod q$. This relation is in the same spirit as the one of Kiayias, Tsiounis and Yung [33], who consider the verifiable encryption of discrete logarithms. While the construction of [33] allow verifiably encrypting discrete-logarithm-type secret keys under the public key of some anonymous TTP, our construction makes it possible to encrypt GPV-type secret keys [23].

$\mathsf{SETUP}_{\mathsf{init}}(1^\lambda)$: This algorithm performs the following:

1. Choose integers $n = \mathcal{O}(\lambda)$, prime $q = \widetilde{\mathcal{O}}(n^4)$, and let $k = \lceil \log_2 q \rceil$, $\bar{m} = nk$ and $m = 2\bar{m} = 2nk$. Choose a B-bounded distribution χ over $\mathbb{Z}$ for some $B = \sqrt{n}\omega(\log n)$.
2. Choose a Gaussian parameter $\sigma = \Omega(\sqrt{n \log q} \log n)$. Let $\beta = \sigma\omega(\log n)$ be the upper bound of samples from $D_{\mathbb{Z},\sigma}$.
3. Select integers $\ell = \ell(\lambda)$ which determines the maximum expected group size 2^ℓ, and $\kappa = \omega(\log \lambda)$ (the number of protocol repetitions).
4. Select a strongly unforgeable one-time signature $\mathcal{OTS} = (\mathsf{Gen}, \mathsf{Sig}, \mathsf{Ver})$. We assume that the verification keys live in $\mathbb{Z}_q^n$.
5. Select public parameters $\mathsf{COM}_{\mathsf{par}}$ for a statistically-hiding commitment scheme like [31]. This commitment will serve as a building block for the zero-knowledge argument system used in $\langle \mathcal{P}, \mathcal{V} \rangle$.
6. Let $\mathsf{FRD} : \mathbb{Z}_q^n \rightarrow \mathbb{Z}_q^{n \times n}$ be the full-rank difference mapping from [1].
7. Pick a random matrix $\mathbf{F} \leftarrow \mathbb{Z}_q^{2n \times n\bar{m}k}$, which will be used to hash users' public keys from $\mathbb{Z}_q^{n \times \bar{m}}$ to $\mathbb{Z}_q^n$.
8. Let $\mathbf{G} \in \mathbb{Z}_q^{n \times \bar{m}}$ be the gadget matrix $\mathbf{G} = \mathbf{I}_n \otimes \left[1\ 2 \ldots 2^{k-1}\right]$ of [43]. Pick matrices $\bar{\mathbf{A}}, \mathbf{U} \leftarrow U(\mathbb{Z}_q^{n \times m})$ and $\mathbf{V} \leftarrow U(\mathbb{Z}_q^{n \times m})$. Looking ahead, $\mathbf{U}$ will be used to encrypt for the receiver while $\mathbf{V}$ will be used to encrypt the user's public key under the OA's public key. As for $\bar{\mathbf{A}}$, it will be used in two instances of the ABB encryption scheme [1].

Output

$$\mathsf{param} = \left\{\lambda, n, q, k, m, B, \chi, \sigma, \beta, \ell, \kappa, \mathcal{OTS}, \mathsf{COM}_{\mathsf{par}}, \mathsf{FRD}, \bar{\mathbf{A}}, \mathbf{G}, \mathbf{F}, \mathbf{U}, \mathbf{V}\right\}.$$

$\langle \mathcal{G}_r, \mathsf{sample}_{\mathcal{R}} \rangle$: Algorithm $\mathcal{G}_r(1^\lambda, 1^n, 1^m)$ proceeds by sampling a random matrix $\mathbf{A}_R \leftarrow U(\mathbb{Z}_q^{n\times m})$ and outputting $(\mathsf{pk}_{\mathcal{R}}, \mathsf{sk}_{\mathcal{R}}) = (\mathbf{A}_R, \varepsilon)$. On input of a public key $\mathsf{pk}_{\mathcal{R}} = \mathbf{A}_R \in \mathbb{Z}_q^{n\times m}$ for the relation $\mathrm{R}_{\mathsf{ISIS}}$, algorithm $\mathsf{sample}_{\mathcal{R}}$ picks $\mathbf{w} \leftarrow U(\{0,1\}^m)$ and outputs a pair $((\mathbf{A}_R, \mathbf{u}_R), \mathbf{w})$, where $\mathbf{u}_R = \mathbf{A}_R \cdot \mathbf{w} \in \mathbb{Z}_q^n$.

$\mathsf{SETUP}_{\mathsf{GM}}(\mathsf{param})$: The GM generates $(\mathsf{sk}_{\mathsf{GM}}, \mathsf{pk}_{\mathsf{GM}}) \leftarrow \mathsf{Keygen}(1^\lambda, q, n, m, \ell, \sigma)$ as a key pair for the SIS-based signature scheme of [37] (as recalled in Appendix A.1). This key pair consists of $\mathsf{sk}_{\mathsf{GM}} := \mathbf{T}_{\mathbf{A}}$ and

$$\mathsf{pk}_{\mathsf{GM}} := \Big(\mathbf{A}, \mathbf{A}_0, \ldots, \mathbf{A}_\ell \in \mathbb{Z}_q^{n\times m},\ \mathbf{D}_0, \mathbf{D}_1 \in \mathbb{Z}_q^{n\times m}, \mathbf{D} \in \mathbb{Z}_q^{n\times \bar{m}}, \mathbf{u} \in \mathbb{Z}_q^n\Big). \quad (12)$$

$\mathsf{SETUP}_{\mathsf{OA}}(\mathsf{param})$: The OA samples a small-norm matrix $\mathbf{T}_{\mathsf{OA}} \leftarrow D_{\mathbb{Z}^m,\sigma}^{\bar{m}}$ in $\mathbb{Z}^{m\times \bar{m}}$ to obtain a statistically uniform $\mathbf{B}_{\mathsf{OA}} = \bar{\mathbf{A}} \cdot \mathbf{T}_{\mathsf{OA}} \in \mathbb{Z}_q^{n\times \bar{m}}$. The OA's key pair consists of $(\mathsf{sk}_{\mathsf{OA}}, \mathsf{pk}_{\mathsf{OA}}) = (\mathbf{T}_{\mathsf{OA}}, \mathbf{B}_{\mathsf{OA}})$.

JOIN: The prospective user U and the GM interact in the following protocol.

1. U first samples $\mathbf{T}_{\mathsf{U}} \leftarrow D_{\mathbb{Z}^m,\sigma}^{\bar{m}}$ in $\mathbb{Z}^{m\times \bar{m}}$ to compute a statistically uniform matrix $\mathbf{B}_{\mathsf{U}} = \bar{\mathbf{A}} \cdot \mathbf{T}_{\mathsf{U}} \in \mathbb{Z}_q^{n\times \bar{m}}$. The prospective user defines his key pair as $(\mathsf{pk}_{\mathsf{U}}, \mathsf{sk}_{\mathsf{U}}) = (\mathbf{B}_{\mathsf{U}}, \mathbf{T}_{\mathsf{U}})$ and sends $\mathsf{pk}_{\mathsf{U}} = \mathbf{B}_{\mathsf{U}}$ to the GM.
2. Upon receiving a public key $\mathsf{pk}_{\mathsf{U}} = \mathbf{B}_{\mathsf{U}} \in \mathbb{Z}_q^{n\times \bar{m}}$ from the user, the GM certifies pk_U via the following steps:
   a. Compute $\mathbf{h}_{\mathsf{U}} = \mathbf{F} \cdot \mathsf{mdec}_{n,\bar{m},q}(\mathbf{B}_{\mathsf{U}}^\top) \in \mathbb{Z}_q^{2n}$ as a hash value of the public key $\mathsf{pk}_{\mathsf{U}} = \mathbf{B}_{\mathsf{U}} \in \mathbb{Z}_q^{n\times \bar{m}}$.
   b. Use the trapdoor $\mathsf{sk}_{\mathsf{GM}} = \mathbf{T}_{\mathbf{A}}$ to generate a signature

$$\mathsf{cert}_{\mathsf{U}} = \big(\tau, \mathbf{d}, \mathbf{r}\big) \in \{0,1\}^\ell \times [-\beta, \beta]^{2m} \times [-\beta, \beta]^m, \quad (13)$$

   satisfying

$$\begin{aligned}&[\mathbf{A} \mid \sum_{j=1}^{\ell} \tau[j]\mathbf{A}_j] \cdot \mathbf{d} \\ &= \mathbf{u} \ + \ \mathbf{D} \cdot \mathsf{vdec}_{n,q-1}(\mathbf{D}_0 \cdot \mathbf{r} + \mathbf{D}_1 \cdot \mathsf{vdec}_{n,q-1}(\mathbf{h}_{\mathsf{U}})) \bmod q, \quad (14)\end{aligned}$$

   where $\tau = \tau[1]\ldots\tau[\ell] \in \{0,1\}^\ell$, as in the scheme of Appendix A.1.

U verifies that $\mathsf{cert}_{\mathsf{U}}$ is tuple of the form (13) satisfying (14) and returns $\perp$ if it is not the case. The GM stores $(\mathsf{pk}_{\mathsf{U}}, \mathsf{cert}_{\mathsf{U}})$ in the user database $\mathsf{database}$ and returns the certificate $\mathsf{cert}_{\mathsf{U}}$ to the new user $\mathcal{U}$.

$\mathsf{ENC}(\mathsf{pk}_{\mathsf{GM}}, \mathsf{pk}_{\mathsf{OA}}, \mathsf{pk}_{\mathsf{U}}, \mathsf{cert}_{\mathsf{U}}, \mathbf{w}, L)$: To encrypt a witness $\mathbf{w} \in \{0,1\}^m$ such that $((\mathbf{A}_R, \mathbf{u}_R), \mathbf{w}) \in \mathrm{R}_{\mathsf{ISIS}}(n, m, q, 1)$ (i.e., $\mathbf{A}_R \cdot \mathbf{w} = \mathbf{u}_R \bmod q$), parse $\mathsf{pk}_{\mathsf{GM}}$ as in (12), $\mathsf{pk}_{\mathsf{OA}}$ as $\mathbf{B}_{\mathsf{OA}} \in \mathbb{Z}_q^{n\times \bar{m}}$, pk_{U} as $\mathbf{B}_{\mathsf{U}} \in \mathbb{Z}_q^{n\times \bar{m}}$ and $\mathsf{cert}_{\mathsf{U}}$ as in (13).

1. Generate a one-time key-pair $(\mathsf{SK}, \mathsf{VK}) \leftarrow \mathsf{Gen}(1^\lambda)$, where $\mathsf{VK} \in \mathbb{Z}_q^n$.
2. Compute a full-rank-difference hash $\mathbf{H}_{\mathsf{VK}} = \mathsf{FRD}(\mathsf{VK}) \in \mathbb{Z}_q^{n\times n}$ of the one-time verification key $\mathsf{VK} \in \mathbb{Z}_q^n$.
3. Encrypt the witness $\mathbf{w} \in \{0,1\}^m$ under U's public key $\mathbf{B}_{\mathsf{U}} \in \mathbb{Z}_q^{n\times \bar{m}}$ using the tag VK by taking the following steps:

a. Sample $\mathbf{s}_{\mathsf{rec}} \leftarrow U(\mathbb{Z}_q^n)$, $\mathbf{R}_{\mathsf{rec}} \leftarrow D_{\mathbb{Z},\sigma}^{m\times\bar{m}}$ and $\mathbf{x}_{\mathsf{rec}}, \mathbf{y}_{\mathsf{rec}} \leftarrow \chi^m$. Compute $\mathbf{z}_{\mathsf{rec}} = \mathbf{R}_{\mathsf{rec}}^\top \cdot \mathbf{y}_{\mathsf{rec}} \in \mathbb{Z}^{\bar{m}}$.

b. Compute

$$\begin{cases} \mathbf{c}_{\mathsf{rec}}^{(1)} = \bar{\mathbf{A}}^\top \cdot \mathbf{s}_{\mathsf{rec}} + \mathbf{y}_{\mathsf{rec}} \bmod q \\ \mathbf{c}_{\mathsf{rec}}^{(2)} = (\mathbf{B}_{\mathsf{U}} + \mathbf{H}_{\mathsf{VK}} \cdot \mathbf{G})^\top \cdot \mathbf{s}_{\mathsf{rec}} + \mathbf{z}_{\mathsf{rec}} \bmod q; \\ \mathbf{c}_{\mathsf{rec}}^{(3)} = \mathbf{U}^\top \cdot \mathbf{s}_{\mathsf{rec}} + \mathbf{x}_{\mathsf{rec}} + \mathbf{w} \cdot \left\lfloor \frac{q}{2} \right\rfloor, \end{cases} \tag{15}$$

and let $\mathbf{c}_{\mathsf{rec}} = \left(\mathbf{c}_{\mathsf{rec}}^{(1)}, \mathbf{c}_{\mathsf{rec}}^{(2)}, \mathbf{c}_{\mathsf{rec}}^{(3)}\right) \in \mathbb{Z}_q^m \times \mathbb{Z}_q^{\bar{m}} \times \mathbb{Z}_q^m$, which forms an ABB ciphertext [1] for the tag $\mathsf{VK} \in \mathbb{Z}_q^n$.

4. Encrypt the decomposition $\mathsf{vdec}_{n,q-1}(\mathbf{h}_{\mathsf{U}}) \in \{0,1\}^m$ of the hashed pk_{U} under the OA's public key $\mathbf{B}_{\mathsf{OA}} \in \mathbb{Z}_q^{n\times\bar{m}}$ w.r.t. the tag $\mathsf{VK} \in \mathbb{Z}_q^n$. Namely, conduct the following steps:

a. Sample $\mathbf{s}_{\mathsf{oa}} \leftarrow U(\mathbb{Z}_q^n)$, $\mathbf{R}_{\mathsf{oa}} \leftarrow D_{\mathbb{Z},\sigma}^{m\times\bar{m}}$, $\mathbf{x}_{\mathsf{oa}} \leftarrow \chi^m, \mathbf{y}_{\mathsf{oa}} \leftarrow \chi^m$. Set $\mathbf{z}_{\mathsf{oa}} = \mathbf{R}_{\mathsf{oa}}^\top \cdot \mathbf{y}_{\mathsf{oa}} \in \mathbb{Z}^{\bar{m}}$.

b. Compute

$$\begin{cases} \mathbf{c}_{\mathsf{oa}}^{(1)} = \bar{\mathbf{A}}^\top \cdot \mathbf{s}_{\mathsf{oa}} + \mathbf{y}_{\mathsf{oa}} \bmod q; \\ \mathbf{c}_{\mathsf{oa}}^{(2)} = (\mathbf{B}_{\mathsf{OA}} + \mathbf{H}_{\mathsf{VK}} \cdot \mathbf{G})^\top \cdot \mathbf{s}_{\mathsf{oa}} + \mathbf{z}_{\mathsf{oa}} \bmod q; \\ \mathbf{c}_{\mathsf{oa}}^{(3)} = \mathbf{V}^\top \cdot \mathbf{s}_{\mathsf{oa}} + \mathbf{x}_{\mathsf{oa}} + \mathsf{vdec}_{n,q-1}(\mathbf{h}_{\mathsf{U}}) \cdot \left\lfloor \frac{q}{2} \right\rfloor, \end{cases} \tag{16}$$

and let $\mathbf{c}_{\mathsf{oa}} = \left(\mathbf{c}_{\mathsf{oa}}^{(1)}, \mathbf{c}_{\mathsf{oa}}^{(2)}, \mathbf{c}_{\mathsf{oa}}^{(3)}\right) \in \mathbb{Z}_q^m \times \mathbb{Z}_q^{\bar{m}} \times \mathbb{Z}_q^m$.

5. Compute a one-time signature $\Sigma = \mathsf{Sig}(\mathsf{SK}, (\mathbf{c}_{\mathsf{rec}}, \mathbf{c}_{\mathsf{oa}}, L))$.

Output the ciphertext

$$\mathbf{\Psi} = (\mathsf{VK}, \mathbf{c}_{\mathsf{rec}}, \mathbf{c}_{\mathsf{oa}}, \Sigma). \tag{17}$$

and the state information $coins_{\mathbf{\Psi}} = \left(\mathbf{s}_{\mathsf{rec}}, \mathbf{R}_{\mathsf{rec}}, \mathbf{x}_{\mathsf{rec}}, \mathbf{y}_{\mathsf{rec}}, \mathbf{s}_{\mathsf{oa}}, \mathbf{R}_{\mathsf{oa}}, \mathbf{x}_{\mathsf{oa}}, \mathbf{y}_{\mathsf{oa}}\right)$.

$\mathsf{DEC}(\mathsf{sk}_{\mathsf{U}}, \mathbf{\Psi}, L)$: The decryption algorithm proceeds as follows:

1. If $\mathsf{Ver}\big(\mathsf{VK}, \Sigma, (\mathbf{c}_{\mathsf{rec}}, \mathbf{c}_{\mathsf{oa}}, L)\big) = 0$, return $\perp$. Otherwise, parse the secret key sk_{U} as $\mathbf{T}_{\mathsf{U}} \in \mathbb{Z}^{m\times\bar{m}}$ and the ciphertext $\mathbf{\Psi}$ as in (17). Define the matrix $\mathbf{B}_{\mathsf{VK}} = \mathbf{B}_{\mathsf{U}} + \mathsf{FRD}(\mathsf{VK}) \cdot \mathbf{G} \in \mathbb{Z}_q^{n\times\bar{m}}$.
2. Decrypt $\mathbf{c}_{\mathsf{rec}}$ using a decryption key for the tag $\mathsf{VK} \in \mathbb{Z}^n$. Namely,

a. Define $\mathbf{B}_{\mathsf{U},\mathsf{VK}} = [\bar{\mathbf{A}}|\mathbf{B}_{\mathsf{VK}}] = [\bar{\mathbf{A}}|\bar{\mathbf{A}} \cdot \mathbf{T}_{\mathsf{U}} + \mathsf{FRD}(\mathsf{VK}) \cdot \mathbf{G}] \in \mathbb{Z}_q^{n\times(m+\bar{m})}$. Using $\mathbf{T}_{\mathsf{U}}$ and the publicly known trapdoor $\mathbf{T}_{\mathbf{G}}$ of $\mathbf{G}$, compute a small-norm matrix $\mathbf{E}_{\mathsf{VK}} \in \mathbb{Z}^{(m+\bar{m})\times m}$ such that $\mathbf{B}_{\mathsf{U},\mathsf{VK}} \cdot \mathbf{E}_{\mathsf{VK}} = \mathbf{U} \bmod q$ by running the $\mathsf{SampleRight}$ algorithm of Lemma 5.

b. Compute

$$\mathbf{w} = \left\lfloor \left(\mathbf{c}_{\mathsf{rec}}^{(3)} - \mathbf{E}_{\mathsf{VK}}^\top \cdot \begin{bmatrix} \mathbf{c}_{\mathsf{rec}}^{(1)} \\ \mathbf{c}_{\mathsf{rec}}^{(2)} \end{bmatrix}\right) / \left\lfloor \frac{q}{2} \right\rfloor \right\rceil \in \mathbb{Z}^m$$

and return the obtained $\mathbf{w} \in \{0,1\}^m$.

$\mathsf{OPEN}(\mathsf{sk_{OA}}, \mathbf{\Psi}, L)$: The opening algorithm proceeds as follows:

1. If $\mathsf{Ver}\big(\mathsf{VK}, \Sigma, (\mathbf{c}_{\mathsf{rec}}, \mathbf{c}_{\mathsf{oa}}), L\big) = 0$, then return $\perp$. Otherwise, parse $\mathsf{sk_{OA}}$ as $\mathbf{T}_{\mathsf{OA}} \in \mathbb{Z}^{m \times \bar{m}}$ and the ciphertext $\mathbf{\Psi}$ as in (17).
2. Decrypt $\mathbf{c}_{\mathsf{oa}}$ using a decryption key for the tag $\mathsf{VK} \in \mathbb{Z}_q^n$ in the same way as in the decryption algorithm. That is, do the following:
   a. Define the matrix $\mathbf{B}_{\mathsf{OA,VK}} = [\bar{\mathbf{A}}|\mathbf{B}_{\mathsf{OA}} + \mathsf{FRD}(\mathsf{VK}) \cdot \mathbf{G}] \in \mathbb{Z}_q^{n \times (m+\bar{m})}$. Use $\mathbf{T}_{\mathsf{OA}}$ to compute a small-norm $\mathbf{E}_{\mathsf{OA,VK}} \in \mathbb{Z}^{(m+\bar{m}) \times m}$ satisfying $\mathbf{B}_{\mathsf{OA,VK}} \cdot \mathbf{E}_{\mathsf{OA,VK}} = \mathbf{V} \bmod q$.
   b. Compute
$$\mathbf{h} = \left\lfloor \left(\mathbf{c}_{\mathsf{oa}}^{(3)} - \mathbf{E}_{\mathsf{OA,VK}}^{\top} \cdot \begin{bmatrix} \mathbf{c}_{\mathsf{oa}}^{(1)} \\ \mathbf{c}_{\mathsf{oa}}^{(2)} \end{bmatrix} \right) / \left\lfloor \frac{q}{2} \right\rfloor \right\rceil \in \{0,1\}^m$$
   and $\mathbf{h}'_{\mathsf{U}} = \mathbf{H}_{2n,q-1} \cdot \mathbf{h} \in \mathbb{Z}_q^{2n}$.
3. Look up database to find a public key $\mathsf{pk_U} = \mathbf{B}_{\mathsf{U}} \in \mathbb{Z}_q^{n \times \bar{m}}$ that hashes to $\mathbf{h}'_{\mathsf{U}} \in \mathbb{Z}_q^{2n}$ (i.e., such that $\mathbf{h}'_{\mathsf{U}} = \mathbf{F} \cdot \mathsf{mdec}_{n,\bar{m},q}(\mathbf{B}_{\mathsf{U}}^{\top})$). If more than one such key exists, return $\perp$. If only one key $\mathsf{pk_U} = \mathbf{B}_{\mathsf{U}} \in \mathbb{Z}_q^{n \times \bar{m}}$ satisfies $\mathbf{h}'_{\mathsf{U}} = \mathbf{F} \cdot \mathsf{mdec}_{n,\bar{m},q}(\mathbf{B}_{\mathsf{U}}^{\top})$, return that key $\mathsf{pk_U}$. In any other situation, return $\perp$.

$\langle \mathcal{P}, \mathcal{V} \rangle$: The common input consists of param and $\mathsf{pk_{GM}}$ as specified above, as well as $(\mathbf{A}_R, \mathbf{u}_R) \in \mathbb{Z}_q^{n \times m} \times \mathbb{Z}_q^n$, $\mathsf{pk_{OA}} = \mathbf{B}_{\mathsf{OA}} \in \mathbb{Z}_q^{n \times \bar{m}}$, and a ciphertext $\mathbf{\Psi}$ as in (17). Both parties compute $\mathbf{B}_{\mathsf{OA,VK}} = [\bar{\mathbf{A}}|\mathbf{B}_{\mathsf{OA}} + \mathsf{FRD}(\mathsf{VK}) \cdot \mathbf{G}]$ as specified above. The prover's secret input consists of a witness $\mathbf{w} \in \{0,1\}^m$, $\mathsf{pk_U} = \mathbf{B}_{\mathsf{U}}$, $\mathsf{cert_U} = (\tau, \mathbf{d}, \mathbf{r}) \in \{0,1\}^\ell \times \mathbb{Z}^{2m} \times \mathbb{Z}^m$, and the random coins $coins_{\mathbf{\Psi}} = \big(\mathbf{s}_{\mathsf{rec}}, \mathbf{R}_{\mathsf{rec}}, \mathbf{x}_{\mathsf{rec}}, \mathbf{y}_{\mathsf{rec}}, \mathbf{s}_{\mathsf{oa}}, \mathbf{R}_{\mathsf{oa}}, \mathbf{x}_{\mathsf{oa}}, \mathbf{y}_{\mathsf{oa}}\big)$ used to generate $\mathbf{\Psi}$.
The prover's goal is to convince the verifier in zero-knowledge that his secret input satisfies the following:

1. $\mathbf{A}_R \cdot \mathbf{w} = \mathbf{u}_R \bmod q$.
2. $\mathbf{h}_{\mathsf{M}} = \mathbf{F} \cdot \mathsf{mdec}_{n,m,q}(\mathbf{M}) \bmod q$.
3. Conditions (13) and (14) hold.
4. Vectors $\mathbf{x}_{\mathsf{rec}}, \mathbf{y}_{\mathsf{rec}}, \mathbf{x}_{\mathsf{oa}}, \mathbf{y}_{\mathsf{oa}}$ have infinity norms bounded by B, and vectors $\mathbf{z}_{\mathsf{rec}}, \mathbf{z}_{\mathsf{oa}}$ have infinity norms bounded by $\beta m B$.
5. Equations in (15) and (16) hold.

To this end $\mathcal{P}$ conducts the following steps.

1. Decompose the matrix $\mathbf{B}_{\mathsf{U}} \in \mathbb{Z}_q^{n \times \bar{m}}$ into $\mathbf{b}_{\mathsf{U}} = \mathsf{mdec}_{n,\bar{m},q}(\mathbf{B}_{\mathsf{U}}^{\top}) \in \{0,1\}^{n\bar{m}k}$ and the vectors $\mathbf{s}_{\mathsf{rec}}, \mathbf{s}_{\mathsf{oa}} \in \mathbb{Z}_q^n$ into $\mathbf{s}_{0,\mathsf{rec}} = \mathsf{vdec}_{n,q-1}(\mathbf{s}_{\mathsf{rec}}) \in \{0,1\}^{nk}$ and $\mathbf{s}_{0,\mathsf{oa}} = \mathsf{vdec}_{n,q-1}(\mathbf{s}_{\mathsf{oa}}) \in \{0,1\}^{nk}$. Combine the first two binary vectors into $\mathbf{z}_{\mathbf{\Psi}} = \mathsf{expand}^{\otimes}(\mathbf{b}_{\mathsf{U}}, \mathbf{s}_{0,\mathsf{rec}}) \in \{0,1\}^{4n\bar{m}k^2}$. Define
$$\mathbf{Q} = \mathbf{H}_{\bar{m},q-1} \cdot \overbrace{[\mathbf{Q}_0|\ldots|\mathbf{Q}_0]}^{n \text{ times}} \in \mathbb{Z}_q^{\bar{m} \times 4n\bar{m}k^2},$$
where $\mathbf{Q}_0 = \mathbf{I}_{\bar{m}k} \otimes \mathbf{g}' \in \mathbb{Z}_q^{\bar{m}k \times 4\bar{m}k^2}$ is the matrix defined as in (7).

2. Generate a zero-knowledge argument of knowledge of

$$\begin{cases} \tau \in \{0,1\}^\ell,\ \mathbf{d} = [\mathbf{d}_1^\top | \mathbf{d}_2^\top]^\top \in [-\beta,\beta]^{2m},\ \mathbf{r} \in [-\beta,\beta]^m \\ \mathbf{t}_\mathsf{U} \in \{0,1\}^m,\ \mathbf{w}_\mathsf{U} \in \{0,1\}^{\bar{m}} \\ \mathbf{b}_\mathsf{U} \in \{0,1\}^{n\bar{m}k},\ \mathbf{s}_{0,\mathsf{rec}} \in \{0,1\}^{nk},\ \mathbf{z}_\Psi = \mathsf{expand}^\otimes(\mathbf{b}_\mathsf{U}, \mathbf{s}_{0,\mathsf{rec}}) \\ \mathbf{x}_\mathsf{rec},\ \mathbf{y}_\mathsf{rec} \in [-B,B]^m,\ \mathbf{z}_\mathsf{rec} \in [-\beta m B, \beta m B]^{\bar{m}},\ \mathbf{w} \in \{0,1\}^m, \\ \mathbf{s}_{0,\mathsf{oa}} \in \{0,1\}^{nk},\ \mathbf{x}_\mathsf{oa},\ \mathbf{y}_\mathsf{oa} \in [-B,B]^m,\ \mathbf{z}_\mathsf{oa} \in [-\beta m B, \beta m B]^{\bar{m}} \end{cases}$$

such that the following system of 10 equations holds:

$$\begin{cases} \mathbf{u} = [\mathbf{A}\,|\,\mathbf{A}_0|\mathbf{A}_1|\ldots|\mathbf{A}_\ell] \cdot \begin{pmatrix} \mathbf{d}_1 \\ \mathbf{d}_2 \\ \tau[1]\cdot\mathbf{d}_2 \\ \vdots \\ \tau[\ell]\cdot\mathbf{d}_2 \end{pmatrix} + (-\mathbf{D})\cdot\mathbf{w}_\mathsf{U} \bmod q, \\ \mathbf{0} = \mathbf{H}_{n,q-1}\cdot\mathbf{w}_\mathsf{U} + (-\mathbf{D}_0)\cdot\mathbf{r} + (-\mathbf{D}_1)\cdot\mathbf{t}_\mathsf{U} \bmod q, \\ \mathbf{0} = \mathbf{H}_{2n,q-1}\cdot\mathbf{t}_\mathsf{U} + (-\mathbf{F})\cdot\mathbf{b}_\mathsf{U} \bmod q, \\ \mathbf{c}_\mathsf{rec}^{(1)} = (\bar{\mathbf{A}}^\top\cdot\mathbf{H}_{n,q-1})\cdot\mathbf{s}_{0,\mathsf{rec}} + \mathbf{I}_m\cdot\mathbf{y}_\mathsf{rec} \bmod q, \\ \mathbf{c}_\mathsf{rec}^{(2)} = \mathbf{Q}\cdot\mathbf{z}_\Psi + (\mathbf{G}^\top\cdot\mathbf{H}_\mathsf{VK}^\top\cdot\mathbf{H}_{n,q-1})\cdot\mathbf{s}_{0,\mathsf{rec}} + \mathbf{I}_{\bar{m}}\cdot\mathbf{z}_\mathsf{rec} \bmod q, \\ \mathbf{c}_\mathsf{rec}^{(3)} = (\mathbf{U}^\top\cdot\mathbf{H}_{n,q-1})\cdot\mathbf{s}_{0,\mathsf{rec}} + \mathbf{I}_m\cdot\mathbf{x}_\mathsf{rec} + (\lfloor\frac{q}{2}\rfloor\cdot\mathbf{I}_m)\cdot\mathbf{w} \bmod q, \\ \mathbf{u}_R = \mathbf{A}_R\cdot\mathbf{w} \bmod q, \\ \mathbf{c}_\mathsf{oa}^{(1)} = (\bar{\mathbf{A}}^\top\cdot\mathbf{H}_{n,q-1})\cdot\mathbf{s}_{0,\mathsf{oa}} + \mathbf{I}_m\cdot\mathbf{y}_\mathsf{oa} \bmod q, \\ \mathbf{c}_\mathsf{oa}^{(2)} = [(\mathbf{B}_\mathsf{OA} + \mathbf{H}_\mathsf{VK}\cdot\mathbf{G})^\top\cdot\mathbf{H}_{n,q-1}]\cdot\mathbf{s}_{0,\mathsf{oa}} + \mathbf{I}_{\bar{m}}\cdot\mathbf{z}_\mathsf{oa} \bmod q, \\ \mathbf{c}_\mathsf{oa}^{(3)} = (\mathbf{V}^\top\cdot\mathbf{H}_{n,q-1})\cdot\mathbf{s}_{0,\mathsf{oa}} + \mathbf{I}_m\cdot\mathbf{x}_\mathsf{oa} + (\lfloor\frac{q}{2}\rfloor\cdot\mathbf{I}_m)\cdot\mathbf{t}_\mathsf{U} \bmod q. \end{cases} \quad (18)$$

Let $\mathbf{w}_1 = \mathbf{b}_\mathsf{U}$, $\mathbf{w}_2 = \mathbf{s}_{0,\mathsf{rec}}$, $\mathbf{w}_3 = \mathbf{z}_\Psi = \mathsf{expand}^\otimes(\mathbf{b}_\mathsf{U}, \mathbf{s}_{0,\mathsf{rec}})$, $\mathbf{w}_4 = \mathbf{w}_\mathsf{U}$, $\mathbf{w}_5 = \mathbf{t}_\mathsf{U}$, $\mathbf{w}_6 = \mathbf{s}_{0,\mathsf{oa}}$, $\mathbf{w}_7 = \mathbf{w}$, $\mathbf{w}_8 = \mathbf{x}_\mathsf{rec}$, $\mathbf{w}_9 = \mathbf{y}_\mathsf{rec}$, $\mathbf{w}_{10} = \mathbf{z}_\mathsf{rec}$, $\mathbf{w}_{11} = \mathbf{r}$, $\mathbf{w}_{12} = \mathbf{x}_\mathsf{oa}$, $\mathbf{w}_{13} = \mathbf{y}_\mathsf{oa}$, $\mathbf{w}_{14} = \mathbf{z}_\mathsf{oa}$ and

$$\mathbf{w}_{15} = \big(\mathbf{d}_1^\top \,\|\, \mathbf{d}_2^\top \,\|\, \tau[1]\cdot\,\mathbf{d}_2^\top \,\|\ldots\|\, \tau[\ell]\cdot\,\mathbf{d}_2^\top\big)^\top.$$

Then system (18) can be rewritten as:

$$\begin{cases} \mathbf{v}_1 = \mathbf{M}_{1,1}\cdot\mathbf{w}_1 + \mathbf{M}_{1,2}\cdot\mathbf{w}_2 + \ldots + \mathbf{M}_{1,15}\cdot\mathbf{w}_{15} \bmod q, \\ \mathbf{v}_2 = \mathbf{M}_{2,1}\cdot\mathbf{w}_1 + \mathbf{M}_{2,2}\cdot\mathbf{w}_2 + \ldots + \mathbf{M}_{2,15}\cdot\mathbf{w}_{15} \bmod q, \\ \ldots\ldots\ldots\ldots\ldots\ldots\ldots\ldots\ldots \\ \mathbf{v}_{10} = \mathbf{M}_{10,1}\cdot\mathbf{w}_1 + \mathbf{M}_{10,2}\cdot\mathbf{w}_2 + \ldots + \mathbf{M}_{10,15}\cdot\mathbf{w}_{15} \bmod q, \end{cases} \quad (19)$$

where $\{\mathbf{M}_{i,j}\}_{(i,j)\in[10]\times[15]}$, $\{\mathbf{v}_i\}_{i\in[10]}$ are public matrices and vectors (which are possibly zero).

The argument system is obtained by invoking the protocol from Sect. 4.2. The protocol is repeated κ times to make the soundness error negligibly small.

5.2 Efficiency and Correctness

EFFICIENCY. It can be seen that the given group encryption scheme can be implemented in polynomial time. We now will evaluate the bit-sizes of keys and ciphertext, as well as the communication cost of the protocol $\langle \mathcal{P}, \mathcal{V} \rangle$.

- The public key of GM, as in (12), has bit-size $\mathcal{O}(\ell n^2 \log^2 q) = \widetilde{\mathcal{O}}(\ell \lambda^2)$.
- The public keys of OA and each user both have bit-size $n\bar{m}\lceil \log_2 q \rceil = \widetilde{\mathcal{O}}(\lambda^2)$.
- The secret key of each party in the scheme is a trapdoor of bit-size $\widetilde{\mathcal{O}}(\lambda^2)$. The user's certificate $\mathsf{cert}_{\mathsf{U}}$ has bit-size $\widetilde{\mathcal{O}}(\lambda)$.
- The ciphertext $\boldsymbol{\Psi}$ consists of $\mathsf{VK} \in \mathbb{Z}_q^n$, two ABB ciphertexts of total size $2(2m+\bar{m})\lceil \log_2 q \rceil$ and a one-time signature Σ. Thus, its bit-size is $\widetilde{\mathcal{O}}(\lambda) + |\Sigma|$.
- The communication cost of the protocol $\langle \mathcal{P}, \mathcal{V} \rangle$ is largely dominated by the bit-size of the witness $\mathbf{z}_{\Psi} = \mathsf{expand}^{\otimes}(\mathbf{b}_{\mathsf{U}}, \mathbf{s}_{0,\mathsf{rec}}) \in \{0,1\}^{4n\bar{m}k^2}$. The total cost is $\kappa \cdot \mathcal{O}(n^2 \log^4 q) = \widetilde{\mathcal{O}}(\lambda^2)$ bits.

CORRECTNESS. The given group encryption scheme is correct with overwhelming probability. We first remark that the scheme parameters are set up so that the two instances of the ABB identity-based encryption [1] are correct. Indeed, during the decryption procedure of $\mathsf{DEC}(\mathsf{sk}_{\mathsf{U}}, \boldsymbol{\Psi}, L)$, we have:

$$\mathbf{c}_{\mathsf{rec}}^{(3)} - \mathbf{E}_{\mathsf{VK}}^{\top} \cdot \begin{bmatrix} \mathbf{c}_{\mathsf{rec}}^{(1)} \\ \mathbf{c}_{\mathsf{rec}}^{(2)} \end{bmatrix} = \mathbf{x}_{\mathsf{rec}} - \mathbf{E}_{\mathsf{VK}}^{\top} \cdot \begin{bmatrix} \mathbf{y}_{\mathsf{rec}} \\ \mathbf{z}_{\mathsf{rec}} \end{bmatrix} + \mathbf{w} \cdot \left\lfloor \frac{q}{2} \right\rfloor.$$

Note that $\|\mathbf{x}_{\mathsf{rec}}\|_\infty$ and $\|\mathbf{y}_{\mathsf{rec}}\|_\infty$ are bounded by B, and $\|\mathbf{z}_{\mathsf{rec}}\|_\infty = \|\mathbf{R}_{\mathsf{rec}}^{\top} \cdot \mathbf{y}_{\mathsf{rec}}\|_\infty \leq \beta m B = \widetilde{\mathcal{O}}(n^2)$. Furthermore, the entries of the discrete Gaussian matrix $\mathbf{E}_{\mathsf{VK}}^{\top}$ are bounded by $\widetilde{\mathcal{O}}(\sqrt{n})$. Hence, the error term $\mathbf{x}_{\mathsf{rec}} - \mathbf{E}_{\mathsf{VK}}^{\top} \cdot \begin{bmatrix} \mathbf{y}_{\mathsf{rec}} \\ \mathbf{z}_{\mathsf{rec}} \end{bmatrix}$ is bounded by $\widetilde{\mathcal{O}}(n^{3.5})$ which is much smaller than $q/4 = \widetilde{\mathcal{O}}(n^4)$. As a result, the decryption algorithm returns $\mathbf{w}$ with overwhelming probability. The correctness of algorithm $\mathsf{OPEN}(\mathsf{sk}_{\mathsf{OA}}, \boldsymbol{\Psi}, L)$ also follows from a similar argument.

Finally, we note that if a certified group user honestly follows all the prescribed algorithms, then he should be able to compute valid witness-vectors to be used in the protocol $\langle \mathcal{P}, \mathcal{V} \rangle$, and he should be accepted by the verifier, thanks to the perfect completeness of the argument system in Sect. 4.2.

Our scheme is proven secure under the SIS and LWE assumptions using classical reduction techniques. The detailed security proofs are given in the full version of the paper.

Acknowledgements. We thank Damien Stehlé for useful discussions and the reviewers for useful comments. The first author was funded by the "Programme Avenir Lyon Saint-Etienne de l'Université de Lyon" in the framework of the programme "Investissements d'Avenir" (ANR-11-IDEX-0007). San Ling, Khoa Nguyen and Huaxiong Wang were supported by the "Singapore Ministry of Education under Research Grant MOE2013-T2-1-041". Huaxiong Wang was also supported by NTU under Tier 1 grant RG143/14.

A Building Blocks

A.1 Signatures Supporting Zero-Knowledge Proofs

We use a signature scheme proposed by Libert, Ling, Mouhartem, Nguyen and Wang [37] who extended the Böhl *et al.* signature [11] (which is itself built upon Boyen's signature [13]) into a signature scheme compatible with zero-knowledge proofs. While the scheme was designed to sign messages comprised of multiple blocks, we only use the single-block version here.

Keygen$(1^\lambda, q, n, m, \ell, \sigma)$: This algorithm takes as input a security parameter $\lambda > 0$ as well as the following parameters: $n = \mathcal{O}(\lambda)$; a prime modulus $q = \widetilde{\mathcal{O}}(n^4)$; dimension $m = 2n\lceil \log q \rceil$; an integer $\ell = \mathsf{poly}(\lambda)$; and Gaussian parameters $\sigma = \Omega(\sqrt{n \log q} \log n)$. It defines the message space as $\{0,1\}^m$.

1. Run $\mathsf{TrapGen}(1^n, 1^m, q)$ to get $\mathbf{A} \in \mathbb{Z}_q^{n \times m}$ and a short basis $\mathbf{T_A}$ of $\Lambda_q^\perp(\mathbf{A})$. This basis allows computing short vectors in $\Lambda_q^\perp(\mathbf{A})$ with a Gaussian parameter σ. Next, choose $\ell + 1$ random $\mathbf{A}_0, \mathbf{A}_1, \ldots, \mathbf{A}_\ell \hookleftarrow U(\mathbb{Z}_q^{n \times m})$.
2. Choose random matrices $\mathbf{D} \hookleftarrow U(\mathbb{Z}_q^{n \times m/2})$, $\mathbf{D}_0, \mathbf{D}_1 \hookleftarrow U(\mathbb{Z}_q^{n \times m})$ as well as a random vector $\mathbf{u} \hookleftarrow U(\mathbb{Z}_q^n)$.

The private key consists of $SK := \mathbf{T_A}$ and the public key is

$$PK := \big(\mathbf{A},\ \{\mathbf{A}_j\}_{j=0}^{\ell},\ \mathbf{D}_0,\ \mathbf{D}_1,\ \mathbf{D},\ \mathbf{u}\big).$$

Sign$\big(SK, \mathfrak{m}\big)$: To sign a message $\mathfrak{m} \in \{0,1\}^m$,

1. Choose a random binary string $\tau \hookleftarrow U(\{0,1\}^\ell)$. Then, using $SK := \mathbf{T_A}$, compute a short delegated basis $\mathbf{T}_\tau \in \mathbb{Z}^{2m \times 2m}$ for the matrix

$$\mathbf{A}_\tau = [\mathbf{A} \mid \mathbf{A}_0 + \sum_{j=1}^{\ell} \tau[j] \mathbf{A}_j] \in \mathbb{Z}_q^{n \times 2m}. \tag{20}$$

2. Choose a discrete Gaussian vector $\mathbf{r} \hookleftarrow D_{\mathbb{Z}^m, \sigma}$. Compute $\mathbf{c}_M \in \mathbb{Z}_q^n$ as a chameleon hash of $\mathfrak{m}$. Namely, compute $\mathbf{c}_M = \mathbf{D}_0 \cdot \mathbf{r} + \mathbf{D}_1 \cdot \mathfrak{m} \in \mathbb{Z}_q^n$, which is used to define $\mathbf{u}_M = \mathbf{u} + \mathbf{D} \cdot \mathsf{vdec}_{n,q-1}(\mathbf{c}_M) \in \mathbb{Z}_q^n$. Using the delegated basis $\mathbf{T}_\tau \in \mathbb{Z}^{2m \times 2m}$, sample a short vector $\mathbf{v} \in \mathbb{Z}^{2m}$ in $D_{\Lambda_q^{\mathbf{u}_M}(\mathbf{A}_\tau), \sigma}$.

Output the signature $sig = (\tau, \mathbf{v}, \mathbf{r}) \in \{0,1\}^\ell \times \mathbb{Z}^{2m} \times \mathbb{Z}^m$.

Verify$\big(PK, \mathfrak{m}, sig\big)$: Given PK, $\mathfrak{m} \in \{0,1\}^m$ and $sig = (\tau, \mathbf{v}, \mathbf{r}) \in \{0,1\}^\ell \times \mathbb{Z}^{2m} \times \mathbb{Z}^m$, return 1 if $\|\mathbf{v}\| < \sigma\sqrt{2m}$, $\|\mathbf{r}\| < \sigma\sqrt{m}$ and

$$\mathbf{A}_\tau \cdot \mathbf{v} = \mathbf{u} + \mathbf{D} \cdot \mathsf{vdec}_{n,q-1}(\mathbf{D}_0 \cdot \mathbf{r} + \mathbf{D}_1 \cdot \mathfrak{m}) \bmod q. \tag{21}$$

Like [11,13], the scheme of [37] was proved secure under the SIS assumption and shown to easily interact with Stern-like protocols when it comes to proving knowledge of a hidden message-signature pair. While such proofs would also be possible using Boyen's signature [13], the number of public matrices $\{\mathbf{A}_j\}_{j=0}^{\ell}$ in the public key can be reduced from $\Theta(n \cdot \log q)$ to $\Theta(\lambda)$ using the scheme of [37].

The above description uses a slightly different variant of [37] where, at step 2 of the signing algorithm, a different binary decomposition of $\mathbf{c}_M$ is used to compute $\mathbf{u}_M$: while [37] uses the standard binary decomposition, we use a non-unique encoding based on the vdec function for convenience. However, the security proof of [37] goes through with this encoding since the function $\mathsf{vdec}_{n,q-1}(.)$ is injective.

Lemma 6 ([37, Theorem 1]). *The above signature scheme is unforgeable under chosen-message attacks if the* SIS *assumption holds.*

A.2 The Agrawal-Boneh-Boyen IBE Scheme

Identity-Based Encryption. An IBE scheme is a tuple of efficient algorithms $(\mathsf{Setup}, \mathsf{Extract}_{\mathsf{PP}}, \mathsf{Encrypt}_{\mathsf{PP}}, \mathsf{Decrypt}_{\mathsf{PP}})$ such that

$\mathbf{Setup}(1^\lambda)$: On security parameter λ, this algorithm outputs public parameters PP and a master secret key msk.

$\mathbf{Extract}_{\mathsf{PP}}(\mathsf{msk}, \mathsf{ID})$: Takes as input a master secret key msk and an identity ID and outputs a secret key $\mathsf{sk}_{\mathsf{ID}}$.

$\mathbf{Encrypt}_{\mathsf{PP}}(\mathsf{ID}, M)$: Given an identity ID and a message M, it outputs a ciphertext C.

$\mathbf{Decrypt}_{\mathsf{PP}}(\mathsf{sk}_{\mathsf{ID}}, C)$: Given a secret key $\mathsf{sk}_{\mathsf{ID}}$ and a ciphertext C, outputs either a decryption error symbol $\perp$, or a message M.

Correctness requires that, for any pair $(\mathsf{PP}, \mathsf{msk}) \leftarrow \mathsf{Setup}(1^\lambda)$, any ID and any message M, we have $\mathsf{Decrypt}_{\mathsf{PP}}\big(\mathsf{Extract}_{\mathsf{PP}}(\mathsf{msk}, \mathsf{ID}), \mathsf{Encrypt}_{\mathsf{PP}}(\mathsf{ID}, M)\big) = M$.

Our proofs rely on the semantic security of the scheme against selective adversaries (IND-sID-CPA) but also on the stronger property of ciphertext pseudorandomness. Informally, this notions demands that the adversary be unable to distinguish an encryption of a message of its choice from a random element of the ciphertext space $\mathcal{C}$. Notice that this property implies IND-sID-CPA security.

Definition 4. *An IBE scheme has pseudo-random-ciphertexts if no PPT adversary $\mathcal{A}$ with access to private key extraction oracle $\mathsf{Extract}_{\mathsf{PP}}(\mathsf{msk},\cdot)$ has non-negligible advantage $\mathbf{Adv}_{\mathcal{A}}^{\mathrm{ROR}}(\lambda) = |\Pr[\mathbf{Expt}_{\mathcal{A}}^{\mathrm{ROR}} = 1] - \frac{1}{2}|$ in this game:*

Experiment $\mathbf{Expt}_{\mathcal{A}}^{\mathrm{ROR}}(\lambda)$

$\mathsf{ID}^\star \leftarrow \mathcal{A}_{id}(\lambda)$; $(\mathsf{PP}, \mathsf{msk}) \leftarrow \mathsf{Setup}(1^\lambda)$;
$M \leftarrow \mathcal{A}_{Ch}^{\mathsf{Extract}_{\mathsf{PP}}(\mathsf{msk},\cdot)}(\mathsf{PP})$;
$b \hookleftarrow U(\{0,1\})$;
if $b = 1$ *then* $C^\star \leftarrow \mathsf{Encrypt}_{\mathsf{PP}}(M, \mathsf{ID}^\star)$ *else* $C^\star \leftarrow U(\mathcal{C})$;
$b' \leftarrow \mathcal{A}_{guess}^{\mathsf{Extract}_{\mathsf{PP}}(\mathsf{msk},\cdot)}(C^\star)$;
if $b = b'$ *then return* 1 *else return* 0;

The ABB System. Agrawal, Boneh and Boyen described [1] a compact IBE scheme in the standard model which allows encrypting multi-bit messages.

$\mathsf{Setup}(1^\lambda)$: Given a security parameter λ, choose parameters q, n, σ, α and define $k = \lfloor \log q \rfloor$, $\bar{m} = nk$, $m = 2\bar{m}$ and choose a noise distribution χ for LWE.

1. Compute $(\bar{\mathbf{A}}, \mathbf{T}_{\bar{\mathbf{A}}}) \leftarrow \mathsf{TrapGen}(1^n, 1^m, q)$.
2. Define $\mathbf{G} = \mathbf{I}_n \otimes [1|2|\ldots|2^{k-1}] \in \mathbb{Z}_q^{n \times \bar{m}}$. Sample matrices $\mathbf{B} \hookleftarrow U(\mathbb{Z}_q^{n \times \bar{m}})$, $\mathbf{U} \hookleftarrow U(\mathbb{Z}_q^{n \times m})$.
3. Let $\mathsf{FRD} : \mathbb{Z}_q^n \to \mathbb{Z}_q^{n \times n}$ be the full-rank difference mapping from [1].

Output $\mathsf{PP} = (\bar{\mathbf{A}}, \mathbf{B}, \mathbf{U})$ and $\mathsf{msk} = \mathbf{T}_{\bar{\mathbf{A}}}$.

$\mathsf{Extract}_{\mathsf{PP}}(\mathsf{msk}, \mathsf{ID})$: Given $\mathsf{msk} = \mathbf{T}_{\bar{\mathbf{A}}}$ and an identity $\mathsf{ID} \in \mathbb{Z}_q^n$, do as follows:

1. Define the matrix $\mathbf{B}_{\mathsf{ID}} = \mathbf{B} + \mathsf{FRD}(\mathsf{ID}) \cdot \mathbf{G} \in \mathbb{Z}_q^{n \times \bar{m}}$.
2. Let $\mathbf{B}_{\mathbf{A},\mathsf{ID}} = [\mathbf{A} \mid \mathbf{B}_{\mathsf{ID}}] \in \mathbb{Z}_q^{n \times (m + \bar{m})}$, use $\mathbf{T}_A$ to compute a delegated basis $\mathbf{T}_{\mathsf{ID}}$ for the lattice $\Lambda^\perp(\mathbf{B}_{\mathbf{A},\mathsf{ID}})$.
3. Use $\mathbf{T}_{\mathsf{ID}}$ to sample a small-norm matrix $\mathbf{E}_{\mathsf{ID}} \in \mathbb{Z}^{(m+\bar{m}) \times m}$ satisfying the equality $\mathbf{B}_{\mathbf{A},\mathsf{ID}} \cdot \mathbf{E}_{\mathsf{ID}} = \mathbf{U} \bmod q$.
4. Output $\mathsf{sk}_{\mathsf{ID}} = \mathbf{E}_{\mathsf{ID}} \in \mathbb{Z}^{(m+\bar{m}) \times m}$.

$\mathsf{\mathbf{Encrypt}}_{\mathsf{PP}}(\mathsf{ID}, \mathbf{m})$: Given identity ID and a message $\mathbf{m} \in \{0,1\}^m$,

1. Compute the matrix $\mathbf{B}_{\mathsf{ID}} = \mathbf{B} + \mathsf{FRD}(\mathsf{ID}) \cdot \mathbf{G} \in \mathbb{Z}_q^{n \times \bar{m}}$. Sample vectors $\mathbf{s} \hookleftarrow U(\mathbb{Z}_q^n), \mathbf{x}, \mathbf{y} \hookleftarrow \chi^m$, $\mathbf{R} \hookleftarrow D_{\mathbb{Z},\sigma}^{m \times \bar{m}}$ and compute $\mathbf{z} = \mathbf{R}^\top \cdot \mathbf{y} \in \mathbf{Z}^m$.
2. Compute

$$\begin{cases} \mathbf{c}^{(1)} = \bar{\mathbf{A}}^\top \cdot \mathbf{s} + \mathbf{y} \bmod q, \\ \mathbf{c}^{(2)} = \mathbf{B}_{\mathsf{ID}}^\top \cdot \mathbf{s} + \mathbf{z} \bmod q, \\ \mathbf{c}^{(3)} = \mathbf{U}^\top \cdot \mathbf{s} + \mathbf{x} + \mathbf{m} \cdot \left\lfloor \frac{q}{2} \right\rfloor. \end{cases} \quad (22)$$

3. Output $\mathbf{c} = (\mathbf{c}^{(1)}, \mathbf{c}^{(2)}, \mathbf{c}^{(3)}) \in \mathbb{Z}_q^m \times \mathbb{Z}_q^{\bar{m}} \times \mathbb{Z}_q^m$.

$\mathsf{\mathbf{Decrypt}}_{\mathsf{PP}}(\mathsf{sk}_{\mathsf{ID}}, \mathbf{c})$: Given $\mathsf{sk}_{\mathsf{ID}} = \mathbf{E}_{\mathsf{ID}}$ and $\mathbf{c} = (\mathbf{c}^{(1)}, \mathbf{c}^{(2)}, \mathbf{c}^{(3)}) \in \mathbb{Z}_q^m \times \mathbb{Z}_q^{\bar{m}} \times \mathbb{Z}_q^m$, compute and output

$$\mathbf{m}' = \left\lfloor \left(\mathbf{c}^{(3)} - \mathbf{E}_{\mathsf{ID}} \cdot \begin{bmatrix} \mathbf{c}^{(1)} \\ \mathbf{c}^{(2)} \end{bmatrix} \right) \cdot \left\lfloor \frac{q}{2} \right\rfloor^{-1} \right\rceil \in \{0,1\}^m.$$

Theorem 2 ([1, Theorem 23]). *The ABB IBE scheme has pseudo-random ciphertexts if the* $\mathsf{LWE}_{n,q,\chi}$ *assumption holds.*

References

1. Agrawal, S., Boneh, D., Boyen, X.: Efficient lattice (H)IBE in the standard model. In: Gilbert, H. (ed.) EUROCRYPT 2010. LNCS, vol. 6110, pp. 553–572. Springer, Heidelberg (2010). doi:10.1007/978-3-642-13190-5_28
2. Aguilar Melchor, C., Bettaieb, S., Boyen, X., Fousse, L., Gaborit, P.: Adapting lyubashevsky's signature schemes to the ring signature setting. In: Youssef, A., Nitaj, A., Hassanien, A.E. (eds.) AFRICACRYPT 2013. LNCS, vol. 7918, pp. 1–25. Springer, Heidelberg (2013). doi:10.1007/978-3-642-38553-7_1

3. Aimani, L., Joye, M.: Toward practical group encryption. In: Jacobson, M., Locasto, M., Mohassel, P., Safavi-Naini, R. (eds.) ACNS 2013. LNCS, vol. 7954, pp. 237–252. Springer, Heidelberg (2013). doi:10.1007/978-3-642-38980-1_15
4. Ajtai, M.: Generating hard instances of the short basis problem. In: Wiedermann, J., Emde Boas, P., Nielsen, M. (eds.) ICALP 1999. LNCS, vol. 1644, pp. 1–9. Springer, Heidelberg (1999). doi:10.1007/3-540-48523-6_1
5. Alwen, J., Peikert, C.: Generating shorter bases for hard random lattices. In: STACS 2009. LIPIcs, vol. 3, pp. 75–86. Schloss Dagstuhl - Leibniz-Zentrum fuer Informatik, Germany (2009)
6. Banaszczyk, W.: New bounds in some transference theorems in the geometry of number. Math. Ann. **296**(1), 625–635 (1993)
7. Bellare, M., Boldyreva, A., Desai, A., Pointcheval, D.: Key-privacy in public-key encryption. In: Boyd, C. (ed.) ASIACRYPT 2001. LNCS, vol. 2248, pp. 566–582. Springer, Heidelberg (2001). doi:10.1007/3-540-45682-1_33
8. Bellare, M., Rogaway, P.: Random oracles are practical: a paradigm for designing efficient protocols. In: CCS 1993, pp. 62–73. ACM Press (1993)
9. Benhamouda, F., Camenisch, J., Krenn, S., Lyubashevsky, V., Neven, G.: Better zero-knowledge proofs for lattice encryption and their application to group signatures. In: Sarkar, P., Iwata, T. (eds.) ASIACRYPT 2014. LNCS, vol. 8873, pp. 551–572. Springer, Heidelberg (2014). doi:10.1007/978-3-662-45611-8_29
10. Benhamouda, F., Krenn, S., Lyubashevsky, V., Pietrzak, K.: Efficient zero-knowledge proofs for commitments from learning with errors over rings. In: Pernul, G., Ryan, P.Y.A., Weippl, E. (eds.) ESORICS 2015. LNCS, vol. 9326, pp. 305–325. Springer, Heidelberg (2015). doi:10.1007/978-3-319-24174-6_16
11. Böhl, F., Hofheinz, D., Jager, T., Koch, J., Striecks, C.: Confined guessing: new signatures from standard assumptions. J. Cryptology **28**(1), 176–208 (2015)
12. Boneh, D., Boyen, X.: Efficient Selective-ID secure identity-based encryption without random oracles. In: Cachin, C., Camenisch, J.L. (eds.) EUROCRYPT 2004. LNCS, vol. 3027, pp. 223–238. Springer, Heidelberg (2004). doi:10.1007/978-3-540-24676-3_14
13. Boyen, X.: Lattice mixing and vanishing trapdoors: a framework for fully secure short signatures and more. In: Nguyen, P.Q., Pointcheval, D. (eds.) PKC 2010. LNCS, vol. 6056, pp. 499–517. Springer, Heidelberg (2010). doi:10.1007/978-3-642-13013-7_29
14. Brakerski, Z., Langlois, A., Peikert, C., Regev, O., Stehlé, D.: On the classical hardness of learning with errors. In: STOC 2013, pp. 575–584. ACM (2013)
15. Camenisch, J., Lysyanskaya, A.: A signature scheme with efficient protocols. In: Cimato, S., Persiano, G., Galdi, C. (eds.) SCN 2002. LNCS, vol. 2576, pp. 268–289. Springer, Heidelberg (2003). doi:10.1007/3-540-36413-7_20
16. Canetti, R., Halevi, S., Katz, J.: Chosen-ciphertext security from identity-based encryption. In: Cachin, C., Camenisch, J.L. (eds.) EUROCRYPT 2004. LNCS, vol. 3027, pp. 207–222. Springer, Heidelberg (2004). doi:10.1007/978-3-540-24676-3_13
17. Cash, D., Hofheinz, D., Kiltz, E., Peikert, C.: Bonsai trees, or how to delegate a lattice basis. In: Gilbert, H. (ed.) EUROCRYPT 2010. LNCS, vol. 6110, pp. 523–552. Springer, Heidelberg (2010). doi:10.1007/978-3-642-13190-5_27
18. Cathalo, J., Libert, B., Yung, M.: Group encryption: non-interactive realization in the standard model. In: Matsui, M. (ed.) ASIACRYPT 2009. LNCS, vol. 5912, pp. 179–196. Springer, Heidelberg (2009). doi:10.1007/978-3-642-10366-7_11
19. Chaum, D., Heyst, E.: Group signatures. In: Davies, D.W. (ed.) EUROCRYPT 1991. LNCS, vol. 547, pp. 257–265. Springer, Heidelberg (1991). doi:10.1007/3-540-46416-6_22

20. Ezerman, M.F., Lee, H.T., Ling, S., Nguyen, K., Wang, H.: A provably secure group signature scheme from code-based assumptions. In: Iwata, T., Cheon, J.H. (eds.) ASIACRYPT 2015. LNCS, vol. 9452, pp. 260–285. Springer, Heidelberg (2015). doi:10.1007/978-3-662-48797-6_12
21. Fiat, A., Shamir, A.: How to prove yourself: practical solutions to identification and signature problems. In: Odlyzko, A.M. (ed.) CRYPTO 1986. LNCS, vol. 263, pp. 186–194. Springer, Heidelberg (1987). doi:10.1007/3-540-47721-7_12
22. Gentry, C.: Fully homomorphic encryption using ideal lattices. In: STOC 2009, pp. 169–178. ACM (2009)
23. Gentry, C., Peikert, C., Vaikuntanathan, V.: Trapdoors for hard lattices and new cryptographic constructions. In: STOC 2008, pp. 197–206. ACM (2008)
24. O. Goldreich, S. Goldwasser, and S. Halevi. Collision-Free Hashing from Lattice Problems. ECCC **3**(42) (1996)
25. Goldwasser, S., Micali, S., Rackoff, C.: The knowledge complexity of interactive proof-systems. In: STOC 1985, pp. 291–304. ACM (1985)
26. Gorbunov, S., Vaikuntanathan, V., Wee, H.: Predicate encryption for circuits from LWE. In: Gennaro, R., Robshaw, M. (eds.) CRYPTO 2015. LNCS, vol. 9216, pp. 503–523. Springer, Heidelberg (2015). doi:10.1007/978-3-662-48000-7_25
27. Gordon, S.D., Katz, J., Vaikuntanathan, V.: A group signature scheme from lattice assumptions. In: Abe, M. (ed.) ASIACRYPT 2010. LNCS, vol. 6477, pp. 395–412. Springer, Heidelberg (2010). doi:10.1007/978-3-642-17373-8_23
28. Groth, J., Sahai, A.: Efficient non-interactive proof systems for bilinear groups. In: Smart, N. (ed.) EUROCRYPT 2008. LNCS, vol. 4965, pp. 415–432. Springer, Heidelberg (2008). doi:10.1007/978-3-540-78967-3_24
29. Izabachène, M., Pointcheval, D., Vergnaud, D.: Mediated traceable anonymous encryption. In: Abdalla, M., Barreto, P.S.L.M. (eds.) LATINCRYPT 2010. LNCS, vol. 6212, pp. 40–60. Springer, Heidelberg (2010). doi:10.1007/978-3-642-14712-8_3
30. Jain, A., Krenn, S., Pietrzak, K., Tentes, A.: Commitments and efficient zero-knowledge proofs from learning parity with noise. In: Wang, X., Sako, K. (eds.) ASIACRYPT 2012. LNCS, vol. 7658, pp. 663–680. Springer, Heidelberg (2012). doi:10.1007/978-3-642-34961-4_40
31. Kawachi, A., Tanaka, K., Xagawa, K.: Concurrently secure identification schemes based on the worst-case hardness of lattice problems. In: Pieprzyk, J. (ed.) ASIACRYPT 2008. LNCS, vol. 5350, pp. 372–389. Springer, Heidelberg (2008). doi:10.1007/978-3-540-89255-7_23
32. Kiayias, A., Tsiounis, Y., Yung, M.: Traceable signatures. In: Cachin, C., Camenisch, J.L. (eds.) EUROCRYPT 2004. LNCS, vol. 3027, pp. 571–589. Springer, Heidelberg (2004). doi:10.1007/978-3-540-24676-3_34
33. Kiayias, A., Tsiounis, Y., Yung, M.: Group encryption. In: Kurosawa, K. (ed.) ASIACRYPT 2007. LNCS, vol. 4833, pp. 181–199. Springer, Heidelberg (2007). doi:10.1007/978-3-540-76900-2_11
34. Kiayias, A., Yung, M.: Group signatures with efficient concurrent join. In: Cramer, R. (ed.) EUROCRYPT 2005. LNCS, vol. 3494, pp. 198–214. Springer, Heidelberg (2005). doi:10.1007/11426639_12
35. Laguillaumie, F., Langlois, A., Libert, B., Stehlé, D.: Lattice-based group signatures with logarithmic signature size. In: Sako, K., Sarkar, P. (eds.) ASIACRYPT 2013. LNCS, vol. 8270, pp. 41–61. Springer, Heidelberg (2013). doi:10.1007/978-3-642-42045-0_3
36. Langlois, A., Ling, S., Nguyen, K., Wang, H.: Lattice-based group signature scheme with verifier-local revocation. In: Krawczyk, H. (ed.) PKC 2014. LNCS, vol. 8383, pp. 345–361. Springer, Heidelberg (2014). doi:10.1007/978-3-642-54631-0_20

37. Libert, B., Ling, S., Mouhartem, F., Nguyen, K., Wang, H.: Signature schemes with efficient protocols and dynamic group signatures from lattice assumptions. In: Cheon, J.H., Takagi, T. (eds.) ASIACRYPT 2016, vol. 10032, pp. 373–403. Springer, Heidelberg (2016)
38. Libert, B., Ling, S., Nguyen, K., Wang, H.: Zero-knowledge arguments for lattice-based accumulators: logarithmic-size ring signatures and group signatures without trapdoors. In: Fischlin, M., Coron, J.-S. (eds.) EUROCRYPT 2016. LNCS, vol. 9666, pp. 1–31. Springer, Heidelberg (2016). doi:10.1007/978-3-662-49896-5_1
39. Libert, B., Yung, M., Joye, M., Peters, T.: Traceable group encryption. In: Krawczyk, H. (ed.) PKC 2014. LNCS, vol. 8383, pp. 592–610. Springer, Heidelberg (2014). doi:10.1007/978-3-642-54631-0_34
40. Ling, S., Nguyen, K., Stehlé, D., Wang, H.: Improved zero-knowledge proofs of knowledge for the ISIS problem, and applications. In: Kurosawa, K., Hanaoka, G. (eds.) PKC 2013. LNCS, vol. 7778, pp. 107–124. Springer, Heidelberg (2013). doi:10.1007/978-3-642-36362-7_8
41. Ling, S., Nguyen, K., Wang, H.: Group signatures from lattices: simpler, tighter, shorter, ring-based. In: Katz, J. (ed.) PKC 2015. LNCS, vol. 9020, pp. 427–449. Springer, Heidelberg (2015). doi:10.1007/978-3-662-46447-2_19
42. Lyubashevsky, V.: Lattice-based identification schemes secure under active attacks. In: Cramer, R. (ed.) PKC 2008. LNCS, vol. 4939, pp. 162–179. Springer, Heidelberg (2008). doi:10.1007/978-3-540-78440-1_10
43. Micciancio, D., Peikert, C.: Trapdoors for lattices: simpler, tighter, faster, smaller. In: Pointcheval, D., Johansson, T. (eds.) EUROCRYPT 2012. LNCS, vol. 7237, pp. 700–718. Springer, Heidelberg (2012). doi:10.1007/978-3-642-29011-4_41
44. Micciancio, D., Vadhan, S.P.: Statistical zero-knowledge proofs with efficient provers: lattice problems and more. In: Boneh, D. (ed.) CRYPTO 2003. LNCS, vol. 2729, pp. 282–298. Springer, Heidelberg (2003). doi:10.1007/978-3-540-45146-4_17
45. Nguyen, P.Q., Zhang, J., Zhang, Z.: Simpler efficient group signatures from lattices. In: Katz, J. (ed.) PKC 2015. LNCS, vol. 9020, pp. 401–426. Springer, Heidelberg (2015). doi:10.1007/978-3-662-46447-2_18
46. Paillier, P.: Public-key cryptosystems based on composite degree residuosity classes. In: Stern, J. (ed.) EUROCRYPT 1999. LNCS, vol. 1592, pp. 223–238. Springer, Heidelberg (1999). doi:10.1007/3-540-48910-X_16
47. Peikert, C.: Public-key cryptosystems from the worst-case shortest vector problem. In: STOC 2009, pp. 333–342. ACM (2009)
48. Peikert, C., Vaikuntanathan, V.: Noninteractive statistical zero-knowledge proofs for lattice problems. In: Wagner, D. (ed.) CRYPTO 2008. LNCS, vol. 5157, pp. 536–553. Springer, Heidelberg (2008). doi:10.1007/978-3-540-85174-5_30
49. Regev, O.: On lattices, learning with errors, random linear codes, and cryptography. In: STOC 2005, pp. 84–93. ACM (2005)
50. Rückert, M.: Lattice-based blind signatures. In: Abe, M. (ed.) ASIACRYPT 2010. LNCS, vol. 6477, pp. 413–430. Springer, Heidelberg (2010). doi:10.1007/978-3-642-17373-8_24
51. Schnorr, C.P.: Efficient identification and signatures for smart cards. In: Quisquater, J.-J., Vandewalle, J. (eds.) EUROCRYPT 1989. LNCS, vol. 434, pp. 688–689. Springer, Heidelberg (1990). doi:10.1007/3-540-46885-4_68
52. Stern, J.: A new paradigm for public key identification. IEEE Trans. Inf. Theory **42**(6), 1757–1768 (1996)
53. Trolin, M., Wikström, D.: Hierarchical group signatures. In: Caires, L., Italiano, G.F., Monteiro, L., Palamidessi, C., Yung, M. (eds.) ICALP 2005. LNCS, vol. 3580, pp. 446–458. Springer, Heidelberg (2005). doi:10.1007/11523468_37

54. Xie, X., Xue, R., Wang, M.: Zero knowledge proofs from ring-LWE. In: Abdalla, M., Nita-Rotaru, C., Dahab, R. (eds.) CANS 2013. LNCS, vol. 8257, pp. 57–73. Springer, Heidelberg (2013). doi:10.1007/978-3-319-02937-5_4

Post Quantum Cryptography

From 5-Pass $\mathcal{MQ}$-Based Identification to $\mathcal{MQ}$-Based Signatures

Ming-Shing Chen[1,2](✉), Andreas Hülsing[3](✉), Joost Rijneveld[4](✉), Simona Samardjiska[5](✉), and Peter Schwabe[4](✉)

[1] Department of Electrical Engineering, National Taiwan University, Taipei, Taiwan
mschen@crypto.tw
[2] Research Center for Information Technology Innovation, Academia Sinica, Taipei, Taiwan
[3] Department of Mathematics and Computer Science, Technische Universiteit Eindhoven, Eindhoven, The Netherlands
andreas@huelsing.net
[4] Digital Security Group, Radboud University, Nijmegen, The Netherlands
joost@joostrijneveld.nl, peter@cryptojedi.org
[5] Faculty of Computer Science and Engineering, "Ss. Cyril and Methodius" University, Skopje, Republic of Macedonia
simona.samardjiska@finki.ukim.mk

Abstract. This paper presents MQDSS, the first signature scheme with a security reduction based on the problem of solving a multivariate system of quadratic equations ($\mathcal{MQ}$ problem). In order to construct this scheme we give a new security reduction for the Fiat-Shamir transform from a large class of 5-pass identification schemes and show that a previous attempt from the literature to obtain such a proof does not achieve the desired goal. We give concrete parameters for MQDSS and provide a detailed security analysis showing that the resulting instantiation MQDSS-31-64 achieves 128 bits of post-quantum security. Finally, we describe an optimized implementation of MQDSS-31-64 for recent Intel processors with full protection against timing attacks and report benchmarks of this implementation.

Keywords: Post-quantum cryptography · Fiat-Shamir · 5-pass identification scheme · Vectorized implementation

1 Introduction

Already since 1997, when Shor published a polynomial-time quantum algorithm for factoring and discrete logarithms, it is known that an attacker equipped with a sufficiently large quantum computer will be able to break essentially

This work was supported by the Netherlands Organization for Scientific Research (NWO) under Veni 2013 project 13114 and by the European Commission through the ICT Programme under contract ICT-645622 PQCRYPTO. Permanent ID of this document: 36edf88b815b75e85fae8684c05ec336. Date: September 6, 2016.

J.H. Cheon and T. Takagi (Eds.): ASIACRYPT 2016, Part II, LNCS 10032, pp. 135–165, 2016.
DOI: 10.1007/978-3-662-53890-6_5

all public-key cryptography in use today. More recently, various statements by physicists and quantum engineers indicate that they may be able to build such a large quantum computer within the next few decades. For example, IBM's Mark Ketchen said in 2012 *"I'm thinking like it's 15 [years] or a little more. It's within reach. It's within our lifetime. It's going to happen."*. In May this year, IBM gave access to their 5-qubit quantum computer to researchers and announced that they are expecting to scale up to 50–100 qubits within one decade [36].

It is still a matter of debate *when* and even *if* we will see a large quantum computer that can efficiently break, for example, RSA-4096 or 256-bit elliptic-curve crypto. However, it becomes more and more clear that cryptography aiming at long-term security can no longer discard the *possibility* of attacks by large quantum computers in the foreseeable future. Consequently, NSA recently updated their Suite B to explicitly emphasize the importance of a migration to *post-quantum* algorithms [41] and NIST announced a call for submissions to a post-quantum competition [40]. Submissions to this competition will be accepted for post-quantum public-key encryption, key exchange, and digital signature. The results presented in this paper fall into the last of these three categories: post-quantum digital signature schemes.

Most experts agree that the most conservative choice for post-quantum signatures are hash-based signatures with tight reductions in the standard model to properties like second-preimage resistance of an underlying cryptographic hash function. Unfortunately, the most efficient hash-based schemes are stateful, a property that makes their use prohibitive in many scenarios [39]. A reasonably efficient stateless construction called SPHINCS was presented at Eurocrypt 2015 [6]; however, eliminating the state in this scheme comes at the cost of decreased speed and increased signature size.

The second direction of research for post-quantum signatures are lattice-based schemes. Various schemes have been proposed with different security and performance properties. The best performance is achieved by BLISS [23] (improved in [22]) whose security reduction relies on the hardness of R-SIS and NTRU, and is non-tight. Furthermore, the performance is achieved at the cost of being vulnerable against cache-attacks as demonstrated in [33]. A more conservative approach is the signature scheme proposed by Bai and Galbraith in [3] with improvements to performance and security in [1,2,17]. The security reduction to LWE in [2] is tight; a variant using the (more efficient) ideal-lattice setting was presented in [1]. However, these schemes either come with enormous key and signature sizes (e.g. sizes in [2] are in the order of megabytes), or sizes are reduced at the cost of switching to assumptions on lattices with additional structure like NTRU, Ring-SIS, or Ring-LWE.

The third large class of post-quantum signature algorithms is based on the hardness of solving large systems of multivariate quadratic equations, the so-called $\mathcal{MQ}$ problem. For random instances this problem is NP-complete [30]. However, all schemes in this class that have been proposed with actual parameters for practical use share two properties that often raise concerns about their security: First, their security arguments are rather ad-hoc; there is no reduction from the hardness of $\mathcal{MQ}$. The reason for this is the second property, namely that these systems require a hidden structure in the system of equations; this implies that their security inherently also relies on the hardness of the so-called isomorphism-of-polynomials (IP) problem [42] (or, more precisely, the Extended IP problem [19] or the similar IP with partial knowledge [51] problem). Time has shown that IP in many of the proposed schemes actually relies on the Min-Rank problem [16,28], and unfortunately, more than often, on an easy instance of this problem. Therefore, many proposed schemes have been broken not by targeting $\mathcal{MQ}$, but by targeting IP (and thus exploiting the structure in the system of equations). Examples of broken schemes include Oil-and-Vinegar [43] (broken in [38]), SFLASH [14] (broken in [21]), MQQ-Sig [31] (broken in [27]), (Enhanced) TTS [57,58] (broken in [52]), and Enhanced STS [53] (broken in [52]). There are essentially only two proposals from the "$\mathcal{MQ}$ + IP" class of schemes that are still standing: HFEv$^-$ variants [44,45] and Unbalanced Oil-and-Vinegar (UOV)variants [20,37]. The literature does not, to the best of our knowledge, describe any instantiation of those schemes with parameters that achieve a conservative *post-quantum* security level.

Contributions of this paper. Obviously what one would want in the realm of $\mathcal{MQ}$-based signatures is a scheme that has a tight reduction to $\mathcal{MQ}$ in the quantum-random-oracle model (QROM) or even better in the standard model, and has small key and signatures sizes and fast signing and verification algorithms when instantiated with parameters that offer 128 bits of post-quantum security. In this paper we make a major step towards such a scheme. Specifically, we present a signature system with a reduction from $\mathcal{MQ}$, a set of parameters that achieves 128 bits of post-quantum security according to our careful post-quantum security analysis, and an optimized implementation of this scheme.

This does not mean that our proposal is going quite all the way to the desired scheme sketched above: our reduction is non-tight and in the ROM. Furthermore, at the 128-bit post-quantum security level, the signature size is 40 952 bytes, which is comparable to SPHINCS [6], but larger than what lattice-based schemes or $\mathcal{MQ}$ + IP schemes achieve. However, the scheme excels in key sizes: it needs only 72 bytes for public keys and 64 bytes for private keys.

The basic idea of our construction is to apply a Fiat-Shamir transform to the $\mathcal{MQ}$-based 5-pass identification scheme (IDS) that was presented by Sakumoto, Shirai, and Hiwatari at Crypto 2011 [48]. In principle, this idea is not new; it already appeared in a 2012 paper by El Yousfi Alaoui, Dagdelen, Véron, Galindo, and Cayrel [24]. In their paper they use the 5-pass IDS from [48] as one example of a scheme with a property they call "n-soundness". According to their proof in the ROM, this property of an IDS guarantees that it can be used in a Fiat-Shamir transform to obtain an existentially unforgeable signature scheme. They give such a transform using the IDS from [48, Sect. 4.2].

One might think that choosing suitable parameters for precisely this transform (and implementing the scheme with those parameters) produces the results we are advertising in this paper. However, we show that not only is the construction from [24, Sect. 4.2] insecure (because it ignores the requirement of an exponentially large challenge space), but also that the proof based on the n-soundness property does not apply to a corrected Fiat-Shamir transform of the 5-pass IDS from [48]. The reason is that the n-soundness property does not hold for this IDS. More than that, we show that any $(2n+1)$-pass scheme for which the n-soundness property holds can trivially be transformed into a 3-pass scheme. This observation essentially renders the results of [24] vacuous, because the declared contribution of that paper is to present *"the first transformation which gives generic security statements for SS derived from* $(2n+1)$*-pass IS"*.

To solve these issues, we present a new proof in the ROM for Fiat-Shamir transforms of a large class of 5-pass IDS, including the 5-pass scheme from [48]. This proof is of independent interest; it applies also, for example, to the IDS from [11,49] and (with minor modifications) to [46]. Equipped with this result, we fix the signature scheme from [24] and instantiate the scheme with parameters for the 128-bit post-quantum security level. We call this signature scheme MQDSS and the concrete instatiation with the proposed parameters MQDSS-31-64. Our optimized implementation of MQDSS-31-64 for Intel Haswell processors takes 8 510 616 cycles for signing and 5 752 612 cycles for verification; key generation takes 1 826 612 cycles. These cycle counts include full protection against timing attacks.

Organization of this paper. We start with some preliminaries in Sect. 2. In Sect. 3, we recall the 5-pass IDS as introduced in [48]. We present our theoretical results in Sect. 4. We discuss the problems with the result from [24] in Subsect. 4.1, and resolve them by providing a new proof in Subsect. 4.3. We present a description of the transformed 5-pass signature scheme and give a security reduction for it in Sect. 5. In Sect. 6 we finally present a concrete instantiation and implementation thereof.

Availability of the software. We place all software described in this paper into the public domain to maximize reusability of our results. The software is available online at https://joostrijneveld.nl/papers/mqdss.

2 Preliminaries

In the following we provide basic definitions used throughout this work.

Digital signatures. The main target of this work are digital signature schemes. These are defined as follows.

Definition 2.1 (Digital signature scheme). *A digital signature scheme* Dss *is a triplet of polynomial time algorithms* $\mathsf{Dss} = (\mathsf{KGen}, \mathsf{Sign}, \mathsf{Vf})$ *defined as:*

- *The key generation algorithm* KGen *is a probabilistic algorithm that on input* 1^k, *where* k *is a security parameter, outputs a key pair* $(\mathsf{sk}, \mathsf{pk})$.
- *The signing algorithm* Sign *is a possibly probabilistic algorithm that on input a secret key* sk *and a message* M *outputs a signature* σ.

- *The verification algorithm* Vf *is a deterministic algorithm that on input a public key* pk, *a message* M *and a signature* σ *outputs a bit* b, *where* $b = 1$ *indicates that the signature is accepted and* $b = 0$ *indicates a reject.*

For correctness of a Dss, we require that for all $k \in \mathbb{N}$, $(\mathsf{sk}, \mathsf{pk}) \leftarrow \mathsf{KGen}(1^k)$, all messages M and all signatures $\sigma \leftarrow \mathsf{Sign}(\mathsf{sk}, M)$, we get $\mathsf{Vf}(\mathsf{pk}, M, \sigma) = 1$, i.e., that correctly generated signatures are accepted.

Existential Unforgeability under Adaptive Chosen Message Attacks. The standard security notion for digital signature schemes is existential unforgeability under adaptive chosen message attacks (EU-CMA) [32] which is defined using the following experiment. By $\mathsf{Dss}(1^k)$ we denote a signature scheme with security parameter k.

Experiment $\mathsf{Exp}^{\mathsf{eu\text{-}cma}}_{\mathsf{Dss}(1^k)}(\mathcal{A})$

$(\mathsf{sk}, \mathsf{pk}) \leftarrow \mathsf{KGen}(1^k)$,

$(M^\star, \sigma^\star) \leftarrow \mathcal{A}^{\mathsf{Sign}(\mathsf{sk},\cdot)}(\mathsf{pk})$, with $\mathcal{A}$'s queries $\{(M_i)\}_1^{Q_s}$.

Return 1 iff $\mathsf{Vf}(\mathsf{pk}, M^\star, \sigma^\star) = 1$ and $M^\star \notin \{M_i\}_1^{Q_s}$.

For the success probability of an adversary $\mathcal{A}$ in the above experiment we write

$$\mathrm{Succ}^{\mathsf{eu\text{-}cma}}_{\mathsf{Dss}(1^k)}(\mathcal{A}) = \Pr\left[\mathsf{Exp}^{\mathsf{eu\text{-}cma}}_{\mathsf{Dss}(1^k)}(\mathcal{A}) = 1\right].$$

A signature scheme is called EU-CMA-secure if any PPT adversary has only negligible success probability:

Definition 2.2 (EU-CMA). *Let* $k \in \mathbb{N}$, *Dss a digital signature scheme as defined above. We call Dss EU-CMA-secure if for all* $Q_s, t = poly(k)$ *the maximum success probability* $\mathrm{InSec}^{\mathsf{eu\text{-}cma}}(\mathsf{Dss}(1^k); t, Q_s)$ *of all possibly probabilistic classical adversaries* $\mathcal{A}$ *running in time* $\leq t$, *making at most* Q_s *queries to* Sign *in the above experiment, is negligible in* k*:*

$$\mathrm{InSec}^{\mathsf{eu\text{-}cma}}\left(\mathsf{Dss}(1^k); t, Q_s\right) \stackrel{def}{=} \max_{\mathcal{A}}\{\mathrm{Succ}^{\mathsf{eu\text{-}cma}}_{\mathsf{Dss}(1^k)}(\mathcal{A})\} = negl(k).$$

Identification Schemes. An identification scheme (IDS) is a protocol that allows a prover $\mathcal{P}$ to convince a verifier $\mathcal{V}$ of its identity. More formally this is covered by the following definition.

Definition 2.3 (Identification scheme). *An identification scheme consists of three probabilistic, polynomial-time algorithms* $\mathsf{IDS} = (\mathsf{KGen}, \mathcal{P}, \mathcal{V})$ *such that:*

- *the key generation algorithm* KGen *is a probabilistic algorithm that on input* 1^k, *where* k *is a security parameter, outputs a key pair* (sk, pk).
- $\mathcal{P}$ *and* $\mathcal{V}$ *are interactive algorithms, executing a common protocol. The prover* $\mathcal{P}$ *takes as input a secret key* sk *and the verifier* $\mathcal{V}$ *takes as input a public key* pk. *At the conclusion of the protocol,* $\mathcal{V}$ *outputs a bit* b *with* $b = 1$ *indicating "accept" and* $b = 0$ *indicating "reject".*

For correctness of the scheme we require that for all $k \in \mathbb{N}$ and all $(\mathsf{pk}, \mathsf{sk}) \leftarrow \mathsf{KGen}(1^k)$ *we have* $\Pr[\langle \mathcal{P}(\mathsf{sk}), \mathcal{V}(\mathsf{pk})\rangle = 1] = 1$, *where* $\langle \mathcal{P}(\mathsf{sk}), \mathcal{V}(\mathsf{pk})\rangle$ *refers to the common execution of the protocol between* $\mathcal{P}$ *with input* sk *and* $\mathcal{V}$ *on input* pk.

In this work we are only concerned with passively secure identification schemes. We define security in terms of two properties: soundness and honest-verifier zero-knowledge.

Definition 2.4 (Soundness (with soundness error κ)). *Let $k \in \mathbb{N}$,* $\mathsf{IDS} = (\mathsf{KGen}, \mathcal{P}, \mathcal{V})$ *an identification scheme. We say that* IDS *is sound with soundness error κ if for every PPT adversary $\mathcal{A}$,*

$$\Pr\left[\begin{matrix}(\mathsf{pk}, \mathsf{sk}) \leftarrow \mathsf{KGen}(1^k) \\ \langle \mathcal{A}(1^k, \mathsf{pk}), \mathcal{V}(\mathsf{pk})\rangle = 1\end{matrix}\right] \leq \kappa + negl(k).$$

Of course, the goal is to obtain an IDS with negligible soundness error. This can be achieved by running r rounds of the protocol for an r that fulfills $\kappa^r = \text{negl}(k)$.

For the following definition we need the notion of a transcript. A transcript of an execution of an identification scheme IDS refers to all the messages exchanged between $\mathcal{P}$ and $\mathcal{V}$ and is denoted by $\mathsf{trans}(\langle \mathcal{P}(\mathsf{sk}), \mathcal{V}(\mathsf{pk})\rangle)$.

Definition 2.5 ((statistical) Honest-verifier zero-knowledge). *Let $k \in \mathbb{N}$,* $\mathsf{IDS} = (\mathsf{KGen}, \mathcal{P}, \mathcal{V})$ *an identification scheme. We say that* IDS *is statistical honest-verifier zero-knowledge if there exists a probabilistic polynomial time algorithm $\mathcal{S}$, called the simulator, such that the statistical distance between the following two distribution ensembles is negligible in k:*

$$\left\{(\mathsf{pk}, \mathsf{sk}) \leftarrow \mathsf{KGen}(1^k) : (\mathsf{sk}, \mathsf{pk}, \mathsf{trans}(\langle \mathcal{P}(\mathsf{sk}), \mathcal{V}(\mathsf{pk})\rangle))\right\}$$
$$\left\{(\mathsf{pk}, \mathsf{sk}) \leftarrow \mathsf{KGen}(1^k) : (\mathsf{sk}, \mathsf{pk}, \mathcal{S}(\mathsf{pk}))\right\}.$$

3 Sakumoto et al. 5-Pass IDS Scheme

In [48], Sakumoto et al. proposed two new identification schemes, a 3-pass and a 5-pass IDS, based on the intractability of the $\mathcal{MQ}$ problem. They showed that assuming existence of a non-interactive commitment scheme that is statistically hiding and computationally binding, their schemes are statistical zero knowledge and argument of knowledge, respectively. They further showed that the parallel composition of their protocols is secure against impersonation under passive attack. Let us quickly recall the basics of the construction.

Let $\mathbf{x} = (x_1, \ldots, x_n)$ and let $\mathcal{MQ}(n, m, \mathbb{F}_q)$ denote the family of vectorial functions $\mathbf{F} : \mathbb{F}_q^n \rightarrow \mathbb{F}_q^m$ of degree 2 over $\mathbb{F}_q$: $\mathcal{MQ}(n, m, \mathbb{F}_q) = \{\mathbf{F}(\mathbf{x}) = (f_1(\mathbf{x}), \ldots, f_m(\mathbf{x})) | f_s(\mathbf{x}) = \sum_{i,j} a_{i,j}^{(s)} x_i x_j + \sum_i b_i^{(s)} x_i, s \in \{1, \ldots, m\}\}$. The function $\mathbf{G}(\mathbf{x}, \mathbf{y}) = \mathbf{F}(\mathbf{x} + \mathbf{y}) - \mathbf{F}(\mathbf{x}) - \mathbf{F}(\mathbf{y})$ is called the polar form of the function $\mathbf{F}$. The $\mathcal{MQ}$ problem $\mathcal{MQ}(\mathbf{F}, \mathbf{v})$ is defined as follows:

Given $\mathbf{v} \in \mathbb{F}_q^m$ *find, if any,* $\mathbf{s} \in \mathbb{F}_q^n$ *such that* $\mathbf{F}(\mathbf{s}) = \mathbf{v}$.

The decisional version of this problem is NP-complete [30]. It is widely believed that the $\mathcal{MQ}$ problem is intractable, i.e., that given

$\mathbf{F} \leftarrow_R \mathcal{MQ}(n, m, \mathbb{F}_q)$, $\mathbf{s} \leftarrow_R \mathbb{F}_q^n$ and $\mathbf{v} = \mathbf{F}(\mathbf{s})$ there does not exist a PPT adversary $\mathcal{A}$ that outputs a solution $\mathbf{s}'$ to the $\mathcal{MQ}(\mathbf{F}, \mathbf{v})$ problem with non-negligible probability.

The novelty of the approach of Sakumoto et al. [48] is that unlike previous public key schemes, their solution provably relies only on the $\mathcal{MQ}$ problem (and the security of the commitment scheme), and not on other related problems in multivariate cryptography such as the Isomorphism of Polynomials (IP) problem [42], the related Extended IP [19] and IP with partial knowledge [51] problems or the MinRank problem [16, 28]. The key for this is the introduction of a technique to split the secret using the polar form $\mathbf{G}(\mathbf{x}, \mathbf{y})$ of a system of polynomials $\mathbf{F}(\mathbf{x})$.

In essence, with their technique, the secret $\mathbf{s}$ is split into $\mathbf{s} = \mathbf{r}_0 + \mathbf{r}_1$, and the public $\mathbf{v} = \mathbf{F}(\mathbf{s})$ can be represented as $\mathbf{v} = \mathbf{F}(\mathbf{r}_0) + \mathbf{F}(\mathbf{r}_1) + \mathbf{G}(\mathbf{r}_0, \mathbf{r}_1)$. In order for the polar form not to depend on both shares of the secret, $\mathbf{r}_0$ and $\mathbf{F}(\mathbf{r}_0)$ are further split as $\alpha\mathbf{r}_0 = \mathbf{t}_0 + \mathbf{t}_1$ and $\alpha\mathbf{F}(\mathbf{r}_0) = \mathbf{e}_0 + \mathbf{e}_1$. Now, due to the linearity of the polar form it holds that $\alpha\mathbf{v} = (\mathbf{e}_1 + \alpha\mathbf{F}(\mathbf{r}_1) + \mathbf{G}(\mathbf{t}_1, \mathbf{r}_1)) + (\mathbf{e}_0 + \mathbf{G}(\mathbf{t}_0, \mathbf{r}_1))$, and from only one of the two summands, represented by $(\mathbf{r}_1, \mathbf{t}_1, \mathbf{e}_1)$ and $(\mathbf{r}_1, \mathbf{t}_0, \mathbf{e}_0)$, nothing can be learned about the secret $\mathbf{s}$. The 5-pass IDS is given in Fig. 1 where $(\mathsf{pk}, \mathsf{sk}) = (\mathbf{v}, \mathbf{s}) \leftarrow \mathsf{KGen}(1^{\mathbf{k}})$.

$\mathcal{P}$		$\mathcal{V}$
$\mathbf{r}_0, \mathbf{t}_0 \leftarrow_R \mathbb{F}_q^n, \mathbf{e}_0 \leftarrow_R \mathbb{F}_q^m$		
$\mathbf{r}_1 \leftarrow \mathbf{s} - \mathbf{r}_0$		
$c_0 \leftarrow Com(\mathbf{r}_0, \mathbf{t}_0, \mathbf{e}_0)$		
$c_1 \leftarrow Com(\mathbf{r}_1, \mathbf{G}(\mathbf{t}_0, \mathbf{r}_1) + \mathbf{e}_0)$	$\xrightarrow{(c_0, c_1)}$	$\alpha \leftarrow_R \mathbb{F}_q$
	$\xleftarrow{\alpha}$	
$\mathbf{t}_1 \leftarrow \alpha\mathbf{r}_0 - \mathbf{t}_0$		
$\mathbf{e}_1 \leftarrow \alpha\mathbf{F}(\mathbf{r}_0) - \mathbf{e}_0$	$\xrightarrow{\mathsf{resp}_1 = (\mathbf{t}_1, \mathbf{e}_1)}$	$\mathsf{ch}_2 \leftarrow_R \{0, 1\}$
	$\xleftarrow{\mathsf{ch}_2}$	
If $\mathsf{ch}_2 = 0$, $\mathsf{resp}_2 \leftarrow \mathbf{r}_0$		
Else $\mathsf{resp}_2 \leftarrow \mathbf{r}_1$	$\xrightarrow{\mathsf{resp}_2}$	If $\mathsf{ch}_2 = 0$, Parse $\mathsf{resp}_2 = \mathbf{r}_0$, check
		$c_0 \stackrel{?}{=} Com(\mathbf{r}_0, \alpha\mathbf{r}_0 - \mathbf{t}_1, \alpha\mathbf{F}(\mathbf{r}_0) - \mathbf{e}_1)$
		Else Parse $\mathsf{resp}_2 = \mathbf{r}_1$, check
		$c_1 \stackrel{?}{=} Com(\mathbf{r}_1, \alpha(\mathbf{v} - \mathbf{F}(\mathbf{r}_1)) - \mathbf{G}(\mathbf{t}_1, \mathbf{r}_1) - \mathbf{e}_1)$

Fig. 1. Sakumoto et al. 5-pass IDS

Sakumoto et al. [48] proved that their 5-pass scheme is statistically zero knowledge when the commitment scheme *Com* is statistically hiding which implies (honest-verifier) zero knowledge. Here we prove the soundness property of the scheme[1].

[1] Sakumoto et al. [48] also sketched a proof that their 5-pass protocol is argument of knowledge when *Com* is computationally binding. Our security arguments rely on the weaker notion of soundness, therefore we include an appropriate proof.

Theorem 3.1. *The 5-pass identification scheme of Sakumoto et al. [48] is sound with soundness error $\frac{1}{2}+\frac{1}{2q}$ when the commitment scheme Com is computationally binding.*

Proof. One can show that there exists an adversary $\mathcal{C}$ that can cheat with probability $\frac{1}{2}+\frac{1}{2q}$ (See the full version [13]). What we want to show now is that there cannot exist a cheater that wins with significantly higher success probability as long as the $\mathcal{MQ}$ problem is hard and the used commitment is computationally binding.

Towards a contradiction, suppose there exists a malicious PPT cheater $\mathcal{C}$ such that it holds that $\epsilon := \Pr[\langle \mathcal{C}(1^k, \mathbf{v}), \mathcal{V}(\mathbf{v})\rangle = 1] - (\frac{1}{2}+\frac{1}{2q}) = \frac{1}{P(k)}$. for some polynomial function $P(k)$. We show that this implies that there exists a PPT adversary $\mathcal{A}$ with access to $\mathcal{C}$ that can either break the binding property of Com or can solve the $\mathcal{MQ}$ problem $\mathcal{MQ}(\mathbf{F}, \mathbf{v})$.

$\mathcal{A}$ can achieve this if she can obtain four accepting transcripts from $\mathcal{C}$ with same internal random tape, equation system $\mathbf{F}$, and public key $\mathbf{v}$, such that for two different α there are two transcripts for each α with different ch_2. This is done by rewinding $\mathcal{C}$ and feeding it with all possible combinations of $\alpha \in [0, q-1]$ and $ch_2 \in \{0,1\}$. That way we obtain $2q$ different transcripts. Now, per assumption $\mathcal{C}$ produces an accepting transcript with probability $\frac{1}{2}+\frac{1}{2q}+\epsilon$. Hence, with non-negligible probability ϵ we get at least $q+2$ accepting transcripts. A simple counting argument gives that there has to be a set of four transcripts fulfilling the above conditions. Let these transcripts be $((c_0, c_1), \alpha^{(i)}, (\mathbf{t}_1^{(i)}, \mathbf{e}_1^{(i)}), \mathsf{ch}_2^{(i)}, \mathsf{resp}_2^{(i)})$, where $\alpha^{(1)} = \alpha^{(2)} \neq \alpha^{(3)} = \alpha^{(4)}$, $\mathbf{t}_1^{(1)} = \mathbf{t}_1^{(2)} \neq \mathbf{t}_1^{(3)} = \mathbf{t}_1^{(4)}$, $\mathbf{e}_1^{(1)} = \mathbf{e}_1^{(2)} \neq \mathbf{e}_1^{(3)} = \mathbf{e}_1^{(4)}$, $\mathsf{ch}_2^{(1)} = \mathsf{ch}_2^{(3)} = 0$, $\mathsf{ch}_2^{(2)} = \mathsf{ch}_2^{(4)} = 1$, $\mathsf{resp}_2^{(1)} = \mathbf{r}_0^{(1)}$, $\mathsf{resp}_2^{(3)} = \mathbf{r}_0^{(3)}$, $\mathsf{resp}_2^{(2)} = \mathbf{r}_1^{(2)}$, $\mathsf{resp}_2^{(4)} = \mathbf{r}_1^{(4)}$. Since the commitment (c_0, c_1) is the same in all four transcripts, we have

$$\begin{aligned} Com(\mathbf{r}_0^{(1)}, \alpha^{(1)}\mathbf{r}_0^{(1)} - \mathbf{t}_1^{(1)}, \alpha^{(1)}\mathbf{F}(\mathbf{r}_0^{(1)}) - \mathbf{e}_1^{(1)}) = \\ Com(\mathbf{r}_0^{(3)}, \alpha^{(3)}\mathbf{r}_0^{(3)} - \mathbf{t}_1^{(3)}, \alpha^{(3)}\mathbf{F}(\mathbf{r}_0^{(3)}) - \mathbf{e}_1^{(3)}) \end{aligned} \tag{1}$$

$$\begin{aligned} Com(\mathbf{r}_1^{(2)}, \alpha^{(2)}(\mathbf{v} - \mathbf{F}(\mathbf{r}_1^{(2)})) - \mathbf{G}(\mathbf{t}_1^{(2)}, \mathbf{r}_1^{(2)}) - \mathbf{e}_1^{(2)}) = \\ Com(\mathbf{r}_1^{(4)}, \alpha^{(4)}(\mathbf{v} - \mathbf{F}(\mathbf{r}_1^{(4)})) - \mathbf{G}(\mathbf{t}_1^{(4)}, \mathbf{r}_1^{(4)}) - \mathbf{e}_1^{(4)}) \end{aligned} \tag{2}$$

If any of the arguments of Com on the left-hand side is different from the one on the right-hand side in (1) or in (2), then we get two different openings of Com, which breaks its computationally binding property.

If they are the same in both (1) and (2), then from (1):

$(\alpha^{(1)} - \alpha^{(3)})\mathbf{r}_0^{(1)} = \mathbf{t}_1^{(1)} - \mathbf{t}_1^{(3)}$ and $(\alpha^{(1)} - \alpha^{(3)})\mathbf{F}(\mathbf{r}_0^{(1)}) = \mathbf{e}_1^{(1)} - \mathbf{e}_1^{(3)}$,

and from (2): $(\alpha^{(2)} - \alpha^{(4)})(\mathbf{v} - \mathbf{F}(\mathbf{r}_1^{(2)})) = \mathbf{G}(\mathbf{t}_1^{(2)} - \mathbf{t}_1^{(4)}, \mathbf{r}_1^{(2)}) + \mathbf{e}_1^{(2)} - \mathbf{e}_1^{(4)}$.

Combining the two,

$$(\alpha^{(2)} - \alpha^{(4)})(\mathbf{v} - \mathbf{F}(\mathbf{r}_1^{(2)})) = (\alpha^{(2)} - \alpha^{(4)})\mathbf{G}(\mathbf{r}_0^{(1)}, \mathbf{r}_1^{(2)}) + (\alpha^{(2)} - \alpha^{(4)})\mathbf{F}(\mathbf{r}_0^{(1)}),$$

and since $\alpha^{(2)} \neq \alpha^{(4)}$ we get $\mathbf{v} = \mathbf{F}(\mathbf{r}_1^{(2)}) + \mathbf{G}(\mathbf{r}_0^{(1)}, \mathbf{r}_1^{(2)}) + \mathbf{F}(\mathbf{r}_0^{(1)})$, i.e., $\mathbf{r}_0^{(1)} + \mathbf{r}_1^{(2)}$ is a solution to the given $\mathcal{MQ}$ problem. □

We will look into the inner workings of the IDS in more detail in Sect. 5, where we also introduce the related 3-pass scheme.

4 Fiat-Shamir for 5-Pass Identification Schemes

For several intractability assumptions, the most efficient IDS are five pass, i.e. IDS where a transcript consists of five messages. Here, efficiency refers to the size of all communication of sufficient rounds to make the soundness error negligible. This becomes especially relevant when one wants to turn an IDS into a signature scheme as it is closely related to the signature size of the resulting scheme.

In [24], the authors present a Fiat-Shamir style transform for $(2n+1)$-pass IDS fulfilling a certain kind of canonical structure. To provide some intuition, a five pass IDS is called canonical in the above sense if $\mathcal{P}$ starts with a commitment com_1, $\mathcal{V}$ replies with a challenge ch_1, $\mathcal{P}$ sends a first response resp_1, $\mathcal{V}$ replies with a second challenge ch_2 and finally $\mathcal{P}$ returns a second response resp_2. Based on this transcript, $\mathcal{V}$ then accepts or rejects. The authors of [24] also present a security reduction for signature schemes derived from such IDS using a security property of the IDS which they call *special n-soundness*. Intuitively, this property says that given two transcripts that agree on all messages but the last challenge and possibly the last response, one can extract a valid secret key.

In this section we first show that any $(2n+1)$-pass IDS that fulfills the requirements of the security reduction in [24] can be converted into a 3-pass IDS by letting $\mathcal{P}$ choose all but the last challenge uniformly at random himself. The main reason this is possible is the *special n-soundness*. On the other hand, we argue that existing 5-pass schemes in the literature do not fulfill *special n-soundness* and prove it for the 5-pass $\mathcal{MQ}$-IDS from [48]. Hence, they can neither be turned into 3-pass schemes, nor does the security reduction from [24] apply. Afterwards we give a security reduction for a less generic class of 5-pass IDS which covers many 5-pass IDS, including [11,46,49]. In particular, it covers the 5-pass $\mathcal{MQ}$ scheme from [48].

4.1 The El Yousfi et al. Proof

Before we can make any statement about IDS that fall into the case of [24] we have to define the target of our analysis. A canonical $(2n+1)$-pass IDS is an IDS where the prover and the verifier exchange n challenges and replies. More formally:

Definition 4.1 (Canonical $(2n+1)$-pass identification schemes). *Let $k \in \mathbb{N}$, $\mathsf{IDS} = (\mathsf{KGen}, \mathcal{P}, \mathcal{V})$ a $(2n+1)$-pass identification scheme with n challenge spaces $\mathsf{C}_j, 0 < j \leq n$. We call IDS a canonical $(2n+1)$-pass identification scheme if the prover can be split into $n+1$ subroutines $\mathcal{P} = (\mathcal{P}_0, \ldots, \mathcal{P}_n)$ and the verifier into $n+1$ subroutines $\mathcal{V} = (\mathsf{ChS}_1, \ldots, \mathsf{ChS}_n, \mathsf{Vf})$ such that*

- $\mathcal{P}_0(\mathsf{sk})$ *computes the initial commitment* com *sent as the first message.*
- $\mathsf{ChS}_j, j \leq n$ *computes the j-th challenge message* $\mathsf{ch}_j \leftarrow_R \mathsf{C}_j$*, sampling a random element from the j-th challenge space.*
- $\mathcal{P}_i(\mathsf{sk}, \mathsf{trans}_{2i}), 0 < i \leq n$ *computes the i-th response of the prover given access to the secret key and* trans_{2i}*, the transcript so far, containing the first 2i messages.*
- $\mathsf{Vf}(\mathsf{pk}, \mathsf{trans})$*, upon access to the public key and the whole transcript outputs* $\mathcal{V}$*'s final decision.*

The definition implies that a canonical $(2n + 1)$-pass IDS is *public coin*. The public coin property just says that the challenges are sampled from the respective challenge spaces using the uniform distribution.

El Yousfi et al. propose a generalized Fiat-Shamir transform that turns a canonical $(2n + 1)$-pass IDS into a digital signature scheme. The algorithms of the obtained signature scheme make use of the IDS algorithms as follows. The key generation is just the IDS key generation. The signature algorithm simulates an execution of the IDS, replacing challenge ch_j by the output of a hash function (that maps into C_j) that takes as input the concatenation of the message to be signed and all $2(j - 1) + 1$ messages that have been exchanged so far. The signature just contains the messages sent by $\mathcal{P}$. The verification algorithm uses the signature and the message to be signed to generate a full transcript, recomputing the challenges using the hash function. Then the verification algorithm runs Vf on the public key and the computed transcript and outputs its result.

El Yousfi et al. give a reduction for the resulting signature scheme if the used IDS is honest-verifier zero-knowledge and fulfills *special n-soundness* defined below. The latter is a generalization of special soundness. Intuitively, special n-soundness says that given two transcripts that agree up to the second-to-last response but disagree on the last challenge, one can extract the secret key.

Definition 4.2 (Special n-soundness). *A canonical* $(2n+1)$*-pass IDS is said to fulfill special n-soundness if there exists a PPT algorithm* $\mathcal{E}$*, called the extractor, that given two accepting transcripts* $\mathsf{trans} = (\mathsf{com}, \mathsf{ch}_1, \mathsf{resp}_1, \ldots, \mathsf{resp}_{n-1}, \mathsf{ch}_n, \mathsf{resp}_n)$ *and* $\mathsf{trans}' = (\mathsf{com}, \mathsf{ch}_1, \mathsf{resp}_1, \ldots, \mathsf{resp}_{n-1}, \mathsf{ch}'_n, \mathsf{resp}'_n)$ *with* $\mathsf{ch}_n \neq \mathsf{ch}'_n$ *as well as the corresponding public key* pk*, outputs a matching secret key* sk *for* pk *with non-negligible success probability.*

The common special soundness for canonical (3-pass) IDS is hence just special 1-soundness. Please note that El Yousfi et al. define special n-soundness for the resulting signature scheme which in turn requires the used IDS to provide special n-soundness. We decided to follow the more common approach, defining the soundness properties for the IDS.

From $(2n+1)$ to three passes. We now show that every canonical $(2n + 1)$-pass IDS that fulfills special n-soundness can be turned into a canonical 3-pass IDS fulfilling special soundness.

Theorem 4.3. *Let* $\mathsf{IDS} = (\mathsf{KGen}, \mathcal{P}, \mathcal{V})$ *be a canonical* $(2n+1)$*-pass IDS that fulfills special* n*-soundness. Then, the following 3-pass IDS* $\mathsf{IDS}' = (\mathsf{KGen}, \mathcal{P}', \mathcal{V}')$ *is canonical and fulfills special soundness.*

IDS' *is obtained from* IDS *by just moving* $\mathsf{ChS}_j, 0 < j < n$, *(i.e. all but the last challenge generation algorithm) from* $\mathcal{V}$ *to* $\mathcal{P}$*:* $\mathcal{P}'$ *computes* $\mathsf{com}' = (\mathsf{com}, \mathsf{ch}_1, \mathsf{resp}_1, \ldots, \mathsf{resp}_{n-1}, \mathsf{ch}_{n-1})$ *using* $\mathcal{P}_0, \ldots, \mathcal{P}_{n-1}$ *and* $\mathsf{ChS}_1, \ldots, \mathsf{ChS}_{n-1}$. *After* $\mathcal{P}'$ *sent* com', $\mathcal{V}'$ *replies with* $\mathsf{ch}'_1 \leftarrow \mathsf{ChS}_n(1^k)$. $\mathcal{P}'$ *computes* $\mathsf{resp}'_1 \leftarrow \mathcal{P}_n(\mathsf{sk}, \mathsf{trans}_{2n})$ *and* $\mathcal{V}'$ *verifies the transcript using* Vf.

Proof. Clearly, IDS' is a canonical 3-pass IDS. It remains to prove that it is honest-verifier zero-knowledge and that it fulfills special soundness. The latter is straight forward as two transcripts for IDS', that fulfill the conditions in the soundness definition, can be turned into two transcripts for IDS fulfilling the conditions in the n-soundness definition, splitting $\mathsf{com}' = (\mathsf{com}, \mathsf{ch}_1, \mathsf{resp}_1, \ldots, \mathsf{resp}_{n-1}, \mathsf{ch}_{n-1})$ into its parts. Consequently, we can use any extractor for IDS as an extractor for IDS' running in the same time and having the exact same success probability.

Showing honest-verifier zero-knowledge is similarly straight forward. A simulator $\mathcal{S}'$ for IDS' can be obtained from any simulator $\mathcal{S}$ for IDS. $\mathcal{S}'$ just runs $\mathcal{S}$ to obtain a transcript and regroups the messages to produce a valid transcript for IDS'. Again, $\mathcal{S}'$ runs in essentially the same time as $\mathcal{S}$ and achieves the exact same statistical distance. □

The Sakumoto et al. 5-pass IDS does not fulfill special n-soundness. The above result raises the question whether this property was overlooked and we can turn all the 5-pass schemes in the literature into 3-pass schemes. This would have the benefit that we could use the classical Fiat-Shamir transform to turn the resulting schemes into signature schemes.

Sadly, this is not the case. The reason is that the extractors for those IDS need more than two transcripts. For example, the extractor for the 5-pass IDS from [48] needs four transcripts such that they all agree on com. The transcripts have to form two pairs such that in a pair the transcripts agree on ch_1 but not on ch_2 and the two pairs disagree on ch_1. The proof given by El Yousfi et al. is flawed. The authors miss that the two secret shares $\mathbf{r}_0$ and $\mathbf{r}_1$ obtained from two different transcripts do not have to be shares of a valid secret key. We now give a formal proof.

Theorem 4.4. *The 5-pass identification scheme from [48] does not fulfill special* n*-soundness if the computational* $\mathcal{MQ}$*-problem is hard.*

Proof. We prove this by showing that there exist pairs of transcripts, fulfilling the special n-soundness criteria that can be generated by an adversary without knowledge of the secret key simulating just two executions of the protocol. As a key pair for the $\mathcal{MQ}$-IDS is a random instance of the $\mathcal{MQ}$ problem, special n-soundness of the 5-pass $\mathcal{MQ}$-IDS would imply that the $\mathcal{MQ}$ problem can be solved in probabilistic polynomial time.

Towards a contradiction, assume there exists a PPT extractor $\mathcal{E}$ against the 5-pass $\mathcal{MQ}$-IDS that fulfills Definition 4.2. We show how to build a PPT solver $\mathcal{A}$ for the $\mathcal{MQ}$ problem. Given an instance of the $\mathcal{MQ}$ problem $\mathbf{v}$, $\mathcal{A}$ sets $\mathsf{pk} = \mathbf{v}$ which is a valid public key for the $\mathcal{MQ}$-IDS. Next, $\mathcal{A}$ computes two transcripts as follows. $\mathcal{A}$ samples a random $\alpha \in \mathbb{F}_q$ and random $\mathbf{s}, \mathbf{r}_0, \mathbf{t}_0 \in \mathbb{F}_q^n$, $\mathbf{e}_0 \in \mathbb{F}_q^m$, and computes $\mathbf{r}_1 \leftarrow \mathbf{s} - \mathbf{r}_0$, and $\mathbf{t}_1 \leftarrow \alpha\mathbf{r}_0 - \mathbf{t}_0$. Then $\mathcal{A}$ simulates two successful protocol executions, one for $\mathsf{ch}_2 = 0$, one for $\mathsf{ch}_2 = 1$. To do so, $\mathcal{A}$ impersonates $\mathcal{P}$ and replaces the first challenge with α and the second with 0 in the first run and 1 in the second run. In addition, $\mathcal{A}$ uses the knowledge of α to compute the commitments as:

$$c_0 \leftarrow Com(\mathbf{r}_0, \mathbf{t}_0, \mathbf{e}_0), \text{ and } c_1 \leftarrow Com(\mathbf{r}_1, \alpha(\mathbf{v} - \mathbf{F}(\mathbf{r}_1)) - \mathbf{G}(\mathbf{t}_1, \mathbf{r}_1) - \alpha\mathbf{F}(\mathbf{r}_0) + \mathbf{e}_0).$$

Then $\mathcal{A}$ computes $\mathbf{e}_1 \leftarrow \alpha\mathbf{F}(\mathbf{r}_0) - \mathbf{e}_0$ and sets the second commitment in both runs to $(\mathbf{t}_1, \mathbf{e}_1)$. For $\mathsf{ch}_2 = 0$, $\mathcal{A}$ sets $\mathsf{resp} = \mathbf{r}_0$, and for $\mathsf{ch}_2 = 1$, $\mathcal{A}$ sets $\mathsf{resp} = \mathbf{r}_1$.

Now, the first transcript (when $\mathsf{ch}_2 = 0$) is valid, since $\mathbf{t}_0 = \alpha\mathbf{r}_0 - \mathbf{t}_1$ and $\mathbf{e}_0 = \alpha\mathbf{F}(\mathbf{r}_0) - \mathbf{e}_1$. The second transcript (when $\mathsf{ch}_2 = 1$) is also valid as a straight forward calculation shows. Finally, $\mathcal{A}$ feeds the transcripts to $\mathcal{E}$ and outputs whatever $\mathcal{E}$ outputs. $\mathcal{A}$ has the same success probability as $\mathcal{E}$ and runs in essentially the same time. As $\mathcal{E}$ is a PPT algorithm per assumption, this contradicts the hardness of the computational $\mathcal{MQ}$ problem. □

Clearly, we can also use $\mathcal{A}$ to deal with a parallel execution of many rounds of the scheme. A similar situation arises for all the 5-pass IDS schemes that we found in the literature.

4.2 A Fiat-Shamir Transform for Most $(2n + 1)$-pass IDS

By now we have established that we are currently lacking security arguments for signature schemes derived from $(2n + 1)$-pass IDS. We now show how to fix this issue for most $(2n + 1)$-pass IDS in the literature. As most of these IDS are 5-pass schemes that follow a certain structure, we restrict ourselves to these cases. There are some generalizations that are straight-forward and possible to deal with, but they massively complicate accessibility of our statements.

We will consider a particular type of 5-pass identification protocols where the length of the two challenges is restricted to q and 2.

Definition 4.5 (q2-Identification scheme). *Let $k \in \mathbb{N}$. A q2-Identification scheme* $\mathsf{IDS}(1^k)$ *is a canonical 5-pass identification scheme where for the challenge spaces* C_1 *and* C_2 *it holds that* $|\mathsf{C}_1| = q$ *and* $|\mathsf{C}_2| = 2$. *Moreover, the probability that the commitment* com *takes a given value is negligible (in k), where the probability is taken over the random choice of the input and the used randomness.*

To keep the security reduction below somewhat generic, we also need a property that defines when an extractor exists for a q2-IDS. As we have seen special n-soundness is not applicable. Hence, we give a less generic definition.

Definition 4.6 (q2-Extractor). *We say that a q2-Identification scheme* $\mathsf{IDS}(1^k)$ *has a* q2-extractor *if there exists a PPT algorithm* $\mathcal{E}$*, the extractor, that given a public key* pk *and four transcripts* $\mathsf{trans}^{(i)} = (\mathsf{com}, \mathsf{ch}_1^{(i)}, \mathsf{resp}_1^{(i)}, \mathsf{ch}_2^{(i)}, \mathsf{resp}_2^{(i)})$, $i \in \{1,2,3,4\}$, *with*

$$\mathsf{ch}_1^{(1)} = \mathsf{ch}_1^{(2)} \neq \mathsf{ch}_1^{(3)} = \mathsf{ch}_1^{(4)}, \mathsf{ch}_2^{(1)} = \mathsf{ch}_2^{(3)} \neq \mathsf{ch}_2^{(2)} = \mathsf{ch}_2^{(4)}, \tag{3}$$

valid with respect to pk*, outputs a matching secret key* sk *for* pk *with non-negligible success probability (in k).*

In what follows, let $\mathsf{IDS}^r = (\mathsf{KGen}, \mathcal{P}^r, \mathcal{V}^r)$ be the parallel composition of r rounds of the identification scheme $\mathsf{IDS} = (\mathsf{KGen}, \mathcal{P}, \mathcal{V})$. As the schemes we are concerned with only achieve a constant soundness error, the construction below uses a polynomial number of rounds to obtain an IDS with negligible soundness error as intermediate step. We denote the transcript of the j-th round by $\mathsf{trans}_j = (\mathsf{com}_j, \mathsf{ch}_{1,j}, \mathsf{resp}_{1,j}, \mathsf{ch}_{2,j}, \mathsf{resp}_{2,j})$.

Construction 4.7 (Fiat-Shamir transform for q2-IDS). *Let* $k \in \mathbb{N}$ *the security parameter,* $\mathsf{IDS} = (\mathsf{KGen}, \mathcal{P}, \mathcal{V})$ *a q2-Identification scheme that achieves soundness with soundness error* κ*. Select* r*, the number of (parallel) rounds of* IDS*, such that* $\kappa^r = negl(k)$*, and that the challenge spaces of the composition* IDS^r*,* $\mathsf{C}_1^r, \mathsf{C}_2^r$ *have exponential size in k. Moreover, select cryptographic hash functions* $H_1 : \{0,1\}^* \rightarrow \mathsf{C}_1^r$ *and* $H_2 : \{0,1\}^* \rightarrow \mathsf{C}_2^r$*. The q2-signature scheme* $q2\text{-}\mathsf{Dss}(1^k)$ *derived from* IDS *is the triplet of algorithms* $(\mathsf{KGen}, \mathsf{Sign}, \mathsf{Vf})$ *with:*

- $(\mathsf{sk}, \mathsf{pk}) \leftarrow \mathsf{KGen}(1^k)$,
- $\sigma = (\sigma_0, \sigma_1, \sigma_2) \leftarrow \mathsf{Sign}(\mathsf{sk}, m)$ *where* $\sigma_0 = \mathsf{com} \leftarrow \mathcal{P}_0^r(\mathsf{sk})$, $h_1 = H_1(m, \sigma_0)$, $\sigma_1 = \mathsf{resp}_1 \leftarrow \mathcal{P}_1^r(\mathsf{sk}, \sigma_0, h_1)$, $h_2 = H_2(m, \sigma_0, h_1, \sigma_1)$, *and* $\sigma_2 = \mathsf{resp}_2 \leftarrow \mathcal{P}_2^r(\mathsf{sk}, \sigma_0, h_1, \sigma_1, h_2)$.
- $\mathsf{Vf}(\mathsf{pk}, m, \sigma)$ *parses* $\sigma = (\sigma_0, \sigma_1, \sigma_2)$*, computes the values* $h_1 = H_1(m, \sigma_0)$, $h_2 = H_2(m, \sigma_0, h_1, \sigma_1)$ *as above and outputs* $\mathcal{V}^r(\mathsf{pk}, \sigma_0, h_1, \sigma_1, h_2, \sigma_2)$.

Correctness of the scheme follows immediately from the correctness of IDS.

4.3 Security of q2-signature Schemes

We now give a security reduction for the above transform in the random oracle model assuming that the underlying q2-IDS is honest-verifier zero-knowledge, achieves soundness with constant soundness error, and has a q2-extractor. More specifically, we prove the following theorem:

Theorem 4.8 (EU-CMA security of q2-signature schemes). *Let* $k \in \mathbb{N}$*,* $\mathsf{IDS}(1^k)$ *a q2-IDS that is honest-verifier zero-knowledge, achieves soundness with constant soundness error* κ *and has a q2-extractor. Then* $q2\text{-}\mathsf{Dss}(1^k)$*, the q2-signature scheme derived applying Construction 4.7 is existentially unforgeable under adaptive chosen message attacks.*

In the following, we model the functions H_1 and H_2 as independent random oracles $\mathcal{O}_1$ and $\mathcal{O}_2$. To proof Theorem 4.8, we proceed in several steps. Our proof builds on techniques introduced by Pointcheval and Stern [47]. As the reduction is far from being tight, we refrain from doing an exact proof as it does not buy us anything but a complicated statement. We first recall an important tool from [47] called the splitting lemma.

Lemma 4.9 (Splitting lemma [47]**).** *Let $A \subset X \times Y$, such that $\Pr[A(x, y)] \geqslant \epsilon$. Then, there exists $\Omega \subset X$, such that*

$$\Pr[x \in \Omega] \geqslant \epsilon/2, \text{ and } \Pr[A(a, y) | a \in \Omega] \geqslant \epsilon/2.$$

We now present a forking lemma for $q2$-signature schemes. The lemma shows that we can obtain four valid signatures which contain four valid transcripts of the underlying IDS, given a successful key-only adversary. Moreover, these four traces fulfill a certain requirement on the challenges (here the related parts of the hash function outputs) that we need later.

Lemma 4.10 (Forking lemma for $q2$-signature schemes). *Let $k \in \mathbb{N}$, $\mathsf{Dss}(1^k)$ a $q2$-signature scheme with security parameter k. If there exists a PPT adversary $\mathcal{A}$ that can output a valid signature message pair (m, σ) with non-negligible success probability, given only the public key as input, then, with non-negligible probability, rewinding $\mathcal{A}$ a polynomial number of times (with same randomness) but different oracles, outputs 4 valid signature message pairs $(m, \sigma = (\sigma_0, \sigma_1^{(i)}, \sigma_2^{(i)})$, $i \in \{1, 2, 3, 4\}$, such that for the associated hash values it holds that*

$$h_{1,j}^{(1)} = h_{1,j}^{(2)} \neq h_{1,j}^{(3)} = h_{1,j}^{(4)}, h_{2,j}^{(1)} = h_{2,j}^{(3)} \neq h_{2,j}^{(2)} = h_{2,j}^{(4)}, \tag{4}$$

for some round $j \in \{1, \ldots, r\}$.

Proof. To prove the Lemma we need to show that we can rewind $\mathcal{A}$ three times and the probability that $\mathcal{A}$ succeeds in forging a (different) signature in all four runs is non-negligible. Moreover, we have to show that the signatures have the additional property claimed in the Lemma, again with non-negligible probability.

Let $\omega \in R_w$ be $\mathcal{A}$'s random tape with R_w the set of allowable random tapes. During the attack $\mathcal{A}$ may ask polynomially many queries (in the security parameter k) $Q_1(k)$ and $Q_2(k)$ to the random oracles $\mathcal{O}_1$ and $\mathcal{O}_2$. Let $q_{1,1}, q_{1,2}, \ldots, q_{1,Q_1}$ and $q_{2,1}, q_{2,2}, \ldots, q_{2,Q_2}$ be the queries to $\mathcal{O}_1$ and $\mathcal{O}_2$, respectively. Moreover, let $(r_{1,1}, r_{1,2}, \ldots, r_{1,Q_1}) \in (\mathsf{C}_1^r)^{Q_1}$ and $(r_{2,1}, r_{2,2}, \ldots, r_{2,Q_2}) \in (\mathsf{C}_2^r)^{Q_2}$ the corresponding answers of the oracles.

Towards proving the first point, we assume that $\mathcal{A}$ also outputs h_1, h_2 with the signature and a signature is considered invalid if those do not match the responses of $\mathcal{O}_1$ and $\mathcal{O}_2$, respectively. This assumption is without loss of generality as we can construct such $\mathcal{A}$ from any $\mathcal{A}'$ that does not output h_1, h_2. $\mathcal{A}$ just runs $\mathcal{A}'$ and given the result queries $\mathcal{O}_1$ and $\mathcal{O}_2$ for h_1, h_2 and outputs everything. Clearly $\mathcal{A}$ succeeds with the same success probability as $\mathcal{A}'$ and runs in essentially the same time, making just one more query to each RO.

Denote by F the event that $\mathcal{A}$ outputs a valid message signature pair $(m, \sigma^{(1)} = (\sigma_0, \sigma_1^{(1)}, \sigma_2^{(1)}))$ with the associated hash values $h_1^{(1)}, h_2^{(1)}$. Per assumption, this event occurs with non-negligible probability, i.e., $\mathsf{Pr}[\mathsf{F}] = \frac{1}{P(k)}$, for some polynomial $P(k)$. In addition, F implies $h_1^{(1)} = \mathcal{O}_1(m, \sigma_0)$ and $h_2^{(1)} = \mathcal{O}_2(m, \sigma_0, h_1^{(1)}, \sigma_1^{(1)})$. As $h_1^{(1)}, h_2^{(1)}$ are chosen uniformly at random from exponentially large sets $\mathsf{C}_1^r, \mathsf{C}_2^r$, the probability that $\mathcal{A}$ did not query $\mathcal{O}_1$ for $h_1^{(1)}$ and $\mathcal{O}_2$ for $h_2^{(1)}$ is negligible. Hence, there exists a polynomial P' such that the event F' that F occurs and $\mathcal{A}$ queried $\mathcal{O}_1$ for $h_1^{(1)}$ and $\mathcal{O}_2$ for $h_2^{(1)}$ has probability $\mathsf{Pr}[\mathsf{F}'] = \dfrac{1}{P'(k)}$.

For the moment only consider the second oracle. As of the previous equation, there exists at least one $\beta \leqslant Q_2$ such that

$$\mathsf{Pr}[\mathsf{F}' \wedge q_{2,\beta} = (m, \sigma_0, h_1^{(1)}, \sigma_1^{(1)})] \geqslant \frac{1}{Q_2(k)P'(k)}$$

where the probability is taken over the random coins of $\mathcal{A}$ and $\mathcal{O}_2$. Informally, the following steps just show that the success of an algorithm with non-negligible success probability cannot be conditioned on an event that occurs only with negligible probability (i.e. the outcome of the $q_{2,\beta}$ query landing in some negligible subset).

Let $\mathcal{B} = \{(\omega, r_{2,1}, r_{2,2}, \ldots, r_{2,Q_2}) | \omega \in R_w \wedge (r_{2,1}, r_{2,2}, \ldots, r_{2,Q_2}) \in (\mathsf{C}_2^r)^{Q_2} \wedge \mathsf{F}' \wedge q_{2,\beta} = (m, \sigma_0, h_1^{(1)}, \sigma_1^{(1)})\}$, i.e., the set of random tapes and oracle responses such that $\mathsf{F}' \wedge q_{2,\beta} = (m, \sigma_0, h_1^{(1)}, \sigma_1^{(1)})$. This implies that there exists a non-negligible set of "good" random tapes $\Omega_\beta \subseteq R_\omega$ for which $\mathcal{A}$ can provide a valid signature and $q_{2,\beta}$ is the oracle query fixing $h_2^{(1)}$. Applying the splitting lemma, we get that

$$\Pr[w \in \Omega_\beta] \geqslant \frac{1}{2Q_2(k)P'(k)}$$
$$\Pr[(\omega, r_{2,1}, r_{2,2}, \ldots, r_{2,Q_2}) \in \mathcal{B} | w \in \Omega_\beta] \geqslant \frac{1}{2Q_2(k)P'(k)}$$

Applying the same reasoning again we can derive from the later probability being non-negligible that there exists a non-negligible subset $\Omega_{\beta,\omega}$ of the "good" oracle responses $(r_{2,1}, r_{2,2}, \ldots, r_{2,\beta-1})$ such that $(\omega, r_{2,1}, r_{2,2}, \ldots, r_{2,Q_2}) \in \mathcal{B}$. Applying the splitting lemma again, we get

$$\Pr[(r_{2,1}, \ldots, r_{2,\beta-1}) \in \Omega_{\beta,\omega}] \geqslant \frac{1}{4Q_2(k)P'(k)}$$
$$\Pr[(\omega, r_{2,1}, \ldots, r_{2,Q_2}) \in \mathcal{B} | (r_{2,1}, \ldots, r_{2,\beta-1}) \in \Omega_{\beta,\omega})] \geqslant \frac{1}{4Q_2(k)P'(k)}$$

This means that rewinding $\mathcal{A}$ to the point where it made query $q_{2,\beta}$ and running it with new, random $r'_{2,\beta}, \ldots, r'_{2,Q_2}$ has a non-negligible probability of $\mathcal{A}$ outputting another valid signature. Therefore, we can use $\mathcal{A}$ to find two valid signature message pairs with associated hash values $(m, \sigma = (\sigma_0, \sigma_1^{(1)}, \sigma_2^{(1)}), h_1^{(1)}, h_2^{(1)})$, $(m, \sigma^{(2)} = (\sigma_0, \sigma_1^{(2)}, \sigma_2^{(2)}), h_1^{(2)}, h_2^{(2)})$, with $h_2^{(1)} \neq h_2^{(2)}$ and such that $(\sigma_0, h_1^{(1)}, \sigma_1^{(1)}) = (\sigma_0, h_1^{(2)}, \sigma_1^{(2)})$, with non-negligible probability.

We now rewind the adversary again using exactly the same technique as above but now considering the queries to $\mathcal{O}_1$ and its responses. In the replay we change the responses of $\mathcal{O}_1$ to obtain a third signature that differs from the previously obtained ones in the first associated hash value. It can be shown that with non-negligible probability $\mathcal{A}$ will output a third signature on m, $\sigma^{(3)} = (\sigma_0, \sigma_1^{(3)}, \sigma_2^{(3)})$, with associated hash values $(h_1^{(3)}, h_2^{(3)})$ such that $h_1^{(3)} \neq h_1^{(2)} = h_1^{(1)}$.

Finally, we rewind the adversary a third time, keeping the responses of $\mathcal{O}_1$ from the last rewind and focusing on $\mathcal{O}_2$ again. Again, with non-negligible probability $\mathcal{A}$ will produce yet another signature on m, $\sigma^{(4)} = (\sigma_0, \sigma_1^{(4)}, \sigma_2^{(4)})$ with associated hash values $(h_1^{(4)}, h_2^{(4)})$ such that $h_1^{(4)} = h_1^{(3)}$ and $h_2^{(4)} \neq h_2^{(3)}$.

Summing up, rewinding the adversary three times, we can find four valid signatures $\sigma^{(1)}, \sigma^{(2)}, \sigma^{(3)}, \sigma^{(4)}$ with the above property on the associated hash values with non-negligible success probability $\frac{1}{P(k)}$ for some polynomial $P(k)$. Let us denote this event by $\mathcal{E}_\sigma$. So we have that $\Pr[\mathcal{E}_\sigma] \geqslant \frac{1}{P(k)}$.

What remains is to show that the obtained signatures satisfy the particular structure from the lemma (Eq. 4) with non-negligible probability.

Let $\mathcal{H}$ be the event that there exists a $j \in \{1, \dots, r\}$ such that (4) is satisfied. We have that

$$\mathsf{Pr}[\mathcal{E}_\sigma \wedge \mathcal{H}] = \mathsf{Pr}[\mathcal{E}_\sigma] - \mathsf{Pr}[\neg\mathcal{H} \wedge \mathcal{E}_\sigma] = \mathsf{Pr}[\mathcal{E}_\sigma] - \mathsf{Pr}[\neg\mathcal{H}|\mathcal{E}_\sigma]\mathsf{Pr}[\mathcal{E}_\sigma] \geqslant \frac{1}{P(k)} - \mathsf{Pr}[\neg\mathcal{H}|\mathcal{E}_\sigma]$$

We will now give a statistical argument why $\mathsf{Pr}[\neg\mathcal{H}|\mathcal{E}_\sigma]$ is negligible.

As argued above, the hash values associated with the signatures must be outcomes of the RO queries of $\mathcal{A}$. During its first run, $\mathcal{A}$ can choose the first hash value $h_1^{(1)}$ from his Q_1 queries to $\mathcal{O}_1$ and the second hash value $h_2^{(1)}$ from his Q_2 queries to $\mathcal{O}_2$. The total number of possible combinations is $Q_1 Q_2$. The hash values associated with the second signature are $h_1^{(2)} = h_1^{(1)}$ (as $\mathcal{E}_\sigma$) and $h_2^{(2)}$. So, the first hash value is fixed and the second is chosen from a set of no more than Q_2 responses from $\mathcal{O}_2$. Following the same arguments, the hash pair associated with the third signature is chosen from a set of size $Q_1 Q_2$ and the one associated with the fourth signature from a set of size Q_2. The oracle outputs are uniformly distributed within C_1^r and C_2^r, respectively. Hence, the set of all possible combinations of hash values that $\mathcal{A}$ could output has size

$$\lambda(k) \leq Q_1 Q_2 \cdot Q_2 \cdot Q_1 Q_2 \cdot Q_2,$$

which is a polynomial in k as Q_1 and Q_2 are.

Recall C_1 has size q and C_2 size 2. The probability that the required pattern did not occur in the four-tupel of challenges derived from random hash values for one internal round j is

$$\Pr[\neg\mathcal{H}_j] = 1 - \Pr[\mathcal{H}_j] = 1 - \frac{q-1}{2^2 q} = \frac{3q+1}{4q}.$$

The last follows from the fact that out of all $2^4 q^2$ 4-tuples $((\alpha_1, \beta_1), (\alpha_1, \beta_2), (\alpha_2, \beta_3), (\alpha_2, \beta_4)) \in (\mathsf{C}_1 \times \mathsf{C}_2)^4$ exactly $2^2 q(q-1)$ satisfy $\alpha_1 \neq \alpha_2$, $\beta_1 \neq \beta_2$, $\beta_3 \neq \beta_4$. Hence, the probability that a random four-tuple of hash values does not have a single internal round that satisfies (4) and hence fulfills $\neg\mathcal{H}$ is

$$\Pr[\neg\mathcal{H}] = \left(\frac{3q+1}{4q}\right)^r = \mathrm{negl}(k).$$

According to Construction 4.7, the number of rounds r must be super-logarithmic (in k), to fulfill C_2^r being exponentially large (in k). Hence, the above is negligible for random hash values.

Finally, we just have to combine the two results. The adversary can at most choose out of a polynomially bounded number of four-tuples of hash pairs. Each of these four-tuples has a negligible probability of fulfilling $\neg\mathcal{H}$. Hence, the probability that all the possible combinations of query responses even contain a four-tuple that does not fulfill $\mathcal{H}$ is negligible. So, $\Pr[\neg\mathcal{H}|\mathcal{E}_\sigma] = \mathrm{negl}(k)$, and hence, the conditions from the lemma are satisfied with non-negligible probability. □

With Lemma 4.10 we can already establish unforgeability under key only attacks:

Corollary 4.11 (Key-only attack resistance). *Let $k \in \mathbb{N}$, $\mathsf{IDS}(1^k)$ a q2-IDS that achieves soundness with constant soundness error κ and has a q2-extractor. Then $q2$ - $\mathsf{Dss}(1^k)$, the q2-signature scheme derived applying Construction 4.7 is unforgeable under key-only attacks.*

A straight forward application of Lemma 4.10 allows to generate the four traces needed to apply the $q2$-extractor. The obtained secret key can then be used to violate soundness.

For EU-CMA security, we still have to deal with signature queries. The following lemma shows that a reduction can produce valid responses to the adversarial signature queries if the identification scheme is honest-verifier zero-knowledge.

Lemma 4.12. *Let $k \in \mathbb{N}$ the security parameter, $\mathsf{IDS}(1^k)$ a q2-IDS that is honest-verifier zero-knowledge. Then any PPT adversary $\mathcal{B}$ against the EU-CMA-security of $q2$ - $\mathsf{Dss}(1^k)$, the q2-signature scheme derived by applying Construction 4.7, can be turned into a key-only adversary $\mathcal{A}$ with the properties described in Lemma 4.10. $\mathcal{A}$ runs in polynomial time and succeeds with essentially the same success probability as $\mathcal{B}$.*

Proof. By construction. We show how to construct an oracle machine $\mathcal{A}^{\mathcal{B},\mathcal{S},\mathcal{O}_1,\mathcal{O}_2}$ that has access to $\mathcal{B}$, an honest-verifier zero-knowledge simulator $\mathcal{S}$, and random oracles $\mathcal{O}_1, \mathcal{O}_2$. $\mathcal{A}$ produces a valid signature for $q2$ - $\mathsf{Dss}(1^k)$ given only a public key running in time polynomial in k and achieving essentially the same success probability (up to a negligible difference) as $\mathcal{B}$.

Upon input of public key pk, $\mathcal{A}$ runs $\mathcal{B}^{\mathcal{O}'_1,\mathcal{O}'_2,\mathsf{Sign}}(\mathsf{pk})$ simulating the random oracles (ROs) $\mathcal{O}'_1, \mathcal{O}'_2$, as well as the signing oracle Sign towards $\mathcal{B}$. When $\mathcal{B}$ outputs a forgery (m^*, σ^*), $\mathcal{A}$ just forwards it.

To simulate the ROs, $\mathcal{A}$ keeps two initially empty tables of query-response pairs, one per oracle. Whenever $\mathcal{B}$ queries $\mathcal{O}'_b$, $\mathcal{A}$ first checks if the table for $\mathcal{O}'_b$ already contains a pair for this query. If such a pair exists, $\mathcal{A}$ just returns the stored response. Otherwise, $\mathcal{A}$ forwards the query to its own $\mathcal{O}_b$.

As IDS is honest-verifier zero-knowledge there exists a PPT simulator $\mathcal{S}$ that upon input of a IDS public key generates a valid transcript that is indistinguishable of the transcripts generated by honest protocol executions. Whenever $\mathcal{B}$ queries the signature oracle with message m, $\mathcal{A}$ runs $\mathcal{S}$ r times, to obtain r valid transcripts. $\mathcal{A}$ combines the transcripts to obtain a valid signature with associated hashes $\sigma = ((\sigma_0, \sigma_1, \sigma_2), h_1, h_2)$. Before outputting σ, $\mathcal{A}$ checks if the table for $\mathcal{O}'_1$ already contains an entry for query (m, σ_0). If so, $\mathcal{A}$ aborts. Otherwise, $\mathcal{A}$ adds the pair $((m, \sigma_0), h_1)$. Then, $\mathcal{A}$ checks the second table for query $(m, \sigma_0, h_1, \sigma_1)$. Again, $\mathcal{A}$ aborts if it finds such an entry and adds $((m, \sigma_0, h_1, \sigma_1), h_2)$, otherwise.

The probability that $\mathcal{A}$ aborts is negligible in k. When answering signature queries, $\mathcal{A}$ verifies that certain queries were not made before. Both queries contain σ_1 which takes any given value only with negligible probability. On the other hand, the total number of queries that $\mathcal{B}$ makes to all its oracles is polynomially bounded. Hence, the probability that one of the two queries was already made before is negligible. If $\mathcal{A}$ does not abort, it perfectly simulates all oracles towards $\mathcal{B}$. Hence, $\mathcal{B}$ – and thereby $\mathcal{A}$ – succeeds with the same probability as in the real EU-CMA game in this case. Hence, $\mathcal{A}$ succeeds with essentially the same probability as $\mathcal{B}$. □

We now got everything we need to prove Theorem 4.8. The proof is a straight forward application of the previous two lemmas.

Proof (of Theorem 4.8*).* Towards a contradiction, assume that there exists a PPT adversary $\mathcal{B}$ against the EU-CMA-security of $q2$-Dss succeeding with non-negligible probability. We show how to construct a PPT impersonator $\mathcal{C}$ breaking the soundness of IDS. Applying Lemma 4.12, $\mathcal{C}$ can construct a PPT key-only forger $\mathcal{A}$, with essentially the same success probability as $\mathcal{B}$. Given a public key for IDS (which is a valid $q2$-Dss public key) $\mathcal{C}$ runs $\mathcal{A}$ as described in Lemma 4.10. That way $\mathcal{C}$ can use $\mathcal{A}$ to obtain four signatures that per (4) lead four transcripts as required by the $q2$-extractor $\mathcal{E}$. Running $\mathcal{E}$, $\mathcal{C}$ can extract a valid secret key that allows to impersonate $\mathcal{P}$ with success probability 1.

$\mathcal{C}$ just runs $\mathcal{A}$ and $\mathcal{E}$, two PPT algorithms. Consequently, $\mathcal{C}$ runs in polynomial time. Also, $\mathcal{A}$ and $\mathcal{E}$ both have non-negligible success probability implying that also $\mathcal{C}$ succeeds with non-negligible probability. □

5 Our Proposal

In the previous sections, we gave security arguments for a Fiat-Shamir transform of 5-pass IDS that contain two challenges, from $\{0, \ldots, q-1\}$ and $\{0, 1\}$

respectively, where $q \in \mathbb{Z}^*$. In this section we apply the transform to the 5-pass IDS from [48] (see Sect. 3). Before discussing the 5-pass scheme, which we dub MQDSS, we first briefly examine the signature scheme obtained by applying the traditional Fiat-Shamir transform to the 3-pass IDS in [48], to obtain a baseline. Then we give a generic description of MQDSS and prove it secure.

The IDS requires an $\mathcal{MQ}$ system $\mathbf{F}$ as input, potentially system-wide. We could simply select one function $\mathbf{F}$ and define it as a system parameter for all users. Instead, we choose to derive it from a unique seed that is included in each public key. This increases the size of pk by k bits, and adds some cost for seed expansion when signing and verifying. However, selecting a single system-wide $\mathbf{F}$ might allow an attacker to focus their efforts on a single $\mathbf{F}$ for all users, and would require whoever selects this system parameter to convince all users of its randomness (which is not trivial [5]). For consistency with literature, we still occasionally refer to $\mathbf{F}$ as the 'system parameter'.

Note that the signing procedure described below is slightly more involved than is suggested by Construction 4.7. Where the transformed construction operates directly on the message m, we first apply what is effectively a randomized hash function. As discussed in [35], this extra step provides resilience against collisions in the hash function at only little extra cost. A similar construction appears e.g. in SPHINCS [6]. The digest (and thus the signature) is still derived from m and sk deterministically.

5.1 Establishing a Baseline Using the 3-Pass Scheme over $\mathbb{F}_2$

In the interest of brevity, we will not go into the details of the derived signature scheme here – instead, we refer to the full version of the paper [13].

For the 3-pass scheme, we select $n = m = 256$ over $\mathbb{F}_2$. This results in signatures of 54.81 KB, and a key pair of 64 bytes per key. We ran benchmarks on a single 3.5 GHz core of an Intel Core i7-4770K CPU, measuring 118 088 992 cycles for signature generation, 8 066 324 cycles for key generation and 82 650 156 cycles for signature verification (or 33.7 ms, 2.30 ms and 23.6 ms, respectively).

5.2 The 5-Pass Scheme over $\mathbb{F}_{31}$

As can be seen from the results above, the plain 3-pass scheme over $\mathbb{F}_2$ is quite inefficient, both in terms of signature size and signing speed. This is a direct consequence of the large number of variables and equations required to achieve 128 bits of post-quantum security using $\mathcal{MQ}$ over $\mathbb{F}_2$, as well as the high number of rounds required (see the full version [13] of the paper for an analysis). Using a 5-pass scheme over $\mathbb{F}_{31}$ allows for a smaller n and m, as well as a smaller number of rounds. One might wonder why we do not consider different fields for the 3-pass scenario, instead. This turns out to be suboptimal: contrary to the 5-pass scheme, this does not result in a knowledge error reduction, but does increase the transcript size per round.

The MQDSS signature scheme. We now explicitly construct the functions KGen, Sign and Vf in accordance with Definition 2.1. Specific values for the parameters that achieve 128 bit post-quantum security are given in the next section. We start by presenting the parameters of the scheme in general.

Parameters. MQDSS is parameterized by a security parameter $k \in \mathbb{N}$, and $m, n \in \mathbb{N}$ such that the security level of the $\mathcal{MQ}$ instance $\mathcal{MQ}(n, m, \mathbb{F}_2) \geq k$. The latter fix the description length of the equation system $\mathbf{F}$, $F_{len} = m \cdot \frac{n \cdot (n+1)}{2}$.

- Cryptographic hash functions $\mathcal{H} : \{0,1\}^* \rightarrow \{0,1\}^k$, $H_1 : \{0,1\}^{2k} \rightarrow {\mathbb{F}_{31}}^r$, and $H_2 : \{0,1\}^{2k} \rightarrow \{0,1\}^r$.
- two string commitment functions $Com_0 : {\mathbb{F}_{31}}^n \times {\mathbb{F}_{31}}^n \times {\mathbb{F}_{31}}^m \rightarrow \{0,1\}^k$ and $Com_1 : {\mathbb{F}_{31}}^n \times {\mathbb{F}_{31}}^m \rightarrow \{0,1\}^k$,
- pseudo-random generators $G_{\mathcal{S}_F} : \{0,1\}^k \rightarrow {\mathbb{F}_{31}}^{F_{len}}$, $G_{SK} : \{0,1\}^k \rightarrow {\mathbb{F}_{31}}^n$, and $G_c : \{0,1\}^{2k} \rightarrow {\mathbb{F}_{31}}^{r \cdot (2n+m)}$.

Key generation. Given the security parameter k, we randomly sample a secret key of k bits $SK \leftarrow_R \{0,1\}^k$ as well as a seed $\mathcal{S}_F \leftarrow_R \{0,1\}^k$. We then select a pseudorandom $\mathcal{MQ}$ system $\mathbf{F}$ from $\mathcal{MQ}(n, m, \mathbb{F}_{31})$ by expanding $\mathcal{S}_F$. In total, we must generate $F_{len} = m \cdot (\frac{n \cdot (n+1)}{2} + n)$ elements for $\mathbf{F}$, to use as coefficients for both the quadratic and the linear monomials. We use the pseudorandom generator $G_{\mathcal{S}_F}$ for this.

In order to compute the public key, we want to use the secret key as input for the $\mathcal{MQ}$ function defined by $\mathbf{F}$. As SK is a k-bit string rather than a sequence of n elements from $\mathbb{F}_{31}$, we instead use it as a seed for a pseudorandom generator as well, deriving $SK_{\mathbb{F}_{31}} = G_{SK}(SK)$. It is then possible to compute $\boldsymbol{PK_v} = \mathbf{F}(SK_{\mathbb{F}_{31}})$. The secret key $\mathsf{sk} = (SK, \mathcal{S}_F)$ and the public key $\mathsf{pk} = (\mathcal{S}_F, \boldsymbol{PK_v})$ require $2 \cdot k$ and $k + 5 \cdot m$ bits respectively, assuming 5 bits per $\mathbb{F}_{31}$ element.

Signing. The signature algorithm takes as input a message $m \in \{0,1\}^*$ and a secret key $\mathsf{sk} = (SK, \mathcal{S}_F)$. Similarly as in the key generation, we derive $\mathbf{F} = G_{\mathcal{S}_F}(\mathcal{S}_F)$. Then, we derive a message-dependent random value $R = \mathcal{H}(SK \parallel m)$, where "$\parallel$" is string concatenation. Using this random value R, we compute the randomized message digest $D = \mathcal{H}(R \parallel m)$. The value R must be included in the signature, so that a verifier can derive the same randomized digest.

As mentioned in Definition 2.4, the core of the derived signature scheme essentially consists of iterations of the IDS. We refer to the number of required iterations to achieve the security level k as r (note that this should not be confused with $\mathbf{r}_0$ and $\mathbf{r}_1$, which are vectors of elements of $\mathbb{F}_{31}$).

Given SK and D, we now compute $G_c(SK, D)$ to produce $(\mathbf{r}_{(0,0)}, \ldots, \mathbf{r}_{(0,r)}, \mathbf{t}_{(0,0)}, \ldots, \mathbf{t}_{(0,r)}, \mathbf{e}_{(0,0)}, \ldots, \mathbf{e}_{(0,r)})$. Using these values, we compute $c_{(0,i)}$ and $c_{(1,i)}$ for each round i, as defined in the IDS. Recall that $\mathbf{G}(\mathbf{x}, \mathbf{y}) = \mathbf{F}(\mathbf{x} + \mathbf{y}) - \mathbf{F}(\mathbf{x}) - \mathbf{F}(\mathbf{y})$, and that Com_0 and Com_1 are string commitment functions:

$$c_{(0,i)} = Com_0(\mathbf{r}_{(0,i)}, \mathbf{t}_{(0,i)}, \mathbf{e}_{(0,i)}) \text{ and } c_{(1,i)} = Com_1(\mathbf{r}_{(1,i)}, \mathbf{G}(\mathbf{t}_{(0,i)}, \mathbf{r}_{(1,i)}) + \mathbf{e}_{(0,i)}).$$

As mentioned in [48], it is not necessary to include all $2r$ commitments in the transcript. Instead, we include a digest over the concatenation of all commitments $\sigma_0 = \mathcal{H}(c_{(0,0)} \| c_{(1,0)} \| \ldots \| c_{(0,r-1)} \| c_{(1,r-1)})$. We derive the challenges[2]

[2] Note that the concatenation of all α_i was previously referred to as ch_1.

$\alpha_i \in \mathbb{F}_{31}$ (for $0 \leq i < r$) by applying H_1 to $h_1 = (D, \sigma_0)$. Using these α_i, the vectors $\mathbf{t}_{(1,i)} = \alpha_i \cdot \mathbf{r}_{(0,i)} - \mathbf{t}_{(0,i)}$ and $\mathbf{e}_{(1,i)} = \alpha_i \cdot \mathbf{F}(\mathbf{r}_{(0,i)}) - \mathbf{e}_{(0,i)}$ can be computed.

Let $\sigma_1 = (\mathbf{t}_{(1,0)} \| \mathbf{e}_{(1,0)} \| \cdots \| \mathbf{t}_{(1,r-1)} \| \mathbf{e}_{(1,r-1)})$. We compute h_2 by applying H_2 to the tuple $(D, \sigma_0, h_1, \sigma_1)$ and use it as r binary challenges $\mathsf{ch}_{2,i} \in \{0,1\}$.

Now we define $\sigma_2 = (\mathbf{r}_{(\mathsf{ch}_{2,i},i)}, \dots, \mathbf{r}_{(\mathsf{ch}_{2,i},r-1)}, c_{1-\mathsf{ch}_{2,i}}, \dots, c_{1-\mathsf{ch}_{2,r-1}})$. Note that here we also need to include the challenges $c_{1-\mathsf{ch}_{2,i}}$ that the verifier cannot recompute. We then output $\sigma = (R, \sigma_0, \sigma_1, \sigma_2)$ as the signature. At 5 bits per $\mathbb{F}_{31}$ element, the size of the signature is $(2+r) \cdot k + 5 \cdot r \cdot (2 \cdot n + m)$ bits.

Verification. The verification algorithm takes as input the message m, the signature $\sigma = (R, \sigma_0, \sigma_1, \sigma_2)$ and the public key $PK = (\mathcal{S}_F, \boldsymbol{PK}_v)$. As above, we use R and m to compute D, and derive $\mathbf{F}$ from $\mathcal{S}_F$ using $G_{\mathcal{S}_F}$. As the signature contains σ_0, we can compose h_1 and, consequentially, the challenge values α_i for all r rounds by using H_1. Similarly, the values $\mathsf{ch}_{2,i}$ are computed by applying H_2 to $(D, \sigma_0, h_1, \sigma_1)$. For each round i, the verifier extracts vectors $\mathbf{t}_i$ and $\mathbf{e}_i$ (which are always $\mathbf{t}_{(1,i)}$ and $\mathbf{e}_{(1,i)}$) from σ_1 and $\mathbf{r}_i$ from σ_2. Depending on $\mathsf{ch}_{2,i}$, half of the commitments can now be computed:

$$\begin{aligned}
&\text{if } \mathsf{ch}_{2,i} = 0 \quad && c_{(0,i)} = Com_0(\mathbf{r}_i, \alpha \cdot \mathbf{r}_i - \mathbf{t}_i, \alpha \cdot \mathbf{F}(\mathbf{r}_i) - \mathbf{e}_i) \\
&\text{if } \mathsf{ch}_{2,i} = 1 \quad && c_{(1,i)} = Com_1(\mathbf{r}_i, \alpha \cdot (\boldsymbol{PK}_v - \mathbf{F}(\mathbf{r}_i)) - \mathbf{G}(\mathbf{t}_i, \mathbf{r}_i) - \mathbf{e}_i)
\end{aligned}$$

Extracting the missing commitments $c_{(1-\mathsf{ch}_{2,i},i)}$ from σ_2, the verifier now computes $\sigma_0' = \mathcal{H}(c_{(0,0)} \| c_{(1,0)} \cdots \| c_{(0,r-1)} \| c_{(1,r-1)})$. For verification to succeed, $\sigma_0' = \sigma_0$ should hold.

5.3 Security of MQDSS

We now give a security reduction for MQDSS in the ROM. As our results from the last section are non-tight we only prove an asymptotic statement. While this does not suffice to make any statement about the security of a specific parameter choice, it provides evidence that the general approach leads a secure scheme. Also, the reduction is in the ROM, not in the QROM, thereby limiting applicability in the post-quantum setting. As already mentioned in the introduction, we consider it important future work to strengthen this statement.

In the remainder of this subsection we prove the following theorem.

Theorem 5.1. MQDSS *is EU-CMA-secure in the random oracle model, if*

- *the search version of the $\mathcal{MQ}$ problem is intractable,*
- *the hash functions $\mathcal{H}$, H_1, and H_2 are modeled as random oracles,*
- *the commitment functions Com_0 and Com_1 are computationally binding, computationally hiding, and the probability that their output takes a given value is negligible in the security parameter,*
- *the pseudorandom generator $G_{\mathcal{S}_F}$ is modeled as random oracle, and*

– *the pseudorandom generators,* G_{SK}*, and* G_c *have outputs computationally indistinguishable from random.*

To prove this theorem we would like to apply Theorem 4.8. However, Theorem 4.8 was formulated for a slightly more generic construction. The point is that we apply an optimization originally proposed in [50]. So, in our actual proposal, the parallel composition of the IDS is slightly different as, instead of the commitments, only the hash of their concatenation is sent. Also, the last message now contains the remaining commitments.

While we could have treated this case in Sect. 4, it would have limited the general applicability of the result, as the above optimization is only applicable to schemes with a certain, less generic, structure. However, it is straightforward to redo the proofs from Sect. 4 for the optimized scheme. When modeling the hash function used to compress the commitments as RO, the arguments are exactly the same with one exception. The proof of Lemma 4.12 uses that the commitment scheme – and thereby the first signature element σ_1 – only takes a given value with negligible probability. Now this statement follows from the same property of the commitment scheme and the randomness of the RO. Altogether this leads to the following corollary:

Corollary 5.2 (EU-CMA security of $q2$-signature schemes). *Let* $k \in \mathbb{N}$*,* $\mathsf{IDS}(1^k)$ *a q2-IDS that is honest-verifier zero-knowledge, achieves soundness with constant soundness error* κ *and has a q2-extractor. Then opt-q2-*$\mathsf{Dss}(1^k)$*, the optimized q2-signature scheme derived by applying Construction 4.7 and the optimization explained above, is existentially unforgeable under adaptive chosen message attacks.*

Based on this corollary we can now prove the above theorem.

Proof (of Theorem 5.1*).* Towards a contradiction, assume there exists an adversary $\mathcal{A}$ that wins the EU-CMA game against MQDSS with non-negligible success probability. We show that this implies the existence of an oracle machine $\mathcal{M}^{\mathcal{A}}$ that solves the $\mathcal{MQ}$ problem, breaks a property of one of the commitment schemes, or distinguishes the outputs of one of the pseudorandom generators from random. We first define a series of games and argue that the difference in success probability of $\mathcal{A}$ between these games is negligible. We assume that $\mathcal{M}$ runs $\mathcal{A}$ in these games.

Game 0: Is the EU-CMA game for MQDSS.

Game 1: Is Game 0 with the difference that $\mathcal{M}$ replaces the outputs of G_{SK} by random bit strings.

Game 2: Is Game 1 with the difference that $\mathcal{M}$ replaces the outputs of G_c by random bit strings.

Game 3: Is Game 2 with the difference that $\mathcal{M}$ takes as additional input a random equation system $\mathbf{F}$. $\mathcal{M}$ simulates $G_{\mathcal{S}_F}$ towards $\mathcal{A}$, programming $G_{\mathcal{S}_F}$ such that it returns the coefficients representing $\mathbf{F}$ upon input of $\mathcal{S}_F$ and uniformly random values on any other input.

Per assumption, $\mathcal{A}$ wins Game 0 with non-negligible success probability. Let's call this ϵ. If the difference in $\mathcal{A}$'s success probability playing Game 0 or Game 1 was non-negligible, we could use $\mathcal{A}$ to distinguish the outputs of G_{SK} from random. The same argument applies for the difference between Game 1 and Game 2, and G_c. Finally, the output distribution of $G_{\mathcal{S}_F}$ in Game 3 is the same as in previous games. Hence, there is no difference for $\mathcal{A}$ between Game 2 and Game 3. Accordingly, $\mathcal{A}$'s success probability in these two games is equal.

Now, Game 3 is exactly the EU-CMA game for the optimized $q2$ signature scheme that is derived from $\mathcal{MQ}$-IDS, the 5-pass IDS from [48]. We obtain the necessary contradiction if we can apply Corollary 5.2. For this, it just remains to be shown that $\mathcal{MQ}$-IDS is a $q2$-IDS that is honest-verifier zero-knowledge, achieves soundness with constant soundness error κ and has a $q2$-extractor. Clearly, $\mathcal{MQ}$-IDS is a $q2$-IDS under the given assumptions on the commitment schemes. Sakumoto et al. [48] show that $\mathcal{MQ}$-IDS is honest-verifier zero-knowledge. Theorem 3.1 shows that $\mathcal{MQ}$-IDS achieves soundness with constant soundness error $\kappa = \frac{q+1}{2q}$. Finally, the proof of Theorem 3.1 provides a construction of a $q2$-extractor. □

6 Instantiating the Scheme

In this section, we provide a concrete instance of MQDSS. We discuss a suitable set of parameters to achieve the desired security level, discuss an optimized software implementation, and present benchmark results.

Parameter choice and security analysis. For the 5-pass scheme, the soundness error κ is affected by the size of q. This motivates a field choice larger than $\mathbb{F}_2$ in order to reduce the number of rounds required. From an implementation point of view, it is beneficial to select a small prime, allowing very cheap multiplications as well as comparatively cheap field reductions. We choose $\mathbb{F}_{31}$ with the intention of storing it in a 16 bit value – the benefits of which become clear in the next subsection, where we discuss the required reductions.

We now consider the choice of $\mathcal{MQ}(n, m, \mathbb{F}_{31})$, i.e. the parameters n and m. There are several known generic classical algorithms for solving systems of quadratic equations over finite fields, such as the F4 algorithm [25] and the F5 algorithm [4,26] using Gröbner basis techniques, the Hybrid Approach [9,10] that is a variant of the F5 algorithm, or the XL algorithm [15,18] and variants [56].

Currently, for fields $\mathbb{F}_q$ where $q \geqslant 4$, the best known technique for solving overdetermined systems of equations over $\mathbb{F}_q$ is combining equation solvers with exhaustive search. The Hybrid Approach [9,10] and the FXL variant of XL [56] use this paradigm. Here we will analyze the complexity using the Hybrid approach. Note that the complexity for the XL family of algorithms is similar [59].

Roughly speaking, for an optimization parameter ℓ, using the Hybrid approach one first fixes ℓ among the n variables, and then computes q^{ℓ} Gröbner bases of the smaller systems in $n - \ell$ variables. Hence, the improvement over the plain F5 algorithm comes from the proper choice of the parameter ℓ. It has been shown in [9] that the best trade-off is achieved when the parameter ℓ is proportional to the number of variables n, i.e. when $\ell = \tau n$.

Let $2 \leqslant \omega \leqslant 3$ be the linear algebra constant. The complexity of computing a Gröbner basis of a system of m equations in n variables, $m \geqslant n$, using the F5 algorithm is given by

$$C_{F5}(n,m) = \mathcal{O}\left(\left(m\binom{n+d_{reg}(n,m)-1}{d_{reg}(n,m)}\right)^{\omega}\right),$$

where $d_{reg}(n,m)$ is the degree of regularity of the system which can be approximated as

$$d_{reg}(n,m) \approx \left(\frac{m}{n} - \frac{1}{2} - \sqrt{\frac{m}{n}\left(\frac{m}{n}-1\right)}\right) + \mathcal{O}\left(n^{1/3}\right).$$

For a fixed $0 < \tau < 1$, the complexity of the Hybrid approach is

$$C_{Hyb}(n,m,\tau,d_{reg}(n(1-\tau),m)) = q^{\tau n} \cdot C_{F5}(n(1-\tau),m,d_{reg,\tau}(n(1-\tau),m)).$$

It is well known (and can be seen from the complexity above) that the F5 algorithm as well as the Hybrid approach perform better when the number of equations is bigger than the number of variables, so from this point of view there is no incentive in choosing $m > n$. On the other hand, if $m < n$, then we can simply fix $n - m$ variables and reduce the problem to a smaller one, with m variables. Therefore, in terms of classical security the best choice is $m = n$.

Following the analysis from [9,10], we calculated the best trade-off for τ for the family of functions $\mathcal{MQ}(n,n,\mathbb{F}_{31})$, when $\omega = 2.3$. Asymptotically, $\tau \to 0.16$, although for smaller values of n (e.g. $n = 32$) we find $\tau = 0.13$.

Since our goal is classical security of at least 128 bits, we need to choose $n \geq 51$, so that for any choice of the linear algebra constant $2 \leqslant \omega \leqslant 3$ the Hybrid approach would need at least 2^{128} operations. Note that if we set the more realistic value of $\omega = 2.3$, the minimum is $n = 45$.

For implementation reasons, we choose $n = 64$. In particular, a multiple of 16 suggests efficient register usage for vectorized implementations. In this case, for $\omega = 2.3$, the complexity of the Hybrid approach is $\approx 2^{177}$ and the best result is obtained for $\tau = 0.14$, which translates to fixing 9 variables in the system.

Regarding post-quantum security, at the moment there is no dedicated quantum algorithm for solving systems of quadratic equations. Instead, we can use Grover's search algorithm [34] to directly attack the $\mathcal{MQ}$ problem, or use Grover's algorithm for the search part in a quantum implementation of the Hybrid method. Note that the later requires an efficient quantum implementation of the F5 algorithm, that we will assume provides no quantum speedup over the classical implementation.

Grover's algorithm searches for an item in a unordered list of size $N = 2^{\mathtt{n}}$ that satisfies a certain condition given in the form of a quantum black-box function $f : \{0,1\}^{\mathtt{n}} \to \{0,1\}$. If the condition is satisfied for the i-th item, then $f(i) = 1$, otherwise $f(i) = 0$. The complexity of Grover's algorithm is $\mathcal{O}(\sqrt{N/M})$, where

M is the number of items in the list that satisfy the condition, i.e. the algorithm provides a quadratic speed-up compared to classical search.

First we will consider a direct application of Grover's algorithm on the $\mathcal{MQ}$ problem in question. In this case, f should provide an answer whether a given n-tuple $\mathbf{x}$ from $\mathbb{F}_{31}^n$ satisfies the system of equations $\mathbf{F}(\mathbf{x}) = \mathbf{v}$. Since the domain is not Boolean, we need to convert it one, so we get a domain of size $n \log 31$.

To estimate the complexity of the algorithm, we need the number of solutions M to the given system of equations. Determining the exact M requires exponential time [54], but it was shown in [29] that the number of solutions of a system of n equations in n variables follows the Poisson distribution with parameter $\lambda = 1$. Therefore the expected value is 1. Furthermore, the probability that there are at least M solutions can be estimated as the tail probability of a Poisson random variable $P[X \geqslant M] \geqslant \frac{(e\lambda)^M}{e^\lambda M^M} = \frac{1}{e}(\frac{e}{M})^M$ which is negligible in M. In practice, we can safely assume that $M \leqslant 4$, since $P[M \geqslant 5] \geqslant 2^{-8}$. In total, Grover's algorithm takes $\mathcal{O}(2^{n \log 31/2}/4) \approx 2^{156}$ operations.

As said earlier, we can also use a quantum version of the Hybrid approach for $m = n$. In this case the complexity will be

$$C_{Hyb,quantum}(n, \tau, d_{reg}(n(1-\tau), n)) = \sqrt{\frac{q^{\tau n}}{M}} \cdot C_{F5}(n(1-\tau), n, d_{reg,\tau}(n(1-\tau), n)).$$

Taking again $M \leqslant 4$, the optimal value for the optimization parameter is $\tau = 0.39$, which means we should fix 25 variables in the system. Hence, the quantum version of the Hybrid method has a time complexity of $\approx 2^{139}$ operations.

To achieve EU-CMA for 128 bits of post-quantum security, we require that $k^r \leq 2^{-256}$, as an adversary could perform a preimage search to effectively control the challenges. As $\kappa = \frac{q+1}{2q}$ with $q = 31$, we need $r = 269$. To complete the scheme, we instantiate the functions $\mathcal{H}$, Com_0 and Com_1 with SHA3-256, and use SHAKE-128 for H_1, H_2, G_{S_F}, G_c, and G_{SK} [7]. In order to convert between the output domain of SHAKE-128 and functions that map to vectors over $\mathbb{F}_{31}$, we simply reject and resample values that are not in $\mathbb{F}_{31}$ (effectively applying an instance of the second TSS08 construction from [55]).

We refer to this instance of the scheme as MQDSS-31-64.

Implementation. The central and most costly computation in this signature scheme is the evaluation of $\mathbf{F}$ (and, by corollary, $\mathbf{G}$). The signing procedure requires one evaluation of each for every round, and the verifier needs to compute either $\mathbf{F}$ (if $\mathsf{ch}_2 = 0$) or both $\mathbf{F}$ and $\mathbf{G}$ (if $\mathsf{ch}_2 = 1$), for each round. Other than these functions, the computational effort is made up of seed expansion, several hash function applications and a small number of additions and subtractions. For SHA3-256 and SHAKE-128, we rely on existing code from the Keccak Code Package [8]. Clearly, the focus for an optimized implementation should be on the $\mathcal{MQ}$ function. Previous work [12] has shown that modern CPUs offer interesting and valuable methods to efficiently implement this primitive, in particular by exploiting the high level of internal parallelism.

Compared to the binary 3-pass scheme, the implementation of the 5-pass scheme over $\mathbb{F}_{31}$ presents more challenges. As $\mathbb{F}_{31}$ does not have closure under regular integer multiplication and addition, results of computations need to be reduced to smaller representations. To avoid having to this too frequently, we generally represent field elements during computation as unsigned 16 bit values. During specific parts of the computation, we vary this representation as needed.

The evaluation of **F** can roughly be divided in two parts: the generation of all monomials, and computation of the resulting polynomials for known monomials. Generating the quadratic monomials based on the given linear monomials requires $n \cdot \frac{n+1}{2}$ multiplications. For the second part, we require $m \cdot (n + n \cdot \frac{n+1}{2})$ multiplications to multiply the coefficients of the system parameter with the quadratic monomials, as well as a number of additions to accumulate all results. As the second part is clearly more computationally intensive, the optimization of this part is our primary concern. We describe an approach for the monomial generation in the full version [13] of the paper.

To efficiently compute all polynomials for a given set of monomials, we keep all required data in registers to avoid the cost of register spilling throughout the computation. Given that $n = m = 64$, for this part of the computation we represent the 64 $\mathbb{F}_{31}$ input values as 8 bit values and the resulting 64 $\mathbb{F}_{31}$ elements as 16 bit values, costing us 2 and 4 `YMM` registers respectively. The coefficients of **F** can be represented as a column major matrix with every column containing all coefficients that correspond to a specific monomial, i.e. one for each output value. That would imply that every row of the matrix represents one polynomial of **F**. In this representation, each result term is computed by accumulating the products of a row of coefficients with each monomial, which is exactly the same as computing the product of the matrix **F** and the vector containing all monomials. This allows us to efficiently accumulate the output terms, minimizing the required output registers.

In order to perform the required multiplications and additions as quickly as possible, we heavily rely on the AVX2 instruction `VPMADDUBSW`. In one instruction, this computes two 8 bit SIMD multiplications and a 16 bit SIMD addition. However, this instruction operates on 8 bit input values that are stored adjacently. This requires a slight variation on the representation of **F** described above: instead, we arrange the coefficients of **F** in a column major matrix with 16 bit elements, each corresponding to two concatenated monomials.

When arranging reductions, we must strike a careful balance between preventing overflow and not reducing more often than necessary. As we make extensive use of `VPMADDUBSW`, which takes both a signed and an unsigned operand to compute the quadratic monomials, we ensure that the input variables for the $\mathcal{MQ}$ function are unsigned values (in particular: $\{0, \ldots, 31\}$). For the coefficients in the system parameter **F**, we can then freely assume the values are in $\{-15, \ldots, 15\}$, as these are the direct result of a pseudo-random generator. It turns out to be efficient to immediately reduce the quadratic monomials back to $\{0, \ldots, 31\}$ when they are computed. When we now multiply such a product with an element from the system parameter and add it to the accumulators, the maximum value of each accumulator word will be at most[3] $64 \cdot 31 \cdot 15 = 29760$.

[3] This follows from the fact that we combine 64 such monomials in two `YMM` registers.

As this does not exceed 32768, we only have to perform reductions on each individual accumulator at the very end.

One should note that [12] approaches this problem from a slightly different angle. In particular, they accumulate each individual output element sequentially, allowing them to keep the intermediate results in the 32 bit representation that is the output of their combined multiplication and addition instructions. This has the natural consequence of also avoiding early reductions.

Benchmark results. The MQDSS-31-64 implementation has been optimized for large Intel processors, supporting AVX2 instructions. Benchmarks were carried out on a single core of an Intel Core i7-4770K CPU, running at 3.5 GHz. *Signature and key sizes.* The signature size of MQDSS-31-64 is considerably smaller than that of the 3-pass scheme. The obvious factor in this is the decreased ratio between the element size (which, in packed form, now require $64 \cdot 5 = 320$ bits each) and the number of rounds, resulting in a signature size of $2 \cdot 256 + 269 \cdot (256 + (5 \cdot 3 \cdot 64)) = 327\,616$ bits, or 40 952 bytes (39.99 KB). The shape of the keys does not change compared to 3-pass scheme, but since a vector of field elements now requires 320 bits, the public key is 72 bytes. The secret key remains 64 bytes.

Performance. As the $\mathcal{MQ}$ function is the most costly part of the computation, parameters are chosen in such a way that its performance is maximized. The required number of multiplications and additions (expressed as functions of n and m) does not change dramatically compared to the 3-pass baseline[4], but the actual values n and m are only a quarter of what they were. As the relation between n and m and the number of multiplications is quadratic for the monomials and cubic for the system parameter masking, and we see only a linear increase in the number of registers needed to operate on, the entire sequence of multiplications and additions becomes much cheaper. This especially impacts operations that involve the accumulators. As the representation allows us to keep reductions out of this innermost repeated loop, we perform (only) $\frac{67 \cdot 4}{2} + 4 = 136$ reductions[5] throughout the main computation and 66 when preparing quadratic monomials. As we were able to arrange the registers in such a way that they do not need to rotate across multiple registers, we greatly reduce the number of rotations required compared to the 3-pass scenario. Furthermore, we note that we use a total of $67 \cdot 16 \cdot 4 = 4288$ `VPMADDUBSW` instructions for the core computations.

For one iteration of the $\mathcal{MQ}$ function $\mathbf{F}$, we measure 6 616 cycles ($\mathbf{G}$ is slightly less costly, at 6 396 cycles). We measure a total of 8 510 616 cycles for the complete signature generation. Key generation costs 1 826 612 cycles, and verification consumes 5 752 612 cycles. On the given platform, that translates to roughly 2.43 ms, 0.52 ms and 1.64 ms, respectively. Verification is expected to require on average $\frac{3}{2}$ calls to an $\mathcal{MQ}$ function per round, whereas signature

[4] A slight difference is introduced by cancellation of the monomials in the $\mathbb{F}_2$ setting.

[5] This follows from the fact that we need a total of $\frac{64+64 \cdot 65}{2 \cdot 32} = 67$ `YMM` registers worth of space to store the monomials and perform 4 reductions after accumulating 2 `YMM` monomials.

generation always requires two. This explains the ratio; note that both signer and verifier incur additional costs besides the $\mathcal{MQ}$ functions, e.g. for seed expansion.

In order to compare these results to the state of the art, we consider the performance figures reported in [12]. In particular, we examine the Rainbow(31, 24, 20, 20) instance, as the 'public map' in this scheme is effectively the $\mathcal{MQ}$ function over $\mathbb{F}_{31}$ with $n = 64$, as used above. The number of equations differs (i.e. $m = 40$ as opposed to $m = 64$), but this can be approximated by normalizing linearly. In [12], the authors report a time measurement of $17.7\,\mu s$, which converts to 50 144 cycles on their 2.833 GHz Intel C2Q Q9550. After normalizing for m, this amounts to 80 230 cycles. Results from the eBACS benchmarking project further show that running the Rainbow verification function from [12] on a Haswell CPU requires approximately 46 520 cycles (and thus 74 432 after normalizing); verification is dominated by the public map. Using their (by now arguably outdated) SSE2-based code to evaluate a public map with $m = 64$ consumes 60 968 cycles on our Intel Core i7-4770K. All of these results provide confidence in the fact that our implementation, which makes extensive use of AVX2 instructions, is performing in line with expectations.

Acknowledgements. The authors would like to thank Marc Fischlin for helpful discussions, the anonymous reviewers for valuable comments, Wen-Ding Li for his contributions to the software, and Arno Mittelbach for the cryptocode package.

References

1. Akleylek, S., Bindel, N., Buchmann, J., Krämer, J., Marson, G.A.: An efficient lattice-based signature scheme with provably secure instantiation. In: Pointcheval, D., Nitaj, A., Rachidi, T. (eds.) AFRICACRYPT 2016. LNCS, vol. 9646, pp. 44–60. Springer, Heidelberg (2016). doi:10.1007/978-3-319-31517-1_3
2. Alkim, E., Bindel, N., Buchmann, J., Dagdelen, O.: TESLA: tightly-secure efficient signatures from standard lattices. Cryptology ePrint Archive (2015)
3. Bai, S., Galbraith, S.D.: An improved compression technique for signatures based on learning with errors. In: Benaloh, J. (ed.) CT-RSA 2014. LNCS, vol. 8366, pp. 28–47. Springer, Heidelberg (2014). doi:10.1007/978-3-319-04852-9_2
4. Bardet, M., Faugère, J., Salvy, B.: On the complexity of the F5 Gröbner basis algorithm. J. Symbolic Comput. **70**, 49–70 (2015)
5. Bernstein, D.J., Chou, T., Chuengsatiansup, C., Hülsing, A., Lange, T., Niederhagen, R., van Vredendaal, C.: How to manipulate curve standards: a white paper for the black hat. Cryptology ePrint Archive (2014)
6. Bernstein, D.J., Hopwood, D., Hülsing, A., Lange, T., Niederhagen, R., Papachristodoulou, L., Schneider, M., Schwabe, P., Wilcox-O'Hearn, Z.: SPHINCS: Practical Stateless Hash-Based Signatures. In: Oswald, E., Fischlin, M. (eds.) EUROCRYPT 2015. LNCS, vol. 9056, pp. 368–397. Springer, Heidelberg (2015). doi:10.1007/978-3-662-46800-5_15
7. Bertoni, G., Daemen, J., Peeters, M., Van Assche, G.: The Keccak reference (2011)
8. Bertoni, G., Daemen, J., Peeters, M., Van Assche, G., Van Keer, R.: Keccak Code Package (2016)
9. Bettale, L., Faugère, J., Perret, L.: Solving polynomial systems over finite fields: improved analysis of the hybrid approach. In: ISSAC 2012, pp. 67–74. ACM (2012)

10. Bettale, L., Faugère, J.-C., Perret, L.: Hybrid approach for solving multivariate systems over finite fields. J. Math. Cryptology, pp. 177–197 (2009)
11. Cayrel, P.-L., Véron, P., Yousfi Alaoui, S.M.: A zero-knowledge identification scheme based on the q-ary syndrome decoding problem. In: Biryukov, A., Gong, G., Stinson, D.R. (eds.) SAC 2010. LNCS, vol. 6544, pp. 171–186. Springer, Heidelberg (2011). doi:10.1007/978-3-642-19574-7_12
12. Chen, A.I.-T., Chen, M.-S., Chen, T.-R., Cheng, C.-M., Ding, J., Kuo, E.L.-H., Lee, F.Y.-S., Yang, B.-Y.: SSE implementation of multivariate PKCs on Modern x86 CPUs. In: Clavier, C., Gaj, K. (eds.) CHES 2009. LNCS, vol. 5747, pp. 33–48. Springer, Heidelberg (2009). doi:10.1007/978-3-642-04138-9_3
13. Chen, M.-S., Hülsing, A., Rijneveld, J., Samardjiska, S., Schwabe, P.: From 5-pass MQ-based identification to MQ-based signatures. Cryptology ePrint Archive (2016)
14. Courtois, N., Goubin, L., Patarin, J.: SFLASH, a fast asymmetric signature scheme for low-cost smartcards - primitive specification and supporting documentation
15. Courtois, N., Klimov, A., Patarin, J., Shamir, A.: Efficient algorithms for solving overdefined systems of multivariate polynomial equations. In: Preneel, B. (ed.) EUROCRYPT 2000. LNCS, vol. 1807, pp. 392–407. Springer, Heidelberg (2000). doi:10.1007/3-540-45539-6_27
16. Courtois, N.T.: Efficient zero-knowledge authentication based on a linear algebra problem minrank. In: Boyd, C. (ed.) ASIACRYPT 2001. LNCS, vol. 2248, pp. 402–421. Springer, Heidelberg (2001). doi:10.1007/3-540-45682-1_24
17. Dagdelen, Ö., Bansarkhani, R., Göpfert, F., Güneysu, T., Oder, T., Pöppelmann, T., Sánchez, A.H., Schwabe, P.: High-speed signatures from standard lattices. In: Aranha, D.F., Menezes, A. (eds.) LATINCRYPT 2014. LNCS, vol. 8895, pp. 84–103. Springer, Heidelberg (2015). doi:10.1007/978-3-319-16295-9_5
18. Diem, C.: The XL-algorithm and a conjecture from commutative algebra. In: Lee, P.J. (ed.) ASIACRYPT 2004. LNCS, vol. 3329, pp. 323–337. Springer, Heidelberg (2004). doi:10.1007/978-3-540-30539-2_23
19. Ding, J., Hu, L., Yang, B.-Y., Chen, J.-M.: Note on design criteria for rainbow-type multivariates. Cryptology ePrint Archive (2006)
20. Ding, J., Schmidt, D.: Rainbow, a new multivariable polynomial signature scheme. In: Ioannidis, J., Keromytis, A., Yung, M. (eds.) ACNS 2005. LNCS, vol. 3531, pp. 164–175. Springer, Heidelberg (2005). doi:10.1007/11496137_12
21. Dubois, V., Fouque, P.-A., Shamir, A., Stern, J.: Practical cryptanalysis of SFLASH. In: Menezes, A. (ed.) CRYPTO 2007. LNCS, vol. 4622, pp. 1–12. Springer, Heidelberg (2007). doi:10.1007/978-3-540-74143-5_1
22. Ducas, L.: Accelerating Bliss: the geometry of ternary polynomials. Cryptology ePrint Archive (2014)
23. Ducas, L., Durmus, A., Lepoint, T., Lyubashevsky, V.: Lattice signatures and bimodal Gaussians. In: Canetti, R., Garay, J.A. (eds.) CRYPTO 2013. LNCS, vol. 8042, pp. 40–56. Springer, Heidelberg (2013). doi:10.1007/978-3-642-40041-4_3
24. Yousfi Alaoui, S.M., Dagdelen, Ö., Véron, P., Galindo, D., Cayrel, P.-L.: Extended security arguments for signature schemes. In: Mitrokotsa, A., Vaudenay, S. (eds.) AFRICACRYPT 2012. LNCS, vol. 7374, pp. 19–34. Springer, Heidelberg (2012). doi:10.1007/978-3-642-31410-0_2
25. Faugère, J.-C.: A new efficient algorithm for computing Gröbner bases (F4). J. Pure Appl. Algebra **139**, 61–88 (1999)
26. Faugère, J.-C.: A new efficient algorithm for computing Gröbner bases without reduction to zero (F5). In: ISSAC 2002, 75–83. ACM (2002)

27. Faugère, J.-C., Gligoroski, D., Perret, L., Samardjiska, S., Thomae, E.: A polynomial-time key-recovery attack on MQQ cryptosystems. In: Katz, J. (ed.) PKC 2015. LNCS, vol. 9020, pp. 150–174. Springer, Heidelberg (2015). doi:10.1007/978-3-662-46447-2_7
28. Faugère, J.-C., Levy-dit-Vehel, F., Perret, L.: Cryptanalysis of minrank. In: Wagner, D. (ed.) CRYPTO 2008. LNCS, vol. 5157, pp. 280–296. Springer, Heidelberg (2008). doi:10.1007/978-3-540-85174-5_16
29. Fusco, G., Bach, E.: Phase transition of multivariate polynomial systems. In: Cai, J.-Y., Cooper, S.B., Zhu, H. (eds.) TAMC 2007. LNCS, vol. 4484, pp. 632–645. Springer, Heidelberg (2007). doi:10.1007/978-3-540-72504-6_58
30. Garey, M.R., Johnson, D.S.: Computers and Intractability: A Guide to the Theory of NP-Completeness. W. H. Freeman and Company, New York (1979)
31. Gligoroski, D., Ødegård, R.S., Jensen, R.E., Perret, L., Faugère, J.-C., Knapskog, S.J., Markovski, S.: MQQ-SIG. In: Chen, L., Yung, M., Zhu, L. (eds.) INTRUST 2011. LNCS, vol. 7222, pp. 184–203. Springer, Heidelberg (2012). doi:10.1007/978-3-642-32298-3_13
32. Goldwasser, S., Micali, S., Rivest, R.L.: A digital signature scheme secure against adaptive chosen-message attacks. SIAM J. Comput. **17**(2), 281–308 (1988)
33. Groot Bruinderink, L., Hülsing, A., Lange, T., Yarom, Y.: Flush, Gauss, and Reload - a cache attack on the BLISS lattice-based signature scheme. Cryptology ePrint Archive (2016)
34. Grover, L.K.: A fast quantum mechanical algorithm for database search. In: STOC 1996, pp. 212–219. ACM (1996)
35. Halevi, S., Krawczyk, H.: Strengthening digital signatures via randomized hashing. In: Dwork, C. (ed.) CRYPTO 2006. LNCS, vol. 4117, pp. 41–59. Springer, Heidelberg (2006). doi:10.1007/11818175_3
36. IBM. IBM makes quantum computing available on IBM cloud to accelerate innovation (2016)
37. Kipnis, A., Patarin, J., Goubin, L.: Unbalanced oil and vinegar signature schemes. In: Stern, J. (ed.) EUROCRYPT 1999. LNCS, vol. 1592, pp. 206–222. Springer, Heidelberg (1999). doi:10.1007/3-540-48910-X_15
38. Gligoroski, D., Ødegård, R.S., Jensen, R.E., Perret, L., Faugère, J.-C., Knapskog, S.J., Markovski, S.: MQQ-SIG. In: Chen, L., Yung, M., Zhu, L. (eds.) INTRUST 2011. LNCS, vol. 7222, pp. 184–203. Springer, Heidelberg (2012). doi:10.1007/978-3-642-32298-3_13
39. McGrew, D., Kampanakis, P., Fluhrer, S., Gazdag, S.-L., Butin, D., Buchmann, J.: State management for hash based signatures. Cryptology ePrint Archive (2016)
40. NIST. Post-quantum cryptography: NIST's plan for the future (2016)
41. NSA. NSA suite B cryptography
42. Patarin, J.: Hidden Fields Equations (HFE) and Isomorphisms of Polynomials (IP): two new families of asymmetric algorithms. In: Maurer, U. (ed.) EUROCRYPT 1996. LNCS, vol. 1070, pp. 33–48. Springer, Heidelberg (1996). doi:10.1007/3-540-68339-9_4
43. Patarin, J.: The Oil and Vinegar signature scheme. In: Dagstuhl Workshop on Cryptography (1997)
44. Patarin, J., Courtois, N., Goubin, L.: QUARTZ, 128-bit long digital signatures. In: Naccache, D. (ed.) CT-RSA 2001. LNCS, vol. 2020, pp. 282–297. Springer, Heidelberg (2001). doi:10.1007/3-540-45353-9_21

45. Petzoldt, A., Chen, M.-S., Yang, B.-Y., Tao, C., Ding, J.: Design principles for HFEv- based multivariate signature schemes. In: Iwata, T., Cheon, J.H. (eds.) ASIACRYPT 2015. LNCS, vol. 9452, pp. 311–334. Springer, Heidelberg (2015). doi:10.1007/978-3-662-48797-6_14
46. Pointcheval, D., Poupard, G.: A new NP-complete problem and public-key identification. Des. Codes Crypt. **28**(1), 5–31 (2003)
47. Pointcheval, D., Stern, J.: Security proofs for signature schemes. In: Maurer, U. (ed.) EUROCRYPT 1996. LNCS, vol. 1070, pp. 387–398. Springer, Heidelberg (1996). doi:10.1007/3-540-68339-9_33
48. Sakumoto, K., Shirai, T., Hiwatari, H.: Public-key identification schemes based on multivariate quadratic polynomials. In: Rogaway, P. (ed.) CRYPTO 2011. LNCS, vol. 6841, pp. 706–723. Springer, Heidelberg (2011). doi:10.1007/978-3-642-22792-9_40
49. Stern, J.: A new identification scheme based on syndrome decoding. In: Stinson, D.R. (ed.) CRYPTO 1993. LNCS, vol. 773, pp. 13–21. Springer, Heidelberg (1994). doi:10.1007/3-540-48329-2_2
50. Stern, J.: A new paradigm for public key identification. IEEE Trans. Inf. Theory **46**(6), 1757–1768 (1996)
51. Thomae, E.: About the Security of Multivariate Quadratic Public Key Schemes. Ph.D. thesis, Ruhr-University Bochum, Germany (2013)
52. Thomae, E., Wolf, C.: Cryptanalysis of Enhanced TTS, STS and All Its Variants, or: why cross-terms are important. In: Mitrokotsa, A., Vaudenay, S. (eds.) AFRICACRYPT 2012. LNCS, vol. 7374, pp. 188–202. Springer, Heidelberg (2012). doi:10.1007/978-3-642-31410-0_12
53. Tsujii, S., Gotaishi, M., Tadaki, K., Fujita, R.: Proposal of a signature scheme based on STS trapdoor. In: Sendrier, N. (ed.) PQCrypto 2010. LNCS, vol. 6061, pp. 201–217. Springer, Heidelberg (2010). doi:10.1007/978-3-642-12929-2_15
54. Valiant, L.G.: The complexity of enumeration and reliability problems. SIAM J. Comput. **8**(3), 410–421 (1979)
55. Weiden, P., Hülsing, A., Cabarcas, D., Buchmann, J.: Instantiating treeless signature schemes. Cryptology ePrint Archive (2013)
56. Yang, B.-Y., Chen, J.-M.: All in the XL Family: theory and practice. In: Park, C., Chee, S. (eds.) ICISC 2004. LNCS, vol. 3506, pp. 67–86. Springer, Heidelberg (2005). doi:10.1007/11496618_7
57. Yang, B.-Y., Chen, J.-M.: Building secure Tame-like multivariate public-key cryptosystems: the new TTS. In: Boyd, C., González Nieto, J.M. (eds.) ACISP 2005. LNCS, vol. 3574, pp. 518–531. Springer, Heidelberg (2005). doi:10.1007/11506157_43
58. Yang, B.-Y., Chen, J.-M., Chen, Y.-H.: TTS: high-speed signatures on a low-cost smart card. In: Joye, M., Quisquater, J.-J. (eds.) CHES 2004. LNCS, vol. 3156, pp. 371–385. Springer, Heidelberg (2004). doi:10.1007/978-3-540-28632-5_27
59. Yeh, J.Y.-C., Cheng, C.-M., Yang, B.-Y.: Operating degrees for XL vs. F_4/F_5 for generic $\mathcal{MQ}$ with number of equations linear in that of variables. In: Fischlin, M., Katzenbeisser, S. (eds.) Number Theory and Cryptography. LNCS, vol. 8260, pp. 19–33. Springer, Heidelberg (2013). doi:10.1007/978-3-642-42001-6_3

Collapse-Binding Quantum Commitments Without Random Oracles

Dominique Unruh(✉)

University of Tartu, Tartu, Estonia
unruh@ut.ee

Abstract. We construct collapse-binding commitments in the standard model. Collapse-binding commitments were introduced in (Unruh, Eurocrypt 2016) to model the computational-binding property of commitments against quantum adversaries, but only constructions in the random oracle model were known.

Furthermore, we show that collapse-binding commitments imply selected other security definitions for quantum commitments, answering an open question from (Unruh, Eurocrypt 2016).

Keywords: Quantum cryptography · Commitments · Hash functions

1 Introduction

Commitment schemes are one of the most fundamental primitives in cryptography. A commitment scheme is a two-party protocol consisting of two phases, the commit and the open phase. The goal of the commitment is to allow the sender to transmit information related to a message m during the commit phase in such a way that the recipient learns nothing about the message (hiding property). But at the same time, the sender cannot change his mind later about the message (binding property). Later, in the open phase, the sender reveals the message m and proves that this was indeed the message that he had in mind earlier (by sending some "opening information" u). Unfortunately, it was shown by [11] that the binding and hiding property of a commitment cannot both hold with statistical (i.e., information-theoretical) security even when using quantum communication. Thus, one typically requires one of them to hold only against computationally-limited adversaries. Since the privacy of data should usually extend far beyond the end of a protocol run, and since we cannot tell which technological advances may happen in that time, we may want the hiding property to hold statistically, and thus are interested in *computationally binding* commitments. Unfortunately, computationally binding commitments turn out to be a subtle issue in the quantum setting. As shown in [1], if we use the natural analogue to the classical definition of computationally binding commitments (called "classical-style binding"),[1] we get a definition that is basically meaningless (the

[1] This definition, called classical-style style binding in [16], roughly states, that it is computationally hard to find a commitment c, two messages $m \neq m'$ and corresponding valid opening informations u, u'.

J.H. Cheon and T. Takagi (Eds.): ASIACRYPT 2016, Part II, LNCS 10032, pp. 166–195, 2016.
DOI: 10.1007/978-3-662-53890-6_6

adversary can open the commitment to whatever message he wishes). [16] suggested a new definition, "collapse binding" commitments, that better captures the idea of computationally binding commitments against quantum adversaries. This definition was shown to perform well in security proofs that use rewinding.[2] (They studied classical non-interactive commitments, i.e., all exchanged messages are classical, but the adversary is quantum.)

We describe basic idea of "collapse-binding" commitments: When committing to a message m using a commitment c, it should be impossible for a quantum adversary to produce a superposition of different messages m that he can open to. Unfortunately, this requirement is too strong to achieve (at least for an statistically hiding commitment).[3] Instead, we require something slightly weaker: Any superposition of different messages m that the adversary can open to should *look like* it is a superposition of only a single message m. Formally, if the adversary produces a classical commitment c, and a superposition of openings m, u in registers M, U, the adversary should not be able to distinguish whether M is measured in the computational basis or not measured. That is, for all quantum-polynomial-time A, B, the circuits (a) and (b) in Fig. 1 are indistinguishable (assuming A only outputs superpositions that contain only valid openings).

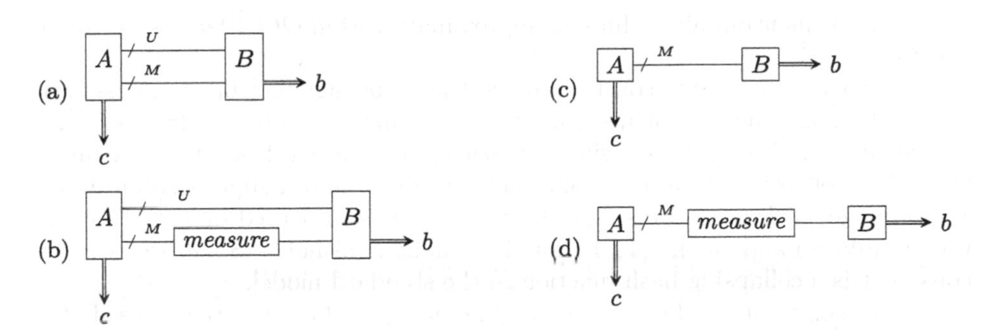

Fig. 1. For collapse-binding commitments, (a) and (b) should be indistinguishable, i.e., $\Pr[b = 1]$ negligibly close in both cases. For collapsing hash functions, (c) and (d) should be indistinguishable.

[16] showed that collapse-binding commitments avoid various problems of other definitions of computationally binding commitments in the quantum setting. In particular, they compose in parallel and are well suited for proofs that involve rewinding (e.g., when constructing zero-knowledge arguments of knowledge).

[2] We do not claim that they will work in every rewinding-based proof, but [16] showed their usefulness in the construction of arguments of knowledge. The proof of their construction did involve the quantum rewinding techniques from [14,17].

[3] The adversary can initialize a register M with the superposition of all messages, run the commit algorithm in superposition, and measure the resulting commitment c. Then M will still be in superposition between many different messages m which the adversary can open c to.

[16] further showed that in the quantum random oracle model, collapse-binding, statistically hiding commitments can be constructed. However, they left open two big questions:

- Can collapse-binding commitments be constructed in the standard model? That is, without the use of random oracles?
- One standard minimum requirement for commitments (called "sum-binding" in [16]) is that for quantum-polynomial-time A, $p_0 + p_1 \leq 1 + \textit{negligible}$ where p_b is the probability that A opens a commitment to b when he learns b only *after* the commit phase. Surprisingly, [16] left it open whether the collapse-binding property implies the sum-binding property.

First contribution: collapse-binding commitments in the standard model. We show that collapse-binding commitments exist in the standard model. More precisely, we construct a non-interactive, classical commitment in the public parameter model (i.e., we assume that some parameters are globally fixed), for arbitrarily long messages (the length of the public parameters and the commitment itself do not grow with the message length), statistically hiding, and collapse-binding. The security assumption is the existence of lossy trapdoor functions [13] with lossiness rate $> \frac{1}{2}$, or alternatively that SIVP and GapSVP are hard for quantum algorithms to approximate within $\tilde{O}(d^c)$ factors for some constant $c > 5$.

The basic idea of our construction is the following: In [16], it was shown that statistically hiding, collapse-binding commitments can be constructed from "collapsing" hash functions (using a classical construction from [6,9]). A function H is collapsing if an adversary that outputs h and a superposition M of H-preimages of h cannot distinguish whether M is measured or not. That is, the circuits (c) and (d) in Fig. 1 should be indistinguishable. So all we need to construct is a collapsing hash function in the standard model.

To do so, we use a lossy trapdoor function (we do not actually need the trapdoor part, though). A lossy function $F_s : A \to B$ is parametrized by a public parameter s. There are two kinds of parameters, which are assumed to be indistinguishable: We call s lossy if $|\mathrm{im}\, F| \ll |A|$, that is, if its image is very sparse. We call s injective if F_s is injective.

If s is injective, then it is easy to see that F_s is collapsing: There can be only one preimage of F_s on register M, so measuring M will not disturb M. But since lossy and injective s are indistinguishable, it follows that F_s is also collapsing for lossy s. Note, however, that F_s is not yet useful on its own, because its range B is much bigger than A, while we want a compressing hash functions (output smaller than input).

However, for lossy s, $|\mathrm{im}\, F_s| \ll |A|$. Let $h_r : B \to C$ be a universal hash function, indexed by r, with $|\mathrm{im}\, F_s| \ll |C| \ll |A|$. We can show that with overwhelming probability, h_r is injective on $\mathrm{im}\, F_s$, for suitable choice of C. Hence h_r is collapsing (on $\mathrm{im}\, F_s$). The composition of two collapsing functions is collapsing, thus $H_{(r,s)} := h_r \circ F_s$ is collapsing for lossy s. (Note that $\mathrm{im}\, F_s$ is not an efficiently decidable set. Fortunately, we can construct all our reductions such that we never need to decide that set.)

Thus far, we have found a collapsing $H_{(r,s)} : A \to C$ that is compressing. But we need something stronger, namely a collapsing hash function $\{0,1\}^* \to C$, i.e., applicable to arbitrary long inputs. A well-known construction (in the classical setting) is the Merkle-Damgård construction, that transforms a compressing collision-resistant function H into a collision-resistant one with domain $\{0,1\}^*$. We prove that the Merkle-Damgård construction also preserves the collapsing property. (This proof is done by a sequence of games that each measure more and more about the hashed message m, each time with a negligible probability of being noticed due to the collapsing property of H_k.) Applying this result to $H_{(r,s)}$, we get a collapsing hash function $\mathrm{MD}_{(r,s)} : \{0,1\}^* \to C$. And from this, we get collapse-binding commitments.

We present our results with concrete security bounds, and our reductions have only constant factors in the runtime, and the security level only has an $O(\mathit{message\ length})$ factor.

We stress that the security proof for the Merkle-Damgård construction has an additional benefit: It shows that existing hash function like SHA-2 [12] are collapsing, assuming that the compression function is collapsing (which in turn is suggested by the random oracle results in [16]). Since we claim that collapsing is a desirable and natural analogue to collision-resistance in the post-quantum setting, this gives evidence for the post-quantum security of SHA-2.

Second contribution: Collapse-binding implies sum-binding. In the classical setting, it relatively straightforward to show that a computationally binding *bit* commitment satisfies the (classical) sum-binding condition. Namely, assume that the adversary breaks sum-binding, i.e., $p_0 + p_1 \geq 1 + \mathit{non\text{-}negligible}$. Then one runs the adversary, lets him open the commitment as $m = 0$ (which succeeds with probability p_0), then rewinds the adversary, and lets him open the same commitment as $m = 1$ (which succeeds with probability p_1). So the probability that both runs succeed is at least $p_0 + p_1 - 1 \geq \mathit{non\text{-}negligible}$, which is a contradiction to the computational binding property.

Since collapse-binding commitments work well with rewinding, one would assume that a similar proof works using the quantum rewinding technique from [14]. Unfortunately, existing quantum rewinding techniques do not seem to work.

To show that a collapse-binding commitment is sum-binding, another proof technique is needed. The basic idea is, instead of simulating two executions of the adversary (opening $m = 0$ and opening $m = 1$) after each other, we perform the two executions in superposition, controlled by a register M, initially in state $|+\rangle$. This entangles M with the execution of the adversary and thus disturbs M. It turns out that the disturbance of M is greater if we measure which bit the adversary opens than if we do not. This allows us to distinguish between measuring and not measuring, breaking the collapse-binding property.

The same proof technique can be used to show that a collapse-binding *string* commitment satisfies the generalization of sum-binding presented in [3]. (In this case we have to use a superposition of a polynomial-number of adversary executions.)

Possibly the technique of "rewinding in superposition" used here might be a special case of a more general new quantum rewinding technique (other than [14,17]), we leave this as an open question.

On the necessity of public parameters. Our commitment scheme assumes the existence of public parameters. This raises the question whether these are necessary. We argue that it would be unlikely to be able to construct non-interactive, statistically hiding, computationally binding commitments without public parameters (not even only classically secure ones) from standard assumptions other than collision-resistant or collapsing hash functions. Namely, such a commitment can always be broken by a *non-uniform* adversary. (Because the adversary could have a commitment and two valid openings hardcoded.) Could there be a such a commitment secure only against uniform adversaries, based on some assumption X? That is, a uniform adversary breaking the commitment could be transformed into an adversary against assumption X. All cryptographic proof techniques that we are aware of would then also transform a *non-uniform* adversary breaking the commitment into a *non-uniform* adversary breaking X. Since a non-uniform adversary breaking the commitment always exists, it follows that X must be an assumption that cannot be secure against non-uniform adversaries. The only such assumptions that we are aware of are (unkeyed) collision-resistant and collapsing hash functions.[4] Thus it is unlikely that there are non-interactive, statistically hiding, computationally binding commitments without public parameters based on standard assumptions different from those two. (We are aware that the above constitutes no proof, but we consider it a strong argument.) We know how to construct such commitments from collapsing hash functions [16]. We leave it as an open problem whether such commitments can be constructed from collision-resistant hash functions.

Of course, it might be possible to have *interactive* statistically-hiding collapse-binding commitments. In fact, our construction can be easily transformed into a two-round scheme by letting the recipient choose the public parameters. This does not affect the collapsing property (because for that property we assume the recipient to be trusted), nor the statistical hiding property (because the proof of hiding did not make any assumptions about the distribution of the public parameters).

Related work. Security definitions for quantum commitments were studied in a number of works: What we call the "sum-binding" definition occurred implicitly and explicitly in different variants in [2,4,7,11]. Of these, [11] showed the impossibility of statistically satisfying that definition (thus breaking [2]). [7] gave a construction of a statistically hiding commitment based on quantum one-way permutations (their commitment sends quantum messages). [4] gives statistically secure commitments in the multi-prover setting. [3] generalizes the sum-binding definition for string commitments, arriving at a computational-binding defini-

[4] By unkeyed hash function, we mean a function that depends only on the security parameter. Such a function might be collision-resistant against uniform adversaries, but not against non-uniform ones.

tion we call CDMS-binding. (Both sum-binding and CDMS-binding are implied by collapse-binding as we show in this paper.) [5] gives another definition of computational-binding (called Q-binding in [16]; see there for a discussion of the differences to collapse-binding commitments). They also show how to construct Q-binding commitments from sigma-protocols. (Both their assumptions and their security definition seem incomparable to ours; finding out how their definition relates to ours is an interesting open problem.) [18] gives a statistical binding definition of commitments sending quantum messages and shows that statistically binding, computationally hiding commitments (sending quantum messages) can be constructed from pseudorandom permutations (and thus from quantum one-way functions, if the results from [10] hold in the quantum setting, as is claimed, e.g., in [19]). [16] gave the collapse-binding definition that we achieve in this paper; they showed how to construct statistically hiding, collapse-binding commitments in the random oracle model. [1] showed that classical-style binding does not exclude that the adversary can open the commitment to any value he chooses. [16] generalized this by showing that this even holds for certain natural constructions based on collision-resistant hash functions.

Organization. In Sect. 2, we give some mathematical preliminaries and cryptographic definitions. In Sect. 3, we recall the notions of collapse-binding commitments and collapsing hash functions, with suitable extensions to model public parameters and to allow for more refined concrete security statements. We also state some known or elementary facts about collapse-binding commitments and collapsing hash functions there. In Sect. 4 we show that the Merkle-Damgård construction allows us to get collapsing hash functions with unbounded input length from collapsing compression functions. In Sect. 5 we show how to construct collapsing hash functions from lossy functions (or from lattice assumptions). Combined with existing results this gives us statistically hiding, collapse-binding commitments for unbounded messages, interactive and non-interactive. In Sect. 6 we show that collapse-binding implies the existing definitions of sum-binding and CDMS-binding. In the full version [15] we give proofs for getting concrete security bounds. Those proofs use the same techniques as the proofs in this paper, but are somewhat less readable due to additional calculations and indices.

2 Preliminaries

Given a function $f : X \to Y$, let $\operatorname{im} f = f(X)$ denote the image of f.

Given a distribution $\mathcal{D}$ on a countable set X, let $\operatorname{supp} \mathcal{D}$ denote the support of $\mathcal{D}$, i.e., the set of all values that have non-zero probability. The statistical distance between two distributions or random variables X, Y with countable range is defined as $\frac{1}{2}\sum_a \big|\Pr[X = a] - \Pr[Y = a]\big|$.

Let λ denote the empty word.

We assume that all algorithms and parameters depend on an integer $\eta > 0$, the security parameter (unless a parameter is explicitly called "constant"). We will keep this dependence implicit (i.e., we write $A(x)$ instead of $A(\eta, x)$ for an algorithm A, and ℓ instead of $\ell(\eta)$ for an integer parameter ℓ). When calling an adversary (quantum-)polynomial-time, we mean that the runtime is polynomial in η.

We do not specify whether our adversaries are uniform or non-uniform. (I.e., whether the adversary's code may depend in an noncomputable way on the security parameter.) All our results hold both in the uniform and in the non-uniform case.

Definition 1 (Universal hash function). *A universal hash function is a function family $h_r : X \to Y$ (with $r \in R$) such that for any $x, x' \in X$ with $x \neq x'$, we have $\Pr[h_r(x) = h_r(x') : r \xleftarrow{\$} R] = 1/|Y|$.*

We define lossy functions, which are like lossy trapdoor functions [13], except that we do not require the existence of a trapdoor.

Definition 2 (Lossy functions). *A collection of (ℓ, k)-*lossy functions *consists of a PPT algorithm S_F and polynomial-time computable deterministic function F_s on $\{0,1\}^\ell$ and a message space M_k such that:*

- *Existence of injective keys: There is a distribution $\mathcal{D}_{inj}$ such that for any $s \in \operatorname{supp} \mathcal{D}_{inj}$ we have that F_s is injective. (We call such a key s injective.)*
- *Existence of lossy keys: There is a distribution $\mathcal{D}_{lossy}$ such that for any $s \in \operatorname{supp} \mathcal{D}_{lossy}$ we have that $|\operatorname{im} F_s| \leq 2^{\ell-k}$. (We call such a key s lossy.)*
- *Hard to distinguish injective from lossy: For any quantum-polynomial-time adversary A, the advantage $\big|\Pr[A(s) = 1 : s \leftarrow \mathcal{D}_{inj}] - \Pr[A(s) = 1 : s \leftarrow \mathcal{D}_{lossy}]\big|$ is negligible.*
- *Hard to distinguish lossy from S: For any quantum-polynomial-time adversary A, the advantage $\big|\Pr[A(s) = 1 : s \leftarrow \mathcal{D}_{lossy}] - \Pr[A(s) = 1 : s \leftarrow S_F]\big|$ is negligible.*

The parameter k is called the lossiness *of F_s.*

This is a weakening of the definition of lossy trapdoor functions from [13]. Our definition does not require the existence of trapdoors, and also does not require that lossy or injective keys can be efficiently sampleable. (We only require that keys that are indistinguishable from both lossy and injective keys can be sampled efficiently using S_F.)

If $k/\ell \geq K$ for some constant K, and $\ell \in \omega(\log \eta)$, we say that the lossy function has *lossiness rate K*.

Any "almost-always lossy trapdoor function" $(S_{\mathrm{ldtf}}, F_{\mathrm{ldtf}}, F^{-1}_{\mathrm{ldtf}})$ in the sense of [13] is a lossy function in the sense of Definition 2.[5]

[5] To see that, let $\mathcal{D}_{inj}$ be the distribution of the first output (i.e., discarding the trapdoor) of the injective key sampler $S_{\mathrm{ldtf}}(\eta, 1)$ conditioned on outputting an injective key. Let $\mathcal{D}_{lossy}$ be the distribution of the first output of the lossy key sampler $S_{\mathrm{ldtf}}(\eta, 0)$ conditioned on outputting a lossy key. Let S_F return the first output of $S_{\mathrm{ldtf}}(\eta, 0)$ (or $S_{\mathrm{ldtf}}(\eta, 1)$). Let $F_k(x) := F_{\mathrm{ldtf}}(k, x)$. For those choices, it is easy to see that (S_F, F_k) satisfies Definition 2.

[13] shows that for any constant $K < 1$, there is an almost-always lossy trapdoor function with lossiness rate K based on the LWE assumption for suitable parameters. [13] further shows that almost-always (ℓ, k)-lossy trapdoor functions with lossiness rate K exist if SIVP and GapSVP are hard for quantum algorithms to approximate within $\tilde{O}(d^c)$ factors, where $c = 2 + \frac{3}{2(1-K)} + \delta$ for any desired $\delta > 0$. The same thus holds for lossy functions in our sense. Furthermore, the construction from [13] has keys that are indistinguishable from uniformly random, hence we can choose S_F to simply return $s \xleftarrow{\$} \{0,1\}^{\ell_s}$ for suitable ℓ_s.[6]

3 Collapse-Binding Commitments and Collapsing Hash Functions

We reproduce the relevant results from [16] here. Note we have extended the definitions in two ways: We include a public parameter $k \leftarrow \mathbf{P}$. And we give additional equivalent definitions for a more refined treatment of the concrete security of commitments.

Commitments. A *commitment scheme* consists of three algorithms $(\mathbf{P}, \mathsf{com}, \mathsf{verify})$. $k \leftarrow \mathbf{P}$ chooses the public parameter. $(c, u) \leftarrow \mathsf{com}(k, m)$ produces a commitment c for a message m, and also returns opening information u to be revealed later. $ok \leftarrow \mathsf{verify}(k, c, m, u)$ checks whether the opening information u is correct for a given commitment c and message m (if so, $ok = 1$, else $ok = 0$).

Definition 3 (Collapse-binding). *For algorithms (A, B), consider the following games:*

$$\begin{aligned} &\mathsf{Game}_1: \; k \leftarrow \mathbf{P}, \; (S, M, U, c) \leftarrow A(k), \; m \leftarrow \mathcal{M}(M), \; b \leftarrow B(S, M, U) \\ &\mathsf{Game}_2: \; k \leftarrow \mathbf{P}, \; (S, M, U, c) \leftarrow A(k), \; b \leftarrow B(S, M, U) \end{aligned}$$

[6] This is not explicitly mentioned in [13], but can be seen as follows: [13] constructs a matrix encryption scheme whose ciphertexts are pairs of matrices $(\mathbf{A}, \mathbf{C}')$ over $\mathbb{Z}_q$ for a suitable prime q. We can see those ciphertexts as a tuple $s' \in \mathbb{Z}_q^n$ for some n. The proof of Lemma 6.2 in [13, full version] shows that the matrix encryption scheme produces ciphertexts that are indistinguishable from uniformly random $s' \xleftarrow{\$} \mathbb{Z}_q^n$.

The lattice-based lossy trapdoor function from [13] uses a ciphertext of that lossy encryption scheme as its key. Thus a key is indistinguishable from $s' \xleftarrow{\$} \mathbb{Z}_q^n$. Hence we can choose S_F to simply return a uniformly random $s' \xleftarrow{\$} \mathbb{Z}_q^n$.

To get an S_F that returns $s \xleftarrow{\$} \{0,1\}^{\ell_s}$ instead, we let S_F choose $s \in \{0, \dots, 2^\ell - 1\}^n$ and set $s_i' := s_i \bmod q$. For sufficiently large ℓ, this changes the distribution of s' only by a negligible amount. Then s can be encoded as an ℓ_s-bitstring with $\ell_s := n\ell$. (Since this way of sampling s_i' is "oblivious", i.e., given s_i' we can efficiently find randomness s_i that leads to that s_i', the security of the lossy function is not affected by outputting s_i as the key instead of s_i'.).

Here S, M, U are quantum registers. $\mathcal{M}(M)$ is a measurement of M in the computational basis.

We call an adversary (A, B) c.b.-valid *for* verify *iff for all k, $\Pr[\mathsf{verify}(k, c, m, u) = 1] = 1$ when we run $(S, M, U, c) \leftarrow A(k)$ and measure M in the computational basis as m, and U in the computational basis as u.*

A commitment scheme is collapse-binding *iff for any quantum-polynomial-time adversary (A, B) that is c.b.-valid for* verify*, $\big|\Pr[b = 1 : \mathsf{Game}_1] - \Pr[b = 1 : \mathsf{Game}_2]\big|$ is negligible.*

The only difference to the definition from [16] is that we have introduced a public parameter k chosen by $\mathbf{P}$. The proofs in [16] are not affected by this change.

For stating concrete security results (i.e., with more specific claims about the runtimes and advantages of adversaries than "polynomial-time" and "negligible"), we could simply call $\big|\Pr[b = 1 : \mathsf{Game}_1] - \Pr[b = 1 : \mathsf{Game}_2]\big|$ the advantage of the adversary (A, B). However, we find that we get stronger results if we directly specify the advantage of an adversary that attacks t commitments simultaneously.[7] This leads to the following definition of advantage. (A reader only interested in asymptotic results may ignore this definition. The main body of this paper will provide statements and proofs with respect to the simpler asymptotic definitions. Concrete security proofs are given in the full version [15].)

Definition 4 (Collapse-binding – concrete security). *For algorithms (A, B), consider the following games:*

$$\begin{aligned}
\mathsf{Game}_1: \ & k \leftarrow \mathbf{P},\ (S, M_1, \dots, M_t, U_1, \dots, U_t, c_1, \dots, c_t) \leftarrow A(k), \\
& m_1 \leftarrow \mathcal{M}(M_1),\ \dots,\ m_t \leftarrow \mathcal{M}(M_t), \\
& b \leftarrow B(S, M_1, \dots, M_t, U_1, \dots, U_t) \\
\mathsf{Game}_2: \ & k \leftarrow \mathbf{P},\ (S, M_1, \dots, M_t, U_1, \dots, U_t, c_1, \dots, c_t) \leftarrow A(k), \\
& b \leftarrow B(S, M_1, \dots, M_t, U_1, \dots, U_t)
\end{aligned}$$

Here $S, M_1, \dots, M_t, U_1, \dots, U_t$ are quantum registers. $\mathcal{M}(M_i)$ is a measurement of M_i in the computational basis.

We call an adversary (A, B) t-c.b.-valid *for* verify *iff for all k, $\Pr[\forall i.\ \mathsf{verify}(k, c_i, m_i, u_i) = 1] = 1$ when we run $(S, M_1, \dots, M_t, U_1, \dots, U_t, c_1, \dots, c_t) \leftarrow A(k)$ and measure all M_i in the computational basis as m_i, and all U_i in the computational basis as u_i.*

For any adversary (A, B), we call $\big|\Pr[b = 1 : \mathsf{Game}_1] - \Pr[b = 1 : \mathsf{Game}_2]\big|$ the collapse-binding-advantage *of (A, B) against $(\mathbf{P}, \mathsf{com}, \mathsf{verify})$.*

[7] We could simply analyze all schemes for adversaries that attack a single commitment at a time, and then invoke the parallel composition theorem from [16] to get the advantage when attacking t commitments. That theorem will then introduce a factor t in the advantage. ([16] states the theorem without concrete security bounds, but they are easily extracted from the proof.) In contrast, a direct analysis for t commitments may give better bounds, since the advantages we get in this paper do not depend on t.

Lemma 5. *A commitment scheme* $(\mathbf{P}, \mathsf{com}, \mathsf{verify})$ *is collapse-binding iff for any polynomially-bounded* t*, and any quantum-polynomial-time adversary* (A, B) *that is* t*-c.b.-valid for* verify*, the collapse-binding-advantage of* (A, B) *against* $(\mathbf{P}, \mathsf{com}, \mathsf{verify})$ *is negligible.*

This follows from the parallel composition theorem from [16].

In [16], two different definitions of collapse-binding were given. The second definition does not require an adversary to be valid (i.e., to output only valid openings) but instead measures whether the adversary's openings are valid. We restate the equivalence here in the public parameter setting, the proof is essentially unchanged.

Lemma 6. (Collapse-binding, alternative characterization). *For a commitment scheme* $(\mathbf{P}, \mathsf{com}, \mathsf{verify})$*, and for algorithms* (A, B)*, consider the following games:*

$$\begin{aligned}
&\mathsf{Game}_1: \; k \leftarrow \mathbf{P}, (S, M, U, c) \leftarrow A(k), ok \leftarrow V_c(M, U), x \leftarrow \mathcal{M}_{ok}(M), b \leftarrow B(S, M, U) \\
&\mathsf{Game}_2: \; k \leftarrow \mathbf{P}, (S, M, U, c) \leftarrow A(k), ok \leftarrow V_c(M, U), \qquad\qquad\qquad\quad b \leftarrow B(S, M, U)
\end{aligned}$$

Here V_c *is a measurement whether* M, U *contains a valid opening. Formally* V_c *is defined through the projector* $\sum_{\substack{m,u \\ \mathsf{verify}(k,c,m,u)=1}} |m\rangle\langle m| \otimes |u\rangle\langle u|$. $\mathcal{M}_{ok}$ *is a measurement of* M *in the computational basis if* $ok = 1$*, and does nothing if* $ok = 0$ *(i.e., it sets* $m := \bot$ *and does not touch the register* M*).*

$(\mathbf{P}, \mathsf{com}, \mathsf{verify})$ *is collapse-binding iff for all polynomial-time adversaries* (A, B)*,* $\big|\Pr[b = 1 : \mathsf{Game}_1] - \Pr[b = 1 : \mathsf{Game}_2]\big|$ *is negligible.*

Hash functions. A *hash function* is a pair $(\mathbf{P}, H_k)$ of a parameter sampler $\mathbf{P}$ and a function $H_k : X \to Y$ for some range X and domain Y. H_k is parametric in the public parameter $k \leftarrow \mathbf{P}$. (Typically, Y consists of fixed length bitstrings, and X consists of fixed length bitstrings or $\{0,1\}^*$.)

Definition 7 (Collapsing). *For algorithms* A*,* B*, consider the following games:*

$$\begin{aligned}
&\mathsf{Game}_1: \quad k \leftarrow \mathbf{P}, \; (S, M, h) \leftarrow A(k), \; m \leftarrow \mathcal{M}(M), \; b \leftarrow B(S, M) \\
&\mathsf{Game}_2: \quad k \leftarrow \mathbf{P}, \; (S, M, h) \leftarrow A(k), \qquad\qquad\qquad\;\; b \leftarrow B(S, M)
\end{aligned}$$

Here S, M *are quantum registers.* $\mathcal{M}(M)$ *is a measurement of* M *in the computational basis.*

For a family of sets $\mathbf{M}_k$*, we call an adversary* (A, B) valid on $\mathbf{M}_k$ *for* H_k *iff for all* k*,* $\Pr[H_k(m) = c \wedge m \in \mathbf{M}_k] = 1$ *when we run* $(S, M, h) \leftarrow A(k)$ *and measure* M *in the computational basis as* m*. If we omit "on* $\mathbf{M}_k$*", we assume* $\mathbf{M}_k$ *to be the domain of* H_k*.*

A function H *is* collapsing *(on* $\mathbf{M}_k$*) iff for any quantum-polynomial-time adversary* (A, B) *that is valid for* H_k *(on* $\mathbf{M}_k$*),* $\big|\Pr[b = 1 : \mathsf{Game}_1] - \Pr[b = 1 : \mathsf{Game}_2]\big|$ *is negligible.*

In contrast to [16] we have added the public parameter k. Furthermore, we have extended the definition to allow to specify the set $\mathbf{M}_k$ of messages the adversary is allowed to use. This extra expressiveness will be needed for stating some intermediate results.

Analogously to case of commitments, we give a definition of advantage for a t-session adversary to get more precise results.

Definition 8 (Collapsing – concrete security). *For algorithms A, B, and an integer t, consider the following games:*

$$\begin{aligned}
\mathsf{Game}_1: \quad & k \leftarrow \mathbf{P},\ (S, M_1, \dots, M_t, h_1, \dots, h_t) \leftarrow A(k), \\
& m_1 \leftarrow \mathcal{M}(M_1),\ \dots,\ m_t \leftarrow \mathcal{M}(M_t), \\
& b \leftarrow B(S, M_1, \dots, M_t) \\
\mathsf{Game}_2: \quad & k \leftarrow \mathbf{P},\ (S, M_1, \dots, M_t, h_1, \dots, h_t) \leftarrow A(k), \\
& b \leftarrow B(S, M_1, \dots, M_t)
\end{aligned}$$

Here $S, M_1, \dots, M_t$ are quantum registers. $\mathcal{M}(M_i)$ is a measurement of M_i in the computational basis.

For a family of sets $\mathbf{M}_k$, we call an adversary (A, B) t-valid on $\mathbf{M}_k$ for H_k *iff for all k, $\Pr[\forall i.\ H_k(m_i) = c_i \wedge m_i \in \mathbf{M}_k] = 1$ when we run $(S, M_1, \dots, M_t, h_1, \dots, h_t) \leftarrow A(k)$ and measure all M_i in the computational basis as m_i. If we omit "on $\mathbf{M}_k$", we assume $\mathbf{M}_k$ to be the domain of H_k.*

We call $adv := \big|\Pr[b = 1 : \mathsf{Game}_1] - \Pr[b = 1 : \mathsf{Game}_2]\big|$ the collapsing-advantage *of (A, B) against $(\mathbf{P}, H_k)$.*

Lemma 9. *A hash function $(\mathbf{P}, H_k)$ is collapsing (on $\mathbf{M}_k$) iff for any polynomially-bounded t, and any quantum-polynomial-time adversary (A, B) that is t-valid for H_k (on $\mathbf{M}_k$), the collapsing-advantage of (A, B) against $(\mathbf{P}, H_k)$ is negligible.*

This follows from the parallel composition theorem for hash functions from [16].

Constructions of commitments. In [16] it was shown that the statistically hiding commitment from Halevi and Micali [9] (which is almost identical to the independently and earlier discovered commitment by Damgård, Pedersen, and Pfitzmann [6]) is collapse-binding, assuming a collapsing hash function. We restate their results with respect to public parameters, the proofs are essentially unchanged.

Definition 10 (Unbounded Halevi-Micali commitment [9]). *Let $(\mathbf{P}, H_k)$ with $H_k : \{0,1\}^* \to \{0,1\}^\ell$ be a hash function. Let $L := 6\ell + 4$. Let $h_r : \{0,1\}^L \to \{0,1\}^\ell$ with $r \in \{0,1\}^{\ell_r}$ be an universal hash function.*

We define the unbounded Halevi-Micali commitment $(\mathbf{P}, \mathsf{com}_{HMu}, \mathsf{verify}_{HMu})$ *as:*

– $\mathbf{P}$ *is the same parameter sampler as in $(\mathbf{P}, H_k)$.*

- $\mathsf{com}_{HMu}(k,m)$: *Pick* $r \in \{0,1\}^{\ell_r}$ *and* $u \in \{0,1\}^L$ *uniformly at random, conditioned on* $h_r(u) = H_k(m)$.[8] *Compute* $h := H_k(u)$. *Let* $c := (h,r)$. *Return commitment* c *and opening information* u.
- $\mathsf{verify}_{HMu}(k,c,m,u)$ *with* $c=(h,r)$: *Check whether* $h_r(u) = H_k(m)$ *and* $h = H_k(u)$. *If so, return* 1.

We define the statistical hiding property in the public parameter model. We use an adaptive definition where the committed message may depend on the public parameter.

Definition 11 (Statistically hiding). *Fix a commitment* $(\mathbf{P}, \mathsf{com}, \mathsf{verify})$ *and an adversary* (A,B). *Let*

$$p_b := \Pr[b'=1 : k \leftarrow \mathbf{P},\ (S,m_0,m_1) \leftarrow A(k),\ (c,u) \leftarrow \mathsf{com}(k,m_b),\ b' \leftarrow B(S,c)].$$

We call $|p_0 - p_1|$ *the* hiding-advantage *of* (A,B). *We call* $(\mathbf{P}, \mathsf{com}, \mathsf{verify})$ statistically hiding *iff for any (possibly unbounded)* (A,B), *the hiding-advantage is negligible.*

Theorem 12 (Security of the unbounded Halevi-Micali commitment). $(\mathbf{P}, \mathsf{com}_{HMu}, \mathsf{verify}_{HMu})$ *is statistically hiding and collapse-binding.*

Miscellaneous facts. These simple facts will be useful throughout the paper.

Lemma 13. *Let* $\mathbf{M}_k$ *be a family of sets. Assume that* $\Pr[H_k$ *is not injective on* $\mathbf{M}_k : k \xleftarrow{\$} \mathbf{P}]$ *is negligible. Then* $(\mathbf{P}, H_k)$ *is collapsing on* $\mathbf{M}_k$.

Lemma 14. *Fix hash functions* $(\mathbf{P}, f_k)$ *and* $(\mathbf{P}, g_k)$ *with the same* $\mathbf{P}$ *and with polynomial-time computable* f_k. *If* $(\mathbf{P}, f_k)$ *is collapsing and* $(\mathbf{P}, g_k)$ *is collapsing on* $\operatorname{im} f_k$, *then* $(\mathbf{P}, g_k \circ f_k)$ *is collapsing.*

Lemma 15. *If* $\mathbf{P}_1$ *and* $\mathbf{P}_2$ *are computationally indistinguishable, and* $(\mathbf{P}_1, H_k)$ *is collapsing, then* $(\mathbf{P}_2, H_k)$ *is collapsing.*

4 Security of Merkle-Damgård Hashes

For this section, fix a hash function $(\mathbf{P}, H_k)$ with $H_k : \{0,1\}^{\ell_{in}} \to \{0,1\}^{\ell_{out}}$ and $\ell_{in} > \ell_{out}$. Let $\ell_{block} := \ell_{in} - \ell_{out}$. Fix some bitstring $iv \in \{0,1\}^{\ell_{out}}$ (may depend on the security parameter). Fix a message space $\mathbf{M}$ with $|\mathbf{M}| \geq 2$ (e.g., $\mathbf{M} = \{0,1\}^*$). Fix a function $pad : \mathbf{M} \to (\{0,1\}^{\ell_{block}})^*$.

Definition 16 (Iterated hash). *We define the* iterated hash $\mathrm{IH}_k : (\{0,1\}^{\ell_{block}})^* \to \{0,1\}^{\ell_{out}}$ *as* $\mathrm{IH}_k(\lambda) := iv$ *for the empty word* λ *and* $\mathrm{IH}_k(\mathbf{m}\|m') := H_k(\mathrm{IH}_k(\mathbf{m})\|m')$ *for* $\mathbf{m} \in (\{0,1\}^{\ell_{block}})^*$ *and* $m' \in \{0,1\}^{\ell_{block}}$.

[8] In general, this can be computationally hard. However, should h_r be a universal hash function where this is hard, one can replace h_r by $h'_{(r,t)}$ defined as $h'_{(r,t)}(x) := t \oplus h_r(x)$. This function is still a universal hash function, and sampling r,t,u is easy.

Definition 17 (Merkle-Damgård). *We call pad a* Merkle-Damgård padding *iff pad is injective and for any* $x, y \in \mathbf{M}$ *with* $x \neq y$, *we have that* $pad(x)$ *is not a suffix of* $pad(y)$ *(in other words,* $pad(\mathbf{M})$ *is a suffix code).*[9]

We define the Merkle-Damgård construction $\mathrm{MD}_k : \mathbf{M} \to \{0,1\}^{\ell_{out}}$ *by* $\mathrm{MD}_k := \mathrm{IH}_k \circ pad$.

Note that IH_k and MD_k depend on the choice of H_k, iv, and pad, but we leave this dependence implicit for brevity.

Lemma 18. (Security of iterated hash). *Let* $\tilde{\mathbf{M}} \subseteq (\{0,1\}^{\ell_{block}})^*$ *be a suffix code with* $|\tilde{\mathbf{M}}| \geq 2$. *If* $(\mathbf{P}, H_k)$ *is a polynomial-time computable collapsing hash function, then* $(\mathbf{P}, \mathrm{IH}_k)$ *is collapsing on* $\tilde{\mathbf{M}}$.

We sketch the idea of the proof: What we have to show is that, if the adversary classically outputs $\mathrm{IH}_k(\mathbf{m})$, we can measure $\mathbf{m}$ on register M without the adversary noticing. We show this by successively measuring more and more information about the message $\mathbf{m}$ on M, each time noting that the additional measurement is not noticed by the adversary. First, measuring $\mathrm{IH}_k(\mathbf{m})$ does not disturb M because $\mathrm{IH}_k(\mathbf{m})$ is already known. Note that $\mathrm{IH}_k(\mathbf{m}) = H_k(\mathrm{IH}_k(\mathbf{m}')\|m)$ for $\mathbf{m} =: \mathbf{m}'\|m$. Thus, we have measured the image of $\mathrm{IH}_k(\mathbf{m}')\|m$ under H_k. Since H_k is collapsing, we know that, once we have measured the hash of a value, we can also measure that value itself without being noticed. Thus we can measure $\mathrm{IH}_k(\mathbf{m}')\|m$ (this value will be called $\mathrm{step}_0(\mathbf{m})$ in the full proof). Now we use the same argument again: $\mathrm{IH}_k(\mathbf{m}') = H_k(\mathrm{IH}_k(\mathbf{m}'')\|m')$ for $\mathbf{m}' =: \mathbf{m}''\|m'$. Since we know classically $\mathrm{IH}_k(\mathbf{m}')$, we can measure $\mathrm{IH}_k(\mathbf{m}'')\|m'$ (this value will be called $\mathrm{step}_1(\mathbf{m})$). Now we already have measured the two last blocks $m'\|m$ of $\mathbf{m}$ without being noticed. We can continue this way, until we have all of $\mathbf{m}$. Since in each step, the adversary did not notice the measurement, he will not notice if we measure all of $\mathbf{m}$.

There is one hidden problem in the above argument: We claimed that given $\mathrm{IH}_k(\mathbf{m}')$, we have that $\mathrm{IH}_k(\mathbf{m}') = H_k(\mathrm{IH}_k(\mathbf{m}'')\|m')$. This is only correct if $\mathbf{m}'$ is not empty! So, the above measurement procedure will implicitly measure whether $\mathbf{m}'$ is empty (and similarly for the values $\mathbf{m}''$ etc. that are measured afterwards). Such a measurement might disturb the state. Here the assumption comes in that $\tilde{\mathbf{M}}$ is a suffix code. Namely, since we know m such that $\mathbf{m} = \mathbf{m}'\|m$, we can tell whether $m \in \tilde{\mathbf{M}}$ (then $\mathbf{m}'$ must be empty) or $m \notin \tilde{\mathbf{M}}$ (then $\mathbf{m}'$ cannot be empty). Thus we already know whether $\mathbf{m}'$ is empty, and measuring this information will not disturb the state. Similarly, we deduce from $m'\|m$ whether $\mathbf{m}''$ is empty, etc.

We now give the formal proof:

[9] Commonly, stronger conditions are placed on pad, see, e.g., [8, Def. 8.7]. However, "suffix-code" and "injective" turns out to be sufficient. For example, the padding using in SHA-256 [12] is a Merkle-Damgård padding for $\mathbf{M} = \{0,1\}^{\leq 2^{64}-1}$ according to our definition.

Proof of Lemma 18. Assume a polynomial-time adversary (A, B) that is valid for IH_k on $\mathbf{M}$. Let Game_1 and Game_2 be the games from Definition 7 for adversary (A, B). Let

$$\varepsilon := \big|\Pr[b = 1 : \mathsf{Game}_1] - \Pr[b = 1 : \mathsf{Game}_2]\big|. \tag{1}$$

We will need to show that ε is negligible.

We have $\lambda \notin \tilde{\mathbf{M}}$. ($\lambda$ denotes the empty word.) Otherwise, we would have $\tilde{\mathbf{M}} = \{\lambda\}$ since $\tilde{\mathbf{M}}$ is a suffix code, which contradicts $|\tilde{\mathbf{M}}| \geq 2$.

For a multi-block message $\mathbf{m} \in (\{0,1\}^{\ell_{block}})^*$, let $|\mathbf{m}|$ denote the number of ℓ_{block}-bit blocks in $\mathbf{m}$. (I.e., $|\mathbf{m}|$ is the bitlength of $\mathbf{m}$ divided by ℓ_{block}.) Let $\mathbf{m}_i$ denote the i-th block of $\mathbf{m}$, and let $\mathbf{m}_{-i}$ denote the i-th block from the end (i.e., $\mathbf{m}_{-i} = \mathbf{m}_{|m|-i+1}$). Let $\mathbf{m}_{\geq -i}$ denote all the blocks in $\mathbf{m}$ starting from $\mathbf{m}_{-i}$ (i.e., $\mathbf{m}_{\geq -i}$ consists of the last i blocks of $\mathbf{m}$). Let $\mathbf{m}_{<-i}$ denote the blocks before $\mathbf{m}_{-i}$. (I.e., $\mathbf{m} = \mathbf{m}_{<-i} \,\|\, \mathbf{m}_{\geq -i}$ for $i \leq |\mathbf{m}|$.)

Let B be a polynomial upper bound on the number of blocks in the message $\mathbf{m}$ output by A on register M.

For a function f, let $\mathcal{M}_f(M)$ denote a measurement that, given a register M that contains values $|\mathbf{m}\rangle$ in superposition, measures $f(\mathbf{m})$, but without measuring more information than that. Formally, $\mathcal{M}_f$ is a projective measurement consisting of projectors P_y ($y \in \operatorname{im} f$) with $P_y = \sum_{\mathbf{m}:f(\mathbf{m})=y} |\mathbf{m}\rangle\langle\mathbf{m}|$.

For $\mathbf{m} \in \tilde{\mathbf{M}}$, we define

$$\mathrm{partial}_i(\mathbf{m}) := \begin{cases} (\bot, \mathbf{m}) & (\text{if } |\mathbf{m}| \leq i) \\ (\mathrm{IH}_k(\mathbf{m}_{<-i}), \mathbf{m}_{\geq -i}) & (\text{if } |\mathbf{m}| > i) \end{cases}$$

(The function $\mathrm{partial}_i$ also depends on k, but we leave that dependence implicit.) Intuitively, $\mathrm{partial}_i(\mathbf{m})$ represents a partial evaluation of $\mathrm{IH}_k(\mathbf{m})$, with the last i blocks not yet processed.

Note that $\mathrm{partial}_i(\mathbf{m})$ always contains enough information to compute $\mathrm{IH}_k(\mathbf{m})$. And the larger i is, the more about $\mathbf{m}$ is revealed. In fact, learning $\mathrm{partial}_0(\mathbf{m})$ is equivalent to learning $\mathrm{IH}_k(\mathbf{m})$, and learning $\mathrm{partial}_B(\mathbf{m})$ is equivalent to learning m as the following easy to verify facts show:

Fact 1 $\mathrm{partial}_0(\mathbf{m}) = (\mathrm{IH}_k(\mathbf{m}), \lambda)$ *for all* $\mathbf{m} \in \tilde{\mathbf{M}}$.

Fact 2 $\mathrm{partial}_B(\mathbf{m}) = (\bot, \mathbf{m})$ *for all* $\mathbf{m} \in \tilde{\mathbf{M}}$ *with* $|\mathbf{m}| \leq B$.

We will need one additional auxiliary function step_i, defined by $\mathrm{step}_i(\mathbf{m}) := \mathrm{IH}_k(\mathbf{m}_{<-(i+1)}) \| \mathbf{m}_{-(i+1)}$ for $|\mathbf{m}| \geq i+1$. (And $\mathrm{step}_i(\mathbf{m}) := \bot$ if $|\mathbf{m}| \leq i$.) Intuitively, $\mathrm{step}_i(\mathbf{m})$ is the input to last call of H_k when computing $\mathrm{partial}_i(\mathbf{m})$. The following facts are again easy to verify using the definition of $\mathrm{partial}_i$, step_i, and IH_k:

Fact 3 *If* $\mathrm{partial}_i(\mathbf{m}) = (h, s)$ *and* $h \neq \bot$, *then* $H_k(\mathrm{step}_i(\mathbf{m})) = h$.

Fact 4 *From* $(\mathrm{partial}_i(\mathbf{m}), \mathrm{step}_i(\mathbf{m}))$ *one can compute* $\mathrm{partial}_{i+1}(\mathbf{m})$ *and vice versa. Formally: there are functions* f, g *such that for all* $m \in \tilde{\mathbf{M}}$, $f(\mathrm{partial}_i(\mathbf{m}), \mathrm{step}_i(\mathbf{m})) = \mathrm{partial}_{i+1}(\mathbf{m})$ *and* $g(\mathrm{partial}_{i+1}(\mathbf{m})) = (\mathrm{partial}_i(\mathbf{m}), \mathrm{step}_i(\mathbf{m}))$.

In a sense, $\text{partial}_i(\mathbf{m})$ interpolates between the knowledge of only $\text{IH}_k(\mathbf{m})$ (case $i = 0$), and full knowledge of $\mathbf{m}$ (case $i = B$). (Cf. Facts 1, 2.) We make this more formal by defining the following hybrid game for $i = 0, \dots, B$:

$$\begin{aligned}\mathsf{Game}_i^{hyb}: \quad & k \leftarrow \mathbf{P},\ (S, M, h) \leftarrow A(k),\\ & (h', s) \leftarrow \mathcal{M}_{\text{partial}_i}(M),\\ & b \leftarrow B(S, M).\end{aligned}$$

(Here $\mathcal{M}_{\text{partial}_i}$ is $\mathcal{M}_f$ as defined above with $f := \text{partial}_i$.)

Consider Game_0^{hyb}. By assumption, (A, B) is valid for IH_k on $\tilde{\mathbf{M}}$, so we have that the register M contains superpositions of states $|\mathbf{m}\rangle$ with $\text{IH}_k(\mathbf{m}) = h_j$ and $\mathbf{m} \in \tilde{\mathbf{M}}$. By Fact 1, this implies that the measurement $\mathcal{M}_{\text{partial}_0}(M)$ will always yield the outcome $(h', s) = (h, \lambda)$. Hence the measurement $\mathcal{M}_{\text{partial}_0}(M)$ has a deterministic outcome. Thus, the probability of $b = 1$ in Game_0^{hyb} does not change if we omit the measurements $y \leftarrow \mathcal{M}_{\text{partial}_i}(M)$. Thus

$$\Pr[b = 1 : \mathsf{Game}_0^{hyb}] = \Pr[b = 1 : \mathsf{Game}_2]. \tag{2}$$

Consider Game_B^{hyb}. By assumption, A outputs only states on M which are superpositions of $|\mathbf{m}\rangle$ with $\mathbf{m} \in \tilde{\mathbf{M}}$ and $|\mathbf{m}| \leq B$. Thus, by Fact 2, $(h', s) \leftarrow \mathcal{M}_{\text{partial}_B}(M)$ is a complete measurement in the computational basis. Hence

$$\Pr[b = 1 : \mathsf{Game}_B^{hyb}] = \Pr[b = 1 : \mathsf{Game}_1]. \tag{3}$$

From (1, 2, 3), we get

$$\big|\Pr[b = 1 : \mathsf{Game}_0^{hyb}] - \Pr[b = 1 : \mathsf{Game}_B^{hyb}]\big| = \varepsilon. \tag{4}$$

For $i = 0, \dots, B$ we now define an adversary (A_i^*, B^*) against H_k. Algorithm $A_i^*(k)$ runs:

- $(S^*, M^*, h^*) \leftarrow A(k)$.
- $(h', s) \leftarrow \mathcal{M}_{\text{partial}_i}(M^*)$.
- Initialize M with $|0^{\ell_{in}}\rangle$.
- If $h' \neq \bot$:
  - Apply U_{step_i} to M^*, M.
  - $h := h'$.
- If $h' = \bot$:
  - Let $h := H_k(0^{\ell_{in}})$.
- Let $S := S^*, M^*, h', i$. (That is, all those registers and classical values are combined into a single register S.)
- Return (S, M, h).

Here U_{step_i} refers to the unitary transformation $|x\rangle|y\rangle \mapsto |x\rangle|y \oplus \text{step}_i(x)\rangle$. See the left dashed box in Fig. 2 for a circuit-representation of A_i^*.

Algorithm $B^*(S, M)$ runs:

- Let $S^*, M^*, h', i := S$.
- If $h' \neq \bot$: apply U_{step_i} to M^*, M.
- Run $b \leftarrow B(S^*, M^*)$.
- Return b.

See the left dashed box in Fig. 2 for a circuit-representation of B^*.

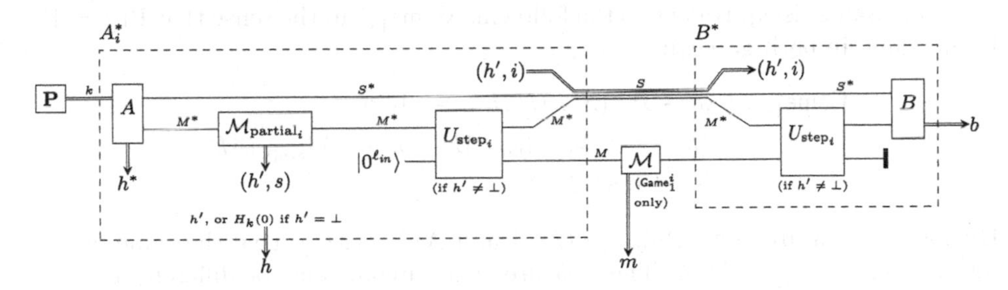

Fig. 2. The adversary (A_i^*, B^*) in games Game_1^i and Game_2^i. Depicted is Game_1^i. Game_2^i is derived by omitting the measurement $\mathcal{M}$ in the middle.

Claim 1. (A_i^*, B^*) is valid.

We show this claim: After the measurement $(h', s) \leftarrow \mathcal{M}_{\mathrm{partial}_i}(M^*)$, we have that M^* contains a superposition of $|\mathbf{m}\rangle$ with $\mathrm{partial}_i(\mathbf{m}) = (h', s)$. If $h' = \bot$, then A_i^* initializes M with $|0^{\ell_{in}}\rangle$ and sets $h := H_k(0^{\ell_{in}})$. Thus in this case, M trivially contains a superposition of $|m\rangle$ with $H_k(m) = h$. If $h' \neq \bot$, then by Fact 3, M^* contains a superposition of $|\mathbf{m}\rangle$ with $H_k(\mathrm{step}_i(\mathbf{m})) = h' = h$. Then A^* initializes M with $|0^{\ell_{in}}\rangle$ and applies U_{step_i} to M^*, M. Thus after that, M is in a superposition of $|m\rangle$ with $H_k(m) = h_j$. Concluding, in both cases M is in a superposition of $|m\rangle$ with $H_k(m) = h$, thus (A_i^*, B^*) is valid and the claim follows.

Let Game_1^i denote Game_1 from Definition 7, but with adversary (A_i^*, B^*) and hash function $(\mathbf{P}, H_k)$. Analogously Game_2^i. Figure 2 depicts both games.

Claim 2. $\Pr[b = 1 : \mathsf{Game}_2^i] = \Pr[b = 1 : \mathsf{Game}_i^{hyb}]$.

We show this claim: In Game_2^i, no measurement occurs between the invocation of U_{step_i} by A_i^* and the invocation of U_{step_i} by B^*. (Cf. Fig. 2.) Since U_{step_i} is an involution, those two invocations cancel out. Thus only the invocations of $\mathbf{P}$, A, $\mathcal{M}_{\mathrm{partial}_i}$, and B remain. This is exactly Game_i^{hyb}. This shows the claim.

Claim 3. $\Pr[b = 1 : \mathsf{Game}_1^i] = \Pr[b = 1 : \mathsf{Game}_{i+1}^{hyb}]$.

We show the claim: Note that in Game_1^i, after the measurement $\mathcal{M}_{\mathrm{partial}_i}$, on the registers M^*, M, we have the following sequence of operations if $h' \neq \bot$:

M is initialized with $|0^{\ell_{in}}\rangle$. U_{step_i} is applied to M^*, M. M is measured in the computational basis (outcome m). U_{step_i} is applied to M^*, M. M is discarded.

This is equivalent to just executing $m \leftarrow \mathcal{M}_{\mathrm{step}_i}(M^*)$.

Furthermore, if $h = \bot$, then the sequence of operations is simply: Initialize M with $|0^{\ell_{in}}\rangle$. Measure M. Discard M. This is equivalent to doing nothing. And doing nothing is equivalent to $m \leftarrow \mathcal{M}_{\mathrm{step}_i}(M^*)$ in case $h' = \bot$. (Because in that case, M^* is in a superposition of $|\mathbf{m}\rangle$ with $|\mathbf{m}| \leq i$, and thus $\mathrm{step}_i(\mathbf{m}) = \bot$, and hence the outcome of $\mathcal{M}_{\mathrm{step}_i}$ is deterministic.)

Thus Game_1^i is equivalent to the following Game_{1*}^i (in the sense that $\Pr[b = 1]$ is the same in both games):

$$\begin{aligned}\mathsf{Game}_{1*}^i: \quad & k \leftarrow \mathbf{P},\ (S^*, M^*, h^*) \leftarrow A(k),\\ & (h', s) \leftarrow \mathcal{M}_{\mathrm{partial}_i}(M^*),\ m \leftarrow \mathcal{M}_{\mathrm{step}_i}(M^*),\\ & b \leftarrow B(S^*, M^*).\end{aligned}$$

By Fact 4, measurements $\mathcal{M}_{\mathrm{partial}_i}(M^*)$ and $\mathcal{M}_{\mathrm{step}_i}(M^*)$ have the same effect on M^* as $\mathcal{M}_{\mathrm{partial}_{i+1}}(M^*)$. (The measurement outcome may be different, but we do not use the measurement outcome in our games.) Thus Game_{1*}^i is equivalent to Game_{1**}^i (in the sense that $\Pr[b = 1]$ is the same in both games):

$$\begin{aligned}\mathsf{Game}_{1**}^i: \quad & k \leftarrow \mathbf{P},\ (S^*, M^*, h^*) \leftarrow A(k),\\ & (h', s) \leftarrow \mathcal{M}_{\mathrm{partial}_{i+1}}(M^*),\\ & b \leftarrow B(S^*, M^*).\end{aligned}$$

But Game_{1**}^i is the same as $\mathsf{Game}_{i+1}^{hyb}$, except that S, M, h are renamed to S^*, M^*, h^*. Hence $\Pr[b = 1]$ is the same in Game_1^i and $\mathsf{Game}_{i+1}^{hyb}$, the claim follows.

Let A^* pick $i \xleftarrow{\$} \{0, \dots, B-1\}$ and then run A_i^*. From Claim 1, it follows that (A^*, B^*) is valid, too. Let Game_1^* denote Game_1 from Definition 7, but with adversary (A^*, B^*) and hash function $(\mathbf{P}, H_k)$. Analogously Game_2^*.

Since $(\mathbf{P}, H_k)$ is collapsing by assumption, and (A^*, B^*) is valid and polynomial-time, we have that $\varepsilon^* := \bigl|\Pr[b = 1 : \mathsf{Game}_1^*] - \Pr[b = 1 : \mathsf{Game}_2^*]\bigr|$ is negligible.

Then we have:

$$\begin{aligned}\varepsilon^* &= \bigl|\Pr[b = 1 : \mathsf{Game}_1^*] - \Pr[b = 1 : \mathsf{Game}_2^*]\bigr|\\ &= \frac{1}{B}\Bigl|\sum_{i=0}^{B-1} \Pr[b = 1 : \mathsf{Game}_1^i] - \sum_{i=0}^{B-1} \Pr[b = 1 : \mathsf{Game}_2^i]\Bigr|\\ &\overset{(*)}{=} \frac{1}{B}\Bigl|\sum_{i=0}^{B-1} \Pr[b = 1 : \mathsf{Game}_{i+1}^{hyb}] - \sum_{i=0}^{B-1} \Pr[b = 1 : \mathsf{Game}_i^{hyb}]\Bigr|\\ &= \frac{1}{B}\Bigl|\Pr[b = 1 : \mathsf{Game}_B^{hyb}] - \Pr[b = 1 : \mathsf{Game}_0^{hyb}]\Bigr| \overset{(4)}{=} \frac{\varepsilon}{B}.\end{aligned}$$

Here $(*)$ follows from Claims 2 and 3.

Since ε^* is negligible, $\varepsilon = B\varepsilon^*$ is negligible. □

Theorem 19 (Security of Merkle-Damgård). *Assume that pad is a polynomial-time computable Merkle-Damgård padding. If* $(\mathbf{P}, H_k)$ *is a polynomial-time computable collapsing hash function,* $(\mathbf{P}, \mathrm{MD}_k)$ *is collapsing.*

A concrete security statement is given in Theorem 20.

Proof. Since *pad* is a Merkle-Damgård padding, we have that *pad* is injective and $\mathrm{im}\, pad$ is a suffix code. Since the domain of *pad* is $\mathbf{M}$, and $|\mathbf{M}| \geq 2$ by assumption, $|\mathrm{im}\, pad| \geq 2$. Thus by Lemma 18, $(\mathbf{P}, \mathrm{IH}_k)$ is collapsing on $\mathrm{im}\, pad$.

Since *pad* is injective, $(\mathbf{P}, pad)$ is collapsing by Lemma 13.

Since $\mathrm{MD}_k = \mathrm{IH}_k \circ pad$, by Lemma 14, $(\mathbf{P}, \mathrm{MD}_k)$ is collapsing. □

Concluding, we also state Theorem 19 in its concrete security variant. Let τ_H denote an upper bound on the time needed for evaluating H_k. Let $\tau_{pad}(\ell)$ denote an upper bound on the time for computing $pad(\mathbf{m})$ for $|\mathbf{m}| \leq \ell$. Let $\ell_{pad}(\ell)$ denote an upper bound on $|pad(\mathbf{m})|$ for $|\mathbf{m}| \leq \ell$. ($|\cdot|$ refers to the length in bits.)

Theorem 20 (Concrete security of Merkle-Damgård). *Assume that pad is a Merkle-Damgård padding.*

Let (A, B) *be a* τ*-time adversary,* t*-valid for* MD_k *on* $\mathbf{M}$*, with collapsing-advantage* ε *against* $(\mathbf{P}, \mathrm{MD}_k)$*.*

Then there is a $(\tau + O(t\tau_{pad}(\ell_A) + t\ell_{pad}(\ell_A)\tau_H/\ell_{block}))$*-time adversary* (A^*, B^*)*,* t*-valid for* H_k*, with collapsing-advantage* $\geq \varepsilon\ell_{block}/\ell_{pad}(\ell_A)$ *against* $(\mathbf{P}, H_k)$*.*

5 Collapsing Hashes in the Standard Model

In the following, let (S_F, F_s) be am (ℓ_{in}, k)-lossy function with $F_s : \{0,1\}^{\ell_{in}} \to \{0,1\}^{\ell_{mid}}$. Let $h_r : \{0,1\}^{\ell_{mid}} \to \{0,1\}^{\ell_{out}}$ be a universal hash function (with key $r \in \{0,1\}^{\ell_{seed}}$). Let $\mathcal{D}_{inj}$ and $\mathcal{D}_{lossy}$ be as in Definition 2.

We will often write $F_{(r,s)}$ and $h_{(r,s)}$ for F_s and h_r to unify notation (one of the parameters will be silently ignored in this case).

Construction 1 (Collapsing compression function). *We define the parameter sampler* $\mathbf{P}_{inj}$ *to return* (r, s) *with* $r \xleftarrow{\$} \{0,1\}^{\ell_{seed}}$*,* $s \leftarrow \mathcal{D}_{inj}$*. We define the parameter sampler* $\mathbf{P}_{lossy}$ *to return* (r, s) *with* $r \xleftarrow{\$} \{0,1\}^{\ell_{seed}}$*,* $s \leftarrow \mathcal{D}_{lossy}$*. We define the parameter sampler* $\mathbf{P}_H$ *to return* (r, s) *with* $r \xleftarrow{\$} \{0,1\}^{\ell_{seed}}$*,* $s \leftarrow S_F$*.*

We define the hash function $H_{(r,s)} : \{0,1\}^{\ell_{in}} \to \{0,1\}^{\ell_{out}}$ *by* $H_{(r,s)} := h_{(r,s)} \circ F_{(r,s)}$*.*

Note that we are mainly interested in the case where $\ell_{out} < \ell_{in}$. Otherwise, $H_{(r,s)}$ could simply be chosen to be an injective function which is always collapsing (Lemma 13).

Furthermore, note that $\mathbf{P}_{inj}$ and $\mathbf{P}_{lossy}$ are not necessarily polynomial-time. The final construction will use $\mathbf{P}_H$, but we need $\mathbf{P}_{inj}$ and $\mathbf{P}_{lossy}$ to state intermediate results.

Lemma 21. *If (S_F, F_s) is a lossy function, then $(\mathbf{P}_{lossy}, F_{(r,s)})$ is collapsing.*

Proof. For $(r,s) \leftarrow \mathbf{P}_{inj}$, $F_{(r,s)}$ is always injective. Hence by Lemma 13, $(\mathbf{P}_{inj}, F_{(r,s)})$ is collapsing.

Since (S_F, F_s) is a lossy function, we have that $\mathcal{D}_{inj}$ and $\mathcal{D}_{lossy}$ are computationally indistinguishable. Hence $\mathbf{P}_{inj}$ and $\mathbf{P}_{lossy}$ are computationally indistinguishable.

Thus by Lemma 15, $(\mathbf{P}_{lossy}, F_{(r,s)})$ is collapsing. □

Lemma 22. *If (S_F, F_s) is a lossy function with lossiness rate K, and if $\ell_{out}/\ell_{in} \geq c > 2 - 2K$ for some constant c, $(\mathbf{P}_{lossy}, h_{(r,s)})$ is collapsing on $\operatorname{im} F_{(r,s)}$.*

Proof. We first compute the probability that $h_{(r,s)}$ is not injective on $\operatorname{im} F_{(r,s)}$.

$$\begin{aligned}
&\Pr[h_{(r,s)} \text{ is not injective on } \operatorname{im} F_{(r,s)} : (r,s) \leftarrow \mathbf{P}_{lossy}] \\
&\overset{(*)}{=} \sum_s \Pr[\mathcal{D}_{lossy} = s] \Pr[h_{(r,s)} \text{ is not injective on } \operatorname{im} F_{(r,s)} : r \xleftarrow{\$} \{0,1\}^{\ell_{seed}}] \\
&\leq \sum_s \Pr[\mathcal{D}_{lossy} = s] \sum_{\substack{x,y \in \operatorname{im} F_s \\ x \neq y}} \Pr[h_{(r,s)}(x) = h_{(r,s)}(y) : r \xleftarrow{\$} \{0,1\}^{\ell_{seed}}] \\
&\overset{(**)}{\leq} \sum_s \Pr[\mathcal{D}_{lossy} = s] \sum_{\substack{x,y \in \operatorname{im} F_s \\ x \neq y}} \frac{1}{2^{\ell_{out}}} \overset{(***)}{\leq} \sum_s \Pr[\mathcal{D}_{lossy} = s] \frac{(2^{\ell_{in}-k})^2}{2^{\ell_{out}}} \\
&= 2^{2\ell_{in} - 2k - \ell_{out}} =: \varepsilon. \qquad (5)
\end{aligned}$$

Here $(*)$ uses the fact that $(r,s) \leftarrow \mathbf{P}_{lossy}$ is the same as $r \xleftarrow{\$} \{0,1\}^{\ell_{seed}}$, $s \leftarrow \mathcal{D}_{lossy}$. And $(**)$ is by definition of universal hash functions. And $(***)$ follows from the fact that for any s in the support of $\mathcal{D}_{lossy}$, $\operatorname{im} F_s = \operatorname{im} F_{(r,s)}$ has size at most $2^{\ell_{in}-k}$ (recall that k is the lossiness of F_s).

Since (S_F, F_s) has lossiness rate K, we have $k \geq K\ell_{in}$ by definition, and ℓ_{in} is superlogarithmic. Remember that $\ell_{out}/\ell_{in} \geq c$. Then

$$\varepsilon = 2^{2\ell_{in} - 2k - \ell_{out}} \leq 2^{2\ell_{in} - 2K\ell_{in} - c\ell_{in}} = 2^{(2-2K)\ell_{in} - c\ell_{in}} = 2^{-d\ell_{in}}$$
$$\text{for } d := c - (2 - 2K).$$

Since by assumption, c and K are constants and $c > 2 - 2K$, we have that $d > 0$ is a constant. Since ℓ_{in} is superlogarithmic, this implies that $\varepsilon \leq 2^{-d\ell_{in}}$ is negligible.

From (5) and Lemma 13, we then have that $(\mathbf{P}_{lossy}, h_{(r,s)})$ is collapsing on $\operatorname{im} F_{(r,s)}$. □

Theorem 23. *If (S_F, F_s) is a polynomial-time computable lossy function with lossiness rate K, and if $\ell_{out}/\ell_{in} \geq c > 2 - 2K$ for some constant c, then $(\mathbf{P}_H, H_{(r,s)})$ is collapsing.*

Proof. By Lemma 21, $(\mathbf{P}_{lossy}, F_{(r,s)})$ is collapsing. By Lemma 22, $(\mathbf{P}_{lossy}, h_{(r,s)})$ is collapsing on $\operatorname{im} F_{(r,s)}$. By Construction 1, $H_{(r,s)} = h_{(r,s)} \circ F_{(r,s)}$. Thus, by Lemma 14, $(\mathbf{P}_{lossy}, H_{(r,s)})$ is collapsing.

Since (S_F, F_s) is a lossy function, $\mathcal{D}_{lossy}$ and S_F are computationally indistinguishable. Hence $\mathbf{P}_{lossy}$ and $\mathbf{P}_H$ are computationally indistinguishable. Hence by Lemma 15, $(\mathbf{P}_H, H_{(r,s)})$ is collapsing. □

Theorem 24. *Assume $\ell_{in} > \ell_{out}$. Let $\mathrm{MD}_{(r,s)}$ be the Merkle-Damgård construction applied to $H_{(r,s)}$ (using a Merkle-Damgård padding pad).*

If (S_F, F_s) is a polynomial-time computable lossy function with lossiness rate K, and h_r is polynomial-time computable, and if $\ell_{out}/\ell_{in} \geq c > 2-2K$ for some constant c, then $(\mathbf{P}_H, \mathrm{MD}_{(r,s)})$ is collapsing.

Proof. By Lemma 23, $(\mathbf{P}_H, H_{(r,s)})$ is collapsing. Then by Theorem 19, $(\mathbf{P}_H, \mathrm{MD}_{(r,s)})$ is collapsing. □

Theorem 25. *Assume $\ell_{in} > \ell_{out}$. Let $\mathrm{MD}_{(r,s)}$ be the Merkle-Damgård construction applied to $H_{(r,s)}$. Let $(\mathsf{com}_{HMu}, \mathsf{verify}_{HMu})$ denote the unbounded Halevi-Micali commitment using $\mathrm{MD}_{(r,s)}$.*

If (S_F, F_s) is a polynomial-time computable lossy function with lossiness rate K, and h_r is polynomial-time computable, and if $\ell_{out}/\ell_{in} \geq c > 2-2K$ for some constant c, then $(\mathbf{P}_H, \mathsf{com}_{HMu}, \mathsf{verify}_{HMu})$ is statistically hiding and collapse-binding.

Proof. By Theorem 24, $(\mathbf{P}_H, \mathrm{MD}_{(r,s)})$ is collapsing. Then by Theorem 12, $(\mathbf{P}_H, \mathsf{com}_{HMu}, \mathsf{verify}_{HMu})$ is statistically hiding and collapse-binding. □

Note that if $K > \frac{1}{2}$, we have $2 - 2K < 1$. Then h_r, c can always be chosen to satisfy the conditions of Theorems 24 and 25 (namely $\ell_{out}/\ell_{in} \geq c > 2 - 2K$ and $\ell_{in} > \ell_{out}$).

For completeness, we now give the concrete security variant of Theorem 25 here. Let τ_F denote the time needed for evaluating $F_{(r,s)}$. Let τ_h denote the time needed for evaluating $h_{(r,s)}$. Let τ'_h denotes an upper bound on the time needed for computing the universal hash function from Definition 10. For a given adversary (A, B), let ℓ_A be a upper bound on the length of each message output by A on the registers M_i (cf. Definition 8).

Theorem 26. *Assume $\ell_{in} > \ell_{out}$. Let $\mathrm{MD}_{(r,s)}$ be the Merkle-Damgård construction applied to $H_{(r,s)}$. Let $(\mathsf{com}_{HMu}, \mathsf{verify}_{HMu})$ denote the unbounded Halevi-Micali commitment using $\mathrm{MD}_{(r,s)}$.*

Then any adversary against $(\mathbf{P}_H, \mathsf{com}_{HMu}, \mathsf{verify}_{HMu})$ has hiding-advantage $\leq 2^{-\ell_{out}-1}$.

Let (A, B) be a τ-time adversary t-c.b.-valid for verify *with collapsing-advantage ε against $(\mathbf{P}_H, \mathsf{com}_{HMu}, \mathsf{verify}_{HMu})$.*

Then there are $(\tau + O(t\tau_{pad}(\ell_A) + t\ell_{pad}(\ell_A)(\tau_F + \tau_h)/(\ell_{in} - \ell_{out}) + \ell_{seed} + t\tau'_h))$-time adversaries $C_1, \dots, C_6$, such that C_1, C_2, C_3 distinguish S_F and $\mathcal{D}_{lossy}$ with some advantages $\varepsilon_1, \varepsilon_2, \varepsilon_3$, and C_4, C_5, C_6 distinguish $\mathcal{D}_{inj}$ and $\mathcal{D}_{lossy}$ with some advantages $\varepsilon_4, \varepsilon_5, \varepsilon_6$, and $\varepsilon \leq (2^{2\ell_{in}-2k-\ell_{out}} + 2\sum_{i=1}^{6} \varepsilon_i) \cdot \frac{\ell_{pad}(\ell_A)}{(\ell_{in} - \ell_{out})}$.

By using existing constructions of lossy functions, we further get:

Theorem 27. *If SIVP and GapSVP are hard for quantum algorithms to approximate within $\tilde{O}(d^c)$ factors for some $c > 5$, then there is a collapsing hash function with domain $\{0,1\}^*$ and codomain $\{0,1\}^{\ell_{out}}$ for some ℓ_{out}, as well as a non-interactive, statistically hiding, collapse-binding commitment schemes with message space $\{0,1\}^*$.*

*Furthermore, the hash function and the commitment scheme can be chosen such that their parameter sampler **P** returns a uniformly random bitstring.*

Proof. [13] shows that almost-always lossy trapdoor functions with lossiness rate $K < 1$ exist if SIVP and GapSVP are hard for quantum algorithms to approximate within $\tilde{O}(d^c)$ factors, where $c = 2 + \frac{3}{2(1-K)} + \delta$ for any desired $\delta > 0$. Almost-always lossy trapdoor functions are in particular lossy functions. If $c > 5$, we can chose some constant $K > \frac{1}{2}$ such that $c = 2 + \frac{3}{2(1-K)} + \delta$ for some $\delta > 0$. Thus there is a lossy function with constant lossiness rate $K > \frac{1}{2}$. Hence by Theorems 24 and 25 there are a collapsing hash function $(\mathbf{P}_H, H_{(r,s)})$ and a non-interactive collapse-binding statistically hiding commitment $(\mathbf{P}_H, \mathsf{com}_{HMu}, \mathsf{verify}_{HMu})$.

$\mathbf{P}_H$ returns (s, r) with $s \leftarrow S_F$ and $r \xleftarrow{\$} \{0,1\}^{\ell_{seed}}$. Furthermore, as discussed after Definition 2, the lossy function (S_F, F_s) can be chosen such that S_F returns uniformly random keys s. In that case $\mathbf{P}_H$ returns a uniformly random bitstring. □

Interactive commitments without public parameters. The above text analyzed non-interactive commitments using public parameters. We refer to the introduction for the reason why it is unlikely that we can get rid of the public parameters in the non-interactive setting. However, in the interactive setting, we get the following result:

Theorem 28. *If lossy function with lossiness rate $K > \frac{1}{2}$ exist, or if SIVP and GapSVP are hard for quantum algorithms to approximate within $\tilde{O}(d^c)$ factors for some $c > 5$, then there is a collapse-binding[10] statistically-hiding commitment scheme with two-round commit phase and non-interactive verification, without public parameters.*

Proof. Let $(\mathbf{P}_H, \mathsf{com}_{HMu}, \mathsf{verify}_{HMu})$ be the commitment scheme analyzed above.

We construct an interactive commitment scheme as follows: To commit to a message m, the recipient runs $k \leftarrow \mathbf{P}_H$ and sends k to the committer. Then the committer computes $(c, u) \leftarrow \mathsf{com}_{HMu}(k, m)$ and sends c. To open to m, the committer sends u, and the verifier checks whether $\mathsf{verify}_{HMu}(k, c, m, u) = 1$.

It is easy to see that if $(\mathbf{P}_H, \mathsf{com}_{HMu}, \mathsf{verify}_{HMu})$ is collapse-binding, so is the resulting interactive scheme. (In the collapse-binding game, the verifier is honest. Hence it is equivalent whether the verifier or $\mathbf{P}_H$ picks k.)

[10] We refer to [16] for the definition of "collapse-binding" for interactive commitments.

In general, having the verifier pick k may break the hiding property of the commitment. However, the proof of the hiding property of $(\mathbf{P}_H, \mathsf{com}_{HMu}, \mathsf{verify}_{HMu})$ (in the full version) reveals that commitment is statistically hiding for any choice of k. Thus the interactive commitment is statistically hiding. □

6 Collapse-Binding Implies Sum-Binding

For the remainder of this section, let $(\mathbf{P}, \mathsf{com}, \mathsf{verify})$ be a commitment scheme with message space $\{0,1\}$. (I.e., a bit commitment.)

A very simple and natural definition of the binding property for bit commitment schemes is the following one (it occurred implicitly and explicitly in different variants in [2–4,7,11]): If an adversary produces a commitment c, and is told only afterwards which bit m he should open it to, then $p_0 + p_1 \leq 1 + \mathit{negligible}$. Here p_0 is the probability that he successfully opens the commitment to $m = 0$, and p_1 analogously. This definition is motivated by the fact that a perfectly binding commitment trivially satisfies $p_0 + p_1 \leq 1 + \mathit{negligible}$.

Definition 29 (Sum-binding). *For any adversary (C_0, C_1) and $m \in \{0,1\}$, let*

$$p_m(C_0, C_1) := \Pr[\mathsf{verify}(k, c, m, u) = 1 : k \leftarrow \mathbf{P}, (S, c) \leftarrow C_0(k),\ u \leftarrow C_1(S, m)].$$

Here S is a quantum register, and c a classical value. We call $adv := p_0 + p_1 - 1$ the sum-binding-advantage *of (C_0, C_1). (With $adv := 0$ if the difference is negative.)*

A commitment is sum-binding *iff for any quantum-polynomial-time (C_0, C_1), adv is negligible.*

Unfortunately, this definition seems too weak to be useful (see [16] for more discussion), but certainly it seems that the sum-binding property is a minimal requirement for a bit commitment scheme. Yet, it was so far not known whether collapse-binding bit commitments are sum-binding. In this section, we will show that collapse-binding bit commitments are sum-binding, thus giving additional evidence that collapse-binding is a sensible definition.

Proof attempt using rewinding. Before we prove our result, we first explain why existing approaches (i.e., rewinding) do not give the required result.

First, the classical case as a warm up. Assume a classical adversary with $p_0 + p_1 = 1 + \varepsilon$ for non-negligible ε. We then break the classical computational-binding property as follows: Run the adversary to get c. Then ask him to provide an opening u for $m = 0$. Then rewind him to the state where he produced c. Then ask him to provide an opening u' for $m = 1$. The probability that u is valid is p_0, the probability that u' is valid is p_1. From the union bound, we get that the probability that both are valid is at least $p_0 + p_1 - 1 = \varepsilon$.[11] But that means

[11] Namely, $\Pr[u \text{ invalid}] = 1 - p_0$, $\Pr[u' \text{ invalid}] = 1 - p_1$. Hence $\Pr[u \text{ invalid or } u' \text{ invalid}] \leq (1 - p_0) + (1 - p_1)$. Thus $\Pr[u, u' \text{ valid}] \geq 1 - ((1 - p_0) + (1 - p_1)) = p_0 + p_1 - 1$.

that the adversary has non-negligible probability ε of finding c, m, m', u, u' with $m \neq m'$ and u, u' being valid openings for m, m'. This contradicts the classical-style binding property.

Now what happens if we try to use rewinding in the quantum case to show that collapse-binding implies sum-binding? If we use the rewinding technique from [14], the basic idea is the following:

Run the adversary to get a commitment c (i.e., $(S, c) \leftarrow C_0(k)$). Run the adversary to get an opening u for $m = 0$ (i.e., run $u \leftarrow C_1(S, 0)$). Here we assume w.l.o.g. that C_1 is unitary. Measure u. Run the inverse of the unitary $C_1(S, 0)$. Run the adversary to get an opening u' for $m = 1$ (i.e., run $u \leftarrow C_1(S, 1)$).

To get a contradiction, we need to show that with non-negligible probability u and u' are both valid openings. While u will be valid with probability p_0, there is nothing we can say about u'. This is because measuring u will disturb the state of the adversary so that $C_1(S, 1)$ may return nonsensical outputs. [14] shows that *if there is only one valid* u, then rewinding works. But there is nothing that guarantees that there is only one valid u.[12] At this point the rewinding-based proof fails.

Collapse-binding implies sum-binding. We now formally state and prove the main result of this section with a technique different from rewinding. (But possibly this is a new rewinding technique under the hood.)

Theorem 30. *If* $(\mathbf{P}, \mathsf{com}, \mathsf{verify})$ *is collapse-binding, then* $(\mathbf{P}, \mathsf{com}, \mathsf{verify})$ *is sum-binding.*

An interesting open question is whether the converse holds. If so, this would immediate give strong results for the parallel composition of sum-binding commitments and their use in rewinding proofs (because all the properties of collapse-binding commitments would carry over).

We give a proof sketch first: As we have seen, running two executions of the adversary sequentially (first opening to $m = 0$, then opening to $m = 1$) via rewinding is problematic because the second execution may not be successful any more. Instead, we will run both executions at the same time in superposition:

Assume an adversary against sum-binding with non-negligible advantage ε. We initialize a qubit M with $|+\rangle = \frac{1}{\sqrt{2}}|0\rangle + \frac{1}{\sqrt{2}}|1\rangle$. Then we let the adversary commit ($(S, c) \leftarrow C_0(k)$), and then we run $C_1(S, 0)$ or $C_1(S, 1)$ in superposition, controlled by the register M. This may entangle M with the rest of the system. And we get openings for $m = 0$ and $m = 1$ in superposition on a register U. Now if we measure whether U contains a valid opening for the message on register M, the answer will be yes with probability $\delta := \frac{p_0 + p_1}{2} = \frac{1+\varepsilon}{2}$ where p_0, p_1 are as in Definition 29 (call this measurement V_c). Now, we either measure the register

[12] Collapse-binding commitments are rewinding-friendly, but this refers only to the case where we wish to measure the opened message m. Roughly, collapse-binding implies that measuring m does disturb the state more than measuring whether the commitment was opened correctly or not, and in that case, the rewinding technique from [14] applies. The [16] for example proofs using this technique.

M in the computational basis or we do not. And finally we apply the inverse of $C_1(S,0)$ or $C_1(S,1)$ in superposition. And finally we measure whether M is still in the state $|+\rangle$ (call this measurement $\mathcal{M}_+$).

We distinguish two cases: If we measure M in the computational basis, then $M = |0\rangle$ or $M = |1\rangle$ afterwards. So the measurement $\mathcal{M}_+$ succeeds with probability $\frac{1}{2}$. Hence the probability that both V_c and $\mathcal{M}_+$ succeed is $\frac{\delta}{2}$.

If we do not measure M in the computational basis, then we have the following situation. The invocation $C_1(S,0)$ or $C_1(S,1)$ in superposition, together with the measurement V_c, together with the uncomputation of $C_1(S,0)$ or $C_1(S,1)$ can be seen as a single binary measurement R_c. Now if we have a measurement that succeeds with high probability, it cannot change the state much. Thus, the higher the success probability δ of R_c, the more likely it is that M is still in state $|+\rangle$ and $\mathcal{M}_+$ succeeds. An exact computation reveals: the probability that both R_c (a.k.a. V_c) and $\mathcal{M}_+$ succeed is δ^2.

Thus the measurement $\mathcal{M}_+$ distinguishes between measuring and not measuring M with non-negligible probability $\frac{\delta}{2} - \delta^2 \geq \frac{\varepsilon}{4}$. This contradicts the collapse-binding property, the theorem follows.

We now give the full proof:

Proof of Theorem 30. Let (C_0, C_1) be an adversary in the sense of Definition 29 (against sum-binding). Let $p_0 := p_0(C_0, C_1)$ and $p_1 := p_1(C_0, C_1)$. We have to show that the advantage $\varepsilon := p_0 + p_1 - 1$ is upper bounded by a negligible function.

Without loss of generality, we can assume that C_1 is unitary. More precisely, $C_1(S,m)$ applies a unitary circuit U_m to S, resulting in two output registers U and E. Then he measures U in the computational basis and returns the outcomes u.

With that notation, we can express the game from Definition 29 as the following circuit (renaming the register S to S' to avoid name clashes later):

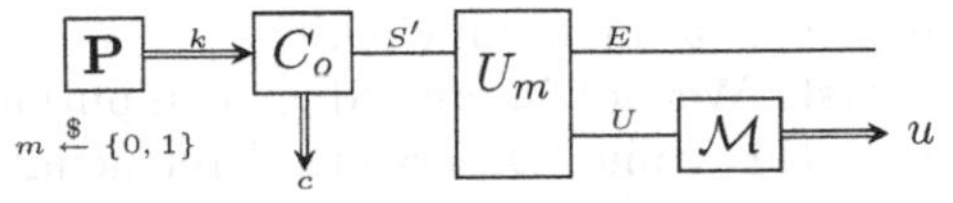

(6)

(Here and in the following, $\mathcal{M}$ denotes a measurement in the computational basis.) In that circuit, $\Pr[\mathsf{verify}(k,c,m,u) = 1] = \delta := \frac{1}{2}(1+\varepsilon)$.

Let M denote a one-qubit quantum register, and define $U_M : |m\rangle_M \otimes |\Psi\rangle_{S'} \mapsto |m\rangle_M \otimes U_m|\Psi\rangle_{S'}$. That is, U_M is a unitary with two input registers M, S', and three output registers M, U, E which is realized by applying U_0 or U_1 to S', depending on whether M is $|0\rangle$ or $|1\rangle$.

Let $\mathcal{M}_+$ be the binary measurement that checks whether register M is in state $|+\rangle = \frac{1}{\sqrt{2}}|0\rangle + \frac{1}{\sqrt{2}}|1\rangle$. Formally, $\mathcal{M}_+$ is defined by the projector $P_+ := |+\rangle\langle+|$ on M.

Recall that V_c from Lemma 6 is the measurement defined by the projector $P_c := \sum_{\substack{m,u \\ \mathsf{verify}(k,c,m,u)=1}} |m\rangle\langle m| \otimes |u\rangle\langle u|$.

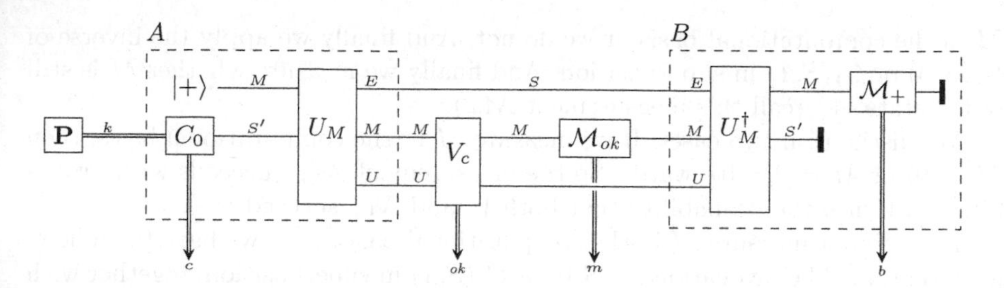

Fig. 3. Circuit describing Game_1. Game_2 can be derived by omitting $\mathcal{M}_{ok}$. The adversary algorithms A and B are depicted in the dashed boxes. (To avoid wires crossing gates, the outgoing wires of U_M are ordered E, M, U, not M, U, E as in the text.)

We define an adversary (A, B) against the collapse-binding property of com (using the alternative definition from Lemma 6). Algorithm $A(k)$ performs the following steps (see also Fig. 3):

- Run $(S', c) \leftarrow C_0(k)$.
- Initialize a register M with $|+\rangle$.
- $(M, U, E) \leftarrow U_M(M, S')$. That is, apply U_M to M, S'.
- $S := E$. (That is, we rename the register E.)
- Return (S, M, U, c).

Algorithm $B(S, M, U)$ performs the following steps (see also Fig. 3):

- $E := S$.
- $(M, S') \leftarrow U_M^\dagger(M, U, E)$.
- $b \leftarrow \mathcal{M}_+(Y)$.
- Return b.

Let $\mathsf{Game}_1, \mathsf{Game}_2$ refer to the games from Lemma 6 with adversary (A, B). Figure 3 depicts those games as a quantum circuit.

We consider Game_1 first. We are interested in computing the probability $p := \Pr[b = 1 \wedge ok = 1]$ in this game. Observe that replacing $\mathcal{M}_{ok}$ by $\mathcal{M}$ (the latter being the measurement in the computational basis, applied even when $ok = 0$) does not change p. (Because $\mathcal{M}_{ok}$ and $\mathcal{M}$ behave differently only when $ok = 0$.) Thus, replacing $\mathcal{M}_{ok}$ on M by $\mathcal{M}$ does not change p. Thus, we get the following circuit:

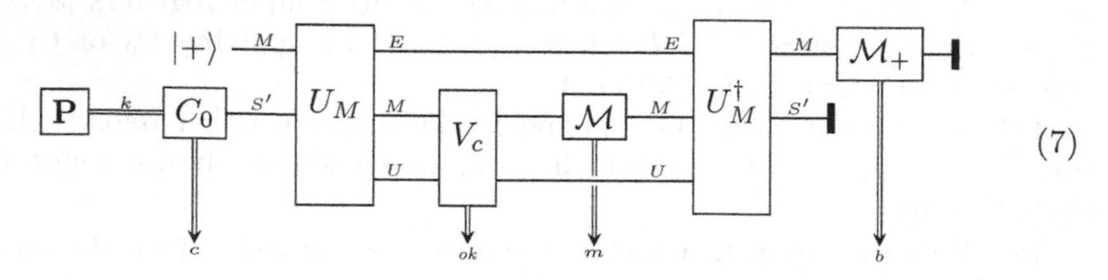

(7)

and have

$$\Pr[b = 1 \wedge ok = 1 : \text{Circuit (7)}] = \Pr[b = 1 \wedge ok = 1 : \mathsf{Game}_1]. \tag{8}$$

Note that $\mathcal{M}$ on M commutes with V_c and U_M. So we can move $\mathcal{M}$ to the beginning (right after initializing M with $|+\rangle$). But measuring $|+\rangle$ in the computational basis yields a uniformly distributed bit m. And furthermore, if M contains $|m\rangle$, then U_M degenerates to U_m on register S', and M stays in state $|m\rangle$ until the measurement $\mathcal{M}_+$. Thus we can simplify (7) as follows:

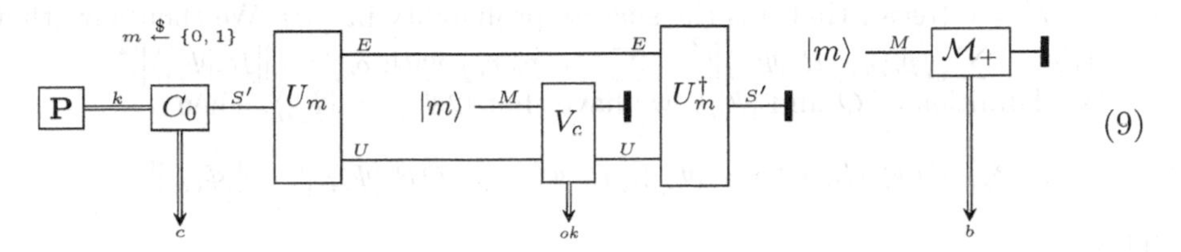

(9)

We thus have

$$\Pr[b = 1 \wedge ok = 1 : \text{Circuit (7)}] = \Pr[b = 1 \wedge ok = 1 : \text{Circuit (9)}]. \tag{10}$$

It is easy to see that

$$\Pr[ok = 1 : \text{Circuit (9)}] = \Pr[\mathsf{verify}(k, c, m, u) = 1 : \text{Circuit (6)}] = \delta.$$

Furthermore, in (9), b is independent of ok, and we have $\Pr[b = 1] = \frac{1}{2}$ by definition of $\mathcal{M}_+$. Thus

$$\Pr[b = 1 \wedge ok = 1 : \mathsf{Game}_1] \overset{(8),(10)}{=} \Pr[b = 1 \wedge ok = 1 : \text{Circuit (9)}] = \frac{\delta}{2}. \tag{11}$$

We now consider Game_2. This game is depicted in Fig. 3 (when omitting the measurement $\mathcal{M}_{ok}$). We are interested in computing the probability $q := \Pr[b = 1 \wedge ok = 1]$ in this game. Recall that P_+, P_c are the projectors describing the measurements $\mathcal{M}_+$, V_c. Thus, $q = \operatorname{tr} \rho$ where ρ is the final state of the following circuit:

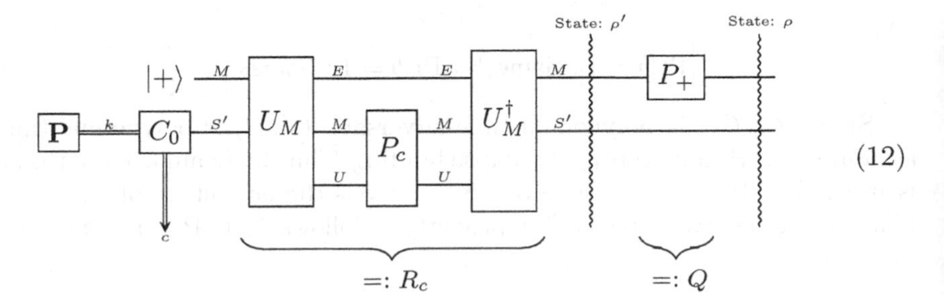

(12)

We abbreviate the product of the operators U_M, P_c, $U_M^\dagger$ with R_c. Note that R_c is a projector since P_c is a projector and U_M is unitary. Let $Q := P_+ \otimes id_{S'}$.

Furthermore, let ρ_c be the state output by C_0 on S' conditioned on classical output c (and let p_c be the probability of that output). We can write ρ_c as $\rho_c = \sum_i p_{ci}|\Psi_{ci}\rangle\langle\Psi_{ci}|$ for some normalized quantum states $|\Psi_{ci}\rangle$ and some probabilities p_{ci} with $\sum_i p_{ci} = 1$. Let $|\Psi'_{ci}\rangle := |+\rangle \otimes |\Psi_{ci}\rangle$. Let $|\Phi_{ci}\rangle := QR_c|\Psi'_{ci}\rangle$. With that notation, we have $\rho = \sum_{c,i} p_c p_{ci}|\Phi_{ci}\rangle\langle\Phi_{ci}|$ and $\sum_{c,i} p_c p_{ci} = 1$. Hence $q = \operatorname{tr}\rho = \sum_{c,i} p_c p_{ci}\big\||\Phi_{ci}\rangle\big\|^2$.

Furthermore, if ρ' is the state in circuit (12) after $U_M^\dagger$, then it is easy to see that $\operatorname{tr}\rho' = \delta$ (recall that δ is the success probability in (6)). We then have that $\delta = \operatorname{tr}\rho' = \sum_{c,i} p_c p_{ci}\big\|R_c|\Psi'_{ci}\rangle\big\|^2 = \sum_{c,i} p_c p_{ci}\delta_{cf}$ with $\delta_{cf} := \big\|R_c|\Psi'_{ci}\rangle\big\|^2$.

By definition of Q and $|\Psi'_{ci}\rangle$, we have that $Q|\Psi'_{ci}\rangle = |\Psi'_{ci}\rangle$. Then

$$\delta_{ci} = \langle\Psi'_{ci}|R_c|\Psi'_{ci}\rangle = \langle\Psi'_{ci}|QR_c|\Psi'_{ci}\rangle \le \big\|QR_c|\Psi'_{ci}\rangle\big\| = \big\||\Phi_{ci}\rangle\big\|.$$

Thus

$$q = \sum_{c,i} p_c p_{ci}\big\||\Phi_{ci}\rangle\big\|^2 \ge \sum_{c,i} p_c p_{ci}\delta_{ci}^2 \overset{(*)}{\ge} \Big(\sum_{c,i} p_c p_{ci}\delta_{ci}\Big)^2 = \delta^2$$

Here $(*)$ uses Jensen's inequality and the fact that $\sum_{c,i} p_c p_{ci} = 1$.

Thus

$$\Pr[b = 1 \wedge ok = 1 : \mathsf{Game}_2] = q \ge \delta^2. \tag{13}$$

Since Game_1 and Game_2 are identical unless $ok = 1$, we have that

$$\Pr[b = 1 \wedge ok \neq 1 : \mathsf{Game}_1] = \Pr[b = 1 \wedge ok \neq 1 : \mathsf{Game}_2]. \tag{14}$$

Thus

$$\begin{aligned}
&\Pr[b = 1 : \mathsf{Game}_2] - \Pr[b = 1 : \mathsf{Game}_1] \\
&= \big(\Pr[b = 1 \wedge ok = 1 : \mathsf{Game}_2] + \Pr[b = 1 \wedge ok \neq 1 : \mathsf{Game}_2]\big) \\
&\quad - \big(\Pr[b = 1 \wedge ok = 1 : \mathsf{Game}_1] + \Pr[b = 1 \wedge ok \neq 1 : \mathsf{Game}_1]\big) \\
&\overset{(14)}{=} \Pr[b = 1 \wedge ok = 1 : \mathsf{Game}_2] - \Pr[b = 1 \wedge ok \neq 1 : \mathsf{Game}_1] \\
&\overset{(11),(13)}{\ge} \delta^2 - \frac{\delta}{2} \ge \frac{\varepsilon}{4}.
\end{aligned}$$

Thus

$$\Big|\Pr[b = 1 : \mathsf{Game}_1] - \Pr[b = 1 : \mathsf{Game}_2]\Big| \ge \frac{\varepsilon}{4}. \tag{15}$$

Since (C_0, C_1) is polynomial-time adversary, (A, B) is polynomial-time. By assumption, $(\mathbf{P}, \mathsf{com}, \mathsf{verify})$ is collapse binding. Thus by Lemma 6, the rhs of (15) is negligible. Hence ε is negligible. Since ε was the advantage of the adversary (C_0, C_1) against the sum-binding property, it follows that $(\mathbf{P}, \mathsf{com}, \mathsf{verify})$ is sum-binding. □

6.1 CDMS-Blinding

For the remainder of this section, let $(\mathbf{P}, \mathsf{com}, \mathsf{verify})$ be a commitment scheme with message space $\{0,1\}^\ell$.

The sum-binding definition is restricted to bit commitments. In [3], a generalization of sum-binding definition is given. Intuitively, for any function f, if the adversary produces a commitment c, then there should be at most one value y such that the adversary can open c to a message m with $f(m) = y$. Slightly more formally, we require that $\sum_y \tilde{p}_y \leq 1 + \mathit{negligible}$ where $\tilde{p}_y$ is the probability that the adversary (who gets y after producing the commitment c) manages to open c to a message m with $f(m) = y$. Again, this definition is motivated by the fact that perfectly binding commitments satisfy $\sum_y \tilde{p}_y \leq 1$. The definition can be parametrized by specifying the set F of allowed functions f.

Definition 31 (CDMS-binding, following [3]). *Let F be a family of functions $\{0,1\}^\ell \to \{0,1\}^\Lambda$.*

For any adversary (C_0, C_1) and any $y \in \{0,1\}^\Lambda$, let

$$\tilde{p}_y(C_0, C_1) := \Pr[\mathsf{verify}(k, c, m, u) = 1 \wedge f(m) = y : \\ k \leftarrow \mathbf{P}, (S, c, f) \leftarrow C_0(k), (m, u) \leftarrow C_1(S, y)].$$

Here S is a quantum register, and c a classical value, and f a function in F (represented as a Boolean circuit).

We call (C_0, C_1) F-CDMS-valid if it only outputs functions $f \in F$. We call $adv := \sum_{y \in \{0,1\}^\Lambda} \tilde{p}_y(C_0, C_1) - 1$ the F-CDMS-advantage *of (C_0, C_1). (With $adv := 0$ if the difference is negative.)*

We call a commitment scheme F-CDMS-binding *iff for all quantum-polynomial time F-CDMS-valid (C_0, C_1), the F-CDMS-advantage of (C_0, C_1) is negligible.*

We have somewhat modified the definition with respect to [3]: Namely, instead of quantifying over all $f \in F$, we let the adversary choose f. This gives the adversary additional power, because f may depend on the public parameter k, but at the same time it also removes some power (because f needs to be efficiently computed in our definition). For non-uniform adversaries, our definition implies the one from [3].

Note that the sum-binding definition is a special case of the CDMS-binding definition: A bit commitment is sum-binding iff it is F-binding where F contains only the identity.

The following theorem is shown using a similar technique as Theorem 30. The main difference is that we have to use a superposition of all possible values y, instead of the superposition $|+\rangle$ of messages 0 and 1. Furthermore, the fact that the adversary has free choice of m, subject to the condition $f(m) = c$ introduces additional technicalities, but these are solved in the full proof.

Theorem 32. *If $(\mathbf{P}, \mathsf{com}, \mathsf{verify})$ is collapse-binding, then $(\mathbf{P}, \mathsf{com}, \mathsf{verify})$ is F-CDMS-binding for any $F \subseteq \{0,1\}^\ell \to \{0,1\}^\Lambda$ with logarithmically-bounded Λ.*

Note the condition that Λ is logarithmically-bounded. This condition is necessary as the following example shows: Let com be a perfectly binding commitment, except that with probability ε the adversary finds a secret that allows him to open the commitment to any message. This small probability ε does not change the fact that the commitment is collapse-binding (and arguably any reasonable definition of computationally binding should tolerate such a negligible error). However, an adversary that commits to 0, then gets $y \in \{0,1\}^\Lambda$, and then tries to open to an arbitrary m with $f(m) = y$ will succeed with probability $\tilde{p}_y = \varepsilon$ for all $y \neq f(0)$, and with probability $\tilde{p}_y = 1$ for $y = f(0)$. Hence $\sum_y \tilde{p}_y = 1+(2^\Lambda - 1)\varepsilon$. If Λ is superlogarithmic, then $(2^\Lambda - 1)\varepsilon$ will not necessarily be negligible. This example shows that collapse-binding cannot imply CDMS-binding for superlogarithmic Λ and also indicates that probably CDMS-binding with superlogarithmic Λ is not a reasonable definition of computationally binding. (Note: in [3], only CDMS-binding with logarithmically-bounded Λ was used and is sufficient for their OT protocol.)

Acknowledgements. We thank Dennis Hofheinz for discussions that led to the commitment protocol from Construction 1. This work was supported by institutional research funding IUT2-1 of the Estonian Ministry of Education and Research and by the Estonian ICT program 2011–2015 (3.2.1201.13-0022).

References

1. Ambainis, A., Rosmanis, A., Unruh, D.: Quantum attacks on classical proof systems (the hardness of quantum rewinding). In: FOCS 2014, pp. 474–483. IEEE (2014). Preprint on IACR ePrint 2014/296
2. Brassard, G., Crépeau, C., Jozsa, R., Langlois, D.: A quantum bit commitment scheme provably unbreakable by both parties. In: FOCS 1993, pp. 362–371. IEEE, Los Alamitos (1993)
3. Crépeau, C., Dumais, P., Mayers, D., Salvail, L.: Computational collapse of quantum state with application to oblivious transfer. In: Naor, M. (ed.) TCC 2004. LNCS, vol. 2951, pp. 374–393. Springer, Heidelberg (2004). doi:10.1007/978-3-540-24638-1_21
4. Crépeau, C., Salvail, L., Simard, J.-R., Tapp, A.: Two provers in isolation. In: Lee, D.H., Wang, X. (eds.) ASIACRYPT 2011. LNCS, vol. 7073, pp. 407–430. Springer, Heidelberg (2011). doi:10.1007/978-3-642-25385-0_22
5. Damgård, I., Fehr, S., Salvail, L.: Zero-knowledge proofs and string commitments withstanding quantum attacks. In: Franklin, M. (ed.) CRYPTO 2004. LNCS, vol. 3152, pp. 254–272. Springer, Heidelberg (2004). doi:10.1007/978-3-540-28628-8_16
6. Damgård, I., Pedersen, T.P., Pfitzmann, B.: On the existence of statistically hiding bit commitment schemes and fail-stop signatures. J. Cryptology **10**(3), 163–194 (1997)
7. Dumais, P., Mayers, D., Salvail, L.: Perfectly concealing quantum bit commitment from any quantum one-way permutation. In: Preneel, B. (ed.) EUROCRYPT 2000. LNCS, vol. 1807, pp. 300–315. Springer, Heidelberg (2000). doi:10.1007/3-540-45539-6_21
8. Goldwasser, S., Bellare, M.: Lecture notes on cryptography. Summer course on cryptography, MIT, 1996–2008 (2008). http://cseweb.ucsd.edu/~mihir/papers/gb.html

9. Halevi, S., Micali, S.: Practical and provably-secure commitment schemes from collision-free hashing. In: Koblitz, N. (ed.) CRYPTO 1996. LNCS, vol. 1109, pp. 201–215. Springer, Heidelberg (1996). doi:10.1007/3-540-68697-5_16
10. Håstad, J., Impagliazzo, R., Levin, L.A., Luby, M.: A pseudorandom generator from any one-way function. SIAM J. Comput. **28**(4), 1364–1396 (1999)
11. Mayers, D.: Unconditionally secure quantum bit commitment is impossible. Phys. Rev. Lett. **78**(17), 3414–3417 (1997). http://arxiv.org/abs/quant-ph/9605044
12. National Institute of Standards and Technology (NIST). Secure hash standard (SHS). FIPS PUBS 180–4 (2015). doi:10.6028/NIST.FIPS.180-4
13. Peikert, C., Waters, B.: Lossy trapdoor functions and their applications. In: STOC, pp. 187–196. ACM, New York (2008). http://ia.cr/2007/279
14. Unruh, D.: Quantum proofs of knowledge. In: Pointcheval, D., Johansson, T. (eds.) EUROCRYPT 2012. LNCS, vol. 7237, pp. 135–152. Springer, Heidelberg (2012). doi:10.1007/978-3-642-29011-4_10
15. Unruh, D.: Collapse-binding quantum commitments without random oracles (2016). IACR ePrint2016/508
16. Unruh, D.: Computationally binding quantum commitments. In: Fischlin, M., Coron, J.-S. (eds.) EUROCRYPT 2016. LNCS, vol. 9666, pp. 497–527. Springer, Heidelberg (2016). doi:10.1007/978-3-662-49896-5_18
17. Watrous, J.: Zero-knowledge against quantum attacks. SIAM J. Comput. **39**(1), 25–58 (2009). https://cs.uwaterloo.ca/~watrous/Papers/ZeroKnowledgeAgainstQuantum.pdf
18. Yan, J., Weng, J., Lin, D., Quan, Y.: Quantum bit commitment with application in quantum zero-knowledge proof (extended abstract). In: Elbassioni, K., Makino, K. (eds.) ISAAC 2015. LNCS, vol. 9472, pp. 555–565. Springer, Heidelberg (2015). doi:10.1007/978-3-662-48971-0_47
19. Mark Zhandry. How to construct quantum random functions. In: FOCS 2013, pages 679–687. IEEE Computer Society, Los Alamitos (2012). IACR ePrint2012/182

Digital Signatures Based on the Hardness of Ideal Lattice Problems in All Rings

Vadim Lyubashevsky(✉)

IBM Research – Zurich, Zurich, Switzerland
vadim.lyubash@gmail.com

Abstract. Many practical lattice-based schemes are built upon the Ring-SIS or Ring-LWE problems, which are problems that are based on the presumed difficulty of finding low-weight solutions to linear equations over polynomial rings $\mathbb{Z}_q[\mathbf{x}]/\langle\mathbf{f}\rangle$. Our belief in the asymptotic computational hardness of these problems rests in part on the fact that there are reduction showing that solving them is as hard as finding short vectors in all lattices that correspond to ideals of the polynomial ring $\mathbb{Z}[\mathbf{x}]/\langle\mathbf{f}\rangle$. These reductions, however, do not give us an indication as to the effect that the polynomial $\mathbf{f}$, which defines the ring, has on the average-case or worst-case problems.

As of today, there haven't been any weaknesses found in Ring-SIS or Ring-LWE problems when one uses an $\mathbf{f}$ which leads to a meaningful worst-case to average-case reduction, but there have been some recent algorithms for related problems that heavily use the algebraic structures of the underlying rings. It is thus conceivable that some rings could give rise to more difficult instances of Ring-SIS and Ring-LWE than other rings. A more ideal scenario would therefore be if there would be an average-case problem, allowing for efficient cryptographic constructions, that is based on the hardness of finding short vectors in ideals of $\mathbb{Z}[\mathbf{x}]/\langle\mathbf{f}\rangle$ for *every* $\mathbf{f}$.

In this work, we show that the above may actually be possible. We construct a digital signature scheme based (in the random oracle model) on a simple adaptation of the Ring-SIS problem which is as hard to break as worst-case problems in every $\mathbf{f}$ whose degree is bounded by the parameters of the scheme. Up to constant factors, our scheme is as efficient as the highly practical schemes that work over the ring $\mathbb{Z}[\mathbf{x}]/\langle\mathbf{x}^n+1\rangle$.

1 Introduction

One of the attractive features of lattice cryptography is that one can construct cryptographic primitives whose security is based on the hardness of worst-case lattice problems [Ajt96]. More concretely, average-case problems such as SIS and LWE are defined in such a way that an adversary who is able to solve these problems could then be used to find short vectors in *any* lattice. While

V. Lyubashevsky—Supported by the SNSF ERC Transfer Grant CRETP2-166734 – FELICITY.

J.H. Cheon and T. Takagi (Eds.): ASIACRYPT 2016, Part II, LNCS 10032, pp. 196–214, 2016.
DOI: 10.1007/978-3-662-53890-6_7

the worst-case to average-case reductions do not help us figure out the exact parameter settings that make SIS and LWE hard, they definitely deserve the credit for leading researchers to the *right* definitions of these problems.

Recent years have seen numerous cryptographic protocols constructed based on SIS and LWE. These schemes, however, are not particularly efficient because SIS and LWE inherently give rise to key sizes and/or outputs which are $\tilde{O}(\lambda^2)$ in the security parameter λ. For this reason, almost all of the practical lattice-based constructions are built upon the average-case problems Ring-SIS and Ring-LWE. The algebraic structure underlying Ring-SIS and Ring-LWE problems are polynomial rings of the form $\mathbb{Z}_q[\mathbf{x}]/\langle \mathbf{f} \rangle$, and it was shown in [PR06, LM06, SSTX09, LPR13] that solving Ring-SIS and Ring-LWE over this ring implies finding short vectors in all ideals of $\mathbb{Z}[\mathbf{x}]/\langle \mathbf{f} \rangle$. Notice that these are somewhat weaker statements than the proof for SIS and LWE because one needs to first pick the ring $\mathbb{Z}[\mathbf{x}]/\langle \mathbf{f} \rangle$ where the worst-case problems are believed to be hard.

As of today, there have not been any attacks on worst-case problems in any ring, nor on the Ring-SIS or Ring-LWE problems in rings for which there exist non-vacuous (i.e. the reduction is not from a problem that is easy) worst-case to average-case reductions. For this reason, most proposals choose to work with cyclotomic rings, such as $\mathbb{Z}[\mathbf{x}]/\langle \mathbf{x}^{2^k} + 1 \rangle$, due to their particularly nice algebraic structure for implementation purposes. Cyclotomics also have the feature that the decision version of the Ring-LWE problem in these rings is hard [LPR13], which makes them even more useful for cryptographic applications.

While the Ring-SIS and Ring-LWE problems remain hard, there have been some recent works that were able to solve other problems in certain rings by taking advantage of the algebraic structure. The work of Cramer et al. [CDPR16], which built on the approach of Campbell et al. [CGS14], showed that the log-unit lattice of cyclotomic rings is efficiently decodable. When combined with a polynomial-time quantum algorithm of Biasse and Song [BS16] (building upon [EHKS14, CGS14]) for finding generators of principal ideals, one obtains a quantum polynomial-time algorithm for finding a $2^{\tilde{O}(\sqrt{n})}$-approximate shortest vector problem in *principal* ideals of cyclotomic rings.

The simultaneous works of Albrecht et al. [ABD16] and Cheon et al. [CJL16] exploited the sub-field structure of number fields to give sub-exponential algorithms for the NTRU problem in which the secret polynomials are very small. This is an approach that is very similar to an early idea mentioned in [GS02, Sect. 6]. While it is interesting to note that none of these attacks say anything about worst-case problems or average-case Ring-SIS and Ring-LWE, they do point out that the choice ring can affect the hardness of problems. For this reason, there have been proposals for using alternative rings (e.g. Bernstein et al. [BCLvV16] suggested using rings $\mathbb{Z}[\mathbf{x}]/\langle \mathbf{x}^p - \mathbf{x} - 1 \rangle$) which do not have the algebraic structure exploited by the aforementioned algorithms. But in the absence of attacks on any of the current constructions, it is of course not clear whether one is more secure than the other.

1.1 Our Result

A more ideal situation would be if one could build efficient cryptographic schemes that are *simultaneously* based on the hardness of average-case (and therefore worst-case) problems in every ring. In this work we show that this indeed may be possible. We construct a digital signature scheme which is up to constant factors, in terms of running time and key/signature sizes, as efficient as the most practical signature schemes [Lyu12, GLP12, DDLL13] (i.e. the key sizes, running time, and output sizes are all $\tilde{O}(\lambda)$), and is based on the hardness of the Ring-SIS problem in *every* ring $\mathbb{Z}[\mathbf{x}]/\langle\mathbf{f}\rangle$, with the obvious restriction that the degree of $\mathbf{f}$ is bounded by the parameters of the scheme.

In the Ring-SIS problem over the ring $\mathbb{Z}_q[\mathbf{x}]/\langle\mathbf{f}\rangle$, called **f**-SIS, one is given k uniformly random polynomials $\mathbf{a}_1, \ldots, \mathbf{a}_k$ and is asked to find elements $\mathbf{z}_1, \ldots, \mathbf{z}_k$ with small coefficients such that $\sum \mathbf{a}_i\mathbf{z}_i = \mathbf{0}$ in the ring $\mathbb{Z}_q[\mathbf{x}]/\langle\mathbf{f}\rangle$. A simple, yet very important, observation is that the input to this problem only very loosely depends on the polynomial $\mathbf{f}$. In particular, for all $\mathbf{f}$ of the same degree, this input has the exact same distribution.

If we then defined a problem over the ring $\mathbb{Z}_q[\mathbf{x}]$ that required finding a combination of the $\mathbf{a}_i$ such that $\sum \mathbf{a}_i\mathbf{z}_i = \mathbf{0}$, then these $\mathbf{z}_i$ would also be a solution to $\sum \mathbf{a}_i\mathbf{z}_i = \mathbf{0} \bmod \mathbf{f}$ for any $\mathbf{f}$. If the degree of $\mathbf{f}$ is larger than the degree of $\mathbf{z}_i$, then as long as one of the $\mathbf{z}_i$ is non-zero in $\mathbb{Z}_q[\mathbf{x}]$, it is also non-zero in $\mathbb{Z}_q[\mathbf{x}]/\langle\mathbf{f}\rangle$.

The intuition for building a digital signature scheme is to let the public key be random polynomials $\mathbf{a}_1, \ldots, \mathbf{a}_k$ in $\mathbb{Z}_q[\mathbf{x}]$ of bounded degree $n-1$, and $\mathbf{t} = \sum \mathbf{a}_i\mathbf{s}_i$ where all operations are performed over $\mathbb{Z}_q[\mathbf{x}]$. We would like to choose the $\mathbf{s}_i$ such that their degree d is somewhat less than n, and also such that the function f defined as $f(\mathbf{s}_1, \ldots, \mathbf{s}_k) = \sum \mathbf{a}_i\mathbf{s}_i$ is compressing. One can then adapt the "Fiat-Shamir with Aborts" technique for Σ-protocols from [Lyu09, Lyu12] to create a signature $(\mathbf{z}_1, \ldots, \mathbf{z}_k)$ that is independent of $\mathbf{s}_i$ and satisfies some linear relation relating $\mathbf{a}_i, \mathbf{t}$ and the "commit" and "challenge" steps of the Σ-protocol.

It can be then shown that an adversary who can break the unforgeability security property of the digital scheme can be used to extract polynomials with small norms $\mathbf{z}_1, \ldots, \mathbf{z}_k$ and $\mathbf{c}$ that satisfy the equation $\sum \mathbf{a}_i\mathbf{z}_i = \mathbf{tc}$ over $\mathbb{Z}_q[\mathbf{x}]$. We then show that a solution to this equation that satisfies certain conditions on the coefficient sizes and degrees of polynomials $\mathbf{z}_i$, $\mathbf{c}$, as well as the polynomials $\mathbf{s}_i$ that were used to construct $\mathbf{t}$, implies a solution to the **f**-SIS problem for any $\mathbf{f}$ whose degree is between $d + \deg(\mathbf{c})$ and n.[1] When combined with the worst-case to average-case reduction from finding short vectors in ideals of $\mathbb{Z}[\mathbf{x}]/\langle\mathbf{f}\rangle$ to the **f**-SIS problem from [LM06], this gives a reduction from worst-case lattice problems in ideals of any ring $\mathbb{Z}[\mathbf{x}]/\langle\mathbf{f}\rangle$ to the hardness of breaking the signature scheme.

---

[1] The lower-bound $d+\deg(\mathbf{c})$ on the degree of $\mathbf{f}$ can be circumvented, but its presence makes the proofs simpler. We also do not think that it's particularly interesting to extend the proofs for $\mathbf{f}$ of very small (compared to n) degree, because those problems will be generally easier than problems over larger rings.

A Note on the Definition of Length. It should be pointed out that the quality of the worst-case to **f**-SIS reduction in [LM06] depends on **f**. If we define the norms of elements in $\mathbb{Z}_q[\mathbf{x}]/\langle\mathbf{f}\rangle$ by computing a standard norm on their coefficients (e.g. the ℓ_∞-norm), then it is possible that a solution to **f**-SIS does not lead to finding short vectors in the lattice. [LM06] defined the "expansion factor" of **f** which determined how much coefficients of polynomial products could grow when multiplied modulo **f**. For some **f**, this growth could be exponential, and one would lose this factor in the reductions, thus making them vacuous. In later works [PR07,LPR13], it was shown that using coefficient sizes is not the most natural way to define the length of elements in $\mathbb{Z}_q[\mathbf{x}]/\langle\mathbf{f}\rangle$. If one instead uses the "canonical embedding" norm whose definition itself depends on **f**, then a lot of the issues concerning the expansion factor disappear, and one can achieve meaningful reductions for all polynomials **f**.

In this current work, though, we cannot use a definition of norm that depends on **f** because there is no **f** in our average-case problem! We therefore need to use the most natural definition for small elements that is independent of any ring. For this, we go back to the definition that simply looks at the coefficients of the polynomials. The reason that we believe that this is most natural is because for many rings, a small coefficient norm implies a small norm in the canonical embedding. Unfortunately, there are rings for which this does not hold true (these are the ones with the large expansion factor), but it seems impossible to define a norm that is independent of **f** in which products of small elements remain small in $\mathbb{Z}_q[\mathbf{x}]/\langle\mathbf{f}\rangle$ for all **f**. We do want to point out that all polynomials that have been proposed for applications such as cyclotomics (of reasonable degree) and others, such as $\mathbf{x}^p - \mathbf{x} - 1$, have small expansion factors. In particular, any polynomial of the form $\mathbf{x}^n + \sum_{i=0}^{\lfloor n/2 \rfloor} a_i\mathbf{x}^i$ where a_i are small, has a relatively small expansion factor [LM06]. Thus the signature scheme in this paper is as hard to break as finding short vectors all such rings $\mathbb{Z}_q[\mathbf{x}]/\langle\mathbf{f}\rangle$, of which there are exponentially many.

1.2 Discussion and Open Problems

While our scheme has keys and ciphertexts which are of size $\tilde{O}(\lambda)$ in the security parameter, just like in signature schemes based on the Ring-SIS and Ring-LWE problems, the concrete instantiations are worse (see Fig. 1) than those of the most practical schemes. Compared to BLISS [DDLL13], the secret key is about 20 times larger, the public key 10 times, and the signature about 30 times. We did not optimize our scheme using the tricks from [GLP12, DDLL13] such as compressing the signature using Huffman codes and altering the random oracle to allow us to output one less polynomial in the signature. A rough estimate shows that these improvements would decrease our signature size by about 20 %, which would still not make it competitive with the best constructions. The biggest contributor to the superiority of the current state-of-the-art schemes is that they are based on Ring-LWE rather than Ring-SIS.

It was shown in [Lyu12] that by creating the public key for the signature scheme based on LWE (or an inhomogeneous version of SIS where there is a unique solution), one can reduce the key/signature sizes by about an order of magnitude. There seems to be a major roadblock to getting a reduction from such problems to those that work over the ring $\mathbb{Z}_q[\mathbf{x}]$, though. As we mentioned in the previous section, one reason that we were able to give a reduction from $\mathbf{f}$-SIS to Ring-SIS over $\mathbb{Z}_q[\mathbf{x}]$ is because the input to $\mathbf{f}$-SIS does not really depend on $\mathbf{f}$. In an inhomogeneous version of $\mathbf{f}$-SIS, however, where one is given $\mathbf{a}_1, \ldots, \mathbf{a}_k$ and $\mathbf{t} = \sum \mathbf{a}_i \mathbf{s}_i \in \mathbb{Z}_q[\mathbf{x}]/\langle \mathbf{f} \rangle$, where $\mathbf{t}$ is *not* statistically-close to uniform in $\mathbb{Z}_q[\mathbf{x}]/\langle \mathbf{f} \rangle$, the value of $\mathbf{t}$ very much depends on $\mathbf{f}$. Thus it is not clear to us how to transform this into an instance that is at the same time independent from $\mathbf{f}$, yet somehow retains pseudo-randomness.

In addition to being able to create more efficient signatures based on the hardness of worst-case problems over all rings, getting such a reduction from $\mathbf{f}$-LWE would then allow for efficient constructions of encryption schemes and a myriad of other primitives with the same hardness guarantees. We therefore believe that finding such a reduction would be truly an outstanding result. A slightly weaker, yet also very interesting achievement, would be to construct schemes which are simultaneously as hard as problems over a few different types of rings. The trivial solution would be to simply combine two schemes over two different rings, so the question here is whether it is possible to get something more efficient than the trivial construction.

Of a more theoretical nature is the direction of trying to understand the real hardness of our new average case problems without relating them to $\mathbb{Z}_q[\mathbf{x}]/\langle \mathbf{f} \rangle$. The average-case problems that we define in this paper operate over the ring $\mathbb{Z}_q[\mathbf{x}]$, so perhaps showing that they are as hard as solving lattice problems over ideals in $\mathbb{Z}[\mathbf{x}]/\langle \mathbf{f} \rangle$ is not the most "natural" reduction. It would therefore be an interesting problem if one could give a reduction to our average-case problem from a different worst-case problem, perhaps more directly related to the ring $\mathbb{Z}_q[\mathbf{x}]$.

1.3 Paper Organization

In Sect. 2 we introduce the notation and definitions that are used throughout the paper. Section 3 presents the new average-case problems defined over the ring $\mathbb{Z}_q[\mathbf{x}]$ and lemmas showing their relation to lattice problems over all polynomial rings. In Sect. 4, we describe a signature scheme and prove its security based on the hardness of our new average-case problems.

2 Preliminaries

2.1 Notation

Throughout the paper, R will denote the polynomial ring $\mathbb{Z}_q[\mathbf{x}]$. We will also assume that all polynomial operations occur in this ring (thus we will not write mod q, as it is implicit). Elements of this ring can be represented by polynomials $\mathbf{a} = \sum_{i=0}^{\infty} a_i \mathbf{x}^i$ where $a_i \in \{-\lfloor \frac{q}{2} \rfloor, \ldots, \lfloor \frac{q-1}{2} \rfloor\}$. For a polynomial $\mathbf{a} \in R$ with a finite degree $\deg(\mathbf{a})$, we denote $\|\mathbf{a}\|_\infty$ to mean $\max_{a_i} |a_i|$ and $\|\mathbf{a}\|_1$ to be $\sum_{i=0}^{\deg(\mathbf{a})-1} |a_i|$.

We will write $R^{<n}$ to mean the set of all polynomials in R of degree less than n, and $R_i^{<n}$ to be polynomials $\mathbf{a} \in R^{<n}$ with $\|\mathbf{a}\|_\infty \leq i$. For a polynomial $\mathbf{a} \in R$ and a monic polynomial $\mathbf{f}$ of degree n, the expression $\mathbf{a} \bmod \mathbf{f}$ denotes the unique polynomial $\mathbf{a}'$ in $R^{<n}$ for which there exists an $\mathbf{r} \in \mathbb{Z}_q[\mathbf{x}]$ such that $\mathbf{a}' + \mathbf{r}\mathbf{f} = \mathbf{a}$.

There is a natural mapping between polynomial in $\mathbb{Z}[\mathbf{x}]$ of degree $n-1$ and vectors in $\mathbb{Z}^n$ that simply maps each coefficient of the polynomial to a vector coordinate. We will make use of this mapping implicitly throughout the paper – that is elements in $\mathbb{Z}^n$ are simultaneously polynomials in $R^{<n}$. If $\mathbf{a}_1, \ldots, \mathbf{a}_k$ are elements in $\mathbb{Z}^n$, then their concatenation $(\mathbf{a}_1 \mid \ldots \mid \mathbf{a}_k)$ is a vector in $\mathbb{Z}^{kn}$.

For a set S, we denote $s \xleftarrow{\$} S$ to mean that s is chosen uniformly at random from S. For a distribution D, we write $s \xleftarrow{\$} D$ to mean that s is chosen according to the distribution D.

2.2 Lattice Problems

Definition 2.1 (Approximate shortest vector problc). *Let Λ be a lattice corresponding to an ideal in the polynomial ring $\mathbb{Z}[\mathbf{x}]/\langle \mathbf{f} \rangle$ and $\gamma \geq 1$ be some real. The* $\mathbf{f}$-SVP$_\gamma(\Lambda)$ *problem asks to find an element $\mathbf{v} \in \Lambda$ such that $\|\mathbf{v}\|_\infty \leq \gamma \cdot \min_{\mathbf{w} \in \Lambda \setminus \{\mathbf{0}\}} (\|\mathbf{w}\|_\infty)$.*

Definition 2.2 (Ring-SIS). *The homogeneous* $\mathbf{f}$-SIS *problem is defined as follows. An instance of the* $\mathbf{f}$-SIS$_{k,q,\beta}$ *problem consists of $\mathbf{a}_1, \ldots, \mathbf{a}_k \xleftarrow{\$} \mathbb{Z}_q[\mathbf{x}]/\langle \mathbf{f} \rangle$. A solution to the problem is k elements $\mathbf{z}_1, \ldots, \mathbf{z}_k$ such that $\|\mathbf{z}_i\|_\infty \leq \beta$ and*

$$\sum_{i=1}^{k} \mathbf{a}_i \mathbf{z}_i = \mathbf{0} \bmod \mathbf{f}.$$

The main result of [LM06] was a connection between the hardness of the $\mathbf{f}$-SVP$_\gamma$ problem for all lattices in $\mathbb{Z}[\mathbf{x}]/\langle \mathbf{f} \rangle$ and the $\mathbf{f}$-SIS$_{k,q,\beta}$ problem. If the length of elements is defined by the $\|\cdot\|_\infty$ function that simply looks at the largest coefficient, then the quality of the reduction has a dependency on a certain property of $\mathbf{f}$ that was called the "expansion factor". This expansion factor explains how much the coefficients of a polynomial in $\mathbb{Z}[\mathbf{x}]$ grow when reduced modulo $\mathbf{f}$.

For the purposes of the theorem, we define the value $\theta_\mathbf{f}$ as

$$\theta_\mathbf{f} = \max_{\mathbf{g} \in \mathbb{Z}[\mathbf{x}], \deg(\mathbf{g}) \leq 3(\deg(\mathbf{f})-1)} \frac{\|\mathbf{g} \bmod \mathbf{f}\|_\infty}{\|\mathbf{g}\|_\infty}.$$

It was shown in [LM06] that for polynomials such as $\mathbf{x}^n + 1$ and $\sum_{i=0}^{p-1} \mathbf{x}^i$, the value of $\theta_\mathbf{f}$ is a small constant (3 and 6 respectively). The paper also showed how to put bounds on the expansion factor of other polynomials. We direct the interested reader to [LM06] for a further discussion of this topic.

Theorem 2.3. *[LM06] For any monic, irreducible (over the integers)* $\mathbf{f}$ *and* $q > 2\theta_\mathbf{f}\beta k n^{1.5} \log n$, *if there is a polynomial-time algorithm that solves the* $\mathbf{f}$-$\mathrm{SIS}_{k,q,\beta}$ *problem with some non-negligible probability, then there is a polynomial-time algorithm that solves the* $\mathbf{f}$-SVP_γ *problem with* $\gamma = 8\theta_\mathbf{f}\beta k n \log^2 n$ *for any lattice* Λ *that corresponds to an ideal in* $\mathbb{Z}[\mathbf{x}]/\langle \mathbf{f} \rangle$.

2.3 The Discrete Normal (Gaussian) Distribution over $\mathbb{Z}^m$

Definition 2.4. *The continuous Normal distribution over* $\mathbb{R}^m$ *centered at* $\mathbf{v}$ *with standard deviation* σ *is defined by the function* $\rho^m_{\mathbf{v},\sigma}(\mathbf{x}) = \left(\frac{1}{\sqrt{2\pi\sigma^2}}\right)^m e^{\frac{-\|\mathbf{x}-\mathbf{v}\|^2}{2\sigma^2}}$

When $\mathbf{v} = 0$, we will just write $\rho^m_\sigma(\mathbf{x})$. The *discrete* Normal distribution over $\mathbb{Z}^m$ is defined as follows:

Definition 2.5. *The discrete Normal distribution over* $\mathbb{Z}^m$ *centered at some* $\mathbf{v} \in \mathbb{Z}^m$ *with standard deviation* σ *is defined as* $D^m_{\mathbf{v},\sigma}(\mathbf{x}) = \rho^m_{\mathbf{v},\sigma}(\mathbf{x})/\rho^m_\sigma(\mathbb{Z}^m)$.

The below is a basic fact about the length of the discrete Gaussian distribution over $\mathbb{Z}$.

Lemma 2.6. *For any* $r > 0$

$$\Pr_{z \leftarrow D^1_\sigma}[|z| > r\sigma] \leq 2e^{-r^2/2}.$$

Lemma 2.7 (Adapted from [Lyu12]**).** *Let* V *be a subset of* $\mathbb{Z}^m$ *in which all elements have norms less than* T, σ *be defined as* $11 \cdot T$, *and* $h : V \to \mathbb{R}$ *be a probability distribution. Then the probability that the following algorithm* $\mathcal{A}$:

1: $\mathbf{v} \xleftarrow{\$} h$
2: $\mathbf{z} \xleftarrow{\$} D^m_{\mathbf{v},\sigma}$
3: output $(\mathbf{z}, \mathbf{v})$ *with probability* $\min\left(\frac{D^m_\sigma(\mathbf{z})}{3 \cdot D^m_{\mathbf{v},\sigma}(\mathbf{z})}, 1\right)$
4: if nothing was output, goto Step 1

terminates within 200 *iterations is greater than* $1 - 2^{-90}$ *(the expected number of iterations is* 3*), and conditioned on its termination, its distribution is within statistical distance* 2^{-95} *of the distribution of the following algorithm* $\mathcal{F}$:

1: $\mathbf{v} \xleftarrow{\$} h$
2: $\mathbf{z} \xleftarrow{\$} D^m_\sigma$
3: output $(\mathbf{z}, \mathbf{v})$

2.4 Digital Signatures

Definition 2.8. *A* signature scheme *consists of a triplet of polynomial-time (possibly probabilistic) algorithms* (G, S, V) *such that for every pair of outputs* (s, v) *of* $G(1^n)$ *and any n-bit message m,*

$$Pr[V(v, m, S(s, m)) = 1] = 1$$

where the probability is taken over the randomness of algorithms S and V.

In the above definition, G is called the key-generation algorithm, S is the signing algorithm, V is the verification algorithm, and s and v are, respectively, the signing and verification keys.

Definition 2.9. *A signature scheme* (G, S, V) *is said to be secure if for every polynomial-time (possibly randomized) forger* $\mathcal{F}$*, the probability that after seeing the public key and* $\{(\mu_1, S(s, \mu_1)), \ldots, (\mu_q, S(s, \mu_q))\}$ *for any q messages* μ_i *of its choosing (where q is polynomial in n),* $\mathcal{F}$ *can produce* $(\mu \neq \mu_i, \sigma)$ *such that* $V(v, \mu, \sigma) = 1$*, is negligibly small. The probability is taken over the randomness of G, S, V, and* $\mathcal{F}$*.*

A stronger notion of security, called *strong unforgeability* requires that in addition to the above, a forger shouldn't even be able to come up with a different signature for a message whose signature he has already seen. The scheme in this paper satisfies this stronger notion.

2.5 Auxiliary Lemmas

Lemma 2.10. *Let* $\mathbf{a}$ *be any monic polynomial in* $\mathbb{Z}[\mathbf{x}]$ *of degree n. If* $\mathbf{b}$ *is a polynomial in* $\mathbb{Z}[\mathbf{x}]$ *of degree m each of whose coefficients is chosen at random modulo q, then the coefficients of* $\mathbf{c} = \mathbf{a} \cdot \mathbf{b} \bmod q$ *corresponding to the terms* $\mathbf{x}^n, \ldots, \mathbf{x}^{m+n}$ *are jointly uniformly random modulo q.*

Proof. If we write $\mathbf{c} = c_0 + c_1\mathbf{x} + \ldots + c_{m+n}\mathbf{x}^{m+n}$, then the coefficient c_{n+m-j} for $0 \leq j \leq m$ is

$$c_{m+n-j} = \sum_{i=0}^{j} a_{n-i} \cdot b_{m-j+i} = b_{m-j} + \sum_{i=1}^{j} a_{n-i} \cdot b_{m-j+i},$$

with the second equality being true because $\mathbf{a}$ is a monic polynomial.

From the above equality, is not hard to see that once we generate the coefficients b_{m-j} through b_m, we will have completely determined the coefficients c_{m+n-j} through c_{m+n} of the product. We can now prove the claim of the lemma by induction. The coefficient $c_{m+n} = b_m$, and is therefore uniformly random modulo q. Now assume that we have already selected the coefficients b_{m-k} through b_m, and therefore completely determined the coefficients of c_{m+n-j} through c_{m+n}, and they are jointly uniformly random modulo q. Once we select the coefficient b_{m-j-1}, we will have $c_{m+n-j-1} = b_{m-j-1} + \sum_{i=1}^{j+1} a_{n-i} \cdot b_{m-j-1+i}$. Because the term b_{m-j-1} was not used to determine c_m through c_{m+n-j}, we have

$$\begin{aligned}
&\Pr\left[c_{m+n-j-1} = \gamma \mid c_m, \ldots, c_{m+n-j}\right] \\
&= \Pr\left[b_{m-j-1} = \gamma - \sum_{i=1}^{j+1} a_{n-i} \cdot b_{m-j-1+i} \mid c_m, \ldots, c_{m+n-j}\right] \\
&= \Pr\left[b_{m-j-1} = \gamma - \sum_{i=1}^{j+1} a_{n-i} \cdot b_{m-j-1+i}\right] = 1/q
\end{aligned}$$

□

Lemma 2.11. *Let $h : X \to Y$ be a deterministic function where X and Y are finite sets and $|X| \geq 2^\lambda |Y|$. If x is chosen uniformly at random from X, then with probability at least $1 - 2^{-\lambda}$, there exists another $x' \in X$ such that $h(x) = h(x')$.*

Proof. There are at most $|Y| - 1$ elements x in X for which there is no x' such that $h(x) = h(x')$. Therefore the probability that a randomly chosen x has a corresponding x' for which $h(x) = h(x')$ is at least $(|X| - |Y| + 1)/|X| = 1 - |Y|/|X| + 1/|X| > 1 - 2^{-\lambda}$. □

3 Ring-SIS over $\mathbb{Z}_q[\mathbf{x}]$

We will now present several average-case problems that are defined over the ring $\mathbb{Z}_q[\mathbf{x}]$ rather than $\mathbb{Z}_q[\mathbf{x}]/\langle\mathbf{f}\rangle$. The first such problem simply asks for a linear combination of the inputs that sum to $\mathbf{0}$ in $\mathbb{Z}_q[\mathbf{x}]$. This is quite similar to the $\mathbf{f}$-SIS problem from Definition 2.2, except that there is no reduction modulo $\mathbf{f}$ and we also limit the degrees of the solution polynomials.

Definition 3.1. *The homogeneous $R^{<n}$-$\mathrm{SIS}_{k,d,\beta}$ problem is defined as follows. An instance of $R^{<n}$-$\mathrm{SIS}_{k,d,\beta}$ consists of $\mathbf{a}_1, \ldots, \mathbf{a}_k \xleftarrow{\$} R^{<n}$ and a solution to the problem is k elements $\mathbf{z}_1, \ldots, \mathbf{z}_k \in R_\beta^{<d}$ such that at least one $\mathbf{z}_i \neq \mathbf{0}$ and $\sum_{i=1}^{k} \mathbf{a}_i \mathbf{z}_i = \mathbf{0}$.*

Notice that if $\deg(\mathbf{f})$ is n, then instances of the $\mathbf{f}$-$\mathrm{SIS}_{k,q,\beta}$ and the $R^{<n}$-$\mathrm{SIS}_{k,d,\beta}$ have exactly the same distributions. Furthermore, it should be clear that if $\mathbf{z}_1, \ldots, \mathbf{z}_k$ is a solution to the instance $\mathbf{a}_1, \ldots, \mathbf{a}_k$ of the $R^{<n}$-$\mathrm{SIS}_{k,d,\beta}$ problem, then it is also a solution to the instance $\mathbf{a}_1, \ldots, \mathbf{a}_k$ of the $\mathbf{f}$-$\mathrm{SIS}_{k,q,\beta}$ problem. The next simple lemma shows that one can also transform instance of the $\mathbf{f}$-$\mathrm{SIS}_{k,q,\beta}$ problem for $d \leq \deg(\mathbf{f}) \leq n$ into instances of the $R^{<n}$-$\mathrm{SIS}_{k,d,\beta}$ problem such that solutions to the latter are still solutions to the former.

Lemma 3.2. *If there is an algorithm that can solve the* $R^{<n}$-$\mathrm{SIS}_{k,d,\beta}$ *problem in time* t *with probability* ϵ*, then there is an algorithm that can solve* $\mathbf{f}$-$\mathrm{SIS}_{k,q,\beta}$ *problem in time* $t + poly(n)$ *with probability* ϵ *as long as* $d \leq \deg(\mathbf{f}) \leq n$.

Proof. Given $\mathbf{a}_1, \ldots, \mathbf{a}_k$ that form an instance of the $\mathbf{f}$-$\mathrm{SIS}_{k,q,\beta}$, we choose polynomials $\mathbf{r}_1, \ldots, \mathbf{r}_k \in R^{<n-\deg(\mathbf{f})}$ and create $\mathbf{a}'_i \leftarrow \mathbf{a}_i + \mathbf{f} \cdot \mathbf{r}_i$. If we write $\mathbf{a}'_i = \sum_{j=0}^{n-1} a_j \mathbf{x}^j$, then Lemma 2.10 states that the coefficients $a_{deg(\mathbf{f})}$ through a_{n-1} are jointly uniformly random modulo q (because they are completely determined by $\mathbf{f} \cdot \mathbf{r}_i$). And since all the $\mathbf{a}_i$ are uniformly random in $R^{<deg(\mathbf{f})}$, we have that all of the $\mathbf{a}'_i = \mathbf{a}_i + \mathbf{f} \cdot \mathbf{r}_i$ are uniformly random in $R^{<n}$.

We feed the instance $\mathbf{a}'_1, \ldots, \mathbf{a}'_k$ to the $R^{<n}$-$\mathrm{SIS}_{k,d,\beta}$ oracle. If he returns a solution $\mathbf{z}_1 \ldots, \mathbf{z}_k \in R_\beta^{<d}$ such that $\sum_{i=1}^{k} \mathbf{a}'_i \mathbf{z}_i = \mathbf{0}$, then we claim that $\mathbf{z}_1, \ldots, \mathbf{z}_k$ is also a solution to the $\mathbf{f}$-$\mathrm{SIS}_{k,q,\beta}$ problem. First observe that

$$\mathbf{0} = \sum_{i=1}^{k} \mathbf{a}'_i \mathbf{z}_i = \sum_{i=1}^{k} (\mathbf{a}_i + \mathbf{r}_i \mathbf{f}) \mathbf{z}_i = \sum_{i=1}^{k} \mathbf{a}_i \mathbf{z}_i + \sum_{i=1}^{k} \mathbf{r}_i \mathbf{f} \mathbf{z}_i = \sum_{i=1}^{k} \mathbf{a}_i \mathbf{z}_i \bmod \mathbf{f}.$$

Furthermore, because $\deg(\mathbf{z}_i) < d \leq \deg(\mathbf{f})$, we have that $\mathbf{z}_i = \mathbf{z}_i \bmod \mathbf{f}$. Thus if at least one of the $\mathbf{z}_i$ is non-zero, so is one of the $\mathbf{z}_i \bmod \mathbf{f}$. □

We next define an approximate inhomogeneous version of the Ring-SIS problem over $\mathbb{Z}_q[\mathbf{x}]$. The exact reasoning for the particular definition is due to the particularities of the signature scheme that we will be constructing in the next section. Intuitively, the inhomogeneous version of Ring-SIS should ask to find a solution $(\mathbf{z}_1, \ldots, \mathbf{z}_k)$ that satisfies $\sum \mathbf{a}_i \mathbf{z}_i = \mathbf{t}$. In our definition below, we additionally specify the distribution that the input $\mathbf{t}$ should have, and also allow an approximate solution to this equation – meaning that the sum $\sum \mathbf{a}_i \mathbf{z}_i$ does not to equal exactly $\mathbf{t}$, but could equal to $\mathbf{tc}$ for some element $\mathbf{c} \in \mathbb{Z}_q[\mathbf{x}]$ with a small ℓ_1 norm.

Definition 3.3. *We define the approximate inhomogeneous Ring-SIS problem as follows. An instance of the* $R^{<n}$-$\mathrm{SIS}_{k,d_1,d_2,s,c,\beta}$ *problem consists of polynomials* $\mathbf{a}_1, \ldots, \mathbf{a}_k \xleftarrow{\$} R^{<n}$ *and a* $\mathbf{t} = \sum_{i=1}^{k} \mathbf{a}_i \mathbf{s}_i$ *where* $\mathbf{s}_i \xleftarrow{\$} R_s^{<d_1}$. *A solution to the problem is* k *elements* $\mathbf{z}_1, \ldots, \mathbf{z}_k \in R_\beta^{<d_2}$ *and a* $\mathbf{c} \in R^{<d_2-d_1+1}$ *with* $0 < \|\mathbf{c}\|_1 \leq c$ *such that*

$$\sum_{i=1}^{k} \mathbf{a}_i \mathbf{z}_i = \mathbf{tc}.$$

The next lemma relates the hardness of solving the inhomogeneous Ring-SIS problem to the homogeneous one. We show that under certain conditions, solving the particular version of the inhomogeneous problem implies being able to solve the homogeneous one.

Lemma 3.4. *Suppose that the following relationships are satisfied:*

1. $d_1 < d_2 \leq n$.
2. $s > 2^{\frac{\lambda}{kd_1}-1} \cdot q^{\frac{n+d_1}{kd_1}}$
3. $sc < q/4$

If there is an algorithm that solves the $R^{<n}$-$\mathrm{SIS}_{k,d_1,d_2,s,c,\beta}$ *problem in time* t *with probability* ϵ*, there is an algorithm that solves the* $R^{<n}$-$\mathrm{SIS}_{k,d_2,\beta+sc}$ *problem with probability at least* $\frac{1}{2} \cdot (\epsilon - 2^{-\lambda})$ *in time* $poly(n) + t$.

Proof. Given an instance $\mathbf{a}_1, \ldots, \mathbf{a}_k$ of an $R^{<n}$-$\mathrm{SIS}_{k,d_2,\beta+sc}$ problem, we select $\mathbf{s}_1, \ldots, \mathbf{s}_k \stackrel{\$}{\leftarrow} R_s^{<d_1}$ and set $\mathbf{t} \leftarrow \sum_{i=1}^{k} \mathbf{a}_i \mathbf{s}_i$. We give the instance $\mathbf{a}_1, \ldots, \mathbf{a}_k, \mathbf{t}$ to the oracle who can solve $R^{<n}$-$\mathrm{SIS}_{k,d_1,d_2,s,c,\beta}$.

Suppose the oracle solves the problem and returns k elements $\mathbf{z}_1, \ldots, \mathbf{z}_k \in R_\beta^{<d_2}$ and a $\mathbf{c} \in R^{<d_2-d_1+1}$ with $\|\mathbf{c}\|_1 \leq c$ such that

$$\sum_{i=1}^{k} \mathbf{a}_i \mathbf{z}_i = \mathbf{tc} = \mathbf{c} \sum_{i=1}^{k} \mathbf{a}_i \mathbf{s}_i,$$

which implies that

$$\sum_{i=1}^{k} \mathbf{a}_i (\mathbf{z}_i - \mathbf{s}_i \mathbf{c}) = \mathbf{0}.$$

Note that $\deg(\mathbf{z}_i - \mathbf{s}_i \mathbf{c}) < d_2$ and

$$\|\mathbf{z}_i - \mathbf{s}_i \mathbf{c}\|_\infty \leq \|\mathbf{z}_i\|_\infty + \|\mathbf{s}_i \mathbf{c}\|_\infty \leq \beta + \|\mathbf{s}_i\|_\infty \cdot \|\mathbf{c}\|_1 \leq \beta + sc.$$

Thus if for some i, $\mathbf{z}_i - \mathbf{s}_i \mathbf{c} \neq \mathbf{0}$, we have a solution for $R^{<n}$-$\mathrm{SIS}_{k,d,\beta+sc}$. If we consider the function $f : (R_s^{<d_1})^k \rightarrow R/^{<n+d_1-1}$ defined as $f(\mathbf{s}_1, \cdots, \mathbf{s}_k) = \sum_{i=1}^{k} \mathbf{a}'_i \mathbf{s}_i$, the domain size of this function is $(2s+1)^{kd_1}$, while the range is of size q^{n+d_1-1}. Because we set $s > 2^{\lambda/(kd_1)-1} \cdot q^{(n+d_1-1)/(kd_1)}$, the size of the domain is greater than $2^\lambda \cdot q^{n+d_1-1}$. By Lemma 2.11, there is probability at least $1-2^{-\lambda}$ that there exists another $\mathbf{s}'_1, \ldots, \mathbf{s}'_k \in R_s^{<d_1}$ such that

$$\mathbf{t} = \sum_{i=1}^{k} \mathbf{a}'_i \mathbf{s}_i = \sum_{i=1}^{k} \mathbf{a}'_i \mathbf{s}'_i.$$

Since it is perfectly indistinguishable whether $\mathbf{s}_1, \ldots, \mathbf{s}_k$ or $\mathbf{s}'_1, \ldots, \mathbf{s}'_k$ were used in creating $\mathbf{t}$ (because both of them have the same posterior probability of having been chosen), the probability of the oracle outputting $\mathbf{z}_1, \ldots, \mathbf{z}_k, \mathbf{c}$ such that

$\mathbf{z}_i - \mathbf{s}_i\mathbf{c} \bmod \mathbf{f} = \mathbf{0}$ is exactly the same if $\mathbf{t}$ were generated as in the reduction, but then after the adversary produced his output, the preimage of $\mathbf{t}$ was chosen at random among all the valid choices. We will now show that $\mathbf{z}_i - \mathbf{s}_i\mathbf{c}$ can only equal $\mathbf{0}$ for all i for at most one of these choices.

If $(\mathbf{s}_1, \ldots, \mathbf{s}_k) \neq (\mathbf{s}'_1, \ldots, \mathbf{s}'_k)$, then there should be at least one $\mathbf{s}_i \neq \mathbf{s}'_i$. For this i, suppose that $\mathbf{z}_i - \mathbf{s}_i\mathbf{c} = \mathbf{0} = \mathbf{z}_i - \mathbf{s}'_i\mathbf{c}$. This implies that $(\mathbf{s}_i - \mathbf{s}'_i)\mathbf{c} = \mathbf{0}$. Since $\mathbb{Z}_q[\mathbf{x}]$ is an integral domain, this can only happen if either $\mathbf{c} = \mathbf{0}$ or if $\mathbf{s}_i = \mathbf{s}'_i$. This is a contradiction. Therefore with probability at least 1/2, some $\mathbf{z}_i - \mathbf{s}_i\mathbf{c} \neq \mathbf{0}$. □

4 The Signature Scheme

We now formally describe our scheme via secret key generation, public key generation, signing, and verification algorithms.

The fixed, public parameters in our scheme are stated below. The values n, k, q, s, d_1, d_2, c are intuitively related to the parametrization of the $R^{<n}$-SIS problem, with the standard deviation σ being related to the parameter β. We furthermore define a cryptographic function H whose range is the set C which consists of bounded-degree polynomials with small ℓ_1 norms.

Fixed Parameters:
- Positive integers $n, k, q, s, d_1, d_2, c, \sigma = 11sc \cdot \sqrt{d_2 k}$
- Ring $R = \mathbb{Z}_q[\mathbf{x}]$
- Set $C = \{\mathbf{c} \in R_1^{<d_2 - d_1 + 1}$ with $\|\mathbf{c}\|_1 \leq c\}$
- Cryptographic hash function $H : \{0,1\}^* \to C$

In Fig. 1, we give some sample parameters with which our scheme can be instantiated. For this, we use the reduction from breaking the signature scheme to the $\mathbf{f}$-SIS problem that is given in the next section. In that section we show that breaking the scheme implies solving the $\mathbf{f}\text{-SIS}_{k,q,\beta}$ problem for $\beta = 2sc + 10\sigma$. Even though there is a reduction from every $\mathbf{f}$ whose degree is between d_2 and n, we instantiate the security based on the hardness of the $\mathbf{f}$-SIS problem for $\mathbf{f}$ whose degree is close to n. Of course if one wants to be more conservative, one could set the parameters so that the scheme is even secure in practice for polynomials whose degrees are closer to d_2.

To set the concrete parameters, we use the standard notion of the Hermite factor defined in [GN08] and the explanation for how to approximate it for the SIS problem given in [MR08].

The key generation algorithm generates $\mathbf{a}_1, \ldots, \mathbf{a}_k \xleftarrow{\$} R^{<n}$ and $\mathbf{s}_1, \ldots, \mathbf{s}_k \xleftarrow{\$} R_s^{<d_1}$, and then outputs $(\mathbf{a}_1, \ldots, \mathbf{a}_k, \mathbf{t} = \sum_{i=1}^{k} \mathbf{a}_i\mathbf{s}_i)$ as the public key. This is, in fact, an instance of the inhomogeneous $R^{<n}$-SIS problem from Definition 3.3.

n	1459
k	6
q	$\approx 2^{30}$
s	1535
d_1	1111
d_2	1285
c	36
σ	$\approx 2^{25.7}$
secret key size	8.8 KB
public key size	9.6 KB
signature size	27 KB
Hermite factor	1.005

Fig. 1. Sample parameters for the signature scheme

To generate a signature of μ, the signer selects "masking" variables $\mathbf{y}_i$ from a particular distribution, computes $\mathbf{c} = \mathrm{H}(\sum \mathbf{a}_i\mathbf{y}_i, \mu)$, and then creates $\mathbf{z}_i = \mathbf{s}_i\mathbf{c} + \mathbf{y}_i$. By the way the parameters were set, each $\mathbf{z}_i$ is in $R^{<d_2}$. Thus the concatenation of the k vectors $\mathbf{z} = (\mathbf{z}_1 \mid \ldots \mid \mathbf{z}_k)$ can be thought of as a vector in $\mathbb{Z}^{kd_2}$. If we similarly define the vector $\mathbf{s} = (\mathbf{s}_1\mathbf{c} \mid \ldots, \mathbf{s}_k\mathbf{c}) \in \mathbb{Z}^{kd_2}$, then we can see that the vector $\mathbf{z}$ is distributed according to the discrete Gaussian distribution $D^{kd_2}_{\mathbf{s},\sigma}$. To get rid of the dependence on $\mathbf{s}$, we use the rejection sampling procedure from [Lyu12] by running the RejectionSample algorithm. By the way the parameters are set, there is a 1/3 probability that the signature will be output, and a 2/3 chance that the signing procedure will need to be restarted. After some $(\mathbf{z}_1, \ldots, \mathbf{z}_k)$ eventually passes the rejection sampling procedure, its distribution will be exactly $D^{kd_2}_{\sigma}$.

Key Generation:

1. Generate $\mathbf{a}_1, \ldots, \mathbf{a}_k \xleftarrow{\$} R^{<n}$
2. Generate $\mathbf{s}_1, \ldots, \mathbf{s}_k \xleftarrow{\$} R_s^{<d_1}$
3. Set $\mathbf{t} \leftarrow \sum_{i=1}^{k} \mathbf{a}_i\mathbf{s}_i$
4. Public Key $\leftarrow (\mathbf{a}_1, \ldots, \mathbf{a}_k, \mathbf{t})$, Secret Key $\leftarrow (\mathbf{s}_1, \ldots, \mathbf{s}_k)$

Because the distribution is being sampled from $\mathbb{Z}^{kn}$, which is an orthogonal lattice, each coefficient of $\mathbf{z}_i$ is distributed according to D^1_{σ}. Thus, by Lemma 2.6, the probability that some coefficient is larger than 5σ in absolute value is less than $2e^{-25/2} < 2^{-17}$. For simplicity, we would like to make sure that all $\mathbf{z}_i$ are small, and so we check that each of their coefficients is less than 5σ. The probability that all kd_2 positions are less than 5σ is at least $1 - kd_2 \cdot 2^{-17}$. In our sample instantiation, $kd_2 < 2^{13}$, and thus the probability that this check is passed is greater than 15/16. So with probabilty at most 1/16, the procedure gets restarted. The signing algorithm finally outputs $(\mathbf{z}_1, \ldots, \mathbf{z}_k, \mathbf{c})$.

Sign(μ, ($\mathbf{a}_1, \ldots, \mathbf{a}_k, \mathbf{t}$), ($\mathbf{s}_1, \ldots, \mathbf{s}_k$)):

1. Generate $\mathbf{y}_1, \ldots, \mathbf{y}_k \in R^{<d_2}$ such that $\mathbf{y}_i \sim D_\sigma^{d_2}$
2. Set $\mathbf{c} = \mathrm{H}\left(\sum_{i=1}^{k} \mathbf{a}_i \mathbf{y}_i, \mu\right)$
3. For $i = 1$ to k, set $\mathbf{z}_i = \mathbf{s}_i \mathbf{c} + \mathbf{y}_i$
4. $b \leftarrow \mathit{RejectionSample}(\mathbf{z}_1, \ldots, \mathbf{z}_k, \mathbf{s}_1, \ldots, \mathbf{s}_k, \mathbf{c}, \sigma, d_2)$
5. If $b = 0$, then goto 1
6. If for some i, $\|\mathbf{z}_i\|_\infty > 5\sigma$, then goto 1
7. Output $(\mathbf{z}_1, \ldots, \mathbf{z}_k, \mathbf{c})$

RejectionSample($\mathbf{z}_1, \ldots, \mathbf{z}_k, \mathbf{s}_1, \ldots, \mathbf{s}_k, \mathbf{c}, \sigma, d_2$):

1. Let $\mathbf{z} \leftarrow (\mathbf{z}_1 \mid \ldots \mid \mathbf{z}_k) \in \mathbb{Z}^{kd_2}$
2. Let $\mathbf{s} \leftarrow (\mathbf{s}_1 \mathbf{c} \mid \ldots \mid \mathbf{s}_k \mathbf{c}) \in \mathbb{Z}^{kd_2}$
3. With probability $D_\sigma^{kd_2}(\mathbf{z})/(3 \cdot D_{\mathbf{s},\sigma}^{kd_2}(\mathbf{z}))$, output 1. Else output 0.

The verification algorithm looks at the signature $(\mathbf{z}_1, \ldots, \mathbf{z}_k, \mathbf{c})$ and accepts if and only if all the coefficients of the $\mathbf{z}_i$ are less than 5σ and $\mathbf{c} = \mathrm{H}\left(\sum_{i=1}^{k} \mathbf{a}_i \mathbf{z}_i - \mathbf{t}\mathbf{c}, \mu\right)$.

Verify(($\mathbf{a}_1, \ldots, \mathbf{a}_k, \mathbf{t}$), ($\mathbf{z}_1, \ldots, \mathbf{z}_k, \mathbf{c}$)):

1. If for some i, $\deg(\mathbf{z}_i) \geq d_2$ or $\|\mathbf{z}_i\|_\infty > 5\sigma$, then Reject
2. If $\mathbf{c} \neq \mathrm{H}\left(\sum_{i=1}^{k} \mathbf{a}_i \mathbf{z}_i - \mathbf{t}\mathbf{c}, \mu\right)$, then Reject
3. Accept

4.1 Security

The main result of this section is a reduction from solving the $R^{<n}$-$\mathrm{SIS}_{k,d_2,2sc+10\sigma}$ problem to forging the signature scheme. We first show how one can simulate the signing algorithm without knowing the secret key $\mathbf{s}_1, \ldots, \mathbf{s}_k$ by programming the random oracle (Lemma 4.1).

We then show in Theorem 4.2 that an adversary who breaks the signature scheme that uses the signing algorithm from Lemma 4.1 can be used to solve either the $R^{<n}$-SIS problem from Definition 3.1 or the one from Definition 3.3. By Lemma 3.4, this implies that the adversary can be used to solve the problem from Definition 3.1, and therefore any instance of the **f**-SIS problem for **f** of degree between d_2 and n. The latter then allows one to solve worst-case lattice problems in the ring $\mathbb{Z}[\mathbf{x}]/\langle \mathbf{f} \rangle$.

Lemma 4.1. *Suppose that the random oracle H is already programmed on v values. Then the statistical distance between the output of the signing procedure and the following Hybrid signing algorithm, which does not take any secret keys* $\mathbf{s}_i$ *as inputs, is at most* $2^{-95} + v(\sqrt{2\pi}\sigma - 1)^{-d_2}$.

HybridSign$(\mu, (\mathbf{a}_1, \ldots, \mathbf{a}_k, \mathbf{t}))$

1. $\mathbf{c} \xleftarrow{\$} C$
2. *Generate* $\mathbf{z}_1, \ldots, \mathbf{z}_k \in R^{<d_2}$ *such that* $\mathbf{z}_i \sim D_\sigma^{d_2}$
3. *If for some* i, $\|\mathbf{z}_i\|_\infty > 5\sigma$, *then goto 1*
4. *Program* $\mathbf{c} = H\left(\sum_{i=1}^{k} \mathbf{a}_i\mathbf{z}_i - \mathbf{tc}, \mu\right)$
5. *Output* $(\mathbf{z}_1, \ldots, \mathbf{z}_k, \mathbf{c})$

Proof. We first define another intermediate signing hybrid algorithm named HybridSign′.

HybridSign′$(\mu, (\mathbf{a}_1, \ldots, \mathbf{a}_k, \mathbf{t}), (\mathbf{s}_1, \ldots, \mathbf{s}_k))$

1. Generate $\mathbf{y}_1, \ldots, \mathbf{y}_k \in R^{<d_2}$ such that $\mathbf{y}_i \sim D_\sigma^{d_2}$
2. $\mathbf{c} \xleftarrow{\$} C$
3. For $i = 1$ to k, set $\mathbf{z}_i = \mathbf{s}_i\mathbf{c} + \mathbf{y}_i$
4. $b \leftarrow$ *RejectionSample*$(\mathbf{z}_1, \ldots, \mathbf{z}_k, \mathbf{s}_1, \ldots, \mathbf{s}_k, \mathbf{c}, \sigma, d_2)$,
5. if $b = 0$, then goto 1
6. If for some i, $\|\mathbf{z}_i\|_\infty > 5\sigma$, hen goto 1
7. Program $\mathbf{c} = \mathrm{H}\left(\sum_{i=1}^{k} \mathbf{a}_i\mathbf{z}_i - \mathbf{tc}, \mu\right)$
8. Output $(\mathbf{z}_1, \ldots, \mathbf{z}_k, \mathbf{c})$

The difference between the real signing procedure and HybridSign′ is that the value of

$$\mathbf{c} = \mathrm{H}\left(\sum_{i=1}^{k} \mathbf{a}_i\mathbf{z}_i - \mathbf{tc}, \mu\right) = \mathrm{H}\left(\sum_{i=1}^{k} \mathbf{a}_i\mathbf{y}_i, \mu\right)$$

gets set uniformly at random in HybridSign′, whereas in the real signature scheme, H would first check whether H was already evaluated on $\left(\sum_{i=1}^{k} \mathbf{a}_i\mathbf{y}_i, \mu\right)$ and only assign it a random value if it wasn't. Therefore HybridSign′ will differ from the real scheme in the case that the value of $\sum_{i=1}^{k} \mathbf{a}_i\mathbf{y}_i$ collides with one of the already-queried values.

For any $\mathbf{w}$,

$$\Pr_{\mathbf{y}_i \xleftarrow{\$} D_\sigma^{d_2}}\left[\sum_i \mathbf{a}_i\mathbf{y}_i = \mathbf{w}\right] \leq \Pr_{\mathbf{z}_1 \xleftarrow{\$} D_\sigma^{d_2}}\left[\mathbf{a}\mathbf{z}_1 = \left(\mathbf{w} - \sum_{i \neq 1} \mathbf{a}_i\mathbf{y}_i\right)\right] < (\sqrt{2\pi}\sigma - 1)^{-d_2},$$

where the last inequality holds because there is at most one possible $\mathbf{z}_1$ that satisfies this equation (because $\mathbb{Z}_q[\mathbf{x}]$ is an integral domain) and because the likeliest element in the discrete Gaussian distribution is $\mathbf{0}$ which has probability less than $(\sqrt{2\pi}\sigma - 1)^{-n}$. Thus if there were already v values of the random oracle that were set, there is less than a $v \cdot (\sqrt{2\pi}\sigma - 1)^{-d_2}$ probability that there would be a collision. In our sample instantiation, for example, σ is approximately 2^{25} and $d_2 > 1200$, and so this probability is extremely small.

We now compare HybridSign′ with Hybrid 2. Lemma 2.7 states that the distribution of the eventual value of $(\mathbf{z}_1, \ldots, \mathbf{z}_k, \mathbf{c})$ after the first 5 steps of HybridSign′ is within statistical distance 2^{-95} of the distribution of $(\mathbf{z}_1, \ldots, \mathbf{z}_k, \mathbf{c})$ after two steps of HybridSign. Since the rest of the steps in both hybrids is identical, their statistical distance is at most 2^{-95}. Thus the statistical distance of the distributions of the output of the real signing algorithm and HybridSign is $2^{-95} + (\sqrt{2\pi}\sigma - 1)^{-d_2}$. □

Theorem 4.2. *Suppose there exists an adversary who makes a total of t queries to the Signing hybrid in Lemma 4.1 and the random oracle H during his attack and succeeds in forging with probability δ. Then there is an algorithm with the same time complexity that solves either the $R^{<n}$-$\mathrm{SIS}_{k,d_1,d_2,s,2c,10\sigma}$ problem or the $R^{<n}$-$\mathrm{SIS}_{k,d_2,10\sigma}$ problem with probability at least*

$$\frac{1}{2} \cdot \left(\delta - \frac{1}{|C|}\right)\left(\frac{\delta - 1/|C|}{t} - \frac{1}{|C|}\right).$$

Proof. Let $(\mathbf{a}_1, \ldots, \mathbf{a}_k, \mathbf{t})$ be an instance of the $R^{<n}$-$\mathrm{SIS}_{k,d_1,d_2,s,2c,10\sigma}$ problem and $(\mathbf{a}'_1, \ldots, \mathbf{a}'_k)$ be an instance of the $R^{<n}$-$\mathrm{SIS}_{k,d_2,10\sigma}$ problem. If we choose $\mathbf{s}'_1, \ldots, \mathbf{s}'_k \xleftarrow{\$} R^{<n}_{d_1}$ and compute $\mathbf{t}' = \sum \mathbf{a}'_i\mathbf{s}'_i$, then the distribution of $(\mathbf{a}_1, \ldots, \mathbf{a}_k, \mathbf{t})$ is exactly the same as that of $(\mathbf{a}'_1, \ldots, \mathbf{a}'_k, \mathbf{t}')$. The simulator then chooses one of those two sets at random and declares it as the public key of the signature scheme. If the adversary produces a forgery on a new message, then we will show that he will solve an instance of the $R^{<n}$-$\mathrm{SIS}_{k,d_1,d_2,s,2c,10\sigma}$ problem. If he produces a signature of a message he has already seen, then he will solve the $R^{<n}$-$\mathrm{SIS}_{k,d_2,10\sigma}$ problem. The simulator's hope is therefore that if he gives the adversary the instance $(\mathbf{a}_1, \ldots, \mathbf{a}_k, \mathbf{t})$, the adversary will forge a signature on a new message, whereas if the simulator gives $(\mathbf{a}'_1, \ldots, \mathbf{a}'_k, \mathbf{t}')$, the adversary will forge on a message he has already seen. It's easy to see that this lowers the success probability of the simulator by a factor of 2.

For simplicity, we will now refer to the public key as $(\mathbf{a}_1, \ldots, \mathbf{a}_k, \mathbf{t})$. During the attack, the adversary may interact with the Simulator in one of three ways. He may ask for a signature of a message μ' for which the Simulator will use Hybrid 2, or query the hash function H on any element in $\{0,1\}^*$, or produce a forgery μ. If the adversary asks for a signature of μ, the Simulator simply returns the output of Hybrid 2. If the adversary queries H on some value, then the Simulator first checks if that value was already assigned and returns it, or otherwise just chooses a random element $\mathbf{c} \in C$ and programs it to be the output of H on the adversary's input.

If the adversary comes up with a signature $(\mathbf{z}_1, \ldots, \mathbf{z}_k, \mathbf{c})$ for a message μ, then this signature satisfies the equality $\mathbf{c} = \mathrm{H}\left(\sum_{i=1}^{k} \mathbf{a}_i\mathbf{z}_i - \mathbf{t}\mathbf{c}, \mu\right)$. If the value for $\mathrm{H}\left(\sum_{i=1}^{k} \mathbf{a}_i\mathbf{z}_i - \mathbf{t}\mathbf{c}, \mu\right)$ has never been programmed during a signing query or a random oracle query, then the adversary has only a $1/|C|$ chance of guessing the $\mathbf{c}$ that equals to $\mathrm{H}\left(\sum_{i=1}^{k} \mathbf{a}_i\mathbf{z}_i - \mathbf{tc}, \mu\right)$. So we will assume that the value for $\mathrm{H}\left(\sum_{i=1}^{k} \mathbf{a}_i\mathbf{z}_i - \mathbf{tc}, \mu\right)$ has already been set.

We will first handle the case where it has been set during a signing query. In this case, the simulator already gave a signature $(\mathbf{z}'_1, \ldots, \mathbf{z}'_k, \mathbf{c})$ for the message μ. In order for $(\mathbf{z}_1, \ldots, \mathbf{z}_k, \mathbf{c})$ to be a valid forgery for μ, some $\mathbf{z}_i$ must be different from $\mathbf{z}'_i$. The adversary's forgery therefore implies that

$$\sum_{i=1}^{k} \mathbf{a}_i\mathbf{z}_i - \mathbf{tc} = \sum_{i=1}^{k} \mathbf{a}_i\mathbf{z}'_i - \mathbf{tc},$$

and therefore

$$\sum_{i=1}^{k} \mathbf{a}_i(\mathbf{z}_i - \mathbf{z}'_i) = \mathbf{0}$$

and at least for one i, $\mathbf{z}_i \neq \mathbf{z}'_i$. Since all $\|\mathbf{z}_i - \mathbf{z}'_i\|_\infty \leq 10\sigma$ and $\deg(\mathbf{z}_i - \mathbf{z}'_i) \leq d_2$, they form a solution to the $R^{<n}\text{-SIS}_{n,q,d_2,10\sigma}$ problem.

We now move to the case where the adversary constructs a signature for a message he has not yet seen. If the adversary comes up with a valid forgery $(\mathbf{z}_1, \ldots, \mathbf{z}_k, \mathbf{c})$ for a new message μ, then $\|\mathbf{z}_i\|_\infty \leq 5\sigma$ and $\mathbf{c} = \mathrm{H}\left(\sum_{i=1}^{k} \mathbf{a}_i\mathbf{z}_i - \mathbf{tc}, \mu\right)$. As before, if the adversary never queried H on $\left(\sum_{i=1}^{k} \mathbf{a}_i\mathbf{z}_i - \mathbf{tc}, \mu\right)$, then he only has at most a $1/|C|$ chance of producing such a forgery. Thus let's assume that the adversary did make such a "winning" query. We then "rewind" the adversary by rerunning him with the same random coins and responding to all the random oracle queries (both his and the ones used in the signing) the same way as before *until* the "winning" query. Starting from the "winning" query, however, we select uniformly random responses to all random oracle queries. Let $\mathbf{c}'$ be the new response to the "winning" query. By the General Forking Lemma of Bellare and Neven [BN06, Lemma 1], the probability that $\mathbf{c} \neq \mathbf{c}'$ and the adversary again forges on the "winning" query is at least

$$\left(\delta - \frac{1}{|C|}\right)\left(\frac{\delta - 1/|C|}{t} - \frac{1}{|C|}\right).$$

With the above probability, then, the Simulator obtains another equation $\mathbf{c}' = \mathrm{H}\left(\sum_{i=1}^{k} \mathbf{a}_i\mathbf{z}'_i - \mathbf{tc}', \mu\right)$ where $\sum_{i=1}^{k} \mathbf{a}_i\mathbf{z}'_i - \mathbf{tc}' = \sum_{i=1}^{k} \mathbf{a}_i\mathbf{z}_i - \mathbf{tc}$ because the query was the same in both runs of the adversary. Therefore

$$\sum_{i=1}^{k} \mathbf{a}_i(\mathbf{z} - \mathbf{z}'_i) = \mathbf{t}(\mathbf{c} - \mathbf{c}')$$

and so $(\mathbf{z}_1 - \mathbf{z}'_1, \ldots, \mathbf{z}_k - \mathbf{z}'_k, \mathbf{c} - \mathbf{c}')$ is a solution to the instance $(\mathbf{a}_1, \ldots, \mathbf{a}_k, \mathbf{t})$ of the $R^{<n}$-SIS$_{k,d_1,d_2,s,2c,10\sigma}$ problem. □

Putting Theorem 4.2, Lemmas 3.4, and 4.1 together, we see that if the signature scheme parameters satisfy the pre-conditions on the public parameters in Lemma 3.4, then an adversary who breaks the signature scheme either solves the $R^{<n}$-SIS$_{k,d_2,10\sigma}$ problem or the $R^{<n}$-SIS$_{k,d_2,2sc+10\sigma}$ problem (the latter is a strictly weaker problem). This implies that an adversary who breaks the signature scheme can be used to break the $\mathbf{f}$-SIS$_{k,q,2sc+10\sigma}$ problem for any polynomial $\mathbf{f}$ of degree between d_2 and n. By Theorem 2.3, this in turn gives a connection between breaking the signature scheme and finding short vectors for any lattice in any polynomial ring $\mathbb{Z}[\mathbf{x}]/\langle\mathbf{f}\rangle$ where the degree of $\mathbf{f}$ is between d_2 and n.

References

[ABD16] Albrecht, M., Bai, S., Ducas, L.: A subfield lattice attack on overstretched NTRU assumptions: cryptanalysis of some FHE and graded encoding schemes. IACR Cryptology ePrint Archive 2016, p. 127 (2016)

[Ajt96] Ajtai, M.: Generating hard instances of lattice problems (extended abstract). In: STOC, pp. 99–108 (1996)

[BCLvV16] Bernstein, D.J., Chuengsatiansup, C., Lange, T., van Vredendaal, C.: NTRU prime. Cryptology ePrint Archive, Report 2016/461 (2016). http://eprint.iacr.org/

[BN06] Bellare, M., Neven, G.: Multi-signatures in the plain public-key model and a general forking lemma. In: ACM Conference on Computer and Communications Security, pp. 390–399 (2006)

[BS16] Jean-François Biasse and Fang Song. Efficient quantum algorithms for computing class groups and solving the principal ideal problem in arbitrary degree number fields. In: SODA, pp. 893–902 (2016)

[CDPR16] Cramer, R., Ducas, L., Peikert, C., Regev, O.: Recovering short generators of principal ideals in cyclotomic rings. In: Fischlin, M., Coron, J.-S. (eds.) EUROCRYPT 2016. LNCS, vol. 9666, pp. 559–585. Springer, Heidelberg (2016). doi:10.1007/978-3-662-49896-5_20

[CGS14] Campbell, P., Groves, M., Shepherd, D.: Soliloquy: a cautionary tale. In: ETSI/IQC 2nd Quantum-Safe Crypto Workshop (2014)

[CJL16] Cheon, J.H., Jeong, J., Lee, C.: An algorithm for NTRU problems and cryptanalysis of the GGH multilinear map without an encoding of zero. IACR Cryptology ePrint Archive (2016)

[DDLL13] Ducas, L., Durmus, A., Lepoint, T., Lyubashevsky, V.: Lattice signatures and bimodal gaussians. In: Canetti, R., Garay, J.A. (eds.) CRYPTO 2013. LNCS, vol. 8042, pp. 40–56. Springer, Heidelberg (2013). doi:10.1007/978-3-642-40041-4_3

[EHKS14] Eisenträger, K., Hallgren, S., Kitaev, A.Y., Song, F.: A quantum algorithm for computing the unit group of an arbitrary degree number field. In: STOC (2014)

[GLP12] Güneysu, T., Lyubashevsky, V., Pöppelmann, T.: Practical lattice-based cryptography: a signature scheme for embedded systems. In: CHES, pp. 530–547 (2012)

[GN08] Gama, N., Nguyen, P.Q.: Predicting lattice reduction. In: Smart, N. (ed.) EUROCRYPT 2008. LNCS, vol. 4965, pp. 31–51. Springer, Heidelberg (2008). doi:10.1007/978-3-540-78967-3_3

[GS02] Gentry, C., Szydlo, M.: Cryptanalysis of the revised NTRU signature scheme. In: Knudsen, L.R. (ed.) EUROCRYPT 2002. LNCS, vol. 2332, pp. 299–320. Springer, Heidelberg (2002). doi:10.1007/3-540-46035-7_20

[LM06] Lyubashevsky, V., Micciancio, D.: Generalized compact knapsacks are collision resistant. In: Bugliesi, M., Preneel, B., Sassone, V., Wegener, I. (eds.) ICALP 2006. LNCS, vol. 4052, pp. 144–155. Springer, Heidelberg (2006). doi:10.1007/11787006_13

[LPR13] Lyubashevsky, V., Peikert, C.: On ideal lattices, learning with errors over rings. J. ACM **60**(6), 43 (2013). Preliminary version appeared in EUROCRYPT 2010

[Lyu09] Lyubashevsky, V.: Fiat-Shamir with aborts: applications to lattice and factoring-based signatures. In: Matsui, M. (ed.) ASIACRYPT 2009. LNCS, vol. 5912, pp. 598–616. Springer, Heidelberg (2009). doi:10.1007/978-3-642-10366-7_35

[Lyu12] Lyubashevsky, V.: Lattice signatures without trapdoors. In: Pointcheval, D., Johansson, T. (eds.) EUROCRYPT 2012. LNCS, vol. 7237, pp. 738–755. Springer, Heidelberg (2012). doi:10.1007/978-3-642-29011-4_43

[MR08] Micciancio, D., Regev, O.: Lattice-based cryptography. In: Bernstein, D.J., Buchmann, J., Dahmen, E. (eds.) Post-Quantum Cryptography, pp. 147–191. Springer, Heidelberg (2009)

[PR06] Peikert, C., Rosen, A.: Efficient collision-resistant hashing from worst-case assumptions on cyclic lattices. In: Halevi, S., Rabin, T. (eds.) TCC 2006. LNCS, vol. 3876, pp. 145–166. Springer, Heidelberg (2006). doi:10.1007/11681878_8

[PR07] Peikert, C., Rosen, A.: Lattices that admit logarithmic worst-case to average-case connection factors. In: STOC, pp. 478–487 (2007)

[SSTX09] Stehlé, D., Steinfeld, R., Tanaka, K., Xagawa, K.: Efficient public key encryption based on ideal lattices. In: Matsui, M. (ed.) ASIACRYPT 2009. LNCS, vol. 5912, pp. 617–635. Springer, Heidelberg (2009). doi:10.1007/978-3-642-10366-7_36

Provable Security

Adaptive Oblivious Transfer and Generalization

Olivier Blazy[1(✉)], Céline Chevalier[2], and Paul Germouty[1]

[1] Université de Limoges, XLim, Limoges, France
olivier.blazy@unilim.fr
[2] CRED, Université Panthéon-Assas, Paris, France

Abstract. Oblivious Transfer (OT) protocols were introduced in the seminal paper of Rabin, and allow a user to retrieve a given number of lines (usually one) in a database, without revealing which ones to the server. The server is ensured that only this given number of lines can be accessed per interaction, and so the others are protected; while the user is ensured that the server does not learn the numbers of the lines required. This primitive has a huge interest in practice, for example in secure multi-party computation, and directly echoes to Symmetrically Private Information Retrieval (SPIR).

Recent Oblivious Transfer instantiations secure in the UC framework suffer from a drastic fallback. After the first query, there is no improvement on the global scheme complexity and so subsequent queries each have a global complexity of $\mathcal{O}(|DB|)$ meaning that there is no gain compared to running completely independent queries. In this paper, we propose a new protocol solving this issue, and allowing to have subsequent queries with a complexity of $\mathcal{O}(\log(|DB|))$ while keeping round optimality, and prove the protocol security in the UC framework with adaptive corruptions and reliable erasures.

As a second contribution, we show that the techniques we use for Oblivious Transfer can be generalized to a new framework we call *Oblivious Language-Based Envelope* (OLBE). It is of practical interest since it seems more and more unrealistic to consider a database with uncontrolled access in access control scenarios. Our approach generalizes Oblivious Signature-Based Envelope, to handle more expressive credentials and requests from the user. Naturally, OLBE encompasses both OT and OSBE, but it also allows to achieve Oblivious Transfer with fine grain access over each line. For example, a user can access a line if and only if he possesses a certificate granting him access to such line.

We show how to generically and efficiently instantiate such primitive, and prove them secure in the Universal Composability framework, with adaptive corruptions assuming reliable erasures. We provide the new UC ideal functionalities when needed, or we show that the existing ones fit in our new framework.

The security of such designs allows to preserve both the secrecy of the database values and the user credentials. This symmetry allows to view our new approach as a generalization of the notion of Symmetrically PIR.

J.H. Cheon and T. Takagi (Eds.): ASIACRYPT 2016, Part II, LNCS 10032, pp. 217–247, 2016.
DOI: 10.1007/978-3-662-53890-6_8

Keywords: Adaptive oblivious transfer · Oblivious signature-based envelope · UC Framework · Private information retrieval

1 Introduction

Oblivious Transfer (OT) is a notion introduced by Rabin in [53]. In its classical 1-out-of-n version, it allows a user $\mathcal{U}$ to access a single line of a database while interacting with the server $\mathcal{S}$ owning the database. The user should be oblivious to the other line values, while the server should be oblivious to which line was indeed received. Oblivious transfer has a fundamental role for achieving secure multi-party computation: It is for example needed for every bit of input in Yao's protocol [59] as well as for Oblivious RAM ([56] for instance), for every AND gate in the Boolean circuit computing the function in [35] or for almost all known garbled circuits [6].

Private Information Retrieval (PIR) schemes [25] allow a user to retrieve information from a database, while ensuring that the database does not learn which data were retrieved. With the increasing need for user privacy, these schemes are quite useful in practice, be they used for accessing records for email repositories, collection of webpages, music... But while protecting the privacy of the user, it is equally important that the user should not learn more information than he is allowed to. This is called database privacy and the corresponding protocol is called a Symmetrically Private Information Retrieval (SPIR), which could be employed in practice, for medical data or biometric information. This notion is closely related to Oblivious Transfer.

Due to their huge interest in practice, it is important to achieve low communication on these Oblivious Transfer protocols. A usual drawback is that the server usually has to send a message equivalent to the whole database each time the user requests a line. If it is logical, in the UC framework, that an OT protocol requires a cost linear in the size of the database for the first line queried. One may then hope to amortize the cost for further queries between the same server and the same user (or even another user, if possible), reducing the efficiency gap between Private Information Retrieval schemes and their stronger equivalent Oblivious Transfer schemes. We thus deal in this paper with a more efficient way, which is to achieve *Adaptive* Oblivious Transfer, in which the user can adaptively ask several lines of the database. In such schemes, the server only sends his database once at the beginning of the protocol, and all the subsequent communication is in $o(n)$, more precisely logarithmic. The linear cost is batched once and for all in this preprocessing phase, achieving then a logarithmic complexity similar to the best PIR schemes.

Smooth Projective Hash Functions (SPHF), used in conjunction with *Commitments* have become the standard way to deal with such secret message transfers. In a commitment scheme, the sender is going to commit to the line required (*i.e.* to give the receiver an analogue of a sealed envelope containing his value i) in such a way that he should not be able to open to a value different from the one he committed to (*binding* property), and that the receiver cannot learn anything about i (*hiding* property) before a potential opening phase. During the opening phase, however, the committer would be asked to reveal i in such a way that the receiver can verify it was indeed i that was contained in the envelope.

But, in our applications, there cannot be an opening phase, due to the *oblivious* requirements on the protocols and the secrecy of the database line i sent. The decommitment (opening phase) will thus be implicit, which means that the committer does not really open its commitment, but rather convinces the receiver that it actually committed to the value it pretended to. We achieve this property thanks to *Smooth Projective Hash Functions* [26,33], which have been widely used in such circumstances (see [1,2,8,9,45] for instance). These hash functions are defined in such a way that their value can be computed in two different ways if the input belongs to a particular subset (the *language*), either using a private hashing key or a public projection key along with a private witness ensuring that the input belongs to the language. The hash value obtained is indistinguishable from random in case the input does not belong to the language (*smoothness*) and in case the input does belong to the language but no witness is known (*pseudo-randomness*).

In a nutshell, to ensure implicit decommitment, the sender will thus simply mask the database line with this hash value computed using the private hashing key. He will then send it along with the public projection key to the user, who will be able to compute the same hash value thanks to the randomness of the commitment of this line he sent in the first place (the randomness is the witness of the membership of the commitment to the language of commitments of this specific line). In order to ensure adaptive security in the universal composability framework, the commitments used are usually required to be both *extractable* (meaning that a simulator can recover the value i committed to thanks to a trapdoor) and *equivocable* (meaning that a simulator can open a commitment to a value i' different from the value i it committed to thanks to a trapdoor).

In order to simplify these commitments, which can be quite technical, we choose here to rely on words in more complex languages rather than on simple line numbers. More precisely, the user will first compute an equivocable commitment on the line number required, which will be his word w in the language. This word will then be encrypted under a CCA encryption scheme, and the SPHF will be constructed for this word (rather than for the line number), which will be simpler. Furthermore, this abstraction consisting in encoding line numbers as words in more complex languages will reveal useful in more general contexts, not only Oblivious Transfer, the simplest of which being Oblivious Signature Based Envelope.

Oblivious Signature-Based Envelope (OSBE) was introduced by Li, Du and Boneh in [49]. OSBE schemes consider the case where Alice (the receiver) is a member of an organization and possesses a certificate produced by an authority attesting she actually belongs to this organization. Bob (the sender) wants to send a private message P to members of this organization. However due to the sensitive nature of the organization, Alice does not want to give Bob neither her certificate nor a proof she belongs to the organization. OSBE lets Bob send an obfuscated version of this message P to Alice, in such a way that she will be able to find P if and only if she is in the required organization. In the process, Bob cannot decide whether Alice does really belong to the organization. We even manage to construct a more general framework to capture many protocols around trust negotiation, where the user receives a message if and only if he possesses some credentials or specific accreditations. As a reference to OSBE, we call this framework *Oblivious Language-Based Envelope* (OLBE).

1.1 Related Work

Since the original paper [53], several instantiations and optimizations of OT protocols have appeared in the literature [23,51], including proposals in the UC framework. More recently, new instantiations have been proposed, trying to reach round-optimality [41], and/or low communication costs [52]. Recent schemes like [1,9] manage to achieve round-optimality while maintaining a small communication cost. Choi *et al.* [24] also propose a generic method and an efficient instantiation secure against adaptive corruptions in the CRS model with erasures, but it is only 1-out-of-2 and it does not scale to 1-out-of-n OT, for $n > 2$. As far as adaptive versions of those protocols are concerned, this problem was first studied by [37,47,50], and more recently UC secure instantiations were proposed, but unfortunately either under the Random Oracle, or under not so standard assumptions such as q-Hidden LRSW or later on q-SDH [17,20,39,43,54], but without allowing adaptive corruptions.

Concerning automated trust negotiation, two frameworks have been proposed to encompass the symmetric protocols (Password-based Authenticated Key-Exchange, Secret Handshakes and Verfier-Based PAKE): The Credential Authenticated Key Exchange [16], and Language-based Authenticated Key Exchange (LAKE) [7], in which two parties establish a common session key if and only if they hold credentials that belong to specific (and possibly independent) languages chosen by the other party. As for OSBE, the authors in [13] improved the security model initially proposed in [49], showing how to use Smooth Projective Hash Functions to do implicit proof of knowledge, and proposed the first efficient instantiation of OSBE, under a standard hypothesis. It fits, as well as Access Controlled Oblivious Transfer [18,19], Priced Oblivious Transfer [4,54]) and Conditional Oblivious Transfer [28], into the generic notion of Conditional Disclosure of Secrets (see for instance [4,5,15,32,34,42,48,58]).

1.2 Contributions

Our first contribution is to give the first round-optimal adaptive Oblivious Transfer protocol secure in the UC framework with adaptive corruptions under standard assumptions (MDDH) and assuming reliable erasures. We show how to instantiate the needed building blocks using standard assumptions, using or extending various basic primitives in order to fit the MDDH framework introduced in [30]. In our scheme, the server first preprocesses its database in a time linear in the length of the database and transfers it to the receiver. After that, the receiver and the sender can run many instances of the protocol on the same database as input and adaptively chosen inputs from the receiver, with a cost sublinear in the database.

It is interesting to note that our resulting adaptive Oblivious Transfer scheme has an amortized complexity in $\mathcal{O}(\log |DB|)$, which is similar to current Private Information Retrieval instantiations [46], that have weaker security prerequisites, and much better than current UC secure Oblivious Transfer under standard assumptions (as they are in $\mathcal{O}(|DB|)$. For a fair comparison it should be stated that the PIR schemes allow this complexity directly from the first query while in our case due to the preprocessing, this amortized cost is only reached after a high number of queries. However, it is interesting to see this convergence in spite of hugely different security models and expectation. Compared to existing versions cited above (either proven in classical security models, or in the UC framework but only with static corruptions and under non-standard assumptions), we manage to prove its security under standard assumptions, like SXDH, and allow UC security with adaptive user corruptions.

As a side result, it is worth noting that we follow some ideas developed in the construction explained in [37] around Blind Identity-Based Encryption and provide techniques in order to transform IBE schemes into blind ones, applying them to revisit the one given in [12], in order to show how we can answer blind user secret key-retrieval, which can be of independent interest.

As a second contribution, we propose our new notion, that we call Oblivious Language-Based Envelope. We provide a security model by giving a UC ideal functionality, and show that this notion supersedes the classical asymmetric automated trust negotiation schemes recalled above such as Oblivious Transfer and Oblivious Signature-Based Envelope. We show how to choose the languages in order to obtain from our framework all the corresponding ideal functionalities, recovering the known ones (such as OT) and providing the new ones (such as OSBE, to the best of our knowledge). We then give a generic construction scheme fulfilling our ideal functionality, which directly gives generic constructions for the specific cases (OT, OSBE). Finally, we show how to instantiate the different simple building blocks in order to recover the standard efficient instantiations of these schemes from our framework. In addition to the two cases most studied (OT, OSBE), we also propose what we call *Conditioned Oblivious Transfer*, which encompasses Access Controlled Oblivious Transfer, Priced Oblivious Transfer and Conditional Oblivious Transfer, and in which the access to each line of the database is hidden behind some possibly secret restriction, be it a credential, a price, or an access policy. The advantage of the OLBE framework on the notion of Conditional Disclosure of Secrets is to allow generic constructions of a large subclass of schemes, as long as two participants are involved. It can be easily applied to any language expressing some new access control policy. Furthermore, those instantiations fit into a global security model, allowing to uniformize (for the better) the security expectations for such schemes. In particular, we allow security in the UC framework with adaptive corruptions for all our constructions (which was already known for some primitives cited above, but not all), and manage to achieve this level of security while staying in the standard model with standard hypothesis.

2 Definitions and Building Blocks

2.1 Notations for Classical Primitives

Throughout this paper, we use the notation $\mathfrak{K}$ for the security parameter.

Digital Signature. A digital signature scheme $\mathcal{S}$ [29,36] allows a signer to produce a verifiable proof that he indeed produced a message. It is described through four algorithms $\sigma = (\mathsf{Setup}, \mathsf{KeyGen}, \mathsf{Sign}, \mathsf{Verify})$. The formal definitions are given in the paper full version [10].

Encryption. An encryption scheme $\mathcal{C}$ is described through four algorithms $(\mathsf{Setup}, \mathsf{KeyGen}, \mathsf{Encrypt}, \mathsf{Decrypt})$. The formal definitions are given in the paper full version [10].

Commitment and Chameleon Hash. Commitments allow a user to commit to a value without revealing it, but without the possibility to later change his mind. It is composed of four algorithms $(\mathsf{Setup}, \mathsf{KeyGen}, \mathsf{Commit}, \mathsf{Decommit})$. Informally, it is extractable if a simulator knowing a certain trapdoor can recover the value committed to, and it is equivocable if a simulator, knowing another trapdoor, can open the commitment to another value than the one it actually committed to. This directly echoes to Chameleon Hashes, traditionally defined by three algorithms $\mathsf{CH} = (\mathsf{KeyGen}, \mathsf{CH}, \mathsf{Coll})$. The formal definitions are given in the paper full version [10].

2.2 Identity-Based Encryption, Identity-Based Key Encapsulation

Identity Based encryption was first introduced by Shamir in [55] who was expecting an encryption scheme where no public key will be needed for sending a message to a precise user, defined by his identity. Thus any user wanting to send a private message to a user only need this user's identity and a master public key. It took 17 years for the cryptographic community to find a way to realize this idea. The first instantiation was proposed in [14] by Boneh and Franklin. It can be described as an identity-based key encapsulation (IBKEM) scheme IBKEM which consists of four algorithms $\mathsf{IBKEM} = (\mathsf{Gen}, \mathsf{USKGen}, \mathsf{Enc}, \mathsf{Dec})$. Every IBKEM can be transformed into an ID-based encryption scheme IBE using a (one-time secure) symmetric cipher.

Definition 1 (Identity-based Key Encapsulation Scheme). *An identity-based key encapsulation scheme* IBKEM *consists of four PPT algorithms* $\mathsf{IBKEM} = (\mathsf{Gen}, \mathsf{USKGen}, \mathsf{Enc}, \mathsf{Dec})$ *with the following properties.*

- $\mathsf{Gen}(\mathfrak{K})$*: returns the (master) public/secret key* $(\mathsf{mpk}, \mathsf{msk})$*. We assume that* mpk *implicitly defines an identity space* $\mathcal{ID}$*, a key space* $\mathcal{KS}$*, and ciphertext space* CS.

- $\mathsf{USKGen}(\mathsf{msk},\mathsf{id})$: *returns the user secret-key* $\mathsf{usk}[\mathsf{id}]$ *for identity* $\mathsf{id} \in \mathcal{ID}$.
- $\mathsf{Enc}(\mathsf{mpk},\mathsf{id})$: *returns the symmetric key* $\mathsf{K} \in \mathcal{KS}$ *together with a ciphertext* $\mathsf{C} \in \mathsf{CS}$ *with respect to identity* id.
- $\mathsf{Dec}(\mathsf{usk}[\mathsf{id}],\mathsf{id},\mathsf{C})$: *returns the decapsulated key* $\mathsf{K} \in \mathcal{KS}$ *or the reject symbol* $\perp$.

For perfect correctness we require that for all $\mathfrak{K} \in \mathbb{N}$, *all pairs* $(\mathsf{mpk},\mathsf{msk})$ *generated by* $\mathsf{Gen}(\mathfrak{K})$, *all identities* $\mathsf{id} \in \mathcal{ID}$, *all* $\mathsf{usk}[\mathsf{id}]$ *generated by* $\mathsf{USKGen}(\mathsf{msk},\mathsf{id})$ *and all* (K,C) *output by* $\mathsf{Enc}(\mathsf{mpk},\mathsf{id})$: $\Pr[\mathsf{Dec}(\mathsf{usk}[\mathsf{id}],\mathsf{id},\mathsf{C}) = \mathsf{K}] = 1$.

The security requirements for an IBKEM we consider here are indistinguishability and anonymity against chosen plaintext and identity attacks (IND-ID-CPA and ANON-ID-CPA). Instead of defining both security notions separately, we define pseudorandom ciphertexts against chosen plaintext and identity attacks (PR-ID-CPA) which means that challenge key and ciphertext are both pseudorandom. Note that PR-ID-CPA trivially implies IND-ID-CPA and ANON-ID-CPA. We define PR-ID-CPA-security of IBKEM formally via the games given in Fig. 1.

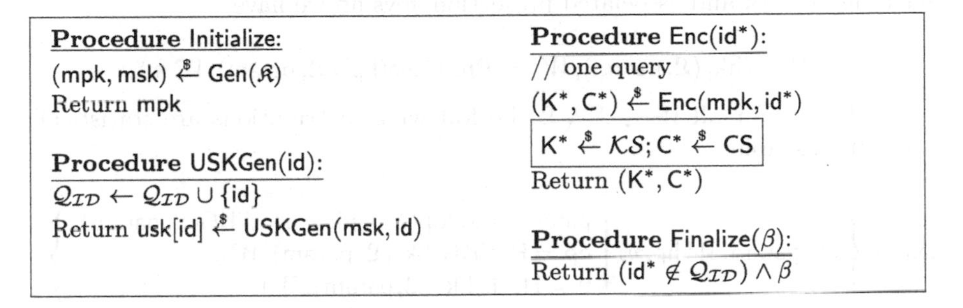

Fig. 1. Security Games $\mathsf{PR\text{-}ID\text{-}CPA}_{\mathsf{real}}$ and $\mathsf{PR\text{-}ID\text{-}CPA}_{\mathsf{rand}}$ (boxed) used for defining PR-ID-CPA-security.

Definition 2 (PR-ID-CPA **Security**). *An ID-based key encapsulation scheme* IBKEM *is* PR-ID-CPA-*secure if for all PPT* $\mathscr{A}$, *the following advantage is negligible:* $\mathsf{Adv}^{\mathsf{pr\text{-}id\text{-}cpa}}_{\mathsf{IBKEM}}(\mathscr{A}) := |\Pr[\mathsf{PR\text{-}ID\text{-}CPA}^{\mathscr{A}}_{\mathsf{real}} \Rightarrow 1] - \Pr[\mathsf{PR\text{-}ID\text{-}CPA}_{\mathsf{rand}}{}^{\mathscr{A}} \Rightarrow 1]|$.

2.3 Smooth Projective Hashing and Languages

Smooth projective hash functions (SPHF) were introduced by Cramer and Shoup in [26] for constructing encryption schemes. A projective hashing family is a family of hash functions that can be evaluated in two ways: using the (secret) hashing key, one can compute the function on every point in its domain, whereas using the (public) *projected* key one can only compute the function on a special subset of its domain. Such a family is deemed *smooth* if the value of the hash function on any point outside the special subset is independent of the projected key. The notion of SPHF has already found applications in various contexts in cryptography (*e.g.* [2,33,44]). A Smooth Projective Hash Function over a language $\mathfrak{L} \subset X$, onto a set $\mathcal{G}$, is defined by five algorithms (Setup, HashKG, ProjKG, Hash, ProjHash):

- $\mathsf{Setup}(1^{\mathfrak{K}})$ where $\mathfrak{K}$ is the security parameter, generates the global parameters param of the scheme, and the description of an $\mathcal{NP}$ language $\mathfrak{L}$;
- $\mathsf{HashKG}(\mathfrak{L}, \mathsf{param})$, outputs a hashing key hk for the language $\mathfrak{L}$;
- $\mathsf{ProjKG}(\mathsf{hk}, (\mathfrak{L}, \mathsf{param}), W)$, derives the projection key hp from the hashing key hk and the word W.
- $\mathsf{Hash}(\mathsf{hk}, (\mathfrak{L}, \mathsf{param}), W)$, outputs a hash value $v \in \mathcal{G}$, using the hashing key hk and the word W.
- $\mathsf{ProjHash}(\mathsf{hp}, (\mathfrak{L}, \mathsf{param}), W, w)$, outputs the hash value $v' \in \mathcal{G}$, using the projection key hp and the witness w that the word $W \in \mathfrak{L}$.

In the following, we assume $\mathfrak{L}$ is a hard-partitioned subset of X, *i.e.* it is computationally hard to distinguish a random element in $\mathfrak{L}$ from a random element in $X \setminus \mathfrak{L}$. An SPHF should satisfy the following properties:

- *Correctness*: Let $W \in \mathfrak{L}$ and w a witness of this membership. Then, for all hashing keys hk and associated projection keys hp we have

$$\mathsf{Hash}(\mathsf{hk}, (\mathfrak{L}, \mathsf{param}), W) = \mathsf{ProjHash}(\mathsf{hp}, (\mathfrak{L}, \mathsf{param}), W, w).$$

- *Smoothness*: For all $W \in X \setminus \mathfrak{L}$ the following distributions are statistically indistinguishable:

$$\Delta_0 = \left\{ (\mathfrak{L}, \mathsf{param}, W, \mathsf{hp}, v) \left| \begin{array}{l} \mathsf{param} = \mathsf{Setup}(1^{\mathfrak{K}}), \mathsf{hk} = \mathsf{HashKG}(\mathfrak{L}, \mathsf{param}), \\ \mathsf{hp} = \mathsf{ProjKG}(\mathsf{hk}, (\mathfrak{L}, \mathsf{param}), W), \\ v = \mathsf{Hash}(\mathsf{hk}, (\mathfrak{L}, \mathsf{param}), W) \end{array} \right. \right\}$$

$$\Delta_1 = \left\{ (\mathfrak{L}, \mathsf{param}, W, \mathsf{hp}, v) \left| \begin{array}{l} \mathsf{param} = \mathsf{Setup}(1^{\mathfrak{K}}), \mathsf{hk} = \mathsf{HashKG}(\mathfrak{L}, \mathsf{param}), \\ \mathsf{hp} = \mathsf{ProjKG}(\mathsf{hk}, (\mathfrak{L}, \mathsf{param}), W), v \stackrel{\$}{\leftarrow} \mathcal{G} \end{array} \right. \right\}.$$

This is formalized by: $\mathsf{Adv}^{\mathsf{smooth}}_{\mathsf{SPHF}}(\mathfrak{K}) = \sum_{V \in \mathbb{G}} |\Pr_{\Delta_1}[v = V] - \Pr_{\Delta_0}[v = V]|$ is negligible.

- *Pseudo-Randomness*: If $W \in \mathfrak{L}$, then without a witness of membership the two previous distributions should remain computationally indistinguishable. For any adversary $\mathcal{A}$ within reasonable time, this advantage is negligible:

$$\mathsf{Adv}^{\mathsf{pr}}_{\mathsf{SPHF},\mathcal{A}}(\mathfrak{K}) = |\Pr_{\Delta_1}[\mathcal{A}(\mathfrak{L}, \mathsf{param}, W, \mathsf{hp}, v) = 1] - \Pr_{\Delta_0}[\mathcal{A}(\mathfrak{L}, \mathsf{param}, W, \mathsf{hp}, v) = 1]|$$

Languages. The language $\mathfrak{L} \subset X$ used in the definition of an SPHF should be a hard-partitioned subset of X, *i.e.* it is computationally hard to distinguish a random element in $\mathfrak{L}$ from a random element not in $\mathfrak{L}$ (see formal definition in [3,33]). The languages used here are more complex and should fulfill the following properties[1]:

[1] We here mainly consider languages which are hard-partitioned subsets, for instance, encryptions of publicly verifiable languages.

- *Publicly Verifiable:* Given a word x in X, anyone should be able to decide in polynomial time whether $x \in \mathfrak{L}$ or not.
- *Self-Randomizable*: Given a word in the language, anyone should be able to sample a new word in the language[2], and the distribution of this resampling should be indistinguishable from an honest distribution. This will be used in order to prevent an adversary, or the authority in charge of distributing the words, to learn which specific form of the word was used by the user.

In case we consider several languages $(\mathfrak{L}_1, \ldots, \mathfrak{L}_n)$, we also assume it is a *Trapdoor Collection of Languages*: It is computationally hard to sample an element in $\mathfrak{L}_1 \cap \cdots \cap \mathfrak{L}_n$, except if one possesses a trapdoor tk (without the knowledge of the potential secret keys)[3]. For instance, if for all i, $\mathfrak{L}_i$ is the language of the equivocable commitments on words in an inner language $\widetilde{\mathfrak{L}}_i = \{i\}$ (as we will consider for OT), the common trapdoor key can be the equivocation trapdoor.

Depending on the applications, we can assume a *Keyed Language*, which means that it is set by a trusted authority, and that it is hard to sample fresh elements from scratch in the language without the knowledge of a secret language key $\mathsf{sk}_{\mathfrak{L}}$. In this case, the authority is also in charge of giving a word in the language to the receiver.

In case the language is keyed, we assume it is also a *Trapdoor Language*: We assume the existence of a trapdoor tk_L allowing a simulator to sample an element in $\mathfrak{L}$ (without the knowledge of the potential secret key $\mathsf{sk}_{\mathfrak{L}}$). For instance, for a language of valid Waters signatures of a message M (as we will consider for OSBE), one can think of $\mathsf{sk}_{\mathfrak{L}}$ as being the signing key, whereas the trapdoor $\mathsf{tk}_{\mathfrak{L}}$ can be the discrete logarithm of h in basis g.[4]

2.4 Security Assumptions

Due to lack of space, instantiations of the primitives recalled above are given in the paper full version [10] and we only give here the security assumptions.

Security Assumption: Pairing groups and Matrix Diffie-Hellman Assumption. Let GGen be a probabilistic polynomial time (PPT) algorithm that on input $1^{\mathfrak{K}}$ returns a description $\mathcal{G} = (p, \mathbb{G}_1, \mathbb{G}_2, \mathbb{G}_T, e, g_1, g_2)$ of asymmetric pairing groups where $\mathbb{G}_1$, $\mathbb{G}_2$, $\mathbb{G}_T$ are cyclic groups of order p for a $\mathfrak{K}$-bit prime

[2] It should be noted that this property is not incompatible with the potential secret key of the language in case it is keyed (see below).

[3] This implicitly means that the languages are compatible, in the sense that one can indeed find a word belonging to all of them.

[4] As another example, one may think of more expressive languages which may not rely directly on generators fixed by the CRS. In this case, one can assume that the CRS contains parameters for an encryption and an associated NIZK proof system. The description of such a language is thus supplemented with an encryption of the language trapdoor, and a non-interactive zero-knowledge proof that the encrypted value is indeed a trapdoor for the said language. Using the knowledge of the decryption key, the simulator is able to recover the trapdoor.

p, g_1 and g_2 are generators of $\mathbb{G}_1$ and $\mathbb{G}_2$, respectively, and $e : \mathbb{G}_1 \times \mathbb{G}_2$ is an efficiently computable (non-degenerated) bilinear map. Define $g_T := e(g_1, g_2)$, which is a generator in $\mathbb{G}_T$.

We use implicit representation of group elements as introduced in [30]. For $s \in \{1, 2, T\}$ and $a \in \mathbb{Z}_p$ define $[a]_s = g_s^a \in \mathbb{G}_s$ as the *implicit representation* of a in $\mathbb{G}_s$. More generally, for a matrix $\boldsymbol{A} = (a_{ij}) \in \mathbb{Z}_p^{n \times m}$ we define $[\boldsymbol{A}]_s$ as the implicit representation of $\boldsymbol{A}$ in $\mathbb{G}_s$:

$$[\boldsymbol{A}]_s := \begin{pmatrix} g_s^{a_{11}} & \ldots & g_s^{a_{1m}} \\ \\ g_s^{a_{n1}} & \ldots & g_s^{a_{nm}} \end{pmatrix} \in \mathbb{G}_s^{n \times m}$$

We will always use this implicit notation of elements in $\mathbb{G}_s$, i.e., we let $[a]_s \in \mathbb{G}_s$ be an element in $\mathbb{G}_s$. Note that from $[a]_s \in \mathbb{G}_s$ it is generally hard to compute the value a (discrete logarithm problem in $\mathbb{G}_s$). Further, from $[b]_T \in \mathbb{G}_T$ it is hard to compute the value $[b]_1 \in \mathbb{G}_1$ and $[b]_2 \in \mathbb{G}_2$ (pairing inversion problem). Obviously, given $[a]_s \in \mathbb{G}_s$ and a scalar $x \in \mathbb{Z}_p$, one can efficiently compute $[ax]_s \in \mathbb{G}_s$. Further, given $[a]_1, [b]_2$ one can efficiently compute $[ab]_T$ using the pairing e. For $\boldsymbol{a}, \boldsymbol{b} \in \mathbb{Z}_p^k$ define $e([\boldsymbol{a}]_1, [\boldsymbol{b}]_2) := [\boldsymbol{a}^\top \boldsymbol{b}]_T \in \mathbb{G}_T$. We recall the definition of the matrix Diffie-Hellman (MDDH) assumption [30].

Definition 3 (Matrix Distribution). *Let $k \in \mathbb{N}$. We call $\mathcal{D}_k$ a matrix distribution if it outputs matrices in $\mathbb{Z}_p^{(k+1) \times k}$ of full rank k in polynomial time.*

Without loss of generality, we assume the first k rows of $\boldsymbol{A} \xleftarrow{\$} \mathcal{D}_k$ form an invertible matrix, we denote this matrix $\overline{\boldsymbol{A}}$, while the last line is denoted $\underline{A}$. The $\mathcal{D}_k$-Matrix Diffie-Hellman problem is to distinguish the two distributions $([\boldsymbol{A}], [\boldsymbol{A}\boldsymbol{w}])$ and $([\boldsymbol{A}], [\boldsymbol{u}])$ where $\boldsymbol{A} \xleftarrow{\$} \mathcal{D}_k$, $\boldsymbol{w} \xleftarrow{\$} \mathbb{Z}_p^k$ and $\boldsymbol{u} \xleftarrow{\$} \mathbb{Z}_p^{k+1}$.

Definition 4 ($\mathcal{D}_k$-MatrixDiffie-HellmanAssumption $\mathcal{D}_k$-MDDH). *Let $\mathcal{D}_k$ be a matrix distribution and $s \in \{1, 2, T\}$. We say that the $\mathcal{D}_k$-Matrix Diffie-Hellman ($\mathcal{D}_k$-MDDH) Assumption holds relative to* GGen *in group $\mathbb{G}_s$ if for all PPT adversaries $\mathcal{D}$,*

$$\mathbf{Adv}_{\mathcal{D}_k, \mathsf{GGen}}(\mathcal{D}) := |\Pr[\mathcal{D}(\mathcal{G}, [\boldsymbol{A}]_s, [\boldsymbol{A}\boldsymbol{w}]_s) = 1] - \Pr[\mathcal{D}(\mathcal{G}, [\boldsymbol{A}]_s, [\boldsymbol{u}]_s) = 1]| = \mathsf{negl}(\lambda),$$

where the probability is taken over $\mathcal{G} \xleftarrow{\$} \mathsf{GGen}(1^\lambda)$, $\boldsymbol{A} \xleftarrow{\$} \mathcal{D}_k$, $\boldsymbol{w} \xleftarrow{\$} \mathbb{Z}_p^k$, $\boldsymbol{u} \xleftarrow{\$} \mathbb{Z}_p^{k+1}$.

For each $k \geq 1$, [30] specifies distributions $\mathcal{L}_k$, $\mathcal{U}_k$, ... such that the corresponding $\mathcal{D}_k$-MDDH assumption is the k-Linear assumption, the k-uniform and others. All assumptions are generically secure in bilinear groups and form a hierarchy of increasingly weaker assumptions. The distributions are exemplified for $k = 2$, where $a_1, \ldots, a_6 \xleftarrow{\$} \mathbb{Z}_p$.

$$\mathcal{L}_2 : \boldsymbol{A} = \begin{pmatrix} a_1 & 0 \\ 0 & a_2 \\ 1 & 1 \end{pmatrix} \quad \mathcal{U}_2 : \boldsymbol{A} = \begin{pmatrix} a_1 & a_2 \\ a_3 & a_4 \\ a_5 & a_6 \end{pmatrix}.$$

It was also shown in [30] that $\mathcal{U}_k$-MDDH is implied by all other $\mathcal{D}_k$-MDDH assumptions.

Lemma 5 (Random self reducibility [30]**).** *For any matrix distribution $\mathcal{D}_k$, $\mathcal{D}_k$-MDDH is random self-reducible. In particular, for any $m \geq 1$ and for all PPT adversaries $\mathcal{D}$ and $\mathcal{D}'$,*

$$\mathbf{Adv}_{\mathcal{D}_k,\mathsf{GGen}}(\mathcal{D}) + \frac{1}{q-1} \geq \mathbf{Adv}^m_{\mathcal{D}_k,\mathsf{GGen}}(\mathcal{D}')$$

where $\mathbf{Adv}^m_{\mathcal{D}_k,\mathsf{GGen}}(\mathcal{D}') := \Pr[\mathcal{D}'(\mathcal{G}, [\boldsymbol{A}], [\boldsymbol{A}\boldsymbol{W}]) \Rightarrow 1] - \Pr[\mathcal{D}'(\mathcal{G}, [\boldsymbol{A}], [\boldsymbol{U}]) \Rightarrow 1]$, *with* $\mathcal{G} \leftarrow \mathsf{GGen}(1^\lambda)$, $\boldsymbol{A} \xleftarrow{\$} \mathcal{D}_k$, $\boldsymbol{W} \xleftarrow{\$} \mathbb{Z}_p^{k\times m}$, $\boldsymbol{U} \xleftarrow{\$} \mathbb{Z}_p^{(k+1)\times m}$.

Remark: It should be noted that $\mathcal{L}_1, \mathcal{L}_2$ are respectively the SXDH and DLin assumptions.

2.5 Security Models

UC Framework. The goal of the UC framework [21] is to ensure that UC-secure protocols will continue to behave in the ideal way even if executed in a concurrent way in arbitrary environments. It is a simulation-based model, relying on the indistinguishability between the real world and the ideal world. In the ideal world, the security is provided by an ideal functionality $\mathcal{F}$, capturing all the properties required for the protocol and all the means of the adversary. In order to prove that a protocol Π emulates $\mathcal{F}$, one has to construct, for any polynomial adversary $\mathscr{A}$ (which controls the communication between the players), a simulator $\mathscr{S}$ such that no polynomial environment $\mathcal{Z}$ can distinguish between the real world (with the real players interacting with themselves and $\mathscr{A}$ and executing the protocol π) and the ideal world (with dummy players interacting with $\mathscr{S}$ and $\mathcal{F}$) with a significant advantage. The adversary can be either *adaptive*, *i.e.* allowed to corrupt users whenever it likes to, or *static*, *i.e.* required to choose which users to corrupt prior to the execution of the session sid of the protocol. After corrupting a player, $\mathscr{A}$ has complete access to the internal state and private values of the player, takes its entire control, and plays on its behalf.

Simple UC Framework. Canetti, Cohen and Lindell formalized a simpler variant in [22], that we use here. This simplifies the description of the functionalities for the following reasons (in a nutshell): All channels are automatically assumed to be authenticated (as if we worked in the $\mathcal{F}_{\text{AUTH}}$-hybrid model); There is no need for *public delayed outputs* (waiting for the adversary before delivering a message to a party), neither for an explicit description of the corruptions. We refer the interested reader to [22] for details.

3 UC-secure Adaptive Oblivious Transfer

As explained in the introduction, the classical OT constructions based on the commitment/SPHF paradigm (with so-called implicit decommitment), among the latest in the UC framework [1,9,24], require the server to send an encryption of the complete database for each line required by the user (thus $O(n)$ each time). We here give a protocol requiring $O(\log(n))$ for each line (except the first one, still in $O(n)$), in the UC framework with adaptive corruptions under classical assumptions (MDDH). This protocol builds upon the more efficient known scheme secure in the UC framework [9] and we use ideas from [37] to make it adaptive.

Using implicit decommitment in the UC framework implies a very strong commitment primitive (formalized as SPHF-friendly commitments in [1]), which is both extractable and equivocable. Our idea is here to split these two properties by using on the one hand an equivocable commitment and on the other hand an (extractable) CCA encryption scheme by generalizing the way to access a line in the database. But this is infeasible with simple line numbers. Indeed, we suggest here not to consider anymore the line numbers as numbers in $\{1, \dots, n\}$ but rather to "encode" them (the exact encoding will depend on the protocol): For every line i, a word W_i in the language $\mathfrak{L}_i$ will correspond to a representation of line i. This representation must be publicly verifiable, in the sense that anyone can associate i to a word W_i. We formalize this in the following definition of oblivious transfer[5], given without loss of generality[6] (the classical notion of OT being easily captured using $\mathfrak{L}_i = \{i\}$).

3.1 Definition and Security Model for Oblivious Transfer

In such a protocol, a server $\mathcal{S}$ possesses a database of n lines $(m_1, \dots, m_n) \in (\{0,1\}^{\mathfrak{K}})^n$. A user $\mathcal{U}$ will be able to recover m_k (in an oblivious way) as soon as he owns a word $W_k \in \mathfrak{L}_k$. The languages $(\mathfrak{L}_1, \dots, \mathfrak{L}_n)$ will be assumed to be a trapdoor collection of languages, publicly verifiable and self-randomizable. As we consider simulation-based security (in the UC framework), we allow a simulated setup SetupT to be run instead of the classical setup Setup in order to allow the simulator to possess some trapdoors. Those two setup algorithms should be indistinguishable.

Definition 6 (Oblivious Transfer). *An* OT *scheme is defined by five algorithms* $(\mathsf{Setup}, \mathsf{KeyGen}, \mathsf{DBGen}, \mathsf{Samp}, \mathsf{Verify})$, *along with an interactive protocol* $\mathsf{Protocol}\langle\mathcal{S},\mathcal{U}\rangle$:

- $\mathsf{Setup}(1^{\mathfrak{K}})$, *where* $\mathfrak{K}$ *is the security parameter, generates the global parameters* param, *among which the number* n;

  or $\mathsf{SetupT}(1^{\mathfrak{K}})$, *where* $\mathfrak{K}$ *is the security parameter, additionally allows the existence*[7] *of a trapdoor* tk *for the collection of languages* $(\mathfrak{L}_1, \dots, \mathfrak{L}_n)$.

---

[5] The adaptive version only implies that the database $(m_1, \dots, m_n)$ is sent only once in the interaction, while the user can query several lines (*i.e.* several words), in an adaptive way.

[6] This formalization furthermore encompasses the variants of OT, such as conditioned OT, where a user accesses a line only if he knows a credential for this line.

[7] The specific trapdoor will depend on the languages and be computed in the KeyGen algorithm.

- KeyGen(param, $\mathfrak{K}$) *generates, for all* $i \in \{1, \ldots, n\}$, *the description of the language* $\mathfrak{L}_i$ *(as well as the language key* $\mathsf{sk}_{\mathfrak{L}_i}$ *if need be). If the parameters* param *were defined by* SetupT, *this implicitly also defines the common trapdoor* tk *for the collection of languages* $(\mathfrak{L}_1, \ldots, \mathfrak{L}_n)$.
- Samp(param) *or* Samp(param, $(\mathsf{sk}_{\mathfrak{L}_i})_{i \in \{1,\ldots,n\}}$) *generates a word* $W_i \in \mathfrak{L}_i$;
- $\mathsf{Verify}_i(W_i, \mathfrak{L}_i)$ *checks whether* W_i *is a valid word in the language* $\mathfrak{L}_i$. *It outputs* 1 *if the word is valid,* 0 *otherwise;*
- $\mathsf{Protocol}\langle \mathcal{S}((\mathfrak{L}_1, \ldots, \mathfrak{L}_n), (m_1, \ldots, m_n)), \mathcal{U}((\mathfrak{L}_1, \ldots, \mathfrak{L}_n), W_i)\rangle$, *which is executed between the server* $\mathcal{S}$ *with the private database* $(m_1, \ldots, m_n)$ *and corresponding languages* $(\mathfrak{L}_1, \ldots, \mathfrak{L}_n)$, *and the user* $\mathcal{U}$ *with the same languages and the word* W_i, *proceeds as follows. If the algorithm* $\mathsf{Verify}_i(W_i, \mathfrak{L}_i)$ *returns* 1, *then* $\mathcal{U}$ *receives* m_k, *otherwise it does not. In any case,* $\mathcal{S}$ *does not learn anything.*

The ideal functionality of an Oblivious Transfer (OT) protocol was given in [1,21,24], and an adaptive version in [38]. We here combine them and rewrite it in simple UC and using our language formalism (instead of directly giving a number line s to the functionality, the user will give it a word $W_s \in \mathfrak{L}_s$). The resulting functionality $\mathcal{F}_{\mathrm{OT}}^{\mathfrak{L}}$ is given in Fig. 2. Recall that there is no need to give an explicit description of the corruptions in the simple version of UC [22].

The functionality $\mathcal{F}_{\mathrm{OT}}^{\mathfrak{L}}$ is parametrized by a security parameter $\mathfrak{K}$ and a set of languages $(\mathfrak{L}_1, \ldots, \mathfrak{L}_n)$ along with the corresponding public verification algorithms $(\mathsf{Verify}_1, \ldots, \mathsf{Verify}_n)$. It interacts with an adversary $\mathscr{S}$ and a set of parties $\mathfrak{P}_1, \ldots, \mathfrak{P}_N$ via the following queries:

- **Upon receiving from party $\mathfrak{P}_i$ an input (NewDataBase, sid, ssid, $\mathfrak{P}_i$, $\mathfrak{P}_j$, $(m_1, \ldots, m_n)$)**, with $m_k \in \{0,1\}^{\mathfrak{K}}$ for all k, record the tuple (sid, ssid, $\mathfrak{P}_i$, $\mathfrak{P}_j$, $(m_1, \ldots, m_n)$) and reveal (Send, sid, ssid, $\mathfrak{P}_i$, $\mathfrak{P}_j$) to the adversary $\mathscr{S}$. Ignore further NewDataBase-message with the same ssid from $\mathfrak{P}_i$.
- **Upon receiving an input (Receive, sid, ssid, $\mathfrak{P}_i$, $\mathfrak{P}_j$, W_k) from party $\mathfrak{P}_j$**: ignore the message if (sid, ssid, $\mathfrak{P}_i$, $\mathfrak{P}_j$, $(m_1, \ldots, m_n)$) is not recorded. Otherwise, reveal (Receive, sid, ssid, $\mathfrak{P}_i$, $\mathfrak{P}_j$) to the adversary $\mathscr{S}$ and send the message (Received, sid, ssid, $\mathfrak{P}_i$, $\mathfrak{P}_j$, m'_k) to $\mathfrak{P}_j$ where $m'_k = m_k$ if $\mathsf{Verify}_k(W_k, \mathfrak{L}_k)$ returns 1, and $m'_k = \perp$ otherwise.

(Non-Adaptive case: Ignore further Receive*-message with the same* ssid *from* $\mathfrak{P}_j$.*)*

Fig. 2. Ideal Functionality for (Adaptive) Oblivious Transfer $\mathcal{F}_{\mathrm{OT}}^{\mathfrak{L}}$

3.2 High Level Idea of the Construction of the Adaptive Oblivious Transfer Scheme

Our construction builds upon the UC-secure OT scheme from [9], with ideas inspired from [37], who propose a neat framework allowing to achieve *adaptive* Oblivious Transfer (but not in the UC framework). Their construction is quite simple: It requires a *blind Identity-Based Encryption*, in other words, an IBE scheme in which there is a way to query for a user key generation without the authority (here the server) learning the targeted identity (here the line in the database). Once such a Blind IBE is defined, one can conveniently obtain an oblivious transfer protocol by asking the database to encrypt (once and for all) each line for an identity (the j-th line being encrypted for the identity j), and having the user do a blind user key generation query for identity i in order to recover the key corresponding to the line i he expects to learn.

This approach is round-optimal: After the database preparation, the first flow is sent by the user as a commitment to the identity i, and the second one is sent by the server with the blinded expected information. But several technicalities arise because of the UC framework we consider here. For instance, the blinded expected information has to be masked, we do this here thanks to an SPHF. Furthermore, instead of using simple line numbers as identities, we have to commit to words in specific languages (so as to ensure extractability and equivocability) as well as to *fragment* the IBE keys into bits in order to achieve $O(\log n)$ in both flows. This allows us to achieve the first UC-secure adaptive OT protocol allowing adaptive corruptions. More details follow in the next sessions.

3.3 Main Building Block: Constructing a Blind Fragmented IBKEM from an IBKEM

Definition and Security Properties of a Blind IBKEM Scheme. Following [12], we recalled in Sect. 2.2 page 6 the definitions, notations and security properties for an IBE scheme, seen as an Identity-Based Key Encapsulation (IBKEM) scheme. We continue to follow the KEM formalism by adapting the definition of a Blind IBE scheme given in [37] to this setting.

Definition 7 (Blind Identity-Based Key Encapsulation Scheme). *A Blind Identity-Based Key Encapsulation scheme* BlindIBKEM *consists of four PPT algorithms* (Gen, BlindUSKGen, Enc, Dec) *with the following properties:*

- Gen, Enc *and* Dec *are defined as for a traditional* IBKEM *scheme.*
- $\mathsf{BlindUSKGen}(\langle(\mathcal{S}, \mathsf{msk})(\mathcal{U}, \mathsf{id}, \ell; \boldsymbol{\rho})\rangle)$ *is an interactive protocol, in which an honest user* $\mathcal{U}$ *with identity* $\mathsf{id} \in \mathcal{ID}$ *obtains the corresponding user secret key* usk[id] *from the master authority* $\mathcal{S}$ *or outputs an error message, while* $\mathcal{S}$*'s output is nothing or an error message* (ℓ *is a label and* $\boldsymbol{\rho}$ *the randomness).*

Defining the security of a BlindIBKEM requires two additional properties, stated as follows (see [37, pages 6 and 7] for the formal security games):

1. **Leak-free Secret Key Generation** (called Leak-free Extract for Blind IBE security in the original paper): A potentially malicious user cannot learn anything by executing the BlindUSKGen protocol with an honest authority which he could not have learned by executing the USKGen protocol with an honest authority; Moreover, as in USKGen, the user must know the identity for which he is extracting a key.

2. **Selective-failure Blindness**: A potentially malicious authority cannot learn anything about the user's choice of identity during the BlindUSKGen protocol; Moreover, the authority cannot cause the BlindUSKGen protocol to fail in a manner dependent on the user's choice.
   For our applications, we only need a weakened property for blindness:[8]
3. **Weak Blindness**: A potentially malicious authority cannot learn anything about the user's choice of identity during the BlindUSKGen protocol.

High-Level Idea of the Transformation. We now show how to obtain a BlindIBKEM scheme from any IBKEM scheme. From a high-level point of view, this transformation mixes two pre-existing approaches.

First, we are going to consider a reverse Naor transform [14,27]: He drew a parallel between Identity-Based Encryption schemes and signature schemes, by showing that a user secret key on an identity can be viewed as the signature on this identity, the verification process therefore being a test that any chosen valid ciphertext for the said identity can indeed be decrypted using the *signature* scheme.

Then, we are going to use Fischlin [31] round-optimal approach to blind signatures, where the whole interaction is done in one pass: First, the user commits to the message, then he recovers a signature linked to his commitment. For sake of simplicity, instead of using a Non-Interactive Zero-Knowledge Proof of Knowledge of a signature, we are going to follow the [11,13] approach, where thanks to an additional term, the user can extract a signature on the identity from a signature on the committed identity.

Omitting technical details described more precisely in the following sections, the main idea of the transformation of the IBKEM scheme in order to blind a user key request is described in Fig. 3.

Generic Transformation of an IBKEM into a Blind IBKEM. It now remains to explain how one can fulfill the idea highlighted in Fig. 3. The technique to blind a user key request uses a smooth projective hash function (see Sect. 2.3), and is often called *implicit decommitment* in recent works: the IBKEM secret key is sent hidden in such a way that it can only be recovered if the user knows how to open the initial commitment on the correct identity. We assume the existence of a labeled CCA-encryption scheme $\mathcal{E} = (\mathsf{Setup}_{\mathrm{cca}}, \mathsf{KeyGen}_{\mathrm{cca}}, \mathsf{Encrypt}^{\ell}_{\mathrm{cca}}, \mathsf{Decrypt}^{\ell}_{\mathrm{cca}})$ compatible with an SPHF defined by $(\mathsf{Setup}, \mathsf{HashKG}, \mathsf{ProjKG}, \mathsf{Hash}, \mathsf{ProjHash})$ onto a set G (where ℓ is a label defined by the global protocol). By "compatible", we mean that the SPHF can be defined over a language $\mathfrak{L}^{c}_{\mathsf{id}} \subset X$, where $\mathfrak{L}^{c}_{\mathsf{id}} = \{C \mid \exists \rho \text{ such that } C = \mathsf{Encrypt}^{\ell}_{\mathrm{cca}}(\mathsf{id}; \rho)\}$. In the KeyGen algorithm, the description of the language $\mathfrak{L}_{\mathsf{id}} = \{\mathsf{id}\}$ thus implicitly defines the language $\mathfrak{L}^{c}_{\mathsf{id}}$ of

[8] Two things to note: First, Selective Failure would be considered as a Denial of Service in the Oblivious Transfer setting. Then, we do not restrict ourselves to schemes where the blindness adversary has access to the generated user keys, as reliable erasures in the OT protocol provide us a way to forget them before being corrupted (otherwise we would need to use a randomizable base IBE).

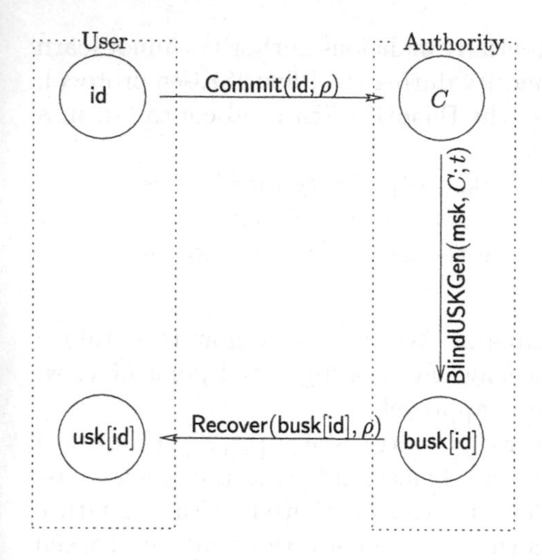

1. A user commits to the targeted identity id using some randomness ρ.
2. The authority possesses an algorithm allowing it to generate keys for committed identities using its master secret key msk, and some randomness t, in order to obtain a blinded user secret key busk[id].
3. The user using solely the randomness used in the initial commitment is able to recover the requested secret key from the authority's generated value.

Fig. 3. Generic Transformation of an IBKEM into a Blind IBKEM (naive approach)

CCA-encryptions of elements of $\mathfrak{L}_{\mathsf{id}}$. We additionally use a key derivation function KDF to derive a pseudo-random bit-string $K \in \{0,1\}^{\mathfrak{K}}$ from a pseudo-random element $v \in G$. One can use the Leftover-Hash Lemma [40], with a random seed defined in param during the global setup, to extract the entropy from v, then followed by a pseudo-random generator to get a long enough bit-string. Many uses of the same seed in the Leftover-Hash Lemma just lead to a security loss linear in the number of extractions. This gives the following protocol for BlindUSKGen, described in Fig. 4.

- The user computes an encryption of the expected identity id and keeps the randomness ρ: $C = \mathsf{Encrypt}^{\ell}_{\mathsf{cca}}(\mathsf{id}; \rho)\}$.
- For every identity id′, the server computes the key usk[id′] along with a pair of (secret, public) hash keys $(\mathsf{hk}_{\mathsf{id}'}, \mathsf{hp}_{\mathsf{id}'})$ for a smooth projective hash function on the language $\mathfrak{L}^{\mathsf{c}}_{\mathsf{id}'}$:
  $\mathsf{hk}_{\mathsf{id}'} = \mathsf{HashKG}(\ell, \mathfrak{L}^{\mathsf{c}}_{\mathsf{id}'}, \mathsf{param})$ and $\mathsf{hp}_{\mathsf{id}'} = \mathsf{ProjKG}(\mathsf{hk}_{\mathsf{id}'}, \ell, (\mathfrak{L}^{\mathsf{c}}_{\mathsf{id}'}, \mathsf{param}), \mathsf{id}')$.
  He also compute the corresponding hash value
  $$H_{\mathsf{id}'} = \mathsf{Hash}(\mathsf{hk}_{\mathsf{id}'}, (\mathfrak{L}^{\mathsf{c}}_{\mathsf{id}'}, \mathsf{param}), (\ell, C)).$$
  Finally, he sends $(\mathsf{hp}_{\mathsf{id}'}, \mathsf{usk}[\mathsf{id}'] \oplus \mathtt{KDF}(H_{\mathsf{id}'}))$ for every id′, where $\oplus$ is a compatible operation.
- Thanks to $\mathsf{hp}_{\mathsf{id}}$, the user is able to compute the corresponding projected hash value $H'_{\mathsf{id}} = \mathsf{ProjHash}(\mathsf{hp}_{\mathsf{id}}, (\mathfrak{L}^{\mathsf{c}}_{\mathsf{id}}, \mathsf{param}), (\ell, C), \rho)$. He then recovers usk[id] for the initially committed identity id since $H_{\mathsf{id}} = H'_{\mathsf{id}}$.

Fig. 4. Summary of the Generic Construction of $\mathsf{BlindUSKGen}(\langle(\mathcal{S}, \mathsf{msk})(\mathcal{U}, \mathsf{id}, \ell; \rho)\rangle)$ for a blind IBE

Theorem 8. *If* IBKEM *is a* PR-ID-CPA*-secure identity-based key encapsulation scheme and* $\mathcal{E}$ *a labeled* CCA*-encryption scheme compatible with an* SPHF*, then* BlindIBKEM *is leak free and weak blind.*

Proof. First, BlindIBKEM satisfies leak-free secret key generation since it relies on the CCA security on the encryption scheme, forbidding a user to open it to another identity than the one initially encrypted. Furthermore, the pseudo-randomness of the SPHF ensures that the blinded user key received for id is indistinguishable from random if he encrypted $\mathsf{id}' \neq \mathsf{id}$. Finally, the weak blindness also relies on the CCA security on the encryption scheme, since an encryption of id is indistinguishable from a encryption of $\mathsf{id}' \neq \mathsf{id}$. □

Using a Blind IBKEM in our Application to Adaptive Oblivious Transfer. The previous approach allows to transform an IBKEM into a Blind IBKEM, but it has a huge drawback in our context: Since we assume an exponential identity space, it requires an exponential number of answers from the authority, which cannot help us to fulfill logarithmic complexity in our application. However, if we focus on the special case of affine IBE with bitwise function[9], a user key can be described as the list $(\mathsf{usk}[0], \mathsf{usk}[0, \mathsf{id}_0], \ldots, \mathsf{usk}[m-1, \mathsf{id}_{m-1}])$ if id_i is the i-th bit of the identity id. One can thus manage to be much more efficient by sending each "bit" evaluation on the user secret key, hidden with a smooth projective hash value on the language "the i-th bit of the identity is a 0 (or 1)", which is common to all identities. We can thus reduce the number of languages from the number of identities (which is exponential) to the length of an identity (which is polynomial). For security reasons, one cannot give directly the evaluation value, but as we are considering the sum of the evaluations for each bit, we simply add a Shamir-like secret sharing, by adding randomness that is going to cancel out at the end.

As a last step, we finally need to make our construction compatible with the UC framework with adaptive corruptions. In this context, interactions should make sense for any possible input chosen by the environment and learnt a posteriori in the simulation during the corruption of an honest party. From the user side, this implies that the last flow should contain enough recoverable information so that a simulator, having sent a commitment to an incorrect identity, can extract the proper user secret key corresponding to the correct identity recovered after the corruption. From the server side, this implies that the IBKEM scheme is defined such as one is able to adapt the user secret keys in order to correspond to the new database learnt a posteriori. Of course, not all schemes allow this property, but this will be the case in the pairing scenario considered in our concrete instantiation.

To deal with corruptions of the user, recall that a simulated server (knowing the secret key of the encryption scheme) is already able to extract the identity

[9] They were defined in [12]. Affine IBE derive their name from the fact that only affine operations are done on the identity bits (no hashing, square rooting, inverting... are allowed).

- The user computes a bit-per-bit encryption of the expected identity id and keeps the randomness ρ: $C = \mathsf{Encrypt}^{\ell}_{\mathsf{cca}}(\mathsf{id}; \rho)\}$.
- The server computes a fragmented version of all the keys $\mathsf{usk}[\mathsf{id}']$, *i.e.* all the values $\mathsf{usk}[i, b]$ for i from 0 up to the length m of the keys and $b \in \{0, 1\}$. He also computes a pair of (secret, public) hash keys $(\mathsf{hk}_{i,b}, \mathsf{hp}_{i,b})$ for a smooth projective hash function on the language $\mathfrak{L}^c_{i,b}$: "The i-th bit of the value encrypted into C is b", *i.e.* $\mathsf{hk}_{i,b} = \mathsf{HashKG}(\ell, \mathfrak{L}^c_{i,b}, \mathsf{param})$ and $\mathsf{hp}_{i,b} = \mathsf{ProjKG}(\mathsf{hk}_{i,b}, \ell, (\mathfrak{L}^c_{i,b}, \mathsf{param}))$. He also computes the corresponding hash value $H_{i,b} = \mathsf{Hash}(\mathsf{hk}_{i,b}, (\mathfrak{L}^c_{i,b}, \mathsf{param}), (\ell, C))$ and chooses random values z_i. Finally, he sends, for each (i, b), $(\mathsf{hp}_{i,b}, \mathsf{busk}[i, b])$, where $\mathsf{busk}[i, b] = \mathsf{usk}[i, b] \oplus \mathtt{KDF}(H_{i,b}) \oplus z_i$, together with $Z = \mathsf{usk}_0 \ominus \left(\bigoplus_i z_i\right)$, where $\oplus$ is a compatible operation and $\ominus$ its inverse.
- Thanks to the $\mathsf{hp}_{i,\mathsf{id}_i}$ for the initially committed identity id, the user is able to compute the corresponding projected hash value
$H'_{i,\mathsf{id}_i} = \mathsf{ProjHash}(\mathsf{hp}_{i,\mathsf{id}_i}, (\mathfrak{L}^c_{i,\mathsf{id}_i}, \mathsf{param}), (\ell, C), \rho)$,
that should be equal to H_{i,id_i} for all i. From the values $\mathsf{busk}[i, \mathsf{id}_i]$, he then recovers $\mathsf{usk}[i, \mathsf{id}_i] \oplus z_i$. Finally, with the operation $\left(\bigoplus_i (\mathsf{usk}[i, \mathsf{id}_i] \oplus z_i)\right) \oplus Z$, he recovers the expected $\mathsf{usk}[\mathsf{id}]$.

Fig. 5. Summary of the Generic Construction of $\mathsf{BlindUSKGen}(\langle(\mathcal{S}, \mathsf{msk})(\mathcal{U}, \mathsf{id}, \ell; \rho)\rangle)$ for a Blind affine IBE

committed to. But we now consider that, for all id, $\mathfrak{L}_{\mathsf{id}}$ is the language of the equivocable commitments on words in the inner language $\widetilde{\mathfrak{L}}_{\mathsf{id}} = \{\mathsf{id}\}$. We assume them to be a *Trapdoor Collection of Languages*, which means that it is computationally hard to sample an element in $\mathfrak{L}_1 \cap \cdots \cap \mathfrak{L}_n$, except for the simulator, who possesses a trapdoor tk (the equivocation trapdoor) allowing it to sample an element in the intersection of languages. This allows a simulated user (knowing this trapdoor) not to really bind to any identity during the commitment phase. The only difference with the algorithm described in Fig. 5 is that the user now encrypts this word W (which is an equivocable commitment on his identity id) rather than directly encrypting his identity id: $C = \mathsf{Encrypt}^{\ell}_{\mathsf{cca}}(W; \rho)$. This technique is also explained as an application of our OLBE framework, in the paper full version [10]. We will directly prove this protocol during the proof of the oblivious transfer scheme.

3.4 Generic Construction of Adaptive OT

We derive from here our generic construction of OT (depicted in Fig. 6). We additionally assume the existence of a Pseudo-Random Generator (PRG) F with input size equal to the plaintext size, and output size equal to the size of the messages in the database and an IND-CPA encryption scheme $\mathcal{E} = (\mathsf{Setup}_{\mathsf{cpa}}, \mathsf{KeyGen}_{\mathsf{cpa}}, \mathsf{Encrypt}_{\mathsf{cpa}}, \mathsf{Decrypt}_{\mathsf{cpa}})$ with plaintext size at least equal to the security parameter. First, the owner of the database generates the keys for such an IBE scheme, and encrypts each line i of the database for the identity i. Then when a user wants to request a given line, he runs the blind user key generation algorithm and recovers the key for the expected given line. This leads to the following security result, proven in the paper full version [10].

Theorem 9. *Assuming that* BlindUSKGen *is constructed as described above, the adaptive Oblivious Transfer protocol described in Fig. 6 UC-realizes the functionality $\mathcal{F}_{OT}^{\mathcal{L}}$ presented in Fig. 2 with adaptive corruptions assuming reliable erasures.*

3.5 Pairing-Based Instantiation of Adaptive OT

Affine Bit-Wise Blind IBE. In [12], the authors propose a generic framework to move from affine Message Authentication Code to IBE, and they propose a

CRS generation:
$\mathsf{crs} \stackrel{\$}{\leftarrow} \mathsf{SetupCom}(1^{\mathfrak{K}}), \mathsf{param}_{\mathsf{cpa}} \stackrel{\$}{\leftarrow} \mathsf{Setup}_{\mathsf{cpa}}(1^{\mathfrak{K}})$.
Database Preparation:

1. Server runs $\mathsf{Gen}(\mathfrak{K})$, to obtain $\mathsf{mpk}, \mathsf{msk}$.
2. For each line t, he computes $(D_t, K_t) = \mathsf{Enc}(\mathsf{mpk}, t)$, and $L_t = K_t \oplus DB(t)$.
3. He also computes $\mathsf{usk}[i, b]$ for all $i = 1 \dots, m$ and $b = 0, 1$ and erases msk.
4. Server generates a key pair $(\mathsf{pk}, \mathsf{sk}) \stackrel{\$}{\leftarrow} \mathsf{KeyGen}_{\mathsf{cpa}}(\mathsf{param}_{\mathsf{cpa}})$ for $\mathcal{E}$, stores sk and completely erases the random coins used by KeyGen.
5. He then publishes $\mathsf{mpk}, \{(D_t, L_t)\}_t, \mathsf{pk}$.

Index query on s**:**

1. User chooses a random value S, computes $R \leftarrow F(S)$ and encrypts S under pk: $c \stackrel{\$}{\leftarrow} \mathsf{Encrypt}_{\mathsf{cpa}}(\mathsf{pk}, S)$
2. User computes $\mathcal{C}$ with the first flow of $\mathsf{BlindUSKGen}(\langle(\mathcal{S}, \mathsf{msk})(\mathcal{U}, s, \ell; \boldsymbol{\rho})\rangle)$ with $\ell = (\mathsf{sid}, \mathsf{ssid}, \mathcal{U}, \mathcal{S})$ (see Figure 5).
3. User stores the random $\boldsymbol{\rho}_s = \{\boldsymbol{\rho}_*\}$ needed to open $\mathcal{C}$ to s, and completely erases the rest, including the random coins used by $\mathsf{Encrypt}_{\mathsf{cpa}}$ and sends $(c, \mathcal{C})$ to the Server

IBE input msk:

1. Server decrypts $S \leftarrow \mathsf{Decrypt}_{\mathsf{cpa}}(\mathsf{sk}, c)$ and computes $R \leftarrow F(S)$
2. Server runs the second flow of $\mathsf{BlindUSKGen}(\langle(\mathcal{S}, \mathsf{msk})(\mathcal{U}, s, \ell; \boldsymbol{\rho})\rangle)$ on $\mathcal{C}$ (see Figure 5).
3. Server erases every new value except $(\mathsf{hp}_{i,b})_{i,b}, (\mathsf{busk}[i, b])_{i,b}, Z \oplus R$ and sends them over a secure channel.

Data recovery:

1. User then using, $\boldsymbol{\rho}_s$ recovers $\mathsf{usk}[s]$ from the values received from the server.
2. He can then recover the expected information with $\mathsf{Dec}(\mathsf{usk}[s], s, D_s) \oplus L_s$ and erases everything else.

Fig. 6. Adaptive UC-Secure 1-out-of-n OT from a Fragmented Blind IBE

tight instantiation of such a MAC, giving an affine bit-wise IBE, which seems like a good candidate for our setting (making it blind and fragmented).

We are thus going to use the family of IBE described in the following picture (Fig. 7), which is their instantiation derived from a Naor-Reingold MAC[10]. In the following, $h_i()$ are injective deterministic public functions mapping a bit to a scalar in $\mathbb{Z}_p$.

A property that was not studied in this paper was the blind user key generation: How to generate and answer blind user secret key queries? We answer to this question by proposing the $k-\mathsf{MDDH}$-based variation presented in Fig. 8. To fit the global framework we are going to consider the equivocable language of each chameleon hash of the identity bits $(\boldsymbol{a}_i, \boldsymbol{b}_{i,m_i})$, and then a Cramer-Shoup like encryption of $\boldsymbol{b}$ into $\boldsymbol{d}$ (more details in the paper full version [10]). We denote this process as Har in the following protocol, and by $\mathfrak{L}_{\mathsf{Har},i,\mathsf{id}_i}$ the language on identity bits. We thus obtain the following security results.

Theorem 10. *This construction achieves both the weak Blindness, and the leak-free secret key generation requirements under the* $k-\mathsf{MDDH}$ *assumption.*

The first one is true under the indistinguishability of the generalized Cramer-Shoup encryption recalled in the paper full version [10], as the server learns nothing about the line requested during the first flow. It should even be noted that because of the inner chameleon hash, a simulator is able to use the trapdoor to do a commitment to every possible words of the set of languages at once, and so can adaptively decide which id he requested. The proof of the second result is delayed to the paper full version [10].

For sake of generality, any bit-wise affine IBE could work (like for example Waters IBE [57]), the additional price paid for tightness here is very small

$\underline{\mathsf{Gen}(\mathfrak{K})}$:	$\underline{\mathsf{Enc}(\mathsf{mpk},\mathsf{id})}$:
$\boldsymbol{A} \stackrel{\$}{\leftarrow} \mathcal{D}_k, \boldsymbol{B} = \overline{\boldsymbol{A}}$	$\boldsymbol{r} \stackrel{\$}{\leftarrow} \mathbb{Z}_p^k$
For $i \in [\![0,\ell]\!] : \boldsymbol{Y}_i \stackrel{\$}{\leftarrow} \mathbb{Z}_p^{k+1}; \boldsymbol{Z}_i = \boldsymbol{Y}_i^\top \cdot \boldsymbol{A} \in \mathbb{Z}_p^k$	$\boldsymbol{c}_0 = \boldsymbol{A}\boldsymbol{r} \in \mathbb{Z}_p^{k+1}$
$\boldsymbol{y}' \stackrel{\$}{\leftarrow} \mathbb{Z}_p^{k+1}; \boldsymbol{z}' = \boldsymbol{y}'^\top \cdot \boldsymbol{A} \in \mathbb{Z}_q^k$	$\boldsymbol{c}_1 = (\boldsymbol{Z}_0 + \sum_{i=1}^{\ell} h_i(\mathsf{id}_i)\boldsymbol{Z}_i) \cdot \boldsymbol{r} \in \mathbb{Z}_p$
$\mathsf{mpk} := (\mathcal{G}, [\boldsymbol{A}]_1, ([\boldsymbol{Z}_i]_1)_{i\in[\![0,\ell]\!]}, [\boldsymbol{z}']_1)$	$K = \boldsymbol{z}' \cdot \boldsymbol{r} \in \mathbb{Z}_p$.
$\mathsf{msk} := (\boldsymbol{Y}_i)_{i\in[\![0,\ell]\!]}, \boldsymbol{y}'$	Return $[K]_T$
Return $(\mathsf{mpk},\mathsf{msk})$	and $\mathsf{C} = ([\boldsymbol{c}_0]_1, [\boldsymbol{c}_1]_1) \in \mathbb{G}_1^{k+1+1}$
$\underline{\mathsf{USKGen}(\mathsf{msk},\mathsf{id})}$:	$\underline{\mathsf{Dec}(\mathsf{usk}[\mathsf{id}],\mathsf{id},\mathsf{C})}$:
$\boldsymbol{s} \stackrel{\$}{\leftarrow} \mathbb{Z}_p^k, \boldsymbol{t} = \boldsymbol{B}\boldsymbol{s}$	Parse $\mathsf{usk}[\mathsf{id}] = ([\boldsymbol{t}]_2, [\boldsymbol{w}]_2)$
$\boldsymbol{w} = (\boldsymbol{Y}_0 + \sum_{i=1}^{\ell} h_i(\mathsf{id}_i)\boldsymbol{Y}_i)\boldsymbol{t} + \boldsymbol{y}' \in \mathbb{Z}_p^{k+1}$	Parse $\mathsf{C} = ([\boldsymbol{c}_0]_1, [\boldsymbol{c}_1]_1)$
Return $\mathsf{usk}[\mathsf{id}] := ([\boldsymbol{t}]_2, [\boldsymbol{w}]_2) \in \mathbb{G}_2^{k+k+1}$	$K = e([\boldsymbol{c}_0]_1, [\boldsymbol{w}]_2) \cdot e([\boldsymbol{c}_1]_1, [\boldsymbol{t}]_2)^{-1}$
	Return $K \in \mathbb{G}_T$

Fig. 7. A fragmentable affine IBKEM.

[10] For the reader familiar with the original result, we combine $x, \boldsymbol{y}$ into a bigger $\boldsymbol{y}$ to lighten the notations, and compact the (x_i', y_i') values into a single y' as this has no impact on their construction.

- First flow: <u>$\mathcal{U}$ starts by computing</u>
  $\boldsymbol{\rho} \stackrel{\$}{\leftarrow} \mathbb{Z}_p^{1+4\times\ell}$,
  $\boldsymbol{a}, \boldsymbol{d} = \mathsf{Har}(\mathsf{id}, \ell; \rho) \in \mathbb{Z}_p^{\ell} \times \mathbb{Z}_p^{2\times(k+3)\ell}$,
  Sends $\mathcal{C} = ([\boldsymbol{a}]_1, [\boldsymbol{d}]_2)$ to S
- Second Flow: <u>$\mathcal{S}$ then proceeds</u>
  $\boldsymbol{s} \stackrel{\$}{\leftarrow} \mathbb{Z}_p^k, \boldsymbol{t} = \boldsymbol{B}\boldsymbol{s}$, $\boldsymbol{f} \stackrel{\$}{\leftarrow} \mathbb{Z}_p^{\ell\times k+1}$,
  For each $i \in [\![1, \lceil \log n \rceil]\!], b \in [\![0, 1]\!]$:
  $\mathsf{hk}_{i,b} = \mathsf{HashKG}(\mathfrak{L}_{\mathsf{Har},i,b}, \boldsymbol{C})$
  $\mathsf{hp}_{i,b} = \mathsf{ProjKG}(\mathsf{hk}_{i,b}, \mathfrak{L}_{\mathsf{Har},i,b}, \boldsymbol{C})$
  $H_{i,b} = \mathsf{Hash}(\mathsf{hk}_{i,b}, \mathfrak{L}_{\mathsf{Har},i,b}, \boldsymbol{C})$
  $\boldsymbol{\omega}_{i,b} = (b\boldsymbol{Y}_i)\boldsymbol{t} + \boldsymbol{f}_i + H_{i,b}$

Then sets $\boldsymbol{w}_0 = \boldsymbol{Y}_0\boldsymbol{t} + \boldsymbol{y}' - \sum_{i=1}^{\ell} \boldsymbol{f}_i \in \mathbb{Z}_p^{k+1}$
Returns $\mathsf{busk} :=$
$([\boldsymbol{t}]_2, [\boldsymbol{w}_0]_2, \{[\boldsymbol{\omega}_{i,b}]_2\}, \{[\mathsf{hp}_{i,b}]_2\})$
- $\mathsf{BlindUSKGen}_3$:
  <u>$\mathcal{U}$ then recovers his key</u>
  For each $i \in [\![1, \ell]\!]$:
  $H_i' =$
  $\mathsf{ProjHash}(\mathsf{hp}_{i,\mathsf{id}_i}, \mathfrak{L}_{\mathsf{Har},i,\mathsf{id}_i}, \boldsymbol{C}, \rho_i)$
  $\boldsymbol{w}_i = \omega_{i,\mathsf{id}_i} - H_i'$
  $\boldsymbol{w} = \boldsymbol{w}_0 + \sum_{i=1}^{\ell} \boldsymbol{w}_i$
  And then recovers $\mathsf{usk}[\mathsf{id}] :=$
  $[\boldsymbol{t}]_2, [\boldsymbol{w}]_2$

Fig. 8. $\mathsf{BlindUSKGen}(\langle(\mathcal{S}, \mathsf{msk})(\mathcal{U}, \mathsf{id}, \ell; \rho)\rangle)$.

and allows to have a better reduction in the proof, but it not required by the framework itself.

Adaptive UC-Secure Oblivious Transfer. We finally get our instantiation by combining this $k - \mathsf{MDDH}$-based blind IBE with a $k - \mathsf{MDDH}$ variant of El Gamal for the CPA encryption needed (see the paper full version [10] for details). The requirement on the IBE blind user secret key generation (being able to adapt the key if the line changes) is achieved assuming that the server knows the discrete logarithms of the database lines. This is quite easy to achieve by assuming that for all line s, $DB(s) = [db(s)]_1$ where $db(s)$ is the real line (thus known). It implies a few more computation on the user's side in order to recover $db(s)$ from $DB(s)$, but this remains completely feasible if the lines belong to a small space. For practical applications, one could imagine to split all 256-bit lines into 8 pieces for a decent/constant trade-off in favor of computational efficiency.

For $k = 1$, so under the classical SXDH assumption, the first flow requires $8 \log |DB|$ elements in $\mathbb{G}_1$ for the CCA encryption part and $\log(|DB| + 1)$ in $\mathbb{G}_2$ for the chameleon one, while the second flow would now require $1 + 4 \log |DB|$ elements in $\mathbb{G}_1$, $1 + 2 \log |DB|$ for the fragmented masked key, and $2 \log |DB|$ for the projection keys.

4 Oblivious Language-Based Envelope

The previous construction opens new efficient applications to the already known Oblivious-Transfer protocols. But what happens when someone wants some additional access control by requesting extra properties, like if the user is only allowed to ask two lines with the same parity bits, the user can only request lines for whose number has been signed by an authority, or even finer control provided through credentials?

In this section we propose to develop a new primitive, that we call Oblivious Language-Based Envelope (OLBE). The idea generalizes that of Oblivious

Transfer and OSBE, recalled right afterwards, for n messages (with n polynomial in the security parameter $\mathfrak{K}$) to provide the best of both worlds.

4.1 Oblivious Signature-Based Envelope

We recall the definition and security requirements of an OSBE protocol given in [13,49], in which a sender $\mathcal{S}$ wants to send a private message $m \in \{0,1\}^{\mathfrak{K}}$ to a recipient $\mathcal{R}$ in possession of a valid certificate/signature on a public message M (given by a certification authority).

Definition 11 (Oblivious Signature-Based Envelope). *An* OSBE *scheme is defined by four algorithms* (Setup, KeyGen, Sign, Verify)*, and one interactive protocol* Protocol$\langle \mathcal{S}, \mathcal{R} \rangle$*:*

- Setup($1^{\mathfrak{K}}$)*, where $\mathfrak{K}$ is the security parameter, generates the global parameters* param*;*
- KeyGen($\mathfrak{K}$) *generates the keys* (vk, sk) *of the certification authority;*
- Sign(sk, M) *produces a signature σ on the input message M, under the signing key* sk*;*
- Verify(vk, M, σ) *checks whether σ is a valid signature on M, w.r.t. the public key* vk*; it outputs 1 if the signature is valid, and 0 otherwise.*
- Protocol$\langle \mathcal{S}(\mathsf{vk}, M, P), \mathcal{R}(\mathsf{vk}, M, \sigma) \rangle$ *between the sender $\mathcal{S}$ with the private message P, and the recipient $\mathcal{R}$ with a certificate σ. If σ is a valid signature under* vk *on the common message M, then $\mathcal{R}$ receives m, otherwise it receives nothing. In any case, $\mathcal{S}$ does not learn anything.*

The authors of [13] proposed some variations to the original definitions from [49], in order to prevent some interference by the authority. Following them, an OSBE scheme should fulfill the following security properties. The formal security games are given in [13]. No UC functionality has already been given, to the best of our knowledge.

- *correct*: the protocol actually allows $\mathcal{R}$ to learn P, whenever σ is a valid signature on M under vk;
- *semantically secure*: the recipient learns nothing about $\mathcal{S}$'s input m if it does not use a valid signature σ on M under vk as input. More precisely, if $\mathcal{S}_0$ owns P_0 and $\mathcal{S}_1$ owns P_1, the recipient that does not use a valid signature cannot distinguish an interaction with $\mathcal{S}_0$ from an interaction with $\mathcal{S}_1$ even if he has eavesdropped on several interactions $\langle \mathcal{S}(\mathsf{vk}, M, P), \mathcal{R}(\mathsf{vk}, M, \sigma) \rangle$ with valid signatures, and the same sender's input P;
- *escrow-free* (*oblivious with respect to the authority*): the authority (owner of the signing key sk), playing as the sender or just eavesdropping, is unable to distinguish whether $\mathcal{R}$ used a valid signature σ on M under vk as input.
- *semantically secure w.r.t. the authority*: after the interaction, the authority (owner of the signing key sk) learns nothing about m from a passive access to a challenge transcript.

4.2 Definition of an Oblivious Language-Based Envelope

In such a protocol, a sender $\mathcal{S}$ wants to send one or several private messages (up to $n_{\max} \leq n$) among $(m_1, \ldots, m_n) \in (\{0,1\}^\ell)^n$ to a recipient $\mathcal{R}$ in possession of a tuple of words $W = (W_{i_1}, \ldots, W_{i_{n_{\max}}})$ such that some of the words W_{i_j} may belong to the corresponding language $\mathfrak{L}_{i_j}$. More precisely, the receiver gets each m_{i_j} as soon as $W_{i_j} \in \mathfrak{L}_{i_j}$ with the requirement that he gets at most $n_{\max}$ messages. In such a scheme, the languages $(\mathfrak{L}_1, \ldots, \mathfrak{L}_n)$ are assumed to be a trapdoor collection of languages, publicly verifiable and self-randomizable (see Sect. 2.3 for the definitions of the properties of the languages).

The collections of words can be a single certificate/signature on a message M (encompassing OSBE, with $n = n_{\max} = 1$), a password, a credential, a line number (encompassing 1-out-of-n oblivious transfer[11], with $n_{\max} = 1$), k line numbers (encompassing k-out-of-n oblivious transfer, with $n_{\max} = k$), etc. (see the paper full version [10] for detailed examples). Following the definitions for OSBE recalled above and given in [13,49], we give the following definition for OLBE. As we consider simulation-based security (in the UC framework), we allow a simulated setup SetupT to be run instead of the classical setup Setup in order to allow the simulator to possess some trapdoors. Those two setup algorithms should be indistinguishable.

Definition 12 (Oblivious Language-Based Envelope). *An* OLBE *scheme is defined by four algorithms* (Setup, KeyGen, Samp, Verify), *and one interactive protocol* Protocol$\langle \mathcal{S}, \mathcal{R} \rangle$*:*

- Setup$(1^{\mathfrak{K}})$, *where $\mathfrak{K}$ is the security parameter, generates the global parameters* param, *among which the numbers n and $n_{\max}$;*
  or SetupT$(1^{\mathfrak{K}})$, *where $\mathfrak{K}$ is the security parameter, additionally allows the existence*[12] *of a trapdoor* tk *for the collection of languages $(\mathfrak{L}_1, \ldots, \mathfrak{L}_n)$.*
- KeyGen(param, $\mathfrak{K}$) *generates, for all $i \in \{1, \ldots, n\}$, the description of the language $\mathfrak{L}_i$ (as well as the language key $\mathsf{sk}_{\mathfrak{L}_i}$ if need be). If the parameters* param *were defined by* SetupT, *this implicitly also defines the common trapdoor* tk *for the collection of languages $(\mathfrak{L}_1, \ldots, \mathfrak{L}_n)$.*
- Samp(param, I) *or* Samp(param, $I, (\mathsf{sk}_{\mathfrak{L}_i})_{i \in I}$) *such that $I \subset \{1, \ldots, n\}$ and $|I| = n_{\max}$, generates a list of words $(W_i)_{i \in I}$ such that $W_i \in \mathfrak{L}_i$ for all $i \in I$;*
- Verify$_i(W_i, \mathfrak{L}_i)$ *checks whether W_i is a valid word in the language $\mathfrak{L}_i$. It outputs 1 if the word is valid, 0 otherwise;*
- Protocol$\langle \mathcal{S}((\mathfrak{L}_1, \ldots, \mathfrak{L}_n), (m_1, \ldots, m_n)), \mathcal{R}((\mathfrak{L}_1, \ldots, \mathfrak{L}_n), (W_i)_{i \in I}) \rangle$, *which is executed between the sender $\mathcal{S}$ with the private messages $(m_1, \ldots, P_n)$ and corresponding languages $(\mathfrak{L}_1, \ldots, \mathfrak{L}_n)$, and the recipient $\mathcal{R}$ with the same languages and the words $(W_i)_{i \in I}$ with $I \subset \{1, \ldots, n\}$ and $|I| = n_{\max}$, proceeds*

[11] Even if, as explained in the former section, we would rather consider equivocable commitments of line numbers than directly line numbers, in order to get adaptive UC security.

[12] The specific trapdoor will depend on the languages and be computed in the KeyGen algorithm.

as follows. For all $i \in I$, if the algorithm $\mathsf{Verify}_i(W_i, \mathfrak{L}_i)$ *returns* 1*, then* $\mathcal{R}$ *receives* m_i*, otherwise it does not. In any case,* $\mathcal{S}$ *does not learn anything.*

4.3 Security Properties and Ideal Functionality of OLBE

Since we aim at proving the security in the universal composability framework, we now describe the corresponding ideal functionality (depicted in Fig. 9). However, in order to ease the comparison with an OSBE scheme, we first list the security properties required, following [13,49]:

- *correct*: the protocol actually allows $\mathcal{R}$ to learn $(m_i)_{i\in I}$, whenever $(W_i)_{i\in I}$ are valid words of the languages $(\mathfrak{L}_i)_{i\in I}$, where $I \subset \{1,\dots,n\}$ and $|I| = n_{\max}$;
- *semantically secure* (*sem*): the recipient learns nothing about the input m_i of $\mathcal{S}$ if it does not use a word in $\mathfrak{L}_i$. More precisely, if $\mathcal{S}_0$ owns $m_{i,0}$ and $\mathcal{S}_1$ owns $m_{i,1}$, the recipient that does not use a word in $\mathfrak{L}_i$ cannot distinguish between an interaction with $\mathcal{S}_0$ and an interaction with $\mathcal{S}_1$ even if the receiver has seen several interactions
$$\langle \mathcal{S}((\mathfrak{L}_1,\dots,\mathfrak{L}_n),(m_1,\dots,m_n)), \mathcal{R}((\mathfrak{L}_1,\dots,\mathfrak{L}_n),(W'_j)_{j\in I})\rangle$$
with valid words $W'_i \in \mathfrak{L}_i$, and the same sender's input m_i;
- *escrow free* (*oblivious with respect to the authority*): the authority corresponding to the language $\mathfrak{L}_i$ (owner of the language secret key $\mathsf{sk}_{\mathfrak{L}_i}$ – if it exists), playing as the sender or just eavesdropping, is unable to distinguish whether $\mathcal{R}$ used a word W_i in the language $\mathfrak{L}_i$ or not. This requirement also holds for anyone holding the trapdoor key tk.
- *semantically secure w.r.t. the authority* (*sem**): after the interaction, the trusted authority (owner of the language secret keys if they exist) learns nothing about the values $(m_i)_{i\in I}$ from the transcript of the execution. This requirement also holds for anyone holding the trapdoor key tk.

Moreover, the Setups should be indistinguishable and it should be infeasible to find a word belonging to two or more languages without the knowledge of tk.

The ideal functionality is parametrized by a set of languages $(\mathfrak{L}_1,\dots,\mathfrak{L}_n)$. Since we show in the following sections that one can see OSBE and OT as special cases of OLBE, it is inspired from the oblivious transfer functionality given in [1, 21,24] in order to provide a framework consistent with works well-known in the literature. As for oblivious transfer (Fig. 2), we adapt them to the simple UC framework for simplicity (this enables us to get rid of `Sent` and `Received` queries from the adversary since the delayed outputs are automatically considered in this simpler framework: We implicitly let the adversary determine if it wants to acknowledge the fact that a message was indeed sent). The first step for the sender (`Send` query) consists in telling the functionality he is willing to take part in the protocol, giving as input his intended receiver and the messages he is willing to send (up to $n_{\max}$ messages). For the receiver, the first step (`Receive` query) consists in giving the functionality the name of the player he intends to receive the messages from, as well as his words. If the word does belong to the

The functionality $\mathcal{F}_{\mathrm{OLBE}}$ is parametrized by a security parameter $\mathfrak{K}$ and a set of languages $(\mathfrak{L}_1, \ldots, \mathfrak{L}_n)$ along with the corresponding public verification algorithms $(\mathsf{Verify}_1, \ldots, \mathsf{Verify}_n)$. It interacts with an adversary $\mathscr{S}$ and a set of parties $\mathfrak{P}_1, \ldots, \mathfrak{P}_N$ via the following queries:

- **Upon receiving from party $\mathfrak{P}_i$ an input of the form $(\mathtt{Send}, \mathsf{sid}, \mathsf{ssid}, \mathfrak{P}_i, \mathfrak{P}_j, (m_1, \ldots, m_n))$**, with $m_k \in \{0,1\}^{\mathfrak{K}}$ for all k: record the tuple $(\mathsf{sid}, \mathsf{ssid}, \mathfrak{P}_i, \mathfrak{P}_j, (m_1, \ldots, m_n))$ and reveal $(\mathtt{Send}, \mathsf{sid}, \mathsf{ssid}, \mathfrak{P}_i, \mathfrak{P}_j)$ to the adversary $\mathscr{S}$. Ignore further $\mathtt{Send}$-message with the same ssid from $\mathfrak{P}_i$.
- **Upon receiving an input of the form $(\mathtt{Receive}, \mathsf{sid}, \mathsf{ssid}, \mathfrak{P}_i, \mathfrak{P}_j, (W_i)_{i \in I})$ with the conditions $I \subset \{1, \ldots, n\}$ and $|I| = n_{\max}$ from party $\mathfrak{P}_j$**: ignore the message if $(\mathsf{sid}, \mathsf{ssid}, \mathfrak{P}_i, \mathfrak{P}_j, (m_1, \ldots, m_n))$ is not recorded. Otherwise, reveal $(\mathtt{Receive}, \mathsf{sid}, \mathsf{ssid}, \mathfrak{P}_i, \mathfrak{P}_j)$ to the adversary $\mathscr{S}$ and send the message $(\mathtt{Received}, \mathsf{sid}, \mathsf{ssid}, \mathfrak{P}_i, \mathfrak{P}_j, (m'_k)_{k \in I})$ to $\mathfrak{P}_j$ where $m'_k = m_k$ if $\mathsf{Verify}_k(W_k, \mathfrak{L}_k)$ returns 1, and $m'_k = \perp$ otherwise. Ignore further $\mathtt{Received}$-message with the same ssid from $\mathfrak{P}_j$.

Fig. 9. Ideal Functionality for Oblivious Language-Based Envelope $\mathcal{F}_{\mathrm{OLBE}}$

language, the receiver recovers the sent message, otherwise, he only gets a special symbol $\perp$.

4.4 Generic UC-Secure Instantiation of OLBE with Adaptive Security

For the sake of clarity, we now concentrate on the specific case where $n_{\max} = 1$. This is the most classical case in practice, and suffices for both OSBE and 1-out-of-n OT. In order to get a generic protocol in which $n_{\max} > 1$, one simply has to run $n_{\max}$ protocols in parallel. This modifies the algorithms Samp and Verify as follows: $\mathsf{Samp}(\mathsf{param}, \{i\})$ or $\mathsf{Samp}(\mathsf{param}, \{i\}, \{\mathsf{sk}_{\mathfrak{L}_i}\})$ generates a word $W = W_i \in \mathfrak{L}_i$ and $\mathsf{Verify}_j(W, \mathfrak{L}_j)$ checks whether W is a valid word in $\mathfrak{L}_j$.

Let us introduce our protocol OLBE: we will call $\mathcal{R}$ the receiver and $\mathcal{S}$ the sender. If $\mathcal{R}$ is an honest receiver, then he knows a word $W = W_i$ in one of the languages $\mathfrak{L}_i$. If $\mathcal{S}$ is an honest sender, then he wants to send up a message among $(m_1, \ldots, m_n) \in (\{0,1\}^{\mathfrak{K}})^n$ to $\mathcal{R}$. We assume the languages $\mathfrak{L}_i$ to be self-randomizable and publicly verifiable. We also assume the collection of languages $(\mathfrak{L}_1, \ldots, \mathfrak{L}_n)$ possess a trapdoor, that the simulator is able to find by programming the common reference string. As recalled in the previous section, this trapdoor enables him to find a word lying in the intersection of the n languages. This should be infeasible without the knowledge of the trapdoor. Intuitively, this allows the simulator to commit to all languages at once, postponing the time when it needs to choose the exact language he wants to bind to. On the opposite, if a user was granted the same possibilities, this would prevent the simulator to extract the chosen language.

We assume the existence of a labeled CCA-encryption scheme $\mathcal{E} = (\mathsf{Setup}_{\mathsf{cca}}, \mathsf{KeyGen}_{\mathsf{cca}}, \mathsf{Encrypt}^{\ell}_{\mathsf{cca}}, \mathsf{Decrypt}^{\ell}_{\mathsf{cca}})$ compatible with an SPHF onto a set G. In the KeyGen algorithm, the description of the languages $(\mathfrak{L}_1, \ldots, \mathfrak{L}_n)$ thus implicitly defines the languages $(\mathfrak{L}_1^c, \ldots, \mathfrak{L}_n^c)$ of CCA-encryptions of elements of the languages $(\mathfrak{L}_1, \ldots, \mathfrak{L}_n)$. We additionally use a key derivation function KDF to derive a pseudo-random bit-string $K \in \{0,1\}^{\mathfrak{K}}$ from a pseudo-random element $v \in G$. One can use the Leftover-Hash Lemma [40], with a random seed defined in param during the global setup, to extract the entropy from v, then followed by a pseudo-random generator to get a long enough bit-string. Many uses of the same seed in the Leftover-Hash Lemma just lead to a security loss linear in the number of extractions. We also assume the existence of a Pseudo-Random Generator (PRG) F with input size equal to the plaintext size, and output size equal to the size of the messages in the database and an IND-CPA encryption scheme $\mathcal{E} = (\mathsf{Setup}_{\mathsf{cpa}}, \mathsf{KeyGen}_{\mathsf{cpa}}, \mathsf{Encrypt}_{\mathsf{cpa}}, \mathsf{Decrypt}_{\mathsf{cpa}})$ with plaintext size at least equal to the security parameter.

We follow the ideas of the oblivious transfer constructions given in [1,9], giving the protocol presented on Fig. 10. For the sake of simplicity, we only give the version for adaptive security, in which the sender generates a public key pk

CRS: $\mathsf{param} \overset{\$}{\leftarrow} \mathsf{Setup}(1^{\mathfrak{K}})$, $\mathsf{param}_{\mathsf{cca}} \overset{\$}{\leftarrow} \mathsf{Setup}_{\mathsf{cca}}(1^{\mathfrak{K}})$, $\mathsf{param}_{\mathsf{cpa}} \overset{\$}{\leftarrow} \mathsf{Setup}_{\mathsf{cpa}}(1^{\mathfrak{K}})$.

Pre-flow:

1. Sender generates a key pair $(\mathsf{pk}, \mathsf{sk}) \overset{\$}{\leftarrow} \mathsf{KeyGen}_{\mathsf{cpa}}(\mathsf{param}_{\mathsf{cpa}})$ for $\mathcal{E}$, stores sk and completely erases the random coins used by KeyGen.
2. Sender sends pk to User.

Flow From the Receiver $\mathcal{R}$:

1. User chooses a random value J, computes $R \leftarrow F(J)$ and encrypts J under pk: $c \overset{\$}{\leftarrow} \mathsf{Encrypt}_{\mathsf{cpa}}(\mathsf{pk}, J)$.
2. User computes $C \overset{\$}{\leftarrow} \mathsf{Encrypt}^{\ell}_{\mathsf{cca}}(W; r)$ with $\ell = (\mathsf{sid}, \mathsf{ssid}, \mathcal{R}, \mathcal{S})$.
3. User completely erases J and the random coins used by $\mathsf{Encrypt}_{\mathsf{cpa}}$ and sends C and c to Sender. He also checks the validity of his words: the receiver only keeps the random coins used by $\mathsf{Encrypt}_{\mathsf{cca}}$ for the j such that $\mathsf{Verify}_j(W, \mathfrak{L}_j) = 1$ (since he knows they will be useless otherwise).

Flow From the Sender $\mathcal{S}$:

1. Sender decrypts $J \leftarrow \mathsf{Decrypt}_{\mathsf{cpa}}(\mathsf{sk}, c)$ and then $R \leftarrow F(J)$.
2. For all $j \in \{1, \ldots, n\}$, sender computes $\mathsf{hk}_j = \mathsf{HashKG}(\ell, \mathfrak{L}_j^c, \mathsf{param})$, $\mathsf{hp}_j = \mathsf{ProjKG}(\mathsf{hk}_j, \ell, (\mathfrak{L}_j^c, \mathsf{param}))$, $v_j = \mathsf{Hash}(\mathsf{hk}_j, (\mathfrak{L}_j^c, \mathsf{param}), (\ell, C))$, $Q_j = m_j \oplus \mathtt{KDF}(v_j) \oplus R$.
3. Sender erases everything except $(Q_j, \mathsf{hp}_j)_{j \in \{1,\ldots,n\}}$ and sends them over a secure channel.

Message recovery:

Upon receiving $(Q_j, \mathsf{hp}_j)_{j \in \{1,\ldots,n\}}$, $\mathcal{R}$ can recover m_i by computing $m_i = Q_i \oplus \mathsf{ProjHash}(\mathsf{hp}_i, (\mathfrak{L}_i^c, \mathsf{param}), (\ell, C), r) \oplus R$.

Fig. 10. UC-Secure OLBE for One Message (Secure Against Adaptive Corruptions)

and ciphertext c to create a somewhat secure channel (they would not be used in the static version).

Theorem 13. *The oblivious language-based envelope scheme described in Fig. 10 is* UC*-secure in the presence of adaptive adversaries, assuming reliable erasures, an* IND-CPA *encryption scheme, and an* IND-CCA *encryption scheme admitting an* SPHF *on the language of valid ciphertexts of elements of* $\mathfrak{L}_i$ *for all* i*, as soon as the languages are self-randomizable, publicly-verifiable and admit a common trapdoor. The proof is given in the paper full version [10].*

4.5 Oblivious Primitives Obtained by the Framework

Classical oblivious primitives such as Oblivious Transfer (both 1-out-of-n and k-out-of-n) or Oblivious Signature-Based Envelope directly lie in this framework and can be seen as examples of Oblivious Language-Based Envelope. We provide in the paper full version [10] details about how to describe the languages and choose appropriate smooth projective hash functions to readily achieve current instantiations of Oblivious Signature-Based Envelope or Oblivious Transfer from our generic protocol. The framework also enables us to give a new instantiation of Access Controlled Oblivious Transfer under classical assumptions. In such a primitive, the user does not automatically gets the line he asks for, but has to prove that he possesses one of the credential needed to access this particular line.

For the sake of simplicity, all the instantiations given are pairing-based but techniques explained in [9] could be used to rely on other families of assumptions, like decisional quadratic residue or even LWE.

Acknowledgments. This work was supported in part by the French ANR EnBid Project (ANR-14-CE28-0003).

References

1. Abdalla, M., Benhamouda, F., Blazy, O., Chevalier, C., Pointcheval, D.: SPHF-friendly non-interactive commitments. In: Sako, K., Sarkar, P. (eds.) ASIACRYPT 2013. LNCS, vol. 8269, pp. 214–234. Springer, Heidelberg (2013). doi:10.1007/978-3-642-42033-7_12
2. Abdalla, M., Chevalier, C., Pointcheval, D.: Smooth projective hashing for conditionally extractable commitments. In: Halevi, S. (ed.) CRYPTO 2009. LNCS, vol. 5677, pp. 671–689. Springer, Heidelberg (2009). doi:10.1007/978-3-642-03356-8_39
3. Abdalla, M., Pointcheval, D.: A scalable password-based group key exchange protocol in the standard model. In: Lai, X., Chen, K. (eds.) ASIACRYPT 2006. LNCS, vol. 4284, pp. 332–347. Springer, Heidelberg (2006). doi:10.1007/11935230_22
4. Aiello, B., Ishai, Y., Reingold, O.: Priced oblivious transfer: how to sell digital goods. In: Pfitzmann, B. (ed.) EUROCRYPT 2001. LNCS, vol. 2045, pp. 119–135. Springer, Heidelberg (2001). doi:10.1007/3-540-44987-6_8

5. Attrapadung, N.: Dual system encryption via doubly selective security: framework, fully secure functional encryption for regular languages, and more. In: Nguyen, P.Q., Oswald, E. (eds.) EUROCRYPT 2014. LNCS, vol. 8441, pp. 557–577. Springer, Heidelberg (2014). doi:10.1007/978-3-642-55220-5_31
6. Bellare, M., Hoang, V.T., Rogaway, P.: Foundations of garbled circuits. In: Yu, T., Danezis, G., Gligor, V.D. (eds.) ACM CCS 12, pp. 784–796. ACM Press, October 2012
7. Ben Hamouda, F., Blazy, O., Chevalier, C., Pointcheval, D., Vergnaud, D.: Efficient UC-secure authenticated key-exchange for algebraic languages. In: Kurosawa, K., Hanaoka, G. (eds.) PKC 2013. LNCS, vol. 7778, pp. 272–291. Springer, Heidelberg (2013). doi:10.1007/978-3-642-36362-7_18
8. Benhamouda, F., Blazy, O., Chevalier, C., Pointcheval, D., Vergnaud, D.: New techniques for SPHFs and efficient one-round PAKE protocols. In: Canetti, R., Garay, J.A. (eds.) CRYPTO 2013. LNCS, vol. 8042, pp. 449–475. Springer, Heidelberg (2013). doi:10.1007/978-3-642-40041-4_25
9. Blazy, O., Chevalier, C.: Generic construction of UC-secure oblivious transfer. In: Malkin, T., Kolesnikov, V., Lewko, A.B., Polychronakis, M. (eds.) ACNS 2015. LNCS, vol. 9092, pp. 65–86. Springer, Heidelberg (2015). doi:10.1007/978-3-319-28166-7_4
10. Blazy, O., Chevalier, C., Germouty, P.: Adaptive oblivious transfer and generalizations. Cryptology ePrint Archive, Report 2016/259 (2016). http://eprint.iacr.org/2016/259
11. Blazy, O., Fuchsbauer, G., Pointcheval, D., Vergnaud, D.: Signatures on randomizable ciphertexts. In: Catalano, D., Fazio, N., Gennaro, R., Nicolosi, A. (eds.) PKC 2011. LNCS, vol. 6571, pp. 403–422. Springer, Heidelberg (2011). doi:10.1007/978-3-642-19379-8_25
12. Blazy, O., Kiltz, E., Pan, J.: (Hierarchical) identity-based encryption from affine message authentication. In: Garay, J.A., Gennaro, R. (eds.) CRYPTO 2014. LNCS, vol. 8616, pp. 408–425. Springer, Heidelberg (2014). doi:10.1007/978-3-662-44371-2_23
13. Blazy, O., Pointcheval, D., Vergnaud, D.: Round-optimal privacy-preserving protocols with smooth projective hash functions. In: Cramer, R. (ed.) TCC 2012. LNCS, vol. 7194, pp. 94–111. Springer, Heidelberg (2012). doi:10.1007/978-3-642-28914-9_6
14. Boneh, D., Franklin, M.: Identity-based encryption from the weil pairing. In: Kilian, J. (ed.) CRYPTO 2001. LNCS, vol. 2139, pp. 213–229. Springer, Heidelberg (2001). doi:10.1007/3-540-44647-8_13
15. Boneh, D., Goh, E.-J., Nissim, K.: Evaluating 2-DNF formulas on ciphertexts. In: Kilian, J. (ed.) TCC 2005. LNCS, vol. 3378, pp. 325–341. Springer, Heidelberg (2005). doi:10.1007/978-3-540-30576-7_18
16. Camenisch, J., Casati, N., Gross, T., Shoup, V.: Credential authenticated identification and key exchange. In: Rabin, T. (ed.) CRYPTO 2010. LNCS, vol. 6223, pp. 255–276. Springer, Heidelberg (2010). doi:10.1007/978-3-642-14623-7_14
17. Camenisch, J., Dubovitskaya, M., Haralambiev, K.: Efficient structure-preserving signature scheme from standard assumptions. In: Visconti, I., Prisco, R. (eds.) SCN 2012. LNCS, vol. 7485, pp. 76–94. Springer, Heidelberg (2012). doi:10.1007/978-3-642-32928-9_5
18. Camenisch, J., Dubovitskaya, M., Neven, G.: Oblivious transfer with access control. In: Al-Shaer, E., Jha, S., Keromytis, A.D. (eds.) ACM CCS 2009, pp. 131–140. ACM Press, November 2009

19. Camenisch, J., Dubovitskaya, M., Neven, G., Zaverucha, G.M.: Oblivious transfer with hidden access control policies. In: Catalano, D., Fazio, N., Gennaro, R., Nicolosi, A. (eds.) PKC 2011. LNCS, vol. 6571, pp. 192–209. Springer, Heidelberg (2011). doi:10.1007/978-3-642-19379-8_12
20. Camenisch, J., Neven, G., shelat, A.: Simulatable adaptive oblivious transfer. In: Naor, M. (ed.) EUROCRYPT 2007. LNCS, vol. 4515, pp. 573–590. Springer, Heidelberg (2007). doi:10.1007/978-3-540-72540-4_33
21. Canetti, R.: Universally composable security: A new paradigm for cryptographic protocols. In: 42nd FOCS, pp. 136–145. IEEE Computer Society Press, October 2001
22. Canetti, R., Cohen, A., Lindell, Y.: A simpler variant of universally composable security for standard multiparty computation. In: Gennaro, R., Robshaw, M. (eds.) CRYPTO 2015. LNCS, vol. 9216, pp. 3–22. Springer, Heidelberg (2015). doi:10.1007/978-3-662-48000-7_1
23. Canetti, R., Lindell, Y., Ostrovsky, R., Sahai, A.: Universally composable two-party and multi-party secure computation. In: 34th ACM STOC, pp. 494–503. ACM Press, May 2002
24. Choi, S.G., Katz, J., Wee, H., Zhou, H.-S.: Efficient, adaptively secure, and composable oblivious transfer with a single, global CRS. In: Kurosawa, K., Hanaoka, G. (eds.) PKC 2013. LNCS, vol. 7778, pp. 73–88. Springer, Heidelberg (2013). doi:10.1007/978-3-642-36362-7_6
25. Chor, B., Goldreich, O., Kushilevitz, E., Sudan, M.: Private information retrieval. In: 36th FOCS, pp. 41–50. IEEE Computer Society Press, October 1995
26. Cramer, R., Shoup, V.: Universal hash proofs and a paradigm for adaptive chosen ciphertext secure public-key encryption. In: Knudsen, L.R. (ed.) EUROCRYPT 2002. LNCS, vol. 2332, pp. 45–64. Springer, Heidelberg (2002). doi:10.1007/3-540-46035-7_4
27. Cui, Y., Fujisaki, E., Hanaoka, G., Imai, H., Zhang, R.: Formal security treatments for signatures from identity-based encryption. In: Susilo, W., Liu, J.K., Mu, Y. (eds.) ProvSec 2007. LNCS, vol. 4784, pp. 218–227. Springer, Heidelberg (2007). doi:10.1007/978-3-540-75670-5_16
28. Crescenzo, G., Ostrovsky, R., Rajagopalan, S.: Conditional oblivious transfer and timed-release encryption. In: Stern, J. (ed.) EUROCRYPT 1999. LNCS, vol. 1592, pp. 74–89. Springer, Heidelberg (1999). doi:10.1007/3-540-48910-X_6
29. Diffie, W., Hellman, M.E.: New directions in cryptography. IEEE Trans. Inf. Theor. **22**(6), 644–654 (1976)
30. Escala, A., Herold, G., Kiltz, E., Ràfols, C., Villar, J.: An algebraic framework for diffie-hellman assumptions. In: Canetti, R., Garay, J.A. (eds.) CRYPTO 2013. LNCS, vol. 8043, pp. 129–147. Springer, Heidelberg (2013). doi:10.1007/978-3-642-40084-1_8
31. Fischlin, M.: Round-optimal composable blind signatures in the common reference string model. In: Dwork, C. (ed.) CRYPTO 2006. LNCS, vol. 4117, pp. 60–77. Springer, Heidelberg (2006). doi:10.1007/11818175_4
32. Gay, R., Kerenidis, I., Wee, H.: Communication complexity of conditional disclosure of secrets and attribute-based encryption. In: Gennaro, R., Robshaw, M. (eds.) CRYPTO 2015. LNCS, vol. 9216, pp. 485–502. Springer, Heidelberg (2015). doi:10.1007/978-3-662-48000-7_24
33. Gennaro, R., Lindell, Y.: A framework for password-based authenticated key exchange. In: Biham, E. (ed.) EUROCRYPT 2003. LNCS, vol. 2656, pp. 524–543. Springer, Heidelberg (2003). doi:10.1007/3-540-39200-9_33

34. Gertner, Y., Ishai, Y., Kushilevitz, E., Malkin, T.: Protecting data privacy in private information retrieval schemes. In: 30th ACM STOC, pp. 151–160. ACM Press, May 1998
35. Goldreich, O., Micali, S., Wigderson, A.: How to play any mental game or A completeness theorem for protocols with honest majority. In: Aho, A. (ed.) 19th ACM STOC, pp. 218–229. ACM Press, May 1987
36. Goldwasser, S., Micali, S., Rivest, R.L.: A digital signature scheme secure against adaptive chosen-message attacks. SIAM J. Comput. **17**(2), 281–308 (1988)
37. Green, M., Hohenberger, S.: Blind identity-based encryption and simulatable oblivious transfer. In: Kurosawa, K. (ed.) ASIACRYPT 2007. LNCS, vol. 4833, pp. 265–282. Springer, Heidelberg (2007). doi:10.1007/978-3-540-76900-2_16
38. Green, M., Hohenberger, S.: Universally composable adaptive oblivious transfer. In: Pieprzyk, J. (ed.) ASIACRYPT 2008. LNCS, vol. 5350, pp. 179–197. Springer, Heidelberg (2008). doi:10.1007/978-3-540-89255-7_12
39. Guleria, V., Dutta, R.: Lightweight universally composable adaptive oblivious transfer. In: Au, M.H., Carminati, B., Kuo, C.-C.J. (eds.) NSS 2014. LNCS, vol. 8792, pp. 285–298. Springer, Heidelberg (2014). doi:10.1007/978-3-319-11698-3_22
40. Håstad, J., Impagliazzo, R., Levin, L.A., Luby, M.: A pseudorandom generator from any one-way function. SIAM J. Comput. **28**(4), 1364–1396 (1999)
41. Horvitz, O., Katz, J.: Universally-composable two-party computation in two rounds. In: Menezes, A. (ed.) CRYPTO 2007. LNCS, vol. 4622, pp. 111–129. Springer, Heidelberg (2007). doi:10.1007/978-3-540-74143-5_7
42. Ishai, Y., Wee, H.: Partial garbling schemes and their applications. In: Esparza, J., Fraigniaud, P., Husfeldt, T., Koutsoupias, E. (eds.) ICALP 2014. LNCS, vol. 8572, pp. 650–662. Springer, Heidelberg (2014). doi:10.1007/978-3-662-43948-7_54
43. Jarecki, S., Liu, X.: Efficient oblivious pseudorandom function with applications to adaptive OT and secure computation of set intersection. In: Reingold, O. (ed.) TCC 2009. LNCS, vol. 5444, pp. 577–594. Springer, Heidelberg (2009). doi:10.1007/978-3-642-00457-5_34
44. Kalai, Y.T.: Smooth projective hashing and two-message oblivious transfer. In: Cramer, R. (ed.) EUROCRYPT 2005. LNCS, vol. 3494, pp. 78–95. Springer, Heidelberg (2005). doi:10.1007/11426639_5
45. Katz, J., Vaikuntanathan, V.: Round-optimal password-based authenticated key exchange. In: Ishai, Y. (ed.) TCC 2011. LNCS, vol. 6597, pp. 293–310. Springer, Heidelberg (2011). doi:10.1007/978-3-642-19571-6_18
46. Kiayias, A., Leonardos, N., Lipmaa, H., Pavlyk, K., Tang, Q.: Optimal rate private information retrieval from homomorphic encryption. PoPETs 2015(2), 222–243 (2015). http://www.degruyter.com/view/j/popets.2015.2015.issue-2/popets-2015-0016/popets-2015-0016.xml
47. Kurosawa, K., Nojima, R., Phong, L.T.: Generic fully simulatable adaptive oblivious transfer. In: Lopez, J., Tsudik, G. (eds.) ACNS 2011. LNCS, vol. 6715, pp. 274–291. Springer, Heidelberg (2011). doi:10.1007/978-3-642-21554-4_16
48. Laur, S., Lipmaa, H.: A new protocol for conditional disclosure of secrets and its applications. In: Katz, J., Yung, M. (eds.) ACNS 2007. LNCS, vol. 4521, pp. 207–225. Springer, Heidelberg (2007). doi:10.1007/978-3-540-72738-5_14
49. Li, N., Du, W., Boneh, D.: Oblivious signature-based envelope. In: Borowsky, E., Rajsbaum, S. (eds.) 22nd ACM PODC, pp. 182–189. ACM, Jul 2003
50. Naor, M., Pinkas, B.: Visual authentication and identification. In: Kaliski, B.S. (ed.) CRYPTO 1997. LNCS, vol. 1294, pp. 322–336. Springer, Heidelberg (1997). doi:10.1007/BFb0052245

51. Naor, M., Pinkas, B.: Efficient oblivious transfer protocols. In: Kosaraju, S.R. (ed.) 12th SODA, pp. 448–457. ACM-SIAM, January 2001
52. Peikert, C., Vaikuntanathan, V., Waters, B.: A framework for efficient and composable oblivious transfer. In: Wagner, D. (ed.) CRYPTO 2008. LNCS, vol. 5157, pp. 554–571. Springer, Heidelberg (2008). doi:10.1007/978-3-540-85174-5_31
53. Rabin, M.O.: How to exchange secrets with oblivious transfer. Technical Report TR81, Harvard University (1981)
54. Rial, A., Kohlweiss, M., Preneel, B.: Universally composable adaptive priced oblivious transfer. In: Shacham, H., Waters, B. (eds.) Pairing 2009. LNCS, vol. 5671, pp. 231–247. Springer, Heidelberg (2009). doi:10.1007/978-3-642-03298-1_15
55. Shamir, A.: Identity-based cryptosystems and signature schemes. In: Blakley, G.R., Chaum, D. (eds.) CRYPTO 1984. LNCS, vol. 196, pp. 47–53. Springer, Heidelberg (1985). doi:10.1007/3-540-39568-7_5
56. Wang, X.S., Huang, Y., Chan, T.H.H., Shelat, A., Shi, E.: SCORAM: Oblivious RAM for secure computation. In: Ahn, G.J., Yung, M., Li, N. (eds.) ACM CCS 14, pp. 191–202. ACM Press, November 2014
57. Waters, B.: Efficient identity-based encryption without random oracles. In: Cramer, R. (ed.) EUROCRYPT 2005. LNCS, vol. 3494, pp. 114–127. Springer, Heidelberg (2005). doi:10.1007/11426639_7
58. Wee, H.: Dual system encryption via predicate encodings. In: Lindell, Y. (ed.) TCC 2014. LNCS, vol. 8349, pp. 616–637. Springer, Heidelberg (2014). doi:10.1007/978-3-642-54242-8_26
59. Yao, A.C.C.: How to generate and exchange secrets (extended abstract). In: 27th FOCS, pp. 162–167. IEEE Computer Society Press, October 1986

Selective Opening Security from Simulatable Data Encapsulation

Felix Heuer(✉) and Bertram Poettering

Horst Görtz Institute for IT Security, Ruhr University Bochum, Bochum, Germany
{felix.heuer,bertram.poettering}@rub.de

Abstract. In the realm of public-key encryption, the confidentiality notion of security against selective opening (SO) attacks considers adversaries that obtain challenge ciphertexts and are allowed to adaptively open them, meaning have the corresponding message and randomness revealed. SO security is stronger than IND-CCA and often required when formally arguing towards the security of multi-user applications. While different ways of achieving SO secure schemes are known, as they generally employ expensive asymmetric building blocks like lossy trapdoor functions or lossy encryption, such constructions are routinely left aside by practitioners and standardization bodies. So far, formal arguments towards the SO security of schemes used in practice (e.g., for email encryption) are not known.

In this work we shift the focus from the asymmetric to the symmetric building blocks of PKE and prove the following statement: If a PKE scheme is composed of a key encapsulation mechanism (KEM) and a blockcipher-based data encapsulation mechanism (DEM), and the DEM has specific combinatorial properties, then the PKE scheme offers SO security in the ideal cipher model. Fortunately, as we show, the required properties hold for popular modes of operation like CTR, CBC and CCM. This paper not only establishes the corresponding theoretical framework of analysis, but also contributes very concretely to practical cryptography by concluding that selective opening security is given for many real-world schemes.

1 Introduction

Public key encryption in the multi-user setting. The most important security notion for public key encryption is indistinguishability under chosen ciphertext attacks (IND-CCA). The modeled setting is as follows: One user generates a key pair, a second users encrypts one out of two messages to her, and the adversary shall find out which one it was. Here, importantly, the adversary controls the distribution of the two messages and may request decryptions of ciphertexts of its choice.

The definition of selective opening (SO) security is more general as it takes into account the fact that the public key setting allows for more than two parties. Concretely, in the SO setting one user generates a key pair, many users encrypt

J.H. Cheon and T. Takagi (Eds.): ASIACRYPT 2016, Part II, LNCS 10032, pp. 248–277, 2016.
DOI: 10.1007/978-3-662-53890-6_9

messages to her key (of course using fresh and independent random coins), and the adversary's goal is to derive any information about any of the messages. Again the adversary controls the message distribution (individually for each participant, but also joint distributions are possible) and may have arbitrary ciphertexts decrypted. On top of that the adversary is allowed to 'open' any subset of ciphertexts, i.e., to corrupt the encrypters, for instance by breaking into their computers, and thereby reveal the messages they encrypted and the random coins they used. (In some applications, like in secure multi-party computation, users even deliberately reveal their messages and randomness to make their computations publicly verifiable.) Selective opening security is provided if in this situation the confidentiality of the remaining 'unopened' ciphertexts is still provided. Intuitively, as all the encryptions occur independently of each other, IND-CCA should imply SO security. Unfortunately, formal analysis reveals that this is not the case.

Notions of Selective Opening security. Formalising suitable notions of SO security has proven to be highly non-trivial. Since encrypted messages may depend on each other, opening some ciphertexts might readily leak information on messages encrypted in other (unopened) ciphertexts. Thus, it is not even clear what it means for unopened messages to remain confidential. Two flavours of SO security have been studied in prior work: notions based on indistinguishability (IND) and notions based on simulatability (SIM). For IND based notions an adversary may open arbitrary ciphertexts and is challenged to tell apart the originally encrypted messages from fresh messages that occur *as likely as the original messages*. One usually restricts the distribution on the messages to be *efficiently conditionally resamplable* to ensure an efficient security game (*weak*-IND-SO). We obtain the security experiment for *full*-IND-SO if arbitrary distributions may occur in the experiment.

In contrast, SIM based notions (capturing semantic security in the SO setting) do not suffer from such a restriction. In a nutshell, a scheme is SIM-SO secure if for every SO adversary there exists a simulator that can compute the same output without seeing any ciphertexts. Importantly, such simulators may corrupt senders to learn the messages they (virtually) encrypted.

Both flavours may be considered for passive (CPA) and active (CCA) adversaries whereby, in contrast to the CPA setting, a CCA adversary has access to a decryption oracle (with the usual restrictions). While any of IND-SO-CPA/CCA and SIM-SO-CPA/CCA implies standard IND-CPA/CCA security, the converse does not hold in general. Only partial results are known for the reverse direction, as discussed below. We give more details on the relations amongst the notions of selective opening security at the end of Sect. 2.

Motivation and contribution. Considering that users in practice may be exposed to the threats modeled in the SO context, and given that the classical indistinguishability notions are formally weaker than notions of SO security, the following question is immediate: Are users 'safe' if they trust in a PKE scheme designed towards the goal of 'only' indistinguishability? At least in theory, if the security proof of the scheme considers exclusively indistinguishability, information about encrypted messages is potentially exposed to the adversary in SO-like attack scenarios. This observation calls for a thorough SO analysis of all encryption schemes covered by international standards. The facts that all PKE schemes that so far were formally confirmed to be SO secure require heavy building blocks like lossy trapdoor functions (except for one work discussed in *Previous work*) and that practitioners systematically avoid such building blocks for reasons of efficiency suggest that likely most practical schemes would not withstand SO attacks. Fortunately, however, in this paper we show that virtually all practical PKE constructions provably do meet SO security.

Our approach is complementary to that of prior works: Instead of analysing the asymmetric building blocks of constructions, we observe that SO security is tightly linked to the security of the symmetric building blocks (i.e., symmetric encryption). We particularly show that in the KEM/DEM paradigm for hybrid encryption certain properties of blockcipher-based DEMs suffice to render the overall PKE scheme SO secure (in the ideal cipher model for the blockcipher) *independently of the properties of the KEM.*

In a nutshell, our result is: We introduce a specific property called *simulatability* for blockcipher-based DEMs that is met by virtually all DEMs used in practice and guarantees that if a corresponding DEM is combined with any IND-CCA secure KEM then the overall hybrid PKE scheme achieves SIM-SO-CCA security (in the ideal cipher model). Intuitively, *simulatable* DEMs can be thought of as some form of non-committing encryption in the realm of symmetric cryptography, while non-committing encryption is usually considered in the public-key setting.

Previous work. The SO problem dates back to [12] where the *selective decommitment problem* was studied for commitment schemes. SO notions for encryption first appeared in [3,6]. The first IND-SO-CPA secure encryption scheme in the standard model was given in [3] and is based on lossy encryption (cf. [29]).

Also *deniable encryption* [7] and techniques from *non-committing encryption* [8,21] already allow for constructing SO secure PKE ([11]). Lots of separation and implication results for SO and standard notions were studied in [2,5,6,26]. While it was known that IND-CPA implies *weak*-IND-SO-CPA when messages are drawn pair-wise independently (cf. [5,12]), the implication does not hold for arbitrary (efficiently conditionally resamplable) distributions as recently reported [25]. The result makes use of heavy machinery as *public-coin differing-inputs obfuscation* and *correlation intractable hash functions.* However, IND-CPA implies *weak*-IND-SO-CPA for low-dependency distributions such as Markov chains [19]. Further, SIM-SO secure constructions in the standard model usually (cf. [28]) suffer in efficiency from bit-wise encryption to ensure *efficient openability.* See [24] for current research. SIM-SO-CCA secure PKE schemes are constructed in [18] employing *extended HPSs* and *cross-authentication codes.* This line of research continued in [28] identifying special properties of a KEM, allowing to construct SIM-SO-CCA secure PKE, when combined with *strengthened cross-authentication codes.*

Note that we only consider SO security under sender corruption. Only recently, security under receiver corruption gained some attention [20] while already defined in [1].

Work analysing the SO security of standardised widely-used encryption schemes appeared only recently (in the random oracle model). Concretely, Heuer *et al.* [22] consider Hashed ElGamal encryption (standardised under the name of DHIES) and RSA-OAEP. Unfortunately, the considered versions of these PKE schemes assume messages that are not longer than the output lengths of the used random oracle, i.e., less than 128 bytes. This severely limits the results of [22] for practical considerations.

Paper organization. In Sect. 2 we recall some important cryptographic notions, including the definition of SO security that we use in this paper. We then, in Sect. 3, identify certain combinatorial properties of DEMs that suffice to achieve SO security of hybrid PKE; more precisely, we expose the central claim of this paper which states that any DEM that has these properties in combination with any KEM results, in the ideal cipher model, in a SIM-SO-CCA secure PKE scheme. In Sect. 3 we also sketch the arguments required for proving this claim. We continue in Sect. 4 with checking whether widely-used DEMs (in particular the NIST standardised: CTR, CBC, CCM) have these properties, and come to the conclusion that they do. We work out the full details of our main claim and its proof in Sect. 5. We conclude in Sect. 6.

In the full version of this paper [23] we further show that also the (NIST standardised) GCM mode of operation possesses the combinatorial properties identified in Sect. 3.

2 Preliminaries

For $n \in \mathbb{N}$ let $[n] := \{1, \ldots, n\}$. We distinguish the following operators for assigning values to variables: We use symbol '$\leftarrow$' when the assigned value results from a constant expression (including the output of a deterministic algorithm), we write '$\leftarrow_U$' when the value is sampled uniformly at random from a finite set, and we write '$\leftarrow_\$$' when the assigned value is the output of a randomised algorithm. If f is a function or a deterministic algorithm that maps elements from a set A to a set B we use notations $f\colon A \to B$ and $A \to f \to B$ interchangeably. If f is a randomised algorithm we correspondingly write $A \to f \to_\$ B$, or simply $f \to_\$ B$ in case the algorithm takes no input. If $A \times B \to f \to C$ is a function then for any $a \in A$ we write $f_a = f(a; \cdot)$ for the partially applied function $B \to f_a \to C$; $b \mapsto f(a, b)$. If R denotes the randomness space of a (randomised) algorithm $A \to f \to_\$ B$, we may write $A \times R \to f \to B$ for its deterministic version. If $A \to f \to B$ is a function or a deterministic algorithm we let $[f] := f(A) \subseteq B$ denote the image of A under f; if $A \to f \to_\$ B$ has randomness space R we correspondingly let $[f] := f(A \times R) \subseteq B$ denote the set of all its possible outputs. When the union $A \cup B$ of two sets A, B is a disjoint union, i.e., if $A \cap B = \emptyset$, we annotate this with $A \mathbin{\dot{\cup}} B$. For a bitstring x of length at least l we write $\mathrm{msb}_l(x)$ for its left-most l bits and $\mathrm{lsb}_l(x)$ for its right-most l bits ('most/least significant bits').

Our security definitions are based on games played between a challenger and an adversary. These games are expressed using program code and terminate when a 'Stop' command is executed; the argument of the latter is the output of the game. We write $\Pr[G \Rightarrow 1]$ for the probability that game G terminates by running into a 'Stop with 1' instruction.

We next define partial permutations and blockciphers. In our proofs, the former play an important role for the abstraction of the latter.

Definition 1 (Permutation, partial permutation, blockcipher). *For a finite domain $\mathcal{D}$ we denote the set of all permutations on $\mathcal{D}$ with $\mathcal{P}(\mathcal{D})$ and the set of all partial permutations on $\mathcal{D}$ with $\mathcal{PP}(\mathcal{D})$. Precisely, a relation $R \subseteq \mathcal{D} \times \mathcal{D}$ is a partial permutation if $\alpha R \beta, \alpha' R \beta \Rightarrow \alpha = \alpha'$ and $\alpha R \beta, \alpha R \beta' \Rightarrow \beta = \beta'$; relation R is a permutation if in addition $|R| = |\mathcal{D}|$ holds. A blockcipher with key space $\mathcal{K}$ and domain $\mathcal{D}$ is a family $(E_k)_{k \in \mathcal{K}}$ of permutations $E_k \in \mathcal{P}(\mathcal{D})$.*

We associate with a partial permutation $R \in \mathcal{PP}(\mathcal{D})$ the partial functions $\mathcal{D} \to R^+ \to \mathcal{D}$ and $\mathcal{D} \to R^- \to \mathcal{D}$ that evaluate R left-to-right and right-to-left, respectively. For instance, if $(\alpha, \beta) \in R$ then $R^+(\alpha) = \beta$ and $R^-(\beta) = \alpha$. We write $\mathrm{Dom}(R)$ and $\mathrm{Rng}(R)$ for the domain and range of R^+, i.e., for the sets $\{\alpha \in \mathcal{D} \mid \exists \beta : (\alpha, \beta) \in R\}$ and $\{\beta \in \mathcal{D} \mid \exists \alpha : (\alpha, \beta) \in R\}$, respectively. If $\alpha \notin \mathrm{Dom}(R)$ and $\beta \notin \mathrm{Rng}(R)$ we denote with $R \leftarrow R \cup \{(\alpha, \beta)\}$ the operation of 'programming' R such that $R^+(\alpha) = \beta$ and $R^-(\beta) = \alpha$ for the updated R, which is again a partial permutation. Note that any partial permutation can be completed to a (full) permutation by adding sufficiently many such pairs (α, β) to it. More importantly, if a partial permutation is selected according to the uniform distribution over some subset of $\mathcal{PP}(\mathcal{D})$, it can be extended to a permutation uniformly distributed in $\mathcal{P}(\mathcal{D})$ by adding random such pairs (α, β) to it.

Our definition of keyed hash functions subsumes both message authentication codes and universal hash functions.

Definition 2 (Keyed hash function). *A keyed hash function for a message space $\mathcal{M}$ consists of a key space $\mathcal{K}$, a tag space $\mathcal{T}$, and an efficient function* khf *of the form $\mathcal{K} \times \mathcal{M} \to \mathrm{khf} \to \mathcal{T}$.*

We proceed with specifying the syntax and functionality of DEMs. As a corresponding notion of authenticity we define integrity of ciphertexts [4]. In a nutshell, a DEM offers this feature if no adversary with access to an encapsulation oracle can find a fresh ciphertext that corresponds to a valid message, i.e., is not rejected by the decapsulation algorithm. Relevant in our work is in particular the corresponding one-time notion where the adversary can pose at most one encapsulation query.

Definition 3 (DEM). *A data encapsulation mechanism (DEM) for a message space $\mathcal{M}$ consists of a finite key space $\mathcal{K}$, a ciphertext space $\mathcal{C}$, and a pair of efficient algorithms* $\mathsf{DEM} = (\mathsf{D.Enc}, \mathsf{D.Dec})$ *of the form*

$$\mathcal{K} \times \mathcal{M} \to \mathsf{D.Enc} \to \mathcal{C} \qquad \mathcal{K} \times \mathcal{C} \to \mathsf{D.Dec} \to \mathcal{M} \cup \{\bot\} ,$$

where symbol '$\bot$' may be used to indicate errors. Correctness requires that for all $k \in \mathcal{K}$ and $m \in \mathcal{M}$, if $\mathsf{D.Enc}(k, m) = c$ then $\mathsf{D.Dec}(k, c) = m$.

Definition 4 (INT-CTXT secure DEM). *A data encapsulation mechanism is (τ, q_d, ϵ)-OT-INT-CTXT secure if all τ-time adversaries $\mathcal{A}$ that interact in the* OT-INT-CTXT *experiment from Fig. 1 and issue at most q_d queries to the* D.DEC *oracle have an advantage of at most ϵ, where we define*

$$\mathbf{Adv}_{\mathcal{A}}^{\mathsf{OT\text{-}INT\text{-}CTXT}} := \Pr[\mathsf{OT\text{-}INT\text{-}CTXT} \Rightarrow 1].$$

This definition can be generalised to $(\tau, q_e, q_d, \epsilon)$-INT-CTXT security by removing line 04 from the experiment and bounding the number of queries to the D.ENC *oracle by q_e.*

Game OT-INT-CTXT	**Oracle** D.ENC(m)	**Oracle** D.DEC(c)
00 $C \leftarrow \emptyset$	04 If $\lvert C \rvert > 0$: Abort	08 If $c \in C$: Abort
01 $k \leftarrow_U \mathcal{K}$	05 $c \leftarrow \mathsf{D.Enc}(k, m)$	09 $m \leftarrow \mathsf{D.Dec}(k, c)$
02 $\mathcal{A}^{\mathrm{D.ENC},\mathrm{D.DEC}}$	06 $C \leftarrow C \cup \{c\}$	10 If $m \neq \bot$:
03 Stop with 0	07 Return c	11 Stop with 1
		12 Return $\bot$

Fig. 1. Security game for defining OT-INT-CTXT security of DEMs. We write 'Abort' as an abbreviation for 'Stop with 0'. Observe that line 04 ensures that the D.ENC oracle is queried at most once.

In most applications a DEM is combined with a KEM to obtain (hybrid) PKE [10]. We recall the concepts of KEMs and PKE below, and include an indistinguishability definition for KEMs.

Definition 5 (KEM). *A key encapsulation mechanism (KEM) for a finite key space $\mathcal{K}$ consists of a public-key space $\mathcal{PK}$, a secret-key space $\mathcal{SK}$, a ciphertext space $\mathcal{C}$, and a triple of efficient algorithms* $\mathsf{KEM} = (\mathsf{K.Gen}, \mathsf{K.Enc}, \mathsf{K.Dec})$ *of the form*

$$\mathsf{K.Gen} \to_\$ \mathcal{PK} \times \mathcal{SK} \qquad \mathcal{PK} \to \mathsf{K.Enc} \to_\$ \mathcal{K} \times \mathcal{C} \qquad \mathcal{SK} \times \mathcal{C} \to \mathsf{K.Dec} \to \mathcal{K} \cup \{\bot\},$$

where symbol '$\bot$' may be used to indicate errors. The randomness space of K.Enc *is typically denoted with $\mathcal{R}$. Correctness requires that for all $(pk, sk) \in [\mathsf{K.Gen}]$, if $(k, c) \in [\mathsf{K.Enc}(pk)]$ then $\mathsf{K.Dec}(sk, c) = k$.*

Definition 6 (IND-CCA secure KEM). *A KEM is (τ, q_d, ϵ)-IND-CCA secure if all τ-time adversaries $\mathcal{A} = (\mathcal{A}_1, \mathcal{A}_2)$ that interact in the* $\mathsf{IND\text{-}CCA}^b$ *experiments from Fig. 2 and issue at most q_d queries to the* K.Dec *oracle have an advantage of at most ϵ, where we define*

$$\mathbf{Adv}^{\mathsf{IND\text{-}CCA}}(\mathcal{A}) := |\Pr[\mathsf{IND\text{-}CCA}^0 \Rightarrow 1] - \Pr[\mathsf{IND\text{-}CCA}^1 \Rightarrow 1]|.$$

Game $\mathsf{IND\text{-}CCA}^b$	**Oracle** K.Dec(c)
00 $C \leftarrow \emptyset$	08 If $c \in C$: Abort
01 $(pk, sk) \leftarrow_\$ \mathsf{K.Gen}$	09 $k \leftarrow \mathsf{K.Dec}(sk, c)$
02 $st \leftarrow_\$ \mathcal{A}_1^{\mathrm{K.Dec}}(pk)$	10 Return k
03 $(k_0^*, c^*) \leftarrow_\$ \mathsf{K.Enc}(pk)$	
04 $k_1^* \leftarrow_U \mathcal{K}$	
05 $C \leftarrow C \cup \{c^*\}$	
06 $b' \leftarrow_\$ \mathcal{A}_2^{\mathrm{K.Dec}}(st, c^*, k_b^*)$	
07 Stop with b'	

Fig. 2. Security games for defining IND-CCA security of KEMs. We write 'Abort' as an abbreviation for 'Stop with 0'.

Definition 7 (PKE). *A scheme for public-key encryption* (PKE) *for a message space $\mathcal{M}$ consists of a public-key space $\mathcal{PK}$, a secret-key space $\mathcal{SK}$, a ciphertext space $\mathcal{C}$, and a triple of efficient algorithms* $\mathsf{PKE} = (\mathsf{P.Gen}, \mathsf{P.Enc}, \mathsf{P.Dec})$ *of the form*

$$\mathsf{P.Gen} \rightarrow_\$ \mathcal{PK} \times \mathcal{SK}, \quad \mathcal{PK} \times \mathcal{M} \rightarrow \mathsf{P.Enc} \rightarrow_\$ \mathcal{C}, \quad \mathcal{SK} \times \mathcal{C} \rightarrow \mathsf{P.Dec} \rightarrow \mathcal{M} \cup \{\bot\},$$

where symbol '$\bot$' may be used to indicate errors. The randomness space of P.Enc *is typically denoted with $\mathcal{R}$. Correctness requires that for all $(pk, sk) \in [\mathsf{P.Gen}]$ and $m \in \mathcal{M}$, if $c \in [\mathsf{P.Enc}(pk, m)]$ then $\mathsf{P.Dec}(sk, c) = m$.*

Construction 1 (Hybrid encryption). *Take a DEM for a message space $\mathcal{M}$ and a KEM for the key space of the DEM. Then the algorithms in Fig. 3 form the hybrid PKE scheme. The randomness space of* P.Enc *coincides with the randomness space of* K.Enc.

We present now the main security definition of this paper: confidentiality under selective opening attacks. Our model is based on works of [6,18] Find a discussion of its details below.

Definition 8 (SIM-SO-CCA secure PKE). *Consider the experiments from Fig. 4. For a function $\epsilon \colon \mathbb{N} \rightarrow \mathbb{R}^{\geq 0}$ we say that a PKE scheme is $(\tau, \tau', q_d, \epsilon)$-SIM-SO-CCA secure if for all τ-time adversaries $\mathcal{A} = (\mathcal{A}_1, \mathcal{A}_2)$ that interact in the* r-SO-CCA *experiment and issue at most q_d decryption queries there exists a*

Proc $\mathsf{P.Gen}(r)$	**Proc** $\mathsf{P.Enc}(pk, m, r)$	**Proc** $\mathsf{P.Dec}(sk, \langle c_1, c_2 \rangle)$
00 $(pk, sk) \leftarrow \mathsf{K.Gen}(r)$	02 $(k, c_1) \leftarrow \mathsf{K.Enc}(pk, r)$	05 $k \leftarrow \mathsf{K.Dec}(sk, c_1)$
01 Return (pk, sk)	03 $c_2 \leftarrow \mathsf{D.Enc}(k, m)$	06 If $k = \perp$: Return $\perp$
	04 Return $\langle c_1, c_2 \rangle$	07 $m \leftarrow \mathsf{D.Dec}(k, c_2)$
		08 Return m

Fig. 3. Hybrid construction of PKE from a KEM and a DEM. We write $\langle c_1, c_2 \rangle$ for the encoding of two ciphertext components into one. For clarity we make the randomness used by P.Gen and P.Enc explicit.

Game $\text{r-SO-CCA}_n^{\mathcal{A}}$	**Game** $\text{i-SO-CCA}_n^{\mathcal{S}}$
00 $\mathcal{I} \leftarrow \emptyset$; $C \leftarrow \emptyset$	15 $\mathcal{I} \leftarrow \emptyset$
01 $(pk, sk) \leftarrow_{\$} \mathsf{P.Gen}$	⋮
02 $(\mathfrak{D}, st) \leftarrow_{\$} \mathcal{A}_1^{\mathrm{P.DEC}}(pk, n)$	16 $(\mathfrak{D}, st) \leftarrow_{\$} \mathcal{S}_1(n)$
03 $(m_1, \ldots, m_n) \leftarrow_{\$} \mathfrak{D}$	17 $(m_1, \ldots, m_n) \leftarrow_{\$} \mathfrak{D}$
04 For $i \leftarrow 1$ to n:	⋮
05 $\quad r_i \leftarrow_U \mathcal{R}$	⋮
06 $\quad c_i \leftarrow \mathsf{P.Enc}(pk, m_i, r_i)$	⋮
07 $\quad C \leftarrow C \cup \{c_i\}$	⋮
08 $out \leftarrow_{\$} \mathcal{A}_2^{\mathrm{OPEN,P.DEC}}(st, c_1, \ldots, c_n)$	18 $out \leftarrow_{\$} \mathcal{S}_2^{\mathrm{OPEN}}(st, \lvert m_1 \rvert, \ldots, \lvert m_n \rvert)$
09 Stop w/ $\mathsf{Pred}(\mathfrak{D}, m_1, \ldots, m_n, \mathcal{I}, out)$	19 Stop w/ $\mathsf{Pred}(\mathfrak{D}, m_1, \ldots, m_n, \mathcal{I}, out)$
Oracle $\mathrm{OPEN}(i)$	**Oracle** $\mathrm{OPEN}(i)$
10 $\mathcal{I} \leftarrow \mathcal{I} \cup \{i\}$	20 $\mathcal{I} \leftarrow \mathcal{I} \cup \{i\}$
11 Return (m_i, r_i)	21 Return m_i
Oracle $\mathrm{P.DEC}(c)$	
12 If $c \in C$: Abort	
13 $m \leftarrow \mathsf{P.Dec}(sk, c)$	
14 Return m	

Fig. 4. Security experiments for defining SIM-SO-CCA security of PKE. With $\mathfrak{D}$ we denote a randomised circuit that induces a distribution over $\mathcal{M}^n$. The randomness space of P.Enc is denoted with $\mathcal{R}$. Oracle OPEN may be called for all $i \in [n]$. We write 'Abort' as an abbreviation for 'Stop with 0'. We show the lines of i-SO-CCA aligned to the ones of r-SO-CCA for easier comparison.

(roughly) τ-time simulator $\mathcal{S} = (\mathcal{S}_1, \mathcal{S}_2)$ that interacts in the i-SO-CCA *experiment such that for all τ'-time predicates $\{0,1\}^* \to \mathsf{Pred} \to_{\$} \{0,1\}$ and all $n \in \mathbb{N}$ the advantage $\mathbf{Adv}_{\mathcal{A},\mathcal{S},\mathsf{Pred}}^{\mathsf{SIM\text{-}SO\text{-}CCA}}(n)$ is at most $\epsilon(n)$, where we define*

$$\mathbf{Adv}_{\mathcal{A},\mathcal{S},\mathsf{Pred}}^{\mathsf{SIM\text{-}SO\text{-}CCA}}(n) := \left|\Pr[\text{r-SO-CCA}_n^{\mathcal{A}} \Rightarrow 1] - \Pr[\text{i-SO-CCA}_n^{\mathcal{S}} \Rightarrow 1]\right|.$$

We give rationale on this formalisation of SO security. The notion compares the information an adversary can deduce about a set of challenge messages in two settings: a real setting (game r-SO-CCA) and an idealised setting (game i-SO-CCA). The real experiment starts with the generation of a key pair. The adversary receives the public key and specifies a message distribution, represented by a randomised circuit $\mathfrak{D}$. Messages $m_1, \ldots, m_n$ are sampled according to this distribution and encrypted using fresh randomnesses $r_1, \ldots, r_n$, and the ciphertexts are given to the adversary which derives some information *out* about the hidden messages. The adversary is supported by two oracles: one that decrypts arbitrary ciphertexts and one that opens honest ciphertexts by revealing the corresponding message and the randomness used to encrypt it (this is meant to model sender corruption).

The ideal experiment is similar but with all the artifacts of public key encryption removed: there is no key generation, no ciphertext generation, and no decryption oracle. Beyond that, the adversary (in this context called 'simulator') performs as above: it specifies a message distribution, adaptively requests openings, and derives some information *out* about unopened messages.

Clearly, in the ideal setting the confidentiality of unopened messages is granted (only their lengths leak in line 18, but this is unavoidable for any practical PKE scheme and implicitly also happens in line 08). We thus deem a public key encryption scheme secure under selective opening attacks if the adversary in the real setting cannot draw more conclusions about unopened messages than can be drawn in the ideal setting. Formally, it is required that for every $\mathcal{A}$ for r-SO-CCA there exists a corresponding $\mathcal{S}$ for i-SO-CCA that derives the same information. This is tested by distinguishing predicate Pred, which also takes further environmental information into account, for instance the recorded opening history $\mathcal{I}$. We proceed with some remarks on the model.

In prior works that give simulation-based definitions of SO security there does not seem to be concensus on the order of quantification of $\mathcal{S}$ and Pred. While most papers (cf. [22,28]) allow for the simulator to depend on the distinguishing predicate, the work of [6] implicitly defines a stronger notion that requires the existence of a simulator that is universal. (Interestingly, many papers that exclusively consider the weaker notion actually *do* construct universal simulators.) We adopt the stronger notion and require the simulator to work for any distinguisher.

In the upcoming sections we construct several PKE schemes that are secure under selective opening attacks. The corresponding proofs will idealise a central building block of the schemes, concretely a blockcipher. By consequence, ideal-cipher oracles have to be added to Fig. 4. There are various options how and where to do this: It is clear that adversary $\mathcal{A}$ should have access to the ideal cipher, but what about $\mathcal{S}$, what about Pred, and what about $\mathfrak{D}$? It seems that each configuration somehow makes sense and gives rise to an individual variant of SIM-SO-CCA security.[1] Each such notion might have particular strengths and weaknesses, so declaring any of them right or wrong is arbitrary. Ultimately, when proving the SO security of our schemes, we decided to go for a model where, besides the relevant algorithms of the encryption scheme itself, only adversary $\mathcal{A}$ gets access to the ideal cipher.

Notions of SO security under active Attacks. As mentioned in the introduction, three notions for SO security under active attacks exist: {*weak*-IND, *full*-IND,

[1] A similar situation emerges with NIZK proofs in the random oracle model: In the corresponding ZK definition, shall the distinguisher have access to the random oracle or not? See [31] for a formal treatment and a comparison of the many possible notions.

SIM}-SO-CCA. Non of them has emerged as a de-facto standard notion, yet. Clearly, *weak*-IND-SO-CCA suffers from the unnatural restriction to efficiently conditionally resamplable message distributions and security implications for practical applications are unclear. While *full*-IND-SO-CCA would provide security for arbitrary underlying message distributions, as of today, no even a *full*-IND-SO-CPA secure scheme is known.

We note that SIM-SO-CCA does not suffer from any of the above disadvantages (there is no resampling involved) and seems to offer a strong security guarantee.

Only few results relating the SO-CCA notions are known; [26] shows that IND-CCA is strictly weaker than *weak*-IND-CCA in general.

3 Simulatable DEMs and Our Main Result

In this section we present our main result on hybrid public key encryption. We define a combinatorial property of a DEM called *simulatability* and show that any KEM and any DEM satisfying standard security notions, if the DEM is in addition simulatable, when composed yield a SIM-SO-CCA secure PKE, in the ideal cipher model [9,17,27].

3.1 Simulatable DEMs

Many practical DEMs are constructed from blockciphers, possibly in combination with further symmetric building blocks like universal hash functions or MACs. We formalise next what it means for a DEM to make use of a blockcipher in a black-box way. Virtually all blockcipher-based DEMs, and in particular those specified by the major standardisation bodies, are of this type. In our definition, $\mathcal{K}$ denotes the key space of the blockcipher and $\mathcal{K}'$ denotes the cartesian product of the key spaces of the remaining cryptographic primitives used by the scheme. For instance, in an encrypt-then-MAC construction, $\mathcal{K}'$ would be the key space of the message authentication code; if the construction requires no further keyed primitive, $\mathcal{K}'$ would be the trivial set containing a single element.

Recall from Definition 1 that $\mathcal{P}(\mathcal{D})$ and $\mathcal{PP}(\mathcal{D})$ denote the sets of all permutations and partial permutations, respectively, on domain $\mathcal{D}$.

Definition 9 (Oracle DEM). *An oracle data encapsulation mechanism (oDEM) for a domain $\mathcal{D}$ and a message space $\mathcal{M}$ consists of a finite key space $\mathcal{K}'$, a ciphertext space $\mathcal{C}$, and efficient algorithms* O.Enc *and* O.Dec *that have oracle access to a permutation on $\mathcal{D}$ (in both directions) and are of the form*

$$\mathcal{K}' \times \mathcal{M} \to \mathsf{O.Enc}^{\pi} \to \mathcal{C} \qquad \mathcal{K}' \times \mathcal{C} \to \mathsf{O.Dec}^{\pi} \to \mathcal{M} \cup \{\bot\},$$

where symbol '$\bot$' may be used to indicate errors. Correctness requires that for all $\pi \in \mathcal{P}(\mathcal{D})$, $k' \in \mathcal{K}'$, and $m \in \mathcal{M}$, if $\mathsf{O.Enc}^{\pi}(k', m) = c$ then $\mathsf{O.Dec}^{\pi}(k', c) = m$.

Definition 10 (Permutation-driven DEM). *A DEM for message space $\mathcal{M}$ with keyspace $\mathcal{K}'' = \mathcal{K} \times \mathcal{K}'$ is* ($\mathcal{K}, \mathcal{D}$)*-permutation-driven if there exists an oracle DEM for $\mathcal{D}$ and $\mathcal{M}$ with algorithms $\mathcal{K}' \times \mathcal{M} \to \mathsf{O.Enc}^{\pi} \to \mathcal{C}$ and $\mathcal{K}' \times \mathcal{C} \to \mathsf{O.Dec}^{\pi} \to \mathcal{M} \cup \{\bot\}$ and a blockcipher $(E_k)_{k \in \mathcal{K}}$ on domain $\mathcal{D}$ such that for all $k' \in \mathcal{K}'$ and $m \in \mathcal{M}$ and $c \in \mathcal{C}$ we have*

$$\mathsf{D.Enc}((k,k'),m) = \mathsf{O.Enc}^{E_k}(k',m) \quad \text{and} \quad \mathsf{D.Dec}((k,k'),c) = \mathsf{O.Dec}^{E_k}(k',c). \tag{1}$$

According to this definition, for any specific permutation-driven DEM multiple corresponding oracle DEMs, i.e., O.Enc and O.Dec algorithms, and blockciphers E might exist. In practice, however, a single canonic specification of these algorithms will stick out. This holds, as we will see, in particular for the standardised DEMs studied in Sect. 4. For the sake of a concise notation, in this paper we thus assume that suitable O.Enc, O.Dec, and E algorithms are always uniquely given.

We next define a combinatorial property called *simulatability* that holds for an oracle DEM if, in principle, the encapsulation algorithm could commit to a ciphertext before seeing the corresponding message; intuitively, this is only possible if the permutation in the oracle is 'flexible enough', i.e., can be 'programmed'. We formalise this idea by splitting the encapsulation routine into two components, Fake and Make. First Fake outputs a ciphertext c without seeing the message m (but it does see the length of m), then Make, on input m, is meant to find a possible (partial) permutation instance $\tilde{\pi}$ under which indeed m would be encapsulated to c. To be useful in our later selective opening related proofs where we want to embed $\tilde{\pi}$ into an ideal cipher, $\tilde{\pi}$ is further required to be uniformly distributed (conditioned on the formulated requirements).

Definition 11 (Simulatable oracle DEM). *Consider an oracle DEM for a domain $\mathcal{D}$ and a message space $\mathcal{M}$ that has an encapsulation algorithm of the form $\mathcal{K}' \times \mathcal{M} \to \mathsf{O.Enc}^{\pi} \to \mathcal{C}$. Consider algorithms* Fake *and* Make *of the form*

$$\mathcal{K}' \times \mathbb{N} \to \mathsf{Fake} \to_{\$} \mathcal{C} \times \Sigma \quad \text{and} \quad \Sigma \times \mathcal{M} \to \mathsf{Make} \to_{\$} \mathcal{PP}(\mathcal{D}),$$

where Σ is a state space shared between the two algorithms. We say that the oracle DEM is ϵ-simulatable *(by* Fake *and* Make*) if for all $k' \in \mathcal{K}'$ and $m \in \mathcal{M}$, for the random variable (defined over the coins of* Fake *and* Make*)*

$$\Pi_{k'}^{m} = \{\tilde{\pi} : (c, st) \leftarrow_{\$} \mathsf{Fake}(k', |m|); \tilde{\pi} \leftarrow_{\$} \mathsf{Make}(st, m)\}$$

we have

(1) *partial permutation $\Pi_{k'}^{m}$ can be extended to a uniformly distributed permutation on $\mathcal{D}$, i.e., by 'filling up' $\Pi_{k'}^{m}$ with random pairs one obtains a permutation uniformly distributed in $\mathcal{P}(\mathcal{D})$;*
(2) *the ciphertext output by* Fake *deviates from the one that would be output by* O.Enc *if invoked with an extension of the partial permutation output by* Make *with probability at most ϵ. More precisely, for any uniformly distributed extension $\pi \in \mathcal{P}(\mathcal{D})$ of $\Pi_{k'}^{m}$ we have $\Pr[c \neq \mathsf{O.Enc}^{\pi}(k', m)] \leq \epsilon$ (where the probability is also taken over the random extension of $\Pi_{k'}^{m}$ to π);*

(3) *the joint running time of* $\mathsf{Fake}(k', |m|)$ *and* $\mathsf{Make}(st, m)$ *does not exceed the running time of* $\mathsf{O.Enc}(k', m)$*, not counting the latter's oracle queries.*

In informal discussions, when we say that a data encapsulation mechanism is simulatable we mean that it is permutation-driven and Fake, Make *algorithms exist for which it is* ϵ*-simulatable with a negligibly small value* ϵ*.*

Concerning the above definition it is important to understand that the random coins of Fake and Make, and the coins used to extend the partial permutation in items (1) and (2), belong to the same probability space. We give an equivalent yet more verbose definition that makes this aspect more explicit in the Appendix of the full version [23].

In line with a comment made above, for all practical DEMs that are simulatable, corresponding specifications for the Fake and Make algorithms emerge canonically. For the sake of notational clarity, from now on we thus assume uniqueness.

Proving Simulatability. We discuss a general technique for proving the simulatability of an oracle DEM. The Fake and Make algorithms are typically explicitly provided in the proof. Fake's strategy is to mimic the behaviour of O.Enc by executing it and answering blockcipher queries with random elements from $\mathcal{D}$. Make constructs a partial permutation $\tilde{\pi}$ that fits this random assignment by starting with the empty relation $\tilde{\pi} = \emptyset$ and iteratively adding pairs $(\alpha, \beta) \in \mathcal{D} \times \mathcal{D}$ to $\tilde{\pi}$ that help meeting the $\mathsf{O.Enc}^{\tilde{\pi}}(k', m) = c$ goal, always taking care that also the $\alpha\tilde{\pi}\beta, \alpha'\tilde{\pi}\beta \Rightarrow \alpha = \alpha'$ and $\alpha\tilde{\pi}\beta, \alpha\tilde{\pi}\beta' \Rightarrow \beta = \beta'$ requirements from Definition 1 are not violated (Make aborts if simultaneously reaching these conditions turns out to be impossible). Simulatability requirement (1) is achieved by ensuring that for each addition of (α, β) to $\tilde{\pi}$ either α or β are uniformly distributed, conditioned on the prior state of $\tilde{\pi}$. Proving the bound from condition (2) typically requires a combinatorial argument that assesses the probability of collisions. Requirement (3) follows by inspection of the specifications of Fake and Make.

3.2 Selective Opening Security from Simulatable DEMs

Our main result is on the SO security of public-key encryption obtained by combining an arbitrary KEM with a permutation-driven DEM. Our analysis is conducted in the ideal cipher model for the blockcipher underlying the DEM. We give an informal version of our main theorem and an outline of the proof. We caution that some technical preconditions are omitted in the statement as we give it here. See Sect. 5 for the full theorem statement and proof.

Theorem 1 (informal). *Combine any KEM and any permutation-driven DEM to obtain a PKE scheme. If the KEM is IND-CCA secure, the DEM is OT-INT-CTXT secure and the corresponding oracle DEM is simulatable, then the combined PKE scheme is SIM-SO-CCA secure, in the ideal cipher model.*

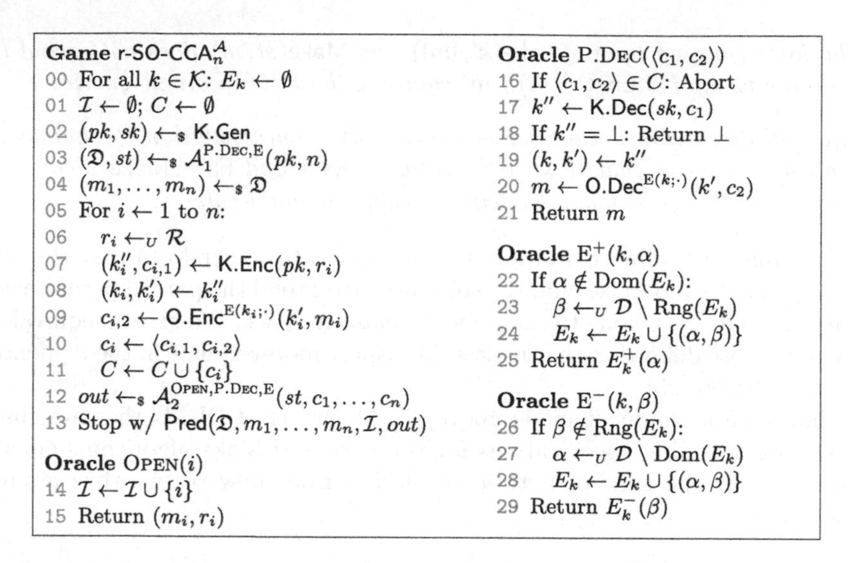

Fig. 5. Game r-SO-CCA adapted towards the analysis of a PKE scheme constructed following the KEM/DEM paradigm using a permutation-driven DEM with corresponding oracle DEM algorithms O.Enc and O.Dec, in the ideal cipher model. We write 'Abort' as an abbreviation for 'Stop with 0'. We further abbreviate the pair E^+, E^- of ideal cipher oracles with just E.

We proceed with the proof outline. The goal is to show that for every adversary $\mathcal{A} = (\mathcal{A}_1, \mathcal{A}_2)$ for the r-SO-CCA game there exists a simulator $\mathcal{S} = (\mathcal{S}_1, \mathcal{S}_2)$ for the i-SO-CCA game that deduces the same information. In Fig. 5 we reproduce the r-SO-CCA game from Fig. 4 with the hybrid construction of the encryption scheme, the oracle DEM underlying the DEM, and the ideal cipher model made explicit. (In the i-SO-CCA game there is nothing to be adapted.) We correspondingly equip adversary $\mathcal{A}$ and the DEM algorithms with oracles E^+ and E^- that implement an ideal blockcipher on domain $\mathcal{D}$. In particular, for each key k, oracles $\text{E}^+(k;\cdot)$ and $\text{E}^-(k;\cdot)$ are inverses of each other. For a concise notation, we typically just write E for the pair consisting of E^+ and E^-. We implement ideal cipher E via lazy sampling and keep track of made assignments using a game internal family $(E_k)_{k \in \mathcal{K}}$ of partial permutations $E_k \in \mathcal{PP}(\mathcal{D})$. Note that we do not also provide the KEM algorithms with access to E, meaning we assume the KEM does not use the same blockcipher as the DEM. See Sect. 5 for a discussion.

When it comes to constructing $\mathcal{S}$ from $\mathcal{A}$, the strategy is to let the former run the latter as a subroutine: Simulator $\mathcal{S}$ converts the own input to an input for $\mathcal{A}$, uses the output of $\mathcal{A}$ as the own output, and answers, and in some cases relays, oracle queries posed by $\mathcal{A}$. We give the footprint of a universal such simulator that leverages on the simulatability of the (permutation-driven) DEM in Fig. 6. For the sake of clarity, we simplified the specifications of algorithms $\mathcal{S}_1$ and $\mathcal{S}_2$ quite a bit, removing many technicalities. While we briefly discuss the missing

$\mathcal{S}_1(n)$	**Oracle** $\text{OPEN}_{\mathcal{A}}(i)$
00 For all $k \in \mathcal{K}$: $E_k \leftarrow \emptyset$	14 $m_i \leftarrow \text{OPEN}_{\mathcal{S}}(i)$
01 $C \leftarrow \emptyset$	15 $\tilde{\pi} \leftarrow_\$ \mathsf{Make}(st_i, m_i)$
02 $(pk, sk) \leftarrow_\$ \mathsf{K.Gen}$	16 $E_{k_i} \leftarrow E_{k_i} \cup \tilde{\pi}$
03 $\mathfrak{D} \leftarrow_\$ \mathcal{A}_1^{\text{P.DEC,E}}(pk, n)$	17 Return (m_i, r_i)
04 Return $\mathfrak{D}$	**Oracle** $\text{P.DEC}(\langle c_1, c_2 \rangle)$ as in Figure 5
$\mathcal{S}_2^{\text{OPEN}_{\mathcal{S}}}(\lvert m_1 \rvert, \ldots, \lvert m_n \rvert)$	
05 For $i \leftarrow 1$ to n:	**Oracle** $\text{E}^+(k, \alpha)$ as in Figure 5
06 $\quad r_i \leftarrow_U \mathcal{R}$	
07 $\quad (k_i'', c_{i,1}) \leftarrow \mathsf{K.Enc}(pk; r_i)$	
08 $\quad (k_i, k_i') \leftarrow k_i''$	**Oracle** $\text{E}^-(k, \beta)$ as in Figure 5
09 $\quad (c_{i,2}, st_i) \leftarrow_\$ \mathsf{Fake}(k_i', \lvert m_i \rvert)$	
10 $\quad c_i \leftarrow \langle c_{i,1}, c_{i,2} \rangle$	
11 $\quad C \leftarrow C \cup \{c_i\}$	
12 $out \leftarrow_\$ \mathcal{A}_2^{\text{OPEN}_{\mathcal{A}},\text{P.DEC,E}}(c_1, \ldots, c_n)$	
13 Return out	

Fig. 6. Simplified version of simulator $\mathcal{S} = (\mathcal{S}_1, \mathcal{S}_2)$, constructed from adversary $\mathcal{A} = (\mathcal{A}_1, \mathcal{A}_2)$. We write $\text{OPEN}_{\mathcal{S}}$ and $\text{OPEN}_{\mathcal{A}}$ for the opening oracles provided to $\mathcal{S}_2$ and $\mathcal{A}_2$, respectively. For simplicity we do not annotate the state information passed from $\mathcal{A}_1$ to $\mathcal{A}_2$ and from $\mathcal{S}_1$ to $\mathcal{S}_2$.

parts below, for the full details of the simulator and a formal analysis we refer to Sect. 5.

We walk the reader through the design principles of our simulator. What above we refered to as 'deduces the same information' formally requires that the inputs $\mathfrak{D}, m_1, \ldots, m_n, \mathcal{I}, out$ of the Pred invocations in the r-SO-CCA and i-SO-CCA games be similar. This is achieved by letting $\mathcal{S}$ simulate for $\mathcal{A}$ the environment of r-SO-CCA in a way such that: $\mathcal{S}_1$ forwards the message distribution $\mathfrak{D}$ obtained from $\mathcal{A}_1$ without modification (this also ensures that the distributions of $m_1, \ldots, m_n$ match), $\mathcal{S}_2$ keeps the index sets $\mathcal{I}$ corresponding to $\mathcal{A}_2$'s and its own OPEN queries consistent (by forwarding the queries), and $\mathcal{S}_2$ forwards $\mathcal{A}_2$'s output out without modification. The lines in Fig. 6 corresponding to these steps are 03,04 and 14 and 12,13, respectively.

Running $\mathcal{A}$ as a subroutine leads to useful results only if $\mathcal{A}$ is exposed to an r-SO-CCA-like environment. Effectively this means that $\mathcal{S}$ has to 'fill all the blank lines' of the i-SO-CCA game in Fig. 4. Concretely this involves (a) generating and providing a public key for $\mathcal{A}_1$, (b) providing ciphertexts to $\mathcal{A}_2$ that correspond to messages $m_1, \ldots, m_n$, (c) providing adequate randomness when processing opening queries of $\mathcal{A}_2$, and (d) handling decryption queries of $\mathcal{A}_1$ and $\mathcal{A}_2$. Further, ideal cipher queries of $\mathcal{A}_1$ and $\mathcal{A}_2$ have to be taken care of. The latter is straight-forward when deploying lazy sampling, i.e., using the mechanisms of the r-SO-CCA version from Fig. 5. Also (a) and (d) are easy to deal with: The public key pk provided to $\mathcal{A}_1$ is a regular KEM key generated by $\mathcal{S}_1$ (lines 02,03); in particular, secret key sk is known to $\mathcal{S}$ and can be used to process decryption queries. Concerning (b), creating ciphertexts $c_1, \ldots, c_n$ for $\mathcal{A}_2$ consists, in principle, of two parts: letting the KEM establish session keys and encapsulating messages with the DEM. Component $\mathcal{S}_2$ of our simulator does the former according to the specification, i.e., by invoking algorithm K.Enc with fresh randomness (lines 06,07), while for the latter, as it cannot invoke D.Enc (or, more precisely, O.Enc) for not knowing the messages it needs to encapsulate, it leverages on the simulatability of the DEM and obtains the corresonding ciphertext from an execution of the Fake algorithm (line 09). How $\mathcal{S}_2$ deals with (c) is now immediate: for each created ciphertext it knows the randomness used, so it can release it in an opening query (line 17). Note, however, that knowledge of this randomness brings $\mathcal{A}_2$ into the position to verify the DEM ciphertext components generated by Fake (e.g., by decapsulating or re-encapsulating them); correspondingly, the OPEN oracle in addition runs the Make algorithm and embeds the partial permutation proposed by it into ideal cipher E (lines 15,16). By the definition of simulatability of a DEM, this fixes the ideal cipher such that overall consistency is established.

As announced earlier, in Fig. 6 we leave out some details of our simulator. These are related to situations in which $\mathcal{S}$ cannot uphold a proper environment for $\mathcal{A}$ and has to abort its execution. This is the case when Fake and Make fail to properly simulate O.Enc (the definition of simulatability considers a small probability of failure), or if the partial permutation output by Make cannot be embedded into the ideal cipher (line 16). The latter condition can result from various actions of adversary $\mathcal{A}$, for instance (explicitly) from queries to the E oracles, or (implicitly) from evaluations of E during the processing of a decryption query. In the full proof given in Sect. 5 we show that if the KEM is IND-CCA secure and the DEM is OT-INT-CTXT secure, then the probability is small that any of these conditions is met. (Very briefly speaking, we use the KEM notion for bounding the probability of explicit queries, and we use the DEM notion for bounding the probability of implicit ones.)

4 Simulatability of Practical DEMs

We prove that three blockcipher-based DEMs that were standardised by NIST are permutation-driven and simulatable. Concretely we analyse the CTR and CBC modes of operation (SP 800-38A [13]), a CBC variant with ciphertext stealing (CTS) (Addendum to SP 800-38A [16]) and the CCM mode (SP 800-38C [14]). The fourth NIST standardised mode of operation, the GCM mode (SP 800-38D [15]), is covered in the full version of this paper [23]. More precisely, as for our results on selective opening security only those DEMs are relevant that offer ciphertext integrity (cf. Definition 4), instead of plain CTR, CBC, and CBC/CTS encryption we actually analyse their encrypt-then-MAC variants, where we assume arbitrary strongly unforgeable MACs. Further, as CCM is an authenticated encryption scheme with associated data (AEAD [30]), we turn it into a DEM by using it with a fixed nonce N_0 and an empty associated data string A_0. As the three named modes follow different design principles, some of which might be incompatible with simulatability, analysing all of them is more than just a matter of diligence. While CTR mode encrypts by XORing blockcipher outputs into the message, CBC mode encrypts by pushing message blocks through the cipher, and CCM combines both approaches is a MAC-then-encrypt design.

In the following we specify the mentioned DEMs in their oracle DEM form, assuming that the underlying blockcipher $(E_k)_{k \in \mathcal{K}}$ is over domain $\mathcal{D} = \{0,1\}^\ell$. We show their simulatability by proposing and analysing corresponding Fake and Make algorithms, following the general strategy suggested at the end of Sect. 3.1.

4.1 CTR-then-MAC

We analyse the DEM obtained by first encrypting the provided message with the CTR0 mode of operation of a blockcipher (counter mode with fixed initial counter value) and then appending a deterministic MAC tag to the ciphertext.

We specify the O.Enc and O.Dec algorithms of CTR0-DEM in Fig. 7, where we assume that $G\colon [1 \,..\, V] \to \mathcal{D}$ denotes a fixed injective function (a 'counter generator') for some sufficiently large value V. The MAC is represented by a keyed hash function $\mathcal{K}' \times \{0,1\}^* \to \text{khf} \to \{0,1\}^T$. The message space of CTR0-DEM is $\mathcal{M} = \{0,1\}^*$ and the ciphertext space is $\mathcal{C} = \{0,1\}^{\geq T}$.

$\mathsf{O.Enc}^\pi(k', m)$	$\mathsf{O.Dec}^\pi(k', c)$
00 Write $\lvert m\rvert$ as $(l-1)\ell + l^*$	12 If $\lvert c\rvert < T$: Return $\perp$
01 Split m into $m_1 \dots m_{l-1} m_l^*$	13 Split c into $\bar{c}t$
02 $m_l \leftarrow m_l^* \,\Vert\, 0^{\ell - l^*}$	14 If $t \neq \text{khf}(k', \bar{c})$:
03 For $i \leftarrow 1$ to l:	15 Return $\perp$
04 $u_i \leftarrow G(i)$	16 Write $\lvert\bar{c}\rvert$ as $(l-1)\ell + l^*$
05 $v_i \leftarrow \pi(u_i)$	17 Split $\bar{c}$ into $c_1 \dots c_{l-1} c_l^*$
06 $c_i \leftarrow m_i \oplus v_i$	18 $c_l \leftarrow c_l^* \,\Vert\, 0^{\ell - l^*}$
07 $c_l^* \leftarrow \text{msb}_{l^*}(c_l)$	19 For $i \leftarrow 1$ to l:
08 $\bar{c} \leftarrow c_1 \dots c_{l-1} c_l^*$	20 $u_i \leftarrow G(i)$
09 $t \leftarrow \text{khf}(k', \bar{c})$	21 $v_i \leftarrow \pi(u_i)$
10 $c \leftarrow \bar{c}t$	22 $m_i \leftarrow c_i \oplus v_i$
11 Return c	23 $m_l^* \leftarrow \text{msb}_{l^*}(m_l)$
	24 $m \leftarrow m_1 \dots m_{l-1} m_l^*$
	25 Return m

Fig. 7. CTR0-DEM. Lines 00 and 16 uniquely identify quantities l and l^* such that $l \in \mathbb{N}^{\geq 1}$ and $0 \leq l^* < \ell$, and $\lvert m\rvert = (l-1)\ell + l^*$ and $\lvert\bar{c}\rvert = (l-1)\ell + l^*$, respectively. Correspondingly, line 01 assumes $\lvert m_1\rvert = \ldots = \lvert m_{l-1}\rvert = \ell$ and $\lvert m_l^*\rvert = l^*$, and line 17 assumes $\lvert c_1\rvert = \ldots = \lvert c_{l-1}\rvert = \ell$ and $\lvert c_l^*\rvert = l^*$. Further, line 13 assumes $\lvert t\rvert = T$.

Lemma 1. *CTR0-DEM is ϵ-simulatable with $\epsilon = (\lceil L/\ell\rceil^2 - \lceil L/\ell\rceil)/2^{\ell+1}$, where L is the maximum message length (in bits).*

Proof. Consider algorithms Fake and Make from Fig. 8. The idea of Fake is to compute intermediate ciphertext $\bar{c}$ on basis of uniformly distributed blockcipher outputs (see how line 01 of Fake replaces l-many iterations of line 06 of O.Enc), but to compute the MAC tag on $\bar{c}$ faithfully. Note that the correct length of $\bar{c}$

is known to Fake as it coincides with the length of m. Inspection shows that, given m, algorithm Make finds a minimal partial permutation $\tilde{\pi}$ such that Fake and Make jointly mimic the behaviour of O.Enc (see here how lines 15–18 of Make arrange the entries of $\tilde{\pi}$ such that they are consistent with lines 05–06 of O.Enc). In some invocations of the algorithms, the described process might fail (lines 16, 17), namely when partial permutation $\tilde{\pi}$ would become inconsistent (i.e., the updated $\tilde{\pi}$ would stop being an element of $\mathcal{PP}$). In such cases Make aborts, outputting the empty partial permutation $\tilde{\pi} = \emptyset$.

We next show that the conditions from Definition 11 are met. Observe that, as Fake picks values $c_1, \ldots, c_l$ uniformly and independently of each other, the same holds for the values $v_1, \ldots, v_l$ computed in line 15. That is, in each iteration of line 18 a value v_i is added to $\mathrm{Rng}(\tilde{\pi})$ that is uniform conditioned on the then current state of $\mathrm{Rng}(\tilde{\pi})$. Thus condition (1) holds. To establish the correctness bound of condition (2) we analyse the probability that Make aborts. By the injectivity of function G the u_i-values from line 14 are pairwise distinct, so the abort condition of line 16 is never met. Further, as values v_i computed in line 15 are uniformly distributed and independent of each other, the abort condition of line 17 is met with probability $\epsilon = (0 + \ldots + (l-1))/|\mathcal{D}| = ((l^2 - l)/2)/|\mathcal{D}|$ (accumulated over all iterations of the loop). Plugging in the maximum value $l = \lceil L/\ell \rceil$ gives the bound claimed in the statement. Condition (3) is clear. □

```
Fake(k', |m|)                        Make(st, m)
00 Write |m| as (l−1)ℓ+l*            08 π̃ ← ∅
01 c_1,...,c_l ←_U D                 09 Write |m| as (l − 1)ℓ + l*
02 c*_l ← msb_{l*}(c_l)              10 Parse st as (c_1,...,c_l)
03 c̄ ← c_1 ... c_{l−1} c*_l          11 Split m into m_1 ... m_{l−1} m*_l
04 t ← khf(k', c̄)                    12 m_l ← m*_l || 0^{ℓ−l*}
05 c ← c̄t                            13 For i ← 1 to l:
06 st ← (c_1,...,c_l)                14    u_i ← G(i)
07 Return c, st                      15    v_i ← m_i ⊕ c_i
                                     16    If u_i ∈ Dom(π̃): Abort
                                     17    If v_i ∈ Rng(π̃): Abort
                                     18    π̃ ← π̃ ∪ {(u_i, v_i)}
                                     19 Return π̃
```

Fig. 8. Fake and Make for CTR0-DEM. We write 'Abort' as an abbreviation for 'Return $\emptyset$'.

4.2 CBC-then-MAC

We consider the DEM obtained by encrypting the message with CBC0 mode (cipher block chaining with initialisation vector zero) and appending a MAC tag to the ciphertext. As a variant we also look at CBC0-CTS (CBC0 with 'ciphertext stealing') that supports a complementary message space.

```
O.Enc^π(k', m)                      O.Dec^π(k', c)
00 Write |m| as lℓ                  10 If |c| < T: Return ⊥
01 Split m into m_1 ... m_l         11 Split c into c̄t
02 c_0 ← 0^ℓ                        12 If t ≠ khf(k', c̄):
03 For i ← 1 to l:                  13    Return ⊥
04    u_i ← m_i ⊕ c_{i-1}           14 Write |c̄| as lℓ
05    c_i ← π(u_i)                  15 Split c̄ into c_1 ... c_l
06 c̄ ← c_1 ... c_l                  16 c_0 ← 0^ℓ
07 t ← khf(k', c̄)                   17 For i ← 1 to l:
08 c ← c̄t                           18    u_i ← π^{-1}(c_i)
09 Return c                         19    m_i ← u_i ⊕ c_{i-1}
                                    20 m ← m_1 ... m_l
                                    21 Return m
```

Fig. 9. CBC-DEM (for multi-block messages). Lines 00 and 14 identify quantity $l \in \mathbb{N}^{\geq 0}$ such that $|m| = l\ell$ and $|\bar{c}| = l\ell$, respectively. Correspondingly, line 01 assumes $|m_1| = \ldots = |m_l| = \ell$ and line 15 assumes $|c_1| = \ldots = |c_l| = \ell$. Further, line 11 assumes $|t| = T$.

```
O.Enc^π(k', m)                           O.Dec^π(k', c)
00 Write |m| as lℓ + l*                  12 If |c| < T: Return ⊥
01 Split m into m_1 ... m_l m*_{l+1}     13 Split c into c̄t
02 m_{l+1} ← m*_{l+1} || 0^{ℓ-l*}        14 If t ≠ khf(k', c̄):
03 c_0 ← 0^ℓ                             15    Return ⊥
04 For i ← 1 to l + 1:                   16 Write |c̄| as lℓ + l*
05    u_i ← m_i ⊕ c_{i-1}                17 Split c̄ into c_1 ... c_{l-1} c*_l c_{l+1}
06    c_i ← π(u_i)                       18 u_{l+1} ← π^{-1}(c_{l+1})
07 c*_l ← msb_{l*}(c_l)                  19 m*_{l+1} ← msb_{l*}(u_{l+1}) ⊕ c*_l
08 c̄ ← c_1 ... c_{l-1} c*_l c_{l+1}      20 c_l ← c*_l || lsb_{ℓ-l*}(u_{l+1})
09 t ← khf(k', c̄)                        21 c_0 ← 0^ℓ
10 c ← c̄t                                22 For i ← 1 to l:
11 Return c                              23    u_i ← π^{-1}(c_i)
                                         24    m_i ← u_i ⊕ c_{i-1}
                                         25 m ← m_1 ... m_l m*_{l+1}
                                         26 Return m
```

Fig. 10. CBC-CTS-DEM (for messages that require padding). Lines 00 and 16 uniquely identify quantities l and l^* such that $l \in \mathbb{N}^{\geq 1}$ and $1 \leq l^* < \ell$, and $|m| = l\ell + l^*$ and $|\bar{c}| = l\ell + l^*$, respectively. Correspondingly, line 01 assumes $|m_1| = \ldots = |m_l| = \ell$ and $|m^*_{l+1}| = l^*$, and line 17 assumes $|c_1| = \ldots = |c_{l-1}| = \ell$ and $|c^*_l| = l^*$ and $|c_{l+1}| = \ell$. Further, line 13 assumes $|t| = T$.

We specify the O.Enc and O.Dec algorithms of CBC-DEM in Fig. 9 and of CBC-CTS-DEM in Fig. 10. Similarly as for CTR0-DEM, the MAC is represented by a keyed hash function of the form $\mathcal{K}' \times \{0,1\}^* \to \text{khf} \to \{0,1\}^T$. The message space of CBC-DEM consists of all messages that have a length that is a multiple of the blocklength ℓ, i.e., $\mathcal{M} = \bigcup_{\lambda \geq \ell, \ell | \lambda} \{0,1\}^\lambda$; the ciphertext space is $\mathcal{C} = \bigcup_{\lambda \geq \ell, \ell | \lambda} \{0,1\}^{\lambda + T}$. In contrast, CBC-CTS-DEM supports all message lengths that are not a multiple of ℓ, with a minimum value of $\ell + 1$; formally,

$\mathcal{M} = \bigcup_{\lambda \geq \ell, \ell \nmid \lambda} \{0,1\}^\lambda$ and $\mathcal{C} = \bigcup_{\lambda \geq \ell, \ell \nmid \lambda} \{0,1\}^{\lambda+T}$. Together, CBC-DEM and CBC-CTS-DEM can handle messages of any length not smaller than ℓ.[2]

Lemma 2. *CBC-DEM is ϵ-simulatable where $\epsilon = ((L/\ell)^2 - (L/\ell))/2^\ell$, and CBC-CTS-DEM is ϵ-simulatable with $\epsilon = (\lfloor L/\ell \rfloor^2 + \lfloor L/\ell \rfloor)/2^\ell$, where L is the maximum message length (in bits).*

```
Fake(k', |m|)                    Make(st, m)
00 Write |m| as lℓ               07 π̃ ← ∅
01 c_1, ..., c_l ←_U D           08 Write |m| as lℓ
02 c̄ ← c_1 ... c_l              09 Parse st as (c_1, ..., c_l)
03 t ← khf(k', c̄)               10 Split m into m_1 ... m_l
04 c ← c̄t                        11 c_0 ← 0^ℓ
05 st ← (c_1, ..., c_l)          12 For i ← 1 to l:
06 Return c, st                  13    u_i ← m_i ⊕ c_{i-1}
                                 14    If u_i ∈ Dom(π̃): Abort
                                 15    If c_i ∈ Rng(π̃): Abort
                                 16    π̃ ← π̃ ∪ {(u_i, c_i)}
                                 17 Return π̃
```

Fig. 11. Fake and Make for CBC-DEM. We write 'Abort' as an abbreviation for 'Return $\emptyset$'.

Proof. The proof is similar to the one of Lemma 1. Consider algorithms Fake and Make from Fig. 11. The idea of Fake is to compute intermediate ciphertext $\bar{c}$ on basis of uniformly distributed blockcipher outputs (see how line 01 of Fake replaces l-many iterations of line 05 of O.Enc), but to compute the MAC tag on $\bar{c}$ faithfully. Note that the correct length of $\bar{c}$ is known to Fake as it coincides with the length of m. Inspection shows that, given m, algorithm Make finds a minimal partial permutation $\tilde{\pi}$ such that Fake and Make jointly mimic the behaviour of O.Enc (see here how lines 13–16 of Make arrange the entries of $\tilde{\pi}$ such that they are consistent with lines 04–05 of O.Enc). In some invocations of the algorithms, the described process might fail (lines 14, 15), namely when partial permutation $\tilde{\pi}$ would become inconsistent. In such cases Make aborts, outputting the empty partial permutation $\tilde{\pi} = \emptyset$.

We next show that the conditions from Definition 11 are met. Observe that, as Fake picks values $c_1, \ldots, c_l$ uniformly and independently of each other, in each iteration of line 16 a value c_i is added to $\mathrm{Rng}(\tilde{\pi})$ that is uniform conditioned on the then current state of $\mathrm{Rng}(\tilde{\pi})$. Thus condition (1) holds. To establish the correctness bound of condition (2) we analyse the probability that Make

[2] Instead of specifying different algorithms for different classes of message length, one could also join them together to a single, more general algorithm. This is usually done in standards [16], but we abstain from doing so in this document to avoid rather obstructing case distinctions in the analysis.

aborts. With values $c_1, \ldots, c_{l-1}$ also the values $u_2, \ldots, u_l$ computed in line 13 are uniformly distributed and independent of each other, so the abort condition of line 14 is met with probability $(0 + \ldots + (l-1))/|\mathcal{D}| = ((l^2 - l)/2)/|\mathcal{D}|$ (accumulated over all iterations of the loop). The same bound holds for line 15. Plugging in the maximum value $l = L/\ell$ gives the bound claimed in the statement. Condition (3) is clear.

Algorithms Fake and Make for CBC-CTS-DEM are given in Fig. 12. The analysis is similar. Here, however, we have $l = \lfloor L/\ell \rfloor$ and for lines 16 and 17 the accumulated probabilities of abort amount to $(0 + \ldots + l)/|\mathcal{D}|$ each. □

```
Fake(k', |m|)                          Make(st, m)
00 Write |m| as lℓ + l*                08 π̃ ← ∅
01 c_1, ..., c_{l+1} ←_U D             09 Write |m| as lℓ + l*
02 c*_l ← msb_{l*}(c_l)                10 Parse st as (c_1, ..., c_{l+1})
03 c̄ ← c_1 ... c_{l-1} c*_l c_{l+1}    11 Split m into m_1 ... m_l m*_{l+1}
04 t ← khf(k', c̄)                      12 m_{l+1} ← m*_{l+1} || 0^{ℓ-l*}
05 c ← c̄t                              13 c_0 ← 0^ℓ
06 st ← (c_1, ..., c_{l+1})            14 For i ← 1 to l+1:
07 Return c, st                        15   u_i ← m_i ⊕ c_{i-1}
                                       16   If u_i ∈ Dom(π̃): Abort
                                       17   If c_i ∈ Rng(π̃): Abort
                                       18   π̃ ← π̃ ∪ {(u_i, c_i)}
                                       19 Return π̃
```

Fig. 12. Fake and Make for CBC-CTS-DEM. We write 'Abort' as an abbreviation for 'Return $\emptyset$'.

4.3 CCM

We analyse the CCM mode of operation ('CTR mode with CBC-MAC') with fixed nonce and associated data field; we call this mode CCM0-DEM. CCM is parameterised by an authentication tag length T, a formatting function $F\colon \mathcal{N} \times \mathcal{A} \times \mathcal{M} \to \mathcal{D}^+$ (where $\mathcal{N}$ and $\mathcal{A}$ denote the nonce space and the associated data space, respectively), and a counter generation function $G\colon \mathcal{N} \times [0\,..\,V] \to \mathcal{D}$, where V is a sufficiently large value. While only one set of instantiations of F and G is suggested in SP 800-38C (and if it is chosen the resulting version of CCM is the one used in wireless encryption standard IEEE 802.11), the specification is explicitly modular in the sense that it works with any F and G that meet certain conditions. Amongst others, the conditions listed in [14] imply that for all $N \in \mathcal{N}$ the function $G(N;\cdot)$ is injective and that for all $(N, A, m) \in \mathcal{N} \times \mathcal{A} \times \mathcal{M}$ and $z_0 \ldots z_r = F(N, A, m)$ we have that $z_0 \notin G(N, [0\,..\,V])$. Now, if we fix any nonce N_0 and any associated data string A_0 (e.g., the all-zero string for N_0 and the empty string for A_0) and define the restrictions $F_0\colon \mathcal{M} \to \mathcal{D}^+$; $m \mapsto F(N_0, A_0, m)$ and $G_0\colon [0\,..\,V] \to \mathcal{D}$; $i \mapsto G(N_0, i)$, then the algorithms of the resulting oracle DEM associated with CCM are given in Fig. 13. The message space of CCM0-DEM is $\mathcal{M} = \{0,1\}^*$ and the ciphertext space is $\mathcal{C} = \{0,1\}^{\geq T}$.

$\mathsf{O.Enc}^\pi(k', m)$	$\mathsf{O.Dec}^\pi(k', c)$
00 $z_0 \ldots z_r \leftarrow F_0(m)$	19 If $\lvert c\rvert < T$: Return $\perp$
01 $y_0 \leftarrow \pi(z_0)$	20 Write $\lvert c\rvert$ as $(l-1)\ell + l^* + T$
02 For $i \leftarrow 1$ to r:	21 Split c into $c_1 \ldots c_{l-1} c_l^* t^*$
03 $\quad x_i \leftarrow z_i \oplus y_{i-1}$	22 $c_l \leftarrow c_l^* \parallel 0^{\ell - l^*}$
04 $\quad y_i \leftarrow \pi(x_i)$	23 For $j \leftarrow 1$ to l:
05 $u_0 \leftarrow G_0(0)$	24 $\quad u_j \leftarrow G_0(j)$
06 $v_0 \leftarrow \pi(u_0)$	25 $\quad v_j \leftarrow \pi(u_j)$
07 $t \leftarrow y_r \oplus v_0$	26 $\quad m_j \leftarrow c_j \oplus v_j$
08 $t^* \leftarrow \mathrm{msb}_T(t)$	27 $m_l^* \leftarrow \mathrm{msb}_{l^*}(m_l)$
09 Write $\lvert m\rvert$ as $(l-1)\ell + l^*$	28 $m \leftarrow m_1 \ldots m_{l-1} m_l^*$
10 Split m into $m_1 \ldots m_{l-1} m_l^*$	29 $z_0 \ldots z_r \leftarrow F_0(m)$
11 $m_l \leftarrow m_l^* \parallel 0^{\ell - l^*}$	30 $y_0 \leftarrow \pi(z_0)$
12 For $j \leftarrow 1$ to l:	31 For $i \leftarrow 1$ to r:
13 $\quad u_j \leftarrow G_0(j)$	32 $\quad x_i \leftarrow z_i \oplus y_{i-1}$
14 $\quad v_j \leftarrow \pi(u_j)$	33 $\quad y_i \leftarrow \pi(x_i)$
15 $\quad c_j \leftarrow m_j \oplus v_j$	34 $u_0 \leftarrow G_0(0)$
16 $c_l^* \leftarrow \mathrm{msb}_{l^*}(c_l)$	35 $v_0 \leftarrow \pi(u_0)$
17 $c \leftarrow c_1 \ldots c_{l-1} c_l^* t^*$	36 $t \leftarrow y_r \oplus v_0$
18 Return c	37 If $t^* \neq \mathrm{msb}_T(t)$: Return $\perp$
	38 Return m

Fig. 13. CCM0-DEM. Lines 09 and 20 uniquely identify quantities l and l^* such that $l \in \mathbb{N}^{\geq 1}$ and $0 \leq l^* < \ell$, and $|m| = (l-1)\ell + l^*$ and $|c| = (l-1)\ell + l^* + T$, respectively. Correspondingly, line 10 assumes $|m_1| = \ldots = |m_{l-1}| = \ell$ and $|m_l^*| = l^*$, and line 21 assumes $|c_1| = \ldots = |c_{l-1}| = \ell$ and $|c_l^*| = l^*$ and $|t^*| = T$.

Lemma 3. *CCM0-DEM is ϵ-simulatable with $\epsilon \leq \lfloor L/\ell \rfloor^2 / 2^{\ell-2}$, where L is the maximum message length (in bits).*

Proof. Consider algorithms Fake and Make from Fig. 14. The idea of Fake is to compute the visible ciphertext components on basis of uniformly distributed blockcipher outputs while completely ignoring the blockcipher invocations of CCM's internal CBC-MAC computation (see how line 07 and l-many iterations of line 15 of O.Enc (in Fig. 13) are replaced by lines 00 and 03 of Fake, while lines 01 and 04 of O.Enc have no counterpart). Inspection shows that, given m, algorithm Make finds a minimal partial permutation $\tilde{\pi}$ such that Fake and Make jointly mimic the behaviour of O.Enc (see here how lines 24–27, 30–33, 35–38, 43–46 of Make arrange the entries of $\tilde{\pi}$ such that they are consistent with lines 01, 04, 06/07, 14/15 of O.Enc). In some invocations of the algorithms, the described process might fail (in lines 25/26, 31/32, 36/37, 44/45), namely when partial permutation $\tilde{\pi}$ would become inconsistent. In such cases Make aborts, outputting the empty partial permutation $\tilde{\pi} = \emptyset$.

We next show that the requirements from Definition 11 are met. To see that condition (1) holds, observe that in Make the values y_0, y_i, v_0, and v_j are uniformly distributed and independent of each other at the point they are added to $\mathrm{Rng}(\tilde{\pi})$ in lines 27, 33, 38, 46. To establish the correctness bound of condition (2) we assess the probability that Make aborts. Using a similar analysis as in the proof of Lemma 1 we obtain the following (accumulated)

$\mathsf{Fake}(k', \lvert m \rvert)$	$\mathsf{Make}(st, m)$	
00 $t \leftarrow_U \mathcal{D}$	20 $\tilde{\pi} \leftarrow \emptyset$	34 $u_0 \leftarrow G_0(0)$
01 $t^* \leftarrow \mathrm{msb}_T(t)$	21 Write $\lvert m \rvert$ as $(l-1)\ell + l^*$	35 $v_0 \leftarrow y_r \oplus t$
02 Write $\lvert m \rvert$ as $(l-1)\ell + l^*$	22 Parse st as $(t, c_1, \ldots, c_l)$	36 If $u_0 \in \mathrm{Dom}(\tilde{\pi})$: Abort
03 $c_1, \ldots, c_l \leftarrow_U \mathcal{D}$	23 $z_0 \ldots z_r \leftarrow F_0(m)$	37 If $v_0 \in \mathrm{Rng}(\tilde{\pi})$: Abort
04 $c_l^* \leftarrow \mathrm{msb}_{l^*}(c_l)$	24 $y_0 \leftarrow_U \mathcal{D}$	38 $\tilde{\pi} \leftarrow \tilde{\pi} \cup \{(u_0, v_0)\}$
05 $c \leftarrow c_1 \ldots c_{l-1} c_l^* t^*$	25 If $z_0 \in \mathrm{Dom}(\tilde{\pi})$: Abort	39 Split m into $m_1 \ldots m_{l-1} m_l^*$
06 $st \leftarrow (t, c_1, \ldots, c_l)$	26 If $y_0 \in \mathrm{Rng}(\tilde{\pi})$: Abort	40 $m_l \leftarrow m_l^* \,\Vert\, 0^{\ell - l^*}$
07 Return c, st	27 $\tilde{\pi} \leftarrow \tilde{\pi} \cup \{(z_0, y_0)\}$	41 For $j \leftarrow 1$ to l:
	28 For $i \leftarrow 1$ to r:	42 $\quad u_j \leftarrow G_0(j)$
	29 $\quad x_i \leftarrow z_i \oplus y_{i-1}$	43 $\quad v_j \leftarrow m_j \oplus c_j$
	30 $\quad y_i \leftarrow_U \mathcal{D}$	44 $\quad$ If $u_j \in \mathrm{Dom}(\tilde{\pi})$: Abort
	31 $\quad$ If $x_i \in \mathrm{Dom}(\tilde{\pi})$: Abort	45 $\quad$ If $v_j \in \mathrm{Rng}(\tilde{\pi})$: Abort
	32 $\quad$ If $y_i \in \mathrm{Rng}(\tilde{\pi})$: Abort	46 $\quad \tilde{\pi} \leftarrow \tilde{\pi} \cup \{(u_j, v_j)\}$
	33 $\quad \tilde{\pi} \leftarrow \tilde{\pi} \cup \{(x_i, y_i)\}$	47 Return $\tilde{\pi}$

Fig. 14. Fake and Make for CCM0-DEM. We write 'Abort' as an abbreviation for 'Return $\emptyset$'.

probabilities: The abort conditions in lines 25 and 26 are never met; for lines 31 and 32 the probabilities are $(1 + \ldots + r)/|\mathcal{D}|$ each; by the properties of CCM's functions F_0 and G_0, for lines 36 and 37 the probabilities are $r/|\mathcal{D}|$ and $(r+1)/|\mathcal{D}|$; for line 44 the probability is $lr/|\mathcal{D}|$; finally, for line 45 the probability is $((r+2)+\ldots+(r+l+1))/|\mathcal{D}|$. If we assume reasonable behaviour of function F_0 and let $r = l$, we obtain quantity $4l^2/|\mathcal{D}|$ as an upper bound for the sum of these probabilities. This establishes the claimed bound. Condition (3) is clear. □

5 A Formal Treatment of Our Main Result

We anticipated the main result of this paper in Sect. 3: Any (hybrid) PKE scheme constructed from a KEM and a permutation-driven DEM offers SIM-SO-CCA security in the ideal cipher model, if the KEM provides confidentiality (IND-CCA), the DEM provides authenticity (OT-INT-CTXT), and the DEM is simulatable. Prerequisites like IND-CCA and OT-INT-CTXT on the KEM and DEM, respectively, are standard for proofs of the IND-CCA security of hybrid encryption, so the important finding is that the added constraint of simulatability suffices to lift security to the stronger notion of SO security.[3]

We discussed an informal version of our result in Sect. 3.2. Recall from the included proof sketch that an important subgoal was bounding the probability of the ideal cipher being evaluated on input a key established by the KEM before a corresponding Open query is posed. (If the cipher is evaluated earlier, the partial permutation found by Fake and Make cannot be smoothly embedded into it any more.) In the following we argue that without putting further restrictions on

[3] We note that a typical proof of IND-CCA security of hybrid PKE requires the DEM to also offer some kind of confidentiality (e.g., OT-IND-CCA). A corresponding notion appears only implicitly in our theorem statement, as it follows from the DEM's simulatability (in the ideal cipher model).

the KEM, bounding this probability to any small value is in general impossible. Indeed, assume for a moment a KEM where K.Enc, before outputting a key k and a ciphertext c, evaluates the blockcipher used by D.Enc on input key k and a value d_0, where the latter is any fixed element $d_0 \in \mathcal{D}$ in the cipher's domain, and assume K.Enc completely ignores the result. Even though this blockcipher evaluation is completely pointless and should not affect security of the overall design, for such a KEM our arguments would not work. Below, in the formal version of our theorem statement, we correspondingly restrict the set of considered KEMs to those that do not evaluate the blockcipher at all. This admittedly is a limitation of our result, but we believe it is a mild one. Indeed, all practical KEMs we are aware of do not (internally) invoke blockcipher operations at all. This holds in particular for Hashed ElGamal, PSEC-KEM, Cramer-Shoup KEM, and RSA-KEM. In the following theorem statement, if E is a blockcipher, we say a KEM is E-independent if no KEM algorithm evaluates E^+ or E^-.

We proceed with the statement and proof of our main theorem.

Theorem 2. *Let* DEM *be a* $(\mathcal{K}, \mathcal{D})$*-permutation-driven DEM with corresponding oracle DEM* oDEM *and blockcipher* E*. Let* KEM *denote an* E*-independent KEM for the key space of the DEM. Let* PKE *denote the hybrid PKE scheme obtained when instantiating Construction 1 in Fig. 3 with* KEM *and* DEM*.*

Let DEM *be* $(\tau, q_d, \epsilon_{ctxt})$*-OT-INT-CTXT secure and* KEM *be* $(\tau, q_d, \epsilon_{cca})$*-IND-CCA secure.*

If oDEM *is* ϵ_{sim}*-simulatable, then* PKE *is* $(\tau, \tau', q_d, q_{ic}, \epsilon)$*-SIM-SO-CCA secure where* ϵ *can be upper-bounded by*

$$\epsilon(n) \leq n \cdot \left(3 \cdot \epsilon_{cca} + \epsilon_{ctxt} + \epsilon_{sim} + 2 \cdot \frac{n + q_{ic} + q_d}{|\mathcal{K}|}\right)$$

and E *is modeled as an ideal cipher.*

See Sect. 3.2 for a proof sketch including the high-level ideas. We proceed with a detailed proof of Theorem 2.

Proof. For the list of n challenge ciphertexts $(\langle c_{1,1}, c_{1,2}\rangle, \ldots, \langle c_{n,1}, c_{n,2}\rangle)$ and $\mathcal{J} \subseteq [n]$ let $C_{\mathcal{J},1}$ denote the set $\{c_{j,1} \mid j \in \mathcal{J}\}$. For the keys $(k_i, k'_i) \leftarrow k''_i$ output by the n iterations of K.Enc, and $\mathcal{J} \subseteq [n]$ let $K_{\mathcal{J}}$ denote the set $\{k_j \mid j \in \mathcal{J}\}$ of blockcipher keys k_i for $i \in \mathcal{J}$. For the family of partial permutations $(E_k)_{k \in \mathcal{K}}$ maintained by $\mathcal{S}$ to implement ideal cipher E, let $\text{supp}(E) := \{k \in \mathcal{K} \mid E_k \neq \emptyset\}$ denote the set of keys $k \in \mathcal{K}$ where partial permutation E_k is not empty.

Fix any SIM-SO-CCA adversary $\mathcal{A}$. We define a simulator $(\mathcal{S}_1, \mathcal{S}_2)$ by giving its pseudocode in Fig. 15. Simulator $\mathcal{S}_1$ consists of lines 00 – 03, $\mathcal{S}_2$ consists of lines 04 – 11. Their code is enhanced by bookkeeping and abort events, while the explicit invocation of $\mathcal{S}_1$, $\mathcal{S}_2$ and their input/output behaviour is merged into the ideal game. Instructions in grey boxes are performed by the ideal game.

We show that $\mathcal{S}$, when run in the ideal game, can simulate the real game for $\mathcal{A}$. To this end we proceed in a sequence of experiments tracing $\mathcal{A}$'s advantage of distinguishing two consecutive games. The sequence interpolates between the

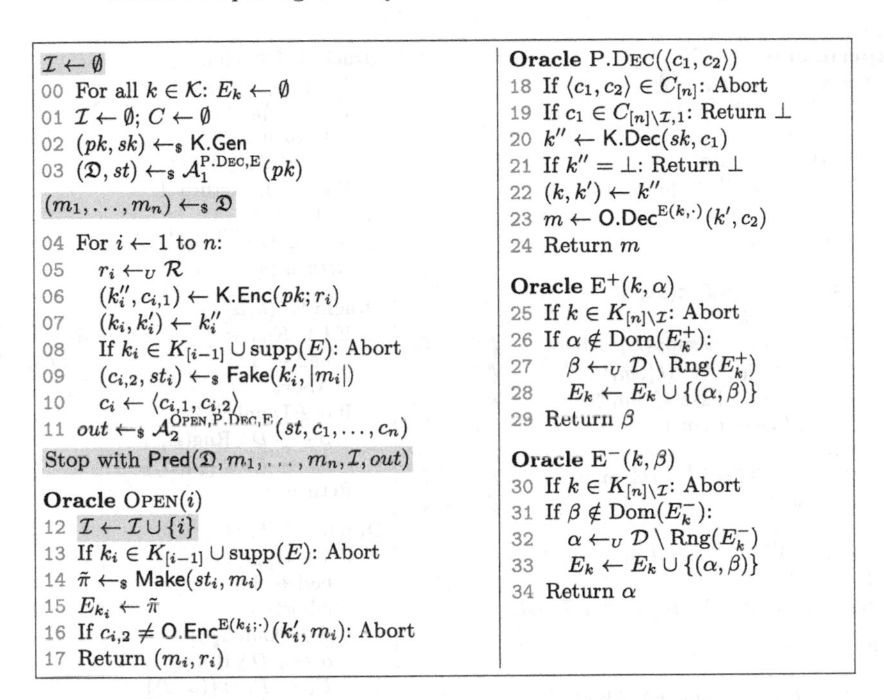

Fig. 15. Proposed simulator $\mathcal{S} = (\mathcal{S}_1, \mathcal{S}_2)$ inlined into the i-SO-CCA experiment. $\mathcal{S}_1$ in lines 00 – 03, $\mathcal{S}_2$ given in lines 04 – 11. Instructions in grey boxes are executed by the ideal experiment. The whole code corresponds to the last game G_6 in our proof. For $\mathcal{J} \subseteq [n]$ we denote $C_{\mathcal{J},1} := \{c_{j,1} \mid j \in \mathcal{J}\}$ and $K_{\mathcal{J}} := \{k_j \mid j \in \mathcal{J}\}$. Further, we denote $\mathrm{supp}(E) := \{k \in \mathcal{K} \mid E_k \neq \emptyset\}$.

real game (G_0 = r-SO-CCA, cf. Fig. 5) and a simulated real game (G_6, cf. Fig. 15) provided by the simulator $\mathcal{S}$ inlined into the ideal game.

The whole sequence of experiments is given in Fig. 16. Lines ending with a range of experiments G_i – G_j (resp. G_i if $j = i$) are only executed when an experiment within the range is run.

Without loss of generality we assume that $\mathcal{A}$ does not make the same opening query twice. We proceed with detailed descriptions of the experiments.

Game G_0. The r-SO-CCA game as given in Fig. 5.

Game G_1. Lines 28 and 29 are added: Any decryption query of the form $\langle c_1, c_2 \rangle$ is answered with $\perp$ if $c_1 \in C_{[n]\setminus\mathcal{I},1}$. That is, there exists $i \in [n]$ such that $c_1 = c_{i,1}$ and $\mathcal{A}$ did not query $\textsc{Open}(i)$.

Claim. There exists an adversary $\mathcal{B}_{cca}$ that $(\tau, q_d, \epsilon_{cca})$-breaks the IND-CCA security of KEM and an adversary $\mathcal{B}_{ctxt}$ that $(\tau, q_d, \epsilon_{ctxt})$-breaks the OT-INT-CTXT security of DEM with $|\Pr[\mathsf{G}_0 \Rightarrow 1] - \Pr[\mathsf{G}_1 \Rightarrow 1]| \leq n \cdot (\epsilon_{cca} + \epsilon_{ctxt})$.

Proof. Games G_0 and G_1 proceed identically, until $\mathcal{A}$ submits a ciphertext $\langle c_1, c_2 \rangle$ to decryption where $c_1 \in C_{[n]\setminus\mathcal{I}}$ and $\mathsf{P.Dec}(sk, \langle c_1, c_2 \rangle) \neq \perp$. We fix some $i \in [n]$

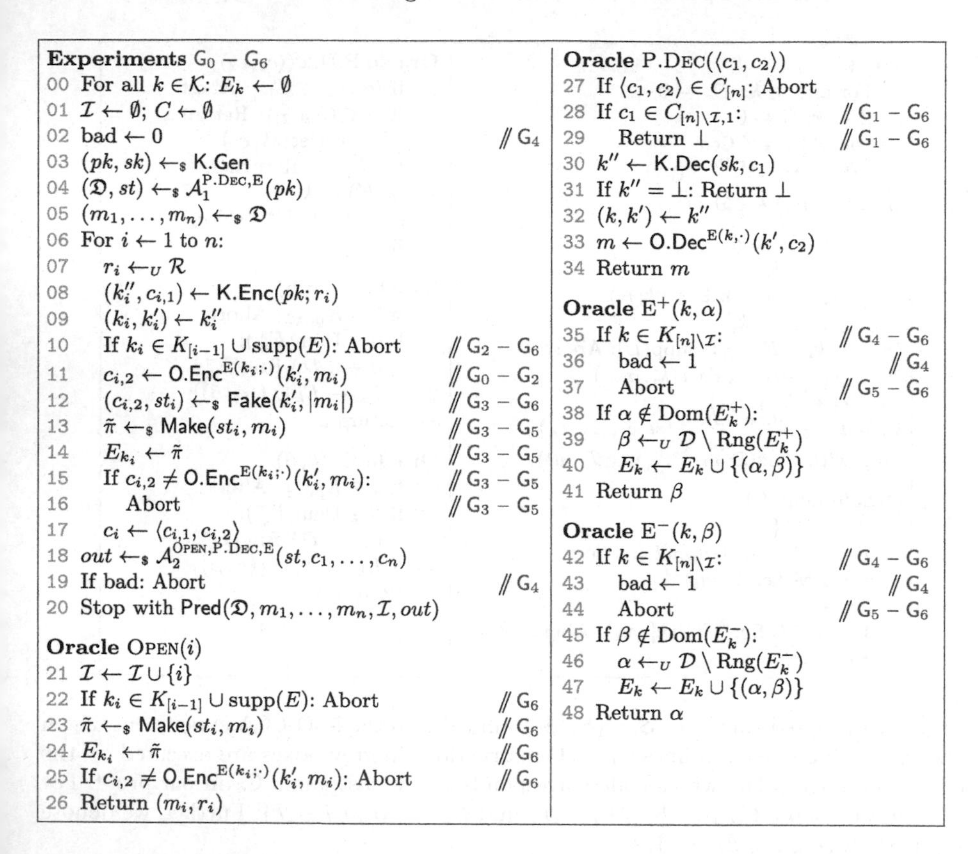

Fig. 16. Experiments $\mathsf{G}_0 - \mathsf{G}_6$ used in the proof of Theorem 2. We write 'Abort' as an abbreviation for 'Stop with 0'.

and analyse the probability that $\mathcal{A}$ submits a ciphertext $\langle c_1, c_2 \rangle$ where $c_1 \in C_{\{i\}\setminus\mathcal{I}}$ and $\mathsf{P.Dec}(sk, \langle c_1, c_2 \rangle) \neq \perp$ we denote this event by '$\langle c_{i,1}, c_2 \rangle \nrightarrow \perp$'.

At first, we replace k''_i as output by the ith invocation of K.Enc with a uniformly random key. We lose an additional summand of ϵ_{cca} in the bound on $\Pr[\langle c_{i,1}, c_2 \rangle \nrightarrow \perp]$ as shown by the following reduction run by adversary $\mathcal{B}_{cca}$: It uses its decapsulation oracle to answer decryption queries from $\mathcal{A}_1$. Receiving (c^*, k^*_b), $\mathcal{B}_{cca}$ parses $(k_b, k'_b) \leftarrow k^*_b$ and computes all ciphertexts faithfully except for $c_i \leftarrow \langle c^*, \mathsf{O.Enc}^{\mathrm{E}(k_b;\cdot)}(k'_b, m_i)\rangle$. Decryption queries $\langle c_1, c_2 \rangle$ by $\mathcal{A}_2$ are answered employing the decapsulation oracle for $c_1 \neq c^*$ and using key k^*_b otherwise.

The reduction perfectly simulates G_1 until $\mathcal{A}$ queries $\textsc{Open}(i)$ which the reduction cannot answer. Yet, to bound the probability of event '$\langle c_{i,1}, c_2 \rangle \nrightarrow \perp$' happening, it suffices to make sure that the reduction 'works' as long as the event can occur. Observe that '$\langle c_{i,1}, c_2 \rangle \nrightarrow \perp$' cannot happen after query $\textsc{Open}(i)$.

We now show how to break the OT-INT-CTXT security of the DEM if '$\langle c_{i,1}, c_2 \rangle \nrightarrow \perp$' happens. We construct $\mathcal{B}_{ctxt}$. The reduction performed by $\mathcal{B}_{ctxt}$ runs K.Gen and starts $\mathcal{A}_1(pk)$. Decryption queries are answered using sk. Once

$\mathcal{A}_1$ outputs $\mathfrak{D}$, $\mathcal{B}_{ctxt}$ samples messages but submits m_i to the D.Enc oracle of its OT-INT-CTXT game to obtain a data encapsulation $c_2^* \leftarrow \mathsf{D.Enc}(k''_\$, m^*)$ under a random key $k''_\$$. Additionally, $\mathcal{B}_{ctxt}$ runs K.Enc to obtain (k, c_1^*) and sends $(c_1, \ldots, c_{i-1}, \langle c_1^*, c_2^* \rangle, \ldots, c_n)$ to $\mathcal{A}$. Adversary $\mathcal{B}_{ctxt}$ answers all further decryption queries on its own, unless the ciphertext is of the form $\langle c_1^*, c_2 \rangle$ where it submits c_2 to the decapsulation oracle of the OT-INT-CTXT experiment. If it receives $\perp$, it returns $\perp$ to $\mathcal{A}_2$.

Clearly, $\mathcal{B}_{ctxt}$ wins the OT-INT-CTXT game when $\mathcal{A}$ submits a ciphertext that causes '$\langle c_{i,1}, c_2 \rangle \nrightarrow \perp$' to happen.

We obtain $\Pr[\langle c_{i,1}, c_2 \rangle \nrightarrow \perp] \leq \epsilon_{cca} + \epsilon_{ctxt}$. The claim follows from the union-bound over all $i \in [n]$. □

The next game hop ensures that (if it is not aborted) the ith invocation of the oracle data encapsulation, i.e., $\mathsf{O.Enc}^{E(k_i;\cdot)}$, has access to an empty partial permutation E_{k_i}. This is a preparational step to ensure that later, when O.Enc is replaced with Fake and Make, the partial permutation output by Make can be embedded into E_{k_i}.

Game G_2. Line 10 is added. That is, G_2 aborts if the ith iteration of O.Enc would have oracle access to a non-empty permutation $E(k_i; \cdot)$.[4]

Claim. There exists an adversary $\mathcal{B}_{cca}$ that $(\tau, q_d, \epsilon_{cca})$-breaks the IND-CCA security of KEM with $|\Pr[\mathsf{G}_1 \Rightarrow 1] - \Pr[\mathsf{G}_2 \Rightarrow 1]| \leq n \cdot (\epsilon_{cca} + (n + q_{ic} + q_d) / |\mathcal{K}|)$.

Proof. We bound $\Pr[k_i \in K_{[i-1]} \cup \mathrm{supp}(E)]$ for fixed $i \in [n]$. Again, we use KEM's IND-CCA security to replace k''_i output by the ith invocation of K.Enc with a uniform key. We construct adversary $\mathcal{B}_{cca}$. It receives pk and starts $\mathcal{A}_1(pk)$. Decryption queries are answered using the decapsulation oracle. When $\mathcal{A}_1$ halts, $\mathcal{B}_{cca}$ requests its IND-CCA challenge (c^*, k_b^*) — let $(k_b, k'_b) \leftarrow k_b^*$ — and runs the For loop 07. In the ith iteration $\mathcal{B}_{cca}$ halts and returns 1 iff $k_b \in K_{[i-1]} \cup \mathrm{supp}(E)$. Clearly, the reduction is perfect until $\mathcal{B}_{cca}$ halts and we have $|\Pr[k_i \in K_{[i-1]} \cup \mathrm{supp}(E)] - \Pr[k_\$ \in K_{[i-1]} \cup \mathrm{supp}(E)]| \leq \epsilon_{cca}$ where $k_\$ \leftarrow_\$ \mathcal{K}$.

Note that each decryption query or query to the ideal cipher oracles adds at most one element to $\mathrm{supp}(E)$, hence $|K_{[i-1]} \cup \mathrm{supp}(E)| \leq n + q_{ic} + q_d$. Thus, we obtain $\Pr[k_\$ \in K_{[i-1]} \cup \mathrm{supp}(E)] \leq (n + q_{ic} + q_d) / |\mathcal{K}|$ and $\Pr[k_i \in K_{[i-1]} \cup \mathrm{supp}(E)] \leq \epsilon_{cca} + (n + q_{ic} + q_d) / (|\mathcal{K}|)$. The claim follows from the union-bound over $i \in [n]$. □

Game G_3. The faithful data encapsulation is replaced by algorithms Fake and Make. More precisely, for each iteration of the For loop (line 06) we replace the invocation $\mathsf{O.Dec}^{E(k_i;\cdot)}(k'_i, m_i)$ (line 11) with running $\mathsf{Fake}(k'_i, |m_i|)$ and

[4] As of now, in the ith iteration of the For loop, we have $K_{[i-1]} \subseteq \mathrm{supp}(E)$ as the invocation of $\mathsf{O.Enc}^{E(k_i;\cdot)}$ adds elements to E_{k_i}. Later, in game G_6, we do not invoke code that (implicitly) adds elements to E_{k_i} and rely on set $K_{[i-1]}$ to detect collisions amongst the (blockcipher) keys.

$\mathsf{Make}(m_i)$ back to back (lines 12,13). E_{k_i} gets assigned partial permutation $\tilde{\pi}$ as output by Make (cf. line 14) and a check is performed whether E_{k_i} has been programmed 'consistently'; if not, experiment G_3 aborts (lines 15, 16).

Claim. $|\Pr[\mathsf{G}_2 \Rightarrow 1] - \Pr[\mathsf{G}_3 \Rightarrow 1]| \leq n \cdot \epsilon_{sim}$.

Proof. Fix $i \in [n]$. Due to the modifications in games G_1 and G_2 partial permutation E_{k_i} is empty at the time of invoking O.Enc. Hence, once we replace O.Enc by Fake and Make, the partial permutation as output by Make can always be embedded into E_{k_i}. Particularly, partial permutations E_{k_i} accessed by O.Enc and $\tilde{\pi}$ output by Make are identically distributed when randomly extended to a full permutation on $\mathcal{D}$. We conclude that the abort in line 16 happens with probability at most ϵ_{sim} as oDEM is ϵ_{sim}-simulatable. The claim follows from the union-bound over all $i \in [n]$. □

Recall from the proof outline that, eventually, Make shall be run as part of the OPEN procedure. The upcoming modifications ensure that partial permutation E_{k_i} remains empty until OPEN(i) is queried.

Game G_4. Line 02 is added to initialise a flag 'bad' as 0. Lines (35, 36) are added to the E^+ oracle, lines (42, 43) are added to the E^- oracle and line 19 is added. That is, if E^+ or E^- is queried on (k_i, z) for any z and $i \notin \mathcal{I}$, 'bad' is set to 1 and the game aborts after the execution of $\mathcal{A}_2$ (in line 19).

Claim. There exists an adversary $\mathcal{B}_{cca}$ that $(\tau, q_d, \epsilon_{cca})$-breaks the IND-CCA security of KEM with $|\Pr[\mathsf{G}_3 \Rightarrow 1] - \Pr[\mathsf{G}_4 \Rightarrow 1]| \leq n \cdot (\epsilon_{cca} + (q_{ic} + q_d)/|\mathcal{K}|)$.

Proof. Fix $i \in [n]$ and let '$k \in K_{\{i\}\setminus\mathcal{I}}$' denote the event that E^+ or E^- is queried on (k, z) where $k \in K_{\{i\}\setminus\mathcal{I}}$. (That is, the condition in lines 35 or 42 holds, even for $K_{\{i\}\setminus\mathcal{I}}$). Again, we replace key k_i'' output in the ith invocation of K.Enc with a uniform key $(k_\$, k_\$') \leftarrow k_\$''$. The reduction run by $\mathcal{B}_{cca}$ proceeds as in the proof to bridge G_0 and G_1. Here, $\mathcal{B}_{cca}$ halts after $\mathcal{A}_2$'s execution and outputs 1 iff bad $= 1$. Clearly $|\Pr[k \in K_{\{i\}\setminus\mathcal{I}}] - \Pr[k \in \{k_\$\} \setminus \mathcal{I}]| \leq \epsilon_{cca}$ for uniform $k_\$ \leftarrow_\$ \mathcal{K}$.

The reduction is perfect unless $\mathcal{A}_2$ queries OPEN(i) which cannot be answered. Note that after query OPEN(i), 'bad' cannot be set to 1 as $K_{\{i\}\setminus\mathcal{I}} = \emptyset$. Similarly to before, it suffices to guarantee the correctness of the simulation as long as the abort in line 19 can potentially happen.

Note that $k_\$$ is uniform from $\mathcal{A}$'s view: Only ciphertext $\langle c_{i,1}, c_{i,2}\rangle$ might contain information on $k_\$$ but $c_{i,1}$ is independent of $k_\$$ as it is sampled after K.Enc output $c_{i,1}$ and data encapsulation $c_{i,2}$ is independent of $k_\$$ as we run $\mathsf{Fake}(k_i', m_i)$ to compute $c_{i,2}$. Thus, $\Pr[k \in \{k_\$\} \setminus \mathcal{I}] \leq (q_{ic} + q_d)/|\mathcal{K}|$ and collecting the probabilities and applying the union-bound gives the desired bound. □

Game G_5. Lines 37 and 44 are added. Instead of aborting after the execution of $\mathcal{A}_2$ if bad $= 1$, game G_5 aborts as soon as bad (as introduced in game G_4) is set to 1. Now obsolete lines 02, 19, 36 and 43 are removed for clarity.

Claim. $\Pr[\mathsf{G}_4 \Rightarrow 1] = \Pr[\mathsf{G}_5 \Rightarrow 1]$.

Proof. The claim follows from observing that game G_5 aborts in lines 37 or 44 if and only if game G_4 aborts in line 19. □

Game G_6. An abort event is added in line 22. The invocation of Make, the embedding of a partial permutation and the consistency check are moved from the For loop in lines 13 – 16 to the OPEN oracle (lines 23 – 24).

Claim. $\Pr[\mathsf{G}_5 \Rightarrow 1] = \Pr[\mathsf{G}_6 \Rightarrow 1]$.

Proof. The abort event in line 22 is solely added for clarity but never met: Assume that line 22 would cause an abort, then the condition in line 10, or lines 35/42 would have been satisfied earlier. Hence, for all $i \in [n]$: a) in game G_5 partial permutation $E_{k_i} \leftarrow \tilde{\pi}$ as output by Make in line 13 is information-theoretically hidden from $\mathcal{A}$ until it queries OPEN and b) in game G_6 partial permutation E_{k_i} remains empty until $\mathcal{A}$ queries OPEN. Thus, embedding partial permutation $\tilde{\pi}$ into E_{k_i} always succeeds. Further, moving the invocation of Make, the embedding and checking to the OPEN oracle is completely oblivious to $\mathcal{A}$. □

We observe that the code as given in game G_6 in Fig. 16 matches the code of the simulator as given in Fig. 15.

The claim of Theorem 2 follows by collecting the probabilities. □

6 Conclusion

The most promising practical approach to public key encryption is through the hybrid KEM/DEM paradigm. Suitable KEMs include Hashed ElGamal, PSEC-KEM, Cramer-Shoup KEM, and RSA-KEM, and candidates for the DEM part are readily derived from the highly efficient encryption modes CTR, CBC, CCM standardised by NIST (to reach CCA security, the former two should be enhanced with a MAC, e.g., CMAC or HMAC). The last NIST standardised mode of operation, GCM, is covered in the full version of this paper [23], too. To compress the contribution of this paper into a single line: We effectively show that if any of these KEMs is combined with any of these DEMs in the sense of hybrid encryption, then the obtained PKE scheme offers a strong notion of selective opening security. Our result holds in the (heuristic) ideal cipher model for the underlying blockcipher. We thus recommend using modern blockciphers like AES as they come closest to meeting such requirements.

Acknowledgements. We thank the reviewers for their helpful feedback. Felix Heuer was funded by the German Research Foundation (DFG) as part of the priority program 1736 Big Data: Scalable Cryptography. Bertram Poettering was supported by ERC Project ERCC (FP7/615074).

References

1. Bellare, M., Dowsley, R., Waters, B., Yilek, S.: Standard security does not imply security against selective-opening. Cryptology ePrint Archive, Report 2011/581 (2011). http://eprint.iacr.org/2011/581
2. Bellare, M., Dowsley, R., Waters, B., Yilek, S.: Standard security does not imply security against selective-opening. In: Pointcheval, D., Johansson, T. (eds.) EUROCRYPT 2012. LNCS, vol. 7237, pp. 645–662. Springer, Heidelberg (2012). doi:10.1007/978-3-642-29011-4_38
3. Bellare, M., Hofheinz, D., Yilek, S.: Possibility and impossibility results for encryption and commitment secure under selective opening. In: Joux, A. (ed.) EUROCRYPT 2009. LNCS, vol. 5479, pp. 1–35. Springer, Heidelberg (2009). doi:10.1007/978-3-642-01001-9_1
4. Bellare, M., Namprempre, C.: Authenticated encryption: relations among notions and analysis of the generic composition paradigm. In: Okamoto, T. (ed.) ASIACRYPT 2000. LNCS, vol. 1976, pp. 531–545. Springer, Heidelberg (2000). doi:10.1007/3-540-44448-3_41
5. Bellare, M., Yilek, S.: Encryption schemes secure under selective opening attack. Cryptology ePrint Archive, Report 2009/101 (2009). http://eprint.iacr.org/2009/101
6. Böhl, F., Hofheinz, D., Kraschewski, D.: On definitions of selective opening security. In: Fischlin, M., Buchmann, J., Manulis, M. (eds.) PKC 2012. LNCS, vol. 7293, pp. 522–539. Springer, Heidelberg (2012). doi:10.1007/978-3-642-30057-8_31
7. Canetti, R., Dwork, C., Naor, M., Ostrovsky, R.: Deniable encryption. In: Kaliski, B.S. (ed.) CRYPTO 1997. LNCS, vol. 1294, pp. 90–104. Springer, Heidelberg (1997). doi:10.1007/BFb0052229
8. Canetti, R., Feige, U., Goldreich, O., Naor, M.: Adaptively secure multi-party computation. In: 28th ACM STOC, pp. 639–648. ACM Press, May 1996
9. Coron, J.-S., Patarin, J., Seurin, Y.: The random oracle model and the ideal cipher model are equivalent. In: Wagner, D. (ed.) CRYPTO 2008. LNCS, vol. 5157, pp. 1–20. Springer, Heidelberg (2008). doi:10.1007/978-3-540-85174-5_1
10. Cramer, R., Shoup, V.: Design and analysis of practical public-key encryption schemes secure against adaptive chosen ciphertext attack. SIAM J. Comput. **33**(1), 167–226 (2003)
11. Dachman-Soled, D.: On minimal assumptions for sender-deniable public key encryption. In: Krawczyk, H. (ed.) PKC 2014. LNCS, vol. 8383, pp. 574–591. Springer, Heidelberg (2014). doi:10.1007/978-3-642-54631-0_33
12. Dwork, C., Naor, M., Reingold, O., Stockmeyer, L.J.: Magic functions. In: 40th FOCS, pp. 523–534. IEEE Computer Society Press, October 1999
13. Dworkin, M.J.: Spp. 800–38A: Recommendation for block cipher modes of operation: Methods and techniques. Technical report, National Institute of Standards & Technology, Gaithersburg, MD, United States (2001)
14. Dworkin, M.J.: Spp. 800–38C: Recommendation for block cipher modes of operation: The CCM mode for authentication and confidentiality. Technical report, National Institute of Standards & Technology, Gaithersburg, MD, United States (2007)
15. Dworkin, M.J.: Spp. 800–38D: Recommendation for block cipher modes of operation: Galois/Counter Mode (GCM) and GMAC. Technical report, National Institute of Standards & Technology, Gaithersburg, MD, United States (2007)

16. Dworkin, M.J.: Addendum to Spp. 800–38A: Recommendation for block cipher modes of operation: Three variants of ciphertext stealing for CBC mode. Technical report, National Institute of Standards & Technology, Gaithersburg, MD, United States (2010)
17. Even, S., Mansour, Y.: A construction of a cipher from a single pseudorandom permutation. In: Imai, H., Rivest, R.L., Matsumoto, T. (eds.) ASIACRYPT 1991. LNCS, vol. 739, pp. 210–224. Springer, Heidelberg (1993). doi:10.1007/3-540-57332-1_17
18. Fehr, S., Hofheinz, D., Kiltz, E., Wee, H.: Encryption schemes secure against chosen-ciphertext selective opening attacks. In: Gilbert, H. (ed.) EUROCRYPT 2010. LNCS, vol. 6110, pp. 381–402. Springer, Heidelberg (2010). doi:10.1007/978-3-642-13190-5_20
19. Fuchsbauer, G., Heuer, F., Kiltz, E., Pietrzak, K.: Standard security does imply security against selective opening for markov distributions. In: Kushilevitz, E., Malkin, T. (eds.) TCC 2016. LNCS, vol. 9562, pp. 282–305. Springer, Heidelberg (2016). doi:10.1007/978-3-662-49096-9_12
20. Hazay, C., Patra, A., Warinschi, B.: Selective opening security for receivers. In: Iwata, T., Cheon, J.H. (eds.) ASIACRYPT 2015. LNCS, vol. 9452, pp. 443–469. Springer, Heidelberg (2015). doi:10.1007/978-3-662-48797-6_19
21. Hemenway, B., Ostrovsky, R., Rosen, A.: Non-committing encryption from Q-hiding. In: Dodis, Y., Nielsen, J.B. (eds.) TCC 2015. LNCS, vol. 9014, pp. 591–608. Springer, Heidelberg (2015). doi:10.1007/978-3-662-46494-6_24
22. Heuer, F., Jager, T., Kiltz, E., Schäge, S.: On the selective opening security of practical public-key encryption schemes. In: Katz, J. (ed.) PKC 2015. LNCS, vol. 9020, pp. 27–51. Springer, Heidelberg (2015). doi:10.1007/978-3-662-46447-2_2
23. Heuer, F., Poettering, B.: Selective opening security from simulatable data encapsulation. Cryptology ePrint Archive, Report 2016/845 (2016). http://eprint.iacr.org/2016/845
24. Hofheinz, D., Jager, T., Rupp, A.: Public-key encryption with simulation-based selective-opening security and compact ciphertexts. Cryptology ePrint Archive, Report 2016/180 (2016). http://eprint.iacr.org/2016/180
25. Hofheinz, D., Rao, V., Wichs, D.: Standard security does not imply indistinguishability under selective opening. Cryptology ePrint Archive, Report 2015/792 (2015). http://eprint.iacr.org/2015/792
26. Hofheinz, D., Rupp, A.: Standard versus selective opening security: separation and equivalence results. In: Lindell, Y. (ed.) TCC 2014. LNCS, vol. 8349, pp. 591–615. Springer, Heidelberg (2014). doi:10.1007/978-3-642-54242-8_25
27. Kilian, J., Rogaway, P.: How to protect DES against exhaustive key search (an analysis of DESX). J. Cryptology **14**(1), 17–35 (2001)
28. Liu, S., Paterson, K.G.: Simulation-based selective opening CCA security for PKE from key encapsulation mechanisms. In: Katz, J. (ed.) PKC 2015. LNCS, vol. 9020, pp. 3–26. Springer, Heidelberg (2015). doi:10.1007/978-3-662-46447-2_1
29. Peikert, C., Waters, B.: Lossy trapdoor functions and their applications. In: Ladner, R.E., Dwork, C. (eds.) 40th ACM STOC, pp. 187–196. ACM Press, May 2008
30. Rogaway, P.: Authenticated-encryption with associated-data. In: Atluri, V. (ed.) ACM CCS 02, pp. 98–107. ACM Press, November 2002
31. Wee, H.: Zero knowledge in the random oracle model, revisited. In: Matsui, M. (ed.) ASIACRYPT 2009. LNCS, vol. 5912, pp. 417–434. Springer, Heidelberg (2009). doi:10.1007/978-3-642-10366-7_25

Selective-Opening Security in the Presence of Randomness Failures

Viet Tung Hoang[1(✉)], Jonathan Katz[2], Adam O'Neill[3], and Mohammad Zaheri[3]

[1] Department of Computer Science, Florida State University, Tallahassee, USA
tvhoang@cs.fsu.edu

[2] Department of Computer Science, University of Maryland, College Park, USA
jkatz@cs.umd.edu

[3] Department of Computer Science, Georgetown University, Washington, D.C., USA
adam@cs.georgetown.edu, mz394@georgetown.edu

Abstract. We initiate the study of public-key encryption (PKE) secure against selective-opening attacks (SOA) in the presence of *randomness failures*, i.e., when the sender may (inadvertently) use low-quality randomness. In the SOA setting, an adversary can adaptively corrupt senders; this notion is natural to consider in tandem with randomness failures since an adversary may target senders by multiple means.

Concretely, we first treat SOA security of *nonce-based PKE*. After formulating an appropriate definition of SOA-secure nonce-based PKE, we provide efficient constructions in the non-programmable random-oracle model, based on lossy trapdoor functions.

We then lift our notion of security to the setting of "hedged" PKE, which ensures security as long as the sender's seed, message, and nonce *jointly* have high entropy. This unifies the notions and strengthens the protection that nonce-based PKE provides against randomness failures even in the non-SOA setting. We lift our definitions and constructions of SOA-secure nonce-based PKE to the hedged setting as well.

1 Introduction

Imagine that an adversary wants to gain access to encrypted communication that various senders are transmitting to a receiver. There are various ways to go about doing this. One is to try to subvert the random-number generator used by the senders. Another is to break-in to the senders' machines, possibly in an adaptive fashion. Encryption schemes resisting the first sort of attack have been studied in the context of *security under randomness failures* [3,7,10,18,23] while resistance to the second sort of attack corresponds to the notion of *security against selective-opening attacks* (SOA) [5,9,11,14–16].[1] However, as far as we are aware, these notions have so far only been considered separately. We initiate the study of

[1] There are two forms of SOA security, called coin-revealing (corresponding to sender corruption) and key-revealing (corresponding to receiver corruption). This paper concerns the first one.

J.H. Cheon and T. Takagi (Eds.): ASIACRYPT 2016, Part II, LNCS 10032, pp. 278–306, 2016.
DOI: 10.1007/978-3-662-53890-6_10

SOA-secure encryption in the presence of randomness failures, providing new definitions and constructions achieving these definitions in the public-key setting.

There are currently three main approaches in the literature to dealing with randomness failures for PKE: (1) *deterministic* PKE [2], which does not use randomness at all but guarantees security only if plaintexts have high entropy, (2) *hedged* PKE, which is randomized and guarantees security as long as plaintexts and the randomness *jointly* have high entropy, and (3) the recently introduced notion of *nonce-based* PKE by Bellare and Tackmann (BT) [10], where each sender uses a uniform seed[2] in addition to a nonce, and security is guaranteed if either the seed is secret and the nonces are unique, or the seed is revealed and the nonces have high entropy. Hedged PKE and nonce-based PKE are incomparable and are useful in different scenarios, and part of our contribution is to unify them into a single primitive. We start by adding consideration of SOA security to nonce-based PKE. We then lift the resulting notions to the setting of hedged PKE (which subsumes deterministic PKE) as well, thereby adding consideration of SOA to a unified primitive with the guarantees of both nonce-based and hedged PKE.

1.1 Our Results

SELECTIVE-OPENING SECURITY FOR NONCE-BASED PKE. As explained above, the first notion we consider for protecting against randomness failures is *nonce-based* PKE, recently introduced by Bellare and Tackmann [10]. For consistency with the definitions of SOA security we introduce for later notions (where new technical challenges arise), we formulate an indistinguishability-based (rather than simulation-based) definition, which we call N-SO-CPA, along the lines of the indistinguishability-based definition of SOA security for standard PKE [9]. Under our definition, the adversary can (i) learn the seeds of some senders, (ii) choose the nonces for all the other senders, as long as nonces of each individual sender do not repeat. Then, *after* seeing the ciphertexts, the adversary can adaptively corrupt some senders to learn their messages *together with* seeds and nonces. The definition asks that the adversary cannot distinguish between the plaintexts of the uncorrupted senders and a *resampling* of these plaintexts conditioned on the revealed plaintexts.

The next question is whether N-SO-CPA security is achievable. Throughout our work, we focus on constructions in the so-called non-programmable random-oracle model (NPROM) [20]. Intuitively, this means that in a security proof, the constructed adversary must honestly answer (i.e., cannot program) the random oracle queries of the assumed adversary. The NPROM is arguably closer to the standard (random oracle devoid) model than the programmable random oracle model (PROM), since real-world hash functions are not programmable. In this

[2] The idea is that because a seed is chosen infrequently, it can be generated using high-quality randomness.

model, we give an efficient construction of N-SO-CPA-secure[3] nonce-based PKE based on any lossy trapdoor function [21]. The idea is to modify the nonce-based PKE scheme of Bellare and Tackmann, which encrypts a message m using public-key pk, seed xk, and nonce N by encrypting m using any standard (randomized) PKE scheme with public key pk and "synthetic" coins derived from a hash of (xk, N, m). Here, we use a *specific* randomized encryption scheme based on any lossy trapdoor function. The security proof of the resulting scheme, which we call NE1, relies on switching to the lossy key-generation algorithm and then using the random oracle to argue that the adversary's choice of which senders to corrupt must be independent of the plaintexts.

SOA+HEDGED SECURITY FOR NONCE-BASED PKE. Unlike nonce-based PKE, *hedged* PKE [3] guarantees security as long as the message and randomness used by the sender *jointly* have high entropy. Indeed, viewing the sender's seed and nonce together as the sender's randomness, nonce-based PKE as defined in [10] *lacks* such a guarantee. To get the best of both worlds, we would like to add such a guarantee to nonce-based PKE. This strengthens the protection provided against randomness failures even in the absence of SOA; however, sticking with the main theme of this work, we aim to achieve it in the SOA setting as well. This leads to a definition that we call HN-SO-CPA, which incorporates both hedged and SOA security into the existing notion of nonce-based PKE.

Modeling SOA in the hedged setting is technically challenging. Indeed, Bellare et al. [4] recently showed that a simulation-based notion of SOA security for deterministic PKE (which is a special case of hedged PKE) is impossible to achieve. They also noted that a natural indistinguishability-based definition is (for different reasons) trivially impossible to achieve, and left open the problem of defining a meaningful (yet achievable) definition. To that end, we introduce a novel "comparison-based" definition of SOA for nonce-based PKE, inspired by the comparison-based definition of SOA for deterministic PKE [2,6] combined with the indistinguishability-based definition of SOA for standard PKE [9]. Roughly, the definition requires that the adversary cannot predict any function of all the plaintexts (i.e., including those of the uncorrupted senders) with much better probability than by computing the same function on a resampling of all the plaintexts conditioned on the revealed plaintexts. For technical reasons, HN-SO-CPA does not protect partial information about the messages depending on the public key, so we still require N-SO-CPA to hold in addition.

We provide two approaches for achieving HN-SO-CPA + N-SO-CPA-secure nonce-based PKE. The first is a generic transform inspired by the "randomized-then-deterministic" transform of [3] in the setting of hedged security. Namely, we propose a "Nonce-then-Deterministic" (NtD) transform in which one obtains a new nonce-based PKE scheme by composing an underlying nonce-based PKE scheme with a deterministic PKE scheme. We require that the underlying deterministic PKE scheme meet a corresponding special case of the HN-SO-CPA definition that we call D-SO-CPA, and achieve it via a scheme DE1 in the NPROM.

[3] In the main body of the paper we treat both CPA and CCA security. For simplicity, we do not discuss CCA here.

Interestingly, the scheme DE1 is exactly the recent construction of Bellare and Hoang [7], except that they assume the hash function is UCE-secure [8] and achieve standard security (not SOA). Again, the analysis is quite involved and deals with subtleties neither present in SOA for randomized PKE nor in prior work on deterministic PKE. Alternatively, we show that the scheme NE1 directly achieves both HN-SO-CPA and N-SO-CPA in the NPROM.

Separation results. Finally, to justify our developing new schemes in the setting of selective-opening security in the presence of randomness failures rather than using existing ones, we show that the N-SO-CPA and D-SO-CPA are not implied by the standard notions (non-SOA) of nonce-based PKE [10] and D-PKE [2], respectively. Our counter-examples rely on the recent result of Hofheinz, Rao, and Wichs (HRW) [15] that separates IND-CCA security from SOA security for randomized PKE. We also show that N-SO-CPA does not imply HN-SO-CPA for nonce-based PKE, meaning the hedged security does strengthen the notion considered for nonce-based PKE in [10].

Open question. We leave obtaining standard-model (versus NPROM) schemes achieving our notions as an open question. Note that our NtD transform is in the standard model, so if we had standard-model instantiations of the underlying primitives we would get a standard-model HN-SO-CPA + N-SO-CPA-secure nonce-based PKE as well.

1.2 Organization

In contrast to the order in which we explained the results above, in the main body of the paper we first present our results on SOA security for deterministic PKE, then move to our results on SOA security for nonce-based PKE, and then finally present our results on hedged security for SOA-secure nonce-based PKE. This is because the results for deterministic PKE constitute the technical core of our work, and form a basis for the results that follow.

2 Preliminaries

Notation and conventions. An adversary is an algorithm or tuple of algorithms. All algorithms may be randomized and are required to be efficient unless otherwise indicated; we let PPT stand for "probabilistic, polynomial time." For an algorithm A we denote by $x \leftarrow\!\!\$\, A(\cdots)$ the experiment that runs A on the elided inputs with uniformly random coins and assigns the output to x, and $x \leftarrow\!\!\$\, A(\cdots; r)$ to denote the same experiment, but under the coins r instead of randomly chosen ones. If A is deterministic we denote this instead by $x \leftarrow A(\cdots)$. We let $[A(\cdots)]$ denote the set of all possible outputs of A when run on the elided arguments. If S is a finite set then $s \leftarrow\!\!\$\, S$ denotes choosing a uniformly random element from S and assigning it to s. We denote by $\Pr[\, P(x) \;:\; \ldots]$ the probability that some predicate P is true of x after executing the elided experiment.

Let $\mathbb{N}$ denote the set of all non-negative integers. For any $n \in \mathbb{N}$ we denote by $[n]$ the set $\{1, \ldots, n\}$. For a vector $\mathbf{x}$, we denote by $|\mathbf{x}|$ its length (number of components) and by $\mathbf{x}[i]$ its i-th component. For a vector $\mathbf{x}$ of length n and any $I \subseteq [n]$, we denote by $\mathbf{x}[I]$ the vector of length $|I|$ such that $\mathbf{x}[I] = (\mathbf{x}[i])_{i \in I}$. For a string X, we let $|X|$ denote its length. For any integer $1 \leq i \leq j \leq |X|$, we write $X[i]$ to denote the ith bit of X, and $X[i, j]$ the substring from the i-th to the j-th bit (inclusive) of X.

PUBLIC-KEY ENCRYPTION. A *public-key encryption scheme* PKE with message space Msg is a tuple of algorithms $(\mathsf{Kg}, \mathsf{Enc}, \mathsf{Dec})$. The key-generation algorithm Kg on input 1^k outputs a public key pk and secret key sk. The encryption algorithm Enc on inputs a public key pk and message $m \in \mathsf{Msg}(k)$ outputs a ciphertext c. The deterministic decryption algorithm Dec on inputs a secret key sk and ciphertext c outputs a message m or $\perp$. We require that for all $(pk, sk) \in [\mathsf{Kg}(1^k)]$ and all $m \in \mathsf{Msg}(1^k)$, the probability that $\mathsf{Dec}(sk, (\mathsf{Enc}(pk, m)) = m$ is 1. We say PKE is *deterministic* if Enc is deterministic.

LOSSY TRAPDOOR FUNCTION. A *lossy trapdoor function* [21] with domain LDom and range LRng is a tuple of algorithms $\mathsf{LT} = (\mathsf{LT.IKg}, \mathsf{LT.LKg}, \mathsf{LT.Eval}, \mathsf{LT.Inv})$ that work as follows. Algorithm $\mathsf{LT.IKg}$ on input a unary encoding of the security parameter 1^k outputs an "injective" evaluation key ek and matching trapdoor td. Algorithm $\mathsf{LT.LKg}$ on input 1^k outputs a "lossy" evaluation key lk. Algorithm $\mathsf{LT.Eval}$ on inputs an (either injective or lossy) evaluation key ek and $x \in \mathsf{LDom}(k)$ outputs $y \in \mathsf{LRng}(1^k)$. Algorithm $\mathsf{LT.Inv}$ on inputs a trapdoor td and a $y' \in \mathsf{LRng}(k)$ outputs $x' \in \mathsf{LDom}(k)$. We require the following properties.

Correctness: For all $k \in \mathbb{N}$ and any $(ek, td) \in [\mathsf{LT.IKg}(1^k)]$, it holds that $\mathsf{Inv}(td, \mathsf{LT.Eval}(ek, x)) = x$ for every $x \in \mathsf{LDom}(k)$.

Key indistinguishability: For every distinguisher D, the advantage $\mathbf{Adv}^{\mathrm{ltdf}}_{\mathsf{LT},D}(k) = \Pr\left[\, D(ek) \Rightarrow 1 \;:\; (ek, td) \leftarrow\!\!\$\; \mathsf{LT.IKg}(1^k)\,\right] - \Pr\left[\, D(lk) \Rightarrow 1 \;:\; lk\,\right] \leftarrow\!\!\$ \; \mathsf{LT.LKg}(1^k)$ is negligible.

Lossiness: The size of the co-domain of $\mathsf{LT.Eval}(lk, \cdot)$ is at most $|\mathsf{LRng}(k)|/2^{\tau(k)}$ for all $k \in \mathbb{N}$ and all $lk \in [\mathsf{LT.LKg}(1^k)]$. We call τ the *lossiness* of LT.

If the function $\mathsf{LT.Eval}(ek, \cdot)$ is a permutation for any $k \in \mathbb{N}$ and any $(ek, td) \in [\mathsf{LT.IKg}(1^k)]$ then we call LT a *lossy trapdoor permutation*. Both RSA and Rabin are lossy trapdoor permutations under appropriate assumptions [19,22].

MESSAGE SAMPLERS. A *message sampler* $\mathcal{M}$ is a PPT algorithm that takes as input 1^k and a string $param \in \{0,1\}^*$, and outputs a vector $\mathbf{m}$ of messages and a vector $\mathbf{a}$ of the same length. Each $\mathbf{a}[i]$ is the auxiliary information that an adversary gains in addition to $\mathbf{m}[i]$, if it breaks into the machine of the sender of $\mathbf{m}[i]$. For example, if each $\mathbf{m}[i]$ is a signature of some string $\mathbf{x}[i]$, then the adversary may be able to obtain even $\mathbf{x}[i]$ in its break-in. We require that $\mathcal{M}$ be associated with functions $v(\cdot)$ and $n(\cdot)$ such that for any $param \in \{0,1\}^*$, for any $k \in \mathbb{N}$, and any $\mathbf{m} \in [\mathcal{M}(1^k, param)]$, we have $|\mathbf{m}| = v(k)$ and $|\mathbf{m}[i]| = n(k)$, for every $i \leq |\mathbf{m}|$.

A message sampler $\mathcal{M}$ is (μ, d)-*entropic* if

- For any $k \in \mathbb{N}$, any $I \subseteq \{1, \ldots, v(k)\}$ such that $|I| \leq d$, any $param \in \{0,1\}^*$, and $(\mathbf{m}, \mathbf{a}) \leftarrow\!\!\$\; \mathcal{M}(1^k, param)$, conditioning on messages $\mathbf{m}[I]$ and

their auxiliary information $\mathbf{a}[I]$ and *param*, each other message $\mathbf{m}[j]$ (with $j \in \{1, \dots, v(k)\} \setminus I$) must have conditional min-entropy at least μ. Note that here $(\mathbf{m}, \mathbf{a})$ is sampled independent of the set I.
- Messages $\mathbf{m}[1], \dots, \mathbf{m}[|\mathbf{m}|]$ must be distinct, for any $param \in \{0,1\}^*$ and any $\mathbf{m} \in [\mathcal{M}(1^k, param)]$.

In this definition d can be ∞, which corresponds to a message sampler in which the conditional distribution of each message, given *param* and all other messages and their corresponding auxiliary information, has at least μ bits of min-entropy.

Resampling. Following [9], let $\mathsf{Coins}[k]$ be the set of coins for $\mathcal{M}(1^k, \cdot)$, and $\mathsf{Coins}[k, \mathbf{m}^*, \mathbf{a}^*, I, param] = \{\omega \in \mathsf{Coins}[k] \mid \mathbf{m}'[I] = \mathbf{m}^* \text{ and } \mathbf{a}'[I] = \mathbf{a}^*, \text{where } (\mathbf{m}', \mathbf{a}') \leftarrow \mathcal{M}(1^k, param; \omega)\}$. Let $\mathsf{Resamp}_{\mathcal{M}}(1^k, I, \mathbf{m}^*, \mathbf{a}^*, param)$ be the algorithm that first samples $r \leftarrow_{\$} \mathsf{Coins}[k, \mathbf{m}^*, \mathbf{a}^*, I, param]$, then runs $(\mathbf{m}', \mathbf{a}') \leftarrow \mathcal{M}(1^k, param; r)$, and then returns $\mathbf{m}'$. (Note that $\mathsf{Resamp}_{\mathcal{M}}$ may run in exponential time.) A *resampling algorithm* of $\mathcal{M}$ is an algorithm Rsmp such that $\mathsf{Rsmp}(1^k, I, \mathbf{m}^*, \mathbf{a}^*, param)$ and $\mathsf{Resamp}_{\mathcal{M}}(1^k, I, \mathbf{m}^*, \mathbf{a}^*, param)$ are identically distributed.[4] A message sampler $\mathcal{M}$ is *fully resamplable* if it admits a PPT resampling algorithm.

Partial resampling. We also introduce a new notion of "partial resampling." Let δ be a function and let $\mathsf{Resamp}_{\mathcal{M},\delta}(1^k, I, \mathbf{m}^*, \mathbf{a}^*, param)$ be the algorithm that samples $r \leftarrow_{\$} \mathsf{Coins}[k, \mathbf{m}^*, \mathbf{a}^*, I, param]$, runs $(\mathbf{m}', \mathbf{a}') \leftarrow \mathcal{M}(1^k, param; r)$, and then returns $\delta(\mathbf{m}', param)$. We say that $\mathcal{M}$ is *δ-partially resamplable* if there is a PT algorithm Rsmp such that $\mathsf{Rsmp}(1^k, I, \mathbf{m}^*, \mathbf{a}^*, param)$ is identically distributed as $\mathsf{Resamp}_{\mathcal{M},\delta}(1^k, I, \mathbf{m}^*, \mathbf{a}^*, param)$. Such an algorithm Rsmp is called a *δ-partial resampling algorithm* of $\mathcal{M}$. If a message sampler is already fully resamplable then it's δ-partially resamplable for any PT function δ.

3 Selective-Opening Security for D-PKE

3.1 Security Notions

Bellare, Dowsley, and Keelveedhi [4] were the first to consider selective-opening security of deterministic PKE (D-PKE). They propose a "simulation-based" semantic security notion, but then show that this definition is unachievable in both the standard model and the non-programmable random-oracle model (NPROM), even if the messages are uniform and independent. To address this, we introduce an alternative, "comparison-based" semantic-security notion that generalizes the original PRIV definition for D-PKE of Bellare, Boldyreva, and O'Neill [2]. In particular, our notion follows the IND-SO-CPA notion of Bellare, Hofheinz, and Yilek (BHY) [9] in the sense that we compare what partial

[4] Here for simplicity, we only consider $\mathcal{M}$ and Rsmp such that the distributions of $\mathsf{Rsmp}(1^k, I, \mathbf{m}^*, \mathbf{a}^*, param)$ and $\mathsf{Resamp}_{\mathcal{M}}(1^k, I, \mathbf{m}^*, \mathbf{a}^*, param)$ are identical. Following [9], one might also consider $\mathcal{M}$ and Rsmp such that the two distributions above are statistically close.

information the adversary learns from the unopened messages, versus messages resampled from the same conditional distribution.

D-SO-CPA1 SECURITY. Let $\mathsf{PKE} = (\mathsf{Kg}, \mathsf{Enc}, \mathsf{Dec})$ be a D-PKE scheme. To a message sampler $\mathcal{M}$ and an adversary $A = (A.\mathrm{pg}, A.\mathrm{cor}, A.\mathrm{g}, A.\mathrm{f})$, we associate the experiment in Fig. 1 for every $k \in \mathbb{N}$. We say that DE is D-SO-CPA1 secure for a class $\mathscr{M}$ of resamplable message samplers and a class $\mathscr{A}$ of adversaries if for every $\mathcal{M} \in \mathscr{M}$ and any $A \in \mathscr{A}$,

$$\mathbf{Adv}^{\text{d-so-cpa1}}_{\mathsf{DE},A,\mathcal{M}}(\cdot)$$
$$= \Pr\left[\text{D-CPA1-REAL}^{A,\mathcal{M}}_{\mathsf{DE}}(\cdot) \Rightarrow 1\right] - \Pr\left[\text{D-CPA1-IDEAL}^{A,\mathcal{M}}_{\mathsf{DE}}(\cdot) \Rightarrow 1\right]$$

is negligible. In these games, the adversary A.pg first creates some parameter *param* to feed the message sampler $\mathcal{M}$. Note that A.pg is not given the public key, and thus messages $\mathbf{m}_1$ created by $\mathcal{M}$ are independent of the public key, a necessary restriction of D-PKE pointed out by Bellare et al. [2]. Next, adversary A.cor will be given both the public key and the ciphertexts $\mathbf{c}$, and decides which set I of indices that it'd like to open $\mathbf{c}[I]$. It then passes its state to adversary A.g. The latter is also given $(\mathbf{m}_1[I], \mathbf{a}[I])$ and has to output some partial information ω of the message vector $\mathbf{m}_1$.

Game D-CPA1-REAL$^{A,\mathcal{M}}_{\mathsf{DE}}$ returns 1 if the string ω above matches the output of $A.\mathrm{f}(\mathbf{m}_1, param)$ which is the partial information of interest to the adversary. On the other hand, game D-CPA1-IDEAL$^{A,\mathcal{M}}_{\mathsf{DE}}$ returns 1 if ω is matches the output of $A.\mathrm{f}(\mathbf{m}_0, param)$, where $\mathbf{m}_0$ is the resampled message vector by $\mathsf{Resamp}_{\mathcal{M}}(1^k, \mathbf{m}_1[I], \mathbf{a}[I], I, param)$. Note that in both games, A.f is not given the public key *pk*, otherwise it can encrypt the messages it receives and output the resulting ciphertexts, while A.g outputs $\mathbf{c}$. Again, this issue is pointed out in [2]: since encryption is deterministic, the ciphertexts themselves are some partial information about the messages. D-PKE can only hope to protect partial information of $\mathbf{m}$ that is independent of *pk*, and A.f is therefore stripped of access to *pk*.

DISCUSSION. For selective-opening attacks against a D-PKE scheme in which an adversary can open d messages, it is clear that the message sampler must be (μ, d)-entropic, where $2^{-\mu(\cdot)}$ is a negligible function, for any meaningful privacy to be achievable. For convenience of discussion, let's say that a scheme is D-SO-CPA1[d] secure if it's D-SO-CPA1 secure for all (μ, d)-entropic, fully resamplable message samplers and all PT adversaries that open at most d ciphertexts, for any μ such that $2^{-\mu(\cdot)}$ is a negligible function. (The resamplability restriction is dropped for $d = 0$.) The D-SO-CPA1[0] security corresponds to the PRIV notion of Bellare et al. [2].[5]

We note that it is unclear if D-SO-CPA1[∞] security implies the classic PRIV security: the latter doesn't allow opening, but it can handle a broader class of

[5] A technical difference is that, to be consistent with [4], we require the "partial information" to be an efficiently computable function of the messages. This formulation can be shown equivalent to a definition in the style of [2] up to a difference of one in the size of the message vectors output by $\mathcal{M}$, following [6, Appendix A].

Game D-CPA1-REAL$_{\mathsf{DE}}^{A,\mathcal{M}}(k)$	**Game** D-CPA1-IDEAL$_{\mathsf{DE}}^{A,\mathcal{M}}(k)$
$param \leftarrow_{\$} A.\mathrm{pg}(1^k)$	$param \leftarrow_{\$} A.\mathrm{pg}(1^k)$; $(pk, sk) \leftarrow_{\$} \mathsf{Kg}(1^k)$
$(pk, sk) \leftarrow_{\$} \mathsf{Kg}(1^k)$	$(\mathbf{m}_1, \mathbf{a}) \leftarrow_{\$} \mathcal{M}(1^k, param)$
$(\mathbf{m}_1, \mathbf{a}) \leftarrow_{\$} \mathcal{M}(1^k, param)$	For $i = 1$ to $\lvert\mathbf{m}\rvert$ do
For $i = 1$ to $\lvert\mathbf{m}\rvert$ do	$\quad \mathbf{c}[i] \leftarrow \mathsf{Enc}(pk, \mathbf{m}_1[i])$
$\quad \mathbf{c}[i] \leftarrow \mathsf{Enc}(pk, \mathbf{m}_1[i])$	$(state, I) \leftarrow_{\$} A.\mathrm{cor}(pk, \mathbf{c}, param)$
$(state, I) \leftarrow_{\$} A.\mathrm{cor}(pk, \mathbf{c}, param)$	$\mathbf{m}_0 \leftarrow_{\$} \mathsf{Resamp}_{\mathcal{M}}(1^k, \mathbf{m}_1[I], \mathbf{a}[I], I, param)$
$\omega \leftarrow_{\$} A.\mathrm{g}(state, \mathbf{m}_1[I], \mathbf{a}[I])$	$\omega \leftarrow_{\$} A.\mathrm{g}(state, \mathbf{m}_1[I], \mathbf{a}[I])$
Return $(\omega = A.\mathrm{f}(\mathbf{m}_1, param))$	Return $(\omega = A.\mathrm{f}(\mathbf{m}_0, param))$

Fig. 1. Games to define D-SO-CPA1 security.

message samplers. Our goal is to find D-PKE schemes that offer D-SO-CPA1$[d]$ security for any value of d, including the important special cases $d = 0$ (PRIV security) and $d = \infty$ (unbounded opening).

SEPARATION. In the full version, we show that the standard PRIV notion of D-PKE doesn't imply D-SO-CPA1. Our construction relies on the recent result of Hofheinz, Rao, and Wichs [15] that separates the standard IND-CPA notion and IND-SO-CPA of randomized PKE. Specifically, we build a contrived D-PKE scheme that is PRIV-secure in the standard model, but subject to the following D-SO-CPA1 attack. The message sampler picks a string $s \leftarrow_{\$} \{0,1\}^{\ell(k)}$ and then secret-share it to $v(k)$ shares $\mathbf{x}[1], \ldots, \mathbf{x}[v(k)]$ such that any $t(k)$ shares reveal no information about the secret s. Let $\mathbf{m}[i] \leftarrow \mathbf{x}[i] \parallel \mathbf{u}[i]$ for every $i \in \{1, \ldots, v(k)\}$, where $\mathbf{u}[i] \leftarrow_{\$} \{0,1\}^{2\ell(k)}$. Since s is uniform, any $t + 1$ shares $\mathbf{x}[i]$ are uniform and independent. Thus, this message sampler is $(3\ell, t)$-entropic. We show that it is also efficiently resamplable. Surprisingly, there is an efficient SOA adversary $(A.\mathrm{cor}, A.\mathrm{g})$ that opens just t ciphertexts and can recover all strings $\mathbf{x}[i]$. Next, $A.\mathrm{g}$ outputs $\mathbf{x}[1] \oplus \cdots \oplus \mathbf{x}[v(k)]$, and $A.\mathrm{f}$ outputs the checksum of the first ℓ bits of the given messages. The adversary A thus wins with advantage $1 - 2^{\ell(k)}$.

D-SO-CPA2 SECURITY. The D-SO-CPA1 security notion only guarantees to protect messages that are fully resamplable. The D-SO-CPA2 notion strengthens that protection, requiring privacy of $\delta(\mathbf{m}, param)$ for any entropic message sampler $\mathcal{M}$ and any δ such that $\mathcal{M}$ is δ-partially resamplable. In Sect. 5, we'll see a concrete use of this extra protection, where (i) we have a sampler $\mathcal{M}$ that is not fully resamplable, but (ii) each message itself is a ciphertext, and there's a function δ such that the plaintexts underneath $\mathbf{m}$ are $\delta(\mathbf{m}, param)$ and $\mathcal{M}$ is δ-partially resamplable. Formally, let

$$\begin{aligned}&\mathbf{Adv}_{\mathsf{DE},A,\mathcal{M},\delta}^{\text{d-so-cpa2}}(\cdot)\\ &= \Pr\left[\text{D-CPA2-REAL}_{\mathsf{DE}}^{A,\mathcal{M},\delta}(\cdot) \Rightarrow 1\right] - \Pr\left[\text{D-CPA2-IDEAL}_{\mathsf{DE}}^{A,\mathcal{M},\delta}(\cdot) \Rightarrow 1\right],\end{aligned}$$

where games D-CPA2-REAL$_{\mathsf{DE}}^{A,\mathcal{M},\delta}$ and D-CPA2-IDEAL$_{\mathsf{DE}}^{A,\mathcal{M},\delta}$ are defined in Fig. 2. In these games, adversary $A.\mathrm{f}$ is given either $\delta(\mathbf{m}_1)$ in the real game,

Game D-CPA2-REAL$_{\mathsf{DE}}^{A,\mathcal{M},\delta}(k)$	Game D-CPA2-IDEAL$_{\mathsf{DE}}^{A,\mathcal{M},\delta}(k)$
$param \leftarrow\!\!\$\ A.\mathrm{pg}(1^k)$	$param \leftarrow\!\!\$\ A.\mathrm{pg}(1^k)$; $(pk, sk) \leftarrow\!\!\$\ \mathsf{Kg}(1^k)$
$(pk, sk) \leftarrow\!\!\$\ \mathsf{Kg}(1^k)$	$(\mathbf{m}, \mathbf{a}) \leftarrow\!\!\$\ \mathcal{M}(1^k, param)$
$(\mathbf{m}, \mathbf{a}) \leftarrow\!\!\$\ \mathcal{M}(1^k, param)$	For $i = 1$ to $\lvert\mathbf{m}\rvert$ do
For $i = 1$ to $\lvert\mathbf{m}\rvert$ do	$\quad \mathbf{c}[i] \leftarrow \mathsf{Enc}(pk, \mathbf{m}[i])$
$\quad \mathbf{c}[i] \leftarrow \mathsf{Enc}(pk, \mathbf{m}[i])$	$(state, I) \leftarrow\!\!\$\ A.\mathrm{cor}(pk, \mathbf{c}, param)$
$(state, I) \leftarrow\!\!\$\ A.\mathrm{cor}(pk, \mathbf{c}, param)$	$z \leftarrow\!\!\$\ \mathsf{Resamp}_{\mathcal{M},\delta}(1^k, \mathbf{m}[I], \mathbf{a}[I], I, param)$
$\omega \leftarrow\!\!\$\ A.\mathrm{g}(state, \mathbf{m}[I], \mathbf{a}[I])$	$\omega \leftarrow\!\!\$\ A.\mathrm{g}(state, \mathbf{m}[I], \mathbf{a}[I])$
Return $(\omega = A.\mathrm{f}(\delta(\mathbf{m}), param))$	Return $(\omega = A.\mathrm{f}(z, param))$

Fig. 2. Games to define D-SO-CPA2 security.

or the output of $\mathsf{Resamp}_{\mathcal{M},\delta}(1^k, \mathbf{m}_1[I], \mathbf{a}[I], I, param)$ in the ideal game. We say that DE is D-SO-CPA2 secure if $\mathbf{Adv}^{\text{d-so-cpa2}}_{\mathsf{DE},A,\mathcal{M},\delta}(\cdot)$ is negligible for any (μ, d)-entropic message sampler $\mathcal{M}$ such that $2^{-\mu}$ is a negligible function, any PT adversary A that opens at most d ciphertexts, and any PT functions δ such that $\mathcal{M}$ is δ-partially resamplable.

Weak Equivalence. Clearly, the D-SO-CPA2 notion implies D-SO-CPA1: the latter is the special case of the former for fully resamplable samplers, and for a specific function $\delta(\mathbf{m}, param)$ that simply returns $\mathbf{m}$. Below, we'll show that if we just restrict to fully resamplable samplers, the D-SO-CPA1 notion actually implies D-SO-CPA2. This is expected, because on an entropic, fully resamplable $\mathcal{M}$, both notions promise to protect any partial information of $\mathbf{m}$ that is independent of the public key.

Proposition 1. Let $\mathcal{M}$ be a fully resamplable sampler, and let δ be a PT function. Then for any adversary A, there is an adversary B such that

$$\mathbf{Adv}^{\text{d-so-cpa2}}_{\mathsf{DE},A,\mathcal{M},\delta}(\cdot) \le \mathbf{Adv}^{\text{d-so-cpa1}}_{\mathsf{DE},B,\mathcal{M}}(\cdot) \ .$$

The adversary B opens as many ciphertexts as A, and its running time is about that of A plus the time to run δ.

Proof. Let B be the adversary that is identical to A, but B.f behaves as follows. When it's given a vector $\mathbf{m}$ and parameter $param$, it'll run $z \leftarrow \delta(\mathbf{m}, param)$ and then outputs $A.\mathrm{f}(z, param)$. Then $\mathbf{Adv}^{\text{d-so-cpa1}}_{\mathsf{DE},B,\mathcal{M}}(\cdot) = \mathbf{Adv}^{\text{d-so-cpa2}}_{\mathsf{DE},A,\mathcal{M},\delta}(\cdot)$. □

In the remainder of the paper, we'll have 6 other notions. Any notion xxx considers an arbitrary message sampler $\mathcal{M}$ with a function δ such that $\mathcal{M}$ is δ-partially resamplable. One can consider a variant xxx1 of xxx, in which the message sampler is fully resamplable and only the specific function $\delta(\mathbf{m}, param) = \mathbf{m}$ is considered, and then establish a weak equivalence between xxx1 and xxx. However, it will lead to a proliferation of 12 definitions. We therefore choose to present just the stronger notion xxx.

$\mathsf{DE.Kg}(1^k)$	$\mathsf{DE.Enc}(pk, m)$	$\mathsf{DE.Dec}(sk, c)$
$(ek, td) \leftarrow_{\$} \mathsf{LT.IKg}(1^k)$	$(hk, ek) \leftarrow pk$	$(hk, td) \leftarrow sk$
$hk \leftarrow_{\$} \{0,1\}^k$	$r \leftarrow H(hk \parallel 0 \parallel m, \mathsf{LT.il}(k))$	$(trap, y) \leftarrow c$
Return $((hk, ek), (hk, td))$	$trap \leftarrow \mathsf{LT.Eval}(ek, r)$	$r \leftarrow \mathsf{LT.Inv}(td, trap)$
	$y \leftarrow H(hk \parallel 1 \parallel r, \lvert m \rvert) \oplus m$	Return $H(hk \parallel 1 \parallel r, \lvert y \rvert) \oplus y$
	Return $(trap, y)$	

Fig. 3. D-PKE scheme $\mathsf{DE1}[H, \mathsf{LT}]$.

CCA EXTENSION. To add a CCA flavor to D-SO-CPA2, a notion which we call D-SO-CCA, one would allow adversaries A.cor and A.g oracle access to $\mathsf{Dec}(sk, \cdot)$ with the restriction that they are forbidden from querying a ciphertext in the given $\mathbf{c}$ to this oracle. Let D-CCA-REAL and D-CCA-IDEAL be the corresponding experiments, and define

$$\mathbf{Adv}^{\text{d-so-cca}}_{\mathsf{DE},A,\mathcal{M},\delta}(\cdot) = \Pr\left[\text{D-CCA-REAL}^{A,\mathcal{M},\delta}_{\mathsf{DE}}(\cdot) \Rightarrow 1\right] - \Pr\left[\text{D-CCA-IDEAL}^{A,\mathcal{M},\delta}_{\mathsf{DE}}(\cdot) \Rightarrow 1\right].$$

We say that DE is D-SO-CCA secure if $\mathbf{Adv}^{\text{d-so-cca}}_{\mathsf{DE},A,\mathcal{M},\delta}(\cdot)$ is negligible for any (μ, d)-entropic message sampler $\mathcal{M}$ such that $2^{-\mu}$ is a negligible function, any PT adversary A that opens at most d ciphertexts, and any PT functions δ such that $\mathcal{M}$ is δ-partially resamplable.

3.2 Achieving D-SO-CPA2 Security

While the simulation-based definition of Bellare et al. [4] is impossible to achieve even in the non-programmable random-oracle model (NPROM), we show that it is possible to build a D-SO-CPA2 secure scheme in the NPROM. A close variant of our scheme is shown to be PRIV-secure in the standard model [7]. Our scheme can handle messages of any length, and is highly efficient: the asymmetric cost is fixed and thus the amortized cost is about as cheap as a symmetric encryption. It's also highly practical on short messages. The only public-key primitive that it uses is a lossy trapdoor function [21], which has practical instantiations, e.g., both Rabin and RSA are lossy [19,22].

ACHIEVING D-SO-CPA2 SECURITY. To handle arbitrary-length messages, we use a hash function H of arbitrary input and output length. On input $(x, \ell) \in \{0,1\}^* \times \mathbb{N}$, the hash returns $y = H(x, \ell) \in \{0,1\}^\ell$. Our scheme $\mathsf{DE1}[H, \mathsf{LT}]$ is shown in Fig. 3, where LT is a lossy trapdoor function with domain $\{0,1\}^{\mathsf{LT.il}}$. Theorem 2 below shows that $\mathsf{DE1}$ is D-SO-CPA2 secure in the NPROM. The proof is in the full version. We stress that for (μ, ∞)-entropic message samplers, our scheme allows the adversary to open as many ciphertexts as it wishes.

Theorem 2. Let LT be a lossy trapdoor function with lossiness τ. Let $\mathcal{M}$ be a (μ, d)-entropic message sampler, and let δ be a function such that $\mathcal{M}$ is δ-partially resamplable. Let $\mathsf{DE1}[H, \mathsf{LT}]$ be as above. In the NPROM, for any adversary A opening at most d ciphertexts, there is an adversary D such that

$$\mathbf{Adv}^{\text{d-so-cpa2}}_{\mathsf{DE1}[H,\mathsf{LT}],A,\mathcal{M},\delta}(k) \le \frac{4q(k)}{2^k} + \frac{4q(k)v(k)}{2^{\mu(k)}} + \frac{v(k)(v(k)+4q(k))}{2^{\tau(k)}} + 2\mathbf{Adv}^{\mathrm{ltdf}}_{\mathsf{LT},D}(k),$$

where $q(k)$ is the total number of random-oracle queries of A and $\mathcal{M}$, and $v(k)$ is the number of messages that $\mathcal{M}$ produces. The running time of D is about that of A plus the time to run δ and an efficient δ-partial resampling algorithm of $\mathcal{M}$ plus the time to run $\mathsf{DE1}[H, \mathsf{LT}]$ to encrypt $\mathcal{M}$'s messages. Adversary D makes at most q random-oracle queries.

Proof ideas. Let $\mathrm{RO}_1, \mathrm{RO}_2, \mathrm{RO}_3$, and RO_4 denote the oracle interface of $(A.\mathrm{pg}, \mathcal{M}), A.\mathrm{cor}, A.\mathrm{g}$, and $A.\mathrm{f}$ respectively. Initially, each interface simply calls RO. In game-based proofs of ROM-based D-PKE constructions, one often considers the event that $A.\mathrm{pg}$ or $\mathcal{M}$ queries $(hk \,\|\, x, \ell)$ to RO_1, and then let the interface lies, instead of calling $\mathrm{RO}(hk \,\|\, x, \ell)$. This allows the coins $\mathbf{r}[i] \leftarrow \mathrm{RO}(hk \,\|\, 0 \,\|\, \mathbf{m}[i], \mathsf{LT.il}(k))$ to be independent of the messages $\mathbf{m}$. The discrepancy due to the lying is tiny, since the chance that $A.\mathrm{pg}$ or $\mathcal{M}$ can make such a query is at most $q(k)/2^k$. However, in the SOA setting, this strategy creates the following subtlety. For the resampling algorithm to behave correctly, one has to give it access to RO_1. Yet the adversary $A.\mathrm{cor}$ can embed some information of hk in I, and therefore it's well possible that the resampling algorithm queries $\mathrm{RO}_1(hk \,\|\, \cdot, \cdot)$. This issue is unique to SOA security of D-PKE: prior papers of SOA security for randomized PKE never have to deal with this. While getting around the subtlety above is not too difficult, it shows that a rigorous proof for Theorem 2 is not as simple as one might expect.

Suppose that $A.\mathrm{pg}$ and $\mathcal{M}$ never query $\mathrm{RO}_1(hk \,\|\, \cdot, \cdot)$. The first step in the proof is to move from an injective key ek of LT to a lossy key lk. Next, recall that the adversary $A.\mathrm{cor}$ is given $\mathsf{LT.Eval}(lk, \mathbf{r}[i])$. Since each synthetic coin $\mathbf{r}[i]$ is uniformly random and LT has lossiness τ, in the view of $A.\mathrm{cor}$, each $\mathbf{r}[i]$ has min-entropy at least $\tau(k)$. Suppose that $A.\mathrm{cor}$ doesn't make any query in $\{hk \,\|\, 0 \,\|\, \mathbf{m}[i], hk \,\|\, 1 \,\|\, \mathbf{r}[i] \mid 1 \le i \le |\mathbf{m}|\}$; this happens with probability at least $1 - q(k)v(k)/2^{\mu(k)} - q(k)v(k)/2^{\tau(k)}$. Then $A.\mathrm{cor}$ knows nothing about $\mathbf{m}$, and thus I is conditionally independent of $\mathbf{m}$, given *param*. Hence in the view of $A.\mathrm{g}$, each $\mathbf{m}[i]$ (for $i \notin I$) still has min-entropy μ, and thus the chance that $A.\mathrm{g}$ can make a query in $\{hk \,\|\, 0 \,\|\, \mathbf{m}[i] \mid i \notin I\}$ is at most $v(k)q(k)/2^{\mu(k)}$.

The core of the proof is to bound the probability that the adversary $A.\mathrm{g}$ can make a query in $\{hk \,\|\, 1 \,\|\, \mathbf{r}[i] \mid i \notin I\}$. Let X_i be the random variable for the number of pre-images of $\mathsf{LT.Eval}(lk, \mathbf{r}[i])$. Although in the view of $A.\mathrm{cor}$, the average conditional min-entropy of each $\mathbf{r}[i]$ is $\tau(k)$, the same claim may *not* hold in the view of $A.\mathrm{g}$. For example, the adversary $A.\mathrm{cor}$ may choose to open all but the ciphertext of $\mathbf{m}[j]$, where j is chosen so that $X_j = \min\{X_1, \ldots, X_{v(k)}\}$: while $\mathbf{E}(1/X_i) \le 2^{-\tau(k)}$ for each fixed $i \in \{1, \ldots, v(k)\}$, the same bound doesn't work for $\mathbf{E}(1/\min\{X_1, \ldots, X_{v(k)}\})$. To get around this, note that the chance that $A.\mathrm{g}$ can make a query in $\{hk \,\|\, 1 \,\|\, \mathbf{r}[i] \mid i \notin I\}$ is at most

$$q(k) \cdot \mathbf{E}\Big(\sum_{i \notin I} \frac{1}{X_i}\Big) \le q(k) \cdot \mathbf{E}\Big(\sum_{i=1}^{v(k)} \frac{1}{X_i}\Big) \le q(k) \cdot \sum_{i=1}^{v(k)} \mathbf{E}\Big(\frac{1}{X_i}\Big) \le \frac{q(k)v(k)}{2^{\tau(k)}}.$$

Finally, if I is conditionally independent of $\mathbf{m}$ given *param*, then the re-sampled string z is identically distributed as $\delta(\mathbf{m}, param)$, even conditioning on hk, I, and *param*.[6] Hence $A.\mathrm{f}$ can query $\mathrm{RO}_4(hk \,\|\, \cdot, \cdot)$ with probability at most $q(k)/2^k$. If all bad events above don't happen then (i) in the joint view of $A.\mathrm{g}$ and $A.\mathrm{f}$, the strings $\delta(\mathbf{m}, param)$ and z are identically distributed, and (ii) the output of $A.\mathrm{f}$ will be conditionally independent of the ciphertexts and the public key, given *param*. This means the d-so-cpa2 advantage of A is 0.

3.3 Achieving D-SO-CCA Security

To achieve D-SO-CCA security, we modify DE1 construction as follows: In the decryption, once we recover the message m, we'll re-encrypt it and return $\perp$ if the resulting ciphertext doesn't match the given one, or the hash image of the message doesn't match the string obtained via inverting the trapdoor function. The resulting construction DE2 is shown in Fig. 4. The scheme $\mathsf{DE} = \mathsf{DE2}[H, \mathsf{LT}]$ is *unique-ciphertext*, as formalized by Bellare and Hoang [7]: for every $k \in \mathbb{N}$, every $(pk, sk) \in [\mathsf{DE.Kg}(1^k)]$, and every $m \in \{0,1\}^*$, there is at most a string c such that $\mathsf{DE.Dec}(sk, c) = m$. Theorem 3 below shows that DE2 is D-SO-CCA secure in the NPROM. The re-encrypting trick for lifting CPA to CCA security in the random-oracle model dates back to a paper of Fujisaki and Okamoto [13], but that work only considers randomized PKE and there's no opening. Still, the proof ideas are quite similar.

Theorem 3. *Let* LT *be a lossy trapdoor function with lossiness τ. Let $\mathcal{M}$ be a (μ, d)-entropic message sampler and let δ be a function such that $\mathcal{M}$ is δ-partially resamplable. Let* $\mathsf{DE2}[H, \mathsf{LT}]$ *be as above. In the NPROM, for any adversary A opening at most d ciphertexts, there is an adversary D such that*

$$\mathbf{Adv}^{\text{d-so-cca}}_{\mathsf{DE2}[H,\mathsf{LT}],A,\mathcal{M},\delta}(k) \le \frac{2p(k)}{2^{\mathsf{LT.il}(k)}} + \frac{10q(k)}{2^k} + \frac{4q(k)v(k)}{2^{\mu(k)}} + \frac{v(k)(v(k)+8q(k))}{2^{\tau(k)}} + 2\mathbf{Adv}^{\mathrm{ltdf}}_{\mathsf{LT},D}(k).$$

where $p(k)$ is the number of decryption-oracle queries of A, $q(k)$ is the total number of random-oracle queries of A and $\mathcal{M}$, and $v(k)$ is the number of messages

[6] Even for the simple case that $\mathcal{M}$ is fully resamplable and outputs empty auxiliary information, and $\delta(\mathbf{m}, param) = \mathbf{m}$, note that if I is correlated to $\mathbf{m}$ then $\mathbf{m}$ and the re-sampled $\mathbf{m}'$ may have completely different distributions. For example, consider $\mathcal{M}$ that outputs (m_1, m_2), with $m_1 \leftarrow\$ \{00, 01\}$ and $m_2 \leftarrow\$ \{10, 11\}$. Since m_1 and m_2 are independent, $\mathcal{M}$ is fully resamplable. Let $I = \{1\}$ if $m_1 = 00$, and $I = \{2\}$ otherwise. Then $\Pr[\mathbf{m}' = (00, 11)] = 3/8$, whereas $\Pr[\mathbf{m} = (00, 11)] = 1/4$.

DE.Kg(1^k)	DE.Enc(pk, m)	DE.Dec(sk, c)
$(ek, td) \leftarrow_\$ \mathsf{LT.IKg}(1^k)$ $hk \leftarrow_\$ \{0,1\}^k$ $pk \leftarrow (hk, ek)$ $sk \leftarrow (hk, ek, td)$ Return (pk, sk)	$(hk, ek) \leftarrow pk$ $r \leftarrow H(hk \,\|\, 0 \,\|\, m, \mathsf{LT.il}(k))$ $trap \leftarrow \mathsf{LT.Eval}(ek, r)$ $y \leftarrow H(hk \,\|\, 1 \,\|\, r, \lvert m \rvert) \oplus m$ Return $(trap, y)$	$(hk, ek, td) \leftarrow sk$ $(trap, y) \leftarrow c$ $r \leftarrow \mathsf{LT.Inv}(td, trap)$ $trap' \leftarrow \mathsf{LT.Eval}(ek, r)$ $m \leftarrow H(hk \,\|\, 1 \,\|\, r, \lvert y \rvert) \oplus y$ $r' \leftarrow H(hk \,\|\, 0 \,\|\, m, \mathsf{LT.il}(k))$ If $r' \neq r$ or $trap' \neq trap$ then Return $\perp$ Return m

Fig. 4. D-PKE scheme $\mathsf{DE} = \mathsf{DE2}[H, \mathsf{LT}]$. If LT is a lossy trapdoor permutation then in the decryption algorithm, the computation of $trap'$ and the check $trap' \neq trap$ can be omitted.

that $\mathcal{M}$ produces. The running time of D is about that of A plus the time to run δ and an efficient δ-partial resampling algorithm of $\mathcal{M}$, plus the time to run $\mathsf{DE2}[H, \mathsf{LT}]$ to encrypt $\mathcal{M}$'s messages and decrypt A's decryption queries. Adversary D makes at most $2q$ random-oracle queries.

Proof. Let Rsmp be an efficient δ-partial resampling algorithm for $\mathcal{M}$. Consider games G_1 and G_2 in Fig. 5. Then

$$\mathbf{Adv}^{\text{d-so-cca}}_{\mathsf{DE2}[H,\mathsf{LT}],A,\mathcal{M}}(\cdot) = 2\Pr[G_1(\cdot) \Rightarrow 1] - 1.$$

Game G_2 is identical to game G_1, except for the following. In procedure $\textsc{Dec}(c)$, instead of using the decryption of $\mathsf{DE2}$ to decrypt c, we maintain the set Dom of the suffixes of random-oracle queries (x, ℓ) that A.cor and A.g make such that $x[1, k+1] = hk \,\|\, 0$ and $\ell = \mathsf{LT.il}(k)$. If there's $m \in \mathsf{Dom}$ such that the corresponding ciphertext of m is c then we return m; otherwise return $\perp$. Wlog, assume that A.cor stores all random-oracle queries/answers in its state; that is, both A.cor and A.g also can track Dom and implement the $\textsc{Dec}$ procedure of game G_2 on their own, without calling the decryption oracle.[7] Let $\mathsf{Range} = \{\mathsf{DE2.Enc}(pk, m) \mid m \in \mathsf{Dom}\}$. On a query $c \in \mathsf{Range}$, the procedures $\textsc{Dec}$ of both games have the same behavior, due to the correctness of the decryption of $\mathsf{DE2}$. Wlog, assume that both A.cor and A.g never query $c \in \mathsf{Range}$ to the decryption oracle. (Adversaries A.cor and A.g are thus assumed to maintain the corresponding ciphertexts of messages in Dom. But this needs additional queries to the random oracle, so the total random-oracle queries of these two adversaries is now at most $2q$.)

[7] This assumption crucially relies on our use of a domain separation in hashing the coins r and the messages m: we employ $H(hk \,\|\, 0 \,\|\, \cdot, \cdot)$ for m, but $H(hk \,\|\, 1 \,\|\, \cdot, \cdot)$ for r. In contrast, BH's variant [7] doesn't use domain separation, and one can't make this assumption anymore: building the corresponding ciphertexts may create additional queries to $H(hk \,\|\, 0 \,\|\, \cdot, \mathsf{LT.il}(k))$, leading to a possible exponential blowup on the number of random-oracle queries.

Games $G_1(k)$, $G_2(k)$
$param \leftarrow_\$ A.\mathrm{pg}^{\mathrm{RO}}(1^k)$; $(\mathbf{m}, \mathbf{a}) \leftarrow_\$ \mathcal{M}^{\mathrm{RO}}(1^k, param)$; $z_1 \leftarrow \delta(\mathbf{m}, param)$
$hk \leftarrow_\$ \{0,1\}^k$; $(ek, td) \leftarrow_\$ \mathsf{LT.IKg}(1^k)$
For $i = 1$ to $|\mathbf{m}|$ do
$\mathbf{r}[i] \leftarrow \mathrm{RO}(hk \,\|\, 0 \,\|\, \mathbf{m}[i], \mathsf{LT.il}(k))$; $trap \leftarrow \mathsf{LT.Eval}(ek, \mathbf{r}[i])$
$y \leftarrow \mathrm{RO}_1(hk \,\|\, 1 \,\|\, \mathbf{r}[i], |\mathbf{m}[i]|) \oplus \mathbf{m}[i]$; $\mathbf{c}[i] \leftarrow (trap, y)$
$\mathsf{Dom} \leftarrow \emptyset$; $(state, I) \leftarrow_\$ A.\mathrm{cor}^{\mathrm{DEC},\mathrm{ROSIM}}((hk, ek), \mathbf{c}, param)$
$\omega \leftarrow_\$ A.\mathrm{g}^{\mathrm{DEC},\mathrm{ROSIM}}(state, \mathbf{m}[I], \mathbf{a}[I])$
$z_0 \leftarrow_\$ \mathsf{Rsmp}^{\mathrm{RO}}(1^k, \mathbf{m}[I], \mathbf{a}[I], I, param)$; $b \leftarrow_\$ \{0,1\}$; $t \leftarrow_\$ A.\mathrm{f}^{\mathrm{RO}}(z_b, param)$
If $(\omega = t)$ then return b else return $1 - b$

Procedure $\mathrm{ROSIM}(x, \ell)$
If $|x| > k+1$ and $x[1, k+1] = hk \,\|\, 0$ then $\mathsf{Dom} \leftarrow \mathsf{Dom} \cup \{x[k+2, |x|]\}$
Return $\mathrm{RO}(x, \ell)$

Procedure $\mathrm{DEC}(c)$ // of game G_1	**Procedure** $\mathrm{DEC}(c)$ // of game G_2
$sk \leftarrow (hk, td)$	For $m \in \mathsf{Dom}$ do
$m \leftarrow \mathsf{DE1}[H, \mathsf{LT}].\mathsf{Dec}(sk, c)$	If $c = \mathsf{DE1}[H, \mathsf{LT}].\mathsf{Enc}(pk, m)$ then
Return m	Return m
	Return $\perp$

Fig. 5. Games G_1 and G_2 of the proof of Theorem 3. Their procedures DEC are in the bottom-left and bottom-right panels, respectively.

Assume that $A.\mathrm{pg}$ and $\mathcal{M}$ never make a random-oracle query (x, ℓ) such that the k-bit suffix of x is hk. This happens with probability at least $1 - q(k)/2^k$. The adversaries can distinguish the games if and only if they can trigger DEC of game G_1 to produce non-$\perp$ output. Let $c = (trap, y)$ be a decryption-oracle query. Let $r = \mathsf{LT.Inv}(td, trap)$ and $m = \mathrm{RO}(hk \,\|\, 1 \,\|\, r) \oplus y$. Due to the unique-ciphertext property of DE2, if this can trigger the DEC procedure of game G_1 to return a non-$\perp$ answer, we must have $m \notin \{\mathbf{m}[1], \ldots, \mathbf{m}[|\mathbf{m}|]\} \cup \mathsf{Dom}$. Then there is no prior random-oracle query $(x, \mathsf{LT.il}(k))$ such that $x = hk \,\|\, 0 \,\|\, m$. Hence procedure DEC of game G_1 will return a non-$\perp$ answer only if $r = \mathrm{RO}(hk \,\|\, 0 \,\|\, m, \mathsf{LT.il}(k))$, which happens with probability $2^{-\mathsf{LT.il}(k)}$. Multiplying for $p(k)$ decryption-oracle queries,

$$\Pr[G_1(k) \Rightarrow 1] - \Pr[G_2(k) \Rightarrow 1] \leq q(k)/2^k + p(k)/2^{\mathsf{LT.il}(k)}.$$

Now in game G_2, the decryption oracle always return $\perp$, and thus wlog, assume that the adversaries never make a decryption query, meaning that they only launch a D-SO-CPA2 attack. Hence

$$2\Pr[G_2(\cdot) \Rightarrow 1] = \mathbf{Adv}^{\text{d-so-cpa2}}_{\mathsf{DE2}[H,\mathsf{LT}],A,\mathcal{M}}(\cdot).$$

But DE2 and DE1 only differ in the decryption algorithms, which doesn't affect the D-SO-CPA security. Hence from Theorem 2, we can construct a distinguisher D of the claimed running time such that

$$\mathbf{Adv}^{\text{d-so-cpa2}}_{\mathsf{DE2}[H,\mathsf{LT}],A,\mathcal{M},\delta}(k) \le \frac{8q(k)}{2^k} + \frac{8q(k)v(k)}{2^{\mu(k)}} + \frac{v(k)(v(k)+8q(k))}{2^{\tau(k)}} + 2\mathbf{Adv}^{\text{ltdf}}_{\mathsf{LT},D}(k) .$$

(Note that the bound above is for adversaries who make at most $2q$ random-oracle queries.) Summing up,

$$\mathbf{Adv}^{\text{d-so-cca}}_{\mathsf{DE2}[H,\mathsf{LT}],A,\mathcal{M},\delta}(k) \le \frac{2p(k)}{2^{\mathsf{LT.il}(k)}} + \frac{10q(k)}{2^k} + \frac{4q(k)v(k)}{2^{\mu(k)}} + \frac{v(k)(v(k)+8q(k))}{2^{\tau(k)}} + 2\mathbf{Adv}^{\text{ltdf}}_{\mathsf{LT},D}(k).$$

□

4 Selective-Opening Security for Nonce-Based PKE

Recall that D-PKE protects only unpredictable messages, but in practice, messages often have very limited entropy [12]. Hedge PKE tries to improve this situation by adding the unpredictability of coins. However, the coins generated by Dual EC are completely determined by Big Brother, and those by the buggy Debian RNG have only about 15 bits of min-entropy. In a recent work, Bellare and Tackmann (BT) [10] propose the notion of nonce-based PKE to address this limitation, supporting arbitrary messages. In this section, we extend the notion of nonce-based PKE for SOA setting, and then show how to achieve this.

4.1 Security Notions

NONCE GENERATORS. A *nonce generator* NG with nonce space $\mathcal{N}$ is an algorithm that takes as input the unary encoding 1^k of the security parameter, a current state St, and a *nonce selector* σ. It then probabilistically produces a nonce $N \in \mathcal{N}$ together with an updated state St. That is, $(N, St) \leftarrow\$ \mathsf{NG}(1^k, St, \sigma)$. A good nonce generator needs to satisfy the following properties: (i) nonces should never repeat, and (ii) each nonce is unpredictable, even if all nonce selectors are adversarially chosen. Formally, let $\mathbf{Adv}^{\text{rp}}_{\mathsf{NG},A}(k) = \Pr[\text{RP}^A_{\mathsf{NG}}(k)]$, where game RP is defined in Fig. 6. We say that NG is RP-secure if for any PT adversary A, $\mathbf{Adv}^{\text{rp}}_{\mathsf{NG},A}(\cdot)$ is a negligible function.

NONCE-BASED PKE. A nonce-based PKE with nonce space $\mathcal{N}$ is a tuple $\mathsf{NE} = (\mathsf{NE.Kg}, \mathsf{NE.Sg}, \mathsf{NE.Enc}, \mathsf{NE.Dec})$. The key generator $\mathsf{NE.Kg}(1^k)$ generates a public key pk and an associated secret key sk. The seed generator $\mathsf{NE.Sg}(1^k)$ produces a sender seed xk. The encryption algorithm NE.Enc takes as input a public key pk, a sender seed xk, a nonce $N \in \mathcal{N}$, and a message m, and then *deterministically* returns a ciphertext c. The decryption algorithm $\mathsf{NE.Dec}(sk, \cdot)$ plays the same role as in traditional randomized PKE; it's not given the nonce or the sender seed.

Nonce-based PKE can be viewed as a way to harden the randomness at the sender side; the receiver is oblivious to this change. Security of nonce-based PKE

Game $\mathrm{RP}^A_{\mathsf{NG}}(k)$	**Procedure** $\mathrm{GEN}(\sigma)$
$St \leftarrow \varepsilon$; $coll \leftarrow \mathsf{false}$	$(N, St) \leftarrow_{\$} \mathsf{NG}(1^k, St, \sigma)$
$\mathsf{Dom} \leftarrow \emptyset$; $N \leftarrow_{\$} A^{\mathrm{GEN}}(1^k)$	If $N \in \mathsf{Dom}$ then $coll \leftarrow \mathsf{true}$
Return $(N \in \mathsf{Dom}) \vee coll$	$\mathsf{Dom} \leftarrow \mathsf{Dom} \cup \{N\}$; Return N

Fig. 6. Game to define security of a nonce generator NG.

should hold when either (i) the seed xk is secret and the nonces are unique, or (ii) the seed is leaked to the adversary, but the nonces are unpredictable to the adversary.[8]

DISCUSSION. To formalize security of nonce-based PKE, BT define two notions, NBP1 and NBP2. Both notions are in the single-sender setting and use nonces generated from a nonce generator NG. The former notion considers the situation when the seed xk is secret, and there's no security requirement from NG, except the uniqueness of nonces. The latter notion considers the case when the seed xk is given to the adversary; now nonces generated from NG have to satisfy RP security.

When we bring SOA extension to nonce-based PKE below, there will be many changes. First, since there are multiple senders and only some of them can keep their seeds secret, one has to merge the SOA variants of NBP1 and NBP2 into a single definition. Next, because the adversary learns the seeds of some senders, the nonce generator NG must be RP-secure. If we let senders whose seeds are secret use unpredictable nonces from NG then our notion will fail to model the possibility that the adversary can corrupt the nonce generator. Therefore, in our notion, for senders whose seeds are secret, we'll let the adversary specify their nonces. We require the adversary to be *nonce-respecting*, meaning that the nonces of every single sender must be distinct.

N-SO-CPA. Let NE be a nonce-based PKE scheme and NG be a nonce generator of the same nonce space $\mathcal{N}$. Let $\mathcal{M}$ be a message sampler, but the generated messages don't have to be distinct or unpredictable. Let δ be a function such that $\mathcal{M}$ is δ-partially resamplable. The game N-SO-CPA defining the N-SO-CPA security is specified in Fig. 7.

Initially, the game picks seed $\boldsymbol{xk}[j] \leftarrow_{\$} \mathsf{NE.Sg}(1^k)$ and sets state $\boldsymbol{st}[j] \leftarrow \varepsilon$ for sender j, with $j = 1, 2, \ldots$. The adversary is then given the public key pk and has to specify the list J of senders that it wishes to get the seeds. It's then granted $\boldsymbol{xk}[J]$ and then has to provide some parameter $param$ for generating $(\mathbf{m}, \mathbf{a}) \leftarrow_{\$} \mathcal{M}(1^k, param)$, together with a vector $\boldsymbol{N}$ of nonces, a map U that

[8] The definition of BT [10] requires that if the seed xk is secret then security should hold as long as the message/nonce pairs are unique. If one directly extends this to the SOA setting, there will be some pesky issue, as the adversary can detect equality within the message vectors by repeating the nonces. Here for simplicity, we only demand that nonces should be unique, which is analogous to nonce-based symmetric encryption. Nevertheless, our constructions are specific instantiations of BT construction, and thus meet their requirement.

Game $\text{N-SO-CPA}_{\mathsf{NE},\mathsf{NG}}^{A,\mathcal{M},\delta}(k)$
For $j = 1, 2, \ldots$ do $\boldsymbol{xk}[j] \leftarrow_{\$} \mathsf{NE.Sg}(1^k)$; $\boldsymbol{st}[j] \leftarrow \varepsilon$
$(pk, sk) \leftarrow_{\$} \mathsf{NE.Kg}(1^k)$; $(J, state) \leftarrow_{\$} A(1^k, pk)$
$(param, \boldsymbol{N}, U, \boldsymbol{\sigma}, state) \leftarrow_{\$} A(state, \boldsymbol{xk}[J])$
$(\mathbf{m}, \mathbf{a}) \leftarrow_{\$} \mathcal{M}(1^k, param)$; $z_1 \leftarrow \delta(\mathbf{m}, param)$
For $i = 1$ to $|\mathbf{m}|$ do
  $j \leftarrow U[i]$
  If $j \in J$ then $(N, \boldsymbol{st}[j]) \leftarrow_{\$} \mathsf{NG}(1^k, \boldsymbol{st}[j], \boldsymbol{\sigma}[i])$; $\boldsymbol{N}[i] \leftarrow N$
  $\mathbf{c}[i] \leftarrow \mathsf{NE.Enc}(pk, \boldsymbol{xk}[j], \boldsymbol{N}[i], \mathbf{m}[i])$
$(I, state) \leftarrow_{\$} A(state, \mathbf{c})$; $b \leftarrow_{\$} \{0, 1\}$
For $i \in I$, $j = 1$ to $|\mathbf{m}|$ do
  If $U[i] = U[j]$ then $I \leftarrow I \cup \{j\}$
$z_0 \leftarrow_{\$} \mathsf{Resamp}_{\mathcal{M},\delta}(1^k, \mathbf{m}[I], \mathbf{a}[I], param)$
$b' \leftarrow_{\$} A(state, \mathbf{m}[I], \mathbf{a}[I], \boldsymbol{N}[I], \boldsymbol{xk}[U[I]], z_b)$; Return $(b = b')$

Fig. 7. Game defining N-SO-CPA security.

specifies message $\mathbf{m}[i]$ belongs to sender $U[i]$, and a vector $\boldsymbol{\sigma}$ of nonce selectors for NG. Note that the messages $\mathbf{m}$ here can depend on the public key. We require that the adversary be *nonce-respecting*, meaning that $(\boldsymbol{N}[1], U[1]), (\boldsymbol{N}[2], U[2]), \ldots$ are distinct.

The game then iterates over $i = 1, \ldots, |\mathbf{m}|$ to encrypt each message $\mathbf{m}[i]$. If $i \in J$ then $\boldsymbol{N}[i]$ is overwritten by a nonce N generated by NG as follows. Let $j \leftarrow U[i]$. The nonce generator NG will read the current state $\boldsymbol{st}[j]$ of sender j and the nonce selector $\boldsymbol{\sigma}[i]$ for the message $\mathbf{m}[i]$, to generate a nonce N and update $\boldsymbol{st}[j]$. The adversary then is given the ciphertexts and has to output a set I to indicate which ciphertexts it wants to open. Note that opening $\mathbf{c}[i]$ returns not only $(\mathbf{m}[i], \mathbf{a}[i])$ but also the associated nonce and sender seed. Moreover, if the adversary opens a message belonging to sender j, then any other messages of this sender are considered open. Finally, the game resamples z_0, and let $z_1 \leftarrow \delta(\mathbf{m}, param)$. It picks $b \leftarrow_{\$} \{0, 1\}$, and gives the adversary z_b and $(\mathbf{m}[I], \mathbf{a}[I], \boldsymbol{xk}[U[I]], \boldsymbol{N}[I])$. The adversary has to guess the challenge bit b. Define

$$\mathbf{Adv}_{\mathsf{NE},\mathsf{NG},A,\mathcal{M},\delta}^{\text{n-so-cpa}}(k) = 2\Pr[\text{N-SO-CPA}_{\mathsf{NE},\mathsf{NG}}^{A,\mathcal{M},\delta}(k)] - 1.$$

We say that NE is N-SO-CPA secure, with respect to NG, if for any message sampler $\mathcal{M}$ and any PT adversary A, and any PT function δ such that $\mathcal{M}$ is δ-partially resamplable, $\mathbf{Adv}_{\mathsf{NE},\mathsf{NG},A,\mathcal{M},\delta}^{\text{n-so-cpa}}(\cdot)$ is a negligible function.

N-SO-CCA. To add a CCA flavor to N-SO-CPA, one would give the adversary oracle access to $\mathsf{Dec}(sk, \cdot)$. Once it's given the ciphertexts $\mathbf{c}$, it's not allowed to query any $\mathbf{c}[i]$ to the decryption oracle. Let N-SO-CCA be the corresponding game, and define $\mathbf{Adv}_{\mathsf{NE},\mathsf{NG},A,\mathcal{M},\delta}^{\text{n-so-cca}}(k) = 2\Pr[\text{N-SO-CCA}_{\mathsf{NE},\mathsf{NG}}^{A,\mathcal{M},\delta}(k)] - 1$.

SIMULATION-BASED SECURITY. One could also define an appropriate simulation-based notion of SOA security for nonce-based PKE, which unlike N-SO-CPA would not require the unrevealed messages to be efficiently resampleable, analogously to the SIM-SOA definition for randomized PKE in [9]. However, we conjecture that such a definition is impossible to achieve. We leave this as an open question. In any case, a simulation-based definition of SOA security for nonce-based PKE will indeed be impossible to achieve later when we lift the primitive to the hedged setting, where an existing impossibility result for a simulation-based notion of SOA security for deterministic PKE [4] applies (because hedged PKE generalizes deterministic PKE).

4.2 Separation

We now show that the standard notions for nonce-based PKE of BT [10] do not imply N-SO-CPA. Our separation is based on the recent result of Hofheinz, Rao, and Wichs (HRW) [15] to show that IND-CCA doesn't imply the notion IND-SO-CPA for randomized PKE.

HRW CONSTRUCTION. Our counterexample is based on the recent (contrived) construction $\mathsf{RE_{bad}} = (\mathsf{RE_{bad}.Kg}, \mathsf{RE_{bad}.Enc}, \mathsf{RE_{bad}.Dec})$ of HRW. The scheme $\mathsf{RE_{bad}}$ is IND-CCA secure, but is vulnerable to the following SOA attack. The message sampler $\mathcal{M}(1^k, param)$ ignores *param*, picks a secret $s \leftarrow_{\$} \{0,1\}^\ell$ and then secret-shares it to $v(k)$ messages $\mathbf{m}[1], \ldots, \mathbf{m}[v(k)]$ so that any $t(k)$ shares reveal no information of the secret s. In other words, it picks $a_0, a_1, \ldots, a_t$ uniformly from $\mathsf{GF}(2^\ell)$, the finite field of size 2^ℓ, and computes $\mathbf{m}[i] \leftarrow f(i)$ for every $i \in \{1, \ldots, v(k)\}$, where $f(x) = a_0 + a_1 x + \cdots + a_t x^t$ is the corresponding polynomial in $\mathsf{GF}(2^\ell)[X]$. Recall that any $t+1$ shares will uniquely determine the polynomial f (via polynomial interpolation), and thus any $t+1$ shares are uniformly and independently random. The auxiliary information is empty. Surprisingly, there's an efficient adversary that opens only t ciphertexts and can recover all messages. We note that HRW's counter-example is based on public-coin differing-inputs obfuscation [17], which is a very strong assumption.

RESULTS. Let H be a hash function. One can model it as a random oracle, or, for a standard-model result, a primitive that BT call *hedged extractor*. BT show that one can build a nonce-based PKE achieving their notions from an arbitrary IND-CCA secure PKE RE as follows. Given seed xk, nonce N, and message m, one uses $H((xk, N, m))$ to extract synthetic coins r, and then encrypt m via RE under coins r. Now, use the scheme $\mathsf{RE_{bad}}$ above to instantiate RE, and let $\mathsf{NE_{bad}}[H, \mathsf{RE_{bad}}]$ be the resulting nonce-based PKE. This $\mathsf{NE_{bad}}[H, \mathsf{RE_{bad}}]$ achieves BT's notions.

We now break the N-SO-CPA security of $\mathsf{NE_{bad}}$. The message sampler $\mathcal{M}$ is as described in HRW attack, and let A be the adversary attacking $\mathsf{RE_{bad}}$ as above. Note that $\mathcal{M}$ is fully resamplable, and let $\delta(\mathbf{m}, param) = \mathbf{m}$. Consider the following adversary B attacking $\mathsf{NE_{bad}}$. It specifies $J = \emptyset$, meaning that it doesn't want to get any sender seed before the opening. It then lets $\boldsymbol{N}[i] = U[i] = i$, for every i. That is, each sender has only a single message. Then, when B gets the

$\mathsf{NE1.Kg}(1^k)$	$\mathsf{NE1.Sg}(1^k)$
$(ek, td) \leftarrow\$ \mathsf{LT.IKg}(1^k)$; $hk \leftarrow\$ \{0,1\}^k$ $pk \leftarrow (hk, ek)$; $sk \leftarrow (hk, td)$; Return (pk, sk)	$xk \leftarrow\$ \{0,1\}^k$; Return xk
$\mathsf{NE1.Enc}(pk, xk, N, m)$	$\mathsf{NE1.Dec}(sk, c)$
$(hk, ek) \leftarrow pk$ $r \leftarrow H(hk \,\|\, 0 \,\|\, (xk, N, m), \mathsf{LT.il}(k))$ $trap \leftarrow \mathsf{LT.Eval}(pk, r)$; $y \leftarrow H(hk \,\|\, 1 \,\|\, r, \lvert m \rvert) \oplus m$ Return $(trap, y)$	$(hk, td) \leftarrow sk$; $(trap, y) \leftarrow c$ $r \leftarrow \mathsf{LT.Inv}(sk, trap)$ $m \leftarrow H(hk \,\|\, 1 \,\|\, r, \lvert y \rvert) \oplus y$ Return m

Fig. 8. Nonce-based PKE scheme $\mathsf{NE1}[H, \mathsf{LT}]$.

ciphertexts $\mathbf{c}$, it runs A on those $\mathbf{c}$. Note that $\mathbf{c}$ are ciphertexts of $\mathsf{RE}_{\mathsf{bad}}$, although the coins are only pseudorandom. Still, adversary A can recover all messages by opening just t ciphertexts. When B is given the messages (real or resampled), it compares that with what A recovers. Then $\mathbf{Adv}^{\text{n-so-cpa}}_{\mathsf{NE},\mathsf{NG},B,\mathcal{M},\delta}(k) \geq 1 - 2^{-\ell(k)}$, where ℓ is the length of each message.

4.3 Achieving N-SO-CPA Security

BT's construction of nonce-based PKE is simple. To encrypt a message m under a seed xk, a nonce N, and public key pk, we hash (xk, N, m) to derive a string r, and then uses a traditional randomized PKE to encrypt m under the synthetic coins r and public key pk. Here we'll use BT's construction, but the underlying randomized PKE is a randomized counterpart of the D-PKE scheme $\mathsf{DE1}$ in Sect. 3.2.

Formally, let H be a hash of arbitrary input and output length, meaning that $H(x, \ell)$ returns an ℓ-bit string. Let LT be a lossy trapdoor function. Our nonce-based PKE $\mathsf{NE1}[H, \mathsf{LT}]$ is described in Fig. 8; it has nonce space $\{0,1\}^*$ and message space $\{0,1\}^*$. Theorem 4 below shows that $\mathsf{NE1}[H, \mathsf{LT}]$ is N-SO-CPA secure in the NPROM; the proof is in the full version.

Theorem 4. *Let LT be a lossy trapdoor function with lossiness τ. Let $\mathcal{M}$ be a message sampler and let δ be a function such that $\mathcal{M}$ is δ-partially resamplable. Let $\mathsf{NE1}[H, \mathsf{LT}]$ be as above, and let NG be a nonce generator. In the NPROM, for any adversary A, there are adversaries B and D such that*

$$\mathbf{Adv}^{\text{n-so-cpa}}_{\mathsf{NE1}[H,\mathsf{LT}],\mathsf{NG},A,\mathcal{M},\delta}(k) \leq 2\mathbf{Adv}^{\text{ltdf}}_{\mathsf{LT},B}(k) + 8q(k)v(k) \cdot \mathbf{Adv}^{\text{rp}}_{\mathsf{NG},D}(k) + \frac{7v(k)(q(k) + v(k))}{2^k} + \frac{12v(k)(q(k) + v(k))}{2^{\tau(k)}},$$

where v is the number of messages that $\mathcal{M}$ generates, and q bounds the total number of random-oracle queries that A and $\mathcal{M}$ make. The running time of B or D is about the time to run game $\text{N-SO-CPA}^{A,\mathcal{M},\delta}_{\mathsf{NE},\mathsf{NG}}$, but using an efficient δ-partial resampling algorithm of $\mathcal{M}$ instead of $\mathsf{Resamp}_{\mathcal{M},\delta}$. Each of B and D makes at most q random-oracle queries.

$\mathsf{NE2.Kg}(1^k)$	$\mathsf{NE2.Sg}(1^k)$
$(ek, td) \leftarrow_{\$} \mathsf{LT.IKg}(1^k)$; $hk \leftarrow_{\$} \{0,1\}^k$ $pk \leftarrow (hk, ek)$; $sk \leftarrow (hk, ek, td)$ Return (pk, sk)	$xk \leftarrow_{\$} \{0,1\}^k$; Return xk
$\mathsf{NE2.Enc}(pk, xk, N, m)$	$\mathsf{NE2.Dec}(sk, c)$
$(hk, ek) \leftarrow pk$ $r \leftarrow H(hk \,\|\, 00 \,\|\, (xk, N, m), \mathsf{LT.il}(k))$ $y \leftarrow H(hk \,\|\, 01 \,\|\, r, \|m\|) \oplus m$ $z \leftarrow H(hk \,\|\, 10 \,\|\, r \,\|\, m, k)$ $trap \leftarrow \mathsf{LT.Eval}(pk, r)$ Return $(trap, y)$	$(hk, ek, td) \leftarrow sk$; $(trap, y, z) \leftarrow c$ $r \leftarrow \mathsf{LT.Inv}(sk, trap)$ $trap' \leftarrow \mathsf{LT.Eval}(pk, r)$ $m \leftarrow H(hk \,\|\, 01 \,\|\, r, \|y\|) \oplus y$ $z' \leftarrow H(hk \,\|\, 10 \,\|\, r \,\|\, m, k)$ If $(z' \neq z) \vee (trap' \neq trap)$ then return $\perp$ Return m

Fig. 9. Nonce-based PKE scheme $\mathsf{NE2}[H, \mathsf{LT}]$.

4.4 Achieving N-SO-CCA Security

To strengthen NE1 with CCA capability, in the encryption, we append to the ciphertext a hash image of $r \,\|\, m$. When we decrypt a ciphertext, we'll recover both r and m, and check if the hash image of $r \,\|\, m$ matches with what's given in the ciphertext. The resulting scheme $\mathsf{NE2}[H, \mathsf{LT}]$ is shown in Fig. 9. The underlying randomized PKE of NE2 is a textbook IND-CCA construction in the ROM (but LT just needs to be an ordinary trapdoor function). Theorem 5 below shows that $\mathsf{NE2}[H, \mathsf{LT}]$ is N-SO-CCA secure in the NPROM; the proof is in the full version.

Theorem 5. Let LT be a lossy trapdoor function with lossiness τ. Let $\mathcal{M}$ be a message sampler and let δ be a function such that $\mathcal{M}$ is δ-partially resamplable. Let $\mathsf{NE2}[H, \mathsf{LT}]$ be as above, and let NG be a nonce generator. In the NPROM, for any adversary A, there are adversaries B and D such that

$$\mathbf{Adv}^{\text{n-so-cca}}_{\mathsf{NE2}[H,\mathsf{LT}],\mathsf{NG},A,\mathcal{M},\delta}(k) \leq 2\mathbf{Adv}^{\text{ltdf}}_{\mathsf{LT},B}(k) + 8v(k)Q(k) \cdot \mathbf{Adv}^{\text{rp}}_{\mathsf{NG},D}(k) + \frac{2p(k)}{2^k} + \frac{7v(k)Q(k)}{2^k} + \frac{12v(k)Q(k)}{2^{\tau(k)}},$$

where v is the number of messages that $\mathcal{M}$ generates, p is the number of A's queries to the decryption oracle, q bounds the total number of random-oracle queries that A and $\mathcal{M}$ make, and $Q = q + 2p + v$. The running time of B or D is about the time to run game $\text{N-SO-CCA}^{A,\mathcal{M},\delta}_{\mathsf{NE},\mathsf{NG}}$, but using an efficient δ-partial resampling algorithm of $\mathcal{M}$ instead of $\mathsf{Resamp}_{\mathcal{M},\delta}$. Each of B and D makes at most $q + 2p$ random-oracle queries.

5 Hedged Security for Nonce-Based PKE

Recall that the security of nonce-based PKE relies on the assumption that the adversary cannot obtain the secret seeds and corrupt the nonce generator simultaneously. Still, this assumption may fail in practice, and it's desirable to retain some security guarantee when seeds and nonces are bad. We capture this via the notion HN-SO-CPA that is a variant of the notion D-SO-CPA2, adapted for the nonce-based setting. A good nonce-based PKE thus has to satisfy both N-SO-CPA and HN-SO-CPA simultaneously. We then extend this treatment to the CCA setting.

5.1 Security Notions

UNPREDICTABLE SAMPLERS. Let $\mathcal{M}$ be a message sampler. We say that $\mathcal{M}$ is (μ, d)-*unpredictable* if for any $param \in \{0,1\}^*$,

(i) For any $(\mathbf{m}, \mathbf{a}) \in [\mathcal{M}(1^k, param)]$, each $\mathbf{a}[i]$ is a tuple (a_i, xk_i, N_i), where xk_i is a seed and N_i is a nonce. Moreover, $(xk_1, N_1, \mathbf{m}[1]), (xk_2, N_2, \mathbf{m}[2]), \ldots$ must be distinct.
(ii) For any $I \subseteq \{1, \ldots, v(k)\}$ such that $|I| \le d$, and any $i \in \{1, \ldots, v(k)\} \backslash I$, for $(\mathbf{m}, \mathbf{a}) \leftarrow_{\$} \mathcal{M}(1^k, param)$ the conditional min-entropy of $(\mathbf{m}[i], xk_i, N_i)$ given $(\mathbf{m}[I], \mathbf{a}[I], param)$ is at least μ, where $v(k)$ is the number of messages that $\mathcal{M}$ produces and xk_i and N_i are the seed and nonce specified by $\mathbf{a}[i]$.

Defining unpredictable samplers allows us to model the situation when the seeds, nonces, and messages are related, and quantify security based on the combined min-entropy of each message with its nonce and seed.

HN-SO-CPA SECURITY. Let NE be a nonce-based PKE scheme, and let $\mathcal{M}$ be an unpredictable message sampler. Let δ be a function such that $\mathcal{M}$ is δ-partially resamplable. Let $A = (A.\text{pg}, A.\text{cor}, A.\text{g}, A.\text{f})$ be an adversary. Define

$$\mathbf{Adv}^{\text{hn-so-cpa}}_{\mathsf{NE},A,\mathcal{M},\delta}(\cdot) = \Pr\left[\text{HN-CPA-REAL}^{A,\mathcal{M},\delta}_{\mathsf{NE}}(\cdot) \Rightarrow 1\right] - \Pr\left[\text{HN-CPA-IDEAL}^{A,\mathcal{M},\delta}_{\mathsf{NE}}(\cdot) \Rightarrow 1\right],$$

where the games are defined in Fig. 10.

HN-SO-CCA SECURITY. To add a CCA flavor to HN-SO-CPA, one would give A.cor and A.g oracle access to $\mathsf{Dec}(sk, \cdot)$. They are not allowed to query any $\mathbf{c}[i]$ to the decryption oracle. Let HN-CCA-REAL and HN-CCA-IDEAL be the corresponding games, and define

$$\mathbf{Adv}^{\text{hn-so-cca}}_{\mathsf{NE},A,\mathcal{M},\delta}(\cdot) = \Pr\left[\text{HN-CCA-REAL}^{A,\mathcal{M},\delta}_{\mathsf{NE}}(\cdot) \Rightarrow 1\right] - \Pr\left[\text{HN-CCA-IDEAL}^{A,\mathcal{M},\delta}_{\mathsf{NE}}(\cdot) \Rightarrow 1\right].$$

SEPARATION. We now show that N-SO-CCA doesn't imply HN-SO-CPA, even if $\mathcal{M}$ picks $\mathbf{m}[i] \leftarrow_{\$} \{0,1\}^k$ and $\mathbf{a}[i] = (i, i, i)$, and there's no opening. Note that $\mathcal{M}$ is fully resamplable, and consider the function δ such that $\delta(\mathbf{m}, param) = param$. Let H be a hash and LT be a lossy trapdoor function. Let $\mathsf{NE}_{\mathsf{bad}}[H, \mathsf{LT}]$ be the following variant of $\mathsf{NE2}[H, \mathsf{LT}]$. To encrypt message m under public key pk,

Game HN-CPA-REAL$_{\mathsf{NE}}^{A,\mathcal{M},\delta}(k)$
$param \leftarrow_\$ A.\mathrm{pg}(1^k)$; $(pk, sk) \leftarrow_\$ \mathsf{Kg}(1^k)$; $(\mathbf{m}, \mathbf{a}) \leftarrow_\$ \mathcal{M}(1^k, param)$
For $i = 1$ to $|\mathbf{m}|$ do $(a, xk, N) \leftarrow \mathbf{a}[i]$; $\mathbf{c}[i] \leftarrow \mathsf{Enc}(pk, xk, N, \mathbf{m}[i])$
$(state, I) \leftarrow_\$ A.\mathrm{cor}(pk, \mathbf{c}, param)$; $\omega \leftarrow_\$ A.\mathrm{g}(state, \mathbf{m}[I], \mathbf{a}[I])$
$z_1 \leftarrow \delta(\mathbf{m}, param)$
Return $(\omega = A.\mathrm{f}(z_1, param))$

Game HN-CPA-IDEAL$_{\mathsf{NE}}^{A,\mathcal{M},\delta}(k)$
$param \leftarrow_\$ A.\mathrm{pg}(1^k)$; $(pk, sk) \leftarrow_\$ \mathsf{Kg}(1^k)$; $(\mathbf{m}, \mathbf{a}) \leftarrow_\$ \mathcal{M}(1^k, param)$
For $i = 1$ to $|\mathbf{m}|$ do $(a, xk, N) \leftarrow \mathbf{a}[i]$; $\mathbf{c}[i] \leftarrow \mathsf{Enc}(pk, xk, N, \mathbf{m}[i])$
$(state, I) \leftarrow_\$ A.\mathrm{cor}(pk, \mathbf{c}, param)$; $\omega \leftarrow_\$ A.\mathrm{g}(state, \mathbf{m}[I], \mathbf{a}[I])$
$z_0 \leftarrow_\$ \mathsf{Resamp}_{\mathcal{M},\delta}(1^k, \mathbf{m}[I], \mathbf{a}[I], I, param)$
Return $(\omega = A.\mathrm{f}(z_0, param))$

Fig. 10. Games to define HN-SO-CPA security.

seed xk and nonce N, instead of hashing (xk, N, m) to derive synthetic coins r, we just hash (xk, N). The proof of Theorem 5 can be recast to justify the N-SO-CCA security $\mathsf{NE}_{\mathsf{bad}}$. However, without even opening, one can trivially break HN-SO-CPA security of $\mathsf{NE}_{\mathsf{bad}}$ as follows. First, adversary A.pg outputs an arbitrary *param*. Next, adversary A.cor stores the ciphertexts and the public key in its state, and outputs $I = \emptyset$. Adversary A.g computes $r \leftarrow H(hk \,\|\, 00 \,\|\, (1,1))$, parses $(trap, y, z) \leftarrow \mathbf{c}[1]$, and outputs $\mathbf{m}[1] = y \oplus H(hk \,\|\, 01 \,\|\, r, |y|)$. Finally, adversary $A.\mathrm{f}(\mathbf{m}^*, param)$ simply outputs $\mathbf{m}^*[1]$. The adversaries win with advantage $1 - 2^{-k}$.

5.2 Achieving HN-SO-CPA Security

NtD TRANSFORM. We first give a transform Nonce-then-Deterministic (NtD). Let DE be a D-SO-CPA2 secure D-PKE and NE be an N-SO-CPA secure nonce-based PKE. Then NtD[NE, DE] achieves both HN-SO-CPA and N-SO-CPA security simultaneously. The resulting nonce-based PKE $\overline{\mathsf{NE}}$ is a double encryption: it first encrypts via NE, and then uses DE to encrypt the resulting ciphertext.[9] The transform NtD is shown in Fig. 11, and Theorem 6 below confirms that it works as claimed.

DISCUSSION. To explain why NtD works, note that using an outer D-PKE on the ciphertext of NE doesn't affect its N-SO-CPA security, and thus $\overline{\mathsf{NE}}$ = NtD[NE, DE] inherits the N-SO-CPA security of NE. For HN-SO-CPA security, there are some subtle points as follows.

[9] For simplicity, we assume that the ciphertext length of $\overline{\mathsf{NE}}$ is the plaintext length of DE. One may also consider a more generalized setting in which the ciphertext length of NE is smaller than the plaintext length of DE. In this case one needs to pad 10^* to the ciphertexts of NE before feeding them to DE.

$\overline{\mathsf{NE}}.\mathsf{Kg}(1^k)$	$\overline{\mathsf{NE}}.\mathsf{Enc}(pk, xk, N, m)$	$\overline{\mathsf{NE}}.\mathsf{Dec}(sk, c)$
$(pk_n, sk_n) \leftarrow_\$ \mathsf{NE.Kg}(1^k)$	$(pk_n, pk_d) \leftarrow pk$	$(sk_n, sk_d) \leftarrow sk$
$(pk_d, sk_d) \leftarrow_\$ \mathsf{DE.Kg}(1^k)$	$m' \leftarrow (m \parallel xk \parallel N)$	$y \leftarrow \mathsf{DE.Dec}(sk_d, c)$
$pk \leftarrow (pk_n, pk_d)$; $sk \leftarrow (sk_n, sk_d)$	$y \leftarrow \mathsf{NE.Enc}(pk_n, xk, N, m')$	$m' \leftarrow \mathsf{NE.Dec}(sk_n, y)$
Return (pk, sk)	$c \leftarrow \mathsf{DE.Enc}(pk_d, y)$	$(m \parallel xk \parallel N) \leftarrow m'$
	Return c	Return $(m \parallel xk \parallel N)$

Fig. 11. Nonce-based PKE scheme $\overline{\mathsf{NE}} = \mathsf{NtD}[\mathsf{NE}, \mathsf{DE}]$. It uses the same seed-generating algorithm as NE.

First, the "messages" for DE are the ciphertexts produced by NE. Now, the D-SO-CPA2 security demands that those "messages" must have good min-entropy, but we only know that the combined min-entropy of each message with its nonce and seed is μ. We need a bound, call it $\mathsf{NE.Guess}(\mu)$, to quantify the min-entropy of the ciphertexts of NE. Therefore, let $\mathsf{NE.Guess}(\mu(k))$ be biggest number that, for any seed xk, any nonce N, any message m, and any random variable X such that the conditional min-entropy of (m, xk, N) given X is at least $\mu(k)$, and $(pk, sk) \leftarrow_\$ \mathsf{NE.Kg}(1^k)$ independent of (m, xk, N, X), the conditional min-entropy of $\mathsf{NE.Enc}(pk, xk, N, m)$ given X is at least $\mathsf{NE.Guess}(\mu(k))$. We say that NE is *entropy-preserving* if for any μ such that $2^{-\mu}$ is negligible, so is $2^{-\mathsf{NE.Guess}(\mu)}$. For example, one can show that $\mathsf{NE1}[H, \mathsf{LT}].\mathsf{Guess}(\mu(k)) \geq \min\{k, \mu(k)/2\} - 1$, by modeling $\mathsf{h}_{hk}(\cdot) = H(hk \parallel 0 \parallel \cdot, \mathsf{LT.il}(k))$ as a universal hash function, and using the Generalized Leftover Hash Lemma [1, Lemma 3.4]. Hence NE1 is entropy-preserving.

Next, we need to build an adversary B attacking DE from an adversary A that attacks $\overline{\mathsf{NE}}$. Then B.pg will run $param \leftarrow_\$ A.\mathrm{pg}(1^k)$, pick $(pk, sk) \leftarrow_\$ \mathsf{NE.Kg}(1^k)$, and outputs $pars = (pk, sk, param)$, asking its sampler $\overline{\mathcal{M}}$ to run $\mathcal{M}$ and encrypt the resulting messages, nonces, and seeds under pk. At some point, A.cor will asks to open some ciphertexts $\mathbf{c}[I]$ to get the corresponding $\mathbf{m}[I], \boldsymbol{xk}[I], \boldsymbol{N}[I]$, but the opened "messages" that B.g receives are $\mathsf{NE.Enc}(pk, \boldsymbol{xk}[i, \boldsymbol{N}[i], \mathbf{m}[i])$. Although B.g knows the secret key sk of NE, if we use NE1 to instantiate NE then one can't recover $(\boldsymbol{N}[i], \boldsymbol{xk}[i])$ from just $c_i = \mathsf{NE.Enc}(pk, \boldsymbol{xk}[i], \boldsymbol{N}[i], \mathbf{m}[i])$ and sk. Thanks to our explicit modeling of the auxiliary information, adversary B does get $(\boldsymbol{xk}[i], \boldsymbol{N}[i])$ when it opens $\mathbf{c}[i]$.

Finally, one has to reason about the resamplability of the constructed sampler $\overline{\mathcal{M}}$. Had we restricted our notions to fully resamplable samplers and the function $\delta(\mathbf{m}, param) = \mathbf{m}$, we would have run into problem here. Why so? The resampling algorithm $\overline{\mathsf{Rsmp}}$ of $\overline{\mathcal{M}}$ has to generate $\mathsf{NE.Enc}(pk, \boldsymbol{xk}'[i], \boldsymbol{N}'[i], \mathbf{m}'[i])$, but it only knows pk and another algorithm Rsmp to generate $\mathbf{m}'$. That is, it's unclear how to resample the seeds $\boldsymbol{xk}'$ and nonces $\boldsymbol{N}'$. Using partial resamplability solves this issue. To justify this, suppose that we need to justify the HN-SO-CPA security of $\overline{\mathsf{NE}}$ with respect to function δ. Then, we'll find *another* function $\overline{\delta}$ such that $\mathbf{Adv}^{\text{hn-so-cpa}}_{\overline{\mathsf{NE}}, A, \mathcal{M}, \delta}(\cdot) \leq \mathbf{Adv}^{\text{d-so-cpa2}}_{\mathsf{DE}, B, \overline{\mathcal{M}}, \overline{\delta}}(\cdot)$, and at the same time, $\overline{\mathcal{M}}$ is $\overline{\delta}$-partially

Algorithm $B(1^k, pk)$ Return $A(1^k, pk)$ **Algorithm** $B(state, \mathbf{c})$ $(pk_{\mathrm{d}}, sk_{\mathrm{d}}) \leftarrow\!\!\$\ \mathsf{DE.Kg}(1^k)$ For $i = 1$ to $\lvert\mathbf{c}\rvert$ do $\overline{\mathbf{c}} \leftarrow \mathsf{DE.Enc}(pk_{\mathrm{d}}, \mathbf{c}[i])$ Return $A(state, \overline{\mathbf{c}})$	**Algorithm** $B(state, \boldsymbol{xk}^*)$ Return $A(state, \boldsymbol{xk}^*)$ **Algorithm** $B(state, \mathbf{m}^*, \mathbf{a}^*, \boldsymbol{N}^*, \boldsymbol{xk}^*)$ Return $A(state, \mathbf{m}^*, \mathbf{a}^*, \boldsymbol{N}^*, \boldsymbol{xk}^*)$

Fig. 12. N-SO-CPA adversary B in the proof of Theorem 6.

resamplable. The function $\overline{\delta}(\mathbf{x}, pars)$ works as follows. It first parses *pars* as $(pk, sk, param)$, runs $\mathbf{m}[i] \leftarrow \mathsf{NE.Dec}(sk, \mathbf{x}[i])$, and then outputs $\delta(\mathbf{m}, param)$. We stress that the NtD transform works in both the standard model and the NPROM. (Of course, this assumes that there are standard-model D-SO-CPA2 secure D-PKE and N-SO-CPA secure entropy-preserving nonce-based PKE.)

Theorem 6. Let NE be a nonce-based PKE, and let DE be a D-PKE scheme such that the ciphertext length of the former is a plaintext length of the latter. Let $\overline{\mathsf{NE}} = \mathsf{NtD}[\mathsf{NE}, \mathsf{DE}]$.

N-SO-CPA security: For any adversary A, any message sampler $\mathcal{M}$, any function δ such that $\mathcal{M}$ is δ-partially resamplable, and any nonce generator NG, there is an adversary B such that

$$\mathbf{Adv}^{\text{n-so-cpa}}_{\overline{\mathsf{NE}},\mathsf{NG},A,\mathcal{M},\delta}(\cdot) \leq \mathbf{Adv}^{\text{n-so-cpa}}_{\mathsf{NE},\mathsf{NG},B,\mathcal{M},\delta}(\cdot).$$

The running time of B is about that of A plus the running time of DE.Kg plus the time to run DE.Enc on the messages that $\mathcal{M}$ produces.

HN-SO-CPA security: For any (μ, d)-unpredictable message sampler $\mathcal{M}$, any function δ such that $\mathcal{M}$ is δ-partially resamplable, and any adversary A, there are an adversary B that opens the same number of ciphertexts, another function $\overline{\delta}$, and another $(\mathsf{NE.Guess}(\mu), d)$-entropic, $\overline{\delta}$-partially resamplable message sampler $\overline{\mathcal{M}}$ such that

$$\mathbf{Adv}^{\text{hn-so-cpa}}_{\overline{\mathsf{NE}},A,\mathcal{M},\delta}(\cdot) \leq \mathbf{Adv}^{\text{d-so-cpa2}}_{\mathsf{DE},B,\overline{\mathcal{M}},\overline{\delta}}(\cdot).$$

The running time of B is about that of A plus the running time of $\overline{\mathsf{NE}}$.Kg plus the time to run $\overline{\mathsf{NE}}$.Dec on v ciphertexts, where v is the number of messages that $\mathcal{M}$ produces. The running time of $\overline{\mathcal{M}}$ is about that of $\mathcal{M}$ plus the time to run NE.Enc on v messages.

Proof. For the first part, consider an arbitrary adversary A. Consider the adversary B in Fig. 12 attacking NE. Then game $\text{N-SO-CPA}^{B,\mathcal{M},\delta}_{\mathsf{NE},\mathsf{NG}}$ coincides with game $\text{N-SO-CPA}^{A,\mathcal{M},\delta}_{\overline{\mathsf{NE}},\mathsf{NG}}$, and thus $\mathbf{Adv}^{\text{n-so-cpa}}_{\overline{\mathsf{NE}},\mathsf{NG},A,\mathcal{M},\delta}(\cdot) = \mathbf{Adv}^{\text{n-so-cpa}}_{\mathsf{NE},\mathsf{NG},B,\mathcal{M},\delta}(\cdot)$.

For the second part, consider an arbitrary adversary A and a message sampler $\mathcal{M}$. Consider the following message sampler $\overline{\mathcal{M}}(1^k, pars)$. It parses *param* as a

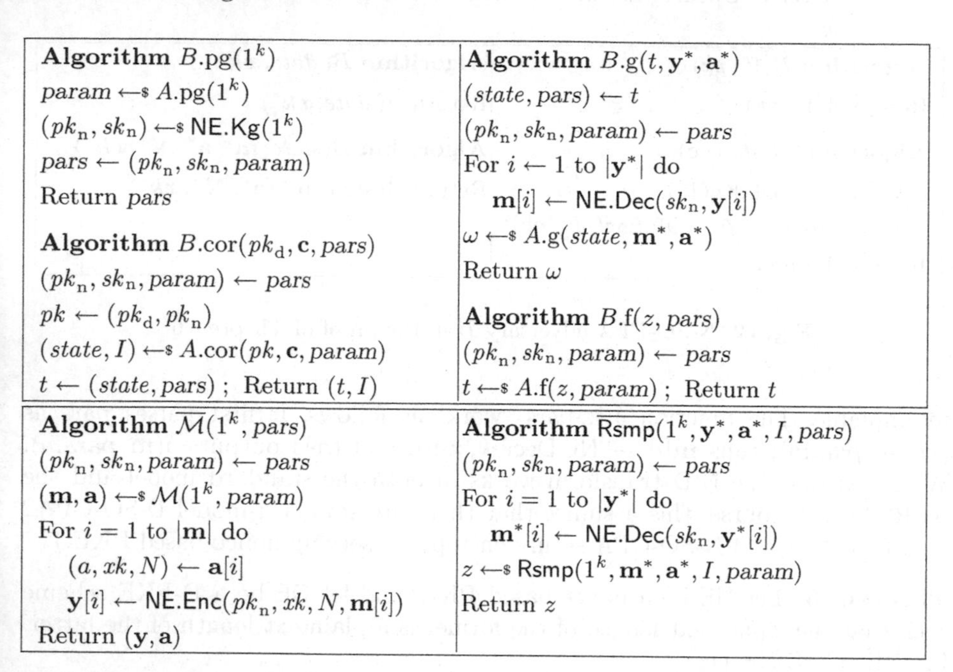
Algorithm B.pg(1^k)
$param \leftarrow_{\$} A$.pg($1^k$)
$(pk_n, sk_n) \leftarrow_{\$}$ NE.Kg(1^k)
$pars \leftarrow (pk_n, sk_n, param)$
Return $pars$

Algorithm B.cor(pk_d, **c**, $pars$)
$(pk_n, sk_n, param) \leftarrow pars$
$pk \leftarrow (pk_d, pk_n)$
$(state, I) \leftarrow_{\$} A$.cor($pk$, **c**, $param$)
$t \leftarrow (state, pars)$; Return (t, I)

Algorithm B.g(t, $\mathbf{y}^*$, $\mathbf{a}^*$)
$(state, pars) \leftarrow t$
$(pk_n, sk_n, param) \leftarrow pars$
For $i \leftarrow 1$ to $|\mathbf{y}^*|$ do
  $\mathbf{m}[i] \leftarrow$ NE.Dec(sk_n, $\mathbf{y}[i]$)
$\omega \leftarrow_{\$} A$.g($state$, $\mathbf{m}^*$, $\mathbf{a}^*$)
Return ω

Algorithm B.f(z, $pars$)
$(pk_n, sk_n, param) \leftarrow pars$
$t \leftarrow_{\$} A$.f($z$, $param$) ; Return t

Algorithm $\overline{\mathcal{M}}(1^k, pars)$
$(pk_n, sk_n, param) \leftarrow pars$
$(\mathbf{m}, \mathbf{a}) \leftarrow_{\$} \mathcal{M}(1^k, param)$
For $i = 1$ to $|\mathbf{m}|$ do
  $(a, xk, N) \leftarrow \mathbf{a}[i]$
  $\mathbf{y}[i] \leftarrow$ NE.Enc(pk_n, xk, N, $\mathbf{m}[i]$)
Return $(\mathbf{y}, \mathbf{a})$

Algorithm $\overline{\mathsf{Rsmp}}(1^k, \mathbf{y}^*, \mathbf{a}^*, I, pars)$
$(pk_n, sk_n, param) \leftarrow pars$
For $i = 1$ to $|\mathbf{y}^*|$ do
  $\mathbf{m}^*[i] \leftarrow$ NE.Dec(sk_n, $\mathbf{y}^*[i]$)
$z \leftarrow_{\$}$ Rsmp(1^k, $\mathbf{m}^*$, $\mathbf{a}^*$, I, $param$)
Return z

Fig. 13. D-SO-CPA2 adversary B, constructed sampler $\overline{\mathcal{M}}$, and its partial resampling algorithm $\overline{\mathsf{Rsmp}}$ in the proof of Theorem 6.

triple $(pk_n, sk_n, param)$, where pk_n and sk_n are public and secret keys for NE. It then runs $\mathcal{M}(1^k, param)$ to generate $(\mathbf{m}, \mathbf{a})$. Since $\mathcal{M}$ is unpredictable, each $\mathbf{a}[i]$ can be parsed as (a_i, xk_i, N_i). Now the "messages" of $\overline{\mathcal{M}}$ is the vector $\mathbf{y}$, where each $\mathbf{y}[i] = \mathsf{NE.Enc}(pk_n, xk_i, N_i, \mathbf{m}[i])$, and the corresponding auxiliary information is still $\mathbf{a}[i]$. The code of $\overline{\mathcal{M}}$ is given in Fig. 13. Since $\mathcal{M}$ is (μ, d)-unpredictable, $\overline{\mathcal{M}}$ is $(\mathsf{NE.Guess}(\mu), d)$-entropic. Let δ be a function such that $\mathcal{M}$ is δ-partially resamplable. Let $\overline{\delta}(\mathbf{y}, pars)$ be the following function. It parses $pars$ as $(pk_n, sk_n, param)$, decrypts $\mathbf{m}[i] \leftarrow \mathsf{NE.Dec}(sk_n, \mathbf{y}[i])$, and then returns $\delta(\mathbf{m}, param)$. Then $\overline{\mathcal{M}}$ is $\overline{\delta}$-partially resamplable: given any δ-partial resampling algorithm Rsmp for $\mathcal{M}$, we can construct a $\overline{\delta}$-partial resampling algorithm $\overline{\mathsf{Rsmp}}$ for $\overline{\mathcal{M}}$ as in Fig. 13.

Now, consider the adversary B attacking DE as given in Fig. 13. It targets message sampler $\overline{\mathcal{M}}$, with respect to function $\overline{\delta}$. Initially, B.pg(1^k) runs $param \leftarrow A(1^k)$, and then generates public and secret keys pk_n and sk_n for NE. It then outputs $pars \leftarrow (pk_n, sk_n, param)$. When B.g receives its "messages" $\mathbf{y}^*$, it extracts the secret key sk_n from its state and decrypts $\mathbf{m}^*[i] \leftarrow \mathsf{NE.Dec}(sk_n, \mathbf{y}^*[i])$, and then gives $\mathbf{m}^*$ to A.g together with the auxiliary information $\mathbf{a}^*$. Then game $\text{HN-CPA-REAL}_{\overline{\mathsf{NE}}}^{A,\mathcal{M},\delta}$ coincides with game $\text{D-CPA2-REAL}_{\mathsf{DE}}^{B,\overline{\mathcal{M}},\overline{\delta}}$. Moreover, game $\text{HN-CPA-IDEAL}_{\overline{\mathsf{NE}}}^{A,\mathcal{M},\delta}$ coincides with $\text{D-CPA2-IDEAL}_{\mathsf{DE}}^{B,\overline{\mathcal{M}},\overline{\delta}}$. Hence $\mathbf{Adv}_{\overline{\mathsf{NE}},A,\mathcal{M},\delta}^{\text{hn-so-cpa}}(\cdot) \le \mathbf{Adv}_{\mathsf{DE},B,\overline{\mathcal{M}},\overline{\delta}}^{\text{d-so-cpa2}}(\cdot)$. □

NE1 ALONE IS ENOUGH. Constructions via NtD transform will be at least twice slower than NE1, because we need to run public primitives twice. But in the NPROM, NE1$[H, \mathsf{LT}]$ alone achieves both N-SO-CPA and HN-SO-CPA security simultaneously. In Theorem 7 below, we'll show that NE1 is HN-SO-CPA secure. See the full version for the proof. We stress that for (μ, ∞)-unpredictable message samplers, NE1 allows the adversary to open as many ciphertexts as it wishes.

Theorem 7. Let LT be a lossy trapdoor function with lossiness τ. Let $\mathcal{M}$ be a (μ, d)-unpredictable resamplable message sampler, and let δ be a function such that $\mathcal{M}$ is δ-partially resamplable. Let NE1$[H, \mathsf{LT}]$ be as above. In the NPROM, for any adversary A opening at most d ciphertexts, there is an adversary D such that

$$\mathbf{Adv}^{\text{hn-so-cpa}}_{\mathsf{NE1}[H,\mathsf{LT}],A,\mathcal{M},\delta}(k) \leq \frac{4q(k)}{2^k} + \frac{4q(k)v(k)}{2^{\mu(k)}} + \frac{v(k)(v(k)+4q(k))}{2^{\tau(k)}} + 2\mathbf{Adv}^{\text{ltdf}}_{\mathsf{LT},D}(k),$$

where $q(k)$ is the total number of random-oracle queries of A and $\mathcal{M}$, and $v(k)$ is the number of messages that $\mathcal{M}$ produces. The running time of D is about that of A plus the time to run δ and an efficient δ-partial resampling algorithm of $\mathcal{M}$ plus the time to run NE1$[H, \mathsf{LT}]$ to encrypt $\mathcal{M}$'s messages. Adversary D makes at most q random-oracle queries.

5.3 Achieving HN-SO-CCA Security

In proving that NtD[NE, DE] achieves HN-SO-CPA security, we don't need any property of the D-PKE scheme DE. This no longer holds for HN-SO-CCA. Indeed, consider a scheme $\mathsf{DE_{bad}}$ such that $\mathsf{DE_{bad}}$.Enc appends 0 to the ciphertexts, and $\mathsf{DE_{bad}}$.Dec ignores the last bit of the ciphertexts. An adversary thus can obtain the plaintexts by modifying the last bits of the ciphertexts, and querying those to the decryption oracle. Hence to obtain HN-SO-CCA, one has to exploit some property of DE. We'll need DE to be *unique-ciphertext*, a property formalized by Bellare and Hoang [7].

Formally, a D-PKE scheme DE is *unique-ciphertext* if for every $k \in \mathbb{N}$, every $(pk, sk) \in [\mathsf{DE.Kg}(1^k)]$, and every $m \in \{0,1\}^*$, there is at most a string c such that $\mathsf{DE.Dec}(sk, c) = m$. The D-PKE scheme $\mathsf{DE_{bad}}$ above is not unique-ciphertext. The unique-ciphertext property of DE ensures that if one modifies a ciphertext of NtD[NE, DE], the underneath ciphertext of NE will be changed.

Bellare and Hoang also show how to efficiently transform a D-PKE scheme DE to a unique-ciphertext one UE: in the decryption, we first recover the message, and then re-encrypt it and return $\perp$ if the newly constructed ciphertext doesn't match the given one. The transform UniqueCtx is given in Fig. 14. Note that this transform doesn't affect the D-SO-CCA security of DE. Indeed, for any message sampler $\mathcal{M}$, any PT adversary A attacking UE = UniqueCtx[DE], it's trivial to construct another PT adversary B attacking DE such that $\mathbf{Adv}^{\text{d-so-cca}}_{\mathsf{UE},B,\mathcal{M}}(\cdot) \leq \mathbf{Adv}^{\text{d-so-cca}}_{\mathsf{DE},A,\mathcal{M}}(\cdot)$.

$\mathsf{UE.Kg}(1^k)$	$\mathsf{UE.Enc}(pk, m)$	$\mathsf{UE.Dec}((pk, sk), c)$

$(pk, sk) \leftarrow_{\$} \mathsf{DE.Kg}(1^k)$
Return $(pk, (pk, sk))$
	$c \leftarrow \mathsf{DE.Enc}(pk, m)$
	Return c
		$m \leftarrow \mathsf{DE.Dec}(sk, c)$
		If $m \neq \bot$ then
		  $c' \leftarrow \mathsf{DE.Enc}(pk, m)$
		  If $c' \neq c$ then return $\bot$
		Return m

Fig. 14. Unique-ciphertext D-PKE scheme $\mathsf{UE} = \mathsf{UniqueCtx}[\mathsf{DE}]$ constructed from D-PKE scheme DE.

Let DE be unique-ciphertext and D-SO-CCA secure D-PKE and NE be an N-SO-CCA secure, entropy-preserving nonce-based PKE. Then, Theorem 8 confirms $\mathsf{NtD}[\mathsf{NE}, \mathsf{DE}]$ achieves both HN-SO-CCA and N-SO-CCA security simultaneously; the proof is in the full version. To instantiate DE, one can either apply the $\mathsf{UniqueCtx}$ transform on a D-SO-CCA secure D-PKE scheme, or directly use our construction DE2 in Sect. 3.3.

Theorem 8. Let NE be a nonce-based PKE as above, and let DE be a unique-ciphertext D-PKE scheme such that the ciphertext length of the former is a plaintext length of the latter. Let $\overline{\mathsf{NE}} = \mathsf{NtD}[\mathsf{NE}, \mathsf{DE}]$.

<u>N-SO-CCA security:</u> For any adversary A, any message sampler $\mathcal{M}$, any function δ such that $\mathcal{M}$ is δ-partially resamplable, and any nonce generator NG, there is an adversary B such that

$$\mathbf{Adv}^{\text{n-so-cca}}_{\overline{\mathsf{NE}},\mathsf{NG},A,\mathcal{M},\delta}(\cdot) \leq \mathbf{Adv}^{\text{n-so-cca}}_{\mathsf{NE},\mathsf{NG},B,\mathcal{M},\delta}(\cdot).$$

The running time of B is about that of A plus the running time of $\mathsf{DE.Kg}$ plus the time to run $\mathsf{DE.Enc}$ on the messages that A produces, and the time to run $\mathsf{DE.Dec}$ on the decryption queries of A. Adversary B makes as many decryption-oracle queries as A.

<u>HN-SO-CCA security:</u> For any adversary A, any (μ, d)-unpredictable message sampler $\mathcal{M}$, and any function δ such that $\mathcal{M}$ is δ-partially resamplable, there are an adversary B that opens the same number of ciphertexts, a function $\overline{\delta}$, and an $(\mathsf{NE.Guess}(\mu), d)$-entropic, $\overline{\delta}$-partially resamplable message sampler $\overline{\mathcal{M}}$ such that

$$\mathbf{Adv}^{\text{hn-so-cca}}_{\overline{\mathsf{NE}},A,\mathcal{M},\delta}(\cdot) \leq \mathbf{Adv}^{\text{d-so-cca}}_{\mathsf{DE},B,\overline{\mathcal{M}},\overline{\delta}}(\cdot).$$

The running time of B is about that of A plus the running time of $\overline{\mathsf{NE}}.\mathsf{Kg}$ plus the time to run $\overline{\mathsf{NE}}.\mathsf{Dec}$ on $v + p$ ciphertexts, where v is the number of messages that $\mathcal{M}$ produces and p is the number of A's decryption-oracle queries. Adversary B makes as many decryption-oracle queries as A. The running time of $\overline{\mathcal{M}}$ is about that of $\mathcal{M}$ plus the time to run $\mathsf{NE.Enc}$ on v messages.

Alternatively, we can use NE2 directly. In Theorem 9 below, we'll show that NE2 is HN-SO-CCA secure. See the full version for the proof.

Theorem 9. Let LT be a lossy trapdoor function with lossiness τ. Let $\mathcal{M}$ be a (μ, d)-unpredictable message sampler, and let δ be a function such that $\mathcal{M}$ is δ-partially resamplable. Let $\mathsf{NE2}[H, \mathsf{LT}]$ be as above. In the NPROM, for any adversary A opening at most d ciphertexts, there is an adversary D such that

$$\mathbf{Adv}^{\text{hn-so-cca}}_{\mathsf{NE2}[H,\mathsf{LT}],A,\mathcal{M},\delta}(k) \leq \frac{6Q(k)}{2^k} + \frac{4Q(k)v(k)}{2^{\mu(k)}} + \frac{v(k)(v(k)+4Q(k))}{2^{\tau(k)}} + 2\mathbf{Adv}^{\text{ltdf}}_{\mathsf{LT},D}(k),$$

where $q(k)$ is the total number of random-oracle queries of A and $\mathcal{M}$, $v(k)$ is the number of messages that $\mathcal{M}$ produces, and $p(k)$ is the number of decryption queries of A, and $Q(k) = q(k) + 2p(k)$. The running time of D is about that of A plus the time to run δ and a δ-partial resampling algorithm of $\mathcal{M}$ plus the time to run $\mathsf{NE2}[H, \mathsf{LT}]$ to encrypt $\mathcal{M}$'s messages. Adversary D makes at most Q random-oracle queries.

Acknowledgments. We thank the Asiacrypt reviewers for their insightful comments. This work was supported in part by NSF awards CNS-1223623, CNS-1423566, and CNS-1553758 (CAREER), as well as the Glen and Susanne Culler Chair. The work of Viet Tung Hoang was done while at the University of Maryland, Georgetown University, and UC Santa Barbara.

References

1. Barak, B., Dodis, Y., Krawczyk, H., Pereira, O., Pietrzak, K., Standaert, F.-X., Yu, Y.: Leftover hash lemma, revisited. In: Rogaway, P. (ed.) CRYPTO 2011. LNCS, vol. 6841, pp. 1–20. Springer, Heidelberg (2011). doi:10.1007/978-3-642-22792-9_1
2. Bellare, M., Boldyreva, A., O'Neill, A.: Deterministic and efficiently searchable encryption. In: Menezes, A. (ed.) CRYPTO 2007. LNCS, vol. 4622, pp. 535–552. Springer, Heidelberg (2007). doi:10.1007/978-3-540-74143-5_30
3. Bellare, M., Brakerski, Z., Naor, M., Ristenpart, T., Segev, G., Shacham, H., Yilek, S.: Hedged public-key encryption: how to protect against bad randomness. In: Matsui, M. (ed.) ASIACRYPT 2009. LNCS, vol. 5912, pp. 232–249. Springer, Heidelberg (2009). doi:10.1007/978-3-642-10366-7_14
4. Bellare, M., Dowsley, R., Keelveedhi, S.: How secure is deterministic encryption? In: Katz, J. (ed.) PKC 2015. LNCS, vol. 9020, pp. 52–73. Springer, Heidelberg (2015). doi:10.1007/978-3-662-46447-2_3
5. Bellare, M., Dowsley, R., Waters, B., Yilek, S.: Standard security does not imply security against selective-opening. In: Pointcheval, D., Johansson, T. (eds.) EUROCRYPT 2012. LNCS, vol. 7237, pp. 645–662. Springer, Heidelberg (2012). doi:10.1007/978-3-642-29011-4_38
6. Bellare, M., Fischlin, M., O'Neill, A., Ristenpart, T.: Deterministic encryption: definitional equivalences and constructions without random oracles. In: Wagner, D. (ed.) CRYPTO 2008. LNCS, vol. 5157, pp. 360–378. Springer, Heidelberg (2008). doi:10.1007/978-3-540-85174-5_20
7. Bellare, M., Hoang, V.T.: Resisting randomness subversion: fast deterministic and hedged public-key encryption in the standard model. In: Oswald, E., Fischlin, M. (eds.) EUROCRYPT 2015. LNCS, vol. 9057, pp. 627–656. Springer, Heidelberg (2015). doi:10.1007/978-3-662-46803-6_21

8. Bellare, M., Hoang, V.T., Keelveedhi, S.: Instantiating random oracles via UCEs. In: Canetti, R., Garay, J.A. (eds.) CRYPTO 2013. LNCS, vol. 8043, pp. 398–415. Springer, Heidelberg (2013). doi:10.1007/978-3-642-40084-1_23
9. Bellare, M., Hofheinz, D., Yilek, S.: Possibility and impossibility results for encryption and commitment secure under selective opening. In: Joux, A. (ed.) EUROCRYPT 2009. LNCS, vol. 5479, pp. 1–35. Springer, Heidelberg (2009). doi:10.1007/978-3-642-01001-9_1
10. Bellare, M., Tackmann, B.: Nonce-based cryptography: retaining security when randomness fails. In: Fischlin, M., Coron, J.-S. (eds.) EUROCRYPT 2016. LNCS, vol. 9665, pp. 729–757. Springer, Heidelberg (2016). doi:10.1007/978-3-662-49890-3_28
11. Böhl, F., Hofheinz, D., Kraschewski, D.: On definitions of selective opening security. In: Fischlin, M., Buchmann, J., Manulis, M. (eds.) PKC 2012. LNCS, vol. 7293, pp. 522–539. Springer, Heidelberg (2012). doi:10.1007/978-3-642-30057-8_31
12. Cash, D., Grubbs, P., Perry, J., Ristenpart, T.: Leakage-abuse attacks against searchable encryption. In CCS, pp. 668–679 (2015)
13. Fujisaki, E., Okamoto, T.: How to enhance the security of public-key encryption at minimum cost. In: Imai, H., Zheng, Y. (eds.) PKC 1999. LNCS, vol. 1560, pp. 53–68. Springer, Heidelberg (1999). doi:10.1007/3-540-49162-7_5
14. Heuer, F., Kiltz, E., Pietrzak, K.: Standard security does imply security against selective opening for markov distributionss. Cryptology ePrint Archive, Report 2015/853 (2015). http://eprint.iacr.org/2015/853
15. Hofheinz, D., Rao, V., Wichs, D.: Standard security does not imply indistinguishability under selective opening. Cryptology ePrint Archive, Report 2015/792 (2015). http://eprint.iacr.org/2015/792
16. Hofheinz, D., Rupp, A.: Standard versus selective opening security: separation and equivalence results. In: Lindell, Y. (ed.) TCC 2014. LNCS, vol. 8349, pp. 591–615. Springer, Heidelberg (2014). doi:10.1007/978-3-642-54242-8_25
17. Ishai, Y., Pandey, O., Sahai, A.: Public-coin differing-inputs obfuscation and its applications. In: Dodis, Y., Nielsen, J.B. (eds.) TCC 2015. LNCS, vol. 9015, pp. 668–697. Springer, Heidelberg (2015). doi:10.1007/978-3-662-46497-7_26
18. Kamara, S., Katz, J.: How to encrypt with a malicious random number generator. In: Nyberg, K. (ed.) FSE 2008. LNCS, vol. 5086, pp. 303–315. Springer, Heidelberg (2008). doi:10.1007/978-3-540-71039-4_19
19. Kiltz, E., O'Neill, A., Smith, A.: Instantiability of RSA-OAEP under chosen-plaintext attack. In: Rabin, T. (ed.) CRYPTO 2010. LNCS, vol. 6223, pp. 295–313. Springer, Heidelberg (2010). doi:10.1007/978-3-642-14623-7_16
20. Nielsen, J.B.: Separating random oracle proofs from complexity theoretic proofs: the non-committing encryption case. In: Yung, M. (ed.) CRYPTO 2002. LNCS, vol. 2442, pp. 111–126. Springer, Heidelberg (2002). doi:10.1007/3-540-45708-9_8
21. Peikert, C., Waters, B.: Lossy trapdoor functions and their applications. In: Ladner, R.E., Dwork, C. (eds.), 40th ACM STOC, pp. 187–196, Victoria, British Columbia, Canada, May 17–20. ACM Press (2008)
22. Seurin, Y.: On the lossiness of the rabin trapdoor function. In: Krawczyk, H. (ed.) PKC 2014. LNCS, vol. 8383, pp. 380–398. Springer, Heidelberg (2014). doi:10.1007/978-3-642-54631-0_22
23. Yilek, S.: Resettable public-key encryption: how to encrypt on a virtual machine. In: Pieprzyk, J. (ed.) CT-RSA 2010. LNCS, vol. 5985, pp. 41–56. Springer, Heidelberg (2010). doi:10.1007/978-3-642-11925-5_4

Efficient KDM-CCA Secure Public-Key Encryption for Polynomial Functions

Shuai Han[1,2], Shengli Liu[1,2,3](✉), and Lin Lyu[1,2]

[1] Department of Computer Science and Engineering, Shanghai Jiao Tong University, Shanghai 200240, China
{dalen17,slliu,lvlin}@sjtu.edu.cn

[2] State Key Laboratory of Cryptology, P.O. Box 5159, Beijing 100878, China

[3] Westone Cryptologic Research Center, Beijing 100070, China

Abstract. KDM[$\mathcal{F}$]-CCA secure public-key encryption (PKE) protects the security of message $f(sk)$, with $f \in \mathcal{F}$, that is computed directly from the secret key, even if the adversary has access to a decryption oracle. An efficient KDM[$\mathcal{F}_{\text{aff}}$]-CCA secure PKE scheme for affine functions was proposed by Lu, Li and Jia (LLJ, EuroCrypt2015). We point out that their security proof cannot go through based on the DDH assumption.

In this paper, we introduce a new concept *Authenticated Encryption with Auxiliary-Input* AIAE and define for it new security notions dealing with related-key attacks, namely *IND-RKA security* and *weak INT-RKA security*. We also construct such an AIAE w.r.t. a set of restricted affine functions from the DDH assumption. With our AIAE,

- we construct the first efficient KDM[$\mathcal{F}_{\text{aff}}$]-CCA secure PKE w.r.t. affine functions with compact ciphertexts, which consist only of a constant number of group elements;
- we construct the first efficient KDM[$\mathcal{F}^d_{\text{poly}}$]-CCA secure PKE w.r.t. polynomial functions of bounded degree d with almost compact ciphertexts, and the number of group elements in a ciphertext is polynomial in d, independent of the security parameter.

Our PKEs are both based on the DDH & DCR assumptions, free of NIZK and free of pairing.

Keywords: Public-key encryption · Key-dependent messages · Chosen-ciphertext security · Authenticated encryption · Related-key attack

1 Introduction

Traditional Chosen-Ciphertext Attack (CCA) security of a public-key encryption (PKE) scheme considers the security of messages chosen by an adversary, even if the adversary obtains the public key pk, challenge ciphertexts of the messages, and has access to a decryption oracle (which provides decryption services to the adversary but refuses to decrypt the challenge ciphertexts). Note that the adversary cannot compute messages directly from secret keys, since it does not possess the secret keys. Therefore, CCA security does not cover the corner, where

J.H. Cheon and T. Takagi (Eds.): ASIACRYPT 2016, Part II, LNCS 10032, pp. 307–338, 2016.
DOI: 10.1007/978-3-662-53890-6_11

messages closely depend on the secret keys, say the secret keys themselves or functions of the secret keys. This issue was first identified in [GM84]. Later the security of key-dependent messages was formalized as KDM-security [BRS02]. KDM-security is an important notion, and has found wide applications, like hard disk encryption [BHHO08], cryptographic protocols [CL01], etc.

KDM-security w.r.t. a set of functions $\mathcal{F}$ is denoted by KDM[$\mathcal{F}$]-security. The larger $\mathcal{F}$ is, the stronger the security is. Roughly speaking, n-KDM[$\mathcal{F}$]-security of PKE considers such a scenario: an adversary is given public keys $(pk_1, pk_2, \cdots, pk_n)$ of n users and an encryption oracle. Whenever the adversary queries a function $f \in \mathcal{F}$, the encryption oracle will always reply with an encryption of a constant say 0, or always reply with an encryption of $f(sk_1, sk_2, \cdots, sk_n)$. If the adversary cannot tell which case it is, the PKE is n-KDM[$\mathcal{F}$]-CPA secure. If the adversary has also access to a decryption oracle in the scenario, then KDM[$\mathcal{F}$]-CPA security is improved to KDM[$\mathcal{F}$]-CCA security. Obviously, KDM-CCA security notion is stronger than KDM-CPA.

KDM[$\mathcal{F}$]-CPA Security. The BHHO scheme [BHHO08] was the first PKE achieving KDM[$\mathcal{F}_{\text{aff}}$]-CPA security based on the Decisional Diffie-Hellman (DDH) assumption, where $\mathcal{F}_{\text{aff}}$ denotes the set of affine functions. It was later generalized by Brakerski and Goldwasser [BG10] to KDM[$\mathcal{F}_{\text{aff}}$]-CPA secure PKE schemes based on the Subgroup Indistinguishability Assumption (including the QR and the DCR assumptions). These schemes have incompact ciphertexts containing $O(\ell)$ group elements, where ℓ denotes the security parameter.

A variant of Regev's scheme [Reg05] was shown to be KDM[$\mathcal{F}_{\text{aff}}$]-CPA secure and has compacter ciphertexts by Applebaum et al. [ACPS09].

Barak et al. [BHHI10] proposed KDM-CPA secure PKE w.r.t. a very large function set, i.e., the function set of boolean circuits of bounded size $p = p(\ell)$. However, their scheme is inflexible and highly impractical, since its encryption algorithm depends on the bound p and the number of users, and the ciphertext contains a garbled circuit of size at least $p = p(\ell)$.

Brakerski et al. [BGK11] amplified the BHHO scheme to KDM[$\mathcal{F}^d_{\text{poly}}$]-CPA security w.r.t. the set of polynomial functions of bounded degree d. However, their ciphertext contains $O(\ell^{d+1})$ group elements.

It is Malkin et al. [MTY11] who designed the first efficient PKE scheme achieving KDM[$\mathcal{F}^d_{\text{poly}}$]-CPA security. Their ciphertext contains only $O(d)$ group elements, thus d can be polynomial in ℓ in their case. The function set $\mathcal{F}^d_{\text{poly}}$ is characterized by a polynomial-size *Modular Arithmetic Circuit* in [MTY11].

KDM[$\mathcal{F}$]-CCA Security. KDM[$\mathcal{F}$]-CCA security of PKE is far more difficult to design than KDM[$\mathcal{F}$]-CPA security. Camenisch et al. [CCS09] gave the first solution, following Naor-Yung's paradigm, which needs a KDM-CPA secure PKE, a CCA-secure PKE and a non-interactive zero-knowledge (NIZK) proving that the two PKEs encrypt the same message.

NIZK is not practical in general, except Groth-Sahai proofs [GS08]. When following [CCS09]'s approach, the only possible way to get an efficient KDM-CCA secure PKE, is using Groth-Sahai proofs together with an efficient

KDM-CPA secure PKE. However, many existing efficient KDM-CPA secure schemes, such as [ACPS09,MTY11], are not based on pairing-friendly groups, thus not compatible with Groth-Sahai's efficient NIZK.

Another work by Galindo et al. [GHV12] is based on the Matrix DDH assumption over pairing-friendly groups. Their scheme has compact ciphertexts, but only obtains a bounded form of KDM-CCA security, i.e., the number of encryption queries is limited to be linear in the size of the secret key.

To get an efficient KDM-CCA secure PKE, Hofheinz [Hof13] proposed another approach, which uses a new tool called "lossy algebraic filter". His work results in the first PKE enjoying both KDM-CCA security and compact ciphertexts (consisting only of a constant number of group elements). However, the function set $\mathcal{F}_{\text{circ}}$ only consists of selection functions $f(sk_1, \cdots, sk_n) = sk_i$ and constant functions.

It is quite challenging to enlarge $\mathcal{F}$ for KDM[$\mathcal{F}$]-CCA security while still keeping PKE efficient. One effort was recently made by Lu, Li and Jia [LLJ15], who proposed the first efficient KDM[$\mathcal{F}_{\text{aff}}$]-CCA secure PKE with compact ciphertexts. We call their construction the LLJ scheme. There is an essential building block called "Authenticated Encryption" ($\overline{\mathsf{AE}}$) in their scheme. The KDM[$\mathcal{F}_{\text{aff}}$]-CCA security heavily relies on a so-called INT-$\mathcal{F}_{\text{aff}}$-RKA security of $\overline{\mathsf{AE}}$. INT-$\mathcal{F}_{\text{aff}}$-RKA security of $\overline{\mathsf{AE}}$ means that a PPT adversary cannot forge a fresh forgery $(f^*, \overline{\mathsf{ae}}.\mathsf{ct}^*)$ such that $\overline{\mathsf{AE}}.\mathsf{Dec}_{f^*(\mathsf{k})}(\overline{\mathsf{ae}}.\mathsf{ct}^*) \neq \perp$, even if the adversary observes multiple outputs of $\overline{\mathsf{AE}}.\mathsf{Enc}_{f_j(\mathsf{k})}(m_j)$ with his choice of (f_j, m_j). Unfortunately, we found that the INT-$\mathcal{F}_{\text{aff}}$-RKA security proof of the specific $\overline{\mathsf{AE}}$ does not go through to the DDH assumption, which in turn affects the KDM[$\mathcal{F}_{\text{aff}}$]-CCA security proof of the LLJ scheme. Our essential observation is that the DDH adversary is not able to employ the fresh forgery from the adversary of $\overline{\mathsf{AE}}$ to solve the DDH problem, since the DDH adversary does not have any trapdoor to convert the computing power (forgery) to a decision bit.

As for KDM[$\mathcal{F}^d_{\text{poly}}$]-CCA security, [CCS09]'s paradigm is the unique path to it up to now. Unfortunately, the only efficient KDM[$\mathcal{F}^d_{\text{poly}}$]-CPA secure scheme [MTY11] does not compose well with Groth-Sahai proofs, so it has to resort to the general NIZK. Other KDM[$\mathcal{F}^d_{\text{poly}}$]-CPA secure schemes either is highly impractical [BHHI10] or has ciphertext containing $O(\ell^{d+1})$ group elements [BGK11], which grows exponentially with the degree d.

Our Contribution. We work on the design of efficient PKE with KDM[$\mathcal{F}_{\text{aff}}$]-CCA security and KDM[$\mathcal{F}^d_{\text{poly}}$]-CCA security.

- We identify the proof flaw in [LLJ15], where an efficient KDM[$\mathcal{F}_{\text{aff}}$]-CCA secure PKE was claimed. We show that for "Authenticated Encryption" ($\overline{\mathsf{AE}}$) used in the LLJ scheme, the INT-$\mathcal{F}_{\text{aff}}$-RKA security reduction to the DDH assumption does not work. This proof flaw directly affects the KDM[$\mathcal{F}_{\text{aff}}$]-CCA security proof of the LLJ scheme.
- We provide the first efficient KDM[$\mathcal{F}_{\text{aff}}$]-CCA secure PKE w.r.t. affine functions with compact ciphertexts. Our scheme has ciphertexts consisting only of a constant number of group elements and is free of NIZK.

– We provide the first efficient KDM[$\mathcal{F}^d_{\text{poly}}$]-CCA secure PKE w.r.t. polynomial functions of bounded degree d with almost compact ciphertexts. Our scheme is free of NIZK. The number of group elements in a ciphertext is polynomial in d, independent of the security parameter ℓ.

We summarize known PKEs either achieving KDM-CCA security or against function set $\mathcal{F}^d_{\text{poly}}$ in Table 1.

Table 1. Comparison between PKEs either achieving KDM-CCA security or against function set $\mathcal{F}^d_{\text{poly}}$. Here ℓ is the security parameter. $\mathcal{F}_{\text{circ}}$, $\mathcal{F}_{\text{aff}}$ and $\mathcal{F}^d_{\text{poly}}$ denote the set of selection functions, the set of affine functions and the set of polynomial functions of bounded degree d, respectively. "CCA" means the scheme is KDM-CCA secure. "Free of Pairing" asks whether the scheme is free of pairing. $|\mathsf{CT}|$ shows the size of ciphertext. $\mathbb{G}$, $\mathbb{Z}_{N^3}$, $\mathbb{Z}_{N^2}$ and $\mathbb{Z}_{\bar{N}}$ are the underlying groups. s can be any integer greater than 1. The symbol "?" means that the security proof is not rigorous.

Scheme	Set	CCA?	Free of Pairing?	$\|\mathsf{CT}\|$	Assumption
[BHHO08] + [CCS09]	$\mathcal{F}_{\text{aff}}$	√	–	$(6\ell+13)\|\mathbb{G}\|$	DDH
[BGK11]	$\mathcal{F}^d_{\text{poly}}$	–	√	$(\ell^{d+1})\|\mathbb{G}\|$	DDH or LWE
[MTY11]	$\mathcal{F}^d_{\text{poly}}$	–	√	$(d+2)\|\mathbb{Z}_{N^s}\|$	DCR
[Hof13]	$\mathcal{F}_{\text{circ}}$	√	–	$6\|\mathbb{Z}_{N^3}\| + 49\|\mathbb{G}\|$	DDH & DCR
[LLJ15]	$\mathcal{F}_{\text{aff}}$	?	√	$3\|\mathbb{Z}_{N^2}\| + 3\|\mathbb{Z}_{N^s}\| + \|\mathbb{Z}_{\bar{N}}\|$	DDH & DCR
Our scheme in Sect. 5	$\mathcal{F}_{\text{aff}}$	√	√	$9\|\mathbb{Z}_{N^2}\| + 9\|\mathbb{Z}_{N^s}\| + 2\|\mathbb{Z}_{\bar{N}}\|$	DDH & DCR
Our scheme in Sect. 6	$\mathcal{F}^d_{\text{poly}}$	√	√	$9\|\mathbb{Z}_{N^2}\| + (8d^9+1)\|\mathbb{Z}_{N^s}\| + 2\|\mathbb{Z}_{\bar{N}}\|$	DDH & DCR

Our Approach. The challenge for KDM[$\mathcal{F}$]-CCA security of PKE lies in the fact that the adversary $\mathcal{A}$ has multiple access to the encryptions of $f(sk)$ and decryption oracle $\mathsf{Dec}(sk,\cdot)$, with $f \in \mathcal{F}$ and sk the secret key. Let us consider only one secret key for simplicity. The information of sk might be leaked completely via encryptions of $f(sk)$.

To solve this problem, we follow a KEM+DEM style and construct our PKE with three building blocks: KEM, $\mathcal{E}$ and AIAE, as shown in Fig. 1.

- We propose a new concept *"Authenticated Encryption with Auxiliary-Input"* (AIAE). We define for it new security notions dealing with related-key attacks, namely *weak INT-$\mathcal{F}'$-RKA security* and IND-$\mathcal{F}'$-RKA security.
- We design the other building blocks KEM and $\mathcal{E}$. KEM.Enc encapsulates a key k for AIAE, and the encapsulation kem.ct serves as an auxiliary input aux for AIAE.Enc. $\mathcal{E}$.Enc encrypts m to get a ciphertext $\mathcal{E}$.ct, which serves as an input for AIAE.Enc.

We show how to achieve KDM[$\mathcal{F}$]-CCA security with our three building blocks.

- $\mathcal{E}$.Enc can behave like an entropy filter (the concept was named in [LLJ15]) for $\mathcal{F}$. That is, through some computationally indistinguishable change, some entropy of sk is always reserved even if multiple encryptions of $f_j(sk)$ are given to $\mathcal{A}$. Here $f_j \in \mathcal{F}$ is chosen by $\mathcal{A}$.
- The fresh keys k_j used by AIAE.Enc can be expressed as functions of a base key k^*, i.e., $\mathsf{k}_j = f'_j(\mathsf{k}^*)$, where $f'_j \in \mathcal{F}'$ for some function set $\mathcal{F}'$. We stress that $\mathcal{F}'$ might be different from $\mathcal{F}$.
- KEM.Enc is able to use the remaining entropy of sk to protect the base key k^*, via some computationally indistinguishable change.
- The weak INT-$\mathcal{F}'$-RKA security of AIAE guarantees: given multiple AIAE ciphertext-auxiliary input pair $(\mathsf{aiae.ct}_j, \mathsf{aux}_j)$ encrypted by $f'_j(\mathsf{k}^*)$, it is infeasible for a PPT algorithm to forge a new $(f', \mathsf{aiae.ct}, \mathsf{aux})$ satisfying (1) $\mathsf{AIAE.Dec}_{f'(\mathsf{k}^*)}(\mathsf{aiae.ct}, \mathsf{aux}) \neq \perp$; (2) if $\mathsf{aux} = \mathsf{aux}_j$ for some j then $f' = f'_j$.
- Decryption oracle can reject all invalid ciphertexts that are not properly generated by the encryption algorithm, via some computationally indistinguishable change. If the invalid ciphertext makes KEM.Dec decapsulate a key $f'(\mathsf{k}^*)$, AIAE.Dec will output $\perp$, due to its weak INT-$\mathcal{F}'$-RKA security. Otherwise, the invalid ciphertext will be rejected by $\mathcal{E}$.Dec or KEM.Dec, due to the remaining entropy of sk. As a result, no extra information about sk is leaked.
- The IND-$\mathcal{F}'$-RKA security of AIAE ensures: given multiple AIAE ciphertext-auxiliary input pair $(\mathsf{aiae.ct}_j, \mathsf{aux}_j)$ with key $f'_j(\mathsf{k}^*)$ encrypting either m_0 or m_1, it is infeasible for a PPT algorithm to distinguish which case it is, even if $f'_j \in \mathcal{F}'$ is submitted by the algorithm.
- By the IND-$\mathcal{F}'$-RKA security of AIAE, the encryption of $\mathcal{E}$.ct can be replaced with an encryption of all zeros. Then the KDM[$\mathcal{F}$]-CCA security follows.

With this approach, we can construct PKEs possessing KDM[$\mathcal{F}_{\text{aff}}$]-CCA and KDM[$\mathcal{F}^d_{\text{poly}}$]-CCA security respectively, by designing specific building blocks.

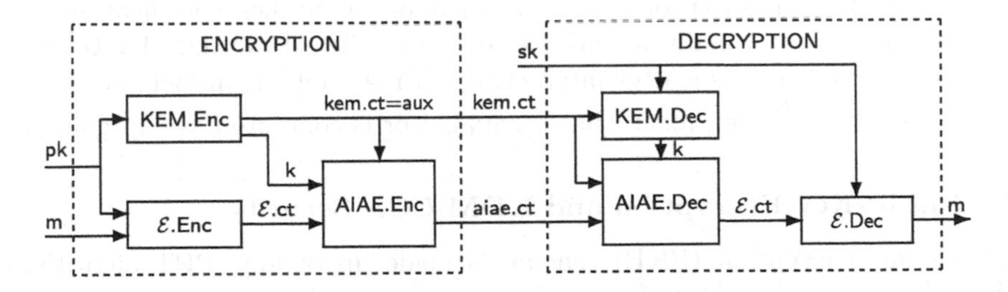

Fig. 1. Our approach of PKE construction. Here KEM and $\mathcal{E}$ share the same public/secret key pair. AIAE.Enc uses k output by KEM to encrypt $\mathcal{E}$.ct with auxiliary input aux := kem.ct, and outputs ciphertext aiae.ct.

Comparison with LLJ. We inherit the idea of utilizing RKA security of AE to achieve KDM security from LLJ. However, our approach deviates from LLJ in three aspects.

1. The structure of our scheme is different from LLJ. It is also possible to explain the LLJ scheme with three components KEM, $\mathcal{E}$ and $\overline{\mathsf{AE}}$. However, their components were composed in a different way. In the LLJ scheme, the output $\mathsf{kem.ct}$ of KEM serves as an additional input for $\mathcal{E}.\mathsf{Enc}$. With their structure, $\mathcal{E}$ is expected to authenticate $\mathsf{kem.ct}$. In our approach, $\mathsf{kem.ct}$ is the auxiliary input of AIAE, thus can be authenticated by AIAE.
2. The syntax and security requirements of our AIAE are different from LLJ's $\overline{\mathsf{AE}}$. Their $\overline{\mathsf{AE}}$ does not support auxiliary input, and the security proof of their $\overline{\mathsf{AE}}$ instantiation has some problem, as shown in Sect. 3.
3. Our KEM and $\mathcal{E}$ are newly designed building blocks which compose well with our AIAE. We give two designs of $\mathcal{E}$ to support KDM[$\mathcal{F}_{\text{aff}}$]-CCA and KDM[$\mathcal{F}^d_{\text{poly}}$]-CCA security respectively.

2 Preliminaries

Let $\ell \in \mathbb{N}$ denote the security parameter. For $i, j \in \mathbb{N}$ with $i < j$, define $[i, j] := \{i, i+1, \cdots, j\}$ and $[j] := \{1, 2, \cdots, j\}$. Denote by $s \leftarrow_\$ \mathcal{S}$ the operation of picking an element s from set $\mathcal{S}$ uniformly at random. For an algorithm $\mathcal{A}$, denote by $y \leftarrow_\$ \mathcal{A}(x; r)$, or simply $y \leftarrow_\$ \mathcal{A}(x)$, the operation of running $\mathcal{A}$ with input x and randomness r and assigning output to y. Let ε denote the empty string. For a primitive XX and a security notion YY, we typically denote the advantage of a PPT adversary $\mathcal{A}$ by $\mathsf{Adv}^{\text{YY}}_{\text{XX},\mathcal{A}}(\ell)$ and define $\mathsf{Adv}^{\text{YY}}_{\text{XX}}(\ell) := \max_{\text{PPT}\mathcal{A}} \mathsf{Adv}^{\text{YY}}_{\text{XX},\mathcal{A}}(\ell)$. Let $2^{-\Omega(\ell)}$ denote the value upper bounded by $2^{-c\cdot\ell}$ for some constant $c > 0$.

Games. Our security proof will be game-based security reductions. A game G starts with an INITIALIZE procedure and ends with a FINALIZE procedure. There are also some optional procedures $\text{PROC}_1, \cdots, \text{PROC}_n$ performing as oracles. All procedures are described using pseudo-code, where initially all variables are empty strings ε and all sets are empty. An adversary $\mathcal{A}$ is executed in game G if it first calls INITIALIZE, obtaining its output. Then the adversary may make arbitrary oracle-queries to procedures PROC_i according to their specification, and obtain their outputs. Finally it makes one single call to FINALIZE. By $\mathsf{G}^{\mathcal{A}} \Rightarrow b$ we means that G outputs b after interacting with $\mathcal{A}$, and b is in fact the output of FINALIZE. By $a \overset{\mathsf{G}}{=} b$ we mean that a equals b or is computed as b in game G.

2.1 Public-Key Encryption and KDM-CCA Security

A public-key encryption (PKE) scheme is made up of four PPT algorithms $\mathsf{PKE} = (\mathsf{Setup}, \mathsf{Gen}, \mathsf{Enc}, \mathsf{Dec})$: $\mathsf{Setup}(1^\ell)$ generates a public parameter prm, which implicitly defines a secret key space $\mathcal{SK}$ and a message space $\mathcal{M}$; $\mathsf{Gen}(\mathsf{prm})$ takes as input the public parameter prm and generates a public/secret key pair $(\mathsf{pk}, \mathsf{sk})$; $\mathsf{Enc}(\mathsf{pk}, m)$ takes as input the public key pk and a message m, and outputs a ciphertext $\mathsf{pke.ct}$; $\mathsf{Dec}(\mathsf{sk}, \mathsf{pke.ct})$ takes as input the secret key sk and a ciphertext $\mathsf{pke.ct}$ and outputs either a message m or a failure symbol $\perp$. The correctness of PKE requires that, for all $\mathsf{prm} \leftarrow_\$ \mathsf{Setup}(1^\ell)$, all $(\mathsf{pk}, \mathsf{sk}) \leftarrow_\$ \mathsf{Gen}(\mathsf{prm})$, all $m \in \mathcal{M}$ and all $\mathsf{pke.ct} \leftarrow_\$ \mathsf{Enc}(\mathsf{pk}, m)$, it holds that $\mathsf{Dec}(\mathsf{sk}, \mathsf{pke.ct}) = m$.

Procedure INITIALIZE:	**Procedure** ENC($f \in \mathcal{F}, i \in [n]$):
$\mathsf{prm} \leftarrow_\$ \mathsf{Setup}(1^\ell)$. For $i \in [n]$ $(\mathsf{pk}_i, \mathsf{sk}_i) \leftarrow_\$ \mathsf{Gen}(\mathsf{prm})$. $\beta \leftarrow_\$ \{0,1\}$. // challenge bit Return $(\mathsf{prm}, \mathsf{pk}_1, \cdots, \mathsf{pk}_n)$.	$m_1 := f(\mathsf{sk}_1, \cdots, \mathsf{sk}_n)$, $m_0 := 0^{\lvert m_1 \rvert}$. $\mathsf{pke.ct} \leftarrow_\$ \mathsf{Enc}(\mathsf{pk}_i, m_\beta)$. $\mathcal{Q}_{\mathcal{ENC}} := \mathcal{Q}_{\mathcal{ENC}} \cup \{(\mathsf{pke.ct}, i)\}$. Return $\mathsf{pke.ct}$.
Procedure FINALIZE(β'):	**Procedure** DEC($\mathsf{pke.ct}, i \in [n]$):
Return $(\beta' = \beta)$.	If $(\mathsf{pke.ct}, i) \in \mathcal{Q}_{\mathcal{ENC}}$, Return $\perp$. Return $\mathsf{Dec}(\mathsf{sk}_i, \mathsf{pke.ct})$.

Fig. 2. n-KDM[$\mathcal{F}$]-CCA security game for PKE.

Let $n \in \mathbb{N}$ and $\mathcal{F}$ be a family of functions from $\mathcal{SK}^n$ to $\mathcal{M}$. We define the n-KDM[$\mathcal{F}$]-CCA security via the security game in Fig. 2.

Definition 1 (KDM[$\mathcal{F}$]-CCA Security for PKE). *Scheme* PKE *is* n-*KDM[$\mathcal{F}$]-CCA secure if for any PPT adversary* $\mathcal{A}$, $\mathsf{Adv}^{kdm\text{-}cca}_{\mathsf{PKE},\mathcal{A}}(\ell) := |\Pr[n\text{-}\mathsf{KDM}[\mathcal{F}]\text{-}\mathsf{CCA}^{\mathcal{A}} \Rightarrow 1] - 1/2|$ *is negligible in* ℓ, *where game* n-KDM[$\mathcal{F}$]-CCA *is specified in Fig. 2.*

2.2 Key Encapsulation Mechanism

A key encapsulation mechanism (KEM) consists of three PPT algorithms $\mathsf{KEM} = (\mathsf{KEM.Gen}, \mathsf{KEM.Enc}, \mathsf{KEM.Dec})$: $\mathsf{KEM.Gen}(1^\ell)$ outputs a public/secret key pair $(\mathsf{pk}, \mathsf{sk})$; $\mathsf{KEM.Enc}(\mathsf{pk})$ uses the public key pk to compute a key k and a ciphertext (or encapsulation) $\mathsf{kem.ct}$; $\mathsf{KEM.Dec}(\mathsf{sk}, \mathsf{kem.ct})$ takes as input the secret key sk and a ciphertext $\mathsf{kem.ct}$, and outputs either a key k or a failure symbol $\perp$. The correctness of KEM requires that, for all $(\mathsf{pk}, \mathsf{sk}) \leftarrow_\$ \mathsf{KEM.Gen}(1^\ell)$ and all $(\mathsf{k}, \mathsf{kem.ct}) \leftarrow_\$ \mathsf{KEM.Enc}(\mathsf{pk})$, it holds that $\mathsf{KEM.Dec}(\mathsf{sk}, \mathsf{kem.ct}) = \mathsf{k}$.

2.3 Authenticated Encryption: One-Time Security and Related-Key Attack Security

Definition 2 (Authenticated Encryption). *An authenticated encryption (AE) scheme* $\mathsf{AE} = (\mathsf{AE.Setup}, \mathsf{AE.Enc}, \mathsf{AE.Dec})$ *consists of three PPT algorithms:*

- $\mathsf{AE.Setup}(1^\ell)$ *outputs a system parameter* $\mathsf{prm}_{\mathsf{AE}}$, *which is an implicit input to* $\mathsf{AE.Enc}$ *and* $\mathsf{AE.Dec}$. *The parameter* $\mathsf{prm}_{\mathsf{AE}}$ *implicitly defines a message space* $\mathcal{M}$ *and a key space* $\mathcal{K}_{\mathsf{AE}}$.
- $\mathsf{AE.Enc}(\mathsf{k}, m)$ *takes as input a key* $\mathsf{k} \in \mathcal{K}_{\mathsf{AE}}$ *and a message* $m \in \mathcal{M}$, *and outputs a ciphertext* $\mathsf{ae.ct}$.
- $\mathsf{AE.Dec}(\mathsf{k}, \mathsf{ae.ct})$ *takes as input a key* $\mathsf{k} \in \mathcal{K}_{\mathsf{AE}}$ *and a ciphertext* $\mathsf{ae.ct}$, *and outputs a message* $m \in \mathcal{M}$ *or a rejection symbol* $\perp$.

Procedure INITIALIZE: $\mathsf{prm_{AE}} \leftarrow_{\$} \mathsf{AE.Setup}(1^\ell)$, $\mathsf{k} \leftarrow_{\$} \mathcal{K}_{\mathsf{AE}}$. $\beta \leftarrow_{\$} \{0,1\}$. // challenge bit Return $\mathsf{prm_{AE}}$. **Procedure** ENC(m_0, m_1): // one query If $|m_0| \neq |m_1|$, Return $\perp$. $\mathsf{ae.ct} \leftarrow_{\$} \mathsf{AE.Enc}(\mathsf{k}, m_\beta)$. Return $\mathsf{ae.ct}$. **Procedure** FINALIZE(β'): Return $(\beta' = \beta)$.	**Procedure** INITIALIZE: $\mathsf{prm_{AE}} \leftarrow_{\$} \mathsf{AE.Setup}(1^\ell)$, $\mathsf{k} \leftarrow_{\$} \mathcal{K}_{\mathsf{AE}}$. Return $\mathsf{prm_{AE}}$. **Procedure** ENC(m): // one query $\mathsf{ae.ct} \leftarrow_{\$} \mathsf{AE.Enc}(\mathsf{k}, m)$. Return $\mathsf{ae.ct}$. **Procedure** FINALIZE$(\mathsf{ae.ct}^*)$: If $\mathsf{ae.ct}^* = \mathsf{ae.ct}$, Return 0. Return $(\mathsf{AE.Dec}(\mathsf{k}, \mathsf{ae.ct}^*) \neq \perp)$.

Fig. 3. Games IND-OT (left) and INT-OT (right) for defining securities of AE.

Correctness of AE *requires that, for all* $\mathsf{prm_{AE}} \leftarrow_{\$} \mathsf{AE.Setup}(1^\ell)$, *all* $\mathsf{k} \in \mathcal{K}_{\mathsf{AE}}$, *all* $m \in \mathcal{M}$ *and all* $\mathsf{ae.ct} \leftarrow_{\$} \mathsf{AE.Enc}(\mathsf{k}, m)$, *it holds that* $\mathsf{AE.Dec}(\mathsf{k}, \mathsf{ae.ct}) = m$.

The security notions for AE include One-time ciphertext-indistinguishability (IND-OT) and One-time ciphertext-integrity (INT-OT). The IND-OT and INT-OT securities of AE are formalized via the security games in Fig. 3.

Definition 3 (One-Time Security for AE). *Scheme* AE *is one-time secure (OT-secure) if it is IND-OT secure and INT-OT secure, i.e., for any PPT adversary* $\mathcal{A}$, *both* $\mathsf{Adv}^{ind\text{-}ot}_{\mathsf{AE},\mathcal{A}}(\ell) := |\Pr[\mathsf{IND\text{-}OT}^{\mathcal{A}} \Rightarrow 1] - 1/2|$ *and* $\mathsf{Adv}^{int\text{-}ot}_{\mathsf{AE},\mathcal{A}}(\ell) := \Pr[\mathsf{INT\text{-}OT}^{\mathcal{A}} \Rightarrow 1]$ *are negligible in* ℓ, *where games* IND-OT *and* INT-OT *are specified in Fig. 3.*

Let $\mathcal{F}$ be a family of functions from $\mathcal{K}_{\mathsf{AE}}$ to $\mathcal{K}_{\mathsf{AE}}$. The $\mathcal{F}$-Related-Key Attack for AE scheme was formalized in [LLJ15], and RKA security notions characterize the ciphertext indistinguishability (IND-$\mathcal{F}$-RKA) and integrity (INT-$\mathcal{F}$-RKA) even if the adversary has multiple access to the encryption oracle and designates a function $f \in \mathcal{F}$ each time such that the encryption oracle uses $f(\mathsf{k})$ as the key.

Definition 4 (IND-RKA and INT-RKA Securities for AE). *Scheme* AE *is IND-$\mathcal{F}$-RKA secure and INT-$\mathcal{F}$-RKA secure, if for any PPT adversary* $\mathcal{A}$, *both* $\mathsf{Adv}^{ind\text{-}rka}_{\mathsf{AE},\mathcal{A}}(\ell) := |\Pr[\mathsf{IND\text{-}}\mathcal{F}\mathsf{\text{-}RKA}^{\mathcal{A}} \Rightarrow 1] - 1/2|$ *and* $\mathsf{Adv}^{int\text{-}rka}_{\mathsf{AE},\mathcal{A}}(\ell) := \Pr[\mathsf{INT\text{-}}\mathcal{F}\mathsf{\text{-}RKA}^{\mathcal{A}} \Rightarrow 1]$ *are negligible in* ℓ, *where games* IND-$\mathcal{F}$-RKA *and* INT-$\mathcal{F}$-RKA *are specified in Fig. 4.*

2.4 DCR, DDH, DL and IV$_d$ Assumptions

Let $\mathsf{GenN}(1^\ell)$ be a PPT algorithm outputting (N, p, q), where p, q are safe primes of ℓ bits and $N = pq$, such that $\bar{N} = 2N+1$ is also a prime. Let $s \in \mathbb{N}$ and $T = 1+N$. Define $\mathbb{QR}_{N^s} := \{a^2 \bmod N^s \mid a \in \mathbb{Z}^*_{N^s}\}$, $\mathbb{SCR}_{N^s} := \{a^{2N^{s-1}} \bmod N^s \mid a \in \mathbb{Z}^*_{N^s}\}$, and $\mathbb{RU}_{N^s} := \{T^r \bmod N^s \mid r \in [N^{s-1}]\}$. Then $\mathbb{SCR}_{N^s}$ is a cyclic group

Procedure INITIALIZE:	**Procedure** INITIALIZE:
$\mathsf{prm_{AE}} \leftarrow_{\$} \mathsf{AE.Setup}(1^\ell)$, $\mathsf{k} \leftarrow_{\$} \mathcal{K}_{\mathsf{AE}}$.	$\mathsf{prm_{AE}} \leftarrow_{\$} \mathsf{AE.Setup}(1^\ell)$, $\mathsf{k} \leftarrow_{\$} \mathcal{K}_{\mathsf{AE}}$.
$\beta \leftarrow_{\$} \{0,1\}$. // challenge bit	Return $\mathsf{prm_{AE}}$.
Return $\mathsf{prm_{AE}}$.	
	Procedure $\mathrm{ENC}(m, f \in \mathcal{F})$:
Procedure $\mathrm{ENC}(m_0, m_1, f \in \mathcal{F})$:	$\mathsf{ae.ct} \leftarrow_{\$} \mathsf{AE.Enc}(f(\mathsf{k}), m)$.
If $\vert m_0\vert \neq \vert m_1\vert$, Return $\perp$.	$\mathcal{Q}_{\mathcal{ENC}} := \mathcal{Q}_{\mathcal{ENC}} \cup \{(f, \mathsf{ae.ct})\}$.
$\mathsf{ae.ct} \leftarrow_{\$} \mathsf{AE.Enc}(f(\mathsf{k}), m_\beta)$.	Return $\mathsf{ae.ct}$.
Return $\mathsf{ae.ct}$.	
	Procedure $\mathrm{FINALIZE}(f^* \in \mathcal{F}, \mathsf{ae.ct}^*)$:
Procedure $\mathrm{FINALIZE}(\beta')$:	If $(f^*, \mathsf{ae.ct}^*) \in \mathcal{Q}_{\mathcal{ENC}}$, Return 0.
Return $(\beta' = \beta)$.	Return $(\mathsf{AE.Dec}(f^*(\mathsf{k}), \mathsf{ae.ct}^*) \neq \perp)$.

Fig. 4. Games IND-$\mathcal{F}$-RKA (left) and INT-$\mathcal{F}$-RKA (right) for defining securities of AE.

of order $\phi(N)/4$, and $\mathbb{QR}_{N^s} = \mathbb{SCR}_{N^s} \otimes \mathbb{RU}_{N^s}$, where $\otimes$ denotes internal direct product. Let $\mathbb{QR}_{\bar{N}} := \{a^2 \bmod \bar{N} \mid a \in \mathbb{Z}_{\bar{N}}\}$, then $\mathbb{QR}_{\bar{N}}$ is a cyclic group of order $N = pq$.

For $X \in \mathbb{RU}_{N^s}$, the discrete logarithm $\mathrm{dlog}_T(X) \in [N^{s-1}]$ can be efficiently computed given only N and X [DJ01]. Note that $\mathbb{Z}_{N^s}^* = \mathbb{Z}_2 \otimes \mathbb{Z}_2' \otimes \mathbb{SCR}_{N^s} \otimes \mathbb{RU}_{N^s}$, hence for any $u = u(\mathbb{Z}_2) \cdot u(\mathbb{Z}_2') \cdot u(\mathbb{SCR}_{N^s}) \cdot T^x \in \mathbb{Z}_{N^s}^*$, $u^{\phi(N)} = T^{x \cdot \phi(N)} \in \mathbb{RU}_{N^s}$ and

$$\mathrm{dlog}_T(u^{\phi(N)})/\phi(N) \bmod N^{s-1} = x. \tag{1}$$

The formal definitions of the Decisional Composite Residuosity (DCR) and the Discrete Logarithm (DL) assumptions are in the full version [HLL16]. The DCR assumption implies the Interactive Vector (IV_d) assumption according to [BG10]. We adopt the version in [LLJ15].

Definition 5 ($\mathbf{IV}_d$ Assumption). *The IV_d Assumption holds w.r.t.* GenN *and group $\mathbb{QR}_{N^s}$ if for any PPT adversary $\mathcal{A}$, the following advantage is negligible in ℓ:*

$$\mathsf{Adv}^{iv_d}_{\mathsf{GenN},\mathcal{A}}(\ell) := \left| \Pr\left[\mathcal{A}^{\mathrm{CHAL}^b_{\mathrm{IV}_d}}(N, g_1, \cdots, g_d) = b\right] - 1/2 \right|,$$

where $(N, p, q) \leftarrow_{\$} \mathsf{GenN}(1^\ell)$, $g_1, \cdots, g_d \leftarrow_{\$} \mathbb{SCR}_{N^s}$, $b \leftarrow_{\$} \{0,1\}$, and the oracle $\mathrm{CHAL}^b_{\mathrm{IV}_d}(\cdot)$ can be queried by $\mathcal{A}$ adaptively. $\mathcal{A}$ submits $(\delta_1, \cdots, \delta_d)$ to the oracle. $\mathrm{CHAL}^b_{\mathrm{IV}_d}(\delta_1, \cdots, \delta_d)$ selects random $r \leftarrow_{\$} [\lfloor N/4 \rfloor]$. If $b = 0$, the oracle returns $(g_1^r, \cdots, g_d^r)$; otherwise it returns $(g_1^r T^{\delta_1}, \cdots, g_d^r T^{\delta_d})$, where $T = 1 + N$.

Definition 6 (DDH Assumption). *The Decisional Diffie-Hellman (DDH) Assumption holds w.r.t.* GenN *and group $\mathbb{QR}_{\bar{N}}$ if for any PPT adversary $\mathcal{A}$, the following advantage is negligible in ℓ:*

$$\mathsf{Adv}^{ddh}_{\mathsf{GenN},\mathcal{A}}(\ell) := \left| \Pr\left[\mathcal{A}(N, p, q, g_1, g_2, g_1^x, g_2^x) = 1\right] - \Pr\left[\mathcal{A}(N, p, q, g_1, g_2, g_1^x, g_2^y) = 1\right] \right|,$$

where $(N, p, q) \leftarrow_{\$} \mathsf{GenN}(1^\ell)$, $g_1, g_2 \leftarrow_{\$} \mathbb{QR}_{\bar{N}}$, $x, y \leftarrow_{\$} \mathbb{Z}_N \setminus \{0\}$.

2.5 Collision Resistant Hashing and Universal Hashing

Definition 7 (Collision Resistant Hashing). *A family of functions* $\mathcal{H} = \{\mathsf{H} : \mathcal{X} \longrightarrow \mathcal{Y}\}$ *is collision-resistant if for any PPT adversary* $\mathcal{A}$*, the following advantage is negligible in* ℓ*:*

$$\mathsf{Adv}^{cr}_{\mathcal{H},\mathcal{A}}(\ell) := \Pr\left[\mathsf{H} \leftarrow_{\$} \mathcal{H},\ (x, x') \leftarrow_{\$} \mathcal{A}(\mathsf{H})\ :\ \mathsf{H}(x) = \mathsf{H}(x')\ \wedge\ x \neq x'\right].$$

Definition 8 (Universal Hashing). *A family of functions* $\mathcal{H} = \{\mathsf{H} : \mathcal{X} \longrightarrow \mathcal{Y}\}$ *is universal, if for all distinct* $x, x' \in \mathcal{X}$*, it follows that*

$$\Pr\left[\mathsf{H} \leftarrow_{\$} \mathcal{H}\ :\ \mathsf{H}(x) = \mathsf{H}(x')\right] \leq 1/|\mathcal{Y}|.$$

3 $\overline{\mathsf{AE}}$ of the LLJ Scheme and Its INT-RKA Security

The LLJ scheme [LLJ15] makes use of an important primitive "Authenticated Encryption" $\overline{\mathsf{AE}}$. Its KDM[$\mathcal{F}_{\text{aff}}$]-CCA security heavily relies on the IND-$\mathcal{F}_{\text{aff}}$-RKA security and INT-$\mathcal{F}_{\text{aff}}$-RKA security of their $\overline{\mathsf{AE}}$. LLJ claimed INT-$\mathcal{F}_{\text{aff}}$-RKA security of their $\overline{\mathsf{AE}}$, however, we point out that their security proof does not go through to the DDH assumption, which in turn affects the KDM[$\mathcal{F}_{\text{aff}}$]-CCA security proof of the LLJ scheme.

Let us briefly review LLJ's $\overline{\mathsf{AE}}$ as follows. The public parameter is $\mathsf{prm}_{\overline{\mathsf{AE}}} = (N, \bar{N}, g)$ where $N = pq$, $\bar{N} = 2N + 1$, and g is a generator of group $\mathbb{QR}_{\bar{N}}$. Let AE be an IND-OT and INT-OT secure authenticated encryption, and H be a 4-wise independent hash function. The secret key space is $\mathbb{Z}_N$.

- $\overline{\mathsf{AE}}.\mathsf{Enc}(k, m)$ computes $u = g^r$ with $r \leftarrow_{\$} \mathbb{Z}_N$, $\kappa = \mathsf{H}(u^k, u)$ and invokes $\chi \leftarrow_{\$} \mathsf{AE}.\mathsf{Enc}(\kappa, m)$. It outputs the ciphertext $\langle u, \chi \rangle$.
- $\overline{\mathsf{AE}}.\mathsf{Dec}(k, \langle u, \chi \rangle)$ computes $\kappa = \mathsf{H}(u^k, u)$ and outputs $m/\perp \leftarrow \mathsf{AE}.\mathsf{Dec}(\kappa, \chi)$.

In the LLJ scheme, $\overline{\mathsf{AE}}$ should have RKA security w.r.t. $\mathcal{F}_{\text{aff}} = \{f : k \longmapsto ak + b \mid a \neq 0\}$. Let us check their security proof. See Table 2. The proof idea is to use the DDH assumption to make sure that each κ_λ, $\lambda \in [Q_e]$, is random to the adversary. Then the INT-OT of AE guarantees that the adversary cannot make a fresh forgery $\left(f^* = (a^*, b^*), \langle u^*, \chi^* \rangle\right)$ such that $\overline{\mathsf{AE}}.\mathsf{Dec}(a^*k + b^*, \langle u^*, \chi^* \rangle) \neq \perp$.

In [LLJ15], the indistinguishability of Game 1.$(i-1)$ and Game 1.i is reduced to the DDH assumption. A PPT algorithm $\mathcal{B}$ is constructed to solve the DDH problem by employing an INT-$\mathcal{F}_{\text{aff}}$-RKA adversary $\mathcal{A}$. Given the challenge (g, g^{r_i}, g^k, Z), $\mathcal{B}$ wants to tell whether $Z = g^{kr_i}$ or $Z = g^{z_i}$ for a random z_i. $\mathcal{B}$ simulates the INT-$\mathcal{F}_{\text{aff}}$-RKA game for $\mathcal{A}$ by computing $\kappa_i = \mathsf{H}(Z^{a_i} g^{r_i b_i}, g^{r_i})$. If $Z = g^{kr_i}$, $\mathcal{B}$ simulates Game 1.$(i-1)$ for $\mathcal{A}$; if $Z = g^{z_i}$, $\mathcal{B}$ simulates Game 1.i.

The problem is now that $\mathcal{B}$ does not know the value of secret key k (it knows g^k). When $\mathcal{A}$ submits a fresh forgery $\left(f^* = (a^*, b^*), \langle u^*, \chi^* \rangle\right)$, $\mathcal{B}$ is not able to see whether $\overline{\mathsf{AE}}.\mathsf{Dec}(a^*k + b^*, \langle u^*, \chi^* \rangle) \neq \perp$ or not without the knowledge of k. More precisely, $\mathcal{B}$ can not compute $\kappa^* = \mathsf{H}(u^{*a^*k+b^*}, u^*) = \mathsf{H}((u^{*k})^{a^*} \cdot u^{*b^*}, u^*)$ from g^k and u^*, unless it is able to compute the CDH value u^{*k} from g^k and u^*. Without κ^*, it is hard for $\mathcal{B}$ to decide whether $\mathsf{AE}.\mathsf{Dec}(\kappa^*, \chi^*) \neq \perp$ or not.

Table 2. INT-$\mathcal{F}_{\text{aff}}$-RKA security proof of $\overline{\mathsf{AE}}$ in the LLJ scheme; we point out a flaw in the security reduction from Game 1.$(i-1)$ to Game 1.i, denoted by "?".

	$\text{Enc}(m_\lambda, f_\lambda = (a_\lambda, b_\lambda))$ oracle, $\lambda \in [Q_e]$, where Q_e is the number of encryption queries	Assumptions
Game 0	$r_\lambda \leftarrow_\$ \mathbb{Z}_N$; $u_\lambda := g^{r_\lambda}$; $\kappa_\lambda := \mathsf{H}(u_\lambda^{(a_\lambda k + b_\lambda)}, u_\lambda)$; $\chi_\lambda \leftarrow_\$ \mathsf{AE.Enc}(\kappa_\lambda, m_\lambda)$; return $\overline{\mathsf{ae}}.\mathsf{ct}_\lambda := \langle u_\lambda, \chi_\lambda \rangle$	–
Game 1	Same as Game 0 except $\kappa_\lambda := \mathsf{H}((g^{kr_\lambda})^{a_\lambda} g^{r_\lambda b_\lambda}, g^{r_\lambda})$	Game 1 = Game 0
Game 1.i	For $\lambda = 1, \cdots, i$, the same as Game 1 except $\kappa_\lambda := \mathsf{H}((g^{z_\lambda})^{a_\lambda} g^{r_\lambda b_\lambda}, g^{r_\lambda})$ with $z_\lambda \leftarrow_\$ \mathbb{Z}_N$; For $\lambda = i+1, \cdots Q_e$, the same as Game 1	DDH (?)
Game 2	Game 2 = Game 1.Q_e	INT-OT of AE

In other words, $\mathcal{B}$ cannot find an efficient (PPT) way to transform the computing power (forgery) of $\mathcal{A}$ into its own decisional power (decision bit) to determine (g, g^{r_i}, g^k, Z) to be a DDH tuple or a random tuple. The failure of the INT-$\mathcal{F}_{\text{aff}}$-RKA security proof results in the failure of the KDM[$\mathcal{F}_{\text{aff}}$]-CCA proof of the LLJ scheme since INT-$\mathcal{F}_{\text{aff}}$-RKA security is used to prevent a KDM[$\mathcal{F}_{\text{aff}}$]-CCA adversary from learning more information about the secret key by querying some invalid ciphertexts for decryption.

4 Authenticated Encryption with Auxiliary-Input

We do not see any hope of successfully fixing the security proof of the LLJ's $\overline{\mathsf{AE}}$ in [LLJ15]. Alternatively, we resort to a different building block, namely AIAE. The intuition is as follows. If LLJ's $\overline{\mathsf{AE}}$ is regarded as (ElGamal + OT-AE), we can design a new AIAE as (Kurosawa-Desmedt [KD04] + OT-AE). But a new problem with our design arises: the secret key of KEM [KD04] consists of several elements, i.e., $\mathsf{k} = (k_1, k_2, k_3, k_4)$. The affine function of k is too complicated to prove the INT-$\mathcal{F}_{\text{aff}}$-RKA security. Fortunately, (a weak) INT-RKA security follows w.r.t. a smaller restricted affine function set $\mathcal{F}_{\text{raff}} = \{f : (k_1, k_2, k_3, k_4) \longmapsto a \cdot (k_1, k_2, k_3, k_4) + (b_1, b_2, b_3, b_4) \mid a \neq 0\}$.

To make AIAE serve KDM-CCA security of our PKE construction in Fig. 1, we have the following requirements.

- AIAE must have auxiliary input aux.
- A weak INT-$\mathcal{F}$-RKA security is defined for AIAE. Compared to INT-$\mathcal{F}$-RKA security, the weak version has an additional special rule for the adversary's forgery $(\mathsf{aux}^*, f^*, \mathsf{aiae.ct}^*)$ to be successful: if the adversary has already queried (m, aux^*, f) to the encryption oracle Enc, it must hold that $f^* = f$.

Next, we introduce the formal definitions of *Authenticated Encryption with Auxiliary-Input*, its *IND-$\mathcal{F}$-RKA Security* and *Weak INT-$\mathcal{F}$-RKA Security*.

4.1 AIAE and Its Related-Key Attack Security

Definition 9 (AIAE). *An auxiliary-input authenticated encryption (AIAE) scheme* $\mathsf{AIAE} = (\mathsf{AIAE.Setup}, \mathsf{AIAE.Enc}, \mathsf{AIAE.Dec})$ *consists of three PPT algorithms:*

- $\mathsf{AIAE.Setup}(1^\ell)$ *outputs a system parameter* $\mathsf{prm}_{\mathsf{AIAE}}$*, which is an implicit input to* $\mathsf{AIAE.Enc}$ *and* $\mathsf{AIAE.Dec}$*. The parameter* $\mathsf{prm}_{\mathsf{AIAE}}$ *implicitly defines a message space* $\mathcal{M}$*, a key space* $\mathcal{K}_{\mathsf{AIAE}}$ *and an auxiliary-input space* $\mathcal{AUX}$*.*
- $\mathsf{AIAE.Enc}(\mathsf{k}, m, \mathsf{aux})$ *takes as input a key* $\mathsf{k} \in \mathcal{K}_{\mathsf{AIAE}}$*, a message* $m \in \mathcal{M}$ *and an auxiliary input* $\mathsf{aux} \in \mathcal{AUX}$*, and outputs a ciphertext* $\mathsf{aiae.ct}$*.*
- $\mathsf{AIAE.Dec}(\mathsf{k}, \mathsf{aiae.ct}, \mathsf{aux})$ *takes as input a key* $\mathsf{k} \in \mathcal{K}_{\mathsf{AE}}$*, a ciphertext* $\mathsf{aiae.ct}$ *and an auxiliary input* $\mathsf{aux} \in \mathcal{AUX}$*, and outputs a message* $m \in \mathcal{M}$ *or a rejection symbol* $\perp$*.*

Correctness of AIAE *requires that, for all* $\mathsf{prm}_{\mathsf{AIAE}} \leftarrow_{\$} \mathsf{AIAE.Setup}(1^\ell)$*, all* $\mathsf{k} \in \mathcal{K}_{\mathsf{AIAE}}$*, all* $m \in \mathcal{M}$*, all* $\mathsf{aux} \in \mathcal{AUX}$ *and all* $\mathsf{aiae.ct} \leftarrow_{\$} \mathsf{AIAE.Enc}(\mathsf{k}, m, \mathsf{aux})$*, we have that* $\mathsf{AIAE.Dec}(\mathsf{k}, \mathsf{aiae.ct}, \mathsf{aux}) = m$*.*

If the auxiliary-input space $\mathcal{AUX} = \emptyset$ for all possible parameters $\mathsf{prm}_{\mathsf{AIAE}}$, the above definition is reduced to traditional AE.

Let $\mathcal{F}$ be a family of functions from $\mathcal{K}_{\mathsf{AIAE}}$ to $\mathcal{K}_{\mathsf{AIAE}}$. We define the related-key security notions for AIAE via Fig. 5.

IND-$\mathcal{F}$-RKA	weak-INT-$\mathcal{F}$-RKA
<u>Procedure INITIALIZE:</u> $\mathsf{prm}_{\mathsf{AIAE}} \leftarrow_{\$} \mathsf{AIAE.Setup}(1^\ell)$, $\mathsf{k} \leftarrow_{\$} \mathcal{K}_{\mathsf{AIAE}}$. $\beta \leftarrow_{\$} \{0,1\}$. // challenge bit Return $\mathsf{prm}_{\mathsf{AIAE}}$. <u>Procedure ENC($m_0, m_1, \mathsf{aux}, f \in \mathcal{F}$):</u> If $\lvert m_0 \rvert \neq \lvert m_1 \rvert$, Return $\perp$. $\mathsf{aiae.ct} \leftarrow_{\$} \mathsf{AIAE.Enc}(f(\mathsf{k}), m_\beta, \mathsf{aux})$. Return $\mathsf{aiae.ct}$. <u>Procedure FINALIZE(β'):</u> Return ($\beta' = \beta$).	<u>Procedure INITIALIZE:</u> $\mathsf{prm}_{\mathsf{AIAE}} \leftarrow_{\$} \mathsf{AIAE.Setup}(1^\ell)$, $\mathsf{k} \leftarrow_{\$} \mathcal{K}_{\mathsf{AIAE}}$. Return $\mathsf{prm}_{\mathsf{AIAE}}$. <u>Procedure ENC($m, \mathsf{aux}, f \in \mathcal{F}$):</u> $\mathsf{aiae.ct} \leftarrow_{\$} \mathsf{AIAE.Enc}(f(\mathsf{k}), m, \mathsf{aux})$. $\mathcal{Q}_{\mathcal{ENC}} := \mathcal{Q}_{\mathcal{ENC}} \cup \{(\mathsf{aux}, f, \mathsf{aiae.ct})\}$. $\mathcal{Q}_{\mathcal{AUXF}} := \mathcal{Q}_{\mathcal{AUXF}} \cup \{(\mathsf{aux}, f)\}$. Return $\mathsf{aiae.ct}$. <u>Procedure FINALIZE($\mathsf{aux}^*, f^* \in \mathcal{F}, \mathsf{aiae.ct}^*$):</u> If $(\mathsf{aux}^*, f^*, \mathsf{aiae.ct}^*) \in \mathcal{Q}_{\mathcal{ENC}}$, Return 0. // Special rule: If there exists $(\mathsf{aux}, f) \in \mathcal{Q}_{\mathcal{AUXF}}$ such that $\mathsf{aux} = \mathsf{aux}^*$ but $f \neq f^*$, Return 0. Return ($\mathsf{AIAE.Dec}(f^*(\mathsf{k}), \mathsf{aiae.ct}^*, \mathsf{aux}^*) \neq \perp$).

Fig. 5. Games IND-$\mathcal{F}$-RKA (left) and weak-INT-$\mathcal{F}$-RKA (right) for defining securities of auxiliary-input authenticated encryption scheme AIAE. We note that the weak INT-$\mathcal{F}$-RKA security needs a special rule to return 0 in FINALIZE as shown in the shadow.

Definition 10 (IND-$\mathcal{F}$-RKA and Weak INT-$\mathcal{F}$-RKA Securities for AIAE). *Scheme* AIAE *is IND-$\mathcal{F}$-RKA secure and weak INT-$\mathcal{F}$-RKA secure, if for any PPT adversary $\mathcal{A}$, both* $\mathsf{Adv}^{ind\text{-}rka}_{\mathsf{AIAE},\mathcal{A}}(\ell) := |\Pr[\mathsf{IND}\text{-}\mathcal{F}\text{-}\mathsf{RKA}^{\mathcal{A}} \Rightarrow 1] - 1/2|$ *and* $\mathsf{Adv}^{weak\text{-}int\text{-}rka}_{\mathsf{AIAE},\mathcal{A}}(\ell) := \Pr[\mathsf{weak}\text{-}\mathsf{INT}\text{-}\mathcal{F}\text{-}\mathsf{RKA}^{\mathcal{A}} \Rightarrow 1]$ *are negligible in ℓ, where games* IND-$\mathcal{F}$-RKA *and* weak-INT-$\mathcal{F}$-RKA *are specified in Fig. 5.*

4.2 AIAE from OT-secure AE and DDH Assumption

Let $\mathsf{AE} = (\mathsf{AE.Setup}, \mathsf{AE.Enc}, \mathsf{AE.Dec})$ be a traditional (without auxiliary-input) authenticated encryption scheme with key space $\mathcal{K}_{\mathsf{AE}}$ and message space $\mathcal{M}$. Let $\mathcal{H}_1 = \{\mathsf{H}_1 : \{0,1\}^* \to \mathbb{Z}_N\}$ and $\mathcal{H}_2 = \{\mathsf{H}_2 : \mathbb{QR}_{\bar{N}} \to \mathcal{K}_{\mathsf{AE}}\}$ be two families of hash functions with $|\mathcal{K}_{\mathsf{AE}}|/|\mathbb{QR}_{\bar{N}}|$ $(= |\mathcal{K}_{\mathsf{AE}}|/N) \leq 2^{-\Omega(\ell)}$. The proposed scheme $\mathsf{AIAE} = (\mathsf{AIAE.Setup}, \mathsf{AIAE.Enc}, \mathsf{AIAE.Dec})$ with key space $\mathcal{K}_{\mathsf{AIAE}} = (\mathbb{Z}_N)^4$, message space $\mathcal{M}$ and auxiliary-input space $\mathcal{AUX} = \{0,1\}^*$ is defined in Fig. 6.

<table>
<tr><td colspan="2">
<u>$\mathsf{prm}_{\mathsf{AIAE}} \leftarrow_\$ \mathsf{AIAE.Setup}(1^\ell)$:</u>
$(N, p, q) \leftarrow_\$ \mathsf{GenN}(1^\ell)$, i.e., pick two ℓ-bit safe primes p and q, such that $2pq+1$ is also a prime, and $N := pq$.
$\bar{N} := 2N + 1 = 2pq + 1$. $g_1, g_2 \leftarrow_\$ \mathbb{QR}_{\bar{N}}$. $\mathsf{H}_1 \leftarrow_\$ \mathcal{H}_1$, $\mathsf{H}_2 \leftarrow_\$ \mathcal{H}_2$.
Return $\mathsf{prm}_{\mathsf{AIAE}} := (N, p, q, \bar{N}, g_1, g_2, \mathsf{H}_1, \mathsf{H}_2)$.
</td></tr>
<tr><td>
<u>$\langle c_1, c_2, \chi \rangle \leftarrow_\$ \mathsf{AIAE.Enc}(\mathsf{k}, m, \mathsf{aux})$:</u>
Parse $\mathsf{k} = (k_1, k_2, k_3, k_4) \in \mathbb{Z}_N^4$.
$w \leftarrow_\$ \mathbb{Z}_N \backslash \{0\}$. $(c_1, c_2) := (g_1^w, g_2^w) \in \mathbb{QR}_{\bar{N}}^2$.
$t := \mathsf{H}_1(c_1, c_2, \mathsf{aux}) \in \mathbb{Z}_N$.
$\kappa := \mathsf{H}_2(c_1^{k_1+k_3t} \cdot c_2^{k_2+k_4t}) \in \mathcal{K}_{\mathsf{AE}}$.
$\chi \leftarrow_\$ \mathsf{AE.Enc}(\kappa, m)$.
Return $\langle c_1, c_2, \chi \rangle$.
</td><td>
<u>$m/\bot \leftarrow \mathsf{AIAE.Dec}(\mathsf{k}, \langle c_1, c_2, \chi \rangle, \mathsf{aux})$:</u>
Parse $\mathsf{k} = (k_1, k_2, k_3, k_4) \in \mathbb{Z}_N^4$.
If $(c_1, c_2) \notin \mathbb{QR}_{\bar{N}}^2 \vee (c_1, c_2) = (1, 1)$,
Return $\bot$.
$t := \mathsf{H}_1(c_1, c_2, \mathsf{aux}) \in \mathbb{Z}_N$.
$\kappa := \mathsf{H}_2(c_1^{k_1+k_3t} \cdot c_2^{k_2+k_4t}) \in \mathcal{K}_{\mathsf{AE}}$.
Return $m/\bot \leftarrow \mathsf{AE.Dec}(\kappa, \chi)$.
</td></tr>
</table>

Fig. 6. Construction of the DDH-based AIAE from AE.

The correctness of AIAE follows from the correctness of AE directly. Note that the factors p, q of N in $\mathsf{prm}_{\mathsf{AIAE}}$ are not needed in the encryption and decryption algorithms of AIAE. Jumping ahead, the factors p, q are necessary when the security of the PKEs presented in Sects. 5 and 6 is reduced to the security of AIAE. We now show the RKA-security of AIAE through the following theorem.

Theorem 1. *If the underlying scheme* AE *is OT-secure, the DDH assumption holds w.r.t.* GenN *and* $\mathbb{QR}_{\bar{N}}$*, $\mathcal{H}_1$ is collision resistant and $\mathcal{H}_2$ is universal, then the resulting scheme* AIAE *in Fig. 6 is IND-$\mathcal{F}_{raff}$-RKA and weak INT-$\mathcal{F}_{raff}$-RKA secure, where the restricted affine function set is defined as* $\mathcal{F}_{raff} := \{f_{(a,\mathsf{b})} : (k_1, k_2, k_3, k_4) \in \mathbb{Z}_N^4 \longmapsto (ak_1 + b_1, ak_2 + b_2, ak_3 + b_3, ak_4 + b_4) \in \mathbb{Z}_N^4 \mid a \in \mathbb{Z}_N^*, \mathsf{b} = (b_1, b_2, b_3, b_4) \in \mathbb{Z}_N^4\}$.

Proof of IND-$\mathcal{F}_{\text{raff}}$-RKA security of AIAE in Theorem 1. The proof proceeds with a sequence of games. Suppose that $\mathcal{A}$ is a PPT adversary against the IND-$\mathcal{F}_{\text{raff}}$-RKA security of AIAE, who makes at most Q_e times of ENC queries. Let $\Pr_i[\cdot]$ (resp., $\Pr_{i'}[\cdot]$) denote the probability of a particular event occurring in game G_i (resp., game G'_i).

- Game G_1: This is the original IND-$\mathcal{F}_{\text{raff}}$-RKA security game. Let Win denote the event that $\beta' = \beta$. Then by definition, $\mathsf{Adv}^{ind\text{-}rka}_{\mathsf{AIAE},\mathcal{A}}(\ell) = \big|\Pr_1[\mathsf{Win}] - \frac{1}{2}\big|$. Denote $\mathsf{prm}_{\mathsf{AIAE}} = (N, p, q, \bar{N}, g_1, g_2, \mathsf{H}_1, \mathsf{H}_2)$ and $\mathsf{k} = (k_1, k_2, k_3, k_4)$. To answer the λ-th ($\lambda \in [Q_e]$) ENC query $(m_{\lambda,0}, m_{\lambda,1}, \mathsf{aux}_\lambda, f_\lambda)$, where $f_\lambda = \langle a_\lambda, \mathsf{b}_\lambda = (b_{\lambda,1}, b_{\lambda,2}, b_{\lambda,3}, b_{\lambda,4})\rangle \in \mathcal{F}_{\text{raff}}$, the challenger proceeds as follows:
  1. pick $w_\lambda \leftarrow_\$ \mathbb{Z}_N \backslash \{0\}$ and compute $(c_{\lambda,1}, c_{\lambda,2}) := (g_1^{w_\lambda}, g_2^{w_\lambda}) \in \mathbb{QR}^2_{\bar{N}}$,
  2. compute a tag $t_\lambda := \mathsf{H}_1(c_{\lambda,1}, c_{\lambda,2}, \mathsf{aux}_\lambda) \in \mathbb{Z}_N$,
  3. compute an encryption key for AE scheme using a related key $f_\lambda(\mathsf{k})$:

$$\kappa_\lambda := \mathsf{H}_2\big(c_{\lambda,1}^{(a_\lambda k_1 + b_{\lambda,1}) + (a_\lambda k_3 + b_{\lambda,3}) t_\lambda} \cdot c_{\lambda,2}^{(a_\lambda k_2 + b_{\lambda,2}) + (a_\lambda k_4 + b_{\lambda,4}) t_\lambda}\big) \in \mathcal{K}_{\mathsf{AE}},$$

  4. invoke $\chi_\lambda \leftarrow_\$ \mathsf{AE.Enc}(\kappa_\lambda, m_{\lambda,\beta})$,

  and returns the challenge ciphertext $\langle c_{\lambda,1}, c_{\lambda,2}, \chi_\lambda\rangle$ to the adversary $\mathcal{A}$.
- Game $\mathsf{G}_{1,i}$, $i \in [Q_e + 1]$: This game is the same as game G_1, except that, the challenger does not use secret key k to answer the λ-th ($\lambda \in [i-1]$) ENC query at all, and instead, it changes steps 1, 3 to steps 1′, 3′ as follows:
  1′. pick $w_{\lambda,1}, w_{\lambda,2} \leftarrow_\$ \mathbb{Z}_N \backslash \{0\}$ and compute $(c_{\lambda,1}, c_{\lambda,2}) := (g_1^{w_{\lambda,1}}, g_2^{w_{\lambda,2}})$,
  3′. choose an encryption key $\kappa_\lambda \leftarrow_\$ \mathcal{K}_{\mathsf{AE}}$ randomly for the AE scheme.

  The challenger still answers the λ-th ($\lambda \in [i, Q_e]$) ENC query as in G_1, i.e., using steps 1, 3.

  Clearly $\mathsf{G}_{1,1}$ is identical to G_1, thus $\Pr_1[\mathsf{Win}] = \Pr_{1,1}[\mathsf{Win}]$.
- Game $\mathsf{G}'_{1,i}$, $i \in [Q_e]$: This game is the same as game $\mathsf{G}_{1,i}$, except that the challenger answers the i-th ENC query using steps 1′, 3 (rather than steps 1, 3 in game $\mathsf{G}_{1,i}$).

  The only difference between $\mathsf{G}_{1,i}$ and $\mathsf{G}'_{1,i}$ is the distribution of $(g_1, g_2, c_{i,1}, c_{i,2})$. In game $\mathsf{G}_{1,i}$, $(g_1, g_2, c_{i,1}, c_{i,2})$ is a DDH tuple, while in game $\mathsf{G}'_{1,i}$, it is a random tuple. It is straightforward to construct a PPT adversary to solve the DDH problem w.r.t. GenN and $\mathbb{QR}_{\bar{N}}$, thus we have that $\big|\Pr_{1,i}[\mathsf{Win}] - \Pr_{1,i'}[\mathsf{Win}]\big| \le \mathsf{Adv}^{ddh}_{\mathsf{GenN}}(\ell)$.

We analyze the difference between $\mathsf{G}'_{1,i}$ and $\mathsf{G}_{1,i+1}$ via the following lemma. Its proof is provided in the full version [HLL16].

Lemma 1. *For all* $i \in [Q_e]$, $\big|\Pr_{1,i'}[\mathsf{Win}] - \Pr_{1,i+1}[\mathsf{Win}]\big| \le \frac{1}{N-1} + 2^{-\Omega(\ell)}$.

- Game G_2: This game is the same as game G_{1,Q_e+1}, except that, to answer the λ-th ($\lambda \in [Q_e]$) ENC query, the challenger changes step 4 to step 4′:
  4′. invoke $\chi_\lambda \leftarrow_\$ \mathsf{AE.Enc}(\kappa_\lambda, 0^{|m_{\lambda,0}|})$.

In game G_{1,Q_e+1}, the challenger computes the AE encryption of $m_{\lambda,\beta}$ under encryption key κ_λ in ENC, while in game G_2 it computes the AE encryption of $0^{|m_{\lambda,0}|}$ in ENC. Both in games G_{1,Q_e+1} and G_2, we have that each κ_λ is chosen uniformly from $\mathcal{K}_{\mathsf{AE}}$ and independent of other parts of the game. Therefore we can reduce the differences between G_{1,Q_e+1} and G_2 to the IND-OT security of AE by a standard hybrid argument, and have that $\big|\Pr_{1,Q_e+1}[\mathsf{Win}] - \Pr_2[\mathsf{Win}]\big| \leq Q_e \cdot \mathsf{Adv}_{\mathsf{AE}}^{ind\text{-}ot}(\ell)$.

Now in game G_2, since the challenger always encrypts the constant message $0^{|m_{\lambda,0}|}$, the challenge bit β is completely hidden. Then $\Pr_2[\mathsf{Win}] = 1/2$.

Taking all things together, the IND-$\mathcal{F}_{\text{raff}}$-RKA security of AIAE follows. ■

Proof of Weak INT-$\mathcal{F}_{\text{raff}}$-RKA security of AIAE in Theorem 1. Again, we prove it through a sequence of games. These games are defined almost the same as those in the previous proof. Suppose that $\mathcal{A}$ is a PPT adversary against the weak INT-$\mathcal{F}_{\text{raff}}$-RKA security of AIAE, who makes at most Q_e times of ENC queries.

- Game G_0: This is the original weak-INT-$\mathcal{F}_{\text{raff}}$-RKA security game.
  Denote $\mathsf{prm}_{\mathsf{AIAE}} = (N, p, q, \bar{N}, g_1, g_2, \mathsf{H}_1, \mathsf{H}_2)$ and $\mathsf{k} = (k_1, k_2, k_3, k_4)$. To answer the λ-th ($\lambda \in [Q_e]$) ENC query $(m_\lambda, \mathsf{aux}_\lambda, f_\lambda)$, the challenger proceeds with steps 1∼4, similar to the previous proof, and returns the challenge ciphertext $\langle c_{\lambda,1}, c_{\lambda,2}, \chi_\lambda\rangle$ to the adversary $\mathcal{A}$. Moreover, the challenger will put $\big(\mathsf{aux}_\lambda, f_\lambda, \langle c_{\lambda,1}, c_{\lambda,2}, \chi_\lambda\rangle\big)$ to a set $\mathcal{Q}_{\mathcal{ENC}}$, put $(\mathsf{aux}_\lambda, f_\lambda)$ to a set $\mathcal{Q}_{\mathcal{AUXF}}$, and put $(c_{\lambda,1}, c_{\lambda,2}, \mathsf{aux}_\lambda, t_\lambda)$ to a set $\mathcal{Q}_{\mathcal{TAG}}$. Finally, the adversary outputs a forgery $\big(\mathsf{aux}^*, f^* = \langle a^*, \mathsf{b}^* = (b_1^*, b_2^*, b_3^*, b_4^*)\rangle, \langle c_1^*, c_2^*, \chi^*\rangle\big)$.
  Let Forge be the event that the following FINALIZE procedure outputs 1:
  - If $\big(\mathsf{aux}^*, f^*, \langle c_1^*, c_2^*, \chi^*\rangle\big) \in \mathcal{Q}_{\mathcal{ENC}}$, Return 0.
  - If there exists $(\mathsf{aux}_\lambda, f_\lambda) \in \mathcal{Q}_{\mathcal{AUXF}}$ such that $\mathsf{aux}_\lambda = \mathsf{aux}^*$ but $f_\lambda \neq f^*$, Return 0.
  - If $(c_1^*, c_2^*) \notin \mathbb{QR}_{\bar{N}}^2 \vee (c_1^*, c_2^*) = (1, 1)$, Return 0.
  - $t^* := \mathsf{H}_1(c_1^*, c_2^*, \mathsf{aux}^*)$, $\kappa^* := \mathsf{H}_2\big(c_1^{*(a^*k_1+b_1^*)+(a^*k_3+b_3^*)t^*} \cdot c_2^{*(a^*k_2+b_2^*)+(a^*k_4+b_4^*)t^*}\big)$.
    Return $(\mathsf{AE.Dec}(\kappa^*, \chi^*) \neq \perp)$.

  By definition, it follows that, $\mathsf{Adv}_{\mathsf{AIAE},\mathcal{A}}^{weak\text{-}int\text{-}rka}(\ell) = \Pr_0[\mathsf{Forge}]$.
- Game G_1: This game is the same as game G_0, except that, the challenger adds the following new rule to the FINALIZE procedure:
  - If there exists $(c_{\lambda,1}, c_{\lambda,2}, \mathsf{aux}_\lambda, t_\lambda) \in \mathcal{Q}_{\mathcal{TAG}}$ such that $t_\lambda = t^*$ but $(c_{\lambda,1}, c_{\lambda,2}, \mathsf{aux}_\lambda) \neq (c_1^*, c_2^*, \mathsf{aux}^*)$, Return 0.

  Since $t_\lambda = \mathsf{H}_1(c_{\lambda,1}, c_{\lambda,2}, \mathsf{aux}_\lambda)$ and $t^* = \mathsf{H}_1(c_1^*, c_2^*, \mathsf{aux}^*)$, any difference between G_0 and G_1 will imply a collision of H_1. Thus $\big|\Pr_0[\mathsf{Forge}] - \Pr_1[\mathsf{Forge}]\big| \leq \mathsf{Adv}_{\mathcal{H}_1}^{cr}(\ell)$.
- Game $\mathsf{G}_{1,i}$, $i \in [Q_e + 1]$: This game is the same as game G_1, except that, the challenger does not use secret key k to answer the λ-th ($\lambda \in [i-1]$) ENC query at all, and instead, it changes the steps 1, 3 to the steps 1′, 3′ respectively, as in the previous proof.
  Clearly $\Pr_1[\mathsf{Forge}] = \Pr_{1,1}[\mathsf{Forge}]$.

- Game $\mathsf{G}'_{1,i}$, $i \in [Q_e]$: This game is the same as game $\mathsf{G}_{1,i}$, except that the challenger answers the i-th ENC query using steps 1′, 3 (rather than steps 1, 3 in game $\mathsf{G}_{1,i}$), as in the previous proof.
  The only difference between $\mathsf{G}_{1,i}$ and $\mathsf{G}'_{1,i}$ is the distribution of $(g_1, g_2, c_{i,1}, c_{i,2})$. In game $\mathsf{G}_{1,i}$, $(g_1, g_2, c_{i,1}, c_{i,2})$ is a DDH tuple, while in game $\mathsf{G}'_{1,i}$, it is a random tuple. It is straightforward to construct a PPT adversary to solve the DDH problem w.r.t. GenN and $\mathbb{QR}_{\bar{N}}$. We stress that the PPT adversary (simulator) can detect the occurrence of event Forge efficiently since it can choose the secret key $\mathsf{k} = (k_1, k_2, k_3, k_4)$ itself. Thus we can reduce the difference between $\mathsf{G}_{1,i}$ and $\mathsf{G}'_{1,i}$ to the DDH assumption smoothly.

Lemma 2. *For all $i \in [Q_e]$, $\big|\Pr_{1,i}[\mathsf{Forge}] - \Pr_{1,i'}[\mathsf{Forge}]\big| \leq \mathsf{Adv}^{ddh}_{\mathsf{GenN}}(\ell)$.*

Proof. We construct a PPT adversary $\mathcal{B}$ to solve the DDH problem. $\mathcal{B}$ is given $(N, p, q, g_1, g_2, g_1^{x_1}, g_2^{x_2})$, where $(N, p, q) \leftarrow_\$ \mathsf{GenN}(1^\ell)$, $g_1, g_2 \leftarrow_\$ \mathbb{QR}_{\bar{N}}$, and aims to distinguish whether $x_1 = x_2 \leftarrow_\$ \mathbb{Z}_N \setminus \{0\}$ or $x_1, x_2 \leftarrow_\$ \mathbb{Z}_N \setminus \{0\}$.

$\mathcal{B}$ will simulate game $\mathsf{G}_{1,i}$ or $\mathsf{G}'_{1,i}$ for adversary $\mathcal{A}$. First, $\mathcal{B}$ picks $\mathsf{H}_1 \leftarrow_\$ \mathcal{H}_1$, $\mathsf{H}_2 \leftarrow_\$ \mathcal{H}_2$ randomly, sets $\mathsf{prm}_{\mathsf{AIAE}} := (N, p, q, \bar{N} = 2N+1, g_1, g_2, \mathsf{H}_1, \mathsf{H}_2)$ and sends $\mathsf{prm}_{\mathsf{AIAE}}$ to $\mathcal{A}$. Then $\mathcal{B}$ generates the secret key $\mathsf{k} = (k_1, k_2, k_3, k_4)$ itself.

To answer the λ-th ($\lambda \in [Q_e]$) ENC query $(m_\lambda, \mathsf{aux}_\lambda, f_\lambda)$, where $f_\lambda = \langle a_\lambda, \mathsf{b}_\lambda = (b_{\lambda,1}, b_{\lambda,2}, b_{\lambda,3}, b_{\lambda,4})\rangle \in \mathcal{F}_{\text{raff}}$, $\mathcal{B}$ proceeds as follows:

- If $\lambda \in [i-1]$, $\mathcal{B}$ proceeds the same as in $\mathsf{G}_{1,i}$ and $\mathsf{G}'_{1,i}$. That is, $\mathcal{B}$ picks $w_{\lambda,1}, w_{\lambda,2} \leftarrow_\$ \mathbb{Z}_N \backslash \{0\}$ randomly and sets $(c_{\lambda,1}, c_{\lambda,2}) := (g_1^{w_{\lambda,1}}, g_2^{w_{\lambda,2}})$. Then $\mathcal{B}$ chooses $\kappa_\lambda \leftarrow_\$ \mathcal{K}_{\mathsf{AE}}$ and invokes $\chi_\lambda \leftarrow_\$ \mathsf{AE.Enc}(\kappa_\lambda, m_\lambda)$.
- If $\lambda \in [i+1, Q_e]$, $\mathcal{B}$ proceeds the same as in $\mathsf{G}_{1,i}$ and $\mathsf{G}'_{1,i}$. That is, $\mathcal{B}$ picks $w_\lambda \leftarrow_\$ \mathbb{Z}_N \backslash \{0\}$ randomly and sets $(c_{\lambda,1}, c_{\lambda,2}) := (g_1^{w_\lambda}, g_2^{w_\lambda})$. Then $\mathcal{B}$ computes $t_\lambda := \mathsf{H}_1(c_{\lambda,1}, c_{\lambda,2}, \mathsf{aux}_\lambda)$, $\kappa_\lambda := \mathsf{H}_2\big(c_{\lambda,1}^{(a_\lambda k_1 + b_{\lambda,1}) + (a_\lambda k_3 + b_{\lambda,3}) t_\lambda} \cdot c_{\lambda,2}^{(a_\lambda k_2 + b_{\lambda,2}) + (a_\lambda k_4 + b_{\lambda,4}) t_\lambda}\big)$, and invokes $\chi_\lambda \leftarrow_\$ \mathsf{AE.Enc}(\kappa_\lambda, m_\lambda)$.
- If $\lambda = i$, $\mathcal{B}$ embedded its DDH challenge to $(c_{i,1}, c_{i,2}) := (g_1^{x_1}, g_2^{x_2})$. Then it computes $t_i := \mathsf{H}_1(c_{i,1}, c_{i,2}, \mathsf{aux}_i)$, $\kappa_i := \mathsf{H}_2\big(c_{i,1}^{(a_i k_1 + b_{i,1}) + (a_i k_3 + b_{i,3}) t_i} \cdot c_{i,2}^{(a_i k_2 + b_{i,2}) + (a_i k_4 + b_{i,4}) t_i}\big)$, and invokes $\chi_i \leftarrow_\$ \mathsf{AE.Enc}(\kappa_i, m_i)$.

$\mathcal{B}$ returns the challenge ciphertext $\langle c_{\lambda,1}, c_{\lambda,2}, \chi_\lambda\rangle$ to $\mathcal{A}$, and puts $\big(\mathsf{aux}_\lambda, f_\lambda, \langle c_{\lambda,1}, c_{\lambda,2}, \chi_\lambda\rangle\big)$ to $\mathcal{Q}_{\mathcal{ENC}}$, $(\mathsf{aux}_\lambda, f_\lambda)$ to $\mathcal{Q}_{\mathcal{AUXF}}$, and $(c_{\lambda,1}, c_{\lambda,2}, \mathsf{aux}_\lambda, t_\lambda)$ to $\mathcal{Q}_{\mathcal{TAG}}$.

In the case of that $(N, p, q, g_1, g_2, g_1^{x_1}, g_2^{x_2})$ is a DDH tuple, i.e., $x_1 = x_2 \leftarrow_\$ \mathbb{Z}_N \setminus \{0\}$, $\mathcal{B}$ simulates game $\mathsf{G}_{1,i}$ perfectly for $\mathcal{A}$; in the case of that $(N, p, q, g_1, g_2, g_1^{x_1}, g_2^{x_2})$ is a random tuple, i.e., $x_1, x_2 \leftarrow_\$ \mathbb{Z}_N \setminus \{0\}$, $\mathcal{B}$ simulates game $\mathsf{G}'_{1,i}$ perfectly for $\mathcal{A}$.

Finally $\mathcal{B}$ receives a forgery $\big(\mathsf{aux}^*, f^*, \langle c_1^*, c_2^*, \chi^*\rangle\big)$ from $\mathcal{A}$, where $f^* = \langle a^*, \mathsf{b}^* = (b_1^*, b_2^*, b_3^*, b_4^*)\rangle \in \mathcal{F}_{\text{raff}}$. $\mathcal{B}$ determines whether or not the FINALIZE procedure outputs 1 using the secret key $\mathsf{k} = (k_1, k_2, k_3, k_4)$. That is,

- If $(\mathsf{aux}^*, f^*, \langle c_1^*, c_2^*, \chi^* \rangle) \in \mathcal{Q}_{\mathcal{ENC}}$, $\mathcal{B}$ outputs 0 (to its DDH challenger).
- If there exists $(\mathsf{aux}_\lambda, f_\lambda) \in \mathcal{Q}_{\mathcal{AUXF}}$ such that $\mathsf{aux}_\lambda = \mathsf{aux}^*$ but $f_\lambda \neq f^*$, $\mathcal{B}$ outputs 0.
- If $(c_1^*, c_2^*) \notin \mathbb{QR}_{\bar{N}}^2 \vee (c_1^*, c_2^*) = (1, 1)$, $\mathcal{B}$ outputs 0.
- $t^* := \mathsf{H}_1(c_1^*, c_2^*, \mathsf{aux}^*)$, $\kappa^* := \mathsf{H}_2(c_1^{*(a^*k_1+b_1^*)+(a^*k_3+b_3^*)t^*} \cdot c_2^{*(a^*k_2+b_2^*)+(a^*k_4+b_4^*)t^*})$.
- If there exists $(c_{\lambda,1}, c_{\lambda,2}, \mathsf{aux}_\lambda, t_\lambda) \in \mathcal{Q}_{\mathcal{TAG}}$ such that $t_\lambda = t^*$ but $(c_{\lambda,1}, c_{\lambda,2}, \mathsf{aux}_\lambda) \neq (c_1^*, c_2^*, \mathsf{aux}^*)$, $\mathcal{B}$ outputs 0.
- Output $(\mathsf{AE.Dec}(\kappa^*, \chi^*) \neq \perp)$.

With the secret key $\mathsf{k} = (k_1, k_2, k_3, k_4)$, $\mathcal{B}$ simulates FINALIZE perfectly, the same as in games $\mathsf{G}_{1,i}$ and $\mathsf{G}'_{1,i}$, and $\mathcal{B}$ outputs 1 to its DDH challenger if and only if FINALIZE outputs 1, i.e., the event Forge occurs.

As a consequence, $\left| \Pr_{1,i}[\mathsf{Forge}] - \Pr_{1,i'}[\mathsf{Forge}] \right| \leq \mathsf{Adv}^{ddh}_{\mathsf{GenN},\mathcal{B}}(\ell)$. ■

We analyze the difference between $\mathsf{G}'_{1,i}$ and $\mathsf{G}_{1,i+1}$ via the following lemma, and the proof is in the full version [HLL16] due to the lack of space.

Lemma 3. *For all* $i \in [Q_e]$, $\Pr_{1,i'}[\mathsf{Forge}] \leq \Pr_{1,i+1}[\mathsf{Forge}] + \mathsf{Adv}^{int\text{-}ot}_{\mathsf{AE}}(\ell) + \frac{1}{(N-1)} + 2^{-\Omega(\ell)}$.

Now in game G_{1,Q_e+1}, the challenger does not use the secret key k to compute κ_λ at all, hence $\mathsf{k} = (k_1, k_2, k_3, k_4)$ is uniformly random to the adversary $\mathcal{A}$. As a result, in the FINALIZE procedure defining the event Forge,

$$\kappa^* = \mathsf{H}_2(\underbrace{g_1^{a^* \cdot ((w_1^* k_1 + w_2^* w k_2) + t^* \cdot (w_1^* k_3 + w_2^* w k_4))} \cdot g_1^{(w_1^* b_1^* + w_2^* w b_2^*) + t^* \cdot (w_1^* b_3^* + w_2^* w b_4^*)}}_{\triangleq Y}),$$

where $w = \mathrm{dlog}_{g_1} g_2 \in \mathbb{Z}_N$ and $(w_1^*, w_2^*) = (\mathrm{dlog}_{g_1} c_1^*, \mathrm{dlog}_{g_2} c_2^*) \in \mathbb{Z}_N^2 \backslash \{(0,0)\}$. The term $(w_1^* k_1 + w_2^* w k_2)$ is uniformly distributed over $\mathbb{Z}_N$. Then as long as $a^* \in \mathbb{Z}_N^*$, Y will be uniformly distributed over $\mathbb{QR}_{\bar{N}}$ and independent of H_2. By the Leftover Hash Lemma, $\kappa^* = \mathsf{H}_2(Y)$ is statistically close to the uniform distribution over $\mathcal{K}_{\mathsf{AE}}$. Thus $\mathsf{AE.Dec}(\kappa^*, \chi^*) \neq \perp$ will hold with probability at most $\mathsf{Adv}^{int\text{-}ot}_{\mathsf{AE}}(\ell)$. Then $\Pr_{1,Q_e+1}[\mathsf{Forge}] \leq \mathsf{Adv}^{int\text{-}ot}_{\mathsf{AE}}(\ell) + 2^{-\Omega(\ell)}$.
Taking all things together, the weak INT-$\mathcal{F}_{\text{raff}}$-RKA security of AIAE follows. ■

Remark. We stress that the problem in the INT-$\mathcal{F}_{\text{aff}}$-RKA security proof of LLJ's $\overline{\mathsf{AE}}$ does not appear here. The weak INT-$\mathcal{F}_{\text{raff}}$-RKA security of our AIAE can be reduced to the DDH assumption smoothly. More precisely, in the security analysis of games $\mathsf{G}_{1,i}$ and $\mathsf{G}'_{1,i}$ (cf. Lemma 2), the simulator chooses the secret key itself and uses it to detect the occurrence of event Forge efficiently. Therefore the simulator can always make use of the difference between $\Pr_{1,i}[\mathsf{Forge}]$ and $\Pr_{1,i'}[\mathsf{Forge}]$ to solve the DDH problem.

5 PKE with n-KDM[$\mathcal{F}_{\text{aff}}$]-CCA Security

Let $\mathsf{AIAE} = (\mathsf{AIAE.Setup}, \mathsf{AIAE.Enc}, \mathsf{AIAE.Dec})$ be the DDH-based auxiliary-input authenticated encryption scheme constructed from OT-secure AE, with key space $(\mathbb{Z}_N)^4$ and a suitable message space $\mathcal{M}$ (cf. Fig. 6). Following our approach in Fig. 1, we have to design the other two building blocks.

KEM: With respect to this AIAE, we design a KEM which can encapsulate a key tuple $(k_1, k_2, k_3, k_4) \in (\mathbb{Z}_N)^4$.

$\mathcal{E}$: With respect to the affine function $\mathcal{F}_{\text{aff}}$, we design a public-key encryption $\mathcal{E}$ such that $\mathcal{E}.\mathsf{Enc}$ can be changed to an entropy filter for affine functions in a computationally indistinguishable way.

The proposed $\mathsf{PKE} = (\mathsf{Setup}, \mathsf{Gen}, \mathsf{Enc}, \mathsf{Dec})$ is defined in Fig. 7, where the shadowed parts describe algorithms of building blocks KEM and $\mathcal{E}$.

<table>
<tr>
<td>

$\mathsf{prm} \leftarrow_{\$} \mathsf{Setup}(1^\ell)$:

$\mathsf{prm}_{\mathsf{AIAE}} \leftarrow_{\$} \mathsf{AIAE.Setup}(1^\ell)$, where
$\mathsf{prm}_{\mathsf{AIAE}} = (N, p, q, \bar{N}, \bar{g}_1, \bar{g}_2, \mathsf{H}_1, \mathsf{H}_2)$,
$N = pq$, $\bar{N} = 2N + 1$, $\bar{g}_1, \bar{g}_2 \in \mathbb{QR}_{\bar{N}}$.
$\mathsf{prm}'_{\mathsf{AIAE}} := (N, \bar{N}, \bar{g}_1, \bar{g}_2, \mathsf{H}_1, \mathsf{H}_2)$.
$g_1, g_2, g_3, g_4, g_5 \leftarrow_{\$} \mathbb{SCR}_{N^s}$.
Return $\mathsf{prm} := (\mathsf{prm}'_{\mathsf{AIAE}}, g_1, g_2, g_3, g_4, g_5)$.

</td>
<td>

$(\mathsf{pk}, \mathsf{sk}) \leftarrow_{\$} \mathsf{Gen}(\mathsf{prm})$:

$x_1, y_1, x_2, y_2, x_3, y_3, x_4, y_4 \leftarrow_{\$} [\lfloor \frac{N^2}{4} \rfloor]$.
$(h_1, h_2, h_3, h_4) := (g_1^{-x_1} g_2^{-y_1}, g_2^{-x_2} g_3^{-y_2}, g_3^{-x_3} g_4^{-y_3}, g_4^{-x_4} g_5^{-y_4}) \bmod N^s$.
$\mathsf{pk} := (h_1, h_2, h_3, h_4)$.
$\mathsf{sk} := (x_1, y_1, x_2, y_2, x_3, y_3, x_4, y_4)$.
Return $(\mathsf{pk}, \mathsf{sk})$.

</td>
</tr>
<tr>
<td>

$\langle \mathsf{aux}, \mathsf{aiae.ct} \rangle \leftarrow_{\$} \mathsf{Enc}(\mathsf{pk}, m)$: $m \in [N^{s-1}]$

// $(\mathsf{k}, \mathsf{aux}) \leftarrow_{\$} \mathsf{KEM.Enc}(\mathsf{pk})$:
$\mathsf{k} = (k_1, k_2, k_3, k_4) \leftarrow_{\$} \mathbb{Z}_N^4$.
$r \leftarrow_{\$} [\lfloor \frac{N}{4} \rfloor]$.
$(u_1, u_2, u_3, u_4, u_5) := (g_1^r, g_2^r, g_3^r, g_4^r, g_5^r) \bmod N^2$.
$(e_1, e_2, e_3, e_4) := (h_1^r T^{k_1}, h_2^r T^{k_2}, h_3^r T^{k_3}, h_4^r T^{k_4}) \bmod N^2$.
$\mathsf{aux} := (u_1, \cdots, u_5, e_1, \cdots, e_4)$.
// $\mathcal{E}.\mathsf{ct} \leftarrow_{\$} \mathcal{E}.\mathsf{Enc}(\mathsf{pk}, m)$:
$\tilde{r}_1, \tilde{r}_2, \tilde{r}_3, \tilde{r}_4 \leftarrow_{\$} [\lfloor \frac{N}{4} \rfloor]$.
$(\tilde{u}_1, \tilde{u}_2, \tilde{u}_3, \tilde{u}_4, \tilde{u}_5, \tilde{u}_6, \tilde{u}_7, \tilde{u}_8) := (g_1^{\tilde{r}_1}, g_2^{\tilde{r}_1}, g_2^{\tilde{r}_2}, g_3^{\tilde{r}_2}, g_3^{\tilde{r}_3}, g_4^{\tilde{r}_3}, g_4^{\tilde{r}_4}, g_5^{\tilde{r}_4}) \bmod N^s$.
$\tilde{e} := h_1^{\tilde{r}_1} h_2^{\tilde{r}_2} h_3^{\tilde{r}_3} h_4^{\tilde{r}_4} T^m \bmod N^s$.
$t := g_1^m \bmod N \in \mathbb{Z}_N$.
$\mathcal{E}.\mathsf{ct} := (\tilde{u}_1, \cdots, \tilde{u}_8, \tilde{e}, t)$.
$\mathsf{aiae.ct} \leftarrow_{\$} \mathsf{AIAE.Enc}(\mathsf{k}, \mathcal{E}.\mathsf{ct}, \mathsf{aux})$.
Return $\langle \mathsf{aux}, \mathsf{aiae.ct} \rangle$.

</td>
<td>

$m/\bot \leftarrow \mathsf{Dec}(\mathsf{sk}, \langle \mathsf{aux}, \mathsf{aiae.ct} \rangle)$:

// $\mathsf{k}/\bot \leftarrow \mathsf{KEM.Dec}(\mathsf{sk}, \mathsf{aux})$:
Parse $\mathsf{aux} = (u_1, \cdots, u_5, e_1, \cdots, e_4)$.
If $e_1 u_1^{x_1} u_2^{y_1}, e_2 u_2^{x_2} u_3^{y_2}, e_3 u_3^{x_3} u_4^{y_3}, e_4 u_4^{x_4} u_5^{y_4} \in \mathbb{RU}_{N^2}$
$(k_1, k_2, k_3, k_4) := (\mathrm{dlog}_T(e_1 u_1^{x_1} u_2^{y_1}), \mathrm{dlog}_T(e_2 u_2^{x_2} u_3^{y_2}), \mathrm{dlog}_T(e_3 u_3^{x_3} u_4^{y_3}), \mathrm{dlog}_T(e_4 u_4^{x_4} u_5^{y_4})) \bmod N$.
$\mathsf{k} := (k_1, k_2, k_3, k_4)$.
Else, Return $\bot$.
$\mathcal{E}.\mathsf{ct}/\bot \leftarrow \mathsf{AIAE.Dec}(\mathsf{k}, \mathsf{aiae.ct}, \mathsf{aux})$.
// $m/\bot \leftarrow \mathcal{E}.\mathsf{Dec}(\mathsf{sk}, \mathcal{E}.\mathsf{ct})$:
Parse $\mathcal{E}.\mathsf{ct} = (\tilde{u}_1, \cdots, \tilde{u}_8, \tilde{e}, t)$.
If $\tilde{e} \tilde{u}_1^{x_1} \tilde{u}_2^{y_1} \tilde{u}_3^{x_2} \tilde{u}_4^{y_2} \tilde{u}_5^{x_3} \tilde{u}_6^{y_3} \tilde{u}_7^{x_4} \tilde{u}_8^{y_4} \in \mathbb{RU}_{N^s}$
$m := \mathrm{dlog}_T(\tilde{e} \tilde{u}_1^{x_1} \tilde{u}_2^{y_1} \tilde{u}_3^{x_2} \tilde{u}_4^{y_2} \tilde{u}_5^{x_3} \tilde{u}_6^{y_3} \tilde{u}_7^{x_4} \tilde{u}_8^{y_4}) \bmod N^{s-1}$.
If $t = g_1^m \bmod N$, Return m.
Else, Return $\bot$.

</td>
</tr>
</table>

Fig. 7. Construction of PKE from AIAE. The shadowed parts describe algorithms of building blocks KEM and $\mathcal{E}$. Here p, q contained in $\mathsf{prm}_{\mathsf{AIAE}}$ are not provided in $\mathsf{prm}'_{\mathsf{AIAE}}$, since they are not necessary in the encryption and decryption algorithms of AIAE.

The correctness of PKE follows from the correctness of AIAE, $\mathcal{E}$ and KEM directly. We now show its KDM-CCA-security through the following theorem.

Theorem 2. *If the underlying scheme* AIAE *is IND-$\mathcal{F}_{raff}$-RKA and weak INT-$\mathcal{F}_{raff}$-RKA secure, the DCR assumption holds w.r.t.* GenN *and group $\mathbb{QR}_{N^s}$, and the DL Assumption holds w.r.t.* GenN *and group $\mathbb{SCR}_{N^s}$, then the resulting scheme* PKE *in Fig. 7 is n-KDM[$\mathcal{F}_{aff}$]-CCA secure.*

Proof of Theorem 2. Suppose that $\mathcal{A}$ is a PPT adversary against the n-KDM[$\mathcal{F}_{\text{aff}}$]-CCA security of PKE, who makes at most Q_e times of ENC queries and Q_d times of DEC queries. We prove the theorem by defining a sequence of games. Before presenting the full detailed proof, we first give a high-level description how n-KDM[$\mathcal{F}_{\text{aff}}$]-CCA security is achieved.

(1) For the n secret key tuples, each tuple can be divided into two parts: for $i \in [n]$, $\mathsf{sk}_i = (x_{i,j}, y_{i,j})_{j=1}^4 = \big((x_{i,j}, y_{i,j})_{j=1}^4 \bmod N, (x_{i,j}, y_{i,j})_{j=1}^4 \bmod \phi(N)/4\big)$.
(2) Each secret key tuple can be generated by adding a random shift $(\overline{x}_{i,j}, \overline{y}_{i,j})_{j=1}^4$ to a fixed base $(x_j, y_j)_{j=1}^4$, i.e., $\mathsf{sk}_i = (x_{i,j}, y_{i,j})_{j=1}^4 := (x_j, y_j)_{j=1}^4 + (\overline{x}_{i,j}, \overline{y}_{i,j})_{j=1}^4$.
(3) Every public key tuple $\mathsf{pk}_i = (h_{i,1}, \cdots, h_{i,4})$ only reveals information about the (mod $\phi(N)/4$) part of the secret key tuple sk_i.
(4) For each encryption query from the adversary (f_λ, i_λ), if the ENC oracle encrypts $f_\lambda(\mathsf{sk}_1, \cdots, \mathsf{sk}_n)$, the ciphertext might reveal information about sk_i through $\mathcal{E}$.ct. We have to change this fact such that the leaked information about sk_i in ENC is bounded.
   - By IV$_d$ assumption, we can change the generation of $\mathcal{E}$.ct by oracle ENC such that it does not reveal any information about $(x_j, y_j)_{j=1}^4 \bmod N$, i.e., the (mod N) part of the base secret key tuple.
   - By IV$_d$ assumption, we can change the generation of kem.ct($=$ aux) by ENC such that it encapsulates a different key, other than the key used in AIAE.Enc. If AIAE.Enc uses key $(r_\lambda k_j^* + s_{\lambda,j})_{j=1}^4$, then KEM.Enc encapsulates $\big(r_\lambda(k_j^* - \alpha_j x_j - \alpha_{j+1} y_j) - r_\lambda(\alpha_j \overline{x}_{i_\lambda,j} + \alpha_{j+1}\overline{y}_{i_\lambda,j}) + s_{\lambda,j}\big)_{j=1}^4 \bmod N$. Thus, $(k_1^*, \cdots, k_4^*)$ is now protected by $(x_j, y_j)_{j=1}^4 \bmod N$.
(5) Oracle DEC might also leak information about $(x_j, y_j)_{j=1}^4 \bmod N$. Therefore, we change how oracle DEC works so that decryption does not use $(x_j, y_j)_{j=1}^4 \bmod N$ any more. Observe that as long as the ciphertext queried by the adversary satisfies $\forall j \in [5], u_j \in \mathbb{SCR}_{N^2}$ and $\forall j \in [8], \tilde{u}_j \in \mathbb{SCR}_{N^s}$, DEC can use $\phi(N)$ and the (mod $\phi(N)/4$) part of secret key for decryption.
   - If $\exists j \in [5], u_j \notin \mathbb{SCR}_{N^2}$ in the ciphertext queried by the adversary, we expect that AIAE.Dec will reject, due to its weak INT-$\mathcal{F}_{\text{raff}}$-RKA security.
   - If $\exists j \in [8], \tilde{u}_j \notin \mathbb{SCR}_{N^s}$ in the ciphertext queried by the adversary, we expect decryption will result in $t \neq g_1^m \bmod N$, so $\mathcal{E}$.Dec will reject.
(6) Consequently, both $(x_j, y_j)_{j=1}^4 \bmod N$ and $(k_1^*, \cdots, k_4^*)$ are random to the adversary, and AIAE.Enc always uses the restricted affine function of $(k_1^*, \cdots, k_4^*)$ for encryption. Then IND-$\mathcal{F}_{\text{raff}}$-RKA security of AIAE implies the n-KDM[$\mathcal{F}_{\text{aff}}$]-CCA security.

In the proof, G_1-G_2 are dedicated to deal with the n-user case; the aim of G_3-G_4 is to eliminate the use of the (mod N) part of $(x_j, y_j)_{j=1}^4$ in Enc; the aim of G_5-G_6 is to use $(x_j, y_j)_{j=1}^4$ mod N to hide the AIAE's base key $(k_1^*, \cdots, k_4^*)$ in Enc, however, Dec may still leak the information about $(x_j, y_j)_{j=1}^4$ mod N; the aim of G_7-G_8 is to eliminate the use of $(x_j, y_j)_{j=1}^4$ mod N in Dec; finally, in G_9-G_{10}, the IND-$\mathcal{F}_{\text{raff}}$-RKA security of AIAE is used to prove the n-KDM[$\mathcal{F}_{\text{aff}}$]-CCA security of PKE, since $(k_1^*, \cdots, k_4^*)$ is perfectly hided by $(x_j, y_j)_{j=1}^4$ mod N.

- Game G_0: This is the original n-KDM[$\mathcal{F}_{\text{aff}}$]-CCA game. Let Win denote the event that $\beta' = \beta$. Then by definition, $\mathsf{Adv}_{\mathsf{PKE},\mathcal{A}}^{kdm\text{-}cca}(\ell) = \left| \Pr_0[\mathsf{Win}] - \frac{1}{2} \right|$. Denote by $\mathsf{pk}_i = (h_{i,1}, \cdots, h_{i,4})$ and $\mathsf{sk}_i = (x_{i,1}, y_{i,1}, \cdots, x_{i,4}, y_{i,4})$ the public and secret keys of the i-th user respectively, $i \in [n]$.
- Game G_1: This game is the same as game G_0, except that, when answering the Dec query $(\langle \mathsf{aux}, \mathsf{aiae.ct} \rangle, i \in [n])$, the challenger outputs $\perp$ if $\langle \mathsf{aux}, \mathsf{aiae.ct} \rangle = \langle \mathsf{aux}_\lambda, \mathsf{aiae.ct}_\lambda \rangle$ for some $\lambda \in [Q_e]$, where $\langle \mathsf{aux}_\lambda, \mathsf{aiae.ct}_\lambda \rangle$ is the challenge ciphertext for the λ-th Enc query (f_λ, i_λ).

  Case 1: $(\langle \mathsf{aux}, \mathsf{aiae.ct} \rangle, i) = (\langle \mathsf{aux}_\lambda, \mathsf{aiae.ct}_\lambda \rangle, i_\lambda)$.

  Dec will output $\perp$ in game G_0 since $(\langle \mathsf{aux}_\lambda, \mathsf{aiae.ct}_\lambda \rangle, i_\lambda)$ is prohibited.

  Case 2: $\langle \mathsf{aux}, \mathsf{aiae.ct} \rangle = \langle \mathsf{aux}_\lambda, \mathsf{aiae.ct}_\lambda \rangle$ but $i \neq i_\lambda$.

  We show that in game G_0, Dec will output $\perp$, due to $e_{\lambda,1} u_{\lambda,1}^{x_{i,1}} u_{\lambda,2}^{y_{i,1}} \notin \mathbb{RU}_{N^2}$, with overwhelming probability. Recall that $u_{\lambda,1} = g_1^{r_\lambda}, u_{\lambda,2} = g_2^{r_\lambda}, e_{\lambda,1} = h_{i_\lambda,1}^{r_\lambda} T^{k_{\lambda,1}}$, so

  $$e_{\lambda,1} u_{\lambda,1}^{x_{i,1}} u_{\lambda,2}^{y_{i,1}} = h_{i_\lambda,1}^{r_\lambda} T^{k_{\lambda,1}} \cdot (g_1^{r_\lambda})^{x_{i,1}} (g_2^{r_\lambda})^{y_{i,1}} = (h_{i_\lambda,1} h_{i,1}^{-1})^{r_\lambda} T^{k_{\lambda,1}} \bmod N^2,$$

  where $h_{i_\lambda,1}$ and $h_{i,1}$ are parts of public key of different users i_λ and i respectively and are uniformly distributed over $\mathbb{SCR}_{N^s}$. So $h_{i_\lambda,1} h_{i,1}^{-1} \neq 1$, hence $e_{\lambda,1} u_{\lambda,1}^{x_{i,1}} u_{\lambda,2}^{y_{i,1}} \notin \mathbb{RU}_{N^2}$, except with probability $2^{-\Omega(\ell)}$.

  By a union bound, G_0 and G_1 are identical except with probability $Q_d \cdot 2^{-\Omega(\ell)}$, therefore $\left| \Pr_0[\mathsf{Win}] - \Pr_1[\mathsf{Win}] \right| \le Q_d \cdot 2^{-\Omega(\ell)}$.
- Game G_2: This game is the same as game G_1, except that, the challenger samples the secret keys $\mathsf{sk}_i = (x_{i,1}, y_{i,1}, \cdots, x_{i,4}, y_{i,4})$, $i \in [n]$, in a different way. First, it chooses random $(x_1, y_1, \cdots, x_4, y_4)$ and $(\bar{x}_{i,1}, \bar{y}_{i,1}, \cdots, \bar{x}_{i,4}, \bar{y}_{i,4})$, $i \in [n]$, from $[\lfloor N^2/4 \rfloor]$, then it computes $(x_{i,1}, y_{i,1}, \cdots, x_{i,4}, y_{i,4}) = (x_1, y_1, \cdots, x_4, y_4) + (\overline{x}_{i,1}, \overline{y}_{i,1}, \cdots, \overline{x}_{i,4}, \overline{y}_{i,4}) \bmod \lfloor N^2/4 \rfloor$ for $i \in [n]$.

  Obviously, the secret keys $\mathsf{sk}_i = (x_{i,1}, y_{i,1}, \cdots, x_{i,4}, y_{i,4})$ are uniformly distributed. Hence G_2 is identical to G_1, and $\Pr_1[\mathsf{Win}] = \Pr_2[\mathsf{Win}]$.
- Game G_3: This game is the same as game G_2, except that, when responding to the adversary's λ-th ($\lambda \in [Q_e]$) Enc query (f_λ, i_λ), instead of using the public keys $\mathsf{pk}_{i_\lambda} = (h_{i_\lambda,1}, \cdots, h_{i_\lambda,4})$, the challenger uses the secret keys $\mathsf{sk}_{i_\lambda} = (x_{i_\lambda,1}, y_{i_\lambda,1}, \cdots, x_{i_\lambda,4}, y_{i_\lambda,4})$ to prepare $(e_{\lambda,1}, \cdots, e_{\lambda,4})$ and $\tilde{e}_\lambda$ as follows:
  - $(e_{\lambda,1}, \cdots, e_{\lambda,4}) := (u_{\lambda,1}^{-x_{i_\lambda,1}} u_{\lambda,2}^{-y_{i_\lambda,1}} T^{k_{\lambda,1}}, \cdots, u_{\lambda,4}^{-x_{i_\lambda,4}} u_{\lambda,5}^{-y_{i_\lambda,4}} T^{k_{\lambda,4}}) \bmod N^2$,
  - $\tilde{e}_\lambda := \tilde{u}_{\lambda,1}^{-x_{i_\lambda,1}} \tilde{u}_{\lambda,2}^{-y_{i_\lambda,1}} \tilde{u}_{\lambda,3}^{-x_{i_\lambda,2}} \tilde{u}_{\lambda,4}^{-y_{i_\lambda,2}} \tilde{u}_{\lambda,5}^{-x_{i_\lambda,3}} \tilde{u}_{\lambda,6}^{-y_{i_\lambda,3}} \tilde{u}_{\lambda,7}^{-x_{i_\lambda,4}} \tilde{u}_{\lambda,8}^{-y_{i_\lambda,4}} T^{m_\beta} \bmod N^s$.

Observe that for $j \in \{1,2,3,4\}$,

$$e_{\lambda,j} \stackrel{\mathsf{G}_2}{=} h_{i_\lambda,j}^{r_\lambda} T^{k_{\lambda,j}} = (g_j^{-x_{i_\lambda,j}} g_{j+1}^{-y_{i_\lambda,j}})^{r_\lambda} T^{k_{\lambda,j}} \stackrel{\mathsf{G}_3}{=} u_{\lambda,j}^{-x_{i_\lambda,j}} u_{\lambda,j+1}^{-y_{i_\lambda,j}} T^{k_{\lambda,j}} \bmod N^2,$$

$$\tilde{e}_\lambda \stackrel{\mathsf{G}_2}{=} h_{i_\lambda,1}^{\tilde{r}_{\lambda,1}} \cdots h_{i_\lambda,4}^{\tilde{r}_{\lambda,4}} T^{m_\beta} = (g_1^{-x_{i_\lambda,1}} g_2^{-y_{i_\lambda,1}})^{\tilde{r}_{\lambda,1}} \cdots (g_4^{-x_{i_\lambda,4}} g_5^{-y_{i_\lambda,4}})^{\tilde{r}_{\lambda,4}} T^{m_\beta}$$

$$\stackrel{\mathsf{G}_3}{=} \tilde{u}_{\lambda,1}^{-x_{i_\lambda,1}} \tilde{u}_{\lambda,2}^{-y_{i_\lambda,1}} \cdots \tilde{u}_{\lambda,7}^{-x_{i_\lambda,4}} \tilde{u}_{\lambda,8}^{-y_{i_\lambda,4}} T^{m_\beta} \bmod N^s.$$

Thus G_3 is identical to G_2, and $\Pr_2[\mathsf{Win}] = \Pr_3[\mathsf{Win}]$.

- Game G_4: This game is the same as game G_3, except that, in the case of the challenge bit $\beta = 1$, to answer the λ-th ($\lambda \in [Q_e]$) Enc query (f_λ, i_λ), the challenger does not use $(x_1, y_1, \cdots, x_4, y_4) \bmod N$ to compute $\tilde{e}_\lambda$ any more, and instead, it computes $(\tilde{u}_{\lambda,1}, \cdots, \tilde{u}_{\lambda,8})$ and $\tilde{e}_\lambda$ as follows:
  - $(\tilde{u}_{\lambda,1}, \cdots, \tilde{u}_{\lambda,8}) := (g_1^{\tilde{r}_{\lambda,1}} T^{\sum_{i=1}^n a_{i,1}}, g_2^{\tilde{r}_{\lambda,1}} T^{\sum_{i=1}^n b_{i,1}}, g_2^{\tilde{r}_{\lambda,2}} T^{\sum_{i=1}^n a_{i,2}}, g_3^{\tilde{r}_{\lambda,2}} \cdot T^{\sum_{i=1}^n b_{i,2}}, g_3^{\tilde{r}_{\lambda,3}} T^{\sum_{i=1}^n a_{i,3}}, g_4^{\tilde{r}_{\lambda,3}} T^{\sum_{i=1}^n b_{i,3}}, g_4^{\tilde{r}_{\lambda,4}} T^{\sum_{i=1}^n a_{i,4}}, g_5^{\tilde{r}_{\lambda,4}} T^{\sum_{i=1}^n b_{i,4}})$,
  - $\tilde{e}_\lambda := h_{i_\lambda,1}^{\tilde{r}_{\lambda,1}} \cdots h_{i_\lambda,4}^{\tilde{r}_{\lambda,4}} T^{\sum_{i=1}^n \sum_{j=1}^4 (a_{i,j}(\bar{x}_{i,j} - \bar{x}_{i_\lambda,j}) + b_{i,j}(\bar{y}_{i,j} - \bar{y}_{i_\lambda,j})) + c} \bmod N^s$,

  where $f_\lambda = (\{a_{i,1}, b_{i,1}, \cdots, a_{i,4}, b_{i,4}\}_{i\in[n]}, c) \in \mathcal{F}_{\mathsf{aff}}$.

  Observe that,

$$\begin{aligned}
\tilde{e}_\lambda &\stackrel{\mathsf{G}_4}{=} \textstyle\prod_{j=1}^4 h_{i_\lambda,j}^{\tilde{r}_{\lambda,j}} \cdot T^{\sum_{i=1}^n \sum_{j=1}^4 (a_{i,j}(\bar{x}_{i,j} - \bar{x}_{i_\lambda,j}) + b_{i,j}(\bar{y}_{i,j} - \bar{y}_{i_\lambda,j})) + c} \\
&= \textstyle\prod_{j=1}^4 h_{i_\lambda,j}^{\tilde{r}_{\lambda,j}} \cdot T^{\sum_{i=1}^n \sum_{j=1}^4 (a_{i,j}(x_{i,j} - x_{i_\lambda,j}) + b_{i,j}(y_{i,j} - y_{i_\lambda,j})) + c} \\
&= \textstyle\prod_{j=1}^4 (g_j^{-x_{i_\lambda,j}} g_{j+1}^{-y_{i_\lambda,j}})^{\tilde{r}_{\lambda,j}} \cdot T^{m_1 - \sum_{i=1}^n \sum_{j=1}^4 (a_{i,j} x_{i_\lambda,j} + b_{i,j} y_{i_\lambda,j})} \\
&= \textstyle\prod_{j=1}^4 (g_j^{\tilde{r}_{\lambda,j}} T^{\sum_{i=1}^n a_{i,j}})^{-x_{i_\lambda,j}} (g_{j+1}^{\tilde{r}_{\lambda,j}} T^{\sum_{i=1}^n b_{i,j}})^{-y_{i_\lambda,j}} \cdot T^{m_1} \\
&= \tilde{u}_{\lambda,1}^{-x_{i_\lambda,1}} \tilde{u}_{\lambda,2}^{-y_{i_\lambda,1}} \cdots \tilde{u}_{\lambda,7}^{-x_{i_\lambda,4}} \tilde{u}_{\lambda,8}^{-y_{i_\lambda,4}} T^{m_1} \bmod N^s,
\end{aligned}$$

  where the third equality follows from $m_1 - \sum_{i=1}^n \sum_{j=1}^4 (a_{i,j} x_{i,j} + b_{i,j} y_{i,j}) + c$. Therefore, $\tilde{e}_\lambda$ can be computed from $(\tilde{u}_{\lambda,1}, \cdots, \tilde{u}_{\lambda,8})$ in the same way as in G_3 and G_4. Hence the only difference between G_3 and G_4 is the distribution of $(\tilde{u}_{\lambda,1}, \cdots, \tilde{u}_{\lambda,8})$ themselves. We analyze the difference via the following lemma, and the proof is presented in the full version [HLL16].

Lemma 4. *There exists a PPT adversary $\mathcal{B}_1$ against the IV_5 assumption w.r.t.* GenN *and $\mathbb{QR}_{N^s}$, such that* $\left|\Pr_3[\mathsf{Win}] - \Pr_4[\mathsf{Win}]\right| \le \mathsf{Adv}^{iv_5}_{\mathsf{GenN},\mathcal{B}_1}(\ell)$.

- Game G_5: This game is the same as game G_4, except that, the challenger chooses random $r^* \in [\lfloor N/4 \rfloor]$ and $\alpha_1, \cdots, \alpha_5 \in \mathbb{Z}_N$ beforehand (in Initialize). In addition, to respond to the λ-th ($\lambda \in [Q_e]$) Enc query (f_λ, i_λ), the challenger computes $(u_{\lambda,1}, \cdots, u_{\lambda,5})$ as follows:
  - $(u_{\lambda,1}, \cdots, u_{\lambda,5}) := ((g_1^{r^*} T^{\alpha_1})^{r_\lambda}, \cdots, (g_5^{r^*} T^{\alpha_5})^{r_\lambda}) \bmod N^2$.

  The only difference between G_4 and G_5 is the distribution of $(u_{\lambda,1}, \cdots, u_{\lambda,5})$. In game G_4, it equals $(g_1^{r_\lambda}, \cdots, g_5^{r_\lambda}) \bmod N^2$, while in game G_5, it equals $((g_1^{r^*} T^{\alpha_1})^{r_\lambda}, \cdots, (g_5^{r^*} T^{\alpha_5})^{r_\lambda}) \bmod N^2$. Similar to the previous lemma, it is straightforward to construct a PPT adversary to solve IV_5 problem by employing the power of adversary $\mathcal{A}$. Thus $\left|\Pr_4[\mathsf{Win}] - \Pr_5[\mathsf{Win}]\right| \le \mathsf{Adv}^{iv_5}_{\mathsf{GenN}}(\ell)$.

- Game G_6: This game is the same as game G_5, except that, the challenger chooses a random tuple $\mathsf{k}^* = (k_1^*, k_2^*, k_3^*, k_4^*)$ beforehand (in INITIALIZE). In addition, to respond to the λ-th ($\lambda \in [Q_e]$) ENC query (f_λ, i_λ), the challenger uses a different way to generate $\mathsf{k}_\lambda = (k_{\lambda,1}, k_{\lambda,2}, k_{\lambda,3}, k_{\lambda,4})$ and $(e_{\lambda,1}, \cdots, e_{\lambda,4})$:
  - pick $\mathsf{s}_\lambda = (s_{\lambda,1}, s_{\lambda,2}, s_{\lambda,3}, s_{\lambda,4}) \leftarrow_\$ \mathbb{Z}_N^4$ and $r_\lambda \leftarrow_\$ [\lfloor N/4 \rfloor]$ uniformly, and compute $\mathsf{k}_\lambda = (k_{\lambda,1}, k_{\lambda,2}, k_{\lambda,3}, k_{\lambda,4}) := (r_\lambda k_1^* + s_{\lambda,1}, \cdots, r_\lambda k_4^* + s_{\lambda,4})$.
  - $(e_{\lambda,1}, \cdots, e_{\lambda,4}) :=$ $(h_{i_\lambda,1}^{r^* r_\lambda} T^{r_\lambda(k_1^* - \alpha_1 x_{i_\lambda,1} - \alpha_2 y_{i_\lambda,1}) + s_{\lambda,1}}, \cdots, h_{i_\lambda,4}^{r^* r_\lambda} T^{r_\lambda(k_4^* - \alpha_4 x_{i_\lambda,4} - \alpha_5 y_{i_\lambda,4}) + s_{\lambda,4}})$.

  Clearly k_λ is uniformly distributed over $\mathbb{Z}_N^4$, as in game G_5. At the same time, observe that for $j \in \{1, 2, 3, 4\}$,

$$\begin{aligned} e_{\lambda,j} &\overset{\mathsf{G}_5}{=} u_{\lambda,j}^{-x_{i_\lambda,j}} u_{\lambda,j+1}^{-y_{i_\lambda,j}} T^{k_{\lambda,j}} = (g_j^{r^*} T^{\alpha_j})^{-r_\lambda \cdot x_{i_\lambda,j}} (g_{j+1}^{r^*} T^{\alpha_{j+1}})^{-r_\lambda \cdot y_{i_\lambda,j}} T^{k_{\lambda,j}} \\ &= (g_j^{-x_{i_\lambda,j}} g_{j+1}^{-y_{i_\lambda,j}})^{r^* r_\lambda} T^{k_{\lambda,j} - r_\lambda \cdot (\alpha_j x_{i_\lambda,j} + \alpha_{j+1} y_{i_\lambda,j})} \\ &\overset{\mathsf{G}_6}{=} h_{i_\lambda,j}^{r^* r_\lambda} T^{r_\lambda \cdot (k_j^* - \alpha_j x_{i_\lambda,j} - \alpha_{j+1} y_{i_\lambda,j}) + s_{\lambda,j}} \bmod N^2. \end{aligned}$$

  Thus G_6 is identical to G_5, and $\Pr_5[\mathsf{Win}] = \Pr_6[\mathsf{Win}]$.
- Game G_7: This game is the same as game G_6, except for a modification to answering the DEC queries $(\langle \mathsf{aux}, \mathsf{aiae.ct} \rangle, i \in [n])$. The challenger uses the i-th user's secret key $\mathsf{sk}_i = (x_{i,1}, y_{i,1}, \cdots, x_{i,4}, y_{i,4})$ together with $\phi(N)$ to compute the decryption of ciphertext $\langle \mathsf{aux}, \mathsf{aiae.ct} \rangle$, where $\mathsf{aux} = (u_1, \cdots, u_5, e_1, \cdots, e_4)$. More precisely, it computes $\mathsf{k} = (k_1, \cdots, k_4)$ and m as follows:
  - $(\alpha_1', \cdots, \alpha_5') := (\mathrm{dlog}_T(u_1^{\phi(N)})/\phi(N), \cdots, \mathrm{dlog}_T(u_5^{\phi(N)})/\phi(N)) \bmod N$, $(\gamma_1', \cdots, \gamma_4') := (\mathrm{dlog}_T(e_1^{\phi(N)})/\phi(N), \cdots, \mathrm{dlog}_T(e_4^{\phi(N)})/\phi(N)) \bmod N$, $\mathsf{k} = (k_1, \cdots, k_4) := (\alpha_1' x_{i,1} + \alpha_2' y_{i,1} + \gamma_1', \cdots, \alpha_4' x_{i,4} + \alpha_5' y_{i,4} + \gamma_4') \bmod N$,
  - $\mathcal{E}.\mathsf{ct} = (\tilde{u}_1, \cdots, \tilde{u}_8, \tilde{e}, t)/\bot \leftarrow \mathsf{AIAE.Dec}(\mathsf{k}, \mathsf{aiae.ct}, \mathsf{aux})$,
  - $(\tilde{\alpha}_1, \cdots, \tilde{\alpha}_8) := (\mathrm{dlog}_T(\tilde{u}_1^{\phi(N)})/\phi(N), \cdots, \mathrm{dlog}_T(\tilde{u}_8^{\phi(N)})/\phi(N)) \bmod N^{s-1}$, $\tilde{\gamma} := \mathrm{dlog}_T(\tilde{e}^{\phi(N)})/\phi(N) \bmod N^{s-1}$, and $m := \tilde{\alpha}_1 x_{i,1} + \tilde{\alpha}_2 y_{i,1} + \tilde{\alpha}_3 x_{i,2} + \tilde{\alpha}_4 y_{i,2} + \tilde{\alpha}_5 x_{i,3} + \tilde{\alpha}_6 y_{i,3} + \tilde{\alpha}_7 x_{i,4} + \tilde{\alpha}_8 y_{i,4} + \tilde{\gamma} \bmod N^{s-1}$.

  According to Eq. (1), for $j \in \{1, 2, 3, 4\}$, we have that

$$\begin{aligned} k_j &\overset{\mathsf{G}_6}{=} \mathrm{dlog}_T(e_j u_j^{x_{i,j}} u_{j+1}^{y_{i,j}}) = \mathrm{dlog}_T((e_j u_j^{x_{i,j}} u_{j+1}^{y_{i,j}})^{\phi(N)})/\phi(N) \bmod N \\ &= \mathrm{dlog}_T(u_j^{\phi(N) \cdot x_{i,j}})/\phi(N) + \mathrm{dlog}_T(u_{j+1}^{\phi(N) \cdot y_{i,j}})/\phi(N) + \mathrm{dlog}_T(e_j^{\phi(N)})/\phi(N) \\ &\overset{\mathsf{G}_7}{=} \underbrace{\mathrm{dlog}_T(u_j^{\phi(N)})/\phi(N)}_{\alpha_j'} \cdot x_{i,j} + \underbrace{\mathrm{dlog}_T(u_{j+1}^{\phi(N)})/\phi(N)}_{\alpha_{j+1}'} \cdot y_{i,j} + \underbrace{\mathrm{dlog}_T(e_j^{\phi(N)})/\phi(N)}_{\gamma_j'}, \end{aligned}$$

$$\begin{aligned} m &\overset{\mathsf{G}_6}{=} \mathrm{dlog}_T(\tilde{e} \tilde{u}_1^{x_{i,1}} \tilde{u}_2^{y_{i,1}} \tilde{u}_3^{x_{i,2}} \tilde{u}_4^{y_{i,2}} \tilde{u}_5^{x_{i,3}} \tilde{u}_6^{y_{i,3}} \tilde{u}_7^{x_{i,4}} \tilde{u}_8^{y_{i,4}}) \bmod N^{s-1} \\ &\overset{\mathsf{G}_7}{=} \underbrace{\mathrm{dlog}_T(\tilde{u}_1^{\phi(N)})/\phi(N)}_{\tilde{\alpha}_1} \cdot x_{i,1} + \cdots + \underbrace{\mathrm{dlog}_T(\tilde{u}_8^{\phi(N)})/\phi(N)}_{\tilde{\alpha}_8} \cdot y_{i,4} + \underbrace{\mathrm{dlog}_T(\tilde{e}^{\phi(N)})/\phi(N)}_{\tilde{\gamma}}. \end{aligned}$$

  These changes are conceptual. So G_7 is identical to G_6, $\Pr_6[\mathsf{Win}] = \Pr_7[\mathsf{Win}]$.

– Game G_8: This game is the same as game G_7, except that, the challenger adds an additional rejection rule when answering DEC queries as follows:
  - if $\alpha'_1 \neq 0 \vee \cdots \vee \alpha'_5 \neq 0 \ \vee \ \tilde{\alpha}_1 \neq 0 \vee \cdots \vee \tilde{\alpha}_8 \neq 0$, return $\perp$.

  That is, the challenger will not output m in DEC unless $\alpha'_1 = \cdots = \alpha'_5 = 0$ and $\tilde{\alpha}_1 = \cdots = \tilde{\alpha}_8 = 0$ holds. Thus the values of $(x_{i,j}, y_{i,j})_{j=1}^4 \bmod N$, in particular $(x_j, y_j)_{j=1}^4 \bmod N$, are not used any more in DEC.

  Let Bad denote the event that $\mathcal{A}$ makes a DEC query $(\langle \mathsf{aux}, \mathsf{aiae.ct}\rangle, i \in [n])$, such that

$$e_1 u_1^{x_{i,1}} u_2^{y_{i,1}}, \cdots, e_4 u_4^{x_{i,4}} u_5^{y_{i,4}} \in \mathbb{RU}_{N^2} \ \wedge \ \mathsf{AIAE.Dec}(\mathsf{k}, \mathsf{aiae.ct}, \mathsf{aux}) \neq \perp \quad (2)$$

$$\wedge \ \tilde{e} \tilde{u}_1^{x_{i,1}} \tilde{u}_2^{y_{i,1}} \tilde{u}_3^{x_{i,2}} \tilde{u}_4^{y_{i,2}} \tilde{u}_5^{x_{i,3}} \tilde{u}_6^{y_{i,3}} \tilde{u}_7^{x_{i,4}} \tilde{u}_8^{y_{i,4}} \in \mathbb{RU}_{N^s} \ \wedge \ t = g_1^m \bmod N \quad (3)$$

$$\wedge \ \big(\alpha'_1 \neq 0 \vee \cdots \vee \alpha'_5 \neq 0 \ \vee \ \tilde{\alpha}_1 \neq 0 \vee \cdots \vee \tilde{\alpha}_8 \neq 0\big).$$

  Clearly, games G_7 and G_8 are the same until Bad happens. Therefore, we have that $\big|\Pr_7[\mathsf{Win}] - \Pr_8[\mathsf{Win}]\big| \leq \Pr_8[\mathsf{Bad}]$.
  To prove that G_7 and G_8 are indistinguishable, we have to show that $\Pr_8[\mathsf{Bad}]$ is negligible. This is not an easy task, and we further divide Bad to two disjoint sub-events:
  * Bad' denotes the event that $\mathcal{A}$ makes a DEC query such that

$$\text{Conditions (2), (3) hold} \ \wedge \ \big(\alpha'_1 \neq 0 \vee \cdots \vee \alpha'_5 \neq 0\big).$$

  * $\widetilde{\mathsf{Bad}}$ denotes the event that $\mathcal{A}$ makes a DEC query such that

$$\text{Conditions (2), (3) hold} \ \wedge \ (\alpha'_1 = \cdots = \alpha'_5 = 0) \ \wedge \ \big(\tilde{\alpha}_1 \neq 0 \vee \cdots \vee \tilde{\alpha}_8 \neq 0\big).$$

  Then $\Pr_8[\mathsf{Bad}] \leq \Pr_8[\mathsf{Bad}'] + \Pr_8[\widetilde{\mathsf{Bad}}]$. We give an upper bound for $\Pr_8[\mathsf{Bad}']$ via the following lemma. See the full version [HLL16] for the proof. The analysis of $\Pr_8[\widetilde{\mathsf{Bad}}]$ is deferred to subsequent games.

Lemma 5. $\Pr_8[\mathsf{Bad}'] \leq 2Q_d \cdot \mathsf{Adv}_{\mathsf{AIAE}}^{weak\text{-}int\text{-}rka}(\ell) + Q_d \cdot 2^{-\Omega(\ell)}$.

– Game G_9: This game is the same as game G_8, except that, the challenger chooses another random tuple $\overline{\mathsf{k}}^* = (\bar{k}_1^*, \bar{k}_2^*, \bar{k}_3^*, \bar{k}_4^*)$ besides $\mathsf{k}^* = (k_1^*, k_2^*, k_3^*, k_4^*)$ in INITIALIZE. In addition, to answer the λ-th ($\lambda \in [Q_e]$) ENC query (f_λ, i_λ), the challenger uses a different key for AIAE to compute $\mathsf{aiae.ct}_\lambda$:
  - set $\overline{\mathsf{k}}_\lambda = (\bar{k}_{\lambda,1}, \bar{k}_{\lambda,2}, \bar{k}_{\lambda,3}, \bar{k}_{\lambda,4}) := (r_\lambda \bar{k}_1^* + s_{\lambda,1}, \cdots, r_\lambda \bar{k}_4^* + s_{\lambda,4})$;
  - invoke $\mathsf{aiae.ct}_\lambda \leftarrow_{\$} \mathsf{AIAE.Enc}(\overline{\mathsf{k}}_\lambda, \mathcal{E}.\mathsf{ct}_\lambda, \mathsf{aux}_\lambda)$.

  But the challenger still uses $\mathsf{k}^* = (k_1^*, k_2^*, k_3^*, k_4^*)$ to compute $(e_{\lambda,1}, \cdots, e_{\lambda,4})$.

  In game G_8, the only place that needs the value of $(x_1, y_1, \cdots, x_4, y_4) \bmod N$ is the computation of $(e_{\lambda,1}, \cdots, e_{\lambda,4})$ in ENC. More precisely, for $j \in \{1, 2, 3, 4\}$,

$$e_{\lambda,j} = h_{i_\lambda,j}^{r^* r_\lambda} T^{r_\lambda \cdot (k_j^* - \alpha_j x_{i_\lambda,j} - \alpha_{j+1} y_{i_\lambda,j}) + s_{\lambda,j}} \bmod N^2$$
$$= h_{i_\lambda,j}^{r^* r_\lambda} T^{r_\lambda \cdot (k_j^* - \alpha_j x_j - \alpha_{j+1} y_j - \alpha_j \bar{x}_{i_\lambda,j} - \alpha_{j+1} \bar{y}_{i_\lambda,j}) + s_{\lambda,j}} \bmod N^2.$$

We stress that the computation of $t_\lambda = g_1^{m_\beta} \bmod N$ in Enc only uses the values of $(x_1, y_1, \cdots, x_4, y_4) \bmod \phi(N)/4$, since the order of $g_1 \in \mathbb{SCR}_{N^s}$ is $\phi(N)/4$. We also note that neither $\mathsf{k}^* = (k_1^*, k_2^*, k_3^*, k_4^*)$ nor $(x_j, y_j)_{j=1}^4 \bmod N$ is involved in Dec since Dec rejects the ciphertext unless $\alpha_1' = \cdots = \alpha_5' = 0$ and $\tilde{\alpha}_1 = \cdots = \tilde{\alpha}_8 = 0$. As a result, $\mathsf{k}^* = (k_1^*, k_2^*, k_3^*, k_4^*)$ is totally hidden by the entropy of $(x_1, y_1, \cdots, x_4, y_4) \bmod N$ and is uniformly random to $\mathcal{A}$.

Thus the challenger can use an independent $\overline{\mathsf{k}}^* = (\bar{k}_1^*, \cdots, \bar{k}_4^*)$ to compute $\overline{\mathsf{k}}_\lambda$, and use $\overline{\mathsf{k}}_\lambda$ to do the encryption of the AIAE scheme in Enc, as in G_9.

Then games G_8 and G_9 are identically distributed from the point of view of $\mathcal{A}$, thus we have $\Pr_8[\mathsf{Win}] = \Pr_9[\mathsf{Win}]$ and $\Pr_8[\widetilde{\mathsf{Bad}}] = \Pr_9[\widetilde{\mathsf{Bad}}]$.

- Game G_{10}: This game is the same as game G_9, except that, to answer the λ-th ($\lambda \in [Q_e]$) Enc query (f_λ, i_λ), the challenger computes $\mathsf{aiae.ct}_\lambda$ as follows:
  - invoke $\mathsf{aiae.ct}_\lambda \leftarrow_{\$} \mathsf{AIAE.Enc}(\overline{\mathsf{k}}_\lambda, 0^{\ell_\mathcal{M}}, \mathsf{aux}_\lambda)$.

  That is, the challenger computes the AIAE encryption of a constant $0^{\ell_\mathcal{M}}$ instead of $\mathcal{E}.\mathsf{ct}_\lambda$ in Enc. Note that in games G_9 and G_{10}, the key $\overline{\mathsf{k}}^* = (\bar{k}_1^*, \bar{k}_2^*, \bar{k}_3^*, \bar{k}_4^*)$ is used only in the computation of the AIAE encryption, where it uses $\overline{\mathsf{k}}_\lambda = r_\lambda \cdot \overline{\mathsf{k}}^* + \mathsf{s}_\lambda$, $\mathsf{s}_\lambda = (s_{\lambda,1}, \cdots, s_{\lambda,4})$, as the encryption key. The difference between G_9 and G_{10} can be reduced to the IND-$\mathcal{F}_{\mathrm{raff}}$-RKA security of the AIAE scheme directly. Thus we have that both $\left| \Pr_9[\mathsf{Win}] - \Pr_{10}[\mathsf{Win}] \right|$, $\left| \Pr_9[\widetilde{\mathsf{Bad}}] - \Pr_{10}[\widetilde{\mathsf{Bad}}] \right| \le \mathsf{Adv}_{\mathsf{AIAE}}^{ind\text{-}rka}(\ell)$.

  Now in G_{10}, the challenger computes the AIAE encryption of a constant $0^{\ell_\mathcal{M}}$ in Enc, thus the challenge bit β is completely hidden. Then $\Pr_{10}[\mathsf{Win}] = \frac{1}{2}$.

  We give an upper bound for $\Pr_{10}[\widetilde{\mathsf{Bad}}]$ via the following lemma, and present its proof in the full version [HLL16].

Lemma 6. $\Pr_{10}[\widetilde{\mathsf{Bad}}] \le (Q_d + 1) \cdot 2^{-\Omega(\ell)} + \mathsf{Adv}_{\mathsf{GenN}}^{dl}(\ell)$.

Taking all things together, the n-KDM[$\mathcal{F}_{\mathrm{aff}}$]-CCA security of PKE follows. ■

6 PKE with n-KDM[$\mathcal{F}_{\mathrm{poly}}^d$]-CCA Security

6.1 The Basic Idea

We consider how to construct a PKE which is n-KDM-CCA secure w.r.t. the set of polynomial functions of bounded degree d, denoted by $\mathcal{F}_{\mathrm{poly}}^d$, where d can be polynomial in security parameter ℓ. We will consider adversaries submitting f in the format of *Modular Arithmetic Circuit* (MAC) [MTY11], i.e., a polynomial-size circuit which computes f. In particular, we do not require a prior bound on the size of circuits, but only require a prior bound d on the degree of the polynomials. Our construction still follows the approach in Fig. 1. In fact, our n-KDM[$\mathcal{F}_{\text{poly}}^d$]-CCA secure PKE shares the same building blocks KEM and AIAE with the previous PKE in Fig. 7 which has n-KDM[$\mathcal{F}_{\text{aff}}$]-CCA security. What we should do is to design a new building block $\mathcal{E}$, which can function as an entropy filter for polynomial functions. Our new $\mathcal{E}$ still share the same secret/public key pair with KEM. Hence for $i \in [n]$, we have $\mathsf{sk}_i = (x_{i,1}, y_{i,1}, \cdots, x_{i,4}, y_{i,4})$ and $\mathsf{pk}_i = (h_{i,1}, \cdots, h_{i,4})$ with $h_{i,1} = g_1^{-x_{i,1}} g_2^{-y_{i,1}}, \cdots, h_{i,4} = g_4^{-x_{i,4}} g_5^{-y_{i,4}} \bmod N^s$.

6.2 Reducing Polynomials of $8n$ Variables to Polynomials of 8 Variables

How to Reduce $8n$-Variable Polynomial f_λ in Enc **($f_\lambda, i_\lambda \in [n]$).** In the n-KDM[$\mathcal{F}_{\text{poly}}^d$]-CCA game, the adversary will submit $(f_\lambda, i_\lambda \in [n])$ to Enc as its λ-th KDM encryption query. Here f_λ is a degree-d polynomial $f_\lambda\big((x_{i,j}, y_{i,j})_{i\in[n],j\in[4]}\big)$ of the n secret keys, which has $8n$ variables. Note that f_λ will contain at most $\binom{8n+d}{8n} = \Theta(d^{8n})$ monomials, which is exponentially large.

To reduce the number of monomials, we can always change the polynomial $f_\lambda\big((x_{i,j}, y_{i,j})_{i\in[n],j\in[4]}\big)$ of $8n$ variables to a polynomial $f'_\lambda\big((x_{i_\lambda,j}, y_{i_\lambda,j})_{j\in[4]}\big)$ of 8 variables as follows. Then f'_λ will contain at most $\binom{8+d}{8} = \Theta(d^8)$ monomials, which is polynomial in ℓ.

In Initialize, the secret keys can be generated with $x_{i,j} := x_j + \bar{x}_{i,j}$ and $y_{i,j} := y_j + \bar{y}_{i,j} \bmod \lfloor N^2/4 \rfloor$ for $i \in [n]$ and $j \in [4]$. Then with the values of $(\bar{x}_{i,j}, \bar{y}_{i,j})_{i\in[n],\, j\in[4]}$, we can represent $(x_{i,j}, y_{i,j})_{i\in[n],\, j\in[4]}$ as shifts of $(x_{i_\lambda,j}, y_{i_\lambda,j})_{j\in[4]}$:

$$x_{i,j} = x_{i_\lambda,j} + \bar{x}_{i,j} - \bar{x}_{i_\lambda,j}, \qquad y_{i,j} = y_{i_\lambda,j} + \bar{y}_{i,j} - \bar{y}_{i_\lambda,j},$$

and reduce the polynomial f_λ in $8n$ variables $(x_{i,j}, y_{i,j})_{i\in[n],\, j\in[4]}$ to a polynomial f'_λ in 8 variables $(x_{i_\lambda,j}, y_{i_\lambda,j})_{j\in[4]}$:

$$f_\lambda\big((x_{i,j}, y_{i,j})_{i\in[n],\, j\in[4]}\big) = f_\lambda\big((\underbrace{x_{i_\lambda,j} + \bar{x}_{i,j} - \bar{x}_{i_\lambda,j}}_{x_{i,j}}, \underbrace{y_{i_\lambda,j} + \bar{y}_{i,j} - \bar{y}_{i_\lambda,j}}_{y_{i,j}})_{i\in[n],\, j\in[4]}\big)$$

$$= f'_\lambda\big((x_{i_\lambda,j}, y_{i_\lambda,j})_{j\in[4]}\big) = \sum_{0 \le c_1 + \cdots + c_8 \le d} a_{(c_1,\cdots,c_8)} \cdot x_{i_\lambda,1}^{c_1} y_{i_\lambda,1}^{c_2} x_{i_\lambda,2}^{c_3} y_{i_\lambda,2}^{c_4} x_{i_\lambda,3}^{c_5} y_{i_\lambda,3}^{c_6} x_{i_\lambda,4}^{c_7} y_{i_\lambda,4}^{c_8}.$$

The resulting polynomial f'_λ is also of degree at most d, and the coefficients $a_{(c_1,\cdots,c_8)}$ are determined by $(\bar{x}_{i,j}, \bar{y}_{i,j})_{i\in[n],j\in[4]}$ completely.

How to Determine Coefficients $a_{(c_1,\cdots,c_8)}$ for f'_λ Efficiently with Only $(\bar{x}_{i,j}, \bar{y}_{i,j})_{i\in[n],\, j\in[4]}$. Repeat choosing values of $(x_{i_\lambda,j}, y_{i_\lambda,j})_{j\in[4]}$ randomly, feeding MAC (which functions as f_λ) with input of $(x_{i_\lambda,j} + \bar{x}_{i,j} - \bar{x}_{i_\lambda,j}, y_{i_\lambda,j} + \bar{y}_{i,j} - \bar{y}_{i_\lambda,j})_{i\in[n],\, j\in[4]}$, where $(\bar{x}_{i,j}, \bar{y}_{i,j})_{i\in[n],j\in[4]}$ always takes the values chosen in Initialize, and recording the output of MAC. After about $\binom{8+d}{8} = \Theta(d^8)$ times, we can extract all $a_{(c_1,\cdots,c_8)}$ by simply solving a system of linear equations (with $a_{(c_1,\cdots,c_8)}$ unknowns):

$$f_\lambda\big((x_{i_\lambda,j}+\bar{x}_{i,j}-\bar{x}_{i_\lambda,j}, y_{i_\lambda,j}+\bar{y}_{i,j}-\bar{y}_{i_\lambda,j})_{i\in[n],\, j\in[4]}\big) = \sum_{0\le c_1+\cdots+c_8\le d} a_{(c_1,\cdots,c_8)} \cdot x_{i_\lambda,1}^{c_1} y_{i_\lambda,1}^{c_2} x_{i_\lambda,2}^{c_3} y_{i_\lambda,2}^{c_4} x_{i_\lambda,3}^{c_5} y_{i_\lambda,3}^{c_6} x_{i_\lambda,4}^{c_7} y_{i_\lambda,4}^{c_8}.$$

This can be done in time polynomial in ℓ.

6.3 How to Design $\mathcal{E}$: A Warmup

Let us first consider a simple case: design $\mathcal{E}$ w.r.t. a specific type of monomials

$$f'_\lambda\big((x_{i_\lambda,j}, y_{i_\lambda,j})_{j\in[4]}\big) = a \cdot x_{i_\lambda,1} y_{i_\lambda,1} x_{i_\lambda,2} y_{i_\lambda,2} x_{i_\lambda,3} y_{i_\lambda,3} x_{i_\lambda,4} y_{i_\lambda,4}.$$

We describe the encryption and decryption algorithms $\mathcal{E}.\mathsf{Enc}$, $\mathcal{E}.\mathsf{Dec}$ in Fig. 8.

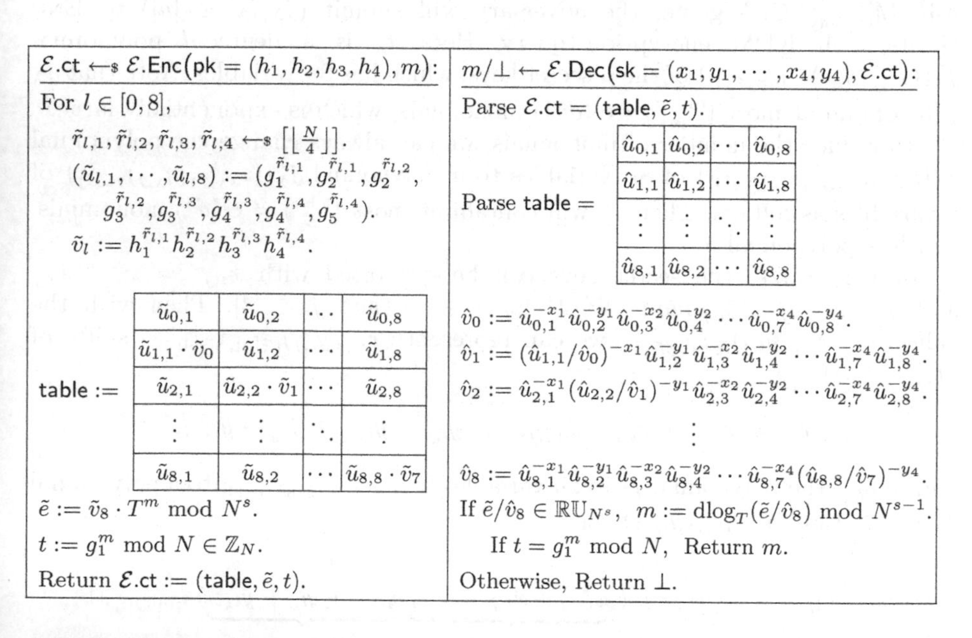

Fig. 8. $\mathcal{E}$ designed for specific monomials $a \cdot x_{i_\lambda,1} y_{i_\lambda,1} x_{i_\lambda,2} y_{i_\lambda,2} x_{i_\lambda,3} y_{i_\lambda,3} x_{i_\lambda,4} y_{i_\lambda,4}$.

Security proof. We can prove KDM-CCA security w.r.t. the specific type of monomials, i.e., $a \cdot x_{i_\lambda,1} y_{i_\lambda,1}\, x_{i_\lambda,2} y_{i_\lambda,2} x_{i_\lambda,3} y_{i_\lambda,3} x_{i_\lambda,4} y_{i_\lambda,4}$, in a similar way as the proof of Theorem 2. The only difference lies in games G_3-G_4, which are related to $\mathcal{E}$. We replace G_3-G_4 with the following three steps (Step 1–Step 3). More precisely, we change the $\mathcal{E}.\mathsf{Enc}$ part of ENC so that it can reserve the entropy of $(x_1, y_1, \cdots, x_4, y_4) \bmod N$, behaving like an entropy filter w.r.t. this specific kind of monomials.

Suppose that the adversary submits $(f_\lambda, i_\lambda \in [n])$ to ENC. Our aim is to reserve the entropy of $(x_j, y_j)_{j=1}^4 \bmod N$ from $\mathcal{E}.\mathsf{Enc}(\mathsf{pk}_{i_\lambda}, f_\lambda((x_{i,j}, y_{i,j})_{i\in[n],\, j\in[4]}))$.

Step 0: In INITIALIZE, the secret keys are generated with $x_{i,j} := x_j + \bar{x}_{i,j}$ and $y_{i,j} := y_j + \bar{y}_{i,j} \bmod \lfloor N^2/4 \rfloor$ for $i \in [n]$, $j \in [4]$. This is the same as G_2 in the proof of Theorem 2.

Step 1: Use $(\bar{x}_{i,j}, \bar{y}_{i,j})_{i\in[n],j\in[4]}$ to re-explain $(f_\lambda, i_\lambda \in [n])$ as $(f'_\lambda, i_\lambda \in [n])$, and determine the coefficient a of the monomial

$$f'_\lambda((x_{i_\lambda,j}, y_{i_\lambda,j})_{j\in[4]}) = a \cdot x_{i_\lambda,1}y_{i_\lambda,1}x_{i_\lambda,2}y_{i_\lambda,2}x_{i_\lambda,3}y_{i_\lambda,3}x_{i_\lambda,4}y_{i_\lambda,4}.$$

Step 2: Use secret key $\mathsf{sk}_{i_\lambda} = (x_{i_\lambda,j}, y_{i_\lambda,j})_{j\in[4]}$ (together with public key $\mathsf{pk}_{i_\lambda} = (h_{i_\lambda,j})_{j\in[4]}$) to implement $\mathcal{E}.\mathsf{Enc}$ (This corresponds to G_3 in the proof of Theorem 2).
- Setup table, just like $\mathcal{E}.\mathsf{Enc}$.
- Compute $\hat{v}_0, \cdots, \hat{v}_8$ from table, just like $\mathcal{E}.\mathsf{Dec}$.
- Use $\hat{v}_8$ instead of $\tilde{v}_8$ to compute $\tilde{e}$ with $\tilde{e} := \hat{v}_8 \cdot T^{f'_\lambda((x_{i_\lambda,j},y_{i_\lambda,j})_{j\in[4]})} \bmod N^s$, and $t := g_1^{f'_\lambda((x_{i_\lambda,j},y_{i_\lambda,j})_{j\in[4]})} \bmod N$.

It is easy to check that $\hat{v}_0, \cdots, \hat{v}_8$ computed from table (via $\mathcal{E}.\mathsf{Dec}$) are identical to $\tilde{v}_0, \cdots, \tilde{v}_8$ that are used to generate table (via $\mathcal{E}.\mathsf{Enc}$). Thus this change is conceptual.

Step 3: This corresponds to G_4 in the proof of Theorem 2.
- table is set up in a similar way as in $\mathcal{E}.\mathsf{Enc}$, but with the following difference. The item of row 1 and column 1 in table now is computed as $\hat{u}_{1,1} = (\tilde{u}_{1,1}T^a) \cdot \tilde{v}_0$ instead of $\hat{u}_{1,1} = \tilde{u}_{1,1} \cdot \tilde{v}_0$. This change is computationally indistinguishable, due to the IV_5 assumption. (We refer to a detailed analysis in the full version [HLL16].)
- Compute $\hat{v}_0, \cdots, \hat{v}_8$ from table, just like $\mathcal{E}.\mathsf{Dec}$.
- $\tilde{e} := \hat{v}_8 \cdot T^{f'_\lambda((x_{i_\lambda,j},y_{i_\lambda,j})_{j\in[4]})} \bmod N^s$, and $t := g_1^{f'_\lambda((x_{i_\lambda,j},y_{i_\lambda,j})_{j\in[4]})} \bmod N$.

It is easy to check that $\hat{v}_0 = \tilde{v}_0, \hat{v}_1 = \tilde{v}_1 \cdot T^{-ax_{i_\lambda,1}}, \hat{v}_2 = \tilde{v}_2 \cdot T^{-ax_{i_\lambda,1}y_{i_\lambda,1}}, \cdots, \hat{v}_8 = \tilde{v}_8 \cdot T^{-ax_{i_\lambda,1}y_{i_\lambda,1}\cdots x_{i_\lambda,4}y_{i_\lambda,4}} = \tilde{v}_8 \cdot T^{-f'_\lambda((x_{i_\lambda,j},y_{i_\lambda,j})_{j\in[4]})}$, thus $\tilde{e} = \hat{v}_8 \cdot T^{f'_\lambda((x_{i_\lambda,j},y_{i_\lambda,j})_{j\in[4]})} = \tilde{v}_8$. Therefore we can also implement Step 3 equivalently as follows.

Step 3 (Equivalent Form):
- table is set up in a similar way as in $\mathcal{E}.\mathsf{Enc}$, but with the following difference. The item of row 1 and column 1 in table is computed as $\hat{u}_{1,1} = (\tilde{u}_{1,1}T^a) \cdot \tilde{v}_0$ instead of $\hat{u}_{1,1} = \tilde{u}_{1,1} \cdot \tilde{v}_0$.
- $\tilde{e} := \tilde{v}_8 \bmod N^s$, and $t := g_1^{f'_\lambda((x_{i_\lambda,j},y_{i_\lambda,j})_{j\in[4]}) \bmod \phi(N)/4} \bmod N$.

In this step, $\mathcal{E}.\mathsf{Enc}$ does not use $(x_1, y_1, \cdots, x_4, y_4) \bmod N$ at all (only uses $(\bar{x}_{i,j}, \bar{y}_{i,j})_{i\in[n],j\in[4]}$ and $(x_1, y_1, \cdots, x_4, y_4) \bmod \phi(N)/4$).

Consequently, through the computationally indistinguishable change, the entropy of $(x_1, y_1, \cdots, x_4, y_4) \bmod N$ is reserved by the $\mathcal{E}.\mathsf{Enc}$ part of ENC.

Similarly, DEC can be changed to do decryptions without $(x_j, y_j)_{j=1}^4 \bmod N$. This can be done with $\phi(N)$ and the (mod $\phi(N)/4$) part of secret key. (This corresponds to G_7-G_8 in the proof of Theorem 2). Use $\phi(N)$ to make sure that all items in table of $\mathcal{E}$.ct belong to $\mathbb{SCR}_{N^s}$. If not, reject immediately. As a result, DEC does not leak any information of $(x_1, y_1, \cdots, x_4, y_4) \bmod N$. This change is computationally indistinguishable, just like the analysis of Pr[Bad] as in the proof of Theorem 2.

6.4 The General $\mathcal{E}$ Designed for $\mathcal{F}_{\text{poly}}^d$

The previous subsection showed how to design $\mathcal{E}$ for a specific type of monomials. A general f'_λ of degree d contains at most $\binom{8+d}{8} = \Theta(d^8)$ monomials. To design a general $\mathcal{E}$ for $\mathcal{F}_{\text{poly}}^d$, we have to consider all possible types of monomials. For each type of non-constant monomial, we create a table and each table is associated with a $\tilde{v}$, which is called a *title*, and those $\tilde{v}$'s are used to hide message in $\tilde{e}$. We describe $\mathcal{E}$.Enc and $\mathcal{E}$.Dec in Fig. 9.

There are totally $\binom{8+d}{8} - 1$ types of non-constant monomials of degree at most d if we neglect the coefficients. Each type of non-constant monomial $x_{i_\lambda,1}^{c_1} y_{i_\lambda,1}^{c_2} x_{i_\lambda,2}^{c_3} y_{i_\lambda,2}^{c_4} x_{i_\lambda,3}^{c_5} y_{i_\lambda,3}^{c_6} x_{i_\lambda,4}^{c_7} y_{i_\lambda,4}^{c_8}$ is associated with a tuple $\mathsf{c} = (c_1, \cdots c_8)$, which determines degrees of each variable. Denote by $\mathcal{S}$ the set containing all such tuples, i.e., $\mathcal{S} := \{\mathsf{c} = (c_1, \cdots c_8) \mid 1 \le c_1 + \cdots + c_8 \le d\}$.

For each $\mathsf{c} = (c_1, \cdots c_8) \in \mathcal{S}$, we generate $\mathsf{table}^{(\mathsf{c})}$ and its title $\tilde{v}^{(\mathsf{c})}$ for monomial $x_{i_\lambda,1}^{c_1} y_{i_\lambda,1}^{c_2} x_{i_\lambda,2}^{c_3} y_{i_\lambda,2}^{c_4} x_{i_\lambda,3}^{c_5} y_{i_\lambda,3}^{c_6} x_{i_\lambda,4}^{c_7} y_{i_\lambda,4}^{c_8}$ via the algorithm TableGen illustrated in Fig. 9. Intuitively, TableGen generates $\mathsf{table}^{(\mathsf{c})}$ of $1+c_1+\cdots+c_8$ rows. The 0-th row of $\mathsf{table}^{(\mathsf{c})}$ is $\tilde{u}_{0,1}, \cdots, \tilde{u}_{0,8}$. The form of other rows are similar to row 0 with a small difference: the next c_1 rows in the 1-st column are multiplied with $\tilde{v}_0$, $\tilde{v}_1, \cdots, \tilde{v}_{c_1-1}$ respectively; the next c_2 rows in the 2-nd column are multiplied with $\tilde{v}_{c_1}, \tilde{v}_{c_1+1}, \cdots, \tilde{v}_{c_1+c_2-1}$ respectively, and so forth. TableGen also generates a title $\tilde{v}^{(\mathsf{c})}$ for $\mathsf{table}^{(\mathsf{c})}$. The product of all the titles, i.e., $\prod_{\mathsf{c}\in\mathcal{S}} \tilde{v}^{(\mathsf{c})}$, is used to hide T^m in $\tilde{e}$.

On the other hand, the title $\hat{v}^{(\mathsf{c})} = \tilde{v}^{(\mathsf{c})}$ can be recovered from $\mathsf{table}^{(\mathsf{c})}$ with secret key $\mathsf{sk} = (x_1, y_1, \cdots, x_4, y_4)$ via the CalculateV algorithm in Fig. 9. Therefore, one can always use the secret key to extract the titles $(\tilde{v}^{(\mathsf{c})})_{\mathsf{c}\in\mathcal{S}}$ from tables $(\mathsf{table}^{(\mathsf{c})})_{\mathsf{c}\in\mathcal{S}}$ one by one with CalculateV and then recover m correctly.

Security proof. The proof of KDM[$\mathcal{F}_{\text{poly}}^d$]-CCA security is similar to that of Theorem 2. But games G_3-G_4 should be replaced with the following three steps (Step 1–Step 3), so that the $\mathcal{E}$.Enc part of ENC can be changed to work as an entropy filter, i.e., reserving the entropy of $(x_1, y_1, \cdots, x_4, y_4) \bmod N$, w.r.t. any polynomial of degree at most d.

Suppose that the adversary submits $(f_\lambda, i_\lambda \in [n])$ to ENC. Our aim is to reserve the entropy of $(x_j, y_j)_{j=1}^4 \bmod N$ from $\mathcal{E}.\mathsf{Enc}(\mathsf{pk}_{i_\lambda}, f_\lambda((x_{i,j}, y_{i,j})_{i\in[n],j\in[4]}))$.

$\mathcal{E}.\mathsf{ct} \leftarrow_{\$} \mathcal{E}.\mathsf{Enc}(\mathsf{pk}, m)$:

For each $\mathsf{c} = (c_1, \cdots, c_8) \in \mathcal{S}$
  $(\mathsf{table}^{(\mathsf{c})}, \tilde{v}^{(\mathsf{c})}) \leftarrow_{\$} \mathsf{TableGen}(\mathsf{pk}, \mathsf{c})$.

$\tilde{e} := \prod_{\mathsf{c} \in \mathcal{S}} \tilde{v}^{(\mathsf{c})} \cdot T^m \bmod N^s$.
$t := g_1^m \bmod N \in \mathbb{Z}_N$.
Return $\mathcal{E}.\mathsf{ct} := \big((\mathsf{table}^{(\mathsf{c})})_{\mathsf{c} \in \mathcal{S}}, \tilde{e}, t\big)$.

$m/\perp \leftarrow \mathcal{E}.\mathsf{Dec}(\mathsf{sk}, \mathcal{E}.\mathsf{ct})$:

Parse $\mathcal{E}.\mathsf{ct} = \big((\mathsf{table}^{(\mathsf{c})})_{\mathsf{c} \in \mathcal{S}}, \tilde{e}, t\big)$.
For each $\mathsf{c} = (c_1, \cdots, c_8) \in \mathcal{S}$
  $\hat{v}^{(\mathsf{c})} \leftarrow \mathsf{CalculateV}(\mathsf{sk}, \mathsf{table}^{(\mathsf{c})}, \mathsf{c})$.
If $\tilde{e} \cdot \big(\prod_{\mathsf{c} \in \mathcal{S}} \hat{v}^{(\mathsf{c})}\big)^{-1} \in \mathbb{RU}_{N^s}$
  $m := \mathrm{dlog}_T\big(\tilde{e} \cdot \big(\prod_{\mathsf{c} \in \mathcal{S}} \hat{v}^{(\mathsf{c})}\big)^{-1}\big) \bmod N^{s-1}$.
  If $t = g_1^m \bmod N$, Return m.
Otherwise, Return $\perp$.

$\mathsf{TableGen}\big(\mathsf{pk} = (h_1, h_2, h_3, h_4), \mathsf{c} = (c_1, \cdots, c_8)\big)$:

For each $l \in \{0, 1, \cdots, \sum_{j=1}^{8} c_j\}$
  $\tilde{r}_{l,1}, \tilde{r}_{l,2}, \tilde{r}_{l,3}, \tilde{r}_{l,4} \leftarrow_{\$} \big[\lfloor \frac{N}{4} \rfloor\big]$.
  $(\tilde{u}_{l,1}, \cdots, \tilde{u}_{l,8}) := (g_1^{\tilde{r}_{l,1}}, g_2^{\tilde{r}_{l,1}}, g_2^{\tilde{r}_{l,2}}, g_3^{\tilde{r}_{l,2}}, g_3^{\tilde{r}_{l,3}}, g_4^{\tilde{r}_{l,3}}, g_4^{\tilde{r}_{l,4}}, g_5^{\tilde{r}_{l,4}})$.
  $\tilde{v}_l := h_1^{\tilde{r}_{l,1}} h_2^{\tilde{r}_{l,2}} h_3^{\tilde{r}_{l,3}} h_4^{\tilde{r}_{l,4}}$.

$\mathsf{table}^{(\mathsf{c})} :=$

$\tilde{u}_{0,1}$	$\tilde{u}_{0,2}$	$\cdots$	$\tilde{u}_{0,8}$	
$\tilde{u}_{1,1} \cdot \tilde{v}_0$	$\tilde{u}_{1,2}$	$\cdots$	$\tilde{u}_{1,8}$	c_1 rows
$\vdots$	$\vdots$	$\ddots$	$\vdots$	
$\tilde{u}_{c_1,1} \cdot \tilde{v}_{c_1-1}$	$\tilde{u}_{c_1,2}$	$\cdots$	$\tilde{u}_{c_1,8}$	
$\tilde{u}_{c_1+1,1}$	$\tilde{u}_{c_1+1,2} \cdot \tilde{v}_{c_1}$	$\cdots$	$\tilde{u}_{c_1+1,8}$	c_2 rows
$\vdots$	$\vdots$	$\ddots$	$\vdots$	
$\tilde{u}_{c_1+c_2,1}$	$\tilde{u}_{c_1+c_2,2} \cdot \tilde{v}_{c_1+c_2-1}$	$\cdots$	$\tilde{u}_{c_1+c_2,8}$	
$\vdots$	$\vdots$	$\vdots$	$\vdots$	
$\tilde{u}_{\sum_{j=1}^{7} c_j+1,1}$	$\tilde{u}_{\sum_{j=1}^{7} c_j+1,2}$	$\cdots$	$\tilde{u}_{\sum_{j=1}^{7} c_j+1,8} \cdot \tilde{v}_{\sum_{j=1}^{7} c_j}$	c_8 rows
$\vdots$	$\vdots$	$\ddots$	$\vdots$	
$\tilde{u}_{\sum_{j=1}^{8} c_j,1}$	$\tilde{u}_{\sum_{j=1}^{8} c_j,2}$	$\cdots$	$\tilde{u}_{\sum_{j=1}^{8} c_j,8} \cdot \tilde{v}_{\sum_{j=1}^{8} c_j-1}$	

Return $(\mathsf{table}^{(\mathsf{c})}, \tilde{v}^{(\mathsf{c})} := \tilde{v}_{\sum_{j=1}^{8} c_j})$.

$\mathsf{CalculateV}\big(\mathsf{sk} = (x_1, y_1, \cdots, x_4, y_4), \mathsf{table}^{(\mathsf{c})}, \mathsf{c} = (c_1, \cdots, c_8)\big)$:

Parse $\mathsf{table}^{(\mathsf{c})} = \left\{ \boxed{\hat{u}_{l,1} \mid \hat{u}_{l,2} \mid \cdots \mid \hat{u}_{l,8}} \right\}_{l \in \{0,1,\cdots,\sum_{j=1}^{8} c_j\}}$.
$\hat{v}_0 := \hat{u}_{0,1}^{-x_1} \hat{u}_{0,2}^{-y_1} \hat{u}_{0,3}^{-x_2} \hat{u}_{0,4}^{-y_2} \hat{u}_{0,5}^{-x_3} \hat{u}_{0,6}^{-y_3} \hat{u}_{0,7}^{-x_4} \hat{u}_{0,8}^{-y_4}$.
For each $l \in \{1, \cdots, c_1\}$
  $\hat{v}_l := (\hat{u}_{l,1}/\hat{v}_{l-1})^{-x_1} \hat{u}_{l,2}^{-y_1} \hat{u}_{l,3}^{-x_2} \hat{u}_{l,4}^{-y_2} \hat{u}_{l,5}^{-x_3} \hat{u}_{l,6}^{-y_3} \hat{u}_{l,7}^{-x_4} \hat{u}_{l,8}^{-y_4}$.
For each $l \in \{c_1 + 1, \cdots, c_1 + c_2\}$
  $\hat{v}_l := \hat{u}_{l,1}^{-x_1} (\hat{u}_{l,2}/\hat{v}_{l-1})^{-y_1} \hat{u}_{l,3}^{-x_2} \hat{u}_{l,4}^{-y_2} \hat{u}_{l,5}^{-x_3} \hat{u}_{l,6}^{-y_3} \hat{u}_{l,7}^{-x_4} \hat{u}_{l,8}^{-y_4}$.

$\vdots$

For each $l \in \{\sum_{j=1}^{7} c_j + 1, \cdots, \sum_{j=1}^{8} c_j\}$
  $\hat{v}_l := \hat{u}_{l,1}^{-x_1} \hat{u}_{l,2}^{-y_1} \hat{u}_{l,3}^{-x_2} \hat{u}_{l,4}^{-y_2} \hat{u}_{l,5}^{-x_3} \hat{u}_{l,6}^{-y_3} \hat{u}_{l,7}^{-x_4} (\hat{u}_{l,8}/\hat{v}_{l-1})^{-y_4}$.
Return $\hat{v}^{(\mathsf{c})} := \hat{v}_{\sum_{j=1}^{8} c_j}$.

Fig. 9. Top: $\mathcal{E}.\mathsf{Enc}$ (left) and $\mathcal{E}.\mathsf{Dec}$ (right) of $\mathcal{E}$ designed for $\mathcal{F}_{\text{poly}}^{d}$; Middle: TableGen, which generates $\mathsf{table}^{(\mathsf{c})}$ together with a title $\tilde{v}^{(\mathsf{c})}$; Bottom: CalculateV, which calculates a title $\hat{v}^{(\mathsf{c})}$ from $\mathsf{table}^{(\mathsf{c})}$ using secret key.

Step 0: In INITIALIZE, the secret keys are generated with $x_{i,j} := x_j + \bar{x}_{i,j}$ and $y_{i,j} := y_j + \bar{y}_{i,j} \bmod \lfloor N^2/4 \rfloor$ for $i \in [n]$, $j \in [4]$. This is the same as G_2 in the proof of Theorem 2.

Step 1: Use $(\bar{x}_{i,j}, \bar{y}_{i,j})_{i \in [n], j \in [4]}$ to re-explain $(f_\lambda, i_\lambda \in [n])$ as $(f'_\lambda, i_\lambda \in [n])$, and determine the coefficients $a_{(c_1,\cdots,c_8)}$ of each monomial of f'_λ, as discussed in Subsect. 6.2. Note that $a_{(c_1,\cdots,c_8)} = 0$ if the associated monomial does not appear in f'_λ. Then

$$f'_\lambda((x_{i_\lambda,j}, y_{i_\lambda,j})_{j \in [4]}) = \sum_{(c_1,\cdots,c_8) \in \mathcal{S}} a_{(c_1,\cdots,c_8)} \cdot x_{i_\lambda,1}^{c_1} y_{i_\lambda,1}^{c_2} x_{i_\lambda,2}^{c_3} y_{i_\lambda,2}^{c_4} x_{i_\lambda,3}^{c_5} y_{i_\lambda,3}^{c_6} x_{i_\lambda,4}^{c_7} y_{i_\lambda,4}^{c_8} + \delta,$$

where $\delta = a_{(0,\cdots,0)}$ denotes the constant term of f'_λ.

Step 2: Use secret key $\mathsf{sk}_{i_\lambda} = (x_{i_\lambda,j}, y_{i_\lambda,j})_{j \in [4]}$ (together with public key $\mathsf{pk}_{i_\lambda} = (h_{i_\lambda,j})_{j \in [4]}$) to implement $\mathcal{E}.\mathsf{Enc}$ (This corresponds to G_3 in the proof of Theorem 2).

- For each $\mathsf{c} = (c_1, \cdots, c_8) \in \mathcal{S}$

  (1) $(\mathsf{table}^{(\mathsf{c})}, \tilde{v}^{(\mathsf{c})}) \leftarrow_\$ \mathsf{TableGen}(\mathsf{pk}_{i_\lambda}, \mathsf{c})$,

  (2) $\hat{v}^{(\mathsf{c})} \leftarrow \mathsf{CalculateV}(\mathsf{sk}_{i_\lambda}, \mathsf{table}^{(\mathsf{c})}, \mathsf{c})$.

- Use $(\hat{v}^{(\mathsf{c})})_{\mathsf{c} \in \mathcal{S}}$ instead of $(\tilde{v}^{(\mathsf{c})})_{\mathsf{c} \in \mathcal{S}}$ to compute $\tilde{e}$ with $\tilde{e} := \prod_{\mathsf{c} \in \mathcal{S}} \hat{v}^{(\mathsf{c})} \cdot T^{f'_\lambda((x_{i_\lambda,j}, y_{i_\lambda,j})_{j \in [4]})} \bmod N^s$, and $t := g_1^{f'_\lambda((x_{i_\lambda,j}, y_{i_\lambda,j})_{j \in [4]})} \bmod N$.

It is easy to check that for each $\mathsf{c} = (c_1, \cdots, c_8) \in \mathcal{S}$, $\hat{v}^{(\mathsf{c})}$ computed from $\mathsf{table}^{(\mathsf{c})}$ via $\mathsf{CalculateV}$ is identical to $\tilde{v}^{(\mathsf{c})}$ associated with $\mathsf{table}^{(\mathsf{c})}$ via $\mathsf{TableGen}$. Thus this change is conceptual.

Step 3: This corresponds to G_4 in the proof of Theorem 2.

- For each $\mathsf{c} = (c_1, \cdots, c_8) \in \mathcal{S}$

  (1) Compute $\mathsf{table}^{(\mathsf{c})}$ via $(\mathsf{table}^{(\mathsf{c})}, \tilde{v}^{(\mathsf{c})}) \leftarrow_\$ \mathsf{TableGen}(\mathsf{pk}_{i_\lambda}, \mathsf{c})$, but with one difference. The item of row 1 and column $j := \min\{i \mid 1 \le i \le 8, c_i \neq 0\}$ in $\mathsf{table}^{(\mathsf{c})}$ now is computed as $\hat{u}_{1,j} = (\tilde{u}_{1,j} T^{a_{(c_1,\cdots,c_8)}}) \cdot \tilde{v}_0$ instead of $\hat{u}_{1,j} = \tilde{u}_{1,j} \cdot \tilde{v}_0$. This change is computationally indistinguishable, due to the IV_5 assumption.

  (2) Invoke $\hat{v}^{(\mathsf{c})} \leftarrow \mathsf{CalculateV}(\mathsf{sk}_{i_\lambda}, \mathsf{table}^{(\mathsf{c})}, \mathsf{c})$ to extract a title $\hat{v}^{(\mathsf{c})}$ from the modified $\mathsf{table}^{(\mathsf{c})}$.

- $\tilde{e} := \prod_{\mathsf{c} \in \mathcal{S}} \hat{v}^{(\mathsf{c})} \cdot T^{f'_\lambda((x_{i_\lambda,j}, y_{i_\lambda,j})_{j \in [4]})}$, and $t := g_1^{f'_\lambda((x_{i_\lambda,j}, y_{i_\lambda,j})_{j \in [4]})} \bmod N$.

Observe that for each $\mathsf{c} = (c_1, \cdots, c_8) \in \mathcal{S}$,

$$\hat{v}^{(\mathsf{c})} = \tilde{v}^{(\mathsf{c})} \cdot T^{-a_{(c_1,\cdots,c_8)} x_{i_\lambda,1}^{c_1} y_{i_\lambda,1}^{c_2} x_{i_\lambda,2}^{c_3} y_{i_\lambda,2}^{c_4} x_{i_\lambda,3}^{c_5} y_{i_\lambda,3}^{c_6} x_{i_\lambda,4}^{c_7} y_{i_\lambda,4}^{c_8}}.$$

Then $\tilde{e} = \prod_{\mathsf{c} \in \mathcal{S}} \hat{v}^{(\mathsf{c})} \cdot T^{f'_\lambda((x_{i_\lambda,j}, y_{i_\lambda,j})_{j \in [4]})}$

$$= \prod_{\mathsf{c} \in \mathcal{S}} \left(\tilde{v}^{(\mathsf{c})} \cdot T^{-a_{(c_1,\cdots,c_8)} x_{i_\lambda,1}^{c_1} y_{i_\lambda,1}^{c_2} x_{i_\lambda,2}^{c_3} \cdots y_{i_\lambda,3}^{c_6} x_{i_\lambda,4}^{c_7} y_{i_\lambda,4}^{c_8}}\right) \cdot T^{f'_\lambda((x_{i_\lambda,j}, y_{i_\lambda,j})_{j \in [4]})}$$

$$= \prod_{\mathsf{c} \in \mathcal{S}} \tilde{v}^{(\mathsf{c})} \cdot T^{\delta},$$

where δ is the constant term of f'_λ. Therefore we can implement Step 3 equivalently as follows.

Step 3 (Equivalent Form):

- For each $\mathsf{c} = (c_1, \cdots, c_8) \in \mathcal{S}$
  Compute $\mathsf{table}^{(\mathsf{c})}$ via $(\mathsf{table}^{(\mathsf{c})}, \tilde{v}^{(\mathsf{c})}) \leftarrow_{\$} \mathsf{TableGen}(\mathsf{pk}_{i_\lambda}, \mathsf{c})$, but with one difference. The item of row 1 and column $j := \min\{i \mid 1 \leq i \leq 8, c_i \neq 0\}$ in $\mathsf{table}^{(\mathsf{c})}$ now is computed as $\hat{u}_{1,j} = (\tilde{u}_{1,j} T^{a(c_1,\cdots,c_8)}) \cdot \tilde{v}_0$ instead of $\hat{u}_{1,j} = \tilde{u}_{1,j} \cdot \tilde{v}_0$.
- $\tilde{e} := \prod_{\mathsf{c} \in \mathcal{S}} \tilde{v}^{(\mathsf{c})} \cdot T^{\delta}$, and $t := g_1^{f'_\lambda((x_{i_\lambda,j}, y_{i_\lambda,j})_{j\in[4]}) \bmod \phi(N)/4} \bmod N$.

In this step, $\mathcal{E}.\mathsf{Enc}$ does not use $(x_1, y_1, \cdots, x_4, y_4) \bmod N$ at all (only uses $(\bar{x}_{i,j}, \bar{y}_{i,j})_{i\in[n], j\in[4]}$ and $(x_1, y_1, \cdots, x_4, y_4) \bmod \phi(N)/4$).

As a result, through the computationally indistinguishable change, the entropy of $(x_1, y_1, \cdots, x_4, y_4) \bmod N$ is reserved by the $\mathcal{E}.\mathsf{Enc}$ part of Enc.

Similarly, Dec can be changed to do decryptions without $(x_j, y_j)_{j=1}^4 \bmod N$, the same argument as in Subsect. 6.3.

Acknowledgments. The authors are supported by the National Natural Science Foundation of China Grant (Nos. 61672346, 61373153 and 61133014). We thank the anonymous reviewers for their comments and suggestions.

References

[ACPS09] Applebaum, B., Cash, D., Peikert, C., Sahai, A.: Fast cryptographic primitives and circular-secure encryption based on hard learning problems. In: Halevi, S. (ed.) CRYPTO 2009. LNCS, vol. 5677, pp. 595–618. Springer, Heidelberg (2009). doi:10.1007/978-3-642-03356-8_35

[BG10] Brakerski, Z., Goldwasser, S.: Circular and leakage resilient public-key encryption under subgroup indistinguishability. In: Rabin, T. (ed.) CRYPTO 2010. LNCS, vol. 6223, pp. 1–20. Springer, Heidelberg (2010). doi:10.1007/978-3-642-14623-7_1

[BGK11] Brakerski, Z., Goldwasser, S., Kalai, Y.T.: Black-box circular-secure encryption beyond affine functions. In: Ishai, Y. (ed.) TCC 2011. LNCS, vol. 6597, pp. 201–218. Springer, Heidelberg (2011). doi:10.1007/978-3-642-19571-6_13

[BHHI10] Barak, B., Haitner, I., Hofheinz, D., Ishai, Y.: Bounded key-dependent message security. In: Gilbert, H. (ed.) EUROCRYPT 2010. LNCS, vol. 6110, pp. 423–444. Springer, Heidelberg (2010). doi:10.1007/978-3-642-13190-5_22

[BHHO08] Boneh, D., Halevi, S., Hamburg, M., Ostrovsky, R.: Circular-secure encryption from decision Diffie-Hellman. In: Wagner, D. (ed.) CRYPTO 2008. LNCS, vol. 5157, pp. 108–125. Springer, Heidelberg (2008). doi:10.1007/978-3-540-85174-5_7

[BRS02] Black, J., Rogaway, P., Shrimpton, T.: Encryption-scheme security in the presence of key-dependent messages. In: Nyberg, K., Heys, H. (eds.) SAC 2002. LNCS, vol. 2595, pp. 62–75. Springer, Heidelberg (2003). doi:10.1007/3-540-36492-7_6

[CCS09] Camenisch, J., Chandran, N., Shoup, V.: A public key encryption scheme secure against key dependent chosen plaintext and adaptive chosen ciphertext attacks. In: Joux, A. (ed.) EUROCRYPT 2009. LNCS, vol. 5479, pp. 351–368. Springer, Heidelberg (2009). doi:10.1007/978-3-642-01001-9_20

[CL01] Camenisch, J., Lysyanskaya, A.: An efficient system for non-transferable anonymous credentials with optional anonymity revocation. In: Pfitzmann, B. (ed.) EUROCRYPT 2001. LNCS, vol. 2045, pp. 93–118. Springer, Heidelberg (2001). doi:10.1007/3-540-44987-6_7

[DJ01] Damgård, I., Jurik, M.: A generalisation, a simplification and some applications of Paillier's probabilistic public-key system. In: Kim, K. (ed.) PKC 2001. LNCS, vol. 1992, pp. 119–136. Springer, Heidelberg (2001). doi:10.1007/3-540-44586-2_9

[GHV12] Galindo, D., Herranz, J., Villar, J.: Identity-based encryption with master key-dependent message security and leakage-resilience. In: Foresti, S., Yung, M., Martinelli, F. (eds.) ESORICS 2012. LNCS, vol. 7459, pp. 627–642. Springer, Heidelberg (2012). doi:10.1007/978-3-642-33167-1_36

[GM84] Goldwasser, S., Micali, S.: Probabilistic encryption. J. Comput. Syst. Sci. **28**(2), 270–299 (1984)

[GS08] Groth, J., Sahai, A.: Efficient non-interactive proof systems for bilinear groups. In: Smart, N. (ed.) EUROCRYPT 2008. LNCS, vol. 4965, pp. 415–432. Springer, Heidelberg (2008). doi:10.1007/978-3-540-78967-3_24

[HLL16] Han, S., Liu, S., Lyu, L.: Efficient KDM-CCA secure public-key encryption for polynomial functions. IACR Cryptology ePrint Archive, Report 2016/829 (2016)

[Hof13] Hofheinz, D.: Circular chosen-ciphertext security with compact ciphertexts. In: Johansson, T., Nguyen, P.Q. (eds.) EUROCRYPT 2013. LNCS, vol. 7881, pp. 520–536. Springer, Heidelberg (2013). doi:10.1007/978-3-642-38348-9_31

[KD04] Kurosawa, K., Desmedt, Y.: A new paradigm of hybrid encryption scheme. In: Franklin, M. (ed.) CRYPTO 2004. LNCS, vol. 3152, pp. 426–442. Springer, Heidelberg (2004). doi:10.1007/978-3-540-28628-8_26

[LLJ15] Lu, X., Li, B., Jia, D.: KDM-CCA security from RKA secure authenticated encryption. In: Oswald, E., Fischlin, M. (eds.) EUROCRYPT 2015. LNCS, vol. 9056, pp. 559–583. Springer, Heidelberg (2015). doi:10.1007/978-3-662-46800-5_22

[MTY11] Malkin, T., Teranishi, I., Yung, M.: Efficient circuit-size independent public key encryption with KDM security. In: Paterson, K.G. (ed.) EUROCRYPT 2011. LNCS, vol. 6632, pp. 507–526. Springer, Heidelberg (2011). doi:10.1007/978-3-642-20465-4_28

[Reg05] Regev, O.: On lattices, learning with errors, random linear codes, and cryptography. In: Gabow, H.N., Fagin, R. (eds.) STOC 2005, pp. 84–93. ACM (2005)

Structure-Preserving Smooth Projective Hashing

Olivier Blazy[1(✉)] and Céline Chevalier[2]

[1] Université de Limoges, XLim, Limoges, France
olivier.blazy@unilim.fr
[2] CRED, Université Panthéon-Assas, Paris, France

Abstract. Smooth projective hashing has proven to be an extremely useful primitive, in particular when used in conjunction with commitments to provide implicit decommitment. This has lead to applications proven secure in the UC framework, even in presence of an adversary which can do adaptive corruptions, like for example Password Authenticated Key Exchange (PAKE), and 1-out-of-m Oblivious Transfer (OT). However such solutions still lack in efficiency, since they heavily scale on the underlying message length.

Structure-preserving cryptography aims at providing elegant and efficient schemes based on classical assumptions and standard group operations on group elements. Recent trend focuses on constructions of structure-preserving signatures, which require message, signature and verification keys to lie in the base group, while the verification equations only consist of pairing-product equations. Classical constructions of Smooth Projective Hash Function suffer from the same limitation as classical signatures: at least one part of the computation (messages for signature, witnesses for SPHF) is a scalar.

In this work, we introduce and instantiate the concept of Structure-Preserving Smooth Projective Hash Function, and give as applications more efficient instantiations for one-round PAKE and three-round OT, and information retrieval thanks to Anonymous Credentials, all UC-secure against adaptive adversaries.

Keywords: Smooth projective hash functions · Structure preserving · Oblivious transfer · Password authenticated key exchange · UC Framework · Credentials

1 Introduction

Smooth Projective Hash Functions (SPHF) were introduced by Cramer and Shoup [30] as a means to design chosen-ciphertext-secure public-key encryption schemes. These hash functions are defined such as their value can be computed in two different ways if the input belongs to a particular subset (the *language*), either using a private hashing key or a public projection key along with a private witness ensuring that the input belongs to the language.

J.H. Cheon and T. Takagi (Eds.): ASIACRYPT 2016, Part II, LNCS 10032, pp. 339–369, 2016.
DOI: 10.1007/978-3-662-53890-6_12

In addition to providing a more intuitive abstraction for their original public-key encryption scheme in [29], the notion of SPHF also enables new efficient instantiations of their scheme under different complexity assumptions such as DLin, or more generally k − MDDH. Due to its usefulness, the notion of SPHF was later extended to several interactive contexts. One of the most classical applications is to combine them with commitments in order to provide *implicit* decommitments.

Commitment schemes have become a central tool used in cryptographic protocols. These two-party primitives (between a committer and a receiver) are divided into two phases. First, in the *commit* phase, the committer gives the receiver an analogue of a sealed envelope containing a value m, while later in the *opening* phase, the committer reveals m in such a way that the receiver can verify whether it was indeed m that was contained in the envelope. In many applications, for example password-based authenticated key-exchange, in which the committed value is a password, one wants the opening to be implicit, which means that the committer does not really open its commitment, but rather convinces the receiver that it actually committed to the value it pretended to.

An additional difficulty arises when one wants to prove the protocols in the universal composability framework proposed in [22]. Skipping the details, when the protocol uses commitments, this usually forces those commitments to be simultaneously *extractable* (meaning that a simulator can recover the committed value m thanks to a trapdoor) and *equivocable* (meaning that a simulator can open a commitment to a value m' different from the committed value m thanks to a trapdoor), which is quite a difficult goal to achieve.

Using SPHF with commitments to achieve an implicit decommitment, the language is usually defined on group elements, with projection keys being group elements, and witnesses being scalars. While in several applications, this has already lead to efficient constructions, the fact that witnesses have to be scalars (and in particular in case of commitments, the randomness used to commit) leads to drastic restrictions when trying to build protocols secure against adaptive corruptions in the UC framework.

This is the classical paradigm of protocol design, where generic primitives used in a modular approach lead to a simple design but quite inefficient constructions, while when trying to move to ad-hoc constructions, the conceptual simplicity is lost and even though efficiency might be gained, a proper security proof gets trickier. Following the same kind of reasoning, [5] introduced the concept of structure-preserving signatures in order to take the best of both worlds. There has been an ongoing series of work surrounding this notion, for instance [3,4,6–8]. This has shown that structure-preserving cryptography indeed provides the tools needed to have simultaneously simple and efficient protocols.

1.1 Related Work

Smooth Projective Hash Functions (SPHF) were introduced by Cramer and Shoup [30] and have been widely used since then, for instance for password-authenticated key exchange (PAKE) [2,14,35,43,44], or oblivious transfer (OT)

[1,28,41], and a classification was introduced separating SPHF into three main kinds, KV-SPHF,CS-SPHF,GL-SPHF depending on how the projection keys are generated and when, the former allowing one-round protocols, while the latter have more efficient communication costs (see Sect. 2.2).

Password-Authenticated Key Exchange (PAKE) protocols were proposed in 1992 by Bellovin and Merritt [12] where authentication is done using a simple password, possibly drawn from a small entropy space subject to exhaustive search. Since then, many schemes have been proposed and studied. SPHF have been extensively used, starting with the work of Gennaro and Lindell [35] which generalized an earlier construction by Katz, Ostrovsky, and Yung [42], and followed by several other works [2,24]. More recently, a variant of SPHF proposed by Katz and Vaikuntanathan even allowed the construction of one-round PAKE schemes [14,44]. The most efficient PAKE scheme so far (using completely different techniques) is the recent Asiacrypt paper [40].

The first ideal functionality for PAKE protocols in the UC framework [22,25] was proposed by Canetti *et al.* [24], who showed how a simple variant of the Gennaro-Lindell methodology [35] could lead to a secure protocol. Though quite efficient, their protocol was not known to be secure against adaptive adversaries, that are capable of corrupting players at any time, and learn their internal states. The first ones to propose an adaptively secure PAKE in the UC framework were Barak *et al.* [10] using general techniques from multi-party computation. Though conceptually simple, their solution results in quite inefficient schemes.

Recent adaptively secure PAKE were proposed by Abdalla *et al.* [1,2], following the Gennaro-Lindell methodology with variation of the Canetti-Fischlin commitment [23]. However their communication size is growing in the size of the passwords, which is leaking information about an upper-bound on the password used in each exchange.

Oblivious Transfer (OT) was introduced in 1981 by Rabin [51] as a way to allow a receiver to get exactly one out of k messages sent by another party, the sender. In these schemes, the receiver should be oblivious to the other values, and the sender should be oblivious to which value was received. Since then, several instantiations and optimizations of such protocols have appeared in the literature, including proposals in the UC framework [26,48].

More recently, new instantiations have been proposed, trying to reach round-optimality [38], or low communication costs [50]. The 1-out-of-2 OT scheme by Choi *et al.* [28] based on the DDH assumption seems to be the most efficient one among those that are secure against adaptive corruptions in the CRS model with erasures. But it does not scale to 1-out-of-m OT, for $m > 2$. [1,17] proposed a generic construction of 1-out-of-m OT secure against adaptive corruptions in the CRS model, however the commitment was still growing in the logarithm of the database length. While this is not so much a security issue for OT as this length is supposed to be fixed at the start of the protocol, this is however a weak spot for the efficiency of the final construction.

1.2 Our Contributions

Similarly to structure-preserving signatures requiring the message, the signature, and the public keys to be group elements, we propose in this paper the notion of structure-preserving Smooth Projective Hash Functions (SP-SPHF), where both words, witnesses and projection keys are group elements, and hash and projective hash computations are doable with simple pairing-product equations in the context of bilinear groups.

This allows, for example, to build Smooth Projective Hash Functions that implicitly demonstrate the knowledge of a Groth Sahai Proof (serving as a witness).

We show how to transform every previously known pairing-less construction of SPHF to fit this methodology, and then propose several applications in which storing a group element as a witness allows to avoid the drastic restrictions that arise when building protocols secure against adaptive corruptions in the UC framework with a scalar as witness. Asking the witness to be a group element enables us to gain more freedom in the simulation (the discrete logarithm of this element and / or real extraction from a commitment). For instance, the simulator can always commit honestly to a random message, since it only needs to modify its witness in the equivocation phase. Furthermore, it allows to avoid bit-per-bit construction. Such design carries similarity with the publicly verifiable MACs from [45], where the pairing operation allows to relax the verification procedure.

A work from Jutla and Roy has appeared on eprint [39] considering a parallel between QA-NIZK and SPHF: Independently from ours, they define a transformation from one to another. Their transformation can then be extended to view QA-NIZK as a special case of SP-SPHF, and so be encompassed by our framework.

As an example, we show that the UC-commitment from [34] (while not fitting with the methodology of traditional SPHF from [1]), is compatible with SP-SPHF and can be used to build UC protocols. As a side contribution, we first generalize this commitment from DLin to the $k-\mathsf{MDDH}$ assumption from [33]. The combination of this commitment and the associated SP-SPHF then enables us to give three interesting applications.

Adaptively secure 1-out-of-m Oblivious Transfer. First, we provide a construction of a three-round UC-secure 1-out-of-m OT. Assuming reliable erasures and a single global CRS, we show in Sect. 5 that our instantiation is UC-secure against adaptive adversaries. Besides having a lesser number of rounds than most recent existing OT schemes with similar security levels, our resulting protocol also has a better communication complexity than the best known solutions so far [1,28] (see Table 1 for a comparison). For ease of readability, we emphasize in this table the SXDH communication cost[1], which is simply k-MDDH for $k = 1$. Our protocol is “nearly optimal” in the sense that it is still linear in the number of lines m, but the constant in front of m is 1.

[1] Our OT and PAKE protocols are described in k-MDDH but one directly obtains the SXDH versions by simply letting $k = 1$ in the commitment presented in Sect. 4.2 (see the paper full version [15] for details).

Table 1. Comparison with existing UC-secure OT schemes

	Flow	Communication Complexity	Assumption	1-out-of
[28]	4	$26\ \mathbb{G} + 7\ \mathbb{Z}_p$	DDH	2
[1]	3	$(m + 8 \log m) \times \mathbb{G}_1 + \log m \times \mathbb{G}_2 + 1 \times \mathbb{Z}_p$	SXDH	m
This paper	3	$(k+3) \times \mathbb{G}_1 + (2 + (3+k)m + k(k+1))$ $\times \mathbb{G}_2 + m \times \mathbb{Z}_p$	k – MDDH	m
This paper	3	$4 \times \mathbb{G}_1 + 12 \times \mathbb{G}_2 + 2 \times \mathbb{Z}_p$	SXDH	m

One-round adaptively secure PAKE. Then, we provide an instantiation of a one-round UC-secure PAKE under any k – MDDH assumption. Once again, we show in Sect. 6 that the UC-security holds against adaptive adversaries, assuming reliable erasures and a single global CRS. Contrarily to most existing one-round adaptively secure PAKE, we show that our scheme enjoys a much better communication complexity while not leaking information about the length of the password used (see Table 2 for a comparison, in particular for the SXDH version). Only [40] achieves a slightly better complexity as ours, but only for SXDH, while ours easily extends to k – MDDH. Furthermore, our construction is an extension to SP-SPHF of well-known classical constructions based on SPHF, which makes it simpler to understand. We omit [17] from the following table, as its contribution is to widen the construction to non-pairing based hypotheses.

Anonymous Credential-Based Message Transmission. Typical credential use involves three main parties. Users need to interact with some authorities to obtain their credentials (assumed to be a set of attributes validated / signed), and then prove to a server that a subpart of their attributes verifies an expect policy. We present a constant-size, round-optimal protocol that allow to use a Credential to retrieve a message without revealing the Anonymous Credentials in a UC secure way, by simply building on the technique proposed earlier in the paper.

Table 2. Comparison with existing UC-secure PAKE schemes where $|\text{password}| = m$

	Adaptive	One-round	Communication complexity	Assumption
[2]	yes	no	$2 \times (2m + 22m\mathfrak{K}) \times \mathbb{G} + \text{OTS}$	DDH
[44]	no	yes	$\approx 2 \times 70 \times \mathbb{G}$	DLIN
[14]	no	yes	$2 \times 6 \times \mathbb{G}_1 + 2 \times 5 \times \mathbb{G}_2$	SXDH
[1]	yes	yes	$2 \times 10m \times \mathbb{G}_1 + 2 \times m \times \mathbb{G}_2$	SXDH
[40]	yes	yes	$4 \times \mathbb{G}_1 + 4 \times \mathbb{G}_2$	SXDH
this paper	yes	yes	$2 \times (k+3) \times \mathbb{G}_1 + 2 \times (k+3$ $+ k(k+1)) \times \mathbb{G}_2$	k-MDDH
this paper	yes	yes	$2 \times 4 \times \mathbb{G}_1 + 2 \times 5 \times \mathbb{G}_2$	SXDH

2 Definitions

2.1 Notations

If $\boldsymbol{x} \in \mathcal{S}^n$, then $|\boldsymbol{x}|$ denotes the length n of the vector, and by default vectors are assumed to be column vectors. Further, $x \stackrel{\$}{\leftarrow} \mathcal{S}$ denotes the process of sampling an element x from the set $\mathcal{S}$ uniformly at random.

2.2 Primitives

Encryption. An encryption scheme $\mathcal{C}$ is described through four algorithms (Setup, KeyGen, Encrypt, Decrypt), defined formally in Appendix A.1.

Commitments. We refer the reader to [1] for formal definitions and results but we give here an informal overview to help the unfamiliar reader with the following. A *non-interactive labelled commitment scheme* $\mathcal{C}$ is defined by three algorithms:

- $\mathsf{SetupCom}(1^{\mathfrak{K}})$ takes as input the security parameter $\mathfrak{K}$ and outputs the global parameters, passed through the CRS ρ to all other algorithms;
- $\mathsf{Com}^{\ell}(x)$ takes as input a label ℓ and a message x, and outputs a pair (C, δ), where C is the commitment of x for the label ℓ, and δ is the corresponding opening data (a.k.a. decommitment information). This is a probabilistic algorithm.
- $\mathsf{VerCom}^{\ell}(C, x, \delta)$ takes as input a commitment C, a label ℓ, a message x, and the opening data δ and outputs 1 (true) if δ is a valid opening data for C, x and ℓ. It always outputs 0 (false) on $x = \perp$.

The basic properties required for commitments are *correctness* (for all correctly generated CRS ρ, all commitments and opening data honestly generated pass the verification VerCom test), the *hiding property* (the commitment does not leak any information about the committed value) and the *binding property* (no adversary can open a commitment in two different ways). More complex properties (equivocability and extractability) are required by the UC framework and described in Appendix A.2 for lack of space.

Smooth Projective Hash Functions. SPHF were introduced by Cramer and Shoup [30] for constructing encryption schemes. A projective hashing family is a family of hash functions that can be evaluated in two ways: using the (secret) hashing key, one can compute the function on every point in its domain, whereas using the (public) *projected* key one can only compute the function on a special subset of its domain. Such a family is deemed *smooth* if the value of the hash function on any point outside the special subset is independent of the projected key. The notion of SPHF has already found numerous applications in various contexts in cryptography (*e.g.* [2,19,35,41]).

Definition 1. Smooth Projective Hashing System. *A Smooth Projective Hash Function over a language $\mathfrak{L} \subset X$, is defined by five algorithms* (Setup, HashKG, ProjKG, Hash, ProjHash)*:*

- $\mathsf{Setup}(1^{\mathfrak{K}})$ *generates the global parameters* param *of the scheme, and the description of an* $\mathcal{NP}$ *language* $\mathfrak{L}$
- $\mathsf{HashKG}(\mathfrak{L}, \mathsf{param})$, *outputs a hashing key* hk *for the language* $\mathfrak{L}$;
- $\mathsf{ProjKG}(\mathsf{hk}, (\mathfrak{L}, \mathsf{param}), W)$, *derives the projection key* hp, *using the hashing key* hk,
- $\mathsf{Hash}(\mathsf{hk}, (\mathfrak{L}, \mathsf{param}), W)$, *outputs a hash value* v, *thanks to the hashing key* hk, *and* W,
- $\mathsf{ProjHash}(\mathsf{hp}, (\mathfrak{L}, \mathsf{param}), W, w)$, *outputs the hash value* v', *thanks to* hp *and the witness* w *that* $W \in \mathfrak{L}$.

In the following, we consider $\mathfrak{L}$ as a hard-partitioned subset of X, *i.e.* it is computationally hard to distinguish a random element in $\mathfrak{L}$ from a random element in $X \setminus \mathfrak{L}$.

A Smooth Projective Hash Function SPHF should satisfy the following properties:

- *Correctness*: Let $W \in \mathfrak{L}$ and w a witness of this membership. Then, for all hashing keys hk and associated projection keys hp we have

$$\mathsf{Hash}(\mathsf{hk}, (\mathfrak{L}, \mathsf{param}), W) = \mathsf{ProjHash}(\mathsf{hp}, (\mathfrak{L}, \mathsf{param}), W, w).$$

- *Smoothness*: For all $W \in X \setminus \mathfrak{L}$ the following distributions are statistically indistinguishable:

$$\Delta_0 = \left\{ (\mathfrak{L}, \mathsf{param}, W, \mathsf{hp}, v) \;\middle|\; \begin{array}{l} \mathsf{param} = \mathsf{Setup}(1^{\mathfrak{K}}), \mathsf{hk} = \mathsf{HashKG}(\mathfrak{L}, \mathsf{param}), \\ \mathsf{hp} = \mathsf{ProjKG}(\mathsf{hk}, (\mathfrak{L}, \mathsf{param}), W), \\ v = \mathsf{Hash}(\mathsf{hk}, (\mathfrak{L}, \mathsf{param}), W) \in \mathbb{G} \end{array} \right\}$$

$$\Delta_1 = \left\{ (\mathfrak{L}, \mathsf{param}, W, \mathsf{hp}, v) \;\middle|\; \begin{array}{l} \mathsf{param} = \mathsf{Setup}(1^{\mathfrak{K}}), \mathsf{hk} = \mathsf{HashKG}(\mathfrak{L}, \mathsf{param}), \\ \mathsf{hp} = \mathsf{ProjKG}(\mathsf{hk}, (\mathfrak{L}, \mathsf{param}), W), v \stackrel{\$}{\leftarrow} \mathbb{G} \end{array} \right\}.$$

A third property called *Pseudo-Randomness*, is implied by the Smoothness on Hard Subset membership languages. If $W \in \mathfrak{L}$, then without a witness of membership the two previous distributions should remain computationally indistinguishable: for any adversary $\mathcal{A}$ within reasonable time the following advantage is negligible

$$\mathsf{Adv}^{\mathsf{pr}}_{\mathsf{SPHF},\mathcal{A}}(\mathfrak{K}) = |\Pr_{\Delta_1}[\mathcal{A}(\mathfrak{L}, \mathsf{param}, W, \mathsf{hp}, v) = 1] - \Pr_{\Delta_0}[\mathcal{A}(\mathfrak{L}, \mathsf{param}, W, \mathsf{hp}, v) = 1]|$$

In [14], the authors introduced a new notation for SPHF: for a language $\mathfrak{L}$, there exist a function Γ and a family of functions Θ, such that $\boldsymbol{u} \in \mathfrak{L}$, if and only if, $\Theta(\boldsymbol{u})$ is a linear combination $\boldsymbol{\lambda}$ of the rows of $\Gamma(\boldsymbol{u})$. We furthermore require that a user, who knows a witness of the membership $\boldsymbol{u} \in \mathfrak{L}$, can efficiently compute the linear combination $\boldsymbol{\lambda}$. The SPHF can now then be described as:

- $\mathsf{HashKG}(\mathfrak{L}, \mathsf{param})$, outputs a hashing key $\mathsf{hk} = \boldsymbol{\alpha}$ for the language $\mathfrak{L}$,
- $\mathsf{ProjKG}(\mathsf{hk}, (\mathfrak{L}, \mathsf{param}), \boldsymbol{u})$, derives the projection key $\mathsf{hp} = \boldsymbol{\gamma}(\boldsymbol{u})$,
- $\mathsf{Hash}(\mathsf{hk}, (\mathfrak{L}, \mathsf{param}), \boldsymbol{u})$, outputs a hash value $H = \Theta(\boldsymbol{u}) \odot \boldsymbol{\alpha}$,
- $\mathsf{ProjHash}(\mathsf{hp}, (\mathfrak{L}, \mathsf{param}), \boldsymbol{u}, \boldsymbol{\lambda})$, outputs the hash value $H' = \boldsymbol{\lambda} \odot \boldsymbol{\gamma}(\boldsymbol{u})$.

In the special case where $\mathsf{hp} = \gamma(\boldsymbol{u}) = \gamma$, we speak about KV-SPHF when the projection key can be given before seeing the word $\boldsymbol{u}$, and of CS-SPHF, when the projection key while independent of the word is given after seeing it. (In reference to [30,44] where those kinds of SPHF were first use). We give in Sect. 3.3 an example of KV-SPHF for Cramer-Shoup encryption, both in classical and new notations.

We will need a third property for our one-round PAKE protocol. This property, called strong pseudo-randomness in [14], is recalled in Appendix A.3 for lack of space.

2.3 Building Blocks

Decisional Diffie-Hellman (DDH). The Decisional Diffie-Hellman hypothesis says that in a multiplicative group $(p, \mathbb{G}, g)$ when we are given $(g^\lambda, g^\mu, g^\psi)$ for unknown random $\lambda, \mu, \psi \overset{\$}{\leftarrow} \mathbb{Z}_p$, it is hard to decide whether $\psi = \lambda \times \mu$.

Pairing groups. Let GGen be a probabilistic polynomial time (PPT) algorithm that on input $1^\mathfrak{K}$ returns a description $\mathcal{G} = (p, \mathbb{G}_1, \mathbb{G}_2, \mathbb{G}_T, e, g_1, g_2)$ of asymmetric pairing groups where $\mathbb{G}_1$, $\mathbb{G}_2$, $\mathbb{G}_T$ are cyclic groups of order p for a $\mathfrak{K}$-bit prime p, g_1 and g_2 are generators of $\mathbb{G}_1$ and $\mathbb{G}_2$, respectively, and $e : \mathbb{G}_1 \times \mathbb{G}_2$ is an efficiently computable (non-degenerated) bilinear map. Define $g_T := e(g_1, g_2)$, which is a generator in $\mathbb{G}_T$.

Matricial Notations. If $\boldsymbol{A} \in \mathbb{Z}_p^{(k+1)\times n}$ is a matrix, then $\overline{\boldsymbol{A}} \in \mathbb{Z}_p^{k\times n}$ denotes the upper matrix of $\boldsymbol{A}$ and $\underline{\boldsymbol{A}} \in \mathbb{Z}_p^{1\times n}$ denotes the last row of $\boldsymbol{A}$. We use classical notations from [36] for operations on vectors (. for the dot product and $\odot$ for the product component-wise). Concatenation of matrices having the same number of lines will be denoted by $\boldsymbol{A}||\boldsymbol{B}$ (where $a||b + c$ should be implicitly parsed as $a||(b + c)$).

We use implicit representation of group elements as introduced in [33]. For $s \in \{1, 2, T\}$ and $a \in \mathbb{Z}_p$ define $[a]_s = g_s^a \in \mathbb{G}_s$ as the *implicit representation* of a in $\mathbb{G}_s$ (we use $[a] = g^a \in \mathbb{G}$ if we consider a unique group). More generally, for a matrix $\boldsymbol{A} = (a_{ij}) \in \mathbb{Z}_p^{n\times m}$ we define $[\boldsymbol{A}]_s$ as the implicit representation of $\boldsymbol{A}$ in $\mathbb{G}_s$:

$$[\boldsymbol{A}]_s := \begin{pmatrix} g_s^{a_{11}} & \dots & g_s^{a_{1m}} \\ \\ g_s^{a_{n1}} & \dots & g_s^{a_{nm}} \end{pmatrix} \in \mathbb{G}_s^{n\times m}$$

We will always use this implicit notation of elements in $\mathbb{G}_s$, i.e., we let $[a]_s \in \mathbb{G}_s$ be an element in $\mathbb{G}_s$. Note that from $[a]_s \in \mathbb{G}_s$ it is generally hard to compute the value a (discrete logarithm problem in $\mathbb{G}_s$). Further, from $[b]_T \in \mathbb{G}_T$ it is hard to compute the value $[b]_1 \in \mathbb{G}_1$ and $[b]_2 \in \mathbb{G}_2$ (pairing inversion problem). Obviously, given $[a]_s \in \mathbb{G}_s$ and a scalar $x \in \mathbb{Z}_p$, one can efficiently compute $[ax]_s \in \mathbb{G}_s$. Further, given $[a]_1, [b]_2$ one can efficiently compute $[ab]_T$ using the pairing e. For $\boldsymbol{a}, \boldsymbol{b} \in \mathbb{Z}_p^k$ define $e([\boldsymbol{a}]_1, [\boldsymbol{b}]_2) := [\boldsymbol{a}^\top \boldsymbol{b}]_T \in \mathbb{G}_T$.

If $a \in \mathbb{Z}_p$, we define the $(k+1)$-vector: $\iota_s(a) := (1_s, \ldots, 1_s, [a]_s)$ (this notion can be implicitly extended to vectors $a \in \mathbb{Z}_p^n$), and the $k+1$ by $k+1$ matrix $\iota_T(a) := \begin{pmatrix} 1 & \ldots & 1 \\ \vdots & \ddots & 1 \\ 1 & 1 & a \end{pmatrix}$.

Assumptions. We recall the definition of the matrix Diffie-Hellman (MDDH) assumption [33].

Definition 2. Matrix Distribution. *Let $k \in \mathbb{N}$. We call $\mathcal{D}_k$ a matrix distribution if it outputs matrices in $\mathbb{Z}_p^{(k+1)\times k}$ of full rank k in polynomial time.*

Without loss of generality, we assume the first k rows of $\mathbf{A} \stackrel{\$}{\leftarrow} \mathcal{D}_k$ form an invertible matrix. The $\mathcal{D}_k$-Matrix Diffie-Hellman problem is to distinguish the two distributions $([\mathbf{A}], [\mathbf{A}\boldsymbol{w}])$ and $([\mathbf{A}], [\boldsymbol{u}])$ where $\mathbf{A} \stackrel{\$}{\leftarrow} \mathcal{D}_k$, $\boldsymbol{w} \stackrel{\$}{\leftarrow} \mathbb{Z}_p^k$ and $\boldsymbol{u} \stackrel{\$}{\leftarrow} \mathbb{Z}_p^{k+1}$.

Definition 3 ($\mathcal{D}_k$-Matrix Diffie-Hellman Assumption $\mathcal{D}_k$-MDDH). *Let $\mathcal{D}_k$ be a matrix distribution and $s \in \{1, 2, T\}$. We say that the $\mathcal{D}_k$-Matrix Diffie-Hellman ($\mathcal{D}_k$-MDDH) Assumption holds relative to GGen in group $\mathbb{G}_s$ if for all PPT adversaries $\mathcal{D}$,*

$$\begin{aligned} \mathbf{Adv}_{\mathcal{D}_k,\mathsf{GGen}}(\mathcal{D}) &:= |\Pr[\mathcal{D}(\mathcal{G}, [\boldsymbol{A}]_s, [\boldsymbol{A}\boldsymbol{w}]_s) = 1] - \Pr[\mathcal{D}(\mathcal{G}, [\boldsymbol{A}]_s, [\boldsymbol{u}]_s) = 1]| \\ &= \mathsf{negl}(\lambda), \end{aligned}$$

where the probability is taken over $\mathcal{G} \stackrel{\$}{\leftarrow} \mathsf{GGen}(1^\lambda)$, $\boldsymbol{A} \stackrel{\$}{\leftarrow} \mathcal{D}_k$, $\boldsymbol{w} \stackrel{\$}{\leftarrow} \mathbb{Z}_p^k$, $\boldsymbol{u} \stackrel{\$}{\leftarrow} \mathbb{Z}_p^{k+1}$.

For each $k \geq 1$, [33] specifies distributions $\mathcal{L}_k$, $\mathcal{U}_k$, ... such that the corresponding $\mathcal{D}_k$-MDDH assumption is the k-Linear assumption, the k-uniform and others. All assumptions are generically secure in bilinear groups and form a hierarchy of increasingly weaker assumptions. The distributions are exemplified for $k = 2$, where $a_1, \ldots, a_6 \stackrel{\$}{\leftarrow} \mathbb{Z}_p$.

$$\mathcal{L}_2 : \boldsymbol{A} = \begin{pmatrix} a_1 & 0 \\ 0 & a_2 \\ 1 & 1 \end{pmatrix} \quad \mathcal{U}_2 : \boldsymbol{A} = \begin{pmatrix} a_1 & a_2 \\ a_3 & a_4 \\ a_5 & a_6 \end{pmatrix}.$$

It was also shown in [33] that $\mathcal{U}_k$-MDDH is implied by all other $\mathcal{D}_k$-MDDH assumptions. In the following, we write $k-\mathsf{MDDH}$ for $\mathcal{D}_k - \mathsf{MDDH}$.

Lemma 4 (Random self reducibility [33]). *For any matrix distribution $\mathcal{D}_k$, $\mathcal{D}_k$-MDDH is random self-reducible. In particular, for any $m \geq 1$,*

$$\mathbf{Adv}_{\mathcal{D}_k,\mathsf{GGen}}(\mathcal{D}) + \frac{1}{q-1} \geq \mathbf{Adv}^m_{\mathcal{D}_k,\mathsf{GGen}}(\mathcal{D}')$$

where $\mathbf{Adv}^m_{\mathcal{D}_k,\mathsf{GGen}}(\mathcal{D}') := \Pr[\mathcal{D}'(\mathcal{G}, [\boldsymbol{A}], [\boldsymbol{A}\boldsymbol{W}]) \Rightarrow 1] - \Pr[\mathcal{D}'(\mathcal{G}, [\boldsymbol{A}], [\boldsymbol{U}]) \Rightarrow 1]$, *with* $\mathcal{G} \leftarrow \mathsf{GGen}(1^\lambda)$, $\boldsymbol{A} \stackrel{\$}{\leftarrow} \mathcal{D}_k$, $\boldsymbol{W} \stackrel{\$}{\leftarrow} \mathbb{Z}_p^{k\times m}$, $\boldsymbol{U} \stackrel{\$}{\leftarrow} \mathbb{Z}_p^{(k+1)\times m}$.

Remark: It should be noted that $\mathcal{L}_1, \mathcal{L}_2$ are respectively the SXDH and DLin assumptions that we recall below for completeness.

Definition 5 (Decisional Linear (*DLin* [20])). *The Decisional Linear hypothesis says that in a multiplicative group* $(p, \mathbb{G}, g)$ *when we are given* $(g^\lambda, g^\mu, g^{\alpha\lambda}, g^{\beta\mu}, g^\psi)$ *for unknown random* $\alpha, \beta, \lambda, \mu \xleftarrow{\$} \mathbb{Z}_p$*, it is hard to decide whether* $\psi = \alpha + \beta$.

Definition 6 (Symmetric External Diffie Hellman (*SXDH* [9])). *This variant of DDH, used mostly in bilinear groups in which no computationally efficient homomorphism exists from* $\mathbb{G}_2$ *in* $\mathbb{G}_1$ *or* $\mathbb{G}_1$ *to* $\mathbb{G}_2$*, states that DDH is hard in both* $\mathbb{G}_1$ *and* $\mathbb{G}_2$.

Labelled Cramer-Shoup Encryption. We present here the well-known encryption schemes based on DDH, and we show in Sect. 4 how to extend it to $\mathcal{D}_k$ − MDDH. We focus on Cramer-Shoup [29] in all the following of the paper, but one easily obtains the same results on El Gamal IND-CPA scheme [32] by simply omitting the corresponding parts. We are going to rely on the IND-CCA property to be able to decrypt queries in the simulation.

VANILLA CRAMER-SHOUP ENCRYPTION. The Cramer-Shoup encryption scheme is an IND-CCA version of the ElGamal Encryption. We present it here as a labeled public-key encryption scheme, the classical version is done with $\ell = \emptyset$.

- Setup$(1^\mathfrak{K})$ generates a group $\mathbb{G}$ of order p, with a generator g
- KeyGen(param) generates $(g_1, g_2) \xleftarrow{\$} \mathbb{G}^2$, $\mathsf{dk} = (x_1, x_2, y_1, y_2, z) \xleftarrow{\$} \mathbb{Z}_p^5$, and sets, $c = g_1^{x_1} g_2^{x_2}$, $d = g_1^{y_1} g_2^{y_2}$, and $h = g_1^z$. It also chooses a Collision-Resistant hash function $\mathfrak{H}_K$ in a hash family $\mathcal{H}$ (or simply a Universal One-Way Hash Function). The encryption key is $\mathsf{ek} = (g_1, g_2, c, d, h, \mathfrak{H}_K)$.
- Encrypt$(\ell, \mathsf{ek}, M; r)$, for a message $M \in \mathbb{G}$ and a random scalar $r \in \mathbb{Z}_p$, the ciphertext is $C = (\ell, \mathbf{u} = (g_1^r, g_2^r), e = M \cdot h^r, v = (cd^\xi)^r)$, where v is computed afterwards with $\xi = \mathfrak{H}_K(\ell, \mathbf{u}, e)$.
- Decrypt(ℓ, dk, C): one first computes $\xi = \mathfrak{H}_K(\ell, \mathbf{u}, e)$ and checks whether $u_1^{x_1+\xi y_1} \cdot u_2^{x_2+\xi y_2} \stackrel{?}{=} v$. If the equality holds, one computes $M = e/(u_1^z)$ and outputs M. Otherwise, one outputs $\perp$.

The security of the scheme is proven under the DDH assumption and the fact the hash function used is a Universal One-Way Hash Function.

In following work [30] they refined the proof, explaining that the scheme can be viewed as a 2-Universal Hash Proof on the language of valid Diffie Hellman tuple.

VANILLA CRAMER-SHOUP ENCRYPTION WITH MATRICIAL NOTATIONS.

- Setup$(1^\mathfrak{K})$ generates a group $\mathbb{G}$ of order p, with a generator g, with an underlying matrix assumption $\mathcal{D}_1$ using a base matrix $[\boldsymbol{A}] \in \mathbb{G}^{2\times 1}$;
- KeyGen(param) generates $\mathsf{dk} = \boldsymbol{t}_1, \boldsymbol{t}_2, \boldsymbol{z} \xleftarrow{\$} \mathbb{Z}_p^2$ (with $\boldsymbol{t}_1 = (x_1, x_2)$, $\boldsymbol{t}_2 = (y_1, y_2)$ and $\boldsymbol{z} = (z, 1)$), and sets $c = \boldsymbol{t}_1\boldsymbol{A}, d = \boldsymbol{t}_2\boldsymbol{A}, h = \boldsymbol{z}\boldsymbol{A}$. It also chooses a hash function $\mathfrak{H}_K$ in a collision-resistant hash family $\mathcal{H}$ (or simply a Universal One-Way Hash Function). The encryption key is $\mathsf{ek} = ([A], [c], [d], [h], \mathfrak{H}_K)$.

- $\mathsf{Encrypt}(\ell, \mathsf{ek}, [m]; r)$, for a message $M = [m] \in \mathbb{G}$ and random scalar $r \stackrel{\$}{\leftarrow} \mathbb{Z}_p$, the ciphertext is $C = (\ell, \mathbf{u} = [\boldsymbol{A}r]), e = [\boldsymbol{h}r + m], v = [(\boldsymbol{c} + \boldsymbol{d} \odot \xi)\boldsymbol{r}]$, where v is computed afterwards with $\xi = \mathfrak{H}_K(\ell, \mathbf{u}, e)$.
- $\mathsf{Decrypt}(\ell, \mathsf{dk}, C)$: one first computes $\xi = \mathfrak{H}_K(\ell, \mathbf{u}, e)$ and checks whether v is consistent with $\boldsymbol{t}_1, \boldsymbol{t}_2$.
  If it is, one computes $M = [e - (\boldsymbol{uz})]$ and outputs M. Otherwise, one outputs $\perp$.

Groth-Sahai Proof System. Groth and Sahai [36] proposed non-interactive zero-knowledge proofs of satisfiability of certain equations over bilinear groups, called *pairing product equations*. Using as witness group elements (and scalars) which satisfy the equation, the prover starts with making commitments on them. To prove satisfiability of an equation (which is the statement of the proof), a Groth-Sahai proof uses these commitments and shows that the committed values satisfy the equation. The proof consists again of group elements and is verified by a pairing equation derived from the statement.

We refer to [36] for details of the Groth-Sahai proof system, and to [33] for the compatibility with the k-MDDH assumptions. More details can be found in the paper full version [15]. We are going to give a rough idea of the technique for SXDH.

To prove that committed variables satisfy a set of relations, the Groth-Sahai techniques require one commitment per variable and one proof element (made of a constant number of group elements) per relation. Such proofs are available for pairing-product relations and for multi-exponentiation equations.

When based on the SXDH assumption, the commitment key is of the form $\mathbf{u}_1 = (u_{1,1}, u_{1,2}), \mathbf{u}_2 = (u_{2,1}, u_{2,2}) \in \mathbb{G}_1^2$ and $\mathbf{v}_1 = (v_{1,1}, v_{1,2}), \mathbf{v}_2 = (v_{2,1}, v_{2,2}) \in \mathbb{G}_2^2$. We write

$$\mathbf{u} = \begin{pmatrix} \mathbf{u}_1 \\ \mathbf{u}_2 \end{pmatrix} = \begin{pmatrix} u_{1,1} & u_{1,2} \\ u_{2,1} & u_{2,2} \end{pmatrix} \quad \text{and} \quad \mathbf{v} = \begin{pmatrix} \mathbf{v}_1 \\ \mathbf{v}_2 \end{pmatrix} = \begin{pmatrix} v_{1,1} & v_{1,2} \\ v_{2,1} & v_{2,2} \end{pmatrix}.$$

The Setup algorithm initializes the parameters as follows: $\mathbf{u}_1 = (g_1, u)$ with $u = g_1^\lambda$ and $\mathbf{u}_2 = \mathbf{u}_1{}^\mu$ with $\lambda, \mu \stackrel{\$}{\leftarrow} \mathbb{Z}_p^*$, which means that $\mathbf{u}$ is a Diffie-Hellman tuple in $\mathbb{G}_1$, since $\mathbf{u}_1 = (g_1, g_1^\lambda)$ and $\mathbf{u}_2 = (g_1^\mu, g_1^{\lambda\mu})$. The TSetup algorithm will use instead $\mathbf{u}_2 = \mathbf{u}_1{}^\mu \odot (1, g_1)^{-1}$: $\mathbf{u}_1 = (g_1, g_1^\lambda)$ and $\mathbf{u}_2 = (g_1^\mu, g_1^{\lambda\mu - 1})$. And it is the same in $\mathbb{G}_2$ for $\mathbf{v}$. Depending on the definition of $\mathbf{u}_2, \mathbf{v}_2$, this commitment can be either perfectly hiding or perfectly binding. The two parameter initializations are indistinguishable under the SXDH assumption.

To commit to $X \in \mathbb{G}_1$, one chooses randomness $s_1, s_2 \in \mathbb{Z}_p$ and sets $\mathcal{C}(X) = (1, X) \odot \mathbf{u}_1^{s_1} \odot \mathbf{u}_2^{s_2} = (1, X) \odot (u_{1,1}^{s_1}, u_{1,2}^{s_1}) \odot (u_{2,1}^{s_2}, u_{2,2}^{s_2}) = (u_{1,1}^{s_1} \cdot u_{2,1}^{s_2}, X \cdot u_{1,2}^{s_1} \cdot u_{2,2}^{s_2})$. Similarly, one can commit to element in $\mathbb{G}_2$ and scalars in $\mathbb{Z}_p$. The committed group elements can be extracted if $\mathbf{u}_2$ is linearly dependant of $\mathbf{u}_1$ by knowing the discrete logarithm x_1 between $\mathbf{u}_{1,1}$ and $\mathbf{u}_{2,2}$: $c_2/(c_1^{x_1}) = X$.

In the following we are going to focus on proof of linear multi-scalar exponentiation in $\mathbb{G}_1$, that is to say we are going to prove equations of the form $\prod_i A_i^{y_i} = A$ where A_i are public elements in $\mathbb{G}_1$ and y_i are going to be scalars committed into $\mathbb{G}_2$.

2.4 Protocols

UC Framework. The goal of this simulation-based model [22] is to ensure that UC-secure protocols will continue to behave in the ideal way even if executed in a concurrent way in arbitrary environments. Due to lack of space, a short introduction to the UC framework is given in the paper full version [15].

Oblivious Transfer and Password-Authenticated Key-Exchange. The security properties for these two protocols are given in terms of ideal functionalities in the paper full version [15].

3 Structure-Preserving Smooth Projective Hashing

3.1 Definition

In this section, we are now going to narrow the classical definition of Smooth Projective Hash Functions to what we are going to name Structure-Preserving Smooth Projective Hash Functions, in which both words, witnesses and projection keys are group elements.

Since witnesses now become group elements, this allows a full compatibility with Groth and Sahai methodology [36], such that for instance possessing a Non-Interactive Zero-Knowledge Proof of Knowledge can become new witnesses of our SP-SPHF, leading to interesting applications, as described later on.

As we are in the context of Structure Preserving cryptography, we assume the existence of a (prime order) bilinear group $(p, \mathbb{G}_1, \mathbb{G}_2, g_1, g_2, \mathbb{G}_T, e)$, and consider Languages (sets of elements) $\mathfrak{L}$ defined over this group. The hash space is usually $\mathbb{G}_T$, the projection key space a group $\mathbb{G}_1^m \times \mathbb{G}_2^n$ and the witness space a group $\mathbb{G}_1^n \times \mathbb{G}_2^m$.

Definition 7 Structure-Preserving Smooth Projective Hash Functions. *A Structure-Preserving Smooth Projective Hash Function over a language* $\mathfrak{L} \subset X$ *onto a set* $\mathcal{H}$ *is defined by 4 algorithms* $(\mathsf{HashKG}, \mathsf{ProjKG}, \mathsf{Hash}, \mathsf{ProjHash})$*:*

- $\mathsf{HashKG}(\mathfrak{L}, \mathsf{param})$, *outputs a hashing key* hk *for the language* $\mathfrak{L}$*;*
- $\mathsf{ProjKG}(\mathsf{hk}, (\mathfrak{L}, \mathsf{param}), W)$, *derives the projection key* hp *thanks to the hashing key* hk*.*
- $\mathsf{Hash}(\mathsf{hk}, (\mathfrak{L}, \mathsf{param}), W)$, *outputs a hash value* $H \in \mathcal{H}$*, thanks to the hashing key* hk*, and* W
- $\mathsf{ProjHash}(\mathsf{hp}, (\mathfrak{L}, \mathsf{param}), W, w)$, *outputs the value* $H' \in \mathcal{H}$*, thanks to* hp *and the witness* w *that* $W \in \mathfrak{L}$*.*

Remark 8. We stress that, contrarily to classical SPHF, both hp, W and more importantly w are base group elements, and so live in the same space.

3.2 Properties

Properties are then inherited by those of classical Smooth Projective Hash Functions.

- *Correctness*: On honest computations with (W, w) compatible with $\mathfrak{L}$, we have $\mathsf{ProjHash}(\mathsf{hp}, (\mathfrak{L}, \mathsf{param}), W, w) = \mathsf{Hash}(\mathsf{hk}, (\mathfrak{L}, \mathsf{param}), W)$.
- *Smoothness*: For all $W \in X \setminus \mathfrak{L}$ the following distributions are statistically indistinguishable:

$$\Delta_0 = \left\{ (\mathfrak{L}, \mathsf{param}, W, \mathsf{hp}, v) \;\middle|\; \begin{array}{l} \mathsf{param} = \mathsf{Setup}(1^{\mathfrak{K}}), \mathsf{hk} = \mathsf{HashKG}(\mathfrak{L}, \mathsf{param}), \\ \mathsf{hp} = \mathsf{ProjKG}(\mathsf{hk}, (\mathfrak{L}, \mathsf{param}), W), \\ v = \mathsf{Hash}(\mathsf{hk}, (\mathfrak{L}, \mathsf{param}), W) \in \mathbb{G}_T \end{array} \right\}$$

$$\Delta_1 = \left\{ (\mathfrak{L}, \mathsf{param}, W, \mathsf{hp}, v) \;\middle|\; \begin{array}{l} \mathsf{param} = \mathsf{Setup}(1^{\mathfrak{K}}), \mathsf{hk} = \mathsf{HashKG}(\mathfrak{L}, \mathsf{param}), \\ \mathsf{hp} = \mathsf{ProjKG}(\mathsf{hk}, (\mathfrak{L}, \mathsf{param}), W), v \stackrel{\$}{\leftarrow} \mathbb{G}_T \end{array} \right\}.$$

This is formalized by

$$\mathsf{Adv}_{\mathsf{SPHF}}^{\mathsf{smooth}}(\mathfrak{K}) = \sum_{V \in \mathbb{G}} \left| \Pr_{\Delta_1}[v = V] - \Pr_{\Delta_0}[v = V] \right| \text{ is negligible.}$$

As usual, a derivative property called *Pseudo-Randomnness*, says the previous distribution are computationally indistinguishable from words in the language while the witnesses remain unknown. This is implied by the Smoothness on Hard Subset membership languages.

3.3 Retro-Compatibility

Constructing SP-SPHF is not that hard of a task. A first naive approach allows to transform every pairing-less SPHF into a SP-SPHF in a bilinear setting. It should be noted that while the resulting Hash/ProjHash values live in the target group, nearly all use cases encourage to use a proper hash function on them before computing anything using their value, hence the communication cost would remain the same. (Only applications where one of the party has to provide an additional proof that the ProjHash was honestly computed might be lost, but besides proof of negativity from [18], this never arises.)

To this goal, simply given a new generator $f \in \mathbb{G}_2$, and a scalar witness vector λ, one generates the new witness vector $\Lambda = [f \odot \boldsymbol{\lambda}]_2$. Words and projection keys belong to $\mathbb{G}_1$, and hash values to $\mathbb{G}_T$. Any SPHF can thus be transformed into an SP-SPHF in the following way:

	SPHF	SP-SPHF
Word $\boldsymbol{u}$	$[\boldsymbol{\lambda} \odot \Gamma(\boldsymbol{u})]_1$	$[\boldsymbol{\lambda} \odot \Gamma(\boldsymbol{u})]_1$
Witness w	$\boldsymbol{\lambda}$	$\boldsymbol{\Lambda} = [f \odot \boldsymbol{\lambda}]_2$
hk	$\boldsymbol{\alpha}$	$\boldsymbol{\alpha}$
$\mathsf{hp} = [\boldsymbol{\gamma}(\boldsymbol{u})]_1$	$[\Gamma(\boldsymbol{u}) \odot \boldsymbol{\alpha}]_1$	$[\Gamma(\boldsymbol{u}) \odot \boldsymbol{\alpha}]_1$
$\mathsf{Hash}(\mathsf{hk}, \boldsymbol{u})$	$[\Theta(\boldsymbol{u}) \odot \boldsymbol{\alpha}]_1$	$[f \odot \Theta(\boldsymbol{u}) \odot \boldsymbol{\alpha}]_T$
$\mathsf{ProjHash}(\mathsf{hp}, \boldsymbol{u}, w)$	$[\boldsymbol{\lambda} \odot \boldsymbol{\gamma}(\boldsymbol{u})]_1$	$[\boldsymbol{\Lambda} \odot \boldsymbol{\gamma}(\boldsymbol{u})]_T$

- *Correctness* is inherited for words in $\mathcal{L}$ as this reduces to computing the same values but in $\mathbb{G}_T$.
- *Smoothness*: For words outside the language, the projection keys, remaining unchanged, do not reveal new information, so that the smoothness will remain preserved.
- *Pseudo-Randomness*: Without any witness, words inside the language are indistinguishable from words outside the language (under the subgroup decision assumption), hence the hash values remain pseudo-random.

It should be noted that in case this does not weaken the subgroup decision assumption (k-MDDH in the following) linked to the original language, one can set $\mathbb{G}_1 = \mathbb{G}_2$.

We give in Fig. 1 two examples of regular Smooth Projective Hash Functions on Diffie-Hellman and Cramer-Shoup encryption of M, where $\alpha = \mathcal{H}(\mathbf{u}, e)$, and their counterparts with SP-SPHF. ElGamal being a simplification of Cramer-Shoup, we skip the description of the associated SP-SPHF. We also give in Fig. 2 the matricial version of Cramer-Shoup encryption, in which we denote by C' the Cramer-Shoup encryption C of M in which we removed M.

	SPHF	SP-SPHF
DH	h^r, g^r	h^r, g^r
Witness w	r	g_2^r
hk	λ, μ	λ, μ
hp	$h^\lambda g^\mu$	$h^\lambda g^\mu$
$\mathsf{Hash}(\mathsf{hk}, \boldsymbol{u})$	$(h^r)^\lambda (g^r)^\mu$	$e((h^r)^\lambda (g^r)^\mu, g_2)$
$\mathsf{ProjHash}(\mathsf{hp}, \boldsymbol{u}, w)$	hp^r	$e(\mathsf{hp}, g_2^r)$
CS(M;r)	$h^r M, f^r, g^r, (cd^\alpha)^r$	$h^r M, f^r, g^r, (cd^\alpha)^r$
Witness w	r	g_2^r
hk	$\lambda_1, \lambda_2, \mu, \nu, \eta$	$\lambda_1, \lambda_2, \mu, \nu, \eta$
hp	$h^{\lambda_1} f^\mu g^\nu c^\eta, h^{\lambda_2} d^\nu$	$h^{\lambda_1} f^\mu g^\nu c^\eta, h^{\lambda_2} d^\nu$
$\mathsf{Hash}(\mathsf{hk}, \boldsymbol{u})$	$H = (h^r)^{\lambda_1 + \alpha\lambda_2} (f^r)^\mu (g^r)^\nu ((cd^\alpha)^r)^\mu$	$e(H, g_2)$
$\mathsf{ProjHash}(\mathsf{hp}, \boldsymbol{u}, w)$ (with $\mathsf{hp} = (\mathsf{hp}_1, \mathsf{hp}_2)$)	$(\mathsf{hp}_1 \mathsf{hp}_2^\alpha)^r$	$e(\mathsf{hp}_1 \mathsf{hp}_2^\alpha, g_2^r)$

Fig. 1. Example of conversion of classical SPHF into SP-SPHF

	SPHF	SP-SPHF
CS(M;r)	$[hr+M, \boldsymbol{A}r, (c+d\alpha)r]$	$[hr+M, \boldsymbol{A}r, (c+d\alpha)r]_1$
$\boldsymbol{B}: \begin{pmatrix} h \\ f \\ g \\ c \end{pmatrix}$	$\left[\boldsymbol{B}r + \begin{pmatrix} 0 \\ 0 \\ 0 \\ d \end{pmatrix} \alpha r + \begin{pmatrix} M \\ 0 \\ 0 \\ 0 \end{pmatrix}\right]$	$\left[\boldsymbol{B}r + \begin{pmatrix} 0 \\ 0 \\ 0 \\ d \end{pmatrix} \alpha r + \begin{pmatrix} M \\ 0 \\ 0 \\ 0 \end{pmatrix}\right]_1$
Witness w	r	$[r]_2$
hk	$\lambda_1, \lambda_2, \mu, \nu, \eta$	$\lambda_1, \lambda_2, \mu, \nu, \eta$
hp	$[\mathsf{hp}_1 = (\lambda_1\ \mu\ \nu\ \eta)\, \boldsymbol{B}]$, $\left[\mathsf{hp}_2 = (\lambda_2\ 0\ 0\ \eta) \begin{pmatrix} h \\ 0 \\ 0 \\ d \end{pmatrix}\right]$	$[\mathsf{hp}_1]_1, [\mathsf{hp}_2]_1$
Hash(hk, $\boldsymbol{u}$)	$[(\lambda_1 + \alpha\lambda_2\ \mu\ \nu\ \eta)\,(C')]$	$[(\lambda_1 + \alpha\lambda_2\ \mu\ \nu\ \eta)\,(C')]_T$
ProjHash(hp, $\boldsymbol{u}$, w)	$[(\mathsf{hp}_1 + \alpha \mathsf{hp}_2)r]$	$[(\mathsf{hp}_1 + \alpha \mathsf{hp}_2)r]_2$

Fig. 2. Example of conversion of SPHF into SP-SPHF (matricial notations)

3.4 Possible Applications

Nearly Constant 1-out-of-m Oblivious Transfer Using FLM. Recent pairing-based constructions [1,28] of Oblivious Transfer use SPHF to mask each line of a database with the hash value of as SPHF on the language corresponding to the first flow being a commitment of the said line.

Sadly, those constructions require special UC commitment on scalars, with equivocation and extraction capacities, leading to very inefficient constructions. In 2011, [34] proposed a UC commitment, whose decommitment operation is done via group elements. In Sect. 5, we are going to show how to combine the existing constructions with this efficient commitment using SP-SPHF, in order to obtain a very efficient round-optimal where there is no longer a growing overhead due to the commitment. As a side result, we show how to generalize the FLM commitment to any MDDH assumption.

Round-Optimal Password Authenticated Key Exchange with Adaptive Corruptions. Recent developments around SPHF-based PAKE have either lead to Round-Optimal PAKE in the BPR model [11], or with static corruptions [14,44]. In order to achieve round-optimality, [1] needs to do a bit-per-bit commitment of the password, inducing a communication cost proportional to the maximum password length.

In the following, we show how to take advantage of the SP-SPHF constructed on the FLM commitment to propose a One-Round PAKE UC secure against adaptive adversaries, providing a constant communication cost.

Using a ZKPK as a witness, Anonymous Credentials. Previous applications allow more efficient instantiations of protocols already using scalar-based SPHF. However, one can imagine additional scenarios, where a scalar based approach may not be possible, due to the inherent nature of the witness used.

For example, one should consider a strong authentication scenario, in which each user possesses an identifier delivered by an authority, and a certification on a commitment to this identifier, together with a proof of knowledge that this commitment is indeed a commitment to this identifier. (Such scenario can be transposed to the delivery of a Social Security Number, where a standalone SSN may not be that useful, but a SSN officially linked to someone is a sensitive information that should be hidden.) In this scenario, a user who wants to access his record on a government service where he is already registered, should give the certificate, and then would use an implicit proof that this corresponds to his identifier. With our technique, the server would neither learn the certificate in the clear nor the user identifier (if he did not possess it earlier), and the user would be able to authenticate only if his certificate is indeed on his committed identifier.

In our scenario, we could even add an additional step, such that Alice does not interact directly with Bob but can instead use a pawn named Carol. She could send to Carol a commitment to the signature on her identity, prove in a black box way that it is a valid signature on an identity, and let Carol do the interaction on her behalf. For example, to allow a medical practitioner to access some subpart of her medical record concerning on ongoing treatment, in this case, Carol would need to anonymously prove to the server that she is indeed a registered medical practitioner, and that Alice has given her access to her data.

4 Encryption and Commitment Schemes Based on k-MDDH

4.1 k-MDDH Cramer-Shoup Encryption

In this paper, we supersede the previous constructions with a k-MDDH based one:

- Setup($1^{\mathfrak{K}}$) generates a group $\mathbb{G}$ of order p, with an underlying matrix assumption using a base matrix $[\boldsymbol{A}] \in \mathbb{G}^{k+1 \times k}$;
- KeyGen(param) generates $\mathsf{dk} = \boldsymbol{t}_1, \boldsymbol{t}_2, \boldsymbol{z} \xleftarrow{\$} \mathbb{Z}_p^{k+1}$, and sets, $\boldsymbol{c} = \boldsymbol{t}_1 \boldsymbol{A} \in \mathbb{Z}_p^k$, $\boldsymbol{d} = \boldsymbol{t}_2 \boldsymbol{A} \in \mathbb{Z}_p^k$, $\boldsymbol{h} = \boldsymbol{z} \boldsymbol{A} \in \mathbb{Z}_p^k$. It also chooses a hash function $\mathfrak{H}_K$ in a collision-resistant hash family $\mathcal{H}$ (or simply a Universal One-Way Hash Function). The encryption key is $\mathsf{ek} = ([\boldsymbol{c}], [\boldsymbol{d}], [\boldsymbol{h}], [\boldsymbol{A}], \mathfrak{H}_K)$.
- Encrypt(ℓ, ek, $[m]$; $\boldsymbol{r}$), for a message $M = [m] \in \mathbb{G}$ and random scalars $\boldsymbol{r} \xleftarrow{\$} \mathbb{Z}_p^k$, the ciphertext is $C = (\mathbf{u} = [\boldsymbol{A}\boldsymbol{r}]), e = [\boldsymbol{h}\boldsymbol{r} + m], v = [(\boldsymbol{c} + \boldsymbol{d} \odot \xi)\boldsymbol{r}]_1$, where v is computed afterwards with $\xi = \mathfrak{H}_K(\ell, \mathbf{u}, e)$.
- Decrypt(ℓ, dk, C): one first computes $\xi = \mathfrak{H}_K(\ell, \mathbf{u}, e)$ and checks whether v is consistent with $\boldsymbol{t}_1, \boldsymbol{t}_2$.
  If it is, one computes $M = [e - (\mathbf{u}\boldsymbol{z})]$ and outputs M. Otherwise, one outputs $\perp$.

Theorem 9. *The k-MDDH Cramer-Shoup Encryption is IND-CCA 2 under k-MDDH assumption and the collision resistance (universal one-wayness) of the Hash Family.*

Proof. To sketch the proof of the theorem, one should remember that the original proof articulate around three main cases noting $\ell, \mathbf{u}, e, v$ the challenge query, and $\ell', \mathbf{u}, e', v'$ the current decryption query:

- $(\ell, \mathbf{u}, e) = (\ell', \mathbf{u}', e')$ but $v \neq v'$. This will fail as v is computed to be the correct checksum, hence we can directly reject the decryption query.
- $(\ell, \mathbf{u}, e) \neq (\ell', \mathbf{u}', e')$ but $\xi = \xi'$, this is a collision on the Hash Function.
- $(\ell, \mathbf{u}, e, v) \neq (\ell, \mathbf{u}, e, v)$ and $\xi \neq \xi'$. This is the argument revolving around the 2-Universality of the Hash Proof system defined by $\boldsymbol{c}, \boldsymbol{d}$. $\boldsymbol{c}, \boldsymbol{d}$ gives $2k$ equations in $2k+2$ variables, hence answering decryption queries always in the same span can give at most 1 more equation leaving at least 1 degree of freedom in the system. □

Structure-Preserving Smooth Projective Hash Function

For ease of readability we are going to set $\boldsymbol{B} = \left[\begin{pmatrix} h \\ \boldsymbol{A} \\ \boldsymbol{c} \end{pmatrix}\right]$ and $\boldsymbol{D} = \left[\begin{pmatrix} 0 \\ \vdots \\ \boldsymbol{d} \end{pmatrix}\right]$, and write $C' = [\boldsymbol{B}\boldsymbol{r} + \xi \boldsymbol{D}\boldsymbol{r}]_1$ the ciphertext without the message M.

- $\mathsf{HashKG}(\mathfrak{L}, \mathsf{param})$, chooses $\Lambda \xleftarrow{\$} \mathbb{Z}_p^{(k+2)\times 1}$, $\lambda \xleftarrow{\$} \mathbb{Z}_p$ and sets

$$\mathsf{hk}_1 = \Lambda, \mathsf{hk}_2 = \begin{pmatrix} \lambda \\ \boldsymbol{0} \\ \Lambda_{k+2} \end{pmatrix};$$

- $\mathsf{ProjKG}(\mathsf{hk}, (\mathfrak{L}, \mathsf{param}), W)$, outputs $\mathsf{hp}_1 = \mathsf{hk}_1^\top \boldsymbol{B}, \mathsf{hp}_2 = \mathsf{hk}_2^\top \begin{pmatrix} h \\ \boldsymbol{0} \\ \boldsymbol{d} \end{pmatrix}$;
- $\mathsf{Hash}(\mathsf{hk}, (\mathfrak{L}, \mathsf{param}), W)$, outputs a hash value $H = [(\mathsf{hk}_1 + \xi \mathsf{hk}_2)^\top C']_T$;
- $\mathsf{ProjHash}(\mathsf{hp}, (\mathfrak{L}, \mathsf{param}), W, w)$, outputs the value $H' = [(\mathsf{hp}_1 + \xi \mathsf{hp}_2)\boldsymbol{r}]_T$.

The Smoothness comes inherently from the fact that we have $2k+2$ unknowns in hk while hp gives at most $2k$ equations. Hence an adversary has a negligible chance to find the real values.

4.2 A Universally Composable Commitment with Adaptive Security Based on MDDH

We first show how to simply generalize FLM's commitment [34] from DLin to **k**-MDDH.

FLM's Commitment on DLin. At Asiacrypt 2011, Fischlin, Libert and Manulis presented a universally composable commitment [34] with adaptive security based on the Decision Linear assumption [20]. We show here how to generalize their scheme to the Matrix Decisional Diffie-Hellman assumption from [33] and recalled in Sect. 2. We first start by recalling their original scheme. Note that sid denotes the session identifier and cid the commitment identifier and that the combination $(\mathsf{sid}, \mathsf{cid})$ is globally unique, as in [34,37].

- **CRS Generation:** $\mathsf{SetupCom}(1^{\mathfrak{K}})$ chooses a bilinear group $(p, \mathbb{G}, \mathbb{G}_T)$ of order $p > 2^{\mathfrak{K}}$, a generator g of $\mathbb{G}$, and sets $g_1 = g^{\alpha_1}$ and $g_2 = g^{\alpha_2}$ with random $\alpha_1, \alpha_2 \in \mathbb{Z}_p^*$. It defines the vectors $\mathbf{g_1} = (g_1, 1, g)$, $\mathbf{g_2} = (1, g_2, g)$ and $\mathbf{g_3} = \mathbf{g_1}^{\xi_1}\mathbf{g_2}^{\xi_2}$ with random $\xi_1, \xi_2 \in \mathbb{Z}_p^*$, which form a Groth-Sahai CRS $\mathbf{g} = (\mathbf{g_1}, \mathbf{g_2}, \mathbf{g_3})$ for the perfect soundness setting. It then chooses a collision-resistant hash function $H : \{0,1\}^* \rightarrow \mathbb{Z}_p$ and generates a public key $\mathsf{pk} = (X_1, \ldots, X_6)$ for the linear Cramer-Shoup encryption scheme. The CRS consists of $\mathsf{crs} = (\mathfrak{K}, \mathbb{G}, \mathbb{G}_T, g, \mathbf{g}, H, \mathsf{pk})$.
- **Commitment algorithm:** $\mathsf{Com}(\mathsf{crs}, M, \mathsf{sid}, \mathsf{cid}, P_i, P_j)$, to commit to message $M \in \mathbb{G}$ for party P_j, party P_i parses crs as $(\mathfrak{K}, \mathbb{G}, \mathbb{G}_T, g, \mathbf{g}, H, \mathsf{pk})$ and conducts the following steps:
  - It chooses random exponents r, s in $\mathbb{Z}_p$ and computes a linear Cramer-Shoup encryption $\psi_{CS} = (U_1, U_2, U_3, U_4, U_5)$ of $M \in \mathbb{G}$ under the label $\ell = P_i \| \mathsf{sid} \| \mathsf{cid}$ and the public key pk.
  - It generates a NIZK proof $\pi_{val-enc}$ that $\psi_{CS} = (U_1, U_2, U_3, U_4, U_5)$ is indeed a valid encryption of $M \in \mathbb{G}$. This requires to commit to exponents r, s and prove that these exponents satisfy the multi-exponentiation equations $U_1 = g_1{}^r$, $U_2 = g_2{}^s$, $U_3 = g^{r+s}$, $U_4/M = X_5{}^r X_6{}^s$ and $U_5 = (X_1 X_3{}^\alpha)^r \cdot (X_2 X_4{}^\alpha)^s$.
  - P_i erases (r, s) after the generation of $\pi_{val-enc}$ but retains the $D_M = \pi_{val-enc}$.

  The commitment is ψ_{CS}.
- **Verification algorithm:** the algorithm $\mathsf{VerCom}(\mathsf{crs}, M, D_M, \mathsf{sid}, \mathsf{cid}, P_i, P_j)$ checks the proof $\pi_{val-enc}$ and ignores the opening if the verification fails.
- **Opening algorithm:** $\mathsf{OpenCom}(\mathsf{crs}, M, D_M, \mathsf{sid}, \mathsf{cid}, P_i, P_j)$ reveals M and $D_M = \pi_{val-enc}$ to P_j.

The extraction algorithm uses Cramer-Shoup decryption algorithm, while the equivocation uses the simulator of the NIZK. It is shown in [1] that the IND-CCA security notion for C and the computational soundness of π make it strongly-binding-extractable, while the IND-CCA security notion and the zero-knowledge property of the NIZK provide the strong-simulation-indistinguishability.

Moving to k-MDDH: We now show how to extend the previous commitment to the k-MDDH assumption. Compared to the original version of the commitment, we split the proof $\pi_{val-enc}$ into its two parts: the NIZK proof denoted here as $[\boldsymbol{\Pi}]_1$ is still revealed during the opening algorithm, while the Groth-Sahai commitment $[\boldsymbol{R}]_2$ of the randomness $\mathbf{r}$ of the Cramer-Shoup encryption is sent during the commitment phase. Furthermore, since the hash value in the Cramer Shoup encryption is used to link the commitment with the session, we include this value $[\boldsymbol{R}]_2$ to the label, in order to ensure that this extra commitment information given with the ciphertext is the original one. We refer the reader to the original security proof in [34, Theorem 1], which remains exactly the same, since this additional commitment provides no information (either computationally or perfectly, depending on the CRS), and since the commitment $[\boldsymbol{R}]_2$ is not modified in the equivocation step (only the value $[\boldsymbol{\Pi}]_1$ is changed).

- **CRS Generation:** algorithm $\mathsf{SetupCom}(1^{\mathfrak{K}})$ chooses a bilinear asymmetric group $(p, \mathbb{G}_1, \mathbb{G}_2, \mathbb{G}_T, e, g_1, g_2)$ of order $p > 2^{\mathfrak{K}}$, and a set of generators $[\boldsymbol{A}]_1$ corresponding to the underlying matrix assumption.
As explained in [33], following their notations, one can define a Groth-Sahai CRS by picking $\boldsymbol{w} \xleftarrow{\$} \mathbb{Z}_p^{k+1}$, and setting $[\boldsymbol{U}]_2 = [\boldsymbol{B}||\boldsymbol{B}\boldsymbol{w}]_2$ for a hiding CRS, and $[\boldsymbol{B}||\boldsymbol{B}\boldsymbol{w} + (0||z)^\top]_2$ otherwise, where $[\boldsymbol{B}]_2$ is an k-MDDH basis, and $\boldsymbol{w}, z$ are the elements defining the challenge vector.
For the Cramer-Shoup like CCA-2 encryption, one additionally picks $\boldsymbol{t}_1, \boldsymbol{t}_2, \boldsymbol{z} \xleftarrow{\$} \mathbb{Z}_p^{k+1}$, and a Universal One-Way Hash Function $\mathcal{H}$ and sets $[\boldsymbol{h}]_1 = [\boldsymbol{z} \cdot \boldsymbol{A}]_1$, $[\boldsymbol{c}]_1 = [\boldsymbol{t}_1 \boldsymbol{A}]_1, [\boldsymbol{d}]_1 = [\boldsymbol{t}_2 \boldsymbol{A}]_1$.
The CRS consists of $\mathsf{crs} = (\mathfrak{K}, p, \mathbb{G}_1, \mathbb{G}_2, \mathbb{G}_T, [\boldsymbol{A}]_1 \in \mathbb{G}_1^{k \times k+1}, [\boldsymbol{U}]_2, [\boldsymbol{h}]_1 \in \mathbb{G}_1^k, [\boldsymbol{c}]_1 \in \mathbb{G}_1^k, [\boldsymbol{d}]_1 \in \mathbb{G}_1^k, \mathcal{H})$.
- **Commitment algorithm:** $\mathsf{Com}(\mathsf{crs}, M, \mathsf{sid}, \mathsf{cid}, P_i, P_j)$, to commit to message $M \in \mathbb{G}_1$ for party P_j, party P_i conducts the following steps:
  - It chooses random exponents $\boldsymbol{r}$ in $\mathbb{Z}_p^k$ and commits to $\boldsymbol{r}$ in $[\boldsymbol{R}]_2$ with randomness $\rho \xleftarrow{\$} \mathbb{Z}_p^{k \times k+1}$, setting $[\boldsymbol{R}]_2 = [\boldsymbol{U}\boldsymbol{\rho} + \iota_2(\boldsymbol{r})]_2 \in \mathbb{G}_2^{k \times k+1}$. It also computes a Cramer-Shoup encryption $\psi_{CS} = [\boldsymbol{C}]_1$ of $M \in \mathbb{G}_1$ under the label $\ell = P_i||\mathsf{sid}||\mathsf{cid}$ and the public key pk:
$$[\boldsymbol{C}]_1 = [\boldsymbol{A}\boldsymbol{r}||\boldsymbol{h}\boldsymbol{r} + M||(\boldsymbol{c} + \boldsymbol{d} \odot \mathcal{H}(\ell||\boldsymbol{C}_1||\boldsymbol{C}_2||\boldsymbol{R}))\boldsymbol{r}]_1 = [\boldsymbol{C}_1||\boldsymbol{C}_2||\boldsymbol{C}_3]_1$$
For simplicity we write $\ell' = \ell||[\boldsymbol{C}_1]_1||[\boldsymbol{C}_2]_1||[\boldsymbol{R}]_2$.
  - It generates a NIZK proof $D_M = [\boldsymbol{\Pi}]_1$ that ψ_{CS} is indeed a valid encryption of $M \in \mathbb{G}_1$ for the committed $\boldsymbol{r}$ in $[\boldsymbol{R}]_2$. This requires to prove that these exponents satisfy the multi-exponentiation equations:
$$[\boldsymbol{C}_1]_1 = [\boldsymbol{A}\boldsymbol{r}]_1, [\boldsymbol{C}_2 - M]_1 = [\boldsymbol{h}\boldsymbol{r}]_1, [\boldsymbol{C}_3 = (\boldsymbol{c} + \boldsymbol{d} \odot \mathcal{H}(\ell'))\boldsymbol{r}]_1$$
The associated proof is then $[\boldsymbol{\Pi}]_1 = [\boldsymbol{\rho}^\top(\boldsymbol{A}||\boldsymbol{h}||\boldsymbol{c} + \boldsymbol{d} \odot \mathcal{H}(\ell'))]_1$.
  - P_i erases r after the generation of $[\boldsymbol{R}]_2$ and $[\boldsymbol{\Pi}]_1$ but retains $D_M = [\boldsymbol{\Pi}]_1$.

  The commitment is $([\boldsymbol{C}]_1, [\boldsymbol{R}]_2)$.
- **Verification algorithm:** the algorithm $\mathsf{VerCom}(\mathsf{crs}, M, D_M, \mathsf{sid}, \mathsf{cid}, P_i, P_j)$ checks the consistency of the proof $\pi_{val-enc}$ with respect to $[\boldsymbol{C}]_1$ and $[\boldsymbol{R}]_2$. and ignores the opening if the verification fails.
- **Opening algorithm:** $\mathsf{OpenCom}(\mathsf{crs}, M, D_M, \mathsf{sid}, \mathsf{cid}, P_i, P_j)$ reveals M and $D_M = [\Pi]_1$ to P_j.

One can easily see that $[\boldsymbol{C}_3]_1$ is the projective hash computation of a 2-universal hash proof on the language "$[\boldsymbol{C}_1]_1$ in the span of $\boldsymbol{A}$", with $[\boldsymbol{C}_2]_1$ being an additional term that uses the same witness to mask the committed message, so that $[\boldsymbol{C}]_1$ is a proper generalization of the Cramer-Shoup CCA-2 encryption. Details on the k-MDDH Groth-Sahai proofs are given in the paper full version [15].

It is thus easy to see that this commitment is indeed a generalization of the FLM non-interactive UC commitment with adaptive corruption under reliable erasures (in which we switched the CRS, the Cramer-Shoup encryption and the Groth-Sahai proof in the k-MDDH setting).

4.3 A Structure-Preserving Smooth Projective Hash Function Associated with This Commitment

Structure-Preserving Smooth Projective Hash Function. We now want to supersede the verification equation of the commitment by a smooth projective hash function providing implicit decommitment, simply using the proof as a witness. We consider the language of the valid encryptions of M using a random r which is committed into $[\boldsymbol{R}]_2$:

$$\mathcal{L}_M = \{[\boldsymbol{C}]_1 \mid \exists r \exists \rho \text{ such that } [\boldsymbol{R}]_2 = [\boldsymbol{U}\boldsymbol{\rho} + \iota_2(\boldsymbol{r})]_2 \text{ and } [\boldsymbol{C}]_1 = [\boldsymbol{A}\boldsymbol{r}||\boldsymbol{h}\boldsymbol{r} + M||(\boldsymbol{c} + \boldsymbol{d} \odot \mathcal{H}(\ell||\boldsymbol{C}_1||\boldsymbol{C}_2||\boldsymbol{R}))\boldsymbol{r}]_1\}$$

The verifier picks a random $\mathsf{hk} = \alpha \stackrel{\$}{\leftarrow} \mathbb{Z}_p^{k+3\times k+1}$ and sets $\mathsf{hp} = [\alpha \odot \boldsymbol{U}]_2$. On one side, the verifier then computes:

$$\mathsf{Hash}(\mathsf{hk}, ([\boldsymbol{C}]_1, [\boldsymbol{R}]_2)) = [\alpha \odot ((\boldsymbol{C}_1||\boldsymbol{C}_2 - M||\boldsymbol{C}_3) - (\boldsymbol{A}||\boldsymbol{h}||\boldsymbol{c} + \boldsymbol{d} \odot \mathcal{H}(\ell')) \cdot \boldsymbol{R})]_T$$

While the prover computes $\mathsf{ProjHash}(\mathsf{hp}, \boldsymbol{\Pi}) = [\boldsymbol{\Pi} \cdot \mathsf{hp}]_T$.

- *Correctness*: comes directly from the previous equations.
- *Smoothness*: on a binding CRS, $[\boldsymbol{U}]_2$'s last column is in the span of the k first (which are simply $[\boldsymbol{B}]_2$), hence as $\mathsf{hk} \in \mathbb{Z}_p^{k+1}$, the k equations given in hp are not enough to determine its value and so it is still perfectly hidden from an information theoretic point of view.
- *Pseudo-Randomness*: Under the MDDH assumption, the subset membership decision is a hard problem, as the generalized Cramer-Shoup is IND-CCA-2, and $[\boldsymbol{R}]_2$ is an IND-CPA commitment to $\boldsymbol{r}$.

Theorem 10. *Under the k-MDDH assumption, the above SP-SPHF is strongly pseudo-random on a perfectly hiding CRS.*

For sake of compactness, the proof is postponed to the paper full version [15].

Efficiency. The rough size of a projection key is $k \times (k+3)$ (number of elements in each proof times number of proofs). It should be noted, that for a $\mathsf{CS\text{-}SPHF}$ (in the case of the oblivious transfer), instead of repeating the projection key $k+3$ times (in order to verify each component of the Cramer-Shoup), one can generate a value $\varepsilon \stackrel{\$}{\leftarrow} \mathbb{Z}_p$, an hp for a single equation, and say that for the other component, one simply uses $\mathsf{hp}^{\varepsilon^i}$, as the trick explained in [1].

5 Application: Nearly Optimal Size 1-out-of-m Oblivious Transfer

5.1 Main Idea of the Construction

Our oblivious transfer scheme builds upon that presented by Abdalla *et al.* at Asiacrypt 2013 [1]. In their scheme, the authors use a SPHF-friendly commitment (which is a notion stronger than a UC commitment) along with its associated SPHF in a now classical way to implicitly open the commitment. They claim that the commitment presented in [34] cannot be used in such an application, since it is not "robust", which is a security notion meaning that one cannot produce a commitment and a label that extracts to x' (possibly $x' = \bot$) such that there exists a valid opening data to a different input x, even with oracle access to the extraction oracle (ExtCom) and to fake commitments (using SCom). Indeed, because of the perfectly-hiding setting of Groth-Sahai proofs, for any ciphertext C and for any message x, there exists a proof Π that makes the verification of C on x. However, we show in this section that in spite of this result, such a commitment can indeed be used in a relatively close construction of oblivious transfer scheme. To this aim, we use our construction of structure-preserving SPHF on FLM's commitment, simply using the decommitment value (a Groth-Sahai proof) as the witness, presented in Sect. 4.3.

It should be noted that the commitment used in [1,2] has the major drawback of leaking the bit-length of the committed message. While in application to Oblivious Transfer this is not a major problem, for PAKE this is a way more sensitive issue, as we show in the next section. Moreover, using FLM's commitment is conceptually simpler, since the equivocation only needs to modify the witness, allowing the user to compute honestly its message in the commitment phase, whereas in the original commitments, a specific flow had to be sent during the commitment phase (with a different computation and more witnesses for the SPHF, than in the honest computation of the commitment).

5.2 A Universally Composable Oblivious Transfer with Adaptive Security Based on MDDH

We denote by DB the database of the server containing $t = 2^m$ lines, and j the line requested by the user in an oblivious way. We assume the existence of a Pseudo-Random Generator (PRG) F with input size equal to the plaintext size, and output size equal to the size of the messages in the database and a IND-CPA encryption scheme $\mathcal{E} = (\mathsf{Setup}_{\mathsf{cpa}}, \mathsf{KeyGen}_{\mathsf{cpa}}, \mathsf{Encrypt}_{\mathsf{cpa}}, \mathsf{Decrypt}_{\mathsf{cpa}})$ with plaintext size at least equal to the security parameter. The commitment used is the variant of [34] described above. It is denoted as Com^{ℓ} in the description of the scheme, with ℓ being a label. Note that sid denotes the session identifier, ssid the subsession identifier and cid the commitment identifier and that the combination (sid, cid) is globally unique, as in [34,37].

We present our construction, in Fig. 3, following the global framework presented in [1], for an easier efficiency comparison (we achieve nearly optimality in the sense that it is linear in the number of lines of the database, but with a constant equal to 1 only).

Theorem 11. *The oblivious transfer scheme described in Fig. 3 is UC-secure in the presence of adaptive adversaries, assuming reliable erasures and authenticated channels.*

The proof is given in the paper full version [15] for completeness.

CRS generation:
$\mathsf{crs} \xleftarrow{\$} \mathsf{SetupCom}(1^{\mathfrak{K}})$, $\mathsf{param}_{\mathsf{cpa}} \xleftarrow{\$} \mathsf{Setup}_{\mathsf{cpa}}(1^{\mathfrak{K}})$.

Pre-flow:

1. Server generates a key pair $(\mathsf{pk}, \mathsf{sk}) \xleftarrow{\$} \mathsf{KeyGen}_{\mathsf{cpa}}(\mathsf{param}_{\mathsf{cpa}})$ for $\mathcal{E}$, stores sk and completely erases the random coins used by KeyGen
2. Server sends pk to User

Index query on j:

1. User chooses a random value J, computes $S \leftarrow F(J)$ and encrypts J under pk: $c \xleftarrow{\$} \mathsf{Encrypt}_{\mathsf{cpa}}(\mathsf{pk}, J)$
2. User computes $([\boldsymbol{C}]_1, [\boldsymbol{R}]_2, [\boldsymbol{\Pi}]_1) \xleftarrow{\$} \mathsf{Com}^{\ell}(\mathsf{crs}, j, \mathsf{sid}, \mathsf{cid}, P_i, P_j)$ with the label $\ell = (\mathsf{sid}, \mathsf{ssid}, P_i, P_j)$
3. User stores $[\boldsymbol{\Pi}]_1$ and completely erases J and the random coins used by Com and $\mathsf{Encrypt}_{\mathsf{cpa}}$ and sends $[\boldsymbol{C}]_1, [\boldsymbol{R}]_2$ and c to Server

Database input $(n_1, \ldots, n_t)$:

1. Server decrypts $J \leftarrow \mathsf{Decrypt}_{\mathsf{cpa}}(\mathsf{sk}, c)$ and computes $S \leftarrow F(J)$
2. For $s = 1, \ldots, t$: Server computes $\mathsf{hk}_s \xleftarrow{\$} \mathsf{HashKG}(\mathcal{L}_s)$, $\mathsf{hp}_s \leftarrow \mathsf{ProjKG}(\mathsf{hk}_s, \mathcal{L}_s)$, $K_s \leftarrow \mathsf{Hash}(\mathsf{hk}_s, (\mathcal{L}_s, (\ell, [\boldsymbol{C}]_1, [\boldsymbol{R}]_2)))$, and $N_s \leftarrow S \oplus K_s \oplus n_s$
3. Server erases everything except $(\mathsf{hp}_s, N_s)_{s=1,\ldots,t}$ and sends them over a secure channel

Data recovery:
Upon receiving $(\mathsf{hp}_s, N_s)_{s=1,\ldots,t}$, User computes
$K_j \leftarrow \mathsf{ProjHash}(\mathsf{hp}_j, (\mathcal{L}_j, \ell, [\boldsymbol{C}]_1, [\boldsymbol{R}]_2), [\boldsymbol{\Pi}]_1)$ and gets $n_j \leftarrow S \oplus K_j \oplus N_j$.

Fig. 3. UC-Secure 1-out-of-t OT from an SPHF-Friendly Commitment (for Adaptive Security)

6 Application: Adaptive and Length-Independent One-Round PAKE

Password-authenticated key exchange (PAKE) protocols allow two players to agree on a shared high entropy secret key, that depends on their own passwords only. Katz and Vaikuntanathan recently came up with the first concrete one-round PAKE protocols [43], where the two players just have to send simultaneous flows to each other. Following their idea, [14] proposed a round-optimal PAKE protocol UC secure against passive corruptions. On the other hand, [2] proposed the first protocol UC secure against adaptive corruptions, and [1] built upon both [43] and [2], to propose the first one-round protocol UC secure against adaptive corruptions. Unfortunately, both of them share a drawback, which is that they use a commitment growing linearly with the length of a password. Besides being an efficiency problem, it is over all a security issue in the UC framework. Indeed, the simulator somehow has to "guess" the length of the password of the player it simulates, otherwise it is unable to equivocate the commitment (since the commitment reveals the length of the password it commits to). Since such a guess is impossible, the apparently only solution to get rid of this limitation seems to give the users an upper-bound on the length of their passwords and to ask them to compute commitments of this length, which leads to costly computations.

In this section, we are now going to present a constant-size, round-optimal, PAKE UC secure against adaptive corruptions. It builds upon the protocol proposed in [1], using the same techniques as in the former section to avoid the apparent impossibility to use FLM's commitment.

It should be noted that we need the classical requirement for extraction capabilities (see for example [16,47] for a detailed explanation), *i.e.* a password pw is assumed to be a bit-string of length bounded by $\log p - 2$, and then one can use a bijective embedding function G mapping $\{0,1\}^{|p|-2}$ in $\mathbb{G}_1$. For the sake of simplicity, we continue to write pw_i in the high level description, but it should be interpreted as a commitment to $G(\mathsf{pw}_i)$.

The language $\mathcal{L}_{\mathsf{pw}_i}$ is then the language of valid Cramer-Shoup encryptions of the embedded password $G(\mathsf{pw}_i)$, consistent with the randomness committed in the second part, and the rest of the label.

Theorem 12. *The Password Authenticated Key Exchange scheme described in Fig. 4 is UC-secure in the presence of adaptive adversaries, assuming reliable erasures and authenticated channels.*

The proof is given in the paper full version [15] for completeness.

CRS: $\mathsf{crs} \overset{\$}{\leftarrow} \mathsf{SetupCom}(1^{\mathfrak{K}})$.
Protocol execution by P_i with pw_i:

1. P_i generates $\mathsf{hk}_i \overset{\$}{\leftarrow} \mathsf{HashKG}(\mathcal{L}_{\mathsf{pw}_i})$, $\mathsf{hp}_i \leftarrow \mathsf{ProjKG}(\mathsf{hk}_i, \mathcal{L}_{\mathsf{pw}_i})$ and erases any random coins used for the generation
2. P_i computes $([\boldsymbol{C}_i]_1, [\boldsymbol{R}_i]_2, [\boldsymbol{\Pi}_i]_1) = \mathsf{Com}^{\ell_i}(\mathsf{crs}, \mathsf{pw}_i, \mathsf{sid}, \mathsf{cid}, P_i, P_j)$ with $\ell_i = (\mathsf{sid}, P_i, P_j, \mathsf{hp}_i)$
3. P_i stores $[\boldsymbol{\Pi}_i]_1$, completely erases random coins used by Com and sends $\mathsf{hp}_i, [\boldsymbol{C}_i]_1, [\boldsymbol{R}_i]_2$ to P_j

Key computation: Upon receiving $\mathsf{hp}_j, [\boldsymbol{C}_j]_1, [\boldsymbol{R}_j]_2$ from P_j

1. P_i computes $H'_i \leftarrow \mathsf{ProjHash}(\mathsf{hp}_j, (\mathcal{L}_{\mathsf{pw}_i}, \ell_i, [\boldsymbol{C}_i]_1, [\boldsymbol{R}_i]_2), [\Pi_i]_1))$ and $H_j \leftarrow \mathsf{Hash}(\mathsf{hk}_i, (\mathcal{L}_{\mathsf{pw}_i}, \ell_j, [\boldsymbol{C}_j]_1, [\boldsymbol{R}_j]_2))$ with $\ell_j = (\mathsf{sid}, P_j, P_i, \mathsf{hp}_j)$
2. P_i computes $\mathsf{sk}_i = H'_i \cdot H_j$ and erases everything else, except pw_i.

Fig. 4. UC-Secure PAKE from the revisited FLM Commitment

7 Application: Anonymous Credential-Based Message Transmission

Anonymous Credential protocols [21,27,31] allow to combine security and privacy. Typical credential use involves three main parties. Users need to interact with some authorities to obtain their credentials (assumed to be a set of attributes validated / signed), and then prove to a server that a subpart of their attributes verifies an expect policy.

In this section, we give another go to Anonymous Credential, this time to allow message recovery. This is between Anonymous Credential but also Conditional Oblivious Transfer [51] and Oblivious Signature-Based Envelope [46].

We present a constant-size, round-optimal protocol that allow to use a Credential to retrieve a message without revealing the Anonymous Credentials in a UC secure way, by simply building on the commitment proposed earlier in the paper.

7.1 Anonymous Credential System

In a Attribute-Based Credential system, we assume that different organization issue credentials to users. A user i possesses a set of credential Cred_i of the form $\{\mathsf{Cred}_{i,j}, \mathsf{vk}_j\}$ where organization j assesses that the user verifies some property. (The DMV will assess that the user is indeed capable of driving, the university that she has a bachelor in Computer Science, while Squirrel Airways that she reached the gold membership, all those authorities don't communicate with each other).

A Server might have an access Policy P requiring some elements (For example being a female, with a bachelor, and capable of driving).

- $\mathsf{Setup}(1^{\mathfrak{K}})$: A probabilistic algorithm that gets a security parameter $\mathfrak{K}$, an upper bound t for the size of attribute sets and returns the public parameters param
- $\mathsf{OKeyGen}(\mathsf{param})$: Generates a pair of signing keys $\mathsf{sk}_j, \mathsf{vk}_j$ for each organization.
- $\mathsf{UKeyGen}(\mathsf{param})$: Generates a pair of keys $\mathsf{sk}_i, \mathsf{vk}_i$ for each use.
- $\mathsf{CredObtain}(\langle U_i, \mathsf{sk}_i \rangle, \langle O_j, \mathsf{sk}_j \rangle)$ Interactive process that allows a user i to obtain some credentials from organization j by providing his public key vk_j and a proof that it belongs to him.
- $\mathsf{CredUse}(\langle U_i, \mathsf{Cred}_i, \mathsf{sk}_i \rangle, \langle S, P, M \rangle)$ Interactive process that allows a user i to access a message guarded by the server S under some policy P by using the already obtained credentials.

An attribute-based anonymous credential system is called secure if it is correct, unforgeable and anonymous.

CRS generation:
$\mathsf{crs} \stackrel{\$}{\leftarrow} \mathsf{SetupCom}(1^{\mathfrak{K}})$, $\mathsf{param}_{\mathsf{cpa}} \stackrel{\$}{\leftarrow} \mathsf{Setup}_{\mathsf{cpa}}(1^{\mathfrak{K}})$.

Pre-flow:

1. Server generates a key pair $(\mathsf{pk}, \mathsf{sk}) \stackrel{\$}{\leftarrow} \mathsf{KeyGen}_{\mathsf{cpa}}(\mathsf{param}_{\mathsf{cpa}})$ for $\mathcal{E}$, stores sk and completely erases the random coins used by KeyGen
2. Server sends pk to User

Credential Use by user i:

1. User chooses a random value J, computes $S \leftarrow F(J)$ and encrypts J under pk: $c \stackrel{\$}{\leftarrow} \mathsf{Encrypt}_{\mathsf{cpa}}(\mathsf{pk}, J)$
2. User computes $([\boldsymbol{C}]_1, [\boldsymbol{R}]_2, [\boldsymbol{\Pi}]_1) \stackrel{\$}{\leftarrow} \mathsf{Com}^{\ell}(\mathsf{crs}, \mathsf{Cred}_i, \mathsf{sid}, \mathsf{cid}, P_i, P_j)$ with $\ell = (\mathsf{sid}, \mathsf{ssid}, P_i, P_j)$
3. User stores $[\boldsymbol{\Pi}]_1$ and completely erases J and the random coins used by Com and $\mathsf{Encrypt}_{\mathsf{cpa}}$ and sends $[\boldsymbol{C}]_1, [\boldsymbol{R}]_2$ and c to Server

Database input M with policy P:

1. Server decrypts $J \leftarrow \mathsf{Decrypt}_{\mathsf{cpa}}(\mathsf{sk}, c)$ and computes $S \leftarrow F(J)$
2. Server computes $\mathsf{hk}_P \stackrel{\$}{\leftarrow} \mathsf{HashKG}(\mathcal{L}_P)$, $\mathsf{hp}_P \leftarrow \mathsf{ProjKG}(\mathsf{hk}_P, \mathcal{L}_P)$, $K_P \leftarrow \mathsf{Hash}(\mathsf{hk}_P, (\mathcal{L}_P, (\ell, [\boldsymbol{C}]_1, [\boldsymbol{R}]_2)))$, and $N_P \leftarrow S \oplus K_P \oplus M$
3. Server erases everything except (hp_P, N_P) and sends them over a secure channel

Data recovery:
Upon receiving (hp_P, N_P), User computes
$K \leftarrow \mathsf{ProjHash}(\mathsf{hp}_P, (\mathcal{L}_P, \ell, [\boldsymbol{C}]_1, [\boldsymbol{R}]_2), [\boldsymbol{\Pi}]_1)$ and gets $M \leftarrow S \oplus K \oplus N_P$.

Fig. 5. UC-Secure Anonymous Credential from an SPSPHF-Friendly Commitment (for Adaptive Security).

7.2 Construction

Smooth Projective Hash Functions have been shown to handle complex languages [2,13], those properties can naturally be extended to Structure Preserving Smooth Projective Hash Function, allowing credentials to be expressive as disjunction / conjunction of sets of credentials, range proofs, or even composition (having a credential from authority A signed by authority B for example).

What is really new with the Structure Preserving part is that now a user can request to have a credential on a witness by requiring a Structure-Preserving signature on it, while before scalars either required to give too much information to the server B or prevented chaining as most signatures requires some sort of Hashing (BLS requires an explicit Hash, while signature *à la* Waters requires to handle a bit per bit version of the message hindering drastically the efficiency of the protocol). This allows more possibilities in both the Credential Generation step and the policy required for accessing messages, while maintaining an efficient construction.

Theorem 13. *The Anonymous Credential Protocol described in Fig. 5 is UC-secure in the presence of adaptive adversaries, assuming reliable erasures and authenticated channels.*

The ideal functionality and a sketch of the proof are given in the paper full version [15] for completeness.

A Commitments and Smooth Projective Hash Functions

A.1 Encryption

An encryption scheme $\mathcal{C}$ is described through four algorithms (Setup, KeyGen, Encrypt, Decrypt):

- Setup($1^{\mathfrak{K}}$), where $\mathfrak{K}$ is the security parameter, generates the global parameters param of the scheme;
- KeyGen(param) outputs a pair of keys, a (public) encryption key pk and a (private) decryption key dk;
- Encrypt(ek, M; ρ) outputs a ciphertext $\mathcal{C}$, on M, under the encryption key pk, with the randomness ρ;
- Decrypt(dk, $\mathcal{C}$) outputs the plaintext M, encrypted in the ciphertext $\mathcal{C}$ or $\perp$.

Such encryption scheme is required to have the following security properties:

- *Correctness*: For every pair of keys (ek, dk) generated by KeyGen, every messages M, and every random ρ, we should have

$$\mathsf{Decrypt}(\mathsf{dk}, \mathsf{Encrypt}(\mathsf{ek}, M; \rho)) = M.$$

- *Indistinguishability under Adaptive Chosen Ciphertext Attack* IND-CCA (see [49,52]): An adversary should not be able to efficiently guess which message has been encrypted even if he chooses the two original plaintexts, and ask several decryption of ciphertexts different from challenge one.
The ODecrypt oracle outputs the decryption of c under the challenge decryption key dk. The input queries (c) are added to the list $\mathcal{CT}$ of decrypted ciphertexts.

$\mathsf{Exp}^{\mathsf{ind\text{-}cca}-b}_{\mathcal{E},\mathcal{A}}(\mathfrak{K})$
1. param $\leftarrow$ Setup($1^{\mathfrak{K}}$)
2. (pk, dk) $\leftarrow$ KeyGen(param)
3. $(M_0, M_1) \leftarrow \mathcal{A}$(FIND : pk, ODecrypt($\cdot$))
4. $c^* \leftarrow$ Encrypt(ek, M_b)
5. $b' \leftarrow \mathcal{A}$(GUESS : c^*, ODecrypt($\cdot$))
6. IF $(c^*) \in \mathcal{CT}$ RETURN 0
7. ELSE RETURN b'

A.2 Commitments

A commitment scheme is said *equivocable* if it has a second setup $\mathsf{SetupComT}(1^{\mathfrak{K}})$ that additionally outputs a trapdoor τ, and two algorithms

- $\mathsf{SimCom}^{\ell}(\tau)$ takes as input the trapdoor τ and a label ℓ and outputs a pair (C, eqk), where C is a commitment and eqk an equivocation key;
- $\mathsf{OpenCom}^{\ell}(\mathsf{eqk}, C, x)$ takes as input a commitment C, a label ℓ, a message x, an equivocation key eqk, and outputs an opening data δ for C and ℓ on x.

such as the following properties are satisfied: *trapdoor correctness* (all simulated commitments can be opened on any message), *setup indistinguishability* (one cannot distinguish the CRS ρ generated by $\mathsf{SetupCom}$ from the one generated by $\mathsf{SetupComT}$) and *simulation indistinguishability* (one cannot distinguish a real commitment (generated by Com) from a fake commitment (generated by SCom), even with oracle access to fake commitments), denoting by SCom the algorithm that takes as input the trapdoor τ, a label ℓ and a message x and which outputs $(C, \delta) \stackrel{\$}{\leftarrow} \mathsf{SCom}^{\ell}(\tau, x)$, computed as $(C, \mathsf{eqk}) \stackrel{\$}{\leftarrow} \mathsf{SimCom}^{\ell}(\tau)$ and $\delta \leftarrow \mathsf{OpenCom}^{\ell}(\mathsf{eqk}, C, x)$.

A commitment scheme $\mathcal{C}$ is said to be *extractable* if it has a second setup $\mathsf{SetupComT}(1^{\mathfrak{K}})$ that additionally outputs a trapdoor τ, and a new algorithm

- $\mathsf{ExtCom}^{\ell}(\tau, C)$ which takes as input the trapdoor τ, a commitment C, and a label ℓ, and outputs the committed message x, or $\perp$ if the commitment is invalid.

such as the following properties are satisfied: *trapdoor correctness* (all commitments honestly generated can be correctly extracted: for all ℓ, x, if $(C, \delta) \stackrel{\$}{\leftarrow} \mathsf{Com}^{\ell}(x)$ then $\mathsf{ExtCom}^{\ell}(C, \tau) = x$), *setup indistinguishability* (as above) and *binding extractability* (one cannot fool the extractor, *i.e.*, produce a commitment and a valid opening data to an input x while the commitment does not extract to x).

A commitment scheme is said *extractable and equivocable* if the indistinguishable setup algorithm outputs a common trapdoor that allows both equivocability and extractability, and the following properties are satisfied: *strong simulation indistinguishability* (one cannot distinguish a real commitment (generated by Com) from a fake commitment (generated by SCom), even with oracle access to the extraction oracle (ExtCom) and to fake commitments (using SCom)) and *strong binding extractability* (one cannot fool the extractor, *i.e.*, produce a commitment and a valid opening data (not given by SCom) to an input x while the commitment does not extract to x, even with oracle access to the extraction oracle (ExtCom) and to fake commitments (using SCom)).

A.3 Smooth Projective Hash Functions Used with Commitments

The strong pseudo-randomness property, from [14], is defined by the experiment $\mathsf{Exp}_{\mathcal{A}}^{\texttt{c-s-ps-rand}}(\mathfrak{K})$ depicted in Fig. 6. It is a strong version of the pseudo-randomness where the adversary is also given the hash value of a commitment

$\mathsf{Exp}_{\mathcal{A}}^{\texttt{c-s-ps-rand-}b}(\mathfrak{K})$
$(\rho, \tau) \stackrel{\$}{\leftarrow} \mathsf{SetupComT}(1^{\mathfrak{K}})$
$(\ell, x, \mathsf{state}) \stackrel{\$}{\leftarrow} \mathcal{A}^{\mathsf{SCom}^{\cdot}(\tau,\cdot),\mathsf{ExtCom}^{\cdot}(\tau,\cdot)}(\rho)$; $C \stackrel{\$}{\leftarrow} \mathsf{SimCom}^{\ell}(\tau)$
$\mathsf{hk} \stackrel{\$}{\leftarrow} \mathsf{HashKG}(L_x)$; $\mathsf{hp} \leftarrow \mathsf{ProjKG}(\mathsf{hk}, L_x, \bot)$
If $(b = 0)$ $H \leftarrow \mathsf{Hash}(\mathsf{hk}, L_x, (\ell, C))$
Else $H \stackrel{\$}{\leftarrow} \Pi$
$(\ell', C', \mathsf{state}) \stackrel{\$}{\leftarrow} \mathcal{A}^{\mathsf{SCom}^{\cdot}(\tau,\cdot),\mathsf{ExtCom}^{\cdot}(\tau,\cdot)}(\mathsf{state}, C, \mathsf{hp}, H)$
If $((\ell', ?, C') \in \Lambda)$ THEN $H' \leftarrow \bot$
Else $H' \leftarrow \mathsf{Hash}(\mathsf{hk}, L_x, (\ell', C')$
Return $\mathcal{A}^{\mathsf{SCom}^{\cdot}(\tau,\cdot),\mathsf{ExtCom}^{\cdot}(\tau,\cdot)}(H')$

Fig. 6. Strong Pseudo-Randomness

of its choice (obviously not generated by SCom or SimCom though, hence the test with Λ which also contains (C, ℓ, x)). This property only makes sense when the projection key does not depend on the word C to be hashed. It thus applies to KV-SPHF, and CS-SPHF only.

References

1. Abdalla, M., Benhamouda, F., Blazy, O., Chevalier, C., Pointcheval, D.: SPHF-friendly non-interactive commitments. In: Sako, K., Sarkar, P. (eds.) ASIACRYPT 2013. LNCS, vol. 8269, pp. 214–234. Springer, Heidelberg (2013). doi:10.1007/978-3-642-42033-7_12
2. Abdalla, M., Chevalier, C., Pointcheval, D.: Smooth projective hashing for conditionally extractable commitments. In: Halevi, S. (ed.) CRYPTO 2009. LNCS, vol. 5677, pp. 671–689. Springer, Heidelberg (2009). doi:10.1007/978-3-642-03356-8_39
3. Abe, M., Chase, M., David, B., Kohlweiss, M., Nishimaki, R., Ohkubo, M.: Constant-size structure-preserving signatures: generic constructions and simple assumptions. In: Wang, X., Sako, K. (eds.) ASIACRYPT 2012. LNCS, vol. 7658, pp. 4–24. Springer, Heidelberg (2012). doi:10.1007/978-3-642-34961-4_3
4. Abe, M., David, B., Kohlweiss, M., Nishimaki, R., Ohkubo, M.: Tagged one-time signatures: tight security and optimal tag size. In: Kurosawa, K., Hanaoka, G. (eds.) PKC 2013. LNCS, vol. 7778, pp. 312–331. Springer, Heidelberg (2013). doi:10.1007/978-3-642-36362-7_20
5. Abe, M., Fuchsbauer, G., Groth, J., Haralambiev, K., Ohkubo, M.: Structure-preserving signatures and commitments to group elements. In: Rabin, T. (ed.) CRYPTO 2010. LNCS, vol. 6223, pp. 209–236. Springer, Heidelberg (2010). doi:10.1007/978-3-642-14623-7_12
6. Abe, M., Groth, J., Haralambiev, K., Ohkubo, M.: Optimal structure-preserving signatures in asymmetric bilinear groups. In: Rogaway, P. (ed.) CRYPTO 2011. LNCS, vol. 6841, pp. 649–666. Springer, Heidelberg (2011). doi:10.1007/978-3-642-22792-9_37
7. Abe, M., Groth, J., Ohkubo, M., Tibouchi, M.: Structure-preserving signatures from type II pairings. In: Garay, J.A., Gennaro, R. (eds.) CRYPTO 2014. LNCS, vol. 8616, pp. 390–407. Springer, Heidelberg (2014). doi:10.1007/978-3-662-44371-2_22

8. Abe, M., Groth, J., Ohkubo, M., Tibouchi, M.: Unified, minimal and selectively randomizable structure-preserving signatures. In: Lindell, Y. (ed.) TCC 2014. LNCS, vol. 8349, pp. 688–712. Springer, Heidelberg (2014). doi:10.1007/978-3-642-54242-8_29
9. Ateniese, G., Camenisch, J., Hohenberger, S., de Medeiros, B.: Practical group signatures without random oracles. Cryptology ePrint Archive, Report 2005/385 (2005). http://eprint.iacr.org/2005/385
10. Barak, B., Canetti, R., Lindell, Y., Pass, R., Rabin, T.: Secure computation without authentication. In: Shoup, V. (ed.) CRYPTO 2005. LNCS, vol. 3621, pp. 361–377. Springer, Heidelberg (2005). doi:10.1007/11535218_22
11. Bellare, M., Pointcheval, D., Rogaway, P.: Authenticated key exchange secure against dictionary attacks. In: Preneel, B. (ed.) EUROCRYPT 2000. LNCS, vol. 1807, pp. 139–155. Springer, Heidelberg (2000). doi:10.1007/3-540-45539-6_11
12. Bellovin, S.M., Merritt, M.: Encrypted key exchange: Password-based protocols secure against dictionary attacks. In: 1992 IEEE Symposium on Security and Privacy, pp. 72–84. IEEE Computer Society Press, May 1992
13. Ben Hamouda, F., Blazy, O., Chevalier, C., Pointcheval, D., Vergnaud, D.: Efficient UC-secure authenticated key-exchange for algebraic languages. In: Kurosawa, K., Hanaoka, G. (eds.) PKC 2013. LNCS, vol. 7778, pp. 272–291. Springer, Heidelberg (2013). doi:10.1007/978-3-642-36362-7_18
14. Benhamouda, F., Blazy, O., Chevalier, C., Pointcheval, D., Vergnaud, D.: New techniques for SPHFs and efficient one-round PAKE protocols. In: Canetti, R., Garay, J.A. (eds.) CRYPTO 2013. LNCS, vol. 8042, pp. 449–475. Springer, Heidelberg (2013). doi:10.1007/978-3-642-40041-4_25
15. Blazy, O., Chevalier, C.: Structure-preserving smooth projective hashing. Cryptology ePrint Archive, Report 2016/258 (2016). http://eprint.iacr.org/2016/258
16. Blazy, O., Chevalier, C., Pointcheval, D., Vergnaud, D.: Analysis and improvement of Lindell's UC-secure commitment schemes. In: Jacobson, M., Locasto, M., Mohassel, P., Safavi-Naini, R. (eds.) ACNS 2013. LNCS, vol. 7954, pp. 534–551. Springer, Heidelberg (2013). doi:10.1007/978-3-642-38980-1_34
17. Blazy, O., Chevalier, C.: Generic construction of uc-secure oblivious transfer. Cryptology ePrint Archive, Report 2015/560 (2015)
18. Blazy, O., Chevalier, C., Vergnaud, D.: Non-interactive zero-knowledge proofs of non-membership. Cryptology ePrint Archive, Report 2015/072 (2015). http://eprint.iacr.org/
19. Blazy, O., Pointcheval, D., Vergnaud, D.: Round-optimal privacy-preserving protocols with smooth projective hash functions. In: Cramer, R. (ed.) TCC 2012. LNCS, vol. 7194, pp. 94–111. Springer, Heidelberg (2012). doi:10.1007/978-3-642-28914-9_6
20. Boneh, D., Boyen, X., Shacham, H.: Short group signatures. In: Franklin, M. (ed.) CRYPTO 2004. LNCS, vol. 3152, pp. 41–55. Springer, Heidelberg (2004). doi:10.1007/978-3-540-28628-8_3
21. Camenisch, J., Lysyanskaya, A.: An efficient system for non-transferable anonymous credentials with optional anonymity revocation. In: Pfitzmann, B. (ed.) EUROCRYPT 2001. LNCS, vol. 2045, pp. 93–118. Springer, Heidelberg (2001). doi:10.1007/3-540-44987-6_7
22. Canetti, R.: Universally composable security: A new paradigm for cryptographic protocols. In: 42nd FOCS, pp. 136–145. IEEE Computer Society Press, October 2001

23. Canetti, R., Fischlin, M.: Universally composable commitments. In: Kilian, J. (ed.) CRYPTO 2001. LNCS, vol. 2139, pp. 19–40. Springer, Heidelberg (2001). doi:10.1007/3-540-44647-8_2
24. Canetti, R., Halevi, S., Katz, J., Lindell, Y., MacKenzie, P.: Universally composable password-based key exchange. In: Cramer, R. (ed.) EUROCRYPT 2005. LNCS, vol. 3494, pp. 404–421. Springer, Heidelberg (2005). doi:10.1007/11426639_24
25. Canetti, R., Krawczyk, H.: Universally composable notions of key exchange and secure channels. In: Knudsen, L.R. (ed.) EUROCRYPT 2002. LNCS, vol. 2332, pp. 337–351. Springer, Heidelberg (2002). doi:10.1007/3-540-46035-7_22
26. Canetti, R., Lindell, Y., Ostrovsky, R., Sahai, A.: Universally composable two-party and multi-party secure computation. In: 34th ACM STOC, pp. 494–503. ACM Press, May 2002
27. Chaum, D.: Showing credentials without identification. In: Pichler, F. (ed.) EUROCRYPT 1985. LNCS, vol. 219, pp. 241–244. Springer, Heidelberg (1986). doi:10.1007/3-540-39805-8_28
28. Choi, S.G., Katz, J., Wee, H., Zhou, H.-S.: Efficient, adaptively secure, and composable oblivious transfer with a single, global CRS. In: Kurosawa, K., Hanaoka, G. (eds.) PKC 2013. LNCS, vol. 7778, pp. 73–88. Springer, Heidelberg (2013). doi:10.1007/978-3-642-36362-7_6
29. Cramer, R., Shoup, V.: A practical public key cryptosystem provably secure against adaptive chosen ciphertext attack. In: Krawczyk, H. (ed.) CRYPTO 1998. LNCS, vol. 1462, pp. 13–25. Springer, Heidelberg (1998). doi:10.1007/BFb0055717
30. Cramer, R., Shoup, V.: Universal hash proofs and a paradigm for adaptive chosen ciphertext secure public-key encryption. In: Knudsen, L.R. (ed.) EUROCRYPT 2002. LNCS, vol. 2332, pp. 45–64. Springer, Heidelberg (2002). doi:10.1007/3-540-46035-7_4
31. Damgård, I.B.: Payment systems and credential mechanisms with provable security against abuse by individuals. In: Goldwasser, S. (ed.) CRYPTO 1988. LNCS, vol. 403, pp. 328–335. Springer, Heidelberg (1990). doi:10.1007/0-387-34799-2_26
32. ElGamal, T.: A public key cryptosystem and a signature scheme based on discrete logarithms. In: Blakley, G.R., Chaum, D. (eds.) CRYPTO 1984. LNCS, vol. 196, pp. 10–18. Springer, Heidelberg (1985). doi:10.1007/3-540-39568-7_2
33. Escala, A., Herold, G., Kiltz, E., Ràfols, C., Villar, J.: An algebraic framework for diffie-hellman assumptions. In: Canetti, R., Garay, J.A. (eds.) CRYPTO 2013. LNCS, vol. 8043, pp. 129–147. Springer, Heidelberg (2013). doi:10.1007/978-3-642-40084-1_8
34. Fischlin, M., Libert, B., Manulis, M.: Non-interactive and re-usable universally composable string commitments with adaptive security. In: Lee, D.H., Wang, X. (eds.) ASIACRYPT 2011. LNCS, vol. 7073, pp. 468–485. Springer, Heidelberg (2011). doi:10.1007/978-3-642-25385-0_25
35. Gennaro, R., Lindell, Y.: A framework for password-based authenticated key exchange. In: Biham, E. (ed.) EUROCRYPT 2003. LNCS, vol. 2656, pp. 524–543. Springer, Heidelberg (2003). doi:10.1007/3-540-39200-9_33
36. Groth, J., Sahai, A.: Efficient non-interactive proof systems for bilinear groups. In: Smart, N. (ed.) EUROCRYPT 2008. LNCS, vol. 4965, pp. 415–432. Springer, Heidelberg (2008). doi:10.1007/978-3-540-78967-3_24
37. Hofheinz, D., Müller-Quade, J.: Universally composable commitments using random oracles. In: Naor, M. (ed.) TCC 2004. LNCS, vol. 2951, pp. 58–76. Springer, Heidelberg (2004). doi:10.1007/978-3-540-24638-1_4

38. Horvitz, O., Katz, J.: Universally-composable two-party computation in two rounds. In: Menezes, A. (ed.) CRYPTO 2007. LNCS, vol. 4622, pp. 111–129. Springer, Heidelberg (2007). doi:10.1007/978-3-540-74143-5_7
39. Jutla, C., Roy, A.: Smooth nizk arguments with applications to asymmetric uc-pake. Cryptology ePrint Archive, Report 2016/233 (2016). http://eprint.iacr.org/
40. Jutla, C.S., Roy, A.: Dual-system simulation-soundness with applications to uc-pake and more. Cryptology ePrint Archive, Report 2014/805 (2014)
41. Kalai, Y.T.: Smooth projective hashing and two-message oblivious transfer. In: Cramer, R. (ed.) EUROCRYPT 2005. LNCS, vol. 3494, pp. 78–95. Springer, Heidelberg (2005). doi:10.1007/11426639_5
42. Katz, J., Ostrovsky, R., Yung, M.: Efficient password-authenticated key exchange using human-memorable passwords. In: Pfitzmann, B. (ed.) EUROCRYPT 2001. LNCS, vol. 2045, pp. 475–494. Springer, Heidelberg (2001). doi:10.1007/3-540-44987-6_29
43. Katz, J., Vaikuntanathan, V.: Smooth projective hashing and password-based authenticated key exchange from lattices. In: Matsui, M. (ed.) ASIACRYPT 2009. LNCS, vol. 5912, pp. 636–652. Springer, Heidelberg (2009). doi:10.1007/978-3-642-10366-7_37
44. Katz, J., Vaikuntanathan, V.: Round-optimal password-based authenticated key exchange. In: Ishai, Y. (ed.) TCC 2011. LNCS, vol. 6597, pp. 293–310. Springer, Heidelberg (2011). doi:10.1007/978-3-642-19571-6_18
45. Kiltz, E., Pan, J., Wee, H.: Structure-preserving signatures from standard assumptions, revisited. In: Gennaro, R., Robshaw, M. (eds.) CRYPTO 2015. LNCS, vol. 9216, pp. 275–295. Springer, Heidelberg (2015). doi:10.1007/978-3-662-48000-7_14
46. Li, N., Du, W., Boneh, D.: Oblivious signature-based envelope. In: Borowsky, E., Rajsbaum, S. (eds.) 22nd ACM PODC, pp. 182–189. ACM, Jul 2003
47. Lindell, Y.: Highly-efficient universally-composable commitments based on the DDH assumption. In: Paterson, K.G. (ed.) EUROCRYPT 2011. LNCS, vol. 6632, pp. 446–466. Springer, Heidelberg (2011). doi:10.1007/978-3-642-20465-4_25
48. Naor, M., Pinkas, B.: Efficient oblivious transfer protocols. In: Kosaraju, S.R. (ed.) 12th SODA, pp. 448–457. ACM-SIAM, January 2001
49. Naor, M., Yung, M.: Public-key cryptosystems provably secure against chosen ciphertext attacks. In: 22nd ACM STOC, pp. 427–437. ACM Press, May 1990
50. Peikert, C., Vaikuntanathan, V., Waters, B.: A framework for efficient and composable oblivious transfer. In: Wagner, D. (ed.) CRYPTO 2008. LNCS, vol. 5157, pp. 554–571. Springer, Heidelberg (2008). doi:10.1007/978-3-540-85174-5_31
51. Rabin, M.O.: How to exchange secrets with oblivious transfer. Technical Report TR81, Harvard University (1981)
52. Rackoff, C., Simon, D.R.: Non-interactive zero-knowledge proof of knowledge and chosen ciphertext attack. In: Feigenbaum, J. (ed.) CRYPTO 1991. LNCS, vol. 576, pp. 433–444. Springer, Heidelberg (1992). doi:10.1007/3-540-46766-1_35

Digital Signature

Signature Schemes with Efficient Protocols and Dynamic Group Signatures from Lattice Assumptions

Benoît Libert[1(✉)], San Ling[2], Fabrice Mouhartem[1], Khoa Nguyen[2], and Huaxiong Wang[2]

[1] École Normale Supérieure de Lyon, Laboratoire LIP, Lyon, France
benoit.libert@ens-lyon.fr

[2] School of Physical and Mathematical Sciences, Nanyang Technological University, Singapore, Singapore

Abstract. A recent line of works – initiated by Gordon, Katz and Vaikuntanathan (Asiacrypt 2010) – gave lattice-based constructions allowing users to authenticate while remaining hidden in a crowd. Despite five years of efforts, known constructions are still limited to static sets of users, which cannot be dynamically updated. This work provides new tools enabling the design of anonymous authentication systems whereby new users can join the system at any time.

Our first contribution is a signature scheme with efficient protocols, which allows users to obtain a signature on a committed value and subsequently prove knowledge of a signature on a committed message. This construction is well-suited to the design of anonymous credentials and group signatures. It indeed provides the first lattice-based group signature supporting dynamically growing populations of users.

As a critical component of our group signature, we provide a simple joining mechanism of introducing new group members using our signature scheme. This technique is combined with zero-knowledge arguments allowing registered group members to prove knowledge of a secret short vector of which the corresponding public syndrome was certified by the group manager. These tools provide similar advantages to those of structure-preserving signatures in the realm of bilinear groups. Namely, they allow group members to generate their own public key without having to prove knowledge of the underlying secret key. This results in a two-message joining protocol supporting concurrent enrollments, which can be used in other settings such as group encryption.

Our zero-knowledge arguments are presented in a unified framework where: (i) The involved statements reduce to arguing possession of a $\{-1, 0, 1\}$-vector $\mathbf{x}$ with a particular structure and satisfying $\mathbf{P} \cdot \mathbf{x} = \mathbf{v} \bmod q$ for some public matrix $\mathbf{P}$ and vector $\mathbf{v}$; (ii) The reduced statements can be handled using permuting techniques for Stern-like protocols. Our framework can serve as a blueprint for proving many other relations in lattice-based cryptography.

Keywords: Lattice-based cryptography · Anonymity · Signatures with efficient protocols · Dynamic group signatures · Anonymous credentials

J.H. Cheon and T. Takagi (Eds.): ASIACRYPT 2016, Part II, LNCS 10032, pp. 373–403, 2016.
DOI: 10.1007/978-3-662-53890-6_13

1 Introduction

Lattice-based cryptography is currently emerging as a promising alternative to traditional public-key techniques. During the last decade, it has received a permanent interest due to its numerous advantages. Not only does it seemingly resist quantum attacks, it also provides a better asymptotic efficiency than its relatives based on conventional number theory. While enabling many advanced functionalities [41,44,45], lattice-based primitives tend to interact with zero-knowledge proofs [43] less smoothly than their counterparts in abelian groups endowed with a bilinear map (see, e.g., [2,18,31,38,49]) or groups of hidden order [6,26,29,30]. Arguably, this partially arises from the fact that lattices have far less algebraic structure than, e.g., pairing-friendly cyclic groups. It is not surprising that the most efficient zero-knowledge proofs for lattice-related languages [15] take advantage of the extra algebraic structure available in the ring setting [64]. A consequence of the scarcity of truly efficient zero-knowledge proofs in the lattice setting is that, in the context of anonymity and privacy-preserving protocols, lattice-based cryptography has undergone significantly slower development than in other areas like functional encryption [44,45]. While natural realizations of ring signatures [70] showed up promptly [22,52] after the seminal work of Gentry, Peikert and Vaikuntanathan (GPV) [42], viable constructions of lattice-based group signatures remained lacking until the work of Gordon, Katz and Vaikuntanathan [46] in 2010. Despite recent advances [14,57,62,66], privacy-preserving primitives remain substantially less practical and powerful in terms of functionalities than their siblings based on traditional number theoretic problems [6,18,38,55] for which solutions even exist outside the random oracle model [10,20,21,48]. For example, we still have no convenient realization of group signature supporting dynamic groups [13,55] or anonymous credentials [28,34].

In this paper, we address the latter two problems by first proposing a lattice-based signature with efficient protocols in the fashion of Camenisch and Lysyanskaya [30]. To ease its use in the design of dynamic group signatures, we introduce a zero-knowledge argument system that allows a user to prove knowledge of a signature on a public key for which the user knows the underlying secret key.

Related Work. Anonymous credentials were first suggested by Chaum [34] and efficiently realized by Camenisch and Lysyanskaya [28,30]. They involve one or more credential issuer(s) and a set of users who have a long-term secret key which constitutes their digital identity and pseudonyms that can be seen as commitments to their secret key. Users can dynamically obtain credentials from an issuer that only knows users' pseudonyms and obliviously certifies users' secret keys as well as (optionally) a set of attributes. Later on, users can make themselves known to verifiers under a different pseudonym and demonstrate possession of the issuer's signature on their secret key without revealing neither the signature nor the key. Anonymous credentials typically consist of a protocol whereby the user obtains the issuer's signature on a committed message, another protocol for proving that two commitments open to the same value (which allows proving that the same secret underlies two distinct pseudonyms) and a protocol for proving possession of a secret message-signature pair.

The first efficient constructions were given by Camenisch and Lysyanskaya under the Strong RSA assumption [28,30] or using bilinear groups [31]. Other solutions were subsequently given with additional useful properties such as non-interactivity [10], delegatability [9] or support for efficient attributes [24] (see [27] and references therein). Anonymous credentials with attributes are often obtained by having the issuer obliviously sign a multi-block message $(\mathfrak{m}_1, \ldots, \mathfrak{m}_N)$, where one block is the secret key while other blocks contain public or private attributes. Note that, for the sake of keeping the scheme compatible with zero-knowledge proofs, the blocks $(\mathfrak{m}_1, \ldots, \mathfrak{m}_N)$ cannot be simply hashed before getting signed using a ordinary, single-block signature.

Group signatures are a central anonymity primitive, introduced by Chaum and van Heyst [35] in 1991, which allows members of a group managed by some authority to sign messages in the name of the entire group. At the same time, users remain accountable for the messages they sign since an opening authority can identify them if they misbehave.

Ateniese, Camenisch, Joye and Tsudik [6] provided the first scalable construction meeting the security requirements that can be intuitively expected from the primitive, although clean security notions were not available yet at that time. Bellare, Micciancio and Warinschi [11] filled this gap by providing suitable security notions for static groups, which were subsequently extended to the dynamic setting[1] by Kiayias and Yung [55] and Bellare, Shi and Zhang [13]. In these models, efficient schemes have been put forth in the random oracle model [38,55] (the ROM) and in the standard model [1,2,48].

Lattice-based group signatures were put forth for the first time by Gordon, Katz and Vaikuntanathan [46] whose solution had linear size signatures in the number of group members. Camenisch, Neven and Rückert [32] extended [46] so as to achieve anonymity in the strongest sense. Laguillaumie *et al.* [56] decreased the signature length to be logarithmic in the number N_{gs} of group members. While asymptotically shorter, their signatures remained space-consuming as, analogously to the Boyen-Waters group signature [20], their scheme encrypts each bit of the signer's identity individually. Simpler and more efficient solutions with $\mathcal{O}(\log N)$ signature size were given by Nguyen, Zhang and Zhang [66] and Ling, Nguyen and Wang [62]. In particular, the latter scheme [62] achieves significantly smaller signatures by encrypting all bits of the signer's identity at once. Benhamouda *et al.* [14] described a hybrid group signature that simultaneously relies on lattice assumptions (in the ring setting) and discrete-logarithm-related assumptions. Recently, Libert, Ling, Nguyen and Wang [60] obtained substantial efficiency improvements via a construction based on Merkle trees which eliminates the need for GPV trapdoors [42]. For the time being, all known group signatures are designed for static groups and analyzed in the model of Bellare,

[1] By "dynamic setting", we refer to a scenario where new group members can register at any time but, analogously to [13,55], we do not consider the orthogonal problem of user revocation here.

Micciancio and Warinschi [11], where no new group member can be introduced after the setup phase. This is somewhat unfortunate given that, in most applications of group signatures (e.g., protecting the privacy of commuters in public transportation), the dynamicity property is arguably what we need. To date, it remains an important open problem to design a lattice-based system that supports dynamically growing population of users in the models of [13,55].

OUR CONTRIBUTIONS. Our first result is a lattice-based signature with efficient protocols for multi-block messages. Namely, we provide a way for a user to obtain a signature on a committed N-block message $(\mathfrak{m}_1, \ldots, \mathfrak{m}_N)$ as well as a protocol for proving possession of a valid message-signature pair. The signature and its companion protocols can serve as a building block for lattice-based anonymous credentials and can potentially find applications in other privacy-preserving protocols (e.g., [25]) based on lattice assumptions.

The main application that we consider in this paper is the design of a lattice-based group signature scheme for dynamic groups. We prove the security of our system in the random oracle model [12] under the Short Integer Solution (SIS) and Learning With Errors (LWE) assumptions. For security parameter λ and for groups of up to N_{gs} members, the scheme features public key size $\widetilde{\mathcal{O}}(\lambda^2) \cdot \log N_{\mathsf{gs}}$, user's secret key size $\widetilde{\mathcal{O}}(\lambda)$, and signature size $\widetilde{\mathcal{O}}(\lambda) \cdot \log N_{\mathsf{gs}}$. As exhibited in Table 1, our scheme achieves a level of efficiency comparable to recent proposals based on standard (i.e., non-ideal) lattices [56,60,62,66] in the static setting [11]. In particular, the cost of moving to dynamic groups is quite reasonable: while using the scheme from [62] as a building block, our construction only lengthens the signature size by a (small) constant factor.

Table 1. Efficiency comparison among recent lattice-based group signatures for static groups and our dynamic scheme. The evaluation is done with respect to 2 governing parameters: security parameter λ and the maximum expected group size N_{gs}. We do not include the earlier schemes [32,46] that have signature size $\widetilde{\mathcal{O}}(\lambda^2) \cdot N_{\mathsf{gs}}$.

Scheme	LLLS [56]	NZZ [66]	LNW [62]	LLNW [60]	Ours
Group PK	$\widetilde{\mathcal{O}}(\lambda^2) \cdot \log N_{\mathsf{gs}}$	$\widetilde{\mathcal{O}}(\lambda^2)$	$\widetilde{\mathcal{O}}(\lambda^2) \cdot \log N_{\mathsf{gs}}$	$\widetilde{\mathcal{O}}(\lambda^2)$	$\widetilde{\mathcal{O}}(\lambda^2) \cdot \log N_{\mathsf{gs}}$
User's SK	$\widetilde{\mathcal{O}}(\lambda^2)$	$\widetilde{\mathcal{O}}(\lambda^2)$	$\widetilde{\mathcal{O}}(\lambda)$	$\widetilde{\mathcal{O}}(\lambda) \cdot \log N_{\mathsf{gs}}$	$\widetilde{\mathcal{O}}(\lambda)$
Signature	$\widetilde{\mathcal{O}}(\lambda) \cdot \log N_{\mathsf{gs}}$	$\widetilde{\mathcal{O}}(\lambda + \log^2 N_{\mathsf{gs}})$	$\widetilde{\mathcal{O}}(\lambda) \cdot \log N_{\mathsf{gs}}$	$\widetilde{\mathcal{O}}(\lambda) \cdot \log N_{\mathsf{gs}}$	$\widetilde{\mathcal{O}}(\lambda) \cdot \log N_{\mathsf{gs}}$

As a stepping stone in the design of our dynamic group signature, we also develop a zero-knowledge argument system allowing a group member to prove knowledge of a secret key (made of a short Gaussian vector) and a membership certificate issued by the group manager on the corresponding public key. Analogously to structure-preserving signatures [2], our signature scheme and zero-knowledge arguments make it possible to sign public keys without hashing them while remaining oblivious of the underlying secret key. They thus enable a round-optimal dynamic joining protocol – which allows the group manager to introduce new group members by issuing a membership certificate on their public key – which does not require any proof of knowledge on behalf of the prospective user. As a result, the interaction is minimal: only one message is sent in each direction between the prospective user and the group manager.[2] Besides being the first lattice-based group signature for dynamic groups, our scheme thus remains secure in the setting advocated by Kiayias and Yung [54], where many users want to join the system at the same time and concurrently interact with the group manager. We believe that, analogously to structure-preserving signatures [1,2], the combination of our signature scheme and zero-knowledge arguments can serve as a building blocks for other primitives, including group encryption [53] or adaptive oblivious transfer [47].

OUR TECHNIQUES. Our signature scheme with efficient protocols builds on the SIS-based signature of Böhl *et al.* [16], which is itself a variant of Boyen's signature [19]. Recall that the latter scheme involves a public key containing matrices $\mathbf{A}, \mathbf{A}_0, \ldots, \mathbf{A}_\ell \in \mathbb{Z}_q^{n \times m}$ and signs an ℓ-bit message $\mathfrak{m} \in \{0,1\}^\ell$ by computing a short $\mathbf{v} \in \mathbb{Z}^{2m}$ such that $[\mathbf{A} \mid \mathbf{A}_0 + \sum_{j=1}^{\ell} \mathfrak{m}[i]\mathbf{A}_j] \cdot \mathbf{v} = \mathbf{0}^n \bmod q$. The variant proposed by Böhl *et al.* [16] only uses a constant number of matrices $\mathbf{A}, \mathbf{A}_0, \mathbf{A}_1 \in \mathbb{Z}_q^{n \times m}$. Each signature is associated with a single-use tag tag (which is only used in one signing query in the proof) and the public key involves an extra matrix $\mathbf{D} \in \mathbb{Z}_q^{n \times m}$ and a vector $\mathbf{u} \in \mathbb{Z}_q^n$. A message Msg is signed by first applying a chameleon hash function $\mathbf{h} = \mathsf{CMHash}(\mathsf{Msg}, \mathbf{s}) \in \{0,1\}^m$ and signing $\mathbf{h}$ by computing a short $\mathbf{v} \in \mathbb{Z}^m$ such that $[\mathbf{A} \mid \mathbf{A}_0 + \mathsf{tag} \cdot \mathbf{A}_1] \cdot \mathbf{v} = \mathbf{u} + \mathbf{D} \cdot \mathbf{h} \bmod q$.

Our scheme extends [16] – modulo the use of a larger number of matrices $(\{\mathbf{A}_j\}_{j=0}^{\ell}, \mathbf{D}, \{\mathbf{D}\}_{k=0}^{N})$ – so that an N-block message $(\mathfrak{m}_1, \ldots, \mathfrak{m}_N) \in (\{0,1\}^L)^N$, for some $L \in \mathbb{N}$, is signed by outputting a tag $\tau \in \{0,1\}^\ell$ and a short $\mathbf{v} \in \mathbb{Z}^{2m}$ such that $[\mathbf{A} \mid \mathbf{A}_0 + \sum_{j=1}^{\ell} \tau[j] \cdot \mathbf{A}_j] \cdot \mathbf{v} = \mathbf{u} + \mathbf{D} \cdot \mathsf{CMHash}(\mathfrak{m}_1, \ldots, \mathfrak{m}_N, \mathbf{s})$, where the chameleon hash function computes $\mathbf{c}_M = \mathbf{D}_0 \cdot \mathbf{s} + \sum_{k=1}^{N} \mathbf{D}_k \cdot \mathfrak{m}_k \bmod q$, for some short vector $\mathbf{s}$, before re-encoding $\mathbf{c}_M$ so as to enable multiplication by $\mathbf{D}$.

In order to obtain a signature scheme akin to the one of Camenisch and Lysyanskaya [30], our idea is to have the tag $\tau \in \{0,1\}^\ell$ play the same role as the prime exponent in Strong-RSA-based schemes [30]. In the security proof of [16], we are faced with two situations: either the adversary produces a signature on a fresh tag $\tau^\star$, or it recycles a tag $\tau^{(i)}$ used by the signing oracle for a new, un-signed message $(\mathfrak{m}_1^\star, \ldots, \mathfrak{m}_N^\star)$. In the former case, the proof can proceed as in Boyen's proof [19]. In the latter case, the reduction must guess upfront which tag $\tau^{(i^\dagger)}$ the adversary will choose to re-use and find a way to properly answer the $i^\dagger$-th signing query without using the vanished trapdoor (for other queries, the Agrawal *et al.* technique [3] applies to compute a suitable $\mathbf{v}$ using a trapdoor hidden in $\{\mathbf{A}_j\}_{j=0}^{\ell}$). Böhl *et al.* [16] solve this problem by "programming" the vector $\mathbf{u} \in \mathbb{Z}_q^n$ in a special way and achieve full security using chameleon hashing.

[2] Note that each signature still requires the user to prove knowledge of his secret key. However, this is not a problem in concurrent settings as the argument of knowledge is made non-interactive via the Fiat-Shamir heuristic.

To adapt this idea in the context of signatures with efficient protocols, we have to overcome several difficulties. The first one is to map $\mathbf{c}_M$ back in the domain of the chameleon hash function while preserving the compatibility with zero-knowledge proofs. To solve this problem, we extend a technique used in [60] in order to build a "zero-knowledge-friendly" chameleon hash function. This function hashes $\mathsf{Msg} = (\mathfrak{m}_1, \ldots, \mathfrak{m}_N)$ by outputting the coordinate-wise binary decomposition $\mathbf{w}$ of $\mathbf{D}_0 \cdot \mathbf{s} + \sum_{k=1}^{N} \mathbf{D}_k \cdot \mathfrak{m}_k$. If we define the "powers-of-2" matrix $\mathbf{H} = \mathbf{I} \otimes [1 \mid 2 \mid \ldots \mid 2^{\lceil \log q \rceil}]$, then we can prove that $\mathbf{w} = \mathsf{CMHash}(\mathfrak{m}_1, \ldots, \mathfrak{m}_N, \mathbf{s})$ by demonstrating the knowledge of short vectors $(\mathfrak{m}_1, \ldots, \mathfrak{m}_N, \mathbf{s}, \mathbf{w})$ such that $\mathbf{H} \cdot \mathbf{w} = \mathbf{D}_0 \cdot \mathbf{s} + \sum_{k=1}^{N} \mathbf{D}_k \cdot \mathfrak{m}_k \bmod q$, which boils down to arguing knowledge of a solution to the ISIS problem [61].

The second problem is to prove knowledge of $(\tau, \mathbf{v}, \mathbf{s})$ and $(\mathfrak{m}_1, \ldots, \mathfrak{m}_N)$ satisfying $[\mathbf{A} \mid \mathbf{A}_0 + \sum_{j=1}^{\ell} \tau[j] \cdot \mathbf{A}_j] \cdot \mathbf{v} = \mathbf{u} + \mathbf{D} \cdot \mathsf{CMHash}(\mathfrak{m}_1, \ldots, \mathfrak{m}_N, \mathbf{s})$, without revealing any of the witnesses. To this end, we provide a framework for proving all the involved statement (and many other relations that naturally arise in lattice-based cryptography) as special cases. We reduce the statements to asserting that a short integer vector $\mathbf{x}$ satisfies an equation of the form $\mathbf{P} \cdot \mathbf{x} = \mathbf{v} \bmod q$, for some public matrix $\mathbf{P}$ and vector $\mathbf{v}$, and belongs to a set VALID of short vectors with a particular structure. While the small-norm property of $\mathbf{x}$ is provable using standard techniques (e.g., [63]), we argue its membership of VALID by leveraging the properties of Stern-like protocols [52,61,72]. In particular, we rely on the fact that their underlying permutations interact well with combinatorial statements pertaining to $\mathbf{x}$, especially $\mathbf{x}$ being a bitstring with a specific pattern. We believe our framework to be of independent interest as it provides a blueprint for proving many other intricate relations in a modular manner.

When we extend the scheme with a protocol for signing committed messages, we need the signer to re-randomize the user's commitment before signing the hidden messages. This is indeed necessary to provide the reduction with a backdoor allowing to correctly answer the $i^{\dagger}$-th query by "programming" the randomness of the commitment. Since we work with integers vectors, a straightforward simulation incurs a non-negligible statistical distance between the simulated distributions of re-randomization coins and the real one (which both have a discrete Gaussian distribution). Camenisch and Lysyanskaya [30] address a similar problem by choosing the signer's randomness to be exponentially larger than that of the user's commitment so as to statistically "drown" the aforementioned discrepancy. Here, the same idea would require to work with an exponentially large modulus q. Instead, we adopt a more efficient solution, inspired by Bai *et al.* [7], which is to apply an analysis based on the Rényi divergence rather than the statistical distance. In short, the Rényi divergence's properties tell us that, if some event E occurs with noticeable probability in some probability space P, so does it in a different probability space Q for which the second order divergence $R_2(P||Q)$ is sufficiently small. In our setting, $R_2(P||Q)$ is precisely polynomially bounded since the two probability spaces only diverge in one signing query.

Our dynamic group signature scheme avoids these difficulties because the group manager only signs known messages: instead of signing the user's secret key as in anonymous credentials, it creates a membership certificate by signing the user's public key. Our zero-knowledge arguments accommodate the requirements of the scheme in the following way. In the joining protocol that dynamically introduces new group members, the user i chooses a membership secret consisting of a short discrete Gaussian vector $\mathbf{z}_i$. This user generates a public syndrome $\mathbf{v}_i = \mathbf{F} \cdot \mathbf{z}_i \mod q$, for some public matrix $\mathbf{F}$, which constitutes his public key. In order to certify $\mathbf{v}_i$, the group manager computes the coordinate-wise binary expansion $\mathsf{bin}(\mathbf{v}_i)$ of $\mathbf{v}_i$. The vector $\mathsf{bin}(\mathbf{v}_i)$ is then signed using our signature scheme. Using the resulting signature $(\tau, \mathbf{v}, \mathbf{s})$ as a membership certificate, the group member is able to sign a message by proving that: (i) He holds a valid signature $(\tau, \mathbf{v}, \mathbf{s})$ on some secret binary message $\mathsf{bin}(\mathbf{v}_i)$; (ii) The latter vector $\mathsf{bin}(\mathbf{v}_i)$ is the binary expansion of some syndrome $\mathbf{v}_i$ of which he knows a GPV pre-image $\mathbf{z}_i$. We remark that condition (ii) can be proved by providing evidence that we have $\mathbf{v}_i = \mathbf{H} \cdot \mathsf{bin}(\mathbf{v}_i) = \mathbf{F} \cdot \mathbf{z}_i \bmod q$, for some short integer vector $\mathbf{z}_i$ and some binary $\mathsf{bin}(\mathbf{v}_i)$, where $\mathbf{H}$ is the "powers-of-2" matrix. Our abstraction of Stern-like protocols [52,61,72] allows us to efficiently argue such statements. The fact that the underlying chameleon hash function smoothly interacts with Stern-like zero-knowledge arguments is the property that maintains the user's capability of efficiently proving knowledge of the underlying secret key.

ORGANIZATION. In the forthcoming sections, we first provide some background in Sect. 2. Our signature with efficient protocols is presented in Sect. 3, where we also give protocols for obtaining a signature on a committed message and proving possession of a message-signature pair. Section 4 uses our signature scheme in the design of a dynamic group signature. The details of the zero-knowledge arguments used in Sect. 3 and Sect. 4 are deferred to Sect. 5, where we present them in a unified framework.

2 Background and Definitions

In the following, all vectors are denoted in bold lower-case letters, whereas bold upper-case letters will be used for matrices. If $\mathbf{b} \in \mathbb{R}^n$, its Euclidean norm and infinity norm will be denoted by $\|\mathbf{b}\|$ and $\|\mathbf{b}\|_\infty$, respectively. The Euclidean norm of matrix $\mathbf{B} \in \mathbb{R}^{m \times n}$ with columns $(\mathbf{b}_i)_{i \le n}$ is denoted by $\|\mathbf{B}\| = \max_{i \le n} \|\mathbf{b}_i\|$. If $\mathbf{B}$ is full column-rank, we let $\widetilde{\mathbf{B}}$ denote its Gram-Schmidt orthogonalization.

When S is a finite set, we denote by $U(S)$ the uniform distribution over S and by $x \hookleftarrow D$ the action of sampling x according to the distribution D.

2.1 Lattices

A (full-rank) lattice L is defined as the set of all integer linear combinations of some linearly independent basis vectors $(\mathbf{b}_i)_{i \le n}$ belonging to some $\mathbb{R}^n$. We work with q-ary lattices, for some prime q.

Definition 1. *Let $m \geq n \geq 1$, a prime $q \geq 2$, $\mathbf{A} \in \mathbb{Z}_q^{n \times m}$ and $\mathbf{u} \in \mathbb{Z}_q^n$, define $\Lambda_q(\mathbf{A}) := \{\mathbf{e} \in \mathbb{Z}^m \mid \exists \mathbf{s} \in \mathbb{Z}_q^n \text{ s.t. } \mathbf{A}^T \cdot \mathbf{s} = \mathbf{e} \bmod q\}$ as well as*

$$\Lambda_q^{\perp}(\mathbf{A}) := \{\mathbf{e} \in \mathbb{Z}^m \mid \mathbf{A} \cdot \mathbf{e} = \mathbf{0}^n \bmod q\}, \quad \Lambda_q^{\mathbf{u}}(\mathbf{A}) := \{\mathbf{e} \in \mathbb{Z}^m \mid \mathbf{A} \cdot \mathbf{e} = \mathbf{u} \bmod q\}$$

For any $\mathbf{t} \in \Lambda_q^{\mathbf{u}}(\mathbf{A})$, $\Lambda_q^{\mathbf{u}}(\mathbf{A}) = \Lambda_q^{\perp}(\mathbf{A}) + \mathbf{t}$ so that $\Lambda_q^{\mathbf{u}}(\mathbf{A})$ is a shift of $\Lambda_q^{\perp}(\mathbf{A})$.

For a lattice L, a vector $\mathbf{c} \in \mathbb{R}^n$ and a real $\sigma > 0$, define the function $\rho_{\sigma,\mathbf{c}}(\mathbf{x}) = \exp(-\pi \|\mathbf{x} - \mathbf{c}\|^2/\sigma^2)$. The discrete Gaussian distribution of support L, parameter σ and center $\mathbf{c}$ is defined as $D_{L,\sigma,\mathbf{c}}(\mathbf{y}) = \rho_{\sigma,\mathbf{c}}(\mathbf{y})/\rho_{\sigma,\mathbf{c}}(L)$ for any $\mathbf{y} \in L$. We denote by $D_{L,\sigma}(\mathbf{y})$ the distribution centered in $\mathbf{c} = \mathbf{0}$. We will extensively use the fact that samples from $D_{L,\sigma}$ are short with overwhelming probability.

Lemma 1 ([8, Le. 1.5]). *For any lattice $L \subseteq \mathbb{R}^n$ and positive real number $\sigma > 0$, we have $\Pr_{\mathbf{b} \hookleftarrow D_{L,\sigma}}[\|\mathbf{b}\| \leq \sqrt{n}\sigma] \geq 1 - 2^{-\Omega(n)}$.*

As shown by Gentry *et al.* [42], Gaussian distributions with lattice support can be sampled efficiently given a sufficiently short basis of the lattice.

Lemma 2 ([23, Le. 2.3]). *There exists a* PPT *(probabilistic polynomial-time) algorithm* GPVSample *that takes as inputs a basis $\mathbf{B}$ of a lattice $L \subseteq \mathbb{Z}^n$ and a rational $\sigma \geq \|\widetilde{\mathbf{B}}\| \cdot \Omega(\sqrt{\log n})$, and outputs vectors $\mathbf{b} \in L$ with distribution $D_{L,\sigma}$.*

Lemma 3 ([4, Th. 3.2]). *There exists a* PPT *algorithm* TrapGen *that takes as inputs 1^n, 1^m and an integer $q \geq 2$ with $m \geq \Omega(n \log q)$, and outputs a matrix $\mathbf{A} \in \mathbb{Z}_q^{n \times m}$ and a basis $\mathbf{T_A}$ of $\Lambda_q^{\perp}(\mathbf{A})$ such that $\mathbf{A}$ is within statistical distance $2^{-\Omega(n)}$ to $U(\mathbb{Z}_q^{n \times m})$, and $\|\widetilde{\mathbf{T_A}}\| \leq \mathcal{O}(\sqrt{n \log q})$.*

Lemma 3 is often combined with the sampler from Lemma 2. Micciancio and Peikert [65] recently proposed a more efficient approach for this combined task, which should be preferred in practice but, for the sake of simplicity, we present our schemes using TrapGen.

We also make use of an algorithm that extends a trapdoor for $\mathbf{A} \in \mathbb{Z}_q^{n \times m}$ to a trapdoor of any $\mathbf{B} \in \mathbb{Z}_q^{n \times m'}$ whose left $n \times m$ submatrix is $\mathbf{A}$.

Lemma 4 ([33, Le. 3.2]). *There exists a* PPT *algorithm* ExtBasis *that takes as inputs a matrix $\mathbf{B} \in \mathbb{Z}_q^{n \times m'}$ whose first m columns span $\mathbb{Z}_q^n$, and a basis $\mathbf{T_A}$ of $\Lambda_q^{\perp}(\mathbf{A})$ where $\mathbf{A}$ is the left $n \times m$ submatrix of $\mathbf{B}$, and outputs a basis $\mathbf{T_B}$ of $\Lambda_q^{\perp}(\mathbf{B})$ with $\|\widetilde{\mathbf{T_B}}\| \leq \|\widetilde{\mathbf{T_A}}\|$.*

In our security proofs, analogously to [16, 19] we also use a technique due to Agrawal, Boneh and Boyen [3] that implements an all-but-one trapdoor mechanism (akin to the one of Boneh and Boyen [17]) in the lattice setting.

Lemma 5 ([3, Th. 19]). *There exists a* PPT *algorithm* SampleRight *that takes as inputs matrices $\mathbf{A}, \mathbf{C} \in \mathbb{Z}_q^{n \times m}$, a low-norm matrix $\mathbf{R} \in \mathbb{Z}^{m \times m}$, a short basis $\mathbf{T_C} \in \mathbb{Z}^{m \times m}$ of $\Lambda_q^{\perp}(\mathbf{C})$, a vector $\mathbf{u} \in \mathbb{Z}_q^n$ and a rational σ such that $\sigma \geq \|\widetilde{\mathbf{T_C}}\| \cdot \Omega(\sqrt{\log n})$, and outputs a short vector $\mathbf{b} \in \mathbb{Z}^{2m}$ such that $\left[\mathbf{A} \mid \mathbf{A} \cdot \mathbf{R} + \mathbf{C}\right] \cdot \mathbf{b} = \mathbf{u} \bmod q$ and with distribution statistically close to $D_{L,\sigma}$ where L denotes the shifted lattice $\Lambda_q^{\mathbf{u}}\left(\left[\mathbf{A} \mid \mathbf{A} \cdot \mathbf{R} + \mathbf{C}\right]\right)$.*

2.2 Computational Problems

The security of our schemes provably relies (in the ROM) on the assumption that both algorithmic problems below are hard, i.e., cannot be solved in polynomial time with non-negligible probability and non-negligible advantage, respectively.

Definition 2. *Let m, q, β be functions of $n \in \mathbb{N}$. The Short Integer Solution problem $\mathsf{SIS}_{n,m,q,\beta}$ is, given $\mathbf{A} \hookleftarrow U(\mathbb{Z}_q^{n\times m})$, find $\mathbf{x} \in \Lambda_q^{\perp}(\mathbf{A})$ with $0 < \|\mathbf{x}\| \leq \beta$.*

If $q \geq \sqrt{n}\beta$ and $m, \beta \leq \mathsf{poly}(n)$, then $\mathsf{SIS}_{n,m,q,\beta}$ is at least as hard as standard worst-case lattice problem SIVP_γ with $\gamma = \widetilde{\mathcal{O}}(\beta\sqrt{n})$ (see, e.g., [42, Se. 9]).

Definition 3. *Let $n, m \geq 1$, $q \geq 2$, and let χ be a probability distribution on $\mathbb{Z}$. For $\mathbf{s} \in \mathbb{Z}_q^n$, let $A_{\mathbf{s},\chi}$ be the distribution obtained by sampling $\mathbf{a} \hookleftarrow U(\mathbb{Z}_q^n)$ and $e \hookleftarrow \chi$, and outputting $(\mathbf{a}, \mathbf{a}^T \cdot \mathbf{s} + e) \in \mathbb{Z}_q^n \times \mathbb{Z}_q$. The Learning With Errors problem $\mathsf{LWE}_{n,q,\chi}$ asks to distinguish m samples chosen according to $\mathcal{A}_{\mathbf{s},\chi}$ (for $\mathbf{s} \hookleftarrow U(\mathbb{Z}_q^n)$) and m samples chosen according to $U(\mathbb{Z}_q^n \times \mathbb{Z}_q)$.*

If q is a prime power, $B \geq \sqrt{n}\omega(\log n)$, $\gamma = \widetilde{\mathcal{O}}(nq/B)$, then there exists an efficient sampleable B-bounded distribution χ (i.e., χ outputs samples with norm at most B with overwhelming probability) such that $\mathsf{LWE}_{n,q,\chi}$ is as least as hard as SIVP_γ (see, e.g., [23,68,69]).

3 A Lattice-Based Signature with Efficient Protocols

Our scheme can be seen as a variant of the Böhl *et al.* signature [16], where each signature is a triple $(\tau, \mathbf{v}, \mathbf{s})$, made of a tag $\tau \in \{0,1\}^\ell$ and integer vectors $(\mathbf{v}, \mathbf{s})$ satisfying $[\mathbf{A} \mid \mathbf{A}_0 + \sum_{j=1}^{\ell} \tau[j] \cdot \mathbf{A}_j] \cdot \mathbf{v} = \mathbf{u} + \mathbf{D} \cdot \mathbf{h} \bmod q$, where matrices $\mathbf{A}, \mathbf{A}_0, \ldots, \mathbf{A}_\ell, \mathbf{D} \in \mathbb{Z}_q^{n\times m}$ are public random matrices and $\mathbf{h} \in \{0,1\}^m$ is a chameleon hash of the message which is computed using randomness $\mathbf{s}$. A difference is that, while [16] uses a short single-use tag $\tau \in \mathbb{Z}_q$, we need the tag to be an ℓ-bit string $\tau \in \{0,1\}^\ell$ which will assume the same role as the prime exponent of Camenisch-Lysyanskaya signatures [30] in the security proof.

We show that a suitable chameleon hash function makes the scheme compatible with Stern-like zero-knowledge arguments [61,62] for arguing possession of a valid message-signature pair. Section 5 shows how to translate such a statement into asserting that a short witness vector $\mathbf{x}$ with a particular structure satisfies a relation of the form $\mathbf{P} \cdot \mathbf{x} = \mathbf{v} \bmod q$, for some public matrix $\mathbf{P}$ and vector $\mathbf{v}$. The underlying chameleon hash can be seen as a composition of the chameleon hash of [33, Section 4.1] with a technique used in [60,67]: on input of a message $(\mathfrak{m}_1, \ldots, \mathfrak{m}_N)$, it outputs the binary decomposition of $\mathbf{D}_0 \cdot \mathbf{s} + \sum_{k=1}^{N} \mathbf{D}_k \cdot \mathfrak{m}_k$, for some discrete Gaussian vector $\mathbf{s}$.

3.1 Description

We assume that messages are vectors of N blocks $\mathsf{Msg} = (\mathfrak{m}_1, \ldots, \mathfrak{m}_N)$, where each block is a $2m$-bit string $\mathfrak{m}_k = \mathfrak{m}_k[1] \ldots \mathfrak{m}_k[2m] \in \{0,1\}^{2m}$ for $k \in \{1, \ldots, N\}$.

For each vector $\mathbf{v} \in \mathbb{Z}_q^L$, we denote by $\mathsf{bin}(\mathbf{v}) \in \{0,1\}^{L\lceil \log q \rceil}$ the vector obtained by replacing each coordinate of $\mathbf{v}$ by its binary representation.

Keygen$(1^\lambda, 1^N)$**:** Given a security parameter $\lambda > 0$ and the number of blocks $N = \mathsf{poly}(\lambda)$, choose the following parameters: $n = \mathcal{O}(\lambda)$; a prime modulus $q = \widetilde{\mathcal{O}}(N \cdot n^4)$; dimension $m = 2n\lceil \log q \rceil$; an integer $\ell = \Theta(\lambda)$; and Gaussian parameters $\sigma = \Omega(\sqrt{n \log q} \log n)$, $\sigma_0 = 2\sqrt{2}(N+1)\sigma m^{3/2}$, and $\sigma_1 = \sqrt{\sigma_0^2 + \sigma^2}$. Define the message space as $(\{0,1\}^{2m})^N$.

1. Run $\mathsf{TrapGen}(1^n, 1^m, q)$ to get $\mathbf{A} \in \mathbb{Z}_q^{n \times m}$ and a short basis $\mathbf{T_A}$ of $\Lambda_q^\perp(\mathbf{A})$. This basis allows computing short vectors in $\Lambda_q^\perp(\mathbf{A})$ with a Gaussian parameter σ. Next, choose $\ell + 1$ random $\mathbf{A}_0, \mathbf{A}_1, \ldots, \mathbf{A}_\ell \hookleftarrow U(\mathbb{Z}_q^{n \times m})$.
2. Choose random matrices $\mathbf{D} \hookleftarrow U(\mathbb{Z}_q^{n \times m})$, $\mathbf{D}_0, \mathbf{D}_1, \ldots, \mathbf{D}_N \hookleftarrow U(\mathbb{Z}_q^{2n \times 2m})$ as well as a random vector $\mathbf{u} \hookleftarrow U(\mathbb{Z}_q^n)$.

The private key consists of $SK := \mathbf{T_A} \in \mathbb{Z}^{m \times m}$ and the public key is

$$PK := (\mathbf{A},\ \{\mathbf{A}_j\}_{j=0}^{\ell},\ \{\mathbf{D}_k\}_{k=0}^{N},\ \mathbf{D},\ \mathbf{u}).$$

Sign(SK, Msg)**:** To sign an N-block message $\mathsf{Msg} = (\mathfrak{m}_1, \ldots, \mathfrak{m}_N) \in \left(\{0,1\}^{2m}\right)^N$,

1. Choose a random string $\tau \hookleftarrow U(\{0,1\}^\ell)$. Then, using $SK := \mathbf{T_A}$, compute with $\mathsf{ExtBasis}$ a short delegated basis $\mathbf{T}_\tau \in \mathbb{Z}^{2m \times 2m}$ for the matrix

$$\mathbf{A}_\tau = [\mathbf{A} \mid \mathbf{A}_0 + \sum_{j=1}^{\ell} \tau[j] \mathbf{A}_j] \in \mathbb{Z}_q^{n \times 2m}. \qquad (1)$$

2. Sample a vector $\mathbf{s} \hookleftarrow D_{\mathbb{Z}^{2m}, \sigma_1}$. Compute $\mathbf{c}_M \in \mathbb{Z}_q^{2n}$ as a chameleon hash of $(\mathfrak{m}_1, \ldots, \mathfrak{m}_N)$: i.e., compute $\mathbf{c}_M = \mathbf{D}_0 \cdot \mathbf{s} + \sum_{k=1}^{N} \mathbf{D}_k \cdot \mathfrak{m}_k \in \mathbb{Z}_q^{2n}$, which is used to define $\mathbf{u}_M = \mathbf{u} + \mathbf{D} \cdot \mathsf{bin}(\mathbf{c}_M) \in \mathbb{Z}_q^n$. Then, using the delegated basis $\mathbf{T}_\tau \in \mathbb{Z}^{2m \times 2m}$, sample a short vector $\mathbf{v} \in \mathbb{Z}^{2m}$ in $D_{\Lambda_q^{\mathbf{u}_M}(\mathbf{A}_\tau), \sigma}$.

Output the signature $sig = (\tau, \mathbf{v}, \mathbf{s}) \in \{0,1\}^\ell \times \mathbb{Z}^{2m} \times \mathbb{Z}^{2m}$.

Verify(PK, Msg, sig)**:** Given PK, $\mathsf{Msg} = (\mathfrak{m}_1, \ldots, \mathfrak{m}_N) \in (\{0,1\}^{2m})^N$ and $sig = (\tau, \mathbf{v}, \mathbf{s}) \in \{0,1\}^\ell \times \mathbb{Z}^{2m} \times \mathbb{Z}^{2m}$, return 1 if $\|\mathbf{v}\| < \sigma\sqrt{2m}$, $\|\mathbf{s}\| < \sigma_1\sqrt{2m}$ and

$$\mathbf{A}_\tau \cdot \mathbf{v} = \mathbf{u} + \mathbf{D} \cdot \mathsf{bin}(\mathbf{D}_0 \cdot \mathbf{s} + \sum_{k=1}^{N} \mathbf{D}_k \cdot \mathfrak{m}_k) \bmod q. \qquad (2)$$

When the scheme is used for obliviously signing committed messages, the security proof follows Bai *et al.* [7] in that it applies an argument based on the Rényi divergence in one signing query. This argument requires to sample $\mathbf{s}$ from a Gaussian distribution whose standard deviation σ_1 is polynomially larger than σ.

We note that, instead of being included in the public key, the matrices $\{\mathbf{D}_k\}_{k=0}^N$ can be part of public parameters shared by many signers. Indeed, only the matrices $(\mathbf{A}, \{\mathbf{A}_i\}_{i=0}^{\ell})$ should be specific to the user who holds $SK = \mathbf{T_A}$. In Sect. 3.3, we use a variant where $\{\mathbf{D}_k\}_{k=0}^N$ belong to public parameters.

3.2 Security Analysis

The security analysis in Theorem 1 requires that $q > \ell$.

Theorem 1. *The signature scheme is secure under chosen-message attacks under the* SIS *assumption.*

Proof (Sketched). To prove the result, we will distinguish three kinds of attacks:

Type I attacks are attacks where, in the adversary's forgery $sig^\star = (\tau^\star, \mathbf{v}^\star, \mathbf{s}^\star)$, $\tau^\star$ did not appear in any output of the signing oracle.

Type II attacks are such that, in the adversary's forgery $sig^\star = (\tau^\star, \mathbf{v}^\star, \mathbf{s}^\star)$, $\tau^\star$ is recycled from an output $sig^{(i^\star)} = (\tau^{(i^\star)}, \mathbf{v}^{(i^\star)}, \mathbf{s}^{(i^\star)})$ of the signing oracle, for some index $i^\star \in \{1, \ldots, Q\}$. However, if $\mathsf{Msg}^\star = (\mathfrak{m}_1^\star, \ldots, \mathfrak{m}_N^\star)$ and $\mathsf{Msg}^{(i^\star)} = (\mathfrak{m}_1^{(i^\star)}, \ldots, \mathfrak{m}_N^{(i^\star)})$ denote the forgery message and the $i^\star$-th signing query, respectively, we have $\mathbf{D}_0 \cdot \mathbf{s}^\star + \sum_{k=1}^N \mathbf{D}_k \cdot \mathfrak{m}_k^\star \neq \mathbf{D}_0 \cdot \mathbf{s}^{(i^\star)} + \sum_{k=1}^N \mathbf{D}_k \cdot \mathfrak{m}_k^{(i^\star)}$.

Type III attacks are those where the adversary's forgery $sig^\star = (\tau^\star, \mathbf{v}^\star, \mathbf{s}^\star)$ recycles $\tau^\star$ from an output $sig^{(i^\star)} = (\tau^{(i^\star)}, \mathbf{v}^{(i^\star)}, \mathbf{s}^{(i^\star)})$ of the signing oracle (i.e., $\tau^{(i^\star)} = \tau^\star$ for some index $i^\star \in \{1, \ldots, Q\}$) and we have the collision

$$\mathbf{D}_0 \cdot \mathbf{s}^\star + \sum_{k=1}^{N} \mathbf{D}_k \cdot \mathfrak{m}_k^\star = \mathbf{D}_0 \cdot \mathbf{s}^{(i^\star)} + \sum_{k=1}^{N} \mathbf{D}_k \cdot \mathfrak{m}_k^{(i^\star)}. \tag{3}$$

Type III attacks imply a collision for the chameleon hash function of Kawachi *et al.* [52]: if (3) holds, a short vector of $\Lambda_q^\perp([\mathbf{D}_0 \mid \mathbf{D}_1 \mid \ldots \mid \mathbf{D}_N])$ is obtained as so that a collision breaks the SIS assumption.

The security against Type I attacks is proved by Lemma 6 which applies the same technique as in [19,65]. In particular, the prefix guessing technique of [50] allows keeping the modulus smaller than the number Q of adversarial queries as in [65]. In order to deal with Type II attacks, we can leverage the technique of [16]. In Lemma 7, we prove that Type II attack would also contradict SIS. □

The following lemmas are proved in the full version of the paper [59].

Lemma 6. *The scheme is secure against Type I attacks if the* $\mathsf{SIS}_{n,m,q,\beta'}$ *assumption holds for* $\beta' = m^{3/2}\sigma^2(\ell + 3) + m^{1/2}\sigma_1$.

Lemma 7. *The scheme is secure against Type II attacks under the* $\mathsf{SIS}_{n,m,q,\beta''}$ *assumption for* $\beta'' = \sqrt{2}(\ell + 2)\sigma^2 m^{3/2} + m^{1/2}$.

3.3 Protocols for Signing a Committed Value and Proving Possession of a Signature

We first show a two-party protocol whereby a user can interact with the signer in order to obtain a signature on a committed message.

In order to prove that the scheme still guarantees unforgeability for obliviously signed messages, we will assume that each message block $\mathfrak{m}_k \in \{0,1\}^{2m}$

is obtained by encoding the actual message $M_k = M_k[1] \dots M_k[m] \in \{0,1\}^m$ as $\mathfrak{m}_k = \mathsf{Encode}(M_k) = (\bar{M}_k[1], M_k[1], \dots, \bar{M}_k[m], M_k[m])$. Namely, each 0 (resp. each 1) is encoded as a pair $(1,0)$ (resp. $(0,1)$). The reason for this encoding is that the proof of Theorem 2 requires that at least one block $\mathfrak{m}_k^\star$ of the forgery message is 1 while the same bit is 0 at some specific signing query. We will show (see Sect. 5) that the correctness of this encoding can be efficiently proved using Stern-like [72] protocols.

To sign committed messages, a first idea is exploit the fact that our signature of Sect. 3.1 blends well with the SIS-based commitment scheme suggested by Kawachi *et al.* [52]. In the latter scheme, the commitment key consists of matrices $(\mathbf{D}_0, \mathbf{D}_1) \in \mathbb{Z}_q^{2n \times 2m} \times \mathbb{Z}_q^{2n \times 2m}$, so that message $\mathfrak{m} \in \{0,1\}^{2m}$ can be committed to by sampling a Gaussian vector $\mathbf{s} \hookleftarrow D_{\mathbb{Z}^{2m},\sigma}$ and computing $\mathbf{C} = \mathbf{D}_0 \cdot \mathbf{s} + \mathbf{D}_1 \cdot \mathfrak{m} \in \mathbb{Z}_q^{2n}$. This scheme extends to commit to multiple messages $(\mathfrak{m}_1, \dots, \mathfrak{m}_N)$ at once by computing $\mathbf{C} = \mathbf{D}_0 \cdot \mathbf{s} + \sum_{k=1}^N \mathbf{D}_k \cdot \mathfrak{m}_k \in \mathbb{Z}_q^{2n}$ using a longer commitment key $(\mathbf{D}_0, \mathbf{D}_1, \dots, \mathbf{D}_N) \in (\mathbb{Z}_q^{2n \times 2m})^{N+1}$. It is easy to see that the resulting commitment remains statistically hiding and computationally binding under the SIS assumption.

In order to make our construction usable in the definitional framework of Camenisch *et al.* [27], we assume common public parameters (i.e., a common reference string) and encrypt all witnesses of which knowledge is being proved under a public key included in the common reference string. The resulting ciphertexts thus serve as statistically binding commitments to the witnesses. To enable this, the common public parameters comprise public keys $\mathbf{G}_0 \in \mathbb{Z}_q^{n \times \ell}$, $\mathbf{G}_1 \in \mathbb{Z}_q^{n \times 2m}$ for multi-bit variants of the dual Regev cryptosystem [42] and all parties are denied access to the underlying private keys. The flexibility of Stern-like protocols allows us to prove that the content of a perfectly hiding commitment $\mathbf{c}_{\mathfrak{m}}$ is consistent with encrypted values.

Global-Setup: Let $B = \sqrt{n}\omega(\log n)$ and let χ be a B-bounded distribution. Let $p = \sigma \cdot \omega(\sqrt{m})$ upper-bound entries of vectors sampled from the distribution $D_{\mathbb{Z}^{2m},\sigma}$. Generate two public keys for the dual Regev encryption scheme in its multi-bit variant. These keys consists of a public random matrix $\mathbf{B} \hookleftarrow U(\mathbb{Z}_q^{n \times m})$ and random matrices $\mathbf{G}_0 = \mathbf{B} \cdot \mathbf{E}_0 \in \mathbb{Z}_q^{n \times \ell}$, $\mathbf{G}_1 = \mathbf{B} \cdot \mathbf{E}_1 \in \mathbb{Z}_q^{n \times 2m}$, where $\mathbf{E}_0 \in \mathbb{Z}^{m \times \ell}$ and $\mathbf{E}_1 \in \mathbb{Z}^{m \times 2m}$ are short Gaussian matrices with columns sampled from $D_{\mathbb{Z}^m,\sigma}$. These matrices will be used to encrypt integer vectors of dimension ℓ and $2m$, respectively. Finally, generate public parameters $CK := \{\mathbf{D}_k\}_{k=0}^N$ consisting of uniformly random matrices $\mathbf{D}_k \hookleftarrow U(\mathbb{Z}_q^{2n \times 2m})$ for a statistically hiding commitment to vectors in $(\{0,1\}^{2m})^N$. Return public parameters consisting of

$$\mathsf{par} := \{\, \mathbf{B} \in \mathbb{Z}_q^{n \times m},\ \mathbf{G}_0 \in \mathbb{Z}_q^{n \times \ell},\ \mathbf{G}_1 \in \mathbb{Z}_q^{n \times 2m},\ CK\}.$$

Issue $\leftrightarrow$ Obtain: The signer S, who has $PK := \{\mathbf{A},\ \{\mathbf{A}_j\}_{j=0}^{\ell},\ \mathbf{D},\ \mathbf{u}\}$ and $SK := \mathbf{T}_{\mathbf{A}}$, interacts with the user U, who has $(\mathfrak{m}_1, \dots, \mathfrak{m}_N)$, as follows.

1. U samples $\mathbf{s}' \hookleftarrow D_{\mathbb{Z}^{2m},\sigma}$ and computes $\mathbf{c}_{\mathfrak{m}} = \mathbf{D}_0 \cdot \mathbf{s}' + \sum_{k=1}^N \mathbf{D}_k \cdot \mathfrak{m}_k \in \mathbb{Z}_q^{2n}$ which is sent to S as a commitment to $(\mathfrak{m}_1, \dots, \mathfrak{m}_N)$. Next, U encrypts

$\{\mathfrak{m}_k\}_{k=1}^N$ and $\mathbf{s}'$ under the key $(\mathbf{B}, \mathbf{G}_1)$ by computing for all $k \in [1, N]$:

$$\begin{aligned} \mathbf{c}_k &= (\mathbf{c}_{k,1}, \mathbf{c}_{k,2}) \\ &= \big(\mathbf{B}^T \cdot \mathbf{s}_k + \mathbf{e}_{k,1},\ \mathbf{G}_1^T \cdot \mathbf{s}_k + \mathbf{e}_{k,2} + \mathfrak{m}_k \cdot \lfloor q/2 \rfloor\big) \in \mathbb{Z}_q^m \times \mathbb{Z}_q^{2m} \end{aligned} \quad (4)$$

for randomly chosen $\mathbf{s}_k \hookleftarrow \chi^n$, $\mathbf{e}_{k,1} \hookleftarrow \chi^m$, $\mathbf{e}_{k,2} \hookleftarrow \chi^{2m}$, and

$$\begin{aligned} \mathbf{c}_{s'} &= (\mathbf{c}_{s',1}, \mathbf{c}_{s',2}) \\ &= \big(\mathbf{B}^T \cdot \mathbf{s}_0 + \mathbf{e}_{0,1},\ \mathbf{G}_1^T \cdot \mathbf{s}_0 + \mathbf{e}_{0,2} + \mathbf{s}' \cdot \lfloor q/p \rfloor\big) \in \mathbb{Z}_q^m \times \mathbb{Z}_q^{2m} \end{aligned} \quad (5)$$

where $\mathbf{s}_0 \hookleftarrow \chi^n$, $\mathbf{e}_{0,1} \hookleftarrow \chi^m$, $\mathbf{e}_{0,2} \hookleftarrow \chi^{2m}$. The ciphertexts $\{\mathbf{c}_k\}_{k=1}^N$ and $\mathbf{c}_{s'}$ are sent to S along with $\mathbf{c}_{\mathfrak{m}}$.
Then, U generates an interactive zero-knowledge argument to convince S that $\mathbf{c}_{\mathfrak{m}}$ is a commitment to $(\mathfrak{m}_1, \ldots, \mathfrak{m}_N)$ with the randomness $\mathbf{s}'$ such that $\{\mathfrak{m}_k\}_{k=1}^N$ and $\mathbf{s}'$ were honestly encrypted to $\{\mathbf{c}_k\}_{i=1}^N$ and $\mathbf{c}_{s'}$, as in (4) and (5). For convenience, this argument system will be described in Sect. 5.3, where we demonstrate that, together with other zero-knowledge protocols used in this work, it can be derived from a Stern-like [72] protocol constructed in Sect. 5.1.

2. If the argument of step 1 properly verifies, S samples $\mathbf{s}'' \hookleftarrow D_{\mathbb{Z}^{2m},\sigma_0}$ and computes a vector $\mathbf{u}_{\mathfrak{m}} = \mathbf{u} + \mathbf{D} \cdot \mathsf{bin}(\mathbf{c}_{\mathfrak{m}} + \mathbf{D}_0 \cdot \mathbf{s}'') \in \mathbb{Z}_q^n$. Next, S randomly picks $\tau \hookleftarrow \{0,1\}^\ell$ and uses $\mathbf{T}_\mathbf{A}$ to compute a delegated basis $\mathbf{T}_\tau \in \mathbb{Z}^{2m \times 2m}$ for the matrix $\mathbf{A}_\tau \in \mathbb{Z}_q^{n \times 2m}$ of (1). Using $\mathbf{T}_\tau \in \mathbb{Z}^{2m \times 2m}$, S samples a short vector $\mathbf{v} \in \mathbb{Z}^{2m}$ in $D^{\mathbf{u}_M}_{\Lambda^\perp(\mathbf{A}_\tau),\sigma}$. It returns the vector $(\tau, \mathbf{v}, \mathbf{s}'') \in \{0,1\}^\ell \times \mathbb{Z}^{2m} \times \mathbb{Z}^{2m}$ to U.
3. U computes $\mathbf{s} = \mathbf{s}' + \mathbf{s}''$ over $\mathbb{Z}$ and verifies that

$$\mathbf{A}_\tau \cdot \mathbf{v} = \mathbf{u} + \mathbf{D} \cdot \mathsf{bin}\big(\mathbf{D}_0 \cdot \mathbf{s} + \sum_{k=1}^N \mathbf{D}_k \cdot \mathfrak{m}_k\big) \bmod q.$$

If so, it outputs $(\tau, \mathbf{v}, \mathbf{s})$. Otherwise, it outputs $\perp$.

Note that, if both parties faithfully run the protocol, the user obtains a valid signature $(\tau, \mathbf{v}, \mathbf{s})$ for which the distribution of $\mathbf{s}$ is $D_{\mathbb{Z}^{2m},\sigma_1}$, where $\sigma_1 = \sqrt{\sigma^2 + \sigma_0^2}$.

The following protocol allows proving possession of a message-signature pair.

Prove: On input of a signature $(\tau, \mathbf{v} = (\mathbf{v}_1^T \mid \mathbf{v}_2^T)^T, \mathbf{s}) \in \{0,1\}^\ell \times \mathbb{Z}^{2m} \times \mathbb{Z}^{2m}$ on the message $(\mathfrak{m}_1, \ldots, \mathfrak{m}_N)$, the user does the following.

1. Using $(\mathbf{B}, \mathbf{G}_0)$ and $(\mathbf{B}, \mathbf{G}_1)$ generate perfectly binding commitments to $\tau \in \{0,1\}^\ell$, $\{\mathfrak{m}_k\}_{k=1}^N$, $\mathbf{v}_1, \mathbf{v}_2 \in \mathbb{Z}^m$ and $\mathbf{s} \in \mathbb{Z}^{2m}$. Namely, compute

$$\begin{aligned} \mathbf{c}_\tau &= (\mathbf{c}_{\tau,1}, \mathbf{c}_{\tau,2}) \\ &= \big(\mathbf{B}^T \cdot \mathbf{s}_\tau + \mathbf{e}_{\tau,1},\ \mathbf{G}_0^T \cdot \mathbf{s}_\tau + \mathbf{e}_{\tau,2} + \tau \cdot \lfloor q/2 \rfloor\big) \in \mathbb{Z}_q^m \times \mathbb{Z}_q^\ell, \\ \mathbf{c}_k &= (\mathbf{c}_{k,1}, \mathbf{c}_{k,2}) \ \in \mathbb{Z}_q^m \times \mathbb{Z}_q^{2m} \\ &= \big(\mathbf{B}^T \cdot \mathbf{s}_k + \mathbf{e}_{k,1},\ \mathbf{G}_1^T \cdot \mathbf{s}_k + \mathbf{e}_{k,2} + \mathfrak{m}_k \cdot \lfloor q/2 \rfloor\big) \qquad \forall k \in [1, N] \end{aligned}$$

where $\mathbf{s}_\tau, \mathbf{s}_k \hookleftarrow \chi^n$, $\mathbf{e}_{\tau,1}, \mathbf{e}_{k,1} \hookleftarrow \chi^m$, $\mathbf{e}_{\tau,2} \hookleftarrow \chi^\ell$, $\mathbf{e}_{k,2} \hookleftarrow \chi^{2m}$, as well as

$$\begin{aligned}
\mathbf{c_v} &= (\mathbf{c}_{\mathbf{v},1}, \mathbf{c}_{\mathbf{v},2}) \\
&= \big(\mathbf{B}^T \cdot \mathbf{s_v} + \mathbf{e}_{\mathbf{v},1},\ \mathbf{G}_1^T \cdot \mathbf{s_v} + \mathbf{e}_{\mathbf{v},2} + \mathbf{v} \cdot \lfloor q/p \rfloor\big) \in \mathbb{Z}_q^m \times \mathbb{Z}_q^{2m} \\
\mathbf{c}_s &= (\mathbf{c}_{s,1}, \mathbf{c}_{s,2}) \\
&= \big(\mathbf{B}^T \cdot \mathbf{s}_0 + \mathbf{e}_{0,1},\ \mathbf{G}_1^T \cdot \mathbf{s}_0 + \mathbf{e}_{0,2} + \mathbf{s} \cdot \lfloor q/p \rfloor\big) \in \mathbb{Z}_q^m \times \mathbb{Z}_q^{2m},
\end{aligned}$$

where $\mathbf{s_v}, \mathbf{s}_0 \hookleftarrow \chi^n$, $\mathbf{e}_{\mathbf{v},1}, \mathbf{e}_{0,1} \hookleftarrow \chi^m$, $\mathbf{e}_{\mathbf{v},2}, \mathbf{e}_{0,2} \hookleftarrow \chi^{2m}$.

2. Prove in zero-knowledge that $\mathbf{c}_\tau$, $\mathbf{c}_s$, $\mathbf{c_v}$, $\{\mathbf{c}_k\}_{k=1}^N$ encrypt a valid message-signature pair. In Sect. 5.4, we show that this involved zero-knowledge protocol can be derived from the statistical zero-knowledge argument of knowledge for a simpler, but more general relation that we explicitly present in Sect. 5.1. The proof system can be made statistically ZK for a malicious verifier using standard techniques (assuming a common reference string, we can use [36]). In the random oracle model, it can be made non-interactive using the Fiat-Shamir heuristic [40].

We require that the adversary be unable to prove possession of a signature of a message $(\mathfrak{m}_1, \ldots, \mathfrak{m}_N)$ for which it did not legally obtain a credential by interacting with the issuer. Note that the messages that are blindly signed by the issuer are uniquely defined since, at each signing query, the adversary is required to supply perfectly binding commitments $\{\mathbf{c}_k\}_{k=1}^N$ to $(\mathfrak{m}_1, \ldots, \mathfrak{m}_N)$.

In instantiations using non-interactive proofs, we assume that these can be bound to a verifier-chosen nonce to prevent replay attacks, as suggested in [27].

The security proof (in Theorem 2) makes crucial use of the Rényi divergence using arguments in the spirit of Bai *et al.* [7]. The reduction has to guess upfront the index $i^\star \in \{1, \ldots, Q\}$ of the specific signing query for which the adversary will re-use $\tau^{(i^\star)}$. For this query, the reduction will have to make sure that the simulation trapdoor of Agrawal *et al.* [3] (used by the SampleRight algorithm of Lemma 5) vanishes: otherwise, the adversary's forgery would not be usable for solving SIS. This means that, as in the proof of [16], the reduction must answer exactly one signing query in a different way, without using the trapdoor. While Böhl *et al.* solve this problem by exploiting the fact that they only need to prove security against non-adaptive forgers, we directly use a built-in chameleon hash function mechanism which is implicitly realized by the matrix $\mathbf{D}_0$ and the vector $\mathbf{s}$. Namely, in the signing query for which the Agrawal *et al.* trapdoor [3] cancels, we assign a special value to the vector $\mathbf{s} \in \mathbb{Z}^{2m}$, which depends on the adaptively-chosen signed message $(\mathsf{Msg}_1^{(i^\star)}, \ldots, \mathsf{Msg}_N^{(i^\star)})$ and some Gaussian matrices $\{\mathbf{R}_k\}_{k=1}^N$ hidden behind $\{\mathbf{D}_k\}_{k=1}^N$.

One issue is that this results in a different distribution for the vector $\mathbf{s} \in \mathbb{Z}^m$. However, we can still view $\mathbf{s}$ as a vector sampled from a Gaussian distribution centered away from $\mathbf{0}^{2m}$. Since this specific situation occurs only once during the simulation, we can apply a result proved in [58] which upper-bounds the Rényi divergence between two Gaussian distributions with identical standard deviations but different centers. By choosing the standard deviation σ_1 of $\mathbf{s} \in \mathbb{Z}^{2m}$ to be polynomially larger than that of the columns of matrices $\{\mathbf{R}_k\}_{k=1}^N$, we can keep the Rényi divergence between the two distributions of $\mathbf{s}$ (i.e., the one of the simulation and the one of the real game) sufficiently small to apply the probability preservation property (which still gives a polynomial reduction since the argument must only be applied on one signing query). Namely, the latter implies that, if the Rényi divergence $R_2(\mathbf{s}^{\mathsf{real}}||\mathbf{s}^{\mathsf{sim}})$ is polynomial, the probability that the simulated vector $\mathbf{s}^{\mathsf{sim}} \in \mathbb{Z}^{2m}$ passes the verification test will only be polynomially smaller than in the real game and so will be the adversary's probability of success.

Another option would have been to keep the statistical distance between $\mathbf{s}^{\mathsf{real}}$ and $\mathbf{s}^{\mathsf{sim}}$ negligible using the smudging technique of [5]. However, this would have implied to use an exponentially large modulus q since σ_1 should have been exponentially larger than the standard deviations of the columns of $\{\mathbf{R}_k\}_{k=1}^N$.

The proofs of the following theorems are given in the full version of the paper.

Theorem 2. *Under the* $\mathsf{SIS}_{n,2m,q,\hat{\beta}}$ *assumption, where* $\hat{\beta} = N\sigma(2m)^{3/2} + 4\sigma_1 m^{3/2}$, *the above protocols are secure protocols for obtaining a signature on a committed message and proving possession of a valid message-signature pair.*

Theorem 3. *The scheme provides anonymity under the* $\mathsf{LWE}_{n,q,\chi}$ *assumption.*

4 A Dynamic Lattice-Based Group Signature

In this section, the signature scheme of Sect. 3 is used to design a group signature for dynamic groups using the syntax and the security model of Kiayias and Yung [55], which is recalled in the full version of the paper.

In the notations hereunder, for any positive integers n, and $q \geq 2$, we define the "powers-of-2" matrix $\mathbf{H}_{\mathsf{n}\times\mathsf{n}\lceil\log q\rceil} = \mathbf{I}_{\mathsf{n}} \otimes [1 \mid 2 \mid 4 \mid \ldots \mid 2^{\lceil\log q\rceil-1}] \in \mathbb{Z}_q^{\mathsf{n}\times\mathsf{n}\lceil\log q\rceil}$. Also, for each vector $\mathbf{v} \in \mathbb{Z}_q^{\mathsf{n}}$, we define $\mathsf{bin}(\mathbf{v}) \in \{0,1\}^{\mathsf{n}\lceil\log q\rceil}$ to be the vector obtained by replacing each entry of $\mathbf{v}$ by its binary expansion. Hence, we have $\mathbf{v} = \mathbf{H}_{\mathsf{n}\times\mathsf{n}\lceil\log q\rceil} \cdot \mathsf{bin}(\mathbf{v})$ for any $\mathbf{v} \in \mathbb{Z}_q^{\mathsf{n}}$.

In our scheme, each group membership certificate is a signature generated by the group manager on the user's public key. Since the group manager only needs to sign known (rather than committed) messages, we can use a simplified version of the signature, where the chameleon hash function does not need to choose the discrete Gaussian vector $\mathbf{s}$ with a larger standard deviation than other vectors.

A key component of the scheme is the two-message joining protocol whereby the group manager admits new group members by signing their public key. The first message is sent by the new user $\mathcal{U}_i$ who samples a membership secret consisting of a short vector $\mathbf{z}_i \hookleftarrow D_{\mathbb{Z}^{4m},\sigma}$ (where $m = 2n\lceil\log q\rceil$), which is used to compute a syndrome $\mathbf{v}_i = \mathbf{F} \cdot \mathbf{z}_i \in \mathbb{Z}_q^{4n}$ for some public matrix $\mathbf{F} \in \mathbb{Z}_q^{4n\times 4m}$. This syndrome $\mathbf{v}_i \in \mathbb{Z}_q^{4n}$ must be signed by $\mathcal{U}_i$ using his long term secret key $\mathsf{usk}[i]$ (as in [13,55], we assume that each user has a long-term key $\mathsf{upk}[i]$ for a digital signature, which is registered in some PKI) and will uniquely identify $\mathcal{U}_i$. In order to generate a membership certificate for $\mathbf{v}_i \in \mathbb{Z}_q^{4n}$, the group manager GM signs its binary expansion $\mathsf{bin}(\mathbf{v}_i) \in \{0,1\}^{4n\lceil\log q\rceil}$ using the scheme of Sect. 3.

Equipped with his membership certificate $(\tau, \mathbf{d}, \mathbf{s}) \in \{0,1\}^\ell \times \mathbb{Z}^{2m} \times \mathbb{Z}^{2m}$, the new group member $\mathcal{U}_i$ can sign a message using a Stern-like protocol for demonstrating his knowledge of a valid certificate for which he also knows the secret key associated with the certified public key $\mathbf{v}_i \in \mathbb{Z}_q^{4n}$. This boils down to providing evidence that the membership certificate is a valid signature on some binary message $\mathsf{bin}(\mathbf{v}_i) \in \{0,1\}^{4n\lceil \log q \rceil}$ for which he also knows a short $\mathbf{z}_i \in \mathbb{Z}^{4m}$ such that $\mathbf{v}_i = \mathbf{H}_{4n \times 2m} \cdot \mathsf{bin}(\mathbf{v}_i) = \mathbf{F} \cdot \mathbf{z}_i \in \mathbb{Z}_q^{4n}$.

Interestingly, the process does not require any proof of knowledge of the membership secret $\mathbf{z}_i$ during the joining phase, which is round-optimal. Analogously to the Kiayias-Yung technique [54] and constructions based on structure-preserving signatures [2], the joining protocol thus remains secure in environments where many users want to register at the same time in concurrent sessions.

4.1 Description of the Scheme

Setup$(1^\lambda, 1^{N_{\mathsf{gs}}})$**:** Given a security parameter $\lambda > 0$ and the maximal expected number of group members $N_{\mathsf{gs}} = 2^\ell \in \mathsf{poly}(\lambda)$, choose lattice parameter $n = \mathcal{O}(\lambda)$; prime modulus $q = \widetilde{\mathcal{O}}(\ell n^3)$; dimension $m = 2n\lceil \log q \rceil$; Gaussian parameter $\sigma = \Omega(\sqrt{n \log q} \log n)$; infinity norm bounds $\beta = \sigma\omega(\log m)$ and $B = \sqrt{n}\omega(\log n)$. Let χ be a B-bounded distribution. Choose a hash function $H : \{0,1\}^* \to \{1,2,3\}^t$ for some $t = \omega(\log n)$, which will be modeled as a random oracle in the security analysis. Then, do the following.

1. Generate a key pair for the signature of Sect. 3.1 for signing single-block messages. Namely, run $\mathsf{TrapGen}(1^n, 1^m, q)$ to get $\mathbf{A} \in \mathbb{Z}_q^{n \times m}$ and a short basis $\mathbf{T_A}$ of $\Lambda_q^\perp(\mathbf{A})$, which allows computing short vectors in $\Lambda_q^\perp(\mathbf{A})$ with Gaussian parameter σ. Next, choose matrices $\mathbf{A}_0, \mathbf{A}_1, \ldots, \mathbf{A}_\ell, \mathbf{D} \hookleftarrow U(\mathbb{Z}_q^{n \times m})$, $\mathbf{D}_0, \mathbf{D}_1 \hookleftarrow U(\mathbb{Z}_q^{2n \times 2m})$ and a vector $\mathbf{u} \hookleftarrow U(\mathbb{Z}_q^n)$.
2. Choose an additional random matrix $\mathbf{F} \hookleftarrow U(\mathbb{Z}_q^{4n \times 4m})$ uniformly. Looking ahead, this matrix will be used to ensure security against framing attacks.
3. Generate a master key pair for the Gentry-Peikert-Vaikuntanathan IBE scheme in its multi-bit variant. This key pair consists of a statistically uniform matrix $\mathbf{B} \in \mathbb{Z}_q^{n \times m}$ and a short basis $\mathbf{T_B} \in \mathbb{Z}^{m \times m}$ of $\Lambda_q^\perp(\mathbf{B})$. This basis will allow us to compute GPV private keys with a Gaussian parameter $\sigma_{\mathrm{GPV}} \geq \|\widetilde{\mathbf{T}}_\mathbf{B}\| \cdot \sqrt{\log m}$.
4. Choose a one-time signature scheme $\Pi^{\mathrm{OTS}} = (\mathcal{G}, \mathcal{S}, \mathcal{V})$ and a hash function $H_0 : \{0,1\}^* \to \mathbb{Z}_q^{n \times 2m}$, that will be modeled as random oracles.

The group public key is defined as

$$\mathcal{Y} := \left(\mathbf{A},\ \{\mathbf{A}_j\}_{j=0}^{\ell},\ \mathbf{B},\ \mathbf{D},\ \mathbf{D}_0,\ \mathbf{D}_1,\ \mathbf{F},\ \mathbf{u},\ \Pi^{\mathrm{OTS}},\ H,\ H_0\right).$$

The opening authority's private key is $\mathcal{S}_{\mathsf{OA}} := \mathbf{T_B}$ and the private key of the group manager consists of $\mathcal{S}_{\mathsf{GM}} := \mathbf{T_A}$. The algorithm outputs $(\mathcal{Y}, \mathcal{S}_{\mathsf{GM}}, \mathcal{S}_{\mathsf{OA}})$.

Join$^{(\mathsf{GM}, \mathcal{U}_i)}$**:** the group manager GM and the prospective user $\mathcal{U}_i$ run the following interactive protocol: $[\mathsf{J}_{\mathsf{user}}(\lambda, \mathcal{Y}), \mathsf{J}_{\mathsf{GM}}(\lambda, St, \mathcal{Y}, \mathcal{S}_{\mathsf{GM}})]$

1. $\mathcal{U}_i$ samples $\mathbf{z}_i \leftarrow D_{\mathbb{Z}^{4m},\sigma}$ and computes $\mathbf{v}_i = \mathbf{F} \cdot \mathbf{z}_i \in \mathbb{Z}_q^{4n}$. He sends the vector $\mathbf{v}_i \in \mathbb{Z}_q^{4n}$, whose binary representation is $\mathsf{bin}(\mathbf{v}_i) \in \{0,1\}^{2m}$, together with an ordinary digital signature $sig_i = \mathrm{Sign}_{\mathsf{usk}[i]}(\mathbf{v}_i)$ to GM.
2. J_{GM} verifies that $\mathbf{v}_i$ was not previously used by a registered user and that sig_i is a valid signature on $\mathbf{v}_i$ w.r.t. $\mathsf{upk}[i]$. It aborts if this is not the case. Otherwise, GM chooses a fresh identifier $\mathrm{id}_i \in \{0,1\}^\ell$ and uses $\mathcal{S}_{\mathsf{GM}} = \mathbf{T_A}$ to certify $\mathcal{U}_i$ as a new group member. To this end, GM defines

$$\mathbf{A}_{\mathrm{id}_i} = \left[\mathbf{A} \;\middle|\; \mathbf{A}_0 + \textstyle\sum_{j=1}^{\ell} \mathrm{id}_i[j] \mathbf{A}_j \right] \in \mathbb{Z}_q^{n \times 2m}. \tag{6}$$

Then, GM runs $\mathbf{T}'_{\mathrm{id}_i} \leftarrow \mathsf{ExtBasis}(\mathbf{A}_{\mathrm{id}_i}, \mathbf{T_A})$ to obtain a short delegated basis $\mathbf{T}'_{\mathrm{id}_i}$ of $\Lambda_q^{\perp}(\mathbf{A}_{\mathrm{id}_i}) \in \mathbb{Z}^{2m \times 2m}$. Finally, GM samples a short vector $\mathbf{s}_i \hookleftarrow D_{\mathbb{Z}^{2m},\sigma}$ and uses the obtained delegated basis $\mathbf{T}'_{\mathrm{id}_i}$ to compute a short vector $\mathbf{d}_i = [\mathbf{d}_{i,1}^T | \mathbf{d}_{i,2}^T]^T \in \mathbb{Z}^{2m}$ such that

$$\begin{aligned} \mathbf{A}_{\mathrm{id}_i} \cdot \mathbf{d}_i &= \left[\mathbf{A} \;\middle|\; \mathbf{A}_0 + \textstyle\sum_{j=1}^{\ell} \mathrm{id}_i[j] \mathbf{A}_j \right] \cdot \mathbf{d}_i \\ &= \mathbf{u} + \mathbf{D} \cdot \mathsf{bin}\big(\mathbf{D}_0 \cdot \mathsf{bin}(\mathbf{v}_i) + \mathbf{D}_1 \cdot \mathbf{s}_i\big) \bmod q. \end{aligned} \tag{7}$$

The triple $(\mathrm{id}_i, \mathbf{d}_i, \mathbf{s}_i)$ is sent to $\mathcal{U}_i$. Then, $\mathsf{J}_{\mathsf{user}}$ verifies that the received $(\mathrm{id}_i, \mathbf{d}_i, \mathbf{s}_i)$ satisfies (7) and that $\|\mathbf{d}_i\|_\infty \leq \beta$, $\|\mathbf{s}_i\|_\infty \leq \beta$. If these conditions are not satisfied, $\mathsf{J}_{\mathsf{user}}$ aborts. Otherwise, $\mathsf{J}_{\mathsf{user}}$ defines the membership certificate as $\mathsf{cert}_i = (\mathrm{id}_i, \mathbf{d}_i, \mathbf{s}_i)$. The membership secret sec_i is defined to be $\mathsf{sec}_i = \mathbf{z}_i \in \mathbb{Z}^{4m}$. J_{GM} stores $\mathsf{transcript}_i = (\mathbf{v}_i, \mathsf{cert}_i, i, \mathsf{upk}[i], sig_i)$ in the database St_{trans} of joining transcripts.

Sign$(\mathcal{Y}, \mathsf{cert}_i, \mathsf{sec}_i, M)$**:** To sign M using $\mathsf{cert}_i = (\mathrm{id}_i, \mathbf{d}_i, \mathbf{s}_i)$, where $\mathbf{d}_i \in \mathbb{Z}^{2m}$ and $\mathbf{s}_i \in \mathbb{Z}^{2m}$, as well as the membership secret $\mathsf{sec}_i = \mathbf{z}_i \in \mathbb{Z}^{4m}$, $\mathcal{U}_i$ generates a one-time signature key pair $(\mathsf{VK}, \mathsf{SK}) \leftarrow \mathcal{G}(n)$ and does the following.

1. Compute $\mathbf{G}_0 = H_0(\mathsf{VK}) \in \mathbb{Z}_q^{n \times 2m}$ and use it as an IBE public key to encrypt $\mathsf{bin}(\mathbf{v}_i) \in \{0,1\}^{2m}$, where $\mathbf{v}_i = \mathbf{F} \cdot \mathbf{z}_i \in \mathbb{Z}_q^{4n}$ is the syndrome of $\mathsf{sec}_i = \mathbf{z}_i \in \mathbb{Z}^{4m}$ for the matrix $\mathbf{F}$. Namely, compute $\mathbf{c}_{\mathbf{v}_i} \in \mathbb{Z}_q^m \times \mathbb{Z}_q^{2m}$ as

$$\mathbf{c}_{\mathbf{v}_i} = (\mathbf{c}_1, \mathbf{c}_2) = \big(\mathbf{B}^T \cdot \mathbf{e}_0 + \mathbf{x}_1,\; \mathbf{G}_0^T \cdot \mathbf{e}_0 + \mathbf{x}_2 + \mathsf{bin}(\mathbf{v}_i) \cdot \lfloor q/2 \rfloor\big) \tag{8}$$

for randomly chosen $\mathbf{e}_0 \hookleftarrow \chi^n$, $\mathbf{x}_1 \hookleftarrow \chi^m, \mathbf{x}_2 \hookleftarrow \chi^{2m}$. Notice that, as in the construction of [62], the columns of $\mathbf{G}_0$ can be interpreted as public keys for the multi-bit version of the dual Regev encryption scheme.
2. Run the protocol in Sect. 5.5 to prove the knowledge of $\mathrm{id}_i \in \{0,1\}^\ell$, vectors $\mathbf{s}_i \in \mathbb{Z}^{2m}, \mathbf{d}_{i,1}, \mathbf{d}_{i,2} \in \mathbb{Z}^m, \mathbf{z}_i \in \mathbb{Z}^{4m}$ with infinity norm bound β; $\mathbf{e}_0 \in \mathbb{Z}^n$, $\mathbf{x}_1 \in \mathbb{Z}^m, \mathbf{x}_2 \in \mathbb{Z}^{2m}$ with infinity norm bound B and $\mathsf{bin}(\mathbf{v}_i) \in \{0,1\}^{2m}, \mathbf{w}_i \in \{0,1\}^m$, that satisfy (8) as well as

$$\mathbf{A} \cdot \mathbf{d}_{i,1} + \mathbf{A}_0 \cdot \mathbf{d}_{i,2} + \sum_{j=1}^{\ell} (\mathrm{id}_i[j] \cdot \mathbf{d}_{i,2}) \cdot \mathbf{A}_j - \mathbf{D} \cdot \mathbf{w}_i = \mathbf{u} \in \mathbb{Z}_q^n \tag{9}$$

and

$$\begin{cases} \mathbf{H}_{2n\times m} \cdot \mathbf{w}_i = \mathbf{D}_0 \cdot \mathsf{bin}(\mathbf{v}_i) + \mathbf{D}_1 \cdot \mathbf{s}_i \in \mathbb{Z}_q^{2n} \\ \mathbf{F} \cdot \mathbf{z}_i = \mathbf{H}_{4n\times 2m} \cdot \mathsf{bin}(\mathbf{v}_i) \in \mathbb{Z}_q^{4n}. \end{cases} \tag{10}$$

The protocol is repeated $t = \omega(\log n)$ times in parallel to achieve negligible soundness error, and then made non-interactive using the Fiat-Shamir heuristic [40] as a triple $\pi_K = (\{\mathsf{Comm}_{K,j}\}_{j=1}^t, \mathsf{Chall}_K, \{\mathsf{Resp}_{K,j}\}_{j=1}^t)$, where $\mathsf{Chall}_K = H(M, \mathsf{VK}, \mathbf{c}_{\mathbf{v}_i}, \{\mathsf{Comm}_{K,j}\}_{j=1}^t) \in \{1,2,3\}^t$

3. Compute a one-time signature $sig = \mathcal{S}(\mathsf{SK}, (\mathbf{c}_{\mathbf{v}_i}, \pi_K))$.

Output the signature that consists of

$$\Sigma = (\mathsf{VK}, \mathbf{c}_{\mathbf{v}_i}, \pi_K, sig). \tag{11}$$

Verify$(\mathcal{Y}, M, \Sigma)$**:** Parse the signature Σ as in (11). Then, return 1 if and only if: (i) $\mathcal{V}(\mathsf{VK}, (\mathbf{c}_{\mathbf{v}_i}, \mathbf{c}_{\mathbf{s}_i}, \mathbf{c}_{\mathrm{id}}, \pi_K), sig) = 1$; (ii) The proof π_K properly verifies.

Open$(\mathcal{Y}, \mathcal{S}_{\mathsf{OA}}, M, \Sigma)$**:** Parse $\mathcal{S}_{\mathsf{OA}}$ as $\mathbf{T}_\mathbf{B} \in \mathbb{Z}^{m\times m}$ and Σ as in (11).

1. Compute $\mathbf{G}_0 = H_0(\mathsf{VK}) \in \mathbb{Z}_q^{n\times 2m}$. Then, using $\mathbf{T}_\mathbf{B}$ to compute a small-norm matrix $\mathbf{E}_{0,\mathsf{VK}} \in \mathbb{Z}^{m\times 2m}$ such that $\mathbf{B} \cdot \mathbf{E}_{0,\mathsf{VK}} = \mathbf{G}_0 \bmod q$.
2. Using $\mathbf{E}_{0,\mathsf{VK}}$, decrypt $\mathbf{c}_{\mathbf{v}_i}$ to obtain a string $\mathsf{bin}(\mathbf{v}) \in \{0,1\}^{2m}$ (i.e., by computing $\lfloor (\mathbf{c}_2 - \mathbf{E}_{0,\mathsf{VK}}^T \cdot \mathbf{c}_1)/(q/2) \rceil$).
3. Determine if the $\mathsf{bin}(\mathbf{v}) \in \{0,1\}^{2m}$ obtained at step 2 corresponds to a vector $\mathbf{v} = \mathbf{H}_{4n\times 2m} \cdot \mathsf{bin}(\mathbf{v}) \bmod q$ that appears in a record $\mathsf{transcript}_i = (\mathbf{v}, \mathsf{cert}_i, i, \mathsf{upk}[i], sig_i)$ of the database St_{trans} for some i. If so, output the corresponding i (and, optionally, $\mathsf{upk}[i]$). Otherwise, output $\perp$.

We remark that the scheme readily extends to provide a mechanism whereby the opening authority can efficiently prove that signatures were correctly opened at each opening operation. The difference between the dynamic group signature models suggested by Kiayias and Yung [55] and Bellare *et al.* [13] is that, in the latter, the opening authority (OA) must be able to convince a judge that the Open algorithm was run correctly. Here, such a mechanism can be realized using the techniques of public-key encryption with non-interactive opening [37]. Namely, since $\mathsf{bin}(\mathbf{v}_i)$ is encrypted using an IBE scheme for the identity VK, the OA can simply reveal the decryption matrix $\mathbf{E}_{0,\mathsf{VK}}$, that satisfies $\mathbf{B} \cdot \mathbf{E}_{0,\mathsf{VK}} = \mathbf{G}_0 \bmod q$ (which corresponds to the verification of a GPV signature) and allows the verifier to perform step 2 of the opening algorithm himself. The resulting construction is easily seen to satisfy the notion of opening soundness of Sakai *et al.* [71].

4.2 Efficiency and Correctness

Efficiency. The given dynamic group signature scheme can be implemented in polynomial time. The group public key has total bit-size $\mathcal{O}(\ell n m \log q) = \widetilde{\mathcal{O}}(\lambda^2) \cdot \log N_{\mathsf{gs}}$. The secret signing key of each user consists of a small constant number of low-norm vectors, and has bit-size $\widetilde{\mathcal{O}}(\lambda)$.

The size of each group signature is largely dominated by that of the non-interactive argument π_K, which is obtained from the Stern-like protocol of Sect. 5.5. Each round of the protocol has communication cost $\widetilde{\mathcal{O}}(m \cdot \log q) \cdot \log N_{\mathsf{gs}}$. Thus, the bit-size of π_K is $t \cdot \widetilde{\mathcal{O}}(m \cdot \log q) \cdot \log N_{\mathsf{gs}} = \widetilde{\mathcal{O}}(\lambda) \cdot \log N_{\mathsf{gs}}$. This is also the asymptotic bound on the size of the group signature.

Correctness. The correctness of algorithm $\mathsf{Verify}(\mathcal{Y}, M, \Sigma)$ follows from the facts that every certified group member is able to compute valid witness vectors satisfying Eqs. (8), (9) and (10), and that the underlying argument system is perfectly complete. Moreover, the scheme parameters are chosen so that the GPV IBE [42] is correct, which implies that algorithm $\mathsf{Open}(\mathcal{Y}, \mathcal{S}_{\mathsf{OA}}, M, \Sigma)$ is also correct.

4.3 Security Analysis

Due to the fact that the number of public matrices $\{\mathbf{A}_j\}_{j=0}^{\ell}$ is only logarithmic in $N_{\mathsf{gs}} = 2^{\ell}$ instead of being linear in the security parameter λ, the proof of security against misidentification attacks (as defined in the full version of this paper and in [53]) cannot rely on the security of our signature scheme in a modular manner. The reason is that, at each run of the Join protocol, the group manager maintains a state and, instead of choosing the ℓ-bit identifier id uniformly in $\{0,1\}^{\ell}$, it chooses an identifier that has not been used yet. Since $\ell \ll \lambda$ (given that $N_{\mathsf{gs}} = 2^{\ell}$ is polynomial in λ), we thus have to prove security from scratch. However, the strategy of the reduction is exactly the same as in the security proof of the signature scheme.

The proofs of the following theorems are given in the full version of the paper.

Theorem 4. *The scheme is secure against misidentification attacks under the* $\mathsf{SIS}_{n,2m,q,\beta'}$ *assumption, for* $\beta' = \mathcal{O}(\ell\sigma^2 m^{3/2})$.

Theorem 5. *The scheme is secure against framing attacks under the* $\mathsf{SIS}_{4n,4m,q,\beta''}$ *assumption, where* $\beta'' = 4\sigma\sqrt{m}$.

Theorem 6. *In the random oracle model, the scheme provides CCA-anonymity if the* $\mathsf{LWE}_{n,q,\chi}$ *assumption holds and if* Π^{OTS} *is a strongly unforgeable one-time signature.*

5 Supporting Zero-Knowledge Argument Systems

This section provides a general framework that allows obtaining zero-knowledge arguments of knowledge (ZKAoK) for many relations appearing in lattice-based cryptography. Since lattice-based cryptosystems are built upon the hardness of the SIS and LWE problems, the relations among objects of the schemes are typically represented by modular linear equations. Thanks to the linearity property, we can often unify the given equations into one equation of the form:

$$\mathbf{P} \cdot \mathbf{x} = \mathbf{v} \bmod q, \tag{12}$$

where (**P**, **v**) are public and **x** is a secret vector (or matrix) that possesses some constraints to be proven in zero-knowledge, e.g., its smallness (like a SIS solution or an LWE noise) or a special arrangement of its entries. Starting from this high-level observation, we look for a tool that handles these constraints well.

Stern's protocol [72], originally proposed in the context of code-based cryptography, appears to be well-suited for our purpose. Stern's main idea is simple, yet elegant: To prove that a binary vector **x** has the fixed-Hamming-weight constraint, simply send the verifier a random permutation $\pi(\mathbf{x})$ which should guarantee that the constraint is satisfied while leaking no additional information about **x**. Ling *et al.* [61] developed this idea to handle the smallness constraint, via a technique called Decomposition-Extension. This technique decomposes a vector with small infinity norm $B \geq 1$ into $\lfloor \log_2 B \rfloor + 1$ vectors with infinity norm 1, and then, extends these vectors into elements of sets that are closed under permutations. Several subsequent works [57,62][60] employed the techniques of [61,72] in different contexts, but did not address the applicability and flexibility of the protocol in an abstract, generalized manner.

In Sect. 5.1, we abstract Stern's protocol to capture many relations that naturally appear in lattice-based cryptography. In particular, the argument systems used in our signature with efficient protocols (Sect. 3) and dynamic group signature (Sect. 4) can all be derived from this abstract protocol, which we will demonstrate in Sects. 5.3, 5.4 and 5.5, respectively.

We note that several works [15,51,73] addressed the problem of proving multiplicative and additive relations among committed linear objects (matrices and vectors over $\mathbb{Z}_q$) in lattice-based cryptography. These results, however, do not yield a simple solution for the relations involved in our schemes. If we were to plug proof systems like [15,51,73] in our relations, we would need to commit to all objects using perfectly binding commitments (which would require very long commitment keys) and express the relations in terms of many multiplications and additions gates before running many instances of the proof systems depending on the circuit. Instead of considering general circuits, our framework aims at a more direct (but still fairly general) solution for a large class of relations that naturally appear in SIS and LWE-based cryptography.

5.1 Abstracting Stern's Protocol

Let $D, L, q \geq 2$ be positive integers let VALID be a subset of $\{-1,0,1\}^L$. Suppose that $\mathcal{S}$ is a finite set such that one can associate every $\pi \in \mathcal{S}$ with a permutation T_π of L elements, satisfying the following conditions:

$$\begin{cases} \mathbf{x} \in \mathsf{VALID} \Longleftrightarrow T_\pi(\mathbf{x}) \in \mathsf{VALID}, \\ \text{If } \mathbf{x} \in \mathsf{VALID} \text{ and } \pi \text{ is uniform in } \mathcal{S}, \text{ then } T_\pi(\mathbf{x}) \text{ is uniform in } \mathsf{VALID}. \end{cases} \tag{13}$$

We aim to construct a statistical ZKAoK for the following abstract relation:

$$\mathrm{R}_{\mathsf{abstract}} = \{(\mathbf{P}, \mathbf{v}), \mathbf{x} \in \mathbb{Z}_q^{D \times L} \times \mathbb{Z}_q^D \times \mathsf{VALID} : \mathbf{P} \cdot \mathbf{x} = \mathbf{v} \bmod q.\}$$

Note that, Stern's original protocol corresponds to the special case when $\mathsf{VALID} = \{\mathbf{x} \in \{0,1\}^L : \mathsf{wt}(\mathbf{x}) = k\}$ (where $\mathsf{wt}(\cdot)$ denotes the Hamming weight and $k < L$ is a given integer), $\mathcal{S} = \mathcal{S}_L$ - hereunder the set of all permutations of L elements, and $T_\pi(\mathbf{x}) = \pi(\mathbf{x})$.

The conditions in (13) play a crucial role in proving in ZK that $\mathbf{x} \in \mathsf{VALID}$: To do so, the prover samples $\pi \hookleftarrow U(\mathcal{S})$ and let the verifier check that $T_\pi(\mathbf{x}) \in \mathsf{VALID}$, while the latter cannot learn any additional information about $\mathbf{x}$ thanks to the randomness of π. Furthermore, to prove in ZK that the linear equation holds, the prover samples a masking vector $\mathbf{r} \hookleftarrow U(\mathbb{Z}_q^L)$, sends $\mathbf{y} = \mathbf{x} + \mathbf{r} \bmod q$, and convinces the verifier instead that $\mathbf{P} \cdot \mathbf{y} = \mathbf{P} \cdot \mathbf{r} + \mathbf{v} \bmod q$.

The interactive protocol between the prover and the verifier with common input $(\mathbf{P}, \mathbf{v})$ and prover's secret input $\mathbf{x}$ is described in Fig. 1. The protocol employs a statistically hiding and computationally binding string commitment scheme COM (e.g., the SIS-based one from [52]).

1. **Commitment:** Prover samples $\mathbf{r} \hookleftarrow U(\mathbb{Z}_q^L)$, $\pi \hookleftarrow U(\mathcal{S})$ and randomness ρ_1, ρ_2, ρ_3 for COM. Then he sends $\mathrm{CMT} = \big(C_1, C_2, C_3\big)$ to the verifier, where

$$C_1 = \mathsf{COM}(\pi, \mathbf{P} \cdot \mathbf{r}; \rho_1), \quad C_2 = \mathsf{COM}(T_\pi(\mathbf{r}); \rho_2), \quad C_3 = \mathsf{COM}(T_\pi(\mathbf{x} + \mathbf{r}); \rho_3).$$

2. **Challenge:** The verifier sends a challenge $Ch \hookleftarrow U(\{1,2,3\})$ to the prover.
3. **Response:** Depending on Ch, the prover sends RSP computed as follows:
   - $Ch = 1$: Let $\mathbf{t}_x = T_\pi(\mathbf{x})$, $\mathbf{t}_r = T_\pi(\mathbf{r})$, and $\mathrm{RSP} = (\mathbf{t}_x, \mathbf{t}_r, \rho_2, \rho_3)$.
   - $Ch = 2$: Let $\pi_2 = \pi$, $\mathbf{y} = \mathbf{x} + \mathbf{r}$, and $\mathrm{RSP} = (\pi_2, \mathbf{y}, \rho_1, \rho_3)$.
   - $Ch = 3$: Let $\pi_3 = \pi$, $\mathbf{r}_3 = \mathbf{r}$, and $\mathrm{RSP} = (\pi_3, \mathbf{r}_3, \rho_1, \rho_2)$.

Verification: Receiving RSP, the verifier proceeds as follows:

- $Ch = 1$: Check that $\mathbf{t}_x \in \mathsf{VALID}$ and $C_2 = \mathsf{COM}(\mathbf{t}_r; \rho_2)$, $C_3 = \mathsf{COM}(\mathbf{t}_x + \mathbf{t}_r; \rho_3)$.
- $Ch = 2$: Check that $C_1 = \mathsf{COM}(\pi_2, \mathbf{P} \cdot \mathbf{y} - \mathbf{v}; \rho_1)$, $C_3 = \mathsf{COM}(T_{\pi_2}(\mathbf{y}); \rho_3)$.
- $Ch = 3$: Check that $C_1 = \mathsf{COM}(\pi_3, \mathbf{P} \cdot \mathbf{r}_3; \rho_1)$, $C_2 = \mathsf{COM}(T_{\pi_3}(\mathbf{r}_3); \rho_2)$.

In each case, the verifier outputs 1 if and only if all the conditions hold.

Fig. 1. A ZKAoK for the relation $\mathrm{R}_{\mathrm{abstract}}$.

The properties of the given protocol are summarized in the following lemma.

Lemma 8. *The protocol in Fig. 1 is a statistical* ZKAoK *for the relation* $\mathrm{R}_{\mathrm{abstract}}$ *with perfect completeness, soundness error* $2/3$*, and communication cost* $\widetilde{\mathcal{O}}(L \log q)$*. In particular:*

- *There exists an efficient simulator that, on input* $(\mathbf{P}, \mathbf{v})$*, outputs an accepted transcript which is statistically close to that produced by the real prover.*
- *There exists an efficient knowledge extractor that, on input a commitment* CMT *and* 3 *valid responses* $(\mathrm{RSP}_1, \mathrm{RSP}_2, \mathrm{RSP}_3)$ *to all* 3 *possible values of the challenge* Ch*, outputs* $\mathbf{x}' \in \mathsf{VALID}$ *such that* $\mathbf{P} \cdot \mathbf{x}' = \mathbf{v} \bmod q$.

The proof of Lemma 8 employs standard simulation and extraction techniques for Stern-like protocols [52,61,62]. It is detailed in the full version of the paper.

5.2 Supporting Notations and Techniques

Below we will describe the notations and techniques, adapted from recent works on Stern-like protocols [39,57,60,61], that we will employ in the next subsections to handle 3 different constraints of the witness vectors.

Let m be an arbitrary dimension, and B be an arbitrary infinity norm bound.

Case 1: $\mathbf{w} \in \{0,1\}^m$. We denote by B_m^2 the set of all vectors in $\{0,1\}^{2m}$ having exactly m coordinates equal to 1. We also let $\mathsf{Ext}_{2m}(\mathbf{w})$ be the algorithm that outputs a vector $\hat{\mathbf{w}} \in \mathsf{B}_m^2$ by appending m suitable coordinates to $\mathbf{w} \in \{0,1\}^m$. Note that, for any permutation $\rho \in \mathcal{S}_{2m}$, we have $\hat{\mathbf{w}} \in \mathsf{B}_m^2 \Leftrightarrow \rho(\hat{\mathbf{w}}) \in \mathsf{B}_m^2$.

Case 2: $\mathbf{w} \in [-B,B]^m$. We define $\delta_B := \lfloor \log_2 B \rfloor + 1$ and denote by $\mathsf{B}_{m\delta_B}^3$ the set of vectors in $\{-1,0,1\}^{3m\delta_B}$ with exactly $m\delta_B$ coordinates equal to j, for every $j \in \{-1,0,1\}$. The Decomposition-Extension technique from [61] consists in transforming $\mathbf{w} \in [-B,B]^m$ to a vector $\mathsf{DecExt}_{m,B}(\mathbf{w}) \in \mathsf{B}_{m\delta_B}^3$, as follows.

Define the sequence $B_1, \ldots, B_{\delta_B}$, where $B_j = \lfloor \frac{B+2^{j-1}}{2^j} \rfloor$ for all $j \in [1, \delta_B]$. As noted in [61], it satisfies $\sum_{j=1}^{\delta_B} B_j = B$, and for any $w \in [-B,B]$, one can efficiently compute $w^{(1)}, \ldots, w^{(\delta_B)} \in \{-1,0,1\}$ such that $\sum_{j=1}^{\delta_B} B_j \cdot w^{(j)} = w$. Next, define the matrix

$$\mathbf{K}_{m,B} = \mathbf{I}_m \otimes [B_1|\ldots|B_{\delta_B}] = \begin{bmatrix} B_1 \ldots B_{\delta_B} & & \\ & \ddots & \\ & & B_1 \ldots B_{\delta_B} \end{bmatrix} \in \mathbb{Z}^{m \times m\delta_B},$$

and its extension $\widehat{\mathbf{K}}_{m,B} = \left[\mathbf{K}_{m,B} \middle| \mathbf{0}^{m \times 2m\delta_B}\right] \in \mathbb{Z}^{m \times 3m\delta_B}$.

If we let $\mathbf{w} = (w_1, \ldots, w_m)^T$, then we can compute

$$\mathbf{w}' = \left(w_1^{(1)}, \ldots, w_1^{(\delta_B)}, \ldots, w_m^{(1)}, \ldots, w_m^{(\delta_B)}\right)^T \in \{-1,0,1\}^{m\delta_B}$$

satisfying $\mathbf{K}_{m,B} \cdot \mathbf{w}' = \mathbf{w}$. By appending $2m\delta_B$ suitable coordinates to $\mathbf{w}'$, we can obtain $\hat{\mathbf{w}} \in \mathsf{B}_{m\delta_B}^3$ satisfying $\widehat{\mathbf{K}}_{m,B} \cdot \hat{\mathbf{w}} = \mathbf{w}$.

Note that for any $\phi \in \mathcal{S}_{3m\delta_B}$, we have $\hat{\mathbf{w}} \in \mathsf{B}_{m\delta_B}^3 \Leftrightarrow \phi(\hat{\mathbf{w}}) \in \mathsf{B}_{m\delta_B}^3$.

Case 3: $\mathbf{w} \in \{0,1\}^{2m}$ is the correct encoding of some $\mathbf{t} \in \{0,1\}^m$.

Recall that the encoding function from Sect. 3.3, hereunder denoted by Encode_m if the input is a binary vector of length m, extends $\mathbf{t} = (t_1, \ldots, t_m)^T$ to $\mathsf{Encode}_m(\mathbf{t}) = (\bar{t}_1, t_1, \ldots, \bar{t}_m, t_m)$. We define $\mathsf{CorEnc}(m) = \{\mathbf{w} = \mathsf{Encode}_m(\mathbf{t}) : \mathbf{t} \in \{0,1\}^m\}$ - the set of all correct encodings of m-bit vectors. To handle the constraint $\mathbf{w} \in \mathsf{CorEnc}(m)$, we adapt the permuting technique from [39,57,60].

For $\mathbf{b} = (b_1, \ldots, b_m)^T \in \{0,1\}^m$, we let $E_\mathbf{b}$ be the permutation transforming vector $\mathbf{w} = (w_1^0, w_1^1, \ldots, w_m^0, w_m^1) \in \mathbb{Z}^{2m}$ to $E_\mathbf{b}(\mathbf{w}) = (w_1^{b_1}, w_1^{\bar{b}_1}, \ldots, w_m^{b_m}, w_m^{\bar{b}_m})$. Note that, $E_\mathbf{b}$ transforms $\mathbf{w} = \mathsf{Encode}_m(\mathbf{t})$ to $E_\mathbf{b}(\mathbf{w}) = \mathsf{Encode}_m(\mathbf{t} \oplus \mathbf{b})$, where $\oplus$ denotes the bit-wise addition modulo 2. Thus, for any $\mathbf{b} \in \{0,1\}^m$, we have

$$\mathbf{w} \in \mathsf{CorEnc}(m) \Leftrightarrow E_\mathbf{b}(\mathbf{w}) \in \mathsf{CorEnc}(m).$$

5.3 Proving the Consistency of Commitments

The argument system used in our protocol for signing a committed value in Sect. 3.3 can be summarized as follows.

Common Input: Matrices $\{\mathbf{D}_k \in \mathbb{Z}_q^{2n\times 2m}\}_{k=0}^N$; $\mathbf{B} \in \mathbb{Z}_q^{n\times m}$; $\mathbf{G}_1 \in \mathbb{Z}_q^{n\times 2m}$; vectors $\mathbf{c}_{\mathfrak{m}} \in \mathbb{Z}_q^{2n}$; $\{\mathbf{c}_{k,1} \in \mathbb{Z}_q^{m}\}_{k=1}^N$; $\{\mathbf{c}_{k,2} \in \mathbb{Z}_q^{2m}\}_{k=1}^N$; $\mathbf{c}_{\mathbf{s}',1} \in \mathbb{Z}_q^{m}$; $\mathbf{c}_{\mathbf{s}',2} \in \mathbb{Z}_q^{2m}$.

Prover's Input: $\mathfrak{m} = (\mathfrak{m}_1^T\|\ldots\|\mathfrak{m}_N^T)^T \in \mathsf{CorEnc}(mN)$; $\{\mathbf{s}_k \in [-B,B]^n, \mathbf{e}_{k,1} \in [-B,B]^m; \mathbf{e}_{k,2} \in [-B,B]^{2m}\}_{k=1}^N$; $\mathbf{s}_0 \in [-B,B]^n$; $\mathbf{e}_{0,1} \in [-B,B]^m; \mathbf{e}_{0,2} \in [-B,B]^{2m}$; $\mathbf{s}' \in [-(p-1),(p-1)]^{2m}$

Prover's Goal: Convince the verifier in ZK that:

$$\begin{cases}\mathbf{c}_{\mathfrak{m}} = \mathbf{D}_0\cdot\mathbf{s}' + \sum_{k=1}^N \mathbf{D}_k\cdot\mathfrak{m}_k \bmod q;\\ \mathbf{c}_{\mathbf{s}',1} = \mathbf{B}^T\cdot\mathbf{s}_0 + \mathbf{e}_{0,1} \bmod q; \mathbf{c}_{\mathbf{s}',2} = \mathbf{G}_1^T\cdot\mathbf{s}_0 + \mathbf{e}_{0,2} + \lfloor q/p\rfloor\cdot\mathbf{s}' \bmod q;\\ \forall k\in[N]: \mathbf{c}_{k,1} = \mathbf{B}^T\cdot\mathbf{s}_k + \mathbf{e}_{k,1}; \mathbf{c}_{k,2} = \mathbf{G}_1^T\cdot\mathbf{s}_k + \mathbf{e}_{k,2} + \lfloor q/2\rfloor\cdot\mathfrak{m}_k.\end{cases} \quad (14)$$

We will show that the above argument system can be obtained from the one in Sect. 5.1. We proceed in two steps.

Step 1: *Transforming the equations in (14) into a unified one of the form* $\mathbf{P}\cdot\mathbf{x} = \mathbf{v} \bmod q$*, where* $\|\mathbf{x}\|_\infty = 1$ *and* $\mathbf{x}\in\mathsf{VALID}$ *- a "specially-designed" set.*

To do so, we first form the following vectors and matrices:

$$\begin{cases}\mathbf{x}_1 = \left(\mathbf{s}_0^T\|\mathbf{e}_{0,1}^T\|\mathbf{e}_{0,2}^T\|\mathbf{s}_1^T\|\mathbf{e}_{1,1}^T\|\mathbf{e}_{1,2}^T\|\ldots\|\mathbf{s}_N^T\|\mathbf{e}_{N,1}^T\|\mathbf{e}_{N,2}^T\right)^T \in [-B,B]^{(n+3m)(N+1)};\\ \mathbf{v} = \left(\mathbf{c}_{\mathfrak{m}}^T\|\mathbf{c}_{\mathbf{s}',1}^T\|\mathbf{c}_{\mathbf{s}',2}^T\|\mathbf{c}_{1,1}^T\|\mathbf{c}_{1,2}^T\|\ldots\|\mathbf{c}_{N,1}^T\|\mathbf{c}_{N,2}^T\right)^T \in \mathbb{Z}_q^{2n+3m(N+1)};\\ \mathbf{P}_1 = \left(\begin{array}{c|c}\mathbf{B}^T & \\ \hline \mathbf{G}_1^T & \mathbf{I}_{3m}\end{array}\right); \mathbf{Q}_2 = \left(\begin{array}{c}\mathbf{0}\\ \hline \lfloor\frac{q}{2}\rfloor\mathbf{I}_{2m}\end{array}\right); \mathbf{Q}_p = \left(\begin{array}{c}\mathbf{0}\\ \hline \lfloor\frac{q}{p}\rfloor\mathbf{I}_{2m}\end{array}\right)\\ \mathbf{M}_1 = \left(\begin{array}{cccc} & \mathbf{0} & & \\ \hline \mathbf{P}_1 & & & \\ & \mathbf{P}_1 & & \\ & & \ddots & \\ & & & \mathbf{P}_1\end{array}\right); \mathbf{M}_2 = \left(\begin{array}{ccc}\mathbf{D}_1|\ldots|\mathbf{D}_N & & \\ \hline & \mathbf{0} & \\ \hline \mathbf{Q}_2 & & \\ & \ddots & \\ & & \mathbf{Q}_2\end{array}\right); \mathbf{M}_3 = \left(\begin{array}{c}\mathbf{D}_0\\ \hline \mathbf{Q}_p\\ \hline \\ \mathbf{0}\end{array}\right).\end{cases}$$

We then observe that (14) can be rewritten as:

$$\mathbf{M}_1\cdot\mathbf{x}_1 + \mathbf{M}_2\cdot\mathfrak{m} + \mathbf{M}_3\cdot\mathbf{s}' = \mathbf{v} \in \mathbb{Z}_q^D, \quad (15)$$

where $D = 2n + 3m(N+1)$. Now we employ the techniques from Sect. 5.2 to convert (15) into the form $\mathbf{P}\cdot\mathbf{x} = \mathbf{v} \bmod q$. Specifically, if we let:

$$\begin{cases}\mathsf{DecExt}_{(n+3m)(N+1),B}(\mathbf{x}_1) \rightarrow \hat{\mathbf{x}}_1 \in \mathsf{B}^3_{(n+3m)(N+1)\delta_B};\\ \mathbf{M}_1' = \mathbf{M}_1\cdot\widehat{\mathbf{K}}_{(n+3m)(N+1),B} \in \mathbb{Z}_q^{D\times 3(n+3m)(N+1)\delta_B};\\ \mathsf{DecExt}_{2m,p-1}(\mathbf{s}') \rightarrow \hat{\mathbf{s}} \in \mathsf{B}^3_{2m\delta_{p-1}}; \mathbf{M}_3' = \mathbf{M}_3\cdot\widehat{\mathbf{K}}_{2m,p-1} \in \mathbb{Z}_q^{D\times 6m\delta_{p-1}},\end{cases}$$

$L = 3(n+3m)(N+1)\delta_B + 2mN + 6m\delta_{p-1}$, and $\mathbf{P} = \left[\mathbf{M}_1'|\mathbf{M}_2|\mathbf{M}_3'\right] \in \mathbb{Z}_q^{D\times L}$, and $\mathbf{x} = \left(\hat{\mathbf{x}}_1^T \| \mathfrak{m}^T \| \hat{\mathbf{s}}^T\right)^T$, then we will obtain the desired equation:

$$\mathbf{P}\cdot\mathbf{x} = \mathbf{v} \bmod q.$$

Having performed the above unification, we now define VALID as the set of all vectors $\mathbf{t} \in \{-1,0,1\}^L$ of the form $\mathbf{t} = \left(\mathbf{t}_1^T\|\mathbf{t}_2^T\|\mathbf{t}_3^T\right)^T$, where $\mathbf{t}_1 \in \mathsf{B}^3_{(n+3m)(N+1)\delta_B}$, $\mathbf{t}_2 \in \mathsf{CorEnc}(mN)$, and $\mathbf{t}_3 \in \mathsf{B}^3_{2m\delta_{p-1}}$. Note that $\mathbf{x} \in$ VALID.

Step 2: *Specifying the set $\mathcal{S}$ and permutations of L elements $\{T_\pi : \pi \in \mathcal{S}\}$ for which the conditions in (13) hold.*

- Define $\mathcal{S} := \mathcal{S}_{3(n+3m)(N+1)\delta_B} \times \{0,1\}^{mN} \times \mathcal{S}_{6m\delta_{p-1}}$.
- For $\pi = (\pi_1, \mathbf{b}, \pi_3) \in \mathcal{S}$, and for vector $\mathbf{w} = \left(\mathbf{w}_1^T\|\mathbf{w}_2^T\|\mathbf{w}_3^T\right)^T \in \mathbb{Z}_q^L$, where $\mathbf{w}_1 \in \mathbb{Z}_q^{3(n+3m)(N+1)\delta_B}$, $\mathbf{w}_2 \in \mathbb{Z}_q^{2mN}$, $\mathbf{w}_3 \in \mathbb{Z}_q^{6m\delta_{p-1}}$, we define:

$$T_\pi = \left(\pi_1(\mathbf{w}_1)^T \| E_{\mathbf{b}}(\mathbf{w}_2)^T \| \pi_3(\mathbf{w}_3)^T\right)^T.$$

By inspection, it can be seen that the properties in (13) are satisfied, as desired. As a result, we can obtain the required argument system by running the protocol in Sect. 5.1 with common input $(\mathbf{P}, \mathbf{v})$ and prover's input $\mathbf{x}$.

5.4 Proving the Possession of a Signature on a Committed Value

We now describe how to derive the protocol for proving the possession of a signature on a committed value, that is used in Sect. 3.3.

Common Input: Matrices $\mathbf{A}, \{\mathbf{A}_j\}_{j=0}^{\ell}, \mathbf{D} \in \mathbb{Z}_q^{n\times m}$; $\{\mathbf{D}_k \in \mathbb{Z}_q^{2n\times 2m}\}_{k=0}^N$; $\mathbf{B} \in \mathbb{Z}_q^{n\times m}$; $\mathbf{G}_1 \in \mathbb{Z}_q^{n\times 2m}$; $\mathbf{G}_0 \in \mathbb{Z}_q^{n\times \ell}$; vectors $\{\mathbf{c}_{k,1}\}_{k=1}^N, \mathbf{c}_{\tau,1}, \mathbf{c}_{\mathbf{v},1}, \mathbf{c}_{s,1} \in \mathbb{Z}_q^m$; $\{\mathbf{c}_{k,2}\}_{k=1}^N, \mathbf{c}_{\mathbf{v},2}, \mathbf{c}_{s,2} \in \mathbb{Z}_q^{2m}$; $\mathbf{c}_{\tau,2} \in \mathbb{Z}_q^{\ell}$; $\mathbf{u} \in \mathbb{Z}_q^n$.

Prover's Input: $\mathbf{v} = \begin{pmatrix}\mathbf{v}_1\\ \mathbf{v}_2\end{pmatrix}$, where $\mathbf{v}_1, \mathbf{v}_2 \in [-\beta,\beta]^m$ and $\beta = \sigma\cdot\omega(\log m)$ - the infinity norm bound of signatures; $\tau \in \{0,1\}^\ell$; $\mathbf{s} \in [-(p-1),(p-1)]^{2m}$; $\mathfrak{m} = (\mathfrak{m}_1^T\|\ldots\|\mathfrak{m}_N^T)^T \in \mathsf{CorEnc}(mN)$; $\{\mathbf{s}_k\}_{k=1}^N$, $\mathbf{s}_{\mathbf{v}}$, $\mathbf{s}_0$, $\mathbf{s}_\tau \in [-B,B]^n$; $\{\mathbf{e}_{k,1}\}_{k=1}^N$, $\mathbf{e}_{\mathbf{v},1}, \mathbf{e}_{0,1}, \mathbf{e}_{\tau,1} \in [-B,B]^m$; $\{\mathbf{e}_{k,2}\}_{k=1}^N, \mathbf{e}_{0,2}, \mathbf{e}_{\mathbf{v},2} \in [-B,B]^{2m}$; $\mathbf{e}_{\tau,2} \in [-B,B]^\ell$.

Prover's Goal: Convince the verifier in ZK that:

$$\mathbf{A}\cdot\mathbf{v}_1 + \mathbf{A}_0\cdot\mathbf{v}_2 + \sum_{i=1}^{\ell}\mathbf{A}_i\cdot\tau[i]\mathbf{v}_2 - \mathbf{D}\cdot\mathsf{bin}(\mathbf{D}_0\cdot\mathbf{s} + \sum_{k=1}^{N}\mathbf{D}_i\cdot\mathfrak{m}_k) = \mathbf{u} \bmod q, \quad (16)$$

and that (modulo q)

$$\begin{cases}\forall k\in[N]: \mathbf{c}_{k,1} = \mathbf{B}^T\cdot\mathbf{s}_k + \mathbf{e}_{k,1}; \mathbf{c}_{k,2} = \mathbf{G}_1^T\cdot\mathbf{s}_k + \mathbf{e}_{k,2} + \lfloor q/2\rfloor\cdot\mathfrak{m}_k;\\ \mathbf{c}_{\mathbf{v},1} = \mathbf{B}^T\cdot\mathbf{s}_{\mathbf{v}} + \mathbf{e}_{\mathbf{v},1};\\ \mathbf{c}_{\mathbf{v},2} = \mathbf{G}_1^T\cdot\mathbf{s}_{\mathbf{v}} + \mathbf{e}_{\mathbf{v},2} + \lfloor\frac{q}{p}\rfloor\cdot\mathbf{v} = \mathbf{G}_1^T\cdot\mathbf{s}_{\mathbf{v}} + \mathbf{e}_{\mathbf{v},2} + \begin{pmatrix}\lfloor\frac{q}{p}\rfloor\mathbf{I}_m\\ \mathbf{0}\end{pmatrix}\cdot\mathbf{v}_1 + \begin{pmatrix}\mathbf{0}\\ \lfloor\frac{q}{p}\rfloor\mathbf{I}_m\end{pmatrix}\cdot\mathbf{v}_2;\\ \mathbf{c}_{s,1} = \mathbf{B}^T\cdot\mathbf{s}_0 + \mathbf{e}_{0,1}; \mathbf{c}_{s,2} = \mathbf{G}_1^T\cdot\mathbf{s}_0 + \mathbf{e}_{0,2} + \lfloor q/p\rfloor\cdot\mathbf{s};\\ \mathbf{c}_{\tau,1} = \mathbf{B}^T\cdot\mathbf{s}_\tau + \mathbf{e}_{\tau,1}; \mathbf{c}_{\tau,2} = \mathbf{G}_0^T\cdot\mathbf{s}_\tau + \mathbf{e}_{\tau,2} + \lfloor q/2\rfloor\cdot\tau.\end{cases} \quad (17)$$

We proceed in two steps.

Step 1: *Transforming the equations in (*16*) and (*17*) into a unified one of the form* $\mathbf{P} \cdot \mathbf{x} = \mathbf{c} \bmod q$*, where* $\|\mathbf{x}\|_\infty = 1$ *and* $\mathbf{x} \in \mathsf{VALID}$ *- a "specially-designed" set.*

Note that, if we let $\mathbf{y} = \mathsf{bin}(\mathbf{D}_0 \cdot \mathbf{s} + \sum_{k=1}^{N} \mathbf{D}_i \cdot \mathfrak{m}_k) \in \{0,1\}^m$, then we have $\mathbf{H}_{2n\times m} \cdot \mathbf{y} = \mathbf{D}_0 \cdot \mathbf{s} + \sum_{k=1}^{N} \mathbf{D}_i \cdot \mathfrak{m}_k \bmod q$, and (16) can be equivalently written as:

$$\begin{pmatrix} \mathbf{A} \\ \mathbf{0} \end{pmatrix} \cdot \mathbf{v}_1 + \begin{pmatrix} \mathbf{A}_0 \\ \mathbf{0} \end{pmatrix} \cdot \mathbf{v}_2 + \sum_{i=1}^{\ell} \begin{pmatrix} \mathbf{A}_i \\ \mathbf{0} \end{pmatrix} \cdot \tau[i]\mathbf{v}_2 + \begin{pmatrix} \mathbf{0} \\ \mathbf{D}_0 \end{pmatrix} \cdot \mathbf{s} + \begin{pmatrix} -\mathbf{D} \\ -\mathbf{H}_{2n\times m} \end{pmatrix} \cdot \mathbf{y}$$
$$+ \begin{pmatrix} \mathbf{0} \\ \mathbf{D}_1|\ldots|\mathbf{D}_N \end{pmatrix} \cdot \mathfrak{m} = \begin{pmatrix} \mathbf{u} \\ \mathbf{0}^{2n} \end{pmatrix} \bmod q.$$

Next, we use linear algebra to combine this equation and (17) into (modulo q):

$$\mathbf{F} \cdot \mathbf{v}_1 + \mathbf{F}_0 \cdot \mathbf{v}_2 + \sum_{i=1}^{\ell} \mathbf{F}_i \cdot \tau[i]\mathbf{v}_2 + \mathbf{M}_1 \cdot \tau + \mathbf{M}_2 \cdot \mathbf{y} + \mathbf{M}_3 \cdot \mathfrak{m} + \mathbf{M}_4 \cdot \mathbf{s} + \mathbf{M}_5 \cdot \mathbf{e} = \mathbf{c}, \quad (18)$$

where, for dimensions $D = \ell + 3n + 7m + 3mN$ and $L_0 = D + nN$,

- Matrices $\mathbf{F}, \mathbf{F}_0, \mathbf{F}_1, \ldots, \mathbf{F}_\ell \in \mathbb{Z}_q^{D\times m}$, $\mathbf{M}_1 \in \mathbb{Z}_q^{D\times \ell}$, $\mathbf{M}_2 \in \mathbb{Z}_q^{D\times m}$, $\mathbf{M}_3 \in \mathbb{Z}_q^{D\times 2mN}$, $\mathbf{M}_4 \in \mathbb{Z}_q^{D\times 2m}$, $\mathbf{M}_5 \in \mathbb{Z}_q^{D\times L_0}$ and vector $\mathbf{c} \in \mathbb{Z}_q^{D}$ are built from the public input.
- Vector $\mathbf{e} = \big(\mathbf{s}_1^T \| \ldots \| \mathbf{s}_N^T \| \mathbf{s}_\mathbf{v}^T \| \mathbf{s}_0^T \| \mathbf{s}_\tau^T \| \mathbf{e}_{1,1}^T \| \ldots \| \mathbf{e}_{N,1}^T \| \mathbf{e}_{\mathbf{v},1}^T \| \mathbf{e}_{0,1}^T \| \mathbf{e}_{\tau,1}^T \| \| \mathbf{e}_{1,2}^T \| \ldots \| \mathbf{e}_{N,2}^T \| \mathbf{e}_{0,2}^T \| \mathbf{e}_{\mathbf{v},2}^T \| \mathbf{e}_{\tau,2}^T\big)^T \in [-B, B]^{L_0}$.

Now we further transform (18) using the techniques from Sect. 5.2. Specifically, we form the following:

$$\begin{cases} \mathsf{DecExt}_{m,\beta}(\mathbf{v}_1) \rightarrow \hat{\mathbf{v}}_1 \in \mathsf{B}^3_{m\delta_\beta}; \mathsf{DecExt}_{m,\beta}(\mathbf{v}_2) \rightarrow \hat{\mathbf{v}}_2 \in \mathsf{B}^3_{m\delta_\beta}; \\ \mathbf{F}' = \big[\mathbf{F} \cdot \widehat{\mathbf{K}}_{m,\beta} | \mathbf{F}_0 \cdot \widehat{\mathbf{K}}_{m,\beta} | \mathbf{F}_1 \cdot \widehat{\mathbf{K}}_{m,\beta} | \ldots | \mathbf{F}_\ell \cdot \widehat{\mathbf{K}}_{m,\beta} | \mathbf{0}^{D\times 3m\delta_\beta \ell}\big] \in \mathbb{Z}_q^{D\times 3m\delta_\beta(2\ell+2)}, \\ \mathsf{Ext}_{2\ell}(\tau) \rightarrow \hat{\tau} = (\tau[1], \ldots, \tau[\ell], \ldots, \tau[2\ell])^T \in \mathsf{B}^2_\ell; \mathbf{M}_1' = [\mathbf{M}_1 | \mathbf{0}^{D\times \ell}] \in \mathbb{Z}_q^{D\times 2\ell}; \\ \mathsf{Ext}_{2m}(\mathbf{y}) \rightarrow \hat{\mathbf{y}} \in \mathsf{B}^2_m; \mathbf{M}_2' = [\mathbf{M}_2 | \mathbf{0}^{D\times m}] \in \mathbb{Z}_q^{D\times 2m}; \\ \mathsf{DecExt}_{2m,p-1}(\mathbf{s}) \rightarrow \hat{\mathbf{s}} \in \mathsf{B}^3_{2m\delta_{p-1}}; \mathbf{M}_4' = \mathbf{M}_4 \cdot \widehat{\mathbf{K}}_{2m,p-1} \in \mathbb{Z}_q^{D\times 6m\delta_{p-1}}; \\ \mathsf{DecExt}_{L_0,B}(\mathbf{e}) \rightarrow \hat{\mathbf{e}} \in \mathsf{B}^3_{L_0\delta_B}; \mathbf{M}_5' = \mathbf{M}_5 \cdot \widehat{\mathbf{K}}_{L_0,B} \in \mathbb{Z}_q^{D\times 3L_0\delta_B}. \end{cases}$$

Now, let $L = 3m\delta_\beta(2\ell+2) + 2\ell + 2m + 2mN + 6m\delta_{p-1} + 3L_0\delta_B$, and construct matrix $\mathbf{P} = \big[\mathbf{F}' | \mathbf{M}_1' | \mathbf{M}_2' | \mathbf{M}_3 | \mathbf{M}_4' | \mathbf{M}_5'\big] \in \mathbb{Z}_q^{D\times L}$ and vector

$$\mathbf{x} = \big(\hat{\mathbf{v}}_1^T \| \hat{\mathbf{v}}_2^T \| \tau[1]\hat{\mathbf{v}}_2^T \| \ldots \| \tau[\ell]\hat{\mathbf{v}}_2^T \| \ldots \| \tau[2\ell]\hat{\mathbf{v}}_2^T \| \hat{\tau}^T \| \hat{\mathbf{y}}^T \| \mathfrak{m}^T \| \hat{\mathbf{s}}^T \| \hat{\mathbf{e}}^T\big)^T,$$

then we will obtain the equation $\mathbf{P} \cdot \mathbf{x} = \mathbf{c} \bmod q$.

Before going on, we define VALID as the set of $\mathbf{w} \in \{-1,0,1\}^L$ of the form:

$$\mathbf{w} = \big(\mathbf{w}_1^T \| \mathbf{w}_2^T \| g_1 \mathbf{w}_2^T \| \ldots \| g_{2\ell} \mathbf{w}_2^T \| \mathbf{g}^T \| \mathbf{w}_3^T \| \mathbf{w}_4^T \| \mathbf{w}_5^T \| \mathbf{w}_6^T\big)^T$$

for some $\mathbf{w}_1, \mathbf{w}_2 \in \mathsf{B}^3_{m\delta_\beta}$, $\mathbf{g} = (g_1, \ldots, g_{2\ell}) \in \mathsf{B}_{2\ell}$, $\mathbf{w}_3 \in \mathsf{B}^2_m$, $\mathbf{w}_4 \in \mathsf{CorEnc}(mN)$, $\mathbf{w}_5 \in \mathsf{B}^3_{2m\delta_{p-1}}$, and $\mathbf{w}_6 \in \mathsf{B}^3_{L_0\delta_B}$. It can be checked that the constructed vector $\mathbf{x}$ belongs to this tailored set VALID.

Step 2: *Specifying the set $\mathcal{S}$ and permutations of L elements $\{T_\pi : \pi \in \mathcal{S}\}$ for which the conditions in (13) hold.*

- Define $\mathcal{S} = \mathcal{S}_{3m\delta_\beta} \times \mathcal{S}_{3m\delta_\beta} \times \mathcal{S}_{2\ell} \times \mathcal{S}_{2m} \times \{0,1\}^{mN} \times \mathcal{S}_{6m\delta_{p-1}} \times \mathcal{S}_{3L_0\delta_B}$.
- For $\pi = (\phi, \psi, \gamma, \rho, \mathbf{b}, \eta, \xi) \in \mathcal{S}$ and $\mathbf{z} = (\mathbf{z}_0^1 \| \mathbf{z}_0^2 \| \mathbf{z}_1 \| \ldots \| \mathbf{z}_{2\ell} \| \mathbf{g} \| \mathbf{t}_1 \| \mathbf{t}_2 \| \mathbf{t}_3 \| \mathbf{t}_4) \in \mathbb{Z}_q^L$, where $\mathbf{z}_0^1, \mathbf{z}_0^2, \mathbf{z}_1, \ldots, \mathbf{z}_{2\ell} \in \mathbb{Z}_q^{3m\delta_\beta}$, $\mathbf{g} \in \mathbb{Z}_q^{2\ell}$, $\mathbf{t}_1 \in \mathbb{Z}_q^{2m}$, $\mathbf{t}_2 \in \mathbb{Z}_q^{2mN}$, $\mathbf{t}_3 \in \mathbb{Z}_q^{6m\delta_{p-1}}$, and $\mathbf{t}_4 \in \mathbb{Z}_q^{3L_0\delta_B}$, we define:

$$T_\pi(\mathbf{z}) = \big(\phi(\mathbf{z}_0^1)^T \| \psi(\mathbf{z}_0^2)^T \| \psi(\mathbf{z}_{\gamma(1)})^T \| \ldots \| \psi(\mathbf{z}_{\gamma(2\ell)})^T \| \gamma(\mathbf{g})^T \| \\ \| \rho(\mathbf{t}_1)^T \| E_{\mathbf{b}}(\mathbf{t}_2)^T \| \eta(\mathbf{t}_3)^T \| \xi(\mathbf{t}_4)^T\big)^T$$

  as the permutation that transforms $\mathbf{z}$ as follows:
  1. It rearranges the order of the 2ℓ blocks $\mathbf{z}_1, \ldots, \mathbf{z}_{2\ell}$ according to γ.
  2. It then permutes block $\mathbf{z}_0^1$ according to ϕ, blocks $\mathbf{z}_0^2$, $\{\mathbf{z}_i\}_{i=1}^{2\ell}$ according to ψ, block $\mathbf{g}$ according to γ, block $\mathbf{t}_1$ according to ρ, block $\mathbf{t}_2$ according to $E_{\mathbf{b}}$, block $\mathbf{t}_3$ according to η, and block $\mathbf{t}_4$ according to ξ.

It can be check that (13) holds. Therefore, we can obtain a statistical ZKAoK for the given relation by running the protocol in Sect. 5.1.

5.5 The Underlying ZKAoK for the Group Signature Scheme

The argument system upon which our group signature scheme is built can be summarized as follows.

Common Input: Matrices $\mathbf{A}, \{\mathbf{A}_j\}_{j=0}^{\ell}, \mathbf{B} \in \mathbb{Z}_q^{n\times m}$, $\mathbf{D}_0, \mathbf{D}_1 \in \mathbb{Z}_q^{2n\times 2m}$, $\mathbf{F} \in \mathbb{Z}_q^{4n\times 4m}$, $\mathbf{H}_{2n\times m} \in \mathbb{Z}_q^{2n\times m}$, $\mathbf{H}_{4n\times 2m} \in \mathbb{Z}_q^{4n\times 2m}$, $\mathbf{G}_0 \in \mathbb{Z}_q^{n\times 2m}$; vectors $\mathbf{u} \in \mathbb{Z}_q^n$, $\mathbf{c}_1 \in \mathbb{Z}_q^m$, $\mathbf{c}_2 \in \mathbb{Z}_q^{2m}$.

Prover's Input: $\mathbf{z} \in [-\beta,\beta]^{4m}$, $\mathbf{y} \in \{0,1\}^{2m}$, $\mathbf{w} \in \{0,1\}^m$, $\mathbf{d}_1, \mathbf{d}_2 \in [-\beta,\beta]^m$, $\mathbf{s} \in [-\beta,\beta]^{2m}$, $\mathrm{id} = (\mathrm{id}[1], \ldots, \mathrm{id}[\ell])^T \in \{0,1\}^\ell$, $\mathbf{e}_0 \in [-B,B]^n$, $\mathbf{e}_1 \in [-B,B]^m$, $\mathbf{e}_2 \in [-B,B]^{2m}$.

Prover's Goal: Convince the verifier in ZK that

$$\begin{cases} \mathbf{F}\cdot\mathbf{z} = \mathbf{H}_{4n\times 2m}\cdot\mathbf{y} \bmod q; \mathbf{H}_{2n\times m}\cdot\mathbf{w} = \mathbf{D}_0\cdot\mathbf{y} + \mathbf{D}_1\cdot\mathbf{s} \bmod q;\\ \mathbf{A}\cdot\mathbf{d}_1 + \mathbf{A}_0\cdot\mathbf{d}_2 + \sum_{j=1}^{\ell}\mathbf{A}_j\cdot(\mathrm{id}[j]\cdot\mathbf{d}_2) - \mathbf{D}\cdot\mathbf{w} = \mathbf{u} \bmod q;\\ \mathbf{c}_1 = \mathbf{B}^T\cdot\mathbf{e}_0 + \mathbf{e}_1 \bmod q; \mathbf{c}_2 = \mathbf{G}_0^T\cdot\mathbf{e}_0 + \mathbf{e}_2 + \lfloor q/2\rfloor\cdot\mathbf{y} \bmod q. \end{cases}$$

Using the same strategy as in Sects. 5.3 and 5.4, we can derive a statistical ZKAoK for the above relation from the protocol in Sect. 5.1. As the transformations are similar to those in Sect. 5.4, we only sketch main points.

In the first step, we combine the given equations to an equation of the form:

$$\mathbf{M}\cdot\begin{pmatrix}\mathbf{d}_1\\ \mathbf{s}\\ \mathbf{z}\end{pmatrix} + \mathbf{M}_0\cdot\mathbf{d}_2 + \sum_{j=1}^{\ell}\mathbf{M}_j(\mathrm{id}[j]\mathbf{d}_2) + \mathbf{M}'\cdot\begin{pmatrix}\mathbf{w}\\ \mathbf{y}\end{pmatrix} + \mathbf{M}''\cdot\begin{pmatrix}\mathbf{e}_0\\ \mathbf{e}_1\\ \mathbf{e}_2\end{pmatrix} = \mathbf{v} \bmod q,$$

where matrices $\mathbf{M}, \mathbf{M}_0, \ldots, \mathbf{M}_\ell, \mathbf{M}', \mathbf{M}''$ and vector $\mathbf{v}$ are built from the input.

We then apply the techniques of Sect. 5.2 for $\mathbf{x}_0 = (\mathbf{d}_1^T \| \mathbf{s}^T \| \mathbf{z}^T)^T \in [-\beta, \beta]^{7m}$, $\mathbf{d}_2 \in [-\beta, \beta]^m$; $\mathbf{x}_1 = (\mathbf{w}^T \| \mathbf{y}^T)^T \in \{0,1\}^{3m}$; and $\mathbf{x}_2 = (\mathbf{e}_0^T \| \mathbf{e}_1^T \| \mathbf{e}_2^T)^T \in [-B, B]^{n+3m}$. This allows us to obtain a unified equation $\mathbf{P} \cdot \mathbf{x} = \mathbf{v} \bmod q$, and to define the sets VALID, $\mathcal{S}$, and permutations $\{T_\pi : \pi \in \mathcal{S}\}$ so that the conditions in (13) hold, in a similar manner as in Sect. 5.4.

Acknowledgements. We thank Damien Stehlé for useful discussions. The first author was funded by the "Programme Avenir Lyon Saint-Etienne de l'Université de Lyon" in the framework of the programme "Investissements d'Avenir" (ANR-11-IDEX-0007). San Ling, Khoa Nguyen and Huaxiong Wang were supported by the "Singapore Ministry of Education under Research Grant MOE2013-T2-1-041". Huaxiong Wang was also supported by NTU under Tier 1 grant RG143/14.

References

1. Abe, M., Chase, M., David, B., Kohlweiss, M., Nishimaki, R., Ohkubo, M.: Constant-size structure-preserving signatures: generic constructions and simple assumptions. In: Wang, X., Sako, K. (eds.) ASIACRYPT 2012. LNCS, vol. 7658, pp. 4–24. Springer, Heidelberg (2012). doi:10.1007/978-3-642-34961-4_3
2. Abe, M., Fuchsbauer, G., Groth, J., Haralambiev, K., Ohkubo, M.: Structure-preserving signatures and commitments to group elements. In: Rabin, T. (ed.) CRYPTO 2010. LNCS, vol. 6223, pp. 209–236. Springer, Heidelberg (2010). doi:10.1007/978-3-642-14623-7_12
3. Agrawal, S., Boneh, D., Boyen, X.: Efficient lattice (H)IBE in the standard model. In: Gilbert, H. (ed.) EUROCRYPT 2010. LNCS, vol. 6110, pp. 553–572. Springer, Heidelberg (2010). doi:10.1007/978-3-642-13190-5_28
4. Alwen, J., Peikert, C.: Generating shorter bases for hard random lattices. In: STACS 2009, pp. 75–86. Schloss Dagstuhl - Leibniz-Zentrum fuer Informatik, Germany (2009)
5. Asharov, G., Jain, A., López-Alt, A., Tromer, E., Vaikuntanathan, V., Wichs, D.: Multiparty computation with low communication, computation and interaction via threshold FHE. In: Pointcheval, D., Johansson, T. (eds.) EUROCRYPT 2012. LNCS, vol. 7237, pp. 483–501. Springer, Heidelberg (2012). doi:10.1007/978-3-642-29011-4_29
6. Ateniese, G., Camenisch, J., Joye, M., Tsudik, G.: A practical and provably secure coalition-resistant group signature scheme. In: Bellare, M. (ed.) CRYPTO 2000. LNCS, vol. 1880, pp. 255–270. Springer, Heidelberg (2000). doi:10.1007/3-540-44598-6_16
7. Bai, S., Langlois, A., Lepoint, T., Stehlé, D., Steinfeld, R.: Improved security proofs in lattice-based cryptography: Using the Rényi divergence rather than the statistical distance. In ASIACRYPT 2015. Springer (2015)
8. Banaszczyk, W.: New bounds in some transference theorems in the geometry of number. Math. Ann. **296**, 625–635 (1993)
9. Belenkiy, M., Camenisch, J., Chase, M., Kohlweiss, M., Lysyanskaya, A., Shacham, H.: Randomizable proofs and delegatable anonymous credentials. In: Halevi, S. (ed.) CRYPTO 2009. LNCS, vol. 5677, pp. 108–125. Springer, Heidelberg (2009). doi:10.1007/978-3-642-03356-8_7

10. Belenkiy, M., Chase, M., Kohlweiss, M., Lysyanskaya, A.: P-signatures and noninteractive anonymous credentials. In: Canetti, R. (ed.) TCC 2008. LNCS, vol. 4948, pp. 356–374. Springer, Heidelberg (2008). doi:10.1007/978-3-540-78524-8_20
11. Bellare, M., Micciancio, D., Warinschi, B.: Foundations of group signatures: formal definitions, simplified requirements, and a construction based on general assumptions. In: Biham, E. (ed.) EUROCRYPT 2003. LNCS, vol. 2656, pp. 614–629. Springer, Heidelberg (2003). doi:10.1007/3-540-39200-9_38
12. Bellare, M., Rogaway, P.: Random oracles are practical: a paradigm for designing efficient protocols. In: ACM-CCS 1993, pp. 62–73. ACM (1993)
13. Bellare, M., Shi, H., Zhang, C.: Foundations of group signatures: the case of dynamic groups. In: Menezes, A. (ed.) CT-RSA 2005. LNCS, vol. 3376, pp. 136–153. Springer, Heidelberg (2005). doi:10.1007/978-3-540-30574-3_11
14. Benhamouda, F., Camenisch, J., Krenn, S., Lyubashevsky, V., Neven, G.: Better zero-knowledge proofs for lattice encryption and their application to group signatures. In: Sarkar, P., Iwata, T. (eds.) ASIACRYPT 2014. LNCS, vol. 8873, pp. 551–572. Springer, Heidelberg (2014). doi:10.1007/978-3-662-45611-8_29
15. Benhamouda, F., Krenn, S., Lyubashevsky, V., Pietrzak, K.: Efficient zero-knowledge proofs for commitments from learning with errors over rings. In: Pernul, G., Ryan, P.Y.A., Weippl, E. (eds.) ESORICS 2015. LNCS, vol. 9326, pp. 305–325. Springer, Heidelberg (2015). doi:10.1007/978-3-319-24174-6_16
16. Böhl, F., Hofheinz, D., Jager, T., Koch, J., Striecks, C.: Confined guessing: New signatures from standard assumptions. J. Cryptology **28**(1), 176–208 (2015)
17. Boneh, D., Boyen, X.: Efficient selective-ID secure identity-based encryption without random oracles. In: Cachin, C., Camenisch, J.L. (eds.) EUROCRYPT 2004. LNCS, vol. 3027, pp. 223–238. Springer, Heidelberg (2004). doi:10.1007/978-3-540-24676-3_14
18. Boneh, D., Boyen, X., Shacham, H.: Short group signatures. In: Franklin, M. (ed.) CRYPTO 2004. LNCS, vol. 3152, pp. 41–55. Springer, Heidelberg (2004). doi:10.1007/978-3-540-28628-8_3
19. Boyen, X.: Lattice mixing and vanishing trapdoors: a framework for fully secure short signatures and more. In: Nguyen, P.Q., Pointcheval, D. (eds.) PKC 2010. LNCS, vol. 6056, pp. 499–517. Springer, Heidelberg (2010). doi:10.1007/978-3-642-13013-7_29
20. Boyen, X., Waters, B.: Compact group signatures without random oracles. In: Vaudenay, S. (ed.) EUROCRYPT 2006. LNCS, vol. 4004, pp. 427–444. Springer, Heidelberg (2006). doi:10.1007/11761679_26
21. Boyen, X., Waters, B.: Full-domain subgroup hiding and constant-size group signatures. In: Okamoto, T., Wang, X. (eds.) PKC 2007. LNCS, vol. 4450, pp. 1–15. Springer, Heidelberg (2007). doi:10.1007/978-3-540-71677-8_1
22. Brakerski, Z., Kalai, Y.T.: A framework for efficient signatures, ring signatures and identity based encryption in the standard model. IACR Cryptology ePrint Arch. **2010**, 86 (2010)
23. Brakerski, Z., Langlois, A., Peikert, C., Regev, O., Stehlé, D.: On the classical hardness of learning with errors. In: STOC 2013, pp. 575–584. ACM (2013)
24. Camenisch, J., Gross, T.: Efficient attributes for anonymous credentials. In: ACM-CCS 2008, pp. 345–356. ACM (2008)
25. Camenisch, J., Hohenberger, S., Lysyanskaya, A.: Compact E-cash. In: Cramer, R. (ed.) EUROCRYPT 2005. LNCS, vol. 3494, pp. 302–321. Springer, Heidelberg (2005). doi:10.1007/11426639_18

26. Camenisch, J., Kiayias, A., Yung, M.: On the portability of generalized schnorr proofs. In: Joux, A. (ed.) EUROCRYPT 2009. LNCS, vol. 5479, pp. 425–442. Springer, Heidelberg (2009). doi:10.1007/978-3-642-01001-9_25
27. Camenisch, J., Krenn, S., Lehmann, A., Mikkelsen, G.L., Neven, G., Pedersen, M.Ø.: Formal treatment of privacy-enhancing credential systems. In: Dunkelman, O., Keliher, L. (eds.) SAC 2015. LNCS, vol. 9566, pp. 3–24. Springer, Heidelberg (2016). doi:10.1007/978-3-319-31301-6_1
28. Camenisch, J., Lysyanskaya, A.: An efficient system for non-transferable anonymous credentials with optional anonymity revocation. In: Pfitzmann, B. (ed.) EUROCRYPT 2001. LNCS, vol. 2045, pp. 93–118. Springer, Heidelberg (2001). doi:10.1007/3-540-44987-6_7
29. Camenisch, J., Lysyanskaya, A.: Dynamic accumulators and application to efficient revocation of anonymous credentials. In: Yung, M. (ed.) CRYPTO 2002. LNCS, vol. 2442, pp. 61–76. Springer, Heidelberg (2002). doi:10.1007/3-540-45708-9_5
30. Camenisch, J., Lysyanskaya, A.: A signature scheme with efficient protocols. In: Cimato, S., Persiano, G., Galdi, C. (eds.) SCN 2002. LNCS, vol. 2576, pp. 268–289. Springer, Heidelberg (2003). doi:10.1007/3-540-36413-7_20
31. Camenisch, J., Lysyanskaya, A.: Signature schemes and anonymous credentials from bilinear maps. In: Franklin, M. (ed.) CRYPTO 2004. LNCS, vol. 3152, pp. 56–72. Springer, Heidelberg (2004). doi:10.1007/978-3-540-28628-8_4
32. Camenisch, J., Neven, G., Rückert, M.: Fully anonymous attribute tokens from lattices. In: Visconti, I., Prisco, R. (eds.) SCN 2012. LNCS, vol. 7485, pp. 57–75. Springer, Heidelberg (2012). doi:10.1007/978-3-642-32928-9_4
33. Cash, D., Hofheinz, D., Kiltz, E., Peikert, C.: Bonsai trees, or how to delegate a lattice basis. In: Gilbert, H. (ed.) EUROCRYPT 2010. LNCS, vol. 6110, pp. 523–552. Springer, Heidelberg (2010). doi:10.1007/978-3-642-13190-5_27
34. Chaum, D.: Security without identification: Transactions ssystem to make big brother obsolete. Commun. ACM **28**(10), 1030–1044 (1985)
35. Chaum, D., Heyst, E.: Group Signatures. In: Davies, D.W. (ed.) EUROCRYPT 1991. LNCS, vol. 547, pp. 257–265. Springer, Heidelberg (1991). doi:10.1007/3-540-46416-6_22
36. Damgård, I.: Efficient concurrent zero-knowledge in the auxiliary string model. In: Preneel, B. (ed.) EUROCRYPT 2000. LNCS, vol. 1807, pp. 418–430. Springer, Heidelberg (2000). doi:10.1007/3-540-45539-6_30
37. Damgård, I., Hofheinz, D., Kiltz, E., Thorbek, R.: Public-key encryption with non-interactive opening. In: Malkin, T. (ed.) CT-RSA 2008. LNCS, vol. 4964, pp. 239–255. Springer, Heidelberg (2008). doi:10.1007/978-3-540-79263-5_15
38. Delerablée, C., Pointcheval, D.: Dynamic fully anonymous short group signatures. In: Nguyen, P.Q. (ed.) VIETCRYPT 2006. LNCS, vol. 4341, pp. 193–210. Springer, Heidelberg (2006). doi:10.1007/11958239_13
39. Ezerman, M.F., Lee, H.T., Ling, S., Nguyen, K., Wang, H.: A provably secure group signature scheme from code-based assumptions. In: Iwata, T., Cheon, J.H. (eds.) ASIACRYPT 2015. LNCS, vol. 9452, pp. 260–285. Springer, Heidelberg (2015). doi:10.1007/978-3-662-48797-6_12
40. Fiat, A., Shamir, A.: How to prove yourself: practical solutions to identification and signature problems. In: Odlyzko, A.M. (ed.) CRYPTO 1986. LNCS, vol. 263, pp. 186–194. Springer, Heidelberg (1987). doi:10.1007/3-540-47721-7_12
41. Gentry, C.: Fully homomorphic encryption using ideal lattices. In: STOC 2009, pp. 169–178. ACM (2009)
42. Gentry, C., Peikert, C., Vaikuntanathan, V.: Trapdoors for hard lattices and new cryptographic constructions. In: STOC 2008, pp. 197–206. ACM (2008)

43. Goldwasser, S., Micali, S., Rackoff, C.: The knowledge complexity of interactive proof-systems. In: STOC 1985, pp. 291–304. ACM (1985)
44. Gorbunov, S., Vaikuntanathan, V., Wee, H.: Attribute-based encryption for circuits. In: STOC 2013, pp. 545–554. ACM (2013)
45. Gorbunov, S., Vaikuntanathan, V., Wee, H.: Predicate encryption for circuits from LWE. In: Gennaro, R., Robshaw, M. (eds.) CRYPTO 2015. LNCS, vol. 9216, pp. 503–523. Springer, Heidelberg (2015). doi:10.1007/978-3-662-48000-7_25
46. Gordon, S.D., Katz, J., Vaikuntanathan, V.: A group signature scheme from lattice assumptions. In: Abe, M. (ed.) ASIACRYPT 2010. LNCS, vol. 6477, pp. 395–412. Springer, Heidelberg (2010). doi:10.1007/978-3-642-17373-8_23
47. Green, M., Hohenberger, S.: Universally composable adaptive oblivious transfer. In: Pieprzyk, J. (ed.) ASIACRYPT 2008. LNCS, vol. 5350, pp. 179–197. Springer, Heidelberg (2008). doi:10.1007/978-3-540-89255-7_12
48. Groth, J.: Fully anonymous group signatures without random oracles. In: Kurosawa, K. (ed.) ASIACRYPT 2007. LNCS, vol. 4833, pp. 164–180. Springer, Heidelberg (2007). doi:10.1007/978-3-540-76900-2_10
49. Groth, J., Sahai, A.: Efficient non-interactive proof systems for bilinear groups. In: Smart, N. (ed.) EUROCRYPT 2008. LNCS, vol. 4965, pp. 415–432. Springer, Heidelberg (2008). doi:10.1007/978-3-540-78967-3_24
50. Hohenberger, S., Waters, B.: Short and stateless signatures from the RSA assumption. In: Halevi, S. (ed.) CRYPTO 2009. LNCS, vol. 5677, pp. 654–670. Springer, Heidelberg (2009). doi:10.1007/978-3-642-03356-8_38
51. Jain, A., Krenn, S., Pietrzak, K., Tentes, A.: Commitments and efficient zero-knowledge proofs from learning parity with noise. In: Wang, X., Sako, K. (eds.) ASIACRYPT 2012. LNCS, vol. 7658, pp. 663–680. Springer, Heidelberg (2012). doi:10.1007/978-3-642-34961-4_40
52. Kawachi, A., Tanaka, K., Xagawa, K.: Concurrently secure identification schemes based on the worst-case hardness of lattice problems. In: Pieprzyk, J. (ed.) ASIACRYPT 2008. LNCS, vol. 5350, pp. 372–389. Springer, Heidelberg (2008). doi:10.1007/978-3-540-89255-7_23
53. Kiayias, A., Tsiounis, Y., Yung, M.: Group encryption. In: Kurosawa, K. (ed.) ASIACRYPT 2007. LNCS, vol. 4833, pp. 181–199. Springer, Heidelberg (2007). doi:10.1007/978-3-540-76900-2_11
54. Kiayias, A., Yung, M.: Group signatures with efficient concurrent join. In: Cramer, R. (ed.) EUROCRYPT 2005. LNCS, vol. 3494, pp. 198–214. Springer, Heidelberg (2005). doi:10.1007/11426639_12
55. Kiayias, A., Yung, M.: Secure scalable group signature with dynamic joins and separable authorities. Int. J. Secur. Netw. **1**(1), 24–45 (2006)
56. Laguillaumie, F., Langlois, A., Libert, B., Stehlé, D.: Lattice-based group signatures with logarithmic signature size. In: Sako, K., Sarkar, P. (eds.) ASIACRYPT 2013. LNCS, vol. 8270, pp. 41–61. Springer, Heidelberg (2013). doi:10.1007/978-3-642-42045-0_3
57. Langlois, A., Ling, S., Nguyen, K., Wang, H.: Lattice-based group signature scheme with verifier-local revocation. In: Krawczyk, H. (ed.) PKC 2014. LNCS, vol. 8383, pp. 345–361. Springer, Heidelberg (2014). doi:10.1007/978-3-642-54631-0_20
58. Langlois, A., Stehlé, D., Steinfeld, R.: GGHLite: more efficient multilinear maps from ideal lattices. In: Nguyen, P.Q., Oswald, E. (eds.) EUROCRYPT 2014. LNCS, vol. 8441, pp. 239–256. Springer, Heidelberg (2014). doi:10.1007/978-3-642-55220-5_14

59. Libert, B., Ling, S., Mouhartem, F., Nguyen, K., Wang, H., Signature schemes with efficient protocols, dynamic group signatures from lattice assumptions. Cryptology ePrint Archive: Report 2016/101 (2016)
60. Libert, B., Ling, S., Nguyen, K., Wang, H.: Zero-knowledge arguments for lattice-based accumulators: logarithmic-size ring signatures and group signatures without trapdoors. In: Fischlin, M., Coron, J.-S. (eds.) EUROCRYPT 2016. LNCS, vol. 9666, pp. 1–31. Springer, Heidelberg (2016). doi:10.1007/978-3-662-49896-5_1
61. Ling, S., Nguyen, K., Stehlé, D., Wang, H.: Improved zero-knowledge proofs of knowledge for the ISIS problem, and applications. In: Kurosawa, K., Hanaoka, G. (eds.) PKC 2013. LNCS, vol. 7778, pp. 107–124. Springer, Heidelberg (2013). doi:10.1007/978-3-642-36362-7_8
62. Ling, S., Nguyen, K., Wang, H.: Group signatures from lattices: simpler, tighter, shorter, ring-based. In: Katz, J. (ed.) PKC 2015. LNCS, vol. 9020, pp. 427–449. Springer, Heidelberg (2015). doi:10.1007/978-3-662-46447-2_19
63. Lyubashevsky, V.: Lattice-based identification schemes secure under active attacks. In: Cramer, R. (ed.) PKC 2008. LNCS, vol. 4939, pp. 162–179. Springer, Heidelberg (2008). doi:10.1007/978-3-540-78440-1_10
64. Lyubashevsky, V., Peikert, C., Regev, O.: On ideal lattices and learning with errors over rings. In: Gilbert, H. (ed.) EUROCRYPT 2010. LNCS, vol. 6110, pp. 1–23. Springer, Heidelberg (2010). doi:10.1007/978-3-642-13190-5_1
65. Micciancio, D., Peikert, C.: Trapdoors for lattices: simpler, tighter, faster, smaller. In: Pointcheval, D., Johansson, T. (eds.) EUROCRYPT 2012. LNCS, vol. 7237, pp. 700–718. Springer, Heidelberg (2012). doi:10.1007/978-3-642-29011-4_41
66. Nguyen, P.Q., Zhang, J., Zhang, Z.: Simpler efficient group signatures from lattices. In: Katz, J. (ed.) PKC 2015. LNCS, vol. 9020, pp. 401–426. Springer, Heidelberg (2015). doi:10.1007/978-3-662-46447-2_18
67. Papamanthou, C., Shi, E., Tamassia, R., Yi, K.: Streaming authenticated data structures. In: Johansson, T., Nguyen, P.Q. (eds.) EUROCRYPT 2013. LNCS, vol. 7881, pp. 353–370. Springer, Heidelberg (2013). doi:10.1007/978-3-642-38348-9_22
68. Peikert, C.: Public-key cryptosystems from the worst-case shortest vector problem. In: STOC 2009, pp. 333–342. ACM (2009)
69. Regev, O.: On lattices, learning with errors, random linear codes, and cryptography. In: STOC 2005, pp. 84–93. ACM (2005)
70. Rivest, R.L., Shamir, A., Tauman, Y.: How to leak a secret. In: Boyd, C. (ed.) ASIACRYPT 2001. LNCS, vol. 2248, pp. 552–565. Springer, Heidelberg (2001). doi:10.1007/3-540-45682-1_32
71. Sakai, Y., Schuldt, J.C.N., Emura, K., Hanaoka, G., Ohta, K.: On the security of dynamic group signatures: preventing signature hijacking. In: Fischlin, M., Buchmann, J., Manulis, M. (eds.) PKC 2012. LNCS, vol. 7293, pp. 715–732. Springer, Heidelberg (2012). doi:10.1007/978-3-642-30057-8_42
72. Stern, J.: A new paradigm for public key identification. IEEE Trans. Inf. Theor. **42**(6), 1757–1768 (1996)
73. Xie, X., Xue, R., Wang, M.: Zero knowledge proofs from ring-LWE. In: Abdalla, M., Nita-Rotaru, C., Dahab, R. (eds.) CANS 2013. LNCS, vol. 8257, pp. 57–73. Springer, Heidelberg (2013). doi:10.1007/978-3-319-02937-5_4

Towards Tightly Secure Lattice Short Signature and Id-Based Encryption

Xavier Boyen and Qinyi Li(✉)

Queensland University of Technology, Brisbane, Australia
qinyi.li@hdr.qut.edu.au

Abstract. Constructing short signatures with tight security from standard assumptions is a long-standing open problem. We present an adaptively secure, short (and stateless) signature scheme, featuring a constant security loss relative to a conservative hardness assumption, Short Integer Solution (SIS), and the security of a concretely instantiated pseudorandom function (PRF). This gives a class of tightly secure short lattice signature schemes whose security is based on SIS and the underlying assumption of the instantiated PRF.

Our signature construction further extends to give a class of tightly and adaptively secure "compact" Identity-Based Encryption (IBE) schemes, reducible with constant security loss from Regev's vanilla Learning With Errors (LWE) hardness assumption and the security of a concretely instantiated PRF. Our approach is a novel combination of a number of techniques, including Katz and Wang signature, Agrawal et al. lattice-based secure IBE, and Boneh et al. key-homomorphic encryption.

Our results, at the first time, eliminate the dependency between the number of adversary's queries and the security of short signature/IBE schemes in the context of lattice-based cryptography. They also indicate that tightly secure PRFs (with constant security loss) would imply tightly, adaptively secure short signature and IBE schemes (with constant security loss).

1 Introduction

Short signatures are useful and desirable for providing data authenticity in low-bandwidth and/or high-throughput applications where many signatures have to be processed very quickly. Most digital signature schemes are based on computationally hard problems on specific algebraic groups, e.g., finite fields, curves, and lattices. A signature is "short" if the signature consists in a (small) constant number of group elements (e.g., field elements or lattice points).

Although bare-bones signatures can be obtained from very weak assumptions (e.g., collision-resistant hash functions), constructing efficient short signatures satisfying standard security requirements (e.g., existential unforgeability under adaptively chosen-message attacks), from reasonable assumptions, appears to be

Xavier Boyen—Research conducted with generous support from the Australian Research Council under Discovery Project grant ARC DP-140103885.

J.H. Cheon and T. Takagi (Eds.): ASIACRYPT 2016, Part II, LNCS 10032, pp. 404–434, 2016.
DOI: 10.1007/978-3-662-53890-6_14

a challenging task. Some of the existing short signature schemes use random oracles, e.g., [10,19,36,48,50], or rely on non-standard computational assumptions (strong, interactive assumptions, and/or q-type parametric assumptions), e.g., [16,26,30,33,34], or require signers to maintain state across signatures, e.g., [45].

The first short signature scheme from a reasonable and non-parametric assumption without random oracles was proposed by Waters [56]. Hohenberger and Waters later proposed a short signature scheme from standard RSA [46]. Lattice-based short signatures from the very mild SIS assumption in the standard model were proposed in [20,51]. Recently, the "confined guessing" technique developed by Böhl et al. [13] has produced short signatures from standard RSA and bilinear-group CDH assumptions, and also from the ring-SIS/SIS assumption in combination with lattice techniques [4,32] with very loose reductions.

Despite these elegant constructions, signature schemes that are *short* and enjoy *tight security* reductions to *standard assumptions* in the *standard model* (without random oracle), remain unknown. Existing tightly secure signature schemes either have large signature size, e.g., [1,11,43], or merely have heuristic security arguments based on random oracles, e.g., [39,48]. We have not been able to ascertain the earliest occurrence of this long-standing folklore problem in cryptography, but here [11] is one recent formulation:

Open Problem #1—**Tightly Secure Short Signatures**
"Construct a tightly secure and short (in the sense that the signature contains constant number of group elements or vectors and the security loss is a constant) signature scheme from standard assumptions." —Blazy, Kakvi, Kiltz, Pan (2015)

1.1 Tight Security

The reductionist approach to cryptographic security algorithms seeks to prove theorems along the lines of: "If a t-time adversary attacks the scheme with successful probability ϵ, then a t'-time algorithm can be constructed to break some computational problem with success probability $\epsilon' = \epsilon/\theta$ and $t' = k \cdot t + o(t)$.". The parameters $\theta \geq 1$ and $k \geq 1$, or more simply the product $k \cdot \theta$, measures how tightly the security of the cryptographic scheme is related to the hardness of the underlying computational problem. Alternatively, when $k \approx 1$ as is the case in many reductions, θ measures the security loss of the security reduction of our cryptographic scheme from the underlying assumption. A cryptographic scheme is *tightly secure* if θ is a small constant that in particular does not depend on parameters under the adversary's control, such as the adversary's own success probability ϵ, the number of queries it chooses to make, and even the scheme's security parameter. The reduction phrases "almost tight security" from the literature refers to the case where θ is a polynomial of the security parameter.

Tight reduction is an elegant notion from a theoretical point of view. A tight reductionist proof (with respect to a well-defined security model) indicates that the security of a cryptographic scheme is (extremely) closely related to the hardness of the underlying hard problem, which is the optimal case we expect from provable security theory. On the other hand, it is also a determinant factor to the practicality of real-world security. Its opposite, loose security, means that in order to realise a desired "real" target security level, one has to increase the "apparent" security level inside the construction to compensate for the loose reduction. This inflates the size of data atoms by some polynomial, with in turn increases the running time of cryptographic operations by another polynomial, combining multiplicatively.

1.2 Identity-Based Encryption with Tight Security

Digital signatures and identity-based encryption (IBE) are closely connected, which suggests that techniques that improve upon the security of signatures might also improve upon the security of IBE. In this work, we also investigate the problem of constructing tightly secure IBE from standard assumptions (without random oracles).

In an IBE system, any random string that uniquely represents a user's identity, such as email address or driver license number, can act as a public key (within a certain domain or realm). Encryption uses this identity, together with some common domain-specific public parameters, to encrypt messages. Users are issued private decryption keys corresponding to their public identities, by a trusted authority (or distributed authorities) called Private Key Generator (PKG) which hold(s) (shares of) the master secret key for a domain. Decryption succeeds if the identity associated with the ciphertext matches the identity associated with the private key, in the same domain.

The strongest, most natural and most widely accepted notion of security for IBE is the *adaptive* security model or *full* security model, formally defined in [17]. In this model, the adversary is able to announce its target (the challenge identity it wants to attack) at any time during the course of its adaptive interaction with the system. Without the luxury of random oracles, an easier security model to achieve was the *selective* security model, where the adversary must announce its target identity at the onset of its interaction with the system.

In the last fifteen years, a great many IBE schemes have been proposed, with varying efficiency, security models, hardness assumptions, and other features. In the standard model (i.e., without random oracles or other idealised oracles), we mention several notable IBE schemes which have been constructed from bilinear maps in the selective model [14,27] and the adaptive model [12,15,29,35,56,57], and from lattices in the adaptive model [2,5,28]. It is fair to say that, by now, the art of selectively secure IBE has been well honed. However, adaptively secure IBE schemes from standard assumptions with tight security (in the sense that the security loss is a small constant) remain unknown. The best known adaptively secure IBE schemes in terms of tight reduction are based on linear assumptions over pairings and achieve almost tight security (e.g., [6,12,29,44]). Waters [56] states this open problem as follows:

Open Problem #2—**Tight Adaptively Secure IBE**
"Construct a tightly, adaptively secure IBE scheme from standard computational hardness assumptions without random oracles." —Waters (2005)

Furthermore, for all known directly constructed adaptively secure IBE scheme from standard post-quantum assumption (specifically the LWE assumption), i.e. [2,5,28], their security loss during reduction depends on the number adversary's of queries. That is there is current no even "almost tightly" secure adaptive IBE scheme based on standard computational problems which are conjectured to be hard under quantum attacks. The following problem is still open.

Open Problem #3—**"Almost" Tight Adaptively Secure, Post-Quantum IBE**
"Construct an "almost" tightly, adaptively secure IBE scheme from standard post-quantum assumptions without random oracles."

1.3 Our Results

Our work uses pseudorandom functions (PRFs). Recall a PRF is a (deterministic) function: $\mathsf{PRF} : \mathcal{K} \times \mathcal{D} \to \mathcal{R}$ with the following security property. For random secret key K from $\mathcal{K}$, $\mathsf{PRF}(K, \cdot)$ is computationally indistinguishable from a random function $\Omega : \mathcal{D} \to \mathcal{R}$, given oracle access to either $\mathsf{PRF}(K, \cdot)$ or $\Omega(\cdot)$. PRFs can be constructed from general assumptions (e.g., the existence of pseudo-random number generators [40]), number-theoretic assumptions (e.g., the DDH/k-LIN assumption [31,47,53]), and lattice assumption LWE [8,9].

Our contribution is a construction of a class of adaptively secure short signature schemes/IBE schemes in the standard model. The schemes' security is tightly related to SIS/LWE and the security of an instantiated PRF PRF in the sense that the security loss is a nearly optimal constant factor. More precisely, let ϵ and ϵ' be the advantage of an adversary in attacking our signature and IBE schemes respectively, ϵ_{SIS} and ϵ_{LWE} be the security level of the SIS and LWE assumptions on which our schemes are based, and ϵ_{PRF} is the security level of the PRF instantiation PRF. Our constructions provide the following: $\epsilon \approx 2(\epsilon_{\mathsf{SIS}} + \epsilon_{\mathsf{PRF}})$, $\epsilon' \approx 2(\epsilon_{\mathsf{LWE}} + \epsilon_{\mathsf{PRF}})$, and the (polynomial) runtime of reduction is approximately the same as attacker's runtime. Depending on the underlying hardness assumption and the reduction of PRF, underlying assumptions and tightness of our signature/IBE scheme vary.

Our work indicates that tightly secure PRFs, which are based on standard assumptions and computable by polynomial size Boolean circuits, are sufficient for us to build tightly, adaptively secure lattice signature/IBE schemes. Ideally, it is better if the PRF instantiations assume weak assumptions and have shallow Boolean circuits implementations. In particular, by instantiating the 'almost" tightly secure PRFs from [8,9], (which are based on LWE assumption with super-polynomial modulus) we obtain the first "almost" tightly secure short signature/IBE schemes from LWE with super-polynomial modulus whose security does not depend on the number of adversarial queries.[1] This, at the first time, eliminates the dependency between the number of adversary's queries and the security of lattice-based short signature scheme/IBE scheme, and allows us to answer the Open Problem #3.

While constructing low-depth (e.g. circuits in NC^1), tightly secure PRFs from standard assumptions with constant security loss in the black-box sense[2] remains an open problem, any progress made in such direction will improve our work toward solving Open Problem #1 and #2 (under SIS/LWE assumption). For instance, if the DDH/k-LIN-based PRFs from [47] achieve security loss $O(\log^2 \lambda)$ for security parameter λ, we obtain signature/IBE schemes enjoy the same security loss under the combined assumptions.

Table 1 provides a comparison between our signature scheme with a LWE-based PRF instantiation (from [9]) and a representative sample of the prominent lattice-based (quantum-safe) signature schemes from the literature. Note, Katz and Wang did not propose a SIS-based signature scheme in [48]. The scheme we refer to is a straightforward application of Katz-Wang's proof technique to GPV'08 signature scheme. Table 2 provides a comparison between our signature scheme with DDH-based PRF instantiation from [47] and the representative signature schemes from traditional number-theoretic assumptions, including (strong) RSA, Dlog and linear assumptions over pairings. Our signature scheme loses a factor of $O(\log^2 \lambda)$ in security proof if the DDH-based PRF instantiation achieves the same security loss. All of those assumptions are not conjectured to be quantum-safe. In each case, the two tables refer to conjectured quantum safe and quantum-unsafe constructions respectively. Table 3 gives a comparison between our IBE scheme (with both direct LWE-based PRF instantiation from [9] and DDH-based instantiation from [47]) and a representative selection of existing IBE schemes from the literature.

It needs to mention that the bit length of PRF secret key determines the number of public matrices in our constructions. In the SIS-based signature scheme from [20] and LWE-based IBE schemes from [2,28], the number of public matrices are determined by the bit length of messages and identities respectively. For the provably secure PRFs, the bit length of secret key is usually significantly larger than the bit length of messages and identities needed in [2,20,28]. So our constructions have larger concrete size of verification key than the signature scheme in [20] and larger concrete size of public parameters than the IBE schemes in [2,28].

Efficiency Consideration. Though we focus on tightness of reduction in the context of short signature and IBE, we do not hide the inefficiency of our schemes, particularly with comparison to the adptively secure lattice-based signature/IBE scheme obained from the "complexity leveraging" [14] of efficient selectively

[1] The (direct) lattice-based PRFs from [8,9] assume LWE assumption with super-polynomial modulus, which makes our schemes rely on LWE assumption for super-polynomial modulus.

[2] The security reduction does not require a priori information about a given adversary.

Table 1. Comparison between signature schemes from quantum-safe (Ring-)SIS assumption

Scheme	Signature size	Security loss	Assumption(s)	Standard model?
KW'03 [48]	$O(1) \times \mathbb{Z}^m$	$O(1)$	SIS, $\beta = \tilde{\Omega}(n^{3/2})$	ROM
GPV'08 [36]	$O(1) \times \mathbb{Z}^m$	$O(q_{\text{hash}})$	SIS, $\beta = \tilde{\Omega}(n^{3/2})$	ROM
Boyen'10 [20]	$O(1) \times \mathbb{Z}^m$	$O(\lambda q_s)$	SIS, $\beta = \tilde{\Omega}(n^{7/2})$	✔
Lyu'12 [50]	$O(1) \times \mathbb{Z}^m$	$O(\lambda q_s)$	SIS, $\tilde{\Omega}(n^{3/2})$	ROM
MP'12 [51]	$O(1) \times \mathbb{Z}^m$	$O(\lambda q_s)$	SIS, $\beta = \tilde{\Omega}(n^{5/2})$	✔
BHJKSS'13 [13]	$O(\log \lambda) \times \mathbb{Z}^m$	$O(\lambda q_s)$	SIS, $\beta = \tilde{\Omega}(n^{5/2})$	✔
DM'14 [32]	$O(1) \times \mathcal{R}_q^{O(\log q)}$	$O(\lambda q_s)$	Ring-SIS, $\beta = \tilde{\Omega}(n^{7/2})$	✔
BKKP'15 [11]	$O(\lambda) \times \mathbb{Z}^m$	$O(1)$	SIS, $\beta = \tilde{\Omega}(n^{3/2})$	✔
Alperin'15 [4]	$O(1) \times \mathbb{Z}^m$	$O(\lambda q_s)$	SIS, $\beta = \tilde{\Omega}(\delta^{2\delta} \cdot n^{11/2})$	✔
Ours	$O(1) \times \mathbb{Z}^m$	$O(\lambda)$	SIS+LWE*, $\beta = \tilde{\Omega}(\ell^{4c} \cdot n^{7/2})$	✔

λ is the security parameter, n is the lattice hardness parameter, m is the lattice dimension, and β is the SIS parameter. q_{hash} is the number of random-oracle queries (if applicable). q_s is the number of signing queries. For DM'14, the ring $\mathcal{R} = \mathbb{Z}_q[X]/(f(X))$ for some cyclotomic polynomial f of degree n and $q \geq \beta\sqrt{n}\omega(\sqrt{\log n})$. For Alperin'15, δ satisfies $2q_s^2/\epsilon < 2^{\lfloor c'^{\delta} \rfloor}$ for attacker's success probability ϵ and arbitrary constant $c' > 1$. Our construction here consider instantiation of the direct LWE-based PRF from [9] which has security loss $O(\lambda)$ and can be computed by a NC^1 circuit with input length ℓ and depth $c \log \ell$ for some constant $c > 1$.
* The security of direct LWE-based PRF construction from [9] relies on LWE assumption with super-polynomial modulus. So LWE here refers to LWE assumption with super-polynomial modulus.

secure lattice-based signature/IBE scheme such as [2]. Although complexity leveraging is not very satisfactory from a theoretical perspective, it indeed often leads to the most practical secure cryptographic schemes. In the context of IBE, we have seen that the adaptively secure IBE scheme leveraged from selective DBDH-based IBE scheme in [14] has higher real-world efficiency than the adaptively secure Waters IBE scheme [56] (as well as the subsequent adaptive IBE schemes from similar standard pairing assumptions without random oracles) for the same security level. This may seem counter-intuitive, but to design adaptively secure IBE schemes one needs to carefully embed some specially crafted complex structures into the scheme, to provide enough freedom for the security reduction. This makes directly constructed adaptive IBE schemes rather bulky and sometimes require even stronger assumptions (in the lattice setting). Therefore, our current results are of more theoretical value. One the other hand, directly constructing adaptively secure schemes from standard assumptions usually requires new proof ideas and techniques which advance the state-of art and lead to further applications. Trying to get tighter reduction for the directly constructed adaptively secure schemes should be always welcome as it remains a very promising way of bridging the efficiency gap.

Table 2. Comparison between signature schemes from various quantum-unsafe assumptions

Scheme	Sig. size	Sec. loss	Assumption(s)	Standard model?
GHR'99 [34]	$O(1) \times \mathbb{Z}_N$	$O(1)$	Strong-RSA + D-I Hash	✔
BLS'01 [19]	$O(1) \times \mathbb{G}$	$O(\lambda q_s)$	CDH	ROM
KW'03 [48]	$O(1) \times \lvert\mathcal{D}\rvert$	$O(1)$	CFP	ROM
BB'04 [16]	$O(1) \times \mathbb{G}$	$O(1)$	q_s-SDH	✔
Waters'05 [56]	$O(1) \times \mathbb{G}$	$O(\lambda q_s)$	CDH	✔
HW'09 [46]	$O(1) \times \mathbb{Z}_N$	$O(\lambda q_s)$	RSA	✔
BHJKSS'13 [13]	$O(1) \times \mathbb{G}$	$O(\lambda q_s)$	DLog	✔
BHJKSS'13 [13]	$O(1) \times \mathbb{Z}_N$	$O(\lambda q_s)$	RSA	✔
ADKMO'13 [1]	$O(\lambda) \times \mathbb{G}$	$O(1)$	DLIN	✔
CW'13 [29]	$O(k) \times \mathbb{G}$	$O(\lambda)$	k-LIN	✔
BKP'14 [12]	$O(k) \times \mathbb{G}$	$O(\lambda)$	k-LIN	✔
BKKP'15 [11]	$O(\lambda) \times \mathbb{G}$	$O(1)$	DLog	✔
BKKP'15 [11]	$O(\lambda) \times \mathbb{Z}_N$	$O(1)$	RSA,FAC	✔
Ours	$O(1) \times \mathbb{Z}^m$	$O(\log^2 \lambda)$	SIS+DDH, $\beta = \tilde{\Omega}(\ell^{4c} \cdot n^{7/2})$	✔

λ is the security parameter, n is the lattice hardness parameter, m is the lattice dimension, q_s the number of signing queries, N is the RSA modulus, m is the lattice dimension, β is the SIS parameter, and k is a non-adversary-query-dependent parameter of the LIN assumption. For GHR'99, D-I hash stands for division-intractable hash. For KW'03, $|\mathcal{D}|$ the domain size of the instantiated claw-free permutation, which is abbreviated as CFP. Our construction here consider instantiating the DDH-based PRF from [47] which has security loss $O(\log^2 \lambda)$ and can be computed by a NC^1 circuit with input length ℓ and depth $c \log \ell$ for some constant $c > 1$.

1.4 Overview of Our Approach

Construction Outline. Our constructions use a PRF $\mathsf{PRF} : \{0,1\}^k \times \{0,1\}^t \rightarrow \{0,1\}$ which takes as input a truly random secret key from $\{0,1\}^k$ and a string from $\{0,1\}^t$, and deterministically outputs a bit which is computationally indistinguishable from a random bit. In our signature scheme, $5 + k$ random matrices are chosen from $\mathbb{Z}_q^{n \times m}$, comprising: a "left" matrix $\mathbf{A}$, two "signature subspace selection" matrices $\mathbf{A}_0, \mathbf{A}_1$, k "PRF secret key" matrices $\{\mathbf{B}_i\}_{i \in [k]}$, and two "message representation" matrices $\mathbf{C}_0, \mathbf{C}_1$. The key generation algorithm further expresses PRF as a NAND Boolean circuit, which serves as a part of the public parameters or perhaps a common reference string. The signing key consists of a "short" basis $\mathbf{T}_{\mathbf{A}}$of $\mathbf{A}$ and a PRF key $K \in \{0,1\}^k$ for PRF.

The signer takes three steps to generate the signature of message $\mathsf{M} = x_1 x_2 \dots x_t \in \{0,1\}^t$. Firstly, it uses the key-homomorphic evaluation algorithm developed from [18,24,38] to compute the unique matrix $\mathbf{A}_{\mathsf{PRF},\mathsf{M}}$ from the circuit

Table 3. Comparison between adaptively secure IBE schemes from various assumptions

Scheme	Security loss	Assumption	Standard model?	Quantum-safe?
BF'01 [17]	$O(q_{\mathsf{id}})$	BDH	ROM	✗
KW'03 [48]	$O(1)$	BDH	ROM	✗
BB'04a [14]	$O(2^{\lambda})$	DBDH, q_{id}-BDHI	✔	✗
BB'04b [15]	$O(\lambda q_{\mathsf{id}})$	DBDH	✔	✗
Waters'05 [56]	$O(\lambda q_{\mathsf{id}})$	DBDH	✔	✗
Gentry'06 [35]	$O(1)$	q_{id}-ABDHE	✔	✗
GPV'08 [36]	$O(q_{\mathrm{hash}})$	LWE	ROM	✔
Waters'09 [57]	$O(q_{\mathsf{id}})$	DBDH	✔	✗
ABB'10 [2]	$O(\lambda q_{\mathsf{id}})$	LWE	✔	✔
CHKP'12 [28]	$O(\lambda q_{\mathsf{id}})$	LWE	✔	✔
LW'12 [49]	$O(q)$	DLIN	✔	✗
CW'13 [29]	$O(\lambda)$	k-LIN	✔	✗
BKP'14 [12]	$O(\lambda)$	k-LIN	✔	✗
Ours	$O(\lambda)$	LWE*	✔	✔
	$O(\log^2(\lambda))$	DDH†+LWE	✔	✗

λ is the security level, q_{id} the number of private key queries and q_{hash} the number of random-oracle queries (if applicable).* Here we instantiate the PRF by direct LWE-based PRF construction from [9] which has $O(\lambda)$ security loss and relies on LWE assumption with super-polynomial modulus. So the LWE here refers to LWE assumption with super-polynomial modulus. The schemes ABB'10 and CHKP'12 assume LWE assumption polynomial modulus.† Here we instantiate the PRF by DDH-based PRF construction from [47] which has (black-box) security loss $O(\log^2(\lambda))$.

of PRF and the $k+t$ matrices $\{\mathbf{B}_i\}_{i\in[k]}$, $\mathbf{C}_{x_1}, \mathbf{C}_{x_2}, \ldots, \mathbf{C}_{x_t}$.[3] Then it computes $b = \mathsf{PRF}(K, \mathsf{M})$ and sets the matrix $\mathbf{F}_{\mathsf{M},1-b} = [\mathbf{A} \mid \mathbf{A}_{1-b} - \mathbf{A}_{\mathsf{PRF},\mathsf{M}}] \in \mathbb{Z}_q^{n\times 2m}$. Finally, it applies the trapdoor $\mathbf{T_A}$ to generate the signature: a low-norm non-zero vector $\mathbf{d}_{\mathsf{M}} \in \mathbb{Z}^{2m}$ such that $\mathbf{F}_{\mathsf{M},1-b} \cdot \mathbf{d}_{\mathsf{M}} = \mathbf{0} \pmod q$. The verification algorithm checks whether the signature is a non-zero vector in $\mathbb{Z}^{2m}$ and has low-norm, and whether $\mathbf{F}_{\mathsf{M},b} \cdot \mathbf{d}_{\mathsf{M}} = \mathbf{0} \pmod q$ or $\mathbf{F}_{\mathsf{M},1-b} \cdot \mathbf{d}_{\mathsf{M}} = \mathbf{0} \pmod q$. If all these conditions are satisfied, the signature is accepted.

Our IBE scheme works as follows. The public parameters contain matrices $\mathbf{A}$, $\mathbf{A}_0$, $\mathbf{A}_1$, $\{\mathbf{B}_i\}_{i\in[k]}$, $\mathbf{C}_0, \mathbf{C}_1$, a secure PRF PRF represented as a NAND Boolean circuit, and a random vector $\mathbf{u} \in \mathbb{Z}_q^n$ which is used to hide messages. The trapdoor basis $\mathbf{T_A}$ and a secret PRF key $K \in \{0,1\}^k$ serve as master secret key. In private key generation for identity $\mathsf{id} = x_1x_2\ldots x_t \in \{0,1\}^t$, the key-homomorphic evaluation algorithm is invoked to compute the unique matrix $\mathbf{A}_{\mathsf{PRF},\mathsf{id}}$ from the circuit of PRF and the $k+t$ matrices

[3] It can be shown that for different massages $\mathsf{M}_0 \neq \mathsf{M}_2$ $\mathbf{A}_{\mathsf{PRF},\mathsf{M}_0} \neq \mathbf{A}_{\mathsf{PRF},\mathsf{M}_1}$ with all but negligible probability. See Sect. 3.3 for details.

$\{\mathbf{B}_i\}_{i\in[k]}, \mathbf{C}_{x_1}, \mathbf{C}_{x_2}, \ldots, \mathbf{C}_{x_t}$. It then sets the "function" matrix to $\mathbf{F}_{\mathsf{id},1-b} = [\mathbf{A} \mid \mathbf{A}_{1-b} - \mathbf{A}_{\mathsf{PRF},\mathsf{id}}] \in \mathbb{Z}_q^{n\times 2m}$ for $b = \mathsf{PRF}(K, \mathsf{M})$, and uses $\mathbf{T}_\mathbf{A}$ to sample a Gaussian vector $\mathbf{d}_{\mathsf{id}} \in \mathbb{Z}^{2m}$ as private identity key where $\mathbf{F}_{\mathsf{id},1-b} \cdot \mathbf{d}_{\mathsf{id}} = \mathbf{u} \pmod q$.

To encrypt a message $\mathsf{Msg} \in \{0,1\}$ with an identity id, the encryptor computes $\mathbf{A}_{\mathsf{PRF},\mathsf{id}}$ and sets two "function" matrices $\mathbf{F}_{\mathsf{id},b} = [\mathbf{A} \mid \mathbf{A}_b - \mathbf{A}_{\mathsf{PRF},\mathsf{id}}]$ and $\mathbf{F}_{\mathsf{id},1-b} = [\mathbf{A} \mid \mathbf{A}_{1-b} - \mathbf{A}_{\mathsf{PRF},\mathsf{id}}]$. It generates two independent GPV-style ciphertexts [36]. The first one uses $\mathbf{F}_{\mathsf{id},b}$:

$$\begin{cases} c_{b,0} = \mathbf{s}_b^\top \mathbf{u} + \nu_{b,0} + \mathsf{Msg} \cdot \lfloor q/2 \rfloor \\ \mathbf{c}_{b,1}^\top = \mathbf{s}_b^\top \mathbf{F}_{\mathsf{id},b} + \boldsymbol{\nu}_{b,1}^\top \end{cases}$$

and the second is based on $\mathbf{F}_{\mathsf{id},1-b}$:

$$\begin{cases} c_{1-b,0} = \mathbf{s}_{1-b}^\top \mathbf{u} + \nu_{1-b,0} + \mathsf{Msg} \cdot \lfloor q/2 \rfloor \\ \mathbf{c}_{1-b,1}^\top = \mathbf{s}_{1-b}^\top \mathbf{F}_{\mathsf{id},1-b} + \boldsymbol{\nu}_{1-b,1}^\top \end{cases}$$

for random vectors $\mathbf{s}_b, \mathbf{s}_{1-b} \xleftarrow{\$} \mathbb{Z}_q^n$, two small noise scalars $\nu_{b,0}, \nu_{1-b,0}$, and two low-norm noise vectors $\boldsymbol{\nu}_{b,1}, \boldsymbol{\nu}_{1-b,1}$.

The decryption algorithm uses $\mathbf{d}_{\mathsf{id}}$ to try both ciphertexts; one of them should work. Here as a technical caveat, we need some redundant information in the messages in order to check whether a recovered message is well-formed. To this end, one option is to apply the standard way of encrypting multiple bits in GPV-style ciphertexts without affecting the security analysis. That is, instead of using just a vector $\mathbf{u} \in \mathbb{Z}_q^n$ in the public key, we use a matrix $\mathbf{U} \in \mathbb{Z}_q^{n\times z}$ allowing us to encrypt z bits. A second option, which costs nothing if hybrid encryption is being used, is to use multi-bit GPV-style encryption to encrypt a symmetric session key without redundancy, again using a matrix $\mathbb{Z}_q^{n\times z}$ and rely on downstream symmetric integrity checks or MACs to weed out the incorrect ciphertexts.

Proof Outline. The security reduction of our signature scheme uses an efficient adversary to solve a of SIS problem instance $\mathbf{A} \in \mathbb{Z}_q^{n\times m}$: a short non-zero vector $\mathbf{e} \in \mathbb{Z}^m$ such that $\mathbf{Ae} = \mathbf{0} \pmod q$. The reduction embeds a randomly picked secret key K for PRF in verification key. More specifically, the reduction selects low-norm matrices $\mathbf{R}_{\mathbf{A}_0}, \mathbf{R}_{\mathbf{A}_1}, \{\mathbf{R}_{\mathbf{B}_i}\}_{i\in[k]}, \mathbf{R}_{\mathbf{C}_0}, \mathbf{R}_{\mathbf{C}_1}$ from $\{1,-1\}^{m\times m}$, a PRF secret key $K = s_1 s_2 \ldots s_k \in \{0,1\}^k$ and sets $\mathbf{A}_0 = \mathbf{A}\mathbf{R}_{\mathbf{A}_0}$, $\mathbf{A}_1 = \mathbf{A}\mathbf{R}_{\mathbf{A}_1} + \mathbf{G}$, $\{\mathbf{B}_i = \mathbf{A}\mathbf{R}_{\mathbf{B}_i} + s_i\mathbf{G}\}_{i\in[k]}$, $\mathbf{C}_0 = \mathbf{A}\mathbf{R}_{\mathbf{C}_0}$ and $\mathbf{C}_1 = \mathbf{A}\mathbf{R}_{\mathbf{C}_1} + \mathbf{G}$. Here, K is completely hidden from adversary's view. For answering a signing query on message M, the reduction computes $\mathbf{A}_{\mathsf{PRF},\mathsf{M}} = \mathbf{AR} + \mathsf{PRF}(K, \mathsf{M})\mathbf{G}$ for some known low-norm $m \times m$ matrix $\mathbf{R}$ that depends on $\mathbf{R}_{\mathbf{A}_0}, \mathbf{R}_{\mathbf{A}_1}, \{\mathbf{R}_{\mathbf{B}_i}\}_{i\in[k]}, \mathbf{R}_{\mathbf{C}_0}, \mathbf{R}_{\mathbf{C}_1}$, K and M. Let $\mathsf{PRF}(K, \mathsf{M}) = b$, the reduction sets $\mathbf{F}_{\mathsf{M},1-b} = [\mathbf{A} \mid \mathbf{A}_{1-b} - \mathbf{A}_{\mathsf{PRF},\mathsf{M}}] = [\mathbf{A} \mid \mathbf{AR} + (1-2b)\mathbf{G}]$ and uses the trapdoor from $\mathbf{G}$ to compute the decryption key. Note, we use PRF to select the matrix $\mathbf{A}_b$ which is the same as the real scheme. For a valid forgery $(\mathsf{M}^*, \mathbf{d}_{\mathsf{M}^*})$, since $b = \mathsf{PRF}(K, \mathsf{M}^*)$ is unpredictable to the adversary, $\mathbf{F}_{\mathsf{M}^*,b} \cdot \mathbf{d}_{\mathsf{M}^*} = \mathbf{0} \pmod q$ happens with essentially probability 1/2 leading to a valid SIS solution.

The security reduction for our IBE scheme is similar to the reduction of the signature scheme. Basically, the reduction answers key generation queries in the same way as answering signing queries in the signature scheme reduction. To construct the challenge ciphertext for a challenge identity id^*, the LWE challenge is embedded in the function matrix $\mathbf{F}_{\mathsf{id}^*,b} = [\mathbf{A} \mid \mathbf{AR}]$ for which the simulator cannot produce private key. Another ciphertext based on $\mathbf{F}_{\mathsf{id}^*,1-b} = [\mathbf{A} \mid \mathbf{AR} + (1-2b)\mathbf{G}]$ is generated as in the real scheme. With essentially half probability, the adversary will choose the ciphertext under $\mathbf{F}_{\mathsf{id}^*,b}$ to attack giving out useful information for solving the LWE challenge.

Related Works. In the related and concurrent work by Brakerski and Vaikuntanathan [25], a similar idea of embedding PRFs into encryption schemes has been used to construct the first semi-adaptively secure attribute-based encryption scheme from lattices supporting an a priori unbounded number of attributes. The recent work by Bai et al. [7] addresses the problem of improving efficiency of lattice-based cryptographic schemes via a different but novel way. Their proposal is about using Rényi divergence instead of statistical distance in the context of lattice-based cryptography which leads to (sometimes simpler) security proofs for more efficient lattice-based schemes.

2 Preliminaries

Notation. 'PPT' abbreviates "probabilistic polynomial-time". If S is a set, we denote by $a \xleftarrow{\$} S$ the uniform sampling of a random element of S. For a positive integer n, we denote by $[n]$ the set of positive integers no greater than n. We use bold lowercase letters (e.g. $\mathbf{a}$) to denote vectors and bold capital letters (e.g. $\mathbf{A}$) to denote matrices. For a positive integer $q \geq 2$, let $\mathbb{Z}_q$ be the ring of integers modulo q. We denote the group of $n \times m$ matrices in $\mathbb{Z}_q$ by $\mathbb{Z}_q^{n\times m}$. Vectors are treated as column vectors. The transpose of a vector $\mathbf{a}$ (resp. a matrix $\mathbf{A}$) is denoted by $\mathbf{a}^\top$ (resp. $\mathbf{A}^\top$). For $\mathbf{A} \in \mathbb{Z}_q^{n\times m}$ and $\mathbf{B} \in \mathbb{Z}_q^{n\times m'}$, let $[\mathbf{A}|\mathbf{B}] \in \mathbb{Z}_q^{n\times(m+m')}$ be the concatenation of $\mathbf{A}$ and $\mathbf{B}$. We denote the Gram-Schmidt ordered orthogonalization of a matrix $\mathbf{A} \in \mathbb{Z}^{m\times m}$ by $\tilde{\mathbf{A}}$. The inner product of two vectors $\mathbf{x}$ and $\mathbf{y}$ is written $\langle \mathbf{x}, \mathbf{y} \rangle$. For a security parameter λ, a function $\mathsf{negl}(\lambda)$ is negligible in λ if it is smaller than all polynomial fractions for a sufficiently large λ.

We recall the following generalisation of left-over hash lemma.

Lemma 1 ([2], Lemma 4). Suppose that $m > (n+1)\log q + \omega(\log n)$ and that $q > 2$ is prime. Let $\mathbf{R}$ be an $m \times k$ matrix chosen uniformly in $\{1, -1\}^{m\times k} \mod q$ where $k = k(n)$ is polynomial in n. Let $\mathbf{A}$ and $\mathbf{B}$ be matrices chosen uniformly in $\mathbb{Z}_q^{n\times m}$ and $\mathbb{Z}_q^{n\times k}$ respectively. Then, for all vectors $\mathbf{w} \in \mathbb{Z}_q^m$, the distribution $(\mathbf{A}, \mathbf{AR}, \mathbf{R}^\top \mathbf{w})$ is statistically close to the distribution $(\mathbf{A}, \mathbf{B}, \mathbf{R}^\top \mathbf{w})$.

For a vector $\mathbf{u}$, we let $\|\mathbf{u}\|$ and $\|\mathbf{u}\|_\infty$ denote its ℓ_2 norm and ℓ_∞ norm, respectively. For a matrix $\mathbf{R} \in \mathbb{Z}^{k\times m}$, we define two matrix norms:

- $\|\mathbf{R}\|$ denotes the ℓ_2 length of the longest column of $\mathbf{R}$.
- $\|\mathbf{R}\|_2$ is the operator norm of $\mathbf{R}$ defined as $\|\mathbf{R}\|_2 = \sup_{\mathbf{x}\in\mathbb{R}^{m+1}} \|\mathbf{R}\cdot\mathbf{x}\|$.

Lemma 2. **([2], Lemma 5).** Let $\mathbf{R}$ be a random chosen matrix from $\{1,-1\}^{m\times m}$, then $\Pr[\|\mathbf{R}\|_2 > 12\sqrt{2m}] < e^{-m}$.

2.1 Lattice Background

Lattice Definitions

Definition 1. *Let a basis* $\mathbf{B} = [\mathbf{b}_1 \mid \ldots \mid \mathbf{b}_m] \in (\mathbb{R}^m)^m$ *of linearly independent vectors. The lattice generated by* $\mathbf{B}$ *is defined as* $\Lambda = \{\mathbf{y} \in \mathbb{R}^m : \exists s_i \in \mathbb{Z}, \mathbf{y} = \sum_{i=1}^m s_i \mathbf{b}_i\}$. *The dual lattice* Λ^* *of* Λ *is defined as* $\Lambda^* = \{\mathbf{z} \in \mathbb{R}^m : \forall \mathbf{y} \in \Lambda, \langle \mathbf{z}, \mathbf{y}\rangle \in \mathbb{Z}\}$.

Definition 2. *For* q *prime,* $\mathbf{A} \in \mathbb{Z}_q^{n\times m}$ *and* $\mathbf{u} \in \mathbb{Z}_q^n$, *we define the* m*-dimensional (full-rank) random integer lattice* $\Lambda_q^\perp(\mathbf{A}) = \{\mathbf{e} \in \mathbb{Z}^m : \mathbf{A}\mathbf{e} = \mathbf{0} \pmod q\}$, *and the "shifted lattice" as the coset* $\Lambda_q^{\mathbf{u}}(\mathbf{A}) = \{\mathbf{e} \in \mathbb{Z}^m : \mathbf{A}\mathbf{e} = \mathbf{u} \pmod q\}$.

Trapdoors of Lattices and Discrete Gaussians. It is shown in [3,51] how to sample a "nearly" uniform random matrix $\mathbf{A} \in \mathbb{Z}^{n\times m}$ along with a trapdoor matrix $\mathbf{T_A} \in \mathbb{Z}^{m\times m}$ which is a short or low-norm basis of the induced lattice $\Lambda_q^\perp(\mathbf{A})$. We refer to this procedure as TrapGen.

Lemma 3. *There is a PPT algorithm TrapGen that takes as input integers* $n \geq 1$, $q \geq 2$ *and a sufficiently large* $m = O(n \log q)$, *outputs a matrix* $\mathbf{A} \in \mathbb{Z}_q^{n\times m}$ *and a trapdoor matrix* $\mathbf{T_A} \in \mathbb{Z}^{m\times m}$, *such that* $\mathbf{A}\cdot\mathbf{T_A} = 0$, *the distribution of* $\mathbf{A}$ *is statistically close to the uniform distribution over* $\mathbb{Z}_q^{n\times m}$ *and* $\|\tilde{\mathbf{T}}_\mathbf{A}\| = O(\sqrt{n \log q})$.

Discrete Gaussians. Let $m \in \mathbb{Z}_{>0}$ be a positive integer and $\Lambda \subset \mathbb{Z}^m$. For any real vector $\mathbf{c} \in \mathbb{R}^m$ and positive parameter $\sigma \in \mathbb{R}_{>0}$, let the Gaussian function $\rho_{\sigma,\mathbf{c}}(\mathbf{x}) = \exp\left(-\pi\|\mathbf{x}-\mathbf{c}\|^2/\sigma^2\right)$ on $\mathbb{R}^m$ with center $\mathbf{c}$ and parameter σ. Define the discrete Gaussian distribution over Λ with center $\mathbf{c}$ and parameter σ as $D_{\Lambda,\sigma} = \rho_{\sigma,\mathbf{c}}(\mathbf{y})/\rho_\sigma(\Lambda)$ for $\forall \mathbf{y} \in \Lambda$, where $\rho_\sigma(\Lambda) = \sum_{\mathbf{x}\in\Lambda} \rho_{\sigma,\mathbf{c}}(\mathbf{x})$. For notational convenience, $\rho_{\sigma,\mathbf{0}}$ and $D_{\Lambda,\sigma,\mathbf{0}}$ are abbreviated as ρ_σ and $D_{\Lambda,\sigma}$.

The following lemma bounds the length of a discrete Gaussian vector with sufficiently large Gaussian parameter.

Lemma 4 **([52]).** For any lattice Λ of integer dimension m with basis $\mathbf{T}$, $\mathbf{c} \in \mathbb{R}^m$ and Gaussian parameter $\sigma \geq \|\tilde{\mathbf{T}}\|\cdot\omega(\sqrt{\log m})$, we have $\Pr[\|\mathbf{x}-\mathbf{c}\| > \sigma\sqrt{m} : \mathbf{x} \leftarrow D_{\Lambda,\sigma,\mathbf{c}}] \leq \mathsf{negl}(n)$.

Smoothing Parameter. We recall the very important notion of smoothing parameter of a lattice Λ. It is the smallest value of s such that the discrete Gaussian $D_{\Lambda,s}$ "behaves" like a continuous Gaussian.

Definition 3 ([52]). For any lattice Λ and positive real tolerance $\epsilon > 0$, the smoothing parameter $\eta_\epsilon(\Lambda)$ is the smallest real $s > 0$ such that $\rho_{1/s}(\Lambda^* \backslash \{\mathbf{0}\}) < \epsilon$.

We will make use of the following lemma, which is a special case of Corollary 3.10 from [55].

Lemma 5 (special case of Corollary 3.10 of [55]). Let $\mathbf{r} \in \mathbb{Z}^m$ be a vector and $r, \alpha > 0$ be reals. Assume that $1/\sqrt{1/r^2 + (\|\mathbf{r}\|/\alpha)^2} \geq \eta_\epsilon(\mathbb{Z}^m)$ for some $\epsilon < 1/2$. Let $\mathbf{y}$ be a vector with distribution $D_{\mathbb{Z}^m,r}$ and e be a scalar with distribution $D_{\mathbb{Z},\alpha}$. The distribution of $\langle \mathbf{r}, \mathbf{y} \rangle + e$ is statistically close to $D_{\mathbb{Z},\sqrt{(r\|\mathbf{r}\|)^2+\alpha^2}}$.

Lattice Sampling Algorithms. Our constructions make use of the "two-sided trapdoor" framework from [2,20] which consists of two sampling algorithms SampleLeft and SampleRight.

$$\textit{Algorithm } \mathsf{SampleLeft}(\mathbf{A}, \mathbf{B}, \mathbf{T_A}, \mathbf{u}, s) \tag{1}$$

Inputs: a full-rank matrix $\mathbf{A} \in \mathbb{Z}_q^{n \times m}$ and a short basis $\mathbf{T_A}$ of $\Lambda_q^\perp(\mathbf{A})$, a matrix $\mathbf{B} \in \mathbb{Z}_q^{n \times m_1}$, a vector $\mathbf{u} \in \mathbb{Z}_q^n$, and a Gaussian parameter s.
Output: Let $\mathbf{F} = [\mathbf{A} \mid \mathbf{B}]$. The algorithm outputs a vector $\mathbf{d} \in \mathbb{Z}^{m+m_1}$ in the set $\Lambda_q^{\mathbf{u}}(\mathbf{F})$.

Theorem 1 ([2,28]). Let $q > 2$, $m > n$ and $s > \|\tilde{\mathbf{T}}_\mathbf{A}\| \cdot \omega(\sqrt{\log(m+m_1)})$. Then the algorithm $\mathsf{SampleLeft}(\mathbf{A}, \mathbf{B}, \mathbf{T_A}, \mathbf{u}, s)$ taking inputs as in (1), outputs a vector $\mathbf{d} \in \mathbb{Z}^{m+m_1}$ distributed statistically close to $D_{\Lambda_q^{\mathbf{u}}(\mathbf{F}),s}$.

$$\textit{Algorithm } \mathsf{SampleRight}(\mathbf{A}, \mathbf{B}, \mathbf{R}, \mathbf{T_B}, \mathbf{u}, s) \tag{2}$$

Inputs: matrices $\mathbf{A} \in \mathbb{Z}_q^{n \times k}$ and $\mathbf{R} \in \mathbb{Z}^{k \times m}$, a full-rank matrix $\mathbf{B} \in \mathbb{Z}_q^{n \times m}$, a short basis $\mathbf{T_B}$ of $\Lambda_q^\perp(\mathbf{B})$, a vector $\mathbf{u} \in \mathbb{Z}_q^n$, and a Gaussian parameter s.
Output: Let $\mathbf{F} = [\mathbf{A} \mid \mathbf{AR} + \mathbf{B}]$; the algorithm outputs a vector $\mathbf{d} \in \mathbb{Z}^{m+m_1}$ in the set $\Lambda_q^{\mathbf{u}}(\mathbf{F})$

Theorem 2 ([2], Theorem 19). Let $q > 2$, $m > n$. Let $s > \|\tilde{\mathbf{T}}_\mathbf{B}\| \cdot \|\mathbf{R}\|_2 \cdot \omega(\sqrt{\log m})$. Then $\mathsf{SampleRight}(\mathbf{A}, \mathbf{B}, \mathbf{R}, \mathbf{T_B}, \mathbf{u}, s)$ taking inputs as in (2), outputs a vector $\mathbf{d} \in \mathbb{Z}^{m+k}$ distributed statistically close to $D_{\Lambda_q^{\mathbf{u}}(\mathbf{F}),s}$.

Gadget Matrix. The "gadget matrix" $\mathbf{G}$ defined in [51]. We recall the following two facts.

Lemma 6 ([51], **Theorem 1**). Let q be a prime, and n, m be integers with $m = n \log q$. There is a fixed full-rank matrix $\mathbf{G} \in \mathbb{Z}_q^{n \times m}$ such that the lattice $\Lambda_q^\perp(\mathbf{G})$ has a publicly known trapdoor matrix $\mathbf{T_G} \in \mathbb{Z}^{n \times m}$ with $\|\tilde{\mathbf{T}}_\mathbf{G}\| \leq \sqrt{5}$.

Lemma 7 ([18], Lemma 2.1). There is a deterministic algorithm, denoted $\mathbf{G}^{-1}(\cdot) : \mathbb{Z}_q^{n\times m} \rightarrow \mathbb{Z}^{m\times m}$, that takes any matrix $\mathbf{A} \in \mathbb{Z}_q^{n\times m}$ as input, and outputs the preimage $\mathbf{G}^{-1}(\mathbf{A})$ of $\mathbf{A}$ such that $\mathbf{G} \cdot \mathbf{G}^{-1}(\mathbf{A}) = \mathbf{A} \pmod q$ and $\|\mathbf{G}^{-1}(\mathbf{A})\| \leq m$.

Computational Assumptions. We recall the two most mainstream and conservative average-case computational assumptions for lattice problems.

The learning with errors problem was first proposed by Regev [55]. For a vector $\mathbf{s} \xleftarrow{\$} \mathbb{Z}_q^n$ and a noise distribution χ over $\mathbb{Z}_q$, let $A_{\mathbf{s},\chi}$ be the distribution over $\mathbb{Z}_q^n \times \mathbb{Z}_q$ by taking $\mathbf{a} \xleftarrow{\$} \mathbb{Z}_q^n$ and $x \leftarrow \chi$, and outputting $(\mathbf{a}, \mathbf{s}^\top \mathbf{a} + x) \pmod q$. Usually, χ is a discrete Gaussian $D_{\mathbb{Z},\alpha q}$ for some $\alpha < 1$, reduced modulo q. We refer to [55] for further details.

Definition 4. *For a security parameter Λ, let a positive integer $n = n(\lambda)$, a prime $q = q(\lambda)$, and a distribution χ over $\mathbb{Z}_q$. The learning with errors problem $LWE_{n,q,\chi}$ is to distinguish the oracle $\mathcal{O}_\mathbf{s}$, which outputs samples from the distribution $A_{\mathbf{s},\chi}$, from the oracle $\mathcal{O}_\$$, which outputs samples from the uniform distribution over $\mathbb{Z}_q^n \times \mathbb{Z}_q$, for an unspecified polynomial number of queries. We define the advantage (in the security parameter λ) of an algorithm $\mathcal{A}$ in solving the $LWE_{n,q,\chi}$ problem as*

$$\mathsf{Adv}_{\mathcal{A}}^{LWE_{n,q,\chi}}(\lambda) = \left|\Pr[\mathcal{A}^{\mathcal{O}_\mathbf{s}}(1^\lambda) = 1] - \Pr[\mathcal{A}^{\mathcal{O}_\$}(1^\lambda) = 1]\right|$$

We say that the $(t, \epsilon_{\mathsf{LWE}})$-$\mathsf{LWE}_{n,q,\chi}$ assumption holds if no t-time algorithm $\mathcal{A}$ that has advantage at least ϵ_{LWE} in solving the $\mathsf{LWE}_{n,q,\chi}$ problem.

For polynomial size q in λ, there are known quantum [55] and classical [22] reductions from the average-case $\mathrm{LWE}_{n,q,\chi}$ assumption to many standard worst-case lattice problems (e.g., GapSVP).[4] Peikert [54] also gave a classic reduction that applies (only) for exponential moduli q in λ. These reductions further strengthen the appeal of the LWE assumption.

The security of our adaptively secure signature scheme is based on the SIS problem, which can be seen as an average-case approximate shortest vector problem on random integer lattices. In a sense, SIS is the computational counterpart to the decisional LWE.

Definition 5. *For a security parameter λ, let $n = n(\lambda)$, $m = m(\lambda)$, and $\beta = \beta(\lambda)$. Let q be a prime integer. The short integer solution problem $\mathsf{SIS}_{n,q,\beta,m}$ is as follows. Given a uniform random matrix $\mathbf{A} \xleftarrow{\$} \mathbb{Z}_q^{n\times m}$, find a non-zero vector $\mathbf{e} \in \mathbb{Z}^m$ such that $\mathbf{A}\mathbf{e} = \mathbf{0} \pmod q$ and $\|\mathbf{e}\| \leq \beta$. We define the advantage (function of the security parameter λ) of an algorithm $\mathcal{A}$ in solving the $\mathsf{SIS}_{n,q,\beta,m}$ problem as*

---

[4] Equivalently, this is to say that many classic worst-case lattice *problems* reduce *to* the average-case LWE *problem*, for suitable parameters.

$$Adv_{\mathcal{A}}^{SIS_{n,q,\beta,m}}(\lambda) = \left[\begin{array}{c} \mathbf{Ae} = \mathbf{0} \pmod q \\ \textit{and } \|\mathbf{e}\| \leq \beta, \\ \textit{and } \mathbf{e} \neq \mathbf{0}. \end{array} : \begin{array}{c} \mathbf{A} \xleftarrow{\$} \mathbb{Z}_q^{n\times m} \\ \mathbf{e} \leftarrow \mathcal{A}(1^\lambda, \mathbf{A}) \end{array} \right]$$

We say the (t, ϵ_{SIS})-$SIS_{n,q,\beta,m}$ assumption holds if no t-time algorithm $\mathcal{A}$ that has advantage at least ϵ_{SIS} in solving the $SIS_{n,q,\beta,m}$ problem.

It has been shown in [52] that solving the average-case instances of the $\mathsf{SIS}_{n,q,\beta,m}$ problem for certain parameters is as hard as solving worst-case instances of the approximate Shortest Independent Vector Problem (SIVP).

2.2 Pseudorandom Functions

Definition 6 (Pseudorandom Functions). *Let $\lambda > 0$ be the security parameter, and let $k = k(\lambda)$, $t = t(\lambda)$ and $l = l(\lambda)$. A pseudorandom function $PRF\colon \{0,1\}^k \times \{0,1\}^t \to \{0,1\}^l$ is an efficiently computable, deterministic two-input function where the first input, denoted by K, is the key. Let Ω be the set of all functions that map t bits strings to l bits strings. We define the advantage (in the security parameter λ) of an adversary $\mathcal{A}$ in attacking the PRF as*

$$Adv_{PRF,\mathcal{A}}(\lambda) = \left| \Pr[\mathcal{A}^{PRF(K,\cdot)}(1^\lambda) = 1] - \Pr[\mathcal{A}^{F(\cdot)}(1^\lambda) = 1] \right|$$

where the probability is taken over a uniform choice of key $K \xleftarrow{\$} \{0,1\}^k$ and $F \xleftarrow{\$} \Omega$, and the randomness of $\mathcal{A}$. We say that PRF is $(t_{PRF}, \epsilon_{PRF})$-secure if for all t_{PRF}-time adversaries $\mathcal{A}$, $Adv_{PRF,\mathcal{A}}(\lambda) \leq \epsilon_{PRF}$.

2.3 Key Homomorphic Evaluation Algorithm

Recall the matrix key-homomorphic evaluation algorithm, which is developed by Gentry et al. [38], Boneh et al. [18] and Brakerski and Vaikuntanathan [24] in the context of fully homomorphic encryption and attribute-based encryption, works generally in the following. Given a fan-in-2 Boolean NAND circuits $C : \{0,1\}^\ell \to \{0,1\}$, ℓ different matrices $\{\mathbf{A}_i = \mathbf{AR}_i + x_i\mathbf{G} \in \mathbb{Z}_q^{n\times m}\}_{i\in[\ell]}$ which correspond to each input wire of C where $\mathbf{A} \xleftarrow{\$} \mathbb{Z}_q^{n\times m}$, $\mathbf{R}_i \xleftarrow{\$} \{1,-1\}^{m\times m}$, $x_i \in \{0,1\}$ and $\mathbf{G} \in \mathbb{Z}_q^{n\times m}$ is the gadget matrix, the key-homomorphic evaluation algorithm deterministically computes $\mathbf{A}_C = \mathbf{AR}_C + C(x_1,\ldots,x_\ell)\mathbf{G} \in \mathbb{Z}_q^{n\times m}$ where $\mathbf{R}_C \in \mathbb{Z}^{m\times m}$ has low norm and $C(x_1,\ldots,x_\ell) \in \{0,1\}$ is the output bit of C on the arguments $x_1,\ldots,x_\ell$. This is done, in general, by inductively evaluating each NAND gate. For a NAND gate $g(u,v;w)$ with input wires u,v and output wire w, matrices $\mathbf{A}_u = \mathbf{AR}_u + x_u\mathbf{G}$ and $\mathbf{A}_v = \mathbf{AR}_v + x_v\mathbf{G}$ where x_u and x_v are input bits of u and v respectively, the evaluation algorithm computes

$$\begin{aligned} \mathbf{A}_w &= \mathbf{G} - \mathbf{A}_u \cdot \mathbf{G}^{-1}(\mathbf{A}_v) \\ &= \mathbf{G} - (\mathbf{AR}_u + x_u\mathbf{G}) \cdot \mathbf{G}^{-1}(\mathbf{AR}_v + x_v\mathbf{G}) \\ &= \mathbf{AR}_g + (1 - x_u x_v)\mathbf{G} \end{aligned}$$

where $1 - x_u x_v \stackrel{\mathsf{def}}{=} \mathsf{NAND}(x_u, x_v)$, and $\mathbf{R}_g = -\mathbf{R}_u \cdot \mathbf{G}^{-1}(\mathbf{A}_v) - x_u \mathbf{R}_v$ has low-norm if $\mathbf{R}_u, \mathbf{R}_v$ have low-norm.

In this paper, we consider evaluating circuits of PRFs. Most of the well-known PRFs from number-theoretic assumptions (e.g. [47,53]) and lattice assumptions (e.g. [8,9]) can be computed by circuits in class NC^1 (i.e. with polynomial size, logarithmic depth $O(\log \ell)$ in input length ℓ and fan-in 2). For circuits in NC^1, by applying above procedure in a general tree-fashion, the norm of $\mathbf{R}_C$ in the matrix $\mathbf{A}_C$ is roughly bounded by $m^{O(\log \ell)}$, which in turn usually results in superpolynomial or sub-exponential LWE/SIS modulus q (in the security parameter) in certain applications.

In [24], Brakerski and Vaikuntanathan observed that the norm of $\mathbf{R}_C$ matrix in above homomorphic evaluation is accumulated in an asymmetric way. They exploited this feature to design a special evaluation algorithm that evaluates NC^1 circuits with moderately increasing the norm of $\mathbf{R}_C$. Specifically, the observation is that any circuit with depth d can be simulated by a length-4^d and width-5 branching program, through the Barrington's theorem. Such a branching program can be computed by multiplying 4^d 5-by-5 permutation matrices. It is showed in [24] that homomorphically evaluating the multiplication of permutation matrices using above homomorphic evaluation procedure and the asymmetrical noise-growth feature only increases the noise by a polynomial factor and, therefore, allows us to use polynomial size LWE/SIS modulus q in the security parameter. Such result has been used to construct efficient ABE scheme for branching programs (with bounded length) from LWE with polynomial modulus [42]. In our constructions, we particularly use the Brakerski and Vaikuntanathan's evaluation algorithm [24] and denote it by $\mathsf{Eval}_{\mathsf{BV}}$.

We recall the Barrington's Theorem.

Theorem 3 (Barrington's Theorem). *Every Boolean NAND circuit C that acts on ℓ inputs and has depth d can be computed by a width-5 permutation branching program Π of length 4^d. Given the description of the circuit Ψ, the description of the branching program C can be computed in poly$(\ell, 4^d)$ time.*

The following theorem follows from the Claim 3.4.2 and Lemma 3.6 of [24] and the Barrington's Theorem.

Lemma 8. *Let $C : \{0,1\}^\ell \rightarrow \{0,1\}$ be a NAND Boolean circuit. Let $\{\mathbf{A}_i = \mathbf{A}\mathbf{R}_i + x_i \mathbf{G} \in \mathbb{Z}_q^{n \times m}\}_{i \in [\ell]}$ be ℓ different matrices correspond to each input wire of C where $\mathbf{A} \stackrel{\$}{\leftarrow} \mathbb{Z}_q^{n \times m}$, $\mathbf{R}_i \stackrel{\$}{\leftarrow} \{1, -1\}^{m \times m}$, $x_i \in \{0,1\}$ and $\mathbf{G} \in \mathbb{Z}_q^{n \times m}$ is the gadget matrix. There is an efficient deterministic algorithm $\mathsf{Eval}_{\mathsf{BV}}$ that takes as input C and $\{\mathbf{A}_i\}_{i \in [\ell]}$ and outputs a matrix $\mathbf{A}_C = \mathbf{A}\mathbf{R}_C + C(x_1, \ldots, x_\ell)\mathbf{G} = \mathsf{Eval}_{\mathsf{BV}}(C, \mathbf{A}_1, \ldots, \mathbf{A}_\ell)$ where $\mathbf{R}_C \in \mathbb{Z}^{m \times m}$ and $C(x_1, \ldots, x_\ell)$ is the output of C on the arguments $x_1, \ldots, x_\ell$. $\mathsf{Eval}_{\mathsf{BV}}$ runs in time poly$(4^d, \ell, n, \log q)$.*

Let $\|\mathbf{R}_{max}\|_2 = max\{\|\mathbf{R}_i\|_2\}_{i\in[\ell]}$, *the norm of* $\mathbf{R}_C$ *in* $\mathbf{A}_C$ *output by* $\mathsf{Eval}_{\mathsf{BV}}$ *can be bounded, with overwhelming probability, by*

$$\begin{aligned}\|\mathbf{R}_C\|_2 &\le O(L\cdot\|\mathbf{R}_{max}\|_2\cdot m)\\ &\le O(L\cdot 12\sqrt{2}\cdot\sqrt{m}\cdot m)\\ &\le O(4^d\cdot m^{3/2})\end{aligned}$$

where L *is the length of the width-5 branching program which simulates* C *and* $\|\mathbf{R}_i\|_2 \le 12\sqrt{2m}$ *for* $i\in[\ell]$ *with overwhelming probability, by Lemma 2.*

Particularly, if C *has depth* $d = c\log\ell$ *for some constant* c, *i.e.* C *is in* NC^1, *we have* $L = 4^d = \ell^{2c}$ *and* $\|\mathbf{R}_C\|_2 \le O(\ell^{2c}\cdot m^{3/2})$.

2.4 Digital Signatures

A digital signature scheme consists of three PPT algorithms: KeyGen, Sign, and Ver. The algorithm KeyGen takes as input a security parameter and generates a public verification key Vk and a private signing key Sk. The signing algorithm Sign takes as input the signing key Sk and a massage M, and outputs the signature Sig of M. The verification algorithm Ver takes as input a signature-message pair (Sig, M) as well as the verification key Vk. It outputs 1 if Sig is valid, or 0 if Sig is invalid.

We review the standard security notion of digital signature schemes. The existential unforgeability under chosen-message attack (EUF-CMA) of a digital signature scheme Π is defined through the following security game between an adversary $\mathcal{A}$ and a challenger $\mathcal{B}$.

Setup. $\mathcal{B}$ runs $\mathsf{Setup}(1^\lambda) \to (\mathsf{Sk}, \mathsf{Vk})$, and passes Vk to $\mathcal{A}$.

Query. $\mathcal{A}$ adaptively selects messages $\mathsf{M}_1, \ldots, \mathsf{M}_{q_s}$ to ask for the corresponding signatures under Vk from $\mathcal{B}$. For the query M_i, $\mathcal{B}$ responds with a signature $\mathsf{Sig}_i \leftarrow \mathsf{Sign}(\mathsf{Sk}, \mathsf{M}_i)$.

Forge. $\mathcal{A}$ outputs a pair $(\mathsf{Sig}^*, \mathsf{M}^*)$ and wins if

1. $\mathsf{M}^* \notin \{\mathsf{M}_1, \ldots, \mathsf{M}_{q_s}\}$, and
2. $\mathsf{Ver}(\mathsf{Vk}, \mathsf{Sig}^*, \mathsf{M}^*) \to 1$.

We refer to such an adversary $\mathcal{A}$ as EUF-CMA adversary. We define the advantage (in the security parameter λ) $\mathsf{Adv}_{\Pi,\mathcal{A}}(\lambda)$ of $\mathcal{A}$ in attacking a digital signature scheme Π to be the probability that $\mathcal{A}$ wins above game.

Definition 7. *For a security parameter* λ, *let* $t = t(\lambda)$, $q_s = q_s(\lambda)$ *and* $\epsilon = \epsilon(\lambda)$. *We say that a digital signature scheme* Π *is* (t, q_s, ϵ)*-EUF-CMA secure if for any* t *time EUF-CMA adversary* $\mathcal{A}$ *that makes at most* q_s *signing queries and has* $\mathsf{Adv}_{\Pi,\mathcal{A}}(\lambda) \le \epsilon$.

2.5 Identity-Based Encryption

An Identity-Based Encryption system (IBE) consists of four PPT algorithms: Setup, KeyGen, Encrypt, and Decrypt. The algorithm Setup takes as input a security parameter and generates public parameters Pub and a master secret key Msk. The algorithm KeyGen uses the master secret key Msk to produce an identity private key $\mathsf{Sk}_{\mathsf{id}}$ corresponding to an identity id. The algorithm Encrypt takes the public parameters Pub to encrypt messages for any given identity id. The algorithm Decrypt decrypts ciphertexts using the identity private key if the identity of the ciphertext matches the identity of the private key.

We review the adaptive (full) security under chosen-plaintext attack (IND-ID-CPA) of IBE system. The IND-ID-CPA security of IBE is defined through the following game between an adversary $\mathcal{A}$ and a challenger $\mathcal{B}$. For a security parameter λ, let $\mathcal{M}_\lambda$ be the message space and $\mathcal{C}_\lambda$ be the ciphertext space.

Setup. $\mathcal{B}$ runs $\mathsf{Setup}(1^\lambda) \rightarrow (\mathsf{Pub}, \mathsf{Msk})$, passes the public parameters Pub to $\mathcal{A}$, and keeps the master secret Msk.

Phase 1. $\mathcal{A}$ adaptively requests keys for any identity id of its choice. $\mathcal{B}$ responds with the corresponding private key $\mathsf{Sk}_{\mathsf{id}}$ by running algorithm KeyGen.

Challenge. When $\mathcal{A}$ decides the Phase 1 is over, it outputs a challenge identity id^*, which is not been queried during Phase 1, and two equal length messages $\mathsf{Msg}_0, \mathsf{Msg}_1 \in \mathcal{M}_\lambda$. $\mathcal{B}$ flips a fair coin $\gamma \xleftarrow{\$} \{0,1\}$ and sets $\mathsf{Ctx}_{\mathsf{id}^*} \leftarrow \mathsf{Encrypt}(\mathsf{Pub}, \mathsf{Msg}_\gamma, \mathsf{id}^*)$. Finally $\mathcal{A}$ passes $\mathsf{Ctx}_{\mathsf{id}^*}$ to $\mathcal{A}$.

Phase 2. $\mathcal{A}$ continues to make key quires for any identity $\mathsf{id} \neq \mathsf{id}^*$.

Guess. $\mathcal{A}$ outputs $\gamma' \in \{0,1\}$ and it wins if $\gamma' = \gamma$.

We refer to such an adversary $\mathcal{A}$ as an IND-ID-CPA adversary. We define the advantage (in the security parameter λ) of $\mathcal{A}$ in attacking an IBE scheme $\mathcal{E}$ as $\mathsf{Adv}_{\mathcal{E},\mathcal{A}}(\lambda) = |\Pr[\gamma' = \gamma] - 1/2|$.

Definition 8. *For a security parameter λ, let $t = t(\lambda)$, $q_{id} = q_{id}(\lambda)$, and $\epsilon = \epsilon(\lambda)$. We say that an IBE system $\mathcal{E}$ is (t, q_{id}, ϵ)-IND-ID-CPA secure if for any t-time IND-ID-CPA adversary $\mathcal{A}$ that makes at most q_{id} private key queries, we have $\mathsf{Adv}_{\mathcal{E},\mathcal{A}}(\lambda) \leq \epsilon$.*

3 Signature Scheme with Tight Security

3.1 Constructions

$\mathsf{KeyGen}(1^\lambda)$ The key generation algorithm does the following.

1. Sample a matrix $\mathbf{A}$ along with a trapdoor basis of lattice $\Lambda_q^\perp(\mathbf{A})$ by TrapGen.
2. Select matrices $\mathbf{A}_0$, $\mathbf{A}_1$, "PRF key" matrices $\mathbf{B}_1, \ldots, \mathbf{B}_k$, and "PRF input" matrices $\mathbf{C}_0$, $\mathbf{C}_1$ from $\mathbb{Z}_q^{n\times m}$ uniformly at random.
3. Select a secure pseudorandom function $\mathsf{PRF} : \{0,1\}^k \times \{0,1\}^t \rightarrow \{0,1\}$, express it as a NAND Boolean circuit C_{PRF} with depth $d = d(\lambda)$, and select a PRF key $K = s_1 s_2 \ldots s_k \xleftarrow{\$} \{0,1\}^k$.

4. Select a Gaussian parameter $s > 0$.
5. Output the verification key and signing key as:

$$\mathsf{Vk} = \left(\mathbf{A}, \{\mathbf{A}_0, \mathbf{A}_1\}, \{\mathbf{B}_i\}_{i\in[k]}, \{\mathbf{C}_0, \mathbf{C}_1\}, s, \mathsf{PRF}, C_{\mathsf{PRF}}\right), \quad \mathsf{Sk} = (\mathbf{T}_{\mathbf{A}}, K)$$

$\mathsf{Sign}(\mathsf{Vk}, \mathsf{Sk}, \mathsf{M})$ The signing algorithm takes as input the public verification key Vk, the signing key Sk and a message $\mathsf{M} = m_1 m_2 \dots m_t \in \{0,1\}^t$. It does:

1. Compute $\mathbf{A}_{C_{\mathsf{PRF}},\mathsf{M}} = \mathsf{Eval}_{\mathsf{BV}}(C_{\mathsf{PRF}}, \{\mathbf{B}_i\}_{i\in[k]}, \mathbf{C}_{m_1}, \mathbf{C}_{m_2}, \dots, \mathbf{C}_{m_t}) \in \mathbb{Z}_q^{n\times m}$.[5]
2. Compute bit value $b = \mathsf{PRF}(K, \mathsf{M})$ and set $\mathbf{F}_{\mathsf{M},1-b} = \left[\mathbf{A} \mid \mathbf{A}_{1-b} - \mathbf{A}_{C_{\mathsf{PRF}},\mathsf{M}}\right]$.
3. Run $\mathsf{SampleLeft}$ to sample $\mathbf{d}_{\mathsf{M}} \in \mathbb{Z}^{2m}$ with distribution $D_{\Lambda_q^{\perp}(\mathbf{F}_{\mathsf{M},1-b}),s}$.
4. Output the signature $\mathsf{Sig} = \mathbf{d}_{\mathsf{M}}$.

$\mathsf{Ver}(\mathsf{Vk}, \mathsf{M}, \mathsf{Sig})$ The verification algorithm takes as input the verification key Vk, message M and the signature of M, verifies as follows:

1. Assume $\mathsf{Sig} = \mathbf{d}$. It checks if $\mathbf{d} \in \mathbb{Z}^{2m}$, $\mathbf{d} \neq \mathbf{0}$, and $\|\mathbf{d}\| \le s\sqrt{2m}$.
2. Compute $\mathbf{A}_{C_{\mathsf{PRF}},\mathsf{M}} = \mathsf{Eval}_{\mathsf{BV}}(C_{\mathsf{PRF}}, \{\mathbf{B}_i\}_{i\in[k]}, \mathbf{C}_{m_1}, \mathbf{C}_{m_2}, \dots, \mathbf{C}_{m_t}) \in \mathbb{Z}_q^{n\times m}$. Check if $\mathbf{F}_{\mathsf{M},b}\mathbf{d} = \left[\mathbf{A} \mid \mathbf{A}_b - \mathbf{A}_{C_{\mathsf{PRF}},\mathsf{M}}\right]\mathbf{d} = \mathbf{0} \pmod q$ for $b = 0$ or 1.
3. If all above verifications pass, accept the signature; otherwise, reject.

3.2 Parameters Selection and Discussion

Let λ be the security parameter, we set $n = n(\lambda)$, let the message length be $t = t(\lambda)$ and the secret key length of PRF be $k = k(\lambda)$. For the most general case, let the circuit depth of C_{PRF} be $d = d(\lambda)$. To ensure we can run $\mathsf{TrapGen}$ in the Lemma 3, we set $m = n^{1+\eta}$ for some η (we assume $n^{\eta} > O(\log q)$). To run $\mathsf{SampleLeft}$ and $\mathsf{SampleRight}$ in the real scheme and simulation per Theorem 2, we set s sufficiently large such that $s > \|\tilde{\mathbf{T}}_{\mathbf{G}}\| \cdot \|\mathbf{R}\|_2 \cdot \omega(\sqrt{\log m})$ for $\mathbf{R} = \mathbf{R}_{\mathbf{A}_b} - \mathbf{R}_{C_{\mathsf{PRF}},\mathsf{M}}$ (see the security proof below). By Lemma 8 we set $s = O(4^d \cdot m^{3/2}) \cdot \omega(\sqrt{\log m})$. For the SIS parameter β, we need $\beta \ge O(4^d \cdot m^{3/2} \cdot s\sqrt{2m})$. So we set $\beta = O(16^d \cdot m^{7/2}) \cdot \omega(\sqrt{\log m})$. To ensure the applicability of the average-case to worst-case reduction for SIS, we need $q \ge \beta \cdot \omega(\sqrt{n \log n})$. So we set $q = O(16^d \cdot m^4) \cdot (\omega(\sqrt{\log m}))^2$.

Particularly, if we choose PRF from the well-known efficient and provably secure candidates of PRFs like the ones from [8,9,31,47,53] can be computed by NC^1 circuits, let $\ell = t + k$ be the input length of PRF (which is a polynomial in the security parameter), the circuit depth of C_{PRF} will be $d = c\log\ell$ for some constant c. In this case we can set $\beta = O(\ell^{4c} \cdot m^{7/2}) \cdot \omega(\sqrt{\log m})$ and $q = O(\ell^{4c} \cdot m^4) \cdot (\omega(\sqrt{\log m}))^2$ which are polynomial in the security parameter.

It needs to mention that if we instantiate PRF by the (direct) LWE-based PRF from [9] or by the LWE-based PRF from [8] whose security relies on LWE assumption with super-polynomial modulus, the security of our signature scheme has to rely on LWE assumption with super-polynomial modulus. Such LWE assumption is stronger than the SIS assumption with polynomial modulus (as we set above) from which we make the proof for the following theorem.

[5] It turns out that if PRF is secure, an efficient SIS algorithm can be tightly reduced to an efficient algorithm that finds $\mathsf{M} \neq \mathsf{M}'$ such that $\mathbf{A}_{C_{\mathsf{PRF}},\mathsf{M}} = \mathbf{A}_{C_{\mathsf{PRF}},\mathsf{M}'}$. We prove this in the Sect. 3.3.

3.3 Security of the Signature Scheme

The security of our signature scheme is stated by the following theorem.

Theorem 4. *Let λ be a security parameter. The parameters n, m, and q are chosen as the Sect. 3.2. If the $(t_{SIS}, \epsilon_{SIS})$-$SIS_{n,q,\beta,m}$ assumption holds and the PRF used in the signature scheme is $(t_{PRF}, \epsilon_{PRF})$-secure, the signature scheme is (t, q_s, ϵ)-EUF-CMA secure where $\epsilon_{SIS} \geq \epsilon/2 - \epsilon_{PRF} - \mathsf{negl}(\lambda)$, for some negligible statistical error $\mathsf{negl}(\lambda)$, and $\max(t_{PRF}, t_{SIS}) \leq t + O(q_s \cdot (T_S + T_E))$ where q_s is the number of signing query, T_S is the maximum running time of SampleRight, and T_E is the maximum running time of $\mathsf{Eval}_{\mathsf{BV}}$ for one input message.*

Proof. Consider the following security game between an adversary $\mathcal{A}$ and a simulator $\mathcal{B}$. Upon receiving a $\mathsf{SIS}_{n,q,\beta,m}$ challenge $\mathbf{A} \in \mathbb{Z}_q^{n \times m}$, the challenger $\mathcal{B}$ prepares Vk as follows:

1. Select $k+4$ matrices $\mathbf{R}_{\mathbf{A}_0}, \mathbf{R}_{\mathbf{A}_1}, \{\mathbf{R}_{\mathbf{B}_i}\}_{i \in [k]}, \mathbf{R}_{\mathbf{C}_0}, \mathbf{R}_{\mathbf{C}_1} \xleftarrow{\$} \{1, -1\}^{m \times m}$.
2. Select a secure pseudorandom function $\mathsf{PRF} : \{0,1\}^k \times \{0,1\}^t \rightarrow \{0,1\}$ and express it as a NAND Boolean circuit C_{PRF} with depth d.
3. Select a PRF key $K = s_1 s_2 \ldots s_k \xleftarrow{\$} \{0,1\}^k$.
4. Set $\mathbf{A}_b = \mathbf{A}\mathbf{R}_{\mathbf{A}_b} + b\mathbf{G}$ and $\mathbf{C}_b = \mathbf{A}\mathbf{R}_{\mathbf{C}_b} + b\mathbf{G}$ for $b = 0, 1$.
5. Set $\mathbf{B}_i = \mathbf{A}\mathbf{R}_{\mathbf{B}_i} + s_i\mathbf{G}$ for $i \in [k]$.
6. Select a Gaussian parameter $s > 0$.
7. Publish $\mathsf{Vk} = \big(\mathbf{A}, \{\mathbf{A}_0, \mathbf{A}_1\}, \{\mathbf{B}_i\}_{i \in [k]}, \{\mathbf{C}_0, \mathbf{C}_1\}, \mathsf{PRF}, C_{\mathsf{PRF}}\big)$.

In the query phase, the adversary $\mathcal{A}$ adaptively issues messages for inquiring the corresponding signatures. Consider a message $\mathsf{M} = m_1 m_2 \ldots m_t \in \{0,1\}^t$. $\mathcal{B}$ does the following to prepare the signature:

1. Compute $\mathbf{A}_{C_{\mathsf{PRF}}} = \mathbf{A}\mathbf{R}_{C_{\mathsf{PRF}},\mathsf{M}} + \mathsf{PRF}(K, \mathsf{M})\mathbf{G} \in \mathbb{Z}_q^{n \times m}$ by $\mathsf{Eval}_{\mathsf{BV}}$ (C_{PRF}, $\{\mathbf{B}_i\}_{i \in [k]}$, $\mathbf{C}_{m_1}$, $\mathbf{C}_{m_2}$, $\ldots$, $\mathbf{C}_{m_t}$).
2. Let $b = \mathsf{PRF}(K, \mathsf{M})$, it sets

$$\begin{aligned}\mathbf{F}_{\mathsf{M},1-b} &= \big[\mathbf{A} \mid \mathbf{A}_{1-b} - \mathbf{A}_{C_{\mathsf{PRF}},\mathsf{M}}\big] \\ &= \big[\mathbf{A} \mid \mathbf{A}(\mathbf{R}_{\mathbf{A}_{1-b}} - \mathbf{R}_{C_{\mathsf{PRF}},\mathsf{M}}) + (1-2b)\mathbf{G}\big]\end{aligned}$$

and runs SampleRight to generate the signature $\mathsf{Sig} = \mathbf{d}_{\mathsf{M}} \sim D_{\Lambda_q^{\perp}(\mathbf{F}_{\mathsf{M},1-b}),s}$.

Finally, $\mathcal{A}$ output a forgery $(\mathbf{d}^*, \mathsf{M}^*)$. Let $\mathsf{PRF}(K, \mathsf{M}^*) = b$. If $\|\mathbf{d}\| > s\sqrt{2m}$ or $\big[\mathbf{A} \mid \mathbf{A}_{1-b} - \mathbf{A}_{C_{\mathsf{PRF}},\mathsf{M}^*}\big]\, \mathbf{d}^* = \mathbf{0} \pmod q$, $\mathcal{B}$ aborts. Otherwise, we have $\big[\mathbf{A} \mid \mathbf{A}_b - \mathbf{A}_{C_{\mathsf{PRF}},\mathsf{M}^*}\big]\, \mathbf{d}^* = \mathbf{0} \pmod q$. Let $\mathbf{d}^* = [\mathbf{d}_1^{\top} \mid \mathbf{d}_2^{\top}]^{\top} \in \mathbb{Z}^{2m}$. $\mathcal{B}$ outputs $\mathbf{e} = \mathbf{d}_1 + (\mathbf{R}_{\mathbf{A}_b} - \mathbf{R}_{C_{\mathsf{PRF}},\mathsf{M}^*})\mathbf{d}_2$ where $\|\mathbf{e}\| \leq \beta$ as a solution for the $\mathsf{SIS}_{n,q,\beta,m}$ problem instance.

We show that Vk output by $\mathcal{B}$ has the correct distribution. In the real scheme, the matrix $\mathbf{A}$ is generated by TrapGen. In the simulation, $\mathbf{A}$ has uniform distribution in $\mathbb{Z}_q^{n \times m}$ as it comes from the SIS challenge. By the Lemma 3, $\mathbf{A}$ generated in the simulation has right distribution except a negligibly small statistical error. Secondly, the matrices $\mathbf{A}$, $\{\mathbf{A}_0, \mathbf{A}_1\}$, $\{\mathbf{B}_i\}_{i \in [k]}$, and $\{\mathbf{C}_0, \mathbf{C}_1\}$ computed in the simulation have distribution that is statistically close to uniform distribution in $\mathbb{Z}_q^{n\times m}$ by the Lemma 1. In particular, the PRF secret key $\{s_i\}_{i\in[k]}$ is information-theoretically concealed by $\{\mathbf{B}_i\}_{i\in[k]}$.

Now we show that given $\{\mathbf{A}_0, \mathbf{A}_1\}$, $\{\mathbf{B}_i\}_{i\in[k]}$, and $\{\mathbf{C}_0, \mathbf{C}_1\}$, it is hard to find two messages $\mathsf{M} \neq \mathsf{M}'$ such that $\mathbf{A}_{C_{\mathsf{PRF}},\mathsf{M}} = \mathbf{A}_{C_{\mathsf{PRF}},\mathsf{M}'}$. Assume an efficient adversary finds $\mathsf{M} \neq \mathsf{M}'$ such that $\mathbf{A}_{C_{\mathsf{PRF}},\mathsf{M}} = \mathbf{A}_{C_{\mathsf{PRF}},\mathsf{M}'}$. With the public parameters set up above, we have

$$\mathbf{A}\mathbf{R}_{C_{\mathsf{PRF}},\mathsf{M}} + \mathsf{PRF}(K,\mathsf{M})\mathbf{G} = \mathbf{A}\mathbf{R}_{C_{\mathsf{PRF}},\mathsf{M}'} + \mathsf{PRF}(K,\mathsf{M}')\mathbf{G}$$

If $\mathsf{PRF}(K,\mathsf{M}) \neq \mathsf{PRF}(K,\mathsf{M}')$, which will happen essentially 1/2 probability if PRF is secure, we have $\mathbf{R}_{C_{\mathsf{PRF}},\mathsf{M}} \neq \mathbf{R}_{C_{\mathsf{PRF}},\mathsf{M}'}$ and $\mathbf{A}(\mathbf{R}_{C_{\mathsf{PRF}},\mathsf{M}} - \mathbf{R}_{C_{\mathsf{PRF}},\mathsf{M}'}) \pm \mathbf{G} = 0 \pmod q$. By Lemma 6 and Algorithm 1, a low-norm vector $\bar{\mathbf{d}} \in \mathbb{Z}^{m\times m}$ can be efficiently found such that $\mathbf{G}\bar{\mathbf{d}} = \mathbf{0} \pmod q$ where $\bar{\mathbf{d}} \neq \mathbf{0}$ and $\|\bar{\mathbf{d}}\| \leq s'\sqrt{m}$ for some Gaussian parameter $s' \geq \sqrt{5}\cdot\omega(\sqrt{\log m})$. Then $(\mathbf{R}_{C_{\mathsf{PRF}},\mathsf{M}} - \mathbf{R}_{C_{\mathsf{PRF}},\mathsf{M}'})\cdot\bar{\mathbf{d}}$ will be a non-zero vector with all but negligible probability and, therefore, a valid the SIS solution for $\mathbf{A}$.

In the query phase, the signatures replied to $\mathcal{A}$ have the correct distribution under the predefined conditions. Indeed, by the Theorem 2, for sufficient large Gaussian parameter s, the the distribution of signatures generated in the simulation by $\mathsf{SampleRight}$ is statistically close to $D_{\Lambda_q^{\perp}(\mathbf{F}_{\mathsf{M},1-b}),s}$ where the distribution of signatures generated in the real scheme by $\mathsf{SampleLeft}$ is also statistically close to $D_{\Lambda_q^{\perp}(\mathbf{F}_{\mathsf{M},1-b}),s}$.

In the forge phase, $\mathcal{A}$ will have at most advantage ϵ_{PRF} in predicting the bit value b with respect to the message it wants to forge. Therefore, if $\mathcal{A}$ can not distinguish PRF from random functions, it will randomly pick either of the matrices $\mathbf{A}_0$ or $\mathbf{A}_1$ to make a forgery. With $\frac{1}{2}$ chance it will pick the one that $\mathcal{B}$ will be able to use to solve the SIS problem. So we have $\epsilon_{\mathsf{SIS}} \geq \epsilon/2 - \epsilon_{\mathsf{PRF}} - \mathsf{negl}(\lambda)$ where $\mathsf{negl}(\lambda)$ stands for negligible statistical error in the simulation.

To argue that $\mathbf{e} = \mathbf{d}_1 + (\mathbf{R}_{\mathbf{A}_1} - \mathbf{R}_{C_{\mathsf{PRF}},\mathsf{M}^*})\mathbf{d}_2$ is a valid solution of the $\mathsf{SIS}_{n,q,\beta,m}$ problem instance, we need to show $\mathbf{e}$ is sufficiently short, and non-zero except with negligible probability. First of all, we have

$$\begin{aligned}\left[\mathbf{A} \mid \mathbf{A}_b - \mathbf{A}_{C_{\mathsf{PRF}},\mathsf{M}^*}\right]\mathbf{d}^* &= \left[\mathbf{A} \mid \mathbf{A}(\mathbf{R}_{\mathbf{A}_b} - \mathbf{R}_{C_{\mathsf{PRF}},\mathsf{M}^*})\right]\mathbf{d}^* \\ &= \mathbf{A}\mathbf{d}_1 + \mathbf{A}(\mathbf{R}_{\mathbf{A}_b} - \mathbf{R}_{C_{\mathsf{PRF}},\mathsf{M}^*})\mathbf{d}_2 \\ &= \mathbf{A}\left(\mathbf{d}_1 + \mathbf{R}\cdot\mathbf{d}_2\right) \\ &= \mathbf{0} \pmod q\end{aligned}$$

where $\mathbf{R} = \mathbf{R}_{\mathbf{A}_b} - \mathbf{R}_{C_{\mathsf{PRF}},\mathsf{M}^*}$. Since $\mathbf{d}_1, \mathbf{d}_2$ have distribution $D_{\mathbb{Z}^m,s}$ with condition $\mathbf{d} \in \Lambda_q^{\perp}(\mathbf{F}_{\mathsf{M},b})$, by the Lemma 4, $\mathbf{d}_1, \mathbf{d}_2 \leq s\sqrt{m}$. By Lemma 8, we have $\|\mathbf{e}\| \leq \|\mathbf{d}_1\| + \|\mathbf{R}\|_2 \cdot \|\mathbf{d}_2\| \leq O(4^d\cdot m^{3/2})\cdot s\sqrt{m}$. Let $\beta \geq O(4^d\cdot m^{3/2})\cdot s\sqrt{m}$ is sufficient.

It remains to show that $\mathbf{e} = \mathbf{d}_1 + \mathbf{R}\cdot\mathbf{d}_2 \neq \mathbf{0}$. Suppose $\mathbf{d}_2 \neq \mathbf{0}$, we have $\mathbf{e} \neq \mathbf{0}$ since $\mathbf{d} \neq \mathbf{0}$. On the other hand, we have $\mathbf{d}_2 = (d_1, \ldots, d_m)^\top \neq \mathbf{0}$ and, thus, at least one coordinate of $\mathbf{d}_2$, say d_j, is not 0. We write $\mathbf{R} = (\mathbf{r}_1, \ldots, \mathbf{r}_m)$ and so

$$\mathbf{R}\cdot\mathbf{d}_2 = \mathbf{r}_j \cdot d_j + \sum_{i=1, i\neq j}^{m} \mathbf{r}_i \cdot d_i$$

Observe that for the fixed message M^* on which $\mathcal{A}$ made the forgery, $\mathbf{R}$ (therefore $\mathbf{r}_j$) depends on the low-norm matrices $\mathbf{R}_{\mathbf{A}_0}, \mathbf{R}_{\mathbf{A}_1}, \{\mathbf{R}_{\mathbf{B}_i}\}_{i\in[k]}, \mathbf{R}_{\mathbf{C}_0}, \mathbf{R}_{\mathbf{C}_1}$ and the secret key of PRF. The only information about $\mathbf{r}_j$ for $\mathcal{A}$ is from the public matrices in Vk, i.e. $\{\mathbf{A}_0, \mathbf{A}_1\}, \{\mathbf{B}_i\}_{i\in[k]}, \{\mathbf{C}_0, \mathbf{C}_1\}$. So by the pigeonhole principle there is a (exponentially) large freedom to pick a value to $\mathbf{r}_j$ which is compatible with $\mathcal{A}$'s view, i.e. $\mathbf{A}\mathbf{r}'_j = \mathbf{A}\mathbf{r}''_j \pmod q$ for admissible (low-norm) $\mathbf{r}'_j, \mathbf{r}''_j$ where $\mathbf{r}'_j \neq \mathbf{r}''_j$. (In fact, here we have more freedom than the case in [20] where $\mathbf{R}$ is picked from $\{1,-1\}^{m\times m}$).

Finally, to answer one signing query, $\mathcal{B}$'s running time is bounded by $O(T_S + T_E)$. So the total running time of $\mathcal{B}$ in the simulation is bounded by $O(q_s(T_S + T_E))$. This concludes the proof. □

4 IBE Scheme with Tight Security

4.1 Construction with CPA Security

Setup(1^λ). The setup algorithm takes as input a security parameter λ and does:

1. Sample a random matrix $\mathbf{A} \in \mathbb{Z}_q^{n\times m}$ along with a trapdoor basis $\mathbf{T}_\mathbf{A} \in \mathbb{Z}^{m\times m}$ of lattice $\Lambda_q^\perp(\mathbf{A})$ by running $\mathsf{TrapGen}$.
2. Select random matrices $\mathbf{A}_0$, $\mathbf{A}_1$, random "PRF key" matrices $\mathbf{B}_1, \ldots, \mathbf{B}_k$, and random "PRF input" matrices $\mathbf{C}_0$, $\mathbf{C}_1$ from $\mathbb{Z}_q^{n\times m}$ uniformly at random.
3. Select a random vector $\mathbf{u} \xleftarrow{\$} \mathbb{Z}_q^n$.
4. Select a secure pseudorandom function $\mathsf{PRF} : \{0,1\}^k \times \{0,1\}^t \to \{0,1\}$, express it as a NAND Boolean circuit C_{PRF} with depth $d = d(\lambda)$, and select a PRF key $K = s_1 s_2 \ldots s_k \xleftarrow{\$} \{0,1\}^k$.
5. Output the public parameters
$$\mathsf{Pub} = \left(\mathbf{A}, \{\mathbf{A}_0, \mathbf{A}_1\}, \{\mathbf{B}_i\}_{i\in[k]}, \{\mathbf{C}_0, \mathbf{C}_1\}, \mathbf{u}, \mathsf{PRF}, C_{\mathsf{PRF}}\right)$$
and the master secret key $\mathsf{Msk} = (\mathbf{T}_\mathbf{A}, K)$.

KeyGen$(\mathsf{Pub}, \mathsf{Msk}, \mathsf{id})$. Upon an input identity $\mathsf{id}{=}x_1x_2\ldots x_t \in \{0,1\}^t$, the key generation algorithm does the following:

1. Compute $b = \mathsf{PRF}(K, \mathsf{id})$.
2. Compute $\mathbf{A}_{C_{\mathsf{PRF}},\mathsf{id}} = \mathsf{Eval}_{\mathsf{BV}}(C_{\mathsf{PRF}}, \{\mathbf{B}\}_{i\in[k]}, \mathbf{C}_{x_1}, \mathbf{C}_{x_2}, \ldots, \mathbf{C}_{x_t}) \in \mathbb{Z}_q^{n\times m}$.
3. Set $\mathbf{F}_{\mathsf{id},1-b} = \left[\mathbf{A} \mid \mathbf{A}_{1-b} - \mathbf{A}_{C_{\mathsf{PRF}},\mathsf{id}}\right] \in \mathbb{Z}_q^{n\times 2m}$.
4. Run $\mathsf{SampleLeft}$ to sample $\mathbf{d}_{\mathsf{id}}$ from the discrete Gaussian distribution $D_{\Lambda_q^{\mathbf{u}}(\mathbf{F}_{\mathsf{id},1-b}),s}$ hence $\mathbf{F}_{\mathsf{id},1-b}\mathbf{d}_{\mathsf{id}} = \mathbf{u} \pmod q$. Output $\mathsf{Sk}_{\mathsf{id}} = \mathbf{d}_{\mathsf{id}}$.

Encrypt$(\mathsf{Pub}, \mathsf{id}, \mathsf{Msg})$. To encrypt a message $\mathsf{Msg} \in \{0,1\}$ with respect to an identity $\mathsf{id} = x_1x_2\ldots x_t \in \{0,1\}^t$:

1. Compute $\mathbf{A}_{C_{\mathsf{PRF}},\mathsf{id}} = \mathsf{Eval}_{\mathsf{BV}}(C_{\mathsf{PRF}}, \{\mathbf{B}_i\}_{i\in[k]}, \mathbf{C}_{x_1}, \mathbf{C}_{x_2}, \ldots, \mathbf{C}_{x_t})$.
2. Set $\mathbf{F}_{\mathsf{id},b} = \left[\mathbf{A} \mid \mathbf{A}_b - \mathbf{A}_{C_{\mathsf{PRF}},\mathsf{id}}\right] \in \mathbb{Z}_q^{n\times 2m}$ for $b = 0, 1$.
3. Select two random vectors $\mathbf{s}_0, \mathbf{s}_1 \xleftarrow{\$} \mathbb{Z}_q^n$.
4. Select two noise scalars $\nu_{0,0}, \nu_{1,0} \leftarrow D_{\mathbb{Z},\sigma_{\mathsf{LWE}}}$ and four noise vectors $\hat{\boldsymbol{\nu}}_{0,1}, \hat{\boldsymbol{\nu}}_{1,1} \leftarrow D_{\mathbb{Z}^m,\sqrt{2}\sigma_{\mathsf{LWE}}}$, $\check{\boldsymbol{\nu}}_{0,1}, \check{\boldsymbol{\nu}}_{1,1} \leftarrow D_{\mathbb{Z}^m,\sigma}$ where σ is sufficiently larger than σ_{LWE}.[6]

[6] For instance we set $\sigma = O(4^d \cdot m^{3/2}) \cdot \omega(\sqrt{\log m}) \cdot \sigma_{\mathsf{LWE}}$.

5. Compute the ciphertext $\mathsf{Ctx}_{\mathsf{id}} = (c_{0,0}, \mathbf{c}_{0,1}, c_{1,0}, \mathbf{c}_{1,1})$ as:

$$\begin{cases} c_{0,0} = \left(\mathbf{s}_0^\top \mathbf{u} + \nu_{0,0} + \mathsf{Msg}\lfloor q/2 \rfloor\right) \bmod q \\ \mathbf{c}_{0,1}^\top = \left(\mathbf{s}_0^\top \mathbf{F}_{\mathsf{id},0} + [\hat{\boldsymbol{\nu}}_{0,1}^\top \mid \check{\boldsymbol{\nu}}_{0,1}^\top]\right) \bmod q \end{cases}$$

$$\begin{cases} c_{1,0} = \left(\mathbf{s}_1^\top \mathbf{u} + \nu_{1,0} + \mathsf{Msg}\lfloor q/2 \rfloor\right) \bmod q \\ \mathbf{c}_{1,1}^\top = \left(\mathbf{s}_1^\top \mathbf{F}_{\mathsf{id},1} + [\hat{\boldsymbol{\nu}}_{1,1}^\top \mid \check{\boldsymbol{\nu}}_{1,1}^\top]\right) \bmod q \end{cases}$$

Decrypt($\mathsf{Pub}, \mathsf{Sk}_{\mathsf{id}}, \mathsf{Ctx}_{\mathsf{id}}$). The decryption algorithm uses the key $\mathbf{d}_{\mathsf{id}}$ to try to decrypt both $(c_{0,0}, \mathbf{c}_{0,1})$ and $(c_{1,0}, \mathbf{c}_{1,1})$[7]. W.l.o.g., assume that $(c_{b,0}, \mathbf{c}_{b,1})$ is the correct ciphertext. The decryption algorithm computes

$$\tau = \left(c_{b,0} - \mathbf{c}_{b,1}^\top \mathbf{d}_{\mathsf{id}}\right) \bmod q$$

View τ as an integer in $(-q/2, q/2]$. If τ is closer to 0 than $\pm q/2$, the output is $\mathsf{Msg} = 0$. Otherwise, it is $\mathsf{Msg} = 1$.

4.2 Correctness

Following the decryption algorithm, let $\mathbf{d}_{\mathsf{id}} = [\mathbf{d}_1^\top \mid \mathbf{d}_2^\top]^\top$. We have

$$\begin{aligned} \tau &= \left(c_{b,0} - \mathbf{c}_{b,1}^\top \mathbf{d}_{\mathsf{id}}\right) \bmod q \\ &= \left(\mathsf{Msg}\lfloor q/2 \rfloor + \nu_{b,0} - \hat{\boldsymbol{\nu}}_{0,1}^\top \mathbf{d}_1 - \check{\boldsymbol{\nu}}_{0,1}^\top \mathbf{d}_2\right) \bmod q \end{aligned}$$

Recall, the norm of $\mathbf{d}_1$ and $\mathbf{d}_2$ is bounded by $s\sqrt{m}$, and the norm of $\hat{\boldsymbol{\nu}}_{b,1}$ and $\check{\boldsymbol{\nu}}_{b,1}$ is bounded by $\sigma_{\mathsf{LWE}}\sqrt{m}$ and $\sigma\sqrt{m}$ respectively, by Lemma 4. To ensure correctness of decryption, we need

$$\begin{aligned} |\tau| &= |c_{b,0} - \hat{\boldsymbol{\nu}}_{b,1}^\top \mathbf{d}_1 - \check{\boldsymbol{\nu}}_{0,1}^\top \mathbf{d}_2| \\ &\le |c_{b,0}| + \|\hat{\boldsymbol{\nu}}_{0,1}\| \cdot \|\mathbf{d}_1\| + \|\hat{\boldsymbol{\nu}}_{0,1}\| \cdot \|\mathbf{d}_2\| \\ &\le O(s \cdot m \cdot (\sigma_{\mathsf{LWE}} + \sigma)) \\ &\le q/4 \end{aligned}$$

Accordingly, it is enough to set q such that $O(s \cdot m \cdot (\sigma_{\mathsf{LWE}} + \sigma)) \le q/4$.

[7] To ensure correct decryption, the message should contain some redundancy to weed out the incorrect ciphertext. It is a standard technique to encrypt multiple bits in GPV-style encryption, by replacing $\mathbf{u}$ with a matrix $\mathbf{U} \in \mathbb{Z}_q^{n \times z}$ in Pub with which we can now independently encrypt $z > 1$ bits without change to the security analysis. If hybrid encryption is used, the multiple bits can be used to encrypt a symmetric key *without* redundancy, deferring the integrity check to the symmetric realm where it can be performed at minimal cost.

4.3 Parameter Selection and Discussion

We now discuss a consistent parameter instantiation that achieves both correctness and security. Let λ be the security parameter, $t = t(\lambda)$ be the identity length, $k = k(\lambda)$ be the secret key length of PRF, and let $\ell = t + k$ be the input length of PRF. Let, for the most general case, the circuit depth of PRF be $d = d(\lambda)$. To ensure we can run $\mathsf{TrapGen}$ in the Lemma 3, we set $m = n^{1+\eta}$ for some $\eta > 0$ (we assume $n^\eta > O(\log q)$). To make sure $\mathsf{SampleLeft}$ in the real scheme and $\mathsf{SampleRight}$ in the simulation algorithm $\mathsf{Sim.KeyGen}$ (see section) 4.4 have the same output distribution per Theorem 2, we set a sufficiently large Gaussian parameter $s = \|\tilde{\mathbf{T}}_{\mathbf{G}}\| \cdot O(4^d \cdot m^{3/2}) \cdot \omega(\sqrt{\log m})$. To ensure the applicability of Regev's [55] and Peikert's [54] LWE reductions from worst-case lattice problems, we set the Gaussian parameter of LWE noise distribution to be $\sigma_{\mathsf{LWE}} = \sqrt{n}$. So the LWE noise distribution is $(D_{\mathbb{Z},\sqrt{n}}) \bmod q$. For the security proof (specifically for the proofs of Lemmas 10 and 16), we set $\sigma = O(4^d \cdot m^{3/2}) \cdot \omega(\sqrt{\log m}) \cdot \sigma_{\mathsf{LWE}}$. Finally, to ensure correctness condition of decryption, we set $q = O(16^d \cdot m^{9/2}) \cdot (\omega\sqrt{\log m})^2$.

As for our signature scheme, if we the PRF can be computed by a NC^1 NAND circuit with depth $d = c \log \ell$ for some constant $c > 1$, we can set the LWE modulus $q = O(\ell^{4c} \cdot m^{9/2}) \cdot (\omega\sqrt{\log m})^2$, which is polynomial in the security parameter λ.

Tight Reduction and Hardness of LWE. It is known that larger modulus results in stronger LWE assumption, if the standard deviation of the noise distribution stays unchanged. More precisely, let B be the maximum magnitude of the LWE noise, and q be the LWE modulus. The hardness of the LWE problem depends on the ratio q/B. The LWE problem becomes easier when this ratio grows. In this regard, the appeal of our tight reduction varies: tight reduction to harder LWE problem is more preferable than tight reduction to easier LWE problem. This is true particularly when one considers the average-case hardness of LWE to worst-case hardness of classic lattice problems, e.g. GapSVP and SIVP, reductions [22,54,55] where ratio q/B is smaller, the solutions for classic lattice problems are better.

One feature of our IBE scheme (and the signature scheme it induces) is that depending on different circuits instantiations, the assumptions we make for our tight reduction may vary. In addition, if we use a LWE-based PRF, our IBE scheme relies on the stronger one of two LWE assumptions: one is made for the PRF and another one is made for our construction, which uses a polynomial modulus q as we chose above. Currently, basing our IBE scheme solely on LWE needs to assume the LWE assumption with super-polynomial modulus. This is because the state-of-art PRFs from LWE (from [8,9]) in terms of efficiency and provable security require super-polynomial LWE modulus.

On the other hand, we believe that our tight reduction is still very valuable even for large ratio q/B. Firstly, it shows that, at the first time, we actually can eliminate the dependency between the number of adversary's queries and the security of lattice-based IBE scheme (as well as *short* lattice signature scheme). This is very important since the number of adversary's queries can be quite large, which will negatively impact the schemes' security seriously. Secondly, the average-case to worst-case reduction does provide some security confidence for the LWE assumption, but this is not the whole story. For certain parameters, many classic lattice problems are NP-hard. However, those parameters have no direct connection to lattice-based cryptography. (There is even evidence that the classic lattice problems with parameters relevant cryptography are not NP-hard.) On the other hand, the LWE problem (with various parameters) could be assured to be a hard problem in its own right. It has shown robustness against various attacks in a relatively long-term period. This has made LWE widely accepted as standard assumption and for use in cryptography. For instance, even for sub-exponentially large ratios $q/B = 2^{O(n^c)}$ where n is the LWE dimension and $0 < c < 1/2$, the LWE problem is still believed to be hard and leads to powerful cryptographic schemes which we were not able to obtain by other means, including fully homomorphic encryption, e.g. [23], attribute-based encryption for circuits, e.g. [18,25,37], and predicate encryption for circuits [41].

4.4 Proof of Security

The security of our IBE scheme with respect to the Definition 8 can be stated by the following theorem.

Theorem 5. *Let λ be a security parameter. The parameters n, q are chosen as the Sect. 4.3. Let χ be the distribution $D_{\mathbb{Z}^m,\sqrt{n}}$. If the $(t_{\mathsf{LWE}}, \epsilon_{\mathsf{LWE}})$-$\mathsf{LWE}_{n,q,\chi}$ assumption holds and the PRF used in the IBE scheme is $(t_{\mathsf{PRF}}, \epsilon_{\mathsf{PRF}})$-secure, then the IBE scheme is (t, q_{id}, ϵ)-IND-ID-CPA secure such that $\epsilon \leq 2(\epsilon_{\mathsf{PRF}} + \epsilon_{\mathsf{LWE}}) + \mathsf{negl}(\lambda)$ for some negligible function $\mathsf{negl}(\lambda)$, and $\max(t_{\mathsf{PRF}}, t_{\mathsf{LWE}}) \leq t + O(q_{id} \cdot (T_S + T_E))$ where T_S is the maximum running time of SampleRight and T_E is the maximum running time of $\mathsf{Eval}_{\mathsf{BV}}$ for one input identity.*

We prove above theorem through a sequence of indistinguishable security games. The first game is identical to the IND-ID-CPA game. In the last game, the adversary has no advantage. We will show that a PPT adversary will not be able to distinguish the neighbouring games which will prove that the adversary has only negligibly small advantage in wining the first (real) game.

Firstly, we define the following simulation algorithms Sim.Setup, Sim.KeyGen and Sim.Encrypt.

Sim.Setup(1^λ). The algorithm does the following:

1. Select matrix $\mathbf{A} \xleftarrow{\$} \mathbb{Z}_q^{n\times m}$.
2. Select $k+4$ random low-norm matrices $\mathbf{R}_{\mathbf{A}_0}$, $\mathbf{R}_{\mathbf{A}_1}$, $\{\mathbf{R}_{\mathbf{B}_i}\}_{i\in[k]}$, $\mathbf{R}_{\mathbf{C}_0}$, $\mathbf{R}_{\mathbf{C}_1}$ from $\{1,-1\}^{m\times m}$.
3. Select a secure pseudorandom function $\mathsf{PRF}: \{0,1\}^k \times \{0,1\}^t \to \{0,1\}$ and express it as a NAND Boolean circuit C_{PRF} with depth $d = d(\lambda)$.
4. Select a uniformly random string $K = s_1 s_2 \dots s_k \xleftarrow{\$} \{0,1\}^k$.
5. Set $\mathbf{A}_b = \mathbf{A}\mathbf{R}_{\mathbf{A}_b} + b\mathbf{G}$ and $\mathbf{C}_b = \mathbf{A}\mathbf{R}_{\mathbf{C}_b} + b\mathbf{G}$ for $b = 0, 1$.
6. Set $\mathbf{B}_i = \mathbf{A}\mathbf{R}_{\mathbf{B}_i} + s_i\mathbf{G}$ for $i \in [k]$.

7. Select vector $\mathbf{u} \xleftarrow{\$} \mathbb{Z}_q^n$.
8. Publish $\mathsf{Pub} = \left(\mathbf{A}, \{\mathbf{A}_0, \mathbf{A}_1\}, \{\mathbf{B}_i\}_{i\in[k]}, \{\mathbf{C}_0, \mathbf{C}_1\}, \mathbf{u}, \mathsf{PRF}, C_{\mathsf{PRF}}\right)$

$\mathsf{Sim.KeyGen}(\mathsf{Pub}, \mathsf{Msk}, \mathsf{id})$. Upon an input identity $\mathsf{id} = x_1 x_2 \dots x_t \in \{0,1\}^t$, the algorithm uses the parameters generated from Sim.Setup to do the following:

1. Compute
   $\mathbf{A}_{\mathsf{PRF},\mathsf{id}} = \mathbf{A}\mathbf{R}_{C_{\mathsf{PRF}},\mathsf{id}} + \mathsf{PRF}(K, \mathsf{id})\mathbf{G} \leftarrow \mathsf{Eval}_{\mathsf{BV}}(C_{\mathsf{PRF}}, \{\mathbf{B}_i\}_{i\in[k]}, \mathbf{C}_{x_1}, \dots, \mathbf{C}_{x_t})$.
2. Let $\mathsf{PRF}(K, \mathsf{id}) = b \in \{0,1\}$. Set

$$\begin{aligned}\mathbf{F}_{\mathsf{id},1-b} &= \left[\mathbf{A} \mid \mathbf{A}_{1-b} - \mathbf{A}_{C_{\mathsf{PRF}},\mathsf{id}}\right] \\ &= \left[\mathbf{A} \mid \mathbf{A}(\mathbf{R}_{\mathbf{A}_{1-b}} - \mathbf{R}_{C_{\mathsf{PRF}},\mathsf{id}}) + (1-2b)\mathbf{G}\right].\end{aligned}$$

3. Run SampleRight to sample $\mathbf{d}_{\mathsf{id}} \in D_{\Lambda_q^{\mathbf{u}}(\mathbf{F}_{\mathsf{id},1-b}),s}$ as the private key $\mathsf{Sk}_{\mathsf{id}}$.

$\mathsf{Sim.Encrypt}(\mathsf{Pub}, \mathsf{id}^*, \mathsf{Msg})$. To encrypt a message $\mathsf{Msg}^* \in \{0,1\}$ with respect to an identity id^*:

1. Compute $b = \mathsf{PRF}(K, \mathsf{id}^*)$.
2. Set

$$\begin{aligned}\mathbf{F}_{\mathsf{id}^*,b} &= \left[\mathbf{A} \mid \mathbf{A}_b - \mathbf{A}_{C_{\mathsf{PRF}},\mathsf{id}^*}\right] \\ &= \left[\mathbf{A} \mid \mathbf{A}(\mathbf{R}_{\mathbf{A}_b} - \mathbf{R}_{C_{\mathsf{PRF}},\mathsf{id}^*})\right]\end{aligned}$$

   and

$$\begin{aligned}\mathbf{F}_{\mathsf{id}^*,1-b} &= \left[\mathbf{A} \mid \mathbf{A}_{1-b} - \mathbf{A}_{C_{\mathsf{PRF}},\mathsf{id}^*}\right] \\ &= \left[\mathbf{A} \mid \mathbf{A}(\mathbf{R}_{\mathbf{A}_{1-b}} - \mathbf{R}_{C_{\mathsf{PRF}},\mathsf{id}^*}) + (1-2b)\mathbf{G}\right].\end{aligned}$$

3. Select random vectors $\mathbf{s}_b, \mathbf{s}_{1-b} \xleftarrow{\$} \mathbb{Z}_q^n$.
4. Select noise scalars $\nu_{b,0}, \nu_{1-b,0} \leftarrow D_{\mathbb{Z},\sigma_{\mathsf{LWE}}}$.
5. Sample noise vectors $\mathbf{x}, \mathbf{y} \leftarrow D_{\mathbb{Z}^m,\sigma_{\mathsf{LWE}}}$ for sufficiently large Gaussian parameter σ_{LWE} ($\sigma_{\mathsf{LWE}} \geq \eta_\varepsilon(\mathbb{Z}^m)$ for some small $\varepsilon > 0$). Set $\hat{\boldsymbol{\nu}}_{b,1} = \mathbf{x} + \mathbf{y}$.
6. Let $\mathbf{R} = \mathbf{R}_{\mathbf{A}_b} - \mathbf{R}_{\mathsf{PRF},\mathsf{id}^*}$ and $\mathbf{r}_i$ be the i-th column of $\mathbf{R}$. We sample the noise vector $\mathbf{z} = (z_1, z_2, \dots, z_m) \in \mathbb{Z}^m$ with $z_i \leftarrow D_{\mathbb{Z},\sigma_{1,i}}$ for the sufficiently large Gaussian parameter $\sigma_{1,i} = \sqrt{\sigma^2 - 2(\|\mathbf{r}_i\| \cdot \sigma_{\mathsf{LWE}})^2}$.[8] Set $\check{\boldsymbol{\nu}}_{b,1} = \mathbf{R}^\top \cdot (\mathbf{x} - \mathbf{y}) + \mathbf{z}$.
7. Select noise vectors $\hat{\boldsymbol{\nu}}_{1-b,1} \leftarrow D_{\mathbb{Z}^m,\sqrt{2}\sigma_{\mathsf{LWE}}}$, $\check{\boldsymbol{\nu}}_{1-b,1} \leftarrow D_{\mathbb{Z}^m,\sigma}$.
8. Set the challenge ciphertext $\mathsf{Ctx}_{\mathsf{id}^*} = (c_{b,0}, \mathbf{c}_{b,1}, c_{1-b,0}, \mathbf{c}_{1-b,1})$ as:

$$\begin{cases} c_{b,0} = \left(\mathbf{s}_b^\top \mathbf{u} + \nu_{b,0} + \mathsf{Msg}\lfloor q/2 \rfloor\right) \bmod q \\ \mathbf{c}_{b,1}^\top = \left(\mathbf{s}_b^\top \mathbf{F}_{\mathsf{id}^*,b} + [\hat{\boldsymbol{\nu}}_{b,1}^\top \mid \check{\boldsymbol{\nu}}_{b,1}^\top]\right) \bmod q \end{cases}$$

$$\begin{cases} c_{1-b,0} = \left(\mathbf{s}_{1-b}^\top \mathbf{u} + \nu_{1-b,0} + \mathsf{Msg}\lfloor q/2 \rfloor\right) \bmod q \\ \mathbf{c}_{1-b,1}^\top = \left(\mathbf{s}_{1-b}^\top \mathbf{F}_{\mathsf{id}^*,1-b} + [\hat{\boldsymbol{\nu}}_{1-b,1}^\top \mid \check{\boldsymbol{\nu}}_{1-b,1}^\top]\right) \bmod q \end{cases}$$

[8] In Sect. 4.3, the σ is set large enough such that $\sigma_{1,i}$ can be larger than $\|\mathbf{R}\| \cdot \eta_\varepsilon(\mathbb{Z}^m)$.

Now we define a series of games and prove that the neighbouring games are either statistically indistinguishable, or computationally indistinguishable.

Game 0. This is the real IND-ID-CPA game from the definition. All the algorithms are the same as the real scheme.

Game 1. This game is the same as **Game 0** except it runs Sim.Setup and Sim.KeyGen instead of Setup and KeyGen.

Game 2. This game is the same as **Game 1** except that the challenge ciphertext is generated by Sim.Encrypt instead of Encrypt.

Game 3. This game is the same as **Game 2** except that during preparation of the challenge ciphertext for identity id*, it samples $(c_{b,0}, \mathbf{c}_{b,1})$ uniformly random from $\mathbb{Z}_q \times \mathbb{Z}_q^{2m}$ for $b = \mathsf{PRF}(K, \mathsf{id}^*)$. Another part of the challenge ciphertext $(c_{1-b,0}, \mathbf{c}_{1-b,1})$ is computed by Sim.Encrypt as in **Game 2**.

Game 4. This game is the same as **Game 3** except for $b = \mathsf{PRF}(K, \mathsf{id}^*)$ it runs real encryption algorithm Encrypt to generate $(c_{1-b,0}, \mathbf{c}_{1-b,1})$ of the challenge ciphertext instead of using Sim.Encrypt.

Game 5. This game is the same as **Game 4** except it runs Setup and KeyGen to generate Pub and private identity keys.

Game 6. This game is the same as **Game 5** except that for $b = \mathsf{PRF}(K, \mathsf{id}^*)$, the challenge ciphertext part $(c_{b,0}, \mathbf{c}_{b,1})$ is generated by Encrypt instead of choosing it randomly, and $(c_{1-b,0}, \mathbf{c}_{1-b,1})$ is chosen randomly.

Game 7. This game is the same as **Game 6** except that it runs Sim.Setup and Sim.KeyGen to generate Pub and private identity keys.

Game 8. This game is the same as **Game 7** except that for the bit value $b = \mathsf{PRF}(K, \mathsf{id}^*)$, it computes the challenge ciphertext $(c_{b,0}, \mathbf{c}_{b,1})$ by Sim.Encrypt.

Game 9. This game is the same as **Game 8** except that the whole challenge ciphertext is sampled uniformly at random from the ciphertext space. Therefore, in **Game 5** the adversary has no advantage in wining the game.

In **Game** i, we let S_i be the event that $\gamma' = \gamma$ at the end of the game. The adversary's advantage in **Game** i is $|\Pr[S_i] - \frac{1}{2}|$. The following lemmas are used to prove Theorem 5. We refer to the full version of this paper ([21]) for the proofs of these lemmas.

Lemma 9. **Game 1** *and* **Game 0** *are statistically indistinguishable, so* $|\Pr[S_0] - \Pr[S_1]| \leq \mathit{negl}(\lambda)$ *for some negligible function* $\mathit{negl}(\lambda)$.

Lemma 10. **Game 2** *and* **Game 1** *are statistically indistinguishable, so* $|\Pr[S_1] - \Pr[S_2]| \leq \mathit{negl}(\lambda)$ *for some negligible function* $\mathit{negl}(\lambda)$.

Lemma 11. *If* $(t, \epsilon_{\mathit{LWE}})$-$\mathit{LWE}_{n,q,\chi}$ *assumption holds where* χ *stands for the distribution* $D_{\mathbb{Z},\sigma_{\mathit{LWE}}}$ *reduced modulo* q, *then* $|\Pr[S_2] - \Pr[S_3]| \leq \epsilon_{\mathit{LWE}}$.

Lemma 12. $|\Pr[S_3] - \Pr[S_4]| = 0$.

Lemma 13. **Game 5** *and* **Game 4** *are statistically indistinguishable, so* $|\Pr[S_4] - \Pr[S_5]| \leq \mathsf{negl}(\lambda)$ *for some negligible function* $\mathsf{negl}(\lambda)$.

Lemma 14. *If the PRF* PRF *is* $(t, \epsilon_{\mathsf{PRF}})$*-secure, then* $|\Pr[S_5] - \Pr[S_6]| \leq 2\epsilon_{\mathsf{PRF}}$.

Lemma 15. **Game 7** *and* **Game 6** *are statistically indistinguishable, so* $|\Pr[S_6] - \Pr[S_7]| \leq \mathsf{negl}(\lambda)$ *for some negligible function* $\mathsf{negl}(\lambda)$.

Lemma 16. **Game 8** *and* **Game 7** *are statistically indistinguishable, so* $|\Pr[S_7] - \Pr[S_8]| \leq \mathsf{negl}(\lambda)$ *for some negligible function* $\mathsf{negl}(\lambda)$.

Lemma 17. *If* $(t, \epsilon_{\mathsf{LWE}})$*-*$\mathsf{LWE}_{n,q,\chi}$ *assumption holds where* χ *stands for the distribution* $D_{\mathbb{Z},\sigma_{\mathsf{LWE}}}$ *reduced modulo* q*, then* $|\Pr[S_8] - \Pr[S_9]| \leq \epsilon_{\mathsf{LWE}}$.

Now we prove the Theorem 5 by the established lemmas.

Proof. Based on the lemmas that show the difference between the sequence of games, we have $\epsilon = |\Pr[S_0] - 1/2| \leq 2(\epsilon_{\mathsf{PRF}} + \epsilon_{\mathsf{LWE}}) + \mathsf{negl}(\lambda)$ for some negligibly small statistical error $\mathsf{negl}(\lambda)$. The running time of $\mathcal{B}$ is dominated by answering q_{id} private key generation queries from $\mathcal{A}$. For answering one such query, $\mathcal{B}$ needs to apply the key-homomorphic algorithm on the circuit of PRF. This requires time T_E. Besides that, $\mathcal{B}$ needs to run $\mathsf{SampleRight}$ to sample Gaussian vectors for constructing the private keys, which requires at most time T_S. Therefore, for one query, $\mathcal{B}$ roughly runs $O(T_S + T_E)$ time. For all q_{id} queries and constructing the challenge ciphertext, the total time is bounded by $O(q_{\mathsf{id}} \cdot (T_S + T_E))$. So if an adversary $\mathcal{A}$ has running time t, $\max(t_{\mathsf{LWE}}, t_{\mathsf{PRF}}) \leq t + O(q_{\mathsf{id}} \cdot (T_S + T_E))$. □

5 Conclusions

In this paper, we propose a short adaptively secure lattice signature scheme and a "compact" adaptively secure IBE scheme in the standard model. Our constructions make use of PRFs in a novel way by combining several recent techniques in the area of lattice-based cryptography. The security of our signature and IBE scheme is tightly related to the conservative lattice assumptions SIS and LWE, respectively, and the security of an instantiated PRF, with a constant loss factor. By instantiating the existing efficient PRFs from lattice and number-theoretic assumptions which can be implemented by shallow circuits, we obtain the first "almost" tightly secure lattice-based short signature/IBE scheme whose security is based on LWE assumption with super-polynomial modulus, and an adaptively secure IBE scheme with the tightest security reduction so far, i.e. with only $O(\log^2 \lambda)$ factor of security loss for the security parameter λ, based on a novel combination of lattice and number-theoretic assumptions.

The problem of constructing a tightly and adaptively secure IBE scheme from standard assumptions (in the sense that the security loss of reduction is a constant) remains open. Our work suggests that constructing tightly secure PRFs, which is another important open problem left by [31,47], would solve it. We leave as a fascinating open problem the question of employing similar (or different) techniques to construct compact and (almost) tightly secure signature and encryption schemes with increased expressiveness, such as hierarchical and attribute-based encryption scheme, or homomorphic signatures. Another interesting open question is to construct an efficient PRF from LWE assumption with polynomial modulus.

Acknowledgements. We would like to thank Jacob Alperin-Sheriff and Josef Pierpzyk as well as the anonymous reviewers for useful comments.

References

1. Abe, M., David, B., Kohlweiss, M., Nishimaki, R., Ohkubo, M.: Tagged one-time signatures: tight security and optimal tag size. In: Kurosawa, K., Hanaoka, G. (eds.) PKC 2013. LNCS, vol. 7778, pp. 312–331. Springer, Heidelberg (2013). doi:10.1007/978-3-642-36362-7_20
2. Agrawal, S., Boneh, D., Boyen, X.: Efficient lattice (H)IBE in the standard model. In: Gilbert, H. (ed.) EUROCRYPT 2010. LNCS, vol. 6110, pp. 553–572. Springer, Heidelberg (2010). doi:10.1007/978-3-642-13190-5_28
3. Ajtai, M.: Generating hard instances of lattice problems (extended abstract). In: STOC 1996, pp. 99–108. ACM (1996)
4. Alperin-Sheriff, J.: Short signatures with short public keys from homomorphic trapdoor functions. In: Katz, J. (ed.) PKC 2015. LNCS, vol. 9020, pp. 236–255. Springer, Heidelberg (2015). doi:10.1007/978-3-662-46447-2_11
5. Apon, D., Fan, X., Liu, F.H.: Fully-secure lattice-based IBE as compact as PKE. Cryptology ePrint Archive, Report 2016/125 (2016)
6. Attrapadung, N., Hanaoka, G., Yamada, S.: A framework for identity-based encryption with almost tight security. In: Iwata, T., Cheon, J.H. (eds.) ASIACRYPT 2015. LNCS, vol. 9452, pp. 521–549. Springer, Heidelberg (2015). doi:10.1007/978-3-662-48797-6_22
7. Bai, S., Langlois, A., Lepoint, T., Stehlé, D., Steinfeld, R.: Improved security proofs in lattice-based cryptography: using the rényi divergence rather than the statistical distance. In: Iwata, T., Cheon, J.H. (eds.) ASIACRYPT 2015. LNCS, vol. 9452, pp. 3–24. Springer, Heidelberg (2015). doi:10.1007/978-3-662-48797-6_1
8. Banerjee, A., Peikert, C.: New and improved key-homomorphic pseudorandom functions. In: Garay, J.A., Gennaro, R. (eds.) CRYPTO 2014. LNCS, vol. 8616, pp. 353–370. Springer, Heidelberg (2014). doi:10.1007/978-3-662-44371-2_20
9. Banerjee, A., Peikert, C., Rosen, A.: Pseudorandom functions and lattices. In: Pointcheval, D., Johansson, T. (eds.) EUROCRYPT 2012. LNCS, vol. 7237, pp. 719–737. Springer, Heidelberg (2012). doi:10.1007/978-3-642-29011-4_42
10. Bellare, M., Rogaway, P.: Random oracles are practical: A paradigm for designing efficient protocols. In: CCS 1993, pp. 62–73. ACM (1993)
11. Blazy, O., Kakvi, S.A., Kiltz, E., Pan, J.: Tightly-secure signatures from chameleon hash functions. In: Katz, J. (ed.) PKC 2015. LNCS, vol. 9020, pp. 256–279. Springer, Heidelberg (2015). doi:10.1007/978-3-662-46447-2_12
12. Blazy, O., Kiltz, E., Pan, J.: (Hierarchical) identity-based encryption from affine message authentication. In: Garay, J.A., Gennaro, R. (eds.) CRYPTO 2014. LNCS, vol. 8616, pp. 408–425. Springer, Heidelberg (2014). doi:10.1007/978-3-662-44371-2_23

13. Böhl, F., Hofheinz, D., Jager, T., Koch, J., Seo, J.H., Striecks, C.: Practical signatures from standard assumptions. In: Johansson, T., Nguyen, P.Q. (eds.) EUROCRYPT 2013. LNCS, vol. 7881, pp. 461–485. Springer, Heidelberg (2013). doi:10.1007/978-3-642-38348-9_28
14. Boneh, D., Boyen, X.: Efficient selective-ID secure identity-based encryption without random oracles. In: Cachin, C., Camenisch, J.L. (eds.) EUROCRYPT 2004. LNCS, vol. 3027, pp. 223–238. Springer, Heidelberg (2004). doi:10.1007/978-3-540-24676-3_14
15. Boneh, D., Boyen, X.: Secure identity based encryption without random oracles. In: Franklin, M. (ed.) CRYPTO 2004. LNCS, vol. 3152, pp. 443–459. Springer, Heidelberg (2004). doi:10.1007/978-3-540-28628-8_27
16. Boneh, D., Boyen, X.: Short signatures without random oracles. In: Cachin, C., Camenisch, J.L. (eds.) EUROCRYPT 2004. LNCS, vol. 3027, pp. 56–73. Springer, Heidelberg (2004). doi:10.1007/978-3-540-24676-3_4
17. Boneh, D., Franklin, M.: Identity-based encryption from the weil pairing. In: Kilian, J. (ed.) CRYPTO 2001. LNCS, vol. 2139, pp. 213–229. Springer, Heidelberg (2001). doi:10.1007/3-540-44647-8_13
18. Boneh, D., Gentry, C., Gorbunov, S., Halevi, S., Nikolaenko, V., Segev, G., Vaikuntanathan, V., Vinayagamurthy, D.: Fully key-homomorphic encryption, arithmetic circuit abe and compact garbled circuits. In: Nguyen, P.Q., Oswald, E. (eds.) EUROCRYPT 2014. LNCS, vol. 8441, pp. 533–556. Springer, Heidelberg (2014). doi:10.1007/978-3-642-55220-5_30
19. Boneh, D., Lynn, B., Shacham, H.: Short signatures from the weil pairing. J. Cryptology **17**(4), 297–319 (2004)
20. Boyen, X.: Lattice mixing and vanishing trapdoors: a framework for fully secure short signatures and more. In: Nguyen, P.Q., Pointcheval, D. (eds.) PKC 2010. LNCS, vol. 6056, pp. 499–517. Springer, Heidelberg (2010). doi:10.1007/978-3-642-13013-7_29
21. Boyen, X., Li, Q.: Towards tightly secure short signature and ibe. Cryptology ePrint Archive, Report 2016/498 (2016)
22. Brakerski, Z., Langlois, A., Peikert, C., Regev, O., Stehlé, D.: Classical hardness of learning with errors. In: STOC 13, pp. 575–584. ACM (2013)
23. Brakerski, Z., Vaikuntanathan, V.: Efficient fully homomorphic encryption from (standard) LWE. In: FOCS 2011. pp. 97–106. IEEE (2011)
24. Brakerski, Z., Vaikuntanathan, V.: Lattice-based FHE as secure as PKE. In: ITCS 2014, pp. 1–12. ACM (2014)
25. Brakerski, Z., Vaikuntanathan, V.: Circuit-ABE from LWE: Unbounded attributes and semi-adaptive security. Cryptology ePrint Archive, Report 2016/118 (2016)
26. Camenisch, J., Lysyanskaya, A.: Signature schemes and anonymous credentials from bilinear maps. In: Franklin, M. (ed.) CRYPTO 2004. LNCS, vol. 3152, pp. 56–72. Springer, Heidelberg (2004). doi:10.1007/978-3-540-28628-8_4
27. Canetti, R., Halevi, S., Katz, J.: A forward-secure public-key encryption scheme. In: Biham, E. (ed.) EUROCRYPT 2003. LNCS, vol. 2656, pp. 255–271. Springer, Heidelberg (2003). doi:10.1007/3-540-39200-9_16
28. Cash, D., Hofheinz, D., Kiltz, E., Peikert, C.: Bonsai trees, or how to delegate a lattice basis. J. Cryptology **25**(4), 601–639 (2012)
29. Chen, J., Wee, H.: Fully, (almost) tightly secure IBE and dual system groups. In: Canetti, R., Garay, J.A. (eds.) CRYPTO 2013. LNCS, vol. 8043, pp. 435–460. Springer, Heidelberg (2013). doi:10.1007/978-3-642-40084-1_25
30. Cramer, R., Shoup, V.: Signature schemes based on the strong RSA assumption. ACM Trans. Inf. Syst. Secur. **3**(3), 161–185 (2000)

31. Döttling, N., Schröder, D.: Efficient pseudorandom functions via on-the-fly adaptation. In: Gennaro, R., Robshaw, M. (eds.) CRYPTO 2015. LNCS, vol. 9215, pp. 329–350. Springer, Heidelberg (2015). doi:10.1007/978-3-662-47989-6_16
32. Ducas, L., Micciancio, D.: Improved short lattice signatures in the standard model. In: Garay, J.A., Gennaro, R. (eds.) CRYPTO 2014. LNCS, vol. 8616, pp. 335–352. Springer, Heidelberg (2014). doi:10.1007/978-3-662-44371-2_19
33. Fischlin, M.: The cramer-shoup strong-rsa signature scheme revisited. In: Desmedt, Y.G. (ed.) PKC 2003. LNCS, vol. 2567, pp. 116–129. Springer, Heidelberg (2003). doi:10.1007/3-540-36288-6_9
34. Gennaro, R., Halevi, S., Rabin, T.: Secure hash-and-sign signatures without the random oracle. In: Stern, J. (ed.) EUROCRYPT 1999. LNCS, vol. 1592, pp. 123–139. Springer, Heidelberg (1999). doi:10.1007/3-540-48910-X_9
35. Gentry, C.: Practical identity-based encryption without random oracles. In: Vaudenay, S. (ed.) EUROCRYPT 2006. LNCS, vol. 4004, pp. 445–464. Springer, Heidelberg (2006). doi:10.1007/11761679_27
36. Gentry, C., Peikert, C., Vaikuntanathan, V.: Trapdoors for hard lattices and new cryptographic constructions. In: STOC 2008, pp. 197–206. ACM (2008)
37. Gorbunov, S., Vaikuntanathan, V., Wee, H.: Attribute-based encryption for circuits. In: STOC 2013, pp. 545–554. ACM (2013)
38. Gentry, C., Sahai, A., Waters, B.: Homomorphic encryption from learning with errors: conceptually-simpler, asymptotically-faster, attribute-based. In: Canetti, R., Garay, J.A. (eds.) CRYPTO 2013. LNCS, vol. 8042, pp. 75–92. Springer, Heidelberg (2013). doi:10.1007/978-3-642-40041-4_5
39. Goh, E.-J., Jarecki, S.: A signature scheme as secure as the diffie-hellman problem. In: Biham, E. (ed.) EUROCRYPT 2003. LNCS, vol. 2656, pp. 401–415. Springer, Heidelberg (2003). doi:10.1007/3-540-39200-9_25
40. Goldreich, O., Goldwasser, S., Micali, S.: How to construct random functions. J. ACM **33**(4), 792–807 (1986)
41. Gorbunov, S., Vaikuntanathan, V., Wee, H.: Predicate encryption for circuits from LWE. In: Gennaro, R., Robshaw, M. (eds.) CRYPTO 2015. LNCS, vol. 9216, pp. 503–523. Springer, Heidelberg (2015). doi:10.1007/978-3-662-48000-7_25
42. Gorbunov, S., Vinayagamurthy, D.: Riding on asymmetry: efficient ABE for branching programs. In: Iwata, T., Cheon, J.H. (eds.) ASIACRYPT 2015. LNCS, vol. 9452, pp. 550–574. Springer, Heidelberg (2015). doi:10.1007/978-3-662-48797-6_23
43. Hofheinz, D., Jager, T.: Tightly secure signatures and public-key encryption. In: Safavi-Naini, R., Canetti, R. (eds.) CRYPTO 2012. LNCS, vol. 7417, pp. 590–607. Springer, Heidelberg (2012). doi:10.1007/978-3-642-32009-5_35
44. Hofheinz, D., Koch, J., Striecks, C.: Identity-based encryption with (almost) tight security in the multi-instance, multi-ciphertext setting. In: Katz, J. (ed.) PKC 2015. LNCS, vol. 9020, pp. 799–822. Springer, Heidelberg (2015). doi:10.1007/978-3-662-46447-2_36
45. Hohenberger, S., Waters, B.: Realizing hash-and-sign signatures under standard assumptions. In: Joux, A. (ed.) EUROCRYPT 2009. LNCS, vol. 5479, pp. 333–350. Springer, Heidelberg (2009). doi:10.1007/978-3-642-01001-9_19
46. Hohenberger, S., Waters, B.: Short and stateless signatures from the RSA assumption. In: Halevi, S. (ed.) CRYPTO 2009. LNCS, vol. 5677, pp. 654–670. Springer, Heidelberg (2009). doi:10.1007/978-3-642-03356-8_38
47. Jager, T.: Tightly-secure pseudorandom functions via work factor partitioning. Cryptology ePrint Archive, Report 2016/121 (2016)

48. Katz, J., Wang, N.: Efficiency improvements for signature schemes with tight security reductions. In: CCS 2003, pp. 155–164. CCS 2003, ACM (2003)
49. Lewko, A., Waters, B.: New proof methods for attribute-based encryption: achieving full security through selective techniques. In: Safavi-Naini, R., Canetti, R. (eds.) CRYPTO 2012. LNCS, vol. 7417, pp. 180–198. Springer, Heidelberg (2012). doi:10.1007/978-3-642-32009-5_12
50. Lyubashevsky, V.: Lattice signatures without trapdoors. In: Pointcheval, D., Johansson, T. (eds.) EUROCRYPT 2012. LNCS, vol. 7237, pp. 738–755. Springer, Heidelberg (2012). doi:10.1007/978-3-642-29011-4_43
51. Micciancio, D., Peikert, C.: Trapdoors for lattices: simpler, tighter, faster, smaller. In: Pointcheval, D., Johansson, T. (eds.) EUROCRYPT 2012. LNCS, vol. 7237, pp. 700–718. Springer, Heidelberg (2012). doi:10.1007/978-3-642-29011-4_41
52. Micciancio, D., Regev, O.: Worst-case to average-case reductions based on gaussian measures. SIAM J. Comput. **37**(1), 267–302 (2007)
53. Naor, M., Reingold, O.: Number-theoretic constructions of efficient pseudo-random functions. J. ACM **51**(2), 231–262 (2004)
54. Peikert, C.: Public-key cryptosystems from the worst-case shortest vector problem: extended abstract. In: STOC 2009, pp. 333–342. ACM (2009)
55. Regev, O.: On lattices, learning with errors, random linear codes, and cryptography. In: STOC 2005, pp. 84–93. ACM (2005)
56. Waters, B.: Efficient identity-based encryption without random oracles. In: Cramer, R. (ed.) EUROCRYPT 2005. LNCS, vol. 3494, pp. 114–127. Springer, Heidelberg (2005). doi:10.1007/11426639_7
57. Waters, B.: Dual system encryption: realizing fully secure ibe and hibe under simple assumptions. In: Halevi, S. (ed.) CRYPTO 2009. LNCS, vol. 5677, pp. 619–636. Springer, Heidelberg (2009). doi:10.1007/978-3-642-03356-8_36

From Identification to Signatures, Tightly: A Framework and Generic Transforms

Mihir Bellare[1(✉)], Bertram Poettering[2], and Douglas Stebila[3]

[1] Department of Computer Science and Engineering, University of California, San Diego, USA
mihir@eng.ucsd.edu
[2] Ruhr University Bochum, Bochum, Germany
bertram.poettering@rub.de
[3] Department of Computing and Software, McMaster University, Hamilton, Canada
stebilad@mcmaster.ca

Abstract. This paper provides a framework to treat the problem of building signature schemes from identification schemes in a unified and systematic way. The outcomes are (1) Alternatives to the Fiat-Shamir transform that, applied to trapdoor identification schemes, yield signature schemes with tight security reductions to standard assumptions (2) An understanding and characterization of existing transforms in the literature. One of our transforms has the added advantage of producing signatures shorter than produced by the Fiat-Shamir transform. Reduction tightness is important because it allows the implemented scheme to use small parameters (thereby being as efficient as possible) while retaining provable security.

1 Introduction

This paper provides a framework to treat the problem of building signature schemes from identification schemes in a unified and systematic way. We are able to explain and characterize existing transforms as well as give new ones whose security proofs give *tight* reductions to *standard* assumptions. This is important so that the implemented scheme can use small parameters, thereby being efficient while retaining provable security. Let us begin by identifying the different elements involved.

ID-TO-SIG TRANSFORMS. Recall that in a three-move identification scheme ID the prover sends a *commitment* Y computed using private randomness y, the verifier sends a random *challenge* c, the prover returns a *response* z computed using y and its secret key *isk*, and the verifier computes a boolean decision from the conversation transcript $Y \| c \| z$ and public key *ivk* (see Fig. 3). We are interested in transforms $\boxed{\textbf{Id2Sig}}$ that take ID and return a signature scheme DS. The transform must be generic, meaning DS is proven to meet some signature security goal $\boxed{\mathrm{P}_{\mathrm{sig}}}$ assuming only that ID meets some identification security goal $\boxed{\mathrm{P}_{\mathrm{id}}}$. This proof is supported by a reduction $\boxed{\mathrm{P}_{\mathrm{sig}} \rightarrow \mathrm{P}_{\mathrm{id}}}$ that may be tight

J.H. Cheon and T. Takagi (Eds.): ASIACRYPT 2016, Part II, LNCS 10032, pp. 435–464, 2016.
DOI: 10.1007/978-3-662-53890-6_15

or loose. Boxing an item here highlights elements of interest and choice in the id-to-sig process.

CANONICAL EXAMPLE. In the most canonical example we have, **Id2Sig** = **FS** is the Fiat-Shamir transform [16] ; $\mathrm{P_{id}}$ = IMP-PA is security against impersonation under passive attack [1,14] ; $\mathrm{P_{sig}}$ = UF is unforgeability under chosen-message attack [20] ; and the reduction $\mathrm{P_{sig}} \rightarrow \mathrm{P_{id}}$ is that of AABN [1], which is loose.

We are going to revisit this to give other choices of the different elements, but first let us recall some more details of the above. In the Fiat-Shamir transform **FS** [16], a signature of a message m is a pair (Y, z) such that the transcript $Y\|c\|z$ is accepting for $c = \mathrm{H}(Y\|m)$, where H is a random oracle. IMP-PA requires that an adversary given transcripts of honest protocol executions still fails to make the honest verifier accept in an interaction where it plays the role of the prover, itself picking Y any way it likes, receiving a random c, and then producing z. The loss in the $\mathrm{P_{sig}} \rightarrow \mathrm{P_{id}}$ reduction of AABN [1] is a factor of the number q of adversary queries to the random oracle H: If $\epsilon_{\mathrm{id}}, \epsilon_{\mathrm{sig}}$ denote, respectively, the advantages in breaking the IMP-PA security of ID and the UF security of DS, then $\epsilon_{\mathrm{sig}} \approx q\,\epsilon_{\mathrm{id}}$.

ALGEBRAIC ASSUMPTION TO ID. Suppose a cryptographer wants to build a signature scheme meeting the definition $\mathrm{P_{sig}}$. The cryptographer would like to base security on some algebraic (or other computational) assumption $\boxed{\mathrm{P_{alg}}}$. This could be factoring, RSA inversion, bilinear Diffie-Hellman, some lattice assumption, or many others. Given an id-to-sig transform as above, the task amounts to designing an identification scheme ID achieving $\mathrm{P_{id}}$ under $\mathrm{P_{alg}}$. (Then one can just apply the transform to ID.) This proof is supported by another reduction $\boxed{\mathrm{P_{id}} \rightarrow \mathrm{P_{alg}}}$ that again may be tight or loose. The tightness of the overall reduction $\mathrm{P_{sig}} \rightarrow \mathrm{P_{alg}}$ thus depends on the tightness of both $\mathrm{P_{sig}} \rightarrow \mathrm{P_{id}}$ and $\mathrm{P_{id}} \rightarrow \mathrm{P_{alg}}$.

CANONICAL EXAMPLE. Continuing with the FS+AABN-based example from above, we would need to build an identification scheme meeting $\mathrm{P_{id}}$ = IMP-PA under $\mathrm{P_{alg}}$. The good news is that a wide swathe of such identification schemes are available, for many choices of $\mathrm{P_{alg}}$ (GQ [23] under RSA, FS [16] under Factoring, Schnorr [36] under Discrete Log, ...). However the reduction $\mathrm{P_{id}} \rightarrow \mathrm{P_{alg}}$ is (very) loose.

Again, we are going to revisit this to give other choices of the different elements, but first let us recall some more details of the above. The practical identification schemes here are typically Sigma protocols (this means they satisfy honest-verifier zero-knowledge and special soundness, the latter meaning that from two accepting conversation transcripts with the same commitment but different challenges, one can extract the secret key) and $\mathrm{P_{alg}}$ = KR ("key recovery") is the problem of computing the secret key given only the public key. To solve this problem, we have to run a given IMP-PA adversary twice and hope for two successes. The analysis exploits the Reset Lemma of [6]. If $\epsilon_{\mathrm{alg}}, \epsilon_{\mathrm{id}}$ denote, respectively, the advantages in breaking the algebraic problem and the IMP-PA security of ID, then it results in $\epsilon_{id} \approx \sqrt{\epsilon_{alg}}$. If ϵ_{sig} is the advantage in breaking UF security of DS, combined with the above, we have $\epsilon_{sig} \approx q\sqrt{\epsilon_{alg}}$.

APPROACH. We see from the above that a tight overall reduction $\mathrm{P_{sig}}{\rightarrow}\mathrm{P_{alg}}$ requires that the $\mathrm{P_{sig}}{\rightarrow}\mathrm{P_{id}}$ and $\mathrm{P_{id}}{\rightarrow}\mathrm{P_{alg}}$ reductions *both* be tight. What we observe is that we have a degree of freedom in achieving this, namely *the choice of the security goal* $\mathrm{P_{id}}$ *for the identification scheme.* Our hope is to pick $\mathrm{P_{id}}$ such that (1) We can give (new) transforms **Id2Sig** for which $\mathrm{P_{sig}}{\rightarrow}\mathrm{P_{id}}$ is tight, and simultaneously (2) We can give identification schemes such that $\mathrm{P_{id}}{\rightarrow}\mathrm{P_{alg}}$ is tight. We view these as two pillars of an edifice and are able to provide both via our definitions of security of identification under constrained impersonation coupled with some new id-to-sig transforms. We first pause to discuss some prior work, but a peek at Fig. 1 gives an outline of the results we will expand on later. Following FS, we work in the random oracle model.

PRIOR WORK. The first proofs of security for **FS**-based signatures [35] reduced UF security of the **FS**-derived signature scheme directly to the hardness of the algebraic problem $\mathrm{P_{alg}}$, assuming H is a random oracle [8]. These proofs exploit forking lemmas [4,5,35]. Modular proofs of the form discussed above, that use identification as an intermediate step, begin with [1,33]. The modular approach has many advantages. One is that since the id-to-sig transforms are generic, we have only to design and analyze identification schemes. Another is the better understanding and isolation of the role of random oracles: they are used by **Id2Sig** but not in the identification scheme. We have accordingly adopted this approach. Note that both the direct (forking lemma based) and the AABN-based indirect (modular) approach result in reductions of the same looseness we discussed above. Our (alternative but still modular) approaches will remove this loss.

Consideration of reduction tightness for signatures begins with BR [9], whose PSS scheme has a tight reduction to the RSA problem. KW [24] give another signature scheme with a tight reduction to RSA, and they and GJ [18] give signature schemes with tight reductions to the Diffie-Hellman problem. GPV [17] give a signature scheme with a tight reduction to the problem of finding short vectors in random lattices.

The lack of tightness of the overall reduction for **FS**-based signatures is well recognized as an important problem and drawback. Micali and Reyzin [30] give a signature scheme, with a tight reduction to factoring, that is obtained from a particular identification scheme via a method they call "swap". ABP [2] say that the method generalizes to other factoring-based schemes. However, "swap" has never been stated as a general transform of an identification scheme into a signature scheme. This lack of abstraction is perhaps due in part to a lack of definitions, and the ones we provide allow us to fill the gap. In Sect. 6.5 we elevate the swap method to a general **Swap** transform, characterize the identification schemes to which it applies, and prove that, when it applies, it gives a tight $\mathrm{P_{sig}}{\rightarrow}\mathrm{P_{id}}$ reduction.

ABP [2] show a tight reduction of **FS**-derived GQ-based signatures to the Φ-hiding assumption of [12]. In contrast, our methods will yield GQ-based signatures with a tight reduction to the standard one-wayness of RSA. AFLT [3] use a slight variant of the Fiat-Shamir transform to turn lossy identification schemes into signature schemes with security based tightly on key indistinguishability, resulting in signature schemes with tight reductions to the decisional short discrete logarithm problem, the shortest vector problem in ideal lattices, and subset sum.

CONSTRAINED IMPERSONATION. Recall our goal is to define a notion of identification security $\mathrm{P_{id}}$ such that (1) We can give transforms **Id2Sig** for which $\mathrm{P_{sig}} \rightarrow \mathrm{P_{id}}$ is tight, and (2) We can give identification schemes such that $\mathrm{P_{id}} \rightarrow \mathrm{P_{alg}}$ is tight. In fact our definitional goal is broader, namely to give a framework that allows us to understand and encompass both old and new transforms, the former including **FS** and **Swap**. We do all this with a definitional framework that we refer to as *constrained impersonation*. It yields four particular definitions denoted CIMP-XY for XY $\in \{\mathrm{CU}, \mathrm{UC}, \mathrm{UU}, \mathrm{CC}\}$. Each, in the role of $\mathrm{P_{id}}$, will be the basis for an id-to-sig transform such that $\mathrm{P_{sig}} \rightarrow \mathrm{P_{id}}$ is tight, and two will allow $\mathrm{P_{id}} \rightarrow \mathrm{P_{alg}}$ to be tight.

In constrained impersonation we continue, as with IMP-PA, to allow a passive attack in which the adversary $\mathcal{A}$ against the identification scheme ID can obtain transcripts $Y_1 \| c_1 \| z_1, Y_2 \| c_2 \| z_2, \ldots$ of interactions between the honest prover and verifier. Then $\mathcal{A}$ tries to impersonate, meaning get the honest verifier to accept. If $\mathrm{X} = \mathrm{C}$ then the *commitment* in this impersonation interaction is adversary-*chosen*, while if $\mathrm{X} = \mathrm{U}$ (*unchosen*) it must be pegged to a commitment from one of the (honest) transcripts. If $\mathrm{Y} = \mathrm{C}$, the *challenge* is adversary-chosen, while if $\mathrm{Y} = \mathrm{U}$ it is as usual picked at random by the verifier. In all cases, multiple impersonation attempts are allowed. The formal definitions are in Sect. 3. CIMP-CU is a multi-impersonation version of IMP-PA, but the rest are novel.

What do any of these notions have to do with identification if one understands the latter as the practical goal of proving one's identity to a verifier? Beyond CIMP-CU, very little. In practice it is unclear how one can constrain a prover to only use, in impersonation, a commitment from a prior transcript. It is even more bizarre to allow a prover to pick the challenge. Our definitions however are not trying to capture any practical usage of identification. They view the latter as an analytic tool, an intermediate land allowing a smooth transition from an algebraic problem to signatures. The constrained impersonation notions work well in this regard, as we will see, both to explain and understand existing work and to obtain new signature schemes with tight reductions.

Relations between the four notions of constrained impersonation are depicted in Fig. 1. An arrow A $\rightarrow$ B is an implication: *Every* identification scheme that is A-secure is also B-secure. A barred arrow A $\not\rightarrow$ B is a separation: *There exists* an identification scheme that is A-secure but *not* B-secure. (For now ignore the boxes around notions.) In particular we see that CIMP-UU is weaker than, and CIMP-UC incomparable to, the more standard CIMP-CU. See Proposition 1 for more precise renditions of the implications.

AUXILIARY DEFINITIONS AND TOOLS. Before we see how to leverage the constrained impersonation framework, we need a few auxiliary definitions and results that, although simple, are, we believe, of independent interest and utility.

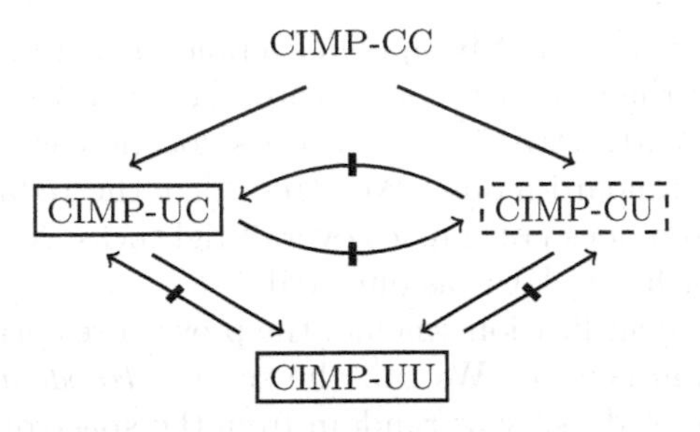

P_{id}	**Id2Sig** Transform	Trapdoor?	P_{sig}-secure **Signature**	**Reductions** $P_{sig} \to P_{id} / P_{id} \to P_{alg}$
CIMP-CU	**FS**	No	$(Y, z) : c = \mathrm{H}(Y \| m)$	Tight/Loose
CIMP-UC	**MdCmt**	Yes	$(c, z) : Y = \mathrm{H}(m)$; $c \leftarrow_\$ \{0,1\}^{\mathsf{ID.cl}}$	Tight/Tight
CIMP-UU	**MdCmtCh**	Yes	$(s, z) : s \leftarrow_\$ \{0,1\}$; $Y = \mathrm{H}_1(m)$ $c = \mathrm{H}_2(m \| s)$	Tight/Tight
CIMP-CC	**MdCh**	No	$(Y, s, z) : s \leftarrow_\$ \{0,1\}^{\mathsf{sl}}$ $c = \mathrm{H}(m \| s)$	Tight/Unknown
CIMP-UC	**Swap**	Yes	$(c, z) : c \leftarrow_\$ \{0,1\}^{\mathsf{ID.cl}}$ $Y = \mathrm{H}(m \| c)$	Tight/Tight

Fig. 1. Top: Relations between notions P_{id} of security for an identification scheme ID under constrained impersonation. Solid arrows denote implications, barred arrows denote separations. A solid box around a notion means a tight $P_{id} \to P_{alg}$ reduction for Sigma protocols; dotted means a loose one; no box means no known reduction. **Bottom:** Transforms of identification schemes into UUF (row 2, 3) or UF (rows 1, 4, 5) signature schemes. The first column is the assumption P_{id} on the identification scheme. The third column indicates whether or not the identification scheme is assumed to be trapdoor. ID.cl is the challenge length and sl is a seed length. In rows 1, 4 the commitment Y is chosen at random. The third transform has the shortest signatures, consisting of a response plus a single bit.

We define a signature scheme to be UUF (Unique Unforgeable) if it is UF with the restriction that a message can be signed at most once. (The adversary is not allowed to twice ask the signing oracle to sign a particular m.) It turns out that some of our id-to-sig transforms naturally achieve UUF, not UF. However there are simple, generic transforms of UUF signature schemes into UF ones — succinctly, UF→UUF— that do not introduce much overhead and have tight reductions. One is to remove randomness, and the other is to add it. In more detail, a well-known method to derandomize a signature scheme is to specify the coins by hashing the secret key along with the message. This has been proved to work in some instances [26, 32] but not in general. We observe that this method has the additional benefit of turning a UUF scheme into a UF one. We call the transform **DR**. Theorem 3 shows that it works. (In particular it shows UF-security of the derandomized scheme in a more general setting than was known before.) The second transform, **AR**, appends a random salt to the message before signing and includes the salt in the signature. Theorem 4 shows that it works. The first transform is attractive because it does not increase signature size. The second does, but is standard-model. We stress that the reductions are tight in both cases, so this step does not impact overall tightness. Now we can take (the somewhat easier to achieve) UUF as our goal.

Recall that in an identification scheme, the prover uses private randomness y to generate its commitment Y. We call the scheme *trapdoor* if the prover can pick the commitment Y directly at random from the space of commitments and then compute the associated private randomness y using its secret key via a prescribed algorithm. The concept is implicit in [30] but does not seem to have been formalized before, so we give a formal definition in Sect. 3. Many existing identification schemes will meet our definition of being trapdoor modulo possibly some changes to the key structure. Thus the GQ scheme of [23] is trapdoor if we add the decryption exponent d to the secret key. With similar changes to the keys, the Fiat-Shamir [16] and Ong-Schnorr [34] identification schemes are trapdoor. The factoring-based identification scheme of [30] is also trapdoor. But not all identification schemes are trapdoor. One that is not is Schnorr's (discrete-log based) scheme [36].

SUMMARY OF RESULTS. For each notion $\mathrm{P_{id}} \in \{\text{CIMP-CU}, \text{CIMP-UC}, \text{CIMP-UU}, \text{CIMP-CC}\}$ we give an id-to-sig transform that turns any given $\mathrm{P_{id}}$-secure identification scheme ID into a $\mathrm{P_{sig}} = \text{UUF}$ signature scheme DS; the transform from CIMP-CC security achieves even a UF signature scheme. *The reduction* $\mathrm{P_{sig}}{\rightarrow}\mathrm{P_{id}}$ *is tight in all four cases.* (To further make the signature schemes UF secure, we can apply the above-mentioned UF→UUF transforms while preserving tightness.) The table in Fig. 1 summarizes the results and the transforms. They are discussed in more detail below and then fully in Sect. 6.

This is one pillar of the edifice, and not useful by itself. The other pillar is the $\mathrm{P_{id}}{\rightarrow}\mathrm{P_{alg}}$ reduction. In the picture at the top of Fig. 1, a solid-line box around $\mathrm{P_{id}}$ means that the reduction $\mathrm{P_{id}}{\rightarrow}\mathrm{P_{alg}}$ is tight, a dotted-line box indicates a reduction is possible but is not tight, and no box means no known reduction. These results assume the identification scheme is a Sigma protocol, as most are, and are discussed in Sect. 4. We see that two points of our framework can be *tightly* obtained from the algebraic problem, so that in these cases the overall $\mathrm{P_{sig}}{\rightarrow}\mathrm{P_{alg}}$ reduction is tight, which was the ultimate goal.

MORE DETAILS ON RESULTS. The id-to-sig transform from CIMP-CU is the classical **FS** one. The reduction is now tight, even though it was not from IMP-PA [1], simply because CIMP-CU is IMP-PA extended to allow multiple impersonation attempts. The result, which we state as Theorem 8, is implicit in [1], but we give a proof to illustrate how simple the proof now is. In this case our framework serves to better understand, articulate and simplify something implicit in the literature, rather than deliver anything particularly new.

For CIMP-UC, we give a transform called **MdCmt**, for "Message-Derived Commitment", where, to sign m, the signer computes the commitment Y as a hash of the message, picks a challenge at random, uses the identification trapdoor to compute the coins y corresponding to Y, uses y and the secret key to compute a response z, and returns the challenge and response as the signature. See Sect. 6.1.

For CIMP-UU, the weakest of the four notions, our transform **MdCmtCh**, for "Message-Derived Commitment and Challenge", has the signer compute the commitment Y as a hash of the message. It then picks a single random bit s and computes the challenge as a hash of the message and seed s, returning as signature the seed and response, the latter computed as before. Beyond a tight reduction, this transform has the added feature of *short signatures*, the signature being a response plus a single bit. (In all other transforms, whether prior or ours, the signature is at least a response plus a challenge, often more.) See Sect. 6.2.

Since CIMP-CC implies CIMP-UC and CIMP-UU (Fig. 1, top), for the former the **MdCmt** and **MdCmtCh** transforms would both work. However, these require the identification scheme to be trapdoor and achieve UUF rather than UF. (The above-mentioned UF→UUF transforms would have to be applied on top to get UF.) We give an alternative transform called **MdCh** ("Message-Derived Challenge") from CIMP-CC that directly achieves UF and works (gives a tight reduction) even if the identification scheme is not trapdoor. It has the signer pick a random commitment, produce the challenge as in **MdCmtCh**, namely as a randomized hash of the message, compute the response, and return the conversation transcript as signature. See Sect. 6.3.

The salient fact is that the reductions underlying all four transforms are tight. To leverage the results we now have to consider achieving CIMP-XY. We do this in Sect. 4. We give reductions $P_{id} \rightarrow P_{alg}$ of the P_{id} = CIMP-XY security of identification schemes that are Sigma protocols to their key-recovery (KR) security, the latter being the problem of recovering the secret key given only the public key, which is typically the algebraic problem P_{alg} whose hardness is assumed. For CIMP-UC and CIMP-UU the $P_{id} \rightarrow P_{alg}$ reduction is tight, as per Theorem 1, making these the most attractive starting points. For CIMP-CU we must use the Reset Lemma [6] so the reduction (cf. Theorem 2) is loose. CIMP-CC is a very strong notion and, as we discuss at the end of Sect. 4, not achieved by Sigma protocols but achievable by other means.

SWAP. As indicated above, our framework allows us to generalize the swap method of [30] into an id-to-sig transform **Swap** and understand and characterize what it does. In Sect. 6.5 we present **Swap** as a generic transform of a trapdoor identification scheme ID to a signature scheme that is just like **MdCmt** (cf. row 2 of the table of Fig. 1) except that the challenge c is included in the input to the hash function (cf. row 5 of the table of Fig. 1). Recall that **MdCmt** turns a CIMP-UC identification scheme into a UUF signature scheme. We can thence get a UF signature scheme by applying the **AR** transform of Sect. 5.2. **Swap** is a shortcut, or optimization, of this two step process: it directly turns a CIMP-UC identification scheme into a UF signature scheme by effectively re-using the randomness of **MdCmt** in **AR**. We note that the composition of our **DR** with our **MdCmtCh** yields a UF signature scheme with shorter signatures than **Swap** while also having a tight reduction to the weaker CIMP-UU assumption, and would thus be superior. However we think **Swap** is of historical interest and accordingly present it. See Sect. 6.5 for details.

INSTANTIATION. As a simple and canonical example, in [7] we apply our framework and transforms to the GQ identification scheme to get signature schemes with tight reductions to RSA. It is also possible to give instantiations based on claw-free permutations [20] and factoring. An intriguing application area to explore for our transforms is in lattice-based cryptography. Here signatures have been obtained via the **FS** transform [28,29]. The underlying lattice-based identification schemes do not appear to be trapdoor, so our transforms would not apply. However, via the techniques of MP [31], one can build lattice-based trapdoor identification schemes to which our transforms apply. Whether there is a performance benefit will depend on the trade-off between the added cost from having the trapdoor and the smaller parameters permitted by the improved security reduction.

DISCUSSION. We measure reduction tightness stringently, in a model where running time, queries and success probability are separate parameters. The picture changes if one considers the expected success ratio, namely the ratio of running time to success probability. Reduction tightness under this metric is considered in PS [35] and the concurrent and independent work of KMP [25].

We establish the classical notion of standard unforgeability (UF) [20]. Our transforms also establish strong unforgeability if the identification scheme has the extra property of unique responses. (For any public key, commitment, and challenge, there exists at most one response that the verifier accepts.)

A reviewer commented that "The signature scheme with the tightest security in this paper is derived from the Swap transform, which makes the result less surprising since the Swap method, first used in [30], has already been found to be generic to some extent by ABP [2]." In response, first, the tightness of the reductions is about the same for **Swap**, $\mathbf{DR} \circ \mathbf{MdCmt}$ and $\mathbf{DR} \circ \mathbf{MdCmtCh}$ (cf. Fig. 14), but the third has shorter signatures, an advantage over **Swap**. Second, while, as indicated above, prior work including ABP [2] did discuss Swap in a broader context than the original MR [30], the discussion was informal and left open to exactly what identification schemes Swap might apply. We have formalized prior intuition using the concept of trapdoor identification, and thus been able to provide a general transform and result for Swap. We view this as a contribution towards understanding the area, making intuition rigorous and providing a result that future work can apply in a blackbox way. Also, as noted above, our framework helps understand Swap, seeing it as an optimized derandomization of the simpler **MdCmt**. We understand, as per what the reviewer says, that our results for **Swap** may not be surprising, but we don't think surprise is the only measure of contribution. Clarifying and formalizing existing intuition, as we have done in this way with **Swap**, puts the area on firmer ground and helps future work.

GK [22] give an example of a 3-move ID protocol where **FS** yields a secure signature scheme in the ROM, but the RO is not instantiable. Their protocol however is not a Sigma protocol, as is assumed for the ones we start with and is true for practical identification schemes. Currently, secure instantiation of the RO, both for **FS** and our transforms, is not ruled out for such identification schemes.

2 Notation and Basic Definitions

NOTATION. We let ε denote the empty string. If X is a finite set, we let $x \leftarrow_{\$} X$ denote picking an element of X uniformly at random and assigning it to x. We use $a_1\|a_2\|\cdots\|a_n$ as shorthand for $(a_1, a_2, \ldots, a_n)$. By $a_1\|a_2\|\cdots\|a_n \leftarrow x$ we mean that x is parsed into its constituents. We use bracket notation for associative arrays, e.g., $T[x] = y$ means that key x is mapped to value y. Algorithms may be randomized unless otherwise indicated. Running time is worst case. If A is an algorithm, we let $y \leftarrow A(x_1, \ldots; r)$ denote running A with random coins r on inputs $x_1, \ldots$ and assigning the output to y. We let $y \leftarrow_{\$} A(x_1, \ldots)$ be the result of picking r at random and letting $y \leftarrow A(x_1, \ldots; r)$. We let $[A(x_1, \ldots)]$ denote the set of all possible outputs of (randomized) A when invoked with inputs $x_1, \ldots$. We use the code based game playing framework of [10]. (See Fig. 2 for an example.) By Pr[G] we denote the event that the execution of game G results in the game returning true. Boolean flags (like bad) in games are assumed initialized to false, and associative arrays empty. We adopt the convention that the running time of an adversary refers to the worst case execution time of the game with the adversary. This means that the time taken for oracles to compute replies to queries is included.

Our treatment of random oracles is more general than usual. In our constructions, we will need random oracles with different ranges. For example we may want one random oracle returning points in a group $\mathbb{Z}_N^*$ and another returning strings of some length l. To provide a single unified definition, we have the procedure H in the games take not just the input x but a description Rng of the set from which outputs are to be drawn at random. Thus $y \leftarrow_{\$} \mathrm{H}(x, \mathbb{Z}_N^*)$ will return a random element of $\mathbb{Z}_N^*$, while $c \leftarrow_{\$} \mathrm{H}(x, \{0,1\}^l)$ will return a random l-bit string, and so on. Sometimes if the range set is understood, it is dropped as an argument.

SIGNATURES. In a signature scheme DS, the signer generates signing key sk and verifying key vk via $(vk, sk) \leftarrow_{\$} \mathsf{DS.Kg}^{\mathrm{H}}$ where H is the random oracle, the latter with syntax as discussed above. Now it can compute a signature $\sigma \leftarrow_{\$} \mathsf{DS.Sig}^{\mathrm{H}}(vk, sk, m)$ on any message $m \in \{0,1\}^*$. A verifier can deterministically compute a boolean $v \leftarrow \mathsf{DS.Vf}^{\mathrm{H}}(vk, m, \sigma)$ indicating whether or not σ is a valid signature of m relative to vk. Correctness as usual requires that $\mathsf{DS.Vf}^{\mathrm{H}}(vk, m, \mathsf{DS.Sig}^{\mathrm{H}}(vk, sk, m)) = \mathsf{true}$ with probability one. Game $\mathbf{G}_{\mathsf{DS}}^{\mathrm{uf}}(\mathcal{A})$ associated to DS and adversary $\mathcal{A}$ as per Fig. 2 captures the classical unforgeability notion of [20] lifted to the ROM as per [8], and we let $\mathbf{Adv}_{\mathsf{DS}}^{\mathrm{uf}}(\mathcal{A}) = \Pr[\mathbf{G}_{\mathsf{DS}}^{\mathrm{uf}}(\mathcal{A})]$ be the UF-advantage of $\mathcal{A}$. The same figure also defines game $\mathbf{G}_{\mathsf{DS}}^{\mathrm{uuf}}(\mathcal{A})$ to capture *unique* unforgeability. The difference is the inclusion of the boxed code, which disallows $\mathcal{A}$ from getting more than one signature on the same message. We let $\mathbf{Adv}_{\mathsf{DS}}^{\mathrm{uuf}}(\mathcal{A}) = \Pr[\mathbf{G}_{\mathsf{DS}}^{\mathrm{uuf}}(\mathcal{A})]$ be the UUF-advantage of $\mathcal{A}$.

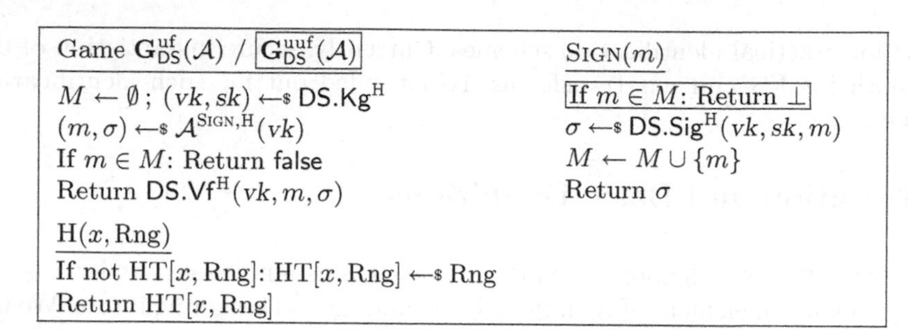
Game $\mathbf{G}^{\mathrm{uf}}_{\mathsf{DS}}(\mathcal{A})$ / $\boxed{\mathbf{G}^{\mathrm{uuf}}_{\mathsf{DS}}(\mathcal{A})}$
$M \leftarrow \emptyset$; $(vk, sk) \leftarrow\!\!\$\ \mathsf{DS.Kg}^{\mathrm{H}}$
$(m, \sigma) \leftarrow\!\!\$\ \mathcal{A}^{\textsc{Sign},\mathrm{H}}(vk)$
If $m \in M$: Return false
Return $\mathsf{DS.Vf}^{\mathrm{H}}(vk, m, \sigma)$

$\mathrm{H}(x, \mathrm{Rng})$
If not $\mathrm{HT}[x, \mathrm{Rng}]$: $\mathrm{HT}[x, \mathrm{Rng}] \leftarrow\!\!\$\ \mathrm{Rng}$
Return $\mathrm{HT}[x, \mathrm{Rng}]$

$\textsc{Sign}(m)$
$\boxed{\text{If } m \in M\text{: Return } \bot}$
$\sigma \leftarrow\!\!\$\ \mathsf{DS.Sig}^{\mathrm{H}}(vk, sk, m)$
$M \leftarrow M \cup \{m\}$
Return σ

Fig. 2. Games defining unforgeability and unique unforgeability of signature scheme DS. Game $\mathbf{G}^{\mathrm{uuf}}_{\mathsf{DS}}(\mathcal{A})$ includes the boxed code and game $\mathbf{G}^{\mathrm{uf}}_{\mathsf{DS}}(\mathcal{A})$ does not.

Of course, UF implies UUF, meaning any signature scheme that is UF secure is also UUF secure. The converse is not true, meaning there exist UUF signature schemes that are not UF secure (we will see natural examples in this paper). In Sect. 5 we give simple, generic and tight ways to turn any given UUF signature scheme into a UF one.

We note that unique unforgeability (UUF) should not be confused with unique signatures as defined in [21, 27]. In a unique signature scheme, there is, for any message, at most one signature the verifier will accept. If a unique signature scheme is UUF then it is also UF. But there are UUF (and UF) schemes that are not unique.

3 Constrained Impersonation Framework

We introduce a framework of definitions of identification schemes secure against constrained impersonation.

IDENTIFICATION. An identification (ID) scheme ID operates as depicted in Fig. 3. First, via $(ivk, isk, itk) \leftarrow\!\!\$\ \mathsf{ID.Kg}$, the prover generates a public *verification key* ivk, private *identification key* isk, and *trapdoor* itk. Via $(Y, y) \leftarrow\!\!\$\ \mathsf{ID.Ct}(ivk)$ it generates *commitment* $Y \in \mathsf{ID.CS}(ivk)$ and corresponding private state y. We refer to $\mathsf{ID.CS}(ivk)$ as the *commitment space* associated to ivk. The verifier sends a random challenge of length $\mathsf{ID.cl}$. The prover's *response* z and the verifier's boolean *decision* v are deterministically computed per $z \leftarrow \mathsf{ID.Rp}(ivk, isk, c, y)$ and $v \leftarrow \mathsf{ID.Vf}(ivk, Y\|c\|z)$, respectively. We assume throughout that identification schemes have perfect correctness. We also assume *uniformly-distributed commitments.* More precisely, the outputs of the following two processes must be identically distributed: the first processes generates $(ivk, isk, itk) \leftarrow\!\!\$\ \mathsf{ID.Kg}$, then lets $(Y, y) \leftarrow\!\!\$\ \mathsf{ID.Ct}(ivk)$ and returns (ivk, Y); the second processes generates $(ivk, isk, itk) \leftarrow\!\!\$\ \mathsf{ID.Kg}$, then lets $Y \leftarrow\!\!\$\ \mathsf{ID.CS}(ivk)$ and returns (ivk, Y). An example ID scheme is GQ [23]; see [7] for a description in our notation. For basic ID schemes, the trapdoor plays no role; its use arises in trapdoor identification, as discussed next.

Prover		Verifier
Input: ivk, isk		Input: ivk
$(Y, y) \leftarrow\$ \mathsf{ID.Ct}(ivk)$	$\xrightarrow{Y}$	
	$\xleftarrow{c}$	$c \leftarrow\$ \{0,1\}^{\mathsf{ID.cl}}$
$z \leftarrow \mathsf{ID.Rp}(ivk, isk, c, y)$	$\xrightarrow{z}$	$v \leftarrow \mathsf{ID.Vf}(ivk, Y\|c\|z)$

Fig. 3. Functioning of an identification scheme ID.

Game $\mathbf{G}_{\mathsf{ID}}^{\text{cimp-xy}}(\mathcal{P})$
$\mathrm{S} \leftarrow \emptyset$; $i \leftarrow 0$; $j \leftarrow 0$
$(ivk, isk, itk) \leftarrow\$ \mathsf{ID.Kg}$
$(k, z) \leftarrow\$ \mathcal{P}^{\mathrm{TR,CH}}(ivk)$
If not $(1 \le k \le j)$: Return false
$T \leftarrow \mathrm{CT}[k]\|z$
Return $\mathsf{ID.Vf}(ivk, T)$

TR()
$i \leftarrow i + 1$
$(Y_i, y_i) \leftarrow\$ \mathsf{ID.Ct}(ivk)$
$c_i \leftarrow\$ \{0,1\}^{\mathsf{ID.cl}}$
$z_i \leftarrow \mathsf{ID.Rp}(ivk, isk, c_i, y_i)$
$\mathrm{S} \leftarrow \mathrm{S} \cup \{(Y_i, c_i)\}$
Return $Y_i\|c_i\|z_i$

CH(l) // xy=uu
If not $(1 \le l \le i)$: Return $\perp$
$j \leftarrow j + 1$; $c \leftarrow\$ \{0,1\}^{\mathsf{ID.cl}}$
$\mathrm{CT}[j] \leftarrow Y_l\|c$; Return c

CH(l, c) // xy=uc
If not $(1 \le l \le i)$: Return $\perp$
If $(c = c_l)$: Return $\perp$
$j \leftarrow j + 1$
$\mathrm{CT}[j] \leftarrow Y_l\|c$; Return c

CH(Y) // xy=cu
$j \leftarrow j + 1$; $c \leftarrow\$ \{0,1\}^{\mathsf{ID.cl}}$
$\mathrm{CT}[j] \leftarrow Y\|c$; Return c

CH(Y, c) // xy=cc
If $(Y, c) \in \mathrm{S}$: Return $\perp$
$j \leftarrow j + 1$
$\mathrm{CT}[j] \leftarrow Y\|c$; Return c

Fig. 4. Games defining security of identification scheme ID against constrained impersonation under passive attack.

TRAPDOOR IDENTIFICATION. We now define what it means for an ID scheme to be *trapdoor*. Namely there is an algorithm $\mathsf{ID.Ct}^{-1}$ that produces y from Y with the aid of the trapdoor *itk*. Formally, the outputs of the following two processes must be identically distributed. Both processes generate (ivk, isk, itk) $\leftarrow\$ \mathsf{ID.Kg}$. The first process then lets $(Y, y) \leftarrow\$ \mathsf{ID.Ct}(ivk)$. The second process picks $Y \leftarrow\$ \mathsf{ID.CS}(ivk)$ and lets $y \leftarrow\$ \mathsf{ID.Ct}^{-1}(ivk, itk, Y)$. (Here $\mathsf{ID.CS}(ivk)$ is the space of commitments associated to ID and *ivk*.) Both processes return (ivk, isk, itk, Y, y).

SECURITY AGAINST IMPERSONATION. Classically, the security goal for an identification scheme ID has been impersonation [1,15]. The framework has two stages. First, the adversary, given *ivk* but not *isk*, attacks the honest, *isk*-using prover. Second, using the information it gathers in the first stage, it engages in an interaction with the verifier, attempting to impersonate the real prover by successfully identifying itself. In the second stage, the adversary, in the role of malicious prover, submits a commitment Y of its choice, receives an honest verifier challenge c, submits a response z of its choice, and wins if $\mathsf{ID.Vf}(ivk, Y\|c\|z) = \mathsf{true}$. A hierarchy of possible first-phase attacks is defined in [6]. In the context of conversion to signatures, the relevant one is the weakest, namely passive attacks, where the adversary is just an eavesdropper and gets honestly-generated protocol transcripts. This is the IMP-PA notion. (Active and even concurrent attacks are relevant in other contexts [6].) We note that in the second stage, the adversary is allowed only one interaction with the honest verifier.

SECURITY AGAINST CONSTRAINED IMPERSONATION. We introduce a new framework of goals for identification that we call *constrained impersonation.* There are two dimensions, the commitment dimension X and the challenge dimension Y, for each of which there are two choices, $\mathrm{X} \in \{\mathrm{C}, \mathrm{U}\}$ and $\mathrm{Y} \in \{\mathrm{C}, \mathrm{U}\}$, where C stands for *chosen* and U for *unchosen.* This results in four notions, CIMP-UU, CIMP-UC, CIMP-CU, CIMP-CC. It works as follows. The adversary is allowed a passive attack, namely the ability to obtain transcripts of interactions between the honest prover and the verifier. The choices pertain to the impersonation, when the adversary interacts with the honest verifier in an attempt to make it accept. When $\mathrm{X} = \mathrm{C}$, the adversary can send the verifier a commitment of its choice, as in classical impersonation. But when $\mathrm{X} = \mathrm{U}$, it cannot. Rather, it is required (constrained) to use a commitment that is from one of the transcripts it obtained in the first phase and thus in particular honestly generated. Next comes the challenge. If $\mathrm{Y} = \mathrm{U}$, this is chosen freshly at random, as in the classical setting, but if $\mathrm{Y} = \mathrm{C}$, the adversary actually gets to pick its own challenge. Regardless of choices made in these four configurations, to win the adversary must finally supply a correct response. And, also regardless of these choices, the adversary can mount *multiple* attempts to convince the verifier, contrasting with the strict two-phase adversary in classical definitions of impersonation security.

For choices $\mathrm{xy} \in \{\mathrm{uu}, \mathrm{uc}, \mathrm{cu}, \mathrm{cc}\}$ of parameters, the formalization considers game $\mathbf{G}_{\mathsf{ID}}^{\mathrm{cimp\text{-}xy}}(\mathcal{P})$ of Fig. 4 associated to identification scheme ID and adversary $\mathcal{P}$. We let

$$\mathbf{Adv}_{\mathsf{ID}}^{\mathrm{cimp\text{-}xy}}(\mathcal{P}) = \Pr[\mathbf{G}_{\mathsf{ID}}^{\mathrm{cimp\text{-}xy}}(\mathcal{P})].$$

The transcript oracle TR returns upon each invocation a transcript of an interaction between the honest prover and verifier, allowing $\mathcal{P}$ to mount its passive attack, and is the same for all four games. The impersonation attempts are mounted through calls to the challenge oracle CH, which creates a partial transcript $\mathrm{CT}[j]$ consisting of a commitment and a challenge, where j is a session id, and it returns the challenge. Multiple impersonation attempts are captured by the adversary being allowed to call CH as often as it wants. Eventually the adversary outputs a session id k and a response z for session k, and wins if the corresponding transcript is accepting. In the UU case, $\mathcal{P}$ would give CH only an index l of an existing transcript already returned by TR, and $\mathrm{CT}[j]$ consists of the commitment from the l-th transcript together with a fresh random challenge. In the UC case, CH takes in addition a challenge c chosen by the adversary. The game requires that it be different from c_l (the challenge in the l-th transcript), and CT[j] then consists of the commitment from the l-th transcript together with this challenge. In CU, the adversary can specify the commitment but the challenge is honestly chosen. In CC, it can specify both, as long as the pair did not occur in a transcript. The adversary can call the oracles as often as it wants and in whatever order it wants.

CIMP-CU is a multi-impersonation extension of the classical IMP-PA notion. The other notions are new, and all will be the basis of transforms of identification to signatures admitting *tight* security reductions. CIMP-CU captures a practical identification security goal. As discussed in Sect. 1, the other notions have no such practical interpretation. However we are not aiming to capture some practical form of identification. We wish to use identification only as an analytic tool in the design of signature schemes. For this purpose, as we will see, our framework and notions are indeed useful, allowing us to characterize past transforms and build new ones.

IMPLICATIONS. Figure 1 shows the relations between the four CIMP-XY notions. The implications are captured by Proposition 1. (The separations will be discussed below.) The bounds in these claims imply some conditions or assumptions for the implications which we did not emphasize before because they hold for typical identification schemes. Namely, CIMP-UC $\rightarrow$ CIMP-UU assumes the identification scheme has large challenge length. CIMP-CC $\rightarrow$ CIMP-UC assumes it has a large commitment space. CIMP-CC $\rightarrow$ CIMP-CU again assumes it has a large challenge length. We remark that in all but one case, the adversary constructed in the proof makes only one CH query, regardless of how many the starting adversary made. The proof of the following is in [7].

Proposition 1. *Let* ID *be an identification scheme. Let*

$$\mathsf{ID.CSS} = \min\{\, |\mathsf{ID.CS}(ivk)| \;:\; (ivk, isk, itk) \in [\mathsf{ID.Kg}]\,\}.$$

Then:

1. [CIMP-UC $\rightarrow$ CIMP-UU] *Given* $\mathcal{P}_{\mathrm{uu}}$ *making* q_c *queries to* CH, *we construct* $\mathcal{P}_{\mathrm{uc}}$, *making one* CH *query, such that* $\mathbf{Adv}^{\text{cimp-uu}}_{\mathsf{ID}}(\mathcal{P}_{\mathrm{uu}}) \le \mathbf{Adv}^{\text{cimp-uc}}_{\mathsf{ID}}(\mathcal{P}_{\mathrm{uc}}) + q_c \cdot 2^{-\mathsf{ID.cl}}$.
2. [CIMP-CU $\rightarrow$ CIMP-UU] *Given* $\mathcal{P}_{\mathrm{uu}}$, *we construct* $\mathcal{P}_{\mathrm{cu}}$ *making as many* CH *queries as* $\mathcal{P}_{\mathrm{uu}}$, *such that* $\mathbf{Adv}^{\text{cimp-uu}}_{\mathsf{ID}}(\mathcal{P}_{\mathrm{uu}}) \le \mathbf{Adv}^{\text{cimp-cu}}_{\mathsf{ID}}(\mathcal{P}_{\mathrm{cu}})$.
3. [CIMP-CC $\rightarrow$ CIMP-UC] *Given* $\mathcal{P}_{\mathrm{uc}}$ *making* q_t *queries to* TR, *we construct* $\mathcal{P}_{\mathrm{cc}}$, *making one* CH *query, such that* $\mathbf{Adv}^{\text{cimp-uc}}_{\mathsf{ID}}(\mathcal{P}_{\mathrm{uc}}) \le \mathbf{Adv}^{\text{cimp-cc}}_{\mathsf{ID}}(\mathcal{P}_{\mathrm{cc}}) + q_t(q_t - 1)/2\mathsf{ID.CSS}$.
4. [CIMP-CC $\rightarrow$ CIMP-CU] *Given* $\mathcal{P}_{\mathrm{cu}}$ *making* q_t *queries to* TR *and* q_c *queries to* CH, *we construct* $\mathcal{P}_{\mathrm{cc}}$, *making one* CH *query, such that* $\mathbf{Adv}^{\text{cimp-cu}}_{\mathsf{ID}}(\mathcal{P}_{\mathrm{cu}}) \le \mathbf{Adv}^{\text{cimp-cc}}_{\mathsf{ID}}(\mathcal{P}_{\mathrm{cc}}) + q_t q_c \cdot 2^{-\mathsf{ID.cl}}$.

In all cases, the constructed adversary makes the same number of TR *queries as the starting adversary and has about the same running time.*

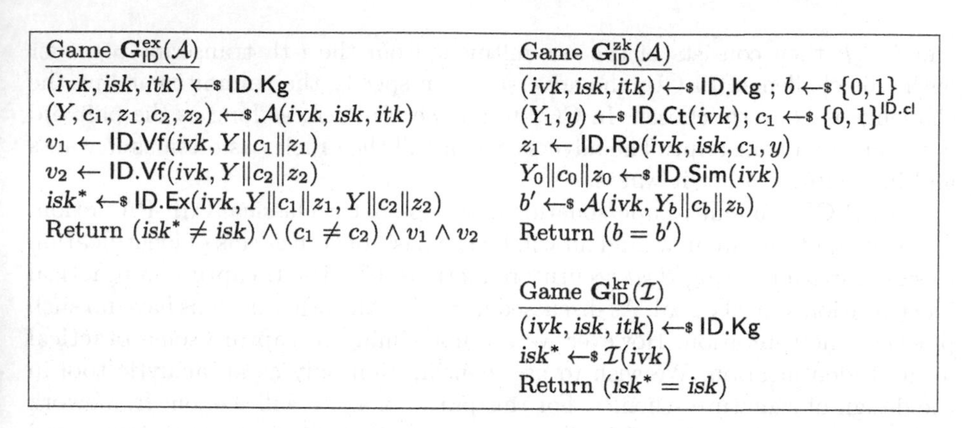

Game $\mathbf{G}^{\mathrm{ex}}_{\mathsf{ID}}(\mathcal{A})$
$(ivk, isk, itk) \leftarrow\$ \mathsf{ID.Kg}$
$(Y, c_1, z_1, c_2, z_2) \leftarrow\$ \mathcal{A}(ivk, isk, itk)$
$v_1 \leftarrow \mathsf{ID.Vf}(ivk, Y\|c_1\|z_1)$
$v_2 \leftarrow \mathsf{ID.Vf}(ivk, Y\|c_2\|z_2)$
$isk^* \leftarrow\$ \mathsf{ID.Ex}(ivk, Y\|c_1\|z_1, Y\|c_2\|z_2)$
Return $(isk^* \neq isk) \wedge (c_1 \neq c_2) \wedge v_1 \wedge v_2$

Game $\mathbf{G}^{\mathrm{zk}}_{\mathsf{ID}}(\mathcal{A})$
$(ivk, isk, itk) \leftarrow\$ \mathsf{ID.Kg}$; $b \leftarrow\$ \{0,1\}$
$(Y_1, y) \leftarrow\$ \mathsf{ID.Ct}(ivk)$; $c_1 \leftarrow\$ \{0,1\}^{\mathsf{ID.cl}}$
$z_1 \leftarrow \mathsf{ID.Rp}(ivk, isk, c_1, y)$
$Y_0\|c_0\|z_0 \leftarrow\$ \mathsf{ID.Sim}(ivk)$
$b' \leftarrow\$ \mathcal{A}(ivk, Y_b\|c_b\|z_b)$
Return $(b = b')$

Game $\mathbf{G}^{\mathrm{kr}}_{\mathsf{ID}}(\mathcal{I})$
$(ivk, isk, itk) \leftarrow\$ \mathsf{ID.Kg}$
$isk^* \leftarrow\$ \mathcal{I}(ivk)$
Return $(isk^* = isk)$

Fig. 5. Games defining the extractability, HVZK and key-recovery security of an identification scheme ID.

SEPARATIONS. We now discuss the separations, beginning with CIMP-CU $\not\Rightarrow$ CIMP-UC. Start with any CIMP-CU scheme. We will modify it so that it remains CIMP-CU-secure but is not CIMP-UC-secure. Distinguish a single challenge $c^* \in \{0,1\}^{\mathsf{ID.cl}}$, e.g., $c^* = 0^{\mathsf{ID.cl}}$. Revise the verifier's algorithm so that it will accept any transcript with challenge c^*. This is still CIMP-CU-secure (as long as ID.cl is large) since, in the CIMP-CU game, challenges are picked uniformly at random for the adversary, so existence of the magic challenge is unlikely to be useful. This is manifestly not CIMP-UC-secure since there the adversary can use any challenge of its choice. CIMP-UU $\not\Rightarrow$ CIMP-UC for the same reason.

We turn to CIMP-UC $\not\Rightarrow$ CIMP-CU. Start with any CIMP-UC scheme. Again we will modify it so that it remains CIMP-UC-secure but is not CIMP-CU-secure. This time, distinguish a single commitment Y^*: one way of doing this is for ID.Kg to sample $Y^* \leftarrow\$ \mathsf{ID.CS}(ivk)$ and include Y^* in the public key ivk; another is to agree for example that $(Y^*, y^*) \leftarrow \mathsf{ID.Ct}(ivk; 0^l)$ where l is the number of random bits required by ID.Ct. Revise the verifier's algorithm so that it will accept any transcript with commitment Y^*. This is still CIMP-UC-secure (assuming $|\mathsf{ID.CS}(ivk)|$ is large) since, in the CIMP-UC game, commitments are generated randomly for the adversary, so existence of a magic commitment is unlikely to be useful. This is manifestly not CIMP-CU-secure since there the adversary can use any commitment of its choice. CIMP-UU $\not\Rightarrow$ CIMP-CU for the same reason.

Finally, CIMP-UC $\not\Rightarrow$ CIMP-CC and CIMP-CU $\not\Rightarrow$ CIMP-CC since otherwise, by transitivity in Fig. 1, we would contradict the separation between CIMP-UC and CIMP-CU.

4 Achieving CIMP-XY Security

Here we show how to obtain identification schemes satisfying our CIMP-XY notions of security. We base CIMP-XY security on the problem of recovering the secret key of the identification scheme given nothing but the public key, which plays the role of the algebraic problem $\mathrm{P}_{\mathrm{alg}}$ in typical identification schemes and corresponds to a standard assumption. (For example for GQ it is one-wayness of RSA.) For CIMP-UC and CIMP-UU, the reductions are tight. For CIMP-CU, the reduction is not tight. CIMP-CC cannot be obtained via these paths, and instead we establish it from signatures. First we need to recall a few standard definitions.

HVZK AND EXTRACTABILITY. We say that an identification scheme ID is *honest verifier zero-knowledge* (HVZK) if there exists an algorithm ID.Sim (called the simulator) that given the verification key, generates transcripts which have the same distribution as honest ones, even given the verification key. Formally, if $\mathcal{A}$ is an adversary, let $\mathbf{Adv}^{\mathrm{zk}}_{\mathsf{ID}}(\mathcal{A}) = 2\Pr[\mathbf{G}^{\mathrm{zk}}_{\mathsf{ID}}(\mathcal{A})] - 1$ where the game is shown in Fig. 5. Then ID is HVZK if $\mathbf{Adv}^{\mathrm{zk}}_{\mathsf{ID}}(\mathcal{A}) = 0$ for all adversaries $\mathcal{A}$ (regardless of the running time of $\mathcal{A}$). We say that an identification scheme ID is *extractable* if there exists an algorithm ID.Ex (called the extractor) which from any two verifying transcripts that have the same commitment but different challenges can recover the secret key. Formally, if $\mathcal{A}$ is an adversary, let $\mathbf{Adv}^{\mathrm{ex}}_{\mathsf{ID}}(\mathcal{A}) = \Pr[\mathbf{G}^{\mathrm{ex}}_{\mathsf{ID}}(\mathcal{A})]$ where the game is shown in Fig. 5. Then ID is extractable if $\mathbf{Adv}^{\mathrm{ex}}_{\mathsf{ID}}(\mathcal{A}) = 0$ for all adversaries $\mathcal{A}$ (regardless of the running time of $\mathcal{A}$). We say that an identification scheme is a Sigma protocol [13] if it is both HVZK and extractable.

SECURITY AGAINST KEY RECOVERY. An identification scheme ID is resilient to *key recovery* if it is hard to recover the secret identification key given nothing but the verification key. This was defined by OO [33]. Formally, if $\mathcal{I}$ is an adversary, let $\mathbf{Adv}^{\mathrm{kr}}_{\mathsf{ID}}(\mathcal{I}) = \Pr[\mathbf{G}^{\mathrm{kr}}_{\mathsf{ID}}(\mathcal{I})]$ where the game is shown in Fig. 5. Security against key recovery is precisely the (standard) assumption $\mathrm{P}_{\mathrm{alg}}$ underlying most identification schemes (e.g., the one-wayness of RSA for the GQ identification scheme, and the factoring assumption for factoring-based schemes).

OBTAINING CIMP-UU AND CIMP-UC. Here we show that for Sigma protocols, CIMP-UU and CIMP-UC security reduce tightly to security under key recovery. The proof of the following is in [7].

Theorem 1. *Let* ID *be an identification scheme that is honest verifier zero-knowledge and extractable. Then for any adversary* $\mathcal{P}$ *against* CIMP-UC *we construct a key recovery adversary* $\mathcal{I}$ *such that*

$$\mathbf{Adv}^{\mathrm{cimp\text{-}uc}}_{\mathsf{ID}}(\mathcal{P}) \leq \mathbf{Adv}^{\mathrm{kr}}_{\mathsf{ID}}(\mathcal{I}). \quad (1)$$

Also for any adversary $\mathcal{P}$ *against* CIMP-UU *that makes* q_c *queries to its* CH *oracle we construct a key recovery adversary* $\mathcal{I}$ *such that*

$$\mathbf{Adv}^{\mathrm{cimp\text{-}uu}}_{\mathsf{ID}}(\mathcal{P}) \leq \mathbf{Adv}^{\mathrm{kr}}_{\mathsf{ID}}(\mathcal{I}) + q_c \cdot 2^{-\mathsf{ID.cl}}. \quad (2)$$

In both cases, the running time of $\mathcal{I}$ *is about that of* $\mathcal{P}$ *plus the time for one execution of* ID.Ex *and the time for a number of executions of* ID.Sim *equal to the number of* TR *queries of* $\mathcal{P}$.

Obtaining CIMP-CU. CIMP-CU security of Sigma protocols can also be established under their key recovery security, but the reduction is not tight.

Theorem 2. *Let* ID *be an identification scheme that is honest verifier zero-knowledge and extractable. For any adversary* $\mathcal{P}$ *against* CIMP-CU *making* q *queries to its* Ch *oracle, we construct a key recovery adversary* $\mathcal{I}$ *such that*

$$\mathbf{Adv}_{\mathsf{ID}}^{\text{cimp-cu}}(\mathcal{P}) \leq q\left(\sqrt{\mathbf{Adv}_{\mathsf{ID}}^{\text{kr}}(\mathcal{I})} + \frac{1}{2^{\mathsf{ID.cl}}}\right). \quad (3)$$

The running time of $\mathcal{I}$ *is about twice that of* $\mathcal{P}$.

To establish Theorem 2, our route will be via standard techniques and known results, and the proof can be found for completeness in [7].

Obtaining CIMP-CC. This is our strongest notion, and is quite different from the rest. Sigma protocols will fail to achieve CIMP-CC because an HVZK identification scheme cannot be CIMP-CC-secure. The attack (adversary) $\mathcal{P}$ showing this is as follows. Assuming ID is HVZK, our adversary $\mathcal{P}$, given the verification key ivk, runs the simulator to get a transcript $Y\|c\|z \leftarrow\$ \mathsf{ID.Sim}(ivk)$. It makes no Tr queries, so the set S in the game is empty. It then makes query $\text{Ch}(Y, c)$ and returns $(1, z)$ to achieve $\mathbf{Adv}_{\mathsf{ID}}^{\text{cimp-cc}}(\mathcal{P}) = 1$.

This doesn't mean CIMP-CC is unachievable. We show in [7] how to achieve it from any UF digital signature scheme.

While this shows CIMP-CC is achievable, and even under standard assumptions, it is not of help for us, since we want to obtain signature schemes from identification schemes and if the latter are themselves built from a signature scheme then nothing has been gained. We consider CIMP-CC nonetheless because our framework naturally gives rise to it and we wish to see the full picture, and also because there may be other ways to achieve CIMP-CC.

5 From UUF to UF

Some of our transforms of identification schemes into signature schemes naturally achieve UUF security rather than UF security. To achieve the latter, one can take our UUF schemes and apply the transforms in this section. The reductions are tight and the costs are low. First we observe that standard derandomization (removing randomness) has the additional benefit (apparently not noted before) of turning UUF into UF. Second, we show that message randomization (adding randomness) is also a natural solution.

5.1 From UUF to UF by Removing Randomness

It is standard to derandomize a signing algorithm by obtaining the coins from a secretly keyed hash function applied to the message. This has been shown to preserve UF security —meaning, if the starting scheme is UF-secure, so is the derandomized scheme— in some cases. One secure instantiation is to use

$\underline{\mathsf{DS}^*.\mathsf{Kg}}$	$\underline{\mathsf{DS}^*.\mathsf{Kg}}$
Return $\mathsf{DS.Kg}$	Return $\mathsf{DS.Kg}$
$\underline{\mathsf{DS}^*.\mathsf{Sig}^{\mathrm{H}}(vk, sk, m)}$	$\underline{\mathsf{DS}^*.\mathsf{Sig}(vk, sk, m)}$
$r \leftarrow \mathrm{H}(sk \| m)$	$s \leftarrow\!\!\$ \{0,1\}^{\mathsf{sl}}$
Return $\mathsf{DS.Sig}(vk, sk, m; r)$	$\sigma \leftarrow\!\!\$ \mathsf{DS.Sig}(vk, sk, m \| s)$
	$\sigma^* \leftarrow \sigma \| s$; Return σ^*
$\underline{\mathsf{DS}^*.\mathsf{Vf}(vk, m, \sigma)}$	
Return $\mathsf{DS.Vf}(vk, m, \sigma)$	$\underline{\mathsf{DS}^*.\mathsf{Vf}(vk, m, \sigma^*)}$
	$\sigma \| s \leftarrow \sigma^*$
	Return $\mathsf{DS.Vf}(vk, m \| s, \sigma)$

Fig. 6. Left: Our construction of deterministic signature scheme $\mathsf{DS}^* = \mathbf{DR}[\mathsf{DS}]$ from a signature scheme DS. By $\mathrm{H}(\cdot)$ we denote $\mathrm{H}(\cdot, \{0,1\}^{\mathsf{DS.rl}})$, which has range $\{0,1\}^{\mathsf{DS.rl}}$. **Right:** Our construction of added-randomness signature scheme $\mathsf{DS}^* = \mathbf{AR}[\mathsf{DS}, \mathsf{sl}]$ from a signature scheme DS and a seed length $\mathsf{sl} \in \mathbb{N}$.

a PRF as the hash function with the PRF key added to the signing secret key [19], however this changes the signing key and can be undesirable in practice. Instead one can hash the signing key with the message using a hash function that one models as a random oracle. This has been proven to work for certain particular choices of the starting signature scheme, namely when this scheme is ECDSA [26]. Such de-randomization is also used in the Ed25519 signature scheme [11]. However, it has not been proven in the general case. This will follow from our results.

The purpose of the method, above, was exactly to derandomize, namely to ensure that the signing process is deterministic, and the starting signature scheme was assumed UF secure. We observe here that the method has an additional benefit which does not seem to have been noted before, namely that it works even if the starting scheme is only UUF secure, meaning it upgrades UUF security to UF security. It is an attractive way to do this because it preserves signature size and verification time, while adding to the signing time only the cost of one hash. We specify a derandomization transform and prove that it turns UUF schemes into UF ones in general, meaning assuming nothing more than UUF security of the starting scheme. In particular, we justify derandomization in a broader context than previous work.

The construction. For a signature scheme DS, let DS.rl denote the length of the randomness (number of coins) used by the signing algorithm DS.Sig. We write $\sigma \leftarrow \mathsf{DS.Sig}(vk, sk, m; r)$ for the execution of DS.Sig on inputs vk, sk, m and coins $r \in \{0,1\}^{\mathsf{DS.rl}}$. Let signature scheme $\mathsf{DS}^* = \mathbf{DR}[\mathsf{DS}]$ be obtained from DS as in Fig. 6. Here, the function $\mathrm{H}(\cdot)$ used to compute r in algorithm $\mathsf{DS}^*.\mathsf{Sig}$ is $\mathrm{H}(\cdot, \{0,1\}^{\mathsf{DS.rl}})$, meaning the range is set to $\{0,1\}^{\mathsf{DS.rl}}$.

While algorithms of the starting scheme DS may invoke the random oracle (and, in the schemes we construct in Sect. 6, they do), it is assumed they do not invoke $\mathrm{H}(\cdot, \{0,1\}^{\mathsf{DS.rl}})$. This can be ensured in a particular case by domain separation. Given this, other calls of the algorithms of the starting scheme to the random oracle can be simulated directly in the proof via the random oracle available to the constructed adversaries. Accordingly in the scheme description of Fig. 6, and proof below, for simplicity, we do not give the algorithms of the starting signature scheme access to the random oracle. That is, think of the starting scheme as being a standard-model one.

UNFORGEABILITY. The following says that the constructed scheme DS^* is UF secure assuming the starting scheme DS was UUF secure, with a tight reduction. The reason a deterministic scheme that is UUF is also UF is clear, namely there is nothing to gain by calling the signing oracle more than once on a particular message, because one just gets back the same thing each time. What the proof needs to ensure is that the method of making the scheme deterministic does not create any weaknesses. The danger is that including the secret key as an input to the hash increases the exposure of the key. The proof says that it might a little, but the advantage does not go up by more than a factor of two. The proof is in [7].

Theorem 3. *Let signature scheme* $\mathsf{DS}^* = \mathbf{DR}[\mathsf{DS}]$ *be obtained from signature scheme* DS *as in Fig. 6. Let* $\overline{\mathcal{A}}$ *be a* UF*-adversary against* DS^* *that makes* q_h *queries to* H *and* q_s *queries to* SIGN. *Then from* $\overline{\mathcal{A}}$ *we can construct* UUF*-adversary* $\mathcal{A}$ *such that*

$$\mathbf{Adv}^{\mathrm{uf}}_{\mathsf{DS}^*}(\overline{\mathcal{A}}) \leq 2 \cdot \mathbf{Adv}^{\mathrm{uuf}}_{\mathsf{DS}}(\mathcal{A}). \quad (4)$$

Adversary $\mathcal{A}$ *makes* q_s *queries to* SIGN. *It has running time about that of* $\overline{\mathcal{A}}$ *plus the time for* q_h *invocations of* DS.Sig *and* DS.Vf.

We remark that adversary $\mathcal{A}_0$ actually violates key recovery security of DS, not just its UUF security.

5.2 From UUF to UF by Adding Randomness

A complementary and natural method for constructing UF signatures from UUF ones is by adding randomness: before being signed, the message is concatenated with a random seed s of some length sl, so even for the same message, the inputs to the UUF signing algorithm are (with high probability) distinct. Compared to derandomization, the drawback of this method is that the signature size increases because the seed must be included in the signature. The potential advantage is that the transform is standard model, not using a random oracle, while preserving the secret key. (Derandomization can be done in the standard model via a PRF, but this requires augmenting the signing key with the PRF key.)

THE CONSTRUCTION. Let signature scheme $\mathsf{DS}^* = \mathbf{AR}[\mathsf{DS}, \mathsf{sl}]$ be obtained from DS as in Fig. 6. As above, DS is for simplicity assumed to be a standard-model scheme, so that its algorithms do not have access to H. The transform itself does not use H.

$\underline{\mathsf{DS.Kg}^{\mathrm{H}}}$	$\underline{\mathsf{DS.Sig}^{\mathrm{H}}(vk, sk, m)}$
$(ivk, isk, itk) \leftarrow\!\!\$\ \mathsf{ID.Kg}$ $vk \leftarrow ivk$; $sk \leftarrow (isk, itk)$ Return (vk, sk) $\underline{\mathsf{DS.Vf}^{\mathrm{H}}(vk, m, \sigma)}$ $ivk \leftarrow vk$; $(c, z) \leftarrow \sigma$ $Y \leftarrow \mathrm{H}(m)$ Return $\mathsf{ID.Vf}(ivk, Y \| c \| z)$	$ivk \leftarrow vk$; $(isk, itk) \leftarrow sk$ $Y \leftarrow \mathrm{H}(m)$ $y \leftarrow\!\!\$\ \mathsf{ID.Ct}^{-1}(ivk, itk, Y)$ $c \leftarrow\!\!\$\ \{0,1\}^{\mathsf{ID.cl}}$ $z \leftarrow \mathsf{ID.Rp}(ivk, isk, c, y)$ $\sigma \leftarrow (c, z)$ Return σ

Fig. 7. The construction of signature scheme $\mathsf{DS} = \mathbf{MdCmt}[\mathsf{ID}]$ from trapdoor identification scheme ID. By $\mathrm{H}(\cdot)$ we denote $\mathrm{H}(\cdot, \mathsf{ID.CS}(ivk))$.

UNFORGEABILITY. The following says that the constructed scheme DS^* is UF secure assuming the starting scheme DS was UUF secure, with a tight reduction. The reason is quite simple, namely that unless seeds collide, the messages being signed are distinct. The proof of the following is in [7].

Theorem 4. *Let signature scheme* $\mathsf{DS}^* = \mathbf{AR}[\mathsf{DS}, \mathsf{sl}]$ *be obtained from signature scheme* DS *and seed length* $\mathsf{sl} \in \mathbb{N}$ *as in Fig. 6. Let* $\overline{\mathcal{A}}$ *be a* UF*-adversary against* DS^* *making* q_s *queries to its* SIGN *oracle. Then from* $\overline{\mathcal{A}}$ *we construct a* UUF *adversary* $\mathcal{A}$ *such that*

$$\mathbf{Adv}^{\mathrm{uf}}_{\mathsf{DS}^*}(\overline{\mathcal{A}}) \leq \mathbf{Adv}^{\mathrm{uuf}}_{\mathsf{DS}}(\mathcal{A}) + \frac{q_s^2}{2^{\mathsf{sl}+1}} .$$

Adversary $\mathcal{A}$ *makes* q_s *queries to its* SIGN *oracle and has about the same running time as* $\overline{\mathcal{A}}$.

6 Signatures from Identification

We specify our three new transforms of identification schemes to signature schemes, namely the ones of rows 2, 3, 4 of the table of Fig. 1. In each case, we give a security proof based on the assumption P_{id} listed in the 1st column of the corresponding row of the table, so that we give transforms from CIMP-UC, CIMP-UU and CIMP-CC. It turns out that these transforms naturally achieve UUF rather than UF, and this is what we prove, with tight reductions of course. The transformation UF→UUF can be done at the level of signatures, not referring to identification, in generic and simple ways, and also with tight reductions, as detailed in Sect. 5. We thus get UF-secure signatures with tight reductions to each of CIMP-UC, CIMP-UU and CIMP-CC. In this section we further study the **FS** transform from [16] and our transform **Swap** which is inspired by the work of MR [30].

6.1 From CIMP-UC Identification to UUF Signatures: MdCmt

MdCmt transforms a CIMP-UC trapdoor identification scheme to a UUF signature scheme using message-dependent commitments.

THE CONSTRUCTION. Let ID be a trapdoor identification scheme and $\mathsf{ID.cl}$ its challenge length. Our **MdCmt** (message-dependent commitment) transform associates to ID the signature scheme $\mathsf{DS} = \mathbf{MdCmt}[\mathsf{ID}]$. The algorithms of DS are defined in Fig. 7. By $\mathrm{H}(\cdot)$ we denote $\mathrm{H}(\cdot, \mathsf{ID.CS}(ivk))$, meaning the range is set to commitment space $\mathsf{ID.CS}(ivk)$. Signatures are effectively identification transcripts, but the commitments are chosen in a particular way. Recall that with trapdoor ID schemes it is the same whether one executes $(Y, y) \leftarrow\$ \mathsf{ID.Ct}$ directly, or samples $Y \leftarrow\$ \mathsf{ID.CS}$ followed by computing $y \leftarrow\$ \mathsf{ID.Ct}^{-1}(Y)$. Our construction exploits this: To each message m it assigns an individual commitment $Y \leftarrow \mathrm{H}(m)$. The signing algorithm, using the trapdoor, completes this commitment to a transcript (Y, c, z) and outputs the pair c, z as the signature. Verification then consists of recomputing Y from m and invoking the verification algorithm of the ID scheme.

UNFORGEABILITY. The following theorem establishes that the (unique) unforgeability of a signature scheme constructed with **MdCmt** tightly reduces to the CIMP-UC security of the underlying ID scheme, in the random oracle model. The proof of the following is in [7].

Theorem 5. *Let signature scheme* $\mathsf{DS} = \mathbf{MdCmt}[\mathsf{ID}]$ *be obtained from trapdoor identification scheme* ID *as in Fig. 7. Let* $\mathcal{A}$ *be a* UUF*-adversary against* DS*. Suppose the number of queries that* $\mathcal{A}$ *makes to its* H *and* SIGN *oracles are* q_h *and* q_s*, respectively. Then from* $\mathcal{A}$ *we construct a* CIMP-UC *adversary* $\mathcal{P}$ *such that*

$$\mathbf{Adv}^{\mathrm{uuf}}_{\mathsf{DS}}(\mathcal{A}) \leq \frac{\mathbf{Adv}^{\mathrm{cimp\text{-}uc}}_{\mathsf{ID}}(\mathcal{P})}{1 - 2^{-\mathsf{ID.cl}}}. \tag{5}$$

Adversary $\mathcal{P}$ *makes* $q_h + q_s + 1$ *queries to* TR *and one query to* CH*. Its running time is about that of* $\mathcal{A}$*.*

The bound of Eq. (5) may be a bit hard to estimate. The following simpler bound is also true and may be easier to use:

$$\mathbf{Adv}^{\mathrm{uuf}}_{\mathsf{DS}}(\mathcal{A}) \leq \mathbf{Adv}^{\mathrm{cimp\text{-}uc}}_{\mathsf{ID}}(\mathcal{P}) + \frac{1}{2^{\mathsf{ID.cl}}}. \tag{6}$$

The justification for Eq. (6) is in [7].

6.2 From CIMP-UU Identification to UUF Signatures: MdCmtCh

MdCmtCh transforms a CIMP-UU trapdoor identification scheme to a UUF signature scheme using message-dependent commitments and challenges.

THE CONSTRUCTION. Our **MdCmtCh** (message-dependent commitment and challenge) transform associates to trapdoor identification scheme ID the signature scheme $\mathsf{DS} = \mathbf{MdCmtCh}[\mathsf{ID}]$ whose algorithms are defined in Fig. 8.

$\underline{\mathsf{DS.Kg}^{\mathrm{H}}}$	$\underline{\mathsf{DS.Sig}^{\mathrm{H}}(vk, sk, m)}$
$(ivk, isk, itk) \leftarrow\$ \mathsf{ID.Kg}$	$s \leftarrow\$ \{0,1\}$
$vk \leftarrow ivk$; $sk \leftarrow (isk, itk)$	$ivk \leftarrow vk$; $(isk, itk) \leftarrow sk$
Return (vk, sk)	$Y \leftarrow \mathrm{H}_1(m)$
$\underline{\mathsf{DS.Vf}^{\mathrm{H}}(vk, m, \sigma)}$	$y \leftarrow\$ \mathsf{ID.Ct}^{-1}(ivk, itk, Y)$
$ivk \leftarrow vk$; $(z, s) \leftarrow \sigma$	$c \leftarrow \mathrm{H}_2(m\|s)$
$Y \leftarrow \mathrm{H}_1(m)$	$z \leftarrow \mathsf{ID.Rp}(ivk, isk, c, y)$
$c \leftarrow \mathrm{H}_2(m\|s)$	$\sigma \leftarrow (z, s)$; Return σ
Return $\mathsf{ID.Vf}(ivk, Y\|c\|z)$	

Fig. 8. Our construction of signature scheme $\mathsf{DS} = \mathbf{MdCmtCh}[\mathsf{ID}]$ from a trapdoor identification scheme ID. By $\mathrm{H}_1(\cdot)$ we denote random oracle $\mathrm{H}(\cdot, \mathsf{ID.CS}(ivk))$ with range $\mathsf{ID.CS}(ivk)$ and by $\mathrm{H}_2(\cdot)$ we denote random oracle $\mathrm{H}(\cdot, \{0,1\}^{\mathsf{ID.cl}})$ with range $\{0,1\}^{\mathsf{ID.cl}}$.

We specify the commitment Y as a hash of the message and use the trapdoor property to allow the signer to obtain $y \leftarrow\$ \mathsf{ID.Ct}^{-1}(ivk, itk, Y)$. We then specify the challenge as a randomized hash of the message. (Unlike in the **FS** transform, the commitment is not hashed along with the message.) The randomization is captured by a one-bit seed s. The construction, and proof below, both use the technique of KW [24].

By $\mathrm{H}_1(\cdot)$ we denote random oracle $\mathrm{H}(\cdot, \mathsf{ID.CS}(ivk))$ with range $\mathsf{ID.CS}(ivk)$ and by $\mathrm{H}_2(\cdot)$ we denote random oracle $\mathrm{H}(\cdot, \{0,1\}^{\mathsf{ID.cl}})$ with range $\{0,1\}^{\mathsf{ID.cl}}$. We assume $\mathsf{ID.CS}(ivk) \neq \{0,1\}^{\mathsf{ID.cl}}$ so that these random oracles are independent. In case $\mathsf{ID.CS}(ivk) = \{0,1\}^{\mathsf{ID.cl}}$, the scheme should be modified to use domain separation, for example prefix a 1 to any query to H_1 and a 0 to any query to H_2.

Notice that the signature consists of a response plus a bit. It is thus shorter than for **MdCmt** (where it is a response plus a challenge) or for **FS** (where it is a response plus a commitment or, in the more compact form, a response plus a challenge). These shorter signatures are a nice feature of **MdCmtCh**.

Unforgeability of our construction. The following shows that unique unforgeability of our signature tightly reduces to the CIMP-UU security of the underlying ID scheme. Standard unforgeability follows immediately (and tightly) by applying one of the UUF-to-UF transforms in Sect. 5.

Theorem 6. *Let signature scheme* $\mathsf{DS} = \mathbf{MdCmtCh}[\mathsf{ID}]$ *be obtained from trapdoor identification scheme* ID *as in Fig. 8. Let* $\mathcal{A}$ *be a* UUF *adversary against* DS. *Suppose the number of queries that* $\mathcal{A}$ *makes to its* H_1 *and* H_2 *oracles is* q_h, *and the number to its* Sign *oracle is* q_s. *Then from* $\mathcal{A}$ *we construct* CIMP-UU *adversary* $\mathcal{P}$ *such that*

$$\mathbf{Adv}^{\mathrm{uuf}}_{\mathsf{DS}}(\mathcal{A}) \leq 2 \cdot \mathbf{Adv}^{\mathrm{cimp\text{-}uu}}_{\mathsf{ID}}(\mathcal{P}) . \quad (7)$$

Adversary $\mathcal{P}$ *makes* $q_h + q_s + 1$ *queries to* Tr *and* $q_h + q_s$ *queries to* Ch. *It has running time about that of* $\mathcal{A}$.

```
Adversary P^{TR,CH}(ivk)                    SIGN(m)
HT1 ← ∅ ; HT2 ← ∅                           If m ∈ M: Return ⊥
M ← ∅ ; i ← 0                               M ← M ∪ {m}
(m, σ) ←$ A^{SIGN,H1,H2}(ivk)               Y ← H1(m) ; l ← Ind[m]
(z, s) ← σ                                  σ ← (z_l, s_l) ; Return σ
Y ← H1(m)
j ← Ind[m]                                  H1(m)
Return (j, z)                               If HT1[m]: Return HT1[m]
                                            i ← i + 1 ; Y_i||c_i||z_i ←$ TR()
                                            HT1[m] ← Y_i ; Ind[m] ← i
                                            s_i ←$ {0,1} ; HT2[m||s_i] ← c_i
                                            HT2[m||s̄_i] ←$ CH(i)
                                            Return HT1[m]

                                            H2(x)
                                            If HT2[x]: Return HT2[x]
                                            m||s ← x ; Y ← H1(m)
                                            Return HT2[x]
```

Fig. 9. Adversary for proof of Theorem 6.

Proof (Theorem 6). Adversary $\mathcal{P}$ is shown in Fig. 9. It executes $\mathcal{A}$, responding to H_1, H_2 and SIGN queries of the latter via the shown procedures, which are subroutines in the code of $\mathcal{P}$. We assume the message m in the forgery (m, σ) returned by $\mathcal{A}$ was not queried to SIGN and is not in the set M, since otherwise $\mathcal{A}$ would automatically lose. The "$Y \leftarrow \mathrm{H}_1(m)$" instructions in the code of SIGN, the code of H_2 and following the execution of $\mathcal{A}$ ensure that $\mathrm{H}_1(m)$ is queried at this point. Each time a new $\mathrm{H}_1(m)$ query is made, a transcript is generated by $\mathcal{P}$ using its TR oracle. The commitment in this transcript is the reply to the $\mathrm{H}_1(m)$ query. Additionally, however, steps are taken to ensure that, if, later, a SIGN(m) query is made, then a signature to return is available. This is done by picking a random one-bit seed s_i and assigning $\mathrm{H}_2(m\|s_i)$ the value c_i. At the time of a signing query, one can use s_i as the seed and use the response of the corresponding transcript to create the signature. To be able to win via the forgery, $\mathrm{H}_2(m\|\overline{s}_i)$ is assigned a challenge via CH, where $\overline{s}_i$ denotes the complement of the bit s_i. Now, when the forgery $(m, (z, s))$ is obtained from $\mathcal{A}$, the associated index j is computed, and then z is returned as a response for that session. Adversary $\mathcal{P}$ will be successful as long as $\mathcal{A}$ is successful and $s = \overline{s}_j$. The events being independent we have

$$\mathbf{Adv}^{\text{cimp-uu}}_{\mathsf{ID}}(\mathcal{P}) \geq \frac{1}{2} \cdot \mathbf{Adv}^{\text{uuf}}_{\mathsf{DS}}(\mathcal{A}).$$

Transposing terms yields Eq. (7). □

6.3 From CIMP-CC Identification to UF Signatures: MdCh

The **MdCmt** and **MdCmtCh** transforms described above rely on the trapdoor property of the underlying identification scheme and achieve UUF rather than UF. The **MdCh** transform we describe here does not have these limitations. (It does not require the identification scheme to be trapdoor, and it directly achieves UF.) However, among the security notions for ID schemes that we defined, **MdCh** assumes the strongest one: CIMP-CC.

THE CONSTRUCTION. Our **MdCh** (message-dependent challenge) transform associates to identification scheme ID and a seed length $\mathsf{sl} \in \mathbb{N}$ the signature scheme $\mathsf{DS} = \mathbf{MdCh}[\mathsf{ID}, \mathsf{sl}]$ whose algorithms are defined in Fig. 10. Signing picks the commitment directly rather than (as in our prior transforms) specifying it as the hash of the message. The challenge is derived as a randomized hash of the message, the randomization being captured by a seed s of length sl. By $\mathrm{H}(\cdot)$ we denote random oracle $\mathrm{H}(\cdot, \{0,1\}^{\mathsf{ID.cl}})$ with range $\{0,1\}^{\mathsf{ID.cl}}$.

UNFORGEABILITY. As we prove below (with tight reduction), the **MdCh** construction yields a UF secure signature scheme if the underlying identification scheme offers CIMP-CC security.

Theorem 7. *Let signature scheme* $\mathsf{DS} = \mathbf{MdCh}[\mathsf{ID}, \mathsf{sl}]$ *be obtained from identification scheme* ID *and seed length* $\mathsf{sl} \in \mathbb{N}$ *as in Fig. 10. Let* $\mathcal{A}$ *be a* UF *adversary against* DS *making* q_h *queries to its* H *oracle and* q_s *queries to its* SIGN *oracle. Then from* $\mathcal{A}$ *we construct a* CIMP-CC *adversary* $\mathcal{P}$ *such that*

$$\mathbf{Adv}^{\mathrm{uf}}_{\mathsf{DS}}(\mathcal{A}) \leq \mathbf{Adv}^{\mathrm{cimp\text{-}cc}}_{\mathsf{ID}}(\mathcal{P}) + \frac{q_h q_s}{2^{\mathsf{ID.cl}}} + \frac{(q_h + q_s)q_s}{2^{\mathsf{sl}}}. \tag{8}$$

Adversary $\mathcal{P}$ *makes* q_s *queries to* TR *and one query to* CH *and has running time about that of* $\mathcal{A}$.

Proof (Theorem 7). Game G_0 of Fig. 11 includes the boxed code, while game G_1 does not. Game G_0 is precisely the UF game of Fig. 2 with the algorithms of DS

$\underline{\mathsf{DS.Kg}^{\mathrm{H}}}$	$\underline{\mathsf{DS.Sig}^{\mathrm{H}}(vk, sk, m)}$
$(ivk, isk, itk) \leftarrow\$ \mathsf{ID.Kg}$	$ivk \leftarrow vk$; $isk \leftarrow sk$
$vk \leftarrow ivk$; $sk \leftarrow isk$	$s \leftarrow\$ \{0,1\}^{\mathsf{sl}}$
Return (vk, sk)	$(Y, y) \leftarrow\$ \mathsf{ID.Ct}(ivk)$
$\underline{\mathsf{DS.Vf}^{\mathrm{H}}(vk, m, \sigma)}$	$c \leftarrow \mathrm{H}(m \| s)$
$ivk \leftarrow vk$; $(Y, s, z) \leftarrow \sigma$	$z \leftarrow \mathsf{ID.Rp}(ivk, isk, c, y)$
$c \leftarrow \mathrm{H}(m \| s)$	$\sigma \leftarrow (Y, s, z)$
Return $\mathsf{ID.Vf}(ivk, Y \| c \| z)$	Return σ

Fig. 10. The construction of signature scheme $\mathsf{DS} = \mathbf{MdCh}[\mathsf{ID}, \mathsf{sl}]$ from identification scheme ID and seed length sl. By $\mathrm{H}(\cdot)$ we denote random oracle $\mathrm{H}(\cdot, \{0,1\}^{\mathsf{ID.cl}})$ with range $\{0,1\}^{\mathsf{ID.cl}}$.

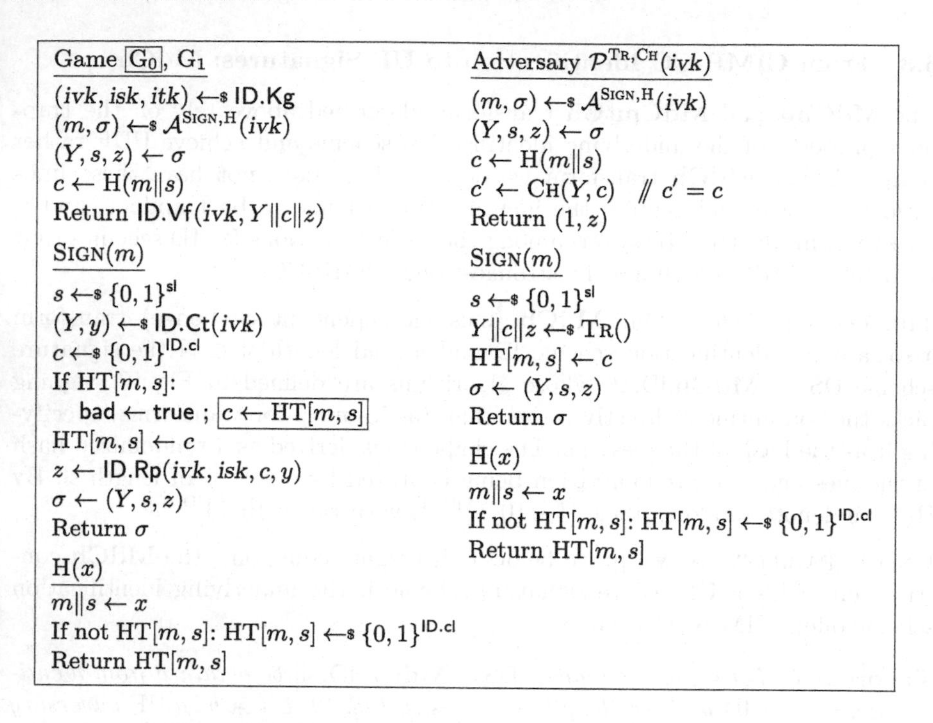

Fig. 11. Games and adversary for proof of Theorem 7.

plugged in. We assume the message m in the forgery (m, σ) returned by $\mathcal{A}$ was not queried to SIGN. Games $\mathrm{G}_0, \mathrm{G}_1$ are identical until bad. By the Fundamental Lemma of Game Playing [10] we have

$$\mathbf{Adv}^{\mathrm{uf}}_{\mathsf{DS}}(\mathcal{A}) = \Pr[\mathrm{G}_0] = \Pr[\mathrm{G}_1] + (\Pr[\mathrm{G}_0] - \Pr[\mathrm{G}_1]) \leq \Pr[\mathrm{G}_1] + \Pr[\mathrm{G}_1 \text{ sets } \mathsf{bad}]$$
$$\leq \Pr[\mathrm{G}_1] + \frac{(q_h + q_s)q_s}{2^{\mathsf{sl}}} .$$

Adversary $\mathcal{P}$ of Fig. 11 executes $\mathcal{A}$, responding to SIGN and H queries of the latter via the shown procedures, which are subroutines in the code of $\mathcal{P}$.

Adversary $\mathcal{P}$ simulates for $\mathcal{A}$ the environment of game G_1. In the execution of game $\mathbf{G}^{\text{cimp-cc}}_{\mathsf{ID}}(\mathcal{P})$ of Fig. 4, let B denote the event that $(Y, c) \in \mathrm{S}$, where Y, c is the argument to the single CH query made by our $\mathcal{P}$. Then

$$\Pr[\mathrm{G}_1] \leq \mathbf{Adv}^{\text{cimp-cc}}_{\mathsf{ID}}(\mathcal{P}) + \Pr[\mathrm{B}] .$$

To complete the proof, it suffices to show that

$$\Pr[\mathrm{B}] \leq \frac{q_h q_s}{2^{\mathsf{ID.cl}}} .$$

We bound Pr[B] by the probability that c is a challenge in one of the transcripts. The message m in the forgery is assumed not one of those signed, so $\mathrm{HT}[m, s]$ was not set by SIGN and is thus independent of the transcript challenges. There are at most q_s transcript challenges and at most q_h queries to H, so $\Pr[\mathrm{B}] \leq q_h q_s / 2^{\mathsf{ID.cl}}$. □

$\mathsf{DS.Kg}^{\mathrm{H}}$	$\mathsf{DS.Sig}^{\mathrm{H}}(vk, sk, m)$
$(ivk, isk, itk) \leftarrow\!\!\$\ \mathsf{ID.Kg}$ $vk \leftarrow ivk\ ;\ sk \leftarrow isk$ Return (vk, sk) $\mathsf{DS.Vf}^{\mathrm{H}}(vk, m, \sigma)$ $ivk \leftarrow vk\ ;\ (Y, z) \leftarrow \sigma$ $c \leftarrow \mathrm{H}(Y \| m)$ Return $\mathsf{ID.Vf}(ivk, Y\|c\|z)$	$ivk \leftarrow vk\ ;\ isk \leftarrow sk$ $(Y, y) \leftarrow\!\!\$\ \mathsf{ID.Ct}(ivk)$ $c \leftarrow \mathrm{H}(Y\|m)$ $z \leftarrow \mathsf{ID.Rp}(ivk, isk, c, y)$ $\sigma \leftarrow (Y, z)$ Return σ

Fig. 12. The construction of signature scheme $\mathsf{DS} = \mathbf{FS}[\mathsf{ID}]$ from identification scheme ID. By $\mathrm{H}(\cdot)$ we denote random oracle $\mathrm{H}(\cdot, \{0,1\}^{\mathsf{ID.cl}})$ with range $\{0,1\}^{\mathsf{ID.cl}}$.

6.4 From CIMP-CU Identification to UF Signatures: FS

The first proofs of UF security of **FS**-based signatures used a Forking Lemma and were quite complex [35]. More modular approaches were given in OO [33] and AABN [1]. AABN reduce UF security of the signature scheme to IMP-PA security of the identification scheme. (The latter is established separately via the Reset Lemma of [6].) The reduction of AABN is not tight.

Our framework allows a tight reduction of the UF security of **FS**-based signatures to the CIMP-CU security of the underlying identification scheme. The reason for this is simple, namely that CIMP-CU is the multi-impersonation version of IMP-PA. The proof is implicit in AABN [1]. We give a proof however for completeness and to illustrate how much simpler this proof is to prior ones.

We note that this tighter reduction does not change overall tightness. That is, in AABN, $\mathrm{P_{sig}}{\rightarrow}\mathrm{P_{id}}$ was not tight, while for us, it is, but the tightness of the overall $\mathrm{P_{sig}}{\rightarrow}\mathrm{P_{alg}}$ reduction remains the same in both cases.

The construction. The **FS** transform [16] associates to identification scheme ID the signature scheme $\mathsf{DS} = \mathbf{FS}[\mathsf{ID}]$ whose algorithms are defined in Fig. 12. Signing picks the commitment directly. The challenge is derived as a hash of the commitment and message. By $\mathrm{H}(\cdot)$ we denote random oracle $\mathrm{H}(\cdot, \{0,1\}^{\mathsf{ID.cl}})$ with range $\{0,1\}^{\mathsf{ID.cl}}$.

Unforgeability. The following theorem says that the **FS** construction yields a UF secure signature scheme if the underlying ID scheme offers CIMP-CU security and the commitment (as generated by the prover) is uniformly distributed over a large space. The latter condition is true for typical identification schemes. The intuition of the proof is that signing queries are answered via transcripts and hash queries are mapped to challenge queries, this failing only if commitments collide. The proof of the following is in [7].

Theorem 8. *Let signature scheme* $\mathsf{DS} = \mathbf{FS}[\mathsf{ID}]$ *be obtained from identification scheme* ID *as in Fig. 12. Let* $\mathsf{ID.CSS} = \min\{\, |\mathsf{ID.CS}(ivk)| \ :\ (ivk, isk, itk) \in [\mathsf{ID.Kg}]\,\}$. *Let* $\mathcal{A}$ *be a* UF *adversary against* DS *making* q_h *queries to its* H *oracle and* q_s *queries to its* Sign *oracle. Then from* $\mathcal{A}$ *we construct a* CIMP-CU *adversary* $\mathcal{P}$ *such that*

$$\mathbf{Adv}^{\mathrm{uf}}_{\mathsf{DS}}(\mathcal{A}) \leq \mathbf{Adv}^{\mathrm{cimp\text{-}cu}}_{\mathsf{ID}}(\mathcal{P}) + \frac{q_s(2q_h + q_s - 1)}{2 \cdot \mathsf{ID.CSS}}. \quad (9)$$

Adversary $\mathcal{P}$ makes q_s queries to TR *and $q_h + 1$ queries to* CH *and has running time about that of $\mathcal{A}$.*

6.5 From CIMP-UC Identification to UF Signatures: Swap

Micali and Reyzin [30] use the term "swap" for a specific construction of a signature scheme that they give with a tight reduction to the hardness of factoring. Folklore, and hints in the literature [2], indicate that researchers understand the method is more general. But exactly how general was not understood or determined before, perhaps for lack of definitions. Our definition of trapdoor identification and the CIMP-XY framework allows us to fill this gap and give a characterization of the swap method and also better understand it.

In this section we define a transform of trapdoor identification schemes to signature schemes that we call **Swap**. We show that it yields UF secure signatures if the identification scheme is CIMP-UC secure.

THE CONSTRUCTION. The **Swap** transform associates to trapdoor identification scheme ID the signature scheme DS = **Swap**[ID] whose algorithms are defined in Fig. 13.

Recall that in Sect. 6.1 we gave the **MdCmt** transform that constructs UUF-secure signatures from CIMP-UC-secure identification. Further, in Sect. 5 we proposed two generic techniques that convert UUF signatures to signatures with full UF security. One of the latter, **AR**, achieves its goal by adding randomness to signed messages as follows: for signing m, it picks a fresh random seed s and signs $m\|s$ instead. The seed is included in the signature. Overall, the combination of **MdCmt** with **AR** yields tightly secure signatures of the form $(c, \mathsf{ID.Rp}(c, \mathsf{ID.Ct}^{-1}(\mathrm{H}(m\|s))), s)$. **Swap** effectively says that it is safe to choose c and s to be identical. Thus it can be viewed as an optimization of **MdCmt**+**AR**, giving up on modularity to achieve more compact UF secure signatures.

We note however that our **MdCmtCh** transform coupled with our UUF-to-UF transform **DR** yields UF signatures that seem superior in every way: they are shorter (response plus a bit as opposed to response plus a challenge), the (tight) reduction is to the weaker CIMP-UU notion, and the efficiency is the same. Thus we would view **Swap** at this point as of mostly historical interest.

UNFORGEABILITY. The following theorem says that the **Swap** construction yields a UF secure signature scheme if the underlying ID scheme offers CIMP-UC security and has sufficiently large challenge length. The proof of the following is in [7].

Theorem 9. *Let signature scheme* DS = **Swap**[ID] *be obtained from trapdoor identification scheme* ID *as in Fig. 13. Let $\mathcal{A}$ be a* UF *adversary against* DS. *Suppose the number of queries that $\mathcal{A}$ makes to its* H *oracle is q_h and the number of queries it makes to* SIGN *is q_s. Then from $\mathcal{A}$ we construct a* CIMP-UC *adversary $\mathcal{P}$ such that*

$\mathsf{DS.Kg}^{\mathrm{H}}$	$\mathsf{DS.Sig}^{\mathrm{H}}(vk, sk, m)$
$(ivk, isk, itk) \leftarrow\$ \mathsf{ID.Kg}$ $vk \leftarrow ivk$; $sk \leftarrow (isk, itk)$ Return (vk, sk)	$ivk \leftarrow vk$; $(isk, itk) \leftarrow sk$ $c \leftarrow\$ \{0,1\}^{\mathsf{ID.cl}}$ $Y \leftarrow \mathrm{H}(m\|c)$ $y \leftarrow\$ \mathsf{ID.Ct}^{-1}(ivk, itk, Y)$ $z \leftarrow \mathsf{ID.Rp}(ivk, isk, c, y)$ $\sigma \leftarrow (c, z)$; Return σ
$\mathsf{DS.Vf}^{\mathrm{H}}(vk, m, \sigma)$	
$ivk \leftarrow vk$; $(c, z) \leftarrow \sigma$ $Y \leftarrow \mathrm{H}(m\|c)$ Return $\mathsf{ID.Vf}(ivk, Y\|c\|z)$	

Fig. 13. The construction of signature scheme $\mathsf{DS} = \mathbf{Swap}[\mathsf{ID}]$ from a trapdoor identification scheme ID. By $\mathrm{H}(\cdot)$ we denote $\mathrm{H}(\cdot, \mathsf{ID.CS}(ivk))$.

Signature scheme DS	$\mathrm{P_{id}}$	Bound on $\mathbf{Adv}^{\mathrm{uf}}_{\mathsf{DS}}(\mathcal{A})$	Sig. size	Equations
DR[**MdCmt**[ID]]	CIMP-UC	$2\epsilon/(1-2^{-l})$	$k+l$	(4),(5),(1)
Swap[ID]	CIMP-UC	$\epsilon + (q_h q_s + q_s^2 + 1)\cdot 2^{-l}$	$k+l$	(10),(1)
DR[**MdCmtCh**[ID]]	CIMP-UU	$4\epsilon + 4(q_h + q_s)\cdot 2^{-l}$	$k+1$	(4),(7),(2)
FS[ID]	CIMP-CU	$(q_h+1)(\sqrt{\epsilon}+2^{-l}) + (2q_h q_s + q_s^2)/2C$	$k+c$	(9),(3)

Fig. 14. UF signature schemes obtained from identification scheme ID. We show bounds on the uf advantage of an adversary $\mathcal{A}$ making q_h queries to H and q_s queries to SIGN. Here $\epsilon = \mathbf{Adv}^{\mathrm{kr}}_{\mathsf{ID}}(\mathcal{I})$ is the kr advantage of an adversary $\mathcal{I}$ of roughly the same running time as $\mathcal{A}$. By l, k, c we denote the lengths of the challenge, response and commitment, respectively. By C we denote the size of the commitment space. By $\mathrm{P_{id}}$ we denote the notion of identification security used in the $\mathrm{P_{sig}} \rightarrow \mathrm{P_{id}}$ reduction.

$$\mathbf{Adv}^{\mathrm{uf}}_{\mathsf{DS}}(\mathcal{A}) < \mathbf{Adv}^{\mathrm{cimp\text{-}uc}}_{\mathsf{ID}}(\mathcal{P}) + \frac{(q_h + q_s)q_s + 1}{2^{\mathsf{ID.cl}}}. \tag{10}$$

Adversary $\mathcal{P}$ makes $q_h + q_s + 1$ queries to TR *and one query to* CH. *Its running time is about that of $\mathcal{A}$.*

6.6 From Identification to UF Signatures: Summary

Figure 14 puts things together. We consider obtaining a UF (not just UUF) signature scheme DS from a given identification scheme ID via the various transforms in this paper. In the first three rows, the identification scheme is assumed to be trapdoor. Whenever a transform achieves (only) UUF, we apply **DR** on top to get UF. We give bounds on the uf advantage $\mathbf{Adv}^{\mathrm{uf}}_{\mathsf{DS}}(\mathcal{A})$ of an adversary $\mathcal{A}$ making q_h queries to H and q_s queries to SIGN. By $l = \mathsf{ID.cl}$ we denote the challenge length of ID, and by $C = \mathsf{ID.CSS}$ the size of the commitment space. We show the full $\mathrm{P_{sig}} \rightarrow \mathrm{P_{alg}}$ reduction, so that the bounds are in terms of the kr advantage $\epsilon = \mathbf{Adv}^{\mathrm{kr}}_{\mathsf{ID}}(\mathcal{I})$ of a kr-adversary $\mathcal{I}$ having about the same running time as $\mathcal{A}$. The bounds are obtained by combining the various relevant theorems, referring to the indicated equations. We show the notion $\mathrm{P_{id}}$ of identification security used as an intermediate point, namely $\mathrm{P_{sig}} \rightarrow \mathrm{P_{id}} \rightarrow \mathrm{P_{alg}}$. Signature size is shown as a function of challenge, response and commitment lengths. In summary, the bounds in the first three rows are tight, but the transform of the third row has the added advantage of shorter signatures and a linear (as opposed to quadratic) additive term in the bound. We do not show the **MdCh** transform from CIMP-CC because the latter is not achieved by Sigma protocols. We note that the bound for **FS** is the same as in [1]. (Our $\mathrm{P_{sig}} \rightarrow \mathrm{P_{id}}$ reduction, unlike theirs, is tight, but there is no change in the tightness of the full $\mathrm{P_{sig}} \rightarrow \mathrm{P_{alg}}$ reduction.)

Acknowledgments. Bellare was supported in part by NSF grant CNS-1526801, a gift from Microsoft corporation and ERC Project ERCC (FP7/615074). Poettering was supported by ERC Project ERCC (FP7/615074). Stebila was supported in part by Australian Research Council (ARC) Discovery Project grant DP130104304 and Natural Sciences and Engineering Research Council (NSERC) of Canada Discovery grant RGPIN-2016-05146.

References

1. Abdalla, M., An, J.H., Bellare, M., Namprempre, C.: From identification to signatures via the Fiat-Shamir transform: minimizing assumptions for security and forward-security. In: Knudsen, L.R. (ed.) EUROCRYPT 2002. LNCS, vol. 2332, pp. 418–433. Springer, Heidelberg (2002). doi:10.1007/3-540-46035-7_28
2. Abdalla, M., Ben Hamouda, F., Pointcheval, D.: Tighter reductions for forward-secure signature schemes. In: Kurosawa, K., Hanaoka, G. (eds.) PKC 2013. LNCS, vol. 7778, pp. 292–311. Springer, Heidelberg (2013). doi:10.1007/978-3-642-36362-7_19
3. Abdalla, M., Fouque, P.-A., Lyubashevsky, V., Tibouchi, M.: Tightly-secure signatures from lossy identification schemes. In: Pointcheval, D., Johansson, T. (eds.) EUROCRYPT 2012. LNCS, vol. 7237, pp. 572–590. Springer, Heidelberg (2012). doi:10.1007/978-3-642-29011-4_34
4. Bagherzandi, A., Cheon, J.H., Jarecki, S.: Multisignatures secure under the discrete logarithm assumption and a generalized forking lemma. In: Ning, P., Syverson, P.F., Jha, S. (eds.) ACM CCS 2008, pp. 449–458. ACM Press, October 2008
5. Bellare, M., Neven, G.: Multi-signatures in the plain public-key model and a general forking lemma. In: Juels, R., Wright, N., Vimercati, S. (eds.) ACM CCS 2006, pp. 390–399. ACM Press, October/November 2006
6. Bellare, M., Palacio, A.: GQ and Schnorr identification schemes: proofs of security against impersonation under active and concurrent attacks. In: Yung, M. (ed.) CRYPTO 2002. LNCS, vol. 2442, pp. 162–177. Springer, Heidelberg (2002). doi:10.1007/3-540-45708-9_11
7. Bellare, M., Poettering, B., Stebila, D.: From identification to signatures, tightly: a framework and generic transforms. Cryptology ePrint Archive, Report 2015/1157 (2015). http://eprint.iacr.org/2015/1157
8. Bellare, M., Rogaway, P.: Random oracles are practical: a paradigm for designing efficient protocols. In: Ashby, V. (ed.) ACM CCS 1993, pp. 62–73. ACM Press, November 1993
9. Bellare, M., Rogaway, P.: The exact security of digital signatures-how to sign with RSA and Rabin. In: Maurer, U. (ed.) EUROCRYPT 1996. LNCS, vol. 1070, pp. 399–416. Springer, Heidelberg (1996). doi:10.1007/3-540-68339-9_34

10. Bellare, M., Rogaway, P.: The security of triple encryption and a framework for code-based game-playing proofs. In: Vaudenay, S. (ed.) EUROCRYPT 2006. LNCS, vol. 4004, pp. 409–426. Springer, Heidelberg (2006). doi:10.1007/11761679_25
11. Bernstein, D.J., Duif, N., Lange, T., Schwabe, P., Yang, B.-Y.: High-speed high-security signatures. In: Preneel, B., Takagi, T. (eds.) CHES 2011. LNCS, vol. 6917, pp. 124–142. Springer, Heidelberg (2011). doi:10.1007/978-3-642-23951-9_9
12. Cachin, C., Micali, S., Stadler, M.: Computationally private information retrieval with polylogarithmic communication. In: Stern, J. (ed.) EUROCRYPT 1999. LNCS, vol. 1592, pp. 402–414. Springer, Heidelberg (1999). doi:10.1007/3-540-48910-X_28
13. Cramer, R.: Modular design of secure, yet practical protocls. Ph.D. thesis, University of Amsterdam (1996)
14. Feige, U., Fiat, A., Shamir, A.: Zero knowledge proofs of identity. In: Aho, A. (ed.) 19th ACM STOC, pp. 210–217. ACM Press, May 1987
15. Feige, U., Fiat, A., Shamir, A.: Zero-knowledge proofs of identity. J. Cryptology **1**(2), 77–94 (1988)
16. Fiat, A., Shamir, A.: How to prove yourself: practical solutions to identification and signature problems. In: Odlyzko, A.M. (ed.) CRYPTO 1986. LNCS, vol. 263, pp. 186–194. Springer, Heidelberg (1987). doi:10.1007/3-540-47721-7_12
17. Gentry, C., Peikert, C., Vaikuntanathan, V.: Trapdoors for hard lattices and new cryptographic constructions. In: Ladner, R.E., Dwork, C. (eds.) 40th ACM STOC, pp. 197–206. ACM Press, May 2008
18. Goh, E.-J., Jarecki, S.: A signature scheme as secure as the Diffie-Hellman problem. In: Biham, E. (ed.) EUROCRYPT 2003. LNCS, vol. 2656, pp. 401–415. Springer, Heidelberg (2003). doi:10.1007/3-540-39200-9_25
19. Goldreich, O.: Two remarks concerning the Goldwasser-Micali-Rivest signature scheme. In: Odlyzko, A.M. (ed.) CRYPTO 1986. LNCS, vol. 263, pp. 104–110. Springer, Heidelberg (1987). doi:10.1007/3-540-47721-7_8
20. Goldwasser, S., Micali, S., Rivest, R.L.: A digital signature scheme secure against adaptive chosen-message attacks. SIAM J. Comput. **17**(2), 281–308 (1988)
21. Goldwasser, S., Ostrovsky, R.: *Invariant* signatures and non-interactive zero-knowledge proofs are equivalent. In: Brickell, E.F. (ed.) CRYPTO 1992. LNCS, vol. 740, pp. 228–245. Springer, Heidelberg (1993). doi:10.1007/3-540-48071-4_16
22. Goldwasser, S., Tauman Kalai, Y.: On the (in)security of the Fiat-Shamir paradigm. In: 44th FOCS, pp. 102–115. IEEE Computer Society Press, October 2003
23. Guillou, L.C., Quisquater, J.-J.: A "Paradoxical" indentity-based signature scheme resulting from zero-knowledge. In: Goldwasser, S. (ed.) CRYPTO 1988. LNCS, vol. 403, pp. 216–231. Springer, Heidelberg (1990). doi:10.1007/0-387-34799-2_16
24. Katz, J., Wang, N.: Efficiency improvements for signature schemes with tight security reductions. In: Jajodia, S., Atluri, V., Jaeger, T. (eds.) ACM CCS 2003, pp. 155–164. ACM Press, October 2003
25. Kiltz, E., Masny, D., Pan, J.: Optimal security proofs for signatures from identification schemes. Cryptology ePrint Archive, Report 2016/191 (2016). http://eprint.iacr.org/2016/191
26. Koblitz, N., Menezes, A.: The random oracle model: a twenty-year retrospective. Cryptology ePrint Archive, Report 2015/140 (2015). http://eprint.iacr.org/2015/140
27. Lysyanskaya, A.: Unique signatures and verifiable random functions from the DH-DDH separation. In: Yung, M. (ed.) CRYPTO 2002. LNCS, vol. 2442, pp. 597–612. Springer, Heidelberg (2002). doi:10.1007/3-540-45708-9_38

28. Lyubashevsky, V.: Fiat-Shamir with aborts: applications to lattice and factoring-based signatures. In: Matsui, M. (ed.) ASIACRYPT 2009. LNCS, vol. 5912, pp. 598–616. Springer, Heidelberg (2009). doi:10.1007/978-3-642-10366-7_35
29. Lyubashevsky, V.: Lattice signatures without trapdoors. In: Pointcheval, D., Johansson, T. (eds.) EUROCRYPT 2012. LNCS, vol. 7237, pp. 738–755. Springer, Heidelberg (2012). doi:10.1007/978-3-642-29011-4_43
30. Micali, S., Reyzin, L.: Improving the exact security of digital signature schemes. J. Cryptology **15**(1), 1–18 (2002)
31. Micciancio, D., Peikert, C.: Trapdoors for lattices: simpler, tighter, faster, smaller. In: Pointcheval, D., Johansson, T. (eds.) EUROCRYPT 2012. LNCS, vol. 7237, pp. 700–718. Springer, Heidelberg (2012). doi:10.1007/978-3-642-29011-4_41
32. M'Raïhi, D., Naccache, D., Pointcheval, D., Vaudenay, S.: Computational alternatives to random number generators. In: Tavares, S., Meijer, H. (eds.) SAC 1998. LNCS, vol. 1556, pp. 72–80. Springer, Heidelberg (1999). doi:10.1007/3-540-48892-8_6
33. Ohta, K., Okamoto, T.: On concrete security treatment of signatures derived from identification. In: Krawczyk, H. (ed.) CRYPTO 1998. LNCS, vol. 1462, pp. 354–369. Springer, Heidelberg (1998). doi:10.1007/BFb0055741
34. Ong, H., Schnorr, C.P.: Fast signature generation with a Fiat Shamir — like scheme. In: Damgård, I.B. (ed.) EUROCRYPT 1990. LNCS, vol. 473, pp. 432–440. Springer, Heidelberg (1991). doi:10.1007/3-540-46877-3_38
35. Pointcheval, D., Stern, J.: Security arguments for digital signatures and blind signatures. J. Cryptology **13**(3), 361–396 (2000)
36. Schnorr, C.-P.: Efficient signature generation by smart cards. J. Cryptology **4**(3), 161–174 (1991)

How to Obtain Fully Structure-Preserving (Automorphic) Signatures from Structure-Preserving Ones

Yuyu Wang[1,2], Zongyang Zhang[2](✉), Takahiro Matsuda[2], Goichiro Hanaoka[2], and Keisuke Tanaka[1,3]

[1] Tokyo Institute of Technology, Tokyo, Japan
wang.y.ar@m.titech.ac.jp, keisuke@is.titech.ac.jp
[2] National Institute of Advanced Industrial Science and Technology (AIST), Tokyo, Japan
zongyang.zhang@gmail.com, {t-matsuda,hanaoka-goichiro}@aist.go.jp
[3] JST CREST, Tokyo, Japan

Abstract. In this paper, we bridge the gap between structure-preserving signatures (SPSs) and fully structure-preserving signatures (FSPSs). In SPSs, all the messages, signatures, and verification keys consist only of group elements, while in FSPSs, even signing keys are required to be a collection of group elements. To achieve our goal, we introduce two new primitives called *trapdoor signature* and *signature with auxiliary key*, both of which can be derived from SPSs. By carefully combining both primitives, we obtain generic constructions of FSPSs from SPSs. Upon instantiating the above two primitives, we get many instantiations of FSPS with unilateral and bilateral message spaces. Different from previously proposed FSPSs, many of our instantiations also have the *automorphic property*, i.e., a signer can sign his own verification key. As by-product results, one of our instantiations has the shortest verification key size, signature size, and lowest verification cost among all previous constructions based on standard assumptions, and one of them is the first FSPS scheme in the *type I* bilinear groups.

Keywords: Signature · Trapdoor signature · Fully structure-preserving · Automorphic

Y. Wang—This author is supported by a JSPS Fellowship for Young Scientists.

Z. Zhang—Corresponding author whose current affiliation is Beihang University. The author is partly supported by the NSFC under No. 61303201. The work is done while the author is a postdoctoral researcher of AIST.

K. Tanaka—A part of this work was supported by a grant of I-System Co. Ltd., NTT Secure Platform Laboratories, Nomura Research Institute, Input Output Hongkong, and MEXT/JSPS KAKENHI 16H01705.

J.H. Cheon and T. Takagi (Eds.): ASIACRYPT 2016, Part II, LNCS 10032, pp. 465–495, 2016.
DOI: 10.1007/978-3-662-53890-6_16

1 Introduction

1.1 Background

Structure-preserving signatures (SPSs). In [3], Abe et al. initiated the study of SPSs which denote pairing-based signatures where all the verification keys, messages, and signatures consist only of group elements and the verification algorithms only make use of pairing product equations (PPEs) to verify signatures.

SPSs are very useful since they can be combined with other structure-preserving (SP) primitives, e.g., ElGamal encryption [19] and Groth-Sahai proofs [29], to obtain efficient cryptographic protocols such as blind signatures [3,23–25], group signatures [3,25,34], homomorphic signatures [33], delegatable anonymous credentials [22], compact verifiable shuffles [17], network coding [6], oblivious transfer [14,37], tightly secure encryption [2,30], and e-cash [7]. Motivated by this, there have been a large deal of works focusing on SPSs (e.g., [1,3]) in the past few years, which provide us with various SPS schemes based on different assumptions and with high efficiency.

Automorphic signatures. In [3], Abe et al. noted that for elaborate applications, the SP property of a signature scheme is not sufficient. In addition, an SPS scheme has to be able to sign its own verification keys, i.e., verification keys have to lie in the message space. They called such kind of SPS *automorphic signature* and gave an instantiation of it, and also provided a generic transformation that converts automorphic signatures for messages of fixed length into ones for messages of arbitrary length.

As argued in [3], since automorphic signatures enable constructions of certification chains (i.e., sequences of verification keys linked by certificates from one key on the next one), they are useful in constructing anonymous proxy signatures and delegatable anonymous credentials. Abe et al. [3] also showed how to combine automorphic signatures with the Groth-Sahai proof system to construct a round-optimal blind signature scheme.

Fully structure-preserving signatures (FSPSs). In [5], Abe et al. introduced FSPSs, where signing keys also consist only of group elements and the correctness of signing keys with respect to verification keys can be verified by PPEs. Since the fully structure-preserving (FSP) property enables efficient signing key extraction, it could help us prevent rogue-key attacks in the public-key infrastructures (PKIs) [36], make anonymous credentials UC-secure [15], achieve privacy in group and ring signatures [10,11,13] in the presence of adversarial keys, and extend delegatable anonymous credentials [8,18,22] with all-or-nothing transferability [16], as noted in [5]. In this paper, we call an automorphic signature scheme that is FSP a *fully automorphic signature (FAS) scheme.*

Abe et al. [5] gave two generic constructions by combining FSPSs unforgeable (UF) against extended random message attacks (xRMA) [1] with other primitives such as one-time SPSs, two-tier SPSs (also called partial one-time SPSs), and trapdoor commitment schemes. Although these constructions are novel and neat, they suffer from three shortcomings due to the use of specific primitives, which make them less generic.

1. As both constructions require a UF-xRMA secure FSPS scheme and one of them also requires a γ-blinding trapdoor commitment scheme, the underlying assumptions and bilinear map of their instantiations are limited. Concretely speaking, all the signature schemes derived from their constructions have to be based on at least the SXDH and XDLIN assumptions and be in the *type III* bilinear group.
2. For the same reason, the efficiency of their instantiations is also potentially limited by the underlying UF-xRMA secure FSPS scheme and the γ-blinding trapdoor commitment scheme. For example, the verification keys and signatures of their most efficient FSPS scheme consist of more than $10n$ group elements in total if messages consist of n^2 group elements.
3. Their instantiations are not automorphic. The reason is that verification keys of the UF-xRMA secure FSPS scheme (which are also verification keys of the resulting schemes) consist of elements in both source groups, while the resulting signature schemes can only sign messages consisting only of elements in one source group.

Note that Abe et al. [5] also gave a variant of their constructions by combining a UF-xRMA secure signature scheme and a trapdoor commitment scheme with SPSs, which can be treated as a generic transformation from SPSs to FSPSs. If the instantiation of SPS is with a bilateral message space (i.e., messages consist of elements in both source groups), then the resulting signature scheme could be automorphic. However, as far as we know, besides the aforementioned shortcomings, all the previously proposed SPS schemes with a bilateral message space require verification keys to consist of elements in both source groups (except for ones that sign messages of "DDH form" [26,27]), which result in very inefficient FSPS schemes, as noted in [5]. The verification keys and signatures (respectively, the verification algorithm) of the most efficient automorphic instantiation that can be derived from their generic construction consist of more than $12n$ group elements in total (respectively, more than $3n$ PPEs) if the messages consist of $2n^2$ group elements.

Following the work of Abe et al. [5], Groth [28] gave an elegant construction of FSPS, which has the shortest verification keys and signatures, and needs the fewest PPEs for verification. Although this FSPS scheme is the most efficient one as far as we know, it is only known to be secure in the generic group model and is not automorphic.

Up until now, a lot of results are devoted to constructing efficient SPSs under different assumptions, while there are very few FSPS schemes. If we can find a generic method to transform existing SPSs into FSPSs or even FASs without directly using specific primitives, it will greatly alleviate the efforts to construct them from scratch.

1.2 Our Results

Generic construction of FSPS. In this paper, we formalize two extensions to ordinary signatures called *trapdoor signatures (TSs)* and *signatures with auxiliary key (AKSs).* We show that any well-formed[1] SPS scheme can be converted into a TS scheme satisfying the signing key structure-preserving (SKSP) property, in which signing keys consist only of group elements and the correctness with respect to verification keys can be verified by PPEs, while messages are not necessarily group elements. Furthermore, it is relatively straightforward to show that any SPS scheme with an algebraic key generation algorithm can be converted into a structure-preserving signature with auxiliary keys (SP-AKS). By combining SKSP-TS with SP-AKS, we obtain a generic construction of FSPS.[2] Our construction implies that for any two SPS schemes, if verification keys of one lie in the message space of the other (which is well-formed), then basically, they can be used to construct an FSPS scheme, without using any other specific primitives or additional assumptions. It also implies that most well-formed SPS schemes with a bilateral message space or unilateral verification key space (i.e., the verification keys consist only of elements in one source group) can be converted into an FSPS scheme.

This generic construction is proved to be secure based on building blocks satisfying different security, which allows us to obtain various instantiations of FSPS based on different assumptions.

Efficient instantiations of FSPSs. By extending the definition of AKSs to two-tier signatures with auxiliary keys (TT-AKSs) and substituting AKSs with TT-AKSs in the above generic construction, we obtain another generic construction, which enables us to obtain more efficient instantiations of FSPS. For instance, by using the TS scheme and TT-AKS scheme adapted from the SPS schemes proposed by Kiltz et al. [31,32], we obtain instantiations of FSPS with unilateral and bilateral message spaces. We give an efficiency comparison between our instantiations and the ones proposed in [5] in Table 1.[3] Note that like the FSPS scheme proposed in [28], a signing key in our instantiations consists of $\Omega(n)$ group elements (concretely, $2n+1$ in [28] and $4n+9$ and $8n+13$ in our results), while that in "AKO+15" consists only of 4 elements. However, in many applications, the

[1] We refer the reader to Definition 10 for details of well-formed SPSs. As far as we know, all the existing SPS schemes are well-formed.

[2] As in [5], we assume the underlying SKSP-TS scheme and SP-AKS scheme share the common setup algorithm.

[3] The second instantiation in Table 1 is derived from the generic construction described in [5, Sect. 6.4], where the underlying SPS scheme is the one with bilateral message space in [31] (based on the SXDH assumption). In this instantiation, we have to add a group element denoting the sequence number to every message block. Furthermore, the underlying two-tier signature schemes of the first and third instantiations have the same efficiency, which makes sure that this comparison is fair. If we allow trusted setup besides the bilinear map generation, the sizes of common parameters $|par|$ in these four schemes are 6, 6, 1, and 2 respectively.

size of a signing key does not have to be "extremely short" since typically, a user generates only one proof for knowing a signing key (e.g., in PKIs and group/ring signatures), while proofs for knowing a signature or a verification key/signature pair are required to be generated for multiple times.[4]

Table 1. Comparison between the most efficient instantiations of FSPS based on standard assumptions derived from the main construction in [5] and the most efficient ones derived from our constructions. Notation (x, y) denotes x elements in $\mathbb{G}_1$ and y elements in $\mathbb{G}_2$. We do not count the two generators in the description of bilinear groups when giving the parameters.

	Security	Assumption	$\|m\|$	$\|pk\| + \|par\|$	$\|\sigma\|$	$\sharp$ PPE
AKO+15 [5]	Full	SXDH, XDLIN	$(n^2, 0)$	$6n + 17$	$4n + 11$	$n + 5$
	Full	SXDH, XDLIN	(n^2, n^2)	$6n + 47$	$13n + 30$	$5n + 6$
Our results	Full	SXDH	$(n^2, 0)$	$2n + 7$	$4n + 8$	$n + 3$
	Full	SXDH	(n^2, n^2)	$4n + 10$	$8n + 12$	$2n + 4$

Our FSPS schemes in Table 1 can also be based on the $\mathcal{U}_k$-matrix Diffie-Hellman (MDDH) assumptions [20] (see the full paper for the definition), while the parameters become $(|m|, |pk| + |par|, |\sigma|, \sharp\text{PPE}) = (n^2, (2nk + 2k + 3 + \mathsf{RE}(\mathcal{D}_k))k + \mathsf{RE}(\mathcal{D}_k), (3k + 1)n + 4 + 3k + \mathsf{RE}(\mathcal{D}_k), kn + 2k + 1)$ and $(|m|, |pk| + |par|, |\sigma|, \sharp\text{PPE}) = (2n^2, (4nk + 3k + 3 + 2\mathsf{RE}(\mathcal{D}_k))k + 2\mathsf{RE}(\mathcal{D})_k, 2(3k + 1)n + 5k + 5 + 2\mathsf{RE}(\mathcal{D}_k), 2kn + 3k + 1)$, where $\mathsf{RE}(\mathcal{D}_k)$ denotes the minimal number of group elements needed to present a matrix sampled from $\mathcal{U}_k$.

Since our constructions only require the underlying schemes to have properties naturally satisfied by SPSs, further improvement on SPS schemes may contribute to the efficiency of FSPSs more via our constructions than the constructions in [5].

FASs. Since we can convert any (well-formed) SPS scheme into an SKSP-TS scheme and an SP-AKS scheme, our generic constructions also derive many instantiations of FAS from various combinations (including the ones in Table 1). As long as verification keys of the underlying TS scheme consist of no more group elements than messages of the underlying AKS scheme in both source groups, the resulting scheme is usually fully automorphic.

We can instantiate our first generic construction with the TS scheme and AKS scheme adapted from the SPS scheme proposed by Groth et al. [28] to obtain our most efficient FAS scheme, while the most efficient one from the

[4] The argument that the signing key size is not as important as verification/signature size does not spoil the motivation for FSPS. FSPS helps avoid extremely heavy key extraction, i.e., extracting a signing key bit by bit (see Introduction in [8]). However, this does not mean we have to make the extraction extremely light. Allowing checking signing keys by using PPEs and keeping the key size linear with message size are enough to achieve the goal.

Table 2. Comparison between the most efficient instantiation of FAS derived from the main construction in [5] and the most efficient one derived from our constructions. Both of them are secure in the generic group model.

	Security	Assumption	$\|m\|$	$\|pk\| + \|par\|$	$\|\sigma\|$	$\sharp$ PPE
AKO+15 [5]	Full	Generic	(n^2, n^2)	$6n + 23$	$6n + 14$	$3n + 6$
Our result	Full	Generic	$(n^2, 0)$	$2n + 1$	$2n + 5$	$n + 3$

generic construction in [5] can be obtained by letting the underlying SPS scheme be the one in [4] and the underlying one-time SPS scheme the one in [26]. For ease of understanding, we give an efficiency comparison in Table 2.

FSPS (FAS) schemes in the symmetric (type I) bilinear map. We also instantiate our generic constructions with the SPS scheme and the tag-based SPS scheme proposed in [2] to obtain the first FSPS and FAS schemes in the type I bilinear map, the most efficient one of which achieves $(|m|, |pk| + |par|, |\sigma|, \sharp\text{PPE}) = (n^2, 6n + 30, 6n + 12, 2n + 7)$.

1.3 High-Level Idea

Our generic construction can be treated as an extension of the well-known EGM paradigm [21]. In this paradigm a signer uses two signature schemes Σ_1 and Σ_2 to sign a message m. It first signs m by using the signing key sk_2 of Σ_2 and then signs the verfication key vk_2 of Σ_2 by using the signing key sk_1 of Σ_1. This paradigm was used to obtain SPSs in [1] and a generic construction of FSPS in [5]. To make sure that the resulting signature scheme is an FSPS scheme, it is natural to require sk_1 to consist only of group elements. This is the reason why Abe et al. [5] instantiated Σ_1 with the xRMA secure signature scheme proposed in [1], which was the only proposed FSPS scheme until then. However, we observe that it is possible to instantiate Σ_1 with all the existing SPS schemes, which also provides us with more options when selecting instantiations of Σ_2 to match Σ_1.

Next, we explain how to choose Σ_1 and Σ_2, and the high level idea of our construction. Roughly speaking, starting from an SPS scheme with a signing key $x \in \mathbb{Z}_p$, we can always derive a signature scheme in which the signing key becomes a group element $X = G^x \in \mathbb{G}$ (where G denotes the generator of $\mathbb{G}$). It is obvious that in this case a message $M \in \mathbb{G}$ cannot be signed by using X since we are not able to compute M^x from X and M. Supposing that $M = G^m$, we can use X to sign m instead of M, i.e., compute X^m instead of M^x when generating a signature. Furthermore, since signatures generated in this way are the same as those generated by the real signing key, and the public key and verification algorithm remain the same, one can verify the signature by using M. We formalize such a signature scheme as a TS scheme. Although such a signature scheme is only "semi"-structure-preserving, we use it to sign the exponent $v \in \mathbb{Z}_p$ of a verification key (called auxiliary keys) of another SPS scheme and use the latter SPS scheme to sign a message $M' \in \mathbb{G}'$. This enables us to obtain an FSPS scheme. We formalize the latter signature scheme which generates auxiliary keys besides verification/signing key pairs as an AKS scheme.

To verify a signature, one only needs to know $V = G^v$ and M', without knowing v. Furthermore, the original signing key x (called trapdoor key) of the TS scheme is never used in the signing process but is necessary as the reduction algorithm in the security proof signs verification keys without knowing the exponent.

Our main contributions lie in two aspects. First, we formalize the notions of TSs and AKSs in order to adapt the EGM paradigm to construct FSPSs. Second, we show that most of existing SPS schemes can be cast as our extended signatures, and consequently we can obtain a number of FSPSs and FASs based on existing SPSs.

Perhaps interestingly, although most of the previously proposed SPS schemes with a unilateral message space are not automorphic (since their verification keys and messages usually consist of elements in different source groups), when some of them are converted into FSPSs using our method, the resulting schemes become automorphic.[5]

Paper organization. We recall several definitions in Sect. 2. Then we formalize TSs and AKSs and show how to instantiate them from any (well-formed) SPS scheme in Sects. 3 and 4 respectively, and give generic constructions of FSPSs based on them in Sect. 5. Finally, we show instantiations of our generic constructions in Sect. 6.

2 Preliminaries

2.1 Notations

In this paper, we let $negl$ be negligible functions, $[n]$ the set $\{1, \dots, n\}$, $\mathbb{N}$ the set of natural numbers, $|X|$ the number of elements in X (where X could be a space, a vector, or a matrix), and $\widetilde{\mathbf{A}}$ the $1 \times mn$ vector $(a_{11}, a_{12}, \dots a_{1n}, a_{21}, a_{22}, \dots a_{2n}, \dots, a_{m1}, a_{m2}, \dots a_{mn})$ where $\mathbf{A}$ denotes the $m \times n$ matrix $(a_{ij})_{i \in [m], j \in [n]}$. If $\mathbf{A} \in \mathbb{Z}_p^{(k+1) \times k}$ lies in the matrix distribution $\mathcal{U}_k$, then we use $\overline{\mathbf{A}}$ to denote the upper square matrix of $\mathbf{A}$. Furthermore, $\vec{a} \in \mathbb{Z}_p^n$ denotes a column vector by default.

2.2 Pairing Group

In this paper, we let $\mathcal{G}$ be an algorithm that takes as input 1^λ and outputs $gk = (p, \mathbb{G}_1, \mathbb{G}_2, \mathbb{G}_T, e, G_1, G_2)$ such that p is a prime satisfying $p = \Theta(2^\lambda)$, $(\mathbb{G}_1, \mathbb{G}_2, \mathbb{G}_T)$ are descriptions of groups of order p, G_1 and G_2 generate $\mathbb{G}_1$

[5] When messages and verification keys of the underlying TS scheme consist of elements in $\mathbb{G}_2$ and $\mathbb{G}_1$ respectively and those of the underlying AKS scheme consist of elements in $\mathbb{G}_1$ and $\mathbb{G}_2$ respectively, verification keys and messages of the resulting FSPS scheme consist of elements only in $\mathbb{G}_1$.

and $\mathbb{G}_2$ respectively, and $e : \mathbb{G}_1 \times \mathbb{G}_2 \rightarrow \mathbb{G}_T$ is an efficiently computable (non-degenerate) bilinear map. Following [28,31], we use the additive notation in [20] such as $e((a+b)[x]_1,[y]_2) = a \cdot e([x]_1,[y]_2) + b \cdot e([x]_1,[y]_2)$ where $[x]_1$ and $[y]_2$ denote G_1^x and G_2^y respectively, and $e([x]_1,[y]_2)$ can be written as $[xy]_T$. Furthermore, $e([\vec{a}]_1^\top,[\vec{b}]_2)$ denotes $\sum_{i=1}^n e([a_i]_1,[b_i]_2)$ where $[\vec{a}]_1 = ([a_1]_1,\ldots,[a_n]_1)^\top$ and $[\vec{b}]_2 = ([b_1]_2,\ldots,[b_n]_2)^\top$, and $e([\mathbf{A}]_1^\top,[\mathbf{B}]_2)$ denotes $(e([\vec{a}_i^\top]_1,[\vec{b}_j]_2))_{i\in[n],j\in[n']}$ where $[\mathbf{A}]_1 = ([\vec{a}_1]_1,\ldots,[\vec{a}_n]_1)$ and $[\mathbf{B}]_2 = ([\vec{b}_1]_2,\ldots,[\vec{b}_{n'}]_2)$.

2.3 Signatures

Definition 1 (Signature). *A* signature scheme *consists of four polynomial-time algorithms* Setup, Gen, Sign, *and* Verify. Setup *takes as input a security parameter* 1^λ *and generates a public parameter par, which determines the message space* $\mathcal{M}$ *and the randomness space* $\mathcal{R}$ *for signing.* Gen *is a randomized algorithm that takes as input a public parameter par and outputs a verification/signing key pair (pk, sk).* Sign *is a randomized algorithm that takes as input a signing key sk and a message m, and returns a signature* σ*.* Verify *is a deterministic algorithm that takes as input a verification key pk, a message M, and a signature* σ*, and returns* 1 *(accept) or* 0 *(reject).*

The correctness is satisfied if we have $\mathsf{Verify}(pk, m, \mathsf{Sign}(sk, m; r)) = 1$ *for all* $\lambda \in \mathbb{N}$, $par \leftarrow \mathsf{Setup}(1^\lambda)$, $(pk, sk) \leftarrow \mathsf{Gen}(par)$, $m \in \mathcal{M}$, *and* $r \in \mathcal{R}$*.*

In [3], Abe et al. firstly defined *SPSs*, in which verification keys, messages, and signatures consist only of group elements in $\mathbb{G}_1$ and $\mathbb{G}_2$, and signatures are verified by evaluating pairing product equations (PPEs), which are of the form $\sum_{ij} a_{ij} e([x_i]_1,[y_j]_2) = [0]_T$, where a_{ij} is an integer constant for all i and j.

Definition 2 (Structure-preserving signature (SPS)). *A signature scheme is said to be* structure-preserving *over a bilinear group generator* $\mathcal{G}$ *if we have (a) a public parameter includes a group description gk generated by* $\mathcal{G}$*, (b) verification keys consist of group elements in* $\mathbb{G}_1$ *and* $\mathbb{G}_2$*, (c) messages consist of group elements in* $\mathbb{G}_1$ *and* $\mathbb{G}_2$*, (d) signatures consist of group elements in* $\mathbb{G}_1$ *and* $\mathbb{G}_2$*, and (e) the verification algorithm consists only of evaluating membership in* $\mathbb{G}_1$ *and* $\mathbb{G}_2$ *and relations described by PPEs.*

SPSs are versatile since they mix well with other pairing-based protocols. Especially, they are compatible with the Groth-Sahai proof system [29]. However, as argued by Abe et al. in [3], Groth-Sahai compatibility of a signature scheme is not sufficient for elaborate applications such as anonymous signatures and delegatable anonymous credentials, which require signatures on verification keys to obtain anonymized certification chains. Abe et al. [3] called an SPS scheme that is able to sign its own verification keys an automorphic signature scheme.

Definition 3 (Automorphic signature). *A signature scheme is said to be an* automorphic signature scheme *over a bilinear group generator* $\mathcal{G}$ *if it is structure-preserving and its (padded) verification keys lie in the message space.*

In [5], Abe et al. introduced FSPSs, which also require a signing key to be group elements in $\mathbb{G}_1$ and $\mathbb{G}_2$ and the correctness of a signing key with respect to a verification key can be verified by PPEs. Such signatures allow efficient key extraction when combined with non-interactive proofs (e.g., the Groth-Sahai proofs), which may help prevent rogue-key attacks [36], build UC-secure privacy preserving protocols [15], strengthen privacy in group and ring signatures [10,11,13] in the presence of adversarial keys, and extend delegatable anonymous credential systems [8,18,22] with all-or-nothing transferability [16].

Definition 4 (Fully structure-preserving signature (FSPS)). *A structure-preserving signature scheme* (Setup, Gen, Sign, Verify) *with the message space $\mathcal{M}$ and randomness space $\mathcal{R}$ for signing is said to be* fully structure-preserving *if we have (a) signing keys consist only of group elements in $\mathbb{G}_1$ and $\mathbb{G}_2$, and additionally, (b) there exists a polynomial-time deterministic algorithm* VerifySK *that takes as input a verification/signing key pair and consists only of evaluating membership in $\mathbb{G}_1$ and $\mathbb{G}_2$ and relations described by PPEs, and it is required that for sufficiently large $\lambda \in \mathbb{N}$, $par \leftarrow \mathsf{Setup}(1^\lambda)$, the following holds:*

- $\mathsf{VerifySK}(pk, sk) = 1$ *if and only if* $\mathsf{Verify}(pk, m, \mathsf{Sign}(sk, m; r)) = 1$ *holds for all $m \in \mathcal{M}$ and $r \in \mathcal{R}$.*

In this paper, we call an automorphic signature scheme which is also FSP a *fully automorphic signature (FAS) scheme.*

Definition 5 (Fully automorphic signature (FAS)). *An automorphic signature scheme is said to be* fully automorphic *if it is also fully structure-preserving.*

Due to space limitation, we recall the UF-CMA, UF-RMA, UF-otCMA, and UF-otRMA security of a signature scheme in the full paper.

3 Trapdoor Signatures

3.1 Definition of Trapdoor Signatures

In this section, we formalize the notion of *γ-trapdoor signature (γ-TS) scheme,* whose instantiations are used as building blocks to obtain FSPSs. Different from standard signatures, a TS scheme verifies the correctness of a signature σ on a message $m \in \mathcal{M}$ by taking as input $(\gamma(m) \in \mathcal{M}_\gamma, \sigma)$ where $\gamma : \mathcal{M} \mapsto \mathcal{M}_\gamma$ is an efficiently computable bijection. Furthermore, there exists a trapdoor key with which we can generate a signature on m if we have $\gamma(m)$ but not m itself.

Definition 6. (γ-Trapdoor signature (γ-TS)). *A* γ-trapdoor signature scheme *consists of five polynomial-time algorithms* Setup, Gen, Sign, Verify, *and* TDSign. Setup *takes as input a security parameter 1^λ and generates a public parameter par, which determines the message space $\mathcal{M}$ for the signing algorithm, the message space $\mathcal{M}_\gamma$ for the verification algorithm, and an efficiently computable bijection $\gamma : \mathcal{M} \mapsto \mathcal{M}_\gamma$.* Gen *is a randomized algorithm that takes* *as input par, and outputs a verification/signing key pair* (pk, sk) *and a trapdoor key* tk. Sign *is a randomized algorithm that takes as input a signing key* sk *and a message* $m \in \mathcal{M}$, *and returns a signature* σ, *where the randomness space is denoted by* $\mathcal{R}$. Verify *is a deterministic algorithm that takes as input a verification key* pk, *a message* $M \in \mathcal{M}_\gamma$, *and a signature* σ, *and returns* 1 *(accept) or* 0 *(reject).* TDSign *takes as input a trapdoor key* tk *and a message* $M \in \mathcal{M}_\gamma$, *and returns a signature* σ. *The randomness space of* TDSign *is also* $\mathcal{R}$.

The correctness is satisfied if for all $\lambda \in \mathbb{N}$, $par \leftarrow \mathsf{Setup}(1^\lambda)$, $((pk, sk), tk) \leftarrow \mathsf{Gen}(par)$, *and* $m \in \mathcal{M}$, *we have (a)* $\mathsf{Verify}(pk, \gamma(m), \mathsf{Sign}(sk, m)) = 1$, *and (b)* $\mathsf{Sign}(sk, m; r) = \mathsf{TDSign}(tk, \gamma(m); r)$ *for all* $r \in \mathcal{R}$.

Key generation algorithm $\mathcal{T}_{\mathsf{Gen}}$. We use $\mathcal{T}_{\mathsf{Gen}}$ to denote an algorithm that runs Gen, which is the key generation algorithm of a TS scheme, in the following way. Taking as input a public parameter *par*, $\mathcal{T}_{\mathsf{Gen}}$ gives *par* to Gen and obtains an output $((pk, sk), tk)$. Then $\mathcal{T}_{\mathsf{Gen}}$ outputs (pk, tk) as a verification/signing key pair.

For a TS scheme $\Sigma = (\mathsf{Setup}, \mathsf{Gen}, \mathsf{Sign}, \mathsf{Verify}, \mathsf{TDSign})$, we denote $(\mathsf{Setup}, \mathcal{T}_{\mathsf{Gen}}, \mathsf{TDSign}, \mathsf{Verify})$ by $\mathcal{T}_\Sigma$. According to the syntax of TS, it is not hard to see that $\mathcal{T}_\Sigma$ forms a standard signature scheme whose message space is $\mathcal{M}_\gamma$.

Now we define SKSP-TSs, in which verification keys, signing keys, and signatures (but not necessarily messages) consist only of group elements, and the correctness of signing keys with respect to verifications keys can be verified by PPEs.

Definition 7 (Signing key structure-preserving (SKSP)). *A* γ*-TS scheme* $\Sigma = (\mathsf{Setup}, \mathsf{Gen}, \mathsf{Sign}, \mathsf{Verify}, \mathsf{TDSign})$ *with message space* $\mathcal{M}$ *is said to be* signing key structure-preserving *over a bilinear group generator* $\mathcal{G}$ *if we have (a)* $\mathcal{T}_\Sigma$ *is an SPS scheme, (b) signing keys (rather than trapdoor keys) consist only of group elements in* $\mathbb{G}_1$ *and* $\mathbb{G}_2$, *and (c)* Σ *satisfies the condition (b) in Definition 4, where* $\mathsf{Verify}(pk, m, \mathsf{Sign}(sk, m; r)) = 1$ *is replaced with* $\mathsf{Verify}(pk, \gamma(m), \mathsf{Sign}(sk, m; r)) = 1$.

Note that different from FSPSs, messages are not required to be group elements in SKSP-TSs.

3.2 Security of Trapdoor Signatures

We now define the UF-CMA security of TSs.

Definition 8 (UF-CMA of TSs). *A* γ*-TS scheme* $(\mathsf{Setup}, \mathsf{Gen}, \mathsf{Sign}, \mathsf{Verify}, \mathsf{TDSign})$ *is said to be* unforgeable against chosen message attacks (UF-CMA) *if for every probabilistic polynomial time (PPT) adversary* $\mathcal{A}$, *we have*

$$\Pr[par \leftarrow \mathsf{Setup}(1^\lambda), ((pk, sk), tk) \leftarrow \mathsf{Gen}(par), (M^*, \sigma^*) \leftarrow \mathcal{A}^{\mathrm{SignO}(\cdot)}(par, pk) : \\ M^* \notin \mathcal{Q}_m \wedge M^* \in \mathcal{M}_\gamma \wedge \mathsf{Verify}(pk, M^*, \sigma^*) = 1] \leq negl(\lambda)$$

where SignO$(\cdot)$ *is the signing oracle that takes as input* $m \in \mathcal{M}$, *runs* $\sigma \leftarrow \mathsf{Sign}(sk, m)$, *adds* $\gamma(m) \in \mathcal{M}_\gamma$ *to* $\mathcal{Q}_m$, *and returns* σ.

Unlike the UF-CMA security of standard signatures, a query m made by an adversary is in $\mathcal{M}$, the signing oracle records $\gamma(m) \in \mathcal{M}_\gamma$, and the message M^* output by the adversary is in $\mathcal{M}_\gamma$.

The UF-CMA security of TSs is similar to the F-unforgeability of standard signatures defined by Belenkiy et al. [9]. Moreover, Libert et al. [35] gave an instantiation of F-unforgeable signatures and combined it with a tagged one-time signature scheme proposed by Abe et al. [2] to obtain a very efficient SPS scheme. However, they neither provided generic constructions nor considered the FSP property.

Now we show the relation between the UF-CMA security of (Setup, Gen, Sign, Verify, TDSign) and that of (Setup, $\mathcal{T}_{\mathsf{Gen}}$, TDSign, Verify) in Theorem 1. We refer the reader to the full paper for the proof.

Theorem 1. *For a γ-TS scheme $\Sigma =$ (Setup, Gen, Sign, Verify, TDSign), if $\mathcal{T}_\Sigma =$ (Setup, $\mathcal{T}_{\mathsf{Gen}}$, TDSign, Verify) is UF-CMA secure, then Σ is UF-CMA secure.*

Now we give the definitions of unforgeability against random message attacks (RMA), one-time chosen message attacks (otCMA), and one-time random message attacks (otRMA) of TSs.

Definition 9 (UF-RMA, UF-otCMA, and UF-otRMA of TSs). *The* UF-RMA *security of TSs is the same as the UF-CMA security of TSs except that to answer a signing query,* SignO$(\cdot)$ *randomly chooses $m \leftarrow \mathcal{M}$ itself, runs $\sigma \leftarrow \mathsf{Sign}(sk, m)$, adds $\gamma(m)$ to $\mathcal{Q}_m$ (initialized with $\emptyset$), and returns (m, σ).*

The UF-otCMA *(respectively,* UF-otRMA*) security is the same as the UF-CMA (respectively, UF-RMA) security of TSs, except that $\mathcal{A}$ is only allowed to make one query to the signing oracle* SignO$(\cdot)$.

3.3 Converting Structure-Preserving Signatures into Signing Key Structure-Preserving Trapdoor Signatures

Before showing our conversion, we define a class of SPSs called well-formed SPSs. Roughly speaking, for a well-formed SPS scheme, it is required that the spaces of randomness and exponents of messages are super-polynomially large in the security parameter, and generating a signature element only involves the group operation, while the scalars of group elements are computed as arithmetic circuits of elements in the signing key and the randomness.

Definition 10 (Well-formed SPS). *For an SPS scheme Σ, let $\mathbb{M}_1 \times \mathbb{M}_2 \times \ldots \times \mathbb{M}_n$ be the space of exponents (with $[1]_1$ and $[1]_2$ for bases) of elements in a message,*[6] *and $\mathbb{R}_1 \times \mathbb{R}_2 \times \ldots \times \mathbb{R}_{n'}$ the randomness space (for signing), where $n, n' \in \mathbb{N}$. Σ is said to be well-formed if (a) for all i, $\mathbb{M}_i, \mathbb{R}_i \subseteq \mathbb{Z}_p$ and $|\mathbb{M}_i|$ and*

[6] We do not count repeated message spaces, e.g., when messages are of the form $([m]_1, [m]_2)$ where $m \in \mathbb{Z}_p$, we have $n = 1$ and $\mathbb{M}_1 = \mathbb{Z}_p$.

$|\mathbb{R}_i|$ are super-polynomial in the security parameter,[7] *and (b) generating a group element $[B]_b$ where $b \in \{1,2\}$ in a signature only involves computing*

$$[B]_b = \sum_i (\prod_j a_{ij}^{c_{ij}})[A_i]_b, \tag{1}$$

where $\{[A_i]_b\}_i$ denotes elements appearing in the public parameters, the message, and the signing key, $\{a_{ij}\}_{ij}$ denotes elements (in $\mathbb{Z}_p$) appearing in the signing key and the randomness for signing, and integer constants, and $\{c_{ij}\}_{ij}$ denotes integer constants. Here, elements in $\{[A_i]_b\}_i$ may represent the same variables, and the same argument is made for $\{a_{ij}\}_{ij}$.[8]

Note that there is no requirement on the distributions of the elements other than the space sizes in the above definition, and as far as we know, all the existing SPSs are well-formed. Now we show that any well-formed SPS scheme can be converted into an SKSP-TS scheme.

Theorem 2. *Any well-formed SPS scheme, the messages of which are supposed to be of the form $([\vec{M}]_1, [\vec{N}]_2)$, can be converted into a γ-SKSP-TS scheme for γ defined by $\gamma(\vec{M}, \vec{N}) = ([\vec{M}]_1, [\vec{N}]_2)$.*

Schwartz-Zippel Lemma. Now we introduce Schwartz-Zippel Lemma [38], based on which we will give the proof of Theorem 2,

Lemma 1 *([38]). Let $P \in F[x]$ be a non-zero polynomial of total degree $d \geq 0$ over a field, S a finite subset of F, and r a randomness uniformly chosen from S. Then, we have*

$$\Pr[P(r) = 0] \leq d/|S|.$$

This lemma indicates that a polynomial of degree d over $\mathbb{Z}_p$ has at most d roots.

Proof (of Theorem 2). We divide the proof of Theorem 2 into two parts. In the first part, we show that any well-formed SPS scheme can be converted into a γ-TS scheme satisfying the conditions (a) and (b) of the SKSP property in Definition 7. In the second part, we prove that the converted TS scheme also satisfies the condition (c).

Part I. Let a group element in a signature be generated as Eq. (1). For all i such that $\{a_{ij}\}_j$ contains a set of variables in the signing key, denoted by $\{s_{ij}\}_j$, we use c'_{ij} to denote the exponent of s_{ij} in Eq. (1), and do the following conversion.

---

[7] It is not hard to see that an SPS scheme whose messages are of the form, e.g., $([m_1]_1, [m_2]_2)$ where $m_1 = m_2 + 1$ and $m_1, m_2 \in \mathbb{Z}_p$, is not well-formed. However, such a scheme can be easily converted to a well-formed one by letting messages be of the form $([m_1]_1, [m_1]_2)$ and compute $[m_1 + 1]_2$ in signing and verification.

[8] For ease of understanding, we give an example here. Supposing that an element in a signature is generated as $(r_1 s_1 + r_2^2 r_1)[U]_1 + s_2^{-1}[M]_1 + [S]_1$, where (r_1, r_2), $(s_1, s_2, [S]_1)$, $[U]_1$, and $[M]_1$ are respectively element(s) in the randomness, signing key, verification key, and message, then we express the formula as $(a_{11}^{c_{11}} a_{12}^{c_{12}})[A_1]_1 + (a_{21}^{c_{21}} a_{22}^{c_{22}})[A_2]_1 + a_{31}^{c_{31}}[A_3]_1 + [A_4]_1$, where $([A_1]_1, [A_2]_1, [A_3]_1, [A_4]_1, a_{11}, a_{12}, a_{21}, a_{22}, a_{31})$ represents $([U]_1, [U]_1, [M]_1, [S]_1, r_1, s_1, r_2, r_1, s_2)$ and $(c_{11}, c_{12}, c_{21}, c_{22}, c_{31}) = (1, 1, 2, 1, -1)$.

- If $[A_i]_b$ is in the message, then we add $[(\prod_j s_{ij}^{c'_{ij}})]_b$ to the signing key.
- Otherwise (i.e., if $[A_i]_b$ is in the signing key or the verification key), then we add $[(\prod_j s_{ij}^{c'_{ij}})A_i]_b$ to the signing key,

For all other group elements in the signature, we execute the same conversions. Then we remove all elements in $\mathbb{Z}_p$, all repeated elements, and elements never used in signing procedures from the original signing key, and set the original signing key as the trapdoor key.

By using the new signing key, we can generate a signature consisting of group elements in the forms of Eq. (1) when taking as input a message consisting of $M_1, M_2, \ldots, N_1, N_2, \ldots \in \mathbb{Z}_p$, which forms the signing algorithm for the resulting γ-TS scheme. Furthermore, taking as input $[M_1]_1, [M_2]_1, \ldots, [N_1]_2, [N_2]_2, \ldots$ and the trapdoor key, we can generate the same signature if the randomness is the same, by using the original signing algorithm. As a result, we have obtained a γ-TS scheme for $\gamma(\vec{M}, \vec{N}) = ([\vec{M}]_1, [\vec{N}]_2)$.

It is straightforward to see that in this γ-TS scheme, the verification keys, signing keys, and signatures consist only of group elements in $\mathbb{G}_1$ and $\mathbb{G}_2$ and the verification consists only of evaluating membership in $\mathbb{G}_1$ and $\mathbb{G}_2$ and relations described by PPEs. This completes the first part of the proof. Here, the verification key size, signature size, and number of PPEs do not change during the conversion, while the signing key size changes depending on the concrete construction of the SPS scheme.[9]

Part II. Next we prove that for the above γ-TS scheme, there exists an algorithm that can check the correctness of signing keys with respect to verification keys by using only PPEs.

Since a group element in the signature is computed as Eq. (1), and a group element in the message $[M]_1$ or $[N]_2$ can be treated as $M[1]_1$ or $N[1]_2$, a PPE in the verification algorithm can be written as

$$\sum_i \Big(\prod_j x_{ij}{}^{d_{ij}}\Big)[X_i]_T = [0]_T, \tag{2}$$

where $\{x_{ij}\}_{ij}$ denotes elements in the randomness, exponents of the message, and integer constants, $\{d_{ij}\}_{ij}$ denotes integer constants, and $\{[X_i]_T\}_i$ denotes pairings between elements in the verification key and the signing key. Here, elements in $\{x_{ij}\}_{ij}$ may represent the same variables, and the same argument is made for $\{[X_i]_T\}_i$.

We now show how to obtain PPEs that check the correctness of signing keys with respect to verification keys as follows. Let $\mathcal{E}$ be the set of all the *distinct* variables in $\{x_{ij}\}_{ij}$ (not including constants). Then for any $x \in \mathcal{E}$, we rewrite Eq. (2) as

$$\sum_i x^{d_i}[Y_i]_T = [0]_T, \tag{3}$$

[9] In the worst case, the resulting signing key size is the total number of elements in all $\{[A_i]_b\}_i$.

where $\{d_i\}_i$ denotes fixed polynomials and $\{[Y_i]_T\}_i$ denotes elements in $\mathbb{G}_T$. Since the SPS scheme is well-formed, the left hand side of Eq. (3) can be treated as a polynomial in x by fixing all $[Y_i]_T$. We rewrite Eq. (3) as

$$[P_0]_T + x^1[P_1]_T + \ldots + x^n[P_n]_T = [0]_T, \tag{4}$$

for some fixed polynomial n, where $[P_k]_T$ denotes the sum of coefficients of x^k. According to the definition of well-formed SPSs, since the space of x is super-polynomial (in the security parameter) and n is a polynomial (in the security parameter), the number of possible values of x must be larger than n for sufficiently large security parameters. As a result, if Eq. (4) holds for all possible value of x, we have

$$[P_0]_T = [0]_T,\ [P_1]_T = [0]_T,\ \ldots,\ [P_n]_T = [0]_T, \tag{5}$$

or the number of roots of Eq. (4) could be larger than n, which is against Schwartz-Zippel Lemma. On the other hand, it is obvious that if PPEs in (5) hold, Eq. (4) holds for any x. For each $[P_i]_T = 0$, we cancel another variable in $\mathcal{E}$ in the same way. Recursively, all the variables in PPEs in the verification algorithm can be cancelled, and we finally obtain a sequence of PPEs of the form

$$\sum_i c_i'[X_i']_T = [0]_T,$$

where $\{c_i'\}_i$ denotes fixed integers, and $\{[X_i']_T\}_i$ denotes pairings between elements in the verification key and the signing key, and elements in $\{[X_i']_T\}_i$ may represent the same variables. Since such collection of PPEs hold *if and only if* PPEs in the verification algorithm holds for all possible randomness and messages, we obtain an algorithm that takes as input verification/signing key pairs and check their correctness using this collection of PPEs.[10]

In conclusion, any well-formed SPS scheme can be converted into an SKSP-TS scheme, completing the proof of Theorem 2. □

Remark. It is not hard to see that the latter half of the proof can also be adopted to show that for a well-formed SPS scheme, if signing keys consist only of group elements, then it is an FSPS scheme.

3.4 Instantiations of Trapdoor Signature

UF-CMA secure TS scheme. Using the conversion described in the proof of Theorem 2, we can convert well-formed SPSs into SKSP-TSs. For ease of understanding, we give an instantiation of γ-TS $\Sigma = (\mathsf{Setup}, \mathsf{Gen}, \mathsf{Sign}, \mathsf{Verify}, \mathsf{TDSign})$ in Fig. 1, which is converted from the SPS scheme (denoted by $\mathcal{T}_\Sigma = (\mathsf{Setup},$

[10] The number of PPEs we finally obtain is smaller than number of elements in $\{[X_i]_T\}_i$ in PPEs of the form Eq. (2).

$\mathcal{T}_{\mathsf{Gen}}, \mathsf{TDSign}, \mathsf{Verify})$) proposed by Kiltz et al. [31]. Here, $\mathcal{T}_\Sigma$ is UF-CMA secure under the $\mathcal{U}_k$-MDDH assumptions and $\gamma : \mathbb{Z}_p^n \mapsto \mathbb{G}_1^n$ is defined by $\gamma(x_1, \ldots, x_n) = ([x_1]_1, \ldots, [x_n]_1)$, where n denotes the number of group elements in a message.

To generate a signature of $\mathcal{T}_\Sigma$, $\sigma_1 = [(1, \vec{m}^\top)]_1\mathbf{K} + \vec{r}^\top[\mathbf{P}_0 + \tau\mathbf{P}_1]_1$ is the only part that needs to be operated by using "$\mathbb{Z}_p$-elements" $\mathbf{K} \in \mathbb{Z}_p^{(n+1)\times(k+1)}$ of the signing key. Following our conversion, we replace $\mathbf{K}$ with $[\mathbf{K}]_1$ in the signing key, and keep the original signing key as the trapdoor key. By using $[\mathbf{K}]_1$, we can compute σ_1 as $(1, \vec{m}^\top)[\mathbf{K}]_1 + \vec{r}^\top[\mathbf{P}_0 + \tau\mathbf{P}_1]_1$. Furthermore, we obtain PPEs that check the correctness of signing keys as follows.

$$e(\sigma_1, [\mathbf{A}]_2) = e([(1, \vec{m}^\top)]_1, [\mathbf{C}]_2) + e(\sigma_2, [\mathbf{C}_0]_2) + e(\sigma_3, [\mathbf{C}_1]_2),$$

$$\Rightarrow e((1, \vec{m}^\top)[\mathbf{K}]_1 + \vec{r}^\top[\mathbf{P}_0 + \tau\mathbf{P}_1]_1, [\mathbf{A}]_2) = e([(1, \vec{m}^\top)]_1, [\mathbf{C}]_2) + e(\vec{r}^\top[\mathbf{B}^\top]_1, [\mathbf{C}_0]_2)$$
$$+ e(\vec{r}^\top[\mathbf{B}^\top\tau]_1, [\mathbf{C}_1]_2), \qquad \text{(Rewrite first equation in Verify)}$$

$$\Rightarrow \begin{cases} e((1, \vec{m}^\top)[\mathbf{K}]_1 + \vec{r}^\top[\mathbf{P}_0]_1, [\mathbf{A}]_2) = e([(1, \vec{m}^\top)]_1, [\mathbf{C}]_2) + e(\vec{r}^\top[\mathbf{B}^\top]_1, [\mathbf{C}_0]_2), \\ e(\vec{r}^\top[\mathbf{P}_1]_1, [\mathbf{A}]_2) = e(\vec{r}^\top[\mathbf{B}^\top]_1, [\mathbf{C}_1]_2), \end{cases} \qquad \text{(Cancelling } \tau)$$

$$\Rightarrow \begin{cases} e((1, \vec{m}^\top)[\mathbf{K}]_1, [\mathbf{A}]_2) = e([(1, \vec{m}^\top)]_1, [\mathbf{C}]_2), \\ e([\mathbf{P}_0]_1, [\mathbf{A}]_2) = e([\mathbf{B}^\top]_1, [\mathbf{C}_0]_2), \\ e([\mathbf{P}_1]_1, [\mathbf{A}]_2) = e([\mathbf{B}^\top]_1, [\mathbf{C}_1]_2), \end{cases} \qquad \text{(Cancelling } \vec{r})$$

$$\stackrel{*}{\Rightarrow} \begin{cases} e([\mathbf{K}]_1, [\mathbf{A}]_2) = e([1]_1, [\mathbf{C}]_2)), \\ e([\mathbf{P}_0]_1, [\mathbf{A}]_2) = e([\mathbf{B}^\top]_1, [\mathbf{C}_0]_2), \\ e([\mathbf{P}_1]_1, [\mathbf{A}]_2) = e([\mathbf{B}^\top]_1, [\mathbf{C}_1]_2). \end{cases} \qquad \text{(Cancelling } \vec{m})$$

Then we rewrite the second equation $e(\sigma_2, \sigma_4) = e(\sigma_3, [1]_2)$ as $e(\vec{r}^\top[\mathbf{B}^\top]_1, [\tau]_2) = e(\vec{r}^\top[\mathbf{B}^\top\tau]_1, [1]_2)$. By cancelling $\vec{r}$ and τ, we obtain $e([\mathbf{B}^\top]_1, [1]_2) = e([\mathbf{B}^\top]_1, [1]_2)$, which is trivial.[11]

Finally, we obtain the algorithm $\mathsf{VerifySK}$ checking correctness of signing keys with respect to verification keys via the above three PPEs (derived from $\stackrel{*}{\Rightarrow}$).

Theorem 3. *The instantiation described in Fig. 1 is a UF-CMA secure γ-SKSP-TS scheme under the $\mathcal{U}_k$-MDDH assumptions.*

The SKSP property of this instantiation is implied by Theorem 2 and the UF-CMA security is implied by Theorem 1. We refer the reader to the full paper for the proof of Theorem 3.

UF-otRMA secure TS scheme. In Fig. 2, we give another instantiation of TS which satisfies the UF-otRMA security under the $\mathcal{U}_k$-MDDH assumptions. This scheme is converted from the UF-otRMA secure SPS scheme in [31]. The proof of correctness is straightforward and the correctness of a signing key with respect to a verification key can be verified by $\mathsf{VerifySK}$ via $e([\mathbf{K}]_1, [\overline{\mathbf{A}}]_2) = e([1]_1, [\mathbf{C}]_2)$.

Unlike the UF-CMA security proved in Theorem 1, the UF-otRMA security of $\Sigma = (\mathsf{Setup}, \mathsf{Gen}, \mathsf{Sign}, \mathsf{Verify}, \mathsf{TDSign})$ is not automatically implied by the UF-otRMA security of $\mathcal{T}_\Sigma = (\mathsf{Setup}, \mathcal{T}_{\mathsf{Gen}}, \mathsf{TDSign}, \mathsf{Verify})$. However, according

[11] Note that for simplicity, we sometimes directly canceled vectors in the above conversion, instead of following the proof of Theorem 2 to cancel elements one by one.

Setup(1^λ): $par = (p, \mathbb{G}_1, \mathbb{G}_2, \mathbb{G}_T, e, [1]_1, [1]_2) \leftarrow \mathcal{G}(1^\lambda)$. For preliminary-fixed $n \in \mathbb{N}$, define $\mathcal{M} = \mathbb{Z}_p^n$ and $\mathcal{M}_\gamma = \mathbb{G}_1^n$. Define γ by $\gamma(m_1, \ldots, m_n) = ([m_1]_1, \ldots [m_n]_1)$. Return par.	Sign$(sk, \vec{m})$: $\vec{r} \leftarrow \mathbb{Z}_p^k$, $\tau \leftarrow \mathbb{Z}_p$, $\sigma_1 = \boxed{(1, \vec{m}^\top)[\mathbf{K}]_1} + \vec{r}^\top [\mathbf{P}_0 + \tau \mathbf{P}_1]_1$, $\sigma_2 = \vec{r}^\top [\mathbf{B}^\top]_1$, $\sigma_3 = \vec{r}^\top [\mathbf{B}^\top \tau]_1$, $\sigma_4 = [\tau]_2 \in \mathbb{G}_2$. Return $(\sigma_1, \sigma_2, \sigma_3, \sigma_4) \in \mathbb{G}_1^{1\times(k+1)} \times \mathbb{G}_1^{1\times(k+1)} \times \mathbb{G}_1^{1\times(k+1)} \times \mathbb{G}_2$.
Gen(par): $\mathbf{A}, \mathbf{B} \leftarrow \mathcal{D}_k$, $\mathbf{K} \leftarrow \mathbb{Z}_p^{(n+1)\times(k+1)}$, $\mathbf{K}_0, \mathbf{K}_1 \leftarrow \mathbb{Z}_p^{(k+1)\times(k+1)}$, $\mathbf{C} = \mathbf{K}\mathbf{A} \in \mathbb{Z}_p^{(n+1)\times k}$, $\mathbf{C}_0 = \mathbf{K}_0\mathbf{A} \in \mathbb{Z}_p^{(k+1)\times k}$, $\mathbf{C}_1 = \mathbf{K}_1\mathbf{A} \in \mathbb{Z}_p^{(k+1)\times k}$, $\mathbf{P}_0 = \mathbf{B}^\top \mathbf{K}_0 \in \mathbb{Z}_p^{k\times(k+1)}$, $\mathbf{P}_1 = \mathbf{B}^\top \mathbf{K}_1 \in \mathbb{Z}_p^{k\times(k+1)}$. $pk = ([\mathbf{C}_0]_2, [\mathbf{C}_1]_2, [\mathbf{C}]_2, [\mathbf{A}]_2)$, $sk = (\boxed{[\mathbf{K}]_1}, [\mathbf{P}_0]_1, [\mathbf{P}_1]_1, [\mathbf{B}]_1)$, $tk = \boxed{(\mathbf{K}, [\mathbf{P}_0]_1, [\mathbf{P}_1]_1, [\mathbf{B}]_1)}$. Return (pk, sk) and tk.	Verify$(pk, [\vec{m}]_1, \sigma)$: Parse $\sigma = (\sigma_1, \sigma_2, \sigma_3, \sigma_4)$, Return 1 if $e(\sigma_1, [\mathbf{A}]_2) = e([(1, \vec{m}^\top)]_1, [\mathbf{C}]_2) + e(\sigma_2, [\mathbf{C}_0]_2) + e(\sigma_3, [\mathbf{C}_1]_2)$ and $e(\sigma_2, \sigma_4) = e(\sigma_3, [1]_2)$. Return 0 otherwise.
VerifySK(pk, sk): Return 1 if $e([\mathbf{K}]_1, [\mathbf{A}]_2) = e([1]_1, [\mathbf{C}]_2)$, $e([\mathbf{P}_0]_1, [\mathbf{A}]_2) = e([\mathbf{B}^\top]_1, [\mathbf{C}_0]_2)$, and $e([\mathbf{P}_1]_1, [\mathbf{A}]_2) = e([\mathbf{B}^\top]_1, [\mathbf{C}_1]_2)$. Return 0 otherwise.	TDSign$(tk, [\vec{m}]_1)$: $\vec{r} \leftarrow \mathbb{Z}_p^k$, $\tau \leftarrow \mathbb{Z}_p$. $\sigma_1 = \boxed{[(1, \vec{m}^\top)]_1 \mathbf{K}} + \vec{r}^\top [\mathbf{P}_0 + \tau \mathbf{P}_1]_1$ $\sigma_2 = \vec{r}^\top [\mathbf{B}^\top]_1$, $\sigma_3 = \vec{r}^\top [\mathbf{B}^\top \tau]_1$, $\sigma_4 = [\tau]_2 \in \mathbb{G}_2$. Return $(\sigma_1, \sigma_2, \sigma_3, \sigma_4) \in \mathbb{G}_1^{1\times(k+1)} \times \mathbb{G}_1^{1\times(k+1)} \times \mathbb{G}_1^{1\times(k+1)} \times \mathbb{G}_2$.

Fig. 1. A UF-CMA secure γ-TS scheme adapted from [31, Sect. 4.2]. The boxes indicate the main differences from the original scheme in [31].

to [31], the proof of the UF-otRMA security of $\mathcal{T}_\Sigma$ remains valid even when an adversary sees the exponents of the messages from the signing oracle, which implies the UF-otRMA security of Σ. We refer the reader to [31] for details of the proof.

Setup(1^λ): $par = (p, \mathbb{G}_1, \mathbb{G}_2, \mathbb{G}_T, e, [1]_1, [1]_2) \leftarrow \mathcal{G}(1^\lambda)$. $\mathcal{M} = \mathbb{Z}_p^n$, $\mathcal{M}_\gamma = \mathbb{G}_1^n$. For preliminary-fixed $n \in \mathbb{N}$, define γ by $\gamma(m_1, \ldots, m_n) = ([m_1]_1, \ldots [m_n]_1)$. Return par.	Sign$(sk, \vec{m})$: $\sigma = \boxed{(1, \vec{m}^\top)[\mathbf{K}]_1}$. Return $\sigma \in \mathbb{G}_1^{1\times k}$.
Gen(par): $\mathbf{A} \leftarrow \mathcal{D}_k$, $\mathbf{K} \leftarrow \mathbb{Z}_p^{(n+1)\times k}$, $\mathbf{C} = \mathbf{K}\overline{\mathbf{A}} \in \mathbb{Z}_p^{(n+1)\times k}$, $pk = ([\mathbf{C}]_2, [\overline{\mathbf{A}}]_2)$, $sk = \boxed{[\mathbf{K}]_1}$, $tk = \boxed{\mathbf{K}}$. Return (pk, sk) and tk.	Verify$(pk, [\vec{m}]_1, \sigma)$: Return 1 if $e(\sigma, [\overline{\mathbf{A}}]_2) = e([(1, \vec{m}^\top)]_1, [\mathbf{C}]_2)$. Return 0 otherwise.
VerifySK(pk, sk): Return 1 if $e([\mathbf{K}]_1, [\overline{\mathbf{A}}]_2) = e([1]_1, [\mathbf{C}]_2)$. Return 0 otherwise.	TDSign$(tk, [\vec{m}]_1)$: $\sigma = \boxed{[(1, \vec{m}^\top)]_1 \mathbf{K}}$. Return $\sigma \in \mathbb{G}_1^{1\times k}$.

Fig. 2. A UF-otRMA secure γ-SKSP-TS scheme adapted from [31, Sect. 5.2]. The boxes indicate the main differences from the original scheme in [31].

4 (Two-tier) Signatures with Auxiliary Key(s)

In this section, we introduce AKSs which are used as building blocks to achieve our generic construction of FSPS. In Sect. 4.1, we give the definition of AKSs, define their properties, and give an instantiation of AKS. In Sect. 4.2, we extend AKS to TT-AKS and give an instantiation of TT-AKS.

4.1 Signature with Auxiliary Key

Definition. Roughly speaking, a γ-AKS scheme is a signature scheme in which the key generation algorithm additionally generates auxiliary keys, and the verification key space and the auxiliary key space have a special (but natural) structure related with γ.

Definition 11 (γ-signature with auxiliary key (γ-AKS)). *A signature scheme $\Sigma = (\mathsf{Setup}, \mathsf{Gen}, \mathsf{Sign}, \mathsf{Verify})$ with verification key space $\mathcal{P}_\gamma$ is said to be a γ-AKS scheme for an efficiently computable bijection $\gamma : \mathcal{P} \mapsto \mathcal{P}_\gamma$ if in addition to the verification/signing key pair (pk, sk), Gen also outputs an auxiliary key $ak \in \mathcal{P}$ such that $pk = \gamma(ak)$.*

Security. The UF-(ot)CMA security and UF-(ot)RMA security of γ-AKSs are exactly the same as those of standard signatures except that Gen addtionally generates ak.

Key generation algorithm $\mathcal{U}_{\mathsf{Gen}}$. Similarly to $\mathcal{T}_{\mathsf{Gen}}$ defined in Sect. 3.1, we use $\mathcal{U}_{\mathsf{Gen}}$ to denote an algorithm that runs Gen, which is the key generation algorithm of a γ-AKS scheme, in the following way.

Taking as input a public parameter par, $\mathcal{U}_{\mathsf{Gen}}$ gives par to Gen and obtains an output $((pk, sk), ak)$. Then $\mathcal{U}_{\mathsf{Gen}}$ outputs (pk, sk) as a verification/signing key pair, without outputting ak. We use $\mathcal{U}_\Sigma$ to denote $(\mathsf{Setup}, \mathcal{U}_{\mathsf{Gen}}, \mathsf{Sign}, \mathsf{Verify})$ when $\Sigma = (\mathsf{Setup}, \mathsf{Gen}, \mathsf{Sign}, \mathsf{Verify})$.

Just like SPSs, we consider γ-AKSs with the SP property.

Definition 12 (γ-SP-AKS). *A γ-AKS scheme Σ is said to be a* γ-SP-AKS scheme *if $\mathcal{U}_\Sigma$ is an SPS scheme.*

Converting SPSs into SP-AKSs. It is straightforward to see that any SPS scheme with an algebraic key generation algorithm, public keys of which are supposed to be of the form $([\vec{u}]_1, [\vec{v}]_2)$, can be converted into a γ-SP-AKS scheme, where γ is defined by $\gamma(\vec{u}, \vec{v}) = ([\vec{u}]_1, [\vec{v}]_2)$, since we can force the setup of any SPS to output no common parameter except for the bilinear map description and let the key generation algorithm additionally output $(\vec{u}_1, \vec{v}_2)$.

We now define the random auxiliary key property for AKSs.

Definition 13 (Random auxiliary key property). *A γ-AKS scheme $(\mathsf{Setup}, \mathsf{Gen}, \mathsf{Sign}, \mathsf{Verify})$ with an auxiliary key space $\mathcal{P}$ is said to satisfy the* random auxiliary key property *if there exists an additional algorithm* AKGen *such that* AKGen *takes as input par and an auxiliary key ak, and outputs a verification/signing key pair (pk, sk) where $\gamma(ak) = pk$. Furthermore, for any PPT adversary $\mathcal{A}$ and all $\lambda \in \mathbb{N}$, we have*

$$|\Pr[par \leftarrow \mathsf{Setup}(1^\lambda) : \mathcal{A}^{\mathrm{GenO}}(par) = 1] - \Pr[par \leftarrow \mathsf{Setup}(1^\lambda) : \mathcal{A}^{\mathrm{AKGenO}}(par) = 1]| \leq negl(\lambda),$$

where GenO *runs* $((pk, sk), ak) \leftarrow \mathsf{Gen}(par)$, *and returns* (pk, sk, ak), *and* AKGenO *uniformly chooses* ak *from* $\mathcal{P}$, *runs* $(pk, sk) \leftarrow \mathsf{AKGen}(par, ak)$, *and returns* (pk, sk, ak).

Instantiation of AKS. Now we give an instantiation of AKS satisfying UF-otCMA security under the $\mathcal{U}_k$-MDDH assumptions (see the full paper) in Fig. 3. This signature scheme is actually the same as the UF-otCMA secure signature scheme in [31] except that Gen additionally generates exponents of a verification key as an auxiliary key. For this instantiation, the bijection γ is defined by $\gamma(\mathbf{X}) = [\mathbf{X}]_2 \in \mathbb{G}_2^{(n+1)\times k} \times \mathbb{G}_2^{(k+1)\times k}$ for n which denotes the length of a message.

We refer the reader to [31] for the proof of the UF-otCMA security of this instantiation.

$\mathsf{Setup}(1^\lambda)$: $par = (p, \mathbb{G}_1, \mathbb{G}_2, \mathbb{G}_T, e, [1]_1, [1]_2) \leftarrow \mathcal{G}(1^\lambda)$. For preliminary-fixed $n \in \mathbb{N}$, define $\mathcal{M} = \mathbb{Z}_p^n$, $\mathcal{M}_\gamma = \mathbb{G}_1^n$, $\mathcal{P} = \mathbb{Z}_p^{(n+1)\times k} \times \mathbb{Z}_p^{(k+1)\times k}$, and $\mathcal{P}_\lambda = \mathbb{G}_2^{(n+1)\times k} \times \mathbb{G}_2^{(k+1)\times k}$. Define γ by $\gamma(\mathbf{X}) = [\mathbf{X}]_2 \in \mathbb{G}_2^{(n+1)\times k} \times \mathbb{G}_2^{(k+1)\times k}$. Return par.	$\mathsf{AKGen}(par, ak)$: Parse $ak = (\mathbf{C}, \mathbf{A})$. Let $\mathbf{A} = \begin{pmatrix} \overline{\mathbf{A}} \\ \vec{a}^\top \end{pmatrix}$, $\vec{k} \leftarrow \mathbb{Z}_p^{n+1}$, $\underline{\mathbf{K}} = (\mathbf{C} - \vec{k}\vec{a}^\top)\overline{\mathbf{A}}^{-1}$, $\mathbf{K} = (\underline{\mathbf{K}}, \vec{k})$. $pk = ([\mathbf{C}]_2, [\mathbf{A}]_2)$, $sk = \mathbf{K}$, $ak = (\mathbf{C}, \mathbf{A})$. Return (pk, sk) and ak.
$\mathsf{Gen}(par)$: $\mathbf{A} \leftarrow \mathcal{U}_k$, $\mathbf{K} \leftarrow \mathbb{Z}_p^{(n+1)\times(k+1)}$, $\mathbf{C} = \mathbf{KA} \in \mathbb{Z}_p^{(n+1)\times k}$. $pk = ([\mathbf{C}]_2, [\mathbf{A}]_2)$, $sk = \mathbf{K}$, and $ak = (\mathbf{C}, \mathbf{A})$. Return (pk, sk) and ak.	$\mathsf{Sign}(sk, [\vec{m}]_1)$: $\sigma = [(1, \vec{m}^\top)]_1 \mathbf{K} \in \mathbb{G}_1^{1\times(k+1)}$. $\mathsf{Verify}(pk, [\vec{m}]_1, \sigma)$: Return 1 if $e(\sigma, [\mathbf{A}]_2) = e([(1, \vec{m}^\top)]_1, [\mathbf{C}]_2)$. Return 0 otherwise.

Fig. 3. A UF-otCMA secure γ-SP-AKS scheme adapted from [31, Sect. 3].

Theorem 4. *The instantiation described in Fig. 3 satisfies the random auxiliary key property.*

This proof follows from the fact that when the distribution of $\mathbf{C}$ is uniform, the distribution of $\underline{\mathbf{K}} = (\mathbf{C} - \vec{k}\vec{a}^\top)\overline{\mathbf{A}}^{-1}$ is uniform as well. We give the proof of Theorem 4 in the full paper due to page limitation.

4.2 Two-Tier Signature with Auxiliary Keys

Definition. Besides AKSs, we also give the definition of (γ_p, γ_s)-TT-AKSs, which is the same as that of two-tier signatures [1,12,32] except that the key generation algorithms additionally generate primary/secondary auxiliary keys. The primary/secondary verification key space and the primary/secondary auxiliary key space have a special (but natural) structure related with γ_p/γ_s. Combining SP-TT-AKSs with SKSP-TSs enables us to obtain more efficient instantiations of FSPS and FAS.

Definition 14 (**(γ_p, γ_s)-TT-AKS**). *A (γ_p, γ_s)-TT-AKS scheme consists of five polynomial-time algorithms* Setup, PGen, SGen, TTSign, *and* TTVerify. Setup *is a randomized algorithm that takes as input 1^λ, and outputs a public parameter par, which determines the message space $\mathcal{M}$, the primary/secondary verification key spaces $\mathcal{P}_\gamma/\mathcal{S}_\gamma$, the primary/secondary auxiliary key spaces $\mathcal{P}/\mathcal{S}$, and the efficiently computable bijections $\gamma_p : \mathcal{P} \mapsto \mathcal{P}_\gamma$ and $\gamma_s : \mathcal{S} \mapsto \mathcal{S}_\gamma$.* PGen *is a randomized algorithm that takes as input par, and outputs a primary verification/signing key pair (Ppk, Psk) where $Ppk \in \mathcal{P}_\gamma$ and a primary auxiliary key $Pak \in \mathcal{P}$.* SGen *is a randomized algorithm that takes as input a primary verification/signing key pair (Ppk, Psk) and a primary auxiliary key Pak, and outputs a secondary verification/signing key pair (opk, osk) where $opk \in \mathcal{S}_\gamma$ and a secondary auxiliary key $oak \in \mathcal{S}$.* TTSign *is a randomized algorithm that takes as input a primary signing key Psk, a secondary signing key osk, and a message m, and returns a signature σ.* TTVerify *is a deterministic algorithm that takes as input a primary verification key Ppk, a secondary verification key opk, a message m, and a signature σ, and returns* 1 *(accept) or* 0 *(reject).*

The correctness is satisfied if for all $\lambda \in \mathbb{N}$, $par \leftarrow$ Setup(1^λ), $((Ppk, Psk), Pak) \leftarrow \mathsf{PGen}(par)$, and $((opk, osk), oak) \leftarrow \mathsf{SGen}(Ppk, Psk, Pak)$, we have (a) $\mathsf{TTVerify}(Ppk, opk, m, \mathsf{TTSign}(Psk, osk, m)) = 1$ for all messages $m \in \mathcal{M}$, and (b) $\gamma_p(Pak) = Ppk$ and $\gamma_s(oak) = opk$.

Unlike the definition of standard two-tier signatures, SGen takes as input (Ppk, Psk, Pak) (instead of (Ppk, Psk)) in the above definition. However, the interface of SGen is not essentially changed since Pak can be treated as part of Psk.

Security. Now we give the definition of unforgeability against two-tier chosen message attacks (UF-TT-CMA).

Definition 15 (**UF-TT-CMA**). *A TT-AKS scheme* (PGen, SGen, TTSign, TTVerify) *is said to be unforgeable against two-tier chosen message attacks if for any PPT adversary $\mathcal{A}$, we have*

$$\begin{aligned}\Pr[par \leftarrow \mathsf{Setup}(1^\lambda), ((Ppk, Psk), Pak) \leftarrow \mathsf{PGen}(par),\\ (i^*, m^*, \sigma^*) \leftarrow \mathcal{A}^{\mathrm{TTSignO}(\cdot)}(Ppk) :\\ (i^*, m) \in \mathcal{TQ}_m \wedge m^* \neq m \wedge \mathsf{TTVerify}(Ppk, opk_{i^*}, m^*, \sigma^*) = 1] \leq negl(\lambda),\end{aligned}$$

where TTSignO$(\cdot)$ *is the signing oracle that takes a message $m \in \mathcal{M}$ as input, runs $i = i + 1$ (initialized with 0), samples $(opk_i, osk_i) \leftarrow \mathsf{SGen}(Ppk, Psk, Pak)$, and computes $\sigma \leftarrow \mathsf{TTSign}(Psk, osk_i, m)$. Then it adds (i, m) to $\mathcal{TQ}_m$ (initialized with $\emptyset$) and returns (opk_i, σ).*

Next we define the SP property of TT-AKS as follows.

Definition 16 (**Structure-preserving TT-AKS (SP-TT-AKS)**). *A TT-AKS scheme is said to be* structure-preserving *over a bilinear group generator $\mathcal{G}$ if we have (a) a public parameter includes a group description gk generated by $\mathcal{G}$, (b) primary and secondary verification keys consist of group elements in*

$\mathbb{G}_1$ and $\mathbb{G}_2$, (c) messages consist of group elements in $\mathbb{G}_1$ and $\mathbb{G}_2$, and (d) the verification algorithm consists only of evaluating membership in $\mathbb{G}_1$ and $\mathbb{G}_2$ and relations described by PPEs.

Converting SP two-tier signatures into SP-TT-AKSs. Like SP-AKSs, an SP two-tier signature scheme, primary and secondary verification keys of which are supposed to be of the form $([\vec{u}]_1, [\vec{v}]_2)$ and $([\vec{u}']_1, [\vec{v}']_2)$ respectively, can be converted into a (γ_p, γ_s)-SP-TT-AKS scheme, where γ_p and γ_s are defined as $\gamma_p(\vec{u}, \vec{v}) = ([\vec{u}]_1, [\vec{v}]_2)$ and $\gamma_s(\vec{u}', \vec{v}') = ([\vec{u}']_1, [\vec{v}']_2)$ respectively, as long as the key generation algorithms are algebraic and primary signing keys consist only of elements in $\mathbb{Z}_p$.[12]

We define the random primary and secondary auxiliary key properties of TT-AKSs as follows.

Definition 17 (Random primary/secondary auxiliary key properties). *A (γ_p, γ_s)-TT-AKS scheme* (Setup, PGen, SGen, TTSign, TTVerify) *is said to satisfy the* random primary auxiliary key property *if there exists an additional polynomial-time algorithm* AKPGen *that takes as input par and a primary auxiliary key Pak, and outputs a primary verification/signing key pair (Ppk, Psk) where $\gamma_p(Pak) = Ppk$. Furthermore, for any PPT adversary $\mathcal{A}$ and all $\lambda \in \mathbb{N}$, we have*

$$|\Pr[par \leftarrow \mathsf{Setup}(1^\lambda) : \mathcal{A}^{\mathrm{PGenO}}(par) = 1] - \Pr[par \leftarrow \mathsf{Setup}(1^\lambda) : \mathcal{A}^{\mathrm{AKPGenO}}(par) = 1]| \leq negl(\lambda),$$

where PGenO *runs $((Ppk, Psk), Pak) \leftarrow \mathsf{PGen}(par)$ and returns $((Ppk, Psk), Pak)$, and* AKPGenO *uniformly chooses Pak from the primary auxiliary key space $\mathcal{P}$, runs $(Ppk, Psk) \leftarrow \mathsf{AKPGen}(par, Pak)$, and returns $((Ppk, Psk), Pak)$.*

Furthermore, it is said to satisfy the random secondary auxiliary key property *if there exists another polynomial-time algorithm* AKSGen *that takes as input a primary verification/signing key pair (Ppk, Psk), a primary auxiliary key Pak, and a secondary auxiliary key oak, and outputs a secondary verification/signing key pair (opk, osk) where $\gamma_s(oak) = opk$. Furthermore, for any PPT adversary $\mathcal{A}$ and all $\lambda \in \mathbb{N}$, we have*

$$|\Pr[par \leftarrow \mathsf{Setup}(1^\lambda) : \mathcal{A}^{\mathrm{SGenO}(\cdot)}(par) = 1] - \Pr[par \leftarrow \mathsf{Setup}(1^\lambda) : \mathcal{A}^{\mathrm{AKSGenO}(\cdot)}(par) = 1]| \leq negl(\lambda),$$

Here, on input a polynomial $n = n(\lambda)$, SGenO$(\cdot)$ *runs $((Ppk, Psk), Pak) \leftarrow \mathsf{PGen}(par)$ and $((opk_i, osk_i), oak_i) \leftarrow \mathsf{SGen}(Ppk, Psk, Pak)$ for $i = 1, \ldots, n$, and returns $(Ppk, Psk, Pak, \{(opk_i, osk_i, oak_i)\}_{i=1}^n)$. On input a polynomial $n = n(\lambda)$,* AKSGenO$(\cdot)$ *runs $((Ppk, Psk), Pak) \leftarrow \mathsf{PGen}(par)$, uniformly chooses oak_i from the secondary auxiliary key space $\mathcal{S}$, runs $(opk_i, osk_i) \leftarrow \mathsf{AKSGen}(Ppk, Psk, Pak, oak_i)$ for $i = 1, \ldots, n$, and returns $(Ppk, Psk, Pak, \{(opk_i, osk_i, oak_i)\}_{i=1}^n)$.*

[12] If a primary signing key consists of group elements, PGen may have trouble in outputting secondary auxiliary keys. However, this can be easily solved by forcing PGen to output the exponents of those group elements as part of a primary signing key.

Instantiation of (γ_p, γ_s)*-SP-TT-AKS.* Now we give an instantiation of (γ_p, γ_s)-SP-TT-AKS satisfying UF-TT-CMA security under the $\mathcal{U}_k$-MDDH assumptions. This signature scheme is the same as the SP two-tier signature scheme in [32] except that PGen and SGen additionally generate the auxiliary keys, and SGen addtionally takes as input the primary auxiliary key. For this instantiation, the bijections (γ_p, γ_s) are defined by $\gamma_p(\mathbf{X}) = [\mathbf{X}]_2 \in \mathbb{G}_2^{n \times k} \times \mathbb{G}_2^{(k+1) \times k}$ and $\gamma_s(\vec{x}) = [\vec{x}]_2 \in \mathbb{G}_2^{1 \times k}$ respectively for some fixed integer n which denotes the length of a message.

Setup(1^λ):
$par = (p, \mathbb{G}_1, \mathbb{G}_2, \mathbb{G}_T, e, [1]_1, [1]_2) \leftarrow \mathcal{G}(1^\lambda)$.
For preliminary-fixed $n \in \mathbb{N}$,
define $\mathcal{M} = \mathbb{Z}_p^n$, $\mathcal{M}_\gamma = \mathbb{G}_1^n$,
$\mathcal{P} = \mathbb{Z}_p^{n \times k} \times \mathbb{Z}_p^{(k+1) \times k}$, $\mathcal{P}_\gamma = \mathbb{G}_2^{n \times k} \times \mathbb{G}_2^{(k+1) \times k}$,
$\mathcal{S} = \mathbb{Z}_p^{1 \times k}$, and $\mathcal{S}_\gamma = \mathbb{G}_2^{1 \times k}$.
Define γ_p by $\gamma_p(\mathbf{X}) = [\mathbf{X}]_2 \in \mathbb{G}_2^{n \times k} \times \mathbb{G}_2^{(k+1) \times k}$
and γ_s by $\gamma_s(\vec{x}) = [\vec{x}]_2 \in \mathbb{G}_2^{1 \times k}$.
Return par.

PGen(par):
$\mathbf{A} \leftarrow \mathcal{U}_k$, $\mathbf{K}' \leftarrow \mathbb{Z}_p^{n \times (k+1)}$, $\mathbf{C}' = \mathbf{K}'\mathbf{A} \in \mathbb{Z}_p^{n \times k}$.
$Ppk = ([\mathbf{C}']_2, [\mathbf{A}]_2))$, $Psk = \mathbf{K}'$, $Pak = (\mathbf{C}', \mathbf{A})$.
Return (Ppk, Psk) and Pak.

SGen(Ppk, Psk, Pak):
$\vec{k} \leftarrow \mathbb{Z}_p^{k+1}$, $\vec{c} = \vec{k}^\top \mathbf{A} \in \mathbb{Z}_p^{1 \times k}$.
$opk = [\vec{c}]_2$, $osk = \vec{k}$, $oak = \vec{c}$.
Return (opk, osk) and oak.

AKPGen(par, Pak):
Parse $Pak = (\mathbf{C}', \mathbf{A})$.
Let $\mathbf{A} = \begin{pmatrix} \overline{\mathbf{A}} \\ \vec{a}^\top \end{pmatrix}$,
$\vec{k}' \leftarrow \mathbb{Z}_p^n$, $\underline{\mathbf{K}'} = (\mathbf{C}' - \vec{k}'\vec{a}^\top)\overline{\mathbf{A}}^{-1} \in \mathbb{Z}_p^{n \times k}$,
$\mathbf{K}' = (\underline{\mathbf{K}'}, \vec{k}') \in \mathbb{Z}_p^{n \times (k+1)}$.
$Ppk = ([\mathbf{C}']_2, [\mathbf{A}]_2)$, $Psk = \mathbf{K}'$, $Pak = (\mathbf{C}', \mathbf{A})$.
Return (Ppk, Psk) and Pak.

AKSGen(Ppk, Psk, Pak, oak):
Parse $Ppk = ([\mathbf{C}']_2, [\mathbf{A}]_2))$, $Psk = \mathbf{K}'$, $Pak = (\mathbf{C}', \mathbf{A})$,
and $oak = \vec{c}$.
Let $\mathbf{A} = \begin{pmatrix} \overline{\mathbf{A}} \\ \vec{a}^\top \end{pmatrix}$, $k \leftarrow \mathbb{Z}_p$, $\vec{k}'^\top = (\vec{c} - k\vec{a}^\top)\overline{\mathbf{A}}^{-1}$, $\vec{k}^\top = (\vec{k}'^\top, k)$.
$opk = [\vec{c}]_2$, $osk = \vec{k}$, $oak = \vec{c}$.
Return (opk, osk) and oak.

TTSign$(Psk, osk, [\vec{m}]_1)$:
$\mathbf{K} = (\vec{k}, \mathbf{K}'^\top)^\top$.
Return $\sigma = [(1, \vec{m}^\top)]_1 \mathbf{K} \in \mathbb{G}_1^{1 \times (k+1)}$.

TTVerify$(Ppk, opk, [\vec{m}]_1, \sigma)$:
$[\mathbf{C}]_2 = ([\vec{c}]_2^\top, [\mathbf{C}']_2^\top)^\top$.
Return 1 if $e(\sigma, [\mathbf{A}]_2) = e([(1, \vec{m}^\top)]_1, [\mathbf{C}]_2)$.
Return 0 otherwise.

Fig. 4. A UF-TT-CMA secure (γ_p, γ_s)-SP-TT-AKS scheme adapted from [32, Sect. 6.1].

Theorem 5. *The instantiation described in Fig. 4 satisfies the random primary and secondary auxiliary key properties.*

This proof follows from the fact that when the distributions of $\mathbf{C}'$ and c are uniform, the distribution of $\underline{\mathbf{K}'} = (\mathbf{C}' - \vec{k}\vec{a}^\top)\overline{\mathbf{A}}^{-1}$ and $\vec{k}' = (\vec{c} - k\vec{a}^\top)\overline{\mathbf{A}}^{-1}$ are uniform as well. We give the proof of Theorem 5 in the full paper due to page limitation.

5 Generic Constructions of Fully Structure-Preserving Signatures (and Fully Automorphic Signatures)

In this section, we give generic constructions of FSPSs and FASs from SKSP-TSs and (TT-)AKSs. Such constructions can be derived from SPSs that are based on various assumptions and with different efficiency performance. In Sects. 5.1, 5.2, and 5.3, we give three generic constructions of UF-CMA secure FSPS schemes respectively. The first two constructions are based on SKSP-TSs and SP-AKSs, and the third one is based on SKSP-TSs and SP-TT-AKSs.

5.1 Generic Construction Sig_1: Trapdoor Signature + Signature with Auxiliary Key

We give a generic construction of FSPSs (and FASs) based on a γ-SKSP-TS scheme and a γ'-SP-AKS scheme, where γ and γ' satisfy a suitable compatibility that we explain shortly.

Let $\Sigma_t = (\mathsf{Setup}, \mathsf{Gen}, \mathsf{Sign}, \mathsf{Verify}, \mathsf{TDSign}, \mathsf{VerifySK})$ be a γ-SKSP-TS scheme with message spaces $\mathcal{M}$ and $\mathcal{M}_\gamma$, and $\Sigma_s = (\mathsf{Setup}, \mathsf{Gen}', \mathsf{Sign}', \mathsf{Verify}')$[13] a γ'-SP-AKS scheme with verification key space $\mathcal{M}_\gamma$, auxiliary key space $\mathcal{M}$, and message space $\mathcal{M}'$, and we have $\gamma'(x) = \gamma(x)$. Then the generic construction of FSPS denoted by $\mathsf{Sig}_1 = (\widehat{\mathsf{Setup}}, \widehat{\mathsf{Gen}}, \widehat{\mathsf{Sign}}, \widehat{\mathsf{Verify}}, \widehat{\mathsf{VerifySK}})$ with message space $\mathcal{M}'$ is described as in Fig. 5.

$\widehat{\mathsf{Setup}}(1^\lambda)$: Run $par \leftarrow \mathsf{Setup}(1^\lambda)$. Determine the message spaces $\mathcal{M}$ and $\mathcal{M}_\gamma$ for Σ_t. Define $\gamma : \mathcal{M} \mapsto \mathcal{M}_\gamma$. Determine the message space $\mathcal{M}'$, verification key space $\mathcal{M}_\gamma$, and auxiliary key space $\mathcal{M}$ for Σ_s. Define $\gamma' : \mathcal{M} \mapsto \mathcal{M}_\gamma$ where $\gamma'(x) = \gamma(x)$. Return par.	$\widehat{\mathsf{Sign}}(sk, M)$: $((pk', sk'), ak') \leftarrow \mathsf{Gen}'(par)$. $\sigma_1 \leftarrow \mathsf{Sign}(sk, ak')$. $\sigma_2 = pk'$. $\sigma_3 \leftarrow \mathsf{Sign}'(sk', M)$. Return $\sigma = (\sigma_1, \sigma_2, \sigma_3)$.
$\widehat{\mathsf{Gen}}(par)$: $((pk, sk), tk) \leftarrow \mathsf{Gen}(par)$. Return (pk, sk). $\widehat{\mathsf{VerifySK}}(pk, sk)$: Return 1 if $\mathsf{VerifySK}(pk, sk) = 1$. Return 0 otherwise.	$\widehat{\mathsf{Verify}}(pk, M, \sigma)$: Parse $\sigma = (\sigma_1, \sigma_2, \sigma_3)$ and $\sigma_2 = pk'$. Return 1 if $\mathsf{Verify}(pk, \sigma_2, \sigma_1) = 1$ and $\mathsf{Verify}'(pk', M, \sigma_3) = 1$. Return 0 otherwise.

Fig. 5. Generic construction Sig_1: TS + AKS (UF-otCMA).

Next we give a theorem for this generic construction.

Theorem 6. *If Σ_t is a UF-CMA secure SKSP-TS scheme, and Σ_s a UF-otCMA secure SP-AKS scheme, then $\mathsf{Sig}_1 = (\widehat{\mathsf{Setup}}, \widehat{\mathsf{Gen}}, \widehat{\mathsf{Sign}}, \widehat{\mathsf{Verify}}, \widehat{\mathsf{VerifySK}})$ is a UF-CMA secure FSPS scheme.*

Proof sketch. The proof of Theorem 6 follows from the fact that if there exists a PPT adversary $\mathcal{A}$ that outputs a successful forgery $(\sigma_1^*, \sigma_2^*, \sigma_3^*)$, where σ_2^* was not queried before (respectively, was queried before), with non-negligible probability,

[13] As in [5], we assume that Σ_t and Σ_s share the common setup algorithm Setup.

then we can construct a PPT adversary $\mathcal{B}_1$ (respectively, $\mathcal{B}_2$) that breaks the UF-CMA security of Σ_t (respectively, the UF-otCMA security of Σ_s). Note that to answer a query from $\mathcal{A}$, $\mathcal{B}_2$ may have to use the signing key of Σ_t to sign an auxiliary key ak' of Σ_s, while it only learns the corresponding verification key pk' from the challenger. In this case, it signs pk' by using the trapdoor key of Σ_t instead. According to the correctness of a TS scheme, $\mathcal{A}$ cannot distinguish such a signature with an honestly generated one, which means that $\mathcal{B}_2$ can perfectly simulate the signing oracle of $\mathcal{A}$. We refer the reader to the full paper for the proof.

UF-RMA secure TSs + UF-otCMA secure AKSs. Now we give another theorem showing that for the generic construction in Fig. 5, the security of the TS scheme can be weakened to the UF-RMA security if the AKS scheme satisfies the random auxiliary key property.

Theorem 7. *If Σ_t is a UF-RMA secure SKSP-TS scheme, and Σ_s a UF-otCMA secure SP-AKS scheme satisfying the random auxiliary key property, then* $\mathsf{Sig}_1 = (\widehat{\mathsf{Setup}}, \widehat{\mathsf{Gen}}, \widehat{\mathsf{Sign}}, \widehat{\mathsf{Verify}}, \widehat{\mathsf{VerifySK}})$ *is a UF-CMA secure FSPS scheme.*

Proof sketch. The proof sketch of Theorem 7 is the same as that of Theorem 6 except that $\mathcal{B}_1$ is against the UF-RMA security of Σ_t instead of the UF-CMA security. To answer a query from $\mathcal{A}$, $\mathcal{B}_1$ makes a query to the signing oracle of Σ_t to obtain a randomly chosen auxiliary key ak' and the corresponding signature σ_1. Then $\mathcal{B}_1$ runs the additional algorithm AKGen (defined in Definition 13) on input (par, ak') to generate a verification/signing key pair (pk', sk'), which is indistinguishable from an honestly generated one according to the random auxiliary key property. Then it lets pk' be σ_2 and use sk' to sign the message. We refer the reader to the full paper for the proof of Theorem 7.

Instantiations of Sig_1. By combining the UF-CMA (respectively, UF-otRMA) secure TS scheme in Fig. 1 (respectively, Fig. 2) with the UF-otCMA secure AKS scheme in Fig. 3 (where $\mathbb{G}_1$ and $\mathbb{G}_2$ are swapped), we obtain an FSPS scheme satisfying UF-CMA (respectively, UF-otCMA) security. We refer the reader to the full paper for the resulting signature schemes.

Furthermore, by converting other previously proposed SPSs into SKSP-TSs and SP-AKSs, we obtain various FSPSs. We list some of them in Table 3 in Sect. 6.

5.2 Variation of $\mathbf{Sig}_1$: Trapdoor Signature + Signature with Auxiliary Key (UF-CMA)

Now we give a variation of the generic construction in Fig. 6 by letting Σ_s be a UF-CMA secure SP-AKS scheme and sign n message blocks with one signing key. Each block is signed with an element indicating its number. This change reduces the signature and verification key sizes from $\Omega(n^2)$ to $\Omega(n)$ when signing n^2 group elements.

Let $\Sigma_t = (\mathsf{Setup}, \mathsf{Gen}, \mathsf{Sign}, \mathsf{Verify}, \mathsf{TDSign}, \mathsf{VerifySK})$ be a γ-SKSP-TS scheme with message spaces $\mathcal{M}$ and $\mathcal{M}_\gamma$, and $\Sigma_s = (\mathsf{Setup}, \mathsf{Gen}', \mathsf{Sign}', \mathsf{Verify}')$[14] a γ-SP-AKS scheme with verification key space $\mathcal{M}_\gamma$, auxiliary key space $\mathcal{M}$, and message space $\mathcal{M}' \times \mathcal{M}_I$, where $\mathcal{M}_I$ is the space for elements indicating numbers of blocks. Then a generic construction of FSPS denoted by $\mathsf{Sig}_1^* = (\widehat{\mathsf{Setup}}, \widehat{\mathsf{Gen}}, \widehat{\mathsf{Sign}}, \widehat{\mathsf{Verify}}, \widehat{\mathsf{VerifySK}})$ with message space $\mathcal{M}'^n$, where n is some fixed integer, is described as in Fig. 6.

$\widehat{\mathsf{Setup}}(1^\lambda)$: Run $par \leftarrow \mathsf{Setup}(1^\lambda)$. Determine the message spaces $\mathcal{M}$ and $\mathcal{M}_\gamma$ for Σ_t. Determine the message space $\mathcal{M}' \times \mathcal{M}_I$, verification key space $\mathcal{M}_\gamma$, and auxiliary key space $\mathcal{M}$ for Σ_s. Define $\gamma : \mathcal{M} \mapsto \mathcal{M}_\gamma$. Return par.	$\widehat{\mathsf{Sign}}(sk, \vec{M})$: Parse $\vec{M} = (M_1, \dots, M_n) \in \mathcal{M}'^n$. $((pk', sk'), ak') \leftarrow \mathsf{Gen}'(par)$. $\sigma_1 \leftarrow \mathsf{Sign}(sk, ak')$. $\sigma_2 = pk'$. $\sigma_{3i} \leftarrow \mathsf{Sign}'(sk', (M_i, I(i)))$ where $I(i) \in \mathcal{M}_I$ for $i = 1, \dots, n$. $\sigma_3 = (\sigma_{31}, \dots, \sigma_{3n})$. Return $\sigma = (\sigma_1, \sigma_2, \sigma_3)$.
$\widehat{\mathsf{Gen}}(par)$: $((pk, sk), tk) \leftarrow \mathsf{Gen}(par)$. Return (pk, sk).	$\widehat{\mathsf{Verify}}(pk, \vec{M}, \sigma)$: Parse $\vec{M} = (M_1, \dots, M_n) \in \mathcal{M}'^n$ and $\sigma = (\sigma_1, \sigma_2, \sigma_3)$. Return 1 if $\mathsf{Verify}(pk, \sigma_2, \sigma_1) = 1$ and $\mathsf{Verify}'(pk', (M_i, I(i)), \sigma_{3i}) = 1$ for all i. Return 0 otherwise.
$\widehat{\mathsf{VerifySK}}(pk, sk)$: Return 1 if $\mathsf{VerifySK}(pk, sk) = 1$. Return 0 otherwise.	

Fig. 6. Generic construction Sig_1^*: TS + AKS (UF-CMA).

For this generic construction, the following two theorems hold.

Theorem 8. *If Σ_t is a UF-CMA secure SKSP-TS scheme, and Σ_s a UF-CMA secure SP-AKS scheme, then $\mathsf{Sig}_1^* = (\widehat{\mathsf{Setup}}, \widehat{\mathsf{Gen}}, \widehat{\mathsf{Sign}}, \widehat{\mathsf{Verify}}, \widehat{\mathsf{VerifySK}})$ is a UF-CMA secure FSPS scheme.*

Theorem 9. *If Σ_t is a UF-RMA secure SKSP-TS scheme, and Σ_s a UF-CMA secure SP-AKS scheme satisfying the random auxiliary key property, then $\mathsf{Sig}_1^* = (\widehat{\mathsf{Setup}}, \widehat{\mathsf{Gen}}, \widehat{\mathsf{Sign}}, \widehat{\mathsf{Verify}}, \widehat{\mathsf{VerifySK}})$ is a UF-CMA secure FSPS scheme.*

We omit the proofs of Theorems 8 and 9 since they are similar to the proofs of Theorems 6 and 7, respectively. We list several instantiations of Sig^* in Table 3 in Sect. 6. Most of them achieve better efficiency than instantiations obtained from Sig_1, and are automorphic.

5.3 Generic Construction Sig_2: Trapdoor Signature + Two-Tier Signature with Auxiliary Keys

In this section, we give another generic construction of FSPS which provides us with FSPSs and FASs based on standard assumptions that have shorter verification keys and signatures.

[14] As in [5], we assume that Σ_t and Σ_s share the common setup algorithm Setup.

Let Σ_t = (Setup, Gen, Sign, Verify, TDSign, VerifySK) be a γ-TS scheme with message spaces $\mathcal{M}_p \times \mathcal{M}_s^n$ and $\mathcal{M}_{\gamma p} \times \mathcal{M}_{\gamma s}^n$, Σ_s = (Setup, PGen, SGen, TTSign, TTVerify)[15] a (γ_p, γ_s)-TT-AKS with primary/secondary verification key spaces $\mathcal{M}_{\gamma p}/\mathcal{M}_{\gamma s}$, auxiliary key spaces $\mathcal{M}_p/\mathcal{M}_s$, and message space $\mathcal{M}'$, where n is some fixed integer and $(\gamma_p(x_1), \gamma_s(x_2), \ldots, \gamma_s(x_{n+1})) = \gamma(x_1, x_2 \ldots, x_{n+1})$. A generic construction of FSPS denoted by $\mathsf{Sig}_2 = (\widehat{\mathsf{Setup}}, \widehat{\mathsf{Gen}}, \widehat{\mathsf{Sign}}, \widehat{\mathsf{Verify}}, \widehat{\mathsf{VerifySK}})$ with message space $\mathcal{M}'^n$ is as described as in Fig. 7.

$\widehat{\mathsf{Setup}}(1^\lambda)$:
Run $par \leftarrow \mathsf{Setup}(1^\lambda)$.
Determine the message spaces $\mathcal{M}_p \times \mathcal{M}_s^n$
 and $\mathcal{M}_{\gamma p} \times \mathcal{M}_{\gamma s}^n$ for Σ_t.
Define $\gamma : \mathcal{M}_p \times \mathcal{M}_s^n \mapsto \mathcal{M}_{\gamma p} \times \mathcal{M}_{\gamma s}^n$.
Determine the message spaces $\mathcal{M}'^n$,
 primary verification key space $\mathcal{M}_{\gamma p}$,
 secondary verification key space $\mathcal{M}_{\gamma s}$,
 primary auxiliary key space $\mathcal{M}_p$,
 and secondary auxiliary key space $\mathcal{M}_s$ for Σ_s.
Define $\gamma_p : \mathcal{M}_p \mapsto \mathcal{M}_{\gamma p}$ and $\gamma_s : \mathcal{M}_s \mapsto \mathcal{M}_{\gamma s}$
where
 $(\gamma_p(x_1), \gamma_s(x_2), \ldots, \gamma_s(x_{n+1})) = \gamma(x_1, x_2 \ldots, x_{n+1})$.
Return public parameter par.

$\widehat{\mathsf{Gen}}(par)$:
$((pk, sk), tk) \leftarrow \mathsf{Gen}(par)$.
Return (pk, sk).

$\widehat{\mathsf{VerifySK}}(pk, sk)$:
Return 1 if $\mathsf{VerifySK}(pk, sk) = 1$.
Return 0 otherwise.

$\widehat{\mathsf{Sign}}(sk, \vec{M})$:
Parse $\vec{M} = (M_1, \ldots, M_n) \in \mathcal{M}'^n$.
$((Ppk, Psk), Pak) \leftarrow \mathsf{PGen}(par)$.
$((opk_i, osk_i), oak_i) \leftarrow \mathsf{SGen}(Ppk, Psk, Pak)$
 for $i = 1, \ldots, n$.
$\sigma_1 \leftarrow \mathsf{Sign}(sk, (Pak, oak_1, \ldots, oak_n))$.
$\sigma_2 = (Ppk, opk_1, \ldots, opk_n)$.
$\sigma_{3i} \leftarrow \mathsf{TTSign}(Psk, osk_i, M_i)$ for $i = 1, \ldots, n$.
$\sigma_3 = (\sigma_{31}, \ldots, \sigma_{3n})$.
Return $\sigma = (\sigma_1, \sigma_2, \sigma_3)$.

$\widehat{\mathsf{Verify}}(pk, \vec{M}, \sigma)$:
Parse $\vec{M} = (M_1, \ldots, M_n) \in \mathcal{M}'^n$
 and $\sigma = (\sigma_1, \sigma_2, \sigma_3)$.
Return 1
 if $\mathsf{Verify}(pk, \sigma_2, \sigma_1) = 1$
 and $\mathsf{TTVerify}(Ppk, opk_i, M_i, \sigma_{3i}) = 1$ for all i.
Return 0 otherwise.

Fig. 7. Generic construction Sig_2: TS + TT-AKS.

For this generic construction, the following two theorems hold.

Theorem 10. *If Σ_t is a UF-CMA secure SKSP-TS scheme, and Σ_s a UF-TT-CMA secure SP-TT-AKS scheme, then* $\mathsf{Sig}_2 = (\widehat{\mathsf{Setup}}, \widehat{\mathsf{Gen}}, \widehat{\mathsf{Sign}}, \widehat{\mathsf{Verify}}, \widehat{\mathsf{VerifySK}})$ *is a UF-CMA secure FSPS scheme.*

Theorem 11. *If Σ_t is a UF-RMA secure SKSP-TS scheme, and Σ_s a UF-TT-CMA secure SP-TT-AKS scheme satisfying the random primary and secondary auxiliary key properties, then* $\mathsf{Sig}_2 = (\widehat{\mathsf{Setup}}, \widehat{\mathsf{Gen}}, \widehat{\mathsf{Sign}}, \widehat{\mathsf{Verify}}, \widehat{\mathsf{VerifySK}})$ *is a UF-CMA secure FSPS scheme.*

The proofs of Theorems 10 and 11 are similar to the proofs of Theorems 6 and 7, respectively. We give them in the full paper.

Instantiations of Sig_2. By combining the UF-CMA (respectively, UF-otRMA) secure TS scheme in Fig. 1 (respectively, Fig. 2) with the UF-TT-CMA secure

[15] As in [5], we assume that Σ_t and Σ_s share the common setup algorithm Setup.

AKS scheme in Fig. 4 (where $\mathbb{G}_1$ and $\mathbb{G}_2$ are swapped), we obtain an FSPS scheme satisfying UF-CMA (respectively, UF-otCMA) security. We refer the reader to the full paper for the resulting signature schemes. Furthermore, we list several instantiations of Sig_2 in Table 3, Sect. 6.

6 Instantiations

In this section, we give several instantiations derived from our generic constructions, which are summarized in Table 3. For notational convenience, we denote these schemes as (A), (B), (C), (D), (E), (F), (G), (H), (I) (see the first column of Table 3) respectively. Many of these instantiations are FAS schemes.[16] It is not hard to see that typically, when signing n^2 group elements, Sig_1 needs $O(n^2)$ verification/signature key elements and $O(1)$ PPEs,[17] while Sig_1^* and Sig_2 need $O(n)$ verification/signature key elements and PPEs.

Besides the UF-CMA secure FSPS schemes in Table 3, we give several UF-otCMA instantiations derived from our generic constructions, which have relatively better efficiency. We refer the reader to the full paper for details.

In Sects. 6.1, 6.2, and 6.3, we give remarks on the instantiations of Sig_1, Sig_1^*, and Sig_2, respectively. Due to page limitation, we refer the reader to the full paper for signing key sizes and numbers of pairings required in verification.

6.1 Sig_1: SKSP-TS + SP-AKS

We give parameters of three instantiations for Sig_1, which are (A), (B), and (C). Especially, (B) is an FSPS scheme in the type I bilinear map and (C) is an FSPS scheme in the generic group model.

The verification key size $|pk|$ of (C) is $(n_1, 0) \leq (n_1^2, 0)$, which makes it automorphic, while its efficiency (considering public parameter size, signature size, and verification cost) is very close to (G) (i.e., the FSPS scheme in [28]). As far as we know, (C) is the most efficient FAS scheme by now. Note that if we follow the definition of basic signatures in [3], which allows no trusted setup except for bilinear group generation, then (C) is not automorphic, and the most efficient FAS scheme becomes (F), in Table 3.

6.2 Sig_1^*: SKSP-TS + SP-AKS (UF-CMA)

We give parameters of two instantiations for Sig_1^*, which are (D) and (E), while (E) is in the type I bilinear map. Both of them are automorphic.

It is obvious that most instantiations derived from Sig_1 have verification key and signature sizes linear in the message size, which makes them less efficient

[16] It is not hard to see that FAS schemes in Table 3 may lose the automorphic property when n_1 (or n_2 or n) is an extremely small number. Furthermore, when k (which is independent with the message size) is a large number, the message size has to be made reasonably large to keep the automorphic property.

[17] There may be exceptions, e.g., (C) in Table 3.

Table 3. Previously proposed FSPSs and FSPSs derived from our work. "Const." is short for "Construction" and "Auto." is short for "Automorphic". We use "`(A)`: `KPW15` [31] (CMA) + `KPW15` [31] (otCMA)" to denote that the underlying TS (respectively, AKS) scheme of `(A)` is adapted from the UF-CMA secure (respectively, UF-otCMA secure) SPS scheme in [31]. We use the same argument for others except that the three FSPSs in the top denote the ones proposed in [5,28]. Especially, "`ADK+13` [2] (TT(TOS))" denotes the tagged one-time signature scheme in [2]. Notation (x, y) denotes x elements in $\mathbb{G}_1$ and y elements in $\mathbb{G}_2$. As noted in Introduction, we do not count the two generators in the bilinear groups in the parameters.

	Const.	Auto.	Assumption	Parameter	♯ Group element (PPE)
`AKO+15` [5]	Generic construction 1	×	SXDH XDLIN	$\lvert m\rvert$ $\lvert pk\rvert + \lvert par\rvert$ $\lvert\sigma\rvert$ ♯ PPE	$(n_1^2, 0)$ (n_1^2+5, n_1^2+11) $(7, 3n_1^2+7)$ $2n_1^2+7$
`AKO+15` [5]	Generic construction 2	×	SXDH XDLIN	$\lvert m\rvert$ $\lvert pk\rvert + \lvert par\rvert$ $\lvert\sigma\rvert$ ♯ PPE	$(n_1^2, 0)$ $(6n_1+13, 4)$ $(2n_1+4, 2n_1+7)$ n_1+5
`Gro15` [28]	FSPS scheme	×	Generic	$\lvert m\rvert$ $\lvert pk\rvert + \lvert par\rvert$ $\lvert\sigma\rvert$ ♯ PPE	$(n_1^2, 0)$ $(2n_1-1, 1)$ (n_1+1, n_1) n_1+1
`(A)`: `KPW15` [31] (CMA) + `KPW15` [31](otCMA)	Sig_1	×	$\mathcal{D}_k$-MDDH $(\mathbb{G}_1, \mathbb{G}_2)$	$\lvert m\rvert$ $\lvert pk\rvert + \lvert par\rvert$ $\lvert\sigma\rvert$ ♯ PPE	$(n_1^2, 0)$ $\left((n_1^2 k+3k+3+\mathsf{RE}(\mathcal{D}_k))k+\mathsf{RE}(\mathcal{D}_k), 0\right)$ $\left(k+2, (n_1^2+4)k+3+\mathsf{RE}(\mathcal{D}_k)\right)$ $3k+1$
`(B)`: `ADK+13` [2] (CMA) + `ADK+13` [2] (CMA)	Sig_1	×	2-Lin $(\mathbb{G}_1 = \mathbb{G}_2)$	$\lvert m\rvert$ $\lvert pk\rvert + \lvert par\rvert$ $\lvert\sigma\rvert$ ♯ PPE	n^2 $4n^2+60$ $2n^2+48$ 14
`(C)`: `Gro15` [28] (CMA) + `Gro15` [28] (CMA)	Sig_1	✓	Generic	$\lvert m\rvert$ $\lvert pk\rvert + \lvert par\rvert$ $\lvert\sigma\rvert$ ♯ PPE	$(n_1^2, 0)$ $(2n_1, 1)$ (n_1+2, n_1+3) n_1+3
`(D)`: `KPW15` [31] (CMA) + `KPW15` [31](CMA)	Sig_1^*	✓	$\mathcal{D}_k$-MDDH $(\mathbb{G}_1, \mathbb{G}_2)$	$\lvert m\rvert$ $\lvert pk\rvert + \lvert par\rvert$ $\lvert\sigma\rvert$ ♯ PPE	$(n_1^2, 0)$ $\left((n_1 k+2k^2+6k+3+\mathsf{RE}(\mathcal{D}_k))k+\mathsf{RE}(\mathcal{D}_k), 0\right)$ $\left(3n_1 k+3n_1+1, (n_1+2k+7)k+n_1+3+\mathsf{RE}(\mathcal{D}_k)\right)$ $(2k+1)(n_1+1)$
`(E)`: `ADK+13` [2] (CMA) + `ADK+13` [2] (CMA)	Sig_1^*	✓	2-Lin $(\mathbb{G}_1 = \mathbb{G}_2)$	$\lvert m\rvert$ $\lvert pk\rvert + \lvert par\rvert$ $\lvert\sigma\rvert$ ♯ PPE	n^2 $4n+64$ $16n+36$ $7(n+1)$
`(F)`: `KPW15` [31] (CMA) + `KPW15` [32] (TT)	Sig_2	✓	$\mathcal{D}_k$-MDDH $(\mathbb{G}_1, \mathbb{G}_2)$	$\lvert m\rvert$ $\lvert pk\rvert + \lvert par\rvert$ $\lvert\sigma\rvert$ ♯ PPE	$(n_1^2, 0)$ $\left((2n_1 k+2k+3+\mathsf{RE}(\mathcal{D}_k))k+\mathsf{RE}(\mathcal{D}_k), 0\right)$ $\left((k+1)n_1+1, 2n_1 k+3k+3+\mathsf{RE}(\mathcal{D}_k)\right)$ kn_1+2k+1
`(G)`: `KPW15` [31] (CMA) (bilateral) + `KPW15` [32] (TT)	Sig_2	✓	$\mathcal{D}_k$-MDDH $(\mathbb{G}_1, \mathbb{G}_2)$	$\lvert m\rvert$ $\lvert pk\rvert + \lvert par\rvert$ $\lvert\sigma\rvert$ ♯ PPE	(n_1^2, n_2^2) $\left((2n_1 k+3k+3+\mathsf{RE}(\mathcal{D}_k))k+\mathsf{RE}(\mathcal{D}_k), (2n_2 k+\mathsf{RE}(\mathcal{D}_k))k+\mathsf{RE}(\mathcal{D}_k)\right)$ $\left((k+1)n_1+2n_2 k+k+2+\mathsf{RE}(\mathcal{D}_k), (k+1)n_2+2n_1 k+4k+3+\mathsf{RE}(\mathcal{D}_k)\right)$ $k(n_1+n_2)+3k+1$
`(H)`: `ADK+13` [2] (CMA) + `ADK+13` [2] (TT(TOS))	Sig_2	✓	2-Lin $(\mathbb{G}_1 = \mathbb{G}_2)$	$\lvert m\rvert$ $\lvert pk\rvert + \lvert par\rvert$ $\lvert\sigma\rvert$ ♯ PPE	n^2 $6n+30$ $6n+12$ $2n+7$
`(I)` `ACD+12` [1] (CMA) + `ACD+12` [1] (TT)	Sig_2	×	SXDH XDLIN	$\lvert m\rvert$ $\lvert pk\rvert + \lvert par\rvert$ $\lvert\sigma\rvert$ ♯ PPE	$(n_1^2, 0)$ $(2n_1+14, 7)$ $(2n_1+4, 2n_1+8)$ n_1+4

and not automorphic (since verification keys have larger size than messages). However, as shown in Table 3, as a variation of Sig_1, Sig_1^* allows us to obtain FSPSs with shorter signatures and verification keys if the underlying SP-AKS scheme is UF-CMA secure. This fact shows that many existing SPSs imply the existence of a corresponding efficient FSPS scheme since any well-formed SPS scheme (respectively, SPS scheme with an algebraic key generation algorithm) can be converted into an SKSP-TS (respectively, SP-AKS) scheme.

6.3 Sig_2: SKSP-TS + SP-TT-AKS

We give parameters of four instantiations for Sig_2, which are `(F)`, `(G)`, `(H)`, and `(I)`, while `(H)` is in the type I bilinear map. The only one that is *not* automorphic among them is `(I)`. Here, `(G)` is achieved by using a UF-CMA secure SKSP-TS scheme to sign auxiliary keys of two SP-TT-AKS schemes with verification keys consisting of elements in $\mathbb{G}_1$ and $\mathbb{G}_2$ respectively, and `(H)` is achieved by using a SKSP-TS scheme to sign auxiliary keys of the tag-based one-time signature scheme in [2]. Tag based one-time signatures can be treated as a special case of two-tier signatures where secondary signing keys are the same as secondary verification keys.

For $k = 1$ (SXDH), we have $(|m|, |pk+par|, |\sigma|, \sharp\text{PPEs}) = (n_1^2, 2n_1+7, 4n_1+8, n_1+3)$ in `(F)`, while the most efficient instantiation given in [5] achieves $(|m|, |pk| + |par|, |\sigma|, \sharp\text{PPEs}) = (n_1^2, 6n_1 + 17, 4n_1 + 11, n_1 + 5)$ and is not automorphic. Furthermore, by sacrificing efficiency, `(F)` can be based on weaker assumptions.

`(G)` achieves $(|m|, |pk| + |par|, |\sigma|, \sharp\text{PPEs}) = (n_1^2, 2n_1 + 2n_2 + 10, 4n_1 + 4n_2 + 12, n_1 + n_2 + 4)$ for $k = 1$ (SDXH), which has the shortest verification key size, signature size, and lowest cost in verification among all FSPS and FAS schemes with a bilateral message space based on standard assumptions by now, as far as we know.

`(H)` is the most efficient FSPS and FAS scheme in the *t*ype I bilinear map, as far as we know.

References

1. Abe, M., Chase, M., David, B., Kohlweiss, M., Nishimaki, R., Ohkubo, M.: Constant-size structure-preserving signatures: generic constructions and simple assumptions. In: Wang, X., Sako, K. (eds.) ASIACRYPT 2012. LNCS, vol. 7658, pp. 4–24. Springer, Heidelberg (2012). doi:10.1007/978-3-642-34961-4_3
2. Abe, M., David, B., Kohlweiss, M., Nishimaki, R., Ohkubo, M.: Tagged one-time signatures: tight security and optimal tag size. In: Kurosawa, K., Hanaoka, G. (eds.) PKC 2013. LNCS, vol. 7778, pp. 312–331. Springer, Heidelberg (2013). doi:10.1007/978-3-642-36362-7_20
3. Abe, M., Fuchsbauer, G., Groth, J., Haralambiev, K., Ohkubo, M.: Structure-preserving signatures and commitments to group elements. In: Rabin, T. (ed.) CRYPTO 2010. LNCS, vol. 6223, pp. 209–236. Springer, Heidelberg (2010). doi:10.1007/978-3-642-14623-7_12

4. Abe, M., Groth, J., Haralambiev, K., Ohkubo, M.: Optimal structure-preserving signatures in asymmetric bilinear groups. In: Rogaway, P. (ed.) CRYPTO 2011. LNCS, vol. 6841, pp. 649–666. Springer, Heidelberg (2011). doi:10.1007/978-3-642-22792-9_37
5. Abe, M., Kohlweiss, M., Ohkubo, M., Tibouchi, M.: Fully structure-preserving signatures and shrinking commitments. In: Oswald, E., Fischlin, M. (eds.) EUROCRYPT 2015. LNCS, vol. 9057, pp. 35–65. Springer, Heidelberg (2015). doi:10.1007/978-3-662-46803-6_2
6. Attrapadung, N., Libert, B., Peters, T.: Computing on authenticated data: new privacy definitions and constructions. In: Wang, X., Sako, K. (eds.) ASIACRYPT 2012. LNCS, vol. 7658, pp. 367–385. Springer, Heidelberg (2012). doi:10.1007/978-3-642-34961-4_23
7. Baldimtsi, F., Chase, M., Fuchsbauer, G., Kohlweiss, M.: Anonymous transferable E-Cash. In: Katz, J. (ed.) PKC 2015. LNCS, vol. 9020, pp. 101–124. Springer, Heidelberg (2015). doi:10.1007/978-3-662-46447-2_5
8. Belenkiy, M., Camenisch, J., Chase, M., Kohlweiss, M., Lysyanskaya, A., Shacham, H.: Randomizable proofs and delegatable anonymous credentials. In: Halevi, S. (ed.) CRYPTO 2009. LNCS, vol. 5677, pp. 108–125. Springer, Heidelberg (2009). doi:10.1007/978-3-642-03356-8_7
9. Belenkiy, M., Chase, M., Kohlweiss, M., Lysyanskaya, A.: P-signatures and noninteractive anonymous credentials. In: Canetti, R. (ed.) TCC 2008. LNCS, vol. 4948, pp. 356–374. Springer, Heidelberg (2008). doi:10.1007/978-3-540-78524-8_20
10. Bellare, M., Micciancio, D., Warinschi, B.: Foundations of group signatures: formal definitions, simplified requirements, and a construction based on general assumptions. In: Biham, E. (ed.) EUROCRYPT 2003. LNCS, vol. 2656, pp. 614–629. Springer, Heidelberg (2003). doi:10.1007/3-540-39200-9_38
11. Bellare, M., Shi, H., Zhang, C.: Foundations of group signatures: the case of dynamic groups. In: Menezes, A. (ed.) CT-RSA 2005. LNCS, vol. 3376, pp. 136–153. Springer, Heidelberg (2005). doi:10.1007/978-3-540-30574-3_11
12. Bellare, M., Shoup, S.: Two-tier signatures, strongly unforgeable signatures, and fiat-shamir without random oracles. In: Okamoto, T., Wang, X. (eds.) PKC 2007. LNCS, vol. 4450, pp. 201–216. Springer, Heidelberg (2007). doi:10.1007/978-3-540-71677-8_14
13. Bender, A., Katz, J., Morselli, R.: Ring signatures: Stronger definitions, and constructions without random oracles. J. Cryptology **22**(1), 114–138 (2009)
14. Camenisch, J., Dubovitskaya, M., Enderlein, R.R., Neven, G.: Oblivious transfer with hidden access control from attribute-based encryption. In: Visconti, I., Prisco, R. (eds.) SCN 2012. LNCS, vol. 7485, pp. 559–579. Springer, Heidelberg (2012). doi:10.1007/978-3-642-32928-9_31
15. Camenisch, J., Krenn, S., Shoup, V.: A framework for practical universally composable zero-knowledge protocols. In: Lee, D.H., Wang, X. (eds.) ASIACRYPT 2011. LNCS, vol. 7073, pp. 449–467. Springer, Heidelberg (2011). doi:10.1007/978-3-642-25385-0_24
16. Camenisch, J., Lysyanskaya, A.: An efficient system for non-transferable anonymous credentials with optional anonymity revocation. In: Pfitzmann, B. (ed.) EUROCRYPT 2001. LNCS, vol. 2045, pp. 93–118. Springer, Heidelberg (2001). doi:10.1007/3-540-44987-6_7
17. Chase, M., Kohlweiss, M., Lysyanskaya, A., Meiklejohn, S.: Malleable proof systems and applications. In: Pointcheval, D., Johansson, T. (eds.) EUROCRYPT 2012. LNCS, vol. 7237, pp. 281–300. Springer, Heidelberg (2012). doi:10.1007/978-3-642-29011-4_18

18. Chase, M., Kohlweiss, M., Lysyanskaya, A., Meiklejohn, S.: Malleable signatures: new definitions and delegatable anonymous credentials. In: CSF 2014, pp. 199–213. IEEE (2014)
19. ElGamal, T.: A public key cryptosystem and a signature scheme based on discrete logarithms. IEEE Trans. Inf. Theory **31**(4), 469–472 (1985)
20. Escala, A., Herold, G., Kiltz, E., Ràfols, C., Villar, J.: An algebraic framework for diffie-hellman assumptions. In: Canetti, R., Garay, J.A. (eds.) CRYPTO 2013. LNCS, vol. 8043, pp. 129–147. Springer, Heidelberg (2013). doi:10.1007/978-3-642-40084-1_8
21. Even, S., Goldreich, O., Micali, S.: On-line/off-line digital signatures. J. Cryptology **9**(1), 35–67 (1996)
22. Fuchsbauer, G.: Commuting signatures and verifiable encryption. In: Paterson, K.G. (ed.) EUROCRYPT 2011. LNCS, vol. 6632, pp. 224–245. Springer, Heidelberg (2011). doi:10.1007/978-3-642-20465-4_14
23. Fuchsbauer, G., Hanser, C., Kamath, C., Slamanig, D.: Practical round-optimal blind signatures in the standard model from weaker assumptions. In: Zikas, V., Prisco, R. (eds.) SCN 2016. LNCS, vol. 9841, pp. 391–408. Springer, Heidelberg (2016). doi:10.1007/978-3-319-44618-9_21
24. Fuchsbauer, G., Hanser, C., Slamanig, D.: Practical round-optimal blind signatures in the standard model. In: Gennaro, R., Robshaw, M. (eds.) CRYPTO 2015. LNCS, vol. 9216, pp. 233–253. Springer, Heidelberg (2015). doi:10.1007/978-3-662-48000-7_12
25. Fuchsbauer, G., Vergnaud, D.: Fair blind signatures without random oracles. In: Bernstein, D.J., Lange, T. (eds.) AFRICACRYPT 2010. LNCS, vol. 6055, pp. 16–33. Springer, Heidelberg (2010). doi:10.1007/978-3-642-12678-9_2
26. Ghadafi, E.: More efficient structure-preserving signatures - or: Bypassing the type-III lower bounds. Cryptology ePrint Archive, Report 2016/255 (2016)
27. Ghadafi, E.: Short structure-preserving signatures. In: Sako, K. (ed.) CT-RSA 2016. LNCS, vol. 9610, pp. 305–321. Springer, Heidelberg (2016). doi:10.1007/978-3-319-29485-8_18
28. Groth, J.: Efficient fully structure-preserving signatures for large messages. In: Iwata, T., Cheon, J.H. (eds.) ASIACRYPT 2015. LNCS, vol. 9452, pp. 239–259. Springer, Heidelberg (2015). doi:10.1007/978-3-662-48797-6_11
29. Groth, J., Sahai, A.: Efficient noninteractive proof systems for bilinear groups. SIAM J. Comput. **41**(5), 1193–1232 (2012)
30. Hofheinz, D., Jager, T.: Tightly secure signatures and public-key encryption. In: Safavi-Naini, R., Canetti, R. (eds.) CRYPTO 2012. LNCS, vol. 7417, pp. 590–607. Springer, Heidelberg (2012). doi:10.1007/978-3-642-32009-5_35
31. Kiltz, E., Pan, J., Wee, H.: Structure-preserving signatures from standard assumptions, revisited. In: Gennaro, R., Robshaw, M. (eds.) CRYPTO 2015. LNCS, vol. 9216, pp. 275–295. Springer, Heidelberg (2015). doi:10.1007/978-3-662-48000-7_14
32. Kiltz, E., Pan, J., Wee, H.: Structure-preserving signatures from standard assumptions, revisited. IACR Cryptology ePrint Archive 2015, 604 (2015)
33. Libert, B., Peters, T., Joye, M., Yung, M.: Linearly homomorphic structure-preserving signatures and their applications. In: Canetti, R., Garay, J.A. (eds.) CRYPTO 2013. LNCS, vol. 8043, pp. 289–307. Springer, Heidelberg (2013). doi:10.1007/978-3-642-40084-1_17
34. Libert, B., Peters, T., Yung, M.: Group signatures with almost-for-free revocation. In: Safavi-Naini, R., Canetti, R. (eds.) CRYPTO 2012. LNCS, vol. 7417, pp. 571–589. Springer, Heidelberg (2012). doi:10.1007/978-3-642-32009-5_34

35. Libert, B., Peters, T., Yung, M.: Short group signatures via structure-preserving signatures: standard model security from simple assumptions. In: Gennaro, R., Robshaw, M. (eds.) CRYPTO 2015. LNCS, vol. 9216, pp. 296–316. Springer, Heidelberg (2015). doi:10.1007/978-3-662-48000-7_15
36. Micali, S., Ohta, K., Reyzin, L.: Accountable-subgroup multisignatures: extended abstract. In: Reiter, M.K., Samarati, P. (eds.) CCS 2001, pp. 245–254. ACM (2001)
37. Rial, A., Kohlweiss, M., Preneel, B.: Universally composable adaptive priced oblivious transfer. In: Shacham, H., Waters, B. (eds.) Pairing 2009. LNCS, vol. 5671, pp. 231–247. Springer, Heidelberg (2009). doi:10.1007/978-3-642-03298-1_15
38. Schwartz, J.T.: Fast probabilistic algorithms for verification of polynomial identities. J. ACM **27**(4), 701–717 (1980)

Functional and Homomorphic Cryptography

Multi-key Homomorphic Authenticators

Dario Fiore[1(✉)], Aikaterini Mitrokotsa[2], Luca Nizzardo[1], and Elena Pagnin[2]

[1] IMDEA Software Institute, Madrid, Spain
{dario.fiore,luca.nizzardo}@imdea.org
[2] Chalmers University of Technology, Gothenburg, Sweden
{aikmitr,elenap}@chalmers.se

Abstract. Homomorphic authenticators (HAs) enable a client to authenticate a large collection of data elements $m_1, \ldots, m_t$ and outsource them, along with the corresponding authenticators, to an untrusted server. At any later point, the server can generate a *short* authenticator vouching for the correctness of the output y of a function f computed on the outsourced data, i.e., $y = f(m_1, \ldots, m_t)$. Recently researchers have focused on HAs as a solution, with minimal communication and interaction, to the problem of delegating computation on outsourced data. The notion of HAs studied so far, however, only supports executions (and proofs of correctness) of computations over data authenticated by a single user. Motivated by realistic scenarios (ubiquitous computing, sensor networks, etc.) in which large datasets include data provided by multiple users, we study the concept of *multi-key homomorphic authenticators*. In a nutshell, multi-key HAs are like HAs with the extra feature of allowing the holder of public evaluation keys to compute on data authenticated under different secret keys. In this paper, we introduce and formally define multi-key HAs. Secondly, we propose a construction of a multi-key homomorphic signature based on standard lattices and supporting the evaluation of circuits of bounded polynomial depth. Thirdly, we provide a construction of multi-key homomorphic MACs based only on pseudorandom functions and supporting the evaluation of low-degree arithmetic circuits. Albeit being less expressive and only secretly verifiable, the latter construction presents interesting efficiency properties.

1 Introduction

The technological innovations offered by modern IT systems are changing the way digital data is collected, stored, processed and consumed. As an example, think of an application where data is collected by some organizations (e.g., hospitals), stored and processed on remote servers (e.g., the Cloud) and finally consumed by other users (e.g., medical researchers) on other devices. On one hand, this computing paradigm is very attractive, particularly as data can be shared and exchanged by multiple users. On the other hand, it is evident that in such scenarios one may be concerned about security: while the users that collect and consume the data may trust each other (up to some extent), trusting the Cloud can be problematic for various reasons. More specifically, two main

J.H. Cheon and T. Takagi (Eds.): ASIACRYPT 2016, Part II, LNCS 10032, pp. 499–530, 2016.
DOI: 10.1007/978-3-662-53890-6_17

security concerns to be addressed are those about the *privacy* and *authenticity* of the data stored and processed in untrusted environments.

While it is widely known that privacy can be solved in such a setting using, e.g., homomorphic encryption [27], in this work we focus on the orthogonal problem of providing authenticity of data during computation. Towards this goal, our contribution is on advancing the study of *homomorphic authenticators* (HAs), a cryptographic primitive that has been the subject of recent work [9,26, 30,32].

Homomorphic Authenticators. Using an homomorphic authenticator (HA) scheme a user Alice can authenticate a collection of data items $m_1, \ldots, m_t$ using her secret key, and send the authenticated data to an untrusted server. The server can execute a program $\mathcal{P}$ on the authenticated data and use a public evaluation key to generate a value $\sigma_{\mathcal{P},y}$ vouching for the correctness of $y = \mathcal{P}(m_1, \ldots, m_t)$. Finally, a user Bob who is given the tuple $(\mathcal{P}, y, \sigma_{\mathcal{P},y})$ and Alice's verification key can use the authenticator to verify the authenticity of y as output of the program $\mathcal{P}$ executed on data authenticated by Alice. In other words, Bob can check that the server did not tamper with the computation's result. Alice's verification key can be either secret or public. In the former case, we refer to the primitive as *homomorphic MACs* [11,26], while in the latter we refer to it as *homomorphic signatures* [9]. One of the attractive features of HAs is that the authenticator $\sigma_{\mathcal{P},y}$ is *succinct*, i.e., much shorter than $\mathcal{P}$'s input size. This means that the server can execute a program on a huge amount of data and convince Bob of its correctness by sending him only a short piece of information. As discussed in previous work (e.g., [5,26,30]), HAs provide a nice solution, with minimal communication and interaction, to the problem of delegating computations on outsourced data, and thus can be preferable to verifiable computation (more details on this comparison appear in Sect. 1.2).

Our Contribution: Multi-key Homomorphic Authenticators. Up to now, the notion of HAs has inherently been single-key, i.e., homomorphic computations are allowed only on data authenticated using the same secret key. This characteristic is obviously a limitation and prevents HA schemes from suiting scenarios where the data is provided (and authenticated) by multiple users. Consider the previously mentioned example of healthcare institutions which need to compute on data collected by several hospitals or even some remote-monitored patients. Similarly, it is often required to compute statistics for time-series data collected from multiple users e.g., to monitor usage data in smart metering, clinical research or to monitor the safety of buildings. Another application scenario is in distributed networks of sensors. Imagine for instance a network of sensors where each sensor is in charge of collecting data about air pollution in a certain area of a city, it sends its data to a Cloud server, and then a central control unit asks the Cloud to compute on the data collected by the sensors (e.g., to obtain the average value of air pollution in a large area).

A trivial solution to address the problem of computing on data authenticated by multiple users is to use homomorphic authenticators in such a way that all data providers share the *same* secret authentication key. The desired functionality is obviously achieved since data would be authenticated using a single secret key. This approach however has several drawbacks. The first one is that users need to coordinate in order to agree on such a key. The second one is that in such a setting there would be no technical/legal way to differentiate between users (e.g., to make each user accountable for his/her duties) as any user can impersonate all the other ones. The third and more relevant reason is that sharing the same key exposes the overall system to way higher risks of attacks and makes disaster recovery more difficult: if a single user is compromised the whole system is compromised too, and everything has to be reinitialized from scratch.

In contrast, this paper provides an innovative solution through the notion of *multi-key homomorphic authenticators* (multi-key HAs). This primitive guarantees that the corruption of one user affects the data of that user only, but does not endanger the authenticity of computations among the other (un-corrupted) users of the system. Moreover, the proposed system is dynamic, in the sense that compromised users can be assigned new keys and be easily reintegrated.

1.1 An Overview of Our Results

Our contribution is mainly threefold. First of all, we elaborate a suitable definition of multi-key HAs. Second, we propose the first construction of a multi-key homomorphic signature (i.e., with public verifiability) which is based on standard lattices and supports the evaluation of circuits of bounded polynomial depth. Third, we present a multi-key homomorphic MAC that is based only on pseudorandom functions and supports the evaluation of low-degree arithmetic circuits. In spite of being less expressive and only secretly verifiable, this last construction is way more efficient than the signature scheme. In what follows, we elaborate more on our results.

Multi-Key Homomorphic Authenticators: What are they? At a high level, multi-key HAs are like HAs with the additional property that one can execute a program $\mathcal{P}$ on data authenticated using different secret keys. In multi-key HAs, Bob verifies using the verification keys of all users that provided inputs to $\mathcal{P}$. These features make multi-key HAs a perfect candidate for applications where multiple users gather and outsource data. Referring to our previous examples, using multi-key HAs each sensor can authenticate and outsource to the Cloud the data it collects; the Cloud can compute statistics on the authenticated data and provide the central control unit with the result along with a certificate vouching for its correctness.

An important aspect of our definition is a mechanism that allows the verifier to keep track of the users that authenticated the inputs of $\mathcal{P}$, i.e., to know which user contributed to each input wire of $\mathcal{P}$. To formalize this mechanism, we build on the model of *labeled data and programs* of Gennaro and Wichs [26] (we refer the reader to Sect. 3 for details). In terms of security, multi-key

HAs allow the adversary to corrupt users (i.e., to learn their secret keys); yet this knowledge should not help the adversary in tampering with the results of programs which involve inputs of honest (i.e., uncorrupted) users only. Our model allows to handle compromised users in a similar way to what occurs with classical digital signatures: a compromised user could be banned by means of a certificate revocation, and could easily be re-integrated via a new key pair.[1] Thinking of the sensor network application, if a sensor in the field gets compromised, the data provided by other sensors remains secure, and a new sensor can be easily introduced in the system with new credentials.

Finally, we require multi-key homomorphic authenticators to be *succinct* in the sense that the size of authenticators is bounded by some fixed polynomial in $(\lambda, n, \log t)$, where λ is the security parameter, n is the number of users contributing to the computation and t is the total number of inputs of $\mathcal{P}$. Although such dependence on n may look undesirable, we stress that it is still meaningful in many application scenarios where n is much smaller than t. For instance, in the application scenario of healthcare institutions a few hospitals can provide a large amount of data from patients.

A Multi-Key Homomorphic Signature for All Circuits. After setting the definition of multi-key homomorphic authenticators, we proceed to construct multi-key HA schemes. Our first contribution is a multi-key homomorphic signature that supports the evaluation of boolean circuits of depth bounded by a fixed polynomial in the security parameter. The scheme is proven secure based on the small integer solution (SIS) problem over standard lattices [36], and tolerates adversaries that corrupt users non-adaptively.[2] Our technique is inspired by the ones developed by Gorbunov, Vaikuntanathan and Wichs [30] to construct a (single-key) homomorphic signature. Our key contribution is on providing a new representation of the signatures that enables to homomorphically compute over them even if they were generated using different keys. Furthermore, our scheme enjoys an additional property, not fully satisfied by [30]: every user can authenticate separately every data item m_i of a collection $m_1, \ldots m_t$, and the correctness of computations is guaranteed even when computing on not-yet-full datasets. Although it is possible to modify the scheme in [30] for signing data items separately, the security would only work against adversaries that query the whole dataset. In contrast, we prove our scheme to be secure under a stronger security definition where the adversary can *adaptively* query the various data items, and it can try to cheat by pretending to possess signatures on data items that it never queried (so-called Type 3 forgeries). We highlight that the scheme in [30] is *not* secure under the stronger definition (with Type 3 forgeries) used in this paper, and we had to introduce new techniques to deal with this scenario. This new property is particularly interesting as it enables users to authenticate and outsource data items in a streaming fashion, without ever having to store

[1] Here we mean that this process does not add more complications than the ones already existing for classical digital signatures (e.g., relying on PKI mechanisms).

[2] Precisely, our "core" scheme is secure against adversaries that make non-adaptive signing queries; this is upgraded to adaptive security via general transformations.

the whole dataset. This is useful in applications where the dataset size can be very large or not fixed a priori.

A MULTI-KEY HOMOMORPHIC MAC FOR LOW-DEGREE CIRCUITS. Our second construction is a multi-key homomorphic MAC that supports the evaluation of arithmetic circuits whose degree d is at most polynomial in the security parameter, and whose inputs come from a small number n of users. For results of such computations the corresponding authenticators have at most size $s = \binom{n+d}{d}$.[3] Notably, the authenticator's size is completely independent of the total number of inputs of the arithmetic circuit. Compared to our multi-key homomorphic signature, this construction is only secretly verifiable (i.e., Bob has to know the secret verification keys of all users involved in the computation) and supports a class of computations that is less expressive; also its succinctness is asymptotically worse. In spite of these drawbacks, our multi-key homomorphic MAC achieves interesting features. From the theoretical point of view, it is based on very simple cryptographic tools: a family of pseudorandom functions. Thus, the security relies only on one-way functions. On the practical side, it is particularly efficient: generating a MAC requires only one pseudo-random function evaluation and a couple of field operations; homomorphic computations boil down to additions and multiplications over a multi-variate polynomial ring $\mathbb{F}_p[X_1, \ldots, X_n]$.

1.2 Related Work

Homomorphic MACs and Signatures. Homomorphic authenticators have received a lot of attention in previous work focusing either on homomorphic signatures (publicly verifiable) or on homomorphic MACs (private verification with a secret key). The notion of homomorphic signatures was originally proposed by Johnson *et al.* [32]. The first schemes that appeared in the literature were homomorphic only for linear functions [8,13–15,23] and found important applications in network coding and proofs of retrievability. Boneh and Freeman [9] were the first to construct homomorphic signatures that can evaluate more than linear functions over signed data. Their scheme could evaluate bounded-degree polynomials and its security was based on the hardness of the SIS problem in ideal lattices in the random oracle model. A few years later, Catalano *et al.* [16] proposed an alternative homomorphic signature scheme for bounded-degree polynomials. Their solution is based on multi-linear maps and bypasses the need for random oracles. More interestingly, the work by Catalano *et al.* [16] contains the first mechanism to verify signatures faster than the running time of the verified function. Recently, Gorbunov *et al.* [30] have proposed the first (leveled) fully homomorphic signature scheme that can evaluate arbitrary circuits of bounded polynomial depth over signed data. Some important advances have been also achieved in the area of homomorphic MACs. Gennaro and Wichs [26] have proposed a fully homomorphic MAC based on fully homomorphic encryption. However, their scheme is not secure in the presence of verification queries. More efficient schemes have been proposed later [5,11,12] that are secure in the

[3] Note that s can be bounded by $poly(n)$ for constant d, or by $poly(d)$ for constant n.

presence of verification queries and are more efficient at the price of supporting only restricted homomorphisms. Finally, we note that Agrawal *et al.* [1] considered a notion of *multi-key* signatures for network coding, and proposed a solution which works for linear functions only. Compared to this work, our contribution shows a full-fledged framework for multi-key homomorphic authenticators, and provides solutions that address a more expressive class of computations.

Verifiable Computation. Achieving correctness of outsourced computations is also the aim of *verifiable delegation of computation* (VC) [6,18,20,25,29,37]. In this setting, a client wants to delegate the computation of a function f on input x to an untrusted cloud-server. If the server replies with y, the client's goal is to verify the correctness of $y = f(x)$ spending less resources than those needed to execute f. As mentioned in previous work (e.g., [26,30]) a crucial difference between verifiable computation and homomorphic authenticators is that in VC the verifier has to know the input of the computation – which can be huge – whereas in HAs one can verify by only knowing the function f and the result y. Moreover, although some results of verifiable computation could be re-interpreted to solve scenarios similar to the ones addressed by HAs, results based on VC would still present several limitations. For instance, using homomorphic authenticators the server can prove correctness of $y = f(x)$ with a single message, without needing any special encoding of f from the delegator. Second, HAs come naturally with a composition property which means that the outputs of some computations on authenticated data (which is already authenticated) can be fed as input for follow-up computations. This feature is of particular interest to parallelize and or distribute computations (e.g., MapReduce). Emulating this composition within VC systems is possible by means of certain non-interactive proof systems [7] but leads to complex statements and less natural realizations. A last advantage is that by using HAs, clients can authenticate various (small) pieces of data independently and without storing previously outsourced data. In contrast, most VC systems require clients to encode the whole input in 'one shot', and often such encoding can be used in a single computation only.

Multi-client Verifiable Computation. Another line of work, closely related to ours is that on *multi-client verifiable computation* [17,31]. This primitive, introduced by Choi *et al.* [17], aims to extend VC to the setting where inputs are provided by multiple users, and one of these users wants to verify the result's correctness. Choi *et al.* [17] proposed a definition and a multi-client VC scheme which generalizes that of Gennaro *et al.* [25]. The solution in [17], however, does not consider malicious or colluding clients. This setting was addressed by Gordon *et al.* in [31], where they provide a scheme with stronger security guarantees against a malicious server or an arbitrary set of malicious colluding clients.

It is interesting to notice that in the definition of multi-client VC all the clients but the one who verifies can encode inputs independently of the function to be later executed on them. One may thus think that the special case in which the verifier provides no input yields a solution similar to the one achieved by multi-key HAs. However, a closer look at the definitions and the existing constructions of multi-client VC reveals three main differences. (1) In multi-client VC, in order to prove the correctness of an execution of a function f, the server has to wait a message from the verifier which includes some encoding of f. This is not necessary in multi-key HAs where the server can directly prove the correctness of f on previously authenticated data with a single message and without any function's encoding. (2) The communication between the server and the verifier is at least linear in the total number of inputs of f: this can be prohibitive in the case of computations on very large inputs (think of TBytes of data). In contrast, with multi-key HAs the communication between the server and the verifier is proportional only to the number of users, and depends only logarithmically on the total number of inputs. (3) In multi-client VC an encoding of one input can be used in a single computation. Thus, if a user wants to first upload data on the server to later execute many functions on it, then the user has to provide as many encodings as the number of functions to be executed. In contrast, multi-key HAs allow one to encode (i.e., authenticate) every input only once and to use it for proving correctness of computations an *unbounded* number of times.

2 Preliminaries

We collect here the notation and basic definitions used throughout the paper.

Notation. The Greek letter λ is reserved for the security parameter of the schemes. A function $\epsilon(\lambda)$ is said to be *negligible* in λ (denoted as $\epsilon(\lambda) = \mathsf{negl}(\lambda)$) if $\epsilon(\lambda) = O(\lambda^{-c})$ for every constant $c > 0$. When a function can be expressed as a polynomial we often write it as *poly*$(\cdot)$. For any $n \in \mathbb{N}$, we refer to $[n]$ as $[n] := \{1, \ldots, n\}$. Moreover, given a set $\mathcal{S}$, the notation $s \xleftarrow{\$} \mathcal{S}$ stays for the process of sampling s uniformly at random from $\mathcal{S}$.

Definition 1 (Statistical Distance). *Let X, Y denote two random variables with support $\mathcal{X}, \mathcal{Y}$ respectively; the statistical distance between X and Y is defined as $\mathbf{SD}(X, Y) := \frac{1}{2}(\sum_{u \in \mathcal{X} \cup \mathcal{Y}} \mid \Pr[X = u] - \Pr[Y = u] \mid)$. If $\mathbf{SD}(X, Y) = \mathsf{negl}(\lambda)$, we say that X and Y are statistically close and we write $X \overset{\mathsf{stat}}{\approx} Y$.*

Definition 2 (Entropy [19]). *The min-entropy of a random variable X is defined as $\boldsymbol{H}_\infty(X) := -\log\big(\max_x \Pr[X = x]\big)$. The (average-) conditional min-entropy of a random variable X conditioned on a correlated variable Y is defined as $\mathbf{H}_\infty(X \mid Y) := -\log\Big(\underset{y \leftarrow Y}{\mathbf{E}}\Big[\max_x \Pr[X = x \mid Y = y]\Big]\Big)$. The optimal probability of an unbounded adversary guessing X when given a correlated value Y is $2^{-\mathbf{H}_\infty(X|Y)}$.*

Lemma 1 ([19]). *Let X, Y be arbitrarily random variables where the support of Y lies in $\mathcal{Y}$. Then $\mathbf{H}_\infty(X \mid Y) \geq \mathbf{H}_\infty(X) - \log(\mid \mathcal{Y} \mid)$.*

3 Multi-key Homomorphic Authenticators

In this section, we present our new notion of *Multi-Key Homomorphic Authenticators* (multi-key HAs). Intuitively, multi-key HAs extend the existing notions of homomorphic signatures [9] and homomorphic MACs [26] in such a way that one can homomorphically compute a program $\mathcal{P}$ over data authenticated using different secret keys. For the sake of verification, in multi-key HAs the verifier needs to know the verification keys of all users that provided inputs to $\mathcal{P}$. Our definitions are meant to be general enough to be easily adapted to both the case in which verification keys are public and the one where verification keys are secret. In the former case, we call the primitive *multi-key homomorphic signatures* whereas in the latter case we call it *multi-key homomorphic MACs*.

As already observed in previous work about HAs, it is important that an authenticator $\sigma_{\mathcal{P},y}$ does not authenticate a value y out of context, but only as the output of a program $\mathcal{P}$ executed on previously authenticated data. To formalize this notion, we build on the model of *labeled data and programs* of Gennaro and Wichs [26]. The idea of this model is that every data item is authenticated under a unique label ℓ. For example, in scenarios where the data is outsourced, such labels can be thought of as a way to index/identify the remotely stored data. A labeled program $\mathcal{P}$, on the other hand, consists of a circuit f where every input wire i has a label ℓ_i. Going back to the outsourcing example, a labeled program is a way to specify on what portion of the outsourced data one should execute a circuit f. More formally, the notion of labeled programs of [26] is recalled below.

Labeled Programs [26]. A labeled program $\mathcal{P}$ is a tuple $(f, \ell_1, \ldots, \ell_n)$, such that $f : \mathcal{M}^n \rightarrow \mathcal{M}$ is a function of n variables (e.g., a circuit) and $\ell_i \in \{0,1\}^*$ is a label for the i-th input of f. Labeled programs can be composed as follows: given $\mathcal{P}_1, \ldots, \mathcal{P}_t$ and a function $g : \mathcal{M}^t \rightarrow \mathcal{M}$, the composed program $\mathcal{P}^*$ is the one obtained by evaluating g on the outputs of $\mathcal{P}_1, \ldots, \mathcal{P}_t$, and it is denoted as $\mathcal{P}^* = g(\mathcal{P}_1, \ldots, \mathcal{P}_t)$. The labeled inputs of $\mathcal{P}^*$ are all the distinct labeled inputs of $\mathcal{P}_1, \ldots \mathcal{P}_t$ (all the inputs with the same label are grouped together and considered as a unique input of $\mathcal{P}^*$). Let $f_{id} : \mathcal{M} \rightarrow \mathcal{M}$ be the identity function and $\ell \in \{0,1\}^*$ be any label. We refer to $\mathcal{I}_\ell = (f_{id}, \ell)$ as the identity program with label ℓ. Note that a program $\mathcal{P} = (f, \ell_1, \ldots, \ell_n)$ can be expressed as the composition of n identity programs $\mathcal{P} = f(\mathcal{I}_{\ell_1}, \ldots, \mathcal{I}_{\ell_n})$.

Using labeled programs to identify users. In our notion of multi-key homomorphic authenticators, one wishes to verify the outputs of computations executed on data authenticated under *different* keys. A meaningful definition of multi-key HAs thus requires that the authenticators are not out of context also with respect to the set of keys that contributed to the computation. To address this issue, we assume that every user has an identity id in some identity space ID, and that the user's keys are associated to id by means of any suitable mechanism (e.g., PKI). Next, in order to distinguish among data of different users and to identify to which input wires a certain user contributed, we assume that the label space contains the set ID. Namely, in our model a data item is assigned a label $\ell := (\mathsf{id}, \tau)$, where id is a user's identity, and τ is a tag; this essentially identifies uniquely a data item of user id with index τ. For compatibility with previous notions of homomorphic authenticators, we assume that data items can be grouped in datasets, and one can compute only on data within the same dataset. In our definitions, a dataset is identified by an arbitrary string Δ.[4]

Definition 3 (Multi-key Homomorphic Authenticator). *A multi-key homomorphic authenticator scheme* MKHAut *consists of a tuple of PPT algorithms (*Setup, KeyGen, Auth, Eval, Ver*) satisfying the following properties:* authentication correctness, evaluation correctness, succinctness *and* security. *The five algorithms work as follows:*

Setup(1^λ). *The setup algorithm takes as input the security parameter λ and outputs some public parameters* pp. *These parameters include (at least) descriptions of a tag space $\mathcal{T}$, an identity space* ID*, the message space $\mathcal{M}$ and a set of admissible functions $\mathcal{F}$. Given $\mathcal{T}$ and* ID*, the label space of the scheme is defined as their cartesian product $\mathcal{L} := \mathsf{ID} \times \mathcal{T}$. For a labeled program $\mathcal{P} = (f, \ell_1, \dots, \ell_t)$ with labels $\ell_i := (\mathsf{id}_i, \tau_i) \in \mathcal{L}$, we use $\mathsf{id} \in \mathcal{P}$ as compact notation for $\mathsf{id} \in \{\mathsf{id}_1, \dots, \mathsf{id}_t\}$. The* pp *are input to all the following algorithms, even when not specified.*

KeyGen(pp). *The key generation algorithm takes as input the public parameters and outputs a triple of keys* (sk, ek, vk)*, where* sk *is a secret authentication key,* ek *is a public evaluation key and* vk *is a verification key which could be either public or private.*[5]

Auth$(\mathsf{sk}, \Delta, \ell, \mathsf{m})$. *The authentication algorithm takes as input an authentication key* sk*, a dataset identifier Δ, a label $\ell = (\mathsf{id}, \tau)$ for the message* m*, and it outputs an authenticator σ.*

Eval$(f, \{(\sigma_i, \mathsf{EKS}_i)\}_{i \in [t]})$. *The evaluation algorithm takes as input a t-input function $f : \mathcal{M}^t \longrightarrow \mathcal{M}$, and a set $\{(\sigma_i, \mathsf{EKS}_i)\}_{i \in [t]}$ where each σ_i is an authenticator and each EKS_i is a set of evaluation keys.*[6]

Ver$(\mathcal{P}, \Delta, \{\mathsf{vk}_{\mathsf{id}}\}_{\mathsf{id} \in \mathcal{P}}, \mathsf{m}, \sigma)$. *The verification algorithm takes as input a labeled program $\mathcal{P} = (f, \ell_1, \dots, \ell_t)$, a dataset identifier Δ, the set of verification keys $\{\mathsf{vk}_{\mathsf{id}}\}_{\mathsf{id} \in \mathcal{P}}$ corresponding to those identities* id *involved in the program $\mathcal{P}$, a message* m *and an authenticator σ. It outputs* 0 *(reject) or* 1 *(accept).*

AUTHENTICATION CORRECTNESS. Intuitively, a Multi-Key Homomorphic Autenticator has authentication correctness if the output of Auth$(\mathsf{sk}, \Delta, \ell, \mathsf{m})$ verifies correctly for m as the output of the identity program $\mathcal{I}_\ell$ over the dataset

[4] Although considering the dataset notion complicates the definition, it also provides some benefits, as we illustrate later in the constructions. For instance, when verifying for the same program $\mathcal{P}$ over different datasets, one can perform some precomputation that makes further verifications cheap.

[5] As mentioned earlier, the generated triple (sk, ek, vk) will be associated to an identity $\mathsf{id} \in \mathsf{ID}$. When this connection becomes explicit, we will refer to (sk, ek, vk) as $(\mathsf{sk}_{\mathsf{id}}, \mathsf{ek}_{\mathsf{id}}, \mathsf{vk}_{\mathsf{id}})$.

[6] The motivation behind the evaluation-keys set EKS_i is that, if σ_i authenticates the output of a labeled program $\mathcal{P}_i$, then $\mathsf{EKS}_i = \{\mathsf{ek}_{\mathsf{id}}\}_{\mathsf{id} \in \mathcal{P}_i}$ should be the set of evaluation keys corresponding to identities involved in the computation of $\mathcal{P}_i$.

Δ. More formally, a scheme MKHAut satisfies authentication correctness if for all public parameters $\mathsf{pp} \leftarrow \mathsf{Setup}(1^\lambda)$, any key triple $(\mathsf{sk}, \mathsf{ek}, \mathsf{vk}) \leftarrow \mathsf{KeyGen}(\mathsf{pp})$, any label $\ell = (\mathsf{id}, \tau) \in \mathcal{L}$ and any authenticator $\sigma \leftarrow \mathsf{Auth}(\mathsf{sk}, \Delta, \ell, \mathsf{m})$, we have $\mathsf{Ver}(\mathcal{I}_\ell, \Delta, \mathsf{vk}, \mathsf{m}, \sigma)$ outputs 1 with all but negligible probability.

Evaluation Correctness. Intuitively, this property says that running the evaluation algorithm on signatures $(\sigma_1, \ldots, \sigma_t)$ such that each σ_i verifies for m_i as the output of a labeled program $\mathcal{P}_i$ over the dataset Δ, it produces a signature σ which verifies for $f(\mathsf{m}_1, \ldots, \mathsf{m}_t)$ as the output of the composed program $f(\mathcal{P}_1, \ldots, \mathcal{P}_t)$ over the dataset Δ. More formally, let us fix the public parameters $\mathsf{pp} \leftarrow \mathsf{Setup}(1^\lambda)$, a set of key triples $\{(\mathsf{sk}_{\mathsf{id}}, \mathsf{ek}_{\mathsf{id}}, \mathsf{vk}_{\mathsf{id}})\}_{\mathsf{id} \in \tilde{\mathsf{ID}}}$ for some $\tilde{\mathsf{ID}} \subseteq \mathsf{ID}$, a dataset Δ, a function $g : \mathcal{M}^t \rightarrow \mathcal{M}$, and any set of program/message/authentica-tor triples $\{(\mathcal{P}_i, m_i, \sigma_i)\}_{i \in [t]}$ such that $\mathsf{Ver}(\mathcal{P}_i, \Delta, \{\mathsf{vk}_{\mathsf{id}}\}_{\mathsf{id} \in \mathcal{P}_i}, \mathsf{m}_i, \sigma_i) = 1$ for all $i \in [t]$. Let $\mathsf{m}^* = g(\mathsf{m}_1, \ldots, \mathsf{m}_t)$, $\mathcal{P}^* = g(\mathcal{P}_1, \ldots, \mathcal{P}_t)$, and $\sigma^* = \mathsf{Eval}(g, \{(\sigma_i, \mathsf{EKS}_i)\}_{i \in [t]})$ where $\mathsf{EKS}_i = \{\mathsf{ek}_{\mathsf{id}}\}_{\mathsf{id} \in \mathcal{P}_i}$. Then, $\mathsf{Ver}(\mathcal{P}^*, \Delta, \{\mathsf{vk}_{\mathsf{id}}\}_{\mathsf{id} \in \mathcal{P}^*}, \mathsf{m}^*, \sigma^*) = 1$ holds with all but negligible probability.

Succinctness. A multi-key HA is said to be *succinct* if the size of every authenticator depends only logarithmically on the size of a dataset. However, we allow authenticators to depend on the number of keys involved in the computation. More formally, let $\mathsf{pp} \leftarrow \mathsf{Setup}(1^\lambda)$, $\mathcal{P} = (f, \ell_1, \ldots, \ell_t)$ with $\ell_i = (\mathsf{id}_i, \tau_i)$, $\{(\mathsf{sk}_{\mathsf{id}}, \mathsf{ek}_{\mathsf{id}}, \mathsf{vk}_{\mathsf{id}}) \leftarrow \mathsf{KeyGen}(\mathsf{pp})\}_{\mathsf{id} \in \mathcal{P}}$, and $\sigma_i \leftarrow \mathsf{Auth}(\mathsf{sk}_{\mathsf{id}_i}, \Delta, \ell_i, \mathsf{m}_i)$ for all $i \in [t]$. A multi-key HA is said to be *succinct* if there exists a fixed polynomial p such that $|\sigma| = p(\lambda, n, \log t)$ where $\sigma = \mathsf{Eval}(g, \{(\sigma_i, \mathsf{ek}_{\mathsf{id}_i})\}_{i \in [t]})$ and $n = |\{\mathsf{id} \in \mathcal{P}\}|$.

Remark 1. Succinctness is one of the crucial properties that make multi-key HAs an interesting primitive. Without succinctness, a trivial multi-key HA construction is the one where Eval outputs the concatenation of all the signatures (and messages) given in input, and the verifier simply checks each message-signature pair and recomputes the function by himself. Our definition of succinctness, where signatures can grow linearly with the number of keys but logarithmically in the total number of inputs, is also non-trivial, especially when considering settings where there are many more inputs than keys (in which case, the above trivial construction would not work). Another property that can make homomorphic signatures an interesting primitive is privacy—context-hiding—as considered in prior work. Intuitively, context-hiding guarantees that signatures do not reveal information on the original inputs. While we leave the study of context-hiding for multi-key HAs for future work, we note that a trivial construction that is context-hiding but not succinct can be easily obtained with the additional help of non-interactive zero-knowledge proofs of knowledge: the idea is to extend the trivial construction above by requiring the evaluator to generate a NIZK proof of knowledge of valid input messages and signatures that yield the public output. In this sense, we believe that succinctness is the most non-trivial property to achieve in homomorphic signatures, and this is what we focus on in this work.

SECURITY. Intuitively, our security model for multi-key HAs guarantees that an adversary, without knowledge of the secret keys, can only produce authenticators that were either received from a legitimate user, or verify correctly on the results of computations executed on the data authenticated by legitimate users. Moreover, we also give to the adversary the possibility of corrupting users. In this case, it must not be able to cheat on the outputs of programs that get inputs from uncorrupted users only. In other words, our security definition guarantees that the corruption of one user affects the data of that user only, but does not endanger the integrity of computations among the other (un-corrupted) users of the system. We point out that preventing cheating on programs that involve inputs of corrupted users is inherently impossible in multi-key HAs, at least if general functions are considered. For instance, consider an adversary who picks the function ($x_1 + x_2 \mod p$) where x_1 is supposed to be provided by user Alice. If the adversary corrupts Alice, it can use her secret key to inject any input authenticated on her behalf and thus bias the output of the function at its will.

The formalization of the intuitions illustrated above is more involved. For a scheme MKHAut we define security via the following experiment between a challenger $\mathcal{C}$ and an adversary $\mathcal{A}$ ($\mathsf{HomUF\text{-}CMA}_{\mathcal{A},\mathsf{MKHAut}}(\lambda)$):

Setup. $\mathcal{C}$ runs $\mathsf{Setup}(1^\lambda)$ to obtain the public parameters pp that are sent to $\mathcal{A}$.

Authentication Queries. $\mathcal{A}$ can adaptively submit queries of the form $(\Delta, \ell, \mathsf{m})$, where Δ is a dataset identifier, $\ell = (\mathsf{id}, \tau)$ is a label in $\mathsf{ID} \times \mathcal{T}$ and $\mathsf{m} \in \mathcal{M}$ are messages of his choice. $\mathcal{C}$ answers as follows:

- If $(\Delta, \ell, \mathsf{m})$ is the first query for the dataset Δ, $\mathcal{C}$ initializes an empty list $L_\Delta = \emptyset$ and proceeds as follows.
- If $(\Delta, \ell, \mathsf{m})$ is the first query with identity id, $\mathcal{C}$ generates keys $(\mathsf{sk}_{\mathsf{id}}, \mathsf{ek}_{\mathsf{id}}, \mathsf{vk}_{\mathsf{id}}) \xleftarrow{\$} \mathsf{KeyGen}(\mathsf{pp})$ (that are implicitly assigned to identity id), gives $\mathsf{ek}_{\mathsf{id}}$ to $\mathcal{A}$ and proceeds as follows.
- If $(\Delta, \ell, \mathsf{m})$ is such that $(\ell, \mathsf{m}) \notin L_\Delta$, $\mathcal{C}$ computes $\sigma_\ell \xleftarrow{\$} \mathsf{Auth}(\mathsf{sk}_{\mathsf{id}}, \Delta, \ell, \mathsf{m})$ (note that $\mathcal{C}$ has already generated keys for the identity id), returns σ_ℓ to $\mathcal{A}$, and updates the list $L_\Delta \leftarrow L_\Delta \cup (\ell, \mathsf{m})$.
- If $(\Delta, \ell, \mathsf{m})$ is such that $(\ell, \cdot) \in L_\Delta$ (which means that the adversary had already made a query $(\Delta, \ell, \mathsf{m}')$ for some message m'), then $\mathcal{C}$ ignores the query.

Verification Queries. $\mathcal{A}$ is also given access to a verification oracle. Namely, the adversary can submit a query $(\mathcal{P}, \Delta, \mathsf{m}, \sigma)$, and $\mathcal{C}$ replies with the output of $\mathsf{Ver}(\mathcal{P}, \Delta, \{\mathsf{vk}_{\mathsf{id}}\}_{\mathsf{id} \in \mathcal{P}}, \mathsf{m}, \sigma)$.

Corruption. The adversary has access to a corruption oracle. At the beginning of the game, the challenger initialises an empty list $L_{\mathsf{corr}} = \emptyset$ of corrupted identities; during the game, $\mathcal{A}$ can adaptively query identities $\mathsf{id} \in \mathsf{ID}$. If $\mathsf{id} \notin L_{\mathsf{corr}}$, then $\mathcal{C}$ replies with the triple $(\mathsf{sk}_{\mathsf{id}}, \mathsf{ek}_{\mathsf{id}}, \mathsf{vk}_{\mathsf{id}})$ (that is generated using KeyGen if not done before) and updates the list $L_{\mathsf{corr}} \leftarrow L_{\mathsf{corr}} \cup \mathsf{id}$. If $\mathsf{id} \in L_{\mathsf{corr}}$, then $\mathcal{C}$ replies with the triple $(\mathsf{sk}_{\mathsf{id}}, \mathsf{ek}_{\mathsf{id}}, \mathsf{vk}_{\mathsf{id}})$ assigned to id before.

Forgery. In the end, $\mathcal{A}$ outputs a tuple $(\mathcal{P}^*, \Delta^*, \mathsf{m}^*, \sigma^*)$. The experiment outputs 1 if the tuple returned by $\mathcal{A}$ is a forgery (defined below), and 0 otherwise.

Definition 4 (Forgery). *Consider an execution of* $\mathsf{HomUF\text{-}CMA}_{\mathcal{A},\mathsf{MKHAut}}(\lambda)$ *where* $(\mathcal{P}^*, \Delta^*, \mathsf{m}^*, \sigma^*)$ *is the tuple returned by the adversary in the end of the experiment, with* $\mathcal{P}^* = (f^*, \ell_1^*, \ldots, \ell_t^*)$. *This is a forgery if* $\mathsf{Ver}(\mathcal{P}^*, \Delta^*, \{\mathsf{vk}_{\mathsf{id}}\}_{\mathsf{id} \in \mathcal{P}^*}, \mathsf{m}^*, \sigma^*) = 1$, *for all* $\mathsf{id} \in \mathcal{P}^*$ *we have that* $\mathsf{id} \notin L_{\mathsf{corr}}$ *(i.e., no identity involved in* $\mathcal{P}^*$ *is corrupted), and either one of the following properties is satisfied:*

Type 1: L_{Δ^*} *has not been initialized during the game (i.e., the dataset* Δ^* *was never queried).*
Type 2: *For all* $i \in [t]$, $\exists(\ell_i^*, \mathsf{m}_i) \in L_{\Delta^*}$, *but* $\mathsf{m}^* \neq f^*(\mathsf{m}_1, \ldots, \mathsf{m}_t)$ *(i.e.,* m^* *is not the correct output of* $\mathcal{P}^*$ *when executed over previously authenticated messages).*
Type 3: *There exists a label* ℓ^* *such that* $(\ell^*, \cdot) \notin L_{\Delta^*}$ *(i.e.,* $\mathcal{A}$ *never made a query with label* ℓ^**).*

We say that a HA scheme MKHAut is *secure* if for every PPT adversary $\mathcal{A}$, its advantage $\mathbf{Adv}^{\mathsf{HomUF\text{-}CMA}}_{\mathsf{MKHAut},\mathcal{A}}(\lambda) = \Pr[\mathsf{HomUF\text{-}CMA}_{\mathcal{A},\mathsf{MKHAut}}(\lambda) = 1]$ is negligible.

Remark 2 (Comparison with previous security definitions). Our security notion can be seen as the multi-key version of the one proposed by Gennaro and Wichs in [26] (in their model our Type 3 forgeries are called 'Type 1' as they do not consider multiple datasets). We point out that even in the special case of a single key, our security definition is stronger than the ones used in previous work [9,16,23,30] (with the only exception of [26]). The main difference lies in our definition of Type 3 forgeries. The intuitive idea of this kind of forgeries is that an adversary who did not receive an authenticated input labeled by a certain ℓ^* cannot produce a valid authenticator for the output of a computation which has ℓ^* among its inputs. In [9,30] these forgeries were not considered at all, as the adversary is assumed to query the dataset always in full. Other works [11,16,23] consider a weaker Type 3 notion, which deals with the concept of "well defined programs", where the input wire labeled by the missing label ℓ^* is required to "contribute" to the computation (i.e., it must change its outcome). The issue with such a Type 3 definition is that it may not be efficient to test if an input contributes to a function, especially if the admissible functions are general circuits. In contrast our definition above is simpler and efficiently testable since it simply considers a Type 3 forgery one where the labeled program $\mathcal{P}^*$ involves an un-queried input.

Multi-key Homomorphic Signatures. As previously mentioned, our definitions are general enough to be easily adapted to either case in which verification is secret or public. The only difference is whether the adversary is allowed to see the verification keys in the security experiment. When the verification is public, we call the primitive *multi-key homomorphic signatures.* More formally:

Definition 5 (Multi-key Homomorphic Signatures). *A multi-key homomorphic signature is a multi-key homomorphic authenticator in which verification keys are also given to the adversary in the security experiment.*

Note that making verification keys public also allows to slightly simplify the security experiment by removing the verification oracle (the adversary can run the verification by itself). In the sequel, when referring to multi-key homomorphic signatures we adapt our notation to the typical one of digital signatures, namely we denote $\mathsf{Auth}(\mathsf{sk}, \Delta, \ell, \mathsf{m})$ as $\mathsf{Sign}(\mathsf{sk}, \Delta, \ell, \mathsf{m})$, and call its outputs *signatures*.

Non-adaptive Corruption Queries. In our work, we consider a relaxation of the security definition in which the adversaries ask for corruptions in a non-adaptive way. More precisely, we say that an adversary $\mathcal{A}$ makes *non-adaptive corruption* queries if for every identity id asked to the corruption oracle, id was not queried earlier in the game to the authentication oracle or the verification oracle. For this class of adversaries, it is easy to see that corruption queries are essentially of no help as the adversary can generate keys on its own. More precisely, the following proposition holds (see the full version [22] for the proof).

Proposition 1. MKHAut *is secure against adversaries that do* not *make corruption queries if and only if* MKHAut *is secure against adversaries that make* non-adaptive *corruption queries.*

Weakly-Adaptive Secure Multi-key HAs. In our work, we also consider a weaker notion of security for multi-key HAs in which the adversary has to declare all the queried messages at the beginning of the experiment. More precisely, we consider a notion in which the adversary declares only the messages and the respective tags that will be queried, for every dataset and identity, without, however, needing to specify the names of the datasets or of the identities. In a sense, the adversary $\mathcal{A}$ is adaptive on identities and dataset names, but not on tags and messages. The definition is inspired by the one, for the single-key setting, of Catalano *et al.* [16].

To define the notion of weakly-adaptive security for multi-key HAs, we introduce here a new experiment $\mathsf{Weak\text{-}HomUF\text{-}CMA}_{\mathcal{A},\mathsf{MKHAut}}$, which is a variation of experiment $\mathsf{HomUF\text{-}CMA}_{\mathcal{A},\mathsf{MKHAut}}$ (Definition 3) as described below.

Definition 6 (Weakly-Secure Multi-key Homomorphic Authenticators). *In the security experiment* $\mathsf{Weak\text{-}HomUF\text{-}CMA}_{\mathcal{A},\mathsf{MKHAut}}$, *before the setup phase, the adversary $\mathcal{A}$ sends to the challenger $\mathcal{C}$ a collection of sets of tags $\mathcal{T}_{i,k} \subseteq \mathcal{T}$ for $i \in [Q_{\mathsf{id}}]$ and $k \in [Q_\Delta]$, where Q_{id} and Q_Δ are, respectively, the total numbers of distinct identities and datasets that will be queried during the game. Associated to every set $\mathcal{T}_{i,k}$, $\mathcal{A}$ also sends a set of messages $\{m_\tau\}_{\tau \in \mathcal{T}_{i,k}}$. Basically the adversary declares, prior to key generation, all the messages and tags that it will query later on; however $\mathcal{A}$ is not required to specify identity and dataset names. Next, the adversary receives the public parameters from $\mathcal{C}$ and can start the query-phase. Verification queries are handled as in* $\mathsf{HomUF\text{-}CMA}_{\mathcal{A},\mathsf{MKHAut}}$.

For authentication queries, $\mathcal{A}$ can adaptively submit pairs (id, Δ) *to* $\mathcal{C}$*. The challenger then replies with a set of authenticators* $\{\sigma_\tau\}_{\tau \in \mathcal{T}_{i,k}}$*, where indices* i, k *are such that* id *is the* i*-th queried identity, and* Δ *is the* k*-th queried dataset.*

An analogous security definition of weakly-secure multi-key homomorphic signatures is trivially obtained by removing a verification oracle.

In the full version of this paper, we present two generic transformations that turn weakly secure multi-key homomorphic authenticator schemes into adaptive secure ones. Our first transformation holds in the standard model and works for schemes in which the tag space $\mathcal{T}$ has polynomial size, while the second one avoids this limitation on the size of $\mathcal{T}$ but holds in the random oracle model.

4 Our Multi-key Fully Homomorphic Signature

In this section, we present our construction of a multi-key homomorphic signature scheme that supports the evaluation of arbitrary circuits of bounded polynomial depth. The scheme is based on the *SIS* problem on standard lattices, a background of which is provided in the next section. Precisely, in Sect. 4.2 we present a scheme that is weakly-secure and supports a single dataset. Later, in Sect. 4.3 we discuss how to extend the scheme to handle multiple datasets, whereas the support of adaptive security can be obtained via the applications of our transformations as shown in [22].

4.1 Lattices and Small Integer Solution Problem

We recall here notation and some basic results about lattices that are useful to describe our homomorphic signature construction.

For any positive integer q we denote by $\mathbb{Z}_q$ the ring of integers modulo q. Elements in $\mathbb{Z}_q$ are represented as integers in the range $(-\frac{q}{2}, \frac{q}{2}]$. The absolute value of any $x \in \mathbb{Z}_q$ (denoted with $|x|$) is defined by taking the modulus q representative of x in $(-\frac{q}{2}, \frac{q}{2}]$, i.e., take $y = x \mod q$ and then set $|x| = |y| \in [0, \frac{q}{2}]$. Vectors and matrices are denoted in bold. For any vector $\mathbf{u} := (u_i, \ldots, u_n) \in \mathbb{Z}_q^n$, its infinity norm is $\|\mathbf{u}\|_\infty := \max_{i \in [n]} |u_i|$, and similarly for a matrix $\mathbf{A} := [a_{i,j}] \in \mathbb{Z}_q^{n \times m}$ we write $\|\mathbf{A}\|_\infty := \max_{i \in [n], j \in [m]} |a_{i,j}|$.

The Small Integer Solution Problem (SIS). For integer parameters n, m, q and β, the $\mathsf{SIS}(n, m, q, \beta)$ problem provides to an adversary $\mathcal{A}$ a uniformly random matrix $\mathbf{A} \in \mathbb{Z}_q^{n \times m}$, and requires $\mathcal{A}$ to find a vector $\mathbf{u} \in \mathbb{Z}_q^n$ such that $\mathbf{u} \neq \mathbf{0}$, $\|\mathbf{u}\|_\infty \leq \beta$, and $\mathbf{A} \cdot \mathbf{u} = \mathbf{0}$. More formally,

Definition 7 (SIS [36]). *Let $\lambda \in \mathbb{N}$ be the security parameter. For values $n = n(\lambda), m = m(\lambda), q = q(\lambda), \beta = \beta(\lambda)$, defined as functions of λ, the* $\mathsf{SIS}(n, m, q, \beta)$ *hardness assumption holds if for any PPT adversary $\mathcal{A}$ we have*

$$\Pr\left[\mathbf{A} \cdot \mathbf{u} = \mathbf{0} \ \wedge \ \mathbf{u} \neq \mathbf{0} \ \wedge \ \|\mathbf{u}\|_\infty \leq \beta : \mathbf{A} \xleftarrow{\$} \mathbb{Z}_q^{n \times m}, \mathbf{u} \leftarrow \mathcal{A}(1^\lambda, \mathbf{A})\right] \leq \mathsf{negl}(\lambda).$$

For standard lattices, the SIS problem is known to be as hard as solving certain worst-case instances of lattice problems [2,33,35,36], and is also implied by the hardness of learning with error (we refer any interested reader to the cited papers for the technical details about the parameters).

In our paper, we assume that for any $\beta = 2^{\mathsf{poly}(\lambda)}$ there are some $n = \mathsf{poly}(\lambda)$, $q = 2^{\mathsf{poly}(\lambda)}$, with $q > \beta$, such that for all $m = \mathsf{poly}(\lambda)$ the $\mathsf{SIS}(n, m, q, \beta)$ hardness assumption holds. This parameters choice assures that hardness of worst-case lattice problems holds with sub-exponential approximation factors.

Trapdoors for Lattices. The SIS problem is hard to solve for a random matrix $\mathbf{A}$. However, there is a way to sample a random $\mathbf{A}$ together with a *trapdoor* such that SIS becomes easy to solve for that $\mathbf{A}$, given the trapdoor. Additionally, it has been shown that there exist "special" (non random) matrices $\mathbf{G}$ for which SIS is easy to solve as well. The following lemma summarizes the above known results (similar to a lemma in [10]):

Lemma 2 ([3,4,28,34]). *There exist efficient algorithms* TrapGen, SamPre, Sam *such that the following holds: given integers $n \geq 1$, $q \geq 2$, there exist some $m^* = m^*(n, q) = O(n \log q)$, $\beta_{\mathsf{sam}} = \beta_{\mathsf{sam}}(n, q) = O(n\sqrt{\log q})$ such that for all $m \geq m^*$ and all k (polynomial in n) we have:*

1. $\mathsf{Sam}(1^m, 1^k, q) \to \mathbf{U}$ *samples a matrix $\mathbf{U} \in \mathbb{Z}_q^{m \times k}$ such that $\|\mathbf{U}\|_\infty \leq \beta_{\mathsf{sam}}$ (with probability 1).*
2. *For $(\mathbf{A}, \mathsf{td}) \leftarrow \mathsf{TrapGen}(1^n, 1^m, q)$, $\mathbf{A}' \xleftarrow{\$} \mathbb{Z}_q^{n \times m}$, $\mathbf{U} \leftarrow \mathsf{Sam}(1^m, 1^k, q)$, $\mathbf{V} := \mathbf{AU}$, $\mathbf{V}' \xleftarrow{\$} \mathbb{Z}_q^{n \times k}$, $\mathbf{U}' \leftarrow \mathsf{SamPre}(\mathbf{A}, \mathbf{V}', \mathsf{td})$, we have the following statistical indistinguishability (negligible in n)*

$$\mathbf{A} \overset{\mathsf{stat}}{\approx} \mathbf{A}' \quad \text{and} \quad (\mathbf{A}, \mathsf{td}, \mathbf{U}, \mathbf{V}) \overset{\mathsf{stat}}{\approx} (\mathbf{A}, \mathsf{td}, \mathbf{U}', \mathbf{V}')$$

and $\mathbf{U}' \leftarrow \mathsf{SamPre}(\mathbf{A}, \mathbf{V}', \mathsf{td})$ always satisfies $\mathbf{AU}' = \mathbf{V}'$ and $\|\mathbf{U}'\|_\infty \leq \beta_{\mathsf{sam}}$.
3. *Given n, m, q as above, there is an efficiently and deterministically computable matrix $\mathbf{G} \in \mathbb{Z}_q^{n \times m}$ and a deterministic polynomial-time algorithm $\mathbf{G}^{-1}$ that on input $\mathbf{V} \in \mathbb{Z}_q^{n \times k}$ (for any integer k) outputs $\mathbf{R} = \mathbf{G}^{-1}(\mathbf{V})$ such that $\mathbf{R} \in \{0,1\}^{m \times k}$ and $\mathbf{GR} = \mathbf{V}$.*

4.2 Our Multi-key Homomorphic Signature for Single Dataset

In this section, we present our multi-key homomorphic signature that supports the evaluation of boolean circuits of bounded polynomial depth. Our construction is inspired by the (single-key) one of Gorbunov *et al.* [30], with the fundamental difference that in our case we enable computations over data signed using different secret keys. Moreover, our scheme is secure against Type 3 forgeries. We achieve this via a new technique which consists into adding to every signature a component that specifically protects against this type of forgeries. We prove the scheme to be weakly-secure under the SIS hardness assumption.

Parameters. Before describing the scheme, we discuss how to set the various parameters involved. Let λ be the security parameter, and let $d = d(\lambda) = \mathsf{poly}(\lambda)$ be the bound on the depth of the circuits supported by our scheme. We define the set of parameters used in our scheme $\mathsf{Par} = \{n, m, q, \beta_{\mathsf{SIS}}, \beta_{\mathsf{max}}, \beta_{\mathsf{init}}\}$ in terms of λ, d and of the parameters required by the trapdoor algorithm in Lemma 2: $m^*, \beta_{\mathsf{sam}}$, where $m^* = m^*(n,q) := O(n \log q)$ and $\beta_{\mathsf{sam}} := O(n\sqrt{\log q})$. More precisely, we set: $\beta_{\mathsf{max}} := 2^{\omega(\log \lambda)d}$; $\beta_{\mathsf{SIS}} := 2^{\omega(\log \lambda)}\beta_{\mathsf{max}}$; $n = \mathsf{poly}(\lambda)$; $q = \mathcal{O}(2^{\mathsf{poly}(\lambda)}) > \beta_{\mathsf{SIS}}$ is a prime (as small as possible) so that the $\mathsf{SIS}(n, m', q, \beta_{\mathsf{SIS}})$ assumption holds for all $m' = \mathsf{poly}(\lambda)$; $m = \max\{m^*, n \log q + \omega(\log(\lambda))\} = \mathsf{poly}(\lambda)$ and, finally, $\beta_{\mathsf{init}} := \beta_{\mathsf{sam}} = \mathsf{poly}(\lambda)$.

Construction. The PPT algorithms $(\mathsf{Setup}, \mathsf{KeyGen}, \mathsf{Sign}, \mathsf{Eval}, \mathsf{Ver})$ which define our construction of Multi-key Homomorphic Signatures work as follows:

$\mathsf{Setup}(1^\lambda)$. The setup algorithm takes as input the security parameter λ and generates the public parameters pp which include: the bound on the circuit depth d (which defines the class $\mathcal{F}$ of functions supported by the scheme, i.e., boolean circuits of depth d), the set $\mathsf{Par} = \{n, m, q, \beta_{\mathsf{SIS}}, \beta_{\mathsf{max}}, \beta_{\mathsf{init}}\}$, the set $\mathcal{U} = \{\mathbf{U} \in \mathbb{Z}_q^{m \times m} : \|\mathbf{U}\|_\infty \le \beta_{\mathsf{max}}\}$, the set $\mathcal{V} = \{\mathbf{V} \in \mathbb{Z}_q^{n \times m}\}$, descriptions of the message space $\mathcal{M} = \{0,1\}$, the tag space $\mathcal{T} = [\mathsf{T}]$, and the identity space $\mathsf{ID} = [\mathsf{C}]$, for integers $\mathsf{T}, \mathsf{C} \in \mathbb{N}$. In this construction, the tag space is of polynomial size, i.e., $\mathsf{T} = \mathsf{poly}(\lambda)$ while the identity space is essentially unbounded, i.e., we set $\mathsf{C} = 2^\lambda$. Also recall that $\mathcal{T}$ and ID immediately define the label space $\mathcal{L} = \mathsf{ID} \times \mathcal{T}$. The final output is $\mathsf{pp} = \{d, \mathsf{Par}, \mathcal{U}, \mathcal{V}, \mathcal{M}, \mathcal{T}, \mathsf{ID}\}$. We assume that these public parameters pp are input of all subsequent algorithms, and often omit them from the input explicitly.

$\mathsf{KeyGen}(\mathsf{pp})$. The key generation algorithm takes as input the public parameters pp and generates a key-triple $(\mathsf{sk}, \mathsf{ek}, \mathsf{vk})$ defined as follows. First, it samples T random matrices $\mathbf{V}_1, \ldots, \mathbf{V}_{\mathsf{T}} \xleftarrow{\$} \mathcal{V}$. Second, it runs $(\mathbf{A}, \mathsf{td}) \leftarrow \mathsf{TrapGen}(1^n, 1^m, q)$ to generate a matrix $\mathbf{A} \in \mathbb{Z}_q^{n \times m}$ along with its trapdoor td. Then, it outputs $\mathsf{sk} = (\mathsf{td}, \mathbf{A}, \mathbf{V}_1, \ldots, \mathbf{V}_{\mathsf{T}})$, $\mathsf{ek} = \mathbf{A}$, $\mathsf{vk} = (\mathbf{A}, \mathbf{V}_1, \ldots, \mathbf{V}_{\mathsf{T}})$. Note that it is possible to associate the key-triple to an identity $\mathsf{id} \in \mathsf{ID}$, when we need to stress this explicitly we write $(\mathsf{sk}_{\mathsf{id}}, \mathsf{ek}_{\mathsf{id}}, \mathsf{vk}_{\mathsf{id}})$. We also observe that the key generation process can be seen as the combination of two independent sub-algorithms[7] KeyGen_1 and KeyGen_2, where $\{\mathbf{V}_1, \ldots \mathbf{V}_{\mathsf{T}}\} \leftarrow \mathsf{KeyGen}_1(\mathsf{pp})$ and $(\mathbf{A}, \mathsf{td}) \leftarrow \mathsf{KeyGen}_2(\mathsf{pp})$.

$\mathsf{Sign}(\mathsf{sk}, \ell, \mathsf{m})$. The signing algorithm takes as input a secret key sk, a label $\ell = (\mathsf{id}, \tau)$ for the message m and it outputs a signature $\sigma := (\mathsf{m}, \mathsf{z}, \mathsf{I}, \mathbf{U}_{\mathsf{id}}, \mathbf{Z}_{\mathsf{id}})$ where $\mathsf{I} = \{\mathsf{id}\}$, $\mathbf{U}_{\mathsf{id}}$ is generated as $\mathbf{U}_{\mathsf{id}} \leftarrow \mathsf{SamPre}(\mathbf{A}, \mathbf{V}_\ell - \mathsf{m}\mathbf{G}, \mathsf{td})$ (using the algorithm SamPre from Lemma 2), $\mathsf{z} = \mathsf{m}$ and $\mathbf{Z}_{\mathsf{id}} = \mathbf{U}_{\mathsf{id}}$. The two latter terms are responsible for protection against Type 3 forgeries. Although they are redundant for fresh signatures, their value will become different from $(\mathsf{m}, \mathbf{U}_{\mathsf{id}})$

[7] This splitting will be used to extend our multi-key homomorphic signature scheme from supporting a single dataset to support multiple datasets. This extension holds in the standard model and is described in Sect. 4.3.

during homomorphic operations, as we clarify later on. More generally, in our construction signatures are of the form $\sigma := (\mathsf{m}, \mathsf{z}, \mathsf{I}, \{\mathbf{U}_{\mathsf{id}}\}_{\mathsf{id}\in\mathsf{I}}, \{\mathbf{Z}_{\mathsf{id}}\}_{\mathsf{id}\in\mathsf{I}})$ with $\mathsf{I} \subseteq \mathsf{ID}$ and $\mathbf{U}_{\mathsf{id}}, \mathbf{Z}_{\mathsf{id}} \in \mathcal{U}, \forall\, \mathsf{id} \in \mathsf{I}$.

$\mathsf{Eval}(f, \{(\sigma_i, \mathsf{EKS}_i)\}_{i\in[t]})$. The evaluation algorithm takes as input a t-input function $f : \mathcal{M}^t \longrightarrow \mathcal{M}$, and a set of pairs $\{(\sigma_i, \mathsf{EKS}_i)\}_{i\in[t]}$ where each σ_i is a signature and each EKS_i is a set of evaluation keys. In our description below we treat f as an arithmetic circuit over $\mathbb{Z}_q$ consisting of addition and multiplication gates.[8] Therefore, we only describe how to evaluate homomorphically a fan-in-2 addition (resp. multiplication) gate as well as a unary multiplication-by-constant gate.

Let g be a fan-in-2 gate with left input $\sigma_{\mathsf{L}} := (\mathsf{m}_{\mathsf{L}}, \mathsf{z}_{\mathsf{L}}, \mathsf{I}_{\mathsf{L}}, \mathbf{U}_{\mathsf{L}}, \mathbf{Z}_{\mathsf{L}})$ and right input $\sigma_{\mathsf{R}} := (\mathsf{m}_{\mathsf{R}}, \mathsf{z}_{\mathsf{R}}, \mathsf{I}_{\mathsf{R}}, \mathbf{U}_{\mathsf{R}}, \mathbf{Z}_{\mathsf{R}})$. To generate the signature $\sigma := (\mathsf{m}, \mathsf{z}, \mathsf{I}, \mathbf{U}, \mathbf{Z})$ on the gate's output one proceeds as follows. First set $\mathsf{I} = \mathsf{I}_{\mathsf{L}} \cup \mathsf{I}_{\mathsf{R}}$. Second, "expand" $\mathbf{U}_{\mathsf{L}} := \{\mathbf{U}_{\mathsf{L}}^{\mathsf{id}}\}_{\mathsf{id}\in\mathsf{I}_{\mathsf{L}}}$ as:

$$\hat{\mathbf{U}}_{\mathsf{L}}^{\mathsf{id}} = \begin{cases} \mathbf{0} & \text{if } \mathsf{id} \notin \mathsf{I}_{\mathsf{L}} \\ \mathbf{U}_{\mathsf{L}}^{\mathsf{id}} & \text{if } \mathsf{id} \in \mathsf{I}_{\mathsf{L}} \end{cases}, \quad \forall\, \mathsf{id} \in \mathsf{I}\,.$$

where $\mathbf{0}$ denotes an $(m \times m)$-matrix with all zero entries. Basically, we extend the set to be indexed over all identities in $\mathsf{I} = \mathsf{I}_{\mathsf{L}} \cup \mathsf{I}_{\mathsf{R}}$ by inserting zero matrices for identities in $\mathsf{I} \setminus \mathsf{I}_{\mathsf{L}}$. The analogous expansion process is applied to $\mathbf{U}_{\mathsf{R}} := \{\mathbf{U}_{\mathsf{R}}^{\mathsf{id}}\}_{\mathsf{id}\in\mathsf{I}_{\mathsf{L}}}$, $\mathbf{Z}_{\mathsf{L}} := \{\mathbf{Z}_{\mathsf{L}}{}^{\mathsf{id}}\}_{\mathsf{id}\in\mathsf{I}_{\mathsf{R}}}$ and $\mathbf{Z}_{\mathsf{R}} := \{\mathbf{Z}_{\mathsf{R}}^{\mathsf{id}}\}_{\mathsf{id}\in\mathsf{I}_{\mathsf{R}}}$, denoting the expanded sets $\{\hat{\mathbf{U}}_{\mathsf{R}}^{\mathsf{id}}\}_{\mathsf{id}\in\mathsf{I}}$, $\{\hat{\mathbf{Z}}_{\mathsf{L}}^{\mathsf{id}}\}_{\mathsf{id}\in\mathsf{I}}$ and $\{\hat{\mathbf{Z}}_{\mathsf{R}}^{\mathsf{id}}\}_{\mathsf{id}\in\mathsf{I}}$ respectively.

Next, depending on whether g is an addition or multiplication gate one proceeds as follows.

Addition gate. If g is additive, compute $\mathsf{m} = \mathsf{m}_{\mathsf{L}} + \mathsf{m}_{\mathsf{R}}$, $\mathsf{z} = \mathsf{z}_{\mathsf{L}} + \mathsf{z}_{\mathsf{R}}$, $\mathbf{U} = \{\mathbf{U}_{\mathsf{id}}\}_{\mathsf{id}\in\mathsf{I}} := \{\hat{\mathbf{U}}_{\mathsf{L}}^{\mathsf{id}} + \hat{\mathbf{U}}_{\mathsf{R}}^{\mathsf{id}}\}_{\mathsf{id}\in\mathsf{I}}$ and $\mathbf{Z} = \{\mathbf{Z}_{\mathsf{id}}\}_{\mathsf{id}\in\mathsf{I}} := \{\hat{\mathbf{Z}}_{\mathsf{L}}^{\mathsf{id}} + \hat{\mathbf{Z}}_{\mathsf{R}}^{\mathsf{id}}\}_{\mathsf{id}\in\mathsf{I}}$.

If we refer to β_{L} and β_{R} as $\|\mathbf{U}_{\mathsf{L}}\|_\infty := \max\{\|\mathbf{U}_{\mathsf{L}}^{\mathsf{id}}\|_\infty : \mathsf{id} \in \mathsf{I}_{\mathsf{L}}\}$ and $\|\mathbf{U}_{\mathsf{R}}\|_\infty := \max\{\|\mathbf{U}_{\mathsf{R}}^{\mathsf{id}}\|_\infty : \mathsf{id} \in \mathsf{I}_{\mathsf{R}}\}$ respectively, then for any fan-in-2 addition gate it holds $\beta := \|\mathbf{U}\|_\infty = \beta_{\mathsf{L}} + \beta_{\mathsf{R}}$. The same noise growth applies to $\mathbf{Z}$.

Multiplication gate. If g is multiplicative, compute $\mathsf{m} = \mathsf{m}_{\mathsf{L}} \cdot \mathsf{m}_{\mathsf{R}}$, $\mathsf{z} = \mathsf{z}_{\mathsf{L}} + \mathsf{z}_{\mathsf{R}}$, define $\mathbf{V}_{\mathsf{L}} = \sum_{\mathsf{id}\in\mathsf{I}_{\mathsf{L}}} \mathbf{A}_{\mathsf{id}} \mathbf{U}_{\mathsf{id}} + \mathsf{m}_{\mathsf{L}} \mathbf{G}$, set

$$\mathbf{U} = \{\mathbf{U}_{\mathsf{id}}\}_{\mathsf{id}\in\mathsf{I}} := \{\mathsf{m}_{\mathsf{R}} \hat{\mathbf{U}}_{\mathsf{L}}^{\mathsf{id}} + \hat{\mathbf{U}}_{\mathsf{R}}^{\mathsf{id}} \cdot \mathbf{G}^{-1}(\mathbf{V}_{\mathsf{L}})\}_{\mathsf{id}\in\mathsf{I}}$$

and $\mathbf{Z} = \{\mathbf{Z}_{\mathsf{id}}\}_{\mathsf{id}\in\mathsf{I}} := \{\hat{\mathbf{Z}}_{\mathsf{L}}^{\mathsf{id}} + \hat{\mathbf{Z}}_{\mathsf{R}}^{\mathsf{id}}\}_{\mathsf{id}\in\mathsf{I}}$.

Letting β_{L} and β_{R} as defined before, then for any fan-in-2 multiplication gate it holds $\beta := \|\mathbf{U}\|_\infty = |\mathsf{m}_{\mathsf{R}}|\beta_{\mathsf{L}} + m\beta_{\mathsf{R}}$, while the noise growth of $\mathbf{Z}$ is the same as in the addition gate.

Multiplication by constant gate. Let g be a unary gate representing a multiplication by a constant $a \in \mathbb{Z}_q$, and let its single input signature

[8] We point out that considering f as an arithmetic circuit over $\mathbb{Z}_q$ is enough to describe any boolean circuits consisting of NAND gates as $\mathsf{NAND}(\mathsf{m}_1, \mathsf{m}_2) = 1 - \mathsf{m}_1 \cdot \mathsf{m}_2$ holds for $\mathsf{m}_1, \mathsf{m}_2 \in \{0, 1\}$.

be $\sigma_R := (m_R, z_R, I_R, \mathbf{U}_R, \mathbf{Z}_R)$. The output $\sigma := (m, z, I, \mathbf{U}, \mathbf{Z})$ is obtained by setting $m = a \cdot m_R \in \mathbb{Z}_q$, $z = z_R$, $I = I_R$, $\mathbf{Z} = \mathbf{Z}_R$, and $\mathbf{U} = \{\mathbf{U}^{id}\}_{id \in I}$ where, for all $id \in I$, $\mathbf{U}^{id} = a \cdot \mathbf{U}_R^{id}$ or, alternatively, $\mathbf{U}^{id} = \mathbf{U}_R^{id} \cdot \mathbf{G}^{-1}(a \cdot \mathbf{G})$. In the first case, the noise parameter becomes $\beta := \|\mathbf{U}\|_\infty = |a|\beta_L$ (thus a needs to be *small*), whereas in the second case it holds $\beta := \|\mathbf{U}\|_\infty \leq m\beta_L$, which is independent of a's size.

$\mathsf{Ver}(\mathcal{P}, \{vk_{id}\}_{id \in \mathcal{P}}, m, \sigma)$. The verification algorithm takes as input a labeled program $\mathcal{P} = (f, \ell_1, \ldots, \ell_n)$, the set of the verification keys $\{vk_{id}\}_{id \in \mathcal{P}}$ of users involved in the program $\mathcal{P}$, a message m and a signature $\sigma = (m, z, I, \mathbf{U}, \mathbf{Z})$. It then performs three main checks and outputs 0 if at least one check fails, otherwise it returns 1.

Firstly, it checks if the list of identities declared in σ corresponds to the ones in the labels of $\mathcal{P}$:

$$I = \{id : id \in \mathcal{P}\} \tag{1}$$

Secondly, from the circuit f (again seen as an arithmetic circuit) and the values $\{\mathbf{V}_{\ell_1}, \ldots, \mathbf{V}_{\ell_t}\}$ contained in the verification keys, it computes two values $\mathbf{V}^*$ and $\mathbf{V}^+$ proceeding gate by gate as follows. Given as left and right input matrices $\mathbf{V}_L^*, \mathbf{V}_R^*$ (resp. $\mathbf{V}_L^+, \mathbf{V}_R^+$), at every *addition gate* one computes $\mathbf{V}^* = \mathbf{V}_L^* + \mathbf{V}_R^*$ (resp. $\mathbf{V}^+ = \mathbf{V}_L^+ + \mathbf{V}_R^+$); at every *multiplication gate* one computes $\mathbf{V}^* = \mathbf{V}_R^* \mathbf{G}^{-1} \mathbf{V}_L^*$ (resp. $\mathbf{V}^+ = \mathbf{V}_L^+ + \mathbf{V}_R^+$). Every gate representing a multiplication by a constant $a \in \mathbb{Z}_q$, on input $\mathbf{V}_R^*$ (resp. $\mathbf{V}_R^+$) outputs $\mathbf{V}^* = a \cdot \mathbf{V}_R^*$ (resp. $\mathbf{V}^+ = \mathbf{V}_R^+$). Note that the computation of $\mathbf{V}^+$ is essentially the computation of a linear function $\mathbf{V}^+ = \sum_{i=1}^t \gamma_i \cdot \mathbf{V}_{\ell_i}$, for some coefficients γ_i that depend on the structure of the circuit f.

Thirdly, the verification algorithm parses $\mathbf{U} = \{\mathbf{U}_{id}\}_{id \in I}$ and $\mathbf{Z} = \{\mathbf{Z}_{id}\}_{id \in I}$ and checks:

$$\|\mathbf{U}\|_\infty \leq \beta_{max} \quad \text{and} \quad \|\mathbf{Z}\|_\infty \leq \beta_{max} \tag{2}$$

$$\sum_{id \in I} \mathbf{A}_{id} \mathbf{U}_{id} + m \cdot \mathbf{G} = \mathbf{V}^* \tag{3}$$

$$\sum_{id \in I} \mathbf{A}_{id} \mathbf{Z}_{id} + z \cdot \mathbf{G} = \mathbf{V}^+ \tag{4}$$

Finally, it is worth noting that the computation of the matrices $\mathbf{V}^*$ and $\mathbf{V}^+$ can be precomputed (or performed offline), prior to seeing the actual signature σ. In the multiple dataset extension of Sect. 4.3 this precomputation becomes particularly beneficial as the same $\mathbf{V}^*, \mathbf{V}^+$ can be re-used every time one wants to verify for the same labeled program $\mathcal{P}$ (i.e., one can verify faster, in an amortized sense, than that of running f).

In the following paragraphs we analyse the correctness, succinctness and the security of the proposed construction. In the following paragraphs, we analyse succinctness and security of the proposed construction; for what regards correctness we just give an intuition and refer the interested reader to the full version of this paper available in [22].

NOISE GROWTH AND SUCCINCTNESS. First we analyse the noise growth of the components $\mathbf{U}, \mathbf{Z}$ in the signatures of our MKHSig construction. In particular we need to show that when starting from "fresh" signatures, in which the noise is bounded by β_{init}, and we apply an admissible circuit, then one ends up with signatures in which the noise is within the allowable amount β_{max}.

An analysis similar to the one of Gorbunov *et al.* [30] is applicable also to our construction whenever the admissible functions are boolean circuits of depth d composed only of NAND gates.

Let us first consider the case of the $\mathbf{U}$ component of the signatures. At every NAND gate, if $\|\mathbf{U}_{\mathsf{L}}\|_\infty, \|\mathbf{U}_{\mathsf{R}}\|_\infty \leq \beta$, the noise of the resulting $\mathbf{U}$ is at most $(m+1)\beta$. Therefore, if the circuit has depth d, the noise of the matrix $\mathbf{U}$ at the end of the evaluation is bounded by $\|\mathbf{U}\|_\infty \leq \beta_{\mathsf{init}} \cdot (m+1)^d \leq 2^{O(\log \lambda)d} \leq \beta_{\mathsf{max}}$. For what regards the computation performed over the matrices $\mathbf{Z}$, we observe that we perform only additions (or identity functions) over them. This means that at every gate of any f, the noise in the $\mathbf{Z}$ component at most doubles. Given that we consider depth-d circuits we have that $\|\mathbf{Z}\|_\infty \leq \beta_{\mathsf{init}} \cdot 2^d \leq 2^{O(\log \lambda)+d} \leq \beta_{\mathsf{max}}$. Finally, by inspection one can see that the size of every signature σ on a computation's output involving n users is at most $(1 + 2^d + n\lambda + 2n\beta_{\mathsf{max}})$ that is $O(n \cdot p(\lambda))$ for some fixed polynomial $p(\cdot)$.

AUTHENTICATION CORRECTNESS. This is rather simple and follows from the noise growth property mentioned above and by observing that equation $\mathbf{A}_{\mathsf{id}}\mathbf{U}_{\mathsf{id}} + \mathsf{m}\mathbf{G} = \mathbf{V}_\ell = \mathbf{V}^*$ holds by construction.

EVALUATION CORRECTNESS. Evaluation correctness follows from two main facts: the noise growth mentioned earlier, and the preservation of the invariant $\sum_{\mathsf{id} \in \mathsf{I}} \mathbf{A}_{\mathsf{id}}\mathbf{U}_{\mathsf{id}} + \mathsf{m}\mathbf{G} = \mathbf{V}^*$. At every gate, it is easy to see that the expansion of $\mathbf{U}$ still preserves the invariant for both left and right inputs. For additive gates, assuming validity of the inputs, i.e., $\mathbf{V}_{\mathsf{L}} = \sum_{\mathsf{id} \in \mathsf{I}_{\mathsf{L}}} \mathbf{A}_{\mathsf{id}}\mathbf{U}_{\mathsf{L}}^{\mathsf{id}} + \mathsf{m}_{\mathsf{L}}\mathbf{G}$ (and similarly $\mathbf{V}_{\mathsf{R}}$) and by construction of $\mathbf{U}_{\mathsf{id}} = \mathbf{U}_{\mathsf{L}}^{\mathsf{id}} + \mathbf{U}_{\mathsf{R}}^{\mathsf{id}}$, one obtains $\sum_{\mathsf{id} \in \mathsf{I}_{\mathsf{L}} \cup \mathsf{I}_{\mathsf{R}}} \mathbf{A}_{\mathsf{id}}\mathbf{U}_{\mathsf{id}} + (\mathsf{m}_{\mathsf{L}} + \mathsf{m}_{\mathsf{R}})\mathbf{G} = \mathbf{V}_{\mathsf{L}} + \mathbf{V}_{\mathsf{R}} = \mathbf{V}^*$. For what regards multiplicative gates, by construction of every $\mathbf{U}_{\mathsf{id}}$ we obtain $\sum_{\mathsf{id} \in \mathsf{I}} \mathbf{A}_{\mathsf{id}}\mathbf{U}_{\mathsf{id}} + \mathsf{m}\mathbf{G} := \sum_{\mathsf{id} \in \mathsf{I}} \mathbf{A}_{\mathsf{id}}(\mathsf{m}_{\mathsf{R}}\hat{\mathbf{U}}_{\mathsf{L}}^{\mathsf{id}} + \hat{\mathbf{U}}_{\mathsf{R}}^{\mathsf{id}}\mathbf{G}^{-1}(\mathbf{V}_{\mathsf{L}})) + (\mathsf{m}_{\mathsf{L}}\mathsf{m}_{\mathsf{R}})\mathbf{G}$. Grouping by m_{R} and applying the definition of $\mathbf{V}_{\mathsf{L}}$, the equation can be rewritten as $\mathsf{m}_{\mathsf{R}}\mathbf{V}_{\mathsf{L}} + \left(\sum_{\mathsf{id} \in \mathsf{I}} \mathbf{A}_{\mathsf{id}} \ \hat{\mathbf{U}}_{\mathsf{R}}^{\mathsf{id}}\right) \mathbf{G}^{-1}(\mathbf{V}_{\mathsf{L}})$. If now we write $\mathsf{m}_{\mathsf{R}}\mathbf{V}_{\mathsf{L}}$ as $\mathsf{m}_{\mathsf{R}}\mathbf{G}\mathbf{G}^{-1}(\mathbf{V}_{\mathsf{L}})$ and we group by $\mathbf{G}^{-1}(\mathbf{V}_{\mathsf{L}})$, we get $\left[\left(\sum_{\mathsf{id} \in \mathsf{I}_{\mathsf{R}}} \mathbf{A}_{\mathsf{id}}\mathbf{U}_{\mathsf{R}}^{\mathsf{id}}\right) + \mathsf{m}_{\mathsf{R}}\mathbf{G}\right] \mathbf{G}^{-1}(\mathbf{V}_{\mathsf{L}}) = \mathbf{V}^*$, where the last equation follows from the definitions of $\mathbf{V}_{\mathsf{R}}$ and $\mathbf{V}^*$. Correctness of computations over the matrices $\mathbf{Z}$ is quite analogous.

Security. The following theorem states the security of the scheme MKHSig.

Theorem 1. *If the* $\mathsf{SIS}(n, m \cdot Q_{\mathsf{id}}, q, \beta_{\mathsf{SIS}})$ *hardness assumption holds,* $\mathsf{MKHSig} = (\mathsf{Setup}, \mathsf{KeyGen}, \mathsf{Sign}, \mathsf{Eval}, \mathsf{Ver})$ *is a multi-key homomorphic signature weakly-*

adaptive secure against adversaries that make signing queries involving at most Q_{id} different identities and that make non-adaptive corruption queries.

Proof. Note that we can deal with corruptions via our generic result of Proposition 1. Therefore it is sufficient to prove the security against adversaries that make no corruptions. Moreover, since this scheme works for a single dataset note that Type 1 forgeries cannot occur.

For the proof let us recall how the weakly-adaptive security experiment (Definition 6) works for our multi-key homomorphic signature scheme MKHSig. This is a game between an adversary $\mathcal{A}$ and a challenger $\mathcal{C}$ that has four main phases:

(1) $\mathcal{A}$ declares an integer Q representing the number of different identities that it will ask in the signing queries. Moreover, for every $i \in [Q]$ $\mathcal{A}$ sends to $\mathcal{C}$ a set $\mathcal{T}_i \subseteq \mathcal{T} := \{\tau_1, \ldots, \tau_T\}$ and a set of pairs $\{(\mathsf{m}_\tau, \tau)\}_{\tau \in \mathcal{T}_i}$.
(2) $\mathcal{C}$ runs $\mathsf{Setup}(1^\lambda)$ to obtain the public parameters and sends them to $\mathcal{A}$.
(3) $\mathcal{A}$ adaptively queries identities $\mathsf{id}_1, \ldots, \mathsf{id}_Q$. When $\mathcal{C}$ receives the query id_i it generates a key-triple $(\mathsf{sk}_{\mathsf{id}_i}, \mathsf{ek}_{\mathsf{id}_i}, \mathsf{vk}_{\mathsf{id}_i})$ by running $\mathsf{KeyGen}(\mathsf{pp})$, and for all labels $\ell = (\mathsf{id}_i, \tau)$ such that $\tau \in \mathcal{T}_i$ it runs $\sigma_\tau^i \leftarrow \mathsf{Sign}(\mathsf{sk}_{\mathsf{id}_i}, \ell, \mathsf{m}_\tau)$. Then $\mathcal{C}$ sends to $\mathcal{A}$: the public keys $\mathsf{vk}_{\mathsf{id}_i} := (\mathbf{A}_{\mathsf{id}_i}, \{\mathbf{V}_\ell\}_{\tau \in \mathcal{T}})$ and $\mathsf{ek}_{\mathsf{id}_i} := (\mathbf{A}_{\mathsf{id}_i})$, and the signatures $\{\sigma_\tau^i\}_{\tau \in \mathcal{T}_i}$.
(4) The adversary produces a forgery consisting of a labeled program $\mathcal{P}^* = (f^*, \ell_1^*, \ldots, \ell_t^*)$ where $f^* \in \mathcal{F}, f^* : \mathcal{M}^t \to \mathcal{M}$, a message m^* and a signature σ^*.

$\mathcal{A}$ wins the non-adaptive security game if $\mathsf{Ver}(\mathcal{P}^*, \{\mathsf{vk}_{\mathsf{id}}\}_{\mathsf{id} \in \mathcal{P}^*}, \mathsf{m}^*, \sigma^*) = 1$ and one of the following conditions holds:

Type 2 Forgery: there exist messages $\mathsf{m}_{\ell_1^*}, \ldots, \mathsf{m}_{\ell_t^*}$ s.t. $\mathsf{m}^* \neq f^*(\mathsf{m}_{\ell_1^*}, \ldots, \mathsf{m}_{\ell_t^*})$ (i.e., m^* is not the correct output of $\mathcal{P}^*$ when executed over previously signed messages).

Type 3 Forgery: there exists at least one label $\ell^* = (\mathsf{id}^*, \tau^*)$ that was not queried by $\mathcal{A}$.

Consider a variation of the above game obtained modifying phase (3) as follows: (3′) $\mathcal{C}$ picks an instance $\mathbf{A} \in \mathbb{Z}_q^{n \times m'}$ of the $\mathsf{SIS}(n, m', q, \beta_{\mathsf{SIS}})$ problem for $m' = m \cdot Q = \mathsf{poly}(\lambda)$, and parse $\mathbf{A} := (\mathbf{A}_{\mathsf{id}_1} | \ldots | \mathbf{A}_{\mathsf{id}_Q}) \in \mathbb{Z}_q^{n \times m'}$ as the concatenation of Q different blocks of $n \times m$ matrices.
Next, when $\mathcal{C}$ receives the i-th query id_i from $\mathcal{A}$, it does the following:

- it samples a matrix $\mathbf{U}_{\mathsf{id}_i, \tau} \xleftarrow{\$} \mathcal{U}$ such that $\|\mathbf{U}_{\mathsf{id}_i, \tau}\| \leq \beta_{\mathsf{init}}$;
- for all $\ell := (\mathsf{id}_i, \tau)$ with $\tau \in \mathcal{T}_i$, $\mathcal{C}$ computes $\mathbf{V}_\ell = \mathbf{A}_{\mathsf{id}_i} \mathbf{U}_{\mathsf{id}_i, \tau} + \mathsf{m}_\tau \cdot \mathbf{G}$;
- for all $\ell := (\mathsf{id}_i, \tau)$ with $\tau \notin \mathcal{T}_i$, $\mathcal{C}$ computes $\mathbf{V}_\ell = \mathbf{A}_{\mathsf{id}_i} \mathbf{U}_{\mathsf{id}_i, \tau} + b_{i,\tau} \cdot \mathbf{G}$, where $b_{i,\tau} \xleftarrow{\$} \{0, 1\}$.
- $\mathcal{C}$ sends to $\mathcal{A}$ the public keys $\mathsf{vk}_{\mathsf{id}_i} := (\mathbf{A}_{\mathsf{id}_i}, \{\mathbf{V}_\ell\}_{\tau \in \mathcal{T}})$ and $\mathsf{ek}_{\mathsf{id}_i} := (\mathbf{A}_{\mathsf{id}_i})$, along with signatures $\{\sigma_\tau^i\}_{\tau \in \mathcal{T}_i}$ where $\sigma_\tau^i := (\mathsf{m}_\tau, \mathsf{m}_\tau, \mathsf{I} := \{\mathsf{id}_i\}, \mathbf{U}_{\mathsf{id}_i, \tau}, \mathbf{U}_{\mathsf{id}_i, \tau})$.

Clearly, if $\mathbf{A}$ is a uniformly random matrix so is each block $\{\mathbf{A}_{\mathsf{id}_i}\}_{i\in[Q]}$.
Due to point (2) of Lemma 2, since $(\mathbf{A}_{\mathsf{id}_i}\mathbf{U}_{\mathsf{id}_i,\tau})$ is statistically indistinguishable from a random matrix, all the matrices $\mathbf{V}_\ell$ generated in (3′) are statistically close to the ones generated in (3). Thus, the two games are statistically indistinguishable. At this point we show that for every PPT adversary $\mathcal{A}$ which produces a forgery in the modified game we can construct a PPT algorithm $\mathcal{B}$ that solves the $\mathsf{SIS}(n, m\cdot Q, q, \beta_{\mathsf{SIS}})$ problem. $\mathcal{B}$ receives an SIS instance $\mathbf{A} := (\mathbf{A}_{\mathsf{id}_1}|\ldots|\mathbf{A}_{\mathsf{id}_Q}) \in \mathbb{Z}_q^{n\times mQ}$ and simulates the modified game to $\mathcal{A}$ by acting exactly as the challenger $\mathcal{C}$ described above. Then, once $\mathcal{A}$ outputs its forgery, according to the forgery's type, $\mathcal{B}$ proceeds as described below.

Type 2 Forgeries. Let $(\mathcal{P}^* := (f^*, \ell_1^*, \ldots, \ell_t^*), \mathsf{m}^*, \sigma^* := (\mathsf{m}^*, \mathsf{z}^*, \mathsf{I}^*, \mathbf{U}^*, \mathbf{Z}^*))$ be a Type 2 forgery produced by $\mathcal{A}$ in the modified game. Moreover let $\sigma = (\mathsf{m}, \mathsf{z}, \mathsf{I}, \mathbf{U}, \mathbf{Z})$ be the signature obtained by honestly applying Eval to the signatures corresponding to labels $\ell_1^*, \ldots, \ell_t^*$ that were given to $\mathcal{A}$. Parse $\mathbf{U} := \{\mathbf{U}_{\mathsf{id}}\}_{\mathsf{id}\in\mathsf{I}}$ and notice that by the correctness of the scheme we have that $\mathsf{m} = f^*(\mathsf{m}_{\ell_1^*}, \ldots, \mathsf{m}_{\ell_t^*})$, $\mathsf{I} = \{\mathsf{id} : \mathsf{id} \in \mathcal{P}^*\}$, and $\sum_{\mathsf{id}\in\mathsf{I}} \mathbf{A}_{\mathsf{id}}\mathbf{U}_{\mathsf{id}} + \mathsf{m}\cdot\mathbf{G} = \mathbf{V}^*$. Moreover, by definition of Type 2 forgery recall that $\mathsf{m}^* \neq f^*(\mathsf{m}_{\ell_1^*}, \ldots, \mathsf{m}_{\ell_t^*})$ and that the tuple satisfies verification. In particular, satisfaction of check (1) implies that $\mathsf{I} = \mathsf{I}^*$, while check (3) means $\sum_{\mathsf{id}\in\mathsf{I}^*} \mathbf{A}_{\mathsf{id}}\mathbf{U}^*_{\mathsf{id}} + \mathsf{m}^*\cdot\mathbf{G} = \mathbf{V}^*$. Combining the two equations above we obtain $\sum_{\mathsf{id}\in\mathsf{I}} \mathbf{A}_{\mathsf{id}}\tilde{\mathbf{U}}_{\mathsf{id}} = \tilde{\mathsf{m}}\cdot\mathbf{G}$, where $\tilde{\mathsf{m}} = \mathsf{m} - \mathsf{m}^* \neq \mathbf{0}$ and, for all $\mathsf{id}\in\mathsf{I}$, $\tilde{\mathbf{U}}_{\mathsf{id}} = \mathbf{U}^*_{\mathsf{id}} - \mathbf{U}_{\mathsf{id}} \in \mathcal{U}$ such that $\|\tilde{\mathbf{U}}_{\mathsf{id}}\|_\infty \le \beta_{\mathsf{max}}$. Notice that there must exist at least one $\bar{\mathsf{id}} \in \mathsf{I}$ for which $\tilde{\mathbf{U}}_{\bar{\mathsf{id}}} \neq \mathbf{0}$.

Moreover, for all $\mathsf{id} \in \{\mathsf{id}_1, \ldots, \mathsf{id}_Q\}\setminus\mathsf{I}$, define $\tilde{\mathbf{U}}_{\mathsf{id}} = \mathbf{0}$ and set $\tilde{\mathbf{U}} = \begin{pmatrix} \tilde{\mathbf{U}}_{\mathsf{id}_1} \\ \vdots \\ \tilde{\mathbf{U}}_{\mathsf{id}_Q} \end{pmatrix} \in \mathbb{Z}_q^{mQ\times m}$. Then, we have $\mathbf{A}\tilde{\mathbf{U}} = \tilde{\mathsf{m}}\cdot\mathbf{G}$.

Next $\mathcal{B}$ samples $\mathbf{r} \xleftarrow{\$} \{0,1\}^{mQ}$, sets $\mathbf{s} = \mathbf{A}\mathbf{r} \in \mathbb{Z}_q^n$, and computes $\mathbf{r}' = \mathbf{G}^{-1}(\tilde{\mathsf{m}}^{-1}\cdot\mathbf{s})$, so that $\mathbf{r}' \in \{0,1\}^m$ and $\tilde{\mathsf{m}}\cdot\mathbf{G}\mathbf{r}' = \mathbf{s}$. Finally, $\mathcal{B}$ outputs $\mathbf{u} = \tilde{\mathbf{U}}\mathbf{r}' - \mathbf{r} \in \mathbb{Z}_q^{mQ}$. We conclude the proof by claiming that the vector $\mathbf{u}$ returned by $\mathcal{B}$ is a solution of the SIS problem for the matrix $\mathbf{A}$. To see this observe that

$$\mathbf{A}(\tilde{\mathbf{U}}\mathbf{r}' - \mathbf{r}) = (\mathbf{A}\tilde{\mathbf{U}})\mathbf{r}' - \mathbf{A}\mathbf{r} = \tilde{\mathsf{m}}\cdot\mathbf{G}\cdot\mathbf{G}^{-1}(\tilde{\mathsf{m}}^{-1}\cdot\mathbf{s}) - \mathbf{s} = \mathbf{0}\,.$$

and $\|\mathbf{u}\|_\infty \le (2m+1)\beta_{\mathsf{max}} \le \beta_{\mathsf{SIS}}$.

It remains to show that $\mathbf{u} \neq \mathbf{0}$. We show that this is the case (i.e., $\tilde{\mathbf{U}}\mathbf{r}' \neq \mathbf{r}$) with overwhelming probability by using an entropy argument (the same argument used in [30]). In particular, this holds for any (worst case) choice of $\mathbf{A}, \tilde{\mathbf{U}}, \tilde{\mathsf{m}}$, and only based on the random choice of $\mathbf{r} \xleftarrow{\$} \{0,1\}^{mQ_{\mathsf{id}}}$. The intuition is that, even if $\mathbf{r}' = \mathbf{G}^{-1}(\mathbf{s}\tilde{\mathsf{m}}^{-1})$ depends on $\mathbf{s} = \mathbf{A}\mathbf{r}$, $\mathbf{s}$ is too small to reveal much information about the random $\mathbf{r}$. More precisely, we have that $\mathbf{H}_\infty(\mathbf{r} \mid \mathbf{r}') \ge \mathbf{H}_\infty(\mathbf{r} \mid \mathbf{A}\mathbf{r})$ because $\mathbf{r}'$ is chosen deterministically based on $\mathbf{s} = \mathbf{A}\mathbf{r}$. Due to the Lemma 1, we have that $\mathbf{H}_\infty(\mathbf{r} \mid \mathbf{A}\mathbf{r}) \ge \mathbf{H}_\infty(\mathbf{r}) - \mathsf{log}(|\mathcal{S}|)$, where $\mathcal{S}$ is the space of all possible $\mathbf{s}$. Since $\mathbf{s} \in \mathbb{Z}_q^n$, $|\mathcal{S}| = q^n$, and then $\mathsf{log}(|\mathcal{S}|) = \log(q^n) = \log((2^{\mathsf{log} q})^n) = n\log((2^{\log q})) = n\log q$. Regarding $\mathbf{H}_\infty(\mathbf{r})$, since $\mathbf{H}_\infty(X) := -\log(\max_x \Pr[X = x])$, we have $\mathbf{H}_\infty(\mathbf{r}) = -\log(2^{-mQ}) = mQ \geq m$. Then, $\mathbf{H}_\infty(\mathbf{r} \mid \mathbf{r}') \geq \mathbf{H}_\infty(\mathbf{r}) - \log(\mathcal{S}) \geq m - n \log q = \omega(\log \lambda)$. Since we know that for random variables X, Y the optimal probability of an unbounded adversary guessing X given the correlated value Y is $2^{-\mathbf{H}_\infty(X|Y)}$, then $\Pr[\mathbf{r} = \tilde{\mathbf{U}}\mathbf{r}'] \leq 2^{-\mathbf{H}_\infty(\mathbf{r}|\mathbf{r}')} \leq 2^{-\omega(\log \lambda)} = \mathsf{negl}(\lambda)$.

Type 3 Forgery. Let $(\mathcal{P}^* := (f^*, \ell_1^*, \ldots, \ell_t^*), \mathsf{m}^*, \sigma^* := (\mathsf{m}^*, \mathsf{z}^*, \mathsf{I}^*, \mathbf{U}^*, \mathbf{Z}^*))$ be a Type 3 forgery produced by $\mathcal{A}$ in the modified game such that there exists (at least) one label $\ell_j^* = (\mathsf{id}^*, \tau^*)$ such that $\mathsf{id}^* = \mathsf{id}_i$ but $\tau^* \notin \mathcal{T}_i$.[9] Actually, without loss of generality we can assume that there is exactly one of such labels; if this is not the case, one could indeed redefine another adversary that makes more queries until it misses only this one. Note that for such a tag $\tau^* \notin \mathcal{T}_i$, $\mathcal{B}$ simulated $\mathbf{V}_{\mathsf{id}_i, \tau^*} = \mathbf{A}\mathbf{U}_{\mathsf{id}_i, \tau^*} + b_{i,\tau^*}\mathbf{G}$ for a randomly chosen bit $b_{i,\tau^*} \xleftarrow{\$} \{0, 1\}$, that is perfectly hidden from $\mathcal{A}$.

By definition of Type 3 forgery, the tuple passes verification, and in particular check (4) $\sum_{\mathsf{id} \in \mathsf{I}^*} \mathbf{A}_{\mathsf{id}} \mathbf{Z}_{\mathsf{id}}^* + \mathsf{z}^* \cdot \mathbf{G} = \mathbf{V}^+ = \sum_{i=1}^{t} \gamma_i \cdot \mathbf{V}_{\ell_i^*}$ where the right hand side of the equation holds by construction of the verification algorithm. Moreover, let $\sigma = (\mathsf{m}, \mathsf{z}, \mathsf{I}, \mathbf{U}, \mathbf{Z})$ be the signature obtained by honestly applying Eval to the signatures corresponding to labels $\ell_1^*, \ldots, \ell_t^*$; in particular for the specific, missing, label ℓ_j^* $\mathcal{B}$ uses the values $\mathbf{U}_{\mathsf{id}_i, \tau^*}, b_{i,\tau^*}$ used to simulate $\mathbf{V}_{\mathsf{id}_i, \tau^*}$. Parsing $\mathbf{Z} := \{\mathbf{Z}_{\mathsf{id}}\}_{\mathsf{id} \in \mathsf{I}}$, notice that by correctness it holds $\mathsf{I} = \{\mathsf{id} : \mathsf{id} \in \mathcal{P}^*\}$ and $\sum_{\mathsf{id} \in \mathsf{I}} \mathbf{A}_{\mathsf{id}} \mathbf{Z}_{\mathsf{id}} + \mathsf{z} \cdot \mathbf{G} = \mathbf{V}^+$ where $\mathsf{z} = \sum_{i=1, i \neq j}^{t} \gamma_i \mathsf{m}_i + \gamma_j b_{i,\tau^*}$. Now, the observation is that every $\gamma_i \leq 2^d < q$, i.e., $\gamma_i \neq 0 \mod q$. Since b_{i,τ^*} is random and perfectly hidden to $\mathcal{A}$ we have that with probability $1/2$ it holds $\mathsf{z} \neq \mathsf{z}^*$.

Thus, if $\mathsf{z} \neq \mathsf{z}^*$, $\mathcal{B}$ combines the equalities on $\mathbf{V}^+$ to come up with an equation $\sum_{\mathsf{id} \in \mathsf{I}} \mathbf{A}_{\mathsf{id}} \tilde{\mathbf{Z}}_{\mathsf{id}} = \tilde{\mathsf{z}} \cdot \mathbf{G}$ where $\tilde{\mathsf{z}} = \mathsf{z} - \mathsf{z}^* \neq 0 \mod q$ and, for all $\mathsf{id} \in \mathsf{I}$, $\tilde{\mathbf{Z}}_{\mathsf{id}} = \mathbf{Z}_{\mathsf{id}}^* - \mathbf{Z}_{\mathsf{id}} \in \mathcal{U}$ such that $\|\tilde{\mathbf{Z}}_{\mathsf{id}}\|_\infty \leq \beta_{\mathsf{max}}$.

Finally, using the same technique as in the case of Type 2 forgeries, $\mathcal{B}$ can compute a vector $\mathbf{u}$ that is a solution of SIS with overwhelming probability, i.e., $\mathbf{A}\mathbf{u} = 0$. Therefore, we have proven that if an adversary $\mathcal{A}$ can break the MKHSig scheme with non negligible probability, then $\mathcal{C}$ can use such an $\mathcal{A}$ to break the SIS assumption for $\mathbf{A}$ with non negligible probability as well.

A Variant with Unbounded Tag Space in the Random Oracle Model. In this section, we show that the construction of multi-key homomorphic signatures of Sect. 4.2 can be easily modified in order to have short public keys and to support an unbounded tag space $\mathcal{T} = \{0, 1\}^*$. Note that once arbitrary tags are allowed, the scheme also allows to handle *multiple datasets* for free. In fact, one can always extend tags to include the dataset name, i.e., simply redefine each tag τ as consisting of two substrings $\tau = (\Delta, \tau')$ where Δ is the dataset name and τ' the actual tag. The idea of modifying the scheme to support an unbounded tag space is simple and was also suggested in [30] for their construction. Instead

[9] It is easy to see that the case in which id^* is new would imply the generation of a new $\mathbf{A}_{\mathsf{id}^*}$, which would make the verification equations hold with negligible probability (over the random choice of $\mathbf{A}_{\mathsf{id}^*}$).

of sampling matrices $\{\mathbf{V}_{\mathsf{id},1}, \ldots \mathbf{V}_{\mathsf{id},\mathsf{T}}\}$ in KeyGen, one can just choose a random string $r_{\mathsf{id}} \xleftarrow{\$} \{0,1\}^{\lambda}$ and define every $\mathbf{V}_{\mathsf{id},\tau} := \hat{\mathsf{H}}(r_{\mathsf{id}}, \tau)$ where $\hat{\mathsf{H}} : \{0,1\}^* \rightarrow \mathcal{V}$ is a hash function chosen in Setup (modeled as a random oracle in the proof). In all the remaining algorithms, every time one needs $\mathbf{V}_{\mathsf{id},\tau}$, this is obtained using $\hat{\mathsf{H}}$.

For this modified scheme, we also provide an idea of how the security proof of Theorem 1 has to be modified to account for these changes. The main change is the simulation of hash queries, which is done as follows.

Before phase (1), where $\mathcal{A}$ declares its queries, $\mathcal{B}$ simply answers every query $\hat{\mathsf{H}}(r, \tau)$ with a randomly chosen $\mathbf{V} \xleftarrow{\$} \mathcal{V}$. Afterwards, once $\mathcal{A}$ has declared all its queries, $\mathcal{B}$ chooses $r_{\mathsf{id}_1}, \ldots, r_{\mathsf{id}_Q} \xleftarrow{\$} \{0,1\}^{\lambda}$ and programs the random oracle so that, for all $\tau \in \mathcal{T}_i$, $\hat{\mathsf{H}}(r_{\mathsf{id}_i}, \tau) = \mathbf{V}_{\mathsf{id}_i,\tau}$ where $\mathbf{V}_{\mathsf{id}_i,\tau}$ is the same matrix generated in the phase (3) of the modified game. On the other hand, for all $\tau \notin \mathcal{T}_i$, $\hat{\mathsf{H}}(r_{\mathsf{id}_i}, \tau) = \mathbf{V}_{\mathsf{id}_i,\tau}$ where $\mathbf{V}_{\mathsf{id}_i,\tau} = \mathbf{A}_{\mathsf{id}}\mathbf{U}_{\mathsf{id}_i,\tau} + b_{i,\tau}\mathbf{G}$ for a randomly chosen $\mathbf{U}_{\mathsf{id}_i,\tau} \xleftarrow{\$} \mathcal{U}$. All other queries $\hat{\mathsf{H}}(r, \tau)$ where $r \neq r_{\mathsf{id}_i}, \forall i \in [Q]$ are answered with random $\mathbf{V} \xleftarrow{\$} \mathcal{V}$. With this simulation, it is not hard to see that, from $\mathcal{A}$'s forgery $\mathcal{B}$ can extract a solution for SIS (except for some negligible probability that $\mathcal{A}$ guesses one of r_{id_i} before seeing it).

4.3 From a Single Dataset to Multiple Datasets

In this section, we present a generic transformation to convert a single-dataset MKHSig scheme into a scheme that supports multiple datasets. The intuition behind this transformation is similar to the one employed in [30] and implicitly used in [13,16], except that here we have to use additional techniques to deal with the multi-key setting. We combine a standard signature scheme NH.Sig (non-homomorphic) with a single dataset multi-key homomorphic signature scheme MKHSig′. The idea is that for every new dataset Δ, every user generates fresh keys of the multi-key homomorphic scheme MKHSig′ and then uses the standard signature scheme NH.Sig to sign the dataset identifier Δ together with the generated public key. More precisely, in our transformation we assume to start with (single-dataset) multi-key homomorphic signature schemes in which the key generation algorithm can be split into two independent algorithms: KeyGen_1 that outputs some public parameters related to the identity id, and KeyGen_2 which outputs the actual keys. Differently than [30], in our scheme the signer does not need to sign the whole dataset at once, nor has to fix a bound N on the dataset size (unless such a bound is already contained in MKHSig′).

In more details, let NH.Sig = (NH.KeyGen, NH.Sign, NH.Ver) be a standard (non-homomorphic) signature scheme, and let MKHSig′ = (Setup′, KeyGen′, Sign′, Eval′, Ver′) be a single-dataset multi-key homomorphic signature scheme. We construct a multi-dataset multi-key homomorphic signature scheme MKHSig = (Setup, KeyGen, Sign, Eval, Ver) as follows.

$\mathsf{Setup}(1^{\lambda})$. The setup algorithm samples parameters of the single-dataset multi-key homomorphic signature scheme, $\mathsf{pp}' \leftarrow \mathsf{Setup}'(1^{\lambda})$, together with a description of a PRF $F : K \times \{0,1\}^* \rightarrow \{0,1\}^{\rho}$, and outputs $\mathsf{pp} = (\mathsf{pp}', F)$.

$\mathsf{KeyGen}(\mathsf{pp})$. The key generation algorithm runs $\mathsf{NH.KeyGen}$ to get $(\mathsf{pk}_{\mathsf{id}}^{\mathsf{NH}}, \mathsf{sk}_{\mathsf{id}}^{\mathsf{NH}})$, a pair of keys for the standard signature scheme. In addition, it runs KeyGen_1 to generate user-specific public parameters $\mathsf{pp}_{\mathsf{id}}$, and chooses a seed K_{id} for the PRF F. The final output is the vector $(\mathsf{sk}_{\mathsf{id}}, \mathsf{ek}_{\mathsf{id}}, \mathsf{vk}_{\mathsf{id}})$: where $\mathsf{sk}_{\mathsf{id}} = (\mathsf{sk}_{\mathsf{id}}^{\mathsf{NH}}, \mathsf{K}_{\mathsf{id}})$, $\mathsf{ek}_{\mathsf{id}} = (\mathsf{pp}_{\mathsf{id}})$ and $\mathsf{vk}_{\mathsf{id}} = (\mathsf{pk}_{\mathsf{id}}^{\mathsf{NH}}, \mathsf{pp}_{\mathsf{id}})$.

$\mathsf{Sign}(\mathsf{sk}_{\mathsf{id}}, \Delta, \ell, \mathsf{m})$. The signing algorithm proceeds as follows. First it samples the keys of the single-dataset multi-key homomorphic signature scheme by feeding randomness $F_{\mathsf{K}_{\mathsf{id}}}(\Delta)$ to KeyGen_2, i.e., it runs $\mathsf{KeyGen}_2(\mathsf{pp}; F_{\mathsf{K}_{\mathsf{id}}}(\Delta))$ to obtain the keys $(\mathsf{sk}_{\mathsf{id}}^{\Delta}, \mathsf{ek}_{\mathsf{id}}^{\Delta}, \mathsf{vk}_{\mathsf{id}}^{\Delta})$.[10] The algorithm then runs $\sigma' \leftarrow \mathsf{Sign}'(\mathsf{sk}_{\mathsf{id}}^{\Delta}, \ell, \mathsf{m})$, and uses the non-homomorphic scheme to sign the concatenation of the public key $\mathsf{vk}_{\mathsf{id}}^{\Delta}$ and the dataset identifier Δ, i.e., $\sigma_{\mathsf{id}}^{\Delta} \leftarrow \mathsf{NH.Sign}(\mathsf{sk}_{\mathsf{id}}^{\mathsf{NH}}, \mathsf{vk}_{\mathsf{id}}^{\Delta}|\Delta)$, The output is the tuple $\sigma := (\mathsf{I} = \{\mathsf{id}\}, \sigma', \mathsf{par}_{\Delta})$ where $\mathsf{par}_{\Delta} = \{(\mathsf{ek}_{\mathsf{id}}^{\Delta}, \mathsf{vk}_{\mathsf{id}}^{\Delta}, \sigma_{\mathsf{id}}^{\Delta})\}$. Note that the use of the PRF allows every signer (having the same K_{id}) to generate the same keys of the scheme MKHSig' on the same dataset Δ.

$\mathsf{Eval}(f, \{(\sigma_i, \mathsf{EKS}_i)\}_{i \in [t]})$. For each $i \in [t]$, the algorithm parses every signature as $\sigma_i := (\mathsf{I}_i, \sigma'_i, \mathsf{par}_{\Delta,i})$ with $\mathsf{par}_{\Delta,i} = \{\mathsf{ek}_{\mathsf{id}}^{\Delta}, \mathsf{vk}_{\mathsf{id}}^{\Delta}, \sigma_{\mathsf{id}}^{\Delta}\}_{\mathsf{id} \in \mathsf{I}_i}$, and sets $\mathsf{EKS}'_i = \{\mathsf{ek}_{\mathsf{id}}^{\Delta}\}_{\mathsf{id} \in \mathsf{I}_i}$. It computes $\sigma' \leftarrow \mathsf{Eval}'(f, \{\sigma'_i, \mathsf{EKS}'_i\}_{i \in [t]})$, defines $\mathsf{I} = \cup_{i=1}^{t} \mathsf{I}_i$ and $\mathsf{par}_{\Delta} = \cup_{i=1}^{t} \mathsf{par}_{\Delta,i}$. The final output is $\sigma = (\mathsf{I}, \sigma', \mathsf{par}_{\Delta})$.

$\mathsf{Ver}(\mathcal{P}, \Delta, \{\mathsf{vk}_{\mathsf{id}}\}_{\mathsf{id} \in \mathcal{P}}, \mathsf{m}, \sigma)$. The verification algorithm begins by parsing the verification keys as $\mathsf{vk}_{\mathsf{id}} := (\mathsf{pk}_{\mathsf{id}}^{\mathsf{NH}}, \mathsf{pp}_{\mathsf{id}})$ for each $\mathsf{id} \in \mathsf{I}$, and also the signature as $\sigma = (\mathsf{I}, \sigma', \mathsf{par}_{\Delta})$ with $\mathsf{par}_{\Delta} = \{(\mathsf{ek}_{\mathsf{id}}^{\Delta}, \mathsf{vk}_{\mathsf{id}}^{\Delta}, \sigma_{\mathsf{id}}^{\Delta})\}_{\mathsf{id} \in \mathsf{I}}$. Then, it proceeds with two main steps. First, for each $\mathsf{id} \in \mathsf{I}$, it verifies the standard signature $\sigma_{\mathsf{id}}^{\Delta}$ on the public key of the single-dataset multi-key homomorphic scheme and the given dataset, i.e., it checks whether $\mathsf{NH.Ver}(\mathsf{pk}_{\mathsf{id}}^{\mathsf{NH}}, \mathsf{vk}_{\mathsf{id}}^{\Delta}|\Delta, \sigma_{\mathsf{id}}^{\Delta}) = 1, \forall \mathsf{id} \in \mathsf{I}$. If at least one of the previous equations is not satisfied, the algorithm returns 0, otherwise it proceeds to the second check and returns the output of $\mathsf{Ver}'(\mathcal{P}, \{\mathsf{pp}_{\mathsf{id}}, \mathsf{vk}_{\mathsf{id}}^{\Delta}\}_{\mathsf{id} \in \mathcal{P}}, \mathsf{m}, \sigma')$.

AUTHENTICATION CORRECTNESS. Correctness of the scheme substantially follows from the correctness of the regular signature scheme $\mathsf{NH.Sig}$, the single-dataset multi-hey homomorphic scheme MKHSig' and the PRF F.

EVALUATION CORRECTNESS. Evaluation correctness follows directly from the correctness of the evaluation algorithm Eval' of the single-dataset MKHSig scheme, the correctness of $\mathsf{NH.Sig}$ and of the PRF.

SECURITY. Intuitively, the security of the scheme follows from two main observations. First, no adversary is able to fake the keys of the single-dataset multi-key homomorphic signature scheme, due to the security of the standard signature scheme and the property of pseudo-random functions. Secondly, no adversary can tamper with the results of Eval for a specific dataset as this would correspond to breaking the security of the single-dataset MKHSig' scheme.

[10] Here we assume that a ρ-bits string is sufficient, otherwise it can always be stretched using a PRG.

Theorem 2. *If F is a secure pseudo-random function,* NH.Sig *is an unforgeable signature scheme and* MKHSig′ *is a secure single-dataset multi-key homomorphic signature scheme, then the* MKHSig *scheme for multiple datasets described in Sect. 4.3 is secure against adversaries that make static corruptions of keys and produce forgeries as in Definition 4.*

The full proof of Theorem 2 is given in [22].

5 Our Multi-key Homomorphic MAC from OWFs

In this section, we describe our construction of a multi-key homomorphic authenticator with private verification keys and supporting the evaluation of low-degree arithmetic circuits. More precisely, for a computation represented by an arithmetic circuit of degree d and involving inputs from n distinct identities, the final authenticator has size $\binom{n+d}{d}$, that is bounded by $poly(n)$ (for constant d) or by $poly(d)$ (for constant n). Essentially, the authenticators of our scheme grow with the degree of the circuit and the number of distinct users involved in the computation, whereas their size remains *independent* of the total number of inputs / users. This property is particularly desirable in contexts that involve a small set of users each of which contributes with several inputs.

Although our multi-key homomorphic MAC supports less expressive computations than our homomorphic signatures of Sect. 4, the scheme comes with two main benefits. First, it is based on a simple, general assumption: it relies on pseudo-random functions and thus is secure only assuming existence of one-way functions (OWF). Second, the scheme is very intuitive and efficient: fresh MACs essentially consist only of two $\mathbb{F}_p$ field elements (where p is a prime of λ bits) and an identity identifier; after evaluation, the authenticators consist of $\binom{n+d}{d}$ elements in $\mathbb{F}_p$, and homomorphic operations are simply additions and multiplications in the multi-variate polynomial ring $\mathbb{F}_p[X_1, \ldots, X_n]$.

We describe the five algorithms of our scheme MKHMac below. We note that our solution is presented for single data set only. However, since it admits labels that are arbitrarily long strings it is straight-forward to extend the scheme for handling multiple data sets: simply redefine each tag τ as consisting of two substrings $\tau = (\Delta, \tau')$ where Δ is the dataset name and τ' the actual tag.

Setup(1^λ). The setup algorithm generates a λ-bit prime p and let the message space be $\mathcal{M} := \mathbb{F}_p$. The set of identities is $\mathsf{ID} = [\mathsf{C}]$ for some integer bound $\mathsf{C} \in \mathbb{N}$, while the tag space consists of arbitrary binary strings, i.e., $\mathcal{T} = \{0,1\}^*$. The set $\mathcal{F}$ of admissible functions is made up of all arithmetic circuits whose degree d is bounded by some polynomial in the security parameter. The setup algorithm outputs the public parameters pp which include the descriptions of $\mathcal{M}, \mathsf{ID}, \mathcal{T}, \mathcal{F}$ as in Sect. 3, as well as the description of a PRF family $F : \mathcal{K} \times \{0,1\}^* \rightarrow \mathbb{F}_p$ with seed space $\mathcal{K}$. The public parameters define also the authenticator space. Each authenticator σ consists of a pair (I, y) where $\mathsf{I} \subseteq \mathsf{ID}$ and y is in the C-variate polynomial ring $\mathbb{F}_p[X_1, \ldots, X_{\mathsf{C}}]$. More precisely, if C is set up as a very large number (e.g., $\mathsf{C} = 2^\lambda$) the polynomials y can still live in some smaller sub-rings of $\mathbb{F}_p[X_1, \ldots, X_{\mathsf{C}}]$.

$\mathsf{KeyGen}(\mathsf{pp})$. The key generation algorithm picks a random $x \xleftarrow{\$} \mathbb{F}_p^*$, a PRF seed $K \xleftarrow{\$} \mathcal{K}$, and outputs $(\mathsf{sk}, \mathsf{ek}, \mathsf{vk})$ where $\mathsf{sk} = \mathsf{vk} = (x, K)$ and ek is void.

$\mathsf{Auth}(\mathsf{sk}, \ell, \mathsf{m})$. In order to authenticate the message m with label $\ell = (\mathsf{id}, \tau) \in \mathsf{ID} \times \mathcal{T}$, the authentication algorithm produces an authenticator $\sigma = (\mathsf{I}, \mathsf{y})$ where $\mathsf{I} \subseteq \mathsf{ID}$ and $\mathsf{y} \in \mathbb{F}_p[X_{\mathsf{id}}] \subset \mathbb{F}_p[X_1, \ldots, X_{\mathsf{C}}]$. The set I is simply $\{\mathsf{id}\}$. The polynomial y is a degree-1 polynomial in the variable X_{id} such that $\mathsf{y}(0) = \mathsf{m}$ and $\mathsf{y}(x_{\mathsf{id}}) = F(K_{\mathsf{id}}, \ell)$. Note that the coefficients of $\mathsf{y}(X_{\mathsf{id}}) = y_0 + y_{\mathsf{id}} X_{\mathsf{id}} \in \mathbb{F}_p[X_{\mathsf{id}}]$ can be efficiently computed with the knowledge of x_{id} by setting $y_0 = \mathsf{m}$ and $y_{\mathsf{id}} = \frac{F(K_{\mathsf{id}}, \ell) - \mathsf{m}}{x_{\mathsf{id}}}$. Moreover, y can be compactly represented by only giving the coefficients $y_0, y_{\mathsf{id}} \in \mathbb{F}_p$.

$\mathsf{Eval}(f, \{\sigma_k\}_{k \in [t]})$. Given a t-input arithmetic circuit $f : \mathbb{F}_p^t \to \mathbb{F}_p$, and the t authenticators $\{\sigma_k := (\mathsf{I}_k, \mathsf{y}_k)\}_k$, the evaluation algorithm outputs $\sigma = (\mathsf{I}, \mathsf{y})$ obtained in the following way. First, it determines all the identities involved in the computation by setting $\mathsf{I} = \cup_{k=1}^t \mathsf{I}_k$. Then every polynomial y_k is "expanded" into a polynomial $\hat{\mathsf{y}}_k$, defined on the variables X_{id} corresponding to all the identities in I. This is done using the canonical embedding $\mathbb{F}_p[X_{\mathsf{id}} : \mathsf{id} \in \mathsf{I}_k] \hookrightarrow \mathbb{F}_p[X_{\mathsf{id}} : \mathsf{id} \in \mathsf{I}]$. It is worth noticing that the terms of $\hat{\mathsf{y}}_k$ that depend on variables in $\mathsf{I} \setminus \mathsf{I}_k$ have coefficient 0. Next, let $\hat{f} : \mathbb{F}_p[X_{\mathsf{id}} : \mathsf{id} \in \mathsf{I}]^t \to \mathbb{F}_p[X_{\mathsf{id}} : \mathsf{id} \in \mathsf{I}]$ be the arithmetic circuit corresponding to the given f, i.e., $\hat{f}$ is the same as f except that additions (resp. multiplications) in $\mathbb{F}_p$ are replaced by additions (resp. multiplications) over the polynomial ring $\mathbb{F}_p[X_{\mathsf{id}} : \mathsf{id} \in \mathsf{I}]$. Finally, y is obtained as $\mathsf{y} = \hat{f}(\hat{\mathsf{y}}_1, \ldots, \hat{\mathsf{y}}_t)$.

$\mathsf{Ver}(\mathcal{P}, \{\mathsf{vk}_{\mathsf{id}}\}_{\mathsf{id} \in \mathcal{P}}, \mathsf{m}, \sigma)$. Let $\mathcal{P} = (f, \ell_1, \ldots, \ell_t)$ be a labeled program where f is a degree-d arithmetic circuit and every label is of the form $\ell_k = (\mathsf{id}_k, \tau_k)$. Let $\sigma = (\mathsf{I}, \mathsf{y})$ where $\mathsf{I} = \{\bar{\mathsf{id}}_1, \ldots, \bar{\mathsf{id}}_n\}$ with $\bar{\mathsf{id}}_i \neq \bar{\mathsf{id}}_j$ for $i \neq j$. The verification algorithm outputs 1 (accept) if and only if the authenticator satisfies the following three checks. Otherwise it outputs 0 (reject).

$$\{\bar{\mathsf{id}}_1, \ldots, \bar{\mathsf{id}}_n\} = \{\mathsf{id} : \mathsf{id} \in \mathcal{P}\}, \tag{5}$$

$$\mathsf{y}(0, \ldots, 0) = \mathsf{m}\ , \tag{6}$$

$$\mathsf{y}(x_{\bar{\mathsf{id}}_1}, \ldots, x_{\bar{\mathsf{id}}_n}) = f(F(K_{\mathsf{id}_1}, \ell_1), \ldots, F(K_{\mathsf{id}_t}, \ell_t))\ . \tag{7}$$

In the remainder of the section, we discuss the efficiency and succinctness of our MKHMac and prove the correctness of our scheme We conclude with the security analysis of the proposed MKHMac scheme.

SUCCINCTNESS. Let us consider the case of an authenticator σ which was obtained after running Eval on a circuit of degree d and taking inputs from n distinct identities. Note that every σ consists of two elements: a set $\mathsf{I} \subseteq [\mathsf{C}]$ and a polynomial $\mathsf{y} \in \mathbb{F}_p[X_{\bar{\mathsf{id}}} : \bar{\mathsf{id}} \in \mathsf{I}]$.

For the set I, it is easy to see that $|\mathsf{I}| = n$ and I can be represented with $n \log \mathsf{C}$ bits. The other part of the authenticator, y, is instead an n-variate polynomial in $\mathbb{F}_p[X_{\bar{\mathsf{id}}_1}, \ldots, X_{\bar{\mathsf{id}}_n}]$ of degree d. Since the circuit degree is d, the maximum number of coefficients of y is $\binom{n+d}{d}$. More precisely, the total size of y depends on the particular representation of the multi-variate polynomial y which is chosen for implementation. In [22] we discuss some possible representations (further details can also be found in [38]). For example, when employing the *sparse representation* of polynomials, the size of y is bounded by $O(nt \log d)$ where t is the number of non-zero coefficients in y (note that in the worst case, a polynomial $\mathsf{y} \in \mathbb{F}_p[X_{\bar{\mathsf{id}}_1}, \ldots, X_{\bar{\mathsf{id}}_n}]$ of degree d has at most $t = \binom{n+d}{d}$ non-zero coefficients). Thus, setting $\log \mathsf{C} \approx \log p \approx \lambda$, we have that the size in bits of the authenticator σ is $|\sigma| \leq \lambda n + \lambda \binom{n+d}{d}$. Ignoring the security parameter, we have that $|\sigma| = \mathsf{poly}(n)$ when d is constant, or $|\sigma| = \mathsf{poly}(d)$ when n is constant.

EFFICIENCY OF Eval. In what follows, we discuss the cost of computing additions and multiplications over authenticators in our MKHMac scheme. Let $\sigma^{(i)} = (\mathsf{I}^{(i)}, \mathsf{y}^{(i)})$, for $i = 1, 2$ be two authenticators and consider the operation $\sigma = \mathsf{Eval}(g, \sigma^{(1)}, \sigma^{(2)})$ where g is a fan-in-2 addition or multiplication gate. In both cases the set I of identities of $\sigma = (\mathsf{I}, \mathsf{y})$ is obtained as the union $\mathsf{I} = \mathsf{I}^{(1)} \cup \mathsf{I}^{(2)}$ that can be computed in time $O(n)$, where $n = |\mathsf{I}|$, assuming the sets $\mathsf{I}^{(1)}, \mathsf{I}^{(2)}$ are ordered. Regarding the computation of y from $\mathsf{y}^{(1)}$ and $\mathsf{y}^{(2)}$, one has to first embed each y_i into the ring $\mathbb{F}_p[X_{\bar{\mathsf{id}}} : \bar{\mathsf{id}} \in \mathsf{I}]$, and then evaluate addition (resp. multiplication) over $\mathbb{F}_p[X_{\bar{\mathsf{id}}} : \bar{\mathsf{id}} \in \mathsf{I}]$. Again, the costs of these operations depend on the adopted representation [24,38].

Using the *sparse representation* of polynomials, expanding a y having t non-zero coefficients into an n-variate polynomial $\hat{\mathsf{y}}$ requires time at most $O(tn)$. To give an idea, such expansion indeed consists simply into inserting zeros in the correct positions of the exponent vectors of every non-zero monomial term of y. On the other hand, the complexity of operations (additions and multiplications) on polynomials using the *sparse representation* is usually estimated in terms of the number of monomial comparisons. The cost of such comparisons depends on the specific monomial ordering chosen, but is usually $O(n \log d)$, where n is the total number of variables and d is the maximum degree. Given two polynomials in sparse representation having t_1 and t_2 non-zero terms respectively, addition costs about $O(t_1 t_2)$ monomial comparisons (if the monomial terms are stored in sorted order the cost of addition drops to $O(t_1 + t_2)$), while multiplication requires to add (merge) t_2 intermediate products of t_1 terms each, and can be performed with $O(t_1 t_2 \log t_2)$ monomial comparisons [24].

Correctness. AUTHENTICATION CORRECTNESS. By construction, each fresh authenticator $\sigma = (\mathsf{I}, \mathsf{y})$ of a message m labeled by $\ell := (\mathsf{id}, \tau)$ is of the form $\mathsf{I} = \{\mathsf{id}\}$ and $\mathsf{y}(X_{\mathsf{id}}) := y_0 + y_{\mathsf{id}} X_{\mathsf{id}} = \mathsf{m} + \frac{F(K_{\mathsf{id}}, \ell) - \mathsf{m}}{x_{\mathsf{id}}} X_{\mathsf{id}}$. Thus the set I satisfies Eq. (5) since $\{\mathsf{id} : \mathsf{id} \in \mathcal{I}_\ell\} = \{\mathsf{id}\}$. The two last verification checks (6) and (7) are automatically granted for the identity program $\mathcal{I}_\ell$ because $\mathsf{y}(0) = y_0 = \mathsf{m}$ and $\mathsf{y}(x_{\mathsf{id}}) = \mathsf{m} + \frac{F(K_{\mathsf{id}}, \ell) - \mathsf{m}}{x_{\mathsf{id}}} x_{\mathsf{id}} = F(K_{\mathsf{id}}, \ell)$.

EVALUATION CORRECTNESS. The correctness of the Eval algorithm essentially comes from the structure of the multi-variate polynomial ring. We provide the detailed proof in the full version of the paper [22].

SECURITY. In what follows we prove the security of our scheme against adversaries that make static corruptions, and produce forgeries according to the following restrictions.

Definition 8 (Weak Forgery). *Consider an execution of the experiment described in Sect. 3,* $\mathsf{HomUF\text{-}CMA}_{\mathcal{A},\mathsf{MKHAut}}(\lambda)$ *where* $(\mathcal{P}^*, \Delta^*, \mathsf{m}^*, \sigma^*)$ *is the tuple returned by the adversary at the end of the experiment, with* $\mathcal{P}^* = (f^*, \ell_1^*, \ldots, \ell_t^*)$, Δ^* *a dataset identifier,* $\mathsf{m}^* \in \mathcal{M}$ *and* σ^* *an authenticator. First, we say that the labeled program* $\mathcal{P}^*$ *is* well-defined *on a list* L *if either one of the following two cases occurs:*

1. *There exists* $i \in [t]$ *such that* $(\ell_i^*, \cdot) \notin L$ *(i.e.,* $\mathcal{A}$ *never made a query with label* ℓ_i^**), and* $f^*(\{\mathsf{m}_j\}_{(\ell_j,\mathsf{m}_j)\in L} \cup \{\tilde{\mathsf{m}}_j\}_{(\ell_j,\cdot)\notin L})$ *outputs the same value for all possible choices of* $\tilde{\mathsf{m}}_j \in \mathcal{M}$*;*
2. L *contains the tuples* $(\ell_1^*, \mathsf{m}_1), \ldots, (\ell_t^*, \mathsf{m}_t)$*, for some messages* $\mathsf{m}_1, \ldots, \mathsf{m}_t$*.*

Then we say that $(\mathcal{P}^*, \Delta^*, \mathsf{m}^*, \sigma^*)$ *is a* weak forgery *if* $\mathsf{Ver}(\mathcal{P}^*, \Delta^*, \{\mathsf{vk}_{\mathsf{id}}\}_{\mathsf{id}\in\mathcal{P}^*}, \mathsf{m}^*, \sigma^*) = 1$ *and either one of the following conditions is satisfied:*

Type 1: L_{Δ^*} *was not initialized during the game (i.e.,* Δ^* *was never queried).*
Type 2: $\mathcal{P}^*$ *is well-defined on* L_{Δ^*} *but* $\mathsf{m}^* \neq f^*(\{\mathsf{m}_j\}_{(\ell_j,\mathsf{m}_j)\in L_{\Delta^*}} \cup \{0\}_{\ell_j \notin L_{\Delta^*}})$ *(i.e.,* m^* *is not the correct output of* $\mathcal{P}^*$ *when executed over previously authenticated messages).*
Type 3: $\mathcal{P}^*$ *is not well-defined on* L_{Δ^*}*.*

Although Definition 8 is weaker than our Definition 4, we stress that the above definition still protects the verifier from adversaries that try to cheat on the output of a computation. In more details, the difference between Definition 8 and Definition 4 is the following: if f^* has an input wire that has never been authenticated during the game (a Type 3 forgery in Definition 4), but f^* is *constant* with respect to such input wire, then the above definition does not consider it a forgery. The intuitive reason why such a relaxed definition still makes sense is that "irrelevant" inputs would not help in any case the adversary to cheat on the output of f^*. Definition 8 is essentially the multi-key version of the forgery definition used in previous (single-key) homomorphic MAC works, e.g., [11]. As discussed in [23] testing whether a program is well-defined may not be doable in polynomial time in the most general case (i.e., every class of functions). However, in [12] it is shown how this can be done efficiently via a probabilistic test in the case of arithmetic circuits of degree d over a finite field of order p such that $d/p < 1/2$. Finally, we notice that for our MKHMac Type 1 forgeries cannot occur as the scheme described here supports only one dataset.[11]

Theorem 3. *If* F *is a pseudo-random function then the multi-key homomorphic MAC described in Sect. 5 is secure against adversaries that make static corruptions of keys and produce forgeries as in Definition 8.*

Note that we can deal with corruptions via our generic result of Proposition 1. Therefore, it is sufficient to prove the security against adversaries that make no

[11] As noted at the beginning of the section the extension to multiple datasets is straightforward given that tags are arbitrary strings. When such extension is applied it is easy to see that Type 1 forgeries are Type 3 ones in the underlying scheme.

corruptions. The proof is done via a chain of games following this (intuitive) path. First, we rule out adversaries that make Type 3 forgeries. Intuitively, this can be done as the adversary has never seen one of the inputs of the computation, and in particular an input which can change the result. Second, we replace every PRF instance with a truly random function. Note that at this point the security of the scheme is information theoretic. Third, we change the way to answer verification queries that are candidates to be Type 2 forgeries. Finally, we observe that in this last game the adversary gains no information on the secret keys x_i and thus has negligible probability of making a Type 2 forgery. Due to space restrictions, the detailed and formal proofs appear in only in the full version [22].

6 Conclusions

In this paper, we introduced the concept of multi-key homomorphic authenticators, a cryptographic primitive that enables an untrusted third party to execute a function f on data authenticated using different secret keys in order to obtain a value certifying the correctness of f's result, which can be checked with knowledge of corresponding verification keys. In addition to providing suitable definitions, we also propose two constructions: one which is publicly verifiable and supports general boolean circuits, and a second one that is secretly verifiable and supports low-degree arithmetic circuits. Although our work does not address directly the problem of privacy, extensions of our results along this direction are possible, and we leave the details to future investigation. A first extension is defining a notion of context-hiding for multi-key HAs. Similarly to the single key setting, this property should guarantee that authenticators do not reveal non-trivial information about the computation's inputs. The second extension has to do with preventing the Cloud from learning the data over which it computes. In this case, we note that multi-key HAs can be executed on top of homomorphic encryption following an approach similar to that suggested in [21]. Finally, an interesting problem left open by our work is to find multi-key HA schemes where authenticators have size independent of the number of users involved in the computation.

Acknowledgements. This work was partially supported by the COST Action IC1306 through a STSM grant to Elena Pagnin, the European Union's H2020 Research and Innovation Programme under grant agreement 688722 (NEXTLEAP), the Spanish Ministry of Economy under project reference TIN2015-70713-R (DEDETIS) and a Juan de la Cierva fellowship to Dario Fiore, and by the Madrid Regional Government under project N-Greens (ref. S2013/ICE-2731). This work was also partially supported by the People Programme (Marie Curie Actions) of the European Union's Seventh Framework Programme (FP7/2007-2013) under REA grant agreement no 608743, the VR grant PRECIS no 621-2014-4845 and the STINT grant Secure, Private & Efficient Healthcare with wearable computing no IB2015-6001.

References

1. Agrawal, S., Boneh, D., Boyen, X., Freeman, D.M.: Preventing pollution attacks in multi-source network coding. In: Nguyen, P.Q., Pointcheval, D. (eds.) PKC 2010. LNCS, vol. 6056, pp. 161–176. Springer, Heidelberg (2010). doi:10.1007/978-3-642-13013-7_10
2. Ajtai, M.: Generating hard instances of lattice problems (extended abstract). In: 28th ACM STOC, pp. 99–108. ACM Press, May 1996
3. Ajtai, M.: Generating hard instances of the short basis problem. In: Wiedermann, J., Emde Boas, P., Nielsen, M. (eds.) ICALP 1999. LNCS, vol. 1644, pp. 1–9. Springer, Heidelberg (1999). doi:10.1007/3-540-48523-6_1
4. Alwen, J., Peikert, C.: Generating shorter bases for hard random lattices. Theory Comput. Syst. **48**(3), 535–553 (2011)
5. Backes, M., Fiore, D., Reischuk, R.M.: Verifiable delegation of computation on outsourced data. In: Sadeghi, A.-R., Gligor, V.D., Yung, M. (eds.) ACM CCS 2013, pp. 863–874. ACM Press, November 2013
6. Benabbas, S., Gennaro, R., Vahlis, Y.: Verifiable delegation of computation over large datasets. In: Rogaway, P. (ed.) CRYPTO 2011. LNCS, vol. 6841, pp. 111–131. Springer, Heidelberg (2011). doi:10.1007/978-3-642-22792-9_7
7. Bitansky, N., Canetti, R., Chiesa, A., Tromer, E.: Recursive composition and bootstrapping for SNARKS and proof-carrying data. In: Boneh, D., Roughgarden, T., Feigenbaum, J. (eds.) 45th ACM STOC, pp. 111–120. ACM Press, June 2013
8. Boneh, D., Freeman, D., Katz, J., Waters, B.: Signing a linear subspace: signature schemes for network coding. In: Jarecki, S., Tsudik, G. (eds.) PKC 2009. LNCS, vol. 5443, pp. 68–87. Springer, Heidelberg (2009). doi:10.1007/978-3-642-00468-1_5
9. Boneh, D., Freeman, D.M.: Homomorphic signatures for polynomial functions. In: Paterson, K.G. (ed.) EUROCRYPT 2011. LNCS, vol. 6632, pp. 149–168. Springer, Heidelberg (2011). doi:10.1007/978-3-642-20465-4_10
10. Boneh, D., Gentry, C., Gorbunov, S., Halevi, S., Nikolaenko, V., Segev, G., Vaikuntanathan, V., Vinayagamurthy, D.: Fully key-homomorphic encryption, arithmetic circuit ABE and compact garbled circuits. In: Nguyen, P.Q., Oswald, E. (eds.) EUROCRYPT 2014. LNCS, vol. 8441, pp. 533–556. Springer, Heidelberg (2014). doi:10.1007/978-3-642-55220-5_30
11. Catalano, D., Fiore, D.: Practical homomorphic MACs for arithmetic circuits. In: Johansson, T., Nguyen, P.Q. (eds.) EUROCRYPT 2013. LNCS, vol. 7881, pp. 336–352. Springer, Heidelberg (2013). doi:10.1007/978-3-642-38348-9_21
12. Catalano, D., Fiore, D., Gennaro, R., Nizzardo, L.: Generalizing homomorphic MACs for arithmetic circuits. In: Krawczyk, H. (ed.) PKC 2014. LNCS, vol. 8383, pp. 538–555. Springer, Heidelberg (2014). doi:10.1007/978-3-642-54631-0_31
13. Catalano, D., Fiore, D., Nizzardo, L.: Programmable hash functions go private: constructions and applications to (homomorphic) signatures with shorter public keys. In: Gennaro, R., Robshaw, M. (eds.) CRYPTO 2015. LNCS, vol. 9216, pp. 254–274. Springer, Heidelberg (2015). doi:10.1007/978-3-662-48000-7_13
14. Catalano, D., Fiore, D., Warinschi, B.: Adaptive pseudo-free groups and applications. In: Paterson, K.G. (ed.) EUROCRYPT 2011. LNCS, vol. 6632, pp. 207–223. Springer, Heidelberg (2011). doi:10.1007/978-3-642-20465-4_13
15. Catalano, D., Fiore, D., Warinschi, B.: Efficient network coding signatures in the standard model. In: Fischlin, M., Buchmann, J., Manulis, M. (eds.) PKC 2012. LNCS, vol. 7293, pp. 680–696. Springer, Heidelberg (2012). doi:10.1007/978-3-642-30057-8_40

16. Catalano, D., Fiore, D., Warinschi, B.: Homomorphic signatures with efficient verification for polynomial functions. In: Garay, J.A., Gennaro, R. (eds.) CRYPTO 2014. LNCS, vol. 8616, pp. 371–389. Springer, Heidelberg (2014). doi:10.1007/978-3-662-44371-2_21
17. Choi, S.G., Katz, J., Kumaresan, R., Cid, C.: Multi-client non-interactive verifiable computation. In: Sahai, A. (ed.) TCC 2013. LNCS, vol. 7785, pp. 499–518. Springer, Heidelberg (2013). doi:10.1007/978-3-642-36594-2_28
18. Chung, K.-M., Kalai, Y., Vadhan, S.: Improved delegation of computation using fully homomorphic encryption. In: Rabin, T. (ed.) CRYPTO 2010. LNCS, vol. 6223, pp. 483–501. Springer, Heidelberg (2010). doi:10.1007/978-3-642-14623-7_26
19. Dodis, Y., Ostrovsky, R., Reyzin, L., Smith, A.: Fuzzy extractors: how to generate strong keys from biometrics and other noisy data. SIAM J. Comput. **38**(1), 97–139 (2008)
20. Fiore, D., Gennaro, R.: Publicly verifiable delegation of large polynomials and matrix computations, with applications. In: Yu, T., Danezis, G., Gligor, V.D. (eds.) ACM CCS 2012, pp. 501–512. ACM Press, October 2012
21. Fiore, D., Gennaro, R., Pastro, V.: Efficiently verifiable computation on encrypted data. In: Ahn, G.-J., Yung, M., Li, N. (eds.) ACM CCS 2014, pp. 844–855. ACM Press, November 2014
22. Fiore, D., Mitrokotsa, A., Nizzardo, L., Pagnin, E.: Multi-key homomorphic authenticators. IACR Cryptology ePrint Archive (2016)
23. Freeman, D.M.: Improved security for linearly homomorphic signatures: a generic framework. In: Fischlin, M., Buchmann, J., Manulis, M. (eds.) PKC 2012. LNCS, vol. 7293, pp. 697–714. Springer, Heidelberg (2012). doi:10.1007/978-3-642-30057-8_41
24. Geddes, K.O., Czapor, S.R., Labahn, G.: Algorithms for Computer Algebra. Kluwer Academic Publishers, Norwell (1992)
25. Gennaro, R., Gentry, C., Parno, B.: Non-interactive verifiable computing: outsourcing computation to untrusted workers. In: Rabin, T. (ed.) CRYPTO 2010. LNCS, vol. 6223, pp. 465–482. Springer, Heidelberg (2010). doi:10.1007/978-3-642-14623-7_25
26. Gennaro, R., Wichs, D.: Fully homomorphic message authenticators. In: Sako, K., Sarkar, P. (eds.) ASIACRYPT 2013, Part II. LNCS, vol. 8270, pp. 301–320. Springer, Heidelberg (2013). doi:10.1007/978-3-642-42045-0_16
27. Gentry, C.: Fully homomorphic encryption using ideal lattices. In: Mitzenmacher, M. (ed.) 41st ACM STOC, pp. 169–178. ACM Press, May/June 2009
28. Gentry, C., Peikert, C., Vaikuntanathan, V.: Trapdoors for hard lattices and new cryptographic constructions. In: Ladner, R.E., Dwork, C. (eds.) 40th ACM STOC, pp. 197–206. ACM Press, May 2008
29. Goldwasser, S., Kalai, Y.T., Rothblum, G.N.: Delegating computation: interactive proofs for muggles. In: Ladner, R.E., Dwork, C. (eds.) 40th ACM STOC, pp. 113–122. ACM Press, May 2008
30. Gorbunov, S., Vaikuntanathan, V., Wichs, D.: Leveled fully homomorphic signatures from standard lattices. In: Servedio, R.A., Rubinfeld, R. (eds.) 47th ACM STOC, pp. 469–477. ACM Press, June 2015
31. Gordon, S.D., Katz, J., Liu, F.-H., Shi, E., Zhou, H.-S.: Multi-client verifiable computation with stronger security guarantees. In: Dodis, Y., Nielsen, J.B. (eds.) TCC 2015. LNCS, vol. 9015, pp. 144–168. Springer, Heidelberg (2015). doi:10.1007/978-3-662-46497-7_6

32. Johnson, R., Molnar, D., Song, D., Wagner, D.: Homomorphic signature schemes. In: Preneel, B. (ed.) CT-RSA 2002. LNCS, vol. 2271, pp. 244–262. Springer, Heidelberg (2002). doi:10.1007/3-540-45760-7_17
33. Micciancio, D.: Almost perfect lattices, the covering radius problem, applications to Ajtai's connection factor. SIAM J. Comput. **34**(1), 118–169 (2004). Preliminary version in STOC 2002
34. Micciancio, D., Peikert, C.: Trapdoors for lattices: simpler, tighter, faster, smaller. In: Pointcheval, D., Johansson, T. (eds.) EUROCRYPT 2012. LNCS, vol. 7237, pp. 700–718. Springer, Heidelberg (2012). doi:10.1007/978-3-642-29011-4_41
35. Micciancio, D., Peikert, C.: Hardness of SIS and LWE with small parameters. In: Canetti, R., Garay, J.A. (eds.) CRYPTO 2013. LNCS, vol. 8042, pp. 21–39. Springer, Heidelberg (2013). doi:10.1007/978-3-642-40041-4_2
36. Micciancio, D., Regev, O.: Worst-case to average-case reductions based on Gaussian measures. In: 45th FOCS, pp. 372–381. IEEE Computer Society Press, October 2004
37. Parno, B., Raykova, M., Vaikuntanathan, V.: How to delegate and verify in public: verifiable computation from attribute-based encryption. In: Cramer, R. (ed.) TCC 2012. LNCS, vol. 7194, pp. 422–439. Springer, Heidelberg (2012)
38. van der Hoeven, J., Lecerf, G.: On the bit-complexity of sparse polynomial and series multiplication. J. Symbolic Comput. **50**, 227–254 (2013)

Multi-input Functional Encryption with Unbounded-Message Security

Vipul Goyal[1](✉), Aayush Jain[2], and Adam O'Neill[3]

[1] Microsoft Research, Bengaluru, India
vipul@microsoft.com

[2] Center for Encrypted Functionalities, University of California, Los Angeles, USA
aayushjainiitd@gmail.com

[3] Georgetown University, Washington, D.C., USA
adam@cs.georgetown.edu

Abstract. Multi-input functional encryption (MIFE) was introduced by Goldwasser *et al.* (EUROCRYPT 2014) as a compelling extension of functional encryption. In MIFE, a receiver is able to compute a joint function of multiple, independently encrypted plaintexts. Goldwasser *et al.* (EUROCRYPT 2014) show various applications of MIFE to running SQL queries over encrypted databases, computing over encrypted data streams, etc.

The previous constructions of MIFE due to Goldwasser *et al.* (EUROCRYPT 2014) based on indistinguishability obfuscation had a major shortcoming: it could only support encrypting an *a priori bounded* number of message. Once that bound is exceeded, security is no longer guaranteed to hold. In addition, it could only support *selective-security*, meaning that the challenge messages and the set of "corrupted" encryption keys had to be declared by the adversary up front.

In this work, we show how to remove these restrictions by relying instead on *sub-exponentially secure* indistinguishability obfuscation. This is done by carefully adapting an alternative MIFE scheme of Goldwasser *et al.* that previously overcame these shortcomings (except for selective security wrt. the set of "corrupted" encryption keys) by relying instead on differing-inputs obfuscation, which is now seen as an implausible assumption. Our techniques are rather generic, and we hope they are useful in converting other constructions using differing-inputs obfuscation to ones using sub-exponentially secure indistinguishability obfuscation instead.

A. Jain–Research supported in part from a DARPA/ARL SAFEWARE award, NSF Frontier Award 1413955, NSF grants 1228984, 1136174, and 1065276, a Xerox Faculty Research Award, a Google Faculty Research Award, an equipment grant from Intel, and an Okawa Foundation Research Grant. This material is based upon work supported by the Defense Advanced Research Projects Agency through the ARL under Contract W911NF-15-C-0205. The views expressed are those of the author and do not reflect the official policy or position of the Department of Defense, the National Science Foundation, or the U.S. Government. Research done in part while visiting Microsoft Research, India.

J.H. Cheon and T. Takagi (Eds.): ASIACRYPT 2016, Part II, LNCS 10032, pp. 531–556, 2016.
DOI: 10.1007/978-3-662-53890-6_18

1 Introduction

In traditional encryption, a receiver in possession of a ciphertext either has a corresponding decryption key for it, in which case it can recover the underlying message, or else it can get no information about the underlying message. *Functional encryption* (FE) [10,21,26,32] is a vast new paradigm for encryption in which the decryption keys are associated to *functions*, whereby a receiver in possession of a ciphertext and a decryption key for a particular function can recover that function of the underlying message. Intuitively, security requires that it learns nothing else. Due to both theoretical appeal and practical importance, FE has gained tremendous attention in recent years.

In particular, this work concerns a compelling extension of FE called *multi-input functional encryption* (MIFE), introduced by Goldwasser *et al.* [25]. In MIFE, decryption operates on *multiple ciphertexts*, such that a receiver with some decryption key is able to recover the associated function applied to all of the underlying plaintexts (*i.e.*, the underlying plaintexts are all arguments to the associated function). MIFE enables an number of important applications not handled by standard (single-input) FE. On the theoretical side, MIFE has interesting applications to non-interactive secure multiparty computation [7]. On the practical side, we reproduce the following example from [25].

Running SQL queries over encrypted data: Suppose we have an encrypted database. A natural goal in this scenario would be to allow a party Alice to perform a certain class of general SQL queries over this database. If we use ordinary functional encryption, Alice would need to obtain a separate secret key for every possible valid SQL query, a potentially exponentially large set. Multi-input functional encryption allows us to address this problem in a flexible way. We highlight two aspects of how Multi-Input Functional Encryption can apply to this example:

- Let f be the function where $f(q, x)$ first checks if q is a valid SQL query from the allowed class, and if so $f(q, x)$ is the output of the query q on the database x. Now, if we give the decryption key corresponding to f *and* the encryption key ek_1 (corresponding to the first input of the function f) to Alice, then Alice can choose a valid query q and encrypt it under her encryption key EK_1 to obtain ciphertext c_1. Then she could use her decryption key on ciphertexts c_1 and c_2, where c_2 is the encrypted database, to obtain the results of the SQL query.
- Furthermore, if our application demanded that multiple users add or manipulate different entries in the database, the most natural way to build such a database would be to have different ciphertexts for each entry in the database. In this case, for a database of size n, we could let f be an $(n+1)$-ary function where $f(q, x_1, \ldots, x_n)$ is the result of a (valid) SQL query q on the database $(x_1, \ldots, x_n)$.

Goldwasser *et al.* [25] discuss various other application of MIFE to non-interactive differentially private data release, delegation of computation, and, computing over encrypted streams, etc. We refer the reader to [25] for a more complete treatment. Besides motivating the notion, Goldwasser *et al.* [25] gave various flavors of definitions for MIFE and its security, as well as constructions based on different forms of program obfuscation. First of all, we note a basic observation about MIFE: in the public-key setting, functions for which one can hope to have any security at all are limited. In particular, a dishonest decryptor in possession of public key PP, a secret key SK_f for (say) a binary function f, and ciphertext CT encrypting message m, can try to learn m by repeatedly choosing some m' and learning $f(m, m')$, namely by encrypting m' under PP to get CT' and decrypting C, C' under SK_f. This means one can only hope for a very weak notion of security in such a case. As a result, in this work we focus on a more general setting where the functions have say a fixed arity n and there are encryption keys $\mathsf{EK}_1, \ldots, \mathsf{EK}_n$ corresponding to each index (*i.e.*, EK_i is used to encrypt a message which can then be used as an i-th argument in any function via decryption with the appropriate key). Only some subset of these keys (or maybe none of them) are known to the adversary. Note that this subsumes both the public key and the secret key setting (in which a much more meaningful notion of security maybe possible). In this setting, [25] presented an MIFE scheme based on indistinguishability obfuscation (iO) [6,21].

Bounded-message security: The construction of Goldwasser *et al.* [25] based on iO has a severe shortcoming namely that it could only support security for an encryption of an *a priori bounded* number of messages[1]. This bound is required to be fixed at the time of system setup and, if exceeded, would result in the guarantee of semantic security not holding any longer. In other words, the number of challenge messages chosen by the adversary in the security game needed to be *a priori* bounded. The size of the public parameters in [25] grows linearly with the number of challenge messages.

Now we go back to the previous example of running SQL queries over encrypted databases where each entry in the database is encrypted individually. This bound would mean that the number of entries in the database would be bounded at the time of the system setup. Also, the number of updates to the database would be bounded as well. Similar restrictions would apply in other applications of MIFE: e.g., while computing over encrypted data streams, the number of data streams would have to be a priori bounded, etc. In addition, the construction of Goldwasser *et al.* [25] could only support *Selective-security:* The challenge messages and the set of "corrupted" encryption keys needed by the adversary is given out at the beginning of the experiment.[2]

Let us informally refer to an MIFE construction that does not have these shortcomings as unbounded-message secure or simply *fully-secure.* In addition

[1] We note that, since we do not work in the public-key setting, there is no generic implication of single-message to multi-message security.

[2] Corruption of encryption keys $\mathsf{EK}_1, \ldots, \mathsf{EK}_n$ is an aspect of MIFE security not present for single-input FE; note that in [25], some subset of these keys could not be requested *adaptively* by the adversary - they were to be chosen even before the setup was done.

to the main construction based on iO, Goldwasser *et al.* [25] also showed a construction of adaptively-secure MIFE (except wrt. the subset of encryption keys given to the adversary, so we still do not call it fully-secure) that relies on a stronger form of obfuscation called *differing-inputs obfuscation* ($di\mathcal{O}$) [1,6,12].[3] Roughly, $di\mathcal{O}$ says that for any two circuits C_0 and C_1 for which it is hard to find an input on which their outputs differ, it should be hard to distinguish their obfuscations, and moreover given such a distinguisher one can extract such a differing input. Unfortunately, due to recent negative results [22], $di\mathcal{O}$ is now viewed as an implausible assumption. The main question we are concerned with in this work is: *Can fully-secure MIFE can be constructed from $i\mathcal{O}$?*

1.1 Our Contributions

Our main result is a fully-secure MIFE scheme from *sub-exponentially secure* $i\mathcal{O}$. More specifically, we use the following primitives: (1) sub-exponentially secure $i\mathcal{O}$, (2) sub-exponentially secure injective one-way functions, and (3) standard public-key encryption (PKE). Here "sub-exponential security" refers to the fact that advantage of any (efficient) adversary should be sub-exponentially small. For primitive (2), this should furthermore hold against adversaries running in sub-exponential time.

A few remarks about these primitives are in order. First, the required security will depend on the function arity, but *not* on the number of challenge messages. Indeed, Goldwasser *et al.* already point out that selective-security (though not bounded-message security, which instead has to do with their use of statistically sound non-interactive proofs) of their MIFE scheme based on $i\mathcal{O}$ can be overcome by standard complexity leveraging. However, in that case the required security level would depend on the the number of challenge messages. As in most applications we expect the number of challenge messages to be orders of magnitude larger than the function arity, this would result in much larger parameters than our scheme. Second, we only use a sub-exponentially secure injective one-way function (*i.e.*, primitive (2)) in our *security proof*, not in the scheme itself. Thus it suffices for such an injective one-way function to simply *exist* for security of our MIFE scheme, even if we do not know an explicit candidate.

1.2 Our Techniques

The starting point of our construction is the fully-secure construction of MIFE based on $di\mathcal{O}$ due to Goldwasser *et al.* [25] mentioned above. In their scheme, the encryption key for an index $i \in [n]$ (where n is the function arity) is a pair of public keys (pk_i^0, pk_i^1) for an underlying PKE scheme, and a ciphertext for index i consists of encryptions of the plaintext under pk_i^0, pk_i^1 respectively, along with a simulation-sound non-interactive zero knowledge proof that the two ciphertexts are well-formed (*i.e.*, both encrypting the same underlying message).

[3] Actually, [25] required even a stronger form of $di\mathcal{O}$ called strong differing-inputs obfuscation or differing-inputs obfuscation secure in presence of an oracle.

The secret key for a function f is an obfuscation of a program that takes as input n ciphertext pairs with proofs $(c_1^0, c_1^1, \pi_1), \ldots, (c_n^0, c_n^1, \pi_n)$, and, if the proofs verify, decrypts the first ciphertext from each pair using the corresponding secret key, and finally outputs f applied to the resulting plaintexts. Note that it is important for the security proof to assume $di\mathcal{O}$, since one needs to argue when the function keys are switched to decrypting the second ciphertext in each pair instead, an adversary who detects the change can be used to extract a false proof.

We will develop modifications that this scheme so that we can instead leverage a result of [12] that any indistinguishability obfuscator is in fact a differing-inputs obfuscator on circuits which differ on polynomially many points. In fact, we we will only need to use this result for circuits which differ on a *single* point. But, we will need to require the extractor to work given an adversary with even exponentially-small distinguishing gap on the obfuscations of two such circuits, due to the exponential number of hybrids in our security proof. Fortunately, [17] showed the result of [12] extends to this case of we start with an indistinguishability obfuscator that is sub-exponentially secure.

Specifically, we need to make the proofs of well-formedness described above *unique* for every ciphertext pair, so that there is only one differing input point in the corresponding hybrids in our security proof. To achieve this, we design novel "special-purpose" proofs built from $i\mathcal{O}$ and punctured pseudorandom functions (PRFs) [11,13,29],[4] which works as follows. We include in the public parameters an obfuscated program that takes as input two ciphertexts and a witness that they are well-formed (*i.e.*, the message and randomness used for both the ciphertexts), and, if this check passes, outputs a (puncturable) PRF evaluation on those ciphertexts. Additionally, the secret key for a function f will now be an obfuscation of a program which additionally has this PRF key hardwired keys and verifies the "proofs" of well-formedness by checking that PRF evaluations are correct. Interestingly, in the security proof, we will switch to doing this check via an injective one-way function applied to the PRF values (*i.e.*, the PRF values themselves are not compared, but rather the outputs of an injective one-way function applied to them). This is so that extracting a differing input at this step in the security proof will correspond to inverting an injective one-way function; otherwise, the correct PRF evaluation would still be hard-coded in the obfuscated function key and we do not know how to argue security.

We now sketch the sequence of hybrids in our security proof. The proof starts from a hybrid where each challenge ciphertext encrypts m_i^0 for $i \in [n]$. Then we switch to a hybrid where each c_i^1 is an encryption of m_i^1 instead. These two hybrids are indistinguishable due to security of the PKE scheme. Let ℓ denote the length of a ciphertext. For each index $i \in [n]$ we define hybrids indexed by x, for all $x \in [2^{2n\ell}]$, in which function key SK_f decrypts the first ciphertext in the pair using SK_i^0 when $(c_1^0, c_1^1, .., c_n^0, c_n^1) < x$ and decrypts the second ciphertext

[4] Due to the number of hybrids in our proof, we will also need the punctured PRFs to be sub-exponentially secure, but this already follows from a sub-exponentially secure injective one-way function.

in the pair using SK_i^1 otherwise. Parse $x = (x_1^0, x_1^1, .., x_n^0, x_n^1)$. Hybrids indexed by x and $x + 1$ can be proven indistinguishable as follows: We first switch to sub-hybrids that puncture the PRF key at $\{x_i^0, x_i^1\}$, changes a function key SK_f to check correctness of an PRF value by applying an injective one-way function as described above, and hard-coded the output of the injective one-way function at the PRF evaluation at the punctured point. Now if the two hybrids differ at an input of the form $(x_1^0, x_1^1, \alpha_1, .., x_n^0, x_n^1, \alpha_n)$ where α_i is some fixed value (a PRF evaluation of (x_i^0, x_i^1)), extracting the differing input can be used to invert the injective one-way function on random input (namely the α_i).

Finally, we note that exponentially many hybrids are indexed by all possible ciphertext vectors that could be input to decryption (i.e., vectors of length the arity of the functionality) and *not* all possible challenge ciphertext vectors. This allows us to handle any unbounded (polynomial) number of ciphertexts for every index.

Our techniques further demonstrate the power of the exponentially-many hybrids technique, together with the iO $\Rightarrow$ one-point-diO, which have also been used recently in works such as [8,17].

1.3 Related Work, Open Problems

In this work we focus on an *indistinguishability-based* security notion for MIFE. This is justified as Goldwasser *et al.* [25] show that an MIFE meeting a stronger simulation-based security definition in general implies black-box obfuscation [6] and hence is impossible. They also point out that in the secret-key setting with small function arity, an MIFE scheme meeting indistinguishability-based security notion can be "compiled" into a simulation-secure one, following the work of De Caro *et al.* [16]; in such a setting we can therefore achieve simulation-based security as well. We note that a main problem left open by our work is whether $i\mathcal{O}$ without sub-exponential security implies MIFE, which would in some sense show these two primitives are equivalent (up to the other primitives used in the construction). Another significant open problem is removing the bound a function's arity in our construction, as well as the bound on the message length, perhaps by building on recent work in the setting of single-input FE [30].

Initial constructions of single-input FE from $i\mathcal{O}$ [21] also had the shortcomings we are concerned with removing for constructions of MIFE in this work, namely selective and bounded-message security. These restrictions were similarly first overcome using differing-inputs obfuscation [1,12], and later removed while only relying on $i\mathcal{O}$ [2,33]. Unfortunately, we have not been able to make the techniques of these works apply to the MIFE setting, which is why we have taken a different route. If they could, this would be a path towards solving the open problem of relying on $i\mathcal{O}$ with standard security mentioned above.

[14] construct an adaptively secure multi-input functional encryption scheme in the secret key setting for any number of ciphertexts from any secret key functional encryption scheme. Their construction builds on a clever observation that function keys of a secret-key function-hiding functional encryption can be used to hide any message. This provides a natural 'arity amplification' procedure

that allows us to go from a t arity secret key MIFE to a $t+1$ arity MIFE. However, because the arity is amplified one by one, it leads to a blow up in the scheme, so the arity of the functions had to be bounded by $O(log(logk))$. [4] builds on similar techniques but considers construction of secret key MIFE from a different view-point (i.e. building iO from functional encryption).

The existence of indistinguishability obfuscation is still a topic of active research. On one hand there has been recent works such as [31] which break many of the existing IO candidates using [20]. However, there have been new/modified constructions which provably resist these attacks under a strengthened model of security [23].

There has also been progress on constructing universal constructions and obfuscation combiners [3,19]. An almost updated list of candidates along with their status can be found here [3]. Since, Multi-Input Functional Encryption implies indistinguishability obfuscation (as shown in [25]) assuming IO is necessary. Finally, we note that the source of trouble in achieving differing-inputs obfuscation is the *auxiliary input* provided to the distinguisher. Another alternative to using differing-inputs obfuscation is *public-coin* diO [28], where this auxiliary input is simply a uniform random string as done in [5] (they however achieve selective security). There are no known implausibility results for public-coin $di\mathcal{O}$, and it is interesting to give an alternative construction of fully-secure MIFE based on it. Our assumption seems incomparable, as we only need $i\mathcal{O}$ but also sub-exponential security.

1.4 Organisation

The rest of this paper is organized as follows: In Sect. 2, we recall some definitions and primitives used in the rest of the paper. In Sect. 3 we formally define MIFE and present our security model. Finally in Sect. 4, we present our construction and a security proof.

2 Preliminaries

In this section we recall various concepts on which the paper is built upon. We assume the familiarity of a reader with concepts such as public key encryption, one way functions and omit formal description in the paper. For the rest of the paper, we denote by $\mathbb{N}$ the set of natural numbers $\{1, 2, 3, ..\}$. Sub-exponential indistinguishability obfuscation and sub-exponentially secure puncturable pseudo-random functions have been used a lot recently such as in the works of [9,15,30]. For completeness, we present these notions below:

2.1 Indistinguisability Obfuscation

The following definition has been adapted from [21]:

Definition 1. *A uniform PPT machine $i\mathcal{O}$ is an indistinguishability obfuscator for a class of circuits $\{\mathcal{C}_n\}_{n\in\mathbb{N}}$ if the following properties are satisfied.*

Correctness: *For every $k \in \mathbb{N}$, for all $\{\mathcal{C}_k\}_{k\in\mathbb{N}}$, we have*

$$Pr[C' \leftarrow i\mathcal{O}(1^k, C) : \forall x, C'(x) = C(x)] = 1$$

Security: *For any pair of functionally equivalent equi-sized circuits $C_0, C_1 \in \mathcal{C}_k$ we have that: For every non uniform PPT adversary $\mathcal{A}$ there exists a negligible function ϵ such that for all $k \in \mathbb{N}$,*

$$| \, Pr[\mathcal{A}(1^n, i\mathcal{O}(1^k, C_0), C_0, C_1, z) = 1] - Pr[\mathcal{A}(1^k, i\mathcal{O}(1^k, C_1), C_0, C_1, z) = 1] \, | \leq \epsilon(k)$$

We additionally say that $i\mathcal{O}$ is sub-exponentially secure if there exists some constant $\alpha > 0$ such that for every non uniform PPT $\mathcal{A}$ the above indistinguishability gap is bounded by $\epsilon(k) = O(2^{-k^\alpha})$.

Definition 2 (Indistinguishability obfuscation for P/poly). *$i\mathcal{O}$ is a secure indistinguishability obfuscator for P/Poly, if it is an indistinguishability obfuscator for the family of circuits $\{\mathcal{C}_k\}_{k\in\mathbb{N}}$ where $\mathcal{C}_k$ is the set of all circuits of size k.*

2.2 Puncturable Psuedorandom Functions

A PRF $F : \mathcal{K}_{k\in\mathbb{N}} \times \mathcal{X} \rightarrow \mathcal{Y}_{k\in\mathbb{N}}$ is a puncturable pseudorandom function if there is an additional key space $\mathcal{K}_p$ and three polynomial time algorithms (F.setup, F.eval, F.puncture) as follows:

- $F.\mathsf{setup}(1^k)$ a randomized algorithm that takes the security parameter k as input and outputs a description of the key space $\mathcal{K}$, the punctured key space $\mathcal{K}_p$ and the PRF F.
- $F.\mathsf{puncture}(K, x)$ is a randomized algorithm that takes as input a PRF key $K \in \mathcal{K}$ and $x \in \mathcal{X}$, and outputs a key $K\{x\} \in \mathcal{K}_p$.
- $F.\mathsf{Eval}(K, x')$ is a deterministic algorithm that takes as input a punctured key $K\{x\} \in \mathcal{K}_p$ and $x' \in \mathcal{X}$. Let $K \in \mathcal{K}$, $x \in \mathcal{X}$ and $K\{x\} \leftarrow F.\mathsf{puncture}(K, x)$.

The primitive satisfies the following properties:

1. **Functionality is preserved under puncturing:** For every $x^* \in \mathcal{X}$,
$$Pr[F.\mathsf{eval}(K\{x^*\}, x) = F(K, x)] = 1$$
here probability is taken over randomness in sampling K and puncturing it.
2. **Psuedo-randomness at punctured point:** For any poly size distinguisher D, there exists a negligible function $\mu(\cdot)$, such that for all $k \in \mathbb{N}$ and $x^* \in \mathcal{X}$,
$$| \, Pr[D(x^*, K\{x^*\}, F(K, x^*)) = 1] - Pr[D(x^*, K\{x^*\}, u) = 1] \, | \leq \mu(k)$$
where $K \leftarrow F.\mathsf{Setup}(1^k)$, $K\{x^*\} \leftarrow F.\mathsf{puncture}(K, x^*)$ and $u \overset{\$}{\leftarrow} \mathcal{Y}_k$

We say that the primitive is sub-exponentially secure if μ is bounded by $O(2^{-k^{c_{PRF}}})$, for some constant $0 < c_{PRF} < 1$. We also abuse the notation slightly and use $F(K, \cdot)$ and $F.\mathsf{Eval}(K, \cdot)$ to mean one and same thing irrespective of whether key is punctured or not.

2.3 Injective One-Way Function

A one-way function with security (s, ϵ) is an efficiently evaluable function $P : \{0,1\}^* \rightarrow \{0,1\}^*$ and $Pr_{x \xleftarrow{\$} \{0,1\}^n}[P(A(P(x))) = P(x)] < \epsilon(n)$ for all circuits A of size bounded by $s(n)$. It is called an injective one-way function if it is injective in the domain $\{0,1\}^n$ for all sufficiently large n.

In this work we require that there *exists*[5] (s, ϵ) injective one-way function with $s(n) = 2^{n^{c_{owp1}}}$ and $\epsilon = 2^{-n^{c_{owp2}}}$ for some constants $0 < c_{owp1}, c_{owp2} < 1$. This assumption is well studied, [27,35] have used $(2^{cn}, 1/2^{cn})$ secure one-way functions and permutations for some constant c.

This is a reasonable assumption due to following result from [24].

Lemma 1. *Fix $s(n) = 2^{n/5}$. For all sufficiently large n, a random permutation π is $(s(n), 1/2^{n/5})$ secure with probability at least $1 - 2^{-2^{n/2}}$.*

Such assumptions have been made and discussed in works of [27,34,35]. In particular, we require the following assumption:

Assumption 1: For any adversary A with running time bounded by $s(n) = O(2^{n^{c_{owp1}}})$, for any apriori bounded polynomial $p(n)$ there exists an injective one-way function P such that,

$$Pr[r_i \xleftarrow{\$} \{0,1\}^n \forall i \in [p], \mathcal{A}^{\mathcal{O}}(P(r_1), .., P(r_p)) = (r_1, .., r_p)] < O(2^{-n^{c_{owp2}}})$$

for some constant $0 < c_{owp_1}, c_{owp_2} < 1$. Here, oracle $\mathcal{O}$ can reveal at most $p - 1$ values out of $r_1, .., r_p$. Note that this assumption follows from the assumption described above with a loss p in the security gap.

2.4 (d, δ)-Weak Extractability Obfuscators

The concept of weak extractability obfuscator was first introduced in [12] where they claimed that if there is an adversary that can distinguish between indistinguishability obfuscations of two circuits that differ on polynomial number of inputs with noticable probability, then there is a PPT extractor that extracts a differing input with overwhelming probability. [17] generalised the notion to what they call (d, δ) weak extractability obfuscator, where they require that if there is any PPT adversary that can distinguish between obfuscations of two circuits (that differ on at most d inputs) with atleast $\epsilon > \delta$ probability, then there is an explicit extractor that extracts a differing input with overwhelming probability and runs in time $poly(1/\epsilon, d, k)$ time. Such a primitive can be constructed from a sub-exponentially secure indistinguishability obfuscation. $(1, 2^{-k})$ weak extractability obfuscation will be crucially used in our construction for our $MIFE$ scheme. We believe that in various applications of differing inputs obfuscation, it may suffice to use this primitive along with other sub-exponentially secure primitives.

[5] We however do not require that the injective one-way function can be sampled efficiently.

Definition 3. *A uniform transformation $we\mathcal{O}$ is a (d, δ) weak extractability obfuscator for a class of circuits $\mathcal{C} = \{\mathcal{C}_k\}$ if the following holds. For every PPT adversary $\mathcal{A}$ running in time $t_\mathcal{A}$ and $1 \geq \epsilon(k) > \delta$, there exists a algorithm E for which the following holds. For all sufficiently large k, and every pair of circuits on n bit inputs, $C_0, C_1 \in \mathcal{C}_k$ differing on at most $d(k)$ inputs, and every auxiliary input z,*

$$| Pr[\mathcal{A}(1^k, we\mathcal{O}(1^k, C_0), C_0, C_1, z) = 1] - Pr[\mathcal{A}(1^k, we\mathcal{O}(1^k, C_1), C_0, C_1, z) = 1] | \geq \epsilon$$

$$\Rightarrow Pr[x \leftarrow E(1^k, C_0, C_1, z) : C_0(x) \neq C_1(x) \geq 1 - \mathsf{negl}(k)$$

and the expected runtime of E is $O(p_E(1/\epsilon, d, t_\mathcal{A}, n, k))$ for some fixed polynomial p_E. In addition, we also require the obfuscator to satisfy correctness.

Correctness: *For every $n \in \mathbb{N}$, for all $\{\mathcal{C}_n\}_{n \in \mathbb{N}}$, we have*

$$Pr[C' \leftarrow we\mathcal{O}(1^n, C) : \forall x, C'(x) = C(x)] = 1$$

We now construct a $(1, 2^{-k})$ input weak extractability obfuscator from sub-exponentially secure indistinguishability obfuscation. Following algorithm describes the obfuscation procedure.

$we\mathcal{O}(1^k, C)$: The procedure outputs $C' \leftarrow i\mathcal{O}(1^{k^{1/\alpha}}, C)$. Here, $\alpha > 0$ is a constant chosen such that any polynomial time adversary against indistinguishability obfuscation has security gap upper bounded by $2^{-k}/4$.

The proof of the following theorem is proven in [17].

Theorem 1. *Assuming sub-exponentially secure indistinguishability obfuscation, there exists $(1, \delta)$ weak obfuscator for P/poly for any $\delta > 2^{-k}$, where k is the size of the circuit.*

In general, assuming sub-exponential security one can construct (d, δ) extractability obfuscator for any $\delta > 2^{-k}$. Our construction is as follows:

$we\mathcal{O}(C)$: Let α be the security constant such that $i\mathcal{O}$ with parameter $1^{k^{1/\alpha}}$ has security gap upper bounded by $O(2^{-3k})$. This can be found due to sub exponential security of indistinguishability obfuscation. The procedure outputs $C' \leftarrow i\mathcal{O}(1^{k^{1/\alpha}}, C)$.

We cite [12] for the proof of the following theorem.

Theorem 2 ([12]). *Assuming sub-exponentially secure indistinguishability obfuscation, there exists (d, δ) weak extractability obfuscator for P/poly for any $\delta > 2^{-k}$.*

3 Multi-input Functional Encryption

Let $\mathcal{X} = \{\mathcal{X}_k\}_{k \in \mathbb{N}}$ and $\mathcal{Y} = \{\mathcal{Y}_k\}_{k \in \mathbb{N}}$ denote ensembles where each $\mathcal{X}_k$ and $\mathcal{Y}_k$ is a finite set. Let $\mathcal{F} = \{\mathcal{F}_k\}_{k \in \mathbb{N}}$ denote an ensemble where each $\mathcal{F}_k$ is a finite collection of n-ary functions. Each $\mathsf{f} \in \mathcal{F}_k$ takes as input n strings $\mathsf{x}_1, .., \mathsf{x}_\mathsf{n}$ where each $\mathsf{x}_\mathsf{i} \in \mathcal{X}_k$ and outputs $\mathsf{f}(\mathsf{x}_1, .., \mathsf{x}_\mathsf{n}) \in \mathcal{Y}_k$. We now describe the algorithms.

- $\mathsf{MIFE.Setup}(1^{\kappa}, \mathsf{n})$: is a PPT algorithm that takes as input the security parameter κ and the function arity n. It outputs n encryption keys $\mathsf{EK}_1, .., \mathsf{EK}_\mathsf{n}$ and a master secret key MSK.
- $\mathsf{MIFE.Enc}(\mathsf{EK}, \mathsf{m})$: is a PPT algorithm that takes as input an encryption key $\mathsf{EK}_\mathsf{i} \in (\mathsf{EK}_1, .., \mathsf{EK}_\mathsf{n})$ and an input message $m \in \mathcal{X}_k$ and outputs a ciphertext CT_i which denotes that the encrypted plaintext constitutes an ith input to a function f.
- $\mathsf{MIFE.Keygen}(\mathsf{MSK}, \mathsf{f})$: is a PPT algorithm that takes as input the master secret key MSK and a $\mathsf{n}-$ary function $\mathsf{f} \in \mathcal{F}_k$ and outputs a corresponding decryption key SK_f.
- $\mathsf{MIFE.Dec}(\mathsf{SK}_\mathsf{f}, \mathsf{CT}_1, .., \mathsf{CT}_\mathsf{n})$: is a deterministic algorithm that takes as input a decryption key SK_f and n ciphertexts $\mathsf{CT}_\mathsf{i}, .., \mathsf{CT}_\mathsf{n}$ and outputs a string $\mathsf{y} \in \mathcal{Y}_k$.

The scheme is said to satisfy *correctness* if for honestly generated encryption and function key and any tuple of honestly generated ciphertexts, decryption of the cipher-texts with function key for f outputs the joint function value of messages encrypted inside the ciphertexts with overwhelming probability.

Definition 4. *Let* $\{\mathsf{f}\}$ *be any set of functions* $f \in \mathcal{F}_\kappa$. *Let* $[n] = \{1, .., n\}$ *and* $I \subseteq [n]$. *Let* $\boldsymbol{X^0}$ *and* $\boldsymbol{X^1}$ *be a pair of input vectors, where* $\boldsymbol{X^b} = \{x^b_{1,j}, .., x^b_{n,j}\}_{j=1}^q$. *We define* $\mathcal{F}$ *and* (X^0, X^1) *to be I-compatible if they satisfy the following property: For every* $f \in \{f\}$, *every* $I' = \{i_1, .., i_t\} \subseteq I$, *every* $j_1, .., j_{n-t} \in [q]$ *and every* $x'_{i_1}, .., x'_{i_t} \in \mathcal{X}_\kappa$,

$$f(< x^0_{i_1,j_1}, .., x^0_{i_{n-t},j_{n-t}}, x'_{i_1}, .., x'_{i_t} >) = f(< x^1_{i_1,j_1}, .., x^1_{i_{n-t},j_{n-t}}, x'_{i_1}, .., x'_{i_t} >)$$

where $< y_{i_1}, ..., y_{i_n} >$ *denotes a permutation of the values* $y_{i_1}, .., y_{i_n}$ *such that the value* y_{i_j} *is mapped to the* l^{th} *location if* y_{i_j} *is the* l^{th} *input (out of* n *inputs) to* f.

IND-Secure MIFE: Security definition in [25] was parameterized by two parameters (t, q) where t denotes the number of encryption keys known to the adversary, and q denotes the number of challenge messages per encryption key. Since, our scheme can handle any unbounded polynomial q and any $t \leq n$, we present a definition independent of these parameters.

Definition 5 (Indistinguishability based security). *We say that a multi-input functional encryption scheme* MIFE *for for* n *ary functions* $\mathcal{F}$ *is fully IND-secure if for every PPT adversary* $\mathcal{A}$, *the advantage of* $\mathcal{A}$ *defined as*

$$\mathsf{Adv}_{\mathcal{A}}^{\mathsf{MIFE,IND}}(1^{\kappa}) = |Pr[\mathsf{IND}_{\mathcal{A}}^{\mathsf{MIFE}}] - 1/2|$$

is $negl(\kappa)$, *where:*

Valid adversaries: In the above experiment, $\mathcal{O}(\mathbf{EK}, \cdot)$ *is an oracle that takes an index* i *and outputs* EK_i. *Let* I *be the set of queries to this oracle.* $\mathcal{E}(\mathbf{EK}, b, \cdot)$ *on a query* $(x^0_{1,j}, .., x^0_{n,j}), (x^1_{1,j}, .., x^1_{n,j})$ *(where* j *denotes the query number) outputs* $\mathsf{CT}_{i,j} \leftarrow \mathsf{MIFE.Enc}(\mathsf{EK}_i, x^b_{i,j})$ $\forall i \in [n]$. *If* q *is the total number of queries to this*

Experiment $\mathsf{IND}^{\mathsf{MIFE}}_{\mathcal{A}}(1^{\kappa})$
$(\boldsymbol{EK}, MSK) \leftarrow \mathsf{MIFE.Setup}(1^{\kappa}, n)$
$b \leftarrow \{0,1\}$
$b' \leftarrow \mathcal{A}^{\mathsf{MIFE.Keygen(MSK,\cdot)}, \mathcal{O}(\mathbf{EK},\cdot), \mathcal{E}(\mathbf{EK},b,\cdot)}(1^{\kappa})$
Output $(b = b')$

Fig. 1. Security game

oracle then let $\boldsymbol{X^l} = \{x^l_{1,j}, .., x^l_{n,j}\}^q_{j=1}$ *and* $l \in \{0,1\}$. *Also, let* $\{f\}$ *denote the entire set of function key queries made by* $\mathcal{A}$. *Then, the challenge message vectors* $\boldsymbol{X^0}$ *and* $\boldsymbol{X^1}$ *chosen by* $\mathcal{A}$ *must be* $I-$*compatible with* $\{f\}$. *The scheme is said to be secure if for any valid adversary* $\mathcal{A}$ *the advantage in the game described above is negligible (Fig. 1).*

4 Our MIFE Construction

Notation: Let k denote the security parameter and $n = n(k)$ denote the bound on arity of the function for which the keys are issued. By $\mathsf{PRF} = (\mathsf{PRF.Setup}, \mathsf{PRF.Puncture}, \mathsf{PRF.Eval})$ denote a sub-exponentially secure puncturable PRF with security constant c_{PRF} and PKE denote a public key encryption scheme. Let P be any one-one function (in the security proof we instantiate with a sub-exponentially secure injective one-way function with security constants c_{owp1} and c_{owp2}). Finally, let $\mathcal{O}$ denote a $(1, 2^{-3nl-k})$ weak extractability obfuscator (here l is the length of the cipher-text of PKE). In particular, for any two equivalent circuits security gap of the obfuscation is bounded by 2^{-3nl-k} (any algorithm that distinguishes obfuscations of two circuits with more than this gap will yield an algorithm that extracts a differing point).

$\mathsf{MIFE.Setup}(1^{\mathsf{k}}, \mathsf{n})$: Sample $K_i \leftarrow \mathsf{PRF.Setup}(1^{\lambda})$ and $\{(PK^b_i, SK^b_i)\}_{b\in\{0,1\}} \leftarrow \mathsf{PKE.Setup}(1^k)$. Let PP_i be the circuit as in Fig. 2. EK_i is declared as the set $EK_i = \{PK^0_i, PK^1_i, \tilde{PP_i} = \mathcal{O}(PP_i), P\}$ and $MSK = \{SK^0_i, SK^1_i, K_i, P\}_{i\in[n]}$. Here injective function P takes as input elements from the co-domain the PRF. λ is set greater than $(3nl+k)^{1/c_{PRF}}$ and so that the length of output of the PRF is at least $max\{(5nl+2k)^{1/c_{owp1}}, (3nl+k)^{1/c_{owp2}}\}$ long.

$\mathsf{MIFE.Enc}(EK_i, m)$: To encrypt a message m, encryptor does the following:

- Compute $c^0_i = \mathsf{PKE.Enc}(PK^0_i, m; r^0)$ and $c^1_i = \mathsf{PKE.Enc}(PK^1_i, m; r^1)$.
- Evaluate $\pi_i \leftarrow \tilde{PP_i}(c^0_i, c^1_i, m, r^0, r^1)$.

Output $CT_i = (c^0_i, c^1_i, \pi_i)$.

$\mathsf{MIFE.KeyGen}(MSK, f)$: Let G^0_f be the circuit described below. Key for f is output as $K_f \leftarrow \mathcal{O}(G^0_f)$.

$\mathsf{MIFE.Decrypt}(K_f, \{c^0_i, c^1_i, \pi_i\}_{i\in[n]})$: Output $K_f(c^0_1, c^1_1, \pi_1, .., c^0_n, c^1_n, \pi_n)$.

Hard-wired: PK_i^0, PK_i^1, K_i.
Input: $c_i^0, c_i^1, m, r_i^0, r_i^1$
The program does the following:

- Check that $c_i^0 = \mathsf{PKE.Enc}(PK_i^0, m; r_i^0)$ and $c_i^1 = \mathsf{PKE.Enc}(PK_i^1, m; r_i^1)$. If the check fails output $\perp$.
- Output $\mathsf{PRF.Eval}(K_i, c_i^0, c_i^1)$

Fig. 2. Program encrypt

Hard-wired: $\{SK_i^0, K_i, P\}_{i \in [n]}$.
Input: $\{c_i^0, c_i^1, \pi_i\}_{i \in [n]}$
The program does the following:

- For all $i \in [n]$, check that $P(\mathsf{PRF.Eval}(K_i, c_i^0, c_i^1)) = P(\pi_i)$. If the check fails output $\perp$.
- Output $f(\mathsf{PKE.Dec}(SK_1^0, c_1^0), .., \mathsf{PKE.Dec}(SK_n^0, c_n^0))$.

Fig. 3. Program G_f^0

Remark

1. We also assume that the circuits are padded appropriately before they are obfuscated.
2. Note that in the scheme, circuit for the key for a function f, G_f^0 is instantiated with any one-one function (denoted by P). In the proofs we replace it with a sub-exponentially secure injective one-way function. We see that the input output behaviour of G_f^0 do not change when it is instantiated with any one-one function, hence we can switch to a hybrid when it is instantiated by sub-exponentially secure injective one way function and due to the security of obfuscation these two hybrids are close.

4.1 Proof Overview

The starting point of our construction is the fully-secure construction of MIFE based on $di\mathcal{O}$ due to Goldwasser *et al.* [25] mentioned above. In their scheme, the encryption key for an index $i \in [n]$ (where n is the function arity) is a pair of public keys (pk_i^0, pk_i^1) for an underlying PKE scheme, and a ciphertext for index i consists of encryptions of the plaintext under pk_i^0, pk_i^1 respectively, along with a simulation-sound non-interactive zero knowledge proof that the two ciphertexts are well-formed (*i.e.*, both encrypting the same underlying message). The secret key for a function f is an obfuscation of a program that takes as input n ciphertext pairs with proofs $(c_1^0, c_1^1, \pi_1), \ldots, (c_n^0, c_n^1, \pi_n)$, and, if the proofs verify, decrypts the first ciphertext from each pair using the corresponding secret key, and finally outputs f applied to the resulting plaintexts. Note that it is important for the security proof to assume $di\mathcal{O}$, since one needs to argue when the function keys are switched to decrypting the second ciphertext in each pair instead, an adversary who detects the change can be used to extract a false proof.

We develop modifications to this scheme so that we can instead leverage a result of [12] that any indistinguishability obfuscator is in fact a differing-inputs obfuscator on circuits which differ on polynomially many points. In fact, we we will only need to use this result for circuits which differ on a *single* point. But, we will need to require the extractor to work given an adversary with even exponentially-small distinguishing gap on the obfuscations of two such circuits, due to the exponential number of hybrids in our security proof. We make use of sub-exponentially secure obfuscation to achieve this.

Specifically, we make the proofs of well-formedness described above *unique* for every ciphertext pair, so that there is only one differing input point in the corresponding hybrids in our security proof. To achieve this, we design novel "special-purpose" proofs built from $i\mathcal{O}$ and punctured pseudorandom functions (PRFs) [11,13,29],[6] which works as follows. We include in the public parameters an obfuscated program that takes as input two cipher-texts and a witness that they are well-formed (*i.e.*, the message and randomness used for both the cipher-texts), and, if this check passes, outputs a (puncturable) PRF evaluation on those ciphertexts. Additionally, the secret key for a function f will now be an obfuscation of a program which additionally has this PRF key hardwired keys and verifies the "proofs" of well-formedness by checking that PRF evaluations are correct. Interestingly, in the security proof, we will switch to doing this check via an injective one-way function applied to the PRF values (*i.e.*, the PRF values themselves are not compared, but rather the outputs of injective one-way function applied to them). This is so that extracting a differing input at this step in the security proof will correspond to inverting a injective one-way function; otherwise, the correct PRF evaluation would still be hard-coded in the obfuscated function key and we do not know how to argue security.

We now sketch the sequence of hybrids in our security proof. The proof starts from a hybrid where each challenge ciphertext encrypts m_i^0 for $i \in [n]$. Then we switch to a hybrid where each c_i^1 is an encryption of m_i^1 instead. These two hybrids are indistinguishable due to security of the PKE scheme. Let ℓ denote the length of a ciphertext. For each index $i \in [n]$ we define hybrids indexed by x, for all $x \in [2^{2n\ell}]$, in which function key SK_f decrypts the first ciphertext in the pair using SK_i^0 when $(c_1^0, c_1^1, .., c_n^0, c_n^1) < x$ and decrypts the second ciphertext in the pair using SK_i^1 otherwise. Parse $x = (x_1^0, x_1^1, .., x_n^0, x_n^1)$. Hybrids indexed by x and $x + 1$ can be proven indistinguishable as follows: We first switch to sub-hybrids that puncture the PRF key at $\{x_i^0, x_i^1\}$, changes a function key SK_f to check correctness of an PRF value by applying an injective one-way function as described above, and hard-coded the output of the injective one-way

[6] Due to the number of hybrids in our proof, we will also need the punctured PRFs to be sub-exponentially secure, but this already follows from sub-exponentially secure injective one-way functions.

function at the punctured point. Now if the two hybrids differ at an input of the form $(x_1^0, x_1^1, \alpha_1, .., x_n^0, x_n^1, \alpha_n)$ where α_i is some fixed value (a PRF evaluation of (x_i^0, x_i^1)), extracting the differing input can be used to invert the injective one-way function on random input (namely the α_i). As in [12], this inverter runs in time inversely proportional to the distinguishing gap between the two consecutive hybrids (which is sub-exponentially small). Hence, we require a sub-exponential secure injective one-way function to argue security.

Finally, we note that exponentially many hybrids are indexed by all possible ciphertext vectors that could be input to decryption (i.e., vectors of length the arity of the functionality) and *not* all possible challenge ciphertext vectors. This allows us to handle any unbounded (polynomial) number of ciphertexts for every index.

4.2 Proof of Security

Theorem 3. *Assuming an existence of a sub-exponentially secure indistinguishability obfuscator, injective one-way function and a polynomially secure public-key encryption scheme there exists a fully IND secure multi-input functional encryption scheme for any polynomially apriori bounded arity n.*

Proof. We start by giving a lemma that will be crucial to the proof.

Lemma 2. *Let X and Y denote two (possibly correlated) random variables from distribution $\mathcal{X}$ and $\mathcal{Y}$, with support $|\mathcal{X}|$ and $|\mathcal{Y}|$, and $U(X,Y)$ denote an event that depends on X, Y. We say that $U(X,Y) = 1$ if the event occurs, and $U(X,Y) = 0$ otherwise. Suppose $\Pr_{(X,Y)\sim\mathcal{X},\mathcal{Y}}[U(X,Y) = 1] = p$. We say that a transcript $\mathbb{X}$ falls in the set 'good' if $Pr_{Y\sim\mathcal{Y}}[U(X,Y|X = \mathbb{X}) = 1] \geq p/2$. Then, $Pr_{X\sim\mathcal{X}}[X \in \mathsf{good}] \geq p/2$.*

Proof. We prove the lemma by contradiction. Suppose $Pr_{X\sim\mathcal{X}}[X \in \mathsf{good}] = c < \frac{p}{2}$. Then,

$$Pr_{(X,Y)\sim(\mathcal{X},\mathcal{Y})}[U(X,Y) = 1] = Pr_{(X,Y)\sim(\mathcal{X},\mathcal{Y})}[U(X,Y) = 1|X \in \mathsf{good}] \cdot \Pr_{X\sim\mathcal{X}}[X \in \mathsf{good}]$$
$$+ Pr_{(X,Y)\sim(\mathcal{X},\mathcal{Y})}[U(X,Y) = 1|X \notin \mathsf{good}] \cdot Pr_{X\sim\mathcal{X}}[X \notin \mathsf{good}]$$

By definition of the set good, $Pr_{(X,Y)\sim(\mathcal{X},\mathcal{Y})}[U(X,Y) = 1|X \notin \mathsf{good}] < \frac{p}{2}$. Then, $p = \Pr[U(X,Y) = 1] < 1 \cdot c + (1-c) \cdot p/2$. Then, if $c < \frac{p}{2}$, we will have that $p < \frac{p}{2} + \frac{p}{2}$, which is a contradiction. This proves our lemma.
We proceed listing hybrids where the first hybrid corresponds to the hybrid where the challenger encrypts message $m_{i,j}^0$ for all $i \in [n]$ and the last hybrid corresponds to the hybrid where the challenger encrypts $m_{i,j}^1$. We then prove that each consecutive hybrid is indistinguishable from each other. Then, we sum up all the advantages between the hybrids and argue that the sum is negligible.

H_0

1. Challenger does setup to compute encryption keys $EK_i \forall i \in [n]$ and MSK as described in the algorithm.

Hard-wired: $\{SK_i^0, SK_i^1, K_i, x, P\}_{i \in [n]}$.
Input: $\{c_i^0, c_i^1, \pi_i\}_{i \in [n]}$
The program does the following:

- For all $i \in [n]$, check that $P(\mathsf{PRF.Eval}(K_i, c_i^0, c_i^1)) = P(\pi_i)$. If the check fails output $\perp$.
- If $(c_1^0, c_1^1, .., c_n^0, c_n^1) < x - 2$, output $f(\mathsf{PKE.Dec}(SK_1^1, c_1^1), .., \mathsf{PKE.Dec}(SK_n^1, c_n^1))$ otherwise output $f(\mathsf{PKE.Dec}(SK_1^0, c_1^0), .., \mathsf{PKE.Dec}(SK_n^0, c_n^0))$.

Fig. 4. Program $G_{f,x}$

2. $\mathcal{A}$ may query for encryption keys EK_i for some $i \in [n]$, function keys for function f and ciphertext queries in an interleaved fashion.
3. If it asks for an encryption key for index i, it is given EK_i.
4. When $\mathcal{A}$ queries keys for n ary function f_j and challenger computes keys honestly using MSK.
5. $\mathcal{A}$ may also ask encryptions of message vectors $M^h = \{(m_{1,j}^h, .., m_{n,j}^h)\}$ where $h \in \{0, 1\}$, where j denotes the encryption query number. The message vectors has to satisfy the constraint as given in the security definition.
6. For all queries j, challenger encrypts $CT_{i,j} \forall i \in [n]$ as follows: $c_{i,j}^0 = \mathsf{PKE.Enc}(PK_i^0, m_{i,j}^0)$ and $c_{i,j}^1 = \mathsf{PKE.Enc}(PK_i^1, m_{i,j}^0)$ and $\pi_{i,j} \leftarrow \mathsf{PRF.Eval}(K_i, c_{i,j}^0, c_{i,j}^1)$. Then the challenger outputs $CT_{i,j} = (c_{i,j}^0, c_{i,j}^1, \pi_{i,j})$.
7. $\mathcal{A}$ can ask for function keys for functions f_j, encryption keys EK_i's and ciphertexts as long as they satisfy the constraint given in the security definition.
8. $\mathcal{A}$ now outputs a guess $b' \in \{0, 1\}$.

H_1: Let q denote the number of cipher-text queries. This hybrid is same as the previous one except that for all indices $i \in [n], j \in [q]$ challenge cipher-text cipher-text component $c_{i,j}^1$ is set as $c_{i,j}^1 = \mathsf{PKE.Enc}(PK_i^1, m_{i,j}^1)$.

$\mathsf{H}_{x \in [2, 2^{2ln}+2]}$: This hybrid is same as the previous one except key for every function query f is generated as an obfuscation of program (Fig. 4) by hard-wiring x (along with SK_i^0, SK_i^1, K_i, P).

$\mathsf{H}_{2^{2ln}+3}$: This hybrid is same as the previous one except that function keys for any function f is generated by obfuscating program (Fig. 5).

$\mathsf{H}_{2^{2ln}+4}$: Let q denote the number of cipher-text queries made by the adversary. This hybrid is same as the previous one except that for all indices $i \in [n], j \in [q]$, challenge cipher-text component $c_{i,j}^0$ is generated as $c_{i,j}^0 = \mathsf{PKE.Enc}(PK_i^0, m_{i,j}^1)$.

$\mathsf{H}_{2^{2ln}+4+x | x \in [2^{2ln}+1]}$: This hybrid is same as the previous one except key for a function f is generated by obfuscating program (Fig. 4) by hard-wiring $2^{2ln}+3-x$ (along with SK_i^0, SK_i^1, K_i, P).

Hard-wired: $\{SK_i^1, K_i, P\}_{i\in[n]}$.
Input: $\{c_i^0, c_i^1, \pi_i\}_{i\in[n]}$
The program does the following:

- For all $i \in [n]$, check that $P(\mathsf{PRF.Eval}(K_i, c_i^0, c_i^1)) = P(\pi_i)$. If the check fails, output $\perp$.
- Output $f(\mathsf{PKE.Dec}(SK_1^1, c_1^1), .., \mathsf{PKE.Dec}(SK_n^1, c_n^1))$.

Fig. 5. Program G_f^1

$\mathsf{H}_{2.2^{2ln}+6}$: This hybrid corresponds to the real security game when $b = 1$.

We now argue indistinguishability by describing following lemmas.

Lemma 3. *For any PPT distinguisher D,* $| Pr[D(\mathsf{H}_0) = 1] - Pr[D(\mathsf{H}_1) = 1] | < \mathsf{negl}(k)$.

Proof. This lemma follows from the security of the encryption scheme PKE. In these hybrids, all function keys only depend on one secret key SK_i^0 for all $i \in [n]$ and SK_i^1 never appears in the hybrids. If there is a distinguisher D that distinguishes between the hybrids then there exists an algorithm $\mathcal{A}$ that breaks the security of the encryption scheme with the same advantage. $\mathcal{A}$ gets set of public keys $PK_1, .., PK_n$ from the encryption scheme challenger and samples public keys $(PK_i^0, SK_i^0)\forall i \in [n]$ himself and sets $PK_i^1 = PK_i \forall i \in [n]$. It also samples PRF keys $K_i \forall i \in [n]$. Using these keys, it generates encryption keys $EK_i \forall i \in [n]$. Then, it invokes D and answers queries for encryption keys EK_i's and function keys. $\mathcal{A}$ generates function keys using only as obfuscation of G_f^0. Finally, D declares $M^b = \{(m_{1,j}^b, .., m_{n,j}^b)\}_{j\in[q]}$. $\mathcal{A}$ sends (M^0, M^1) to the encryption challenger and gets $c_{i,j} \forall i \in [n], j \in [q]$ from the challenger. $\mathcal{A}$ computes $c_{i,j}^0 \leftarrow \mathsf{PKE.Enc}(\mathsf{PK_i^0}, \mathsf{m_{i,j}^0})$. Then evaluates $\pi_{i,j} \leftarrow \mathsf{PRF.Eval}(K_i, c_{i,j}^0, c_{i,j})$. Then it sets, $CT_{i,j} = (c_{i,j}^0, c_{i,j}, \pi_{i,j})$ and sends it to D. After that D may query keys for functions and encryption keys and the response is given as before. D now submits a guess b' which is also output by $\mathcal{A}$ as its guess for the encryption challenge. If $c_{i,j}$ is an encryption of $m_{i,j}^0$ then D's view is identical to the view in H_1 otherwise its view is identical to the view in H_2. Hence, distinguishing advantage of D in distinguishing hybrids is less than the advantage of $\mathcal{A}$ in breaking the security of the encryption scheme.

Lemma 4. *For any PPT distinguisher D,* $| Pr[D(\mathsf{H}_1) = 1] - Pr[D(\mathsf{H}_2) = 1] | < \mathsf{negl}(k)$.

Proof. For simplicity, we consider the case when there is only single function key query f. General case can be argued by introducing v many intermediate hybrids where v is the number of keys issued to the adversary. Indistinguishability of these hybrids follows from the fact that circuit G_f^0 and $G_{f,x=2}$ are functionally equivalent. Hence, due to the security of indistinguishability obfuscation property of the weak extractability obfuscator the lemma holds. For completeness, we describe the reduction. Namely, we construct an adversary $\mathcal{A}$ that uses D to break the security of weak extractability obfuscator. $\mathcal{A}$ invokes D and does setup (by sampling PKE encryption key pairs and PRF keys for all indices) and answers cipher-text queries as in the previous hybrid H_1. On query f from D, it sends G_f^0 and $G_{f,x}$ to the obfuscation challenger. It receives K_f and sends it to $\mathcal{A}$. $\mathcal{A}$ sends it to D. It replies to the encryption key queries to D using the sampled PKE keys and PRF keys. Then it outputs whatever D outputs. Note that view of D is identical to the view in H_1 (if K_f is an obfuscation of G_f^0) or H_2 (if K_f is an obfuscation of $G_{f,x=2}$). Hence, advantage of $\mathcal{A}$ is at least the advantage of D in distinguishing hybrids. Due to security of obfuscation claim holds.

Lemma 5. *For any PPT distinguisher D, $\mid Pr[D(\mathsf{H}_{2^{2ln}+2}) = 1] - Pr[D(\mathsf{H}_{2^{2ln}+3}) = 1] \mid < \mathsf{negl}(k)$.*

Proof. This follows from the indistinguishability obfuscator $\mathcal{O}$. For any function f, G_f^1 is functionally equivalent to $G_{f,x=2^{2ln}+2}$. Proof of the lemma is similar to the proof of Lemma 4.

Lemma 6. *For any PPT distinguisher D, $\mid Pr[D(\mathsf{H}_{2^{2ln}+3}) = 1] - Pr[D(\mathsf{H}_{2^{2ln}+4}) = 1] \mid < \mathsf{negl}(k)$.*

Proof. This follows from the security of encryption scheme PKE. Note that in both the hybrids SK_i^0 is not used anywhere. Proof is similar to the proof of Lemma 3.

Lemma 7. *For any PPT distinguisher D, $\mid Pr[D(\mathsf{H}_{2^{2ln}+4}) = 1] - Pr[D(\mathsf{H}_{2^{2ln}+5}) = 1] \mid < \mathsf{negl}(k)$.*

Proof. This follows from the security of indistinguishability obfuscator $\mathcal{O}$. Proof is similar to the proof of Lemma 4.

Lemma 8. *For any PPT distinguisher D, $\mid Pr[D(\mathsf{H}_{2.2^{2ln}+5}) = 1] - Pr[D(\mathsf{H}_{2.2^{2ln}+6}) = 1] \mid < \mathsf{negl}(k)$.*

Proof. This follows from the security of indistinguishability obfuscator $\mathcal{O}$. Proof is similar to the proof of Lemma 4.

Lemma 9. *For any PPT distinguisher D and $x \in [2, 2^{2ln} + 1]$, $\mid Pr[D(\mathsf{H}_x) = 1] - Pr[D(\mathsf{H}_{x+1}) = 1] \mid < O(v \cdot 2^{-2ln-k})$ for some polynomial v.*

Proof. We now list following sub hybrids and argue indistinguishability between these hybrids.

$\mathsf{H}_{x,1}$

1. Challenger samples key pairs $(PK_i^0, SK_i^0), (PK_i^1, SK_i^1)$ for each $i \in [n]$.
2. Parses $x - 2 = (x_1^0, x_1^1, .., x_n^0, x_n^1)$ and computes $(a_i^0, a_i^1) \leftarrow (\mathsf{PKE.Dec}(SK_i^0, x_i^0), \mathsf{PKE.Dec}(SK_i^1, x_i^1))$.
3. Samples puncturable PRF's keys $K_i \forall i \in [n]$.
4. Denote by set $Z \subset [n]$ such that $i \in Z$ if $a_i^0 \neq a_i^1$. Computes $\alpha_i \leftarrow \mathsf{PRF.Eval}(K_i, x_i^0, x_i^1)$ and derives punctured keys $K_i' \leftarrow \mathsf{PRF.Puncture}(K_i, x_i^0, x_i^1)$ for all $i \in [n]$.
5. If $\mathcal{A}$ queries for encryption keys for any index i, for any i in Z, $\tilde{PP_i}$ is generated as an obfuscation of circuit in Fig. 2 instantiated with the punctured key K_i' (α_i will never be accessed by the circuit PP_i in this case). For all other indices i, $\tilde{PP_i}$ is constructed by using the punctured key K_i' and hard-coding the value α_i (for input (x_i^0, x_i^1)) as done in Fig. 6. These $\tilde{PP_i}$ are used to respond to the queries for EK_i.
6. If $\mathcal{A}$ queries keys for n ary function f_j and challenger computes keys honestly as in H_x using MSK.
7. If $\mathcal{A}$ releases message vectors $M^h = \{(m_{1,j}^h, .., m_{n,j}^h)\}$ where $h \in \{0, 1\}$, challenger encrypts $CT_{i,j} \forall i \in [n], j \in [q]$ as follows: $c_{i,j}^0 = \mathsf{PKE.Enc}(PK_i^0, m_{i,j}^0)$ and $c_{i,j}^1 = \mathsf{PKE.Enc}(PK_i^1, m_{i,j}^1)$. If $(c_{i,j}^0, c_{i,j}^1) = (x_i^0, x_i^1)$ set $\pi_{i,j} = \alpha_i$ otherwise set $\pi_{i,j} \leftarrow \mathsf{PRF.Eval}(K_i, c_{i,j}^0, c_{i,j}^1)$. Then the challenger outputs $CT_{i,j} = (c_{i,j}^0, c_{i,j}^1, \pi_{i,j})$. Here q denotes the total number of encryption queries.
8. Challenger can ask for function keys for functions f_j and encryption keys EK_i as long as they satisfy the constraint with the message vectors.
9. $\mathcal{A}$ now outputs a guess $b' \in \{0, 1\}$.

Hard-wired: $PK_i^0, PK_i^1, K_i', \alpha_i, x_i^0, x_i^1$.
Input: $c_i^0, c_i^1, m, r_i^0, r_i^1$
The program does the following:

- Checks that $c_i^0 = \mathsf{PKE.Enc}(PK_i^0, m; r_i^0)$ and $c_i^1 = \mathsf{PKE.Enc}(PK_i^1, m; r_i^1)$. If the check fails output $\perp$.
- If $(c_i^0, c_i^1) = (x_i^0, x_i^1)$ output α_i otherwise output $\mathsf{PRF.Eval}(K_i', c_i^0, c_i^1)$

Fig. 6. Program Encrypt*

$\mathsf{H}_{x,2}$: This hybrid is similar to the previous one except that function key for any function f is generated as an obfuscation of program (Fig. 7) by hard-wiring $(SK_i^0, SK_i^1, K_i^{'}, P, P(\alpha_i), x_i^0, x_i^1) \forall i \in [n]$.

$\mathsf{H}_{x,3}$ This hybrid is similar to the previous hybrid except that for all $i \in [n]$, α_i is chosen randomly from the domain of the injective one way function P.

$\mathsf{H}_{x,4}$: This hybrid is similar to the previous hybrid except that the function key is generated as an obfuscation program (Fig. 7) initialised $x + 1$.

Hard-wired: $\{SK_i^0, SK_i^1, K_i^{'}, P, P(\alpha_i), x_i^0, x_i^1\}_{i \in [n]}$.
Input: $\{c_i^0, c_i^1, \pi_i\}_{i \in [n]}$
The program does the following:

- For any $i \in [n]$, if $(c_i^0, c_i^1) = (x_i^0, x_i^1)$ check that $P(\alpha_i) = P(\pi_i)$. If the check fails output $\perp$.
- Otherwise, for $i \in [n]$, check that $P(\mathsf{PRF.Eval}(K_i, c_i^0, c_i^1)) = P(\pi_i)$. If the check fails output $\perp$.
- If $(c_1^0, c_1^1, .., c_n^0, c_n^1) < x - 2$, output $f(\mathsf{PKE.Dec}(SK_1^1, c_1^1), .., \mathsf{PKE.Dec}(SK_n^1, c_n^1))$ otherwise output $f(\mathsf{PKE.Dec}(SK_1^0, c_1^0), .., \mathsf{PKE.Dec}(SK_n^0, c_n^0))$.

Fig. 7. Program $G^*_{f,x}$

$\mathsf{H}_{x,5}$: This hybrid is the same as the previous one except that $\alpha_i \forall i \in [n]$ is chosen as actual PRF values at (x_i^0, x_i^1) using the key K_i.

$\mathsf{H}_{x,6}$: This hybrid is the same as the previous one except that key for the function f, keys are generated as obfuscation of program (Fig. 4) initialised with $x + 1$.

$\mathsf{H}_{x,7}$: This hybrid is the same as the previous one except for all $i \in [n]$, $\tilde{PP}_i$ is generated as an obfuscation of (Fig. 2) initialised with genuine PRF key K_i. This hybrid is identical to the hybrid H_{x+1}.

Claim. For any PPT distinguisher D, $\mid Pr[D(\mathsf{H}_x) = 1] - Pr[D(\mathsf{H}_{x,1}) = 1] \mid < O(n \cdot 2^{-3nl-k})$.

Proof. This claim follows from the indistinguishability security of weak extractability obfuscator. We have that circuits for $i \in Z$, circuit in Fig. 2 initialised with regular PRF key K_i is functionally equivalent to when it is initialised with punctured key K_i'. This is because for $i \in Z$, (x_i^0, x_i^1) never satisfies the check and the PRF is never evaluated at this point and also the fact the punctured key outputs correctly at all points except the point at which the PRF is punctured. For $i \in [n] \backslash Z$, program in Fig. 2 initialised with K_i is functionally equivalent to the program in Fig. 6 initialised with (K_i', α_i).

From the above observation, we can prove the claim by at most n intermediate hybrids where we switch one by one obfuscation $\tilde{PP}_i$ to use the punctured key and each intermediate hybrid is indistinguishable due to the security of obfuscation.

Claim. For any PPT distinguisher D, $\mid Pr[D(\mathsf{H}_{x,1}) = 1] - Pr[D(\mathsf{H}_{x,2}) = 1] \mid < O(p(k) \cdot 2^{-3nl-k})$. Here, $p(k)$ is some polynomial.

Proof. This follows from the indistinguishability obfuscation property of the weak extractability obfuscator $\mathcal{O}$. The proof follows by at most p intermediate hybrids where each queried K_f is switched to an obfuscation of program

(Fig. 4) (with hard-wired values $SK_i^0, SK_i^1, K_i, x, P$) to an obfuscation of program (Fig. 7) (with hard-wired values $SK_i^0, SK_i^1, K_i', P, P(\alpha_i), x$). Note that in this hybrids, both these programs are functionally equivalent. This reduction is straight forward and we omit details.

Claim. For any PPT distinguisher D, $\mid Pr[D(\mathsf{H}_{x,2}) = 1] - Pr[D(\mathsf{H}_{x,3}) = 1] \mid < O(n \cdot 2^{-2nl-k})$.

Proof. This claim follows from the property that puncturable PRF's value is psuedo-random at punctured point given the punctured key (sub-exponential security of the puncturable PRF). This proof goes through by a sequence of at most n hybrids where for each index $i \in [n]$, $(K_i', \alpha_i = \mathsf{PRF.Eval}(K_i, x_i^0, x_i^1))$ is replaced with $(K_i', \alpha_i \leftarrow \mathcal{R})$ for all $i \in [n]$. This can be done because in both these hybrids, function keys and the encryption keys use only the punctured keys and a the value of the PRF at the punctured point. Here $\mathcal{R}$ is the co-domain of the PRF, which is equal to the domain of the injective one way function P. Since, PRF is sub exponentially secure with parameter c_{PRF} (c_{PRF} be the security constant of the PRF) when PRF is initialised with parameter greater than $(2nl+k)^{1/c_{PRF}}$, distinguishing advantage between each intermediate hybrid is bounded by $O(2^{-2nl-k})$. The reduction is straight forward and we omit the details.

Claim. For any PPT distinguisher D, $\mid Pr[D(\mathsf{H}_{x,3}) = 1] - Pr[D(\mathsf{H}_{x,4}) = 1] \mid < O(p(k).2^{-2nl-k})$. for some polynomial $p(k)$.

Proof. We prove this claim for a simplified case when only one function key is queried. The general case by considering a sequence of intermediate hybrids where function keys are changed one by one, hence the factor $p(k)$. Assume that there is a PPT algorithm D such that $\mid Pr[D(\mathsf{H}_{x,3}) = 1] - Pr[D(\mathsf{H}_{x,4}) = 1] \mid > \epsilon > 2^{-2nl-k}$. Note that these hybrids are identical upto the point the adversary asks for a key for a function f. We argue indistinguishability according to following cases.

1. **Case 0:** Circuit given in Fig. 7 initialised with x is functionally equivalent to circuit Fig. 7 initialised with $x + 1$.
2. **Case 1:** This is the case in which the two circuits described above are not equivalent.

Let Q denote the random variable and $Q = 0$ if adversary is in case 0, otherwise $Q = 1$. By $\epsilon_{Q=b}$ denote the value $\mid Pr[D(\mathsf{H}_{x,3}) = 1/Q = b] - Pr[D(\mathsf{H}_{x,4}) = 1/Q = b] \mid$. It is known that $Pr[Q = 0]\epsilon_{Q=0} + Pr[Q = 1]\epsilon_{Q=1} > \epsilon$.

Now we analyse both these cases:

$\mathbf{Pr[Q = 0]\epsilon_{Q=0} < 2^{-2nl-k}}$: This claim follows due to the indistinguishability security of $(1, 2^{-3nl-k})$ weak extractability obfuscator. Consider an adversary D with $Q = 0$ and challenger C, we construct an algorithm $\mathcal{A}$ that uses D and breaks the indistinguishability obfuscation of the weak extractability obfuscator with the same advantage. $\mathcal{A}$ works as follows: $\mathcal{A}$ invokes C that invokes D. C does the setup as in the hybrid and responds to the queries of D. D outputs f. $\mathcal{A}$ gives $G^*_{f,x}$ and $G^*_{f,x+1}$ to the obfuscation challenger and gets back K_f in return which is given to D. D's queries are now answered by C. $\mathcal{A}$ outputs whatever D outputs. $\mathcal{A}$ breaks the indistinguishability obfuscation security of the weak extractability obfuscator with advantage at least $\epsilon_{Q=0}$ as the view of D is identical to $\mathsf{H}_{x,3}$ if $G^*_{f,x}$ was obfuscated and it is identical to $\mathsf{H}_{x,4}$ otherwise.

$\mathbf{Pr[Q = 1]\epsilon_{Q=1} < 2^{-2nl-k}}$: The only point at which the two circuits $G^*_{f,x}$ and $G^*_{f,x+1}$ in this case may differ is $(x^0_1, x^1, \alpha_1, ..., x^0_n, x^n, \alpha_n)$ where α_i is the inverse of a fixed injective one way function value $P(\alpha_i)$. In this case, due to security of weak extractability obfuscator the claim holds. Assume to the contrary $Pr[Q = 1]\epsilon_{Q=1} > \delta > 2^{-2nl-k}$. In this case, let τ be the transcript (including the randomness to generate PKE keys, PRF keys along with chosen α_i's) between the challenger and the adversary till the point function key for function f is queried. We denote $\tau \in \mathsf{good}$ if conditioned on τ, $\epsilon_{\tau,Q=1} > \epsilon_{Q=1}/2$. Then, using Lemma 2, one can show that $Pr[\tau \in \mathsf{good}] > \epsilon_{Q=1}/2$.

Now, let us denote by set Z a set that contains indices in $i \in [n]$ such that $a^0_i \neq a^1_i$. Note that α_i can be requested by the adversary in one of the two following ways: $a^0_i = a^1_i$ and adversary queries for EK_i or adversary queries for an encryption of (a^0_i, a^1_i) and challenger sends encryption as (x^0_i, x^1_i, α_i) with some probability. Let E denote the set of indices for which α_i's queried by the adversary through first method and S denote the set queried through second method. Then it holds that $S \cup E \neq [n]$. This is because adversary cannot query for such cipher-texts and encryption keys in these hybrids since $Q = 1$ and in particular it holds that $f(< \{a^0_i\}_{i\in S}, \{a^0_i\}_{i\in E} >) \neq f(< \{a^1_i\}_{i\in S}, \{a^0_i\}_E >)$. Here $<,>$ denotes the permutation which sends a variable with subscript i to index i.

Now we let $T \subsetneq [n]$ denote the set of α_i for $i \in [n]$ requested by D (either by querying cipher-text or by querying for EK_i such that $a^0_i = a^1_i$). We know that conditioned on τ (randomness upto the point f is queried),

$$| Pr[D(\mathsf{H}_{x,3}) = 1/Q = 1, \tau] - Pr[D(\mathsf{H}_{x,4}) = 1/Q = 1, \tau] \,| > \epsilon_{Q=1}/2$$

For all $t \subsetneq Z$,

$$\Sigma_t \,| Pr[D(\mathsf{H}_{x,3}) = 1 \cap T = t/Q = 1, \tau] - Pr[D(\mathsf{H}_{x,4}) = 1 \cap T = t/Q = 1, \tau] \,| > \epsilon_{Q=1}/2$$

Since number of proper subsets of $[n]$ is bounded by 2^n, there exists a set t such that

$$| Pr[D(\mathsf{H}_{x,3}) = 1 \cap T = t/Q = 1, \tau] - Pr[D(\mathsf{H}_{x,4}) = 1 \cap T = t/Q = 1, \tau] \,| > \epsilon_{Q=1}/2^{n+1}$$

Now we construct an adversary $\mathcal{A}$ that breaks the security of injective one way function with probability $Pr[Q = 1]\epsilon_{Q=1}/2^{n+1}$ that runs in time $O(2^{2n}/\epsilon^2_{Q=1})$. $\mathcal{A}$ runs as follows:

1. $\mathcal{A}$ invokes D. Then it does setup and generates PKE keys and punctured PRF keys K_i' for all indices in $[n]$ according to hybrid $\mathsf{H}_{x,3}$.

2. $\mathcal{A}$ gets injective one way function values from the injective one way function challenger $(P, P(\alpha_1), .., P(\alpha_n))$.
3. $\mathcal{A}$ now guesses a random proper subset $t \subset [n]$.
4. For all indices in $i \in t$ it gets α_i from the injective one way function challenger.
5. If EK_i is asked for any $i \in t \cup Z$, it is generated as in $\mathsf{H}_{x,3}$ and given out. Otherwise, $\mathcal{A}$ aborts. We call the transcript till here τ.
6. When D asks for a key for f. If f is such that $Q = 0$, $\mathcal{A}$ outputs $\perp$. $\mathcal{A}$ now constructs a distinguisher $\mathcal{B}$ of obfuscation of circuits $G^*_{f,x}$ and $G^*_{f,x+1}$ as follows:
   - $\mathcal{A}$ gets as a challenge obfuscation $\tilde{C}_f$ which is an obfuscation $G^*_{f,x}$ or $G^*_{f,x+1}$.
   - $\mathcal{A}$ gives this obfuscation to $\mathcal{B}$ which invokes D from the point of the transcript τ and gives this obfuscation to D.
   - When D asks for a cipher-text, if the queries are such that $\mathcal{B}$ can generate it using $\alpha_i \forall i \in t$ then answer the cipher-text query. Otherwise, it outputs 0.
   - If EK_i is asked by D for any $i \in t \cup Z$, it is generated as in $\mathsf{H}_{x,3}$ and given out. If any other encryption key is queried, it outputs 0.
   - If set of indices for which α_i's used to generate response to the queries (in the transcript τ and the queries asked by D when run by $\mathcal{B}$) equals t it outputs whatever D outputs otherwise, $\mathcal{B}$ outputs 0.
7. If t is correctly guessed as t^*, it is easy to check that $| \ Pr[\mathcal{B}(G^*_{f,x}, G^*_{f,x+1}, \mathcal{O}(G^*_{f,x}), aux) = 1] - Pr[\mathcal{B}(G^*_{f,x}, G^*_{f,x+1}, \mathcal{O}(G^*_{f,x+1}), aux) = 1] \ | > \epsilon_{Q=1}/2^{n+1}$. (Here aux is the information with $\mathcal{A}$ required to run $\mathcal{B}$ including $\alpha_i \forall i \in t$, $P(\alpha_i), PK^0_i, PK^1_i, SK^0_i, SK^1_i, K'_i \forall i \in [n]$ and transcript τ till point 4). This is because,

$$| \ Pr[\mathcal{B}(G^*_{f,x}, G^*_{f,x+1}, \mathcal{O}(G^*_{f,x}), aux) = 1] - Pr[\mathcal{B}(G^*_{f,x}, G^*_{f,x+1}, \mathcal{O}(G^*_{f,x+1}), aux) = 1] \ | =$$

$$| \ Pr[D(\mathsf{H}_{x,3}) = 1 \cap T = t/Q = 1, \tau] - Pr[D(\mathsf{H}_{x,4}) = 1 \cap T = t/Q = 1, \tau] \ | > \epsilon_{Q=1}/2^{n+1}$$

8. We finally run the extractor E of the weak extractability obfuscator using $\mathcal{B}$ to extract a point $(x^0_1, x^1_1, \alpha_1, .., x^0_n, x^1_n, \alpha_n)$. (This extraction can be run as long as $\epsilon_{Q=1}/2^{n+1} > 2^{-3nl}$ implying $\epsilon_{Q=1} > 2^{-2nl-k}$ as otherwise there is nothing to prove and claim trivially goes through). This extractor runs in time $O(t_D.2^{2n}/\epsilon^2_{Q=1})$. Probability of success of this extraction is

$$Pr[Q = 1] \cdot Pr[\tau \text{ is good}] \cdot Pr[\text{ t is guessed correctly}] > Pr[Q = 1] \cdot \epsilon_{Q=1}/2^{n+1}$$

Let μ be the input length for injective one way function. We note the following cases:

Case 0: If $Pr[Q = 1]\epsilon_{Q=1} < O(2^{-2nl-k})$, in this case the claim goes through.

Case 1: If $Pr[Q = 1]\epsilon_{Q=1}/2^{n+1} < O(2^{-\mu^{c_{owp2}}})$, in this case the claim goes through if μ is set to be greater than $(3nl + k)^{1/c_{owp2}}$.

Case 2: If case 1 does not occur, then we must have that $2^{2n}/\epsilon^2_{Q=1} > 2^{\mu^{c_{owp1}}}$, implying that if μ is greater than $(5nl + 2k)^{1/c_{owp1}}$ the claim holds (due to the security of injective one way function P).

Hence, if $\mu > max\{(3nl + k)^{1/c_{owp2}}, (5nl + 2k)^{1/c_{owp1}}\}$, $Pr[Q = 1]\epsilon_{Q=1} < 2^{-2nl-k}$ and the claim holds.

Claim. For any PPT distinguisher D, $\mid Pr[D(\mathsf{H}_{x,4}) = 1] - Pr[D(\mathsf{H}_{x,5}) = 1] \mid < O(n \cdot 2^{-2nl-k})$.

Proof. This claim follows from the security of the puncturable PRF's. This is similar to the proof of the Claim 4.2.

Claim. For any PPT distinguisher D, $\mid Pr[D(\mathsf{H}_{x,5}) = 1] - Pr[D(\mathsf{H}_{x,6}) = 1] \mid < O(p(k) \cdot 2^{-2nl-k})$. Here $p(\cdot)$ is a some polynomial.

Proof. This claim follows from the indistinguishability obfuscation security of the weak extractability obfuscator. This proof is similar the proof of the Claim 4.2.

Claim. For any PPT distinguisher D, $\mid Pr[D(\mathsf{H}_{x,6}) = 1] - Pr[D(\mathsf{H}_{x,7}) = 1] \mid < O(n \cdot 2^{-2nl-k})$.

Proof. This claim follows from the indistinguishability obfuscation security of the weak extractability obfuscator $\mathcal{O}$. This proof is similar the proof of the Claim 4.2.

Combining all the claims above, we prove the lemma.

Lemma 10. *For any PPT distinguisher* D *and* $x \in [2^{2ln}]$, $\mid Pr[D(\mathsf{H}_{2^{2ln}+4+x}) = 1] - Pr[D(\mathsf{H}_{2^{2ln}+5+x}) = 1] \mid < O(v(k) \cdot 2^{-2nl-k})$ *for some polynomial* $v(k)$.

Proof. Proof of this lemma is similar to the proof of Lemma 9.
Combining all these lemmas above, we get that for any PPT D,

$$\mid Pr[D(\mathsf{H}_0) = 1] - Pr[D(\mathsf{H}_{2.2^{2ln}+6}) = 1] \mid < \mathsf{negl}(k) + 2.2^{2nl} O(v(k) \cdot 2^{-2nl-k}) < \mathsf{negl}(k).$$

References

1. Ananth, P., Boneh, D., Garg, S., Sahai, A., Zhandry, M.: Differing-inputs obfuscation and applications. IACR Cryptology ePrint Archive 2013, p. 689 (2013). http://eprint.iacr.org/2013/689
2. Ananth, P., Brakerski, Z., Segev, G., Vaikuntanathan, V.: The trojan method in functional encryption: From selective to adaptive security, generically. IACR Cryptology ePrint Archive 2014, p. 917 (2014). http://eprint.iacr.org/2014/917
3. Ananth, P., Jain, A., Naor, M., Sahai, A., Yogev, E.: Universal obfuscation and witness encryption: Boosting correctness and combining security. IACR Cryptology ePrint Archive 2016, p. 281 (2016). http://eprint.iacr.org/2016/281
4. Ananth, P., Jain, A.: Indistinguishability obfuscation from compact functional encryption. In: Gennaro, R., Robshaw, M. (eds.) CRYPTO 2015. LNCS, vol. 9215, pp. 308–326. Springer, Heidelberg (2015). doi:10.1007/978-3-662-47989-6_15
5. Badrinarayanan, S., Gupta, D., Jain, A., Sahai, A.: Multi-input functional encryption for unbounded arity functions. In: Iwata, T., Cheon, J.H. (eds.) ASIACRYPT 2015. LNCS, vol. 9452, pp. 27–51. Springer, Heidelberg (2015). doi:10.1007/978-3-662-48797-6_2

6. Barak, B., Goldreich, O., Impagliazzo, R., Rudich, S., Sahai, A., Vadhan, S., Yang, K.: On the (Im)possibility of obfuscating programs. In: Kilian, J. (ed.) CRYPTO 2001. LNCS, vol. 2139, pp. 1–18. Springer, Heidelberg (2001). doi:10.1007/3-540-44647-8_1
7. Beimel, A., Gabizon, A., Ishai, Y., Kushilevitz, E., Meldgaard, S., Paskin-Cherniavsky, A.: Non-interactive secure multiparty computation. IACR Cryptology ePrint Archive 2014, p. 960 (2014). http://eprint.iacr.org/2014/960
8. Bitansky, N., Paneth, O., Rosen, A.: On the cryptographic hardness of finding a nash equilibrium. In: Electronic Colloquium on Computational Complexity (ECCC) vol. 22, p. 1 (2015). http://eccc.hpi-web.de/report/2015/001
9. Bitansky, N., Vaikunthanathan, V.: Indistinguishability obfuscation from functional encryption. IACR Cryptology ePrint Archive 2013 (2015). http://eprint.iacr.org/2015/163
10. Boneh, D., Sahai, A., Waters, B.: Functional encryption: definitions and challenges. In: Ishai, Y. (ed.) TCC 2011. LNCS, vol. 6597, pp. 253–273. Springer, Heidelberg (2011). doi:10.1007/978-3-642-19571-6_16
11. Boneh, D., Waters, B.: Constrained pseudorandom functions and their applications. IACR Cryptology ePrint Archive 2013, p. 352 (2013). http://eprint.iacr.org/2013/352
12. Boyle, E., Chung, K.-M., Pass, R.: On extractability obfuscation. In: Lindell, Y. (ed.) TCC 2014. LNCS, vol. 8349, pp. 52–73. Springer, Heidelberg (2014). doi:10.1007/978-3-642-54242-8_3
13. Boyle, E., Goldwasser, S., Ivan, I.: Functional signatures and pseudorandom functions. IACR Cryptology ePrint Archive 2013, p. 401 (2013). http://eprint.iacr.org/2013/401
14. Brakerski, Z., Komargodski, I., Segev, G.: From single-input to multi-input functional encryption in the private-key setting. IACR Cryptology ePrint Archive 2015, p. 158 (2015). http://eprint.iacr.org/2015/158
15. Canetti, R., Lin, H., Tessaro, S., Vaikuntanathan, V.: Obfuscation of probabilistic circuits and applications. In: Dodis, Y., Nielsen, J.B. (eds.) TCC 2015. LNCS, vol. 9015, pp. 468–497. Springer, Heidelberg (2015). doi:10.1007/978-3-662-46497-7_19
16. Caro, A., Iovino, V., Jain, A., O'Neill, A., Paneth, O., Persiano, G.: On the achievability of simulation-based security for functional encryption. In: Canetti, R., Garay, J.A. (eds.) CRYPTO 2013. LNCS, vol. 8043, pp. 519–535. Springer, Heidelberg (2013). doi:10.1007/978-3-642-40084-1_29
17. Chandran, N., Goyal, V., Jain, A., Sahai, A.: Functional encryption: decentralised and delegatable. IACR Cryptology ePrint Archive (2015)
18. Dodis, Y., Nielsen, J.B. (eds.): TCC 2015. LNCS, vol. 9015. Springer, Heidelberg (2015). doi:10.1007/978-3-662-46497-7
19. Fischlin, M., Herzberg, A., Noon, H.B., Shulman, H.: Obfuscation combiners. IACR Cryptology ePrint Archive 2016, p. 289 (2016). http://eprint.iacr.org/2016/289
20. Garg, S., Gentry, C., Halevi, S.: Candidate multilinear maps from ideal lattices. In: Johansson, T., Nguyen, P.Q. (eds.) EUROCRYPT 2013. LNCS, vol. 7881, pp. 1–17. Springer, Heidelberg (2013). doi:10.1007/978-3-642-38348-9_1
21. Garg, S., Gentry, C., Halevi, S., Raykova, M., Sahai, A., Waters, B.: Candidate indistinguishability obfuscation and functional encryption for all circuits. In: 54th Annual IEEE Symposium on Foundations of Computer Science (FOCS 2013), 26–29 , Berkeley, CA, USA. pp. 40–49. IEEE Computer Society (2013). http://dx.doi.org/10.1109/FOCS.2013.13

22. Garg, S., Gentry, C., Halevi, S., Wichs, D.: On the implausibility of differing-inputs obfuscation and extractable witness encryption with auxiliary input. In: Garay, J.A., Gennaro, R. (eds.) CRYPTO 2014. LNCS, vol. 8616, pp. 518–535. Springer, Heidelberg (2014). doi:10.1007/978-3-662-44371-2_29
23. Garg, S., Mukherjee, P., Srinivasan, A.: Obfuscation without the vulnerabilities of multilinear maps. IACR Cryptology ePrint Archive 2016, p. 390 (2016). http://eprint.iacr.org/2016/390
24. Gennaro, R., Gertner, Y., Katz, J., Trevisan, L.: Bounds on the efficiency of generic cryptographic constructions. SIAM J. Comput. **35**(1), 217–246 (2005). http://dx.doi.org/10.1137/S0097539704443276
25. Goldwasser, S., et al.: Multi-input functional encryption. In: Nguyen, P.Q., Oswald, E. (eds.) EUROCRYPT 2014. LNCS, vol. 8441, pp. 578–602. Springer, Heidelberg (2014). doi:10.1007/978-3-642-55220-5_32
26. Goldwasser, S., Kalai, Y.T., Popa, R.A., Vaikuntanathan, V., Zeldovich, N.: Reusable garbled circuits and succinct functional encryption. In: Boneh, D., Roughgarden, T., Feigenbaum, J. (eds.) Symposium on Theory of Computing Conference (STOC 2013), Palo Alto, CA, USA, June 1–4, pp. 555–564. ACM (2013). http://doi.acm.org/10.1145/2488608.2488678
27. Holenstein, T.: Pseudorandom generators from one-way functions: a simple construction for any hardness. In: Halevi, S., Rabin, T. (eds.) TCC 2006. LNCS, vol. 3876, pp. 443–461. Springer, Heidelberg (2006). doi:10.1007/11681878_23
28. Ishai, Y., Pandey, O., Sahai, A.: Public-coin differing-inputs obfuscation and its applications. In: Dodis, Y., Nielsen, J.B. (eds.) TCC 2015. LNCS, vol. 9015, pp. 668–697. Springer, Heidelberg (2015). doi:10.1007/978-3-662-46497-7_26
29. Kiayias, A., Papadopoulos, S., Triandopoulos, N., Zacharias, T.: Delegatable pseudorandom functions and applications. In: Sadeghi, A., Gligor, V.D., Yung, M. (eds.) 2013 ACM SIGSAC Conference on Computer and Communications Security (CCS 2013), Berlin, Germany, November 4–8, pp. 669–684. ACM (2013). http://doi.acm.org/10.1145/2508859.2516668
30. Koppula, V., Lewko, A.B., Waters, B.: Indistinguishability obfuscation for turing machines with unbounded memory. In: Proceedings of the Forty-Seventh Annual ACM on Symposium on Theory of Computing (STOC 2015), Portland, OR, USA, June 14–17, pp. 419–428 (2015). http://doi.acm.org/10.1145/2746539.2746614
31. Miles, E., Sahai, A., Zhandry, M.: Annihilation attacks for multilinear maps: cryptanalysis of indistinguishability obfuscation over GGH13. In: Robshaw, M., Katz, J. (eds.) CRYPTO 2016. LNCS, vol. 9815, pp. 629–658. Springer, Heidelberg (2016). doi:10.1007/978-3-662-53008-5_22
32. Sahai, A., Waters, B.: Fuzzy identity-based encryption. In: Cramer, R. (ed.) EUROCRYPT 2005. LNCS, vol. 3494, pp. 457–473. Springer, Heidelberg (2005). doi:10.1007/11426639_27
33. Waters, B.: A punctured programming approach to adaptively secure functional encryption. IACR Cryptology ePrint Archive 2014, p. 588 (2014). http://eprint.iacr.org/2014/588
34. Wee, H.: On obfuscating point functions. In: Gabow, H.N., Fagin, R. (eds.) Proceedings of the 37th Annual ACM Symposium on Theory of Computing, Baltimore, MD, USA, May 22–24, pp. 523–532. ACM (2005). http://doi.acm.org/10.1145/1060590.1060669
35. Wee, H.: One-way permutations, interactive hashing and statistically hiding commitments. In: Vadhan, S.P. (ed.) TCC 2007. LNCS, vol. 4392, pp. 419–433. Springer, Heidelberg (2007). doi:10.1007/978-3-540-70936-7_23

Verifiable Functional Encryption

Saikrishna Badrinarayanan[1(✉)], Vipul Goyal[2], Aayush Jain[1], and Amit Sahai[1]

[1] Center for Encrypted Functionalities,
University of California, Los Angeles, USA
{saikrishna,sahai}@cs.ucla.edu, aayushjainiitd@gmail.com
[2] Microsoft Research, Bengaluru, India
vipul@microsoft.com

Abstract. In light of security challenges that have emerged in a world with complex networks and cloud computing, the notion of functional encryption has recently emerged. In this work, we show that in several applications of functional encryption (even those cited in the earliest works on functional encryption), the formal notion of functional encryption is actually *not* sufficient to guarantee security. This is essentially because the case of a malicious authority and/or encryptor is not considered. To address this concern, we put forth the concept of *verifiable functional encryption*, which captures the basic requirement of output correctness: even if the ciphertext is maliciously generated (and even if the setup and key generation is malicious), the decryptor is still guaranteed a meaningful notion of correctness which we show is crucial in several applications.

We formalize the notion of verifiable function encryption and, following prior work in the area, put forth a simulation-based and an indistinguishability-based notion of security. We show that simulation-based verifiable functional encryption is unconditionally impossible even in the most basic setting where there may only be a single key and a single ciphertext. We then give general positive results for the indistinguishability setting: a general compiler from any functional encryption scheme into a verifiable functional encryption scheme with the only additional assumption being the Decision Linear Assumption over Bilinear Groups (DLIN). We also give a generic compiler in the secret-key setting for functional encryption which maintains both message privacy and function privacy. Our positive results are general and also apply to other simpler settings such as Identity-Based Encryption, Attribute-Based Encryption and Predicate Encryption. We also give an application of verifiable functional encryption to the recently introduced primitive

A. Sahai—Research supported in part from a DARPA/ARL SAFEWARE award, NSF Frontier Award 1413955, NSF grants 1228984, 1136174, and 1065276, a Xerox Faculty Research Award, a Google Faculty Research Award, an equipment grant from Intel, and an Okawa Foundation Research Grant. This material is based upon work supported by the Defense Advanced Research Projects Agency through the ARL under Contract W911NF-15-C-0205. The views expressed are those of the author and do not reflect the official policy or position of the Department of Defense, the National Science Foundation, or the U.S. Government.

J.H. Cheon and T. Takagi (Eds.): ASIACRYPT 2016, Part II, LNCS 10032, pp. 557–587, 2016.
DOI: 10.1007/978-3-662-53890-6_19

of functional commitments. Finally, in the context of indistinguishability obfuscation, there is a fundamental question of whether the correct program was obfuscated. In particular, the recipient of the obfuscated program needs a guarantee that the program indeed does what it was intended to do. This question turns out to be closely related to verifiable functional encryption. We initiate the study of verifiable obfuscation with a formal definition and construction of verifiable indistinguishability obfuscation.

1 Introduction

Encryption has traditionally been seen as a way to ensure confidentiality of a communication channel between a unique sender and a unique receiver. However, with the emergence of complex networks and cloud computing, recently the cryptographic community has been rethinking the notion of encryption to address security concerns that arise in these more complex environments.

In particular, the notion of functional encryption (FE) was introduced [29,30], with the first comprehensive formalizations of FE given in [13,26]. In FE, there is an authority that sets up public parameters and a master secret key. Encryption of a value x can be performed by any party that has the public parameters and x. Crucially, however, the master secret key can be used to generate limited "function keys." More precisely, for a given allowable function f, using the master secret key, it is possible to generate a function key SK_f. Applying this function key to an encryption of x yields only $f(x)$. In particular, an adversarial entity that holds an encryption of x and SK_f learns nothing more about x than what is learned by obtaining $f(x)$. It is not difficult to imagine how useful such a notion could be – the function f could enforce access control policies, or more generally only allow highly processed forms of data to be learned by the function key holder.

Our work: The case of dishonest authority and encryptor. However, either implicitly or explicitly, almost[1] all known prior work on FE has not considered the case where either the authority or the encryptor, or both, could be dishonest. This makes sense historically, since for traditional encryption, for example, there usually isn't a whole lot to be concerned about if the receiver that chooses the public/secret key pair is herself dishonest. However, as we now illustrate with examples, there are simple and serious concerns that arise for FE usage scenarios when the case of a dishonest authority and encryptor is considered:

- **Storing encrypted images:** Let us start with a motivating example for FE given in the paper of Boneh, Sahai, and Waters [13] that initiated the systematic study of FE. Suppose that there is a cloud service on which customers

[1] One of the few counter-examples to this that we are aware of is the following works [19,20,28] on Accountable Authority IBE that dealt with the very different problem of preventing a malicious authority that tries to sell decryption boxes.

store encrypted images. Law enforcement may require the cloud to search for images containing a particular face. Thus, customers would be required to provide to the cloud a restrictive decryption key which allows the cloud to decrypt images containing the target face (but nothing else). Boneh et al. argued that one could use functional encryption in such a setting to provide these restricted decryption keys.

However, we observe that if we use functional encryption, then law enforcement inherently has to trust the customer to be honest, because the customer is acting as both the authority and the encryptor in this scenario. In particular, suppose that a malicious authority could create malformed ciphertexts and "fake" decryption keys that in fact do not provide the functionality guarantees required by law enforcement. Then, for example, law enforcement could be made to believe that there are no matching images, when in fact there might be several matching images.

A similar argument holds if the cloud is storing encrypted text or emails (and law enforcement would like to search for the presence of certain keywords or patterns).

- **Audits:** Next, we consider an even older example proposed in the pioneering work of Goyal, Pandey, Sahai, and Waters [21] to motivate Attribute-Based Encryption, a special case of FE. Suppose there is a bank that maintains large encrypted databases of the transactions in each of its branches. An auditor is required to perform a financial audit to certify compliance with various financial regulations such as Sarbanes-Oxley. For this, the auditor would need access to certain types of data (such as logs of certain transactions) stored on the bank servers. However the bank does not wish to give the auditors access to the entire data (which would leak customer personal information, etc.). A natural solution is to have the bank use functional encryption. This would enable it to release a key to the auditor which selectively gives him access to only the required data.

  However, note that the entire purpose of an audit is to provide assurances even in the setting where the entity being audited is not trusted. What if either the system setup, or the encryption, or the decryption key generation is maliciously done? Again, with the standard notions of FE, all bets are off, since these scenarios are simply not considered.

Surprisingly, to the best of our knowledge, this (very basic) requirement of adversarial correctness *has not been previously captured* in the standard definitions of functional encryption. Indeed, it appears that many previous works overlooked this correctness requirement while envisioning applications of (different types of) functional encryption. The same issue also arises in the context of simpler notions of functional encryption such as identity based encryption (IBE), attribute based encryption (ABE), and predicate encryption (PE), which have been studied extensively [11,17,18,21,23,29,32].

In order to solve this problem, we define the notion of *Verifiable Functional Encryption*[2] (VFE). Informally speaking, in a VFE scheme, regardless of how the system setup is done, for each (possibly maliciously generated) ciphertext C that passes a publicly known verification procedure, there must exist a unique message m such that: for any allowed function description f and function key SK_f that pass another publicly known verification procedure, it must be that the decryption algorithm given C, SK_f, and f is guaranteed to output $f(m)$. In particular, this also implies that if two decryptions corresponding to functions f_1 and f_2 of the same ciphertext yield y_1 and y_2 respectively, then there must exist a single message m such that $y_1 = f_1(m)$ and $y_2 = f_2(m)$.

We stress that even the public parameter generation algorithm can be corrupted. As illustrated above, this is critical for security in many applications. The fact that the public parameters are corrupted means that we cannot rely on the public parameters to contain an honestly generated Common Random String or Common Reference String (CRS). This presents the main technical challenge in our work, as we describe further below.

1.1 Our Contributions for Verifiable Functional Encryption

Our work makes the following contributions with regard to VFE:

- We formally define verifiable functional encryption and study both indistinguishability and simulation-based security notions. Our definitions can adapt to all major variants and predecessors of FE, including IBE, ABE, and predicate encryption.
- We show that simulation based security is unconditionally impossible to achieve by constructing a one-message zero knowledge proof system from any simulation secure verifiable functional encryption scheme. Interestingly, we show the impossibility holds even in the most basic setting where there may only be a single key and a single ciphertext that is queried by the adversary (in contrast to ordinary functional encryption where we know of general positive results in such a setting from minimal assumptions [27]). Thus, in the rest of our work, we focus on the indistinguishability-based security notion.
- We give a generic compiler from any public-key functional encryption scheme to a verifiable public-key functional encryption scheme, with the only additional assumption being Decision Linear Assumption over Bilinear Groups (DLIN). Informally, we show the following theorem.

Theorem 1. *(Informal) Assuming there exists a secure public key functional encryption scheme for the class of functions $\mathcal{F}$ and DLIN is true, there exists an explicit construction of a secure verifiable functional encryption scheme for the class of functions $\mathcal{F}$.*

[2] A primitive with the same name was also defined in [8]. However, their setting is entirely different to ours. They consider a scenario where the authority as well as the encryptor are honest. Their goal is to convince a weak client that the decryption (performed by a potentially malicious cloud service provider) was done correctly using the actual ciphertext and function secret key.

Table 1. Our Results for Verifiable FE

Verifiable Functionality	Assumptions Needed
Verifiable IBE	BDH+Random Oracle [11]
Verifiable IBE	BDH+DLIN [32]
Verifiable ABE for NC^1	DLIN [25,31]
Verifiable ABE for all Circuits	LWE + DLIN [12,17]
Verifiable PE for all Circuits	LWE + DLIN [18]
Verifiable FE for Inner Product Equality	DLIN [25,31]
Verifiable FE for Inner Product	DLIN [1]
Verifiable FE for Bounded Collusions	DLIN [16,27]
Verifiable FE for Bounded Collusions	LWE + DLIN [15]
Verifiable FE for all Circuits	iO + Injective OWF [14]

IBE stands for identity-based encryption, ABE for attribute-based encryption and PE for predicate encryption. The citation given in the assumption column shows a relevant paper that builds ordinary FE without verifiability for the stated function class.

In the above, the DLIN assumption is used only to construct non-interactive witness indistinguishable (NIWI) proof systems. We show that NIWIs are necessary by giving an explicit construction of a NIWI from any verifiable functional encryption scheme. This compiler gives rise to various verifiable functional encryption schemes under different assumptions. Some of them have been summarized in Table 1.

- We next give a generic compiler for the *secret-key* setting. Namely, we convert from any secret-key functional encryption scheme to a verifiable secret-key functional encryption scheme with the only additional assumption being DLIN. Informally, we show the following theorem:

Theorem 2. *(Informal) Assuming there exists a message hiding and function hiding secret-key functional encryption scheme for the class of functions $\mathcal{F}$ and DLIN is true, there exists an explicit construction of a message hiding and function hiding verifiable secret-key functional encryption scheme for the class of functions $\mathcal{F}$.*

An Application: Non-Interactive Functional Commitments: In a traditional non-interactive commitment scheme, a committer commits to a message m which is revealed entirely in the decommitment phase. Analogous to the evolution of functional encryption from traditional encryption, we consider the notion of *functional commitments* which were recently studied in [24] as a natural generalization of non-interactive commitments. In a functional commitment scheme, a committer commits to a message m using some randomness r. In the decommitment phase, instead of revealing the entire message m, for any function f agreed upon by both parties, the committer outputs a pair of values (a, b) such that using b and the commitment, the receiver can verify that $a = f(m)$ where m was the committed value. Similar to a traditional commitment scheme, we require the properties of hiding and binding. Roughly, hiding states that for any pair of messages (m_0, m_1), a commitment of m_0 is indistinguishable to a commitment of m_1 if $f(m_0) = f(m_1)$ where f is the agreed upon function. Informally, binding states that for every commitment c, there is a unique message m committed inside c.

We show that any verifiable functional encryption scheme directly gives rise to a non-interactive functional commitment scheme with no further assumptions.

Verifiable iO: As shown recently [3,4,10], functional encryption for general functions is closely tied to indistinguishability obfuscation [6,14]. In obfuscation, aside from the security of the obfuscated program, there is a fundamental question of whether the *correct* program was obfuscated. In particular, the recipient of the obfuscated program needs a guarantee that the program indeed does what it was intended to do.

Indeed, if someone hands you an obfuscated program, and asks you to run it, your first response might be to run away. After all, you have no idea what the obfuscated program does. Perhaps it contains backdoors or performs other problematic behavior. In general, before running an obfuscated program, it makes sense for the recipient to wait to be convinced that the program behaves in an appropriate way. More specifically, the recipient would want an assurance that only certain specific secrets are kept hidden inside it, and that it uses these secrets only in certain well-defined ways.

In traditional constructions of obfuscation, the obfuscator is assumed to be honest and no correctness guarantees are given to an honest evaluator if the obfuscator is dishonest. To solve this issue, we initiate a formal study of verifiability in the context of indistinguishability obfuscation, and show how to convert any iO scheme into a usefully verifiable iO scheme.

We note that verifiable iO presents some nontrivial modeling choices. For instance, of course, it would be meaningless if a verifiable iO scheme proves that a specific circuit C is being obfuscated – the obfuscation is supposed to hide exactly which circuit is being obfuscated. At the same time, of course every obfuscated program does correspond to some Boolean circuit, and so merely proving that there exists a circuit underlying an obfuscated program would be trivial. To resolve this modeling, we introduce a public predicate P, and our definition will require that there is a public verification procedure that takes both P and any maliciously generated obfuscated circuit $\tilde{C}$ as input. If this verification procedure is satisfied, then we know that there exists a circuit C equivalent to $\tilde{C}$ such that $P(C) = 1$. In particular, P could reveal almost everything about C, and only leave certain specific secrets hidden. (We also note that our VFE schemes can also be modified to also allow for such public predicates to be incorporated there, as well.)

iO requires that given a pair (C_0, C_1) of equivalent circuits, the obfuscation of C_0 should be indistinguishable from the obfuscation of C_1. However, in our construction, we must restrict ourselves to pairs of circuits where this equivalence can be proven with a short witness. In other words, there should be an NP language L such that $(C_0, C_1) \in L$ implies that C_0 is equivalent to C_1. We leave removing this restriction as an important open problem. However, we note that, to the best of our knowledge, all known applications of iO in fact only consider pairs of circuits where proving equivalence is in fact easy given a short witness[3].

1.2 Technical Overview

At first glance, constructing verifiable functional encryption may seem easy. One naive approach would be to just compile any functional encryption (FE) system with NIZKs to achieve verifiability. However, note that this doesn't work, since if the system setup is maliciously generated, then the CRS for the NIZK would also be maliciously generated, and therefore soundness would not be guaranteed to hold.

Thus, the starting point of our work is to use a relaxation of NIZK proofs called non-interactive witness indistinguishable proof (NIWI) systems, that do guarantee soundness even without a CRS. However, NIWIs only guarantee witness indistinguishability, not zero-knowledge. In particular, if there is only one valid witness, then NIWIs do not promise any security at all. When using NIWIs, therefore, it is typically necessary to engineer the possibility of multiple witnesses.

A failed first attempt and the mismatch problem: Two parallel FE schemes. A natural initial idea would be to execute two FE systems in parallel and prove using a NIWI that at least one of them is fully correct: that is, its setup was generated correctly, the constituent ciphertext generated using this system was computed correctly and the constituent function secret key generated using this system was computed correctly. Note that the NIWI computed for proving correctness of the ciphertext will have to be separately generated from the NIWI computed for proving correctness of the function secret key.

This yields the *mismatch problem*: It is possible that in one of the FE systems, the ciphertext is maliciously generated, while in the other, the function secret key is! Then, during decryption, if either the function secret key or the ciphertext is malicious, all bets are off. In fact, several known FE systems [14,15] specifically provide for programming either the ciphertext or the function secret key to force a particular output during decryption.

Could we avoid the mismatch problem by relying on majority-based decoding? In particular, suppose we have three parallel FE systems instead of two. Here, we run into the following problem: If we prove that at least two of the three ciphertexts are honestly encrypting *the same* message, the NIWI may not hide this message at all: informally speaking, the witness structure has too few

[3] For instance, suppose that C_0 uses an ordinary GGM PRF key, but C_1 uses a punctured GGM PRF key. It is easy to verify that these two keys are equivalent by simply verifying each node in the punctured PRF tree of keys by repeated application of the underlying PRG.

"moving parts", and it is not known how to leverage NIWIs to argue indistinguishability. On the other hand, if we try to relax the NIWI and prove only that at least two of the three ciphertexts are honestly encrypting *some* (possibly different) message, each ciphertext can no longer be associated with a unique message, and the mismatch problem returns, destroying verifiability.

Let's take a look at this observation a bit more in detail in the context of functional commitments, which is perhaps a simpler primitive. Consider a scheme where the honest committer commits to the same message m thrice using a non-interactive commitment scheme. Let $\mathsf{Z}_1, \mathsf{Z}_2, \mathsf{Z}_3$ be these commitments. Note that in the case of a malicious committer, the messages being committed m_0, m_1, m_2, may all be potentially different. In the decommitment phase, the committer outputs a and a NIWI proving that two out of the three committed values (say m_i and m_j) are such that $a = f(m_i) = f(m_j)$. With such a NIWI, it is possible to give a hybrid argument that proves the hiding property (which corresponds to indistinguishability in the FE setting). However, binding (which corresponds to verifiability) is lost: One can maliciously commit to m_0, m_1, m_2 such that they satisfy the following property: there exists functions f, g, h for which it holds that $f(m_0) = f(m_1) \neq f(m_2)$, $g(m_0) \neq g(m_1) = g(m_2)$ and $h(m_0) = h(m_2) \neq h(m_1)$. Now, if the malicious committer runs the decommitment phase for these functions separately, there is no fixed message bound by the commitment.

As mentioned earlier, one could also consider a scheme where in the decommitment phase, the committer outputs $f(m)$ and a NIWI proving that two out of the three commitments correspond to the same message m (i.e. there exists i, j such that $m_i = m_j$) and $f(m_i) = a$. The scheme is binding but does not satisfy hiding any more. This is because there is no way to move from a hybrid where all three commitments correspond to message m_0^* to one where all three commitments correspond to message m_1^*, since at every step of the hybrid argument, two messages out of three must be equal.

This brings out the reason why verifiability and security are two conflicting requirements. Verifiability seems to demand a majority of some particular message in the constituent ciphertexts whereas in the security proof, we have to move from a hybrid where the majority changes (from that of m_0^* to that of m_1^*). Continuing this way it is perhaps not that hard to observe that having any number of systems will not solve the problem. Hence, we have to develop some new techniques to solve the problem motivated above. This is what we describe next.

Our solution: Locked trapdoors. Let us start with a scheme with five parallel FE schemes. Our initial idea will be to *commit to the challenge constituent ciphertexts* as part of the public parameters, but we will need to introduce a twist to make this work, that we will mention shortly. Before we get to the twist, let's first see why having a commitment to the challenge ciphertext doesn't immediately solve the problem. Let's introduce a trapdoor statement for the relation used by the NIWI corresponding to the VFE ciphertexts. This trapdoor statement states that two of the constituent ciphertexts are encryptions of the same message and all the constituent ciphertexts are committed in the public parameters. Initially, the NIWI in the challenge ciphertext uses the fact that the trapdoor statement is correct with the indices 1 and 2 encrypting the same message m_0^*. The NIWIs in the function secret keys use the fact that the first four indices are secret keys for the same function. Therefore, this leaves the fifth index free (not being part of the NIWI in any function secret key or challenge ciphertext) and we can switch the fifth constituent challenge ciphertext to be an encryption of m_1^*. We can switch the indices used in the NIWI for the function secret keys (one at a time) appropriately to leave some other index free and transform the challenge ciphertext to encrypt m_0^* in the first two indices and m_1^* in the last three. We then switch the proof in the challenge ciphertext to use the fact that the last two indices encrypt the same message m_1^*. After this, in the same manner as above, we can switch the first two indices (one by one) of the challenge ciphertext to also encrypt m_1^*. This strategy will allow us to complete the proof of indistinguishability security.

Indeed, such an idea of committing to challenge ciphertexts in the public parameters has been used in the FE context before, for example in [14]. However, observe that if we do this, then verifiability is again lost, because recall that even the public parameters of the system are under the adversary's control! If a malicious authority generates a ciphertext using the correctness of the trapdoor statement, he could encrypt the tuple (m, m, m_1, m_2, m_3) as the set of messages in the constituent ciphertexts and generate a valid NIWI. Now, for some valid function secret key, decrypting this ciphertext may not give rise to a valid function output. The inherent problem here is that any ciphertext for which the NIWI is proved using the trapdoor statement and any honestly generated function secret key need not agree on a majority (three) of the underlying systems.

To overcome this issue, we introduce the idea of a *guided locking mechanism*. Intuitively, we require that the system cannot have both valid ciphertexts that use the correctness of the trapdoor statement and valid function secret keys. Therefore, we introduce a new "lock" in the public parameters. The statement being proved in the function secret key will state that this lock is a commitment of 1, while the trapdoor statement for the ciphertexts will state that the lock is a commitment of 0. Thus, we cannot simultaneously have valid ciphertexts that use the correctness of the trapdoor statement and valid function secret keys. This ensures verifiability of the system. However, while playing this cat and mouse game of ensuring security and verifiability, observe that we can no longer prove that the system is secure! In our proof strategy, we wanted to switch the challenge ciphertext to use the correctness of the trapdoor statement which would mean that no valid function secret key can exist in the system. But, the adversary can of course ask for some function secret keys and hence the security proof wouldn't go through.

We handle this scenario by introducing another trapdoor statement for the relation corresponding to the function secret keys. This trapdoor statement is similar to the honest one in the sense that it needs four of the five constituent function secret keys to be secret keys for the same function. Crucially, however, additionally, it states that if you consider the five constituent ciphertexts committed to in the public parameters, decrypting each of them with the corresponding constituent function secret key yields the same output. Notice that for any function secret key that uses the correctness of the trapdoor statement and any ciphertext generated using the correctness of its corresponding trapdoor statement, verifiability is not lost. This is because of the condition that all corresponding decryptions yield the same output. Indeed, for any function secret key that uses the correctness of the trapdoor statement and any ciphertext generated using the correctness of its non-trapdoor statement, verifiability is maintained. Thus, this addition doesn't impact the verifiability of the system.

Now, in order to prove security, we first switch every function secret key to be generated using the correctness of the trapdoor statement. This is followed by changing the lock in the public parameter to be a commitment of 1 and then switching the NIWI in the ciphertexts to use their corresponding trapdoor statement. The rest of the security proof unravels in the same way as before. After the challenge ciphertext is transformed into an encryption of message m_1^*, we reverse the whole process to switch every function secret key to use the real statement (and not the trapdoor one) and to switch the challenge ciphertext to use the corresponding real statement. Notice that the lock essentially guides the sequence of steps to be followed by the security proof as any other sequence is not possible. In this way, the locks *guide* the hybrids that can be considered in the security argument, hence the name "guided" locking mechanism for the technique. In fact, using these ideas, it turns out that just having four parallel systems suffices to construct verifiable functional encryption in the public key setting.

In the secret key setting, to achieve verifiability, we also have to commit to all the constituent master secret keys in the public parameters. However, we need an additional system (bringing the total back to five) because in order to switch a constituent challenge ciphertext from an encryption of m_0^* to that of m_1^*, we need to puncture out the corresponding master secret key committed in the public parameters. We observe that in the secret key setting, ciphertexts and function secret keys can be seen as duals of each other. Hence, to prove function hiding, we introduce indistinguishable modes and a switching mechanism. At any point in time, the system can either be in function hiding mode or in message hiding mode but not both. At all stages, verifiability is maintained using similar techniques.

Organisation: In Sect. 2 we define the preliminaries used in the paper. In Sect. 3, we give the definition of a verifiable functional encryption scheme. This is followed by the construction and proof of a verifiable functional encryption scheme in Sect. 4. In Sect. 5, we give the construction of a secret key verifiable functional encryption scheme. Section 6 is devoted to the study of verifiable obfuscation. An application of verifiable functional encryption is in achieving functional commitments. Due to lack of space, this has been discussed in the full version [5].

2 Preliminaries

Throughout the paper, let the security parameter be λ and let PPT denote a probabilistic polynomial time algorithm. We assume that reader is familiar with the concept of public key encryption and non-interactive commitment schemes.

2.1 One Message WI Proofs

We will be extensively using one message witness indistinguishable proofs NIWI as provided by [22].

Definition 1. *A pair of PPT algorithms* $(\mathcal{P}, \mathcal{V})$ *is a* NIWI *for an NP relation* $\mathcal{R}_{\mathcal{L}}$ *if it satisfies:*

1. *Completeness: for every* $(x, w) \in \mathcal{R}_{\mathcal{L}}$, $Pr[\mathcal{V}(x, \pi) = 1 : \pi \leftarrow \mathcal{P}(x, w)] = 1$.
2. *(Perfect) Soundness: Proof system is said to be perfectly sound if there for every* $x \notin L$ *and* $\pi \in \{0,1\}^*$
   $Pr[\mathcal{V}(x, \pi) = 1] = 0$.
3. *Witness indistinguishability: for any sequence* $\mathcal{I} = \{(x, w_1, w_2) : w_1, w_2 \in \mathcal{R}_{\mathcal{L}}(x)\}$
   $\{\pi_1 : \pi_1 \leftarrow \mathcal{P}(x, w_1)\}_{(x,w_1,w_2)\in\mathcal{I}} \approx_c \{\pi_2 : \pi_2 \leftarrow \mathcal{P}(x, w_2)\}_{(x,w_1,w_2)\in\mathcal{I}}$

[22] provides perfectly sound one message witness indistinguishable proofs based on the decisional linear (DLIN) assumption. [7] also provides perfectly sound proofs (although less efficient) under a complexity theoretic assumption, namely that Hitting Set Generators against co-non deterministic circuits exist. [9] construct NIWI from one-way permutations and indistinguishability obfuscation.

3 Verifiable Functional Encryption

In this section we give the definition of a (public-key) verifiable functional encryption scheme. Let $\mathcal{X} = \{\mathcal{X}_\lambda\}_{\lambda\in\mathbb{N}}$ and $\mathcal{Y} = \{\mathcal{Y}_\lambda\}_{\lambda\in\mathbb{N}}$ denote ensembles where each $\mathcal{X}_\lambda$ and $\mathcal{Y}_\lambda$ is a finite set. Let $\mathcal{F} = \{\mathcal{F}_\lambda\}_{\lambda\in\mathbb{N}}$ denote an ensemble where each $\mathcal{F}_\lambda$ is a finite collection of functions, and each function $f \in \mathcal{F}_\lambda$ takes as input a string $x \in \mathcal{X}_\lambda$ and outputs $f(x) \in \mathcal{Y}_\lambda$. A verifiable functional encryption scheme is similar to a regular functional encryption scheme with two additional algorithms $(\mathsf{VerifyCT}, \mathsf{VerifyK})$. Formally, $\mathsf{VFE} = (\mathsf{Setup}, \mathsf{Enc}, \mathsf{KeyGen}, \mathsf{Dec}, \mathsf{VerifyCT}, \mathsf{VerifyK})$ consists of the following polynomial time algorithms:

- $\mathsf{Setup}(1^\lambda)$. The setup algorithm takes as input the security parameter λ and outputs a master public key-secret key pair $(\mathsf{MPK}, \mathsf{MSK})$.
- $\mathsf{Enc}(\mathsf{MPK}, x) \to \mathsf{CT}$. The encryption algorithm takes as input a message $x \in \mathcal{X}_\lambda$ and the master public key MPK. It outputs a ciphertext CT.
- $\mathsf{KeyGen}(\mathsf{MPK}, \mathsf{MSK}, f) \to \mathsf{SK}_f$. The key generation algorithm takes as input a function $f \in \mathcal{F}_\lambda$, the master public key MPK and the master secret key MSK. It outputs a function secret key SK_f.

- $\mathsf{Dec}(\mathsf{MPK}, f, \mathsf{SK}_f, \mathsf{CT}) \rightarrow y$ or $\perp$. The decryption algorithm takes as input the master public key MPK, a function f, the corresponding function secret key SK_f and a ciphertext CT. It either outputs a string $y \in \mathcal{Y}$ or $\perp$. Informally speaking, MPK is given to the decryption algorithm for verification purpose.
- $\mathsf{VerifyCT}(\mathsf{MPK}, \mathsf{CT}) \rightarrow 1/0$. Takes as input the master public key MPK and a ciphertext CT. It outputs 0 or 1. Intuitively, it outputs 1 if CT was correctly generated using the master public key MPK for some message x.
- $\mathsf{VerifyK}(\mathsf{MPK}, f, \mathsf{SK}) \rightarrow 1/0$. Takes as input the master public key MPK, a function f and a function secret key SK_f. It outputs either 0 or 1. Intuitively, it outputs 1 if SK_f was correctly generated as a function secret key for f.

The scheme has the following properties:

Definition 2. *(Correctness) A verifiable functional encryption scheme* VFE *for* $\mathcal{F}$ *is correct if for all* $f \in \mathcal{F}_\lambda$ *and all* $x \in \mathcal{X}_\lambda$

$$\Pr\left[\begin{array}{c}(\mathsf{MPK},\mathsf{MSK}) \leftarrow \mathsf{Setup}(1^\lambda) \\ \mathsf{SK}_f \leftarrow \mathsf{KeyGen}(\mathsf{MPK}, \mathsf{MSK}, f) \\ \mathsf{Dec}(\mathsf{MPK}, f, \mathsf{SK}_f, \mathsf{Enc}(\mathsf{MPK}, x)) = f(x)\end{array}\right] = 1$$

Definition 3. *(Verifiability) A verifiable functional encryption scheme* VFE *for* $\mathcal{F}$ *is verifiable if, for all* $\mathsf{MPK} \in \{0,1\}^*$, *for all* $\mathsf{CT} \in \{0,1\}^*$, *there exists* $x \in \mathcal{X}$ *such that for all* $f \in \mathcal{F}$ *and* $\mathsf{SK} \in \{0,1\}^*$, *if*

$$\mathsf{VerifyCT}(\mathsf{MPK}, \mathsf{CT}) = 1 \text{ and } \mathsf{VerifyK}(\mathsf{MPK}, f, \mathsf{SK}) = 1$$

then

$$\Pr\left[\mathsf{Dec}(\mathsf{MPK}, f, \mathsf{SK}, \mathsf{CT}) = f(x)\right] = 1$$

Remark. Intuitively, verifiability states that each ciphertext (possibly associated with a maliciously generated public key) should be associated with a unique message and decryption for a function f using any possibly maliciously generated key SK should result in $f(x)$ for that unique message $f(x)$ and nothing else (if the ciphertext and keys are verified by the respective algorithms).

We also note that a verifiable functional encryption scheme should satisfy perfect correctness. Otherwise, a non-uniform malicious authority can sample ciphertexts/keys from the space where it fails to be correct. Thus, the primitives that we will use in our constructions are assumed to have perfect correctness. Such primitives have been constructed before in the literature.

3.1 Indistinguishability Based Security

The indistinguishability based security for verifiable functional encryption is similar to the security notion of a functional encryption scheme. For completeness, we define it below. We also consider a {full/selective} CCA secure variant where the adversary, in addition to the security game described below, has access to a decryption oracle which takes a ciphertext and a function as input and decrypts the ciphertext with an honestly generated key for that function and returns the output. The adversary is allowed to query this decryption oracle for all ciphertexts of his choice except the challenge ciphertext itself.

We define the security notion for a verifiable functional encryption scheme using the following game ($\mathsf{Full-IND}$) between a challenger and an adversary.

Setup Phase: The challenger $(\mathsf{MPK}, \mathsf{MSK}) \leftarrow \mathsf{vFE.Setup}(1^\lambda)$ and then hands over the master public key MPK to the adversary.

Key Query Phase 1: The adversary makes function secret key queries by submitting functions $\mathsf{f} \in \mathcal{F}_\lambda$. The challenger responds by giving the adversary the corresponding function secret key $\mathsf{SK_f} \leftarrow \mathsf{vFE.KeyGen}(\mathsf{MPK}, \mathsf{MSK}, \mathsf{f})$.

Challenge Phase: The adversary chooses two messages $(\mathsf{m_0}, \mathsf{m_1})$ of the same size (each in $\mathcal{X}_\lambda$)) such that for all queried functions f in the key query phase, it holds that $\mathsf{f}(\mathsf{m_0}) = \mathsf{f}(\mathsf{m_1})$. The challenger selects a random bit $\mathsf{b} \in \{0, 1\}$ and sends a ciphertext $\mathsf{CT} \leftarrow \mathsf{vFE.Enc}(\mathsf{MPK}, \mathsf{m_b})$ to the adversary.

Key Query Phase 2: The adversary may submit additional key queries $\mathsf{f} \in \mathcal{F}_\lambda$ as long as they do not violate the constraint described above. That is, for all queries f, it must hold that $\mathsf{f}(\mathsf{m_0}) = \mathsf{f}(\mathsf{m_1})$.

Guess: The adversary submits a guess b' and wins if $b' = b$. The adversary's advantage in this game is defined to be $2 * |\Pr[\mathsf{b} = \mathsf{b}'] - 1/2|$.

We also define the *selective* security game, which we call ($\mathsf{sel-IND}$) where the adversary outputs the challenge message pair even before seeing the master public key.

Definition 4. *A verifiable functional encryption scheme* VFE *is* $\{$ selective, fully$\}$ *secure if all polynomial time adversaries have at most a negligible advantage in the* $\{\mathsf{Sel-IND}, \mathsf{Full-IND}\}$ *security game.*

Functional Encryption: In our construction, we will use functional encryption as an underlying primitive. Syntax of a functional encryption scheme is defined in [14]. It is similar to the syntax of a verifiable functional encryption scheme except that it doesn't have the $\mathsf{VerifyCT}$ and $\mathsf{VerifyK}$ algorithms, the KeyGen algorithm does not take as input the master public key and the decryption algorithm does not take as input the master public key and the function. Other than that, the security notions and correctness are the same. However, in general any functional encryption scheme is not required to satisfy the verifiability property.

3.2 Simulation Based Security

Many variants of simulation based security definitions have been proposed for functional encryption. In general, simulation security (where the adversary can request for keys arbitrarily) is shown to be impossible [2]. We show that even the weakest form of simulation based security is impossible to achieve for verifiable functional encryption.

Theorem 3. *There exists a family of functions, each of which can be represented as a polynomial sized circuit, for which there does not exist any simulation secure verifiable functional encryption scheme.*

Proof. Let L be a NP complete language. Let $\mathcal{R}$ be the relation for this language. $\mathcal{R} : \{0,1\}^* \times \{0,1\}^* \rightarrow \{0,1\}$, takes as input a string x and a polynomial sized (in the length of x) witness w and outputs 1 iff $x \in L$ and w is a witness to this fact. For any security parameter λ, let us define a family of functions $\mathcal{F}_\lambda$ as a family indexed by strings $y \in \{0,1\}^\lambda$. Namely, $\mathcal{F}_\lambda = \{\mathcal{R}(y, \cdot) \; \forall y \in \{0,1\}^\lambda\}$.

Informally speaking, any verifiable functional encryption scheme that is also simulation secure for this family implies the existence of one message zero knowledge proofs for L. The proof system is described as follows: the prover, who has the witness for any instance x of length λ, samples a master public key and master secret key pair for a verifiable functional encryption scheme with security parameter λ. Using the master public key, it encrypts the witness and samples a function secret key for the function $\mathcal{R}(x, \cdot)$. The verifier is given the master public key, the ciphertext and the function secret key. Informally, simulation security of the verifiable functional encryption scheme provides computational zero knowledge while perfect soundness and correctness follow from verifiability. A formal proof is can be found in the fullversion [5].

In a similar manner, we can rule out even weaker simulation based definitions in the literature where the simulator also gets to generate the function secret keys and the master public key. Interestingly, IND secure VFE for the circuit family described in the above proof implies one message witness indistinguishable proofs(NIWI) for NP and hence it is intuitive that we will have to make use of NIWI in our constructions.

Theorem 4. *There exists a family of functions, each of which can be represented as a polynomial sized circuit, for which (selective) IND secure verifiable functional encryption implies the existence of one message witness indistinguishable proofs for* NP *(*NIWI*).*

We prove the theorem in the full version [5].

The definition for verifiable secret key functional encryption and verifiable multi-input functional encryption can be found in the full version [5].

4 Construction of Verifiable Functional Encryption

In this section, we give a compiler from any Sel − IND secure public key functional encryption scheme to a Sel − IND secure verifiable public key functional encryption scheme. The techniques used in this construction have been elaborated upon in Sect. 1.2. The resulting verifiable functional encryption scheme has the same security properties as the underlying one - that is, the resulting scheme is q-query secure if the original scheme that we started out with was q-query secure and so on, where q refers to the number of function secret key queries that the adversary is allowed to make. We prove the following theorem:

Theorem 5. *Let $\mathcal{F} = \{\mathcal{F}_\lambda\}_{\lambda \in \mathbb{N}}$ be a parameterized collection of functions. Then, assuming there exists a* $\mathsf{Sel-IND}$ *secure public key functional encryption scheme* FE *for the class of functions $\mathcal{F}$, a non-interactive witness indistinguishable proof system, a non-interactive perfectly binding and computationally hiding commitment scheme, the proposed scheme* VFE *is a* $\mathsf{Sel-IND}$ *secure verifiable functional encryption scheme for the class of functions $\mathcal{F}$ according to Definition 3.*

Notation: Without loss of generality, let's assume that every plaintext message is of length λ where λ denotes the security parameter of our scheme. Let $(\mathsf{Prove}, \mathsf{Verify})$ be a non-interactive witness-indistinguishable (NIWI) proof system for NP, $\mathsf{FE} = (\mathsf{FE.Setup}, \mathsf{FE.Enc}, \mathsf{FE.KeyGen}, \mathsf{FE.Dec})$ be a $\mathsf{Sel-IND}$ secure public key functional encryption scheme, Com be a statistically binding and computationally hiding commitment scheme. Without loss of generality, let's say Com commits to a string bit-by-bit and uses randomness of length λ to commit to a single bit. We denote the length of ciphertexts in FE by $\mathsf{c\text{-}len} = \mathsf{c\text{-}len}(\lambda)$. Let $\mathsf{len} = 4 \cdot \mathsf{c\text{-}len}$.
Our scheme $\mathsf{VFE} = (\mathsf{VFE.Setup}, \mathsf{VFE.Enc}, \mathsf{VFE.KeyGen}, \mathsf{VFE.Dec}, \mathsf{VFE.VerifyCT}, \mathsf{VFE.VerifyK})$ is as follows:

- **Setup** $\mathsf{VFE.Setup}(1^\lambda)$:
  The setup algorithm does the following:
  1. For all $i \in [4]$, compute $(\mathsf{MPK}_i, \mathsf{MSK}_i) \leftarrow \mathsf{FE.Setup}(1^\lambda; s_i)$ using randomness s_i.
  2. Set $\mathsf{Z} = \mathsf{Com}(0^{\mathsf{len}}; u)$ and $\mathsf{Z}_1 = \mathsf{Com}(1; u_1)$ where u,u_1 represent the randomness used in the commitment.
     The master public key is $\mathsf{MPK} = (\{\mathsf{MPK}_i\}_{i \in [4]}, \mathsf{Z}, \mathsf{Z}_1)$.
     The master secret key is $\mathsf{MSK} = (\{\mathsf{MSK}_i\}_{i \in [4]}, \{s_i\}_{i \in [4]}, u, u_1)$.
- **Encryption** $\mathsf{VFE.Enc}(\mathsf{MPK}, m)$:
  To encrypt a message m, the encryption algorithm does the following:
  1. For all $i \in [4]$, compute $\mathsf{CT}_i = \mathsf{FE.Enc}(\mathsf{MPK}_i, m; r_i)$.
  2. Compute a proof $\pi \leftarrow \mathsf{Prove}(y, w)$ for the statement that $y \in L$ using witness w where:
     $y = (\{\mathsf{CT}_i\}_{i \in [4]}, \{\mathsf{MPK}_i\}_{i \in [4]}, \mathsf{Z}, \mathsf{Z}_1)$,
     $w = (m, \{r_i\}_{i \in [4]}, 0, 0, 0^{|u|}, 0^{|u_1|})$.
     L is defined corresponding to the relation R defined below.
- **Relation**R:
  Instance: $y = (\{\mathsf{CT}_i\}_{i \in [4]}, \{\mathsf{MPK}_i\}_{i \in [4]}, \mathsf{Z}, \mathsf{Z}_1)$
  Witness: $w = (m, \{r_i\}_{i \in [4]}, i_1, i_2, u, u_1)$
  $R_1(y, w) = 1$ if and only if either of the following conditions hold:
  1. All 4 constituent ciphertexts encrypt the same message. That is,
     $\forall i \in [4]$, $\mathsf{CT}_i = \mathsf{FE.Enc}(\mathsf{MPK}_i, m; r_i)$
     (OR)
  2. 2 constituent ciphertexts (corresponding to indices i_1, i_2) encrypt the same message, Z is a commitment to all the constituent ciphertexts and Z_1 is a commitment to 0. That is,

(a) $\forall i \in \{\imath_1, i_2\}$, $\mathsf{CT}_i = \mathsf{FE.Enc}(\mathsf{MPK}_i, m; r_i)$.
(b) $\mathsf{Z} = \mathsf{Com}(\{\mathsf{CT}_i\}_{i \in [4]}; u)$.
(c) $\mathsf{Z}_1 = \mathsf{Com}(0; u_1)$.
The output of the algorithm is the ciphertext $\mathsf{CT} = (\{\mathsf{CT}_i\}_{i \in [4]}, \pi)$.
π is computed for statement 1 of relation R.

- **Key Generation** $\mathsf{VFE.KeyGen}(\mathsf{MPK}, \mathsf{MSK}, f)$**:**
To generate the function secret key K^f for a function f, the key generation algorithm does the following:
  1. $\forall i \in [4]$, compute $\mathsf{K}_i^f = \mathsf{FE.KeyGen}(\mathsf{MSK}_i, f; r_i)$.
  2. Compute a proof $\gamma \leftarrow \mathsf{Prove}(y, w)$ for the statement that $y \in L_1$ using witness w where:
  $y = (\{\mathsf{K}_i^f\}_{i \in [4]}, \{\mathsf{MPK}_i\}_{i \in [4]}, \mathsf{Z}, \mathsf{Z}_1)$,
  $w = (f, \{\mathsf{MSK}_i\}_{i \in [4]}, \{s_i\}_{i \in [4]}, \{r_i\}_{i \in [4]}, 0^3, 0^{|u|}, u_1)$.
  L_1 is defined corresponding to the relation R_1 defined below.
- **Relation**R_1:
Instance: $y = (f, \{\mathsf{K}_i^f\}_{i \in [4]}, \{\mathsf{MPK}_i\}_{i \in [4]}, \mathsf{Z}, \mathsf{Z}_1)$.
Witness: $w = (\{\mathsf{MSK}_i\}_{i \in [4]}, \{s_i\}_{i \in [4]}, \{r_i\}_{i \in [4]}, i_1, i_2, i_3, u, u_1)$
$R_1(y, w) = 1$ if and only if either of the following conditions hold:
  1. Z_1 is a commitment to 1, all 4 constituent function secret keys are secret keys for the same function and are constructed using honestly generated public key-secret key pairs.
     (a) $\forall i \in [4]$, $\mathsf{K}_i^f = \mathsf{FE.KeyGen}(\mathsf{MSK}_i, f; r_i)$.
     (b) $\forall i \in [4]$, $(\mathsf{MPK}_i, \mathsf{MSK}_i) \leftarrow \mathsf{FE.Setup}(1^\lambda; s_i)$.
     (c) $\mathsf{Z}_1 = \mathsf{Com}(1; u_1)$.
     (OR)
  2. 3 of the constituent function secret keys (corresponding to indices i_1, i_2, i_3) are keys for the same function and are constructed using honestly generated public key-secret key pairs, Z is a commitment to a set of ciphertexts CT such that each constituent ciphertext in CT when decrypted with the corresponding function secret key gives the same output. That is,
     (a) $\forall i \in \{i_1, i_2, i_3\}$, $\mathsf{K}_i^f = \mathsf{FE.KeyGen}(\mathsf{MSK}_i, f; r_i)$.
     (b) $\forall i \in \{i_1, i_2, i_3\}$, $(\mathsf{MPK}_i, \mathsf{MSK}_i) \leftarrow \mathsf{FE.Setup}(1^\lambda; s_i)$.
     (c) $\mathsf{Z} = \mathsf{Com}(\{\mathsf{CT}_i\}_{i \in [4]}; u)$.
     (d) $\exists x \in \mathcal{X}_\lambda$ such that $\forall i \in [4]$, $\mathsf{FE.Dec}(\mathsf{CT}_i, \mathsf{K}_i^f) = x$
     The output of the algorithm is the function secret key $\mathsf{K}^f = (\{\mathsf{K}_i^f\}_{i \in [4]}, \gamma)$.
     γ is computed for statement 1 of relation R_1.
- **Decryption** $\mathsf{VFE.Dec}(\mathsf{MPK}, f, \mathsf{K}^f, \mathsf{CT})$**:** This algorithm decrypts the ciphertext $\mathsf{CT} = (\{\mathsf{CT}_i\}_{i \in [4]}, \pi)$ using function secret key $\mathsf{K}^f = (\{\mathsf{K}_i^f\}_{i \in [4]}, \gamma)$ in the following way:
  1. Let $y = (\{\mathsf{CT}_i\}_{i \in [4]}, \{\mathsf{MPK}_i\}_{i \in [4]}, \mathsf{Z}, \mathsf{Z}_1)$ be the statement corresponding to proof π. If $\mathsf{Verify}(y, \pi) = 0$, then stop and output $\perp$. Else, continue to the next step.
  2. Let $y_1 = (f, \{\mathsf{K}_i^f\}_{i \in [4]}, \{\mathsf{MPK}_i\}_{i \in [4]}, \mathsf{Z}, \mathsf{Z}_1)$ be the statement corresponding to proof γ. If $\mathsf{Verify}(y_1, \gamma) = 0$, then stop and output $\perp$. Else, continue to the next step.

3. For $i \in [4]$, compute $m_i = \mathsf{FE.Dec}(\mathsf{CT}_i, \mathsf{K}_i^f)$. If at least 3 of the m_i's are equal (let's say that value is m), output m. Else, output $\perp$.

- **VerifyCT** $\mathsf{VFE.VerifyCT}(\mathsf{MPK}, \mathsf{CT})$**:** Given a ciphertext $\mathsf{CT} = (\{\mathsf{CT}_i\}_{i\in[4]}, \pi)$, this algorithm checks whether the ciphertext was generated correctly using master public key MPK. Let $y = (\{\mathsf{CT}_i\}_{i\in[4]}, \{\mathsf{MPK}_i\}_{i\in[4]}, \mathsf{Z}, \mathsf{Z}_1)$ be the statement corresponding to proof π. If $\mathsf{Verify}(y, \pi) = 1$, it outputs 1. Else, it outputs 0.
- **VerifyK** $\mathsf{VFE.VerifyK}(\mathsf{MPK}, f, \mathsf{K})$**:** Given a function f and a function secret key $\mathsf{K} = (\{\mathsf{K}_i\}_{i\in[4]}, \gamma)$, this algorithm checks whether the key was generated correctly for function f using the master secret key corresponding to master public key MPK. Let $y = (f, \{\mathsf{K}_i\}_{i\in[4]}, \{\mathsf{MPK}_i\}_{i\in[4]}, \mathsf{Z}, \mathsf{Z}_1)$ be the statement corresponding to proof γ. If $\mathsf{Verify}(y, \gamma) = 1$, it outputs 1. Else, it outputs 0.

Correctness: Correctness follows directly from the correctness of the underlying FE scheme, correctness of the commitment scheme and the completeness of the NIWI proof system.

4.1 Verifiability

Consider any master public key MPK and any ciphertext $\mathsf{CT} = (\{\mathsf{CT}_i\}_{i\in[4]}, \pi)$ such that
$\mathsf{VFE.VerifyCT}(\mathsf{MPK}, \mathsf{CT}) = 1$. Now, there are two cases possible for the proof π.

1. Statement 1 of relation R is correct:
   Therefore, there exists $m \in \mathcal{X}_\lambda$ such that $\forall i \in [4]$, $\mathsf{CT}_i = \mathsf{FE.Enc}(\mathsf{MPK}_i, m; r_i)$ where r_i is a random string. Consider any function f and function secret key $\mathsf{K} = (\{\mathsf{K}_i\}_{i\in[4]}, \gamma)$ such that $\mathsf{VFE.VerifyK}(\mathsf{MPK}, f, \mathsf{K}) = 1$. There are two cases possible for the proof γ.
   (a) Statement 1 of relation R_1 is correct:
   Therefore, $\forall i \in [4]$, K_i is a function secret key for the same function - f. That is, $\forall i \in [4]$, $\mathsf{K}_i = \mathsf{FE.KeyGen}(\mathsf{MSK}_i, f; r_i')$ where r_i' is a random string. Thus, for all $i \in [4]$, $\mathsf{FE.Dec}(\mathsf{CT}_i, \mathsf{K}_i) = f(m)$. Hence, $\mathsf{VFE.Dec}(\mathsf{MPK}, f, \mathsf{K}, \mathsf{CT}) = f(m)$.
   (b) Statement 2 of relation R_1 is correct:
   Therefore, there exists 3 indices i_1, i_2, i_3 such that $\mathsf{K}_{i_1}, \mathsf{K}_{i_2}, \mathsf{K}_{i_3}$ are function secret keys for the same function - f. That is, $\forall i \in \{i_1, i_2, i_3\}$, $\mathsf{K}_i = \mathsf{FE.KeyGen}(\mathsf{MSK}_i, f; r_i')$ where r_i' is a random string Thus, for all $i \in \{i_1, i_2, i_3\}$, $\mathsf{FE.Dec}(\mathsf{CT}_i, \mathsf{K}_i) = f(m)$. Hence, $\mathsf{VFE.Dec}(\mathsf{MPK}, f, \mathsf{K}, \mathsf{CT}) = f(m)$.
2. Statement 2 of relation R is correct:
   Therefore, $\mathsf{Z}_1 = \mathsf{Com}(0; u_1)$ and $\mathsf{Z} = \mathsf{Com}(\{\mathsf{CT}_i\}_{i\in[4]}; u)$ for some random strings u, u_1. Also, there exists 2 indices i_1, i_2 and a message $m \in \mathcal{X}_\lambda$ such that for $i \in \{i_1, i_2\}$, $\mathsf{CT}_i = \mathsf{FE.Enc}(\mathsf{MPK}_i, m; r_i)$ where r_i is a random string. Consider any function f and function secret key $\mathsf{K} = (\{\mathsf{K}_i\}_{i\in[4]}, \gamma)$ such that $\mathsf{VFE.VerifyK}(\mathsf{MPK}, f, \mathsf{K}) = 1$. There are two cases possible for the proof γ.

(a) Statement 1 of relation R_1 is correct:
Then, it must be the case that $\mathsf{Z}_1 = \mathsf{Com}(1; u'_1)$ for some random string u'_1. However, we already know that $\mathsf{Z}_1 = \mathsf{Com}(0; u_1)$ and Com is a perfectly binding commitment scheme. Thus, this scenario isn't possible. That is, both $\mathsf{VFE.VerifyCT}(\mathsf{MPK}, \mathsf{CT})$ and $\mathsf{VFE.VerifyK}(\mathsf{MPK}, f, \mathsf{K})$ can't be equal to 1.

(b) Statement 2 of relation R_1 is correct:
Therefore, there exists 3 indices i'_1, i'_2, i'_3 such that $\mathsf{K}_{i'_1}, \mathsf{K}_{i'_2}, \mathsf{K}_{i'_3}$ are function secret keys for the same function - f. That is, $\forall i \in \{i'_1, i'_2, i'_3\}$, $\mathsf{K}_i = \mathsf{FE.KeyGen}(\mathsf{MSK}_i, f; r'_i)$ where r'_i is a random string. Thus, by pigeonhole principle, there exists $i^* \in \{i'_1, i'_2, i'_3\}$ such that $i^* \in \{i_1, i_2\}$ as well. Also, $\mathsf{Z} = \mathsf{Com}(\{\mathsf{CT}_i\}_{i\in[4]}; u)$ and $\forall i \in [4]$, $\mathsf{FE.Dec}(\mathsf{CT}_i, \mathsf{K}_i)$ is the same. Therefore, for the index i^*, $\mathsf{FE.Dec}(\mathsf{CT}_{i^*}, \mathsf{K}_{i^*}) = f(m)$. Hence, $\forall i \in [4]$, $\mathsf{FE.Dec}(\mathsf{CT}_i, \mathsf{K}_i) = f(m)$. Therefore, $\mathsf{VFE.Dec}(\mathsf{MPK}, f, \mathsf{K}, \mathsf{CT}) = f(m)$.

4.2 Security Proof

We now prove that the proposed scheme VFE is Sel − IND secure. We will prove this via a series of hybrid experiments $\mathsf{H}_1, \ldots, \mathsf{H}_{16}$ where H_1 corresponds to the real world experiment with challenge bit $b = 0$ and H_{16} corresponds to the real world experiment with challenge bit $b = 1$. The hybrids are summarized below in Table 2.

We briefly describe the hybrids below. A more detailed description can be found in the full version [5].

- **Hybrid** H_1: This is the real experiment with challenge bit $b = 0$. The master public key is $\mathsf{MPK} = (\{\mathsf{MPK}_i\}_{i\in[4]}, \mathsf{Z}, \mathsf{Z}_1)$ such that $\mathsf{Z} = \mathsf{Com}(0^{\mathsf{len}}; u)$ and $\mathsf{Z}_1 = \mathsf{Com}(1; u_1)$ for random strings u, u_1. The challenge ciphertext is $\mathsf{CT}^* = (\{\mathsf{CT}^*_i\}_{i\in[4]}, \pi^*)$, where for all $i \in [4]$, $\mathsf{CT}^*_i = \mathsf{FE.Enc}(\mathsf{MPK}_i, m_0; r_i)$ for some random string r_i. π^* is computed for statement 1 of relation R.
- **Hybrid** H_2: This hybrid is identical to the previous hybrid except that Z is computed differently. $\mathsf{Z} = \mathsf{Com}(\{\mathsf{CT}^*_i\}_{i\in[4]}; u)$.
- **Hybrid** H_3: This hybrid is identical to the previous hybrid except that for every function secret key K^f, the proof γ is now computed for statement 2 of relation R_1 using indices $\{1, 2, 3\}$ as the set of 3 indices $\{i_1, i_2, i_3\}$ in the witness. That is, the witness is $w = (\mathsf{MSK}_1, \mathsf{MSK}_2, \mathsf{MSK}_3, 0^{|\mathsf{MSK}_4|}, s_1, s_2, s_3, 0^{|s_4|}, r_1, r_2, r_3, 0^{|r_4|}, 1, 2, 3, u, 0^{|u_1|})$.
- **Hybrid** H_4: This hybrid is identical to the previous hybrid except that Z_1 is computed differently. $\mathsf{Z}_1 = \mathsf{Com}(0; u_1)$.
- **Hybrid** H_5: This hybrid is identical to the previous hybrid except that the proof π^* in the challenge ciphertext is now computed for statement 2 of relation R using indices $\{1, 2\}$ as the 2 indices $\{i_1, i_2\}$ in the witness. That is, the witness is $w = (m, r_1, r_2, 0^{|r_3|}, 0^{|r_4|}, 1, 2, u, u_1)$.
- **Hybrid** H_6: This hybrid is identical to the previous hybrid except that we change the fourth component CT^*_4 of the challenge ciphertext to be an encryption of the challenge message m_1 (as opposed to m_0). That is,

Table 2. Here, (m_0, m_0, m_0, m_0) indicates the messages that are encrypted to form the challenge ciphertext $\{\mathsf{CT}_i^*\}_{i\in[4]}$. Similarly for the column $\{\mathsf{K}_i^f\}_{i\in[4]}$. The column π^* (and γ) denote the statement proved by the proof in relation R (and R_1). The text in red indicates the difference from the previous hybrid. The text in blue denotes the indices used in the proofs π^* and γ. That is, the text in blue in the column $(\{\mathsf{CT}_i^*\}_{i\in[4]})$ denotes the indices used in the proof π^* and the text in blue in the column $(\{\mathsf{K}_i^f\}_{i\in[4]})$ denotes the indices used in the proof γ for every function secret key K^f corresponding to function f. In some cases, the difference is only in the indices used in the proofs π^* or γ and these are not reflected using red.

Hybrid	$(\{\mathsf{CT}_i^*\}_{i\in[4]})$	π^*	$\{\mathsf{K}_i^f\}_{i\in[4]}$	γ	Z	Z_1	Security
H_1	(m_0, m_0, m_0, m_0)	1	(f, f, f, f)	1	$\mathsf{Com}(0)$	$\mathsf{Com}(1)$	-
H_2	(m_0, m_0, m_0, m_0)	1	(f, f, f, f)	1	$\mathsf{Com}(\{\mathsf{CT}_i^*\}_{i\in[4]})$	$\mathsf{Com}(1)$	Com-Hiding
H_3	(m_0, m_0, m_0, m_0)	1	(f, f, f, f)	2	$\mathsf{Com}(\{\mathsf{CT}_i^*\}_{i\in[4]})$	$\mathsf{Com}(1)$	NIWI
H_4	(m_0, m_0, m_0, m_0)	1	(f, f, f, f)	2	$\mathsf{Com}(\{\mathsf{CT}_i^*\}_{i\in[4]})$	$\mathsf{Com}(0)$	Com-Hiding
H_5	(m_0, m_0, m_0, m_0)	2	(f, f, f, f)	2	$\mathsf{Com}(\{\mathsf{CT}_i^*\}_{i\in[4]})$	$\mathsf{Com}(0)$	NIWI
H_6	(m_0, m_0, m_0, m_1)	2	(f, f, f, f)	2	$\mathsf{Com}(\{\mathsf{CT}_i^*\}_{i\in[4]})$	$\mathsf{Com}(0)$	IND-secure FE
H_7	(m_0, m_0, m_0, m_1)	2	(f, f, f, f)	2	$\mathsf{Com}(\{\mathsf{CT}_i^*\}_{i\in[4]})$	$\mathsf{Com}(0)$	NIWI
H_8	(m_0, m_0, m_1, m_1)	2	(f, f, f, f)	2	$\mathsf{Com}(\{\mathsf{CT}_i^*\}_{i\in[4]})$	$\mathsf{Com}(0)$	IND-secure FE
H_9	(m_0, m_0, m_1, m_1)	2	(f, f, f, f)	2	$\mathsf{Com}(\{\mathsf{CT}_i^*\}_{i\in[4]})$	$\mathsf{Com}(0)$	NIWI
H_{10}	(m_0, m_1, m_1, m_1)	2	(f, f, f, f)	2	$\mathsf{Com}(\{\mathsf{CT}_i^*\}_{i\in[4]})$	$\mathsf{Com}(0)$	IND-secure FE
H_{11}	(m_0, m_1, m_1, m_1)	2	(f, f, f, f)	2	$\mathsf{Com}(\{\mathsf{CT}_i^*\}_{i\in[4]})$	$\mathsf{Com}(0)$	NIWI
H_{12}	(m_1, m_1, m_1, m_1)	2	(f, f, f, f)	2	$\mathsf{Com}(\{\mathsf{CT}_i^*\}_{i\in[4]})$	$\mathsf{Com}(0)$	IND-secure FE
H_{13}	(m_1, m_1, m_1, m_1)	1	(f, f, f, f)	2	$\mathsf{Com}(\{\mathsf{CT}_i^*\}_{i\in[4]})$	$\mathsf{Com}(0)$	NIWI
H_{14}	(m_1, m_1, m_1, m_1)	1	(f, f, f, f)	2	$\mathsf{Com}(\{\mathsf{CT}_i^*\}_{i\in[4]})$	$\mathsf{Com}(1)$	Com-Hiding
H_{15}	(m_1, m_1, m_1, m_1)	1	(f, f, f, f)	1	$\mathsf{Com}(\{\mathsf{CT}_i^*\}_{i\in[4]})$	$\mathsf{Com}(1)$	NIWI
H_{16}	(m_1, m_1, m_1, m_1)	1	(f, f, f, f)	1	$\mathsf{Com}(0)$	$\mathsf{Com}(1)$	Com-Hiding

$\mathsf{CT}_4^* = \mathsf{FE.Enc}(\mathsf{MPK}_4, m_1; r_4)$ for some random string r_4. Note that the proof π^* is unchanged and is still proven for statement 2 of relation R.

- **Hybrid** H_7: This hybrid is identical to the previous hybrid except that for every function secret key K^f, the proof γ is now computed for statement 2 of relation R_1 using indices $\{1, 2, 4\}$ as the set of 3 indices $\{i_1, i_2, i_3\}$ in the witness. That is, the witness is $w = (\mathsf{MSK}_1, \mathsf{MSK}_2, 0^{|\mathsf{MSK}_3|}, \mathsf{MSK}_4, s_1, s_2, 0^{|s_3|}, s_4, r_1, r_2, 0^{|r_3|}, r_4, 1, 2, 4, u, 0^{|u_1|})$.
- **Hybrid** H_8: This hybrid is identical to the previous hybrid except that we change the third component CT_3^* of the challenge ciphertext to be an encryption of the challenge message m_1 (as opposed to m_0). That is, $\mathsf{CT}_3^* = \mathsf{FE.Enc}(\mathsf{MPK}_3, m_1; r_3)$ for some random string r_3.
  Note that the proof π^* is unchanged and is still proven for statement 2 of relation R.
- **Hybrid** H_9: This hybrid is identical to the previous hybrid except that the proof π^* in the challenge ciphertext is now computed for statement 2 of relation R using message m_1 and indices $\{3, 4\}$ as the 2 indices $\{i_1, i_2\}$ in the witness. That is, the witness is $w = (m_1, 0^{|r_1|}, 0^{|r_2|}, r_3, r_4, 3, 4, u, u_1)$.

Also, for every function secret key K^f, the proof γ is now computed for statement 2 of relation R_1 using indices $\{1, 3, 4\}$ as the set of 3 indices $\{i_1, i_2, i_3\}$ in the witness. That is, the witness is $w = (\mathsf{MSK}_1, 0^{|\mathsf{MSK}_2|}, \mathsf{MSK}_3, \mathsf{MSK}_4, s_1, 0^{|s_2|}, s_3, s_4, r_1, 0^{|r_2|}, r_3, r_4, 1, 3, 4, u, 0^{|u_1|})$.

- **Hybrid** H_{10}: This hybrid is identical to the previous hybrid except that we change the second component CT_2^* of the challenge ciphertext to be an encryption of the challenge message m_1 (as opposed to m_0). That is, $\mathsf{CT}_2^* = \mathsf{FE.Enc}(\mathsf{MPK}_2, m_1; r_2)$ for some random string r_2.
  Note that the proof π^* is unchanged and is still proven for statement 2 of relation R.
- **Hybrid** H_{11}: This hybrid is identical to the previous hybrid except that for every function secret key K^f, the proof γ is now computed for statement 2 of relation R_1 using indices $\{2, 3, 4\}$ as the set of 3 indices $\{i_1, i_2, i_3\}$ in the witness. That is, the witness is $w = (0^{|\mathsf{MSK}_1|}, \mathsf{MSK}_2, \mathsf{MSK}_3, \mathsf{MSK}_4, 0^{|s_1|}, s_2, s_3, s_4, 0^{|r_1|}, r_2, r_3, r_4, 2, 3, 4, u, 0^{|u_1|})$.
- **Hybrid** H_{12}: This hybrid is identical to the previous hybrid except that we change the first component CT_1^* of the challenge ciphertext to be an encryption of the challenge message m_1 (as opposed to m_0). That is, $\mathsf{CT}_1^* = \mathsf{FE.Enc}(\mathsf{MPK}_1, m_1; r_1)$ for some random string r_1. Note that the proof π^* is unchanged and is still proven for statement 2 of relation R.
- **Hybrid** H_{13}: This hybrid is identical to the previous hybrid except that the proof π^* in the challenge ciphertext is now computed for statement 1 of relation R. The witness is $w = (m_1, \{r_i\}_{i \in [4]}, 0, 0, 0^{|u|}, 0^{|u_1|})$.
- **Hybrid** H_{14}: This hybrid is identical to the previous hybrid except that Z_1 is computed differently. $\mathsf{Z}_1 = \mathsf{Com}(1; u_1)$.
- **Hybrid** H_{15}: This hybrid is identical to the previous hybrid except that for every function secret key K^f, the proof γ is now computed for statement 1 of relation R_1. The witness is $w = (\{\mathsf{MSK}_i\}_{i \in [4]}, \{s_i\}_{i \in [4]}, \{r_i\}_{i \in [4]}, 0^3, 0^{|u|}, u_1)$.
- **Hybrid** H_{16}: This hybrid is identical to the previous hybrid except that Z is computed differently. $\mathsf{Z} = \mathsf{Com}(0^{\mathsf{len}}; u)$. This hybrid is identical to the real experiment with challenge bit $b = 1$.

Below we will prove that $(\mathsf{H}_1 \approx_c \mathsf{H}_2)$ and $(\mathsf{H}_5 \approx_c \mathsf{H}_6)$. The indistinguishability of other hybrids will follow along the same lines and is described in the full version [5].

Lemma 1 $(\mathsf{H}_1 \approx_c \mathsf{H}_2)$. *Assuming that* Com *is a (computationally) hiding commitment scheme, the outputs of experiments* H_1 *and* H_2 *are computationally indistinguishable.*

Proof. The only difference between the two hybrids is the manner in which the commitment Z is computed. Let's consider the following adversary $\mathcal{A}_{\mathsf{Com}}$ that interacts with a challenger $\mathcal{C}$ to break the hiding of the commitment scheme. Also, internally, it acts as the challenger in the security game with an adversary $\mathcal{A}$ that tries to distinguish between H_1 and H_2. $\mathcal{A}_{\mathsf{Com}}$ executes the hybrid H_1 except that it does not generate the commitment Z on it's own. Instead, after receiving the challenge messages (m_0, m_1) from $\mathcal{A}$, it computes $\mathsf{CT}^* = (\{\mathsf{CT}_i^*\}_{i \in [4]}, \pi^*)$ as an encryption of message m_0 by following the honest encryption algorithm as in H_1 and H_2. Then, it sends two strings, namely (0^{len}) and $(\{\mathsf{CT}_i^*\}_{i\in[4]})$ to the outside challenger $\mathcal{C}$. In return, $\mathcal{A}_{\mathsf{Com}}$ receives a commitment Z corresponding to either the first or the second string. It then gives this to $\mathcal{A}$. Now, whatever bit b $\mathcal{A}$ guesses, $\mathcal{A}_{\mathsf{Com}}$ forwards the same guess to the outside challenger $\mathcal{C}$. Clearly, $\mathcal{A}_{\mathsf{Com}}$ is a polynomial time algorithm and breaks the hiding property of Com unless $\mathsf{H}_1 \approx_c \mathsf{H}_2$.

Lemma 2. *($\mathsf{H}_5 \approx_c \mathsf{H}_6$). Assuming that* FE *is a* $\mathsf{Sel-IND}$ *secure functional encryption scheme, the outputs of experiments* H_5 *and* H_6 *are computationally indistinguishable.*

Proof. The only difference between the two hybrids is the manner in which the challenge ciphertext is created. More specifically, in H_5, the fourth component of the challenge ciphertext CT_4^* is computed as an encryption of message m_0, while in H_6, CT_4^* is computed as an encryption of message m_1. Note that the proof π^* remains same in both the hybrids.

Let's consider the following adversary $\mathcal{A}_{\mathsf{FE}}$ that interacts with a challenger $\mathcal{C}$ to break the security of the underlying FE scheme. Also, internally, it acts as the challenger in the security game with an adversary $\mathcal{A}$ that tries to distinguish between H_5 and H_6. $\mathcal{A}_{\mathsf{FE}}$ executes the hybrid H_5 except that it does not generate the parameters $(\mathsf{MPK}_4, \mathsf{MSK}_4)$ itself. It sets (MPK_4) to be the public key given by the challenger $\mathcal{C}$. After receiving the challenge messages (m_0, m_1) from $\mathcal{A}$, it forwards the pair (m_0, m_1) to the challenger $\mathcal{C}$ and receives a ciphertext CT which is either an encryption of m_0 or m_1 using public key MPK_4. $\mathcal{A}_{\mathsf{FE}}$ sets $\mathsf{CT}_4^* = \mathsf{CT}$ and computes $\mathsf{CT}^* = (\{\mathsf{CT}_i^*\}_{i\in[4]}, \pi^*)$ as the challenge ciphertext as in H_5. Note that proof π^* is proved for statement 2 of relation R. It then sets the public parameter $\mathsf{Z} = \mathsf{Com}(\{\mathsf{CT}_i^*\}_{i\in[4]}; u)$ and sends the master public key MPK and the challenge ciphertext CT^* to $\mathcal{A}$.

Now, whatever bit b $\mathcal{A}$ guesses, $\mathcal{A}_{\mathsf{FE}}$ forwards the same guess to the outside challenger $\mathcal{C}$. Clearly, $\mathcal{A}_{\mathsf{FE}}$ is a polynomial time algorithm and breaks the security of the functional encryption scheme FE unless $\mathsf{H}_5 \approx_c \mathsf{H}_6$.

5 Construction of Verifiable Secret Key Functional Encryption

In this section, we give a compiler from any $\mathsf{Sel-IND}$ secure message hiding and function hiding secret key functional encryption scheme to a $\mathsf{Sel-IND}$ secure message hiding and function hiding verifiable secret key functional encryption scheme. The resulting verifiable functional encryption scheme has the same security properties as the underlying one - that is, the resulting scheme is q-query secure if the original scheme that we started out with was q-query secure and so on, where q refers to the number of function secret key queries (or encryption queries) that the adversary is allowed to make. We prove the following theorem.

Theorem 6. *Let $\mathcal{F} = \{\mathcal{F}_\lambda\}_{\lambda \in \mathbb{N}}$ be a parameterized collection of functions. Then, assuming there exists a* $\mathsf{Sel-IND}$ *secure message hiding and function hiding secret key functional encryption scheme* FE *for the class of functions $\mathcal{F}$, a non-interactive witness indistinguishable proof system, a non-interactive perfectly binding and computationally hiding commitment scheme, the proposed scheme* VFE *is a* $\mathsf{Sel-IND}$ *secure message hiding and function hiding verifiable secret key functional encryption scheme for the class of functions $\mathcal{F}$ according to Definition 3.*

Notation: Without loss of generality, let's assume that every plaintext message is of length λ where λ denotes the security parameter of our scheme and that the length of every function in $\mathcal{F}_\lambda$ is the same. Let $(\mathsf{Prove}, \mathsf{Verify})$ be a non-interactive witness-indistinguishable (NIWI) proof system for NP, $\mathsf{FE} = (\mathsf{FE.Setup}, \mathsf{FE.Enc}, \mathsf{FE.KeyGen}, \mathsf{FE.Dec})$ be a $\mathsf{Sel-IND}$ secure message hiding and function hiding secret key functional encryption scheme, Com be a statistically binding and computationally hiding commitment scheme. Without loss of generality, let's say Com commits to a string bit-by-bit and uses randomness of length λ to commit to a single bit. We denote the length of ciphertexts in FE by $\mathsf{c\text{-}len} = \mathsf{c\text{-}len}(\lambda)$. Let the length of every function secret key in FE be $\mathsf{k\text{-}len} = \mathsf{k\text{-}len}(\lambda)$. Let $\mathsf{len}_{CT} = 5 \cdot \mathsf{c\text{-}len}$ and $\mathsf{len}_f = 5 \cdot \mathsf{k\text{-}len}$.
Our scheme $\mathsf{VFE} = (\mathsf{VFE.Setup}, \mathsf{VFE.Enc}, \mathsf{VFE.KeyGen}, \mathsf{VFE.Dec}, \mathsf{VFE.VerifyCT}, \mathsf{VFE.VerifyK})$ is as follows:

- **Setup** $\mathsf{VFE.Setup}(1^\lambda)$ **:**
  The setup algorithm does the following:
  1. For all $i \in [5]$, compute $(\mathsf{MSK}_i) \leftarrow \mathsf{FE.Setup}(1^\lambda; p_i)$ and $\mathsf{S}_i = \mathsf{Com}(\mathsf{MSK}_i; s_i)$ using randomness s_i.
  2. Set $\mathsf{Z_{CT}} = \mathsf{Com}(0^{\mathsf{len}}_{CT}; a)$ and $\mathsf{Z_f} = \mathsf{Com}(0^{\mathsf{len}}_f; b)$ where a, b represents the randomness used in the commitments.
  3. For all $i \in [3]$, set $\mathsf{Z}_i = \mathsf{Com}(1; u_i)$ where u_i represents the randomness used in the commitment. Let's denote $\mathsf{u\text{-}len} = |u_1| + |u_2| + |u_3|$.

  The public parameters are $\mathsf{PP} = (\{\mathsf{S}_i\}_{i\in[5]}, \mathsf{Z_{CT}}, \mathsf{Z_f}, \{\mathsf{Z}_i\}_{i\in[3]})$.
  The master secret key is $\mathsf{MSK} = (\{\mathsf{MSK}_i\}_{i\in[5]}, \{p_i\}_{i\in[5]}, \{s_i\}_{i\in[5]}, a, b, \{u_i\}_{i\in[3]})$.
- **Encryption** $\mathsf{VFE.Enc}(\mathsf{PP}, \mathsf{MSK}, m)$ **:**
  To encrypt a message m, the encryption algorithm does the following:
  1. For all $i \in [5]$, compute $\mathsf{CT}_i = \mathsf{FE.Enc}(\mathsf{MSK}_i, m; r_i)$.
  2. Compute a proof $\pi \leftarrow \mathsf{Prove}(y, w)$ for the statement that $y \in L$ using witness w where:
     $y = (\{\mathsf{CT}_i\}_{i\in[5]}, \mathsf{PP})$,
     $w = (m, \mathsf{MSK}, \{r_i\}_{i\in[5]}, 0^2, 5, 0)$.
     L is defined corresponding to the relation R defined below.
- **Relation** R:
  Instance: $y = (\{\mathsf{CT}_i\}_{i\in[5]}, \mathsf{PP})$
  Witness: $w = (m, \mathsf{MSK}, \{r_i\}_{i\in[5]}, i_1, i_2, j, k)$
  $R_1(y, w) = 1$ if and only if either of the following conditions hold:

1. 4 out of the 5 constituent ciphertexts (except index j) encrypt the same message and are constructed using honestly generated secret keys. Also, Z_1 is a commitment to 1. That is,
   (a) $\forall i \in [5]/\{j\}$, $\mathsf{CT}_i = \mathsf{FE.Enc}(\mathsf{MSK}_i, m; r_i)$.
   (b) $\forall i \in [5]/\{j\}$, $\mathsf{S}_i = \mathsf{Com}(\mathsf{MSK}_i; s_i)$ and $\mathsf{MSK}_i \leftarrow \mathsf{FE.Setup}(1^\lambda; p_i)$
   (c) $\mathsf{Z}_1 = \mathsf{Com}(1; u_1)$
   (OR)
2. 2 constituent ciphertexts (corresponding to indices i_1, i_2) encrypt the same message and are constructed using honestly generated secret keys. Z_{CT} is a commitment to all the constituent ciphertexts, Z_2 is a commitment to 0 and Z_3 is a commitment to 1. That is,
   (a) $\forall i \in \{\imath_1, i_2\}$, $\mathsf{CT}_i = \mathsf{FE.Enc}(\mathsf{MSK}_i, m; r_i)$.
   (b) $\forall i \in \{\imath_1, i_2\}$, $\mathsf{S}_i = \mathsf{Com}(\mathsf{MSK}_i; s_i)$ and $\mathsf{MSK}_i \leftarrow \mathsf{FE.Setup}(1^\lambda; p_i)$
   (c) $\mathsf{Z}_{\mathsf{CT}} = \mathsf{Com}(\{\mathsf{CT}_i\}_{i \in [5]}; a)$.
   (d) $\mathsf{Z}_2 = \mathsf{Com}(0; u_2)$.
   (e) $\mathsf{Z}_3 = \mathsf{Com}(1; u_3)$.
   (OR)
3. 4 out of 5 constituent ciphertexts (except for index k) encrypt the same message and are constructed using honestly generated secret keys. Z_f is a commitment to a set of function secret keys K such that each constituent function secret key in K when decrypted with the corresponding ciphertext gives the same output . That is,
   (a) $\forall i \in [5]/\{k\}$, $\mathsf{CT}_i = \mathsf{FE.Enc}(\mathsf{MSK}_i, m; r_i)$.
   (b) $\forall i \in [5]/\{k\}$, $\mathsf{S}_i = \mathsf{Com}(\mathsf{MSK}_i; s_i)$ and $\mathsf{MSK}_i \leftarrow \mathsf{FE.Setup}(1^\lambda; p_i)$
   (c) $\mathsf{Z}_\mathsf{f} = \mathsf{Com}(\{\mathsf{K}_i\}_{i \in [5]}; b)$.
   (d) $\exists x \in \mathcal{X}_\lambda$ such that $\forall i \in [5]$, $\mathsf{FE.Dec}(\mathsf{CT}_i, \mathsf{K}_i) = x$

The output of the algorithm is the ciphertext $\mathsf{CT} = (\{\mathsf{CT}_i\}_{i \in [5]}, \pi)$
π is computed for statement 1 of relation R.

- **Key Generation** $\mathsf{VFE.KeyGen}(\mathsf{PP}, \mathsf{MSK}, f)$:
  To generate the function secret key K^f for a function f, the key generation algorithm does the following:
  1. $\forall i \in [5]$, compute $\mathsf{K}^f_i = \mathsf{FE.KeyGen}(\mathsf{MSK}_i, f; r_i)$.
  2. Compute a proof $\gamma \leftarrow \mathsf{Prove}(y, w)$ for the statement that $y \in L_1$ using witness w where:
     $y = (\{\mathsf{K}^f_i\}_{i \in [5]}, \mathsf{PP})$,
     $w = (f, \mathsf{MSK}, \{r_i\}_{i \in [5]}, 0^3, 5, 0)$.
     L_1 is defined corresponding to the relation R_1 defined below.
- **Relation** R_1:
  Instance: $y = (\{\mathsf{K}^f_i\}_{i \in [5]}, \mathsf{PP})$.
  Witness: $w = (f, \mathsf{MSK}, \{r_i\}_{i \in [5]}, i_1, i_2, j, k)$
  $R_1(y, w) = 1$ if and only if either of the following conditions hold:
  1. 4 out of 5 constituent function secret keys (except index j) are keys for the same function and are constructed using honestly generated secret keys. Also, Z_2 is a commitment to 1. That is,
     (a) $\forall i \in [5]/\{j\}$, $\mathsf{K}^f_i = \mathsf{FE.KeyGen}(\mathsf{MSK}_i, f; r_i)$.

   (b) $\forall i \in [5]/\{j\}$, $\mathsf{S}_i = \mathsf{Com}(\mathsf{MSK}_i; s_i)$ and $\mathsf{MSK}_i \leftarrow \mathsf{FE.Setup}(1^\lambda; p_i)$
   (c) $\mathsf{Z}_2 = \mathsf{Com}(1; u_1)$
   (OR)
2. 4 out of 5 constituent function secret keys (except index k) are keys for the same function and are constructed using honestly generated secret keys. Z_{CT} is a commitment to a set of ciphertexts CT such that each constituent ciphertext in CT when decrypted with the corresponding function secret key gives the same output . That is,
   (a) $\forall i \in [5]/\{k\}$, $\mathsf{K}_i^f = \mathsf{FE.KeyGen}(\mathsf{MSK}_i, f; r_i)$.
   (b) $\forall i \in [5]/\{k\}$, $\mathsf{S}_i = \mathsf{Com}(\mathsf{MSK}_i; s_i)$ and $\mathsf{MSK}_i \leftarrow \mathsf{FE.Setup}(1^\lambda; p_i)$
   (c) $\mathsf{Z}_{\mathsf{CT}} = \mathsf{Com}(\mathsf{CT}; a)$.
   (d) $\exists x \in \mathcal{X}_\lambda$ such that $\forall i \in [5]$, $\mathsf{FE.Dec}(\mathsf{CT}_i, \mathsf{K}_i^f) = x$
   (OR)
3. 2 constituent function secret keys (corresponding to indices i_1, i_2) are keys for the same function and are constructed using honestly generated secret keys. Z_f is a commitment to all the constituent function secret keys, Z_1 is a commitment to 0 and Z_3 is a commitment to 0. That is,
   (a) $\forall i \in \{\imath_1, i_2\}$, $\mathsf{K}_i^f = \mathsf{FE.KeyGen}(\mathsf{MSK}_i, f; r_i)$.
   (b) $\forall i \in \{\imath_1, i_2\}$, $\mathsf{S}_i = \mathsf{Com}(\mathsf{MSK}_i; s_i)$ and $\mathsf{MSK}_i \leftarrow \mathsf{FE.Setup}(1^\lambda; p_i)$
   (c) $\mathsf{Z}_\mathsf{f} = \mathsf{Com}(\{\mathsf{K}_i^f\}_{i\in[5]}; b)$.
   (d) $\mathsf{Z}_1 = \mathsf{Com}(0; u_1)$.
   (e) $\mathsf{Z}_3 = \mathsf{Com}(0; u_3)$.

The output of the algorithm is the function secret key $\mathsf{K}^f = (\{\mathsf{K}_i^f\}_{i\in[5]}, \gamma)$. γ is computed for statement 1 of relation R_1.

- **Decryption** $\mathsf{VFE.Dec}(\mathsf{PP}, \mathsf{K}^f, \mathsf{CT})$:
  This algorithm decrypts the ciphertext $\mathsf{CT} = (\{\mathsf{CT}_i\}_{i\in[5]}, \pi)$ using function secret key $\mathsf{K}^f = (\{\mathsf{K}_i^f\}_{i\in[5]}, \gamma)$ in the following way:
  1. Let $y = (\{\mathsf{CT}_i\}_{i\in[5]}, \mathsf{PP})$ be the statement corresponding to proof π. If $\mathsf{Verify}(y, \pi) = 0$, then stop and output $\perp$. Else, continue to the next step.
  2. Let $y_1 = (\{\mathsf{K}_i^f\}_{i\in[5]}, \mathsf{PP})$ be the statement corresponding to proof γ. If $\mathsf{Verify}(y_1, \gamma) = 0$, then stop and output $\perp$. Else, continue to the next step.
  3. For $i \in [5]$, compute $m_i = \mathsf{FE.Dec}(\mathsf{CT}_i, \mathsf{K}_i^f)$. If at least 3 of the m_i's are equal (let's say that value is m), output m. Else, output $\perp$.
- **VerifyCT** $\mathsf{VFE.VerifyCT}(\mathsf{PP}, \mathsf{CT})$:
  Given a ciphertext $\mathsf{CT} = (\{\mathsf{CT}_i\}_{i\in[5]}, \pi)$, this algorithm checks whether the ciphertext was generated correctly using the master secret key corresponding to the public parameters PP. Let $y = (\{\mathsf{CT}_i\}_{i\in[5]}, \mathsf{PP})$ be the statement corresponding to proof π. If $\mathsf{Verify}(y, \pi) = 1$, it outputs 1. Else, it outputs 0.
- **VerifyK** $\mathsf{VFE.VerifyK}(\mathsf{PP}, \mathsf{K})$:
  Given a function secret key $\mathsf{K} = (\{\mathsf{K}_i\}_{i\in[5]}, \gamma)$, this algorithm checks whether the key was generated correctly for some function using the master secret key corresponding to public parameters PP. Let $y = (\{\mathsf{K}_i\}_{i\in[5]}, \mathsf{PP})$ be the statement corresponding to proof γ. If $\mathsf{Verify}(y, \gamma) = 1$, it outputs 1. Else, it outputs 0.

Correctness: Correctness follows directly from the correctness of the underlying FE scheme, correctness of the commitment scheme and the completeness of the NIWI proof system.

The proofs for verifiability and security can be found in the full version [5].

Verifiable Multi-Input Functional Encryption: We also study verifiability in the case of multi-input functional encryption. The construction (and proofs) of a verifiable multi-input functional encryption scheme are given in the full version [5].

6 Verifiable Indistinguishability Obfuscation

In this section, we first we recall the notion of indistinguishability obfuscation that was first proposed by [6] and then define the notion of verifiable indistinguishability obfuscation. For indistinguishability obfuscation, intuitively, we require that for any two circuits C_0 and C_1 that are "functionally equivalent" (i.e. for all inputs x in the domain, $C_0(x) = C_1(x)$, the obfuscation of C_0 must be computationally indistinguishable from the obfuscation of C_1. Below, we present the formal definition following the syntax of [14].

Definition 5 *(Indistinguishability Obfuscation). A uniform PPT machine* $\mathsf{i}\mathcal{O}$ *is called an indistinguishability obfuscator for a circuit class* $\{\mathcal{C}_\lambda\}_{\lambda \in \mathbb{N}}$ *if the following conditions are satisfied:*

- *Functionality:*
  For every $\lambda \in \mathbb{N}$*, every* $C \in \mathcal{C}_\lambda$ *, every input* x *to* C*:*

$$Pr[(\mathsf{i}\mathcal{O}(C))(x) \neq C(x)] <= negl(|C|),$$

  where the probability is over the coins of $\mathsf{i}\mathcal{O}$*.*
- *Polynomial Slowdown:*
  There exists a polynomial q *such that for every* $\lambda \in \mathbb{N}$ *and every* $C \in \mathcal{C}_\lambda$*, we have that* $|\mathsf{i}\mathcal{O}(C)| <= q(|C|)$*.*
- *Indistinguishability:*
  For all PPT distinguishers D*, there exists a negligible function* α *such that for every* $\lambda \in \mathbb{N}$*, for all pairs of circuits* $C_0, C_1 \in \mathcal{C}_\lambda$*, we have that if* $C_0(x) = C_1(x)$ *for all inputs* x*, then*

$$|Pr[D(\mathsf{i}\mathcal{O}(C_0))] - Pr[D(\mathsf{i}\mathcal{O}(C_1))]| <= \alpha(\lambda).$$

Definition 6 *($(L, \mathcal{C})-$Restricted Verifiable Indistinguishability Obfuscation). Let* $\mathcal{C} = \{\mathcal{C}_\lambda\}_{\lambda \in \mathbb{N}}$ *denote an ensemble where each* $\mathcal{C}_\lambda$ *is a finite collection of circuits. Let* L *be any language in* NP *defined by a relation* R *satisfying the following two properties:*

1. *For any two circuits $C_0, C_1 \in \mathcal{C}$, if there exists a string w such that $R(C_0, C_1, w) = 1$, then C_0 is equivalent to C_1.*
2. *For any circuit $C \in \mathcal{C}$, $R(C, C, 0) = 1$.*

Let $\mathcal{X}_\lambda$ be the ensemble of inputs to circuits in $\mathcal{C}_\lambda$. Let $\mathcal{P} = \{\mathcal{P}_\lambda\}_{\lambda \in \mathbb{N}}$ be an ensemble where each $\mathcal{P}_\lambda$ is a collection of predicates and each predicate $\mathsf{P} \in \mathcal{P}_\lambda$ takes as input a circuit $C \in \mathcal{C}_\lambda$ and outputs a bit. A verifiable indistinguishability obfuscation scheme consists of the following algorithms:

- $\mathsf{vi}\mathcal{O}(1^\lambda, C, P \in \mathcal{P}_\lambda) \rightarrow \widehat{C}$. *$\mathsf{vi}\mathcal{O}$ is a PPT algorithm that takes as input a security parameter λ, a circuit $C \in \mathcal{C}_\lambda$ and a predicate P in $\mathcal{P}_\lambda$. It outputs an obfuscated circuit $\widehat{C}$.*
- $\mathsf{Eval}(\widehat{C}, x, P \in \mathcal{P}_\lambda) \rightarrow y$. *$\mathsf{Eval}_P$ is a deterministic algorithm that takes as input an obfuscation $\widehat{C}$, an input x and a predicate P in $\mathcal{P}_\lambda$. It outputs a string y.*

The scheme must satisfy the following properties:

- *Functionality:*
  For every $\lambda \in \mathbb{N}$, every $C \in \mathcal{C}_\lambda$, every $\mathsf{P} \in \mathcal{P}_\lambda$ such that $\mathsf{P}(C) = 1$ and every input x to C:
  $$Pr[\mathsf{Eval}(\mathsf{vi}\mathcal{O}(\lambda, C, P), x, P) \neq C(x)] = 0,$$
  where the probability is over the coins of $\mathsf{vi}\mathcal{O}$.
- *Polynomial Slowdown:*
  There exists a polynomial q such that for every $\lambda \in \mathbb{N}$, every $C \in \mathcal{C}_\lambda$ and every $\mathsf{P} \in \mathcal{P}_\lambda$, we have that $|\mathsf{vi}\mathcal{O}(\lambda, C, \mathsf{P})| <= q(|C| + |\mathsf{P}| + \lambda)$. We also require that the running time of Eval on input $(\widehat{C}, x, \mathsf{P})$ is polynomially bounded in $|P| + \lambda + |\widehat{C}|$
- *Indistinguishability:*
  We define indistinguishability with respect to two adversaries $\mathcal{A} = (\mathcal{A}_1, \mathcal{A}_2)$. We place no restriction on the running time of $\mathcal{A}_1$. On input 1^λ $\mathcal{A}_1$ outputs two equivalent circuits (C_0, C_1) in $\mathcal{C}_\lambda$, such that $(C_0, C_1) \in L$. For all PPT distinguishers $\mathcal{A}_2$, there exists a negligible function α such that for every $\lambda \in N$, for pairs of circuits (C_0, C_1) and for all predicates $\mathsf{P} \in \mathcal{P}_\lambda$, we have that if $C_0(x) = C_1(x)$ for all inputs x and $\mathsf{P}(C_0) = \mathsf{P}(C_1)$, then
  $$|Pr[\mathcal{A}_2(\mathsf{vi}\mathcal{O}(\lambda, C_0, P))] - Pr[\mathcal{A}_2(\mathsf{vi}\mathcal{O}(\lambda, C_1, P))]| \leq \mathsf{negl}(\lambda)$$
- *Verifiability:*
  In addition to the above algorithms, there exists an additional deterministic polynomial time algorithm **VerifyO** *that takes as input a string in $\{0,1\}^*$ and a predicate $\mathsf{P} \in \mathcal{P}_\lambda$. It outputs 1 or 0. We say that the obfuscator $\mathsf{vi}\mathcal{O}$ is verifiable if: For any $\mathsf{P} \in \mathcal{P}_\lambda$ and $\widehat{C} \in \{0,1\}^*$, if* **VerifyO**$(\widehat{C}, \mathsf{P}) = 1$, *then there exists a circuit $C \in \mathcal{C}_\lambda$ such that $P(C) = 1$ and for all $x \in \mathcal{X}_\lambda$,* $\mathsf{Eval}(\widehat{C}, x, \mathsf{P}) = C(x)$.

6.1 Construction

Let $\mathcal{C} = \{\mathcal{C}\}_\lambda$ be the set of all polynomial sized circuits and let L_{eq} be an NP language given by some relation R_{eq}.

Relation R_{eq}:
Instance: C', D'
Witness: γ
$\mathsf{R}_{eq}(C', D', \pi) = 1$ implies that:

1. $C' = D' \in \mathcal{C}_\lambda$ for some $\lambda \in \mathbb{N}$. That is, both circuits are equal. (OR)
2. $C', D' \in \mathcal{C}_\lambda$, and there exists a witness γ of size $\mathsf{poly}(|C'|, |D'|)$ proving that C' is functionally equivalent to D'.

We now construct an $(L_{eq}, \mathcal{C})$−restricted verifiable indistinguishability obfuscation scheme. Let $\mathsf{i}\mathcal{O}$ be a perfectly correct indistinguishability obfuscator and $(\mathsf{Prove}, \mathsf{Verify})$ be a NIWI for NP. Formally, we prove the following theorem:

Theorem 7. *Assuming* NIWI *is a witness indistinguishable proof system and* $\mathsf{i}\mathcal{O}$ *is a secure indistinguishability obfuscator for* $\mathcal{C}_\lambda$*, the proposed scheme* $\mathsf{vi}\mathcal{O}$ *is a secure* $(L_{eq}, \mathcal{C})$*−restricted verifiable indistinguishability obfuscator.*

$\underline{\mathsf{vi}\mathcal{O}(\mathbf{1}^\lambda, \mathbf{C}, \mathsf{P})}$: The obfuscator does the following.

- Compute $C^i = \mathsf{i}\mathcal{O}(C; r_i)\ \forall i \in [3]$.
- Compute a NIWI proof π for the following statement $(\mathsf{P}, C^1, C^2, C^3) \in L$ using witness $(1, 2, C, C, r_1, r_2, 0)$ where L is an NP language defined by the following relation R_1 where
  Relation R_1
  Instance: $y = (\mathsf{P}, C^1, C^2, C^3)$
  Witness: $w = (i, j, C_i, C_j, r_i, r_j, \gamma)$
  $\mathsf{R}_1(y, w) = 1$ if and only if:
  1. $C^i = \mathsf{i}\mathcal{O}(C_i; r_i)$ and $C^j = \mathsf{i}\mathcal{O}(C_j; r_j)$ where $i \neq j$ and $i, j \in [3]$. (AND)
  2. $\mathsf{P}(C^i) = \mathsf{P}(C^j) = 1$ (AND)
  3. $R_{eq}(C_i, C_j, \gamma) = 1$.
- Output (C^1, C^2, C^3, π) as the obfuscation.

$\underline{\mathsf{Eval}(\mathcal{O} = (\mathbf{C^1}, \mathbf{C^2}, \mathbf{C^3}, \pi), \mathbf{x}, \mathsf{P})}$: To evaluate:

- Verify the proof π. Output $\perp$ if the verification fails.
- Otherwise, output the majority of $\{C^i(x)\}_{i \in [3]}$.

We now investigate the properties of this scheme.

Correctness: By completeness of NIWI and correctness of the obfuscator $\mathsf{i}\mathcal{O}$ it is straightforward to see that our obfuscator is correct.

Verifiability: We now present the algorithm **VerifyO**. It takes as input an obfuscation (C^1, C^2, C^3, π) and a predicate P. It outputs 1 if π verifies and 0 otherwise. Note that if π verifies then there are two indices $i, j \in [3]$ such that C^i (C^j) is an $\mathsf{i}\mathcal{O}$ obfuscation of some circuit C_i (C_j) and it holds that

$\mathsf{P}(C_i) = \mathsf{P}(C_j) = 1$. Also, either $C_i = C_j$ or C_i is equivalent to C_j (due to the soundness of NIWI). Hence, the evaluate algorithm always outputs $C_i(x)$ on any input x due to perfect correctness of $\mathsf{i}\mathcal{O}$.

Security Proof: Let P be a predicate and (C_0, C_1) be any two equivalent circuits in $\mathcal{C}_\lambda$ such that $P(C_0) = P(C_1) = 1$ and there exists a string γ_1 such that $\mathsf{R}_{\mathsf{eq}}(C_0, C_1, \gamma_1) = 1$. Let (C^1, C^2, C^3, π) be the challenge obfuscated circuit. We now define indistinguishable hybrids such that the first hybrid (Hybrid_0) corresponds to the real world security game where the challenger obfuscates C_0 and the final hybrid (Hybrid_5) corresponds to the security game where the challenger obfuscates C_1.

- $\mathbf{Hybrid_0}$: In this hybrid, $C^i = \mathsf{i}\mathcal{O}(C_0; r_i)\ \forall i \in [3]$ and $(1, 2, C_0, C_0, r_1, r_2, 0)$ is used as a witness to compute π.
- $\mathbf{Hybrid_1}$: This hybrid is same as the previous hybrid except that C^3 is computed as $C^3 = \mathsf{i}\mathcal{O}(C_1; r_3)$.
- $\mathbf{Hybrid_2}$: This hybrid is same as the previous hybrid except that the witness used to compute π is$(1, 3, C_0, C_1, r_1, r_2, \gamma_1)$ where γ_1 is the witness for the statement $(C_0, C_1) \in L_{eq}$.
- $\mathbf{Hybrid_3}$: This hybrid is identical to the previous hybrid except that C^2 is computed as $C^2 = \mathsf{i}\mathcal{O}(C_1; r_2)$.
- $\mathbf{Hybrid_4}$: This hybrid is same as the previous hybrid except that the witness used to compute π is $(2, 3, C_1, C_1, r_1, r_2, 0)$.
- $\mathbf{Hybrid_5}$: This hybrid is identical to the previous hybrid except that $C^1 = \mathsf{i}\mathcal{O}(C_1; r_1)$. This hybrid corresponds to the real world security game where the challenger obfuscates C_1.

Now, we prove indistinguishability of the hybrids.

Lemma 3. *Assuming* $\mathsf{i}\mathcal{O}$ *is a secure indistinguishability obfuscator for* $\mathcal{C}_\lambda$, Hybrid_0 *is computationally indistinguishable from* Hybrid_1.

Proof. Note that the only difference between Hybrid_0 and Hybrid_1 is the way C^3 is generated. In Hybrid_0, it is generated as an obfuscation of C_0, while in Hybrid_1 it is generated as an obfuscation of C_1. Since C_0 and C_1 are equivalent the lemma now follows from the security of $\mathsf{i}\mathcal{O}$.

Lemma 4. *Assuming* NIWI *is a witness indistinguishable proof system,* Hybrid_1 *is computationally indistinguishable from* Hybrid_2.

Proof. Note that the only difference between Hybrid_1 and Hybrid_2 is the way in which π is generated. In Hybrid_1 it uses $(1, 2, C_0, C_0, r_1, r_2, 0)$ as its witness while in Hybrid_2 it uses $(1, 3, C_0, C_1, r_1, r_2, \gamma_1)$ as its witness where γ_1 is the witness for the instance (C_0, C_1) satisfying the relation R_{eq}. The lemma now follows due to the witness indistinguishability of NIWI.

Lemma 5. *Assuming* $\mathsf{i}\mathcal{O}$ *is a secure indistinguishability obfuscator for* $\mathcal{C}_\lambda$, Hybrid_2 *is computationally indistinguishable from* Hybrid_3.

Proof. The only difference between the two hybrids is that C^2 is generated as an obfuscation of C_0 in Hybrid_2 and as an obfuscation of C_1 in Hybrid_3. Since C_0 and C_1 are equivalent the lemma now follows from the security of $i\mathcal{O}$.

Lemma 6. *Assuming* NIWI *is a witness indistinguishable proof system,* Hybrid_3 *is computationally indistinguishable from* Hybrid_4.

Proof. Note that the only difference between Hybrid_3 and Hybrid_4 is the way π is generated. In Hybrid_3 it uses $(1, 3, C_0, C_1, r_1, r_3, \gamma_1)$ as its witness while in Hybrid_4 it uses $(2, 3, C_1, C_1, r_2, r_3, 0)$ as its witness where γ_1 is the witness for the instance (C_0, C_1) satisfying the relation $\mathbf{R_{eq}}$. The lemma now follows due to the witness indistinguishability of NIWI.

Lemma 7. *Assuming* $i\mathcal{O}$ *is a secure indistinguishability obfuscator for* $\mathcal{C}_\lambda$, Hybrid_4 *is computationally indistinguishable from* Hybrid_5.

Proof. The only difference between the two hybrids is that C^1 is generated as an obfuscation of C_0 in Hybrid_4 and as an obfuscation of C_1 in Hybrid_5. Since C_0 and C_1 are equivalent the lemma now follows from the security of $i\mathcal{O}$.

References

1. Abdalla, M., Raykova, M., Wee, H.: Multi-input inner-product functional encryption from pairings. IACR Cryptology ePrint Archive 2016, 425 (2016). http://eprint.iacr.org/2016/425
2. Agrawal, S., Gorbunov, S., Vaikuntanathan, V., Wee, H.: Functional encryption: new perspectives and lower bounds. In: Canetti, R., Garay, J.A. (eds.) CRYPTO 2013. LNCS, vol. 8043, pp. 500–518. Springer, Heidelberg (2013). doi:10.1007/978-3-642-40084-1_28
3. Ananth, P., Jain, A.: Indistinguishability obfuscation from compact functional encryption. In: Gennaro, R., Robshaw, M. (eds.) CRYPTO 2015. LNCS, vol. 9215, pp. 308–326. Springer, Heidelberg (2015). doi:10.1007/978-3-662-47989-6_15
4. Ananth, P., Jain, A., Sahai, A.: Achieving compactness generically: Indistinguishability obfuscation from non-compact functional encryption. IACR Cryptology ePrint Archive 2015, 730 (2015). http://eprint.iacr.org/2015/730
5. Badrinarayanan, S., Goyal, V., Jain, A., Sahai, A.: Verifiable functional encryption. IACR Cryptology ePrint Arch. **2016**, 629 (2016)
6. Barak, B., Goldreich, O., Impagliazzo, R., Rudich, S., Sahai, A., Vadhan, S., Yang, K.: On the (im)possibility of obfuscating programs. In: Kilian, J. (ed.) CRYPTO 2001. LNCS, vol. 2139, pp. 1–18. Springer, Heidelberg (2001). doi:10.1007/3-540-44647-8_1
7. Barak, B., Ong, S.J., Vadhan, S.P.: Derandomization in cryptography. SIAM J. Comput. **37**(2), 380–400 (2007). http://dx.doi.org/10.1137/050641958
8. Barbosa, M., Farshim, P.: Delegatable homomorphic encryption with applications to secure outsourcing of computation. In: Dunkelman, O. (ed.) CT-RSA 2012. LNCS, vol. 7178, pp. 296–312. Springer, Heidelberg (2012). doi:10.1007/978-3-642-27954-6_19

9. Bitansky, N., Paneth, O.: ZAPs and non-interactive witness indistinguishability from indistinguishability obfuscation. In: Dodis, Y., Nielsen, J.B. (eds.) TCC 2015. LNCS, vol. 9015, pp. 401–427. Springer, Heidelberg (2015). doi:10.1007/978-3-662-46497-7_16
10. Bitansky, N., Vaikuntanathan, V.: Indistinguishability obfuscation from functional encryption. In: FOCS (2015)
11. Boneh, D., Franklin, M.K.: Identity-based encryption from the weil pairing. SIAM J. Comput. **32**(3), 586–615 (2003)
12. Boneh, D., Gentry, C., Gorbunov, S., Halevi, S., Nikolaenko, V., Segev, G., Vaikuntanathan, V., Vinayagamurthy, D.: Fully key-homomorphic encryption, arithmetic circuit abe and compact garbled circuits. In: Nguyen, P.Q., Oswald, E. (eds.) EUROCRYPT 2014. LNCS, vol. 8441, pp. 533–556. Springer, Heidelberg (2014). doi:10.1007/978-3-642-55220-5_30
13. Boneh, D., Sahai, A., Waters, B.: Functional encryption: definitions and challenges. In: Ishai, Y. (ed.) TCC 2011. LNCS, vol. 6597, pp. 253–273. Springer, Heidelberg (2011). doi:10.1007/978-3-642-19571-6_16
14. Garg, S., Gentry, C., Halevi, S., Raykova, M., Sahai, A., Waters, B.: Candidate indistinguishability obfuscation and functional encryption for all circuits. In: FOCS (2013)
15. Goldwasser, S., Kalai, Y.T., Popa, R.A., Vaikuntanathan, V., Zeldovich, N.: Reusable garbled circuits and succinct functional encryption. In: STOC 2013 (2013)
16. Gorbunov, S., Vaikuntanathan, V., Wee, H.: Functional encryption with bounded collusions via multi-party computation. In: Safavi-Naini, R., Canetti, R. (eds.) CRYPTO 2012. LNCS, vol. 7417, pp. 162–179. Springer, Heidelberg (2012). doi:10.1007/978-3-642-32009-5_11
17. Gorbunov, S., Vaikuntanathan, V., Wee, H.: Attribute-based encryption for circuits. In: STOC 2013 (2013)
18. Gorbunov, S., Vaikuntanathan, V., Wee, H.: Predicate encryption for circuits from LWE. In: Gennaro, R., Robshaw, M. (eds.) CRYPTO 2015. LNCS, vol. 9216, pp. 503–523. Springer, Heidelberg (2015). doi:10.1007/978-3-662-48000-7_25
19. Goyal, V.: Reducing trust in the PKG in identity based cryptosystems. In: Menezes, A. (ed.) CRYPTO 2007. LNCS, vol. 4622, pp. 430–447. Springer, Heidelberg (2007). doi:10.1007/978-3-540-74143-5_24
20. Goyal, V., Lu, S., Sahai, A., Waters, B.: Black-box accountable authority identity-based encryption. In: CCS (2008)
21. Goyal, V., Pandey, O., Sahai, A., Waters, B.: Attribute-based encryption for fine-grained access control of encrypted data. In: CCS (2006)
22. Groth, J., Ostrovsky, R., Sahai, A.: Non-interactive zaps and new techniques for NIZK. In: Dwork, C. (ed.) CRYPTO 2006. LNCS, vol. 4117, pp. 97–111. Springer, Heidelberg (2006). doi:10.1007/11818175_6
23. Katz, J., Sahai, A., Waters, B.: Predicate encryption supporting disjunctions, polynomial equations, and inner products. In: Smart, N. (ed.) EUROCRYPT 2008. LNCS, vol. 4965, pp. 146–162. Springer, Heidelberg (2008). doi:10.1007/978-3-540-78967-3_9
24. Libert, B., Ramanna, S., Yung, M.: Functional commitment schemes: From polynomial commitments to pairing-based accumulators from simple assumptions. In: ICALP 2016 (2016)
25. Okamoto, T., Takashima, K.: Fully secure functional encryption with general relations from the decisional linear assumption. In: Rabin, T. (ed.) CRYPTO 2010. LNCS, vol. 6223, pp. 191–208. Springer, Heidelberg (2010). doi:10.1007/978-3-642-14623-7_11

26. O'Neill, A.: Definitional issues in functional encryption. IACR Cryptology ePrint Archive 2010, 556 (2010). http://eprint.iacr.org/2010/556
27. Sahai, A., Seyalioglu, H.: Worry-free encryption: functional encryption with public keys. In: CCS (2010)
28. Sahai, A., Seyalioglu, H.: Fully secure accountable-authority identity-based encryption. In: Catalano, D., Fazio, N., Gennaro, R., Nicolosi, A. (eds.) PKC 2011. LNCS, vol. 6571, pp. 296–316. Springer, Heidelberg (2011). doi:10.1007/978-3-642-19379-8_19
29. Sahai, A., Waters, B.: Fuzzy identity-based encryption. In: Cramer, R. (ed.) EUROCRYPT 2005. LNCS, vol. 3494, pp. 457–473. Springer, Heidelberg (2005). doi:10.1007/11426639_27
30. Sahai, A., Waters, B.: Slides on functional encryption, powerpoint presentation (2008). http://www.cs.utexas.edu/~bwaters/presentations/files/functional.ppt
31. Takashima, K.: Expressive attribute-based encryption with constant-size ciphertexts from the decisional linear assumption. In: Abdalla, M., Prisco, R. (eds.) SCN 2014. LNCS, vol. 8642, pp. 298–317. Springer, Heidelberg (2014). doi:10.1007/978-3-319-10879-7_17
32. Waters, B.: Efficient identity-based encryption without random oracles. In: Cramer, R. (ed.) EUROCRYPT 2005. LNCS, vol. 3494, pp. 114–127. Springer, Heidelberg (2005). doi:10.1007/11426639_7

ABE and IBE

Dual System Encryption Framework in Prime-Order Groups via Computational Pair Encodings

Nuttapong Attrapadung(✉)

National Institute of Advanced Industrial Science and Technology (AIST), Tokyo, Japan
n.attrapadung@aist.go.jp

Abstract. We propose a new generic framework for achieving fully secure attribute based encryption (ABE) in *prime-order* bilinear groups. Previous generic frameworks by Wee (TCC'14) and Attrapadung (Eurocrypt'14) were given in *composite-order* bilinear groups. Both provide abstractions of dual-system encryption techniques introduced by Waters (Crypto'09). Our framework can be considered as a prime-order version of Attrapadung's framework and works in a similar manner: it relies on a main component called *pair encodings*, and it generically compiles any secure pair encoding scheme for a predicate in consideration to a fully secure ABE scheme for that predicate. One feature of our new compiler is that although the resulting ABE schemes will be newly defined in prime-order groups, we require essentially the same security notions of pair encodings as before. Beside the security of pair encodings, our framework assumes only the Matrix Diffie-Hellman assumption (Escala *et al.*, Crypto'13), which includes the Decisional Linear assumption as a special case.

Recently and independently, prime-order frameworks are proposed also by Chen *et al.* (Eurocrypt'15), and Agrawal and Chase (TCC'16-A). The main difference is that their frameworks can deal only with *information-theoretic* encodings, while ours can also deal with *computational* ones, which admit wider applications. We demonstrate our applications by obtaining the first fully secure prime-order realizations of ABE for regular languages, ABE for monotone span programs with short-ciphertext, short-key, or completely unbounded property, and ABE for branching programs with short-ciphertext, short-key, or unbounded property.

Keywords: Attribute-based encryption · Full security · Prime-order groups

1 Introduction

Attribute based encryption (ABE), initiated by Sahai and Waters [40], is an emerging paradigm that extends beyond normal public-key encryption. In an ABE scheme for predicate $R : \mathbb{X} \times \mathbb{Y} \to \{0, 1\}$, a ciphertext is associated with a ciphertext attribute, say, $Y \in \mathbb{Y}$, while a key is associated with a key attribute,

J.H. Cheon and T. Takagi (Eds.): ASIACRYPT 2016, Part II, LNCS 10032, pp. 591–623, 2016.
DOI: 10.1007/978-3-662-53890-6_20

say, $X \in \mathbb{X}$, and the decryption is possible if and only if $R(X, Y) = 1$.[1] In Key-Policy (KP) type, $\mathbb{X}$ is a set of Boolean functions (often called *policies*), while $\mathbb{Y}$ is a set of inputs to functions, and we define $R(f, x) = f(x)$. Ciphertext-Policy (CP) type is the dual of KP where the roles of $\mathbb{X}$ and $\mathbb{Y}$ are swapped (that is, policies are associated to ciphertexts). Besides direct applications of fine-grained access control [21], ABE is also known to imply verifiable computation outsourcing [38].

The standard security requirement for ABE is *full security*, where an adversary is allowed to adaptively query keys for any attribute X as long as $R(X, Y) = 0$, where Y is an adversarially chosen attribute for a challenge ciphertext. *Dual system encryption techniques* introduced by Waters [44] have been successful approaches for constructing fully secure ABE systems that are based on bilinear groups. Despite being versatile as they can be applied to ABE systems for many predicates, until only recently, however, there were no known generic frameworks that can use the techniques in a black-box and modular manner. Wee [46] and Attrapadung [3] recently proposed such generic frameworks that abstract the dual system techniques by decoupling what seem to be essential underlying primitives and characterizing their sufficient conditions so as to obtain fully-secure ABE automatically via generic constructions. However, their frameworks are inherently constructed over bilinear groups of *composite-order*. Although composite-order bilinear groups are more intuitive to work with, especially in the case of dual system techniques, *prime-order* bilinear groups are more preferable as they provide more efficient and compact instantiations. This has been motivated already in a line of research [18,22,24,28,29,34,36,41]. More concretely, group elements in composite-order groups are more than 12 times larger than those in prime-order groups for the same security level (3072 bits or 3248 bits for composite-order vs 256 bits for prime-order in case of 128-bit security, according to NIST or ECRYPT II recommendations [22]). Regarding time performances, Guillevic [22] reported that bilinear pairings are 254 times slower in composite-order than in prime-order groups for the same 128-bit security. Moreover, exponentiations are also more than 200 times slower [22, Table 6]. In this work, our goal is to propose a generic framework for dual-system encryption in prime-order groups.

The generic frameworks of [3,46] work similarly but with the difference that the latter [3] captures also dual system techniques with *computational approaches*, which are generalized from techniques implicitly used in the ABE of Lewko and Waters [32]. (The former [46] only captures the traditional dual systems, which implicitly use information-theoretic approaches). Using computational approaches, the framework of [3] is able to obtain the first fully secure schemes for many ABE primitives for which only selectively secure constructions were known before, including KP-ABE for regular languages [45],

[1] Traditionally, ABE refers to only ABE for *Boolean formulae* predicate [21]. In this paper, however, we use the term ABE for arbitrary predicate R. Indeed, it corresponds to the "public-index predicate encryption" class of functional encryption, as per [12].

KP-ABE for Boolean formulae[2] with constant-size ciphertexts [9], and (completely) unbounded KP-ABE for Boolean formulae [31,39]. Moreover, Attrapadung and Yamada [10] recently show that, within the framework of [3], we can generically convert ABE to its *dual* scheme, *i.e.*, key-policy to ciphertext-policy type, and vice versa. They also show a conversion to its *dual-policy* [8] type, which is the conjunctive of KP and CP. Many instantiations were then obtained in [10], including the first CP-ABE for formulae with short keys. We therefore choose to build upon [3].

1.1 Our Contributions on Framework

New Framework. We present a new generic framework for achieving fully secure ABE in *prime-order groups*. It is generic in the sense that it can be applied to ABE for *arbitrary* predicate. Our framework extends the framework of [3], which was constructed in composite-order groups, and works in a similar manner as follows. First, the main component is a primitive called *pair encoding* scheme defined for a predicate. Second, we provide a generic construction that compiles any secure pair encoding scheme for a predicate R to a fully secure ABE scheme for the same predicate R. The *security* requirement for the underlying encoding scheme is exactly the same as that in the framework of [3]; in particular, our framework can deal with both information-theoretic and computational encodings. On the other hand, we restrict the *syntax* of encodings into a class we call *regular encodings*, via some simple requirements. This confinement, however, seems natural and does not affect any concrete pair encoding schemes proposed so far [3,10,46]. Beside the security of pair encodings, our framework assumes only the Matrix Diffie-Hellman assumption [17], which includes the Decisional Linear assumption as a special case.

Conceptually, since our framework uses the same security requirement for pair encodings as in the composite-order framework of [3], we can view it as an automatic way for translating ABE from composite-order to prime-order settings.

Prime-order frameworks are recently and independently proposed by Chen, Gay, and Wee [14] and Agrawal and Chase [2], albeit they can deal only with information-theoretic encodings. We compare them later in Sect. 1.4. As a side result, we also simplify our scheme using a simpler basis from [14] in Sect. 8.

1.2 Our Contributions on Instantiations

New Instantiations (the First in Prime-order Settings). By using exactly the same encoding instantiations in [3,10], we automatically obtain fully secure ABE schemes, *for the first time in prime-order groups*, for various predicates:

- KP-ABE and CP-ABE for regular languages,

[2] Or more precisely, ABE for monotone span programs, which implies ABE for Boolean formulae [21]. We will use both terms interchangeably.

Table 1. Composite-order ABE, positioned by properties (for comparing to Table 2)

Predicate	Properties		Unbounded		KP	CP	DP
	Security	Universe	Input	Multi-use			
ABE-PDS	full	-	-	-	A14 [3]	AY15 [10]	AY15 [10]
Unbounded ABE-MSP	selective	large	yes	yes	LW11 [32],	sub	sub
	full	small	yes	yes	sub	LW12 [33]	sub
	full	large	yes	no	sub	sub	sub
	full	large	yes	yes	A14 [3]	AY15 [10]	AY15 [10]
Short-Cipher ABE-MSP	selective	large	no	yes	sub	sub	open
	semi	large	no	yes	sub	AC16 [2]	open
	full	large	no	yes	A14 [3]	open	open
Short-Key ABE-MSP	selective	large	no	yes	sub	sub	open
	full	large	no	yes	open	AY15 [10]	open
(Bounded) ABE-MSP	selective	large	no	yes	sub	sub	sub
	full	small	no	no	LOS+10 [34], A14 [3], W14 [47]	LOS+10 [34], A14 [3], W14 [47]	AY15 [10]
	full	large	no	no	A14 [3],	A14 [3]	AY15 [10]
ABE-RL	selective	small	-	-	sub	sub	sub
	full	large	-	-	A14 [3]	A14 [3]	AY15 [10]

Acronym: "ABE-PDS" = ABE for policy over doubly-spatial relations, "ABE-MSP" = ABE for monotone span programs, "ABE-RL" = ABE for regular languages, "ABE-BP" = ABE for branching programs. "KP" = key-policy. "CP" = ciphertext-policy. "DP" = dual-policy. "sub" = subsumed (no previous work but is subsumed by another system with stronger properties such as full security or prime-order). "open" = was open problem (before our work and subsequent work that uses ours). "-" = undefined. "Unbounded input" = unbounded size of attribute set size per ciphertext in KP-ABE-MSP, attribute set size per key in CP-ABE-MSP, and input string in ABE-BP. "Unbounded Multi-use" = unbounded multi-use of attributes in a policy in ABE-MSP, and in a branching program in ABE-BP. "semi" = semi-adaptive security.

- KP-ABE for monotone span programs with constant-size ciphertexts,
- CP-ABE for monotone span programs with constant-size keys,
- Completely unbounded KP-ABE and CP-ABE for monotone span programs.

The assumptions for respective encodings are the same as those in [3] (albeit with a minor syntactic change to prime-order groups); some are parameterized assumptions (or often called q-type), as in [3]. Moreover, via the dual-policy conversion of [10], we also obtain their respective dual-policy variants.

We give their detailed comparisons in Tables 5, 6 in Sect. 7. Here, for high-level overview, we position our instantiations in Table 2, which show prime-order schemes by their properties. In Table 2, our instantiations that are the first such schemes for given predicates and properties are specified by **New**. Our new instantiations that are not the first of a kind are specified by **New**′. Table 1 provides composite-order schemes for comparison.

Table 2. Prime-order ABE schemes, positioned by properties

Predicate	Properties		Unbounded		KP	CP	DP
	Security	Universe	Input	Multi-use			
ABE-PDS	full	-	-	-	**New**$_1$	**New**$_2$	**New**$_3$
Unbounded ABE-MSP	selective	large	yes	yes	RW13 [40]	RW13 [40]	sub
	full	small	yes	yes	sub	LW12 [33]	sub
	full	large	yes	no	OT12 [38]	OT12 [38]	sub
	full	large	yes	yes	**New**$_4$	**New**$_5$	**New**$_6$
Short-Cipher ABE-MSP	selective	large	no	yes	ALP11 [9]	sub	sub
	semi	large	no	yes	CW14,T14 [17,43]	AC16 [2]	sub
	full	large	no	yes	**New**$_7$	AHY15 [7]*	**Newer**$_{28}$
Short-Key ABE-MSP	selective	large	no	yes	BGG+14 [12]†	sub	sub
	full	large	no	yes	AHY15 [7]*	**New**$_8$	**Newer**$_{29}$
(Bounded) ABE-MSP	selective	large	no	yes	GPSW06 [22]	W11 [44]	AI09 [8]
	full	small	no	no	CGW15 [15], **New**$'_9$	CGW15 [15], **New**$'_{10}$	**New**$_{11}$
	full	large	no	no	OT10 [37], **New**$'_{12}$	OT10 [37], **New**$'_{13}$	**New**$_{14}$
ABE-RL	selective	small	-	-	W12 [46]	sub	sub
	full	large	-	-	**New**$_{15}$	**New**$_{16}$	**New**$_{17}$
Unbounded ABE-BP	full	-	yes	yes	**New**$_{18}$	**New**$_{19}$	**New**$_{20}$
Short-Cipher ABE-BP	full	-	no	yes	**New**$_{21}$	**Newer**$_{27}$	**Newer**$_{30}$
Short-Key ABE-BP	selective	-	no	yes	GV15 [21]†	sub	sub
	full	-	no	yes	**Newer**$_{26}$	**New**$_{22}$	**Newer**$_{31}$
(Bounded) ABE-BP	selective	-	no	yes	GVW13 [20]†	sub	sub
	full	-	no	no	CGW15 [15], **New**$'_{23}$	CGW15 [15], **New**$'_{24}$	**New**$_{25}$

Acronym: "**New**$_i$" = new instantiations from our framework that are the first such schemes for given predicates and properties. The subscript i is the scheme numbering. "**Newer**$_i$" = newer instantiations (that are the first of a kind) obtained here using a subsequent work to our work, namely [7]. "**New**$'_i$" = new instantiations but not the first of a kind. † refers to a solution based on LWE. * refers to subsequent work that essentially uses our work as their building block. Also refer to the acronym of Table 1.

First Realizations. We also obtain the first-ever realizations of ABE for some predicates, namely,

- Unbounded KP-ABE and CP-ABE for branching programs (BP),
- KP-ABE for branching programs with constant-size ciphertexts,
- CP-ABE for branching programs with constant-size keys.

Unbounded ABE-BP refers to a system that allows an encryptor to associate a ciphertext with an input string of any length (in the case of KP). All of our above ABE-BP schemes are the first such schemes for respective variants even among composite-order or selectively secure schemes. Comparing to the previous schemes, KP-ABE-BP of [14,19,25] are of bounded type and require linear-size ciphertexts and keys[3], while (selective) KP-ABE-BP of [20] achieves short keys. We obtain our above ABE-BP schemes by invoking the theorem stating a generic implication from ABE for monotone span programs (MSP) to ABE-BP (see Remark 6 for further discussion on this theorem).

Update after Subsequent Work. Subsequent to our work, Attrapadung et al. [7] present various conversions for ABE. By applying their conversions to some of our instantiations, they obtain CP-ABE with short ciphertexts and KP-ABE with short keys for (non-)monotone span programs. Now, by applying the ABE-MSP-to-ABE-BP conversion back to their instantiations, we obtain further (fully secure) schemes not explicitly achievable before, namely:

- KP-ABE for branching programs with constant-size keys,
- CP-ABE for branching programs with constant-size ciphertexts.

Moreover, we can combine KP-ABE and CP-ABE both with short keys to DP-ABE with short keys. The same goes for short ciphertexts. We mark the schemes after this update as **Newer**$_i$ in Table 2. Interestingly, all of our results complete the whole Table 2, which had been otherwise filled with open problems before.

1.3 Our Techniques

Due to the lack of space, we defer a more detailed discussion on our techniques to the full version [4]. We provide only a summary here.

Background on [3]**.** We first briefly review the framework of [3]. In the generic construction of [3], a ciphertext CT encrypting M, and a key SK take the forms:

$$\mathsf{CT} = (\boldsymbol{C}, C_0) = (g_1^{\boldsymbol{c}(\boldsymbol{s},\boldsymbol{h})}, Me(g_1, g_2)^{\alpha s_0}), \qquad \mathsf{SK} = g_2^{\boldsymbol{k}(\alpha,\boldsymbol{r},\boldsymbol{h})}$$

where $\boldsymbol{c}$ and $\boldsymbol{k}$ are *encodings* of attributes Y and X associated to a ciphertext and a key, respectively. Here, g_1, g_2 are generators of subgroups of order p_1 of $\mathbb{G}_1, \mathbb{G}_2$, which are asymmetric bilinear groups of composite order $N = p_1p_2p_3$ with bilinear map $e : \mathbb{G}_1 \times \mathbb{G}_2 \to \mathbb{G}_T$. The bold fonts denote vectors. Intuitively, α plays the role of a master key, $\boldsymbol{h}$ represents common variables (or called parameters). These define a public key $\mathsf{PK} = (g_1^{\boldsymbol{h}}, e(g_1, g_2)^{\alpha})$. $\boldsymbol{s}, \boldsymbol{r}$ represents randomness in the ciphertext and the key, respectively, with s_0 being the first element in $\boldsymbol{s}$. The pair $(\boldsymbol{c}, \boldsymbol{k})$ form a *pair encoding* scheme for predicate R. Informally, the main theorem of [3] states that if the pair encoding is secure and subgroup decision assumptions hold, then the ABE scheme (with CT, SK as above) is fully secure.

Our Approach. Towards translating to a new prime-order based framework, we identify a set of features consisting of *element representations*, *procedures*, *properties*, and *assumptions* that are required by the framework of [3]. We list up the first three categories in Sect. 4.

[3] Note that we consider only *Boolean* branching programs here as in [19], in contrast with [14,25], where *arithmetic* branching programs are also considered.

As for assumptions, our goal is to use the security definition of pair encoding "as is", since this will allow us to instantly instantiate the encoding schemes already proposed and proved secure in [3]. If we can leave encoding "as is", we will only have to replace subgroup decision assumptions provided by composite-order groups with some mechanisms from prime-order groups that mimic them.

Candidate Techniques. There are two candidate tools for simulating subgroup decision in prime-order groups: *Dual Pairing Vector Space (DPVS)* [28,35,36] and *Prime-order Dual System Group (PDSG)* [15]. We argue (in the full version [4]) that DPVS would require modifying one of the encoding (in the pair encoding) to an "orthogonal form" in order to enable inner-product spaces, which seems essential in this approach. This, however, would violate our goal to use encoding "as is". We thus turn to use the other tool: PDSG. Although PDSG was devised for specific predicates such as HIBE in the first place [15], it seems compatible to the pair encoding syntax in terms of *element representations* since, roughly speaking, it provides one-to-one translation of elements. (This itself is although implicit in [15]). Intuitively, each $\mathbb{Z}_N$ element in $\boldsymbol{s}, \boldsymbol{r}, \boldsymbol{h}$ is mapped to elements of vector spaces over $\mathbb{Z}_p$ (such as vectors or matrices), and subgroup assumptions are emulated by some subspace assumptions.

Difficulties and Our Solutions. We argue that the out-of-the-box formulation of PDSG [15] is, however, not sufficient for applying to the framework of [3], mainly due to the following four issues.

First, out-of-the-box PDSG does not allow a direct exponentiation procedure that is required by [3], such as $g_1^{\boldsymbol{h}}$. This is since translated elements involve matrices, of which multiplication is not commutative. We solve this by properly re-ordering translated elements in multiplicative terms in encoding, and enabling exponentiation via *left multiplication* of matrices (in exponents). See Sect. 4.

Second, and more importantly, subgroup decision-like assumptions provided by PDSG would guarantee indistinguishability for elements that have *only one element of randomness* in the encoding. On the other hand, pair encodings in the framework of [3] are formulated to deal with *arbitrary number of randomness elements*, that is, $\boldsymbol{s}, \boldsymbol{r}$ can be of any length. We solve this by introducing a new technique that uses *random self-reducibility* of the Matrix-DH assumption. We also note that this technique becomes possible only after our re-formulation, designed for solving the first issue. We depict this in the proof of lemma 2 in Sect. 6.

Third, the syntax of pair encodings [3] allows multiplication such as $h_k h_{k'}$ (and implicitly uses commutativity: $h_k h_{k'} = h_{k'} h_k$), when encodings are paired. However, these elements would translate to matrices, which do not commute. We solve this by restricting the syntax of pair encodings so that such multiplication is not allowed (and using only the associativity property [15]). It turns out that, however, all available pair encodings still satisfy these new restriction; hence, our new framework applies to them. We define this as Rule 1 of *regularity* in Sect. 3.1.

The fourth issue is perhaps the most important since it is unique to our new framework. In order to achieve our goal of using *computational* security of

Table 3. High-level conceptual comparison among generic dual-system frameworks

Framework	Settings	Applicable encodings	Restrictions on encodings	Additional features
W14 [46]	Composite	Info.-theoretic	-	-
A14 [3]	Composite	Info.-theoretic, computational	-	Tighter reduction
CGW15 [14]	Composite, prime	Info.-theoretic	One unit of randomness	Weak attribute-hiding
AC16 [2]	Composite, prime	Info.-theoretic	Rule 1 of our Regularity	Relaxed perfect security
This work	Prime	Info.-theoretic, computational	Regularity	Tighter reduction

encodings "as is", we need to establish a reduction from the new "matrix-form" of encodings, exponentiated over prime-order group elements, to the original encodings, in the security proof. This was not a problem in the original composite-order framework of [3] since the original hybrid proof uses exactly the same form of original encodings. Also, it was not a problem for (prime-order) frameworks using *information-theoretic* encodings [2,14] since, intuitively, information-theoretic properties will preserve regardless of whether their elements are in the exponents. We resolve this issue, for the case of *computational* encodings, by identifying which terms will be needed in the aforementioned reduction and enforcing them to be given out explicitly in encodings *by definition.* We define this as Rule 2–4 of *regularity* in Sect. 3.1. We provide more intuition on this at the end of Sect. 4.

1.4 Independent Works and Their Comparisons

Independently, Chen et al. [14] recently proposed a generic dual-system framework in prime-order groups. The main difference is that our framework can deal with *computationally secure encodings*, while theirs can deal only with information-theoretic ones. As motivated in [3], computational approaches have an advantage in that they are applicable to ABE for predicates where information-theoretic theoretic argument seems insufficient. These include ABE with some *unbounded* properties, or *constant-size* ciphertexts (or keys). We compare some instantiations of [14] that are relevant to ours in Table 2. Another difference is that the syntax of encoding in [14] seems more restricted in the sense that it can deal with only one element of randomness, while our syntax can deal with arbitrary many elements. On one hand, one unit of randomness is shown to suffice for all known information-theoretic encodings in [14]. On the other hand, multi-unit randomness seems essential in more esoteric predicates such as ABE for regular languages (of which information-theoretic encodings are not known). An extension with weak attribute-hiding property is also given in [14] (although currently applicable to small predicate classes such as HIBE, inner-product). Moreover, a simpler basis of PDSG is proposed in [14]. Although our main construction is based upon the original basis of [15], it is possible to use the simplified basis by [14]. We provide this simplification in Sect. 8.

In another concurrent[4] and independent work, Agrawal and Chase [2] also presented a prime-order dual system framework. As in [14], their work consider only information-theoretic encodings, albeit with a useful extension that allows to relax perfect encodings, which yields CP-ABE with short ciphertexts.

In the conceptual view, both frameworks [2,14] unify both composite-order and prime-order groups into one generic construction. Contrastingly, we focus solely on the prime-order generic construction.[5] We compare them in Table 3. A feature of our framework, inherited from [3], is that it enjoys tighter reduction, of which the cost does not depend on the number of post-challenge queries.

Some technical difficulties we pointed out in Sect. 1.3 have been addressed in these frameworks [2,14]. For instance, the loss of commutativity is coped by restricting encodings (differently in [14], but similarly in [2]). Also, the random self-reducibility is implicitly utilized in [2]. On the other hand, the technique that is all unique to ours is our solution in accommodating *computational* encodings.

We comment that although computational encodings enjoy much wider applications than information-theoretic ones, they come with a drawback that some encodings, especially for esoteric predicates, often use parameterized (q-type) assumptions. Some plausible future research directions to reduce them to simpler assumptions may include extending the recent Deja-q method [13,47], or relaxing encodings analogously to [2], but in computational settings.

Some recent subsequent works that use some of our instantiations include ABE with parameter tradeoffs [5] and ABE for range attributes [6].

2 Preliminaries

2.1 Definitions of Attribute Based Encryption

Predicate Family. We consider a predicate family $R = \{R_\kappa\}_{\kappa \in \mathbb{N}^c}$, for some constant $c \in \mathbb{N}$, where a relation $R_\kappa : \mathbb{X}_\kappa \times \mathbb{Y}_\kappa \to \{0,1\}$ is a predicate function that maps a pair of key attribute in a space $\mathbb{X}_\kappa$ and ciphertext attribute in a space $\mathbb{Y}_\kappa$ to $\{0,1\}$. The family index $\kappa = (n_1, n_2, \ldots)$ specifies the description of a predicate from the family. We will often neglect κ for simplicity of exposition.

Attribute Based Encryption Syntax. An ABE scheme for predicate family R consists of the following algorithms. Let $\mathcal{M}$ be the message space.

- $\mathsf{Setup}(1^\lambda, \kappa) \to (\mathsf{PK}, \mathsf{MSK})$: takes as input a security parameter 1^λ and a family index κ of predicate family R, and outputs a master public key PK and a master secret key MSK.

[4] A preliminary version of our full version [4] has been made available before that of [2].

[5] Nevertheless, since we use the same notion of pair encoding as in the composite-order framework of [3], it can be said that our framework together with [3] provide a unified framework albeit with two generic constructions.

- $\mathsf{Encrypt}(Y, M, \mathsf{PK}) \to \mathsf{CT}$: takes as input a ciphertext attribute $Y \in \mathbb{Y}_\kappa$, a message $M \in \mathcal{M}$, and public key PK. It outputs a ciphertext CT.
- $\mathsf{KeyGen}(X, \mathsf{MSK}, \mathsf{PK}) \to \mathsf{SK}$: takes as input a key attribute $X \in \mathbb{X}_\kappa$ and the master key MSK. It outputs a secret key SK.
- $\mathsf{Decrypt}(\mathsf{CT}, \mathsf{SK}) \to M$: given a ciphertext CT with its attribute Y and the decryption key SK with its attribute X, it outputs a message M or $\perp$.

Correctness. Consider all indexes κ, all $M \in \mathcal{M}$, $X \in \mathbb{X}_\kappa$, $Y \in \mathbb{Y}_\kappa$ such that $R_\kappa(X, Y) = 1$. If $\mathsf{Encrypt}(Y, M, \mathsf{PK}) \to \mathsf{CT}$ and $\mathsf{KeyGen}(X, \mathsf{MSK}, \mathsf{PK}) \to \mathsf{SK}$ where $(\mathsf{PK}, \mathsf{MSK})$ is generated from $\mathsf{Setup}(1^\lambda, \kappa)$, then $\mathsf{Decrypt}(\mathsf{CT}, \mathsf{SK}) \to M$.

We use the standard security definition for ABE and refer to the full version [4].

2.2 Bilinear Groups, Notations, and Assumptions

In our framework, for maximum generality and clarity, we consider asymmetric bilinear groups $(\mathbb{G}_1, \mathbb{G}_2, \mathbb{G}_T)$ of prime order p, with an efficiently computable bilinear map $e : \mathbb{G}_1 \times \mathbb{G}_2 \to \mathbb{G}_T$. The symmetric version of our framework can be obtained by just setting $\mathbb{G}_1 = \mathbb{G}_2$. We define a bilinear group generator $\mathcal{G}(\lambda)$ that takes as input a security parameter λ and outputs $(\mathbb{G}_1, \mathbb{G}_2, \mathbb{G}_T, e, p)$. We recall that e has the bilinear property: $e(g_1^a, g_2^b) = e(g_1, g_2)^{ab}$ for any $g_1 \in \mathbb{G}_1, g_2 \in \mathbb{G}_2$, $a, b \in \mathbb{Z}$ and the non-degeneration property: $e(g_1, g_2) \neq 1 \in \mathbb{G}_T$ whenever $g_1 \neq 1 \in \mathbb{G}_1, g_2 \neq 1 \in \mathbb{G}_2$.

Notation for Matrix in the Exponents. Vectors will be treated as either row or column matrices. When unspecified, we shall let it be a row vector. Let $\mathbb{G}$ be a group. Let $\boldsymbol{a} = (a_1, \ldots, a_n)$ and $\boldsymbol{b} = (b_1, \ldots, b_n) \in \mathbb{G}^n$. We denote $\boldsymbol{a} \cdot \boldsymbol{b} = (a_1 \cdot b_1, \ldots, a_n \cdot b_n)$, where '$\cdot$' is the group operation of $\mathbb{G}$. For $g \in \mathbb{G}$ and $\boldsymbol{c} = (c_1, \ldots, c_n) \in \mathbb{Z}^n$, we denote $g^{\boldsymbol{c}} = (g^{c_1}, \ldots, g^{c_n})$. We denote by $\mathbb{GL}_{p,n}$ the group of invertible matrices (the general linear group) in $\mathbb{Z}_p^{n \times n}$. Consider $\boldsymbol{M} \in \mathbb{Z}_p^{d \times n}$ (the set of all $d \times n$ matrices in $\mathbb{Z}_p$). We denote the transpose of $\boldsymbol{M}$ as $\boldsymbol{M}^\top$. Denote $\boldsymbol{M}^{-\top} = (\boldsymbol{M}^\top)^{-1}$. Denote by $g^{\boldsymbol{M}}$ the matrix in $\mathbb{G}^{d \times n}$ of which its (i, j) entry is $g^{\boldsymbol{M}_{i,j}}$, where $\boldsymbol{M}_{i,j}$ is the (i, j) entry of $\boldsymbol{M}$. For $\boldsymbol{Q} \in \mathbb{Z}_p^{\ell \times d}$, we denote $(g^{\boldsymbol{Q}})^{\boldsymbol{M}} = g^{\boldsymbol{Q}\boldsymbol{M}}$. Note that from $\boldsymbol{M}$ and $g^{\boldsymbol{Q}} \in \mathbb{G}^{\ell \times d}$, we can compute $g^{\boldsymbol{Q}\boldsymbol{M}}$ without knowing $\boldsymbol{Q}$, since its (i, j) entry is $\prod_{k=1}^{d} (g^{\boldsymbol{Q}_{i,k}})^{\boldsymbol{M}_{k,j}}$. The same can be said about $g^{\boldsymbol{M}}$ and $\boldsymbol{Q}$. For $\boldsymbol{X} \in \mathbb{Z}_p^{r \times c_1}$ and $\boldsymbol{Y} \in \mathbb{Z}_p^{r \times c_2}$, denote its pairing as:

$$e(g_1^{\boldsymbol{X}}, g_2^{\boldsymbol{Y}}) = e(g_1, g_2)^{\boldsymbol{Y}^\top \boldsymbol{X}} \in \mathbb{G}_T^{c_2 \times c_1}.$$

Projection Maps. $\binom{\boldsymbol{I}_d}{0}$ denotes the $(d+1) \times d$ matrix where the first d rows comprise the identity matrix while the last row is zero. It functions as a left-projection map. That is, $X \binom{\boldsymbol{I}_d}{0} \in \mathbb{Z}_p^{(d+1) \times d}$ is the matrix consisting of all left d columns of X for any $X \in \mathbb{Z}_p^{(d+1) \times (d+1)}$. Similarly, $\binom{0}{1}$ is the $(d+1) \times 1$ matrix where the last row is 1; it functions as a right-projection map.

Matrix-DH Assumptions [17]. We call $\mathcal{D}_d$ a matrix distribution if it outputs (in poly time, with overwhelming probability) matrices in $\mathbb{Z}_p^{(d+1)\times(d+1)}$ of the form:

$$\boldsymbol{T} = \begin{array}{c} \\ d \\ 1 \end{array}\!\!\begin{array}{c} \begin{array}{cc} d & 1 \end{array} \\ \begin{pmatrix} \boldsymbol{M} & \boldsymbol{0} \\ \boldsymbol{c} & 1 \end{pmatrix} \end{array} \xleftarrow{\$} \mathcal{D}_d. \tag{1}$$

such that $\boldsymbol{M}$ is an invertible matrix in $\mathbb{Z}_p^{d\times d}$ (*i.e.*, $\boldsymbol{M} \in \mathbb{GL}_{p,d}$) and $\boldsymbol{c} \in \mathbb{Z}_p^{1\times d}$. We say that the $\mathcal{D}_d$-*Matrix Diffie-Hellman Assumption* for $\mathcal{G}$ holds in $\mathbb{G}_1$ if for all ppt adversaries $\mathcal{A}$, the advantage $\mathsf{Adv}_{\mathcal{A}}^{\mathcal{D}_d\text{-MatDH}}(\lambda) :=$

$$\left| \Pr\left[\mathcal{A}(\mathbb{G}, g_1^{\boldsymbol{T}}, g_1^{\boldsymbol{T}\binom{\boldsymbol{y}}{0}}) = 1\right] - \Pr\left[\mathcal{A}(\mathbb{G}, g_1^{\boldsymbol{T}}, g_1^{\boldsymbol{T}\binom{\boldsymbol{y}}{\hat{y}}}) = 1\right] \right| \tag{2}$$

is negligible in λ, where the probability is taken over $(\mathbb{G}_1, \mathbb{G}_2, \mathbb{G}_T, e, p) \xleftarrow{\$} \mathcal{G}(\lambda)$, $g_1 \xleftarrow{\$} \mathbb{G}_1$, $g_2 \xleftarrow{\$} \mathbb{G}_2$, $\boldsymbol{T} \xleftarrow{\$} \mathcal{D}_d$, $\boldsymbol{y} \xleftarrow{\$} \mathbb{Z}_p^{d\times 1}$, $\hat{y} \xleftarrow{\$} \mathbb{Z}_p$, and the randomness of $\mathcal{A}$. Denote $\mathbb{G} = (\mathbb{G}_1, \mathbb{G}_2, \mathbb{G}_T, e, p, g_1, g_2)$.

Remark 1. We remark that the assumption is progressively weaker as d increases. In symmetric bilinear groups, we require that $d \geq 2$ (otherwise, it is trivially broken [17]), while in asymmetric bilinear groups, we can choose also $d = 1$. The most well-known special case of the $\mathcal{D}_d$-Matrix-DH Assumption is the Decision d-Linear Assumption, for which $\boldsymbol{M}$ are restricted to random diagonal matrices and $\boldsymbol{c}$ is fixed as the vector with all 1's. The SXDH assumption is a special case of the Matrix-DH when $d = 1$ (hence, operates in asymmetric bilinear groups).

Our scheme will use arbitrary $\mathcal{D}_d$ for maximal generality. One can directly tradeoff the weakness of assumption and the sizes of ciphertexts and keys by d.

Random Self Reducibility of Matrix-DH Assumptions. The $\mathcal{D}_d$-Matrix-DH Assumption is random self reducible, as shown in [17]: the problem instance defined by $(\boldsymbol{T}, \binom{\boldsymbol{y}}{\hat{y}})$ can be randomized to another instance defined by $(\boldsymbol{T}, \binom{\boldsymbol{y}'}{\hat{y}'})$. This is done by choosing $\boldsymbol{\delta} \xleftarrow{\$} \mathbb{Z}_p^{d\times 1}, \hat{\delta} \xleftarrow{\$} \mathbb{Z}_p$ and setting $g_1^{\boldsymbol{T}\binom{\boldsymbol{y}'}{\hat{y}'}} = g_1^{\boldsymbol{T}\binom{\boldsymbol{y}}{\hat{y}}\hat{\delta}} g_1^{\boldsymbol{T}\binom{\boldsymbol{\delta}}{0}}$, and observe that $y = 0$ iff $y' = 0$. We can gather each new instance $\binom{\boldsymbol{y}'}{\hat{y}'}$ into columns of a matrix and consider the *m-fold* $\mathcal{D}_d$-Matrix-DH Assumption for which the advantage is defined as $\mathsf{Adv}_{\mathcal{A}}^{m,\mathcal{D}_d\text{-MatDH}}(\lambda) :=$

$$\left| \Pr\left[\mathcal{A}(\mathbb{G}, g_1^{\boldsymbol{T}}, g_1^{\boldsymbol{T}\binom{\boldsymbol{Y}}{\boldsymbol{0}}}) = 1\right] - \Pr\left[\mathcal{A}(\mathbb{G}, g_1^{\boldsymbol{T}}, g_1^{\boldsymbol{T}\binom{\boldsymbol{Y}}{\hat{\boldsymbol{y}}}}) = 1\right] \right|, \tag{3}$$

where the probability is taken over $(\mathbb{G}_1, \mathbb{G}_2, \mathbb{G}_T, e, p) \xleftarrow{\$} \mathcal{G}(\lambda)$, $g_1 \xleftarrow{\$} \mathbb{G}_1$, $g_2 \xleftarrow{\$} \mathbb{G}_2$, $\boldsymbol{T} \xleftarrow{\$} \mathcal{D}_d$, $\boldsymbol{Y} \xleftarrow{\$} \mathbb{Z}_p^{d\times m}$, $\hat{\boldsymbol{y}} \xleftarrow{\$} \mathbb{Z}_p^{1\times m}$, and the randomness of $\mathcal{A}$. Again, we denote $\mathbb{G} = (\mathbb{G}_1, \mathbb{G}_2, \mathbb{G}_T, e, p, g_1, g_2)$. Due to the random self-reducibility, the reduction to the m-fold variant is tight.

Proposition 1 ([17]). *For any integer m, for all ppt adversary $\mathcal{A}$, there exists a ppt algorithm $\mathcal{A}'$ such that* $\mathsf{Adv}_{\mathcal{A}'}^{m,\mathcal{D}_d\text{-MatDH}}(\lambda) = \mathsf{Adv}_{\mathcal{A}}^{\mathcal{D}_d\text{-MatDH}}(\lambda)$.

3 Definition of Pair Encoding

We recall the definition of pair encoding schemes as given in [3]. A pair encoding scheme for predicate family R consists of four deterministic algorithms given by $\mathsf{P} = (\mathsf{Param}, \mathsf{Enc1}, \mathsf{Enc2}, \mathsf{Pair})$ as follows:

- $\mathsf{Param}(\kappa) \to n$. It takes as input an index κ and outputs an integer n, which specifies the number of *common variables* in Enc1, Enc2. For the default notation, let $\boldsymbol{h} = (h_1, \ldots, h_n)$ denote the the list of common variables.
- $\mathsf{Enc1}(X) \to (\boldsymbol{k} = (k_1, \ldots, k_{m_1}); m_2)$. It takes as inputs $X \in \mathbb{X}_\kappa$, and outputs a sequence of polynomials $\{k_i\}_{i\in[1,m_1]}$ with coefficients in $\mathbb{Z}_p$, and $m_2 \in \mathbb{N}$. We require that each polynomial k_i is a *linear combination of monomials* $\alpha, r_j, h_k r_j$, where $\alpha, r_1, \ldots, r_{m_2}, h_1, \ldots, h_n$ are variables. More precisely, it outputs a set of coefficients $\{b_i\}_{i\in[1,m_1]}$, $\{b_{i,j}\}_{i\in[1,m_1],j\in[1,m_2]}$, $\{b_{i,j,k}\}_{i\in[1,m_1],j\in[1,m_2],k\in[1,n]}$ that defines the following sequence of polynomials, where we denote $\boldsymbol{r} = (r_1, \ldots, r_{m_2})$:

$$\boldsymbol{k}(\alpha, \boldsymbol{r}, \boldsymbol{h}) = \left\{ b_i\alpha + \left(\sum_{j\in[1,m_2]} b_{i,j} r_j \right) + \left(\sum_{\substack{j\in[1,m_2]\\ k\in[1,n]}} b_{i,j,k} h_k r_j \right) \right\}_{i\in[1,m_1]} . \tag{4}$$

- $\mathsf{Enc2}(Y) \to (\boldsymbol{c} = (c_1, \ldots, c_{w_1}); w_2)$. It takes as inputs $Y \in \mathbb{Y}_\kappa$, and outputs a sequence of polynomials $\{c_i\}_{i\in[1,w_1]}$ with coefficients in $\mathbb{Z}_p$, and $w_2 \in \mathbb{N}$. We require that each polynomial c_i is a *linear combination of monomials* $s_j, h_k s_j$, where $s_0, s_1, \ldots, s_{w_2}, h_1, \ldots, h_n$ are variables. Denote $\boldsymbol{s} = (s_0, s_1, \ldots, s_{w_2})$. Indeed, it outputs $\{a_{i,j}\}_{i\in[1,w_1],j\in[0,w_2]}$, $\{a_{i,j,k}\}_{i\in[1,w_1],j\in[0,w_2],k\in[1,n]}$ which is a set of coefficients that defines the following sequence of polynomials:

$$\boldsymbol{c}(\boldsymbol{s}, \boldsymbol{h}) = \left\{ \left(\sum_{j\in[0,w_2]} a_{i,j} s_j \right) + \left(\sum_{\substack{j\in[0,w_2]\\ k\in[1,n]}} a_{i,j,k} h_k s_j \right) \right\}_{i\in[1,w_1]} . \tag{5}$$

- $\mathsf{Pair}(X, Y) \to \boldsymbol{E}$. It takes as inputs X, Y, and output $\boldsymbol{E} \in \mathbb{Z}_p^{m_1\times w_1}$.

Correctness. The correctness requirement is defined as follows. Let $(\boldsymbol{k}; m_2) \leftarrow \mathsf{Enc1}(X)$, $(\boldsymbol{c}; w_2) \leftarrow \mathsf{Enc2}(Y)$, and $\boldsymbol{E} \leftarrow \mathsf{Pair}(X, Y)$. We have that if $R(X, Y) = 1$, then $\boldsymbol{k}\boldsymbol{E}\boldsymbol{c}^\top = \alpha s_0$, where the equality holds symbolically.

Note that since $\boldsymbol{k}\boldsymbol{E}\boldsymbol{c}^\top = \sum_{i\in[1,m_1],j\in[1,w_1]} E_{i,j} k_i c_j$, the correctness amounts to check if there is a linear combination of $k_i c_j$ terms summed up to αs_0.

3.1 Regular Pair Encoding

Towards proving the security of our framework in prime-order groups, we require new properties for pair encoding. We formalize them as *regularity*. This would generally confine the class of encoding schemes that the new framework can deal with from the previous framework by [3]. Nonetheless, the confinement seems natural since all the pair encoding schemes proposed so far [3, 10, 46] turn out to be regular, and hence are not affected. Below, we use notation: $[m] = \{1, \ldots, m\}$.

Definition 1 (Regular Pair Encoding). *We call a pair encoding* regular *if the following hold:*

1. *For all* $(i, i') \in [m_1] \times [w_1]$ *such that there is* $(j, k, j', k') \in [m_2] \times [n] \times [w_2] \times [n]$ *where* $b_{i,j,k} \neq 0$ *and* $a_{i',j',k'} \neq 0$*, we require that* $E_{i,i'} = 0$.
2. *If* $r_j \notin \mathbf{k}$,[6] *then* $b_{i,j,k} = 0$ *for all* $i \in [m_1], k \in [n]$.
3. *If* $s_j \notin \mathbf{c}$,[6] *then* $a_{i,j,k} = 0$ *for all* $i \in [w_1], k \in [n]$.
4. $s_0 \in \mathbf{c}$. *Wlog, we always let* $\mathbf{c} = (s_0, \ldots)$*, that is,* s_0 *is the first entry of* $\mathbf{c}$.

Explaining the Definition. The first restriction basically states that the multiplication of $(h_k r_j)$ and $(h_{k'} s_{j'})$ will not be allowed when pairing. The reason to do so is that the parameter $h_k, h_{k'}$ will be translated to matrices, and the matrix multiplication does not commute; hence, the multiplication procedure would not be mimicked correctly (from the composite-order setting) if it were to be allowed (see Eq. (9)). This restriction is quite natural since the product $r_j h_k, h_{k'} s_{j'}$ can be implemented by grouping $h_{k''} = h_k h_{k'}$, and just using associativity $(r_j h_{k''}) s_{j'} = r_j (h_{k''} s_{j'})$ instead; therefore, the multiplication of $(h_k r_j)$ and $(h_{k'} s_{j'})$ will not be needed in the first place.

The second restriction basically states that a term $h_k r_j$ is allowed in the key encoding only if r_j is given out explicitly in the key encoding. The third is similar but for the ciphertext encoding.

These restrictions are also natural since intuitively to cancel out $h_k r_j$ (so that the bilinear combination would give only the term αs_0 and no others), one would need r_j to multiply with, say $h_k s_{j'}$ (since we cannot do the multiplication concerning two parameters, as depicted above). The meaning of the fourth is clear: s_0 must be given out in the encoding.

These latter three restrictions will be used for the security proofs in hybrid games that are based on the security of encodings. We explain the intuition why we require them at the end of Sect. 4.

3.2 Security Definitions for Pair Encodings

The security notions of pair encoding schemes are given in [3], with a refinement regarding the number of queries in [10]. We describe almost the same definitions here and remark slight differences from [3, 10] below.

[6] For a polynomial u, we say that $u \in \boldsymbol{v} = (v_1, \ldots, v_q)$, if $u = v_i$ for some $i \in [q]$.

(Perfect Security). The pair encoding scheme P is *perfectly master-key hiding* (PMH) if the following holds. Suppose $R(X,Y)=0$. Let $n \leftarrow \mathsf{Param}(\kappa)$, $(\boldsymbol{k};m_2) \leftarrow \mathsf{Enc1}(X)$, $(\boldsymbol{c};w_2) \leftarrow \mathsf{Enc2}(Y)$, then the following two distributions are identical:

$$\{\boldsymbol{c}(\boldsymbol{s},\boldsymbol{h}),\ \boldsymbol{k}(0,\boldsymbol{r},\boldsymbol{h})\} \qquad \text{and} \qquad \{\boldsymbol{c}(\boldsymbol{s},\boldsymbol{h}),\ \boldsymbol{k}(\alpha,\boldsymbol{r},\boldsymbol{h})\},$$

where the probability is taken over $\boldsymbol{h} \xleftarrow{\$} \mathbb{Z}_p^n, \alpha \xleftarrow{\$} \mathbb{Z}_p, \boldsymbol{r} \xleftarrow{\$} \mathbb{Z}_p^{m_2}, \boldsymbol{s} \xleftarrow{\$} \mathbb{Z}_p^{(w_2+1)}$.

(Computational Security). We define two flavors for computational security notions: *selectively* and *co-selectively secure master-key hiding* (SMH, CMH) in a bilinear group generator $\mathcal{G}$. We first define the following game template, denoted as $\mathsf{Exp}_{\mathcal{G},\mathsf{P},\mathsf{G},b,\mathcal{A},t_1,t_2}(\lambda)$, for pair encoding P, a flavor $\mathsf{G} \in \{\mathsf{CMH},\mathsf{SMH}\}$, $b \in \{0,1\}$, and $t_1,t_2 \in \mathbb{N}$. It takes as input the security parameter λ and does the experiment with the adversary $\mathcal{A}=(\mathcal{A}_1,\mathcal{A}_2)$, and outputs b' (as a guess of b). Denote by st a state information by $\mathcal{A}$. The game is defined as:

$$\begin{aligned}\mathsf{Exp}_{\mathcal{G},\mathsf{G},b,\mathcal{A},t_1,t_2}(\lambda) : (\mathbb{G}_1,\mathbb{G}_2,\mathbb{G}_T,e,p) \leftarrow \mathcal{G}(\lambda);\ g_1 \xleftarrow{\$} \mathbb{G}_1, g_2 \xleftarrow{\$} \mathbb{G}_2,\\ \alpha \xleftarrow{\$} \mathbb{Z}_p, n \leftarrow \mathsf{Param}(\kappa), \boldsymbol{h} \xleftarrow{\$} \mathbb{Z}_p^n;\\ \mathsf{st} \leftarrow \mathcal{A}_1^{\mathcal{O}^1_{\mathsf{G},b,\alpha,\boldsymbol{h}}(\cdot)}(g_1,g_2);\ b' \leftarrow \mathcal{A}_2^{\mathcal{O}^2_{\mathsf{G},b,\alpha,\boldsymbol{h}}(\cdot)}(\mathsf{st}),\end{aligned}$$

where each oracle $\mathcal{O}^1,\mathcal{O}^2$ *can be queried at most* t_1,t_2 *times respectively*, and is defined as follows.

- **Selective Security**
  - $\mathcal{O}^1_{\mathsf{SMH},b,\alpha,\boldsymbol{h}}(Y)$: Run $(\boldsymbol{c};w_2) \leftarrow \mathsf{Enc2}(Y); \boldsymbol{s} \xleftarrow{\$} \mathbb{Z}_p^{(w_2+1)}$; return $\boldsymbol{U} \leftarrow g_1^{\boldsymbol{c}(\boldsymbol{s},\boldsymbol{h})}$.
  - $\mathcal{O}^2_{\mathsf{SMH},b,\alpha,\boldsymbol{h}}(X)$: If $R(X,Y)=1$ for some queried Y, then return $\perp$. Else, run $(\boldsymbol{k};m_2) \leftarrow \mathsf{Enc1}(X); \boldsymbol{r} \xleftarrow{\$} \mathbb{Z}_p^{m_2}$; return $\boldsymbol{V} \leftarrow g_2^{\boldsymbol{k}(b\alpha,\boldsymbol{r},\boldsymbol{h})}$.
- **Co-selective Security**
  - $\mathcal{O}^1_{\mathsf{CMH},b,\alpha,\boldsymbol{h}}(X)$: Run $(\boldsymbol{k};m_2) \leftarrow \mathsf{Enc1}(X); \boldsymbol{r} \xleftarrow{\$} \mathbb{Z}_p^{m_2}$; return $\boldsymbol{V} \leftarrow g_2^{\boldsymbol{k}(b\alpha,\boldsymbol{r},\boldsymbol{h})}$.
  - $\mathcal{O}^2_{\mathsf{CMH},b,\alpha,\boldsymbol{h}}(Y)$: If $R(X,Y)=1$ for some queried X, then return $\perp$. Else, run $(\boldsymbol{c};w_2) \leftarrow \mathsf{Enc2}(Y); \boldsymbol{s} \xleftarrow{\$} \mathbb{Z}_p^{(w_2+1)}$; return $\boldsymbol{U} \leftarrow g_1^{\boldsymbol{c}(\boldsymbol{s},\boldsymbol{h})}$.

We define the advantage of $\mathcal{A}$ against the pair encoding scheme P in the security game $\mathsf{G} \in \{\mathsf{SMH},\mathsf{CMH}\}$ for bilinear group generator $\mathcal{G}$ with the bounded number of queries (t_1,t_2) as

$$\mathsf{Adv}_{\mathcal{A}}^{(t_1,t_2)\text{-}\mathsf{G}(\mathsf{P})}(\lambda) := |\Pr[\mathsf{Exp}_{\mathcal{G},\mathsf{P},\mathsf{G},0,\mathcal{A},t_1,t_2}(\lambda)=1] - \Pr[\mathsf{Exp}_{\mathcal{G},\mathsf{P},\mathsf{G},1,\mathcal{A},t_1,t_2}(\lambda)=1]|$$

We say that P is (t_1,t_2)-*selectively master-key hiding* in $\mathcal{G}$ if $\mathsf{Adv}_{\mathcal{A}}^{(t_1,t_2)\text{-}\mathsf{SMH}(\mathsf{P})}(\lambda)$ is negligible for all polynomial time attackers $\mathcal{A}$. Analogously, P is (t_1,t_2)-*co-selectively master-key hiding* in $\mathcal{G}$ if $\mathsf{Adv}_{\mathcal{A}}^{(t_1,t_2)\text{-}\mathsf{CMH}(\mathsf{P})}(\lambda)$ is negligible for all polynomial time attackers $\mathcal{A}$.

Poly-many Queries. We also consider the case where t_i is *not a-priori bounded* and hence the corresponding oracle can be queried polynomially many times. In such a case, we denote t_i as poly.

Remark 2. The original notions considered in [3] are $(1, \mathsf{poly})$-SMH, $(1,1)$-CMH for selective and co-selective master-key hiding security, respectively. The refinement with (t_1, t_2) is done recently in [10]. An advantage of this refinement is that we can have a "dual" conversion that converts between $(1,1)$-CMH and $(1,1)$-SMH for dual predicate [10].

Remark 3. The definition of computational security for encoding here is slightly different from that in [3,10] in that here we define it in *asymmetric* and *prime-order* groups, while it was defined in *symmetric* and *prime-order subgroup of composite-order* groups in [3,10]. We use asymmetric groups for the purpose of generality, one can obtain schemes in symmetric groups by just setting $\mathbb{G}_1 = \mathbb{G}_2$. Hence, we can use all the proposed encodings in [3,10] by working on the symmetric group version of our framework. For the latter issue, the difference of definitions between prime-order groups and prime-order subgroups are merely *syntactic*. This is since although the original definition was defined in prime-order subgroups, the hardness of factorization was not assumed (*i.e.*, generators of each subgroup or even factors of composites N can be given out to the adversary). Hence, the encoding schemes in [3,10] are secure in our definition under the security proofs in their present forms.

4 Approach for Translation to Prime-Order Groups

Before describing our prime-order framework, we intuitively describe how we translate elements, procedures, and properties from the composite-order group setting to the prime-order group setting, following the intuition overview in Sect. 1.3.

• **Generators.** In composite-order groups $(\mathbb{C}_1, \mathbb{C}_2, \mathbb{C}_T)$ of order $N = p_1p_2p_3$, we consider generators $c_1 \in \mathbb{C}_{1,p_1}$, $\hat{c}_1 \in \mathbb{C}_{1,p_2}$, $c_2 \in \mathbb{C}_{2,p_1}$, $\hat{c}_2 \in \mathbb{C}_{2,p_2}$, where $\mathbb{C}_{i,p_j}$ is the subgroup of $\mathbb{C}_i$ of order p_j. In prime-order groups $(\mathbb{G}_1, \mathbb{G}_2, \mathbb{G}_T)$ with generators $g_1 \in \mathbb{G}_1, g_2 \in \mathbb{G}_2$, we use the following elements to mimic generators $c_1, \hat{c}_1, c_2, \hat{c}_2$, respectively:

$$g_1^{\boldsymbol{B}\binom{\boldsymbol{I}_d}{0}} \in \mathbb{G}_1^{(d+1)\times d}, \qquad g_1^{\boldsymbol{B}\binom{0}{1}} \in \mathbb{G}_1^{(d+1)\times 1},$$
$$g_2^{\boldsymbol{Z}\binom{\boldsymbol{I}_d}{0}} \in \mathbb{G}_2^{(d+1)\times d}, \qquad g_2^{\boldsymbol{Z}\binom{0}{1}} \in \mathbb{G}_2^{(d+1)\times 1}.$$

where we let $(\boldsymbol{B}, \boldsymbol{Z}) \xleftarrow{\$} \mathbb{S}_d$ where the distribution $\mathbb{S}_d$ does as follows: sample $\boldsymbol{B} \xleftarrow{\$} \mathbb{GL}_{p,d+1}$, $\tilde{\boldsymbol{D}} \xleftarrow{\$} \mathbb{GL}_{p,d}$ and set $\boldsymbol{Z} := \boldsymbol{B}^{-\top}\boldsymbol{D}$ where $\boldsymbol{D} := \begin{pmatrix} \tilde{\boldsymbol{D}} & 0 \\ 0 & 1 \end{pmatrix} \in \mathbb{GL}_{p,d+1}$.

• **Variables.** The role of parameter h_k (in $\boldsymbol{h}$) in the composite-order setting will be played by a matrix $\boldsymbol{H}_k \in \mathbb{Z}_p^{(d+1)\times(d+1)}$. The role of randomness s_j, r_j (in $\boldsymbol{s}, \boldsymbol{r}$)

to be exponentiated over c_1, c_2 in the composite-order setting for a ciphertext and a key will be played by vectors $\boldsymbol{s}_j, \boldsymbol{r}_j \in \mathbb{Z}_p^{d\times 1}$, respectively, in the prime-order setting. The role of randomness $\hat{s}_j, \hat{r}_j$ (in $\hat{\boldsymbol{s}}, \hat{\boldsymbol{r}}$) to be exponentiated over $\hat{c}_1, \hat{c}_2$ will be used as it is (a scalar in $\mathbb{Z}_p$) in the prime-order setting.

• **Exponentiation by parameter.** To mimic exponentiation $c_1^{h_k}, \hat{c}_1^{\hat{h}_k}, c_2^{h_k}, \hat{c}_2^{\hat{h}_k}$ in the composite-order setting, we do the following in the prime-order setting:

$$g_1^{\boldsymbol{H}_k \boldsymbol{B}\binom{\boldsymbol{I}_d}{0}} \in \mathbb{G}_1^{(d+1)\times d}, \qquad g_1^{\boldsymbol{B}\binom{\boldsymbol{0}}{1}\hat{h}_k} \in \mathbb{G}_1^{(d+1)\times 1},$$

$$g_2^{\boldsymbol{H}_k^\top \boldsymbol{Z}\binom{\boldsymbol{I}_d}{0}} \in \mathbb{G}_2^{(d+1)\times d}, \qquad g_2^{\boldsymbol{Z}\binom{\boldsymbol{0}}{1}\hat{h}_k} \in \mathbb{G}_2^{(d+1)\times 1}.$$

• **Exponentiation by randomness.** To mimic exponentiation $c_1^{s_j}, \hat{c}_1^{\hat{s}_j}, c_2^{r_j}, \hat{c}_2^{\hat{r}_j}$, in the composite-order setting, we do the following in the prime-order setting:

$$g_1^{\boldsymbol{B}\binom{\boldsymbol{I}_d}{0}\boldsymbol{s}_j} = g_1^{\boldsymbol{B}\binom{\boldsymbol{s}_j}{0}} \in \mathbb{G}_1^{(d+1)\times 1}, \qquad g_1^{\boldsymbol{B}\binom{\boldsymbol{0}}{1}\hat{s}_j} = g_1^{\boldsymbol{B}\binom{\boldsymbol{0}}{\hat{s}_j}} \in \mathbb{G}_1^{(d+1)\times 1},$$

$$g_2^{\boldsymbol{Z}\binom{\boldsymbol{I}_d}{0}\boldsymbol{r}_j} = g_2^{\boldsymbol{Z}\binom{\boldsymbol{r}_j}{0}} \in \mathbb{G}_2^{(d+1)\times 1}, \qquad g_2^{\boldsymbol{Z}\binom{\boldsymbol{0}}{1}\hat{r}_j} = g_2^{\boldsymbol{Z}\binom{\boldsymbol{0}}{\hat{r}_j}} \in \mathbb{G}_2^{(d+1)\times 1}.$$

• **Exponentiation by randomness over parameter.** To mimic $(c_1^{h_k})^{s_j}$, $(\hat{c}_1^{\hat{h}_k})^{\hat{s}_j}$, $(c_2^{h_k})^{r_j}$, $(\hat{c}_2^{\hat{h}_k})^{\hat{r}_j}$, in the composite-order setting, we do as follows:

$$g_1^{\boldsymbol{H}_k\boldsymbol{B}\binom{\boldsymbol{I}_d}{0}\boldsymbol{s}_j} = g_1^{\boldsymbol{H}_k\boldsymbol{B}\binom{\boldsymbol{s}_j}{0}} \in \mathbb{G}_1^{(d+1)\times 1}, \quad g_1^{\boldsymbol{B}\binom{\boldsymbol{0}}{1}\hat{h}_k\hat{s}_j} = g_1^{\boldsymbol{B}\binom{\boldsymbol{0}}{\hat{h}_k\hat{s}_j}} \in \mathbb{G}_1^{(d+1)\times 1},$$

$$g_2^{\boldsymbol{H}_k^\top\boldsymbol{Z}\binom{\boldsymbol{I}_d}{0}\boldsymbol{r}_j} = g_2^{\boldsymbol{H}_k^\top\boldsymbol{Z}\binom{\boldsymbol{r}_j}{0}} \in \mathbb{G}_2^{(d+1)\times 1}, \quad g_2^{\boldsymbol{Z}\binom{\boldsymbol{0}}{1}\hat{h}_k\hat{r}_j} = g_2^{\boldsymbol{Z}\binom{\boldsymbol{0}}{\hat{h}_k\hat{r}_j}} \in \mathbb{G}_2^{(d+1)\times 1}.$$

• **Evaluating Pair Encoding with Vectors/Matrices.** We can evaluate the ciphertext attribute encoding $\boldsymbol{c}(\boldsymbol{s}, \boldsymbol{h})$, defined in Eq.(5), with each s_j being substituted by a vector $\boldsymbol{x}_j \in \mathbb{Z}_p^{(d+1)\times 1}$ and each h_k being substituted by a matrix $\boldsymbol{H}_k \in \mathbb{Z}_p^{(d+1)\times(d+1)}$. Let $\boldsymbol{X} = (\boldsymbol{x}_0, \ldots, \boldsymbol{x}_{w_2}) \in \mathbb{Z}_p^{(d+1)\times(w_2+1)}$ and $\mathbb{H} = (\boldsymbol{H}_1, \ldots, \boldsymbol{H}_n)$. We define

$$\boldsymbol{c}(\boldsymbol{X}, \mathbb{H}) := \left\{ \left(\sum_{j\in[0,w_2]} a_{i,j}\boldsymbol{x}_j \right) + \left(\sum_{\substack{j\in[0,w_2]\\ k\in[1,n]}} a_{i,j,k}\boldsymbol{H}_k\boldsymbol{x}_j \right) \right\}_{i\in[1,w_1]} . \tag{6}$$

Similarly for the key attribute encoding $\boldsymbol{k}(\alpha, \boldsymbol{r}, \boldsymbol{h})$, defined in Eq. (4), we replace each r_j with a vector $\boldsymbol{y}_j \in \mathbb{Z}_p^{(d+1)\times 1}$ and α with $\boldsymbol{\alpha} \in \mathbb{Z}_p^{(d+1)\times 1}$. Let $\boldsymbol{Y} = (\boldsymbol{y}_1, \ldots, \boldsymbol{y}_{m_2}) \in \mathbb{Z}_p^{(d+1)\times m_2}$. We define

$$\boldsymbol{k}(\boldsymbol{\alpha}, \boldsymbol{Y}, \mathbb{H}) := \left\{ b_i \boldsymbol{\alpha} + \left(\sum_{j \in [1, m_2]} b_{i,j} \boldsymbol{y}_j \right) + \left(\sum_{\substack{j \in [1, m_2] \\ k \in [1, n]}} b_{i,j,k} \boldsymbol{H}_k^\top \boldsymbol{y}_j \right) \right\}_{i \in [1, m_1]} . \tag{7}$$

• **Associativity.** In the composite-order setting, we have that $e(t_1^{h_k s_j}, t_2^{r_i}) = e(t_1^{s_j}, t_2^{h_k r_i})$, for any $t_1 \in \mathbb{C}_1, t_2 \in \mathbb{C}_2$. In the prime-order setting, we have

$$e(g_1^{\boldsymbol{H}_k \boldsymbol{B} \binom{\boldsymbol{s}_j}{\hat{s}_j}}, g_2^{\boldsymbol{Z} \binom{\boldsymbol{r}_i}{\hat{r}_i}}) = e(g_1^{\boldsymbol{B} \binom{\boldsymbol{s}_j}{\hat{s}_j}}, g_2^{\boldsymbol{H}_k^\top \boldsymbol{Z} \binom{\boldsymbol{r}_i}{\hat{r}_i}}). \tag{8}$$

as $\left((\boldsymbol{r}_i^\top \ \hat{r}_i) \boldsymbol{Z}^\top \right) \left(\boldsymbol{H}_k \boldsymbol{B} \binom{\boldsymbol{s}_j}{\hat{s}_j} \right) = \left((\boldsymbol{r}_i^\top \ \hat{r}_i) \boldsymbol{Z}^\top \boldsymbol{H}_k \right) \left(\boldsymbol{B} \binom{\boldsymbol{s}_j}{\hat{s}_j} \right)$.

• **Unavailable Commutativity.** We also give an intuition why commutativity does not preserve to prime-order settings. In the composite-order setting, we allow for any $t_1 \in \mathbb{C}_1, t_2 \in \mathbb{C}_2$, $e(t_1^{h_k s_j}, t_2^{h_{k'} r_i}) = e(t_1^{h_{k'} s_j}, t_2^{h_k r_i})$. However, when translating to our prime-order setting using our rules so far, an analogous mechanism would not hold as we can see that:

$$e(g_1^{\boldsymbol{H}_k \boldsymbol{B} \binom{\boldsymbol{s}_j}{\hat{s}_j}}, g_2^{\boldsymbol{H}_{k'}^\top \boldsymbol{Z} \binom{\boldsymbol{r}_i}{\hat{r}_i}}) \neq e(g_1^{\boldsymbol{H}_{k'} \boldsymbol{B} \binom{\boldsymbol{s}_j}{\hat{s}_j}}, g_2^{\boldsymbol{H}_k^\top \boldsymbol{Z} \binom{\boldsymbol{r}_i}{\hat{r}_i}}), \tag{9}$$

as $\left((\boldsymbol{r}_i^\top \ \hat{r}_i) \boldsymbol{Z}^\top \boldsymbol{H}_{k'} \right) \left(\boldsymbol{H}_k \boldsymbol{B} \binom{\boldsymbol{s}_j}{\hat{s}_j} \right) \neq \left((\boldsymbol{r}_i^\top \ \hat{r}_i) \boldsymbol{Z}^\top \boldsymbol{H}_k \right) \left(\boldsymbol{H}_{k'} \boldsymbol{B} \binom{\boldsymbol{s}_j}{\hat{s}_j} \right)$, due to the fact that the matrix multiplication is not commutative. This is exactly why we will *not* use this commutativity-based computation in our framework by disallowing exactly this kind of multiplication to occur. We enable this with the first rule of *regular encoding*, which exactly prevents multiplying $h_k s_j$ with $h_{k'} r_{j'}$.

• **Parameter-Hiding.** In composite-order groups, we have that: given $c_1^{h_k}, c_2^{h_k}, c_1, \hat{c}_1, c_2, \hat{c}_2, p_1, p_2$; $h_k \bmod p_2$ is information-theoretically hidden (due to the Chinese Remainder Theorem). In prime-order settings, we have Lemma 1.

Lemma 1. *Let* $(\boldsymbol{B}, \boldsymbol{Z}) \xleftarrow{\$} \mathbb{S}_d$. *For any* $\boldsymbol{H}_k \in \mathbb{Z}_p^{(d+1)\times(d+1)}$, *we have that, given* $\boldsymbol{H}_k \boldsymbol{B} \binom{\boldsymbol{I}_d}{0}$ *and* $\boldsymbol{H}_k^\top \boldsymbol{Z} \binom{\boldsymbol{I}_d}{0}$, *along with* $\boldsymbol{B}, \boldsymbol{Z}$, *the quantity of the entry at* $(d+1, d+1)$ *of the matrix* $\boldsymbol{B}^{-1} \boldsymbol{H}_k \boldsymbol{B}$ *is information-theoretically hidden.*

Proof. Write $\boldsymbol{B}^{-1} \boldsymbol{H}_k \boldsymbol{B} = \begin{pmatrix} M_1 & M_2 \\ M_3 & \delta \end{pmatrix}$ where $M_1 \in \mathbb{Z}_p^{d \times d}$, $M_2 \in \mathbb{Z}_p^{d \times 1}$, $M_3 \in \mathbb{Z}_p^{1 \times d}$, and $\delta \in \mathbb{Z}_p$. We have

$$\boldsymbol{H}_k \boldsymbol{B} \binom{\boldsymbol{I}_d}{0} = \boldsymbol{B} \begin{pmatrix} M_1 & M_2 \\ M_3 & \delta \end{pmatrix} \binom{\boldsymbol{I}_d}{0} = \boldsymbol{B} \binom{M_1}{M_3},$$
$$\boldsymbol{H}_k^\top \boldsymbol{Z} \binom{\boldsymbol{I}_d}{0} = \boldsymbol{H}_k^\top \boldsymbol{B}^{-\top} \begin{pmatrix} \tilde{\boldsymbol{D}} & 0 \\ \boldsymbol{0} & 1 \end{pmatrix} \binom{\boldsymbol{I}_d}{0} = \boldsymbol{B}^{-\top} \begin{pmatrix} M_1^\top & M_3^\top \\ M_2^\top & \delta \end{pmatrix} \binom{\tilde{\boldsymbol{D}}}{\boldsymbol{0}} = \boldsymbol{B}^{-\top} \binom{M_1^\top \tilde{\boldsymbol{D}}}{M_2^\top \tilde{\boldsymbol{D}}},$$

where in the second line, we use the fact that $\boldsymbol{B}^\top \boldsymbol{H}_k^\top \boldsymbol{B}^{-\top} = \begin{pmatrix} M_1^\top & M_3^\top \\ M_2^\top & \delta \end{pmatrix}$. We can see that both $\boldsymbol{H}_k \boldsymbol{B} \binom{\boldsymbol{I}_d}{0}, \boldsymbol{H}_k^\top \boldsymbol{Z} \binom{\boldsymbol{I}_d}{0}$ do not contain information on δ. □

• **Using Security of Encodings in Hybrid Games.** In the composite-order setting, intuitively, we embed the security of encodings *as it is* in one hybrid game in the proof of the scheme. That is, we simply invoke a trivial implication:

$$\hat{C}_2^{\boldsymbol{k}(0,\hat{\boldsymbol{r}},\hat{\boldsymbol{h}})} \approx_c \hat{C}_2^{\boldsymbol{k}(\hat{\alpha},\hat{\boldsymbol{r}},\hat{\boldsymbol{h}})} \implies \hat{C}_2^{\boldsymbol{k}(0,\hat{\boldsymbol{r}},\hat{\boldsymbol{h}})} \approx_c \hat{C}_2^{\boldsymbol{k}(\hat{\alpha},\hat{\boldsymbol{r}},\hat{\boldsymbol{h}})}$$

where we refer the left-hand side as the security of encoding and the right-hand side as the hybrid in the proof of the scheme. Also, $\approx_c$ denotes computational indistinguishability (informally). In the prime-order setting, contrastingly, we will need to prove the following reduction: (stated informally here)

$$g_2^{\boldsymbol{k}(0,\hat{\boldsymbol{r}},\hat{\boldsymbol{h}})} \approx_c g_2^{\boldsymbol{k}(\hat{\alpha},\hat{\boldsymbol{r}},\hat{\boldsymbol{h}})} \implies g_2^{\boldsymbol{k}(\boldsymbol{0},\boldsymbol{Z}\hat{\boldsymbol{R}},\mathbb{H})} \approx_c g_2^{\boldsymbol{k}(\hat{\boldsymbol{\alpha}},\boldsymbol{Z}\hat{\boldsymbol{R}},\mathbb{H})}$$

where the left-hand side refers to the same security of encodings as before, so that we can achieve our goal of using security of encoding "as is".[7] Now, however, the right-hand side, which refers to one hybrid in our scheme[8], is of a *different* form, as it contains the matrix-based definition of encodings in Eq. (7). To this end, we will relate both sides as follows. First, we implicitly define $\hat{\boldsymbol{\alpha}}$ from $\hat{\alpha}$, and $\hat{\boldsymbol{R}}$ from $\hat{\boldsymbol{r}}$.[9] Second, we invoke the parameter-hiding property to implicitly replace each $\boldsymbol{H}_k$ with $\boldsymbol{H}_k + \boldsymbol{B}\begin{pmatrix} 0 & 0 \\ 0 & \hat{h}_k \end{pmatrix}\boldsymbol{B}^{-1}$ in $\mathbb{H}$. Our novelty here then lies in identifying the following sufficient condition: (stated informally here)

$$g_2^{\boldsymbol{k}(\hat{\boldsymbol{\alpha}},\boldsymbol{Z}\hat{\boldsymbol{R}},\mathbb{H})} \text{ can be fully simulated by } g_2^{\boldsymbol{k}(\hat{\alpha},\hat{\boldsymbol{r}},\hat{\boldsymbol{h}})} \text{ and } (g_2^{\hat{r}_j})^{b_{i,j,k}} \text{ (for all } i,j,k),$$

where $\hat{\alpha}, \hat{\boldsymbol{r}} = (\hat{r}_1, \ldots, \hat{r}_{m_2}), \hat{\boldsymbol{h}} = (\hat{h}_1, \ldots, \hat{h}_n)$ are unknown, and $b_{i,j,k}$ is defined by the encoding (Eq.(4)). We note that this is quite surprising in the first place, since we might expect that only $g_2^{\boldsymbol{k}(\hat{\alpha},\hat{\boldsymbol{r}},\hat{\boldsymbol{h}})}$ would suffice to simulate $g_2^{\boldsymbol{k}(\hat{\boldsymbol{\alpha}},\boldsymbol{Z}\hat{\boldsymbol{R}},\mathbb{H})}$ (intuitively due to one-to-one translation of elements into matrix forms). Now, to establish the reduction, we require the availability of the latter term $(g_2^{\hat{r}_j})^{b_{i,j,k}}$, which was not a-priori guaranteed. We simply resolve this by observing that it is only available if either $\hat{r}_j$ is given out in the definition of $\boldsymbol{k}(\hat{\alpha}, \hat{\boldsymbol{r}}, \hat{\boldsymbol{h}})$ or $b_{i,j,k} = 0$. This is why we thus define this to be exactly one of the rules for regular encodings (Rule 2 of Definition 1). The case for the encoding $\boldsymbol{c}$ can be argued analogously.

5 Our Generic Construction for Fully Secure ABE

We are now ready to describe our generic construction in prime-order groups. It is obtained by translating the composite-order scheme of [3], recapped also in the full version [4], to the prime-order setting using the above rules of Sect. 4.

[7] The only difference is that now it is defined in prime-order groups, instead of prime-order subgroups of composite-order groups.

[8] Looking ahead, it corresponds to the hybrid game between type 1 and 2 keys (*cf.* Eqs. (20), (21)).

[9] Details can be found in the proof for the hybrid between the games $\mathsf{G}_{i,1}$ and $\mathsf{G}_{i,2}$, deferred to [4].

We use the distribution $\mathcal{S}_d$ defined in Sect. 4. From a pair encoding scheme P for a predicate R, we construct an ABE scheme for R, denoted $\mathsf{ABE}(\mathsf{P})$, as follows.

- $\mathsf{Setup}(1^\lambda, \kappa)$: Run $(\mathbb{G}_1, \mathbb{G}_2, \mathbb{G}_T, e, p) \xleftarrow{\$} \mathcal{G}(\lambda)$. Pick generators $g_1 \xleftarrow{\$} \mathbb{G}_1$ and $g_2 \xleftarrow{\$} \mathbb{G}_2$. Run $n \leftarrow \mathsf{Param}(\kappa)$. Pick $\mathbb{H} = (\boldsymbol{H}_1, \ldots, \boldsymbol{H}_n) \xleftarrow{\$} (\mathbb{Z}_p^{(d+1)\times(d+1)})^n$ and $\boldsymbol{\alpha} \xleftarrow{\$} \mathbb{Z}_p^{(d+1)\times 1}$. Sample $(\boldsymbol{B}, \boldsymbol{Z}) \xleftarrow{\$} \mathcal{S}_d$. Note that $\boldsymbol{B}, \boldsymbol{Z} \in \mathbb{Z}_p^{(d+1)\times(d+1)}$. Output

$$\begin{aligned} \mathsf{PK} &= \left(e(g_1, g_2)^{\boldsymbol{\alpha}^\top \boldsymbol{B}\binom{\boldsymbol{I}_d}{0}}, g_1^{\boldsymbol{B}\binom{\boldsymbol{I}_d}{0}}, g_1^{\boldsymbol{H}_1\boldsymbol{B}\binom{\boldsymbol{I}_d}{0}}, \ldots, g_1^{\boldsymbol{H}_n\boldsymbol{B}\binom{\boldsymbol{I}_d}{0}} \right), \\ \mathsf{MSK} &= \left(g_2^{\boldsymbol{\alpha}}, \quad g_2^{\boldsymbol{Z}\binom{\boldsymbol{I}_d}{0}}, g_2^{\boldsymbol{H}_1^\top\boldsymbol{Z}\binom{\boldsymbol{I}_d}{0}}, \ldots, g_2^{\boldsymbol{H}_n^\top\boldsymbol{Z}\binom{\boldsymbol{I}_d}{0}} \right). \end{aligned} \tag{10}$$

- $\mathsf{Encrypt}(Y, M, \mathsf{PK})$: Upon input $Y \in \mathbb{Y}$, run $(\boldsymbol{c}; w_2) \leftarrow \mathsf{Enc2}(Y)$. Randomly pick $\boldsymbol{s}_0, \boldsymbol{s}_1, \ldots, \boldsymbol{s}_{w_2} \xleftarrow{\$} \mathbb{Z}_p^{d\times 1}$. Let $\boldsymbol{S} := (\binom{\boldsymbol{s}_0}{0}, \binom{\boldsymbol{s}_1}{0}, \ldots, \binom{\boldsymbol{s}_{w_2}}{0}) \in \mathbb{Z}_p^{(d+1)\times(w_2+1)}$. Output the ciphertext as $\mathsf{CT} = (\boldsymbol{C}, C_0)$:

$$\begin{aligned} \boldsymbol{C} &= g_1^{\boldsymbol{c}(\boldsymbol{BS}, \mathbb{H})} && \in (\mathbb{G}_1^{(d+1)\times 1})^{w_1}, \\ C_0 &= e(g_1, g_2)^{\boldsymbol{\alpha}^\top \boldsymbol{B}\binom{\boldsymbol{s}_0}{0}} \cdot M && \in \mathbb{G}_T. \end{aligned} \tag{11}$$

- $\mathsf{KeyGen}(X, \mathsf{MSK})$: Upon input $X \in \mathbb{X}$, run $(\boldsymbol{k}; m_2) \leftarrow \mathsf{Enc1}(X)$. Randomly pick $\boldsymbol{r}_1, \ldots, \boldsymbol{r}_{m_2} \xleftarrow{\$} \mathbb{Z}_p^{d\times 1}$. Let $\boldsymbol{R} := (\binom{\boldsymbol{r}_1}{0}, \ldots, \binom{\boldsymbol{r}_{m_2}}{0}) \in \mathbb{Z}_p^{(d+1)\times m_2}$. Output

$$\mathsf{SK} = g_2^{\boldsymbol{k}(\boldsymbol{\alpha}, \boldsymbol{ZR}, \mathbb{H})} \qquad \in (\mathbb{G}_2^{(d+1)\times 1})^{m_1}. \tag{12}$$

- $\mathsf{Decrypt}(\mathsf{CT}, \mathsf{SK})$: Obtain Y, X from CT, SK. Suppose $R(X, Y) = 1$. Run $\boldsymbol{E} \leftarrow \mathsf{Pair}(X, Y)$. Compute the mask

$$e(g_1, g_2)^{\boldsymbol{\alpha}^\top \boldsymbol{B}\binom{\boldsymbol{s}_0}{0}} \leftarrow \prod_{i\in[1,m_1], j\in[1,w_1]} e(\boldsymbol{C}[j], \mathsf{SK}[i])^{E_{i,j}}. \tag{13}$$

  where we denote by $\boldsymbol{C}[j] \in \mathbb{G}_1^{(d+1)\times 1}$ the j-th vector in $\boldsymbol{C}$, and $\mathsf{SK}[i] \in \mathbb{G}_2^{(d+1)\times 1}$ the i-th vector in SK. Finally, remove this mask from C_0 to get M.

Remark on Computability. We note that $\boldsymbol{C}$ can be computed from PK since

$$\boldsymbol{c}(\boldsymbol{BS}, \mathbb{H}) = \left\{ \left(\sum_{j\in[0,w_2]} a_{i,j} \boldsymbol{B}\binom{\boldsymbol{s}_j}{0} \right) + \left(\sum_{\substack{j\in[0,w_2] \\ k\in[1,n]}} a_{i,j,k} \boldsymbol{H}_k \boldsymbol{B}\binom{\boldsymbol{s}_j}{0} \right) \right\}_{i\in[1,w_1]} \tag{14}$$

and thanks to the identity relation $\left(\boldsymbol{X}\left(\begin{smallmatrix}\boldsymbol{I}_d\\0\end{smallmatrix}\right)\right)\boldsymbol{y} = \boldsymbol{X}\left(\begin{smallmatrix}\boldsymbol{y}\\0\end{smallmatrix}\right)$ for any $\boldsymbol{X} \in \mathbb{Z}_p^{(d+1)\times(d+1)}$, $\boldsymbol{y} \in \mathbb{Z}_p^{d\times 1}$. Similarly, SK can be computed from MSK since

$$\boldsymbol{k}(\boldsymbol{\alpha}, \boldsymbol{ZR}, \mathbb{H}) =$$
$$\left\{ b_i\boldsymbol{\alpha} + \left(\sum_{j\in[1,m_2]} b_{i,j}\boldsymbol{Z}\left(\begin{smallmatrix}\boldsymbol{r}_j\\0\end{smallmatrix}\right)\right) + \left(\sum_{\substack{j\in[1,m_2]\\k\in[1,n]}} b_{i,j,k}\boldsymbol{H}_k^\top\boldsymbol{Z}\left(\begin{smallmatrix}\boldsymbol{r}_j\\0\end{smallmatrix}\right)\right)\right\}_{i\in[1,m_1]}. \quad (15)$$

Correctness. We would like to prove that if $R(X,Y) = 1$ then

$$\boldsymbol{\alpha}^\top\boldsymbol{B}\left(\begin{smallmatrix}\boldsymbol{s}_0\\0\end{smallmatrix}\right) = \sum_{i\in[1,m_1],j\in[1,w_1]} E_{i,j}\cdot\left(\boldsymbol{k}(\boldsymbol{\alpha}, \boldsymbol{ZR}, \mathbb{H})[i]\right)^\top\cdot\boldsymbol{c}(\boldsymbol{BS}, \mathbb{H})[j].$$

This is implied from the correctness of the pair encoding which states that: if $R(X,Y) = 1$, then $\alpha s_0 = \sum_{i\in[1,m_1],j\in[1,w_1]} E_{i,j}\cdot\boldsymbol{k}(\alpha,\boldsymbol{r},\boldsymbol{h})[i]\cdot\boldsymbol{c}(\boldsymbol{s},\boldsymbol{h})[j]$. Intuitively, since we translate to the prime-order setting by substituting variables and procedures while preserving their properties as in Sect. 4, this relation should also translate to the above equation. In particular, we use associativity but not use commutativity, as clarified in Sect. 4. We verify the correctness more formally in the full version [4].

6 Security Theorems and Proofs

We obtain three security theorems for the generic construction. The first one is the main theorem and is for the case when the pair encoding is $(1,\mathsf{poly})$-SMH and $(1,1)$-CMH, where we achieve tighter reduction cost, $O(q_1)$. The other two are for the case of PMH and the pair of $(1,1)$-SMH, $(1,1)$-CMH, where we obtain normal reduction cost, $O(q_{\mathrm{all}})$. We postpone the latter two to [4].

Theorem 1. *Suppose that a pair encoding scheme* P *for predicate* R *is* $(1,\mathsf{poly})$*-selectively and* $(1,1)$*-co-selectively master-key hiding in* $\mathcal{G}$*, and the Matrix-DH Assumption holds in* $\mathcal{G}$*. Then the construction* $\mathsf{ABE}(\mathsf{P})$ *in* $\mathcal{G}$ *is fully secure. More precisely, for any PPT adversary* $\mathcal{A}$*, let* q_1 *denote the number of queries in phase 1, there exist PPT algorithms* $\mathcal{B}_1, \mathcal{B}_2, \mathcal{B}_3$*, whose running times are the same as* $\mathcal{A}$ *plus some polynomial times, such that for any* λ*,*

$$\mathsf{Adv}_{\mathcal{A}}^{\mathsf{ABE}}(\lambda) \le (2q_1+3)\mathsf{Adv}_{\mathcal{B}_1}^{\mathcal{D}_d\text{-}\mathsf{MatDH}}(\lambda) + q_1\mathsf{Adv}_{\mathcal{B}_2}^{(1,1)\text{-}\mathsf{CMH}}(\lambda) + \mathsf{Adv}_{\mathcal{B}_3}^{(1,\mathsf{poly})\text{-}\mathsf{SMH}}(\lambda).$$

Semi-functional Algorithms. We define semi-functional algorithms which will be used in the security proof. These are also translated from semi-functional algorithms from the framework of [3] (also recapped in [4]).

- $\mathsf{SFSetup}(1^\lambda, \kappa) \to (\mathsf{PK}, \mathsf{MSK}, \widehat{\mathsf{PK}}, \widehat{\mathsf{MSK}}_{\mathsf{base}}, \widehat{\mathsf{MSK}}_{\mathsf{aux}})$: This is exactly the same as Setup albeit it additionally outputs also $\widehat{\mathsf{PK}}, \widehat{\mathsf{MSK}}_{\mathsf{base}}, \widehat{\mathsf{MSK}}_{\mathsf{aux}}$ defined as

$$\widehat{\mathsf{PK}} = \left(e(g_1, g_2)^{\boldsymbol{\alpha}^\top \boldsymbol{B}\binom{\mathbf{0}}{1}}, g_1^{\boldsymbol{B}\binom{\mathbf{0}}{1}}, g_1^{\boldsymbol{H}_1 \boldsymbol{B}\binom{\mathbf{0}}{1}}, \ldots, g_1^{\boldsymbol{H}_n \boldsymbol{B}\binom{\mathbf{0}}{1}} \right), \tag{16}$$

$$\widehat{\mathsf{MSK}}_{\mathsf{base}} = g_2^{\boldsymbol{Z}\binom{\mathbf{0}}{1}}, \qquad \widehat{\mathsf{MSK}}_{\mathsf{aux}} = \left(g_2^{\boldsymbol{H}_1^\top \boldsymbol{Z}\binom{\mathbf{0}}{1}}, \ldots, g_2^{\boldsymbol{H}_n^\top \boldsymbol{Z}\binom{\mathbf{0}}{1}} \right). \tag{17}$$

- $\mathsf{SFEncrypt}(Y, M, \mathsf{PK}, \widehat{\mathsf{PK}}) \to \mathsf{CT}$: Run $(\boldsymbol{c}; w_2) \leftarrow \mathsf{Enc2}(Y)$. Pick $\boldsymbol{S}$ as in Encrypt. Pick $\hat{s}_0, \hat{s}_1, \ldots, \hat{s}_{w_2} \xleftarrow{\$} \mathbb{Z}_p$. Let $\hat{\boldsymbol{S}} := \left(\binom{\mathbf{0}}{\hat{s}_0}, \binom{\mathbf{0}}{\hat{s}_1}, \ldots, \binom{\mathbf{0}}{\hat{s}_{w_2}} \right) \in \mathbb{Z}_p^{(d+1)\times(w_2+1)}$. Output the ciphertext as $\mathsf{CT} = (\boldsymbol{C}, C_0)$:

$$\begin{aligned} \boldsymbol{C} &= g_1^{\boldsymbol{c}(\boldsymbol{BS}, \mathbb{H}) + \boldsymbol{c}(\boldsymbol{B}\hat{\boldsymbol{S}}, \mathbb{H})} = g_1^{\boldsymbol{c}(\boldsymbol{B}(\boldsymbol{S}+\hat{\boldsymbol{S}}), \mathbb{H})} && \in (\mathbb{G}_1^{(d+1)\times 1})^{w_1}, \\ C_0 &= e(g_1, g_2)^{\boldsymbol{\alpha}^\top \boldsymbol{B}\binom{\boldsymbol{s}_0}{\hat{s}_0}} \cdot M. && \in \mathbb{G}_T. \end{aligned} \tag{18}$$

- $\mathsf{SFKeyGen}(X, \mathsf{MSK}, \widehat{\mathsf{MSK}}_{\mathsf{base}}, \widehat{\mathsf{MSK}}_{\mathsf{aux}}, \mathsf{t} \in \{0,1,2,3\}, \beta \in \mathbb{Z}_p) \to \mathsf{SK}$: Run $(\boldsymbol{k}; m_2) \leftarrow \mathsf{Enc1}(X)$. Pick $\boldsymbol{R}$ as in KeyGen. Pick $\hat{r}_1, \ldots, \hat{r}_{m_2} \xleftarrow{\$} \mathbb{Z}_p$. $\hat{\boldsymbol{R}} := \left(\binom{\mathbf{0}}{\hat{r}_1}, \ldots, \binom{\mathbf{0}}{\hat{r}_{m_2}} \right) \in \mathbb{Z}_p^{(d+1)\times m_2}$ Output the secret key SK:

$$\mathsf{SK} = \begin{cases} g_2^{\boldsymbol{k}(\boldsymbol{\alpha}, \boldsymbol{ZR}, \mathbb{H})} & \text{if } \mathsf{t} = 0 \ (19) \\ g_2^{\boldsymbol{k}(\boldsymbol{\alpha}, \boldsymbol{ZR}, \mathbb{H}) + \boldsymbol{k}(\mathbf{0}, \boldsymbol{Z}\hat{\boldsymbol{R}}, \mathbb{H})} = g_2^{\boldsymbol{k}(\boldsymbol{\alpha}, \boldsymbol{Z}(\boldsymbol{R}+\hat{\boldsymbol{R}}), \mathbb{H})} & \text{if } \mathsf{t} = 1 \ (20) \\ g_2^{\boldsymbol{k}(\boldsymbol{\alpha}, \boldsymbol{ZR}, \mathbb{H}) + \boldsymbol{k}(\boldsymbol{Z}\binom{\mathbf{0}}{\beta}, \boldsymbol{Z}\hat{\boldsymbol{R}}, \mathbb{H})} = g_2^{\boldsymbol{k}(\boldsymbol{\alpha}+\boldsymbol{Z}\binom{\mathbf{0}}{\beta}, \boldsymbol{Z}(\boldsymbol{R}+\hat{\boldsymbol{R}}), \mathbb{H})} & \text{if } \mathsf{t} = 2 \ (21) \\ g_2^{\boldsymbol{k}(\boldsymbol{\alpha}, \boldsymbol{ZR}, \mathbb{H}) + \boldsymbol{k}(\boldsymbol{Z}\binom{\mathbf{0}}{\beta}, \mathbf{0}, \mathbf{0})} = g_2^{\boldsymbol{k}(\boldsymbol{\alpha}+\boldsymbol{Z}\binom{\mathbf{0}}{\beta}, \boldsymbol{ZR}, \mathbb{H})} & \text{if } \mathsf{t} = 3 \ (22) \end{cases}$$

We call t the type of semi-functional keys. Note that

– In computing type 0, 3, $\widehat{\mathsf{MSK}}_{\mathsf{aux}}$ is not required as input (and no $\hat{\boldsymbol{R}}$ needed).

– In computing type 0, 1, β is not required as input.

Proof (of Theorem 1). We use a sequence of games in the following order:

$\mathsf{G}_{\mathrm{real}} \to \mathsf{G}_0 \to \mathsf{G}_{1,1} \to \cdots \to \mathsf{G}_{i-1,3} \to \mathsf{G}_{i,1} \to \mathsf{G}_{i,2} \to \mathsf{G}_{i,3} \to \cdots \to \mathsf{G}_{q_1,3} \to \mathsf{G}_{q_1+1} \to \mathsf{G}_{q_1+2} \to \mathsf{G}_{q_1+3} \to \mathsf{G}_{\mathrm{final}}$

MatDH (between $\mathsf{G}_{\mathrm{real}}$ and G_0); MatDH, CMH, MatDH (between $\mathsf{G}_{i-1,3}$, $\mathsf{G}_{i,1}$, $\mathsf{G}_{i,2}$, $\mathsf{G}_{i,3}$); MatDH, SMH, MatDH, = (between $\mathsf{G}_{q_1,3}$, G_{q_1+1}, G_{q_1+2}, G_{q_1+3}, $\mathsf{G}_{\mathrm{final}}$)

where each game is defined as follows.[10] $\mathsf{G}_{\mathrm{real}}$ is the actual security game. Each of the following game is defined exactly as *its previous game* in the sequence except the specified modification as follows. For notational purpose, let $\mathsf{G}_{0,3} := \mathsf{G}_0$.

[10] For formality and ease of viewing, we depict these game definitions in Fig. 1 in [4].

- G_0: We modify the challenge ciphertext to be semi-functional type.
- $\mathsf{G}_{i,t}$ where $i \in [1, q_1]$, $t \in \{1, 2, 3\}$: We modify the i-th queried key to be semi-functional of type-t. We use fresh β for each key (for type $t = 2, 3$).
- G_{q_1+t} where $t \in \{1, 2, 3\}$: We modify all keys in phase 2 to be semi-functional of type-t at once. We use the same β for all these keys (for type $t = 2, 3$).
- $\mathsf{G}_{\mathsf{final}}$: We modify the challenge to encrypt a random message.

In the final game, the advantage of $\mathcal{A}$ is trivially 0. We prove the indistinguishability between all these adjacent games (under the underlying assumptions as written in the diagram). Due to the lack of space, we defer most of them to [4] and show only the proof of the indistinguishability between $\mathsf{G}_{\mathsf{real}}$ and G_0 under MatDH here below (Lemma 2). Other MatDH-based transitions can be done similarly. On the other hand, the transitions based on the security of encodings (namely, CMH and SMH), although are a bit more involved, will basically follow the intuition explained at the end of Sect. 4. In particular, we will be able to establish the reduction to the security of encodings thanks to the restriction for regular encodings (Rule 2–4) and the parameter-hiding lemma. From these, we obtain Theorem 1. □

Lemma 2 ($\mathsf{G}_{\mathsf{real}}$ **to** G_0)**.** *For any adversary $\mathcal{A}$ against* ABE, *there exists an algorithm $\mathcal{B}$ that breaks the $\mathcal{D}_d$-Matrix-DH with* $|\mathsf{G}_{\mathsf{real}}\mathsf{Adv}_{\mathcal{A}}^{\mathsf{ABE}}(\lambda) - \mathsf{G}_0\mathsf{Adv}_{\mathcal{A}}^{\mathsf{ABE}}(\lambda)| \leq \mathsf{Adv}_{\mathcal{B}}^{\mathcal{D}_d\text{-}\mathsf{MatDH}}(\lambda)$. *(Denote $\mathsf{G}_j\mathsf{Adv}_{\mathcal{A}}^{\mathsf{ABE}}(\lambda)$ as the advantage of $\mathcal{A}$ in the game G_j.)*

Proof (of Lemma 2). $\mathcal{B}$ obtains an input $(\mathbb{G}, g_1^{\boldsymbol{T}}, g_1^{\boldsymbol{T}\binom{\boldsymbol{y}}{\hat{y}}})$ from the $\mathcal{D}_d$-Matrix DH Assumption where either $\hat{y} = 0$ or $\hat{y} \xleftarrow{\$} \mathbb{Z}_p$, and $\boldsymbol{T} \xleftarrow{\$} \mathcal{D}_d$, $\boldsymbol{y} \xleftarrow{\$} \mathbb{Z}_p^{d \times 1}$.

Setup. $\mathcal{B}$ runs Setup except that it uses $\mathbb{G}$ from its input, and that it will set $(\boldsymbol{B}, \boldsymbol{Z})$ in an *implicit* manner as follows. $\mathcal{B}$ chooses $\tilde{\boldsymbol{B}} \xleftarrow{\$} \mathbb{GL}_{p,d+1}$, $\boldsymbol{J} \xleftarrow{\$} \mathbb{GL}_{p,d}$ and sets

$$\boldsymbol{B} = \tilde{\boldsymbol{B}}\boldsymbol{T}, \qquad \boldsymbol{Z} = \tilde{\boldsymbol{B}}^{-\top}\tilde{\boldsymbol{Z}} := (\tilde{\boldsymbol{B}}^{-\top}) \begin{array}{c} \\ d \\ 1 \end{array}\!\!\overset{\begin{array}{cc} d & \quad 1 \end{array}}{\begin{pmatrix} \boldsymbol{J} & -\boldsymbol{M}^{-\top}\boldsymbol{c}^{\top} \\ \boldsymbol{0} & 1 \end{pmatrix}},$$

where we recall that $\boldsymbol{T} = \left(\begin{smallmatrix} \boldsymbol{M} & \boldsymbol{0} \\ \boldsymbol{c} & 1 \end{smallmatrix}\right)$ from Eq. (1). We can see that $(\boldsymbol{B}, \boldsymbol{Z})$ are properly distributed as from $\mathcal{S}_d$ as follows.

- $\boldsymbol{B}$ is properly distributed due to uniformly random $\tilde{\boldsymbol{B}}, \boldsymbol{T} \in \mathbb{GL}_{p,d+1}$.
- $\boldsymbol{Z}$ is properly distributed as we observe that $\boldsymbol{D} = \boldsymbol{B}^{\top}\boldsymbol{Z}$ is

$$\boldsymbol{D} = \boldsymbol{B}^{\top}\boldsymbol{Z} = (\boldsymbol{T}^{\top}\tilde{\boldsymbol{B}}^{\top})(\tilde{\boldsymbol{B}}^{-\top}\tilde{\boldsymbol{Z}}) = \boldsymbol{T}^{\top}\tilde{\boldsymbol{Z}}$$

$$= \begin{pmatrix} \boldsymbol{M}^{\top} & \boldsymbol{c}^{\top} \\ \boldsymbol{0} & 1 \end{pmatrix} \begin{pmatrix} \boldsymbol{J} & -\boldsymbol{M}^{-\top}\boldsymbol{c}^{\top} \\ \boldsymbol{0} & 1 \end{pmatrix} = \begin{pmatrix} \boldsymbol{M}^{\top}\boldsymbol{J} & \boldsymbol{0} \\ \boldsymbol{0} & 1 \end{pmatrix},$$

where the last equality holds since $(\boldsymbol{M}^{\top})(-\boldsymbol{M}^{-\top}\boldsymbol{c}^{\top}) + (\boldsymbol{c}^{\top})(1) = \boldsymbol{0}$ (for the upper right block). We can see that $\boldsymbol{D}$ is properly distributed due to uniformly random $\boldsymbol{M}^{\top}, \boldsymbol{J} \in \mathbb{GL}_{p,d}$.

$\mathcal{B}$ can then compute $g_1^{\boldsymbol{B}} = g_1^{\tilde{\boldsymbol{B}}\boldsymbol{T}}$ and $g_2^{\boldsymbol{Z}\binom{\boldsymbol{I}_d}{0}} = g_2^{\tilde{\boldsymbol{B}}^{-\top}\binom{\boldsymbol{J}}{0}}$. Here, the first term is computable from $g_1^{\boldsymbol{T}}$, while in the second term, the unknown last column of $\boldsymbol{Z}$ vanishes through the left projection map, $\binom{\boldsymbol{I}_d}{0}$. From these two terms, $\mathcal{B}$ can compute $\mathsf{PK}, \mathsf{MSK}$. The public key PK is given to $\mathcal{A}$.

Phase 1, 2. $\mathcal{B}$ answer all key queries to $\mathcal{A}$ using KeyGen (with the known MSK).

Challenge. The adversary $\mathcal{A}$ outputs $M_0, M_1 \in \mathbb{G}_T$ and a target $Y^\star$. $\mathcal{B}$ runs $(\boldsymbol{c}; w_2) \leftarrow \mathsf{Enc2}(Y^\star)$ as usual. Using *random self reducibility*, $\mathcal{B}$ extends the Matrix-DH Assumption to (w_2+1)-fold and obtains $(g_1^{\boldsymbol{T}}, g_1^{\boldsymbol{T}\binom{\boldsymbol{Y}}{\hat{\boldsymbol{y}}}})$ where either $\hat{\boldsymbol{y}} = \boldsymbol{0}$ or $\hat{\boldsymbol{y}} \xleftarrow{\$} \mathbb{Z}_p^{1\times(w_2+1)}$ with $\boldsymbol{T} \xleftarrow{\$} \mathcal{D}_d$, $\boldsymbol{Y} \xleftarrow{\$} \mathbb{Z}_p^{d\times(w_2+1)}$. $\mathcal{B}$ chooses $b \xleftarrow{\$} \{0,1\}$ and uses $g_1^{\boldsymbol{T}\binom{\boldsymbol{Y}}{\hat{\boldsymbol{y}}}}$ to compute $\mathsf{CT}^\star = (\boldsymbol{C}^\star, C_0^\star)$ as

$$\boldsymbol{C}^\star = g_1^{\boldsymbol{c}\left(\tilde{\boldsymbol{B}}\boldsymbol{T}\binom{\boldsymbol{Y}}{\hat{\boldsymbol{y}}}, \mathbb{H}\right)}, \qquad C_0^\star = e(g_1, g_2)^{\boldsymbol{\alpha}^\top \tilde{\boldsymbol{B}}\boldsymbol{T}\binom{\boldsymbol{y}_0}{\hat{y}_0}} \cdot M_b,$$

where we let $\binom{\boldsymbol{y}_0}{\hat{y}_0}$ be the first column of $\binom{\boldsymbol{Y}}{\hat{\boldsymbol{y}}}$. This can be done since $\mathcal{B}$ possesses $\boldsymbol{\alpha}, \mathbb{H}, \tilde{\boldsymbol{B}}$. From this setting, we have

- If $\hat{\boldsymbol{y}} = \boldsymbol{0}$, then $\mathsf{CT}^\star$ is exactly a normal ciphertext as in Eq. (11) with $\boldsymbol{S} = \binom{\boldsymbol{Y}}{\boldsymbol{0}}$.
- If $\hat{\boldsymbol{y}} \xleftarrow{\$} \mathbb{Z}_p^{1\times(w_2+1)}$, then $\mathsf{CT}^\star$ is semi-functional as in Eq. (18) with $\boldsymbol{S}+\hat{\boldsymbol{S}} = \binom{\boldsymbol{Y}}{\hat{\boldsymbol{y}}}$.

Guess. The algorithm $\mathcal{B}$ has properly simulated $\mathsf{G}_{\mathsf{real}}$ if $\hat{y} = 0$ and G_0 if $\hat{y} \xleftarrow{\$} \mathbb{Z}_p$. Hence, $\mathcal{B}$ can use the output of $\mathcal{A}$ to break the Matrix DH Assumption. $\square$

7 Concrete Predicates and Our New Instantiations

In this section, we briefly describe the definitions of considering predicates and our new instantiations for them. Regarding the instantiations, their specifications are completely defined in Table 4, where we provide what pair encoding scheme to be instantiated for each scheme.

Dual, Conjunctive, and Dual-policy. We first define basic operations on predicates. For a predicate $R : \mathbb{X} \times \mathbb{Y} \rightarrow \{0,1\}$, its *dual* predicate is defined by $\bar{R} : \bar{\mathbb{X}} \times \bar{\mathbb{Y}} \rightarrow \{0,1\}$ where $\bar{\mathbb{X}} = \mathbb{Y}, \bar{\mathbb{Y}} = \mathbb{X}$ and $\bar{R}(X,Y) := R(Y,X)$. Let $R_1 : \mathbb{X}_1 \times \mathbb{Y}_1 \rightarrow \{0,1\}$, $R_2 : \mathbb{X}_2 \times \mathbb{Y}_2 \rightarrow \{0,1\}$ be two predicates. We define the *conjunctive* predicate of R_1, R_2 as $[R_1 \wedge R_2] : \tilde{\mathbb{X}} \times \tilde{\mathbb{Y}} \rightarrow \{0,1\}$ where $\tilde{\mathbb{X}} = \mathbb{X}_1 \times \mathbb{X}_2$, $\tilde{\mathbb{Y}} = \mathbb{Y}_1 \times \mathbb{Y}_2$ and $[R_1 \wedge R_2]((X_1,X_2),(Y_1,Y_2)) = 1$ iff $R_1(X_1,Y_1) = 1$ *and* $R_2(X_2,Y_2) = 1$. For predicate R, we define its *dual-policy* predicate (DP) [8,10] as the conjunctive of itself and its dual predicate, $\bar{R}$. Generic dual and conjunctive conversions (and hence also dual-policy conversion) for pair encodings are recently given in [10]. We mostly use this conjunctive conversion to obtain dual-policy variants. It is indicated by '+' in Table 4.

ABE for Policy over Doubly-Spatial Relation (ABE-PDS). This predicate was defined in [3] as a generalization that captures doubly-spatial encryption [23] and ABE for monotone span programs (and hence Boolean formulae)

Table 4. Our instantiations

Instantiation	Scheme	Obtained from what encoding
$\mathbf{New}_1$	KP-ABE-PDS	[3, Scheme 6]
$\mathbf{New}_2$	CP-ABE-PDS	[10, Scheme 2]
$\mathbf{New}_3$	DP-ABE-PDS	[3, Scheme 6] + [10, Scheme 2]
$\mathbf{New}_4$	Completely unbounded KP-ABE-MSP	[3, Scheme 4]
$\mathbf{New}_5$	Completely unbounded CP-ABE-MSP	[10, Scheme 3]
$\mathbf{New}_6$	Completely unbounded DP-ABE-MSP	[10, Scheme 4]
$\mathbf{New}_7$	KP-ABE-MSP with constant-size ciphertexts	[3, Scheme 5]
$\mathbf{New}_8$	CP-ABE-MSP with constant-size keys	[10, Scheme 5]
$\mathbf{New}'_9$	KP-ABE-MSP with small universe	[3, Scheme 9]
$\mathbf{New}'_{10}$	CP-ABE-MSP with small universe	[3, Scheme 11]
$\mathbf{New}_{11}$	DP-ABE-MSP with small universe	[3, Scheme 9] + [3, Scheme 11]
$\mathbf{New}'_{12}$	KP-ABE-MSP with large universe	[3, Scheme 12]
$\mathbf{New}'_{13}$	CP-ABE-MSP with large universe	[3, Scheme 13]
$\mathbf{New}_{14}$	DP-ABE-MSP with large universe	[3, Scheme 12] + [3, Scheme 13]
$\mathbf{New}_{15}$	KP-ABE-RL	[3, Scheme 3]
$\mathbf{New}_{16}$	CP-ABE-RL	[3, Scheme 7]
$\mathbf{New}_{17}$	DP-ABE-RL	[3, Scheme 3] + [3, Scheme 7]
$\mathbf{New}_{18}$	Unbounded KP-ABE-BP	$\mathbf{New}_4$ & Theorem 2
$\mathbf{New}_{19}$	Unbounded CP-ABE-BP	$\mathbf{New}_5$ & Theorem 2
$\mathbf{New}_{20}$	Unbounded DP-ABE-BP	$\mathbf{New}_6$ & Theorem 2
$\mathbf{New}_{21}$	KP-ABE-BP with constant-size ciphertexts	$\mathbf{New}_7$ & Theorem 2
$\mathbf{New}_{22}$	CP-ABE-BP with constant-size keys	$\mathbf{New}_8$ & Theorem 2
$\mathbf{New}'_{23}$	Bounded KP-ABE-BP	$\mathbf{New}'_9$ & Theorem 2
$\mathbf{New}'_{24}$	Bounded CP-ABE-BP	$\mathbf{New}'_{10}$ & Theorem 2
$\mathbf{New}_{25}$	Bounded DP-ABE-BP	$\mathbf{New}_{11}$ & Theorem 2
$\mathbf{Newer}_{26}$	KP-ABE-BP with constant-size keys	KP-ABE-MSP with short keys of [7] & Theorem 2
$\mathbf{Newer}_{27}$	CP-ABE-BP with constant-size ciphertexts	CP-ABE-MSP with short ciphertexts of [7] & Theorem 2
$\mathbf{Newer}_{28}$	DP-ABE-MSP with constant-size ciphertexts	CP-ABE-MSP with short ciphertexts of [7] + $\mathbf{New}_7$
$\mathbf{Newer}_{29}$	DP-ABE-MSP with constant-size keys	KP-ABE-MSP with short keys of [7] + $\mathbf{New}_8$
$\mathbf{Newer}_{30}$	DP-ABE-BP with constant-size ciphertexts	$\mathbf{New}_{28}$ & Theorem 2
$\mathbf{Newer}_{31}$	DP-ABE-BP with constant-size keys	$\mathbf{New}_{29}$ & Theorem 2

'+' refers to the conjunctive conjunction given in [10].

into one primitive. We refer the definition to [3]. By using exactly the same encodings as in [3,10], we automatically obtain the first fully-secure prime-order KP-ABE-PDS, CP-ABE-PDS, DP-ABE-PDS schemes ($\mathbf{New}_1$-$\mathbf{New}_3$).

ABE for Monotone Span Programs (ABE-MSP). Let $\mathcal{U}$ be the universe of attributes. If $|\mathcal{U}|$ is of super-polynomial size, it is called large universe [21,39], otherwise, it is small universe. In ABE-MSP [21], a policy is specified by a monotone span program (A, ρ) where A is an integer matrix of dimension $m \times k$ for some m, k, and ρ is a map $\rho : [1, m] \to \mathcal{U}$. For a set of attributes $S \subseteq \mathcal{U}$, let $A|_S$ be the sub-matrix of A that takes all the rows j such that $\rho(j) \in S$. We say that (A, ρ) accepts S if $(1, 0, \ldots, 0) \in \text{rowspan}(A|_S)$. ABE-MSP is the most popular predicate studied in the literature since it is known to imply ABE for Boolean formulae [21]. Let $t := |S|$. Some schemes specifies bounds on maximum allowed sizes of t, m, k (we denote these bounds as T, M, K). Some may restrict the maximum number, denoted by R, of attribute multi-use in one policy (that is, the number of distinct i for the same $\rho(i)$). We call a large-universe scheme without any bounds a *completely unbounded* ABE scheme.

By using the same encodings as in [3,10], we obtain the first fully-secure, prime-order ABE-MSP with various properties: completcly unbounded KP/CP/DP-ABE, and short-ciphertext KP-ABE, short-key CP-ABE ($\mathbf{New}_4$-$\mathbf{New}_8$). By using encodings in [3] for bounded schemes, we also obtain some bounded schemes $\mathbf{New}'_9$-$\mathbf{New}_{14}$; these latter encodings are perfectly master-key hiding, hence the resulting schemes rely solely on the Matrix-DH assumption. Furthermore, we also observe that, by using also new encodings in [7] (which is then a subsequent work based on our work), we further obtain the first DP-ABE with short ciphertexts ($\mathbf{Newer}_{28}$), or short keys ($\mathbf{Newer}_{29}$).

For concreteness, we explicitly give the description for one of our instantiations, $\mathbf{New}_4$, in the full version [4].

Performances of Our ABE-MSP Schemes. We compare performances of our KP-ABE-MSP, CP-ABE-MSP to others in the literature in Tables 5 and 6, respectively. For clarity of comparison, we augment schemes in the literature which were proposed for one-use, to multi-use (with bound R) by using the transformation in [33]. Available pair encodings in [3,10] were proved secure in symmetric groups, hence to be able to use them as they are, we will evaluate our construction at $d = 2$, which yields the most efficient instantiations in symmetric settings. In such a case, schemes can rely on DLIN (See also Remark 5).

The numbers of group elements in our schemes for SK, CT are 3 times as large as their composite-order counterparts in A14, AY15 [3,10]. But since composite-order elements are 12 times larger than prime-order ones [22], we achieve improvements of 25 % size reduction. More importantly, time performance is significantly improved. We recall that pairing is 250 times slower in composite-order groups than in prime-order ones [22]. In unbounded ABE ($\mathbf{New}_4$, $\mathbf{New}_5$), the dominant operation is pairing, and the numbers of pairings in decryption are 3 times as large as their composite-order counterparts in [3,10]. As a result, our decryption is about 80 times faster. In constant-size ABE ($\mathbf{New}_7$, $\mathbf{New}_8$), the numbers of pairing are constant, and exponentiation may dominate

Table 5. Performance by each KP-ABE for monotone span programs

	Scheme	|PK|	|SK|	|CT|	Decryption complexity			Sec.	Assumptions	Reduction cost
					Pairing	Exp$\mathbb{G}$	Exp$\mathbb{G}_T$			
Composite-order schemes	LW11 [32]	5	$4m$	$3t+1$	$4m$	0	m	sel.	SD	$O(q_{\text{all}})$
	A14 [3, Scheme 4]	8	$3m+3$	$2t+4$	$3m+3$	0	m	full	SD, $(1,t)$-EDHE3, $(1,m,k)$-EDHE4	$O(q_1)$ 1 $O(q_1)$
	A14 [3, Scheme 5]	$T+8$	$Tm+3m$ $+3$	6	6	$Tm+3m$	0	full	SD, $(T+1,1)$-EDHE3, $(T+1,m,k)$-EDHE4	$O(q_1)$ 1 $O(q_1)$
	CW14 [17]	$U+1$	$Um+m$	2	$2m$	U	m	semi	3DHsub SD	$O(U)$ $O(1)$
	L+10 [34]	$UR+1$	$2m$	$tR+1$	$2m$	0	m	full	SD	$O(q_{\text{all}})$
	A14 [3, Scheme 9]	$UR+1$	$m+1$	$tR+1$	2	$2m$	0	full	SD	$O(q_{\text{all}})$
	W14 [47]	$UR+1$	$m+1$	$tR+1$	2	$2m$	0	full	SD	$O(q_{\text{all}})$
	A14 [3, Scheme 12]	$16(M+TR)^2$ $\times \log(UR)$	$m+1$	$tR+1$	2	$2m$	0	full	SD	$O(q_{\text{all}})$
	KL15 [28]	$2\log(UR)+1$	$3m$	$3tR$	$3m$	0	m	full	DLIN, SD	$O(URq_{\text{all}})$ $O(q_{\text{all}})$
Prime-order schemes	RW13 [40]	4	$3m$	$2t+1$	$3m$	0	m	sel.	t-RW2	1
	OT12 [38]	99	$14m+5$	$14tR+5$	$14m+5$	0	m	full	DLIN	$O(t^2R^2q_{\text{all}})$
	New$_4$	42	$9m+9$	$6t+12$	$9m+9$	0	m	full	DLIN, $(1,t)$-EDHE3p, $(1,m,k)$-EDHE4p	$O(q_1)$ 1 $O(q_1)$
	ALP11 [9]	$T+1$	$Tm+m$	3	3	$Tm+m$	0	sel.	T-DBDHE	1
	T14 [43]	$12T^2+15$	$6Tm+6T$	17	17	$6Tm+6T$	0	semi	DLIN	$O(T)$
	New$_7$	$6T+42$	$3Tm+9m$ $+9$	18	18	$3Tm+9m$	0	full	DLIN, $(T+1,1)$-EDHE3p, $(T+1,m,k)$-EDHE4p	$O(q_1)$ 1 $O(q_1)$
	GPSW06 [22]	$T+3$	$2m$	$t+1$	$2m$	0	m	sel.	DBDH	1
	CGW15 [15]	$6UR+6$	$3m+3$	$3tR+3$	6	$6m$	0	full	DLIN	$O(q_{\text{all}})$
	New$'_9$	$6UR+6$	$3m+3$	$3tR+3$	6	$6m$	0	full	DLIN	$O(q_{\text{all}})$
	OT10 [37]	$21TR+15$	$7m+5$	$7tR+5$	$7m+5$	0	m	full	DLIN	$O(q_{\text{all}})$
	New$'_{12}$	$96(M+TR)^2$ $\times \log(UR)$	$3m+3$	$3tR+3$	6	$6m$	0	full	DLIN	$O(q_{\text{all}})$
	KL15 [28]	$24\log^2(UR)$ $+48\log(UR)$	$3m\log U R$ $+6m$	$3tR\log UR$ $+6tR$	$3m\log UR$ $+6m$	0	m	full	DLIN	$O(URq_{\text{all}})$

[1] Variables:
- t is the attribute set size; T is the maximum size for t (if bounded).
- $m \times k$ is the dimension of the matrix for the span program (the policy); M, K are the maximum sizes for m, k (if bounded).
- U is the size of the attribute universe (if bounded small-universe).
- R is the maximum number of attribute multi-use in one policy (if bounded).
- q_1 is the number of key queries in phase 1 (before the challenge). q_{all} is the number of all key queries.

[2] |PK|, |SK|, |CT| depict the number of source group elements ($\mathbb{G}_1$ or $\mathbb{G}_2$) in public key, secrete key, and ciphertext, respectively. Composite-order group elements are about 12 times larger than prime-order group elements [23]. We omit target group elements ($\mathbb{G}_T$): in PK, all the schemes above have at most 3 elements; in CT, all schemes contain 1 element.

[3] In Decryption complexity, 'Pairing' = the number of pairings, 'Exp$\mathbb{G}$' = the number of exponentiations in source groups ($\mathbb{G}_1$ or $\mathbb{G}_2$), 'Exp$\mathbb{G}_T$' = the number of exponentiations in the target group ($\mathbb{G}_T$).

[4] Sec. is for security. 'sel.'= selective; 'full'= full security. 'semi'= semi-adaptive security [17,43] (an intermediate of selective/full).

[5] We refer assumptions to corresponding papers. Particularly, SD refers to some subgroup decision assumptions in composite-order groups [31,34].

[6] The reduction cost refers to the security factor loss to the corresponding assumption in the same line in the table. The security of each scheme relies on all assumptions for it combined.

Table 6. Performance by each CP-ABE for monotone span programs

	Scheme	|PK|	|SK|	|CT|	Decryption complexity			Sec.	Assumptions	Reduction cost
					Pairing	ExpG	Exp$\mathbb{G}_T$			
Composite-order schemes	LW12 [33]	$U+3$	$t+3$	$2m+2$	$2m+2$	0	m	full	SD, 3DHsub, $\max\{m,k\}$-SPBDHE	$O(q_{\text{all}})$, $O(q_1)$, $O(q_2)$
	AY15 [10, Scheme 3]	10	$2t+6$	$3m+5$	$3m+5$	0	m	full	SD, $(1,t)$-EDHE3, $(1,m,k)$-EDHE4dual	$O(q_1)$, $O(q_1)$, 1
	AY15 [10, Scheme 5]	$T+10$	8	$Tm+3m+5$	8	$Tm+3m$	0	full	SD, $(T+1,1)$-EDHE3, $(T+1,m,k)$-EDHE4dual	$O(q_1)$, $O(q_1)$, 1
	L+10 [34]	$UR+2$	$tR+2$	$2m+1$	$2m+1$	0	m	full	SD	$O(q_{\text{all}})$
	A14 [3, Scheme 11]	$UR+2$	$tR+2$	$m+2$	3	$2m$	0	full	SD	$O(q_{\text{all}})$
	W14 [47]	$UR+2$	$tR+2$	$m+2$	3	$2m$	0	full	SD	$O(q_{\text{all}})$
	A14 [3, Scheme 13]	$16(M+TR)^2 \times \log(UR)$	$tR+2$	$m+2$	3	$2m$	0	full	SD	$O(q_{\text{all}})$
	AC16 [2]	$M(K+T)+M$	$M^2(T+1)+M(K+t-T)$	2	2	$M^2(K+T)$	0	semi	SD	$O((M+K)q_{\text{all}})$
Prime-order schemes	RW13 [40]	5	$2t+2$	$3m+1$	$3m+1$	0	m	sel.	$\max\{m,k\}$-RW1	1
	LW12 [33]	$24U+12$	$6t+6$	$6m+6$	$6m+9$	0	m	full	DLIN, 3DH, $\max\{m,k\}$-SPBDHEp	$O(q_{\text{all}})$, $O(q_1)$, $O(q_2)$
	OT12 [38]	99	$14tR+5$	$14m+5$	$14m+5$	0	m	full	DLIN	$O(t^2R^2q_{\text{all}})$
	New$_5$	54	$6t+18$	$9m+15$	$9m+15$	0	m	full	DLIN, $(1,t)$-EDHE3p, $(1,m,k)$-EDHE4dualp	$O(q_1)$, $O(q_1)$, 1
	New$_8$	$6T+54$	24	$3Tm+9m+15$	24	$3Tm+9m$	0	full	DLIN, $(T+1,1)$-EDHE3p, $(T+1,m,k)$-EDHE4dualp	$O(q_1)$, $O(q_1)$, 1
	W11 [44]	$U+2$	$t+2$	$2m+1$	$2m+1$	0	m	sel.	$\max\{m,k\}$-PDBDH	1
	CGW15 [15]	$6UR+12$	$3tR+6$	$3m+3$	6	$6m$	0	full	DLIN	$O(q_{\text{all}})$
	New$'_{10}$	$6UR+12$	$3tR+6$	$3m+6$	9	$6m$	0	full	DLIN	$O(q_{\text{all}})$
	OT10 [37]	$21TR+15$	$7tR+5$	$7m+5$	$7m+5$	0	m	full	DLIN	$O(q_{\text{all}})$
	New$'_{13}$	$96(M+TR)^2 \times \log(UR)$	$3tR+6$	$3m+6$	9	$6m$	0	full	DLIN	$O(q_{\text{all}})$
	AC16 [2]	$6M(K+T)+6M$	$3M^2(T+1)+3M(K+t-T)$	6	6	$3M^2(K+T)$	0	semi	DLIN	$O((M+K)q_{\text{all}})$

[1] q_2 is the number of queries in phase 2 (after the challenge).
[2] We refer for the remaining parameters to the note under Table 5.

(depending on m, T), but the improvement is similar, since exponentiation (in $\mathbb{G}_1, \mathbb{G}_2$) can be more than 200 times faster in prime-order groups [22, Table 6].

Remark 4. The underlying pair encodings of our schemes **New**$_4$, **New**$_7$ are those proposed in [3, Sect. 7.1, 7.2], of which security rely on parameterized assumptions, namely, EDHE3, EDHE4, also given in [3]. We indeed use *prime-order group* versions, hence denoted as EDHE3p, EDHE4p, instead of *prime-order subgroup in composite-order group* as defined in [3]. These are defined exactly the same as the original except only that the group generator $\mathbb{G}$ outputs a prime-order group instead of a composite-order group (see [3, Defininition 6, 7]). For self-containment, we recapture them in the full version [4]. This modification is merely syntactic, see Remark 3.

Remark 5. As mentioned above, we use $d = 2$ so that the security and assumptions for available pair encoding schemes can be argued in the present form. On the other hand, if we are willing to modify the assumptions and security proofs of pair encodings in [3,10] to asymmetric groups, we can also instantiate at $d = 1$, where we can rely on the SXDH assumption (for framework). This yields even more efficient construction.

The modification for assumptions (such as EDHE3p, EDHE4p) to asymmetric settings can be done straightforwardly by defining all elements in both groups $\mathbb{G}_1, \mathbb{G}_2$ (instead of $\mathbb{G}$ in symmetric settings). The proof can be modified by using $\mathbb{G}_1$ for all elements of ciphertexts, and $\mathbb{G}_2$ for all elements of keys, as defined in our construction. To optimize the size of assumptions (which is otherwise two times larger than the original), we can use automated tools of [1].

ABE for Regular Languages (ABE-RL). In ABE-RL [45], a policy is a deterministic finite automata (DFA) M, and an input to policy is a string w, and $R(M, w) = 1$ if the automata M accepts the string w. We defer the detailed definition to [3,4]. We obtain the first fully-secure prime-order KP-ABE, CP-ABE, DP-ABE for regular languages ($\mathbf{New}_{15}$-$\mathbf{New}_{17}$).

ABE for Branching Programs (ABE-BP). In ABE-BP [19], a policy is associated to a branching program Γ, which is a directed acyclic graph in which every non-terminal node has exactly two outgoing edges labeled $(i, 0)$ and $(i, 1)$ for some $i \in \mathbb{N}$. For an edge j, denote its label as ℓ_j. Moreover, there is a distinguished terminal node called accept node. We can also assume wlog that there is exactly one start node. We can assume wlog that there is at most only one edge connecting any two nodes in Γ (See [19]).

An input to policy is a binary string w. Every input binary string w induces a subgraph Γ_w that contains exactly all the edges labeled (i, w_i) for $i \in [1, |w|]$, where we write $w = (w_1, \ldots, w_{|w|})$ as the binary representation of w. We say that Γ accepts w if there is a path from the start node to the accept node in Γ_w. If the allow length of w is bounded, we say that it is a *bounded* ABE-BP, otherwise, it is an *unbounded* scheme. In the latter, a label (i, b) has no bound on i.

We invoke the following theorem, which holds unconditionally.

Theorem 2. *Large-universe ABE-MSP implies ABE-BP.*

Remark 6. Karchmer and Wigderson proved in 1993 [26] that $\mathbf{SL} \subseteq \mathbf{PSP}$ (Symmetric Logspace $\subseteq$ Poly-size Span Program). Thus, the ABE-MSP-to-ABE-BP implication can be inferred from this. (We thank an anonymous reviewer for pointing this out.) Nevertheless, to the best of our knowledge, there is no explicit use of this theorem in the context of ABE, as ABE-MSP and ABE-BP were often studied separately. For self-containment and independent interest, we offer our alternative proof for this ABE-MSP-to-ABE-BP implication in the full version [4].

Our proof for this implication in [4] is constructive and the conversion preserves efficiency and the unbounded property (if satisfied) of the original ABE-MSP. Therefore, by using our instantiated ABE-MSP, we obtain the first schemes for the following schemes of ABE-BP: unbounded, short-ciphertext, short-key for all KP/CP/DP variants of ABE-BP (See Table 4). Our schemes are the first such schemes for each given property, not to mention that they are fully-secure and prime-order schemes. (This is with the only exception to the *selectively*-secure short-key KP-ABE-BP of [20]).

8 Generic Construction from Simpler Basis

Our main construction in Sect. 5 is based upon the original basis of PDSG in [15], where both $\boldsymbol{B}, \boldsymbol{B}^{-\top}$ are required for setup. Chen et al. [14] proposed a simpler basis where the inverse matrix is not required. This substantially simplifies the proofs for subgroup decision-like assumptions provided by PDSG. In this section, we provide a simplification of our scheme using the basis from [14].

Simpler Basis from CGW [14]. Let $\mathcal{W}_d$ be an efficiently samplable distribution of pair $(\boldsymbol{A}, \boldsymbol{a}^\perp)$ over $\mathbb{Z}_p^{(d+1)\times d} \times \mathbb{Z}_p^{(d+1)\times 1}$ so that $(\boldsymbol{a}^\perp)^\top \boldsymbol{A} = \boldsymbol{0}$ and $\boldsymbol{a}^\perp \neq \boldsymbol{0}$. A useful property of $\mathcal{W}_d$ is the Basis Lemma [14], which we also recap in [4].

Our Simplified Construction. From a pair encoding scheme P, our simplified generic construction, denoted $\mathsf{SimplerABE}(\mathsf{P})$, can be described as follows. The correctness, the security theorem, and the security proof are similar to our main construction and are deferred to [4].

- $\mathsf{Setup}(1^\lambda, \kappa)$: Run $(\mathbb{G}_1, \mathbb{G}_2, \mathbb{G}_T, e, p) \xleftarrow{\$} \mathcal{G}(\lambda)$. Pick generators $g_1 \xleftarrow{\$} \mathbb{G}_1$ and $g_2 \xleftarrow{\$} \mathbb{G}_2$. Run $n \leftarrow \mathsf{Param}(\kappa)$. Pick $\mathbb{H} = (\boldsymbol{H}_1, \ldots, \boldsymbol{H}_n) \xleftarrow{\$} (\mathbb{Z}_p^{(d+1)\times(d+1)})^n$. Sample $(\boldsymbol{A}, \boldsymbol{a}^\perp) \xleftarrow{\$} \mathcal{W}_d$ and $(\boldsymbol{B}, \boldsymbol{b}^\perp) \xleftarrow{\$} \mathcal{W}_d$. Choose $\boldsymbol{\alpha} \xleftarrow{\$} \mathbb{Z}_p^{(d+1)\times 1}$. Output

$$\begin{aligned} \mathsf{PK} &= \Big(e(g_1, g_2)^{\boldsymbol{\alpha}^\top \boldsymbol{A}}, \; g_1^{\boldsymbol{A}}, g_1^{\boldsymbol{H}_1 \boldsymbol{A}}, \ldots, g_1^{\boldsymbol{H}_n \boldsymbol{A}} \Big), \\ \mathsf{MSK} &= \Big(\quad g_2^{\boldsymbol{\alpha}}, \quad g_2^{\boldsymbol{B}}, g_2^{\boldsymbol{H}_1^\top \boldsymbol{B}}, \ldots, g_2^{\boldsymbol{H}_n^\top \boldsymbol{B}} \Big). \end{aligned} \tag{23}$$

- $\mathsf{Encrypt}(Y, M, \mathsf{PK})$: Upon input $Y \in \mathbb{Y}$, run $(\boldsymbol{c}; w_2) \leftarrow \mathsf{Enc2}(Y)$. Randomly pick $\boldsymbol{S} := (\boldsymbol{s}_0, \boldsymbol{s}_1, \ldots, \boldsymbol{s}_{w_2}) \xleftarrow{\$} \mathbb{Z}_p^{d\times(w_2+1)}$. Output the ciphertext as $\mathsf{CT} = (\boldsymbol{C}, C_0)$:

$$\begin{aligned} \boldsymbol{C} &= g_1^{\boldsymbol{c}(\boldsymbol{AS}, \mathbb{H})} && \in (\mathbb{G}_1^{(d+1)\times 1})^{w_1}, \\ C_0 &= e(g_1, g_2)^{\boldsymbol{\alpha}^\top \boldsymbol{A} \boldsymbol{s}_0} \cdot M && \in \mathbb{G}_T. \end{aligned} \tag{24}$$

- $\mathsf{KeyGen}(X, \mathsf{MSK})$: Upon input $X \in \mathbb{X}$, run $(\boldsymbol{k}; m_2) \leftarrow \mathsf{Enc1}(X)$. Randomly pick $\boldsymbol{R} := (\boldsymbol{r}_1, \ldots, \boldsymbol{r}_{m_2}) \xleftarrow{\$} \mathbb{Z}_p^{d\times m_2}$. Output

$$\mathsf{SK} = g_2^{\boldsymbol{k}(\boldsymbol{\alpha}, \boldsymbol{BR}, \mathbb{H})} \qquad \in (\mathbb{G}_2^{(d+1)\times 1})^{m_1}. \tag{25}$$

- $\mathsf{Decrypt}(\mathsf{CT}, \mathsf{SK})$: Obtain Y, X from CT, SK. Suppose $R(X, Y) = 1$. Run $\boldsymbol{E} \leftarrow \mathsf{Pair}(X, Y)$. Compute $e(g_1, g_2)^{\boldsymbol{\alpha}^\top \boldsymbol{A} \boldsymbol{s}_0} = \prod_{i\in[1,m_1], j\in[1,w_1]} e(\boldsymbol{C}[j], \mathsf{SK}[i])^{E_{i,j}}$. Finally, remove this mask from C_0 to get M.

References

1. Abe, M., Groth, J., Ohkubo, M., Tango, T.: Converting cryptographic schemes from symmetric to asymmetric bilinear groups. In: Garay, J.A., Gennaro, R. (eds.) CRYPTO 2014. LNCS, vol. 8616, pp. 241–260. Springer, Heidelberg (2014). doi:10.1007/978-3-662-44371-2_14
2. Agrawal, S., Chase, M.: A study of pair encodings: predicate encryption in prime order groups. In: Kushilevitz, E., Malkin, T. (eds.) TCC 2016. LNCS, vol. 9563, pp. 259–288. Springer, Heidelberg (2016). doi:10.1007/978-3-662-49099-0_10
3. Attrapadung, N.: Dual system encryption via doubly selective security: framework, fully secure functional encryption for regular languages, and more. In: Nguyen, P.Q., Oswald, E. (eds.) EUROCRYPT 2014. LNCS, vol. 8441, pp. 557–577. Springer, Heidelberg (2014). doi:10.1007/978-3-642-55220-5_31. Full version available at Cryptology ePrint Archive: Report 2014/428
4. Attrapadung, N.: Dual system encryption framework in prime-order groups via computational pair encodings. Full version of this paper. Cryptology ePrint Archive: Report 2015/390 (2015)
5. Attrapadung, N., Hanaoka, G., Matsumoto, T., Teruya, T., Yamada, S.: Attribute based encryption with direct efficiency tradeoff. In: Manulis, M., Sadeghi, A.-R., Schneider, S. (eds.) ACNS 2016. LNCS, vol. 9696, pp. 249–266. Springer, Heidelberg (2016). doi:10.1007/978-3-319-39555-5_14
6. Attrapadung, N., Hanaoka, G., Ogawa, K., Ohtake, G., Watanabe, H., Yamada, S.: Attribute-based encryption for range attributes. In: Zikas, V., Prisco, R. (eds.) SCN 2016. LNCS, vol. 9841, pp. 42–61. Springer, Heidelberg (2016). doi:10.1007/978-3-319-44618-9_3
7. Attrapadung, N., Hanaoka, G., Yamada, S.: Conversions among several classes of predicate encryption and applications to ABE with various compactness tradeoffs. In: Iwata, T., Cheon, J.H. (eds.) ASIACRYPT 2015. LNCS, vol. 9452, pp. 575–601. Springer, Heidelberg (2015). doi:10.1007/978-3-662-48797-6_24
8. Attrapadung, N., Imai, H.: Dual-policy attribute based encryption. In: Abdalla, M., Pointcheval, D., Fouque, P.-A., Vergnaud, D. (eds.) ACNS 2009. LNCS, vol. 5536, pp. 168–185. Springer, Heidelberg (2009). doi:10.1007/978-3-642-01957-9_11
9. Attrapadung, N., Libert, B., Panafieu, E.: Expressive key-policy attribute-based encryption with constant-size ciphertexts. In: Catalano, D., Fazio, N., Gennaro, R., Nicolosi, A. (eds.) PKC 2011. LNCS, vol. 6571, pp. 90–108. Springer, Heidelberg (2011). doi:10.1007/978-3-642-19379-8_6
10. Attrapadung, N., Yamada, S.: Duality in ABE: converting attribute based encryption for dual predicate and dual policy via computational encodings. In: Nyberg, K. (ed.) CT-RSA 2015. LNCS, vol. 9048, pp. 87–105. Springer, Heidelberg (2015). doi:10.1007/978-3-319-16715-2_5. Full version available at Cryptology ePrint Archive: Report 2015/157
11. Boneh, D., Gentry, C., Gorbunov, S., Halevi, S., Nikolaenko, V., Segev, G., Vaikuntanathan, V., Vinayagamurthy, D.: Fully key-homomorphic encryption, arithmetic circuit ABE and compact garbled circuits. In: Nguyen, P.Q., Oswald, E. (eds.) EUROCRYPT 2014. LNCS, vol. 8441, pp. 533–556. Springer, Heidelberg (2014). doi:10.1007/978-3-642-55220-5_30
12. Boneh, D., Sahai, A., Waters, B.: Functional encryption: definitions and challenges. In: Ishai, Y. (ed.) TCC 2011. LNCS, vol. 6597, pp. 253–273. Springer, Heidelberg (2011). doi:10.1007/978-3-642-19571-6_16

13. Chase, M., Meiklejohn, S.: Déjà Q: using dual systems to revisit q-type assumptions. In: Nguyen, P.Q., Oswald, E. (eds.) EUROCRYPT 2014. LNCS, vol. 8441, pp. 622–639. Springer, Heidelberg (2014). doi:10.1007/978-3-642-55220-5_34
14. Chen, J., Gay, R., Wee, H.: Improved dual system ABE in prime-order groups via predicate encodings. In: Oswald, E., Fischlin, M. (eds.) EUROCRYPT 2015. LNCS, vol. 9057, pp. 595–624. Springer, Heidelberg (2015). doi:10.1007/978-3-662-46803-6_20
15. Chen, J., Wee, H.: Fully, (Almost) tightly secure ibe and dual system groups. In: Canetti, R., Garay, J.A. (eds.) CRYPTO 2013. LNCS, vol. 8043, pp. 435–460. Springer, Heidelberg (2013). doi:10.1007/978-3-642-40084-1_25
16. Chen, J., Wee, H.: Semi-adaptive attribute-based encryption and improved delegation for boolean formula. In: Abdalla, M., Prisco, R. (eds.) SCN 2014. LNCS, vol. 8642, pp. 277–297. Springer, Heidelberg (2014). doi:10.1007/978-3-319-10879-7_16
17. Escala, A., Herold, G., Kiltz, E., Ràfols, C., Villar, J.: An algebraic framework for Diffie-Hellman assumptions. In: Canetti, R., Garay, J.A. (eds.) CRYPTO 2013. LNCS, vol. 8043, pp. 129–147. Springer, Heidelberg (2013). doi:10.1007/978-3-642-40084-1_8
18. Freeman, D.M.: Converting pairing-based cryptosystems from composite-order groups to prime-order groups. In: Gilbert, H. (ed.) EUROCRYPT 2010. LNCS, vol. 6110, pp. 44–61. Springer, Heidelberg (2010). doi:10.1007/978-3-642-13190-5_3
19. Gorbunov, S., Vaikuntanathan, V., Wee, H.: Attribute-based encryption for circuits. In: STOC 2013 (2013)
20. Gorbunov, S., Vinayagamurthy, D.: Riding on asymmetry: efficient ABE for branching programs. In: Iwata, T., Cheon, J.H. (eds.) ASIACRYPT 2015. LNCS, vol. 9452, pp. 550–574. Springer, Heidelberg (2015). doi:10.1007/978-3-662-48797-6_23
21. Goyal, V., Pandey, O., Sahai, A., Waters, B.: Attribute-based encryption for fine-grained access control of encrypted data. In: ACM CCS 2006, pp. 89–98 (2006)
22. Guillevic, A.: Comparing the pairing efficiency over composite-order and prime-order elliptic curves. In: Jacobson, M., Locasto, M., Mohassel, P., Safavi-Naini, R. (eds.) ACNS 2013. LNCS, vol. 7954, pp. 357–372. Springer, Heidelberg (2013). doi:10.1007/978-3-642-38980-1_22
23. Hamburg, M.: Spatial Encryption. Cryptology. ePrint Archive: Report 2011/389
24. Herold, G., Hesse, J., Hofheinz, D., Ràfols, C., Rupp, A.: Polynomial spaces: a new framework for composite-to-prime-order transformations. In: Garay, J.A., Gennaro, R. (eds.) CRYPTO 2014. LNCS, vol. 8616, pp. 261–279. Springer, Heidelberg (2014). doi:10.1007/978-3-662-44371-2_15
25. Ishai, Y., Wee, H.: Partial garbling schemes and their applications. In: Esparza, J., Fraigniaud, P., Husfeldt, T., Koutsoupias, E. (eds.) ICALP 2014. LNCS, vol. 8572, pp. 650–662. Springer, Heidelberg (2014). doi:10.1007/978-3-662-43948-7_54
26. Karchmer, M., Wigderson, A.: On span programs. In: Structure in Complexity Theory Conference (1993)
27. Kowalczyk, L., Lewko, A.B.: Bilinear entropy expansion from the decisional linear assumption. In: Gennaro, R., Robshaw, M. (eds.) CRYPTO 2015. LNCS, vol. 9216, pp. 524–541. Springer, Heidelberg (2015). doi:10.1007/978-3-662-48000-7_26. Report 2014/754 (retrieved version: Sep. 4, 2015)
28. Lewko, A.: Tools for simulating features of composite order bilinear groups in the prime order setting. In: Pointcheval, D., Johansson, T. (eds.) EUROCRYPT 2012. LNCS, vol. 7237, pp. 318–335. Springer, Heidelberg (2012). doi:10.1007/978-3-642-29011-4_20

29. Lewko, A., Meiklejohn, S.: A profitable sub-prime loan: obtaining the advantages of composite order in prime-order bilinear groups. In: Katz, J. (ed.) PKC 2015. LNCS, vol. 9020, pp. 377–398. Springer, Heidelberg (2015). doi:10.1007/978-3-662-46447-2_17
30. Lewko, A., Waters, B.: New techniques for dual system encryption and fully secure HIBE with short ciphertexts. In: Micciancio, D. (ed.) TCC 2010. LNCS, vol. 5978, pp. 455–479. Springer, Heidelberg (2010). doi:10.1007/978-3-642-11799-2_27
31. Lewko, A., Waters, B.: Unbounded HIBE and attribute-based encryption. In: Paterson, K.G. (ed.) EUROCRYPT 2011. LNCS, vol. 6632, pp. 547–567. Springer, Heidelberg (2011). doi:10.1007/978-3-642-20465-4_30
32. Lewko, A., Waters, B.: New proof methods for attribute-based encryption: achieving full security through selective techniques. In: Safavi-Naini, R., Canetti, R. (eds.) CRYPTO 2012. LNCS, vol. 7417, pp. 180–198. Springer, Heidelberg (2012). doi:10.1007/978-3-642-32009-5_12
33. Lewko, A., Okamoto, T., Sahai, A., Takashima, K., Waters, B.: Fully secure functional encryption: attribute-based encryption and (hierarchical) inner product encryption. In: Gilbert, H. (ed.) EUROCRYPT 2010. LNCS, vol. 6110, pp. 62–91. Springer, Heidelberg (2010). doi:10.1007/978-3-642-13190-5_4
34. Meiklejohn, S., Shacham, H., Freeman, D.M.: Limitations on transformations from composite-order to prime-order groups: the case of round-optimal blind signatures. In: Abe, M. (ed.) ASIACRYPT 2010. LNCS, vol. 6477, pp. 519–538. Springer, Heidelberg (2010). doi:10.1007/978-3-642-17373-8_30
35. Okamoto, T., Takashima, K.: Hierarchical predicate encryption for inner-products. In: Matsui, M. (ed.) ASIACRYPT 2009. LNCS, vol. 5912, pp. 214–231. Springer, Heidelberg (2009). doi:10.1007/978-3-642-10366-7_13
36. Okamoto, T., Takashima, K.: Fully secure functional encryption with general relations from the decisional linear assumption. In: Rabin, T. (ed.) CRYPTO 2010. LNCS, vol. 6223, pp. 191–208. Springer, Heidelberg (2010). doi:10.1007/978-3-642-14623-7_11
37. Okamoto, T., Takashima, K.: Fully secure unbounded inner-product and attribute-based encryption. In: Wang, X., Sako, K. (eds.) ASIACRYPT 2012. LNCS, vol. 7658, pp. 349–366. Springer, Heidelberg (2012). doi:10.1007/978-3-642-34961-4_22
38. Parno, B., Raykova, M., Vaikuntanathan, V.: How to delegate and verify in public: verifiable computation from attribute-based encryption. In: Cramer, R. (ed.) TCC 2012. LNCS, vol. 7194, pp. 422–439. Springer, Heidelberg (2012). doi:10.1007/978-3-642-28914-9_24
39. Rouselakis, Y., Waters, B..: Practical constructions and new proof methods for large universe attribute-based encryption. In: ACM CCS 2013, pp. 463–474 (2013)
40. Sahai, A., Waters, B.: Fuzzy identity-based encryption. In: Cramer, R. (ed.) EUROCRYPT 2005. LNCS, vol. 3494, pp. 457–473. Springer, Heidelberg (2005). doi:10.1007/11426639_27
41. Seo, J.H., Cheon, J.H.: Beyond the limitation of prime-order bilinear groups, and round optimal blind signatures. In: Cramer, R. (ed.) TCC 2012. LNCS, vol. 7194, pp. 133–150. Springer, Heidelberg (2012). doi:10.1007/978-3-642-28914-9_8
42. Takashima, K.: Expressive attribute-based encryption with constant-size ciphertexts from the decisional linear assumption. In: Abdalla, M., Prisco, R. (eds.) SCN 2014. LNCS, vol. 8642, pp. 298–317. Springer, Heidelberg (2014). doi:10.1007/978-3-319-10879-7_17

43. Waters, B.: Ciphertext-policy attribute-based encryption: an expressive, efficient, and provably secure realization. In: Catalano, D., Fazio, N., Gennaro, R., Nicolosi, A. (eds.) PKC 2011. LNCS, vol. 6571, pp. 53–70. Springer, Heidelberg (2011). doi:10.1007/978-3-642-19379-8_4
44. Waters, B.: Dual system encryption: realizing fully secure IBE and HIBE under simple assumptions. In: Halevi, S. (ed.) CRYPTO 2009. LNCS, vol. 5677, pp. 619–636. Springer, Heidelberg (2009). doi:10.1007/978-3-642-03356-8_36
45. Waters, B.: Functional encryption for regular languages. In: Safavi-Naini, R., Canetti, R. (eds.) CRYPTO 2012. LNCS, vol. 7417, pp. 218–235. Springer, Heidelberg (2012). doi:10.1007/978-3-642-32009-5_14
46. Wee, H.: Dual system encryption via predicate encodings. In: Lindell, Y. (ed.) TCC 2014. LNCS, vol. 8349, pp. 616–637. Springer, Heidelberg (2014). doi:10.1007/978-3-642-54242-8_26
47. Wee, H.: Déjà Q: encore! un petit IBE. In: Kushilevitz, E., Malkin, T. (eds.) TCC 2016. LNCS, vol. 9563, pp. 237–258. Springer, Heidelberg (2016). doi:10.1007/978-3-662-49099-0_9

Efficient IBE with Tight Reduction to Standard Assumption in the Multi-challenge Setting

Junqing Gong[1], Xiaolei Dong[2(✉)], Jie Chen[3,4,5(✉)], and Zhenfu Cao[2(✉)]

[1] Department of Computer Science and Engineering, Shanghai Jiao Tong University, Shanghai, China
gongjunqing@126.com

[2] Shanghai Key Lab for Trustworthy Computing, East China Normal University, Shanghai, China
{dongxiaolei,zfcao}@sei.ecnu.edu.cn

[3] School of Computer Science and Software Engineering, East China Normal University, Shanghai, China
S080001@e.ntu.edu.sg

[4] École Normale Supérieure de Lyon, Laboratoire LIP, Lyon, France

[5] College of Information Science and Technology, Jinan University, Guangzhou, China
http://www.jchen.top

Abstract. In 2015, Hofheinz *et al.* [PKC, 2015] extended Chen and Wee's almost-tight reduction technique for identity based encryptions (IBE) [CRYPTO, 2013] to the multi-instance, multi-ciphertext (MIMC, or multi-challenge) setting, where the adversary is allowed to obtain multiple challenge ciphertexts from multiple IBE instances, and gave the first almost-tightly secure IBE in this setting using composite-order bilinear groups. Several prime-order realizations were proposed lately. However there seems to be a dilemma of high system performance (involving ciphertext/key size and encryption/decryption cost) or weak/standard security assumptions. A natural question is: can we achieve high performance without relying on stronger/non-standard assumptions?

In this paper, we answer the question in the affirmative by describing a prime-order IBE scheme with the same performance as the most efficient solutions so far but whose security still relies on the *standard* k-linear (k-Lin) assumption. Our technical start point is Blazy *et al.*'s almost-tightly secure IBE [CRYPTO, 2014]. We revisit their concrete IBE scheme and associate it with the framework of nested dual system group. This allows us to extend Blazy *et al.*'s almost-tightly secure IBE to the MIMC setting using Gong *et al.*'s method [PKC, 2016]. We emphasize that, when instantiating our construction by the Symmetric eXternal Diffie-Hellman assumption (SXDH = 1-Lin), we obtain the most efficient concrete IBE scheme with almost-tight reduction in the MIMC setting, whose performance is even comparable to the most efficient IBE in the classical model (i.e., the single-instance, single-ciphertext setting). Besides pursuing high performance, our IBE scheme also achieves a weaker form of anonymity pointed out by Attrapadung *et al.* [AsiaCrypt, 2015].

J.H. Cheon and T. Takagi (Eds.): ASIACRYPT 2016, Part II, LNCS 10032, pp. 624–654, 2016.
DOI: 10.1007/978-3-662-53890-6_21

Keywords: Identity based encryption · Tight security · Multi-challenge setting · Nested dual system group · Prime-order bilinear group · Groth-Sahai proof system · (Weak) Anonymity

1 Introduction

1.1 Background and Motivation

The notion of identity based encryption (IBE) was proposed by Shamir [32] in 1984 and realized by Boneh and Franklin [7] in 2001 using bilinear groups. In an IBE system, an authority publishes a set of public parameters and issues secret keys for users according to their identities, the encryption requires the public parameters and receiver's identity (for example, his/her e-mail address). As an advantage over traditional PKI-based cryptosystems, users in an IBE system only need to authenticate and store the system-level public parameter once and for all, while users' identities are always self-explained and thus easy to validate.

Since Boneh and Franklin's work [7], a series of constructions [5,6,13,33] appeared making trade-off between several features such as security model, strength of complexity assumption, and public key size. In 2009, Waters [34] proposed a novel proof technique, called *dual system encryption*, and showed the first adaptively secure IBE scheme with constant-size public key and polynomially related to the k-linear (k-Lin) assumption, a standard assumption, in the standard model. Nowadays the dual system technique has become a regular and powerful tool for achieving adaptive security of attribute based encryptions (ABE) and inner-product encryption (IPE) (and more general primitives) in the standard model [21,22,26–28]. More importantly, under the framework of dual system encryption, we have obtained a clean, deep, and uniform understanding on the construction of a branch of encryption systems, including IBE, ABE, IPE and so on [1,2,8,35].

The classical adaptive security model for IBE [7] requires that the challenge ciphertext for the challenge identity reveals nothing even when the adversary has held secret keys for other identities. The dual system technique [34] generally works as follows. There are two forms of secret keys and ciphertexts, *normal* and *semi-functional* form. The normal ciphertexts/keys are used in the real system, while the semi-functional ciphertexts/keys are often constructed by introducing extra entropy into normal ones and will only be used for the security proof. We say normal object is in the normal space and the extra entropy is in the semi-functional space and require that they are independent in some sense. The proof follows the hybrid argument method. One first transforms the challenge ciphertext from normal to its semi-functional form. Next, one converts secret keys from normal to semi-functional form in an one-by-one fashion. Finally, one can immediately prove the security utilizing the extra entropy we have introduced in the semi-functional space.

Tight Security. Clearly, the reduction described above suffers from a security loss proportional to the number of secret keys the adversary held. Due to the generality of such a loss, a natural question is whether such a security loss is inherent for IBE in the standard model under standard assumptions? In practical point of view, a tightly secure IBE allows practitioners to implement this system in a smaller group, which always leads to shorter ciphertexts/keys and faster encryption/decryption operations in the real world.

Fortunately, Chen and Wee [9] answered the question in the negative. They proposed the first almost-tightly secure IBE in the standard model based on the k-Lin assumption. Here the so-called *almost-tight* means the security loss is proportional to the security parameter instead of the amount of secret keys revealed to the adversary. Technically, they combined the high-level idea of dual system encryption with the proof technique of Naor and Reingold [25]. In the next year, Blazy *et al.* showed an almost-tightly secure IBE with higher space and time efficiency. In fact, they proved that an adaptively secure IBE can be generically constructed from affine message authentication code (MAC) and Groth-Sahai non-interactive zero-knowledge (NIZK) proof [15], and offered us a realization of affine MAC based on Naor and Reingold's proof technique [25]. Roughly speaking, their high-level strategy is still identical to Chen and Wee's [9].

Let us take a look at Chen and Wee's idea [9]. Essentially, they borrowed the proof strategy from Naor and Reingold [25] in order to introduce entropy into semi-functional space more quickly. After converting normal ciphertext to semi-functional form, one may conceptually introduce a truly random function RF to all secret keys and challenge ciphertext whose domain is just $\{\epsilon\}$, i.e., unrelated to the identity. Relying on the binary encoding of the identities in secret keys, one can increase the dependency of RF on the identity, from 0-bit prefix to 1-bit prefix, 2-bit prefix, ..., and finally the entire identity. They called such a property *nested hiding*. At this moment, $\mathsf{RF}(\text{ID})$ is revealed to adversary through secret key for ID while $\mathsf{RF}(\text{ID}^*)$ for the challenge identity ID^* is still unpredictable since adversary is not allowed to hold its secret key. This feature is sufficient for proving the security. It is worth noting that for an identity space $\{0,1\}^n$, we just need n steps to construct such a random function RF and just arise $\mathcal{O}(n)$ security loss.

Multi-instance, Multi-ciphertext Setting. The classical security model for IBE [7] requires that the *single* challenge ciphertext from the *single* challenge identity should leak nothing about the corresponding message even with secret keys for adversarially-chosen identities. In 2015, Hofheinz *et al.* [18] considered a more realistic security model, called adaptive security in the *multi-instance, multi-ciphertext setting* (MIMC, or multi-challenge setting), which ensures the security of *multiple* challenge ciphertexts for *multiple* challenge identities in *multiple* IBE instances. In general, an IBE scheme secure in the classical single-instance, single-ciphertext (SISC) model must be secure in the MIMC setting. However the implication is *not* tightness-preserving. Assuming the number of IBE instances and challenge ciphertexts per instance are μ and Q, the general reduction from MIMC to SISC will arise a multiplicative security loss $\mathcal{O}(Q\mu)$.

Hofheniz *et al.* [18] extended Chen and Wee's tight reduction technique [9] and gave the first almost-tight secure IBE in the MIMC setting. Technically, the ηth *nested hiding* step in Chen and Wee's proof procedure requires that the ηth bit of all challenge identities should be identical. It is the case in the SISC setting but is not necessarily hold in the MIMC setting. To overcome this difficulty, they introduced another semi-functional space. Now the original semi-functional space may be called $\wedge$-semi-functional space and the new-comer may be named $\sim$-semi-functional space. They also employed two independent random functions $\widehat{\mathsf{RF}}$ and $\widetilde{\mathsf{RF}}$ for them, respectively, acting the same role of RF in Chen and Wee's proof. As the preparation for the ηth nested hiding, they transfer the entropy in $\wedge$-semi-functional space to $\sim$-semi-functional space for all challenge ciphertexts whose identity has 1 on its ηth bit. At this moment, we reach the configuration that, in every semi-functional spaces, the challenge identities indeed share the same ηth bit, and nested hiding can be done as Chen and Wee did but in each of two semi-functional spaces *independently.*

However their construction was built in composite-order bilinear groups. Attrapadung *et al.* [3] and Gong *et al.* [14] gave prime-order solutions independently. Attrapadung *et al.* [3] provided a generic framework building almost-tight secure IBE from *broadcast encoding* which is compatible with both composite-order and prime-order bilinear groups. Utilizing the power of broadcast encoding, they proposed not only ordinary IBE scheme but also IBE with other features such as sublinear-size master public key. Gong *et al.* [14] followed the line of extended nested dual system groups (ENDSG) [18] and proposed two constructions from more general assumptions, the second of which is an improved version based on the first one. In this paper, we do not consider additional feature and name Attrapadung *et al.*'s basic IBE in the prime-order group (i.e., $\Phi_{\mathsf{cc}}^{\mathsf{prime}}$) [3] as AHY, while name Gong *et al.*'s two constructions [14] as GCDCT and $\mathsf{GCDCT+}$.

Motivation. Among existing prime-order IBE constructions with almost-tight reduction in the MIMC model, there is a trade-off between the efficiency and strength of complexity assumption. On one hand, GCDCT was proven secure based on the k-Lin assumption but less efficient in terms of both ciphertext/key size and encryption/decryption cost. On the other hand, $\mathsf{GCDCT+}$ and AHY were more efficient but relied on the k-linear assumption with auxiliary input (k-LinAI) in asymmetric bilinear groups and the decisional linear assumption (sDLIN) in symmetric bilinear groups, respectively, which are stronger and less general than the k-Lin assumption. Therefore it is still an interesting and non-trivial problem to find a solution with some real improvements instead of just a trade-off. More concretely, we ask the following question:

QUESTION: Can we find a tightly secure IBE scheme in the MIMC setting, which is (at least) as efficient as $\mathsf{GCDCT+}$ and AHY but still proven secure under the standard k-Lin assumption as GCDCT?

1.2 Our Main Result

In this paper, we answer the question in the affirmative by proposing an IBE scheme using prime-order bilinear groups in the MIMC setting. The adaptive security of the construction is almost-tightly based on the k-Lin assumption as GCDCT. At the same time, its performance is better than GCDCT and is identical to GCDCT+ and AHY for corresponding parameter.

We compare existing almost-tightly secure IBE in prime-order groups with ours in detail in Table 1. The comparison involves the complexity assumption, the sizes of master public key, secret keys and ciphertexts, and encryption/decryption cost. As a base line, we also investigate almost-tightly secure prime-order IBE by Chen and Wee [9], denoted by CW, and Blazy *et al.* [4], denoted by BKP, both of which are adaptively secure in the SISC setting.

- All schemes take $\{0,1\}^n$ as identity space.
- "DLIN" and "sDLIN" in Column "Sec." stand for decisional linear assumption in asymmetric and symmetric bilinear groups, respectively.
- Column $|\text{MPK}|$, $|\text{SK}|$, and $|\text{CT}|$ present numbers of group elements in master public keys, secret keys and ciphertexts, respectively. Here G refers to the source group of symmetric bilinear groups; G_1, G_2 are those of asymmetric bilinear groups; G_T stands for the target group for both cases.
- Column T_{Enc} and T_{Dec} give numbers of costly operations required during encryption and decryption procedures. E_1, E and E_T refer to exponentiation

Table 1. Comparison among almost-tight IBE schemes in the prime-order group.

Scheme	Sec.	$\|\text{MPK}\|$		$\|\text{SK}\|$	$\|\text{CT}\|$		T_{Enc}		T_{Dec}	MIMC
		G_1/G	G_T	G_2/G	G_1/G	G_T	E_1/E	E_T	P	
CW	k-Lin	$2k^2(2n+1)$	k	$4k$	$4k$	1	$4k^2$	k	$4k$	✗
	DLIN	$16n+8$	2	8	8	1	16	2	8	
	SXDH	$4n+2$	1	4	4	1	4	1	4	
BKP	k-Lin	$k^2(2n+1)+k$	k	$2k+1$	$2k+1$	1	$2k^2+1$	k	$2k+1$	✗
	DLIN	$8n+6$	2	5	5	1	9	2	5	
	SXDH	$2n+2$	1	3	3	1	3	1	3	
GCDCT	k-Lin	$3k^2(2n+1)$	k	$6k$	$6k$	1	$6k^2$	k	$6k$	✓
	DLIN	$24n+12$	2	12	12	1	24	2	12	
	SXDH	$6n+3$	1	6	6	1	6	1	6	
GCDCT+	k-LinAI	$2k^2(2n+1)$	k	$4k$	$4k$	1	$4k^2$	k	$4k$	✓
	XDLIN	$16n+8$	2	8	8	1	16	2	8	
AHY	sDLIN	$16n+8$	2	8	8	1	16	2	8	✓
Ours	k-Lin	$k^2(2n+3)$	k	$4k$	$4k$	1	$4k^2$	k	$4k$	✓
	DLIN	$8n+12$	2	8	8	1	16	2	8	
	SXDH	$2n+3$	1	4	4	1	4	1	4	

on the first source group of asymmetric bilinear groups, the only source group of symmetric bilinear groups, and target group in both cases, respectively. P is for pairing operation for both cases.

Benefit of Standard k-Lin. Compared with k-Lin, the k-LinAI assumption (used by GCDCT+) is not well-understood[1] and the sDLIN assumption (used by AHY) is stronger especially in the case of AHY[2]. Without doubt k-Lin is the best choice. However we want to emphasize that achieving the same performance (as GCDCT+ and AHY) under the k-Lin assumption is not just advantageous to theorist, since we can indeed derive a strictly more efficient instantiation than all previous solutions. We note that, AHY is based on the sDLIN assumption and no related generalization was given, while the k-LinAI assumption, on which GCDCT+ is built, is not well-defined[3] for $k = 1$. In contrast, our construction can be naturally instantiated by $k = 1$ and yield an IBE scheme based on SXDH (see Sect. 6), whose performance is shown in the last row (in gray) of the table. Clearly, it has the shortest secret key/ciphertext and the most efficient encryption/decryption algorithm. Compared with BKP under the SXDH assumption, the cost we pay for stronger and more practical MIMC security is quite small: just one more group element is added to secret keys and ciphertexts, and just one more exponentiation and pairing operation are added to encryption and decryption procedure, respectively.

(Weak) Anonymity. Apart from the concern on performance, our main construction achieves anonymity as BKP and AHY. However the notion here is weaker than the standard anonymity, which was first pointed out by Attrapadung *et al.* [3]. All of them are proven to be anonymous under the restriction that all secret keys for the same identity must be created using the same random coin. It's reported in [3] that this can be fulfilled by generating the random coin using a PRF from each identity. A subtlety here is the newly introduced PRF itself should be tightly secure otherwise our effort pursuing tight security will finally come to nothing. In the paper we continue working in this restricted model and neglect this subtlety to keep a clean exposition.

1.3 Our Method

All of AHY, GCDCT, and GCDCT+ are extended from Chen and Wee's construction [9] or its recent development by Chen *et al.* [8]. However, from Table 1, we can see that BKP, Blazy *et al.*'s almost-tightly secure IBE in the SISC model [4], is more efficient in terms of both space and time efficiency. Therefore our idea is

[1] The k-LinAI assumption is an extended (and stronger) version of k-Lin. However only its generic security has been investigated in [14].

[2] One may convert AHY into an asymmetric bilinear group and the security now relies on the XDLIN assumption, which is stronger than 2-Lin. Furthermore it's of course stronger than k-Lin for $k > 2$.

[3] The improvement technique behind GCDCT+ does not work for the special case $k = 1$ since two semi-functional spaces are 1-dimension and too small to compress.

to extend BKP to the MIMC setting and we hope that the resulting construction inherits its high performance and could become a solution to the problem we posed in Sect. 1.1.

Although Blazy *et al.* essentially followed the dual system technique, their concrete realization relied on the Groth-Sahai NIZK proof system [15], which is very different from constructions in [8,9], the common bases of AHY, GCDCT, and GCDCT+. The existing extension strategy seemingly can not be directly applied to updating BKP to the MIMC setting.

To circumvent the difficulty, we reconsider BKP and observe a surprising connection between BKP and Chen *et al.*'s (non-tight) IBE [8]. This allows us to study and manipulate BKP in the framework of nested dual system groups (NDSG) [9] which is much easier to understand and also more feasible to extend towards the MIMC setting [14,18] with existing techniques. We provide the reader with a technical overview in Sect. 3 covering our basic observation and sketching our two technical results which formally treat the observation.

1.4 Related Work

In 2013, Jutla and Roy [19] investigated the notion of quasi-adpative NIZK (QANIZK) and developed an IBE scheme from their SXDH based QANIZK. Both this work and Blazy *et al.*'s work [4] realized the dual system technique using NIZK proof and the idea is actually quite similar. Blazy *et al.* focused on generic frameworks from affine MAC to IBE, while Jutla and Roy considered many other applications of newly proposed QANIZK. A series of work [29–31] extended Jutla and Roy's IBE constructions to more complex functionality.

Since being introduced in 2013, Chen and Wee's technique of almost-tight reduction [9] has been applied to other primitives such as public key encryption against chosen-ciphertext attack and a signature [23] and QANIZK with unbounded simulation soundness [24]. Recently, Hofheinz [16,17] proposed a series of novel techniques based on Chen and Wee's [9] and achieved constant-size parameters and better efficiency for public key encryptions with chosen-ciphertext security and signatures. In the pairing-free setting, Gay *et al.* [12] provided more efficient CCA secure PKE with tight reduction and applied their basic idea to NIZK proof system.

Roadmap. We review necessary preliminary background in Sect. 2. Section 3 is an overview with more technical detail. Sections 4 and 5 present our two technical results. We show our main result (from k-Lin assumption) and its concrete instantiation under SXDH assumption in Sect. 6.

2 Preliminaries

Notation. We use $a \leftarrow A$ to denote the process of uniformly sampling an element from set A and assigning it to variable a. We employ $\{x_i\}_{i \in I}$ to denote a family (or list) of objects with index set I. The abbreviation $\{x_i\}$ will be used when index set is clear in the context. Let G be a group of order p. Given two vectors $\mathbf{a} = (a_1, \ldots, b_n) \in G^n$ and $\mathbf{b} = (b_1, \ldots, b_n) \in G^n$, we let $\mathbf{a} \cdot \mathbf{b} = (a_1 b_1, \ldots, a_n b_n) \in G^n$. For $\mathbf{c} = (c_1, \ldots, c_n) \in \mathbb{Z}_p$ and $g \in G$, we define $g^{\mathbf{c}} = (g^{c_1}, \ldots, g^{c_n}) \in G^n$. For any matrix $\mathbf{A} \in \mathbb{Z}_p^{m \times n}$ with $m > n$, we use $\overline{\mathbf{A}}$ to refer to the square matrix consisting of the first n rows of $\mathbf{A}$ and let $\underline{\mathbf{A}}$ be the sub-matrix consisting of the remaining $m - n$ rows. For any square matrix $\mathbf{A} \in \mathbb{Z}_p^{m \times m}$, we define $\mathbf{A}^* = (\mathbf{A}^\top)^{-1}$. We use $(\mathbf{A}|\mathbf{B})$ to denote the matrix formed by concatenating columns of matrix $\mathbf{A}$ and $\mathbf{B}$ in order.

2.1 Prime-Order Bilinear Group

Let GrpGen be a prime-order bilinear group generator which takes as input security parameter 1^λ and outputs group description $\mathcal{G} = (G_1, G_2, G_T, p, e, g_1, g_2)$. Here G_1, G_2 and G_T are finite cyclic groups of prime order p and $|p| = \Theta(\lambda)$. $e : G_1 \times G_2 \to G_T$ is an admissible (non-degenerated and efficiently computable) bilinear map. g_1, g_2 and $g_T = e(g_1, g_2)$ are respective generators of G_1, G_2, G_T. We employ the *implicit representation* of group elements [11]. For any $a \in \mathbb{Z}_p$ and any $s \in \{1, 2, T\}$, we define $[a]_s = g_s^a \in G_s$. For any matrix $\mathbf{A} = (a_{i,j}) \in \mathbb{Z}_p^{m \times n}$, we define $[\mathbf{A}]_s = ([a_{i,j}]_s) \in G_s^{m \times n}$ and let $e([\mathbf{A}]_1, [\mathbf{B}]_2) = [\mathbf{A}^\top \mathbf{B}]_T$ when $\mathbf{A}^\top \mathbf{B}$ is well-defined.

The security of our construction relies on the *Matrix Decisional Diffie-Hellman* (MDDH) Assumption introduced in [11].

Definition 1 (Matrix Distribution [11]). *For any $\ell, k \in \mathbb{N}$ with $\ell > k$, we let $\mathcal{D}_{\ell,k}$ be a matrix distribution over all full-rank matrices in $\mathbb{Z}_p^{\ell \times k}$. Furthermore, we assume the first k rows of the output matrix form an invertible matrix.*

Assumption 1 ($\mathcal{D}_{\ell,k}$-Matrix Diffie-Hellman Assumption [11]). *Let $\mathcal{D}_{\ell,k}$ be a matrix distribution and $s \in \{1, 2, T\}$. For any p.p.t. adversary $\mathcal{A}$ against* GrpGen, *the following advantage function is negligible in λ.*

$$\mathsf{Adv}_{\mathcal{A}}^{\mathcal{D}_{\ell,k}}(\lambda) = |\Pr[\mathcal{A}(\mathcal{G}, [\mathbf{A}]_s, [\mathbf{Au}]_s) = 1] - \Pr[\mathcal{A}(\mathcal{G}, [\mathbf{A}]_s, [\mathbf{v}]_s) = 1]|$$

where $\mathcal{G} \leftarrow \mathsf{GrpGen}(1^\lambda)$, $\mathbf{A} \leftarrow \mathcal{D}_{\ell,k}$, $\mathbf{u} \leftarrow \mathbb{Z}_p^k$, $\mathbf{v} \leftarrow \mathbb{Z}_p^\ell$.

The matrix distribution $\mathcal{D}_{k+1,k}$ will extensively appear in the paper. For simplicity, we take $\mathcal{D}_k$ as its abbreviation. As in [8], we let $\mathcal{D}_k$ output an additional vector $\mathbf{a}^\perp \in \mathbb{Z}_p^{k+1}$ satisfying $\mathbf{A}^\top \mathbf{a}^\perp = \mathbf{0}$ and $\mathbf{a}^\perp \neq \mathbf{0}$. The notable *$k$-Linear* ($k$-Lin) Assumption is a special case of the $\mathcal{D}_k$-MDDH assumption with

$$\mathbf{A} = \begin{pmatrix} a_1 & & \\ & \ddots & \\ & & a_k \\ 1 & \cdots & 1 \end{pmatrix} \in \mathbb{Z}_p^{(k+1) \times k} \quad \text{and} \quad \mathbf{a}^\perp = \begin{pmatrix} a_1^{-1} \\ \vdots \\ a_k^{-1} \\ -1 \end{pmatrix} \in \mathbb{Z}_p^{k+1}$$

where $a_1, \ldots, a_k \leftarrow \mathbb{Z}_p$. We describe a lemma similar to that shown in [8].

Lemma 1. *With probability* $1-1/p$ *over* $(\mathbf{A}, \mathbf{a}^\perp) \leftarrow \mathcal{D}_k$ *and* $\mathbf{b} \leftarrow \mathbb{Z}_p^{k+1}$, *we have*

$$\mathbf{b} \notin \mathsf{Span}(\mathbf{A}) \quad \textit{and} \quad \mathbf{b}^\top \mathbf{a}^\perp \neq 0.$$

We will heavily use the uniform matrix distribution $\mathcal{U}_{\ell,k}$, which uniformly samples a matrix over $\mathbb{Z}_p^{\ell \times k}$. Similarly, we let $\mathcal{U}_k$ be the short form of $\mathcal{U}_{k+1,k}$. A direct observation is "$\mathcal{D}_k$-MDDH $\Rightarrow \mathcal{U}_k$-MDDH" with constant security loss, since any $\mathcal{D}_k$-MDDH instance can be disguised as a $\mathcal{U}_k$-MDDH instance using a random square matrix (c.f. [11,12]). Besides, we have the following lemma.

Lemma 2 **($\mathcal{U}_k \Rightarrow \mathcal{U}_{\ell,k}$, $\ell > k$ [12]).** *For any p.p.t. adversary* $\mathcal{A}$, *there exists an adversary* $\mathcal{B}$ *with* $\mathsf{T}(\mathcal{B}) \approx \mathsf{T}(\mathcal{A}) + k^2\ell \cdot \mathsf{poly}(\lambda)$ *and*

$$\mathsf{Adv}_{\mathcal{A}}^{\mathcal{U}_{\ell,k}\text{-MDDH}}(\lambda) \leq \mathsf{Adv}_{\mathcal{B}}^{\mathcal{U}_k\text{-MDDH}}(\lambda).$$

The observation and the lemma lead to the fact that $\mathcal{U}_{\ell,k}$-MDDH with $\ell > k$ is *constantly* implied by the well-known k-Lin assumption. In the paper, we utilize the following structural lemma [12].

Lemma 3. *For a fixed full-rank* $\mathbf{A} \in \mathbb{Z}_p^{3k \times k}$, *with probability at least* $1 - 2k/p$ *over* $\widehat{\mathbf{A}}, \widetilde{\mathbf{A}} \leftarrow \mathcal{U}_{3k,k}$, *we have* $\mathsf{Span}\big((\mathbf{A}|\widehat{\mathbf{A}}|\widetilde{\mathbf{A}})\big) = \mathbb{Z}_p^{3k}$, *in which case it holds that*

$$\mathsf{Span}(\mathbf{A}^\perp) = \mathsf{Ker}\big((\mathbf{A}|\widehat{\mathbf{A}})^\top\big) \oplus \mathsf{Ker}\big((\mathbf{A}|\widetilde{\mathbf{A}})^\top\big).$$

and $\widehat{\mathbf{A}}^\top \widehat{\mathbf{A}}^* \in \mathbb{Z}_p^{k \times k}$ *is invertible if* $\widehat{\mathbf{A}}^*$ *forms a basis of* $\mathsf{Ker}\big((\mathbf{A}|\widetilde{\mathbf{A}})^\top\big)$.

For $Q \in \mathbb{N}$, we recall the Q-fold $\mathcal{U}_{\ell,k}$-MDDH assumption [11] as follows. One may view it as Q independent instances of the basic $\mathcal{U}_{\ell,k}$-MDDH problem.

Assumption 2 **(Q-fold $\mathcal{U}_{\ell,k}$-MDDH [11]).** *Let* $\mathcal{U}_{\ell,k}$ *be the uniform matrix distribution and* $s \in \{1, 2, T\}$. *For any p.p.t. adversary* $\mathcal{A}$ *against* GrpGen, *the following advantage function is negligible in* λ.

$$\mathsf{Adv}_{\mathcal{A},Q}^{\mathcal{U}_{\ell,k}}(\lambda) = |\Pr[\mathcal{A}(\mathcal{G}, [\mathbf{A}]_s, [\mathbf{AU}]_s) = 1] - \Pr[\mathcal{A}(\mathcal{G}, [\mathbf{A}]_s, [\mathbf{V}]_s) = 1]|$$

where $\mathcal{G} \leftarrow \mathsf{GrpGen}(1^\lambda)$, $\mathbf{A} \leftarrow \mathcal{U}_{\ell,k}$, $\mathbf{U} \leftarrow \mathbb{Z}_p^{k \times Q}$, $\mathbf{V} \leftarrow \mathbb{Z}_p^{\ell \times Q}$.

It would be direct to prove "$\mathcal{U}_{\ell,k}$-MDDH $\Rightarrow$ Q-fold $\mathcal{U}_{\ell,k}$-MDDH" with a security loss Q. The *Random Self-reducibility Lemma* by Escala *et al.* [11] (see below) provided us with a tighter reduction, the security loss solely depends on the property of matrix $\mathbf{A}$ instead of Q. Namely one can deal with *unbounded* number of instances simultaneously with *constant* security loss for a fixed $\mathbf{A}$.

Lemma 4 (Random Self-reducibility [11]). *Assume* $Q > \ell - k$. *For any uniform matrix distribution* $\mathcal{U}_{\ell,k}$ *and any p.p.t. adversary* $\mathcal{A}$, *there exists an adversary* $\mathcal{B}$ *such that*

$$\mathsf{Adv}_{\mathcal{A},Q}^{\mathcal{U}_{\ell,k}}(\lambda) \leqslant (\ell - k) \cdot \mathsf{Adv}_{\mathcal{B}}^{\mathcal{U}_{\ell,k}}(\lambda) + 1/(p-1)$$

and $\mathsf{T}(\mathcal{B}) \approx \mathsf{T}(\mathcal{A}) + \ell^2 k \cdot \mathsf{poly}(\lambda)$ *where* $\mathsf{poly}(\lambda)$ *is independent of* $\mathsf{T}(\mathcal{A})$.

2.2 Identity Based Encryption

Algorithms. An *Identity Based Encryption* (IBE) in the multi-instance setting [3,14,18] consists of five p.p.t. algorithms:

- $\mathsf{Param}(1^\lambda, \text{SYS}) \to \text{GP}$. The *parameter generation algorithm* takes as input a security parameter $\lambda \in \mathbb{Z}^+$ and a system-level parameter SYS, and outputs a global parameter GP.
- $\mathsf{Setup}(\text{GP}) \to (\text{MPK}, \text{MSK})$. The *setup algorithm* takes as input a global parameter GP, and outputs a master public/secret key pair (MPK, MSK).
- $\mathsf{KeyGen}(\text{MPK}, \text{MSK}, \text{ID}) \to \text{SK}_{\text{ID}}$. The *key generation algorithm* takes as input a master public key MPK, a master secret key MSK and an identity ID, and outputs a secret key SK_{ID}.
- $\mathsf{Enc}(\text{MPK}, \text{ID}, \text{M}) \to \text{CT}_{\text{ID}}$. The *encryption algorithm* takes as input a master public key MPK, an identity ID and a message M, outputs a ciphertext CT_{ID}.
- $\mathsf{Dec}(\text{MPK}, \text{SK}, \text{CT}) \to \text{M}$. The *decryption algorithm* takes as input a master public key MPK, a secret key SK and a ciphertext CT, outputs message M or $\perp$.

If the IBE scheme in question is in the classical single-instance setting, we may merge the first two algorithms into a single Setup algorithm for clarity. The merged Setup algorithm takes 1^λ and SYS as inputs and creates a master public/secret key pair (MPK, MSK).

Correctness. For any parameter $\lambda \in \mathbb{N}$, any SYS, any $\text{GP} \in [\mathsf{Param}(1^\lambda, \text{SYS})]$, any $(\text{MPK}, \text{MSK}) \in [\mathsf{Setup}(\text{GP})]$, any identity ID and any message M, it holds that

$$\Pr\left[\mathsf{Dec}(\text{MPK}, \text{SK}, \text{CT}) = \text{M} \,\middle|\, \begin{array}{c} \text{SK} \leftarrow \mathsf{KeyGen}(\text{MPK}, \text{MSK}, \text{ID}) \\ \text{CT} \leftarrow \mathsf{Enc}(\text{MPK}, \text{ID}, \text{M}) \end{array}\right] \geqslant 1 - 2^{-\Omega(\lambda)}.$$

Security Definition. We investigate both ciphertext indistinguishability and anonymity under chosen identity and plaintext attacks in the multi-instance, multi-ciphertext setting. We define the advantage function

$$\mathsf{Adv}_{\mathcal{A}}^{\mathsf{IBE}}(\lambda) = \left| \Pr\left[\beta = \beta' \,\middle|\, \begin{array}{c} \mu \leftarrow \mathcal{A}(),\ \text{GP} \leftarrow \mathsf{Param}(1^\lambda, \text{SYS}),\ \beta \leftarrow \{0,1\} \\ (\text{MPK}_1, \text{MSK}_1), \ldots, (\text{MPK}_\mu, \text{MSK}_\mu) \leftarrow \mathsf{Setup}(\text{GP}) \\ \beta' \leftarrow \mathcal{A}^{\mathsf{O}_\beta^{\mathsf{Enc}}, \mathsf{O}^{\mathsf{KeyGen}}}(\text{MPK}_1, \ldots, \text{MPK}_\mu) \end{array} \right] - \frac{1}{2} \right|$$

where oracles $\mathsf{O}_\beta^{\mathsf{Enc}}$ and $\mathsf{O}^{\mathsf{KeyGen}}$ work as follows

- $\mathsf{O}_\beta^{\mathsf{Enc}}$: Given $(\iota_0^*, \text{ID}_0^*, \iota_1^*, \text{ID}_1^*, \text{M}_0^*, \text{M}_1^*)$, return $\mathsf{Enc}(\text{MPK}_{\iota_\beta^*}, \text{ID}_\beta^*, \text{M}_\beta^*)$ and update $\mathcal{Q}_C = \mathcal{Q}_C \cup \{(\iota_0^*, \text{ID}_0^*), (\iota_1^*, \text{ID}_1^*)\}$.
- $\mathsf{O}^{\mathsf{KeyGen}}$: Given (ι, ID), return $\mathsf{KeyGen}(\text{MPK}_\iota, \text{MSK}_\iota, \text{ID})$ and update $\mathcal{Q}_K = \mathcal{Q}_K \cup \{(\iota, \text{ID})\}$.

An identity based encryption scheme is *adaptively secure* and *anonymous* in the multi-instance, multi-ciphertext setting if for all p.p.t. adversary $\mathcal{A}$ the advantage function $\mathsf{Adv}_{\mathcal{A}}^{\mathsf{IBE}}(\lambda)$ is negligible in λ and $\mathcal{Q}_K \cap \mathcal{Q}_C = \emptyset$.

As a special case, the adaptive security and anonymity in the single-instance, single-ciphertext setting can be derived by setting two restrictions: (1) There is only one master public/secret key pair, i.e., we set $\mu = 1$ and all $\iota_0^*, \iota_1^*, \iota$ submitted to oracles are restricted to be 1. (2) There is only one challenge ciphertext, i.e., $\mathcal{A}$ can send only one query to oracle $\mathsf{O}_\beta^{\mathsf{Enc}}$.

3 A Technical Overview

3.1 Revisiting BKP

A Short Overview of BKP. Let $(G_1, G_2, G_T, p, e, g_1, g_2) \leftarrow \mathsf{GrpGen}(1^\lambda)$, let's review BKP, i.e., $\mathsf{IBE}[\mathsf{MAC_{NR}}[\mathcal{D}_k], \mathcal{D}_k]$ in [4], which is derived from the affine MAC based on Naor-Reingold PRF. The affine MAC can be described as follows.

$$\begin{array}{|lcl|} \hline \mathrm{SK}_{\mathsf{MAC}} & : & \mathbf{x}_{1,0},\ \mathbf{x}_{1,1},\ \ldots,\ \mathbf{x}_{n,0},\ \mathbf{x}_{n,1},\ x \\ \mathrm{TAG}_m & : & [\mathbf{t}]_2,\quad \left[\sum_{i=1}^n \mathbf{x}_{i,m[i]}^\top \mathbf{t} + x\right]_2 \\ \hline \end{array}$$

Here $\mathbf{x}_{i,b} \leftarrow \mathbb{Z}_p^k$ for $(i,b) \in [n] \times \{0,1\}$ and $x \leftarrow \mathbb{Z}_p$, random coin $\mathbf{t} \in \mathbb{Z}_p^k$ is uniformly sampled for each tag and $m[i]$ represents the ith bit of message $m \in \{0,1\}^n$. It's beneficial to define *randomized verification key* for m^* as

$$\mathrm{VK}_{m^*} \quad : \quad [h]_1,\quad \left[h \cdot \sum_{i=1}^n \mathbf{x}_{i,m^*[i]}\right]_1,\quad [h \cdot x]_T$$

where $h \leftarrow \mathbb{Z}_p$. Blazy *et al.* can prove that a verification key for m^* is pseudorandom for any p.p.t adversary holding tags for $m_1, \ldots, m_q \neq m^*$ under k-Lin assumption with $\mathcal{O}(n)$ security loss.

In a nutshell, the IBE scheme is obtained as follows: master secret key MSK is $\mathrm{SK}_{\mathsf{MAC}}$; master public key MPK consists of perfectly hiding commitments to $\mathrm{SK}_{\mathsf{MAC}}$; a secret key SK for $\mathrm{ID} \in \{0,1\}^n$ is composed of a tag TAG for ID and a Groth-Sahai NIZK proof [15] showing that TAG is correct under $\mathrm{SK}_{\mathsf{MAC}}$; a ciphertext under ID and decryption algorithm are derived from verification method of the NIZK proof system. A more detailed description is given below.

$$\begin{array}{|lcll|} \hline \mathrm{MPK} & : & [\mathbf{A}]_1, [\mathbf{Z}_{1,0}]_1, [\mathbf{Z}_{1,1}]_1, \ldots, [\mathbf{Z}_{n,0}]_1, [\mathbf{Z}_{n,1}]_1, [\mathbf{z}]_1 & (\text{commitment to } \mathrm{SK}_{\mathsf{MAC}}) \\ \mathrm{SK}_{\mathrm{ID}} & : & [\mathbf{k}_0]_2,\ [k_1]_2 = \left[\sum_{i=1}^n \mathbf{x}_{i,\mathrm{ID}[i]}^\top \mathbf{k}_0 + x\right]_2 & (\text{MAC tag TAG for ID}) \\ & & [\mathbf{k}_2]_2 = \left[\sum_{i=1}^n \mathbf{Y}_{i,\mathrm{ID}[i]}^\top \mathbf{k}_0 + \mathbf{y}^\top\right]_2 & (\text{proving validity of TAG}) \\ \mathrm{CT}_{\mathrm{ID}} & : & [\mathbf{As}]_1,\quad \left[\sum_{i=1}^n \mathbf{Z}_{i,\mathrm{ID}[i]}\mathbf{s}\right]_1,\quad [\mathbf{zs}]_T \cdot \mathrm{M} & \\ \hline \end{array}$$

Here $\mathbf{A} \leftarrow \mathcal{D}_k$ is commitment key, $\mathbf{Z}_{i,b} = (\mathbf{Y}_{i,b}|\mathbf{x}_{i,b})\mathbf{A}$ is a commitment to $\mathbf{x}_{i,b}$ with random coin $\mathbf{Y}_{i,b} \leftarrow \mathbb{Z}_q^{k \times k}$ for $(i,b) \in [\ell] \times \{0,1\}$, and $\mathbf{z} = (\mathbf{y}|x)\mathbf{A}$ is a commitment to x with random coin $\mathbf{y} \leftarrow \mathbb{Z}_q^{1 \times k}$. To prove the security of BKP, one first transform the challenge ciphertext $\mathrm{CT}_{\mathrm{ID}^*}$ into the form

$$[\mathbf{As} + \boxed{h} \cdot \mathbf{e}_{k+1}]_1,\ \left[\sum_{i=1}^n \mathbf{Z}_{i,\mathrm{ID}^*[i]}\mathbf{s} + \boxed{h \cdot \sum_{i=1}^n \mathbf{x}_{i,\mathrm{ID}^*[i]}}\right]_1,\ [\mathbf{zs} + \boxed{h \cdot x}]_T \cdot \mathrm{M}$$

in which the boxed terms in fact form a verification key of ID^*. Then we may rewrite the proof part $[\mathbf{k}_2]_2$ of SK_{ID} as

$$\mathbf{k}_2 = \overline{\mathbf{A}}^* \cdot \left(\textstyle\sum_{i=1}^n \mathbf{Z}_{i,\text{ID}[i]}^\top \mathbf{k}_0 + \mathbf{z}^\top - k_1 \underline{\mathbf{A}}^\top\right).$$

Here we use the following relation

$$\begin{aligned}\mathbf{Z}_{i,b} = (\mathbf{Y}_{i,b}|\mathbf{x}_{i,b})\mathbf{A} &\Leftrightarrow \mathbf{Y}_{i,b} = \mathbf{Z}_{i,b}\overline{\mathbf{A}}^{-1} - \mathbf{x}_{i,b}\underline{\mathbf{A}}\overline{\mathbf{A}}^{-1}, \quad (i,b) \in [n] \times \{0,1\} \\ \mathbf{z} = (\mathbf{y}|x)\mathbf{A} &\Leftrightarrow \mathbf{y} = \mathbf{z}\overline{\mathbf{A}}^{-1} - x\underline{\mathbf{A}}\overline{\mathbf{A}}^{-1}.\end{aligned}$$

From the standpoint of NIZK proof system, we have replaced the real proof with a *simulated* proof. An observation is that we do not need $\mathbf{Y}_{i,b}$ (resp. $\mathbf{y}$) and $\mathbf{Z}_{i,b}$ (resp. $\mathbf{z}$) and $\mathbf{x}_{i,b}$ (resp. x) are distributed *independently* by the property of perfectly hiding commitment. In this case we can reduce the adaptive security and anonymity of BKP to the property of underlying affine MAC we just mentioned.

BKP in the Dual-system Lens. Although Blazy *et al.*'s proof [4] is in the framework of dual system encryption [9,34], from their exposition, it's seemingly difficult to identify normal space and semi-functional space, which may guide us to a better understanding and has been formulated via dual system group (DSG) [10] and NDSG [9] (as well as ENDSG [14,18]). Fortunately, ciphertexts and keys used in the proof (c.f. paragraph **A Short Overview of BKP**) give us the following (informal) observations:

- the commitments $\mathbf{Z}_{i,b}$ and $\mathbf{z}$ lie in the normal space;
- the values being committed to, $\mathbf{x}_{i,b}$ and x, lie in the semi-functional space.

Now we try to put the structure into the real system instead of in the proof. For simplicity, we ignore the master secret (i.e., $\mathbf{z}$, $\mathbf{y}$ and x). From the relation in the previous paragraph, we readily have the following representation:

$$\begin{pmatrix}\mathbf{Y}_{i,b}^\top \\ \mathbf{x}_{i,b}^\top\end{pmatrix} = \begin{pmatrix}\overline{\mathbf{A}}^* & -\overline{\mathbf{A}}^*\underline{\mathbf{A}}^\top \\ \mathbf{0}_{1\times k} & 1\end{pmatrix}\begin{pmatrix}\mathbf{Z}_{i,b}^\top \\ \mathbf{x}_{i,b}^\top\end{pmatrix}, \quad \forall\ (i,b) \in [n] \times \{0,1\}.$$

We find that the transformation matrix above actually forms the dual basis of $(\mathbf{A}|\mathbf{e}_{k+1}) = \begin{pmatrix}\overline{\mathbf{A}} & \mathbf{0}_k \\ \underline{\mathbf{A}} & 1\end{pmatrix}$. A simple substitution results in secret keys (without master secret) in the following form:

$$[\mathbf{k}_0]_2, \qquad \begin{bmatrix}\mathbf{k}_2 \\ k_1\end{bmatrix}_2 = \left[\sum_{i=1}^n (\mathbf{A}|\mathbf{e}_{k+1})^* \begin{pmatrix}\mathbf{Z}_{i,\text{ID}[i]}^\top \\ \mathbf{x}_{i,\text{ID}[i]}^\top\end{pmatrix}\mathbf{k}_0\right]_2.$$

As we have observed, $\mathbf{Y}_{i,b}$ is not needed when creating secret keys and ciphertexts *in the real system* and $\mathbf{Z}_{i,b}$ and $\mathbf{x}_{i,b}$ are distributed independently. Therefore we may sample them *directly* instead of through $\mathbf{Y}_{i,b}$. In particular, we sample $\mathbf{W}_{i,b} \leftarrow \mathbb{Z}_p^{k\times(k+1)}$ for all $(i,b) \in [n] \times \{0,1\}$ and define $\mathbf{Z}_{i,b}$ and $\mathbf{x}_{i,b}$ such that

$$\mathbf{W}_{i,b}^\top = (\mathbf{A}|\mathbf{e}_{k+1})^* \begin{pmatrix}\mathbf{Z}_{i,b}^\top \\ \mathbf{x}_{i,b}^\top\end{pmatrix}$$

or equivalently define $\mathbf{Z}_{i,b} = \mathbf{W}_{i,b}\mathbf{A}$ and $\mathbf{x}_{i,b} = \mathbf{W}_{i,b}\mathbf{e}_{k+1}$. This allows us to simplify BKP (without considering master secret key and payload) as follows:

MPK	$: [\mathbf{A}]_1, [\mathbf{W}_{1,0}\mathbf{A}]_1, [\mathbf{W}_{1,1}\mathbf{A}]_1, \ldots, [\mathbf{W}_{n,0}\mathbf{A}]_1, [\mathbf{W}_{n,1}\mathbf{A}]_1$	
$\mathrm{CT}_{\mathrm{ID}}$	$: [\mathbf{As}]_1, \left[\sum_{i=1}^{n} \mathbf{W}_{i,\mathrm{ID}[i]}\mathbf{As}\right]_1$	$\in G_1^{k+1} \times G_1^k$
$\mathrm{SK}_{\mathrm{ID}}$	$: [\mathbf{k}_0]_2, \left[\sum_{i=1}^{n} \mathbf{W}_{i,\mathrm{ID}[i]}^{\top}\mathbf{k}_0\right]_2$	$\in G_2^k \times G_2^{k+1}$

which is surprisingly close to Chen *et al.*'s structure [8].

Remark 1. The structure presented here also appeared in a quasi-adaptive NIZK (QA-NIZK) recently proposed by Gay *et al.* [12]. They obtained this structure from their pairing-free designated-verifier QA-NIZK. In fact, we can alternatively derive their QA-NIZK from the basic QA-NIZK with no support to simulation soundness in [20] (see their Introduction) and a randomized PRF underlying the above structure (following the semi-general method of reaching unbounded simulation soundness in [20]).

3.2 Technical Result 1: Generalizing NDSG

The similarity between Chen *et al.*'s structure [8] and simplified BKP suggests that one may study simplified BKP under the framework of NDSG [9]. However Chen and Wee's NDSG [9] is not sufficient for our purpose and a series of adjustments are seemingly necessary.

Informally, NDSG defines an abstract bilinear group $(\mathbb{G}, \mathbb{H}, \mathbb{G}_T, e)$ equipped with a collection of algorithms sampling group elements. In the generic construction of IBE, a ciphertext (excluding the payload) consists of elements from $\mathbb{G}$ while a secret key is composed of elements in $\mathbb{H}$. However both ciphertexts and keys in the above observation involve elements from two distinct groups, i.e., G_1^{k+1} and G_1^k for $\mathrm{CT}_{\mathrm{ID}}$ and G_2^k and G_2^{k+1} for $\mathrm{SK}_{\mathrm{ID}}$. We generalize Chen and Wee's NDSG [9] in the following aspects:

- replace $\mathbb{G}$ with $\mathbb{G}_0$ and $\mathbb{G}$;
- replace $\mathbb{H}$ with $\mathbb{H}_0$ and $\mathbb{H}$;
- replace e with e and e_0 which map $\mathbb{G} \times \mathbb{H}_0$ and $\mathbb{G}_0 \times \mathbb{H}$ to $\mathbb{G}_T$, respectively.

The first two points are straightforward while the last one is motivated by the decryption procedure where only two vectors of the same dimensions, i.e., either k or $k+1$ dimension, can be paired together and the results should lie in $\mathbb{G}_T$ in both case. Of course, more fine-tunings are required for other portions of NDSG (including making SampH private as in [14], see Sect. 4 for more detail).

Furthermore, following Chen *et al.* [8], we also upgrade NDSG (with all above generalization) to support *weak* anonymity. In particular, we define an additional requirement, called $\mathbb{G}$-uniformity, which is a combination of $\mathbb{H}$-hiding and a weakened $\mathbb{G}$-uniformity in [8]. This allows us to implement its computational version (we will discuss it later) in a tighter fashion.

It's not hard to verify that our generalized NDSG implies an almost-tightly secure IBE in the SISC setting with weaker anonymity [3]. Motivated by our simplified BKP, we can provide a prime-order instantiation of our generalized NDSG. All computational requirements (i.e., left-subgroup and nested-hiding indistinguishability) are proved under the k-Lin assumption based on [4,8].

3.3 Technical Result 2: Towards MIMC Setting

All previous informal discussion and formal treatment are preparations for moving from SISC towards MIMC settings. Having a generalized NDSG with a prime-order instantiation, we can now apply the extension technique proposed in [14,18]. This finally results in a *generalized* extended NDSG (ENDSG) [14,18] and its prime-order instantiation, which immediately gives us an almost-tightly secure and *weakly* anonymous IBE in the MIMC setting, i.e., our main result (c.f. Sects. 1.2 and 6).

Apart from regular extension procedure [14,18] introducing new algorithms and requirements, we also update the $\mathbb{G}$-uniformity (in our generalized NDSG) to its computational version. It's direct to check that the computational $\mathbb{G}$-uniformity gives to our generalized ENDSG the power of reaching *weak* anonymity [3] in the MIMC setting.

The prime-order instantiation of generalized ENDSG and its proofs are obtained from those for the generalized NDSG following the extension strategy by Gong *et al.* [14] and its recent refinement from Gay *et al.* [12]. In particular, the most important extensions must be:

- We let the bases of normal, $\wedge$-semi-functional, and $\sim$-semi-functional space be $\mathbf{A}$, $\widehat{\mathbf{A}}$, and $\widetilde{\mathbf{A}}$, respectively, all of which are sampled from uniform matrix distribution over $\mathbb{Z}_p^{3k\times k}$. The size of matrix $\mathbf{W}$ randomizing bases are extended from $k \times (k+1)$ to $k \times 3k$ accordingly.
- Random functions $\widehat{\mathsf{RF}}_i$ and $\widetilde{\mathsf{RF}}_i$ map an binary string (say, the i-bit prefix of an identity) to a random element in $\mathsf{Span}(\widehat{\mathbf{A}}^*)$ and $\mathsf{Span}(\widetilde{\mathbf{A}}^*)$, respectively. Here we let $\widehat{\mathbf{A}}^*$ (resp. $\widetilde{\mathbf{A}}^*$) be a basis of $\mathsf{Ker}((\mathbf{A}|\widetilde{\mathbf{A}})^\top)$ (resp. $\mathsf{Ker}((\mathbf{A}|\widehat{\mathbf{A}})^\top)$) following Gay *et al.*'s method [12].

This prime-order instantiation derives an IBE (i.e., our main result) with ciphertexts of size $(3k + \boxed{k})|G_1| = 4k|G_1|$ and secret keys of size $(\boxed{k} + 3k)|G_2| = 4k|G_2|$. We highlight that, with the above extension,

- all $\mathbf{W}_{i,b}\mathbf{A}$ are still of size $k \times k$ (see the first boxed term);
- the random coin $\mathbf{r}$ for key is still k dimensional (see the second boxed term).

Namely *not all components in ciphertexts and secret keys swell in our extension procedure* which seemingly benefits from Blazy *et al.*'s structure [4]. More importantly, we gain this feature without relying on the technique presented in [14] which compresses both two semi-functional spaces and thus has to turn to a non-standard assumption.

3.4 Discussion and Perspective

Besides acting as the cornerstone of Technical Result 2, we believe Technical Result 1 may be of independent interest due to its clean description and proofs. For instance, it allows us to explain why BKP can be more efficient than CW, which is not quite obvious before. As a matter of fact, through Technical Result 1, we can compare CW with BKP in the same framework and perceive two differences between them which make BKP more efficient.

Firstly, the secret keys in CW contain a structure supporting *parameter-hiding* which is not found in BKP's secret keys. It is previously used to achieve *right subgroup indistinguishability* in Chen and Wee's prime-order instantiation of DSG [10] but is actually not needed when proving almost-tight adaptive security using Chen and Wee's technique [9].

Secondly, the proof of *nested-hiding indistinguishability* is stronger such that corresponding structure on the key side in BKP are much simpler than in CW. We highlight this point in our proof (in Sect. 4.3) via a lemma (Lemma 5) extracted from Blazy *et al.*'s proof. We specially describe it in the same flavor as Chen and Wee's *Many Tuple Lemma* [9]. One can think of it as a stronger version of *Many Tuple Lemma* [9] since it just involves a secret *vector* instead of a *matrix* which costs less space to hide.

4 Blazy-Kiltz-Pan Almost-Tightly Secure IBE, Revisited

4.1 Generalized Nested Dual System Group

Keeping our informal discussion in Sect. 3 in mind, we generalize the notion of nested dual system group (NDSG) [9] in this section. The formal definition is followed by remarks illustrating main differences with the original one.

Algorithms. Our generalized NDSG consists of five p.p.t. algorithms as follows:

- SampP$(1^\lambda, n)$: Output (PP, SP) where:
  - PP contains group $(\mathbb{G}_0, \mathbb{G}, \mathbb{H}_0, \mathbb{H}, \mathbb{G}_T)$ and admissible bilinear maps
$$e_0 : \mathbb{G}_0 \times \mathbb{H} \to \mathbb{G}_T \quad \text{and} \quad e : \mathbb{G} \times \mathbb{H}_0 \to \mathbb{G}_T,$$
  an efficient linear map μ defined on $\mathbb{H}$, and public parameters for SampG;
  - SP contains $h^* \in \mathbb{H}$ and secret parameters for SampH, $\widehat{\mathsf{SampG}}$.
- SampGT: $\mathrm{Im}(\mu) \to \mathbb{G}_T$.
- SampG(PP): Output $\mathbf{g} = (g_0;\ g_1,\ \ldots,\ g_n) \in \mathbb{G}_0 \times \mathbb{G}^n$.
- SampH(PP, SP): Output $\mathbf{h} = (h_0;\ h_1,\ \ldots,\ h_n) \in \mathbb{H}_0 \times \mathbb{H}^n$.
- $\widehat{\mathsf{SampG}}$(PP, SP): Output $\widehat{\mathbf{g}} = (\widehat{g}_0;\ \widehat{g}_1,\ \ldots,\ \widehat{g}_n) \in \mathbb{G}_0 \times \mathbb{G}^n$.

We employ SampG_0 (resp., $\widehat{\mathsf{SampG}}_0$) to indicate the first element $g_0 \in \mathbb{G}_0$ (resp., $\widehat{g}_0 \in \mathbb{G}_0$) in the output of SampG (resp., $\widehat{\mathsf{SampG}}$). We simply view the outputs of the last three algorithms as vectors but use a semicolon to emphasize the first element and all remaining ones belong to distinct groups.

Correctness. For all $\lambda, n \in \mathbb{Z}^+$ and all $(\text{PP}, \text{SP}) \in [\mathsf{SampP}(1^\lambda, n)]$, we require:

(projective) For all $h \in \mathbb{H}$ and coin s, $\mathsf{SampGT}(\mu(h); s) = e_0(\mathsf{SampG}_0(\text{PP}; s), h)$.
(associative) For all $(g_0;\ g_1, \ldots, g_n) \in [\mathsf{SampG}(\text{PP})]$ and $(h_0;\ h_1, \ldots, h_n) \in [\mathsf{SampH}(\text{PP}, \text{SP})]$, $e_0(g_0, h_i) = e(g_i, h_0)$ for all $i \in [n]$.

Security. For all $\lambda, n \in \mathbb{Z}^+$ and $(\text{PP}, \text{SP}) \leftarrow \mathsf{SampP}(1^\lambda, n)$, we require:

(orthogonality) $\mu(h^*) = 1$.
(non-degeneracy) With overwhelming probability when $\widehat{g}_0 \leftarrow \widehat{\mathsf{SampG}_0}(\text{PP}, \text{SP})$, the value $e_0(\widehat{g}_0, h^*)^\alpha$ is uniformly distributed over $\mathbb{G}_T$ where $\alpha \leftarrow \mathbb{Z}_{\mathsf{ord}(\mathbb{H})}$.
($\mathbb{H}$-subgroup) The output of $\mathsf{SampH}(\text{PP}, \text{SP})$ is uniformly distributed over some subgroup of $\mathbb{H}_0 \times \mathbb{H}^n$.
(left subgroup indistinguishability) For any p.p.t. adversary $\mathcal{A}$, the following advantage function is negligible in λ.

$$\mathsf{Adv}_{\mathcal{A}}^{\text{LS}}(\lambda, q) = \left|\Pr[\mathcal{A}(\text{PP}, \{\mathbf{h}_j\}_{j\in[q]}, \boxed{\mathbf{g}}) = 1] - \Pr[\mathcal{A}(\text{PP}, \{\mathbf{h}_j\}_{j\in[q]}, \boxed{\mathbf{g}\cdot\widehat{\mathbf{g}}}) = 1]\right|$$

where $\mathbf{g} \leftarrow \mathsf{SampG}(\text{PP})$, $\widehat{\mathbf{g}} \leftarrow \widehat{\mathsf{SampG}}(\text{PP}, \text{SP})$, $\mathbf{h}_j \leftarrow \mathsf{SampH}(\text{PP}, \text{SP})$.
(nested-hiding indistinguishability) For all $\eta \in [n]$ and any p.p.t. adversary $\mathcal{A}$, the following advantage function is negligible in λ.

$$\mathsf{Adv}_{\mathcal{A}}^{\text{NH}(\eta)}(\lambda, q) = |\Pr[\mathcal{A}(D, T_0) = 1] - \Pr[\mathcal{A}(D, T_1) = 1]|,$$

where $D = \left(\text{PP}, h^*, \widehat{\mathbf{g}}_{-\eta}, \{\mathbf{h}'_j\}_{j\in[q]}\right)$,

$$T_0 = \left\{\mathbf{h}_j\right\}_{j\in[q]}, \qquad T_1 = \left\{\mathbf{h}_j \cdot \boxed{(1_{\mathbb{H}_0}; (h^*)^{\gamma_j \mathbf{e}_\eta})}\right\}_{j\in[q]}$$

and $\widehat{\mathbf{g}} \leftarrow \widehat{\mathsf{SampG}}(\text{PP}, \text{SP})$, $\mathbf{h}_j, \mathbf{h}'_j \leftarrow \mathsf{SampH}(\text{PP}, \text{SP})$, $\gamma_j \leftarrow \mathbb{Z}_{\mathsf{ord}(\mathbb{H})}$, $\widehat{\mathbf{g}}_{-\eta}$ refers to $(\widehat{g}_0; \widehat{g}_1, \ldots, \widehat{g}_{\eta-1}, \widehat{g}_{\eta+1}, \ldots, \widehat{g}_n)$, $\mathbf{e}_\eta$ is an n-dimension identity vector with a 1 on the ηth position. We can define $\mathsf{Adv}_{\mathcal{A}}^{\text{NH}}(\lambda, q) = \max_{\eta\in[n]}\left\{\mathsf{Adv}_{\mathcal{A}}^{\text{NH}(\eta)}(\lambda, q)\right\}$.
($\mathbb{G}$-uniformity) The statistical distance between the following two distributions is bounded by $2^{-\Omega(\lambda)}$.

$$\left\{\text{PP}, h^*, \left\{\mathbf{h}_j \cdot (1_{\mathbb{H}_0}; (h^*)^{\widehat{\mathbf{v}}_j})\right\}_{j\in[q]}, \boxed{\mathbf{g}\cdot\widehat{\mathbf{g}}}\right\} \quad \text{and}$$
$$\left\{\text{PP}, h^*, \left\{\mathbf{h}_j \cdot (1_{\mathbb{H}_0}; (h^*)^{\widehat{\mathbf{v}}_j})\right\}_{j\in[q]}, \boxed{\mathbf{g}\cdot\widehat{\mathbf{g}}\cdot(1_{\mathbb{G}_0}; (g')^{\mathbf{1}_n})}\right\}$$

where $\mathbf{h}_j \leftarrow \mathsf{SampH}(\text{PP}, \text{SP})$, $\mathbf{g} \leftarrow \mathsf{SampG}(\text{PP})$, $\widehat{\mathbf{g}} \leftarrow \widehat{\mathsf{SampG}}(\text{PP}, \text{SP})$, $\widehat{\mathbf{v}}_j \leftarrow \mathbb{Z}^n_{\mathsf{ord}(\mathbb{H})}$, $g' \leftarrow \mathbb{G}$, $\mathbf{1}_n$ is a vector of n 1's.

One can construct an IBE scheme from generalized NDSG following Chen and Wee's generic construction [9]. The master public/secret key pair is

$$\text{MPK} = (\text{PP}, \mu(\text{MSK}_0)) \quad \text{and} \quad \text{MSK} = (\text{MSK}_0, \text{SP}).$$

where $(\text{PP}, \text{SP}) \leftarrow \mathsf{SampP}(1^\lambda, 2n)$ and $\text{MSK}_0 \leftarrow \mathbb{H}$. A secret key for ID is

$$\text{SK}_{\text{ID}} = \left(K_0 = h_0,\ K_1 = \text{MSK}_0 \cdot \prod_{i \in [n]} h_{2i-\text{ID}[i]}\right) \in \mathbb{H}_0 \times \mathbb{H}.$$

where $(h_0; h_1, \ldots, h_{2n}) \leftarrow \mathsf{SampH}(\text{PP}, \text{SP})$. A ciphertext for M under ID is

$$\text{CT}_{\text{ID}} = \left(C_0 = g_0,\ C_1 = \prod_{i \in [n]} g_{2i-\text{ID}[i]},\ C_2 = g'_T \cdot \text{M}\right) \in \mathbb{G}_0 \times \mathbb{G} \times \mathbb{G}_T.$$

where $(g_0; g_1, \ldots, g_{2n}) \leftarrow \mathsf{SampG}(\text{PP}; s)$ and $g'_T = \mathsf{SampGT}(\mu(\text{MSK}_0); s)$ for random coin s. The message can be recovered by $\text{M} = C_2 \cdot e(C_1, K_0)/e_0(C_0, K_1)$.

Remark 2 (group structure). We generalized SampG, $\widehat{\mathsf{SampG}}$ and SampH such that elements they outputs may come from two different groups. Of course, the new groups $\mathbb{G}_0$ and $\mathbb{H}_0$ are generated via SampP and described in PP. Motivated by the decryption procedure (see the graph below), we require two bilinear maps e_0 and e, denoted by dash line and solid line, respectively, in the graph.

$$\begin{array}{llll} \text{CT}_{\text{ID}}: & C_0 \in \mathbb{G}_0 & C_1 \in \mathbb{G} & C_2 \in \mathbb{G}_T \\ \text{SK}_{\text{ID}}: & K_0 \in \mathbb{H}_0 & K_1 \in \mathbb{H} & \end{array}$$

It's worth noting that both maps share the same range $\mathbb{G}_T$, which helps us to preserve the *associative* property and thus the correctness of IBE scheme.

Remark 3 (private SampH*).* We make the algorithm SampH private as in [14]. One should run SampH with SP besides PP. Therefore *left subgroup* and *nested-hiding indistinguishability* are modified accordingly [14] since adversary now cannot run SampH by itself.

Remark 4 ($\mathbb{G}$-uniformity and anonymity). The $\mathbb{G}$-uniformity property is used to achieve the anonymity. Our definition could be viewed as a direct combination of $\mathbb{H}$-hiding and $\mathbb{G}$-uniformity described by Chen *et al.* in [8] with a tiny relaxation. In particular, we require the last n elements in $\mathbf{g} \cdot \widehat{\mathbf{g}}$ to be hidden by *one* random element from $\mathbb{G}$ instead of n *i.i.d.* random elements in $\mathbb{G}$ as in [8]. One can check that our definition is sufficiently strong to prove the *weak* anonymity [3] (c.f. Sect. 2.2) of our generic IBE scheme.

4.2 A Prime-Order Instantiation Motivated by BKP

We provide an instantiation of our generalized NDSG in the prime-order bilinear group. This formulates our (informal) observation in Sect. 3.1.

– $\mathsf{SampP}(1^\lambda, n)$: Run $\mathcal{G} = (G_1, G_2, G_T, p, e, g_1, g_2) \leftarrow \mathsf{GrpGen}(1^\lambda)$. Define

$$\mathbb{G}_0 = G_1^{k+1}, \quad \mathbb{G} = G_1^k, \quad \mathbb{H}_0 = G_2^k, \quad \mathbb{H} = G_2^{k+1}$$

and bilinear map e_0 and e are natural extensions of e (given in $\mathcal{G}$) to $(k+1)$-dim and k-dim, respectively. Sample $(\mathbf{A}, \mathbf{a}^\perp) \leftarrow \mathcal{D}_k$ and $\mathbf{b} \leftarrow \mathbb{Z}_p^{k+1}$. For each $\mathbf{k} \in \mathbb{Z}_p^{k+1}$, define $\mu : G_2^{k+1} \to G_T^k$ by

$$\mu([\mathbf{k}]_2) = e([\mathbf{A}]_1, [\mathbf{k}]_2) = [\mathbf{A}^\top \mathbf{k}]_T.$$

Let $h^* = [\mathbf{a}^\perp]_2 \in G_2^{k+1}$. Pick $\mathbf{W}_i \leftarrow \mathbb{Z}_p^{k\times(k+1)}$ for all $i \in [n]$ and output

$$\text{PP} = \big([\mathbf{A}]_1,\ [\mathbf{W}_1\mathbf{A}]_1,\ \ldots,\ [\mathbf{W}_n\mathbf{A}]_1\big), \quad \text{SP} = \big(\mathbf{a}^\perp,\ \mathbf{b},\ \mathbf{W}_1,\ \ldots,\ \mathbf{W}_n\big).$$

- $\mathsf{SampGT}([\mathbf{p}]_T)$: Sample $\mathbf{s} \leftarrow \mathbb{Z}_p^k$ and output $[\mathbf{s}^\top\mathbf{p}]_T \in G_T$ for $\mathbf{p} \in \mathbb{Z}_p^k$.
- $\mathsf{SampG}(\text{PP})$: Sample $\mathbf{s} \leftarrow \mathbb{Z}_p^k$ and output

$$\big([\mathbf{As}]_1;\ [\mathbf{W}_1\mathbf{As}]_1,\ \ldots,\ [\mathbf{W}_n\mathbf{As}]_1\big) \in G_1^{k+1} \times (G_1^k)^n.$$

- $\mathsf{SampH}(\text{PP}, \text{SP})$: Sample $\mathbf{r} \leftarrow \mathbb{Z}_p^k$ and output

$$\big([\mathbf{r}]_2;\ [\mathbf{W}_1^\top\mathbf{r}]_2,\ \ldots,\ [\mathbf{W}_n^\top\mathbf{r}]_2\big) \in G_2^k \times (G_2^{k+1})^n.$$

- $\widehat{\mathsf{SampG}}(\text{PP}, \text{SP})$: Sample $\widehat{s} \leftarrow \mathbb{Z}_p$ and output

$$\big([\mathbf{b}\widehat{s}]_1;\ [\mathbf{W}_1\mathbf{b}\widehat{s}]_1,\ \ldots,\ [\mathbf{W}_n\mathbf{b}\widehat{s}]_1\big) \in G_1^{k+1} \times (G_1^k)^n.$$

We only describe formal proof for *nested-hiding indistinguishability* for the lack of space. The remaining requirements can be proved following [8,12].

4.3 Nested-Hiding Indistinguishability

We may rewrite the advantage function $\mathsf{Adv}_{\mathcal{A}}^{\mathrm{NH}(\eta)}(\lambda, q)$ using

$$\begin{aligned}
\text{PP} &= \big([\mathbf{A}]_1,\ [\mathbf{W}_1\mathbf{A}]_1,\ \ldots,\ [\mathbf{W}_n\mathbf{A}]_1\big);\\
h^* &= [\mathbf{a}^\perp]_2;\\
\widehat{\mathbf{g}} &= \big([\mathbf{b}\widehat{s}]_1;\ [\mathbf{W}_1\mathbf{b}\widehat{s}]_1, \ldots, [\mathbf{W}_n\mathbf{b}\widehat{s}]_1\big), \quad \widehat{s} \leftarrow \mathbb{Z}_p;\\
\mathbf{h}'_j &= \big([\mathbf{r}'_j]_2;\ [\mathbf{W}_1^\top\mathbf{r}'_j]_2,\ \ldots,\ [\mathbf{W}_n^\top\mathbf{r}'_j]_2\big), \quad \mathbf{r}'_j \leftarrow \mathbb{Z}_p^k
\end{aligned}$$

and the challenge term $\{\mathbf{h}_j \cdot (1_{\mathbb{H}_0}; (h^*)^{\gamma_j\mathbf{e}_\eta})\}$ may be written as

$$\big([\mathbf{r}_j]_2;\ [\mathbf{W}_1^\top\mathbf{r}_j]_2, \ldots, [\mathbf{W}_\eta^\top\mathbf{r}_j + \mathbf{a}^\perp\gamma_j]_2, \ldots, [\mathbf{W}_n^\top\mathbf{r}_j]_2\big), \quad \mathbf{r}_j \leftarrow \mathbb{Z}_p^k,$$

where either $\gamma_j \leftarrow \mathbb{Z}_p$ or $\gamma_j = 0$.

Before we proceed, we first prove a lemma implicitly used in Blazy *et al.*'s proof [4], which looks like the *Many Tuple Lemma* by Chen and Wee [9].

Lemma 5. *Given $Q \in \mathbb{N}$, group G of prime order p, $[\mathbf{M}] \in G^{(k+1)\times k}$ and $[\mathbf{T}] = [\mathbf{t}_1|\cdots|\mathbf{t}_Q] \in G^{(k+1)\times Q}$ (Here $[\cdot]$ is the implicit representation on G.) where either $\mathbf{t}_i \leftarrow \mathsf{Span}(\mathbf{M})$ or $\mathbf{t}_i \leftarrow \mathbb{Z}_p^{k+1}$, one can efficiently compute*

$$[\mathbf{Z}], \quad [\mathbf{vZ}], \quad \{[\boldsymbol{\tau}_j], [\tau_j]\}_{j\in[Q]}$$

where $\mathbf{Z} \in \mathbb{Z}_p^{k\times k}$ is full-rank, $\mathbf{v} \in \mathbb{Z}_p^{1\times k}$ is a secret row vector, $\boldsymbol{\tau}_j \leftarrow \mathbb{Z}_p^k$, either $\tau_j = \mathbf{v}\boldsymbol{\tau}_j$ (when $\mathbf{t}_j \leftarrow \mathsf{Span}(\mathbf{M})$) or $\tau_j \leftarrow \mathbb{Z}_p$ (when $\mathbf{t}_j \leftarrow \mathbb{Z}_p^{k+1}$).

Proof. Given Q, G, $[\mathbf{M}]$, $[\mathbf{T}] = [\mathbf{t}_1|\cdots|\mathbf{t}_Q]$, the algorithm works as follows:

Programming $[\mathbf{Z}]$ and $[\mathbf{vZ}]$. Define $\mathbf{Z} = \overline{\mathbf{M}}$. Pick $\mathbf{m} = (m_1, \ldots, m_k, m_{k+1}) \leftarrow \mathbb{Z}_p^{1\times(k+1)}$ and implicitly define $\mathbf{v} \in \mathbb{Z}_p^{1\times k}$ such that

$$\mathbf{vZ} = \mathbf{v}\overline{\mathbf{M}} = \mathbf{mM}.$$

One can compute $[\mathbf{Z}]$ and $[\mathbf{vZ}]$ using $[\mathbf{M}]$ and $\mathbf{m}$.

Generating Q tuples. For all $j \in [Q]$, we compute

$$[\boldsymbol{\tau}_j] = [\overline{\mathbf{t}}_j] \quad \text{and} \quad [\tau_j] = [\mathbf{mt}_j].$$

Here $\overline{\mathbf{t}}_j$ indicates the first k entries of $\mathbf{t}_j$.

Observe that: if $\mathbf{t}_j = \mathbf{Mu}_j$ for some $\mathbf{u}_j \in \mathbb{Z}_p^k$, we have that $\boldsymbol{\tau}_j = \overline{\mathbf{M}}\mathbf{u}_j$ and $\tau_j = \mathbf{mMu}_j = \mathbf{v}\overline{\mathbf{M}}\mathbf{u}_j = \mathbf{v}\boldsymbol{\tau}_j$; if $\mathbf{t}_j \leftarrow \mathbb{Z}_p^{k+1}$, we can see that

$$\begin{pmatrix} \boldsymbol{\tau}_j \\ \tau_j \end{pmatrix} = \begin{pmatrix} 1 & & \\ & \ddots & \\ & & 1 \\ m_1 & \cdots & m_k \; m_{k+1} \end{pmatrix} \mathbf{t}_j$$

is uniformly distributed over $\mathbb{Z}_p^{k+1}$. This readily proves the lemma. □

We now prove the following lemma for all $\eta \in [n]$.

Lemma 6. *For any p.p.t. adversary $\mathcal{A}$, there exists an adversary $\mathcal{B}$ such that*

$$\mathsf{Adv}_{\mathcal{A}}^{\mathrm{NH}(\eta)}(\lambda, q) \leqslant \mathsf{Adv}_{\mathcal{B},q}^{\mathcal{D}_k}(\lambda)$$

where $\mathsf{T}(\mathcal{B}) \approx \mathsf{T}(\mathcal{A}) + k^2 \cdot q \cdot \mathsf{poly}(\lambda, n)$ *and* $\mathsf{poly}(\lambda, n)$ *is independent of* $\mathsf{T}(\mathcal{A})$.

Proof. Given $[\mathbf{M}]_2 \in G_2^{(k+1)\times k}$ and $[\mathbf{T}]_2 = [\mathbf{t}_1|\cdots|\mathbf{t}_q]_2 \in G_2^{(k+1)\times q}$ where $\mathbf{t}_j \leftarrow \mathsf{Span}(\mathbf{M})$ or $\mathbf{t}_j \leftarrow \mathbb{Z}_p^{k+1}$, $\mathcal{B}$ proceeds as follows:

Generating q tuples. We invoke the algorithm described in Lemma 5 on input $(q, G_2, [\mathbf{M}]_2, [\mathbf{T}]_2)$ and obtain $\big([\mathbf{Z}]_2, [\mathbf{vZ}]_2, \{[\boldsymbol{\tau}_j]_2, [\tau_j]_2\}_{j\in[q]}\big)$.

Simulating pp and h^*. Sample $(\mathbf{A}, \mathbf{a}^\perp) \leftarrow \mathcal{D}_k$ and define $h^* = [\mathbf{a}^\perp]_2$. Sample $\mathbf{W}_i \leftarrow \mathbb{Z}_p^{k\times(k+1)}$ for all $i \in [n] \setminus \{\eta\}$. Pick $\bar{\mathbf{W}}_\eta \leftarrow \mathbb{Z}_p^{k\times(k+1)}$ and implicitly set

$$\mathbf{W}_\eta = \bar{\mathbf{W}}_\eta + \mathbf{v}^\top {\mathbf{a}^\perp}^\top.$$

Therefore we can simulate all entries in PP with the observation

$$\mathbf{W}_\eta\mathbf{A} = \big(\bar{\mathbf{W}}_\eta + \mathbf{v}^\top {\mathbf{a}^\perp}^\top\big)\mathbf{A} = \bar{\mathbf{W}}_\eta\mathbf{A},$$

where the secret vector $\mathbf{v}$ has been eliminated by the fact $\mathbf{A}^\top\mathbf{a}^\perp = \mathbf{0}$.

Simulating $\widehat{\mathbf{g}}_{-\eta}$. Sample $\mathbf{b} \leftarrow \mathbb{Z}_p^{k+1}$. We can directly simulate $\widehat{\mathbf{g}}_{-\eta}$ since we know $\mathbf{W}_i$ for all $i \in [n] \setminus \{\eta\}$. Note that we do not know $\mathbf{W}_\eta$ where there is a secret vector $\mathbf{v}$, but it is not needed here.

Simulating $\mathbf{h}'_j$. Sample $\bar{\mathbf{r}}_j \leftarrow \mathbb{Z}_p^k$ and implicitly define

$$\mathbf{r}'_j = \mathbf{Z}\bar{\mathbf{r}}_j \quad \text{for all } j \in [q].$$

We are ready to produce $[\mathbf{r}'_j]_2$ and $[\mathbf{W}_i^\top \mathbf{r}'_j]_2$ for $i \in [n] \setminus \{\eta\}$. Observe that

$$\mathbf{W}_\eta^\top \mathbf{r}'_j = \left(\bar{\mathbf{W}}_\eta + \mathbf{v}^\top \mathbf{a}^{\perp\top}\right)^\top \mathbf{Z}\bar{\mathbf{r}}_j = \bar{\mathbf{W}}_\eta^\top \mathbf{Z}\bar{\mathbf{r}}_j + \mathbf{a}^\perp (\mathbf{v}\mathbf{Z})\,\bar{\mathbf{r}}_j.$$

The entry $[\mathbf{W}_\eta^\top \mathbf{r}'_j]_2$ can be simulated with $\bar{\mathbf{W}}_\eta$, $\mathbf{a}^\perp$, $\bar{\mathbf{r}}_j$ and $[\mathbf{Z}]_2, [\mathbf{v}\mathbf{Z}]_2$.

Simulating the challenge. For all $j \in [q]$, we produce the challenge as

$$\left([\boldsymbol{\tau}_j]_2, [\mathbf{W}_1^\top \boldsymbol{\tau}_j]_2, \ldots, [\bar{\mathbf{W}}_\eta^\top \boldsymbol{\tau}_j + \mathbf{a}^\perp \tau_j]_2, \ldots, [\mathbf{W}_n^\top \boldsymbol{\tau}_j]_2\right).$$

Here we implicitly set $\mathbf{r}_j = \boldsymbol{\tau}_j$. Observe that, when $\mathbf{t}_j \leftarrow \mathsf{Span}(\mathbf{M})$, we have $\tau_j = \mathbf{v}\boldsymbol{\tau}_j$, the challenge is identical to $\{\mathbf{h}_j\}$, i.e., $\gamma_j = 0$; when $\mathbf{t}_j \leftarrow \mathbb{Z}_p^{k+1}$, we have $\tau_j \leftarrow \mathbb{Z}_p$, the challenge is identical to $\{\mathbf{h}_j \cdot (1_{\mathbb{H}_0}; (h^*)^{\gamma_j \mathbf{e}_\eta})\}$ where $\gamma_j = \tau_j - \mathbf{v}\boldsymbol{\tau}_j$ is uniformly distributed over $\mathbb{Z}_p$. This proves the lemma. □

5 Towards Tight Security in MIMC Setting

5.1 A Generalization of Extended Nested Dual System Group

Applying Gong *et al.*'s idea of extending NDSG [14], a variant of Hofheinz *et al.*'s method [18], to our generalization described in Sect. 4.1, we obtain a generalization of extended nested dual system group (ENDSG).

Algorithms. Our ENDSG consists of eight p.p.t. algorithms defined as follows:

- $\mathsf{SampP}(1^\lambda, n)$: Output (PP, SP) where:
  - PP contains group description $(\mathbb{G}_0, \mathbb{G}, \mathbb{H}_0, \mathbb{H}, \mathbb{G}_T)$ and two admissible bilinear maps

    $$e_0 : \mathbb{G}_0 \times \mathbb{H} \to \mathbb{G}_T \quad \text{and} \quad e : \mathbb{G} \times \mathbb{H}_0 \to \mathbb{G}_T,$$

    an efficient linear map μ defined on $\mathbb{H}$, and public parameters for SampG;
  - SP contains secret parameters for SampH, $\widehat{\mathsf{SampG}}$, $\widetilde{\mathsf{SampG}}$, $\widehat{\mathsf{SampH}}^*$, and $\widetilde{\mathsf{SampH}}^*$.
- SampGT: $\mathrm{Im}(\mu) \to \mathbb{G}_T$.
- $\mathsf{SampG}(\text{PP})$: Output $\mathbf{g} = (g_0;\ g_1,\ \ldots,\ g_n) \in \mathbb{G}_0 \times \mathbb{G}^n$.
- $\mathsf{SampH}(\text{PP}, \text{SP})$: Output $\mathbf{h} = (h_0;\ h_1,\ \ldots,\ h_n) \in \mathbb{H}_0 \times \mathbb{H}^n$.
- $\widehat{\mathsf{SampG}}(\text{PP}, \text{SP})$: Output $\widehat{\mathbf{g}} = (\widehat{g}_0;\ \widehat{g}_1,\ \ldots,\ \widehat{g}_n) \in \mathbb{G}_0 \times \mathbb{G}^n$.
- $\widetilde{\mathsf{SampG}}(\text{PP}, \text{SP})$: Output $\widetilde{\mathbf{g}} = (\widetilde{g}_0;\ \widetilde{g}_1,\ \ldots,\ \widetilde{g}_n) \in \mathbb{G}_0 \times \mathbb{G}^n$.
- $\widehat{\mathsf{SampH}}^*(\text{PP}, \text{SP})$: Output $\widehat{h}^* \in \mathbb{H}$.

– $\widetilde{\mathsf{SampH}}^*(\text{PP}, \text{SP})$: Output $\widetilde{h}^* \in \mathbb{H}$.

We employ SampG_0 (resp., $\widehat{\mathsf{SampG}}_0$, $\widetilde{\mathsf{SampG}}_0$) to indicate the first element $g_0 \in \mathbb{G}_0$ (resp., $\widehat{g}_0 \in \mathbb{G}_0$, $\widetilde{g}_0 \in \mathbb{G}_0$) in the output of SampG (resp., $\widehat{\mathsf{SampG}}$, $\widetilde{\mathsf{SampG}}$).

Correctness and Security. The correctness requirement is exactly the same as our generalized NDSG including *projective* and *associative* (c.f. Sect. 4.1). For all $\lambda, n \in \mathbb{Z}^+$ and $(\text{PP}, \text{SP}) \leftarrow \mathsf{SampP}(1^\lambda, n)$, the security requirement involves:

(orthogonality) For all $\widehat{h}^* \in [\widehat{\mathsf{SampH}}^*(\text{PP}, \text{SP})]$ and all $\widetilde{h}^* \in [\widetilde{\mathsf{SampH}}^*(\text{PP}, \text{SP})]$, (1) $\mu(\widehat{h}^*) = \mu(\widetilde{h}^*) = 1$; (2) $e_0(\widehat{g}_0, \widetilde{h}^*) = 1$ for all $\widehat{g}_0 \in [\widehat{\mathsf{SampG}}_0(\text{PP}, \text{SP})]$; (3) $e_0(\widetilde{g}_0, \widehat{h}^*) = 1$ for all $\widetilde{g}_0 \in [\widetilde{\mathsf{SampG}}_0(\text{PP}, \text{SP})]$.

($\mathbb{H}$-subgroup) The output of $\mathsf{SampH}(\text{PP}, \text{SP})$ is uniformly distributed over some subgroup of $\mathbb{H}_0 \times \mathbb{H}^n$, while those of $\widehat{\mathsf{SampH}}^*(\text{PP}, \text{SP})$ and $\widetilde{\mathsf{SampH}}^*(\text{PP}, \text{SP})$ are uniformly distributed over some subgroup of $\mathbb{H}$, respectively.

(left subgroup indistinguishability 1) For any p.p.t. adversary $\mathcal{A}$, the following advantage function is negligible in λ.

$$\mathsf{Adv}_{\mathcal{A}}^{\text{LS1}}(\lambda, q, q') := |\Pr[\mathcal{A}(D, T_0) = 1] - \Pr[\mathcal{A}(D, T_1) = 1]|,$$

where $D = \left(\text{PP}, \{\mathbf{h}_j\}_{j \in [q']}\right)$,

$$T_0 = \{\mathbf{g}_j\}_{j \in [q]}, \quad T_1 = \left\{\mathbf{g}_j \cdot \boxed{\widehat{\mathbf{g}}_j \cdot \widetilde{\mathbf{g}}_j}\right\}_{j \in [q]}$$

and $\mathbf{g}_j \leftarrow \mathsf{SampG}(\text{PP})$, $\widehat{\mathbf{g}}_j \leftarrow \widehat{\mathsf{SampG}}(\text{PP}, \text{SP})$, $\widetilde{\mathbf{g}}_j \leftarrow \widetilde{\mathsf{SampG}}(\text{PP}, \text{SP})$, $\mathbf{h}_j \leftarrow \mathsf{SampH}(\text{PP}, \text{SP})$.

(left subgroup indistinguishability 2) For any p.p.t. adversary $\mathcal{A}$, the following advantage function is negligible in λ.

$$\mathsf{Adv}_{\mathcal{A}}^{\text{LS2}}(\lambda, q, q') = |\Pr[\mathcal{A}(D, T_0) = 1] - \Pr[\mathcal{A}(D, T_1) = 1]|,$$

where $D = \left(\text{PP}, \{\widehat{h}_j^* \cdot \widetilde{h}_j^*\}_{j \in [q+q']}, \{\mathbf{g}'_j \cdot \widehat{\mathbf{g}}'_j \cdot \widetilde{\mathbf{g}}'_j\}_{j \in [q]}, \{\mathbf{h}_j\}_{j \in [q']}\right)$,

$$T_0 = \left\{\mathbf{g}_j \cdot \widehat{\mathbf{g}}_j \cdot \boxed{\widetilde{\mathbf{g}}_j}\right\}_{j \in [q]}, \quad T_1 = \{\mathbf{g}_j \cdot \widehat{\mathbf{g}}_j\}_{j \in [q]},$$

and $\widehat{h}_j^* \leftarrow \widehat{\mathsf{SampH}}^*(\text{PP}, \text{SP})$, $\widetilde{h}_j^* \leftarrow \widetilde{\mathsf{SampH}}^*(\text{PP}, \text{SP})$, $\mathbf{g}_j, \mathbf{g}'_j \leftarrow \mathsf{SampG}(\text{PP})$, $\widehat{\mathbf{g}}_j, \widehat{\mathbf{g}}'_j \leftarrow \widehat{\mathsf{SampG}}(\text{PP}, \text{SP})$, $\widetilde{\mathbf{g}}_j, \widetilde{\mathbf{g}}'_j \leftarrow \widetilde{\mathsf{SampG}}(\text{PP}, \text{SP})$, $\mathbf{h}_j \leftarrow \mathsf{SampH}(\text{PP}, \text{SP})$.

(left subgroup indistinguishability 3) For any p.p.t. adversary $\mathcal{A}$, the following advantage function is negligible in λ.

$$\mathsf{Adv}_{\mathcal{A}}^{\text{LS3}}(\lambda, q, q') = |\Pr[\mathcal{A}(D, T_0) = 1] - \Pr[\mathcal{A}(D, T_1) = 1]|,$$

where $D = \left(\text{PP}, \{\widehat{h}_j^* \cdot \widetilde{h}_j^*\}_{j \in [q+q']}, \{\mathbf{g}'_j \cdot \widehat{\mathbf{g}}'_j\}_{j \in [q]}, \{\mathbf{h}_j\}_{j \in [q']}\right)$,

$$T_0 = \left\{\mathbf{g}_j \cdot \boxed{\widehat{\mathbf{g}}_j} \cdot \widetilde{\mathbf{g}}_j\right\}_{j \in [q]}, \quad T_1 = \{\mathbf{g}_j \cdot \widetilde{\mathbf{g}}_j\}_{j \in [q]},$$

and $\widehat{h}_j^* \leftarrow \widehat{\mathsf{SampH}}^*(\text{PP}, \text{SP})$, $\widetilde{h}_j^* \leftarrow \widetilde{\mathsf{SampH}}^*(\text{PP}, \text{SP})$, $\mathbf{g}_j, \mathbf{g}'_j \leftarrow \mathsf{SampG}(\text{PP})$, $\widehat{\mathbf{g}}_j, \widehat{\mathbf{g}}'_j \leftarrow \widehat{\mathsf{SampG}}(\text{PP}, \text{SP})$, $\widetilde{\mathbf{g}}_j \leftarrow \widetilde{\mathsf{SampG}}(\text{PP}, \text{SP})$, $\mathbf{h}_j \leftarrow \mathsf{SampH}(\text{PP}, \text{SP})$.

(nested-hiding indistinguishability) For all $\eta \in [\lfloor n/2 \rfloor]$ and any p.p.t. adversary $\mathcal{A}$, the following advantage function is negligible in λ.

$$\mathsf{Adv}_{\mathcal{A}}^{\mathrm{NH}(\eta)}(\lambda, q, q') = |\Pr[\mathcal{A}(D, T_0) = 1] - \Pr[\mathcal{A}(D, T_1) = 1]|,$$

where $D = \left(\mathrm{PP}, \{\widehat{h}_j^*, \widetilde{h}_j^*\}_{j\in[q+q']}, \{(\widehat{\mathbf{g}}_j)_{-(2\eta-1)}, (\widetilde{\mathbf{g}}_j)_{-2\eta}\}_{j\in[q]}, \{\mathbf{h}'_j\}_{j\in[q']}\right)$,

$$T_0 = \{\mathbf{h}_j\}_{j\in[q']}, \quad T_1 = \left\{\mathbf{h}_j \cdot \boxed{(1_{\mathbb{H}_0}; (\widehat{h}_j^{**})^{\mathbf{e}_{2\eta-1}}) \cdot (1_{\mathbb{H}_0}; (\widetilde{h}_j^{**})^{\mathbf{e}_{2\eta}})}\right\}_{j\in[q']}$$

and $\widehat{\mathbf{g}}_j \leftarrow \widehat{\mathsf{SampG}}(\mathrm{PP}, \mathrm{SP})$, $\widetilde{\mathbf{g}}_j \leftarrow \widetilde{\mathsf{SampG}}(\mathrm{PP}, \mathrm{SP})$, $\widehat{h}_j^*, \widehat{h}_j^{**} \leftarrow \widehat{\mathsf{SampH}}^*(\mathrm{PP}, \mathrm{SP})$, $\widetilde{h}_j^*, \widetilde{h}_j^{**} \leftarrow \widetilde{\mathsf{SampH}}^*(\mathrm{PP}, \mathrm{SP})$, $\mathbf{h}_j, \mathbf{h}'_j \leftarrow \mathsf{SampH}(\mathrm{PP}, \mathrm{SP})$. We may further define $\mathsf{Adv}_{\mathcal{A}}^{\mathrm{NH}}(\lambda, q, q') = \max_{\eta\in[\lfloor n/2\rfloor]}\{\mathsf{Adv}_{\mathcal{A}}^{\mathrm{NH}(\eta)}(\lambda, q, q')\}$.

(non-degeneracy) For any p.p.t. adversary $\mathcal{A}$, the following advantage function is negligible in λ.

$$\mathsf{Adv}_{\mathcal{A}}^{\mathrm{ND}}(\lambda, q, q', q'') = |\Pr[\mathcal{A}(D, T_0) = 1] - \Pr[\mathcal{A}(D, T_1) = 1]|,$$

where $D = \left(\mathrm{PP}, \{\widehat{h}_j^* \cdot \widetilde{h}_j^*, \ \mathbf{h}_j\}_{j\in[q']}, \{\widehat{\mathbf{g}}_{j,j'} = (\widehat{g}_{0,j,j'}; \ldots)\}_{j\in[q], j'\in[q'']}\right)$,

$$T_0 = \left\{e_0(\widehat{g}_{0,j,j'}, \widehat{h}_j^{**})\right\}_{j\in[q],j'\in[q'']}, \; T_1 = \left\{e_0(\widehat{g}_{0,j,j'}, \widehat{h}_j^{**}) \cdot \boxed{R_{j,j'}}\right\}_{j\in[q],j'\in[q'']}$$

and $\widehat{\mathbf{g}}_{j,j'} \leftarrow \widehat{\mathsf{SampG}}(\mathrm{PP}, \mathrm{SP})$, $\widetilde{h}_j^* \leftarrow \widetilde{\mathsf{SampH}}^*(\mathrm{PP}, \mathrm{SP})$, $\widehat{h}_j^*, \widehat{h}_j^{**} \leftarrow \widehat{\mathsf{SampH}}^*(\mathrm{PP}, \mathrm{SP})$, $\mathbf{h}_j \leftarrow \mathsf{SampH}(\mathrm{PP}, \mathrm{SP})$, and $R_{j,j'} \leftarrow \mathbb{G}_T$.

($\mathbb{G}$-uniformity) For any p.p.t. adversary $\mathcal{A}$, the following advantage function is negligible in λ.

$$\mathsf{Adv}_{\mathcal{A}}^{\mathbb{G}\text{-uni}}(\lambda, q, q') = |\Pr[\mathcal{A}(D, T_0) = 1] - \Pr[\mathcal{A}(D, T_1) = 1]|,$$

where $D = \left(\mathrm{PP}, \{\mathbf{h}_j \cdot (1_{\mathbb{H}_0}; \ \widehat{h}_{1,j}^*, \ldots, \widehat{h}_{n,j}^*), \ \widehat{h}_j^*, \ \widetilde{h}_j^*\}_{j\in[q']}\right)$,

$$T_0 = \{\mathbf{g}_j \cdot \widehat{\mathbf{g}}_j\}_{j\in[q]}, \quad T_1 = \left\{\mathbf{g}_j \cdot \widehat{\mathbf{g}}_j \cdot \boxed{(1_{\mathbb{G}_0}; (g'_j)^{\mathbf{1}_n})}\right\}_{j\in[q]}$$

and $\mathbf{h}_j \leftarrow \mathsf{SampH}(\mathrm{PP}, \mathrm{SP})$, $\mathbf{g}_j \leftarrow \mathsf{SampG}(\mathrm{PP})$, $\widehat{\mathbf{g}}_j \leftarrow \widehat{\mathsf{SampG}}(\mathrm{PP}, \mathrm{SP})$, $\widetilde{h}_j^* \leftarrow \widetilde{\mathsf{SampH}}^*(\mathrm{PP}, \mathrm{SP})$, $\widehat{h}_j^*, \widehat{h}_{1,j}^*, \ldots, \widehat{h}_{n,j}^* \leftarrow \widehat{\mathsf{SampH}}^*(\mathrm{PP}, \mathrm{SP})$, $g'_j \leftarrow \mathbb{G}$.

The generic IBE in the multi-instance setting is similar to the IBE scheme in Sect. 4.1 except that we take $(\mathrm{PP}, \mathrm{SP}) \leftarrow \mathsf{SampP}(1^\lambda, 2n)$ as the global parameter GP and master secret $\mathrm{MSK}_0 \in \mathbb{H}$ will be picked for each instance (in algorithm Setup).

5.2 An Instantiation in the Prime-Order Group

The generalized ENDSG described above can be implemented by extending the construction in Sect. 4.2. In particular, we follow the extension technique by Gong *et al.* [14] and Gay *et al.* [12] (c.f. Sect. 3.3).

– $\mathsf{SampP}(1^\lambda, n)$: Run $\mathcal{G} = (G_1, G_2, G_T, p, e, g_1, g_2) \leftarrow \mathsf{GrpGen}(1^\lambda)$. Define
$$\mathbb{G}_0 = G_1^{3k}, \quad \mathbb{G} = G_1^k, \quad \mathbb{H}_0 = G_2^k, \quad \mathbb{H} = G_2^{3k}$$
and bilinear map e_0 and e are natural extension of e (given in $\mathcal{G}$) to $3k$-dim and k-dim, respectively. Sample $\mathbf{A}, \widehat{\mathbf{A}}, \widetilde{\mathbf{A}} \leftarrow \mathcal{U}_{3k,k}$ and randomly pick $\widehat{\mathbf{A}}^*, \widetilde{\mathbf{A}}^* \in \mathbb{Z}_p^{3k\times k}$ as respective bases of $\mathsf{Ker}\big((\mathbf{A}|\widetilde{\mathbf{A}})^\top\big)$ and $\mathsf{Ker}\big((\mathbf{A}|\widehat{\mathbf{A}})^\top\big)$. For each $\mathbf{k} \in \mathbb{Z}_p^{3k}$, define $\mu : G_2^{3k} \to G_T^k$ by $\mu([\mathbf{k}]_2) = e([\mathbf{A}]_1, [\mathbf{k}]_2) = [\mathbf{A}^\top\mathbf{k}]_T$. Sample $\mathbf{W}_i \leftarrow \mathbb{Z}_p^{k\times 3k}$ for all $i \in [n]$ and output
$$\mathrm{PP} = \big([\mathbf{A}]_1, [\mathbf{W}_1\mathbf{A}]_1, \ldots, [\mathbf{W}_n\mathbf{A}]_1\big), \quad \mathrm{SP} = \big(\widehat{\mathbf{A}}, \widetilde{\mathbf{A}}, \widehat{\mathbf{A}}^*, \widetilde{\mathbf{A}}^*, \mathbf{W}_1, \ldots, \mathbf{W}_n\big).$$
– $\mathsf{SampGT}([\mathbf{p}]_T)$: Sample $\mathbf{s} \leftarrow \mathbb{Z}_p^k$ and output $[\mathbf{s}^\top\mathbf{p}]_T$ for $\mathbf{p} \in \mathbb{Z}_p^k$.
– $\mathsf{SampG}(\mathrm{PP})$: Sample $\mathbf{s} \leftarrow \mathbb{Z}_p^k$ and output
$$\big([\mathbf{As}]_1;\ [\mathbf{W}_1\mathbf{As}]_1,\ \ldots,\ [\mathbf{W}_n\mathbf{As}]_1\big) \in G_1^{3k} \times (G_1^k)^n.$$
– $\mathsf{SampH}(\mathrm{PP}, \mathrm{SP})$: Sample $\mathbf{r} \leftarrow \mathbb{Z}_p^k$ and output
$$\big([\mathbf{r}]_2;\ [\mathbf{W}_1^\top\mathbf{r}]_2,\ \ldots,\ [\mathbf{W}_n^\top\mathbf{r}]_2\big) \in G_2^k \times (G_2^{3k})^n.$$
– $\widehat{\mathsf{SampG}}(\mathrm{PP}, \mathrm{SP})$: Sample $\widehat{\mathbf{s}} \leftarrow \mathbb{Z}_p^k$ and output
$$\big([\widehat{\mathbf{A}}\widehat{\mathbf{s}}]_1;\ [\mathbf{W}_1\widehat{\mathbf{A}}\widehat{\mathbf{s}}]_1,\ \ldots,\ [\mathbf{W}_n\widehat{\mathbf{A}}\widehat{\mathbf{s}}]_1\big) \in G_1^{3k} \times (G_1^k)^n.$$
– $\widetilde{\mathsf{SampG}}(\mathrm{PP}, \mathrm{SP})$: Sample $\widetilde{\mathbf{s}} \leftarrow \mathbb{Z}_p^k$ and output
$$\big([\widetilde{\mathbf{A}}\widetilde{\mathbf{s}}]_1;\ [\mathbf{W}_1\widetilde{\mathbf{A}}\widetilde{\mathbf{s}}]_1,\ \ldots,\ [\mathbf{W}_n\widetilde{\mathbf{A}}\widetilde{\mathbf{s}}]_1\big) \in G_1^{3k} \times (G_1^k)^n.$$
– $\widehat{\mathsf{SampH}}^*(\mathrm{PP}, \mathrm{SP})$: Sample $\widehat{\mathbf{r}} \in \mathbb{Z}_p^k$ and output $\big[\widehat{\mathbf{A}}^*\widehat{\mathbf{r}}\big]_2 \in G_2^{3k}$.
– $\widetilde{\mathsf{SampH}}^*(\mathrm{PP}, \mathrm{SP})$: Sample $\widetilde{\mathbf{r}} \in \mathbb{Z}_p^k$ and output $\big[\widetilde{\mathbf{A}}^*\widetilde{\mathbf{r}}\big]_2 \in G_2^{3k}$.

For the lack of space, we only show that our instantiation satisfies *Left Subgroup Indistinguishability 2* and *3*, *Nested-hiding Indistinguishability* and $\mathbb{G}$-*uniformity* in the next several subsections.

5.3 Left Subgroup Indistinguishability 2 and 3

We rewrite the advantage function $\mathsf{Adv}_{\mathcal{A}}^{\mathrm{LS2}}(k, q, q')$ using

$$\begin{aligned}
\mathrm{PP} &= \big([\mathbf{A}]_1,\ [\mathbf{W}_1\mathbf{A}]_1,\ \ldots,\ [\mathbf{W}_n\mathbf{A}]_1\big);\\
\widehat{h}_j^* \cdot \widetilde{h}_j^* &= \big[\widehat{\mathbf{A}}^*\widehat{\mathbf{r}}_j + \widetilde{\mathbf{A}}^*\widetilde{\mathbf{r}}_j\big]_2,\ \widehat{\mathbf{r}}_j, \widetilde{\mathbf{r}}_j \leftarrow \mathbb{Z}_p^k;\\
\mathbf{g}_j' \cdot \widehat{\mathbf{g}}_j' \cdot \widetilde{\mathbf{g}}_j' &= \big([\mathbf{s}_j']_1;\ [\mathbf{W}_1\mathbf{s}_j']_1,\ \ldots,\ [\mathbf{W}_n\mathbf{s}_j']_1\big),\ \mathbf{s}_j' \leftarrow \mathbb{Z}_p^{3k};\\
\mathbf{h}_j &= \big([\mathbf{r}_j]_2;\ [\mathbf{W}_1^\top\mathbf{r}_j]_2,\ \ldots,\ [\mathbf{W}_n^\top\mathbf{r}_j]_2\big),\ \mathbf{r}_j \leftarrow \mathbb{Z}_p^k;\\
\mathbf{g}_j \cdot \widehat{\mathbf{g}}_j &= \big([\mathbf{As}_j + \widehat{\mathbf{A}}\widehat{\mathbf{s}}_j]_1;\ [\mathbf{W}_1(\mathbf{As}_j + \widehat{\mathbf{A}}\widehat{\mathbf{s}}_j)]_1,\ \ldots,\ [\mathbf{W}_n(\mathbf{As}_j + \widehat{\mathbf{A}}\widehat{\mathbf{s}}_j)]_1\big),\\
&\qquad \mathbf{s}_j, \widehat{\mathbf{s}}_j \leftarrow \mathbb{Z}_p^k;\\
\mathbf{g}_j \cdot \widehat{\mathbf{g}}_j \cdot \widetilde{\mathbf{g}}_j &= \big([\mathbf{s}_j]_1;\ [\mathbf{W}_1\mathbf{s}_j]_1,\ \ldots,\ [\mathbf{W}_n\mathbf{s}_j]_1\big),\ \mathbf{s}_j \leftarrow \mathbb{Z}_p^{3k}.
\end{aligned}$$

Note that the distribution here is identical to the original one except that $\mathbf{A}$, $\widehat{\mathbf{A}}$, $\widetilde{\mathbf{A}}$ fail to span the entire space $\mathbb{Z}_p^{3k}$ whose probability is bounded by $2k/p$ (c.f. Lemma 3). We prove the following lemma.

Lemma 7. *For any p.p.t. adversary $\mathcal{A}$, there exists an adversary $\mathcal{B}$ such that*

$$\mathsf{Adv}_{\mathcal{A}}^{\mathrm{LS2}}(\lambda, q, q') \leqslant \mathsf{Adv}_{\mathcal{B},q}^{\mathcal{U}_{3k,k}}(\lambda) + 2^{-\Omega(\lambda)}$$

where $\mathsf{T}(\mathcal{B}) \approx \mathsf{T}(\mathcal{A}) + k^2 \cdot (q+q') \cdot \mathsf{poly}(\lambda, n)$ *and* $\mathsf{poly}(\lambda, n)$ *is independent of* $\mathsf{T}(\mathcal{A})$.

Proof. Given $[\widehat{\mathbf{A}}]_1 \in G_1^{3k\times k}$ and $[\mathbf{T}]_1 = [\mathbf{t}_1|\cdots|\mathbf{t}_q]_1 \in G_1^{3k\times q}$, $\mathcal{B}$ works as follows:

Simulating pp. Sample $\mathbf{A} \leftarrow \mathcal{U}_{3k,k}$ and $\mathbf{W}_i \leftarrow \mathbb{Z}_p^{k\times 3k}$ for all $i \in [n]$. We can then simulate PP directly.

Simulating $\widehat{h}_j^* \cdot \widetilde{h}_j^*$. Calculate $\mathbf{A}^\perp \in \mathbb{Z}_p^{3k\times 2k}$ from $\mathbf{A} \in \mathbb{Z}_p^{3k\times k}$ and one may simulate $\widehat{h}_j^* \cdot \widetilde{h}_j^*$ by sampling $\widehat{h}_j^* \cdot \widetilde{h}_j^* \leftarrow \mathsf{Span}([\mathbf{A}^\perp]_2)$ by Lemma 3.

Simulating $\mathbf{g}_j' \cdot \widehat{\mathbf{g}}_j' \cdot \widetilde{\mathbf{g}}_j'$ and $\mathbf{h}_j$. We can simply simulate each $\mathbf{g}_j' \cdot \widehat{\mathbf{g}}_j' \cdot \widetilde{\mathbf{g}}_j'$ (resp. $\mathbf{h}_j$) using $\mathbf{W}_i$ for all $i \in [n]$ and a freshly chosen $\mathbf{s}_j' \leftarrow \mathbb{Z}_p^{3k}$ for all $j \in [q]$ (resp. $\mathbf{r}_j \in \mathbb{Z}_p^k$ for all $j \in [q']$).

Simulating the Challenge. Sample $\bar{\mathbf{s}}_j \leftarrow \mathbb{Z}_p^k$ for all $j \in [q]$. We simulate the challenge as

$$\big([\mathbf{A}\bar{\mathbf{s}}_j + \mathbf{t}_j]_1; [\mathbf{W}_1(\mathbf{A}\bar{\mathbf{s}}_j + \mathbf{t}_j)]_1, \ldots, [\mathbf{W}_n(\mathbf{A}\bar{\mathbf{s}}_j + \mathbf{t}_j)]_1\big) \quad \text{for all } j \in [q].$$

Observe that: when $\mathbf{t}_j \leftarrow \mathsf{Span}(\widehat{\mathbf{A}})$ for all $j \in [q]$, the challenge equals $\{\mathbf{g}_j \cdot \widehat{\mathbf{g}}_j\}$; when $\mathbf{t}_j \leftarrow \mathbb{Z}_p^{3k}$ for all $j \in [q]$, the challenge is identical to $\{\mathbf{g}_j \cdot \widehat{\mathbf{g}}_j \cdot \widetilde{\mathbf{g}}_j\}$ (we described above). This proves the lemma. ⊔

We can prove a similar lemma for $\mathsf{Adv}_{\mathcal{A}}^{\mathrm{LS3}}(k, q, q')$. The proof is almost the same as above with the exception that $\mathcal{B}$ controls $\mathbf{A}$ and $\widehat{\mathbf{A}}$ this time, and embeds q-fold $\mathcal{U}_{3k,k}$-MDDH instance through $\widetilde{\mathbf{A}}$. More concretely, one may simulate PP, $\{\widehat{h}_j^* \cdot \widetilde{h}_j^*\}$, $\{\mathbf{h}_j\}$ and the challenge with $\mathbf{A}$ and $\widehat{\mathbf{A}}$ as before, while the simulation of $\{\mathbf{g}_j' \cdot \widehat{\mathbf{g}}_j'\}$ needs the help of $\widehat{\mathbf{A}}$.

5.4 Nested-Hiding Indistinguishability

For all $\eta \in [\lfloor n/2 \rfloor]$, we rewrite the advantage function $\mathsf{Adv}_{\mathcal{A}}^{\mathrm{NH}(\eta)}(\lambda, q, q')$ using

$$\begin{aligned}
\text{PP} &= \big([\mathbf{A}]_1,\ [\mathbf{W}_1\mathbf{A}]_1,\ \ldots,\ [\mathbf{W}_n\mathbf{A}]_1\big);\\
\widehat{h}_j^* &= [\widehat{\mathbf{A}}^*\widehat{\mathbf{r}}_j']_2,\ \widehat{\mathbf{r}}_j' \leftarrow \mathbb{Z}_p^k; \qquad \widetilde{h}_j^* = [\widetilde{\mathbf{A}}^*\widetilde{\mathbf{r}}_j']_2,\ \widetilde{\mathbf{r}}_j' \leftarrow \mathbb{Z}_p^k;\\
\widehat{\mathbf{g}}_j &= \big([\widehat{\mathbf{A}}\widehat{\mathbf{s}}_j]_1;\ [\mathbf{W}_1\widehat{\mathbf{A}}\widehat{\mathbf{s}}_j]_1,\ \ldots,\ [\mathbf{W}_n\widehat{\mathbf{A}}\widehat{\mathbf{s}}_j]_1\big),\ \widehat{\mathbf{s}}_j \leftarrow \mathbb{Z}_p^k;\\
\widetilde{\mathbf{g}}_j &= \big([\widetilde{\mathbf{A}}\widetilde{\mathbf{s}}_j]_1;\ [\mathbf{W}_1\widetilde{\mathbf{A}}\widetilde{\mathbf{s}}_j]_1,\ \ldots,\ [\mathbf{W}_n\widetilde{\mathbf{A}}\widetilde{\mathbf{s}}_j]_1\big),\ \widetilde{\mathbf{s}}_j \leftarrow \mathbb{Z}_p^k;\\
\mathbf{h}_j' &= \big([\mathbf{r}_j']_2;\ [\mathbf{W}_1^\top\mathbf{r}_j']_2,\ \ldots,\ [\mathbf{W}_n^\top\mathbf{r}_j']_2\big),\ \mathbf{r}_j' \leftarrow \mathbb{Z}_p^k
\end{aligned}$$

and the challenge term $\mathbf{h}_j \cdot (1_{\mathbb{H}_0}; (\widehat{h}_j^{**})^{\mathbf{e}_{2\eta-1}}) \cdot (1_{\mathbb{H}_0}; (\widetilde{h}_j^{**})^{\mathbf{e}_{2\eta}})$ equals

$$([\mathbf{r}_j]_2;\ [\mathbf{W}_1^\top \mathbf{r}_j]_2,\ \ldots,\ [\mathbf{W}_{2\eta-1}^\top \mathbf{r}_j + \widehat{\mathbf{A}}^* \widehat{\mathbf{r}}_j]_2,\ [\mathbf{W}_{2\eta}^\top \mathbf{r}_j + \widetilde{\mathbf{A}}^* \widetilde{\mathbf{r}}_j]_2,\ \ldots,\ [\mathbf{W}_n^\top \mathbf{r}_j]_2)$$

where $\mathbf{r}_j \leftarrow \mathbb{Z}_p^k$, either $\widehat{\mathbf{r}}_j, \widetilde{\mathbf{r}}_j \leftarrow \mathbb{Z}_p^k$ or $\widehat{\mathbf{r}}_j = \widetilde{\mathbf{r}}_j = \mathbf{0}_k$. We prove the lemma below.

Lemma 8. *For any p.p.t. adversary $\mathcal{A}$, there exists an adversary $\mathcal{B}$ such that*

$$\mathsf{Adv}_{\mathcal{A}}^{\mathrm{NH}(\eta)}(\lambda, q, q') \leqslant \mathsf{Adv}_{\mathcal{B},q'}^{\mathcal{U}_{3k,k}}(\lambda)$$

where $\mathsf{T}(\mathcal{B}) \approx \mathsf{T}(\mathcal{A}) + k^2 \cdot (q + q') \cdot \mathsf{poly}(\lambda, n)$ and $\mathsf{poly}(\lambda, n)$ is independent of $\mathsf{T}(\mathcal{A})$.

Before we prove the lemma, we describe and prove an extension of Lemma 5.

Lemma 9. *Given $Q \in \mathbb{N}$, group G of prime order p, $[\mathbf{M}] \in G^{3k \times k}$ and $[\mathbf{T}] = [\mathbf{t}_1 | \cdots | \mathbf{t}_Q] \in G^{3k \times Q}$ where either $\mathbf{t}_i \leftarrow \mathsf{Span}(\mathbf{M})$ or $\mathbf{t}_i \leftarrow \mathbb{Z}_p^{3k}$, one can efficiently compute*

$$[\mathbf{Z}],\quad [\mathbf{V}_0\mathbf{Z}],\quad [\mathbf{V}_1\mathbf{Z}],\quad \{[\boldsymbol{\tau}_j], [\boldsymbol{\tau}_{0,j}], [\boldsymbol{\tau}_{1,j}]\}_{j \in [Q]}$$

where $\mathbf{Z} \in \mathbb{Z}_p^{k \times k}$ is full-rank, $\mathbf{V}_0, \mathbf{V}_1 \in \mathbb{Z}_p^{k \times k}$ are secret matrices, $\boldsymbol{\tau}_j \leftarrow \mathbb{Z}_p^k$ and either $\boldsymbol{\tau}_{0,j} = \mathbf{V}_0 \boldsymbol{\tau}_j$, $\boldsymbol{\tau}_{1,j} = \mathbf{V}_1 \boldsymbol{\tau}_j$ (when $\mathbf{t}_j \leftarrow \mathsf{Span}(\mathbf{M})$) or $\boldsymbol{\tau}_{0,j}, \boldsymbol{\tau}_{1,j} \leftarrow \mathbb{Z}_p^k$ (when $\mathbf{t}_j \leftarrow \mathbb{Z}_p^{3k}$).

Proof. Given Q, G, $[\mathbf{M}]$, $[\mathbf{T}] = [\mathbf{t}_1 | \cdots | \mathbf{t}_Q]$, the algorithm works as follows:

Programming $[\mathbf{Z}], [\mathbf{V}_0\mathbf{Z}], [\mathbf{V}_1\mathbf{Z}]$. Define $\mathbf{Z} = \overline{\mathbf{M}}$. Randomly pick $\mathbf{M}_0, \mathbf{M}_1 \leftarrow \mathbb{Z}_p^{k \times 3k}$ and implicitly define $\mathbf{V}_0, \mathbf{V}_1 \in \mathbb{Z}_p^{k \times k}$ such that

$$\mathbf{V}_0\mathbf{Z} = \mathbf{V}_0\overline{\mathbf{M}} = \mathbf{M}_0\mathbf{M} \quad \text{and} \quad \mathbf{V}_1\mathbf{Z} = \mathbf{V}_1\overline{\mathbf{M}} = \mathbf{M}_1\mathbf{M}.$$

One can generate $[\mathbf{Z}]$ along with $[\mathbf{V}_0\mathbf{Z}], [\mathbf{V}_1\mathbf{Z}]$ using $[\mathbf{M}]$ and $\mathbf{M}_0, \mathbf{M}_1$.

Generating Q tuples. For all $j \in [Q]$, we compute

$$[\boldsymbol{\tau}_j] = [\overline{\mathbf{t}}_j],\quad [\boldsymbol{\tau}_{0,j}] = [\mathbf{M}_0\mathbf{t}_j],\quad [\boldsymbol{\tau}_{1,j}] = [\mathbf{M}_1\mathbf{t}_j].$$

Here $\overline{\mathbf{t}}_j$ indicates the first k entries of $\mathbf{t}_j$.

Observe that: if $\mathbf{t}_j = \mathbf{M}\mathbf{u}_j$ for some $\mathbf{u}_j \leftarrow \mathbb{Z}_p^k$, we have that $\boldsymbol{\tau}_j = \overline{\mathbf{M}}\mathbf{u}_j$ and

$$\boldsymbol{\tau}_{0,j} = \mathbf{M}_0\mathbf{M}\mathbf{u}_j = \mathbf{V}_0\overline{\mathbf{M}}\mathbf{u}_j = \mathbf{V}_0\boldsymbol{\tau}_j, \qquad \boldsymbol{\tau}_{1,j} = \mathbf{M}_1\mathbf{M}\mathbf{u}_j = \mathbf{V}_1\overline{\mathbf{M}}\mathbf{u}_j = \mathbf{V}_1\boldsymbol{\tau}_j;$$

if $\mathbf{t}_j \leftarrow \mathbb{Z}_p^{3k}$, we can see that

$$\begin{pmatrix} \boldsymbol{\tau}_j \\ \boldsymbol{\tau}_{0,j} \\ \boldsymbol{\tau}_{1,j} \end{pmatrix} = \begin{pmatrix} \mathbf{I}_{k \times 3k} \\ \mathbf{M}_0 \\ \mathbf{M}_1 \end{pmatrix} \mathbf{t}_j$$

is uniformly distributed over $\mathbb{Z}_p^{3k}$ where the left-most k columns of $\mathbf{I}_{k \times 3k}$ form an identity matrix and remaining columns are zero vectors. □

We are ready to prove Lemma 8 by extending the strategy proving Lemma 6.

Proof. Given $[\mathbf{M}]_2 \in G_2^{3k \times k}$ and $[\mathbf{T}]_2 = [\mathbf{t}_1 | \cdots | \mathbf{t}_{q'}]_2 \in G_2^{3k \times q'}$ where either $\mathbf{t}_j \leftarrow \mathsf{Span}(\mathbf{M})$ or $\mathbf{t}_j \leftarrow \mathbb{Z}_p^{3k}$, $\mathcal{B}$ proceeds as follows:

Generating q' tuples. We invoke the algorithm described in Lemma 9 on input $(q', G_2, [\mathbf{M}]_2, [\mathbf{T}]_2)$ and obtain

$$\big([\mathbf{Z}]_2, [\mathbf{V}_0\mathbf{Z}]_2, [\mathbf{V}_1\mathbf{Z}]_2, \{[\boldsymbol{\tau}_j]_2, [\boldsymbol{\tau}_{0,j}]_2, [\boldsymbol{\tau}_{1,j}]_2\}_{j\in[q']}\big).$$

Simulating pp. Sample $\mathbf{A}, \widehat{\mathbf{A}}, \widetilde{\mathbf{A}} \leftarrow \mathcal{U}_{3k,k}$ and randomly pick $\widehat{\mathbf{A}}^*$ and $\widetilde{\mathbf{A}}^*$, the respective bases of $\mathsf{Ker}\big((\mathbf{A}|\widetilde{\mathbf{A}})^\top\big)$ and $\mathsf{Ker}\big((\mathbf{A}|\widehat{\mathbf{A}})^\top\big)$. Select $\bar{\mathbf{W}}_{2\eta-1}, \bar{\mathbf{W}}_{2\eta} \leftarrow \mathbb{Z}_p^{k\times 3k}$ and define

$$\mathbf{W}_{2\eta-1} = \bar{\mathbf{W}}_{2\eta-1} + \mathbf{V}_1^\top \cdot (\widehat{\mathbf{A}}^*)^\top \quad \text{and} \quad \mathbf{W}_{2\eta} = \bar{\mathbf{W}}_{2\eta} + \mathbf{V}_0^\top \cdot (\widetilde{\mathbf{A}}^*)^\top.$$

Then we sample $\mathbf{W}_i \leftarrow \mathbb{Z}_p^{k\times 3k}$ for all $i \in [n] \setminus \{2\eta-1, 2\eta\}$. We can simulate PP using the following observation:

$$\mathbf{W}_{2\eta-1}\mathbf{A} = \big(\bar{\mathbf{W}}_{2\eta-1} + \mathbf{V}_1^\top \cdot (\widehat{\mathbf{A}}^*)^\top\big)\mathbf{A} = \bar{\mathbf{W}}_{2\eta-1}\mathbf{A},$$
$$\mathbf{W}_{2\eta}\mathbf{A} = \big(\bar{\mathbf{W}}_{2\eta} + \mathbf{V}_0^\top \cdot (\widetilde{\mathbf{A}}^*)^\top\big)\mathbf{A} = \bar{\mathbf{W}}_{2\eta}\mathbf{A}.$$

Simulating $\widehat{h}_j^*$ and $\widetilde{h}_j^*$. It is direct to simulate all $\widehat{h}_j^*$ and $\widetilde{h}_j^*$ using $\widehat{\mathbf{A}}^*$ and $\widetilde{\mathbf{A}}^*$.
Simulating $(\widehat{\mathbf{g}}_j)_{-(2\eta-1)}$ and $(\widetilde{\mathbf{g}}_j)_{-2\eta}$. We can simulate $(\widehat{\mathbf{g}}_j)_{-(2\eta-1)}$ following the fact that

$$\mathbf{W}_{2\eta}\widehat{\mathbf{A}} = \big(\bar{\mathbf{W}}_{2\eta} + \mathbf{V}_0^\top \cdot (\widetilde{\mathbf{A}}^*)^\top\big)\widehat{\mathbf{A}} = \bar{\mathbf{W}}_{2\eta}\widehat{\mathbf{A}}.$$

Similarly, we can also simulate $(\widetilde{\mathbf{g}}_j)_{-2\eta}$ because

$$\mathbf{W}_{2\eta-1}\widetilde{\mathbf{A}} = \big(\bar{\mathbf{W}}_{2\eta-1} + \mathbf{V}_1^\top \cdot (\widehat{\mathbf{A}}^*)^\top\big)\widetilde{\mathbf{A}} = \bar{\mathbf{W}}_{2\eta}\widetilde{\mathbf{A}}.$$

Although $\mathbf{W}_{2\eta-1}\widehat{\mathbf{A}}$ and $\mathbf{W}_{2\eta}\widetilde{\mathbf{A}}$ contain secret matrices and are unknown to $\mathcal{B}$ due to Lemma 3, they are not necessary in our simulation.
Simulating $\mathbf{h}_j'$. Sample $\bar{\mathbf{r}}_j \leftarrow \mathbb{Z}_p^k$ and implicitly define $\mathbf{r}_j' = \mathbf{Z}\bar{\mathbf{r}}_j$ for all $j \in [q']$. We can simply produce $[\mathbf{r}_j']_2$ and $[\mathbf{W}_i^\top\mathbf{r}_j']_2$ for $i \in [n]\setminus\{2\eta-1, 2\eta\}$ while the remaining two entries are simulated following the fact

$$\mathbf{W}_{2\eta-1}^\top\mathbf{r}_j' = \big(\bar{\mathbf{W}}_{2\eta-1} + \mathbf{V}_1^\top \cdot (\widehat{\mathbf{A}}^*)^\top\big)^\top\mathbf{Z}\bar{\mathbf{r}}_j = \bar{\mathbf{W}}_{2\eta-1}^\top\mathbf{Z}\bar{\mathbf{r}}_j + \widehat{\mathbf{A}}^* \cdot (\mathbf{V}_1\mathbf{Z}) \cdot \bar{\mathbf{r}}_j,$$
$$\mathbf{W}_{2\eta}^\top\mathbf{r}_j' = \big(\bar{\mathbf{W}}_{2\eta} + \mathbf{V}_0^\top \cdot (\widetilde{\mathbf{A}}^*)^\top\big)^\top\mathbf{Z}\bar{\mathbf{r}}_j = \bar{\mathbf{W}}_{2\eta}^\top\mathbf{Z}\bar{\mathbf{r}}_j + \widetilde{\mathbf{A}}^* \cdot (\mathbf{V}_0\mathbf{Z}) \cdot \bar{\mathbf{r}}_j,$$

because $[\mathbf{Z}]_2$, $[\mathbf{V}_0\mathbf{Z}]_2$ and $[\mathbf{V}_1\mathbf{Z}]_2$ are known to $\mathcal{B}$.
Simulating the challenge. For all $j \in [q']$, we compute the challenge as

$$([\boldsymbol{\tau}_j]_2, [\mathbf{W}_1^\top\boldsymbol{\tau}_j]_2, \ldots, [\bar{\mathbf{W}}_{2\eta-1}^\top\boldsymbol{\tau}_j + \widehat{\mathbf{A}}^*\boldsymbol{\tau}_{1,j}]_2, [\bar{\mathbf{W}}_{2\eta}^\top\boldsymbol{\tau}_j + \widetilde{\mathbf{A}}^*\boldsymbol{\tau}_{0,j}]_2, \ldots, [\mathbf{W}_n^\top\boldsymbol{\tau}_j]_2).$$

Observe that, when $\mathbf{t}_j \leftarrow \mathsf{Span}(\mathbf{M})$, we have that $\boldsymbol{\tau}_{0,j} = \mathbf{V}_0\boldsymbol{\tau}_j$ and $\boldsymbol{\tau}_{1,j} = \mathbf{V}_1\boldsymbol{\tau}_j$, the challenge is identical to $\{\mathbf{h}_j\}$, i.e., $\widehat{\mathbf{r}}_j = \widetilde{\mathbf{r}}_j = \mathbf{0}_k$; when $\mathbf{t}_j \leftarrow \mathbb{Z}_p^{3k}$, we have $\boldsymbol{\tau}_{0,j}, \boldsymbol{\tau}_{1,j} \leftarrow \mathbb{Z}_p^k$, the challenge is identical to $\{\mathbf{h}_j\cdot(1_{\mathbb{H}_0}; (\widehat{h}_j^{**})^{\mathbf{e}_{2\eta-1}})\cdot(1_{\mathbb{H}_0}; (\widetilde{h}_j^{**})^{\mathbf{e}_{2\eta}})\}$ where $\widehat{\mathbf{r}}_j = \boldsymbol{\tau}_{1,j} - \mathbf{V}_1\boldsymbol{\tau}_j$ and $\widetilde{\mathbf{r}}_j = \boldsymbol{\tau}_{0,j} - \mathbf{V}_0\boldsymbol{\tau}_j$ are uniformly distributed over $\mathbb{Z}_p^k$. This proves the lemma. □

5.5 $\mathbb{G}$-Uniformity

We rewrite the advantage function $\mathsf{Adv}_{\mathcal{A}}^{\mathbb{G}\text{-uni}}(\lambda, q, q')$ using

$$\text{PP} = ([\mathbf{A}]_1, [\mathbf{W}_1\mathbf{A}]_1, \ldots, [\mathbf{W}_n\mathbf{A}]_1); \quad \widehat{h}_j^* = [\widehat{\mathbf{A}}^*\widehat{\mathbf{r}}_j]_2; \quad \widetilde{h}_j^* = [\widetilde{\mathbf{A}}^*\widetilde{\mathbf{r}}_j]_2$$

where $\widehat{\mathbf{r}}_j, \widetilde{\mathbf{r}}_j \leftarrow \mathbb{Z}_p^k$ and $\mathbf{h}_j \cdot (1_{\mathbb{H}_0}; \widehat{h}_{1,j}^*, \ldots, \widehat{h}_{n,j}^*)$ equals

$$([\mathbf{r}_j]_2; [\mathbf{W}_1^\top \mathbf{r}_j + \widehat{\mathbf{A}}^*\widehat{\mathbf{r}}_{1,j}]_2, \ldots, [\mathbf{W}_n^\top \mathbf{r}_j + \widehat{\mathbf{A}}^*\widehat{\mathbf{r}}_{n,j}]_2), \; \mathbf{r}_j, \widehat{\mathbf{r}}_{1,j}, \ldots, \widehat{\mathbf{r}}_{n,j} \leftarrow \mathbb{Z}_p^k;$$

and the challenge term $\mathbf{g}_j \cdot \widehat{\mathbf{g}}_j \cdot (1_{\mathbb{G}_0}; (g_j')^{\mathbf{1}_n})$ equals

$$([\mathbf{A}\mathbf{s}_j + \widehat{\mathbf{A}}\widehat{\mathbf{s}}_j]_1; [\mathbf{W}_1(\mathbf{A}\mathbf{s}_j + \widehat{\mathbf{A}}\widehat{\mathbf{s}}_j) + \mathbf{s}_j']_1, \ldots, [\mathbf{W}_n(\mathbf{A}\mathbf{s}_j + \widehat{\mathbf{A}}\widehat{\mathbf{s}}_j) + \mathbf{s}_j']_1)$$

where $\mathbf{s}_j, \widehat{\mathbf{s}}_j \leftarrow \mathbb{Z}_p^k$, either $\mathbf{s}_j' \leftarrow \mathbb{Z}_p^k$ or $\mathbf{s}_j' = \mathbf{0}_k$. We prove the following lemma using essentially the same method as in [3].

Lemma 10. *For any p.p.t. adversary $\mathcal{A}$, there exists an adversary $\mathcal{B}$ such that*

$$\mathsf{Adv}_{\mathcal{A}}^{\mathbb{G}\text{-uni}}(\lambda, q, q') \leqslant \mathsf{Adv}_{\mathcal{B},q}^{\mathcal{U}_{2k,k}}(\lambda)$$

where $\mathsf{T}(\mathcal{B}) \approx \mathsf{T}(\mathcal{A}) + k^2 \cdot (q + q') \cdot \mathsf{poly}(\lambda, n)$ and $\mathsf{poly}(\lambda, n)$ is independent of $\mathsf{T}(\mathcal{A})$.

We describe a simple extension of Lemma 5 without proof which is basically identical to *Generalized Many-Tuple Lemma* in [14].

Lemma 11. *Given $Q \in \mathbb{N}$, group G of prime order p, $[\mathbf{M}] \in G^{2k \times k}$ and $[\mathbf{T}] = [\mathbf{t}_1 | \cdots | \mathbf{t}_Q] \in G^{2k \times Q}$ where either $\mathbf{t}_i \leftarrow \mathsf{Span}(\mathbf{M})$ or $\mathbf{t}_i \leftarrow \mathbb{Z}_p^{2k}$, one can efficiently compute $[\mathbf{Z}]$, $[\mathbf{VZ}]$ and Q tuples $([\boldsymbol{\tau}_j], [\boldsymbol{\tau}_j'])_{j \in [Q]}$ where $\mathbf{Z} \in \mathbb{Z}_p^{k \times k}$ is full-rank, $\mathbf{V} \in \mathbb{Z}_p^{k \times k}$ is a secret matrix, $\boldsymbol{\tau}_j \leftarrow \mathbb{Z}_p^k$, either $\boldsymbol{\tau}_j' = \mathbf{V}\boldsymbol{\tau}_j$ (when $\mathbf{t}_j \leftarrow \mathsf{Span}(\mathbf{M})$) or $\boldsymbol{\tau}_j' \leftarrow \mathbb{Z}_p^k$ (when $\mathbf{t}_j \leftarrow \mathbb{Z}_p^{2k}$).*

We are ready to prove Lemma 10.

Proof. Given $[\mathbf{M}]_1 \in G_1^{2k \times k}$ and $[\mathbf{T}]_1 = [\mathbf{t}_1 | \cdots | \mathbf{t}_q]_1 \in G_1^{3k \times q}$ where either $\mathbf{t}_j \leftarrow \mathsf{Span}(\mathbf{M})$ or $\mathbf{t}_j \leftarrow \mathbb{Z}_p^{2k}$, $\mathcal{B}$ proceeds as follows:

Generating q tuples. We invoke the algorithm described in Lemma 11 on input $(q, G_1, [\mathbf{M}]_1, [\mathbf{T}]_1)$ and obtain $([\mathbf{Z}]_1, [\mathbf{VZ}]_1, \{[\boldsymbol{\tau}_j]_1, [\boldsymbol{\tau}_j']_1\}_{j \in [q]})$.

Simulating pp. Sample $\mathbf{A}, \widehat{\mathbf{A}}, \widetilde{\mathbf{A}} \leftarrow \mathcal{U}_{3k,k}$ and randomly pick $\widehat{\mathbf{A}}^*$ and $\widetilde{\mathbf{A}}^*$, the respective bases of $\mathsf{Ker}((\mathbf{A}|\widetilde{\mathbf{A}})^\top)$ and $\mathsf{Ker}((\mathbf{A}|\widehat{\mathbf{A}})^\top)$. For all $i \in [n]$, pick $\bar{\mathbf{W}}_i \leftarrow \mathbb{Z}_p^{k \times 3k}$ and implicitly define

$$\mathbf{W}_i = \bar{\mathbf{W}}_i + \bar{\mathbf{V}} \cdot (\widehat{\mathbf{A}}^*)^\top$$

where $\bar{\mathbf{V}} = \mathbf{V}((\widehat{\mathbf{A}}^*)^\top \widehat{\mathbf{A}})^{-1} \in \mathbb{Z}_p^{k \times k}$. We can simulate PP from the observation

$$\mathbf{W}_i\mathbf{A} = (\bar{\mathbf{W}}_i + \bar{\mathbf{V}} \cdot (\widehat{\mathbf{A}}^*)^\top)\mathbf{A} = \bar{\mathbf{W}}_i\mathbf{A}.$$

Simulating $\widehat{h}_j^*$ and $\widetilde{h}_j^*$. It is direct to simulate all $\widehat{h}_j^*$ and $\widetilde{h}_j^*$ using $\widehat{\mathbf{A}}^*$ and $\widetilde{\mathbf{A}}^*$.
Simulating $\mathbf{h}_j \cdot (1_{\mathbb{H}_0};\ \widehat{h}_{1,j}^*, \ldots, \widehat{h}_{n,j}^*)$. Observe that

$$\mathbf{W}_i^\top \mathbf{r}_j + \widehat{\mathbf{A}}^* \widehat{\mathbf{r}}_{i,j} = \bar{\mathbf{W}}_i^\top \mathbf{r}_j + \widehat{\mathbf{A}}^* (\bar{\mathbf{V}}^\top \mathbf{r}_j + \widehat{\mathbf{r}}_{i,j}) \text{ for all } i \in [n], j \in [q'].$$

We can alternatively simulate $\mathbf{h}_j \cdot (1_{\mathbb{H}_0};\ \widehat{h}_{1,j}^*, \ldots, \widehat{h}_{n,j}^*)$ as $\bar{\mathbf{W}}_i^\top \mathbf{r}_j + \widehat{\mathbf{A}}^* \widehat{\mathbf{r}}_{i,j}$ for all $i \in [n], j \in [q']$ where $\mathbf{r}_j, \widehat{\mathbf{r}}_{i,j} \leftarrow \mathbb{Z}_p^k$ without secret matrix $\mathbf{V}$.
Simulating the challenge. Observe that

$$\mathbf{W}_i \widehat{\mathbf{A}} = (\bar{\mathbf{W}}_i + \bar{\mathbf{V}} \cdot (\widehat{\mathbf{A}}^*)^\top) \widehat{\mathbf{A}} = \bar{\mathbf{W}}_i \widehat{\mathbf{A}} + \mathbf{V}.$$

We can sample $\bar{\mathbf{s}}_j \leftarrow \mathbb{Z}_p^k$ and simulate the challenge as

$$\left([\mathbf{A}\bar{\mathbf{s}}_j + \widehat{\mathbf{A}}\boldsymbol{\tau}_j]_1, [\bar{\mathbf{W}}_1 \mathbf{A}\bar{\mathbf{s}}_j + \bar{\mathbf{W}}_1 \widehat{\mathbf{A}}\boldsymbol{\tau}_j + \boldsymbol{\tau}_j']_1, \ldots, [\bar{\mathbf{W}}_n \mathbf{A}\bar{\mathbf{s}}_j + \bar{\mathbf{W}}_n \widehat{\mathbf{A}}\boldsymbol{\tau}_j + \boldsymbol{\tau}_j']_1\right).$$

Observe that, when $\mathbf{t}_j \leftarrow \mathsf{Span}(\mathbf{M})$, we have $\boldsymbol{\tau}_j' = \mathbf{V}\boldsymbol{\tau}_j$, the challenge is identical to $\{\mathbf{g}_j \cdot \widehat{\mathbf{g}}_j\}$; when $\mathbf{t}_j \leftarrow \mathbb{Z}_p^{2k}$, we have $\boldsymbol{\tau}_j' \leftarrow \mathbb{Z}_p^k$, the challenge is identical to $\{\mathbf{g}_j \cdot \widehat{\mathbf{g}}_j \cdot (1_{\mathbb{G}_0}; (g_j')^{\mathbf{1}_n})\}$ where $\mathbf{s}_j' = \boldsymbol{\tau}_j' - \mathbf{V}\boldsymbol{\tau}_j$ is uniformly distributed over $\mathbb{Z}_p^k$. This proves the lemma. □

6 Concrete Constructions

We present our main result in Fig. 1 whose adaptive security and anonymity in the MIMC setting is almost-tightly based on the k-Lin assumption.

Figure 2 presents a concrete instantiation of our main result based on SXDH (1-Lin) assumption by setting $k = 1$. Our description below only involves vectors and scalars.

$\underline{\mathsf{Param}(1^\lambda, n)}$	$\underline{\mathsf{KeyGen}(\text{MPK}, \text{MSK}, \text{ID})}$
$\mathbf{A} \leftarrow \mathcal{U}_{3k,k}$ **for** $(i,b) \in [n] \times \{0,1\}$ **do** $\quad \mathbf{W}_{i,b} \leftarrow \mathbb{Z}_p^{k \times 3k}$, $\mathbf{Z}_{i,b} = \mathbf{W}_{i,b}\mathbf{A} \in \mathbb{Z}_p^{k \times k}$ $\text{GP} = ([\mathbf{A}]_1, \{[\mathbf{Z}_{i,b}]_1, [\mathbf{W}_{i,b}]_2\})$ **return** GP	$\mathbf{r} \leftarrow \mathbb{Z}_p^k$ $\text{SK} = ([\mathbf{r}]_2, [\boldsymbol{\alpha} + \sum_{i=1}^n \mathbf{W}_{i,\text{ID}[i]}^\top \mathbf{r}]_2) \in G_2^{4k}$ **return** SK $\underline{\mathsf{Enc}(\text{MPK}, \text{ID}, \text{M})}$ $\mathbf{s} \leftarrow \mathbb{Z}_p^k$
$\underline{\mathsf{Setup}(\text{GP})}$ $\boldsymbol{\alpha} \leftarrow \mathbb{Z}_p^{3k}$ $\text{MPK} = ([\mathbf{A}]_1, \{[\mathbf{Z}_{i,b}]_1\}, [\mathbf{A}^\top \boldsymbol{\alpha}]_T)$ $\text{MSK} = ([\boldsymbol{\alpha}]_2, \{[\mathbf{W}_{i,b}]_2\})$ **return** MPK, MSK	$\text{CT}' = ([\mathbf{A}\mathbf{s}]_1, [\sum_{i=1}^n \mathbf{Z}_{i,\text{ID}[i]}\mathbf{s}]_1) \in G_1^{4k}$ $\text{KEY} = [\mathbf{s}^\top \mathbf{A}^\top \boldsymbol{\alpha}]_T \in G_T$ **return** $\text{CT} = (\text{CT}', \text{KEY} \cdot \text{M})$ $\underline{\mathsf{Dec}(\text{MPK}, \text{SK} = (\mathbf{k}_0, \mathbf{k}_1), \text{CT} = (\mathbf{c}_0, \mathbf{c}_1, c_2))}$ **return** $\text{M} = c_2 \cdot e(\mathbf{c}_1, \mathbf{k}_0)/e(\mathbf{c}_0, \mathbf{k}_1)$

Fig. 1. Main result: a concrete IBE scheme based on the k-Lin assumption.

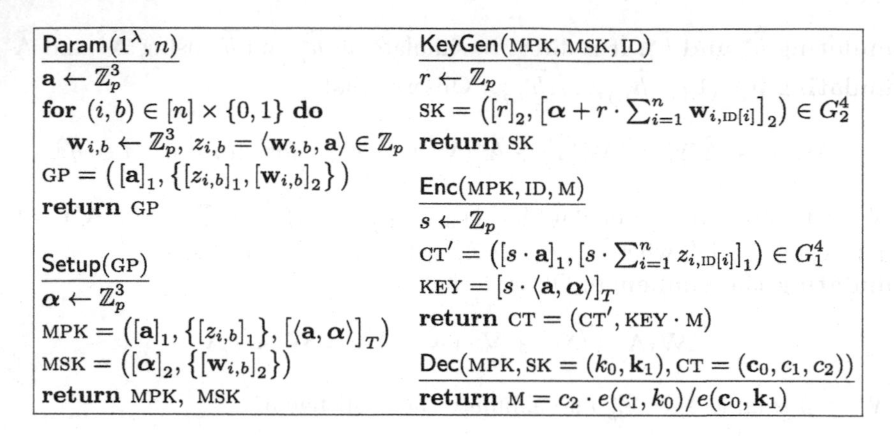

Fig. 2. A concrete IBE scheme based on SXDH ($k = 1$). Here we let $\langle \mathbf{x}, \mathbf{y} \rangle$ be the inner product of $\mathbf{x}$ and $\mathbf{y}$ of the same length and $e([\mathbf{x}]_1, [\mathbf{y}]_2) = [\langle \mathbf{x}, \mathbf{y} \rangle]_T$ in this case.

Acknowledgement. We want to thank Hoeteck Wee for his constructive suggestion and Kai Zhang for his useful advice. We also thank all anonymous reviewers of AsiaCrypt 2016 for their helpful comments. This work was supported by the National Natural Science Foundation of China (Grant No. 61373154, 61371083, 61472142, 61632012, 61672239, 61602180), the Prioritized Development Projects through the Specialized Research Fund for the Doctoral Program of Higher Education of China (Grant No. 20130073130004), Shanghai High-tech field project (Grant No. 16511101400), Science and Technology Commission of Shanghai Municipality (Grant No. 14YF1404200), and the "Programme Avenir Lyon Saint-Etienne de l'Université de Lyon" in the framework of the programme "Investissements d'Avenir" (ANR-11-IDEX-0007).

References

1. Attrapadung, N.: Dual system encryption via doubly selective security: framework, fully secure functional encryption for regular languages, and more. In: Nguyen, P.Q., Oswald, E. (eds.) EUROCRYPT 2014. LNCS, vol. 8441, pp. 557–577. Springer, Heidelberg (2014). doi:10.1007/978-3-642-55220-5_31
2. Attrapadung, N.: Dual system encryption framework in prime-order groups. IACR Cryptology ePrint Archive (2015)
3. Attrapadung, N., Hanaoka, G., Yamada, S.: A framework for identity-based encryption with almost tight security. In: Iwata, T., Cheon, J.H. (eds.) ASIACRYPT 2015. LNCS, vol. 9452, pp. 521–549. Springer, Heidelberg (2015). doi:10.1007/978-3-662-48797-6_22
4. Blazy, O., Kiltz, E., Pan, J.: (Hierarchical) Identity-based encryption from affine message authentication. In: Garay, J.A., Gennaro, R. (eds.) CRYPTO 2014. LNCS, vol. 8616, pp. 408–425. Springer, Heidelberg (2014). doi:10.1007/978-3-662-44371-2_23
5. Boneh, D., Boyen, X.: Efficient selective-ID secure identity-based encryption without random oracles. In: Cachin, C., Camenisch, J.L. (eds.) EUROCRYPT 2004. LNCS, vol. 3027, pp. 223–238. Springer, Heidelberg (2004). doi:10.1007/978-3-540-24676-3_14

6. Boneh, D., Boyen, X.: Secure identity based encryption without random oracles. In: Franklin, M. (ed.) CRYPTO 2004. LNCS, vol. 3152, pp. 443–459. Springer, Heidelberg (2004). doi:10.1007/978-3-540-28628-8_27
7. Boneh, D., Franklin, M.: Identity-based encryption from the weil pairing. In: Kilian, J. (ed.) CRYPTO 2001. LNCS, vol. 2139, pp. 213–229. Springer, Heidelberg (2001). doi:10.1007/3-540-44647-8_13
8. Chen, J., Gay, R., Wee, H.: Improved dual system ABE in prime-order groups via predicate encodings. In: Oswald, E., Fischlin, M. (eds.) EUROCRYPT 2015. LNCS, vol. 9057, pp. 595–624. Springer, Heidelberg (2015). doi:10.1007/978-3-662-46803-6_20
9. Chen, J., Wee, H.: Fully, (almost) tightly secure IBE and dual system groups. In: Canetti, R., Garay, J.A. (eds.) CRYPTO 2013. LNCS, vol. 8043, pp. 435–460. Springer, Heidelberg (2013). doi:10.1007/978-3-642-40084-1_25
10. Chen, J., Wee, H.: Dual system groups and its applications - compact HIBE and more. IACR Cryptology ePrint Archive 2014, p. 265 (2014)
11. Escala, A., Herold, G., Kiltz, E., Ràfols, C., Villar, J.: An algebraic framework for Diffie-Hellman assumptions. In: Canetti, R., Garay, J.A. (eds.) CRYPTO 2013. LNCS, vol. 8043, pp. 129–147. Springer, Heidelberg (2013). doi:10.1007/978-3-642-40084-1_8
12. Gay, R., Hofheinz, D., Kiltz, E., Wee, H.: Tightly CCA-secure encryption without pairings. In: Fischlin, M., Coron, J.-S. (eds.) EUROCRYPT 2016. LNCS, vol. 9665, pp. 1–27. Springer, Heidelberg (2016). doi:10.1007/978-3-662-49890-3_1
13. Gentry, C.: Practical identity-based encryption without random oracles. In: Vaudenay, S. (ed.) EUROCRYPT 2006. LNCS, vol. 4004, pp. 445–464. Springer, Heidelberg (2006). doi:10.1007/11761679_27
14. Gong, J., Chen, J., Dong, X., Cao, Z., Tang, S.: Extended nested dual system groups, revisited. In: Cheng, C.-M., Chung, K.-M., Persiano, G., Yang, B.-Y. (eds.) PKC 2016. LNCS, vol. 9614, pp. 133–163. Springer, Heidelberg (2016). doi:10.1007/978-3-662-49384-7_6
15. Groth, J., Sahai, A.: Efficient noninteractive proof systems for bilinear groups. SIAM J. Comput. **41**(5), 1193–1232 (2012)
16. Hofheinz, D.: Adaptive partitioning. IACR Cryptology ePrint Archive 2016, p. 373 (2016)
17. Hofheinz, D.: Algebraic partitioning: fully compact and (almost) tightly secure cryptography. In: Kushilevitz, E., Malkin, T. (eds.) TCC 2016. LNCS, vol. 9562, pp. 251–281. Springer, Heidelberg (2016). doi:10.1007/978-3-662-49096-9_11
18. Hofheinz, D., Koch, J., Striecks, C.: Identity-based encryption with (almost) tight security in the multi-instance, multi-ciphertext setting. In: Katz, J. (ed.) PKC 2015. LNCS, vol. 9020, pp. 799–822. Springer, Heidelberg (2015). doi:10.1007/978-3-662-46447-2_36
19. Jutla, C.S., Roy, A.: Shorter quasi-adaptive NIZK proofs for linear subspaces. In: Sako, K., Sarkar, P. (eds.) ASIACRYPT 2013. LNCS, vol. 8269, pp. 1–20. Springer, Heidelberg (2013). doi:10.1007/978-3-642-42033-7_1
20. Kiltz, E., Wee, H.: Quasi-adaptive NIZK for linear subspaces revisited. In: Oswald, E., Fischlin, M. (eds.) EUROCRYPT 2015. LNCS, vol. 9057, pp. 101–128. Springer, Heidelberg (2015). doi:10.1007/978-3-662-46803-6_4
21. Lewko, A., Okamoto, T., Sahai, A., Takashima, K., Waters, B.: Fully secure functional encryption: attribute-based encryption and (hierarchical) inner product encryption. In: Gilbert, H. (ed.) EUROCRYPT 2010. LNCS, vol. 6110, pp. 62–91. Springer, Heidelberg (2010). doi:10.1007/978-3-642-13190-5_4

22. Lewko, A., Waters, B.: New proof methods for attribute-based encryption: achieving full security through selective techniques. In: Safavi-Naini, R., Canetti, R. (eds.) CRYPTO 2012. LNCS, vol. 7417, pp. 180–198. Springer, Heidelberg (2012). doi:10.1007/978-3-642-32009-5_12
23. Libert, B., Joye, M., Yung, M., Peters, T.: Concise multi-challenge CCA-secure encryption and signatures with almost tight security. In: Sarkar, P., Iwata, T. (eds.) ASIACRYPT 2014. LNCS, vol. 8874, pp. 1–21. Springer, Heidelberg (2014). doi:10.1007/978-3-662-45608-8_1
24. Libert, B., Peters, T., Joye, M., Yung, M.: Compactly hiding linear spans. In: Iwata, T., Cheon, J.H. (eds.) ASIACRYPT 2015. LNCS, vol. 9452, pp. 681–707. Springer, Heidelberg (2015). doi:10.1007/978-3-662-48797-6_28
25. Naor, M., Reingold, O.: Number-theoretic constructions of efficient pseudo-random functions. J. ACM **51**(2), 231–262 (2004)
26. Okamoto, T., Takashima, K.: Fully secure functional encryption with general relations from the decisional linear assumption. In: Rabin, T. (ed.) CRYPTO 2010. LNCS, vol. 6223, pp. 191–208. Springer, Heidelberg (2010). doi:10.1007/978-3-642-14623-7_11
27. Okamoto, T., Takashima, K.: Adaptively attribute-hiding (hierarchical) inner product encryption. In: Pointcheval, D., Johansson, T. (eds.) EUROCRYPT 2012. LNCS, vol. 7237, pp. 591–608. Springer, Heidelberg (2012). doi:10.1007/978-3-642-29011-4_35
28. Okamoto, T., Takashima, K.: Fully secure unbounded inner-product and attribute-based encryption. In: Wang, X., Sako, K. (eds.) ASIACRYPT 2012. LNCS, vol. 7658, pp. 349–366. Springer, Heidelberg (2012). doi:10.1007/978-3-642-34961-4_22
29. Ramanna, S.C.: More efficient constructions for inner-product encryption. In: Manulis, M., Sadeghi, A.-R., Schneider, S. (eds.) ACNS 2016. LNCS, vol. 9696, pp. 231–248. Springer, Heidelberg (2016). doi:10.1007/978-3-319-39555-5_13
30. Ramanna, S.C., Sarkar, P.: Efficient adaptively secure IBBE from standard assumptions. IACR Cryptology ePrint Archive 2014, p. 380 (2014)
31. Ramanna, S.C., Sarkar, P.: Efficient (anonymous) compact HIBE from standard assumptions. In: Chow, S.S.M., Liu, J.K., Hui, L.C.K., Yiu, S.M. (eds.) ProvSec 2014. LNCS, vol. 8782, pp. 243–258. Springer, Heidelberg (2014). doi:10.1007/978-3-319-12475-9_17
32. Shamir, A.: Identity-based cryptosystems and signature schemes. In: Blakley, G.R., Chaum, D. (eds.) CRYPTO 1984. LNCS, vol. 196, pp. 47–53. Springer, Heidelberg (1985). doi:10.1007/3-540-39568-7_5
33. Waters, B.: Efficient identity-based encryption without random oracles. In: Cramer, R. (ed.) EUROCRYPT 2005. LNCS, vol. 3494, pp. 114–127. Springer, Heidelberg (2005). doi:10.1007/11426639_7
34. Waters, B.: Dual system encryption: realizing fully secure IBE and HIBE under simple assumptions. In: Halevi, S. (ed.) CRYPTO 2009. LNCS, vol. 5677, pp. 619–636. Springer, Heidelberg (2009). doi:10.1007/978-3-642-03356-8_36
35. Wee, H.: Dual system encryption via predicate encodings. In: Lindell, Y. (ed.) TCC 2014. LNCS, vol. 8349, pp. 616–637. Springer, Heidelberg (2014). doi:10.1007/978-3-642-54242-8_26

Déjà Q All Over Again: Tighter and Broader Reductions of q-Type Assumptions

Melissa Chase[1(✉)], Mary Maller[2], and Sarah Meiklejohn[2]

[1] Microsoft Research Redmond, Redmond, USA
melissac@microsoft.com
[2] University College London, London, UK
{mary.maller.15,s.meiklejohn}@ucl.ac.uk

Abstract. In this paper, we demonstrate that various cryptographic constructions—including ones for broadcast, attribute-based, and hierarchical identity-based encryption—can rely for security on only the static subgroup hiding assumption when instantiated in composite-order bilinear groups, as opposed to the dynamic q-type assumptions on which their security previously was based. This specific goal is accomplished by more generally extending the recent Déjà Q framework (Chase and Meiklejohn, Eurocrypt 2014) in two main directions. First, by teasing out common properties of existing reductions, we expand the q-type assumptions that can be covered by the framework; i.e., we demonstrate broader classes of assumptions that can be reduced to subgroup hiding. Second, while the original framework applied only to asymmetric composite-order bilinear groups, we provide a reduction to subgroup hiding that works in symmetric (as well as asymmetric) composite-order groups. As a bonus, our new reduction achieves a tightness of $\log(q)$ rather than q.

1 Introduction

In cryptography, the provable security paradigm crucially relies on the existence of hard mathematical problems. To prove the security of a candidate cryptographic construction, one must demonstrate that any adversary that can break its security can be used to construct another adversary that can break the underlying mathematical problem; if the problem is assumed to be hard, then it logically follows that the construction is secure.

To be confident in the security of a construction, we must therefore also be confident in the underlying assumption; i.e., the assumption that the given mathematical problem is hard. Cryptographic assumptions come in many forms, and confidence in them can be gained through various means: one can perform cryptanalysis on the problem and attempt to break it, prove its security in the generic group model [40], or generalize multiple assumptions using a construct like the *uber-assumption* [11,15] to provide general lower bounds on security.

As a field, cryptography has in the past decade become increasingly tolerant of assumptions that are new, not particularly well understood, and in some cases even "hard to untangle from the constructions which utilize them" [26]. While

J.H. Cheon and T. Takagi (Eds.): ASIACRYPT 2016, Part II, LNCS 10032, pp. 655–681, 2016.
DOI: 10.1007/978-3-662-53890-6_22

there are of course good reasons for doing so (e.g., driving the state of the art forward), and it is demonstrably impossible to reduce every construction to a simple assumption like DDH, the growth in the volume and complexity of new assumptions nevertheless provides an opportunity to revisit this landscape of assumptions and attempt to simplify and systematize it where possible.

Our specific focus in this paper is the class of *q-type assumptions*, in which the assumption is not static, but rather can grow dynamically; e.g., the decisional q-wBDHI (weak Bilinear Diffie-Hellman Inversion) assumption [11] says that given $(g, g^c, g^b, g^{b^2}, \dots, g^{b^q})$, it should be hard to distinguish $e(g,g)^{b^{q+1}c}$ from random. These assumptions are closely tied to the schemes that rely on them for security, as the value q is often equal to the number of oracle calls that can be made in a reduction; e.g., in identity-based encryption (IBE), a distinct value from the assumption is used within the reduction to respond to each of q key extraction queries. Moreover, q-type assumptions become stronger as q grows, and the time to recover the discrete logarithm scales inversely with q [22].

In a recent paper [18], Chase and Meiklejohn demonstrated the potential to move away from q-type assumptions by demonstrating that certain types of q-type assumptions (under the umbrella of the uber-assumption) were implied by the static subgroup hiding assumption [13] in asymmetric composite-order groups. Specifically, they demonstrated a reduction—with looseness q—to the subgroup hiding assumption from all q-type assumptions that either (1) gave out functions on only one side of the pairing and asked the adversary to distinguish elements in the source group or (2) gave out functions on both sides of the pairing and asked the adversary to compute an element in the source group. Following Wee [44], we dub their set of techniques and results the "Déjà Q framework."

1.1 Our Contributions

In this paper, we seek to expand the applicability of the Déjà Q framework to encompass wider classes of assumptions and to apply to settings that are used more commonly in cryptographic constructions. In particular, we provide the following three main contributions:

Broader classes of assumptions. In terms of specific schemes and assumptions, the original Déjà Q framework implied that the Dodis-Yampolskiy PRF [23] and the q-SDH assumption [10] could be reduced to subgroup hiding. To broaden not only the class of assumptions but also the concrete applicability of the framework, we capture computational and decisional uber-assumptions in the target group, including commonly used q-type assumptions such as q-BDHE [11] and q-wBDHI. We also demonstrate techniques for translating concrete schemes—in particular, the BGW broadcast encryption scheme [12], the BBG hierarchical identity-based encryption scheme [11], the Waters attribute-based encryption scheme [41], and the ACF identity-based key encapsulation mechanism [1]—that rely on the symmetric versions of these assumptions for security into asymmetric composite-order bilinear groups, where they can then be reduced to subgroup hiding.

Tighter reductions. We provide a new reduction from both computational and decisional uber-assumptions in the target group to subgroup hiding. Our new reduction requires adding at least one additional prime to the factorization of N, but it achieves logarithmic—rather than linear—tightness. These results can then be applied to any scheme based on these assumptions, including the ones mentioned above, which directly gives a tightly (or almost tightly, depending on ones preferred terminology) secure instantiation, albeit in a somewhat inefficient setting.

Symmetric and asymmetric groups. The original Déjà Q framework could operate only in asymmetric composite-order bilinear groups (or composite-order groups where no pairing existed), of which only one construction is known [14,37]. Our new proof works in both symmetric and asymmetric settings, thus allowing us to consider the more "usual" instantiations of composite-order bilinear groups.

1.2 Our Techniques

In terms of the techniques we use, our proof in Sect. 3 that computational and decisional uber-assumptions in the target group can be reduced to subgroup hiding is closely based on the proof in the original Déjà Q framework for computational uber-assumptions in the source group. To achieve this, we observe that reductions frequently treat group generators in separate ways; i.e., separate sets of generators are used to answer separate types of queries, and the reduction crucially relies on this separation to ensure that the adversary can't test the relationships between different objects as they (separately) incorporate additional randomness or otherwise shift in value. By explicitly acknowledging this usage in our statement of the uber-assumption, we can treat the separate generators in different ways in our reductions and thus extend the results to the target group. To further demonstrate how to securely move symmetric constructions into the asymmetric setting, where they can then be covered by these results, we rely on a recent set of techniques due to Abe et al. [3] for doing automated symmetric-to-asymmetric translations.

Next, in Sect. 4, we consider a modified version of this proof strategy, where in each game hop we double the amount of randomness included in the assumption. To do this, we require three subgroups instead of two, meaning we can write $G = G_1 \times G_2 \times G_3$. As in the original Déjà Q framework, we start by shifting the variables used in the q-type assumption from G_1 into G_2 and G_3, which following the usual dual-system technique we can argue goes unnoticed by *subgroup hiding* [13]. We then change the variables in G_2 and G_3 to take on entirely new values, which again following the dual-system technique we can argue goes unnoticed by *parameter hiding* [29]. Now, however, instead of continuing to shift the same variables from G_1 into G_2 and change them one by one, we shift the new variables from G_3 into G_2, so that G_2 has effectively doubled the number of new variables it contains. By repeating this process of shifting all the variables from G_2 into G_3, changing them, and shifting them back, we achieve the same outcome as the original framework of having ℓ sets of variables in G_2, but using $\log_2(\ell)$ game transitions instead of ℓ.

While one additional subgroup suffices to achieve this tighter reduction in asymmetric bilinear groups, our reduction relies on the use of subgroup generators that would break subgroup hiding in symmetric groups. To address this, our new reduction brings in certain aspects of the more traditional application of the dual-system technique to constructions (rather than assumptions) [9,20,29,31,32], and in particular a recent result due to Wee [44] that used an adaption of the Déjà Q framework to reduce both an IBE scheme and a broadcast encryption scheme to subgroup hiding. We thus demonstrate that by folding in random values from a fourth subgroup, we can sufficiently "mask" the subgroups to push through the same reduction in symmetric groups. Thus, while our results in Sect. 3 apply to versions of concrete constructions translated into the asymmetric setting (but otherwise unmodified), our results in Sect. 4 provide tighter reductions for the (original) symmetric versions in which additional randomness is incorporated when instantiated in groups with two additional subgroups, or for asymmetric versions with an additional subgroup (but no additional randomness).

1.3 Related Work

Our work closely builds on the Déjà Q framework due to Chase and Meiklejohn [18]. In order to go beyond the original set of contributions, we draw on certain aspects of the dual-system technique [31,32,42], the notion of parameter hiding [29,30], and the general notion of subgroup hiding [8]. For our results in the symmetric setting, we draw on ideas in a recent work by Wee [44], who extended the original Déjà Q framework but focused specifically on constructions for broadcast encryption and IBE.

The search for tight reductions goes back to the paper of Bellare and Rogaway [7], and the results are extensive. To compare with the results most similar to ours, we focus on results for pairing-based primitives, where much related work has provided (almost) tight reductions for various primitives, including identity-based encryption [1,9,21,28,35], inner product encryption [38], authenticated key exchange [6], and quasi-adaptive non-interactive zero-knowledge proofs [24,36]. Each of these results focuses on a specific construction, and employs a specific set of techniques to achieve tight security. (One exception is a paper by Attrapadung, Hanoaka, and Yamada [5] that gives an abstraction from which several different IBE variants can be constructed. This work, however, is still focused on IBE and on a particular construction approach.) By presenting our results at the level of assumptions, we can instead prove tight security for an entire class of constructions; i.e., constructions that are instantiated in appropriate groups and have been previously proved secure under an appropriate class of q-type assumptions. To the best of our knowledge, we are thus the first to use the dual-system technique to provide a tightly secure reduction in a more general setting. Finally, we note that while much of the previous work has focused on reductions whose running time is linear in the security parameter, our reduction is linear in $\log(q)$, which in practice may be a much smaller number.

2 Definitions and Notation

2.1 Preliminaries

If x is a binary string then $|x|$ denotes its bit length. If S is a finite set then $|S|$ denotes its size and $x \xleftarrow{\$} S$ denotes sampling a member uniformly from S and assigning it to x. $\lambda \in \mathbb{N}$ denotes the security parameter and 1^λ denotes its unary representation. $[n]$ denotes the set $\{1, \ldots, n\}$.

Algorithms are randomized unless explicitly noted otherwise. "PT" stands for "polynomial-time." By $y \leftarrow A(x_1, \ldots, x_n; R)$ we denote running algorithm A on inputs $x_1, \ldots, x_n$ and random coins R and assigning its output to y. By $y \xleftarrow{\$} A(x_1, \ldots, x_n)$ we denote $y \leftarrow A(x_1, \ldots, x_n; R)$ for coins R sampled uniformly at random. By $[A(x_1, \ldots, x_n)]$ we denote the set of values that have positive probability of being output by A on inputs $x_1, \ldots, x_n$. Adversaries are algorithms.

We use games in definitions of security and in proofs. A game G has a MAIN procedure whose output is the output of the game. $\Pr[\mathsf{G}]$ denotes the probability that this output is true.

2.2 Basic Bilinear Groups

A *bilinear group* is a tuple $\mathbb{G} = (N, G, H, G_T, e)$, where N is either prime or composite, $|G| = |H| = kN$ and $|G_T| = \ell N$ for $k, \ell \in \mathbb{N}$, all elements of G, H, and G_T are of order at most N, and $e : G \times H \to G_T$ is a *bilinear map*: it is efficiently computable, satisfies $e(A^x, B^y) = e(A, B)^{xy}$ for all $A \in G$, $B \in H$, and $x, y \in \mathbb{Z}/N\mathbb{Z}$ (bilinearity), and if $e(A, B) = 1$ for all $B \in H$ then $A = 1$ and vice versa if this holds for all $A \in G$ (non-degeneracy). We use $\mathsf{BilinearGen}$ to denote the algorithm by which bilinear groups are generated.

When G and H are cyclic, the description of the group may include their respective generators g and h. If the groups can be decomposed as $G = G_1 \times G_2$ and $H = H_1 \times H_2$, the description of the group may include information about these subgroups and their generators; additionally, the number of cyclic subgroups may be provided as an argument n to $\mathsf{BilinearGen}$.

2.3 Subgroup Hiding and Parameter Hiding

We highlight two structural properties of bilinear groups—*subgroup hiding* and *parameter hiding*—that are essential to the Déjà Q framework, using adapted versions of the definitions given by Chase and Meiklejohn [18].

Assumption 2.1 (Subgroup hiding). *For $n \in \mathbb{N}$ and a bilinear group generation algorithm* $\mathsf{BilinearGen}(\cdot,\cdot)$*, define* $\mathbf{Adv}^{sgh}_{\mathcal{A}}(\lambda) = 2Pr[SGH^{\mathcal{A}}_{\mu}(\lambda)] - 1$*, where* $SGH^{\mathcal{A}}_{\mu}(\lambda)$ *is defined as follows:*

$$\begin{array}{l} \underline{\text{MAIN } SGH^{\mathcal{A}}_{\mu}(\lambda)} \\ b \xleftarrow{\$} \{0,1\};\ (N, G, H, G_T, e, \mu) \xleftarrow{\$} \mathsf{BilinearGen}(1^{\lambda}, n) \\ \textit{if } (b = 0) \textit{ then } w \xleftarrow{\$} G \\ \textit{if } (b = 1) \textit{ then } w \xleftarrow{\$} G_1 \\ b' \xleftarrow{\$} \mathcal{A}(N, G, H, G_T, e, \mu, w) \\ \textit{return } (b' = b) \end{array}$$

Then subgroup hiding *holds in G_1 with auxiliary information μ if for all PT adversaries $\mathcal{A}$ there exists a negligible function $\nu(\cdot)$ such that* $\mathbf{Adv}^{sgh}_{\mathcal{A}}(\lambda) < \nu(\lambda)$.

Subgroup hiding is defined analogously for G_2, $G_{1,T}$, and $G_{2,T}$ (where $G_{1,T}$ and $G_{2,T}$ are cyclic subgroups of G_T), and the auxiliary information μ is designed to capture additional subgroup generators that may also be given out (with the observation that revealing certain subgroup generators might allow one to trivially distinguish subgroups when using a canceling pairing, so one must be careful with what μ contains). If we switch between different subgroups rather than one subgroup and the full group—e.g., between G_2 and G_{23}, as we do in Sect. 4—then we say subgroup hiding holds *between* the subgroups.

To elaborate on the point about μ, subgroup hiding can be trivially broken if the adversary has knowledge of certain generators; e.g., if an adversary is given a value w and asked to determine if it is in G or G_1, knowledge of the generator h_2 allows it to check if $e(w, h_2) = 1$ and trivially break subgroup hiding. To avoid this, the many variants of subgroup hiding used in the literature often specify which subgroup elements the adversary can see [16,25,27,33,34,39], and the rules about which generators can be given out have been codified in the *general subgroup decision* assumption due to Bellare, Waters, and Yilek [8]. The variants of subgroup hiding that we use in Sects. 3 and 4 are specific instantiations of this general assumption.

Definition 2.1 (Extended parameter hiding). *For $m, n \in \mathbb{N}$ and a bilinear group $(N, G, H, G_T, e, \mu) \in [\mathsf{BilinearGen}(1^{\lambda}, n)]$, we say* extended parameter hiding *holds with respect to a family of functions $\mathcal{F}$, auxiliary information* aux*, and a pair of subgroups (G_{i_1}, G_{i_2}) if for all $g_{i_1} \in G_{i_1}$ and $g_{i_2} \in G_{i_2}$, the distribution* $\{g_{i_1}^{f(\vec{x})} g_{i_2}^{f(\vec{x})}, a(\vec{x})\}_{f \in \mathcal{F}, a \in \mathsf{aux}}$ *is identical to* $\{g_{i_1}^{f(\vec{x})} g_{i_2}^{f(\vec{x}')}, a(\vec{x})\}_{f \in \mathcal{F}, a \in \mathsf{aux}}$ *for* $\vec{x}, \vec{x}' \xleftarrow{\$} (\mathbb{Z}/N\mathbb{Z})^m$.

Chase and Meiklejohn proved [18, Lemma 5.2] that their original definition of extended parameter hiding (which used $n = 2$) holds in composite-order bilinear groups with respect to all polynomial functions and the version of aux that we require in Sect. 3. In Sect. 4, however, we consider a group with $n > 2$ subgroups and we want parameter hiding to hold across subgroups beyond G_1 and G_2. We thus prove that parameter hiding still holds in this setting as long as the orders of G_{i_1} and G_{i_2} have no primes in common and the auxiliary information is not in G_{i_2}.

Lemma 2.1. *For all $m, n \in \mathbb{N}$, and for all bilinear groups $(N, G, H, G_T, e) \in [\mathsf{BilinearGen}(1^\lambda, n)]$ where $N = p_1 \cdot \ldots \cdot p_n$, (i_1, i_2) such that $1 \le i_1, i_2 \le n$, and for the class $\mathcal{F}$ of all polynomials $f(\cdot)$ over $\mathbb{Z}/N\mathbb{Z}$, if $\gcd(p_{i_1}, p_{i_2}) = 1$ and if for all $a \in \mathsf{aux}$, $a(\cdot) \in A$ such that $\gcd(|A|, p_{i_2}) = 1$, then the distribution over $\{g_{i_1}^{f(\vec{x})} g_{i_2}^{f(\vec{x})}, a(\vec{x})\}_{f \in \mathcal{F}, a \in \mathsf{aux}}$ is identical to the distribution over $\{g_{i_1}^{f(\vec{x})} g_{i_2}^{f(\vec{x}')}, a(\vec{x})\}_{f \in \mathcal{F}, a \in \mathsf{aux}}$ for $\vec{x}, \vec{x}_1' \xleftarrow{\$} (\mathbb{Z}/N\mathbb{Z})^m$.*

Proof. For any polynomial $f(\cdot)$, one can compute $g_{i_1}^{f(\vec{x})}$ knowing just the value of $x_j \bmod p_{i_1}$ for all j, $1 \le j \le m$, and can similarly compute $g_{i_2}^{f(\vec{x})}$ knowing just the value of $x_j \bmod p_{i_2}$ for all j, $1 \le j \le m$. If $\gcd(p_{i_1}, p_{i_2}) = 1$ and the functions in aux reveal no information about $x_j \bmod p_{i_2}$, then by the Chinese Remainder theorem the values of $x_j \bmod p_{i_2}$ are independent of all the other values, so this is identical to using an independent x_j' for the g_{i_2} values. □

3 Uber-Assumptions in the Target Group

In this section, we consider how to capture new classes of assumptions within the Déjà Q framework [18]. In particular, we first prove in Sect. 3.1 that decisional and computational uber-assumptions in the target group are implied—through the repeated application of subgroup hiding and parameter hiding—by assumptions with significant amounts of randomness folded into particular subgroups. (The framework previously covered only computational assumptions in the source group, which are implied by computational assumptions in the target group, or "one-sided" decisional assumptions in the source group; i.e., assumptions where meaningful functions could be given out on only one side of the pairing.)

Next, in Sect. 3.2, we show that the computational variant of the transitioned uber-assumption is so weak that it holds by a statistical argument; thus, the computational uber-assumption can be implied solely by subgroup hiding. By relying on an additional mild subgroup hiding assumption in the target, we can show the same results for decisional variants as well; i.e., we can show that the decisional uber-assumption is implied by three variants of subgroup hiding.

Finally, in Sect. 3.3, we observe that many examples of uber-assumptions (including widely used q-type assumptions) have been used only in symmetric bilinear groups to date, making it difficult to cover them directly with our analysis. (In Sect. 4, we do provide ways to cover the symmetric setting, but this requires an extra prime in the order of the group.) We thus demonstrate how to convert popular symmetric assumptions into asymmetric variants using techniques due to Abe et al. [3]. All of our converted symmetric schemes—e.g., the BGW broadcast encryption scheme [12] and the Waters attribute-based encryption scheme [41]—rely for security on q-type decisional uber-assumptions of the appropriate form, so our results demonstrate the security of these schemes when instantiated in groups where subgroup hiding holds.

3.1 Reducing Asymmetric Assumptions to Weaker Variants

In the uber-assumption [18, Assumption 4.1], the adversary is given three sets of values with respect to a set of c variables $\vec{x}$: a generator $g \in G$ raised to a set of functions $R(\vec{x})$, a generator $h \in H$ raised to a set of functions $S(\vec{x})$, and the value $e(g,h)$ raised to a set of functions $T(\vec{x})$ (where $g^{R(\vec{x})}$ is used as shorthand for $\{g^{\rho_i(\vec{x})}\}_{i=1}^r$ for $R = \langle \rho_1(\vec{x}), \ldots, \rho_r(\vec{x})\rangle$, and similarly for S and T). The adversary is then asked to either compute $e(g,h)^{f(\vec{x})}$ (in the computational assumption in the target group) or distinguish it from random.

This definition captures a broad range of q-type assumptions, but in some cases it may be instructive to explicitly identify the qualities of the assumption that are used in the reduction. In particular, constructions that use the dual-system technique must add noise into group elements in such a way that valuable information is hidden but one can nevertheless continue to correctly perform operations (e.g., decryption) without noticing the added noise. This is often accomplished by using two separate generators that are primarily used for separate operations—e.g., in the case of identity-based encryption, one generator is used to create the parameters and the other to form the challenge ciphertext—and this separation is acknowledged in the assumption. For example, the (symmetric) q-BDHE assumption [11] says that given $(g, g^s, \{g^{a^i}\}_{i\in[2q], i\neq q+1})$, it should be hard to distinguish $e(g,g)^{a^{q+1}s}$ from random.

We thus modify slightly the original definition of the uber-assumption to (1) make explicit the role of two generators h and $\hat{h}$, the former of which we move into a subgroup to provide the necessary correctness and the latter of which we keep in the full group to provide the necessary hiding guarantee, and (2) combine computational and decisional assumptions into the same definition so we can cover them both in our main theorem.

Assumption 3.1 (Uber-assumption). *Define the advantage of an adversary* $\mathcal{A}$ *by* $\mathbf{Adv}_{\mathcal{A}}^{comp\text{-}uber}(\lambda) = Pr[comp\text{-}UBER_{c,R,S,T,f}^{\mathcal{A}}(\lambda)]$ *in the computational case and* $\mathbf{Adv}_{\mathcal{A}}^{dec\text{-}uber}(\lambda) = 2Pr[dec\text{-}UBER_{c,R,S,T,f}^{\mathcal{A}}(\lambda) - 1$ *in the decisional case, where* $\mathsf{type}\text{-}UBER_{c,R,S,T,f}^{\mathcal{A}}(\lambda)$ *is defined as follows for* $\mathsf{type} \in \{comp, dec\}$*:*

MAIN $\mathsf{type}\text{-}UBER_{c,R,S,T,f}^{\mathcal{A}}(\lambda)$

$(N, G, H, G_T, e) \xleftarrow{\$} \mathsf{BilinearGen}(1^\lambda, 2)$; $g \xleftarrow{\$} G$, $h, \hat{h} \xleftarrow{\$} H$
$x_1, \ldots, x_c \xleftarrow{\$} \mathbb{Z}/N\mathbb{Z}$
$\mathsf{inputs} \leftarrow (N, G, H, G_T, e, g, \hat{h}, g^{R(\vec{x})}, h^{S(\vec{x})}, e(g,h)^{T(\vec{x})})$
$\mathsf{chal} \leftarrow e(g, \hat{h})^{f(\vec{x})}$
return type-PLAY$(\lambda, \mathsf{inputs}, \mathsf{chal})$

comp-PLAY$(\lambda, \mathsf{inputs}, \mathsf{chal})$

$y \xleftarrow{\$} \mathcal{A}(1^\lambda, \mathsf{inputs})$
return $(y = \mathsf{chal})$

$dec\text{-PLAY}(\lambda, \mathsf{inputs}, \mathsf{chal})$

$b \xleftarrow{\$} \{0,1\}$
$\mathit{if}\ (b=0)\ \mathit{then}\ y \xleftarrow{\$} G_T$
$\mathit{if}\ (b=1)\ \mathit{then}\ y \leftarrow \mathsf{chal}$
$b' \xleftarrow{\$} \mathcal{A}(1^\lambda, \mathsf{inputs}, y)$
$\mathit{return}\ (b' = b)$

Then the uber-assumption in the target group holds if for all PT algorithms $\mathcal{A}$ there exists a negligible function $\nu(\cdot)$ such that $\mathbf{Adv}^{uber}_{\mathcal{A}}(\lambda) < \nu(\lambda)$.

We now proceed to prove a theorem analogous to the one in the original Déjà Q framework [19, Theorem 4.8], but which treats these different bases in H in different ways. For ease of exposition, we make explicit the original assumption used in this proof, which (with our additional generator $\hat{h}$ added) is as follows:

Assumption 3.2. *For a bilinear group $\mathbb{G} = (N, G, H, G_T, e) \in [\mathsf{BilinearGen}(1^\lambda, 2)]$, $\ell \in \mathbb{N}$, and classes of functions R, S, T, and f (as defined in the uber-assumption in Assumption 3.1), given*

$$\mathsf{inputs} = (\mathbb{G}, g_1 g_2^{\sum_{i=1}^{\ell} r_i}, \hat{h}, \{g_1^{\rho_k(\vec{x})} g_2^{\sum_{i=1}^{\ell} r_i \rho_k(\vec{x}_i)}\}_{k=1}^{r}, h_1^{S(\vec{x})}, e(g_1, h_1)^{T(\vec{x})})$$

for $g_1 \xleftarrow{\$} G_1$, $g_2 \xleftarrow{\$} G_2 \backslash \{1\}$, $\hat{h} \xleftarrow{\$} H$, $h_1 \xleftarrow{\$} H_1 \backslash \{1\}$, and $r_1, \ldots, r_\ell, \xleftarrow{\$} \mathbb{Z}/N\mathbb{Z}$, $\vec{x}, \vec{x}_1, \ldots, \vec{x}_\ell \xleftarrow{\$} (\mathbb{Z}/N\mathbb{Z})^c$, no PT adversary has more than negligible advantage when playing $\mathsf{type}\text{-PLAY}(\lambda, \mathsf{inputs}, e(g_1, \hat{h})^{f(\vec{x})} e(g_2, \hat{h})^{\sum_{i=1}^{\ell} r_i f(\vec{x}_i)})$.

Theorem 3.3. *For a bilinear group $\mathbb{G} = (N, G, H, G_T, e) \in [\mathsf{BilinearGen}(1^\lambda, 2)]$, consider the uber-assumption in the target group parameterized by (c, R, S, T, f). Then this is implied by Assumption 3.2 if*

1. *subgroup hiding holds in G_1 with $\mu = \{g_2, h_1\}$;*
2. *subgroup hiding holds in H_1 with $\mu = \{g_1\}$; and*
3. *extended parameter hiding holds with respect to $\mathcal{F} = R \cup \{f\}$ and* $\mathsf{aux} = \{h_1^{\sigma(\cdot)}\}_{\sigma \in S \cup T}$ *for all $h_1 \in H_1$.*

In particular, for $\ell \in \mathbb{N}$ we have that

$$\mathbf{Adv}^{uber}_{\mathcal{A}}(\lambda) \leq \mathbf{Adv}^{sgh}_{\mathcal{B}_0}(\lambda) + \mathbf{Adv}^{sgh}_{\mathcal{C}_0}(\lambda) + \ell \mathbf{Adv}^{sgh}_{\mathcal{B}_i}(\lambda) + \mathbf{Adv}^{3.2}_{\mathcal{A}}(\lambda).$$

A proof of this theorem can be found in the full version of the paper [17]. Intuitively, the outline is similar to that of the original proof: to start, all elements in G are first shifted into G_1, and elements using h as the base are shifted into H_1. Elements using $\hat{h}$ remain in the full group H (this is our main point of divergence from the original Déjà Q proof). We argue that both of these changes go unnoticed by subgroup hiding. Then, the elements in G_1 are added into G_2, which we again argue goes unnoticed by subgroup hiding. The elements in G_2 are then switched to use a new set of variables $\vec{x}_1$, which we argue is identical by parameter hiding. Now, we repeat this process of adding the original elements from G_1 into G_2 and switching them to a new set of variables, until—after ℓ transitions—we end up with ℓ sets of variables in G_2.

3.2 Reducing Asymmetric Assumptions to Subgroup Hiding

We now deal separately with the case of computational and decisional assumptions, as decisional assumptions require an extra assumption on the indistinguishability of random elements in $G_{2,T}$ and random elements in G_T (we use $G_{i,T}$ to denote the ith subgroup of G_T). For both, however, we first recall two relevant components from the Déjà Q framework: the matrix V defined as

$$V = \begin{bmatrix} 1 & \rho_1(\vec{x}_1) & \rho_2(\vec{x}_1) & \cdots & \rho_q(\vec{x}_1) & f(\vec{x}_1) \\ 1 & \rho_1(\vec{x}_2) & \rho_2(\vec{x}_2) & \cdots & \rho_q(\vec{x}_2) & f(\vec{x}_2) \\ \vdots & \vdots & & \ddots & \vdots & \vdots \\ & & & & \ddots & \\ 1 & \rho_1(\vec{x}_\ell) & \rho_2(\vec{x}_\ell) & \cdots & \rho_q(\vec{x}_\ell) & f(\vec{x}_\ell) \end{bmatrix} \quad (1)$$

and a lemma that relates the linear independence of the polynomials with the invertibility of V as follows:

Lemma 3.1. [18] *For all $\lambda \in \mathbb{N}$, if the functions in $R \cup \{f\}$ are linearly independent and of maximum degree poly(λ), $\ell = q + 2$ for $q =$ poly(λ), and $N = p_1 \cdot \ldots \cdot p_n$ for $n =$ poly(λ) distinct primes $p_1, \ldots, p_n \in \Omega(2^{poly(\lambda)})$, then with all but negligible probability the matrix V is invertible.*

We also make explicit the argument used in the Déjà Q framework concerning the multiplication of this matrix with a random vector.

Lemma 3.2. *If V is invertible, then the distribution over $\vec{r} \cdot V$ for $r_1, \ldots, r_{q+2} \xleftarrow{\$} \mathbb{Z}/N\mathbb{Z}$ is uniformly random.*

Proof. Define $\vec{y} \leftarrow \vec{r} \cdot V$, and consider the set of all vectors of length $q + 2$ over $\mathbb{Z}/N\mathbb{Z}$. Since $\vec{r}$ and $\vec{y}$ are both members of this set, multiplication by V maps the set to itself; as V is furthermore invertible, it is a permutation over this set. Thus, sampling $\vec{r}$ uniformly at random and multiplying by V yields a vector $\vec{y}$ that is also distributed uniformly at random. □

Computational Assumptions. For computational assumptions, we can now argue directly that, by transitioning to Assumption 3.2, we reach an assumption so weak that it holds by a statistical argument. Thus, the computational uber-assumption reduces directly to subgroup hiding.

Proposition 3.1. *For a bilinear group $\mathbb{G}$ of order N, the computational uber-assumption parameterized by (c, R, S, T, f) holds in the target group if*

1. *subgroup hiding holds in G_1 with $\mu = \{g_2, h_1\}$;*
2. *subgroup hiding holds in H_1 with $\mu = \{g_1\}$;*
3. *extended parameter hiding holds with respect to $\mathcal{F} = R \cup f$ and $\mathsf{aux} = \{h_1^{\sigma(\cdot)}\}_{\forall \sigma \in S \cup T}$ for all $h_1 \in H_1$;*
4. *$N = p_1 \cdot \ldots \cdot p_n$ for distinct primes $p_1, \ldots, p_n \in \Omega(2^{poly(\lambda)})$; and*

5. *the polynomials in $R \cup f$ are linearly independent and have maximum degree poly(λ).*

Proof. By requirements (1)–(3), Theorem 3.3 tells us that the original assumption is implied by the computational variant of Assumption 3.2. We make the problem strictly easier if we assume that g_1 and $\vec{x}$ are public, in which case $g_1^{R(\vec{x})}$, $h_1^{S(\vec{x})}$, and $e(g_1,h_1)^{T(\vec{x})}$ provide no additional information, and $\mathcal{A}$ can compute the $G_{1,T}$ component of chal directly.

We thus consider a problem where $\mathcal{A}$ is given $g_2^{\sum r_i}$ and $\{g_2^{\sum_{i=1}^{q+2} r_i \rho_k(\vec{x}_i)}\}_{k=0}^{r}$ and we must argue that it is hard for it to compute $e(g_2,\hat{h})^{\sum_{i=1}^{q+2} r_i f(\vec{x}_i)}$. If we let $\ell = q+2$, requirements (4)–(5) and Lemma 3.1 imply that V is invertible with all but negligible probability, and Lemma 3.2 then tells us that the distribution over $\vec{y} \leftarrow \vec{r} \cdot V$ is uniformly random. As $\mathcal{A}$ is given values in G_2 raised to the first $q+1$ entries of $\vec{y}$ and is asked to compute $e(g_2,\hat{h})$ raised to the last, it is thus given uniformly random values and asked to compute something uniformly random, which it has at most negligible probability in doing. □

Decisional Assumptions. Finally, to enable an argument about the decisional assumption in the target, we introduce the following assumption:

Assumption 3.4. *For $\ell \in \mathbb{N}$ and a bilinear group $\mathbb{G} = (N,G,H,G_T,e) \in [\mathsf{BilinearGen}(1^\lambda,2)]$, consider the inputs given to $\mathcal{A}$ in Assumption 3.2. Given the same set of inputs, it is difficult to distinguish $e(g_1,\hat{h})^{f(\vec{x})} e(g_2,\hat{h})^{\sum_{i=1}^{\ell} r_i f(\vec{x}_i)}$ from $e(g_1,\hat{h})^{f(\vec{x})} \cdot R$ for $R \xleftarrow{\$} G_{2,T}$.*

We now prove the following lemma:

Lemma 3.3. *If subgroup hiding holds in $G_{2,T}$ with $\mu = \{g_1, g_2, h_1\}$, then Assumption 3.2 is implied by Assumption 3.4.*

MAIN $\mathsf{G}^{\mathsf{A}}_{3.2}(\lambda)$ / $\mathsf{G}^{\mathsf{A}}_0(\lambda)$ / $\mathsf{G}^{\mathsf{A}}_1(\lambda)$	
if $(b=0)$ then $\mathsf{chal} \xleftarrow{\$} G_T$	// $\mathsf{G}^{\mathsf{A}}_{3.2}(\lambda)$
if $(b=0)$ then $R \xleftarrow{\$} G_T$; $\mathsf{chal} \leftarrow \boxed{e(g_1,\hat{h})^{f(\vec{x})}} \cdot R$	// $\mathsf{G}^{\mathsf{A}}_0(\lambda)$
if $(b=0)$ then $\boxed{R \xleftarrow{\$} G_{2,T}}$; $\mathsf{chal} \leftarrow e(g_1,\hat{h})^{f(\vec{x})} \cdot R$	// $\mathsf{G}^{\mathsf{A}}_1(\lambda)$

Fig. 1. Games for the proof of Lemma 3.3. Each game introduces the boxed code on its corresponding line.

Proof. Let $\mathcal{A}$ be a PT adversary playing game $\mathsf{G}_{3.2}^{\mathcal{A}}(\lambda)$, and let $\mathbf{Adv}_{\mathcal{A}}^{3.4}(\lambda)$ denote its advantage in the game specified in Assumption 3.4. We build a PT adversary $\mathcal{B}$ such that

$$\mathbf{Adv}_{\mathcal{A}}^{3.2}(\lambda) \leq \mathbf{Adv}_{\mathcal{B}}^{\mathrm{sgh}}(\lambda) + \mathbf{Adv}_{\mathcal{A}}^{3.4}(\lambda)$$

for all $\lambda \in \mathbb{N}$, from which the theorem follows. To do this, we build $\mathcal{B}$ such that

$$\Pr[\mathsf{G}_{3.2}^{\mathcal{A}}(\lambda)] - \Pr[\mathsf{G}_{0}^{\mathcal{A}}(\lambda)] = 0 \quad (2)$$

$$\Pr[\mathsf{G}_{0}^{\mathcal{A}}(\lambda)] - \Pr[\mathsf{G}_{1}^{\mathcal{A}}(\lambda)] \leq \mathbf{Adv}_{\mathcal{B}}^{\mathrm{sgh}}(\lambda) \quad (3)$$

$$\Pr[\mathsf{G}_{1}^{\mathcal{A}}(\lambda)] = \mathbf{Adv}_{\mathcal{A}}^{3.4}(\lambda). \quad (4)$$

We then have that

$$\begin{aligned}\mathbf{Adv}_{\mathcal{A}}^{3.2}(\lambda) &= \Pr[\mathsf{G}_{3.2}^{\mathcal{A}}(\lambda)] \\ &= (\Pr[\mathsf{G}_{3.2}^{\mathcal{A}}(\lambda)] - \Pr[\mathsf{G}_{0}^{\mathcal{A}}(\lambda)]) + (\Pr[\mathsf{G}_{0}^{\mathcal{A}}(\lambda)] - \Pr[\mathsf{G}_{1}^{\mathcal{A}}(\lambda)]) + \Pr[\mathsf{G}_{1}^{\mathcal{A}}(\lambda)] \\ &\leq \mathbf{Adv}_{\mathcal{B}}^{\mathrm{sgh}}(\lambda) + \mathbf{Adv}_{\mathcal{A}}^{3.4}(\lambda).\end{aligned}$$

We follow the game hops presented in Figure 1.

Equation 2 : $\mathsf{G}_{3.2}^{\mathcal{A}}(\lambda)$ to $\mathsf{G}_{0}^{\mathcal{A}}(\lambda)$

This follows trivially, as the values $\mathsf{chal} \cdot A$ and chal are identically distributed for $\mathsf{chal} \xleftarrow{\$} G_T$ and $A \in G_{1,T}$.

Equation 3 : $\mathsf{G}_{0}^{\mathcal{A}}(\lambda)$ to $\mathsf{G}_{1}^{\mathcal{A}}(\lambda)$

$\mathcal{B}$ behaves as follows:

$\underline{\mathcal{B}(1^\lambda, N, G, H, G_T, e, g_1, g_2, h_1, w)}$
$b \xleftarrow{\$} \{0,1\}$
$\vec{x}, \vec{x}_1, \ldots, \vec{x}_\ell \xleftarrow{\$} (\mathbb{Z}/N\mathbb{Z})^c,\ r_1, \ldots, r_\ell \xleftarrow{\$} \mathbb{Z}/N\mathbb{Z}$
$v_k \leftarrow g_1^{\rho_k(\vec{x})} g_2^{\sum_{j=1}^{\ell} r_j \rho_k(\vec{x}_j)}\ \forall k \in [r]$ (Here we define $\rho_0 = 1$.)
$y_k \leftarrow h_1^{\sigma_k(\vec{x})}\ \forall k \in [s]$
$z_k \leftarrow e(g_1, h_1)^{\tau_k(\vec{x})}\ \forall k \in [t]$
$\mathsf{inputs} \leftarrow (N, G, H, G_T, e, \hat{h}, v_0, \ldots, v_r, y_1, \ldots, y_s, z_1, \ldots, z_t)$
if $(b = 0)$ then $\mathsf{chal} \leftarrow e(g_1, \hat{h})^{f(\vec{x})} \cdot w$
if $(b = 1)$ then $\mathsf{chal} \leftarrow e(g_1, \hat{h})^{f(\vec{x})} e(g_2, \hat{h})^{\sum_{j=1}^{\ell} r_j f(\vec{x}_j)}$
$b' \xleftarrow{\$} \mathcal{A}(1^\lambda, \mathsf{inputs}, \mathsf{chal})$
return $(b' = b)$

If $w \xleftarrow{\$} G_T$, then this is identical to $\mathsf{G}_{0}^{\mathcal{A}}(\lambda)$. If $w \xleftarrow{\$} G_{2,T}$, then this is identical to $\mathsf{G}_{1}^{\mathcal{A}}(\lambda)$. □

Proposition 3.2. *For a bilinear group $\mathbb{G}$ of order N, the decisional uber-assumption parameterized by (c, R, S, T, f) holds in the target group if*

1. *subgroup hiding holds in G_1 with $\mu = \{g_2, h_1\}$;*

2. *subgroup hiding holds in* H_1 *with* $\mu = \{g_1\}$;
3. *subgroup hiding holds in* $G_{2,T}$ *with* $\mu = \{g_1, g_2, h_1\}$;
4. *extended parameter hiding holds with respect to* $\mathcal{F} = R \cup f$ *and* $\mathsf{aux} = \{h_1^{\sigma(\cdot)}\}_{\forall \sigma \in S \cup T}$ *for all* $h_1 \in H_1$;
5. $N = p_1 \cdot \ldots \cdot p_n$ *for distinct primes* $p_1, \ldots, p_n \in \Omega(2^{poly(\lambda)})$; *and*
6. *the polynomials in* $R \cup f$ *are linearly independent and have maximum degree* $poly(\lambda)$.

Proof. By requirements (1)–(4), Theorem 3.3 and Lemma 3.3 tell us that the original assumption is implied by Assumption 3.4. We make the problem strictly easier if we assume that g_1 and $\vec{x}$ is public, in which case $g_1^{R(\vec{x})}$, $h_1^{S(\vec{x})}$, and $e(g_1, h_1)^{T(\vec{x})}$ provide no additional information, and $\mathcal{A}$ can compute the $G_{1,T}$ component of chal directly (which is the same in either case).

We thus consider a problem where $\mathcal{A}$ is given $g_2^{\sum r_i}$ and $\{g_2^{\sum_{i=1}^{q+2} r_i \rho_k(\vec{x}_i)}\}_{k=0}^{r}$ and we must argue that it is hard for it to distinguish $e(g_2, \hat{h})^{\sum_{i=1}^{q+2} r_i f(\vec{x}_i)}$ from random. If we let $\ell = q+2$, requirements (5)–(6) and Lemmas 3.1 and 3.2 imply that the distribution over $\vec{y} \leftarrow \vec{r} \cdot V$ is uniformly random with all but negligible probability. As $\mathcal{A}$ is given values in G_2 raised to the first $q+1$ entries of $\vec{y}$ and is asked to distinguish $e(g_2, \hat{h})$ raised to the last from random, it is thus given uniformly random values and asked to distinguish two uniformly random things, which it has at most negligible advantage in doing. □

3.3 Converting Symmetric Uber-Assumptions

As mentioned earlier, most schemes that rely on q-type assumptions do so in the symmetric setting, whereas our analysis above works only in the asymmetric setting. To nevertheless capture these useful examples of q-type assumptions, we use the technique of Abe et al. [3] to convert the assumptions from the symmetric to the asymmetric setting so that they can be covered by our analysis.

To perform this conversion, we must of course do so in a way that respects the underlying reduction; i.e., we must ensure that the asymmetric variant of the scheme can still be proved secure under the asymmetric variant of the assumption. The main technique for doing this revolves around the idea of *dependency graphs* that reflect the usage of all values in the source groups and how they interact with each other and with the pairing. Thus, all of the dependencies in both the scheme and its security reduction are represented in a directed graph Γ, with pairings represented by two nodes (one for each side of the pairing). To find an asymmetric variant that respects these dependencies, one must search for a *valid split* of Γ into Γ_0 and Γ_1; this is defined as a split in which

- No nodes or edges are lost; i.e., merging Γ_0 and Γ_1 recovers Γ,
- For every pair of pairing nodes, if one node is in Γ_0, the other node is exclusively in Γ_1, and
- For every node X in each split graph, the ancestor subgraph of X in Γ is included in the same graph.

For more details on this technique and the process of automating it, we refer to the original paper of Abe et al. or to a paper by Akinyele et al. [4] that proposes a tool, AutoGroup+, that improves on the tool developed by Abe et al. and applies the technique to additional schemes.

To demonstrate the coverage of our analysis, we have identified four influential schemes that rely on symmetric uber-assumptions and demonstrated their conversion to asymmetric variants that fit into the class of uber-assumptions our analysis can cover. These are:

- The general construction of the Boneh-Gentry-Waters broadcast encryption scheme [12], based on the q-BDHE assumption;
- the Boneh-Boyen-Goh hierarchical identity-based encryption scheme with constant-sized ciphertexts [11], based on the q-wBDHI assumption;
- the version of Waters' attribute-based encryption scheme [41] that uses the q-BDHE assumption (as opposed to the more efficient construction that uses the q-parallel BDHE assumption [43], which we cannot cover); and
- the Abdalla-Catalano-Fiore identity-based key encapsulation mechanism [1], based on the q-wBDHI assumption.

These schemes are given in Table 1, along with the assumptions they rely on for security, and the number of elements in both the symmetric and the asymmetric variants of the public key. As an example of our analysis, we include in Fig. 2 the dependency graph for the Boneh-Boyen-Goh HIBE. In the graph, the shape of the node indicates which side of the split each element goes on: triangle nodes are in G, inverted triangle nodes are in H, and diamond nodes are replicated across G and H. Pairing equations are denoted by $pn[i]$, where $n \in \mathbb{N}$ indicates a particular usages of the pairing and $i \in \{0, 1\}$ indicates the side of the pairing in which the element is used. The nodes with an i included represent multiple (related) values; e.g., the node yi represents $\{g^{\alpha^i}\}_i$.

The original q-wBDHI assumption states that given $(g, g^c, g^\alpha, g^{\alpha^2}, \ldots, g^{\alpha^q})$, it should be hard to distinguish $e(g,g)^{\alpha^{q+1}c}$ from random. Looking at the graph

Table 1. Examples of schemes whose reductions are compatible with the desired conversion from symmetric to asymmetric assumptions, along with the assumptions they rely on and the numbers of group elements in both the symmetric and asymmetric variants of the public key. The value A refers to the number of parallel instances of the system being run in the BGW scheme, and the value U refers to the maximum number of system attributes in Waters' scheme.

Scheme	Assumption	Elements in public key	
		symmetric	asymmetric
BGW [12]	q-BDHE	$2q + A$	$4q + A$
BBG [11]	q-wBDHI	$q + 4$	$2q + 7$
Waters [41]	q-BDHE	$3 + U$	$5 + 2U$
ACF [1]	q-wBDHI	$2 + 2q$	$3 + 2q$

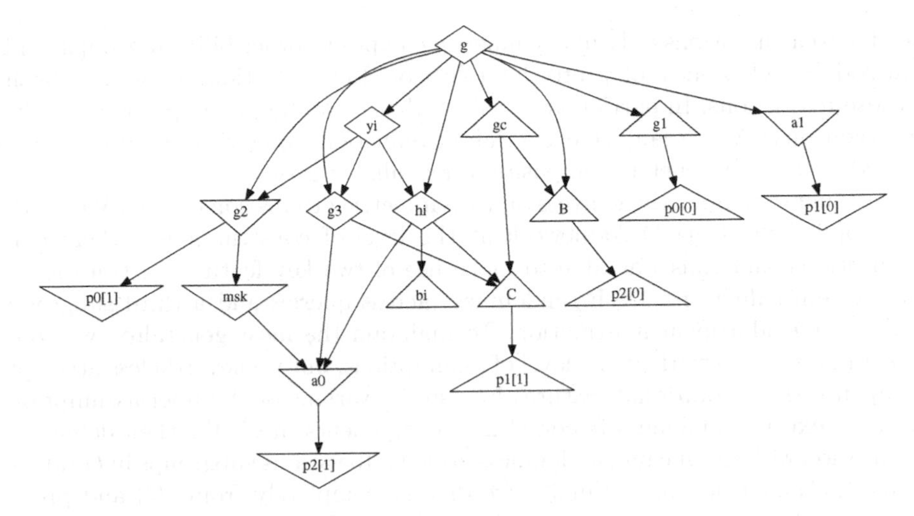

Fig. 2. Dependency graph for the BBG HIBE scheme [11]. The public key consists of g, $g1$, $g2$, $g3$, and hi, the master secret key is denoted msk, and the secret keys consist of $a0$, $a1$, and bi. Encryption uses the pairing $p0$ and produces B and C, and decryption uses the pairings $p1$ and $p2$. In the reduction, yi and gc are derived from the q-wBDHI assumption.

in Fig. 2, in which these quantities are represented by yi and gc, we see that the yi nodes must be replicated across G and H but gc can remain in only one source group. Writing h^c as $\hat{h}$, the asymmetric q-wBDHI assumption thus states that given $(g, h, \hat{h}, g^{\alpha}, h^{\alpha}, \ldots, g^{\alpha^q}, h^{\alpha^q})$, it should be hard to distinguish $e(g, \hat{h})^{\alpha^{q+1}}$ from random. This same converted version of the assumption also works for the Abdalla-Catalano-Fiore IB-KEM (whose dependency graph is included in Appendix A).

A similar analysis works for the schemes that rely on the q-BDHE assumption (whose dependency graphs are also included in Appendix A), which states that given $(g, g^c, \{g^{\alpha^i}\}_{i\in[2q],i\neq q+1})$, it should be hard to distinguish $e(g, g)^{\alpha^{q+1}c}$ from random. Here we find that the asymmetric variant states that—again, rewriting h^c as $\hat{h}$—given $(g, h, \hat{h}, \{g^{\alpha^i}, h^{\alpha^i}\}_{i\in[2q],i\neq q+1}$, it should be hard to distinguish $e(g, \hat{h})^{\alpha^{q+1}}$ from random.

As each of the converted assumptions fits the set of requirements for the uber-assumption needed for Proposition 3.2, we thus obtain as a corollary that, when instantiated in asymmetric composite-order bilinear groups, the security of each of these schemes can rely solely on (three variants of) the subgroup hiding assumption.

4 Tighter Reductions in (A)symmetric Groups

The results in the previous section already demonstrate a broader application of the Déjà Q framework, but two fundamental restrictions remain: it can be applied directly to assumptions only in asymmetric composite-order bilinear groups, and it introduces a looseness of q into the reduction. In this section, we address both of these restrictions. In particular, we show that by adding more primes into the factorization of N, we can achieve a tighter reduction—one with $\log(q)$ looseness instead of q—in symmetric composite-order bilinear groups.

Our inspiration for the conversion to symmetric groups comes from Wee [44], who applied the Déjà Q framework at the level of constructions rather than assumptions, and thus was able to make use of two key features of traditional dual-system reductions: fresh randomness across queries and a third subgroup used to hide additional information. To maintain the most generality, we continue in Sect. 4.1 to work at the level of assumptions, but we nevertheless attempt to capture these additional features by using a variant of the uber-assumption in which extra randomness is added into components in G. We then define an assumption with significant randomness added into various subgroups in G (analogous to Assumption 3.2). Finally, we diverge completely from [44] and prove that—in only a logarithmic number of game hops—this assumption implies these additionally randomized computational and decisional uber-assumptions in the target group.

Next, in Sect. 4.2, we show—in a manner almost completely analogous to that in Sect. 3.2—that the computational variant of the transitioned uber-assumption is so weak that it holds by a statistical argument; thus the computational randomized uber-assumption is implied by two variants of subgroup hiding. In the case of the decisional uber-assumption, we transition to an assumption analogous to Assumption 3.4 and show that it is implied by three variants of subgroup hiding.

Finally, in Sect. 4.3, we briefly discuss the implications of our results for the concrete schemes presented in Sect. 3.3. Although our discussion here is not as formal as our symmetric-to-asymmetric conversions, we nevertheless suggest ways to transform existing schemes to provide them with tight reductions to subgroup hiding.

4.1 Reducing Randomized Assumptions to Weaker Variants

We begin by formalizing the *randomized* uber-assumption as follows:

Assumption 4.1 (Randomized uber-assumption). *Define the advantage of an adversary $\mathcal{A}$ by $\mathbf{Adv}_{\mathcal{A}}^{comp\text{-}r\text{-}uber}(\lambda) = Pr[comp\text{-}RandUBER_{c,R,S,T,f}^{\mathcal{A}}(\lambda)]$ in the computational case and $\mathbf{Adv}_{\mathcal{A}}^{dec\text{-}r\text{-}uber}(\lambda) = 2Pr[dec\text{-}RandUBER_{c,R,S,T,f}^{\mathcal{A}}(\lambda)]-1$ in the decisional case, where for* $\mathsf{type} \in \{comp, dec\}$, type-*$RandUBER_{c,R,S,T,f}^{\mathcal{A}}(\lambda)$ is defined as follows (with the omitted end games comp-*PLAY *and dec-*PLAY *the same as in Assumption 3.1):*

MAIN type-$RandUBER^{\mathcal{A}}_{c,R,S,T,f}(\lambda)$

$(N, G, H, G_T, e) \xleftarrow{\$} \mathsf{BilinearGen}(1^\lambda, 4);\ g \xleftarrow{\$} G,\ g_4 \xleftarrow{\$} G_4,\ h_{123}, \hat{h} \xleftarrow{\$} H_{123}$
$x_1, \ldots, x_c, \chi_1, \ldots, \chi_r \xleftarrow{\$} \mathbb{Z}/N\mathbb{Z}$
$\mathsf{inputs} \leftarrow (N, G, H, G_T, e, g, g_4, \hat{h}, g^{R(\vec{x})} g_4^{\vec{\chi}}, h_{123}^{S(\vec{x})}, e(g, h_{123})^{T(\vec{x})})$
$\mathsf{chal} \leftarrow e(g, \hat{h})^{f(\vec{x})}$
return type-PLAY(λ, inputs, chal)

The randomized uber-assumption in the target group holds if for PT algorithms $\mathcal{A}$ there exists a negligible function $\nu(\cdot)$ such that $\mathbf{Adv}^{r\text{-}uber}_{\mathcal{A}}(\lambda) < \nu(\lambda)$.

The main difference from the regular uber-assumption is the additional randomness in G_4 (hence the name), and the fact that h and $\hat{h}$ are now sampled from the subgroup H_{123} rather than the full group H. As discussed further in Sect. 4.3, this latter change is needed to balance out the former, as the canceling property of the pairing means that we can still obtain meaningful values in G_T (i.e., values without added randomness) by pairing an element with a random G_4 component with an element in H_{123}. To maintain full generality, we also continue to write G and H separately, but in a symmetric pairing they would be the same group.

Assumption 4.2. *For $\mathbb{G} = (N, G, H, G_T, e) \in [\mathsf{BilinearGen}(1^\lambda, 4)]$ a bilinear group, $\ell \in \mathbb{N}$, and classes of functions R, S, T, and f (as defined in the uber-assumption in Assumption 4.1), given*

$$\mathsf{inputs} = (\mathbb{G}, g_1 g_2^{\sum_{i=1}^{\ell} r_i} g_4^{\chi}, g_4, \hat{h}, \{g_1^{\rho_k(\vec{x})} g_2^{\sum_{i=1}^{\ell} r_i \rho_k(\vec{x}_i)} g_4^{\chi_k}\}_{k=1}^{r}, h_1^{S(\vec{x})}, e(g_1, h_1)^{T(\vec{x})}),$$

for $g_1 \xleftarrow{\$} G_1$, $g_2 \xleftarrow{\$} G_2 \backslash \{1\}$, $g_4 \xleftarrow{\$} G_4$, $h_1 \xleftarrow{\$} H_1 \backslash \{1\}$, $\hat{h} \xleftarrow{\$} H_{123}$; $\vec{x}, \ldots, \vec{x}_\ell \xleftarrow{\$} (\mathbb{Z}/N\mathbb{Z})^c$, $r_1, \ldots, r_\ell, \chi, \chi_1, \ldots, \chi_r \xleftarrow{\$} \mathbb{Z}/N\mathbb{Z}$, there does not exist a PT adversary with better than negligible advantage when playing the game type-PLAY(λ, inputs, $e(g_1, \hat{h})^{f(\vec{x})} e(g_2, \hat{h})^{\sum_{i=1}^{\ell} r_i f(\vec{x}_i)}$).

In addition to the extra subgroups, our new reduction also makes use of a different class of functions for extended parameter hiding. In particular, our old proof added variables into G_2 one at a time, which allowed us to fold in a freshly random coefficient r_j in this step. As we now add many variables at a time, however, the extra randomness added by the subgroup hiding transition is not sufficient, so we instead use parameter hiding to argue that the randomness can be "freshened up" in the new subgroup instead. In the main parameter hiding step, we thus want to transition the quantity $\sum_j r_j \rho_k(\vec{x}_j)$ to $\sum_j r'_j \rho_k(\vec{x}'_j)$, which we accomplish using the set of functions defined as

$$\mathcal{F} = \left\{ p'(y_1, \vec{y}_1, \ldots, y_m, \vec{y}_m) = \sum_{i=1}^{m} y_m p(\vec{y}_m) \right\}_{p \in R \cup \{f\}}. \tag{5}$$

Theorem 4.3. *For a bilinear group $(N, G, H, G_T, e) \in [\mathsf{BilinearGen}(1^\lambda, 4)]$, consider the randomized uber-assumption parameterized by (c, R, S, T, f). Then this is implied by Assumption 4.2 if*

1. *subgroup hiding holds between H_1 and H_{123} with $\mu = \{g_4, h_{123}\}$;*
2. *subgroup hiding holds between G_{24} and G_{34} with $\mu = \{g_1, g_{24}, g_4, h_1, h_{123}\}$;*
3. *extended parameter hiding holds with respect to $R \cup \{f\}$, with respect to* $\mathsf{aux} = \{g_3^{\rho(\cdot)}, h_1^{\sigma(\cdot)}\}_{\rho \in R \cup \{f\}, \sigma \in S \cup T}$ *for all $g_3 \in G_3$ and $h_1 \in H_1$, and subgroups (G_1, G_2);*
4. *extended parameter hiding holds with respect to $R \cup \{f\}$, with respect to* $\mathsf{aux} = \{h_1^{\sigma(\cdot)}\}_{\sigma \in S \cup T}$ *for all $h_1 \in H_1$, and subgroups (G_1, G_3); and*
5. *extended parameter hiding holds with respect to the $\mathcal{F}$ defined in Eq. 5,* $\mathsf{aux} = \emptyset$*, and subgroups (G_2, G_3).*

In particular, we have that

$$\mathbf{Adv}_{\mathcal{A}}^{r\text{-}uber}(\lambda) \leq \mathbf{Adv}_{\mathcal{C}_0}^{sgh}(\lambda) + \mathbf{Adv}_{\mathcal{C}_1}^{sgh}(\lambda) + \log_2(\ell)(\mathbf{Adv}_{\mathcal{B}_i}^{sgh}(\lambda) + \mathbf{Adv}_{\mathcal{B}_{i+1}}^{sgh}(\lambda)) + \mathbf{Adv}_{\mathcal{A}}^{4.2}(\lambda).$$

Our two subgroup hiding variants are valid instantiations of the general subgroup decision assumption [8] discussed in Sect. 2. Similarly, we proved in Lemma 2.1 that in composite-order groups extended parameter hiding holds for all polynomials and the aux and subgroups that we use here, so the three variants all hold and are listed separately solely for insight into the reduction.

A proof of this theorem can be found in the full version of the paper [17]. To start, all elements using h as the base are shifted into the H_1 subgroup, but elements using $\hat{h}$ or in G remain unchanged. Using the first two variants of parameter hiding, we now switch the variables in G_2 to $\vec{x}'$ and in G_3 to $\vec{x}''$, and—using subgroup hiding—fold the $\vec{x}''$ elements into G_2. At this point we now have the original variables $\vec{x}$ in G_1, two new sets of variables in G_2, nothing in G_3, and random values in G_4.

Our reduction now proceeds by exploiting this "semi-functional" subgroup G_3 and the masking effect provided by the randomness in G_4. First, a shadow copy of *all of the variables* in G_2 is added to G_3, which we argue goes unnoticed by subgroup hiding. Second, the variables in G_3 are changed to a new set of variables, which is identical by the third variant of parameter hiding. Finally, we fold all of the new variables back into G_2, which we again argue goes unnoticed by subgroup hiding. By working with all of the variables at once—as opposed to the one-at-a-time approach of the original Déjà Q framework—we double the number of new variables in the G_2 subgroup after each iteration, so after only $\log_2(\ell)$ transitions we end up with ℓ sets of variables in the G_2 subgroup.

As described, we move new variables from G_3 to G_2 while using the generator g_2 to compute the existing variables in the G_2 subgroup. In symmetric groups with a canceling pairing, however, one could use knowledge of this generator to violate subgroup hiding by checking if $e(g_2, w) = 1$. The G_4 subgroup is thus needed to mask this transition, so in symmetric groups we transition from G_{34} to G_{24} instead, and argue that the randomness in G_4 "absorbs" the variables that are added there. In an asymmetric setting, however, knowledge of g_2 does not provide the ability to distinguish G_2 and G_3, so the masking effect of G_4 is unnecessary and the same reduction goes through without it. We thus state the simplified version of Theorem 4.3 for asymmetric groups as the following corollary:

Corollary 4.1. *For $(N, G, H, G_T, e) \in [\mathsf{BilinearGen}(1^\lambda, 3)]$ an asymmetric bilinear group, consider the uber-assumption parameterized by (c, R, S, T, f). Then this is implied by a version of Assumption 3.2 (using $\mathsf{BilinearGen}(1^\lambda, 3)$) if*

1. *subgroup hiding holds between H and H_1 with $\mu = \{\ \}$;*
2. *subgroup hiding holds between G_2 and G_3 with $\mu = \{g_1, g_2, h_1\}$;*
3. *extended parameter hiding holds with respect to $R \cup \{f\}$, with respect to $\mathsf{aux} = \{g_3^{\rho(\cdot)}, h_1^{\sigma(\cdot)}\}_{\rho \in R \cup \{f\}, \sigma \in S \cup T}$ for all $g_3 \in G_3$ and $h_1 \in H_1$, and subgroups (G_1, G_2);*
4. *extended parameter hiding holds with respect to $R \cup \{f\}$, with respect to $\mathsf{aux} = \{h_1^{\sigma(\cdot)}\}_{\sigma \in S \cup T}$ for all $h_1 \in H_1$, and subgroups (G_1, G_3); and*
5. *extended parameter hiding holds with respect to the $\mathcal{F}$ defined in Eq. 5, $\mathsf{aux} = \emptyset$, and subgroups (G_2, G_3).*

In particular, we have that

$$\mathbf{Adv}^{uber}_{\mathcal{A}}(\lambda) \leq \mathbf{Adv}^{sgh}_{\mathcal{C}_0}(\lambda) + \mathbf{Adv}^{sgh}_{\mathcal{C}_1}(\lambda) + \log_2(\ell)(\mathbf{Adv}^{sgh}_{\mathcal{B}_i}(\lambda) + \mathbf{Adv}^{sgh}_{\mathcal{B}_{i+1}}(\lambda)) + \mathbf{Adv}^{3.2}_{\mathcal{A}}(\lambda).$$

Thus, under the conditions in Propositions 3.1 and 3.2, we get tight reductions in the asymmetric setting with $N = p_1 p_2 p_3$.

For the rest of this section we will focus on the symmetric setting.

4.2 Reducing Randomized Assumptions to Subgroup Hiding

As in Sect. 3, we now treat computational and decisional assumptions separately.

Computational Assumptions. Our argument that the computational randomized uber-assumption holds is nearly identical to our previous argument that the (regular) computational uber-assumption holds.

Proposition 4.1. *For a bilinear group $\mathbb{G}$ of order N, the computational uber-assumption parameterized by (c, R, S, T, f) holds in the target group if*

1. *subgroup hiding holds between H_1 and H_{123} with $\mu = \{g_4, h_{123}\}$;*
2. *subgroup hiding holds between G_{34} and G_{24} with $\mu = \{g_1, g_{24}, g_4, h_1, h_{123}\}$;*
3. *extended parameter hiding holds with respect to $R \cup \{f\}$, with respect to $\mathsf{aux} = \{g_3^{\rho(\cdot)}, h_1^{\sigma(\cdot)}\}_{\rho \in R \cup \{f\}, \sigma \in S \cup T}$ for all $g_3 \in G_3$ and $h_1 \in H_1$, and subgroups (G_1, G_2);*
4. *extended parameter hiding holds with respect to $R \cup \{f\}$, with respect to $\mathsf{aux} = \{h_1^{\sigma(\cdot)}\}_{\sigma \in S \cup T}$ for all $h_1 \in H_1$, and subgroups (G_1, G_3);*
5. *extended parameter hiding holds with respect to the $\mathcal{F}$ defined in Eq. 5, $\mathsf{aux} = \emptyset$, and subgroups (G_2, G_3);*

6. *$N = p_1 \cdot \ldots \cdot p_n$ for distinct primes $p_1, \ldots, p_n \in \Omega(2^{poly(\lambda)})$; and*
7. *the polynomials in $R \cup f$ are linearly independent and have maximum degree $poly(\lambda)$.*

Proof. By requirements (1)–(5), Theorem 4.3 tells us that the computational uber-assumption is implied by the computational variant of Assumption 4.2. We make the problem strictly easier if we assume that g_1, g_4, $\vec{x}$, and $\vec{\chi}$ are public, in which case $g_1^{R(\vec{x})}$, $g_4^{\vec{\chi}}$, $h_1^{S(\vec{x})}$ and $e(g_1, h_1)^{T(\vec{x})}$ provide no additional information. In this case $\mathcal{A}$ can also compute the $G_{1,T}$ component of chal directly, so we need only to argue that it is hard for it to compute $e(g_2, \hat{h})^{\sum_{i=1}^{q+2} r_i f(\vec{x}_i)}$. The rest of the argument can thus proceed as in the proof of Proposition 3.1. □

Decisional Assumptions. To enable an argument about the decisional assumption in the target group, we introduce an assumption analogous to Assumption 3.4.

Assumption 4.4. *For a bilinear group $(N, G, H, G_T, e) \in [\mathsf{BilinearGen}(1^\lambda, 4)]$, $\ell \in \mathbb{N}$, consider the values given to $\mathcal{A}$ in Assumption 4.2. Given the same set of values, it is difficult to distinguish $e(g_1, \hat{h}_1)^{f(\vec{x})} e(g_2, \hat{h}_2)^{\sum_{i=1}^{\ell} r_i f(\vec{x}_i)}$ from $e(g_1, \hat{h}_1)^{f(\vec{x})} \cdot R$ for $R \xleftarrow{\$} G_{2,T}$.*

We now prove the following lemma:

Lemma 4.1. *If subgroup hiding holds in $G_{2,T}$ with $\mu = \{g_1, g_2, g_4, h_1, h_{123}\}$, then Assumption 4.2 is implied by Assumption 4.4.*

Proof. Let $\mathcal{A}$ be a PT adversary playing game $\mathsf{G}_{4.2}^{\mathcal{A}}(\lambda)$, and let $\mathbf{Adv}_{\mathcal{A}}^{4.2}(\lambda)$ denote its advantage in the game specified in Assumption 4.2. We build a PT adversary $\mathcal{B}$ such that

$$\mathbf{Adv}_{\mathcal{A}}^{4.2}(\lambda) \leq \mathbf{Adv}_{\mathcal{B}}^{\mathrm{sgh}}(\lambda) + \mathbf{Adv}_{\mathcal{A}}^{4.4}(\lambda)$$

for all $\lambda \in \mathbb{N}$, from which the theorem follows. To do this, we build $\mathcal{B}$ such that

$$\Pr[\mathsf{G}_{4.2}^{\mathcal{A}}(\lambda)] - \Pr[\mathsf{G}_{4.4}^{\mathcal{A}}(\lambda)] \leq \mathbf{Adv}_{\mathcal{B}}^{\mathrm{sgh}}(\lambda) \qquad (6)$$

We then have that

$$\mathbf{Adv}_{\mathcal{A}}^{4.2}(\lambda) \leq \mathbf{Adv}_{\mathcal{B}}^{\mathrm{sgh}}(\lambda) + \mathbf{Adv}_{\mathcal{A}}^{4.2}(\lambda).$$

<u>Equation 6 : $\mathsf{G}_{4.2}^{\mathcal{A}}(\lambda)$ to $\mathsf{G}_{4.4}^{\mathcal{A}}(\lambda)$</u>
$\mathcal{B}$ behaves as follows (again assuming $\rho_0 = 1$):

$\underline{\mathcal{B}(1^\lambda, N, G, H, G_T, e, g_1, g_2, g_4, h_1, w)}$
$b \xleftarrow{\$} \{0,1\}$
$\vec{x}, \vec{x}_1, \ldots, \vec{x}_\ell \xleftarrow{\$} (\mathbb{Z}/N\mathbb{Z})^c, r_1, \ldots, r_\ell, \chi_1, \ldots, \chi_r \xleftarrow{\$} \mathbb{Z}/N\mathbb{Z}$
$v_k \leftarrow g_1^{\rho_k(\vec{x})} g_2^{\sum_{j=1}^{\ell} r_j \rho_k(\vec{x}_j)} g_4^{\chi_k} \; \forall k \in [r]$
$y_k \leftarrow h_1^{\sigma_k(\vec{x})} \; \forall k \in [s]$
$z_k \leftarrow e(g_1, h_1)^{\tau_k(\vec{x})} \; \forall k \in [t]$
$\mathsf{inputs} \leftarrow (N, G, H, G_T, e, g_4, \hat{h}, v_0, \ldots, v_r, y_1, \ldots, y_s, z_1, \ldots, z_t)$
if $(b = 0)$ then $\mathsf{chal} \leftarrow e(g_1, \hat{h})^{f(\vec{x})} \cdot w$
if $(b = 1)$ then $\mathsf{chal} \leftarrow e(g_1, \hat{h})^{f(\vec{x})} e(g_2, \hat{h})^{\sum_{j=1}^{\ell} r_j f(\vec{x}_j)}$
$b' \xleftarrow{\$} \mathcal{A}(1^\lambda, \mathsf{inputs}, \mathsf{chal})$
return $(b' = b)$

If $w \xleftarrow{\$} G_T$, then this is identical to $\mathsf{G}_{4.2}^{\mathcal{A}}(\lambda)$. If $w \xleftarrow{\$} G_{2,T}$, then this is identical to $\mathsf{G}_{4.4}^{\mathcal{A}}(\lambda)$. □

Proposition 4.2. *For a bilinear group $\mathbb{G}$ of order N, the decisional uber-assumption parameterized by (c, R, S, T, f) holds in the target group if*

1. *subgroup hiding holds between H_{123} and H_1 with $\mu = \{g_4, h_{123}\}$;*
2. *subgroup hiding holds between G_{24} and G_{34} with $\mu = \{g_1, g_{24}, g_4, h_1, h_{123}\}$;*
3. *subgroup hiding holds in $G_{2,T}$ with $\mu = \{g_1, g_2, g_4, h_1, h_{123}\}$;*
4. *extended parameter hiding holds with respect to $R \cup \{f\}$, with respect to* $\mathsf{aux} = \{g_3^{\rho(\cdot)}, h_1^{\sigma(\cdot)}\}_{\rho \in R \cup \{f\}, \sigma \in S \cup T}$ *for all $g_3 \in G_3$ and $h_1 \in H_1$, and subgroups (G_1, G_2);*
5. *extended parameter hiding holds with respect to $R \cup \{f\}$, with respect to* $\mathsf{aux} = \{h_1^{\tau(\cdot)}\}_{\sigma \in S \cup T}$ *for all $h_1 \in H_1$, and subgroups (G_1, G_3);*
6. *extended parameter hiding holds with respect to the $\mathcal{F}$ defined in Eq. 5,* $\mathsf{aux} =$ *$\emptyset$, and subgroups (G_2, G_3);*
7. *$N = p_1 \cdot \ldots \cdot p_n$ for distinct primes $p_1, \ldots, p_n \in \Omega(2^{poly(\lambda)})$; and*
8. *the polynomials in $R \cup f$ are linearly independent and have maximum degree $poly(\lambda)$.*

Proof. By requirements (1)–(6), Theorem 4.3 and Lemma 4.1 tell us that the original assumption is implied by Assumption 4.2. We make the problem strictly easier if we assume that g_1, g_4, $\vec{x}$, and $\vec{\chi}$ are public, in which case $g_1^{R(\vec{x})}$, $g_4^{\vec{\chi}}$, $h_1^{S(\vec{x})}$ and $e(g_1, h_1)^{T(\vec{x})}$ provide no additional information. In this case $\mathcal{A}$ can also compute the $G_{1,T}$ component of chal directly (which is the same in either case), so we need only to argue that it is hard for it to distinguish $e(g_2, \hat{h})^{\sum_{i=1}^{q+2} r_i f(\vec{x}_i)}$ from random. The rest of the argument can thus proceed as in the proof of Proposition 3.2. □

4.3 Application to Existing Schemes

In Sect. 3.3, we demonstrated how to convert schemes that rely on symmetric version of the uber-assumption to work in asymmetric groups and thus be covered by our overall results in Sect. 3. Here, we briefly demonstrate how to convert schemes to be covered by our results in this section as well.

Suppose we have a scheme and corresponding reduction that work in asymmetric groups and performs only group operations, pairings, and equality tests between group elements. We can then modify both the scheme and reduction as follows: instead of sampling elements from H we sample them from H_{123}; when we multiply any elements in G we also include a freshly random element in G_4; and when we compare two elements g and g' in G for equality, rather than return $(g = g')$ we return $(e(g, h_{123}) = e(g', h_{123}))$. In particular, this last alteration—combined with the fact that $e(g_4, h_{123}) = 1$ and an asymmetric scheme only ever pairs elements of G with elements of H—allows us to preserve the functionality of the original scheme despite the fact that additional randomness is added into the G_4 subgroup.

If the original assumption relied on for security is a case of the uber-assumption (Assumption 3.1), then the resulting assumption is a case of the randomized uber-assumption (Assumption 4.1). Thus, the concrete schemes presented in Sect. 3.3 can be instantiated either in asymmetric groups of order $N = p_1p_2p_3$ under the asymmetric variants of their original (symmetric) assumptions, or in symmetric groups of order $N = p_1p_2p_3p_4$ under the randomized variants. In either case, the results of Theorem 4.3 and Corollary 4.1 imply a tight reduction to the appropriate variants of the subgroup hiding assumption.

Acknowledgments. Mary Maller is supported by a scholarship from Microsoft Research and Sarah Meiklejohn is supported in part by EPSRC Grant EP/M029026/1.

A Dependency Graphs from Sect. 3.3

In this section, we include the rest of the dependency graphs (Figs. 3, 4 and 5) for the converted schemes in Table 1. As a reminder from Sect. 3.3 (in which we included the graph for the Boneh-Boyen-Goh HIBE), the shape of the node indicates which side of the split each element goes on: triangle nodes are in G, inverted triangle nodes are in H, and diamond nodes are replicated across G and H. Pairing equations are denoted by $pn[i]$, where $n \in \mathbb{N}$ indicates a particular usage of the pairing and $i \in \{0, 1\}$ indicates the side of the pairing in which the element is used. The nodes with an i included represent multiple (related) values; e.g., the node gi represents $\{g^{\alpha^i}\}_i$.

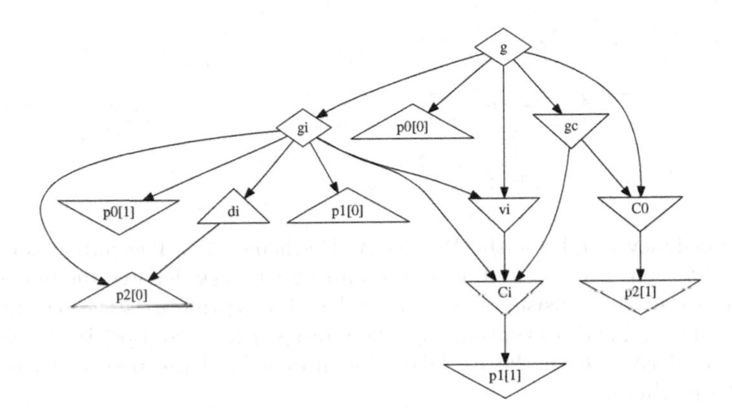

Fig. 3. Dependency graph for the BGW broadcast encryption scheme [12]. The public key consists of g, gi and vi, and the secret key of di. Encryption uses the pairing $p0$ and produces $C0$ and Ci, and decryption uses the pairings $p1$ and $p2$. In the reduction, gi are derived from the q-BDHE assumption.

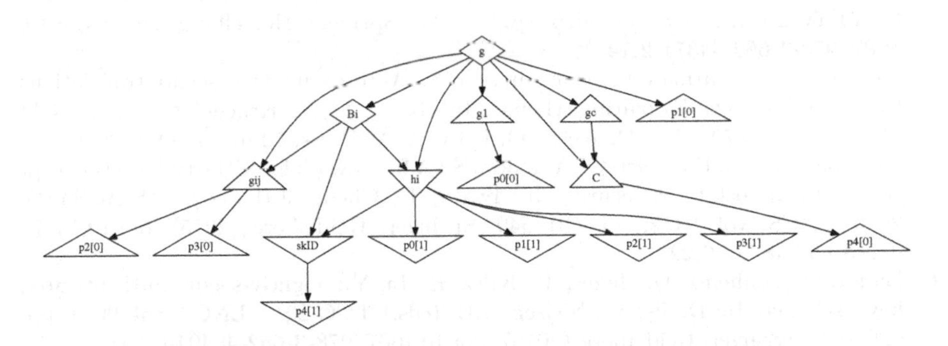

Fig. 4. Dependency graph for the ACF IB-KEM [2]. The master public key consists of g, gij, and $g1$. Secret key derivation uses hi as the auxiliary information and $skID$ as the secret key for identity ID. The pairing $p0$ and the ciphertext C are used in the encapsulation process, decapsulation uses the pairings $p1$, $p2$, and $p3$, and the key is calculated from the encapsulation using $p4$. In the reduction, Bi and gc are derived from the q-wBDHI assumption.

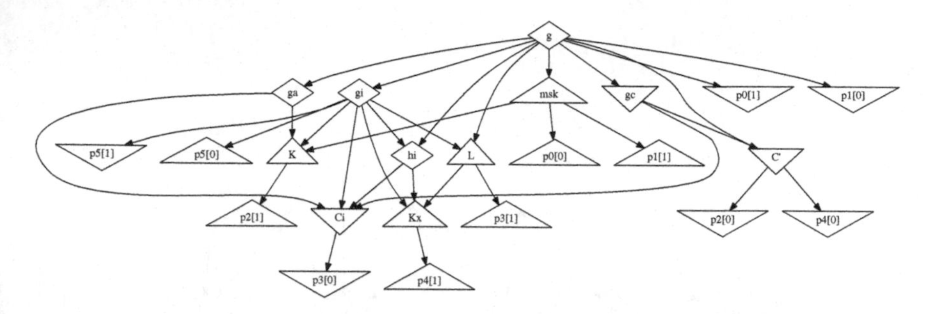

Fig. 5. Dependency graph for the Waters ABE scheme [41]. The public key consists of g, ga, and hi, and is computed using the pairing $p0$. msk denotes the master secret key and the secret key consists of K, Kx, and L. Encryption uses the pairing $p1$ and produces C' and Ci, and decryption uses the pairings $p2$, $p3$, and $p4$. In the reduction, gi and gc are derived from the q-BDHE assumption, and the pairing $p5$ is used to simulate the pairing $p0$.

References

1. Abdalla, M., Catalano, D., Fiore, D.: Verifiable random functions from identity-based key encapsulation. In: Joux, A. (ed.) EUROCRYPT 2009. LNCS, vol. 5479, pp. 554–571. Springer, Heidelberg (2009). doi:10.1007/978-3-642-01001-9_32
2. Abdalla, M., Catalano, D., Fiore, D.: Verifiable random functions: relations to identity-based key encapsulation and new constructions. J. Cryptology **27**(3), 544–593 (2014)
3. Abe, M., Groth, J., Ohkubo, M., Tango, T.: Converting cryptographic schemes from symmetric to asymmetric bilinear groups. In: Garay, J.A., Gennaro, R. (eds.) CRYPTO 2014. LNCS, vol. 8616, pp. 241–260. Springer, Heidelberg (2014). doi:10.1007/978-3-662-44371-2_14
4. Akinyele, J.A., Garman, C., Hohenberger, S., Automating fast, secure translations from type-I to type-III pairing schemes. In: Ray, I., Li, N., Kruegel, C. (eds.), ACM CCS 15, pp. 1370–1381, Denver, CO, USA. ACM Press, 12–16 October 2015
5. Attrapadung, N., Hanaoka, G., Yamada, S.: A framework for identity-based encryption with almost tight security. In: Iwata, T., Cheon, J.H. (eds.) ASIACRYPT 2015. LNCS, vol. 9452, pp. 521–549. Springer, Heidelberg (2015). doi:10.1007/978-3-662-48797-6_22
6. Bader, C., Hofheinz, D., Jager, T., Kiltz, E., Li, Y.: Tightly-secure authenticated key exchange. In: Dodis, Y., Nielsen, J.B. (eds.) TCC 2015. LNCS, vol. 9014, pp. 629–658. Springer, Heidelberg (2015). doi:10.1007/978-3-662-46494-6_26
7. Bellare, M., Rogaway, P.: The exact security of digital signatures-how to sign with RSA and Rabin. In: Maurer, U. (ed.) EUROCRYPT 1996. LNCS, vol. 1070, pp. 399–416. Springer, Heidelberg (1996). doi:10.1007/3-540-68339-9_34
8. Bellare, M., Waters, B., Yilek, S.: Identity-based encryption secure against selective opening attack. In: Ishai, Y. (ed.) TCC 2011. LNCS, vol. 6597, pp. 235–252. Springer, Heidelberg (2011). doi:10.1007/978-3-642-19571-6_15
9. Blazy, O., Kiltz, E., Pan, J.: (Hierarchical) identity-based encryption from affine message authentication. In: Garay, J.A., Gennaro, R. (eds.) CRYPTO 2014. LNCS, vol. 8616, pp. 408–425. Springer, Heidelberg (2014). doi:10.1007/978-3-662-44371-2_23

10. Boneh, D., Boyen, X.: Short signatures without random Oracles. In: Cachin, C., Camenisch, J.L. (eds.) EUROCRYPT 2004. LNCS, vol. 3027, pp. 56–73. Springer, Heidelberg (2004). doi:10.1007/978-3-540-24676-3_4
11. Boneh, D., Boyen, X., Goh, E.-J.: Hierarchical identity based encryption with constant size ciphertext. In: Cramer, R. (ed.) EUROCRYPT 2005. LNCS, vol. 3494, pp. 440–456. Springer, Heidelberg (2005). doi:10.1007/11426639_26
12. Boneh, D., Gentry, C., Waters, B.: Collusion resistant broadcast encryption with short ciphertexts and private keys. In: Shoup, V. (ed.) CRYPTO 2005. LNCS, vol. 3621, pp. 258–275. Springer, Heidelberg (2005). doi:10.1007/11535218_16
13. Boneh, D., Goh, E.-J., Nissim, K.: Evaluating 2-DNF formulas on ciphertexts. In: Kilian, J. (ed.) TCC 2005. LNCS, vol. 3378, pp. 325–341. Springer, Heidelberg (2005). doi:10.1007/978-3-540-30576-7_18
14. Boneh, D., Rubin, K., Silverberg, A.: Finding ordinary composite order elliptic curves using the Cocks-Pinch method. J. Number Theory **131**(5), 832–841 (2011)
15. Boyen, X.: The uber-assumption family. In: Galbraith, S.D., Paterson, K.G. (eds.) Pairing 2008. LNCS, vol. 5209, pp. 39–56. Springer, Heidelberg (2008). doi:10.1007/978-3-540-85538-5_3
16. Boyen, X., Waters, B.: Anonymous hierarchical identity-based encryption (Without Random Oracles). In: Dwork, C. (ed.) CRYPTO 2006. LNCS, vol. 4117, pp. 290–307. Springer, Heidelberg (2006). doi:10.1007/11818175_17
17. Chase, M., Maller, M., Meiklejohn, S.: Déjà Q all over again: tighter and broader reductions of q-type assumptions. Cryptology ePrint Archive, Report 2016/840 (2016). https://eprint.iacr.org/2016/840
18. Chase, M., Meiklejohn, S.: Déjà Q: using dual systems to revisit q-type assumptions. In: Nguyen, P.Q., Oswald, E. (eds.) EUROCRYPT 2014. LNCS, vol. 8441, pp. 622–639. Springer, Heidelberg (2014). doi:10.1007/978-3-642-55220-5_34
19. Chase, M., Meiklejohn, S.: Déjà Q: using dual systems to revisit q-type assumptions. Cryptology ePrint Archive, Report 2014/570 (2014). http://eprint.iacr.org/2014/570
20. Chen, J., Gay, R., Wee, H.: Improved dual system ABE in prime-order groups via predicate encodings. In: Oswald, E., Fischlin, M. (eds.) EUROCRYPT 2015. LNCS, vol. 9057, pp. 595–624. Springer, Heidelberg (2015). doi:10.1007/978-3-662-46803-6_20
21. Chen, J., Wee, H.: Fully, (Almost) tightly secure IBE and dual system groups. In: Canetti, R., Garay, J.A. (eds.) CRYPTO 2013. LNCS, vol. 8043, pp. 435–460. Springer, Heidelberg (2013). doi:10.1007/978-3-642-40084-1_25
22. Cheon, J.H.: Security analysis of the strong Diffie-Hellman problem. In: Vaudenay, S. (ed.) EUROCRYPT 2006. LNCS, vol. 4004, pp. 1–11. Springer, Heidelberg (2006). doi:10.1007/11761679_1
23. Dodis, Y., Yampolskiy, A.: A verifiable random function with short proofs and keys. In: Vaudenay, S. (ed.) PKC 2005. LNCS, vol. 3386, pp. 416–431. Springer, Heidelberg (2005). doi:10.1007/978-3-540-30580-4_28
24. Gay, R., Hofheinz, D., Kiltz, E., Wee, H.: Tightly CCA-secure encryption without pairings. In: Fischlin, M., Coron, J.-S. (eds.) EUROCRYPT 2016. LNCS, vol. 9665, pp. 1–27. Springer, Heidelberg (2016). doi:10.1007/978-3-662-49890-3_1
25. Gerbush, M., Lewko, A., O'Neill, A., Waters, B.: Dual Form Signatures: an approach for proving security from static assumptions. In: Wang, X., Sako, K. (eds.) ASIACRYPT 2012. LNCS, vol. 7658, pp. 25–42. Springer, Heidelberg (2012). doi:10.1007/978-3-642-34961-4_4

26. Goldwasser, S., Tauman Kalai, Y.: Cryptographic Assumptions: a position paper. In: Kushilevitz, E., Malkin, T. (eds.) TCC 2016. LNCS, vol. 9562, pp. 505–522. Springer, Heidelberg (2016). doi:10.1007/978-3-662-49096-9_21
27. Hemenway, B., Libert, B., Ostrovsky, R., Vergnaud, D.: Lossy Encryption: constructions from general assumptions and efficient selective opening chosen ciphertext security. In: Lee, D.H., Wang, X. (eds.) ASIACRYPT 2011. LNCS, vol. 7073, pp. 70–88. Springer, Heidelberg (2011). doi:10.1007/978-3-642-25385-0_4
28. Hofheinz, D., Jager, T.: Tightly secure signatures and public-key encryption. In: Safavi-Naini, R., Canetti, R. (eds.) CRYPTO 2012. LNCS, vol. 7417, pp. 590–607. Springer, Heidelberg (2012). doi:10.1007/978-3-642-32009-5_35
29. Lewko, A.: Tools for simulating features of composite order bilinear groups in the prime order setting. In: Pointcheval, D., Johansson, T. (eds.) EUROCRYPT 2012. LNCS, vol. 7237, pp. 318–335. Springer, Heidelberg (2012). doi:10.1007/978-3-642-29011-4_20
30. Lewko, A., Meiklejohn, S.: A Profitable Sub-prime Loan: obtaining the advantages of composite order in prime-order bilinear groups. In: Katz, J. (ed.) PKC 2015. LNCS, vol. 9020, pp. 377–398. Springer, Heidelberg (2015). doi:10.1007/978-3-662-46447-2_17
31. Lewko, A., Okamoto, T., Sahai, A., Takashima, K., Waters, B.: Fully Secure Functional Encryption: attribute-based encryption and (hierarchical) inner product encryption. In: Gilbert, H. (ed.) EUROCRYPT 2010. LNCS, vol. 6110, pp. 62–91. Springer, Heidelberg (2010). doi:10.1007/978-3-642-13190-5_4
32. Lewko, A., Waters, B.: New techniques for dual system encryption and fully secure HIBE with short ciphertexts. In: Micciancio, D. (ed.) TCC 2010. LNCS, vol. 5978, pp. 455–479. Springer, Heidelberg (2010). doi:10.1007/978-3-642-11799-2_27
33. Lewko, A., Waters, B.: Decentralizing attribute-based encryption. In: Paterson, K.G. (ed.) EUROCRYPT 2011. LNCS, vol. 6632, pp. 568–588. Springer, Heidelberg (2011). doi:10.1007/978-3-642-20465-4_31
34. Lewko, A., Waters, B.: Unbounded HIBE and attribute-based encryption. In: Paterson, K.G. (ed.) EUROCRYPT 2011. LNCS, vol. 6632, pp. 547–567. Springer, Heidelberg (2011). doi:10.1007/978-3-642-20465-4_30
35. Libert, B., Joye, M., Yung, M., Peters, T.: Concise multi-challenge CCA-secure encryption and signatures with almost tight security. In: Sarkar, P., Iwata, T. (eds.) ASIACRYPT 2014. LNCS, vol. 8874, pp. 1–21. Springer, Heidelberg (2014). doi:10.1007/978-3-662-45608-8_1
36. Libert, B., Peters, T., Joye, M., Yung, M.: Compactly hiding linear spans. In: Iwata, T., Cheon, J.H. (eds.) ASIACRYPT 2015. LNCS, vol. 9452, pp. 681–707. Springer, Heidelberg (2015). doi:10.1007/978-3-662-48797-6_28
37. Meiklejohn, S., Shacham, H.: New trapdoor projection maps for composite-order bilinear groups. Cryptology ePrint Archive, Report 2013/657 (2013). http://eprint.iacr.org/2013/657
38. Okamoto, T., Takashima, K.: Efficient (hierarchical) inner-product encryption tightly reduced from the decisional linear assumption. IEICE Trans. **96–A**(1), 42–52 (2013)
39. Peikert, C., Waters, B.: Lossy trapdoor functions and their applications. In: Ladner, R.E., Dwork, C. (eds.) 40th ACM STOC, pp. 187–196, Victoria, British Columbia, Canada. ACM Press, 17–20 May 2008
40. Shoup, V.: Lower bounds for discrete logarithms and related problems. In: Fumy, W. (ed.) EUROCRYPT 1997. LNCS, vol. 1233, pp. 256–266. Springer, Heidelberg (1997). doi:10.1007/3-540-69053-0_18

41. Waters, B.: Ciphertext-policy attribute-based encryption: an expressive, efficient, and provably secure realization. Cryptology ePrint Archive, Report 2008/290 (2008). http://eprint.iacr.org/2008/290
42. Waters, B.: Dual System Encryption: realizing fully secure IBE and HIBE under simple assumptions. In: Halevi, S. (ed.) CRYPTO 2009. LNCS, vol. 5677, pp. 619–636. Springer, Heidelberg (2009). doi:10.1007/978-3-642-03356-8_36
43. Waters, B.: Ciphertext-Policy Attribute-Based Encryption: an expressive, efficient, and provably secure realization. In: Catalano, D., Fazio, N., Gennaro, R., Nicolosi, A. (eds.) PKC 2011. LNCS, vol. 6571, pp. 53–70. Springer, Heidelberg (2011). doi:10.1007/978-3-642-19379-8_4
44. Wee, H.: Déjà Q: Encore! Un Petit IBE. In: Kushilevitz, E., Malkin, T. (eds.) TCC 2016. LNCS, vol. 9563, pp. 237–258. Springer, Heidelberg (2016). doi:10.1007/978-3-662-49099-0_9

Partitioning via Non-linear Polynomial Functions: More Compact IBEs from Ideal Lattices and Bilinear Maps

Shuichi Katsumata[1,2] and Shota Yamada[2(✉)]

[1] The University of Tokyo, Tokyo, Japan
shuichi_katsumata@it.k.u-tokyo.ac.jp
[2] National Institute of Advanced Industrial Science and Technology (AIST), Tokyo, Japan
yamada-shota@aist.go.jp

Abstract. In this paper, we present new adaptively secure identity-based encryption (IBE) schemes. One of the distinguishing properties of the schemes is that it achieves shorter public parameters than previous schemes. Both of our schemes follow the general framework presented in the recent IBE scheme of Yamada (Eurocrypt 2016), employed with novel techniques tailored to meet the underlying algebraic structure to overcome the difficulties arising in our specific setting. Specifically, we obtain the following:

- Our first scheme is proven secure under the ring learning with errors (RLWE) assumption and achieves the best asymptotic space efficiency among existing schemes from the same assumption. The main technical contribution is in our new security proof that exploits the ring structure in a crucial way. Our technique allows us to greatly weaken the underlying hardness assumption (e.g., we assume the hardness of RLWE with a fixed polynomial approximation factor whereas Yamada's scheme requires a super-polynomial approximation factor) while improving the overall efficiency.

- Our second IBE scheme is constructed on bilinear maps and is secure under the 3-computational bilinear Diffie-Hellman exponent assumption. This is the first IBE scheme based on the hardness of a computational/search problem, rather than a decisional problem such as DDH and DLIN on bilinear maps with sub-linear public parameter size.

1 Introduction

Background. Identity-based encryption (IBE) is a generalization of public key encryption (PKE) where the public key of a user can be any arbitrary string such as an e-mail address. The concept of IBE was first proposed by Shamir [Sha85] in 1984, but it took nearly two decades for the first realizations of IBE [SOK00, BF01, Coc01] to appear. Since then, the construction of IBE has been one of the central topics in cryptography. Nowadays, we have constructions of IBEs from assumptions on bilinear maps [BF01, BB04a, BB04b, Wat05, Gen06, Wat09], the

J.H. Cheon and T. Takagi (Eds.): ASIACRYPT 2016, Part II, LNCS 10032, pp. 682–712, 2016.
DOI: 10.1007/978-3-662-53890-6_23

quadratic residue assumption [Coc01, BGH07], and from the learning with error (LWE) assumption [GPV08, CHKP10, ABB10] whose hardness is implied by the worst case reductions to certain lattice problems [Reg05].

One of the most standard security definitions for IBE is the adaptive security, or often called full security. While it is not quite hard to obtain the adaptive security for an IBE in the random oracle model [BF01, Coc01, GPV08], the realization in the standard model is much harder. Roughly speaking, currently there are two general techniques in achieving adaptive security in the standard model: the partitioning technique [BB04b, Wat05] and the dual system encryption methodology [Wat09, LW10]. The latter is very attractive, because it allows us to construct very efficient IBE schemes [CW13, JR13] and even more advanced cryptosystems such as attribute-based encryptions [LOS+10] with adaptive security. However, it inherently relies on *decisional assumptions* on bilinear maps (e.g., SXDH and DLIN) and cannot be extended to the proofs based on *computational assumptions* on bilinear maps (e.g., computational bilinear Diffie-Hellman (CBDH) assumption) or assumptions on lattices. On the other hand, the application of the former technique is wider. We can construct adaptively secure IBE from the CBDH assumption (by the straightforward combination of the Goldreich-Levin bit [GL89] and Waters IBE [Wat05]) and from the LWE assumption [CHKP10, ABB10, Boy10]. However, IBE schemes constructed from the former approach typically requires larger parameters due to the use of the Waters' hash [Wat05] or the admissible hash [BB04b, CHKP10]. Very recently, Yamada [Yam16] constructed IBE schemes from lattices based on the partitioning technique with novel ideas that are different from the Waters' hash or the admissible hash. His schemes achieve asymptotically shorter public parameters than previous works. One of the drawbacks of the schemes is that they require super-polynomial size modulus for LWE. As a result, their ciphertexts are longer than those of previous works by a rather large super-constant factor. In addition, they have to assume the hardness of the LWE problem for *all polynomial* (i.e., $O(n^c)$ for *all* $c \in \mathbb{N}$) or the more aggressive *super-polynomial* approximation factor. Though their assumption is plausible, it is much stronger than those used in the previous works where the hardness of the LWE problem for some *fixed polynomial* approximation factor (i.e., $O(n^c)$ for *some* $c \in \mathbb{N}$) is assumed. Furthermore, since he used fully homomorphic computations of trapdoors [BGG+14], a technique unique to the lattice setting, it is a highly non-trivial task to construct analogous schemes in other settings such as bilinear maps.

Our Contribution. In this paper, we focus on the constructions of adaptively secure IBE in these settings where dual system encryption methodology is unavailable. In particular, we propose IBE schemes with shorter public parameters from ring/ideal lattices and from a certain computational assumption (rather than a decisional assumption) on bilinear groups, by extending and adding twists to the techniques of [Yam16]. Specifically, we obtain the following results. See Tables 1 and 2 for the overview.

- We propose an anonymous and adaptively secure IBE scheme from the ring LWE (RLWE) assumption with *fixed polynomial* approximation factors, which is further reduced to certain worst case problems on ideal lattices. Note that

simply instantiating Yamada's scheme using ideal lattices[1] will still require the RLWE assumption for *all polynomial* approximation factors, which is a much stronger assumption than what we use. As for the efficiency, the size of the public parameters, private keys, and ciphertexts in our scheme are $O(n\kappa^{1/d}\log n)$, $O(n\log n)$, and $O(n\log n)$, respectively. Here, n is the dimension of the ring elements, κ is the length of the identities, and d is a flexible constant that can be set arbitrary, but will affect the reduction cost exponentially. We note that all of them achieve the best efficiency among the other adaptively secure IBE from the RLWE assumption in an asymptotic sense. Compared to the ring version of Yamada's scheme, we managed to reduce the poly-logarithmic factors contained in the public parameters, private keys, and ciphertexts.
- We propose a (non anonymous and) adaptively secure IBE scheme from the 3-computational bilinear Diffie-Hellman exponent (3-CBDHE) assumption. The 3-CBDHE assumption is a weaker variant of the n-decisional bilinear Diffie-Hellman exponent (n-DBDHE) assumption [BBG05, BGW05, BH08]. The former seems to be much a weaker assumption than the latter in two aspects. First, the former is a computational assumption whereas the latter is a decisional assumption. Second, the former is not a parameterized assumption, in the sense that the size of the problem instance only depends on the security parameter. As for the efficiency, the public parameters, private keys, and ciphertexts in our scheme require $O(\sqrt{\kappa})$ group elements. Here, κ is the length of the identities. This is the first adaptively secure IBE scheme from a computational assumption on bilinear groups with public parameters consisting of sub-linear number of group elements in the length of the identities. However, we note that the sizes of the ciphertexts and private keys of our scheme are larger than the previous schemes.

We emphasize that our result for the lattice based construction cannot be obtained through the simple switch to the ring setting in Yamada's scheme. Their proof will still require a super-polynomial-size modulus to work, whereas our new technique allows for a polynomial-size modulus. In addition, the security proof of our scheme requires new ideas that did not appear in [Yam16]. It exploits the commutative properties of the underlying ring elements in an essential way, involves a more generalized partitioning argument, and a careful analysis of the Gaussian error. Refer Sect. 2 for the technical overview. We note that the public parameter of our second scheme could be further reduced to $O(\kappa^{1/d})$ assuming the $d+1$-CBDHE assumption. However, it would come at the cost of even longer ciphertexts and complicated description of the scheme. This is beyond the scope of our work. We finally remark that the reduction costs for both of our schemes are inadmissible as was in the case of [Yam16]. In fact, the reduction loss for the first scheme is worse than [Yam16]. Improving them is left as an open problem.

Related Works. One way to reduce the size of the public parameters in Waters' hash and its analogue is to use Naccache's approach [Nac07, SRB12]. However,

[1] Note that he does not describe nor mention the ring variant of the scheme. However, we can convert his scheme into a ring variant in a straightforward manner as is the case in most previous works [CHKP10, ABB10, Boy10].

with this approach, we are only allowed to reduce the size of public parameters up to logarithmic factor. Ducas et al. [DLP14] constructed efficient IBE over NTRU lattices in the random oracle model. Gentry [Gen06] proposed adaptively secure IBE with compact parameters from a parameterized (or q-type) assumption on bilinear maps. Galindo [Gal10] and Chen et al. [CCZ11] proposed selectively secure CCA-secure IBE schemes from the CBDH assumption.

Note on Recent Works. Here, we mention two important recent related works.

Apon et al. [AFL16] proposed an adaptively secure IBE scheme from lattices whose parameters are very compact, using collision resistant hash function with output-length $\kappa = \omega(\log \lambda)$. Here, λ is the security parameter. While their scheme is more efficient than our scheme, we clarify that they implicitly assume exponential security on the collision resistant hash function, which is a stronger assumption than what we use. To demonstrate this, let us set $\kappa = \log^2 \lambda$. If there is no better attack than the birthday attack against the hash function, no PPT adversary can find a collision with more than negligible probability. On the other hand, the existence of even a sub-exponential time attack would compromise the security of the IBE. For example, assume that there exists an attack that finds a collision in time $2^{\sqrt{\kappa}}$. Then, the collision for the hash can be found in linear time in λ, since $2^{\sqrt{\kappa}} = 2^{\log \lambda} = \lambda$.

In their very recent work, Zhang et al. [ZCZ16] constructed an IBE scheme with poly-logarithmic public parameters. While their scheme achieves better asymptotic space efficiency than our scheme, their scheme is Q-bounded, in the sense that the security of the scheme is not guaranteed any more if the adversary obtains more than Q private keys. This restriction cannot be removed by just making Q super-polynomial, because the running time of the encryption algorithm in their scheme is at least linear in Q. We note that our scheme is secure against an unbounded collusion.

2 Overview of Our Techniques

2.1 Construction from Ring and Ideal Lattices

The Yamada IBE. We briefly review the Yamada IBE [Yam16], for our proposed IBE scheme follows the framework of theirs and overcomes some of the major problems posed by their construction. Their construction follows the general framework of constructing lattice-based IBE schemes that associates to each identity ID the matrix $[\mathbf{A}|\mathsf{H}(\mathsf{ID})] \in \mathbb{Z}_q^{n\times 2m}$. In previous IBE constructions [ABB10, CHKP10], the function $\mathsf{H}(\mathsf{ID})$ was computed by using the rather long κ public matrices $\{\mathbf{B}_i\}_{i\in[\kappa]}$, where $\kappa = O(n)$ is the length of the identities. The main technical contribution of the Yamada IBE was in reducing the size of the public matrices to $\kappa^{1/d}$ for any constant d and hence reducing the size of the public parameters by incorporating a primitive called fully homomorphic trapdoor functions. Hereafter, we consider the case $d = 2$ for simplicity. In detail, they used an injective map $S : \{0,1\}^\kappa \to 2^{[\ell]\times[\ell]}$ that maps an identity to a subset of the set $[\ell] \times [\ell]$ where $\ell = \lceil \kappa^{1/2} \rceil$, and computed the function $\mathsf{H}(\mathsf{ID})$ as

$$\mathsf{H}(\mathsf{ID}) = \mathbf{B}_0 + \sum_{(i,j)\in S(\mathsf{ID})} \mathbf{B}_{1,i} \cdot \mathbf{G}^{-1}(\mathbf{B}_{2,j}) \tag{1}$$

where the number of public matrices $\mathbf{B}_0, \{\mathbf{B}_{i,j}\}_{(i,j)\in[2]\times[\ell]}$ are now reduced to $O(\kappa^{1/2})$. Here, $\mathbf{G}$ is a special gadget matrix whose trapdoor is publicly known [MP12] and $\mathbf{G}^{-1}$ is viewed as a deterministic function rather than a matrix, that maps a matrix $\mathbf{V} \in \mathbb{Z}_q^{n\times m}$ to a matrix $\mathbf{U} \in \{0,1\}^{m\times m}$ such that $\mathbf{G}\cdot\mathbf{U} = \mathbf{V} \mod q$.

During the security proof, the reduction algorithm first prepares random integers $y_0, \{y_{i,j}\}_{(i,j)\in[2]\times[\ell]} \in \mathbb{Z}_q$ from certain domains whose size grows linear in the number of key extraction query Q of the adversary. Then after sampling $\mathbf{R}_0, \{\mathbf{R}_{i,j}\}_{i\in[2],j\in[\ell]} \in \mathbb{Z}^{m\times m}$ with small spectral norm, the reduction algorithm prepares the public parameters as

$$\mathbf{B}_0 = \mathbf{A}\mathbf{R}_0 + y_0\mathbf{G}, \quad \mathbf{B}_{i,j} = \mathbf{A}\mathbf{R}_{i,j} + y_{i,j}\mathbf{G}$$

for $(i,j) \in [2]\times[\ell]$. Then during the security reduction the hash value for identity ID Eq. (1) is computed as

$$\begin{aligned}
\mathsf{H}(\mathsf{ID}) &= (\mathbf{A}\mathbf{R}_0 + y_0\mathbf{G}) + \sum_{(i,j)\in S(\mathsf{ID})} (\mathbf{A}\mathbf{R}_{1,i} + y_{1,i}\mathbf{G})\cdot\mathbf{G}^{-1}(\mathbf{B}_{2,j}) \\
&= (\mathbf{A}\mathbf{R}_0 + y_0\mathbf{G}) + \sum_{(i,j)\in S(\mathsf{ID})} (\mathbf{A}\mathbf{R}_{1,i}\mathbf{G}^{-1}(\mathbf{B}_{2,j}) + y_{1,i}\mathbf{B}_{2,j}) \\
&= (\mathbf{A}\mathbf{R}_0 + y_0\mathbf{G}) + \sum_{(i,j)\in S(\mathsf{ID})} (\mathbf{A}\mathbf{R}_{1,i}\mathbf{G}^{-1}(\mathbf{B}_{2,j}) + y_{1,i}(\mathbf{A}\mathbf{R}_{2,j} + y_{2,j}\mathbf{G})) \\
&= (\mathbf{A}\mathbf{R}_0 + y_0\mathbf{G}) + \sum_{(i,j)\in S(\mathsf{ID})} (\mathbf{A}\mathbf{R}_{1,i}\mathbf{G}^{-1}(\mathbf{B}_{2,j}) + \mathbf{A}(y_{1,i}\mathbf{R}_{2,j}) + y_{1,i}y_{2,j}\mathbf{G}) \\
&= \mathbf{A}\underbrace{\left(\mathbf{R}_0 + \sum_{(i,j)\in S(\mathsf{ID})} \left(\mathbf{R}_{1,i}\mathbf{G}^{-1}(\mathbf{B}_{2,j}) + y_{1,i}\mathbf{R}_{2,j}\right)\right)}_{:=\mathbf{R}_{\mathsf{ID}},\ \text{which is ``small''}} + \underbrace{\left(y_0 + \sum_{(i,j)\in S(\mathsf{ID})} y_{1,i}y_{2,j}\right)}_{:=\mathsf{F}_{\mathbf{y}}(\mathsf{ID})}\cdot\mathbf{G} \\
&= \mathbf{A}\mathbf{R}_{\mathsf{ID}} + \mathsf{F}_{\mathbf{y}}(\mathsf{ID})\mathbf{G}.
\end{aligned} \tag{2}$$

Observe that we implicitly relied on the fact that $\mathbf{A}$ and $y_{1,i}$ commutes. Therefore, the reduction algorithm is able to sample a secret key for ID using the trapdoor of $\mathbf{G}$ if and only if $\mathsf{F}_{\mathbf{y}}(\mathsf{ID}) \neq 0 \mod q$. Hence, the simulation succeeds when the adversary queries on secret keys for ID satisfying $\mathsf{F}_{\mathbf{y}}(\mathsf{ID}) \neq 0 \mod q$, and queries for a challenge ciphertext for $\mathsf{ID}^\star$ satisfying $\mathsf{F}_{\mathbf{y}}(\mathsf{ID}^\star) = 0 \mod q$ in which case the reduction algorithm can embed its LWE challenge.

Overview of the Construction and Security Proof. The major drawback of the Yamada IBE is that they require the modulus size q to be super-polynomial. This stems from the fact that the size of $y_0, y_{i,j} \in \mathbb{Z}_q$ must grow linearly in the number of adversarial key extraction query Q for the security proof to be meaningful, i.e., $\Pr_{\mathbf{y}}[\mathsf{F}_{\mathbf{y}}(\mathsf{ID}^\star) = 0 \wedge \mathsf{F}_{\mathbf{y}}(\mathsf{ID}_1) \neq 0 \wedge \cdots \wedge \mathsf{F}_{\mathbf{y}}(\mathsf{ID}_Q) \neq 0]$ is noticeable in n. However, since the size of the $\mathbf{G}$-trapdoor $\mathbf{R}_{\mathsf{ID}}$ used during simulation grows proportionally to the size of $y_{1,i}$ (check above Eq. (2) to see how $\mathbf{R}_{\mathsf{ID}}$ was created), thereby growing proportional to $Q = \mathsf{poly}(n)$, we need to set the modulus size q to be at least super-polynomial in n for the trapdoor to operate properly. Therefore, if we try to restrict ourselves to a polynomial sized modulus q, it seems the best we can achieve is a scheme where we have to set a bound on the number of adversarial key extraction queries before instantiation, i.e., a Q-bounded scheme.

In our work, we combine several ideas in a novel way to circumvent the above seemingly inevitable problem. The first idea is to extend the elements $y_0, y_{i,j} \in \mathbb{Z}_q$ to matrices $\mathbf{Y}_0, \mathbf{Y}_{i,j} \in \mathbb{Z}_q^{n\times n}$ so that instead of increasing the size of the element $y \in \mathbb{Z}_q$, we can "pack" small elements in the entries of the matrix $\mathbf{Y} \in \mathbb{Z}_q^{n\times n}$. Namely, since the matrix has n^2 entries, if the number of key extraction query is $Q = n^c$ for some constant c, we can always set up the matrix so that c of the entries are packed by elements of size $O(n)$. Since there are n^2 entries in total, this allows us to pack the matrix with small entries (e.g., $O(n)$) for arbitrary $Q = \mathsf{poly}(n)$ without the need of increasing the modulus size q. However, this simple idea alone does not work, since during the security proof to obtain Eq. (2), we crucially relied on the fact that $\mathbf{A}$ and $y_{1,i}$ commutes. For our idea to work we need the two matrices $\mathbf{A}$ and $\mathbf{Y}_{1,i}$ to commute, however, in general this does not hold.

To overcome this problem, we introduce our second idea of using the ring structure of ideal lattices. Concretely, we use the special polynomial ring $R = \mathbb{Z}[X]/(X^n + 1)$ to construct our scheme for n a power of 2. The construction itself is exactly the same as the ring analogue of the Yamada IBE, however, our new security proof relies crucially on the underlying ring structure. In detail, the reduction algorithm prepares the public parameters as

$$\boldsymbol{b}_0 = \boldsymbol{a}\boldsymbol{R}_0 + y_0\boldsymbol{g}, \quad \boldsymbol{b}_{i,j} = \boldsymbol{a}\boldsymbol{R}_{i,j} + y_{i,j}\boldsymbol{g}$$

for $(i,j) \in [2] \times [\ell]$, where $\boldsymbol{a}, \boldsymbol{b}_0, \boldsymbol{b}_{i,j} \in R_q^k$, $\boldsymbol{R} \in R_q^{k\times k}$, $y_0, y_{i,j} \in R_q$ and $\boldsymbol{g} \in R_q^k$ is the ring analogue of the $\mathbf{G}$-trapdoor. Observe that $y_0, y_{i,j}$ are now elements in R_q instead of $\mathbb{Z}_q$. Although this y is not quite a matrix, this is actually more than enough for us to use the packing technique described above. This can be seen by first noticing the natural isomorphism between $R_q \cong \mathbb{Z}_q^n$ induced by the coefficient embedding and viewing $y \in R_q$ as a vector in $\mathbb{Z}_q^n$. Since y has n entries when viewed as vectors, it can support up to n^n queries by packing each entry with small elements of size $O(n)$. Furthermore, the second part of the problem addressed above is naturally resolved, since now that we are working in a ring we get the commutativity of $\boldsymbol{a}$ and $y_{1,j}$ for free. This key role in the commutativity for rings is somewhat reminiscent to the signature scheme of [DM14]. We note that the technique used by [Alp15] (which has also been used in [Xag13]) to extend the results of [DM14] to matrices seems to be inapplicable in our setting. This is because in our setting we need to commute the LWE challenge matrix $\mathbf{A}$ instead of the gadget matrix $\mathbf{G}$ whose associating trapdoor is known. To summarize, by incorporating our second idea, we obtain the ring variant of Eq. (2) and the trapdoor operates as specified. We note that one might be tempted to pack the entries of y with constant size elements, since 2^n is still exponential in n and hence $Q(n) < 2^n$. However, the security proof relies heavily on the fact that the density (i.e., the number of entries that are packed) of y is bounded by some constant. Therefore, we must choose the size of the packed elements with care to make the overall scheme secure.

The final idea is carefully crafting a properly distributed challenge ciphertext. To be precise, the main issue is in the difficulty of creating a ciphertext that has errors that are properly distributed. This problem of generating a properly distributed challenge ciphertext was addressed in [Yam16] as well, however, they used the standard technique called the "smudging" or "noise flooding" technique which came at the cost of making the modulus size q super-polynomial in n. This was not a problem for them, since as we pointed out earlier, their scheme inherently needed a super-polynomial sized modulus to work. However, this tactic is inapplicable to our setting since we want to restrict ourselves to the polynomial sized modulus. To overcome this we devise a way to carefully craft the error term; a technique reminiscent of [GPV08, ACPS09]. First, assume we have $\mathsf{F}(\mathsf{ID}^\star) = 0$ for the challenge identity $\mathsf{ID}^\star$ and thus $\mathsf{H}(\mathsf{ID}) = \mathbf{A}\mathbf{R}_{\mathsf{ID}^\star}$. Note that for ease of understanding we explain the technique in the matrix form instead of the ring form. To prove security, we have to embed the LWE challenge $\mathbf{A}$ and $\mathbf{v}$ into the challenge ciphertext, where $\mathbf{v} = \mathbf{s}\mathbf{A} + \mathbf{x}$ or $\mathbf{v}$ a random vector. One natural way is to set

$$\mathbf{x}_1 = \mathbf{x}, \quad \mathbf{x}_2 = \mathbf{x}\mathbf{R}_{\mathsf{ID}^\star} \tag{3}$$

and compute the challenge ciphertext as

$$\mathbf{s}[\mathbf{A}|\mathsf{H}(\mathsf{ID}^\star)] + [\mathbf{x}_1|\mathbf{x}_2] = [\mathbf{v}|\mathbf{v}\mathbf{R}_{\mathsf{ID}^\star}].$$

However, one can not simply use the standard generalized leftover hash lemma for lattices presented in [ABB10]; a technique often used in proving such forms. This is because $\mathbf{R}_{\mathsf{ID}^\star}$ is not uniformly sampled as in the case of [ABB10], but instead highly correlated to the values of y, $\{y_{i,j}\}$ used during the simulation. Alternatively, we present a noise rerandomization technique and add a small extra noise to Eq. (3) and statistically hide $\mathbf{R}_{\mathsf{ID}}$. Namely, we sample noises $\mathbf{e}_1$ and $\mathbf{e}_2$ from a particular Gaussian distribution with variance computed from $\mathbf{R}_{\mathsf{ID}^\star}$ and set

$$\mathbf{x}_1 = \mathbf{x} + \mathbf{e}_1, \quad \mathbf{x}_2 = \mathbf{x}\mathbf{R}_{\mathsf{ID}^\star} + \mathbf{e}_2.$$

Thus the challenge ciphertext is created as above by further adding the new noise terms. Although the general idea of this technique has been around since [Reg05, GPV08] and has been used in contexts elsewhere, as far as we know, we believe this is a nice application for rerandomizing the noise without the need of adding a huge (super-polynomial sized) noise.

An Additional Idea. Working in the ring setting introduces some subtle yet crucial obstacles, which we did not have to address before. Namely, for q a prime and n a power of 2, the domain $R_q = \mathbb{Z}[X]/(q, X^n + 1)$ we work in is no longer a field as in the case of $\mathbb{Z}_q$. Additionally, if we use a modulus q such that $q \equiv 1 \mod 2n$ as in [LPR10,LPR13], the ring R_q completely splits into n fields. In such a ring, each field only contains $q = \mathsf{poly}(n)$ elements so the Schwartz-Zippel lemma during our security proof can not be applied. We get around this by using a modulus q such that $q \equiv 3 \mod 8$ where it is known to split into only two fields. Then, since each field now contains $q^{n/2}$ elements and R_q acts roughly as a field, we are able to apply our proof techniques. As for the purpose of completeness, we prove the hardness of LWE over such rings by the straightforward combination of previous results. We finally note that we also obtain a nice regularity lemma over such rings which helps us attain better parameters for the scheme.

We also employ some ideas to further optimize the sizes of the public parameters, secret keys and ciphertexts. Namely, we use the (ring version of the) $\mathbf{G}$-trapdoor where the base is set as n^η for some positive constant η. We use $\eta = \frac{1}{4}$ for our concrete parameter selection. By incorporating this idea, we can further reduce the size of the parameters by a factor of $\log n$. However, this comes at the cost of making the scheme less efficient, since the function $\mathbf{G}^{-1}(\cdot)$ has a slower running time for a larger base.

2.2 Construction from Bilinear Maps

Here, we explain our IBE scheme from bilinear maps. We start with a slightly modified version of Waters IBE [Wat05] and gradually modify it to obtain our scheme. Let us consider a group $\mathbb{G}$ with prime order p whose generator is g. The group is equipped with a efficiently computable bilinear map $e : \mathbb{G} \times \mathbb{G} \to \mathbb{G}_T$. The public parameters of the scheme contains rather long $\kappa + 3$ group elements $\{g^{w_i}\}_{i\in[0,\kappa]}$, g^α, g^β, and a randomness $\mathsf{rand} \in \{0,1\}^{|\mathbb{G}_T|}$ that is used to derive the Goldreich-Levin hardcore bit function $\mathsf{GL} : \{0,1\}^{|\mathbb{G}_T|} \times \{0,1\}^{|\mathbb{G}_T|} \to \{0,1\}$. The form of the ciphertexts and private keys in the scheme are as follows:

$$C = \left(g^s,\ g^{s\mathsf{H}(\mathsf{ID})},\ \mathsf{GL}\left(e(g^\alpha, g^\beta)^s, \mathsf{rand}\right) \oplus \mathsf{M} \right), \quad \mathsf{sk}_{\mathsf{ID}} = \left(g^{\alpha\beta} \cdot g^{r\mathsf{H}(\mathsf{ID})},\ g^{-r} \right)$$

where $\mathsf{M} \in \{0,1\}$ is the message to be encrypted, and s and r are random elements in $\mathbb{Z}_p$ that are picked during the encryption and key generation algorithms, respectively.

Here, $\mathsf{H} : \{0,1\}^\kappa \to \mathbb{Z}_p$ is defined as $\mathsf{H}(\mathsf{ID}) = w_0 + \sum_{\mathsf{ID}_i=1} w_i$ where ID_i is the i-th bit of ID. The reason why we use the hardcore bit function is to base the security of the scheme on the *computational* bilinear Diffie-Hellman (CBDH) assumption, rather than the stronger *decisional* bilinear Diffie-Hellman (DBDH) assumption which was used to prove the security of the original Waters IBE.

Next, we try to reduce the size of the public parameters using the idea of the Yamada IBE. A natural way to do this would be to introduce the injective map $S : \{0,1\}^\kappa \to 2^{[\ell]\times[\ell]}$ with $\ell = \lceil \kappa^{1/2} \rceil$, change the public parameters to be $g^{w_0}, \{g^{w_{i,j}}\}_{(i,j)\in[2]\times[\ell]}$, and modify the function H as

$$\mathsf{H}(\mathsf{ID}) = w_0 + \sum_{(i,j)\in S(\mathsf{ID})} w_{1,i} w_{2,j}.$$

Through this change, we can reduce the size of the public parameters from $O(\kappa)$ group elements to $O(\sqrt{\kappa})$, just in as [Yam16]. However, we come across an immediate problem: We cannot efficiently compute $g^{s\mathsf{H}(\mathsf{ID})}$ from the public parameters! A straightforward solution to this problem is to put "helper" terms $\{g^{w_{1,i}w_{2,j}}\}$ into the public parameters. However, this makes the size of the public parameters large again.

Our solution to this problem is to rely on the Boneh-Boyen technique [BB04a] to compute something similar to the problematic term. Namely, we compute

$$g^{s\mathsf{H}(\mathsf{ID})+\sum_{j\in S(\mathsf{ID})}\tilde{t}_j w_{2,j}}, \qquad \{\ g^{\tilde{t}_j}\ \}_{j\in[\ell]} \tag{4}$$

instead of computing only $g^{s\mathsf{H}(\mathsf{ID})}$. Here, $\{\tilde{t}_j\}$ are additional randomness introduced by the encryption algorithm. Accordingly, we change the form of the ciphertexts and private keys of our scheme as follows:

$$C = \left(\ g^s,\ g^{s\mathsf{H}(\mathsf{ID})+\sum_{j\in[\ell]}\tilde{t}_j w_{2,j}},\ \{g^{\tilde{t}_j}\}_{j\in[\ell]},\ \mathsf{GL}\left(e(g^\alpha,g^\beta)^s,\mathsf{rand}\right)\oplus\mathsf{M}\ \right),$$
$$\mathsf{sk}_{\mathsf{ID}} = \left(\ g^{\alpha\beta}\cdot g^{r\mathsf{H}(\mathsf{ID})},\ g^{-r},\ \{g^{rw_{2,j}}\}_{j\in[\ell]}\ \right). \tag{5}$$

Note that although the size of the public parameters is smaller than the original scheme, the sizes of the ciphertexts and private keys are larger due to the additional terms. We now show that one can efficiently compute the ciphertext. In particular, we show that it is possible to generate the terms in Eq. (4). To see this, let us introduce the variables $\{t_j\}$ such that

$$\tilde{t}_j := t_j - s\left(\sum_{i\in\{i\in[1,\ell]\,|\,(i,j)\in S(\mathsf{ID})\}} w_{1,i}\right). \tag{6}$$

Then, we have

$$\begin{aligned}
&s\mathsf{H}(\mathsf{ID}) + \sum_{j\in[\ell]}\tilde{t}_j w_{2,j}\\
&= s\mathsf{H}(\mathsf{ID}) + \sum_{j\in[\ell]} w_{2,j}\left(t_j - s\left(\sum_{i\in\{i\in[1,\ell]\,|\,(i,j)\in S(\mathsf{ID})\}} w_{1,i}\right)\right)\\
&= s\mathsf{H}(\mathsf{ID}) + \sum_{j\in[\ell]} w_{2,j}t_j - s\sum_{j\in[\ell]}\left(\sum_{i\in\{i\in[1,\ell]\,|\,(i,j)\in S(\mathsf{ID})\}} w_{1,i}w_{2,j}\right)\\
&= sw_0 + s\cancel{\sum_{(i,j)\in S(\mathsf{ID})} w_{1,i}w_{2,j}} + \sum_{j\in[\ell]} w_{2,j}t_j - s\cancel{\sum_{(i,j)\in S(\mathsf{ID})} w_{1,i}w_{2,j}}\\
&= sw_0 + \sum_{j\in[\ell]} w_{2,j}t_j.
\end{aligned} \tag{7}$$

Since Eqs. (6) and (7) are linear in w_0, $w_{i,j}$, it can be seen that the terms in Eq. (4) can be computed efficiently, as desired.

By substituting $\tilde{t}_j$ in Eq. (5) with the right-hand side of Eq. (4), we obtain our final scheme. As for the security, we can prove the adaptive security of the scheme from the 3-computational bilinear Diffie-Hellman exponent (3-CBDHE) assumption. We need to rely on this stronger assumption than the standard CBDH assumption, because of the different algebraic structure incorporated by the modified Waters IBE.

3 Preliminaries

Due to the space limitation, most of the proofs for the lemmas presented in this paper are omitted. For the full proof refer to our full version.

Notations. We use non-italic bold lowercase letters (e.g., $\mathbf{v}$) for vectors with entries in $\mathbb{R}$ and italic bold lowercase letters (e.g., $\boldsymbol{v}$) for vectors with entries in rings or number fields. We view vectors in the row form stated otherwise. Matrices are denoted by uppercase bold letters analogously. For a vector $\mathbf{v} \in \mathbb{R}^n$, denote $\|\mathbf{v}\|_p$ as the L_p-norm, where $p = 2$ is the standard Euclidean norm. For a matrix $\mathbf{R} \in \mathbb{R}^{n \times n}$, denote $\|\mathbf{R}\|_{\mathsf{GS}}$ as the longest column of the Gram-Schmidt orthogonalization of $\mathbf{R}$ and denote $s_1(\mathbf{R})$ as the largest singular value (spectral norm). We denote $[\cdot|\cdot]$ (resp. $[\cdot;\cdot]$) as the horizontal (resp. vertical) concatenation of vectors and matrices. We denote $[a, b]$ as the set $\{a, a+1, \ldots, b-1, b\}$ for any integers $a, b \in \mathbb{N}$ satisfying $a \leq b$, and for simplicity write $[b]$ for the special case $a = 1$. For a (quotient) polynomial ring R over $\mathbb{Z}$, we denote $[-b, b]_R \subseteq R$ as the set of elements in R with all coefficients in the interval $[-b, b]$. Statistical distance between two random variables X and Y with support Ω is defined as $\Delta(X; Y) = \frac{1}{2} \sum_{s \in \Omega} |\Pr[X = s] - \Pr[Y = s]|$. A function $f : \mathbb{N} \to \mathbb{R}_{\geq 0}$ is said to be negligible, if for all c, there exists λ_0 such that $f(\lambda) < 1/\lambda^c$ for all $\lambda > \lambda_0$. We denote by $\mathsf{negl}(\lambda)$ a negligible function in λ.

3.1 Identity-Based Encryption

We use the standard syntax of IBE [BF01]. We briefly recall the security notion of IBEs and refer the exact definition to the full version. In our paper, we define two security notions: *adaptive security* and *adaptively-anonymous security*. The former adaptive security is the standard notion for IBEs as in [Wat05]. The latter adaptively-anonymous security is a notion that additionally requires the ciphertext to be indistinguishable from random. The term anonymous captures the fact the the ciphertext does not reveal the identity for which it was sent to. Furthermore, we use two random variables coin and $\widehat{\mathsf{coin}}$ in $\{0, 1\}$ for defining the security for IBEs. coin refers to the random value chosen by the challenger at the beginning of the security game and $\widehat{\mathsf{coin}}$ refers to the random value outputted by the adversary at the end of the game. We provide a general statement concerning coin and $\widehat{\mathsf{coin}}$ in Sect. 3.4.

3.2 Lattices and Gaussian Distributions

An n-dimensional (full rank) lattice $\Lambda \subseteq \mathbb{R}^n$ is the set of all integer linear combinations of some set of n linearly independent basis vectors $\mathbf{B} = \{\mathbf{b}_1, \dots, \mathbf{b}_n\} \subseteq \mathbb{R}^n$, $\Lambda = \{\sum_{i \in [n]} z_i \mathbf{b}_i | \mathbf{z} \in \mathbb{Z}^n\}$. For positive integers q, n, m, a matrix $\mathbf{A} \in \mathbb{Z}_q^{n \times m}$ and a vector $\mathbf{u} \in \mathbb{Z}_q^n$, the m-dimensional "shifted" integer lattice is defined as $\Lambda_{\mathbf{u}}^{\perp}(\mathbf{A}) = \{\mathbf{z} \in \mathbb{Z}^m | \mathbf{A}\mathbf{z}^T = \mathbf{u}^T \mod q\}$. We simply write $\Lambda^{\perp}(\mathbf{A})$ in case $\mathbf{u} = \mathbf{0}$.

For $s > 0$, the n-dimensional Gaussian function $\rho_s : \mathbb{R}^n \to (0, 1]$ is defined as $\rho_s(\mathbf{x}) = \exp(-\pi \|\mathbf{x}\|_2^2 / s^2)$. The (spherical) continuous Gaussian distribution D_s over $\mathbb{R}^n$ is the distribution with density function proportional to ρ_s. When the dimension n is not clear from context, we explicitly write it as D_s^n. More generally, for any matrix $\mathbf{B} \in \mathbb{R}^{n \times m}$, denote $D_{\mathbf{B}}$ as the distribution of $\mathbf{x}\mathbf{B}^T$ where $\mathbf{x}$ is distributed as D_1^m. A well known fact is that for any two matrices $\mathbf{B}_1, \mathbf{B}_2$, the sum of an independent sample from $D_{\mathbf{B}_1}$ and $D_{\mathbf{B}_2}$ is distributed as $D_{\mathbf{C}}$ where $\mathbf{C} = (\mathbf{B}_1\mathbf{B}_1^T + \mathbf{B}_2\mathbf{B}_2^T)^{1/2}$.

For a n-dimensional lattice Λ and a vector in $\mathbf{u} \in \mathbb{R}^n$, the discrete Gaussian distribution $D_{\Lambda+\mathbf{u},s}$ over the coset $\Lambda + \mathbf{u}$ is defined as $D_{\Lambda+\mathbf{u},s}(\mathbf{x}) = \rho_s(\mathbf{x}) / \rho_s(\Lambda + \mathbf{u})$ for all $\mathbf{x} \in \Lambda + \mathbf{u}$. We also define the discrete Gaussian distribution $D_{\Lambda+\mathbf{u},r}^{\mathsf{coeff}}$ over a (quotient) polynomial ring R in X over $\mathbb{R}$. The discrete Gaussian distribution $D_{\Lambda+\mathbf{u},r}^{\mathsf{coeff}}$ is the distribution of $a = \sum_{i=0}^{n-1} \alpha_i X^i \in R$ where the coefficient vector $[\alpha_0, \dots, \alpha_{n-1}] \in \mathbb{R}^n$ is sampled from the discrete Gaussian distribution $D_{\Lambda+\mathbf{u},r}$. This definition naturally extends to vectors $\boldsymbol{a} \in R^k$ in case of nk-dimensional lattices.

The following lemma on noise rerandomization plays an important role in the security proof of our scheme when creating a properly distributed challenge ciphertext. This allows us to simulate the challenge ciphertext without resorting to the noise flooding technique as in [Yam16]. Namely, during simulation we set $\ell = 2m$, $\mathbf{V} = [\mathbf{I}_m | \mathbf{R}_{\mathsf{ID}}]$ and $\mathbf{b} + \mathbf{x}$ as the LWE challenge (note that we view the LWE challenge in a slightly different way than usual).

Lemma 1 (Noise Rerandomization). *Let q, ℓ, m be positive integers and r a positive real satisfying $r > \max\{\omega(\sqrt{\log m}), \omega(\sqrt{\log \ell})\}$. Let $\mathbf{b} \in \mathbb{Z}_q^m$ be arbitrary and $\mathbf{x}$ chosen from $D_{\mathbb{Z}^m,r}$. Then for any $\mathbf{V} \in \mathbb{Z}^{m \times \ell}$ and positive real $\sigma > s_1(\mathbf{V})$, there exists a PPT algorithm* $\mathsf{ReRand}(\mathbf{V}, \mathbf{b}+\mathbf{x}, r, \sigma)$ *that outputs $\mathbf{b}' = \mathbf{b}\mathbf{V} + \mathbf{x}' \in \mathbb{Z}_q^{\ell}$ where $\mathbf{x}'$ is distributed statistically close to $D_{\mathbb{Z}^{\ell},2r\sigma}$.*

3.3 Rings and Ideal Lattices

We try to provide a minimum exposition of rings and ideal lattices to keep it self-contained. For further detail see the full version or refer to other works [LPR10, LPR13].

Preparation. Let n be a power of 2 and set $m = 2n$. Define the ring $R = \mathbb{Z}[X]/(\Phi_m(X))$, where $\Phi_m(X) = X^n + 1$ is the mth cyclotomic polynomial. For an integer q, denote R_q as $R/qR = \mathbb{Z}[X]/(q, \Phi_m(X))$. By viewing the elements in R as $n-1$ degree polynomials in $\mathbb{Z}[X]$, we can consider a natural coefficient embedding of R onto the integer lattice $\mathbb{Z}^n$. Namely, we define the coefficient embedding $\phi : R \to \mathbb{Z}^n$ that maps $a = \sum_{i=0}^{n-1} \alpha_i X^i \in R$ to $[\alpha_0, \alpha_1, \dots, \alpha_{n-1}] \in \mathbb{Z}^n$. We extend the coefficient embedding naturally to vectors and matrices. On the other hand, we can also identify R as the subring of anti-circulant matrices in $\mathbb{Z}^{n\times n}$ by viewing each ring element $a \in R$ as a linear transformation $r \to a{\cdot}r$ of R. Concretely, we define the ring homomorphism rot : $R \to \mathbb{Z}^{n\times n}$ that sends $a \in R$ to a matrix in $\mathbb{Z}^{n\times n}$ such that the i-th row is $\phi(a \cdot X^{i-1} \mod \Phi_m(X)) \in \mathbb{Z}^n$. Note that the first row of rot(a) is $\phi(a)$. Similarly to above, the definition of the map rot naturally extends to vectors and matrices.

Norms in R. We define the Euclidean length for an element $a \in R$ and a vector $\boldsymbol{v} \in R^k$ by identifying R with $\mathbb{Z}^n$ through the coefficient embedding.[2] Therefore, when we say a vector $\boldsymbol{v}$ in R^k is "short", we mean that $\|\phi(\boldsymbol{v})\|_2$ is small. We also define the largest singular value of a matrix $\boldsymbol{R} \in R^{s\times t}$ by identifying the ring R with $\mathbb{Z}^{n\times n}$ through the map rot.[3] Namely, $s_1(\boldsymbol{R}) := \max_{\|\mathbf{z}\|_2=1} \|\mathbf{z} \cdot \text{rot}(\boldsymbol{R})\|_2$. Note that this definition allows us to consider singular values of an element in R as well.

Properties for Elements in R. As with matrices with entries in $\mathbb{R}$, we have similar singular value bounds for matrices with elements in R. Namely, we can bound the singular value of a random matrix chosen from $[-b, b]_R^{s\times t}$. Recall that an element of $[-b, b]_R$ is an element in R with all of its coefficients in the interval $[-b, b]$.

Lemma 2 ([DM15], **Special case of Fact 1**). *Let b be a positive integer and $\boldsymbol{R}$ be a $s \times t$ matrix chosen uniformly at random from $[-b, b]_R^{s\times t}$. Then, there exists a universal constant $C(\approx 1/\sqrt{2\pi})$ such that*

$$\Pr[s_1(\boldsymbol{R}) \geq C \cdot b\sqrt{n} \cdot (\sqrt{s} + \sqrt{t} + \omega(\sqrt{\log n}))] = \mathsf{negl}(n)$$

We note that similarly to matrices with entries in $\mathbb{R}$, we have $s_1(\boldsymbol{R}_1\boldsymbol{R}_2) \leq s_1(\boldsymbol{R}_1)s_1(\boldsymbol{R}_2)$ for all $\boldsymbol{R}_1, \boldsymbol{R}_2 \in R^{k\times k}$, which follows from the fact that rot is a ring homomorphism. Furthermore, it also holds when $\boldsymbol{R}_1$ is replaced by an element a in R.

Regularity Lemma. The former Lemma shows that there exists a quotient ring $R_q = R/(q, \Phi_m(X))$ that acts roughly as a field, or in other words, R_q has exponentially many invertible elements. The latter Lemma is a ring analogue of the standard lattice regularity lemma.

Lemma 3. *Let q be a prime such that $q \equiv 3 \mod 8$ and n be a power of 2. Then, $\Phi_{2n}(X) = X^n + 1$ splits as $X^n + 1 \equiv t_1 t_2 \mod q$ for two irreducible polynomials $t_1 = X^{n/2} + uX^{n/4} - 1$ and $t_2 = X^{n/2} - uX^{n/4} - 1$ in $\mathbb{Z}_q[X]$ where $u^2 \equiv -2 \mod q$. Furthermore, all $x \in R_q$ satisfying $\|\phi(x)\|_2 < \sqrt{q}$ are invertible, i.e., $x \in R_q^*$.*

[2] We could have identified the Euclidean length by the *canonical* embedding as done in other works. However, for our special case where n is power of 2, the lengths are equivalent up to a factor of $\sqrt{n}$.

[3] For the special case where n is a power of 2, $s_1(\boldsymbol{R})$ defined by the coefficient and canonical embeddings are both equivalent to the one defined by the map rot.

Lemma 4 (Regularity Lemma). *Let n be a power of 2, q be a prime larger than $4n$ such that $q \equiv 3 \mod 8$, and ℓ, k', k, ρ be positive integers satisfying $\ell, k' \geq 1$, $k \geq 2$, $\rho < \frac{1}{2}\sqrt{q/n}$. Define the family of hash functions $\mathcal{H} = \{h_{\boldsymbol{A}}(\boldsymbol{x}) : [-\rho,\rho]_R^k \to R_q^{k'}\}$, where $h_{\boldsymbol{A}}(\boldsymbol{x}) = \boldsymbol{A}\boldsymbol{x}$ for $\boldsymbol{A} \in R_q^{k' \times k}$, $\boldsymbol{x} \in R_q^{k \times 1}$. Then, $\mathcal{H}$ is a universal hash family. Furthermore, for $\boldsymbol{A} \xleftarrow{\$} R_q^{k' \times k}$ and $\boldsymbol{X} \xleftarrow{\$} [-\rho,\rho]_R^{k \times \ell}$, we have*

$$\Delta((\boldsymbol{A}, \boldsymbol{A}\boldsymbol{X}) \ ; \ (\boldsymbol{A}, U(R_q^{k' \times \ell}))) \leq \frac{\ell}{2} \cdot \sqrt{\left(\frac{q^{k'}}{(2\rho+1)^k}\right)^n}.$$

Ring Learning with Errors. The ring LWE problem was introduced by Lyubashevsky et al. [LPR10]. They showed that solving it on the average is as hard as (quantumly) solving several standard problems on ideal lattices in the worst case.

Definition 1 (RLWE). *For positive integers $n = n(\lambda)$, $k = k(n)$, a prime integer $q = q(n) > 2$, an error distribution $\chi = \chi(n)$ over R_q, and an PPT algorithm $\mathcal{A}$, an advantage for the RLWE problem $\mathsf{RLWE}_{n,k,q,\chi}$ of $\mathcal{A}$ is defined as follows:*

$$\mathsf{Adv}_{\mathcal{A}}^{\mathsf{RLWE}_{n,k,q,\chi}} = |\Pr[\mathcal{A}(\{(a_i, v_i)\}_{i=1}^k) \to 1] - \Pr[\mathcal{A}(\{(a_i, a_i s + e_i)\}_{i=1}^k) \to 1]|$$

where $a_1, \ldots, a_k, v_1, \ldots, v_k, s \xleftarrow{\$} R_q$ and $e_1, \ldots, e_k \xleftarrow{\$} \chi$. We say that $\mathsf{RLWE}_{n,k,q,\chi}$ assumption holds if $\mathsf{Adv}_{\mathcal{A}}^{\mathsf{RLWE}_{n,k,q,\chi}}$ is negligible for all PPT $\mathcal{A}$.

Theorem 1. *Let α be a positive real, m be a power of 2, ℓ be an integer, $\Phi_m(X) = X^n + 1$ be the mth cyclotomic polynomial where $m = 2n$, and $R = \mathbb{Z}[X]/(\Phi_m(X))$. Let $q \equiv 3 \mod 8$ be a (polynomial size) prime such that there is another prime $p \equiv 1 \mod m$ satisfying $p \leq q \leq 2p$ and $\alpha q \geq n^{3/2} k^{1/4} \omega(\log^{9/4} n)$. Then, there is a probabilistic polynomial-time quantum reduction from $\tilde{O}(\sqrt{n}/\alpha)$-approximate SIVP (or SVP) to $\mathsf{RLWE}_{n,k,q,\chi}$ with $\chi = D_{\mathbb{Z}^n,\alpha q}^{\mathsf{coeff}}$.*

The proof is obtained by a straightforward combination of previous results [LPR10,LS15]. Due to the Linnik's theorem and Dirichlet's theorem on arithmetic progressions, we have that there are sufficiently many primes p and q satisfying the assumption of the theorem.

Trapdoors for Rings. Define the gadget matrix $\boldsymbol{g}_b = [1|b|\cdots|b^{k'-1}|\boldsymbol{0}] \in R_q^k$, where b is a positive integer and $k \geq k' = \lceil \log_b q \rceil$. When $k = k'$ and $b = 2$, this corresponds to the matrix representation of the gadget matrix $\mathbf{G} \in \mathbb{Z}_q^{n \times nk}$ often used in the literatures by properly rearranging the rows and columns of $\mathrm{rot}(\boldsymbol{g}_2)$. The following algorithms are simple modification of traditional lattice based algorithms.

Lemma 5. *Let n be a power of 2, q be a prime larger than $4n$ such that $q \equiv 3 \mod 8$, and b, ρ be a positive integer satisfying $\rho < \frac{1}{2}\sqrt{q/n}$. Furthermore, define $\log_1(\cdot) := \log_2(\cdot)$. Then, there exist polynomial time algorithms with the properties below:*

- $\mathsf{TrapGen}(1^n, 1^k, q, \rho) \to (\boldsymbol{a}, \boldsymbol{T_a})$ ([MP12], Lemma 5.3): *a randomized algorithm that, when $k \geq 2\log_\rho q$, outputs a vector $\boldsymbol{a} \in R_q^k$ and a matrix $\boldsymbol{T_a} \in R^{k\times k}$, where $\mathrm{rot}(\boldsymbol{a}^T)^T \in \mathbb{Z}_q^{n\times nk}$ is a full-rank matrix and $\mathrm{rot}(\boldsymbol{T_a}) \in \mathbb{Z}^{nk\times nk}$ is a basis for $\Lambda^\perp(\mathrm{rot}(\boldsymbol{a}^T)^T)$ such that $\boldsymbol{a}$ is $\mathsf{negl}(n)$-close to uniform and $\|\mathrm{rot}(\boldsymbol{T_a})\|_{\mathsf{GS}} = O(b\rho \cdot \sqrt{n \log_\rho q})$.*[4]
- $\mathsf{SampleLeft}(\boldsymbol{a}, \boldsymbol{b}, u, \boldsymbol{T_a}, \sigma) \to \boldsymbol{e}$ ([CHKP10]): *a randomized algorithm that, given vectors $\boldsymbol{a}, \boldsymbol{b} \in R_q^k$ where $\mathrm{rot}(\boldsymbol{a}^T)^T, \mathrm{rot}(\boldsymbol{b}^T)^T \in \mathbb{Z}_q^{n\times nk}$ are full-rank, an element $u \in R_q$, a matrix $\boldsymbol{T_a} \in R^{k\times k}$ such that $\mathrm{rot}(\boldsymbol{T_a}) \in \mathbb{Z}^{nk\times nk}$ is a basis for $\Lambda^\perp(\mathrm{rot}(\boldsymbol{a}^T)^T)$, and a Gaussian parameter $\sigma > \|\mathrm{rot}(\boldsymbol{T_a})\|_{\mathsf{GS}} \cdot \omega(\sqrt{\log nk})$, outputs a vector $\boldsymbol{e} \in R^{2k}$ sampled from a distribution which is $\mathsf{negl}(n)$-close to $D^{\mathsf{coeff}}_{\Lambda^\perp_{\phi(u)}([\mathrm{rot}(\boldsymbol{a}^T)^T|\mathrm{rot}(\boldsymbol{b}^T)^T]),\sigma}$, i.e., $[\boldsymbol{a}|\boldsymbol{b}]\boldsymbol{e}^T = u$ and $\phi(\boldsymbol{e}) \in \mathbb{Z}^{2nk}$ is distributed according to $D_{\Lambda^\perp_{\phi(u)}([\mathrm{rot}(\boldsymbol{a}^T)^T|\mathrm{rot}(\boldsymbol{b}^T)^T]),\sigma}$.*
- $\mathsf{SampleRight}(\boldsymbol{a}, \boldsymbol{g}_b, \boldsymbol{R}, y, u, \boldsymbol{T}_{\boldsymbol{g}_b}, \sigma) \to \boldsymbol{e}$ *where* $\boldsymbol{b} = \boldsymbol{aR} + y\boldsymbol{g}_b$ ([ABB10]): *a randomized algorithm that, given vectors $\boldsymbol{a}, \boldsymbol{g}_b \in R_q^k$ such that $\mathrm{rot}(\boldsymbol{a}^T)^T, \mathrm{rot}(\boldsymbol{g}_b)$*[5] *$\in \mathbb{Z}_q^{n\times nk}$ are full-rank matrices, elements $y \in R_q^*, u \in R_q$, a matrix $\boldsymbol{R} \in R^{k\times k}$, a matrix $\boldsymbol{T}_{\boldsymbol{g}_b} \in R^{k\times k}$ such that $\mathrm{rot}(\boldsymbol{T}_{\boldsymbol{g}_b}) \in \mathbb{Z}^{nk\times nk}$ is a basis for $\Lambda^\perp(\mathrm{rot}(\boldsymbol{g}_b))$, and a Gaussian parameter $\sigma > s_1(\boldsymbol{R}) \cdot \|\mathrm{rot}(\boldsymbol{T}_{\boldsymbol{g}_b})\|_{\mathsf{GS}} \cdot \omega(\sqrt{\log nk})$, outputs a vector $\boldsymbol{e} \in R^{2k}$ sampled from a distribution which is $\mathsf{negl}(n)$-close to $D^{\mathsf{coeff}}_{\Lambda^\perp_{\phi(u)}([\mathrm{rot}(\boldsymbol{a}^T)^T|\mathrm{rot}(\boldsymbol{b}^T)^T]),\sigma}$, i.e., $[\boldsymbol{a}|\boldsymbol{b}]\boldsymbol{e}^T = u$ and $\phi(\boldsymbol{e}) \in \mathbb{Z}^{2nk}$ is distributed according to $D_{\Lambda^\perp_{\phi(u)}([\mathrm{rot}(\boldsymbol{a}^T)^T|\mathrm{rot}(\boldsymbol{b}^T)^T]),\sigma}$.*
- ([MP12]:) *Let $k \geq \lceil \log_b q \rceil$. There exists a publicly known matrix $\boldsymbol{T}_{\boldsymbol{g}_b}$ such that $\mathrm{rot}(\boldsymbol{T}_{\boldsymbol{g}_b}) \in \mathbb{Z}^{nk\times nk}$ is a basis for the lattice $\Lambda^\perp(\mathrm{rot}(\boldsymbol{g}_b))$ and $\|\mathrm{rot}(\boldsymbol{T}_{\boldsymbol{g}_b})\|_{\mathsf{GS}} \leq \sqrt{b^2+1}$. Furthermore, there exists a deterministic polynomial time algorithm $\boldsymbol{g}_b^{-1}$ which takes input $\boldsymbol{u} \in R_q^k$ and outputs $\boldsymbol{R} = \boldsymbol{g}_b^{-1}(\boldsymbol{u})$ such that $\boldsymbol{R} \in [-b, b]_R^{k\times k}$ and $\boldsymbol{g}_b\boldsymbol{R} = \boldsymbol{u}$.*

Note that we abuse the notation $\boldsymbol{g}_b^{-1}$ by viewing it as a function rather than a vector. Namely, for any $\boldsymbol{u} \in R_q^k$ there are many choices for $\boldsymbol{R} \in R^{k\times k}$ such that $\boldsymbol{g}_b\boldsymbol{R} = \boldsymbol{u}$, and $\boldsymbol{g}_b^{-1}(\boldsymbol{u})$ is a function that deterministically outputs a particular short matrix from the possible candidates. Since we have $s_1(\boldsymbol{R}) \leq b \cdot nk$ for any $\boldsymbol{R} \in [-b, b]_R^{k\times k}$, $s_1(\boldsymbol{g}_b^{-1}(\boldsymbol{u})) \leq bnk$ holds for arbitrary $\boldsymbol{u} \in R_q^k$.

Homomorphic Computation. Let d be a natural number. We introduce the function $\mathsf{PubEval}_d : (R_q^k)^d \to R_q^k$ as in [Yam16], which takes a set of vectors $\boldsymbol{b}_1, \boldsymbol{b}_2, \ldots, \boldsymbol{b}_d \in R_q^k$ as inputs and outputs a vector in R_q^k. This function will be used to hash identities to R_q^k in our lattice-based IBE construction. The function is defined recursively as follows:

$$\mathsf{PubEval}_d(\boldsymbol{b}_1, \ldots, \boldsymbol{b}_d) = \begin{cases} \boldsymbol{b}_1 & \text{if } d = 1 \\ \boldsymbol{b}_1 \cdot \boldsymbol{g}_b^{-1}(\mathsf{PubEval}_{d-1}(\boldsymbol{b}_2, \ldots, \boldsymbol{b}_d)) & \text{if } d \geq 2. \end{cases}$$

[4] We combine several lemmas from [MP12] and the regularity lemma (Lemma 4) to show correctness of TrapGen. See the full version for further detail. Further, the unusual lattice $\Lambda^\perp(\mathrm{rot}(\boldsymbol{a}^T)^T)$ is used only to be consistent with the other algorithms. Namely, we could have instead defined the trapdoor for the lattice $\Lambda^\perp(\mathrm{rot}(\boldsymbol{a}))$.

[5] We have $\mathrm{rot}(\boldsymbol{g}_b^T)^T = \mathrm{rot}(\boldsymbol{g}_b)$ since all the entries of $\boldsymbol{g}_b$ are integers.

Lemma 6. *Let $y_1, \ldots, y_d$ be elements in R, $\boldsymbol{a}, \boldsymbol{b}_1, \ldots, \boldsymbol{b}_d$ be vectors in R_q^k and $\boldsymbol{R}_1, \ldots, \boldsymbol{R}_d$ be matrices in $R^{k \times k}$ such that $\boldsymbol{b}_i = \boldsymbol{a}\boldsymbol{R}_i + y_i \boldsymbol{g}_b$ for $i \in [d]$. Furthermore, we assume that $s_1(\boldsymbol{R}_i) \le B, \|\phi(y_i)\|_1 \le \delta$ for $i \in [d]$. Then, there exists an efficient algorithm $\mathsf{TrapEval}_d$ that takes $\boldsymbol{R}_1, \ldots, \boldsymbol{R}_d$, $y_1, \ldots, y_d$ as inputs and outputs $\boldsymbol{R}' \in R^{k \times k}$ such that*

$$\mathsf{PubEval}_d(\boldsymbol{b}_1, \ldots, \boldsymbol{b}_d) = \boldsymbol{a}\boldsymbol{R}' + y_1 \cdots y_d \boldsymbol{g}_b \in R_q^k$$

and $s_1(\boldsymbol{R}') \le B\delta^{d-1} + Bbnk\left(\frac{\delta^{d-1}-1}{\delta-1}\right)$.

3.4 Other Facts

Lemma 7 (Expansion of Coefficients). *Let $c_1, c_2, B_1, B_2 \in \mathbb{N}$. Let also $u = u_0 + u_1 X + \cdots u_{c_1-1} X^{c_1-1} \in R$ and $v = v_0 + v_1 X + \cdots v_{c_2-1} X^{c_2-1} \in R$ be ring elements. We further assume that $c_1 + c_2 < n$ and $\|\phi(u)\|_\infty < B_1$ and $\|\phi(v)\|_\infty < B_2$. Then we have $\|\phi(uv)\|_\infty \le \min\{c_1, c_2\} \cdot B_1 B_2$.*

The following Lemma addresses a general statement for bounding the success probability of an adversary engaging with the security game of IBE. In more detail, when the partitioning technique is used to prove security, the guess returned by the adversary is correlated with the key extraction queries it has made. Therefore, we need to argue with care to obtain a meaningful bound on the success probability that holds for arbitrary key extraction queries.

Lemma 8 (Implicit in [BR09,Yam16]). *Let us consider an IBE scheme and an adversary $\mathcal{A}$ that breaks adaptive security (adaptively-anonymous security) with advantage ϵ. Let us also consider a map γ that maps a sequence of identities to a value in $[0,1]$. We consider the following experiment. We first execute the security game for $\mathcal{A}$. Let $\mathsf{ID}^\star$ be the challenge identity and $\mathsf{ID}_1, \ldots, \mathsf{ID}_Q$ be the identities for which key extraction queries were made. We denote $\mathbb{ID} = (\mathsf{ID}^\star, \mathsf{ID}_1, \ldots, \mathsf{ID}_Q)$. At the end of the game, we set $\mathsf{coin}' \in \{0,1\}$ as $\mathsf{coin}' = \widehat{\mathsf{coin}}$ with probability $\gamma(\mathbb{ID})$ and $\mathsf{coin}' \xleftarrow{\$} \{0,1\}$ with probability $1 - \gamma(\mathbb{ID})$. Then, the following holds.*

$$\left| \Pr[\mathsf{coin}' = \mathsf{coin}] - \frac{1}{2} \right| \ge \gamma_{\min} \cdot \epsilon - \frac{\gamma_{\max} - \gamma_{\min}}{2}$$

where $\gamma_{\min}$ (resp. $\gamma_{\max}$) is the maximum (resp. minimum) of $\gamma(\mathbb{ID})$ taken over all possible $\mathbb{ID}$.

Injective map. Let d and κ be some integers. Furthermore, let ℓ be $\ell = \lceil \kappa^{1/d} \rceil$. Then, an element of $[1, \kappa]$ can be written as an element of $[1, \ell]^d$ using some canonical map. Furthermore, it is also possible to write a subset of $[1, \kappa]$ as a subset of $[1, \ell]^d$ by naturally extending the canonical map. By identifying a bit string in $\{0,1\}^\kappa$ with a subset of $[1, \kappa]$ (for example, by regarding the former as the indicator vector of a subset of $[1, \kappa]$), we can define an efficiently computable injective map S that maps a bit string $\mathsf{ID} \in \{0,1\}^\kappa$ to a subset $S(\mathsf{ID})$ of $[1, \ell]^d$.

3.5 Core Lemma for Our Partitioning

We make a general statement concerning the partitioning technique for IBEs, which we use during the security analysis for both our lattice and bilinear map based constructions. Namely, we use the following Lemma in order to argue that the probability of the hash value for identities corresponding to the key extraction queries being invertible and the hash value for the challenge identity being zero is non-negligible.

Lemma 9. *Let $\nu, \mu, d, Q \geq 1$ be any integers. Let Φ be a ring and $\Omega_1, \ldots, \Omega_\nu$ be a set of fields equipped with homomorphisms $\pi_j : \Phi \to \Omega_j$ for $j \in [\nu]$. Assume that the map Π defined as $\Pi : \Phi \ni y \mapsto (\pi_1(y), \ldots, \pi_\nu(y)) \in \Omega_1 \times \cdots \times \Omega_\nu$ is an isomorphism. Let S_0 and S_1 be subsets of Φ with finite cardinality. Let us consider a set of multivariate polynomials $f_i(Y_1, \ldots, Y_\mu) \in \Phi[Y_1, \ldots, Y_\mu]$ for $i \in [0, Q]$ We further assume the following properties:*

1. *The map π_j is injective on S_1 for all $j \in [\nu]$.*
2. *We have $\pi_j(f_0) - \pi_j(f_i)$ is a non-zero polynomial with degree d for all $i \in [Q]$ and $j \in [\nu]$. Here π_j is extended to $\pi_j : \Phi[X] \to \Omega_j[X]$ in a natural way.*
3. *We have $S_0 \supseteq \cup_{i \in [0,Q]} \{-f_i(y_1, \ldots, y_\mu) | y_1, \ldots, y_\mu \in S_1\}$.*

Then, for $y_0 \overset{\$}{\leftarrow} S_0$ and $y_1, \ldots, y_\mu \overset{\$}{\leftarrow} S_1$, we have

$$\frac{1}{|S_0|}\left(1 - \frac{d\nu Q}{|S_1|}\right) \leq \gamma \leq \frac{1}{|S_0|}$$

where we denote

$$\gamma = \Pr_{y_0, \boldsymbol{y}'}[\, y_0 + f_0(\boldsymbol{y}') = 0 \ \wedge\ y_0 + f_1(\boldsymbol{y}') \in \Phi^* \ \wedge \cdots \wedge\ y_0 + f_Q(\boldsymbol{y}') \in \Phi^*],$$

$\boldsymbol{y}' = (y_1, \ldots, y_\mu)$, and $\Phi^* = \Pi^{-1}(\Omega_1^* \times \cdots \times \Omega_\nu^*)$.

4 Construction from RLWE

In this section, we show our IBE scheme from the RLWE assumption. Let d be a (flexible) constant number. In addition, let the identity space of the scheme be $\mathcal{ID} = \{0,1\}^\kappa$ for some $\kappa \in \mathbb{N}$ and the message space be $\{0,1\}^n \subset R$.[6] For our construction, we consider an efficiently computable injective map S that maps an identity $\mathsf{ID} \in \{0,1\}^\kappa$ to a subset $S(\mathsf{ID})$ of $[1, \ell]^d$, where $\ell = \lceil \kappa^{1/d} \rceil$. Such a map can be constructed easily as we explained in Sect. 3.4. Let $n := n(\lambda)$, $b := b(n)$, $\rho := \rho(n)$, $m := 2n$, $k := k(n)$, $q := q(n)$, $\ell := \ell(n)$, $\alpha := \alpha(n)$, $\alpha' := \alpha'(n)$, and $\sigma := \sigma(n)$ be parameters that are specified later. Let also $\Phi_m(X) = X^n + 1$ be the mth cyclotomic polynomial and $R = \mathbb{Z}[X]/(\Phi_m(X))$.

[6] Note that we regard m as an elements in R via $\phi^{-1} : \mathbb{Z}^n \to R$ (the inversion of coefficient embedding).

Setup(1^λ): On input 1^λ, it first runs $(\boldsymbol{a}, \boldsymbol{T_a}) \xleftarrow{\$} \mathsf{TrapGen}(1^n, 1^k, q, \rho)$ to obtain $\boldsymbol{a} \in R_q^k$ and $\boldsymbol{T_a} \in R^{k\times k}$. It also picks $u \xleftarrow{\$} R_q$, $\boldsymbol{b}_0, \boldsymbol{b}_{i,j} \xleftarrow{\$} R_q^k$ for $(i,j) \in [d]\times[\ell]$ and outputs

$$\mathsf{mpk} = (\boldsymbol{a}, \boldsymbol{b}_0, \{\boldsymbol{b}_{i,j}\}_{(i,j)\in[d]\times[\ell]}, u) \quad \text{and} \quad \mathsf{msk} = \boldsymbol{T_a}.$$

In the following, we use a deterministic function $\mathsf{H} : \mathcal{ID} \to R_q^k$ defined as

$$\mathsf{H}(\mathsf{ID}) = \boldsymbol{b}_0 + \sum_{(j_1,\ldots,j_d)\in S(\mathsf{ID})} \mathsf{PubEval}_d(\boldsymbol{b}_{1,j_1}, \boldsymbol{b}_{2,j_2}, \ldots, \boldsymbol{b}_{d,j_d}) \in R_q^k.$$

KeyGen(mpk, msk, ID): It first computes H(ID) and picks $\boldsymbol{e} \in R^{2k}$ such that

$$[\boldsymbol{a}|\mathsf{H}(\mathsf{ID})] \cdot \boldsymbol{e}^T = u$$

using $\mathsf{SampleLeft}(\boldsymbol{a}, \mathsf{H}(\mathsf{ID}), u, \mathbf{T}_{\boldsymbol{a}}, \sigma) \to \boldsymbol{e}$. It returns $\mathsf{sk_{ID}} = \boldsymbol{e}$.

Encrypt(mpk, ID, M): To encrypt a message $\mathsf{M} \in \{0,1\}^n \subset R$, it first picks $s \xleftarrow{\$} R_q$, $x_0 \xleftarrow{\$} D^{\mathsf{coeff}}_{\mathbb{Z}^n,\alpha q}$, $\boldsymbol{x}_1, \boldsymbol{x}_2 \xleftarrow{\$} (D^{\mathsf{coeff}}_{\mathbb{Z}^n,\alpha'})^k$. Then it computes

$$c_0 = su + x_0 + \lfloor q/2 \rceil \cdot \mathsf{M}, \quad \boldsymbol{c}_1 = s[\boldsymbol{a}|\mathsf{H}(\mathsf{ID})] + [\boldsymbol{x}_1|\boldsymbol{x}_2].$$

Finally, it outputs the ciphertext $C = (c_0, \boldsymbol{c}_1) \in R_q \times R_q^{2k}$.

Decrypt($\mathsf{mpk}, \mathsf{sk_{ID}}, C$): To decrypt a ciphertext $C = (c_0, \boldsymbol{c}_1)$ using a private key $\mathsf{sk_{ID}} = \boldsymbol{e}$, it computes $\left(\lfloor (2/q) \cdot \phi(c_0 - \boldsymbol{c}_1\boldsymbol{e}^T) \rceil \mod 2\right) = m$. Here, the rounding function $\lfloor \cdot \rceil$ is applied componentwise.

4.1 Correctness and Parameter Selection

The following lemma addresses the correctness of the scheme.

Lemma 10 (Correctness). *Assume $\alpha q \omega(\sqrt{\log n}) + \sqrt{nk}\alpha'\sigma\omega(\sqrt{\log nk}) \le q/5$ holds with over whelming probability. Then the above scheme has negligible decryption error.*

Parameter selection. We refer the precise requirements for the parameter selection to the full version. One concrete selection for the parameters is as follows:

$$k = 8d + 12, \qquad q = n^{2d+3}, \qquad b = \rho = n^{\frac{1}{4}},$$
$$\sigma = n^d \cdot \omega(\log n), \quad \alpha = n^{-2d-\frac{3}{8}} \cdot \omega(\log^2 n)^{-1}, \quad \alpha' = n^{d+\frac{5}{2}} \cdot \omega(\log^{\frac{3}{4}} n)^{-1},$$

where d is a (flexible) constant which may be set very small (e.g., $d = 2$ or 3) in a typical setting and the length κ of the identities ID is set as n. This specific instantiation is denoted as the Type 2 IBE scheme in Sect. 6. Table 1. Furthermore, the other concrete instantiation provided only in the full version, where we set $b = 2$ and $k = O(\log n)$, is denoted as the Type 1 IBE scheme.

4.2 Security Proof for the Scheme

The following theorem addresses the security of the scheme. The proof proceeds in a similar manner as in [Yam16], but we incorporate several novel ideas as we explained in Sect. 2.

Theorem 2. *The above IBE scheme is adaptively-anonymous secure assuming* $\mathsf{RLWE}_{n,k+1,q,D^{\mathsf{coeff}}_{\mathbb{Z}^n,\alpha q}}$ *is hard, where the ciphertext space is* $\mathcal{C} = R_q \times R_q^{2k}$.

Proof. Let $\mathcal{A}$ be a PPT adversary that breaks the adaptively-anonymous security of the scheme. In addition, let $\epsilon = \epsilon(n)$ and $Q = Q(n)$ be its advantage and the upper bound of the number of key extraction queries, respectively.

Since $\mathcal{A}$ is PPT and λ and n are polynomially related (namely, $n = O(\lambda^\delta)$ for some constant δ), there exists a constant number $c_1 \in \mathbb{N}$ such that $4(dQ+1) \leq n^{c_1}$ for all n that are sufficiently large. Similarly, since $\mathcal{A}$ breaks the security of the scheme, there exists $c_2 \in \mathbb{N}$ such that $2\epsilon \geq n^{-c_2}$ holds for infinitely many n. By setting $c = c_1 + c_2$, we have that

$$4dQ \leq n^c \text{ for all } n \in \mathbb{N} \quad \text{and} \quad \frac{\epsilon}{2(dQ+1)} \geq \frac{1}{n^c} \quad \text{for infinitely many } n \in \mathbb{N}. \qquad (8)$$

In the proof, we will assume $d(c-1) < n$. Since both c and d are constant numbers, this holds for sufficiently large n.

We show the security of the scheme via the following games. In each game, a value $\mathsf{coin}' \in \{0,1\}$ is defined. While it is set $\mathsf{coin}' = \mathsf{coin}$ in the first game, these values might be different in the later games. In the following, we define X_i to be the event that $\mathsf{coin}' = \mathsf{coin}$.

Game_0: This is the real security game. In the challenge phase, the challenge ciphertext is set as $C^\star = (c_0, \mathbf{c}_1) \xleftarrow{\$} R_q \times R_q^{2k}$ if $\mathsf{coin} = 1$. Otherwise, it is set as $C^\star \leftarrow \mathsf{Encrypt}(\mathsf{mpk}, \mathsf{ID}, \mathsf{M})$, where M is the message chosen by $\mathcal{A}$. At the end of the game, $\mathcal{A}$ outputs a guess $\widehat{\mathsf{coin}}$ for coin. Finally, the challenger sets $\mathsf{coin}' = \widehat{\mathsf{coin}}$. By definition, we have

$$\left|\Pr[X_0] - \frac{1}{2}\right| = \left|\Pr[\mathsf{coin}' = \mathsf{coin}] - \frac{1}{2}\right| = \left|\Pr[\widehat{\mathsf{coin}} = \mathsf{coin}] - \frac{1}{2}\right| = \epsilon.$$

Game_1: For integers $t_0, t_1 \in \mathbb{Z}$ such that $t_0 \leq t_1$ and positive integer $c \in \mathbb{N}$, let us denote $[t_0, t_1]_{R,c}$ as

$$[t_0, t_1]_{R,c} := \left\{ \sum_{i=0}^{c-1} a_i X^i \;\middle|\; a_i \in [t_0, t_1] \text{ for all } i \in [0, c-1] \right\} \subseteq R.$$

In words, $[t_0, t_1]_{R,c}$ denotes the set of polynomials of degree less then $c-1$ with all of its coefficients in the interval $[t_0, t_1]$. Note that c is the constant defined in Eq. (8). In this game, we change Game_0 so that the challenger performs the following additional step at the end of the game. First, the challenger picks $\mathbf{y} = (y_0, \{y_{i,j}\}_{(i,j)\in[d,\ell]})$ as

$$y_0 \xleftarrow{\$} [-\kappa(cn)^d, -1]_{R,(c-1)d+1} \qquad \text{and} \qquad y_{i,j} \xleftarrow{\$} [1, n]_{R,c} \tag{9}$$

for $(i,j) \in [d] \times [\ell]$. Recall κ is the length of the identities. We then define a function $\mathsf{F}_{\boldsymbol{y}} : \mathcal{ID} \to R_q$ as follows:

$$\mathsf{F}_{\boldsymbol{y}}(\mathsf{ID}) = y_0 + \sum_{(j_1,\ldots,j_d)\in S(\mathsf{ID})} y_{1,j_1} \cdots y_{d,j_d}.$$

Then the challenger checks whether the following condition holds:

$$\mathsf{F}_{\boldsymbol{y}}(\mathsf{ID}^\star) = 0 \ \wedge \ \mathsf{F}_{\boldsymbol{y}}(\mathsf{ID}_1) \in R_q^* \ \wedge \ \cdots \ \wedge \ \mathsf{F}_{\boldsymbol{y}}(\mathsf{ID}_Q) \in R_q^*, \tag{10}$$

where $\mathsf{ID}^\star$ is the challenge identity, and $\mathsf{ID}_1, \ldots, \mathsf{ID}_Q$ are identities for which $\mathcal{A}$ has made key extraction queries. If it does not hold, the challenger ignores the output $\widehat{\mathsf{coin}}$ of $\mathcal{A}$, and sets $\mathsf{coin}' \xleftarrow{\$} \{0,1\}$. In this case, we say that the challenger aborts. If condition (10) holds, the challenger sets $\mathsf{coin}' = \widehat{\mathsf{coin}}$. As we will show in Lemma 11, we have

$$\left|\Pr[X_1] - \frac{1}{2}\right| \geq \frac{1}{(\kappa c^d n^d)^{(c-1)d+1}} \left(\frac{\epsilon}{2} - \frac{dQ}{n^c}\right).$$

So as not to interrupt the proof of Theorem 2, we intentionally skip the proof for the time being.

Game_2: In this game, we change the way $\boldsymbol{b}_0$ and $\boldsymbol{b}_{i,j}$ are chosen. At the beginning of the game, the challenger picks $\boldsymbol{R}_0, \boldsymbol{R}_{i,j} \xleftarrow{\$} [-\rho, \rho]_R^{k\times k}$ for $(i,j) \in [d] \times [\ell]$. It also picks $\boldsymbol{y}$ as in Game_1. Then, $\boldsymbol{a}$, $\boldsymbol{b}_0$, and $\boldsymbol{b}_{i,j}$ are defined as

$$\boldsymbol{b}_0 = \boldsymbol{a}\boldsymbol{R}_0 + y_0\boldsymbol{g}_b, \qquad \boldsymbol{b}_{i,j} = \boldsymbol{a}\boldsymbol{R}_{i,j} + y_{i,j}\boldsymbol{g}_b, \tag{11}$$

for $(i,j) \in [d] \times [\ell]$. The rest of the game is the same as in Game_1.
Now, we bound $|\Pr[X_2] - \Pr[X_1]|$. By Lemma 4, the distributions

$$\left(\boldsymbol{a}, \boldsymbol{a}\boldsymbol{R}_0 + y_0\boldsymbol{g}_b, \{\boldsymbol{a}\boldsymbol{R}_{i,j} + y_{i,j}\boldsymbol{g}_b\}_{(i,j)\in[d]\times[\ell]}\right) \quad \text{and} \quad \left(\boldsymbol{a}, \boldsymbol{b}_0, \{\boldsymbol{b}_{i,j}\}_{(i,j)\in[d]\times[\ell]}\right)$$

are $\mathsf{negl}(n)$-close, where $\boldsymbol{b}_0, \boldsymbol{b}_{i,j} \xleftarrow{\$} R_q^k$. Thus, we have $|\Pr[X_1] - \Pr[X_2]| = \mathsf{negl}(n)$.

Game_3: Recall that in the previous game, the challenger aborts at the end of the game if condition (10) is not satisfied. In this game, we change the game so that the challenger aborts as soon as the abort condition becomes true. Since this is only a conceptual change, we have $\Pr[X_2] = \Pr[X_3]$.

Before describing the next game, we define $\boldsymbol{R}_{\mathsf{ID}} \in R^{k\times k}$ for an identity $\mathsf{ID} \in \mathcal{ID}$ as

$$\boldsymbol{R}_{\mathsf{ID}} = \boldsymbol{R}_0 + \sum_{(j_1,\ldots,j_d)\in S(\mathsf{ID})} \mathsf{TrapEval}_d(\boldsymbol{R}_{1,j_1}, \ldots, \boldsymbol{R}_{d,j_d}, y_{1,j_1}, \ldots, y_{d,j_d}). \tag{12}$$

Note that by the definition of $\boldsymbol{R}_{\mathsf{ID}}$, $\mathsf{H}(\mathsf{ID})$, $\mathsf{PubEval}$ and $\mathsf{TrapEval}$ (Lemma 6) we have

$$\begin{aligned}\mathsf{H}(\mathsf{ID}) &= \boldsymbol{b}_0 + \sum_{(j_1,\ldots,j_d)\in S(\mathsf{ID})} \mathsf{PubEval}_d(\boldsymbol{b}_{1,j_1}\boldsymbol{b}_{2,j_2}, \ldots, \boldsymbol{b}_{d,j_d}) \\ &= \boldsymbol{a}\boldsymbol{R}_{\mathsf{ID}} + \mathsf{F}_{\boldsymbol{y}}(\mathsf{ID})\boldsymbol{g}_b.\end{aligned} \tag{13}$$

Since $\boldsymbol{R}_0, \boldsymbol{R}_{i,j} \xleftarrow{\$} [-\rho, \rho]_R^{k \times k}$, from Lemma 2 we have $s_1(\boldsymbol{R}_0), s_1(\boldsymbol{R}_{i,j}) \leq B$ with all but negligible probability where $B = C' \cdot \rho\sqrt{n}(\sqrt{k} + \omega(\sqrt{\log n}))$ for some positive absolute constant C'. Furthermore, we have $\|y_{i,j}\|_1 \leq cn$ from Eq. (9). Therefore by Lemma 6, we have

$$\begin{aligned} s_1(\boldsymbol{R}_{\mathsf{ID}}) &\leq s_1(\boldsymbol{R}_0) + \sum_{(j_1,\ldots,j_d) \in S(\mathsf{ID})} s_1(\mathsf{TrapEval}_d(\boldsymbol{R}_{1,j_1}, \ldots, \boldsymbol{R}_{d,j_d}, y_{1,j_1}, \ldots, y_{d,j_d})) \\ &\leq B\left(1 + \kappa(cn)^{d-1} + \kappa bnk\frac{(cn)^{d-1} - 1}{cn - 1}\right), \end{aligned} \tag{14}$$

for any $\mathsf{ID} \in \mathcal{ID}$ with all but negligible probability.

Game_4: In this game, we change the way the vector $\boldsymbol{a}$ is sampled. Namely, Game_4 challenger picks $\boldsymbol{a} \xleftarrow{\$} R_q^k$ instead of generating it with a trapdoor. By Lemma 5, this makes only negligible difference. Furthermore, we also change the way the key extraction queries are answered. When $\mathcal{A}$ makes a key extraction query for an identity ID, the challenger first computes $\boldsymbol{R}_{\mathsf{ID}}$ as in Eq. (12). It aborts if $\mathsf{F}_{\boldsymbol{y}}(\mathsf{ID}) \notin R_q^*$ as in the previous game and runs

$$\mathsf{SampleRight}(\boldsymbol{a}, \boldsymbol{g}_b, \boldsymbol{R}_{\mathsf{ID}}, \mathsf{F}_{\mathbf{y}}(\mathsf{ID}), u, \mathbf{T}_{\boldsymbol{g}_b}, \sigma) \to \boldsymbol{e},$$

otherwise. Note that in the previous game the private key was sampled as

$$\mathsf{SampleLeft}(\boldsymbol{a}, \mathsf{H}(\mathsf{ID}), u, \mathbf{T}_{\boldsymbol{a}}, \sigma) \to \boldsymbol{e}.$$

By Eq. (14) and for our choice of σ, the output distribution of SampleRight is $\mathsf{negl}(n)$-close to $D^{\mathsf{coeff}}_{\Lambda^{\perp}_{\phi(u)}([\mathrm{rot}(\boldsymbol{a}^T)^T | \mathrm{rot}(\mathsf{H}(\mathsf{ID})^T)^T]), \sigma}$. Furthermore, by the choice of σ, this distribution is $\mathsf{negl}(n)$-close to the output distribution of SampleLeft. Therefore, the above change alters the view of $\mathcal{A}$ only negligibly. Thus, we have $|\Pr[X_3] - \Pr[X_4]| = \mathsf{negl}(n)$.

Game_5: In this game, we change the way the challenge ciphertext is created when $\mathsf{coin} = 0$. Recall in the previous games when $\mathsf{coin} = 0$, we created a valid challenge ciphertext as in the real scheme. If $\mathsf{coin} = 0$ and $\mathsf{F}_{\boldsymbol{y}}(\mathsf{ID}^\star) = 0$ (i.e., if it does not abort), to create the challenge ciphertext Game_5 challenger first picks $s \xleftarrow{\$} R_q$ and $\boldsymbol{x} \xleftarrow{\$} (D^{\mathsf{coeff}}_{\mathbb{Z}^n, \alpha q})^k$ and computes $\boldsymbol{v} = s\boldsymbol{a} + \boldsymbol{x} \in R^k$. It then runs the algorithm

$$\mathsf{ReRand}\left(\mathrm{rot}([\boldsymbol{I}_k | \boldsymbol{R}_{\mathsf{ID}^\star}]), \phi(\boldsymbol{v}), \alpha q, \frac{\alpha'}{2\alpha q}\right) \to \mathbf{c} \in \mathbb{Z}_q^{2nk}$$

from Lemma 1, where $\boldsymbol{I}_k \in R^{k \times k}$ is the identity matrix of size $k \times k$. Finally, it picks $x_0 \xleftarrow{\$} D^{\mathsf{coeff}}_{\mathbb{Z}^n, \alpha q}$ and sets the challenge ciphertext as

$$C^\star = \left(\ c_0 = v_0 + \lfloor q/2 \rceil \cdot \mathsf{M},\ \boldsymbol{c}_1 = \phi^{-1}(\mathbf{c})\ \right) \in R_q \times R_q^{2k}, \tag{15}$$

where $v_0 = su + x_0$ and M is the message chosen by $\mathcal{A}$. We claim that this change alters the view of $\mathcal{A}$ only negligibly. To show this, observe that the input to ReRand is $\mathrm{rot}([\boldsymbol{I}_k | \boldsymbol{R}_{\mathsf{ID}^\star}]) \in \mathbb{Z}_q^{nk \times 2nk}$ and

$$\phi(\boldsymbol{v}) = \phi(s\boldsymbol{a} + \boldsymbol{x}) = \phi(s)\mathrm{rot}(\boldsymbol{a}) + \phi(\boldsymbol{x}) \in \mathbb{Z}_q^{nk},$$

where $\phi(\boldsymbol{x})$ is distributed as $\phi(\boldsymbol{x}) \xleftarrow{\$} D_{\mathbb{Z}^{nk},\alpha q}$. Therefore, by the property of ReRand and our choice of α and α', the output $\mathbf{c} \in \mathbb{Z}_q^{2nk}$ is

$$\begin{aligned}\mathbf{c} &= \Big(\phi(s)\mathrm{rot}(\boldsymbol{a})\Big) \cdot \mathrm{rot}\big([\boldsymbol{I}_k|\boldsymbol{R}_{\mathsf{ID}^\star}]\big) + \mathbf{x}' \\ &= \phi(s) \cdot \mathrm{rot}\big([\boldsymbol{a}|\mathsf{H}(\mathsf{ID}^\star)]\big) + \mathbf{x}' \\ &= \phi\big(s\big[\boldsymbol{a}|\mathsf{H}(\mathsf{ID}^\star)\big]\big) + \mathbf{x}',\end{aligned}$$

where the distribution of $\mathbf{x}'$ is within negligible distance from $\mathbf{x}' \xleftarrow{\$} D_{\mathbb{Z}^{2nk},\alpha'}$ due to Lemma 1. Here, we use the fact that $\mathsf{H}(\mathsf{ID}^\star) = \boldsymbol{a}\boldsymbol{R}_{\mathsf{ID}^\star}$ holds since $\mathsf{F}_{\boldsymbol{y}}(\mathsf{ID}^\star) = 0$. It can be readily seen that the distribution of $\boldsymbol{c}_1 = \phi^{-1}(\mathbf{c})$ in Game_5 is statistically close to that in Game_4. Therefore, we conclude that $|\Pr[X_4] - \Pr[X_5]| = \mathsf{negl}(n)$.

Game_6: In this game, we change the way the challenge ciphertext is created when $\mathsf{coin} = 0$. If $\mathsf{coin} = 0$ and the abort condition is not satisfied, to create the challenge ciphertext for identity $\mathsf{ID}^\star$ and message M, Game_6 challenger first picks $v_0 \xleftarrow{\$} R_q$, $\boldsymbol{v}' \xleftarrow{\$} R_q^k$ and $\boldsymbol{x} \xleftarrow{\$} (D^{\mathsf{coeff}}_{\mathbb{Z}^n,\alpha q})^k$, and runs

$$\mathsf{ReRand}\left(\mathrm{rot}\big([\boldsymbol{I}_k|\boldsymbol{R}_{\mathsf{ID}^\star}]\big), \phi(\boldsymbol{v}), \alpha q, \frac{\alpha'}{2\alpha q}\right) \to \mathbf{c} \in \mathbb{Z}_q^{2nk}, \tag{16}$$

where $\boldsymbol{v} = \boldsymbol{v}' + \boldsymbol{x}$. Then, the challenge ciphertext is set as in Eq. (15). As we will show in Lemma 12, assuming $\mathsf{RLWE}_{n,k+1,q,D^{\mathsf{coeff}}_{\mathbb{Z}^n,\alpha q}}$ is hard, we have $|\Pr[X_5] - \Pr[X_6]| = \mathsf{negl}(n)$.

Game_7: In this game, we further change the way the challenge ciphertext is created. When $\mathsf{coin} = 0$ and the abort condition is not satisfied, the challenge ciphertext for $\mathsf{ID}^\star$ is created as

$$C^\star = (\ c_0 = v_0 + \lfloor q/2 \rceil \cdot \mathsf{M},\ \boldsymbol{c}_1 = [\boldsymbol{v}'|\boldsymbol{v}'\boldsymbol{R}_{\mathsf{ID}^\star}] + [\boldsymbol{x}_1|\boldsymbol{x}_2]\ ) \in R_q \times R^{2k},$$

where $v_0 \xleftarrow{\$} R_q$, $\boldsymbol{v}' \xleftarrow{\$} R_q^k$ and $\boldsymbol{x}_1, \boldsymbol{x}_2 \xleftarrow{\$} (D^{\mathsf{coeff}}_{\mathbb{Z}^n,\alpha'})^k$.

We claim that this change alters the view of $\mathcal{A}$ only negligibly. This can be seen by a similar argument to that we made in the step from Game_3 to Game_4. We first observe that in Game_6 the input to ReRand is $\mathrm{rot}\big([\boldsymbol{I}_k|\boldsymbol{R}_{\mathsf{ID}^\star}]\big) \in \mathbb{Z}_q^{nk \times 2nk}$ and

$$\phi(\boldsymbol{v}) = \phi(\boldsymbol{v}' + \boldsymbol{x}) = \phi(\boldsymbol{v}') + \phi(\boldsymbol{x}) \in \mathbb{Z}_q^{nk}, \tag{17}$$

where $\phi(\boldsymbol{x})$ is distributed as $D_{\mathbb{Z}^{nk},\alpha q}$. Therefore, the output $\mathbf{c} \in \mathbb{Z}_q^{2nk}$ of ReRand is

$$\mathbf{c} = \phi(\boldsymbol{v}') \cdot \mathrm{rot}\big([\boldsymbol{I}_k|\boldsymbol{R}_{\mathsf{ID}^\star}]\big) + \mathbf{x}' = \phi\big([\boldsymbol{v}'|\boldsymbol{v}'\boldsymbol{R}_{\mathsf{ID}^\star}]\big) + \mathbf{x}',$$

where the distribution of $\mathbf{x}'$ is within negligible distance from $\mathbf{x}' \xleftarrow{\$} D_{\mathbb{Z}^{2nk},\alpha'}$ due to Lemma 1. Hence, the distribution of $\boldsymbol{c}_1 = \phi^{-1}(\mathbf{c})$ in Game_6 is statistically close to that in Game_7. Therefore, we have $|\Pr[X_6] - \Pr[X_7]| = \mathsf{negl}(n)$.

Game_8: In this game, we change the way the key extraction queries are answered. Instead of running $\mathsf{SampleLeft}$ or $\mathsf{SampleRight}$, the (possibly inefficient) challenger directly picks a secret key $\mathsf{sk_{ID}}$ for identity ID as $\mathsf{sk_{ID}} \xleftarrow{\$} D^{\mathsf{coeff}}_{\Lambda^{\perp}_{\phi(u)}([\mathrm{rot}(\boldsymbol{a}^T)^T|\mathrm{rot}(\mathsf{H}(\mathsf{ID})^T)^T]),\sigma}$ without using $\boldsymbol{R}_{\mathsf{ID}}$. Similarly to the change from Game_3 to Game_4, by the choice of σ and Eq. (14), this alters the view of $\mathcal{A}$ only negligibly. Therefore, we have $|\Pr[X_7] - \Pr[X_8]| = \mathsf{negl}(n)$. Note that this is only a conceptual game in order to get rid of any (negligible) correlation between the secret key and $\boldsymbol{R}_{\mathsf{ID}}$ so as not to interfere with the statistical argument using $\boldsymbol{R}_{\mathsf{ID}^\star}$ in the following game.

Game_9: In this game, we change the challenge ciphertext to be a random vector, regardless of whether $\mathsf{coin} = 0$ or $\mathsf{coin} = 1$. Namely, Game_9 challenger generates the challenge ciphertext $C^\star = (c_0, \boldsymbol{c}_1)$ as

$$c_0 \xleftarrow{\$} R_q, \qquad \text{and} \qquad \boldsymbol{c}_1 \xleftarrow{\$} R_q^{2k}.$$

We now proceed to bound $|\Pr[X_8] - \Pr[X_9]|$. Since Game_8 and Game_9 differ only in the creation of the challenge ciphertext when $\mathsf{coin} = 0$, we focus on this case. First, it is easy to see that c_0 is uniformly random over R_q in both of Game_8 and Game_9. Therefore, we only need to show that the distribution of $\boldsymbol{c}_1$ in Game_8 is $\mathsf{negl}(n)$-close to the uniform distribution over R_q^{2k}. To see this, it suffices to show that $[\boldsymbol{v}'|\boldsymbol{v}'\boldsymbol{R}_{\mathsf{ID}^\star}]$ is distributed statistically close to the uniform distribution over R_q^{2k}. First, observe that the following distributions are $\mathsf{negl}(n)$-close:

$$(\boldsymbol{a}, \boldsymbol{a}\boldsymbol{R}_0, \boldsymbol{v}', \boldsymbol{v}'\boldsymbol{R}_0) \approx (\boldsymbol{a}, \boldsymbol{a}', \boldsymbol{v}', \boldsymbol{v}'') \approx (\boldsymbol{a}, \boldsymbol{a}\boldsymbol{R}_0, \boldsymbol{v}', \boldsymbol{v}''), \tag{18}$$

where $\boldsymbol{a}, \boldsymbol{a}' \xleftarrow{\$} R_q^k$, $\boldsymbol{R}_0 \xleftarrow{\$} [-\rho, \rho]_R^{k\times k}$, $\boldsymbol{v}', \boldsymbol{v}'' \xleftarrow{\$} R_q^k$. It can be seen that the first and the second distributions are $\mathsf{negl}(n)$-close, by applying Lemma 4 for $[\boldsymbol{a}; \boldsymbol{v}'] \in R_q^{2\times k}$ and $\boldsymbol{R}_0$. It can also be seen that the second and the third distributions are $\mathsf{negl}(n)$-close, by applying the same lemma for $\boldsymbol{a}$ and $\boldsymbol{R}_0$. From the above, the following distributions are statistically close:

$$\begin{aligned}
&(\boldsymbol{a}, \boldsymbol{a}\boldsymbol{R}_0, \boldsymbol{v}', \boldsymbol{v}'\boldsymbol{R}_{\mathsf{ID}^\star}) \\
&= \left(\boldsymbol{a}, \boldsymbol{a}\boldsymbol{R}_0, \boldsymbol{v}', \boldsymbol{v}'\left(\boldsymbol{R}_0 + \boldsymbol{R}'_{\mathsf{ID}^\star}\right)\right) \\
&\approx \left(\boldsymbol{a}, \boldsymbol{a}\boldsymbol{R}_0, \boldsymbol{v}', \boldsymbol{v}'' + \boldsymbol{v}'\boldsymbol{R}'_{\mathsf{ID}^\star}\right) \\
&\approx (\boldsymbol{a}, \boldsymbol{a}\boldsymbol{R}_0, \boldsymbol{v}', \boldsymbol{v}'')
\end{aligned}$$

where $\boldsymbol{a}, \boldsymbol{a}' \xleftarrow{\$} R_q^k$, $\boldsymbol{R}_0 \xleftarrow{\$} [-\rho, \rho]_R^{k\times k}$, $\boldsymbol{v}', \boldsymbol{v}'' \xleftarrow{\$} R_q^k$, and

$$\boldsymbol{R}'_{\mathsf{ID}^\star} := \sum_{(j_1,\ldots,j_d)\in S(\mathsf{ID})} \mathsf{TrapEval}_d(\boldsymbol{R}_{1,j_1}, \ldots, \boldsymbol{R}_{d,j_d}, y_{1,j_1}, \ldots, y_{d,j_d}).$$

The second and the third distributions above are $\mathsf{negl}(n)$-close by Eq. (18). Note that we intentionally ignored all the $\boldsymbol{a}\boldsymbol{R}_{i,j}$ terms to keep the argument simple, since focusing on the $\boldsymbol{a}\boldsymbol{R}_0$ term is enough to prove randomness of $[\boldsymbol{v}'|\boldsymbol{v}'\boldsymbol{R}_{\mathsf{ID}^\star}]$. Therefore, we conclude that $|\Pr[X_8] - \Pr[X_9]| = \mathsf{negl}(n)$.

Analysis. From the above, we have

$$\begin{aligned}\left|\Pr[X_9] - \frac{1}{2}\right| &= \left|\Pr[X_1] - \frac{1}{2} + \sum_{i=1}^{8} (\Pr[X_{i+1}] - \Pr[X_i])\right| \\ &\geq \left|\Pr[X_1] - \frac{1}{2}\right| - \sum_{i=1}^{8} |\Pr[X_{i+1}] - \Pr[X_i]| \\ &\geq \frac{1}{(\kappa c^d n^d)^{(c-1)d+1}} \left(\frac{\epsilon}{2} - \frac{dQ}{n^c}\right) - \mathsf{negl}(n) \\ &= \frac{1}{\mathsf{poly}(n)} \left(\frac{\epsilon}{2} - \frac{dQ}{n^c}\right) - \mathsf{negl}(n) \end{aligned} \tag{19}$$

where the last equality follows from the facts that c and d are constants and $\kappa = \mathsf{poly}(n)$. Since the challenge ciphertext is independent from the value of coin in Game_9, we have $\Pr[X_9] = 1/2$ and thus $|\Pr[X_9] - 1/2| = 0$. Therefore, we have that $\epsilon/2 - dQ/n^c$ is negligible. However, by Eq. (8),

$$\frac{\epsilon}{2} - \frac{dQ}{n^c} \geq \frac{dQ+1}{n^c} - \frac{dQ}{n^c} = \frac{1}{n^c}$$

holds for infinitely many n, which is a contradiction.

To complete the proof of Theorem 2, it remains to prove Lemmas 11 and 12.

Lemma 11. *For any PPT adversary $\mathcal{A}$, we have*

$$\left|\Pr[X_1] - \frac{1}{2}\right| \geq \frac{1}{(\kappa c^d n^d)^{(c-1)d+1}} \left(\frac{\epsilon}{2} - \frac{dQ}{n^c}\right).$$

Proof. For a sequence of identities $\mathbb{ID} = (\mathsf{ID}^\star, \mathsf{ID}_1, \ldots, \mathsf{ID}_Q) \in \mathcal{ID}^{Q+1}$, we define $\gamma(\mathbb{ID})$ as

$$\gamma(\mathbb{ID}) = \Pr_{\boldsymbol{y}}[\mathsf{F}_{\boldsymbol{y}}(\mathsf{ID}^\star) = 0 \wedge \mathsf{F}_{\boldsymbol{y}}(\mathsf{ID}_1) \neq 0 \wedge \mathsf{F}_{\boldsymbol{y}}(\mathsf{ID}_2) \neq 0 \wedge \cdots \wedge \mathsf{F}_{\boldsymbol{y}}(\mathsf{ID}_Q) \neq 0]$$

where the probability is taken over $\boldsymbol{y} = (y_0, \{y_{i,j}\}_{(i,j)\in[d,\ell]})$, which is chosen as specified in Game_1. Then, it suffices to show

$$\frac{1}{(\kappa c^d n^d)^{(c-1)d+1}} \left(1 - \frac{2dQ}{n^c}\right) \leq \gamma(\mathbb{ID}) \leq \frac{1}{(\kappa c^d n^d)^{(c-1)d+1}} \tag{20}$$

since by Lemma 8, this implies

$$\begin{aligned}&\left|\Pr[X_1] - \frac{1}{2}\right| \\ &\geq \frac{\epsilon}{(\kappa c^d n^d)^{(c-1)d+1}} \left(1 - \frac{2dQ}{n^c}\right) - \frac{1}{2(\kappa c^d n^d)^{(c-1)d+1}} \left(1 - \left(1 - \frac{2dQ}{n^c}\right)\right) \\ &= \frac{1}{(\kappa c^d n^d)^{(c-1)d+1}} \left(\epsilon \left(1 - \frac{2dQ}{n^c}\right) - \frac{dQ}{n^c}\right) \\ &\geq \frac{1}{(\kappa c^d n^d)^{(c-1)d+1}} \left(\frac{\epsilon}{2} - \frac{dQ}{n^c}\right)\end{aligned}$$

where the last inequality follows from Eq. (8). In the following, we will prove Eq. (20) by applying Lemma 9. We set

$$\begin{aligned} &\nu = 2,\ \mu = d\ell & &\Phi = R_q, \\ &\Omega_j = R_q/\langle t_j\rangle, & &\pi_j : R_q \to R_q/\langle t_j\rangle, \qquad \text{for } j \in [2], \\ &S_0 = [-\kappa(cn)^d, -1]_{R,(c-1)d+1}, & &S_1 = [1, n]_{R,c} \end{aligned}$$

where π_j is a natural homomorphism and t_1, t_2 are elements in R_q as defined in Lemma 3. Therefore, the map $\Pi : \Phi \ni y \mapsto (\pi_1(y), \pi_2(y)) \in \Omega_1 \times \Omega_2$ is an isomorphism. We define $f_i(\{Y_{j,j'}\}_{(j,j')\in[d]\times[\ell]})$ for $i \in [0, Q]$ as

$$f_i\left(\{Y_{j,j'}\}_{(j,j')\in[d]\times[\ell]}\right) = \sum_{(j'_1,\ldots,j'_d)\in S(\mathsf{ID}_i)} Y_{1,j'_1} Y_{2,j'_2} \cdots Y_{d,j'_d}$$

where we define $\mathsf{ID}_0 := \mathsf{ID}^\star$. Note that we have $\mathsf{F}_{\boldsymbol{y}}(\mathsf{ID}_i) = y_0 + f_i(\{y_{i,j}\}_{(i,j)\in[d]\times[\ell]})$. We now check that the three conditions for Lemma 9 hold.

- We prove that π_j is injective on S_1 for $j \in \{1, 2\}$. Assume for contradiction that there are $a_1, a_2 \in S_1$ with $a_1 \neq a_2$ and $\pi_j(a_1) = \pi_j(a_2) \Leftrightarrow \pi_j(a_1 - a_2) = 0$. We then have $a_1 - a_2 \notin R_q^*$. On the other hand, we have $\|\phi(a_1 - a_2)\|_2 \leq \sqrt{c}n < \sqrt{q}$. However, this contradicts Lemma 3.
- For $i \in [1, Q]$, we have

$$\begin{aligned} &f_0\left(\{Y_{j,j'}\}\right) - f_i\left(\{Y_{j,j'}\}\right) \\ &= \sum_{(j'_1,\ldots,j'_d)\in S(\mathsf{ID}^\star)} Y_{1,j'_1} Y_{2,j'_2} \cdots Y_{d,j'_d} - \sum_{(j'_1,\ldots,j'_d)\in S(\mathsf{ID}_i)} Y_{1,j'_1} Y_{2,j'_2} \cdots Y_{d,j'_d}. \end{aligned}$$

  Since $\mathsf{ID}^\star \neq \mathsf{ID}_i$ and S is an injective map, we have $S(\mathsf{ID}^\star) \neq S(\mathsf{ID}_i)$. Therefore, there exists $(j_1^\star, \ldots, j_d^\star) \in [\ell]^d$ such that $(j_1^\star, \ldots, j_d^\star) \in S(\mathsf{ID}^\star) \triangle S(\mathsf{ID}_i)$, where $S(\mathsf{ID}^\star) \triangle S(\mathsf{ID}_i)$ denotes the symmetric difference of $S(\mathsf{ID}^\star)$ and $S(\mathsf{ID}_i)$. Thus, the above polynomial is a non-zero polynomial with degree d. Since the coefficients of $f_0 - f_i$ are all in $\{-1, 0, 1\}$ and $\pi_j(\pm 1) = \pm 1$, $\pi_j(f_0 - f_i)$ is a non-zero polynomial for $j \in \{1, 2\}$ as well.
- We prove $S_0 \supseteq \{-f_i(\{y_{j,j'}\}_{(j,j')\in[d]\times[\ell]}) | y_{1,1}, \ldots, y_{d,\ell} \in S_1\}$ for all $i \in [0, Q]$. By our assumption $d(c-1) < n$ and by regarding elements $y_{j,j'}$ as polynomials in $\mathbb{Z}[X]/(X^n + 1)$ with degree $c - 1$, we have $f_i(\{y_{j,j'}\})$ are all in $[*, *]_{R,d(c-1)+1}$ where $*$ represents some integer. It then suffices to show $\|\phi(f_i(\{y_{j,j'}\}_{(j,j')\in[d]\times[\ell]}))\|_\infty \leq \kappa(cn)^d$. For any $\{y_{j,j'}\}_{(j,j')\in[d]\times[\ell]}$, we have

$$\begin{aligned} &\|\phi(f_i(\{y_{j,j'}\}_{(j,j')\in[d]\times[\ell]}))\|_\infty \\ &= \left\| \phi\left(\sum_{(j'_1,\ldots,j'_d)\in S(\mathsf{ID}_i)} y_{1,j'_1} y_{2,j'_2} \cdots y_{d,j'_d} \right) \right\|_\infty \end{aligned} \tag{21}$$

$$= \left\| \sum_{(j'_1,\ldots,j'_d)\in S(\mathsf{ID}_i)} \phi(y_{1,j'_1} y_{2,j'_2} \cdots y_{d,j'_d}) \right\|_\infty \quad (22)$$

$$\leq \sum_{(j'_1,\ldots,j'_d)\in S(\mathsf{ID}_i)} \left\| \phi(y_{1,j'_1} y_{2,j'_2} \cdots y_{d,j'_d}) \right\|_\infty \quad (23)$$

$$\leq \kappa(cn)^d \quad (24)$$

where Eq. (21) follows from the definition, Eq. (22) holds because ϕ^{-1} is a homomorphism, Eq. (23) is from the triangle inequality, and Eq. (24) is from Lemma 7 and the fact that $\|y_{j,j'}\|_\infty \leq n$.

This completes the proof of Lemma 11.

Lemma 12. *For any PPT adversary $\mathcal{A}$, there exists another PPT adversary $\mathcal{B}$ such that*

$$|\Pr[X_5] - \Pr[X_6]| \leq \mathsf{Adv}_{\mathcal{B}}^{\mathsf{RLWE}_{n,k+1,q,D^{\mathsf{coeff}}_{\mathbb{Z}^n,\alpha q}}}.$$

In particular, we have $|\Pr[X_5] - \Pr[X_6]| = \mathsf{negl}(n)$ *under the* $\mathsf{RLWE}_{n,k+1,q,D^{\mathsf{coeff}}_{\mathbb{Z}^n,\alpha q}}$ *assumption,.*

We omit the proof here. It is a standard proof where we convert the adversary distinguishing Game_5 from Game_6 into another adversary against the RLWE assumption. This is accomplished by noticing that the trapdoor information for $\boldsymbol{a}$ nor (secret) randomness used to create the ciphertext is no longer required to simulate the challengers in Game_5 and Game_6.

5 Construction from Bilinear Maps

In the following, we present our IBE scheme from bilinear maps. Here, for simplicity, we present the scheme with only single-bit message space. A variant of our scheme that can deal with longer message space will appear in the full version. Let the identity space of the scheme be $\mathcal{ID} = \{0,1\}^\kappa$ for some $\kappa \in \mathbb{N}$. For our construction, we consider an efficiently computable injective map S that maps an identity $\mathsf{ID} \in \{0,1\}^\kappa$ to a subset $S(\mathsf{ID})$ of $[1,\ell] \times [1,\ell]$, where $\ell = \lceil \sqrt{\kappa} \rceil$. We would typically set $\kappa = O(\lambda)$, and thus $\ell = O(\sqrt{\lambda})$ in such a case. We also use $\mathsf{GL}(\mathsf{K}, \mathsf{rand})$ to denote the Goldreich-Levin hardcore bit [GL89] of K using randomness rand. Recall that $\mathsf{GL}(\mathsf{K}, \mathsf{rand})$ is the bitwise inner product between K and rand.

$\mathsf{Setup}(1^\lambda)$: On input 1^λ, it chooses an asymmetric bilinear group $\mathbb{G}_1, \mathbb{G}_2, \mathbb{G}_T$ with efficiently computable map $e : \mathbb{G}_1 \times \mathbb{G}_2 \to \mathbb{G}_T$ of prime order $p = p(\lambda)$. Let g and h be generators of $\mathbb{G}_1$ and $\mathbb{G}_2$ respectively. It then picks $w_0, w_{1,1}, \ldots, w_{1,\ell}, w_{2,1}, \ldots, w_{2,\ell}, \alpha, \beta \stackrel{\$}{\leftarrow} \mathbb{Z}_p$ and $\mathsf{rand} \stackrel{\$}{\leftarrow} \{0,1\}^{|\mathbb{G}_T|}$. It finally outputs

$\mathsf{mpk} = (g, W_0 = g^{w_0}, \{W_{1,i} = g^{w_{1,i}}\}_{i=1}^\ell, \{W_{2,i} = g^{w_{2,i}}\}_{i=1}^\ell, g^\alpha, h^\beta, \mathsf{rand})$ and
$\mathsf{msk} = (h, \alpha, \beta, w_0, w_{1,1}, \ldots, w_{1,\ell}, w_{2,1}, \ldots, w_{2,\ell})$

In the following, we use a deterministic function $\mathsf{H} : \mathcal{ID} \to \mathbb{Z}_p$ that is defined as follows.

$$\mathsf{H}(\mathsf{ID}) = w_0 + \sum_{(i,j)\in S(\mathsf{ID})} w_{1,i}w_{2,j} \in \mathbb{Z}_p.$$

$\mathsf{KeyGen}(\mathsf{mpk}, \mathsf{msk}, \mathsf{ID})$: It first computes $\mathsf{H}(\mathsf{ID})$ using msk and picks $r \stackrel{\$}{\leftarrow} \mathbb{Z}_p$. It then returns

$$\mathsf{sk}_{\mathsf{ID}} = (\ A_1 = h^{\alpha\beta + r\cdot \mathsf{H}(\mathsf{ID})},\ A_2 = h^{-r},\ \{B_j = h^{rw_{2,j}}\}_{j=1}^{\ell}\ ).$$

$\mathsf{Encrypt}(\mathsf{mpk}, \mathsf{ID}, \mathsf{M})$: To encrypt a message $\mathsf{M} \in \{0,1\}$, it picks $s, t_1, \dots, t_\ell \stackrel{\$}{\leftarrow} \mathbb{Z}_p$ and computes

$$C_0 = \mathsf{M} \oplus \mathsf{GL}\big(e(g^\alpha, h^\beta)^s, \mathsf{rand}\big), \quad C_1 = g^s, \quad C_2 = W_0^s \cdot \prod_{j\in[1,\ell]} W_{2,j}^{t_j},$$

$$D_j = g^{t_j} \cdot \left(\prod_{i\in\{i\in[1,\ell] \mid (i,j)\in S(\mathsf{ID})\}} W_{1,i} \right)^{-s} \quad \text{for} \quad j \in [1,\ell]$$

Finally, it returns the ciphertext $C = (C_0, C_1, C_2, \{D_j\}_{j=1}^{\ell})$.

$\mathsf{Decrypt}(\mathsf{mpk}, \mathsf{sk}_{\mathsf{ID}}, C)$: To decrypt a ciphertext $C = (C_0, C_1, C_2, \{D_j\}_{j=1}^{\ell})$ using a private key $\mathsf{sk}_{\mathsf{ID}} = (A_1, A_2, \{B_j\}_{j=1}^{\ell})$, it first computes

$$e(C_1, A_1) \cdot e(C_2, A_2) \cdot \prod_{j\in[1,\ell]} e(D_j, B_j) = e(g,h)^{s\alpha\beta}.$$

Then it retrieves the message by $C_0 \oplus \mathsf{GL}(e(g,h)^{s\alpha\beta}, \mathsf{rand})$.

The correctness of the scheme will be shown by a simple calculation.

Definition 2 (3-Computational Bilinear Diffie-Hellman Exponent (3-CBDHE) Assumption). *We say that 3-CBDHE holds on* $(\mathbb{G}_1, \mathbb{G}_2, \mathbb{G}_T)$ *if*

$$\Pr[\mathcal{A}(g, g^s, g^a, g^{a^2}, h, h^a, h^{a^2}) \to e(g,h)^{sa^3}]$$

is negligible for any PPT adversary $\mathcal{A}$ *where* $g \stackrel{\$}{\leftarrow} \mathbb{G}_1$, $h \stackrel{\$}{\leftarrow} \mathbb{G}_2$, $s, a \stackrel{\$}{\leftarrow} \mathbb{Z}_p$.

The following theorem addresses the security of the scheme.

Theorem 3. *The above IBE scheme is adaptively secure assuming the 3-CBDHE assumption.*

6 Comparisons and Discussions

In this section, we compare our IBE schemes obtained in Sects. 4 and 5 with previous schemes. Throughout this section, $|\mathsf{mpk}|$, $|C|$, and $|\mathsf{sk}_{\mathsf{ID}}|$ denote the sizes of the master public keys, ciphertexts, and private keys, respectively. We denote by κ the length of the identity, which corresponds to the output length of the collision resistant hash if we choose to hash the bit string representing an identity.

Ideal Lattice Based IBE. In Sect. 4. we proposed a new ideal lattice based IBE scheme. By changing the base b of the $\boldsymbol{g}_b$-trapdoor, we obtain two types of instantiation offering tradeoffs. Namely, by setting $b = 2$ we obtain the Type 1 IBE scheme presented in the full version and by setting $b = n^{\frac{1}{4}}$ we obtain the Type 2 IBE scheme presented in Sect. 4.1. The Type 2 IBE allows for a more compact size parameters compared to the Type 1 IBE, whereas the Type 1 IBE allows for a more efficient sampling procedure due to the smaller Gaussian width. Note that the technique of changing the base b is applicable for other existing IBE schemes as well, offering a similar tradeoff presented above. Both of our schemes achieve the best efficiency among existing adaptively secure IBE schemes assuming the fixed polynomial approximation of the RLWE problem. This is illustrated in Table 1. We point out that the largest improvement from the Yamada's IBE is that we greatly weakened the underlying hardness assumption while improving the overall efficiency of the scheme.

Bilinear Map Based IBE. Here, we compare our scheme in Sect. 5 with other adaptively secure IBE schemes based on the hardness of computational/search

Table 1. Comparison of Lattice-Base IBEs in the standard model.

Schemes	$\|\mathsf{mpk}\|$	$\|C\|$, $\|\mathsf{sk}_{\mathsf{ID}}\|$	$1/\alpha$ for LWE Assumption	Anonymous?
[CHKP10]	$O(n\kappa \log^2 n)$	$O(n\kappa \log^2 n)$	Fixed $\mathsf{poly}(n)$	Yes
[ABB10]+[Boy10]*	$O(n\kappa \log^2 n)$	$O(n \log^2 n)$	Fixed $\mathsf{poly}(n)$	Yes
[Yam16]: Scheme 1	$O(n\kappa^{\frac{1}{d}} \log^4 n)$	$O(n \log^4 n)$	$n^{\omega(1)}$	Yes
[Yam16]: Scheme 2	$O(n\kappa^{\frac{1}{d}} \log^4 n)$	$O(n \log^4 n)$	All $\mathsf{poly}(n)$	No
Ours: Sect. 4. Type 1.	$O(n\kappa^{\frac{1}{d}} \log^2 n)$	$O(n \log^2 n)$	Fixed $\mathsf{poly}(n)$	Yes
Ours: Sect. 4. Type 2.	$O(n\kappa^{\frac{1}{d}} \log n)$	$O(n \log n)$	Fixed $\mathsf{poly}(n)$	Yes

All parameters presented in the table are obtained by instantiating the schemes in the ring setting. $d \in \mathbb{N}$ is a flexible constant, which can be set to be any value. "$1/\alpha$" for LWE assumption refers to the underlying LWE assumption used in the security reduction. "Fixed $\mathsf{poly}(n)$"means that the corresponding scheme is proven secure under the LWE assumption with $1/\alpha$ being some fixed polynomial (e.g., n^3). "All $\mathsf{poly}(n)$" mean that we have to assume the LWE assumption for all polynomial.

* In the security proof for the adaptively secure variant of IBE in [ABB10], we have a restriction that $q > Q$. Namely, only bounded form of the security is proven. This restriction is removed in the refined analysis due to Boyen [Boy10].

Table 2. Comparison of IBE from bilinear maps in the standard model.

Schemes	$\|\mathsf{mpk}\|$	$\|C\|$, $\|\mathsf{sk}_{\mathsf{ID}}\|$	Assumption
[Wat05] + Hardcore bit	$O(\kappa)$	2	CBDH
[Nac07] + Hardcore bit	$O(\kappa/\log(\lambda)) = O(\kappa/\log(\kappa))$	2	CBDH
Ours: Sect. 5	$O(\sqrt{\kappa})$	$O(\sqrt{\kappa})$	3-CBDHE

problems on bilinear maps in the standard model. To base the security of IBE schemes on such problems, we have to mask the message using the Goldreich-Levin hardcore bit [GL89]. To the best of our knowledge, there are only two IBE schemes that we can apply this modification: Waters IBE [Wat05] and Naccache IBE [Nac07]. As shown in Table 2, our scheme achieves asymptotically shorter master public key size than these schemes. We note that to compare the efficiency, we count the number of group elements. However our method comes at the cost of increasing the ciphertext and private key size and we further have to rely on a stronger assumption than theirs.

Acknowledgement. We would like to thank anonymous reviewers of Asiacrypt 2016 for helpful comments. We also thank the members of Shin-Akarui-Angou-Benkyoukai for their helpful discussions and comments. This research was partially supported by CREST, JST. The second author is supported by JSPS KAKENHI Grant Number 16K16068.

References

[ABB10] Agrawal, S., Boneh, D., Boyen, X.: Efficient Lattice (H)IBE in the standard model. In: Gilbert, H. (ed.) EUROCRYPT 2010. LNCS, vol. 6110, pp. 553–572. Springer, Heidelberg (2010). doi:10.1007/978-3-642-13190-5_28

[ACPS09] Applebaum, B., Cash, D., Peikert, C., Sahai, A.: Fast cryptographic primitives and circular-secure encryption based on hard learning problems. In: Halevi, S. (ed.) CRYPTO 2009. LNCS, vol. 5677, pp. 595–618. Springer, Heidelberg (2009). doi:10.1007/978-3-642-03356-8_35

[Alp15] Alperin-Sheriff, J.: Short signatures with short public keys from homomorphic trapdoor functions. In: Katz, J. (ed.) PKC 2015. LNCS, vol. 9020, pp. 236–255. Springer, Heidelberg (2015). doi:10.1007/978-3-662-46447-2_11

[AFL16] Apon, D., Fan, X., Liu, F.: Fully-secure lattice-based IBE as compact as PKE. In: IACR Cryptology ePrint Archive 2016, p. 125 (2016)

[BB04a] Boneh, D., Boyen, X.: Efficient selective-ID secure identity-based encryption without random oracles. In: Cachin, C., Camenisch, J.L. (eds.) EUROCRYPT 2004. LNCS, vol. 3027, pp. 223–238. Springer, Heidelberg (2004). doi:10.1007/978-3-540-24676-3_14

[BB04b] Boneh, D., Boyen, X.: Secure identity based encryption without random oracles. In: Franklin, M. (ed.) CRYPTO 2004. LNCS, vol. 3152, pp. 443–459. Springer, Heidelberg (2004). doi:10.1007/978-3-540-28628-8_27

[BBG05] Boneh, D., Boyen, X., Goh, E.-J.: Hierarchical identity based encryption with constant size ciphertext. In: Cramer, R. (ed.) EUROCRYPT 2005. LNCS, vol. 3494, pp. 440–456. Springer, Heidelberg (2005). doi:10.1007/11426639_26

[BF01] Boneh, D., Franklin, M.: Identity-based encryption from the weil pairing. In: Kilian, J. (ed.) CRYPTO 2001. LNCS, vol. 2139, pp. 213–229. Springer, Heidelberg (2001). doi:10.1007/3-540-44647-8_13

[BGG+14] Boneh, D., Gentry, C., Gorbunov, S., Halevi, S., Nikolaenko, V., Segev, G., Vaikuntanathan, V., Vinayagamurthy, D.: Fully key-homomorphic encryption, arithmetic circuit ABE and compact garbled circuits. In: Nguyen, P.Q., Oswald, E. (eds.) EUROCRYPT 2014. LNCS, vol. 8441, pp. 533–556. Springer, Heidelberg (2014). doi:10.1007/978-3-642-55220-5_30

[BGH07] Boneh, D., Gentry, C., Hamburg, M.: Space-efficient identity based encryption without pairings. In: FOCS, pp. 647–657 (2007)

[BGW05] Boneh, D., Gentry, C., Waters, B.: Collusion resistant broadcast encryption with short ciphertexts and private keys. In: Shoup, V. (ed.) CRYPTO 2005. LNCS, vol. 3621, pp. 258–275. Springer, Heidelberg (2005). doi:10.1007/11535218_16

[BH08] Boneh, D., Hamburg, M.: Generalized identity based and broadcast encryption schemes. In: Pieprzyk, J. (ed.) ASIACRYPT 2008. LNCS, vol. 5350, pp. 455–470. Springer, Heidelberg (2008). doi:10.1007/978-3-540-89255-7_28

[Boy10] Boyen, X.: Lattice Mixing and Vanishing Trapdoors: a framework for fully secure short signatures and more. In: Nguyen, P.Q., Pointcheval, D. (eds.) PKC 2010. LNCS, vol. 6056, pp. 499–517. Springer, Heidelberg (2010). doi:10.1007/978-3-642-13013-7_29

[BR09] Bellare, M., Ristenpart, T.: Simulation without the Artificial Abort: simplified proof and improved concrete security for Waters' IBE scheme. In: Joux, A. (ed.) EUROCRYPT 2009. LNCS, vol. 5479, pp. 407–424. Springer, Heidelberg (2009). doi:10.1007/978-3-642-01001-9_24

[CHKP10] Cash, D., Hofheinz, D., Kiltz, E., Peikert, C.: Bonsai Trees, or how to delegate a lattice basis. In: Gilbert, H. (ed.) EUROCRYPT 2010. LNCS, vol. 6110, pp. 523–552. Springer, Heidelberg (2010). doi:10.1007/978-3-642-13190-5_27

[CCZ11] Chen, Y., Chen, L., Zhang, Z.: CCA secure IB-KEM from the computational bilinear Diffie-Hellman assumption in the standard model. In: Kim, H. (ed.) ICISC 2011. LNCS, vol. 7259, pp. 275–301. Springer, Heidelberg (2012). doi:10.1007/978-3-642-31912-9_19

[Coc01] Cocks, C.: An identity based encryption scheme based on quadratic residues. In: Honary, B. (ed.) Cryptography and Coding 2001. LNCS, vol. 2260, pp. 360–363. Springer, Heidelberg (2001). doi:10.1007/3-540-45325-3_32

[CW13] Chen, J., Wee, H.: Fully, (almost) tightly secure IBE and dual system groups. In: Canetti, R., Garay, J.A. (eds.) CRYPTO 2013. LNCS, vol. 8043, pp. 435–460. Springer, Heidelberg (2013). doi:10.1007/978-3-642-40084-1_25

[DM14] Ducas, L., Micciancio, D.: Improved short lattice signatures in the standard model. In: Garay, J.A., Gennaro, R. (eds.) CRYPTO 2014. LNCS, vol. 8616, pp. 335–352. Springer, Heidelberg (2014). doi:10.1007/978-3-662-44371-2_19

[DLP14] Ducas, L., Lyubashevsky, V., Prest, T.: Efficient identity-based encryption over NTRU lattices. In: Sarkar, P., Iwata, T. (eds.) ASIACRYPT 2014. LNCS, vol. 8874, pp. 22–41. Springer, Heidelberg (2014). doi:10.1007/978-3-662-45608-8_2

[DM15] Ducas, L., Micciancio, D.: FHEW: bootstrapping homomorphic encryption in less than a second. In: Oswald, E., Fischlin, M. (eds.) EUROCRYPT 2015. LNCS, vol. 9056, pp. 617–640. Springer, Heidelberg (2015). doi:10.1007/978-3-662-46800-5_24

[Gal10] Galindo, D.: Chosen-ciphertext secure identity-based encryption from computational bilinear Diffie-Hellman. In: Joye, M., Miyaji, A., Otsuka, A. (eds.) Pairing 2010. LNCS, vol. 6487, pp. 367–376. Springer, Heidelberg (2010). doi:10.1007/978-3-642-17455-1_23

[Gen06] Gentry, C.: Practical identity-based encryption without random oracles. In: Vaudenay, S. (ed.) EUROCRYPT 2006. LNCS, vol. 4004, pp. 445–464. Springer, Heidelberg (2006). doi:10.1007/11761679_27

[GL89] Goldreich, O., Levin, L.: A hard-core predicate for all one-way functions. In: STOC, pp. 25–32 (1989)

[GPV08] Gentry, C., Peikert, C., Vaikuntanathan, V.: Trapdoors for hard lattices and new cryptographic constructions. In: STOC, pp. 197–206 (2008)

[JR13] Jutla, C.S., Roy, A.: Shorter Quasi-Adaptive NIZK proofs for linear subspaces. In: Sako, K., Sarkar, P. (eds.) ASIACRYPT 2013. LNCS, vol. 8269, pp. 1–20. Springer, Heidelberg (2013). doi:10.1007/978-3-642-42033-7_1

[LOS+10] Lewko, A., Okamoto, T., Sahai, A., Takashima, K., Waters, B.: Fully Secure Functional Encryption: attribute-based encryption and (hierarchical) inner product encryption. In: Gilbert, H. (ed.) EUROCRYPT 2010. LNCS, vol. 6110, pp. 62–91. Springer, Heidelberg (2010). doi:10.1007/978-3-642-13190-5_4

[LPR10] Lyubashevsky, V., Peikert, C., Regev, O.: On ideal lattices and learning with errors over rings. In: Gilbert, H. (ed.) EUROCRYPT 2010. LNCS, vol. 6110, pp. 1–23. Springer, Heidelberg (2010). doi:10.1007/978-3-642-13190-5_1

[LPR13] Lyubashevsky, V., Peikert, C., Regev, O.: A toolkit for ring-LWE cryptography. In: Johansson, T., Nguyen, P.Q. (eds.) EUROCRYPT 2013. LNCS, vol. 7881, pp. 35–54. Springer, Heidelberg (2013). doi:10.1007/978-3-642-38348-9_3

[LS15] Langlois, A., Stehlé, D.: Worst-case to average-case reductions for module lattices. DES **75**(3), 565–599 (2015)

[LW10] Lewko, A., Waters, B.: New techniques for dual system encryption and fully secure HIBE with short ciphertexts. In: Micciancio, D. (ed.) TCC 2010. LNCS, vol. 5978, pp. 455–479. Springer, Heidelberg (2010). doi:10.1007/978-3-642-11799-2_27

[MP12] Micciancio, D., Peikert, C.: Trapdoors for Lattices: simpler, tighter, faster, smaller. In: Pointcheval, D., Johansson, T. (eds.) EUROCRYPT 2012. LNCS, vol. 7237, pp. 700–718. Springer, Heidelberg (2012). doi:10.1007/978-3-642-29011-4_41

[Nac07] Naccache, D.: Secure and practical identity-based encryption. IET Inf. Sec. **1**(2), 59–64 (2007)

[Reg05] Regev, O.: On lattices, learning with errors, random linear codes, and cryptography. In: STOC, pp. 84–93. ACM Press (2005)

[Sha85] Shamir, A.: Identity-based cryptosystems and signature schemes. In: Blakley, G.R., Chaum, D. (eds.) CRYPTO 1984. LNCS, vol. 196, pp. 47–53. Springer, Heidelberg (1985). doi:10.1007/3-540-39568-7_5

[SOK00] Sakai, R., Ohgishi, K., Kasahara, M.: Cryptosystems based on pairings. In: SCIS (2000). (In Japanese)

[SRB12] Singh, K., Pandu Rangan, C., Banerjee, A.K.: Adaptively secure efficient Lattice (H)IBE in standard model with short public parameters. In: Bogdanov, A., Sanadhya, S. (eds.) SPACE 2012. LNCS, vol. 7644, pp. 153–172. Springer, Heidelberg (2012). doi:10.1007/978-3-642-34416-9_11

[Wat05] Waters, B.: Efficient identity-based encryption without random oracles. In: Cramer, R. (ed.) EUROCRYPT 2005. LNCS, vol. 3494, pp. 114–127. Springer, Heidelberg (2005). doi:10.1007/11426639_7

[Wat09] Waters, B.: Dual System Encryption: realizing fully secure IBE and HIBE under simple assumptions. In: Halevi, S. (ed.) CRYPTO 2009. LNCS, vol. 5677, pp. 619–636. Springer, Heidelberg (2009). doi:10.1007/978-3-642-03356-8_36

[Xag13] Xagawa, K.: Improved (hierarchical) inner-product encryption from lattices. In: Kurosawa, K., Hanaoka, G. (eds.) PKC 2013. LNCS, vol. 7778, pp. 235–252. Springer, Heidelberg (2013). doi:10.1007/978-3-642-36362-7_15

[Yam16] Yamada, S.: Adaptively secure identity-based encryption from lattices with asymptotically shorter public parameters. In: Fischlin, M., Coron, J.-S. (eds.) EUROCRYPT 2016. LNCS, vol. 9666, pp. 32–62. Springer, Heidelberg (2016). doi:10.1007/978-3-662-49896-5_2

[ZCZ16] Zhang, J., Chen, Y., Zhang, Z.: Programmable Hash Functions from Lattices: short signatures and IBEs with small key sizes. In: Robshaw, M., Katz, J. (eds.) CRYPTO 2016. LNCS, vol. 9816, pp. 303–332. Springer, Heidelberg (2016). doi:10.1007/978-3-662-53015-3_11

Foundation

How to Generate and Use Universal Samplers

Dennis Hofheinz[1], Tibor Jager[2], Dakshita Khurana[3](✉), Amit Sahai[3], Brent Waters[4], and Mark Zhandry[5]

[1] Karlsruher Institut Für Technologie, Karlsruhe, Germany
Dennis.Hofheinz@kit.edu

[2] Ruhr-Universität Bochum, Bochum, Germany
Tibor.Jager@rub.de

[3] Center for Encrypted Functionalities, UCLA, Los Angeles, USA
{dakshita,sahai}@cs.ucla.edu

[4] Center for Encrypted Functionalities, University of Texas at Austin, Austin, USA
bwaters@cs.utexas.edu

[5] Princeton University, Princeton, USA
mzhandry@princeton.edu

Abstract. A random oracle is an idealization that allows us to model a hash function as an oracle that will output a uniformly random string given any input. We introduce the notion of a *universal sampler* scheme that extends the notion of a random oracle, to a method of sampling securely from *arbitrary* distributions.

We describe several applications that provide a natural motivation for this notion; these include generating the trusted parameters for many schemes from just a single trusted setup. We further demonstrate the versatility of universal samplers by showing how they give rise to simple constructions of identity-based encryption and multiparty key exchange. In particular, we construct adaptively secure non-interactive multiparty key exchange in the random oracle model based on indistinguishability

D. Hofheinz—Supported by DFG grants GZ HO 4534/2-1 and GZ HO 4534/4-1.
D. Khurana and A. Sahai—Research supported in part from a DARPA/ARL SAFEWARE award, NSF Frontier Award 1413955, NSF grants 1228984, 1136174, and 1065276, a Xerox Faculty Research Award, a Google Faculty Research Award, an equipment grant from Intel, and an Okawa Foundation Research Grant. This material is based upon work supported by the Defense Advanced Research Projects Agency through the ARL under Contract W911NF-15-C-0205. The views expressed are those of the author and do not reflect the official policy or position of the Department of Defense, the National Science Foundation, or the U.S. Government.
B. Waters—Supported by NSF CNS-0915361 and CNS-0952692, CNS-1228599 DARPA through the U.S. Office of Naval Research under Contract N00014-11-1-0382, DARPA SAFEWARE Award, Google Faculty Research award, the Alfred P. Sloan Fellowship, Microsoft Faculty Fellowship, and Packard Foundation Fellowship.
M. Zhandry—Work done while a graduate student at Stanford University supported by the DARPA PROCEED program. Any opinions, findings and conclusions or recommendations expressed in this material are those of the author(s) and do not necessarily reflect the views of DARPA.

J.H. Cheon and T. Takagi (Eds.): ASIACRYPT 2016, Part II, LNCS 10032, pp. 715–744, 2016.
DOI: 10.1007/978-3-662-53890-6_24

obfuscation; obtaining the first known construction of adaptively secure NIKE without complexity leveraging.

We give a solution that shows how to transform any random oracle into a universal sampler scheme, based on indistinguishability obfuscation. At the heart of our construction and proof is a new technique we call "delayed backdoor programming" that we believe will have other applications.

1 Introduction

Many cryptographic systems rely on the trusted generation of common parameters to be used by participants. There may be several reasons for using such parameters. For example, many cutting edge cryptographic protocols rely on the generation of a common reference string.[1] Constructions for other primitives such as aggregate signatures [10] or batch verifiable signatures [15] require all users to choose their public keys using the same algebraic group structure. Finally, common parameters are sometimes used for convenience and efficiency—such as when generating an EC-DSA public signing key, one can choose the elliptic curve parameters from a standard set and avoid the cost of completely fresh selection.

In most of these systems it is extremely important to make sure that the parameters were indeed generated in a trustworthy manner, and failure to do so often results in total loss of security. In cryptographic protocols that explicitly create a common reference string it is obvious how and why a corrupt setup results in loss of security. In other cases, security breaks are more subtle. The issue of trust is exemplified by the recent concern over NSA interference in choosing public parameters for cryptographic schemes [2,27,30].

Given these threats it is important to establish a trusted setup process that engenders the confidence of all users, even though users will often have competing interests and different trust assumptions. Realizing such trust is challenging and requires a significant amount of investment. For example, we might try to find a single trusted authority to execute the process. Alternatively, we might try to gather different parties that represent different interests and have them jointly execute a trusted setup algorithm using secure multiparty computation. For instance, one could imagine gathering disparate parties ranging from the Electronic Frontier Foundation, to large corporations, to national governments.

Pulling together such a trusted process requires a considerable investment. While we typically measure the costs of cryptographic processes in terms of computational and communication costs, the organizational overhead of executing

[1] Several cryptographic primitives (e.g. NIZKs) are realizable using only a common *random* string and thus only need access to a trusted random source for setup. However, many cutting edge constructions need to use a common *reference* string that is setup by some private computation. For example, the NIZKs in Sahai-Waters [32] and the recent two-round MPC protocol of Garg et al. [19] uses a trusted setup phase that generates public parameters drawn from a nontrivial distribution, where the randomness underlying the specific parameter choice needs to be kept secret.

a trusted setup may often be the most significant barrier to adoption of a new cryptographic system. Given the large number of current and future cryposystems, it is difficult to imagine that a carefully executed trusted setup can be managed for each one of these. We address this problem by asking an ambitious question:

Can a single trusted setup output a set of trusted parameters, which can (securely) serve all cryptographic protocols?

In this work, we address this question by introducing a new primitive that we call Universal Samplers, and we show how to achieve a strong adaptive notion of security for universal samplers in the random oracle model, using indistinguishability obfuscation (iO). To obtain our result, we introduce a new construction and proof technique called *delayed backdoor programming*. There are only a small handful of known high-level techniques for leveraging iO, and we believe delayed backdoor programming will have other applications in the future.

Universal Sampler Schemes. We want a cryptographic primitive that allows us to (freshly) sample from an *arbitrary* distribution, without revealing the underlying randomness used to generate that sample. We call such a primitive a universal sampler scheme. In such a system there will exist a function, `Sample`, which takes as input a polynomial-size circuit description, d, and outputs a sample $p = d(x)$ for a randomly chosen x. Intuitively, p should "look like" it was freshly sampled from the distribution induced by the function d. That is from an attack algorithm's perspective it should look like a call to the `Sample` algorithm induces a fresh sample by first selecting a random string x and then outputting $d(x)$, but keeping x hidden. (We will return to a formal definition shortly.)

Perhaps the most natural comparison of our notion is to the random oracle model put forth in the seminal work of Bellare and Rogaway [5]. In the random oracle model, a function H is modeled as an oracle that when called on a certain input will output a fresh sample of a random string x. The random oracle model has had a tremendous impact on the development of cryptography and several powerful techniques such as "programming" and "rewinding" have been used to leverage its power. However, functions modeled as random oracles are inherently limited to sampling random strings. Our work explores the power of a primitive that is "smarter" and can do this for any distribution.[2] Indeed, our main result is a *transformation*: we show how to transform any ordinary random oracle into a universal sampler scheme, by making use of indistinguishability obfuscation

[2] We note that random oracles are often used as a tool to help sample from various distributions. For example, we might use them to select a prime. In RSA full domain hash signatures [6], they are used to select a group element in $\mathbb{Z}_N^*$. This sampling occurs as a two step process. First, the function H is used to sample a fresh string x *which is completely visible to the attacker*. Then there is some post processing phase such as taking $x \pmod N$ to sample an integer mod N. In the literature this is often described as one function for the sake of brevity. However, the distinction between sampling with a universal sampler scheme and applying post processing to a random oracle output is very important.

applied to a function that *interacts* with the outputs of a random oracle – our construction does not obfuscate a random oracle itself, which would be problematic to model in a theoretically reasonable way.

On Random Oracles, Universal Samplers and Instantiation. We view universal samplers as the next generation of the random oracle model. Universal samplers are an intuitive yet powerful tool: they capture the idea of a trusted box in the sky that can sample from arbitrary user-specified distributions, and provide consistent samples to every user - including providing multiple samples from the same user-specified distribution. Such a trusted box is at least as strong as a random oracle, which is a box in the sky that samples from just the uniform distribution. Our notion formalizes a conversion process in the other direction, from a random oracle to a universal sampler that can sample from arbitrary (possibly adaptively chosen) distributions.

An important issue is how to view universal samplers, given that our strongest security model requires a random oracle for realization. We again turn to the history of the random oracle model for perspective. The random oracle model itself is a well-defined and rigorous model of computation. While it is obvious that a hash function cannot actually be a random oracle, a cryptographic primitive that utilizes a hash function in place of the random oracle, and is analyzed in the random oracle model, might actually lead to a secure realization of that primitive. While it is possible to construct counterexamples [16], there are no natural cryptographic schemes designed in the random oracle model that are known to break when utilizing a cryptographic hash function in place of a random oracle.

In fact, the random oracle model has historically served two roles: (1) for efficiency, and (2) for initial feasibility results. We focus exclusively on the latter role. Our paper shows that for achieving feasibility results, by assuming iO, one can bootstrap the random oracle model to the Universal Sampler Model. And just as random oracle constructions led to standard model constructions in the past, most notably for Identity-Based Encryption, we expect the Universal Sampler Model to be a gateway to new standard model constructions. Indeed, the random-oracle IBE scheme of Boneh-Franklin [9] led to the standard model IBE schemes of Canetti-Halevi-Katz [17], Boneh-Boyen [8], and beyond. It is uncontroverted that these latter constructions owe a lot to Boneh-Franklin [9], even though completely new ideas were needed to remove the random oracle.

Similarly, we anticipate that future standard model constructions will share intuition from universal sampler constructions, but new ideas will be needed as well. Indeed, since the initial publication of our work, this has already happened: for the notion of universal signature aggregators [25], an initial solution was obtained using our universal samplers, and then a standard model notion was obtained using additional ideas, but building upon the intuition conceived in the Universal Sampler Model. We anticipate many other similar applications to arise from our work. Indeed, identifying specific distributions that do not require the full power of iO may allow one to avoid both the random oracle model and iO. But our work would provide the substrate for this exploration.

We stress that unlike the random oracle model, where heuristic constructions of cryptographic hash functions preceded the random oracle model, before our work there were not even heuristic constructions of universal samplers. Our work goes further, and gives a candidate whose security can be rigorously analyzed in the random oracle model. Moreover, just as iO and UCEs (universal computational extractors) [4] have posited achievable standard-model notions related to ideal models like VBB and random oracles, we anticipate that future work will do so for universal samplers. Our work lays the foundation for this; indeed our bounded-secure notion of universal samplers is already a realizable notion in the standard model, that can be a starting point for such work.

Our work and subsequent work give examples of the power of the universal sampler model. For example, prior to our work obtaining even weak notions of adaptivity for NIKE required extremely cumbersome schemes and proofs, whereas universal samplers give an extremely simple and intuitive solution, detailed in the full version of our paper. Thus, we argue that having universal samplers in the toolkit facilitates the development of new primitives by allowing for very intuitive constructions (as evidenced in subsequent works [7,21,24,25]).

Last, but not least, in settings where only a bounded number of secure samples are required (including a subsequent work [28]), universal samplers are a useful tool for obtaining standard model solutions.

1.1 Our Technical Approach

We now describe our approach. We begin with a high level overview of the definition we wish to satisfy; details of the definition are in Sect. 3. In our system there is a universal sampler parameter generation algorithm, Setup, which is invoked with security parameter 1^λ and randomness r. The output of this algorithm are the universal sampler parameters U. In addition, there is a second algorithm Sample which takes as input the parameters U and the (circuit) description of a setup algorithm, d, and outputs the induced parameters p_d.

We model security as an ideal/real game. In the real game an attacker will receive the parameters U produced from the universal parameter generation algorithm. Next, it will query an oracle on multiple setup algorithm descriptions $d_1, \ldots, d_q$ and iteratively get back $p_i = \texttt{Sample}(U, d_i)$ for $i = 1, 2, \ldots, q$.

In the ideal world, the attacker will first get the universal sampler parameters U, as before. Now, when the adversary queries on d_i, a unique true random string r_i is chosen for each distinct d_i, and the adversary gets back $p_i = d_i(r_i)$, as if obtaining a freshly random sample from d_i.

A scheme is secure if no poly-time attacker can distinguish between the real and ideal game with non-negligible advantage after observing their transcripts. Since p_i is a deterministic function of d_i, this strong definition is only achievable in the random oracle model. This strongest definition is formalized in Sect. 3.2.

To make progress toward our eventual solution we begin with a relaxed security notion, which is in fact realizable in the standard model, without random oracles. We relax the definition in two ways: (1) we consider a setting where the attacker makes only a single query to the oracle and (2) he commits to the query statically (a.k.a. selectively) before seeing the sampler parameters U. While this security notion is too weak for our long term goals, developing a solution will serve as step towards our final solution and provide insights.

In the selective setting, in the ideal world, it will be possible to program U to contain the output corresponding to the attacker's query. Given this insight, it is straightforward to obtain the selective and bounded notion of security by using indistinguishability obfuscation and applying punctured programming [32] techniques. In our construction we consider setup programs to all come from a polynominal circuit family of size $\ell(\lambda)$, where each setup circuit d takes in input $m(\lambda)$ bits and outputs parameters of $k(\lambda)$ bits. The polynomials of ℓ, m, k are fixed for a class of systems; we often will drop the dependence on λ when it is clear from context.

The `Setup` algorithm will first choose a puncturable pseudo random function (PRF) key K for function F where $F(K, \cdot)$ takes as input a circuit description d and outputs coins $x \stackrel{\$}{\leftarrow} \{0,1\}^m$. The universal sampler parameters are created as an obfuscation of a program that on input d computes and outputs $d(F(K,d))$. To prove security we perform a hybrid argument between the real and ideal games in the 1-bounded and selective model. First, we puncture out d^*, the single program that the attacker queried on, from K to get the punctured key $K(d^*)$. We change the parameters to be an obfuscation of the program which uses $K(d^*)$ to compute the program for any $d \neq d^*$. And for $d = d^*$ we simply hardwire in the output z where $z = d(F(K,d))$. This computation is functionally equivalent to the original program—thus indistinguishability of this step from the previous follows from indistinguishability obfuscation. In this next step, we change the hardwired value to $d(r)$ for freshly chosen randomness $r \in \{0,1\}^m$. This completes the transition to the ideal game.

Achieving Adaptive Security. We now turn our attention to achieving our original goal of universal sampler generation for adaptive security. While selective security might be sufficient in some limited situations, the adaptive security notion covers many plausible real world attacks. For instance, suppose a group of people perform a security analysis and agree to use a certain cryptographic protocol and its corresponding setup algorithm. However, for any one algorithm there will be a huge number of functionally equivalent implementations. In a real life setting an attacker could choose one of these implementations based on the universal sampler parameters and might convince the group to use this one. A selectively secure system is not necessarily secure against such an attack, while this is captured by the adaptive model.

Obtaining a solution in the adaptive unbounded setting will be significantly more difficult. Recall that we consider a setting where a random oracle may be augmented by a program to obtain a universal sampler scheme for arbitrary distributions[3]. Indeed, for uniformly distributed samples, our universal sampler scheme will imply a programmable random oracle.

A tempting idea is to simply replace the puncturable PRF call from our last construction with a call to a hash function modeled as a programmable random oracle. This solution is problematic: what does it mean to obfuscate an oracle-aided circuit? It is not clear how to model this notion without yielding an impossibility result *even within the random oracle model*, since the most natural formulation of indistinguishability obfuscation for random-oracle-aided circuits would yield VBB obfuscation, a notion that is known to be impossible to achieve [3]. In particular, Goldwasser and Rothblum [23] also showed a family of random-oracle-aided circuits that are provably impossible to indistinguishably obfuscate. However, these impossibilities only show up when we try to obfuscate circuits that make random oracle calls. Therefore we need to obtain a solution where random oracle calls are only possible outside of obfuscated programs. This complicates matters considerably, since the obfuscated program then has no way of knowing whether a setup program d is connected to a particular hash output.

A new proof technique: delayed backdoor programming. To solve this problem we develop a novel way of allowing what we call "delayed backdoor programming" using a random oracle. In our construction, users will be provided with universal sampler parameters which consist of an obfuscated program U (produced from `Setup`) as well as a hash function H modeled as a random oracle. Users will use these overall parameters to determine the induced samples. We will use the notion of "hidden triggers" [32] that loosely corresponds to information hidden in an otherwise pseudorandom string, that can only be recovered using a secret key.

Let's begin by seeing how `Setup` creates a program, P, that will be obfuscated to create U. The program takes an input w (looking ahead, this input w will be obtained by a user as a result of invoking the random oracle on his input distribution d). The program consists of two main stages. In the first stage, the program checks to see if w encodes a "hidden trigger" using secret key information. If it does, this step will output the "hidden trigger" $x \in \{0,1\}^n$, and the program P will simply output x. However, for a uniformly randomly chosen string w, this step will fail to decode with very high probability, since trigger values are encoded sparsely. Moreover, without the secret information it will be difficult to distinguish an input w containing a hidden trigger value from a uniformly sampled string.

If decoding is unsuccessful, P will move into its second stage. It will compute randomness $r = F(K, w)$ for a puncturable PRF F. Now instead of directly computing the induced samples using r, we add a level of indirection. The program will run the Setup algorithm for a 1-bounded universal parameter generation scheme using randomness r—in particular the program P could call the

[3] Note that once the universal sampler parameters of a fixed polynomial size are given out, it is not possible for a standard model proof to make an unbounded number of parameters consistent with the already-fixed universal sampler parameters.

1-bounded selective scheme we just illustrated above[4]. The program P then outputs the 1-bounded universal sampler parameters U_w.

In order to generate an induced sample by executing $\texttt{Sample}(U, d)$ on an input distribution d, the algorithm first calls the random oracle to obtain $H(d) = w$. Next, it runs the program U to obtain output program $U_w = U(w)$. Finally, it obtains the induced parameters by computing $p_d = U_w(d)$. The extra level of indirection is critical to our proof of security.

We now give an overview of the proof of security. At the highest level the goal of our proof is to construct a sequence of hybrids where parameter generation is "moved" from being directly computed by the second stage of U (as in the real game) to where the parameters for setup algorithm d are being programmed in by the first stage hidden trigger mechanism via the input $w = H(d)$. Any poly-time algorithm $\mathcal{A}$ will make at most a polynomial number $Q = Q(\lambda)$ (unique) queries $d_1, \ldots, d_Q$ to the random oracle with RO outputs $w_1, \ldots, w_Q$. We perform a hybrid of Q outer steps where at outer step i we move from using U_{w_i} to compute the induced parameters for d_i, to having the induced parameter for d_i being encoded in w_i itself.

Let's zoom in on the ith transition for input distribution d_i. The first hybrid step uses punctured programming techniques to replace the normal computation of the 1-time universal sampler parameters U_{w_i} inside the program, with a hard-wired and randomly sampled value $U_{w_i} = U'$. These techniques require making changes to the universal sampler parameter U. Since U is published *before* the adversary queries the random oracle on distribution d_i, note that we cannot "program" U to specialize to d_i.

The next step[5] involves a "hand-off" operation where we move the source of the one time parameters U' to the trigger that will be hidden inside the random oracle output w_i, instead of using the hardwired value U' inside the program. This step is critical to allowing an unbounded number of samples to be programmed into the universal sampler scheme via the random oracle. Essentially, we first choose U' independently and then set w_i to be a hidden trigger encoding of U'. At this point on calling $U(w_i)$ the program will get $U_{w_i} = U'$ from the Stage 1 hidden trigger detection and never proceed to Stage 2. Since the second stage is no longer used, we can use iO security to return to the situation where U' is no longer hardwired into the program—thus freeing up the a-priori-bounded "hardwiring resources" for future outer hybrid steps.

Interestingly, all proof steps to this point were independent of the actual program d_i. We observe that this fact is essential to our proof since the reduction was able to choose and program the one-time parameters U' ahead of time into U which had to be published well before d_i was known. However, now $U_{w_i} = U'$ comes programmed in to the random oracle output w_i obtained as a result of the

[4] In our construction of Sect. 5 we directly use our 1-bounded scheme inside the construction. However, we believe our construction can be adapted to work for any one bounded scheme.

[5] This is actually performed by a sequence of smaller steps in our proof. We simplify to bigger steps in this overview.

call to $H(d_i)$. At this point, the program U' needs to be constructed only *after* the oracle call $H(d_i)$ has been made and thus d_i is known to the challenger. We can now use our techniques from the selective setting to force $U'(d_i)$ to output the ideally generated parameters $d_i(r)$ for distribution d_i.

We believe our "delayed backdoor programming" technique may be useful in other situations where an unbounded number of backdoors are needed in a program of fixed size.

1.2 Applications of Universal Samplers

Universal setup. Our notion of arbitrary sampling allows for many applications. For starters let's return to the problem of providing a master setup for all cryptographic protocols. Using a universal sampler scheme this is quite simple. One will simply publish the universal sampler $U \leftarrow \texttt{Setup}(1^\lambda)$, for security parameter λ. Then if subsequently a new scheme is developed that has a trusted setup algorithm d, everyone can agree to use $p = \texttt{Sample}(U, d)$ as the scheme's parameters.

We can also use universal sampler schemes as a technical tool to build applications as varied as identity-based encryption (IBE), non-interactive key exchange (NIKE), and broadcast encryption (BE) schemes. We note that our goal is not to claim that our applications below are the "best" realizations of such primitives, but more to demonstrate the different and perhaps surprising ways a universal sampler scheme can be leveraged.

From the public-key to the identity-based setting. As a warmup, we show how to transport cryptographic schemes from the public-key to the identity-based setting using universal samplers. For instance, consider a public-key encryption (PKE) scheme $\mathsf{PKE} = (\mathsf{PKGen}, \mathsf{PKEnc}, \mathsf{PKDec})$. Intuitively, to obtain an IBE scheme IBE from PKE, we use one PKE instance for each identity id of IBE.

A first attempt to do so would be to publish a description of U as the master public key of IBE, and then to define a public key pk_{id} for identity id as $pk_{id} = \texttt{Sample}(U, d_{id})$, where d_{id} is the algorithm that first generates a PKE key pair $(pk, sk) \leftarrow \mathsf{PKGen}(1^\lambda)$ and then outputs pk. (Furthermore, to distinguish the keys for different identities, d_{id} contains id as a fixed constant that is built into its code, but not used.) This essentially establishes a "virtual" public-key infrastructure in the identity-based setting.

Encryption to an identity id can then be performed using PKEnc under public key pk_{id}. However, at this point, it is not clear how to derive individual secret keys sk_{id} that would allow to decrypt these ciphertexts. (In fact, this first scheme does not appear to have any master secret key to begin with.)

Hence, as a second attempt, we add a "master PKE public key" pk' from a chosen-ciphertext secure PKE scheme to IBE's master public key. Furthermore, we set $(pk_{id}, c'_{id}) = \texttt{Sample}(U, d_{id})$ for the algorithm d_{id} that first samples $(pk, sk) \leftarrow \mathsf{PKGen}(1^\lambda)$, then encrypts sk under pk' via $c' \leftarrow \mathsf{PKEnc}'(pk', sk)$, and finally outputs (pk, c'). This way, we can use sk' as a "master secret key" to extract sk from c'_{id} – and thus extract individual user secret keys.

We show that this construction yields a selectively-secure IBE scheme once the used universal sampler scheme is selectively secure and the underlying PKE schemes are secure. Intuitively, during the analysis, we substitute the user public key pk_{id^*} for the challenge identity id^* with a freshly generated PKE public key, and we substitute the corresponding c'_{id^*} with a random ciphertext. This allows to embed an externally given PKE public key pk^*, and thus to use PKE's security.

Non-interactive key exchange and broadcast encryption. We provide a very simple construction of a multiparty non-interactive key exchange (NIKE) scheme. In an n-user NIKE scheme, a group of n parties wishes to agree on a shared random key k without any communication. User i derives k from its own secret key and the public keys of the other parties. (Since we are in the public-key setting, each party chooses its key pair and publishes its public key.) Security demands that k look random to any party not in the group.

We construct a NIKE scheme from a universal sampler scheme and a PKE scheme $\mathsf{PKE} = (\mathsf{PKGen}, \mathsf{PKEnc}, \mathsf{PKDec})$ as follows: the public parameters are the universal samplers U. Each party chooses a keypair $(pk, sk) \leftarrow \mathsf{PKGen}(1^\lambda)$. A shared key K among n parties with public keys from the set $S = \{pk_1, \ldots, pk_n\}$ is derived as follows. First, each party computes $(c_1, \ldots, c_n) = \mathtt{Sample}(U, d_S)$, where d_S is the algorithm that chooses a random key k, and then encrypts it under each pk_i to c_i (i.e., using $c_i \leftarrow \mathsf{PKEnc}(pk_i, k)$). Furthermore, d_S contains a description of the set S, e.g., as a comment. (This ensures that different sets S imply different algorithms d_S and thus different independently random $\mathtt{Sample}$ outputs.) Obviously, the party with secret key sk_i can derive k from c_i. On the other hand, we show that k remains hidden to any outsiders, even in an adaptive setting, assuming the universal sampler scheme is adaptively secure, and the encryption scheme is (IND-CPA) secure.

We also give a variant of the protocol that has no setup at all. Roughly, we follow Boneh and Zhandry [12] and designate one user as the "master party" who generates and publishes the universal sampler parameters along with her public key. Unfortunately, as in [12], the basic conversion is totally broken in the adaptive setting. However, we make a small change to our protocol so that the resulting no-setup scheme *does* have adaptive security. This is in contrast to [12], which required substantial changes to the scheme, achieved only a weaker *semi-static* security, and only obtained security though complexity leveraging.

Not only is our scheme the first adaptively secure multiparty NIKE without any setup, but it is the first to achieve adaptive security even among schemes with trusted setup, and it is the first to achieve *any* security beyond static security without relying on complexity leveraging. Subsequent to our work, Rao [31] gave an adaptive multi-party non-interactive key exchange protocol under adaptive assumptions on multilinear maps. One trade-off is that our scheme is only proved secure in the random oracle model, whereas [12,31] are proved secure in the standard model. Nevertheless, we note that adaptively secure NIKE with polynomial loss to underlying assumptions is not known to be achievable outside of the random oracle model unless one makes very strong adaptive (non-falsifiable) assumptions [31].

Finally, using an existing transformation of Boneh and Zhandry [12], we obtain a new adaptive distributed broadcast encryption from our NIKE scheme.

1.3 Subsequent Work Leveraging Universal Sampler Schemes

After the initial posting of our paper, a few other papers have applied universal sampler schemes. Hohenberger, Koppula and Waters [25] used universal samplers to achieve adaptive security without complexity leveraging for a new notion they called universal signature aggregators. Hofheinz, Kamath, Koppula and Waters [24] showed how to build adaptively secure constrained PRFs [11,14,26], for any circuits, using universal parameters as a key ingredient. All previous constructions were only selectively secure, or required complexity leveraging.

Our adaptively secure universal sampler scheme in the random oracle model, also turns out to be a key building block in the construction of proof of human-work puzzles of Blocki and Zhou [7]. Again, the abstraction of universal samplers proved useful for constructing NIKE schemes based on polynomially-hard functional encryption [21].

Another paper that appeared subsequent to ours [18], introduced the notion of explainability compilers and used them to obtain adaptively secure, universally composable MPC in constant rounds based on indistinguishability obfuscation and one-way functions. We note that explainability compilers are related to our notion of selectively secure universal samplers.

1.4 Organization of the Paper

We give an overview of indistinguishability obfuscation and puncturable PRFs, the main technical tools required for our constructions, in Sect. 2. In Sect. 3, we define our notion of universal sampler schemes. We give a realization and proof of security for a 1-bounded selectively secure scheme in Sect. 4. In Sect. 5, we give the construction and security overview for our main notion of an unbounded adaptively secure scheme. The full proof of security of the adaptive unbounded universal sampler scheme is in the full version. Applications of Universal Samplers to IBE and NIKE are also detailed in the full version.

2 Preliminaries

2.1 Indistinguishability Obfuscation and PRFs

In this section, we define indistinguishability obfuscation, and variants of pseudo-random functions (PRFs) that we will make use of. All variants of PRFs that we consider can be constructed from one-way functions.

Indistinguishability Obfuscation. The definition below is adapted from [20]:

Definition 1 (Indistinguishability Obfuscator (iO)). *A uniform PPT machine iO is called an* indistinguishability obfuscator *for circuits if the following conditions are satisfied:*

- *For all security parameters $\lambda \in \mathbb{N}$, for all circuits C, for all inputs x, we have that*

$$\Pr[C'(x) = C(x) : C' \leftarrow iO(\lambda, C)] = 1$$

- *For any (not necessarily uniform) PPT adversaries Samp, D, there exists a negligible function α such that the following holds: if $\Pr[|C_0| = |C_1| \text{ and } \forall x, C_0(x) = C_1(x) : (C_0, C_1, \sigma) \leftarrow Samp(1^\lambda)] > 1 - \alpha(\lambda)$, then we have:*

$$\Big|\Pr\big[D(\sigma, iO(\lambda, C_0)) = 1 : (C_0, C_1, \sigma) \leftarrow Samp(1^\lambda)\big]$$
$$-\Pr\big[D(\sigma, iO(\lambda, C_1)) = 1 : (C_0, C_1, \sigma) \leftarrow Samp(1^\lambda)\big]\Big| \leq \alpha(\lambda)$$

We will sometimes omit λ from the notation whenever convenient and clear from context.

Such indistinguishability obfuscators for circuits were constructed under novel algebraic hardness assumptions in [20].

PRF variants. We first consider some simple types of constrained PRFs [11,14,26], where a PRF is only defined on a subset of the usual input space. We focus on *puncturable* PRFs, which are PRFs that can be defined on all bit strings of a certain length, except for any polynomial-size set of inputs:

Definition 2. *A* puncturable *family of PRFs F is given by a triple of Turing Machines* Key_F, $\mathrm{Puncture}_F$, *and* Eval_F, *and a pair of computable functions $n(\cdot)$ and $m(\cdot)$, satisfying the following conditions:*

- ***[Functionality preserved under puncturing]****. For every PPT adversary A such that $A(1^\lambda)$ outputs a set $S \subseteq \{0,1\}^{n(\lambda)}$, then for all $x \in \{0,1\}^{n(\lambda)}$ where $x \notin S$, we have that:*

$$\Pr\big[\mathrm{Eval}_F(K, x) = \mathrm{Eval}_F(K_S, x) : K \leftarrow \mathrm{Key}_F(1^\lambda), K_S = \mathrm{Puncture}_F(K, S)\big] = 1$$

- ***[Pseudorandom at punctured points]****. For every PPT adversary (A_1, A_2) such that $A_1(1^\lambda)$ outputs a set $S \subseteq \{0,1\}^{n(\lambda)}$ and state σ, consider an experiment where $K \leftarrow \mathrm{Key}_F(1^\lambda)$ and $K_S = \mathrm{Puncture}_F(K, S)$. Then we have*

$$\Big|\Pr\big[A_2(\sigma, K_S, S, \mathrm{Eval}_F(K, S)) = 1\big] - \Pr\big[A_2(\sigma, K_S, S, U_{m(\lambda)\cdot|S|}) = 1\big]\Big| = negl(\lambda)$$

where $\mathrm{Eval}_F(K, S)$ denotes the concatenation of $\mathrm{Eval}_F(K, x_1)), \ldots, \mathrm{Eval}_F(K, x_k))$ where $S = \{x_1, \ldots, x_k\}$ is the enumeration of the elements of S in lexicographic order, $negl(\cdot)$ is a negligible function, and U_ℓ denotes the uniform distribution over ℓ bits.

For ease of notation, we write $F(K,x)$ *to represent* $\text{Eval}_F(K,x)$. *We also represent the punctured key* $\text{Puncture}_F(K,S)$ *by* $K(S)$.

The GGM tree-based construction of PRFs [22] from one-way functions are easily seen to yield puncturable PRFs, as recently observed by [11,14,26]. Thus:

Theorem 1. [11,14,22,26] *If one-way functions exist, then for all efficiently computable functions* $n(\lambda)$ *and* $m(\lambda)$, *there exists a puncturable PRF family that maps* $n(\lambda)$ *bits to* $m(\lambda)$ *bits.*

3 Definitions

In this section, we describe our definitional framework for universal sampler schemes. The essential property of a universal sampler scheme is that given the sampler parameters, and given any program d that generates samples from randomness (subject to certain size constraints, see below), it should be possible for any party to use the sampler parameters and the description of d to obtain induced samples that look like the samples that d would have generated given uniform and independent randomness.

We will consider two definitions – a simpler definition promising security for a single arbitrary but fixed protocol, and a more complex definition promising security in a strong adaptive sense against many protocols chosen after the sampler parameters are fixed. All our security definitions follow a "Real World" vs. "Ideal World" paradigm. Before we proceed to our definitions, we will first set up some notation and conventions:

- We will consider programs d that are bounded in the following ways: Note that we will use d to refer to both the program, and the description of the program. Below, $\ell(\lambda), m(\lambda)$, and $k(\lambda)$ are all computable polynomials. The description of d is as an $\ell(\lambda)$-bit string describing a circuit[6] implementing d. The program d takes as input $m(\lambda)$ bits of randomness, and outputs samples of length $k(\lambda)$ bits. Without loss of generality, we assume that $\ell(\lambda) \geq \lambda$ and $m(\lambda) \geq \lambda$. When context is clear, we omit the dependence on the security parameter λ. The quantities (ℓ, m, k) are bounds that are set during the setup of the universal sampler scheme.
- We enforce that every ℓ-bit description of d yields a circuit mapping m bits to k bits; this can be done by replacing any invalid description with a default circuit satisfying these properties.
- We will sometimes refer to the program d that generates samples as a "protocol". This is to emphasize that d can be used to generate arbitrary parameters for some protocol.

A universal parameter scheme consists of two algorithms:

[6] Note that if we assume iO for Turing Machines, then we do not need to restrict the size of the description of d. Candidates for iO for Turing Machines were given by [1,13].

(1) The first randomized algorithm `Setup` takes as input a security parameter 1^λ and outputs sampler parameters U.
(2) The second algorithm `Sample` takes as input sampler parameters U and a circuit d of size at most ℓ, and outputs induced samples p_d.

Intuition. Before giving formal definitions, we will now describe the intuition behind our definitions. We want to formulate security definitions that guarantee that induced samples are indistinguishable from honestly generated samples to an arbitrary interactive system of adversarial and honest parties.

We first consider an "ideal world," where a trusted party, on input a program description d, simply outputs $d(r_d)$ where r_d is independently chosen true randomness, chosen once and for all for each given d. In other words, if F is a truly random function, then the trusted party outputs $d(F(d))$. In this way, if any party asks for samples corresponding to a specific program d, they are all provided with the same honestly generated value. This corresponds precisely to the shared trusted public parameters model in which protocols are typically constructed.

In the real world, however, all parties would only have access to the trusted *sampler* parameters. Parties would use the sampler parameters to derive induced samples for any specific program d. Following the ideal/real paradigm, we would like to argue that for any adversary that exists in the real world, there should exist an equivalently successful adversary in the ideal world. However, the general scenario of an interaction between multiple parties, some malicious and some honest, interacting in an arbitrary security game would be cumbersome to model in a definition. To avoid this, we note that the only way that honest parties ever use the sampler parameters is to execute the sample derivation algorithm using the sampler parameters and some program descriptions d (corresponding to the protocols in which they participate) to obtain derived samples, which these honest parties then use in their interactions with the adversary.

Thus, instead of modeling these honest parties explicitly, we can "absorb" them into the adversary, as we now explain: We will require that for every real-world adversary $\mathcal{A}$, there exists a simulator $\mathcal{S}$ that can provide simulated sampler parameters U to the adversary such that these simulated sampler parameters U actually induce the completely honestly generated samples $d(F(d))$ created by the trusted party: in other words, that $\texttt{Sample}(U, d) = d(F(d))$. Note that since honest parties are instructed to simply honestly compute induced samples, this ensures that honest parties in the ideal world would obtain these completely honestly generated samples $d(F(d))$. Thus, we do not need to model the honest parties explicitly – the adversary $\mathcal{A}$ can internally simulate any (set of) honest parties. By the condition we impose on the simulation, these honest parties would have the correct view in the ideal world.

Selective (and bounded) vs. Adaptive (and unbounded) Security. We explore two natural formulations of the simulation requirement. The simpler variant is the *selective* case, where we require that the adversary declare at the start a single program d^* on which it wants the ideal world simulator to enforce equality between the honestly generated samples $d^*(F(d^*))$ and the induced samples $\texttt{Sample}(U, d^*)$. This simpler variant has two advantages: First, it is achievable in the standard model. Second, it is achieved by natural and simple construction based on indistinguishability obfuscation.

However, ideally, we would like our security definition to capture a scenario where sampler parameters U are set, and then an adversary can potentially adaptively choose a program d for generating samples for some adaptively chosen application scenario. For example, there may be several plausible implementations of a program to generate samples, and an adversary could influence which specific program description d is used for a particular protocol. Note, however, that such an adaptive scenario is trivially impossible to achieve in the standard model: there is no way that a simulator can publish sampler parameters U of polynomial size, and then with no further interaction with the adversary, force $\texttt{Sample}(U, d^*) = d^*(F(d^*))$ for a d^* chosen after U has already been declared. This impossibility is very similar to the trivial impossibility for reusable non-interactive non-committing public-key encryption [29] in the plain model. Such causality problems can be addressed, however, in the random-oracle model. As discussed in the introduction, the sound use of the random oracle model together with obfuscation requires care: we do not assume that the random oracle itself can be obfuscated, which presents an intriguing technical challenge.

Furthermore, we would like our sampler parameters to be useful to obtain induced samples for an unbounded number of other application scenarios. We formulate and achieve such an adaptive unbounded definition of security in the random oracle model.

3.1 Selective One-Time Universal Samplers

We now formally define a selective one-time secure universal sampler scheme.

Definition 3 (Selectively-Secure One-Time Universal Sampler Scheme). *Let $\ell(\lambda)$, $m(\lambda)$, $k(\lambda)$ be efficiently computable polynomials. A pair of efficient algorithms* ($\texttt{Setup}, \texttt{Sample}$) *where* $\texttt{Setup}(1^\lambda) \to U, \texttt{Sample}(U, d) \to p_d$, *is a selectively-secure one-time universal sampler scheme if there exists an efficient algorithm* $\texttt{SimUGen}$ *such that:*

- *There exists a negligible function $negl(\cdot)$ such that for all circuits d of length ℓ, taking m bits of input, and outputting k bits, and for all strings $p_d \in \{0,1\}^k$, we have that:*

$$\Pr[\texttt{Sample}(\texttt{SimUGen}(1^\lambda, d, p_d), d) = p_d] = 1 - negl(\lambda)$$

- *For every efficient adversary $\mathcal{A} = (\mathcal{A}_1, \mathcal{A}_2)$, where $\mathcal{A}_2$ outputs one bit, there exists a negligible function $negl(\cdot)$ such that*

$$\left|\Pr[\texttt{Real}(1^\lambda) = 1] - \Pr[\texttt{Ideal}(1^\lambda) = 1]\right| = negl(\lambda) \tag{1}$$

where the experiments $\texttt{Real}$ *and* $\texttt{Ideal}$ *are defined below (σ denotes auxiliary information).*

The experiment $\texttt{Real}(1^\lambda)$ *is as follows:*	*The experiment* $\texttt{Ideal}(1^\lambda)$ *is as follows:*
– $(d^*, \sigma) \leftarrow \mathcal{A}_1(1^\lambda)$	– $(d^*, \sigma) \leftarrow \mathcal{A}_1(1^\lambda)$
– *Output* $\mathcal{A}_2(\texttt{Setup}(1^\lambda), \sigma)$	– *Choose* r *uniformly from* $\{0,1\}^m$
	– *Let* $p_d = d^*(r)$
	– *Output* $\mathcal{A}_2(\texttt{SimUGen}(1^\lambda, d^*, p_d), \sigma)$

3.2 Adaptively Secure Universal Samplers

We now define universal sampler schemes for the adaptive setting in the random oracle model, handling an unbounded number of induced samples simultaneously. We *do not assume* obfuscation of circuits that call the random oracle. Thus, we allow the random oracle to be used only outside of obfuscated programs.

We consider an adversary that uses a universal sampler to obtain samples on (adaptively chosen) distributions of his choice. We want to guarantee that for any distribution specified by the adversary, the output samples he obtains are indistinguishable from externally generated parameters from the same distribution. In other words, there must exist a simulator that can force the adversary to obtain the externally generated parameters as output of the universal sampler.

Converting this intuition into an actual formal definition turns out to be somewhat complicated. The reason is that in the real world, the adversary must be able to generate samples on his own, using the universal sampler provided to him. However, the simulator which is required to force the external parameters cannot learn the adversary's queries to the sampler program. Such a simulator must observe all of the adversary's queries to the random oracle, and use them to program the output of the samplers, without knowing any of the adversary's actual queries to the sampler program.

Definition 4 (Adaptively-Secure Universal Sampler Scheme). *Let $\ell(\lambda)$, $m(\lambda)$, $k(\lambda)$ be efficiently computable polynomials. A pair of efficient oracle algorithms* $(\texttt{Setup}, \texttt{Sample})$ *where* $\texttt{Setup}^{\mathcal{H}}(1^\lambda) \rightarrow U, \texttt{Sample}^{\mathcal{H}}(U, d) \rightarrow p_d$ *is an adaptively-secure universal sampler scheme if there exist efficient interactive Turing Machines* $\texttt{SimUGen}, \texttt{SimRO}$ *such that for every efficient admissible adversary $\mathcal{A}$, there exists a negligible function negl$(\cdot)$ such that:*

$$\left|\Pr[\texttt{Real}(1^\lambda) = 1] - \Pr[\texttt{Ideal}(1^\lambda) = 1]\right| = negl(\lambda)$$

where admissible adversaries, the experiments $\texttt{Real}$ *and* $\texttt{Ideal}$ *and our (non-standard) notion of the* $\texttt{Ideal}$ *experiment aborting, are described below.*

– *An* admissible *adversary $\mathcal{A}$ is an efficient interactive Turing Machine that outputs one bit, with the following input/output behavior:*
  • *$\mathcal{A}$ initially takes input security parameter λ and sampler parameters U.*
  • *$\mathcal{A}$ can send a message (RO, x) corresponding to a random oracle query. In response, $\mathcal{A}$ receives the output of the random oracle on input x.*

- *$\mathcal{A}$ can send a message* (sample, d), *where d is a circuit of length ℓ, taking m bits of input, and outputting k bits. $\mathcal{A}$ does not expect any response to this message. Instead, upon sending this message, $\mathcal{A}$ is required to* **honestly compute** *$p_d = \texttt{Sample}(U, d)$, making use of any additional RO queries, and append (d, p_d) to an auxiliary tape.*

  Remark. Intuitively, (sample, d) corresponds to an honest party seeking a sample generated by program d. Recall that $\mathcal{A}$ is meant to internalize the behavior of honest parties that compute parameters by correctly querying the random oracle and recording the sampler's output[7].

- *The experiment* $\texttt{Real}(1^\lambda)$ *is as follows:*
  1. *Throughout this experiment, a random oracle $\mathcal{H}$ is implemented by assigning random outputs to each unique query made to $\mathcal{H}$.*
  2. $U \leftarrow \texttt{Setup}^{\mathcal{H}}(1^\lambda)$
  3. *$\mathcal{A}(1^\lambda, U)$ is executed, where every message of the form* (RO, x) *receives the response $\mathcal{H}(x)$.*
  4. *The output of the experiment is the final output of the execution of $\mathcal{A}$(which is a bit $b \in \{0, 1\}$).*
- *The experiment* $\texttt{Ideal}(1^\lambda)$ *is as follows:*
  1. *A truly random function F that maps ℓ bits to m bits is implemented by assigning random m-bit outputs to each unique query made to $\mathcal{F}$[8]. Throughout this experiment, a Samples Oracle $\mathcal{O}$ is implemented as follows: On input d, where d is a circuit of length ℓ, taking m bits of input, and outputting k bits, $\mathcal{O}$ outputs $d(F(d))$.*
  2. $(U, \tau) \leftarrow \texttt{SimUGen}(1^\lambda)$. *Here,* $\texttt{SimUGen}$ *can make arbitrary queries to the Samples Oracle $\mathcal{O}$.*
  3. $\texttt{SimRO}$ *corresponds to the output of a programmable random oracle in the ideal world.*
  4. *$\mathcal{A}(1^\lambda, U)$ and* $\texttt{SimRO}(\tau)$ *begin simultaneous execution. Messages for $\mathcal{A}$ or* $\texttt{SimRO}$ *are handled as:*
     - *Whenever $\mathcal{A}$ sends a message of the form* (RO, x), *this is forwarded to* $\texttt{SimRO}$, *which produces a response to be sent back to $\mathcal{A}$.*
     - $\texttt{SimRO}$ *can make any number of queries to the Samples Oracle $\mathcal{O}$*[9].
     - *Finally, after $\mathcal{A}$ sends a message of the form* (sample, d), *the auxiliary tape of $\mathcal{A}$ is examined until $\mathcal{A}$ adds an entry of the form (d, p_d) to it. At this point, if $p_d \neq d(F(d))$, the experiment aborts and we say that an "Honest Sample Violation" has occurred. Note that this*

---

[7] Note that proving security against such admissible adversaries suffices to capture the intuition behind a universal sampler and in particular suffices for all our applications. This is because honest parties will still use the correctly generated output, and we would like to guarantee that no malicious adversary will be able to distinguish the samples used by **honest parties** from externally generated samples.

[8] $\mathcal{A}$ does not have direct access to $\mathcal{F}$, in fact $\mathcal{A}$ will only have access to $\texttt{SimRO}$ which we define later to model the output of a programmable random oracle.

[9] Looking ahead, in our proof, $\texttt{SimRO}$ will use the output of queries to $\mathcal{O}$ to generate a programmed output of the Random Oracle.

corresponds to a correctness requirement in the ideal world, and is the only way that the experiment `Ideal` *can abort*[10]*. In this case, if the adversary itself "aborts", we consider this to be an output of zero by the adversary, not an abort of the experiment itself.*

5. *The output of the experiment is the final output of the execution of* $\mathcal{A}$ *(which is a bit* $b \in \{0, 1\}$*).*

Remark 1. We note that indistinguishability of the real and ideal worlds also implies that: $\Pr[\mathtt{Ideal}(1^\lambda) \text{ aborts}] < negl(\lambda)$

4 Selective One-Time Universal Samplers

In this section, we show the following:

Theorem 2 (Selective One-Time Universal Samplers). *If indistinguishability obfuscation and one-way functions exist, then there exists a selectively secure one-time universal sampler scheme, according to Definition 3.*

The required Selective One-Time Universal Sampler Scheme consists of programs `Setup` and `Sample`.

- `Setup`(1^λ) first samples the key K for a PRF that takes ℓ bits as input and outputs m bits. It then sets Sampler Parameters U to be an indistinguishability obfuscation of the program[11] Selective-Single-Samples in Figure 1. It outputs U.
- `Sample`(U, d) runs the program U on input d to generate and output $U(d)$.

Selective-Single-Samples

Constant: PRF key K.
Input: Program description d.

1. Output $d(F(K, d))$.
Recall that d is a program description which outputs k bits.

Fig. 1. Program Selective-Single-Samples

[10] Recall that an admissible adversary only honestly computes samples and adds them to its tape – i.e., an admissible adversary always writes $p_d = \mathtt{Sample}^{\mathcal{H}}(U, d)$ as the honest output of the sampler program. Thus, an honest sample violation in the ideal world indicates that the simulator did not force the correct samples $d(F(d))$ obtained externally from a trusted party, into the output of the sampler program.

[11] Appropriately padded to the maximum of the size of itself and Program Selective-Single-Samples: 2 in Fig. 2.

4.1 Overview of Security Proof

The proof follows straightforwardly from the puncturing techniques of [32] and we give a brief overview before giving the full proof. In the real world, the adversary commits to his input d^* and then the challenger gives the Selective-Single-Samples program to the adversary. In the first hybrid, we puncture the PRF key K at value d^*, and hardwire the output $f^* = d^*(PRF(K, d^*))$ into the program, arguing security by iO of the functionally equivalent programs. In the next hybrid, $PRF(K, d^*)$ can be replaced with a random value x, setting $f^* = d^*(x)$ and arguing security because of the puncturable PRF. Finally, the value f^* can be replaced with the external sample p_d.

4.2 Hybrids

We prove security by a sequence of hybrids, starting with the original experiment Hybrid_0 in the Real World and replacing the output at d^* with an external sample in the final hybrid (Ideal World). Each hybrid is an experiment that takes as input 1^λ. The output of each hybrid is the adversary's output when it terminates. We denote changes between subsequent hybrids using red underlined font.

Hybrid_0:

- The adversary picks protocol description d^* and sends it to the challenger.
- The challenger picks PRF key K and sends the adversary an iO of the program[12] Selective-Single-Samples in Fig. 1.
- The adversary queries the program on input d^* to obtain the sample.

Hybrid_1:

- The adversary picks protocol description d^* and sends it to the challenger.
- The challenger picks PRF key K, sets $f^* = d^*(F(K, d^*))$, punctures K at d^* and sends the adversary an iO of the program[13] Selective-Single-Samples: 2 in Fig. 2.
- The adversary queries the program on input d^* to obtain the sample.

Hybrid_2:

- The adversary picks protocol description d^* and sends it to the challenger.
- The challenger picks PRF key K, picks $x \leftarrow \{0,1\}^m$, sets $f^* = d^*(x)$, punctures K at d^* and sends the adversary an iO of the program[14] Selective-Single-Samples: 2 in Fig. 2.
- The adversary queries the program on input d^* to obtain the sample.

[12] Padded to the maximum of the size of itself and Selective-Single-Samples: 2.
[13] Padded to the maximum of the size of itself and Selective-Single-Samples.
[14] Padded to the maximum of the size of itself and Selective-Single-Samples.

Selective-Single-Samples: 2

Constant: PRF key $K\{d^*\}$, d^*, f^*.
Input: Program description d.

1. If $d = d^*$ output f^*.
2. Else output $d(F(K, d))$. Recall that d is a program description which outputs k bits.

Fig. 2. Program Selective-Single-Samples: 2

Hybrid_3:

- This hybrid describes how `SimUGen` works.
- The adversary picks protocol description d^* and sends it to the challenger.
- The challenger executes $\mathtt{SimUGen}(1^\lambda, d^*)$, which does the following: It picks PRF key K, sets $f^* = p_d$ for externally obtained sample p_d, punctures K at d^* and outputs an iO of the program[15] Selective-Single-Samples: 2 in Fig. 2. This is then sent to the adversary.
- The adversary queries the program on input d^* to obtain the sample.

4.3 Indistinguishability of the Hybrids

To prove Theorem 2, it suffices to prove the following claims,

Claim. $\mathsf{Hybrid}_0(1^\lambda)$ and $\mathsf{Hybrid}_1(1^\lambda)$ are computationally indistinguishable.

Proof. Hybrid_0 and Hybrid_1 are indistinguishable by security of iO, since the programs Selective-Single-Samples and Selective-Single-Samples: 2 are functionally equivalent. Suppose not, then there exists a distinguisher $\mathcal{D}_1$ that distinguishes between the two hybrids. This can be used to break security of the iO via the following reduction to distinguisher $\mathcal{D}$.

$\mathcal{D}$ acts as challenger in the experiment of Hybrid_0. He activates the adversary $\mathcal{D}_1$ to obtain input d^*, and computes $f^* = d^*(F(K, d^*))$, to obtain circuits C_0 = Selective-Single-Samples according to Fig. 1 and C_1 = Selective-Single-Samples: 2 according to Fig. 2 with inputs d^*, f^*. He gives C_0, C_1 to the iO challenger.

The iO challenger pads these circuits in order to bring them to equal size. It is easy to see that these circuits are functionally equivalent. Next, the iO challenger gives circuit $C_x = iO(C_0)$ or $C_x = iO(C_1)$ to $\mathcal{D}$.

$\mathcal{D}$ continues the experiment of Hybrid_1 except that he sends the obfuscated circuit C_x instead of the obfuscation of Selective-Single-Samples to the adversary $\mathcal{D}_1$. Since $\mathcal{D}_1$ has significant distinguishing advantage, there exists a polynomial $\mathsf{p}(\cdot)$ such that, $\left|\Pr[\mathcal{D}_1(\mathsf{Hybrid}_0) = 1] - \Pr[\mathcal{D}_1(\mathsf{Hybrid}_1) = 1]\right| \geq 1/\mathsf{p}(\lambda)$.

[15] Padded to the maximum of the size of itself and Selective-Single-Samples.

We note that Hybrid_0 and Hybrid_1 correspond exactly to C_x being C_0 and C_1 respectively, thus we can just have $\mathcal{D}$ echo the output of $\mathcal{D}_1$ such that the following is true, for $\alpha(\cdot) = 1/\mathsf{p}(\cdot)$

$$\left|\Pr\left[\mathcal{D}(\sigma, iO(n, C_0)) = 1\right] - \Pr\left[\mathcal{D}(\sigma, iO(n, C_1)) = 1\right]\right| \geq \alpha(\lambda)$$

Claim. $\mathsf{Hybrid}_1(1^\lambda)$ and $\mathsf{Hybrid}_2(1^\lambda)$ are computationally indistinguishable.

Proof. Hybrid_1 and Hybrid_2 are indistinguishable by security of the punctured PRF $K\{d^*\}$. Suppose they are not, then consider an adversary $\mathcal{D}_2$ who distinguishes between these hybrids with significant advantage.

This adversary can be used to break *selective* security of the punctured PRF K via the following reduction algorithm to distinguisher $\mathcal{D}$, that first gets the protocol d^* after activating the distinguisher $\mathcal{D}_2$. The PRF challenger gives the punctured PRF K along with challenge a to the PRF attacker $\mathcal{D}$, which is either the output of the PRF at d^* or is set uniformly at random in $\{0,1\}^m$. $\mathcal{D}$ sets $f^* = d^*(a)$ and continues the experiment of Hybrid_1 against $\mathcal{D}_2$. Then, $\left|\Pr\left[\mathcal{D}_2(\mathsf{Hybrid}_1) = 1\right] - \Pr\left[\mathcal{D}_2(\mathsf{Hybrid}_2) = 1\right]\right| \geq 1/p(\lambda)$ for some polynomial $p(\cdot)$.

If a is the output of the punctured PRF K at d^*, then we are in Hybrid_1. If a was chosen uniformly at random, then we are in Hybrid_2. Therefore, we can just have $\mathcal{D}$ echo the output of $\mathcal{D}_2$ such that

$$\left|\Pr\left[\mathcal{D}(F(K\{d^*\}, d^*)) = 1\right] - \Pr\left[\mathcal{D}(y \leftarrow \{0,1\}^n) = 1\right]\right| \geq 1/\mathsf{p}(\lambda).$$

Claim. $\mathsf{Hybrid}_2(1^\lambda)$ and $\mathsf{Hybrid}_3(1^\lambda)$ are identical.

Proof. These are identical since x is sampled uniformly at random in $\{0,1\}^n$.

Claim. $\Pr[\mathtt{Sample}(\mathtt{SimUGen}(1^\lambda, d, p_d), d) = p_d] = 1$

Proof. It follows from inspection of our construction that the program always outputs the external samples in the ideal world, therefore condition (1) in Definition 3 is fulfilled.

5 Adaptively Secure Universal Samplers

Theorem 3 (Adaptively Secure Universal Samplers). *If indistinguishability obfuscation and one way functions exist, then there exists an adaptively secure universal sampler scheme, according to Definition 4, in the Random Oracle Model.*

Our scheme consists of algorithms `Setup` and `Sample`, defined below. We rely on injective PRGs and indistinguishability obfuscation.

- `Setup`$(1^\lambda, r)$ first samples PRF keys K_1, K_2, K_2' and then sets Sampler Parameters U to be an indistinguishability obfuscation of the program Adaptive-Samples [16], Figure 3. The first three steps in the program look for "hidden triggers" and extract an output if a trigger is found, the final step represents the normal operation of the program (when no triggers are found).

  The program takes as input a value u, where $|u| = n^2$ and v where $|v| = n$, such that $u||v$ is obtained as the output of a random oracle $\mathcal{H}$ on input d. Here, n is the size of an iO of program[17] P_{K_3}(Figure 4). As such, n will be some fixed polynomial in the security parameter λ. The key to our proof is to instantiate the random oracle $\mathcal{H}$ appropriately to generate the sample for any input protocol description d.

  Denote by $F_1^{(n)} = \{F_1^{1,0}, F_1^{1,1}, F_1^{2,0}, F_1^{2,1} \ldots F_1^{n,0}, F_1^{n,1}\}$ a sequence of $2n$ puncturable PRF's that each take n-bit inputs and output n bits. For some key sequence $\{K_1^{1,0}, K_1^{1,1}, K_1^{2,0}, K_1^{2,1} \ldots K_1^{n,0}, K_1^{n,1}\}$, denote the combined key by $K_1^{(n)}$. Then, on a n-bit input v_1, denote the combined output of the function $F_1^{(n)}$ using key $K_1^{(n)}$ by $F_1^{(n)}(K_1^{(n)}, v_1)$. Note that the length of this combined output is $2n^2$. Denote by F_2 a puncturable PRF that takes inputs of $(n^2 + n)$ bits and outputs n_1 bits, where n_1 is the size of the key K_3 for the program P_{K_3} in Fig. 4. In particular, $n_1 = \lambda$. Denote by F_2' another puncturable PRF that takes inputs of $(n^2 + n)$ bits and outputs n_2 bits, where n_2 is the size of the randomness r used by the iO given the program P_{K_3} in Fig. 4. Denote by F_3 another puncturable PRF that takes inputs of ℓ bits and outputs m bits. Denote by PRG an *injective length-doubling* pseudo-random generator that takes inputs of n bits and outputs $2n$ bits.

  Here m is the size of uniform randomness accepted by $d(\cdot)$, k is the size of samples generated by $d(\cdot)$.
- `Sample`(U, d) queries the random oracle $\mathcal{H}$ to obtain $(u, v) = \mathcal{H}(d)$. It then runs the program U generated by `Setup`(1^λ) on input (u, v) to obtain as output the obfuscated program P. It now runs this program P on input d to obtain the required samples.

5.1 Overview of the Security Game and Hybrids

We convert any admissible adversary $\mathcal{A}$ - that is allowed to send *any* message (RO, x) or (params, d) - and construct a modified adversary, such that whenever $\mathcal{A}$ sends message (params, d), our modified adversary sends message (RO, d) and then sends message (params, d). It suffices to prove the security of our scheme with respect to such modified adversaries because this modified adversary is functionally equivalent to the admissible adversary. Because the modified adversary always provides protocol description d to the random oracle, our proof will

[16] This program must be padded appropriately to maximum of the size of itself and other corresponding programs in various hybrids, as described in the next section.

[17] Appropriately padded to the maximum of the size of itself and P'_{K_3, p_j^*, d_j^*} in future hybrids.

Adaptive-Samples

Constants: PRF keys $K_1^{(n)}$, K_2, K_2'.
Input: Program hash $u = u[1], \dots, u[n]$, v.

1. Compute $F_1(K_1^{(n)}, v) = (y_{1,0}, y_{1,1}), \dots, (y_{n,0}, y_{n,1})$.
2. For $i = 1, \dots, n$, if $u[i] = y_{i,0}$ set $x_i = 0$ else if $u[i] = y_{i,1}$ set $x_i = 1$ else set $x_i = \perp$
3. If $x \in \{0,1\}^n$ (i.e. no $\perp$s), output x.
4. Else set $K_3 = F_2(K_2, u|v)$, $r = F_2(K_2', u|v)$. Output $P = iO(P_{K_3}; r)$ of the program [a] P_{K_3} of Figure 4.

[a] Appropriately padded to the maximum of the size of itself and P'_{K_3, p_j^*, d_j^*} in future hybrids

Fig. 3. Program Adaptive-Samples

P_{K_3}

Constant: PRF key K_3.
Input: Program description d.

1. Output $d(F_3(K_3, d))$. Recall that d is a program description which outputs k bits.

Fig. 4. Program P_{K_3}

not directly deal with messages of the form (params, d) and it will suffice to handle only messages (RO, d) sent by the adversary.

We prove via a sequence of hybrids, that algorithms `Setup` and `Sample` satisfy the security requirements of Definition 4 in the Random Oracle Model. Hybrid_0 corresponds to the real world in the security game described above. Suppose the adversary makes $q(\lambda)$ queries to the random oracle $\mathcal{H}$, for some polynomial $q(\cdot)$. The argument proceeds via the sequence $\mathsf{Hybrid}_0, \mathsf{Hybrid}_{1,1}, \mathsf{Hybrid}_{1,2}, \dots \mathsf{Hybrid}_{1,13}$, $\mathsf{Hybrid}_{2,1}$, $\dots \mathsf{Hybrid}_{2,13} \dots \mathsf{Hybrid}_{q(\lambda),13}$, each of which we prove to be indistinguishable from the previous one. We define $\mathsf{Hybrid}_0 \equiv \mathsf{Hybrid}_{0,13}$ for convenience. The final hybrid $\mathsf{Hybrid}_{q(\lambda),13}$ corresponds to the ideal world in the security game described above, and contains (implicitly) descriptions of `SimUGen`, `SimRO` as required in Definition 4. For brevity, we only describe Hybrid_0 and $\mathsf{Hybrid}_{s,13}$ for a generic $s \in q(\lambda)$ in this section. We also give a short overview of how the sequence of hybrids progresses. The complete sequence of hybrids along with complete indistinguishability arguments, beginning with Hybrid_0 and then $\mathsf{Hybrid}_{s,1}$, $\mathsf{Hybrid}_{s,2}, \dots \mathsf{Hybrid}_{s,13}$ for a generic $s \in [q(\lambda)]$, can be found in the next sections.

In the following experiments, the challenger chooses PRF keys $K_1^{(n)}$, K_2 and K_2' for PRFs $F_1^{(n)}$, F_2 and F_2'. Each hybrid is an experiment that takes input 1^λ. The output of any hybrid experiment denotes the output of the adversary upon termination. Changes between hybrids are denoted using red underlined font.

Hybrid_0:

- The challenger pads the program Adaptive-Samples in Fig. 3 to be the maximum of the size of itself and all corresponding programs (Adaptive-Samples: 2, Adaptive-Samples: 3) in other hybrids. Next, he sends the obfuscation of the program in Fig. 3 to the adversary.
- Set $j = 0$. While the adversary queries the RO, increment j and repeat:
  1. Let the adversary query the random oracle on protocol description d_j^*.
  2. The challenger sets the output of the RO, $(u_j^*, v_j^*) \leftarrow \{0,1\}^{n^2+n}$.
- The adversary then outputs a single bit b'.

$\mathsf{Hybrid}_{s,13}$:

- The challenger pads the program Adaptive-Samples in Fig. 5 appropriately [18] and sends an iO of the program to the adversary.
- Set $j = 0$. While the adversary queries the RO, increment j and repeat:
  1. Let the adversary query the random oracle on protocol description d_j^*.
  2. If $j \leq s$, the challenger sets the output of the random oracle, $v_j^* \leftarrow \{0,1\}^n$. He sets $K_3 \leftarrow \{0,1\}^n, e' \leftarrow \{0,1\}^n$. He queries the oracle to obtain the sample p_j^* and sets $g = iO(P'_{K_3,p_j^*,d_j^*}, e')$ (See Fig. 7). For all $b \in \{0,1\}$ and $i \in [1,n]$, he sets $(y_{1,0}^*, y_{1,1}^*), \ldots, (y_{n,0}^*, y_{n,1}^*) = F_1(K_1^{(n)}, v_j^*)$, $u_j^*[i] = y_{i,g_i}^*$, where g_i is the ith bit of g.
  3. If $j > s$, challenger sets the RO output, $(u_j^*, v_j^*) \leftarrow \{0,1\}^{n^2+n}$.
- The adversary then outputs a single bit b'.

Note that $\mathsf{Hybrid}_{q(\lambda),13}$ is the Ideal World and it describes how `SimUGen` and `SimRO` work in the first and second bullet points above, respectively.

From $\mathsf{Hybrid}_{s-1,13}$ to $\mathsf{Hybrid}_{s,13}$.

We now outline a series of sub-hybrids from $\mathsf{Hybrid}_{s-1,13}$ to $\mathsf{Hybrid}_{s,13}$ for a generic $s \in [1, q]$, where we program the universal sampler U to output external parameters on the sth query of the adversary. Our proof comprises of two main steps: the first step consists in hardwiring a fresh single-use program into the random oracle output for the sth query – this is done by first hardwiring values into the obfuscated program, then changing the output of the random oracle, and then un-hardwiring these values from the obfuscated program.

Once this is done, the second step comprises of hardwiring the external parameters into this single-use program. The complete hybrids and indistinguishability arguments are in the next subsection.

[18] To the maximum of the size of itself and all corresponding programs in the other hybrids.

Adaptive-Samples

Constants: PRF keys $K_1^{(n)}$, K_2, K_2'.
Input: Program hash $u = u[1], \ldots, u[n]$, v.

1. Compute $F_1(K_1^{(n)}, v) = (y_{1,0}, y_{1,1}), \ldots, (y_{n,0}, y_{n,1})$.
2. For $i = 1, \ldots, n$, if $u[i] = y_{i,0}$ set $x_i = 0$ else if $u[i] = y_{i,1}$ set $x_i = 1$ else set $x_i = \perp$
3. If $x \in \{0,1\}^n$ (i.e. no $\perp$s), output x.
4. Else set $K_3 = F_2(K_2, u|v)$, $r = F_2(K_2', u|v)$. Output $iO(P_{K_3}; r)$ of the program[a] P_{K_3} of Figure 6.

[a] Appropriately padded to the maximum size of itself and P'_{K_3, p_j^*, d_j^*}

Fig. 5. Program Adaptive-Samples

P_{K_3}

Constant: PRF key K_3. **Input**: Program description d.

1. Output $d(F_3(K_3, d))$.

Fig. 6. Program P_{K_3}

P'_{K_3, p_j^*, d_j^*}

Constants: PRF key $K_3\{d_j^*\}, d_j^*, p_j^*$. **Input**: Program description d.

1. If $d = d_j^*$ output p_j^*.
2. Else output $d(F_3(K_3, d))$.

Fig. 7. Program P'_{K_3, p_j^*, d_j^*}

First step. $\mathsf{Hybrid}_{s,1}$: Let the sth random oracle query of the adversary be on input d_s^*. We first use punctured programming to hardwire computation corresponding to input d_s^* into the Adaptive-Samples program.

To do this, in $\mathsf{Hybrid}_{s,1}$ the challenger picks v_s^* uniformly at random as the output of the random oracle on input d_s^*. He sets $(y_{1,0}^*, y_{1,1}^*, \cdots y_{n,0}^*, y_{n,1}^*) = F_1(K_1^{(n)}, v_s^*)$. Then, for all $b \in \{0,1\}, i \in [n]$ he sets $z_{i,b}^* = \mathsf{PRG}(y_{i,b}^*)$. Next, he adds a check at the beginning of the main program such that for $v = v_s^*$, if $u[i] = z_{i,b}^*$, the program sets $x_i = b$. The program Adaptive-Samples of $\mathsf{Hybrid}_{s-1,13}$ is replaced by the program Adaptive-Samples: 2 illustrated in Fig. 8. This is indistinguishable from the previous hybrid by the security of indistinguishability obfuscation, because the programs Adaptive-Samples and Adaptive-Samples: 2 are functionally equivalent.

Adaptive-Samples: 2

Constants: v_s^*, PRF key $K_1^{(n)}\{v_s^*\}$, K_2, K_2', $z_{i,b}^*$ for $i \in [1, n]$ and $b \in \{0, 1\}$
Input: Program hash $u = u[1], \ldots, u[n]$, v.

1. If $v = v_s^*$ then for $i = 1, \ldots, n$ do
   If $\mathsf{PRG}(u[i]) = z_{i,0}^*$ let $x_i = 0$, if $\mathsf{PRG}(u[i]) = z_{i,1}^*$ $x_i = 1$, else $x_i = \perp$.
   Go to step 4.
2. Compute $F_1(K_1^{(n)}, v) = (y_{1,0}, y_{1,1}), \ldots, (y_{n,0}, y_{n,1})$.
3. For $i = 1, \ldots, n$, if $u[i] = y_{i,0}$ set $x_i = 0$ else if $u[i] = y_{i,1}$ set $x_i = 1$ else set $x_i = \perp$
4. If $x \in \{0, 1\}^n$ (i.e. no $\perp$s), output x.
5. Else set $K_3 = F_2(K_2, u|v)$, $r = F_2(K_2', u|v)$. Output $iO(P_{K_3}; r)$ of the program[a] P_{K_3} of Figure 6.

[a] Appropriately appended to the maximum of the size of itself and P'_{K_3, p_j^*, d_j^*}

Fig. 8. Program Adaptive-Samples: 2

$\mathsf{Hybrid}_{s,2}$: In $\mathsf{Hybrid}_{s,2}$, the output of PRF F_1 on input v_s^* is replaced with random. That is for all $b \in \{0, 1\}, i \in [n]$, he sets $y_{i,b} \xleftarrow{\$} \{0, 1\}^n$. This hybrid is indistinguishable from $\mathsf{Hybrid}_{s,1}$ by security of the puncturable PRF.

$\mathsf{Hybrid}_{s,3}$: Next, the string z^* is set uniformly at random. That is, for each $i \in [n], b \in \{0, 1\}$, instead of setting $z_{i,b}^* = \mathsf{PRG}(y_{i,b}^*)$, the challenger sets $z_{i,b}^* \xleftarrow{\$} \{0, 1\}^{2\lambda}$. This hybrid is indistinguishable from $\mathsf{Hybrid}_{s,2}$ by security of the PRG. Note that this step "deactivates" the extra check we had added in $\mathsf{Hybrid}_{s,1}$, because with overwhelming probability, z^* will lie outside the image of the PRG.

$\mathsf{Hybrid}_{s,4}$: Once this is done, for u_s^* and v_s^* both fixed uniformly at random as random oracle response to query d_s^*, in $\mathsf{Hybrid}_{s,4}$ the challenger sets $e = F_2(K_2, u_s^*|v_s^*)$, $e' = F_2'(K_2', u_s^*|v_s^*)$, $g = iO(P_e, e')$ and adds an initial check in the main program: if input $u = u_s^*$ and $v = v_s^*$, then output g and exit. Simultaneously, the challenger punctures the keys K_2 and K_2' in the main program. The modified program Adaptive-Samples: 3 is depicted in Fig. 9. At this point, we have hardwired Adaptive-Samples: 3 to output g on input values (u_s^*, v_s^*), obtained from the RO on input d_s^*. This is indistinguishable from $\mathsf{Hybrid}_{s,3}$ by the security of indistinguishability obfuscation, because the programs Adaptive-Samples: 3 and Adaptive-Samples: 2 are functionally equivalent.

$\mathsf{Hybrid}_{s,5}$: In this hybrid, the challenger generates e uniformly at random instead of the output of the punctured PRF F_2.

Adaptive-Samples: 3

Constants: v_s^*, u_s^*, g, PRF keys $K_1^{(n)}\{v_s^*\}$, $K_2\{u_s^*|v_s^*\}$, $K_2'\{u_s^*|v_s^*\}$, $z_{i,b}^*$ for $i \in [1,n]$ and $b \in \{0,1\}$
Input: Program hash $u = u[1], \ldots, u[n]$, v.

1. If $u = u_s^*$ and $v = v_s^*$ output g and stop.
2. If $v = v_s^*$ then for $i = 1, \ldots, n$ do
   If $\mathsf{PRG}(u[i]) = z_{i,0}^*$ let $x_i = 0$, if $\mathsf{PRG}(u[i]) = z_{i,1}^*$ let $x_i = 1$, else $x_i = \perp$.
   Go to step 4.
3. Compute $F_1(K_1^{(n)}, v) = (y_{1,0}, y_{1,1}), \ldots, (y_{n,0}, y_{n,1})$.
4. For $i = 1, \ldots, n$, if $u[i] = y_{i,0}$ set $x_i = 0$ else if $u[i] = y_{i,1}$ set $x_i = 1$ else set $x_i = \perp$
5. If $x \in \{0,1\}^n$ (i.e. no $\perp$s), output x.
6. Else set $K_3 = F_2(K_2, u|v)$, $r = F_2(K_2', u|v)$. Output $iO(P_{K_3}; r)$ of the program[a] P_{K_3} of Figure 6.

[a] Appropriately appended to the maximum of the size of itself and P'_{K_3, p_j^*, d_j^*}

Fig. 9. Program Adaptive-Samples: 3

$\mathsf{Hybrid}_{s,6}$: In this hybrid, the challenger generates e' uniformly at random instead of the output of the punctured PRF F_2'. This will be needed in the next few hybrids when we start programming the single-use parameters.

$\mathsf{Hybrid}_{s,7}$: Since the (bounded size) program Adaptive-Samples: 3 must remain programmable for an unbounded number of samples, we now move the hardwired single-use paramters g from the Adaptive-Samples: 3 program to a hidden trigger encoding in the output of the random oracle, u_s^*. Specifically, this is done by setting for all $i \in [1,n], z_{i,g_i}^* = \mathsf{PRG}(u_s^*[i])$ in $\mathsf{Hybrid}_{s,7}$. This is made possible also by injectivity of the PRG. Once u_s^* has been programmed appropriately to encode the value g, hardwiring g into the program becomes redundant, and it is possible to replace Adaptive-Samples: 3 with the previous program Adaptive-Samples: 2.

At this point, we can *seal* back the punctured keys, un-hardwire g from the program and return to the original program Adaptive-Samples in a sequence of hybrids, $\mathsf{Hybrid}_{s,8}$ to $\mathsf{Hybrid}_{s,10}$ which reverse our sequence of operations from $\mathsf{Hybrid}_{s,1}$ to $\mathsf{Hybrid}_{s,3}$. More specifically, $\mathsf{Hybrid}_{s,8}$ involves generating $z_{i,b}^*$ for all $i \in [n], b \in \{0,1\}$ as outputs of a PRG, and this is indistinguishable by security of the PRG. Then $\mathsf{Hybrid}_{s,9}$ involves generating $(y_{1,0}^*, y_{1,1}^* \cdots y_{n,0}^*, y_{n,1}^*$ as the output of $F_1(K_1^{(n)}, v_s^*)$, and this is indistinguishable by security of the puncturable PRF.

At this point, hardwiring the z^* values becomes redundant, and it is possible to go back to program Adaptive-Samples, in $\mathsf{Hybrid}_{s,10}$ arguing indistinguishability via indistinguishability obfuscation.

Now, $\mathsf{Hybrid}_{s,10}$ becomes identical to $\mathsf{Hybrid}_{s-1,13}$ except for a *trapdoor* that has been programmed into the random oracle output u_s^*, which outputs specific selective single-use parameters.

Second Step. Now, it is straightforward (following the same sequence of hybrids as the selective single-use case) to force the single-use parameters that were programmed into u_s^* to output external parameters p_s^*, in hybrids $\mathsf{Hybrid}_{s,11}$ through $\mathsf{Hybrid}_{s,13}$. Please refer to the full version for a more detailed proof.

No honest sample violations. At this point, in the final hybrid, whenever the adversary queries $\mathcal{H}$ on any input d, in the final hybrid we set $(u, v) = \mathcal{H}(d)$ to output the externally specified samples p_s^*. Thus, the correctness requirement in the ideal world is always met, and there are no honest sample violations according to Definition 4.

Acknowledgements. The authors would like to thank the anonymous Asiacrypt 2016 reviewers for their helpful comments, and in particular for pointing out the contents of Remark 1.

References

1. Ananth, P., Boneh, D., Garg, S., Sahai, A., Zhandry, M.: Differing-inputs obfuscation and applications. IACR Cryptology ePrint Archive 2013, p. 689 (2013)
2. Ball, J., Borger, J., Greenwald, G.: Revealed: how US and UK spy agencies defeat internet privacy and security. The Guardian (2013). http://www.theguardian.com/world/2013/sep/05/nsa-gchq-encryption-codes-security
3. Barak, B., Goldreich, O., Impagliazzo, R., Rudich, S., Sahai, A., Vadhan, S., Yang, K.: On the (Im)possibility of obfuscating programs. In: Kilian, J. (ed.) CRYPTO 2001. LNCS, vol. 2139, pp. 1–18. Springer, Heidelberg (2001). doi:10.1007/3-540-44647-8_1
4. Bellare, M., Hoang, V.T., Keelveedhi, S.: Instantiating random Oracles via UCEs. In: Canetti, R., Garay, J.A. (eds.) CRYPTO 2013. LNCS, vol. 8043, pp. 398–415. Springer, Heidelberg (2013). doi:10.1007/978-3-642-40084-1_23
5. Bellare, M., Rogaway, P.: Random Oracles are practical: a paradigm for designing efficient protocols. In: CCS 1993, Proceedings of the 1st ACM Conference on Computer and Communications Security, Fairfax, Virginia, USA, 3–5 November 1993, pp. 62–73 (1993)
6. Bellare, M., Rogaway, P.: The exact security of digital signatures - how to sign with RSA and Rabin. In: Proceeding Advances in Cryptology - EUROCRYPT 1996, International Conference on the Theory and Application of Cryptographic Techniques, Saragossa, Spain, 12–16 May 1996, pp. 399–416 (1996)
7. Blocki, J., Zhou, H.: Designing proof of human-work puzzles for cryptocurrency and beyond. IACR Cryptology ePrint Archive 2016, p. 145 (2016). http://eprint.iacr.org/2016/145
8. Boneh, D., Boyen, X.: Efficient selective identity-based encryption without random oracles. J. Cryptology **24**(4), 659–693 (2011). http://dx.doi.org/10.1007/s00145-010-9078-6

9. Boneh, D., Franklin, M.: Identity-based encryption from the weil pairing. In: Kilian, J. (ed.) CRYPTO 2001. LNCS, vol. 2139, pp. 213–229. Springer, Heidelberg (2001). doi:10.1007/3-540-44647-8_13
10. Boneh, D., Gentry, C., Lynn, B., Shacham, H.: Aggregate and verifiably encrypted signatures from bilinear maps. In: Biham, E. (ed.) EUROCRYPT 2003. LNCS, vol. 2656, pp. 416–432. Springer, Heidelberg (2003). doi:10.1007/3-540-39200-9_26
11. Boneh, D., Waters, B.: Constrained pseudorandom functions and their applications. IACR Cryptology ePrint Archive 2013, p. 352 (2013)
12. Boneh, D., Zhandry, M.: Multiparty key exchange, efficient traitor tracing, and more from indistinguishability obfuscation. In: Garay, J.A., Gennaro, R. (eds.) CRYPTO 2014. LNCS, vol. 8616, pp. 480–499. Springer, Heidelberg (2014). doi:10.1007/978-3-662-44371-2_27
13. Boyle, E., Chung, K.-M., Pass, R.: On extractability obfuscation. In: Lindell, Y. (ed.) TCC 2014. LNCS, vol. 8349, pp. 52–73. Springer, Heidelberg (2014). doi:10.1007/978-3-642-54242-8_3
14. Boyle, E., Goldwasser, S., Ivan, I.: Functional signatures and pseudorandom functions. IACR Cryptology ePrint Archive 2013, p. 401 (2013)
15. Camenisch, J., Hohenberger, S., Pedersen, M.Ø.: Batch verification of short signatures. J. Cryptology **25**(4), 723–747 (2012)
16. Canetti, R., Goldreich, O., Halevi, S.: The random oracle methodology, revisited. J. ACM **51**(4), 557–594 (2004)
17. Canetti, R., Halevi, S., Katz, J.: A forward-secure public-key encryption scheme. J. Cryptology **20**(3), 265–294 (2007)
18. Dachman-Soled, D., Katz, J., Rao, V.: Adaptively secure, universally composable, multiparty computation in constant rounds. In: Dodis, Y., Nielsen, J.B. (eds.) TCC 2015. LNCS, vol. 9015, pp. 586–613. Springer, Heidelberg (2015). doi:10.1007/978-3-662-46497-7_23
19. Garg, S., Gentry, C., Halevi, S., Raykova, M.: Two-round secure MPC from indistinguishability obfuscation. In: Lindell, Y. (ed.) TCC 2014. LNCS, vol. 8349, pp. 74–94. Springer, Heidelberg (2014). doi:10.1007/978-3-642-54242-8_4
20. Garg, S., Gentry, C., Halevi, S., Raykova, M., Sahai, A., Waters, B.: Candidate indistinguishability obfuscation and functional encryption for all circuits. In: FOCS (2013)
21. Garg, S., Pandey, O., Srinivasan, A., Zhandry, M.: Breaking the sub-exponential barrier in obfustopia. IACR Cryptology ePrint Archive 2016, p. 102 (2016). http://eprint.iacr.org/2016/102
22. Goldreich, O., Goldwasser, S., Micali, S.: How to construct random functions (extended abstract). In: FOCS, pp. 464–479 (1984)
23. Goldwasser, S., Rothblum, G.N.: On best-possible obfuscation. In: Vadhan, S.P. (ed.) TCC 2007. LNCS, vol. 4392, pp. 194–213. Springer, Heidelberg (2007). doi:10.1007/978-3-540-70936-7_11
24. Hofheinz, D., Kamath, A., Koppula, V., Waters, B.: Adaptively secure constrained pseudorandom functions. IACR Cryptology ePrint Archive 2014, p. 720 (2014)
25. Hohenberger, S., Koppula, V., Waters, B.: Universal signature aggregators. In: Oswald, E., Fischlin, M. (eds.) EUROCRYPT 2015. LNCS, vol. 9057, pp. 3–34. Springer, Heidelberg (2015). doi:10.1007/978-3-662-46803-6_1
26. Kiayias, A., Papadopoulos, S., Triandopoulos, N., Zacharias, T.: Delegatable pseudorandom functions and applications. IACR Cryptology ePrint Archive 2013, p. 379 (2013)

27. Larson, J., Perlroth, N., Shane, S.: Revealed: The NSA's secret campaign to crack, undermine internet security. Pro-Publica (2013). http://www.propublica.org/article/the-nsas-secret-campaign-to-crack-undermine-internet-encryption
28. Liang, B., Li, H., Chang, J.: The generic transformation from standard signatures to identity-based aggregate signatures. In: Lopez, J., Mitchell, C.J. (eds.) ISC 2015. LNCS, vol. 9290, pp. 21–41. Springer, Heidelberg (2015). doi:10.1007/978-3-319-23318-5_2
29. Nielsen, J.B.: Separating Random Oracle Proofs from Complexity Theoretic Proofs: the non-committing encryption case. In: Yung, M. (ed.) CRYPTO 2002. LNCS, vol. 2442, pp. 111–126. Springer, Heidelberg (2002). doi:10.1007/3-540-45708-9_8
30. Perlroth, N., Larson, J., Shane, S.: N.S.A. able to foil basic safeguards of privacy on web. Internation New York Times (2013). http://www.nytimes.com/2013/09/06/us/nsa-foils-much-internet-encryption.html
31. Rao, V.: Adaptive multiparty non-interactive key exchange without setup in the standard model. Cryptology ePrint Archive, Report 2014/910 (2014). http://eprint.iacr.org/
32. Sahai, A., Waters, B.: How to use indistinguishability obfuscation: deniable encryption, and more. In: STOC, pp. 475–484 (2014)

Iterated Random Oracle: A Universal Approach for Finding Loss in Security Reduction

Fuchun Guo[1(✉)], Willy Susilo[1], Yi Mu[1], Rongmao Chen[1,2], Jianchang Lai[1], and Guomin Yang[1]

[1] Centre for Computer and Information Security Research, School of Computing and Information Technology, University of Wollongong, Wollongong, NSW 2522, Australia
{fuchun,wsusilo,ymu,rc517,jl967,gyang}@uow.edu.au

[2] College of Computer, National University of Defense Technology, Changsha, China

Abstract. The indistinguishability security of a public-key cryptosystem can be reduced to a computational hard assumption in the random oracle model, where the solution to a computational hard problem is hidden in one of the adversary's queries to the random oracle. Usually, there is a *finding loss* in finding the correct solution from the query set, especially when the decisional variant of the computational problem is also hard. The problem of finding loss must be addressed towards tight(er) reductions under this type. In EUROCRYPT 2008, Cash, Kiltz and Shoup proposed a novel approach using a trapdoor test that can solve the finding loss problem. The simulator can find the correct solution with overwhelming probability 1, if there exists a trapdoor test for the adopted hard problem. The proposed approach is efficient and can be used for many Diffie-Hellman computational assumptions. The only limitation is the requirement of a trapdoor test that must be found for the adopted computational assumptions.

In this paper, we introduce a universal approach for finding loss, namely *Iterated Random Oracle*, which can be applied to all computational assumptions. The *finding loss* in our proposed approach is very small. For 2^{60} queries to the random oracle, the success probability of finding the correct solution from the query set will be as large as $1/64$ compared to $1/2^{60}$ by a random pick. We show how to apply the iterated random oracle for security transformation from key encapsulation mechanism with one-way security to normal encryption with indistinguishability security. The security reduction is very tight due to a small finding loss. The transformation does not expand the ciphertext size. We also give the application of the iterated random oracle in the key exchange.

Keywords: Random oracle · Indistinguishability security under computational assumptions · Finding loss

---

This work was partially supported by ARC Discovery Grant DP130101383.

J.H. Cheon and T. Takagi (Eds.): ASIACRYPT 2016, Part II, LNCS 10032, pp. 745–776, 2016.
DOI: 10.1007/978-3-662-53890-6_25

1 Introduction

Security reduction is a kind of reduction techniques in cryptography where we construct a simulator that uses an adversary's attack to solve a mathematically hard problem. According to the type of attack and the type of hard problem, cryptosystems have the following two popular types of security reduction.

- Unforgeability security based on a computational hard problem (UF-CHP). This type of security reduction has been used to prove the security of digital signature schemes. We construct a simulator that uses a forged signature from the adversary to solve a computational hard problem.
- Indistinguishability security based on a decisional hard problem (IND-DHP). This type of security reduction has been used to prove the security of encryption schemes. We construct a simulator that uses the guess of random message in the challenge ciphertext from the adversary to decide whether a solution in a given instance is correct or incorrect.

Roughly speaking, a computational problem is to find a correct solution to a given instance, while a decisional problem is to decide whether or not a solution in a given instance is correct. A computational hard problem is always harder than its decisional variant. However, without any additional assumption, it seems impossible to carry out a security reduction for a cryptosystem with indistinguishability security based on a computational hard problem. We call this type of reduction IND-CHP for short. This is because the guess from the adversary only has two answers: 0 or 1, which cannot provide sufficient information to find a correct solution. Fortunately, IND-CHP reduction becomes possible with the help of random oracles. Random oracles were first introduced by Bellare and Rogaway in [5] for designing efficient protocols. In the random oracle model, at least one hash function namely H is treated as a random oracle where responses on queries are assumed to be uniformly distributed. Anyone especially the adversary has no advantage in guessing the hash value of an input before querying the input to the random oracle. With the help of this "magical" property, many cryptosystems such as asymmetric encryption and key exchange can achieve IND-CHP security reduction.

The IND-CHP security reduction is programmed as follows. Suppose the simulator aims to compute $\mathcal{C}[I, P]$ as the solution to a given instance I under a computational hard problem P. The simulator who controls the random oracle programs the simulation using the instance I. In the simulation, the adversary must make a set of queries including a challenge query denoted by $\overline{\mathcal{Q}}^*$ to the random oracle to break the security, and the solution $\mathcal{C}[I, P]$ can be extracted from this challenge query. Different from UF-CHP and IND-DHP security reductions, the simulator solves the hard problem using the adversary's query set to random oracles instead of the adversary's forgery or guess. This distinctive security reduction arises a very important and interesting question:

How to find the correct solution from the adversary's query set?

We call this problem as a **finding problem** and the reduction has a **finding loss**, if the simulator can only succeed in finding the correct solution from the query set with a probability less than 1. When the decisional variant of the computational hard problem P is easy, there is no finding loss by verifying all solutions extracted from each query. However, when the decisional variant is also hard, it seems finding loss cannot be avoided. In this work, we focus on the non-trivial case that the decisional variant of P is also hard.

1.1 Finding Loss in Previous Approaches

In the IND-CHP security reduction, when the adversary can break a scheme simulated using an instance I, the challenge query will appear in the adversary query set and contain the solution $\mathcal{C}[I, P]$ to the instance I. The reduction after disclosing the simulation is equivalent to that the adversary who is given an instance I will make a set of queries including a challenge query $\overline{\mathcal{Q}}^* = \mathcal{C}[I, P]$. Using this disclosed reduction, we can use the following theories to describe how the finding problem is addressed.

The traditional approach in the literature is described in Theory 1. It has been applied to many cryptosystems such as [8] for IND-CHP security reductions.

Theory 1 (Traditional Approach). *Suppose an adversary, who is given an instance I generated by the simulator, must make a set of queries $\mathbb{Q}$ ($|\mathbb{Q}| = q$) including a challenge query $\overline{\mathcal{Q}}^* = \mathcal{C}[I, P]$ to the random oracle. We can construct a simulator who controls the random oracle to solve the hard problem P using the query set $\mathbb{Q}$ in $O(1)$ time with success probability $1/q$.*

It is easy to construct such a simulator. Given an instance I, the simulator forwards the instance to the adversary. Then, the challenge query is equal to the solution for the simulator. A random pick from the query set with q number of queries therefore has the success probability $1/q$.

In the security reduction, the adversary can make a polynomial number of queries to the random oracle. The query number q can be as large as $q = 2^{60}$, and hence the success probability of finding the correct solution is $1/2^{60}$. It means that all cryptosystems using this traditional approach in reduction will have at least 60-bit security loss. In the concrete security of group-based cryptosystems, we must expand the corresponding group size with 60-bit more security to compensate the security loss. This compensation at least requires 120-bit length more of security parameter in group choice, and it is therefore accompanied with inefficient group operation and large group representation.

In EUROCRYPT 2008, Cash, Kiltz and Shoup [10] introduced the first novel approach for finding loss. They proposed a new computational problem called the twin Diffie-Hellman problem. This new problem is as hard as the Computational Diffie-Hellman (CDH) problem even given access to a corresponding decision oracle. The heart of their approach is a trapdoor test, which allows the simulator to simulate an effective decision oracle without knowing any of the corresponding discrete logarithms. Their approach can be summarized using a theory described as follows.

Theory 2 (Cash-Kiltz-Shoup). *Suppose an adversary, who is given instances* (I_1, I_2) *generated by the simulator, must make a set of queries* $\mathbb{Q}$ $(|\mathbb{Q}| = q)$ *including a challenge query* $\overline{\mathcal{Q}}^* = \mathcal{C}[I_1, P] \,||\, \mathcal{C}[I_2, P]$ *to the random oracle. We can construct a simulator who controls the random oracle to solve the hard problem* P *using the query set* $\mathbb{Q}$ *in* $O(q)$ *time with nearly success probability* 1, *if there exists a trapdoor test on solutions to a given instance and a created instance under the hard problem* P.

The simulator can be constructed as follows. Given an instance I, the simulator sets $I_1 = I$, Then, it randomly chooses a trapdoor and creates the second instance I_2 from I_1 and the trapdoor. The trapdoor test holds with the property that a query $\overline{\mathcal{Q}} = Q_1 \,||\, Q_2$ can pass the trapdoor test run by the simulator if and only if $Q_1 = \mathcal{C}[I_1, P]$ and $Q_2 = \mathcal{C}[I_2, P]$ except with a negligible probability. Therefore, only the challenge query can pass the test and the simulator can successfully find the correct solution $\mathcal{C}[I, P]$ without any finding loss after all queries are tested.

Based on this theory, Cash, Kiltz and Shoup [10] proposed many twin schemes based on original schemes using two key pairs, whose IND-CHP security reductions are tight(er) without any finding loss. The price to pay for an encryption scheme is two times less efficient in terms of key size and computations compared to the original one, but the size of ciphertext is not changed. However, this theory has a limitation. It can only be applied to those cryptosystems whose underling computational assumptions have a corresponding trapdoor test. The trapdoor test proposed in [10] is a very special construction and it can be adopted by some computational Diffie-Hellman hard problems only.

1.2 Our Contribution

We propose a completely new approach for finding loss, namely *iterated random oracle*, which can be applied to all computational hard problems. Instead of using a trapdoor test to find the correct solution, the simulator in our approach can remove most of useless queries such that a random pick from remaining queries will merely have a small finding loss only. The corresponding theory is described as follows.

Theory 3 (Iterated Random Oracle). *Let* H *be a random oracle. Suppose an adversary, who is given instances* $(I_1, I_2, \cdots, I_n)$ *generated by the simulator, must make a set of queries* $\mathbb{Q}$ $(|\mathbb{Q}| = q)$ *including a challenge query* $\overline{\mathcal{Q}}^* = \overline{\mathcal{Q}}_*^{(n)}$ *to the random oracle, where* $\overline{\mathcal{Q}}_*^{(n)}$ *is defined as*

$$\overline{\mathcal{Q}}_*^{(i)} = H(\overline{\mathcal{Q}}_*^{(i-1)}) \,||\, \mathcal{C}[I_i, P] \,||\, i: \quad i \in [1, n], \quad H(\overline{\mathcal{Q}}_*^{(0)}) = 0_\epsilon \text{ is an empty string.}$$

We can construct a simulator who controls the random oracle to solve the hard problem P *using the query set* $\mathbb{Q}$ *in* $O(n)$ *time with success probability at least* $1/(nq^{\frac{1}{n}})$.

The simulator construction and probability analysis are given in Sect. 3. We give an example in the next subsection to overview the simulator construction and the probability analysis. When this theory holds, the success probability is $1/640$ for $q = 2^{60}$ and $n = 10$. We can further increase the success probability to $1/64$ by repeating hash operations for ten times. In comparison with the traditional approach with success probability $1/2^{60}$ only, our approach significantly improves the success probability even with a small integer n. We compare the different approaches for finding loss in Table 1.

We show how to apply the iterated random oracle in encryption and key exchange for tight(er) reduction. In the application to encryption, we show how to use a key encapsulation mechanism with one-way security to construct an encryption scheme with indistinguishability security against a chosen-plaintext attack and a chosen-ciphertext attack. The security transformation from one-way security to indistinguishability security will only have a small finding loss. Notice that the security reduction for encapsulation mechanism with one-way security does not have the finding loss because the adversary must return the encapsulation key, which can be programmed as the solution to the computational hard problem in the reduction. Therefore, our security transformation is equivalent to a provably secure encryption under IND-CHP security reduction with a small finding loss. The transformation is n times ($n = 10$) less efficient in terms of key size and computations. However, the transformation does not expand the ciphertext size when the generation of key encapsulation is independent of public key. Many encryption schemes such as the ElGamal encryption [21] and BF-IBE [8] can be modified into key encapsulation mechanisms capturing this property. We also study the application of the iterated random oracle in an identity-based non-interactive key exchange protocol and other key exchange protocols.

Table 1. Comparison of different approaches for finding loss. The finding efficiency refers to the time cost of picking a query from the query set. The query efficiency refers to the time cost of generating the challenge query. Here, q is the size of query set including the challenge query and n is the maximum iteration time.

	Theory 1	Theory 2	Theory 3 (Ours)
For all problems	✓	×	✓
Success probability	$\frac{1}{q}$	1	$\frac{1}{n\cdot q^{\frac{1}{n}}}$
Finding efficiency	$O(1)$	$O(q)$	$O(n)$
Query efficiency	1	2	$O(n)$

1.3 Overview of the Approach

For simplicity, we use the concrete CDH problem as an example to describe the overview of the approach in the iterated random oracle. Suppose an adversary, who is given instances $I_i = (g,\ g^{a_i},\ g^b)$ for all $i \in [1, n]$ generated by the simulator, must make a query set $\mathbb{Q}$ ($|\mathbb{Q}| = q$) including a challenge query $\overline{\mathcal{Q}}^* = A_n$ to the random oracle, where A_n is defined as

$$A_i = H(A_{i-1})\ ||\ g^{a_i b}\ ||\ i: \quad i \in [1, n], \quad H(A_0) = 0_\epsilon \text{ is an empty string.}$$

We can construct a simulator to solve the CDH problem using the query set $\mathbb{Q}$ with success probability at least $1/(nq^{\frac{1}{n}})$. Given as input an instance (g, g^a, g^b) under a cyclic group $\mathbb{G}$ of prime order p, the aim of the simulator is to find g^{ab} from the query set generated by the adversary. This reduction is mainly composed of two tasks: (1) how to generate the instances $I_i = (g,\ g^{a_i},\ g^b), i \in [1, n]$ for the adversary, namely *instance generation* and (2) how to pick the query from the adversary's query set, namely *query selection*.

Instance Generation. The simulator randomly chooses $d \in [1, n], a_1, a_2, \cdots, a_{d-1},\ a_{d+1}, \cdots, a_n \in \mathbb{Z}_p$ and sets $a_d = a$. Then, it gives $I_i = (g,\ g^{a_i},\ g^b)$ for all $i \in [1, n]$ to the adversary who is required to make a query set including A_n. It requires that the adversary does not know d. Since all instances are chosen randomly, this requirement holds trivially. In the instances given to the adversary, the simulator can compute $g^{a_i b} = (g^b)^{a_i}$ for all $i \in [d+1, n]$ by itself since all related a_i are known. This is very important in the query selection for a small finding loss.

Query Selection. In this phase, a query is defined as either a candidate query or a useless query. The simulator will randomly pick a query from candidate queries, after all useless queries are removed. Before introducing what are useless queries and how to remove them, we first introduce what all iterated queries look like.

The query $\overline{\mathcal{Q}} = H(\overline{\mathcal{Q}}')\ ||\ Q\ ||\ i$ in the iterated random oracle is an iterated query, composed of an oracle response, a weight (the solution will appear here) and an iteration time. All iterated queries to the random oracle can be depicted in an arbitrary tree, where a node denotes a response on a query and an edge denotes a query. The root is an empty string. The edge $\overline{\mathcal{Q}} = H(\overline{\mathcal{Q}}')\ ||\ Q\ ||\ i$ starts from the node $H(\overline{\mathcal{Q}}')$ and ends at the node $H(\overline{\mathcal{Q}})$, which is depicted at the level i. When the maximum iteration time is n, the height of this arbitrary tree is n. For example, the two queries $\overline{\mathcal{Q}}_{1,2}^{(1)} = 0_\epsilon\ ||Q_{1,2}^{(1)}\ ||\ 1$ and $\overline{\mathcal{Q}}_{2,1}^{(2)} = H(\overline{\mathcal{Q}}_{1,2}^{(1)})\ ||Q_{1,2}^{(2)}\ ||\ 2$ can be depicted in a path from the root to a leaf shown in Fig. 1.

According to the property of random oracle, if $\overline{\mathcal{Q}}^* = A_n$ appears in the query set, all queries $A_1, A_2, \cdots, A_n$ must appear in the query set. Now, we can roughly describe what are useless queries. First, all queries with iteration time which is not equal to d are useless queries. Second, a query $\overline{\mathcal{Q}}$ with iteration time equal to d is a useless query if there is no valid path from the node $H(\overline{\mathcal{Q}})$ to a leaf node at the level n. Here, a valid path is the path where all edges for $i \in [d+1, n]$ in this path are valid queries whose weights are equal to $g^{a_i b}$. The simulator can verify whether a path is valid or not, because $g^{a_i b} = (g^b)^{a_i}$ for all $i \in [d+1, n]$ are computable using a_i. All queries with iteration time equal to n are candidate queries.

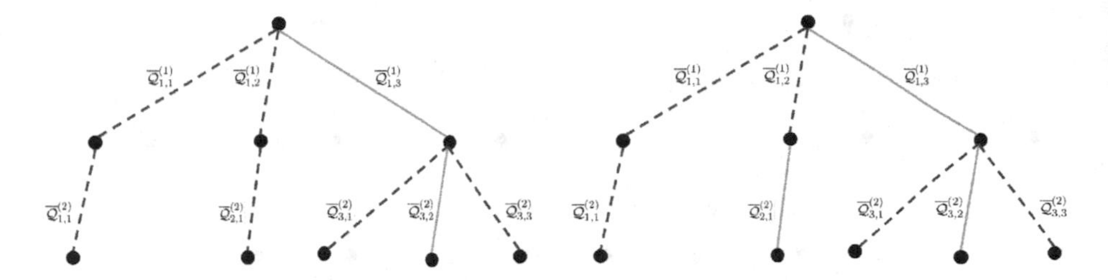

Fig. 1. Example 1

Fig. 2. Example 2

Probability Analysis. Based on the above instance generation and query selection, we can prove there must exist an integer $i^* \in [1, n]$ satisfying the minimum probability $1/q^{\frac{1}{n}}$. Precisely, for those queries with iteration time i^*, the success probability of picking a valid query from candidate queries is $1/q^{\frac{1}{n}}$. The integer i^* is adaptively decided by the adversary in query set generation, while the integer d is randomly chosen by the simulator. When $d = i^*$ (i.e. the simulator happens to embed the solution in this level), all useless queries with iteration time i^* will be removed and the corresponding success probability is $1/q^{\frac{1}{n}}$. Therefore, we yield the success probability result by

$$\Pr[suc] = \sum_{i=1}^{n} \Pr[suc|d = i] \Pr[d = i] \geq \Pr[suc|d = i^*] \Pr[d = i^*] = \frac{1}{nq^{\frac{1}{n}}}.$$

We now give four simple examples where $n = 2$ and $q = 8$ to analyze the above result. The corresponding success probability of $\Pr[suc|d = i^*]$ for some i^* should be at least $1/\sqrt{8}$. We use a solid line to denote a query at the level i if it has a valid weight equal to $g^{a_i b}$. Otherwise, we denote the query with a dashed line. In this arbitrary tree, $\overline{\mathcal{Q}}^{(i)}$ denotes a query at the level i. Notice that all queries from the same node have at most one query with a valid weight, but all queries at the same level i could have more than one valid query whose weights are all valid and equal to $g^{a_i b}$.

In these examples, if the adversary only makes two queries at the first level, we immediately have $\Pr[suc|d = 1] = \frac{1}{2} \geq \frac{1}{\sqrt{8}}$ when $d = 1$. Therefore, in the following examples, the adversary is assumed to make three queries at the first level.

- Suppose the query set can be depicted as the tree in Fig. 1. When $d = 1$, the two queries $\overline{\mathcal{Q}}^{(1)}_{1,1}, \overline{\mathcal{Q}}^{(1)}_{1,2}$ will be removed because their nodes do not have a valid path such that only one query is remained at this level. Therefore, we have $\Pr[suc|d = 1] = 1 \geq \frac{1}{\sqrt{8}}$.
- Suppose the query set can be depicted as the tree in Fig. 2. When $d = 1$, the query $\overline{\mathcal{Q}}^{(1)}_{1,1}$ will be removed because this node does not have a valid path such that two queries are remained at this level. Therefore, we have $\Pr[suc|d = 1] = \frac{1}{2} \geq \frac{1}{\sqrt{8}}$.

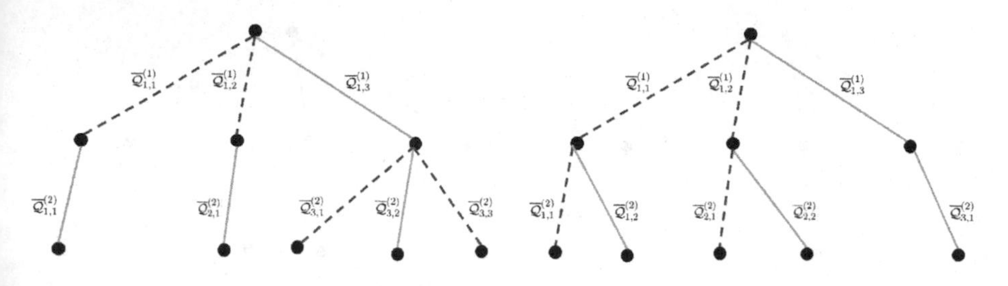

Fig. 3. Example 3 **Fig. 4.** Example 4

- Suppose the query set can be depicted as the tree in Fig. 3. When $d = 2$, it is easy to see that $\Pr[suc|d = 2] = \frac{3}{5} \geq \frac{1}{\sqrt{8}}$.
- Suppose the query set can be depicted as the tree in Fig. 4. The result is exactly the same as Fig. 3, where $\Pr[suc|d = 2] = \frac{3}{5} \geq \frac{1}{\sqrt{8}}$.

1.4 Other Related Work

The UF-CHP security reduction with a tight reduction for digital signatures has been studied in [1,2,6,13,14,24–26]. A tight reduction requires no abortion in signature simulation and enables to solve a hard problem from the forged signature. With the help of random oracles, it seems easier to achieve a tight reduction by adding a random bit after the message to be signed. In this reduction, the simulator uses the bit to control the hash values of messages to be signed and to be forged, such that the probability of abortion is very small.

The IND-DHP security reduction with a tight reduction for encryption has been studied in [4,7,9,15,16,22,23,26–28]. To achieve a tight reduction, the simulator must be able to simulate decryption queries for CCA security and private key queries for identity-based encryption and its variants. It also requires the simulator to program the challenge ciphertext into a one-time pad or an indistinguishable ciphertext depending on the given instance. We note that the approaches for tight reduction are different. This is because there is no general technique enabling a tight reduction for encryption, especially without random oracles.

The IND-CHP security reduction is a special reduction requiring the help of random oracles, where the simulator solves a hard problem using the adversary's queries instead of its direct attack. How to find the correct solution from the adversary's query set is necessary to achieve a tight reduction. The problem of finding loss only exists in this reduction type especially when the decisional variant is also hard. The traditional approach for finding loss is via a random pick, which results in a huge finding loss. The first non-trivial approach was introduced by Cash, Kiltz and Shoup [10] in EUROCRYPT 2008. The proposed trapdoor test can be used to solve finding loss during the corresponding IND-CHP security reductions. They had shown that the proposed approach can be applied to Diffie-Hellman key exchange [17], Cramer-Shoup encryption [16], BF-IBE [8] and password-authenticated key exchange [3] to achieve the tightness of security reduction. This approach, however, requires that the computational hard problem can be embedded with a trapdoor test on solutions to a given instance and a created instance. This work has been extended and applied in [11,12] but they still have the same restriction. There is no efficient approach for finding loss in the IND-CHP security reduction without any restriction on the adopted computational hard assumptions.

The rest of this paper is organized as follows. We use an example to introduce how the IND-CHP security reduction works in Sect. 2. The generalization of computational hard problems is also given and discussed. In Sect. 3, we prove the correctness of Theory 3. Then, we show how to apply the iterated random oracle for encryption in Sect. 4 and key exchange towards tight(er) security reduction in Sect. 5.

2 IND-CHP Security Reduction and Generalized Problems

2.1 An Example of IND-CHP Security Reduction

Let $\mathbb{G}$ be a cyclic group of the prime order p and g be a generator. Let $H : \{0,1\}^* \rightarrow \{0,1\}^n$ be a one-way hash function. Considering the following bare ciphertext CT without a public/secret key pair, where $x, y \in \mathbb{Z}_p$ and $coin \in \{0,1\}$ are chosen randomly and secretly.

$$CT = (c_1, c_2, c_3) = \Big(g^x,\ g^y,\ H(g^{xy}) \oplus m_{coin}\Big)$$

Suppose there exists an adversary who can distinguish the message $m_{coin} \in \{m_0, m_1\}$ in CT with a non-negligible advantage ϵ in a polynomial time, where the two messages $\{m_0, m_1\} \in \{0,1\}^n$ are adaptively chosen by the adversary. We can construct a simulator to solve the CDH problem in the random oracle model, where H is set as a random oracle controlled by the simulator.

Before we introduce how to program the security reduction, we first introduce the nice feature of using random oracle in security reduction. In the random oracle model, the message is encrypted with $H(g^{xy})$, which is a random string from $\{0,1\}^n$ and is independent of its hash input g^{xy} and (g^x, g^y). Without making a query on g^{xy} to the random oracle, the ciphertext CT is a one-time pad encryption on m_{coin} because $H(g^{xy})$ is random and independent of (g^x, g^y) in the ciphertext. Then, the success probability of guessing the encrypted message is $\frac{1}{2}$ only. According to the assumption, the adversary can distinguish the encrypted message with probability $\frac{1}{2} + \epsilon$. This assumption indicates that the adversary ever queried g^{xy} to the random oracle with probability 2ϵ [8]. That is, one of queries in the adversary's query set is equal to g^{xy}. This query is called challenge query, which is used to break the security of cryptosystem.

The security reduction works as follows. Given (g, g^a, g^b), the simulator aims to compute g^{ab}. Upon receiving $m_0, m_1 \in \{0,1\}^n$ from the adversary, the simulator creates the challenge ciphertext as $CT = (c_1, c_2, c_3) = (g^a,\ g^b,\ R)$, where R is a random string from $\{0,1\}^n$. What the simulator will do is to wait for queries from the adversary. Notice that if the adversary does not make a query on g^{ab} to the simulator, the adversary cannot either distinguish the message with a non-negligible advantage or distinguish the simulation ciphertext from the real ciphertext. According to the assumption, the group element g^{ab} will appear in one of queries with probability 2ϵ. Suppose the adversary made q queries to the random oracle in total. The simulator randomly picks one of queries as the solution to the CDH problem. We have the randomly picked element is equal to g^{ab} with probability $\frac{2\epsilon}{q}$. That is, the simulator will solve the hard problem with probability $\frac{2\epsilon}{q}$ in the corresponding security reduction. This completes the description of security reduction. This reduction has a finding loss whose corresponding success probability is in the linear of hash query number q.

We note that the above bare ciphertext cannot be decrypted by anyone when the CDH problem is hard. However, in the real encryption scheme, the encryptor and the decryptor know more information than the bare ciphertext. When treating g^x as the public key and y is the chosen random number by the encryptor, we have that the bare ciphertext is equivalent to the hashed ElGamal encryption scheme, where the encryptor knows y and the decryptor knows the secret key x such that the ciphertext can be created and decrypted respectively. Roughly speaking, a secure encryption scheme is constructed in the way that a computational hard problem can be easily solved by the encryptor and decryptor with an additional secret, while outsiders (adversaries) without knowing a secret must solve the computational hard problem in order to break the scheme.

2.2 Generalized Computational Hard Problems

We generalize all computational hard problems into the following description.

I: The input arbitrary string (also known as instance)

P: The computational problem

$\mathcal{C}[I,P]$: The solution to the instance I under the computational problem P.

For example, given an instance $I = (g, g^a, g^b) \in \mathbb{G}$, based on different problems P, the solution can be

$$\mathcal{C}[I,P_1] = g^{ab}, \qquad \mathcal{C}[I,P_2] = g^{\frac{b}{a}}.$$

The generalized computational hard problem is defined as

$$\Pr\Big[\mathcal{A}(I,P) = \mathcal{C}[I,P]\Big] \leq \epsilon,$$

where no adversary who is given (I,P) can find a solution $\mathcal{C}[I,P]$ with a non-negligible advantage ϵ. Here, ϵ is a function of the security parameter in the generation of the instance I.

For the computational hard problem (I,P), anyone can verify whether a solution is correct or not if the decisional variant of this problem is easy. However, if the decisional variant is also hard, it seems no one can verify the correctness of a solution. However, this observation is not correct because the instance generator, who generates the instance, can generate the instance in the way that it knows its correct solution. Taking the CDH problem in a cyclic group as an example where the DDH problem is also hard. The instance generator can randomly choose $a, b \in \mathbb{Z}_p$ and set the instance to be (g, g^a, g^b), where the solution g^{ab} is computable by the instance generator. Hence, for the computational hard problem P, we assume the instance generator enables to generate an instance I such that $\mathcal{C}[I, P]$ can be efficiently computed. This assumption is necessary to support the definition of computational hard problems whose decisional variants are also hard. We emphasize the importance of this property here because the simulator in the iterated random oracle requires generating some instances indistinguishable from the challenge instance, such that the simulator can compute solutions to all self-generated instances under the challenge hard problem P.

3 Iterated Random Oracle and Its Proof

In the iterated random oracle, each query will be programmed using iterations, and hence it will be called as an iterated query. An iterated query is composed of an oracle response, a weight (the solution to a hard problem will appear here) and an iteration time. They are put together using a concatenation symbol "||". Given a hash list recording all iterated queries and their responses, we can depict all queries in the hash list using an arbitrary tree. The height of this arbitrary tree is n, where n is the maximum time of iteration. The details are described in the following subsections.

3.1 Iterated Query and Tree Representation

Iterated Query. We define an iterated query $\overline{\mathcal{Q}}$ to the random oracle as

$$\overline{\mathcal{Q}} = \text{Response} \ || \ \text{Weight} \ || \ \text{Iteration Time} = \overline{\mathcal{R}} \ || \ Q \ || \ i,$$

where $\overline{\mathcal{R}}$ is a response on a query from the random oracle H (an empty string 0_ϵ is assumed as the initialized response), Q is a weight (any arbitrary string) chosen by the adversary and i is the iteration time. The iteration time denotes the minimum time for making such an iterated query. If $i = 1$, it means the adversary can immediately make such a query. Otherwise, for example, given $\overline{\mathcal{Q}}_1 = 0_\epsilon||Q_1||1$ and $\overline{\mathcal{Q}}_2 = H(\overline{\mathcal{Q}}_1)||Q_2||2$, it requires the adversary to query $\overline{\mathcal{Q}}_1$ first before $\overline{\mathcal{Q}}_2$. We will use the following symbols associated with queries and responses in the following representations.

- $\overline{\mathcal{Q}}^{(i)}$ is an iterated query with the iteration time i.
- $Q^{(i)}_{j,k}$ is the weight in the iterated query $\overline{\mathcal{Q}}^{(i)}_{j,k}$.
- $\mathbb{Q}$ is the set of all queries made by the adversary.
- $\mathbb{Q}^{(i)}$ is the set of all iterated queries whose iteration time are all equal to i.
- $H(\overline{\mathcal{Q}}^{(i)})$ is the response from the random oracle on the query $\overline{\mathcal{Q}}^{(i)}$.

Tree Representation. Suppose the adversary only makes the above iterated queries to the random oracle, and an empty hash list $\mathcal{L}$ is used to record all queries and responses. We can depict all queries and corresponding responses using an artitrary tree (such as Fig. 5), where the root is the empty string 0_ϵ.

- All edges denote iterated queries and their end nodes denote their corresponding responses.
- The query $\overline{\mathcal{Q}}_{j,k}^{(i)} = H(\overline{\mathcal{Q}}^{(i-1)}) \,||\, Q_{j,k}^{(i)} \,||\, i$ is the edge with connection between the node $H(\overline{\mathcal{Q}}^{(i-1)})$ and the node $H(\overline{\mathcal{Q}}_{j,k}^{(i)})$ at the level i. Here, j in this query represents that $\overline{\mathcal{Q}}^{(i-1)}$ is the j-th query at the level $i-1$ counted from left to right, and k in this query represents that $H(\overline{\mathcal{Q}}_{j,k}^{(i)})$ is the k-th child of $H(\overline{\mathcal{Q}}^{(i-1)})$ counted from left to right.
- The height of the arbitrary tree is the maximum time of iteration in all iterated queries.

The hash list and the tree representation have the following connections. First, this is an arbitrary tree because the adversary can make any number of iterated queries $\overline{\mathcal{Q}} = \overline{\mathcal{R}} \,||\, Q \,||\, i$ with the same $\overline{\mathcal{R}}$ and i. Second, all edges starting from the same node are the depiction of queries with the same $\overline{\mathcal{R}}$ and i but distinct weights Q. Third, all iterated queries are different such that all nodes are distinct, but the weights in those queries (edges) from different nodes could be the same. For example, the weight $Q_{3,1}^{(2)}$ must be different from $Q_{3,2}^{(2)}$ because the queries $\overline{\mathcal{Q}}_{3,1}^{(2)}, \overline{\mathcal{Q}}_{3,2}^{(2)}$ already have the same oracle response and iteration time. However, $Q_{3,2}^{(2)}$ could be equal to $Q_{1,1}^{(2)}$ in Fig. 5. This observation is very important in the analysis of success probability for the iterated random oracle. Finally, the total query number is equal to the total number of edges in this arbitrary tree, if all queries are iterated queries.

In the random oracle model, the adversary can make any arbitrary string as a query chosen by itself. However, we focus on the defined iterated queries only. We emphasize that our focus does not compromise any problem because all other queries that cannot be described in this arbitrary tree must be not the challenge query and will be removed from the query set before selection.

3.2 Proof of Theory 3

It is complicated to prove this theory directly especially the analysis of success probability. We split the proof for this theory into the following steps.

Simulator Construction. Given as input an instance I and the problem P, the simulator aims to compute $\mathcal{C}[I, P]$. The simulator generates $(I_1, I_2, \cdots, I_n)$ for the adversary as follows.

- Randomly choose $d \in [1, n]$ and set $I_d = I$. We have $\mathcal{C}[I_d, P] = \mathcal{C}[I, P]$.
- Choose random instances $I_1, I_2, \cdots, I_{d-1}, I_{d+1}, \cdots, I_n$ under the problem P such that $\mathcal{C}[I_i, P]$ for all $i \in [1, n]/\{d\}$ are known by the simulator.

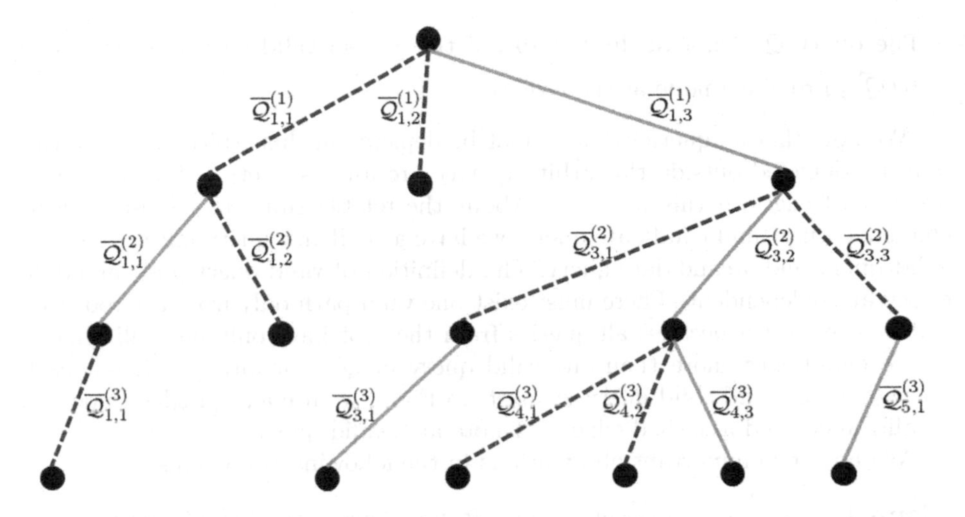

Fig. 5. An example of arbitrary tree generated from iterated queries and responses.

– Set and give $(I_1, I_2, \cdots, I_n)$ to the adversary.

According to the assumption, the adversary will make a query set $\mathbb{Q}$ to the random oracle including a challenge query $\overline{\mathcal{Q}}^* \in \mathbb{Q}$, where $\overline{\mathcal{Q}}^* = \overline{\mathcal{Q}}_*^{(n)}$. According to the definition of $\overline{\mathcal{Q}}^{(i)}$ and the property of random oracles, the adversary must ever make all challenge queries $\overline{\mathcal{Q}}_*^{(1)}, \overline{\mathcal{Q}}_*^{(2)}, \cdots, \overline{\mathcal{Q}}_*^{(n)}$ to the random oracle. Otherwise, the adversary cannot generate $\overline{\mathcal{Q}}^* \in \mathbb{Q}$. Notice that $\mathcal{C}[I, P]$ exists in $\overline{\mathcal{Q}}_*^{(d)} \in \mathbb{Q}^{(d)}$. The simulator will solve the hard problem by removing all useless queries in $\mathbb{Q}^{(d)}$, picking a random query from the remaining set $\mathbb{Q}^{(d)}$ and extracting the weight from the picked query as the solution to the hard problem. The success probability of finding the correct solution will be the one given in our theory.

Further Tree Representation. We further define queries and weights in order to clarify how to remove all useless queries from $\mathbb{Q}^{(d)}$.

– The query $\overline{\mathcal{Q}}_{j,k}^{(i)}$ is a **challenge query** if $\overline{\mathcal{Q}}_{j,k}^{(i)} = \overline{\mathcal{Q}}_*^{(i)}$.
– The weight $Q_{j,k}^{(i)}$ is a **valid weight** if $Q_{j,k}^{(i)} = \mathcal{C}[I_i, P]$.
– The query $\overline{\mathcal{Q}}_{j,k}^{(i)}$ is a **valid query** if it has a valid weight.
– A path from a node to a leaf is a **valid path** if all edges in this path are valid queries.
– The query $\overline{\mathcal{Q}}_{j,k}^{(i)}$ is a **child query** of $\overline{\mathcal{Q}}^{(i-1)}$ if $\overline{\mathcal{Q}}_{j,k}^{(i)} = H(\overline{\mathcal{Q}}^{(i-1)})||Q_{j,k}^{(i)}||i$.
– The query $\overline{\mathcal{Q}}_{j,k}^{(i)}$ is a **candidate query** if there exists a valid path from the node $H(\overline{\mathcal{Q}}_{j,k}^{(i)})$ to a leaf node at the level n. All queries in $\mathbb{Q}^{(n)}$ are defined as candidate queries.

– The query $\overline{\mathcal{Q}}_{j,k}^{(i)}$ is a **useless query** if there is no valid path from the node $H(\overline{\mathcal{Q}}_{j,k}^{(i)})$ to a leaf node at the level n.

We note that all queries that cannot be depicted in this arbitrary tree or can only be depicted outside this arbitrary tree are useless queries. The maximum number of edges in this tree is q. About the relationship among valid query, challenge query and candidate query, we have a challenge query must be both a valid query and a candidate query. The definition of valid query and candidate query are independent. There must exist one valid path only from the root to a leaf at the level n because all queries from the root have only one valid query. There could exist more than one valid query in $\mathbb{Q}^{(i)}$ for any $i \geq 2$, but each query has one valid child query at most. In Fig. 5, we use a solid edge to denote a valid query and a dashed edge to denote an invalid query.

We have two important observations in the following two claims.

Claim 1. *If $\overline{\mathcal{Q}}^{(i)}$ is a candidate query, it must have a valid child query.*

According to the definition of candidate query, there exists a valid path from the node $H(\overline{\mathcal{Q}}^{(i)})$ to a leaf node at the level n. The first edge in this valid path is a valid query comprising of the response $H(\overline{\mathcal{Q}}^{(i)})$. This is the valid child query of $\overline{\mathcal{Q}}^{(i)}$.

Claim 2. *If $\overline{\mathcal{Q}}^{(i)}$ is a candidate query and its child query denoted by $\overline{\mathcal{Q}}^{(i+1)}$ is a valid query, we have that $\overline{\mathcal{Q}}^{(i+1)}$ is also a candidate query.*

We prove by contradiction. According to the first claim and the tree representation, there exists only one valid child query of $\overline{\mathcal{Q}}^{(i)}$ denoted by $\overline{\mathcal{Q}}^{(i+1)}$. All paths starting from the node $H(\overline{\mathcal{Q}}^{(i)})$ through invalid child queries of $\overline{\mathcal{Q}}^{(i)}$ must be invalid paths. If all paths starting from the node $H(\overline{\mathcal{Q}}^{(i)})$ through the edge $\overline{\mathcal{Q}}^{(i+1)}$ are invalid paths either, there is no valid path from the node $H(\overline{\mathcal{Q}}^{(i)})$ to a leaf node. Hence $\overline{\mathcal{Q}}^{(i)}$ is not a candidate query. Therefore, the assumption is incorrect and there should exist a valid path starting from the node $H(\overline{\mathcal{Q}}^{(i)})$ through the edge $\overline{\mathcal{Q}}^{(i+1)}$, which implies that $\overline{\mathcal{Q}}^{(i+1)}$ is also a candidate query.

Lemma 1. *If the following rate*

$$R^{(i)} = \frac{\textit{The number of valid queries in } \mathbb{Q}^{(i)}}{\textit{The number of candidate queries in } \mathbb{Q}^{(i)}} < \frac{1}{q^{\frac{1}{n}}}$$

holds for all $i \in [1, n]$, the adversary must make more than q candidate queries.

Proof. Let $N = q^{\frac{1}{n}}$. All queries in $\mathbb{Q}^{(i)}$ must be either valid or invalid. Let VQ_i denote the number of valid queries at the level i of tree. Let IQ_i denote the number of invalid queries at the level i of tree. If the rate $R^{(i)}$ holds for all $i \in [1, n]$, we have the following deduction from the first level to the last level based on the above two claims, where only candidate queries are counted.

- **Level 1.** All queries are from the root and there is one valid query only, which is also a candidate query. That is, $VQ_1 = 1$. To make sure the rate is less than $1/N$, the adversary must make $IQ_1 \geq (N-1) \cdot VQ_1 + 1$ invalid queries that are also candidate queries. The total number of candidate queries in this level therefore is $VQ_1 + IQ_1$. Hence, according to the Claim 1, the total number of valid queries in the next level is $VQ_1 + IQ_1$.
- **Level 2.** According to the result in the level 1, the number of valid queries is $VQ_2 = VQ_1 + IQ_1$. According to Claim 2, these valid queries are also candidate queries. To make sure the rate is less than $1/N$, the adversary must make $IQ_2 \geq (N-1) \cdot VQ_2 + 1$ invalid queries that are also candidate queries. The total number of candidate queries in this level therefore is $VQ_2 + IQ_2$. Hence, according to Claim 1, the total number of valid queries in the next level is $VQ_2 + IQ_2$.
- **Level 3.** According to the result in the level 2, the number of valid queries is $VQ_3 = VQ_2 + IQ_2$. According to Claim 2, these valid queries are also candidate queries. To make sure the rate is less than $1/N$, the adversary must make $IQ_3 \geq (N-1) \cdot VQ_3 + 1$ invalid queries that are also candidate queries. The total number of candidate queries in this level therefore is $VQ_3 + IQ_3$. Hence, according to Claim 1, the total number of valid queries in the next level is $VQ_3 + IQ_3$.
- The result in the level i is the same as the previous analysis.
- **Level** $n-1$**.** According to the result in the level $n-2$, the number of valid queries is $VQ_{n-1} = VQ_{n-2} + IQ_{n-2}$. According to Claim 2, these valid queries are also candidate queries. To make sure the rate is less than $1/N$, the adversary must make $IQ_{n-1} \geq (N-1) \cdot VQ_{n-1} + 1$ invalid queries that are also candidate queries. The total number of candidate queries in this level therefore is $VQ_{n-1} + IQ_{n-1}$. Hence, according to Claim 1, the total number of valid queries in the next level is $VQ_{n-1} + IQ_{n-1}$.
- **Level** n**.** According to the result in the level $n-1$, the number of valid queries is $VQ_n = VQ_{n-1} + IQ_{n-1}$. To make sure the rate is less than $1/N$, the adversary must make $IQ_n \geq (N-1) \cdot VQ_n + 1$ invalid queries. The total query number in this level therefore is $VQ_n + IQ_n$. All queries are treated as candidate queries.

From the above analysis, we obtain the following results for all $i \in [1, n]$.

$$\begin{aligned}
VQ_1 + IQ_1 &= N + 1 \\
VQ_i + IQ_i &\geq VQ_i + (N-1) \cdot VQ_i + 1 \\
&= N \cdot VQ_i + 1 \\
&> N \cdot VQ_i \\
&= N \cdot (VQ_{i-1} + IQ_{i-1}).
\end{aligned}$$

Then, we yield

$$\sum_{i=1}^{n} (VQ_i + IQ_i) > (VQ_n + IQ_n) > N^{n-1}(VQ_1 + IQ_1) > N^n = q.$$

This completes the proof of Lemma 1. □

Based on the above definitions and explanations, we are ready to give the proof of Theory 3.

Proof of Theory 3. In the simulation, the number d is randomly chosen by the simulator and all instances $(I_1, I_2, \cdots, I_n)$ are indistinguishable. The adversary therefore does not know d. The query set $\mathbb{Q}$ generated by the adversary hence is independent of d.

According to Lemma 1, if the adversary makes q queries at most, there must exist an integer $i^* \in [1, n]$ satisfying

$$R^{(i^*)} = \frac{\text{The number of valid queries in } \mathbb{Q}^{(i^*)}}{\text{The number of candidate queries in } \mathbb{Q}^{(i^*)}} \geq \frac{1}{q^{\frac{1}{n}}}.$$

When $d = i^*$, the simulator can remove all useless queries in $\mathbb{Q}^{(i^*)}$ because $\mathcal{C}[I_i, P]$ for all $i \in [d+1, n]$ are computable by the simulator. Then, the success probability of picking a valid query from all candidate queries is at least $1/q^{\frac{1}{n}}$. The success probability $\Pr[suc]$ given in Theory 3 holds because

$$\begin{aligned} \Pr[suc] &= \sum_{i=1}^{n} \Pr[suc|d = i] \Pr[d = i] \\ &\geq \Pr[suc|d = i^*] \Pr[d = i^*] \\ &= \frac{1}{q^{\frac{1}{n}}} \cdot \frac{1}{n}. \end{aligned}$$

This completes the proof of Theory 3. □

3.3 Variant

The success probability given in Theory 3 is the lower bound probability because the probability $\Pr[suc|d = i] > 0$ holds for all $i \neq i^*$. We can repeat hash operations in the iterated random oracle to obtain a larger lower bound success probability.

Theory 4 (Improved Iterated Random Oracle). *Let H be a random oracle. Suppose an adversary, who is given instances $(I_1, I_2, \cdots, I_n)$ generated by the simulator, must make a set of queries $\mathbb{Q}$ ($|\mathbb{Q}| = q$) including a challenge query $\overline{\mathcal{Q}}^* = H^{k-1}(\overline{\mathcal{Q}}_*^{(n)})$ to the random oracle, where $\overline{\mathcal{Q}}_*^{(n)}$ is defined as*

$$\overline{\mathcal{Q}}_*^{(i)} = H^k(\overline{\mathcal{Q}}_*^{(i-1)}) \,||\, \mathcal{C}[I_i, P] \,||\, i: \quad i \in [1, n], \quad H(\overline{\mathcal{Q}}_*^{(0)}) = 0_\epsilon \text{ is an empty string.}$$

We can construct a simulator who controls the random oracle to solve the hard problem P using the query set $\mathbb{Q}$ with success probability at least $k/(nq^{\frac{1}{n}})$. Here, $H^i(\overline{\mathcal{Q}})$ is to repeat hash operation on $\overline{\mathcal{Q}}$ for i times. $H^i(\overline{\mathcal{Q}}) = H\Big(H^{i-1}(\overline{\mathcal{Q}})\Big)$ and $H^0(\overline{\mathcal{Q}}) = \overline{\mathcal{Q}}$.

$H^{k-1}(\overline{\mathcal{Q}}_*^{(n)})$	$H^{k-2}(\overline{\mathcal{Q}}_*^{(n)})$	$\cdots$	$H(\overline{\mathcal{Q}}_*^{(n)})$	$\overline{\mathcal{Q}}_*^{(n)}$
$H^{k-1}(\overline{\mathcal{Q}}_*^{(n-1)})$	$H^{k-2}(\overline{\mathcal{Q}}_*^{(n-1)})$	$\cdots$	$H(\overline{\mathcal{Q}}_*^{(n-1)})$	$\overline{\mathcal{Q}}_*^{(n-1)}$
$\cdots$	$\cdots$	$\cdots$	$\cdots$	$\cdots$
$H^{k-1}(\overline{\mathcal{Q}}_*^{(2)})$	$H^{k-2}(\overline{\mathcal{Q}}_*^{(2)})$	$\cdots$	$H(\overline{\mathcal{Q}}_*^{(2)})$	$\overline{\mathcal{Q}}_*^{(2)}$
$H^{k-1}(\overline{\mathcal{Q}}_*^{(1)})$	$H^{k-2}(\overline{\mathcal{Q}}_*^{(1)})$	$\cdots$	$H(\overline{\mathcal{Q}}_*^{(1)})$	$\overline{\mathcal{Q}}_*^{(1)}$

In this theory, the adversary must make $k \cdot n$ queries to obtain the challenge query $\overline{\mathcal{Q}}^* = H^{k-1}(\overline{\mathcal{Q}}_*^{(n)}) \in \mathbb{Q}$.

The query on $H^i(\overline{\mathcal{Q}})$ requires the adversary to make a query on $H^{i-1}(\overline{\mathcal{Q}})$ first. In particular, the query on $\overline{\mathcal{Q}}_*^{(j)}$ requires the adversary to make a query on $H^{k-1}(\overline{\mathcal{Q}}_*^{(j-1)})$ to obtain $H^k(\overline{\mathcal{Q}}_*^{(j-1)})$ to compose $\overline{\mathcal{Q}}_*^{(j)}$. The proof of this theory is based on a slightly different lemma where the rate is $k/q^{\frac{1}{n}}$. This is because the total number of queries in each level is $k\cdot(VQ_i+IQ_i)$ instead of (VQ_i+IQ_i). The other analysis is similar and we omit them here without redundancy. Therefore we have the success probability shown in the theory.

3.4 Comparison of Success Probability

We compare the success probability of finding the solution from the query set among the traditional approach, the Cash-Kiltz-Shoup approach and the iterated random oracle, where concrete integers $n = 10$ and $k = 10$ are chosen. The result is given in Table 2. It shows that the iterated random oracle has a very small finding loss compared to the traditional approach even the iteration time n is very small. With a proper hash repeating time k, it further improves the success probability. Notice that the Cash-Kiltz-Shoup's approach is the most efficient approach, but it is not a universal approach for any computational hard problem.

Table 2. Comparison of success probability.

	$q = 2^{40}$	$q = 2^{50}$	$q = 2^{60}$
Traditional approach	$\frac{1}{2^{40}}$	$\frac{1}{2^{50}}$	$\frac{1}{2^{60}}$
Cash-Kiltz-Shoup [10,11]	1	1	1
Iterated random oracle with $n = 10, k = 1$	$\frac{1}{160}$	$\frac{1}{320}$	$\frac{1}{640}$
Iterated random oracle with $n = 10, k = 10$	$\frac{1}{16}$	$\frac{1}{32}$	$\frac{1}{64}$

3.5 Comparison of Query Efficiency and Finding Efficiency

The price to pay for small finding loss from the iterated random oracle is the efficiency loss in the generation of challenge query. Recall that the challenge query is associated with one instance computation in the traditional approach (Theory 1) and two instance computations in the Cash-Kiltz-Shoup [10,11] approach (Theory 2). The challenge query in the iterated random oracle is associated with n instance computations and n queries or $n \cdot k$ queries. The efficiency loss is in the linear of n. Fortunately, n can be as small as 10 in the iteration. Furthermore, when the efficiency is mainly dominated by the computation of $\mathcal{C}[I_i, P]$, they can be performed in parallel because all computations are independent.

In the iterated random oracle, the simulator needs to compute $\mathcal{C}[I_i, P]$ for all $i \in [d+1, n]$, where d is randomly chosen from $[1, n]$[1], in order to remove all useless queries. Then, the simulator randomly picks one query from all candidate queries. Hence, the time cost of finding a solution is mainly dominated by instance computations and the time complexity is $O(n)$. In comparison with the other two approaches, the simulator in the traditional approach (Theory 1) directly picks one solution in a random way and the time complexity is $O(1)$. The simulator in the Cash-Kiltz-Shoup [10,11] approach (Theory 2) has to test each query until it finds the correct solution. Therefore, their time complexity is (q) more expensive than the iterated random oracle.

3.6 Remarks of Simulation Based on Theories

The introduced three theories for finding loss can be described as follows in a general summary. Suppose an adversary, who is given an instance $I_{\mathcal{A}}$ generated by the simulator, must make a set of queries $\mathbb{Q}$ ($|\mathbb{Q}| = q$) including a challenge query $\overline{\mathcal{Q}}^* = \mathcal{C}[I_{\mathcal{A}}, P_{\mathcal{A}}]$ to the random oracle. Here, $\mathcal{C}[I_{\mathcal{A}}, P_{\mathcal{A}}]$ is the solution to the instance $I_{\mathcal{A}}$ under the computational hard problem $P_{\mathcal{A}}$ which is defined by the simulator. We aim to construct a simulator to solve a hard problem P using the query set $\mathbb{Q}$.

In the corresponding simulator construction, the simulator is given an instance I under the hard problem P and aims to solve it with the help of the adversary. The simulator should construct an instance $I_{\mathcal{A}}$ for the adversary using the given instance I and define the hard problem $P_{\mathcal{A}}$ such that $\mathcal{C}[I, P]$ will appear in the query set. The resulting results $(I_{\mathcal{A}}, P_{\mathcal{A}})$ from the traditional approach, Cash-Kiltz-Shoup's approach and our approach are different. We remark that the successful construction of such a simulator is not the end of simulation. It merely introduces the approach of how to find the correct solution from the adversary's query set. To complete the reduction, the simulator must enable to use the created instance $I_{\mathcal{A}}$ to simulate the proposed cryptosystem and make sure the challenge query including $\mathcal{C}[I_{\mathcal{A}}, P_{\mathcal{A}}]$ will appear in the query set. This is required in the security reduction because the adversary is not to solve a hard problem for the simulator but is going to break a cryptosystem.

[1] When $d = n$, the simulator does not need to compute any $\mathcal{C}[I, P]$.

4 Tight Security in Security Transformation for Encryption

The principle application of the iterated random oracle is the security transformation from a key encapsulation mechanism with one-way security to an encryption with indistinguishability security, whose reduction is tight. In this section, we show how to achieve such a security transformation without expanding ciphertext size.

A key encapsulation mechanism (KEM) is an asymmetric encryption whose encryption algorithm will generate a random key (a.k.a. the encapsulation key), together with a corresponding ciphertext (a.k.a. the encapsulation). The random key is then used for symmetric encryption while the encapsulation forms part of the message ciphertext to deliver the random key in an asymmetric manner. Any receiver who owns a valid secret key can decapsulate the random key from the encapsulation. In the definition of one-way security for KEM, the challenger generates a challenge ciphertext CT^* for the adversary and the aim of the adversary is to return the corresponding challenge random key.

We observe that any KEM with one-way security does not have a security loss in finding a correct solution, if the random key is the solution to a computational hard problem in security reduction. This is because the adversary only returns one answer to the simulator, which is the correction solution to a hard problem. However, in the IND-CHP security reduction with the help of random oracles, the correct solution is hidden in a large query set made by the adversary. In this section, we show how to fill this gap by using the iterated random oracle.

Our security transformation is based on the KEM of functional encryption, namely functional key encapsulation mechanism (FKEM). The functional encryption can be seen as a generalized asymmetric encryption including public key encryption, identity-based encryption and attribute-based encryption. We adopt the FKEM because the iterated random oracle is a general approach fitting for all asymmetric encryptions.

Our security transformation can be applied to any FKEM. However, this generic transformation could be accompanied with a long ciphertext under iterated random oracles. This is because the challenge ciphertext must be associated with n different instances using the iterated random oracle, where the adversary is required to compute n solutions to different instances. To obtain a short ciphertext after transformation, these n instances must have shared input parameters. We extract one special type of FKEM from all FKEM with the following two properties.

- Firstly, global system parameters Param will be defined for FKEM, where many master key pairs $(\widetilde{\mathsf{mpk}}_1, \widetilde{\mathsf{msk}}_1), (\widetilde{\mathsf{mpk}}_2, \widetilde{\mathsf{msk}}_2), \cdots, (\widetilde{\mathsf{mpk}}_n, \widetilde{\mathsf{msk}}_n)$ can be generated with this global parameters. We note that these global system parameters are very common in an asymmetric encryption. It could include the definitions of pairing group, chosen generator and hash functions. All of these parameters are shared and used by different users or authorities.

- Secondly, the ciphertext encapsulation is computed without the input of master public keys, which will be the shared input parameters for all generated master key pairs. We note that many asymmetric encryptions fall into this type, such as the ElGamal public-key encryption scheme [21], the Boneh-Franklin identity-based encryption scheme [8] and the Waters identity-based encryption scheme [30]. One instantiation is given at the end of this section.

In the remaining of this section, we first give the definition of FKEM under our chosen type, and then show how to transform the FKEM with one-way security to a functional encryption with indistinguishability security against a chosen-plaintext attack (CPA) and a chosen-ciphertext attack (CCA).

4.1 Functional Key Encapsulation Mechanism

The functional key encapsulation mechanism (FKEM) is defined as follows.

Functional Key Encapsulation Mechanism

- $\mathsf{Param} \xleftarrow{\$} \mathsf{SysGen}(1^\lambda)$. This algorithm takes as input the security parameter λ and outputs the global system parameters Param.
- $(\mathsf{mpk}, \mathsf{msk}) \xleftarrow{\$} \mathsf{Setup}(\mathsf{Param})$. This algorithm takes as input Param and outputs the master key pair $(\mathsf{mpk}, \mathsf{msk})$.
- $\mathsf{usk} \xleftarrow{\$} \mathsf{KeyGen}(\mathsf{Param}, \mathsf{mpk}, \mathsf{msk}, \mathsf{upk})$. This algorithm takes as input Param, the master key pair $(\mathsf{mpk}, \mathsf{msk})$ and upk (*will be explained later*) and outputs the user secret key usk.
- $(\mathsf{C}, \mathsf{K}) \xleftarrow{\$} \mathsf{Encap}(\mathsf{Param}, \mathsf{mpk}, \mathsf{str}, \mathsf{r})$. This algorithm consists of two sub-algorithms.
  - $\mathsf{C} \xleftarrow{\$} \underline{\mathsf{Encap_c}(\mathsf{Param}, \mathsf{str}, \mathsf{r})}$. This sub-algorithm takes as input Param, a string str (*will be explained later*), a randomness r and outputs the encapsulation C. The encapsulation generation is independent of mpk.
  - $\mathsf{K} \xleftarrow{\$} \mathsf{Encap_k}(\mathsf{Param}, \mathsf{mpk}, \mathsf{str}, \mathsf{r})$. This sub-algorithm takes as input $(\mathsf{Param}, \mathsf{mpk}, \mathsf{str}, \mathsf{r})$ and outputs the encapsulation key K.
- $\mathsf{K} \xleftarrow{\$} \mathsf{Decap}(\mathsf{Param}, \mathsf{mpk}, \mathsf{upk}, \mathsf{usk}, \mathsf{C})$. This algorithm takes as input $(\mathsf{Param}, \mathsf{mpk}, \mathsf{upk}, \mathsf{usk})$ and the encapsulation C, and outputs the encapsulation key K or $\perp$.

Definition 1 (Correctness). *For any* $(\mathsf{C}, \mathsf{K}) \xleftarrow{\$} \mathsf{Encap}(\mathsf{Param}, \mathsf{mpk}, \mathsf{str}, \mathsf{r})$ *and* $\mathsf{usk} \xleftarrow{\$} \mathsf{KeyGen}(\mathsf{Param}, \mathsf{mpk}, \mathsf{msk}, \mathsf{upk})$*, we have that,*

$$\mathsf{Decap}(\mathsf{Param}, \mathsf{mpk}, \mathsf{upk}, \mathsf{usk}, \mathsf{C}) = \begin{cases} \mathsf{K} & F(\mathsf{upk}, \mathsf{str}) = 1, \\ \perp & \textit{otherwise}, \end{cases}$$

where $\mathsf{Param} \xleftarrow{\$} \mathsf{SysGen}(1^\lambda)$ *and* $(\mathsf{mpk}, \mathsf{msk}) \xleftarrow{\$} \mathsf{Setup}(\mathsf{Param})$. *The function* F *evaluates the relationship between the* upk *and the string* str.

The key pair $(\mathsf{mpk}, \mathsf{msk}), (\mathsf{upk}, \mathsf{usk})$, the string str and the function F have different representations in specified asymmetric encryptions. For example,

- In a public-key encryption, $\mathsf{mpk} = \mathsf{upk}$ is the public key while $\mathsf{msk} = \mathsf{usk}$ is the corresponding secret. str is also a public key and the function $F(\mathsf{upk}, \mathsf{str}) = 1$ if and only if $\mathsf{str} = \mathsf{mpk} = \mathsf{upk}$.
- In an identity-based encryption, upk is the identity of user and str is the identity of receiver. The function $F(\mathsf{upk}, \mathsf{str}) = 1$ if and only if $\mathsf{str} = \mathsf{upk}$.
- In an identity-based broadcast encryption, upk is the identity of user and str is the identity set of receivers. The function $F(\mathsf{upk}, \mathsf{str}) = 1$ if and only if upk is one of identities in the identity set str.
- In a ciphertext-policy attribute-based encryption, upk is an attribute set of a user while str is an access policy. The function $F(\mathsf{upk}, \mathsf{str}) = 1$ if and only if the access policy str accepts the attribute set upk.
- In a key-policy attribute-based encryption, upk is an access policy for a user while str is an attribute set. $F(\mathsf{upk}, \mathsf{str}) = 1$ if and only if the access policy upk accepts the attribute set str.
- In an inner-product encryption, both upk and str are vectors. $F(\mathsf{upk}, \mathsf{str}) = 1$ if and only if the inner product $\mathsf{upk} \cdot \mathsf{str} = 0$.

Definition 2 (One-Way FKEM). *A functional key encapsulation mechanism* $(\mathsf{SysGen}, \mathsf{Setup}, \mathsf{KeyGen}, \mathsf{Encap}, \mathsf{Decap})$ *is one-way secure if for any PPT adversary* $\mathcal{A}$,

$$\mathsf{Adv}^{\mathsf{OW}}_{\mathcal{A},\mathsf{FKEM}}(\lambda) = \Pr\left[\mathsf{K}' = \mathsf{K}^* : \begin{array}{l} \mathsf{Param} \xleftarrow{\$} \mathsf{SysGen}(1^\lambda); \\ (\mathsf{mpk}^*, \mathsf{msk}^*) \xleftarrow{\$} \mathsf{Setup}(\mathsf{Param}); \\ \mathsf{str}^* \leftarrow \mathcal{A}^{\mathcal{O}_K(\cdot)}(\mathsf{Param}, \mathsf{mpk}^*); \\ (\mathsf{C}^*, \mathsf{K}^*) \xleftarrow{\$} \mathsf{Encap}(\mathsf{Param}, \mathsf{mpk}, \mathsf{str}^*, \mathsf{r}^*); \\ \mathsf{K}' \leftarrow \mathcal{A}^{\mathcal{O}_K(\cdot)}(\mathsf{Param}, \mathsf{mpk}, \mathsf{str}^*, \mathsf{C}^*) \end{array}\right] \leq \mathsf{negl}(\lambda),$$

where $\mathcal{O}_K(\cdot)$ *is a key generation oracle that on input of any* upk, *returns* $\mathsf{usk} \xleftarrow{\$} \mathsf{KeyGen}(\mathsf{Param}, \mathsf{mpk}, \mathsf{msk}, \mathsf{upk})$ *on the condition that* $F(\mathsf{upk}, \mathsf{str}^*) \neq 1$.

The definition of function encryption is similar with the FKEM except the encryption algorithm and the decryption algorithm. The encryption algorithm additionally takes as input a message and returns a ciphertext for the message directly. While the decryption algorithm directly returns the message or outputs failure. The corresponding security model under indistinguishability against a chosen-plaintext attack and a chosen-ciphertext attack is also similar except that the adversary outputs $\mathsf{str}^*, \mathsf{m}_0, \mathsf{m}_1$ for challenge and the challenge ciphertext is encrypted with a random message from $\{m_0, m_1\}$ chosen by the simulator. We define IND-CCA for FE in the following definition. The definition of IND-CPA is the same as IND-CCA except that the adversary cannot access the decryption oracle in the security model.

Definition 3 (IND-CCA FE). *A functional encryption (*SysGen, Setup, *KeyGen,* Encrypt, Decrypt*) is IND-CCA secure if for any PPT adversary $\mathcal{A}$,*

$$\mathsf{Adv}^{\mathsf{IND-CCA}}_{\mathcal{A},\mathsf{FKEM}}(\lambda) = \Pr\left[\begin{array}{c}\mathsf{coin}' \\ = \\ \mathsf{coin}\end{array} : \begin{array}{l}\mathsf{Param} \xleftarrow{\$} \mathsf{SysGen}(1^\lambda); \\ (\mathsf{mpk}^*, \mathsf{msk}^*) \xleftarrow{\$} \mathsf{Setup}(\mathsf{Param}); \\ (\mathsf{str}^*, \mathsf{m}_0, \mathsf{m}_1) \leftarrow \mathcal{A}^{\mathcal{O}_K(\cdot),\mathcal{O}_D(\cdot)}(\mathsf{Param}, \mathsf{mpk}^*); \\ \mathsf{coin} \xleftarrow{R} \{0,1\} \\ \mathsf{CT}^* \xleftarrow{\$} \mathsf{Encypt}(\mathsf{Param}, \mathsf{mpk}, \mathsf{str}^*, \mathsf{r}^*, \mathsf{m}_{\mathsf{coin}}); \\ \mathsf{coin}' \leftarrow \mathcal{A}^{\mathcal{O}_K(\cdot),\mathcal{O}_D(\cdot)}(\mathsf{Param}, \mathsf{mpk}, \mathsf{str}^*, \mathsf{CT}^*)\end{array}\right] \leq \mathsf{negl}(\lambda),$$

where $\mathcal{O}_K(\cdot)$ *is a key generation oracle that on input of any* upk*, returns* $\mathsf{usk} \xleftarrow{\$} \mathsf{KeyGen}(\mathsf{Param}, \mathsf{mpk}, \mathsf{msk}, \mathsf{upk})$ *on the condition that* $F(\mathsf{upk}, \mathsf{str}^*) \neq 1$ *and* $\mathcal{O}_D(\cdot)$ *is a decryption oracle that on input of any* str, CT*, returns* $\{\mathsf{m}, \perp\} \xleftarrow{\$} \mathsf{Decrypt}(\mathsf{Param}, \mathsf{mpk}, \mathsf{upk}, \mathsf{usk}, \mathsf{CT})$ *on the condition that* $\mathsf{str} \neq \mathsf{str}^*$ *or* $\mathsf{CT} \neq \mathsf{CT}^*$.

4.2 Generic Conversion from OW-FKEM to IND-CPA-FE with Tight Reduction

Let $\mathsf{Param}_{\mathsf{OW}}$ be the global system parameters of FKEM with one-way security. Let $(\widetilde{\mathsf{mpk}_i}, \widetilde{\mathsf{msk}_i})$ for all $i \in [1, n]$ be n master key pairs of FKEM and $\widetilde{\mathsf{usk}_i}$ be the secret key of upk generated from $(\widetilde{\mathsf{mpk}_i}, \widetilde{\mathsf{msk}_i})$. Here, n can be as small as $n = 10$ depending on the choice of security loss. We choose n pairs in order to compute a different encapsulation key under each key pair, such that all n encapsulation keys can be iterated together following the iterated random oracle approach to generate the final encapsulation key. The functional encryption with IND-CPA security is constructed as follows.

SysGen: Choose a secure one-way hash function $H : \{0,1\}^* \to \{0,1\}^\ell$, where the message space is $\{0,1\}^\ell$. The global system parameters of FE are

$$\mathsf{Parama} = (\mathsf{Param}_{\mathsf{OW}}, H).$$

Setup: Set the master public key mpk and the master secret key msk of FE as

$$\mathsf{mpk} = (\widetilde{\mathsf{mpk}_1}, \widetilde{\mathsf{mpk}_2}, \cdots, \widetilde{\mathsf{mpk}_n}), \quad \mathsf{msk} = (\widetilde{\mathsf{msk}_1}, \widetilde{\mathsf{msk}_2}, \cdots, \widetilde{\mathsf{msk}_n}).$$

KeyGen: Taking as input Param, mpk, msk, upk, run the key generation algorithm $\mathsf{KeyGen}(\mathsf{Param}, \widetilde{\mathsf{mpk}_i}, \widetilde{\mathsf{msk}_i}, \mathsf{upk}) \xrightarrow{\$} \widetilde{\mathsf{usk}_i}$ for all $i \in [1, n]$ and output the private key usk for upk as

$$\mathsf{usk} = (\widetilde{\mathsf{usk}_1}, \widetilde{\mathsf{usk}_2}, ..., \widetilde{\mathsf{usk}_n}).$$

Encrypt: Taking as input Param, mpk, str and a message $\mathsf{m} \in \{0,1\}^\ell$, create the ciphertext CT for upk as follows

- Choose a random r for the Encap algorithm.
- Compute $\mathsf{C}_1 = \mathsf{Encap_c}(\mathsf{Param_{OW}}, \mathsf{str}, \mathsf{r})$.
- Compute $\mathsf{K_i} = \mathsf{Encap_k}(\mathsf{Param_{OW}}, \widetilde{\mathsf{mpk}}_\mathsf{i}, \mathsf{str}, \mathsf{r})$ for all $i \in [1, n]$.
- Compute the iteration as

$$A_i = H(A_{i-1}) \,||\, \mathsf{K_i} \,||\, i: \quad i \in [1, n], \quad \text{where } H(A_0) = 0_\epsilon.$$

- Set $\mathsf{C}_2 = H(A_n) \oplus \mathsf{m}$.

The output ciphertext is $\mathsf{CT} = (\mathsf{C}_1, \mathsf{C}_2)$.

Decrypt: Taking as input $\mathsf{Param}, \mathsf{mpk}, \mathsf{str}, \mathsf{upk}, \mathsf{usk}$ and a ciphertext $\mathsf{CT} = (\mathsf{C}_1, \mathsf{C}_2)$, decrypt the message as

- Compute $\mathsf{K_i} = \mathsf{Decap}(\mathsf{Param_{OW}}, \widetilde{\mathsf{mpk}}_\mathsf{i}, \mathsf{upk}, \widetilde{\mathsf{usk}}_\mathsf{i}, \mathsf{C}_1)$ for all $i \in [1, n]$.
- Compute the iteration as

$$B_i = H(B_{i-1}) || \mathsf{K_i} || i: \quad i \in [1, n], \quad \text{where } H(B_0) = 0_\epsilon.$$

- Compute $\mathsf{m} = \mathsf{C}_2 \oplus H(B_n)$.

This completes the description of FE construction. Without counting the size of the encrypted message, the ciphertext size is the same as FKEM. That is, the generic conversion from OW-FKEM to IND-CPA-FE does not expand the ciphertext size. This conversion without expanding ciphertext requires that the encapsulation is independent of the master public key $\widetilde{\mathsf{mpk}}$. Otherwise, the ciphertext is composed of n number of distinct C_1 generated under a different $\widetilde{\mathsf{mpk}}_i$. In the following theorem, we prove that the IND-CPA security of FE can be tightly reduced to one-way security of an FKEM.

Theorem 1. *Let H be a random oracle. If there exists an adversary $\mathcal{A}$ who makes q queries to H has an advantage ϵ in the IND-CPA security model against the constructed encryption scheme, then we can construct a simulator $\mathcal{B}$ that has advantage $\mathsf{Adv}^{\mathsf{OW}}_{\mathcal{B},\mathsf{FKEM}}(\lambda) = \frac{2\epsilon}{nq^{\frac{1}{n}}}$ in breaking the underlying FKEM in the one-way security model.*

Proof. Suppose there exists an adversary $\mathcal{A}$ who can break the above encryption scheme with an advantage ϵ. We construct a simulator $\mathcal{B}$ to break the one-way security of the underlying FKEM. The reduction works as follows.

Setup: $\mathcal{B}$ first obtains $(\mathsf{Param_{OW}}, \widetilde{\mathsf{mpk}}^*)$ from FKEM. It then picks a random $d \in [1, n]$, and runs the setup algorithm $\mathsf{Setup}(\mathsf{Param_{OW}}) \xrightarrow{\$} (\widetilde{\mathsf{mpk}}_\mathsf{i}, \widetilde{\mathsf{msk}}_\mathsf{i})$ for all $i \in [1, n] \backslash d$ to generate master key pairs. Finally, it sets $\mathsf{mpk} = (\widetilde{\mathsf{mpk}}_1, ..., \widetilde{\mathsf{mpk}}_\mathsf{n})$ where

$$\widetilde{\mathsf{mpk}}_\mathsf{i} = \begin{cases} \widetilde{\mathsf{mpk}}_\mathsf{i} & \text{if } i \neq d, \\ \widetilde{\mathsf{mpk}}^* & \text{otherwise} \end{cases},$$

and $\mathsf{Param} = \mathsf{Param}_{\mathsf{OW}}$ where H is treated as a random oracle controlled by the simulator. Finally, the simulator returns $(\mathsf{Param}, \mathsf{mpk})$ to $\mathcal{A}$.

H-Query: $\mathcal{B}$ maintains a hash list L to record all queries to the random oracle H. If a query $\overline{\mathcal{Q}}$ has been made and appears in the list $(\overline{\mathcal{Q}}, \overline{\mathcal{R}})$, $\mathcal{B}$ responds with the same response $\overline{\mathcal{R}}$. Otherwise, the simulator randomly chooses $\overline{\mathcal{R}}$ from $\{0,1\}^\ell$ as the response $\overline{\mathcal{R}} = H(\overline{\mathcal{Q}})$ and adds $(\overline{\mathcal{Q}}, \overline{\mathcal{R}})$ into the list.

Phase 1: $\mathcal{A}$ requests the secret key of upk in this phase, which is adaptively chosen by the adversary. The simulator $\mathcal{B}$ first queries upk to the key generation oracle $\mathcal{O}_K(\cdot)$ which returns $\widetilde{\mathsf{usk}}$ and sets $\widetilde{\mathsf{usk}_\mathsf{d}} = \widetilde{\mathsf{usk}}$. For all other $i \in [1,n]\backslash d$, $\mathcal{B}$ runs $\mathsf{KeyGen}(\mathsf{Param}, \mathsf{mpk}_\mathsf{i}, \mathsf{msk}_\mathsf{i}, \mathsf{upk}) \xrightarrow{\$} \widetilde{\mathsf{usk}_\mathsf{i}}$ by itself to compute $\widetilde{\mathsf{usk}_i}$. Finally, it sets $\mathsf{usk} = (\widetilde{\mathsf{usk}_1}, \widetilde{\mathsf{usk}_2}, ..., \widetilde{\mathsf{usk}_\mathsf{n}})$ and returns usk to $\mathcal{A}$ as the query response.

Challenge: $\mathcal{A}$ outputs two distinct challenge messages m_0, m_1 from $\{0,1\}^n$ and a challenge string str^* with the restriction that for any upk queried in the **Phase 1**, $F(\mathsf{upk}, \mathsf{str}^*) \neq 1$. $\mathcal{B}$ then forwards str^* to FKEM and obtains the challenge encapsulation ciphertext C^*. Finally, $\mathcal{B}$ randomly chooses $\mathsf{R} \in \{0,1\}^\ell$ and sets the challenge ciphertext as

$$\mathsf{CT}^* = (\mathsf{C}^*, \mathsf{R}).$$

Phase 2: $\mathcal{A}$ issues more secret key queries on any chosen upk such that $F(\mathsf{upk}, \mathsf{str}^*) \neq 1$. $\mathcal{B}$ responds the same as in the **Phase 1**.

Output: Finally, $\mathcal{A}$ outputs its guess $coin' \in \{0,1\}$. $\mathcal{B}$ then follows the approach in Theory 3 to find the underlying key K^* from the recorded hash list L to break the FKEM.

This completes the description of simulation and solution. All master key pairs are generated from the setup algorithm of FKEM. They are therefore indistinguishable from the view of the adversary, such that the adversary has no advantage in guessing d. The random oracle is simulated using truly random string, and hence the simulator performs a correct simulation on the random oracle. Let $\mathsf{C}^* = \mathsf{Encap}_\mathsf{c}(\mathsf{Param}_{\mathsf{OW}}, \mathsf{str}^*, \mathsf{r}^*)$. Since C^* is generated from the encapsulation algorithm and R^* is randomly chosen, the challenge ciphertext is a one-time pad unless the adversary queries A_n^*, which is defined as

$$A_i^* = H(A_{i-1}^*) \,||\, \mathsf{Encap}_\mathsf{k}(\mathsf{Param}_{\mathsf{OW}}, \widetilde{\mathsf{mpk}_\mathsf{i}}, \mathsf{str}^*, \mathsf{r}^*) \,||\, i: \quad i \in [1,n], \quad H(A_0^*) = 0_\epsilon.$$

We have

$$K^* = \mathsf{Encap}_\mathsf{k}(\mathsf{Param}_{\mathsf{OW}}, \widetilde{\mathsf{mpk}_\mathsf{d}}, \mathsf{str}^*, \mathsf{r}^*) = \mathsf{Encap}_\mathsf{k}(\mathsf{Param}_{\mathsf{OW}}, \widetilde{\mathsf{mpk}}^*, \mathsf{str}^*, \mathsf{r}^*),$$

in A_d^* is the solution to the FKEM. The approach of finding the correct encapsulation key exactly falls into Theory 3 where the simulator can successfully pick a valid query with probability $1/(nq^{\frac{1}{n}})$. According to the definition of advantage, the adversary will make such a query with probability 2ϵ to the random oracle. We therefore yield Theorem 1. □

4.3 Generic Conversion from OW-FKEM to IND-CCA-FE

Given an FKEM composed of

$$\mathsf{C} = \mathsf{Encap_c}(\mathsf{Param_{OW}}, \mathsf{str}, \mathsf{r}),$$
$$\mathsf{K_i} = \mathsf{Encap_k}(\mathsf{Param_{OW}}, \widetilde{\mathsf{mpk}}_\mathsf{i}, \mathsf{str}, \mathsf{r}) : \; i \in [1, n],$$

we have shown how to construct an IND-CPA FE via

$$\mathsf{CT} = \Big(\mathsf{Encap_c}(\mathsf{Param_{OW}}, \mathsf{str}, \mathsf{r}), \quad H(A_n) \oplus \mathsf{m}\Big),$$

where $A_i = H(A_{i-1}) \;||\; \mathsf{K_i} \;||\; i : \quad i \in [1, n], \;\; \text{and}\, H(A_0) = 0_\epsilon$.

We can further transfer the conversion from FKEM to FE with IND-CCA security by applying the Fujisaki-Okamoto transformation approach [19,20]. This approach requires two more one-way secure hash functions H_1, H_2 in the global system parameters and they are also treated as random oracles in the security proof. The first hash function H_1 has the same output space as the randomness r and the second one H_2 has the same output space as H.

Taking as input $\mathsf{Param}, \mathsf{mpk}, \mathsf{str}$ and a message $\mathsf{m} \in \{0,1\}^\ell$, the encryption algorithm for IND-CCA security works as follows.

- Choose a random string $\sigma \in \{0,1\}^\ell$ and compute $\mathsf{r} = \mathsf{H_1}(\sigma, \mathsf{m})$.
- Run the IND-CPA encryption algorithm using the randomness r to encrypt σ, which returns

$$(\mathsf{C_1}, \mathsf{C_2}) = \Big(\mathsf{Encap_c}\Big(\mathsf{Param_{OW}}, \mathsf{str}, \mathsf{H_1}(\sigma, \mathsf{m})\Big), \quad H(A_n) \oplus \sigma\Big).$$

- Set $\mathsf{C_3} = H_2(\sigma) \oplus \mathsf{m}$.

The output ciphertext is

$$\mathsf{CT} = (\mathsf{C_1}, \mathsf{C_2}, \mathsf{C_3}) = \Big(\mathsf{Encap_c}(\mathsf{Param_{OW}}, \mathsf{str}, \mathsf{H_1}(\sigma, \mathsf{m})), \quad \mathsf{H}(\mathsf{A_n}) \oplus \sigma, \quad \mathsf{H_2}(\sigma) \oplus \mathsf{m}\Big).$$

In the corresponding decryption algorithm, the decryptor first runs the IND-CPA decryption algorithm to obtain σ, and then it computes $H_2(\sigma) \oplus \mathsf{C_3}$ to obtain m. Finally, it outputs the message m if $\mathsf{C_1}$ is the generation using the randomness $H_1(\sigma, \mathsf{m})$. Otherwise, it simply returns $\perp$.

It is not hard to obtain the security proof based on the proposed security reduction for CPA security and the Fujisaki-Okamoto transformation. First, all key queries will be generated the same as the proof in Theorem 1; Second, all decryption queries will be responded using the Fujisaki-Okamoto transformation approach. Finally, the challenge ciphertext is simulated using $(\mathsf{C}^*, \mathsf{R_1}, \mathsf{R_2})$, where C^* is the challenge encapsulation from FKEM and $\mathsf{R_1}, \mathsf{R_2}$ are random strings. From the view of the adversary, if the adversary has an advantage in distinguishing the encrypted message, it must make a query on σ. The probability of obtaining σ is bounded by a random guess which is negligible for a large ℓ and bounded by breaking the IND-CPA construction. While the probability of breaking the IND-CPA construction is bounded by making a query on A_n. Therefore, if σ appears in the query list with probability 2ϵ, the probability of querying A_n is nearly 2ϵ. The simulator then is able to break the underlying FKEM with probability $2\epsilon/(nq^{1/n})$ by applying the approach in Theory 3.

4.4 Identity-Based Key Encapsulation Mechanism

At the end of this section, we give an instantiation using the Park-Lee identity-based encryption [28], which can be modified to a key encapsulation mechanism satisfying the requirement for short ciphertext in transformation. We choose this scheme as an example because there is no security loss during the private key simulation. By using the iterated random oracle, the corresponding encryption with indistinguishability security can be tightly reduced to solve the Bilinear Diffie-Hellman problem.

SysGen: This algorithm takes as input the security parameter λ. It selects a pairing group $\mathbb{PG} = (\mathbb{G}, \mathbb{G}_T, g, p, e)$ and one secure one-way hash function $H_1 : \{0,1\}^* \rightarrow \mathbb{G}$. Then it randomly chooses $u \in \mathbb{G}$. The global system parameters Param are

$$\mathsf{Param} = (\mathbb{PG}, u, H_1).$$

Setup: It randomly chooses $\alpha \in \mathbb{Z}_p$ and computes $e(g,g)^\alpha$. The algorithm returns a master public/secret key pair (mpk, msk) as

$$\mathsf{mpk} = e(g,g)^\alpha, \quad \mathsf{msk} = \alpha.$$

KeyGen: The key generation algorithm takes as input Param, an identity $ID \in \{0,1\}^*$ and the master key pair (mpk, msk). It randomly chooses $s, t_k \in \mathbb{Z}_p$ and creates the private key as

$$d_{ID} = (d_0, d_1, d_2, d_3) = \left(t_k,\ g^s,\ \left(H_1(ID)u^{t_k}\right)^s, g^\alpha u^s\right).$$

It requires that the random number t_k for ID is the same in the private key generation.

Encap: The encryption algorithm takes as input Param, the master public key mpk and an identity ID. It randomly chooses $r, t_c \in \mathbb{Z}_p$ and creates the ciphertext and the encapsulation key as follows.

$$\mathsf{C} = \mathsf{Encap_c}(\mathsf{Param}, \mathsf{ID}, \mathsf{r}) = \left(t_c,\ (H_1(ID)u^{t_c})^r,\ g^r\right),$$
$$\mathsf{K} = \mathsf{Encap_k}(\mathsf{Param}, \mathsf{mpk}, \mathsf{ID}, \mathsf{r}) = e(g,g)^{\alpha \cdot r}.$$

Decap: The decryption algorithm takes as input Param, the master public key mpk, an identity ID, a private key d_{ID} and a ciphertext $\mathsf{C} = (\mathsf{C}_0, \mathsf{C}_1, \mathsf{C}_2)$. It computes the random key as

$$\mathsf{K} = e(d_3, \mathsf{C}_2) \cdot \left(\frac{e(\mathsf{C}_1, d_1)}{e(\mathsf{C}_2, d_2)}\right)^{-\frac{1}{t_c - t_k}}.$$

The correctness of the decapsulation is showed as follows.

$$\begin{aligned}
\mathsf{K} &= e(d_3, \mathsf{C}_2) \cdot \left(\frac{e(\mathsf{C}_1, d_1)}{e(\mathsf{C}_2, d_2)}\right)^{-\frac{1}{t_c - t_k}} \\
&= e(g^{\alpha} u^s, g^r) \cdot \left(\frac{e\Big((H_1(ID)u^{t_c})^r, g^s\Big)}{e\left(g^r, \Big(H_1(ID)u^{t_k}\Big)^s\right)}\right)^{-\frac{1}{t_c - t_k}} \\
&= e(g,g)^{\alpha r} \cdot e(u,g)^{rs} \cdot \Big(e(u,g)^{rs(t_c - t_k)}\Big)^{-\frac{1}{t_c - t_k}} \\
&= e(g,g)^{\alpha r}.
\end{aligned}$$

Theorem 2. *Let H_1 be a random oracle. If there exists an adversary who can break the Park-Lee identity-based key encapsulation mechanism with (t, q_1, q_k, ϵ) in the one-way security model, where the adversary makes q_1 queries to H_1 and q_k numbers of private keys, then we can construct a simulator to solve the BDH problem with $(t + T_s, \epsilon)$ where T_s denotes the time cost of simulation.*

Proof. Suppose there exists an adversary $\mathcal{A}$ who can break the identity-based encryption scheme. We can construct a simulator $\mathcal{B}$ to solve the BDH problem. Given as input the instance (g, g^a, g^b, g^c) in the pairing group $\mathbb{PG}$, the simulator aims to compute $e(g,g)^{abc}$. $\mathcal{B}$ interacts with the adversary as follows.

Setup: $\mathcal{B}$ picks a random $z \in \mathbb{Z}_p$, sets $u = g^{z-a}, \alpha = ab$ and computes $e(g,g)^{\alpha} = e(g^a, g^b)$. Then, it gives $\mathsf{Param} = (\mathbb{PG}, u)$ and $\mathsf{mpk} = e(g,g)^{\alpha}$ except H_1 to the adversary, where H_1 is treated as a random oracle controlled by the simulator.

H-Query: $\mathcal{B}$ maintains a hash list L_1 to record all queries to the random oracle H_1. If a query ID_i has been made and $(ID_i, x_i, y_i, H_1(ID_i))$ is in the list, $\mathcal{B}$ responds with $H_1(ID_i)$. Otherwise, $\mathcal{B}$ randomly chooses $x_i, y_i \in \mathbb{Z}_p$, sets $H_1(ID_i) = g^{x_i a + y_i}$ and adds $(ID_i, x_i, y_i, H_1(ID_i))$ into the hash list.

Phase 1: $\mathcal{A}$ requests private keys of identities in this phase. For the query on ID, $\mathcal{B}$ first runs the H_1 query to get the corresponding $(ID, x, y, H_1(ID))$, randomly chooses $s \in \mathbb{Z}_p$ and computes the private key as

$$\begin{aligned}
d_0 &= t_k = x \\
d_1 &= g^{b+s} \\
d_2 &= g^{b(y+xz)+s(y+xz)} \\
d_3 &= g^{zs+zb-sa},
\end{aligned}$$

which can be computed by the simulator. Let $s' = b + s$ and $t_k = x$. We have

$$\begin{aligned}
(d_1, d_2, d_3) &= \left(g^{s'},\ \ (H_1(ID)u^{t_k})^{s'},\ \ g^{\alpha} u^{s'}\right) \\
&= \left(g^{b+s},\ \ (g^{xa+y} g^{x(z-a)})^{b+s},\ \ g^{ba} g^{(z-a)(b+s)}\right) \\
&= \left(g^{b+s},\ \ g^{b(y+xz)+s(y+xz)},\ \ g^{zs+zb-sa}\right).
\end{aligned}$$

Therefore, $d_{ID} = (d_0, d_1, d_2, d_3)$ is a valid private key of ID.

Challenge: The adversary $\mathcal{A}$ outputs an identity ID^* for challenge, where the adversary never requested the private key of ID^*. Let the query response of ID in the random oracle be $(ID^*, x^*, y^*, H_1(ID^*))$. The simulator $\mathcal{B}$ sets the challenge encapsulation as

$$\mathsf{C}^* = \left(x^*,\ g^{(y+x^*z)c},\ g^c\right).$$

Ler $r = c$ and $t_c = x^*$. We have

$$\begin{aligned}\mathsf{C}^* &= \left(t_c,\ (H_1(ID^*)u^{t_c})^r,\ g^r\right)\\ &= \left(x^*,\ (g^{x^*a+y^*}g^{x^*(z-a)})^c,\ g^c\right)\\ &= \left(x^*,\ g^{(y+x^*z)c},\ g^c\right).\end{aligned}$$

Therefore, C^* is a valid challenge encapsulation whose corresponding key K^* is

$$e(g,g)^{\alpha r} = e(g,g)^{abc}.$$

Output: Finally, $\mathcal{A}$ outputs K^* and the simulator outputs K^* as the solution to the BDH problem.

This completes the simulation and solution. We have that a, z are chosen randomly and independently such that both Param and mpk are indistinguishable from the real scheme. x, y are chosen randomly and independently such that the random oracle simulation is correctly performed. x^*, c are chosen randomly and independently such that the challenge ciphertext is indistinguishable from the real scheme. According to the definition of advantage and the assumption, we have the adversary will output K^* with probability ϵ and the simulator will solve the BDH problem with probability ϵ. This completes the proof of Theorem 2. □

5 Tight Reduction for Key Exchange

The iterated random oracle can also be applied in the key exchange for tight(er) reduction in the IND-CHP security reduction. However, we observe that the application is a little complicated due to many different definitions of key exchange protocols. In this section, we discuss how to apply the iterated random oracle for this cryptographic primitive and what will occur during the applications.

Identity-Based Non-Interactive Key Exchange (IB-NIKE). In the Sakai-Ohgishi-Kasahara IB-NIKE protocol [29], the private key of ID is $d_{ID} = H_1(ID)^\alpha$, where $\alpha \in \mathbb{Z}_p$ is the master secret key and $H_1 : \{0,1\}^* \rightarrow \mathbb{G}$ is a collision-resistant hash function. Here, the IB-NIKE is constructed over a pairing group. The NIKE between ID_A and ID_B is defined as

$$\begin{aligned} K &= H\Big(e(d_{ID_A}, H_1(ID_B))\Big) \\ &= H\Big(e(d_{ID_B}, H_1(ID_A))\Big) \\ &= H\Big(e\Big(H_1(ID_A), H_1(ID_B)\Big)^{\alpha}\Big), \end{aligned}$$

where $H : \{0,1\}^* \rightarrow \{0,1\}^{\ell}$ is another secure one-way hash function.

The above IB-NIKE protocol is provably secure in the random oracle model (assuming H_1, H are random oracles) under the BDH assumption. The finding loss exists because the simulator cannot decide which query in the adversary's query set is the correct solution to the BDH problem. We can apply the iterated random oracle by iterating the section keys as follows.

- Compute the private key d_{ID} of ID as

$$d_{ID} = \Big(H_1(ID,1)^{\alpha},\;\; H_1(ID,2)^{\alpha},\;\; \cdots, H_1(ID,n)^{\alpha}\Big).$$

- Compute the i-th intermediate key between ID_A and ID_B as

$$K_i = e\Big(H_1(ID_A,i), H_1(ID_B,i)\Big)^{\alpha}.$$

- The final section key between ID_A and ID_B is $H(EK_n)$ where

$$EK_i = H(EK_{i-1}) \mid\mid K_i \mid\mid i: \quad i \in [1,n], \;\; \text{where } H(EK_0) = 0_{\epsilon}.$$

It is not hard to prove its security when the simulator can simulate all private keys except $H(ID_A,d)^{\alpha}$ and $H(ID_B,d)^{\alpha}$ where $e\Big(H_1(ID_A,d), H_1(ID_B,d)\Big)^{\alpha}$ is programmed as the solution to the BDH problem. By applying Theory 3, we have the final security reduction will have a very small finding loss.

In comparison with the original scheme, ours gives a tighter reduction. We admit that our scheme requires each user to store n private keys. Although n can be as small as 10, the final key length is still longer compared to the length of original scheme by expending group size for security loss. Therefore, this construction is somewhat theoretically interesting only for short length. However, when parallel computation is allowed, all pairing computations and hash group operations in our scheme can be completed in parallel within a group. Our scheme will reduce the time cost because there is no need to expand group size for security loss.

Other (Authenticated) Key Exchange. Similarly, we can utilize the above approach to solve the finding loss in other key exchange protocols by generating n keys for each user instead of one. Only the d-th sub-key can be programmed to solve a hard problem while the others can be simulated or computed by the simulator. However, it seems that we still have to resort to the help of decision oracle [18] because the simulator cannot simulate some section keys for the adversary. Let $\mathsf{upk_A}$ and $\mathsf{upk_B}$ be the challenge public keys. In the security model for key exchange, the adversary is allowed to launch section key query between for example $\mathsf{upk_A}$ and a corrupted user namely $\mathsf{upk_C}$. Notice that the secret key of $\mathsf{usk_A}$ is unknown (only the d-th sub-key is programmed as unknown). When the secret key of $\mathsf{usk_C}$ is also unknown, the simulator cannot simulate the section keys correctly for the adversary especially on the random oracle without the help of decision oracle. If the assumption still needs a decision oracle, there is no finding loss in security reduction because the simulator can use the decision oracle to find the correct solution.

We emphasize that there is still a benefit of applying the iterated random oracle for key exchange, whose security assumption is a strong computational assumption with a decision oracle. Notice that the iterated random oracle will exponentially consume the hash queries from the adversary if it wants to hide the challenge query. Then, the simulator can make less number of queries to the decision oracle especially when the simulator wants to simulate the section key and find the correct solution. That is, by applying the iterated random oracle, we can adopt a strong computational assumption where the access time to the decision oracle is bounded with a small number. This assumption is better than the assumption with q times access to the decision oracle.

6 Conclusion

Finding loss is a common security loss in those security reductions for indistinguishability security under computational hard assumptions, when their decisional variants are also hard. This security loss will result in a significant loose reduction by a random pick because the number of queries can be as large as 2^{60}. The novel Cash-Kiltz-Shoup's approach is efficient without any finding loss, but can only be applied to a computational hard problem with a trapdoor test. We proposed a completely new approach, namely the iterated random oracle, as a universal approach for finding loss, which can be applied to any computational hard problem without any restriction on the adopted hard problem. The finding loss in this approach is very small. The corresponding success probability is $\frac{1}{64}$ compared to $\frac{1}{2^{60}}$ by a random pick. This approach has been applied to achieve a security transformation for encryption and key exchange towards tight(er) reductions.

References

1. Abdalla, M., Ben Hamouda, F., Pointcheval, D.: Tighter reductions for forward-secure signature schemes. In: Kurosawa, K., Hanaoka, G. (eds.) PKC 2013. LNCS, vol. 7778, pp. 292–311. Springer, Heidelberg (2013). doi:10.1007/978-3-642-36362-7_19

2. Abdalla, M., Fouque, P.-A., Lyubashevsky, V., Tibouchi, M.: Tightly-secure signatures from lossy identification schemes. In: Pointcheval, D., Johansson, T. (eds.) EUROCRYPT 2012. LNCS, vol. 7237, pp. 572–590. Springer, Heidelberg (2012). doi:10.1007/978-3-642-29011-4_34
3. Abdalla, M., Pointcheval, D.: Simple password-based encrypted key exchange protocols. In: Menezes, A. (ed.) CT-RSA 2005. LNCS, vol. 3376, pp. 191–208. Springer, Heidelberg (2005). doi:10.1007/978-3-540-30574-3_14
4. Attrapadung, N., Furukawa, J., Gomi, T., Hanaoka, G., Imai, H., Zhang, R.: Efficient identity-based encryption with tight security reduction. In: Pointcheval, D., Mu, Y., Chen, K. (eds.) CANS 2006. LNCS, vol. 4301, pp. 19–36. Springer, Heidelberg (2006). doi:10.1007/11935070_2
5. Bellare, M., Rogaway, P.: Random oracles are practical: a paradigm for designing efficient protocols. In: Denning, D.E., Pyle, R., Ganesan, R., Sandhu, R.S., Ashby, V. (eds.) CCS 1993, pp. 62–73. ACM (1993)
6. Blazy, O., Kakvi, S.A., Kiltz, E., Pan, J.: Tightly-secure signatures from chameleon hash functions. In: Katz, J. (ed.) PKC 2015. LNCS, vol. 9020, pp. 256–279. Springer, Heidelberg (2015). doi:10.1007/978-3-662-46447-2_12
7. Boneh, D., Boyen, X.: Efficient selective-ID secure identity-based encryption without random oracles. In: Cachin, C., Camenisch, J.L. (eds.) EUROCRYPT 2004. LNCS, vol. 3027, pp. 223–238. Springer, Heidelberg (2004). doi:10.1007/978-3-540-24676-3_14
8. Boneh, D., Franklin, M.: Identity-based encryption from the weil pairing. In: Kilian, J. (ed.) CRYPTO 2001. LNCS, vol. 2139, pp. 213–229. Springer, Heidelberg (2001). doi:10.1007/3-540-44647-8_13
9. Boyen, X.: Miniature CCA2 PK encryption: tight security without redundancy. In: Kurosawa, K. (ed.) ASIACRYPT 2007. LNCS, vol. 4833, pp. 485–501. Springer, Heidelberg (2007). doi:10.1007/978-3-540-76900-2_30
10. Cash, D., Kiltz, E., Shoup, V.: The twin Diffie-Hellman problem and applications. In: Smart, N. (ed.) EUROCRYPT 2008. LNCS, vol. 4965, pp. 127–145. Springer, Heidelberg (2008). doi:10.1007/978-3-540-78967-3_8
11. Cash, D., Kiltz, E., Shoup, V.: The twin Diffie-Hellman problem and applications. J. Cryptology **22**(4), 470–504 (2009)
12. Chen, L., Chen, Y.: The n-Diffie-Hellman problem and its applications. In: Lai, X., Zhou, J., Li, H. (eds.) ISC 2011. LNCS, vol. 7001, pp. 119–134. Springer, Heidelberg (2011). doi:10.1007/978-3-642-24861-0_9
13. Chevallier-Mames, B.: An efficient CDH-based signature scheme with a tight security reduction. In: Shoup, V. (ed.) CRYPTO 2005. LNCS, vol. 3621, pp. 511–526. Springer, Heidelberg (2005). doi:10.1007/11535218_31
14. Chevallier-Mames, B., Joye, M.: A practical and tightly secure signature scheme without hash function. In: Abe, M. (ed.) CT-RSA 2007. LNCS, vol. 4377, pp. 339–356. Springer, Heidelberg (2006). doi:10.1007/11967668_22
15. Coron, J.: A variant of boneh-franklin IBE with a tight reduction in the random oracle model. Des. Codes Cryptography **50**(1), 115–133 (2009)
16. Cramer, R., Shoup, V.: A practical public key cryptosystem provably secure against adaptive chosen ciphertext attack. In: Krawczyk, H. (ed.) CRYPTO 1998. LNCS, vol. 1462, pp. 13–25. Springer, Heidelberg (1998). doi:10.1007/BFb0055717
17. Diffie, W., Hellman, M.E.: New directions in cryptography. IEEE Trans. Inf. Theory **22**(6), 644–654 (1976)
18. Fiore, D., Gennaro, R.: Making the Diffie-Hellman protocol identity-based. In: Pieprzyk, J. (ed.) CT-RSA 2010. LNCS, vol. 5985, pp. 165–178. Springer, Heidelberg (2010). doi:10.1007/978-3-642-11925-5_12

19. Fujisaki, E., Okamoto, T.: Secure integration of asymmetric and symmetric encryption schemes. In: Wiener, M. (ed.) CRYPTO 1999. LNCS, vol. 1666, pp. 537–554. Springer, Heidelberg (1999). doi:10.1007/3-540-48405-1_34
20. Fujisaki, E., Okamoto, T.: Secure integration of asymmetric and symmetric encryption schemes. J. Cryptology **26**(1), 80–101 (2013)
21. ElGamal, T.: A public key cryptosystem and a signature scheme based on discrete logarithms. In: Blakley, G.R., Chaum, D. (eds.) CRYPTO 1984. LNCS, vol. 196, pp. 10–18. Springer, Heidelberg (1985). doi:10.1007/3-540-39568-7_2
22. Gay, R., Hofheinz, D., Kiltz, E., Wee, H.: Tightly CCA-secure encryption without pairings. In: Fischlin, M., Coron, J.-S. (eds.) EUROCRYPT 2016. LNCS, vol. 9665, pp. 1–27. Springer, Heidelberg (2016). doi:10.1007/978-3-662-49890-3_1
23. Gentry, C.: Practical identity-based encryption without random oracles. In: Vaudenay, S. (ed.) EUROCRYPT 2006. LNCS, vol. 4004, pp. 445–464. Springer, Heidelberg (2006). doi:10.1007/11761679_27
24. Goh, E., Jarecki, S., Katz, J., Wang, N.: Efficient signature schemes with tight reductions to the diffie-hellman problems. J. Cryptology **20**(4), 493–514 (2007)
25. Hofheinz, D., Jager, T.: Tightly secure signatures and public-key encryption. In: Safavi-Naini, R., Canetti, R. (eds.) CRYPTO 2012. LNCS, vol. 7417, pp. 590–607. Springer, Heidelberg (2012). doi:10.1007/978-3-642-32009-5_35
26. Katz, J., Wang, N.: Efficiency improvements for signature schemes with tight security reductions. In: Jajodia, S., Atluri, V., Jaeger, T. (eds.) CCS 2003, pp. 155–164. ACM (2003)
27. Kurosawa, K., Takagi, T.: Some RSA-based encryption schemes with tight security reduction. In: Laih, C.-S. (ed.) ASIACRYPT 2003. LNCS, vol. 2894, pp. 19–36. Springer, Heidelberg (2003). doi:10.1007/978-3-540-40061-5_2
28. Park, J.H., Lee, D.H.: An efficient ibe scheme with tight security reduction in the random oracle model. Des. Codes Crypt. **79**(1), 63–85 (2016)
29. Sakai, R., Ohgishi, K., Kasahara, M.: Cryptosystems based on pairing. In: The 2000 Symposium on Cryptography and Information Security, vol. 45, pp. 26–28 (2000)
30. Waters, B.: Efficient identity-based encryption without random oracles. In: Cramer, R. (ed.) EUROCRYPT 2005. LNCS, vol. 3494, pp. 114–127. Springer, Heidelberg (2005). doi:10.1007/11426639_7

NIZKs with an Untrusted CRS: Security in the Face of Parameter Subversion

Mihir Bellare[1](✉), Georg Fuchsbauer[2], and Alessandra Scafuro[3]

[1] Department of Computer Science and Engineering, University of California, San Diego, San Diego, USA
mihir@eng.ucsd.edu

[2] Inria, Ecole Normale Supérieure, CNRS and PSL Research University, Paris, France
georg.fuchsbauer@ens.fr

[3] Department of Computer Science, North Carolina State University, Raleigh, USA
ascafur@ncsu.edu

Abstract. Motivated by the subversion of "trusted" public parameters in mass-surveillance activities, this paper studies the security of NIZKs in the presence of a maliciously chosen common reference string. We provide definitions for subversion soundness, subversion witness indistinguishability and subversion zero knowledge. We then provide both negative and positive results, showing that certain combinations of goals are unachievable but giving protocols to achieve other combinations.

1 Introduction

The summer of 2013 brought shocking news of mass surveillance being conducted by the NSA and its counter-parts in other countries. The documents revealed new ways in which the adversary compromises security, ways not covered by standard models and definitions in cryptography. This opens up a new research agenda, namely to formalize security goals that defend against these novel attacks, and study the achievability of these goals. This agenda is being pursued along several fronts. The front we pursue here is *parameter subversion*, namely the compromise of security by the malicious creation of supposedly trusted public parameters for cryptographic systems. The representative example is the Dual EC random number generator (RNG).

DUAL EC. Dual EC is an NSA-designed, elliptic-curve-based random number generator, standardized as NIST SP 800-90 and ANSI X9.82. BLN [14] say that its story is "one of the most interesting in modern cryptography." The RNG includes two points P, Q on an elliptic curve that function as public parameters for the algorithm. At the Crypto 2007 rump session, Shumow and Ferguson noted that anyone who knew the discrete logarithm of P to base Q, meaning a scalar s such that $P = sQ$, could predict generator outputs. In a Wired Magazine article the same year, Schneier warned against Dual EC because it "just might contain

J.H. Cheon and T. Takagi (Eds.): ASIACRYPT 2016, Part II, LNCS 10032, pp. 777–804, 2016.
DOI: 10.1007/978-3-662-53890-6_26

a backdoor for the NSA." The NSA's response was that they had "generated P, Q in a secure, classified way." But the Snowden revelations (documents from project Bullrun and SIGINT) show that Dual EC was part of a systematic NSA effort to subvert standards. And in 2014, CNEGLRBMSF [24] showed the practical effectiveness of the subversion by demonstrating how the backdoor could be exploited to break TLS.

Two things are remarkable. The first is that the "trusted" public parameters were in fact subverted. The second is the effort put into ensuring that the subverted parameters were standardized and used. NSA-based pressure and lobbying not only lead to Dual EC remaining a US standard but even to its being in an international standard, ISO 18031:2005. In 2013 Reuters reported that the NSA paid RSA corporation $10 million to make Dual EC the default method for random number generation in their BSafe library.

<u>Cryptography resistant to parameter subversion.</u> The lesson to take away is that a cryptographic system that relies on public parameters assumed to have been honestly generated, say by some "trusted" party, is at great practical risk from the possibility that the parameters were in fact maliciously generated with intent to subvert security of their use. We suggest that in response we should develop cryptography that is resistant to parameter subversion. This means that it should provide its usual security with trusted parameters, but retain as much security as possible when the parameters are maliciously generated.

Parameters arise in many places in cryptography, but a prominent one that springs to mind are non-interactive zero-knowledge (NIZK) systems, where the common reference string (CRS) is assumed to be honestly generated. NIZKs are not only important in their own right but used in a wide variety of applications, so their security under parameter subversion has far-reaching effects. This paper provides a treatment of resistance to parameter subversion for NIZKs, with definitions, negative results and positive results.

<u>NIZKs.</u> Non-interactive zero-knowledge systems originate with BFM [17] and BDMP [16] and have since seen an explosion in constructions and applications. The Groth-Sahai framework for efficient NIZKs [44] is widely utilized and we are seeing not only efficient NIZKs but also their implementation in systems [12, 13, 31, 39, 44]. Structure-preserving cryptography [1, 2, 40] was developed to allow these NIZKs to be used for efficient applications.

The NIZK model postulates a common reference string (CRS) that has been honestly generated according to some distribution. The pragmatics of how this is done receives little explicit attention. Some early works talk of using digits of π and others speak whimsically of "a random string in the sky," but for the most part the understanding is that a trusted party will generate, and make public, the CRS. In light of the above, however, we must be concerned that the CRS is in fact maliciously generated. This is the issue addressed by our work.

An immediate avenue of attack that may come to mind is the following. NIZK security requires that there is a simulator that generates a simulated CRS (indistinguishable from the honest one) together with a trapdoor allowing the simulator to generate proofs without knowing the witness. What if the subvertor generates the CRS via the simulator, so that it knows the trapdoor? Since this CRS is indistinguishable from an honestly generated one, the subversion will not be detected. Now, what does the subvertor gain? This seems to depend on the particular system and its properties. For example, the subvertor may be able to generate proofs of *false* statements and violate soundness. In some cases the trapdoor permits extraction of witnesses from honest proofs, in which case the subvertor would be able to violate zero knowledge. What we see here is that features built into the standard notions and constructions of NIZKs turn out to be potential liabilities in the face of subversion. Put another way, current NIZKs have the possibility of subversion effectively built into the security requirement because the simulator works by "subverting" the CRS.

Two remarks with regard to the above. (1) First, if it is unclear what is going on, or what conclusion to draw, there is a good reason, namely that we are trying to think or talk about what subversion does in the absence of a clear understanding of the subversion-resistance goal, effectively jumping the gun. To be able to effectively assess security we first need precise definitions of the new goal(s) underlying resistance to CRS subversion. Providing such definitions is the first contribution of this paper. (2) Second, while the above discussion may lead one to be pessimistic, we will see that in fact a surprising amount of security can be retained even under a maliciously generated CRS.

NIZK SECURITY, NOW. To discuss the new goals in subversion-resistant NIZKs we first back up to recall the standard goals in the current model where the CRS is trusted and assumed to be honestly generated. We distinguish three standard goals for a non-interactive (NI) system Π relative to an **NP** relation R defining the language $L(\mathsf{R}) \in \mathbf{NP}$. The formalizations are recalled in Sect. 4.

SND: (Soundness) It is hard for an adversary, given an honestly generated *crs*, to find an $x \notin L(\mathsf{R})$ together with a valid proof π (meaning one that the verification algorithm $\Pi.\mathsf{V}$ accepts) for x relative to *crs*.

WI: (Witness indistinguishability) Assuming *crs* is honestly generated, an adversary can't tell under which of two valid witnesses an honest proof (i.e., generated by the prover algorithm $\Pi.\mathsf{P}$ under *crs*) for an instance x was created, and this even holds for multiple, adaptively chosen instances depending on *crs*.

ZK: (Zero-knowledge) There is a simulator $\Pi.\mathsf{Sim.crs}$ returning a simulated CRS crs_0 and associated trapdoor *std*, and an accomplice simulator $\Pi.\mathsf{Sim.pf}$ taking an instance $x \in L(\mathsf{R})$ and *std* and returning a proof, such that an adversary given crs_b cannot tell whether a proof it receives was created honestly (with the honest prover algorithm, an honest crs_1 and a witness; the $b = 1$ case) or via $\Pi.\mathsf{Sim.pf}$ (the $b = 0$ case). Moreover this holds even for multiple, adaptively chosen instances depending on crs_b.

NIZK SECURITY UNDER SUBVERSION. The key change in our model is that the adversary generates the CRS. It can retain, via its coins r, some kind of "backdoor" related to this CRS. In Sect. 4 we formalize the following goals:

S-SND: (Subversion soundness) It is hard for the adversary to generate a (malicious) CRS *crs* together with an instance $x \notin L(\mathsf{R})$ and a valid proof π for x relative to *crs*. (The goal of the subvertor here is to create a CRS that allows it to give proofs of false statements.)

S-WI: (Subversion witness indistinguishability) Even if the adversary creates *crs* maliciously and retains the corresponding coins r, it can't tell under which of two valid witnesses an honest proof (meaning one generated by the prover algorithm $\Pi.\mathsf{P}$ under the subverted *crs*) for an instance x was created, and moreover this holds even for multiple, adaptively chosen instances depending on *crs*.

S-ZK: (Subversion zero knowledge) For any adversary X creating a malicious CRS crs_1 using coins r_1, there is a simulator $\mathsf{S.crs}$ returning not only a simulated CRS crs_0 and associated trapdoor *std but also simulated coins* r_0, and an accomplice simulator $\mathsf{S.pf}$ taking an instance $x \in L(\mathsf{R})$ and *std* and returning a proof, such that an adversary A given crs_b, r_b cannot tell whether a proof it receives was created honestly (with $\Pi.\mathsf{P}$ using crs_1 and a witness; the $b = 1$ case) or via $\mathsf{S.pf}$ (the $b = 0$ case). Moreover this holds even for multiple, adaptively chosen instances depending on crs_b, r_b.

The right side of Fig. 1 may help situate the notions. It shows the obvious relations: S-X implies X; ZK implies WI and S-ZK implies S-WI.

<u>ACHIEVABILITY.</u> Is subversion resistance achievable? This question first needs to be meaningfully posed. The subversion resistance goals are easy to achieve *in isolation*. For example, S-SND is achieved for any **NP** relation by having the prover send the witness, but this is not ZK. S-ZK is achieved by having the prover send the empty string as the proof and having the verifier always accept, but this is not SND. Such trivial constructions are un-interesting. The interesting question is whether meaningful combinations of the goals are simultaneously achievable. A pragmatic viewpoint is that we already have systems achieving SND+WI+ZK. We want to "upgrade" these to get some resistance to subversion. While retaining SND, WI and ZK, what can be added from the list S-SND, S-WI, S-ZK? Can we have them all? Are things so bad that we can have none? We will be able to completely categorize what is achievable and what is not and will see that the truth is somewhere between these extremes and on the whole the news is perhaps more positive than we might have expected. Our core results are summarized in the table on the left side of Fig. 1. In any row, we are considering simultaneously achieving the notions indicated by the bullets. The last column indicates whether or not it is possible. We now discuss these results, beginning with the negative result of the first row.

<u>NEGATIVE RESULT.</u> We first ask whether we can achieve S-SND (soundness for a malicious CRS) while retaining what we have now, namely SND, WI and ZK. Result **N** (the first row of Fig. 1) indicates that we cannot. It says that

	Standard			Subversion resistant			Achievable?
	SND	ZK	WI	S-SND	S-ZK	S-WI	
N		•		•			✗ Thm. 1
P1	•	•	•		•	•	✓ Thm. 3
P2	•		•	•		•	✓ Thm. 5
P3	•	•	•			•	✓ Thm. 6

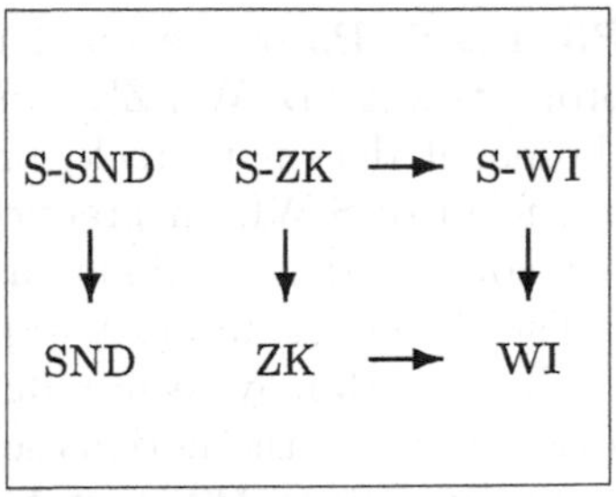

S-SND   S-ZK → S-WI
↓   ↓   ↓
SND   ZK → WI

Fig. 1. Left: Achievability chart showing our negative result **N** and positive results **P1**, **P2**, **P3**. In a row we refer to simultaneously achieving all selected notions. **Right:** Relations.

there is no NI system that achieves both ZK and S-SND. (More precisely, this is only possible for trivial **NP**-relations, i.e., where verifiers can check if $x \in L(\mathsf{R})$ themselves.) We stress that ZK here is the standard notion where the CRS is honest. We are not asking for S-ZK but only to retain ZK. The proof of Theorem 1 establishing this uses the paradigm of GO [36] of using the simulator to break soundness.

<u>Positive results.</u> Figure 1 lists three positive results that we discuss in turn:

P1: The most desirable target is S-ZK. By result **N** it cannot be achieved in combination with S-SND. The next best thing would be to get it in combination with SND. We show in Theorem 3 that this is possible. Since S-ZK implies ZK, S-WI and WI, this yields result **P1** of the table of Fig. 1, showing we can simultaneously achieve all notions but S-SND. Theorem 3 is based on a knowledge-of-exponent assumption (KEA) in a group equipped with a bilinear map. The assumption is certainly strong, but (1) this is to be expected since our goal implies certain forms of 2-move interactive ZK that have themselves only been achieved under extractability assumptions [15], (2) similar assumptions have been made before [39], and (3) unlike other knowledge assumptions [15], our assumption is not ruled out assuming indistinguishability obfuscation. See the beginning of Sect. 6.1 for a high-level description of the ideas of our construction.

P2: The question left open by **P1** is whether there is some meaningful way to achieve S-SND. (It is the one item missing in row **P1**.) We know from result **N** that we cannot do this in combination with ZK. Result **P2** of the table of Fig. 1 says that we can do the best possible given this limitation. Namely we can simultaneously achieve both S-SND and S-WI (and thus SND and WI). Theorem 5 establishing this is under a standard assumption, namely the decision-linear assumption (DLin). It follows easily from the existence of a SND and WI NI system with trivial CRS under DLin [42] and the observation (Lemma 4) that any such system is obviously also S-SND and S-WI.

P3: Result **P3** of the Fig. 1 represents "hedging." The system has the desired properties (SND, WI, ZK) under an honest CRS. When the CRS is maliciously chosen, it does not break completely; it retains witness indistinguishability in the form of S-WI. In practice this offers quite a bit of protection. Our hedging construction combines a PRG with a zap. (A zap is a 2-move witness-indistinguishable interactive protocol [30].)

Result **P3** may seem redundant; isn't it implied by **P1**? (Indeed it selects a strict subset of the notions selected by **P1**.) While **P1** uses strong (extractability) assumptions, **P3** is established in Theorem 6 under the minimal assumption that some SND+WI+ZK NI system exists. Our hedging thus adds no extra assumptions. This is because a zap can be built from any SND+ZK NI system [30].

FULL ACHIEVABILITY PICTURE. The broad question we have asked is, which combinations of the six notions SND, WI, ZK, S-SND, S-WI, S-ZK are simultaneously achievable? Fig. 1 looks at four combinations. But there are in principle 2^6 combinations about which one could ask. In the full version [6] we go systematically over *all* combinations and evaluate achievability. We are able to give the answer in all cases. Briefly, Fig. 1 covers the interesting cases, which is why we have focused on those here, and other cases are dealt with relatively easily.

OTHER NOTIONS. We have been selective rather than exhaustive with regard to which notions to consider in this setting, focusing on the basic soundness, witness indistinguishability and zero knowledge. There are many other notions in this area that could be considered including robustness, simulation soundness and extractability [26,28,38,41] but it seems fairly apparent that these stronger notions will be subject to commensurately strong negative results with regard to security under CRS subversion. For example, extractability asks that the simulator can create a CRS such that, with a trapdoor it withholds, it can extract the witness from a valid proof. But if so, a subvertor can create the CRS like the simulator so that it has the trapdoor and can also extract the witness.

2 Discussion and Related Work

RELATION TO 2-MOVE PROTOCOLS. There is a natural connection between NI systems and 2-move interactive protocols in which NI system Π corresponds to the protocol 2MV in which the verifier first sends the CRS and the prover sends the proof in the second move. We can then think of the following correspondence of notions for Π and 2MV: S-WI ↔ ZAP; ZK ↔ honest-verifier ZK; S-ZK ↔ full (cheating-verifier) ZK. This analogy provides intuition and insight and opens up connections we exploit for both positive and negative results, but one must be wary that the analogy is not fully accurate in either direction. We look separately at this for negative and positive results.

On the negative side, many forms of 2-move ZK are impossible [4,36]. This does not directly imply that S-ZK is impossible because S-ZK does not imply these particular forms of 2-move ZK. For example, S-ZK does not incorporate auxiliary inputs and thus does not imply auxiliary-input 2-move ZK, so the fact that the latter is ruled out [36] does not mean the former is ruled out. (Why does our definition of S-ZK not incorporate auxiliary inputs? One reason was exactly to avoid the impossibility results. But also, an important reason to introduce auxiliary inputs in the interactive case was to be able to prove that ZK for multiple instances is provided, by sequential composition. But our S-ZK formulation already and directly requires security for multiple, adaptively chosen instances, removing the main motivation for auxiliary inputs.)

On the positive side, some forms of 2-move ZK are possible [4,5,15,50]. A natural question is whether one can obtain S-ZK+SND (the goal of **P1**) from them by the obvious transformation, namely to make the verifier's move the CRS. Unfortunately, this does not in general achieve S-ZK. In particular the simulation requirement for S-ZK is stronger than for ZK because the simulated CRS must be produced upfront without knowing the instance, and then the simulator must be able to adaptively produce simulated proofs for multiple instances.

So 2-move ZK as claimed and proven by [4,5,15] does not directly yield S-ZK. The next natural question is whether the protocols of these papers can, nonetheless, be directly shown to have the stronger properties needed to obtain S-ZK. This appears to be the case for the protocols of [4,15,50], because the verifier's first message does not depend on the instance. Starting from BLV [4], the assumption would be that Micali's conjecture [48] (there exist CS proofs or two-round universal arguments) is true. Starting from BCPR [15], the assumption would be the existence of privately verifiable P-delegation, 1-hop FHE, and a complexity-leveraging commitment scheme. In this light, we have chosen to present our knowledge of exponent based **P1** construction as a concrete, self-contained illustration of one simple route to S-ZK+SND from a plausible assumption, but other routes are possible. We do note that BLV [4] themselves view their assumption as so strong that they hesitate to call their result a positive one, instead referring to it as "a negative result on negative results."

BP [5] build one-message ZK arguments, but the simulation is super polynomial time. (This is also true of the construction of Pass [50].) These would thus yield S-ZK with super-polynomial-time simulation. But we require simulation for S-ZK to be polynomial time. This is in keeping with the intuition behind zero-knowledge that the entity running the verifier in the protocol should be able to run the simulator to produce a similar view.

Finally, in the bare public-key model of [21], Wee [56] constructs a weak non-uniform non-interactive zero-knowledge argument. This can be turned into a NI system by using the verifier's public key as the CRS. However this form of ZK allows a super-polynomial simulator whose size depends on the size of the distinguisher and the distinguishing gap, and this is weaker than S-ZK. Also Wee's [56] construction is only proved for one instance, while in S-ZK we require security for multiple, adaptively-chosen instances.

<u>Context.</u> Resistance of NIZKs to parameter subversion may not be of *immediate* practical relevance but we believe it is an important long-term consideration for this technology. The foundational tradition has always had as its stated goal

to model and capture realistic, practical attacks and then investigate theoretically whether or not security can be achieved. Parameter subversion is such a realistic attack not previously considered, and it leads us to revisit the foundations of NIZKs to bring it into the picture. We are seeing large efforts in the creation of efficient NIZKs and their implementation in systems towards eventual applications [11–13, 31, 39, 44]. For security, parameter subversion must be kept in mind from the start.

A standard suggestion to protect against CRS subversion is to generate the CRS via a multi-party computation protocol so that no particular party controls the outcome. This is pursued in [11]. The effectiveness and practicality of this solution are not very clear. What parties would perform this task, and why can we trust *any* of them? The Snowden revelations indicate that corporations cooperate with the NSA toward subversion, either willingly or due to court orders. NIZKs with built-in resistance to subversion, as we define and achieve, provide greater protection.

One might note that in some applications, such as the use of NIZKs for signatures [7, 23, 28] and IND-CCA encryption [29, 49], users can pick their own CRS and be confident of its quality. However this blows up key sizes and increases system complexity. It would be more convenient if there were a single, global CRS, in which case resistance to subversion matters.

CPs [22] study UC-secure computation in a model where the CRS is drawn from a distribution that is adversarially chosen subject to several restrictions, including that it has high min-entropy and is efficiently sampleable via an algorithm known to the simulator. They do not consider NIZKs, and in their model the CRS is not chosen fully maliciously, with no restrictions, as in our model. GO [41] studied the "multi-CRS" model where the adversary can substitute t out of m CRSs, GGJS [33] consider replacing a single trusted setup in UC with multiple, untrusted ones and KKZZ [46] consider distributing the setup for UC-secure multi-party computation. Concern with trust in a CRS is exhibited in the context of elections by KZZ [47], who have the CRS generated by the election authority using the voter's coins.

Algorithm-substitution attacks, studied in [3, 9], are another form of subversion, going back to the broader framework of kleptography [57, 58]. Back-doored blockciphers were studied in [51–53]. DGGJR [27] provide a formal treatment of back-dooring of PRGs in response to the Dual EC debacle. The cliptography framework [54] aims to capture many forms of subversion.

3 Notation

The empty string is denoted by ε. If x is a (binary) string then $|x|$ is its length. If S is a finite set then $|S|$ denotes its size and $s \leftarrow\!\!\$\ S$ denotes picking an element uniformly from S and assigning it to s. We denote by $\lambda \in \mathbb{N}$ the security parameter and by 1^λ its unary representation. Algorithms are randomized unless otherwise indicated. "PT" stands for "polynomial time", whether for randomized or deterministic algorithms. By $y \leftarrow A(x_1, \ldots; r)$ we denote the operation of running A on inputs $x_1, \ldots$ and coins r and letting y denote the output. By $y \leftarrow_{\$} A(x_1, \ldots)$, we denote letting $y \leftarrow A(x_1, \ldots; r)$ for random r. We denote by $[A(x_1, \ldots)]$ the set of points that have positive probability of being output by A on inputs $x_1, \ldots$ Adversaries are algorithms. Complexity is uniform throughout: scheme algorithms and adversaries are Turing Machines, not circuit families.

For our security definitions and some proofs we use the code-based game playing framework of [10]. A game G (e.g. Fig. 2) usually depends on some scheme and executes one or more adversaries. It defines oracles for the adversaries as procedures. The game eventually returns a boolean. We let Pr[G] denote the probability that G returns true.

4 Security of NIZKs Under CRS Subversion

We first recall and discuss standard notions of NIZK security in the setting used until now where the CRS is trusted. We then formulate new notions of NIZK security in the setting where the CRS is subverted, starting with the syntax.

4.1 NP Relations and NI Systems

NP RELATIONS. Proofs pertain to membership in an **NP** language defined by an **NP** relation, and we begin with the latter. Suppose R: $\{0,1\}^* \times \{0,1\}^* \rightarrow \{\mathsf{true}, \mathsf{false}\}$. For $x \in \{0,1\}^*$ we let $\mathsf{R}(x) = \{\, w : \mathsf{R}(x, w) = \mathsf{true} \,\}$ be the *witness set* of x. We say that R is an **NP** relation if it is PT and there is a polynomial $\mathsf{R.wl}$: $\mathbb{N} \rightarrow \mathbb{N}$ called the maximum witness length such that every w in $\mathsf{R}(x)$ has length at most $\mathsf{R.wl}(|x|)$ for all $x \in \{0,1\}^*$. We let $L(\mathsf{R}) = \{\, x : \mathsf{R}(x) \neq \emptyset \,\}$ be the *language* associated to R. The fact that R is an **NP** relation means that $L(\mathsf{R}) \in \mathbf{NP}$. We now go on to security properties, first giving formal definitions and then discussions.

NI SYSTEMS. A non-interactive (NI) system specifies the syntax of the proof system. We can then consider various security attributes, including soundness, zero knowledge and witness indistinguishability. Formally, a NI system Π for R specifies the following PT algorithms. Via $crs \leftarrow_{\$} \Pi.\mathsf{Pg}(1^\lambda)$ one generates a common reference string crs. Via $\pi \leftarrow_{\$} \Pi.\mathsf{P}(1^\lambda, crs, x, w)$ the honest prover, given x and $w \in \mathsf{R}(x)$, generates a proof π that $x \in L(\mathsf{R})$. Via $d \leftarrow \Pi.\mathsf{V}(1^\lambda, crs, x, \pi)$ a verifier can produce a decision $d \in \{\mathsf{true}, \mathsf{false}\}$ indicating whether π is a valid proof that $x \in L(\mathsf{R})$. We require (perfect) completeness, namely $\Pi.\mathsf{V}(1^\lambda, crs, x, \Pi.\mathsf{P}(1^\lambda, crs, x, w)) = \mathsf{true}$ for all $\lambda \in \mathbb{N}$, all $crs \in [\Pi.\mathsf{Pg}(\lambda)]$, all $x \in L(\mathsf{R})$ and all $w \in \mathsf{R}(x)$. We also require that $\Pi.\mathsf{V}$ returns false if any of its arguments is $\perp$.

4.2 Notions for Honest CRS: SND, WI and ZK

SOUNDNESS. Soundness asks that it be hard to create a valid proof for $x \notin L(\mathsf{R})$. Formally, we say that Π is sound for R, abbreviated SND, if $\mathbf{Adv}^{\mathrm{snd}}_{\Pi,\mathsf{R},\mathsf{A}}(\cdot)$ is negligible for all PT adversaries A, where $\mathbf{Adv}^{\mathrm{snd}}_{\Pi,\mathsf{R},\mathsf{A}}(\lambda) = \Pr[\mathrm{SND}_{\Pi,\mathsf{R},\mathsf{A}}(\lambda)]$ and game SND is specified in Fig. 2. This is a computational soundness requirement as opposed to a statistical one, as is sufficient for applications.

WI. This notion [32] requires that a PT adversary, which chooses two witnesses, cannot tell which one was used to create a proof. Formally, we say that Π is witness-indistinguishable (WI) for R, if $\mathbf{Adv}^{\mathrm{wi}}_{\Pi,\mathsf{R},\mathsf{A}}(\cdot)$ is negligible for all PT adversaries A, where $\mathbf{Adv}^{\mathrm{wi}}_{\Pi,\mathsf{R},\mathsf{A}}(\lambda) = 2\Pr[\mathrm{WI}_{\Pi,\mathsf{R},\mathsf{A}}(\lambda)] - 1$ and game WI is specified in Fig. 2. In this game, an adversary A can request a proof for x under one of two witnesses w_0, w_1. It is returned an honestly generated proof under w_b where b is the challenge bit. It can adaptively request and obtain many such proofs before outputting a guess b' for b. The game returns true if this guess is correct.

ZK. We say that Π is zero-knowledge for R, abbreviated ZK, if Π specifies additional PT algorithms $\Pi.\mathsf{Sim.crs}$ and $\Pi.\mathsf{Sim.pf}$ such that $\mathbf{Adv}^{\mathrm{zk}}_{\Pi,\mathsf{R},\mathsf{A}}(\cdot)$ is negligible for all PT adversaries A, where $\mathbf{Adv}^{\mathrm{zk}}_{\Pi,\mathsf{R},\mathsf{A}}(\lambda) = 2\Pr[\mathrm{ZK}_{\Pi,\mathsf{R},\mathsf{A}}(\lambda)] - 1$ and game ZK is specified in Fig. 2. Adversary A can adaptively request proofs by supplying an instance and a valid witness for it. The proof is produced either by the honest prover using the witness, or by the proof simulator $\Pi.\mathsf{Sim.pf}$ using a trapdoor *std*. The adversary outputs a guess b' as to whether the proofs were real or simulated.

DISCUSSION. The classical definitions of soundness and zero knowledge for proof systems [37] were in what we will call the complexity-theoretic style. The soundness condition said that for all $x \notin L(\mathsf{R})$, the probability that a dishonest prover could convince the honest verifier to accept was low. Zero knowledge, similarly, looked at distributions associated to a fixed $x \in L(\mathsf{R})$ and then at ensembles over x. The first definition for NIZK was similar [16]. But over time, NIZK definitions have adapted to what we call a cryptographic style [26,43]. This is the style we use because it seems more prevalent now and it works better for applications. Here x is not quantified but chosen by an adversary. The definitions directly capture proofs for multiple, related statements. All adversaries are PT, meaning all metrics are computational.

One consequence of the complexity-theoretic style was a need for non-uniform complexity for adversaries and assumptions [35,37]. In [34] Goldreich made a case for uniform complexity. The cryptographic style we adopt is in this vein, and in our setting all complexity (adversaries, algorithms, assumptions) is uniform.

4.3 Notions for Subverted CRS: S-SND, S-WI and S-ZK

A core assumption in NIZKs is that the CRS is honestly generated. In light of subversion of parameters in other contexts as part of the mass-surveillance revelations, we ask what would happen if the CRS were maliciously generated. We will define subversion-resistance analogues S-SND, S-WI and S-ZK of the SND, WI, ZK goals above. The key difference is that the CRS is selected by an adversary rather than via the CRS-generation algorithm $\Pi.\mathsf{Pg}$ prescribed by Π.

Game $\mathrm{SND}_{\Pi,R,A}(\lambda)$	Game $\mathrm{S\text{-}SND}_{\Pi,R,A}(\lambda)$
$crs \leftarrow_{\$} \Pi.\mathsf{Pg}(1^\lambda)$ $(x,\pi) \leftarrow_{\$} \mathsf{A}(1^\lambda, crs)$ Return $(x \notin L(\mathsf{R})$ and $\Pi.\mathsf{V}(1^\lambda, crs, x, \pi))$	$(crs, x, \pi) \leftarrow_{\$} \mathsf{A}(1^\lambda)$ Return $(x \notin L(\mathsf{R})$ and $\Pi.\mathsf{V}(1^\lambda, crs, x, \pi))$

Game $\mathrm{WI}_{\Pi,R,A}(\lambda)$	Game $\mathrm{S\text{-}WI}_{\Pi,R,A}(\lambda)$
$b \leftarrow_{\$} \{0,1\}$ $crs \leftarrow_{\$} \Pi.\mathsf{Pg}(1^\lambda)$ $b' \leftarrow_{\$} \mathsf{A}^{\textsc{Prove}}(1^\lambda, crs)$ Return $(b = b')$	$b \leftarrow_{\$} \{0,1\}$ $(crs, st) \leftarrow_{\$} \mathsf{A}(1^\lambda)$ $b' \leftarrow_{\$} \mathsf{A}^{\textsc{Prove}}(1^\lambda, crs, st)$ Return $(b = b')$
$\textsc{Prove}(x, w_0, w_1)$ If $\mathsf{R}(x, w_0) = \mathsf{false}$ or $\mathsf{R}(x, w_1) = \mathsf{false}$ then Return $\perp$ $\pi \leftarrow_{\$} \Pi.\mathsf{P}(1^\lambda, crs, x, w_b)$ Return π	$\textsc{Prove}(x, w_0, w_1)$ If $\mathsf{R}(x, w_0) = \mathsf{false}$ or $\mathsf{R}(x, w_1) = \mathsf{false}$ then Return $\perp$ $\pi \leftarrow_{\$} \Pi.\mathsf{P}(1^\lambda, crs, x, w_b)$ Return π

Game $\mathrm{ZK}_{\Pi,R,A}(\lambda)$	Game $\mathrm{S\text{-}ZK}_{\Pi,R,X,S,A}(\lambda)$
$b \leftarrow_{\$} \{0,1\}$ $crs_1 \leftarrow_{\$} \Pi.\mathsf{Pg}(1^\lambda)$ $(crs_0, std) \leftarrow_{\$} \Pi.\mathsf{Sim.crs}(1^\lambda)$ $b' \leftarrow_{\$} \mathsf{A}^{\textsc{Prove}}(1^\lambda, crs_b)$ Return $(b = b')$	$b \leftarrow_{\$} \{0,1\}$ $r_1 \leftarrow_{\$} \{0,1\}^{\mathsf{X.rl}(\lambda)}$; $crs_1 \leftarrow \mathsf{X}(1^\lambda; r_1)$ $(crs_0, r_0, std) \leftarrow_{\$} \mathsf{S.crs}(1^\lambda)$ $b' \leftarrow_{\$} \mathsf{A}^{\textsc{Prove}}(1^\lambda, crs_b, r_b)$ Return $(b = b')$
$\textsc{Prove}(x, w)$ If $\mathsf{R}(x, w) = \mathsf{false}$ then Return $\perp$ If $b = 1$ then $\pi \leftarrow_{\$} \Pi.\mathsf{P}(1^\lambda, crs_1, x, w)$ Else $\pi \leftarrow_{\$} \Pi.\mathsf{Sim.pf}(1^\lambda, crs_0, std, x)$ Return π	$\textsc{Prove}(x, w)$ If $\mathsf{R}(x, w) = \mathsf{false}$ then Return $\perp$ If $b = 1$ then $\pi \leftarrow_{\$} \Pi.\mathsf{P}(1^\lambda, crs_1, x, w)$ Else $\pi \leftarrow_{\$} \mathsf{S.pf}(1^\lambda, crs_0, std, x)$ Return π

Fig. 2. Games defining standard (left) and subversion (right) security of NI system Π. Top to bottom: Soundness, witness indistinguishability, zero knowledge.

<u>Subversion soundness.</u> Subversion soundness asks that if a subvertor creates a CRS in any way it likes, it will still be unable to prove false statements under that CRS. Formally, we say that Π is subversion-sound (abbreviated S-SND) for R if $\mathbf{Adv}^{\text{s-snd}}_{\Pi,R,A}(\cdot)$ is negligible for all PT adversaries A, where $\mathbf{Adv}^{\text{s-snd}}_{\Pi,R,A}(\lambda) = \Pr[\mathrm{S\text{-}SND}_{\Pi,R,A}(\lambda)]$ and game S-SND is specified in Fig. 2. Compared to the honest-CRS game SND to the left of it, the adversary now not only generates x and π, but itself supplies crs, modeling a malicious choice of the latter.

<u>Subversion WI.</u> Subversion WI asks that if a subvertor creates a CRS in any way it likes then it will still be unable to tell which of two witnesses was used to create a proof, even given both witnesses. Formally, we say that Π is subversion

witness-indistinguishable (S-WI) for R if $\mathbf{Adv}^{\text{s-wi}}_{\Pi,\mathsf{R},\mathsf{A}}(\cdot)$ is negligible for all PT adversaries A, where $\mathbf{Adv}^{\text{s-wi}}_{\Pi,\mathsf{R},\mathsf{A}}(\lambda) = 2\Pr[\text{S-WI}_{\Pi,\mathsf{R},\mathsf{A}}(\lambda)] - 1$ and game S-WI is specified in Fig. 2. Compared to the honest-CRS game WI, the CRS *crs* is now generated by the adversary in a first stage, along with state information *st* passed to its second stage. In the latter, via its PROVE oracle, it adaptively obtains proofs for instances of its choice under a challenge witness, and outputs a guess b' for the challenge b. The state can contain the coins of A or any trapdoor associated to *crs* that A chooses to put there helping its distinguishing task.

SUBVERSION ZK. Subversion ZK asks that for any CRS subvertor X creating a CRS in any way it likes there is a simulator able to produce the full view of the CRS subvertor, including its coins and proofs corresponding to adaptively chosen instances, without knowing the witnesses. Formally, a simulator S for X specifies PT algorithms $\mathsf{S.crs}$ and $\mathsf{S.pf}$. Now consider game S-ZK of Fig. 2 associated to $\Pi, \mathsf{R}, \mathsf{X}, \mathsf{S}$ and an adversary A. We let $\mathbf{Adv}^{\text{s-zk}}_{\Pi,\mathsf{R},\mathsf{X},\mathsf{S},\mathsf{A}}(\lambda) = 2\Pr[\text{S-ZK}_{\Pi,\mathsf{R},\mathsf{X},\mathsf{S},\mathsf{A}}(\lambda)] - 1$. We say that Π is subversion zero-knowledge (S-ZK) for R if for all PT CRS subvertors X there is a PT simulator S such that for all PT A the function $\mathbf{Adv}^{\text{s-zk}}_{\Pi,\mathsf{R},\mathsf{X},\mathsf{S},\mathsf{A}}(\cdot)$ is negligible.

In this game, if the challenge bit b is 1 then the CRS crs_1 is generated via X with the coins r_1 made explicit. Otherwise, if $b = 0$, the first stage $\mathsf{S.crs}$ of the simulator is run to produce simulated versions crs_0, r_0 not only of the CRS but *also of the coins of* X. Alongside, $\mathsf{S.crs}$ produces a simulation trapdoor *std* as in ZK to allow its second stage to simulate proofs. Now, A gets to request its PROVE oracle for proofs of instances of its choice. If $b = 1$, these are produced by the honest prover with the given witness; but if $b = 0$, they are produced via the second stage $\mathsf{S.pf}$ of the simulator using the simulation trapdoor *std* and no witness. Adversary A produces its guess b' and wins of $b' = b$.

The definition reflects that X here is like a cheating verifier in classical ZK [37]. The simulator thus needs to produce its coins as well as the transcript of its interaction with its oracle. But also, to reflect the ZK requirement of *non-interactive* systems above, more is required, namely that the simulator must first produce the simulated CRS and coins, and then, in its second stage, be able to produce simulated proofs. The definition is thus quite demanding. Note that the simulator can depend (in a non-blackbox way) on X, but not on A. The latter is important to ensure that S-ZK implies ZK.

4.4 2-Move Protocols

We will have many occasions to refer to and use 2-move interactive protocols, so we fix a syntax for them. A 2-move protocol 2MV for **NP** relation R specifies PT algorithms $\mathsf{2MV.V}$, $\mathsf{2MV.P}$, $\mathsf{2MV.D}$. Via $(m_1, st) \leftarrow\$\ \mathsf{2MV.V}(1^\lambda, x)$ the honest verifier generates the first move message m_1 on input x, retaining associated state information st. Via $m_2 \leftarrow\$\ \mathsf{2MV.P}(1^\lambda, x, w, m_1)$ the honest prover generates a reply computed from x, a witness $w \in \mathsf{R}(x)$ and the first move message m_1. Deterministic decision algorithm $\mathsf{2MV.D}$ takes x, m_1, m_2, st and returns a boolean decision. Security notions will be discussed as needed.

5 Negative Result: ZK and S-SND Are Not Compatible

All the different forms of subversion security (S-SND, S-WI, S-ZK) are easy to achieve in isolation. For example sending the witness as the proof achieves S-SND (but this is not ZK). Having the verification algorithm always accept and sending the empty string as the proof achieves S-ZK (but not SND). These kinds of results are not interesting. We want to study the simultaneous achievability of meaningful combinations of the notions, meaning some kind of soundness together with some kind of zero knowledge or witness indistinguishability.

We already have NI systems that are SND+ZK and we do not want to degrade this. If now the CRS is subverted, what more can we have without losing the initial properties? The first question we ask is, can we up the ante for soundness, meaning add S-SND? That is, we want subversion soundness while retaining ZK. We will show that this is not possible.

An impossibility result in this domain means no NI system satisfying the conditions exists unless the relation R is trivial. Roughly, trivial means that the verification algorithm can decide membership in $L(\mathsf{R})$ on its own. Impossibility results of this type begin with Goldreich and Oren (GO) [36]. Their definition of R being trivial was simple, namely that it is in **BPP**. This will not suffice here, so we begin with a more precise definition of relation triviality and an explanation of why it is needed.

GAME $\mathrm{DEC}_{\mathsf{IG},\mathsf{R},\mathsf{M}}(\lambda)$
$(x, w) \leftarrow\$ \mathsf{IG}(1^\lambda)$; $d_1 \leftarrow \mathsf{R}(x, w)$
If ($x \in L(\mathsf{R})$ and $d_1 = \mathsf{false}$) then return false
$d_0 \leftarrow\$ \mathsf{M}(1^\lambda, x)$; return $(d_0 \neq d_1)$

Fig. 3. Game defining language triviality

RELATION TRIVIALITY. The definition of a relation R being trivial if $L(\mathsf{R}) \in$ **BPP** works when the formulations of ZK and soundness are in the complexity-theoretic style, meaning the conditions refer to universally quantified inputs. As discussed in Sect. 4.2 however, our formulations, following modern treatments of NI systems in the literature, are in the cryptographic style, which is better suited for applications. Here the only instances that come into play are those that can be generated by PT algorithms, and the only positive instances that come into play are those generated with witnesses. In this setting, **BPP** will not work as a definition of triviality because membership in standard complexity classes like **BPP** refers to arbitrary inputs, not merely ones that one can generate in PT. For our purposes we thus give a definition of a language (actually an **NP** relation) being trivial, which can be seen as defining a cryptographic version of **BPP**.

Let R be an **NP** relation. An *instance generator* is a PT algorithm that on input 1^λ returns a pair (x, w). Here x is a challenge instance that may or may not be in $L(\mathsf{R})$, and w should be in $\mathsf{R}(x)$ if $x \in L(\mathsf{R})$. Let M be an algorithm (decision procedure) taking $1^\lambda, x$ and returning a boolean representing whether or not it thinks x is in $L(\mathsf{R})$. Consider game DEC of Fig. 3 associated to $\mathsf{IG}, \mathsf{R}, \mathsf{M}$ and let $\mathbf{Adv}^{\mathrm{dec}}_{\mathsf{IG},\mathsf{R},\mathsf{M}}(\lambda) = \Pr[\mathrm{DEC}_{\mathsf{IG},\mathsf{R},\mathsf{M}}(\lambda)]$. We say that algorithm M decides R if for every PT IG the function $\mathbf{Adv}^{\mathrm{dec}}_{\mathsf{IG},\mathsf{R},\mathsf{M}}(\cdot)$ is negligible. We say that R is trivial if there is a PT algorithm M that decides R. Intuitively, in game DEC, think of IG as an adversary trying to make M fail. The game returns true when IG succeeds, meaning M returns the wrong decision. A technical point is that if IG generates a positive instance x, the game forces it to lose if the witness w is not valid. Thus we are asking that M is able to decide membership in PT for instances that can be efficiently generated with valid witnesses if the instance is positive. But this does not mean it can decide membership on all instances. Thus if $L(\mathsf{R}) \in \mathbf{BPP}$ then R is certainly trivial, but the converse need not be true.

<u>Result.</u> We show that ZK and subversion soundness (S-SND) cannot co-exist, meaning only trivial relations will have NI systems with both attributes. We stress that we are not asking here for subversion ZK but just plain ZK.

Theorem 1. *Let Π be a NI system satisfying zero knowledge (ZK) and subversion soundness (S-SND) for an* **NP** *relation* R*. Then* R *is trivial.*

The proof follows the basic paradigm of GO [36]. We use the simulator to build a cheating prover that violates soundness. In our case this works if soundness holds relative to a simulated CRS, but S-SND guarantees this.

Proof. (Theorem 1). Define the following decision procedure M:

<u>Algorithm $\mathsf{M}(1^\lambda, x)$</u>
$(crs_0, std_0) \leftarrow_\$ \Pi.\mathsf{Sim.crs}(1^\lambda)$; $\pi \leftarrow_\$ \Pi.\mathsf{Sim.pf}(1^\lambda, crs_0, std_0, x)$
Return $\Pi.\mathsf{V}(1^\lambda, crs_0, x, \pi)$

Thus, to decide if $x \in L(\mathsf{R})$, algorithm M runs the simulator to get a simulated CRS and simulation trapdoor, uses the latter to generate a simulated proof, and decides that $x \in L(\mathsf{R})$ if this proof is valid. Let IG be any PT instance generator. We will show below that $\mathbf{Adv}^{\mathrm{dec}}_{\mathsf{IG},\mathsf{R},\mathsf{M}}(\cdot)$ is negligible. This shows that R is trivial.

To show $\mathbf{Adv}^{\mathrm{dec}}_{\mathsf{IG},\mathsf{R},\mathsf{M}}(\cdot)$ is negligible, below we will define PT adversaries A, B such that

$$\mathbf{Adv}^{\mathrm{dec}}_{\mathsf{IG},\mathsf{R},\mathsf{M}}(\lambda) \leq \mathbf{Adv}^{\mathrm{zk}}_{\Pi,\mathsf{R},\mathsf{A}}(\lambda) + \mathbf{Adv}^{\mathrm{s\text{-}snd}}_{\Pi,\mathsf{R},\mathsf{B}}(\lambda) \qquad (1)$$

for all $\lambda \in \mathbb{N}$. By assumption, Π satisfies ZK and S-SND for R, so the functions $\mathbf{Adv}^{\mathrm{zk}}_{\Pi,\mathsf{R},\mathsf{A}}(\cdot)$ and $\mathbf{Adv}^{\mathrm{s\text{-}snd}}_{\Pi,\mathsf{R},\mathsf{B}}(\cdot)$ are both negligible. Thus Eq. (1) implies that $\mathbf{Adv}^{\mathrm{dec}}_{\mathsf{IG},\mathsf{R},\mathsf{M}}(\cdot)$ is negligible, as desired.

Consider games $\mathrm{G}_0, \mathrm{G}_1, \mathrm{G}_2$ of Fig. 4. Game G_0 is defined ignoring the box, while game $\boxed{\mathrm{G}_1}$ includes it. Games G_0 and G_1 split up the decision process depending on whether or not $x \in L(\mathsf{R})$. Game G_2 switches to a real CRS and proofs, which it can do since the instance generator provided a witness.

Game DEC returns true iff $((x \notin L(\mathsf{R}))$ AND $(d_0 = \mathsf{true}))$ OR $((x \in L(\mathsf{R}))$ AND $(d_1 = \mathsf{true})$ AND $(d_0 = \mathsf{false}))$. The first condition in the OR is when game

Games G_0, $\boxed{\mathrm{G}_1}$	Game G_2
$(x, w) \leftarrow\$ \mathsf{IG}(1^\lambda)$; $d_1 \leftarrow \mathsf{R}(x, w)$	$(x, w) \leftarrow\$ \mathsf{IG}(1^\lambda)$; $d_1 \leftarrow \mathsf{R}(x, w)$
$(crs, std) \leftarrow\$ \Pi.\mathsf{Sim.crs}(1^\lambda)$	$crs \leftarrow\$ \Pi.\mathsf{Pg}(1^\lambda)$
$\pi \leftarrow\$ \Pi.\mathsf{Sim.pf}(1^\lambda, crs, std, x)$	$\pi \leftarrow\$ \Pi.\mathsf{P}(1^\lambda, crs, x, w)$
$d_0 \leftarrow \Pi.\mathsf{V}(1^\lambda, crs, x, \pi)$	$d_0 \leftarrow \Pi.\mathsf{V}(1^\lambda, crs, x, \pi)$
$b \leftarrow ((x \notin L(\mathsf{R})) \wedge (d_0 = \mathsf{true}))$	$b \leftarrow ((d_1 = \mathsf{true}) \wedge (d_0 = \mathsf{false}))$
$\boxed{b \leftarrow ((d_1 = \mathsf{true}) \wedge (d_0 = \mathsf{false}))}$	Return b
Return b	

Fig. 4. Games for proof of Theorem 1

G_0 returns true. The second condition in the OR is equivalent to $((d_1 = \mathsf{true})$ AND $(d_0 = \mathsf{false}))$, which is the condition under which game G_1 returns true. Furthermore the conditions are mutually exclusive. We thus have

$$\mathbf{Adv}^{\mathrm{dec}}_{\mathsf{IG},\mathsf{R},\mathsf{M}}(\lambda) = \Pr[\mathrm{G}_0] + \Pr[\mathrm{G}_1] = \Pr[\mathrm{G}_0] + \Pr[\mathrm{G}_2] + (\Pr[\mathrm{G}_1] - \Pr[\mathrm{G}_2]) \quad (2)$$

Notice that by completeness of Π we have

$$\Pr[\mathrm{G}_2] = 0 \,. \quad (3)$$

Now we specify the adversaries A, B as follows:

Adversary $\mathsf{A}^{\mathrm{PROVE}}(1^\lambda, crs)$

$(x, w) \leftarrow\$ \mathsf{IG}(1^\lambda)$; $d_1 \leftarrow \mathsf{R}(x, w)$
$\pi \leftarrow\$ \mathrm{PROVE}(x, w)$; $d_0 \leftarrow \Pi.\mathsf{V}(1^\lambda, crs, x, \pi)$
If $((d_1 = \mathsf{true}) \wedge (d_0 = \mathsf{false}))$ then $b' \leftarrow 0$
Else $b' \leftarrow 1$
Return b'

Adversary $\mathsf{B}(1^\lambda)$

$(x, w) \leftarrow\$ \mathsf{IG}(1^\lambda)$
$(crs, std) \leftarrow\$ \Pi.\mathsf{Sim.crs}(1^\lambda)$
$\pi \leftarrow\$ \Pi.\mathsf{Sim.pf}(1^\lambda, crs, std, x)$
Return (crs, x, π)

Then we have

$$\Pr[\mathrm{G}_0] \leq \mathbf{Adv}^{\text{s-snd}}_{\Pi,\mathsf{R},\mathsf{B}}(\lambda) \quad (4)$$

$$\Pr[\mathrm{G}_1] - \Pr[\mathrm{G}_2] \leq \mathbf{Adv}^{\mathrm{zk}}_{\Pi,\mathsf{R},\mathsf{A}}(\lambda) \,. \quad (5)$$

Putting together Eqs. (2), (3), (4) and (5) we get Eq. (1). □

6 Positive Results

We already have NI systems that are SND+ZK, or SND+WI. We ask, if the CRS is subverted, what more can we have without losing the initial properties? Can we add S-ZK? In Sect. 6.1 we answer positively to this question (result **P1**), showing a protocol that is SND+S-ZK under a knowledge-of-exponent assumption (KEA) in a group equipped with a bilinear map. In light of negative result **N**, this is the best we can achieve if we want to retain ZK in presence of CRS subversion.

Can we add S-SND? In light of **N**, we know that we cannot have S-SND and any form of ZK together. The best we can achieve while retaining S-SND is S-WI. In Sect. 6.2 we show that there exist NI systems that are S-SND+S-WI (result **P2**).

Result **P1** provides S-ZK but requires KEA. A natural question is, if we relax the requirement of S-ZK and aim to retain S-WI, can we achieve it from weaker assumptions? In Sect. 6.3 we show that there exists a NI system that is SND, ZK and S-WI under the weaker assumption that one-way functions and zaps exist.

6.1 Soundness and Subversion ZK

<u>OVERVIEW.</u> To achieve S-ZK, a simulator must be able to simulate proofs under a CRS output by a subvertor. As opposed to ZK, the simulator thus cannot embed a trapdoor in the CRS, nor can it extract one from the subvertor by rewinding, as there is no interaction with it. We will instead rely on a knowledge assumption, stating that an algorithm can only produce a certain output if it knows underlying information. This is formalized by requiring that there exists an extractor that extracts the information from the algorithm. We will use this information as the simulation trapdoor, which we can extract from a subvertor outputting a CRS. For soundness, a minimal requirement is that it is hard for the adversary to obtain the trapdoor from an honestly generated CRS.

The knowledge-of-exponent assumption (KEA) for a group $\mathbb{G}$, generated by g, states that from any algorithm which given a random element $h \leftarrow_{\$} \mathbb{G}$ returns a pair of the form (g^s, h^s) one can efficiently extract s. A possible approach for a NI system is to define the CRS as a pair (g^s, h^s), for random s, and define a proof for $x \in L$ to prove that either $x \in L$ or one knows the value s in the CRS. By extracting s, the simulator in the S-ZK game can simulate proofs, while the adversary in the soundness game must supposedly use a witness for x, since it does not know s.

There are two problems with this approach: who chooses the group $\mathbb{G}$ and who chooses the element h used to prove knowledge of s? We address the first problem by letting the group $\mathbb{G}$ be part of the scheme specification. As for the choice of h, it cannot be chosen at CRS setup, since if the subvertor knows $\eta = \log_g h$, it can produce a CRS (S_1, S_2) *without* knowing s by randomly picking $S_1 \leftarrow_{\$} \mathbb{G}$ and setting $S_2 \leftarrow S^{\eta}$. Fixing h and letting it also be also part of the scheme description is problematic, since again, what guarantees that the subvertor does not know its logarithm and can thereby break KEA? We overcome this issue by defining a new type of KEA, stating that in order to produce elements $(h = g^{\eta}, g^s, h^s)$, one has to *either* know s *or* η. As tuples of this form are Diffie-Hellman tuples, we call the assumption DH-KEA.

We define a CRS as a tuple $(g^{s_0}, g^{s_1}, g^{s_0 s_1})$ and let a proof for a statement x prove that either there is a witness for x or one knows s_0 or s_1. We prove knowledge by adding a ciphertext C and use a perfectly sound witness-indistinguishable NI proof ζ with trivial CRS (a.k.a. a non-interactive zap) to prove that either $x \in L$ or C encrypts s_0 or s_1. (Using linear encryption for C

and the NI system by GOS [42], both IND-CPA of C, as well as WI of ζ, follow from the decision-linear assumption (Dlin) [18].)

The sketched scheme is ZK since by encrypting the trapdoor s_0 (or s_1) proofs can be simulated, and by IND-CPA of C and WI of ζ they are indistinguishable from real ones. But we defined the CRS to allow even more: by DH-KEA, from a CRS subvertor we can *extract* either s_0 or s_1, which should yield S-ZK. Not quite, since the subvertor could simply output random group elements (S_0, S_1, S_2), from which we cannot extract. Since the GOS NI system requires a *bilinear* group, we can use its pairing to check CRS well-formedness. The prove (and verification) algorithm can then reject a malformed CRS, which together with simulatability under a well-formed CRS yields S-ZK.

Soundness intuitively holds because, by soundness of ζ, a proof for a wrong statement must contain an encryption of s_0 or s_1, which should be infeasible to obtain from an honestly generated CRS if computing discrete logarithms (DL) is hard. (Given a DL challenge S, one can randomly set S_0 or S_1 to S and with probability $\frac{1}{2}$, the proof contains an encryption of $\log S$.) To formally prove soundness, the reduction must recover s from C. We could include in the CRS a public key under which C is to be encrypted: the reduction sets up the CRS, knows the decryption key and can obtain s. Alas, this would break S-ZK: an adversary that created the CRS could also decrypt C and thereby distinguish real proofs from simulated ones.

We therefore include the linear-encryption key $pk = (g^u, g^v)$ in the proof rather than the CRS. But how would the soundness reduction then retrieve s? Could we use KEA again? Since we can only extract one of two possible logarithms, we do the following. The proof contains *two* public keys $pk_0 = (g^{u_0}, g^{v_0})$ and $pk_1 = (g^{u_1}, g^{v_1})$ and s is encrypted under both of them. Additionally, the proof contains elements $g^{u_0 u_1}, g^{u_0 v_1}, g^{v_0 u_1}, g^{v_0 v_1}$, whose consistency can be verified via the pairing. By DH-KEA, there exists an extractor which from $(g^{u_0}, g^{u_1}, g^{u_0 u_1})$ extracts either u_0 or u_1, another extractor that from $(g^{u_0}, g^{v_1}, g^{u_0 v_1})$ extracts u_0 or v_1, and so on. Together these four extractors either yield (u_0, v_0) or (u_1, v_1), thus one of the secret keys corresponding to pk_0 and pk_1. This way the soundness reduction can extract the value s encrypted in a proof for a false statement. At the same time we show that S-ZK still holds.

In our actual scheme we use the CDH assumption (defined below and implied by DLin) instead of DL. The reason is that CDH solutions are group elements, which can be efficiently encrypted using linear encryption. The trapdoor is then a solution to a CDH instance in the CRS. Besides 14 group elements, the most costly component of our proofs is the GOS NI proof ζ. It uses a circuit representation of the NP relation R and shows that (a) either $\mathsf{R}(x, w)$ for some w, or (b) the simulation trapdoor was encrypted (see Eq. (6)). The GOS system [42] was further developed by Groth and Sahai [44] yielding very efficient proofs for algebraic statements, and we could replace GOS by GS. As the clause (b) that we added has precisely this algebraic form, the overhead for turning a proof that is merely WI into one that is S-ZK would be quite modest.

<u>DISCUSSION.</u> Our scheme specification includes the bilinear group, so one might ask whether we have not just shifted the subversion risk from the CRS to the choice of the group. Since the group generation algorithm is deterministic and public, anyone can run the algorithm to re-obtain the group; moreover, different entities can implement it independently if they think that some standardized implementation was subverted, as a check. With the CRS, the situation is different. There is no easy way to check that it was properly generated, at least without compromising security. Perhaps a vocabulary that speaks to this is that the group is *reproducible*, whereas the CRS is not. Someone is trusted to produce it and one cannot easily check that they did it honestly.

Still, one must ask whether the algorithms used allow embedding of backdoors. Here we must look at the specific algorithms. Thus, while one could use a bilinear group in which the discrete-log problem is easy, leading to an insecure scheme, we know it is possible to publicly specify good algorithms. The specifications, given for example in research papers, may be used by anyone to re-produce the results of the algorithms with some faith that there are no backdoors, in the case (as here) that these algorithms are deterministic.

<u>GAME $\mathrm{KE}_{\mathsf{dGG},\mathsf{M},\mathsf{E}}(\lambda)$</u>

$(p, \mathbb{G}, \mathbb{G}_T, \mathbf{e}, g) \leftarrow \mathsf{dGG}(1^\lambda)$; $h_0, h_1 \leftarrow\!\!\$\ \mathbb{G}$; $r \leftarrow\!\!\$\ \{0,1\}^{\mathsf{M.rl}(\lambda)}$

$(S_0, S_1, S_2) \leftarrow \mathsf{M}(1^\lambda, h_0, h_1; r)$; $s \leftarrow\!\!\$\ \mathsf{E}(1^\lambda, h_0, h_1, r)$

Return $\big(\mathbf{e}(S_0, S_1) = \mathbf{e}(g, S_2)$ and $g^s \neq S_0$ and $g^s \neq S_1\big)$

<u>GAME $\mathrm{CDH}_{\mathsf{dGG},\mathsf{A}}(\lambda)$</u>

$(p, \mathbb{G}, \mathbb{G}_T, \mathbf{e}, g) \leftarrow \mathsf{dGG}(1^\lambda)$; $s, t \leftarrow\!\!\$\ \mathbb{Z}_p$; $C \leftarrow\!\!\$\ \mathsf{A}(1^\lambda, g^s, g^t)$

Return $(C = g^{st})$

<u>GAME $\mathrm{DLin}_{\mathsf{dGG},\mathsf{A}}(\lambda)$</u>

$b \leftarrow\!\!\$\ \{0,1\}$; $(p, \mathbb{G}, \mathbb{G}_T, \mathbf{e}, g) \leftarrow \mathsf{dGG}(1^\lambda)$

$u, v, s, t, \xi \leftarrow\!\!\$\ \mathbb{Z}_p$; $b' \leftarrow\!\!\$\ \mathsf{A}(1^\lambda, g^u, g^v, g^{us}, g^{vt}, g^{s+t+b\cdot\xi})$

Return $(b = b')$

Fig. 5. Games defining the knowledge-of-exponent assumption, the CDH assumption and the DLin assumption.

Speaking broadly, we cannot (and do not claim to) prevent all possible subversion. This is not possible. Our goal is to put in defenses that make the most obvious paths harder, one of which is subversion of the CRS.

<u>BILINEAR GROUPS.</u> Our construction is based on bilinear groups for which we introduce a new type of knowledge-of-exponent assumption. A bilinear-group generator GGen is a PT algorithm that takes input a security parameter 1^λ and outputs a description of a bilinear group $(p, \mathbb{G}, \mathbb{G}_T, \mathbf{e}, g)$, where p is a prime of length λ, $\mathbb{G}$ and $\mathbb{G}_T$ are groups of order p, g generates $\mathbb{G}$ and $\mathbf{e}\colon \mathbb{G} \times \mathbb{G} \to \mathbb{G}_T$ is a bilinear map that is non-degenerate (i.e. $\langle \mathbf{e}(g, g) \rangle = \mathbb{G}_T$).

While in the cryptographic literature bilinear groups are often assumed to be probabilistically generated, real-world pairing-based schemes are defined for groups that are fixed for every λ. We reflect this by defining the group generator as a deterministic PT algorithm dGG. An advantage of doing so is that every entity in the scheme can compute the group from the security parameter and no party must be trusted with generating the group.

KEA. The knowledge-of-exponent assumption (KEA) [8,25,45] in a group $\mathbb{G}$ states that an algorithm M that is given two random generators g, h of $\mathbb{G}$ and outputs (g^c, h^c) must know c. This is formalized by requiring that there exists an extractor for M which when given M's coins outputs c. Generalizations of KEA were used in the bilinear-group setting in [39]. We introduce a new type of KEA in bilinear groups, which we call DH-KEA, where we assume that if M outputs a Diffie-Hellman (DH) tuple g^s, g^t, g^{st} then it must either know s or t. This should also be the case when M is given two additional random generators h_0, h_1. We note that while an adversary may produce one group element without knowing its discrete logarithm by hashing into the elliptic curve [19,20,55], it seems hard to produce a DH tuple without knowing at least one of the logarithms.

Formally, let $\mathbf{Adv}^{\mathrm{ke}}_{\mathsf{dGG},\mathsf{M},\mathsf{E}}(\lambda) = \Pr[\mathrm{KE}_{\mathsf{dGG},\mathsf{M},\mathsf{E}}(\lambda)]$, where game KE is defined in Fig. 5. The DH-KEA assumption holds for dGG if for every PT M there exists a PT E s.t. $\mathbf{Adv}^{\mathrm{ke}}_{\mathsf{dGG},\mathsf{M},\mathsf{E}}(\cdot)$ is negligible.

We note that due to deterministic group generation the assumption does not hold for non-uniform machines M, as their advice for inputs 1^λ could simply be a DH tuple (S_0, S_1, S_2) w.r.t. the group output by $\mathsf{dGG}(1^\lambda)$. However, we follow Goldreich [34] and only consider uniform machines. As a sanity check, we show that DH-KEA holds in the generic-group model. To reflect hashing into elliptic curves, we provide the adversary with an additional generic operation: it can create new group elements without knowing their discrete log. In the full version [6] we show the following.

Theorem 2. *DH-KEA, as defined above, holds in the generic-group model with hashing into the group.*

CDH. The computational Diffie-Hellman assumption in a group $\mathbb{G}$ states that given g^s and g^t for a random s, t, it should be hard to compute g^{st}. Formally, the CDH assumption holds for dGG if $\mathbf{Adv}^{\mathrm{cdh}}_{\mathsf{dGG},\mathsf{A}}(\cdot)$ is negligible for all PT adversaries A, where $\mathbf{Adv}^{\mathrm{cdh}}_{\mathsf{dGG},\mathsf{A}}(\lambda) = \Pr[\mathrm{CDH}_{\mathsf{dGG},\mathsf{A}}(\lambda)]$ and game CDH is specified in Fig. 5.

DLin. The decision linear (DLIN) assumption [18] in a group $\mathbb{G}$ states that given $(g^u, g^v, g^{us}, g^{vt})$ for random u, v, s, t, the element g^{s+t} is indistinguishable from a random group element. Formally, the DLin assumption holds for dGG if $\mathbf{Adv}^{\mathrm{dlin}}_{\mathsf{dGG},\mathsf{A}}(\cdot)$ is negligible for all PT adversaries A, where $\mathbf{Adv}^{\mathrm{dlin}}_{\mathsf{dGG},\mathsf{A}}(\lambda) = 2\Pr[\mathrm{DLin}_{\mathsf{dGG},\mathsf{A}}(\lambda)] - 1$ and game DLin is defined in Fig. 5.

We will make use of the fact that DLin is self-reducible. This means that given a tuple (U, V, S, T, X) one can produce a new tuple (U', V', S', T', X') so that if the original tuple was linear then the new tuple is so too, but with fresh u, v, s and t; and if X is random then (U', V', S', T', X') are all independently random as well. In particular, consider the following algorithm that takes input a DLin challenge $(U, V, S, T, X) \in \mathbb{G}^5$:

Algorithm $\mathsf{Rnd}(1^\lambda, (U, V, S, T, X))$

$(p, \mathbb{G}, \mathbb{G}_T, \mathbf{e}, g) \leftarrow \mathsf{dGG}(1^\lambda)$; $z, a, b, c, d \leftarrow_{\$} \mathbb{Z}_p$
$U' \leftarrow U^c$; $V' \leftarrow V^d$; $S' \leftarrow S^{cz} U^{ca}$; $T' \leftarrow T^{dz} V^{db}$; $X' \leftarrow X^z g^a g^b$
Return (U', V', S', T', X')

Let s, t, ξ be such that $S = U^s, T = V^t, X = g^\xi$. Define $s' := sz + a$ and $t' := tz + b$ and note that they are both uniformly random. We have $S' = (U')^{s'}$, $T' = (V')^{t'}$ and $X' = g^{\xi z + a + b} = g^{(\xi - s - t)z + sz + tz + a + b} = g^{(\xi - s - t)z + s' + t'}$. Thus, if the original challenge was a linear tuple (i.e., $\xi = s + t$) then the new tuple is also linear with new randomness uc, vd, s', t', whereas otherwise (i.e., $\xi - s - t \neq 0$) U', V', S', T' and X' are independently random.

THE SCHEME. Our S-ZK scheme is based on a bilinear-group generator dGG, for which we define *linear commitments* to messages $M \in \mathbb{G}$ as follows:

$\mathsf{Ln.C}(M; (\boldsymbol{u}, \boldsymbol{t}))$
$\boldsymbol{C} \leftarrow (g^{u_0}, g^{u_1}, g^{u_0 t_0}, g^{u_1 t_1}, g^{t_0 + t_1} \cdot M)$
Return $\boldsymbol{C}$

$\mathsf{Ln.D}(\boldsymbol{u}, (C_2, C_3, C_4))$
$M \leftarrow C_4 \cdot C_2^{-1/u_0} \cdot C_3^{-1/u_1}$
Return M

Commitments are hiding under DLin. Since (C_2, C_3, C_4) is a linear encryption under public key (C_0, C_1), the logarithms of the latter let one recover the message via $\mathsf{Ln.D}$.

We also use a statistically sound NI system with trivial CRS (also called "non-interactive zap" by GOS [42]) $\mathsf{Z} = (\mathsf{Z.P}, \mathsf{Z.V})$ for the following relation:

$\mathsf{R_Z}((x, S_0, S_1, h, \boldsymbol{C}_0, \boldsymbol{C}_1), ((w, (s, \boldsymbol{u}_0, \boldsymbol{u}_1, \boldsymbol{t}_0, \boldsymbol{t}_1)))$
If $\mathsf{R}(x, w) = \mathsf{true}$ then return true
If $(g^s = S_0 \text{ or } g^s = S_1)$ and $\boldsymbol{C}_0 = \mathsf{Ln.C}(h^s; (\boldsymbol{u}_0, \boldsymbol{t}_0))$ and $\boldsymbol{C}_1 = \mathsf{Ln.C}(h^s; (\boldsymbol{u}_1, \boldsymbol{t}_1))$
then return true
Return false (6)

The NI proof system Z can for example be instantiated by the construction from [42], which does not require a CRS, is perfectly sound and WI under the DLin assumption. Our NIZK system $\Pi[\mathsf{R}, \mathsf{dGG}]$ is given in Fig. 6.

Theorem 3. *Let* R *be an* **NP** *relation and let* dGG *be a bilinear-group generator. Then* $\Pi[\mathsf{R}, \mathsf{dGG}]$*, defined in Fig. 6, satisfies (1) soundness under DH-KEA and CDH; and (2) subversion zero knowledge under DH-KEA and DLin.*

Below we give some intuition. A proof can be found in the full version [6].

$\underline{\Pi.\mathsf{Pg}(1^\lambda)}$
$(p, \mathbb{G}, \mathbb{G}_T, \mathbf{e}, g) \leftarrow \mathsf{dGG}(1^\lambda)$; $t, s_0, s_1 \leftarrow_\$ \mathbb{Z}_p$; $h \leftarrow g^t$
$S_0 \leftarrow g^{s_0}$; $S_1 \leftarrow g^{s_1}$; $S_2 \leftarrow g^{s_0 s_1}$; $crs \leftarrow (S_0, S_1, S_2, h)$; Return crs

$\underline{\Pi.\mathsf{V}(1^\lambda, (S_0, S_1, S_2, h), x, \pi)}$
$(p, \mathbb{G}, \mathbb{G}_T, \mathbf{e}, g) \leftarrow \mathsf{dGG}(1^\lambda)$; $(\boldsymbol{C}_0, \boldsymbol{C}_1, \boldsymbol{D}_0, \boldsymbol{D}_1, \zeta) \leftarrow \pi$
If $\mathbf{e}(S_0, S_1) \neq \mathbf{e}(g, S_2)$ then return false
For $i, j = 0, 1$ do
  If $\mathbf{e}(C_{0,i}, C_{1,j}) \neq \mathbf{e}(g, D_{i,j})$ then return false
Return $\mathsf{Z.V}((x, S_0, S_1, h, \boldsymbol{C}_0, \boldsymbol{C}_1), \zeta)$

$\underline{\Pi.\mathsf{P}(1^\lambda, (S_0, S_1, S_2, h), x, w)}$
If $\mathsf{R}(x, w) =$ false then return $\perp$
$(p, \mathbb{G}, \mathbb{G}_T, \mathbf{e}, g) \leftarrow \mathsf{dGG}(1^\lambda)$
If $\mathbf{e}(S_0, S_1) \neq \mathbf{e}(g, S_2)$ then return $\perp$
$C_{0,0}, \ldots, C_{0,4}, C_{1,2}, C_{1,3}, C_{1,4} \leftarrow_\$ \mathbb{G}$; $u_0, u_1 \leftarrow_\$ \mathbb{Z}_p$; $C_{1,0} \leftarrow g^{u_0}$; $C_{1,1} \leftarrow g^{u_1}$
For $i, j = 0, 1$ do $D_{i,j} \leftarrow C_{0,i}^{u_j}$
$\zeta \leftarrow_\$ \mathsf{Z.P}((x, S_0, S_1, h, \boldsymbol{C}_0, \boldsymbol{C}_1), (w, \perp))$; $\pi \leftarrow (\boldsymbol{C}_0, \boldsymbol{C}_1, \boldsymbol{D}_0, \boldsymbol{D}_1, \zeta)$
Return π

Fig. 6. NIZK scheme $\Pi[\mathsf{R}, \mathsf{dGG}]$ satisfying SND and S-ZK

Soundness. Assume an adversary A outputs a proof $\pi = (\boldsymbol{C}_0, \boldsymbol{C}_1, \boldsymbol{D}_0, \boldsymbol{D}_1, \zeta)$ for a false statement. Since there does not exist a witness w, by statistical soundness of the proof ζ, R_Z must return 1 in the second line in Eq. (6), meaning $\boldsymbol{C}_0$ and $\boldsymbol{C}_1$ are commitments to either $h^{\log S_0}$ or $h^{\log S_1}$; intuitively, the adversary has thus broken the CDH assumption either for challenge (S_0, h) or (S_1, h).

To make this formal, we construct an algorithm B that on input (g^s, h) outputs h^s with probability close to $\frac{1}{2}$. We first construct four machines $\mathsf{M}_{i,j}$, $0 \leq i, j \leq 1$ that are given given (S, h), set $S_b \leftarrow S$ for a random b, complete this to a CRS, on which they run A; when A returns π, $\mathsf{M}_{i,j}$ outputs $(C_{0,i}, C_{1,j}, D_{i,j})$. By DH-KEA there exist four extractors $\mathsf{E}_{i,j}$ which on input (S, h) and $\mathsf{M}_{i,j}$'s coins (which include A's coins) return either $u_{0,i} = \log C_{0,i}$ or $u_{1,j} = \log C_{1,j}$.

Using $\mathsf{M}_{0,0}, \mathsf{M}_{0,1}, \mathsf{M}_{1,0}, \mathsf{M}_{1,1}$, we define B: given a CDH challenge (S, h), it picks coins $\bar{r}$ and uses $\bar{r}$ to pick $b \leftarrow_\$ \{0, 1\}$, $s' \leftarrow_\$ \mathbb{Z}_p$ and coins r for A; it sets $S_b \leftarrow S$, $S_{1-b} \leftarrow g^{s'}$ and $S_2 \leftarrow S^{s'}$ and runs A on input (S_0, S_1, S_2, h) and coins r to get π containing $(\boldsymbol{C}_0, \boldsymbol{C}_1, \boldsymbol{D}_0, \boldsymbol{D}_1)$; it then runs all $\mathsf{E}_{i,j}$ on input $(S, h, \bar{r})$, which each returns either $u_{0,i} = \log C_{0,i}$ or $u_{1,j} = \log C_{1,j}$. This implies that for some i, B obtains both $u_{i,0}$ and $u_{i,1}$. Using this, B recovers $T \leftarrow \mathsf{Ln.D}((u_{i,0}, u_{i,1}), (C_{i,2}, C_{i,3}, C_{i,4}))$, which it outputs. By soundness of ζ, we have either $T = h^{\log S_0}$ or $T = h^{\log S_1}$. Since A has no information on where the challenge S was embedded, B solves CDH with probability $\frac{1}{2}$.

$\underline{\Pi.\mathsf{Pg}(1^\lambda)}$
$\sigma \leftarrow\!\!{\scriptstyle\$}\ \{0,1\}^{2\lambda}$; $m_1 \leftarrow\!\!{\scriptstyle\$}\ \mathsf{Z.V}(1^\lambda)$; Return $crs \leftarrow (\sigma, m_1)$

$\underline{\Pi.\mathsf{P}(1^\lambda, (\sigma, m_1), x, w)}$
$m_2 \leftarrow\!\!{\scriptstyle\$}\ \mathsf{Z.P}(1^\lambda, (\sigma, x), (\perp, w), m_1)$; $\pi \leftarrow m_2$; Return π

$\underline{\Pi.\mathsf{V}(1^\lambda, (\sigma, m_1), x, \pi)}$
Return $\mathsf{Z.D}(1^\lambda, (\sigma, x), m_1, \pi)$

Fig. 7. NIZK scheme $\Pi[\mathsf{R}, \mathsf{dGG}]$ satisfying SND and S-ZK

Subversion zero knowledge. By DH-KEA, for every X that outputs a CRS of the form $(g^{s_0}, g^{s_1}, g^{s_0 s_1}, h)$ there exists an algorithm E that extracts either s_0 or s_1. To show S-ZK we first construct a simulator S. Its first part $\mathsf{S.crs}$ picks r, runs $crs \leftarrow \mathsf{X}(1^\lambda, r)$ and sets $s \leftarrow\!\!{\scriptstyle\$}\ \mathsf{E}(1^\lambda, r)$ if crs is correctly formed and $s \leftarrow \perp$ otherwise, and outputs crs, r and the trapdoor $std \leftarrow s$. It is immediate that crs_1 output by X on coins r_1 is indistinguishable from crs_0, r_0 output by $\mathsf{S.crs}$.

We next construct a proof simulator $\mathsf{S.pf}$ for statements x under $crs = (S_0, S_1, S_2, h)$ using trapdoor s. Like $\Pi.\mathsf{P}$ it returns $\perp$ if crs is malformed. Else, it chooses $\boldsymbol{u}_0, \boldsymbol{t}_0, \boldsymbol{u}_1, \boldsymbol{t}_1$ and defines $\boldsymbol{C}_0$ and $\boldsymbol{C}_1$ as commitments to h^s and computes the corresponding elements $D_{i,j} \leftarrow g^{u_{0,i} u_{1,j}}$. Since either $g^s = S_0$ or $g^s = S_1$, $\mathsf{S.pf}$ has thus a witness for the statement $(x, S_0, S_1, h, \boldsymbol{C}_0, \boldsymbol{C}_1) \in \mathsf{R}_Z$, which it uses to compute a proof ζ. The simulated proof is $\pi \leftarrow (\boldsymbol{C}_0, \boldsymbol{C}_1, \boldsymbol{D}_0, \boldsymbol{D}_1, \zeta)$, which we now argue is indistinguishable from a real proof output by $\Pi.\mathsf{P}$ under DLin by a series of game hops.

We first note that when constructing ζ, instead of witness $(s, \boldsymbol{u}_0, \boldsymbol{u}_1, \boldsymbol{t}_0, \boldsymbol{t}_1)$ we could use w; this is indistinguishable under WI, which for the GOS system follows from DLin. In the next game hop, we replace $\boldsymbol{C}_0$ by a random quintuple and construct the $D_{i,j}$'s as in $\Pi.\mathsf{P}$; this is indistinguishable under DLin. In the final game hop we replace $\boldsymbol{C}_1$ by a random quintuple. This is also reduced to DLin using the fact that we can compute the $D_{i,j}$'s using the logarithms of $\boldsymbol{C}_0$. The result is a proof π that is distributed like one output by $\Pi.\mathsf{P}$.

6.2 Subversion SND and Subversion WI

In this section we prove result **P2**: there exists an NI system that is simultaneously SND, WI, S-SND and S-WI. We call Π an NI system with *trivial* CRS if $crs = \varepsilon$ and $\Pi.\mathsf{P}$ and $\Pi.\mathsf{V}$ ignore input crs. In Lemma 4 we observe that if such a Π is SND and WI then it is also S-SND and S-WI. (Intuitively, if the CRS is ignored then there's no harm in subverting it.) In Theorem 5 we then notice that an NI system with trivial CRS exists [42] which is SND and WI under the DLin assumption in bilinear groups (defined on p. 19). As in this instantiation the group is chosen by the prover (rather than fixed as for **P1**), it needs to be *verifiable* [42] (that is, one can efficiently check that it is a bilinear group).

Lemma 4. *Let* R *be an* **NP** *relation. Let* Π *be an NI system with trivial CRS for* R*. If* Π *is SND and WI then it is also S-SND and S-WI.*

Proof. Let A be an S-SND adversary. Define B against SND: on input $(1^\lambda, \varepsilon)$, run $(crs, x, \pi) \leftarrow\!\!\$\ \mathsf{A}(1^\lambda)$ and return (x, π). Since $\Pi.\mathsf{V}(1^\lambda, \varepsilon, x, \pi) = \Pi.\mathsf{V}(1^\lambda, crs, x, \pi)$, we have $\Pr[\mathrm{SND}_{\Pi,\mathsf{R},\mathsf{B}}(\lambda)] = \Pr[\text{S-SND}_{\Pi,\mathsf{R},\mathsf{A}}(\lambda)]$. Thus, if Π is SND, it is S-SND.

Let A be a WI adversary. Define B against S-WI: on input $(1^\lambda, \varepsilon)$, run $(crs, st) \leftarrow\!\!\$\ \mathsf{A}(1^\lambda)$; $b' \leftarrow\!\!\$\ \mathsf{A}^{\mathrm{PROVE}}(1^\lambda, crs, st)$ and return b'; forward A's queries to own oracle (this simulates A's oracle since $\Pi.\mathsf{P}(1^\lambda, \varepsilon, x, w_b) = \Pi.\mathsf{P}(1^\lambda, crs, x, w_b)$). We have $\Pr[\mathrm{WI}_{\Pi,\mathsf{R},\mathsf{B}}(\lambda)] = \Pr[\text{S-WI}_{\Pi,\mathsf{R},\mathsf{A}}(\lambda)]$. Thus, if Π is WI, it is S-WI. □

Theorem 5. *Let* R *be an* **NP** *relation. If the decision-linear assumption holds for a verifiable bilinear group then there exists an NI system* Π *for* R *that is S-SND and S-WI.*

Proof. Let Π be the NI system presented in [42]. Π is an NI system with trivial CRS satisfying SND and WI under the DLin assumption. By Lemma 4 it follows that Π is also S-SND and S-WI. □

6.3 Soundness, ZK and Subversion WI

We prove result **P3** by presenting an NI system that is SND, ZK, and S-WI.

<u>ZAPS.</u> A zap [30] for a relation R is a 2-move protocol (cf. Sect. 4.4), where the first move is *public-coin* and is generated *independently of the statement* to be proved. Zaps retain soundness and witness-indistinguishability even if the statements are chosen adaptively after the first move m_1 is fixed. Consequently, the same m_1 can be reused for many proofs. We denote zaps by

$$m_1 \leftarrow\!\!\$\ \mathsf{Z.V}(1^\lambda)\ ;\ m_2 \leftarrow\!\!\$\ \mathsf{Z.P}(1^\lambda, x, w, m_1)\ ;\ b \leftarrow \mathsf{Z.D}(x, m_1, m_2)\ .$$

Dwork and Naor [30] show that zaps can be constructed from any NIZK in the shared random string model. Concretely, zaps can be based on any family of doubly-enhanced trapdoor permutations, when the underlying NIZK is instantiated with the system of FLS [32].

<u>THE SCHEME.</u> The CRS of our scheme consists of a random bit string σ of length 2λ and the first move m_1 of a zap. A proof consists of the second move of the zap for statement (x, σ), proving that either $x \in L$ or s is the pre-image of σ under a PRG G. The formal description of Π follows.

Let $\mathsf{G}\colon \{0,1\}^\lambda \to \{0,1\}^{2\lambda}$ be a pseudorandom generator and let Z be a zap for the following relation R_Z:

<u>$\mathsf{R}_Z((\sigma, x), (s, w))$</u>
If $\sigma = \mathsf{G}(s)$ then return true
Return $\mathsf{R}(x, w)$

Then NI system $\Pi[\mathsf{G}, \mathsf{Z}]$ is given in Fig. 7.

Theorem 6. *Let* R *be an* **NP** *relation. Let* G *be a length-doubling function and* Z *be a zap for relation* R_Z*. If* G *is pseudorandom and* Z *is sound and witness-indistinguishable then* $\Pi[\mathsf{G},\mathsf{Z}]$ *is SND, ZK and S-WI.*

Proof. Soundness of Π follows from the soundness of the zap and the fact that the probability that a randomly sampled string σ is in the range of the PRG G is negligible. ZK follows as in [32]: The ZK simulator picks $s \leftarrow_{\$} \{0,1\}^\lambda$, sets the CRS to be $\sigma \leftarrow \mathsf{G}(s)$ and $m_1 \leftarrow_{\$} \mathsf{Z.V}(1^\lambda)$. When the simulator is challenged to prove a theorem x, it has a witness for $(\sigma, x) \in \mathsf{R}_Z$ and can therefore compute $\pi \leftarrow_{\$} \mathsf{Z.P}(1^\lambda, (\sigma, x), (s, \bot), m_1)$. Indistinguishability of the simulated CRS and proofs follows from the pseudorandomness of G and zap-WI (defined below).

To show S-WI, we prove that from an adversary A winning game $\text{S-WI}_{\Pi,\mathsf{R},\mathsf{X},\mathsf{A}}$ we can construct an adversary B winning the WI game of the underlying zap for relation R_Z. We denote this game by $\text{Z-WI}_{\mathsf{Z},\mathsf{R}_Z,\mathsf{B}}$ and define it in Fig. 8. Note that it reflects the stronger notion of WI where the verifier can obtain several proofs, for theorems of her choice, computed using the same first move m_1.

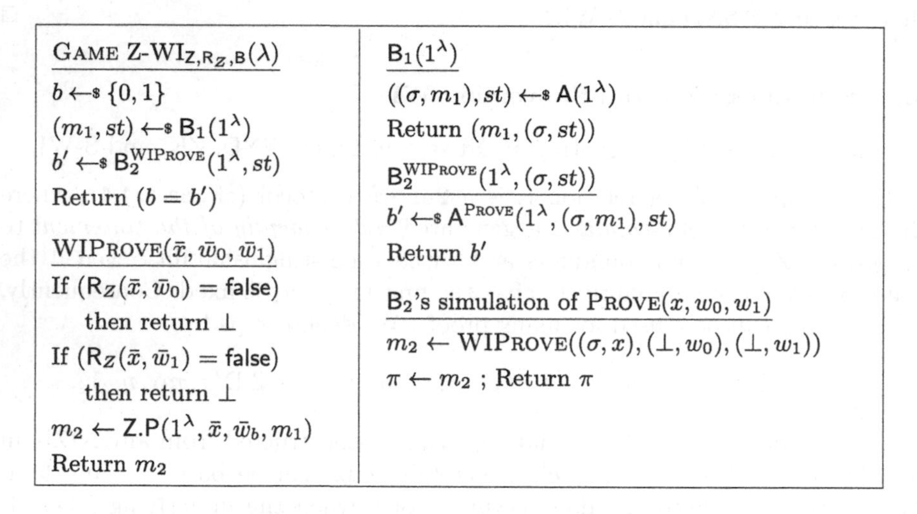

Fig. 8. Game defining WI for zaps (left) and adversary in proof of S-WI of Π

In its first stage B runs A to obtain a CRS consisting of σ and the first message m_1 and returns m_1. B then simulates oracle $\text{Prove}(x, w_0, w_1)$ for A by accessing its own oracle WIProve. Figure 8 specifies adversary B. Plugging its description into game $\text{Z-WI}_{\mathsf{Z},\mathsf{R}_Z,\mathsf{B}}$, we obtain

Game Z-WI$_{\mathsf{Z},\mathsf{R}_Z,\mathsf{B}}(\lambda)$
$b \leftarrow_{\$} \{0,1\}$
$((\sigma, m_1), st) \leftarrow_{\$} \mathsf{A}(1^\lambda)$
$b' \leftarrow_{\$} \mathsf{A}^{\text{Prove}}(1^\lambda, (\sigma, m_1), st)$
Return $(b = b')$

Prove(x, w_0, w_1)
If $\mathsf{R}_Z((\sigma, x), (\bot, w_0)) = \mathsf{false}$ then return $\bot$
If $\mathsf{R}_Z((\sigma, x), (\bot, w_1)) = \mathsf{false}$ then return $\bot$
$m_2 \leftarrow \mathsf{Z.P}(1^\lambda, (\sigma, x), (\bot, w_b), m_1)$
Return m_2

As this is precisely the description of game $\text{S-WI}_{\Pi,R,A}$, we have

$$\Pr[\text{Z-WI}_{Z,R_Z,B}(\lambda)] = \Pr[\text{S-WI}_{\Pi,R,A}(\lambda)] \ . \tag{7}$$

Since Z is zap-WI, $2\Pr[\text{Z-WI}_{Z,R_Z,B}(\cdot)] - 1$ is negligible and thus by Eq. (7) $\mathbf{Adv}^{\text{s-wi}}_{\Pi,R,A}(\cdot)$ is negligible, which proves the theorem. □

Acknowledgments. Bellare was supported in part by NSF grants CNS-1228890 and CNS-1526801, ERC Project ERCC FP7/615074 and a gift from Microsoft corporation. Fuchsbauer was supported in part by the European Research Council under the European Communitys Seventh Framework Programme (FP7/2007-2013 Grant Agreement no. 339563 CryptoCloud). This work was done in part while Bellare and Scafuro were visiting the Simons Institute for the Theory of Computing, supported by the Simons Foundation and by the DIMACS/Simons Collaboration in Cryptography through NSF grant CNS-1523467. We thank Yuval Ishai for helpful discussions and information.

References

1. Abe, M., Fuchsbauer, G., Groth, J., Haralambiev, K., Ohkubo, M.: Structure-preserving signatures and commitments to group elements. In: Rabin, T. (ed.) CRYPTO 2010. LNCS, vol. 6223, pp. 209–236. Springer, Heidelberg (2010). doi:10.1007/978-3-642-14623-7_12
2. Abe, M., Groth, J., Ohkubo, M., Tibouchi, M.: Structure-preserving signatures from type II pairings. In: Garay, J.A., Gennaro, R. (eds.) CRYPTO 2014. LNCS, vol. 8616, pp. 390–407. Springer, Heidelberg (2014). doi:10.1007/978-3-662-44371-2_22
3. Ateniese, G., Magri, B., Venturi, D.: Subversion-resilient signature schemes. In: Ray, I., Li, N., Kruegel, C. (eds.) ACM CCS 15, pp. 364–375. ACM Press, October 2015
4. Barak, B., Lindell, Y., Vadhan, S.P.: Lower bounds for non-black-box zero knowledge. In: 44th FOCS, pp. 384–393. IEEE Computer Society Press, October 2003
5. Barak, B., Pass, R.: On the possibility of one-message weak zero-knowledge. In: Naor, M. (ed.) TCC 2004. LNCS, vol. 2951, pp. 121–132. Springer, Heidelberg (2004). doi:10.1007/978-3-540-24638-1_7
6. Bellare, M., Fuchsbauer, G., Scafuro, A.: NIZKs with an untrusted CRS: Security in the face of parameter subversion. Cryptology ePrint Archive, Report 2016/372 (2016). http://eprint.iacr.org/2016/372
7. Bellare, M., Goldwasser, S.: New paradigms for digital signatures and message authentication based on non-interative zero knowledge proofs. In: Brassard, G. (ed.) CRYPTO 1990. LNCS, vol. 435, pp. 194–211. Springer, Heidelberg (1990)
8. Bellare, M., Palacio, A.: The knowledge-of-exponent assumptions and 3-round zero-knowledge protocols. In: Franklin, M. (ed.) CRYPTO 2004. LNCS, vol. 3152, pp. 273–289. Springer, Heidelberg (2004). doi:10.1007/978-3-540-28628-8_17
9. Bellare, M., Paterson, K.G., Rogaway, P.: Security of symmetric encryption against mass surveillance. In: Garay, J.A., Gennaro, R. (eds.) CRYPTO 2014. LNCS, vol. 8616, pp. 1–19. Springer, Heidelberg (2014). doi:10.1007/978-3-662-44371-2_1
10. Bellare, M., Rogaway, P.: The security of triple encryption and a framework for code-based game-playing proofs. In: Vaudenay, S. (ed.) EUROCRYPT 2006. LNCS, vol. 4004, pp. 409–426. Springer, Heidelberg (2006). doi:10.1007/11761679_25

11. Ben-Sasson, E., Chiesa, A., Green, M., Tromer, E., Virza, M.: Secure sampling of public parameters for succinct zero knowledge proofs. In: 2015 IEEE Symposium on Security and Privacy (SP), pp. 287–304. IEEE (2015)
12. Ben-Sasson, E., Chiesa, A., Tromer, E., Virza, M.: Scalable zero knowledge via cycles of elliptic curves. In: Garay, J.A., Gennaro, R. (eds.) CRYPTO 2014. LNCS, vol. 8617, pp. 276–294. Springer, Heidelberg (2014). doi:10.1007/978-3-662-44381-1_16
13. Ben-Sasson, E., Chiesa, A., Tromer, E., Virza, M.: Succinct non-interactive zero knowledge for a Von Neumann architecture. In: 23rd USENIX Security Symposium (USENIX Security 14), pp. 781–796 (2014)
14. Bernstein, D.J., Lange, T., Niederhagen, R., Dual, E.C.: A standardized back door. Cryptology ePrint Archive, Report 2015/767 (2015). http://eprint.iacr.org/2015/767
15. Bitansky, N., Canetti, R., Paneth, O., Rosen, A.: On the existence of extractable one-way functions. In: Shmoys, D.B. (ed.) 46th ACM STOC, pp. 505–514. ACM Press, May/June 2014
16. Blum, M., De Santis, A., Micali, S., Persiano, G.: Noninteractive zero-knowledge. SIAM J. Comput. **20**(6), 1084–1118 (1991)
17. Blum, M., Feldman, P., Micali, S.: Non-interactive zero-knowledge and its applications (extended abstract). In: 20th ACM STOC, pp. 103–112. ACM Press, May 1988
18. Boneh, D., Boyen, X., Shacham, H.: Short group signatures. In: Franklin, M. (ed.) CRYPTO 2004. LNCS, vol. 3152, pp. 41–55. Springer, Heidelberg (2004). doi:10.1007/978-3-540-28628-8_3
19. Boneh, D., Franklin, M.: Identity-based encryption from the weil pairing. In: Kilian, J. (ed.) CRYPTO 2001. LNCS, vol. 2139, pp. 213–229. Springer, Heidelberg (2001). doi:10.1007/3-540-44647-8_13
20. Brier, E., Coron, J.-S., Icart, T., Madore, D., Randriam, H., Tibouchi, M.: Efficient indifferentiable hashing into ordinary elliptic curves. In: Rabin, T. (ed.) CRYPTO 2010. LNCS, vol. 6223, pp. 237–254. Springer, Heidelberg (2010). doi:10.1007/978-3-642-14623-7_13
21. Canetti, R., Goldreich, O., Goldwasser, S., Micali, S.: Resettable zero-knowledge (extended abstract). In: 32nd ACM STOC, pp. 235–244. ACM Press, May 2000
22. Canetti, R., Pass, R., Shelat, A.: Cryptography from sunspots: how to use an imperfect reference string. In: 48th FOCS, pp. 249–259. IEEE Computer Society Press, October 2007
23. Chase, M., Lysyanskaya, A.: On signatures of knowledge. In: Dwork, C. (ed.) CRYPTO 2006. LNCS, vol. 4117, pp. 78–96. Springer, Heidelberg (2006). doi:10.1007/11818175_5
24. Checkoway, S., Fredrikson, M., Niederhagen, R., Everspaugh, A., Green, M., Lange, T., Ristenpart, T., Bernstein, D.J., Maskiewicz, J., Shacham, H.: On the practical exploitability of Dual EC in TLS implementations. In: USENIX Security (2014)
25. Damgård, I.: Towards practical public key systems secure against chosen ciphertext attacks. In: Feigenbaum, J. (ed.) CRYPTO 1991. LNCS, vol. 576, pp. 445–456. Springer, Heidelberg (1992). doi:10.1007/3-540-46766-1_36
26. Santis, A., Crescenzo, G., Ostrovsky, R., Persiano, G., Sahai, A.: Robust non-interactive zero knowledge. In: Kilian, J. (ed.) CRYPTO 2001. LNCS, vol. 2139, pp. 566–598. Springer, Heidelberg (2001). doi:10.1007/3-540-44647-8_33

27. Dodis, Y., Ganesh, C., Golovnev, A., Juels, A., Ristenpart, T.: A formal treatment of backdoored pseudorandom generators. In: Oswald, E., Fischlin, M. (eds.) EUROCRYPT 2015. LNCS, vol. 9056, pp. 101–126. Springer, Heidelberg (2015). doi:10.1007/978-3-662-46800-5_5
28. Dodis, Y., Haralambiev, K., López-Alt, A., Wichs, D.: Efficient public-key cryptography in the presence of key leakage. In: Abe, M. (ed.) ASIACRYPT 2010. LNCS, vol. 6477, pp. 613–631. Springer, Heidelberg (2010). doi:10.1007/978-3-642-17373-8_35
29. Dolev, D., Dwork, C., Naor, M.: Nonmalleable cryptography. SIAM J. Comput. **30**(2), 391–437 (2000)
30. Dwork, C., Naor, M.: Zaps and their applications. In: 41st FOCS, pp. 283–293. IEEE Computer Society Press, November 2000
31. Escala, A., Groth, J.: Fine-tuning groth-sahai proofs. In: Krawczyk, H. (ed.) PKC 2014. LNCS, vol. 8383, pp. 630–649. Springer, Heidelberg (2014). doi:10.1007/978-3-642-54631-0_36
32. Feige, U., Lapidot, D., Shamir, A.: Multiple non-interactive zero knowledge proofs based on a single random string (extended abstract). In: 31st FOCS, pp. 308–317. IEEE Computer Society Press, October 1990
33. Garg, S., Goyal, V., Jain, A., Sahai, A.: Bringing people of different beliefs together to do UC. In: Ishai, Y. (ed.) TCC 2011. LNCS, vol. 6597, pp. 311–328. Springer, Heidelberg (2011). doi:10.1007/978-3-642-19571-6_19
34. Goldreich, O.: A uniform-complexity treatment of encryption and zero-knowledge. J. Cryptology **6**(1), 21–53 (1993)
35. Goldreich, O., Micali, S., Wigderson, A.: Proofs that yield nothing but their validity or all languages in NP have zero-knowledge proof systems. J. ACM **38**(3), 691–729 (1991)
36. Goldreich, O., Oren, Y.: Definitions and properties of zero-knowledge proof systems. J. Cryptology **7**(1), 1–32 (1994)
37. Goldwasser, S., Micali, S., Rackoff, C.: The knowledge complexity of interactive proof systems. SIAM J. Comput. **18**(1), 186–208 (1989)
38. Groth, J.: Simulation-sound NIZK proofs for a practical language and constant size group signatures. In: Lai, X., Chen, K. (eds.) ASIACRYPT 2006. LNCS, vol. 4284, pp. 444–459. Springer, Heidelberg (2006). doi:10.1007/11935230_29
39. Groth, J.: Short pairing-based non-interactive zero-knowledge arguments. In: Abe, M. (ed.) ASIACRYPT 2010. LNCS, vol. 6477, pp. 321–340. Springer, Heidelberg (2010). doi:10.1007/978-3-642-17373-8_19
40. Groth, J.: Efficient fully structure-preserving signatures for large messages. In: Iwata, T., Cheon, J.H. (eds.) ASIACRYPT 2015. LNCS, vol. 9452, pp. 239–259. Springer, Heidelberg (2015). doi:10.1007/978-3-662-48797-6_11
41. Groth, J., Ostrovsky, R.: Cryptography in the multi-string model. In: Menezes, A. (ed.) CRYPTO 2007. LNCS, vol. 4622, pp. 323–341. Springer, Heidelberg (2007). doi:10.1007/978-3-540-74143-5_18
42. Groth, J., Ostrovsky, R., Sahai, A.: Non-interactive zaps and new techniques for NIZK. In: Dwork, C. (ed.) CRYPTO 2006. LNCS, vol. 4117, pp. 97–111. Springer, Heidelberg (2006). doi:10.1007/11818175_6
43. Groth, J., Ostrovsky, R., Sahai, A.: Perfect non-interactive zero knowledge for NP. In: Vaudenay, S. (ed.) EUROCRYPT 2006. LNCS, vol. 4004, pp. 339–358. Springer, Heidelberg (2006). doi:10.1007/11761679_21
44. Groth, J., Sahai, A.: Efficient non-interactive proof systems for bilinear groups. In: Smart, N. (ed.) EUROCRYPT 2008. LNCS, vol. 4965, pp. 415–432. Springer, Heidelberg (2008). doi:10.1007/978-3-540-78967-3_24

45. Hada, S., Tanaka, T.: On the existence of 3-round zero-knowledge protocols. In: Krawczyk, H. (ed.) CRYPTO 1998. LNCS, vol. 1462, pp. 408–423. Springer, Heidelberg (1998). doi:10.1007/BFb0055744
46. Katz, J., Kiayias, A., Zhou, H.-S., Zikas, V.: Distributing the setup in universally composable multi-party computation. In: Halldórsson, M.M., Dolev, S. (eds.) 33rd ACM PODC, pp. 20–29. ACM, July 2014
47. Kiayias, A., Zacharias, T., Zhang, B.: DEMOS-2: scalable E2E verifiable elections without random oracles. In: Ray, I., Li, N., Kruegel, C. (eds.) ACM CCS 15, pp. 352–363. ACM Press, October 2015
48. Micali, S.: CS proofs (extended abstracts). In: 35th FOCS, pp. 436–453. IEEE Computer Society Press, November 1994
49. Naor, M., Yung, M.: Public-key cryptosystems provably secure against chosen ciphertext attacks. In: 22nd ACM STOC, pp. 427–437. ACM Press, May 1990
50. Pass, R.: On deniability in the common reference string and random oracle model. In: Boneh, D. (ed.) CRYPTO 2003. LNCS, vol. 2729, pp. 316–337. Springer, Heidelberg (2003). doi:10.1007/978-3-540-45146-4_19
51. Patarin, J., Goubin, L.: Asymmetric cryptography with S-Boxes Is it easier than expected to design efficient asymmetric cryptosystems? In: Han, Y., Okamoto, T., Qing, S. (eds.) ICICS 1997. LNCS, vol. 1334, pp. 369–380. Springer, Heidelberg (1997). doi:10.1007/BFb0028492
52. Paterson, K.G.: Imprimitive permutation groups and trapdoors in iterated block ciphers. In: Knudsen, L. (ed.) FSE 1999. LNCS, vol. 1636, pp. 201–214. Springer, Heidelberg (1999). doi:10.1007/3-540-48519-8_15
53. Rijmen, V., Preneel, B.: A family of trapdoor ciphers. In: Biham, E. (ed.) FSE 1997. LNCS, vol. 1267, pp. 139–148. Springer, Heidelberg (1997). doi:10.1007/BFb0052342
54. Russell, A., Tang, Q., Yung, M., Zhou, H.-S.: Cliptography: clipping the power of kleptographic attacks. Cryptology ePrint Archive, Report 2015/695 (2015). http://eprint.iacr.org/2015/695
55. Shallue, A., Woestijne, C.E.: Construction of rational points on elliptic curves over finite fields. In: Hess, F., Pauli, S., Pohst, M. (eds.) ANTS 2006. LNCS, vol. 4076, pp. 510–524. Springer, Heidelberg (2006). doi:10.1007/11792086_36
56. Wee, H.: Lower bounds for non-interactive zero-knowledge. In: Vadhan, S.P. (ed.) TCC 2007. LNCS, vol. 4392, pp. 103–117. Springer, Heidelberg (2007). doi:10.1007/978-3-540-70936-7_6
57. Young, A., Yung, M.: The dark side of "Black-Box" cryptography or: should we trust capstone? In: Koblitz, N. (ed.) CRYPTO 1996. LNCS, vol. 1109, pp. 89–103. Springer, Heidelberg (1996). doi:10.1007/3-540-68697-5_8
58. Young, A., Yung, M.: Kleptography: using cryptography against cryptography. In: Fumy, W. (ed.) EUROCRYPT 1997. LNCS, vol. 1233, pp. 62–74. Springer, Heidelberg (1997). doi:10.1007/3-540-69053-0_6

Cryptographic Protocol

Universal Composition with Responsive Environments

Jan Camenisch[1], Robert R. Enderlein[1,2], Stephan Krenn[3], Ralf Küsters[4](✉), and Daniel Rausch[4]

[1] IBM Research – Zurich, Rüschlikon, Switzerland
jca@zurich.ibm.com
[2] Department of Computer Science, ETH Zürich, Zürich, Switzerland
asiacrypt2016@e7n.ch
[3] AIT Austrian Institute of Technology GmbH, Vienna, Austria
stephan.krenn@ait.ac.at
[4] University of Trier, Trier, Germany
{kuesters,rauschd}@uni-trier.de

Abstract. In universal composability frameworks, adversaries (or environments) and protocols/ideal functionalities often have to exchange meta-information on the network interface, such as algorithms, keys, signatures, ciphertexts, signaling information, and corruption-related messages. For these purely modeling-related messages, which do not reflect actual network communication, it would often be very reasonable and natural for adversaries/environments to provide the requested information immediately or give control back to the protocol/functionality immediately after having received some information. However, in none of the existing models for universal composability is this guaranteed. We call this the *non-responsiveness problem*. As we will discuss in the paper, while formally non-responsiveness does not invalidate any of the universal composability models, it has many disadvantages, such as unnecessarily complex specifications and less expressivity. Also, this problem has often been ignored in the literature, leading to ill-defined and flawed specifications. Protocol designers really should not have to care about this problem at all, but currently they have to: giving the adversary/environment the option to not respond immediately to modeling-related requests does not translate to any real attack scenario.

This paper solves the non-responsiveness problem and its negative consequences completely, by avoiding this artificial modeling problem altogether. We propose the new concepts of responsive environments and adversaries. Such environments and adversaries must provide a valid response to modeling-related requests before any other protocol/functionality is activated. Hence, protocol designers do no longer have to worry about artifacts resulting from such requests not being answered promptly. Our concepts apply to all existing models for universal composability, as exemplified for the UC, GNUC, and IITM models,

This project was in part funded by the European Commission through grant agreements n°s 321310 (PERCY) and 644962 (PRISMACLOUD), and by the *Deutsche Forschungsgemeinschaft* (DFG) through Grant KU 1434/9-1.

J.H. Cheon and T. Takagi (Eds.): ASIACRYPT 2016, Part II, LNCS 10032, pp. 807–840, 2016.
DOI: 10.1007/978-3-662-53890-6_27

with full definitions and proofs (simulation relations, transitivity, equivalence of various simulation notions, and composition theorems) provided for the IITM model.

Keywords: Universal composability · Protocol design · Cryptographic security proofs · Responsive environments

1 Introduction

One of the most demanding tasks when designing a cryptographic protocol is to define its intended security guarantees, and to then prove that it indeed satisfies them. In the best case, these proofs should guarantee the security of the protocol in arbitrary contexts, i.e., also when composed with other, potentially insecure, protocols. This would allow one to split complex protocols into smaller components, which can then be separately analyzed one by one and once and for all, thus allowing a modular security analysis. Over the past two decades, many models to achieve this goal have been proposed [3,8–10,19,22,24,27–29], with Canetti's UC model being one of the first and most prominent ones.

All these models have in common that the designer first needs to specify an *ideal functionality* $\mathcal{F}$ defining the intended security and functional properties of the protocol. Informally, a *real protocol* realizes $\mathcal{F}$ if no efficient distinguisher (the *environment*) can decide whether it is interacting with the ideal functionality and a *simulator*, or with the real world protocol and an *adversary*.

Urgent requests/messages. In the specifications of such real protocols and ideal functionalities, it is often required for the adversary (and the environment) to provide some meta-information via the network interface to the protocol or the functionality, such as cryptographic algorithms, cryptographic values of signatures, ciphertexts, and keys, or corruption-related messages. Conversely, protocols/functionalities often have to provide the adversary with meta-information, for example, signaling information (e.g., the existence of machines) or again corruption-related messages. Such meta-information does not correspond to any real network messages, but is merely used for modeling purposes. Typically, giving the adversary/environment the option to not respond immediately to such modeling-specific messages does not translate to any real attack scenario. Hence, often it is natural for protocol designers to expect that the adversary/environment (answers and) returns control back to the protocol/functionality immediately when the adversary is requested to provide meta-information or when the adversary receives meta-information from the protocol/functionality. In the following, we call such messages from protocols/ideal functionalities on the network interface *urgent messages* or *urgent requests*.

Urgent requests occur in many functionalities and protocols from the literature, see, e.g., [1,4,5,8,11–13,15,16,21,25,26,31]. This is not surprising as the exchange of meta-information between the adversary/environment and the protocols/functionalities is an important mechanism for protocol designs in any

UC-like model. For example, one can specify the behaviour of cryptographic values or algorithms by an ideal functionality in a natural manner without having to worry about how these values are generated or the parameters for the algorithms are set up, e.g., using a CRS. Also, protocols should be able to provide the adversary with meta-information in situations where it is not intended to give control to the adversary, such as certain information leaks (e.g., honest-but-curious corruption) or signaling messages. In general, it seems impossible to dispense with urgent requests altogether, and certainly, such requests are very convenient and widely used in the literature (see also Sect. 3).

The non-responsiveness problem. In the existing universal composability models, it currently is not guaranteed that urgent requests are answered immediately by the adversary: when receiving an urgent request on the network interface, adversaries and environments can freely activate protocols and ideal functionalities in between, on network and I/O interfaces, without answering the request. In what follows, we refer to this problem as the *problem of non-responsive adversaries/environments* or the *non-responsiveness problem.*

This problem formally does not invalidate any of the UC-style models. It, however, often makes the specification of protocols and functionalities much harder and the models less expressive (see below). Most disturbingly, as mentioned, the non-responsiveness problem is really an artificial problem: urgent requests do not correspond to any real messages, and the adversary not responding promptly to such requests does not reflect any real attack scenario. Hence, non-responsiveness forces protocol designers to take care of artificial adversarial behavior that was unintended in the first place and is merely a modeling artifact.

In particular, protocol designers currently have to deal with various delicate problems: (i) While waiting for a response to an urgent request, a protocol/ideal functionality might receive other requests, and hence, protocol designers have to take care of interleaving and dangling requests. (ii) While a protocol/ideal functionality is waiting for an answer from the adversary to an urgent request, other parties and parts of the protocol/ideal functionality can be activated in the meantime (via the network or the I/O interface), which might change their state, even their corruption status, and which in turn might lead to race conditions (see Sect. 3 for examples from the literature).

This, as further discussed in the paper, makes it difficult to deal with the non-responsiveness problem and results in unnecessarily complex and artificial specifications of protocols and ideal functionalities, which, in addition, are then hard to re-use. In some cases, one might not even be able to express certain desired properties. As explained in Sect. 3, there is no generic and generally applicable way to deal with the non-responsiveness problem, and hence, one has to resort to solutions specifically tailored to the protocols at hand.

Importantly, the non-responsiveness problem propagates to higher-level protocols as they might not get responses from their subprotocols as expected. The security proofs also become more complex because one, again, has to deal with runs having various dangling and interleaving requests as well as unexpected and unintuitive state changes, which do not translate into anything in the real world, but are just an artifact of the modeling.

Clearly, in the context of actual network messages, one has to deal with many of the above problems in the specifications of protocols and ideal functionalities too. But, in contrast to the non-responsiveness problem, dealing with the asynchronous nature of networks has a real counterpart, and these two types of interactions with the adversary should not be confused.

In the literature, urgent requests and the non-responsiveness problem occur in many protocols and functionalities. Nevertheless, protocol designers frequently ignore this problem (see, e.g., [1,4,5,13,14,18,21,25,26,30,31]), i.e., they seem to implicitly assume that urgent request *are* answered immediately, probably, at least as far as ideal functionalities are concerned, because their simulators promptly respond to these kinds of requests. As a result, protocols and ideal functionalities are underspecified and/or expose unexpected behavior, and thus, are not usable in other (hybrid) protocols, or security proofs of hybrid protocols are flawed (see Sect. 3).

Our contribution. In this paper, we propose a universal composability framework with the new concept of *responsive environments* and *adversaries*, which should be applicable to all existing UC-style models (see below). This framework completely avoids and, by this, solves the non-responsiveness problem as it guarantees that urgent requests *are* answered immediately. This really is the most obvious and most natural solution to the problem: there is no reason that protocol designers should have to take care of the non-responsiveness problem and its many negative consequences.

More specifically, the main idea behind our framework is as follows. When a protocol/ideal functionality sends what we call a *restricting* message to the adversary/environment on the network interface, then the adversary/environment is forced to be responsive, i.e., to reply with a valid response before sending any other message to the protocol. This requires careful definitions and non-trivial proofs to ensure that all properties and features that are expected in models for universal composition are lifted to the setting with responsive environments and adversaries.

By using our framework and concepts, protocols and ideal functionalities can be modeled in a very natural way: protocol designers can simply declare urgent requests to be restricting messages, which hence have to be answered immediately. This elegantly and completely solves the non-responsiveness problem. In particular, protocol designers no longer have to worry about this problem, and specifications of protocols and ideal functionalities are greatly simplified, as one can dispense with artificial solutions. In fact, as illustrated in Sect. 6, with our concepts we can easily fix existing specifications from the literature in which the non-responsiveness problem has not been dealt with properly or has simply been ignored as protocol designers often implicitly assumed responsiveness for urgent messages. In some cases, we can now even express certain functionalities in a natural and elegant way that could not be expressed before (see Sects. 3.2.2 and 6). Of course, with simplified and more natural functionalities and protocols, also security proofs become easier because the protocol designer does not have to consider irrelevant and unrealistic adversarial behavior and execution orders.

We emphasize that protocol designers must exercise discretion when using restricting messages: *such messages should be employed for meta-information used for modeling purposes only, and not for real network traffic, where immediate answers cannot be guaranteed in reality.*

We illustrate that our framework and concepts apply to existing models for universal composability. This is exemplified for three prominent models: UC [8], GNUC [19], and IITM [22,24]. In the full version of this paper [6], we provide full proofs for the IITM model. In particular, we define all common notions of simulatability, including UC, dummy UC, strong simulatability, and blackbox simulatability with respect to responsive environments and adversaries, and show that all of these notions are equivalent. This result can be seen as a sanity check of our concepts, as it has been a challenge in previous models (see, e.g., the discussions in [20,24]). We also prove in detail that all composition theorems from the original IITM model carry over to the IITM model with our concepts.

Related work. The concept of responsive adversary and environments is new and has not been considered before.

In [2], composition for restricted classes of environments is studied, motivated by impossibility results in UC frameworks and to weaken the requirements on realizations of ideal functionalities. In this setting, environments are restricted in that they may send only certain sequences of messages to the I/O interfaces of protocols and functionalities. These restrictions cannot express that urgent requests are answered immediately and also do not restrict adversaries in any way. Hence, this approach cannot be used to solve the non-responsiveness problem, which anyway was not the intention of the work in [2].

In the first version of his seminal work [8], Canetti introduced *immediate functionalities*. According to the definition (cf. page 35 of the 2001 version), an immediate functionality uses an *immediate adversary* to guarantee that messages are delivered immediately between the functionality and its dummy. To be more precise, an immediate functionality may force an immediate adversary to deliver a message to a specific dummy party within a single activation. This construct was necessary as in the initial version of Canetti's model, the ideal functionality could not directly pass an output to its dummy but had to rely on the adversary instead. In current versions of UC, the problem addressed by immediate adversaries has vanished completely because ideal functionalities can directly communicate with their dummies. Clearly, immediate adversaries do not address, let alone solve, the non/responsiveness problem, which is about immediate answers for certain request to the adversary on the network interface rather than between a functionality and its dummies.

Outline. In Sect. 2, we briefly recall basic terminology and notation. We observe in Sect. 3 that the non-responsiveness problem affects many protocols from the literature, with many papers ignoring the problem altogether, resulting in underspecified and ill-specified protocols and functionalities, that are thus hard to re-use. Furthermore, that section shows that properly taking this problem into consideration is quite difficult and does not have a simple and generally applicable solution. Our universal composability framework with responsive environments and adversaries is then presented in Sect. 4. This section is kept quite model independent to highlight the main new concepts and the fact that these concepts are not restricted to specific models. Section 5 then illustrates how our concepts can be implemented in the UC, GNUC, and IITM models. Section 6 shows how the problems with non-responsive environment and adversaries discussed in Sect. 3 can be avoided elegantly with our restricting messages and responsive environments/adversaries. We conclude in Sect. 7. Further details can be found in the full version of this paper [6]. In particular, as mentioned before, we provide full details for the IITM model with responsive environments and adversaries in the full version.

2 Preliminaries

In this section, we briefly recap the basic concepts of universal composability and fix some notation and terminology. The description is independent of the model being used and can easily be mapped to any concrete model, such as UC, GNUC, or IITM. For now, we ignore runtime issues as they are handled differently in the models and only implicitly assume that all systems run in polynomial time in the security parameter and the length of the external input (if any). Runtime issues are discussed in detail in Sect. 5.

Universal composability models use *machines* to model programs. Each machine may have I/O and network tapes/interfaces. These machines are then used as blueprints to create *instances* which execute the machine code while having their own local state. Machines can be combined into a *system* $\mathcal{S}$. In a run of $\mathcal{S}$, multiple instances of machines may be generated and different instances can communicate by sending messages via I/O or network interfaces. Given two systems $\mathcal{R}$ and $\mathcal{Q}$, we define the system $\{\mathcal{R}, \mathcal{Q}\}$ which contains all machines from $\mathcal{R}$ and $\mathcal{Q}$.

There are three different kinds of entities, which can themselves be considered as systems and which can be combined to one system: *protocols*, *adversaries*, and *environments*. One distinguishes *real* and *ideal* protocols, where ideal protocols are often called *ideal functionalities*. An ideal protocol can be thought of as the specification of a task, whereas a real protocol models an actual protocol that is supposed to realize the ideal protocol (cf. Definition 2.1). These protocols have an I/O interface to communicate with the environment and a network interface to communicate with the adversary. An *adversary* controls the network communication of protocols and can also interact with the environment. *Environments* connect to the I/O interface of protocols and may communicate with the adversary, cf. Fig. 1 for an illustration of how environments, adversaries, and protocols are connected.

Environments try to distinguish whether they run with a real protocol and an adversary or an ideal protocol and an adversary (then often called a simulator or ideal adversary). An environment may get some external input to start a run. It is expected to end the run by outputting a single bit.

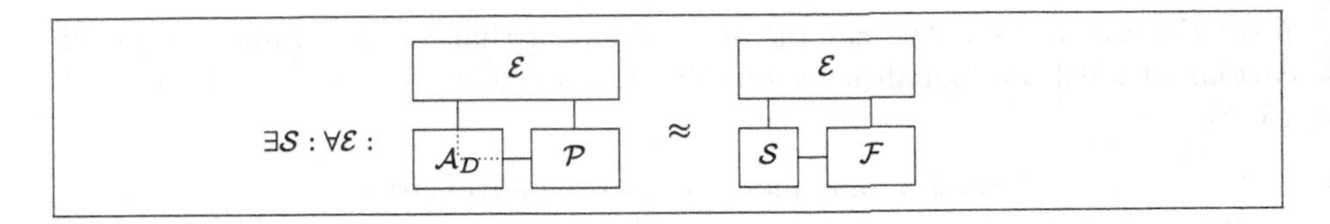

Fig. 1. A real protocol $\mathcal{P}$ realizing an ideal functionality $\mathcal{F}$; $\mathcal{A}_D$ denotes the dummy adversary which just forwards messages to and from the environment $\mathcal{E}$.

Given an environment $\mathcal{E}$, an adversary $\mathcal{A}$, and a protocol $\mathcal{P}$, we denote both the combined system and the output distribution of the environment by $\{\mathcal{E}, \mathcal{A}, \mathcal{P}\}$. We use the binary operator $\equiv$ to denote two output distributions that are negligibly close in the security parameter η (and the external input, if any).

Now, in models for universal composability, the realization of an ideal protocol by a real protocol is defined as follows.

Definition 2.1 (Realization Relation). *Let $\mathcal{P}$ and $\mathcal{F}$ be protocols, the* real *and* ideal protocol, *respectively. Then, $\mathcal{P}$* realizes *$\mathcal{F}$ ($\mathcal{P} \leq \mathcal{F}$) if for every adversary $\mathcal{A}$, there exists an ideal adversary $\mathcal{S}$ such that $\{\mathcal{E}, \mathcal{A}, \mathcal{P}\} \equiv \{\mathcal{E}, \mathcal{S}, \mathcal{F}\}$ for every environment $\mathcal{E}$.*

We note that, in the definition above and in all reasonable models, instead of quantifying over all adversaries, it suffices to consider only the dummy adversary $\mathcal{A}_D$ which forwards all network messages between $\mathcal{P}$ and $\mathcal{E}$. Intuitively, this is true because $\mathcal{A}$ can be subsumed by $\mathcal{E}$. Hence, we have that $\mathcal{P} \leq \mathcal{F}$ iff there exists an ideal adversary $\mathcal{S}$ such that $\{\mathcal{E}, \mathcal{A}_D, \mathcal{P}\} \equiv \{\mathcal{E}, \mathcal{S}, \mathcal{F}\}$ for every environment $\mathcal{E}$.

The main result in any universal composability model is a composition theorem. Informally, once a protocol $\mathcal{P}$ has been proven to realize an ideal protocol $\mathcal{F}$, one can securely replace (all instances of) $\mathcal{F}$ by $\mathcal{P}$ in arbitrary higher-level systems without affecting the security of the higher-level system.

3 The Non-responsiveness Problem and Its Consequences: Examples from the Literature

We have already introduced and discussed the non-responsiveness problem and sketched its consequences in Sect. 1. In this section, we illustrate this problem and its consequences by examples from the literature. We also point to concrete cases in which this problem has been ignored (i.e., immediate answers to urgent requests were assumed implicitly) and where this has led to ill-defined protocols and functionalities as well as invalid proofs and statements.

3.1 Underspecified and Ill-Defined Protocols and Functionalities

In many papers, the non-responsiveness problem is ignored in the specifications of both ideal functionalities and (higher-level) protocols (see, e.g., [1,4,5,13,14, 18,21,25,26,30,31]). We discuss a number of typical cases in the following.

Ideal Functionalities. An example of a statement that one often finds in specifications of ideal functionalities is one like the following (see, e.g., [1,4,13,18,21, 25,26]):

$$\text{“send <some message> to the adversary;} \\ \text{upon receiving <some answer> from the adversary do <something>”,} \tag{1}$$

where the message sent to the adversary, in our terminology, is an urgent request, i.e., as explained in Sect. 1, some meta-information provided to the adversary or a request for some meta-information the adversary is supposed to provide. For example, ideal functionalities might ask for cryptographic material (cryptographic algorithms and keys, ciphertexts, or signatures), ask whether the adversary wants to corrupt a party, or simply signal their existence.

In specifications containing formulations as in (1) it is not specified what happens if the adversary does not respond immediately, but, for example, other requests on the I/O interface are received; intermediate state changes in other parts might also occur, which might require different actions. There does not seem to exist a generic solution to handle such problems (see Sects. 3.2.1 and 3.2.3). It rather seems to be necessary to find solutions tailored to the specific protocol and ideal functionality at hand, making it even more important to precisely specify the behavior in case the adversary does not respond immediately to urgent requests.

Many research papers on universal composability focus on proposing new functionalities and realizations thereof, including proofs that a realization actually realizes a functionality; to a lesser extent the functionalities are then used in higher-level protocols. In realization proofs, one might not notice that formulations as that in (1) are problematic because for such proofs an ideal functionality $\mathcal{F}$ runs alongside a (friendly) simulator and this simulator might provide answers to urgent requests immediately (see also Fig. 1). However, if used in a hybrid protocol (see Fig. 2), an ideal functionality $\mathcal{F}$ runs alongside a (hostile) adversary/environment. In this case, it is important that specifications capture the case that urgent requests are not answered immediately. If this is ignored or not handled correctly, it yields (i) underspecified protocols, with the problem that they cannot be re-used in hybrid protocols, which in turn defeats the purpose of universal composability frameworks, and (ii) possibly false statements.

To illustrate these points by a concrete example, we consider the “signature of knowledge” functionality $\mathcal{F}_{\text{sok}}(L)$ proposed by Chase and Lysyanskaya [14]. This functionality contains a Setup instruction (reproduced in Fig. 3), where the adversary provides the keys and algorithms, and signing and verification instructions that then use those keys and algorithms without requiring interaction with the adversary - a very common mechanism in the literature (see, e.g., [1,5,12,15,16,30]). This functionality is explicitly intended to be used in a hybrid setting to realize delegatable credentials.

If the adversary does not respond to the first ($\texttt{Setup}$, sid) request, all subsequent requests (e.g., a Setup request by a different party) will cause the functionality to use or output the undefined Sign and Verify algorithms, which is a

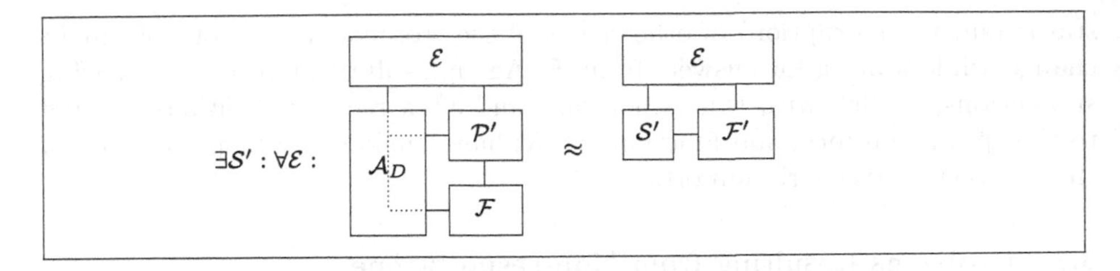

Fig. 2. An $\mathcal{F}$-hybrid protocol $\mathcal{P}'$ realizing some ideal functionality $\mathcal{F}'$.

Upon receiving a value (`Setup`, *sid*) from any party P, verify that $sid = (M_L, sid')$ for some sid'. If not, then ignore the request. Else, if this is the first time that (`Setup`, *sid*) was received, **hand** (`Setup`, *sid*) **to the adversary; upon receiving** (`Algorithms`, *sid*, Verify, Sign, Simsign, Extract) **from the adversary**, where Sign, Simsign, Extract are descriptions of PPT TMs, and Verify is a description of a deterministic polytime TM, store these algorithms. Output the stored (Algorithms, *sid*, Sign, Verify) to P.

Fig. 3. The Setup instruction of $\mathcal{F}_{\text{sok}}(L)$ from [14].

problem: Chase and Lysyanskaya provide a protocol in the $\mathcal{F}_{\text{sok}}(L)$-hybrid model that can be used for realizing delegated credentials, i.e., an ideal functionality for signatures on a signature. They then prove that this protocol realizes the functionality. They, however, missed the fact that $\mathcal{F}_{\text{sok}}(L)$ may interact with a non-responsive adversary in the hybrid world. Such an adversary can force $\mathcal{F}_{\text{sok}}(L)$ to use undefined algorithms, and their simulator does not handle that situation in the ideal world. It is thus easy to distinguish the real from the ideal world. Hence, their proof is flawed, and in fact it seems that the statement cannot be proven.

(Higher-Level) Protocols. As already mentioned in the introduction, real protocols often also send urgent requests to the adversary (e.g., signaling their existence or asking whether the adversary wants to corrupt). In addition, one often finds protocol specifications containing formulations of the following form to make requests to subprotocols (see, e.g., [5,14,30,31]):

"send <some message> to $\mathcal{F}$;
upon receiving <some answer> from $\mathcal{F}$ do <something>." (2)

Intuitively, $\mathcal{F}$ might indeed model some non-interactive functionality, such as signature functionalities. However, because of the use of urgent requests in such functionalities, even when completely uncorrupted, $\mathcal{F}$ might not return answers right away. So, again, formulations as the one in (2) are greatly underspecified. What happens if other requests are received at the network or I/O interface? Should they be ignored? Or may they be queued somehow? Also, the state and status (such as corruption) of other parts of the protocol or subprotocols might change while waiting for answers from $\mathcal{F}$. Again, as illustrated in the following subsections, dealing with this is not easy and often requires solutions tailored to the specific protocol and functionality at hand, making a full specification of the behavior particularly important.

3.2 Problems Resulting from Non-responsiveness

We now discuss challenges resulting from the non-responsiveness problem (when actually taken into account, rather than ignored) and illustrate them by examples from the literature.

3.2.1 Unintended State Changes and Race Conditions

As mentioned before, a general problem one has to take care of when dealing with the non-responsiveness problem is that while a protocol is waiting for answers to urgent requests, the adversary might cause changes in the state of other parts of the protocol/functionality and of subprotocols, which in turn influences the behavior of the protocol. Keeping track of the actual current overall state might be tricky, and race conditions are possible.

The following is a simple example which illustrates that the problem can occur already locally within a single functionality. It can often become even trickier in higher-level protocols which use urgent requests themselves and where possibly several subroutines use urgent requests.

We consider the dual-authentication certification functionality $\mathcal{F}_{\text{D-Cert}}$ [31]. In this functionality, the adversary needs to be contacted when verifying a signature (a common mechanism to verify cryptographic values that is also used in many other functionalities [7,13,21]). Such requests are urgent as this is supposed to model local computations. However, the adversary may not answer immediately.

More specifically, Fig. 4 shows the Verify instruction of $\mathcal{F}_{\text{D-Cert}}$. Assume now that S' has received a message m and a signature σ for this message, which supposedly was created by an honest party P with SID *sid*. Now, if the signature actually was not created by P, the verification should fail as P is not

Upon receiving a value $(\texttt{Verify}, sid, m, \sigma)$ **from some party** S', **hand** $(\texttt{Verify}, sid, m, \sigma)$ **to the adversary. Upon receiving** $(\texttt{Verified}, sid, m, \phi)$ **from the adversary**, do:

1. If $(m, \sigma, 1)$ is recorded then set $f = 1$.
2. Else, if the signer is not corrupted, and no entry $(m, \sigma', 1)$ for any σ' is recorded, then set $f = 0$ and record the entry $(m, \sigma, 0)$.
3. Else, if there is an entry (m, σ, f') recorded, then set $f = f'$.
4. Else, set $f = \phi$, and record the entry (m, σ, ϕ).

Output $(\texttt{Verified}, sid, m, f)$ to S'.

Fig. 4. The Verify instruction of $\mathcal{F}_{\text{D-Cert}}$ from [31].

corrupted. However, as the adversary gets activated during this allegedly local task, it could corrupt the signer during the verification process, return $\phi = 1$, and therefore let the functionality *accept* σ. This behavior is certainly unexpected and counterintuitive.

Such a functionality also considerably complicates the security analysis of any higher-level application that uses $\mathcal{F}_{\text{D-Cert}}$ as a subroutine, as one has to also consider the possibility of a party getting corrupted during the invocation of a subroutine modeling a local task, which, even worse, in that case returns unexpected answers.

3.2.2 Problems Expressing the Desired Properties

The following is an example where the authors struggled with the non-responsiveness problem in that it finally led to a functionality that, as the authors acknowledge, is not completely satisfying. This functionality, denoted $\mathcal{F}_{\text{NIKE}}$, is supposed to model a non-interactive key exchange and was proposed by Freire et al. [17]. Figure 5 shows a central part of this functionality, namely, the actual key exchange. A party P_i may ask for the key that is shared between the parties P_i and P_j. If this session of P_i and P_j is considered corrupted, namely, because one of the parties is corrupted, and no key has been recorded for this session yet, the adversary is allowed to freely choose the key that is shared between the two parties. The functionality uses an urgent request to model this, i.e., it directly sends a message to the adversary if she is allowed to choose a key.

Upon input $(\texttt{init}, P_i, P_j)$ from P_i, if $P_j \notin \Lambda_{\text{ref}}$, return $(P_i, P_j, \perp)$ to P_i. If $P_j \in \Lambda_{\text{reg}}$, we consider two cases:

- Corrupted session mode: if there exists an entry $(\{P_i, P_j\}, K_{i,j})$ in Λ_{keys}, set $key = K_{i,j}$. Else, **send $(\texttt{init}, P_i, P_j)$ to the adversary. After receiving $(\{P_i, P_j\}, K_{i,j})$ from the adversary**, set $key = K_{i,j}$ and add $(\{P_i, P_j\}, K_{i,j})$ to Λ_{keys}.
- Honest session mode: if there exists an entry $(\{P_i, P_j\}, K_{i,j})$ in Λ_{keys}, set $key = K_{i,j}$, else choose $key \leftarrow \{0,1\}^k$ and add $(\{P_i, P_j\}, K_{i,j})$ to Λ_{keys}.

Return (P_i, P_j, key) to P_i.

Fig. 5. The init instruction of $\mathcal{F}_{\text{NIKE}}$ from [17].

As the authors state, they would have liked to also model "immediateness" of the functionality, i.e., a higher-level protocol that requests a key should be able to expect an answer without the adversary being able to interfere with the protocol in the meantime. This indeed would be expected and natural because $\mathcal{F}_{\text{NIKE}}$ models a *non-interactive* key exchange. However, this is in conflict with allowing the adversary to choose the key of a corrupted session. The authors suggest that one option to also model immediateness might be to let the adversary choose an algorithm upon setup, which is then used to compute the keys for corrupt parties. Nevertheless, they chose the non-immediate modeling because the other solution would lead to "technical complications"; it would also limit the adaptiveness of the adversary and might add other problems. Indeed, code upload constructs (see also Sect. 3.2.3), in general, do not solve the non-responsiveness problem.

As a consequence of the formulation chosen in $\mathcal{F}_{\text{NIKE}}$, the adversary can now, e.g., block requests, which again also needs to be considered in any higher-level protocol using $\mathcal{F}_{\text{NIKE}}$ as a subroutine, even though in the real world the honest party would always obtain some key because of the non-interactivity of the primitive.

More generally, ideal functionalities that use urgent messages (which in current models are not answered immediately) might have weaker security guarantees than their realizations, in particular when the functionality is supposed to model a non-interactive task, because the realization might not give control to the adversary. So for hybrid protocols one might not be able to prove certain properties when using an ideal functionality, whereas the same protocol using the realization of the ideal functionality instead might enjoy such properties.

This is in contrast to one of the goals of universal composability models, namely, reducing the complexity of security analyses by enabling the use of conceptually simpler ideal functionalities as subroutines.

3.2.3 The Reentrance Problem

As already mentioned in Sect. 1, a protocol designer has to specify the behavior of protocols and ideal functionalities upon receiving another input (on the I/O interface) while they are waiting for a response to an urgent request on the network. In other words, protocols and ideal functionalities have to be *reentrant*. Note that, as pointed out, a protocol has to be reentrant not only when it uses urgent requests itself, but also if a subroutine uses such messages.

As explained next, dealing with the reentrance problem can be difficult. Approaches to solve this problem complicate the specifications of protocols and ideal functionalities, and none of them is sufficiently general to be applicable in every case.

We now illustrate this by an example ideal functionality. However, similar issues occur in specifications for real and hybrid protocols. Let $\mathcal{F}$ be any ideal functionality which sends an urgent request to the adversary upon its first creation, say, to retrieve some modeling-related information. This is a common situation. For example, ideal functionalities often require some cryptographic material such as keys and algorithms from the adversary before they can continue their execution (e.g., functionalities for digital signatures or public-key encryption). We also assume that $\mathcal{F}$ is meant to be realized by a real protocol consisting of two independent parties/roles A and B (e.g., signer and verifier). We further assume that both of these parties also send an urgent request to the adversary upon their first activation and expect an answer before they can continue with their computation. Again, this is a common situation as, for example, real protocols often ask for their corruption status or notify the adversary of their creation.[1] While the above is only one illustrative example, it already describes a large and common class of real and ideal protocols often encountered in the literature.

We now present several approaches to make $\mathcal{F}$ reentrant in the above sense, i.e., to deal with I/O requests while waiting for a response to an urgent request on the network. We show that the obvious approaches in general cannot be used. In particular, with most of these approaches $\mathcal{F}$ cannot even be realized by A and B in the setting outlined above. This in turn shows that solutions that are tailored to the specific functionality at hand and even the envisioned realization are required, which is very unsatisfactory, as this leads to more complex and yet less general functionalities and protocols.

Ignore Requests. After sending an urgent request to the adversary, the most straightforward approach would be to ignore all incoming messages until a response from the adversary is received.[2] This, however, is not only an unexpected behavior in many cases – for example, why should a request silently fail if the ideal functionality models a local computation? – but the ideal functionality in fact might no longer be realizable by some real protocols:

If $\mathcal{F}$, in our example functionality, would simply ignore incoming messages, an environment can distinguish $\mathcal{F}$ (with a simulator) from the realization A and B (with the dummy adversary). It first sends a message to A which, as we assume, then in turn sends an urgent request to the dummy adversary and hence to the environment. Now the environment, which does not have to respond to urgent requests immediately, sends a message to B which in turn also sends an urgent request to the adversary and hence to the environment. Consider the behavior of the ideal world in this case: After receiving the message for A, $\mathcal{F}$ will send an urgent request to the simulator. The simulator, however, cannot answer this urgent request because it has to simulate A by sending an urgent request to the environment. (This might be the case because the simulator first has to consult the environment before answering the urgent request by $\mathcal{F}$ or because $\mathcal{F}$ does not return control to the simulator after receiving an answer to the urgent request.) The environment then sends the second message (for B) to $\mathcal{F}$, which is ignored because $\mathcal{F}$ still waits for an answer to its urgent request. This behavior is different from the real world, and thus, the environment can distinguish the real world from the ideal one.

This illustrates that an ideal functionality that simply blocks *all* requests while waiting for a response to an urgent request can in general not be realized by two or more *independent* parties that also send urgent requests to the adversary. Instead one needs to adjust the blocking approach to the specific protocols at

[1] The latter is, for example, required by the definition of "subroutine respecting protocols" in the 2013 version of UC [8]. While prompt responses by the adversary are formally not required, they would be very convenient for all of the reasons discussed in Sect. 3.2.

[2] Alternatively, one could send error messages as response to intermediate requests. However, the exact same problems discussed for the approach of ignoring requests occur.

hand. For example, often it might be possible to block messages that would be processed by a single party in the real protocol, while messages for other parties are still processed. But this does not work if, for instance, $\mathcal{F}$ cannot process messages for any party before receiving a response to its urgent requests, e.g., because $\mathcal{F}$ first needs to receive cryptographic material (algorithms, keys, etc.). Thus, in this case yet another workaround is required.

Queuing of Intermediate Requests. Another potential general approach to deal with the reentrance problem is to store all incoming messages to process them later on. The simplest implementation of this approach would be the following: Upon receiving another input while still waiting for a response to an urgent request, the ideal functionality stores the input in a queue and then ends its activation. After receiving a response from the adversary, the ideal functionality processes the messages stored in the queue.

This approach is vulnerable to the same attack as the previous approaches: if the environment executes this attack in the real world, it will eventually receive an urgent request from B. This, however, cannot be simulated in the ideal world. The simulator does not get control when B is activated as the ideal functionality simply ends its activation after queuing the input for B.

Another problem with this approach is that in all current universal composability models, a machine is allowed to send only one message per activation. Hence, the ideal functionality will never be able to catch up with the inputs that have been stored. Every time it is activated by another input, it will have to process both the new input and several older inputs that are still stored in the queue. But it can only answer one of these messages at a time. This observation leads to another approach based on the queuing of unanswered requests which we discuss in the full version of this paper [6]. This approach, which does not seem to have been used in the literature so far, is, however, very complex and weakens the security of the ideal functionality to an extent that for some tasks is unacceptable: it allows the adversary to determine the order in which requests are processed by an ideal functionality.

Further Approaches. In the full version of this paper [6], we discuss several alternative approaches, namely, *default answers* and *code uploads*, which, however, can merely help reduce the use of urgent requests, but do not solve the reentrance problem, let alone the general non-responsiveness problem.

3.2.4 Unnatural Specifications of Higher-Level Protocols

Higher-level protocols have to deal with the non-responsiveness problem for two reasons. First, they might use urgent requests themselves. Second, subprotocols might use urgent requests, and hence, if requests are sent to subprotocols (even for those that intuitively should model non-interactive primitives), the adversary might get control. In both cases, higher-level protocols have to deal with the problem that while waiting for answers, the state of other parts of them and of any of their subprotocols might change and new requests (from the network or I/O interface) might have to be processed. This can lead to unnecessarily complex and often unnatural specifications, if the non-responsiveness problem is actually taken into account rather than being ignored (which in turn would result in underspecified, and hence, unusable protocols).

We illustrate this by a joint state realization, which represents one form of a higher-level protocol: Consider a digital signature functionality $\mathcal{F}_{\text{sig}}$. Let us assume that $\mathcal{F}_{\text{sig}}$ is specified in such a way that at the beginning it asks the adversary for signing and verification algorithms and keys before it answers other requests; as already mentioned, this is a very common design pattern. Because the adversary might not answer requests for the cryptographic material right away (non-responsiveness), $\mathcal{F}_{\text{sig}}$ might receive further requests while waiting for the answer. Let us assume that $\mathcal{F}_{\text{sig}}$ ignores/drops all such requests (this seems to be the option mainly used in the literature, see, e.g., [4,23]).[3]

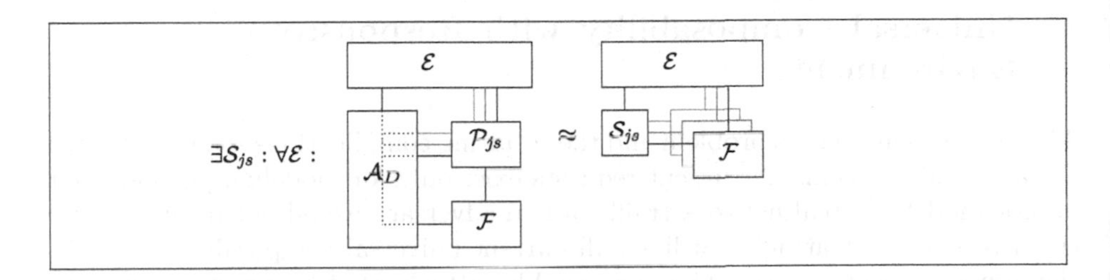

Fig. 6. Joint state realization.

In a joint state realization of $\mathcal{F}_{\text{sig}}$, one instance of $\mathcal{F}_{\text{sig}}$ (per party) is used to realize all sessions of $\mathcal{F}_{\text{sig}}$ (for one party) in the ideal world (see also Fig. 6). The idea behind the joint state realization is that if in session *sid* a message m is to be signed/verified, then one would instead sign/verify the message (sid, m). In this way, messages of different sessions cannot interfere. In the realization proof, a simulator would provide an instance $\mathcal{F}_{\text{sig}}$ in session *sid* with a signing and verification algorithm that exactly mimics the behavior of $\mathcal{F}_{\text{sig}}$ in session *sid* (i.e., signing/verifying prefix messages with *sid*). Unfortunately, because of the non-responsiveness problem, the joint state realization is more complex than that, even if, for the purpose of the discussion, we ignore the handling of corruption. To see this, assume that the environment sends a signing request for some message m in session *sid*. The joint state realization would now invoke $\mathcal{F}_{\text{sig}}$ with (sid, m). Before $\mathcal{F}_{\text{sig}}$ can answer, $\mathcal{F}_{\text{sig}}$ asks the adversary for the cryptographic material. Hence, the adversary/environment gets activated again, and the environment can send a new, say, signing request for message m' in session sid'. As $\mathcal{F}_{\text{sig}}$ is still waiting for the adversary to provide the cryptographic material, this later request

[3] As explained in Sect. 3.2.3, this approach, just as all other approaches discussed in Sect. 3.2.3, does not work in general, e.g., when the signer and verifier are independent and send urgent requests to the adversary upon first activation. It really depends on the details of $\mathcal{F}_{\text{sig}}$ and its realization.

will be ignored by $\mathcal{F}_{\text{sig}}$ and hence will never be answered. To mimic this behavior in the ideal world, the simulator should not provide the cryptographic material to the instance of $\mathcal{F}_{\text{sig}}$ in session sid' (otherwise, $\mathcal{F}_{\text{sig}}$ in session sid' would return a signature for m'). But then, this instance of $\mathcal{F}_{\text{sig}}$ is blocked completely. Hence, in turn, the joint state realization also has to block all further requests for session sid'. That is, it has to store all SIDs for which it received requests while waiting for $\mathcal{F}_{\text{sig}}$ to respond, and all future requests for all such SIDs have to be dropped.

This is very unnatural and certainly would not correspond to anything one would do in actual implementations: there one would simply prefix messages with SIDs, but one would never block requests for certain SIDs. This is just an artifact of the non-responsiveness problem, i.e., the fact that, in current models, urgent requests (in this case the request for cryptographic material by $\mathcal{F}_{\text{sig}}$) might not be answered immediately.

4 Universal Composability with Responsive Environments

The non-responsiveness problem and the resulting complications shown in Sect. 3 are artificial problems. As urgent requests exist only for modeling purposes but do not model any real network traffic, a real adversary would not be able to use them to carry out attacks. Still, in all current universal composability models, the non-responsiveness of adversaries enables attacks that do not correspond to anything in reality. If we could force the adversary to answer urgent requests immediately, which, as already mentioned before, would be the natural and expected behavior, there would not be any need for coming up with workarounds that try to solve the non-responsiveness problem in the specifications of protocols and functionalities and one would not have to consider such artificial attacks in security proofs.

In this section, we present our framework which extends universal composability models by allowing protocol designers to specify messages that have to be answered immediately by (responsive) environments and adversaries. We first give a brief overview of our approach, then define in more detail responsive environments, responsive adversaries and the realization relation in this setting, and finally prove that the composition theorems still hold for our extension. As our framework and concepts can be used by any universal composability model and to highlight the new concepts, we keep this section independent of specific models. In particular, we mostly ignore runtime considerations. In Sect. 5, we then discuss in detail how our framework can be adapted to specific models.

4.1 Overview

To avoid the non-responsiveness problem altogether, we introduce the concept of *responsive environments* and *responsive adversaries*. In a nutshell, when these environments and adversaries receive specific messages from the network (we call these messages *restricting*) then they have to respond to these messages immediately, i.e., without activating other parts of the protocol before sending an answer. Furthermore, depending on the restricting message, they may send an answer from a specific set of messages only. Restricting messages and the possible answers can be specified by the protocol designer; they are not hardwired into the framework. More specifically, restricting messages and the possible responses are specified by a binary relation $R \subseteq \{0,1\}^+ \times \{0,1\}^+$ over non-empty messages, called a *restriction.* If $(m, m') \in R$, then m is a restricting message and m' a possible answer to m. That is, if an environment/adversary receives m on its network interface, then it has to answer immediately with some m' such that $(m, m') \in R$.

This allows a protocol designer to specify all urgent requests as restricting messages by defining a restriction R appropriately; such requests are then answered not only immediately but also with an expected answer. Therefore the adversary can no longer interfere with the protocol run in an unintended way by activating other parts of the protocol or sending unexpected inputs before answering an urgent request.

Note that this concept is very powerful and needs to be handled with care: While, as motivated above, it does not weaken security results if one models urgent requests as restricting messages, one must not use such messages when modeling real network traffic, as real network messages are not guaranteed to be answered immediately in reality.

4.2 Defining Responsiveness

To define responsive environments and responsive adversaries, we first precisely define the notion of a restriction. As mentioned, restrictions are used to define both restricting messages, which have to be answered immediately by the environment/adversary, and possible answers to each restricting message.

Definition 4.1. *A* restriction *R is a set of pairs of non-empty messages, i.e., $R \subseteq \{0,1\}^+ \times \{0,1\}^+$, such that, given a pair of messages (m, m'), it is efficiently decidable whether R allows m' as an answer to m. We define $R[0] := \{m | \exists m' : (m, m') \in R\}$. A message $m \in R[0]$ is called a* restricting message.

The idea is that if an environment/adversary receives m on the network interface, there are two cases: If m is not a restricting message, i.e., $m \notin R[0]$, then the environment/adversary is not restricted in any way. Otherwise, if $m \in R[0]$, then the first message (if any) sent back to the protocol (both on the network and I/O interface of the protocol) has to be some message m' with $(m, m') \in R$. This message has to be sent on the network interface of the same machine that issued the request m, without any other message being sent to another machine of the protocol (see also Definition 4.2).

By requiring efficient decidability we ensure that environments are able to check whether some answer is allowed by the restriction; this is necessary, e.g., for Lemma 4.4. We refer to Sect. 5 for the exact definitions of "efficiently decidable", which depend on the runtime definitions of the underlying models.

As mentioned in Sect. 4.1, only urgent requests should be defined as restricting messages via a restriction. For example, upon creation of a new instance by receiving a message m, instances of protocols are often expected to first ask the adversary whether they are corrupted before they process the message m. An adversary can be forced to answer such a request immediately by the following restriction:

$$R := \{(m, m')|m = \texttt{AmICorrupted?}, m' = (\texttt{Corruption}, b), b \in \{\mathsf{false}, \mathsf{true}\}\}.$$

We now formalize the responsiveness property of environments and adversaries.

Definition 4.2 (Responsive Environments). *An environment $\mathcal{E}$ is called* responsive *for a system of machines $\mathcal{Q}$ with respect to a restriction R if in an overwhelming set of runs of $\{\mathcal{E}, \mathcal{Q}\}$ every restricting message from $\mathcal{Q}$ on the network is answered correctly, i.e., for any restricting message $m \in R[0]$ sent by $\mathcal{Q}$ on the network, the first message m' that $\mathcal{Q}$ receives afterwards (be it on the network interface or the I/O interface of $\mathcal{Q}$), if any, is sent by $\mathcal{E}$ on the network interface of $\mathcal{Q}$ to the same machine of $\mathcal{Q}$ that sent m and m' satisfies $(m, m') \in R$. By $\mathsf{Env}_R(\mathcal{Q})$ we denote the set of responsive environments for $\mathcal{Q}$.*

In the above definition, "same machine" typically means the same *instance* of a machine. So if an instance of a machine of $\mathcal{Q}$ sent a restricting message m on the network interface to the environment, the first message m' received by any instance of $\mathcal{Q}$ (on the network or I/O interface), including all currently running instances of $\mathcal{Q}$ and an instance that might be created as a result of m', has to be sent back on the network interface to the same instance of $\mathcal{Q}$ which sent m, and m' has to satisfy $(m, m') \in R$. The exact definition of "same machine" depends on the model under consideration (see Sect. 5).

The system $\mathcal{Q}$ usually is either $\{\mathcal{A}_D, \mathcal{P}\}$, where $\mathcal{P}$ is a real protocol and $\mathcal{A}_D$ is the dummy adversary, or $\{\mathcal{S}, \mathcal{F}\}$, where $\mathcal{S}$ is an ideal adversary and $\mathcal{F}$ is an ideal protocol.

Responsive adversaries have to provide the same guarantees as responsive environments; however, they have to do so only when running in combination with a responsive environment. In other words, they can use the responsiveness property of the environment to ensure their own responsiveness property.

Definition 4.3 (Responsive Adversaries). *Let $\mathcal{Q}$ be a system and let $\mathcal{A}$ be an adversary that controls the network interface of $\mathcal{Q}$. Then, $\mathcal{A}$ is called a* responsive adversary *if, for all $\mathcal{E} \in \mathsf{Env}_R(\{\mathcal{A}, \mathcal{Q}\})$, in an overwhelming set of runs of $\{\mathcal{E}, \mathcal{A}, \mathcal{Q}\}$ every restricting message from $\mathcal{Q}$ on the network is immediately answered (in the sense of Definition 4.2). We denote the set of all such adversaries for a protocol $\mathcal{Q}$ by $\mathrm{Adv}_R(\mathcal{Q})$.*

We note that the dummy adversary $\mathcal{A}_D$ is responsive.

Also note that the definitions of both responsive environments and responsive adversaries depend on a specific system, i.e., an environment which is responsive for a system $\mathcal{Q}$ is not necessarily responsive for a system $\mathcal{Q}'$. If we required environments to be responsive for *every* system, we would also have to require this from simulators (ideal adversaries). This in turn would needlessly complicate security proofs. Let us elaborate on this. Many theorems and lemmas in UC-like models, such as transitivity of the realization relation (cf. Lemma 4.7) and the composition theorems (cf. Theorems 4.8 and 4.9), are proven by simulating (some instances of) adversaries/simulators and protocols within the environment. In such proofs, we need to make sure that if an environment is responsive, then it is still responsive if we move a simulator (ideal adversary) into the environment, i.e., run the simulator within the environment. Now, if we require strong responsiveness (i.e., responsiveness for all systems), then moving a simulator into a responsive environment might result in an environment that is no longer responsive (in the strong sense), unless we require from the simulator that it is responsive in the strong sense as well. However, imposing such a strong requirement on simulators seems unreasonable. Simulators are constructed in security proofs to work with exactly one protocol. So a protocol designer should only have to care about runs with this specific protocol, not with arbitrary systems that might try to actively violate the responsiveness property of the simulator. This is why we require responsiveness for specific systems only and this indeed is sufficient.

In fact, for security proofs, there are two important properties that should be fulfilled and for which we now show that they are. The first says that if an environment is responsive for one system, then it is also responsive for any system indistinguishable from that system. The second property says that a responsive environment can internally simulate a responsive adversary/simulator without losing its responsiveness property. In other words, we can move a responsive adversary/simulator into a responsive environment without losing the responsiveness property of the environment. As mentioned before, this is necessary, for example, for the transitivity of the realization relation and the composition theorems.

Lemma 4.4. *Let R be a restriction. Let $\mathcal{Q}$ and $\mathcal{Q}'$ be two systems of machines such that $\{\mathcal{E}, \mathcal{Q}\} \equiv \{\mathcal{E}, \mathcal{Q}'\}$ for all $\mathcal{E} \in \mathsf{Env}_R(\mathcal{Q})$. Then, $\mathsf{Env}_R(\mathcal{Q}) = \mathsf{Env}_R(\mathcal{Q}')$.*

For the proof of this lemma, we refer the reader to the full version of this paper [6].

Lemma 4.5. *Let R be a restriction. Let $\mathcal{Q}$ be a system, $\mathcal{A} \in \mathrm{Adv}_R(\mathcal{Q})$ be a responsive adversary, and $\mathcal{E} \in \mathsf{Env}_R(\{\mathcal{A}, \mathcal{Q}\})$ be a responsive environment. Let $\mathcal{E}'$ denote the environment that internally simulates the system $\{\mathcal{E}, \mathcal{A}\}$. Then, $\mathcal{E}' \in \mathsf{Env}_R(\mathcal{Q})$.*

For the proof of this lemma, we refer the reader to the full version of this paper [6].

4.3 Realization Relation for Responsive Environments

We can now define the realization relation for responsive environments. The definition is analogous to the one for general environments and adversaries (see Definition 2.1), but restricts these entities to being responsive.

Definition 4.6 (Realizing Protocols with Responsive Environments). *Let $\mathcal{P}$ and $\mathcal{F}$ be protocols, the* real *and* ideal protocol, *respectively, and R be a restriction. Then, $\mathcal{P}$* realizes *$\mathcal{F}$* with respect to responsive environments *($\mathcal{P} \leq_R \mathcal{F}$) if for every responsive adversary $\mathcal{A} \in \mathrm{Adv}_R(\mathcal{P})$, there exists an (ideal) responsive adversary $\mathcal{S} \in \mathrm{Adv}_R(\mathcal{F})$ such that $\{\mathcal{E}, \mathcal{A}, \mathcal{P}\} \equiv \{\mathcal{E}, \mathcal{S}, \mathcal{F}\}$ for every environment $\mathcal{E} \in \mathsf{Env}_R(\{\mathcal{A}, \mathcal{P}\})$.*

Just as in the case of Definition 2.1, we have that instead of quantifying over all responsive adversaries, it suffices to consider only the dummy adversary $\mathcal{A}_D$, which forwards all network messages between $\mathcal{P}$ and $\mathcal{E}$ (we provide a formal proof in the full version of this paper [6]). As already mentioned, $\mathcal{A}_D$ is always responsive. This means that in security proofs, one has to construct only one responsive simulator $\mathcal{S}$ for $\mathcal{A}_D$.

As mentioned before Lemma 4.5, the responsiveness of $\mathcal{S}$ is necessary for the transitivity of $\leq_R$. While the responsiveness of $\mathcal{S}$ is a property a protocol designer has to ensure, this property is easy to check and guarantee: upon receiving a restricting message from the protocol, it either answers immediately and correctly or sends only restricting messages to the environment until it can provide a correct answer to the original restricting message from the protocol. In such a situation, the simulator should not send a non-restricting message to the environment because, if it does so, it cannot make sure that it gets back an answer immediately from the environment and that the environment does not invoke the protocol in between. In the full version of this paper [6], we specify and provide a formal proof of this intuition.

We also note that Definition 4.6 is a generalization of Definition 2.1: with $R := \emptyset$, we obtain Definition 2.1.

We now prove that the realization relation with responsive environments is reflexive and transitive. This is crucial for the modular and step-wise design of protocols: once we have proven $\mathcal{P} \leq_R \mathcal{P}'$ and $\mathcal{P}' \leq_R \mathcal{P}''$, we want to conclude immediately that $\mathcal{P} \leq_R \mathcal{P}''$.

Lemma 4.7. *The $\leq_R$ relation is reflexive and transitive.*

For the proof of this lemma, we refer the reader to the full version of this paper [6].

4.4 Composition Theorems

The core of every universal composability model is the composition theorems. We now present a first composition theorem that handles concurrent composition of any (fixed) number of potentially different protocols.

Theorem 4.8. *Let R be a restriction. Let $k \geq 1$, $\mathcal{Q}$ be a protocol, and $\mathcal{P}_1, \ldots, \mathcal{P}_k$, $\mathcal{F}_1, \ldots, \mathcal{F}_k$ be protocols such that for all $j \leq k$ it holds true that $\mathcal{P}_j \leq_R \mathcal{F}_j$.*

Then, $\{\mathcal{Q}, \mathcal{P}_1, \ldots, \mathcal{P}_k\} \leq_R \{\mathcal{Q}, \mathcal{F}_1, \ldots, \mathcal{F}_k\}$.

Proof. In what follows, we take the (equivalent) formulation of $\leq_R$ with the dummy adversary $\mathcal{A}_D$.

It suffices to prove the theorem for the case $k = 1$. The argument can then be iterated to obtain the theorem for $k > 1$ using transitivity of the $\leq_R$ relation. Let $\mathcal{S} \in \mathrm{Adv}_R(\mathcal{F}_1)$ be the simulator from the definition of $\mathcal{P}_1 \leq_R \mathcal{F}_1$. Define the simulator $\mathcal{S}'$ to forward messages between the environment and $\mathcal{Q}$, while internally simulating $\mathcal{S}$ for messages between the environment and $\mathcal{F}_1$. Now let $\mathcal{E} \in \mathsf{Env}_R(\{\mathcal{A}_D, \mathcal{Q}, \mathcal{P}_1\})$. For convenience, in what follows, we split $\mathcal{A}_D$ into $\mathcal{A}_D^{\mathcal{Q}}$ and $\mathcal{A}_D^{\mathcal{P}_1}$ where $\mathcal{A}_D^{\mathcal{Q}}$ forwards all communication betweeen $\mathcal{E}$ and $\mathcal{Q}$ and $\mathcal{A}_D^{\mathcal{P}_1}$ forwards all communication betweeen $\mathcal{E}$ and $\mathcal{P}_1$.

We first prove that $\{\mathcal{E}, \mathcal{A}_D, \mathcal{Q}, \mathcal{P}_1\} \equiv \{\mathcal{E}, \mathcal{S}', \mathcal{Q}, \mathcal{F}_1\}$. Suppose that this is not the case. Then we can define a new environment $\mathcal{E}'$ that distinguishes $\{\mathcal{A}_D^{\mathcal{P}_1}, \mathcal{P}_1\}$ from $\{\mathcal{S}, \mathcal{F}_1\}$. The environment $\mathcal{E}'$ internally simulates $\{\mathcal{E}, \mathcal{A}_D^{\mathcal{Q}}, \mathcal{Q}\}$, and hence, distinguishes with the same probability as $\mathcal{E}$. Now observe that $\mathcal{E}'$ is responsive for $\{\mathcal{A}_D^{\mathcal{P}_1}, \mathcal{P}_1\}$: All network messages from $\{\mathcal{A}_D^{\mathcal{P}_1}, \mathcal{P}_1\}$ in $\{\mathcal{E}, \mathcal{A}_D, \mathcal{Q}, \mathcal{P}_1\}$ are handled by $\mathcal{E}$ only, not by $\mathcal{Q}$. Moreover, as $\mathcal{E}$ is responsive for $\{\mathcal{A}_D, \mathcal{Q}, \mathcal{P}_1\}$, we have that these messages are answered correctly (in the sense of Definition 4.2), implying the responsiveness of $\mathcal{E}'$ for $\{\mathcal{A}_D^{\mathcal{P}_1}, \mathcal{P}_1\}$. This contradicts the assumption that $\mathcal{P}_1 \leq_R \mathcal{F}_1$, and hence $\{\mathcal{E}, \mathcal{A}_D, \mathcal{Q}, \mathcal{P}_1\} \equiv \{\mathcal{E}, \mathcal{S}', \mathcal{Q}, \mathcal{F}_1\}$ must be true.

We still have to show the responsiveness property of $\mathcal{S}'$, that is, $\mathcal{S}' \in \mathrm{Adv}_R(\{\mathcal{Q}, \mathcal{F}_1\})$. Let $\mathcal{E} \in \mathsf{Env}_R(\{\mathcal{S}', \mathcal{Q}, \mathcal{F}_1\})$. We have to show that all restricting network messages from $\mathcal{Q}$ and $\mathcal{F}_1$ to $\mathcal{E}$ and $\mathcal{S}'$ are answered correctly (in the sense of Definition 4.2). Suppose that there is a non-negligible set of runs of $\{\mathcal{E}, \mathcal{S}', \mathcal{Q}, \mathcal{F}_1\}$ in which a restricting network message from $\{\mathcal{Q}, \mathcal{F}_1\}$ is not answered correctly. As $\mathcal{S}'$ only forwards network messages from $\mathcal{Q}$ to the environment and the environment is responsive for $\{\mathcal{S}', \mathcal{Q}, \mathcal{F}_1\}$, we have that with overwhelming probability these messages are answered correctly. Hence, there must be a non-negligible set of runs in which network messages from $\mathcal{F}_1$ are not answered correctly. Now consider $\mathcal{E}'$ from above. Then there also is a non-negligible set of runs of $\{\mathcal{E}', \mathcal{S}, \mathcal{F}_1\}$ in which restricting messages on the network from $\mathcal{F}_1$ are answered incorrectly because, by construction of $\mathcal{E}'$, the behavior of the system $\{\mathcal{E}', \mathcal{S}, \mathcal{F}_1\}$ coincides with $\{\mathcal{E}, \mathcal{S}', \mathcal{Q}, \mathcal{F}_1\}$. We already know that $\mathcal{E}' \in \mathsf{Env}_R(\{\mathcal{A}_D^{\mathcal{P}_1}, \mathcal{P}_1\})$ from above. Also, by assumption, we have that $\{\mathcal{E}'', \mathcal{A}_D^{\mathcal{P}_1}, \mathcal{P}_1\} \equiv \{\mathcal{E}'', \mathcal{S}, \mathcal{F}_1\}$ for all $\mathcal{E}'' \in \mathsf{Env}_R(\{\mathcal{A}_D^{\mathcal{P}_1}, \mathcal{P}_1\})$. Now, by Lemma 4.4, it follows that $\mathsf{Env}_R(\{\mathcal{A}_D, \mathcal{P}_1\}) = \mathsf{Env}_R(\{\mathcal{S}, \mathcal{F}_1\})$, and hence $\mathcal{E}' \in \mathsf{Env}_R(\{\mathcal{S}, \mathcal{F}_1\})$. This contradicts the responsiveness property of $\mathcal{S}$. □

The following composition theorem guarantees the secure composition of an unbounded number of instances of the same protocol system. To state this theorem, we consider single-session (responsive) environments, i.e., environments that invoke a single session of a protocol only. In universal composability models, instances of protocol machines have IDs that consist of party IDs and session IDs. Instances with the same session ID form a session. Instances from different sessions may not directly interact with each other. A *single-session environment* may invoke machines with the same session ID only. We denote the set of single-session environments for a system $\mathcal{Q}$ by $\mathsf{Env}_{R,\mathsf{single}}(\mathcal{Q})$. We say that $\mathcal{P}$

single-session realizes $\mathcal{F}$ ($\mathcal{P} \leq_{R,\mathsf{single}} \mathcal{F}$) if there exists a simulator $\mathcal{S} \in \mathrm{Adv}_R(\mathcal{F})$ such that $\{\mathcal{E}, \mathcal{A}_D, \mathcal{P}\} \equiv \{\mathcal{E}, \mathcal{S}, \mathcal{F}\}$ for all $\mathcal{E} \in \mathsf{Env}_{R,\mathsf{single}}(\{\mathcal{A}_D, \mathcal{P}\})$. Now, the composition theorem states that if a single session of a real protocol $\mathcal{P}$ realizes a single session of an ideal protocol $\mathcal{F}$, then multiple sessions of $\mathcal{P}$ realize multiple sessions of $\mathcal{F}$.

Theorem 4.9. *Let* R *be a restriction, and let* $\mathcal{P}$ *and* $\mathcal{F}$ *be protocols. Then,* $\mathcal{P} \leq_{R,\mathsf{single}} \mathcal{F}$ *implies* $\mathcal{P} \leq_R \mathcal{F}$.

Proof. Let $\mathcal{S}$ be the simulator for $\mathcal{P} \leq_{R,\mathsf{single}} \mathcal{F}$. A new simulator $\mathcal{S}'$ for arbitrary responsive environments can be constructed just as in the original (non-responsive) composition theorem, i.e., $\mathcal{S}'$ internally keeps one copy of $\mathcal{S}$ per session and uses these copies to answer messages from/to the corresponding sessions.

The proof has two main steps: The first step shows indistinguishability of $\{\mathcal{A}_D, \mathcal{P}\}$ and $\{\mathcal{S}', \mathcal{F}\}$ for every responsive environment $\mathcal{E} \in \mathsf{Env}_R(\{\mathcal{A}_D, \mathcal{P}\})$. The second step shows the responsiveness property of the simulator.

The first part uses a hybrid argument in which one builds a series of single-session environments $\mathcal{E}_i, i \geq 1$, which internally simulate $\mathcal{E}$ such that all messages to the first $i-1$ sessions are sent to internally simulated instances of $\{\mathcal{S}, \mathcal{F}\}$, messages to the i-th session are sent to the (external) system $\{\mathcal{A}_D, \mathcal{P}\}$ or $\{\mathcal{S}, \mathcal{F}\}$, respectively, and the remaining messages are sent to internally simulated instances of $\{\mathcal{A}_D, \mathcal{P}\}$. As different sessions of a protocol do not directly interact with each other, it is easy to see that $\{\mathcal{E}_1, \mathcal{A}_D, \mathcal{P}\}$ behaves just as $\{\mathcal{E}, \mathcal{A}_D, \mathcal{P}\}$ (*), and $\{\mathcal{E}_n, \mathcal{S}, \mathcal{F}\}$ behaves just as $\{\mathcal{E}, \mathcal{S}', \mathcal{F}\}$, where $n \in \mathbb{N}$ is an upper bound of the number of sessions created by $\mathcal{E}$ (note that n is a polynomial in the security parameter and the length of the external input given to the environment, if any). Hence, the distinguishing advantage of $\mathcal{E}$ is bounded by the sum of the advantages of $\mathcal{E}_1, \ldots, \mathcal{E}_n$, i.e., it is sufficient to show that the advantages of $\mathcal{E}_1, \ldots, \mathcal{E}_n$ are bounded by the *same* negligible function[4] to show that $\mathcal{E}$ cannot distinguish $\{\mathcal{A}_D, \mathcal{P}\}$ from $\{\mathcal{S}', \mathcal{F}\}$. In what follows, to show the existence of a single negligible function, we consider environments with external input because the argument is simpler in that case. Nevertheless, using sampling of runs, the argument also works without external input, i.e., in the uniform case (see the full version of this paper [6] for details).

To show that such a bound exists, it is first necessary to prove that there is a (single) negligible function f that, for every $i \leq n$, bounds the probability of $\mathcal{E}_i$ of violating the responsiveness property in runs of $\{\mathcal{A}_D, \mathcal{P}\}$ or $\{\mathcal{S}, \mathcal{F}\}$, respectively. Let $\hat{C}_i^{\{\mathcal{A}_D,\mathcal{P}\}}$ be the event that in runs of $\{\mathcal{E}_i, \mathcal{A}_D, \mathcal{P}\}$ the environment $\mathcal{E}$, which is internally simulated by $\mathcal{E}_i$, answers a restricting message of the external system $\{\mathcal{A}_D, \mathcal{P}\}$ *or* one of the internally simulated instances of $\{\mathcal{A}_D, \mathcal{P}\}$ and $\{\mathcal{S}, \mathcal{F}\}$ incorrectly; $\hat{C}_i^{\{\mathcal{S},\mathcal{F}\}}$ is defined analogously. Because $\mathcal{E} \in \mathsf{Env}_R(\{\mathcal{A}_D, \mathcal{P}\})$

[4] It is not sufficient to show that the advantage of every environment $\mathcal{E}_i$ is bounded by a negligible function f_i, which is actually rather easy to show. The negligible functions f_i might be different and then their sum $f_1 + \cdots + f_n$ might not be negligible.

and because of (*), we have that $\hat{C}_1^{\{\mathcal{A}_D,\mathcal{P}\}}$ is negligible. It also holds true that (**) there exists a single negligible function that bounds $|\Pr\left[\hat{\mathcal{C}}_i^{\{\mathcal{A}_D,\mathcal{P}\}}\right]-\Pr\left[\hat{\mathcal{C}}_i^{\{\mathcal{S},\mathcal{F}\}}\right]|$ for all $i \geq 1$. This is because one can define a single-session responsive environment $\mathcal{E}'$ that gets i as external input and then simulates $\mathcal{E}_i$; $\mathcal{E}'$ aborts and outputs 1 as soon as a restricting message is about to be answered incorrectly, and 0 otherwise. Note that because the restriction R can be decided efficiently, $\mathcal{E}'$ can perform the task described. Also, by construction, $\mathcal{E}'$ is a single-session environment (it invokes a single external session only) and it is responsive (it stops the execution before the responsiveness requirement would be violated). As $\mathcal{E}'$ distinguishes $\{\mathcal{A}_D,\mathcal{P}\}$ and $\{\mathcal{S},\mathcal{F}\}$ only based on the events $\hat{C}_i^{\{\mathcal{A}_D,\mathcal{P}\}}$ and $\hat{C}_i^{\{\mathcal{S},\mathcal{F}\}}$, and both systems are indistinguishable for every single session responsive environment, statement (**) holds true. Finally, observe that, for all $i \geq 2$, the systems $\{\mathcal{E}_{i-1},\mathcal{S},\mathcal{F}\}$ and $\{\mathcal{E}_i,\mathcal{A}_D,\mathcal{P}\}$ behave exactly the same, and hence $\Pr\left[\hat{\mathcal{C}}_i^{\{\mathcal{A}_D,\mathcal{P}\}}\right]-\Pr\left[\hat{\mathcal{C}}_i^{\{\mathcal{S},\mathcal{F}\}}\right]$. This implies that there is a single negligible function that bounds $\Pr\left[\hat{\mathcal{C}}_i^{\{\mathcal{A}_D,\mathcal{P}\}}\right]$ for all $1 \leq i \leq n$ (here we need that n is polynomially bounded).[5] In particular, we have that the probability that $\mathcal{E}_i$ is not responsive for the system $\{\mathcal{A}_D,\mathcal{P}\}$ is bounded by a single negligible function independently of $i \leq n$.

We can now conclude the indistinguishability argument by showing that the advantages of $\mathcal{E}_i$, $1 \leq i \leq n$, in distinguishing $\{\mathcal{A}_D,\mathcal{P}\}$ from $\{\mathcal{S},\mathcal{F}\}$ are bounded by the same negligible function. For this, we construct another single-session responsive environment $\mathcal{E}''$ analogously to $\mathcal{E}'$. The system $\mathcal{E}''$ expects $1 \leq i \leq n$ as external input (and otherwise stops) and then exactly simulates $\mathcal{E}_i$. Importantly, $\mathcal{E}''$ is responsive for $\{\mathcal{A}_D,\mathcal{P}\}$ because we have shown that every $\mathcal{E}_i$ violates responsiveness with at most the same negligible probability, i.e., the same bound also holds for $\mathcal{E}''$ for every input. As $\mathcal{E}''$ is a single-session responsive environment, its distinguishing advantage for the systems $\{\mathcal{A}_D,\mathcal{P}\}$ or $\{\mathcal{S},\mathcal{F}\}$ is negligible for every possible input. Moreover, with external input i, its distinguishing advantage is the same as that for $\mathcal{E}_i$. Hence, the same negligible function that bounds the advantage of $\mathcal{E}''$ also bounds all advantages of $\mathcal{E}_i, i \leq n$. As mentioned at the beginning of the proof, this implies indistinguishability of $\{\mathcal{A}_D,\mathcal{P}\}$ and $\{\mathcal{S}',\mathcal{F}\}$ for every responsive environment $\mathcal{E} \in \mathsf{Env}_R(\{\mathcal{A}_D,\mathcal{P}\})$.

Having proved indistinguishability, it remains to show that $\mathcal{S}'$ is responsive, i.e., $\mathcal{S}' \in \mathrm{Adv}_R(\mathcal{F})$. Let $\mathcal{E} \in \mathsf{Env}_R(\{\mathcal{S}',\mathcal{F}\})$. We have to show that the probability that all restricted messages from $\mathcal{F}$ in runs of $\{\mathcal{E},\mathcal{S}',\mathcal{F}\}$ are answered correctly (in the sense of Definition 4.2) is overwhelming. For this, consider the following single-session environment $\mathcal{E}'$ that is meant to run with $\{\mathcal{S},\mathcal{F}\}$: The system $\mathcal{E}'$ first flips $r \leq n$, with n as above, and then internally simulates $\mathcal{E}$ and several sessions of $\{\mathcal{S},\mathcal{F}\}$ such that messages from $\mathcal{E}$ to the r-th session are sent to the external session, whereas all other messages are processed by the internally simulated sessions. Note that $\{\mathcal{E}',\mathcal{S},\mathcal{F}\}$ behaves just as $\{\mathcal{E},\mathcal{S}',\mathcal{F}\}$, and hence,

[5] Note that it also follows that $\Pr\left[\hat{\mathcal{C}}_i^{\{\mathcal{S},\mathcal{F}\}}\right]$ is bounded for all $1 \leq i \leq n$. However, we do not need this result in the following.

because $\mathcal{E} \in \mathsf{Env}_R(\{\mathcal{S}', \mathcal{F}\})$, by Lemma 4.4 we have that $\mathcal{E}'$ is responsive for $\{\mathcal{S}, \mathcal{F}\}$. Because $\mathcal{S}$ is a responsive adversary, this implies that there is only a negligible set of runs of $\{\mathcal{E}', \mathcal{S}, \mathcal{F}\}$ in which a restricting message of $\mathcal{F}$ is answered incorrectly (by $\mathcal{E}'$ or $\mathcal{S}$). Hence, the probability for this to happen is bounded by some negligible function f. From this and the fact that there are only polynomially many sessions, it follows that the probability that a restricting message from some session of $\mathcal{F}$ is answered incorrectly is negligible. Hence, $\mathcal{S}'$ is a responsive adversary. □

We note that Theorems 4.8 and 4.9 can be combined to obtain increasingly complex protocols. For example, one can first show that a single session of a real protocol $\mathcal{P}$ realizes a single session of an ideal protocol $\mathcal{F}$. Using the two theorems, it then follows, for example, that a protocol $\mathcal{Q}$ using multiple sessions of $\mathcal{P}$ realizes $\mathcal{Q}$ using multiple sessions of $\mathcal{F}$.

To conclude this section, we note that all of our lemmas and theorems have been proven using a single restriction R. Hence, formally, a protocol designer would have to use the same restriction in all of her security proofs in order to be able to use our results. However, as we show in the full version of this paper [6], this is actually not the case because it is very easy to extend and combine different restrictions while still retaining all security results. Also, as discussed in Sect. 6, there is in fact one generic restriction that would suffice for all purposes.

5 Responsive Environments in Concrete Models

In the preceding section, we have presented our universal composability framework with responsive environments in a rather model-independent way. In this section, we outline how to implement this framework in the prominent UC, GNUC, and IITM models to exemplify that our framework and concepts are sufficiently general to be applicable to any universal composability model. While these three models follow the same general idea, they differ in several details which affect the concrete implementation of our concepts in these models (see, e.g., [19,24] for a discussion of these differences). The main differences and details to be considered concern runtime definitions and the mechanism for addressing (instances of) machines.

To instantiate our universal composability framework with responsive environments for the models mentioned, we mainly have to concretize the definitions in Sect. 4.2 for these models, that is, the definitions of restrictions as well as of the responsive environments and adversaries. For some models we also have to adjust their runtime notions slightly. Before presenting the details for the specific models, let us briefly explain the central points to be taken care of:

Runtime. In the GNUC and IITM models, the runtime of systems/protocols is required to be polynomially bounded only for a certain class of environments. As we now want to consider responsive environments, we should restrict the class of environments considered in the GNUC and IITM models to those that are responsive. This also has some technical advantages. To see this, let $\mathcal{R}$ and $\mathcal{R}'$ be two systems/protocols. For example, $\mathcal{R}$ and $\mathcal{R}'$ could be the systems $\{\mathcal{E}, \mathcal{A}_D, \mathcal{Q}, \mathcal{P}\}$ and $\{\mathcal{E}, \mathcal{S}, \mathcal{Q}, \mathcal{I}\}$ as considered in the composition theorem (Theorem 4.8) when we want to prove that $\{\mathcal{Q}, \mathcal{P}\}$ realizes $\{\mathcal{Q}, \mathcal{I}\}$. We often face the situation that we know that, say, $\mathcal{R}$ satisfies the model's runtime bound for all environments in a certain class and that $\mathcal{R}$ and $\mathcal{R}'$ are indistinguishable for every *responsive* environment $\mathcal{E}$ (in this class). This implies that $\mathcal{R}'$ also has to satisfy the runtime notion, but only for all responsive environments of the class. Hence, one cannot necessarily use $\mathcal{R}'$, with any environment, in another system as it does not satisfy the model's runtime notion (for non-responsive environments $\mathcal{E}$, the runtime of $\mathcal{R}'$ might not be polynomial). Hence, also from a technical point of view, it makes sense to relax the runtime notions in these models in that the runtime of systems/protocols should only be required to be polynomially bounded for responsive environments.

Definition of restrictions. According to Definition 4.1, we require that restrictions are "efficiently decidable". As mentioned, the exact definition depends on the model at hand. The important property this definition should satisfy is the following: An environment $\mathcal{E}'$ which internally simulates another environment $\mathcal{E}$ should be able to decide whether the output $\mathcal{E}$ produces is a correct answer (according to the restriction) when receiving some message as input. That is, $\mathcal{E}'$ must be able to check whether the input message was restricting at all, and if it was, $\mathcal{E}'$ must be able to check whether the response was valid. We often use such simulations in proofs. Depending on the model under consideration, we might not yet (at this point of the proof) have guarantees about the length of the restricting message sent to $\mathcal{E}$. A model-dependent definition of an efficiently decidable restriction should take this into account.

Definition of responsive environments. In the definition of responsive environments (Definition 4.2), we require that an answer to a restricting message be sent back to the same machine and we already explained that "same machine" typically means the same instance from which the restricting message has been received. This has to be specified for the different models.

Definition of responsive adversaries. Depending on the restriction R considered, in some models, in particular UC and GNUC, Definition 4.3 can be too restrictive, and, for example, the dummy adversary in these models might not satisfy the definition. The dummy adversary in these models is required to perform multiplexing. When it receives a message from an instance of the protocol and forwards this message to the environment, it has to prefix the message with the ID of that instance to tell the environment where the message came from. This alters the message, and the resulting message might no longer be restricting, depending on the definition of the restriction R. Hence, the environment would no longer be obliged to answer directly, and thus the (dummy) adversary would not be responsive. One way to fix this is to require a certain closure property of restrictions, namely that adding IDs at the beginning of restricting messages still yields restricting messages and that these message permit the same answers. But this is quite cumbersome. For example, by recursively applying this constraint one would have to require that R be closed under arbitrarily long prefixes of sequences of IDs. A more elegant solution that would still allow simple and natural restrictions would redefine what it means for a message from an adversary to the environment to be restricting. This is what we suggest for the UC and GNUC models (see below).

In what follows, we sketch how to adjust and concretize the runtime notions and the definitions for the UC, GNUC, and IITM models. As mentioned in the introduction, we have carried out the implementation of responsive environments in this model in full detail for the IITM model.

5.1 UC

For the UC model, we do not have to change the runtime definition because the runtime of a protocol is not defined w.r.t. a class of environments, but simply bounded by a fixed polynomial (see also below).

Definition of Restrictions. For UC we require both R and $R[0]$ to be decidable in polynomial time in the length of the input. Because of UC's strict runtime definition, this is sufficient to satisfy the requirement mentioned above, namely, that an environment $\mathcal{E}'$ simulating another environment $\mathcal{E}$ can check whether a restricting message received by $\mathcal{E}$ is answered correctly by $\mathcal{E}$. To see this, recall that every machine in UC is required to be parameterized with a polynomial. At every point in the run, the runtime of every instance of a machine is bounded by this polynomial, where the polynomial is in $n := n_I - n_O$, with n_I being the number of bits received so far on the I/O interface from higher-level machines and n_O being the number of bits sent on the I/O interface to lower level machines. Environment machines have to satisfy this condition as well, where n_I is the number of bits of the external input (which contains the security parameter η). Hence, as protocols will receive only a polynomial number of input bits from the environment, they can send messages of polynomial length in the length of the external input plus η only. Therefore, given some message m that was received by an environment and a response m' to this message, the message pair (m, m') has at most polynomial length in the external input plus η, and an environment is able to decide within its runtime bound whether m' is a correct answer to m if we use the above definition of effectively decidable restrictions.

Definition of Responsive Environments. We require that a response to a restricting message be sent back to the *instance* of the machine that sent the restricting message. This is possible because every instance in UC is assigned a globally unique ID, which is then used to specify the sender and the recipient of a message.

Definition of Responsive Adversaries. As explained above, messages from the adversary to the environment and vice versa may contain a prefix (typically an ID). For reasons explained above, we say in UC that such a prefix is ignored for the sake of checking whether a message is restricting and whether the answer is correct. To be more specific, a message $m = (pre, \bar{m})$ from the adversary to the environment is restricting iff $\bar{m} \in R[0]$. Also, if m is restricting (in this sense), an answer $m' = (pre', \bar{m}')$ from the environment is allowed if $(\bar{m}, \bar{m}') \in R$ and $pre' = pre$. Using this definition, it is easy to see that the dummy adversary in UC, which adds some prefix to messages from a protocol to the environment and strips off a prefix from messages from the environment to a protocol, is responsive.

5.2 GNUC

The changes necessary for the GNUC model are similar to those for the UC model. However, the runtime notion has to be modified:

Runtime. Let us first recall the relevant parts of the runtime definition of GNUC.[6] In this model, the runtime definition depends on the entity considered. For an environment $\mathcal{E}$, there has to exist a polynomial p that bounds the runtime of the environment in runs with *every* system where p gets as input the number of bits of all messages that have been received by the environment during the run, including the external input, plus the security parameter η. For a protocol $\mathcal{P}$, there has to exist a polynomial q such that the runtime of $\mathcal{P}$ is bounded by q in runs with any environment and the dummy adversary where q gets as input the number of bits that are output by the environment (to both the adversary and the protocol). This definition has to be changed such that the runtime of a protocol needs to be bounded only for all environments (in the sense of GNUC) that in addition are responsive.

Definition of Restrictions. Analogously to UC, we require R and $R[0]$ to be decidable in polynomial time in the length of the input. This is sufficient to satisfy the described requirement ($\mathcal{E}'$ simulating $\mathcal{E}$) as the runtime of environments in GNUC depends on the number of bits received from a protocol. Hence, an environment is always able to read a potentially restricting message m entirely, whereas the length of an answer m' is bounded by the runtime bound of the environment.

Definition of Responsive Environments. Just as for UC, we require that responses to restricting messages be sent to the same *instance* of a machine. This is possible in GNUC because, again, all machines have globally unique IDs to address instances.

Definition of Responsive Adversaries. Just as for UC, the adversary in GNUC might (have to) add IDs as prefixes or remove such prefixes, therefore these prefixes are ignored in the definition of responsive adversaries.

[6] Note that there are several additional requirements, such as bounds on the number of bits that are sent by the environment as well as so-called invited messages. These details, however, are not relevant here.

5.3 IITM

Just as for the other models, we now outline how to adjust and concretize the runtime notion and the definitions from Sect. 4 for the IITM model. As mentioned, in the full version of this paper [6], we provide full details for the IITM model with responsive environments, with a brief summary of the results presented at the end of this subsection.

Runtime. In the IITM model, the runtime depends on the type of entity. For an environment $\mathcal{E}$, it is required that there exists a polynomial p (in the length of the external input, if any, plus the security parameter) such that for *every* system running with $\mathcal{E}$ the runtime of $\mathcal{E}$ with this system is bounded by p. For a protocol $\mathcal{P}$, it is merely required that it be environmentally bounded, i.e., for every environment $\mathcal{E}$ there is a polynomial q (again, in the length of the external input plus the security parameter) that bounds the overall runtime of runs of $\{\mathcal{E}, \mathcal{P}\}$ (except for at most a negligible set of runs).[7] Given a protocol $\mathcal{P}$, for an adversary $\mathcal{A}$ for $\mathcal{P}$ it is required only that $\{\mathcal{A}, \mathcal{P}\}$ be environmentally bounded. (Clearly, the dummy adversary is environmentally bounded.) To adjust the runtime notions for the setting with responsive environments, instead of quantifying over all environments in the definition of environmentally bounded protocols/adversaries, one should now quantify over responsive environments only, as motivated at the beginning of Sect. 5.

Definition of Restrictions. We require that a restriction R is *efficiently decidable in the second component*, i.e., there is an algorithm A which expects pairs (m, m') of messages as input and which runs in polynomial time in $|m'|$ in order to decide whether m' is a correct answer to m according to R (see the full version of this paper [6] for a formal definition). This stronger definition is necessary to obtain the property described, namely, that an environment $\mathcal{E}'$ internally simulating another environment $\mathcal{E}$ can check that answers of $\mathcal{E}$ to restricting messages are correct. Owing to the very liberal runtime notion for protocols used in the IITM model, in proofs (e.g., of the composition theorem) we sometimes have to establish that a system is environmentally bounded. Therefore, we do not know a priori that the length of the message m is polynomially bounded. Hence, the environment might not be able to read m completely. Conversely, the length of m' is guaranteed to be polynomially bounded as it is output by the environment $\mathcal{E}$, which, by definition, is polynomially bounded. With R being efficiently decidable in the second component, $\mathcal{E}'$ can then efficiently decide whether m' is a correct answer to m. Compared with the definition of restrictions for the UC and GNUC models presented above, this formally is more restricted. It is, however, sufficient for all practical purposed, as discussed in Sect. 6, as one has to consider one generic restriction only and this restriction is efficiently decidable in the second component.

[7] Here $\mathcal{E}$ may directly connect to $\mathcal{P}$'s network interface. Equivalently one could have $\mathcal{E}$ communicate with $\mathcal{P}$ on the network interface via a dummy adversary.

Definition of Responsive Environments. Unlike the UC and GNUC models, the IITM model does not hardwire a specific addressing mechanism for instances of machines and specific IDs for such instances into the model. Instead, it supports a flexible addressing mechanism which allows a protocol designer to specify how machine instances are addressed and what they consider to be their ID. More specifically, the IITM model allows a protocol designer to specify an algorithm run by machine instances that decides whether the message received is accepted by the instance or not. Therefore, in the IITM model, we can require only that responses to restricting messages be sent to the same machine, but not necessarily the same machine *instance*. This, however, is indeed sufficient. A protocol designer, can specify that a (protocol) machine accepts a message iff it is prefixed by a certain ID (the one seen in the first activation of the instance) as typically done in the IITM model. This ID can then be considered to be the ID of this machine instance, and messages output by this machine would also be prefixed by this ID. Now, a protocol designer can use restrictions to manually enforce that the same instance receives a response. Such a restriction would contain message pairs of the form $((id, m), (id, m'))$. By this, it is guaranteed that if a restricting message has been sent by a protocol machine instance with ID id, then the response is returned to this instance, as the response is prefixed with id.

Definition of Responsive Adversaries. For the IITM model, we do not have to change the definition of responsive adversaries. Adversaries in the IITM model do not have to add prefixes to messages, and hence, do not have to modify restricting messages. In particular, the dummy adversary simply forwards messages between the environment and the protocol without changing messages.

Detailed Results for the IITM Model. In the full version of this paper [6] we provide full details of the IITM model with responsive environments. That is, we adjust the runtime notion of the IITM model accordingly, and provide full definitions of restrictions, responsive environments and adversaries. Based on these definitions we define the various security notions for realization relations considered in the literature (now with responsive environments), namely, (dummy) UC, black-box simulatability, strong simulatability, and reactive simulatability. These new and adjusted notions have been carefully developed in order to be general and to preserve central properties. In particular, we show that all the notions mentioned for realization relations are equivalent (for reactive simulatability, this requires environments with external input). We also prove that these relations are reflexive and transitive. We finally prove the composition theorems for responsive environments. As should be clear from the proof sketches in Sect. 4, the proofs are more involved than those without responsive environments because one always has to ensure that the constructed environments and simulators are responsive. The full proofs are even more intricate and non-trivial because they take all model-specific details, such as the liberal runtime notions, into account. We note, however, that this is a once and for all effort. Protocol designers no longer have to perform such proofs. They can simply use the results. That is, responsive environments do not put any burden on the protocol designer. On the contrary, as explained, they greatly simplify the specification and analysis of protocols.

6 Applying Our Concepts to the Literature

Our new concepts of restricting messages and responsive environments and adversaries allow protocol designers to avoid the non-responsiveness problem elegantly and completely. As mentioned, urgent requests can simply be declared to be restricting messages, causing the adversary/environment to reply with a valid response before sending any other message to the protocol. This indeed seems to be the most reasonable and natural solution to the non-responsiveness problem. We now show that our approach indeed easily solves all the problems mentioned in Sects. 1 and 3.

The frequently encountered formulations of the form (1) mentioned in Sect. 3.1 can now actually be used without causing confusion and flawed specifications, if the message sent to the adversary is declared to be restricting: there will now in fact be an immediate answer to this message. Similarly, ideal functionalities which are intended to be non-interactive can now be made non-interactive (at least if uncorrupted; but, if desired and realistic, also in the corrupted case) just like their realizations, which solves the problems discussed in Sect. 3.2.2 (lack of expressivity), and also makes it possible to use the, again, often encountered specifications of the form (2): if such ideal functionalities have to send urgent requests to the adversary, such requests can be made restricting, and hence, prompt replies are guaranteed, i.e., if the (responsive) adversary/environment contacts the protocol at all again, it first has to answer the request. Clearly, the other problems caused by urgent requests not being answered immediately discussed in Sect. 3.2, namely, unintended state changes and race conditions, the reentrance problem, and unnatural specifications of higher-level protocols, vanish also; again, because urgent request now *are answered immediately.*

Two ways of defining restrictions. We note that there are two approaches to define restrictions R.

Tailored Restrictions. One approach is to define restrictions tailored to specific protocols and functionalities. For example, for $\mathcal{F}_{\text{D-Cert}}$ the restriction could be defined as follows:

$$\{((\texttt{Verify}, sid, m, \sigma), (\texttt{Verified}, sid, m, \phi)) : sid, m, \sigma \in \{0,1\}^*, \phi \in \{0,1\}\}$$

Now, whenever the adversary is asked to verify some σ, the next message sent to the ideal functionality is guaranteed to be the expected response. This directly resolves the issues discussed in Sect. 3.2.1. Similarly, one could, for example, define restrictions for $\mathcal{F}_{\text{NIKE}}$ and $\mathcal{F}_{\text{sok}}$.[8]

[8] Note that to show that the respective real protocols realize their ideal functionalities, according to Definition 4.6, one needs to prove that there exists a *responsive*

We note that the above approach of defining a separate restriction for each protocol is general in the sense that it can be used independently of the underlying model for universal composition, and is thus applicable, e.g., to the UC, GNUC, and IITM models. Furthermore, this solution allows one to fix many ideal functionalities and their realizations found in the literature without any modifications to the specifications, including all examples mentioned in this document. However, since the composition theorems and the transitivity property assume one restriction, different restrictions have to be combined into a single one. This is always possible as shown in the full version of this paper [6]. Nevertheless, the following solution seems preferable.

Generic Restriction. Alternatively to employing tailored restrictions, one can use the following generic restriction:

$$R_G := \{(m, m') \mid m = (\texttt{Respond}, m''), m', m'' \in \{0,1\}^*\}.$$

This means that messages prefixed with `Respond` are considered to be restricting, and hence protocol designers can declare a message to be restricting by simply prefixing it by `Respond`. According to the definition of R_G, the adversary/environment can respond with any message to these messages, but protocols or ideal functionalities can be defined in such a way that they repeat their requests until they receive the expected answer: for instance, in the case of $\mathcal{F}_{\mathrm{sok}}$, it can repeatedly send $m'' = (\texttt{Setup}, sid)$ to the adversary until it receives the expected algorithms. In this way, the adversary is forced to eventually provide an expected answer (if she wants the protocol to proceed).

Using this fixed multi-purpose restriction has the advantage that, in contrast to the former approach, there is no need to combine different restrictions. Also, in protocol specifications, the prefixing immediately makes clear which messages are considered to be restricting.

The main reasons we did not hardwire the generic restriction into our framework are twofold. First, this is not required to prove our results, but makes our framework only more general, and the flexibility might become useful in some situations. Second, as protocols and ideal functionalities have to send several requests until they get the expected answer, depending on the runtime notions used, they might run out of resources. In the IITM model, however, this is not an issue, and hence the generic restriction can be used.

7 Conclusion

In this paper, we highlighted the non-responsiveness problem, the fact that it has often been ignored in the literature, and its many negative consequences.

---

simulator. However, it is easy to verify that the simulators constructed in [14,17,31] for the functionalities mentioned already are responsive, and thus these realizations can be used unalteredly also in a responsive setting.

We have proposed a framework that completely avoids this problem. It enables protocol designers to declare urgent requests to be restricting messages, causing such requests to be answered immediately by (responsive) environments/adversaries. This, in particular, allows protocols and ideal functionalities to be defined in the expected and natural way. It also avoids unnecessarily complex and artificial specifications, unintended state changes and race conditions while waiting for responses to urgent requests, the reentrance problem, the lack of expressivity when modeling non-interactive tasks, and the propagation of such problems to higher-level protocols and proofs. We discussed how our concepts can be adopted by existing models for universal composition, as exemplified in this work for the UC, GNUC, and IITM models. In the full version of this paper [6], we also provide full details for the IITM model, showing that our concepts can seamlessly be integrated into the existing model without losing any of the properties of the setting without responsive environments: all security notions for the realization relations are formulated, shown to (still) be equivalent, and enjoy reflexivity and transitivity; the composition theorems also carry over to the setting with responsive environments.

References

1. Abe, M., Ohkubo, M.: A framework for universally composable non-committing blind signatures. In: Matsui, M. (ed.) ASIACRYPT 2009. LNCS, vol. 5912, pp. 435–450. Springer, Heidelberg (2009). doi:10.1007/978-3-642-10366-7_26
2. Backes, M., Dürmuth, M., Hofheinz, D., Küsters, R.: Conditional reactive simulatability. Int. J. Inf. Secur. (IJIS) **7**(2), 155–169 (2008)
3. Backes, M., Pfitzmann, B., Waidner, M.: The reactive simulatability (RSIM) framework for asynchronous systems. Inf. Comput. **205**(12), 1685–1720 (2007)
4. Backes, M., Hofheinz, D.: How to break and repair a universally composable signature functionality. In: Zhang, K., Zheng, Y. (eds.) ISC 2004. LNCS, vol. 3225, pp. 61–72. Springer, Heidelberg (2004). doi:10.1007/978-3-540-30144-8_6
5. Camenisch, J., Dubovitskaya, M., Haralambiev, K., Kohlweiss, M.: Composable & modular anonymous credentials: definitions and practical constructions. In: ASIACRYPT 2015 (2015)
6. Camenisch, J., Enderlein, R.R., Krenn, S., Küsters, R., Rausch, D.: Universal composition with responsive environments. Technical report, Cryptology ePrint Archive, Report 2016/034 (2016). http://eprint.iacr.org/2016/034
7. Canetti, R.: Universally composable signature, certification, and authentication. In: CSFW 2004, pp. 219–233. IEEE (2004)
8. Canetti, R.: Universally composable security: a new paradigm for cryptographic protocols. In: 42nd FOCS, pp. 136–145. IEEE Computer Society Press, October 2001. For full and previous versions https://eprint.iacr.org/2000/067.pdf
9. Canetti, R., Cheung, L., Kaynar, D., Liskov, M., Lynch, N., Pereira, O., Segala, R.: Time-bounded task-PIOAs: a framework for analyzing security protocols. In: Dolev, S. (ed.) DISC 2006. LNCS, vol. 4167, pp. 238–253. Springer, Heidelberg (2006). doi:10.1007/11864219_17

10. Canetti, R., Dodis, Y., Pass, R., Walfish, S.: Universally composable security with global setup. In: Vadhan, S.P. (ed.) TCC 2007. LNCS, vol. 4392, pp. 61–85. Springer, Heidelberg (2007). doi:10.1007/978-3-540-70936-7_4
11. Canetti, R., Halevi, S., Katz, J.: Adaptively-secure, non-interactive public-key encryption. In: Kilian, J. (ed.) TCC 2005. LNCS, vol. 3378, pp. 150–168. Springer, Heidelberg (2005). doi:10.1007/978-3-540-30576-7_9
12. Canetti, R., Krawczyk, H., Nielsen, J.B.: Relaxing chosen-ciphertext security. In: Boneh, D. (ed.) CRYPTO 2003. LNCS, vol. 2729, pp. 565–582. Springer, Heidelberg (2003). doi:10.1007/978-3-540-45146-4_33
13. Canetti, R., Shahaf, D., Vald, M.: Universally composable authentication and key-exchange with global PKI. Cryptology ePrint Archive, Report 2014/432 (2014)
14. Chase, M., Lysyanskaya, A.: On signatures of knowledge. In: Dwork, C. (ed.) CRYPTO 2006. LNCS, vol. 4117, pp. 78–96. Springer, Heidelberg (2006). doi:10.1007/11818175_5
15. Damgård, I., Hofheinz, D., Kiltz, E., Thorbek, R.: Public-key encryption with non-interactive opening. In: Malkin, T. (ed.) CT-RSA 2008. LNCS, vol. 4964, pp. 239–255. Springer, Heidelberg (2008). doi:10.1007/978-3-540-79263-5_15
16. Dowsley, R., Müller-Quade, J., Otsuka, A., Hanaoka, G., Imai, H., Nascimento, A.C.A.: Universally composable and statistically secure verifiable secret sharing scheme based on pre-distributed data. IEICE Trans. **94–A**(2), 725–734 (2011)
17. Freire, E.S.V., Hesse, J., Hofheinz, D.: Universally composable non-interactive key exchange. In: Abdalla, M., Prisco, R. (eds.) SCN 2014. LNCS, vol. 8642, pp. 1–20. Springer, Heidelberg (2014). doi:10.1007/978-3-319-10879-7_1
18. Hazay, C., Venkitasubramaniam, M.: On black-box complexity of universally composable security in the CRS model. In: Iwata, T., Cheon, J.H. (eds.) ASIACRYPT 2015. LNCS, vol. 9453, pp. 183–209. Springer, Heidelberg (2015). doi:10.1007/978-3-662-48800-3_8
19. Hofheinz, D., Shoup, V.: GNUC: a new universal composability framework. Cryptology ePrint Archive, Report 2011/303 (2011)
20. Hofheinz, D., Unruh, D., Müller-Quade, J.: Polynomial runtime and composability. J. Cryptology **26**(3), 375–441 (2013)
21. Kurosawa, K., Furukawa, J.: Universally composable undeniable signature. In: Aceto, L., Damgård, I., Goldberg, L.A., Halldórsson, M.M., Ingólfsdóttir, A., Walukiewicz, I. (eds.) ICALP 2008. LNCS, vol. 5126, pp. 524–535. Springer, Heidelberg (2008). doi:10.1007/978-3-540-70583-3_43
22. Küsters, R.: Simulation-based security with inexhaustible interactive turing machines. In: CSFW 2006, pp. 309–320. IEEE (2006)
23. Küsters, R., Tuengerthal, M.: Joint state theorems for public-key encryption and digital signature functionalities with local computation. In: Proceedings of the 21st IEEE Computer Security Foundations Symposium (CSF 2008), pp. 270–284. IEEE Computer Society (2008)
24. Küsters, R., Tuengerthal, M.: The IITM model: a simple and expressive model for universal composability. Cryptology ePrint Archive, Report 2013/025 (2013)
25. Laud, P., Ngo, L.: Threshold homomorphic encryption in the universally composable cryptographic library. Cryptology ePrint Archive, Report 2008/367 (2008)
26. Matsuo, T., Matsuo, S.: On universal composable security of time-stamping protocols. In: IWAP 2005, pp. 169–181 (2005)

27. Maurer, U.: Constructive cryptography – a new paradigm for security definitions and proofs. In: Mödersheim, S., Palamidessi, C. (eds.) TOSCA 2011. LNCS, vol. 6993, pp. 33–56. Springer, Heidelberg (2012). doi:10.1007/978-3-642-27375-9_3
28. Maurer, U., Renner, R.: Abstract cryptography. In: ICS 2011, pp. 1–21. Tsinghua University Press (2011)
29. Pfitzmann, B., Waidner, M.: Composition and integrity preservation of secure reactive systems. In: ACM CCS 2000, pp. 245–254. ACM Press (2000)
30. Tian, Y., Peng, C.: Universally composable secure group communication. Cryptology ePrint Archive, Report 2014/647 (2014). http://eprint.iacr.org/
31. Zhao, S., Zhang, Q., Qin, Y., Feng, D.: Universally composable secure tnc protocol based on IF-T binding to TLS. In: Au, M.H., Carminati, B., Kuo, C.-C.J. (eds.) NSS 2014. LNCS, vol. 8792, pp. 110–123. Springer, Heidelberg (2014). doi:10.1007/978-3-319-11698-3_9

A Shuffle Argument Secure in the Generic Model

Prastudy Fauzi([✉]), Helger Lipmaa, and Michał Zając

University of Tartu, Tartu, Estonia
prastudy.fauzi@gmail.com

Abstract. We propose a new random oracle-less NIZK shuffle argument. It has a simple structure, where the first verification equation ascertains that the prover has committed to a permutation matrix, the second verification equation ascertains that the same permutation was used to permute the ciphertexts, and the third verification equation ascertains that input ciphertexts were "correctly" formed. The new argument has 3.5 times more efficient verification than the up-to-now most efficient shuffle argument by Fauzi and Lipmaa (CT-RSA 2016). Compared to the Fauzi-Lipmaa shuffle argument, we (i) remove the use of knowledge assumptions and prove our scheme is sound in the generic bilinear group model, and (ii) prove standard soundness, instead of culpable soundness.

Keywords: Common reference string · Bilinear pairings · Generic bilinear group model · Mix-net · Shuffle argument · Zero knowledge

1 Introduction

A typical application of mix-nets is in e-voting, where each voter (assume that there are n of them) encrypts his ballot by using an additively homomorphic public-key cryptosystem, and sends it to the bulletin board. After the vote casting period has ended, the bulletin board (considered to be the 0th, non-mixing, mix-server) forwards all encrypted ballots to the first mix-server. A small number (say, M) of mix-servers are ordered sequentially. The kth mix-server obtains a tuple $\mathfrak{v}$ of input ciphertexts from the $(k-1)$th mix-server, shuffles them, and sends a tuple $\mathfrak{v}'$ of output ciphertexts to the $(k+1)$th mix-server. Shuffling means that the kth mix-server generates a random permutation $\sigma \leftarrow_r S_n$ and a vector $\boldsymbol{s}$ of randomizers, and sets $\mathfrak{v}'_i = \mathfrak{v}_{\sigma(i)} \cdot \mathsf{enc}_{\mathsf{pk}}(0; s_i)$. The last mix-server (the $(M+1)$th one, usually implemented by using multi-party computation) is again a non-mixing server, who instead decrypts the results.

A mix-net clearly preserves the anonymity of voters, if at least one of the participating mix-servers is honest. To achieve security against an active attack (where some of the shuffles were not done correctly) is more difficult. In a nutshell, each server should prove in zero knowledge [24] that her shuffle was done correctly, i.e., prove that there exists a permutation σ and a vector $\boldsymbol{s}$, such that $\mathfrak{v}'_i = \mathfrak{v}_{\sigma(i)} \cdot \mathsf{enc}_{\mathsf{pk}}(0; s_i)$ for each i. The resulting zero-knowledge proof is usually called a *(zero-knowledge) shuffle argument*.

J.H. Cheon and T. Takagi (Eds.): ASIACRYPT 2016, Part II, LNCS 10032, pp. 841–872, 2016.
DOI: 10.1007/978-3-662-53890-6_28

Moreover, to obtain active security of the whole mix-net, it is important that the outputs of incorrect shuffles are ignored. This means that each mix-server (including the $(M+1)$th one) has to verify the correctness of each previous mix-server, and only apply her own shuffle to the output of the (multi-)shuffle where each previous server has been correct. Intuitively, this means that the verification time is the real bottleneck of mix-nets.

Substantial amount of work has been done on interactive zero-knowledge shuffle arguments. Random oracle model shuffle arguments are already quite efficient, see, e.g., [25]. However, an ever-growing amount of research [6,12,23,36] has provided evidence that the random oracle model yields properties that are impossible to achieve in the standard model. (See [14] for recent progress on NIZK arguments in the random oracle model.)

Much less is known about shuffle arguments in the common reference string (CRS, [7]) model, without using random oracles. Based on earlier work [28,33], Fauzi and Lipmaa recently proposed a shuffle argument in the CRS model [19]. Assuming that basic group operations are as efficient in both cases, and that a pairing is about 8 times slower than a group exponentiation (both assumptions should be taken with a caveat), the Fauzi-Lipmaa shuffle is about two times less efficient for the prover than the most efficient known shuffle argument in the random oracle model [25], while its verification is about 25 times less efficient.

The security of the Fauzi-Lipmaa shuffle argument is proven under a knowledge assumption [15] (PKE, [26]) and three computational assumptions (PCDH, TSDH, PSP). Knowledge assumptions are non-falsifiable [35], and their validity has to be very carefully checked in each application [5]. Moreover, the PSP assumption of Fauzi and Lipmaa [19] is novel (albeit closely related to SP, an earlier assumption of Groth and Lu [28]), and its security is proven in the generic bilinear group model [8,34,38].

The Fauzi-Lipmaa shuffle differs from the shuffle of Lipmaa and Zhang [33] in its security model. Briefly, in the security proof of the Lipmaa-Zhang shuffle argument it is assumed that the adversary obtains — by using knowledge assumptions — not only the secrets of the possibly malicious mix-server, but also the plaintexts and randomizers computed by all voters. This model was called *white-box soundness* by Fauzi and Lipmaa [19], where it was also criticized. Moreover, in the Lipmaa-Zhang shuffle argument [32], the plaintexts have to be small for the soundness proof to go through; for this, all voters should use efficient CRS-model range proofs [13,20,31].

On the other hand, the Fauzi-Lipmaa shuffle is proven culpably sound [28] though also under knowledge assumptions. Intuitively, culpable soundness means that if a cheating adversary produces an invalid (yet acceptable) shuffle together with the secret key, then one can break one of the underlying knowledge or computational assumptions.

Our Contribution. In all three results mentioned above [19,28,33], the authors based the soundness of their shuffle argument on some novel hardness assumptions, and then proved that the assumptions are secure in the generic bilinear group model (GBGM). It seems to be an obvious question whether one can obtain some efficiency benefit by bypassing the intermediate assumption and proving the soundness of the shuffle argument directly in the GBGM. We show this is indeed the case. We improve on the efficiency of the previous CRS-based shuffle arguments by proving the security of our protocol in the GBGM and *without* using knowledge assumptions. Due to the use of GBGM, we must first define a sensible security model.

First, recall that in the GBGM, the adversary inputs some group elements $\mathfrak{G}_i = \mathfrak{g}^{\chi_i}$, where $\mathfrak{g}$ is a group generator and χ_i are various (not necessarily independent) random values. One assumes that each group element $\mathfrak{H}_j$ output by the adversary is of the form $\mathfrak{H}_j = \mathfrak{g}_z^{F_j(\boldsymbol{\chi})}$, where $F_j(\boldsymbol{X})$ are known linear polynomials and $\mathfrak{g}_z$ is a generator of the group $\mathbb{G}_z$, $z \in \{1, 2\}$. (Within this paper, $\boldsymbol{\chi}$ is a concrete instantiation of the indeterminate $\boldsymbol{X}$.) We call such values *admissible.* (In addition, elements output from the target group can also use the bilinear map, but in the current paper, we do not use this fact.)

One philosophical question when using the GBGM is what exactly is the input of the adversary. In our intended usage cases, the shuffle argument is a part of a mix-net. Clearly, the mix-net should remain secure against coalitions between parties (in the case of e-voting, either voters, or some of the mix-servers themselves) that create the input ciphertexts, and parties who perform the shuffling. It is a common practice to model such coalitions as a single adversary. In the GBGM, it is natural to model this single adversary — who may corrupt everybody who has produced any part of the input to the verifier — as a generic adversary. This means that an adversary, who has generated a (say, ILin [18]) ciphertext $\mathfrak{v}_i = (\mathfrak{v}_{i1}, \mathfrak{v}_{i2}, \mathfrak{v}_{i3})$, knows polynomials $V_{ij}(\boldsymbol{X})$ and $V'_{ij}(\boldsymbol{X})$, such that $\log \mathfrak{v}_{ij} = V_{ij}(\boldsymbol{\chi})$ and $\log \mathfrak{v}'_{ij} = V'_{ij}(\boldsymbol{\chi})$. This is somewhat similar to the approach taken in [33] who used knowledge assumptions to then obtain the random variables — more precisely, plaintexts and randomizers — hidden in $\mathfrak{v}$.

We will assume that the mix-net is structured as follows. First, the encrypters (e.g., voters) prove that their ciphertexts (e.g., *encrypted* ballots) are admissible. More precisely, by using a *validity argument*, a voter proves that each component (e.g., an ILin [18] ciphertext consists of three group elements) of her ciphertext is equal to $\mathfrak{g}_1^{F(\boldsymbol{\chi})}$, where the polynomial $F(\boldsymbol{X})$ has specific form. The validity argument guarantees that the input ciphertexts to the first mix-server have been computed only from certain, "allowed", elements of the CRS.

Each mix-server first verifies the validity of original (unshuffled) ciphertexts and the soundness of each previous shuffle argument. After that the mix-server produces her shuffle $(\mathfrak{v}'_i)_{i=1}^n$ together with her shuffle argument π_{sh}. This means that we consider shuffling a part of the shuffle argument.

Our generic approach in the shuffle argument is as follows. We first let the prover (a mix-server) choose a permutation matrix and then commit separately to its every row. The prover then proves that the committed matrix is a permutation matrix, by proving that each row is 1-sparse (i.e., it has at most one non-zero element) as in [33], while computing the last row explicitly, see Sect. 5. The 1-sparsity argument is based loosely on Square Span Programs [16]. Basically, to show that a vector $\boldsymbol{a}$ is 1-sparse, we construct $n+1$ polynomials $(P_i(X))_{i=0}^n$ that interpolate a certain matrix (and a certain vector) connected to the definition of 1-sparsity, and then commit to $\boldsymbol{a}$ by using a "polynomial" version of the extended Pedersen commitment scheme, $\mathfrak{c} \leftarrow \mathfrak{g}_2^{\sum a_i P_i(\chi)+r\varrho}$, for random secrets χ and ϱ.

To obtain the full shuffle argument, we use the same underlying idea as [19,28,33]. Namely, we construct a specific *consistency* verification equation that ensures that $(\mathfrak{v}_i)_{i=1}^n$ is permuted to $(\mathfrak{v}'_i)_{i=1}^n$ by using the same permutation matrix that was used to permute $(\mathfrak{g}_2^{P_i(\chi)})_{i=1}^n$ to $(\mathfrak{A}_{i2})_{i=1}^n$. This is done by using a pairing equation of type $\prod \hat{e}(\mathfrak{v}'_i, \mathfrak{g}_2^{P_i(\chi)}) / \prod \hat{e}(\mathfrak{v}_i, \mathfrak{A}_{i2}) = \mathfrak{R}$, where $\mathfrak{R}$ is a value that takes care of the rerandomization (i.e., it depends on the values $\boldsymbol{s}$ used to rerandomize $\mathfrak{v}$, but not on σ).

Both [19,28] had an additional problem here, namely it can be the case that a maliciously created $\mathfrak{v}'_i$ depends on $P_j(X)$ (in [28], one has $P_j(X_1, \ldots, X_n) = X_j$, where X_j are independent random variables) so $\log_{\mathfrak{g}_T} \hat{e}(\mathfrak{v}'_i, \mathfrak{g}_2^{P_i(\chi)})$ can depend on $P_j(X)P_i(X)$, for arbitrary i and j. In this case, this equation is not sufficient for soundness, since $\{P_i(X)P_j(X)\}_{i,j\in[1\,..\,n]}$ is not linearly independent (e.g., an adversary can cancel out $P_j(X)P_i(X)$ easily with $-P_i(X)P_j(X)$). Therefore, they had to go through additional complicated steps — that reduced the efficiency of their arguments — to achieve (culpable) soundness even in this case.

In our case, such complications are not needed, due to the validity argument. Since the validity argument guarantees that $\mathfrak{v}_i$ and $\mathfrak{v}'_i$ do not depend on $P_i(X)$, it means that the values $\log_{\mathfrak{g}_T} \hat{e}(\mathfrak{v}'_i, \mathfrak{g}_2^{P_i(\chi)})$ and $\log_{\mathfrak{g}_T} \hat{e}(\mathfrak{v}_i, \mathfrak{A}_{i2})$ do not depend on $P_i(X)P_j(X)$, which removes the problem evident in both [19,28]. On the other hand, [19,28] solved this problem by proving culpable soundness only, while we prove that the new argument satisfies the standard soundness property.

We emphasize that the full GBGM soundness proof of the new shuffle argument is quite intricate. In particular, the verification of the permutation matrix argument results in a system of more than 20 polynomial equations. As some other recent papers like [1,3], we use computer-based tools to solve the latter system. More precisely, we use a computer algebra system to find its Gröbner basis [11], and then continue to find solutions from there on. It is interesting that a simple shuffle argument has such a complicated security proof. On the other hand, both researchers and practitioners can write their own computer algebra code to verify the security proof; this is not possible in many other arguments.

We further optimize the verification by the use of batching techniques [4], thus replacing many pairings with less costly exponentiations. Batching has not been used before in the context of pairing-based shuffle arguments.

Table 1 compares our work and known NIZK shuffle arguments in the CRS model. However, differently from other papers, [28] uses symmetric pairings, and thus its computational and communication complexity is not directly comparable. The prover's computational complexity and the communication includes the computation and sending of the ciphertexts themselves. (This is fair, since different shuffle arguments use different public-key cryptosystems that incur different overhead to these complexity measures.) The highlighted cells in each row

Table 1. A comparison of different NIZK shuffle arguments. We always consider shuffling to be a part of the communication and prover's computation. Units (the main parameter, a weighted sum of other parameters) are defined in Sect. 9.

	Groth-Lu	Lipmaa-Zhang	Fauzi-Lipmaa	Current Work
Type of pairings	Symmetric	Asymmetric		
$\lvert\text{CRS}\rvert$ in $(\mathbb{G}_1, \mathbb{G}_2, \mathbb{G}_T)$	$2n+8$	$(2n+2, 5n+4, 0)$	$(6n+8, 2n+8, 1)$	$(2n+6, n+7, 1)$
Communication	$18n+120$	$(8n+6, 4n+5, 0)$	$(7n+2, 2n, 0)$	$(4n+1, 3n+2, 0)$
Prover's computation				
Exp. in $(\mathbb{G}_1, \mathbb{G}_2)$	$54n+246$	$(16n+6, 12n+5)$	$(14n+3, 4n)$	$(9n+2, 9n+3)$
Units		36	19.8	24.3
Verifier's computation				
Exp. in $(\mathbb{G}_1, \mathbb{G}_2, \mathbb{G}_T)$		—	—	$(11n+5, 3n+6, 1)$
Pairings	$75n+282$	$28n+18$	$18n+6$	$3n+6$
Units		196	126	36.3
Knowl. assumpt-s.	No	Yes	Yes	No
Relying on GBGM	PP, SP	Knowledge	Knowl., PSP	Complete
Random oracle	No			
Soundness	Culpable	Full	Culpable	Full

are the values with best efficiency, or best security properties. A more precise efficiency comparison is given in Sect. 9.

Finally, each of the CRS-model shuffle arguments relies substantially on the GBGM. The Groth-Lu and Fauzi-Lipmaa shuffles rely on the GBGM to prove security of complicated computational assumptions. The Lipmaa-Zhang shuffle relies on the GBGM to prove security of non-falsifiable knowledge assumptions. The current paper gives the full shuffle soundness proof in the GBGM. See Sect. 10 for a more thorough discussion of the GBGM security proof versus using knowledge assumptions.

2 Preliminaries

Let S_n be the symmetric group on n elements. For a (Laurent) polynomial or a rational function f and its monomial μ, denote by $\mathrm{coeff}_\mu(f)$ the coefficient of μ in f. We write $f(\kappa) \approx_\kappa g(\kappa)$, if $f(\kappa) - g(\kappa)$ is negligible as a function of κ.

Bilinear Maps. Let κ be the security parameter. Let q be a prime of length $O(\kappa)$ bits. Assume we use a secure bilinear group generator $\mathsf{genbp}(1^\kappa)$ that returns $\mathsf{gk} = (q, \mathbb{G}_1, \mathbb{G}_2, \mathbb{G}_T, \hat{e})$, where $\mathbb{G}_1$, $\mathbb{G}_2$, and $\mathbb{G}_T$ are three multiplicative groups of order q, and $\hat{e} : \mathbb{G}_1 \times \mathbb{G}_2 \to \mathbb{G}_T$. Within this paper, we denote the elements of $\mathbb{G}_1$, $\mathbb{G}_2$, and $\mathbb{G}_T$ as in $\mathfrak{g}_1$ (i.e., by using the Fraktur typeface). It is required that $\hat{e}$ is bilinear (i.e., $\hat{e}(\mathfrak{g}_1^a, \mathfrak{g}_2^b) = \hat{e}(\mathfrak{g}_1, \mathfrak{g}_2)^{ab}$), efficiently computable, and non-degenerate. We define $\hat{e}((\mathfrak{A}_1, \mathfrak{A}_2, \mathfrak{A}_3), \mathfrak{B}) = (\hat{e}(\mathfrak{A}_1, \mathfrak{B}), \hat{e}(\mathfrak{A}_2, \mathfrak{B}), \hat{e}(\mathfrak{A}_3, \mathfrak{B}))$ and $\hat{e}(\mathfrak{B}, (\mathfrak{A}_1, \mathfrak{A}_2, \mathfrak{A}_3)) = (\hat{e}(\mathfrak{B}, \mathfrak{A}_1), \hat{e}(\mathfrak{B}, \mathfrak{A}_2), \hat{e}(\mathfrak{B}, \mathfrak{A}_3))$. Assume that $\mathfrak{g}_i$ is a generator of $\mathbb{G}_i$ for $i \in \{1, 2\}$, and set $\mathfrak{g}_T \leftarrow \hat{e}(\mathfrak{g}_1, \mathfrak{g}_2)$.

For $\kappa = 128$, the current recommendation is to use an optimal (asymmetric) Ate pairing over a subclass of Barreto-Naehrig curves. In that case, at security level of $\kappa = 128$, an element of $\mathbb{G}_1/\mathbb{G}_2/\mathbb{G}_T$ can be represented in respectively 256/512/3072 bits.

Zero Knowledge. A NIZK argument for a group-dependent language $\mathcal{L}$ consists of four algorithms, setup, gencrs, pro and ver. The setup algorithm setup takes as input 1^κ and n (the input length), and outputs the group description gk. The CRS generation algorithm gencrs takes as input gk and outputs the prover's CRS crs_p, the verifier's CRS crs_v, and a trapdoor td. The distinction between crs_p and crs_v is only important for efficiency. The prover pro takes as input gk and crs_p, a statement u, and a witness w, and outputs an argument π. The verifier ver takes as input gk and crs_v, a statement u, and an argument π, and either accepts or rejects.

Some of the properties of an argument are: (i) *perfect completeness* (honest verifier always accepts honest prover's argument), (ii) *perfect zero knowledge* (there exists an efficient simulator that can, given u, $(\mathsf{crs}_p, \mathsf{crs}_v)$ and td, output an argument that comes from the same distribution as the argument produced by the prover), (iii) *adaptive computational soundness* (if $u \notin \mathcal{L}$, then an arbitrary non-uniform probabilistic polynomial time prover has negligible probability of success in creating a satisfying argument), and (iv) *adaptive computational culpable soundness* [28,29] (if $u \notin \mathcal{L}$, then an arbitrary NUPPT prover has negligible success in creating a satisfying argument together with a witness that $u \notin \mathcal{L}$). An argument is an *argument of knowledge*, if from an accepting argument it follows that the prover knows the witness. See Appendix A for formal definitions.

Generic Bilinear Group Model. We will prove the soundness of the new shuffle argument in the generic bilinear group model (GBGM, [8,34,38]). Our description of the GBGM is based on [34].

We start by picking a random asymmetric bilinear group $\mathsf{gk} := (q, \mathbb{G}_1, \mathbb{G}_2, \mathbb{G}_T, \hat{e}) \leftarrow \mathsf{genbp}(1^\kappa)$. Consider a black box $\mathbf{B}$ that can store values from groups $\mathbb{G}_1, \mathbb{G}_2, \mathbb{G}_T$ in internal state variables $\mathsf{cell}_1, \mathsf{cell}_2, \ldots$, where for simplicitly we allow the storage space to be infinite (this only increases the power of a generic adversary). The initial state consists of some values $(\mathsf{cell}_1, \mathsf{cell}_2, \ldots, \mathsf{cell}_{|inp|})$, which are set according to some probability distribution. Each state variable cell_i has an accompanying type $\mathsf{type}_i \in \{1, 2, T, \perp\}$. We assume initially $\mathsf{type}_i = \perp$ for $i > |inp|$. The black box allows computation operations on internal state variables and queries about the internal state. No other interaction with $\mathbf{B}$ is possible.

Let Π be the allowed set of computation operations. A computation operation consists of selecting a (say, t-ary) operation $f \in \Pi$ together with $t + 1$ indices $i_1, i_2, \ldots, i_{t+1}$. Assuming inputs have the correct type, $\mathbf{B}$ computes $f(\mathsf{cell}_{i_1}, \ldots, \mathsf{cell}_{i_t})$ and stores the result in $\mathsf{cell}_{i_{t+1}}$. For a set Σ of relations, a query consists of selecting a (say, t-ary) relation $\varrho \in \Sigma$ together with t indices

$i_1, i_2, \ldots, i_t$. Assuming inputs have the correct type, **B** replies to the query with $\varrho(\mathsf{cell}_{i_1}, \ldots, \mathsf{cell}_{i_t})$.

In the GBGM, we define $\Pi = \{\cdot, \hat{e}\}$ and $\Sigma = \{=\}$, where

1. On input $(\cdot, i_1, i_2, i_3)$: if $\mathsf{type}_{i_1} = \mathsf{type}_{i_2} \neq \perp$ then set $\mathsf{cell}_{i_3} \leftarrow \mathsf{cell}_{i_1} \cdot \mathsf{cell}_{i_2}$ and $\mathsf{type}_{i_3} \leftarrow \mathsf{type}_{i_1}$.
2. On input $(\hat{e}, i_1, i_2, i_3)$: if $\mathsf{type}_{i_1} = 1$ and $\mathsf{type}_{i_2} = 2$ then set $\mathsf{cell}_{i_3} \leftarrow \hat{e}(\mathsf{cell}_{i_1}, \mathsf{cell}_{i_2})$ and $\mathsf{type}_{i_3} \leftarrow T$.
3. On input $(=, i_1, i_2)$: if $\mathsf{type}_{i_1} = \mathsf{type}_{i_2} \neq \perp$ and $\mathsf{cell}_{i_1} = \mathsf{cell}_{i_2}$ then return 1. Otherwise return 0.

Since we are proving lower bounds, we will give a generic adversary adv additional power. We assume that all relation queries are for free. We also assume that adv is successful if after τ operation queries, he makes an equality query $(=, i_1, i_2)$, $i_1 \neq i_2$, that returns 1; at this point adv quits. Thus, if $\mathsf{type}_i \neq \perp$, then $\mathsf{cell}_i = F_1(\mathsf{cell}_1, \ldots, \mathsf{cell}_{|inp|})$ for a polynomial F_i known to adv.

The GBGM has proved itself to be very fruitful since its introduction, [8]. In particular, the generic (bilinear) group model is amenable to computerized analysis, and as such, has proven itself to be very useful say in the area of structure-preserving signature schemes [3]; see also [1].

Finally, Fischlin [21] and Dent [17] have pointed out that there exist constructions that are secure in (Shoup's version of) the generic group model but cannot be instantiated given any efficient instantiation of the group encoding. However, their constructions are utterly artificial; e.g., Dent constructed a signature scheme that under certain conditions outputs the signing key as a part of the signature.

Cryptosystems. A public-key cryptosystem Π is a triple $(\mathsf{genpkc}, \mathsf{enc}, \mathsf{dec})$ of efficient algorithms. The key generation algorithm $\mathsf{genpkc}(1^\kappa)$ returns a fresh public and secret key pair $(\mathsf{pk}, \mathsf{sk})$. The encryption algorithm $\mathsf{enc}_{\mathsf{pk}}(m; r)$, given a public key pk, a message m, and a randomizer r (from some randomizer space $\mathcal{R}$), returns a ciphertext. The decryption algorithm $\mathsf{dec}_{\mathsf{sk}}(c)$, given a secret key sk and a ciphertext c, returns a plaintext m. It is required that for each $(\mathsf{pk}, \mathsf{sk}) \in \mathsf{genpkc}(1^\kappa)$ and each m, r, it holds that $\mathsf{dec}_{\mathsf{sk}}(\mathsf{enc}_{\mathsf{pk}}(m; r)) = m$. Informally, Π is *IND-CPA secure*, if the distributions of ciphertexts corresponding to any two plaintexts are computationally indistinguishable.

We will use the ILin cryptosystem from [18]; it is distinguished from other well-known cryptosystems like the BBS cryptosystem [9] by having shorter secret and public keys. Consider group $\mathbb{G}_k$, $k \in \{1, 2\}$. In this cryptosytem, where the secret key is $\mathsf{sk} = \gamma \leftarrow_r \mathbb{Z}_q \setminus \{0, -1\}$, the public key is $\mathsf{pk}_k \leftarrow (\mathfrak{g}_k, \mathfrak{h}_k) = (\mathfrak{g}_k, \mathfrak{g}_k^\gamma)$, and the encryption of a small $m \in \mathbb{Z}_q$ is

$$\mathsf{enc}_{\mathsf{pk}_k}(m; \boldsymbol{s}) := (\mathfrak{h}_k^{s_1}, (\mathfrak{g}_k \mathfrak{h}_k)^{s_2}, \mathfrak{g}_k^m \mathfrak{g}_k^{s_1 + s_2})$$

for $\boldsymbol{s} \leftarrow_r \mathbb{Z}_q^{1 \times 2}$. Denote $\mathfrak{P}_{k1} := (\mathfrak{h}_k, \mathfrak{1}_k, \mathfrak{g}_k)$ and $\mathfrak{P}_{k2} := (\mathfrak{1}_k, \mathfrak{g}_k \mathfrak{h}_k, \mathfrak{g}_k)$, thus $\mathsf{enc}_{\mathsf{pk}_k}(m; \boldsymbol{s}) = (\mathfrak{1}_k, \mathfrak{1}_k, \mathfrak{g}_k^m) \cdot \mathfrak{P}_{k1}^{s_1} \mathfrak{P}_{k2}^{s_2}$. Given $\mathfrak{v} \in \mathbb{G}_k^3$, the decryption sets

$$\mathsf{dec}_{\mathsf{sk}}(\mathfrak{v}) := \log_{\mathfrak{g}_k}(\mathfrak{v}_3 \mathfrak{v}_2^{-1/(\gamma+1)} \mathfrak{v}_1^{-1/\gamma}) \ ,$$

Decryption succeeds since $\mathfrak{v}_3\mathfrak{v}_2^{-1/(\gamma+1)}\mathfrak{v}_1^{-1/\gamma} = \mathfrak{g}_k^m\mathfrak{g}_k^{s_1+s_2} \cdot (\mathfrak{g}_k\mathfrak{h}_k)^{-s_2/(\gamma+1)} \cdot \mathfrak{h}_k^{-s_1/\gamma} = \mathfrak{g}_k^m\mathfrak{g}_k^{s_1+s_2} \cdot \mathfrak{g}_k^{-s_2/(\gamma+1)}\mathfrak{g}_k^{-s_2\cdot\gamma/(\gamma+1)} \cdot \mathfrak{g}_k^{-s_1} = \mathfrak{g}_k^m$. This cryptosystem is CPA-secure under the 2-Incremental Linear (2-ILin) assumption, see [18]. The ILin cryptosystem is *blindable*, $\mathsf{enc}_{\mathsf{pk}_k}(m; \boldsymbol{s}) \cdot \mathsf{enc}_{\mathsf{pk}_k}(0; \boldsymbol{s}') = \mathsf{enc}_{\mathsf{pk}}(m; \boldsymbol{s} + \boldsymbol{s}')$.

We use a variant of the ILin cryptosystem where each plaintext is encrypted twice, in group $\mathbb{G}_1$ and in $\mathbb{G}_2$ (but by using the same secret key an the same randomizer $\boldsymbol{s}$ in both). For technical reasons (relevant to the shuffle argument but not to the ILin cryptosystem), in group $\mathbb{G}_1$ we will use an auxiliary generator $\hat{\mathfrak{g}}_1 = \mathfrak{g}_1^{\varrho/\beta}$ instead of $\mathfrak{g}_1$, for $(\varrho, \beta) \leftarrow_r (\mathbb{Z}_q \setminus \{0\})^2$; both encryption and decryption are done as before but just using the secret key $\mathsf{sk} = (\varrho, \beta, \gamma)$ and the public key $\mathsf{pk}_1 = (\hat{\mathfrak{g}}_1, \mathfrak{h}_1 = \hat{\mathfrak{g}}_1^\gamma)$; this also redefines $\mathfrak{P}_{k1}$. That is, $\mathsf{enc}_{\mathsf{pk}}(m; \boldsymbol{s}) = (\mathsf{enc}_{\mathsf{pk}_1}(m; \boldsymbol{s}), \mathsf{enc}_{\mathsf{pk}_2}(m; \boldsymbol{s}))$, where $\mathsf{pk}_1 = (\hat{\mathfrak{g}}_1, \mathfrak{h}_1 = \hat{\mathfrak{g}}_1^\gamma)$, and $\mathsf{pk}_2 = (\mathfrak{g}_2, \mathfrak{h}_2 = \mathfrak{g}_2^\gamma)$, and $\mathsf{dec}_{\mathsf{sk}}(\boldsymbol{\mathfrak{v}}) := \log_{\hat{\mathfrak{g}}_1}(\mathfrak{v}_3\mathfrak{v}_2^{-1/(\gamma+1)}\mathfrak{v}_1^{-1/\gamma}) = \log_{\mathfrak{g}_1}(\mathfrak{v}_3\mathfrak{v}_2^{-1/(\gamma+1)}\mathfrak{v}_1^{-1/\gamma})/(\varrho/\beta)$ for $\boldsymbol{\mathfrak{v}} \in \mathbb{G}_1^3$. We call this the *validity-enhanced* ILin cryptosystem.

In this case we denote the ciphertext in group k by $\boldsymbol{\mathfrak{v}}_k$, and its jth component by $\mathfrak{v}_{kj}$. In the case when we have many ciphertexts, we denote the ith ciphertext by $\boldsymbol{\mathfrak{v}}_i$ and the jth component of the ith ciphertext in group k by $\mathfrak{v}_{ikj}$.

3 Shuffle Argument

In the current section, we will give a full description of the new shuffle argument, followed by its efficiency analysis. Intuition behind its soundness will be given in Sect. 4. The full soundness proof is long, and postponed to Sects. 5, 6, and 7. Its zero knowledge property will be proven in Sect. 8.

Let $\Pi = (\mathsf{genpkc}, \mathsf{enc}, \mathsf{dec})$ be an additively homomorphic cryptosystem with randomizer space R; we assume henceworth that one uses the validity-enhanced ILin cryptosystem. Assume that $\boldsymbol{\mathfrak{v}}_i$ and $\boldsymbol{\mathfrak{v}}'_i$ are valid ciphertexts of Π. In a shuffle argument, the prover aims to convince the verifier in zero-knowledge that given $(\mathsf{pk}, (\boldsymbol{\mathfrak{v}}_i, \boldsymbol{\mathfrak{v}}'_i)_{i=1}^n)$, he knows a permutation $\sigma \in S_n$ and randomizers s_{ij}, $i \in [1 \mathinner{..} n]$ and $j \in [1 \mathinner{..} 2]$, such that $\boldsymbol{\mathfrak{v}}'_i = \boldsymbol{\mathfrak{v}}_{\sigma(i)} \cdot \mathsf{enc}_{\mathsf{pk}}(0; \boldsymbol{s}_i)$ for $i \in [1 \mathinner{..} n]$. More precisely, we define the group-specific binary relation $\mathcal{R}_{sh,n}$ exactly as in [28,33]:

$$\mathcal{R}_{sh,n} := \begin{pmatrix} (\mathsf{gk}, (\mathsf{pk}, \boldsymbol{\mathfrak{v}}_i, \boldsymbol{\mathfrak{v}}'_i)_{i=1}^n), (\sigma, \boldsymbol{s})) : \\ \sigma \in S_n \wedge \boldsymbol{s} \in R^{n\times 2} \wedge \left(\forall i : \boldsymbol{\mathfrak{v}}'_i = \boldsymbol{\mathfrak{v}}_{\sigma(i)} \cdot \mathsf{enc}_{\mathsf{pk}}(0; \boldsymbol{s}_i)\right) \end{pmatrix} .$$

See Protocol 1 for the full description of the new shuffle argument.

We note that in the real mix-net, (γ, ϱ, β) is handled differently (in particular, γ — and possibly ϱ/β — will be known to the decrypting party while (ϱ, β) does not have to be known to anybody) than the real trapdoor (χ, α) that enables one to simulate the argument and thus cannot be known to anybody. Moreover, $(\mathfrak{g}_1, \mathfrak{g}_2)^{\sum P_i(\chi)}$ is in the CRS only to optimize computation. A precise efficiency analysis of this argument is given in Sect. 9.

In the rest of this section, we will explain the notion of batching and define non-batched versions (that are easier to read and analyse in the soundness proof) of the verification equations. We then state the main security theorem.

$\mathsf{gencrs}(1^\kappa, n \in \mathrm{poly}(\kappa))$: Call $\mathsf{gk} = (q, \mathbb{G}_1, \mathbb{G}_2, \mathbb{G}_T, \hat{e}) \leftarrow \mathsf{genbp}(1^\kappa)$. Let $P_i(X)$ for $i \in [0 \,..\, n]$ be polynomials, chosen in Sect. 5. Set $\boldsymbol{\chi} = (\chi, \alpha, \varrho, \beta, \gamma) \leftarrow_r \mathbb{Z}_q^2 \times (\mathbb{Z}_q \setminus \{0\})^2 \times (\mathbb{Z}_q \setminus \{0, -1\})$. Let enc be the ILin cryptosystem with the secret key γ, and let $(\mathsf{pk}_1, \mathsf{pk}_2)$ be its public key. Set

$$\mathsf{crs} \leftarrow \begin{pmatrix} \mathsf{gk}, (\mathfrak{g}_1^{P_i(\chi)})_{i=1}^n, \mathfrak{g}_1^{\varrho}, \mathfrak{g}_1^{\alpha + P_0(\chi)}, \mathfrak{g}_1^{P_0(\chi)}, (\mathfrak{g}_1^{((P_i(\chi)+P_0(\chi))^2-1)/\varrho})_{i=1}^n, \\ \mathsf{pk}_1 = (\hat{\mathfrak{g}}_1 = \mathfrak{g}_1^{\varrho/\beta}, \mathfrak{h}_1 = \hat{\mathfrak{g}}_1^{\gamma}), \\ (\mathfrak{g}_2^{P_i(\chi)})_{i=1}^n, \mathfrak{g}_2^{\varrho}, \mathfrak{g}_2^{-\alpha + P_0(\chi)}, \mathsf{pk}_2 = (\mathfrak{g}_2, \mathfrak{h}_2 = \mathfrak{g}_2^{\gamma}), \mathfrak{g}_2^{\beta}, \\ \hat{e}(\mathfrak{g}_1, \mathfrak{g}_2)^{1-\alpha^2}, (\mathfrak{g}_1, \mathfrak{g}_2)^{\sum_{i=1}^n P_i(\chi)} \end{pmatrix}.$$

and $\mathsf{td} \leftarrow (\chi, \varrho)$. Return $(\mathsf{crs}, \mathsf{td})$.

$\mathsf{pro}(\mathsf{crs}; \mathfrak{v} \in (\mathbb{G}_1 \times \mathbb{G}_2)^{3n}; \sigma \in S_n, \boldsymbol{s} \in \mathbb{Z}_q^{n \times 2})$:

1. For $i = 1$ to $n - 1$:
   (a) Set $r_i \leftarrow_r \mathbb{Z}_q$. Set $(\mathfrak{A}_{i1}, \mathfrak{A}_{i2}) \leftarrow (\mathfrak{g}_1, \mathfrak{g}_2)^{P_{\sigma^{-1}(i)}(\chi) + r_i \varrho}$.
2. Set $r_n \leftarrow -\sum_{i=1}^{n-1} r_i$.
3. Set $(\mathfrak{A}_{n1}, \mathfrak{A}_{n2}) \leftarrow (\mathfrak{g}_1, \mathfrak{g}_2)^{\sum_{i=1}^n P_i(\chi)} / \prod_{i=1}^{n-1} (\mathfrak{A}_{i1}, \mathfrak{A}_{i2})$.
4. For $i = 1$ to n: `/* Sparsity, for permutation matrix: */`
   (a) Set $\pi_{\mathsf{1sp}:i} \leftarrow (\mathfrak{A}_{i1} \mathfrak{g}_1^{P_0(\chi)})^{2r_i} (\mathfrak{g}_1^{\varrho})^{-r_i^2} \mathfrak{g}_1^{((P_{\sigma^{-1}(i)}(\chi)+P_0(\chi))^2-1)/\varrho}$.
5. For $i = 1$ to n: `/* Shuffling itself */`
   (a) Set $(\mathfrak{v}'_{i1}, \mathfrak{v}'_{i2}) \leftarrow (\mathfrak{v}_{\sigma(i)1}, \mathfrak{v}_{\sigma(i)2}) \cdot (\mathsf{enc}_{\mathsf{pk}_1}(0; \boldsymbol{s}_i), \mathsf{enc}_{\mathsf{pk}_2}(0; \boldsymbol{s}_i))$.
6. Set `/* Consistency */`
   (a) For $k = 1$ to 2: Set $r_{s:k} \leftarrow_r \mathbb{Z}_q$. Set $\pi_{\mathsf{c1}:k} \leftarrow \mathfrak{g}_2^{\sum_{i=1}^n s_{ik} P_i(\chi) + r_{s:k} \varrho}$.
   (b) $$(\boldsymbol{\pi}_{\mathsf{c2}:1}, \boldsymbol{\pi}_{\mathsf{c2}:2}) \leftarrow \prod_{i=1}^n (\mathfrak{v}_{i1}, \mathfrak{v}_{i2})^{r_i} \cdot (\mathsf{enc}_{\mathsf{pk}_1}(0; \boldsymbol{r}_s), \mathsf{enc}_{\mathsf{pk}_2}(0; \boldsymbol{r}_s)).$$
7. Return $\pi_{sh} \leftarrow (\mathfrak{v}', (\mathfrak{A}_{i1}, \mathfrak{A}_{i2})_{i=1}^{n-1}, (\pi_{\mathsf{1sp}:i})_{i=1}^n, \pi_{\mathsf{c1}:1}, \pi_{\mathsf{c1}:2}, \boldsymbol{\pi}_{\mathsf{c2}:1}, \boldsymbol{\pi}_{\mathsf{c2}:2})$.

$\mathsf{ver}(\mathsf{crs}; \mathfrak{v}; \mathfrak{v}', (\mathfrak{A}_{i1}, \mathfrak{A}_{i2})_{i=1}^{n-1}, (\pi_{\mathsf{1sp}:i})_{i=1}^n, \pi_{\mathsf{c1}:1}, \pi_{\mathsf{c1}:2}, \boldsymbol{\pi}_{\mathsf{c2}:1}, \boldsymbol{\pi}_{\mathsf{c2}:2})$:

1. Set $(\mathfrak{A}_{n1}, \mathfrak{A}_{n2}) \leftarrow (\mathfrak{g}_1, \mathfrak{g}_2)^{\sum_{i=1}^n P_i(\chi)} / \prod_{i=1}^{n-1} (\mathfrak{A}_{i1}, \mathfrak{A}_{i2})$.
2. Set $(p_{1i}, p_{2j}, p_{3ij}, p_{4j})_{i \in [1 \,..\, n], j \in [1 \,..\, 3]} \leftarrow_r \mathbb{Z}_q^{4n+6}$.
3. Check that `/* Permutation matrix: */`
   $$\prod_{i=1}^n \hat{e}\left((\mathfrak{A}_{i1} \mathfrak{g}_1^{\alpha + P_0(\chi)})^{p_{1i}}, \mathfrak{A}_{i2} \mathfrak{g}_2^{-\alpha + P_0(\chi)}\right) = \hat{e}\left(\prod_{i=1}^n \pi_{\mathsf{1sp}:i}^{p_{1i}}, \mathfrak{g}_2^{\varrho}\right) \cdot \hat{e}(\mathfrak{g}_1, \mathfrak{g}_2)^{(1-\alpha^2) \sum_{i=1}^n p_{1i}}.$$
4. Check that `/* Validity: */`
   $$\hat{e}\left(\mathfrak{g}_1^{\varrho}, \prod_{j=1}^3 \pi_{\mathsf{c2}:2j}^{p_{2j}} \cdot \prod_{i=1}^n \prod_{j=1}^3 (\mathfrak{v}'_{i2j})^{p_{3ij}}\right) = \hat{e}\left(\prod_{j=1}^3 \pi_{\mathsf{c2}:1j}^{p_{2j}} \cdot \prod_{i=1}^n \prod_{j=1}^3 (\mathfrak{v}'_{i1j})^{p_{3ij}}, \mathfrak{g}_2^{\beta}\right).$$
5. Set $\mathfrak{R} \leftarrow \hat{e}(\hat{\mathfrak{g}}_1, \pi_{\mathsf{c1}:2}^{p_{42}} (\pi_{\mathsf{c1}:1} \pi_{\mathsf{c1}:2})^{p_{43}}) \cdot \hat{e}(\mathfrak{h}_1, \pi_{\mathsf{c1}:1}^{p_{41}} \pi_{\mathsf{c1}:2}^{p_{42}}) / \hat{e}\left(\prod_{j=1}^3 \pi_{\mathsf{c2}:1j}^{p_{4j}}, \mathfrak{g}_2^{\varrho}\right)$.
6. Check that `/* Consistency: */`
   $$\prod_{i=1}^n \hat{e}\left(\prod_{j=1}^3 (\mathfrak{v}'_{i1j})^{p_{4j}}, \mathfrak{g}_2^{P_i(\chi)}\right) / \prod_{i=1}^n \hat{e}\left(\prod_{j=1}^3 \mathfrak{v}_{i1j}^{p_{4j}}, \mathfrak{A}_{i2}\right) = \mathfrak{R}.$$

Protocol 1: The new shuffle argument

3.1 Batching

We assume that verifier checks that the batched version [4] of the equations (given in Protocol 1) hold. However, for soundness we need that the individual (non-batched) verification equations hold. We will show that we still have soundness even if the verifier checks batched versions of the equations.

We first prove the following lemma. We state it in the case where $f_i(\boldsymbol{X})$ are polynomials, but one can obviously transform it to the case where $f_i(\boldsymbol{X})$ are Laurent polynomials or even rational functions.

Lemma 1. *Assume $(p_i)_{i \in [1\,..\,k]}$ are values chosen uniformly random from $\mathbb{Z}_q^k$. Assume χ are values chosen uniformly at random from $\mathbb{Z}_q$. Assume f_i are some polynomials of degree* poly(κ). *If the equation $\prod_{i=1}^k \hat{e}(\mathfrak{g}_1, \mathfrak{g}_2)^{f_i(\chi) p_i} = 1_T$ holds, then with probability $\geq 1 - 1/q$ the k pairing equations $\hat{e}(\mathfrak{g}_1, \mathfrak{g}_2)^{f_i(\chi)} = 1_T$, $i \in [1\,..\,k]$ also hold.*

Proof. As the pairing is non-degenerate, $\prod_{i=1}^k \hat{e}(\mathfrak{g}_1, \mathfrak{g}_2)^{f_i(\chi) p_i} = 1_T$ iff $\sum_{i=1}^k f_i(\chi) p_i = 0$. By the Schwartz-Zippel lemma [37,39], with probability $\geq 1 - 1/q$ this means $\sum_{i=1}^k f_i(\chi) Y_i = 0$ as a polynomial, where $(Y_i)_{i \in [1\,..\,k]}$ are random variables corresponding to p_i. Hence all individual coefficients of Y_i must be zero, i.e., $f_i(\chi) = 0$ for $i \in [1\,..\,k]$. But then we have for $i \in [1\,..\,k]$ that $\hat{e}(\mathfrak{g}_1, \mathfrak{g}_2)^{f_i(\chi)} = \hat{e}(\mathfrak{g}_1, \mathfrak{g}_2)^0 = 1_T$, as desired. □

The following corollary follows immediately from Lemma 1.

Corollary 1. *Assume $\chi = (\chi, \alpha, \varrho, \beta, \gamma)$ is chosen uniformly random from $\mathbb{Z}_q^2 \times (\mathbb{Z}_q \setminus \{0\})^2 \times (\mathbb{Z}_q \setminus \{0, -1\})$. Assume $(p_{1i}, p_{2j}, p_{3ij}, p_{4j})_{i \in [1\,..\,n], j \in [1\,..\,3]}$ are values chosen uniformly random from $\mathbb{Z}_q^{4n+6}$. Consider the verification steps in Protocol 1.*

- *If the verification on Step 3 accepts, then (with probability $\geq 1 - 1/q$) for $i \in [1\,..\,n]$,*

$$\hat{e}\left(\mathfrak{A}_{i1} \mathfrak{g}_1^{\alpha + P_0(\chi)}, \mathfrak{A}_{i2} \mathfrak{g}_2^{-\alpha + P_0(\chi)}\right) = \hat{e}\left(\pi_{1\mathsf{sp}:i}, \mathfrak{g}_2^{\varrho}\right) \hat{e}(\mathfrak{g}_1, \mathfrak{g}_2)^{1-\alpha^2} \; . \tag{1}$$

- *If the verification on Step 4 accepts, then with probability $\geq 1 - 1/q$,*

$$\hat{e}(\mathfrak{g}_1^{\varrho}, \pi_{\mathsf{c}2:2i}) = \hat{e}(\pi_{\mathsf{c}2:1i}, \mathfrak{g}_2^{\beta}) \; , \quad i \in [1\,..\,3] \; , \tag{2}$$

$$\hat{e}\left(\mathfrak{g}_1^{\varrho}, \mathfrak{v}'_{i2j}\right) = \hat{e}(\mathfrak{v}'_{i1j}, \mathfrak{g}_2^{\beta}) \; , \quad i \in [1\,..\,n], j \in [1\,..\,3] \; . \tag{3}$$

- *If the verification on Step 6 accepts, then with probability $\geq 1 - 1/q$,*

$$\prod_{i=1}^{n} \hat{e}\left(\mathfrak{v}'_{i1}, \mathfrak{g}_2^{P_i(\chi)}\right) / \prod_{i=1}^{n} \hat{e}\left(\mathfrak{v}_{i1}, \mathfrak{A}_{i2}\right) = \hat{e}\left(\mathfrak{P}_{11}, \pi_{\mathsf{c}1:1}\right) \hat{e}\left(\mathfrak{P}_{12}, \pi_{\mathsf{c}1:2}\right) / \hat{e}\left(\pi_{\mathsf{c}2:1}, \mathfrak{g}_2^{\varrho}\right) \; . \tag{4}$$

Proof. If the verification on Step 3 accepts, then we get that

$$\prod_{i=1}^{n}\left(\hat{e}\left(\mathfrak{A}_{i1}\mathfrak{g}_1^{\alpha+P_0(\chi)},\mathfrak{A}_{i2}\mathfrak{g}_2^{-\alpha+P_0(\chi)}\right)/\left(\hat{e}(\pi_{1\mathsf{sp}:i},\mathfrak{g}_2^{\varrho})\hat{e}(\mathfrak{g}_1,\mathfrak{g}_2)^{1-\alpha^2}\right)\right)^{p_{1i}}$$
$$=\prod_{i=1}^{n}\hat{e}\left((\mathfrak{A}_{i1}\mathfrak{g}_1^{\alpha+P_0(\chi)})^{p_{1i}},\mathfrak{A}_{i2}\mathfrak{g}_2^{-\alpha+P_0(\chi)}\right)/\prod_{i=1}^{n}\left(\hat{e}(\pi_{1\mathsf{sp}:i},\mathfrak{g}_2^{\varrho})\hat{e}(\mathfrak{g}_1,\mathfrak{g}_2)^{1-\alpha^2}\right)^{p_{1i}}$$
$$=1_T.$$

By Lemma 1, with probability $\geq 1-1/q$ we get

$$\hat{e}\left(\mathfrak{A}_{i1}\mathfrak{g}_1^{\alpha+P_0(\chi)},\mathfrak{A}_{i2}\mathfrak{g}_2^{-\alpha+P_0(\chi)}\right)/\left(\hat{e}(\pi_{1\mathsf{sp}:i},\mathfrak{g}_2^{\varrho})\hat{e}(\mathfrak{g}_1,\mathfrak{g}_2)^{1-\alpha^2}\right)=1_T,$$

for $i\in[1\,..\,n]$. Simplifying, this is precisely Eq. (1). The other cases are similar. □

This means that with probability $\geq 1-3/q$, checking the batched version of verification equations (as in Protocol 1) is equivalent to the checking of individual verification equations (as in Corollary 1).

We note that Corollary 1 also holds when χ is chosen according to the distribution, stipulated in Protocol 1.

3.2 Statement of Security

Theorem 1 (Shuffle Security). *The shuffle argument from Protocol 1 is perfectly complete, computationally sound in the GBGM, and perfectly zero knowledge. More precisely, any generic adversary attacking the soundness of the new shuffle argument requires $\Omega(\sqrt{q/n})$ computation.*

Proof. Completeness: we deal with other verifications in later sections. Currently we only show that if the prover and the verifier are honest, then Eq. (4) (and thus also, the verification on step 6 in Protocol 1) accepts. Really, let $\mathfrak{v}'_{ik}=\mathfrak{v}_{\sigma(i)k}\cdot\mathsf{enc}_{\mathsf{pk}_k}(0;\boldsymbol{s}_i)$ and $\mathsf{pk}_1=(\hat{\mathfrak{g}}_1,\mathfrak{h}_1)$ for some $\boldsymbol{s}_i\in\mathbb{Z}_q^{1\times 2}$. Then,

$$\prod_{i=1}^{n}\hat{e}\left(\mathfrak{v}'_{i1},\mathfrak{g}_2^{P_i(\chi)}\right)=\prod_{i=1}^{n}\hat{e}\left(\mathfrak{v}_{\sigma(i)1}\cdot\mathsf{enc}_{\mathsf{pk}_1}(0;\boldsymbol{s}_i),\mathfrak{g}_2^{P_i(\chi)}\right)$$
$$=\prod_{i=1}^{n}\hat{e}\left(\mathfrak{v}_{\sigma(i)1},\mathfrak{g}_2^{P_i(\chi)}\right)\cdot\prod_{i=1}^{n}\hat{e}\left(\mathsf{enc}_{\mathsf{pk}_1}(0;\boldsymbol{s}_i),\mathfrak{g}_2^{P_i(\chi)}\right)$$
$$=\prod_{i=1}^{n}\hat{e}\left(\mathfrak{v}_{i1},\mathfrak{g}_2^{P_{\sigma^{-1}(i)}(\chi)}\right)\cdot\prod_{i=1}^{n}\hat{e}\left(\mathfrak{P}_{11}^{s_1}\mathfrak{P}_{12}^{s_2},\mathfrak{g}_2^{P_i(\chi)}\right)$$
$$=\prod_{i=1}^{n}\hat{e}\left(\mathfrak{v}_{i1},\mathfrak{g}_2^{P_{\sigma^{-1}(i)}(\chi)}\right)\cdot\hat{e}\left(\mathfrak{P}_{11},\mathfrak{g}_2^{\sum_{i=1}^{n}s_{i1}P_i(\chi)}\right)\cdot\hat{e}\left(\mathfrak{P}_{12},\mathfrak{g}_2^{\sum_{i=1}^{n}s_{i2}P_i(\chi)}\right)$$

and

$$\prod_{i=1}^{n} \hat{e}(\mathfrak{v}_{i1}, \mathfrak{A}_{i2}) = \prod_{i=1}^{n} \hat{e}\left(\mathfrak{v}_{i1}, \mathfrak{g}_2^{P_{\sigma^{-1}(i)}(\chi)+r_i\varrho}\right)$$
$$= \prod_{i=1}^{n} \hat{e}\left(\mathfrak{v}_{i1}, \mathfrak{g}_2^{P_{\sigma^{-1}(i)}(\chi)}\right) \cdot \hat{e}\left(\prod_{i=1}^{n} \mathfrak{v}_{i1}^{r_i}, \mathfrak{g}_2^{\varrho}\right) .$$

Hence, as needed,

$$\prod_{i=1}^{n} \hat{e}\left(\mathfrak{v}'_{i1}, \mathfrak{g}_2^{P_i(\chi)}\right) / \prod_{i=1}^{n} \hat{e}(\mathfrak{v}_{i1}, \mathfrak{A}_{i2})$$
$$= \hat{e}\left(\mathfrak{P}_{11}, \mathfrak{g}_2^{\sum_{i=1}^{n} s_{i1} P_i(\chi)}\right) \cdot \hat{e}\left(\mathfrak{P}_{12}, \mathfrak{g}_2^{\sum_{i=1}^{n} s_{i2} P_i(\chi)}\right) / \hat{e}\left(\prod_{i=1}^{n} \mathfrak{v}_{i1}^{r_i}, \mathfrak{g}_2^{\varrho}\right)$$
$$= \hat{e}\left(\mathfrak{P}_{11}, \mathfrak{g}_2^{\sum_{i=1}^{n} s_{i1} P_i(\chi)+r_{s:1}\varrho}\right) \cdot$$
$$\hat{e}\left(\mathfrak{P}_{12}, \mathfrak{g}_2^{\sum_{i=1}^{n} s_{i2} P_i(\chi)+r_{s:2}\varrho}\right) / \hat{e}\left(\mathfrak{P}_{11}^{r_{s:1}} \mathfrak{P}_{12}^{r_{s:2}} \cdot \prod_{i=1}^{n} \mathfrak{v}_{i1}^{r_i}, \mathfrak{g}_2^{\varrho}\right)$$
$$= \hat{e}(\mathfrak{P}_{11}, \pi_{c1:1})\, \hat{e}(\mathfrak{P}_{12}, \pi_{c1:2}) / \hat{e}(\boldsymbol{\pi}_{c2:1}, \mathfrak{g}_2^{\varrho}) .$$

SOUNDNESS. Intuition behind soundness will be given in Sect. 4. Soundness of this argument will be proven in Sects. 5, 6, and 7.

ZERO-KNOWLEDGE: The zero-knowledge property will be proven in Sect. 8. □

Since we work in the GBGM, where the adversary knows how all values were computed, Protocol 1 is actually an argument of knowledge.

4 Intuition Behind Soundness

Throughout this paper, we use a variation of the polynomial commitment scheme of type $\mathsf{com}_j(\boldsymbol{a}; r) := \mathfrak{h}^{\sum_{i=1}^{n} a_i P_i(\chi)+r\varrho}$, where $\mathfrak{h}$ is a generator of $\mathbb{G}_j$, χ and ϱ are random values from $\mathbb{Z}_q$, and $P_i(X)$ are well-chosen polynomials. (The choice of $P_i(X)$ is fixed by the 1-sparsity argument, see Sect. 5.1.) Several variants of this commitment scheme are well-known to be perfectly hiding and computationally binding (under a suitable computational assumption, security of which is usually proved in the GBGM, [26,30]). However, since we only rely on the security of this commitment scheme within the GBGM soundness proof of the shuffle, we will state neither the concrete assumption nor the security requirements (like hiding and binding) of a commitment scheme.

On the last three steps, see Protocol 1, the verifier executes four different verifications, restated in an easier to read format in Corollary 1. Each of these verifications has an intuitive meaning, resulting in a different subargument. However, since all of them have to use the same CRS and the soundness proof is in the GBGM, the subarguments interact strongly.

Our soundness proof in the GBGM uses the following idea. An adversary can only produce group elements from $\mathbb{G}_1$ or $\mathbb{G}_2$ that are products of the elements of the same group given in the CRS; elements of $\mathbb{G}_T$ can also be output by the pairing operation. Let $\boldsymbol{\chi} = (\chi, \alpha, \varrho, \beta, \gamma)$ be concrete (randomly chosen) values from $\mathbb{Z}_q$ and $\boldsymbol{X} = (X, X_\alpha, X_\varrho, X_\beta, X_\gamma)$ be the corresponding random variables. E.g., if $\mathcal{F}(\boldsymbol{X}) = \{F_i(\boldsymbol{X})\}$ is the set of all rational functions such that $\mathfrak{g}_1^{\mathcal{F}(\boldsymbol{\chi})} = \{\mathfrak{g}_1^{F_i(\boldsymbol{\chi})}\}$ is equal to the set of all CRS values in $\mathbb{G}_1$, then any value that the adversary creates in $\mathbb{G}_1$ must be of the form $\mathfrak{g}_1^{A(\boldsymbol{\chi})}$, where $A(\boldsymbol{X}) \in \operatorname{span} \mathcal{F}(\boldsymbol{X})$.

In this way, after taking a discrete logarithm, each verification equation can be written in the form $\mathcal{V}(\boldsymbol{\chi}) = 0$ for some polynomial $\mathcal{V}(\boldsymbol{X})$ known to the adversary. However, since the values in $\boldsymbol{\chi}$ were chosen uniformly random, from the Schwartz-Zippel lemma [37,39] we can conclude that $\mathcal{V}(\boldsymbol{X}) = 0$ as a polynomial (or a rational function), except with negligible probability $O(n)/q$. From $\mathcal{V}(\boldsymbol{X}) = 0$, we deduce that all the coefficients of terms $X_\alpha^{i_1} X_\varrho^{i_2} X_\beta^{i_3} X_\gamma^{i_4}$ in $\mathcal{V}(\boldsymbol{X}) \cdot \mathcal{V}^*(\boldsymbol{X})$ (where $\mathcal{V}^*(\boldsymbol{X})$ is the denominator of $\mathcal{V}(\boldsymbol{X})$) are zero, giving us several equations related to the adversary's chosen values. From these equations and the linear independence of polynomials $P_i(X)$, we can deduce that the adversary's chosen values must be of a certain form, except with negligible probability $O(n)/q$.

More precisely, for symbolic values T and t, define (by following the definition of the CRS in Protocol 1)

$$\begin{aligned}\mathsf{crs}_1(\boldsymbol{X}, T, t) =& t(X) + T_\varrho X_\varrho + T_\alpha \cdot (X_\alpha + P_0(X)) + T_0 P_0(X) + \frac{t^\dagger(X) Z(X)}{X_\varrho} + \\ & \frac{T_{\varrho\beta} X_\varrho}{X_\beta} + \frac{T_\gamma X_\varrho X_\gamma}{X_\beta} \enspace , \\ \mathsf{crs}_2(\boldsymbol{X}, T, t) =& t(X) + T_\varrho X_\varrho + T_\alpha \cdot (-X_\alpha + P_0(X)) + T_1 + T_\gamma X_\gamma + T_\beta X_\beta \enspace ,\end{aligned}$$

where $t^\dagger(X)$ is in the span of $\{((P_i(X) + P_0(X)) - 1)^2 / Z(X)\}_{i=1}^n$ and $t(X)$ is in the span of $\{P_i(X)\}_{i=1}^n$. We will follow the same notation in the rest of the paper. In particular, all "daggered" polynomials (e.g., $b^\dagger(X)$) are in the span of $\{((P_i(X) + P_0(X)) - 1)^2 / Z(X)\}_{i=1}^n$. Since $\deg Z(X) = n + 1$, $\deg t^\dagger(X) \leq n - 1$, and $\deg t(X) \leq n$, then $\deg(\mathsf{crs}_1(\boldsymbol{X}, T, t) \cdot X_\varrho X_\beta) \leq (n - 1) + (n + 1) - 1 + 2 = 2n + 1$. (Multiplication with $X_\varrho X_\beta$ is needed to make $\mathsf{crs}_1(\boldsymbol{X}, T, t)$ a polynomial.) Analogously, $\deg \mathsf{crs}_2(\boldsymbol{X}, T, t) \leq n$. Importantly, $\{P_i(X)\}_{i=0}^n$ is linearly independent. In particular, $P_0(X)$ is linearly independent to all other polynomials present in $\mathsf{crs}_1(\boldsymbol{X})$ and $\mathsf{crs}_2(\boldsymbol{X})$, except the "daggered" polynomial $t^\dagger(X)$.

Since the shuffle argument adversary is a GBGM adversary (and one uses ILin encryption), she knows the following polynomials (in the case of crs_2-functions), Laurent polynomials (in the case of crs_1-functions) or rational functions (in the case of $M_{ij}(\boldsymbol{X})$, $M'_{ij}(\boldsymbol{X})$, and $M_{E:j}(\boldsymbol{X})$), where $\hat{\mathfrak{g}}_2 = \mathfrak{g}_2$:

$$\begin{aligned}
A(\boldsymbol{X}) &= \mathsf{crs}_1(\boldsymbol{X}, A, a) && \text{s.t. } \mathfrak{A}_1 = \mathfrak{g}_1^{A(\chi)} ,\\
B(\boldsymbol{X}) &= \mathsf{crs}_2(\boldsymbol{X}, B, b) && \text{s.t. } \mathfrak{A}_2 = \mathfrak{g}_2^{B(\chi)} ,\\
C(\boldsymbol{X}) &= \mathsf{crs}_1(\boldsymbol{X}, C, c) && \text{s.t. } \pi_{1\mathsf{sp}} = \mathfrak{g}_1^{C(\chi)} ,\\
D_j(\boldsymbol{X}) &= \mathsf{crs}_2(\boldsymbol{X}, D_j, d_j) && \text{s.t. } \pi_{\mathsf{c}1:j} = \mathfrak{g}_2^{D_j(\chi)} ,\\
E_{kj}(\boldsymbol{X}) &= \mathsf{crs}_k(\boldsymbol{X}, E_{kj}, e_{kj}) && \text{s.t. } \pi_{\mathsf{c}2:kj} = \mathfrak{g}_k^{E_{kj}(\chi)} ,\\
V_{ikj}(\boldsymbol{X}) &= \mathsf{crs}_k(\boldsymbol{X}, V_{ikj}, v_{ikj}) && \text{s.t. } \mathfrak{v}_{ikj} = \hat{\mathfrak{g}}_k^{V_{ikj}(\chi)} ,\\
V'_{ikj}(\boldsymbol{X}) &= \mathsf{crs}_k(\boldsymbol{X}, V'_{ikj}, v'_{ikj}) && \text{s.t. } \mathfrak{v}'_{ikj} = \hat{\mathfrak{g}}_k^{V'_{ikj}(\chi)} ,\\
M_{ij}(\boldsymbol{X}) &= V_{ij3}(\boldsymbol{X}) - V_{ij2}(\boldsymbol{X})/(X_\gamma + 1) - V_{ij1}/X_\gamma && \text{s.t. } \mathsf{dec}_{\mathsf{sk}}(\mathfrak{v}_{ij}) = M_{ij}(\chi) ,\\
M'_{ij}(\boldsymbol{X}) &= V'_{ij3}(\boldsymbol{X}) - V'_{ij2}(\boldsymbol{X})/(X_\gamma + 1) - V'_{ij1}/X_\gamma && \text{s.t. } \mathsf{dec}_{\mathsf{sk}}(\mathfrak{v}'_{ij}) = M'_{ij}(\chi) ,\\
M_{E:j}(\boldsymbol{X}) &= E_{j3}(\boldsymbol{X}) - E_{j2}(\boldsymbol{X})/(X_\gamma + 1) - E_{j1}/X_\gamma && \text{s.t. } \mathsf{dec}_{\mathsf{sk}}(\boldsymbol{\pi}_{\mathsf{c}2:j}) = M_{E:j}(\chi) .
\end{aligned} \tag{5}$$

We are now almost ready to explain the meaning of each individual verification equation. Before doing so, we emphasize that a major obstacle in proving soundness in the GBGM is that all subarguments must use the same CRS. In particular, a subargument that is sound by itself might stop being sound due to the elements in the CRS that are added because of other subarguments. Intuitively, we tackle this problem by introducing random variables α (that is only needed in Eq. (1)) and β (that is needed in Eqs. (2) and (3)).

Briefly, the verifier makes three checks. Equation (1), the "permutation matrix argument", guarantees that the prover has committed to a permutation matrix corresponding to some permutation σ. Equations (2) and (3), the "validity argument", guarantee that the ciphertexts have not been formed in a devious way that would make the consistency argument to be unsound. Equation (4), the "consistency argument", guarantees that the prover has used the same permutation σ to shuffle the ciphertexts.

Permutation Matrix Argument. Consider the subargument of Protocol 1, where the verifier just computes $(\mathfrak{A}_{n1}, \mathfrak{A}_{n2})$ and then performs the verification Eq. (1) for each $i = 1$ to n. We will call it the *permutation matrix argument.* In Sect. 5 we motivate this name, by showing that after the permutation matrix argument only, the verifier is convinced that $(\mathfrak{A}_{11}, \dots, \mathfrak{A}_{n1})$ commits to a permutation matrix. For this, we first prove the security of its subargument — the 1-sparsity argument [33] — where the verifier performs the verification Eq. (1) for exactly one i.

To prove the security of permutation matrix argument, we have to solve a quite complicated system of polynomial equations. We do it by using a computer algebra system, see Sect. 5 for more details.

Validity Argument. As a subroutine in our argument, we make the verifier check the validity of all ciphertexts. This is done by checking Eq. (3) (and Eq. (2)). The main goal of the validity check is to show that the prover did not use "forbidden" terms $\mathfrak{g}_k^{P_i(\chi)}$ and $\mathfrak{g}_i^{\varrho}$ when computing the ciphertexts $\mathfrak{v}'_{ik}$ and $\boldsymbol{\pi}_{\mathsf{c}2:k}$. In the case of the Elgamal cryptosystem, the validity argument provides a proof that both $\mathfrak{v}'_{i1}$ and $\mathfrak{v}'_{i2}$ decrypt to a plaintext of form $M_i(\boldsymbol{X}) = \sum M_{ij} f_{ij}(\boldsymbol{X})$, for known coefficients M_{ij} and polynomials $f_{ij}(\boldsymbol{X})$, where none of the rational functions f_{ij} depends on either X or X_ϱ. (See Eq. (12).) Similar assurance is provided about the plaintext hidden in $\boldsymbol{\pi}_{\mathsf{c}2:k}$. Employing validity subarguments allows the consistency subargument to be more efficient than in [19,28].

Consistency Argument. Finally, we show that performing all checks guarantees that $\mathsf{dec}_{\mathsf{sk}}(\mathfrak{v}'_i) = \mathsf{dec}_{\mathsf{sk}}(\mathfrak{v}_{\sigma(i)}) \neq \perp$ for some permutation $\sigma \in S_n$. The main observation is that a permutation of ciphertexts (without rerandomization) is invariant under multiplication: without rerandomizing the ciphertexts, the (non-batched) verification Eq. (4) would just be the identity $\hat{e}(\mathfrak{v}'_{i1}, \mathfrak{g}_2^{P_i(\chi)}) = \hat{e}(\mathfrak{v}_{i1}, \mathfrak{g}_2^{P_{\sigma^{-1}(i)}(\chi)})$, for all i. However, this trivially leaks the permutation σ, and hence is not secure. To ensure privacy, $\mathfrak{v}'_{i1}$ must be rerandomized, and $\mathfrak{g}_2^{P_{\sigma^{-1}(i)}(\chi)}$ must be replaced by a commitment to the unit vector $\boldsymbol{e}_{\sigma^{-1}(i)}$. This makes the final verification slightly more complicated, as we need extra terms to adjust it to the added random values.

A version of Eq. (4) was also used in [19,28,33]. However, the shuffle arguments from [19,28] need to execute two versions of Eq. (4), once with $P_i(X)$ and once with different carefully chosen polynomials $\hat{P}_i(X)$ in $\mathbb{G}_2$. (See [19,28] for an explanation.) In addition, one must prove that those two versions are consistent between each other (by providing a same-message argument, in the terminology of [19]). This makes the arguments of [19,28] quite complicated.

Similarly to [33], we avoid this complication by having a validity argument on the ciphertexts. Since valid ciphertexts are not dependent of $P_i(X)$, it suffices for the verifier to execute just one version of Eq. (4).

5 Permutation Matrix Argument

In this section, we show that a subargument of Protocol 1, where the verifier only computes $\mathfrak{A}_{n1}$ as shown and then executes verification at Eq. (1) (for each $i \in [1 \mathinner{..} n]$) gives us a permutation matrix argument. This argument will be by far the most complex subargument that we use.

5.1 New 1-Sparsity Argument

In a 1-sparsity argument [33], the prover aims to convince the verifier that he knows how to open a commitment $\mathfrak{A}_1$ to $(\boldsymbol{a}, r)$, such that *at most* one coefficient a_I is non-zero. If, in addition, $a_I = 1$, then we have a unit vector argument [19]. A 1-sparsity argument can be constructed by using square span programs [16], an especially efficient variant of the quadratic span programs of [22]. We prove its security in the GBGM and therefore use a technique similar to that of [27], and this introduces some complications as we will demonstrate below. While we start using ideas behind the unit vector argument of [19], we only obtain a 1-sparsity argument. Then, in Sect. 5, we show how to obtain an efficient permutation matrix argument from it.

Clearly, $\boldsymbol{a} \in \mathbb{Z}_q^n$ is a unit vector iff the following $n+1$ conditions hold [19]:

- $a_i \in \{0,1\}$ for $i \in [1 \mathinner{..} n]$ (i.e., $\boldsymbol{a}$ is Boolean), and
- $\sum_{i=1}^n a_i = 1$.

Let $\{0,2\}^{n+1}$ denote the set of $(n+1)$-dimensional vectors where every coefficient is from $\{0,2\}$, let $\circ$ denote the Hadamard (entry-wise) product of two vectors, let $V := \begin{pmatrix} 2 \cdot I_{n \times n} \\ \mathbf{1}_n^\top \end{pmatrix} \in \mathbb{Z}_q^{(n+1) \times n}$ and $\boldsymbol{b} := \begin{pmatrix} \mathbf{0}_n \\ 1 \end{pmatrix} \in \mathbb{Z}_q^{n+1}$. Clearly, the above $n+1$ conditions hold iff $V\boldsymbol{a} + \boldsymbol{b} \in \{0,2\}^{n+1}$, i.e.,

$$(V\boldsymbol{a} + \boldsymbol{b} - \mathbf{1}_{n+1}) \circ (V\boldsymbol{a} + \boldsymbol{b} - \mathbf{1}_{n+1}) = \mathbf{1}_{n+1} \ . \tag{6}$$

Let ω_i, $i \in [1 \mathinner{..} n+1]$ be $n+1$ different values. Let

$$Z(X) := \prod_{i=1}^{n+1} (X - \omega_i)$$

be the unique degree $n+1$ monic polynomial, such that $Z(\omega_i) = 0$ for all $i \in [1 \mathinner{..} n+1]$. Let the ith Lagrange basis polynomial

$$\ell_i(X) := \prod_{j \in [1 \mathinner{..} n+1], j \neq i} ((X - \omega_j)/(\omega_i - \omega_j))$$

be the unique degree n polynomial, s.t. $\ell_i(\omega_i) = 1$ and $\ell_i(\omega_j) = 0$ for $j \neq i$.

For $i \in [1 \mathinner{..} n]$, let $P_i(X)$ be the polynomial that interpolates the ith column of the matrix V. That is,

$$P_i(X) = 2\ell_i(X) + \ell_{n+1}(X)$$

for $i \in [1 \mathinner{..} n]$. Let

$$P_0(X) = \ell_{n+1}(X) - 1$$

be the polynomial that interpolates $\boldsymbol{b} - \mathbf{1}_{n+1}$. In the rest of this paper, we will heavily use the following simple result.

Lemma 2. *$\{P_i(X)\}_{i=0}^n$ is linearly independent.*

Proof. Assume that $\sum_{i=0}^n b_i P_i(X) = 0$ for some constants b_i. Thus, $\sum_{i=0}^n b_i P_i(\omega_k) = 0$ for each k. Consider any $k \in [1 \mathinner{..} n]$. Then, $0 = b_0 P_0(\omega_k) + \sum_{i=1}^n b_i P_i(\omega_k) = b_0(\ell_{n+1}(\omega_k) - 1) + \sum_{i=1}^n b_i (2\ell_i(\omega_k) + \ell_{n+1}(\omega_k)) = -b_0 + 2b_k$. Thus, $b_k = b_0/2$ for $k \in [1 \mathinner{..} n]$. Consider now the case $k = n+1$, then $0 = b_0 P_0(\omega_{n+1}) + \sum_{i=1}^n b_i P_i(\omega_{n+1}) = b_0(\ell_{n+1}(\omega_{n+1}) - 1) + \sum_{i=1}^n b_i(2\ell_i(\omega_{n+1}) + \ell_{n+1}(\omega_{n+1})) = \sum_{i=1}^n b_i = n/2 \cdot b_0$. Thus $b_k = 0$ for $k \in [0 \mathinner{..} n]$. □

We arrive at the polynomial $Q(X) = (\sum_{i=1}^{n} a_i P_i(X) + P_0(X))^2 - 1 = (P_I(X) + P_0(X))^2 - 1$ (here, we used the fact that $\boldsymbol{a} = \boldsymbol{e_I}$ for some $I \in [1 \,..\, n]$), such that $\boldsymbol{a}$ is a unit vector iff $Z(X) \mid Q(X)$. As in [27], to obtain privacy, we now add randomness $A_\varrho X_\varrho$ to $Q(X)$, arriving at the degree $2n$ polynomial

$$Q_{wi}(X, X_\varrho) = (P_I(X) + P_0(X) + A_\varrho X_\varrho)^2 - 1 \ . \tag{7}$$

Here, X_ϱ is a special independent random variable, and $A_\varrho \leftarrow_r \mathbb{Z}_q$. This means that we will use an instantiation of the polynomial commitment scheme (see Sect. 4) with $P_i(X)$ defined as in the current subsection.

The new 1-sparsity argument is the subargument of the shuffle argument on Protocol 1, where the verifier only executes verification step Eq. (1) for one concrete value of i.

Theorem 2. *Consider $i \in [1 \,..\, n]$. The 1-sparsity argument is perfectly complete. The following holds in the GBGM, given that the generic adversary works in polynomial time. If the honest verifier accepts on Step 3 for this i, then there exists $I \in [1 \,..\, n]$, such that*

$$\mathfrak{A}_{i1} = \mathfrak{g}_1^{a(\chi) + A_\varrho \varrho + A_\alpha(\alpha + P_0(\chi))} \ , \tag{8}$$

where $a(X) = (1 + A_\alpha) P_I(X)$ for some constant A_α.

Proof. Completeness: For an honest prover, $\mathfrak{A}_{i1} = \mathfrak{g}_1^{A(\chi)}$, $\mathfrak{A}_{i2} = \mathfrak{g}_2^{B(\chi)}$, and $\pi_{\mathsf{1sp}:i} = \mathfrak{g}_1^{C(\chi)}$, where $A(\boldsymbol{X}) = B(\boldsymbol{X}) = P_I(X) + A_\varrho X_\varrho$ and $C(\boldsymbol{X}) = 2A_\varrho \cdot (A(\boldsymbol{X}) + P_0(X)) - A_\varrho^2 X_\varrho + Q_{wi}(X, X_\varrho)/X_\varrho$. Write

$$\mathcal{V}_{1sp}(\boldsymbol{X}) := (A(\boldsymbol{X}) + X_\alpha + P_0(X)) \cdot (B(\boldsymbol{X}) - X_\alpha + P_0(X)) - C(\boldsymbol{X}) \cdot X_\varrho - (1 - X_\alpha^2) \ . \tag{9}$$

The verification equation Eq. (1) assesses that $\mathcal{V}_{1sp}(\chi) = 0$. This simplifies to $\mathcal{V}_{1sp}(\boldsymbol{X}) = (A_\varrho X_\varrho + P_I(X) + P_0(X))^2 - 1 - Q_{wi}(X, X_\varrho)$. Hence for an honest prover, it follows from Eq. (7) that $\mathcal{V}_{1sp}(\chi) = 0$.

Soundness: Assume that the verifier has accepted inputs $\mathsf{cell}_{i_1} = A(\boldsymbol{X})$, $\mathsf{cell}_{i_2} = B(\boldsymbol{X})$, and $\mathsf{cell}_{i_3} = C(\boldsymbol{X})$, for some polynomials $A(\boldsymbol{X})$, $B(\boldsymbol{X})$, and $C(\boldsymbol{X})$. In the GBGM, the adversary knows all coefficients. (This corresponds to $\mathfrak{A}_{i1} = \mathfrak{g}_1^{A(\chi)}, \mathfrak{A}_{i2} = \mathfrak{g}_2^{B(\chi)}, \pi_{\mathsf{1sp}:i} = \mathfrak{g}_1^{C(\chi)}$.) Let $\mathcal{V}_{1sp}(\boldsymbol{X})$ then be as in Eq. (9) with $A(\boldsymbol{X})$, $B(\boldsymbol{X})$, and $C(\boldsymbol{X})$ as in Eq. (5). Let $\mathcal{V}_{1sp}^*(\boldsymbol{X})) := X_\varrho X_\beta$. Clearly, $\mathcal{V}_{1sp}(\boldsymbol{X}) \cdot \mathcal{V}_{1sp}^*(\boldsymbol{X})$ is a polynomial, with $\deg(\mathcal{V}_{1sp}(\boldsymbol{X}) \cdot \mathcal{V}_{1sp}^*(\boldsymbol{X})) \leq 3n + 1$. Since the verifier accepts, $\mathcal{V}_{1sp}(\boldsymbol{X}) = 0$ as a polynomial.

In Table 2, we enlist all the coefficients of $\mu(\boldsymbol{i}) = X_\alpha^{i_1} X_\varrho^{i_2} X_\beta^{i_3} X_\gamma^{i_4}$ in $\mathcal{V}_{1sp}(\boldsymbol{X}) \cdot \mathcal{V}_{1sp}^*(\boldsymbol{X})$. We remark that we found those polynomials by using a computer algebra system[1], but they can be verified manually.

---

[1] In the concrete case, Mathematica 9.0, but any other reasonably powerful system can be used. See [1] for references on the prior use of computer algebra systems to prove security in the generic (bilinear) group model.

Table 2. $\mathrm{coeff}_{\mu(i)}(\mathcal{V}_{1sp}(\boldsymbol{X}) \cdot \mathcal{V}^*_{1sp}(\boldsymbol{X}))$, where $\mu(i) = X_\alpha^{i_1} X_\varrho^{i_2} X_\beta^{i_3} X_\gamma^{i_4}$

$\{i_1, \ldots, i_4\}$	$\mathrm{coeff}_{\mu(i)}(\mathcal{V}_{1sp}(\boldsymbol{X}) \cdot \mathcal{V}^*_{1sp}(\boldsymbol{X}))$
$\{1,2,1,0\}$	$-A_\varrho(B_\alpha+1)+(A_\alpha+1)B_\varrho - C_\alpha$
$\{1,2,0,1\}$	$-A_\gamma(B_\alpha+1)$
$\{1,2,0,0\}$	$-A_{\varrho\beta}(B_\alpha+1)$
$\{1,1,2,0\}$	$(A_\alpha+1)B_\beta$
$\{1,1,1,1\}$	$(A_\alpha+1)B_\gamma$
$\{1,1,1,0\}$	$-a(X)(B_\alpha+1)+(A_\alpha+1)(b(X)+B_1) - A_0(B_\alpha+1)P_0(X)$
$\{1,0,1,0\}$	$-(B_\alpha+1)Z(X)a^\dagger(X)$
$\{0,3,1,0\}$	$A_\varrho B_\varrho - C_\varrho$
$\{0,3,0,1\}$	$A_\gamma B_\varrho - C_\gamma$
$\{0,3,0,0\}$	$A_{\varrho\beta} B_\varrho - C_{\varrho\beta}$
$\{0,2,2,0\}$	$A_\varrho B_\beta$
$\{0,2,1,1\}$	$A_\gamma B_\beta + A_\varrho B_\gamma$
$\{0,2,1,0\}$	$a(X)B_\varrho + A_\varrho(b(X)+B_1) + A_{\varrho\beta}B_\beta +$ $P_0(X)(A_\varrho(B_\alpha+1)+(A_\alpha+A_0+1)B_\varrho - C_\alpha - C_0) - c(X)$
$\{0,2,0,2\}$	$A_\gamma B_\gamma$
$\{0,2,0,1\}$	$A_\gamma(b(X)+B_1) + A_{\varrho\beta}B_\gamma + A_\gamma(B_\alpha+1)P_0(X)$
$\{0,2,0,0\}$	$A_{\varrho\beta}(b(X)+B_1) + A_{\varrho\beta}(B_\alpha+1)P_0(X)$
$\{0,1,2,0\}$	$a(X)B_\beta + (A_\alpha+A_0+1)B_\beta P_0(X)$
$\{0,1,1,1\}$	$a(X)B_\gamma + (A_\alpha+A_0+1)B_\gamma P_0(X)$
$\{0,1,1,0\}$	$-Z(X)c^\dagger(X) + P_0(X)(a(X)(B_\alpha+1)+(A_\alpha+A_0+1)(b(X)+B_1)) +$ $a(X)(b(X)+B_1) + (A_\alpha+A_0+1)(B_\alpha+1)P_0(X)^2 - 1 +$ $B_\varrho Z(X)a^\dagger(X)$
$\{0,0,2,0\}$	$B_\beta Z(X)a^\dagger(X)$
$\{0,0,1,1\}$	$B_\gamma Z(X)a^\dagger(X)$
$\{0,0,1,0\}$	$Z(X)(b(X)+B_1)a^\dagger(X) + (B_\alpha+1)P_0(X)Z(X)a^\dagger(X)$

Consider now each monomial of $\mathrm{coeff}_{\mu(i)}(\mathcal{V}_{1sp}(\boldsymbol{X}) \cdot \mathcal{V}^*_{1sp}(\boldsymbol{X})) = 0$ as a polynomial $F_i^*(\boldsymbol{Y})$ of formal variables $\boldsymbol{Y} := (a(X), A_\varrho, A_\alpha, \ldots, C_\gamma)$ (i.e., in all coefficients of $A(\boldsymbol{X})$, $B(\boldsymbol{X})$, and $C(\boldsymbol{X})$). We can now set $F_i^*(\boldsymbol{Y}) = 0$ for each monomial, and the solution set of this system of polynomial equations gives us all possible ways of "cheating" the adversary can do. However, the resulting polynomial equation system is too complicated, and moreover, it contains some formal variables that are *not* linearly independent, like $a(X)$ and $a^\dagger(X)$.

We hence execute two additional steps. First, we take into account (by using Lemma 2) that $P_0(X)$ is linearly independent of all other polynomials except "daggered" polynomials $a^\dagger(X)$ and $c^\dagger(X)$. This allows us to simplify some of the coefficients and gives some more polynomial equations. After that step, we get a new polynomial equation system $\{F_i(\boldsymbol{Y}) = 0\}$ for some polynomials F_i.

Second, we use a computer algebra system to derive a Gröbner basis [11] in variables in $\boldsymbol{Y}$ for the system $\{F_i(\boldsymbol{Y}) = 0\}$. By using lexicographic order (more precisely, we used the function `GroebnerBasis` of Mathematica, with parameters `MonomialOrder -> Lexicographic` and `Method -> "Buchberger"`), we get the Gröbner basis $\{\mathcal{B}_i(\boldsymbol{Y})\}$ on Fig. 1.

$$\begin{pmatrix}
C_\gamma \\
C_{\varrho\beta} \\
4C_\varrho - C_0(C_0 + 2C_\alpha) \\
c(X)^2 + 2(C_0 + C_\alpha)P_0(X)c(X) + (C_0 + C_\alpha)^2\left(P_0(X)^2 - Z(X)c^\dagger(X) - 1\right) \\
B_\beta \\
B_\gamma \\
2B_\varrho - (B_\alpha + 1)(C_0 + 2C_\alpha) \\
(b(X) + B_1)(C_0 + C_\alpha) - c(X)(B_\alpha + 1) \\
b(X)c(X) + (B_1 + 2(B_\alpha + 1)P_0(X))c(X) + \\
\qquad (B_\alpha + 1)(C_0 + C_\alpha)\left(P_0(X)^2 - Z(X)c^\dagger(X) - 1\right) \\
-(B_\alpha + 1)^2 - B_\alpha(B_\alpha + 2)Z(X)c^\dagger(X) - Z(X)c^\dagger(X) + \\
\qquad (b(X) + B_1 + (B_\alpha + 1)P_0(X))^2 \\
A_\gamma \\
A_{\varrho\beta} \\
Z(X)a^\dagger(X) \\
A_0 \\
B_\alpha + A_\alpha(B_\alpha + 1) \\
2A_\varrho - (A_\alpha + 1)C_0 \\
a(X) - (A_\alpha + 1)^2(b(X) + B_1)
\end{pmatrix}$$

Fig. 1. Gröbner basis $\{\mathcal{B}_i(\boldsymbol{Y})\}$

The system of polynomial equations $\{\mathcal{B}_i(\boldsymbol{Y}) = 0\}$ can be solved manually. First, we simplify this system by setting $C_\gamma = 0$, $C_{\varrho\beta} = 0$, $B_\beta = 0$, $B_\gamma = 0$, $A_\gamma = 0$, $A_{\varrho\beta} = 0$, $a^\dagger(X) = 0$, $A_0 = 0$, $B_\alpha = -A_\alpha/(A_\alpha + 1)$, $C_0 = 2A_\varrho/(A_\alpha + 1)$, $b(X) = a(X)/(A_\alpha + 1)^2 - B_1$, $C_\varrho = (A_\alpha + 1)B_\varrho((A_\alpha + 1)B_\varrho - C_\alpha)$. Then, we get a new system of polynomial equations, with the Gröbner basis $\{\mathcal{B}'_i(\boldsymbol{Y}) = 0\}$ as given on Fig. 2.

We can further simplify this system by noting that $C_\alpha = (A_\alpha + 1)B_\varrho - A_\varrho/(A_\alpha + 1)$ and thus $c(X) = a(X)\left(A_\varrho/(A_\alpha + 1)^2 + B_\varrho\right)$. By inserting those two values to the Gröbner basis $\{\mathcal{B}'_i(\boldsymbol{Y})\}$, we get that the resulting system of polynomial equations has the following simple Gröbner basis $\{\mathcal{B}''_i(\boldsymbol{Y})\}$:

$$\left((A_\alpha + 1)^2\left(-Z(X)c^\dagger(X) + P_0(X)^2 - 1\right) + 2a(X)(A_\alpha + 1)P_0(X) + a(X)^2\right)$$

$$\begin{pmatrix} -\left(C_\alpha - 2(A_\alpha+1)B_\varrho\right)^2 \left(-Z(X)c^\dagger(X) + P_0(X)^2 - 1\right) + \\ 2c(X)P_0(X)\left(C_\alpha - 2(A_\alpha+1)B_\varrho\right) - c(X)^2 \\ \frac{\left(C_\alpha - 2(A_\alpha+1)B_\varrho\right)^2\left(-Z(X)c^\dagger(X)+P_0(X)^2-1\right)+2c(X)P_0(X)\left(2(A_\alpha+1)B_\varrho - C_\alpha\right)+c(X)^2}{A_\alpha+1} \\ (A_\alpha+1)\left(C_\alpha - (A_\alpha+1)B_\varrho\right) + A_\varrho \\ (A_\alpha+1)\Big(2(A_\alpha+1)B_\varrho\left(-Z(X)c^\dagger(X)+P_0(X)^2-1\right) + \\ C_\alpha\left(Z(X)c^\dagger(X) - P_0(X)^2+1\right) + 2c(X)P_0(X)\Big) + a(X)c(X) \\ -(A_\alpha+1)\left(c(X) - 2a(X)B_\varrho\right) - a(X)C_\alpha \\ a(X)\left(\frac{C_\alpha}{A_\alpha+1} - 2B_\varrho\right) + c(X) \\ (A_\alpha+1)^2\left(-Z(X)c^\dagger(X)+P_0(X)^2-1\right) + 2a(X)(A_\alpha+1)P_0(X) + a(X)^2 \end{pmatrix} .$$

Fig. 2. Gröbner basis $\{\mathcal{B}'_i(\boldsymbol{Y})\}$

By solving $\mathcal{B}''_i(\boldsymbol{Y}) = 0$, we get

$$c^\dagger(X) = \frac{\left(\frac{a(X)}{A_\alpha+1} + P_0(X)\right)^2 - 1}{Z(X)} ,$$

which is a witness that $a(X)/(A_\alpha+1) = P_I(X)$ for some I.

Hence, if verification Step 3 in Protocol 1 succeeds for $j = i$, then, after replacing all coefficients with values derived in this proof, we get

$$\begin{aligned} A(\boldsymbol{X}) =& a(X) + A_\varrho X_\varrho + A_\alpha\left(X_\alpha + P_0(X)\right) , \\ B(\boldsymbol{X}) =& \frac{a(X)}{(A_\alpha+1)^2} + B_\varrho X_\varrho + \frac{A_\alpha(X_\alpha - P_0(X))}{A_\alpha+1} . \end{aligned}$$

Hence, Eq. (8) holds. □

5.2 Permutation Matrix Argument

Assume we explicitly compute $\mathfrak{A}_{n1} = \mathfrak{g}_1^{\sum_{i=1}^n P_i(X)} / \prod_{j=1}^{n-1} \mathfrak{A}_{j1}$ as in Protocol 1, and then apply the 1-sparsity argument to each $\mathfrak{A}_{i1}$, $i \in [1 .. n]$. Then, as in [33], we get that $(\mathfrak{A}_{11}, \dots, \mathfrak{A}_{n1})$ commits to a permutation matrix. More precisely, according to Eq. (8), the ith commitment is represented by the polynomial

$$A_i(\boldsymbol{X}) = a_i(X) + A_{\varrho i}X_\varrho + A_{\alpha i}\cdot\left(X_\alpha + P_0(X)\right) ,$$

where $a_i(X)/(1 + A_{\alpha i}) = P_{f(i)}(X)$ for some f. Since $\sum_i A_i(\boldsymbol{X}) = \sum_i P_i(X)$, we get in particular that $\sum_i (A_{\alpha i}+1)P_{f(i)}(X) = \sum_i P_i(X)$. Since due to Lemma 2, $\{P_i(X)\}_{i=0}^n$ is linearly independent, it means that $A_{\alpha i} = 0$ for each i, and $f = \sigma^{-1}$ is a permutation.

Theorem 3. *The described permutation matrix argument is perfectly complete. The following holds in the GBGM, assuming that the generic adversary works in polynomial time. If the honest verifier accepts Eq. (1) for all $i \in [1 .. n]$, and $(\mathfrak{A}_{n1}, \mathfrak{A}_{n2})$ is explicitly computed as in Protocol 1, then there exists a permutation $\sigma \in S_n$ and randomizers $A_{\varrho i}$, such that*

$$\mathfrak{A}_{i1} = \mathfrak{g}_1^{P_{\sigma^{-1}(i)}(\chi) + A_{\varrho i}\varrho} \quad \textit{for all } i \in [1 .. n] \ . \tag{10}$$

6 Validity Argument

The shuffle argument employs validity arguments for $(\boldsymbol{\pi}_{c2:1}, \boldsymbol{\pi}_{c2:2})$ and for each $(\mathfrak{v}'_{i1}, \mathfrak{v}'_{i2})$. We outline this argument for $(\boldsymbol{\pi}_{c2:1}, \boldsymbol{\pi}_{c2:2})$, the argument is the same for $(\mathfrak{v}'_{i1}, \mathfrak{v}'_{i2})$. More precisely, in the validity argument for $(\boldsymbol{\pi}_{c2:1}, \boldsymbol{\pi}_{c2:2})$, the verifier checks that $\hat{e}(\mathfrak{g}_1^{\varrho}, \pi_{c2:2j}) = \hat{e}(\pi_{c2:1j}, \mathfrak{g}_2^{\beta})$ for $j \in [1 .. 3]$. Thus, for

$$\mathcal{V}_{val:j}(\boldsymbol{X}) = E_{1j}(\boldsymbol{X}) X_\beta - X_\varrho E_{2j}(\boldsymbol{X}) \ ,$$

this argument guarantees that in the GBGM, $\mathcal{V}_{val:j}(\boldsymbol{X}) = 0$ for $j \in [1 .. 3]$.

Table 3. $coeff_{\mu(\boldsymbol{i})}(\mathcal{V}_{val:j}(\boldsymbol{X}) \cdot X_\varrho X_\beta)$, where $\mu(\boldsymbol{i}) = X_\alpha^{i_1} X_\varrho^{i_2} X_\beta^{i_3} X_\gamma^{i_4}$

$\{i_1, \dots, i_4\}$	$\mathrm{coeff}_{\mu(\boldsymbol{i})}(\mathcal{V}_{val:j}(\boldsymbol{X}) \cdot X_\varrho X_\beta)$
$\{1, 2, 1, 0\}$	$E_{2j,\alpha}$
$\{1, 1, 2, 0\}$	$E_{1j,\alpha}$
$\{0, 3, 1, 0\}$	$-E_{2j,\varrho}$
$\{0, 2, 2, 0\}$	$E_{1j,\varrho} - E_{2j,\beta}$
$\{0, 2, 1, 1\}$	$E_{1j,\gamma} - E_{2j,\gamma}$
$\{0, 2, 1, 0\}$	$E_{1j,\varrho\beta} - e_{2j}(X) - E_{2j,\alpha} P_0(X) - E_{2j,1}$
$\{0, 1, 2, 0\}$	$e_{1j}(X) + (E_{1j,\alpha} + E_{1j,0}) P_0(X)$
$\{0, 0, 2, 0\}$	$Z(X) e_{1j}^{\dagger}(X)$

In this case, it is much easier to solve the resulting polynomial system of equations than it was in Sect. 5. First, we find the coefficients of $\mu(\boldsymbol{i}) = X_\alpha^{i_1} X_\varrho^{i_2} X_\beta^{i_3} X_\gamma^{i_4}$ in $\mathcal{V}_{val:j}(\boldsymbol{X}) \cdot X_\varrho X_\beta$, see Table 3. Taking into account (see Lemma 2) that $\{P_i(X)\}_{i=0}^n$ are linearly independent and that $1 \notin \mathrm{span}\{P_i(X)\}_{i=1}^n$, we get from solving this polynomial system of equations that

$$\begin{aligned}
E_{1j}(\boldsymbol{X}) &= (E_{1j,\varrho\beta} + E_{2j,\beta} X_\beta + E_{2j,\gamma} X_\gamma) X_\varrho / X_\beta \ , \\
E_{2j}(\boldsymbol{X}) &= E_{1j,\varrho\beta} + E_{2j,\beta} X_\beta + E_{2j,\gamma} X_\gamma \ , \quad \text{and thus} \\
M_E(\boldsymbol{X}) &= M_{E:1}(\boldsymbol{X}) = M_{E:2}(\boldsymbol{X}) = E_{23}(\boldsymbol{X}) - E_{22}(\boldsymbol{X})/(X_\gamma + 1) - E_{21}(\boldsymbol{X})/X_\gamma \\
&= M_{E:1} + M_{E:2} X_\beta + M_{E:3} X_\gamma + \frac{M_{E:4}}{X_\gamma} + \frac{M_{E:5} X_\beta}{X_\gamma} + \frac{M_{E:6}}{X_\gamma + 1} + \frac{M_{E:7} X_\beta}{X_\gamma + 1}
\end{aligned} \tag{11}$$

for some coefficients $M_{E:j}$ known to the adversary. Here, say $M_{E:2}(\boldsymbol{X}) = E_{23}(\boldsymbol{X}) - E_{22}(\boldsymbol{X})/(X_\gamma + 1) - E_{21}(\boldsymbol{X})/X_\gamma$; we will call such an operation a "generic decryption' in group $\mathbb{G}_k$.

Theorem 4. *The validity argument for* $(\pi_{c2:1}, \pi_{c2:2})$ *is perfectly complete. The following holds in the GBGM, assuming that the generic adversary works in polynomial time. If the honest verifier accepts Eq.* (2)*, then the generic adversary knows coefficients* $M_{E:j}$*, s.t.* $\mathsf{dec}_{\mathsf{sk}}(\pi_{c2}) = M_E(\chi)$ *where* $M_E(\boldsymbol{X})$ *is as in Eq.* (11)*.*

Assuming similarly that also validity of $V_{i1j}(\boldsymbol{X})$, $V_{i2j}(\boldsymbol{X})$, $V'_{i1j}(\boldsymbol{X})$, and $V'_{i2j}(\boldsymbol{X})$ is checked, we get that

$$\begin{aligned}
&V_{i1j}(\boldsymbol{X}) = (V_{i1j,\varrho\beta} + V_{i2j,\beta}X_\beta + V_{i2j,\gamma}X_\gamma)X_\varrho/X_\beta \ , \\
&V_{i2j}(\boldsymbol{X}) = V_{i1j,\varrho\beta} + V_{i2j,\beta}X_\beta + V_{i2j,\gamma}X_\gamma \ , \\
&V'_{i1j}(\boldsymbol{X}) = (V'_{i1j,\varrho\beta} + V'_{i2j,\beta}X_\beta + V'_{i2j,\gamma}X_\gamma)X_\varrho/X_\beta \ , \\
&V'_{i2j}(\boldsymbol{X}) = V'_{i1j,\varrho\beta} + V'_{i2j,\beta}X_\beta + V'_{i2j,\gamma}X_\gamma \ , \quad \text{and thus} \\
M_i(\boldsymbol{X}) =& M_{i1}(\boldsymbol{X}) = M_{i2}(\boldsymbol{X}) = V_{i23}(\boldsymbol{X}) - V_{i22}(\boldsymbol{X})/(X_\gamma + 1) - V_{i21}(\boldsymbol{X})/X_\gamma \\
=& M_{i1} + M_{i2}X_\beta + M_{i3}X_\gamma + \frac{M_{i4}}{X_\gamma} + \frac{M_{i5}X_\beta}{X_\gamma} + \frac{M_{i6}}{X_\gamma + 1} + \frac{M_{i7}X_\beta}{X_\gamma + 1} \\
M'_i(\boldsymbol{X}) =& M'_{i1}(\boldsymbol{X}) = M'_{i2}(\boldsymbol{X}) = V'_{i23}(\boldsymbol{X}) - V'_{i22}(\boldsymbol{X})/(X_\gamma + 1) - V'_{i21}(\boldsymbol{X})/X_\gamma \\
=& M'_{i1} + M'_{i2}X_\beta + M'_{i3}X_\gamma + \frac{M'_{i4}}{X_\gamma} + \frac{M'_{i5}X_\beta}{X_\gamma} + \frac{M'_{i6}}{X_\gamma + 1} + \frac{M'_{i7}X_\beta}{X_\gamma + 1} \qquad (12)
\end{aligned}$$

for some coefficients M_{ik}, $k \in [1 \mathinner{..} 3]$, known to the adversary.

Corollary 2. *The validity argument for* $(\mathfrak{v}'_{i1}, \mathfrak{v}'_{i2})$ *is perfectly complete. The following holds in the GBGM, assuming that the generic adversary works in polynomial time. If the honest verifier accepts Eq.* (3) *for some* $i \in [1 \mathinner{..} n]$*, then the generic adversary knows coefficients* M'_{ij}*, s.t.* $\mathsf{dec}_{\mathsf{sk}}(\mathfrak{v}'_i) = M'_i(\chi)$ *where* $M'_i(\boldsymbol{X})$ *is as in Eq.* (12)*.*

7 Consistency Argument

We call the subargument of Protocol 1, where the verifier only executes the last verification (namely, Eq. (4)), the *consistency argument.* Intuitively, the consistency argument guarantees that the ciphertexts have been permuted by using the same permutation according to which the elements $\mathfrak{g}_k^{P_i(\chi)}$ were permuted inside the commitments $\mathfrak{A}_{i1}$.

According to Sects. 5 and 6, the permutation matrix argument and validity arguments are "sound". In what follows, we show that if the verifier executes all verification steps in Protocol 1, then this shuffle argument is sound in the GBGM. Now, we are finally able to finish the soundness proof of Theorem 1.

Proof (Of Theorem 1). Since all the batch verifications in Protocol 1 accept, by Corollary 1 we have that with probability $\geq 1 - 3/q$ all individual equations also hold. Since the permutation matrix argument and ciphertext validity are sound, Eq. (8) and Eq. (12) hold with overwhelming probability for all i.

Since we have a generic adversary, from the verification equation Eq. (4) we get that $\mathcal{V}_{cons:j}(\boldsymbol{X}) = 0$ for $j \in [1\,..\,3]$, where

$$\begin{aligned}
\mathcal{V}_{cons:1}(\boldsymbol{X}) =& \sum (V'_{i11}(\boldsymbol{X}) - V_{\sigma(i)11}(\boldsymbol{X}))P_i(X) - \sum V_{i11}(\boldsymbol{X}) r_i X_\varrho \\
&- D_1(\boldsymbol{X}) X_\gamma X_\varrho / X_\beta + E_{11}(\boldsymbol{X}) X_\varrho \ , \\
\mathcal{V}_{cons:2}(\boldsymbol{X}) =& \sum (V'_{i12}(\boldsymbol{X}) - V_{\sigma(i)12}(\boldsymbol{X}))P_i(X) - \sum V_{i12}(\boldsymbol{X}) r_i X_\varrho \\
&- D_2(\boldsymbol{X})(X_\gamma + 1) X_\varrho / X_\beta + E_{12}(\boldsymbol{X}) X_\varrho \ , \\
\mathcal{V}_{cons:3}(\boldsymbol{X}) =& \sum (V'_{i13}(\boldsymbol{X}) - V_{\sigma(i)13}(\boldsymbol{X}))P_i(X) - \sum V_{i13}(\boldsymbol{X}) r_i X_\varrho \\
&- D_1(\boldsymbol{X}) X_\varrho / X_\beta - D_2(\boldsymbol{X}) X_\varrho / X_\beta + E_{13}(\boldsymbol{X}) X_\varrho
\end{aligned}$$

are rational functions. By doing a "generic decryption", define $M_i(\boldsymbol{X})$, $M'_i(\boldsymbol{X})$, and $M_E(\boldsymbol{X})$ as in Eq. (5). Then we get that $\mathcal{V}_{cons}(\boldsymbol{X}) = 0$, where

$$\begin{aligned}
\mathcal{V}_{cons}(\boldsymbol{X}) =& \frac{\mathcal{V}_{cons:3}(\boldsymbol{X}) X_\beta}{X_\varrho} - \frac{\mathcal{V}_{cons:2}(\boldsymbol{X}) X_\beta}{X_\varrho (X_\gamma + 1)} - \frac{\mathcal{V}_{cons:1}(\boldsymbol{X}) X_\beta}{X_\varrho X_\gamma} \\
=& \sum_i \left(M'_i(\boldsymbol{X}) - M_{\sigma(i)}(\boldsymbol{X}) \right) P_i(X) - \left(\sum_i M_i(\boldsymbol{X}) r_i - M_E(\boldsymbol{X}) \right) X_\varrho = 0
\end{aligned}$$

is again a "generic decryption". Clearly, the last equality holds clearly independently of the shape of $D_j(\boldsymbol{X})$.

Now, since the validity argument is sound, $M_E(\boldsymbol{X})$ is as in Eq. (11) and $M_i(\boldsymbol{X})$ and $M'_i(\boldsymbol{X})$ are as in Eq. (12). Denote $\mathcal{V}^*_{cons}(\boldsymbol{X}) := X_\gamma(X_\gamma + 1)$. Inserting the obtained representations of $M_i(\boldsymbol{X})$, $M'_i(\boldsymbol{X})$, and $M_E(\boldsymbol{X})$ to $\mathcal{V}_{cons}(\boldsymbol{X})$, we find the coefficients of $\mathcal{V}_{cons}(\boldsymbol{X}) \cdot \mathcal{V}^*_{cons}(\boldsymbol{X})$, as given in Table 4.

Since $\{P_i(X)\}$ is linearly independent, this directly gives us $M_{\sigma(i)j} = M'_{ij}$, for each $j \in [2\,..\,5]$, and hence also for $j = 7$, as needed. In addition, we get that $M_{\sigma(i)1} + M_{\sigma(i)3} = M'_{i1} + M'_{i3}$ (and hence $M_{\sigma(i)1} = M'_{i1}$) and $M_{\sigma(i)1} + M_{\sigma(i)4} + M_{\sigma(i)6} = M'_{i1} + M'_{i4} + M'_{i6}$ (and hence $M_{\sigma(i)6} = M'_{i6}$). Hence, we have proven that $M_{\sigma(i)}(\boldsymbol{X}) = M'_i(\boldsymbol{X})$ as a polynomial, which gives us soundness of the new shuffle argument in the GBGM.

Let us now compute a lower bound to the efficiency of a generic adversary. Assume that after some τ steps, the adversary has made a successful equality query $(=, i_1, i_2)$, i.e., $\mathsf{cell}_{i_1} = \mathsf{cell}_{i_2}$ for $i_1 \neq i_2$. Hence, she has found a collision $B_1(\chi) = B_2(\chi)$ such that $B_1(\boldsymbol{X}) \neq B_2(\boldsymbol{X})$. If $\mathsf{type}_{i_1} \in \{1, T\}$ (this is not needed for group $\mathbb{G}_2$, since we do not have rational functions there), then redefine $B_j(\boldsymbol{X}) := B_j(\boldsymbol{X}) \cdot X_\varrho X_\beta$, this guarantees $B_j(\boldsymbol{X})$ is a polynomial. Thus,

$$B_1(\chi) - B_2(\chi) \equiv 0 \pmod{q} \ . \tag{13}$$

Note that

- If $\mathsf{type}_{i_1} = 1$, then $\deg B_j(\boldsymbol{X}) \leq 2n + 1 =: d_1$,

Table 4. $\operatorname{coeff}_{\mu(i)}(\mathcal{V}_{cons}(\boldsymbol{X}) \cdot \mathcal{V}^*_{cons}(\boldsymbol{X}))$, where $\mu(\boldsymbol{i}) = X_\alpha^{i_1} X_\varrho^{i_2} X_\beta^{i_3} X_\gamma^{i_4}$

$\{i_1, \ldots, i_4\}$	$\operatorname{coeff}_{\mu(i)}(\mathcal{V}_{cons}(\boldsymbol{X}) \cdot \mathcal{V}^*_{cons}(\boldsymbol{X}))$
$\{0,1,1,2\}$	$M_{E:2} - \sum M_{i2} r_i$
$\{0,1,0,3\}$	$M_{E:3} - \sum M_{i3} r_i$
$\{0,1,0,0\}$	$M_{E:4} - \sum M_{i4} r_i$
$\{0,1,1,0\}$	$M_{E:5} - \sum M_{i5} r_i$
$\{0,1,1,1\}$	$M_{E:2} + M_{E:5} + M_{E:7} - \sum (M_{i2} + M_{i5} + M_{i7}) r_i$
$\{0,1,0,2\}$	$M_{E:1} + M_{E:3} - \sum (M_{i1} + M_{i3}) r_i$
$\{0,1,0,1\}$	$M_{E:1} + M_{E:4} + M_{E:6} - \sum (M_{i1} + M_{i4} + M_{i6}) r_i$
$\{0,0,1,2\}$	$\sum(M'_{i2} - \sum M_{\sigma(i)2}) P_i(X)$
$\{0,0,0,3\}$	$\sum(M'_{i3} - \sum M_{\sigma(i)3}) P_i(X)$
$\{0,0,0,0\}$	$\sum(M'_{i4} - \sum M_{\sigma(i)4}) P_i(X)$
$\{0,0,1,0\}$	$\sum(M'_{i5} - \sum M_{\sigma(i)5}) P_i(X)$
$\{0,0,0,2\}$	$\sum \left((M'_{i1} + M'_{i3}) - (M_{\sigma(i)1} + M_{\sigma(i)3})\right) P_i(X)$
$\{0,0,0,1\}$	$\sum \left((M'_{i1} + M'_{i4} + M'_{i6}) - (M_{\sigma(i)1} + M_{\sigma(i)4} + M_{\sigma(i)6})\right) P_i(X)$
$\{0,0,1,1\}$	$\sum \left((M'_{i2} + M'_{i5} + M'_{i7}) - (M_{\sigma(i)2} + M_{\sigma(i)5} + M_{\sigma(i)7})\right) P_i(X)$

- If $\mathsf{type}_{i_1} = 2$, then $\deg B_j(\boldsymbol{X}) \leq n =: d_2$, and thus
- If $\mathsf{type}_{i_1} = T$, then $\deg B_j(\boldsymbol{X}) \leq (2n+1) + n = 3n + 1 =: d_T$.

Due to the Schwartz-Zippel lemma, since χ is chosen uniformly random from $\mathbb{Z}_q^2 \times (\mathbb{Z}_q \setminus \{0\})^2 \times (\mathbb{Z}_q \setminus \{0, -1\})$, and since $B_1(\boldsymbol{X}) \neq B_2(\boldsymbol{X})$ as a polynomial, Eq. (13) holds with probability at most $\deg B_j(\boldsymbol{X})/(q-2) \leq d_{\mathsf{type}_{i_1}}/(q-2)$. Clearly, an adversary working in time τ can generate up to τ new group elements. Then the probability that there exists a collision between any two of those group elements is upper bounded by $\binom{\tau}{2} \cdot \deg B_j(\boldsymbol{X})/(q-2) \leq \binom{\tau}{2} \cdot d_{\mathsf{type}_{i_1}}/(q-2) \leq \tau^2/2 \cdot d_{\mathsf{type}_{i_1}}/(q-2)$. Thus, a successful adversary on average requires time at least

$$\tau^2 \geq 2(q-2)/d_{\mathsf{type}_{i_1}} \geq 2(q-2)/d_T = 2(q-2)/(3n+1)$$

to produce a collision. Simplifying, we get $\tau \in \Omega(\sqrt{q/n})$. □

8 Zero-Knowledge

Theorem 5. *The new shuffle argument is perfectly zero knowledge.*

Proof. Consider the simulator Sim that, given the CRS crs, the trapdoor $\mathsf{td} = (\chi, \varrho)$, and input $(\mathfrak{v}, \mathfrak{v}')$, simulates the prover in the shuffle argument. If the simulator can create an accepting argument with correct distribution for *any* $(\mathfrak{v}, \mathfrak{v}')$, this means that an accepting argument provides no information on $(\mathfrak{v}, \mathfrak{v}')$ or the relation between the two sets of ciphertext.

The complete simulator construction is as follows:

1. For $i = 1$ to $n - 1$:
   (a) Set $r_i \leftarrow_r \mathbb{Z}_q$. Compute $(\mathfrak{A}_{i1}, \mathfrak{A}_{i2}) \leftarrow (\mathfrak{g}_1, \mathfrak{g}_2)^{P_i(\chi)+r_i\varrho}$.
2. Set $r_n \leftarrow -\sum_{i=1}^{n-1} r_i$.
3. Set $(\mathfrak{A}_{n1}, \mathfrak{A}_{n2}) \leftarrow (\mathfrak{g}_1, \mathfrak{g}_2)^{\sum_{i=1}^n P_i(\chi)} / \prod_{i=1}^{n-1}(\mathfrak{A}_{i1}, \mathfrak{A}_{i2})$.
4. For $i = 1$ to n: Compute

$$\pi_{\mathsf{1sp}:i} \leftarrow \left(\mathfrak{A}_{i1}\mathfrak{g}_1^{P_0(\chi)}\right)^{2r_i} (\mathfrak{g}_1^{\varrho})^{-r_i^2} \mathfrak{g}_1^{((P_i(X)+P_0(X))^2-1)/\varrho} .$$

5. Set $\boldsymbol{r}_s \leftarrow_r \mathbb{Z}_q^2$. Set $\pi_{\mathsf{c1}:1} \leftarrow \mathfrak{g}_2^{r_{s:1}\varrho}$, $\pi_{\mathsf{c1}:2} \leftarrow \mathfrak{g}_2^{r_{s:2}\varrho}$. (I.e., they commit to $\mathbf{0}$.)
6. Compute $\boldsymbol{\pi}_{\mathsf{c2}:1} \leftarrow \prod_{i=1}^n (\mathfrak{v}_{i1}^{r_i+P_i(\chi)/\varrho}/(\mathfrak{v}'_{i1})^{P_i(\chi)/\varrho}) \cdot \mathsf{enc}_{\mathsf{pk}_1}(0; \boldsymbol{r}_s)$.
   Compute $\boldsymbol{\pi}_{\mathsf{c2}:2} \leftarrow \prod_{i=1}^n (\mathfrak{v}_{i2}^{r_i+P_i(\chi)/\varrho}/(\mathfrak{v}'_{i2})^{P_i(\chi)/\varrho}) \cdot \mathsf{enc}_{\mathsf{pk}_2}(0; \boldsymbol{r}_s)$.
7. Return $\pi_{sh} := (\boldsymbol{\mathfrak{v}}', (\mathfrak{A}_{i1}, \mathfrak{A}_{i2})_{i=1}^{n-1}, (\pi_{\mathsf{1sp}:i})_{i=1}^n, \pi_{\mathsf{c1}:1}, \pi_{\mathsf{c1}:2}, \boldsymbol{\pi}_{\mathsf{c2}:1}, \boldsymbol{\pi}_{\mathsf{c2}:2})$.

The simulator calculates all values $(\mathfrak{A}_{i1}, \mathfrak{A}_{i2})_{i=1}^{n-1}, (\pi_{\mathsf{1sp}:i})_{i=1}^n, \boldsymbol{\pi}_{\mathsf{c1}}$ exactly as an honest prover would have when $\sigma = \mathrm{Id}$, and hence these values will have the same distribution as the same values computed by an honest prover. Since the commitment scheme is obviously perfectly hiding, these values have the same distribution independently of the choice of σ. Moreover, there is a unique pair of values $\boldsymbol{\pi}_{\mathsf{c2}:1}, \boldsymbol{\pi}_{\mathsf{c2}:2}$ that satisfy Eqs. (2) and (4). (Computing $\boldsymbol{\pi}_{\mathsf{c2}:1}$ and $\boldsymbol{\pi}_{\mathsf{c2}:2}$ is the only place in the simulation where one needs the trapdoor $\mathsf{td} = (\chi, \varrho)$.) Thus we are left to show that our chosen values satisfy these two equations.

But assuming the ciphertexts are valid, Eq. (2) trivially holds. We get $\hat{e}(\mathfrak{P}_{11}^{r_{s:1}}, \mathfrak{g}_2^{\varrho}) = \hat{e}(\mathfrak{P}_{11}, \pi_{\mathsf{c1}:1})$ and $\hat{e}(\mathfrak{P}_{12}^{r_{s:2}}, \mathfrak{g}_2^{\varrho}) = \hat{e}(\mathfrak{P}_{12}, \pi_{\mathsf{c1}:2})$. Hence,

$$\begin{aligned}
&\prod_{i=1}^n \hat{e}\left(\mathfrak{v}'_{i1}, \mathfrak{g}_2^{P_i(\chi)}\right) / \prod_{i=1}^n \hat{e}\left(\mathfrak{v}_{i1}, \mathfrak{A}_{i2}\right) \\
&\qquad = \prod_{i=1}^n \hat{e}\left((\mathfrak{v}'_{i1})^{P_i(\chi)}, \mathfrak{g}_2\right) / \prod_{i=1}^n \hat{e}\left(\mathfrak{v}_{i1}^{P_i(\chi)+r_i\varrho}, \mathfrak{g}_2\right) \\
&\qquad = \hat{e}\left(\prod_{i=1}^n ((\mathfrak{v}'_{i1})^{P_i(\chi)/\varrho}/\mathfrak{v}_{i1}^{P_i(\chi)/\varrho+r_i}), \mathfrak{g}_2^{\varrho}\right) \\
&\qquad = \hat{e}\left(\mathfrak{P}_{11}, \pi_{\mathsf{c1}:1}\right) \hat{e}\left(\mathfrak{P}_{12}, \pi_{\mathsf{c1}:2}\right) / \hat{e}\left(\boldsymbol{\pi}_{\mathsf{c2}:1}, \mathfrak{g}_2^{\varrho}\right) ,
\end{aligned}$$

so the verifier will accept the shuffle argument. As the simulator did not know anything about the honest prover's permutation σ, the shuffle argument is thus perfectly zero knowledge. □

9 Efficiency

We use exponentiation speed records from [10] and pairing speed records from [2] for Barreto-Naehrig curves. According to Table 4 in [10], a pairing, exponentiation in $\mathbb{G}_1$, exponentiation in $\mathbb{G}_2$, and exponentiation in $\mathbb{G}_T$ take respectively 7.0, 0.9, 1.8, and 3.1 million clock cycles on the Core i7-3520M CPU. This does *not* take into account the possible speed-ups by employing fast fixed-based exponentiation or multi-exponentiation algorithms. Thus, all following comparisons are imprecise, and just to give a gut feeling about the difference. They also depend on the known speed records on implementing pairings and exponentiations.

Prover's Computation:

- Step 1: $n-1$ exponentiations in $\mathbb{G}_1$, $n-1$ exponentiations in $\mathbb{G}_2$.
- Step 4: $2n$ exponentiations in $\mathbb{G}_1$.
- Step 5a: $3n$ exponentiations in $\mathbb{G}_1$, $3n$ exponentiations in $\mathbb{G}_2$.
- Step 6a: $2n+2$ exponentiations in $\mathbb{G}_2$.
- Step 6b: $3n+3$ exponentiations in $\mathbb{G}_1$, $3n+3$ exponentiations in $\mathbb{G}_2$.

Hence, the prover executes $9n+2$ exponentiations in $\mathbb{G}_1$ and $9n+4$ exponentiations in $\mathbb{G}_2$. Here, *all* costly (i.e., at least n-wide) exponentiations can be written as either multi-exponentiations or fixed-base exponentiations — e.g., Step 5a — and are hence relatively cheap. The only exception is the computation of general exponentiation in $(\mathfrak{A}_{i1}\mathfrak{g}_1^{P_0(X)})^{2r_i}$ for $i \in [1\,..\,n]$. Taking n million clock cycles as the basic unit (and *not* taking into account possible speed-ups by employing fast multi- exponentiation and fixed-base exponentiation algorithms), the prover's computation is dominated by $9 \cdot 0.9 + 9 \cdot 1.8 = 24.3$ units.

Verifier's Computation: by using batching techniques [4,31], we reduced the number of pairings by introducing a number of exponentiations either in $\mathbb{G}_1$, $\mathbb{G}_2$, or $\mathbb{G}_T$. The verifier does:

- Step 3: $2n$ exponentiations in $\mathbb{G}_1$, 1 exponentiation in $\mathbb{G}_T$, and $n+1$ pairings.
- Step 4: $3n+3$ exponentiations in $\mathbb{G}_1$, $3n+3$ exponentiations in $\mathbb{G}_2$, and 2 pairings.
- Step 5: 2 exponentiations in $\mathbb{G}_1$, 3 exponentiations in $\mathbb{G}_2$ ($\pi_{c1:2}^{p42}$ is reused), and 3 pairings.
- Step 6: $3n+3n=6n$ exponentiations in $\mathbb{G}_1$, and $n+n=2n$ pairings.

In total, the verifier has to do $11n+5$ exponentiations in $\mathbb{G}_1$, $3n+6$ exponentiations in $\mathbb{G}_2$, 1 exponentiation in $\mathbb{G}_T$, and $3n+6$ pairings. Taking n million clock cycles as the basic unit, the verifier's computation is dominated by $11 \cdot 0.9 + 3 \cdot 1.8 + 3 \cdot 7.0 = 36.3$ units; around 58 % (21 units) of this is the cost of pairings. Also here, most of the exponentiations are multi-exponentiations or fixed-base exponentiations.

Communication: $3n+(n-1)+n+3 = 5n+2$ elements from $\mathbb{G}_1$ and $3n+(n-1)+2+3 = 4n+4$ elements from $\mathbb{G}_2$, that is, $9n+6$ group elements.
CRS length (excluding gk): $2n+6$ elements from $\mathbb{G}_1$, $n+6$ elements from $\mathbb{G}_2$, and 1 element from $\mathbb{G}_T$, that is, $3n+13$ group elements.

Comparison with Prior Work. To compare, the verifier's computation in [33] (resp., [19]) is dominated by $28 \cdot 7.0 = 196$ (resp., $18 \cdot 7.0 = 126$) units. Hence, the verification of the new shuffle is effectively about 5.4 (resp., 3.5) times faster than that of the Lipmaa-Zhang (resp., Fauzi-Lipmaa) shuffle. Since verification is a bottleneck of mix-nets, this constitutes of a major improvement.

In [33], the prover's computation is dominated by $28n + 11$ exponentiations, $16n + 6$ in $\mathbb{G}_1$ and $12n + 5$ in $\mathbb{G}_2$ (this also includes reshuffling) which yields $16 \cdot 0.9 + 12 \cdot 1.8 = 36$ units. In [19], the prover's computation is dominated by $18n+3$ exponentiations, $14n+3$ in $\mathbb{G}_1$ and $4n$ in $\mathbb{G}_2$ (this also includes reshuffling) which yields $14 \cdot 0.9 + 4 \cdot 1.8 = 19.8$ units. Hence, in the new protocol, the prover is about 1.5 times more efficient compared to [33], but about 1.2 times less efficient compared to [19].

As mentioned above, the most efficient shuffle scheme up to now [25] works in the random oracle model which allows to obtain better computational complexity both for the prover ($6 \cdot 0.9 = 5.4$ units) and verifier ($6 \cdot 0.9 = 5.4$ units), assuming that computation is done in $\mathbb{G}_1$. In reality, non pairing-friendly groups have usually somewhat faster arithmetic than pairing-friendly groups. Hence, there is still a significant gap.

10 On GBGM Versus Knowledge Assumptions

A knowledge assumption guarantees that if an adversary, given an input (that includes the CRS and some auxiliary input), outputs some values then there exists an extractor running on the same input that outputs the same values together with some witness. Following [15], each input to the knowledge assumption has a well-defined knowledge component. Apart from that, the precise definition of a knowledge assumption is left to the imagination of its proposers. However, it is known that knowledge assumptions are unacceptable if the auxiliary input is not well chosen [5], and hence special care has to be taken when defining them.

In contrast, in the GBGM, the adversary can compute output values as a product or pairing of given inputs (and other previously computed values), so it is assumed that she knows a polynomial relationship between the discrete logarithms of its outputs and inputs. There is little need for imagination of how to define the GBGM, since this has been done before in sufficient detail [34,38]. The known impossibility results about the generic (bilinear) group model [17,21] use quite contrived constructions.

We think that GBGM is preferable to knowledge assumptions, hence Table 1 has a highlighted cell for arguments that do not use knowledge assumptions. The validity of knowledge assumptions can and should be proven in the GBGM anyhow; indeed, one should be very suspicious of knowledge assumptions that cannot be proven in the GBGM. However, this should be done very carefully, taking into account the precise shape of the CRS and the adversary's auxiliary input. To guarantee correct use of a knowledge assumption, we think that it is prudent that one proves in the GBGM the security of the knowledge assumption given the auxiliary string the adversary gets in the concrete application. This seems to hint that one should reprove in the GBGM the security of all used knowledge assumptions in each individual paper.

Instead of proving the security of non-falsifiable knowledge assumptions (on top of several novel computational assumptions like PP and SP [28] or PSP [19]) in the GBGM, and then using such assumptions in the security proof, we think it is more reasonable to work directly in the GBGM. Moreover, GBGM model arguments tend to be more efficient, in particular since there is a reduced need to compute the knowledge components.

In fact, most of the known knowledge assumptions make a very specific use of the power of the GBGM. E.g., reinterpreting the knowledge assumptions used in say [19] in the language of GBGM, one assumes for a *specific (unique!)* random variable X_k the following holds: for any polynomial F, if $F(X_1, \ldots, X_k, \ldots, X_m) = 0$ and $F(X_1, \ldots, X_k, \ldots, X_m)$ has μ_k as a coefficient of X_k then $\mu_k = 0$. It is questionable why this concrete coefficient is handled differently from all other coefficients; in the GBGM, from $F(X_1, \ldots, X_m) = 0$ one can derive that *all* coefficients $\mu_{i_1,\ldots,i_m}$ of $F(X_1, \ldots, X_m) = \sum \mu_{i_1,\ldots,i_m} X_1^{i_1} \ldots X_m^{i_m}$ are equal to 0.

Acknowledgment. We would like to thank Jens Groth for useful discussion. The authors were supported by the European Union's Horizon 2020 research and innovation programme under grant agreement No 653497 (project PANORAMIX), and by institutional research funding IUT2-1 of the Estonian Ministry of Education and Research.

A Preliminaries: Zero Knowledge

Let $\mathcal{R} = \{(u, w)\}$ be an efficiently computable binary relation with $|w| = \text{poly}(|u|)$. Here, u is a statement, and w is a witness. Let $\mathcal{L} = \{u : \exists w, (u, w) \in \mathcal{R}\}$ be an **NP**-language. Let $n = |u|$ be the input length. For fixed n, we have a relation $\mathcal{R}_n$ and a language $\mathcal{L}_n$. Here, as in [28], since we argue about group elements, both $\mathcal{L}_n$ and $\mathcal{R}_n$ are group-dependent and thus we add gk as an input to $\mathcal{L}_n$ and $\mathcal{R}_n$. Let $\mathcal{R}_n(\mathsf{gk}) := \{(u, w) : (\mathsf{gk}, u, w) \in \mathcal{R}_n\}$.

A *non-interactive argument* for a group-dependent relation family $\mathcal{R}$ consists of four PPT algorithms: a setup algorithm setup, a common reference string (CRS) generator gencrs, a prover pro, and a verifier ver. For $\mathsf{gk} \leftarrow \mathsf{setup}(1^\kappa, n)$ (where n is the input length) and $(\mathsf{crs} = (\mathsf{crs}_p, \mathsf{crs}_v), \mathsf{td}) \leftarrow \mathsf{gencrs}(\mathsf{gk})$ (where td is not accessible to anybody but the simulator), $\mathsf{pro}(\mathsf{crs}_p; u, w)$ produces an argument π, and $\mathsf{ver}(\mathsf{crs}_v; u, \pi)$ outputs either 1 (accept) or 0 (reject). Here, crs_p (resp., crs_v) is the part of the CRS given to the prover (resp., the verifier). Distinction between crs_p and crs_v is not important from the security point of view, but in many cases crs_v is significantly shorter.

A non-interactive argument Ψ is *perfectly complete*, if for all $n = \text{poly}(\kappa)$,

$$\Pr\left[\begin{array}{l}\mathsf{gk} \leftarrow \mathsf{setup}(1^\kappa, n), ((\mathsf{crs}_p, \mathsf{crs}_v), \mathsf{td}) \leftarrow \mathsf{gencrs}(\mathsf{gk}), (u, w) \leftarrow \mathcal{R}_n(\mathsf{gk}) : \\ \mathsf{ver}(\mathsf{gk}, \mathsf{crs}_v; u, \mathsf{pro}(\mathsf{gk}, \mathsf{crs}_p; u, w)) = 1\end{array}\right] = 1 .$$

Ψ is adaptively *computationally sound* for $\mathcal{L}$, if for all $n = \text{poly}(\kappa)$ and non-uniform probabilistic polynomial-time adv,

$$\Pr\left[\begin{array}{l}\mathsf{gk} \leftarrow \mathsf{setup}(1^\kappa, n), ((\mathsf{crs}_p, \mathsf{crs}_v), \mathsf{td}) \leftarrow \mathsf{gencrs}(\mathsf{gk}), \\ (u, \pi) \leftarrow \mathsf{adv}(\mathsf{gk}, \mathsf{crs}_p, \mathsf{crs}_v) : (\mathsf{gk}, u) \notin \mathcal{L}_n \wedge \mathsf{ver}(\mathsf{gk}, \mathsf{crs}_v; u, \pi) = 1\end{array}\right] \approx_\kappa 0 .$$

We recall that in situations where the inputs have been committed by using a computationally binding trapdoor commitment scheme, the notion of computational soundness does not make sense (since the commitments could be to any input messages). Instead, one should either proof culpable soundness or the argument of knowledge property.

Ψ is adaptively *computationally culpably sound* [28,29] for $\mathcal{L}$, if for all $n = \text{poly}(\kappa)$, for all polynomial-time decidable binary relations $\mathcal{R}^{\mathsf{guilt}} = \{\mathcal{R}_n^{\mathsf{guilt}}\}$ consisting of elements from $\bar{\mathcal{L}}$ and witnesses w^{guilt}, and for all non-uniform probabilistic polynomial-time adv,

$$\Pr\left[\begin{array}{l} \mathsf{gk} \leftarrow \mathsf{setup}(1^\kappa, n), ((\mathsf{crs}_p, \mathsf{crs}_v), \mathsf{td}) \leftarrow \mathsf{gencrs}(\mathsf{gk}), \\ (u, \pi, w^{\mathsf{guilt}}) \leftarrow \mathsf{adv}(\mathsf{gk}, \mathsf{crs}_p, \mathsf{crs}_v) : \\ (\mathsf{gk}, u, w^{\mathsf{guilt}}) \in \mathcal{R}_n^{\mathsf{guilt}} \wedge \mathsf{ver}(\mathsf{gk}, \mathsf{crs}_v; u, \pi) = 1 \end{array}\right] \approx_\kappa 0 \ .$$

For algorithms adv and X_{adv}, we write $(y; y') \leftarrow (\mathsf{adv}||X_{\mathsf{adv}})(\chi)$ if adv on input χ outputs y, and X_{adv} on the same input (including the random tape of adv) outputs y'.

Ψ is *an argument of knowledge*, if for all $n = \text{poly}(\kappa)$ and every non-uniform probabilistic polynomial-time adv, there exists a non-uniform probabilistic polynomial-time extractor X, s.t. for every auxiliary input $\mathsf{aux} \in \{0,1\}^{\text{poly}(\kappa)}$,

$$\Pr\left[\begin{array}{l} \mathsf{gk} \leftarrow \mathsf{setup}(1^\kappa, n), ((\mathsf{crs}_p, \mathsf{crs}_v), \mathsf{td}) \leftarrow \mathsf{gencrs}(\mathsf{gk}), \\ ((u, \pi); w) \leftarrow (\mathsf{adv}||X_{\mathsf{adv}})(\mathsf{crs}_p, \mathsf{crs}_v; \mathsf{aux}) : \\ (u, w) \notin \mathcal{R} \wedge \mathsf{ver}(\mathsf{crs}_v; u, \pi) = 1 \end{array}\right] \approx_\kappa 0 \ .$$

Here, aux can be seen as the common auxiliary input to adv and X_{adv} that is generated by using benign auxiliary input generation [5].

Ψ is *perfectly zero-knowledge*, if there exists a probabilistic polynomial-time simulator $\mathcal{X}_\gamma$, such that for all stateful adversaries adv and $n = \text{poly}(\kappa)$,

$$\Pr\left[\begin{array}{l} \mathsf{gk} \leftarrow \mathsf{setup}(1^\kappa, n), \\ ((\mathsf{crs}_p, \mathsf{crs}_v), \mathsf{td}) \leftarrow \mathsf{gencrs}(\mathsf{gk}), \\ (u, w) \leftarrow \mathsf{adv}(\mathsf{gk}, \mathsf{crs}_p, \mathsf{crs}_v), \\ \pi \leftarrow \mathsf{pro}(\mathsf{gk}, \mathsf{crs}_p; u, w) : \\ (\mathsf{gk}, u, w) \in \mathcal{R}_n \wedge \mathsf{adv}(\mathsf{gk}, \pi) = 1 \end{array}\right] = \Pr\left[\begin{array}{l} \mathsf{gk} \leftarrow \mathsf{setup}(1^\kappa, n), \\ ((\mathsf{crs}_p, \mathsf{crs}_v); \mathsf{td}) \leftarrow \mathsf{gencrs}(\mathsf{gk}), \\ (u, w) \leftarrow \mathsf{adv}(\mathsf{gk}, \mathsf{crs}_p, \mathsf{crs}_v), \\ \pi \leftarrow \mathcal{X}_\gamma(\mathsf{gk}, \mathsf{crs}_p, \mathsf{crs}_v; u, \mathsf{td}) : \\ (\mathsf{gk}, u, w) \in \mathcal{R}_n \wedge \mathsf{adv}(\mathsf{gk}, \pi) = 1 \end{array}\right]$$

Here, the prover and the simulator use the same CRS. That is, we have *same-string zero knowledge*. A same-string statistical zero knowledge argument stay secure even when using the CRS an unbounded number of times.

References

1. Ambrona, M., Barthe, G., Schmidt, B.: Automated unbounded analysis of cryptographic constructions in the generic group model. In: Fischlin, M., Coron, J.-S. (eds.) EUROCRYPT 2016. LNCS, vol. 9666, pp. 822–851. Springer, Heidelberg (2016). doi:10.1007/978-3-662-49896-5_29

2. Aranha, D.F., Barreto, P.S.L.M., Longa, P., Ricardini, J.E.: The realm of the pairings. In: Lange, T., Lauter, K., Lisoněk, P. (eds.) SAC 2013. LNCS, vol. 8282, pp. 3–25. Springer, Heidelberg (2014). doi:10.1007/978-3-662-43414-7_1
3. Barthe, G., Fagerholm, E., Fiore, D., Scedrov, A., Schmidt, B., Tibouchi, M.: Strongly-optimal structure preserving signatures from type II pairings: synthesis and lower bounds. In: Katz, J. (ed.) PKC 2015. LNCS, vol. 9020, pp. 355–376. Springer, Heidelberg (2015). doi:10.1007/978-3-662-46447-2_16
4. Bellare, M., Garay, J.A., Rabin, T.: Batch verification with applications to cryptography and checking. In: Lucchesi, C.L., Moura, A.V. (eds.) LATIN 1998. LNCS, vol. 1380, pp. 170–191. Springer, Heidelberg (1998). doi:10.1007/BFb0054320
5. Bitansky, N., Canetti, R., Paneth, O., Rosen, A.: On the existence of extractable one-way functions. In: STOC 2014, pp. 505–514 (2014)
6. Bitansky, N., Dachman-Soled, D., Garg, S., Jain, A., Kalai, Y.T., López-Alt, A., Wichs, D.: Why "fiat-shamir for proofs" lacks a proof. In: Sahai, A. (ed.) TCC 2013. LNCS, vol. 7785, pp. 182–201. Springer, Heidelberg (2013). doi:10.1007/978-3-642-36594-2_11
7. Blum, M., Feldman, P., Micali, S.: Non-interactive zero-knowledge and its applications. In: STOC 1988, pp. 103–112 (1988)
8. Boneh, D., Boyen, X., Goh, E.-J.: Hierarchical identity based encryption with constant size ciphertext. In: Cramer, R. (ed.) EUROCRYPT 2005. LNCS, vol. 3494, pp. 440–456. Springer, Heidelberg (2005). doi:10.1007/11426639_26
9. Boneh, D., Boyen, X., Shacham, H.: Short group signatures. In: Franklin, M. (ed.) CRYPTO 2004. LNCS, vol. 3152, pp. 41–55. Springer, Heidelberg (2004). doi:10.1007/978-3-540-28628-8_3
10. Bos, J.W., Costello, C., Naehrig, M.: Exponentiating in pairing groups. In: Lange, T., Lauter, K., Lisoněk, P. (eds.) SAC 2013. LNCS, vol. 8282, pp. 438–455. Springer, Heidelberg (2014). doi:10.1007/978-3-662-43414-7_22
11. Buchberger, B.: An Algorithm for Finding the Basis Elements of the Residue Class Ring of a Zero Dimensional Polynomial Ideal. Ph.D. thesis, University of Innsbruck (1965)
12. Canetti, R., Goldreich, O., Halevi, S.: The random oracle methodology, revisited. In: STOC 1998, pp. 209–218 (1998)
13. Chaabouni, R., Lipmaa, H., Zhang, B.: A non-interactive range proof with constant communication. In: Keromytis, A.D. (ed.) FC 2012. LNCS, vol. 7397, pp. 179–199. Springer, Heidelberg (2012). doi:10.1007/978-3-642-32946-3_14
14. Ciampi, M., Persiano, G., Siniscalchi, L., Visconti, I.: A transform for NIZK almost as efficient and general as the fiat-shamir transform without programmable random oracles. In: Kushilevitz, E., Malkin, T. (eds.) TCC 2016. LNCS, vol. 9563, pp. 83–111. Springer, Heidelberg (2016). doi:10.1007/978-3-662-49099-0_4
15. Damgård, I.: Towards practical public key systems secure against chosen ciphertext attacks. In: Feigenbaum, J. (ed.) CRYPTO 1991. LNCS, vol. 576, pp. 445–456. Springer, Heidelberg (1992). doi:10.1007/3-540-46766-1_36
16. Danezis, G., Fournet, C., Groth, J., Kohlweiss, M.: Square span programs with applications to succinct NIZK arguments. In: Sarkar, P., Iwata, T. (eds.) ASIACRYPT 2014. LNCS, vol. 8873, pp. 532–550. Springer, Heidelberg (2014). doi:10.1007/978-3-662-45611-8_28
17. Dent, A.W.: Adapting the weaknesses of the random oracle model to the generic group model. In: Zheng, Y. (ed.) ASIACRYPT 2002. LNCS, vol. 2501, pp. 100–109. Springer, Heidelberg (2002). doi:10.1007/3-540-36178-2_6

18. Escala, A., Herold, G., Kiltz, E., Ràfols, C., Villar, J.: An algebraic framework for diffie-hellman assumptions. In: Canetti, R., Garay, J.A. (eds.) CRYPTO 2013. LNCS, vol. 8043, pp. 129–147. Springer, Heidelberg (2013). doi:10.1007/978-3-642-40084-1_8
19. Fauzi, P., Lipmaa, H.: Efficient culpably sound NIZK shuffle argument without random oracles. In: Sako, K. (ed.) CT-RSA 2016. LNCS, vol. 9610, pp. 200–216. Springer, Heidelberg (2016). doi:10.1007/978-3-319-29485-8_12
20. Fauzi, P., Lipmaa, H., Zhang, B.: Efficient modular NIZK arguments from shift and product. In: Abdalla, M., Nita-Rotaru, C., Dahab, R. (eds.) CANS 2013. LNCS, vol. 8257, pp. 92–121. Springer, Heidelberg (2013). doi:10.1007/978-3-319-02937-5_6
21. Fischlin, M.: A note on security proofs in the generic model. In: Okamoto, T. (ed.) ASIACRYPT 2000. LNCS, vol. 1976, pp. 458–469. Springer, Heidelberg (2000). doi:10.1007/3-540-44448-3_35
22. Gennaro, R., Gentry, C., Parno, B., Raykova, M.: Quadratic span programs and succinct NIZKs without PCPs. In: Johansson, T., Nguyen, P.Q. (eds.) EUROCRYPT 2013. LNCS, vol. 7881, pp. 626–645. Springer, Heidelberg (2013). doi:10.1007/978-3-642-38348-9_37
23. Goldwasser, S., Kalai, Y.T.: On the (In)security of the Fiat-Shamir Paradigm. In: FOCS 2003, pp. 102–113 (2003)
24. Goldwasser, S., Micali, S., Rackoff, C.: The knowledge complexity of interactive proof-systems. In: STOC 1985, pp. 291–304 (1985)
25. Groth, J.: A verifiable secret shuffle of homomorphic encryptions. J. Cryptology **23**(4), 546–579 (2010)
26. Groth, J.: Short pairing-based non-interactive zero-knowledge arguments. In: Abe, M. (ed.) ASIACRYPT 2010. LNCS, vol. 6477, pp. 321–340. Springer, Heidelberg (2010). doi:10.1007/978-3-642-17373-8_19
27. Groth, J.: On the size of pairing-based non-interactive arguments. In: Fischlin, M., Coron, J.-S. (eds.) EUROCRYPT 2016. LNCS, vol. 9666, pp. 305–326. Springer, Heidelberg (2016). doi:10.1007/978-3-662-49896-5_11
28. Groth, J., Lu, S.: A non-interactive shuffle with pairing based verifiability. In: Kurosawa, K. (ed.) ASIACRYPT 2007. LNCS, vol. 4833, pp. 51–67. Springer, Heidelberg (2007). doi:10.1007/978-3-540-76900-2_4
29. Groth, J., Ostrovsky, R., Sahai, A.: New techniques for noninteractive zero-knowledge. J. ACM **59**(3), 1–35 (2012)
30. Lipmaa, H.: Progression-free sets and sublinear pairing-based non-interactive zero-knowledge arguments. In: Cramer, R. (ed.) TCC 2012. LNCS, vol. 7194, pp. 169–189. Springer, Heidelberg (2012). doi:10.1007/978-3-642-28914-9_10
31. Lipmaa, H.: Prover-efficient commit-and-prove zero-knowledge SNARKs. In: Pointcheval, D., Nitaj, A., Rachidi, T. (eds.) AFRICACRYPT 2016. LNCS, vol. 9646, pp. 185–206. Springer, Heidelberg (2016). doi:10.1007/978-3-319-31517-1_10
32. Lipmaa, H., Zhang, B.: A more efficient computationally sound non-interactive zero-knowledge shuffle argument. In: Visconti, I., Prisco, R. (eds.) SCN 2012. LNCS, vol. 7485, pp. 477–502. Springer, Heidelberg (2012). doi:10.1007/978-3-642-32928-9_27
33. Lipmaa, H., Zhang, B.: A more efficient computationally sound non-interactive zero-knowledge shuffle argument. J. Comput. Secur. **21**(5), 685–719 (2013)
34. Maurer, U.: Abstract models of computation in cryptography. In: Smart, N.P. (ed.) Cryptography and Coding 2005. LNCS, vol. 3796, pp. 1–12. Springer, Heidelberg (2005). doi:10.1007/11586821_1

35. Naor, M.: On cryptographic assumptions and challenges. In: Boneh, D. (ed.) CRYPTO 2003. LNCS, vol. 2729, pp. 96–109. Springer, Heidelberg (2003). doi:10.1007/978-3-540-45146-4_6
36. Nielsen, J.B.: Separating random oracle proofs from complexity theoretic proofs: the non-committing encryption case. In: Yung, M. (ed.) CRYPTO 2002. LNCS, vol. 2442, pp. 111–126. Springer, Heidelberg (2002). doi:10.1007/3-540-45708-9_8
37. Schwartz, J.T.: Fast probabilistic algorithms for verification of polynomial identities. J. ACM **27**(4), 701–717 (1980)
38. Shoup, V.: Lower bounds for discrete logarithms and related problems. In: Fumy, W. (ed.) EUROCRYPT 1997. LNCS, vol. 1233, pp. 256–266. Springer, Heidelberg (1997). doi:10.1007/3-540-69053-0_18
39. Zippel, R.: Probabilistic algorithms for sparse polynomials. In: Ng, E.W. (ed.) Symbolic and Algebraic Computation. LNCS, vol. 72, pp. 216–226. Springer, Heidelberg (1979). doi:10.1007/3-540-09519-5_73

Efficient Public-Key Distance Bounding Protocol

Handan Kılınç(✉) and Serge Vaudenay

EPFL, Lausanne, Switzerland
handan.kilinc@epfl.ch

Abstract. Distance bounding protocols become more and more important because they are the most accurate solution to defeat relay attacks. They consist of two parties: a verifier and a prover. The prover shows that (s)he is close enough to the verifier. In some applications such as payment systems, using public-key distance bounding protocols is practical as no pre-shared secret is necessary between the payer and the payee. However, public-key cryptography requires much more computations than symmetric key cryptography. In this work, we focus on the efficiency problem in public-key distance bounding protocols and the formal security proofs of them. We construct two protocols (one without privacy, one with) which require fewer computations on the prover side compared to the existing protocols, while keeping the highest security level. Our construction is generic based on a key agreement model. It can be instantiated with only one resp. three elliptic curve computations for the prover side in the two protocols, respectively. We proved the security of our constructions formally and in detail.

Keywords: Distance bounding · RFID · NFC · Relay attack · Key agreement · Mafia fraud · Distance fraud · Distance hijacking

1 Introduction

Nowadays, various technologies, such as contactless payment (e.g. NFC), access control in a building, remote keyless system (e.g. car keys) are part of our lives since they provide us efficient usage of time and accessibility. However, these applications are exposed to simple but dangerous attacks such as relay attacks. A malicious person can abuse all these technologies by just relaying messages.

Distance bounding (DB) is a solution to detect the relay attacks. The detection of the attack is simpler, cheaper and more practical than preventing it because prevention could require a special hardware equipment [4]. The first DB protocol is introduced by Brands and Chaum [9]. Basically in DB, the verifying party measures the physical distance of the proving party by sending the challenges and receiving the responses (they are generally 1 or 2 bit(s)). In the end, if too many rounds have too long round trip times or too many incorrect responses, the verifier rejects the proving party since he may be exposed to a relay attack.

J.H. Cheon and T. Takagi (Eds.): ASIACRYPT 2016, Part II, LNCS 10032, pp. 873–901, 2016.
DOI: 10.1007/978-3-662-53890-6_29

Threats for DB is not limited to only relay attacks. The other threats are the following:

Distance Fraud (DF): A malicious, far-away prover tries to prove that (s)he is close enough.

Mafia Fraud (MiM) [13]: A man-in-the-middle (MiM) adversary between a verifier and a far-away honest prover tries to make the verifier accept.

Terrorist fraud (TF) [13]: A far-away malicious prover, with the help of the adversary, tries to make the verifier accept, but without giving any advantage to the adversary to later pass the protocol alone.

Distance Hijacking (DH) [12]: A far-away malicious prover takes advantage of some honest and active provers who are close to the verifier to make the verifier grant privileges to the far-away prover.

Privacy threat: An adversary tries to learn any useful information such as the identity of a prover. In *strong privacy*, the adversary tries to identify the identity of a prover with access to the prover's secret (e.g. by corruption).

DB protocols are categorized as symmetric DB protocols (the verifier and the prover share a secret) [5–8,16,23–25,34] and public-key DB protocols (the verifier and the prover only know the public key of each other) [9,10,17,20,35,37,38].

In some applications, we cannot assume that the prover and the verifier have established a secret. For example, in a payment system, it is not realistic to assume that the payment terminal and the customer share a secret. We can mention as an instance of a payment protocol the EMV standard [1] which now uses the public-key DB protocol PaySafe from [11]. However, this protocol sends nonces of several bits through the time-critical channel. Normally, a time-critical exchange should only take a few nanoseconds to reach a distance bound

Table 1. The review of the existing public-key DB protocols. ✓ means that it is secure for corresponding threat model and × means it is not. ✓* means that it is secure against the adversaries that cannot relay the messages close to the speed of light. EC is elliptic curve, ZK is zero knowledge, NIZK is non-interactive zero knowledge, AKA is authenticated key agreement. Public key (PK) computations are counted only on prover side. n is the number of rounds in the challenge phase and s is the security parameter.

Protocol	MiM	DF	DH	TF	Privacy	Strong privacy	PK computations for the prover
Brands-Chaum [9]	✓	✓	×	×	×	×	1 commitment, 1 signature
HPO [20]	✓	✓	×	×	✓	×	4 EC multiplications
GOR [17]	✓	✓	×	×	×	×	4 EC multiplications, 1 encryption, 1 NIZK proof
PaySafe [11]	✓*	×	×	×	×	×	1 signature
PrivDB [37]	✓	✓	✓	×	✓	✓	1 signature, 1 IND-CCA encryption
ProProx [38]	✓	✓	✓	✓	×	×	$n+1$ commitments, n ZK proofs
eProProx [35]	✓	✓	✓	✓	✓	✓	1 encryption, s hashing, $n+1$ commitments, n ZK proofs
Simp-pkDB	✓	✓	×	×	×	×	1 IND-CCA decryption
Eff-pkDB	✓	✓	✓	×	×	×	1 AKA protocol
Eff-pkDB^p	✓	✓	✓	×	✓	✓	1 IND-CCA Encryption, 1 AKA protocol

of meters with the speed of light, but sending a string of several bits typically takes microseconds. This is why usual DB protocols only exchange single bits through the time-critical phases. Actually, the protocol from [11] does not protect against adversaries running computations at the speed of light but only against adversaries using standard equipment which induce natural delays.

Although public-key distance bounding protocols are useful, it can cause **considerable energy consumption** on the prover side since public-key cryptography needs heavier computations than symmetric-key cryptography. Energy constraints on most of the powerless devices using RFID and NFC technologies cause very limited computation resources. One of the solutions could be to add more computational power to these devices but it increases their costs.

In this paper, we construct new protocols called Eff-pkDB, Eff-pkDB^p, and Simp-pkDB (Eff-pkDB^p is the privacy-preserving variant of Eff-pkDB).

Table 1 shows the security and the efficiency properties of previous protocols and our protocols. We can see that most of the previous public-key DB protocols [9,10,17,35,37,38] do not concentrate on this efficiency problem, except HPO [20]. So far, HPO is the most efficient one among them since it requires only 4 elliptic curve (EC) multiplications on the prover side, but it is not strong private [36] and it is not secure against DH [22] and TF. In addition to this, its security is based on several ad-hoc assumptions [20] which are not so well studied: "OMDL", "Conjecture 1", "extended ODH" and "XL".

GOR [17] was constructed to have strong privacy, but it has been shown in [36] that it is neither strong private nor private.

ProProx [38] satisfies all the security properties except privacy. Its version eProProx [35] is secure against all threat models and strong private. However, both ProProx and eProProx suffer from heavy cryptographic operations as zero-knowledge (ZK) proofs. These are the only TF-secure protocols, but we can see that their cost is unreasonable.

PrivDB [37] and our new protocol Eff-pkDB^p have the same security properties. However, PrivDB is a bit less efficient on the prover side than Eff-pkDB^p and it has no light privacy-less variant, contrarily to Eff-pkDB^p.

Our lighter protocol Eff-pkDB and our first attempt Simp-pkDB in Appendix B are the most efficient public-key DB protocols as seen in Table 1. Eff-pkDB is secure against DF, MF, DH but it is not private. Simp-pkDB is secure only against DF, MiM and not private. It is more efficient than the Brand-Chaum protocol which has the same security level with Simp-pkDB. We focus on Eff-pkDB in the rest of the paper since it gives higher security level. Eff-pkDB's variant Eff-pkDB^p uses one extra encryption and it is strong private. We propose an instance of these protocols based on the Gap Diffie-Hellman (GDH) problem [30] in EC with a random oracle. The detailed efficiency analysis is presented in Sect. 6.

PaySafe [11] is very efficient but we do not compare it with the other protocols and our protocols since it assumes weaker adversarial model. It is only secure against MiM. It is not secure against DF, DH and TF because the response of the prover in the time critical phase which is a nonce picked by the prover does not depend on any message of the verifier. It also does not protect the privacy of the prover.

Our contributions are:

- We design two public-key DB protocols. The first protocol is secure against **DF, MF and DH** but it is not private. It uses **only one public key related operation** on the prover side. Basically, this protocol can be used in applications not requiring privacy in a very efficient way. Then, we modify this protocol by adding a public-key encryption to make it **strong private**. Both protocols are **quite efficient compared with the previous protocols**. Our constructions are generic based on a key agreement protocol, a weakly-secure symmetric DB protocols, and a cryptosystem. We formally prove the security following the model of Boureanu-Vaudenay [8] which was adapted to public-key DB in Vaudenay [37].
- We define a new key agreement (KA) security game (D-AKA). In literature, the extended Canetti-Krawczyk (eCK) security model [27] is widely accepted for KA. However, **a weaker security model (D-AKA) is sufficient** for the security of our new public-key DB protocols since we care both the efficiency and the security. Finally, we design a D-AKA secure key agreement protocol (Nonce-DH) based on the hardness of the GDH problem and a random oracle. The Nonce-DH key agreement protocol can be used in our DB constructions.

We show in Appendix B another reasonable protocol Simp-pkDB which was our first attempt to construct an efficient and a secure protocol. Although this protocol is quite efficient and does not require any public-key of a verifier, it fails in DH-security. This shows that it is hard to make a protocol which is secure for MiM, DF, and DH at the same time. Adding privacy in protocols is yet another challenge. Strong privacy cannot be achieved so easily as shown in Sect. 5.2. HPO and GOR failed to on this.

Organization of the paper: In Sect. 2, we give the formal definitions for the notion of DB and all necessary security definitions we are considering in our new protocols. In Sect. 3, we describe one time DB protocol OTDB [37] and give new security results on this protocol. OTDB and all the results about OTDB can be employed by Eff-pkDB or Eff-pkDB^p in a very efficient way. In Sect. 4, we introduce our new and weaker KA security model (D-AKA). Then, we construct a new KA protocol Nonce-DH which is D-AKA secure. We have Nonce-DH to show that both Eff-pkDB and Eff-pkDB^p can employ it and to make more precise efficiency analysis on these protocols. In Sect. 5, we introduce Eff-pkDB and Eff-pkDB^p with all security and privacy proofs. Finally, in Sect. 6, we do the efficiency and security analyses of all previous public-key DB protocols in detail.

2 Definitions

The formalism in DB started by Avoine et al. [2]. Then, the first complete model was introduced by Dürholz et al. [15] where the threat models are defined according to the number of tainted time critical phase. The SKI model by Boureanu et al. [5–7] is another formal model which includes a clear communication model between parties in DB. The last model BV model [8] by Boureanu and Vaudenay is a more natural multi-party security model.

In this section, we give the definitions from the literature that we use in our security proofs.

2.1 Public Key Distance Bounding

Definition 1 (Public key DB Protocol [37]). *A public key distance bounding protocol is a two-party probabilistic polynomial-time (PPT) protocol and it consists of a tuple $(\mathcal{K}_P, \mathcal{K}_V, V, P, B)$. Here, $(\mathcal{K}_P, \mathcal{K}_V)$ are the key generation algorithms of P and V, respectively. The output of $\mathcal{K}_P$ is a secret/public key pair $(\mathsf{sk}_P, \mathsf{pk}_P)$ and similarly the output of $\mathcal{K}_V$ is a secret/public key pair $(\mathsf{sk}_V, \mathsf{pk}_V)$. P is the proving algorithm, V is the verifying algorithm where the inputs of P and V are from $\mathcal{K}_P$ and $\mathcal{K}_V$. B is the distance bound. $P(\mathsf{sk}_P, \mathsf{pk}_P, \mathsf{pk}_V)$ and $V(\mathsf{sk}_V, \mathsf{pk}_V)$ interact with each other. At the end of the protocol, $V(\mathsf{sk}_V, \mathsf{pk}_V)$ outputs a final message Out_V and have pk_P as a private output. If $\mathsf{Out}_V = 1$, then V accepts. If $\mathsf{Out}_V = 0$, then V rejects.*

A public-key DB protocol is correct if and only if under honest execution, whenever a verifier $\mathcal{V}$ and a close (to $\mathcal{V}$) prover P run the protocol, then $\mathcal{V}$ always outputs $\mathsf{Out}_V = 1$ and pk_P.

Remark that this definition combines identification with DB: pk_P is not an input of the algorithm V, but it is an output. So, V learns the identity of P during the protocol.

We formalize the security notions of DB protocols. In the setting below, we have parties called provers, verifiers and other actors. Each party has instances and each instance I has its own location. It is called *close* to the instance J, if $d(I, J) \leq B$ and *far* from J, if $d(I, J) > B$ where d is a distance function.

An instance of an honest prover runs the algorithm denoted by $P(\mathsf{sk}_P, \mathsf{pk}_P, \mathsf{pk}_V)$. An instance of a malicious prover runs an arbitrary algorithm denoted by P^*. The verifier is always honest and its instances run $V(\mathsf{sk}_V, \mathsf{pk}_V)$. Without loss of generality, we say that the other actors are malicious. They may run any algorithm.

The locations of the participants are elements of a metric space. We summarize the *communication and adversarial model* (See [5] for the details):

DB protocols run in natural communication settings. There is a notion of time, e.g. time-unit, a notion of measurable distance and a location. Besides, timed communication follows the laws of physics, e.g., communication cannot be faster than the speed of light. An adversary can see all messages (whenever they reach him). He can change the destination of a message subject to constraints.

This communication and adversarial model will only play a role in the DF and MiM security (defined below) but we will not have to deal with it. Indeed, we will start from an existing weakly secure symmetric DB protocol (such as OTDB [37]) and reduce the DF and MiM security of our protocol to the security of that protocol. So, we do not need to formalize more this model.

Now, we explain the security games for the distance fraud, mafia fraud and distance hijacking from [37].

Definition 2 (Distance fraud [37]). *The game begins by running the key setup algorithm* $\mathcal{K}_V$ *which outputs* $(\mathsf{sk}_V, \mathsf{pk}_V)$. *The game includes a verifier instance* $\mathcal{V}$ *and instances of an adversary. Given* pk_V, *the adversary generates* $(\mathsf{sk}_P, \mathsf{pk}_P)$ *with an arbitrary key setup algorithm* $\mathcal{K}^*(\mathsf{pk}_V)$ *(instead of* $\mathcal{K}_P$*). There is no participant close to* $\mathcal{V}$. *The adversary wins if* $\mathcal{V}$ *outputs* $\mathsf{Out}_V = 1$ *and* pk_P. *A DB protocol is DF-secure, if for any such game, the adversary wins with negligible probability.*

Definition 3 (Mafia fraud (MiM security) [37]). *The game begins by running the key setup algorithms* $\mathcal{K}_V$ *and* $\mathcal{K}_P$ *which output* $(\mathsf{sk}_V, \mathsf{pk}_V)$ *and* $(\mathsf{sk}_P, \mathsf{pk}_P)$, *respectively. The adversary receives* pk_V *and* pk_P. *The game consists of several verifier instances including a distinguished one* $\mathcal{V}$, *an honest prover* P *with its instances which are far away from* $\mathcal{V}$ *and an adversary with its instances at any location. The adversary wins if* $\mathcal{V}$ *outputs* $\mathsf{Out}_V = 1$ *and* pk_P. *A DB protocol is MiM-secure if for any such game, the probability of an adversary to win is negligible.*

Definition 4 (Distance hijacking [37]). *The game consists of several verifier instances* $\mathcal{V}, V_1, V_2, ...$, *a far away adversary* P, *and also honest prover instances* $\mathsf{P}', \mathsf{P}'_1, \mathsf{P}'_2 ....$ *A DB protocol* $(\mathcal{K}_P, \mathcal{K}_V, V, P, B)$ *having an initialization, a challenge and a verification phases is DH-secure if for all PPT algorithms* $\mathcal{K}_P^*$ *and* $\mathcal{A}$, *the probability of* P *to win the following game is negligible.*

- $\mathcal{K}_V \to (\mathsf{sk}_V, \mathsf{pk}_V)$, $\mathcal{K}_{P'} \to (\mathsf{sk}_{P'}, \mathsf{pk}_{P'})$.
- $\mathcal{K}_P^*(\mathsf{pk}_{P'}, \mathsf{pk}_V) \to (\mathsf{sk}_P, \mathsf{pk}_P)$ *and if* $\mathsf{pk}_P = \mathsf{pk}_{P'}$, *the game aborts. Then, instances of* P *run* $\mathcal{A}(\mathsf{sk}_P, \mathsf{pk}_P, \mathsf{pk}_V, \mathsf{pk}_{P'})$, $\mathsf{P}', \mathsf{P}'_1, \mathsf{P}'_2, ...$ *run* $P(\mathsf{sk}_{p'}, \mathsf{pk}_V)$, $\mathcal{V}, V_1, V_2, ...$ *run* $V(\mathsf{sk}_V, \mathsf{pk}_V)$.
- P *interacts with* $\mathsf{P}', \mathsf{P}'_1, \mathsf{P}'_2, ...$ *and* $\mathcal{V}, V_1, V_2, ...$ *during the initialization phase of* $\mathcal{V}$ *and* P' *concurrently.*
- P' *and* $\mathcal{V}$ *continue interacting with each other in their challenge phase and* P *remains passive even though he sees the exchanged messages.*
- P *interacts with* $\mathsf{P}', \mathsf{P}'_1, \mathsf{P}'_2, ...$ *and* $\mathcal{V}, V_1, V_2, ...$ *in the verification phase concurrently.*

The adversary wins if $\mathcal{V}$ *outputs* $\mathsf{Out}_V = 1$ *and* pk_P.

The notion of initialization/challenge/verification phase is arbitrary but the notion of DH-security depends on this. To make it correspond to the notion in [12], the challenge phase must correspond to the time critical part where the verifier and the prover exchange challenge/response so fast that responses from far away would be rejected.

Definition 5 (HPVP Privacy Game [19]). *The privacy game is the following: Pick* $b \in \{0, 1\}$ *and let the adversary* $\mathcal{A}$ *play with the following oracles:*

- $\texttt{CreateP}(ID) \to P_i$: *It creates a new prover identity of* ID *and returns its identifier* P_i.
- $\texttt{Launch}() \to \pi$: *It launches a new protocol with the verifier* V_j *and returns the session identifier* π.
- $\texttt{Corrupt}(P_i)$: *It returns the current state of* P_i. *Current state means the all the values in* P_i*'s current memory. It does not include volatile memory.*
- $\texttt{DrawP}(P_i, P_j) \to vtag$: *It draws either* P_i *(if* $b = 0$*) or draws* P_j *(if* $b = 1$*) and returns the virtual tag reference* $vtag$. *If one of the provers was already an input of* $\texttt{DrawP} \to vtag'$ *query and* $vtag'$ *has not been released, then it outputs* $\emptyset$.
- $\texttt{Free}(vtag)$: *It releases* $vtag$ *which means* $vtag$ *can no longer be accessed.*
- $\texttt{SendP}(vtag, m) \to m'$: *It sends the message* m *to the drawn prover and returns the response* m' *of the prover. If* $vtag$ *was not drawn or was released, nothing happens.*
- $\texttt{SendV}(\pi, m) \to m'$: *It sends the message* m *to the verifier in the session* π *and returns the response* m' *of the verifier. If* π *was not launched, nothing happens.*
- $\texttt{Result}(\pi) \to b'$: *It returns a bit that shows if the session* π *is accepted by the verifier (i.e. the message* Out_V*).*

In the end of the game, the adversary outputs a bit g. *If* $g = b$, *then* $\mathcal{A}$ *wins. Otherwise, it loses.*

A DB protocol is strong private *if for all PPT adversaries, the advantage of winning the privacy game is negligible.*

We distinguish strong and weak privacy [33]. The weak privacy game does not include any '`Corrupt`' oracle. The other kind of classification is *wide* and *narrow* private. Wide privacy game is allowing to use the '`Result`' oracle while the narrow privacy game does not. In this paper, we implicitly consider wide privacy by making Out_V a protocol message, which means we always obtain this bit without using '`Result`' oracle.

2.2 Symmetric Distance Bounding

In this section, we give the useful definitions about the symmetric distance bounding that we need to use for our public key distance bounding protocols. Therefore, we do not explain all security notions for symmetric DB protocols.

Definition 6 (Symmetric DB Protocol [37]). *A symmetric distance bounding protocol is a two-party PPT protocol and it consists of a tuple* $(\mathcal{K}, V, P, B)$. *Here,* $\mathcal{K}$ *is the key generation algorithm,* P *is the proving algorithm and* V *is the verifying algorithm. The inputs of* P *and* V *is the output* s *of* $\mathcal{K}$. B *is the distance bound.* $P(s)$ *and* $V(s)$ *interact with each other. At the end of the protocol,* $V(s)$ *outputs a final message* Out_V. *If* $\mathsf{Out}_V = 1$, *then* V *accepts. If* $\mathsf{Out}_V = 0$, *then* V *rejects.*

A symmetric DB protocol is correct if and only if under honest execution, whenever a verifier $\mathcal{V}$ *and a close (to* $\mathcal{V}$*) prover* P *run the protocol, then* $\mathcal{V}$ *always outputs* $\mathsf{Out}_V = 1$.

Definition 7 (One Time DF (OT-DF) [37]). *The game begins by running a malicious key setup algorithm K^* which outputs s. It consists of a single verifier instance $\mathcal{V}$ running $V(s)$ and instances of an adversary P^*. P^* receives s. There is no participant close to $\mathcal{V}$. The adversary wins if $\mathcal{V}$ outputs $\mathsf{Out}_V = 1$. A symmetric DB protocol is OT-DF-secure, if for any such game, the adversary wins with negligible probability.*

Definition 8 (One Time MiM (OT-MiM) [37]). *The game begins by running the key setup algorithm $\mathcal{K}$ which outputs s. It consists of a single verifier instance $\mathcal{V}$ running $V(s)$, a single far away prover instance P running $P(s)$ and instances of an adversary. The adversary wins if $\mathcal{V}$ outputs $\mathsf{Out}_V = 1$. A symmetric DB protocol is OT-MiM-secure, if for any such game, the probability that the adversary wins is negligible.*

Multi-verifier OT-MiM: The OT-MiM game with more than one verifier instance is called as *multi-verifier OT-MiM-security.* We defined this new notion to be able to have the result in Theorem 1 which helps us to prove the security of our constructions.

Definition 9 (One Time DH (OT-DH) [37]). *The game consists of a verifier instance $\mathcal{V}$, a far away adversary P, and also honest (and close) prover instance P'. A symmetric DB protocol $(\mathcal{K}, V, P, B)$ having an initialization, a challenge and a verification phases is OT-DH-secure if for all PPT algorithms $\mathcal{A}, \mathcal{K}^*$, the probability of P to win the following game is negligible.*

- $\mathcal{K}^* \to s$, $\mathcal{K} \to s'$. *Then,* P' *runs* $P(s')$, $\mathcal{V}$ *runs* $V(s)$ *and* P *runs* $\mathcal{A}(s)$.
- P *interacts with* P' *and* $\mathcal{V}$ *in their initialization phase concurrently.*
- P' *and* $\mathcal{V}$ *continue interacting with each other in their challenge phase and* P *remains passive even though he sees the exchanged messages.*
- P *interacts with* P' *and* $\mathcal{V}$ *in their verification phase concurrently.*

The adversary wins if $\mathcal{V}$ *outputs* $\mathsf{Out}_V = 1$.

Definition 10 (Multi-verifier Impersonation Fraud (IF) [3]). *The game begins by running the key setup algorithm $\mathcal{K}$ which outputs s. It consists of verifier instances running $V(s)$ and an adversary with no inputs. The adversary wins if any verifier instance outputs $\mathsf{Out}_V = 1$. A distance bounding protocol is multi-verifier IF-secure, if for any such game, the probability of an adversary to win is negligible.*

The above definition is with several verifiers, contrarily to others, because we will only use multi-verifier IF security.

MiM-security covers multi-verifier IF-security. So, if a DB protocol is MiM-secure, then it is multi-verifier IF-secure.

We will see in Theorem 2 that OT-MiM-security also implies multi-verifier IF-security for a DB following the canonical structure.

Definition 11 (Canonical Structure [37]). *A symmetric DB protocol $(\mathcal{K}, V, P, B)$ follows the canonical structure, if there exist an initialization/challenge/verification phases,* P *does not use* s *during the initialization phase,* $\mathcal{V}$ *does not use* s *at all except for computing the final* Out_V*, and the verification phase is not interactive.*

Remark that the notion of phase is used in DH and OT-DH security.

3 OTDB

As an example of one-time secure protocol, we can give the protocol OTDB by Vaudenay [37] which is a symmetric DB adapted from Hancke-Kuhn protocol [18]. The OTDB protocol follows the canonical structure (See Definition 11), only requires one xor operation before the challenge phase on the prover side and it is OT-DF, OT-MiM, multi-verifier OT-MiM and OT-DH secure [37]. (See Fig. 1.) We complement these known results by showing multi-verifier OT-MiM security and multi-verifier IF-security.

Theorem 1. *OTDB is multi-verifier OT-MiM secure.*

Proof. Γ_0: In this game, an adversary $\mathcal{A}$ plays multi-verifier OT-MiM game. Here, we have a distinguished verifier instance $\mathcal{V}$ with other instances $\{V_1, ..., V_k\}$ and one prover instance P. The success probability of Γ_0 is p_0.

Γ_1 : We reduce Γ_0 to Γ_1 where at most one verifier instance outputs 1. Let's say E is an event in Γ_0 where at least two verifier instances output 1 ($\mathsf{Out}_V = 1$). To reduce Γ_0 to Γ_1, we show that $\Pr[E]$ is negligible.

First, we define hybrid games $\Gamma_{i,j}$'s to analyze $\Pr[E]$. $\Gamma_{i,j}$ is similar to Γ_0 except the game stops right after V_i and V_j have sent their final outputs and all Out_V is replaced by 0 except V_i and V_j. The adversary wins the game if $\mathsf{Out}_{V_i} = \mathsf{Out}_{V_j} = 1$.

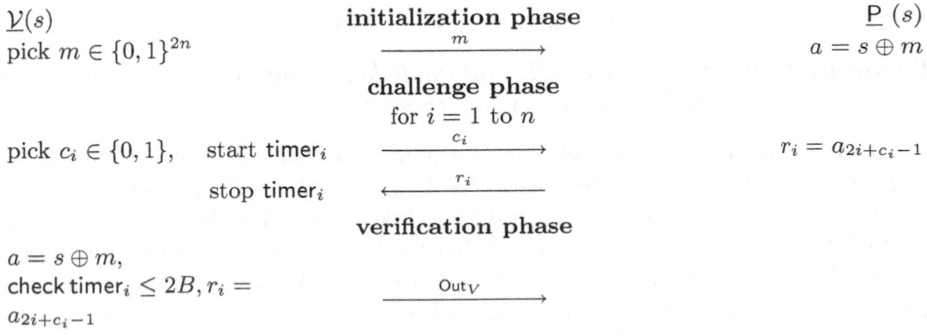

$\underline{\mathcal{V}}(s)$		**initialization phase**	$\underline{\mathsf{P}}\ (s)$
pick $m \in \{0,1\}^{2n}$		$\xrightarrow{m}$	$a = s \oplus m$
		challenge phase	
		for $i = 1$ to n	
pick $c_i \in \{0,1\}$,	start timer_i	$\xrightarrow{c_i}$	$r_i = a_{2i+c_i-1}$
	stop timer_i	$\xleftarrow{r_i}$	
		verification phase	
$a = s \oplus m$, check $\mathsf{timer}_i \leq 2B, r_i = a_{2i+c_i-1}$		$\xrightarrow{\mathsf{Out}_V}$	

Fig. 1. OTDB

In $\Gamma_{i,j}$, we define three kinds of arrays for the challenges. The first array C_{V_i} includes the challenges sent by V_i, the second array C_{V_j} includes the challenges sent by V_j and the third array C_P includes the challenges seen by P. The bits in C_{V_i} and C_{V_j} are independent. We also define a response function $\mathsf{resp}_k(c) = a_{2k+c-1}$ for each round k. Since the bits of the secret s are independent, the bits of $\{\mathsf{resp}_k(0)||\mathsf{resp}_k(1)\}_{k=1}^n$ are independent as well. If $C_{V_i}[k] \neq C_{V_j}[k]$, then the adversary could have taken $C_P[k] = c$ where c is equal either $C_{V_i}[k]$ or $C_{V_j}[k]$ and learned $\mathsf{resp}_k(c)$. So, he responds correctly to either V_i or V_j for sure, but to the other instance with probability $\frac{1}{2}$. We define an event $E_{ij,k}$ where the responses are correct for V_i and V_j in round k. Clearly, all events $\{E_{ij,k}\}_{k=1}^n$ are independent. So, $\Gamma_{i,j} = \prod_k \Pr[E_{ij,k}]$. Hence,

$$\Pr[E_{ij,k}] \leq \Pr[C_{V_i}[k] = C_{V_j}[k]] + \Pr[E_{ij,k}|C_{V_i}[k] \neq C_{V_j}[k]] \\ \times \Pr[C_{V_i}[k] \neq C_{V_j}[k]] \leq \frac{3}{4}$$

So, the adversary wins $\Gamma_{i,j}$ with the probability $(\frac{3}{4})^n$ which is negligible. Now, we can analyze E.

$$\Pr[E] \leq \sum_{i,j} \Pr[\Gamma_{i,j}] = \mathsf{negl}(n)$$

Since E happens with the negligible probability, we can reduce Γ_0 to Γ_1 and conclude $p_1 - p_0$ is negligible. For Γ_1 to succeed, only $\mathcal{V}$ must produce $\mathsf{Out}_\mathcal{V} = 1$.

Γ_2 : We reduce Γ_1 to Γ_2 where we simulate all verifier instances except $\mathcal{V}$. We can do this simulation because the messages but Out_V sent by a verifier does not depend on the secret. Since $\mathsf{Out}_\mathsf{V} = 0$ for all verifier instance except $\mathcal{V}$ in the winning case (only $\mathcal{V}$ can output 1), $p_1 \leq p_2$.

Now in Γ_2, we are in OT-MiM game where there is only one verifier instance $\mathcal{V}$ and one prover instance P. By using the OT-MiM-security result of OTDB [37], we deduce p_2 is negligible so p_0 is negligible. □

We prove the following result which will be used in Theorem 6.

Theorem 2. *If a (symmetric) DB protocol following the canonical structure is OT-MiM secure, then it is multi-verifier IF-secure.*

Proof. We take an adversary $\mathcal{M}$ playing the multi-verifier IF game. $\mathcal{M}$ interacts with polynomially many verifier instances V_j's. We define adversaries $\mathcal{A}_i$'s playing the OT-MiM game. $\mathcal{A}_i$ simulates $\mathcal{M}$ and takes the verifier instance V_i as $\mathcal{V}$ in the OT-MiM game. Concretely, we number the V_j's by their order of appearance during the simulation of $\mathcal{M}$. When $\mathcal{M}$ queries $V_1, ..., V_{i-1}, V_{i+1}, ..., V_k$ (where k is the total number of verifier instances), $\mathcal{A}_i$ just simulates them (this is possible since the protocol follows the canonical structure. So, no message from the verifier except Out_V depends on s). If Out_V needs to be returned to $\mathcal{M}$, $\mathcal{A}_i$ returns 0. When $\mathcal{M}$ queries V_i, $\mathcal{A}_i$ relays it to $\mathcal{V}$ and sends the response of $\mathcal{V}$ to $\mathcal{M}$.

Let E_i be the event in the multi-verifier IF game which is $\mathsf{Out}_{V_i} = 1$ and all previously released Out_V are equal to 0. Clearly, we have $\Pr[\mathcal{M}\,\mathsf{wins}] = \sum_{i\geq 1} \Pr[\mathcal{M}\,\mathsf{wins} \wedge E_i]$. On the other hand, $\Pr[\mathcal{M}\,\mathsf{wins} \wedge E_i] \leq \Pr[\mathcal{A}_i\,\mathsf{wins}]$ because for all coins making $\mathcal{M}$ win the multi-verifier IF-game and E_i occur at the same time, we have $\mathsf{Out}_{V_j} = 0$ for all $j < i$ and $\mathsf{Out}_{V_i} = 1$ so the same coins make $\mathcal{A}_i$ win the OT-MiM game. So, $\Pr[\mathcal{M}\,\mathsf{wins}] \leq \sum_{i\geq 1} \Pr[\mathcal{A}_i\,\mathsf{wins}]$. Due to OT-MiM security, $\Pr[\mathcal{A}_i\,\mathsf{wins}]$ is negligible for every i. So, $\Pr[\mathcal{M}\,\mathsf{wins}]$ is negligible. So, we have multi-verifier IF-security. □

Thanks to Theorem 2, OTDB is multi-verifier IF-secure.

4 Authenticated Key Agreement (AKA) Protocols

In this section, we show our new KA security model and some preliminaries about the AKA protocols. The security models in this section are used *to construct secure and private public-key DB protocols* in Sect. 5.

We note that the DB protocols we constructed in Sect. 5 can employ any eCK-secure [27] key agreement protocol to have the same security properties. However, eCK-security is stronger than we need in our protocols. Therefore, we define a weaker notion **to have simpler, more efficient and secure public-key DB.** Table 3 in Appendix A shows that Nonce-DH which is secure in our weaker model is more efficient than the previous KA protocols.

Definition 12 (AKA in one-pass). *A one-pass AKA protocol is a tuple* $(\mathsf{Gen}_A, \mathsf{Gen}_B, D, A, B)$ *of PPT algorithms. Let A and B be the two parties. A and B generate secret/public key pairs* $(\mathsf{sk}_A, \mathsf{pk}_A)$ *and* $(\mathsf{sk}_B, \mathsf{pk}_B)$ *with the algorithms* $\mathsf{Gen}_A(1^n)$ *and* $\mathsf{Gen}_B(1^n)$*, respectively where n is the security parameter. B picks N from the sampling algorithm D and runs* $B(\mathsf{sk}_B, \mathsf{pk}_B, \mathsf{pk}_A, N)$ *which outputs the session key s. Then, (s)he sends N and finally, A gets the session key s by running* $A(\mathsf{sk}_A, \mathsf{pk}_A, \mathsf{pk}_B, N)$ *(See Fig. 2). We say that AKA is correct, if A and B obtain the same s at the end of the protocol.*

$\underline{\mathbf{A}}(\mathsf{sk}_A, \mathsf{pk}_A, \mathsf{pk}_B)$		$\underline{\mathbf{B}}(\mathsf{sk}_B, \mathsf{pk}_B, \mathsf{pk}_A)$
		$N \leftarrow D(1^n)$
$A(\mathsf{sk_A}, \mathsf{pk_A}, \mathsf{pk_B}, \mathsf{N}) \rightarrow s$	$\xleftarrow{\mathsf{N}}$	$B(\mathsf{sk}_B, \mathsf{pk}_B, \mathsf{pk}_A, N) \rightarrow s$

Fig. 2. The structure of an authenticated key agreement (AKA) protocols in one pass.

Definition 13 (Decisional-Authenticated Key Agreement (D-AKA) security). *We define two oracles set up with* $\mathsf{sk}_A, \mathsf{pk}_A, \mathsf{sk}_B, \mathsf{pk}_B$.

$\underline{\mathcal{O}_A(.,.):}$
return $A(\mathsf{sk}_A, \mathsf{pk}_A, ., .)$

$\underline{\mathcal{O}_B(.):}$
$N' \leftarrow D(1^n)$
$s' \leftarrow B(\mathsf{sk}_B, \mathsf{pk}_B, ., N')$
return s', N'

Given $b \in \{0,1\}$ *and the oracles* $\mathcal{O}_A(.,.), \mathcal{O}_B(.)$, *the game* $\mathsf{KA}^{d-aka}_{b,\mathcal{A}(n)}$ *is:*

1. *Challenger executes* $\mathsf{Gen}_A(1^n) \to (\mathsf{sk}_A, \mathsf{pk}_A), \mathsf{Gen}_B(1^n) \to (\mathsf{sk}_B, \mathsf{pk}_B)$, *sets up the oracles, calls* $\mathcal{O}_B(\mathsf{pk}_A) \to (s_0, N)$ *and picks* $s_1 \in \{0,1\}^n$. *Then, he sends* $s_b, N, \mathsf{pk}_B, \mathsf{pk}_A$ *to the adversary* $\mathcal{A}$.
2. $\mathcal{A}$ *has access to the oracle* $\mathcal{O}_B(.)$ *and* $\mathcal{O}_A(.,.)$ *under the condition of not querying the oracle* $\mathcal{O}_A$ *with the input* (pk_B, N). *Eventually,* $\mathcal{A}$ *outputs* b'.
3. *The advantage of the game is*

$$\mathsf{Adv}(\mathsf{KA}^{d-aka}_{\mathcal{A}(n)}) = \mathsf{Pr}[\mathsf{KA}^{d-aka}_{0,\mathcal{A}(n)} = 1] - \mathsf{Pr}[\mathsf{KA}^{d-aka}_{1,\mathcal{A}(n)} = 1].$$

A KA protocol $(\mathsf{Gen}_A(1^n), \mathsf{Gen}_B(1^n), D, A, B)$ *is D-AKA secure if for all PPT algorithms* $\mathcal{A}$, $\mathsf{Adv}(\mathsf{KA}^{d-aka}_{\mathcal{A}(n)})$ *is negligible.*

We show that eCK-security implies D-AKA security in Theorem 8 in Appendix A. It means that our new public-key DB protocols can employ eCK-secure key agreement protocols as well.

Note that as a result of Lemma 1 in Appendix A, the probability that the same nonce is picked by the oracle B is negligible when we have D-AKA security.

Definition 14 (D-AKA^p privacy). *Given* $b \in \{0,1\}$ *and the oracle* $\mathcal{O}_A(.,.)$ *(defined in Definition 13), the game* $\mathsf{KA}^{d-aka^p}_{b,\mathcal{A}(n)}$ *is:*

1. *Challenger runs* $\mathsf{Gen}_A(1^n) \to (\mathsf{sk}_A, \mathsf{pk}_A)$ *and* $\mathsf{Gen}_B(1^n) \to (\mathsf{sk}_{B_1}, \mathsf{pk}_{B_1})$, *sets up the oracle and gives* $\mathsf{pk}_A, \mathsf{pk}_{B_1}$ *and* sk_{B_1} *to* $\mathcal{A}$.
2. $\mathcal{A}$ *selects* sk_{B_0} *and* pk_{B_0} *and sends them to the challenger.*
3. *Challenger executes* $D(1^n) \to N$, $B(\mathsf{sk}_{B_b}, \mathsf{pk}_{B_b}, \mathsf{pk}_A^{\mathsf{sk}_{B_b}}, N) \to s$. *Then, he sends* s *to the adversary* $\mathcal{A}$.
4. $\mathcal{A}$ *has access to the oracle* $\mathcal{O}_A$. *Eventually,* $\mathcal{A}$ *outputs* b'. *(Remark that* $\mathcal{A}$ *does not know* N.*)*
5. *The advantage of the game is*

$$\mathsf{Adv}(\mathsf{KA}^{d-aka^p}_{\mathcal{A}(n)}) = \mathsf{Pr}[\mathsf{KA}^{d-aka^p}_{0,\mathcal{A}(n)} = 1] - \mathsf{Pr}[\mathsf{KA}^{d-aka^p}_{1,\mathcal{A}(n)} = 1].$$

A KA protocol $(\mathsf{Gen}_A(1^n), \mathsf{Gen}_B(1^n), D, A, B)$ *is D-AKA^p private if for all PPT algorithms* $\mathcal{A}$, $\mathsf{Adv}(\mathsf{KA}^{d-aka^p}_{\mathcal{A}(n)})$ *is negligible.*

A One-Pass AKA Protocol (Nonce-DH): We construct a D-AKA secure protocol (Nonce-DH) based on the Diffie-Hellman (DH) [14] as in Fig. 3. Here g is a generator of a group of prime order q. g and q depend on a security parameter. The parties know each others' public keys beforehand where $\mathsf{pk}_A = g^{\mathsf{sk}_A}$ and $\mathsf{pk}_B = g^{\mathsf{sk}_B}$ and sk_A and sk_B are the corresponding secret keys which are uniformly picked in $\mathbb{Z}_q$.

The party B has input $(\mathsf{sk}_B, \mathsf{pk}_B, \mathsf{pk}_A)$. He randomly picks N from $\{0,1\}^\ell$ and computes $B(\mathsf{sk}_B, \mathsf{pk}_B, \mathsf{pk}_A, N) = H(g, \mathsf{pk}_B, \mathsf{pk}_A, \mathsf{pk}_A^{\mathsf{sk}_B}, N)$ to get s. The party A computes $A(\mathsf{sk}_A, \mathsf{pk}_A, \mathsf{pk}_B, N) = H(g, \mathsf{pk}_B, \mathsf{pk}_A, \mathsf{pk}_B^{\mathsf{sk}_A}, N)$ and gets s. Here, H is a deterministic function.

Clearly, Nonce-DH is correct since H is deterministic.

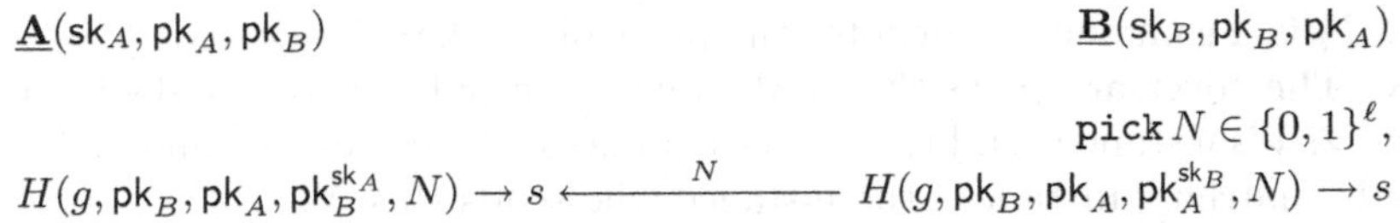

$\underline{\mathbf{A}}(\mathsf{sk}_A, \mathsf{pk}_A, \mathsf{pk}_B)$ $\underline{\mathbf{B}}(\mathsf{sk}_B, \mathsf{pk}_B, \mathsf{pk}_A)$

pick $N \in \{0,1\}^\ell$,

$H(g, \mathsf{pk}_B, \mathsf{pk}_A, \mathsf{pk}_B^{\mathsf{sk}_A}, N) \to s \xleftarrow{N} H(g, \mathsf{pk}_B, \mathsf{pk}_A, \mathsf{pk}_A^{\mathsf{sk}_B}, N) \to s$

Fig. 3. The Nonce-DH key agreement protocol.

Theorem 3. *Assuming that the Gap Diffie-Hellman problem [30] is hard and $\ell = \Omega(n)$, Nonce-DH is D-AKA secure and D-AKA^p private in the random oracle model.*

The proof is in Appendix C.

5 Efficient Public Key Distance Bounding Protocol

In this section, we first introduce Eff-pkDB which is secure against DF, MF and DH and then Eff-pkDB^p a variant of it preserving the strong privacy as well.

5.1 Eff-PkDB

Eff-pkDB (Fig. 4) is constructed on an AKA in one-pass and a symmetric DB protocol. P and $\mathcal{V}$ first agree on a secret key s using an AKA protocol. Then, they together run a symmetric key DB protocol (symDB) by using s. Using OTDB as symDB and Using Nonce-DH as an AKA protocol will appear to be enough for its security.

Theorem 4. *If symDB is OT-DF-secure, then Eff-pkDB is DF-secure.*

Proof sketch: The malicious and far away prover with its instances play the DF game. We can easily reduce it to the game where $\mathcal{V}$ and the adversary receive the same s' from outside (even if maliciously selected). Since symDB is OT-DF-secure, the prover passes the protocol with negligible probability. □

Theorem 5. *If symDB is multi-verifier OT-MiM-secure and the key agreement protocol with the algorithms $\mathsf{Gen}_A, \mathsf{Gen}_B, A, B, D$ is D-AKA secure then Eff-pkDB is MiM-secure.*

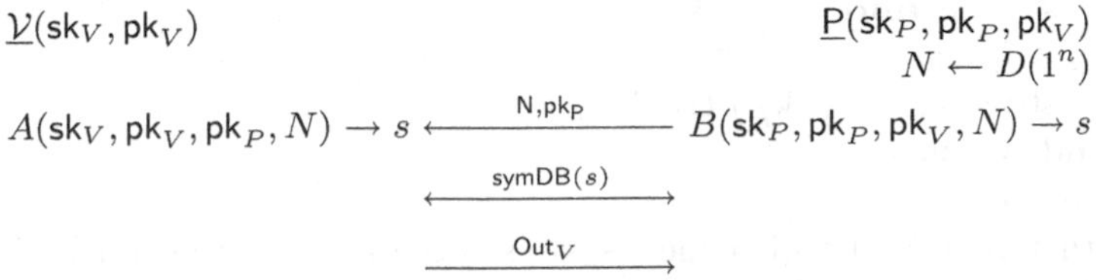

$\underline{\mathcal{V}}(\mathsf{sk}_V, \mathsf{pk}_V)$ $\underline{\mathsf{P}}(\mathsf{sk}_P, \mathsf{pk}_P, \mathsf{pk}_V)$

$N \leftarrow D(1^n)$

$A(\mathsf{sk}_V, \mathsf{pk}_V, \mathsf{pk}_P, N) \to s \xleftarrow{\mathsf{N}, \mathsf{pk}_\mathsf{P}} B(\mathsf{sk}_P, \mathsf{pk}_P, \mathsf{pk}_V, N) \to s$

$\xleftrightarrow{\mathsf{symDB}(s)}$

$\xrightarrow{\mathsf{Out}_V}$

Fig. 4. Public-key DB protocol based on D-AKA secure KA (Eff-pkDB)

Proof. Γ_i is a game and p_i denotes the probability that Γ_i succeeds.

Γ_0 : The adversary plays the MiM game in Eff-pkDB with the distinguished verifier $\mathcal{V}$, $\mathcal{V}$'s instances and the prover instances. $\mathcal{V}$ receives pk_P and a given N. We call "matching instance" the instance who sends this N.

Γ_1 : We reduce Γ_0 to Γ_1 where no nonce produced by any prover instance is duplicated or equal to any nonce received by any a verifier instance before. Thanks to Lemma 1 in Appendix A, $p_1 - p_0$ is negligible. So, the matching instance (if any) is unique and sets N before it is sent to $\mathcal{V}$.

Γ_2 : We simulate the prover instances and $\mathcal{V}$ as below in this game. Basically, in Γ_2, the prover and the verifier do not use the secret generated by the oracles $\mathcal{O}_B$ and $\mathcal{O}_A$, respectively.

<u>$P(.)$</u> (in Γ_2)
  run $\mathcal{O}_B(\mathsf{pk}_V) \rightarrow (s_0, N')$
  send N', pk_P
  pick s_1
  store (N', s_1, pk_P) **in** T
  run symDB(s_1)

<u>$V(.)$</u> (in Γ_2)
  receive N', pk_P
  if $(N', ., \mathsf{pk}_P) \in T$
    retrieve s from T
    where $(N', s, \mathsf{pk}_P) \in T$
  else:
    $s \leftarrow \mathcal{O}_A(\mathsf{pk}_P, N')$
  run symDB(s)

With the reduction from Γ_1 to Γ_2, we show that the secret generated by A and B are indistinguishable from the randomly picked secret. The reduction is showed below:

We define the hybrid games $\Gamma_{2,t}$ to show $p_2 - p_1$ is negligible. Here, $t \in \{0, 1, 2, ..., k\}$ and k is the number of prover instances bounded by a polynomial.

$\Gamma_{2,i}$: $\mathcal{V}$ is simulated as in Γ_2 and the jth instance of P is simulated as in Γ_2 for $j \leq i$ and as in Γ_1 for $j > i$. Clearly, $\Gamma_{2,0} = \Gamma_1$ and $\Gamma_{2,k} = \Gamma_2$.

First, we show that $\Gamma_{2,i}$ and $\Gamma_{2,i+1}$ are indistinguishable. For this, we use an adversary $\mathcal{B}$ that plays the D-AKA game. $\mathcal{B}$ receives $\mathsf{pk}_A, \mathsf{pk}_B, s_b, N$ from the D-AKA challenger and simulates against the adversary $\mathcal{A}$ which distinguishes $\Gamma_{2,i}$ and $\Gamma_{2,i+1}$. $\mathcal{B}$ assigns $\mathsf{pk}_V = \mathsf{pk}_A$ and $\mathsf{pk}_P = \mathsf{pk}_B$. $\mathcal{B}$ simulates each prover P_j as described below.

<u>$P_j(.)$</u>
  if $j \neq i + 1$
    $\mathcal{O}_B(\mathsf{pk}_V) \rightarrow (s', N')$
    if $j \leq i$
      pick s'
  else:
    $s' \leftarrow s_b$ **and** $N' \leftarrow N$
  if $j \leq i + 1$
    store (N', s', pk_P) **to** T
  send N', pk_P
  run symDB(s')

Note that if $b = 0$ which means s_b is generated by the oracle B then $\mathcal{B}$ simulates the game $\Gamma_{2,i}$. Otherwise, he simulates $\Gamma_{2,i+1}$.

For the verifier simulation, $\mathcal{B}$ first checks, if $(N', ., \mathsf{pk}_P)$ is stored by himself as $\mathcal{V}$ in Γ_2. Otherwise, he sends (pk_P, N') to the oracle $\mathcal{O}_A$ and receives s'. Since (N, s_b, pk_P) is always stored in T, (pk_P, N) is not queried to $\mathcal{O}_A$ oracle. In the end of the game, $\mathcal{A}$ sends his decision. If $\mathcal{A}$ outputs i, then $\mathcal{B}$ outputs 0. If $\mathcal{A}$ outputs $i+1$, then $\mathcal{B}$ outputs 1. Clearly, the advantage of $\mathcal{B}$ is $p_{2,i} - p_{2,i+1}$. Due to the D-AKA security, we obtain that $p_{2,i} - p_{2,i+1}$ is negligible. From the hybrid theorem, we can conclude that $p_{2,0} - p_{2,k}$ is negligible where $p_{2,0} = p_1$ and $p_{2,k} = p_2$.

Γ_3 : We simulate the prover instances as below so that they do not run the oracle $\mathcal{O}_B$ to have N. The only change in this game is the generation of the nonce. Since the prover in Γ_3 picks the nonce from the same distribution that $\mathcal{O}_B$ picks, $p_3 = p_2$. This game shows that the prover generates N' (and also s_1) independently from $\mathcal{O}_B$.

<u>$P(.)$</u> (in Γ_3)
 pick $N' \in D(1^n)$
 send N', pk_P
 pick s_1
 store (N', s_1, pk_P) **to** T
 run symDB(s_1)

Γ_4 : We reduce Γ_3 to the multi-verifier OT-MiM-security game Γ_4 where there is only matching instance and the other instances are simulated. With this final reduction, we show that the adversary has to break the multi-verifier OT-MiM-security of symDB in order to break the MiM-security of Eff-pkDB.

The reduction is the following. $\mathcal{A}^3$ plays the Γ_3 game. We construct an adversary $\mathcal{A}_i^4$ in Γ_4. $\mathcal{A}_i^4$ receives N from the matching prover in Γ_4. $\mathcal{A}_i^4$ takes P_i as a matching prover in Γ_3 where $i \in \{1, ..., k\}$. $\mathcal{A}_i^4$ simulates all of the provers except P_i against $\mathcal{A}^3$. For P_i, $\mathcal{A}_i^4$ just sends (pk_P, N). In the end, if P_i is the matching instance in Γ_3 and $\mathcal{A}^3$ wins then $\mathcal{A}_i^4$ wins. Therefore $p_3 < \sum_i p_{4,i}$ where $p_{4,i}$ is the probability that $\mathcal{A}_i^4$ wins. Due to multi-verifier OT-MiM-security, all $p_{4,i}$'s are negligible. So, p_3 is negligible. Hence, p_0 is negligible. □

Theorem 6. *If symDB is OT-MiM-secure, OT-DH-secure and follows the canonical structure and if the key agreement protocol with the algorithms* $\mathsf{Gen}_A, \mathsf{Gen}_B, A, B, D$ *is D-AKA secure then Eff-pkDB is DH-secure.*

Proof. Γ_i is a game and p_i denotes the probability that Γ_i succeeds.

Γ_0 : The adversary P with its instances plays the DH-security game in Eff-pkDB with the distinguished verifier $\mathcal{V}$ and its instances and an honest prover P'. The probability that the adversary succeeds in Γ_0 is p_0.

Γ_1 and Γ_2 : These games are like in the proof of Theorem 5 except that P_j is replaced by P'_j. The reduction from Γ_0 to Γ_1 and Γ_2 is similar to the proof of Theorem 5. So we can conclude that $p_2 - p_0$ is negligible.

We let N be the nonce produced by the instance of P' and s_1 be its key which is playing a role during the challenge phase of $\mathcal{V}$ in the DH game.

We reduce Γ_2 to Γ_3 in which all Out_V from a verifier instance who receives pk_P and N is replaced by 0 during the initialization phase. Intuitively, in this case, Out_V cannot be equal 1 because if it is 1, it means P' impersonates P. The reduction is as follows: During the initialization game, P' sends messages which do not depend on s_1 because of the canonical structure, and which can be simulated. So, we can reduce this phase to the multi-verifier IF game and use Theorem 2 to show that $p_3 - p_2$ is negligible. This reduction shows that the DH-adversary P cannot win the game with sending pk_P and N generated by P'.

We reduce Γ_3 to Γ_4 where the game stops after the challenge phase for $\mathcal{V}$. Since the verification phase which is after the challenge phase is non-interactive and Out_V is determined at the end of the challenge phase, $p_4 = p_3$.

We reduce Γ_4 to Γ_5 which is OT-DH game. In Γ_4, s_1 has never been used so s (the key of $\mathcal{V}$ which is given by the adversary) is independent from s_1. In this case, P' and $\mathcal{V}$ run symDB with independent secrets. So, $p_5 = p_4$. Because of the OT-DH security of symDB, p_5 is negligible. □

5.2 Eff-pkDB^p

Eff-pkDB is not strong private as the public key of the prover is sent in clear. Adding one encryption operation to Eff-pkDB is enough to have strong privacy.

Eff-pkDB^p in Fig. 5 is the following: The prover and the verifier generate their secret/public key pairs by running the algorithms $\mathsf{Gen}_P(1^n)$ and $\mathsf{Gen}_V(1^n)$, respectively. We denote $(\mathsf{sk}_P, \mathsf{pk}_P)$ for the secret/public key pair of the prover and $(\mathsf{sk}_V, \mathsf{pk}_V)$ for the secret/public key pair of the verifier where $\mathsf{sk}_V = (\mathsf{sk}_{V_1}, \mathsf{sk}_{V_2})$ and $\mathsf{pk}_V = (\mathsf{pk}_{V_1}, \mathsf{pk}_{V_2})$ and the first key is used for the encryption and the second key is used for the AKA protocol. The prover picks N from the sampling algorithm D and generates s with the algorithm $B(\mathsf{sk}_P, \mathsf{pk}_P, \mathsf{pk}_{V_2}, N)$. Then, he encrypts pk_P and N with pk_{V_1}. After, he sends the ciphertext e to the verifier. The verifier decrypts e with sk_{V_1} and learns N and pk_P which helps him to understand who is interacting with him. Next, the verifier runs $A(\mathsf{sk}_{V_2}, \mathsf{pk}_{V_2}, \mathsf{pk}_P, N)$ and gets s. Finally, the prover and verifier run a symmetric DB protocol symDB protocol with s.

Assuming that the AKA protocol is D-AKA secure and symDB is OT-X secure symmetric key DB protocol for all $X \in \{DF, MiM, DH\}$ and follows

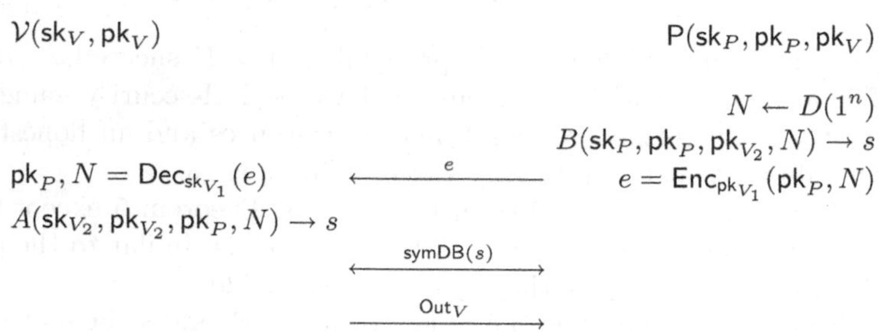

$\mathcal{V}(\mathsf{sk}_V, \mathsf{pk}_V)$		$\mathsf{P}(\mathsf{sk}_P, \mathsf{pk}_P, \mathsf{pk}_V)$
		$N \leftarrow D(1^n)$
		$B(\mathsf{sk}_P, \mathsf{pk}_P, \mathsf{pk}_{V_2}, N) \rightarrow s$
$\mathsf{pk}_P, N = \mathsf{Dec}_{\mathsf{sk}_{V_1}}(e)$	$\xleftarrow{\quad e \quad}$	$e = \mathsf{Enc}_{\mathsf{pk}_{V_1}}(\mathsf{pk}_P, N)$
$A(\mathsf{sk}_{V_2}, \mathsf{pk}_{V_2}, \mathsf{pk}_P, N) \rightarrow s$		
	$\xleftrightarrow{\quad \mathsf{symDB}(s) \quad}$	
	$\xrightarrow{\quad \mathsf{Out}_V \quad}$	

Fig. 5. Eff-pkDB^p: private variant of Eff-pkDB

canonical structure, we can easily show that Eff-pkDB^p is X-secure from Theorems 4 to 6. To prove this, we start from an adversary playing the X-security game against Eff-pkDB^p. We construct an adversary playing the same game against Eff-pkDB to whom we give sk_{V_1}. The simulation is straightforward.

Theorem 7. *Assuming the key agreement protocol is D-AKA^p secure and the cryptosystem is IND-CCA secure, then the Eff-pkDB^p is strong private in the HPVP model (Definition 5).*

Proof. Γ_i is a game and p_i denotes the probability that Γ_i succeeds.

Γ_0 : The adversary $\mathcal{A}$ plays the HPVP privacy game.

Γ_1 : The verifiers skip the decryption when they receive a ciphertext produced by any prover and continue with the values encrypted by the prover. Because of the correctness of the encryption scheme $p_1 = p_0$.

Γ_2 : This game is the same with Γ_1 except the provers encrypt a random string instead of pk_P, N. The verifier retrieves e and s from the table T so that it does not decrypt any ciphertext that comes from a prover as in Γ_1. Thanks to the IND-CCA security (Verifiers are simulated using a decryption oracle due to our Γ_1 reduction. The use of this oracle is valid in IND-CCA game), $p_2 - p_1$ is negligible. So, P and $\mathcal{V}$ works as follows:

<u>$P(.)$</u> (in Γ_2)
pick $N \in D(1^n)$
$s \leftarrow B(\mathsf{sk}_P, \mathsf{pk}_P, \mathsf{pk}_{V_2}, N)$
pick r
$e \leftarrow \mathsf{Enc}_{\mathsf{pk}_{V_1}}(r)$
store (e, s) **to** T
send e
run symDB(s)

<u>$V(.)$</u> (in Γ_2)
receive e
if $(e, .) \in T$
  retrieve s from T
  where $(e, s) \in T$
else:
  $(\mathsf{pk}', N) \leftarrow \mathsf{Dec}_{\mathsf{sk}_{V_1}}(e)$
  $s \leftarrow A(\mathsf{sk}_{V_2}, \mathsf{pk}_{V_2}, \mathsf{pk}', N)$
run symDB(s)

This reduction shows that the adversary cannot retrieve pk_P and N from the encryption.

Γ_3 : It is the same with Γ_3 except that we simulate the prover as below. In this game, s is generated independently from sk_P and pk_P.

<u>$P(.)$</u> (in Γ_3)
$(\mathsf{sk}, \mathsf{pk}) \leftarrow \mathsf{Gen}_B(1^n)$
pick $N \in D(1^n)$
run $s \leftarrow B(\mathsf{sk}, \mathsf{pk}, \mathsf{pk}_{\mathsf{V}_2}, \mathsf{N})$
pick r
$e \leftarrow \mathsf{Enc}_{\mathsf{pk}_{V_1}}(r)$
store (e, s) **to** T
send e
run symDB(s)

We defined the hybrid games $\Gamma_{3,t}$ to show $p_3 - p_2$ is negligible. Here, $t \in \{0, 1, 2, ..., k\}$ and k is the number of prover instances bounded by a polynomial.

$\Gamma_{3,i}$: $\mathcal{V}$ is simulated as in Γ_3 and the jth instance of P is simulated as in Γ_3 if $j \leq i$ and as in Γ_2 if $j > i$.

First, we show that $\Gamma_{3,i}$ and $\Gamma_{3,i+1}$ are indistinguishable. For this, we use an adversary $\mathcal{B}$ that plays D-AKA^p game. $\mathcal{B}$ receives $\mathsf{pk}_A, \mathsf{pk}_{B_1}$ and sk_{B_1} from the D-AKA^p challenger, picks $(\mathsf{sk}_{B_0}, \mathsf{pk}_{B_0})$ and sends them to the challenger. Finally, $\mathcal{B}$ receives s. After, he begins simulating against the adversary $\mathcal{A}$ that wants to distinguish $\Gamma_{3,i}$ and $\Gamma_{3,i+1}$.

$\underline{P_{i+1}(.)}$
pick r
$e \leftarrow \mathsf{Enc}_{\mathsf{pk}_V}(r)$
store (e, s) **to** T
send e
run symDB(s)

$\mathcal{B}$ assigns $\mathsf{pk}_V = \mathsf{pk}_A$ and $\mathsf{pk}_P = \mathsf{pk}_{B_1}$. For all of the prover simulations, if $j \neq i+1$, P_j is simulated normally. $\mathcal{V}$ is simulated using the $\mathcal{O}_A$ oracle. Corrupt can be simulated since sk_{B_1} is available.

Note that if s is generated from $B(\mathsf{sk}_{B_0}, \mathsf{pk}_{B_0}, \mathsf{pk}_V, N)$ then $\mathcal{B}$ simulates $\Gamma_{3,i+1}$ and if it is generated from $B(\mathsf{sk}_{B_1}, \mathsf{pk}_{B_1}, \mathsf{pk}_V, N)$ then $\mathcal{B}$ simulates $\Gamma_{3,i}$.

For the verifier simulation, $\mathcal{B}$ first checks if $(e, .)$ is stored by himself as $\mathcal{V}$ in Γ_3. Otherwise, he decrypts e and sends (pk_{P_j}, N) to the oracle $\mathcal{O}_A(\mathsf{pk}_P, N)$ and receives s. In the end of the game, $\mathcal{A}$ sends his decision. If $\mathcal{A}$ outputs i, then $\mathcal{B}$ outputs 1. If $\mathcal{A}$ outputs $i+1$, then $\mathcal{B}$ outputs 0. Clearly, the advantage of $\mathcal{B}$ is $p_{3,i} - p_{3,i+1}$ which is negligible because of the D-AKA^p assumption. From the hybrid theorem, we can conclude that $p_{3,0}$ and $p_{3,k}$ is negligible where $p_{3,0} = p_2$ and $p_{3,k} = p_3$.

Now, in Γ_3, no identity is used by the provers. Hence, $\mathcal{A}$ does not have any advantage to guess the prover which means $p_3 = \frac{1}{2}$. As a result of it, $p_0 - \frac{1}{2}$ is negligible.

Consequently, if we use D-AKA secure and D-AKA^p private key agreement protocol in Eff-pkDB^p, then we have DF, MF, DH secure and strong private public-key DB protocol. For instance, Nonce-DH key agreement protocol is a good candidate for Eff-pkDB^p.

Difficulties of having strong privacy: The strong privacy is the hardest privacy notion to achieve in DB protocols. Sending all prover messages with an IND-CCA secure encryption is not always enough to have strong privacy. We exemplify our argument as follows: Clearly, Eff-pkDB protocol is still DF-MiM and DH-secure, if we replace the nonce selection by a counter. So, we can make a new version of Eff-pkDB^p based on the counter version of Eff-pkDB where the prover sends his public key and the counter by an IND-CCA encryption. However, clearly, it does not give strong privacy because when an adversary calls Corrupt oracle, he learns the counter of two drawn provers. Since the adversary knows the corresponding secret keys for both of them, he can easily differentiate the drawn provers based on the counter. This attack is not possible in Eff-pkDB^p which uses a nonce instead of a counter because the nonce is in the volatile memory. So, the adversary does not learn it with the Corrupt oracle.

Table 2. The review of the existing public-key DB protocols.

Protocol	Security	Privacy	PK operations	Number of computations
Brands-Chaum [9]	MiM, DF	No Privacy	1 commitment, 1 signature	1 EC multiplication, 2 hashings, 1 mapping, 1 modular inversion, 1 random string selection
HPO [20]	MiM, DF	Weak Private		4 EC multiplications, 2 random string selections, 2 mappings
PrivDB [37]	MiM, DF, DH	Strong Private	1 signature, 1 IND-CCA encryption	3 EC multiplications, 2 hashings, 2 random string selection, 1 modular inversion, 1 mapping, 1 symmetric key encryption, 1 MAC
Simp-pkDB	MiM, DF	No Privacy	1 decryption	1 EC multiplication, 1 hashing, 1 symmetric key decryption, MAC
Eff-pkDB	MiM, DF, DH	No privacy	1 D-AKA secure KA protocol	1 EC multiplication, 1 hashing, 1 random string selection
Eff-pkDB^p	MiM, DF, DH	Strong Private	1 IND-CCA Encryption, 1 D-AKA secure KA protocol	3 EC multiplications, 2 hashings, 2 random string selections, 1 symmetric key encryption, 1 MAC

6 Conclusion

Our main purpose in this work was to design an efficient and a secure public-key DB protocol. First, we designed Eff-pkDB which is secure against DF, MiM and DH. We did not consider privacy in this one because privacy is not the main concern of some applications. Therefore, Eff-pkDB can be employed by the applications that do not need privacy. Eff-pkDB is one of the most efficient public key DB protocols compared to the previous ones (See Table 2).

Second, we added strong privacy to the Eff-pkDB protocol and obtained Eff-pkDB^p. We succeeded it by adding one public-key IND-CCA secure encryption. In this case, the protocol is not as efficient as before but still one of the most efficient ones with the same security and privacy properties.

In Table 2, we give the security properties of existing public-key DB protocols along with the number of computations done on prover side. We use the number of elliptic curve multiplications and hashing as a metric in our efficiency analysis. We exclude GOR, ProProx and eProProx (in Table 1) since they clearly require a lot more computation than the other public-key DB protocols. In our counting for the number of computations in Table 2, 1 commitment is counted as 1 hashing operation. For the signature, we prefer an efficient and existentially unforgeable under chosen-message attacks resistant signature scheme ECDSA [21]. ECDSA requires 1 EC multiplication, 1 mapping, 1 hashing, 1 modular inversion and 1 random string selection. For the IND-CCA encryption scheme, we use ECIES [31] which requires 2 EC multiplications, 1 KDF, 1 symmetric key encryption, 1 MAC and 1 random string selection. For the D-AKA secure key agreement protocol, we use Nonce-DH which requires 1 EC multiplication, 1 hashing and 1 random string selection.

We first compare the protocols considering the security and the efficiency trade-off. Eff-pkDB and Simp-pkDB are the most efficient ones. However, Simp-pkDB is secure only against MiM and DF. After Eff-pkDB, the second most efficient protocol is Brands-Chaum protocol [9] but this protocol is only secure against MiM and DF while Eff-pkDB is secure against DH as well.

Now, we compare the protocols considering security, privacy and efficiency trade-off. In this case, HPO requires 4 EC multiplications while PrivDB and Eff-pkDB^p require 3 EC multiplications and 1 hashing. Hashing is more efficient than elliptic curve multiplication so it looks like PrivDB and Eff-pkDB^p are more efficient. However, HPO has an advantage in efficiency if it is used in a dedicated hardware allowing only EC operations. On the other hand, Eff-pkDB^p and PrivDB are secure against MiM, DF, DH and strong private while HPO is only MiM and DF secure and only private.

Eff-pkDB^p and PrivDB have the same security and privacy properties and almost the same efficiency level. However, if we analyze the efficiency with more metrics, we see that PrivDB requires extra 1 modular inversion and 1 mapping. More importantly, Eff-pkDB^p has lighter version Eff-pkDB which can be used efficiently in the applications which do not need privacy.

One of the important useful property of Eff-pkDB is that it can employ any D-AKA secure key agreement protocol to satisfy DF, MiM and DH security.

Acknowledgements. This work was partly sponsored by the ICT COST Action IC1403 Cryptacus in the EU Framework Horizon 2020.

A More Results About D-AKA Security Model

The Extended Canetti-Krawczyk (eCK) Security Model [27]. The eCK security model consists of t parties with their certificated public keys. The key exchange protocol is executed between two parties A and B. When A starts a key exchange protocol with B, it is called as a session and A is the owner of the session and B is the peer. A (initiator) starts the protocol by sending a message $\mathsf{M_A}$, then B (responder) responds with a message $\mathsf{M_B}$. The session id sid corresponds to an instance of A or B.

There is a probabilistic polynomial time (PPT) adversary $\mathcal{A}$ controlling all communication and some instances. The activation of the parties starts by `Send`(A, B,message) (or `Send`(B, A,message)). Besides `Send`, $\mathcal{A}$ can do following queries:

- `Long-Term Key Reveal`(A): Outputs the long term public-key of A.
- `Ephemeral Key Reveal`(sid): Outputs an ephemeral key of a session sid.
- `Reveal`(sid): Outputs the session key of a completed session sid.
- `Test`(sid): If sid is clean then outputs $s \leftarrow$ `Reveal`(sid) if $b = 1$, outputs $s \leftarrow \{0,1\}^\lambda$ if $b = 0$ (λ is the size of the session key).
  The advantage is the difference of the probability that $\mathcal{A}$ gives 1 for $b = 0$ and $b = 1$.

Table 3. Existing KA protocols with their security and efficiency. Efficiency column shows the number of exponentiation done by per party.

KA Protocol	Efficiency	Security
MQV [29]	2.5	unproven
HMQV [26]	2.5	CK
KEA+ [28]	3	CK
NAXOS [27]	4	eCK
CMQV [32]	3	eCK
Nonce-DH	1	D-AKA

A clean session is basically a session where winning the game for $\mathcal{A}$ is not trivial. See [27] for more details.

Theorem 8. *If a key agreement protocol is eCK secure [27], then it is D-AKA secure.*

Proof. Let's assume that there is an adversary $\mathcal{A}$ playing D-AKA game. We construct an adversary $\mathcal{B}$ simulating the D-AKA game and playing the eCK game. $\mathcal{B}$ receives all the public keys in the eCK game. $\mathcal{B}$ first picks two parties A and B. Then, he creates a session sid between them by sending the query Send(A,B, message) and he assigns the ephemeral public key of B as a nonce N. Then, he sends the query Test(sid) and receives s_b. Finally, he sends $s_b, N, \mathsf{pk}_B, \mathsf{pk}_A$ to $\mathcal{A}$. Whenever $\mathcal{A}$ calls the oracle $\mathcal{O}_B(\mathsf{pk}_{A'})$, $\mathcal{B}$ creates a new session sid' with A' on behalf of B as explained above. Similarly, he assigns the ephemeral public key of B as a nonce N'. After, he sends the query Reveal(sid') and receives the session key s'. As a response of $\mathcal{O}_B(\mathsf{pk}_{A'})$, he sends s', N' to $\mathcal{A}$. In addition, whenever $\mathcal{A}$ calls the oracle $\mathcal{O}_A(\mathsf{pk}_{B'}, N'')$, first, $\mathcal{B}$ checks if $(\mathsf{pk}_{B'}, N'')$ equals (pk_B, N). If it is not equal, he creates a new session sid'' on behalf of B' with the ephemeral public key N'' and calls the oracle Reveal(sid'') to receive the session key s''. Then, he responds to $\mathcal{A}$ with s''. In the end, $\mathcal{B}$ outputs whatever $\mathcal{A}$ outputs. The simulation of D-AKA game is perfect. So the advantage of $\mathcal{B}$ equals to the advantage of $\mathcal{A}$. Therefore, since the advantage of $\mathcal{B}$ is negligible, the advantage of $\mathcal{A}$ is negligible as well. □

As a result of Theorem 8, we can conclude any eCK secure key agreement protocol can be used in Eff-pkDB. However, we suggest using D-AKA secure key agreement protocols since they may require less public-key operations.

Lemma 1. *We consider D-AKA secure key agreement protocol* $(\mathsf{Gen}_A, \mathsf{Gen}_B, D, A, B)$. *We define the random variables* $(\mathsf{sk}_A, \mathsf{pk}_A) \leftarrow \mathsf{Gen}_A(1^n)$, $(\mathsf{sk}_B, \mathsf{pk}_B) \leftarrow \mathsf{Gen}_B(1^n)$, *and* $(s, N) \leftarrow \mathcal{O}_B(\mathsf{pk}_A)$ *and* $(s', N') \leftarrow \mathcal{O}_B(\mathsf{pk}_A)$. *We have that* $\Pr[N = N']$ *is negligible. Furthermore, for all values u which could depend on* $\mathsf{sk}_A, \mathsf{pk}_A, \mathsf{sk}_B, \mathsf{pk}_B$, $\Pr[N = u]$ *is negligible.*

Proof. We define an adversary $\mathcal{A}$ playing the D-AKA game as follows:

$\underline{\mathcal{A}}$

receive $s_b, N, \mathsf{pk}_B, \mathsf{pk}_A$
$(s', N') \leftarrow \mathcal{O}_B(\mathsf{pk}_A)$
if $N' = N$
  if $s' = s_b$
    output 0
  else:
    output 1
else:
  output $b' \leftarrow_r \{0,1\}$

In this strategy, $\mathcal{A}$ wins if $N = N'$ (except $s_1 = s_0$ and $b = 1$). Otherwise, he wins with $\frac{1}{2}$ probability.

$$\begin{aligned}\Pr[\mathcal{A}\,\mathsf{win}] =& \frac{1}{2}(1 - \Pr[N = N']) + \Pr[N = N'] - \Pr[N = N', s_1 = s_0, b = 1]\\ =& \frac{1}{2} + \frac{1}{2}\Pr[N = N'] - \Pr[N = N', s_1 = s_0, b = 1]\end{aligned}$$

We know from the D-AKA security that $\Pr[\mathcal{A}\,\mathsf{win}] - \frac{1}{2}$ is negligible. $\Pr[s_1 = s_0] = 2^{-n}$ is negligible as well. So, $\Pr[N = N']$ is negligible. Now, we need to show that it holds for all values u.

Let v be the most probable value for N. We have

$$\begin{aligned}\Pr[N = N'] =& \sum_w \Pr[N = N' = w]\\ =& \sum_w \Pr[N = w]^2\\ \geq& \Pr[N = v]^2\end{aligned}$$

So, we have the following inequality in the end:

$$\Pr[N = u] \leq \Pr[N = v] \leq \sqrt{\Pr[N = N']}$$

We know that $\Pr[N = N']$ is negligible so $\Pr[N = u]$ is negligible.

□

B Mafia and Distance Fraud Secure Public Key DB

We consider the Simp-pkDB protocol in Fig. 6. In Simp-pkDB the prover P selects a nonce $N \in \{0,1\}^n$ where n is security parameter and sends it to the verifier together with pk. Then verifier V selects a secret $s \in \{0,1\}^n$, encrypts it with N by the public key pk of the prover and sends the encryption e to P. After receiving e, P decrypts it with the secret key sk and gets s, N. If the N is the nonce by P, then they run one-time secure symDB(s).

We show that this protocol is MiM-secure but not DH-secure. Simp-pkDB requires only one operation which is IND-CCA decryption.

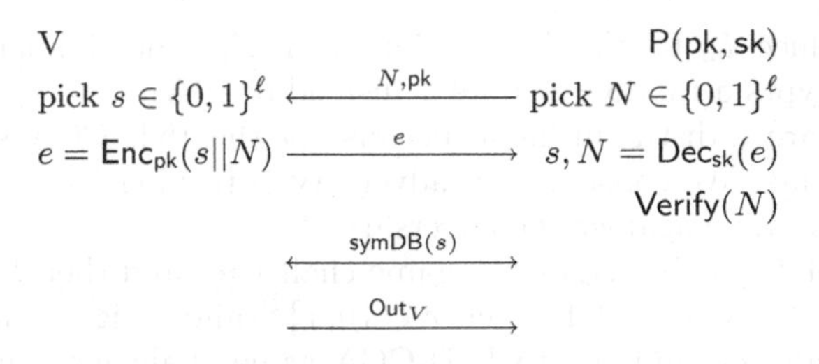

Fig. 6. Simp-pkDB

Theorem 9. *If symDB is DF-secure then Simp-pkDB is DF-secure.*

Proof. It is trivial.

Theorem 10. *If symDB is one-time MiM-secure and the cryptosystem resists chosen-ciphertext attacks (IND-CCA secure) then Simp-pkDB is MiM-secure.*

Proof. Γ_i is a game and p_i denotes the probability that Γ_i succeeds.

Γ_0 : Adversary plays MiM game in the protocol in Fig. 6 with the verifier with its instances, the prover with its instances and other actors. Let's assume that the number of prover instances is k where k is polynomially bounded.

Let s, pk, N and e be the values seen by the distinguished instance $\mathcal{V}$ of the verifier. Here $e = \mathsf{Enc}_{\mathsf{pk}}(s||N)$. We group the prover's instances as the following:

1. The provers seeing N and e,
2. The provers seeing e but another nonce N'.
3. The provers not seeing e (see a ciphertext e' which is not e).
   The probability that an adversary succeeds in Γ_0 is p_0.

Γ_1 : We reduce Γ_0 to Γ_1 where the first group has up to one prover instance P. We call $\mathcal{V}$ and P the matching instances. The probability that more than one prover picks same N is bounded by $\binom{k}{2}2^{-\ell}$ which is negligible. So, $p_1 - p_0$ is negligible.

Γ_2 : We reduce Γ_1 to Γ_2 where the matching P receives e after $\mathcal{V}$ has released e which means that e which is encryption of $s||N$ is only sent by the verifier. In Γ_1, the probability that $\mathcal{V}$ selects s after P has received e so that $\mathsf{Dec}_{\mathsf{sk}}(e) = s$ is $\frac{1}{2^\ell}$ which means that $p_2 - p_1$ is negligible.

Γ_3 : We reduce Γ_2 to Γ_3 where the provers are simulated as below:

The prover in the first group after receiving e run symDB(s) without decrypting e. Since e was released before, the value of s is already defined. The provers in the second group, abort the protocol after receiving e. The provers in the third group, call decryption oracle $\mathsf{Dec}_{\mathsf{sk}}(.)$ after receiving e' and check if the nonce is the same nonce that was chosen by them. Then they run symDB(s') with s' obtained from the decryption oracle.

The simulation gives identical result so the success probabilities in Γ_3 and Γ_2 are the same.

Γ_4 : We reduce Γ_3 to Γ_4. We simulate $\mathcal{V}$ in Γ_4. The simulation of $\mathcal{V}$ after selecting s encrypts a random plaintext instead of $s||N$.

Γ_3 and Γ_4 are indistinguishable because of the IND-CCA security of the encryption scheme. We construct an adversary $\mathcal{B}$ playing IND-CCA game and simulating MiM game against the adversary $\mathcal{A}$.

$\mathcal{B}$ receives pk from the IND-CCA game challenger and then $\mathcal{B}$ forwards it to $\mathcal{A}$. Firstly, $\mathcal{B}$ picks $N, s \in \{0,1\}^\ell$ and $r \in \{0,1\}^{2\ell}$ and assigns $m_0 = s||N, m_1 = r$. Then he sends m_0 and m_1 to IND-CCA game challenger and receives the response e_b where $e_b = \mathsf{Enc}_{\mathsf{pk}}(m_0)$ or $\mathsf{Enc}_{\mathsf{pk}}(m_1)$. If $\mathcal{A}$ interacts with $\mathcal{V}$ then $\mathcal{B}$ sends e_b, if $\mathcal{A}$ interacts with P, then $\mathcal{B}$ sends N. For the simulation of other prover instances P' (controlled by $\mathcal{A}$), when P' asks for the decryption of e', $\mathcal{B}$ sends e' to IND-CCA game challenger and receives decryption of e' to send P'. In the end, if $\mathcal{A}$ succeeds then $\mathcal{B}$ outputs 0, otherwise he outputs 1. If $\mathcal{A}$ succeeds given $b = 0$, then it means that he succeeds Γ_3 and if $\mathcal{A}$ succeeds given $b = 1$ then it means that he succeeds Γ_4. Therefore we have the following success probability of $\mathcal{B}$.

$$\mathsf{Adv}(\mathcal{B}) = \Pr[\mathcal{B} \to 1 | b = 0] + \Pr[\mathcal{B} \to 1 | b = 1] = p_3 - p_4$$

Since we know that the advantage of $\mathcal{B}$ is negligible, we can deduce that $p_3 - p_4$ is negligible (if we multiply negligible function with a polynomial we still have a negligible function).

Γ_5 : Now in Γ_5 we have at most two matching instances and they both run $\mathsf{symDB}(s)$ with the same and fresh random s. In Γ_5, The rest of the game (including the selection of pk and sk and the the decryption oracle $\mathsf{Dec}_{\mathsf{sk}}(.)$) is simulated by the adversary, Γ_4 and Γ_5 work the same. So $p_4 = p_5$. So they run symDB(s). The success probability p_5 of Γ_5 is negligible because of the security of OT-MiM-security of symDB.

As a conclusion, since $p_1 - p_0 = \mathsf{negl}$, $p_2 - p_1 = \mathsf{negl}$, $p_2 - p_3 = 0$, $p_4 - p_3 = \mathsf{negl}$, $p_5 - p_4 = 0$ and $p_5 = \mathsf{negl}$, we deduce that p_0 is negligible.

DH-Security: The protocol in Fig. 6 is not secure against DH because of the attack in Fig. 7. In this attack, the malicious and far away prover P uses honest and close prover P' so that in the end V accepts P.

Basically, P chooses the same nonce that P' chose. Then V encrypts $s||N$ with the public key pk_P of P and then sends it to P. P decrypts e with his own secret key sk_P and then behaves as if he is the verifier and prepares encryption $e' = \mathsf{Enc}_{\mathsf{pk}_{\mathsf{P}'}}$ with using P''s public key $\mathsf{pk}_{\mathsf{P}'}$ and sends it to P'. Since e' is valid encryption for P', he continues by executing $\mathsf{symDB}(s)$ with V. In the end of the protocol, V accepts P since V has the P's public key. P' is used by P only to be able to pass the distance bounding phase of $\mathsf{symDB}(s)$ protocol.

C Security of Nonce-DH

Definition 15 (Gap Diffie-Hellman (GDH) [30]). *Let $\mathbb{G}$ be a prime order group and $g \in \mathbb{G}$ be a generator. We have the following problems:*

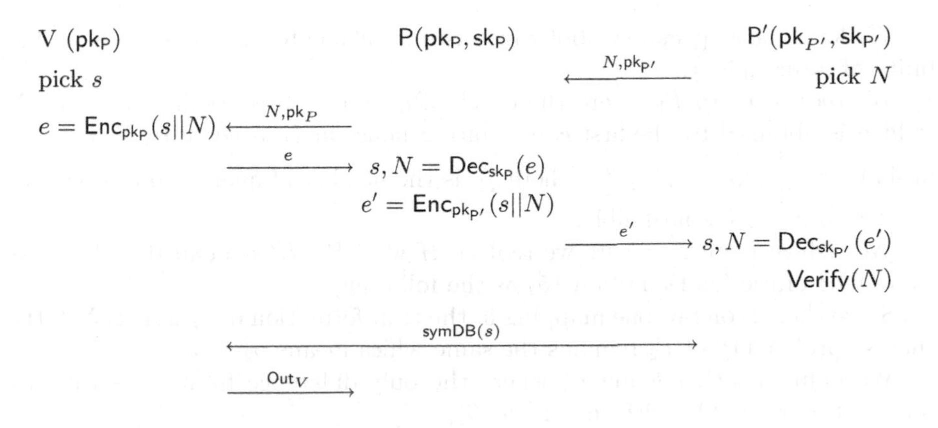

Fig. 7. DH attack on Simp-pkDB.

- ***Computational Diffie-Hellman Problem (CDH):*** *Given* $g, X, Y \in \mathbb{G}$ *compute* $Z = g^{log_g X . log_g Y}$.
- ***Decisional Diffie-Hellman Problem (DDH):*** *Given* $g, X, Y, Z \in \mathbb{G}$, *decide if* $Z = g^{log_g X . log_g Y}$ *or* $Z = g^r$ *where* r *is a random element.*

The GDH problem is solving the CDH given (g, X, Y) *with the help of a DDH oracle which answers whether a given quadruple is a Diffie-Hellman quadruple.*

Theorem 11. *Assuming that the GDH problem is hard and* $\ell = \Omega(n)$, *Nonce-DH is D-AKA secure in the random oracle model.*

Proof. The game Γ_0 is the D-AKA game. The challenger works as follows: He picks q and g as described in Nonce-DH. He randomly picks $\mathsf{sk}_A, \mathsf{sk}_B \in \mathbb{Z}_q$, and computes $\mathsf{pk}_A = g^{\mathsf{sk}_A}$, $\mathsf{pk}_B = g^{\mathsf{sk}_B}$. He picks randomly $s_1 \in \{0,1\}^n$ and then he assigns $(s_0, N) \leftarrow \mathcal{O}_B(\mathsf{pk}_A)$. Then, he picks $b \in \{0,1\}$ and gives $g, q, \mathsf{pk}_A, \mathsf{pk}_B, N, s_b$ to the adversary $\mathcal{A}$. $\mathcal{A}$ has access to the oracle H, $\mathcal{O}_A(.,.)$ (with the restriction not asking for pk_B, N) and $\mathcal{O}_B(.)$ defined below.

$\underline{\mathcal{O}_A(.,.)}$
Input: pk'_B, N'
  if (pk'_B, N') **equals** (pk_B, N)
  send $\perp$
  else:
  $s \leftarrow H(g, \mathsf{pk}'_B, \mathsf{pk}_A, \mathsf{pk}_B'^{\mathsf{sk}_A}, N')$
  send s

$\underline{\mathcal{O}_B(.)}$
Input: pk'_A
  pick $N' \in \{0,1\}^\ell$
  $s \leftarrow H(g, \mathsf{pk}_B, \mathsf{pk}'_A, \mathsf{pk}_A'^{\mathsf{sk}_B}, N')$
  send (s, N')

$\underline{\mathrm{H}(.)}$
Input: U
  if $(U, .) \in T$
  send V where $(U, V) \in T$
  else:
  pick $V \in \{0,1\}^n$
  save (U, V) to T
  send V

$\underline{\mathrm{H}'(.)}$
Input: (w, x, y, z, N')
  if $w = g$ **and** $1 \leftarrow \mathsf{DDH}(g, x, y, z)$
  $z \leftarrow \perp$
  send $\mathrm{H}(w, x, y, z, N')$

We let $\perp$ be a special symbol which is unavailable to $\mathcal{A}$. The success probability of $\mathcal{A}$ in Γ_0 is p_0.

We reduce Γ_0 to Γ_1 where the oracle $\mathcal{O}_B$ never selects again the nonce N (which is obtained by the first call). Since a nonce in Γ_0 is equal to N with the probability $\frac{1}{2^\ell}$, $|p_1 - p_0| \leq \frac{q_B}{2^\ell}$ where q_B is the number of queries to $\mathcal{O}_B$. Due to $\ell = \Omega(n)$, $p_1 - p_0$ is negligible.

We reduce Γ_1 to Γ_2 where we replace H with H'. H' is defined with access to a DDH oracle (as Definition 15) as the following:

Since there is one-to-one mapping in the transformation of (g, x, y, z, N'), the success probability of Γ_2 remains the same which means $p_2 = p_1$.

We define another game Γ_3 where the only difference from Γ_2 is that we replace the oracle $\mathcal{O}_B$ with the oracle $\mathcal{O}'_B$.

$\underline{\mathcal{O}'_B(.)}$
Input: pk'_A
  pick $N' \in \{0,1\}^\ell$
  $s \leftarrow H(g, \mathsf{pk}_B, \mathsf{pk}'_A, \perp, N')$
  send (s, N')

Note that $\mathcal{O}'_B$ queries H instead of H' and $N' \neq N$ due to the reduction to Γ_1. Γ_3 is exactly same with Γ_2 so the success probabilities p_3 and p_2 are the same as well.

Now in Γ_3, sk_B is used only by the DDH oracle.

We reduce Γ_3 to Γ_4 where $\mathcal{A}$ does not make the query $H'(g, \mathsf{pk}_B, \mathsf{pk}_A, z, N)$ with $z = \mathsf{pk}_A^{\mathsf{sk}_B}$. Indeed, any such query can be filtered using the DDH oracle and stopped to solve the GDH problem. Since the GDH problem is hard, $\mathcal{A}$ in Γ_3 selects $z = \mathsf{pk}_A^{\mathsf{sk}_B}$ given $(\mathsf{pk}_A, \mathsf{pk}_B)$ with negligible probability. Therefore, $p_4 - p_3$ is negligible.

In Γ_4, $H(g, \mathsf{pk}_B, \mathsf{pk}_A, \perp, N)$ is queried only once and this query is done by the challenger. Lastly, we reduce Γ_4 to Γ_5 where the challenger picks a random s_0 instead of picking $s_0 = H(g, \mathsf{pk}_B, \mathsf{pk}_A, \perp, N)$.

Γ_4 and Γ_5 are the same because if $(g, \mathsf{pk}_B, \mathsf{pk}_A, \perp, N)$ is never being queried again, it is not necessary that H stores $((g, \mathsf{pk}_B, \mathsf{pk}_A, \perp, N), s_0)$ in T. So, $p_4 = p_5$.

In Γ_5, s_0 and s_1 play a symmetric role and could be erased with b from the game after s_b is released. So, the state of the game after erasure of b, s_0 and s_1 are independent from b. Hence, $p_5 = \frac{1}{2}$ leading to $p_0 - \frac{1}{2}$ is negligible. □

Theorem 12. *Assuming that $\ell = \Omega(n)$, Nonce-DH is D-AKA^p private in the random oracle model.*

Proof. The game Γ_0 is D-AKA^p game. The challenger works as follows: He picks q and g as described in Nonce-DH. He selects $\mathsf{sk}_A, \mathsf{sk}_{B_1} \in \mathbb{Z}_q$, and computes $\mathsf{pk}_A = g^{\mathsf{sk}_A}$ and $\mathsf{pk}_{B_1} = g^{\mathsf{sk}_{B_1}}$. Then, he sends $\mathsf{pk}_A, \mathsf{pk}_{B_1}$ and sk_{B_1} to $\mathcal{A}$. $\mathcal{A}$ selects sk_{B_0} and pk_{B_0} and sends them to the challenger. Next, the challenger picks $b \in \{0,1\}$, $N \in \{0,1\}^\ell$, queries $(g, \mathsf{pk}_{B_b}, \mathsf{pk}_A, \mathsf{pk}_A^{\mathsf{sk}_{B_b}}, N)$ to H and receives s.

He sends s to $\mathcal{A}$. $\mathcal{A}$ has access to the oracle H as defined in the proof of Theorem 11, and to the oracle $\mathcal{O}_A(.,.)$.

We reduce Γ_0 to Γ_1 where $\mathcal{A}$ never selects the same nonce with N in the query of the oracle H or $\mathcal{O}_A$. The probability that he selects N is $\frac{1}{2^\ell}$ so $p_2 - p_1$ is negligible.

We reduce Γ_1 to Γ_2 where $\mathcal{O}_B$ picks s at random instead of a response from H. Since, the query $(g, \mathsf{pk}_{B_b}, \mathsf{pk}_A, \mathsf{pk}_A^{\mathsf{sk}_{B_b}}, N)$ by the challenger is never done again, we have $p_1 = p_2$. Now, b is never used in Γ_2. It means that s is independent from b, so $p_2 = \frac{1}{2}$. Therefore, $p_0 - \frac{1}{2}$ is negligible.

□

References

1. EMVCo version 2.6 in book c-2 kernel 2 specification
2. Avoine, G., Bingöl, M.A., Kardaş, S., Lauradoux, C., Martin, B.: A framework for analyzing RFID distance bounding protocols. J. Comput. Secur. **19**(2), 289–317 (2011)
3. Avoine, G., Tchamkerten, A.: An efficient distance bounding RFID authentication protocol: balancing false-acceptance rate and memory requirement. In: Samarati, P., Yung, M., Martinelli, F., Ardagna, C.A. (eds.) ISC 2009. LNCS, vol. 5735, pp. 250–261. Springer, Heidelberg (2009). doi:10.1007/978-3-642-04474-8_21
4. Bengio, S., Brassard, G., Desmedt, Y.G., Goutier, C., Quisquater, J.-J.: Secure implementation of identification systems. J. Cryptology **4**(3), 175–183 (1991)
5. Boureanu, I., Mitrokotsa, A., Vaudenay, S.: Secure and lightweight distance-bounding. In: Avoine, G., Kara, O. (eds.) LightSec 2013. LNCS, vol. 8162, pp. 97–113. Springer, Heidelberg (2013). doi:10.1007/978-3-642-40392-7_8
6. Boureanu, I., Mitrokotsa, A., Vaudenay, S.: Towards secure distance bounding. In: Moriai, S. (ed.) FSE 2013. LNCS, vol. 8424, pp. 55–67. Springer, Heidelberg (2014). doi:10.1007/978-3-662-43933-3_4
7. Boureanu, I., Mitrokotsa, A., Vaudenay, S.: Practical and Provably Secure Distance-Bounding. IOS Press, Amsterdam (2015)
8. Boureanu, I., Vaudenay, S.: Optimal proximity proofs. In: Lin, D., Yung, M., Zhou, J. (eds.) Inscrypt 2014. LNCS, vol. 8957, pp. 170–190. Springer, Heidelberg (2015). doi:10.1007/978-3-319-16745-9_10
9. Brands, S., Chaum, D.: Distance-bounding protocols. In: Helleseth, T. (ed.) EUROCRYPT 1993. LNCS, vol. 765, pp. 344–359. Springer, Heidelberg (1994). doi:10.1007/3-540-48285-7_30
10. Bussard, L., Bagga, W.: Distance-bounding proof of knowledge to avoid real-time attacks. In: Sasaki, R., Qing, S., Okamoto, E., Yoshiura, H. (eds.) SEC 2005. IAICT, vol. 181, pp. 223–238. Springer, Heidelberg (2005). doi:10.1007/0-387-25660-1_15
11. Chothia, T., Garcia, F.D., Ruiter, J., Breekel, J., Thompson, M.: Relay cost bounding for contactless EMV payments. In: Böhme, R., Okamoto, T. (eds.) FC 2015. LNCS, vol. 8975, pp. 189–206. Springer, Heidelberg (2015). doi:10.1007/978-3-662-47854-7_11
12. Cremers, C., Rasmussen, K.B., Schmidt, B., Capkun, S.: Distance hijacking attacks on distance bounding protocols. In: SP, pp. 113–127 (2012)

13. Desmedt, Y.: Major security problems with the unforgeable (Feige-) Fiat-Shamir proofs of identity and how to overcome them. In: SECURICOM, pp. 147–159 (1988)
14. Diffie, W., Hellman, M.E.: New directions in cryptography. IEEE Trans. Inf. Theory **22**(6), 644–654 (1976)
15. Dürholz, U., Fischlin, M., Kasper, M., Onete, C.: A formal approach to distance-bounding RFID protocols. In: Lai, X., Zhou, J., Li, H. (eds.) ISC 2011. LNCS, vol. 7001, pp. 47–62. Springer, Heidelberg (2011). doi:10.1007/978-3-642-24861-0_4
16. Fischlin, M., Onete, C.: Terrorism in distance bounding: modeling terrorist-fraud resistance. In: Jacobson, M., Locasto, M., Mohassel, P., Safavi-Naini, R. (eds.) ACNS 2013. LNCS, vol. 7954, pp. 414–431. Springer, Heidelberg (2013). doi:10.1007/978-3-642-38980-1_26
17. Gambs, S., Onete, C., Robert, J.-M.: Prover anonymous and deniable distance-bounding authentication. In: ASIA CCS, ACM Symposium, pp. 501–506 (2014)
18. Hancke, G.P., Kuhn, M.G.: An RFID distance bounding protocol. In: SecureComm 2005, pp. 67–73. IEEE (2005)
19. Hermans, J., Pashalidis, A., Vercauteren, F., Preneel, B.: A new RFID privacy model. In: Atluri, V., Diaz, C. (eds.) ESORICS 2011. LNCS, vol. 6879, pp. 568–587. Springer, Heidelberg (2011). doi:10.1007/978-3-642-23822-2_31
20. Hermans, J., Peeters, R., Onete, C.: Efficient, secure, private distance bounding without key updates. In: WiSec, pp. 207–218 (2013)
21. Johnson, D., Menezes, A., Vanstone, S.: The elliptic curve digital signature algorithm (ecdsa). Int. J. Inf. Secur. **1**(1), 36–63 (2001)
22. Kılınç, H., Vaudenay, S.: Comparison of public-key distance bounding protocols, under submission
23. Kılınç, H., Vaudenay, S.: Optimal proximity proofs revisited. In: Malkin, T., Kolesnikov, V., Lewko, A.B., Polychronakis, M. (eds.) ACNS 2015. LNCS, vol. 9092, pp. 478–494. Springer, Heidelberg (2015). doi:10.1007/978-3-319-28166-7_23
24. Kim, C.H., Avoine, G.: RFID distance bounding protocol with mixed challenges to prevent relay attacks. In: Garay, J.A., Miyaji, A., Otsuka, A. (eds.) CANS 2009. LNCS, vol. 5888, pp. 119–133. Springer, Heidelberg (2009). doi:10.1007/978-3-642-10433-6_9
25. Kim, C.H., Avoine, G., Koeune, F., Standaert, F.-X., Pereira, O.: The swiss-knife RFID distance bounding protocol. In: Lee, P.J., Cheon, J.H. (eds.) ICISC 2008. LNCS, vol. 5461, pp. 98–115. Springer, Heidelberg (2009). doi:10.1007/978-3-642-00730-9_7
26. Krawczyk, H.: HMQV: a high-performance secure Diffie-Hellman protocol. In: Shoup, V. (ed.) CRYPTO 2005. LNCS, vol. 3621, pp. 546–566. Springer, Heidelberg (2005). doi:10.1007/11535218_33
27. LaMacchia, B., Lauter, K., Mityagin, A.: Stronger security of authenticated key exchange. In: Susilo, W., Liu, J.K., Mu, Y. (eds.) ProvSec 2007. LNCS, vol. 4784, pp. 1–16. Springer, Heidelberg (2007). doi:10.1007/978-3-540-75670-5_1
28. Lauter, K., Mityagin, A.: Security analysis of KEA authenticated key exchange protocol. In: Yung, M., Dodis, Y., Kiayias, A., Malkin, T. (eds.) PKC 2006. LNCS, vol. 3958, pp. 378–394. Springer, Heidelberg (2006). doi:10.1007/11745853_25
29. Law, L., Menezes, A., Qu, M., Solinas, J., Vanstone, S.: An efficient protocol for authenticated key agreement. Des. Codes Crypt. **28**(2), 119–134 (2003)
30. Okamoto, T., Pointcheval, D.: The gap-problems: a new class of problems for the security of cryptographic schemes. In: Kim, K. (ed.) PKC 2001. LNCS, vol. 1992, pp. 104–118. Springer, Heidelberg (2001). doi:10.1007/3-540-44586-2_8
31. Shoup, V.: A proposal for an ISO standard for public key encryption (2.0) (2001)

32. Ustaoglu, B.: Obtaining a secure and efficient key agreement protocol from (H)MQV and NAXOS. Des. Codes Crypt. **46**(3), 329–342 (2008)
33. Vaudenay, S.: On privacy models for RFID. In: Kurosawa, K. (ed.) ASIACRYPT 2007. LNCS, vol. 4833, pp. 68–87. Springer, Heidelberg (2007). doi:10.1007/978-3-540-76900-2_5
34. Vaudenay, S.: On modeling terrorist frauds. In: Susilo, W., Reyhanitabar, R. (eds.) ProvSec 2013. LNCS, vol. 8209, pp. 1–20. Springer, Heidelberg (2013). doi:10.1007/978-3-642-41227-1_1
35. Vaudenay, S.: On privacy for RFID. In: Au, M.-H., Miyaji, A. (eds.) ProvSec 2015. LNCS, vol. 9451, pp. 3–20. Springer, Heidelberg (2015). doi:10.1007/978-3-319-26059-4_1
36. Vaudenay, S.: Privacy failure in the public-key distance-bounding protocol. IET Inf. Secur. **10**(4), 188–193 (2015)
37. Vaudenay, S.: Private and secure public-key distance bounding. In: Böhme, R., Okamoto, T. (eds.) FC 2015. LNCS, vol. 8975, pp. 207–216. Springer, Heidelberg (2015). doi:10.1007/978-3-662-47854-7_12
38. Vaudenay, S.: Sound proof of proximity of knowledge. In: Au, M.-H., Miyaji, A. (eds.) ProvSec 2015. LNCS, vol. 9451, pp. 105–126. Springer, Heidelberg (2015). doi:10.1007/978-3-319-26059-4_6

Indistinguishable Proofs of Work or Knowledge

Foteini Baldimtsi[1(✉)], Aggelos Kiayias[2], Thomas Zacharias[2], and Bingsheng Zhang[3]

[1] George Mason University, Fairfax, USA
foteini@gmu.edu
[2] University of Edinburgh, Edinburgh, UK
{akiayias,tzachari}@inf.ed.ac.uk
[3] Security Lancaster Research Centre, Lancaster University, Lancaster, UK
b.zhang2@lancaster.ac.uk

Abstract. We introduce a new class of protocols called *Proofs of Work or Knowledge* (PoWorKs). In a PoWorK, a prover can convince a verifier that she has either performed work or that she possesses knowledge of a witness to a public statement *without* the verifier being able to distinguish which of the two has taken place. We formalize PoWorK in terms of three properties, completeness, f-soundness and indistinguishability (where f is a function that determines the tightness of the proof of work aspect) and present a construction that transforms 3-move HVZK protocols into 3-move public-coin PoWorKs. To formalize the work aspect in a PoWorK protocol we define cryptographic puzzles that adhere to certain uniformity conditions, which may also be of independent interest. We instantiate our puzzles in the random oracle (RO) model as well as via constructing "dense" versions of suitably hard one-way functions.

We then showcase PoWorK protocols by presenting a number of applications. We first show how non-interactive PoWorKs can be used to *reduce spam email* by forcing users sending an e-mail to either prove to the mail server they are approved contacts of the recipient or to perform computational work. As opposed to previous approaches that applied proofs of work to this problem, our proposal of using PoWorKs is privacy-preserving as it hides the list of the receiver's approved contacts from the mail server. Our second application, shows how PoWorK can be used to *compose* cryptocurrencies that are based on proofs of work ("Bitcoin-like") with cryptocurrencies that are based on knowledge relations (these include cryptocurrencies that are based on "proof of stake", and others). The resulting PoWorK-based cryptocurrency inherits the robustness properties of the underlying two systems while PoWorK-indistinguishability ensures a uniform population of miners. Finally, we show that PoWorK protocols imply straight-line quasi-polynomial simulatable arguments of knowledge and based on our construction we obtain an efficient straight-line concurrent 3-move statistically quasi-polynomial simulatable argument of knowledge.

F. Baldimtsi et al.—Work performed while at the National and Kapodistrian University of Athens. Research supported by ERC project CODAMODA, #259152 and H2020 Project Panoramix #653497. Baldimtsi also did part of this work while at Boston University supported by NSF #1012910.

J.H. Cheon and T. Takagi (Eds.): ASIACRYPT 2016, Part II, LNCS 10032, pp. 902–933, 2016.
DOI: 10.1007/978-3-662-53890-6_30

Keywords: Proof of Work · Cryptographic puzzle · Concurrent zero-knowledge · Dense one-way functions · Cryptocurrencies

1 Introduction

We introduce a new class of prover verifier protocols where the prover wishes to convince the verifier that it is either in possession of a witness to a publicly known statement or that it has invested a certain amount of computational effort. A *Proof of Work or Knowledge* (PoWorK) enables the prover to achieve this objective while at the same time ensuring that the verifier is incapable of distinguishing which way the prover has followed: performing the work or exploiting her knowledge of the witness.

At an intuitive level a PoWorK protocol is a disjunction of a *proof of work* and a *proof of knowledge*. Proofs of knowledge are a fundamental notion in cryptography [GMR85] with a very wide array of applications in the design of cryptographic protocols. They have been studied extensively, both in terms of efficient constructions, e.g., [Sch89], as well as in terms of their composability with themselves or within larger protocols, see e.g., [CDS94, DNS98, CGGM00, Can01, CF01, Pas03, Pas04]. Proofs of work on the other hand, were first introduced in [DN92], further studied in [RSW96, Bac97, JB99, DGN03, CMSW09], and were primarily applied as a denial of service network or spam protection mechanism; recently they have also found important applications in building decentralized cryptocurrencies (notably Bitcoin [Nak08] but also many others).

In an interactive proof protocol, we are interested primarily in two basic properties, soundness and zero-knowledge, that represent the adversarial objectives of the prover and the verifier respectively: the prover must not be able to convince the verifier of false statements while the verifier should not extract any knowledge from interacting with the prover beyond what can be inferred by the public statement. An important class of prover verifier protocols is the 3-move honest-verifier zero knowledge (HVZK) protocols. They are three-move protocols that are "public-coin", i.e., the verifier in the second move merely selects a random value (that is drawn independently to the statement of the prover's first move) and submits it to the prover. 3-move HVZK protocols capture a very wide class of practical proofs of knowledge (including Schnorr's identification scheme [Sch89]) but also all languages in $\mathcal{NP}$ can be shown with a (computational) HVZK protocol via reduction to e.g., the Hamilton cycle protocol [Blu87]. The class of Σ-protocols possesses very useful properties including being closed under conjunction and disjunction operations [CDS94].

Given the above, one may construct a PoWorK protocol for a language $\mathcal{L}$ as follows: the verifier samples a cryptographic puzzle, puz, and submits it to the prover. The prover provides a commitment ψ and shows that she either possesses a witness w showing that the statement x belongs to $\mathcal{L}$ or that the commitment ψ contains a solution to puz. It is easy to prove that this is a general four-move protocol that implements a PoWorK for any language $\mathcal{L}$ and any cryptographic puzzle. On the other hand, it is known that for zero-knowledge proofs, two-round protocols do not exist for non-trivial languages [GO94] and this result remains true even if the zero-knowledge property is relaxed to $O(\lambda^{\log^c(\lambda)})$-simulatability [Pas03], in the sense that only languages decidable in quasi-polynomial time may have two-round quasi-polynomial-time simulatable protocols.

1.1 Our Results

We define and construct efficient *three-move* PoWorK protocols as well as relevant cryptographic puzzles. Morerover, we demonstrate how PoWorK can instantiate systems that reduce email spam while preserving user privacy, how they are useful in composition of cryptocurrency systems and how they can give rise to concurrent simulatable protocols. In more details:

Definition of PoWorKs. Our formalization entails two definitions, f-soundness and (statistical) indistinguishability. In f-soundness we require that any prover that has running time (in number of steps) less than a specified parameter calibrated according to the function f of the running time of the puzzle solver, it is guaranteed to lead to a knowledge extractor. The importance of the function f is to provide a safe running time upper bound under which the complete protocol execution is successful only via an (a-priori) knowledge of the witness. Indistinguishability on the other hand, ensures that a malicious verifier is incapable of discerning whether the prover performs the proof of work or possesses the knowledge of the witness. We note that timing issues are not taken into account in our model (i.e., we assume that the prover always takes the same amount of time to finish no matter which one of the two strategies it follows). What we do care about though, is that the prover who performs a proof of work spends at least a certain amount of computational resources. Note that indistinguishability easily implies witness indistinguishability [FS90], and thus any PoWorK is also a witness indistinguishable protocol.

PoWorK Constructions. We present a three-move public-coin protocol instantiating a PoWorK given any 3-move HVZK protocol with special soundness. Our protocol transformation preserves the structure and round complexity of the given 3-move HVZK protocol. Observe that the verifier cannot simply provide a puzzle challenge since this would violate the public-coin characteristic of the protocol. To achieve our construction we require puzzle generation algorithms that have a suitable uniformity characteristics, specifically, we require that the domain of puzzles (the "puzzle space") and the challenge space of the 3-move HVZK protocol are statistically very close (in terms of the distributions induced by the puzzle sample algorithm and the verifier in the protocol). Given such suitable puzzle distribution we present a protocol where the prover is capable of generating a puzzle solution on the fly (utilizing the verifier's public coins) and solve it, if she wishes. To establish the practicality of our approach we also construct puzzles that are "dense" within $\{0,1\}^l$ and hence consistent with the challenge space of many natural 3-move HVZK protocols. Our dense puzzle based PoWorK construction has the characteristic that is *black-box* with respect to the underlying puzzle system (which is suitable for puzzles whose security is argued, say, in the RO model).

Definition and Instantiations of Puzzles. We give formal definitions of cryptographic puzzle systems PuzSys that are easy to generate, hard to solve, and easy to verify. We define additional properties like density and amortization resistance and we give two instantiations. Our first instantiation utilizes the random oracle model [BR93] while the second relies on complexity assumptions. More specifically, we use *Universal One Way Hash Function* families (UOWHF) [NY89] to build extractors with special properties, invoking a variant of leftover hash lemma [Dod05]. We then combine this special extractor with suitably hard one-way functions to obtain our second puzzle instantiation; we present an instantiation of this methodology for the discrete-logarithm problem. As an intermediate result, which may be of independent interest, we show how to convert any arbitrary oneway function to a "dense" oneway function over $\{0,1\}^{\ell(\lambda)}$ for some $\ell(\cdot)$ and security parameter $\lambda \in \mathbb{Z}^+$ (cf. Theorem 3).

Our puzzle definitions are close in spirit to previous formalizations [RSW96, WJHF04, CMSW09, MMV11, BGJ+16] with the following distinctions. In [CMSW09] the hardness of a puzzle is defined as a monotonically increasing function that maps the running time of an adversary to the success rate of solving the puzzle. Contrary to this, our definition, motivated by our proof of knowledge application, imposes a sharp time threshold, below which the success rate of solving a puzzle becomes negligible. Also, contrary to time-lock puzzles [RSW96, WJHF04, MMV11, BGJ+16], we do not restrict the parallelizability of our puzzles as such feature does not hurt (and may even be desirable) in the PoWorK context. Parallelizable puzzles, like the ones we are focusing on here, have become very popular by their applications to cryptocurrencies. The requirement there is that the puzzle solver should spend a minimum of computational resources to find a solution to the puzzle (and may or may not choose to parallelize).

Applications. Generally speaking, PoWorKs can be used in applications where we would like to allow access to either "registered" or "approved" users (who know a witness) or to every user who is willing to invest computational effort. The key property of PoWorKs is that they enhance privacy since they do not leak the type of user (i.e. approved or not) to the entity that verifies access. A nice illustration of this type of application of PoWorKs is in regard to *reducing spam email.* Dwork and Naor proposed using proofs of work to control spam e-mails [DN92]. The gist of the idea is that every non-approved contact of a receiver would have to perform some work (i.e. invest computational effort) in order to send her an email. A downside of the method is that the mail server has to maintain an updated list of "approved-contacts" for every user; this can be a privacy concern for the users (not to mention the cost of updating the approved contacts database). We show how by using PoWorK's, one can still enforce the non-approved senders to perform work while preserving user privacy, since the mail server (who acts as a PoWorK verifier) will not be able to distinguish between approved and non-approved contacts because of PoWorK indistinguishability property.

Our second application is related to cryptocurrencies based on blockchains to maintain the ledger of transactions. These systems can be naturally divided by the mechanism they use to produce the next block in the blockchain as follows: first there are "puzzle-based" ones, (e.g., Bitcoin [Nak08] and many others that followed[1]), and then there are "knowledge-based" ones, that include those[2] that use "proof-of-stake", "proof-of-activity" or other type of consensus mechanism that relies e.g., on a public-key infrastructure, e.g., [BLMR14, DM16, Maz15]). We demonstrate how given two cryptocurrencies $\mathsf{C}_1, \mathsf{C}_2$ of each type, one can use PoWorK to fuse them into a single cryptocurrency C with the following properties: (i) in C, the miners that perform C_1-type of mining are indistinguishable from those that perform C_2-type of mining, (ii) C would reach consensus in the sense of persistence of transactions in the ledger under the conjunction of the conditions that systems $\mathsf{C}_1, \mathsf{C}_2$ would do, (iii) C would satisfy liveness under the disjunction of the conditions that systems $\mathsf{C}_1, \mathsf{C}_2$ would do.[3] PoWorK-based cryptocurrencies that fuse the knowledge-based and the puzzle-based approach have novel features in the context of cryptocurrencies: for instance, by composing a regular Bitcoin-like cryptocurrency C_1 with a centralized cryptocurrency C_2 supported by a single authority, we get a cryptocurrency C that resembles Bitcoin but has a trusted authority with a trapdoor that enables it to regulate and normalize the block production rate. Such systems may offer a more attractive solution for nation-states or central banks that wish to issue centralized cryptocurrencies, however they do not want to be constantly involved with block production and they prefer to leave ledger maintenance to the public, while retaining the ability to issue blocks in case of an emergency situation (e.g., many miners go offline due to a software problem). The PoWorK indistinguishability property is critically useful in this setting, since it enables the regulation of the block production rate made by the trusted party to be indistinguishable to everyone, thus ensuring that the trusted party's involvement will be unnoticed and hence will have no impact to the economy that the cryptocurrency supports.

Our third application relates to zero-knowledge protocols and concerns quasi-polynomial time straight-line simulatable arguments of knowledge. This class of protocols was introduced by [Pas03] and was motivated by the construction of concurrent zero-knowledge proofs in the plain model (as opposed to using a "setup" assumption). In [Pas03] a four-move argument of knowledge was presented that is quasi-polynomial time simulatable. We show that any suitable PoWorK protocol (see Theorem 1 for the precise formulation) implies

[1] E.g., Litecoin, Dogecoin, Ethereum, Dashcoin, etc.
[2] E.g., Peercoin, NXT, Nushares, Faircoin etc.
[3] For definitions of properties like liveness and persistence of the ledger we refer to e.g., [GKL15, BMC+15].

quasi-polynomial time straight-line simulatable arguments of knowledge. Given our 3-move PoWorK construction, this immediately yields a 3-round protocol in this setting which is optimal in terms of efficiency (round complexity is optimal and computational overhead is just two exponentiations for prover and verifier in total when using the elliptic curves from [BHKL13]); we note that a similar result in terms of rounds can be obtained via a different route, specifically, via the efficient OR composition with an input-delayed Σ-protocol as recently observed in [CPS+16], however the resulting complexity overhead would be at least 5 exponentiations for prover and verifier in total when instantiated using discrete logarithms.

Roadmap. The rest of this paper is organized as follows. In Sect. 2, we provide basic notation, and formalize cryptographic puzzles, the additional properties of dense samplable puzzles and the property of amortization resistance, as well as the notion of PoWorKs by defining completeness, f-soundness and indistinguishability. In Sect. 3, we present our efficient dense puzzle based construction built upon an arbitrary 3-move special sound HVZK protocol for a language $\mathcal{L}$ and some puzzle system, and prove that our construction achieves f-soundness and indistinguishability. In the same section, we present two dense puzzle instantiations. Finally, in Sect. 4, we describe the applications of PoWorKs. Namely, (i) a method to reduce the amount of spam email while preserving the privacy of the receiver, (ii) the composition of knowledge-based and puzzle-based cryptocurrencies that gives rise to PoWorK-based cryptocurrencies, (iii) an efficient 3-move straight-line concurrent statistically $\lambda^{\mathsf{poly}(\log\lambda)}$-simulatable argument of knowledge as defined in [Pas03, Pas04].

Alternative PoWorK Constructions. In the full version of this work [BKZZ15] we provide a second PoWorK construction based on the Lapidot-Shamir 3-move special sound computationally special HVZK protocol [LS90], which is less efficient than the dense puzzle based construction but works for all puzzle systems; note that this construction is not black-box with respect to the puzzle and depending on the puzzle may not be public-coin. A third way to construct PoWorK's can be derived from the recent efficient OR composition technique that was introduced in [CPS+16] that can be used with "input-delayed" Σ-protocols, where the statement need not be determined ahead of time. It is easy to see that in the case a puzzle accepts an "input-delayed" Σ proof of knowledge of the puzzle solution (e.g., a puzzle based on discrete-logarithms), a third possible construction method for PoWorK's is facilitated. We stress however that these alternative methods for constructing PoWorK's do not combine well with puzzles based on hash functions and thus may be of only theoretical interest in the context of our primitive.

2 Definitions

We start by setting the notation to be used in the rest of the paper. By λ we denote the security parameter and by $\mathsf{negl}(\cdot)$ the property that a function is negligible in some parameter. Let $z \xleftarrow{\$} \mathcal{Z}$ denote the uniformly at random selection of z from space $\mathcal{Z}$ and $\Delta[\mathbf{X}, \mathbf{Y}]$ the statistical distance of random variables (or distributions) $\mathbf{X}, \mathbf{Y}$. Composition of functions is denoted by $\circ$.

Let $\langle \mathcal{P}(y) \leftrightarrow \mathcal{V} \rangle(x, z)$ denote the interaction between a prover $\mathcal{P}$ and a verifier $\mathcal{V}$ on common input x, auxiliary input z, and $\mathcal{P}$'s private input y. For an algorithm $\mathcal{B}$ that is part of an interactive protocol let $view_{\mathcal{B}}$ and $output_{\mathcal{B}}$ denote the views and the output of $\mathcal{B}$ respectively. Let $\mathsf{Steps}_{\mathcal{B}}(x)$ be the number of steps (i.e. machine/operation cycles) executed by algorithm $\mathcal{B}$ on input x, and $\mathsf{Steps}_{\mathcal{P}}(\langle \mathcal{P}(y) \leftrightarrow \mathcal{V} \rangle(x, z))$ be the number of steps of $\mathcal{P}$, when interacting on inputs x, y, z[4]. If $R_{\mathcal{L}}$ is a witness relation for the language $\mathcal{L} \in \mathcal{NP}$ (i.e. $R_{\mathcal{L}}$ polynomial-time-decidable and $(x, w) \in R_L$ implies that $|w| \leq \text{poly}(|x|)$), we define the set of witnesses for the membership $x \in L$ as $R_L(x) = \{w : (x, w) \in R_L\}$.

2.1 Cryptographic Puzzles

Roughly speaking, a cryptographic puzzle should be easy to generate, hard to solve, and easy to verify. Given a specific security parameter λ, we denote the puzzle space as $\mathcal{PS}_\lambda$, the solution space as $\mathcal{SS}_\lambda$, and the hardness space as $\mathcal{HS}_\lambda$. We first define puzzles with a minimum set of properties, and then add extra properties that are useful in our constructions.

Definition 1. *A puzzle system* $\mathsf{PuzSys} = (\mathsf{Sample}, \mathsf{Solve}, \mathsf{Verify})$ *consists of the following four algorithms:*

- $\mathsf{Sample}(1^\lambda, h)$ *is a probabilistic puzzle instance sampling algorithm. On input the security parameter* 1^λ *and a hardness factor* $h \in \mathcal{HS}_\lambda$*, it outputs a puzzle instance* $\mathsf{puz} \in \mathcal{PS}_\lambda$.
- $\mathsf{Solve}(1^\lambda, h, \mathsf{puz})$ *is a probabilistic puzzle solving algorithm. On input the security parameter* 1^λ*, a hardness factor* $h \in \mathcal{HS}_\lambda$ *and a puzzle instance* $\mathsf{puz} \in \mathcal{PS}_\lambda$*, it outputs a potential solution* $\mathsf{soln} \in \mathcal{SS}_\lambda$.
- $\mathsf{Verify}(1^\lambda, h, \mathsf{puz}, \mathsf{soln})$ *is a deterministic puzzle verification algorithm. On input the security parameter* 1^λ*, a hardness factor* $h \in \mathcal{HS}_\lambda$*, a puzzle instance* $\mathsf{puz} \in \mathcal{PS}_\lambda$ *and a potential solution* $\mathsf{soln} \in \mathcal{SS}_\lambda$ *it outputs* true *or* false.

Subsequently, we define the following properties for a puzzle system.

Completeness: We say that a puzzle system PuzSys is *complete*, if for every $h \in \mathcal{HS}_\lambda$:

$$\Pr\left[\begin{matrix}\mathsf{puz} \leftarrow \mathsf{Sample}(1^\lambda, h); \mathsf{soln} \leftarrow \mathsf{Solve}(1^\lambda, h, \mathsf{puz}) : \\ \mathsf{Verify}(1^\lambda, h, \mathsf{puz}, \mathsf{soln}) = \mathsf{false}\end{matrix}\right] = \mathsf{negl}(\lambda).$$

Note that the number of steps that Solve takes to run is monotonically decreasing in the hardness factor h and may exponentially depend on λ, while Verify should run in polynomial time in λ.

[4] In this work we focus on parallelizable puzzles so counting in number steps as opposed to actual running time is more intuitive.

g-Hardness: We say that a puzzle system PuzSys is *g-hard* for some function g, if for every adversary $\mathcal{A}$, for every auxiliary tape $z \in \{0,1\}^*$ and for every $h \in \mathcal{HS}_\lambda$:

$$\Pr\left[\begin{array}{l}\mathsf{puz} \leftarrow \mathsf{Sample}(1^\lambda, h); \mathsf{soln} \leftarrow \mathcal{A}(z, 1^\lambda, h, \mathsf{puz}) : \\ \mathsf{Verify}(1^\lambda, h, \mathsf{puz}, \mathsf{soln}) = \mathsf{true} \wedge \\ \wedge \mathsf{Steps}_{\mathcal{A}}(z, 1^\lambda, h, \mathsf{puz}) \leq g(\mathsf{Steps}_{\mathsf{Solve}}(1^\lambda, h, \mathsf{puz}))\end{array}\right] = \mathsf{negl}(\lambda).$$

Dense Samplable Puzzles. In addition to the standard puzzle definition, for our PoWorK construction in Sect. 3 we need puzzles that can be sampled by just generating random strings (i.e. the puzzle instances should be "dense" over $\{0,1\}^{\ell(\lambda,h)}$ for some function ℓ and $\lambda, h \in \mathbb{Z}^+$). Formally it holds that for some function ℓ in λ and h,

$$\Delta[\mathsf{Sample}(1^\lambda, h), \mathbf{U}_{\ell(\lambda,h)}] = \mathsf{negl}(\lambda),$$

where $\mathbf{U}_{\ell(\lambda,h)}$ stands for the uniform distribution over $\{0,1\}^{\ell(\lambda,h)}$. For such puzzles we will require some additional properties. First there should be a puzzle sampler that outputs a valid solution together with puz:

- $\mathsf{SampleSol}(1^\lambda, h)$ is a probabilistic solved puzzle instance sampling algorithm. On input the security parameter 1^λ and a hardness factor $h \in \mathcal{HS}_\lambda$, it outputs a puzzle instance and solution pair $(\mathsf{puz}, \mathsf{soln}) \in \mathcal{PS}_\lambda \times \mathcal{SS}_\lambda$.

Correctness of Sampling: We say that a puzzle system PuzSys is *correct* with respect to sampling, if for every $h \in \mathcal{HS}_\lambda$, we have that:

$$\Pr\left[(\mathsf{puz}, \mathsf{soln}) \leftarrow \mathsf{SampleSol}(1^\lambda, h) : \mathsf{Verify}(1^\lambda, h, \mathsf{puz}, \mathsf{soln}) = \mathsf{false}\right] = \mathsf{negl}(\lambda).$$

Efficiency of Sampling: We say $\mathsf{SampleSol}$ is *efficient* with respect to the puzzle g-hardness, if for every $\lambda \in \mathbb{Z}^+$, $h \in \mathcal{HS}_\lambda$ and $\mathsf{puz} \in \mathcal{PS}_\lambda$, we have that:

$$\mathsf{Steps}_{\mathsf{SampleSol}}(1^\lambda, h) < g(\mathsf{Steps}_{\mathsf{Solve}}(1^\lambda, h, \mathsf{puz})).$$

Statistical Indistinguishability: We define the following two probability distributions

$$\mathbf{D}_{s,\lambda,h} \stackrel{def}{=} \{(\mathsf{puz}, \mathsf{soln}) \leftarrow \mathsf{SampleSol}(1^\lambda, h)\} \quad \text{and}$$

$$\mathbf{D}_{p,\lambda,h} \stackrel{def}{=} \{\mathsf{puz} \leftarrow \mathsf{Sample}(1^\lambda, h), \mathsf{soln} \leftarrow \mathsf{Solve}(1^\lambda, h, \mathsf{puz}) : (\mathsf{puz}, \mathsf{soln})\}.$$

We say a PuzSys is *statistically indistinguishable*, if for every $\lambda \in \mathbb{Z}^+$ and $h \in \mathcal{HS}_\lambda$:

$$\Delta[\mathbf{D}_{s,\lambda,h}, \mathbf{D}_{p,\lambda,h}] = \mathsf{negl}(\lambda).$$

(τ, k)-Amortization Resistance. For certain applications it is important that the puzzle is not amenable to amortization. We say that a g-hard puzzle system, PuzSys, is *(τ, k)-amortization resistant* if for every adversary $\mathcal{A}$, for every auxiliary tape $z \in \{0,1\}^*$ and for every $h \in \mathcal{HS}_\lambda$:

$$\Pr\left[\begin{array}{l}\forall 1 \le i \le k : \mathsf{puz}_i \leftarrow \mathsf{Sample}(1^\lambda, h);\\ \{\mathsf{soln}_1, \ldots, \mathsf{soln}_k\} \leftarrow \mathcal{A}(z, 1^\lambda, h, \{\mathsf{puz}_1, \ldots, \mathsf{puz}_k\}) :\\ (\forall 1 \le i \le k : \quad \mathsf{Verify}(1^\lambda, h, \mathsf{puz}_i, \mathsf{soln}_i) = \mathsf{true} \quad)\wedge\\ \wedge\Big(\mathsf{Steps}_{\mathcal{A}}(z, 1^\lambda, h, \{\mathsf{puz}_1\}_{i=1}^k) \le \tau\big(\sum_{i=1}^k g(\mathsf{Steps}_{\mathsf{Solve}}(1^\lambda, h, \mathsf{puz}_i))\big)\Big)\end{array}\right] = \mathsf{negl}(\lambda).$$

Informally, (τ, k)-amortization resistance implies a lower bound on the hardness preservation against adversaries that attempt to benefit from solving vectors of puzzles of length k.

2.2 Definition of PoWorK

In a PoWorK, the prover $\mathcal{P}$ may interact with the verifier $\mathcal{V}$ by running in either of the two following modes: (a) the *Proof of Knowledge (PoK)* mode, where $\mathcal{P}$ convinces $\mathcal{V}$ that she knows a witness for some statement x, or (b) the *Proof of WorK (PoW)* mode, where $\mathcal{P}$ makes calls to the puzzle solving algorithm to solve a certain puzzle. For some language in $\mathcal{NP}$ and a fixed puzzle system PuzSys, we define PoWorK to satisfy: (i) completeness, (ii) f-soundness (for some "computation-scaling" function f) and (iii) indistinguishability, as follows:

Definition 2 (PoWorK). *Let $\mathcal{L}$ be a language in $\mathcal{NP}$ and $R_{\mathcal{L}}$ be a witness relation for $\mathcal{L}$. Let* PuzSys = (Sample, Solve, Verify) *be a puzzle system anf f be a function. We say that $(\mathcal{P}, \mathcal{V})$ is an f-sound* Proof of Work or Knowledge (PoWorK) *for $\mathcal{L}$ and* PuzSys, *if the following properties are satisfied:*

(i). ***Completeness:*** *for every $x \in \mathcal{L} \cap \{0,1\}^{\mathrm{poly}(\lambda)}$, $w \in R_{\mathcal{L}}(x)$, $z \in \{0,1\}^*$ and every hardness factor $h \in \mathcal{HS}_\lambda$, it holds that*
(i.a) $\Pr[out_{\mathcal{V}} \leftarrow \langle \mathcal{P}(w) \leftrightarrow \mathcal{V}\rangle(x, z, h) : out_{\mathcal{V}} = \text{accept}] > 1 - 1/\text{poly}(\lambda)$ *and*
(i.b) $\Pr[out_{\mathcal{V}} \leftarrow \langle \mathcal{P}^{\mathsf{Solve}(1^\lambda, h, \cdot)} \leftrightarrow \mathcal{V}\rangle(x, z, h) : out_{\mathcal{V}} = \text{accept}] > 1 - 1/\text{poly}(\lambda)$.

(ii). ***f-Soundness:*** *For every $x \in \{0,1\}^{\mathrm{poly}(\lambda)}$, $y, z \in \{0,1\}^*$, every hardness factor $h \in \mathcal{HS}_\lambda$ and prover $\mathcal{P}^*$ define by $\pi_{x,y,z,h,\lambda}$ the probability*

$$\Pr\left[\begin{array}{l}\mathsf{puz} \leftarrow \mathsf{Sample}(1^\lambda, h); out_{\mathcal{V}} \leftarrow \langle \mathcal{P}^*(y) \leftrightarrow \mathcal{V}\rangle(x, z, h) : (out_{\mathcal{V}} = \text{accept})\\ \wedge \mathsf{Steps}_{\mathcal{P}^*}(\langle \mathcal{P}^*(y) \leftrightarrow \mathcal{V}\rangle(x, z, h)) \le f(\mathsf{Steps}_{\mathsf{Solve}}(1^\lambda, h, \mathsf{puz}))\end{array}\right].$$

f-Soundness holds if there are non-negligible functions s, q such that for any $\mathcal{P}^*$, there exists a PPT witness-extraction algorithm $\mathcal{K}$ such that for any $\lambda \in \mathbb{N}, x \in \{0,1\}^{\mathrm{poly}(\lambda)}, y, z \in \{0,1\}^*, h \in \mathcal{HS}_\lambda$, if $\pi_{x,y,z,h,\lambda} \ge s(\lambda)$ (representing the knowledge error), then

$$\Pr[\mathcal{K}^{\mathcal{P}^*}(x, y, z, h) \in R_{\mathcal{L}}(x)] \ge q(\lambda) .$$

(iii) ***Statistical (resp. Computational) Indistinguishability:*** *for every $x \in \mathcal{L} \cap \{0,1\}^{\mathrm{poly}(\lambda)}$, $w \in R_{\mathcal{L}}(x)$, $z \in \{0,1\}^*$, for every hardness factor $h \in \mathcal{HS}_\lambda$ and for every verifier (resp. PPT verifier) $\mathcal{V}^*$, the following two random variables are statistically (resp. computationally) indistinguishable:*

$$\mathbf{D}_{PoK}^{\mathcal{V}^*} \overset{def}{=} \{view_{\mathcal{V}^*} \leftarrow \langle \mathcal{P}(w) \leftrightarrow \mathcal{V}^*\rangle(x, z, h)\}$$
$$\mathbf{D}_{PoW}^{\mathcal{V}^*} \overset{def}{=} \Big\{view_{\mathcal{V}^*} \leftarrow \langle \mathcal{P}^{\mathsf{Solve}(1^\lambda, h, \cdot)} \leftrightarrow \mathcal{V}^*\rangle(x, z, h)\Big\}.$$

Intuitively, soundness is related to the hardness of solving a presumably hard cryptographic puzzle. The hardness threshold T is set to be the (probabilistic) computational complexity (in number of steps) of the puzzle solver, when the latter is provided some output of the puzzle sampling algorithm, scaled to some function f. According to Definition 2, any prover who does not know a witness, cannot convince the verifier in less than $f(T)$ steps with some good probability. Observe that in the definition of f-soundness, the convincing capability of the prover is limited by the hardness of solving puzzle challenges. This implies that in an f-sound protocol, provers who do not know (per the knowledge extractor) are forced to "work" in order to convince the verifier. The indistinguishability property of PoWorKs implies that a (potentially malicious) verifier cannot distinguish the running mode (PoK or PoW) that $\mathcal{P}$ follows.

3 The Dense Puzzle Based PoWorK Construction

In this section, we show how to transform an arbitrary 3-move, public coin, special sound, honest verifier zero-knowledge (SS-HVZK) into a 3-move public-coin PoWorK. Our construction is lightweight and requires dense samplable puzzle systems that we formalized in Sect. 2.1. In our full version [BKZZ15] we provide a second construction which is less efficient, non-black-box on the puzzle, but it works for all puzzle systems and may not be public-coin (depending on the puzzle). For both constructions, we consider a puzzle system PuzSys that achieves completeness and g-hardness for some function $g : \mathbb{N} \longrightarrow \mathbb{R}^+$. In addition, for dense samplable puzzle systems, we require correctness, efficient samplability, and statistical indistinguishability.

3.1 Preliminaries

The puzzle, solution and hardness spaces are denoted by $\mathcal{PS}_\lambda, \mathcal{SS}_\lambda, \mathcal{HS}_\lambda$, as in Sect. 2.1. Our PoWorK protocols are interactive proofs between a prover $\mathcal{P}$ and a verifier $\mathcal{V}$, denoted by $(\mathcal{P}, \mathcal{V})$.

The challenge space of our dense puzzle based construction $(\mathcal{P}, \mathcal{V})$, denoted by $\mathcal{CS}_\lambda$, is determined by the security parameter λ. From an algebraic point of view, $\mathcal{CS}_\lambda$ is set to be a group with operation $\oplus$, where performing $\oplus$ and inverting an element should be efficient. For the first construction, we require that $\mathcal{PS}_\lambda \subseteq \mathcal{CS}_\lambda$. For instance, we may set $\mathcal{CS}_\lambda$ as the group $\left(\mathbb{GF}(2^{\ell(\lambda)}, \oplus\right)$, where $\ell(\lambda)$ is the length of the challenges and $\oplus$ is the bitwise XOR operation. Of course, one may select a different setting which could be tailor made to the algebraic properties of the underlying primitives.

Let ChSampler be the algorithm that samples a challenge from $\mathcal{CS}_\lambda$. For a fixed security parameter, we define the following random variables (r.v.):

- The challenge sampling r.v. $\mathbf{C}_{\lambda,h} \stackrel{def}{=} \mathsf{ChSampler}(1^\lambda, h)$.
- The puzzle sampling r.v. $\mathbf{P}_{\lambda,h} \stackrel{def}{=} \{\mathsf{puz} \leftarrow \mathsf{Sample}(1^\lambda, h) : \mathsf{puz}\}$.

Finally, we denote by $x \oplus \mathbf{D}$ (resp. $\mathbf{D}^{\mathsf{Inv}}$) the r.v. of performing $\oplus$ on some fixed $x \in \mathcal{CS}_\lambda$ and an element y sampled from r.v. $\mathbf{D}$ (resp. inverting an element sampled from $\mathbf{D}$). The r.v. $\mathbf{D} \oplus x$ is defined similarly. Formally,

$$x \oplus \mathbf{D} \stackrel{def}{=} \{y \leftarrow \mathbf{D} : x \oplus y\},\ \mathbf{D} \oplus x \stackrel{def}{=} \{y \leftarrow \mathbf{D} : y \oplus x\},\ \mathbf{D}^{\mathsf{Inv}} \stackrel{def}{=} \{y \leftarrow \mathbf{D} : -y\}.$$

3.2 The Dense Puzzle Based Compiler

We now provide a detailed description of our protocol $(\mathcal{P}, \mathcal{V})$, which can be viewed as a compiler that can transform a SS-HVZK protocol $\Pi = (\mathsf{P1}_\Pi, \mathsf{P2}_\Pi, \mathsf{Ver}_\Pi)$ for $\mathcal{L} \in \mathcal{NP}$ and a g-hard puzzle system PuzSys into a 3-move PoWorK. The resulting PoWorK protocol achieves $\Theta(g)$-hardness and statistical indistiguishability. From a syntax point of view, our compiler will set the challenge space of the PoWorK $\mathcal{CS}_\lambda$ to be equal to $\mathcal{CS}_\Pi$. We denote by Sim_Π the HVZK simulator of Π.

The protocol $(\mathcal{P}, \mathcal{V})$ can be executed in either of the two following modes:

1. **Proof of Knowledge (PoK) mode:** $\mathcal{P}$ has a witness $w \in \mathcal{R}_\mathcal{L}(x)$ as private input. In order to prove knowledge of w to $\mathcal{V}$, $\mathcal{P}$ runs $\mathsf{P1}_\Pi$ and $\mathsf{P2}_\Pi$ as described by the original SS-HVZK protocol, with the difference that instead of providing $\mathsf{P2}_\Pi$ with the challenge c from $\mathcal{V}$ directly, $\mathcal{P}$ runs the puzzle sampler algorithm to receive a pair of a puzzle and its solution, $(\mathsf{puz}, \mathsf{soln})$, computes the value $\tilde{c} = c \oplus \mathsf{puz}$ and runs $\mathsf{P2}_\Pi$ with challenge $\tilde{c}$.
2. **Proof of Work (PoW) mode:** $\mathcal{P}$ has no private input and tries to convince $\mathcal{V}$ that it has performed a minimum amount of computational "work" (i.e. at least some expected number of steps). To achieve this, $\mathcal{P}$ runs Sim_Π to simulate a transcript of the original SS-HVZK protocol. Then, it receives the challenge c from $\mathcal{V}$ and computes the value $\mathsf{puz} = (-c) \oplus \tilde{c}$. It runs the Solve algorithm on input puz, and if puz is a puzzle in $\mathcal{PS}_\lambda$ (which, as we argue later, must occur with high probability), then it obtains a solution soln of puz, except for some negligible error.

The verification mechanism, must be the same for both modes, so that indistinguishability can be achieved. Namely, the verifier checks that: (i) the relation $\tilde{c} = c \oplus \mathsf{puz}$ holds, (ii) the transcript of the SS-HVZK protocol is accepting and (iii) the prover has output a correct pair of a puzzle puz and some solution soln of puz. The protocol $(\mathcal{P}, \mathcal{V})$ is presented in detail in Fig. 1.

3.3 Security of the Dense Puzzle Based Construction

In order to prove that our protocol satisfies soundness and indistinguishability, we need to assume that the challenge and puzzle distributions satisfy some plausible properties and that the presumed g-hardness of the puzzle system dominates the step complexity of the group operation and challenge sampling algorithms. In detail, we require that:

(A). The challenge and puzzle sampling distributions are statistically close.

Statement: $x \in \mathcal{L} \cap \{0,1\}^{\text{poly}(\lambda)}$. **Prover's private input:** $w \in R_{\mathcal{L}}(x)$. $\mathcal{P}$: $(\tilde{a}, \phi_1) \leftarrow \mathsf{P1}_{\Pi}(w, x)$. $\mathcal{P} \rightarrow \mathcal{V}$: $\tilde{a}$. $\mathcal{P} \leftarrow \mathcal{V}$: $c \leftarrow \mathsf{ChSampler}(1^\lambda, h)$; $\mathcal{P}$: • sample a puzzle-solution pair $(\mathsf{puz}, \mathsf{soln}) \leftarrow \mathsf{SampleSol}(1^\lambda, h)$; • set $\tilde{c} = c \oplus \mathsf{puz}$; • execute $\tilde{r} \leftarrow \mathsf{P2}_{\Pi}(\phi_1, \tilde{c})$; $\mathcal{P} \rightarrow \mathcal{V}$: $\tilde{c}, \tilde{r}, \mathsf{puz}, \mathsf{soln}$. **Verification:** 1. $\tilde{c} = c \oplus \mathsf{puz}$. 2. $\mathsf{Ver}_{\Pi}(x, \tilde{a}, \tilde{c}, \tilde{r}) = 1$. 3. $\mathsf{Verify}(1^\lambda, h, \mathsf{puz}, \mathsf{soln}) = \mathsf{true}$.	**Statement:** $x \in \mathcal{L} \cap \{0,1\}^{\text{poly}(\lambda)}$. **Prover's private input:** – $\mathcal{P}$: • execute $(\tilde{a}, \tilde{c}, \tilde{r}) \leftarrow \mathsf{Sim}_{\Pi}(x)$; $\mathcal{P} \rightarrow \mathcal{V}$: $\tilde{a}$. $\mathcal{P} \leftarrow \mathcal{V}$: $c \leftarrow \mathsf{ChSampler}(1^\lambda, h)$; $\mathcal{P}$: • set $\mathsf{puz} = (-c) \oplus \tilde{c}$; • compute a puzzle solution $\mathsf{soln} \leftarrow \mathsf{Solve}(1^\lambda, h, \mathsf{puz})$; $\mathcal{P} \rightarrow \mathcal{V}$: $\tilde{c}, \tilde{r}, \mathsf{puz}, \mathsf{soln}$. **Verification:** 1. $\tilde{c} = c \oplus \mathsf{puz}$. 2. $\mathsf{Ver}_{\Pi}(x, \tilde{a}, \tilde{c}, \tilde{r}) = 1$. 3. $\mathsf{Verify}(1^\lambda, h, \mathsf{puz}, \mathsf{soln}) = \mathsf{true}$.
(a) Knowing the witness (PoK)	(b) Doing work (PoW)

Fig. 1. The Dense Puzzle Based PoWorK Construction for fixed security parameter λ and pre-determined hardness factor $h \in \mathcal{HS}_\lambda$, given a 3-move-SS-HVZK protocol Π for language $\mathcal{L}$ and a dense samplable puzzle system PuzSys satisfying that $\mathcal{PS}_\lambda \subseteq \mathcal{CS}_\lambda = \mathcal{CS}_\Pi$; ChSampler is the challenge sampling algorithm over $\mathcal{CS}_\lambda$.

(B). The challenge sampling distribution is (statistically) *invariant* to any group operation, i.e. (a) inverting a challenge sampled from $\mathcal{CS}_\lambda$ and (b) performing $\oplus$ operations on some element x in $\mathcal{CS}_\lambda = \mathcal{CS}_\Pi$ and a sampled challenge. Observe that these two assumptions imply that the puzzle sampling distribution is also (statistically) $\oplus$-invariant.

(C). With high probability, the number of steps needed for $\mathsf{Steps}_{\mathsf{Solve}}(1^\lambda, h, \mathsf{puz})$ to solve a g-hard puzzle puz according to $\mathbf{P}_{\lambda,h}$, scaled to the puzzle hardness function g, is more than the number of steps of performing group operations (inversion and $\oplus$ operation), or sampling from $\mathcal{CS}_\lambda$.

The assumptions described are stated formally in Fig. 2. Assumptions (**A**) and (**B**) can be met for meaningful distributions, widely used in cryptographic protocols. For example, when $\mathbf{C}_{\lambda,h}$ and $\mathbf{P}_{\lambda,h}$ are close to uniform, it is straightforward that assumption (**A**) holds. Moreover, since the uniform distribution is invariant under group operations, we have that assumption (**B**) also holds. The assumption (**C**) is expected to hold for any meaningful cryptographic puzzle construction. Indeed, if solving a puzzle is believed to be hard (on average) within a bounded amount of steps T, then performing efficient tasks, such as group operations or sampling a challenge in the space where this puzzle belongs must be feasible in a number of steps much less than T.

(A). For every hardness factor $h \in \mathcal{HS}_\lambda$, the r.v. $\mathbf{C}_{\lambda,h}$ and $\mathbf{P}_{\lambda,h}$ are ϵ_1-statistically close, where $\epsilon_1(\cdot)$ is a negligible function.

(B). For every $x \in \mathcal{CS}_\lambda$ and hardness factor $h \in \mathcal{HS}_\lambda$, the r.v. $\mathbf{C}_{\lambda,h}$ is ϵ_2-statistically close to the r.v. $x \oplus \mathbf{C}_{\lambda,h}$, $\mathbf{C}_{\lambda,h} \oplus x$ and $\mathbf{C}^{\mathsf{Inv}}_{\lambda,h}$, where $\epsilon_2(\cdot)$ is a negligible function.

(C). There exists a constant $\kappa < 1$ and a negligible function $\epsilon_3(\cdot)$ s.t. for every hardness factor $h \in \mathcal{HS}_\lambda$ and every $r, r' \in \mathcal{CS}_\lambda$

$$\Pr[\mathsf{puz} \leftarrow \mathsf{Sample}(1^\lambda, h) : \kappa \cdot g(\mathsf{Steps}_{\mathsf{Solve}}(1^\lambda, h, \mathsf{puz})) > \\ > \mathsf{Steps}_{\mathsf{ChSampler}}(1^\lambda, h) + \mathsf{Steps}_{\mathsf{Inv}}(r) + \mathsf{Steps}_{\oplus}(r, r')] \geq 1 - \epsilon_3(\lambda),$$

where $\mathsf{Steps}_{\mathsf{Inv}}$, $\mathsf{Steps}_{\oplus}$ denote the number of steps needed for inversion and group operation in $\mathcal{CS}_\lambda$.

Fig. 2. Assumptions for our Dense Puzzle Based PoWorK Construction, where $\mathbf{C}_{\lambda,h}$ and $\mathbf{P}_{\lambda,h}$ are the challenge sampling and the puzzle sampling distributions respectively.

We prove that our dense puzzle based construction is a PoWorK, assuming (**A**), (**B**) and (**C**), the g-hardness of PuzSys and the soundness and ZK properties of the original SS-HVZK protocol. The soundness of our protocol is in constant relation with the hardness of PuzSys.

Theorem 1. *Let $\mathcal{L}$ be a language in $\mathcal{NP}$ and let $\Pi = (\mathsf{P1}_\Pi, \mathsf{P2}_\Pi, \mathsf{Ver}_\Pi)$ be a special-sound 3-move statistical HVZK protocol for $\mathcal{L}$, where the challenge sampling distribution is uniform. Let* PuzSys = (Sample, SampleSol, Solve, Verify) *be a dense samplable puzzle system that satisfies g-hardness for some function g. Define $(\mathcal{P}, \mathcal{V})$ as the protocol described in Fig. 1 when built upon Π,* PuzSys *and assume that (**A**), (**B**), (**C**) in Fig. 2 hold. Then, $(\mathcal{P}, \mathcal{V})$ is a $((1-\kappa)/2) \cdot g$-sound PoWorK for $\mathcal{L}$ and* PuzSys *with statistical indistiguishability, where κ is the constant defined in assumption (**C**).*

Proof. **Completeness:** By the completeness of Π and the correctness of PuzSys, the dense puzzle based PoWorK construction is complete in the case that $\mathcal{P}$ executes the PoK mode of the protocol. Regarding the PoW mode, an honest execution of PuzSys is incorrect, only if either of the two following cases is true:

(i). $\mathsf{puz} = (-c) \oplus \tilde{c} \in \mathcal{CS}_\lambda \setminus \mathcal{PS}_\lambda$, i.e. puz is not a puzzle. By assumptions (**A**), (**B**) in Fig. 2, this happens with negligible probability, since

$$\Delta[\,\mathbf{P}_{\lambda,h}, \mathbf{C}_{\lambda,h}] \leq \epsilon_1(\lambda) \wedge \Delta[\mathbf{C}_{\lambda,h}, \mathbf{C}^{\mathsf{Inv}}_{\lambda,h} \oplus \tilde{c}] \leq 2 \cdot \epsilon_2(\lambda) \Rightarrow \\ \Rightarrow \Delta[\mathbf{P}_{\lambda,h}, \mathbf{C}^{\mathsf{Inv}}_{\lambda,h} \oplus \tilde{c}] \leq \epsilon_1(\lambda) + 2 \cdot \epsilon_2(\lambda),$$

where we applied (**B**) two times (one for inversion and one for $\oplus$ operation).

(ii). puz is a puzzle, but the puzzle solver algorithm Solve does not output a solution for puz. Namely, we have that $\mathsf{Verify}(1^\lambda, h, \mathsf{puz}, \mathsf{soln}) = \mathsf{false}$. By the completeness property of PuzSys, this also happens with negligible probability.

Therefore, $(\mathcal{P}, \mathcal{V})$ achieves completeness with high probability, as required in Definition 2.

$\big((1-\kappa)/2\big) \cdot g$**-Soundness.** First, we make use of the special soundness PPT extractor K_Π of Π to construct a knowledge extractor $\mathcal{K}$ that on input (x, y, z, h) and given the code of an arbitrary prover $\hat{\mathcal{P}}$, executes the following steps:

1. By applying standard rewinding, $\mathcal{K}$ interacts with $\hat{\mathcal{P}}(y)$ for statement x and auxiliary input z, using two challenges c_1, c_2 sampled from $\mathbf{C}_{\lambda,h}$ and receives two protocol transcripts $\langle \tilde{a}_1, c_1, (\tilde{c}_1, \tilde{r}_1, \mathsf{puz}_1, \mathsf{soln}_1)\rangle$ and $\langle \tilde{a}_1, c_2, (\tilde{c}_2, \tilde{r}_2, \mathsf{puz}_2, \mathsf{soln}_2)\rangle$.
2. $\mathcal{K}$ runs $\mathcal{K}_\Pi$ on input $(x, \langle \tilde{a}_1, \tilde{c}_1, \tilde{r}_1\rangle, \langle \tilde{a}_1, \tilde{c}_2, \tilde{r}_2\rangle)$.
3. $\mathcal{K}$ returns the output of $\mathcal{K}_\Pi$.

Since K_Π is a PPT algorithm, $\mathcal{K}$ also runs in polynomial time.

Assume that for some $x \in \{0,1\}^{\mathrm{poly}(\lambda)}$, $y \in \{0,1\}^*, z \in \{0,1\}^*$, $h \in \mathcal{HS}_\lambda$, there exists a prover $\mathcal{P}^*$ and a non-negligible function $s(\cdot)$ s.t

$$\Pr[\mathsf{puz} \leftarrow \mathsf{Sample}(1^\lambda, h); out_{\mathcal{V}} \leftarrow \langle \mathcal{P}^*(y) \leftrightarrow \mathcal{V}\rangle(x, z, h) : (out_{\mathcal{V}} = \text{accept}) \wedge$$
$$\wedge\, \mathsf{Steps}_{\mathcal{P}^*}(\langle \mathcal{P}^*(y) \leftrightarrow \mathcal{V}\rangle(x, z, h)) \leq \big((1-\kappa)/2\big) \cdot g(\mathsf{Steps}_{\mathsf{Solve}}(1^\lambda, h, \mathsf{puz}))] \geq s(\lambda).$$

We construct an algorithm $\mathcal{W}$ that makes use of $\mathcal{P}^*$ to break the g-hardness of PuzSys. The input that $\mathcal{W}$ receives is $\langle (x, y, z), 1^\lambda, h, \mathsf{puz}\rangle$, where (x, y, z) is the auxiliary input and puz sampled from $\mathsf{Sample}(1^\lambda, h)$. Then, $\mathcal{W}$ executes the following steps:

1. It samples c_1 by running $\mathsf{ChSampler}(1^\lambda, h)$.
2. It interacts with $\mathcal{P}^*(y)$ for statement x, auxiliary input z, hardness factor h and challenge c_1. It receives the transcript $\langle \tilde{a}_1, c_1, (\tilde{c}_1, \tilde{r}_1, \mathsf{puz}_1, \mathsf{soln}_1)\rangle$.
3. It computes the inverse of puz, denoted by $(-\mathsf{puz})$.
4. It computes $c_2 = \tilde{c}_1 \oplus (-\mathsf{puz})$.
5. It rewinds $\mathcal{P}^*$ at the challenge phase and provides $\mathcal{P}^*$ with challenge c_2. It receives a second transcript $\langle \tilde{a}_1, c_2, (\tilde{c}_2, \tilde{r}_2, \mathsf{puz}_2, \mathsf{soln}_2)\rangle$.
6. It returns the value soln_2.

By the assumption for $\mathcal{P}^*$ and the splitting Lemma, we have that when $\mathcal{P}^*$ is challenged with two honestly selected c_1, c_2, it outputs two accepting transcripts by running in no more than $\big((1-\kappa)/2\big) \cdot g(\mathsf{Steps}_{\mathsf{Solve}}(1^\lambda, h, \mathsf{puz}))$ steps with at least $(s(\lambda)/2)^2$ probability. By Equal we denote the event that this happens and $\tilde{c}_1 = \tilde{c}_2$ holds. Obviously, either Equal, or $\neg$Equal will occur with probability at least $(s(\lambda)/2)^2/2 = s(\lambda)^2/8$.

Assume that Equal happens with at least $s(\lambda)^2/8$ probability. We will show that this case leads to a contradiction; namely, $\mathcal{W}$ will output a solution of puz while running in no more than $g(\mathsf{Steps}_{\mathsf{Solve}}(1^\lambda, h, \mathsf{puz}))$ steps, hence breaking the g-hardness of PuzSys.

We observe that for any puz, if both transcripts generated by the interaction with $\mathcal{P}^*$ are accepting and the values $\tilde{c}_1, \tilde{c}_2$ are equal, then we have that

$$(c_2 = \tilde{c}_1 \oplus (-\mathsf{puz})) \wedge (\tilde{c}_2 = c_2 \oplus \mathsf{puz}_2) \wedge (\tilde{c}_1 = \tilde{c}_2) \Rightarrow \mathsf{puz}_2 = \big(-(-\mathsf{puz})\big) = \mathsf{puz},$$

where the second equality holds due to verification step 1. Therefore, it holds that

$$\mathsf{Verify}(1^\lambda, h, \mathsf{puz}_2, \mathsf{soln}_2) = \mathsf{true} \Leftrightarrow \mathsf{Verify}(1^\lambda, h, \mathsf{puz}, \mathsf{soln}_2) = \mathsf{true}. \tag{1}$$

By the assumptions (**A**), (**B**) in Fig. 2, we have that there are negligible functions $\epsilon_1(\lambda), \epsilon_2(\lambda)$ s.t. for any $\tilde{c}_1$ that $\mathcal{P}^*$ returns,

$$\Delta[\tilde{c}_1 \oplus \mathbf{C}^{\mathsf{Inv}}_{\lambda,h}, \tilde{c}_1 \oplus \mathbf{P}^{\mathsf{Inv}}_{\lambda,h}] < 2\epsilon_1(\lambda) \quad \text{and} \quad \Delta[\mathbf{C}_{\lambda,h}, \tilde{c}_1 \oplus \mathbf{C}^{\mathsf{Inv}}_{\lambda,h}] < 2\epsilon_2(\lambda),$$

where in the first and second inequality, we applied assumptions (**A**) and (**B**) respectively two times (one for inversion and one for $\oplus$ operation). Therefore, by the triangular inequality we have that

$$\Delta[\mathbf{C}_{\lambda,h}, \tilde{c}_1 \oplus \mathbf{P}^{\mathsf{Inv}}_{\lambda,h}] < 2\epsilon_1(\lambda) + 2\epsilon_2(\lambda). \tag{2}$$

Eq. (2) implies that the probability distribution of $c_2 = \tilde{c}_1 \oplus (-\mathsf{puz})$ that $\mathcal{W}$ computes is $[2\epsilon_1(\cdot) + 2\epsilon_2(\cdot)]$-statistically close to the challenge sampling distribution of $\mathcal{V}$.

By construction, the running time of $\mathcal{W}$ (in number of steps) is at most

$$2 \cdot \mathsf{Steps}_{\mathcal{P}^*}(\langle \mathcal{P}^*(y) \leftrightarrow \mathcal{V}\rangle(x, z, h)) + \mathsf{Steps}(((-\mathsf{puz})))+ \\ +\mathsf{Steps}(\tilde{c}_1 \oplus (-\mathsf{puz})) + \mathsf{Steps}_{\mathsf{ChSampler}}(1^\lambda, h).$$

By assumption (**C**) in Fig. 2, there is a negligible function $\epsilon_3(\cdot)$ and a constant $\kappa < 1$ s.t.

$$\Pr[\mathsf{puz} \leftarrow \mathsf{Sample}(1^\lambda, h) : \kappa \cdot g(\mathsf{Steps}_{\mathsf{Solve}}(1^\lambda, h, \mathsf{puz})) < \mathsf{Steps}_{\mathsf{ChSampler}}(1^\lambda, h)+ \\ +\mathsf{Steps}((-\mathsf{puz})) + \mathsf{Steps}(\tilde{c}_1 \oplus (-\mathsf{puz}))] \leq \epsilon_3(\lambda). \tag{3}$$

When Equal occurs, then it holds that

$$\mathsf{Steps}_{\mathcal{P}^*}(\langle \mathcal{P}^*(y) \leftrightarrow \mathcal{V}\rangle(x, z, h)) \leq \big((1-\kappa)/2\big) \cdot g(\mathsf{Steps}_{\mathsf{Solve}}(1^\lambda, h, \mathsf{puz})),$$

hence by the assumption for $\mathcal{P}^*$ and Eqs. (2) and (3), the probability that the running time of $\mathcal{W}$ is bounded by

$$\begin{aligned}
&\mathsf{Steps}_{\mathcal{W}}(1^\lambda, (x, y, z), h, \mathsf{puz}) \leq \\
&\leq 2 \cdot \mathsf{Steps}_{\mathcal{P}^*}(\langle \mathcal{P}^*(y) \leftrightarrow \mathcal{V}\rangle(x, z, h)) + \kappa \cdot g(\mathsf{Steps}_{\mathsf{Solve}}(1^\lambda, h, \mathsf{puz})) \leq \\
&\leq (2 \cdot ((1-\kappa)/2)) \cdot g(\mathsf{Steps}_{\mathsf{Solve}}(1^\lambda, h, \mathsf{puz})) + \kappa \cdot g(\mathsf{Steps}_{\mathsf{Solve}}(1^\lambda, h, \mathsf{puz})) = \\
&= g(\mathsf{Steps}_{\mathsf{Solve}}(1^\lambda, h, \mathsf{puz})),
\end{aligned}$$

is at least Pr[Equal] $-\big(2\epsilon_1(\lambda)+2\epsilon_2(\lambda)+\epsilon_3(\lambda)\big)$. By Eqs. (1), (2) and (3), and the assumption $\Pr[\text{Equal}] \geq s(\lambda)^2/8$, we have that for auxiliary tape (x, y, z) and hardness factor h:

$$\Pr\left[\begin{array}{l}\mathsf{puz} \leftarrow \mathsf{Sample}(1^\lambda, h);\\ \mathsf{soln}_* \leftarrow \mathcal{W}(1^\lambda,(x,y,z),h,\mathsf{puz}):\\ \mathsf{Verify}(1^\lambda,h,\mathsf{puz},\mathsf{soln}_*)=\mathsf{true} \quad \wedge\\ \wedge \mathsf{Steps}_{\mathcal{W}}(1^\lambda,(x,y,z),h,\mathsf{puz})\\ \leq g(\mathsf{Steps}_{\mathsf{Solve}}(1^\lambda,h,\mathsf{puz}))\end{array}\right] \geq s(\lambda)^2/8 - \big(2\epsilon_1(\lambda)+2\epsilon_2(\lambda)+\epsilon_3(\lambda)\big),$$

which contradicts to the g-hardness of PuzSys, as $s(\lambda)^2/8 - \big(2\epsilon_1(\lambda)+2\epsilon_2(\lambda)+\epsilon_3(\lambda)\big)$ is a non-negligible function. Therefore, it holds that $\Pr[\text{Equal}] \leq s(\lambda)^2/8$ which implies

$$\Pr[\neg\text{Equal}] \geq s(\lambda)^2/8. \tag{4}$$

By the construction of $\mathcal{K}$ and the special soundness property of Π, we have that $\mathcal{K}$ will return a witness for x whenever $\mathcal{K}_\Pi$ is provided with different $\tilde{c}_1, \tilde{c}_2$. Define $q(\lambda) = s(\lambda)^2/8$. By Eq. (4), when $\mathcal{K}$ is given oracle access to $\mathcal{P}^*$ it holds that

$$\Pr[\mathcal{K}^{\mathcal{P}^*}(x,y,z,h) \in R_{\mathcal{L}}(x)] = \Pr[\neg\text{Equal}] \geq q(\lambda).$$

Thus, we conclude that our protocol is $\big((1-\kappa)/2\big)\cdot g$-sound.

Statistical Indistinguishability. Assume that the protocol described in Fig. 1 does not satisfy the PoWorK indistinguishability property in Definition 2. Then, for some (x, z, h) there exists a verifier $\mathcal{V}^*$ that w.l.o.g. outputs a single bit and can distinguish between:

$$\mathbf{D}^{\mathcal{V}^*}_{PoK} = \{view_{\mathcal{V}^*} \leftarrow \langle \mathcal{P}(w) \leftrightarrow \mathcal{V}^*\rangle(x,z,h)\} \quad \text{and}$$
$$\mathbf{D}^{\mathcal{V}^*}_{PoW} = \left\{view_{\mathcal{V}^*} \leftarrow \langle \mathcal{P}^{\mathsf{Solve}(1^\lambda,h,\cdot)} \leftrightarrow \mathcal{V}^*\rangle(x,z,h)\right\}.$$

with non-negligible advantage $\eta(\lambda)$.

In the following, we will show that if such a $\mathcal{V}^*$ exists, then we can construct an adversary $\mathcal{B}$ who breaks the statistical (auxiliary input) HVZK property of the underlying 3-move protocol $\Pi = (\mathsf{P1}_\Pi, \mathsf{P2}_\Pi, \mathsf{Ver}_\Pi)$. This means that $\mathcal{B}$ can distinguish between:

$$\mathbf{D}_\Pi = \left\{(\tilde{a},\phi_1) \leftarrow \mathsf{P1}_\Pi(w,x); \tilde{c} \xleftarrow{\$} \mathcal{CS}_\Pi; \tilde{r} \leftarrow \mathsf{P2}_\Pi(\phi_1,\tilde{c}) : (\tilde{a},\tilde{c},\tilde{r})\right\} \text{ and}$$
$$\mathbf{D}_{\mathsf{Sim}} = \{(\tilde{a},\tilde{c},\tilde{r}) \leftarrow \mathsf{Sim}_\Pi(x,(z,h)) : (\tilde{a},\tilde{c},\tilde{r})\}$$

with some non-negligible advantage $\eta'(\lambda)$, where (z, h) is the auxiliary input. Namely, $\mathcal{B}$ takes as input $(x, (z, h), (\tilde{a}, \tilde{c}, \tilde{r}))$, and works as follows:

1. Invokes $\mathcal{V}^*$ with input x, z, h and first move message $\tilde{a}$.
2. $\mathcal{V}^*$ responds back with his challenge c.
3. $\mathcal{B}$ computes $\mathsf{puz} = (-c) \oplus \tilde{c}$ and runs Solve on input $(1^\lambda, h, \mathsf{puz})$ to receive back soln.

4. $\mathcal{B}$ sends $(\tilde{c}, \tilde{r}, \mathsf{puz}, \mathsf{soln})$ to $\mathcal{V}^*$.
5. $\mathcal{B}$ returns $\mathcal{V}^*$'s output b^*.

By construction of $\mathcal{B}$, what is left to argue is that $\mathsf{puz} = (-c) \oplus \tilde{c}$ and $\mathsf{soln} \leftarrow \mathsf{Solve}(1^\lambda, h, \mathsf{puz})$ are indistinguishable from a pair $(\mathsf{puz}', \mathsf{soln}')$ that was picked by $\mathsf{SampleSol}(1^\lambda, h)$. We stusy the following two cases:

1. *$\mathcal{B}$'s input is sampled according to $\mathbf{D}_\Pi$*: By the assumption (**B**) in Fig. 2 and for any c returned by $\mathcal{V}^*$, we have that:
$$\Delta[\mathbf{C}_{\lambda,h}, \mathbf{C}^{\mathsf{Inv}}_{\lambda,h} \oplus \tilde{c}] < 2\epsilon_2(\lambda),$$
where we applied (**B**) two times (one for inversion and one for $\oplus$ operation). By assumption (**A**), we have that
$$\Delta[\mathbf{C}_{\lambda,h}, \mathbf{P}_{\lambda,h}] < \epsilon_1(\lambda).$$
By the triangular inequality, we have that for the distribution of $\mathsf{puz} = (-c) \oplus \tilde{c}$, it holds that
$$\Delta[\mathbf{P}_{\lambda,h}, \mathbf{C}^{\mathsf{Inv}}_{\lambda,h} \oplus \tilde{c}] < \epsilon_1(\lambda) + 2\epsilon_2(\lambda).$$
By the statistical indistinguishability property of PuzSys (Definition 1), we have that the distribution $\{\mathsf{soln} \leftarrow \mathsf{Solve}(1^\lambda, h, \mathsf{puz}) : \mathsf{soln}\}$ is $\epsilon_4(\lambda)$-statistically close to the distribution $\{(\mathsf{soln}', \mathsf{puz}') \leftarrow \mathsf{SampleSol}(1^\lambda, h) : \mathsf{soln}'\}$, for some negligible function ϵ_4. Consequently, the probability distribution of puz that $\mathcal{B}$ computes is $[\epsilon_1(\lambda) + 2\epsilon_2(\lambda) + \epsilon_4(\lambda)]$-statistically close to the puzzle sampling distribution.
2. *$\mathcal{B}$'s input is sampled according to $\mathbf{D}_{\mathsf{Sim}}$*: in this case, it is straightforward that $\mathcal{B}$ simulates perfectly the *PoW* mode of the PoWorK protocol.

By the above and given that the probability of success of $\mathcal{V}^*$ is at least $\eta(\lambda)$, we have that

$$\begin{aligned}
&\Big|\Pr[(\tilde{a}, \tilde{c}, \tilde{r}) \leftarrow \mathbf{D}_\Pi : \mathcal{B}(x, (z,h), \tilde{a}, \tilde{c}, \tilde{r}) = 1] - \\
&\qquad - \Pr[(\tilde{a}, \tilde{c}, \tilde{r}) \leftarrow \mathbf{D}_{\mathsf{Sim}} : \mathcal{B}(x, (z,h), \tilde{a}, \tilde{c}, \tilde{r}) = 1]\Big| \geq \\
&\geq \Big|\Big(\Pr[view_{\mathcal{V}^*} \leftarrow \mathbf{D}^{\mathcal{V}^*}_{PoK} : \mathcal{V}^*(view_{\mathcal{V}^*}) = 1] - (\epsilon_1(\lambda) + 2\epsilon_2(\lambda) + \epsilon_4(\lambda))\Big) - \\
&\qquad - \Pr[view_{\mathcal{V}^*} \leftarrow \mathbf{D}^{\mathcal{V}^*}_{PoW} : \mathcal{V}^*(view_{\mathcal{V}^*}) = 1]\Big| \geq \\
&\geq \Big|\Pr[view_{\mathcal{V}^*} \leftarrow \mathbf{D}^{\mathcal{V}^*}_{PoK} : \mathcal{V}^*(view_{\mathcal{V}^*}) = 1] - \\
&\qquad - \Pr[view_{\mathcal{V}^*} \leftarrow \mathbf{D}^{\mathcal{V}^*}_{PoW} : \mathcal{V}^*(view_{\mathcal{V}^*}) = 1]\Big| - (\epsilon_1(\lambda) + 2\epsilon_2(\lambda) + \epsilon_4(\lambda))) \geq \\
&\geq \eta(\lambda) - (\epsilon_1(\lambda) + 2\epsilon_2(\lambda) + \epsilon_4(\lambda)).
\end{aligned}$$

Therefore, $\mathcal{B}$ is successful in breaking the statistical HVZK property of the underlying 3-move SS-HVZK protocol with non-negligible advantage $\eta'(\lambda) = \eta(\lambda) - (\epsilon_1(\lambda) + 2\epsilon_2(\lambda) + \epsilon_4(\lambda))$. This leads us to the conclusion that the protocol in Fig. 1 is a PoWorK with statistical indistinguishability. $\square$

Remark. Theorem 1 can be extended to encompass the case where the protocol Π to be compiled in the construction described in Fig. 1 achieves $T(\lambda)$-*computational HVZK*, i.e. it is HVZK for every verifier $\mathcal{B}$ which runs in $T(\lambda)$ steps. Specifically, in the indistinguishability proof the running time of the HVZK adversary $\mathcal{B}$ is (in number of steps) bounded by:

$$\mathsf{Steps}_{\mathcal{V}^*}(\langle(\mathsf{P1}_\Pi, \mathsf{P2}_\Pi)(w), \mathsf{Ver}_\Pi(\tilde{c})\rangle(x, z, h)) + \\ +\mathsf{Steps}_{\mathsf{Inv}}(c) + \mathsf{Steps}_{\oplus}((-c), \tilde{c}) + \mathsf{Steps}_{\mathsf{Solve}}(1^\lambda, h, \mathsf{puz}).$$

Therefore, we can prove that if $T(\lambda)$ is an asymptotically larger function than the time of the puzzle solving algorithm, then our dense puzzle based construction achieves computational indistinguishability.

3.4 Dense Puzzle Instantiation in the Random Oracle Model

We now instantiate a dense puzzle system in the random oracle model. For a given security parameter λ, let $\mathcal{O} : \{0,1\}^* \mapsto \{0,1\}^m$ be a random oracle, where $m \geq \lambda/2$. Our dense puzzle system is described in Fig. 3.

Theorem 2. *Let $\lambda \in \mathbb{Z}^+$ be the security parameter. Define $\mathcal{PS}_\lambda = \{0,1\}^\lambda$, $\mathcal{SS}_\lambda = \{0,1\}^\lambda$, and $\mathcal{HS}_\lambda = [\log^2 \lambda, \lambda/4]$. Let $\mathcal{O}$ be a random oracle mapping from $\{0,1\}^*$ to $\{0,1\}^m$, where $m \geq \lambda/2$. For any $h \in \mathcal{HS}_\lambda$, the puzzle system* PuzSys *described in Fig. 3 is correct, complete with* Solve*'s running time $2^{h+2\log\lambda}$, efficiently samplable, statistically indistinguishable, and g-hard, where $g(T) = T^{1/c}$, for any constant $c > 2$. In addition, for any k that is $O(2^{\lambda/8})$,* PuzSys *is $(\mathsf{id}(\cdot), k)$-amortization resistant, where $\mathsf{id}(\cdot)$ is the identity function.*

Proof. Please see the full version [BKZZ15].

Define $\mathcal{PS}_\lambda = \{0,1\}^\lambda$, $\mathcal{SS}_\lambda = \{0,1\}^\lambda$, and $\mathcal{HS}_\lambda = [\log^2 \lambda, \lambda/4]$. Let $H(\cdot) := \mathsf{LSB}_{\lambda/2}(\mathcal{O}(\cdot))$, where LSB_k stands for k least significant bits.

- $\mathsf{Sample}(1^\lambda, h)$: Return $\mathsf{puz} \leftarrow \{0,1\}^\lambda$.
- $\mathsf{SampleSol}(1^\lambda, h)$: Pick random $x \leftarrow \{0,1\}^\lambda$ and $y \leftarrow \{0,1\}^{\lambda/2}$. Return $\mathsf{puz} = (H(x,y), y)$ and $\mathsf{soln} = x$.
- $\mathsf{Solve}(1^\lambda, h, \mathsf{puz})$:
  - Parse puz to (z, y); set $\mathsf{soln} = \perp$ and initialize an empty set X.
  - For $ctr = \{1, \ldots, 2^{h+2\log\lambda}\}$:
    Randomly pick $x \leftarrow \{0,1\}^\lambda \setminus X$, and add x to X. Set $\mathsf{soln} = x$ if $\mathsf{LSB}_h(z) = \mathsf{LSB}_h(H(x,y))$.
  - Return soln.
- $\mathsf{Verify}(1^\lambda, h, \mathsf{puz}, \mathsf{soln})$: Parse puz to (z, y). Return true if and only if $\mathsf{LSB}_h(z) = \mathsf{LSB}_h(H(\mathsf{soln}, y))$.

Fig. 3. The Dense Puzzle system from the random oracle $\mathcal{O}$.

3.5 Dense Puzzle Instantiation from Complexity Assumptions

In this section, we show how to construct a puzzle system whose puzzle instance distribution is statistically close to the uniform distribution (over $\{0,1\}^{m(\lambda)}$) without random oracles. The main challenge is, given an arbitrary oneway function $\psi : \mathcal{X} \mapsto \mathcal{Y}$, to build another oneway function with uniform output distribution (on random inputs) while still maintaining its onewayness. As an intuition, we would like to first map the output of the given oneway function from $\mathcal{Y}$ to $\{0,1\}^{\ell}$ using an efficient injective map (which is usually the bit representation of $y \in \mathcal{Y}$), and then apply a strong extractor on it. Let $\mathsf{Ext} : \{0,1\}^{\ell} \times \{0,1\}^{d} \mapsto \{0,1\}^{m}$ be a strong extractor as defined at Definition 3.

Definition 3. *Function* $\mathsf{Ext} : \{0,1\}^{\ell} \times \{0,1\}^{d} \mapsto \{0,1\}^{m}$ *is* (t, ϵ)*-strong extractor if for any* t*-source* X *(over* $\{0,1\}^{\ell}$*), we have* $\Delta[(S, \mathsf{Ext}(X, S)), (S, \mathbf{U}_m)] \leq \epsilon$*, where* $S \leftarrow \{0,1\}^{d}$ *and* $\mathbf{U}_m \leftarrow \{0,1\}^{m}$ *are drawn uniformly and independently of* X.

The new oneway function $\psi^U : \mathcal{X} \times \{0,1\}^{d} \mapsto \{0,1\}^{m} \times \{0,1\}^{d}$ is defined as $\psi^U(x, s) = (\mathsf{Ext}(\psi(x), s), s)$. According to LHL [HILL93], if $H_{\infty}(x) \geq m + 2\log(1/\epsilon)$, then the output of ψ^U is at most ϵ-far from the uniform distribution over $\{0,1\}^{m+d}$. However, in order to maintain its onewayness, we need an extra property of the strong extractor – *Target Collision Resistance* (TCR), i.e. given x and s, it is computationally infeasible to find x' such that $x \neq x'$ and $\mathsf{Ext}(x, s) = \mathsf{Ext}(x', s)$. We construct TCR strong extractors from *regular universal oneway hash functions* (UOWHFs), initially proposed by Naor and Yung [NY89]. We first formally define the TCR property for a strong extractor in Definition 4.

Definition 4. *Let* $\mathsf{Ext} : \{0,1\}^{\ell(\lambda)} \times \{0,1\}^{d(\lambda)} \mapsto \{0,1\}^{m(\lambda)}$ *be a strong extractor. We say* Ext *is target collision resistant if for all PPT adversary* $\mathcal{A}$*, the following probability:*

$$\Pr\left[\begin{array}{l} x \leftarrow \mathcal{A}(1^{\lambda}); s \leftarrow \{0,1\}^{d(\lambda)} : x' \leftarrow \mathcal{A}(s) : \\ x, x' \in \{0,1\}^{\ell(\lambda)} \wedge x \neq x' \wedge \mathsf{Ext}(x, s) = \mathsf{Ext}(x', s) \end{array}\right] = \mathsf{negl}(\lambda).$$

A stronger notion, *collision resistant extractors*, was introduced by Dodis [Dod05]. Collision resistant extractors were applied to construct *perfectly oneway probabilistic hash functions* proposed [CMR98] in 2005. The construction of such collision resistant extractors relies on a variant of leftover hash lemma proved by Dodis and Smith [DS05]. Our observation is that in the same way that [Dod05] employ regular collision resistant hash functions (CRHF) to derive collision resistant strong extractors, we can use regular universal oneway hash function (UOWHF), to obtain TCR strong extractor. The notion of UOWHF was initially proposed by Naor and Yung [NY89] where they showed that UOWHFs can be constructed by composing oneway permutations with (weakly) pairwise independent hash functions. Since then, many constructions of UOWHFs have been proposed, assuming the existence of regular oneway functions [SY90] or any oneway functions [Rom90,HHR+10].[5]

We would like to use $\mathcal{H}_{2n} = \{H_{(a,b)}(x) = ax + b | \forall a \neq 0, a, b \in \mathbb{GF}(2^n)\}$ as the family of pairwise independent permutations and a regular UOWHF family $\mathcal{F}_\lambda$ to construct our TCR strong extractors. Define $\hat{F}_i(\cdot) := (F_i(\cdot), i)$, where $F_i \in \mathcal{F}_\lambda$. Our TCR strong extractor is constructed as $\mathsf{Ext}(x, (i, s)) = \hat{F}_i \circ H_s(x)$. Note that regularity of the UOWHFs is important to ensure that the output distribution of such strong extractors is close to the uniform distribution, as $F_i(U_{\ell_1(\lambda)}) \equiv U_{\ell_2(\lambda)}$. On the other hand, some UOWHF constructions give regular UOWHFs by default (i.e., the UOWHFs constructed by the oneway permutation based approach [NY89]).

Dense Oneway Functions and Dense Puzzles from Complexity Assumptions. We apply a TCR strong extractor for our construction. The key to the construction will be a "dense" oneway function: a oneway function is ϵ-*dense* oneway if its output distribution is at most ϵ-far from $\mathbf{U}_m$ for some $m \in \mathbb{Z}^+$. We now present a transformation of a one-way function to a dense one-way function via the application of a TCR-strong extractor. The TCR property will ensure that any attempt to invert the dense one-way function will result to an inversion of the underlying one-way function. Formally we prove the following.

Theorem 3. *Let $\lambda_1, \lambda_2 \in \mathbb{Z}^+$ be the security parameters. Let $\psi_{\lambda_1} : \mathcal{X}_{\lambda_1} \mapsto \mathcal{Y}_{\lambda_1}$ be an arbitrary oneway function, and define $H_{\lambda_1} = H_\infty(\psi_{\lambda_1}(X))$ for random variable X drawn uniformly from $\mathcal{X}_{\lambda_1}$. Assume there exists an efficient injective map $\zeta_{\lambda_1} : \mathcal{Y}_{\lambda_1} \mapsto \{0,1\}^{\ell(\lambda_2)}$. If*

$$\mathsf{Ext}_{\lambda_2}(x, (s_1, s_2)) : \{0,1\}^{\ell(\lambda_2)} \times \{0,1\}^{\lambda_2 + 2\cdot\ell(\lambda_2)} \mapsto \{0,1\}^{H_{\lambda_1} - 2\log(1/\epsilon) - 1}$$

is a $(H_{\lambda_1}, \epsilon)$-TCR strong extractor, then

$$\psi^U_{\lambda_1,\lambda_2}(x, s_1, s_2) = (\mathsf{Ext}_{\lambda_2}(\zeta_{\lambda_1}(\psi_{\lambda_1}(x)), (s_1, s_2)), s_2)$$

is an ϵ-dense oneway function with range $\{0,1\}^{2\cdot\ell(\lambda_2) + H_{\lambda_1} - 2\log(1/\epsilon) - 1}$ and domain $\mathcal{X}_{\lambda_1} \times \{0,1\}^{\lambda_2 + 2\cdot\ell(\lambda_2)}$.

Proof. Please see the full version [BKZZ15].

The above result paves the way for constructing dense puzzles from complexity assumptions. Essentially, given a function with moderately hard characteristics making it suitable for a puzzle, it is possible to transform it to a dense puzzle by applying a suitably hard TCR extractor ("suitable" here means that breaking the TCR property should be harder than solving the puzzle). We now illustrate this methodology by applying it to the discrete logarithm problem. More generally this methodology transforms any puzzle in the sense of Definition 1 to a dense puzzle (assuming again a suitably hard TCR extractor).

[5] We note that, on the contrary, CR strong extractors cannot be built from arbitrary oneway functions, since Simon [Sim98] gave a black-box separation between CRHFs and oneway functions.

The DLP Based Puzzle and Calibrating Its Hardness. Consider the discrete logarithm problem (DLP) as the candidate oneway function for our puzzle. Let $\mathbb{G} = \langle G \rangle$ be some (multiplicative) cyclic group where the DLP is hard, and G is a generator with order p, which is a λ_1-bit prime. The oneway function $\psi_G : \mathbb{Z}_p \mapsto \mathbb{G}$ is defined as $\psi_G(x) = G^x$. It is shown by Shoup [Sho97] that any probabilistic algorithm takes $\Omega(\sqrt{p})$ steps to solve the DLP over generic groups. Analogously, [GJKY13] shows any probabilistic algorithm must take at least $\sqrt{2p\epsilon}$ steps to solve DLP with probability ϵ in the generic group model. To build a puzzle, we would like to calibrate the hardness of the DLP by revealing the most significant bits of the pre-image. For example, for a puzzle with hardness factor$h \leq \lfloor \frac{\lambda_1 - 1}{2} \rfloor$, we pick $x \in \{0,1\}^h$ and $y \in \{0,1\}^{\lfloor (\lambda_1 - 1)/2 \rfloor}$ uniformly at random, and set the puzzle as $(\mathsf{Ext}_{\lambda_2}(\psi_G(x + 2^h \cdot y), (s_1, s_2)), s_2, y)$. We assume the calibrated DLP is still moderately hard with respect to the min-entropy of x. Note that a similar assumption was used by Gennaro to construct a more efficient pseudo-random generator [Gen00]. It is easy to see that this assumption holds for DLP in generic groups, i.e. given $\psi_G(x + 2^h \cdot y)$ and y, the best generic algorithm must take at least $\sqrt{2^{h+1}\epsilon}$ steps to solve DLP with probability ϵ. We note that this problem is closely related to leakage-resilient cryptography [AM11, ADVW13].

On the other hand, due to the out-layer extractor, we cannot directly adopt any known (generic) DLP algorithms, such as [GTY07, GPR13]. Instead, our puzzle solver just exhaustively searches for a valid solution. There is a subtle caveat, namely the expected running time of solving a puzzle with hardness factor$h$, i.e. $x \leftarrow \{0,1\}^h$ is designed to be 2^h, whereas the TCR property of UOWHF is only guaranteed against PPT adversaries with respect to λ_2 (the security parameter of the UOWHF). To address this issue, we introduce an additional assumption, that is the expected running time of any adversary $\mathcal{A}$ (in number of steps) can break the TCR property of the underlying UOWHF with non-negligible probability on $x \leftarrow \{0,1\}^h$ is $\omega(2^{h/2})$, (i.e. breaking TCR is expected to happen after the birthday paradox bound). The dense puzzle system from DLP (combining with TCR strong extractors) is depicted in Fig. 4.

Theorem 4. *Let $\lambda \in \mathbb{Z}^+$ be the security parameter and $h \in [\log^4 \lambda + \log^2 \lambda + 1, \log^5 \lambda]$ be the hardness factor. Let $\mathsf{Ext}_\lambda : \{0,1\}^\lambda \times \{0,1\}^{3\lambda} \mapsto \{0,1\}^{\lambda + \log^4 \lambda}$ be a TCR strong extractor such that the expected running time of any adversary $\mathcal{A}$ that breaks its TCR property with non-negligible probability on $x \leftarrow \{0,1\}^h$ is $\omega(2^{h/2})$. Assume $\psi_G : \mathbb{Z}_p \mapsto \mathbb{G}$ is a hard DLP in generic groups such that the best generic algorithm must take at least $\sqrt{2^{h+1}\varepsilon}$ steps to solve it with probability ε. The puzzle system* $\mathsf{PuzSys} = (\mathsf{Sample}, \mathsf{SampleSol}, \mathsf{Solve}, \mathsf{Verify})$ *described in Fig. 4 is correct, complete with* Solve*'s running time 2^h, efficiently samplable, statistically indistinguishable, and g-hard, where $g(T) = T^{1/c}$ for any constant $c > 2$. In addition, for any k that is $O(2^{\log^3 \lambda})$,* PuzSys *is $(\mathsf{id}(\cdot), k)$-amortization resistant, where $\mathsf{id}(\cdot)$ is the identity function.*

Proof. Please see the full version [BKZZ15].

Define $\mathcal{PS}_\lambda = \{0,1\}^{7\lambda/2+\log^4\lambda}$, $\mathcal{SS}_\lambda = \{0,1\}^{\log^4\lambda}$, and $\mathcal{HS}_\lambda = [\log^4\lambda + \log^2\lambda + 1, \log^5\lambda]$. For the given λ, select a pre-defined $\mathsf{Ext}_\lambda : \{0,1\}^\lambda \times \{0,1\}^{3\lambda} \mapsto \{0,1\}^{\lambda+\log^4\lambda}$. Set the DLP $\psi_G : \mathbb{Z}_p \mapsto \mathbb{G}$ over the pre-defined elliptic curve, where p is λ-bit prime such that there exists an efficient injective map $\zeta : \mathbb{G} \mapsto \{0,1\}^\lambda$. (We will omit this map ζ in the rest of the description for notation simplicity.)

- $\mathsf{Sample}(1^\lambda, h)$: Return $\mathsf{puz} \leftarrow \{0,1\}^{7\lambda/2+\log^4\lambda}$.
- $\mathsf{SampleSol}(1^\lambda, h)$:
  - Pick random $s_1 \leftarrow \{0,1\}^\lambda$, $s_2 \leftarrow \{0,1\}^{2\lambda}$, $x \leftarrow \{0,1\}^h$ and $y \leftarrow \{0,1\}^{\lambda/2}$.
  - Return $\mathsf{puz} = (\mathsf{Ext}_\lambda(\psi_G(x + 2^h \cdot y), (s_1, s_2)), s_2, y)$ and $\mathsf{soln} = x$.
- $\mathsf{Solve}(1^\lambda, h, \mathsf{puz})$:
  - Parse puz to (z, s_1, s_2, y); set $\mathsf{soln} = \bot$ and initialize an empty set X.
  - For $ctr = \{1, \ldots, 2^h\}$:
    - ∘ Randomly pick $x \leftarrow \{0,1\}^h \setminus X$, and add x to X.
    - ∘ Set $\mathsf{soln} = x$ if $z = \mathsf{Ext}_\lambda(\psi_G(x + 2^h \cdot y), (s_1, s_2))$.
  - Return soln.
- $\mathsf{Verify}(1^\lambda, h, \mathsf{puz}, \mathsf{soln})$: Parse puz to (z, s_1, s_2, y). Return true if and only if $z = \mathsf{Ext}_\lambda(\psi_G(\mathsf{soln} + 2^h \cdot y), (s_1, s_2))$.

Fig. 4. The Dense Puzzle system From DLP.

Remark. For notation simplicity, we let the puzzle space "independent" of the hardness factor h, therefore we have to limit h within a small interval to ensure (i) $\psi_G(x + 2^h \cdot y)$ has enough entropy and (ii) it is infeasible to break the TCR property of the underlying UOWHF within $2^{h/2}$ steps. In practice, for any desired h, we can always pick a suitable $\mathsf{Ext}_\lambda : \{0,1\}^\lambda \times \{0,1\}^{3\lambda} \mapsto \{0,1\}^{\lambda+h-\log^2\lambda-1}$.

3.6 Instantiation of the Dense Puzzle Based PoWorK

We instantiate our PoWorK protocol as described in Fig. 1 by building it upon the Schnorr identification scheme [Sch89] and the dense puzzle system instantiation in the RO model[6] (see Sect. 3.4). The description of our instantiation is presented in the full version of this work [BKZZ15].

4 Applications

Below we present some practical and theoretical applications of our PoWorK. When using PoWorK in practice we must ensure that the verifier cannot distinguish between the two types of provers based on their response time. In Sect. 2.2 we argued that for our indistinguishability proofs, $\mathcal{P}(w)$ (i.e. the prover who knows the witness) should perform some idle steps so that his running time will be lower bounded by the time that one would need to solve the puzzle. However,

[6] The construction using the DLP based puzzle system is similar. We chose to employ the RO instantiation for simplicity in presentation.

enforcing a real user to wait is not ideal. Luckily though, the time needed for a prover who solves a puzzle (i.e., does not know the witness) depends on his total computational power and on whether the puzzle is parallelizable or not. Provers who own specialized hardware (e.g., based on ASICs) or that have access to powerful computer clusters (in case that a puzzle is parallelizable) might be able to solve the puzzle very fast – paying of course the relevant computation cost. Thus, when applying PoWorK in practice, the time that takes a prover to respond to a challenge is not a distinguishing factor: the prover might have as well solved the puzzle in constant time by fully parallelizing its computation or alternatively, for the case of non-interactive PoWorK's the receiver may not know when the prover started proof computation. Finally note that in any case, we do care that the prover has paid the corresponding computational cost and he is not able to amortize a previous solution of a puzzle to solve a new one.

4.1 Email Spam Application

Using proofs of work to reduce the amount of spam email was suggested back in 1992 by Dwork and Naor [DN92]. Their idea can be summarized in the following:

> "If I don't know you and you want to send me a message, then you must prove that you spent, say, ten seconds of CPU time, just for me and just for this message" [DN92].

In their proposal there exists some special software[7] that operates on behalf of the receiver and checks whether the sender has properly computed the proof of work or the sender is an *approved* (by the receiver) *contact*. The reason that this approach helps to reduce spam is mainly economic: in order for spammers to send high volumes of emails they would have to invest in powerful computational resources which makes spamming non cost-effective.

A disadvantage of the method described above is that the list of the approved contacts (i.e. email addresses) of the receiver has to be given to this special software/mail server in order to check whether the sender belongs in this list or not - in which case she will have to perform additional computation. This violates the privacy of the receiver who needs to reveal which of her contacts she considers to be approved and thus allows them to send emails "for free". Adopting our PoWorK protocol would give a *privacy preserving* solution to the spam problem: given the indistinguishability feature of PoWorK, the software/verifier does not need to know the approved list of contacts, in fact it does not even need to know whether the incoming email is from an approved contact or a non-approved user who successfully fulfilled the computational work.

Non-interactive PoWorKs. Sending an email should not require any extra communication between the sender and the mail server. Our 3-move PoWorK is public-coin, thus can be turned into non-interactive by applying the Fiat-Shamir transformation [FS86]. Namely, the prover, instead of receiving a challenge from

[7] This special software could for example run on the receiver's mail server or be an independent program running on the receiver's side.

the verifier, hashes the first move message **a** together with the context of the email and the email address of the receiver into **c**, and provides the verifier with the whole proof, π, which includes $(\mathbf{a}, \mathbf{c}, \mathbf{r})$ and the context of the email, in one round.

Multi-witness Hard Relation. In order for a user to approve a list of contacts she will have to provide each one of them with a unique witness for the same statement (in order to ensure indistinguishability). Let $R_{\mathcal{L}}$ be a multi-witness hard relation with a trapdoor for a language $\{x \mid \exists w : (x, w) \in R_{\mathcal{L}}\}$. A relation is said to be hard if for $(x, w) \in R_{\mathcal{L}}$, a PPT adversary given x can only output w' s.t. $(x, w') \in R_{\mathcal{L}}$ with negligible probability. A multi-witness hard relation *with a trapdoor* is described by the following algorithms: (a) a trapdoor generation algorithm sets a pair of a statement x and associated trapdoor t: $(x, t) \leftarrow \mathsf{GenT}(R_{\mathcal{L}})$, (b) an efficient algorithm GenW that on input $x \in \mathcal{L}$ and a trapdoor t outputs a witness w such that $(x, w) \in R_{\mathcal{L}}$ and, (c) a verification algorithm $1/0 \leftarrow \mathsf{Ver}(R_{\mathcal{L}}, x, w)$ outputs 1 if $(x, w) \in R_{\mathcal{L}}$ and 0 otherwise[8].

PoWorK Based Spam Reducing System. Consider a PoWorK scheme as presented in Fig. 1 for a security parameter λ, a puzzle system PuzSys and a multi-witness hard relation with a trapdoor $R_{\mathcal{L}}$ as described above. A spam reducing system SRS consists of the following algorithms:

- *MailServerSetup*(1^λ): the mail server $\mathcal{S}_{mail}$ on input the security parameter, λ, selects the hardness of the puzzle system $h \in \mathcal{HS}_\lambda$.
- *ReceiverSetup*$(1^\lambda, h)$: user $\mathcal{R}$ (i.e. the receiver) runs $(x, t) \leftarrow \mathsf{GenT}(R_{\mathcal{L}}$ and sends x and her email address $ad_{\mathcal{R}}$ to the mail server (potentially signed together). The trapdoor t is secretly stored by $\mathcal{R}$.
- *ApproveContact* (t, x): in order for $\mathcal{R}$ to approve a sender $\mathcal{S}$, it will run $w \leftarrow \mathsf{GenW}(t, x)$ and will give $w \in R_{\mathcal{L}}(x)$ to the sender (unique witnesses allow for revocation). From now on, $\mathcal{S}$ can use w to send emails to $\mathcal{R}$.
- *SendEMail*(w, h, x): a sender $\mathcal{S}$ with input the public parameters v, statement $x \in \mathcal{L}$ and with a private input $w \in R_{\mathcal{L}}(x) \cup \{\perp\}$, prepares a PoWorK proof $\pi = (\mathbf{a}, \mathbf{c}, \mathbf{r})$. If $\mathcal{S}$ is an approved contact of $\mathcal{R}$, then she will use the witness w to perform the PoK side of PoWorK, while if $\mathcal{R}$ is not an approved contact (i.e. $w = \perp$) she will have to execute the PoW side. To compute π non-interactively she will fix c to be $H(\mathbf{a}, m)$, where $\mathbf{a}$ is the first message of PoWorK, m stands for the body of the email[9], and H is a hash. The rest of PoWorKis computed as before.

[8] Examples of multi-witness hard relations with trapdoors are (a) the DL representation problem [Bra94, BF99] over prime order groups, (b) the representation problem in composite modular groups [ACJT00] which has constant size parameters in the number of adversarial parties.

[9] We can assume that the email body also contains a time-stamp (or that the time-stamp is added later by the mail server) and also includes $(ad_{\mathcal{S}}, ad_{\mathcal{R}})$ which are the sender/receiver email addresses.

- *ApproveEMail*(h, x, π): is run by the mail server $\mathcal{S}_{mail}$ who verifies π and outputs 0/1. If proof is π valid, then $\mathcal{S}_{mail}$ forwards the enclosed email to $\mathcal{R}$.

Note that our proposal, similar to [DN92, DGN03], requires to implement additional protocols between the sender and the recipient (i.e. a change in the internet mail standards would be required). In the full version of this work [BKZZ15] we discuss some interesting extensions of our protocol that address revocation, prevention of witness sharing and solving "useful" puzzles.

Security. Although a formal definition and description of properties of an email system is out of the scope of this paper, we do define and prove *spam resistance* and *privacy*. Briefly, spam resistance guarantees that the mail server will allow an email message to reach the recipient if and only if a valid proof (of work or knowledge) has been attached. At the same time for a non-approved contact the number of valid proofs of work prepared should not affect the time required to prepare a new one (similar to puzzle amortization property). Privacy implies that the mail server cannot distinguish whether the sender of a message is an approved contact of the recipient or not.

Definition 5. *Let* SRS *be a spam reducing system built upon a PoWorK* $(\mathcal{P}, \mathcal{V})$ *for a language* $\mathcal{L} \in \mathcal{NP}$ *and a puzzle system* PuzSys $=$ (Sample, Solve, Verify). *We define spam resistance and privacy of* SRS *as follows:*

(i). (σ, k)**-Spam Resistance:** *We say that* SRS *is* (σ, k)-spam resistant *if there exists a PPT witness-extraction algorithm* $\mathcal{K}$*, such that for every hardness factor* $h \in \mathcal{HS}_\lambda$*, auxiliary tape* $z \in \{0,1\}^*$ *and every adversary* $\mathcal{A}$*, if for non-negligible functions* $\alpha_1(\cdot), \alpha_2(\cdot)$*:*

$$\Pr\left[\begin{array}{l} (t,x) \leftarrow \mathit{ReceiverSetup}(1^\lambda, h); \forall 1 \le i \le k : \mathsf{puz}_i \leftarrow \mathsf{Sample}(1^\lambda, h); \\ \{\pi_i = (\mathbf{a}_i, \mathbf{c}_i, \mathbf{r}_i)\}_{i \in [k]} \leftarrow \mathcal{A}(z, 1^\lambda, h, x) : \\ (\forall 1 \le i \le k : \ \mathit{ApproveEMail}(h, x, \pi_i) = 1) \wedge \\ \wedge (\forall i \neq j \in [k] : \pi_i \neq \pi_j) \wedge \\ \wedge \Big(\mathsf{Steps}_{\mathcal{A}}(z, 1^\lambda, h, x) \le \sigma\big(\sum_{i=1}^{k} \mathsf{Steps}_{\mathsf{Solve}}(1^\lambda, h, \mathsf{puz}_i)\big)\Big) \end{array}\right] = \alpha_1(\lambda),$$

then $\Pr[\mathcal{K}^{\mathcal{A}}(z, 1^\lambda, h, x) \in R_{\mathcal{L}}(x)] = \alpha_2(\lambda)$.

(ii). **Privacy:** *We say that* SRS *is* private, *if for every hardness factor* $h \in \mathcal{HS}_\lambda$*, auxiliary tape* $z \in \{0,1\}^*$ *and every adversarial mail server* $\mathcal{A}$*, it holds that:*

$$\left| \Pr\left[\begin{array}{c} (t,x) \leftarrow \mathit{ReceiverSetup}(1^\lambda, h); w \leftarrow \mathit{ApproveContact}(t, x); \\ \pi \leftarrow \mathit{SendEMail}(w, h, x) : \mathcal{A}(z, h, x, \pi) = 1 \end{array}\right] - \right.$$
$$\left. - \Pr\left[\begin{array}{c} (t,x) \leftarrow \mathit{ReceiverSetup}(1^\lambda, h); \\ \pi \leftarrow \mathit{SendEMail}(\bot, h, x) : \mathcal{A}(z, h, x, \pi) = 1 \end{array}\right] \right| = \mathsf{negl}(\lambda).$$

We prove the following theorem for a *private spam reducing* email system:

Theorem 5. *Let* SRS *be a spam reducing system built upon dense puzzle-based PoWorK* $(\mathcal{P}, \mathcal{V})$ *for a* g*-hard and* (τ, k)*-amortization resistant dense puzzle system* PuzSys = (Sample, Solve, Verify)*, where* k *is polynomial in* λ*,* τ *is an increasing function and* g *is a subadditive function. Let* H *be a hash function with output domain equal to challenge sampling space* $\mathcal{CS}_\lambda$ *modeled as a random oracle. Assume that the worst-case running time of* $\mathsf{Solve}(1^\lambda, \cdot, \cdot)$ *is* $o(|\mathcal{CS}_\lambda|)$ *and that* $(\sqrt{\tau \circ g}(\mathsf{Solve}(1^\lambda, \cdot, \cdot))$ *is super-polynomial in* λ*. Then, the email system described above is* private *and* $(\sqrt{\tau \circ g}, k)$-spam resistant.

Proof. Please see the full version [BKZZ15].

Intuitively, the privacy holds because of the indistinguishability of PoWorK. The $(\sqrt{\tau \circ g}, k)$*-spam resistance* property holds because of the soundness of PoWorK and the amortization resistance of the underlying PuzSys.

4.2 PoWorK-Based Cryptocurrencies

Proofs of work is the basic primitive used in achieving the type of distributed consensus required in cryptocurrencies, notably Bitcoin [Nak08] and many others that use the same approach. The main idea is that a proof of work operation can be used to calibrate the ability of parties to build a hash chain that contains transaction records, commonly referred to as the blockchain.

An important feature of a blockchain is its decentralized nature. Given the view of a participant (commonly referred to as a miner) that includes its view of the blockchain, a fresh instance of a puzzle of a specified difficulty is created (which itself may depend on the blockchain) and has to be solved in order to add another block in the chain. Formally, the operation of a PoW-based miner as used in Bitcoin and numerous other cryptocurrencies (such as Litecoin, Namecoin, Dogecoin) is as shown in Fig. 5.

Let $\langle B_1, \ldots, B_n \rangle$ be the current blockchain where B_i is a tuple (t_i, T_i, u_i, π_i) with t_i a time-stamp, T_i a set of transactions, $u_i = H(B_{i-1})$ (for a hash function H) and π_i is such that $\mathsf{Verify}(1^\lambda, h_i, H(B_i), \pi_i) = \mathsf{true}$. The hardness h_i is calculated via a function operating on the time-stamps as follows $h_i = \mathsf{HC}(t_1, \ldots, t_{i-1})$. A new block B_{n+1} is created as follows.

1. **Collect transactions into a vector T_{n+1}.**
2. **Calculate $h_{n+1} = \mathsf{HC}(t_1, \ldots, t_n)$.**
3. **Set puz $= H(t_{n+1}, T_{n+1})$ where t_{n+1} is a current timestamp and run $\mathsf{Solve}(1^\lambda, h, \mathsf{puz})$ to produce a soln $= \pi_{n+1}$.**
4. **If the above step is successful, broadcast $B_{n+1} = (t_{n+1}, T_{n+1}, u_{n+1}, \pi_{n+1})$.**

Fig. 5. Miner operation in a puzzle-based cryptocurrency (using a puzzle PuzSys = (Sample, Solve, Verify) that is dense). $\mathsf{HC}(\cdot)$ is the puzzle hardness calculation function which depends on the timestamps of the blocks of the current blockchain.

Under certain assumptions about the network synchronicity and the hardness of the proof, the above mechanism has been shown to be robust in the sense of satisfying two properties, *persistence* (transactions remain stable in the "ledger") and *liveness* (all transactions are eventually inserted in the ledger) assuming that the honest parties are above majority [GKL15]. Puzzle-based cryptocurrencies have also drawn a lot of criticism due to the fact that they require a lot of natural resources (e.g., in [OM14] it is reported that Bitcoin mining in 2014 already consumed as much energy as the needs of the country of Ireland for electricity).

This lead to the development of a number of systems that circumvent puzzles (including, [DM16,BLMR14,Maz15] as well as Peercoin, DasHCoin, NXT, Nushares, ACHCoin, Faircoin and others). These systems maintain a blockchain as well, however they rely on different mechanisms for producing blocks. We call them, generically, "knowledge-based cryptocurrencies" since the production of a block is associated with the production of a witness for a public-relation relation $\mathcal{R}$ which parameterizes the system. Formally, we present the miner[10] operation in Fig. 6.

Let $\langle B_1, \ldots, B_n\rangle$ be the current blockchain where B_i is a tuple (t_i, T_i, u_i, π_i), for t_i, T_i, u_i defined as in Figure 5 and π_i being a NIZK that shows $x_i \in \{x \mid \exists w : (x, w) \in \mathcal{R}\}$, where $x_i = V(B_1, \ldots, B_{i-1}, t_i, T_i)$ for $i = 1, \ldots, n$. The miner, equipped with secret-key sk, produces the next block as follows.

1. **Collect transactions into a vector T_{n+1}.**
2. **Calculate the pair $(x_{n+1}, \mathsf{aux}) \leftarrow V(B_1, \ldots, B_n, t_{n+1}, T_{n+1})$ where t_{n+1} is the current time. Then calculate $W_{sk}(x_{n+1}, \mathsf{aux}) = w_{n+1}$. If $w_{n+1} \neq \perp$ it holds that $(x_{n+1}, w_{n+1}) \in \mathcal{R}$.**
3. **If the above step is successful, compute a NIZK proof π_{n+1} for x_{n+1} using witness w_{n+1}.**
4. **Broadcast $B_{n+1} = (t_{n+1}, T_{n+1}, u_{n+1}, \pi_{n+1})$.**

Fig. 6. Miner operation in a knowledge-based cryptocurrency parameterized by relation $\mathcal{R}$. The function $V(\cdot)$, given the blockchain information, the current set of transactions and the time-stamp produces a statement x, while the function $W_{sk}(\cdot)$ given a statement produces a witness w so that $(x, w) \in \mathcal{R}$.

A trivial way to construct a knowledge-based cryptocurrency would be to have a a single trusted authority with a public and secret key pair, (pk, sk), acting as the sole miner.[11] At a time-step $n+1$, the function $V(\cdot)$ would set simply $x_{n+1} = (t_{n+1}, T_{n+1}, u_{n+1})$ and $W_{sk}(x_{n+1})$ would produce a signature on x_{n+1} that would serve as π_{n+1} (there is no need for a NIZK). Another example of a knowledge-based cryptocurrency is NXT. On a high level, in this system each miner (called forger) has a digital signature public and secret key, (pk, sk), associated with her account. The function $V(B_1, \ldots, B_n, t_{n+1}, T_{n+1})$ (run by each

[10] Note that we use the term "miner" for symmetry. Miners are associated with puzzle based cryptocurrencies and thus different terminology has been introduced in knowledge-based systems including "mintettes", "forgers" and others.

[11] For instance, this would be a single "mintette" instantiation of [DM16].

miner), operates as follows: it parses T_{n+1} to recover the public pk of the miner (note that it is always present in the transaction collecting the fees). Then, based on the public-key pk and the blockchain $B_1, \ldots, B_n$ it determines how much currency is associated with the account that corresponds to the public-key pk; this results in a time-window $d \in \mathbb{R}^+$ whose expectation is proportionate to the amount of currency in the account (the more currency, the shorter the expectation of d is; we omit the exact dependency in this high level description). The function $V(\cdot)$ returns (x_{n+1}, aux) with $x_{n+1} = (t_{n+1}, T_{n+1}, u_n)$ and $\mathsf{aux} = d$. The procedure $W_{sk}(x_{n+1}, d)$, will produce a signature w on the message (t_{n+1}, T_{n+1}, u_n) if $t_{n+1} \geq t_n + d$; else, it produces $\perp$. Note that in this system no NIZK is employed, one may just set $\pi_{n+1} = w$; however, the system would operate similarly if a NIZK was employed to establish knowledge of a signature w on the message (t_{n+1}, T_{n+1}, u_n).

We now show how to construct a PoWorK-based cryptocurrency derived from a knowledge-based cryptocurrency C_1 and a puzzle-based cryptocurrency C_2 for a dense puzzle, see Fig. 7. The construction is straightforward: a new block can be added to the blockchain by someone who can efficiently compute a proof π_i using some secret key or by someone who is computing a π_i by performing computational work.

The properties of the composition are informally stated in the following (meta)-theorem; the proof of the theorem follows from the properties of PoWorK and is similar in spirit to the proof of Theorem 5. The formal statement and proof of the theorem (that should also include a formalization of all relevant underlying properties of cryptocurrencies, both in the puzzle-based and knowledge-based setting, e.g., in the sense of [GKL15]) is out of scope for the present exposition.

Let $\langle B_1, \ldots, B_n \rangle$ be the current blockchain where B_i is a tuple (t_i, T_i, u_i, π_i), for t_i, T_i, u_i defined as in Figure 5 and π_i being a non-interactive PoWorKthat demonstrates either the solution of the puzzle puz $= H(t_i, T_i)$ with hardness $h_i = R(t_1, \ldots, t_{i-1})$ or that $x_i \in \{x \mid \exists w : (x, w) \in \mathcal{R}\}$ where $x_i = V(B_1, \ldots, B_{i-1}, t_i, T_i)$.

1. **Collect transactions into a vector T_{n+1}.**
2. **If a secret-key sk is available, perform steps 2-3 of Figure 6 and follow the PoK direction of PoWorK(cf. Figure 1), using the $H(\cdot)$ to compute the challenge of the verifier.**
3. **Else, perform steps 2-3 of Figure 5 and follow the PoW direction of PoWorK(cf. Figure 1) using the $H(\cdot)$ to compute the challenge of the verifier.**
4. **Broadcast $B_{n+1} = (t_{n+1}, T_{n+1}, u_{n+1}, \pi_{n+1})$.**

Fig. 7. Miner operation in a PoWorK-based cryptocurrency parameterized by relation $\mathcal{R}$ and $\mathsf{PuzSys} = (\mathsf{Sample}, \mathsf{Solve}, \mathsf{Verify})$. The functions $V(\cdot), W_{sk}(\cdot)$ are as in Fig. 5 and the function $C(\cdot)$ is as in Fig. 6.

Theorem 6. *(informally stated) The cryptocurrency* C *of Fig. 7 is the composition of a knowledge-based cryptocurrency* C_1 *and a puzzle-based cryptocurrency*

C_2 so that (i) the population of miners of $\mathsf{C}_1, \mathsf{C}_2$ becomes a single set that is indistinguishable to any adversary that controls a subset of miners of C, (ii) the persistence property of C is upheld as long as the conditions for persistence of $\mathsf{C}_1, \mathsf{C}_2$ hold in conjunction. (iii) the liveness property of C is upheld as long as the conditions for liveness of $\mathsf{C}_1, \mathsf{C}_2$ hold in disjunction.

4.3 PoWorKs as 3-Move Straight-Line Concurrent Simulatable Arguments of Knowledge

In this section, we present a theoretical application of PoWorKs. Namely, we show that any PoWorK protocol that satisfies a couple of reasonable assumptions, implies straight-line concurrent ($\lambda^{\text{poly}(\log \lambda)}$)-simulatable arguments of knowledge. Our application is described at length in our full version [BKZZ15]. Here, we provide the statement of our main result.

Theorem 7. *Let $\mathcal{L}$ be a language in $\mathcal{NP}$ and let PuzSys be a puzzle system. Let $(\mathcal{P}, \mathcal{V})$ be a 3-move f-sound PoWorK for $\mathcal{L}$ and PuzSys with statistical indistinguishability such that for every hardness factor $h \in \mathcal{HS}_\lambda$, it holds that:*

(i). $\Pr[\mathsf{puz} \leftarrow \mathsf{Sample}(1^\lambda, h) : f(\mathsf{Steps}_{\mathsf{Solve}}(1^\lambda, h, \mathsf{puz})) \leq \lambda^{\log \lambda}] = \mathsf{negl}(\lambda)$.
(ii). The worst-case running time of $\mathsf{Solve}(1^\lambda, h, \cdot)$ is $\lambda^{\text{poly}(\log \lambda)}$ and $\mathcal{P}$ is a polynomial time algorithm that makes oracle calls to $\mathsf{Solve}(1^\lambda, h, \cdot)$.

Then, $(\mathcal{P}, \mathcal{V})$ is a 3-move straight-line concurrent statistically $\lambda^{\text{poly}(\log \lambda)}$-simulatable argument of knowledge.

Remark. In practice, we can instantiate the dense puzzle with a DL function over a dense elliptic curve [BHKL13] (without the need of an extractor). This means that we can transform a 3-move proof/argument of knowledge to a concurrent one with minimal computational overhead – 1 exponentiation for the prover and 1 exponentiation for the verifier. (cf. Fig. 1(a).) Note that a similar result in terms of rounds and with similar assumptions (i.e. DL) can be obtained via the efficient OR composition with an input-delayed Σ-protocol as recently observed in [CPS+16], however the resulting complexity overhead would be at least 3 exponentiations for the prover and 2 exponentiations for the verifier when the underlying Chameleon Σ-protocol is instantiated from Schnorr's protocol.

References

[ACJT00] Ateniese, G., Camenisch, J., Joye, M., Tsudik, G.: A practical and provably secure coalition-resistant group signature scheme. In: Bellare, M. (ed.) CRYPTO 2000. LNCS, vol. 1880, pp. 255–270. Springer, Heidelberg (2000). doi:10.1007/3-540-44598-6_16

[ADVW13] Agrawal, S., Dodis, Y., Vaikuntanathan, V., Wichs, D.: On continual leakage of discrete log representations. In: Sako, K., Sarkar, P. (eds.) ASIACRYPT 2013. LNCS, vol. 8270, pp. 401–420. Springer, Heidelberg (2013). doi:10.1007/978-3-642-42045-0_21

[AM11] Aggarwal, D., Maurer, U.: The leakage-resilience limit of a computational problem is equal to its unpredictability entropy. In: Lee, D.H., Wang, X. (eds.) ASIACRYPT 2011. LNCS, vol. 7073, pp. 686–701. Springer, Heidelberg (2011). doi:10.1007/978-3-642-25385-0_37

[Bac97] Back, A.: Hashcash (1997). http://www.cypherspace.org/hashcash

[BF99] Boneh, D., Franklin, M.: An efficient public key traitor tracing scheme. In: Wiener, M. (ed.) CRYPTO 1999. LNCS, vol. 1666, pp. 338–353. Springer, Heidelberg (1999). doi:10.1007/3-540-48405-1_22

[BGJ+16] Bitansky, N., Goldwasser, S., Jain, A., Paneth, O., Vaikuntanathan, V., Waters, B.: Time-lock puzzles from randomized encodings. In: ITCS (2016)

[BHKL13] Bernstein, D.J., Hamburg, M., Krasnova, A., Lange, T.: Elligator: elliptic-curve points indistinguishable from uniform random strings. In: CCS (2013)

[BKZZ15] Baldimtsi, F., Kiayias, A., Zacharias, T., Zhang, B.: Indistinguishable proofs of work or knowledge. Cryptology ePrint Archive, Report 2015/1230 (2015). http://eprint.iacr.org/2015/1230

[BLMR14] Bentov, I., Lee, C., Mizrahi, A., Rosenfeld, M.: Proof of activity: extending bitcoin's proof of work via proof of stake [extended abstract]. SIGMETRICS Perform. Eval. Rev. **42**(3), 34–37 (2014)

[Blu87] Blum, M.: How to prove a theorem so no one else can claim it. In: Proceedings of the International Congress of Mathematicians, pp. 1444–1451 (1987)

[BMC+15] Bonneau, J., Miller, A., Clark, J., Narayanan, A., Kroll, J.A., Felten, E.W.: Sok: research perspectives and challenges for bitcoin and crypto currencies. In: IEEE Symposium on Security and Privacy (2015)

[BR93] Bellare, M., Rogaway, P.: Random oracles are practical: a paradigm for designing efficient protocols. In: CCS (1993)

[Bra94] Brands, S.: An efficient off-line electronic cash system based on the representation problem. In: CWI Technical Report CS-R9323 (1994)

[Can01] Canetti, R.: Universally composable security: a new paradigm for cryptographic protocols. In: FOCS (2001)

[CDS94] Cramer, R., Damgård, I., Schoenmakers, B.: Proofs of partial knowledge and simplified design of witness hiding protocols. In: Desmedt, Y.G. (ed.) CRYPTO 1994. LNCS, vol. 839, pp. 174–187. Springer, Heidelberg (1994). doi:10.1007/3-540-48658-5_19

[CF01] Canetti, R., Fischlin, M.: Universally composable commitments. In: Kilian, J. (ed.) CRYPTO 2001. LNCS, vol. 2139, pp. 19–40. Springer, Heidelberg (2001). doi:10.1007/3-540-44647-8_2

[CGGM00] Canetti, R., Goldreich, O., Goldwasser, S., Micali, S.: Resettable zero-knowledge (extended abstract). In: STOC (2000)

[CMR98] Canetti, R., Micciancio, D., Reingold, O.: Perfectly one-way probabilistic hash functions (preliminary version). In: STOC (1998)

[CMSW09] Chen, L., Morrissey, P., Smart, N.P., Warinschi, B.: Security notions and generic constructions for client puzzles. In: Matsui, M. (ed.) ASIACRYPT 2009. LNCS, vol. 5912, pp. 505–523. Springer, Heidelberg (2009). doi:10.1007/978-3-642-10366-7_30

[CPS+16] Ciampi, M., Persiano, G., Scafuro, A., Siniscalchi, L., Visconti, I.: Improved OR-composition of sigma-protocols. In: Kushilevitz, E., Malkin, T. (eds.) TCC 2016. LNCS, vol. 9563, pp. 112–141. Springer, Heidelberg (2016). doi:10.1007/978-3-662-49099-0_5

[DGN03] Dwork, C., Goldberg, A., Naor, M.: On memory-bound functions for fighting spam. In: Boneh, D. (ed.) CRYPTO 2003. LNCS, vol. 2729, pp. 426–444. Springer, Heidelberg (2003). doi:10.1007/978-3-540-45146-4_25

[DM16] Danezis, G., Meiklejohn, S.: Centrally banked cryptocurrencies. In: NDSS (2016)

[DN92] Dwork, C., Naor, M.: Pricing via processing or combatting junk mail. In: Brickell, E.F. (ed.) CRYPTO 1992. LNCS, vol. 740, pp. 139–147. Springer, Heidelberg (1993). doi:10.1007/3-540-48071-4_10

[DNS98] Dwork, C., Naor, M., Sahai, A.: Concurrent zero-knowledge. In: STOC (1998)

[Dod05] Dodis, Y.: On extractors, error-correction and hiding all partial information. In: ITW (2005)

[DS05] Dodis, Y., Smith, A.: Correcting errors without leaking partial information. In: STOC (2005)

[FS86] Fiat, A., Shamir, A.: How To Prove Yourself: practical solutions to identification and signature problems. In: Odlyzko, A.M. (ed.) CRYPTO 1986. LNCS, vol. 263, pp. 186–194. Springer, Heidelberg (1987). doi:10.1007/3-540-47721-7_12

[FS90] Feige, U., Shamir, A.: Witness indistinguishable and witness hiding protocols. In: STOC (1990)

[Gen00] Gennaro, R.: An improved pseudo-random generator based on discrete log. In: Bellare, M. (ed.) CRYPTO 2000. LNCS, vol. 1880, pp. 469–481. Springer, Heidelberg (2000). doi:10.1007/3-540-44598-6_29

[GJKY13] Garay, J.A., Johnson, D.S., Kiayias, A., Yung, M.: Resource-based corruptions and the combinatorics of hidden diversity. In: ITCS (2013)

[GKL15] Garay, J., Kiayias, A., Leonardos, N.: The Bitcoin Backbone Protocol: analysis and applications. In: Oswald, E., Fischlin, M. (eds.) EUROCRYPT 2015. LNCS, vol. 9057, pp. 281–310. Springer, Heidelberg (2015). doi:10.1007/978-3-662-46803-6_10

[GMR85] Goldwasser, S., Micali, S., Rackoff, C.: The knowledge complexity of interactive proof-systems (extended abstract). In: STOC (1985)

[GO94] Goldreich, O., Oren, Y.: Definitions and properties of zero-knowledge proof systems. J. Cryptology **7**(1), 1–32 (1994)

[GPR13] Galbraith, S.D., Pollard, J.M., Ruprai, R.S.: Computing discrete logarithms in an interval. Math. Comput. **82**(282), 1181–1195 (2013)

[GTY07] Gopalakrishnan, K., Thériault, N., Yao, C.Z.: Solving discrete logarithms from partial knowledge of the key. In: Srinathan, K., Rangan, C.P., Yung, M. (eds.) INDOCRYPT 2007. LNCS, vol. 4859, pp. 224–237. Springer, Heidelberg (2007). doi:10.1007/978-3-540-77026-8_17

[HHR+10] Haitner, I., Holenstein, T., Reingold, O., Vadhan, S., Wee, H.: Universal one-way hash functions via inaccessible entropy. In: Gilbert, H. (ed.) EUROCRYPT 2010. LNCS, vol. 6110, pp. 616–637. Springer, Heidelberg (2010). doi:10.1007/978-3-642-13190-5_31

[HILL93] Håstad, J., Impagliazzo, R., Levin, L.A., Luby, M.: Construction of a pseudo-random generator from any one-way function. SIAM J. Comput. **28**, 12–24 (1993)

[JB99] Juels, A., Brainard, J.G.: Client puzzles: A cryptographic countermeasure against connection depletion attacks. In: NDSS (1999)

[LS90] Lapidot, D., Shamir, A.: Publicly verifiable non-interactive zero-knowledge proofs. In: Menezes, A.J., Vanstone, S.A. (eds.) CRYPTO 1990. LNCS, vol. 537, pp. 353–365. Springer, Heidelberg (1991). doi:10.1007/3-540-38424-3_26

[Maz15] Mazieres, D.: The stellar consensus protocol: A federated model for internet-levelconsensus (2015).https://www.stellar.org/papers/stellar-consensus-protocol.pdf

[MMV11] Mahmoody, M., Moran, T., Vadhan, S.: Time-lock puzzles in the random oracle model. In: Rogaway, P. (ed.) CRYPTO 2011. LNCS, vol. 6841, pp. 39–50. Springer, Heidelberg (2011). doi:10.1007/978-3-642-22792-9_3

[Nak08] Nakamoto, S.: Report (2008). https://bitcoin.org/bitcoin.pdf. Accessed 7 Oct. 2015

[NY89] Naor, M., Yung, M.: Universal one-way hash functions and their cryptographic applications. In: STOC (1989)

[OM14] ODwyer, K.J., Malone, D.: Bitcoin mining and its energy footprint. In: ISSC 2014/CIICT 2014 (2014)

[Pas03] Pass, R.: Simulation in quasi-polynomial time, and its application to protocol composition. In: Biham, E. (ed.) EUROCRYPT 2003. LNCS, vol. 2656, pp. 160–176. Springer, Heidelberg (2003). doi:10.1007/3-540-39200-9_10

[Pas04] Pass, R.: Alternative variants of zero-knowledge proofs. In: Licentiate (Master's) Thesis, ISBN 91-7283-933-3 (2004)

[Rom90] Rompel, J.: One-way functions are necessary and sufficient for secure signatures. In: STOC (1990)

[RSW96] Rivest, R., Shamir, A., Wagner, D.: Time-lock puzzles and timed-release crypto. Technical report (1996)

[Sch89] Schnorr, C.P.: Efficient identification and signatures for smart cards. In: Quisquater, J.-J., Vandewalle, J. (eds.) EUROCRYPT 1989. LNCS, vol. 434, pp. 688–689. Springer, Heidelberg (1990). doi:10.1007/3-540-46885-4_68

[Sho97] Shoup, V.: Lower bounds for discrete logarithms and related problems. In: Fumy, W. (ed.) EUROCRYPT 1997. LNCS, vol. 1233, pp. 256–266. Springer, Heidelberg (1997). doi:10.1007/3-540-69053-0_18

[Sim98] Simon, D.R.: Finding collisions on a one-way street: can secure hash functions be based on general assumptions? In: Nyberg, K. (ed.) EUROCRYPT 1998. LNCS, vol. 1403, pp. 334–345. Springer, Heidelberg (1998). doi:10.1007/BFb0054137

[SY90] Santis, A., Yung, M.: On the design of provably-secure cryptographic hash functions. In: Damgård, I.B. (ed.) EUROCRYPT 1990. LNCS, vol. 473, pp. 412–431. Springer, Heidelberg (1991). doi:10.1007/3-540-46877-3_37

[WJHF04] Waters, B., Juels, A., Halderman, J.A., Felten, E.W.: New client puzzle outsourcing techniques for dos resistance. In: CCS (2004)

Multi-party Computation

Size-Hiding Computation for Multiple Parties

Kazumasa Shinagawa[1,2(✉)], Koji Nuida[2,3], Takashi Nishide[1], Goichiro Hanaoka[2], and Eiji Okamoto[1]

[1] University of Tsukuba, Tsukuba, Ibaraki, Japan
shinagawa@cipher.risk.tsukuba.ac.jp
[2] AIST, Koto-ku, Tokyo, Japan
[3] Japan Science and Technology Agency (JST) PRESTO Researcher, Chiyoda-ku, Tokyo, Japan

Abstract. Lindell, Nissim, and Orlandi (ASIACRYPT 2013) studied feasibility and infeasibility of general two-party protocols that hide not only the contents of the inputs of parties, but also some sizes of the inputs and/or the output. In this paper, we extend their results to n-party protocols for $n \geq 2$, and prove that it is infeasible to securely compute every function while hiding two or more (input or output) sizes. Then, to circumvent the infeasibility, we naturally extend the communication model in a way that any adversary can learn neither the contents of the messages nor the numbers of bits exchanged among honest parties. We note that such "size-hiding"computation is never a trivial problem even by using our "size-hiding"channel, since size-hiding computation of some function remains infeasible as we show in the text. Then, as our main result, we give a necessary and sufficient condition for feasibility of size-hiding computation of an arbitrary function, in terms of which of the input and output sizes must be hidden from which of the n parties. In particular, it is now possible to let each input/output size be hidden from some parties, while the previous model only allows the size of at most one input to be hidden. Our results are based on a security model slightly stronger than the honest-but-curious model.

Keywords: Secure multiparty computation · Size-hiding

1 Introduction

Secure multiparty computation (MPC) protocols enable parties to compute a function while hiding the contents of the inputs from each other. Goldreich, Micali, and Wigderson [GMW87] first constructed a general MPC protocol in the presence of semi-honest and malicious adversaries. Here, we say that a protocol is general when it can securely compute *every* efficient function.

Most of the previous MPC protocols (implicitly) assume that the input sizes of parties may be revealed. However, the input sizes may be confidential in some settings. Let us consider the following situation: A police department has a list of suspected terrorists and each company has its customers' list. The police wants

J.H. Cheon and T. Takagi (Eds.): ASIACRYPT 2016, Part II, LNCS 10032, pp. 937–966, 2016.
DOI: 10.1007/978-3-662-53890-6_31

to know the intersection of the lists without revealing any information. However, if we straightforwardly utilize the standard MPC, there is no guarantee that the number of terrorists (i.e., input size) will be protected against companies, and this might cause a serious problem since the number of terrorists is often sensitive information. We may also consider the case where the police wants to hide the number of terrorists in customers' lists (i.e., output size) from companies. For resolving these issues, we require MPC that hides input and output sizes. This type of MPC is called *size-hiding computation.*

Currently, several size-hiding protocols have been proposed [MRK03,IP07, ACT11,CV12], but these protocols can compute only specific functionalities such as set intersection, homomorphic evaluation for branching programs, and database commitments. In 2013, Lindell, Nissim, and Orlandi [LNO13] exhaustively investigated feasibility and infeasibility of general size-hiding *two-party* protocols. They showed that, when the output size is not hidden, every efficient function can be securely computed while hiding one size (i.e., the input size of one party). Furthermore, they also proved that there is an efficient function that *cannot* be securely computed while hiding two sizes (i.e., either the input sizes of both parties, or the input size of one party and the output size). Recently, Chase, Ostrovsky, and Visconti [COV15] further strengthened the feasibility result of Lindell et al. by constructing a general size-hiding two-party protocol in the presence of malicious adversaries while hiding the input size of one party. However, these existing works investigated only the two-party setting, and therefore, feasibility and infeasibility of size-hiding n-party computation for $n > 2$ are still not clear.

1.1 Our Results

In this paper, we study general size-hiding $n(\geq 2)$-party protocols in the presence of static and semi-honest adversaries corrupting up to $n-1$ of the n parties. For a technical reason, our semi-honest model is slightly stronger than the standard honest-but-curious model. (See the last paragraph in this Section and Appendix.) To clarify our results, we classify size-hiding computations as *size-hiding classes* according to which of the input sizes and the output size must be hidden from which of the n parties. We note that, as in the previous work on two-party cases, we assume that every party wants to compute a *common* function. To study generalized settings is a future research plan.

Our results in the secure channel model. We extend the two-party results [LNO13] into multiparty settings in the *secure channel model*, in which an adversary cannot learn the contents of messages exchanged among honest parties, but may learn the number of bits of the messages. In the multiparty setting, the inputs and the output sizes can be hidden from a subset[1] of parties. See Table 1 (part corresponding to secure channel) for a summary. As the feasibility

[1] An input or the output size cannot be hidden from *all* parties because an input size is known to the holding party and we assume that at least one party obtains the output value.

results, when at most one input size is hidden from some parties, every efficient function can be securely computed (Lemma 3). The computation is also possible when the output size is hidden from some parties but then the input sizes are not hidden. On the other hand, when two or more sizes are hidden from some parties, there exists a function that *cannot* be securely computed (Lemmas 4 and 5).

Table 1. Our results (Sects. 4 and 5)

∘ Secure channel model			
	# of hidden input sizes	Output size	Feasible?
Lemma 3	≤ 1	known	yes
Lemma 4	≥ 2	known	no
(Trivial)	0	hidden	yes
Lemma 5	≥ 1	hidden	no
∘ Strong secure channel model			
	Condition for hidden sizes	Output size	Feasible?
Lemma 6	(A)	known	yes
Lemma 7	(A)	known	no
Lemma 8	(B)	hidden	yes
Lemma 9	(B)	hidden	no

(A) When all parties may learn the output size; for every pair of parties P_i and P_j, there is a party P_k (possibly $P_k = P_i$ or $P_k = P_j$) who may learn both input sizes of P_i and P_j. (B) When some parties must not learn the output size; for every party P_i who must not learn the output size, P_i may learn all the input sizes, and some other party may learn the input size of P_i and the output size.

For example, if two of n parties must hide their input sizes from each other, then a general size-hiding protocol is infeasible even when the other $n-2$ parties can support the computation. Our result assumes the existence of threshold fully homomorphic encryption (threshold FHE), which is, for example, derived by combining MPC with ordinary FHE; see Appendix A of [LNO13]. The above result shows that *almost all sizes of inputs and the output must be revealed in the standard setting of MPC.*

Our results in the strong secure channel model. In order to circumvent the aforementioned infeasibility, we introduce a new communication model, a *strong secure channel model* such that an adversary cannot learn even the number of bits exchanged among honest parties. We note that this model is justified from steganographic techniques [Cachin04, HAL09], i.e., if communications are hidden from other parties using steganography, an adversary cannot learn the number of communication bits between uncorrupted parties. Moreover, secure steganography is implied by one-way functions, thus, our new model requires no additional assumption inherently. (However, it should also be noted that straightforward implementation of steganography requires large computational and communication cost.)

We show that the feasibility of size-hiding computations is dramatically improved in the strong secure channel model. See Table 1 (part corresponding to strong secure channel) for a summary of our main result. We prove that, in the strong secure channel model, a general size-hiding protocol exists if either the condition (A) holds when the output size is known to all parties (Lemma 6) or the condition (B) holds when the output size is hidden from some parties (Lemma 8). (Unlike our results in the secure channel model, these conditions depend on *what sizes a party may learn.*) We also prove the reverse direction, i.e., there is a function that cannot be securely computed if a given size-hiding class does not satisfy the conditions above (Lemmas 7 and 9). Therefore, it is a necessary and sufficient condition for a general size-hiding protocol.

Surprisingly, in contrast to the standard secure channel model, we show that each input/output size can be hidden from some parties, while the previous model only allows the size of at most one input to be hidden. For example, let us consider the case of three parties where P_1 hides $|x_1|$ from P_2 (but not P_3), P_2 hides $|x_2|$ from P_3 (but not P_1), and P_3 hides $|x_3|$ from P_1 (but not P_2), where $|x_i|$ denotes the size of the input of P_i. Now the number of hidden sizes (three) is beyond the limitation in the previous model mentioned above, but our new model allows computation of a general function even in this case. By generalizing this observation, we see that there are concrete cases where it is possible to hide *all* input and output sizes for any $n > 2$.

The honest-but-randomness-controlling model. In the two-party setting, [LNO13] classified size-hiding classes in terms of feasibility in the honest-but-curious (HBC) model. Recently, [LNO13] (uploaded on IACR ePrint Archive on 01-Apr-2016) revisited that some of their infeasibility results in fact holds in the *honest-but-deterministic* (HBD) model, proposed by Hubácek and Wichs [HW15], rather than the HBC model. In light of the revision, we have also to modify the model since some of our results are based on the results in [LNO13]. However, there is an issue that the HBD model is likely to be incomparable with the HBC model. Alternatively, we introduce a new model, the *honest-but-randomness-controlling* (HBRC) model, where an adversary can use any string as its random tape. We believe that the HBRC model would be a reasonable security model by the following reasons. First, the HBRC model is stronger than the HBC model, i.e., the security in the HBRC model implies the security in the HBC model (see Appendix). Moreover, almost all of the previous standard protocols in the HBC model are also secure in the HBRC model. In particular, all (in)feasibility results in the two-party setting [LNO13] still hold in the HBRC model by an easy observation. We left it as an open problem to give a complete feasibility characterization of both two-party and multiparty settings in the HBC model.

1.2 Our New Techniques

In this section, we clarify the most technical part of size-hiding multiparty computations, and introduce the basic idea for our main results (Sect. 5).

First, we recall the general *two-party* computation [LNO13]. In their protocols, at least one of the parties *always* learns *all* sizes (i.e., the input size of the other party and the output size). This party can correctly compute any function of input x_1 and x_2 by using FHE. However, in the multiparty setting, we cannot assume the existence of such a party who may learn all sizes, and otherwise, almost all sizes of inputs cannot be protected.

For circumventing the above problem, we develop new techniques which guarantee the computation of the correct output even under the situation where no party knows all of sizes. Our techniques are based on a novel way to use a threshold FHE, and we consider this is the main non-trivial part of this work. More specifically, we propose two independent techniques which handle the following two different cases: **(1)** all parties may learn the output size, and **(2)** some parties must not learn the output sizes. In the rest of this subsection, we explain them more in detail.

(1) The case of public output size. Suppose that parties wish to compute a function while hiding some input sizes, but do not need to hide the output size. In addition, we assume that for every pair of parties P_i and P_j, at least one of n parties (including P_i and P_j) may learn both input sizes of P_i and P_j. In this setting, we call the party who has a longest input a *server*, and the other parties *clients*. In the protocol, all parties perform in the same way as the server since nobody (even the server itself) knows who is the server. We overview the protocol and show the idea behind it as follows.

First, all parties invoke a threshold key generation protocol of FHE. Next, each pair of parties P_i and P_j exchange ciphertexts of their inputs with the support from P_k who may learn both $|x_i|$ and $|x_j|$ as follows. Without loss of generality, we can assume $|x_i| \geq |x_j|$. First, P_i and P_j send ciphertexts $c_i = \mathsf{Enc}_{pk}(1||x_i)$ and $c_j = \mathsf{Enc}_{pk}(1||x_j)$ to the party P_k, respectively. Then, P_k computes a ciphertext $c_{(j,i)} = \mathsf{Enc}_{pk}(0^{|x_i|-|x_j|})||c_j$ and a ciphertext of zeroes $c_{(i,j)} = \mathsf{Enc}_{pk}(0^{|x_j|+1})$, and sends $c_{(j,i)}$ to P_i and $c_{(i,j)}$ to P_j, respectively. We call the former ciphertext a *valid ciphertext*, and the latter all-zero ciphertext a *dummy ciphertext*. Note that nobody except P_k knows whether a ciphertext is the valid one or the dummy one, due to the security of FHE. Next, parties attempt to obtain the output value using homomorphic computation. However, each party cannot estimate the output size since he/she does not know all of input sizes. Thus, a circuit which computes the output value cannot be constructed (the number of output wires is unknown). To avoid the problem, parties first obtain the output size and then compute the desired function as follows. Each party P_i constructs a circuit that takes $x'_1, \cdots, x'_n$ (x'_j is either $0^{|x_i|-|x_j|}||1||x_j$ or $0^{|x_i|+1}$) as inputs, and if one of inputs is all-zero, then outputs all-zero string, otherwise, outputs a representation of the size of the function value $|f(\vec{x})|$ of appropriate length. (For example, $(\log \kappa)^2$ bits, where κ is security parameter, can be used since it must hold $|f(\vec{x})| < 2^{(\log \kappa)^2}$ for sufficiently large κ. See also the discussion at the end of Sect. 4.2.) Each party P_i computes ciphertext c_i^{size} by homomorphic evaluation of the circuit from the ciphertexts $c_{(1,i)}^{\mathsf{in}}, \cdots, c_{(n,i)}^{\mathsf{in}}$. Then, each party P_i sends c_i^{size} to all parties. Each party computes c^{size} by homomorphic evaluation of a max function from $c_1^{\mathsf{size}}, \cdots, c_n^{\mathsf{size}}$. The underlying message of c^{size} is the output size since one of the party (specifically, the server) correctly computed the encrypted output size. All parties invoke threshold decryption protocol for the ciphertext c^{size} and obtain the output size. Now we have the output size and thus can construct the circuit that computes the function. In the similar way, all parties can compute c^{out} from which the output value is decrypted. The full description appears in Protocol 2 (Sect. 5.3).

(2) The case of private output size. Suppose that parties wish to compute a function while hiding some input sizes, and some parties must not learn the output size. In this setting, we call parties who must not learn the output size *servers*, and the other parties *clients*. In addition, we assume that every server may learn all input sizes of parties, and each server may tell its input size to some client (we call such a client a *partner*). We overview the protocol and show the idea behind it as follows.

First, all *clients* execute a threshold key generation protocol of FHE. The reason why servers are not involved in the threshold key generation is that the clients need to be able to decrypt an output ciphertext (whose plaintext length is related to the output size) without servers. Then, every party computes secret shares of its own input (the number of shares is the number of servers), and sends a ciphertext of a share to each server. All servers securely compute ciphertexts c^{size}, whose message is the output size, and c^{out}, whose message is the output, over the FHE. Here, the plaintext for the ciphertext c^{out} is padded zeroes up to L bits, where L is an upper bound of the output size. Note that for every polynomial-time computable function f, there exists a polynomial $p(\cdot)$ such that $|f(x'_1, \cdots, x'_n)| < p(\max(|x'_1|, \cdots, |x'_n|))$ for all $x'_i \in \{0,1\}^*$. Thus, a server can compute the bound $L = p(\max(|x_1|, \cdots, |x_n|))$, since every server knows all input sizes. Next, one server attempts to send c^{size} and c^{out} to all clients, but the length of c^{out} is related to $p(\max(|x_1|, \cdots, |x_n|))$ and it may reveal the maximum input size (possibly private size) to clients. To avoid this, for each client, the server sends the ciphertext whose length only depends on sizes which the client may learn. Let σ_i be the maximum size which P_i may learn. The server sends the truncated ciphertext c^{out} of length $p(\sigma_i)$ to P_i. If a server has the longest input, then the partner learn the maximum input size. Otherwise, it is trivial that there is a client who learns the maximum input size. In any case, at least one of the clients has the ciphertext c^{out} which is not truncated. Then, all clients collaboratively decrypt c^{size} and obtain the output size ℓ. Finally, all clients decrypt ℓ-bit ciphertexts of c^{out} and obtain the output value. The full description appears in Protocol 3 (Sect. 5.5).

1.3 Related Works

As earlier results relevant to size-hiding computation, Micali, Rabin, and Kilian [MRK03] provided a zero-knowledge set, which is a commitment to a set S that hides also the cardinality of S, where the committer can prove $x \in S$ or $x \notin S$ for any string x. Ishai and Paskin [IP07] constructed a public key encryption scheme that can evaluate any branching program in such a way that the size of the program is hidden. These works concentrated on efficient realization of specific functionalities, while our work aims at clarifying the (in)feasibility of general size-hiding computation depending on a given size-hiding class. On the other hand, Chase and Visconti [CV12] showed the first size-hiding protocol in the presence of malicious adversaries for a specific task, secure database commitments. Chase, Ostrovsky and Visconti [COV15] strengthened the feasibility results of [CV12,LNO13] by constructing a general size-hiding two-party protocol in the presence of malicious adversaries while hiding input size of one party. In contrast, we only consider the honest-but-randomness-controlling adversaries in this paper. We leave constructions of size-hiding MPC protocols against malicious adversaries as future work.

2 Preliminaries

We review the basic notations and the definition of threshold FHE.

2.1 Basic Notations

Throughout this paper, we use the following notations: "$\mathbb{N}$" denotes the set of natural numbers, i.e., $\mathbb{N} = \{1, 2, 3, \cdots\}$. "$\log x$" denotes the logarithm of x with the base two, i.e., $\log_2 x$. "$x||y$" denotes the concatenation of x and y. "$|x|$" denotes the bit length of x. "$\emptyset$" denotes an empty set. If S is a finite set, then "$x \stackrel{\mathrm{U}}{\leftarrow} S$" denotes that x is chosen uniformly at random from S. If $\vec{v}$ is a vector, "$\vec{v}[i]$" denotes the i-th element of the vector. If $m = m_1 m_2 \cdots m_\ell \in \{0,1\}^\ell$ is a plaintext and Enc_{pk} is an encryption algorithm for 1-bit message, "$c = \mathsf{Enc}_{pk}(m)$" denotes a vector of ℓ ciphertexts $(c_1, c_2, \cdots, c_\ell)$, where c_i is a ciphertext $c_i = \mathsf{Enc}_{pk}(m_i)$. If $I = \{i_1, \cdots, i_t\}$ is a subset of $\mathbb{N}$, "x_I" denotes the set $x_I = \{x_{i_1}, \cdots, x_{i_t}\}$. If $I = (i_1, \cdots, i_t)$ is an element of $\mathbb{N}^t$, "x_I" denotes the vector $x_I = (x_{i_1}, \cdots, x_{i_t})$. If $\Phi = \{\Phi(x, \kappa)\}_{x,\kappa}$ and $\Psi = \{\Psi(x, \kappa)\}_{x,\kappa}$ are probability distributions indexed by $\kappa \in \mathbb{N}$ and $x \in X_\kappa$ where X_κ is an auxiliary parameter set indexed by κ, then we say that Φ and Ψ are computationally indistinguishable, denoted by "$\Phi \stackrel{\mathrm{c}}{\equiv} \Psi$", if for every non-uniform probabilistic polynomial-time (PPT) algorithm $\mathcal{D}$ and every (positive) polynomial p, there exists a number $\kappa_0 \in \mathbb{N}$ with the property that $\left|\Pr[\mathcal{D}(\Phi(x,\kappa)) = 1] - \Pr[\mathcal{D}(\Psi(x,\kappa)) = 1]\right| < 1/p(\kappa)$ for any $\kappa \in \mathbb{N}$ with $\kappa > \kappa_0$ and any $x \in X_\kappa$.

2.2 Threshold Fully Homomorphic Encryption

We present a definition of threshold FHE. Asharov et al. [AJL+12] constructed an efficient threshold FHE scheme from the learning with error assumption, whose threshold key generation and threshold decryption protocols have only one round. In general, the threshold version of FHE is implied from an ordinary FHE scheme [LNO13].

Definition 1 (Threshold FHE). *We say that a tuple of protocols and algorithms* ($\mathsf{ThrGen}, \mathsf{Enc}, \mathsf{Eval}, \mathsf{ThrDec}$) *is a* threshold FHE scheme *if* ($\mathsf{Gen}, \mathsf{Enc}, \mathsf{Dec}$) *is a public-key encryption with message space* $\{0, 1\}$*, that is secure under chosen-plaintext attacks, and the protocols* ThrGen *and* ThrDec *with parties* $P_1, \cdots, P_n$ *realize the following functionalities and the following conditions:*

Threshold Key Generation: *The functionality of* ThrGen *takes security parameter* 1^κ *from* $P_1, \cdots, P_n$*, computes* $(pk, sk) \leftarrow \mathsf{Gen}(1^\kappa)$ *and chooses uniformly random values* $sk_1, \cdots, sk_{n-1} \in \{0, 1\}^{|sk|}$*. Then, the functionality outputs* (pk, sk_i) *to each* P_i $(i = 1, \cdots, n)$*, where* $sk_n = sk_1 \oplus \cdots \oplus sk_{n-1} \oplus sk$.

Threshold Decryption: *For a subset* $I \subset \{1, \cdots, n\}$*, the functionality of* ThrDec_I *takes security parameter* 1^κ*, a ciphertext* c *and shares of secret key* $sk_1, \cdots, sk_n$ *from* $P_1, \cdots, P_n$*, computes* $m = \mathsf{Dec}_{sk_1 \oplus \cdots \oplus sk_n}(c)$*, and outputs* m *to each* P_i $(i \in I)$*. If it holds* $I = \{1, \cdots, n\}$*, we omit the index* I.

Correctness: *For every polynomial-size circuit* $\mathcal{C}$ *that takes* n *inputs, and every inputs of the circuit* $m_1, \cdots, m_n \in \{0, 1\}$*:*

$$\Pr\left[\mathsf{Dec}_{sk}(\mathsf{Eval}_{pk}(\mathcal{C}, \mathsf{Enc}_{pk}(m_1), \cdots, \mathsf{Enc}_{pk}(m_n))) = \mathcal{C}(m_1, \cdots, m_n)\right] = 1,$$

where the probability is taken over the random coins of all the algorithms ($\mathsf{Gen}, \mathsf{Enc}, \mathsf{Eval}, \mathsf{Dec}$).

Security of the Threshold Key Generation: *There exists a PPT* $\mathcal{S}_{\mathsf{ThrGen}}$ *such that for every* $I \subsetneq \{1, \cdots, n\}$*, the view in a real execution of* ThrGen *with security parameter* κ *is computationally indistinguishable from the output of* $\mathcal{S}_{\mathsf{ThrGen}}$ *with inputs* $I, 1^\kappa$ *and keys obtained by* P_i $(i \in I)$.

Security of the Threshold Decryption: *There exists a PPT* $\mathcal{S}_{\mathsf{ThrDec}}$ *such that for every* $I \subsetneq \{1, \cdots, n\}$*, the view in a real execution of* ThrDec *with security parameter* κ *is computationally indistinguishable from the output of* $\mathcal{S}_{\mathsf{ThrDec}}$ *with inputs a subset* I*, keys, the ciphertext and the decrypted value.*

3 Size-Hiding Computation

In this section, first, we give a definition of size-hiding classes and provide their graphical representations in Sect. 3.1. Second, as an extension of [LNO13] to n-party settings, we give definitions of polynomial-time protocols and the security of size-hiding protocols in Sect. 3.2. Next, for later references, we review the previous two-party results [LNO13] using our graphical representation in Sect. 3.3. Finally, we introduce tools for proving lemmas in the later section, *protocol compilers*, that can derive a size-hiding protocol from another protocol in Sect. 3.4.

3.1 Classes of Size-Hiding

We provide a definition of a class of size-hiding that specifies *what sizes a party may learn* in an execution of a protocol. A size-hiding class can be represented by $(G, \vec{v})$, where G is a directed graph which specifies how input sizes are hidden (more precisely, which input size may be known to which party), and $\vec{v}$ is a vector which specifies how the output size may be known to each party. A directed graph G with n vertices is called an *input size graph with n vertices*, a vector $\vec{v}$ with n elements is called an *output size vector with n elements*, and a tuple $(G, \vec{v})$ is called a *size-hiding class with n parties.*

An input size graph with n vertices has a set of vertices $V(G) = \{1, 2, \cdots, n\}$ and a set of edges $E(G)$. Each vertex $i \in V(G)$ corresponds to the party P_i. If there is an edge $(j, i) \in E(G)$ directed from j to i, the party P_i *may* learn $|x_j|$, which is the input size x_j of P_j in a protocol execution. If there is no edge (j, i), the party P_i *must not* learn any partial information of $|x_j|$ except trivial information which can be computed from other information that P_i obtained legally. From now on, we assume that any input size graph with n vertices has edges $(1,1), (2,2), \cdots, (n,n)$ since P_i always knows its own input size $|x_i|$.

An output size vector with n elements is a member of $\{\perp, |\mathsf{f}|, \mathsf{f}\}^n$, where $\perp$, $|\mathsf{f}|$ and f are symbols that represent how to receive the output information. The i-th element $\vec{v}[i]$ specifies how P_i receives the output information. If $\vec{v}[i] = \perp$, the party P_i *must not* receive any partial information of the output $f(\vec{x})$ (except trivial information which can be computed efficiently). If $\vec{v}[i] = |\mathsf{f}|$, the party P_i *may* learn the output size $|f(\vec{x})|$ but *must not* receive $f(\vec{x})$ beyond the size information $|f(\vec{x})|$ (except trivial information). If $\vec{v}[i] = \mathsf{f}$, the party P_i *must* learn $f(\vec{x})$. From now on, we assume that any output size vector $\vec{v}$ contains at least one f since if there is no f in $\vec{v}$, nobody obtains the output $f(\vec{x})$ even though the protocol aims at computing the function f.

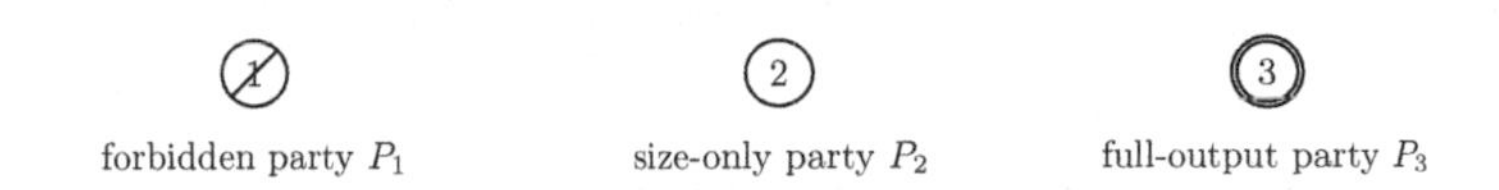

Fig. 1. Graphical representation of parties

We provide a graphical representation of a size-hiding class $(G, \vec{v})$. We use a circle to denote a vertex of G, and an arrow $i \to j$ to denote an edge $(i, j) \in E(G)$. For simplicity, we omit arrows $1 \to 1, 2 \to 2, \cdots, n \to n$ since the edges $(1,1), (2,2), \cdots, (n,n) \in E(G)$ always exist. We also use three types of circles to denote the output size vector $\vec{v}$ as follows. For a vertex i, we use a double circle to denote $\vec{v}[i] = \mathsf{f}$, a normal circle to denote $\vec{v}[i] = |\mathsf{f}|$, and a forbidden circle to denote $\vec{v}[i] = \perp$; see Fig. 1.

Figure 2 is an example of a size-hiding class $(G, \vec{v})$ as follows: The input size graph with 3 vertices G has a set of vertices $V(G) = \{1, 2, 3\}$ and a set of edges $E(G) = \{(1,2), (2,3), (1,3), (1,1), (2,2), (3,3)\}$. The output size vector

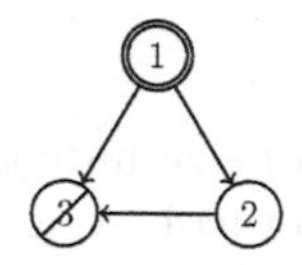

Fig. 2. Example of a size-hiding class

with 3 elements is a vector $\vec{v} = (\mathsf{f}, |\mathsf{f}|, \perp)$. The size-hiding class $(G, \vec{v})$ means the following: The party P_1 may learn $|x_1|$, must learn $f(\vec{x})$, and must not learn $|x_2|$ nor $|x_3|$. The party P_2 may learn $|x_1|, |x_2|$ and $|f(\vec{x})|$, and must not learn $|x_3|$ nor $f(\vec{x})$. The party P_3 may learn $|x_1|, |x_2|$, and must not learn $|f(\vec{x})|$.

Throughout this paper, we use the following terminology.

Public size is a size which all parties may learn. Formally, we say that an input size $|x_i|$ is public if there are edges $(i, 1), (i, 2), \cdots, (i, n) \in E(G)$, and the output size is public if the output size vector $\vec{v}$ is an element of $\{|\mathsf{f}|, \mathsf{f}\}^n$.

Private size is a size which some parties must not learn. Formally, we say that an input size $|x_i|$ is private if there is a vertex $j \in V(G)$ such that $(i, j) \notin E(G)$, and the output size is private if there is an index i such that $\vec{v}[i] = \perp$.

Forbidden party is a party who must not learn the output size. "$I_\perp$" denotes all indices of the forbidden parties, i.e., $I_\perp = \{I \mid \vec{v}[i] = \perp\} \subset \{1, \cdots, n\}$.

Size-only party is a party who may learn the output size but must not learn the exact output value. "$I_{|\mathsf{f}|}$" denotes all indices of size-only parties, i.e., $I_{|\mathsf{f}|} = \{I \mid \vec{v}[i] = |\mathsf{f}|\} \subset \{1, \cdots, n\}$.

Full-output party is a party who must learn the output. "I_f" denotes all indices of full-output parties, i.e., $I_\mathsf{f} = \{I \mid \vec{v}[i] = \mathsf{f}\} \subset \{1, \cdots, n\}$.

Permitted party is a party who may learn the output size. "I_p" denotes all indices of permitted parties, i.e., $I_\mathsf{p} = I_{|\mathsf{f}|} \cup I_\mathsf{f}$. It holds $I_\perp \cup I_\mathsf{p} = I_\perp \cup I_{|\mathsf{f}|} \cup I_\mathsf{f} = \{1, \cdots, n\}$.

3.2 Basic Notions for Size-Hiding Multiparty Protocols

Our definitions of notions for size-hiding n-party protocols follow the two-party version of [LNO13]. Let $(G, \vec{v})$ be a size-hiding class with n parties, and let f be an n-ary polynomial-time computable function $f : (\{0,1\}^*)^n \to \{0,1\}^*$. Let π be an n-party protocol with parties $P_1, \cdots, P_n$, and let $\kappa \in \mathbb{N}$ be a security parameter of π. Each party P_i has an input $x_i \in \{0,1\}^*$, which may be polynomially unbounded. We denote by $\mathrm{TIME}^{\pi}_{P_i}(\vec{x}, \kappa)$ the running time of P_i in π for the inputs $\vec{x} = (x_1, \cdots, x_n)$. We denote by $\mathrm{OUTPUT}^{(G,\vec{v},f)}_i(\vec{x})$ the P_i's output specified by $(G, \vec{v})$, e.g., for the example of Fig. 2, we have that $\mathrm{OUTPUT}^{(G,\vec{v},f)}_1(\vec{x}) = (f(\vec{x}))$, $\mathrm{OUTPUT}^{(G,\vec{v},f)}_2(\vec{x}) = (1^{|f(\vec{x})|}, 1^{|x_1|})$ and $\mathrm{OUTPUT}^{(G,\vec{v},f)}_3(\vec{x}) = (1^{|x_1|}, 1^{|x_2|})$. Now we are ready to define a polynomial-time protocol for $(G, \vec{v}, f)$.

Definition 2 (Polynomial-time protocol). *Let $(G, \vec{v})$ be a size-hiding class with n parties, let f be an n-ary function, and let π be an n-party protocol. We say that π is a polynomial-time protocol for $(G, \vec{v}, f)$ if there exists a polynomial $p(\cdot)$ such that for every $\kappa \in \mathbb{N}$, every $\vec{x} \in (\{0,1\}^*)^n$ and every $i \in \{1, \cdots, n\}$,*

$$\mathrm{TIME}^{\pi}_{P_i}(\vec{x}, \kappa) \leq p(|x_i| + |\mathrm{OUTPUT}^{(G,\vec{v},f)}_i(\vec{x})| + \kappa)$$

Next, we define the security of protocols against *honest-but-randomness-controlling* (HBRC) adversaries in the secure channel model (See Sect. 1.1 and Appendix for details of the HBRC model). In the HBRC model, a simulator must simulate a transcript on given random tapes which is produced by a *randomness producer.* It is a PPT algorithm that chooses corrupted parties' random tapes. Formally, we say that a PPT $\mathcal{R}$ is a *randomness producer* if $\mathcal{R}(1^\kappa, I)$ outputs a vector of strings $\vec{r}_I = (r_{i_1}, \cdots, r_{i_t}) \in (\{0,1\}^*)^{|I|}$ for all $I = \{i_1, \cdots, i_t\} \subsetneq \{1, \cdots, n\}$.

We denote by $\mathrm{MSIZE}^{\pi}(\vec{x})$ the numbers of all bits exchanged among $P_1, \cdots, P_n$ in an execution of π with inputs $\vec{x}$, expressed by unary expression such as $1^{|m|}$. The view of the party P_i during an execution of π with inputs $\vec{x}$ is defined as $\mathrm{view}^{\pi}_i(\vec{x}) = (x_i, r_i, m_{i_1}, \cdots, m_{i_t})$, where r_i is his internal coin tosses and m_{i_j} is the j-th message that was received by P_i in the protocol execution. We also use $\mathrm{view}^{\pi}_i(\vec{x})|_{r_i} = (x_i, m_{i_1}, \cdots, m_{i_t})$ to denote $\mathrm{view}^{\pi}_i(\vec{x})$ on given randomness r_i. Here, if the length of r_i is shorter than the length of its internal randomness, its internal randomness is $r_i||0^k$ for appropriate $k \in \mathbb{N}$.

Definition 3 (Security in the secure channel model). *Let $(G, \vec{v})$ be a size-hiding class with n parties, let f be an n-ary function, and let π be a polynomial-time protocol for $(G, \vec{v}, f)$. We say that π correctly computes $(G, \vec{v}, f)$ if for every $\kappa \in \mathbb{N}$, and every $\vec{x} \in (\{0,1\}^*)^n$, all full-output parties output $f(\vec{x})$ at the end of the execution of π with the input $\vec{x}$ and security parameter κ. We say that π realizes $(G, \vec{v}, f)$ in the* secure channel model *if π correctly computes $(G, \vec{v}, f)$ and for every randomness producer $\mathcal{R}$, there exists a PPT $\mathcal{S}$ such that for every $I \subsetneq \{1, \cdots, n\}$, every polynomials $q_1, q_2, \cdots, q_n$,*

$$\{\mathcal{S}(1^\kappa, I, \vec{x}_I, \mathrm{OUTPUT}^{(G,\vec{v},f)}_I(\vec{x}), \vec{r}_I \leftarrow \mathcal{R}(1^\kappa, I))\}_{\kappa,\vec{x}} \stackrel{c}{\equiv} \{(\mathrm{view}^{\pi}_I(\vec{x})|_{\vec{r}_I}, \mathrm{MSIZE}^{\pi}(\vec{x}))\}_{\kappa,\vec{x}}$$

where $x_1 \in \{0,1\}^{q_1(\kappa)}, \cdots, x_n \in \{0,1\}^{q_n(\kappa)}$.

In this paper, we focus on which size-hiding class has a general protocol. For a size-hiding class $(G, \vec{v})$, we say that $(G, \vec{v})$ is *feasible* if for every polynomial-time computable function f, there exists a protocol π that realizes $(G, \vec{v}, f)$ in the secure channel model. On the other hand, we say that $(G, \vec{v})$ is infeasible if it is not feasible.

3.3 Overview of the Two-Party Results

We overview the results in the two-party setting shown by Lindell, Nissim and Orlandi [LNO13] using our graphical representation. Later, we use them in order to prove infeasibility results in multiparty settings. We note that their original paper shows their feasibility and infeasibility against honest-but-curious adversaries. However, very recently, they (implicitly) revised the infeasibility of class 1.d is in fact holds against honest-but-randomness-controlling (HBRC) adversaries[2] rather than honest-but-curious adversaries. Since all of their protocol can be easily modified to the HBRC setting, the following results are based on the HBRC model.

They defined three classes of size-hiding: (class 0) the input sizes of both parties are revealed, (class 1) the input size of one party is revealed and the other is hidden, (class 2) the input sizes of both parties are hidden. In addition, they define five subclasses of class 1, and three subclasses of class 2.

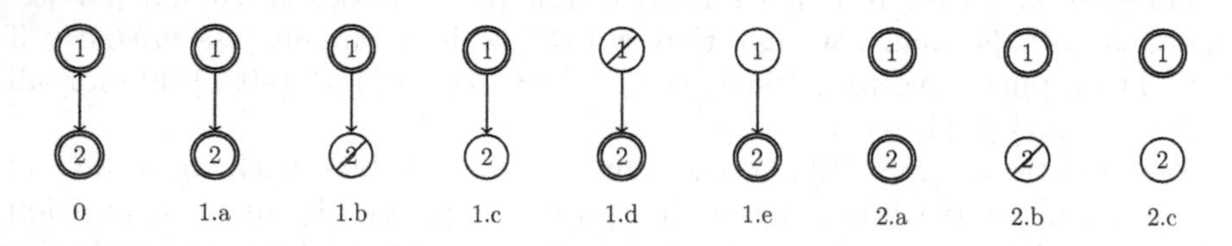

Fig. 3. Graphical representations of subclasses in the two-party setting

Let $(G_0, \vec{v}_0), (G_{1.a}, \vec{v}_{1.a}), \cdots, (G_{2.c}, \vec{v}_{2.c})$ be size-hiding classes $0, 1.a, \cdots, 2.c$ in Fig. 3, respectively. They (implicitly) showed that size-hiding classes $(G_0, \vec{v}_0)$, $(G_{1.a}, \vec{v}_{1.a}), (G_{1.c}, \vec{v}_{1.c})$ and $(G_{1.e}, \vec{v}_{1.e})$ are feasible while the other classes are infeasible in the HBRC model. Later in this paper, we use the following results.

- There is a two-ary function f such that the functionality $(G_{1.b}, \vec{v}_{1.b}, f)$ cannot be realized. An example of the f is the oblivious transfer; see Sect. 4.3.
- There is a two-ary function f such that the functionality $(G_{1.d}, \vec{v}_{1.d}, f)$ cannot be realized. An example of the f is an oblivious multi-input pseudorandom function evaluation omprf introduced in [LNO13].
- There is a two-ary function f with constant output length such that the functionality $(G_{2.a}, \vec{v}_{2.a}, f)$ cannot be realized. An example of the f is the binary inner product $\{0,1\}^* \times \{0,1\}^* \to \{0,1\}$; see [LNO13].

3.4 Tools for Infeasibility – Protocol Compilers

Here we introduce auxiliary algorithms used in the proofs of our infeasibility results, which we call *protocol compilers*. Namely, to give a proof by contradiction, we start with an n-party protocol for a given size-hiding class whose existence is assumed, and convert it by the protocol compilers into a two-party size-hiding protocol for some size-hiding class, where the existence of the latter protocol has been denied by the result of [LNO13]. Below we give two kinds of protocol compilers, which we call a *reduction compiler* and a *wrapping compiler*.

[2] Their original revision states that it holds against HBD adversaries. But it also holds in the HBRC model.

Reduction Compiler. A *reduction compiler* takes as inputs an MPC protocol with $P_1, \cdots, P_n$ and two subsets I_1, I_2 that is a partition of $\{1, \cdots, n\}$, and outputs a two-party protocol with P'_1 and P'_2, where P'_i ($i \in \{1,2\}$) behaves in the same way as $\{P_j\}_{j \in I_i}$. More concretely, if P_i computes/sends/receives messages in π, then P'_j ($i \in I_j$) behaves in the same way as P_i. At the end of the compiled protocol, if P_i outputs $f(\vec{x})$ in π, then P'_j ($i \in I_j$) outputs $f(\vec{x})$ in π'. The reason why we call it a "reduction" compiler is that it *reduces* the number of parties, and we use it in a *reduction* to prove infeasibility results.

Lemma 1. *Let f' be a two-ary function, let f be an n-ary function such that $f(x_1, x_2, \cdots, x_n) = f'(x_1, x_2)$, let $(G, \vec{v}, f)$ be a functionality for n parties, and let π be a protocol that realizes $(G, \vec{v}, f)$. Let I_1 and I_2 be non-empty subsets of $\{1, \cdots, n\}$ such that $1 \in I_1$, $2 \in I_2$, $I_1 \cap I_2 = \emptyset$ and $I_1 \cup I_2 = \{1, 2, \cdots, n\}$. There exists a protocol π' that realizes a functionality $(G', \vec{v}', f')$ as follows:*

- *The party P'_1 has the same input x_1 of P_1, and the party P'_2 has the same input x_2 of P_2.*
- *The input size graph with two parties G' has a set of edges $E(G')$ as follows. The edge $(1, 2)$ exists in $E(G')$ if and only if an edge $(1, i)$ exists in $E(G)$ such that $i \in I_2$. Similarly, the edge $(2, 1)$ exists in $E(G')$ if and only if an edge $(2, i)$ exists in $E(G)$ such that $i \in I_1$.*
- *$\vec{v}'$ is an output size vector with two elements such that $\vec{v}'[i] = \max_{j \in I_i}(\vec{v}[j])$, for an order $\perp < |\mathsf{f}| < \mathsf{f}$.*

Proof. Based on a simulator $\mathcal{S}$ of the protocol π, we construct a simulator $\mathcal{S}'$ of the protocol π'. By the symmetry, it suffices to show the simulator when P'_1 is corrupted. Given 1^κ, $I' = \{1\}$, the input x_1, the output $f(x_1, x_2)$ if there is an full-output party P_i ($i \in I_1$), and a random tape r_1 produced by a randomness producer, $\mathcal{S}'$ invokes $\mathcal{S}$ on the same inputs except I_1 instead of I'. Since the simulator $\mathcal{S}'$ works correctly, the protocol π' securely computes $(G', \vec{v}', f')$. $\square$

Wrapping Compiler. For a subset $I \subset \{1, \cdots, n\}$, we say that a protocol π is I-independent[3] if there exists a polynomial p such that for every $\kappa \in \mathbb{N}$ and every $\vec{x} \in (\{0,1\}^*)^n$, the output size and the number of bits, exchanged among all parties in an execution of π with κ and $\vec{x}$, are upper bounded by $p(\kappa, |x_{j_1}|, \cdots, |x_{j_t}|)$ except negligible probability, where $\{j_1, \cdots, j_t\} \cap I = \emptyset$. A *wrapping compiler* takes an I-independent protocol π (it is not necessary for π to be secure) that computes $f(\vec{x})$, and outputs a size-hiding protocol π' that computes $f(\vec{x})$ while hiding the inputs $|x_i|$ ($i \in I$) from all parties. It is used in the proof of Lemma 4. The following lemma is the security of a protocol that is compiled by the wrapping compiler.

Lemma 2. *Let I be a non-empty subset of $\{1, \cdots, n\}$, and let π be an I-independent protocol computing $f(\vec{x})$ with $P_1, \cdots, P_n$. Assume that threshold FHE exists. There exists a protocol π' that realizes a functionality $(G', \vec{v}', f)$ as follows:*

[3] It is an generalization of *size independent protocol*; see Sect. 4.3 in [LNO13].

- *The party* P'_i $(i \in \{1, \cdots, n\})$ *has the same input* x_i *of* P_i.
- G' *is an input size graph with* n *parties, where* $\{|x_i|\}_{i \in I}$ *are private sizes and the others are public sizes.*
- $\vec{v}'$ *is any element of* $\{|\mathsf{f}|, \mathsf{f}\}^n$.

Proof. Given an I-independent protocol π that computes $f(\vec{x})$, the compiled protocol π' with $P'_1, \cdots, P'_n$ proceeds as follows. First, parties execute a threshold key generation protocol of threshold FHE, and each party encrypts own input under the public key. Second, every party P'_j $(j \notin I)$ sends $|x_j|$ to all parties. On receiving $|x_{j_1}|, \cdots, |x_{j_t}|$ (they are not independent sizes), parties compute the upper bound $B = p(|x_{j_1}|, \cdots, |x_{j_t}|)$. Since the communication complexity is bounded by B, each party P_i can construct a circuit that can produce the next messages of P_i. More concretely, for each k round (k is also bounded by B), the circuit takes as inputs previous messages that are received by P_i at $1, 2, \cdots, k-1$ rounds and P_i's input x_i, and outputs the next messages for each party. Using these circuits, all parties homomorphically evaluate the protocol π. Finally, parties obtain an output ciphertext (whose message is of length B), invoke a threshold decryption protocol, and obtain the output value. (It is easy to obtain a protocol for *any* $\vec{v}' \in \{|\mathsf{f}|, \mathsf{f}\}^n$ by specifying parties who can obtain the output appropriately.)

Now we show the above protocol π' realizes $(G', \vec{v}', f)$ in the secure channel model. In order to prove the security of π', we construct a simulator $\mathcal{S}$ that can generate views of corrupted parties. Given 1^κ, $I \subsetneq \{1, \cdots, n\}$, the inputs $\vec{x}_I$, the output $f(\vec{x})$ (or the output size $|f(\vec{x})|$), all input sizes which are not independent sizes $\{1^{|x_{j_1}|}, \cdots, 1^{|x_{j_t}|}\}$, and random tapes $\vec{r}_I$ produced by a randomness producer, the simulator $\mathcal{S}$ first computes the upper bound $B = p(\kappa, |x_{j_1}|, \cdots, |x_{j_t}|)$. Second, $\mathcal{S}$ simulates a threshold key generation protocol, and computes ciphertexts of x_i for all $i \in I$. Next, $\mathcal{S}$ simulates messages sent by P_i to P_j $(i \in \{1, \cdots, n\}$ and $j \in I)$ as follows. If P_i is corrupted, $\mathcal{S}$ does the same as P_i. Otherwise, $\mathcal{S}$ computes a ciphertext for zero string of appropriate length. At the end of the protocol π', $\mathcal{S}$ simulates a threshold decryption protocol. Finally, $\mathcal{S}$ computes message sizes, and outputs views of corrupted parties and message sizes generated as above. The views generated by $\mathcal{S}$ are indistinguishable from the views in a real execution of the protocol due to the IND-CPA security of FHE and the security of the threshold protocols. Thus, the protocol π' securely computes $(G', \vec{v}', f)$ in the secure channel model. □

4 Results in the Secure Channel Model

In this section, we show that every function can be realized while hiding one (input or output) size in the secure channel model. On the other hand, we also prove that there exists a function that cannot be realized while hiding two or more (input or output) sizes in the secure channel model. Our result shows that, in the secure channel model, a general size-hiding protocol exists only in the case where parties wish to hide at most one of $n+1$ (n inputs and the output) sizes.

In Sect. 4.1, we give a formal statement of our result in Theorem 1, and show examples of (feasible or infeasible) classes. Then, we show the feasibility part of the theorem in Sect. 4.2, and the infeasibility part of the theorem in Sect. 4.3.

4.1 Our Result

Our result in the secure channel model is as follows.

Theorem 1. *Let* $(G, \vec{v})$ *be a size-hiding class with n parties. Assume that threshold FHE exists. The class* $(G, \vec{v})$ *is feasible in the secure channel model if and only if the number of private sizes of* $(G, \vec{v})$ *is at most* 1.

Examples. Examples of feasible size-hiding classes are shown in Fig. 4. The number of private sizes of them is just one. On the other hand, classes shown in Fig. 5 are infeasible. The number of private sizes of the left and the center graphs is two, and of the right graph is three.

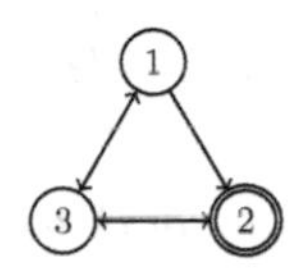

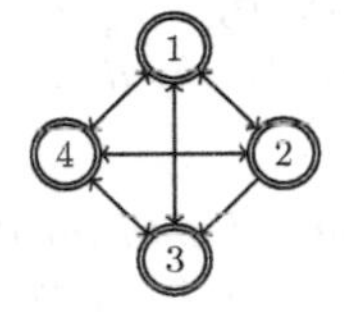

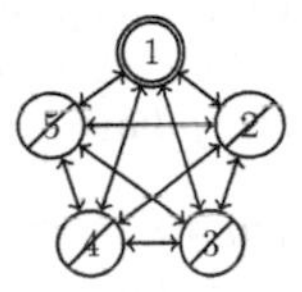

Fig. 4. Examples of feasible classes

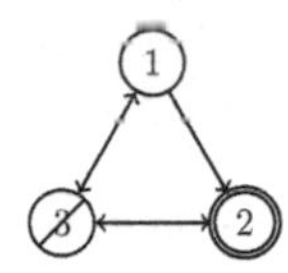

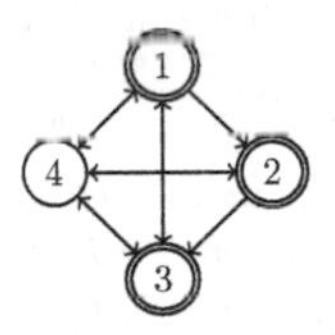

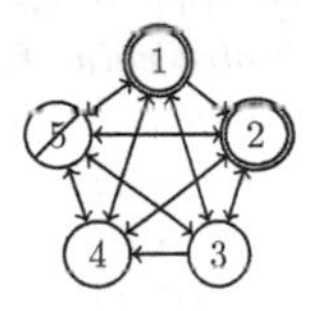

Fig. 5. Examples of infeasible classes

4.2 Protocol Hiding One Size

We construct a general size-hiding MPC protocol that can hide one (input or output) size, in order to show the feasibility part of Theorem 1. The case where only the output size is private is an easy application of ordinary MPC. Indeed, since now the input sizes are public, the output size also has a public and efficient upper bound derived from the complexity of the function f, therefore the output size can be hidden from the forbidden parties by a naive padding technique.

From now on, we consider the case where the output size is public. Let "server" denote the unique party who wants to hide its own input size, and let "clients" denote the other parties. The outline of the protocol construction, which is a natural extension of the two-party results for classes 1.a, 1.c, and 1.e in [LNO13], is explained as follows. Each client sends an FHE ciphertext of its own input to the server, which can be freely performed since their input sizes are public. Given these ciphertexts, the server seems to be able to compute the encrypted output of the function using homomorphic evaluation, which is then decrypted for the full-output parties by the threshold decryption. However, the ciphertext of the output value may have a length longer than the actual output size since the precise inputs are not known at the homomorphic evaluation, and the difference from the actual output size may reveal some non-trivial information on the server's input size. To avoid the problem, the server first homomorphically computes the ciphertext of the *actual output length* ℓ, and the parties know ℓ via the threshold decryption. Then the server generates the ciphertext of the output value where the length is exactly set to ℓ, which prevents the leak of the server's input size mentioned above.

The full description of the protocol appears in Protocol 1. In the following argument, we assume by symmetry that P_1 is the server.

Protocol 1. *Suppose that parties $P_1, P_2, \cdots, P_n$ have inputs $x_1, x_2, \cdots, x_n$, respectively, and all sizes are public except the input size $|x_1|$ of the party P_1. The protocol proceeds as follows.*

1. *All parties invoke a* ThrGen *protocol with inputs 1^κ, and each party P_i obtains a public key pk and a share of the secret key sk_i.*
2. *Each party P_i computes $c_i^{\mathsf{in}} = \mathsf{Enc}_{pk}(x_i)$, and sends c_i^{in} to P_1.*
3. *P_1 constructs a circuit C_{size}, which takes $\vec{x}$ as inputs and outputs $|f(\vec{x})|$ padded with zeroes up to $(\log \kappa)^2$ bits. Then, P_1 computes $c^{\mathsf{size}} \leftarrow \mathsf{Eval}_{pk}(C_{\mathsf{size}}, c_1^{\mathsf{in}}, \cdots, c_n^{\mathsf{in}})$ and sends c^{size} to all parties.*
4. *All parties invoke a* ThrDec *protocol with the ciphertext c^{size}, and obtain the decrypted value ℓ.*
5. *P_1 computes $c^{\mathsf{out}} \leftarrow \mathsf{Eval}_{pk}(C_{\mathsf{out}}, c_1^{\mathsf{in}}, \cdots, c_n^{\mathsf{in}})$, where the circuit C_{out} computes $f(x_1, \cdots, x_n)$ of length ℓ, and sends c^{out} to all parties.*
6. *All parties invoke a $\mathsf{ThrDec}_{I_{\mathsf{f}}}$ protocol with the ciphertext c^{out} as only full-output parties obtain the decrypted value $z \in \{0,1\}^\ell$. Then, all full-output parties output z, and the other parties output nothing. The protocol terminates.*

Lemma 3 (Security of Protocol 1). *Let $(G, \vec{v}, f)$ be a functionality with n parties, where $|x_1|$ is private and the other sizes are public. Assume that threshold FHE exists. Then, Protocol 1 realizes the functionality $(G, \vec{v}, f)$ in the secure channel model.*

Proof. In order to prove the security, we construct a simulator $\mathcal{S}$ that, given inputs, outputs, and random tapes of corrupted parties, generates their view in the protocol. We note that it suffices to only consider the most difficult case that $|x_1|$ is hidden from *all* other parties. Given 1^κ, $I = \{i_1, \cdots, i_t\}$, the inputs $\vec{x}_I$, public sizes $(1^{|x_2|}, \cdots, 1^{|x_n|}, 1^{|f(\vec{x})|})$, the output $f(\vec{x})$ if $I \cap I_{\mathsf{f}} \neq \emptyset$, and random tapes $\vec{R}_I = (r_{i_1}, \cdots, r_{i_t})$ produced by a randomness producer, the simulator $\mathcal{S}$ works as follows. (In the following probabilistic computation, $\mathcal{S}$ uses a string

$r_i||000\cdots$ as P_i's random tape.) First, $\mathcal{S}$ computes $(pk, sk) \leftarrow \mathsf{Gen}(1^\kappa)$, chooses $sk_i \xleftarrow{U} \{0,1\}^{|sk|}$ for all $i \in I$, and simulates a threshold key generation protocol under the keys. If P_1 is corrupted, $\mathcal{S}$ computes $c_i^{\mathsf{in}} = \mathsf{Enc}_{pk}(x_i)$ $(i \in I)$, $c_i^{\mathsf{in}} = \mathsf{Enc}_{pk}(0^{|x_i|})$ $(i \notin I)$, and evaluates c^{size} and c^{out} from these ciphertexts. Otherwise, $\mathcal{S}$ computes $c^{\mathsf{size}} = \mathsf{Enc}_{pk}(0^{(\log\kappa)^2})$ and $c^{\mathsf{out}} = \mathsf{Enc}_{pk}(0^{|f(\vec{x})|})$. Next, $\mathcal{S}$ simulates threshold decryption protocols for c^{size} and c^{out}. Then, $\mathcal{S}$ computes message sizes $\mathrm{MSIZE}^\pi(\vec{x})$. ($\mathcal{S}$ can compute them since all sizes of messages are only dependent on the public sizes $|x_2|, \cdots, |x_n|$ and $|f(\vec{x})|$.) Finally, $\mathcal{S}$ outputs views of corrupted parties and message sizes generated as above.

Let us observe the difference between the view generated in a real execution and the view generated by $\mathcal{S}$. The views of threshold key generation and threshold decryption protocols generated by $\mathcal{S}$ are indistinguishable from them in a real execution due to the security of these protocols. The ciphertexts generated by $\mathcal{S}$ are indistinguishable from them in a real execution due to the IND-CPA security of the underlying FHE scheme. Therefore, the above protocol realizes the functionality in the secure channel model. □

Fully avoiding upper bounding of input sizes. In the same way as the protocols in [LNO13], Protocol 1 above assumes that all input sizes are bounded by $2^{(\log\kappa)^2} = \kappa^{\log\kappa}$. From the viewpoint of security, this restriction causes no problems, since now the input sizes are polynomially bounded and thus the bound above indeed holds asymptotically. However, it may cause a problem from the viewpoint of correctness, since now the polynomial bounds for input sizes do not exist and the correctness should be satisfied at every parameter κ rather than just asymptotically. To resolve the issue, we show the assumption $|x_i| < 2^{(\log\kappa)^2}$ can be avoided by using a *flag technique*. A flag function $\mathsf{flag}_\ell : \{0,1\}^\ell \to \{0,1\}^\ell$ takes $x = x_\ell \cdots x_2 x_1 \in \{0,1\}^\ell$ as an input, and outputs $z = 0^{\ell-i}||1^i$, where i is an index such that $i = \max(j-1 \text{ s.t. } x_j = 1)$. For example, a flag function flag_{10} with an input $x = 0010000001$ outputs $z = 0001111111$. Next we explain how to use the flag function in Protocol 1. Let p be a polynomial such that $|f(x_1', \cdots, x_n')| < p(|x_1'|, \cdots, |x_n'|)$ for all $x_i' \in \{0,1\}^*$. In step 3, the party P_1 first computes $B = \log_2 p(|x_1|, \cdots, |x_n|)$, and then constructs a circuit C_{size}, which takes $\vec{x}$ as inputs and outputs $|f(\vec{x})|$ padded with zeroes up to $B = \log_2 p(|x_1|, \cdots, |x_n|)$, and a circuit C_{flag}, which takes $x \in \{0,1\}^B$ as an input and outputs a string $\mathsf{flag}_B(x)$. Then, P_1 computes $c^{\mathsf{size}} \leftarrow \mathsf{Eval}_{pk}(C_{\mathsf{size}}, c_1^{\mathsf{in}}, \cdots, c_n^{\mathsf{in}})$ and $c^{\mathsf{flag}} \leftarrow \mathsf{Eval}_{pk}(C_{\mathsf{flag}}, c^{\mathsf{size}})$. For $i = 1, 2, 3, \cdots$, P_1 sends $c^{\mathsf{flag}}[i]$ to all parties, and parties decrypt it. If the decrypted value equals zero, then P_1 sends $(c^{\mathsf{size}}[j])_{1\le j\le i}$ to all parties, otherwise, continue the loop. Now $(c^{\mathsf{size}}[j])_{1\le j\le i}$ indeed involves the whole information of $|f(\vec{x})|$ by the definition of the flag function, and thus we can avoid the upper bound of input sizes. The flag technique can also be applied to all of our protocols and previous two-party protocols [LNO13].

4.3 Infeasibility for Hiding Two Sizes

Unfortunately, in the secure channel model, there is no general size-hiding MPC protocol that can hide two or more (input or output) sizes. The rest of this subsection is devoted to proving the infeasibility part of Theorem 1. In particular, we prove the infeasibility when two input sizes are hidden (Lemma 4), and the infeasibility when one input and the output sizes are hidden (Lemma 5).

We first prove the infeasibility when two input sizes are hidden. In this case, the infeasibility of n-party protocol can be reduced to the infeasibility of two-party protocol when both input sizes are hidden (class 2). First, assume by contradiction, there exists a protocol π that realizes an n-ary function. Then, using a reduction compiler and a wrapping compiler, we compile the protocol π into a two-party protocol π' that realizes an impossible functionality. By the contradiction, we conclude that there exists a function while hiding two input sizes. The formal statement and the proof are as follows.

Lemma 4 (Hiding two input sizes). *Let $(G, \vec{v})$ be a size-hiding class for n parties, such that two input sizes are private, and the others are public. Assuming the existence of threshold FHE, there exists a function f such that the functionality $(G, \vec{v}, f)$* cannot *be realized in the secure channel model.*

Proof. Without loss of generality, we can assume the private input sizes are $|x_1|$ and $|x_2|$. Let f' be a two-ary function such that its range is a constant size, and $(G_{2a}, \vec{v}_{2a}, f')$ cannot be realized in the secure channel model. (The existence of such a function is shown by [LNO13].) Let f be an n-ary function such that $f(x_1, \cdots, x_n) = f'(x_1, x_2)$. Assume by contradiction that there exists an n-party protocol π with $P_1, \cdots, P_n$ that realizes $(G, \vec{v}, f)$ in the secure channel model.

Let $T(\kappa, \vec{x})$ be a random variable representing the number of bits exchanged among all parties when running π with inputs $\vec{x}$ and security parameter κ. In this case, by the argument similar to [LNO13], there exists a polynomial p such that $T(\kappa, \vec{x}) < p(\kappa)$ for all large enough κ. Let us consider the simulator $\mathcal{S}$ for the protocol π corrupting $P_2, \cdots, P_n$. For a fixed output value α, let x_2^* be the smallest string for which there exists x_1 such that $f'(x_1, x_2^*) = \alpha$. At this time, there exists a polynomial p_α such that the running time of the simulator $\mathcal{S}$ is bounded by $p_\alpha(|x_2^*|, |\alpha|, \kappa)$, and there exists a polynomial p'_α such that $p'_\alpha(\kappa) = p_\alpha(|x_2^*|, |\alpha|, \kappa)$ since $|x_2^*|$ and $|\alpha|$ are constant sizes. We claim that, for every (x_1, x_2) such that $f'(x_1, x_2) = \alpha$, the length of the transcript with input (x_1, x_2) is upper bounded by $p'_\alpha(\kappa)$ except negligible probability. Otherwise, it contradicts the security of π. (For example, the simulator $\mathcal{S}$, corrupting P_3 only, cannot compute the message size since $\mathcal{S}$ does not know P_2's input is x_2^* or not.) Since the number of possible output value is constant, there exists a polynomial p such that $T(\kappa, \vec{x}) < p(\kappa)$ for every $\vec{x}$ except negligible probability. Therefore, the protocol π is I-independent for any I (especially, $I = \{1, 2\}$).

Now we are ready to derive the contradiction. We first construct a two-party π' with $P'_1 = \{P_1\}$ and $P'_2 = \{P_2, \cdots, P_n\}$ that is compiled by a reduction compiler from the protocol π. Note that the protocol π' is also I-independent

($I = \{1, 2\}$) since the communication complexity and the computation complexity are the same as π. Then, we construct a protocol π'' that is compiled by a wrapping compiler from the protocol π'. From Lemma 2, the protocol π'' realizes $(G_{2.a}, \vec{v}_{2.a}, f')$, in contradiction to the infeasibility of f'. Now we have that the functionality $(G, \vec{v}, f)$ cannot be realized in the secure channel model. □

Next, we prove the infeasibility when one input and the output sizes are hidden. In order to prove this, we introduce a new function, a *truncated oblivious multi-input pseudorandom function* tomprf defined as follows. Let F be a pseudorandom function $F : \{0,1\}^\kappa \times \{0,1\}^\kappa \to \{0,1\}^\kappa$. A *truncated oblivious multi-input pseudorandom function* tomprf_n is an n-party functionality (but ignoring inputs $x_4, \cdots, x_n$) that takes as inputs a vector of arbitrary length $x_1 = (a_1, \cdots, a_m) \in (\{0,1\}^\kappa)^m$ from P_1, a κ-bit string $x_2 \in \{0,1\}^\kappa$ from P_2, and a key for the pseudorandom function $x_3 \in \{0,1\}^\kappa$ from P_3. The functionality outputs to P_1 $(F_{x_3}(a_1), \cdots, F_{x_3}(a_\ell))$, where $\ell = \min(x_2, m)$. Now we are ready to prove the following lemma.

Lemma 5 (Hiding an input and the output sizes). *Let $(G, \vec{v})$ be a size-hiding class for n parties, such that an input and the output sizes are private, and the others are public. Assume that one-way functions exist. There exists a function f such that $(G, \vec{v}, f)$ cannot be realized in the secure channel model.*

Proof. Without loss of generality, we can assume the private input size is $|x_1|$. Essentially, there are three settings regarding who must not learn $|x_1|$ and $|f(\vec{x})|$:

1. The party P_2 must not learn both of $|x_1|$ and $|f(\vec{x})|$.
2. The party P_2 must not learn $|x_1|$, and the party P_1 must not learn $|f(\vec{x})|$.
3. The party P_2 must not learn $|x_1|$, and the party P_3 must not learn $|f(\vec{x})|$.

First, let us consider the case where P_2 must not learn both of $|x_1|$ and $|f(\vec{x})|$. Let f be an n-ary function ignoring $x_3, \cdots, x_n$ such that $f(\vec{x}) = f'(x_1, x_2)$, where the functionality $(G_{1.d}, \vec{v}_{1.d}, f')$ cannot be realized in the secure channel model. Assume by contradiction that there exists an n-party protocol π with $P_1, \cdots, P_n$ that securely computes $(G, \vec{v}, f)$ in the secure channel model. We can construct a two-party protocol π' with $P'_1 = \{P_2\}$ and $P'_2 = \{P_1, P_3, \cdots, P_n\}$ that is compiled by a reduction compiler from π. From Lemma 1, the protocol π' realizes $(G_{1.d}, \vec{v}_{1.d}, f')$, in contradiction to the infeasibility of f'. Now in this case we obtain a function f such that $(G, \vec{v}, f)$ *cannot* be realized in the secure channel model.

Second, let us consider the case where P_2 must not learn $|x_1|$, and P_1 must not learn $|f(\vec{x})|$. An oblivious transfer OT is a two-party function that takes $x_1 = (s_0, s_1)$ from P_1, where s_0 and s_1 are strings of arbitrary length, and $x_2 \in \{0,1\}$ from P_2 as inputs, and outputs a string s_{x_2} to only the party P_2. Let f be an n-ary function such that $f(x_1, \cdots, x_n) = \mathsf{OT}(x_1, x_2)$. Now we show that the function f cannot be realized in the secure channel model by the technique similar to [LNO13]. Assume by contradiction that there exists an n-party protocol π with $P_1, \cdots, P_n$ that realizes $(G, \vec{v}, f)$ in the secure channel model.

We denote the inputs $\vec{x}$ by $\vec{x} = ((s_0, s_1), x_2)$ since inputs $x_3, \cdots, x_n$ are ignored. Let $T(\kappa, \vec{x})$ be a random variable representing the number of bits exchanged among $P_1, \cdots, P_n$ when running π with inputs $\vec{x}$ and security parameter κ. For inputs $\vec{x}^* = ((0,0),0)$, there exists a polynomial p such that $T(\kappa, \vec{x}^*) < p(\kappa)$ for all large enough κ since π is a polynomial-time protocol. Let s' be a random string whose length is $\omega(p(\kappa))$, and let $\vec{x}'_0 = ((0, s'), 0)$ and $\vec{x}'_1 = ((0, s'), 1)$. It must hold that $T(\kappa, \vec{x}'_0) < p(\kappa)$, otherwise P_2 can distinguish the other input of P_1 is 0 or s'. And it must hold $T(\kappa, \vec{x}'_1) < p(\kappa)$, otherwise P_1 can distinguish that P_2 obtains 0 or s'. However, in the case of $\vec{x}'_1 = ((0, s'), 1)$, the party P_2 must compute s', the random string of length $\omega(p(\kappa))$, from a transcript of length less than $p(\kappa)$. This contradicts to the incompressibility of a random string. Thus, in this case, there is a function f such that $(G, \vec{v}, f)$ *cannot* be realized in the secure channel model.

Finally, let us consider the case where P_2 must not learn $|x_1|$, and P_3 must not learn $|f(\vec{x})|$. Let f be a truncated oblivious multi-input pseudorandom function tomprf_n with a pseudorandom function $F : \{0,1\}^\kappa \times \{0,1\}^\kappa \to \{0,1\}^\kappa$. Assume by contradiction that there exists an n-party protocol π with $P_1, \cdots, P_n$ that realizes $(G, \vec{v}, f)$ in the secure channel model. Let $T(\kappa, \vec{x})$ be a random variable representing the number of bits exchanged among $P_1, \cdots, P_n$ when running π with inputs $\vec{x}$ and security parameter κ. There exists a polynomial p such that $T(\kappa, (\emptyset, 0, x_3)) < p(\kappa)$ for all large enough κ since π is a polynomial-time protocol. For any x_1^* of the cardinality $\omega(p(\kappa))$, it must hold $T(\kappa, (x_1^*, 0, x_3)) < p(\kappa)$ for all large enough κ, otherwise P_2 can distinguish that P_1 has $\emptyset$ or x_1^*, although P_2 must not learn $|x_1|$. (Note that since $\mathsf{tomprf}_n(x_1^*, 0, x_3) = \emptyset$, the party P_2, who may learn the output, must not learn any partial information of the size of P_1.) It must also hold $T(\kappa, (x_1^*, 2^\kappa - 1, x_3)) < p(\kappa)$ for all large enough κ, otherwise P_3 can distinguish that the output size is 0 or $\omega(p(\kappa))$. Now we construct an algorithm $\mathcal{D}$ that distinguishes between outputs of the pseudorandom function $F_{x_3}(a_1), \cdots, F_{x_3}(a_m) \in \{0,1\}^\kappa$, and truly random values $r_1, \cdots, r_m \in \{0,1\}^\kappa$, using a simulator $\mathcal{S}$ for a randomness producer $\mathcal{R}(1^\kappa) = 0$. The distinguisher $\mathcal{D}$ invokes $\mathcal{S}$ with inputs $(1^\kappa, x_i, \vec{z}, 0)$ where x_i is the input of P_i, $\vec{z}$ is either $(F_{x_3}(a_1), \cdots, F_{x_3}(a_m))$ or $(r_1, \cdots, r_m)$ (here, we omit a set of indices I and the input sizes). If $\vec{z}$ is the pseudorandom values, the simulator $\mathcal{S}$ outputs a transcript of length less than $p(\kappa)$, otherwise $\mathcal{S}$ cannot output consistent transcript due to the incompressibility of a random string. The distinguisher $\mathcal{D}$ should output 1 if $\mathcal{S}$ outputs consistent transcript. $\mathcal{D}$ distinguishes pseudorandom values and random values[4], in contradiction to the pseudorandomness of F. Thus, in this case, assuming the existence of one-way functions, there is a function f such that $(G, \vec{v}, f)$ *cannot* be realized in the secure channel model. □

Theorem 1 is proven by Lemmas 3, 4 and 5.

[4] The above strategy works even for any randomness producer whose output size is bounded. However, in the HBC model, the proof does not work since a simulator can generate a transcript with a *long random tape*.

5 Results in the Strong Secure Channel Model

In previous section, we show that, in the secure channel model, a general size-hiding protocol cannot hide two or more (input or output) size information. In order to circumvent the infeasibility, we introduce a new communication model, a *strong secure channel model* such that an adversary cannot learn the number of bits exchanged among honest parties. We show that, in the strong secure channel model, a general size-hiding protocol exists even hiding all sizes of inputs and output from some parties, while the secure channel model only allows the size of at most one input to be hidden. Furthermore, we also prove that some functions still remain infeasible even in the strong secure channel model. More specifically, we give a sufficient and necessary condition under which a general size-hiding protocol can exist. Because the condition depends on whether the output size is public or private, our result is stated in Theorem 2 (when the output size is public) and Theorem 3 (when the output size is private).

In Sect. 5.1, we introduce the strong secure channel model. In Sect. 5.2, we give our main results, Theorem 2 and Theorem 3, and some examples of (feasible or infeasible) classes. We show the feasibility part of Theorem 2 in Sect. 5.3, and the infeasibility part of the theorem in Sect. 5.4. We show the feasibility part of Theorem 3 in Sect. 5.5, and the infeasibility part of the theorem in Sect. 5.6.

5.1 Strong Secure Channel Model

One of the standard communication model is the secure channel model, which is an abstraction of secure communication. In the secure channel model, an adversary cannot learn messages exchanged among honest parties, but can learn the number of bits of them. The model is very powerful and used in various works, however, in the context of size-hiding computations, there are strong infeasibility results. In order to circumvent the infeasibility, we introduce a new communication model, a strong secure channel model such that an adversary can learn neither messages nor the number of bits exchanged among honest parties, At first glance, the existence of such a communication channel seems to be suspicious, but we emphasize that the strong secure channel model can be instantiated by using steganographic techniques.

We provide a security definition of the strong secure channel model. The only difference from the secure channel model is that a simulator does not have to create message sizes in the strong secure channel model. Thus, the security in the secure channel model implies the security in the strong secure channel model. The security of protocols in the model is formally defined as follows.

Definition 4 (Security in the strong secure channel model). *Let $(G, \vec{v}, f)$ be a functionality for n parties and let π be a protocol that correctly computes $(G, \vec{v}, f)$. We say that π realizes $(G, \vec{v}, f)$ in the* strong secure channel model *if for every randomness producer $\mathcal{R}$, there exists a PPT $\mathcal{S}$ such that for every $I \subsetneq \{1, \cdots, n\}$, every polynomials $q_1, \cdots, q_n$,*

$$\left\{\mathcal{S}(1^\kappa, I, \vec{x}_I, \mathrm{OUTPUT}_I^{(G,\vec{v},f)}(\vec{x}), \vec{r}_I \leftarrow \mathcal{R}(1^\kappa, I))\right\}_{\kappa,\vec{x}} \stackrel{c}{\equiv} \left\{\mathrm{view}_I^\pi(\vec{x})_{\vec{r}_I}\right\}_{\kappa,\vec{x}}$$

where $x_1 \in \{0,1\}^{q_1(\kappa)}, \cdots, x_n \in \{0,1\}^{q_n(\kappa)}$.

5.2 Our Main Results

In the strong secure channel model, it is possible to realize any functionality while hiding two or more sizes. The condition under which any functionality can exist is different depending on the case where the output size is public and the case where the output size is private.

The main theorem for the case where the output size is public is as follows.

Theorem 2 (Public output size). *Let $(G, \vec{v})$ be a size-hiding class with n parties, where the output size is public. Assume that threshold FHE exists. A class of size-hiding $(G, \vec{v})$ is feasible in the strong secure channel model if and only if for every two distinct vertices $i, j \in V(G)$, there exists a vertex k such that $(i, k) \in E(G)$ and $(j, k) \in E(G)$.*

Examples. Suppose parties $P_1, \cdots, P_5$ wish to compute a function while hiding their input sizes, but each party thinks it is permitted to leak its own size information to the neighboring parties; see the right most graph in Fig. 6. In this case, the parties can securely compute every function, since every two distinct parties have a party who may learn both input sizes of them. (For example, a pair of parties P_1 and P_4 has the party P_5 who may learn input sizes of P_1 and P_4.) Thus, in such a pentagon case, a general size-hiding protocol exists in the strong secure channel model. Similarly, the triangle and the square cases also have a general size-hiding protocol. On the other hand, there is no general size-hiding protocol in the hexagon case; see the right most graph in Fig. 7. This is due to the fact that the pair of parties P_1 and P_4 do not have a party who may learn both input sizes of them. Other feasible and infeasible classes are shown in Figs. 6 and 7.

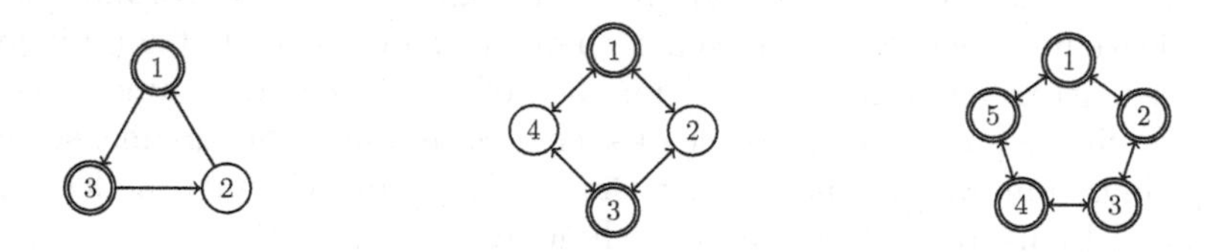

Fig. 6. Feasible classes with public output size

The main theorem for that case where the output size is private is as follows.

Theorem 3 (Private output size). *Let $(G, \vec{v})$ be a size-hiding class with n parties, where the output size is private. Assume that threshold FHE exists. A class of size-hiding $(G, \vec{v})$ is feasible in the strong secure channel model if and only if any vertex $i \in V(G)$ such that $\vec{v}[i] = \perp$ satisfies the both conditions:*

1. *For all vertices $j \in V(G)$, there exists an edge $(j, i) \in E(G)$.*
2. *There exists an edge $(i, j) \in E(G)$ such that $\vec{v}[j] \neq \perp$.*

Fig. 7. Ineasible classes with public output size

Examples. See the center graph in Fig. 8. Suppose two clients (P_3 and P_4) wish to compute a function while hiding their input sizes, with the help of servers (P_1 and P_2, they also have input data). Furthermore, suppose clients want to hide the output size from servers. In this case, if servers may learn all input sizes of clients and each server has a client who may learn the server's input size, every function can be realized while meeting the demand. Every feasible class with private output size is interpreted as such a client-server situation. On the other hand, there is no general size-hiding protocol in classes of Fig. 9.

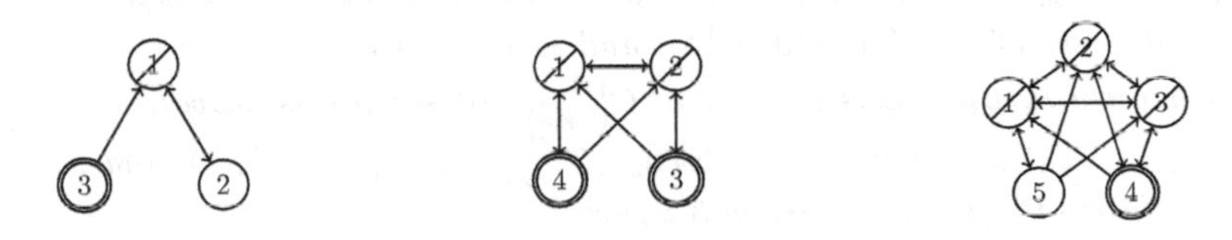

Fig. 8. Feasible classes with private output size

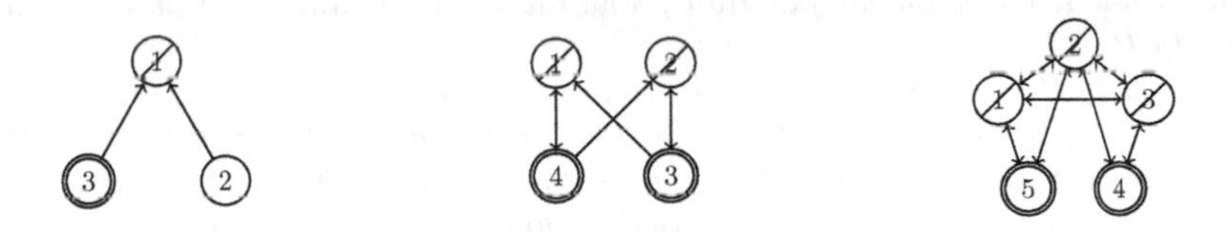

Fig. 9. Infeasible classes with private output size

5.3 Protocol with Public Output Size

We show that, in the strong secure channel model, the feasibility of size-hiding computations is dramatically improved compared to the secure channel model. In this subsection, we construct a size-hiding protocol where all parties may learn the output size. (The case of private output size is described in Sect. 5.5.) In particular, we construct a general size-hiding protocol if every pair of parties has a party who may learn input sizes of them. The condition includes the case where each input size is hidden from some parties, i.e., the number of hidden sizes is the number of parties. The protocol idea is explained in Introduction; see Sect. 1.2. The full description of the protocol appears in Protocol 2.

Building blocks – circuits for homomorphic evaluation. Let f be a function $f : (\{0,1\}^*)^n \to \{0,1\}^*$ and let ℓ be an integer. We can construct the following circuit, denoted by $C^{\mathsf{Sac}}_{f,\ell}$. On receiving a string $x'_1, x'_2, \cdots, x'_n \in \{0,1\}^*$ as inputs, if there is an all-zero string in inputs, then it outputs 0^ℓ, otherwise, parse as $x'_i = 00\cdots0||1||x_i$ $(i = 1, \cdots, n)$, and outputs $f(x_1, \cdots, x_n)$ padded with zeroes up to ℓ.

Protocol 2. *Suppose that parties $P_1, P_2, \cdots, P_n$ have inputs $x_1, x_2, \cdots, x_n$, respectively, and each party is either a full-output party or a size-inly party. The protocol proceeds as follows.*

1. *All parties invoke a* ThrGen *protocol with inputs 1^κ, and each party P_i obtains a public key pk and a share of the secret key sk_i.*
2. *For all edges $(i,j) \in E(G)$, P_i sends the size information $|x_i|$ to P_j.*
3. *For all two vertices $i, j \in V(G)$, the party P_i sends ciphertexts $c^{\mathsf{in}}_{(i,j)}$ to P_j as follows. Let P_k be a party who may learn both of $|x_i|$ and $|x_j|$. The party P_i computes $c_i = \mathsf{Enc}_{pk}(1||x_i)$, and sends to P_k. (If P_i and P_k are the same party, then P_i computes only.) If it holds $|x_j| \geq |x_i|$, then P_k computes a ciphertext $c^{\mathsf{in}}_{(i,j)} = \mathsf{Enc}_{pk}(0^{|x_j|-|x_i|})||c_i$, and sends it to P_j. Otherwise, P_k computes a ciphertext of zeroes $c^{\mathsf{in}}_{(i,j)} = \mathsf{Enc}_{pk}(0^{|x_j|+1})$, and sends it to P_j.*
4. *Each party P_i constructs the circuit $C^{|f|}_{(\log\kappa)^2}$ (described as above), where $|f| = |f(x_1, \cdots, x_n)|$, computes $c^{\mathsf{size}}_i \leftarrow \mathsf{Eval}_{pk}(C^{|f|}_{(\log\kappa)^2}, c^{\mathsf{in}}_{(1,i)}, \cdots, c^{\mathsf{in}}_{(n,i)})$, and sends c^{size}_i to the party P_1 (or other designated party).*
5. *The party P_1 computes c^{size} by homomorphic evaluation of a max function from $c^{\mathsf{size}}_1, \cdots, c^{\mathsf{size}}_n$, and sends it to all parties. Then, all parties invoke a* ThrDec *protocol with the ciphertext c^{size}, and obtain the decrypted value ℓ.*
6. *Each party P_i computes $c^{\mathsf{out}}_i \leftarrow \mathsf{Eval}_{pk}(C^f_\ell, c^{\mathsf{in}}_{(1,i)}, \cdots, c^{\mathsf{in}}_{(n,i)})$, where C^f_ℓ is a circuit described as above for the function f and the integer ℓ, and then sends c^{out}_i to the party P_1.*
7. *The party P_1 computes c^{out} by homomorphic evaluation of a max function from $c^{\mathsf{out}}_1, \cdots, c^{\mathsf{out}}_n$, and sends it to all parties. Then, all parties invoke a $\mathsf{ThrDec}_{I_\mathsf{f}}$ protocol with the ciphertext c^{out}, and all full-output parties obtain the decrypted value z. All full-output parties output z, and the other parties output nothing. The protocol terminates.*

Lemma 6 (Security of Protocol 2). *Let $(G, \vec{v})$ be a size-hiding class with n parties, which holds the conditions stated in Theorem 2. Let $f(x_1, \cdots, x_n)$ be any n-ary polynomial-time computable function. Assuming the existence of threshold FHE, Protocol 2 realizes $(G, \vec{v}, f)$ in the strong secure channel model.*

Proof. Given 1^κ, I, inputs $\vec{x}_I$, input sizes $\{1^{|x_j|} | (j,i) \in E(G)\}_{i \in I}$, the output $f(\vec{x})$ if $I \cap I_\mathsf{f} \neq \emptyset$, and random tapes $\vec{R}_I$ produced by a randomness producer, the simulator $\mathcal{S}$ works as follows. First, $\mathcal{S}$ computes $(pk, sk) \leftarrow \mathsf{Gen}(1^\kappa)$, chooses $sk_i \overset{\mathrm{U}}{\leftarrow} \{0,1\}^{|sk|}$ for every $i \in I$, and simulates ThrGen protocol with keys $(pk, sk_i)_{i \in I}$. Next, for every $i \in \{1, \cdots, n\}$ and every $j \in I$, $\mathcal{S}$ computes $c^{\mathsf{in}}_{(i,j)} = \mathsf{Enc}_{pk}(0^{|x_j|-|x_i|}||1||x_i)$ if it holds $i \in I$ and $|x_j| \geq |x_i|$, otherwise, $c^{\mathsf{in}}_{(i,j)} = \mathsf{Enc}_{pk}(0^{|x_j|+1})$. Then, $\mathcal{S}$ computes c^{size}_i and c^{out}_i for every $i \in I$, and evaluates c^{size} and c^{out}. $\mathcal{S}$ simulates threshold decryption protocols for c^{size} and c^{out}. Finally, $\mathcal{S}$ outputs views of corrupted parties generated as above. The views generated by $\mathcal{S}$ are indistinguishable from the views in a real execution of the protocol due to the IND-CPA security of threshold FHE and the security of threshold key generation and decryption protocols. □

5.4 Infeasibility Result with Public Output Size

We show that, when all parties may learn the output size, the condition "every pair has a party who may learn input sizes of the pair" is a necessary and sufficient condition under which a general size-hiding protocol can exist. In this subsection, in order to prove this, we show that the condition is not satisfied, a general size-hiding protocol does not exist.

Lemma 7. *Let $(G, \vec{v})$ be a size-hiding class with n parties, where the output size is public. The size-hiding class $(G, \vec{v})$ is infeasible in the strong secure channel model if there exists two distinct vertices $i^*, j^* \in V(G)$ such that there is no vertex $k \in V(G)$ such that $(i^*, k) \in E(G)$ and $(j^*, k) \in E(G)$.*

Proof. Let $(G, \vec{v})$ be a size-hiding class that satisfies the conditions as above. Without loss of generality, we can assume $i^* = 1$ and $j^* = 2$. Let $P_1' = \{P_i | (2, i) \notin E(G)\}$ and let $P_2' = \{P_1, \cdots, P_n\} \setminus P_1'$. The parties P_1' must not learn $|x_2|$ from the definition, and the parties P_2' must not learn $|x_1|$, otherwise, it contradicts to the condition. (Note that $P_1 \in P_1'$ and $P_2 \in P_2'$.) Let f' be a two-ary function such that $(G_{2.a}, \vec{v}_{2.a}, f')$ cannot be realized in the (strong) secure channel model[5], and let f be a function such that $f(x_1, x_2, \cdots, x_n) = f'(x_1, x_2)$. Assume by contradiction that there exists an n-party protocol π that realizes $(G, \vec{v}, f)$ in the strong secure channel model. We can construct a two-party protocol π' with P_1' and P_2', that is compiled by a reduction compiler from the protocol π. From Lemma 1, the protocol π' realizes the functionality $(G_{2.a}, \vec{v}_{2.a}, f')$, in contradiction to the assumption. Therefore, the size-hiding class $(G, \vec{v})$ is infeasible in the strong secure channel model. □

Theorem 2 is proven by Lemmas 6 and 7.

5.5 Protocol with Private Output Size

In this subsection, we construct a size-hiding protocol where some parties must not learn the output size; see Protocol 3. (The case of public output size is described in Sect. 5.3.) Note that it is not superior to Protocol 2 since these size-hiding conditions are different. Interestingly, the underlying idea of the protocol is completely different from Protocol 2; see Sect. 1.2.

---

[5] Note that, in the two-party setting, the strong secure channel model and the secure channel model are essentially the same.

Building block – GMW protocol. Goldreich et al. [GMW87] constructed a general MPC protocol that is secure in the presence of HBRC adversaries[6] corrupting up to $n-1$ of n parties, and showed that it can be compiled to a protocol that is secure in the presence of malicious adversaries. Our protocol uses the HBRC protocol in order to compute the desired function by all servers. For simplicity, we use *GMW protocol* to denote a protocol that realizes the following functionality.

- Input: Each party is given a secret share of x_j for all $j = 1, \cdots, n$.
- Output: All parties output $f(x_1, \cdots, x_n)$.

Protocol 3. *Suppose that parties $P_1, P_2, \cdots, P_n$ have inputs $x_1, x_2, \cdots, x_n$, respectively, and there are some forbidden parties. The function f to be computed has a polynomial p s.t. $|f(x'_1, \cdots, x'_n)| < p(|x'_1|, \cdots, |x'_n|)$ for all $x'_i \in \{0,1\}^*$. Let p' be a polynomial such that $p'(x) = p(x, x, \cdots, x)$. For a ciphertext $c = (c_1, \cdots, c_\ell)$ (each c_i is a ciphertext of 1-bit message), $[c^{\mathsf{out}}]_{\ell'}$ ($\ell' \leq \ell$) denotes $c' = (c_1, \cdots, c_{\ell'})$. Without loss of generality, we can assume $1 \in I_\perp$ and $2 \in I_{\mathsf{p}}$. The protocol proceeds as follows.*

1. *All permitted parties invoke a ThrGen protocol with inputs 1^κ, and then each permitted party P_i obtains the public key pk and a share of the secret key sk_i. Then, P_2 sends the public key pk to all forbidden parties.*
2. *Each party P_i computes shares of additive secret sharing $\{r_{i,j}\}_{j \in I_\perp}$, where $\{R_{i,j} \in \{0,1\}^{|x_i|}$, whose secret is x_i, i.e., $r_{i_1,j} \oplus r_{i_2,j} \oplus \cdots \oplus r_{i_t,j} = x_i$ ($\{i_1, \cdots, i_t\} = I_\perp$). Then, P_i computes $c^{\mathsf{in}}_{(i,j)} = \mathsf{Enc}_{pk}(r_{i,j})$, and sends $c^{\mathsf{in}}_{(i,j)}$ to all forbidden parties.*
3. *The forbidden parties homomorphically evaluate a GMW protocol computing $f(\vec{x})$ of length $L = p'(\max(|x_1|, \cdots, |x_n|))$ using FHE, i.e., all messages in the execution are encrypted by FHE and all computations are done by homomorphic evaluation. As the output of the protocol, they obtain a ciphertext c^{out}.*
4. *The party P_1 constructs a circuit C_{size} that takes $x \in \{0,1\}^L$ as an input, and outputs $|x| \in \{0,1\}^{(\log \kappa)^2}$. Then, P_1 computes $c^{\mathsf{size}} \leftarrow \mathsf{Eval}_{pk}(C_{\mathsf{size}}, c^{\mathsf{out}})$, and sends c^{size} to all permitted parties.*
5. *Let $\sigma_i = \max\{|x_j| | (j,i) \in E(G)\}$, and let $L_i = p'(\sigma_i)$. (It must hold $L_i \leq L$ for all i by the definition.) For every $i \in I_{\mathsf{p}}$, the party P_1 homomorphically evaluates a max function for $[c^{\mathsf{out}}]_{L_i}$ and $\mathsf{Enc}_{pk}(0^{L_i})$, and obtains a ciphertext c^{out}_i. Then, P_1 sends c^{out}_i to P_i for every $i \in I_{\mathsf{p}}$.*
6. *All permitted parties invoke a ThrDec protocol with inputs 1^κ and c^{size}, and obtain the decrypted value ℓ.*
7. *Each permitted party P_i sends $[c^{\mathsf{out}}_i]_\ell$ to the party P_2. (If the length of the ciphertext is less than ℓ, then the party uses a padding with zero ciphertexts up to ℓ.) The party P_2 computes c^{out} by homomorphic evaluation of a max function from $[c^{\mathsf{out}}_1]_\ell, \cdots, [c^{\mathsf{out}}_n]_\ell$, and sends it to all permitted parties.*
8. *All permitted parties invoke a $\mathsf{ThrDec}_{I_{\mathsf{f}}}$ protocol with inputs 1^κ and c^{out}, and all full-output parties obtain the decrypted value $z \in \{0,1\}^\ell$. All full-output parties output z, and the other parties output nothing. The protocol terminates.*

[6] It is well known that the protocol is secure against HBC adversaries. However, it is also secure against HBRC adversaries since the simulation algorithm does not have to choose random tapes by itself.

Lemma 8 (Security of Protocol 3). *Let $(G, \vec{v})$ be a size-hiding class with n parties, which satisfies the conditions stated in Theorem 3. Let $f(x_1, \cdots, x_n)$ be any n-ary polynomial-time computable function. Assuming the existence of threshold FHE, Protocol 3 realizes $(G, \vec{v}, f)$ in the strong secure channel model.*

Proof. Given 1^κ, I, inputs $\vec{x}_I$, input sizes $\{1^{|x_j|}|(j,i) \in E(G)\}_{i \in I}$, the output $f(\vec{x})$ if $I \cap I_{\mathsf{f}} \neq \emptyset$, and random tapes $\vec{R}_I$ produced by a randomness producer, the simulator $\mathcal{S}$ works as follows. First, $\mathcal{S}$ computes $(pk, sk) \leftarrow \mathsf{Gen}(1^\kappa)$, chooses $sk_i \overset{\text{U}}{\leftarrow} \{0,1\}^{|sk|}$ for every $i \in I$, and simulates ThrGen protocol with keys $(pk, sk_i)_{i \in I}$. Next, for every $i \in \{1, \cdots, n\}$ and every $j \in I$, $\mathcal{S}$ computes $c^{\mathsf{in}}_{(i,j)} = \mathsf{Enc}_{pk}(0^{|x_j|-|x_i|}||1||x_i)$ if it holds $i \in I$ and $|x_j| \geq |x_i|$, otherwise, $c^{\mathsf{in}}_{(i,j)} = \mathsf{Enc}_{pk}(0^{|x_j|+1})$. Then, $\mathcal{S}$ computes c^{size}_i and c^{out}_i for every $i \in I$, and evaluates c^{size} and c^{out}. $\mathcal{S}$ simulates threshold decryption protocols for c^{size} and c^{out}. Finally, $\mathcal{S}$ outputs views of corrupted parties generated as above. The views generated by $\mathcal{S}$ are indistinguishable from the views in a real execution of the protocol due to the IND-CPA security of threshold FHE and the security of threshold key generation and decryption protocols. □

Proof. Let I_1 and I_2 be a partition of $I = I_1 \cup I_2$ such that $I_1 \subset I_\perp$ and $I_2 \subset I_{\mathsf{p}}$. We consider the following cases: (1) $I_1 \subsetneq I_\perp$ and $I_2 = I_{\mathsf{p}}$, (2) $I_1 = \emptyset$ and $I_2 = I_{\mathsf{p}}$, (3) $I_1 = I_\perp$ and $I_2 \subsetneq I_{\mathsf{p}}$, (4) $I_1 \subsetneq I_\perp$ and $I_2 \subsetneq I_{\mathsf{p}}$, (5) $I_1 = \emptyset$ and $I_2 \subsetneq I_{\mathsf{p}}$, (6) $I_1 = I_\perp$ and $I_2 = \emptyset$, and (7) $I_1 \subsetneq I_\perp$ and $I_2 = \emptyset$. We show the simulator in the cases of (1) and (3). It is easy to adapt the proof to the other cases.

We construct the simulator $\mathcal{S}$ in the case of (1), where all clients and some servers are corrupted. Given 1^κ, I, inputs $\vec{x}_I$, all input sizes $\{1^{|x_1|}, \cdots, 1^{|x_n|}\}$, the output $f(\vec{x})$, and random tapes $\vec{r}_I$ produced by a randomness producer, the simulator $\mathcal{S}$ works as follows. First, $\mathcal{S}$ invokes a threshold key generation protocol, and obtains a public key pk and all shares of the secret key. Second, $\mathcal{S}$ computes secret shares of x_i for $i \in I$ and $0^{|x_i|}$ for $i \notin I$, and encrypts them by threshold FHE. Next, $\mathcal{S}$ computes $c^{\mathsf{out}} = \mathsf{Enc}_{pk}(f(\vec{x}))$ padded with zeroes up to appropriate length and simulates an encrypted GMW protocol on the input ciphertexts and c^{out}. Then, $\mathcal{S}$ invokes threshold decryption protocols. Finally, $\mathcal{S}$ outputs views of corrupted parties generated as above. The view generated by $\mathcal{S}$ and the view in a real execution are indistinguishable due to the security of the GMW protocol.

Next we construct the simulator $\mathcal{S}$ in the case of (3), where all servers and some clients are corrupted. Given 1^κ, $I = I_1 \cup I_2$, inputs $\vec{x}_I$, all input sizes $\{1^{|x_1|}, \cdots, 1^{|x_n|}\}$, the output $f(\vec{x})$, and random tapes $\vec{r}_I$ produced by a randomness producer, the simulator $\mathcal{S}$ works as follows. First, $\mathcal{S}$ computes $(pk, sk) \leftarrow \mathsf{Gen}(1^\kappa)$, chooses $sk_i \overset{\text{U}}{\leftarrow} \{0,1\}^{|sk|}$ for all $i \in I_2$, and simulates a threshold key generation protocol with the keys. Second, $\mathcal{S}$ computes secret shares of x_i for $i \in I$ and $0^{|x_i|}$ for $i \notin I$, and encrypts them by threshold FHE. Next, $\mathcal{S}$ homomorphically executes GMW protocol, and obtains the output ciphertext c^{out}. Then, $\mathcal{S}$ computes c^{size} honestly, and simulates threshold decryption protocols. Finally, $\mathcal{S}$ outputs views of corrupted parties generated as above. The view generated by $\mathcal{S}$ and the view in a real execution are indistinguishable due to the IND-CPA security of FHE and the security of threshold protocols. □

5.6 Infeasibility Result with Private Output Size

We show that, when some parties must not learn the output size, the condition stated in Theorem 3 is a necessary and sufficient condition under which a general size-hiding protocol can exist. In this subsection, in order to prove this, we show that if the condition is not satisfied, a general size-hiding protocol does not exist.

Lemma 9. *Let $(G, \vec{v})$ be a size-hiding class with n parties, where the output size is private. The size-hiding class $(G, \vec{v})$ is infeasible in the strong secure channel model if there exists a vertex $i^* \in V(G)$ such that $\vec{v}[i^*] = \bot$, which satisfies one of the following conditions:*

1. *There exists a vertex $j^* \in V(G)$ such that $(j^*, i^*) \notin E(G)$.*
2. *There is no edge $(i^*, j) \in E(G)$ such that $\vec{v}[j] \neq \bot$.*

Proof. Let $(G, \vec{v})$ be a size-hiding class that satisfies the former condition. Without loss of generality, we can assume $i^* = 1$ and $j^* = 2$, i.e., $\vec{v}[1] = \bot$ and $(2, 1) \notin E(G)$. Let f' be a two-ary function such that the functionality $(G_{1.d}, \vec{v}_{1.d}, f')$ cannot be realized in the (strong) secure channel model, and let f be an n-ary function such that $f(x_1, x_2, \cdots, x_n) = f'(x_1, x_2)$. Assume by contradiction that there exists n-party protocol π that realizes $(G, \vec{v}, f)$ in the strong secure channel model. Now we construct a two-party protocol π' with $P'_1 = \{P_1\}$ and $P'_2 = \{P_2, \cdots, P_n\}$ that is compiled by a reduction compiler from the protocol π. Since P'_1 must not know both the output size and the input size $|x_2|$, the protocol π' realizes $(G_{1.d}, \vec{v}_{1.d}, f')$ in the (strong) secure channel model, in contradiction to the infeasibility of f'. Therefore, the size-hiding class $(G, \vec{v})$ is infeasible in the strong secure channel model.

Let $(G, \vec{v})$ be a size-hiding class that satisfies the latter condition. Let P'_1 be a subset of parties $P'_1 = \{P_i | \vec{v}[i] = \bot\}$, and let $P'_2 = \{P_1, \cdots, P_n\} \setminus P'_1$. Without loss of generality, we can assume $i^* = 1$ and $P_2 \in P'_2$. Let f' be a two-ary function such that $(G_{1.b}, \vec{v}_{1.b}, f')$ cannot be realized in the (strong) secure channel model, and let f be an n-ary function such that $f(x_1, x_2, \cdots, x_n) = f'(x_1, x_2)$. Assume by contradiction that there exists n-party protocol π that realizes $(G, \vec{v}, f)$ in the strong secure channel model. Now we construct a two-party protocol π' with P'_1 and P'_2, that is compiled by a reduction compiler from the protocol π. Since P'_1 must not learn the output size, and P'_2 must not learn the input size $|x_1|$, the protocol π' realizes $(G_{1.b}, \vec{v}_{1.b}, f')$ in the (strong) secure channel model, in contradiction to the infeasibility of f'. Therefore, the size-hiding class $(G, \vec{v})$ is infeasible in the strong secure channel model. □

Theorem 3 is proven by Lemmas 8 and 9.

Acknowledgment. The authors would like to thank members of the study group "Shin-Akarui-Angou-Benkyou-Kai" for the valuable discussions and helpful comments, and thank the anonymous reviewers for their comments. This work was supported in part by JSPS KAKENHI Grant Number 26330151 and JSPS A3 Foresight Program.

Appendix

Honest-But-Randomness-Controlling Model

In this section, we show that the honest-but-randomness-controlling (HBRC) model is truly stronger than the honest-but-curious (HBC) model. We also explain the relation between the HBRC model and the honest-but-deterministic (HBD) model proposed by [HW15][7]. To clarify the difference among them, we describe them in a standard setting (not size-hiding settings).

The view of the party P_i during an execution of π with inputs $\vec{x}$ is defined as $\text{view}_i^\pi(\vec{x}) = (x_i, r_i, m_{i_1}, \cdots, m_{i_t})$, where r_i is its internal coin tosses and m_{i_j} is the j-th message that was received by P_i in the protocol execution. We also use $\text{view}_i^\pi(\vec{x})|_{r_i} = (x_i, m_{i_1}, \cdots, m_{i_t})$ to denote $\text{view}_i^\pi(\vec{x})$ on given randomness r_i. Here, if the length of r_i is shorter than the length of its internal randomness, its internal randomness is $r_i||0^k$ for appropriate $k \in \mathbb{N}$.

Definition 5 (HBC). *Let f be a polynomial-time computable n-ary function. We say that π securely computes f in the HBC model if there exists a PPT $\mathcal{S}$ such that for every $I \subsetneq \{1, \cdots, n\}$, every polynomials $q_1, q_2, \cdots, q_n$, $\{\mathcal{S}(1^\kappa, I, \vec{x}_I, \vec{y}_I)\}_{\kappa,\vec{x}} \stackrel{c}{\equiv} \{\text{view}_I^\pi(\vec{x})\}_{\kappa,\vec{x}}$, where $x_1 \in \{0,1\}^{q_1(\kappa)}, \cdots, x_n \in \{0,1\}^{q_n(\kappa)}$.*

Definition 6 (HBRC Model). *Let f be a polynomial-time computable n-ary function. We say that a PPT $\mathcal{R}$ is a* randomness producer *if $\mathcal{R}(1^\kappa, I)$ outputs a vector of strings $\vec{r}_I = (r_{i_1}, \cdots, r_{i_t}) \in (\{0,1\}^*)^{|I|}$ for all $I = \{i_1, \cdots, i_t\} \subsetneq \{1, \cdots, n\}$. We say that π securely computes f in the HBRC model if for every randomness producer $\mathcal{R}$ there exists a PPT $\mathcal{S}$ such that for every $I \subsetneq \{1, \cdots, n\}$, every polynomials $q_1, \cdots, q_n$, $\{\mathcal{S}(1^\kappa, I, \vec{x}_I, \vec{y}_I, \vec{r}_I \leftarrow \mathcal{R}(1^\kappa, I))\}_{\kappa,\vec{x}} \stackrel{c}{\equiv} \{\text{view}_I^\pi(\vec{x})|_{\vec{r}_I}\}_{\kappa,\vec{x}}$, where $x_1 \in \{0,1\}^{q_1(\kappa)}, \cdots, x_n \in \{0,1\}^{q_n(\kappa)}$.*

Definition 7 (HBD Model). *Let f be a polynomial-time computable n-ary function. We say that π securely computes f in the HBD model if there exists a PPT $\mathcal{S}$ such that for every $I \subsetneq \{1, \cdots, n\}$, every polynomials $q_1, q_2, \cdots, q_n$, $\{\mathcal{S}(1^\kappa, I, \vec{x}_I, \vec{y}_I)\}_{\kappa,\vec{x}} \stackrel{c}{\equiv} \{\text{view}_I^\pi(\vec{x})|_0\}_{\kappa,\vec{x}}$, where $x_1 \in \{0,1\}^{q_1(\kappa)}, \cdots, x_n \in \{0,1\}^{q_n(\kappa)}$.*

It is trivial that the security in the HBRC model implies the security in the HBD model. Moreover, the HBRC model implies the HBC model by the following Theorem.

Theorem 4. *Let f be a polynomial-time computable n-ary function. If a protocol π securely computes f in the HBRC model then it also securely computes f in the HBC model.*

---

[7] In original definition in [HW15], the model captures precomputation settings. Our formalization does not include a precomputation for simplicity.

Proof. For simplicity, we consider the case of $n = 2$ and that one party is corrupted. We note that the general case $n > 2$ can be proven in the same way.

Assume that a protocol π is secure in the HBRC model. We construct a PPT $\mathcal{S}'$ that produces $\text{view}^{\pi}(\vec{x})$ given an input $(1^{\kappa}, x, y)$. $\mathcal{S}'$ computes $T = p(\kappa, |x|, |y|)$ and generates an uniformly random string $r \in \{0,1\}^T$. Then, $\mathcal{S}'$ invokes $\mathcal{S}$ on inputs $(1^{\kappa}, x, y, r)$ and outputs the same output as $\mathcal{S}$. Therefore, if a protocol is secure in the HBRC model then it is also secure in the HBC model. □

References

[ACT11] Ateniese, G., Cristofaro, E., Tsudik, G.: (If) size matters: size-hiding private set intersection. In: Catalano, D., Fazio, N., Gennaro, R., Nicolosi, A. (eds.) PKC 2011. LNCS, vol. 6571, pp. 156–173. Springer, Heidelberg (2011). doi:10.1007/978-3-642-19379-8_10

[AJL+12] Asharov, G., Jain, A., López-Alt, A., Tromer, E., Vaikuntanathan, V., Wichs, D.: Multiparty computation with low communication, computation and interaction via threshold FHE. In: Pointcheval, D., Johansson, T. (eds.) EUROCRYPT 2012. LNCS, vol. 7237, pp. 483–501. Springer, Heidelberg (2012). doi:10.1007/978-3-642-29011-4_29

[Cachin04] Cachin, C.: An information-theoretic model for steganography. Inf. Comput. **192**, 41–56 (2004)

[COV15] Chase, M., Ostrovsky, R., Visconti, I.: Executable proofs, input-size hiding secure computation and a new ideal world. In: Oswald, E., Fischlin, M. (eds.) EUROCRYPT 2015. LNCS, vol. 9057, pp. 532–560. Springer, Heidelberg (2015). doi:10.1007/978-3-662-46803-6_18

[CV12] Chase, M., Visconti, I.: Secure database commitments and universal arguments of quasi knowledge. In: Safavi-Naini, R., Canetti, R. (eds.) CRYPTO 2012. LNCS, vol. 7417, pp. 236–254. Springer, Heidelberg (2012). doi:10.1007/978-3-642-32009-5_15

[GMW87] Goldreich, O., Micali, S., Wigderson, A.: How to play any mental game or a completeness theorem for protocols with honest majority. In: STOC, pp. 218–229 (1987)

[HAL09] Hopper, N.J., von Ahn, L., Langford, J.: Provably secure steganography. IEEE Trans. Comput. **58**(5), 662–676 (2009)

[HW15] Hubácek, P., Wichs, D.: On the communication complexity of secure function evaluation with long output. In: ITCS, pp. 163–172 (2015)

[IP07] Ishai, Y., Paskin, A.: Evaluating branching programs on encrypted data. In: Vadhan, S.P. (ed.) TCC 2007. LNCS, vol. 4392, pp. 575–594. Springer, Heidelberg (2007). doi:10.1007/978-3-540-70936-7_31

[LNO13] Lindell, Y., Nissim, K., Orlandi, C.: Hiding the input-size in secure two-party computation. In: Sako, K., Sarkar, P. (eds.) ASIACRYPT 2013. LNCS, vol. 8270, pp. 421–440. Springer, Heidelberg (2013). doi:10.1007/978-3-642-42045-0_22

[MRK03] Micali, S., Rabin, M.O., Kilian, J.: Zero-knowledge sets. In: FOCS, pp. 80–91 (2003)

How to Circumvent the Two-Ciphertext Lower Bound for Linear Garbling Schemes

Carmen Kempka, Ryo Kikuchi(✉), and Koutarou Suzuki

NTT Secure Platform Laboratories, Tokyo, Japan
{kempka.carmen,kikuchi.ryo}@lab.ntt.co.jp

Abstract. At EUROCRYPT 2015, Zahur et al. argued that all linear, and thus, efficient, garbling schemes need at least two k-bit elements to garble an AND gate with security parameter k. We show how to circumvent this lower bound, and propose an efficient garbling scheme which requires less than two k-bit elements per AND gate for most circuit layouts. Our construction slightly deviates from the linear garbling model, and constitutes no contradiction to any claims in the lower-bound proof. With our proof of concept construction, we hope to spur new ideas for more practical garbling schemes.

Our construction can directly be applied to semi-private function evaluation by garbling XOR, XNOR, NAND, OR, NOR and AND gates in the same way, and keeping the evaluator oblivious of the gate function.

Keywords: Garbled circuits · Lower bound on linear garbling schemes · Semi-private function evaluation

1 Introduction

Yao's *garbled circuit* technique [28], modeled as a stand-alone primitive by Bellare et al. [4], is one of the most important techniques to achieve secure two-party computation. In this technique, one of the parties, the *garbler*, creates an encrypted form of a circuit, a so-called *garbled circuit*, which the other party, the *evaluator*, can evaluate without being able to learn anything other than the output of the computed function. Malkhi et al. demonstrated practical feasibility of Yao's technique with their implementation Fairplay [21].

Continued research on Yao's technique has improved its efficiency in terms of computational as well as communication cost. After Yao's original proposal, which needed four ciphertexts to garble a single gate, several techniques have been proposed which reduce the number of ciphertexts in a garbled circuit. The most important works achieve a reduction to a factor roughly between 0.25 and 0.75. Naor et al. [22] pointed out that the number of ciphertexts needed per gate can be reduced from four to three, by setting one of them to the all-zero string. Kolesnikov and Schneider [15] showed how to garble XOR gates "for free", by setting their output keys to be the XOR of their input keys. Pinkas et al. [26] use polynomial interpolation to garble gates with only two ciphertexts per gate. Their technique is not compatible with the free XOR technique.

J.H. Cheon and T. Takagi (Eds.): ASIACRYPT 2016, Part II, LNCS 10032, pp. 967–997, 2016.
DOI: 10.1007/978-3-662-53890-6_32

Recently, Zahur et al. [29] observed that all of these garbling schemes mentioned above share a structure which they model as *linear garbling schemes*. Basically, garbler and evaluator use only XOR operations in the field $GF(2^k)$, and calls to a random oracle, to process the circuit. Zahur et al. showed that garbling an AND gate in this linear structure requires at least two ciphertexts. They further proposed a garbling scheme, the *half gate* construction, which matches this lower bound, and is compatible with the free XOR technique. They concluded that to require less ciphetexts, one needs to employ non-linear, and thus, presumably inefficient techniques. This gives the impression that the optimum we can achieve concerning communication cost in the semi-honest case has already been reached. In this work, we show that this is not necessarily the case.

Our Contribution: We propose an efficient garbling scheme which requires strictly less than two k-bit ciphertexts per AND gate. Our construction is easy to understand and implement, its computational cost comparable to existing practical schemes. Evaluation looks the same for XOR, XNOR, AND, NAND, OR, and NOR gates, so our technique can be applied to secure function evaluation of semi-private functions (SPF-SFE) [25], where the evaluator knows the circuit topology, but not the gate functions. If the positions of XOR gates are known, the number of ciphertext can be further reduced. We prove that our garbling scheme achieves simulation-based privacy [4] in the random oracle model.

Our construction requires only a single k-bit ciphertext for AND gates of which at least one input wire is a circuit input wire. This already seems contradictory to the lower bound, which considers a single AND gate, rather than a whole circuit. All other (inner) AND gates need one additional k-bit ciphertext for adjustment. Thus, general circuits require $1 \leq s < 2$ k-bit ciphertexts[1] per AND gate. In circuits with fan-out one, at least half of the gates are input gates, so we require $1 \leq s \leq 1.5$ ciphertexts per gate. Even though, we do not break the lower bound. We circumvent it by slightly deviating from the linear garbling model, and we do need $5 > 2$ ciphertexts for circuit input gates, and $6 > 2$ ciphertexts for inner gates. But four of them have the length of merely 2 bit.

We demonstrate how we circumvent the lower bound, and hope that our observations sow new ideas for further improvement. We further show that there is at least one other garbling scheme which circumvents the lower bound in a very similar way: a secret-sharing based construction introduced by Kolesnikov in 2005 [13] garbles AND gates with zero ciphertexts. Kolesnikov's technique produces a large blow-up of the input key size, and is impractical for large circuits. It is nonetheless interesting to look at in order to find directions for more efficient constructions.

Idea of our Construction: The linear garbling model performs all operations in $GF(2^k)$. It allows only XOR operations (denoted by $\oplus$) and random oracle calls. In contrast, we also use $(\mathbb{Z}_{2^k}, +)$, where $+$ denotes standard addition, in cases where we need $d + d \neq 0$ for some value d.

---

[1] The case $s = 2$ can only happen in circuits which have no input.

Consider a hash function H and an AND gate with input wires A and B, to which we want to assign input wire labels K_A^0, K_A^1 and K_B^0, K_B^1, respectively, as well as output wire labels K_i^0 and K_i^1. We exploit a similar relation as the free XOR technique, but in $\mathbb{Z}_{2^k}$ rather than $GF(2^k)$: if $K_A^1 = K_A^0 + d$ and $K_B^1 = K_B^0 + d$ for some $d \in \mathbb{Z}_{2^k}$, we have

$$K_A^0 + K_B^1 = K_A^1 + K_B^0 = K_A^0 + K_B^0 + d.$$

The garbler can then set the output wire label K_i^0 to either

$$K_i^0 := H(K_A^0 + K_B^0) \quad \text{or} \quad K_i^0 := H(K_A^0 + K_B^0 + d),$$

each with probability $\frac{1}{2}$, and include the single ciphertext

$$G := H(K_A^0 + K_B^0) \oplus H(K_A^0 + K_B^0 + d)$$

in the garbled circuit. If we further set

$$K_i^1 := H(K_A^0 + K_B^0 + 2d) = H(K_A^1 + K_B^1),$$

the evaluator simply needs to hash his input keys, and XOR the hash value with the ciphertext G if necessary.

Obviously, this construction is not yet secure, since the ciphertext is never used if the gate's output truth value is 1. Therefore, in the case of input $(1, 1)$, with probability $\frac{1}{2}$, we just let the evaluator use the ciphertext anyway, by setting

$$K_i^1 = H(K_A^0 + K_B^0 + 2d) \oplus b_1 G$$

for a random bit $b_1 \in \{0, 1\}$. This way, we need to provide only a single k-bit ciphertext G for security parameter k. The evaluator needs to use G with probability $\frac{1}{2}$ in any case, and learns nothing about the actual input.

Additionally, we need four 2-bit ciphertexts[2] to communicate whether the k-bit ciphertext is to be used or not. Also, the difference d is not preserved for the output wire labels, so for inner gates, we need one additional k-bit ciphertext for adjustment. For the same reason, our construction is not compatible with free XOR. However, XOR gates of which at least one input wire does not depend on the output of an AND gate, can use the free XOR technique and need 0 ciphertexts, while inner XOR gates can be garbled with only one k-bit ciphertext.

How we bypass the lower bound: In all known linear garbling schemes, the operation the evaluator needs to perform, for example, which ciphertext to use, depends on wire-specific permute bits. Changing even one permute bit assigns a differrent operation to the output truth value 1. The lower-bound proof strongly depends on this fact, and on the assumption that all ciphertexts are elements of $GF(2^k)$. However, 2-bit values can be masked with 2-bit ciphertexts[3].

[2] One bit in each ciphertexts contains the actual choice bit whether to use the ciphertext, and the other contains the color bit of the output label.

[3] More precisely, these ciphertexts need only 2 bits of entropy, and can be represented with a bitstring of length 2.

Our scheme can be divided into a k-bit part *not dependent* on any permute bits, and a 2-bit part which depends on permute bits. For the k-bit part, the *same* operation using the *same* ciphertext might be performed by the evaluator for two different inputs, which might even lead to different output truth values. Thus, arguments of the lower-bound proof do not apply[4] to our k-bit part. However, we communicate which operation to perform via several 2-bit ciphertexts, which depend on permute bits in the standard way, and for which all arguments in the lower-bound proof hold — we do need more than two of them per AND gate.

Some Remarks: Our construction offers significant improvements for semi-private functions, where the gate function needs to be hidden and free XOR cannot be used anyway. If the gate functions are known to the evaluator, whether our construction actually performs better than the half gate construction strongly depends on the circuit layout. It might offer significant improvement for circuits with fan-out one consisting mostly of odd gates like AND, NAND, OR and NOR. However, for most interesting circuits, the actual practical improvement might be insignificant or non-existent in the non-semi-private case.

One could argue that, since each circuit input is known either by the garbler or by the evaluator, all input gates can be garbled as half gates, which require only one ciphertext. This would make the half gate construction [29] strictly better than ours in the case of known gate functions. However, this approach has several problems. When used with the cut and choose technique, check circuits would reveal part of the garbler's input if generator half gates on input level are opened. In addition, inputs need to be known at the time of garbling, which makes this approach incompatible with reactive garbling [24], and prevents proponong the garbling process to an offline phase. Compliance with simulation based privacy is unclear, since the simulator does not know the inputs. In addition, this approach seems to contradict the lower bound introduced in the same paper. Nontheless, we introduce an optimization in Appendix A, which combines our scheme with this idea, such that the first *two* gate levels require only one k-bit ciphertexts per AND gate for fortunate circuit layouts.

Other Related Work: There are at least two garbling schemes [12,13] which do not need to communicate any k-bit ciphertexts at all, if the garbled circuit has fan-out one, by garbling the circuit backwards from output gate to input gates. Both schemes produce larger input keys, and when garbling general circuits, require additional ciphertexts. One is the information theoretically secure construction by Kolesnikov [13]. Output keys are secret-shared into the input keys, and no ciphertexts are required at all. However, the secret sharing produces a blow-up in the input key size which is quadratic in the circuit depth. The other, introduced by Kempka et al. [12], creates ciphertexts by hashing public data, sparing the need to communicate them. Fitting decryption keys are then determined by the garbler, who uses a secret trapdoor to invert the ciphertexts with

[4] As we will see, this is also the case in Kolesnikov's scheme [13], where permute bits are only assigned to one input wire per gate.

an inverse trapdoor one-way permutation. Due to the asymmetric primitive, the construction requires a larger security parameter.

Huang et al. [8] garble AND gates as generator half gates with one ciphertext, to realize a permutation network. Paus et al. [25] eliminate constant inputs to reduce the circuit size. Both techniques might be used before applying ours. However, the benefits do not necessarily add up, because they reduce the number of input wires. Compliance with simulation-based privacy [4] is unclear since the simulator does not know the garbler's input.

Secure function evaluation is called *semi-private* (SPF-SFE), if the topology of the circuit is known to the evaluator, but the gate functions are kept secret. As pointed out by Paus et al. [25], Yao's original construction [28] already hides the gate function, and can directly be used for SPF-SFE. The same holds for the three-row reduction (GRR3) [22]. Both constructions allow using free XOR in circuit parts which are known to the evaluator. Paus et al. implement circuits with privately programmable blocks by garbling several functions (sub-circuits) with Yao's construction and multiplexing their output. Their construction can easily be combined with our technique, giving up free XOR for non-private parts, but reducing the garbled circuit size significantly for the private part of the circuit. One limitation here is that we cannot realize left-or-right wire choosing (multiplexing), or constant gates within a single gate. Therefore, the multiplexer-subcircuit by Paus et al. still needs to be realized using Yao[5] circuits. The half gate approach [29] hides which odd gate (AND, NAND, OR, NOR) is evaluated. However, the positions of XOR gates need to be known to the evaluator. The same holds for the GRR2-techniques of Pinkas et al. [26] and Gueron et al. [7]. SPF-SFE is also covered by works on private function evaluation, which additionally hide the circuit topology. Naturally, hiding the topology comes with a lager overhead in the circuit size. Constructions using universal circuits require $O(l \cdot log(l))$ [27] or $O(l \cdot log^2(l))$ [16] additional gates, where l is the number of gates of the original circuit. The LEGO-like construction of Katz and Malka [11] produces less overhead, but requires asymmetric primitives, in particular, one-time homomorphic encryption.

Another line of research focuses on security against malicious adversaries [1,5,6,9,10,17,19,20,23]. This work focuses on the semi-honest case.

2 Preliminaries

2.1 Notation

We use the following notations. By $x \stackrel{U}{\leftarrow} X$, we denote that x is randomly selected from the set X according to uniform distribution, $x \leftarrow \mathsf{Algo}$ denotes that x is the output of a probabilistic algorithm Algo, $A := B$ denotes that A is defined by B, and $[S]_x$ denotes the x-th bit of bitstring S. Our security parameter is k.

[5] It is easy to see that we can use Yao's garbling technique, GRR3 and our technique in the same circuit, and even adjust the difference of output wire labels in gates garbled with Yao's technique or GRR3 on the way.

2.2 Garbling Scheme

In this section, we recall the definition of garbling schemes and the notion of *simulation-based privacy* of Bellare et al. [4].

A circuit is described as $f = (n, m, l, A, B, g)$. Here, $n \geq 2$ is the number of circuit input wires, $m \geq 1$ is the number of circuit output wires, and $l \geq 1$ is the number of gates (and their output wires). Let $W = \{1, ..., n + l\}$ be the set of all wires, $W_{input} = \{1, ..., n\}$ the set of circuit input wires, $W_{output} = \{n+l-m+1, ..., n+l\}$ the set of circuit output wires, and $W_{gate} = \{n+1, ..., n+l\}$ the set of gates (and their output wires). The functions $A : W_{gate} \to W \setminus W_{output}$ and $B : W_{gate} \to W \setminus W_{output}$ specify the first input wire $A(i)$ and the second input wire $B(i)$ of each gate i, respectively. We require $A(i) < B(i) < i$ for all $i \in W_{gate}$. The function $g : W_{gate} \times \{0,1\}^2 \to \{0,1\}$ specifies the gate function $g(i, \cdot, \cdot) = g_i(\cdot, \cdot)$ of each gate i. We leave out the parameter i if it is clear from context. We define the notion of garbling schemes as follows.

Definition 1 (Garbling Scheme). *A garbling scheme for a family of circuits $\mathcal{F} = \{\mathcal{F}_n\}_{n \in \mathbb{N}}$, where n is a polynomial in a security parameter k, consists of probabilistic polynomial-time algorithms* $\mathsf{GC} = (\mathsf{Garble}, \mathsf{Encode}, \mathsf{Eval}, \mathsf{Decode})$ *defined as follows.*

- Garble *takes as input security parameter 1^k and circuit $f \in \mathcal{F}_n$, and outputs garbled circuit F, encoding information e, and decoding information d, i.e.,* $(F, e, d) \leftarrow \mathsf{Garble}(1^k, f)$.
- Encode *takes as input encoding information e and circuit input $x \in \{0,1\}^n$, and outputs garbled input X, i.e.,* $X \leftarrow \mathsf{Encode}(e, x)$.
- Eval *takes as input garbled circuit F and garbled input X, and outputs garbled output Y, i.e.,* $Y \leftarrow \mathsf{Eval}(F, X)$
- Decode *takes as input decoding information d and garbled output Y, and outputs circuit output y, i.e.,* $y \leftarrow \mathsf{Decode}(d, Y)$.

A garbling scheme should have the following correctness *property: for all security parameters k, circuits $f \in \mathcal{F}_n$, and input values $x \in \{0,1\}^n$,* $(F, e, d) \leftarrow \mathsf{Garble}(1^k, f)$, $X \leftarrow \mathsf{Encode}(e, x)$, $Y \leftarrow \mathsf{Eval}(F, X)$, $y \leftarrow \mathsf{Decode}(d, Y)$, *it holds that $y = f(x)$.*

We then define *simulation-based privacy* of garbling schemes as follows. We adapt the notion of Bellare et al. [4] slightly to allow the adversary access to a random oracle H. We denote by $\Phi(f)$ the information about circuit f that is allowed to be leaked by the garbling scheme, e.g., size $\Phi_{size}(f) = (n, m, l)$, topology $\Phi_{topo}(f) = (n, m, l, A, B)$, or the entire information $\Phi_{circ}(f) = (n, m, l, A, B, g)$ of circuit $f = (n, m, l, A, B, g)$.

Definition 2 (Simulation-based Privacy). *For a garbling scheme* $\mathsf{GC} = (\mathsf{Garble}, \mathsf{Encode}, \mathsf{Eval}, \mathsf{Decode})$, *function $f \in \mathcal{F}_n$, input values $x \in \{0,1\}^n$, simulator* Sim, *and random oracle H, the advantage of the adversary $\mathcal{A}$ is defined as* $Adv^{\mathsf{prv.sim}}_{\mathsf{GC},\mathsf{Sim},\Phi,\mathcal{A}}(k) :=$

$$|\Pr[s \leftarrow \mathcal{A}^H(1^k), (F, e, d) \leftarrow \mathsf{Garble}(1^k, f), X \leftarrow \mathsf{Encode}(e, x) : \mathcal{A}^H(s, F, X, d) = 1]$$
$$- \Pr[s \leftarrow \mathcal{A}^H(1^k), (F, X, d) \leftarrow \mathsf{Sim}(1^k, \Phi(f), f(x)) : \mathcal{A}^H(s, F, X, d) = 1]|.$$

A garbling scheme GC = (Garble, Encode, Eval, Decode) *is* private, *if there exists a probabilistic polynomial-time simulator* Sim, *such that for any function* $f \in \mathcal{F}_n$*, input values* $x \in \{0,1\}^n$*, and probabilistic polynomial-time adversary* $\mathcal{A}$*, the advantage* $Adv^{\mathrm{prv.sim}}_{\mathsf{GC},\mathsf{Sim},\Phi,\mathcal{A}}(k)$ *is negligible.*

3 A Garbling Scheme Which Circumvents the Lower Bound

We first describe our basic garbling scheme considering only AND gates, in the semi-honest model. In Sect. 3.2, we describe how to garble other gate types, and application to semi-private functions. Our scheme is not compatible with free XOR, but Sect. 3.3 shows that we can garble XOR gates with 0 or 1 k-bit ciphertexts, and sometimes even inner AND gates can be garbled with 1 k-bit ciphertext. Section 3.4 briefly discusses the malicious case. We estimate efficiency in Sect. 4, and prove that our scheme achieves simulation-based privacy as defined by Bellare et al. [4] in the semi-honest setting in Sect. 5.

3.1 Our Construction

We use the following notation. Let k be our security parameter. With the $+$ symbol, we denote addition in $\mathbb{Z}_{2^k}$. The operation $\oplus$ performs a bitwise XOR on bitstrings. Elements in $\mathbb{Z}_{2^k}$ are interpreted as bitstrings when used with the $\oplus$ operation. The function $\mathsf{lsb}(x)$ returns the least significant bit of its input x, and the function $\mathsf{lsb}_2(x)$ returns the two least significant bits of x.

We assign to each wire i two labels $K_i^0, K_i^1 \in \mathbb{Z}_{2^k}$, where K_i^b represents the truth value $b \in \{0,1\}$ on that wire. To each wire i, we assign a random permute bit λ_i known only to the garbler. Each wire label K_i^b has assigned a bit $c_i^b = \lambda_i \oplus b$, which we call the *color bit* or the *color* of a wire label, in the style of previous work, and to avoid confusion with other choice bits which we describe below. So far, this is no different from most existing garbling schemes. However, jumping ahead, to circumvent the lower bound, the actual operation to compute a gate's output label needs to be somewhat detached from the color bits and the permute bits. To achieve this, we use three additional kinds of choice bits. Their exact role, and their relations among each other as well as to the permute and color bits, will become clear in the scheme description. We provide a brief overview here. In the garbling process, the garbler chooses two random bits b_0 and b_1 for each gate. These bits define by which operation the gate's output labels are computed. The bits b_0 and b_1 are independent of all color bits c_i^b and permute bits λ_i. They need to remain secret, but define a single choice bit $\gamma^{(a,b)}$ for each gate input $(a,b) \in \{0,1\}^2$. The appropriate $\gamma^{(a,b)}$ needs to be communicated to the evaluator. We use the color bits c_i^a, c_i^b of the gate's input wires to point to the correct encryption of the corresponding choice bit $\gamma \in \{\gamma^{(a,b)}\}_{(a,b)\in\{0,1\}^2}$, which then points to the correct operation to compute the gate's output label.

Our garbling algorithm is described in Fig. 1. Encoding of inputs and evaluation are described in Figs. 2 and 3. Decoding consists of XORing the color bits of the circuit output wires with the corresponding permute bits, as specified in Fig. 4. To prevent attacks similar to the one described by Bellare et al. [3], we include a second parameter in our hash function H: a unique tweak j, incremented before each (evaluator's) call to the hash function. This is also done in the half gate construction for similar reasons. We denote this in the same way, using a stateful procedure nextindex(), which increments an internal counter and returns it. For the sake of readability, we leave out this tweak in the following informal description of our garbling scheme.

Let i be an AND gate with input wires A and B. Similar to the free XOR technique, we exploit commutativity of the + operation: if $K_A^1 = K_A^0 + d_i$ and $K_B^1 = K_B^0 + d_i$, we have

$$K_A^0 + K_B^1 = K_A^1 + K_B^0 = K_A^0 + K_B^0 + d_i.$$

We further arrange the output wire labels to be either the hash of the input keys, or the k-bit ciphertext included in the garbled gate XOR this hash value.

In more detail, to garble an AND gate, the garbler chooses two random bits $b_0, b_1 \in \{0,1\}$, sets the output wire label K_i^0 assigned to truth value 0 to

$$K_i^0 := H(K_A^0 + K_B^0 + b_0 d_i),$$

and includes in the garbled circuit the *single* ciphertext

$$G := H(K_A^0 + K_B^0) \oplus H(K_A^0 + K_B^0 + d_i).$$

The garbler further sets the output wire label K_i^1 assigned to truth value 1 to

$$K_i^1 := H(K_A^0 + K_B^0 + 2d_i) \oplus b_1 G = H(K_A^1 + K_B^1) \oplus b_1 G.$$

To evaluate an AND gate, given the input wire labels K_A and K_B, the evaluator needs to compute either $H(K_A + K_B)$ or $H(K_A + K_B) \oplus G$. We let the evaluator know whether he needs to use the ciphertext G via a choice bit γ, which he can compute using his input keys as described below. The choice bit γ does not reveal any information about the input, since for any input combination $(a,b) \in \{0,1\}^2$, the evaluator needs to use the ciphertext with probability $\frac{1}{2}$.

Before we continue our description, let us point out that so far, we have not used any permute bits or color bits. In fact, whether the evaluator needs to use the ciphertext G, only depends on the input, and the bits b_0 and b_1, which are independent of any wire-specific permute bits. This fact plays an important role in circumventing the lower bound on garbling schemes [29]. Details on this can be found in Sect. 6. Arguments in the lower-bound proof show us that to circumvent the lower bound, we need to avoid a direct dependency between permute bits assigned to the input wires and the choice bit γ, which implies that γ cannot be computed by the evaluator as a function of the color bits. Instead, we include in the garbled circuit the four 1-bit ciphertexts

$$b_{(a,b)}^{\gamma} := \mathsf{lsb}(H(K_A^a || K_B^b)) \oplus \gamma^{(a,b)},$$

which encrypt the correct choice bit $\gamma^{(a,b)}$ for each possible input combination $(a,b) \in \{0,1\}^2$. The choice bits $\gamma^{(a,b)}$ only depend on b_0 and b_1, we have

$$\gamma^{(0,0)} = b_0, \gamma^{(0,1)} = \gamma^{(1,0)} = 1 - b_0, \gamma^{(1,1)} = b_1.$$

However, we order the four ciphertext $b^{\gamma}_{(a,b)}$ according to the permute bits λ_A and λ_B of the input wires, so the evaluator can choose the correct ciphertext using his color bits $c^a_A = \lambda_A \oplus a$ and $c^b_B = \lambda_B \oplus b$ as usual.

We still need to describe how the evaluator learns the color bits. For the circuit input wires, we can use the least significant bit of the input wire labels as usual. However, we have little freedom in choosing output wire labels, and thus cannot guarantee their least significant bits to be different. Instead, as for γ, we include in the garbled circuit four additional 1-bit ciphertexts,

$$b^c_{(a,b)} := \mathsf{lsb}(H(K^a_A || K^b_B)) \oplus g_i(a,b) \oplus \lambda_i,$$

among which the evaluator chooses using the color bits of the input wire labels, so together with the four ciphertexts encrypting γ, we would have eight 1-bit ciphertexts in total. To reduce the number of oracle calls, we use the two least significant bits of the hash output, denoted by $\mathsf{lsb}_2(H(.))$, and create the four 2-bit ciphertexts

$$b^c_{(a,b)} || b^{\gamma}_{(a,b)} := \mathsf{lsb}_2(H(K^a_A || K^b_B)) \oplus ((g_i(a,b) \oplus \lambda_i) || \gamma^{(a,b)})$$

instead. This way we avoid having to evaluate the hash function twice on the same input values but with different tweaks.

Unfortunately, we cannot have a global difference d such that $K^1_i = K^0_i + d$ for each wire i. Since the labels of circuit input wires can be chosen freely, they can be given the same difference. However, this difference is not preserved and cannot be controlled in non-input wires. In the next circuit level, gate i's input wires A and B will thus have wire labels (K^0_A, K^1_A) and (K^0_B, K^1_B) with $K^1_A - K^0_A \neq K^1_B - K^0_B$ with high probability. We provide one additional ciphertext to adjust the difference: let λ_B the permute bit on wire B, and the difference d' used for this gate $d' := K^1_A - K^0_A$. Then we set $K^{\lambda_B}_{B'} := K^{\lambda_B}_B$, $K^{1-\lambda_B}_{B'} := K^{\lambda_B}_{B'} + d'$ and include a second k-bit ciphertext $E := K^{1-\lambda_B}_{B'} + K^{1-\lambda_B}_B$ in the garbled circuit. This is why we need two ciphertexts for inner[6] AND gates. The complete garbling algorithm is described in Fig. 1. For better readability, we only describe AND gates in the main algorithm. A discussion about arbitrary gates and semi-private function evaluation can be found in Sect. 3.2.

We cannot use a field with characteristic two to compute addition, since we require $2d \neq 0$ for all differences d occurring in the garbled circuit. Therefore, we perform addition in $\mathbb{Z}_{2^k}$, which gives us a small error probability: there is one element $d_0 \in \mathbb{Z}_{2^k}$ with order 2. Since $K + 2d_0 = K$ for all $K \in \mathbb{Z}_{2^k}$, garbling a

[6] The situation changes when an inner AND gate has only XOR gates as predecessors. In this case, we can use the freedom of key choices in the XOR gates to adjust the difference of the AND gate's input keys.

Garbling algorithm $\mathsf{Garble}(1^k, f)$

Input: Security parameter k, Circuit $f = (n, m, l, A, B, g)$
Algorithm: 1. Initialize empty arrays $F[]$, $e[]$, $d[]$ with $|F| = l$, $|e| = n$ and $|d| = m$.
2. Garbling the gates: For $i := n + 1$ to $l + n$ do:
(a) Set $A := A(i)$ and $B := B(i)$
(b) If undefined, choose permute bits $\lambda_A, \lambda_B \in \{0, 1\}$ at random.
For all $(a, b) \in \{0, 1\}^2$, if undefined, set $c_A^a := \lambda_A \oplus a$, $c_B^b := \lambda_B \oplus b$.
(c) Input keys:
– If A's and B's labels are defined:
Set $d_i := K_A^1 - K_A^0$,
$K_{B'}^{\lambda_B} := K_B^{\lambda_B}$, and
$K_{B'}^{1-\lambda_B} := K_{B'}^{\lambda_B} + (-1)^{\lambda_B} d_i$.
Set $j_E := \mathsf{nextindex}()$,
$E := H(K_B^{1-\lambda_B}, j_E) \oplus K_{B'}^{1-\lambda_B}$.
– If A's labels are defined and B's labels are undefined (vice versa analog), set $d_i := K_A^1 - K_A^0$, choose K_B^0 at random and set $K_B^1 := K_B^0 + d_i$, $K_{B'}^0 := K_B^0$, $K_{B'}^1 := K_B^1$.
– If A's and B's labels are undefined,
choose K_A^0, K_B^0 and d_i at random and
set $K_A^1 := K_A^0 + d_i$, $K_B^1 := K_B^0 + d_i$, $K_{B'}^0 := K_B^0$, $K_{B'}^1 := K_B^1$.
(d) GarbleAND:
Set $j_L := \mathsf{nextindex}()$.
Set $G := H(K_A^0 + K_{B'}^0, j_L) \oplus H(K_A^0 + K_{B'}^0 + d_i, j_L)$.
Choose random bits $b_0, b_1 \in \{0, 1\}$.
Set $K_i^0 := H(K_A^0 + K_{B'}^0 + b_0 d_i, j_L)$.
Set $K_i^1 := H(K_A^0 + K_{B'}^0 + 2d_i, j_L) \oplus b_1 G$.
Set $\gamma^{(0,0)} := b_0$, $\gamma^{(0,1)} := \gamma^{(1,0)} := 1 - b_0$, $\gamma^{(1,1)} := b_1$.
(e) Encrypt choice bits $\gamma^{(a,b)}$ and color bits of output wire:
Set $j_{c,\gamma} := \mathsf{nextindex}()$.
Choose random permute bit $\lambda_i \in \{0, 1\}$.
For all $(a, b) \in \{0, 1\}^2$,
$b^{c,\gamma}_{2c_A^a + c_B^b} := \mathsf{lsb}_2(H(K_A^a || K_{B'}^b, j_{c,\gamma})) \oplus ((g_i(a, b) \oplus \lambda_i) || \gamma^{(a,b)})$,
(f) Set $F[i] := (b_0^{c,\gamma}, b_1^{c,\gamma}, b_2^{c,\gamma}, b_3^{c,\gamma}, G, E)$, if E is defined.
Set $F[i] := (b_0^{c,\gamma}, b_1^{c,\gamma}, b_2^{c,\gamma}, b_3^{c,\gamma}, G)$, otherwise.
(g) If $j \in \{A, B\}$ is a circuit input wire ($j \in W_{input}$),
set $e[j] := (K_j^0 || c_j^0, K_j^1 || c_j^1)$.
If $i \in W_{output}$, set $d[i - (n + l) + m] := \lambda_i$.
Output: Garbled circuit F, encoding e, decoding $d = (\lambda_{n+l-m+1}, \ldots, \lambda_{n+l})$

Fig. 1. The proposed garbling algorithm.

Encoding algorithm $\mathsf{Encode}(e, x)$

Inputs: Garbled input keys e, input x
Algorithm: Parse x to $x = x_1 \ldots x_n$
For $i = 1$ to n do:
Parse $e[i] = (e_0, e_1)$
$X[i] := e_{x_i}$
Return X

Fig. 2. The function Encode.

Evaluation algorithm $\mathsf{Eval}(F, X)$

Inputs: Garbled circuit F, garbled input X
Algorithm: 1. For $j = 1$ to n do
$K_j || c_j := X[j]$
2. Compute wire labels:
For $i := n + 1$ to $l + n$ do
- Set $A := A(i)$ and $B := B(i)$.
- Set $x := 2c_A + c_B$.
- Parse $F[i] = (b_0^{c,\gamma}, b_1^{c,\gamma}, b_2^{c,\gamma}, b_3^{c,\gamma}, G, E)$
- If E is defined:
  - Set $j_E :=$ nextindex().
  - If $c_B = 0$, set $K_B := E \oplus H(K_B, j_E)$.
- Set $j_L :=$ nextindex(), $j_{c,\gamma} :=$ nextindex().
- Compute $c_i || \gamma := \mathsf{lsb}_2(H(K_A || K_B, j_{c,\gamma})) \oplus b_x^{c,\gamma}$.
- Set $K_i := H(K_A + K_B, j_L) \oplus \gamma G$.
3. Return $Y := (c_{n+l-m+1}, \ldots, c_{n+l})$.

Fig. 3. The evaluation algorithm.

Decoding algorithm $\mathsf{Decode}(d, Y)$

Inputs: Decoding d, evaluation output Y
Algorithm: Parse $Y = (c_1, \ldots, c_m)$
Parse $d = (\lambda_1, \ldots, \lambda_m)$
Return $f(x) := (c_1 \oplus \lambda_1, \ldots, c_m \oplus \lambda_m)$

Fig. 4. The function Decode.

Garbling other gate types

GarbleOR:
Set $j_L :=$ nextindex().
Set $G := H(K_A^0 + K_{B'}^0 + d_i, j_L) \oplus H(K_A^0 + K_{B'}^0 + 2d_i, j_L)$.
Choose random bits $b_0, b_1 \in \{0, 1\}$.
Set $K_i^0 := H(K_A^0 + K_{B'}^0, j_L) \oplus b_1 G$.
Set $K_i^1 := H(K_A^0 + K_{B'}^0 + d_i + b_0 d_i, j_L)$.
Set $\gamma^{(0,0)} := b_1$, $\gamma^{(0,1)} := \gamma^{(1,0)} := b_0$, $\gamma^{(1,1)} := 1 - b_0$.

GarbleXOR:
Set $j_L :=$ nextindex().
Set $G := H(K_A^0 + K_{B'}^0, j_L) \oplus H(K_A^0 + K_{B'}^0 + 2d_i, j_L)$.
Choose random bits $b_0, b_1 \in \{0, 1\}$.
Set $K_i^0 := H(K_A^0 + K_{B'}^0 + b_0 2d_i, j_L)$.
Set $K_i^1 := H(K_A^0 + K_{B'}^0 + d_i, j_L) \oplus b_1 G$.
Set $\gamma^{(0,0)} := 1 - b_0$, $\gamma^{(0,1)} := \gamma^{(1,0)} := b_1$, $\gamma^{(1,1)} := b_0$.

Fig. 5. Garbling OR and XOR gates. Garbling NAND, NOR and XNOR can be done by swapping K_i^0 and K_i^1 in the AND, OR and XOR description, respectively.

gate with input wire labels differing by d_0 produces identical output wire labels for this gate, or labels differing by G. However, the error probability is negligible, and the garbler can detect it and start over with different randomness. We need to take care of this in the malicious case, as discussed in Sect. 3.4.

3.2 Arbitrary Gates and Semi-private Function Evaluation

We can garble other odd gates like NAND, OR and NOR, as well as the even gates XOR and XNOR, in a very similar way, by substituting the GarbleAND part (Step (d) in Fig. 1) with the appropriate one in Fig. 5. Evaluation is the same as for AND gates, so the evaluator only needs to know the circuit topology. This makes our construction directly applicable to semi-private functions [25]. To our knowledge, the best construction in previous work which garbles XOR and odd gates in the same way is Yao's original construction with GRR3, which needs three ciphertexts per garbled gate, while our construction needs one k-bit ciphertext for each input gate and two k-bit ciphertexts for each inner gate.

The reason we can easily garble odd and even gates in the same way is the shared additive difference d in $(\mathbb{Z}_{2^k}, +)$ of the gate input wires. In most garbling schemes, a function $F(\cdot,\cdot)$ is applied to the two input wire labels to compute the output labels in some way. Often F is a hash function or a key derivation function. The mapping $F(K_A, K_B)) \mapsto K_i$ has a different input/output pattern for odd and even gates in most garbling schemes (see Table 1): leaving out free-XOR, $F(K_A^a, K_B^b))$ usually has a different value for each of the four gate inputs $(a, b) \in \{0,1\}^2$. In odd gates, three of them are mapped to a value v, and one is mapped to $1-v$, where v depends on the gate type, we call this a 3/1 pattern. In the even gates XOR and XNOR, the two values $F(K_A^0, K_B^0))$ and $F(K_A^1, K_B^1))$ are mapped to a value v, and the other two to $1 - v$, producing the even 2/2 pattern. In our construction, $F(K_A^a, K_B^b)) = H(K_A^a + K_B^b)$. We only have the three values $H(K_A^0 + K_B^0)$, $H(K_A^0 + K_B^0 + d)$, and $H(K_A^0 + K_B^0 + 2d)$. In each gate, two of them are mapped to a value v, and one is mapped to $1-v$, creating a 2/1 pattern for both odd and even gates (see Table 2).

Table 1. Usual output patterns

input	odd gates		even gates
	(N)AND	(N)OR	X(N)OR
$F(K_A^0, K_B^0)$	v	$1-v$	v
$F(K_A^0, K_B^1)$	v	v	$1-v$
$F(K_A^1, K_B^0)$	v	v	$1-v$
$F(K_A^1, K_B^1)$	$1-v$	v	v
Pattern:	3/1	3/1	2/2

Table 2. Output patterns in our construction

input	odd gates		even gates
	(N)AND	(N)OR	X(N)OR
$F(K_A^0 + K_B^0)$	v	$1-v$	v
$F(K_A^0 + K_B^0 + d)$	v	v	$1-v$
$F(K_A^0 + K_B^0 + 2d)$	$1-v$	v	v
Pattern:	2/1	2/1	2/1

3.3 More Efficient Handling of XOR Gates

Our wire labels do not share a global difference Δ with $K_i^1 = K_i^0 \oplus \Delta$ for each wire i. Thus, we cannot use the free XOR technique directly. We can still incorporate its idea in our garbling scheme to save ciphertexts.

Free XOR and 1-Ciphertext-XOR. Input XOR gates can be garbled with zero ciphertexts. An XOR gate with only circuit input wires as input can simply be garbled as in the free XOR technique. Now assume an XOR gate i with input wires A and B with labels $K_A^0, K_A^1, K_B^0, K_B^1$, where B is a circuit input wire, and the labels for A are already defined. We can set $\Delta_i := K_A^0 \oplus K_A^1$, choose K_B^0 at random, set $K_B^1 := K_B^0 \oplus \Delta_i$, $K_i^0 := K_A^0 \oplus K_B^0$, and $K_i^1 := K_i^0 \oplus \Delta_i$.

Inner XOR gates can be garbled using one ciphertext to adjust the difference between the input wire labels in the same way as for the AND gates, but in $GF(2^k)$ rather than $\mathbb{Z}_{2^k}$. Alternatively, one could use the FleXOR technique [14] or the technique by Gueron et al. [7] for inner XOR gates.

Backward Construction for Inner Gates with Preceding XOR Gates. If all paths of an input wire of an inner gate to circuit input wires consist only of XOR gates, we can sometimes adjust these preceding XOR gates in a backward manner, such that we can garble an inner XOR gate for free, or garble an inner AND gate with one ciphertext as if it were an input gate.

As an example, consider the circuit $w_A := w_1 \oplus w_2$, $w_B := w_3 \oplus w_4$, and $w_O := w_A \wedge w_B$ where w_1, w_2, w_3, w_4 are the circuit input wires, w_A and w_B are the left and right input wires of the AND gate, and w_O is the circuit output wire. Using the following construction, we only need 0, 0, and 1 ciphertexts for the left XOR gate, right XOR gate, and AND gate, respectively.

1. Construct the left XOR gate with 0 ciphertexts as in the usual free XOR technique, using some random difference Δ_0. This defines the labels K_A^0, K_A^1 for the left input wire w_A of the AND gate.
2. Define the additive difference $d := K_A^1 - K_A^0$. Select random K_B^0, set $K_B^1 := K_B^0 + d$ for the right input wire w_B of the AND gate. The AND gate can now be garbled with 1 ciphertext.
3. Define the XOR difference $\Delta := K_B^1 \oplus K_B^0$. Select random K_3^0 and K_4^0, set $K_3^1 := K_3^0 \oplus \Delta$ and $K_4^1 := K_4^0 \oplus \Delta$ for the input wires w_3 and w_4 of the right XOR gate. No ciphertexts are needed for this gate.

Using intelligent difference adjustment like this, we can save adjustment ciphertexts for inner gates.

3.4 Security Against Malicious Adversaries

To achieve security against malicious adversaries, we can combine our construction with standard cut and choose [18]. Additional care needs to be taken that a malicious garbler cannot violate correctnes by choosing input wire labels K_A^0, K_A^1

and K_B^0, K_B^1 with a difference d with order 2 in $\mathbb{Z}_{2^k}$. Otherwise, he could set the labels of the output wire to identical values $K_i^0 = K_i^1 := H(K_A^0 + K_B^0) = H(K_A^0 + K_B^0 + 2d)$, or even make the circuit output the same for any input. A standard cut and choose check as in Lindell et al. [18] can prevent this, too.

4 Efficiency

In this efficiency estimation, we focus on the communication cost of our garbling scheme. Computational cost is comparable to existing practical constructions. A comparison of the number of calls to the hash function is listed in Table 3, where we consider plain SFE and handling of XOR gates as described in Sect. 3.3.

Table 3. Number of oracle calls per gate in plain SFE

Technique	Garbler		Evaluator	
	XOR	AND	XOR	AND
classical [2]	4	4	4	4
row reduction (GRR3) [22]	4	4	1	1
row reduction [26]	4	4	1	1
free XOR + GRR3 [15]	0	4	1	1
fleXOR [14]	$\{0,2,4\}$	4	$\{0,1,2\}$	1
half gates [29]	0	4	0	2
this work	$\{0,1\}$	$\{4,5\}$	$\{0,1\}$	$\{2,3\}$

We estimate efficiency in three settings: Plain secure function evaluation (SFE) in which the evaluator knows all gate functions, SPF-SFE in which he only knows the circuit topology, and SFE with semi-private sub-circuits. Garbled odd gates do not differ in size, so it is sufficient to consider AND and XOR gates.

4.1 Efficiency in Plain SFE

First, we estimate efficiency assuming the evaluator knows all gate functions. We call a gate with at least one circuit input wire as input wire an *input gate*, and an *inner gate* is a gate which is not an input gate. Let l_{A} denote the number of AND gates, $l_{\text{A,in}}$ the number of AND gates which are input gates, and $l_{\text{A,mid}} = l_{\text{A}} - l_{\text{A,in}}$ the number of inner AND gates. Similarly, l_{X} denotes the number of XOR gates, $l_{\text{X,in}}$ the number of XOR gates which are input gates, and $l_{\text{X,mid}} = l_{\text{X}} - l_{\text{X,in}}$ the number of inner XOR gates. We have $l = l_{\text{A}} + l_{\text{X}} = l_{\text{A,in}} + l_{\text{A,mid}} + l_{\text{X,in}} + l_{\text{X,mid}}$.

We consider handling XOR gates as described in Sect. 3.3, without the optimization for inner gates preceded by XOR gates. We compare the size of our garbled circuits with several garbling schemes in Table 4. In our construction, an XOR gate requires 0 or 1 k-bit elements, and an AND gate requires 1 or 2 k-bit

Table 4. Size of garbled circuit in the plain SFE

technique	k-bit elements/gate		total bits of garbled circuit
	XOR	AND	
classical [2]	4	4	$4(k+1)l$
row reduction (GRR3) [22]	3	3	$3(k+1)l$
row reduction (GRR2) [26]	2	2	$2(k+1)l$
free XOR + GRR3 [15]	0	3	$3(k+1)l_{\mathrm{A}}$
fleXOR [14]	$\{0,1,2\}$	2	x *s.t.* $2l_{\mathrm{A}}(k+1) \leq x \leq 2(k+1)l$
half gates [29]	0	2	$2l_{\mathrm{A}}(k+1)$
this work	$\{0,1\}$	$\{1,2\}$	$8l_{\mathrm{A}} + k(l_{\mathrm{A,in}} + 2l_{\mathrm{A,mid}} + l_{\mathrm{X,mid}})$

elements, depending on the gate's position in the circuit. The other constructions use the least significant bits of wire labels to communicate color bits. This reduces security by one bit, so $(k+1)$-bit elements are needed to achieve the same security parameter k. Our construction requires 8 bits per gate in addition to the k-bit elements, $8l + k(l_{\mathrm{A,in}} + 2l_{\mathrm{A,mid}} + l_{\mathrm{X,mid}})$ bits in total.

Our construction generates smaller garbled circuits than the half gate construction when $k(l_{\mathrm{A,in}} - l_{\mathrm{X,mid}}) - 8l > 0$, i.e., when there are more input AND gates than inner XOR gates. Although our construction circumvents the lower bound and generates smaller garbled circuits in some cases, the half gate construction may still be the most efficient garbling scheme for most realistic circuits in plain SFE.

4.2 Efficiency in SPF-SFE

Second, we consider semi-private functions, where the evaluator is only allowed to learn the circuit topology. We assume that the garbler knows the function before garbling, and circuits consist of AND, NAND, OR, NOR, XOR and XNOR gates[7]. In the SPF-SFE setting, we garble XOR gates according to Fig. 5 to make them indistinguishable from other gate types. Therefore, the size of a gate does not depend on its type. Let l_{in} denote the number of input gates, l_{mid} the number of inner gates, and $l = l_{\mathrm{in}} + l_{\mathrm{mid}}$ the total number of gates. We compare our construction to other garbling schemes compatible with SPF-SFE in Table 5. We omit GRR2, free XOR + GRR3 and fleXOR in this comparison, since they are less efficient than the half gates approach, and require the evaluator to know the positions of XOR gates. The same is true for the half gates approach, so for SPF-SFE, the circuit has to be transformed into one without XOR gates, which can be done by replacing each XOR gate with two odd gates. Therefore, effectively four ciphertexts are required for an XOR gate in the half gate approach. Note that

[7] Circuits containing multiplexers (for example to realize programmable blocks), or gates with constant output, could use GRR3 only for these gates.

Table 5. Size of garbled circuit in SPF-SFE

technique	k-bit elements/gate	total bits of garbled circuit
classical [2]	4	$4(k+1)l$
GRR3 [22]	3	$3(k+1)l$
half gates [29]	$\{2,4\}$	$(k+1)(2l_{\mathrm{A}}+4l_{\mathrm{X}})$
this work	$\{1,2\}$	$8l+k(l_{\mathrm{in}}+2l_{\mathrm{mid}})$

free XOR cannot be used, because the gate types need to be indistinguishable. As shown in Table 5, our construction is the most efficient one in this setting.

4.3 Efficiency in SFE with Semi-private Sub-circuits

Finally, we consider an evaluator who knows the gate function in some parts of the circuit, and only the topology in the other parts. Let $l^{(\mathrm{pub})}$ be the number of gates of which the evaluator knows the gate function, and $l^{(\mathrm{prv})}$ be the number of the other, "private" gates. Let $l^{(\mathrm{pub})}_{\mathrm{A,in}}$, $l^{(\mathrm{pub})}_{\mathrm{X,in}}$, and $l^{(\mathrm{prv})}_{\mathrm{in}}$ denote the public/private part of $l_{\mathrm{A,in}}$, $l_{\mathrm{X,in}}$, and l_{in}, respectively.

We observe that GRR3 and the half gate construction can be combined easily. The difference between the half gate construction and ours is $k(l^{(\mathrm{pub})}_{\mathrm{A,in}} - l^{(\mathrm{pub})}_{\mathrm{X,mid}}) - 8l^{(\mathrm{pub})}$ in the public part. The difference between GRR3 and our technique is $2k(l^{(\mathrm{prv})} + l^{(\mathrm{prv})}_{\mathrm{in}}) - 5l^{(\mathrm{prv})}$ in the private part. Therefore, our construction generates smaller garbled circuits when $k(l^{(\mathrm{pub})}_{\mathrm{A,in}} + 2l^{(\mathrm{prv})} + 2l^{(\mathrm{prv})}_{\mathrm{in}} - l^{(\mathrm{pub})}_{\mathrm{X,mid}}) - 5l - 3l^{(\mathrm{pub})} > 0$. Which construction is the most efficient depends on how much of the circuit is private, and on the number of inner XOR gates in the public part.

5 Proof of Security

5.1 Correctness

Correctness of our garbling scheme clearly holds. In the case of AND gates, correctness follows from the following equations:

$$((g_i(a,b)\oplus\lambda_i)||\gamma^{(a,b)}) = b^{c,\gamma}_{2c^a_A+c^b_B} \oplus \mathsf{lsb}_2(H(K^a_A||K^b_{B'}, j_{c,\gamma}))$$

and

$$H(K^0_A+K^0_B, j_L)\oplus\gamma^{(0,0)}G = H(K^0_A+K^0_B, j_L)\oplus b_0G = H(K^0_A+K^0_B+b_0d_i, j_L),$$

$$H(K^0_A+K^1_B, j_L)\oplus\gamma^{(0,1)}G = H(K^0_A+K^0_B+d_i, j_L)\oplus(1-b_0)G = H(K^0_A+K^0_B+b_0d_i, j_L),$$

$$H(K^1_A+K^0_B, j_L)\oplus\gamma^{(1,0)}G = H(K^0_A+K^0_B+d_i, j_L)\oplus(1-b_0)G = H(K^0_A+K^0_B+b_0d_i, j_L),$$

$$H(K^1_A+K^1_B, j_L)\oplus\gamma^{(1,1)}G = H(K^0_A+K^0_B+2d_i, j_L)\oplus b_1G.$$

Correctness of the other gate types can be shown analogously.

5.2 Simulation-Based Privacy of Semi-private Functions

By *active* labels we denote labels which are used in an actual evaluation. An *inactive* label is a label which is not active. For example, if the actual truth value on wire i is v_i, $K_i^{v_i}$ is an active label and $K_i^{1-v_i}$ is an inactive one.

In the proof, the simulator can obtain only active labels, and cannot obtain inactive labels and differences d_i. It means that the simulator can compute only one of $\{H(K_A^0 + K_B^0 + bd_A, j)\}_{b\in\{0,1,2\}}$, and one of $\{H(K_A^0 + ad_A||K_B^0 + bd_A, j)\}_{a,b\in\{0,1\}}$.

To simulate the hash values of inactive labels without knowledge of d_i's, the simulator uses the following 10 oracles for a given hash function $H : \{0,1\}^* \longrightarrow \mathbb{Z}_{2^k}$,

- $\mathsf{Corr}_{d_i}^{(1)}(K, b, j) = H(K + bd_i, j)$ where $K \in \mathbb{Z}_{2^k}$ and $b \in \{-2, -1, 1, 2\}$.
- $\mathsf{Corr}_{d_i}^{(2)}(K, b_1, b_2, j) = H(K + b_1d_i, j) \oplus H(K + b_2d_i, j)$ where $K \in \mathbb{Z}_{2^k}$ and $(b_1, b_2) \in \{(-1, -2), (1, 2), (-1, 1)\}$.
- For each (K, j), one can query $\mathsf{Corr}_{d_i}^{(1)}(K, b, j)$ or $\mathsf{Corr}_{d_i}^{(2)}(K, b_1, b_2, j)$ only once (one cannot query both).
- $\mathsf{Corr}_{d_i}^{(3)}(K_1, K_2, a, b, j) = H(K_1 + ad_i||K_2 + bd_i, j)$ where $K_1, K_2 \in \mathbb{Z}_{2^k}$ and $a, b \in \{-1, 0, 1\}$.
- $\mathsf{Corr}_{d_i,d_j}^{(4)}(K, a, b, j') = H(K + ad_j, j') \oplus (K + bd_i)$ where $K \in \mathbb{Z}_{2^k}$ and $a, b \in \{-1, 1\}$.
- $\mathsf{Corr}_{d_i,d_j}^{(5)}(i, j, a, b) = ad_i + bd_j$ where $(a, b) \in \{(1, -1), (-1, 1)\}$.
- $\mathsf{Rand}_{d_i}^{(1)}(K, b, j), \mathsf{Rand}_{d_i}^{(2)}(K, j), \mathsf{Rand}_{d_i}^{(3)}(K_1, K_2, a, b, j)$ and $\mathsf{Rand}_{d_i,d_j}^{(4)}(K, a, b, j)$ output a random value in $\mathbb{Z}_{2^k}$.
- $\mathsf{Rand}_{d_i,d_j}^{(5)}(i, j, a, b)$ chooses $\hat{d}_i$ and $\hat{d}_j$ at random and outputs $a\hat{d}_i + b\hat{d}_j$.

We use $\mathsf{Corr}_{d_A}^{(1)}$, $\mathsf{Corr}_{d_A}^{(2)}$ and $\mathsf{Corr}_{d_A}^{(3)}$ for obtaining $H(K_A^0 + K_B^0 + bd_i, j)$, $H(K_A^0 + K_B^0 + b_1d_i, j) \oplus H(K_A^0 + K_B^0 + b_2d_i, j)$ and $H(K_A^0 + ad_i||K_B^0 + bd_i, j)$, which are used for simulating G and $b^{c,\gamma}$. For simulating E, we use $\mathsf{Corr}_{d_A,d_B}^{(4)}$ and $\mathsf{Corr}_{d_A,d_B}^{(5)}$ to obtain $H(K_B^{1-v_i}, j) \pm d_A$ and $\pm d_A \mp d_B$, respectively.

In the random oracle model, it is clear that $\mathsf{Corr}_{d_i}^{(1)}$, $\mathsf{Corr}_{d_i}^{(2)}$, $\mathsf{Corr}_{d_i}^{(3)}$ and $\mathsf{Corr}_{d_i,d_j}^{(4)}$ output a uniformly random distribution. In addition, each d_i is uniformly random since d_i is either chosen uniformly at random or the difference of two hash values. Therefore, $\mathsf{Corr}_{d_i,d_j}^{(5)}$ and $\mathsf{Rand}_{d_i,d_j}^{(5)}$ output an identical distribution. In our sequence of games, we replace the Corr oracles with Rand oracles to move from the real game to the simulation. The proposed garbling scheme is simulation-based private for $\Phi = \Phi_{topo}$ in the random oracle model.

Theorem 1 (Simulation-based Privacy of Semi-private Functions). *The proposed garbling scheme described in Sect. 3 satisfies simulation-based privacy, for* $\Phi = \Phi_{topo} = (n, m, l, A, B)$ *of ciruit* $f = (n, m, l, A, B, g)$*, of*

$$\mathcal{S}(1^k, \Phi_{topo}(f), f(x))$$

Input: Security parameter k, Circuit $f = (n, m, l, A, B, g)$, and output $f(x)$.

Algorithm: 1. Initialize empty arrays $F[]$, $X[]$ and d with $|F| = l$, $|X| = n$ and $d = m$.

2. Garbling the gates: For $i := n+1$ to $l+n$ do:
   (a) Set $A := A(i)$ and $B := B(i)$
   (b) If undefined, choose permute bits $\lambda_A, \lambda_B \in \{0,1\}$ at random.
   For all $(a,b) \in \{0,1\}^2$, if undefined, set $c_A^a := \lambda_A \oplus a$, $c_B^b := \lambda_B \oplus b$.
   (c) Input keys:
   – If A's and B's labels are defined, set $j_E :=$ nextindex(). If $0 = \lambda_B$, set $K_{B'}^0 := K_B^0$ and $E := \mathsf{Rand}^{(4)}_{d_A, d_B}(K_B^0, 1, 1, j_E)$. Otherwise, set $K_{B'}^0 := K_B^0 + \mathsf{Rand}^{(5)}_{d_A, d_B}(A, B, 1, -1)$ and $E := H(K_B^0, j_E) \oplus K_{B'}^0$.
   – If A's labels are defined and B's labels are undefined (vice versa analog), choose K_B^0 at random and set $K_{B'}^0 := K_B^0$.
   – If A's and B's labels are undefined, choose K_A^0 and K_B^0 at random and set $K_{B'}^0 := K_B^0$.
   (d) Output Keys:
   Choose random bit $b_0, b_1, \in \{0,1\}$ and set $j_L :=$ nextindex().
   Set $\gamma^{(0,0)} := b_0$, $\gamma^{(0,1)} := \gamma^{(1,0)} := 1 - b_0$, $\gamma^{(1,1)} := b_1$.
   Choose random G and $K_i^0 := H(K_A^0 + K_{B'}^0, j_L) \oplus \gamma^{(0,0)} G$.
   (e) Encrypt choice bit:
   Set $j_{c,\gamma} :=$ nextindex(), choose $\lambda_i \in \{0,1\}$ at random.
   – If $(a,b) = (0,0)$,
   $b^{c,\gamma}_{2c_A^0 + c_B^0} := \mathsf{lsb}_2(H(K_A^0 || K_{B'}^0, j_{c,\gamma})) \oplus ((g_i(0,0) \oplus \lambda_i) || \gamma^{(0,0)})$.
   Exception: if $i = l+n$, set
   $b^{c,\gamma}_{2c_A^0 + c_B^0} := \mathsf{lsb}_2(H(K_A^0 || K_{B'}^0, j_{c,\gamma})) \oplus ((g_i(0,0) \oplus \lambda_i \oplus f(x)) || \gamma^{(0,0)})$.
   – Otherwise,
   $b^{c,\gamma}_{2c_A^a + c_B^b} := \mathsf{lsb}_2(\mathsf{Rand}^{(3)}_{d_A}(K_A^0, K_{B'}^0, a, b, j_{c,\gamma})) \oplus ((g_i(a,b) \oplus \lambda_i) || \gamma^{(a,b)})$.
   (f) Set $F[i] := (b_0^{c,\gamma}, b_1^{c,\gamma}, b_2^{c,\gamma}, b_3^{c,\gamma}, G, E)$, if E is defined.
   Set $F[i] := (b_0^{c,\gamma}, b_1^{c,\gamma}, b_2^{c,\gamma}, b_3^{c,\gamma}, G)$, otherwise.
   (g) If $j \in \{A, B\}$ is a circuit input wire, set $X[j] := K_j^0 || c_j^0$.
   If $i \in W_{output}$, set $d[i - (n+l) + m] := \lambda_i$.

Output: Garbled circuit F, garbled input X, decoding d.

Fig. 6. The simulator $\mathcal{S}$.

Definition 2, if we assume that H is a random oracle. More precisely, for any adversary $\mathcal{A}$ there exists an adversary $\mathcal{B}$ such that

$$Adv^{\mathsf{prv.sim}}_{\mathsf{GC},\mathsf{Sim},\Phi_{topo},\mathcal{A}}(k) \leq \frac{l}{2^{k-2}} + \frac{lq}{2^k}$$

where k is the length of keys, l is the number of gates, and q is the number of random oracle queries.

Proof. We consider the simulator $\mathcal{S}$ in the simulated experiment of Definition 2, and the games $\mathcal{G}_0$, $\mathcal{G}_1$, $\mathcal{G}_2^{\mathcal{O}^{(1)}_{d_i}, \mathcal{O}^{(2)}_{d_i}, \mathcal{O}^{(3)}_{d_i}, \mathcal{O}^{(4)}_{d_i,d_j}, \mathcal{O}^{(5)}_{d_i,d_j}}$, $\mathcal{G}_3^{\mathsf{Corr}^{(1)}_{d_i}, \mathsf{Corr}^{(2)}_{d_i}, \mathsf{Corr}^{(3)}_{d_i}, \mathsf{Corr}^{(4)}_{d_i,d_j}, \mathsf{Corr}^{(5)}_{d_i,d_j}}$, and $\mathcal{G}_{real}$. We explain the simulator and the games in the following. For simplicity, we only consider AND, OR, and XOR gates. For NAND, NOR, and XNOR gates, we can swap K_i^0 and K_i^1 in $\mathcal{S}$, $\mathcal{G}_0$, $\mathcal{G}_1$, $\mathcal{G}_2$, $\mathcal{G}_3$, and $\mathcal{G}_{real}$.

$\mathcal{G}_0(1^k, \Phi_{circ}(f), f(x))$

Input: Security parameter k, Circuit $f = (n, m, l, A, B, g)$, and output $f(x)$.
Algorithm: 1. Initialize empty arrays $F[]$, $X[]$ and $d[]$ with $|F| = l$, $|X| = n$ and $|d| = m$.
2. Compute $f(0)$ and $v_i \in \{0,1\}$ that is the actual value on wire i for $x = 0$.
3. Garbling the gates: For $i := n+1$ to $l+n$ do:
(a) Set $A := A(i)$ and $B := B(i)$
(b) If undefined, choose permute bits $\lambda_A, \lambda_B \in \{0,1\}$ at random.
For all $(a,b) \in \{0,1\}^2$, if undefined, set $c_A^a := \lambda_A \oplus a$, $c_B^b := \lambda_B \oplus b$.
(c) Input keys:
Same as $\mathcal{S}$.
(d) Output Keys:
Choose random bit $b_0, b_1, \in \{0,1\}$ and set $j_L :=$ nextindex().
– AND gate case: Set $\gamma^{(0,0)} := b_0$, $\gamma^{(0,1)} := \gamma^{(1,0)} := 1 - b_0$, $\gamma^{(1,1)} := b_1$.
Choose random G and $K_i^{v_i} := H(K_A^{v_A} + K_{B'}^{v_B}, j_L) \oplus \gamma^{(v_A, v_B)} G$.
– OR gate case: Set $\gamma^{(0,0)} := b_1$, $\gamma^{(0,1)} := \gamma^{(1,0)} := b_0$, $\gamma^{(1,1)} := 1 - b_0$.
Choose random G and $K_i^{v_i} := H(K_A^{v_A} + K_{B'}^{v_B}, j_L) \oplus \gamma^{(v_A, v_B)} G$.
– XOR gate case: Set $\gamma^{(0,0)} := 1 - b_0$, $\gamma^{(0,1)} := \gamma^{(1,0)} := b_1$, $\gamma^{(1,1)} := b_0$.
Choose random G and $K_i^{v_i} := H(K_A^{v_A} + K_{B'}^{v_B}, j_L) \oplus \gamma^{(v_A, v_B)} G$.
(e) Encrypt choice bit:
Set $j_{c,\gamma} :=$ nextindex(), choose $\lambda_i \in \{0,1\}$ at random.
– If $(a,b) = (v_A, v_B)$,
$b^{c,\gamma}_{2c_A^a + c_B^b} := \mathsf{lsb}_2(H(K_A^{v_A} || K_{B'}^{v_B}, j_{c,\gamma})) \oplus ((g_i(a,b) \oplus \lambda_i) || \gamma^{(a,b)})$.
Exception: if $i = l + n$, set
$b^{c,\gamma}_{2c_A^a + c_B^b} := \mathsf{lsb}_2(H(K_A^a || K_{B'}^b, j_{c,\gamma})) \oplus ((g_i(a,b) \oplus \lambda_i \oplus f(x)) || \gamma^{(a,b)})$.
– Otherwise,
$b^{c,\gamma}_{2c_A^a + c_B^b} := \mathsf{lsb}_2(\mathsf{Rand}^{(3)}_{d_A}(K_A^{v_A}, K_{B'}^{v_B}, a - v_A, b - v_B, j_{c,\gamma})) \oplus ((g_i(a,b) \oplus \lambda_i) || \gamma^{(a,b)})$.
(f) Set $F[i] := (b_0^{c,\gamma}, b_1^{c,\gamma}, b_2^{c,\gamma}, b_3^{c,\gamma}, G, E)$, if E is defined.
Set $F[i] := (b_0^{c,\gamma}, b_1^{c,\gamma}, b_2^{c,\gamma}, b_3^{c,\gamma}, G)$, otherwise.
(g) If $j \in \{A, B\}$ is a circuit input wire, set $X[j] := K_j^0 || c_j^0$.
If $i \in W_{output}$, set $d[i - (n+l) + m] := \lambda_i$.
Output: Garbled circuit F, garbled input X, decoding d.

Fig. 7. The game $\mathcal{G}_0$ in which the output keys are generated as in the real scheme.

$\mathcal{S}(1^k, \Phi_{topo}(f), f(x))$**:** $\mathcal{S}$ given in Fig. 6 generates the garbled circuit and garbled input (F, X, d) without knowledge of x. $\mathcal{S}$ generates only labels corresponding to truth value 0, chooses G, E uniformly at random. $\mathcal{S}$ chooses $b^c_{2c_A^0 + c_B^0}$ so that Eval and Decode output $f(x)$.

$\mathcal{G}_0(1^k, \Phi_{circ}(f), f(x))$**:** $\mathcal{G}_0$ generates the garbled circuit and garbled input (F, X, d) as described in Fig. 7. In this game, the actual value v_i for each wire i for input $x = 0$ is computed and known to the simulator. $\mathcal{G}_0$ chooses $b^c_{2c_A^a + c_B^b}$ so that Eval and Decode output $f(x)$.

$\mathcal{G}_1(1^k, \Phi_{circ}(f), x)$**:** $\mathcal{G}_0$ generates the garbled circuit and garbled input (F, X, d) as described in Fig. 8. In this game, the actual value v_i for each wire i for input x is computed and known to the simulator. $\mathcal{G}_0$ generates the output label in one of two ways, which depends on v_i.

$\mathcal{G}_2^{\mathcal{O}^{(1)}_{d_i}, \mathcal{O}^{(2)}_{d_i}, \mathcal{O}^{(3)}_{d_i}, \mathcal{O}^{(4)}_{d_i,d_j}, \mathcal{O}^{(5)}_{d_i,d_j}}(1^k, \Phi_{circ}(f), x)$**:** In the game, (F, X, d) is generated as described in Fig. 9, with three oracles without knowledge of d_i's. The oracles

$\mathcal{G}_1(1^k, \Phi_{circ}(f), x)$

Input: Security parameter k, Circuit $f = (n, m, l, A, B, g)$, and input x.
Algorithm: 1. Initialize empty arrays $F[]$, $X[]$ and $d[]$ with $|F| = l$, $|X| = n$ and $|d| = m$.
2. Compute $f(x)$ and $v_i \in \{0,1\}$ that is the actual value on wire i.
3. Garbling the gates: For $i := n+1$ to $l+n$ do:
   (a) Set $A := A(i)$ and $B := B(i)$
   (b) If undefined, choose permute bits $\lambda_A, \lambda_B \in \{0,1\}$ at random.
   For all $(a,b) \in \{0,1\}^2$, if undefined, set $c_A^a := \lambda_A \oplus a$, $c_B^b := \lambda_B \oplus b$.
   (c) Input keys:
   Same as $\mathcal{S}$.
   (d) Output Keys:
   Choose random bit $b_0, b_1, \in \{0,1\}$ and set $j_L :=$ nextindex().
   – AND gate case: Set $\gamma^{(0,0)} := b_0$, $\gamma^{(0,1)} := \gamma^{(1,0)} := 1 - b_0$, $\gamma^{(1,1)} := b_1$.
   Choose random G and $K_i^{v_i} := H(K_A^{v_A} + K_{B'}^{v_B}, j_L) \oplus \gamma^{(v_A,v_B)}G$.
   – OR gate case: Set $\gamma^{(0,0)} := b_1$, $\gamma^{(0,1)} := \gamma^{(1,0)} := b_0$, $\gamma^{(1,1)} := 1 - b_0$.
   Choose random G and $K_i^{v_i} := H(K_A^{v_A} + K_{B'}^{v_B}, j_L) \oplus \gamma^{(v_A,v_B)}G$.
   – XOR gate case: Set $\gamma^{(0,0)} := 1 - b_0$, $\gamma^{(0,1)} := \gamma^{(1,0)} := b_1$, $\gamma^{(1,1)} := b_0$.
   Choose random G and $K_i^{v_i} := H(K_A^{v_A} + K_{B'}^{v_B}, j_L) \oplus \gamma^{(v_A,v_B)}G$.
   (e) Encrypt choice bit:
   Set $j_{c,\gamma} :=$ nextindex(), choose $\lambda_i \in \{0,1\}$ at random.
   – If $(a,b) = (v_A, v_B)$,
   $b^{c,\gamma}_{2c_A^a + c_B^b} := \mathsf{lsb}_2(H(K_A^{v_A} || K_{B'}^{v_B}, j_{c,\gamma})) \oplus ((g_i(a,b) \oplus \lambda_i) || \gamma^{(a,b)})$.
   – Otherwise,
   $b^{c,\gamma}_{2c_A^a + c_B^b} := \mathsf{lsb}_2(\mathsf{Rand}^{(3)}_{d_A}(K_A^{v_A}, K_{B'}^{v_B}, a - v_A, b - v_B, j_{c,\gamma})) \oplus ((g_i(a,b) \oplus \lambda_i) || \gamma^{(a,b)})$.
   (f) Set $F[i] := (b_0^{c,\gamma}, b_1^{c,\gamma}, b_2^{c,\gamma}, b_3^{c,\gamma}, G, E)$, if E is defined.
   Set $F[i] := (b_0^{c,\gamma}, b_1^{c,\gamma}, b_2^{c,\gamma}, b_3^{c,\gamma}, G)$, otherwise.
   (g) If $j \in \{A, B\}$ is a circuit input wire, set $X[j] := K_j^0 || c_j^0$.
   If $i \in W_{output}$, set $d[i - (n+l) + m] := \lambda_i$.

Output: Garbled circuit F, garbled input X, decoding d.

Fig. 8. The game $\mathcal{G}_1$ in which the output keys are generated as in the real scheme.

queried are either $(\mathsf{Rand}^{(1)}_{d_i}, \mathsf{Rand}^{(2)}_{d_i}, \mathsf{Rand}^{(3)}_{d_i}, \mathsf{Rand}^{(4)}_{d_i,d_j}, \mathsf{Rand}^{(5)}_{d_i,d_j})$ or the oracles $(\mathsf{Corr}^{(1)}_{d_i}, \mathsf{Corr}^{(2)}_{d_i}, \mathsf{Corr}^{(3)}_{d_i}, \mathsf{Corr}^{(4)}_{d_i,d_j}, \mathsf{Corr}^{(5)}_{d_i,d_j})$. In $\mathcal{G}_1$, the active labels $K_i^{v_i}$ and oracle outputs $H(K_A^0 + K_B^0 + (v_A + v_B)d, j)$ are computed similar to the real scheme.

For simulating the ciphertext G, $H(K_A^0 + K_B^0, j_L) + H(K_A^0 + K_B^0 + d, j_L)$ has to be computed. The simulator makes oracle query $\mathcal{O}^{(1)}_{d_A}(K_A^{v_A} + K_B^{v_B}, b - (v_A + v_B), j_L)$ if $(v_A, v_B) \neq (1,1)$, and $\mathcal{O}^{(2)}_{d_A}(K_A^{v_A} + K_B^{v_B}, j_L)$ if $(v_A, v_B) = (1,1)$.
For simulating the ciphertexts $b^c_{2c_A^a + c_B^b} || b^\gamma_{2c_A^a + c_B^b}$, the simulator makes oracle query $\mathcal{O}^{(3)}_{d_A}(K_A^{v_A}, K_B^{v_B}, a - v_A, b - v_B, j_{c,\gamma})$, obtains $H(K_A^0 + ad_A || K_B^0 + bd_B, j_{c,\gamma})$, and computes $b^c_{2c_A^a + c_B^b} || b^\gamma_{2c_A^a + c_B^b}$.
The ciphertext E is computed as

$$\begin{aligned} E &= H(K_B^{1-\lambda_B}, j_e) + K_{B'}^{1-\lambda_B} \\ &= \begin{cases} H(K_B^{v_B} + (1 - 2v_B)d_B, j_e) + K_B^{v_B} + (1 - 2v_B)d_A & \text{if } v_B = \lambda_B \\ H(K_B^{v_B}, j_e) + K_B^{v_B} + (2v_B - 1)d_B + (1 - 2v_B)d_A & \text{otherwise} \end{cases} . \end{aligned}$$

$$\mathcal{G}_2^{\mathcal{O}^{(1)}_{d_i}, \mathcal{O}^{(2)}_{d_i}, \mathcal{O}^{(3)}_{d_i}, \mathcal{O}^{(4)}_{d_i,d_j}, \mathcal{O}^{(5)}_{d_i,d_j}}(1^k, \Phi_{circ}(f), x)$$

Input: Security parameter k, Circuit $f = (n, m, l, A, B, g)$, and input x.
Algorithm: 1. Initialize empty arrays $F[]$, $X[]$ and $d[]$ with $|F| = l$, $|X| = n$ and $|d| = m$.

2. Compute $f(x)$ and $v_i \in \{0,1\}$ that is the actual value on wire i.
3. Garbling the gates: For $i := n+1$ to $l+n$ do:
   (a) Set $A := A(i)$ and $B := B(i)$
   (b) If undefined, choose permute bits $\lambda_A, \lambda_B \in \{0,1\}$ at random.
   For all $(a,b) \in \{0,1\}^2$, if undefined, set $c_A^a := \lambda_A \oplus a$, $c_B^b := \lambda_B \oplus b$.
   (c) Input keys:
   - If A's and B's labels are defined, set $j_E :=$ nextindex(). If $v_B = \lambda_B$, set $K_{B'}^{v_B} := K_B^{v_B}$ and $E := \mathcal{O}^{(4)}_{d_A,d_B}(K_B^{v_B}, 1-2v_B, 1-2v_B, j_E)$.
   Otherwise, set $K_{B'}^{v_B} := K_B^{v_B} + \mathcal{O}^{(5)}_{d_A,d_B}(A, B, 1-2v_B, 2v_B - 1)$ and $E := H(K_B^{v_B}, j_E) \oplus K_{B'}^{v_B}$.
   - If A's labels are defined and B's labels are undefined (vice versa analog), choose $K_B^{v_B}$ at random and set $K_{B'}^{v_B} := K_B^{v_B}$.
   - If A's and B's labels are undefined, choose $K_A^{v_A}$ and $K_B^{v_B}$ at random and set $K_{B'}^{v_B} := K_B^{v_B}$.

   (d) Output Keys:
   Choose random bit $b_0, b_1, \in \{0,1\}$ and set $j_L :=$ nextindex().
   - AND gate case: Set $\gamma^{(0,0)} := b_0$, $\gamma^{(0,1)} := \gamma^{(1,0)} := 1 - b_0$, $\gamma^{(1,1)} := b_1$.
     - If $(v_A, v_B) \neq (1,1)$,
     $G := H(K_A^{v_A} + K_{B'}^{v_B}, j_L) \oplus \mathcal{O}^{(1)}_{d_A}(K_A^{v_A} + K_{B'}^{v_B}, b, j_L)$ where $b = 0$ if $v_A + v_B = 1$, $b = 1$ if $v_A + v_B = 0$,
     $K_i^{v_i} := H(K_A^{v_A} + K_{B'}^{v_B}, j_L) \oplus \gamma^{(v_A,v_B)} G$.
     - If $(v_A, v_B) = (1,1)$,
     $G := \mathcal{O}^{(2)}_{d_A}(K_A^{v_A} + K_{B'}^{v_B}, -1, -2, j_L)$,
     $K_i^{v_i} := H(K_A^{v_A} + K_{B'}^{v_B}, j_L) \oplus \gamma^{(v_A,v_B)} G$.
   - OR gate case: Set $\gamma^{(0,0)} := b_1$, $\gamma^{(0,1)} := \gamma^{(1,0)} := b_0$, $\gamma^{(1,1)} := 1 - b_0$.
     - If $(v_A, v_B) \neq (0,0)$,
     $G := H(K_A^{v_A} + K_{B'}^{v_B}, j_L) \oplus \mathcal{O}^{(1)}_{d_A}(K_A^{v_A} + K_{B'}^{v_B}, b, j_L)$ where $b = 1$ if $v_A + v_B = 2$, $b = 2$ if $v_A + v_B = 1$,
     $K_i^{v_i} := H(K_A^{v_A} + K_{B'}^{v_B}, j_L) \oplus \gamma^{(v_A,v_B)} G$.
     - If $(v_A, v_B) = (0,0)$,
     $G := \mathcal{O}^{(2)}_{d_A}(K_A^{v_A} + K_{B'}^{v_B}, 1, 2, j_L)$,
     $K_i^{v_i} := H(K_A^{v_A} + K_{B'}^{v_B}, j_L) \oplus \gamma^{(v_A,v_B)} G$.
   - XOR gate case: Set $\gamma^{(0,0)} := 1 - b_0$, $\gamma^{(0,1)} := \gamma^{(1,0)} := b_1$, $\gamma^{(1,1)} := b_0$.
     - If $(v_A, v_B) \neq (0,1), (1,0)$,
     $G := H(K_A^{v_A} + K_{B'}^{v_B}, j_L) \oplus \mathcal{O}^{(1)}_{d_A}(K_A^{v_A} + K_{B'}^{v_B}, b, j_L)$ where $b = 0$ if $v_A + v_B = 2$, $b = 2$ if $v_A + v_B = 0$,
     $K_i^{v_i} := H(K_A^{v_A} + K_{B'}^{v_B}, j_L) \oplus \gamma^{(v_A,v_B)} G$.
     - If $(v_A, v_B) = (0,1), (1,0)$,
     $G := \mathcal{O}^{(2)}_{d_A}(K_A^{v_A} + K_{B'}^{v_B}, -1, 1, j_L)$,
     $K_i^{v_i} := H(K_A^{v_A} + K_{B'}^{v_B}, j_L) \oplus \gamma^{(v_A,v_B)} G$.

   (e) Encrypt choice bit:
   Set $j_{c,\gamma} :=$ nextindex(), choose $\lambda_i \in \{0,1\}$ at random.
   - If $(a,b) = (v_A, v_B)$,
   $b^{c,\gamma}_{2c_A^a + c_B^b} := \mathsf{lsb}_2(H(K_A^{v_A} || K_{B'}^{v_B}, j_{c,\gamma})) \oplus ((g_i(a,b) \oplus \lambda_i) || \gamma^{(a,b)})$.
   - Otherwise,
   $b^{c,\gamma}_{2c_A^a + c_B^b} := \mathsf{lsb}_2(\mathcal{O}^{(3)}_{d_A}(K_A^{v_A}, K_{B'}^{v_B}, a - v_A, b - v_B, j_{c,\gamma})) \oplus ((g_i(a,b) \oplus \lambda_i) || \gamma^{(a,b)})$.

   (f) Set $F[i] := (b_0^{c,\gamma}, b_1^{c,\gamma}, b_2^{c,\gamma}, b_3^{c,\gamma}, G, E)$, if E is defined.
   Set $F[i] := (b_0^{c,\gamma}, b_1^{c,\gamma}, b_2^{c,\gamma}, b_3^{c,\gamma}, G)$, otherwise.
   (g) If $j \in \{A, B\}$ is a circuit input wire, set $X[j] := K_j^{v_j} || c_j^{v_j}$.
   If $i \in W_{output}$, set $d[i - (n+l) + m] := \lambda_i$.

Output: Garbled circuit F, garbled input X, decoding d.

Fig. 9. The game $\mathcal{G}_2$ in which garbling with actual values.

$$\mathcal{G}_3^{\mathsf{Corr}^{(1)}_{d_i},\mathsf{Corr}^{(2)}_{d_i},\mathsf{Corr}^{(3)}_{d_i},\mathsf{Corr}^{(4)}_{d_i,d_j},\mathsf{Corr}^{(5)}_{d_i,d_j}}(1^k, \Phi_{circ}(f), x)$$

Input: Security parameter k, Circuit $f = (n, m, l, A, B, g)$, and input x.
Algorithm: 1. Initialize empty arrays $F[]$, $X[]$ and $d[]$ with $|F| = l$, $|X| = n$ and $|d| = m$.

2. Compute $f(x)$ and $v_i \in \{0,1\}$ that is the actual value in wire i.
3. Garbling the gates: For $i := n+1$ to $l+n$ do:
   (a) Set $A := A(i)$ and $B := B(i)$
   (b) If undefined, choose permute bits $\lambda_A, \lambda_B \in \{0,1\}$ at random. For all $(a,b) \in \{0,1\}^2$, if undefined, set $c_A^a := \lambda_A \oplus a$, $c_B^b := \lambda_B \oplus b$.
   (c) Input keys: Same as $\mathcal{G}_2^{\mathsf{Corr}^{(1)}_{d_i},\mathsf{Corr}^{(2)}_{d_i},\mathsf{Corr}^{(3)}_{d_i},\mathsf{Corr}^{(4)}_{d_i,d_j},\mathsf{Corr}^{(5)}_{d_i,d_j}}$ except setting $K_A^{1-v_A} := (1-2v_A)d_A + K_A^{v_A}$, $K_B^{1-v_B} := (1-2v_B)d_A + K_B^{v_B}$, $K_{B'}^{1-v_B} := (1-2v_B)d_A + K_{B'}^{v_B}$ if undefined.
   (d) Output Keys: Same as $\mathcal{G}_2^{\mathsf{Corr}^{(1)}_{d_i},\mathsf{Corr}^{(2)}_{d_i},\mathsf{Corr}^{(3)}_{d_i,d_j},\mathsf{Corr}^{(4)}_{d_i,d_j},\mathsf{Corr}^{(5)}_{d_i,d_j}}$ except adding the following in the last step:
      - AND gate case:
        If $v_i = 0$, set $K_i^{1-v_i} := H(K_A^{v_A} + K_{B'}^{v_B} + 2d_i, j_L) \oplus b_1 G$.
        If $v_i = 1$, set $K_i^{1-v_i} := H(K_A^{v_A} + K_{B'}^{v_B} + b_0 d_i, j_L)$.
      - OR gate case:
        If $v_i = 0$, set $K_i^{1-v_i} := H(K_A^{v_A} + K_{B'}^{v_B} + d_i + b_0 d_i, j_L)$.
        If $v_i = 1$, set $K_i^{1-v_i} := H(K_A^{v_A} + K_{B'}^{v_B}, j_L) \oplus b_1 G$.
      - XOR gate case:
        If $v_i = 0$, set $K_i^{1-v_i} := H(K_A^{v_A} + K_{B'}^{v_B} + d_i, j_L) \oplus b_1 G$.
        If $v_i = 1$, set $K_i^{1-v_i} := H(K_A^{v_A} + K_{B'}^{v_B} + b_0 2d_i, j_L)$.
   (e) Encrypt choice bit: Same as $\mathcal{G}_2^{\mathsf{Corr}^{(1)}_{d_i},\mathsf{Corr}^{(2)}_{d_i},\mathsf{Corr}^{(3)}_{d_i},\mathsf{Corr}^{(4)}_{d_i,d_j},\mathsf{Corr}^{(5)}_{d_i,d_j}}$.
   (f) Set $F[i] := (b_0^{c,\gamma}, b_1^{c,\gamma}, b_2^{c,\gamma}, b_3^{c,\gamma}, G, E)$, if E is defined. Set $F[i] := (b_0^{c,\gamma}, b_1^{c,\gamma}, b_2^{c,\gamma}, b_3^{c,\gamma}, G)$, otherwise.
   (g) If $j \in \{A, B\}$ is a circuit input wire, set $X[j] := K_j^{v_j} || c_j^{v_j}$. If $i \in W_{output}$, set $d[i-(n+l)+m] := \lambda_i$.

Output: Garbled circuit F, garbled input X, decoding d.

Fig. 10. The game $\mathcal{G}_3$ in which including inactive keys.

The simulator makes oracle query $\mathcal{O}^{(4)}_{d_A,d_B}(K_B^{v_B}, 1-2v_B, 1-2v_B, j_e)$ instead of computing $H(K_B^{v_B} + (1-2v_B)d_B, j_e) + (1-2v_B)d_A$, and oracle query $\mathcal{O}^{(5)}_{d_A,d_B}(A, B, 2v_B-1, 1-2v_B)$ instead of computing $(2v_B-1)d_B + (1-2v_B)d_A$.

$\mathcal{G}_3^{\mathsf{Corr}^{(1)}_{d_i},\mathsf{Corr}^{(2)}_{d_i},\mathsf{Corr}^{(3)}_{d_i},\mathsf{Corr}^{(4)}_{d_i,d_j},\mathsf{Corr}^{(5)}_{d_i,d_j}}(1^k, \Phi_{circ}(f), x)$: This game, described in Fig. 10, is almost identical to $\mathcal{G}_1$, except that inactive labels are defined. In this game, the simulator knows d_i's.

$\mathcal{G}_{real}$: This is the real experiment of Definition 2.

Now we prove the indistinguishability between the simulators and the real protocol. We use the following chain of simulators and hybrid games.

1. $\mathcal{S} \equiv \mathcal{G}_0$: The only difference between the two games is that K^0 is used in $\mathcal{S}$ but K^v is used in $\mathcal{G}_0$ and there are 3 cases of AND, OR, and XOR in $\mathcal{G}_0$. However,

the distributions of simulation are identical, since the inactive $\gamma^{(a,b)}$'s are masked by random oracle Rand.

2. $\mathcal{G}_0 \equiv \mathcal{G}_1$: The only difference between the two games is that v_i for input $x = 0$ is used and output $f(x)$ is embedded in $\mathcal{G}_0$ but v_i for input x is used in $\mathcal{G}_1$. However, the distributions of simulation are identical, since the inactive $\gamma^{(a,b)}$'s are masked by random oracle Rand and embedded $f(x)$ is masked by random λ_i.
3. $\mathcal{G}_1 \equiv \mathcal{G}_2^{\mathsf{Rand}^{(1)}_{d_i},\mathsf{Rand}^{(2)}_{d_i},\mathsf{Rand}^{(3)}_{d_i},\mathsf{Rand}^{(4)}_{d_i,d_j},\mathsf{Rand}^{(5)}_{d_i,d_j}}$: The only difference between them is how G is generated. However, G is uniformly distributed in both games and therefore these two games are indistinguishable.
4. $\mathcal{G}_2^{\mathsf{Rand}^{(1)}_{d_i},\mathsf{Rand}^{(2)}_{d_i},\mathsf{Rand}^{(3)}_{d_i},\mathsf{Rand}^{(4)}_{d_i,d_j},\mathsf{Rand}^{(5)}_{d_i,d_j}} \equiv \mathcal{G}_2^{\mathsf{Corr}^{(1)}_{d_i},\mathsf{Corr}^{(2)}_{d_i},\mathsf{Corr}^{(3)}_{d_i},\mathsf{Corr}^{(4)}_{d_i,d_j},\mathsf{Corr}^{(5)}_{d_i,d_j}}$:
The Rand oracles are replaced with Corr oracles. These games are indistinguishable since the hash function is a random oracle except if $d_i = 0$ or $2d_i = 0$ for some i. However, l d_i's are chosen uniformly at random and then the probability of the event is bounded by $\Pr[\exists i \text{ s.t. } (d_i = 0) \vee (2d_i = 0)] = 1 - (1 - 2/2^k)^l \leq l/2^{k-2}$.
5. $\mathcal{G}_2^{\mathsf{Corr}^{(1)}_{d_i},\mathsf{Corr}^{(2)}_{d_i},\mathsf{Corr}^{(3)}_{d_i},\mathsf{Corr}^{(4)}_{d_i,d_j},\mathsf{Corr}^{(5)}_{d_i,d_j}} \equiv \mathcal{G}_3^{\mathsf{Corr}^{(1)}_{d_i},\mathsf{Corr}^{(2)}_{d_i},\mathsf{Corr}^{(3)}_{d_i},\mathsf{Corr}^{(4)}_{d_i,d_j},\mathsf{Corr}^{(5)}_{d_i,d_j}}$: The only difference between these two games is adding the inactive labels. However, the inactive labels are not used, so the distribution is unchanged. In $\mathcal{G}_1^{\mathsf{Corr}^{(1)}_{d_i},\mathsf{Corr}^{(2)}_{d_i},\mathsf{Corr}^{(3)}_{d_i},\mathsf{Corr}^{(4)}_{d_i,d_j},\mathsf{Corr}^{(5)}_{d_i,d_j}}$, if the adversary correctly guess one of l d_i's, and ask a key unknown to the simulator for random oracle H among q oracle queries, the simulator fails to simulate random oracle H since the simulator does not know d_i's. The probability of this simulation failure is $lq/2^k$.
6. $\mathcal{G}_3^{\mathsf{Corr}^{(1)}_{d_i},\mathsf{Corr}^{(2)}_{d_i},\mathsf{Corr}^{(3)}_{d_i},\mathsf{Corr}^{(4)}_{d_i,d_j},\mathsf{Corr}^{(5)}_{d_i,d_j}} \equiv \mathcal{G}_{real}$: In $\mathcal{G}_2$, we first define $K_i^{v_i}$ and then define $K_i^{1-v_i}$ as $K_i^{1-v_i} := (1 - 2v_i)d_i + K_i^{v_i}$ if either or both input wires is a circuit input wire. In $\mathcal{G}_{real}$, we first define K_i^0 and then $K_i^1 := d_i + K_i^0$. In addition, we define $K_i^{1-v_i} := H(K_A^{v_A} + K_{B'}^{v_B} + (2-(v_A+v_B))d_A, j_L) + b_1 G$ or $K_i^{1-v_i} := H(K_A^{v_A} + K_{B'}^{v_B} + (b_0 - 2)d_A, j_L)$, depending on v_i and b_0 for inner wires in $\mathcal{G}_2$. In $\mathcal{G}_{real}$, we define $K_i^0 := H(K_A^0 + K_{B'}^0 + 2d_A, j_L) + b_1 G$ and $K_i^1 := H(K_A^0 + K_{B'}^0 + b_0 d_A, j_L) + b_1 G$. Although the steps to generate the labels are changed, the outputs are unchanged. Therefore, these changes do not affect the distribution.

Consequently, the simulated experiment is indistinguishable from the real experiment except negligible probability $l/2^{k-2} + lq/2^k$. □

6 On the Lower Bound of Linear Garbling Schemes

Zahur et al. [29] observed that many practical garbling schemes share a common structure, which they formalize in their model of linear garbling schemes. They proved that in this model, garbling a single AND gate requires at least two rows. They concluded that to garble an odd gate with significantly[8] less than $2k$ bits, an inherently different, non-linear structure is needed. Our garbling scheme contradicts this, while maintaining a computational efficiency comparable to previous work. In this chapter, we first provide an intuition of how our construction circumvents the lower bound. In Sect. 6.2, we provide an outline of the lower-bound proof. Then, we state our garbling scheme in the linear garbling model, and show how it exploits loopholes in the lower-bound proof more formally. Since this chapter mostly discusses a single AND gate, we denote the color bits of a gate's input wires by α and β for better readability.

6.1 How We Circumvent the Lower Bound: An Intuition

Intuitively, the arguments in the lower-bound proof should also hold for our "almost linear" garbling scheme. We now show where our construction exploits loopholes. The proof relies on two assumptions which hold for most linear constructions, but are not needed for linearity in an algebraic sense:

Assumption (1): The linear operations to compute a gate's output labels *directly* depend on permute bits assigned to the gate's input wires.
Assumption (2): Each ciphertext/row consists of k bit.

Let us first take a closer look at Assumption (1). The lower-bound proof strongly depends on the fact that changing the permute bits assigned to the two input wires of a gate changes the operation the evaluator needs to perform when processing this gate. This is true for most existing garbling schemes, but not for ours. In existing schemes, the evaluator usually uses two color bits α and β to choose one out of four options. In Yao's original scheme [28], when used with the point and permute technique [2], the four options are four ciphertexts. In the three-row reduction [22,26], the options are three ciphertexts and the zero string. In the interpolation-based two-row reduction [26], the options are four x-coordinates. The half gate construction [29] has the options ciphertext or zero string for each half gate, so four possible options per garbled AND gate. The common way to let the evaluator choose the correct option is letting the options depend on permute bits, and communicating corresponding color bits to the evaluator, which keeps him oblivious of the actual input. All of the above mentioned schemes use this technique. As a side-effect, in all these constructions, changing even one permute bit inevitably changes the assignment of options to input values, in particular, which option is assigned to the input leading to output truth value 1. In our scheme, the evaluator has only two options: to use the given k-bit ciphertext or not. Neither this ciphertext itself nor its usage depends on any permute bit. In fact, it might happen that the *same* operation using this *same* ciphertext might be performed by the evaluator for two possible inputs $(x_a, x_b) \neq (x'_a, x'_b)$, with $g(x_a, x_b) \neq g(x'_a, x'_b)$. Since we have only two options, we need only one choice bit, which does not depend on permute bits.

[8] As they point out, one can only prove a lower-bound of at least $2k$ minus some bits, as one can always take away a few bits and maintain asymptotic security.

To communicate this choice bit, we include four 2-bit ciphertexts in the garbled circuit, which do depend on permute bits. And this is where we exploit the second loophole in the lower-bound proof: Assumption (2), which says that each row has the length of k bit, is neither used nor needed[9] in any arguments in the proof. Since $M < k$ bit of information can perfectly be masked with a M-bit ciphertext, we can instead "fill the necessary rows" with our 2-bit ciphertexts.

6.2 How We Circumvent the Lower Bound: Formal Discussion

Let us first briefly summarize the linear garbling model. All elements considered in the model are in $GF(2^k)$, and the only operations allowed are XOR operations and calls to a random oracle, which outputs elements in $GF(2^k)$. The model considers garbling a *single* AND gate. Let r and h be constants, and let $\langle \cdot, \cdot \rangle$ denote the scalar product of two vectors. The garbler chooses $R_1, \ldots, R_r \in GF(2^k)$ at random. Using linear combinations of these as inputs to a random oracle, he obtains oracle responses $Q_1, \ldots, Q_h$. Let $S := (R_1, \ldots, R_r, Q_1, \ldots, Q_h)$. The garbler applies linear functions on S to obtain input wire labels A_0, A_1, B_0, B_1, output wire labels C_0, C_1 and ciphertexts $G_1, \ldots, G_m$. The function to obtain the ciphertexts can be written as a matrix $\mathbb{G}_{\lambda_a,\lambda_b}$ with $S \cdot \mathbb{G}_{\lambda_a,\lambda_b} = (G_1, \ldots, G_m)$, and can depend on the permute bits λ_a and λ_b of the input wires.

The evaluator obtains as input the wire labels $K_A \in \{A_0, A_1\}$ and $K_B \in \{B_0, B_1\}$, and color bits α and β. He makes several oracle queries using this input to obtain a vector T, which consists of his input, the oracle responses and the ciphertexts $G_1, \ldots, G_m$. He computes a linear function on T, denoted by a vector $V_{\alpha,\beta}$, to compute the output wire label $C_{(\lambda_a \oplus \alpha) \wedge (\lambda_b \oplus \beta)} = \langle V_{\alpha,\beta}, T \rangle$.

The Lower-Bound Proof. We recap the parts of the lower-bound proof [29] which are important for a more formal discussion. The proof argues that the matrix $\mathbb{G}_{\lambda_a,\lambda_b}$ must have at least two rows, and thus creates at least two ciphertexts. This is based on a chain of claims, of which we circumvent the first one: it says that the $\mathbb{G}_{\lambda_a,\lambda_b}$ are all distinct. The claim is argued for as follows. The output wire label $C_{(\lambda_a \oplus \alpha) \wedge (\lambda_b \oplus \beta)}$, computed by the evaluator as $C_{(\lambda_a \oplus \alpha) \wedge (\lambda_b \oplus \beta)} = \langle V_{\alpha,\beta}, T \rangle$, can be written as

$$C_{(\lambda_a \oplus \alpha) \wedge (\lambda_b \oplus \beta)} = \left\langle V^{pub}_{\alpha,\beta}, \mathbb{M}_{\alpha,\beta} \times S^T \right\rangle + \left\langle V^{prv}_{\alpha,\beta}, \mathbb{G}_{\lambda_a,\lambda_b} \times S^T \right\rangle, \quad (1)$$

for an appropriate matrix $\mathbb{M}_{\alpha,\beta}$, where $V_{\alpha,\beta}$ is divided into a public part $V^{pub}_{\alpha,\beta}$, independent of permute bits, and a private part $V^{prv}_{\alpha,\beta}$, which depends on λ_a and λ_b. For only *one* input $((\lambda_a \oplus \alpha), (\lambda_b \oplus \beta))$, it holds that $(\lambda_a \oplus \alpha) \wedge (\lambda_b \oplus \beta) = 1$, and thus $C_{(\lambda_a \oplus \alpha) \wedge (\lambda_b \oplus \beta)} = C_1$, the label assigned to truth value 1. When changing a permute bit, a different combination of (α, β) is assigned to C_1. However, since all other values in Eq. 1 do not depend on permute bits, only $\mathbb{G}_{\lambda_a,\lambda_b}$ can change

[9] More precisely, elements in $GF(2^k)$ or $\mathbb{Z}_{2^k}$ do not necessarily have k bits of entropy, and those with less entropy can be represented using shorter strings.

when changing λ_a or λ_b. Thus, all $\mathbb{G}_{\lambda_a,\lambda_b}$ must be distinct. Basic algebra then implies that all $\mathbb{G}_{\lambda_a,\lambda_b}$ must have at least two rows. If all rows have k bit, this implies the lower bound of $2k$ bits per gate. In our garbing scheme, we divide $\mathbb{G}_{\lambda_a,\lambda_b}$ into a k-bit(-entropy) part and a 2-bit(-entropy) part. Our k-bit part does not depend on permute bits and has only one row. The 2-bit part has four rows and thus does not contradict the arguments in the lower-bound proof.

Our Construction and the Model of Linear Garbling Schemes. We now compare our scheme to the linear garbling model, and explain how it bypasses the lower bound more formally. Since the lower-bound proof only considers a single AND gate, the labels of both input wires can be chosen freely, and we can leave out the ciphertext for difference adjustment in the following discussion.

Our construction does not perfectly fit into the model in two points. The first point is that we use $\mathbb{Z}_{2^k}$ rather than $\mathbb{F}_{2^k}$, simply because we need $+$ and $\oplus$ to be different operations. The second point is that the linear garbling model considers only k-bit values. In contrast, we use oracles with k-bit output and with 2-bit output. The 2-bit oracle is implemented by using the lsb_2 function on the k-bit oracle output. Similarly, we have k-bit ciphertexts and 2-bit ciphertexts.

A garbling algorithm in the linear garbling model consists of five steps. We describe our scheme in these steps, using the same enumeration as in [29]. We omit the tweaks implemented by nextindex() in the calls to the random oracle H.

1. The garbler chooses several random k-bit values. The only 1-bit randomness considered in the model are the permute bits. In our scheme, the garbler chooses the random k-bit values K_A^0, K_B^0, and d, and the random bits b_0 and b_1. So we allow 1-bit randomness here, which is only a technical issue.
2. The garbler makes several oracle queries, using the random values from Step 1 as input. The random values and oracle responses form a vector S, on which all following linear operations are performed. In our construction, we have k-bit queries and 2-bit queries. We divide S into the two vectors S_k, containing k-bit values, and S_2, containing 2-bit values. We have:
   k-bit queries:
   $Q_1 := H(K_A^0 + K_B^0)$, $Q_2 := H(K_A^0 + K_B^0 + d)$, $Q_3 := H(K_A^0 + K_B^0 + 2d)$,
   $\Rightarrow S_k = (K_A^0, K_B^0, d, Q_1, Q_2, Q_3)$.
   2-bit queries Q_{4-7}:
   $Q_{2(\lambda_a \oplus a)+(\lambda_B \oplus b)+1} = \mathsf{lsb}_2(H(K_A^0 + ad || K_B^0 + bd))$ for all $(a, b) \in \{0,1\}^2$.
3. The random permute bits λ_A, λ_B and λ_C are chosen.
4. Linear operations are performed on S to compute the input wire labels A_0, A_1, B_0, B_1 and the output wire labels C_0, C_1. The latter can be written as $C_i = \langle C_{\lambda_a,\lambda_b,i}, S\rangle$, $i \in \{0,1\}$, for appropriate vectors $C_{\lambda_a,\lambda_b,i}$, which can depend on permute bits. In our case, we have:

$$(A_0, B_0, A_1, B_1, C_0, C_1) = \begin{pmatrix} 1 & 0 & 0 & 0 & 0 & 0 \\ 0 & 1 & 0 & 0 & 0 & 0 \\ 1 & 0 & 1 & 0 & 0 & 0 \\ 0 & 1 & 1 & 0 & 0 & 0 \\ 0 & 0 & 0 & 1-b_0 & b_0 & 0 \\ 0 & 0 & 0 & b_1 & b_1 & 1 \end{pmatrix} S_k$$

The rows $C_{\lambda_a,\lambda_b,0} = (0,0,0,1-b_0,b_0,0)$ and $C_{\lambda_a,\lambda_b,1} = (0,0,0,b_1,b_1,1)$ define the output labels. They depend on b_0 and b_1, but not on λ_a and λ_b.

5. Ciphertexts are computed: for $i \in [m]$, $G_i = \left\langle G^{(i)}_{\lambda_a,\lambda_b}, S \right\rangle$ for appropriate $G^{(i)}_{\lambda_a,\lambda_b}$, where m is the number of ciphertexts included in the garbled circuit. In our scheme, we have k-bit and 2-bit ciphertexts:

$$G_1^{k-bit} = \left\langle G^{(1)}_{\lambda_a,\lambda_b}, S_k \right\rangle, G_i^{2-bit} = \left\langle G^{(i)}_{\lambda_a,\lambda_b}, S_2 \right\rangle, i = 2, \ldots, 5.$$

Let $\mathbb{G}_{\lambda_a,\lambda_b}$ be the matrix consisting of the rows $G^{(i)}_{\lambda_a,\lambda_b}$ for $i \in [m]$. We divide $\mathbb{G}_{\lambda_a,\lambda_b}$ into a k-bit part and a 2-bit part. The k-bit part has only one row:

$$\mathbb{G}^{k-bit}_{\lambda_a,\lambda_b} = G^{(1)}_{\lambda_a,\lambda_b} = (0,0,0,1,1,0)$$

for all $(\lambda_a, \lambda_b) \in \{0,1\}^2$. So our k-bit ciphertext is

$$G_1^{k-bit} := \langle (0,0,0,1,1,0), S_k \rangle = (H(K_A^0 + K_B^0) \oplus H(K_A^0 + K_B^0 + d)).$$

As we can see, $\mathbb{G}^{k-bit}_{\lambda_a,\lambda_b}$ does not depend on λ_a and λ_b, so changing the permute bits *cannot* change $\mathbb{G}^{k-bit}_{\lambda_a,\lambda_b}$. This seems like a contradiction to the lower-bound proof, which argues that changing the permute bits must change $\mathbb{G}_{\lambda_a,\lambda_b}$ in order to assign a different pair (α, β) to the label C_1. However, in our construction, the labels C_0 and C_1 only depend on the random bits b_0 and b_1 chosen by the garbler. Thus, the assignment of color bits to output truth values is irrelevant for the computation of the labels on the garbler's side. However, to allow the evaluator to compute the correct output label, a choice bit γ needs to be communicated using ciphertexts which do depend on α and β. These are computed by the rows of the 2-bit part $\mathbb{G}^{2-bit}_{\lambda_a,\lambda_b}$, which takes care of the dependence on permute bits this way. Thus, the k-bit part $\mathbb{G}^{k-bit}_{\lambda_a,\lambda_b}$ can stay unchanged for changing permute bits (and thus consist of only one row), without causing a contradiction to the lower-bound proof.

To enable the evaluator to compute the color bit $(\alpha \oplus \lambda_A)(\beta \oplus \lambda_B) \oplus \lambda_C$ of the output wire, we define the four rows $G^{(2)-(5)}_{\lambda_a,\lambda_b}$ of the 2-bit part such that

$$b^c_{2\alpha+\beta} || b^\gamma_{2\alpha+\beta} = \left\langle G^{(2+2\alpha+\beta)}_{\lambda_A,\lambda_B}, S_2 \right\rangle$$

$$= \mathsf{lsb}_2(H(K_A^{\alpha \oplus \lambda_A} + K_B^{\beta \oplus \lambda_B})) \oplus (((\alpha \oplus \lambda_A)(\beta \oplus \lambda_B) \oplus \lambda_C) || \gamma^{(\alpha,\beta)})$$

for all $(\alpha, \beta) \in \{0,1\}^2$, where

$$\gamma^{(\lambda_A,\lambda_B)} = b_0, \gamma^{(1\oplus\lambda_A,\lambda_B)} = \gamma^{(\lambda_A,1\oplus\lambda_B)} = 1 - b_0, \gamma^{(1\oplus\lambda_A,1\oplus\lambda_B)} = b_1.$$

The order of the rows in $\mathbb{G}^{2-bit}_{\lambda_a,\lambda_b}$ depends on λ_a and λ_b. Thus, in compliance with the lower-bound proof, $\mathbb{G}^{2-bit}_{\lambda_a,\lambda_b}$ is different for each choice of (λ_a, λ_b).

6.3 Further Analyzing the Lower Bound

In 2005, Kolesnikov [13] introduced an information-theoretically secure secret sharing based garbling scheme which requires zero ciphertexts, using only XOR operations. Kolesnikov's construction produces an exponential blow-up of the input key size, so the comparison is slightly unfair. It does not fit into the model for the simple reason that the wire labels of one input wire are not elements in $GF(2^k)$, but have the form $B_L||B_R$ for $B_L, B_R \in GF(2^k)$. Regardless, Kolesnikov's construction seems like a contradiction to the lower bound. We analyze how this "almost linear" construction circumvents the lower bound.

Kolesnikov also introduces an optimization which reduces the blow-up to approximately $\sum d_j^2$, where d_j is the depth of the jth leaf of the circuit. However, the optimization is irrelevant for our analysis.

Outline of Kolesnikov's Construction. Kolesnikov's construction works by garbling circuits backwards from output gates to input gates. Consider garbling an AND gate i with yet undefined input wire labels $K_A^0, K_A^1, K_B^0, K_B^1$, and given output wire labels K_i^0 and K_i^1. The labels K_i^0 and K_i^1 are secret-shared in the following way: Assign a *single* random permute bit λ_A to input wire A, no permute bit is assigned to the second input wire B. Choose the input labels K_A^0 and K_A^1 at random, append λ_A to K_A^0, and $1 - \lambda_A$ to K_A^1. The labels K_B^0 and K_B^1 each consist of two entries, which are permuted according to λ_A: $K_B^0 := K_i^0 \oplus K_A^{\lambda_A}||K_i^0 \oplus K_A^{1-\lambda_A}$, and $K_B^1 := K_i^{\lambda_A} \oplus K_A^{\lambda_A}||K_i^{1-\lambda_A} \oplus K_A^{1-\lambda_A}$. To evaluate gate i, the evaluator XORs his label K_A with the entry of his label K_B which is indicated by the color bit appended to K_A.

Kolesnikov's Construction and the Lower Bound. Kolesnikov's construction is linear in an algebraic sense. It circumvents the lower bound in a way similar to our scheme: the operations performed by the evaluator do not depend on two permute bits. Kolesnikov's construction only assigns permute bits to "A-wires" to indicate which part of the "B-wire" to use. "B-wires" are not assigned any permute or color bit. Thus we have only one bit assigned to four possible input combinations, making claim one in the lower-bound proof meaningless. And in fact, similar to the k-bit part of our construction, the same linear operation is performed for different truth values on the output wire.

6.4 Conclusion

If less than two rows imply less than two possible operations, only one or no choice bit is needed, making claim one in the lower-bound proof meaningless. It is left for future work whether our observations can be used to break the lower bound and omit even the small ciphertexts altogether without input key blow-up. It would be interesting whether garbling schemes with less than two k-bit rows can be constructed without sacrifising free XOR.

Acknowledgements. We thank the reviewers for their helpful and constructive comments.

A Combining our Construction with Half Gates

By combining our technique with the half gate construction [29], we can garble the first two gate levels in a circuit with only one ciphertext per AND gate and 0 ciphertexts per XOR gate, if the circuit layout is fortunate. Each circuit input is known either by the garbler or the evaluator, so one could argue that all input gates can be garbled as half gates. A similar technique is used by Huang et al. [8], who use generator half gates as input gates. We need to modify the half gate technique such that output wires i which are used as inputs for AND gates get an additive global difference d such that $K_i^1 = K_i^0 + d$, and output wires j which are used as input to an XOR gate get a global difference Δ such that $K_j^1 = K_j^0 \oplus \Delta$. We cannot do both at the same time, so this only saves ciphertexts when most output wires of input gates go to either an AND gate or an XOR gate, but not both. If this is the case, the next level can be garbled with our construction using one ciphertext for most AND gates, and zero ciphertexts for most XOR gates. A half gate can produce an additive difference in its output wire using the following modification: A generator half gate with input a known to the garbler produces the ciphertexts $H(K_B) \oplus K_i^0$ and $H(K_B \oplus \Delta) \oplus (K_i^0 + ad)$, of which one is set to the zero string as in the original scheme. It is evaluated as in the original half gate construction. An evaluator half gate, where the evaluator knows input a, and gets input labels K_A^a and K_B, consists of the ciphertexts $G_1 = H(K_A^0) \oplus K_C^0$, which is set to the zero string by setting $K_C^0 = H(K_A^0)$, and $G_2 = H(K_A^1) \oplus (K_C^0 - B)$. In the evaluator half gate, we require an additive difference $K_B^1 - K_B^0 = d$ for the labels of input wire B. If $a = 0$, it is evaluated by computing $H(K_A^0) \oplus G_1$. If $a = 1$, the evaluator computes $(G_2 \oplus H(K_A^1)) + K_B$.

References

1. shelat, A., Shen, C.: Two-output secure computation with malicious adversaries. In: Paterson, K.G. (ed.) EUROCRYPT 2011. LNCS, vol. 6632, pp. 386–405. Springer, Heidelberg (2011). doi:10.1007/978-3-642-20465-4_22
2. Beaver, D., Micali, S., Rogaway, P.: The round complexity of secure protocols (extended abstract). In: STOC, pp. 503–513 (1990)
3. Bellare, M., Hoang, V.T., Keelveedhi, S., Rogaway, P.: Efficient garbling from a fixed-key blockcipher. In: IEEE Symposium on Security and Privacy, pp. 478–492. IEEE Computer Society (2013)
4. Bellare, M., Hoang, V.T., Rogaway, P.: Foundations of garbled circuits. In: ACM CCS, pp. 784–796. ACM (2012)

5. Brandão, L.T.A.N.: Secure two-party computation with reusable bit-commitments, via a cut-and-choose with forge-and-lose technique. In: Sako, K., Sarkar, P. (eds.) ASIACRYPT 2013. LNCS, vol. 8270, pp. 441–463. Springer, Heidelberg (2013). doi:10.1007/978-3-642-42045-0_23
6. Frederiksen, T.K., Jakobsen, T.P., Nielsen, J.B., Nordholt, P.S., Orlandi, C.: MiniLEGO: efficient secure two-party computation from general assumptions. In: Johansson, T., Nguyen, P.Q. (eds.) EUROCRYPT 2013. LNCS, vol. 7881, pp. 537–556. Springer, Heidelberg (2013). doi:10.1007/978-3-642-38348-9_32
7. Gueron, S., Lindell, Y., Nof, A., Pinkas, B.: Fast garbling of circuits under standard assumptions. In: Ray, I., Li, N., Kruegel, C. (eds.) ACM CCS, pp. 567–578. ACM (2015)
8. Huang, Y., Evans, D., Katz, J.: Private set intersection: Are garbled circuits better than custom protocols? In: NDSS. The Internet Society (2012)
9. Huang, Y., Katz, J., Evans, D.: Efficient Secure two-party computation using symmetric cut-and-choose. In: Canetti, R., Garay, J.A. (eds.) CRYPTO 2013. LNCS, vol. 8043, pp. 18–35. Springer, Heidelberg (2013). doi:10.1007/978-3-642-40084-1_2
10. Huang, Y., Katz, J., Kolesnikov, V., Kumaresan, R., Malozemoff, A.J.: Amortizing garbled circuits. In: Garay, J.A., Gennaro, R. (eds.) CRYPTO 2014. LNCS, vol. 8617, pp. 458–475. Springer, Heidelberg (2014). doi:10.1007/978-3-662-44381-1_26
11. Katz, J., Malka, L.: Constant-round private function evaluation with linear complexity. In: Lee, D.H., Wang, X. (eds.) ASIACRYPT 2011. LNCS, vol. 7073, pp. 556–571. Springer, Heidelberg (2011). doi:10.1007/978-3-642-25385-0_30
12. Kempka, C., Kikuchi, R., Kiyoshima, S., Suzuki, K.: Garbling scheme for formulas with constant size of garbled gates. In: Iwata, T., Cheon, J.H. (eds.) ASIACRYPT 2015. LNCS, vol. 9452, pp. 758–782. Springer, Heidelberg (2015). doi:10.1007/978-3-662-48797-6_31
13. Kolesnikov, V.: Gate evaluation secret sharing and secure one-round two-party computation. In: Roy, B. (ed.) ASIACRYPT 2005. LNCS, vol. 3788, pp. 136–155. Springer, Heidelberg (2005). doi:10.1007/11593447_8
14. Kolesnikov, V., Mohassel, P., Rosulek, M.: FleXOR: flexible garbling for XOR gates that beats free-XOR. In: Garay, J.A., Gennaro, R. (eds.) CRYPTO 2014. LNCS, vol. 8617, pp. 440–457. Springer, Heidelberg (2014). doi:10.1007/978-3-662-44381-1_25
15. Kolesnikov, V., Schneider, T.: Improved garbled circuit: free XOR gates and applications. In: Aceto, L., Damgård, I., Goldberg, L.A., Halldórsson, M.M., Ingólfsdóttir, A., Walukiewicz, I. (eds.) ICALP 2008. LNCS, vol. 5126, pp. 486–498. Springer, Heidelberg (2008). doi:10.1007/978-3-540-70583-3_40
16. Kolesnikov, V., Schneider, T.: A practical universal circuit construction and secure evaluation of private functions. In: Tsudik, G. (ed.) FC 2008. LNCS, vol. 5143, pp. 83–97. Springer, Heidelberg (2008). doi:10.1007/978-3-540-85230-8_7
17. Lindell, Y.: Fast cut-and-choose based protocols for malicious and covert adversaries. In: Canetti, R., Garay, J.A. (eds.) CRYPTO 2013. LNCS, vol. 8043, pp. 1–17. Springer, Heidelberg (2013). doi:10.1007/978-3-642-40084-1_1
18. Lindell, Y., Pinkas, B.: An efficient protocol for secure two-party computation in the presence of malicious adversaries. In: Naor, M. (ed.) EUROCRYPT 2007. LNCS, vol. 4515, pp. 52–78. Springer, Heidelberg (2007). doi:10.1007/978-3-540-72540-4_4
19. Lindell, Y., Pinkas, B.: Secure two-party computation via cut-and-choose oblivious transfer. In: Ishai, Y. (ed.) TCC 2011. LNCS, vol. 6597, pp. 329–346. Springer, Heidelberg (2011). doi:10.1007/978-3-642-19571-6_20

20. Lindell, Y., Riva, B.: Cut-and-choose yao-based secure computation in the online/offline and batch settings. In: Garay, J.A., Gennaro, R. (eds.) CRYPTO 2014. LNCS, vol. 8617, pp. 476–494. Springer, Heidelberg (2014). doi:10.1007/978-3-662-44381-1_27
21. Malkhi, D., Nisan, N., Pinkas, B., Sella, Y.: Fairplay - a secure two-party computation system. In: USENIX Security Symposium, pp. 287–302 (2004)
22. Naor, M., Pinkas, B., Sumner, R.: Privacy preserving auctions and mechanism design. In: ACM conference on Electronic commerce, pp. 129–139 (1999)
23. Nielsen, J.B., Orlandi, C.: LEGO for two-party secure computation. In: Reingold, O. (ed.) TCC 2009. LNCS, vol. 5444, pp. 368–386. Springer, Heidelberg (2009). doi:10.1007/978-3-642-00457-5_22
24. Nielsen, J.B., Ranellucci, S.: Foundations of reactive garbling schemes. Cryptology ePrint Archive, Report 2015/693 (2015). http://eprint.iacr.org/2015/693
25. Paus, A., Sadeghi, A.-R., Schneider, T.: Practical secure evaluation of semi-private functions. In: Abdalla, M., Pointcheval, D., Fouque, P.-A., Vergnaud, D. (eds.) ACNS 2009. LNCS, vol. 5536, pp. 89–106. Springer, Heidelberg (2009). doi:10.1007/978-3-642-01957-9_6
26. Pinkas, B., Schneider, T., Smart, N.P., Williams, S.C.: Secure two-party computation is practical. In: Matsui, M. (ed.) ASIACRYPT 2009. LNCS, vol. 5912, pp. 250–267. Springer, Heidelberg (2009). doi:10.1007/978-3-642-10366-7_15
27. Valiant, L.G.: Universal circuits (preliminary report). In: STOC, pp. 196–203. ACM (1976)
28. Yao, AC.-C.: How to generate and exchange secrets (extended abstract). In: FOCS, pp. 162–167 (1986)
29. Zahur, S., Rosulek, M., Evans, D.: Two halves make a whole. In: Oswald, E., Fischlin, M. (eds.) EUROCRYPT 2015. LNCS, vol. 9057, pp. 220–250. Springer, Heidelberg (2015). doi:10.1007/978-3-662-46803-6_8

Constant-Round Asynchronous Multi-Party Computation Based on One-Way Functions

Sandro Coretti[1(✉)], Juan Garay[2], Martin Hirt[3], and Vassilis Zikas[4]

[1] New York University, New York City, USA
corettis@nyu.edu
[2] Yahoo Research, Sunnyvale, USA
garay@yahoo-inc.com
[3] Department of Computer Science, ETH Zurich, Zurich, Switzerland
hirt@inf.ethz.ch
[4] Department of Computer Science, RPI, Troy, USA
vzikas@cs.rpi.edu

Abstract. Secure multi-party computation (MPC) allows several mutually distrustful parties to securely compute a joint function of their inputs and exists in two main variants: In *synchronous* MPC parties are connected by a synchronous network with a global clock, and protocols proceed in *rounds* with strong delivery guarantees, whereas *asynchronous* MPC protocols can be deployed even in networks that deliver messages in an arbitrary order and impose arbitrary delays on them.

The two models—synchronous and asynchronous—have to a large extent developed in parallel with results on both feasibility and asymptotic efficiency improvements in either track. The most notable gap in this parallel development is with respect to round complexity. In particular, although under standard assumptions on a synchronous communication network (availability of secure channels and broadcast), synchronous MPC protocols with (exact) constant rounds have been constructed, to the best of our knowledge, thus far no constant-round asynchronous MPC protocols based on standard assumptions are known, with the best protocols requiring a number of rounds that is linear in the multiplicative depth of the arithmetic circuit computing the desired function.

In this work we close this gap by providing the first constant-round asynchronous MPC protocol that is optimally resilient (i.e., it tolerates up to $t < n/3$ corrupted parties), adaptively secure, and makes black-box use of a pseudo-random function. It works under the standard network

The full version of this paper can be found on the Cryptology ePrint Archive [18]

S. Coretti—Work supported by the Swiss NSF project no. 200020-132794.

J. Garay and V. Zikas—Work done in part while the author was visiting the Simons Institute for the Theory of Computing, supported by the Simons Foundation and by the DIMACS/Simons Collaboration in Cryptography through NSF grant #CNS-1523467.

V. Zikas—Work supported in part by the Swiss NSF Ambizione grant PZ00P2_142549.

V. Zikas and S. Correti—Work done in part while these authors were at ETH Zurich

J.H. Cheon and T. Takagi (Eds.): ASIACRYPT 2016, Part II, LNCS 10032, pp. 998–1021, 2016.
DOI: 10.1007/978-3-662-53890-6_33

assumptions for protocols in the asynchronous MPC setting, namely, a complete network of point-to-point (secure) asynchronous channels with eventual delivery and asynchronous Byzantine agreement (aka consensus). We provide formal definitions of these primitives and a proof of security in the Universal Composability framework.

1 Introduction

In secure multi-party computation (MPC), a set of n parties $p_1, \ldots, p_n$, each holding some private input, wish to jointly compute a function on these inputs in a fashion such that even up to t colluding adversarial parties are unable to obtain any information beyond what they can extract from their inputs and outputs or to affect the computation in any way other than contributing their desired inputs. The problem of MPC has been studied in the two important settings of *synchronous* and *asynchronous* networks, respectively.

MPC protocols for the synchronous setting assume a network in which parties proceed in rounds, with the guarantee that messages sent by any party in any given round are delivered to all recipients in the same round. Consequently, in all such protocols the parties are assumed to be (at least partially) synchronized, i.e., to be in the same round at all times.

In real-world networks, such as the Internet, this synchrony assumption corresponds to assuming that the parties have (partially) synchronized clocks and communicate over channels with a known upper-bounded latency. The synchronous structure is then imposed by "timeouts," i.e., in each round the parties wait for an amount T of time defined by their estimate of when other parties send their messages and the bound on the network latency. If their estimate is accurate and their clocks are indeed synchronized, this will ensure that parties receive all messages sent to them from honest senders before the end of the round (timeout). Thus, after time T has passed, they can safely assume that if a message was expected for the current round but has not been received, then the sender must be adversarial. The security of synchronous protocols heavily relies on this assumption. In fact, many of them would become completely insecure if there is even a single delayed message. As a result, the round length T must typically be set much higher than the average transmission time.

A natural question is therefore to study the security one can obtain if no synchrony assumption is made but merely under the assumption that messages sent by honest parties are *eventually* delivered. [1] In particular, messages sent by parties can be reordered arbitrarily and delayed by arbitrary (albeit finite) amounts of time in such an asynchronous network. Note that one could consider even more pessimistic networks where the adversary can block messages sent by honest parties; this is for example the case in the base network assumed in Canetti's UC framework [13]. In such networks, however, protocols cannot be

[1] The eventual-delivery assumption is supported by the fact that whenever a message is dropped or delayed for too long, Internet protocols typically resend that message.

guaranteed to (eventually) terminate as the adversary can delay the computation indefinitely.

In asynchronous MPC protocols parties do not wait until a round times out. Rather, as soon as a party has received enough messages to compute its next message[2], it computes that message, sends it, and moves on. In that sense, asynchronous MPC protocols are "opportunistic" and terminate essentially as quickly as the network allows. Hence, they can be much faster than their synchronous counterparts depending on the network latency. [3]

In this work, unless explicitly stated otherwise, whenever we refer to the *asynchronous (communication) model*, we mean the above asynchronous model with eventual delivery.

On the importance of round complexity. The inherent need for waiting until each round times out clearly makes round complexity an important consideration for the performance of *synchronous* MPC protocols. Indicatively, Schneider and Zohner [34] have shown that as the latency between machines increases, the cost of each round becomes more and more significant.

Despite their opportunistic nature, round complexity is just as important a consideration for *asynchronous* protocols, since a protocol's round complexity can be a more relevant efficiency metric than, for example, its bit complexity. Indeed, at the conceptual/theoretical level, having constant-round protocols allows us to use them as sub-routines in a higher level protocol without blowing up (asymptotically) the round complexity of the calling protocol, while at the practical level, communication time is often dominated by the round-trip times in the network and not by the size of the messages. For example, it takes about the same amount of time to transmit a byte and a megabyte, while sending a message from A to B over many intermediate nodes, computing something at B, and sending an answer back to A may take a (comparatively) long time.

Our contributions. In this paper, we first formalize the asynchronous model with *eventual delivery* in the universal composability (UC) framework [13], introduce a suitable formal notion of *asynchronous* round complexity, and formulate the basic communication resources (such as asynchronous secure channel and asynchronous Byzantine agreement [A-BA]) as ideal functionalities in that model.[4] (See Sect. 3.) We then present the—to the best of our knowledge—first *constant-round* MPC protocol for this asynchronous setting (i.e., a protocol whose round complexity is independent of the multiplicative depth of the evaluated circuit and the number n of parties) based on standard assumptions, namely, the existence

[2] What "enough" means is concretely specified by the party's protocol.

[3] This speed up, however, does not come for free, as it inevitably allows the adversary to exclude some of the honest parties' inputs from being considered in the computation.

[4] Note that while the UC framework already is asynchronous, asynchronous communication *with eventual delivery* has not been modeled in it so far; moreover, standard (asynchronous) UC protocols do not achieve achieve eventual termination/delivery (cf. Sect. 3).

of pseudo-random functions (PRFs).[5] The protocol is UC-secure in the secure-channels model with A-BA, and makes *black-box* use of the underlying PRF, tolerating a computationally bounded, *adaptive* adversary actively corrupting up to $t < n/3$ parties, which is optimal for this setting.[6]

At a high level, here is how we construct our constant round protocol. First, we devise a constant-depth circuit for computing the keys, masked values, and (shares of the) garbled gates needed for a distributed evaluation of a Yao garbled circuit that encodes the function the parties wish to compute. This circuit is then evaluated by means of a linear-round (in the depth of the circuit and in n) asynchronous protocol. However, this circuit is Boolean whereas all existing asynchronous protocols evaluate arithmetic circuits. To deal with this mismatch we devise an asynchronous protocol for computing Boolean circuits by appropriately adapting the protocol by Ben-Or, Kelmer, and Rabin [8]. Any party who receives the output from the evaluation of the Boolean circuit uses it to encrypt shares of each garbled gate, which it sends to all other parties. Finally, each party *locally* evaluates the (distributed) garbled circuit by decrypting incoming encrypted shares of each gate and reconstructing the function table of the gate as soon as sufficiently many consistent shares have arrived until all gates are evaluated. Once all gates are evaluated in this fashion, the party is in possession of the output. The protocol and its analysis are presented in Sect. 4.

Related work. Beaver, Micali, and Rogaway [2] were the first to provide a constant-round MPC protocol in the synchronous stand-alone model. (Refer to Appendix A for a more detailed and historical account of the development of MPC protocols in both the synchronous and asynchronous settings, together with the tools that are used in each setting.) Their protocol is secure in the computational setting and tolerates an adaptive adversary who actively corrupts up to $t < n/2$ parties. The complexity of [2] was improved by Damgård and Ishai [19], who provided the first constant-round protocol making black-box use of the underlying cryptographic primitive (a pseudo-random generator). Importantly, both [2] and [19] assume a broadcast channel, an assumption essential for obtaining constant-round MPC. Indeed, as proved in [20,22], it is impossible to implement such a broadcast channel from point-to-point communication in a constant number of rounds, and although expected constant-round broadcast protocols exist in the literature (e.g., [21,30]), using them to instantiate calls within the constructions of [2] or [19] would not yield an expected constant-round protocol [6]. The intuitive reason—formally argued by Ben-Or and El-Yaniv [6]—is that the process of running n such broadcast protocols (even in parallel) does not terminate in an expected constant number of rounds.

The model of *asynchronous* communication with eventual delivery was considered early on in seminal works on fault-tolerant distributed computing

[5] An approach based on threshold fully homomorphic encryption was recently proposed by Cohen [17]; see the discussion in the related work section below.

[6] The necessity of this bound in the asynchronous setting is also discussed in related work below.

(e.g., [23]), although to our knowledge this paper is the first to formalize this capability in the UC framework. The study of optimally resilient MPC in this type of asynchronous networks was initiated by Ben-Or, Canetti, and Goldreich [5], who proved that any function can be computed by a perfectly secure asynchronous protocol if and only if at most $t < n/4$ parties are corrupted. Following that result, Ben-Or, Kelmer, and Rabin [8] showed that if a negligible error probability is allowed, the bound $t < n/3$ is necessary and sufficient for asynchronous MPC.[7] More recently, Hirt *et al.* [27,28] provided computationally secure solutions (i.e., protocols tolerating a computationally bounded adversary) and Beerliová and Hirt [3] perfectly secure solutions with improved communication complexity.

The above asynchronous protocols are secure—according to simpler, stand-alone security definitions—if one assumes point-to-point communication and an A-BA protocol. Similarly to their synchronous counterparts, all the above protocols—even assuming an A-BA primitive—have round complexity linear in the multiplicative depth of the arithmetic circuit that computes the function, as they follow the standard gate-by-gate evaluation paradigm.

Concurrently and independently, Cohen [17] recently put forth an asynchronous MPC protocol that is secure against a computationally bounded attacker statically corrupting up to $t < n/3$ parties and in which all parties run in constant time. Notably, the protocol from [17] relies on fully homomorphic encryption, thus leaving the question of constant-round MPC from standard assumptions open, which is answered in this work.

We note in passing that although in the synchronous setting BA implies broadcast, this is not the case in the asynchronous setting. Indeed, Canetti and Rabin [15] provide an asynchronous BA protocol tolerating $t < n/3$ malicious parties, which if every honest party terminates at the latest after a poly-logarithmic number of rounds, securely implements asynchronous BA except with negligible probability. A broadcast protocol with similar guarantees is provably impossible [23], and existence of an asynchronous BA protocol which terminates in a strict constant number of rounds would contradict the impossibility from [20,22]. Similarly to the synchronous case, although solutions for asynchronous BA with expected constant number of rounds exist [11,15], using them in the above asynchronous protocol to replace invocations to asynchronous BA would not yield an expected constant-round MPC protocol [6].[8]

---

[7] The necessity of the $t < n/3$ bound follows from the result by Canetti *et al.* [5,12], who argue that this bound is necessary for fail-stop adversaries; it also applies to computational security and assuming A-BA. Moreover, note that in the asynchronous setting, all feasibility bounds are worse by an additive term of t compared to the synchronous setting. Intuitively, this stems from the fact that honest parties cannot distinguish between messages by other honest parties being delayed and messages by corrupted parties not being sent. Thus, in particular, perfectly secure asynchronous MPC is possible only if $t < n/4$.

[8] Nonetheless, [6] does describe an alternative way of obtaining several asynchronous BA protocols that are guaranteed to all terminate in expected constant number of rounds.

2 Model and Building Blocks

We denote the player set by $\mathcal{P} = \{p_1, \ldots, p_n\}$ and consider a computationally bounded *adaptive t-adversary*, i.e., the adversary gets to corrupt up to t parties dynamically during the protocol execution and depending on its protocol view. The most common network model for the execution of asynchronous protocols is secure channels with eventual delivery, where the adversary is allowed to delay the delivery of any message by an arbitrary but finite amount of time, i.e., he is not able to block messages sent among honest parties. Moreover, as argued in the introduction, existing asynchronous protocols rely on an additional resource, namely, an asynchronous version of Byzantine agreement (A-BA) instead of a broadcast channel, and such a resource is even necessary to obtain an (exact) constant number of rounds. We formalize this model and formulate the ideal functionalities corresponding to these communication resources separately in Sect. 3.

We now present some basic tools we use in our protocol.

Secret sharing. Our construction makes use of Shamir's secret sharing scheme [35], which allows to encode a secret into n shares such that any subset of t shares gives no information about the secret and such that from any subset of $t+1$ shares the secret can be reconstructed.

For a sharing of a secret s, let $[s]_i$ denote the ith share. A set of shares are called *t-consistent* if they lie on a polynomial of degree at most t. For a tuple of secrets $s = (s_1, \ldots, s_\ell)$, denote—in slight abuse of notation—by $[s]_i := ([s_1]_i, \ldots, [s_\ell]_i)$ the tuple of the ith shares of the values and refer to it as the ith share of s. A set of such tuples is called *t-consistent* if the property holds for all components.

A linear-round asynchronous MPC protocol. In [8], Ben-Or, Kelmer, and Rabin constructed a protocol, call it π_{BKR}, that computes an arbitrary n-party function f in an asynchronous environment assuming asynchronous point-to-point (secure) channels and asynchronous BA.[9] The protocol follows the gate-by-gate evaluation paradigm [7,16,26], where the function to be evaluated is represented as an arithmetic circuit over a sufficiently large finite field, and the computation proceeds by evaluating sequentially the gates of depth one, then the gates of depth two, and so on. The evaluation of each gate requires a constant number of (asynchronous) rounds,[10] thus making the round complexity of the overall protocol linear in the depth of the circuit.

π_{BKR} was designed for a simpler, stand-alone security definition, which only ensures sequential composition. In the next section we show how it can be cast in our eventual-delivery model to give UC-security guarantees.

---

[9] [8] also assumes A-Cast to get a more efficient solution, but as argued in the introduction, A-Cast can be easily reduced to asynchronous BA in two rounds.

[10] Note that in each such round the parties might invoke the asynchronous BA resource.

3 A UC Model for Asynchronous Computation with Eventual Message Delivery

In this section we formalize the asynchronous network model *with eventual message delivery* in the UC framework. We start with the basic task of point-to-point communication and proceed with asynchronous SFE and BA. Note that the asynchronous model with evenutal delivery has previously been informally treated only in the stand-alone model without composition. Thus, although at first read one might consider our treatment pedantic, providing a UC proof of asynchronous MPC protocol with eventual termination/delivery requires the design of appropriate UC functionalities that can be used as hybrids. Indeed, while the plain UC framework is inherently asynchronous, the adversary has full control over message delivery and may even choose to delete messages sent between uncorrupted parties (by delaying them indefinitely). Hence, without the extensions in this section, the UC model does *not* capture eventual delivery.[11]

Asynchronous communication with eventual delivery. Our formulation of communication with eventual delivery within the UC framework builds on ideas from [31]. In particular, we capture such communication by allowing the parties access to multi-use bilateral secure channels, where a sender $p_i \in \mathcal{P}$ can input a messages to be delivered to some recipient $p_j \in \mathcal{P}$; messages are buffered and delivered in an order specified by the adversary.

To ensure that when p_s and p_r are honest, the adversary cannot delay the delivery of submitted messages arbitrarily, we make the following modifications: We first turn the UC secure channels functionality to work in a "fetch message" mode, where the channel delivers the message to its intended recipient p_j if and only if p_j asks to receive it by issuing a special "fetch" command. If the adversary wishes to delay the delivery of some message, he needs to submit to the channel functionality an integer value T—the *delay*. This will have the effect of the channel ignoring the first T fetch attempts following the reception of the sender's message. Importantly, we require the adversary send the delay T in unary notation; this will ensure that the delay will be bounded by the adversary's running time,[12] and thus a polynomial environment will be able to observe the transmission through its completion. To allow the adversary freedom in scheduling delivery of messages, we allow him to input delays more than once, which are added to the current delay amount. (If the adversary wants to deliver the message in the next activation, all he needs to do is submit a negative delay.)

[11] Standard UC constant-round protocols in the plain (fully asynchronous) UC framework do not work in this setting as, in these protocols, a party waits for all his r-round messages before proceeding to round $r+1$, which would allow the adversary to make honest parties wait indefinitely (for messages from corrupted parties), thereby preventing them from terminating. Instead, in asynchronous protocols with eventual delivery, parties need to proceed to the next round as soon as they have sufficiently many (but not necessarily all) their messages for the current round.

[12] We refer to [13] for a formal definition of running time in the UC framework.

The detailed specification of secure channels with eventual delivery, denoted $\mathcal{F}_{\text{A-SMT}}$, is given in Fig. 1. In the description, we denote by $\boldsymbol{M}$ a vector of strings. We also use $||$ to denote the operation which adds a string to $\boldsymbol{M}$; concretely, if $\boldsymbol{M} = (m_1, \ldots, m_\ell)$, then $\boldsymbol{M}||m := (m_1, \ldots, m_\ell, m)$ and $m||\boldsymbol{M} = (m, m_1, \ldots, m_\ell)$.

Functionality $\mathcal{F}_{\text{A-SMT}}(p_s, p_r)$

Initialize $D := 0$ and $\boldsymbol{M} := (\texttt{eom})$, where `eom` is a special "end-of-messages" symbol.

- Upon receiving a message $(\texttt{send}, m)$ from p_s set $D := D + 1$ and $\boldsymbol{M} := (m, \mathit{mid})||\boldsymbol{M}$, where mid is a unique message ID, and send (D, mid) to the adversary.
- Upon receiving a message $(\texttt{fetch})$ from p_r:
  1. Set $D := D - 1$.
  2. If $D = 0$ and $\boldsymbol{M} = ((m_1, \mathit{mid}_1), \ldots, (m_\ell, \mathit{mid}_\ell), \texttt{eom})$ then set $\boldsymbol{M} := ((m_2, \mathit{mid}_2), \ldots, (m_\ell, \mathit{mid}_\ell), \texttt{eom})$, and send the message m_1 to p_r (otherwise, no message is sent and the activation is given back to the environment, as defined in the UC framework).
- Upon receiving a message $(\texttt{delay}, T)$ from the adversary, if T is a valid delay (i.e., it encodes an integer in unary notation), set $D := \max\{1, D + T\}$; otherwise, ignore the message.
- Upon receiving a message $(\texttt{permute}, \pi)$ from the adversary, if $\pi : [|\boldsymbol{M}| - 1] \to [|\boldsymbol{M}| - 1]$ is a permutation over $[|\boldsymbol{M}| - 1]$, then set $\boldsymbol{M} := \boldsymbol{M}' = ((m_{\pi(1)}, \mathit{mid}_{\pi(1)}), \ldots, (m_{\pi(\ell)}, \mathit{mid}_{\pi(\ell)}), \texttt{eom})$.
- *(Adaptive message replacement)* Upon receiving a message $(p_s, ((m_1, \mathit{mid}_1), \ldots, (m_{\ell'}, \mathit{mid}_{\ell'})), T')$ from $\mathcal{A}$, if p_s is corrupted and $D > 0$ and T' is a valid delay, then set $D = \max\{1, T'\}$ and set $\boldsymbol{M} := ((m_1, \mathit{mid}_1), \ldots, (m_{\ell'}, \mathit{mid}_{\ell'}), \texttt{eom})$.

Fig. 1. Asynchronous secure channel with eventual delivery

We refer to the model in which every two parties p_i and p_j in $\mathcal{P}$ have access to an independent instance of $\mathcal{F}_{\text{A-SMT}}(p_i, p_j)$ as the $\mathcal{F}_{\text{A-SMT}}$*-hybrid model.* An asynchronous protocol in such a model proceeds as follows: Whenever a party p_j gets activated, if its current protocol instructions include sending some message m to some other party p_j, then the party inputs $(\texttt{send}, m)$ to $\mathcal{F}_{\text{A-SMT}}(p_i, p_j)$; otherwise, p_i sends a `fetch` message to every channel $\mathcal{F}_{\text{A-SMT}}(p_j, p_i)$, $j \in [n]$ in a round-robin fashion, i.e., if in the previous activation it sent a $(\texttt{fetch})$ message to $\mathcal{F}_{\text{A-SMT}}(p_j, p_i)$, then it sends a $(\texttt{fetch})$ message to $\mathcal{F}_{\text{A-SMT}}(p_{(j \bmod n)+1}, p_i)$.

Remark 1 (On permuting messages). Our formulation of an asynchronous channel captures the somewhat pessimistic view of asynchronous communication, implicit in many works on asynchronous distributed protocols, where the adversary has full scheduling power and can, in particular, reorder the messages sent by any party as he wishes. One could attempt to emulate a network which does not allow for reordering of the messages—the so-called *first-in-first-out (FIFO)* channel—by adding appropriate (publicly known) message identifiers and instructing the parties to wait until a specific identifier is delivered before outputting messages with other identifiers. However, we note that such an emulating protocol would be distinguished from the original when, for example, we consider an adversary that introduces no delay and an environment that inputs two messages in a row and corrupts the receiver as soon as the first message is supposed to have been delivered.

Functionality $\mathcal{F}^f_{\text{A-SFE}}(\mathcal{P})$

$\mathcal{F}^f_{\text{A-SFE}}$ proceeds as follows, given a function $f : (\{0,1\}^* \cup \{\bot\})^n \times R \rightarrow (\{0,1\}^*)^n$ and a player set $\mathcal{P}$. For each $i \in \mathcal{P}$, initialize variables x_i and y_i to a default value $\bot$ and a current delay $D_i := 0$. Additionally, set $\mathcal{I} := \mathcal{H}$. (Recall that $\mathcal{H}$ denotes the set of honest parties)

- Upon receiving message ($\texttt{no-input}, \mathcal{P}'$) from the adversary, if $|\mathcal{P}'| \leq |\mathcal{P} \setminus \mathcal{H}|$ and no party has received an output ($\texttt{output}, y$) yet, then set $\mathcal{I} = \mathcal{H} \setminus \mathcal{P}'$; otherwise ignore this message.
- Upon receiving input ($\texttt{input}, v$) from party $p_i \in \mathcal{P}$ (or from the adversary in case p_i is corrupted), do the following:
  - If some party (or the adversary) has received an output ($\texttt{output}, y$), then ignore this message; otherwise, set $x_i := v$.
  - If $x_i \neq \bot$ for every $p_i \in \mathcal{I}$, then compute $(y_1, \ldots, y_n) = f((x'_1, \ldots, x'_n), r)$ for a uniformly random r, where $x'_i = x_i$ for $p_i \in \mathcal{I} \cup (\mathcal{P} \setminus \mathcal{H})$ and $x'_i = \bot$ for all other parties.
  - Send ($\texttt{input}, i$) to the adversary.
- Upon receiving ($\texttt{delay}, p_i, T$), from the adversary, set $D_i := D_i + T$.
- Upon receiving message ($\texttt{fetch}$) from party $p_i \in \mathcal{P}$, if y_i has not yet been set (i.e., $y_i = \bot$) then ignore this message, else do:
  - Set $D_i := D_i - 1$
  - If $D_i = 0$, send ($\texttt{output}, y_i$) to p_i.

Fig. 2. Asynchronous SFE with eventual delivery

Asynchronous secure function evaluation (SFE). In an asynchronous environment, it is impossible to get guaranteed (eventual) termination and input completeness, i.e., take into account all inputs in the computation of the function (cf. [31] and early work on fault-tolerant distributed computing). The reason is that if honest parties wait until the inputs of all parties are delivered, then the adversary can make them wait indefinitely for corrupted parties to give their inputs (honest parties have no way of distinguishing between an honest sender whose message is delayed and a corrupt sender who did not send a message). Thus, to ensure eventual termination, the parties cannot afford to wait for input from more than $n-t$ parties, as the t remaining parties might be the corrupted ones. Therefore, protocols for asynchronous computation of a multi-party function f on inputs $x_1, \ldots, x_n$ from parties $p_1, \ldots, p_n$ end up computing the function $f|_{\mathcal{P}'}(x_1, \ldots, x_n) = f(x'_1, \ldots, x'_n)$ for some $\mathcal{P}' \subseteq \mathcal{P}$ with $|\mathcal{P}'| = t$, where $x'_i = x_i$ if $p_i \notin \mathcal{P}'$, and otherwise a default value (denoted $\perp$).

Moreover, by being able to schedule the delivery of messages from honest parties, the adversary can (in worst-case scenarios) choose exactly the set $\mathcal{P}'$. Therefore, the ideal functionality corresponding to asynchronous SFE with eventual termination needs to allow the simulator to choose this set depending on the adversary's strategy. Moreover, the simulator should be allowed to schedule delivery of the outputs depending on the adversary's strategy, but not allowed to delay them arbitrarily. This last requirement can be achieved, as in the case of $\mathcal{F}_{\text{A-SMT}}$, by turning the SFE functionality into a "fetch message"-mode functionality and allowing the simulator to specify a delay on the delivery to every party.

The SFE functionality with the above properties is described in Fig. 2. In the description, we use $\mathcal{H} \subseteq \mathcal{P}$ to denote the set of honest parties; note that $\mathcal{H}$ is dynamically updated as the adversary corrupts new parties. Moreover, we use $\mathcal{I}$ to denote the set of honest parties whose input is allowed to be considered in the computation, and require that $|\mathcal{I}| \geq n-2t$. We provide a generic description of the functionality for an arbitrary number t of corruptions; however, and as implied by classical impossibility results, we are only able to realize it for $t < n/3$ [5].

Asynchronous BA with eventual delivery. The last primitive we describe is (UC) asynchronous BA with eventual message delivery. In such a BA primitive, every party has an input, and we want to ensure the following properties: All honest parties (eventually) output the same value y (consistency), and if all honest parties have the same input x, then this output is $y = x$. Intuitively, asynchronous BA can be cast as a version of asynchronous SFE for the function that looks at the set of received inputs and, if all inputs contributed by honest parties are identical[13] and equal to some x, sets the output to x for every party; otherwise, it sets the output to the input of some corrupted party (for example, the first in any ordering, e.g., lexicographic). The formal definition of $\mathcal{F}_{\text{A-BA}}$ is given in Fig. 3.

We will refer to the setting where every two parties p_i and p_j in $\mathcal{P}$ have access to an independent instance of $\mathcal{F}_{\text{A-SMT}}(p_i, p_j)$ and, additionally, the parties have access to independent instances of $\mathcal{F}_{\text{A-BA}}(\mathcal{P})$ as the $\{\mathcal{F}_{\text{A-SMT}}, \mathcal{F}_{\text{A-BA}}\}$*-hybrid model*. The execution in the $\{\mathcal{F}_{\text{A-SMT}}, \mathcal{F}_{\text{A-BA}}\}$-hybrid model is analogous to the execution in the $\mathcal{F}_{\text{A-SMT}}$-hybrid model: Whenever a party p_i gets activated, if its current protocol instructions include sending some message m to some other party p_j or inputing a message m' to $\mathcal{F}_{\text{A-BA}}(\mathcal{P})$, then the party inputs $(\texttt{send}, m)$ to $\mathcal{F}_{\text{A-SMT}}(p_i, p_j)$ or m' to $\mathcal{F}_{\text{A-BA}}(\mathcal{P})$, respectively; otherwise, p_i keeps sending

[13] Similarly to the SFE case, the adversary might prevent some of the honest parties from providing an input.

Functionality $\mathcal{F}_{\text{A-BA}}(\mathcal{P})$

For each $i \in \mathcal{P}$, initialize variables x_i and y_i to a default value $\perp$ and a current delay $D_i := 0$. Additionally, set $\mathcal{I} := \mathcal{H}$. (Recall that $\mathcal{H}$ denotes the set of honest parties)

- Upon receiving message $(\texttt{no-input}, \mathcal{P}')$ from the adversary, if $|\mathcal{P}'| \leq |\mathcal{P} \setminus \mathcal{H}|$ and no party has received an output $(\texttt{output}, y)$ yet, then set $\mathcal{I} = \mathcal{H} \setminus \mathcal{P}'$; otherwise ignore this message.
- Upon receiving input $(\texttt{input}, v)$ from party $p_i \in \mathcal{P}$ (or from the adversary in case p_i is corrupted), do the following:
  - If some party (or the adversary) has received an output $(\texttt{output}, y)$, then ignore this message; otherwise, set $x_i := v$.
  - If $x_i \neq \perp$ for every $p_i \in \mathcal{I}$, then set $(y_1, \ldots, y_n) := (y, \ldots, y)$, where if there exists $x \neq \perp$ such that $x_i = x$ for every $p_i \in \mathcal{I}$, then $y = x$; otherwise $y = x_j$, where p_j is the party in $\mathcal{P} \setminus \mathcal{H}$ with the smallest index.
  - Send $(\texttt{input}, i)$ to the adversary.
- Upon receiving $(\texttt{delay}, p_i, T)$, from the adversary, set $D_i := D_i + T$.
- Upon receiving message $(\texttt{fetch})$ from party $p_i \in \mathcal{P}$, if y_i has not yet been set (i.e., $y_i = \perp$) then ignore this message, else do:
  - Set $D_i := D_i - 1$
  - If $D_i = 0$, send $(\texttt{output}, y_i)$ to p_i.

Fig. 3. Asynchronous BA with eventual delivery

(with each activation) a $\texttt{fetch}$ to every channel $\mathcal{F}_{\text{A-SMT}}(p_i, p_j)$, $j \in [n]$ and then to $\mathcal{F}_{\text{A-BA}}(\mathcal{P})$ in a round-robin fashion.

Asynchronous rounds. We now briefly elaborate on the notion of rounds in an asynchronous environment. Unlike the situation in the synchronous case, where rounds are well specified by the protocol, the definition of rounds in an asynchronous setting requires a bit more care. Intuitively, two messages m_i and m_j sent by some party p_i in an asynchronous protocol are considered to be sent in rounds i and j, $j > i$, if m_j is generated by computation which takes as input a message received after p_i sent m_i. Following the above intuition, we define for each p_i and for each point in the protocol execution, the current round in which p_i is to be the number of times p_i alternated between sending $(\texttt{send}, m)$ to some channel $\mathcal{F}_{\text{A-SMT}}(p_i, p_j)$, $p_j \in \mathcal{P}$ (or to the asynchronous BA functionality $\mathcal{F}_{\text{A-BA}}(\mathcal{P})$) and sending $(\texttt{fetch})$ to some $\mathcal{F}_{\text{A-SMT}}(p_k, p_i)$, $p_k \in \mathcal{P}$ or to $\mathcal{F}_{\text{A-BA}}(\mathcal{P})$. That is, every round (except for the first one)[14] starts by sending a $(\texttt{send}, m)$ to some $\mathcal{F}_{\text{A-SMT}}(p_i, p_j)$ or to $\mathcal{F}_{\text{A-BA}}(\mathcal{P})$ after some $(\texttt{fetch})$ was sent by p_i and finishes with the first $(\texttt{fetch})$ command that p_i sends. The round complexity of the protocol is the maximum (over all honest parties) number of rounds that an honest party uses in the protocol execution.

[14] The first round starts as soon as the party receives its protocol input from the environment.

We note in passing that, similarly to [31], the above formulation allows for any party to send several messages in each round: the party buffers the messages and while the buffer is not empty, in each activation the party pops the next message and sends it to its intended recipient.

A UC-secure linear-round MPC protocol with eventual delivery. Finally, we argue the security of protocol π_{BKR} mentioned in Sect. 2 in our model. π_{BKR} is information-theoretic and proved simulation-based secure, where the simulation is in fact black-box (i.e., the simulator uses the corresponding adversary in a black-box manner) and straight-line (the simulator does not rewind the adversary). Moreover, the protocol tolerates any adaptive t-adversary where $t < n/3$, a bound which is also tight [5]. Thus, by casting π_{BKR} in our UC $\{\mathcal{F}_{\text{A-SMT}}, \mathcal{F}_{\text{A-BA}}\}$-hybrid model—where every bilateral message exchange is implemented by the sender p_i and the receiver p_j using (an instance of the) channel $\mathcal{F}_{\text{A-SMT}}(p_i, p_j)$ and every call to asynchronous BA done by invocation of $\mathcal{F}_{\text{A-BA}}(\mathcal{P})$—we obtain a protocol for UC securely evaluating $\mathcal{F}^{f}_{\text{A-SFE}}(\mathcal{P})$, which is linear in the depth of the circuit computing f. More formally:

Theorem 1 ([8]). *Let f be an n-ary function and C be an arithmetic circuit for computing f by parties in $\mathcal{P}$. Then there exists a protocol, π_{BKR}, which UC-securely realizes $\mathcal{F}^{f}_{\text{A-SFE}}$ in the $\{\mathcal{F}_{\text{A-SMT}}, \mathcal{F}_{\text{A-BA}}\}$-hybrid model tolerating an adaptive t-adversary in a linear (in the depth of the circuit) number of rounds, provided $t < n/3$.*

4 Constant-Round Asynchronous SFE

In this section we present our asynchronous SFE protocol and prove its security and round complexity.

4.1 Description of the Protocol

Let Circ be a Boolean circuit that is to be evaluated in a multi-party computation. In our protocol for securely evaluating the function that Circ computes, denoted $\pi_{\text{A-SFE}}(\mathsf{Circ}, \mathcal{P})$, the parties first jointly compute a garbled version of Circ (along the lines of [2,4,37]); every party then evaluates this garbled circuit locally to obtain the output of the computation. Computing the garbled circuit takes place in two phases: First, the parties evaluate a function $f^{\mathsf{Circ}}_{\mathrm{PREP}}$ (described below) which is represented by a *constant-depth* arithmetic circuit over a finite field using a (non-constant-round) asynchronous MPC protocol. Given the outputs of this function, the parties can then complete the computation of the garbled circuit within one additional asynchronous round.[15] Since the evaluation of the garbled circuit takes place locally and $f^{\mathsf{Circ}}_{\mathrm{PREP}}$ is computed via a constant-depth circuit, the entire protocol is a constant-round protocol.

[15] Refer to Sect. 2 for a definition of asynchronous round complexity.

Function $f_{\text{PREP}}^{\mathsf{Circ}}((b_{\omega_{11}},\ldots,b_{\omega_{1L_1}}),\ldots,(b_{\omega_{n1}},\ldots,b_{\omega_{nL_n}}))$

The preparation function is parameterized by a Boolean circuit Circ describing the function to be computed. The wires of Circ are labeled by values $\omega \in \mathbb{N}$. We use Greek letters $\alpha, \beta, \gamma, \omega$ for referring to the wire labels.

INPUT. For every input wire ω, b_ω denotes the corresponding input bit.

CREATE RANDOM VALUES. For each wire ω do:

1. For each $i \in [n]$ choose a random sub-key $k_{\omega,0}^i \in \mathbb{F}^n$. Set $k_{\omega,0} := (k_{\omega,0}^1,\ldots,k_{\omega,0}^n)$.
2. For each $i \in [n]$ choose a random sub-key $k_{\omega,1}^i \in \mathbb{F}^n$. Set $k_{\omega,1} := (k_{\omega,1}^1,\ldots,k_{\omega,1}^n)$.
3. Choose random mask $m_\omega \in \{0,1\}$.

INPUT WIRES. For every input wire ω do:

1. Compute masked value $z_\omega := b_\omega \oplus m_\omega$.
2. Choose corresponding key $k_\omega := k_{\omega,z_\omega}$.

COMPUTE MASKED FUNCTION TABLES. For every gate g with wires α, β, γ do:

1. For every $x, y \in \{0,1\}$ do:
   (a) Compute masked value $z_\gamma^{xy} := ((x \oplus m_\alpha)\ \texttt{NAND}\ (y \oplus m_\beta)) \oplus m_\gamma$.
   (b) Choose corresponding key $k_\gamma^{xy} := k_{\gamma,z_\gamma^{xy}}$.
   (c) Set $t_g^{xy} := (z_\gamma^{xy}, k_\gamma^{xy})$ and $T_g := (t_g^{00}, t_g^{01}, t_g^{10}, t_g^{11})$.
2. Compute a Shamir sharing of T_g (i.e., of every entry).

OUTPUT. Proceed as follows:

(Public outputs) Output the following values to *all* players:

1. For every *input* wire ω: the masked value z_ω and the corresponding key k_ω.

(Private outputs) Output the following values to every $p_i \in \mathcal{P}$:

1. For every wire ω: the subkeys $k_{\omega,0}^i$ and $k_{\omega,1}^i$.
2. For every gate g: the ith share $[T_g]_i$ of T_g.
3. For every output wire ω: the mask m_ω if p_i is to learn that output.

Fig. 4. The description of function $f_{\text{PREP}}^{\mathsf{Circ}}$ corresponding to the distributed version of circuit garbling.

We define and analyze our protocol in the $\{\mathcal{F}_{\text{A-SFE}}^{f_{\text{PREP}}^{\mathsf{Circ}}}, \mathcal{F}_{\text{A-SMT}}\}$-hybrid model. Furthermore we provide a protocol for UC securely realizing the functionality $\mathcal{F}_{\text{A-SFE}}^{f_{\text{PREP}}^{\mathsf{Circ}}}$ from asynchronous secure channels and BA with eventual delivery based on π_{BKR} [8] (cf. Lemma 1).

Circuit garbling. Before elaborating on the protocol, we describe what the garbled version of Circ looks like.[16] Boolean circuit Circ consists of wires and NAND gates.[17] In the garbled version, every wire ω of Circ has a corresponding (secret) random mask m_ω, which is used to hide the real value on that wire. Consequently, every gate g, with input wires α and β and output wire γ, has a special function table T_g that works on masked values. It contains four entries z_γ^{xy}, corresponding to the masked value on the outgoing wire γ under the four possible combinations of masked inputs $x, y \in \{0,1\}$ on wires α and β. Each entry is obtained by unmasking the inputs, applying the gate function (NAND), and re-masking the result with the mask of the outgoing wire. That is,

$$z_\gamma^{xy} = ((x \oplus m_\alpha) \ \mathtt{NAND}\ (y \oplus m_\beta)) \oplus m_\gamma,$$

for $x, y \in \{0,1\}$.

The entries of each function table need to be protected so that only the one entry necessary to evaluate the circuit can be accessed. To that end, for each wire ω there are two (secret) keys $k_{\omega,0}$ and $k_{\omega,1}$. In the function tables T_g, each entry z_γ^{xy} is now augmented by the corresponding key $k_{\gamma,z_\gamma^{xy}}$ of the outgoing wire γ. The pair $t_g^{xy} := (z_\gamma^{xy}, k_{\gamma,z_\gamma^{xy}})$ is encrypted with $k_{\alpha,x}$ and $k_{\beta,y}$ under the "tweak" (g, x, y). The resulting ciphertexts

$$c_g^{xy} := \mathsf{Enc}_{k_{\alpha,x},k_{\beta,y}}^{g,x,y}\left(t_g^{xy}\right) = \mathsf{Enc}_{k_{\alpha,x},k_{\beta,y}}^{g,x,y}\left(z_\gamma^{xy}, k_{\gamma,z_\gamma^{xy}}\right)$$

make up the garbled function table

$$C_g := (c_g^{00}, c_g^{01}, c_g^{10}, c_g^{11}),$$

where $(\mathsf{Enc}_{k_1,k_2}^T(\cdot), \mathsf{Dec}_{k_1,k_2}^T(\cdot))$ is a tweakable dual-key cipher. A suitable such cipher can be realized using a PRF [4].[18]

In order to be compatible with the garbled function tables, inputs to the circuit must be garbled as well. That is, for the input bit b_ω on input wire ω, the garbled input is $z_\omega := b_\omega \oplus m_\omega$ and the corresponding key is $k_\omega := k_{\omega,z_\omega}$.

With the garbled inputs and function tables, any party can (locally) evaluate the circuit as follows: Given the masked values and the corresponding keys of the incoming wires of some gate, the party can decrypt the corresponding row, obtaining the masked value on the outgoing wire and the corresponding key. In the end, the values on the output wires can be unblinded if the corresponding masks are known.

---

[16] Note that $f_{\text{PREP}}^{\mathsf{Circ}}$ actually computes a "distributed" version of the garbled circuit (described below).

[17] Any (arithmetic or Boolean) circuit can be efficiently transformed into such a circuit.

[18] The security required from such a cipher is roughly semantic security even if one of the keys is known (see [4] for more details). Moreover, we assume a canonical injective mapping of triples (g, x, y) to the tweak space of the cipher.

Distributed encryption. Given the input bits b_ω for all input wires ω, computing the garbled circuit could be described by a constant-depth circuit, since the garbled function tables can be computed in parallel after choosing the wire masks and keys. This circuit, however, would be rather large since it entails evaluating the cipher. Therefore, to avoid evaluating the cipher within the asynchronous MPC, the parties use the distributed-encryption technique by Damgård and Ishai [19]: Instead of computing $\mathsf{Enc}^T_{k_1,k_2}(m)$ for a message m, two keys k_1 and k_2, and a tweak T, the parties first jointly choose $2n$ subkeys $k_1^1,\ldots,k_1^n$ and $k_2^1,\ldots,k_2^n$, compute a Shamir sharing of $m = ([m]_1,\ldots,[m]_n)$, open $[m]_i$ as well as k_1^i and k_2^i to p_i for every i, and then each party encrypts its share $[m]_i$ of m using its two subkeys k_1^i and k_2^i and sends the resulting ciphertext $\mathsf{Enc}^T_{k_1^1,k_2^1}([m]_i)$ to all parties.

In order to decrypt, a party in possession of all keys recovers the shares by decrypting the ciphertexts received from other players and waits until it has $2t+1$ t-consistent shares, which it uses to reconstruct m.[19]

Asynchronously evaluating Boolean circuits. Protocol π_{BKR} [8], which we wish to use to realize $\mathcal{F}^{f^{\mathsf{Circ}}_{\mathrm{PREP}}}_{\mathrm{A\text{-}SFE}}$, evaluates arithmetic circuits over fields with more than two elements; the circuit representing $f^{\mathsf{Circ}}_{\mathrm{PREP}}$, however, is Boolean. Thus, in order to evaluate it via π_{BKR} we transform it into an (arithmetic) circuit over an extension field F of GF(2), denoted $C^F_{f^{\mathsf{Circ}}_{\mathrm{PREP}}}$, by having every `NAND` gate with inputs $x, y \in \{0,1\}$ replaced by the computation $1 - xy$, which can be implemented using addition and multiplication over the extension field F. The above transformation, however, works only if all the inputs to the circuit corresponding to bits in the Boolean circuit are either 0 or 1 in the corresponding field F. For the honest parties this is easy to enforce: they encode a 0 (resp., 1) input bit into the 0 (resp., 1) element of F. The adversary, however, might try to cheat by giving "bad" inputs, namely, inputs in $F \setminus \{0,1\}$. We now show an explicit construction to ensure that the adversary cannot give any value other than 0 or 1, resulting in a simple adaptation of protocol π_{BKR}.[20]

Before describing the solution we recall the reader how π_{BKR} evaluates a given circuit. We omit several low-level details and keep the description at the level which is appropriate for a formal description of our adaptation; the interested reader is referred to [8] for further details. π_{BKR} follows the *gate-by-gate evaluation paradigm* [7,26]: The circuit is evaluated in a gate-by-gate fashion, where the invariant is that after the evaluation of each gate (in fact, of each bulk of gates that are at the same multiplicative depth), the output of the gates is verifiably shared among all the parties. In fact, [8] defines the notion of *Ultimate Secret*

[19] Our protocol ensures that each party eventually receives these many encrypted shares (see below).

[20] In principle, the arithmetic circuit "re-compiler" technique by Genkin *et al.* [24] could also be used for this purpose, although it is not shown to work for π_{BKR} nor be constant-depth. In addition, the functionality of the re-compiler is richer, as it allows to restrict possible malicious strategies during the evaluation of the circuit, which is not needed here.

Protocol $\pi_{\text{A-SFE}}(\mathsf{Circ}, \mathcal{P})$: Code for p_i

First, mark all gates as *unevaluated*. Initialize empty variables z_ω and k_ω for every wire ω and m_ω for every output wire ω accessible to p_i. Initialize $\phi := 0$ (phase indicator). Then, proceed as follows:

- Upon first activation with input b, input b to $\mathcal{F}_{\text{A-SFE}}^{f_{\text{PREP}}^{\mathsf{Circ}}}$.
- Upon later activations:
  - If $\phi = 0$, check if output from $\mathcal{F}_{\text{A-SFE}}^{f_{\text{PREP}}^{\mathsf{Circ}}}$ received. If not, output (`output`) to $\mathcal{F}_{\text{A-SFE}}^{f_{\text{PREP}}^{\mathsf{Circ}}}$ and become inactive. Otherwise, encrypt every gate g, with wires α, β, γ, as follows:
    1. Output by functionality includes:
       (a) Subkeys $k_{\alpha,0}^i$ and $k_{\alpha,1}^i$ as well as $k_{\beta,0}^i$ and $k_{\beta,1}^i$.
       (b) Function table share $[T_g]_i = ([t_g^{00}]_i, [t_g^{01}]_i, [t_g^{10}]_i, [t_g^{11}]_i)$.
    2. For $x, y \in \{0,1\}$, compute $c_g^{xy,i} := \mathsf{Enc}_{k_{\alpha,x}^i, k_{\beta,y}^i}^{g,x,y}([t_g^{xy}]_i)$.
    3. Send $C_g^i := (c_g^{00,i}, c_g^{01,i}, c_g^{10,i}, c_g^{11,i})$ to all parties by invocation of $\mathcal{F}_{\text{A-SMT}}(p_i, p_j), j \in [n]$.

    Further, for all input wires ω, set z_ω and k_ω, to the values output by $\mathcal{F}_{\text{A-SFE}}^{f_{\text{PREP}}^{\mathsf{Circ}}}$. Similarly, set the masks m_ω for the (accessible) output wires to the values output by $\mathcal{F}_{\text{A-SFE}}^{f_{\text{PREP}}^{\mathsf{Circ}}}$. Set $\phi := 1$.
  - If $\phi = 1$, upon reception of any encryption, proceed as follows for every unevaluated gate g, with incoming wires α and β and outgoing wire γ:
    1. Let z_α, z_β, and z_γ be the masked bits and k_α, k_β, and k_γ the keys of the incoming wires α and β and of the outgoing wire γ. If z_α and z_β, are not defined yet, skip this gate; else:
       (a) For each ciphertext $C_g^j = (c_g^{00,j}, c_g^{01,j}, c_g^{10,j}, c_g^{11,j})$ from a party p_j, decrypt $t_g^{xy,j} := \mathsf{Dec}_{k_\alpha^j, k_\beta^j}^{g,x,y}(c_g^{xy,j})$ for $x = z_\alpha$ and $y = z_\beta$, thereby recovering j^{th} shares of z_γ and of every entry of $k_\gamma = (k_\gamma^1, \ldots, k_\gamma^n)$.
       (b) Check if z_γ and the entries of k_γ can be safely computed by interpolation, i.e., if there are at least $2t+1$ t-consistent shares for each value. If not, skip this gate. Otherwise, interpolate and mark g as *evaluated*.

    If all gates have been evaluated, output $z_\omega \oplus m_\omega$ for all (accessible) output wires ω.

Fig. 5. Our constant-round asynchronous SFE protocol in the $\{\mathcal{F}_{\text{A-SFE}}^{f_{\text{PREP}}^{\mathsf{Circ}}}, \mathcal{F}_{\text{A-SMT}}\}$-hybrid model

Sharing (USS) which is a version of VSS that is appropriate for asynchronous computation with $t < n/3$;[21] More concretely, the first step is to process all

[21] USS is an adaptation of the bivariate-polynomial sharing technique [7,33] to the asynchronous setting.

input gates in parallel (i.e., receive inputs from all parties); this step finishes with every party holding a share of the input of each party p_i. As already mentioned, due to asynchrony, the inputs of some, up to t, (honest) parties might not be considered. The set Core of these parties whose inputs are considered (the so-called *core set* [5,8]) is decided by π_{BKR} (and agreed upon by all parties) during the evaluation of the input gates, while the input of the parties not in the core set is set to a default value, in our case to 0 (i.e., a default USS of 0 is adopted as a sharing of these parties' inputs [8]). Once any party has agreed on the core set parties giving input, it goes on to the evaluation of the next gate (in fact, of all gates which are one level deeper in the circuit in parallel).

Our modification to π_{BKR} is as follows. For any party p_j, as soon as p_j has processed all input gates (i.e., holds shares of inputs of all parties in the core set and default shares for the remaining parties), and before any other gate of the arithmetic circuit is computed, p_j does the following: For each party we denote by x'_i the value that is (eventually) shared as p_i's input when all parties have evaluated the corresponding input gate, and denote by $[x'_i]_j$ p_j's share of this value.[22]

Now, instead of continuing to process the original circuit $C^F_{f^{\mathsf{Circ}}_{\text{PREP}}}$, we use the following trick from [9], which will allow us to enforce zero/one inputs. Each party uses the shared values x'_i to compute the circuit evaluating the following function: output $\mathbf{c} = (c_1, \ldots, c_n)$, where $c_i = x'_i - {x'_i}^2$ for each p_i. Each party that received the output $\mathbf{c}$ does the following:[23] For each p_i, if $c_i \neq 0$, then the parties replace the sharing of x'_i with a default sharing of 0. (That is, as soon as p_j receives the vector $\mathbf{c}$, for each i with $c_i \neq 0$ p_i replaces his share $[x_i]_j$ of x'_i with a default sharing of 0.) Once a party has completed this step (and replaced his local shares), he continues the execution of π_{BKR} with the modified shares to compute the remainder of the circuit $C^F_{f^{\mathsf{Circ}}_{\text{PREP}}}$.

We denote the above modification of protocol π_{BKR} (in the $\mathcal{F}_{\text{A-BA}}(\mathcal{P})$-hybrid world where calls to A-BA are replaced by invocations of $\mathcal{F}_{\text{A-BA}}(\mathcal{P})$) with the above mechanism for enforcing inputs in $\{0,1\}$ by $\pi^{0/1}_{\text{BKR}}$. In order to evaluate the (Boolean) circuit for $f^{\mathsf{Circ}}_{\text{PREP}}$, the parties execute $\pi^{0/1}_{\text{BKR}}$ encoding their inputs and outputs with the following trivial encoding: An input-bit 0 (resp., 1) is encoded as the 0 (resp., 1) element in F, and output 0 (resp., 1) in F is decoded back to the bit 0 (resp., 1). The following lemma states the achieved security.

Lemma 1. *Protocol $\pi^{0/1}_{\text{BKR}}$ for evaluating the circuit $C^F_{f^{\mathsf{Circ}}_{\text{PREP}}}$ with the above trivial encoding UC-securely realizes $\mathcal{F}^{f^{\mathsf{Circ}}_{\text{PREP}}}_{\text{A-SFE}}$.*

Proof. (sketch). First note that if the inputs of all (honest and corrupted) parties are 0 or 1 (in the arithmetic field F), then the (decoded) output of $C^F_{f^{\mathsf{Circ}}_{\text{PREP}}}$ is the

[22] By the USS property, at this point p_i is committed to x'_i but the adversary has no information on it, i.e., the adversary holds random shares of a USS of x_i.

[23] Observe that the eventual delivery property ensures that every party will eventually receive the output.

same as the output of the (Boolean) circuit for $f_{\text{PREP}}^{\mathsf{Circ}}$ since all NAND gates with inputs $x, y \in \{0,1\}$ are computed by $1 - xy$. Next, we argue that $\pi_{\text{BKR}}^{0/1}$ forces the inputs of the adversary to be 0 or 1 and does not modify the inputs of honest parties. Indeed, an honest party p_i in the core set will share inputs $x'_i \in \{0,1\}$ and therefore $c_i = 0$, which means that his input sharing is not modified by $\pi_{\text{BKR}}^{0/1}$. The same holds for any corrupted party that shares $x'_i = 0$ or $x'_i = 1$. On the other hand, any corrupted party sharing a value other than 0 or 1 will result into an output $c_i \neq 0$ (since the non-zero elements in F form a multiplicative group of order $|F| - 1$) and its input will be set to 0.

Note that the eventual termination of π_{BKR} ensures that all parties will eventually receive the output vector c and will therefore resume the computation of the original circuit $C'^{F}_{f_{\text{PREP}}^{\mathsf{Circ}}}$, which (also due to the eventual termination of π_{BKR}) will terminate. The simulation of $\pi_{\text{BKR}}^{0/1}$ is easily reduced to the simulation of π_{BKR}: The evaluation of the extra component that computes the c_i's can be easily simulated as they are random sharings of 0 for all honest parties in the core set, and for corrupted parties they are functions of the sharing of x'_i that the adversary creates in the input-processing phase, which for corrupted parties is fully simulatable. For the rest of the simulation, the simulator simply uses the π_{BKR} simulator. Thus the indistinguishability of the simulation follows from the security of π_{BKR}. □

Putting things together. The detailed description of protocol $\pi_{\text{A-SFE}}(\mathsf{Circ}, \mathcal{P})$ is presented in Fig. 5. As already said, we describe the protocol in the $\mathcal{F}_{\text{A-SFE}}^{f_{\text{PREP}}^{\mathsf{Circ}}}$-hybrid model, where $\mathcal{F}_{\text{A-SFE}}^{f_{\text{PREP}}^{\mathsf{Circ}}}$ can be replaced with $\pi_{\text{BKR}}^{0/1}$ using Lemma 1 and the universal composition theorem. At a high-level, the protocol proceeds as follows: In the first phase, the parties send their inputs to the functionality $\mathcal{F}_{\text{A-SFE}}^{f_{\text{PREP}}^{\mathsf{Circ}}}$. The (randomized) function $f_{\text{PREP}}^{\mathsf{Circ}}$ chooses the random masks, the subkeys, and computes Shamir sharings of the masked function tables (which are the values that need to be encrypted). Moreover, based on the inputs, it computes the masked value and the corresponding key of every input wire. The formal specification of $f_{\text{PREP}}^{\mathsf{Circ}}$ can be found in Fig. 4. The fact that $\mathcal{F}_{\text{A-SFE}}^{f_{\text{PREP}}^{\mathsf{Circ}}}$ can be evaluated by a constant-depth circuit is illustrated in Fig. 6, which provides a diagram describing the structure of such a circuit. Each of the rectangles corresponds to a collection of independent constant-depth circuits that are evaluated in parallel.

In the second phase of the protocol, as soon as a party receives output from $\mathcal{F}_{\text{A-SFE}}^{f_{\text{PREP}}^{\mathsf{Circ}}}$, it encrypts all the shares obtained using the appropriate subkeys and sends the resulting ciphertexts to all parties, as shown in Fig. 5. Then, it proceeds to locally evaluate the gates. For each gate, the party waits for ciphertexts from other parties and decrypts them. For a specific entry in the function table, the party has to wait until it has $2t+1$ t-consistent shares of that entry (see again Fig. 5).[24] Note that since all of the at least $2t+1$ honest parties are guaranteed to obtain an output from $\mathcal{F}_{\text{A-SFE}}^{f_{\text{PREP}}^{\mathsf{Circ}}}$, they will all properly encrypt their function tables

[24] Note that using the Berlekamp-Welch algorithm, this can be achieved efficiently.

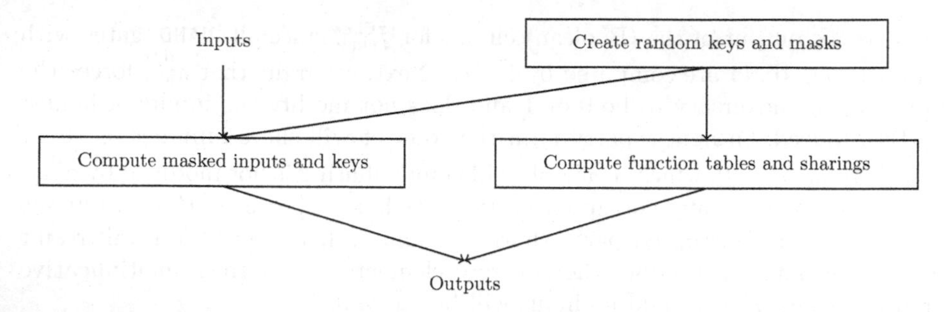

Fig. 6. Bird's-eye view of the arithmetic circuit Prep computing function $f^{\mathsf{Circ}}_{\text{PREP}}$. Each box represents a constant-depth circuit.

and send out the resulting ciphertexts. Therefore, the wait for $2t+1$ t-consistent shares is finite.

4.2 Analysis of the Protocol

Theorem 2. *Let* Circ *be a given boolean circuit and* f_{Circ} *be the n-party function computed by* Circ. *Protocol* $\pi_{\text{A-SFE}}(\mathsf{Circ}, \mathcal{P})$ *securely realizes* $\mathcal{F}^{f_{\mathsf{Circ}}}_{\text{A-SFE}}$ *in the* $\{\mathcal{F}^{f^{\mathsf{Circ}}_{\text{PREP}}}_{\text{A-SFE}}, \mathcal{F}_{\text{A-SMT}}\}$*-hybrid model tolerating an adaptive adversary who corrupts up to* $t < n/3$ *of the parties and making black-box use of a PRF.*

A full proof of Theorem 2 can be found in the full version of this paper [18]. Here we only provide a high-level sketch.

Proof (sketch). The output of each party from the evaluation of $\mathcal{F}^{f^{\mathsf{Circ}}_{\text{PREP}}}_{\text{A-SFE}}$ contains (among other things) a t-out-of-n sharing of the garbled circuit for computing function f_{Circ}. After receiving the output from $\mathcal{F}^{f^{\mathsf{Circ}}_{\text{PREP}}}_{\text{A-SFE}}$ and encrypting as described in Fig. 5, the only time the parties have to wait is for the encryptions of $2t+1$ t-consistent shares of garbled function-table entries from other parties. Since all of the at least $2t+1$ honest parties are guaranteed to obtain an output from $\mathcal{F}^{f^{\mathsf{Circ}}_{\text{PREP}}}_{\text{A-SFE}}$, they will all properly encrypt their function tables and send out the resulting ciphertexts at some point. Therefore, the wait for $2t+1$ t-consistent shares is finite.

Moreover, the adversary cannot make an honest party accept a wrong value for any entry of the garbled gate: Observe that in any set of $2t+1$ shares that a party receives, at least $t+1$ are from honest parties. These $t+1$ shares uniquely define the degree-t sharing polynomial F and, therefore, they can only be combined with correct shares (as output by $\mathcal{F}^{f^{\mathsf{Circ}}_{\text{PREP}}}_{\text{A-SFE}}$). This implies that wrong shares sent by the adversary cannot make any honest party choose any other polynomial than F.

The simulator $\mathcal{S}$ for an adversary $\mathcal{A}$ roughly proceeds as follows: It emulates towards $\mathcal{A}$ the behavior of $\mathcal{F}^{f^{\mathsf{Circ}}_{\text{PREP}}}_{\text{A-SFE}}$ and the channels $\mathcal{F}_{\text{A-SMT}}$. The security of the circuit-garbling technique and that of Shamir sharings allows $\mathcal{S}$ to perfectly simulate the entries of the garbled function tables that one would decrypt during a local evaluation of the garbled circuit, even *without* knowing the actual inputs. Moreover, the security of the double-key cipher ensures that the remaining entries are hidden, and can thus be replaced by dummy values (which can again be done without knowing the inputs) without causing a noticeable difference in the view of $\mathcal{A}$. □

We now turn to the analysis of the protocol's round complexity. It is straightforward to verify that our protocol (assuming hybrid $\mathcal{F}_{\text{A-SFE}}^{f_{\text{PREP}}^{\mathsf{Circ}}}$) needs only two communication rounds for each party p_i: one round in which p_i sends its inputs to the functionality and receives its outputs, and one round in which p_i sends all its corresponding encryptions and receives the encryptions of other parties. Moreover, the function $f_{\text{PREP}}^{\mathsf{Circ}}$ can be represented by an arithmetic circuit Prep over a finite field $\mathbb{F}$ with constant (multiplicative) depth: The players first jointly generate the subkeys and the masks. A straightforward method for generating a random field element (such as the subkeys) is to take a random input from every party and computing the sum. Generating a random bit $b \in \{0,1\} \subseteq \mathbb{F}$ can be reduced to generating random field elements as shown by Genkin *et al.* [24]. Given the masks and the subkeys, computing the function table and a Shamir sharing thereof can clearly be done in constant depth and, most importantly, in parallel for every gate.

Combining the above and Theorem 2 with Theorem 1 yields the following corollary:

Corollary 1. *Let* Circ *be a given boolean circuit and* f_{Circ} *be the n-party function computed by* Circ*. There exists a constant-round protocol which securely realizes* $\mathcal{F}_{\text{A-SFE}}^{f_{\mathsf{Circ}}}$ *in the* $\{\mathcal{F}_{\text{A-DA}}, \mathcal{F}_{\text{A-SMT}}\}$*-hybrid model tolerating an adaptive adversary who corrupts up to* $t < n/3$ *of the parties and making black-box use of a PRF.*

A A History and Related Work (cont'd)

Here we provide a fuller account of related work and put our results in perspective. To give a more complete picture, we start by discussing the development of MPC protocols in the synchronous setting, and then contrast it with the development in the asynchronous setting. Along the way we also discuss the tools (e.g., setup assumptions and communication resources) that are used in each setting.

Starting with Yao's seminal paper [36], which introduced the problem of MPC and provided the first solution, a long line of interesting results proved feasibility bounds for *synchronous* networks in various adversarial settings. Goldreich, Micali, and Wigderson [25,26] proved that under computational assumptions (the existence of enhanced trapdoor permutations), any n-party function can be securely computed if and only if up to $t < n$ parties are corrupted passively or up to $t < n/2$ actively. Corresponding bounds for information-theoretic security were shown by Ben-Or, Goldwasser, and Wigderson [7], who proved that perfect security is possible if and only if the adversary corrupts up to $t < n/2$ parties passively or up to $t < n/3$ actively. Similar bounds where concurrently proved by Chaum, Crépeau, and Damgård [16] for the case where a negligible error-probability is allowed and were later improved by Rabin and Ben-Or [33] to achieve optimal resiliency $t < n/2$.

The above works assume point-to-point secure communication and a broadcast channel and, under these assumptions, are secure even against an adaptive adversary [14]. However, in [25,26] both these resources can be implemented with adaptive security assuming a public-key infrastructure and non-committing encrpytion [14,29]. Similarly, the broadcast channel in [7,16,33] can be emulated by an adaptively secure broadcast protocol [29].[25] The round complexity of all the above protocols in the malicious multi-party setting —even with the assumption of a broadcast channel—is linear in the multiplicative depth of the arithmetic circuit corresponding to the function the parties aim to compute.

Beaver, Micali, and Rogaway [2] were the first to provide a constant-round MPC protocol in the synchronous stand-alone model. Their protocol is secure in the computational setting and tolerates an adaptive adversary who actively corrupts up to $t < n/2$ parties. The complexity of [2] was improved by Damgård and Ishai [19], who provided the first constant-round protocol making black-box use of the underlying cryptographic primitive (a pseudo-random generator). Importantly, both [2] and [19] assume a broadcast channel, an assumption essential for obtaining constant-round MPC. Indeed, as proved in [20,22], it is impossible to implement such a broadcast channel from point-to-point communication in a constant number of rounds, and although expected constant-round broadcast protocols exist in the literature (e.g., [21,30]), using them to instantiate calls within the constructions of [2] or [19] would not yield an expected constant-round protocol [6]. The intuitive reason—formally argued by Ben-Or and El-Yaniv [6]—is that the process of running n such broadcast protocols (even in parallel) does not terminate in an expected constant number of rounds.

The model of *asynchronous* communication with eventual delivery was considered early on in seminal works on fault-tolerant distributed computing (e.g., [23]). The study of optimally resilient MPC in such an asynchronous network was initiated by Ben-Or, Canetti, and Goldreich [5], who proved that any function can be computed by a perfectly secure asynchronous protocol if and only if at most $t < n/4$ parties are corrupted. Following that result, Ben-Or, Kelmer, and Rabin [8] showed that if a negligible error probability is allowed, the bound $t < n/3$ is necessary and sufficient for asynchronous MPC.[26] More recently, Hirt *et al.* [27,28] provided computationally secure solutions

[25] Because [33] tolerates even $n/3 \leq t < n/2$ corrupted parties, the emulation of broadcast would require an additional setup of information-theoretic pseudo-signatures [32].

[26] The necessity of the $t < n/3$ bound follows from the result by Canetti *et al.* [5,12], who argue that this bound is necessary for fail-stop adversaries; it also applies to computational security and assuming A-BA. Moreover, note that in the asynchronous setting, all feasibility bounds are worse by an additive term of t compared to the synchronous setting. Intuitively, this stems from the fact that honest parties cannot distinguish between messages by other honest parties being delayed and messages by corrupted parties not being sent. Thus, in particular, perfectly secure asynchronous MPC is possible only if $t < n/4$.

(i.e., protocols tolerating a computationally bounded adversary) and Beerliová and Hirt [3] perfectly secure solutions with improved communication complexity.

The above asynchronous protocols are secure if one assumes point-to-point communication and an A-BA protocol. Similarly to their synchronous counterparts, all the above protocols—even assuming an A-BA primitive—have round complexity linear in the multiplicative depth of the arithmetic circuit that computes the function, as they follow the standard gate-by-gate evaluation paradigm.

We note in passing that although in the synchronous setting BA implies broadcast, this is not the case in the asynchronous setting. Indeed, Canetti and Rabin [15] provide an asynchronous BA protocol tolerating $t < n/3$ malicious parties, which if every honest party terminates at the latest after a poly-logarithmic number of rounds, securely implements asynchronous BA except with negligible probability. A broadcast protocol with similar guarantees is provably impossible [23], and existence of an asynchronous BA protocol which terminates in a strict constant number of rounds would contradict the impossibility from [20,22]. Similarly to the synchronous case, although solutions for asynchronous BA with expected constant number of rounds exist [11,15], using them in the above asynchronous protocol to replace invocations to asynchronous BA would not yield an expected constant-round MPC protocol [6].[27]

Finally, if one gives up the requirement that the broadcast protocol (eventually) terminates when the sender is corrupted (this results in a primitive known as *A-Cast* [10]), then one can implement it even in a constant number of rounds. (In fact, A-Cast can be easily reduced to asynchronous BA by having the sender send his input to all parties, who then forward this input as soon as it is received to the asynchronous BA primitive).

References

1. Simon, J.: Proceedings of the Twentieth Annual ACM Symposium on Theory of Computing. ACM, Chicago (1988)
2. Beaver, D., Micali, S., Rogaway, P.: The round complexity of secure protocols (extended abstract). In: STOC, pp. 503–513. ACM (1990)
3. Beerliová-Trubíniová, Z., Hirt, M.: Simple and efficient perfectly-secure asynchronous MPC. In: Kurosawa, K. (ed.) ASIACRYPT 2007. LNCS, vol. 4833, pp. 376–392. Springer, Heidelberg (2007). doi:10.1007/978-3-540-76900-2_23
4. Bellare, M., Hoang, V.T., Rogaway, P.: Foundations of garbled circuits. In: the ACM Conference on Computer and Communications Security, CCS 2012, Raleigh, NC, USA, October 16–18, 2012, pp. 784–796 (2012). http://doi.acm.org/10.1145/2382196.2382279
5. Ben-Or, M., Canetti, R., Goldreich, O.: Asynchronous secure computation. In: STOC, pp. 52–61 (1993)
6. Ben-Or, M., El-Yaniv, R.: Resilient-optimal interactive consistency in constant time. Distrib. Comput. **16**(4), 249–262 (2003). http://dx.doi.org/10.1007/s00446-002-0083-3

[27] Nonetheless, [6] does describe an alternative way of obtaining several asynchronous BA protocols that are guaranteed to all terminate in expected constant number of rounds.

7. Ben-Or, M., Goldwasser, S., Wigderson, A.: Completeness theorems for non-cryptographic fault-tolerant distributed computation (extended abstract). In: STOC [1], pp. 1–10
8. Ben-Or, M., Kelmer, B., Rabin, T.: Asynchronous secure computations with optimal resilience (extended abstract). In: PODC, pp. 183–192 (1994)
9. Boneh, D., Goh, E.-J., Nissim, K.: Evaluating 2-DNF formulas on ciphertexts. In: Kilian, J. (ed.) TCC 2005. LNCS, vol. 3378, pp. 325–341. Springer, Heidelberg (2005). doi:10.1007/978-3-540-30576-7_18
10. Bracha, G.: An asynchronou [(n-1)/3]-resilient consensus protocol. In: Probert, R.L., Lynch, N.A., Santoro, N. (eds.) 3rd ACM PODC. pp. 154–162. ACM Press, Vancouver, British Columbia, Canada (Aug 27–29, 1984)
11. Cachin, C., Kursawe, K., Shoup, V.: Random oracles in Constantinople: Practical asynchronous byzantine agreement using cryptography. J. Cryptology **18**(3), 219–246 (2005)
12. Canetti, R.: Studies in Secure Multiparty Computation and Applications. Ph.D. thesis, Weizmann Institute of Technology. http://www.wisdom.weizmann.ac.il/oded/PSX/ran-phd.pdf. 6 1995
13. Canetti, R.: Universally composable security: A new paradigm for cryptographic protocols. In: 42nd FOCS, pp. 136–145. IEEE Computer Society Press, Las Vegas, Nevada, USA (Oct 14–17, 2001)
14. Canetti, R., Feige, U., Goldreich, O., Naor, M.: Adaptively secure multi-party computation. In: 28th ACM STOC. pp. 639–648. ACM Press, Philadephia, Pennsylvania, USA (May 22–24, 1996)
15. Canetti, R., Rabin, T.: Fast asynchronous byzantine agreement with optimal resilience. In: 25th ACM STOC, pp. 42–51. ACM Press, San Diego, California, USA (May 16–18, 1993)
16. Chaum, D., Crépeau, C., Damgård, I.: Multiparty unconditionally secure protocols (extended abstract). In: STOC [1], pp. 11–19
17. Cohen, R.: Asynchronous secure multiparty computation in constant time. In: Cheng, C.-M., Chung, K.-M., Persiano, G., Yang, B.-Y. (eds.) PKC 2016. LNCS, vol. 9615, pp. 183–207. Springer, Heidelberg (2016). doi:10.1007/978-3-662-49387-8_8
18. Coretti, S., Garay, J., Hirt, M., Zikas, V.: Constant-round asynchronous multi-party computation. Cryptology ePrint Archive, Report 2016/208 (2016). http://eprint.iacr.org/2016/208
19. Damgård, I., Ishai, Y.: Constant-round multiparty computation using a black-box pseudorandom generator. In: Shoup, V. (ed.) CRYPTO 2005. LNCS, vol. 3621, pp. 378–394. Springer, Heidelberg (2005). doi:10.1007/11535218_23
20. Dolev, D., Strong, H.R.: Authenticated algorithms for byzantine agreement. SIAM J. Comput. **12**(4), 656–666 (1983). http://dx.doi.org/10.1137/0212045
21. Feldman, P., Micali, S.: Optimal algorithms for byzantine agreement. In: 20th ACM STOC, pp. 148–161. ACM Press, Chicago, Illinois, USA (May 2–4, 1988)
22. Fischer, M.J., Lynch, N.A.: A lower bound for the time to assure interactive consistency. Inf. Process. Lett. **14**(4), 183–186 (1982). http://dx.doi.org/10.1016/0020-0190(82)90033-3
23. Fischer, M.J., Lynch, N.A., Paterson, M.: Impossibility of distributed consensus with one faulty process. In: Fagin, R., Bernstein, P.A. (eds.) Proceedings of the Second ACM SIGACT-SIGMOD Symposium on Principles of Database Systems, March 21–23, 1983, Colony Square Hotel, Atlanta, Georgia, USA. pp. 1–7. ACM (1983). http://doi.acm.org/10.1145/588058.588060

24. Genkin, D., Ishai, Y., Prabhakaran, M., Sahai, A., Tromer, E.: Circuits resilient to additive attacks with applications to secure computation. In: Symposium on Theory of Computing, STOC 2014, New York, NY, USA, May 31 - June 03, 2014. pp. 495–504 (2014). http://doi.acm.org/10.1145/2591796.2591861
25. Goldreich, O.: Foundations of Cryptography: Basic Applications, vol. 2. Cambridge University Press, Cambridge (2004)
26. Goldreich, O., Micali, S., Wigderson, A.: How to play any mental game or a completeness theorem for protocols with honest majority. In: STOC, pp. 218–229. ACM (1987)
27. Hirt, M., Nielsen, J.B., Przydatek, B.: Cryptographic asynchronous multi-party computation with optimal resilience. In: Cramer, R. (ed.) EUROCRYPT 2005. LNCS, vol. 3494, pp. 322–340. Springer, Heidelberg (2005). doi:10.1007/11426639_19
28. Hirt, M., Nielsen, J.B., Przydatek, B.: Asynchronous multi-party computation with quadratic communication. In: Aceto, L., Damgård, I., Goldberg, L.A., Halldórsson, M.M., Ingólfsdóttir, A., Walukiewicz, I. (eds.) ICALP 2008. LNCS, vol. 5126, pp. 473–485. Springer, Heidelberg (2008). doi:10.1007/978-3-540-70583-3_39
29. Hirt, M., Zikas, V.: Adaptively secure broadcast. In: Gilbert, H. (ed.) EUROCRYPT 2010. LNCS, vol. 6110, pp. 466–485. Springer, Heidelberg (2010). doi:10.1007/978-3-642-13190-5_24
30. Katz, J., Koo, C.-Y.: On expected constant-round protocols for byzantine agreement. In: Dwork, C. (ed.) CRYPTO 2006. LNCS, vol. 4117, pp. 445–462. Springer, Heidelberg (2006). doi:10.1007/11818175_27
31. Katz, J., Maurer, U., Tackmann, B., Zikas, V.: Universally composable synchronous computation. In: Sahai, A. (ed.) TCC 2013. LNCS, vol. 7785, pp. 477–498. Springer, Heidelberg (2013). doi:10.1007/978-3-642-36594-2_27
32. Pfitzmann, B., Waidner, M.: Unconditional Byzantine agreement for any number of faulty processors. In: Finkel, A., Jantzen, M. (eds.) STACS 1992. LNCS, vol. 577, pp. 337–350. Springer, Heidelberg (1992). doi:10.1007/3-540-55210-3_195
33. Rabin, T., Ben-Or, M.: Verifiable secret sharing and multiparty protocols with honest majority (extended abstract). In: 21st ACM STOC. pp. 73–85. ACM Press, Seattle, Washington, USA (May 15–17, 1989)
34. Schneider, T., Zohner, M.: GMW vs. yao? efficient secure two-party computation with low depth circuits. In: Sadeghi, A.-R. (ed.) FC 2013. LNCS, vol. 7859, pp. 275–292. Springer, Heidelberg (2013). doi:10.1007/978-3-642-39884-1_23
35. Shamir, A.: How to share a secret. Commun. ACM **22**(11), 612–613 (1979)
36. Yao, A.C.C.: Protocols for secure computations (extended abstract). In: FOCS, pp. 160–164. IEEE (1982)
37. Yao, A.C.C.: How to generate and exchange secrets (extended abstract). In: FOCS, pp. 162–167. IEEE (1986)

Reactive Garbling: Foundation, Instantiation, Application

Jesper Buus Nielsen(✉) and Samuel Ranellucci

Aarhus University, Aarhus, Denmark
{jbn,samuel}@cs.au.dk

Abstract. Garbled circuits is a cryptographic technique, which has been used among other things for the construction of two and three-party secure computation, private function evaluation and secure outsourcing. Garbling schemes is a primitive which formalizes the syntax and security properties of garbled circuits. We define a generalization of garbling schemes called *reactive garbling schemes*. We consider functions and garbled functions taking multiple inputs and giving multiple outputs. Two garbled functions can be linked together: an encoded output of one garbled function can be transformed into an encoded input of the other garbled function without communication between the parties. Reactive garbling schemes also allow partial evaluation of garbled functions even when only some of the encoded inputs are provided. It is possible to further evaluate the linked garbled functions when more garbled inputs become available. It is also possible to later garble more functions and link them to the ongoing garbled evaluation. We provide rigorous definitions for reactive garbling schemes. We define a new notion of security for reactive garbling schemes called confidentiality. We provide both simulation based and indistinguishability based notions of security. We also show that the simulation based notion of security implies the indistinguishability based notion of security. We present an instantiation of reactive garbling schemes. We finally present an application of reactive garbling schemes to reactive two-party computation secure against a malicious, static adversary.

1 Introduction

Garbled circuits is a technique originating in the work of Yao and later formalised by Bellare, Hoang and Rogaway [2], who introduced the notion of a garbling scheme along with an instantiation. Garbling schemes have found a wide range of applications. However, many of these applications are using specific constructions of garbled circuits instead of the abstract notion of a garbling scheme. One possible explanation is that the notion of a garbling scheme falls short of capturing many of the current uses. In the notion of a garbling scheme, the constructed garbled function can only be used for a single evaluation and the garbled function has no further use. In contrast, many of the most interesting current applications of garbled circuits have a more granular look at garbling,

J.H. Cheon and T. Takagi (Eds.): ASIACRYPT 2016, Part II, LNCS 10032, pp. 1022–1052, 2016.
DOI: 10.1007/978-3-662-53890-6_34

where several components are garbled, dynamically glued together and possibly evaluated at different points in time. We now give a few examples of this.

In the standard cut-and-choose paradigm for two-party computation, Alice sends s copies of a garbled function to Bob. Half of the garblings (chosen by Bob) are opened to check that they were correctly constructed. This guarantees that the majority of the remaining instances were correctly constructed. Alice and Bob then use the remaining garblings for evaluation. Bob takes the majority output of these evaluations as his output. Although conceptually simple, this introduces a number of problems: Bob must ensure that Alice uses consistent inputs. It is also required that the probability that Bob aborts does not depend on his choice of input. Previous protocols solve these problems by doing white-box modifications of the underlying garbling scheme. We will show how to solve these problems by using reactive garbling schemes in a black-box manner.

In [18], Lindell presents a very efficient protocol for achieving active secure two-party computation from garbled circuits. In the scheme of Lindell, first s circuits are sent. Then a random subset of them are opened up to test that they were correctly constructed and the rest, the so-called evaluation circuits, are then evaluated in parallel. If the evaluations don't all give the same output, then the evaluator can construct a certificate of cheating which can be fed into a small corrective garbled circuit. Another example is a technique introduced simultaneously by Krater, shelat and Shen [16] and Frederiksen, Jakobsen and Nielsen [6], where a part of the circuit which checks the so-called input consistency of one of the parties is constructed *after* the main garbled circuit has been constructed and *after* Alice has given her input. We use a similar technique in our example application, showing that this trick can be applied to (reactive) garbling schemes in general. Another example is the work of Huang, Katz, Kolesnikov, Kumaresan and Malozemoff [14] on amortising garbled circuits, where one of the analytic challenges is a setting where many circuits are garbled prior to inputs being given. Our security notion allows this behaviour and this part of their protocol could therefore be cast as using a general (reactive) garbling scheme. Another example is the work of Huang, Evans, Katz and Malka [13] on fast secure two-party computation using garbled circuits, where they use pipelining: the circuit is garbled and evaluated in blocks for efficiency. Finally, we remark that sometimes the issue of garbling many circuits and gluing them together and having them interact with other security components can also lead to subtle insecurity problems, as demonstrated by the notion of a garbled RAM as introduced by Lu and Ostrovsky in [19], where the construction was later proven to be insecure by Gentry, Halevi, Lu, Ostrovsky, Raykova and Wichs [10]. We believe that having well founded abstract notions of partial garbling and gluing will make it harder to overlook security problems.

Our goal is to introduce a notion of reactive garbling schemes, which is general enough to capture the use of garbled circuits in most of the existing applications and which will hopefully form a foundation for many future applications of garbling schemes. Reactive garbling schemes generalize garbling schemes in several ways. First of all, we allow a garbled evaluation to save a state and use it in further computations. Specifically, when garbling a function f one can link it to a previous garbling of some function g and as a result get a garbling of $f \circ g$. Even more, given two independent garblings of f and g, it is possible to do a linking which will produce a garbling of $f \circ g$ or $g \circ f$. The linking depends only on the output encoding and input encoding of the linked garblings. We also allow garbling of a single function which allows partial evaluation and which allows dynamic input selection based on partial outputs. This can be mixed with linking, so that the choice of which functions to garble and link can be based on partial outputs. This can be important in *reactive* secure computation which allows inputs to arrive gradually and allows branching based on public partial outputs. We introduce the syntax and security definitions for this notion. We give an instantiation of reactive garbling schemes in the random oracle model. We also construct a reactive, maliciously UC secure two-party computation protocol based on reactive garbling schemes in a black-box manner.

1.1 Discussion and Motivation

In this section, we describe the purpose of our framework and why certain design choices were made for the framework in this paper.

One of the main goals of garbling schemes was to define a primitive that would be used in constructions without relying on the underlying instantiation. Unfortunately, most secure two-party computation protocols still rely on garbled circuits to provide security. In some sense, the notion of garbling schemes is not able to achieve this goal for the given task. One way of thinking of our result is to note that many techniques that previously only worked for garbled circuits, now work for reactive garbling schemes.

More precisely, to achieve reactive secure computation, the protocol for reactive computation shows how three issues which typically are solved using the underlying instantiation of garbled circuits in cut-and-choose protocols can be solved using reactive garbling schemes. These issues are Alice's input consistency, selective failure attacks and how to run the simulator against a corrupted Bob. We solve these three issues by using the notion of reactive garbling schemes. This means that many protocols in the literature can easily be modified to achieve security by only relying on the properties of reactive garbling schemes.

We now discuss why certain design choices were made. In particular, why we included notions such as linking multiple output wires to a single input wire, partial evaluation and output encoding. The reason that we allow multiple output wires to link to a single input wire is that otherwise we would exclude important constructions such as Minilego [7] and Lindell's reduced circuit optimization [18].

Output encodings are important for many reasons. First, it provides a method for defining linking. Roughly because of this notion, it is easy to define a linking as information which allows an encoded output to be converted into an encoded input. Secondly, in certain cases, constructions based on garbling schemes require a special property of the encoded output which otherwise cannot be described. This is the case of [11] where the encoded input has to be the same size as the encoded output. It is also useful for output reuse, covers pipelining and has applications to protocols where the receiver can use a proof of cheating to extract the sender's input.

We included partial evaluation for two main reasons, first we consider that it can be an important feature for reactive computation, secure outsourcing and secure computation where a partial output would be valuable. A partial output could be used to determine what future computation to run on data. In addition, we could garble blocks of functions and decide to link certain blocks together based on partial outputs.

In addition, many schemes in the literature inherently allow partial evaluation and not allowing partial evaluation imposes artificial restrictions on the constructions. For example, fine-grained privacy in [1] cannot be realized by standard schemes precisely because those schemes give out partial outputs.

1.2 Recasting Previous Constructions

The concept of using output encoding and linking has been implicitly used in many previous works. In particular, in cut-and-choose protocols, it has been used in [5,6,17,23] to enforce sender input consistency (ensure that the sender uses the same input in each instance) and to prevent selective failure attacks (an attack that works by having the probability that the receiver aborts depend on his choice of input). These concepts have also been used for different optimizations. Pipelining [13,16] and output reuse [11,20] are examples of direct optimizations. Linking has also been employed to reduce the number of circuits that need to be sent in protocols that apply cut-and-choose at the circuit level [4,18]. This is done by adding a phase where a receiver can extract the input of a cheating sender. Another example is gate soldering [7,21]. This technique works by employing cut-and-choose at the gate level. The gates are then randomly split among different buckets and soldered together. This optimization reduces the replication factor for a security $\Theta(s)$ to $\Theta(\frac{s}{\log(n)})$ where n is the number of non-xor gates. There are many applications that benefit from output encoding and linking in garbling schemes. In addition, if we allow sequences where the input is chosen as a function of the garbling, reactive garbling schemes are also adaptive. The constructions of [9,12] require adaptive garbling.

1.3 Structure of the Paper

In Sect. 2, we give the preliminaries. In Sect. 3, we define the syntax and security of reactive garbling schemes. In Sect. 4, we describe an instantiation of a reactive garbling scheme. In Sect. 4.1, we give a full description of the reactive garbling scheme. In Sect. 5, we give an intuitive description of the reactive two-party computation protocol based on reactive garbling schemes. In Sect. 5.1, we provide a full description of the reactive two-party computation protocol. We note that the techniques that we introduce in Sect. 5 can be applied to previous secure two-party computation protocol to convert them into constructions that only use reactive garbling schemes in a black-box manner. There is a full version of the paper with more details. [22] In the full version there is a detailed simulation proof that our reactive computation protocol is secure, a proof of security of our reactive garbling scheme using the indistinguishability based notion of security, we recast Lindell's construction using reactive garbling schemes, we describe Minilego's garbling and soldering as a reactive garbling scheme, and we prove security of our garbling scheme using the simulation based definition of confidentiality. We also show that simulation based definition implies the indistinguishability based definition of security.

2 Preliminaries

Let $\mathbb{N}$ be the set of natural numbers. For $n \in \mathbb{N}$, let $\{0,1\}^n$ be the set of n-bit strings. Let $\{0,1\}^* = \bigcup_{n \in \mathbb{N}} \{0,1\}^n$. We use $\top$ and $\bot$ as the syntax for *true* and *false* and we assume that $\top, \bot \notin \{0,1\}^*$. We use $()$ to denote the empty sequence. For a sequence σ, we use $x \in \sigma$ to denote that x is in the sequence. When we iterate over $x \in \sigma$ in a for-loop, we do it from left to right. For a sequence σ and an element x we use $\sigma \parallel x$ to denote that we append x to σ. We use $\parallel$ to denote concatenation of sequences. When unambiguous, we also use juxtaposition for concatenating and appending. We use $x \xleftarrow{\$} X$ to denote sampling a uniformly random x from a finite set X. We use $[A]$ to denote the possible legal outputs of an algorithm A. This is just the set of possible outputs, with $\bot$ removed.

We prove security of protocols in the UC framework and we assume that the reader is familiar with the framework. When we specify entities for the UC framework, ideal functionalities, parties in protocol, adversaries and simulators we give them by a set of rules of the form EXAMPLE (which sends (x_1, x_2) to the adversary in its last line). In Fig. 1, we give an example of a rule. A line of the form "**send** m **to** $\mathcal{F}$.R", where $\mathcal{F}$ is another entity and R the name of a rule, the entity will send (R, id, m) to $\mathcal{F}$, where id is a unique identifier of the rule that is sending, including the session and sub-session identifier, in case many copies of the same rule are currently in execution. We then give $(\mathrm{R}, id, ?)$ to the adversary and let the adversary decide when to deliver the message. Here ? is just a special reserved string indicating that the real input has been removed. When a message of the form (R, id, m) arrives from an entity A, the receiver stores $(\mathrm{R}, \mathsf{A}, id, m)$ in a pool of pending messages and turns the activation over to the adversary. A line of the form "**on** P **from** A" executed in a rule named R running with identifier id and where P is a pattern, is executed as follows. The entity executing the rule stores $(\mathrm{R}, \mathsf{A}, id, P)$ in a pool of pending receives and turns over the activation to the adversary. We say that a pending message $(\mathrm{R}, \mathsf{A}, id, m)$ matches pending receive $(\mathrm{R}, \mathsf{A}, id, P)$ if m can be parsed on the form P. Whenever an entity turns over the activation to the adversary it sends along $(\mathrm{R}, \mathsf{A}, id, ?)$ for all matched $(\mathrm{R}, \mathsf{A}, id, P)$, where ? is just a special

rule Example
on $(7, x_1)$ **from** A
on x_2 **from** B
$x \leftarrow ()$
$x \leftarrow x \parallel x_1 \parallel x_2$
$z \leftarrow 0$
for $y \in (1,2,4)$ **do**
  if $z \geq y$ **then abort**
  $z \leftarrow z + y$
send x **to** $\mathcal{A}$

Fig. 1. A rule

reserved bit-string. There is a special procedure INITIALIZE which is executed once, when the entity is created. All other rules begin with an **on**-command. The rule is considered *ready* for *id* if the first line is of the form "**on** P **from** A" and $(\mathrm{R}, \mathsf{A}, id, P)$ is matched and the rule was never executed with identifier *id*. In that case $(\mathrm{R}, \mathsf{A}, id, P)$ is considered to be in the set of pending receives. If the adversary sends $(\mathrm{R}, \mathsf{A}, id, ?)$ to an entity that has some pending receive $(\mathrm{R}, \mathsf{A}, id, P)$ matched by some pending message $(\mathrm{R}, \mathsf{A}, id, m)$, then the entity parses m using P and starts executing right after the line "**on** P **from** A" which added $(\mathrm{R}, \mathsf{A}, id, P)$ to the list of pending receives. A line of the form "**await** P" where P is a predicate on the state of the entity works like the **on**-command. The line turns activation over to the adversary along with an identifier, and the entity will report to the adversary which predicates have become true. The adversary can instruct the entity to resume execution right after any "**await** P" where P is true on the state of the entity. If an entity executes a rule which terminates, it turns the activation over to the adversary. The keyword **abort** makes an entity terminate and ignore all future inputs. A line of the form "**verify** P" makes the entity abort if P is not true on the state of the entity. We use $\mathcal{A}$ to denote the adversary and $\mathcal{Z}$ to denote the environment. A line of the form "**on** P" is equivalent to "**on** P **from** $\mathcal{Z}$". When specifying ideal functionalities we use Corrupt to denote the set of corrupted parties.

We define security of cryptographic schemes via code-based games [3]. The game is given by a set of procedures. There is a special procedure INITIALIZE which is called once, as the first call. There is another special procedure FINALIZE which may be called by the adversary. The output is true or false, $\top$ or $\bot$, where $\top$ indicates that the adversary won the game. In between INITIALIZE and FINALIZE, the adversary might call the other procedures at will. The other procedures might also output $\bot$ or $\top$ at which point the game ends with that output. Other outputs go back to the adversary.

3 Syntax and Security of Reactive Garbling Schemes

Section overview. We will start by defining the notion of gradual function, this will allow us to describe the type of functions that can be garbled. The functions that we define, in contrast to standard garbling schemes allow multiple inputs and outputs as well as partial evaluation.

Next, we will define the syntax of a reactive garbling scheme in the same way that a garbling scheme was described before. We will describe tags, a way of assigning identities to garbled functions, so that we can refer to them later. We will then describe different algorithms: how to encode inputs, decode outputs, link garblings together and other algorithms. Next, we will define correctness. The work of [2] defined the notion of correctness by comparing it to a plaintext evaluation. We define the notion of garbling sequences which is the equivalent of plaintext evaluation but for reactive garbling. Some garbling sequences don't make sense, for example producing an encoded input for a function that has not been defined. As a result, we will define the concept of legal garbling sequences to avoid sequences that are nonsensical. Finally, we can define correctness by comparing the plaintext evaluation of a garbling sequence with the evaluation of a garbling sequence by applying the algorithms define before. We then use the notion of garbling sequence to define the side-information function for reactive garbling. This is necessary to describe our notion of security which we call confidentiality.

Gradual Functions. We first define the notion of a gradual function. A gradual function is an extension of the usual notion of a function $f : A_1 \times \cdots \times A_n \rightarrow B_1 \times \cdots \times B_m$, where we allow to partially evaluate the function on a subset of the input components. Some output components might become available before all input components have arrived. We require that when an output component has become available, it cannot become unavailable or change as more input components arrive. We also require that the set of available outputs depends only on which inputs are ready, not on the exact value of the inputs. In our framework, we only allow garblings of gradual functions. This allows us to define partial evaluation and to avoid issues such as circular evaluation and determining when outputs are defined. These issues would make our framework more complex. The access function will be the function describing which outputs are available when a given set of inputs is ready. We will use $\perp$ to denote that an input is not yet specified and that an output is not yet available. We therefore require that $\perp$ is not a usual input or output of the function. We now formalize these notions. For a function $f : A_1 \times \cdots \times A_n \rightarrow B_1 \times \cdots \times B_m$ we use the following notation. $f.n := n$ and $f.m := m$, $f.A := A_1 \times \cdots \times A_n$, $f.B := B_1 \times \cdots \times B_m$, and $f.A_i := A_i$ and $f.B_i := B_i$.

Definition 1. *We use* component *to denote a set* $C = \{0,1\}^\ell \cup \{\perp\}$ *for some* $\ell \in \mathbb{N}$, *where* $\perp \notin \{0,1\}^*$. *We call* ℓ *the length of* C *and we write* $\text{len}(C) = \ell$. *Let* $C_1, \ldots, C_n$ *be components and let* $x', x \in C_1 \times \cdots \times C_n$.

- *We say that* x' *is an* extension *of* x, *written* $x \sqsubset x'$ *if* $x_i \neq \perp$ *implies that* $x_i = x'_i$ *for* $i = 1, \ldots, n$.
- *We say that* x *and* x' *are* equivalently undefined, *written* $x \bowtie x'$, *if for all* $i = 1, \ldots, n$ *it holds that* $x_i = \perp$ *iff* $x'_i = \perp$.

Definition 2 (Gradual Function). *Let* $A_1, \ldots, A_n, B_1, \ldots, B_m$ *be components and let* $f : A_1 \times \cdots \times A_n \rightarrow B_1 \times \cdots \times B_m$. *We say that* f *is a* gradual function *if it is monotone and variable defined.*

- *It is* monotone *if for all* $x, x' \in A_1 \times \cdots \times A_m$ *it holds that* $x \sqsubset x'$ *implies that* $f(x) \sqsubset f(x')$.
- *It is* variable defined *if* $x \bowtie x'$ *then* $f(x) \bowtie f(x')$.

We say that an algorithm *computes a gradual function* $f : A_1 \times \cdots \times A_n \rightarrow B_1 \times \cdots \times B_m$ if on all inputs $x \in A_1 \times \cdots \times A_m$ it accepts with output $f(x)$ and on all other inputs it rejects. We define a notion of access function which specifies which outputs components will be available given that a given subset of input components are available.

Definition 3 (Access Function). *The access function of a gradual function* $f : A_1 \times \cdots \times A_n \to B_1 \times \cdots \times B_m$ *is a function* $\mathsf{access}(f) : \{\bot, \top\}^n \to \{\bot, \top\}^m$ *defined as follows. For* $j = 1, \ldots, m$, *let* $q_j : B_j \to \{\bot, \top\}$ *be the function where* $q_j(\bot) = \bot$ *and* $q_j(y) = \top$ *otherwise. Let* $q : B_1 \times \cdots \times B_m \to \{\bot, \top\}^m$ *be the function* $(y_1, \ldots, y_m) \mapsto (q_1(y_1), \ldots, q_m(y_m))$. *For* $i = 1, \ldots, n$, *let* $p_i : \{\bot, \top\} \to A_i$ *be the function with* $p_i(\bot) = \bot$ *and* $p_i(\top) = 0^{\mathrm{len}(A_i)}$. *Let* $p : \{\bot, \top\}^n \to A_1 \times \cdots \times A_n$ *be the function* $(x_1, \ldots, x_n) \mapsto (p_1(x_1), \ldots, p_n(x_n))$. *Then* $\mathsf{access}(f) = q \circ f \circ p$.

Definition 4 (Gradual functional similarity). *Let* f,g *be gradual functions. We say that* f *is* similar *to* g ($f \sim g$) *if* $f.n = g.n$, $f.m = g.m$, $f.A = g.A$, $f.B = g.B$ *and* $\mathsf{access}(f) = \mathsf{access}(g)$.

In the following, if we use a function at a place where a gradual function is expected and nothing else is explicitly mentioned, we extend it to be a gradual function by adding $\bot$ to all input and output components and letting all outputs be undefined until all inputs are defined.

Syntax of Algorithms. A *reactive garbling scheme* consists of seven algorithms $\mathcal{G} = (\mathsf{St}, \mathsf{Gb}, \mathsf{En}, \mathsf{li}, \mathsf{Ev}, \mathsf{ev}, \mathsf{De})$. The algorithms St, Gb and Li are randomized and the other algorithms are deterministic. Gradual functions are described by strings f. We call f the *original gradual function.* For each such description, we require that $\mathsf{ev}(f, \cdot)$ computes some gradual function $\mathsf{ev}(f, \cdot) : A_1 \times \cdots \times A_n \to B_1 \times \cdots \times B_m$. This is the function that f describes. We often use f also to denote the gradual function $\mathsf{ev}(f, \cdot)$.

- On input of a security parameter $k \in \mathbb{N}$, the *setup algorithm* outputs a pair of parameters $(\mathsf{sps}, \mathsf{pps}) \leftarrow \mathsf{St}(1^k)$, where $\mathsf{sps} \in \{0,1\}^*$ is the *secret parameters* and $\mathsf{pps} \in \{0,1\}^*$ is the *public parameters*. All other algorithms will also receive 1^k as their first input, but we will stop writing that explicitly.
- On input f, a tag[1] $t \in \{0,1\}^*$ and the secret parameters sps the *garbling algorithm* Gb produces as output a quadruple of strings (F, e, o, d), where F is the *garbled function*, e is the *input encoding function*, d is the *output decoding function*, which is of the form $d = (d_1, \ldots, d_m)$, and o is the *output encoding function.* When $(F, e, o, d) \leftarrow \mathsf{Gb}(\mathsf{sps}, f, t)$ we use F_t to denote F, we use $d_{t,i}$ to denote the ith entry of d, and similarly for the other components. This naming is unique by the *function-tag uniqueness* and *garble-tag uniqueness* conditions described later.
- The *encoding algorithm* En takes input (e, t, i, x) and produces *encoded input* $X_{t,i}$.
- The *linking algorithm* li takes input of the form $(t_1, i_1, t_2, i_2, o, e)$ and produces an output L_{t_1,i_1,t_2,i_2} called the *encoded linking information.* Think of this as information which allows to take an encoded output Y_{t_1,i_1} for F_{t_1} and turn it into an encoded input X_{t_2,i_2} for F_{t_2}. In other words, we link the output wire

[1] Some of the algorithms will take as input values output by other algorithms. To identify where these inputs originate from we use tags.

with index i_1 of the garbling with tag t_1 to the input wire with index i_2 of the garbling with tag t_2.
- The *garbled evaluation algorithm* Ev takes as input a set $\mathcal{F}$ of pairs (t, F_t) where t is a tag and F_t a garbled function (let T be the set of tags t occurring in $\mathcal{F}$), a set $\mathcal{X}$ of triples $(t, i, X_{t,i})$ where $t \in T$, $i \in [F_t.n]$ and $X_{i,j} \neq \perp$ is an encoded input, and a set $\mathcal{L}$ of tuples $(t_1, i_1, t_2, i_2, L_{t_1,i_1,t_2,i_2})$ with $t_1, t_2 \in T$ and $i_1 \in [F_{t_1}.m]$ and $i_2 \in [F_{t_2}.n]$ and $L_{t_1,i_1,t_2,i_2} \neq \perp$ an encoded linking information. It outputs a set $\mathcal{Y} = \{(t, i, Y_{t,i})\}_{t \in T, i \in [F_t.m]}$, where each $Y_{t,i}$ is an *encoded output*. It might be that $Y_{t,i} = \perp$ if the corresponding output is not ready.
- The *decoding algorithm* takes input $(t, i, d_{t,i}, Y_{t,i})$, and produces a *final output* $y_{t,i}$. We require that $\mathsf{De}(\cdot,\cdot,\cdot,\perp) = \perp$. The reason for this is that $Y_{t,i} = \perp$ is used to signal that the encoded output cannot be computed yet, and we want this to decode to $y_{t,i} = \perp$. We extend the decoding algorithm to work on sets of decoding functions and sets of encoded outputs, by simply decoding each encoded output for which the corresponding output decoding function is given, as follows. For a set δ, called the *overall decoding function*, consisting of triples of the form $(t, i, d_{t,i})$, and a set $\mathcal{Y}$ of triples of the form $(t, i, Y_{t,i})$, we let $\mathsf{De}(\delta, \mathcal{Y})$ output the set of $(t, i, \mathsf{De}(t, i, d_{t,i}, Y_{t,i}))$ for which $(t, i, d_{t,i}) \in \delta$ and $(t, i, Y_{t,i}) \in \mathcal{Y}$.

Basic requirements. We require that $f.n$ and $f.m$ can be computed in linear time from a function description f. We require that $\mathrm{len}(f.A_i)$ and $\mathrm{len}(f.B_j)$ can be computed in linear time for $i = 1, \ldots, n$ and $j = 1, \ldots, m$. We require that the same numbers can be computed in linear time from any garbling F of f. We finally require that one can compute $\mathsf{access}(f)$ in polynomial time given a garbling F of f. We do not impose the length condition and the non-degeneracy condition from [2], i.e., e and d might depend on f. Our security definitions ensure that the dependency does not leak unwarranted information (Fig. 2).

Projective Schemes. Following [2], we call a scheme projective (on input component i) if all $X \in \{\ En(e, t, i, x) \mid x \in \{0,1\}^n\ \}$ are of the form $\{X_{1,0}, X_{1,1}\} \times$

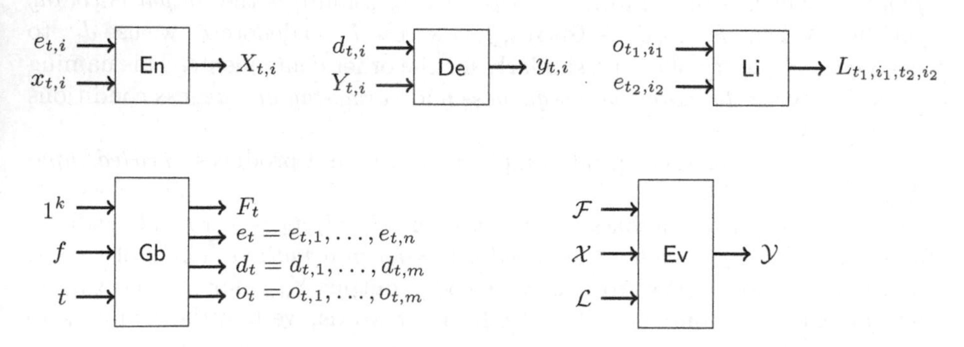

Fig. 2. Input-output behaviour of the central algorithms of a reactive garbling scheme.

$\ldots \times \{X_{c,0}, X_{c,1}\}$, where $c = \text{len}(f.X_i)$, and $En(e,t,i,x) = (X_{1,x[1]}, \ldots, X_{c,x[c]})$. This should hold for all k, f, t, ℓ, $x \in \{0,1\}^c$ and $(\mathsf{sps}, \mathsf{pps}) \in [\mathsf{St}(1^k)]$ and $(F,e,o,d) \in [\mathsf{Gb}(\mathsf{sps}, f, t, \ell)]$. As in [2] being projective is defined only relative to the input encodings. One can define a similar notion for output decodings. Having projective output decodings is needed for capturing some applications using reactive garbling scheme, for instance [18].

Correctness. To define correctness, we need a notion of calling the algorithms of a garbling scheme in a meaningful order. For this purpose, we define a notion of *garbling sequence* σ. A garbling sequence is a sequence of *garbling commands*, each command has one of the following forms: $(\texttt{Func}, f, t)$, $(\texttt{Link}, t_1, i_1, t_2, i_2)$, $(\texttt{Input}, t, i, x)$, $(\texttt{Output}, t, i)$, $(\texttt{Garble}, t)$. In the rest of the paper, we will use σ to refer to a garbling sequence. A garbling sequence is called *legal* if the following conditions hold.

Function uniqueness: σ does not contain distinct commands $(\texttt{Func}, f_1, t)$ and $(\texttt{Func}, f_2, t)$.

Garble uniqueness: Each command $(\texttt{Garble}, t)$ occurs at most once in σ.

Garble legality: If $(\texttt{Garble}, t)$ occurs in σ, it is preceded by $(\texttt{Func}, \cdot, t)$.

Linkage legality: If the command $(\texttt{Link}, t_1, i_1, t_2, i_2)$ occurs in σ, then the command is preceded by commands of the forms $(\texttt{Func}, f_1, t_1)$, $(\texttt{Garble}, t_1)$, $(\texttt{Func}, f_2, t_2)$ and $(\texttt{Garble}, t_2)$, and $1 \leq i_1 \leq f_1.m$, $1 \leq i_2 \leq f_2.n$ and $f_1.B_{i_1} = f_2.A_{i_2}$.

Input legality: If $(\texttt{Input}, t, i, x)$ occurs in σ it is preceded by $(\texttt{Func}, f, t)$ and $(\texttt{Garble}, f)$ and $x \in f.A_i \setminus \{\perp\}$.

Output legality: If $(\texttt{Output}, t, i)$ occurs in a sequence it is preceded by $(\texttt{Func}, f, t)$ and $(\texttt{Garble}, t)$ and $1 \leq i \leq f.m$.

Note that if a sequence is legal, then so is any prefix of the sequence. We call a garbling sequence *illegal* if it is not legal. Since we allow to link several output components onto the same input component we have to deal with the case where they carry different values. We consider this an error, and to catch it, we use the following safe assignment operator.

proc $\mathsf{eval}(\sigma \in \mathbb{L})$
for $(\texttt{Func}, t, f) \in \sigma$, **do**
  $f_t \leftarrow f$
  for $i = 1, \ldots, f_t.n$ **do** $x_{t,i} \leftarrow \perp$
  for $j = 1, \ldots, f_t.m$ **do** $y_{t,j} \leftarrow \perp$
for $(\texttt{Input}, t, i, x) \in \sigma$ **do** $x_{t,i} \hookleftarrow x$
$T \leftarrow \emptyset$
repeat
  $U \leftarrow T$
  for $(\texttt{Func}, t, f) \in \sigma$ **do**
    $(y_{t,1}, \ldots, y_{t,f_t.m}) \leftarrow f_t(x_{t,1}, \ldots, x_{t,f_t.n})$
    for $(\texttt{Link}, t, i_1, t_2, i_2) \in \sigma$ **do** $x_{t_2,i_2} \hookleftarrow y_{t,i_1}$
  $T \leftarrow \{(t, i, y_{t,i}) \mid t \in \text{Tags}(\sigma), i = 1, \ldots, f_t.m\}$
until $T = U \vee (\cdot, \cdot, \texttt{Error}) \in T$
return T

Fig. 3. Plaintext evaluation

$$(u \hookleftarrow v) := \begin{cases} u \leftarrow \texttt{Error} & \text{if } v = \texttt{Error} \\ u \leftarrow u & \text{if } v = \bot \\ u \leftarrow v & \text{if } u = \bot \vee u = v \\ u \leftarrow \texttt{Error} & \text{otherwise} \end{cases}$$

We now define an algorithm eval, which takes as input a legal garbling sequence σ and outputs a set of tuples $(t, i, y_{t,i})$, one for each command $(\texttt{Output}, t, i)$, where possibly $y_{t,i} = \bot$. The values are computed by taking the least fix point of the evaluation of all the gradual functions, see Fig. 3. We call this the *plain evaluation* of σ. We extend the definition of a legal sequence to include the requirement that

Input uniqueness $(\cdot, \cdot, \texttt{Error}) \notin \mathsf{eval}(\sigma)$.

Therefore the use of the safe assignment in eval is only to conveniently define the notion of legal sequence. In the rest of the paper we assume that all inputs to eval are legal. The values $y_{t,i} \neq \bot$ are by definition the values that are *ready* in σ, i.e., $\mathsf{ready}(\sigma) = \{(t, i) | \exists (t, i, y_{t,i}) \in \mathsf{eval}(\sigma)(y_{t,i} \neq \bot)\}$. Note that since the gradual functions are variable defined, which outputs are ready does not depend on the values of the inputs, except via whether they are $\bot$ or not.

The procedure Eval in Fig. 4 demonstrates how a legal garbling sequence is intended to be translated into calls to the algorithms of the garbling scheme. We call the procedure executed by Eval *garbled evaluation* of σ.

proc $\mathsf{Eval}(\sigma \in \mathbb{L})$
for $c \in \sigma$ **do**
  if $c = (\texttt{Func}, t, f)$ **then** $f_t \leftarrow f$;
  if $c = (\texttt{Garble}, t)$ **then**
    $(F_t, e_t, o_t, d_t) \leftarrow \mathsf{Gb}(\mathsf{sps}, f_t, t)$
    $\mathcal{F} \leftarrow \mathcal{F} \| (t, F_t)$
  if $c = (\texttt{Input}, t, i, x)$ **then**
    $X_{t,i} \leftarrow \mathsf{En}(e_t, t, i, x)$
    $\mathcal{X} \leftarrow \mathcal{X} \| (t, i, X_{t,i})$
  if $c = (\texttt{Link}, t_1, i_1, t_2, i_2)$ **then**
    $L_{t_1,i_1,t_2,i_2} \leftarrow \mathsf{li}(t_1, i_1, t_2, i_2, o_{t_1}, e_{t_2})$
    $\mathcal{L} \leftarrow \mathcal{L} \| (t_1, i_1, t_2, i_2, L_{t_1,i_1,t_2,i_2})$
  if $c = (\texttt{Output}, t, i)$ **then**
    $\delta \leftarrow \delta \| (t, i, d_{t,i})$
return $\mathsf{De}(\delta, \mathsf{Ev}(\mathcal{F}, \mathcal{X}, \mathcal{L}))$

Fig. 4. Garbled evaluation

Lemma 1. *For a function description f, let $T(f)$ be the worst case running time of $\mathsf{ev}(f, \cdot)$. The algorithm eval will terminate in time $\mathrm{poly}(T|\sigma|(n + m))$, where $n = \max_{(\texttt{Func},t,f)\in\sigma} f.n$, $m = \max_{(\texttt{Func},t,f)\in\sigma} f.m$, and $T = \max_{(\texttt{Func},t,f)\in\sigma} T(f)$.*

Proof. By monotonicity, if the loop in eval does not terminate, another variable $y_{t,i}$ has changed from $\bot$ to $\neq \bot$ and can never change value again. This bounds the number of iterations as needed.

Side-Information Functions. We use the same notion of side-information functions as in [2]. A side information function Φ maps function descriptions f into the side information $\Phi = \Phi(f) \in \{0, 1\}^*$. Intuitively, a garbling of f should not leak more than $\Phi(f)$. The exact meaning of the side information functions are given by our security definition. We extend a side information function Φ to the set of garbling sequences. For the empty sequence $\sigma = ()$ we let $\Phi(\sigma) = ()$.

For a sequence σ, we define the side-information as $\Phi(\sigma) := \Phi_\sigma(\sigma)$ where for a sequence $\bar{\sigma}$ and a command c: $\Phi_\sigma(\bar{\sigma} \,\|\, c) = \Phi_\sigma(\bar{\sigma}) \,\|\, \Phi_\sigma(c)$, where $\Phi_\sigma(\texttt{Func}, t, f) = (\texttt{Func}, t, \Phi(f))$, $\Phi_\sigma(\texttt{Link}, t_1, i_1, t_2, i_2) = (\texttt{Link}, t_1, i_1, t_2, i_2)$, $\Phi_\sigma(\texttt{Input}, t, i, x) = (\texttt{Input}, t, i, |x|)$, $\Phi_\sigma(\texttt{Garble}, t) = (\texttt{Garble}, t)$ and $\Phi_\sigma(\texttt{Output}, t, i) = (\texttt{Output}, t, i, y_{t,i})$, where $y_{t,i}$ is defined by $\mathsf{eval}(\sigma)$.

Legal Sequence Classes. We define the notion of a *legal sequence class* $\mathbb{L}$ (relative to a given side-information function Φ). It is a subset of the legal garbling sequences which additionally has these five properties:

Monotone: If $\sigma' \,\|\, \sigma'' \in \mathbb{L}$, then $\sigma' \in \mathbb{L}$.
Input independent: If $\sigma' \,\|(\texttt{Input}, t, i, x) \,\|\, \sigma'' \in \mathbb{L}$, then $\sigma' \,\|(\texttt{Input}, t, i, x') \,\|\, \sigma'' \in \mathbb{L}$ for all $x' \in \{0,1\}^{|x|}$.
Function independent: If $\sigma' \,\|(\texttt{Func}, t, f) \,\|\, \sigma'' \in \mathbb{L}$, then $\sigma' \,\|(\texttt{Func}, t, f') \,\|\, \sigma'' \in \mathbb{L}$ for all f with $\Phi(f') = \Phi(f)$.
Name invariant: If $\sigma \in \mathbb{L}$ and σ' is σ with all tags t replaced by $t' = \pi(t)$ for an injection π, then $\sigma' \in \mathbb{L}$.
Efficient: Finally, the language $\mathbb{L}$ should be in P, i.e., in polynomial time.

It is easy to see that the set of all legal garbling sequences is a legal sequence class.

Definition 5 (Correctness). *For a legal sequence class* $\mathbb{L}$ *and a reactive garbling scheme* $\mathcal{G}$ *we say that* $\mathcal{G}$ *is* $\mathbb{L}$-correct *if for all* $\sigma \in \mathbb{L}$, *it holds that* $\mathsf{De}(\mathsf{Eval}(\sigma)) \subseteq \mathsf{eval}(\sigma)$ *for all choices of randomness by the randomized algorithms.*

Function Individual Garbled Evaluation. The garbled evaluation function Ev just takes as input sets of garbled functions, inputs and linking information and then somehow produces a set of garbled outputs. It is often convenient to have more structure to the garbled evaluation than this.

We say that garbled evaluation is *function individual* if each garbled function F is evaluated on its own. Specifically there exist deterministic poly-time algorithms Evl and Li called the *individual garbled evaluation algorithm* and the *garbled linking algorithm*. The input to Evl is a garbled function and some garbled inputs. For each fixed garbled function F with $n = F.n$ and $m = F.m$ the algorithm computes a gradual

proc $\mathsf{Ev}(\mathcal{F}, \mathcal{X}, \mathcal{L})$
for $(t, F) \in \mathcal{F}$ **do**
  $F_t \leftarrow F$
  for $i = 1, \ldots, F_t.n$ **do** $X_{t,i} \leftarrow \perp$
for $(t, i, X) \in \mathcal{X}$ **do** $X_{t,i} \leftarrow X$
$T \leftarrow \emptyset$
repeat
  $U \leftarrow T$
  for $(t, F_t) \in \mathcal{F}$ **do**
    $(Y_{t,1}, \ldots, Y_{t,F_t.m}) \leftarrow \mathsf{Evl}(F_t, (X_{t,1}, \ldots, X_{t,F_t.n}))$
    for $(t, i_1, t_2, i_2, L) \in \mathcal{L}$ **do** $X_{t_2,i_2} \leftarrow \mathsf{Li}(L, Y_{t,i_1})$
    $T \leftarrow \{(t, i, Y_{t,i}) \mid t \in \mathrm{Tags}(\sigma) \wedge i = 1, \ldots, F_t.m\}$
until $T = U$
return T

Fig. 5. Function individual evaluation

function $\mathsf{Evl}(F) : A_1 \times \cdots \times A_n \to B_1 \times \cdots \times B_m$ and $(X_1, \ldots, X_n) \mapsto \mathsf{Evl}(F, X_1, \ldots, X_n)$, with $\mathsf{access}(\mathsf{Evl}(F)) = \mathsf{access}(f)$, where f is the function garbled by F. We denote the output by $(Y_1, \ldots, Y_m) = \mathsf{Evl}(F, X_1, \ldots, X_n)$. The intention is that the Y_j are garbled outputs (or $\perp$). To say that Ev has individual garbling we then require that it is defined from Evl and Li as in Fig. 5.

Security of Reactive Garbling. We define a notion of security that we call confidentiality, which unifies privacy and obliviousness as defined in [2]. Obliviousness says that if the evaluator is given a garbled function and garbled inputs but no output decoding function it can learn a garbled output of the function but learns no information on the plaintext value of the output. Privacy says that if the evaluator is given a garbled function, garbled inputs and the output decoding function it can learn the plaintext value of the function, but no other information, like intermediary values from the evaluation. It is necessary to synthesise these properties as we envision protocols where the receiver of the garbled functions might receive the output decoding function for *some* of the output components but *not all* of them. Obliviousness does not cover this case, since the adversary has some of the decoding keys. It is not covered by privacy either, as the receiver should not gain any information about outputs for which he does not have a decoding function.

In the confidentiality (indistinguishability) game, the adversary feeds two sequences σ_0 and σ_1 to the game, which produces a garbling of one of the two sequences, σ_b for a uniform bit b. The adversary wins if it can guess which sequence was garbled. It is required that the two sequences are not trivially distinguishable. For instance, the two commands at position i in the two sequences should have the same type, the side information of functions at the same positions in the sequences should be the same, and all outputs produced by the sequences should be the same. This is formalized by requiring that the side information of the sequences are the same. This is done by checking that $\Phi(\sigma_0) = \Phi(\sigma_1)$ in the rule FINALIZE. If one considers garbling sequences with only one function command, one garbling command, one input command per input component, no linking and where no output command is given, then confidentiality implies obliviousness. If in addition an output command is given for each output component, then confidentiality implies privacy.

In the confidentiality (simulation) game, the adversary feeds a sequence σ to the game. The game samples a uniform bit b. If $b = 0$, then the game uses the reactive garbling scheme to produce values for the sequence. Otherwise, if the bit $b = 1$, the game feeds the output of the side-information function to the simulator and forwards any response to the adversary. The simulation-based notion of confidentiality implies the indistinguishability-based notion of indistinguishability [22].

Definition 6 (Confidentiality). *For a legal sequence class $\mathbb{L}$ relative to side-information function Φ and a reactive garbling scheme $\mathcal{G}$, we say that $\mathcal{G}$ is $(\mathbb{L}, \Phi)$-*confidential *if for all PPT $\mathcal{A}$ it holds that $\mathbf{Adv}^{\text{adp.ind.con}}_{\mathcal{G},\mathbb{L}',\Phi,\mathcal{A}}(1^k)$ is negligible, where*

$\mathbf{Adv}_{\mathcal{G},\mathbb{L}',\Phi,\mathcal{A}}^{\text{adp.ind.con}}(1^k) = \Pr[\mathbf{Game}_{\mathcal{G},\mathbb{L}',\Phi,\mathcal{A}}^{\text{adp.ind.con}}(1^k) = \top] - \frac{1}{2}$ *and* $\mathbf{Game}_{\mathcal{G},\mathbb{L}',\Phi}^{\text{adp.ind.con}}$ *is given in Fig. 6.*

Notice that this security definition is indistinguishability based, which is known to be very weak in some cases for garbling (cf. [2]). Consider for instance garbling a function f where the input x is secret and $y = f(x)$ is made a public output. The security definition then only makes a requirement on the garbling scheme in the case where the adversary inputs two sequences where in sequence one the input is x_1 and in sequence two the input is x_2 and where $f(x_1) = f(x_2)$. Consider then what happens if f is collision resistant. Since no adversary can compute such x_1 and x_2 where $x_1 \neq x_2$, it follows that $x_1 = x_2$ in all pairs of sequences that the adversary can submit to the game. It can then be seen that it would be secure to "garble" collision resistant functions f by simply sending f in plaintext. Despite this weak definition, we later manage to prove that it is sufficient for building secure two-party computation. Looking ahead, when we need to securely compute f, we will garble a function f' which takes an additional input p which is the same length as the output of f and where $f'(x,p) = p \oplus f(x)$ and ask the party that supplies p to always let p be the all-zero string. Our techniques for ensuring active security in general is used to enforce that even a corrupted party does this. Correctness is thus preserved. Clearly f' is not collision resistant even if f is collision resistant. This prevents a secure garbling scheme from making insecure garblings of f'. In fact, note that this trick ensures that f' has the efficient invertibility property defined by [2], which means that the indistinguishability and simulation based security coincide.

4 Instantiating a Confidential Reactive Garbling Scheme

We show that the instantiation of garbling schemes in [2] can be extended to a reactive garbling scheme in the random oracle (RO) model. We essentially implement the dual-key cipher construction from [2] using the RO. To link a wire with 0-token T_0 and 1-token T_1 to an input wire with tokens I_0 and I_1, we provide the linking information $L_0 = RO(T_0) \oplus I_0$ and $L_1 = RO(T_1) \oplus I_1$ in a random order with each value tagged by the permutation bits of their corresponding input wires and output wires. Evaluation is done using function individual evaluation. Evaluation of a single garbled circuit is done as in [2]. Evaluation of a linking is: given T_b and a permutation bit, the bit is used to retrieve L_b from which $I_b = L_b \oplus RO(T_b)$ is computed. We provide the details in Sect. 4.1. We use the RO because reactive garbling schemes run into many of the same subtle security problems as adaptive garbling schemes [1], which are conveniently handled by being able to program the RO. We leave as an open problem the construction of (efficient) reactive garbling schemes in the standard model.

4.1 A Reactive Garbling Scheme

We will now give the details of the construction of a confidential reactive garbling scheme based on a random oracle. The protocol is inspired by the construction

```
proc INITIALIZE()                      proc FUNC(f_0, f_1, t)
b <-$- {0,1}                           for c ∈ {0,1} do
σ_0 <- ∅                               σ_c <- σ_c || (Func, f_c, t)
σ_1 <- ∅                               if f_0 ≁ f_1 then return ⊥

proc OUTPUT(t, i)                      proc INPUT(t, i, x_0, x_1)
for c ∈ {0,1} do                       for c ∈ {0,1} do
σ_c <- σ_c || (Output, t, i)           σ_c <- σ_c || (Input, t, i, x_c)
return d_{t,i}                         return En(e_t, t, i, x_b)

proc LINK(t_1, i_1, t_2, i_2)          proc FINALIZE(b')
for c ∈ {0,1} do                       if b = b' ∧ Φ(σ_0) = Φ(σ_1) ∧ σ_0 ∈ 𝕃
σ_c <- σ_c || (Link, t_1, i_1, t_2, i_2)   then return ⊤
return li(t_1, i_1, t_2, i_2, o_{t_1,i_1}, e_{t_2,i_2})   else return ⊥

proc GARBLE(t)
for c ∈ {0,1} do σ_c <- σ_c || (Garble, t)
(F_t, e_t, o_t, d_t) <- Gb(sps, f_t, t)
return F_t
```

Fig. 6. The game $\mathbf{Game}^{\text{adp.ind.con}}_{\mathcal{G},\mathbb{L},\Phi}(1^k)$ defining *adaptive indistinguishability confidentiality*. In FINALIZE we check that $\sigma_0 \in \mathbb{L}$ and the adversary loses if this is not the case. It is easy to see that when $\mathbb{L}$ is a legal sequence class and $\Phi(\sigma_0) = \Phi(\sigma_1)$, then $\sigma_0 \in \mathbb{L}$ iff $\sigma_1 \in \mathbb{L}$. We can therefore by monotonicity assume that the game returns $\perp$ as soon as it happens that $\sigma_c \notin \mathbb{L}$. We use a number of notational conventions from above. Tags are used to name objects relative to σ_c, which is assumed to be legal. As an example, in Garble(t), the function f_t refers to the function f_c occurring in the command (Func, $f_c, t)$ which was added to σ_c in FUNC by **Garble Legality**. For another example, the $d_{t,i}$ in OUTPUT(t,i) refers to the ith component of the d_t component output by Gb(sps, $f_t, t)$ in the execution of GARBLE(t,π) which must have been executed by **Output Legality**.

of garbling schemes from dual-key ciphers presented in [2]. The pseudocode for our reactive garbling scheme is shown in Figs. 7 and 8.

To simplify notation, we define lsb as the least significant bit, slsb as the second least significant bit. The operation Root removes the last two bits of a string. The symbol H denotes the random oracle.

We use the notation of [2] to represent a circuit. A circuit is a 6-tuple $f = (n, m, q, A, B, G)$. Here $n \geq 2$ is the number of inputs, $m \geq 1$ is the number of outputs and $q \geq 1$ is the number of gates. We let $r = n + q$ be the number of wires. We let Inputs $= \{1, \ldots, n\}$, Wire $= \{1, \ldots, n+q\}$, OutputWires $= \{n+q-m+1, \ldots, n+q\}$ and Gates $= \{n, \ldots, n+q\}$. Then A : Gates $\rightarrow$ Wires $\setminus$ OutputWires is a function to identify each gate's first incoming wire and B : Gates $\rightarrow$ Wires $\setminus$ OutputWires is a function to identify each gate's second incoming wire. Finally, G : Gates $\times \{0,1\}^2 \rightarrow \{0,1\}$ is a function that determines the functionality of each gate. We require that $A(g) < B(g) < g$ for all $g \in$ Gates.

Our protocol will also follow the approach of [2]. To garble a circuit, two tokens are selected for each wire, one denoted by $X_{t,i,0}$ which shall encode the

proc $\mathsf{Gb}(f_t, t)$
$(n, m, q, A, B, G) \leftarrow f_t$
for $i \in \{1, \ldots, n + q - m\}$ **do**
  $c \xleftarrow{\$} \{0,1\}$ // Type of the zero-encoding
  $X_{t,i,0} \leftarrow \{0,1\}^{k-1} \,\|\, c$
  $X_{t,i,1} \leftarrow \{0,1\}^{k-1} \,\|\, 1 - c$
for $i \in \{1, \ldots, m\}$ **do**
  $c \xleftarrow{\$} \{0,1\}, r_i \xleftarrow{\$} \{0,1\}$ // Type and mask of zero-encoding
  $Y_{t,i,0} \leftarrow \{0,1\}^{k-2} \,\|\, r_i \,\|\, c$
  $Y_{t,i,1} \leftarrow \{0,1\}^{k-2} \,\|\, 1 - r_i \,\|\, 1 - c$
  $X_{t,n+q-m+i,0} \leftarrow Y_{t,i,0}$
  $X_{t,n+q-m+i,1} \leftarrow Y_{t,i,1}$
for $(i, u, v) \in \{n + 1, \ldots, n + q\} \times \{0,1\} \times \{0,1\}$ **do**
  $a \leftarrow A(i), b \leftarrow B(i)$ // Left wire, right wire
  // Left-wire encoding of u and its type.
  $A \leftarrow \mathsf{root}(X_{t,a,u}), \mathfrak{a} \leftarrow \mathsf{lsb}(X_{t,a,u})$
  // Right-wire encoding of v and its type.
  $B \leftarrow \mathsf{root}(X_{t,b,v}), \mathfrak{b} \leftarrow \mathsf{lsb}(X_{t,b,v})$
  // Unique tag
  $T \leftarrow t \,\|\, i \,\|\, \mathfrak{a} \,\|\, \mathfrak{b}$
  // Row of Garbled table associated to gate i and input (u, v)
  $P[i, \mathfrak{a}, \mathfrak{b}] \leftarrow \mathsf{H}(T \,\|\, A \,\|\, B) \oplus Y_{t,i,G(i,u,v)}$
$F_t \leftarrow (n, m, q, A, B, P)$
$e_t \leftarrow ((X_{1,0}, X_{1,1}), \ldots, (X_{n,0}, X_{n,1}))$
$o_t \leftarrow ((Y_{1,0}, Y_{1,1}), \ldots, (Y_{m,0}, Y_{m,1}))$
$d_t \leftarrow \{r_1, \ldots, r_m\}$
return (F_t, e_t, o_t, d_t)

proc $\mathsf{En}(t, i, x)$
$X_{t,i} \leftarrow e_{t,i,x}$
return $X_{t,i}$

proc $\mathsf{De}(t, i, Y_{t,i}, d_{t,i})$
$y_{t,i} \leftarrow \mathsf{slsb}(Y_{t,i}) \oplus d_{t,i}$
return $y_{t,i}$

Fig. 7. Reactive garbling scheme

value 0 and the other denoted by $X_{t,i,1}$ which will encode the value 1, we refer to this mapping as the semantic of a token.

The encoding of an input for a value x is simply the token of the given wire with semantic x. The decoding of an output is the mask for that wire. We decouple the decoding from the linking to simplify the proof of security. The simulator will be able to produce linking without having to worry about the semantics of the output encoding.

For each wire, the two associated tokens will be chosen such that the least significant bit (the type of a token) will differ. It is important to note that the

```
proc li(o_{t1,i1}, e_{t2,i2})
// Type of zero-encoding
c ← lsb(o_{t1,i1,0})
K_0 ← root(o_{t1,i1,0})
K_1 ← root(o_{t1,i1,1})
T ← (t_1, i_1, t_2, i_2)
// Encryption of encoded input whose
   associated output encoding has
   type 0
U_0 ← H(T || k_c) ⊕ e_{t2,i2,c}
// Encryption of input encoding whose
   associated output encoding has
   type 1
U_1 ← H(T || k_{1⊕c}) ⊕ e_{t2,i2,1⊕c}
L_{t1,i1,t2,i2} ← (U_0, U_1)
return L_{t1,i1,t2,i2}

proc Li(L_{t1,i1,t2,i2}, Y_{t1,i1})
r ← lsb(Y_{t1,i1})
K ← root(Y_{t1,i1})
T ← (t_1, i_1, t_2, i_2)
X_{t,i} ← H(T || k) ⊕ L_{t1,i1,t2,i2,r}
return X_{t,i}

proc Evl(F_t, X_1, ..., X_n)
(n, m, q, A, B, P) ← F_t
for i ← n + 1 to n + q do
    a ← A(i), b ← B(i)
    A ← X_{t,a}, B ← X_{t,b}
    if A ≠ ⊥ ∧ B ≠ ⊥ then
        𝔞 ← lsb(A), 𝔟 ← lsb(B)
        T ← t || i || 𝔞 || 𝔟
        X_g ← P[g, 𝔞, 𝔟] ⊕ H(T || A || B)
(Y_{t,i}, ..., Y_{t,m}) ← (X_{n+q−m+1}, ..., X_{n+q})
return (Y_{t,1}, ..., Y_{t.m})
```

Fig. 8. Reactive garbling scheme (continued)

semantics and type of a token are independent. The second least significant bit is called the mask and will have a special meaning later when the tokens are output tokens. We use $\mathsf{root}(X)$ to denote the part of a token that is not the type bit or the mask bit.

Each gate g will be garbled by producing a garbled table. A garbled table will consist of four ciphertexts $p[g, a, b]$ where $a, b \in \{0, 1\}$, The ciphertext $P[g, a, b]$ will be produced in the following way: first find the token associated to the left input wire (i_1) with type a, denote the semantic of this token as x. Secondly, find the token associated to the right input wire (i_2) with type b, denote the semantic of this token as y. The ciphertext will be an encryption of the token of

$z \leftarrow G(g, x, y)$. We will denote $T \leftarrow t \,\|\, g \,\|\, a \,\|\, b$. The encryption will be $P[g, a, b] \leftarrow \mathsf{H}(T \,\|\, \mathsf{root}(X_{t,i_1,x}) \,\|\, \mathsf{root}(X_{t,i_2,y})) \oplus (X_{t,i,z})$

For each non-output wire, the token with semantic 0 will be chosen randomly and the token with semantic 1 will be chosen uniformly at random except for the last bit which will be chosen to be the negation of the least significant bit of the token with semantic 0 for the same wire.

For each output wire, the first token will also be chosen uniformly at random. The token with semantic 0 will be chosen randomly and the token with semantic 1 will be chosen uniformly at random except for the least significant bit and the second least significant bit. For both of these positions, the second token will be chosen so that they differ from the value in the 0-token for the same position. We refer to the second least significant bit of the 0-token of an output token as the mask of an output wire.

A linking between output (t_1, i_1) and input (t_2, i_2) consists of two ciphertexts: let c be the type of the 0-token for the output wire. In this case, we set $T = t_1 \,\|\, i_1 \,\|\, t_2 \,\|\, i_2$. The linking is simply

$$L \leftarrow (E^T_{\mathsf{root}(Y_{t_1,i_1,c})}(X_{t_2,i_2,c}), E_{\mathsf{root}(Y_{t_1,i_1,1-c})}(X_{t_2,i_2,1-c}))$$

where $E^T_k(z) = \mathsf{H}(T||k) \oplus z$. Converting an encoded output into an encoded input follows naturally.

In [22] we prove the following theorem.

Theorem 1. *Let $\mathbb{L}$ be the set of all legal garbling sequence, let Φ denote the circuit topology of a function. Then RGS is $(\mathbb{L}, \Phi)$-confidential in the random oracle model.*

5 Application to Secure Reactive Two-Party Computation

We now show how to implement reactive two-party computation secure against a malicious, static adversary using a projective reactive garbling scheme. For simplicity we assume that $\mathbb{L}$ is the set of all legal sequences. It can, however, in general consist of a set of sequences closed under the few augmentations we do of the sequence in the protocol. The implementation could be optimized using contemporary tricks for garbling based protocols, but we have chosen to not do this, as the purpose of this section is to demonstrate the use of our security definition, not efficiency.

We implement the ideal functionality in Fig. 9. The inputs to the parties will be a garbling sequence. The commands are received one-by-one, to have a well defined sequence, but can be executed in parallel. We assume that at any point in time the input sequence received by a party is a prefix or suffix of the input sequence of the other parties, except that when a party receives a secret input by receiving input $(\texttt{Input}, t, i, x)$, then the other party receives $(\texttt{Input}, t, i, ?)$, to not leak the secret x, where we use ? to denote a special reserved input indicating that the real input has been removed. We also assume that the sequence of

rule INITIALIZE
$\sigma \leftarrow \{\}$

rule INPUTA
on $(\texttt{Input}, t, i, x)$ **from** A
on $(\texttt{Input}, t, i, ?)$ **from** B
await $(\texttt{Garble}, t) \in \sigma$
on $(\texttt{Input}, t, i, x')$ **from** $\mathcal{S}$
if $\mathsf{A} \in \mathsf{Corrupt}$ **then** $x \leftarrow x'$
send $(\texttt{Input}, t, i, \texttt{done})$ **to** A
send $(\texttt{Input}, t, i, \texttt{done})$ **to** B
$\sigma \leftarrow \sigma \,\|\, (\texttt{Input}, t, i, x)$

rule LINK
await $(\texttt{Garble}, t) \in \sigma)$
on $(\texttt{Link}, t_1, i_1, t_2, i_2)$ **from** A
on $(\texttt{Link}, t_1, i_1, t_2, i_2)$ **from** B
await $(\texttt{Garble}, t) \in \sigma)$
send $(\texttt{Link}, t_1, i_1, t_2, i_2, \texttt{done})$ **to** A
send $(\texttt{Link}, t_1, i_1, t_2, i_2, \texttt{done})$ **to** B
$\sigma \leftarrow \sigma \,\|\, (\texttt{Link}, t_1, i_1, t_2, i_2)$

rule OUTPUT
on $(\texttt{Output}, t, i)$ **from** A
on $(\texttt{Output}, t, i)$ **from** B
await $\exists(t, i, y_{t,i} \neq \perp) \in \mathsf{eval}(\sigma)$
send $(\texttt{Output}, t, i, \texttt{done})$ **to** A
send $(\texttt{Output}, t, i, y_{t,i})$ **to** B
$\sigma \leftarrow \sigma \,\|\, (\texttt{Output}, t, i)$

rule FUNC
on $(\texttt{Func}, t, f)$ **from** A
on $(\texttt{Func}, t, f)$ **from** B
$\sigma \leftarrow \sigma \,\|\, (\texttt{Func}, t, f)$

rule INPUTB
on $(\texttt{Input}, t, i, ?)$ **from** A
on $(\texttt{Input}, t, i, x)$ **from** B
await $(\texttt{Garble}, t) \in \sigma)$
on $(\texttt{Input}, t, i, x')$ **from** $\mathcal{S}$
if $\mathsf{B} \in \mathsf{Corrupt}$ **then** $x \leftarrow x'$
send $(\texttt{Input}, t, i, \texttt{done})$ **to** A
send $(\texttt{Input}, t, i, \texttt{done})$ **to** B
$\sigma \leftarrow \sigma \,\|\, (\texttt{Input}, t, i, x)$

rule GARBLE
on $(\texttt{Garble}, t)$ **from** A
on $(\texttt{Garble}, t)$ **from** B
await $(\texttt{Func}, t, f) \in \sigma$
send $(\texttt{Garble}, t, \texttt{done})$ **to** A
send $(\texttt{Garble}, t, \texttt{done})$ **to** B
$\sigma \leftarrow \sigma \,\|\, (\texttt{Garble}, t)$

Fig. 9. Ideal Functionality $\mathcal{F}_{\text{R2PC}}^{\mathbb{L},\Phi}$ (only suitable for static security). For each line of the form, "**on** c **from** P" for a command c and a party P, when the activation is given to the adversary the ideal functionality sends along $(\Phi(c), \mathsf{P})$.

inputs given to any party is in $\mathbb{L}$. If not, the ideal functionality will simply stop operating. We only specify an ideal functionality for static security. To correctly handle adaptive security a party should sometimes be allowed to replace its input when becoming adaptively corrupted. Since we only prove static security, we chose to not add these complication to the specification.

The implementation will be based on the idea of a watchlist [15]. Alice and Bob will run many instances of a base protocol where Alice is the garbler and Bob is the evaluator. Alice will in each instance provide Bob with garbled functions, linking information, encoded inputs for Alice's inputs and encoded inputs for Bob's inputs, and decoding information. For all Bob's input bits, Alice computes encodings of both 0 and 1, and Bob uses an oblivious transfer to pick the encoding he wants. For a given input bit, the same oblivious transfer instance is used to choose the appropriate encodings in all the instances. This forces Bob to use the same input in all instances. Bob then does a garbled evaluation and decodes to get a plaintext output. Bob therefore gets one possible value of the output from each instance. If Alice cheats by sending incorrect garblings or using different inputs in different instances, the outputs might be different. We combat this by using a watchlist. For a random subset of the instances, Bob will learn all the randomness used by Alice to run the algorithms of the garbling scheme and Bob can therefore check whether Alice is sending the expected values in these instances. The instances inspected by Bob are called the *watchlist instances*. The other instances are called the *evaluation instances*. The watchlist is random and unknown to Alice. The number of instances and the size of the watchlist is set up such that except with negligible probability, either a majority of the evaluation instances are correct or Bob will detect that Alice cheated without leaking information about his input. Bob can therefore take the output value that appears the most often among the evaluation instances as his output. There are several issues with this general approach that must be handled.

1. We cannot allow Bob to learn the encoded inputs of Alice in watchlist instances, as Bob also knows the input encoding functions for the watchlist instances. This is handled by letting Alice send her random tape r_i for each instance i to Bob in an oblivious transfer, where the other message is a key that will be used by Alice to encrypt the encodings of her input. That way Bob can choose to *either* make instance i a watchlist instance, by choosing r_i, or learn the encoded inputs of Alice, but not both.
2. Alice might not send correct input encodings of her own inputs, in which case correctness is not guaranteed. This is not caught by the watchlist mechanism as Bob does not learn Alice's input encodings for the watchlist instances. To combat this attack, Alice must for all input bits of Alice, in all instances, commit to both the encoding of 0 and 1, in a random order, and send along with her input encodings an opening of one of the commitments. The randomness used to commit is picked from the random tape that Bob knows in the watchlist instances. That way Bob can check in the watchlist instances that the commitments were computed correctly, and hence the check in the evaluation position that the encoding sent by Alice opens one of the commitments will ensure that most evaluation instances were run with correct input encodings, except with negligible probability.
3. We have to ensure that Alice uses the same input for herself in all instances. For the same reason as item 2, this cannot be caught by the watchlist mechanism. Instead, it is done by revealing in all instances a privacy-preserving message digest of Alice's input. Bob can then check that this digest is the same in all instances. For efficiency, the digest is computed using a two-universal hash function. This is a common trick by now, see [6,8,23]. However, all previous work used garbled circuits in a white box manner to make this trick work. We can do it by a black box use of reactive garbling, as follows. First Alice garbles the function f to be evaluated producing the garbling F where

Alice is to provide some input component x. Then Alice garbles the function g which takes as input a mask m, an index c for a family h of two-universal hash functions and an input x for the hash function and which outputs x and $y = h_c(x) \oplus m$. Alice then randomly samples a mask m and then sends encodings of m and x to Bob as well as the output decoding function for y. Bob then samples an index c at random and makes it public. Then Alice sends the encoding of c to Bob. Alice then links the output component x of G into the input component x of F. This lets Bob compute y and an encoding X of the input x of f.

4. As usual Alice can mount a selective attack by for example offering Bob a correct encoding of 0 and an incorrect encoding of 1 in one of the OTs used for picking Bob's input. This will not be caught by the watchlist mechanism if Bob's input is 0. As usual this is combated by encoding Bob's input and instead using the encoding as input. The encoding is such that any s positions are uniformly random and independent of the input of Bob. Hence if Alice learns up to s bits of the encoding, it gives her no information on the input of Bob, and if she mounts more than s selective attacks, she will get caught except with probability 2^{-s}. This is again a known trick used in a white box manner in previous works, and again we use linking to generalize this technique to (reactive) garbling schemes. First, Alice will garble an identity function for which Bob will get an encoding of a randomly chosen input x' via OT. Then Bob selects a random hash function h from a two-universal family of hash functions such that $h(x') = x$ where x is Bob's real input. Bob sends h to Alice. Alice then garbles the hash function and links the output of the identity function to the input of the hash function and she links the output of the hash function to the encoded function which Bob is providing an input for.

With the above augmentations which solves obvious security problems, along with an augmentation described below, addressing a problem with simulation, the protocol is UC secure against a static adversary. We briefly sketch how to achieve simulation security.

Simulating corrupted Alice is easy. The simulator can cheat in the OTs used to set up the watchlist and learn both the randomness r_i and the input encodings of Alice in all the evaluation instances. The mechanisms described above ensure that in a majority of evaluation instances Alice correctly garbled and also used the same correct input encoding. Since the input encoding is projective, the input x of Alice can be computed from the input encoding function and her garbled input. By correctness of the garbling scheme, it follows that all correct evaluation instances would give the same output z consistent with x. Hence the simulator can use x as the input of Alice in the simulation.

As usual simulating corrupted Bob is more challenging. To get a feeling for the problem, assume that Alice has to send a garbled circuit F of the function f to be computed before Bob gives inputs. When Bob then gives input, the input y of Bob can be extracted in the simulation by cheating in the OTs and inspecting the choice bits used by Bob. The simulator then inputs y to the ideal functionality and gets back the output $z = f(x, y)$ that Bob is to learn. However, the simulator then in addition has to make F output z in the simulated execution of the protocol. This in general would require finding an input x' of Alice such that $z = f(x', y)$, which could be computationally hard. Previous papers have used white-box modifications of the garbled circuit or the output decoding function to facilitate enough cheating to make F hit z without having to compute x'. We show how to do it in a very simple and elegant way in a black-box manner from any reactive garbling scheme which can garble the exclusive-or function. In our protocol Alice will not send to Bob the decoding key for the encoded output Z. Instead, she garbles a masking function ($\psi(z, m) = z \oplus m$) and links the output of the function f to the first argument of the masking function. Then she produces an encoding M of the all-zero string for m and sends M to Bob along with the output decoding function for ψ. Bob can then compute and decode from Z and M the value $z \oplus 0 = z$. In the simulation, the simulator of corrupted Bob knows the watchlist and can hence behave honestly in the watchlist instances and use the freedom of m to make the output $z \oplus m$ hit the desired output from the ideal functionality in the evaluation positions. This will be indistinguishable from the real world because of the confidentiality property. Since this trick does not require modifying the garbled function, our protocol will only require a projective garbling scheme which is confidential. It will work for any side-information function. Earlier protocols required that the side-information be the topology of the circuit to hide the modification of the function f needed for simulation, or they needed to do white box modifications of the output decoding function to make the needed cheating occur as part of the output decoding.

5.1 Details of the Reactive 2PC Protocol

We now give more details on the protocol. The different instances will be indexed by $j \in I = \{1, \ldots, s\}$. The watchlist is given by $\boldsymbol{w} = (w_1, \ldots, w_s) \in \{0, 1\}^s$, where $w_j = 1$ iff j is a watchlist instance. In the protocol s instances are run in parallel. When a copy of a variable v is used in each instance, the copy used in instance j is denoted by v^j. In most cases the code for an instance does not depend on j explicitly but only on whether the instance is on the watchlist or the evaluation list, in which case we will write the code generically using the variable name v. The convention is that all s copies $v^1, \ldots, v^s$ are manipulated the same way, in single instruction multiple data program style. For instance, $w = 1$ will mean $w^j = 1$, such that $w = 1$ is true iff the instance is in the watchlist.

We will use commitments and oblivious transfer within the protocol. We work in the OT hybrid model. We use $\mathsf{OT.send}(m_0, m_1)$ to mean that Alice sends two messages via the oblivious-transfer functionality and we use the notation $\mathsf{OT.choose}(b)$ to say that Bob chooses to receive m_b. We use a perfect binding and computationally hiding commitment scheme. If a public key is needed, it could be generated by Alice and sent to Bob in initialization. A commitment to a message m produced with randomness r is denoted by $\mathsf{com}(m; r)$, sending (m, r) constitutes an opening of the commitment.

rule A.INITIALIZE
`// Sample watchlist key and an evaluation key`
$\mathsf{wk}, \mathsf{ek} \xleftarrow{\$} \{0,1\}^k$
$\mathsf{OT.send}(\mathsf{ek}, \mathsf{wk})$
$\sigma \leftarrow ()$

rule A.FUNC
on $(\mathtt{Func}, t, f)$
$\sigma \leftarrow \sigma \,\|\, (\mathtt{Func}, t, f)$

rule A.GARBLE
on $(\mathtt{Garble}, t)$
await $\exists f\colon (\mathtt{Func}, t, f) \in \sigma$
$(F_t, e_t, o_t, d_t) \leftarrow \mathsf{Gb}(f, t; r)$
$E \leftarrow \mathsf{E}_{\mathsf{wk}}(r)$
send F_t, E **to** B
$\sigma \leftarrow \sigma \,\|\, (\mathtt{Garble}, t)$

rule A.LINK
on $(\mathtt{Link}, t_1, i_1, t, i_2)$
await $(\mathtt{Garble}, t) \in \sigma$
await $(\mathtt{Garble}, t_1) \in \sigma$
send $\mathsf{li}(o_{t_1,i_1}, e_{t,i_2})$ **to** B
$\sigma \leftarrow \sigma \,\|\, (\mathtt{Link}, t_1, i_1, t, i_2)$

rule B.INITIALIZE
`// Learn either the watchlist key or the evaluation key`
$w \xleftarrow{\$} \{0,1\}$
$k \leftarrow \mathsf{OT.choose}(w)$
$\sigma \leftarrow ()$

rule B.FUNC
on $(\mathtt{Func}, t, f)$
$\sigma \leftarrow \sigma \,\|\, (\mathtt{Func}, t, f)$

rule B.GARBLE
on $(\mathtt{Garble}, t)$
await $\exists f\colon (\mathtt{Func}, t, f) \in \sigma$
on F', E **from** A
if $w = 1$ **then**
  $r \leftarrow \mathsf{D}_{\mathsf{wk}}(E)$
  $(F_t, e_t, o_t, d_t) \leftarrow \mathsf{Gb}(f, t; r)$
  verify $F' = F_t$
$F_t \leftarrow F'$
$\sigma \leftarrow \sigma \,\|\, (\mathtt{Garble}, t)$

rule B.LINK
on $(\mathtt{Link}, t_1, i_1, t, i_2)$
await $(\mathtt{Garble}, t) \in \sigma$
await $(\mathtt{Garble}, t_1) \in \sigma$
on $\bar{L}$ **from** A
$\mathcal{L} \leftarrow \mathcal{L} \,\|\, (t_1, i_1, t, i_2, \bar{L})$
if $w = 1$ **then verify**
$\bar{L} = \mathsf{li}(o_{t_1,i_1}, e_{t,i_2})$
$\sigma \leftarrow \sigma \,\|\, (\mathtt{Link}, t_1, i_1, t, i_2)$

Fig. 10. Protocol (INITIALIZE, GARBLE, LINK)

If we write $A(x; r)$ for a randomized algorithm, where r is not bound before, then it means that we make a random run of A on input x and that we use r in the following to denote the randomness used by A. If we send a set $\{x, y\}$, then it is sent as a vector with the bit strings x and y sorted lexicographically, such that all information extra to the elements is removed before sending. When rules are called, tags t are provided. It follows from the input sequences being legal that these tags are unique, except when referring to a legal previous occurrence. We further assume that all tags provided as inputs are of the form $0\|\{0,1\}^*$, which allows us to use tags of the form $1\|\{0,1\}^*$ for internal book keeping. Tags for internal use will be derived from the tags given as input and the name of the rule creating the new tag. For a garbling scheme $\mathcal{G}$, a commitment scheme com and an encryption scheme $\mathcal{E}$, we use $\pi_{\mathcal{G},\mathsf{com},\mathcal{E}}$ to denote protocol given by the set of rules in Figs. 10, 11, 12 13, 14 and 15. We add a few remarks to the figures.

rule A.INPUT_A
on $(\texttt{Input}, t, i, x)$
await $(\texttt{Garble}, t) \in \sigma$
$\bar{t} \leftarrow 1 \| (\texttt{Input}, t, i) \| 0$
$\ell_1 \leftarrow \text{len}(f_t.A_i)$
$\ell_2 \leftarrow \text{len}(g_{\ell_1}.A_2)$
$\ell_3 \leftarrow \text{len}(g_{\ell_1}.A_3)$
$m \xleftarrow{\$} \{0,1\}^{\ell_2}$
`// Garble auxiliary function` g
$(G_{\bar{t}}, e_{\bar{t}}, d_{\bar{t}}, o_{\bar{t}}) \leftarrow \mathsf{Gb}(g_{\ell_1}, \bar{t}; r)$
`// Watchlist encryption of garbled auxilary function's randomness`
$E \leftarrow E_{\mathsf{wk}}(r)$
send $(G_{\bar{t}}, d_{\bar{t},2}, E)$ **to** B
for $u \in \{1, \ldots, \ell_1\}$ **do**
  $X_{u,0} \leftarrow \mathsf{En}(e_{\bar{t},1,u}, 0)$
  $X_{u,1} \leftarrow \mathsf{En}(e_{\bar{t},1,u}, 1)$
  $r_{u,0}, r_{u,1} \xleftarrow{\$} \{0,1\}^k$
  `// Commit to tokens`
  $S_{u,1} \leftarrow \{\mathsf{com}(X_{u,0}; r_{u,0}), \mathsf{com}(X_{u,1}; r_{u,1})\}$
  `// Watchlist encryption of tokens`
  $E_{u,1} \leftarrow \mathsf{E}_{\mathsf{wk}}((X_{u,0}, X_{u,1}))$
  `// Watchlist encryption of commitment's randomness`
  $E_{u,2} \leftarrow \mathsf{E}_{\mathsf{wk}}((r_{u,0}, r_{u,1}))$
  `// Evaluation encryption of tokens for Alice's choice of input`
  $E_{u,3} \leftarrow \mathsf{E}_{\mathsf{ek}}((X_{u,x_{i,u}}, r_{u,x_{i,u}}))$
  `// Linking` G `to` F_t
  $L_u \leftarrow \mathsf{li}(o_{\bar{t},1,u}, e_{t,i,u})$
  send $(S_{u,1}, E_{u,1}, E_{u,2}, E_{u,3}, L_u)$ **to** B
for $u \in \{1, \ldots, \ell_2\}$ **do**
  $M_{u,0} \leftarrow \mathsf{En}(e_{\bar{t},2,u}, 0)$
  $M_{u,1} \leftarrow \mathsf{En}(e_{\bar{t},2,u}, 1)$
  $r'_{u,0}, r'_{u,1} \xleftarrow{\$} \{0,1\}^k$
  $S_{u,2} \leftarrow \{\mathsf{com}(M_{u,0}; r'_{u,0}), \mathsf{com}(M_{u,1}; r'_{u,1})\}$
  $E_{u,4} \leftarrow \mathsf{E}_{\mathsf{wk}}((M_{u,0}, M_{u,1}))$
  $E_{u,5} \leftarrow \mathsf{E}_{\mathsf{wk}}((r'_{u,0}, r'_{u,1}))$
  $E_{u,6} \leftarrow \mathsf{E}_{\mathsf{ek}}((M_{u,m_{i,u}}, r'_{u,m_u}))$
  send $(S_{u,2}, E_{u,4}, E_{u,5}, E_{u,6})$ **to** B
`// Auxiliary input from Bob`
on c **from** B
`// Encoding of auxiliary input`
for $u \in \{1, \ldots, \ell_3\}$ **do send** C_{u,c_u} **to** B
$\sigma \leftarrow \sigma \| (\texttt{Input}, t, i, \top)$

Fig. 11. INPUT_A

rule B.$\textsc{Input}_\mathsf{A}$
on $(\texttt{Input}, t, i, ?)$
await $(\texttt{Garble}, t) \in \sigma$
$\bar{t} \leftarrow 1 \| (\texttt{Input}, t, i) \| 0$
$c \xleftarrow{\$} \{0,1\}^{\ell_3}$
on $G'_{\bar{t}}, d'_{\bar{t},2}, E$ **from** A
for $u \in \{1, \ldots, \ell_1\}$ **do** **on** $(S_{u,1}, E_{u,1}, E_{u,2}, E_{u,3}, L_u)$ **from** A
for $u \in \{1, \ldots, \ell_2\}$ **do** **on** $(S_{u,2}, E_{u,4}, E_{u,5}, E_{u,6})$ **from** A
send c **to** A
for $u \in \{1, \ldots, \ell_3\}$ **do** **on** C_{u,c_u} **from** A
// Use watchlist key to verify correctness of garbling and commitments.
if $w = 1$ **then**
    $r \leftarrow \mathsf{D}_{\mathsf{wk}}(E)$, $(G_{\bar{t}}, e_{\bar{t}}, d_{\bar{t}}, o_{\bar{t}}) \leftarrow \mathsf{Gb}(g_{\ell_1}, \bar{t}; r)$
    for $u \in \{1, \ldots, \ell_1\}$ **do**
        $X_{u,0} \leftarrow \mathsf{En}(e_{\bar{t},1,u}, 0)$
        $X_{u,1} \leftarrow \mathsf{En}(e_{\bar{t},1,u}, 1)$
    for $u \in \{1, \ldots, \ell_2\}$ **do**
        $M_{u,0} \leftarrow \mathsf{En}(e_{\bar{t},2,u}, 0)$
        $M_{u,1} \leftarrow \mathsf{En}(e_{\bar{t},2,u}, 1)$
    for $u \in \{1, \ldots, \ell_3\}$ **do**
        $C_{u,0} \leftarrow \mathsf{En}(e_{\bar{t},3,u}, 0)$
        $C_{u,1} \leftarrow \mathsf{En}(e_{\bar{t},3,u}, 1)$
    for $u \in \{1, \ldots, \ell_1\}$ **do**
        $(r_{u,0}, r_{u,1}) \leftarrow \mathsf{D}_{\mathsf{wk}}(E_{u,2})$
        verify $\mathsf{D}_{\mathsf{wk}}(E_{u,1}) = (X_{u,0}, X_{u,1})$
        verify $S_{u,1} = \{\mathsf{com}(X_{u,0}; r_{u,0}), \mathsf{com}(X_{u,1}; r_{u,1})\}$
        verify $L_u = \mathsf{li}(o_{\bar{t},1,u}, e_{t,i,u})$
    for $u \in \{1, \ldots, \ell_2\}$ **do**
        $(r'_{u,0}, r'_{u,1}) \leftarrow \mathsf{D}_{\mathsf{wk}}(E_{u,5})$
        verify $\mathsf{D}_{\mathsf{wk}}(E_{u,3}) = (M_{u,0}, M_{u,1})$
        verify $S_{u,2} = \{\mathsf{com}(M_{u,0}; r_{u,0}), \mathsf{com}(M_{u,1}, r_{u,1})\}$
    for $u \in \{1, \ldots, \ell_3\}$ **do**
        verify $C_{u,c_u} = \mathsf{En}(e_{\bar{t},3,u}, c_u)$
else
// Use evaluation key to extract tokens for Alice's choice of input
    for $u \in \{1, \ldots, \ell_1\}$ **do**
        $(X_{u,x_{i,u}}, r_{u,x_{i,u}}) \leftarrow \mathsf{D}_{\mathsf{ek}}(E_{u,3})$
    for $u \in \{1, \ldots, \ell_2\}$ **do**
        $(M_{u,x_{i,u}}, r'_{u,x_{i,u}}) \leftarrow \mathsf{D}_{\mathsf{ek}}(E_{u,6})$
    // Verify commitments of tokens for Alice's choice of input
    verify $\forall u \in \{1, \ldots, \ell_1\}\ (\mathsf{com}(X_{u,x_{i,u}}; r_{u,x_{i,u}}) \in S_{u,1})$
    verify $\forall u \in \{1, \ldots, \ell_2\}\ (\mathsf{com}(M_{u,m_u}; r'_{u,m_u}) \in S_{u,2})$
    $\bar{\mathcal{X}} \leftarrow \{(\bar{t}, 1, X_x), (\bar{t}, 2, M_m), (\bar{t}, 3, C_c)\}$
    $\bar{Y} \leftarrow \mathsf{Ev}(\{(\bar{t}, G_{\bar{t}})\}, \bar{\mathcal{X}})$
    $y_2 \leftarrow \mathsf{De}(d_2, \bar{Y}_2)$
    // Verify that auxiliary outputs are the same in each instance
    verify $\forall j, j'\ (y_2^j = y_2^{j'})$
    $\mathcal{X} \leftarrow \mathcal{X} \| \bar{\mathcal{X}}$
    $\mathcal{F} \leftarrow \mathcal{F} \| (\bar{t}, G_{\bar{t}})$
    $\mathcal{L} \leftarrow \mathcal{L} \| (\bar{t}, 1, t, i, L)$
    $\sigma \leftarrow \sigma \| (\texttt{Input}, t, i, \top)$

Fig. 12. $\textsc{Input}_\mathsf{A}$ (continued)

In the INITIALIZE-rules Alice and Bob setup the watchlist. They use a (symmetric) encryption scheme $\mathcal{E} = (\mathsf{E}, \mathsf{D})$ with k-bit keys. For each instance j, Alice sends two keys via the oblivious transfer functionality, the watchlist key wk^j and the evaluation key ek^j. Alice will later encrypt and send the information

rule $\mathsf{A}.\textsc{Input}_\mathsf{B}$
on $(\texttt{Input}, t, i, ?)$
await $(\texttt{Garble}, t) \in \sigma$
$\ell \leftarrow \mathrm{len}(f_t.A_i)$
$\ell_1 \leftarrow \ell + 2s + 1$
$\bar{t} \leftarrow 1 \| (\texttt{Input}, t, i) \| 0$
$t' \leftarrow 1 \| (\texttt{Input}, t, i) \| 1$
`// Garble the identity function`
$(\mathsf{Id}_{\bar{t}}, e_{\bar{t}}, o_{\bar{t}}, d_{\bar{t}}) \leftarrow \mathsf{Gb}(\mathsf{id}_{\ell_1}, \bar{t}; r)$
`// Send to Bob the garbled identity function and the watchlist encryption of its randomness to Bob`
send $E \leftarrow \mathsf{E}_{\mathsf{wk}}(r), \mathsf{Id}_t$ **to** B
for $u \in \{1, \ldots, \ell_1\}$ **do**
  $X_{u,0} \leftarrow \mathsf{En}(e_{\bar{t},u}, 0)$
  $X_{u,1} \leftarrow \mathsf{En}(e_{\bar{t},u}, 1)$
  `// Oblivious Transfer of Bob's input tokens`
  $\mathsf{OT.send}(\{X^j_{u,0}\}_{j \in \{1,\ldots,s\}}, \{X^j_{u,1}\}_{j \in \{1,\ldots,s\}})$
`// Await universal hash function`
on h **from** B
`// Garble universal hash function`
$(H_{t'}, e_{t'}, o_{t'}, d_{t'}) \leftarrow \mathsf{Gb}(h, t'; r')$
`// Send garbled hash function and the watchlist encryption of its randomness to Bob`
send $H_{t'}, \mathsf{E}_{\mathsf{wk}}(r')$ **to** B
for $u \in \{1, \ldots, \ell_1\}$ **do**
  `// Link` $\mathsf{Id}_{\bar{t}}$ `to` $H_{t'}$
  send $\bar{L}_u \leftarrow \mathsf{li}(o_{\bar{t},u}, e_{t',u})$ **to** B
  `// Link` $H_{t'}$ `to` F_t
  send $L_u \leftarrow \mathsf{li}(o_{t',u}, e_{t,i,u})$ **to** B
$\sigma \leftarrow \sigma \| (\texttt{Input}, t, i, \top)$

Fig. 13. $\textsc{Input}_\mathsf{B}$

Bob is to learn for watchlist (evaluation) instances with the key wk (ek). In the FUNC-rules they simply associate a function to a tag. In the GARBLE-rules Alice garbles the function and sends the garbling to Bob, she also sends an encryption using the watchlist key of the randomness used to produce this garbling. This allows Bob, for the watchlist positions to check that Alice produced a correct garbling and to store the result of garbling. This knowledge will be used in other rules. In the LINK-rules Alice sends linking information. Bob can for all watchlist positions check that the information is correct, since he knows the randomness used to garble. In the OUTPUT-rules Alice awaits that she has sent to Bob the encoded inputs and linkings to produce the encoded output associated to this rule. She produces a garbling of ψ. She will link the output to ψ and produce an encoding of the zero-string for the second component, she also sends an encryption of the randomness used to produce the garbling of ψ to Bob. Bob awaits

```
rule B.INPUT_B
on (Input, t, i, x)
await (Garble, t) ∈ σ
$\bar{t} \leftarrow 1\|(\text{Input}, t, i)\|0$
$t' \leftarrow 1\|(\text{Input}, t, i)\|1$
// sample a random string $\bar{x}$
$\bar{x} \xleftarrow{\$} \{0,1\}^{\ell_1}$
// Sample a random universal hash function $h$ such that $h(\bar{x}) = x$
$h \xleftarrow{\$} \{ \bar{h} \in \mathcal{H}_\ell \mid \bar{h}(\bar{x}) = x \}$
// Await a garbled identity function from Alice
on $E, \mathsf{Id}'_{\bar{t}}$ from A
// Obliviously learn tokens for $\bar{x}$
for $u \in \{1, \ldots, \ell_1\}$ do
    $\{\bar{X}^j_{u,\bar{x}_u}\}_{j \in \{1,\ldots,s\}} \leftarrow \mathsf{OT.choose}(\bar{x}_u)$
$\bar{X}_{\bar{t},\bar{x}} \leftarrow (\bar{X}_{1,\bar{x}_1}, \ldots, \bar{X}_{\ell_1,\bar{x}_{\ell_1}})$
if $w = 1$ then
    /* Verify garbled identity function and the correctness of
       received tokens using the watchlist encryption of the
       randomness                                              */
    $r \leftarrow \mathsf{D}_{\mathsf{wk}}(E)$
    $(\mathsf{Id}_{\bar{t}}, e_{\bar{t}}, o_{\bar{t}}, d_{\bar{t}}) \leftarrow \mathsf{Gb}(\mathsf{id}_{\ell_1}, \bar{t}; r)$
    verify $\mathsf{Id}_{\bar{t}} = \mathsf{Id}'_{\bar{t}}$
    verify $\forall u \in \{1, \ldots, \ell_1\} : \bar{X}_{\bar{t},u} = \mathsf{En}(e_{\bar{t},u}, \bar{x}_u)$
else
    $\mathcal{X} \leftarrow \mathcal{X} \|(\bar{t}, \bar{X}_{\bar{t},\bar{x}})$
send $h$ to A
on $H', E'$ from A
for $u \in \{1, \ldots, \ell_1\}$ do
    on $\bar{L}_u$ from A
    on $L_u$ from A
if $w = 1$ then
    /* Verify garbled hash function using the watchlist encryption
       of the randomness                                       */
    $r' \leftarrow \mathsf{D}_{\mathsf{wk}}(E')$
    $(H_{t'}, e_{t'}, o_{t'}, d_{t'}) \leftarrow \mathsf{Gb}(h, t'; r')$
    verify $H_{t'} = H'$
    // Verify linking information
    for $u \in \{1, \ldots, \ell_1\}$ do
        verify $\bar{L}_u = \mathsf{li}(o_{\bar{t},u}, e_{t',u})$
        verify $L_u = \mathsf{li}(o_{t',u}, e_{t,i,u})$
else
    $\mathcal{F} \leftarrow \mathcal{F} \|(\bar{t}, Id)$
    $\mathcal{F} \leftarrow \mathcal{F} \|(t', H)$
    $\mathcal{X} \leftarrow \mathcal{X} \|(\bar{t}, \bar{X}_{\bar{t},\bar{x}})$
    $\mathcal{L} \leftarrow \mathcal{L} \|(t', 1, t, i, L)$
    for $u \in \{1, \ldots, \ell_1\}$ do
        $\mathcal{L} \leftarrow \mathcal{L} \|(\bar{t}, u, t', u, \bar{L}_u)$
$\sigma \leftarrow \sigma \|(\text{Input}, t, i, \top)$
```

Fig. 14. INPUT$_\mathsf{B}$ (continued)

rule A.OUTPUT
on (Output, t, i)
await $(t, i) \in \mathsf{ready}(\sigma)$
$\bar{t} \leftarrow 1 \| (\text{Output}, t, i)$
`// Garble` ψ
$(\Psi, e_{\bar{t}}, d_{\bar{t}}, o_{\bar{t}}) \leftarrow \mathsf{Gb}(\psi, \bar{t}; r)$
$L \leftarrow \mathsf{li}(o_{t,i}, e_{\bar{t},1})$
$E \leftarrow \mathsf{E}_{\mathsf{wk}}(r)$
`// Encode all zero-string`
$X_{\bar{t},0} \leftarrow \mathsf{En}(e_{\bar{t},2}, 0)$
send $(L, E, \Psi, X_{\bar{t},0}, d_{\bar{t}})$ **to** B

rule B.OUTPUT
on (Output, t, i)
await $(t, i, \top) \in \mathsf{ready}(\sigma)$
$\bar{t} \leftarrow 1 \| (\text{Output}, t, i)$
on $(\bar{L}, \bar{E}, \bar{\Psi}, \bar{X}_{\bar{t},0}, \bar{d}_{\bar{t}})$ **from** A
if $w = 1$ **then**
  $r \leftarrow \mathsf{D}_{\mathsf{wk}}(\bar{E})$
  $(\Psi, e_{\bar{t}}, d_{\bar{t}}, o_{\bar{t}}) \leftarrow \mathsf{Gb}(\psi, \bar{t}; r)$
  $L \leftarrow \mathsf{li}(o_{t,i}, e_{\bar{t},1})$
  `/* Verify:`
  `1)` $\bar{\Psi}$ `is the garbling of` ψ
  `2) Linking is correct`
  `3) Encoding of the all zero-string was sent`
  `4) Correct output decoding was sent */`
  verify $\bar{L} = L \wedge \bar{\Psi} = \Psi$
  verify $\bar{X}_{\bar{t},0} = \mathsf{En}(e_{\bar{t},2}, 0) \wedge \bar{d}_{\bar{t},1} = d_{\bar{t},1}$
else
  $\mathcal{F} \leftarrow \mathcal{F} \| (\bar{t}, \Psi)$
  $\mathcal{X} \leftarrow \mathcal{X} \| (\bar{t}, 2, \bar{X}_{\bar{t},0})$
  $\mathcal{L} \leftarrow \mathcal{L} \| (t, i, \bar{t}, 1, \bar{L})$
  $\delta \leftarrow \delta \| (\bar{t}, 1, \bar{d}_{\bar{t},1})$
  await $\exists(\bar{t}, 1, Y_{t,1}) \in \mathsf{Ev}(\mathcal{F}, \mathcal{X}, \mathcal{L})$
  $y^j_{t,i} \leftarrow \mathsf{De}(\bar{d}_{\bar{t},1}, Y_{\bar{t},1})$
  `// Apply majority decoding`
  $y_{t,i} \leftarrow \mathsf{maj}(y^1_{t,i}, \dots, y^1_{t,i})$

Fig. 15. Protocol (OUTPUT)

that he has received the garbling, linking and encoding to produce the encoded output in question. For each instance of the watchlist, he uses the randomness to check that the linking was done correctly, that ψ was garbled correctly and that an encoding of an all zero-string was sent for the second component of ψ. He then evaluates each instance in the evaluation set and takes the majority value as his output.

In the $\mathtt{Input_A}$-rules Alice commits to both her input encodings and encrypts the openings of the commitments using the watchlist key. The opening of Alice's input encoding will be encrypted using the evaluation key. To verify Alice's input, we first pass Alice's input through an auxiliary function which combines the identity function with an additional verification function which forces Alice to use the same input in different instances. We then link the output of the identity function to the appropriate input. We denoted the auxiliary function by $g_l : A_1 \times A_2 \times A_3 \rightarrow B_1 \times B_2$ and $g_\ell(x, m, c) = (x, v_\ell(x, m, c))$ where $A_1 = A_2 = B_1 = \{0,1\}^\ell \cup \{\perp\}$ and $v_\ell : A_1 \times A_2 \times A_3 \rightarrow B_2$. Efficient such functions with the properties needed for the security of the protocol can be based on universal hash functions, see for instance [6,23].

In the $\mathtt{Input_B}$-rules Alice first garbles the identity function. Bob then randomly samples a value x' and gets an encoding of that value via oblivious transfer for the garbled identity function. Then Bob samples uniformly at random a function h from a two-universal family of hash functions such that $h(x') = x$ where x is the input of Bob. Alice will then garble the hash function. She will link the garbling of the identity function to the garbling of the hash function. She will then link the garbled hash function to the garbled function. We will denote by $\mathcal{H}_\ell$ a two-universal family of hash functions $h : \{0,1\}^{\ell+2s+1} \rightarrow \{0,1\}^\ell$. We use $\mathsf{id} : A \rightarrow A$ to denote the identify function on A.

In [22] we prove the following theorem.

Theorem 2. *Let* $\mathbb{L}$ *be the set of all legal sequences and let* Φ *be a side-information function. Let* $\mathcal{G}$ *be a reactive garbling scheme. Let* com *be a commitment scheme and* $\mathcal{E}$ *an encryption scheme. If* $\mathcal{G}$ *is* $\mathbb{L}$*-correct and* $(\mathbb{L}, \Phi)$*-confidential and* com *is computationally hiding and perfect binding and* $\mathcal{E}$ *is IND-CPA secure, then* $\pi_{\mathcal{G},\mathsf{com},\mathcal{E}}$ *UC securely realizes* $\mathcal{F}_{\mathrm{R2PC}}^{\mathbb{L},\Phi}$ *in the* $\mathcal{F}_{\mathrm{OT}}$*-hybrid model against a static, malicious adversary.*

References

1. Bellare, M., Hoang, V.T., Rogaway, P.: Adaptively secure garbling with applications to one-time programs and secure outsourcing. In: Wang, X., Sako, K. (eds.) ASIACRYPT 2012. LNCS, vol. 7658, pp. 134–153. Springer, Heidelberg (2012). doi:10.1007/978-3-642-34961-4_10
2. Bellare, M., Hoang, V.T., Rogaway, P.: Foundations of garbled circuits. In: Yu, T., Danezis, G., Gligor, V.D. (eds.) The ACM Conference on Computer and Communications Security, CCS 2012, Raleigh, NC, USA, 16–18 October 2012, pp. 784–796. ACM (2012)
3. Bellare, M., Rogaway, P.: The security of triple encryption and a framework for code-based game-playing proofs. In: Vaudenay, S. (ed.) EUROCRYPT 2006. LNCS, vol. 4004, pp. 409–426. Springer, Heidelberg (2006). doi:10.1007/11761679_25
4. Brandão, L.T.A.N.: Secure two-party computation with reusable bit-commitments, via a cut-and-choose with forge-and-lose technique. In: Sako, K., Sarkar, P. (eds.) ASIACRYPT 2013. LNCS, vol. 8270, pp. 441–463. Springer, Heidelberg (2013). doi:10.1007/978-3-642-42045-0_23

5. Carter, H., Lever, C., Traynor, P.: Whitewash: Outsourcing garbled circuit generation for mobile devices (2014)
6. Frederiksen, T.K., Jakobsen, T.P., Nielsen, J.B.: Faster maliciously secure two-party computation using the GPU. In: Abdalla, M., Prisco, R. (eds.) SCN 2014. LNCS, vol. 8642, pp. 358–379. Springer, Heidelberg (2014). doi:10.1007/978-3-319-10879-7_21
7. Frederiksen, T.K., Jakobsen, T.P., Nielsen, J.B., Nordholt, P.S., Orlandi, C.: MiniLEGO: efficient secure two-party computation from general assumptions. In: Johansson, T., Nguyen, P.Q. (eds.) EUROCRYPT 2013. LNCS, vol. 7881, pp. 537–556. Springer, Heidelberg (2013). doi:10.1007/978-3-642-38348-9_32
8. Frederiksen, T.K., Nielsen, J.B.: Fast and maliciously secure two-party computation using the GPU. IACR Cryptology ePrint Archive 2013, p. 46 (2013)
9. Gennaro, R., Gentry, C., Parno, B.: Non-interactive verifiable computing: outsourcing computation to untrusted workers. In: Rabin, T. (ed.) CRYPTO 2010. LNCS, vol. 6223, pp. 465–482. Springer, Heidelberg (2010). doi:10.1007/978-3-642-14623-7_25
10. Gentry, C., Halevi, S., Lu, S., Ostrovsky, R., Raykova, M., Wichs, D.: Garbled RAM revisited. In: Nguyen, P.Q., Oswald, E. (eds.) EUROCRYPT 2014. LNCS, vol. 8441, pp. 405–422. Springer, Heidelberg (2014). doi:10.1007/978-3-642-55220-5_23
11. Gentry, C., Halevi, S., Vaikuntanathan, V.: i-hop homomorphic encryption and rerandomizable yao circuits. In: Rabin, T. (ed.) CRYPTO 2010. LNCS, vol. 6223, pp. 155–172. Springer, Heidelberg (2010). doi:10.1007/978-3-642-14623-7_9
12. Goldwasser, S., Kalai, Y.T., Rothblum, G.N.: One-time programs. In: Wagner, D. (ed.) CRYPTO 2008. LNCS, vol. 5157, pp. 39–56. Springer, Heidelberg (2008). doi:10.1007/978-3-540-85174-5_3
13. Huang, Y., Evans, D., Katz, J., Malka, L.: Faster secure two-party computation using garbled circuits. In: 20th USENIX Security Symposium, Proceedings, San Francisco, CA, USA, 8–12 August 2011. USENIX Association (2011)
14. Huang, Y., Katz, J., Kolesnikov, V., Kumaresan, R., Malozemoff, A.J.: Amortizing garbled circuits. In: Garay, J.A., Gennaro, R. (eds.) CRYPTO 2014. LNCS, vol. 8617, pp. 458–475. Springer, Heidelberg (2014). doi:10.1007/978-3-662-44381-1_26
15. Ishai, Y., Prabhakaran, M., Sahai, A.: Founding cryptography on oblivious transfer – efficiently. In: Wagner, D. (ed.) CRYPTO 2008. LNCS, vol. 5157, pp. 572–591. Springer, Heidelberg (2008). doi:10.1007/978-3-540-85174-5_32
16. Kreuter, B., Shelat, A., Shen, C.H.: Billion-gate secure computation with malicious adversaries. Cryptology ePrint Archive, Report 2012/179 (2012). http://eprint.iacr.org/
17. Kreuter, B., Shelat, A., Shen, C.-H.: Billion-gate secure computation with malicious adversaries. In: USENIX Security Symposium, pp. 285–300 (2012)
18. Lindell, Y.: Fast cut-and-choose based protocols for malicious and covert adversaries. In: Canetti, R., Garay, J.A. (eds.) CRYPTO 2013. LNCS, vol. 8043, pp. 1–17. Springer, Heidelberg (2013). doi:10.1007/978-3-642-40084-1_1
19. Lu, S., Ostrovsky, R.: How to garble RAM programs? In: Johansson, T., Nguyen, P.Q. (eds.) EUROCRYPT 2013. LNCS, vol. 7881, pp. 719–734. Springer, Heidelberg (2013). doi:10.1007/978-3-642-38348-9_42
20. Mood, B., Gupta, D., Butler, K., Feigenbaum, J.: Reuse it or lose it: more efficient secure computation through reuse of encrypted values. In: Proceedings of the 2014 ACM SIGSAC Conference on Computer and Communications Security, pp. 582–596. ACM (2014)

21. Nielsen, J.B., Orlandi, C.: LEGO for two-party secure computation. In: Reingold, O. (ed.) TCC 2009. LNCS, vol. 5444, pp. 368–386. Springer, Heidelberg (2009). doi:10.1007/978-3-642-00457-5_22
22. Nielsen, J.B., Ranellucci, S.: Foundations of reactive garbling schemes. IACR Cryptology ePrint Archive 2015, p. 693 (2015)
23. Shelat, A., Shen, C.: Fast two-party secure computation with minimal assumptions. In: 2013 ACM SIGSAC Conference on Computer and Communications Security, CCS 2013, Berlin, Germany, 4–8 November 2013, pp. 523–534 (2013)

Author Index

Printed in the United States
By Bookmasters